# PRINCIPLES OF TISSUE ENGINEERING

**Third Edition**

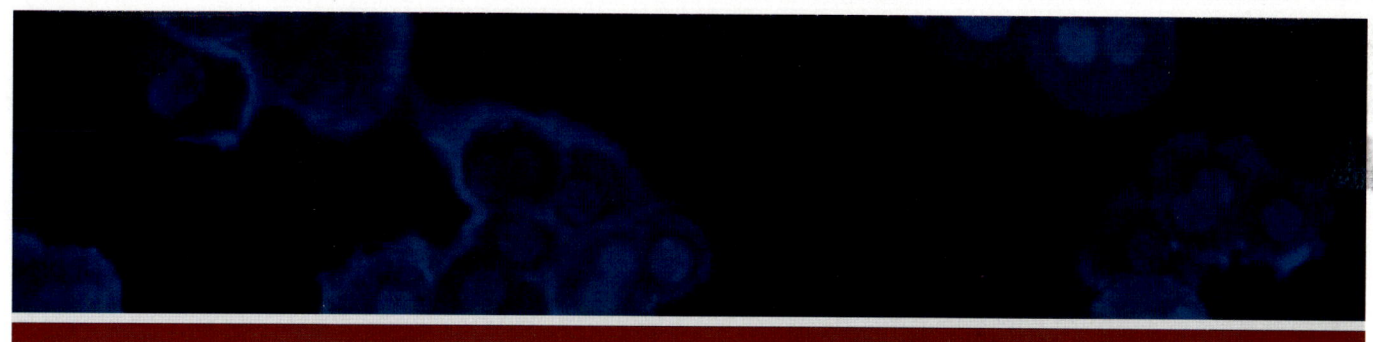

# PRINCIPLES OF TISSUE ENGINEERING

## Third Edition

*Edited by*

**Robert Lanza**
VP Research and Scientific Development, Advanced Cell Technology
Worcester, Massachusetts
Adjunct Professor, Institute of Regenerative Medicine
Wake Forest University School of Medicine
Winston-Salem, North Carolina


**Robert Langer**
Institute Professor
Massachusetts Institute of Technology
Cambridge, Massachusetts


**Joseph Vacanti**
John Homans Professor of Surgery
Director, Laboratory for Tissue Engineering and Organ Fabrication
Harvard Medical School and Massachusetts General Hospital
Boston, Massachusetts

AMSTERDAM • BOSTON • HEIDELBERG • LONDON
NEW YORK • OXFORD • PARIS • SAN DIEGO
SAN FRANCISCO • SINGAPORE • SYDNEY • TOKYO
Academic Press is an imprint of Elsevier

ELSEVIER

*On the cover:* Deborah Odum Hutchinson is an award winning and nationally published medical artist. After working seven years for the College of Medicine at Texas. A&M University, she has been doing freelance art for 19 years. She is also known for her work in watercolors as a fine artist. Her work is a sensitive creation of the delicate anatomical aspects of the body as well as the artistic aspect of the figure in her final illustrations.

Elsevier Academic Press
30 Corporate Drive, Suite 400, Burlington, MA 01803, USA
525 B Street, Suite 1900, San Diego, California 92101-4495, USA
84 Theobald's Road, London WC1X 8RR, UK

This book is printed on acid-free paper. ∞

Library of Congress Cataloging-in-Publication Data
Principles of tissue engineering / editors, Robert Lanza, Robert Langer,
Joseph P. Vacanti. — 3rd ed.
    p. cm.
  ISBN-13: 978-0-12-370615-7 (hardcover : alk. paper)
  ISBN-10: 0-12-370615-7 (hardcover : alk. paper)   1. Animal cell
biotechnology.   2. Tissue engineering.   3. Transplantation of organs, tissues,
etc.   I. Lanza, R. P. (Robert Paul), 1956–   II. Langer, Robert S.   III. Vacanti, Joseph.
  TP248.27.A53P75 2007
  612'.028 — dc22

                                                                    2007002966

British Library Cataloguing in Publication Data
A catalogue record for this book is available from the British Library

ISBN 13: 978-0-12-370615-7
ISBN 10: 0-12-370615-7

For all information on all Elsevier Academic Press publications
visit our Web site at www.books.elsevier.com

Printed in China
09  10    9  8  7  6  5  4  3  2

# CONTENTS IN BRIEF

# CONTENTS

# CONTRIBUTORS

Jon D. Ahlstrom
Section of Molecular and Cellular Biology
University of California, Davis
Davis, CA 95616

Julie Albon
School of Optometry and Vision Sciences
Cardiff University
CF10 3NB Cardiff
UK

Richard A. Altschuler
Kresge Hearing Research Institute
University of Michigan
Department of Otolaryngology and Department of Anatomy &
Cell Biology
Ann Arbor, MI 48109-0506

A. Amendola
Department of Orthopedics
University of Iowa College of Medicine
Iowa City, IA 52242

David J. Anderson
Kresge Hearing Research Institute
Department of Electrical Engineering & Computer Sciences
Department of Biomedical Engineering & Kresge Hearing
Research Institute
University of Michigan
Ann Arbor, MI 48109-0506

Piero Anversa
Cardiovascular Research Institute
Department of Medicine
New York Medical College
Valhalla, NY 10595

Anthony Atala
Wake Forest Institute for Regenerative Medicine
Wake Forest University School of Medicine
Winston-Salem, NC 27157

Kyriacos A. Athanasiou
Department of Bioengineering
Rice University
Houston, TX 77251-1892

François A. Auger
Laboratoire d'Organogénèse Expérimentale
Québec, Qc, G1S 4L8
Canada

Debra T. Auguste
Division of Engineering and Applied Sciences
Massachusetts Institute of Technology
Cambridge, MA 02139

Claudia Bearzi
Cardiovascular Research Institute
Department of Medicine
New York Medical College
Valhalla, NY 10595

Daniel Becker
International Center for Spinal Cord Injury
Kennedy-Krieger Institute
Baltimore, MD 21205

Francisco J. Bedoya
Centro Andaluz de Biología Molecular y Medicina
Regenerativa (Cabimer)
C/Américo Vespucio, s/n
41092 Isla de la Cartuja, Seville
Spain

Eugene Bell
TEI Biosciences Inc.
Department of Biology
Boston, MA 02127

Timothy Bertram
Tengion Inc.
Winston-Salem, NC 27103

Valérie Besnard
Division of Pulmonary Biology
Cincinnati Children's Hospital Medical Center
Cincinnati, OH 45229-3039

Christopher J. Bettinger
Department of Materials Science and Engineering
Massachusetts Institute of Technology
Cambridge, MA 02142

Sangeeta N. Bhatia
Harvard-M.I.T. Division of Health Sciences and Technology/
Electrical Engineering and Computer Science
Laboratory for Multiscale Regenerative Technologies
Massachusetts Institute of Technology
Cambridge, MA 02139

Paolo Bianco
Dipartimento di Medicina Sperimentale e Patologia
Universita "La Sapienza"
324-00161 Rome
Italy

Anne E. Bishop
Stem Cells & Regenerative Medicine,
Section on Experimental Medicine & Toxicology
Imperial College Faculty of Medicine
Hammersmith Campus
W12 ONN London
UK

C. Clare Blackburn
MRC/JDRF Centre Development in Stem Cell Biology
Institute for Stem Cell Research
University of Edinburgh
EH9 3JQ Edinburgh
UK

Michael P. Bohrer
New Jersey Center for Biomaterials
Rutgers, The State University of New Jersey
Piscataway, NJ 08854

Roberto Bolli
Institute of Molecular Cardiology
University of Louisville
Louisville, KY 40292

Lawrence J. Bonassar
Department of Biomedical Engineering
Sibley School of Mechanical and Aerospace Engineering
Cornell University
Ithaca, NY 14853

Jeffrey T. Borenstein
Biomedical Engineering Center
Charles Stark Draper Laboratory
Cambridge, MA, 02139

Michael E. Boulton
AMD Center
Department of Ophthalmology & Visual Sciences
The University of Texas Medical Branch
Galveston, TX 77555-1106

Amy D. Bradshaw
Gazes Cardiac Research Institute
Medical University of South Carolina
Charleston, SC 29425

Christopher Breuer
Department of Pediatric Surgery
Yale University School of Medicine
New Haven, CT 06510

Luke Brewster
Department of Surgery
Loyola University Medical Center
Maywood, IL 60153

Eric M. Brey
Department of Biomedical Engineering
Illinois Institute of Technology
Chicago, IL 60616
and
Hines VA Hospital
Hines, IL 60141

Mairi Brittan
Institute of Cell & Molecular Science
Queen Mary's University of London
E1 2AT London
UK

T. Brown
Department of Orthopedics
University of Iowa College of Medicine,
Iowa City, IA 52242

Scott P. Bruder
Johnson & Johnson Regenerative Therapeutics
Raynham, MA 02767

Joseph A. Buckwalter
Department of Orthopedics
University of Iowa College of Medicine
Iowa City, IA 52242

Christopher Cannizzaro
Harvard-M.I.T. Division for Health Sciences and Technology
Massachusetts Institute of Technology
Cambridge, MA 02139

Yilin Cao
Shanghai Ninth People's Hospital
Shanghai Jiao Tong University, School of Medicine
200011 Shanghai
P.R. China

Lamont Cathey
Department of General Surgery
Carolinas Medical Center
Charlotte, NC 28232

Thomas M. S. Chang
Department of Physiology
McGill University
Montréal, PQ, H3G 1Y6
Canada

Yunchao Chang
Division of Molecular Oncology
The Scripps Research Institute
La Jolla, CA 92037

Robert G. Chapman
National Research Council
Institute for Nutrisciences and Health
Charlottetown, PE, C1A 4P3
Canada

Alice A. Chen
Harvard-M.I.T. Division of Health Sciences and Technology
Massachusetts Institute of Technology
Cambridge, MA 02139

Faye H. Chen
Cartilage Biology and Orthopaedics Branch
National Institute of Arthritis, and Musculoskeletal and
Skin Diseases
National Institutes of Health
Bethesda, MD 20892-8022

Una Chen
Stem Cell Therapy Program
Medical Microbiology, AG Chen
University of Giessen
D-35394 Giessen
Germany

Richard A.F. Clark
Departments of Biomedical Engineering, Dermatology and Medicine
Health Sciences Center
State University of New York
Stony Brook, NY 11794-8165

Clark K. Colton
Department of Chemical Engineering
Massachusetts Institute of Technology
Cambridge, MA 02139

George Cotsarelis
Department of Dermatology
University of Pennsylvania School of Medicine
Philadelphia, PA 19104

Stephen C. Cowin
Department of Mechanical Engineering
The City College
New York, NY 10031

Ronald Crystal
Department of Genetic Medicine
Weill Medical College of Cornell University
New York, NY 10021

Gislin Dagnelie
Lions Vision Center
Johns Hopkins University School of Medicine
Baltimore, MD 21205-2020

Jeffrey M. Davidson
Department of Medical Pathology
Vanderbilt University
Nashville, TN 37235-1604
and
Research Service
VA Tennessee Valley Healthcare System
Nashville, TN 37212-2637

Thomas F. Deuel
Departments of Molecular and Experimental Medicine and Cell Biology
The Scripps Research Institute
La Jolla, CA 92037

Elizabeth Deweerd
Department of Ophthalmology
Novartis Institutes for Biomedical Research
Cambridge, MA 02143

Gregory R. Dressler
Department of Pathology
University of Michigan
Ann Arbor, MI 48109

George C. Engelmayr, Jr.
Harvard-M.I.T. Division of Health Sciences and Technology
Massachusetts Institute of Technology
Cambridge, MA 02139

Carol A. Erickson
Department of Molecular and Cellular Biology
University of California, Davis
Davis, CA 95616

Thomas Eschenhagen
Institute of Experimental and Clinical Pharmacology
University Medical Center Hamburg-Eppendorf
D-20246 Hamburg
Germany

Vincent Falanga
Boston University School of Medicine
Department of Dermatology and Skin Surgery
Roger Williams Medical Center
Boston, MA 02118

Katie Faria
Organogenesis Inc.
Canton, MA 02021

Denise L. Faustman
Immunobiology Laboratory
Massachusetts General Hospital
Harvard Medical School
Boston, MA 02129

Dario O. Fauza
Children's Hospital Boston
Harvard Medical School
Boston, MA 02115

Lino da Silva Ferreira
Department of Chemical Engineering
Massachusetts Institute of Technology
Cambridge, MA 02139
and
Center of Neurosciences and Cell Biology
University of Coimbra
3004-517 Coimbra
Portugal
and
Biocant Centro de Inovação em Biotecnologia
3060-197 Cantanhede
Portugal

Hanson K. Fong
Department of Materials Science and Engineering
College of Engineering
University of Washington
Seattle, WA 98195

Peter Fong
Department of Biomedical Engineering
Yale University
New Haven, CT 06510

Lisa E. Freed
Harvard-M.I.T. Division of Health Sciences and Technology
Massachusetts Institute of Technology
Cambridge, MA 02139

R.I. Freshney
Centre for Oncology and Applied Pharmacology
University of Glasgow
G12 8QQ Glasgow
UK

Mark E. Furth
Wake Forest Institute for Regenerative Medicine
Wake Forest University Health Sciences
Winston-Salem, NC 27101

Jeffrey Geesin
Johnson & Johnson Regenerative Therapeutics
Raynham, MA 02767

Sharon Gerecht
Department of Chemical and Biomolecular Engineering
The Johns Hopkins University
Baltimore, MD 21218

Lucie Germain
Laboratoire d'Organogénèse Expérimentale
Québec, Qc, G1S 4L8
Canada

Kaustabh Ghosh
Department of Biomedical Engineering
Health Sciences Center
State University of New York
Stony Brook, NY 11794-8165

William V. Giannobile
Michigan Center for Oral Health Research
University of Michigan School of Dentistry
Ann Arbor, MI 48106

Francine Goulet
Laboratoire d'Organogénèse Expérimentale
Québec, Qc, G1S 4L8
Canada

Maria B. Grant
Pharmacology & Therapeutics
University of Florida
Gainesville, FL 32610-0267

Warren Grayson
Department of Biomedical Engineering
Columbia University
New York, NY 10027

Howard P. Greisler
Department of Surgery and Department of Cell Biology,
Neurobiology and Anatomy
Loyola University Medical Center
Maywood, IL 60153
and
Hines VA Hospital
Hines, IL 60141

Farshid Guilak
Departments of Surgery, Biomedical Engineering, and
Mechanical Engineering & Materials Science
Duke University Medical Center
Durham, NC 27710

Craig Halberstadt
Department of General Surgery
Carolinas Medical Center
Cannon Research Center
Charlotte, NC 28232-2861

Brendan Harley
Department of Mechanical Engineering
Massachusetts Institute of Technology
Cambridge, MA 02139

Kiki B. Hellman
The Hellman Group, LLC
Clarksburg, MD 20871

Abdelkrim Hmadcha
Centro Andaluz de Biología Molecular y Medicina
Regenerativa (Cabimer)
C/Américo Vespucio, s/n
41092 Isla de la Cartuja, Seville
Spain

Steve J. Hodges
Department of Urology
Wake Forest University School of Medicine
Winston-Salem, NC 27157

Walter D. Holder
The Polyclinic
Seattle, WA 98122

Chantal E. Holy
Johnson & Johnson Regenerative Therapeutics
Raynham, MA 02767-0650

Toru Hosoda
Cardiovascular Research Institute
Department of Medicine
New York Medical College
Valhalla, NY 10595

Jeffrey A. Hubbell
Laboratory for Regenerative Medicine and Pharmacobiology
Institute of Bioengineering
Ecole Polytechnique Fédérale de Lausanne (EPFL)
CH-1015 Lausanne
Switzerland

H. David Humes
Department of Internal Medicine
Division of Nephrology
University of Michigan Medical School
Ann Arbor, MI 48109

Donald E. Ingber
Vascular Biology Program
Departments of Pathology & Surgery
Children's Hospital and Harvard Medical School
Boston, MA 02115

Ana Jaklenec
Department of Molecular Pharmacology and Biotechnology
Brown University
Providence, RI 02912

Xingyu Jiang
National Center for NanoScience and Technology
100080 Beijing
China

Hee-Sook Jun
Rosalind Franklin Comprehensive Diabetes Center
Chicago Medical School
North Chicago, IL 60064

Jan Kajstura
Cardiovascular Research Institute
Department of Medicine
New York Medical College
Valhalla, NY 10595

Ravi S. Kane
Department of Chemical and Biological Engineering
Rensselaer Polytechnic Institute
Troy, NY 12180

Jeffrey M. Karp
Department of Chemical Engineering
Massachusetts Institute of Technology
Cambridge MA 02139

John Kay
Isotis, Inc.
Irvine, CA 92618

Ali Khademhosseini
Harvard-M.I.T. Division of Health Sciences and Technology
Brigham and Women's Hospital
Harvard Medical School
Cambridge, MA 02139

Salman R. Khetani
Harvard-M.I.T. Division of Health Sciences and Technology
Massachusetts Institute of Technology
Cambridge, MA 02139

Joachim Kohn
Department of Chemistry and Chemical Biology
Rutgers, The State University of New Jerscy
Piscataway, NJ 08854

Shaun M. Kunisaki
Department of Surgery
Massachusetts General Hospital
Boston, MA 02114

Matthew D. Kwan
Stanford University School of Medicine
Department of Surgery
Stanford, CA 94305-5148

Themis R. Kyriakides
Department of Pathology
Yale University School of Medicine
New Haven, CT 06519

Eric Lagasse
McGowan Institute for Regenerative Medicine
Department of Pathology
University of Pittsburgh
Pittsburgh, PA 15219-3130

Robert Langer
Department of Chemical Engineering
Massachusetts Institute of Technology
Cambridge, MA 02142

Douglas A. Lauffenburger
Department of Chemical Engineering
Massachusetts Institute of Technology
Cambridge, MA 02139

Kuen Yong Lee
Department of Bioengineering
Hanyang University
133-791 Seoul
South Korea

Annarosa Leri
Cardiovascular Research Institute
Department of Medicine
New York Medical College
Valhalla, NY 10595

David W. Levine
Genzyme
Cambridge, MA 02142

Amy S. Lewis
Department of Chemical Engineering
Massachusetts Institute of Technology
Cambridge, MA 02139

Wan-Ju Li
Cartilage Biology and Orthopaedics Branch
National Institute of Arthritis, and Musculoskeletal and Skin Diseases
National Institutes of Health
Bethesda, MD 20892-8022

Wei Liu
Shanghai Ninth People's Hospital
Shanghai Jiao Tong University, School of Medicine
200011 Shanghai
P.R. China

Michael T. Longaker
Stanford University School of Medicine
Department of Surgery
Stanford, CA 94305-5148

Ying Luo
Department of Chemical Engineering
Massachusetts Institute of Technology
Cambridge, MA 02139-4307

Michael J. Lysaght
Department of Molecular Pharmacology and Biotechnology
Brown University
Providence, RI 02912

Nancy Ruth Manley
Department of Genetics
University of Georgia
Athens, GA 30602

Jonathan Mansbridge
Tecellact LLC
La Jolla, CA 92037

J.L. Marsh
Department of Orthopaedics
University of Iowa College of Medicine
Iowa City, IA 52242

David C. Martin
Macromolecular Science and Engineering Center
University of Michigan
Ann Arbor, MI 48109-2136

J.A. Martin
Department of Orthopaedics
University of Iowa College of Medicine
Iowa City, IA 52242

Manuela Martins-Green
Department of Cell Biology & Neuroscience
University of California at Riverside
Riverside, CA 92521

Koichi Masuda
Department of Orthopedic Surgery and Biochemistry
Rush Medical College
Chicago, IL 60612

Robert L. Mauck
Department of Orthopaedic Surgery
University of Pennsylvania School of Medicine
Philadelphia, PA 19104

John W. McDonald, III
International Center for Spinal Cord Injury
Kennedy Krieger Institute
Baltimore, MD 21205

Antonios G. Mikos
Department of Bioengineering
Rice University
Houston, TX 77251-1892

Josef M. Miller
Kresge Hearing Research Institute
Department of Otolaryngology
University of Michigan
Ann Arbor, MI 48109-0506

David J. Mooney
Division of Engineering and Applied Sciences
Harvard University
Boston, MA 02138

Malcolm A.S. Moore
Department of Cell Biology
Memorial Sloan-Kettering Cancer Center
New York, NY 10021

Matthew B. Murphy
Department of Bioengineering
Rice University
Houston, TX 77251-1892

Robert M. Nerem
Georgia Institute of Technology
Parker H. Petit Institute for Bioengineering & Bioscience
Atlanta, GA 30332-0363

William Nikovits, Jr.
Division of Oncology
Stanford University School of Medicine
Stanford, CA 94305

Craig Scott Nowell
MRC/JDRF Centre Development in Stem Cell Biology
Institute for Stem Cell Research
University of Edinburgh
EH9 3JQ Edinburgh
UK

Bojana Obradovic
Department of Chemical Engineering
Faculty of Technology and Metallurgy
University of Belgrade
11000 Belgrade
Serbia

Bjorn R. Olsen
Department of Developmental Biology
Harvard School of Dental Medicine
Boston, MA 02115

James M. Pachence
Veritas Medical Technologies, Inc.
Princeton, NJ 08540-5799

Hyoungshin Park
Harvard-M.I.T. Division of Health Sciences and Technology
Massachusetts Institute of Technology
Cambridge, MA 02139

Jason Park
Department of Biomedical Engineering
Yale University School of Medicine
New Haven, CT 06510

M. Petreaca
Department of Cell Biology & Neuroscience
University of California at Riverside
Riverside, CA 92521

Julia M. Polak
Department of Chemical Engineering, Tissue Engineering and
Regenerative Medicine
Imperial College
South Kensington Campus
SW7 2AZ London
UK

A. Robin Poole
Joint Diseases Laboratory
Shiners Hospital for Crippled Children
Montréal, Qc, H3G 1A6
Canada

Christopher S. Potten
EpiStem Ltd.
M13 9XX Manchester
UK

Ales Prokop
Department of Chemical Engineering
Vanderbilt University
Nashville, TN 37235-1604

Milica Radisic
Institute of Biomaterials and Biomedical Engineering
Department of Chemical Engineering and Applied Chemistry
University of Toronto
Toronto, ON, M5S 3E5
Canada

Yehoash Raphael
Kresge Hearing Research Institute
Department of Otolaryngology
University of Michigan
Ann Arbor, MI 48109-0648

A. Hari Reddi
Ellison Center for Tissue Regeneration
University of California, Davis
UC Davis Health System
Sacramento, CA 95817

Herrmann Reichenspurner
Department of Cardiovascular Surgery
University Medical Center Hamburg-Eppendorf
D-20246 Hamburg
Germany

Ellen Richie
MD Anderson Cancer Center
University of Texas
Smithville, TX 78957

Pamela G. Robey
NIH/NIDCR
Bethesda, MD 20817-4320

Marcello Rota
Cardiovascular Research Institute
Department of Medicine
New York Medical College
Valhalla, NY 10595

Jeffrey W. Ruberti
Department of Mechanical and Industrial Engineering
Northeastern University
Boston, MA 02115

Alan J. Russell
McGowan Institute for Regenerative Medicine
University of Pittsburgh
Pittsburgh, PA 15219

E. Helene Sage
Hope Heart Program
The Benaroya Research Institute at Virginia Mason
Seattle, WA 98101

Rajiv Saigal
Medical Engineering
Harvard-M.I.T. Division of Health Sciences and Technology
Massachusetts Institute of Technology
Cambridge, MA 02139

W. Mark Saltzman
Department of Biomedical Engineering
Yale University
New Haven, CT 06520-8267

Athanassios Sambanis
Georgia Institute of Technology
School of Chemical & Biomolecular Engineering
Atlanta, GA 30332-0100

Jochen Schacht
Kresge Hearing Research Institute
Department of Otolaryngology and Department of Biochemistry
University of Michigan
Ann Arbor, MI 48109-0506

Lori A. Setton
Departments of Biomedical Engineering and Surgery
Duke University
Durham, NC 27708-0281

Upma Sharma
Department of Bioengineering
Rice University
Houston, TX 77251-1892

Paul T. Sharpe
Department of Craniofacial Development
Dental Institute
Kings College London
Guy's Hospital, London Bridge
SE1 9RT London
UK

Jonathan M.W. Slack
Stem Cell Institute
Minneapolis, MN 55455

Anthony J. Smith
School of Dentistry
University of Birmingham
B4 6NN Birmingham
UK

Martha J. Somerman
School of Dentistry
University of Washington
Seattle, WA 98195

Lin Song
Cartilage Biology and Orthopedics Branch
National Institute of Arthritis, and Musculoskeletal and
Skin Diseases
National Institutes of Health
Bethesda, MD 20892-8022
and
Stryker Orthopaedics
Mahwah, NJ 07430

Bernat Soria
Centro Andaluz de Biología Molecular y Medicina
Regenerativa (Cabimer)
C/Américo Vespucio, s/n
41092 Isla de la Cartuja, Seville
Spain

Frank E. Stockdale
Stanford University School of Medicine
Stanford Cancer Center
Department of Medicine
Division of Oncology
Stanford, CA 94305-5826

Lorenz Studer
Developmental Biology Program
Memorial Sloan-Kettering Cancer Center
New York, NY 10021

Shuichi Takayama
Department of Biomedical Engineering
The University of Michigan
Ann Arbor, MI 48109-2099

Juan R. Tejedo
Centro Andaluz de Biología Molecular y Medicina
Regenerativa (Cabimer)
C/Américo Vespucio, s/n
41092 Isla de la Cartuja, Seville
Spain

Vickery Trinkaus-Randall
Department of Biochemisty
Department of Ophthalmology
Boston University
Boston, MA 02118

Alan Trounson
Monash Immunology and Stem Cell Laboratories
Australian Stem Cell Centre
Monash University
Clayton, Victoria 3800
Australia

Rocky S. Tuan
Cartilage Biology and Orthopaedics Branch National Institute of
Arthritis, and Musculoskeletal and Skin Diseases
National Institutes of Health
Bethesda, MD 20892-8022

Gregory H. Underhill
Harvard-M.I.T. Division of Health Sciences and Technology
Massachusetts Institute of Technology
Cambridge, MA 02139

Konrad Urbanek
Cardiovascular Research Institute
Department of Medicine
New York Medical College
Valhalla, NY 10595

Charles A. Vacanti
Harvard Medical School
Brigham and Women's Hospital
Boston, MA 02114

Joseph Vacanti
Harvard Medical School
Massachusetts General Hospital
Boston, MA 02114

F. Jerry Volenec
Johnson & Johnson Regenerative Therapeutics
Raynham, MA 02767

Gordana Vunjak-Novakovic
Department of Biomedical Engineering
Columbia University
New York, NY 10027

Lars U. Wahlberg
NsGene A/S
2750 Ballerup
Denmark

Derrick C. Wan
Stanford University School of Medicine
Department of Surgery
Stanford, CA 94305-5148

George M. Whitesides
Department of Chemistry and Chemical Biology
Harvard University
Cambridge, MA 02138

Jeffrey A. Whitsett
Division of Pulmonary Biology
Cincinnati Children's Hospital Medical Center
Cincinnati, OH 45229-3039

James W. Wilson
EpiStem Ltd.
M13 9XX Manchester
UK

Stefan Worgall
Department of Pediatrics
Weill Medical College of Cornell University
New York, NY 10021

Mark E.K. Wong
Department of Oral and Maxillofacial Surgery
University of Texas Health Science Center — Houston
Houston, TX 77030

Nicholas A. Wright
Institute of Cell & Molecular Science
Queen Mary's University of London
E1 2AT London
UK

Ioannis V. Yannas
Division of Biological Engineering and Mechanical Engineering
Massachusetts Institute of Technology
Cambridge, MA 02139

Ji-Won Yoon
Rosalind Franklin Comprehensive Diabetes Center
Chicago Medical School
North Chicago, IL 60064

Simon Young
Department of Bioengineering
Rice University
Houston, TX 77251-1892

Hai Zhang
Department of Restorative Dentistry
School of Dentistry
University of Washington
Seattle, WA 98195

Wenjie Zhang
Shanghai Ninth People's Hospital
Shanghai Jiao Tong University, School of Medicine
200011 Shanghai
P.R. China

Beth A. Zielinski
The Department of Molecular Pharmacology, Physiology and Biotechnology
Brown University
Providence, RI, 02912
and
Biotechnology Manufacturing Program
Biotechnology and Clinical Laboratory Science Programs
Department of Cell and Molecular Biology
University of Rhode Island
Feinstein College of Continuing Education
Providence, RI 02903

James D. Zieske
Schepen's Eye Research Institute
and
Department of Opthalmology
Harvard Medical School
Boston, MA 02114

Wolfram-Hubertus Zimmermann
Institute of Experimental and Clinical Pharmacology
University Medical Center Hamburg-Eppendorf
D-20246 Hamburg
Germany

Laurie Zoloth
Center for Bioethics, Science and Society
Northwestern University
Feinberg School of Medicine
Chicago, IL 60611

# FOREWORD

*Robert Langer*

Since the mid-1980s, tissue engineering has moved from a concept to a very significant field. Already we are at the point where numerous tissues, such as skin, cartilage, bone, liver, blood vessels, and others, are in the clinic or even approved by regulatory authorities. Many other tissues are being studied. In addition, the advent of human embryonic stem cells has brought forth new sources of cells that may be useful in a variety of areas of tissue engineering.

This third edition of *Principles of Tissue Engineering* examines a variety of important areas. In the introductory section, an important overview on the history of tissue engineering and the movement of engineered tissues into the clinic is examined. This is followed by an analysis of important areas in cell growth and differentiation, including aspects of molecular biology, extracellular matrix interactions, cell morphogenesis, and gene expression and differentiation. Next, *in vitro* and *in vivo* control of tissue and organ development is examined. Important aspects of tissue culture and bioreactor design are covered, as are aspects of cell behavior and control by growth factors and cell mechanics. Models for tissue engineering are also examined. This includes mathematical models that can be used to predict important phenomena in tissue engineering and related medical devices. The involvement of biomaterials in tissue engineering is also addressed. Important aspects of polymers, extracellular matrix, materials processing, novel polymers such as biodegradable polymers as well as micro- and nano-fabricated scaffolds and three-dimensional scaffolds are discussed. Tissue and cell transplantation, including methods of immunoisolation, immunomodulation, and even transplantation in the fetus, are analyzed.

As mentioned earlier, stem cells have become an important part of tissue engineering. As such, important coverage of embryonic stem cells, adult stem cells, and postnatal stem cells is included. Gene therapy is another important area, and both general aspects of gene therapy as well as intracellular delivery of genes and drugs to cells and tissues are discussed. Various important engineered tissues, including breast-tissue engineering, tissues of the cardiovascular systems, such as myocardium, blood vessels, and heart valves, endocrine organs, such as the pancreas and the thymus, are discussed, as are tissues of the gastrointestinal system, such as liver and the alimentary tract. Important aspects of the hematopoietic system are analyzed, as is the engineering of the kidney and genitourinary system.

Much attention is devoted to the muscular skeletal system, including bone and cartilage regeneration and tendon and ligament placement. The nervous system is also discussed, including brain implants and the spinal cord. This is followed by a discussion of the eye, where corneal replacement and vision enhancement systems are examined. Oral and dental applications are also discussed, as are the respiratory system and skin. The concluding sections of the book cover clinical experience in such areas as cartilage, bone, skin, and cardiovascular systems as well as the bladder. Finally, regulatory and ethical considerations are examined.

In sum, the 86 chapters of this third edition of *Principles of Tissue Engineering* examine the important advances in the burgeoning field of tissue engineering. This volume will be very useful for scientists, engineers, and clinicians engaging in this important new area of science and medicine.

# PREFACE

The third edition of *Principles of Tissue Engineering* attempts to incorporate the latest advances in the biology and design of tissues and organs and simultaneously to connect the basic sciences — including new discoveries in the field of stem cells — with the potential application of tissue engineering to diseases affecting specific organ systems. While the third edition furnishes a much-needed update of the rapid progress that has been achieved in the field since the turn of the century, we have retained those facts and sections that, while not new, will assist students and general readers in understanding this exciting area of biology.

The third edition of *Principles* is divided into 22 parts plus an introductory section and an Epilogue. The organization remains largely unchanged from previous editions, combining the prerequisites for a general understanding of tissue growth and development, the tools and theoretical information needed to design tissues and organs, and a presentation by the world's experts on what is currently known about each specific organ system. As in previous editions, we have striven to create a comprehensive book that, on one hand, strikes a balance among the diversity of subjects that are related to tissue engineering, including biology, chemistry, materials science, and engineering, while emphasizing those research areas likely to be of clinical value in the future.

No topic in the field of tissue engineering is left uncovered, including basic biology/mechanisms, biomaterials, gene therapy, regulation and ethics, and the application of tissue engineering to the cardiovascular, hematopoietic, musculoskeletal, nervous, and other organ systems. While we cannot describe all of the new and updated material of the third edition, we can say that we have expanded and given added emphasis to stem cells, including adult and embryonic stem cells, and progenitor populations that may soon lead to new tissue-engineering therapies for heart disease, diabetes, and a wide variety of other diseases that afflict humanity. This up-to-date coverage of stem cell biology and other emerging technologies is complemented by a series of new chapters on recent clinical experience in applying tissue engineering. The result is a comprehensive book that we believe will be useful to students and experts alike.

*Robert Lanza*
*Robert Langer*
*Joseph Vacanti*

# PREFACE TO THE SECOND EDITION

The first edition of this textbook, published in 1997, was rapidly recognized as the comprehensive textbook of tissue engineering. This edition is intended to serve as a comprehensive text for the student at the graduate level or the research scientist/physician with a special interest in tissue engineering. It should also function as a reference text for researchers in many disciplines. It is intended to cover the history of tissue engineering and the basic principles involved, as well as to provide a comprehensive summary of the advances in tissue engineering in recent years and the state of the art as it exists today.

Although many reviews had been written on the subject and a few textbooks had been published, none had been as comprehensive in its defining of the field, description of the scientific principles and interrelated disciplines involved, and discussion of its applications and potential influence on industry and the field of medicine in the future as the first edition.

When one learns that a more recent edition of a textbook has been published, one has to wonder if the base of knowledge in that particular discipline has grown sufficiently to justify writing a revised textbook. In the case of tissue engineering, it is particularly conspicuous that developments in the field since the printing of the first edition have been tremendous. Even experts in the field would not have been able to predict the explosion in knowledge associated with this development. The variety of new polymers and materials now employed in the generation of engineered tissue has grown exponentially, as evidenced by data associated with specialized applications. More is learned about cell/biomaterials interactions on an almost daily basis. Since the printing of the last edition, recent work has demonstrated a tremendous potential for the use of stem cells in tissue engineering. While some groups are working with fetal stem cells, others believe that each specialized tissue contains progenitor cells or stem cells that are already somewhat committed to develop into various specialized cells of fully differentiated tissue.

Parallel to these developments, there has been a tremendous "buy in" concerning the concepts of tissue engineering not only by private industry but also by practicing physicians in many disciplines. This growing interest has resulted in expansion of the scope of tissue engineering well beyond what could have been predicted five years ago and has helped specific applications in tissue engineering to advance to human trials.

The chapters presented in this text represent the results of the coordinated research efforts of several hundred scientific investigators internationally. The development of this text in a sense parallels the development of the field as a whole and is a true reflection of the scientific cooperation expressed as this field evolves.

*Robert Lanza*
*Robert Langer*
*Joseph Vacanti*

# PREFACE TO THE FIRST EDITION

Although individual papers on various aspects of tissue engineering abound, no previous work has satisfactorily integrated this new interdisciplinary subject area. *Principles of Tissue Engineering* combines in one volume the prerequisites for a general understanding of tissue growth and development, the tools and theoretical information needed to design tissues and organs, as well as a presentation of applications of tissue engineering to diseases affecting specific organ system. We have striven to create a comprehensive book that, on the one hand, strikes a balance among the diversity of subjects that are related to tissue engineering, including biology, chemistry, materials science, engineering, immunology, and transplantation among others, while, on the other hand, emphasizing those research areas that are likely to be of most value to medicine in the future.

The depth and breadth of opportunity that tissue engineering provides for medicine is extraordinary. In the United States alone, it has been estimated that nearly half-a-trillion dollars are spent each year to care for patients who suffer either tissue loss or end-stage organ failure. Over four million patients suffer from burns, pressure sores, and skin ulcers, over twelve million patients suffer from diabetes, and over two million patients suffer from defective or missing supportive structures such as long bones, cartilage, connective tissue, and intervertebral discs. Other potential applications of tissue engineering include the replacement of worn and poorly functioning tissues as exemplified by aged muscle or cornea; replacement of small caliber arteries, veins, coronary, and peripheral stents; replacement of the bladder, ureter, and fallopian tube; and restoration of cells to produce necessary enzymes, hormones, and other bioactive secretory products.

*Principles of Tissue Engineering* is intended not only as a text for biomedical engineering students and students in cell biology, biotechnology, and medical courses at advanced undergraduate and graduate levels, but also as a reference tool for research and clinical laboratories. The expertise required to generate this text far exceeded that of its editors. It represents the combined intellect of more than eighty scholars and clinicians whose pioneering work has been instrumental to ushering in this fascinating and important field. We believe that their knowledge and experience have added indispensable depth and authority to the material presented in this book and that in the presentation, they have succeeded in defining and capturing the sense of excitement, understanding, and anticipation that has followed from the emergence of this new field, tissue engineering.

*Robert Lanza*
*Robert Langer*
*William Chick*

# Introduction to Tissue Engineering

# Chapter One

# The History and Scope of Tissue Engineering

*Joseph Vacanti and Charles A. Vacanti*

## I. INTRODUCTION

The dream is as old as humankind. Injury, disease, and congenital malformation have always been part of the human experience. If only damaged bodies could be restored, life could go on for loved ones as though tragedy had not intervened. In recorded history, this possibility first was manifested through myth and magic, as in the Greek legend of Prometheus and eternal liver regeneration. Then legend produced miracles, as in the creation of Eve in Genesis or the miraculous transplantation of a limb by the saints Cosmos and Damien. With the introduction of the scientific method came new understanding of the natural world. The methodical unraveling of the secrets of biology was coupled with the scientific understanding of disease and trauma. Artificial or prosthetic materials for replacing limbs, teeth, and other tissues resulted in the partial restoration of lost function. Also, the concept of using one tissue as a replacement for another was developed. In the 16th century, Tagliacozzi of Bologna, Italy, reported in his work *Decusorum Chirurgia per Insitionem* a description of a nose replacement that he constructed from a forearm flap. With the 19th-century scientific understanding of the germ theory of disease and the introduction of sterile technique, modern surgery emerged. The advent of anesthesia by the mid-19th century enabled the rapid evolution of many surgical techniques. With patients anesthetized, innovative and courageous surgeons could save lives by examining and treating internal areas of the body: the thorax, the abdomen, the brain, and the heart. Initially the surgical techniques were primarily extirpative, for example, removal of tumors, bypass of the bowel in the case of intestinal obstruction, and repair of life-threatening injuries. Maintenance of life without regard to the crippling effects of tissue loss or the psychosocial impact of disfigurement, however, was not an acceptable end goal. Techniques that resulted in the restoration of function through structural replacement became integral to the advancement of human therapy.

Now whole fields of reconstructive surgery have emerged to improve the quality of life by replacing missing function through rebuilding body structures. In our current era, modern techniques of transplanting tissue and organs from one individual into another have been revolutionary and lifesaving. The molecular and cellular events of the immune response have been elucidated sufficiently to suppress the response in the clinical setting of transplantation and to produce prolonged graft survival and function in patients. In a sense, transplantation can be viewed as the most extreme form of reconstructive surgery, transferring tissue from one individual into another.

As with any successful undertaking, new problems have emerged. Techniques using implantable foreign body materials have produced dislodgment, infection at the foreign body/tissue interface, fracture, and migration over time. Techniques moving tissue from one position to another have produced biologic changes because of the abnormal interaction of the tissue at its new location. For example, diverting urine into the colon can produce fatal colon cancers 20–30 years later. Making esophageal tubes from the skin can result in skin tumors 30 years later. Using intestine for urinary tract replacement can result in severe scarring and obstruction over time.

Transplantation from one individual into another, although very successful, has severe constraints. The major problem is accessing enough tissue and organs for all of the patients who need them. Currently, 92,587 people are on transplant waiting lists in the United States, and many will die waiting for available organs. Also, problems with the immune system produce chronic rejection and destruction over time. Creating an imbalance of immune surveillance from immunosuppression can cause new tumor formation. The constraints have produced a need for new solutions to provide needed tissue.

It is within this context that the field of tissue engineering has emerged. In essence, new and functional living tissue is fabricated using living cells, which are usually associated, in one way or another, with a matrix or scaffolding to guide tissue development. New sources of cells, including many types of stem cells, have been identified in the past several years, igniting new interest in the field. In fact, the emergence of stem cell biology has led to a new term, *regenerative medicine*. Scaffolds can be natural, man-made, or a composite of both. Living cells can migrate into the implant after implantation or can be associated with the matrix in cell culture before implantation. Such cells can be isolated as fully differentiated cells of the tissue they are hoped to recreate, or they can be manipulated to produce the desired function when isolated from other tissues or stem cell sources. Conceptually, the application of this new discipline to human health care can be thought of as a refinement of previously defined principles of medicine. The physician has historically treated certain disease processes by supporting nutrition, minimizing hostile factors, and optimizing the environment so that the body can heal itself. In the field of tissue engineering, the same thing is accomplished on a cellular level. The harmful tissue is eliminated; the cells necessary for repair are then introduced in a configuration optimizing survival of the cells in an environment that will permit the body to heal itself. Tissue engineering offers an advantage over cell transplantation alone in that organized three-dimensional functional tissue is designed and developed. This chapter summarizes some of the challenges that must be resolved before tissue engineering can become part of the therapeutic armamentarium of physicians and surgeons. Broadly speaking, the challenges are scientific and social.

## II. SCIENTIFIC CHALLENGES

As a field, tissue engineering has been defined only since the mid-1980s. As in any new undertaking, its roots are firmly implanted in what went before. Any discussion of when the field began is inherently fuzzy. Much still needs to be learned and developed to provide a firm scientific basis for therapeutic application. To date, much of the progress in this field has been related to the development of model systems, which have suggested a variety of approaches. Also, certain principles of cell biology and tissue development have been delineated. The field can draw heavily on the explosion of new knowledge from several interrelated, well-established disciplines and in turn may promote the coalescence of relatively new, related fields to achieve their potential. The rate of new understanding of complex living systems has been explosive since the 1970s. Tissue engineering can draw on the knowledge gained in the fields of cell and stem cell biology, biochemistry, and molecular biology and apply it to the engineering of new tissues. Likewise, advances in materials science, chemical engineering, and bioengineering allow the rational application of engineering principles to living systems. Yet another branch of related knowledge is the area of human therapy as applied by surgeons and physicians. In addition, the fields of genetic engineering, cloning, and stem cell biology may ultimately develop hand in hand with the field of tissue engineering in the treatment of human disease, each discipline depending on developments in the others.

We are in the midst of a biologic renaissance. Interactions of the various scientific disciplines can elucidate not only the potential direction of each field of study, but also the right questions to address. The scientific challenge in tissue engineering lies both in understanding cells and their mass transfer requirements and the fabrication of materials to provide scaffolding and templates.

## III. CELLS

If we postulate that living cells are required to fabricate new tissue substitutes, much needs to be learned with regard to their behavior in two normal circumstances: normal development in morphogenesis and normal wound healing. In both of these circumstances, cells create or recreate functional structures using preprogrammed information and signaling. Some approaches to tissue engineering rely on guided regeneration of tissue using materials that serve as templates for ingrowth of host cells and tissue. Other approaches rely on cells that have been implanted as part of an engineered device. As we gain understanding of normal developmental and wound-healing gene programs and cell behavior, we can use them to our advantage in the rational design of living tissues.

Acquiring cells for creation of body structures is a major challenge, the solution of which continues to evolve. The ultimate goal in this regard — the large-scale fabrication of structures — may be to create large cell banks composed of universal cells that would be immunologically transparent to an individual. These universal cells could be differentiated cell types that could be accepted by an individual or could be stem cell reservoirs, which could respond to signals to differentiate into differing lineages for specific structural applications. Much is already known about stem cells and cell lineages in the bone marrow and blood. Studies suggest that progenitor cells for many differentiated tissues exist within the marrow and blood and may very well be ubiquitous. Our knowledge of the existence and behavior of such cells in various mesenchymal tissues (muscle, bone, and cartilage), endodermally derived tissues (intestine and liver), or ectodermally derived tissues (nerves, pancreas, and skin) expands on a daily basis. A new area of stem cell biology involving embryonic stem cells holds promise for tissue engineering. The calling to the scientific community is to understand the principles of stem and progenitor cell biology and then to apply that understanding to tissue engineering. The development of immunologically inert universal cells may come from advances in genetic manipulation as well as stem cell biology.

As intermediate steps, tissue can be harvested as allograft, autograft, or xenograft. The tissues can then be dissociated and placed into cell culture, where proliferation of cells can be initiated. After expansion to the appropriate cell number, the cells can then be transferred to templates, where further remodeling can occur. Which of these strategies are practical and possibly applicable in humans remains to be explored.

Large masses of cells for tissue engineering need to be kept alive, not only *in vitro* but also *in vivo*. The design of systems to accomplish this, including *in vitro* flow bioreactors and *in vivo* strategies for maintenance of cell mass, presents an enormous challenge, in which significant advances have been made. The fundamental biophysical constraint of mass transfer of living tissue needs to be understood and dealt with on an individual basis as we move toward human application.

## IV. MATERIALS

There are so many potential applications to tissue engineering that the overall scale of the undertaking is enormous. The field is ripe for expansion and requires training of a generation of materials scientists and chemical engineers.

The optimal chemical and physical configurations of new biomaterials as they interact with living cells to produce tissue-engineered constructs are under study by many research groups. These biomaterials can be permanent or biodegradable. They can be naturally occurring materials, synthetic materials, or hybrid materials. They need to be developed to be compatible with living systems or with living cells *in vitro* and *in vivo*. Their interface with the cells and the implant site must be clearly understood so that the interface can be optimized. Their design characteristics are major challenges for the field and should be considered at a molecular chemical level. Systems can be closed, semipermeable, or open. Each design should factor into the specific replacement therapy considered. Design of biomaterials can also incorporate the biologic signaling that the materials may offer. Examples include release of growth and differentiation factors, design of specific receptors and anchorage sites, and three-dimensional site specificity using computer-assisted design and manufacture techniques. New nanotechnologies have been incorporated to design systems of extreme precision. Combining computational models with nanofabrication can produce microfluidic circulations to nourish and oxygenate new tissues.

## V. GENERAL SCIENTIFIC ISSUES

As new scientific knowledge is gained, many conceptual issues need to be addressed. Related to mass transfer is the fundamental problem associated with nourishing tissue of large mass as opposed to tissue with a relatively high ratio of surface area to mass. Also, functional tissue equivalents necessitate the creation of composites containing different cell types. For example, all tubes in the body are laminated tubes composed of a vascularized mesodermal element, such as smooth muscle, cartilage, or fibrous tissue. The inner lining of the tube, however, is specific to the organ system. Urinary tubes have a stratified transitional epithelium. The trachea has a pseudostratified columnar epithelium. The esophagus has an epithelium that changes along the gradient from mouth to stomach. The intestine has an enormous, convoluted surface area of columnar epithelial cells that migrate from a crypt to the tip of the villus. The colonic epithelium is, again, different for the purposes of water absorption and storage.

Even the well-developed manufacture of tissue-engineered skin used only the cellular elements of the dermis for a long period of time. Attention is now focusing on creating new skin consisting of both the dermis and its associated fibroblasts as well as the epithelial layer, consisting of keratinocytes. Obviously, this is a significant advance. But for truly "normal" skin to be engineered, all of the cellular elements should be contained so that the specialized appendages can be generated as well. These "simple" composites will indeed prove to be quite complex and require intricate designs. Thicker structures with relatively high ratios of surface area to mass, such as liver, kidney, heart, breast, and the central nervous system, will offer engineering challenges.

Currently, studies for developing and designing materials in three-dimensional space are being developed utilizing both naturally occurring and synthetic molecules. The applications of computer-assisted design and manufacture techniques to the design of these matrices are critically important. Transformation of digital information obtained from magnetic resonance scanning or computerized tomography scanning can then be developed to provide appropriate templates. Some tissues can be designed as universal tissues that will be suitable for any individual, or they may be custom-developed tissues specific to one patient. An important area for future study is the entire field of neural regeneration, neural ingrowth, and neural function toward end organ tissues such as skeletal or smooth muscle. Putting aside the complex architectural structure of these tissues, the cells contained in them have a very high metabolic requirement. As such, it is exceedingly difficult to isolate a large number of viable cells. An alternate approach may be the use of less mature progenitor cells, or stem cells, which not only would have a higher rate of survival as a result of their lower metabolic demand but also would be more able to survive the insult and hypoxic environment of transplantation. As stem cells develop and require more oxygen, their differentiation may stimulate the development of a vascular complex to nourish them. The understanding of and solutions to these problems are fundamentally important to the success of any replacement tissue that needs ongoing neural interaction for maintenance and function.

It has been shown that some tissues can be driven to completion *in vitro* in bioreactors. However, optimal incubation times will vary from tissue to tissue. Even so, the new tissue will require an intact blood supply at the time of implantation for successful engraftment and function.

Finally, all of these characteristics need to be understood in the fourth dimension, time. If tissues are implanted in a growing individual, will the tissues grow at the same rate? Will cells taken from an older individual perform as young cells in their new "optimal" environment? How will the biochemical characteristics change over time after implantation? Can the strength of structural support tissues such as bone, cartilage, and ligaments be improved in a bioreactor in which force vectors can be applied? When is the optimal timing of this transformation? When does tissue strength take over the biochemical characteristics as the material degrades?

## VI. SOCIAL CHALLENGES

If tissue engineering is to play an important role in human therapy, in addition to scientific issues, fundamental issues that are economic, social, and ethical in nature will arise. Something as simple as a new vocabulary will need to be developed and uniformly applied. A universal problem is funding. Can philanthropic dollars be accessed for the purpose of potential new human therapies? Will industry recognize the potential for commercialization and invest heavily? If this occurs, will the focus be changed from that of a purely academic endeavor? What role will governmental agencies play as the field develops? How will the field be regulated to ensure its safety and efficacy prior to human application? Is the new tissue to be considered transplanted tissue and, therefore, not be subject to regulation, or is it a pharmaceutical that must be subjected to the closest scrutiny by regulatory agencies? If lifesaving, should the track be accelerated toward human trials?

There are legal ramifications of this emerging technology as new knowledge is gained. What becomes proprietary through patents? Who owns the cells that will be sourced to provide the living part of the tissue fabrication?

In summary, one can see from this brief overview that the challenges in the field of tissue engineering remain significant. All can be encouraged by the progress that has been made in the past few years, but much discovery lies ahead. Ultimate success will rely on the dedication, creativity, and enthusiasm of those who have chosen to work in this exciting but still unproved field. Quoting from the epilogue of the previous edition: "At any given instant in time, humanity has never known so much about the physical world and will never again know so little."

## VII. REFERENCES

Langer, R., and Vacanti, J. P. (1993). Tissue engineering. *Science* **260**, 920–926.

Lanza, R. P., Langer, R., and Vacanti, J. P. (2000). "Principles of Tissue Engineering," 2nd ed., p. 929.

Nerem, R. M. (2006). Tissue engineering: the hope, the hype and the future. *Tissue Eng.* **12**, 1143–1150.

Vacanti, C. A. (2006). History of tissue engineering and a glimpse into its future. *Tissue Eng.* **12**, 1137–1142.

Vacanti, J. P., and Langer, R. (1999). Tissue engineering: the design and fabrication of living replacement devices for surgical reconstruction and transplantation. *Lancet* **354**, SI32–34.

# Chapter TWO

# The Challenge of Imitating Nature

*Robert M. Nerem*

## I. INTRODUCTION

Tissue engineering, through the imitation of nature, has the potential to confront the transplantation crisis caused by the shortage of donor tissues and organs and also to address other important, but yet unmet, patient needs. If we are to be successful in this, a number of challenges need to be faced. In the area of cell technology, these include cell sourcing, the manipulation of cell function, and the effective use of stem cell technology. Next are those issues that are part of what is called here *construct technology*. These include the design and engineering of tissuelike constructs and/or delivery vehicles and the manufacturing technology required to provide off-the-shelf availability to the clinician. Finally, there are those issues associated with the integration of cells or a construct into the living system, where the most critical issue may be the engineering of immune acceptance. Only if we can meet the challenges presented by these issues and only if we can ultimately address the tissue engineering of the most vital of organs will it be possible to achieve success in confronting the crisis in transplantation.

An underlying premise of this is that the utilization of the natural biology of the system will allow for greater success in developing therapeutic strategies aimed at the replacement, maintenance, and/or repair of tissue and organ function. Another way of saying this is that, just maybe, the great creator, in whatever form one believes he or she exists, knows something that we mere mortals do not, and if we can only tap into a small part of this knowledge base, if we can only imitate nature in some small way, then we will be able to achieve greater success in our efforts to address patient needs in this area. It is this challenge of imitating nature that has been accepted by those who are providing leadership to this new area of technology called *tissue engineering* (Langer and Vacanti, 1993; Nerem and Sambanis, 1995). To imitate nature requires that we first understand the basic biology of the tissues and organs of interest, including developmental biology; with this we then can develop methods for the control of these biologic processes; and based on the ability to control, we finally can develop strategies either for the engineering of living tissue substitutes or for the fostering of tissue repair or regeneration.

The initial successes have been for the most part substitutes for skin, a relatively simple tissue, at least by comparison with most other targets of opportunity. In the long term, however, tissue engineering has the potential for creating vital organs, such as the kidney, the liver, and the pancreas. Some even believe it will be possible to tissue engineer an entire heart. In addressing the repair, replacement, and/or regeneration of such vital organs, tissue engineering has the potential literally to confront the transplantation crisis, i.e., the shortage of donor tissues and organs available for transplantation. It also has the potential

*Principles of Tissue Engineering, 3rd Edition*
*ed. by Lanza, Langer, and Vacanti*

to develop strategies for the regeneration of nerves, another important and unmet patient need.

Although research in what we now call tissue engineering started more than a quarter of a century ago, the term *tissue engineering* was not coined until 1987, when Professor Y. C. Fung, from the University of California, San Diego, suggested this name at a National Science Foundation meeting. This led to the first meeting called "tissue engineering," held in early 1988 at Lake Tahoe, California (Skalak and Fox, 1988). More recently the term *regenerative medicine* has come into use. For some this is a code word for stem cell technology, while for others regenerative medicine is the broader term, with tissue engineering representing only replacement, not repair or regeneration. Still others use the terms *tissue engineering* and *regenerative medicine* interchangeably. What is important is that it is a more biologic approach that has the potential to lead to new patient therapies and treatments, where in some cases none is currently available.

It should be noted that the concept of a more biologic approach dates back to 1938 (Carrel and Lindbergh, 1938). Since then there has been a large expansion in research efforts in this field and a considerable recognition of the enormous potential that exists. With this hope, there also has been a lot of hype; however, the future long term remains bright (Nerem, 2006). As the technology has become further developed, an industry has begun to emerge. This industry is still very much a fledgling one, with only a few companies possessing product income streams (Ahsan and Nerem, 2005). A study based on 2002 data documents a total of 89 companies active in the field, with $500 million annually in industrial research and development taking place (Lysaght and Hazlehurst, 2004). Although this study will soon be updated, based on the 2002 data, 80% of the new firms were in the stem cell area and 40% were located outside of the United States.

Tissue engineering is literally at the interface of the traditional medical implant industry and the biological revolution (Galletti *et al.*, 1995). By harnessing the advances of this revolution, we can create an entirely new generation of tissue and organ implants as well as strategies for repair and regeneration. Already we are seeing increased investments in this field by the large medical device companies. A part of this is a convergence of biologics and devices, which is recognized by the medical implant industry. It is from this that the short-term successes in tissue engineering will come; however, long term it is the potential for a literal revolution in medicine and in the medical device/implant industry that must be realized.

This revolution will only occur, however, if we successfully meet the challenge of imitating nature. Thus, in the remainder of this chapter the critical issues involved in this are addressed. This is done by first discussing those issues associated with cell technology, i.e., issues important in cell sourcing and in the achievement of the functional characteristics required of the cells to be employed. Next to be discussed are those issues associated with construct technology. These include the organization of cells into a three-dimensional architecture that functionally mimics tissue, the development of vehicles for the delivery of genes, cells, and proteins, and the technologies required to manufacture such products and provide them off the shelf to the clinician. Finally, issues involved in the integration of a living cell construct into, or the fostering of remodeling within, the living system is discussed. These range from the use of appropriate animal models to the issues of biocompatibility and immune acceptance. Success in tissue engineering will only be achieved if issues at these three different levels, i.e., cell technology, construct technology, and the technology for integration into the living system, can be addressed.

## II. CELL TECHNOLOGY

The starting point for any attempt to engineer a tissue or organ substitute is a consideration of the cells to be employed. Not only will one need to have a supply of sufficient quantity and one that can be ensured to be free of pathogens and contamination of any type whatsoever, but one will need to decide whether the source to be employed is to be autologous, allogeneic, or xenogeneic. As indicated in Table 2.1, each of these has both advantages and disadvantages; however, it should be noted that one important consideration for any product or treatment strategy is its off-the-shelf availability. This is obviously required for surgeries that must be carried out on short notice. However, even when the time for surgery is elective, it is only with off-the-shelf availability that the product and strategy will be used for the wide variety of patients who are in need and who are being treated throughout the entire health care system, including those in community hospitals.

With regard to the use of autologous cells, there are a number of potential sources. These include both differenti-

| Table 2.1. | Cell source |
|---|---|
| *Type* | *Comments* |
| Autologous | Patient's own cells; immune acceptable, but does not lend itself to off-the-shelf availability unless recruited from the host |
| Allogeneic | Cells from other human sources; lends itself to off-the-shelf availability, but may require engineering immune acceptance |
| Xenogeneic | From different species; not only requires engineering immune acceptance, but must be concerned with animal virus transmission |

ated cells and adult stem/progenitor cells. It is only, however, if we can recruit the host's own cells, e.g., to an acellular implant, that we can have off-the-shelf availability, and it is only by moving to off-the-shelf availability for the clinician that routine use becomes possible.

The skin substitutes developed by Organogenesis (Canton, MA) and Advanced Tissue Sciences (La Jolla, CA) represented the first living-cell, tissue-engineered products, and these in fact use allogeneic cells. The Organogenesis product, Apligraf™, is a bilayer model of skin involving fibroblasts and keratinocytes that are obtained from donated human foreskin (Parenteau, 1999). Apligraf™ is approved by the Food and Drug Administration (FDA); however, the first tissue-engineered products approved by the FDA were acellular. These included Integra™, based on a polymeric template approach (Yannas *et al.*, 1982), and the Advanced Tissue Sciences product, TransCyte™. Approved initially for third-degree burns, TransCyte™ is made by seeding dermal fibroblasts in a polymeric scaffold; however, once cryopreserved it becomes a nonliving wound covering. Advanced Tissue Sciences also has a living-cell product, called Dermagraft™. It is a dermis model, also with dermal fibroblasts obtained from donated human foreskin (Naughton, 1999). Even though the cells employed by both Organogenesis and Advanced Tissue Sciences are allogeneic, immune acceptance did not have to be engineered because both the fibroblast and the keratinocyte do not constitutively express major histocompatibility complex (MHC) II antigens.

The next generation of tissue-engineered products will involve other cell types, and the immune acceptance of allogeneic cells will be a critical issue in many cases. As an example, consider a blood vessel substitute that employs both endothelial cells and smooth muscle cells. Although there is some unpublished data that suggest allogeneic smooth muscle cells may be immune acceptable, allogeneic endothelial cells certainly would not be. Thus, for the latter, one either uses autologous cells or else engineers the immune acceptance of allogeneic cells, as is discussed in a later section. Undoubtedly the first human clinical trials will be done using autologous endothelial cells; however, it appears that the use of such cells would severely limit the availability of a blood vessel substitute, unless the host's own endothelial cells are recruited.

Once one has selected the cell type(s) to be employed, then the next issue relates to the manipulation of the functional characteristics of a cell so as to achieve the behavior desired. This can be done either by (1) manipulating a cell's microenvironment, e.g., its matrix, the mechanical stresses to which it is exposed, or its biochemical environment, or by (2) manipulating a cell's genetic program. With regard to the latter, the manipulation of a cell's genetic program could be used as an ally to tissue engineering in a variety of ways. A partial list of possibilities includes the alteration of matrix synthesis; inhibition of the immune response; enhance-

ment of nonthrombogenicity, e.g., through increased synthesis of antithrombotic agents; engineering the secretion of specific biologically active molecules, e.g., a specific insulin secretion rate in response to a specific glucose concentration; and the alteration of cell proliferation.

Much of the foregoing is in the context of creating a cell-seeded construct that can be implanted as a tissue or organ substitute; however, the fostering of the repair or remodeling of tissue also represents tissue engineering as defined here. Here a critical issue is how to deliver the necessary biologic cues in a spatially and temporally controlled fashion so as to achieve a "healing" environment. In the repair and/or regeneration of tissue, the use of genetic engineering might take a form that is more what we would call *gene therapy*. An example of this would be the introduction of growth factors to foster the repair of bone defects. In using a gene therapy approach to tissue engineering it should be recognized that in many cases only a transient expression will be required. Because of this, the use of gene therapy as a strategy in tissue engineering may become viable prior to its employment in treating genetically related diseases.

Returning to the issue of cell selection, there is considerable interest in the use of stem cells as a primary source for cell-based therapies, ones ranging from replacement to repair and/or regeneration. This interest includes both adult stem cells and progenitor cells as well as embryonic stem cells (Ahsan and Nerem, in press; Vats *et al.*, 2005). With regard to the latter, the excitement about stem cells reached a new height in the late 1990s with articles reporting the isolation of the first lines of human embryonic stem cells (Thomson *et al.*, 1998; Solter and Gearhart, 1999; Vogel, 1999). Since then considerable progress has been made; however, the hype continues to outpace the progress. This reached an unfortunate crescendo in the latter part of 2005 with the revelation that the major advances reported by the Korean scientist Woo Suk Hwang were based on the fabrication of results (Normile and Vogel, 2005; Normile *et al.*, 2005, 2006). This was compounded by ethical issues and by the inclusion of Dr. Gerald Schatten from the University of Pittsburgh as a senior author (Guterman, 2006). Korea must be credited with launching a full investigation that led to Dr. Hwang's losing his position. The University of Pittsburgh also conducted an investigation and found Dr. Schatten guilty of "research misbehavior," a term not fully understood by the scientific community (Holden, 2006). The unfortunate thing is that this all happened at a time of considerable ethical and political controversy surrounding human embryonic stem cell research. From this we must all learn (Cho *et al.*, 2006), and, in spite of this setback in the public arena, research in the human embryonic stem cell area continues to hold considerable promise for the future.

There is in fact a variety of different stem cells, and several comprehensive reviews of a general nature have recently appeared (Vats *et al.*, 2005; Ahsan and Nerem, in press). It is the adult stem cells and progenitor cells that are being and will be used first clinically; however, long term there is considerable interest in embryonic stem cells. These cells are pluripotent, i.e., capable of differentiating into many cell types, even totipotent, i.e., capable of developing into all cell types. Although we are quite a long way from being able to use embryonic stem cells, a number of companies are working with stem cells in the context of tissue engineering and regenerative medicine. It needs to be recognized, however, that immunogenicity issues may be associated with the use of embryonic stem cells. Furthermore, different embryonic stem cell lines, even when in a totally undifferentiated state, can be significantly different. This is illustrated by the results of Rao *et al.* (2004) in a comparison of the transcriptional profile of two different embryonic stem cell lines. This difference should not be considered surprising, since the lines were derived from different embryos and undoubtedly cultured under different conditions.

To take full advantage of stem cell technology, however, it will be necessary to understand more fully how a stem cell differentiates into a tissue-specific cell. This requires knowledge not just about the molecular pathways of differentiation, but, even more importantly, about the identification of the combination of signals leading to a stem cell's becoming a specific type of differentiated tissue cell. As an example, with the recognized differences between large-vessel endothelial cells and valvular endothelial cells (Butcher *et al.*, 2004), what are the signals that will drive the differentiation toward one type of endothelial cell versus the other? Only with this type of knowledge will we be able to realize the full potential of stem cells. In addition, however, we will need to develop the technologies necessary to expand a cell population to the number necessary for clinical application, to do this in a controlled, reproducible manner, and to deliver cells at the right place and at the time required.

## III. CONSTRUCT TECHNOLOGY

With the selection of a source of cells, the next challenge in imitating nature is to develop an organized three-dimensional architecture (with functional characteristics such that a specific tissue is mimicked) and/or a delivery vehicle for the cells. In this it is important to recognize the importance of a cell's microenvironment in determining its function. *In vivo* a cell's function is orchestrated by a symphony of signals. This symphony includes soluble molecules, the mechanical environment, i.e., physical forces, to which the cell is exposed, and the extracellular matrix. These are all part of the symphony. And if we want the end result to replicate the characteristics of native tissue, attention must be given to each of these components of a cell's microenvironment.

The design and engineering of a tissuelike substitute are challenges in their own right. If the approach is to seed cells into a scaffold, then a basic issue is the type of scaffold that will allow the cells to make their own matrix. There are, of course, many possible approaches. One of these is a cell-seeded polymeric scaffold, an approach pioneered by Langer and his collaborators (Langer and Vacanti, 1993; Cima *et al.*, 1991). This is the technology that was used by Advanced Tissue Sciences, and many consider this the classic tissue-engineering approach. There are other approaches, however, with one of these being a cell-seeded collagen gel. This approach was pioneered by Bell in the late 1970s and early 1980s (Bell *et al.*, 1979; Weinberg and Bell, 1986), and this is used by Organogenesis in their skin substitute, Apligraf™.

A rather intriguing approach is that of Auger and his group in Quebec, Canada (Auger *et al.*, 1995; Heureux *et al.*, 1998). Auger refers to this as *cell self-assembly*, and it involves a layer of cells secreting their own matrix, which over a period of time becomes a sheet. Originally developed as part of the research on skin substitutes by Auger's group, it has been extended to other applications. For example, the blood vessel substitute developed in Quebec involves rolling up one of these cell self-assembled sheets into a tube. One can in fact make tubes of multiple layers so as to mimic the architecture of a normal blood vessel.

An equally intriguing approach is that pioneered by the Campbells in Australia and their collaborators (Chue *et al.*, 2004). In this they literally use the peritoneal cavity as an *in vivo* bioreactor to grow a blood vessel substitute. The concept is that a free-floating body in the peritoneal cavity initiates an inflammatory response and becomes encapsulated with cells. This is an autologous-cell approach, and it is also one where the cells make their own matrix.

Any discussion of different approaches to the creation of a three-dimensional, functional tissue equivalent would be remiss if acellular approaches were not included. Although in tissue engineering the end result should include functional cells, there are those who are employing a strategy whereby the implant is without cells, i.e., acellular, and the cells are then recruited from the recipient or host. A number of laboratories and companies are developing this approach. Examples include the products Integra™ and TransCyte™, already noted, and the development of SIS, i.e., small intestine submucosa (Badylak *et al.*, 1999; Lindberg and Badylak, 2001). One result of this approach, in effect, is to bypass the cell-sourcing issue and replace this with the issue of cell recruitment, i.e., the recruiting of cells from the host in order to populate the construct. Because these are the patient's own cells, there is no need for any engineering of immune acceptance.

Whatever is done, an objective in imitating nature must be to create a healing environment, one that will foster remodeling and ultimately repair. To do this requires delivering the appropriate, necessary cues in a controlled spatial and temporal fashion. This is needed whether the goal is replacement or repair or regeneration. Whatever the approach, the engineering of an architecture and of functional characteristics that allow one to mimic a specific tissue is critical to achieving any success and to meeting the challenge of imitating nature. In fact, because of the interrelationship of structure and function in cells and tissues, it would be unlikely to have the appropriate functional characteristics without the appropriate three-dimensional architecture. Thus, many of the chapters in this book describe in some detail the approach being taken in the design and engineering of constructs for specific tissues and organs, and any further discussion of this is left to those chapters.

The challenge of imitating nature, however, does not stop with the design and engineering of a specific tissuelike substitute or a delivery vehicle. This is because the patient need that exists cannot be met by making one construct at a time on a benchtop in some research laboratory. Accepting the challenge of imitating nature must include the development of cost-effective manufacturing processes. These must allow for a scale-up from making one at a time to a production quantity of 100 or 1000 per week. Anything significantly less would not be cost effective; and if a product cannot be manufactured in large quantities and cost effectively, then it will not be widely available for routine use.

Much of the research on manufacturing technology has focused on bioreactor technology. A bioreactor simply represents a controlled environment — both chemically and mechanically — in which a tissuelike construct can be grown (Freed *et al.*, 1993; Neitzel *et al.*, 1998; Saini and Wick, 2003). The design of a bioreactor involves a number of critical issues. The list starts with the configuration of the bioreactor, its mass transport characteristics, and its scaleability. Then, if it is to be used in a manufacturing process, it is desirable to minimize the number of aseptic operations while maximizing automation. Reliability and reproducibility obviously will be critical, and it needs to be user friendly.

Although it is generally recognized that a construct, once implanted in the living system, will undergo remodeling, it is equally true that the environment of a bioreactor can be tailored to induce the *in vitro* remodeling of a construct so as to enhance characteristics critical to the success to be achieved when it is implanted (Seliktar *et al.*, 1998). Thus, the manufacturing process can be used to influence directly the final product and is part of the overall process leading to the imitation of nature. An important issue in developing a substitute for replacement, however, is how much of the maturation of a substitute is done *in vitro* in a bioreactor as compared to what is done *in vivo* through the remodeling that takes place within the body itself, i.e., in the body's own bioreactor environment. As pointed out by Dr. Frederick Schoen (private communication), in this one needs to recognize that the rate at which remodeling *in vivo* takes place will be extremely different from individual to individual. It is equally true that the extent of remodeling also will be different. Thus, the degree of maturation

that occurs *in vivo* will be highly variable, depending on the host response.

Once a product is manufactured, a critical issue will be how it is delivered and made available to the clinician. The Organogenesis product, Apligraf™, is delivered fresh and originally had a 5-day shelf life at room temperature (Parenteau, 1999), although recently this has been extended. On the other hand, Dermagraft™, the skin substitute developed by Advanced Tissue Sciences, is cryopreserved and shipped and stored at −70°C (Naughton, 1999). This provides for a much more extended shelf life but introduces other issues that one must address. Ultimately, the clinician will want off-the-shelf availability, and one way or another this will need to be provided if a tissue-engineered product or strategy is to have wide use. Although cryobiology is a relatively old field and most cell types can be cryopreserved, there is much that still needs to be learned if we are successfully to cryopreserve three-dimensional tissue-engineered products.

## IV. INTEGRATION INTO THE LIVING SYSTEM

The final challenge to imitating nature is presented by moving a tissue-engineering concept into the living system. Here one starts with animal experiments, and there is a lack of good animal models for use in the evaluation of a tissue-engineered implant or strategy. This is despite the fact that a variety of animal models have been developed for the study of different diseases. Unfortunately, these models are still somewhat unproved, at least in many cases, when it comes to their use in evaluating the success of a tissue-engineering concept.

In addition, there is a significant need for the development of methods to evaluate quantitatively the performance of an implant, and a number of concepts are being advanced (Guldberg *et al.*, 2003; Stabler *et al.*, 2005). This is not only the case for animal studies, but is equally true for human clinical trials. With regard to the latter, it may not be enough to show efficacy and long-term patency; it may also be necessary to demonstrate the mechanism(s) that lead to the success of the strategy. Furthermore, it is not just clinical trials that have a need for more quantitative tools for assessment; it also would be desirable to have available the technologies necessary to assess periodically the continued viability and functionality of a tissue substitute or strategy after implantation into a patient.

Also, one cannot state that one has successfully met the challenge of imitating nature unless the implanted construct is biocompatible. Even if the implant is immune acceptable, there can still be an inflammatory response. This response can be considered separate from the immune response, although obviously interactions between these two might occur. In addition to any inflammatory response, for some types of tissue-engineered substitutes thrombosis will be an issue. This is certainly an important part of the biocompatibility of a blood vessel substitute.

Finally, important to the success of any tissue-engineering approach is the immune response and that it be immune acceptable. This comes naturally with the use of autologous cells; however, if one moves to nonautologous cell systems (as this author believes we must, at least in many cases, if we are to make the products of tissue engineering widely available for routine use), then the challenge of engineering immune acceptance is critical to our achieving success in the imitation of nature. Today we have immunosuppressive drugs, e.g., cyclosporine; however, transplant patients treated this way face a lifetime where their entire immune system is affected, placing them at risk of infection and other problems.

It should be recognized that the issues surrounding the immune acceptance of an allogeneic cell-seeded implant are no different than those associated with a transplanted human tissue or organ. Both represent allogeneic cell transplantation, and this means that much of what is being learned in the field of transplant immunology can help us understand implant immunology and the engineering of immune acceptance for tissue-engineered substitutes. For example, it is now known that to have immune rejection there must not only be a recognition by the host of a foreign body, but there also must be present what is called the *costimulatory signal*, or sometimes simply signal 2. It has been demonstrated that, with donated allogeneic tissue, if one can block the costimulatory signal, one can extend survival of the transplant considerably (Larsen *et al.*, 1996). There also is the chimeric approach, where one transplants into the patient from the donor both the specific tissue/organ and bone marrow. This suggests that perhaps in the future one will be able to use a stem cell–based chimeric approach. As an example, if one were to differentiate an embryonic stem cell both into the tissue-specific cells needed and into the cells required for implantation into the bone marrow, then from a single cell source one would create the chimerism desired.

Another approach is that of therapeutic cloning. Here a patient's DNA is transferred into an embryonic stem cell, which in turn is differentiated into the cells needed for a particular tissue-engineering approach. As attractive as this approach appears, many think it is unrealistic, simply because of the scarcity of eggs and embryonic stem cells. Furthermore, as our knowledge of immunology continues to advance, other approaches might make the need for therapeutic cloning disappear (Brown, 2006). Thus, strategies are under development, and these may provide greater opportunities in the future for the use of allogeneic cells.

## V. CONCLUDING DISCUSSION

If we are to meet the challenge of imitating nature, there are a variety of issues. These have been divided here into three different categories. The issue of cell technology includes cell sourcing, the manipulation of cell function, and the use of stem cell technology. Construct technology includes the engineering of a tissuelike construct as a substitute or delivery vehicle and the manufacturing technology required to provide the product and ensure its off-the-shelf availability. Finally is the issue of integration into living systems. This has several important facets, with the most critical one being the engineering of immune acceptance.

Much of the discussion here has focused on the challenge of engineering tissuelike constructs for implantation. As noted earlier, however, equally important to tissue engineering are strategies for the fostering of remodeling and ultimately the repair and enhancement of function. As the field moves to the more complex biological tissues, e.g., ones that require innervation and vascularization, it may well be that a strategy of repair and/or regeneration is preferable to one of replacement.

As one example, consider a damaged, failing heart. Should the approach be to tissue engineer an entire heart, or should the strategy be to foster the repair of the myocardium? In this latter case, it may be possible to return the heart to relatively normal function through the implantation of a myocardial patch or even through the introduction of growth factors, angiogenic factors, or other biologically active molecules. Which strategy has the highest potential for success? Which approach will have the greatest public acceptance?

Even though short-term successes in tissue engineering may come from the convergence of biologics and devices, long term it is the generation of totally biologic products and strategies that must be envisioned. These will result in advances that include, for example, the following: *in vitro* models for the study of basic biology and for use in drug discovery; blood cells derived from stem cells and expanded *in vitro*, thus reducing the need for blood donors; an insulin-secreting, glucose-responsive bioartificial pancreas; and heart valves that when implanted into an infant grow as the child grows. In addition, the repair/regeneration of the central nervous system will become a reality. Furthermore, as one thinks about the future, medicine will move to being more predictive, more personalized, and, where possible, more preventive. It is entirely possible that we will be able to diagnose disease at a preclinical stage. In that event, the concept of inducing biological repair prior to the appearance of the clinical manifestations of disease becomes even more attractive.

Thus, the strategy being evolved in Atlanta, Georgia, by the Georgia Tech/Emory Center for the Engineering of Living Tissues, an engineering research center funded by the National Science Foundation, is one that more and more is placing the emphasis on repair and/or regeneration. It is moving beyond replacement that may provide the best opportunity to meet the challenge of imitating nature. Fundamental to this is understanding the basic biology, including developmental biology, even though the biological mechanisms involved in adult tissue repair/regeneration are far different from those involved in fetal development. Furthermore, to translate a basic biological understanding into a technology that reaches the patient bedside will require a multidisciplinary, even an interdisciplinary, effort, one involving life scientists, engineers, and clinicians. Only with such teams will we be able to meet the challenge of imitating nature, and only then can the existing patient need be addressed and will we as a community be able to confront the transplantation crisis.

## VI. ACKNOWLEDGMENTS

The author acknowledges with thanks the support by the National Science Foundation of the Georgia Tech/Emory Center for the Engineering of Living Tissues and the many discussions with GTEC's faculty and student colleagues and with the representatives of the center's industrial partners.

## VII. REFERENCES

Ahsan, T., and Nerem, R. M. (2005). Bioengineered tissues: the science, the technology, and the industry. *Ortho. Cranofacial Res.* **8**, 134–140.

Ahsan, T., and Nerem, R. M. (in press). Stem cell research in regenerative medicine. *In* "Principles of Regenerative Medicine" (A. Atala, J. A. Thomson, R. M. Nerem, and R. Lanza, eds.). Elsevier Academic Press, Boston, MA.

Auger, P. A., Lopez Valle, C. A., Guignard, R., Tremblay, N., Noel, B., Goulet, F., and Germain, L. (1995). Skin equivalent produced with human collagen. *In Vitro Cell. Dev. Biol.* **31**, 432–439.

Badylak, S., *et al.* (1999). Naturally occurring extracellular matrix as a scaffold for musculoskeletal repair. *Clin. Ortho. Related Res.* **3675**, 333–343.

Bell, E., Ivarsson, B., and Merrill, C. (1979). Production of a tissue-like structure by contraction of collagen lattices by human fibroblasts of different proliferative potential in vitro. *Proc. Natl. Acad. Sci. U.S.A.* **76**, 1274–1278.

Brown, P. (2006). Do we even need eggs? *Nature* **439**(7077), 655–657.

Butcher, J. T., *et al.* (2004). Unique morphology and focal adhesion development of valvular endothelial cells in static and fluid flow environments. *Arterioscler. Thromb. Vasc. Biol.* **24**, 1429–1434.

Carrel, A., and Lindbergh, C. (1938). "The Culture of Organs." Paul B. Hoeber Inc., Harper Brothers, New York.

Chue, W. L., *et al.* (2004). Dog peritoneal and pleural cavities as bioreactors to grow autologous vascular grafts. *J. Vasc. Surg.* **39**(4), 859–867.

Cho, M. K., McGee, G., and Magnus, D. (2006). Lessons of the stem cell scandal. *Science* **311**, 614–615.

Cima, L. G., Langer, R., and Vacanti, J. P. (1991). Polymers for tissue and organ culture. *Bioact. Compat. Polym.* **6**, 232–239.

Freed, L. E., Vunjak, G., and Langer, R. (1993). Cultivation of cell-polymer cartilage implants in bioreactors. *J. Cell. Biochem.* **51**, 257–264.

Galletti, P. M., Aebischer, P., and Lysaght, M. J. (1995). The dawn of biotechnology in artificial organs. *Am. Soc. Artif. Intern. Organs* **41**, 49–57.

Guldberg, R. E., *et al.* (2003). Microcomputed tomography imaging and analysis of bone, blood vessels, and biomaterials. *IEEE Eng. Med. Biol. Mag.* **22**(5), 77–83.

Guterman, L. (2006). A silent scientist under fire. *Chron. Higher Ed.* **LII**(22), A15, A18–A19.

Heureux, N. L., Paquet, S., Labbe, R., Germain, L., and Auger, R. A. (1998). A completely biological tissue-engineered human blood vessel. *FASEB J.* **12**, 47–56.

Holden, C. (2006). Schatten: Pitt panel finds "misbehavior" but not misconduct. *Science* **311**, 928.

Langer, R., and Vacanti, J. P. (1993). Tissue engineering. *Science* **260**, 920–926.

Larsen, C. P., Elwood, E. T., Alexander, D. Z., Ritchie, S. C, Hendrix, R., Tucker-Burden, C., Cho, H. R., Aruffo, A., Hollenbaugh, D., Unsley, P. S., Wmn, K. J., and Pearson, T. C. (1996). Long-term acceptance of skin and cardiac allografts by blockade of the CD40 and CD28 pathways. *Nature (London)* **381**, 434–438.

Lindberg, K., and Badylak, S. (2001). Small intestine submucosa (SIS): a bioscaffold supporting *in vitro* primary epidermal cell differentiation and synthesis of basement membrane proteins. *Burns* **27**, 254–256.

Lysaght, M. J., and Hazlehurst, A. L. (2004). Tissue engineering: the end of the beginning. *Tissue Eng.* **10**(1–2), 309–320.

Naughton, G. (1999). Skin: The first tissue-engineered products — the advanced tissue sciences story. *Sci. Am.* **280**(4), 84–85.

Neitzel, G. P., *et al.* (1998). Cell function and tissue growth in bioreactors: fluid mechanical and chemical environments. *J. Jpn. Soc. Microgravity Appl.* **15**(S-11), 602–607.

Nerem, R. M. (2006). Tissue engineering: the hope, the hype, and the future. *Tissue Eng.* **12**, 1143–1150.

Nerem, R. M., and Sambanis, A. (1995). Tissue engineering: from biology to biological substitutes. *Tissue Eng.* **1**, 3–13.

Normile, D., and Vogel, G. (2005). Korean university will investigate cloning paper. *Science* **310**, 1748–1749.

Normile, D., Vogel, G., and Holden, C. (2005). Cloning researcher says work is flawed but claims results stand. *Science* **310**, 1886–1887.

Normile, D., Vogel, G., and Couzin, J. (2006). South Korean team's remaining human stem cell claim demolished. *Science* **311**, 156–157.

Parenteau, N. (1999). Skin: the first tissue-engineered products — the organogenesis story. *Sci. Am.* **280**(4), 83–84.

Rao, R. R., *et al.* (2004). Comparative transcriptional profiling of two human embryonic stem cell lines. *Biotechnol. Bioeng.* **88**(3), 273–286.

Saini, S., and Wick, T.M. (2003). Concentric cylinder bioreactor for production of tissue engineered cartilage: effect of seeding density and hydrodynamic loading on construct development. *Biotechnol. Prog.* **19**, 510–521.

Seliktar, D., Black, R. A., and Nerem, R. M. (1998). Use of a cyclic strain bioreactor to precondition a tissue-engineered blood vessel substitute. *Ann. Biomed. Eng.* **26**(Suppl. 1), S-137.

Skalak, R., and Fox, C., ed. (1998). "NSF Workshop, UCLA Symposia on Molecular and Cellular Biology." Alan R. Liss, New York.

Solter, D., and Gearhart, J. (1999). Putting stem cells to work. *Science* **283**, 1468–1470.

Stabler, C. L., *et al.* (2005). *In vivo* noninvasive monitoring of a tissue-engineered construct using 1H NMR spectroscopy. *Cell Transplant.* **14**, 139–149.

Thomson, J. A., *et al.* (1998). Embryonic stem cell lines derived from human blastocysts. *Science* **282**, 1145–1147.

Vats, A., *et al.* (2005). Stem cells. *Lancet* **366**, 592–602.

Vogel, G. (1999). Harnessing the power of stem cells. *Science* **283**, 1432–1434.

Weinberg, C. B., and Bell, E. (1986). A blood vessel model constructed from collagen and cultured vascular cells. *Science* **231**, 397–399.

Yannas, I. V., *et al.* (1982). Wound tissue can utilize a polymeric template to synthesize a functional extension of skin. *Science* **215**, 174–176.

# Chapter Three

# Moving into the Clinic

*Alan J. Russell and Timothy Bertram*

## I. INTRODUCTION

In the early 1930s Charles Lindbergh, who was better known for his aerial activities, went to Rockefeller University and began to study the culture of organs. After the publication of his book about the culturing of organs *ex vivo* in order to repair or replace damaged or diseased organs, the field lay dormant for many years. Indeed, delivering respite to failing organs with devices or total replacement (transplant) became far more fashionable. Transplantation medicine has been a dramatic success. But in the late 1980s scientists, engineers, and clinicians began to conceptualize how *de novo* tissue generation might be used to address the tragic shortage of donated organs. The approach they proposed was as simple as it was dramatic. Biodegradable materials would be seeded with cells and cultured outside the body for a period of time before exchanging this artificial bioreactor for a natural bioreactor by implanting the seeded material into a patient. These early pioneers believed that the cells would degrade the material, and after implantation the cell-material construct would become a vascularized native tissue. Tissue engineering, as this approach came to be known, can be accomplished once we understand which materials and cells to use, how to culture these together *ex vivo*, and how to integrate the resulting construct into the body.

Most major medical advances take decades to progress from the laboratory to broad clinical implementation. Tissue engineering was such a compelling concept that the process moved much faster. As we discuss later, the speed at which tissue-engineering solutions can be implemented is inherently faster than traditional drug development strategies. For this reason, coupled with what was probably unexplained exuberance, business investors saw an immediate role for industry in delivering tissue-engineered products to patients. Traditionally, new fields are seeded with foundational research, development, and engineering prior to implementation, but an apparent alignment of interests caused many to believe that companies could deliver products immediately and that the traditional foundational aspects could wait.

The race to clinical implementation of a tissue-engineered medical product began with the incorporation of Advanced Tissue Science (ATS) in 1987. ATS and Organogenesis, an early competitor, began their quest by focusing on seeding biodegradable matrices with human foreskin fibroblasts. In the early days, Organogenesis, Integra, and Ortec focused on bovine-derived scaffolds, while ATS focused on human-derived scaffolds. Other companies focused on developing tissue-engineering products using scaffold alone or cells alone. The path to implementation has been very different for each class of company, as is summarized later.

With hindsight, one might say that the choice of living-skin equivalents as a first commercial product was probably driven by the willingness of the FDA to regulate them as Class III devices rather than biologics. This attractive feature of the products was supplemented by large predicted market sizes and the ease of culturing skin cells. As these products moved from the laboratory to the clinic, development issues such as which cells, materials, and bioreactors were supplanted with industrial challenges such as scale-up and immunocompatibility.

At the same time that ATS and Organogenesis were rapidly growing, the view emerged that delivery of tissue-engineered products to patients would require an allogeneic off-the-shelf solution having ease of use and long storage life in the United States — and persists today. This business-driven decision predicated development of allogeneic, cell-based therapies. Interestingly, in a study sponsored by the National Science Foundation, the World Technology Evaluation Center discovered that non-U.S. investors were focused on autologous cell therapies resulting from a belief that allogeneic therapy would be unsuccessful because of the need to suppress the patient's immune system. This difference in emphasis between U.S. and non-U.S. investors continues today. Both approaches have genuine advantages and disadvantages. However, once it became clear that cell-seeded scaffolds could trigger dramatic changes in natural wound healing, thereby inducing *de novo* tissue formation and function, clinical implementation through industry progressed from skin to a wide array of tissues. In addition, pockets of excellence arose at major medical centers, where new innovations were tested clinically in relatively small numbers of patients.

So one is left with several questions: What have we learned from these early adoptions of tissue engineering? How can these lessons drive sustainable innovation that will both heal and generate a return on investment? Is broad clinical implementation of tissue engineering limited by the nature and structure of regulatory bodies? This chapter seeks to answer these questions by looking historically at selected high-profile clinical tissue-engineering programs and looking forward with a suggested generic approach to rapid clinical translation in this new era of advanced medical therapies.

## II. HISTORY OF CLINICAL TISSUE ENGINEERING

### What Is Clinical Tissue Engineering?

As mentioned earlier, in the early 1990s the term *tissue engineering* was generally used to describe the combination of biomaterials and cells *ex vivo* to provide benefit once implanted *in vivo*. What emerged over the next decade, however, were biomaterials designed to alter the natural wound-healing response and cell-only therapies. Lessons learned in the development of each led to the fusion of these tools under the rubric of tissue engineering.

Today, there remains confusion about what the term *tissue engineering* truly encompasses. A related term, *regenerative medicine*, has emerged recently. The boundaries of what falls under each of these terms are unclear. We do not seek to provide a definitive answer in this chapter, but herein we discuss the use of biomaterials and cell-seeded biomaterials. We exclude the use of cell-only therapies. Thus, we define *clinical tissue engineering* as "the use of a synthetic or natural biodegradable material, which has been seeded with living cells when necessary, to regenerate the form and/or function of a damaged or diseased tissue or organ in a human patient." We see clinical tissue engineering as a set of tools that can be used to perform regenerative medicine, but not all regenerative medicine has to be done with that set of tools.

### Two-Dimensional Clinical Tissue Engineering

The earliest clinical applications of tissue engineering revolved around the use of essentially flat materials designed to stimulate wound care. Tissue-engineered skin substitutes dominated the market for almost a decade. Another small and slim tissue that found a clinical application was cartilage. Later in the 1990s thin sheets of cells were produced in culture and then applied to patients using a powerful cell-sheet technology. In both the applications, engineered tissue equivalent is relatively easy to culture *ex vivo* because oxygen and nutrient delivery to thin, essentially two-dimensional, materials is not challenging. In addition, once the construct has been cultured *ex vivo*, integration into the body is not an insurmountable barrier for thin materials.

#### Tissue-Engineered Skin Substitutes

Since the inception of tissue engineering there has been a focus on the regeneration of skin. A number of drivers led to this early focus, not least of which was the mistaken assumption that skin is simple to reconstitute *in vitro*. Skin cells proliferate readily without signs of senescence. Indeed, fibroblasts and keratinocytes have been cultured *in vitro* for many years with ease. Interestingly, other highly regenerative tissues, such as the liver, are populated with cells that cannot be proliferated *in vitro*. The clinical need for effective skin wound healing was also a major driver. One in seven Medicare dollars is spent on treating diabetes-induced disease in the United States. The largest component of that cost goes toward treating diabetic ulcers. This attractiveness drew many tissue-engineering efforts into the wound-care market.

#### Regenerative Biomaterials

For almost two decades scientists have explored the use of processed natural materials as biodegradable scaffolds that induce improved healing from skin wounds. One of the first products to market was the INTEGRA® Dermal Regeneration Template. INTEGRA® is an acellular scaffold designed to provide an environment for healing using the patient's own cells. The INTEGRA® label describes the product as follows:

> INTEGRA® Dermal Regeneration Template is a bilayer membrane system for skin replacement. The dermal replacement layer is made of a porous matrix of fibers of cross-linked bovine tendon collagen and a glycosamino-glycan (chondroitin-6-sulfate) that is manufactured with a controlled porosity and defined degradation rate. The temporary epidermal substitute layer is made of synthetic polysiloxane polymer (silicone) and functions to control moisture loss from the wound. The collagen dermal replacement layer serves as a matrix for the infiltration of fibroblasts, macrophages, lymphocytes, and capillaries derived from the wound bed. As healing progresses an endogenous collagen matrix is deposited by fibroblasts, simultaneously the dermal layer of INTEGRA® Dermal Regeneration Template is degraded. Upon adequate vas-cularization of the dermal layer and availability of donor autograft tissue, the temporary silicone layer is removed and a thin, meshed layer of epidermal autograft is placed over the "neodermis." Cells from the epidermal autograft grow and form a confluent stratum corneum, thereby closing the wound, reconstituting a functional dermis and epidermis.

INTEGRA® is now one of many processed natural materials used to stimulate healing. Since the material is not vascularized at point of use, it is best used in thin (two-dimensional) applications. INTEGRA® is an FDA-approved tissue-engineering material widely used in patients today. INTEGRA® does not, however, contain biological factors that are released during the tissue-remodeling process.

Another class of products, the thin extracellular matrix-based materials, does release natural factors as the material degrades, and these factors serve to reset the natural tissue-remodeling process, thereby producing a healing outcome. The most common ECM-based material is derived from the submucosal layer of pig small intestine. The Cook OASIS® Wound Matrix label describes the product as follows:

> The OASIS® Wound Matrix is a biologically derived extra-cellular matrix–based wound product that is compatible with human tissue. Unlike other collagen-based wound care materials, OASIS is unique because it is a complex scaffold that provides an optimal environment for a favor-able host tissue response, a response characterized by restoration of tissue structure and function. OASIS is comprised of porcine-derived acellular small intestine submucosa. The OASIS Wound Matrix is indicated for use in all partial- and full-thickness wounds and skin loss injuries as well as superficial and second-degree burns.

Regenerative biomaterials, or materials designed to alter and enhance the natural tissue-remodeling process, are being used in hundreds of thousands of patients world-wide. These materials recruit a patient's own cells into the healing process postimplantation, and their relative simplicity makes them compelling clinical tools for indica-tions where a thin, essentially two-dimensional material will achieve the desired result. In an elegant series of accomplishments, natural matrices have been applied for skin wounds and many other tissue-replacement therapies.

### Cell-Seeded Scaffolds (Fig. 3.1)

Cultured skin-substitute products, where cells are seeded onto a biodegradable matrix and cultured *ex vivo* prior to shipment and use, have been extraordinarily diffi-cult to market. Given this reality, it is interesting that the purveyors of the two leading skin equivalents are the true early pioneers of tissue engineering. Both Advanced Tissue Sciences and Organogenesis engaged in a valiant effort to use human fibroblasts and biomaterials to regenerate skin. They were challenged by a changing regulatory landscape, an ongoing struggle with reimbursement issues, and the highly complex need to manufacture and ship a living product. A full case study of ATS or Organogenesis would be of tremendous value to the next generation of tissue-engineering companies but is beyond the scope of this chapter. Dermagraft®, the Advanced Tissue Science product now manufactured by Smith & Nephew, uses skin cells

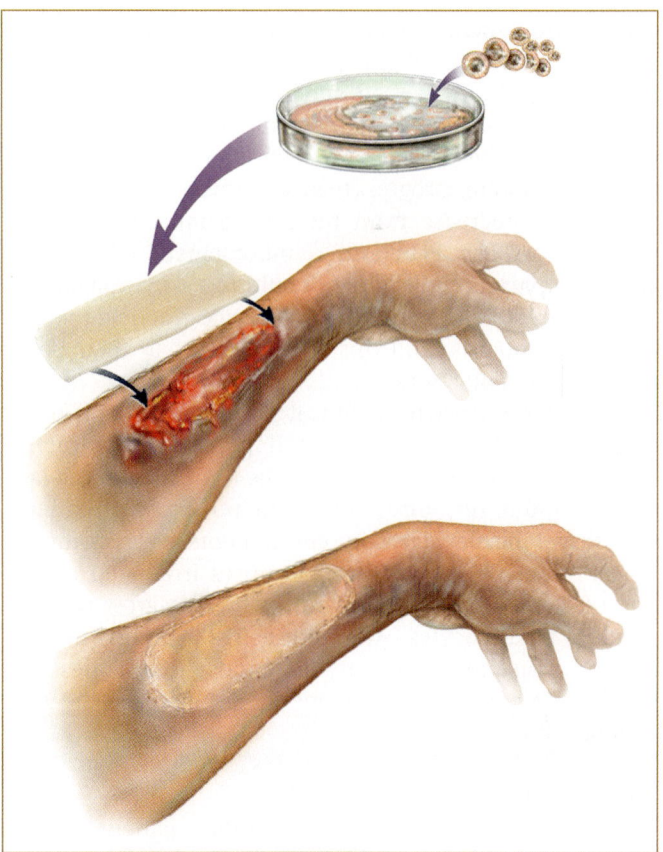

**FIG. 3.1.**

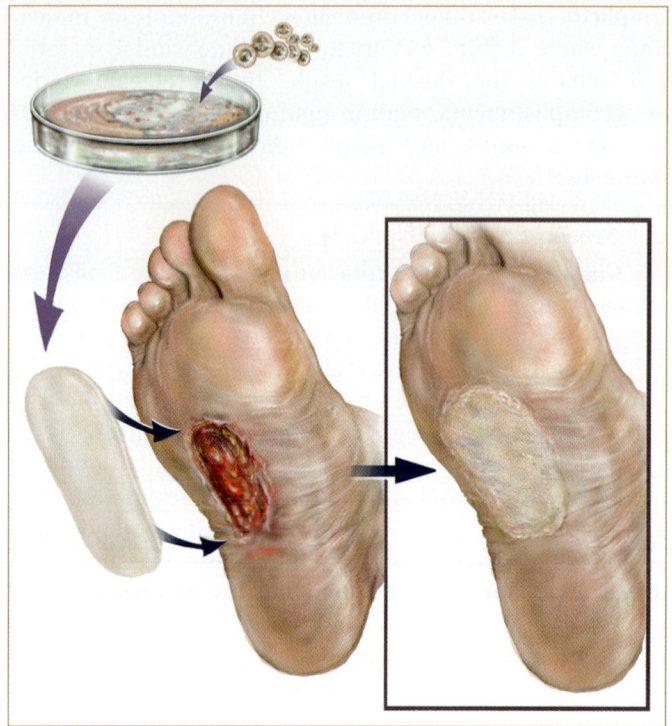

**FIG. 3.2.**

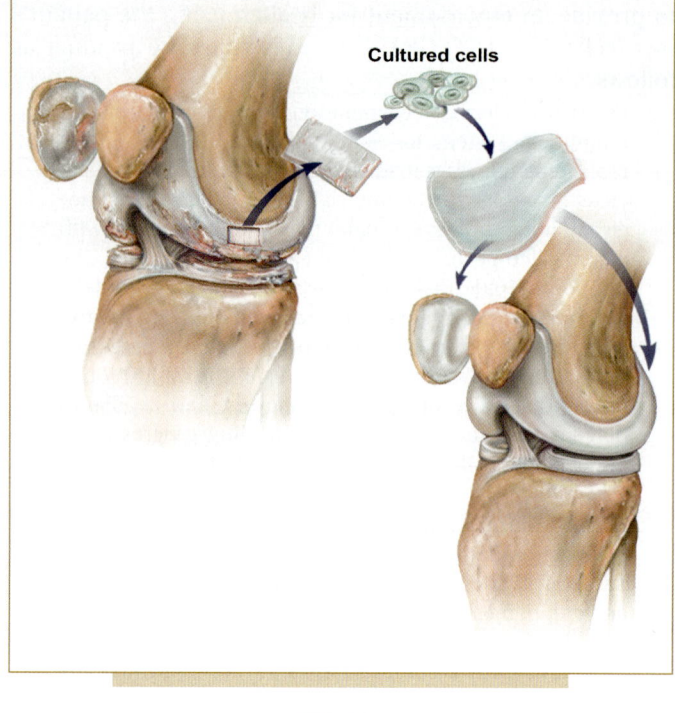

**FIG. 3.3.**

isolated from neonatal foreskins prior to seeding onto a polymeric scaffold. Apligraf®, Organogenesis' product, used similar technology to seed and culture cells on collagen-based scaffolds. Tissue-engineered skin equivalents continue to be developed. As technology improves and multilayer systems progress to broad clinical use, the market should also increase from today's anemic levels of $15 million/year. The FDA and reimbursement issues have greatly impacted clinical use of cultured skin equivalents. The FDA treated these products as devices yet held them to biologic standards, and this led inevitably to their being reimbursed as biologics.

Another approach to clinical skin remodeling is using autologous cell–based therapies (Fig. 3.2). One attractive feature of using a patient's own cells is, of course, the lack of an immune response, but the manufacture of patient-specific yet inexpensive skin replacements is very complex. Epicel® from Genzyme Biosurgery uses irradiated mouse fibroblasts as a feeder layer from which to grow patient-specific keratinocytes. Co-culture with animal-derived cells may raise regulatory and infectious disease questions requiring manufacturing practices that increase the cost of goods.

## Cartilage (Fig. 3.3)

In 1995, Genzyme began expanding patient-specific chondrocytes. Small biopsies were sent to Genzyme, where they were cultured and returned to the surgeon for implantation. The product, Carticel®, was approved as a biologic by

the FDA in 1997. At time of treatment, a patient typically receives 10 million to 15 million cells after five weeks of custom *ex vivo* culturing. As with skin remodeling, a number of companies have focused on the use of acellular regenerative materials. Approved products are currently sold in many countries around the world that are based on collagen and/or extracellular matrixes (ECMs). Thousands of patients around the world have benefited from orthobiologic approaches to cartilage replacement. Although patient-specific cartilage replacement therapy has also provided benefit, it is a good example of the difficulty of delivering individualized therapies while deriving a profit. The considerable infrastructure required to culture tissue safely in this manner presents unique challenges for the manufacturer to overcome. Many research groups around the world have sought to improve on the efficacy of Carticel®, focusing on cell-based and regenerative material–based approaches. Although cartilage segments *in vivo* and *in vitro* are generally small and non-vascularized, the biomechanical properties of those tissue-engineered cartilage products overall have not achieved the standards required for clinical application.

## Corneal Cell Sheets (Fig. 3.4)

Okano at Tokyo Women's Hospital has invented a remarkable technology that produces intact cell sheets for clinical application. In general, when human cells are cultured *in vitro* they adhere to their culture dish substrate. Traditional culturing techniques extract cells by adding enzymes and other materials that digest cell–surface and

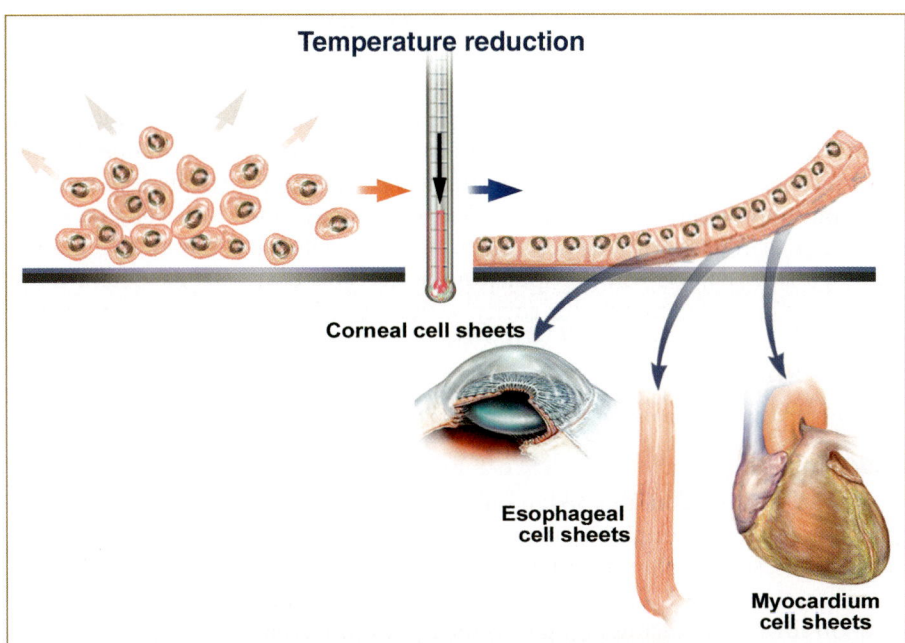

**FIG. 3.4.**

cell–cell contacts. Cells processed in this manner are delivered as single cells for clinical application. Okano envisioned an alternative for removing cells that has had a dramatic clinical impact. Okano covalently bonds a layer of *N*-isopropylamide to the surface of the culture dish prior to adding cells and has shown that, under normal growth temperatures, cells adhere but that when slightly chilled, the entire sheet of cells is repelled from the dish without disrupting the cell–cell contacts and can be lifted from the surface rather like a Post-it note®. For a first clinical application, Okano's team cultured corneal epithelial cells and used the resulting sheets to replace the damaged corneal epithelia of dozens of patients. Okano has reported significant success and, although the number of patients in need of corneal epithelial replacement is limited, this cell-sheet technology has real potential for broader clinical tissue-engineering application.

### Encapsulated Pancreatic Islets

The use of biomaterials to immunoisolate pancreatic islets of Langerhans has been studied since the mid-1990s. If one could build a cage that surrounded the islet and had a mesh size small enough to prevent the approach of antibodies to the islet but large enough to enable nutrient diffusion, it may be possible to diminish a patient's dependence on insulin posttransplant. Alginate-encapsulated islets have been studied for many years, and an ongoing clinical trial (Novocell) is using interfacially polymerized PEG-encapsulated islets (Fig. 3.5). The success of these trials is not yet known, but porcine islets have already been shown to be protectable in a short-term discordant xenotransplantation model. Interestingly, our own work has shown that even a molecular-scale PEG cage can immunoisolate islets,

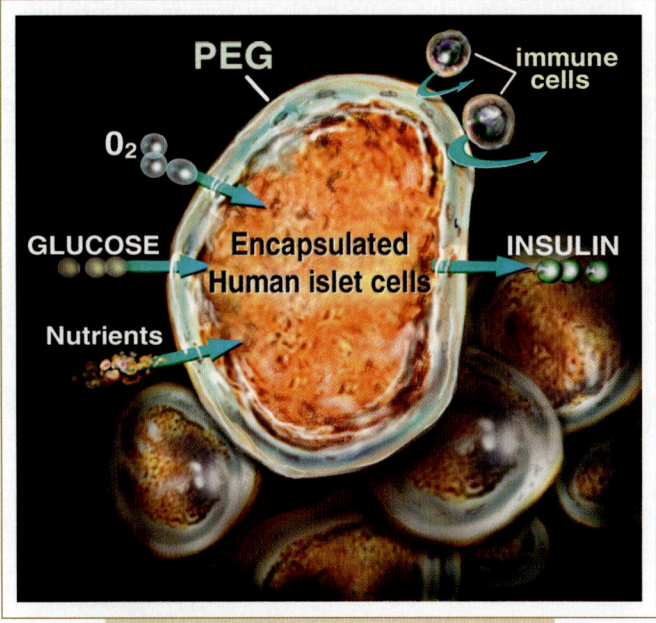

**FIG. 3.5.**

and this has now been shown to eliminate insulin dependence in diabetic animals.

## Three-Dimensional Clinical Tissue Engineering

### Bone Regeneration

Since the turn of the century, Dr. Yilin Cao has led a remarkable clinical tissue-engineering approach to craniofacial reconstruction in Shanghai, China. Regeneration of craniofacial bone in patients has now been reported by using demineralized bone and autologous cells. Using tissue engineering to rebuild lost bone is novel, but it is not the only regenerative medicine approach being applied to the challenge. Peptide-based therapy is an established treatment for stimulating bone formation. Bone morphogenetic protein (BMP) is the most common drug currently employed to induce bone growth. In a novel application of BMP, Medtronic developed a spine fusion device containing a collagen sponge infused with the peptide that is now in clinical use (Fig. 3.6). Deployed in the spine, the device and BMP induce native bone to fill the cavity within the device. Although not often identified as such, this combination of a biodegradable material and a tissue-formation-inducing biologic molecule is clinical tissue engineering at its best.

### Bladder

Early successes in the tissue-engineering field were gained in relatively simple tissue structures, organizations, or functions, such as chondrocytes, or two-dimensional cellular structures with limited organ function required. Tengion advanced a technology pioneered by Anthony Atala to augment or replace failing three-dimensional internal organs and tissues, requiring functionality and a vascularization platform using autologous progenitor cells, isolated and cultured *ex vivo*, and seeded onto a degradable biomaterial optimized for the body tissue it is intended to augment or replace. This cell-seeded neo-organ construct is implanted into the patient for final regeneration of the neo-organ

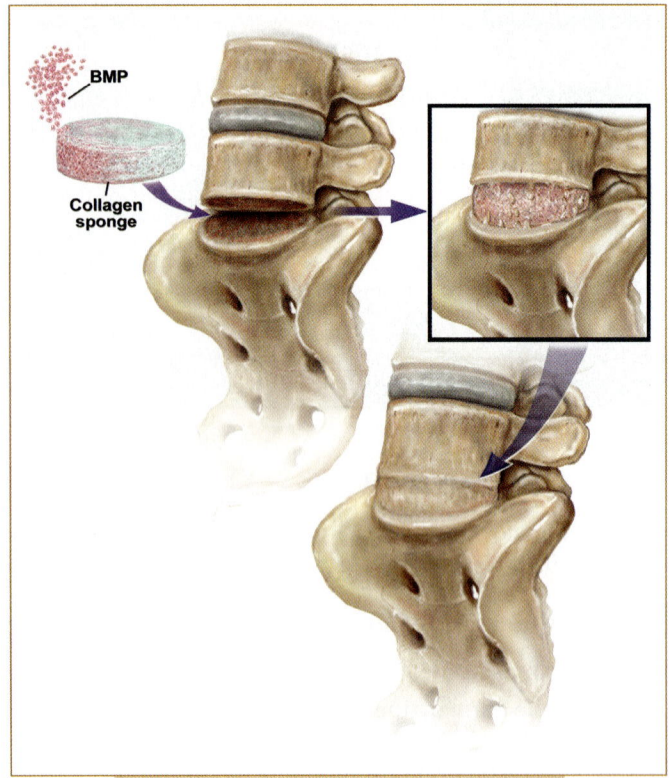

**FIG. 3.6.**

(Fig. 3.7). Using the neo-organ construct as a template, the body regenerates healthy tissue, restoring function to the patient's failing organ. This autologous organ and tissue regeneration avoids many of the negative implications of traditional donor transplantation techniques, such as requisite immunosuppression and limited donor supply.

Tengion's initial focus on the genitourinary system was based on a bladder augmentation and ultimately an organ replacement for patients who have undergone radical cystectomy, or removal of the bladder. Tengion developed a robust focus on manufacturing capabilities to support neo-organ construct production in accordance with regulatory standards.

### Blood Vessel

At Tokyo Women's Hospital, Dr. Toshi Shin'Oka has used patient-specific tissue engineering to replace malformations of pediatric pulmonary arteries. Working with a biodegradable matrix designed by one of the "fathers" of biomaterials, Dr. Ikada, Shin'Oka seeded a tubular material with the patient's own bone marrow cells at the time of vessel reconstruction (Fig. 3.8). In a series of clinical experiments, Shin'Oka demonstrated that biodegradable scaffold's strength during the degradation period was sufficient to allow complete natural vessel replacement without rupture. This first successful clinical replacement of a pediatric blood vessel with a tissue-engineered construct designed to become as natural as the patient's own vasculature was performed in almost 50 patients.

As one considers these historical events and the advances in tissue engineering over the past 80 years, we are seeing that tissue engineering is moving toward the regeneration and repair of increasingly complex tissues and even whole-organ replacement. This field holds the realistic promise of regenerating damaged tissues and organs *in vivo* (in the living body) through reparative techniques that stimulate previously irreparable organs into healing themselves. Regenerative medicine also empowers scientists to grow tissues and organs *in vitro* (in the laboratory) and safely implant them when the body is unable to be prompted into healing itself. We have the technological potential to develop therapies for previously untreatable diseases and conditions. Examples of diseases regenerative medicine could cure include diabetes, heart disease, renal failure, and spinal cord injuries. Virtually any disease that results from malfunctioning, damaged, or failing tissues may be potentially cured through regenerative medicine therapies. Having these tissues available to treat sick patients creates the concept of *tissues for life* (U.S. Department of Health and Human Services, 2005).

## III. STRATEGIES TO ADVANCE TOWARD THE CLINIC

Since the mid-1980s we have had many opportunities to learn how one might quickly convert tissue-engineering

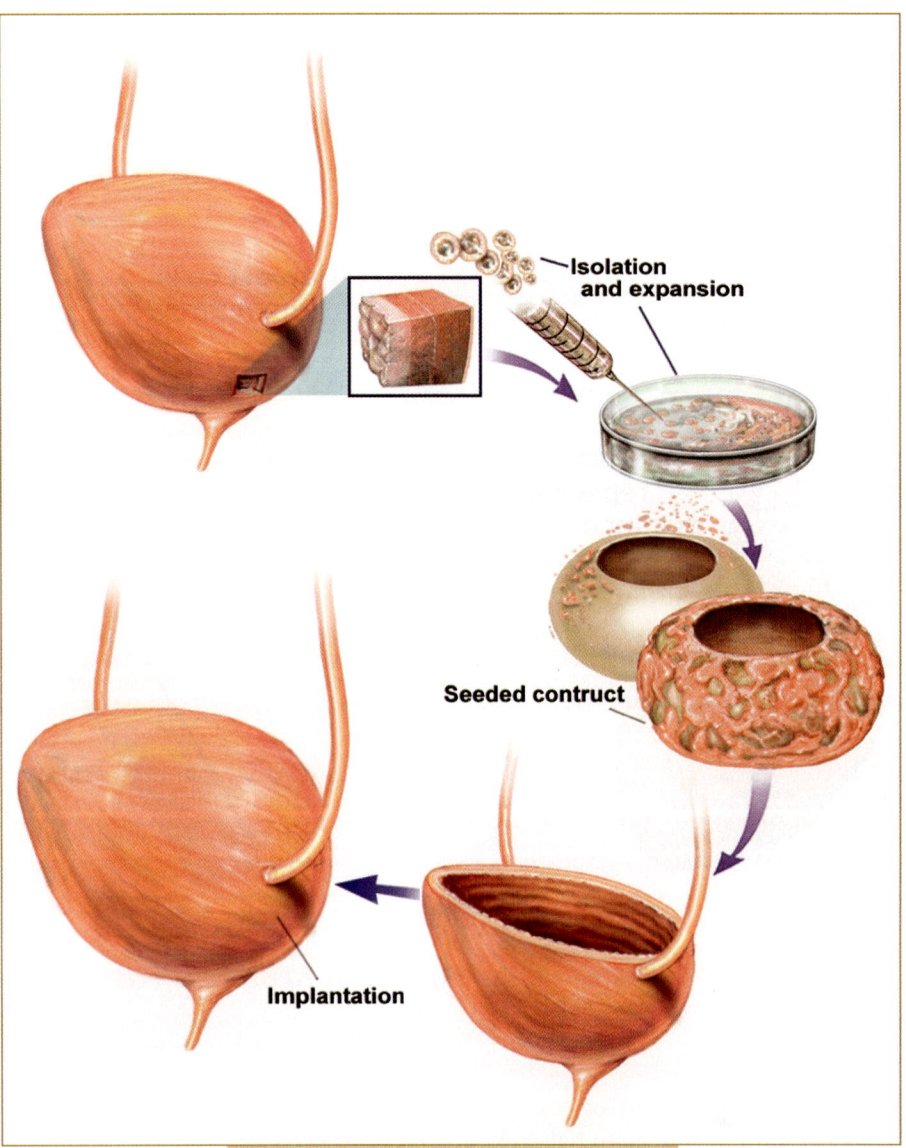

Isolation and expansion

Seeded contruct

Implantation

**FIG. 3.7.**

technology into regenerative medical products from the bench to the bedside. Establishing a plan to move toward clinical testing rests on a strategy of defining the unmet medical need (patient population), determining the intended use of the tissue-engineered/regenerative medical product (TERMP) that addresses the need, and defining the processes necessary to ensure that the product can be reproducibly manufactured to be both safe and effective once it is placed into the patient. A requisite scientific basis for partial or complete structural and/or functional replacement of a diseased organ or tissue requires a definition of what constitutes a successful outcome (i.e., primary clinical endpoint). Ultimately, any clinical testing will require the application of existing regulatory guidelines for testing and manufacturing a product prior to use in a human subject. With this information in hand, initial steps into clinical testing phases can be contemplated. As we have already seen, sound scientific strategies are not effective in clinical translation unless there is a balancing sustainable business strategy. Naturally, the scientific and business strategies must be woven together, in terms of both specific outcomes and timelines. As we discuss in detail later, the significant differences between traditional drug therapies and tissue-engineering therapies actually offer the opportunity to accelerate the bench-to-bedside process. We take the position that instead of a 10- to 15-year development cycle, tissue-engineering therapies can be brought to market in 8–10 years.

In considering the exploratory clinical testing phase with a scientifically based program, final product characteristics must be defined as well as standardizing the production processes and anticipating what justifications will indicate readiness for entering into the next phase of clinical testing (confirmatory studies). Table 3.1 presents an overview of a prototypical product development process.

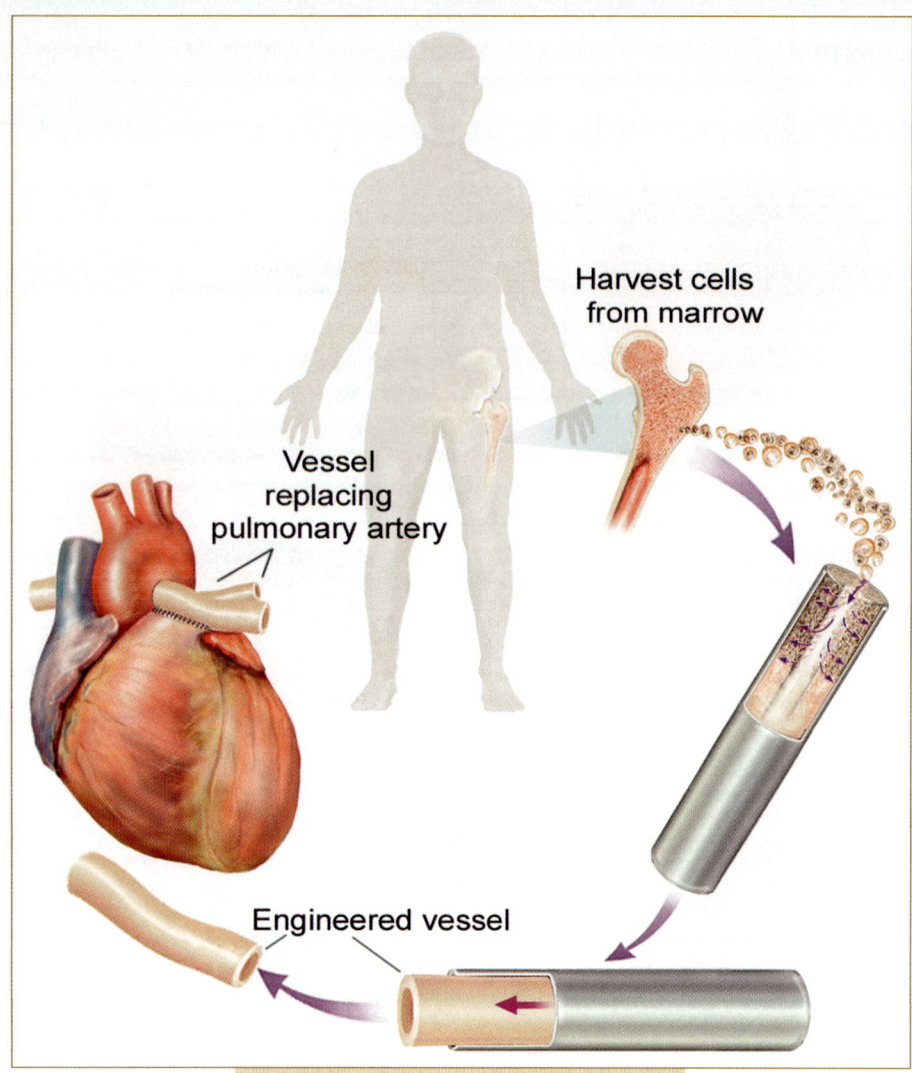

**Harvest cells from marrow**

**Vessel replacing pulmonary artery**

**Engineered vessel**

**FIG. 3.8.**

Transitioning into an initial exploratory clinical evaluation rests on understanding the objectives for the first regulatory review as they relate to the specific product characteristic, the process to make the product, translational medical study results, and how preclinical information demonstrates the desired clinical outcome (Preti, 2005; Weber, 2004). In a rapidly changing field where regulatory agencies are still maturing in their decision-making processes, the decisions made during the initial clinical evaluation phase can have far-reaching impact. Table 3.2 provides an overview of data that will be needed prior to entering into an exploratory clinical trial.

Regulatory considerations affect the types of data required and process technologies that must be in place prior to initiating clinical trials. Indeed, the regulatory environment is much more defined today than in the early days of tissue engineering because of scientific advances and insights gained from various attempts to commercially develop tissue-engineering and regenerative medical products. Once again, understanding the regulatory environ-

ment is foundational to implementing successfully the development plan and using the scientific objectives laid out in Tables 3.1 and 3.2.

Several regulatory considerations have significant impact on the development plan necessary to bring a TERMP to clinical testing: extent of cellular manipulation, cell source and use, and scaffold characteristics (Table 3.3). With a scientific foundation, an established product characterization, and application of appropriate regulatory considerations established, three additional considerations come into play for a particular technology to be transitioned from the bench to the bedside: raw materials testing, manufacturing process controls testing, and translational medicine.

## Raw Materials Testing

### Cells

Cellular components of a TERMP are raw materials encompassing viable cells from the patient (autologous), other donors (allogeneic), or animals (xenogeneic). Standards

**Table 3.1.** Overview of a potential testing program to support clinical entry of a prototypical tissue-engineered/regenerative medical product (TERMP)

*Cellular/chemistry manufacturing control*

- Define product production and early manufacturing processes
- Establish cell, tissue, and biomaterial sourcing for good manufacturing practices (GMPs)
- Validate product processing and final product testing scheme
- Characterize adventitious agents and impurities for each element
- Define lot-to-lot consistency criteria
- Validate quality control procedures

*Translational medical studies*

- Complete *in vitro* and *in vivo* testing
- Define toxicity testing of raw materials composing the TERMP
- Evaluate biomaterial biocompatibility
- Establish immunogenic and inflammatory responses to each component
- Develop rationale for animal and *in vitro* models to test product effectiveness
- Define endpoints for establishing TERMP durability

*Clinical trials*

- Develop rationale for safety and clinical benefit (risk/benefit analysis)
- Design exploratory and confirmatory trials
- Select patient population and define inclusion/exclusion criteria
- Identify investigational comparators and control treatments
- Establish primary and secondary study endpoints
- Consider options for data analysis and potential labeling claims

**Table 3.2.** Information needed prior to evaluating a TERMP in clinical studies*

|  | In vitro | In vivo |
|---|---|---|
| Raw materials supply | | |
| Cells* | + | ± |
| Scaffold | + | ± |
| Manufacturing process controls — in-process and potency | | |
| Cellular processing* | + | − |
| Biomaterial processing | + | − |
| Final combination product* | + | + |
| Translational medical studies | | |
| Safety and efficacy | + | + |
| Endpoint selection | + | + |
| Translation into clinical design | ± | + |

*For products composed solely of acellular scaffold material, evaluation of cellular components is not needed.

for cellular quality have been extensively reviewed and considered by regulatory bodies and generally focus on controlling introduction of infectious diseases and cross-contamination from other patients. These standards also consider potential for environmental contamination from the facility and equipment and the introduction of infectious agents from materials used to process cells (e.g., bovine-derived material that may contain infectious agents).

For TERMPs that have cells placed onto a scaffold, scientific and regulatory considerations focus on ensuring that both the raw materials comprising the scaffold and its three-dimensional characteristics are biocompatible (FDA, 1995). Biocompatibility testing involves evaluation of the scaffold's potential cytotoxicity to cells being seeded, potential toxic-

ity that may be inflicted on the recipient's tissues once implanted, as well as the consequences of immune and inflammatory responses to the TERMP after implantation. Biocompatibility extends through the *in vivo* regeneration process; therefore, biocompatibility should be evaluated in parallel with demonstrating that the scaffold maintains the necessary biomechanical properties to support new tissue or organ growth.

## Scaffold

Synthetic, natural, or semisynthetic materials are readily available from various commercial sources, but the quality control of a material varies substantially between medical and research grades. As testing of a potential TERMP moves from research bench to clinical testing, scaffold composition and designs must be controlled for reproducibility of production and product characterization. Final production must consider quality management and organization, device design, production-facility environmental controls, equipment, component handling, production and process controls, packaging and labeling control, distribution and shipping, complaint handling, and records management, as outlined in 21 CFR 820 (FDA, 2005b). However, during the exploratory phase and transition from bench to clinical testing, the most relevant of these guidelines are process validation and design controls.

Typically, a design input phase is a continuum beginning with feasibility and formal input requirements and continuing through early physical design activities. Engineering input on final prototype specifications follow the initial design input phase and establish the design reviews and qualification. For a combination TERMP, defining quality for the chemical polymer (e.g., PGA) or natural

**Table 3.3.** Regulatory considerations for the development of a tissue-engineered/regenerative medical product

| | Description | Impact |
|---|---|---|
| Manipulation of cells | For structural and nonstructural tissues, manipulation is minimal if it involves centrifugation, separation, cutting, grinding and shaping, sterilization, lyophilizing, or freezing (e.g., cells are removed and reintroduced in a single procedure).<br><br>Manipulation is not minimal if cells are expanded during culture or growth factors are used to activate cells to divide or differentiate.<br><br>Defined in 21 CFR 1271.3(f) — see also FDA (2005a) | More extensive regulatory requirements applied to TERMP when manipulation is more than minimal. |
| Cell source and application | Homologous use is interpreted as the augmentation tissue using cells of the same cellular origin. Examples include applying bone cells to skeletal defects and using acellular dermis as a urethral sling.<br><br>Nonhomologous-use examples include using cartilage to treat bladder incontinence or hematopoietic cells to treat cardiac defects.<br><br>Defined in 21 CFR 1271.3(c) — see also FDA (2001). | Nonhomologous use triggers additional requirements for entering clinical trials. |
| Scaffold characterization | Final scaffold composition and design determine whether the TERMP is characterized as a device, a biologic, or a combination product.<br><br>Defined in Quality Systems Regulations (QSRs) in 21 CFR 820 (FDA, 2005b) — see also FDA (1999). | Devices are held to the QSRs in 21 CFR 820 (FDA, 2005b), biologics are required to comply with good manufacturing processes (GMPs) (FDA, 1991), and combination products are often required to comply with both sets of regulations and guidelines. |

material (e.g., collagen), including any residues introduced during machine processing (e.g., mineral oil), can require QSR integration into a product that would otherwise be regulated as a biologic. Since many TERMPs are combination products, testing of scaffold, cells, and the cell-seeded scaffold (i.e., construct) are required to ensure that, in exploratory clinical trials, the product is sterile, potent, fit for use, and composed of the appropriate raw materials to function properly following *in vivo* placement.

## Manufacturing Process Controls and Testing

### Cellular Processing

In-process controls generally focus on sterility, viability, and functional analysis of cells from isolation, through expan-

sion and before they are placed on the scaffold. Release criteria generally ensure that cells remain viable and functioning properly after being attached to the scaffold. Functional evaluations of cells and potency assessments of their "fitness for use" are performed after cells are combined with (or seeded onto) a scaffold. Taken together, these tests determine whether the final product can be released from the production facility for surgical implantation in the clinical setting.

### Biomaterial Process and Testing

The focus of biomaterial process testing is to evaluate the *in vivo* behavior of the scaffold material following implantation. Characterizing the scaffold degradation profile ensures that breakdown time and other degradation attributes will support the regenerating tissue long enough for it to acquire the appropriate functional and structural integrity as the scaffold material degrades. Defining scaffold-breakdown products identifies the biochemical factors that may impact reparative, inflammatory, immunologic, and regenerative processes once the product is placed into the body. Measuring biomechanical properties such as stress–strain relationships, Young's modulus, and other characteristics ensures that the scaffold portion of the combination product will perform properly during the *in vivo* regenerative phase.

Final Combination Product Testing

Analytical methods for final product testing vary substantially, depending on the composition of the TERMP. In general, any product intended for customization to individual patients (e.g., autologous products) requires confirmation that release and potency standards are met via nondestructive test methods. Such test methods are typically novel and specific to each product type and are frequently based on a battery or "matrix" of tests that evaluate cellular function and physical parameters of the scaffold. In contrast, lot-testing strategies, statistical sampling, and more routine analytical methods are available for scaffold-only products and cell-based products produced in large lots (e.g., allogeneic and xenogeneic cellular products). In the future we may see allogeneic therapies that are customized for patient-specific needs. Naturally, such innovations will require a combination of analytic approaches.

## Translational Medicine

Safety and efficacy evaluation of a TERMP is conducted in animals, and the findings are foundational to designing the first clinical trial protocol. These translational studies are the basis for safely transitioning a potential product into clinical testing. Since the regenerative process invoked by components product involve multiple homeostatic (e.g., metabolic), defense (e.g., immune), and healing (e.g., inflammation) pathways, animal studies provide an approach to understanding the inherent function of the TERMP (i.e., if the product contains cells, it can be considered a living "tissue") and the inherent response of the body to a product composed of biomaterials with or without cells. Animal studies are a regulatory necessity, but we must also remind ourselves that many therapies function effectively in animals and fail in humans. The reverse is not often discussed. Some therapies may fail preclinical testing and never enter clinical trials, but this is not to say that some of those therapies would not be excellent when applied in humans. An interesting example of this conundrum is emerging in artificial-blood therapies. The U.S. Army received approval for clinical use of a natural blood substitute in trauma applications during war. In a postapproval attempt to understand why the product worked so effectively in humans it was tested in a porcine model of hemorrhagic shock. The pigs did not do well with the therapy. The investigators proceeded to test the product in rodents and demonstrated that the rodents died when injected with the blood substitute that worked so effectively in the clinic. If preclinical animal testing had been performed first, this excellent product would never have been submitted for clinical evaluation.

Translational medical studies can be conducted in large animals (e.g., dog) or small (e.g., rat). Selection of the correct animal model should be based on the similarities of the pathophysiology, physiology, and structural components intended for treatment in the clinical setting. Exploratory clinical trials for most medical products utilize normal human volunteers as the first line of clinical testing. However, some products can be tested outside of the intended clinical population. As such, the animal model employed in translational medicine should resemble the human condition as closely as possible — immune status, inflammatory response, and healing pathways as well as the medical approaches used to treat the human condition (e.g., surgical procedure) and monitoring methods to follow a clinical benefit or risk (e.g., imaging).

Many pivotal preclinical experiments are performed in academia, and the importance of complying with the good laboratory, manufacturing, and tissue practice regulatory standards is critical at this phase. There is no such thing as GLP/GMP/GTP-light, and many academic animal facilities are not compliant to the degree needed by the FDA. This issue will increase in significance as the FDA increases its post-approval auditing of preclinical compliance.

Since the regenerative response starts at the moment of TERMP implantation and concludes with the final functioning neo-tissue or neo-organ, animal studies provide an understanding of how to evaluate the early body responses as well as longer-term outcomes reflecting the desired benefit — an augmented or replaced tissue or organ. Since most products are surgically implanted for the life of the patient, the duration of a translational study would extend to the time when final clinical outcome is achieved. Regulatory agencies have given considerable thought to the duration of translational studies, and many are of long duration — months to years. Nonetheless, since the final outcome is frequently achieved in a shorter period of time, the potential to conduct shorter-duration studies based on final patient outcome may present a rational solution to testing clinical utility in the shortest possible time while ensuring a high benefit:risk outcome.

Understanding which endpoints are available and appropriate for clinical testing is achieved through translational studies. Standards are provided for the proper safety evaluation of TERMPs, whether they are regulated as a device (FDA, 1997) or a biological product (e.g., 351 or 361). Although the optimal testing strategy will typically be product specific (FDA, 2001), some basic guidelines for testing device-like products can be found in the ISO10993-1 guidance document. These testing guidelines cover a number of *in vitro* and *in vivo* assays (Table 3.4).

A scaffold-only product that is similar to an already-tested material or medical device can be accelerated through the testing process using a 510K approach under an existing PMA (Rice and Lowery, 1995). Appropriate translational testing approaches will follow the biocompatibility flow-chart for the selection of toxicity tests for 510(k)s (FDA, 1995).

If the device requires an IDE/PMA level of testing, then the translational studies will be more extensive and influenced by the length of time that the TERMP is in contact with the body of the recipient. As already mentioned, many regenerative medical devices or the tissues that replace the initial implant are in bodily contact for longer than 30 days and are therefore considered permanent devices. These products require a full range of *in vitro* and *in vivo* testing approaches prior to clinical testing.

If the TERMP is cell based or the primary mode of action is mediated through the cellular constituents of a scaffold–cell combination product, then the device requires an IND/BLA. The testing approach for these products will usually involve an assortment of studies that evaluate scaffold and cellular components through appropriate endpoint selection and experimental design for both *in vitro* and *in vivo* translational studies. An example of a preclinical development program for a cell-based product is presented in Table 3.5.

Although specific testing approaches are not defined absolutely, the scope and testing approaches for a specific TERMP can frequently be predicted by evaluating the development approach used for related technology platforms. A number of tissue-engineering technologies can bridge from bench to clinical application. Testing a TERMP prior to moving into clinical evaluation is based on (i) scientific information demonstrating that the potential clinical product can invoke a response in the body of potential therapeutic benefit; (ii) demonstration of a controlled and reproducible manufacturing process; and (iii) demonstration of the safety of each component and the final product. This stage in the development of a prototypical clinical product is typically the first point of regulatory authority and governance body interactions and an area where procedural approaches for establishing controls are frequently reviewed and clarified.

## IV. BRINGING TECHNOLOGY PLATFORMS TO THE CLINICAL SETTING (Fig. 3.9)

### General

Technology platforms that intend to recapitulate a tissue (e.g., skeletal muscle, bone, cardiac muscle) or an organ may address a range of unmet medical needs, from simple cosmetic defects in the body (tissue-focused technologies) to life-threatening maladies (organ and organ system replacement). Bringing a TERMP technology to clinical testing may rest on the scope of unmet medical need and the availability of alternative therapies. The array of available alternative therapies influences the early testing strategy of a particular product by determining comparable products to be evaluated, selection of animal models, appropriate endpoints and amount of preclinical information needed to enter into clinical testing. Ultimately, the safety and efficacy of the prototype product are balanced by a risk:benefit analysis versus other available products, which directly influences the ability to test it in human trials.

### Tissue-Focused Technologies

Tissue-focused technologies, such as bone and tendon repair, may move into clinical testing through routes that have been established by previous successes (e.g., Depuy's Restore®). If animal models and alternative therapeutic approaches are established, comparing the benefit of a proposed product to an existing therapy may be an appropriate approach to potential clinical testing. Ultimately, comparing the benefit of the TERMP versus the "gold standard"

| **Table 3.4.** | **Test categories described in ISO10993-1** |
|---|---|
| *In vitro assays* | *In vivo assays* |
| Cytotoxicity | Irritation |
| Pyrogenicity | Sensitization |
| Hemocompatibility | Acute systemic toxicity |
| Genotoxicity/genetic tests | Subchronic toxicity |
| | Local tolerance |

| **Table 3.5.** | **General translational medical testing paradigm for a cell-based tissue-engineered/regenerative medical product** | |
|---|---|---|
| *Cellular component* | *Scaffold* | *Combination* |
| Phenotype characterization | Early stage (acute) | Early stage (acute toxicity) |
| Genetic stability | Late stage (chronic) | Late stage (chronic toxicity) |
| | Biocompatibility | |
| | Biomechanical properties | |
| | Degradation profile | |

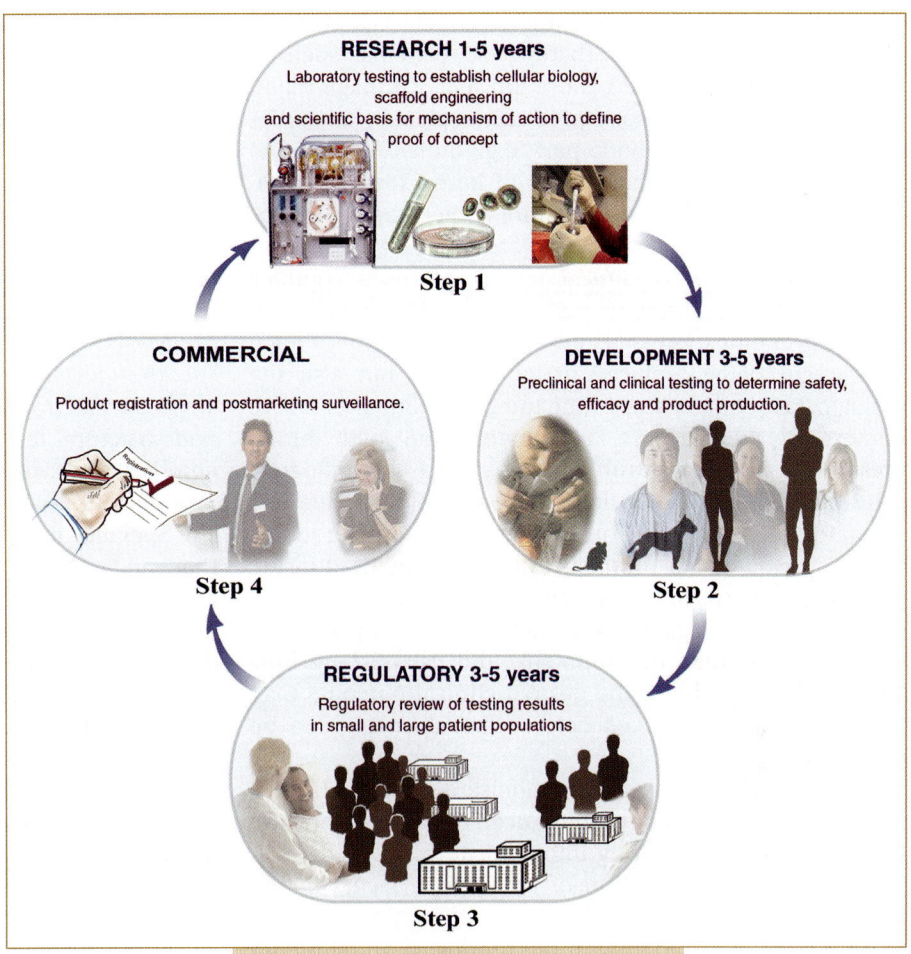

**FIG. 3.9.**

commercial product or surgical therapy is the foundational rationale to evaluate potential human use.

Depending on the raw materials composing the TERMP, the primary mode of action may drive the testing strategy for novel products. The primary mode of action is defined by the scientific studies demonstrating the range of bodily responses invoked by the product and the range of long-term outcomes. Products that elicit an immune response (e.g., allogeneic, xenogeneic, or genetically modified cells) will need to include an evaluation of immunotoxicity, immunomodulation, and/or potential for rendering the recipient sensitive to infectious diseases. Those products whose production employs animal materials will require testing for adventitious infectious agents or the use of materials from certified sources. Testing for potential endogenous infectious agents prior to clinical testing is especially relevant for products that contain, or whose production process includes, xenogeneic cells. Products using scaffold material for which there is little or no previous human testing will require testing that follows established FDA Guidelines (see G95-1) (FDA, 1995). Biodegradable scaffolds have a testing paradigm similar to that used for a nonbiodegradable material, with additional requirements for defining the degrada-

tion profile, breakdown products released, route of excretion, and response of the body to the material as it breaks down. Ultimately, the final safety/efficacy testing strategy may rest with the regulatory pathway selected through a process established by the "Office of Combination Products."

## Organ-Based Technologies

Tissue-engineered/regenerative medical technologies offer the promise of alleviating the vast organ shortage that exists worldwide. In spite of this great promise, the pathway to clinical testing with a product that replaces an entire organ is the least clearly defined. Although the clinical benefit of such a TERMP may be definitive, the endpoints readily discernible, and the animal models established, the delivery mechanism, procedures for connecting the neo-organ to other parts of the body, may pose substantial development hurdles and actually preclude clinical testing.

The complexity of whole-organ replacement by these types of products spans defining what is actually being replaced through defining what ancillary products may be needed if all organ functions are not included in the product characteristics. Traditional therapeutic approaches have generally focused on one pathway or target (e.g., pharmaceutical) or possibly two therapeutic benefits, such as structural and functional restoration (e.g., cartilage repair products). However, those products replacing an entire organ (e.g., kidney) or body part (e.g., limbs) will need to consider broad functional testing of both exocrine/excretory and endocrine functions before clinical testing can be considered.

The transition to clinical testing of more complex TERMPs will have commensurate preclinical testing requirements to demonstrate not only the functionality of each component being replaced or augmented, but also the biological responsiveness of the integrated organ to native homestatic mechanisms (e.g., integration with blood pressure or glucose control). Matters such as percutaneous conduits, skin infections, and controlling biofilms may be substantial development hurdles for the use of products outside the body. For products intended to be used inside the body, solutions for vascular connections, waste product release pathways (e.g., urinary tract and GI), clinical monitoring of neo-organ development, and establishing how long it takes to achieve the desired clinical outcome may all need to be established before clinical testing can be considered.

Biosensors and integration of biosensors with TERMPs replacing whole or major portions of an organ's function are becoming a reality. Moving into clinical testing with such products requires definition of *recovery pathways* in the event of product failure; definition of alternative therapies to be used in association with the product if not all organ functions are replaced; understanding TERMP longevity and how to replace the product if the product/neo-organ wears out; and understanding the rate of product failure for proper clinical management.

In spite of these hurdles, the lure of replacing an entire organ is considerable. The benefit to society of replacing a kidney or pancreas is unimaginable. As scientific advances in *in vitro* organ growth are made and regenerative templates for entire organs are pioneered (e.g., through such technologies as organ printing), the potential to replace, regenerate, repair, and restore entire organ systems is being considered. Tissue-engineering approaches may yield solutions for some of the most devastating human conditions, including congenital agenesis, cancer, degenerative disorders, and infectious diseases. However, entry into clinical testing with such products has not yet been defined.

## V. TRANSITION TO CLINICAL TESTING

### Defining and Testing a Prototype

Prior to beginning a clinical testing program, the TERMP's specific characteristics must be defined to the point that the product can be repeatedly and reproducibly manufactured for *in vitro* and *in vivo* testing as defined earlier. Product characteristics should be sufficiently stable to allow for data-driven demonstration of their clinical utility. Once a prototype is defined, its characteristics are evaluated in a series of tests to define the limits of the initial design criteria that allow for durability testing of the product design by establishing failure points, limits of TERMP application, and the achievement of design criteria.

Anticipating the clinical conditions, complications, and untoward events that may arise during clinical testing also establishes a prototype's potential for clinical utility. A TERMP is seldom introduced as a final functioning neotissue; therefore, characterizing the pharmacological responsiveness, electrophysiological parameters, and phenotypic and structural features of the neotissue or neo-organ that emerges following implantation is key to demonstrating the product's ultimate clinical benefit.

Specific design elements of a final TERMP prototype that will be tested in humans are the culmination of a series of biological, physical, and chemical evaluations obtained during the prototyping phase. This characterization also defines sourcing and control of raw materials, assembly processes (aka: in-process testing) and release criteria. Additionally, the product's shelf life, shipping conditions (temperature, humidity, nutrients, etc.), stability, sterility, and method of use are established before clinical testing. Any unique surgical procedures, clinical management practices during and after implantation, and recovery times are estimated based on the translational medical results using the final prototype product with the fully embodied characteristics.

### Extending Existing Technology

Using previously tested technology platforms can accelerate the entry of any TERMP into clinical testing. Most products are combination products based on multiple technology platforms. Using one or more already-approved scaffold materials, cell-processing methods, culture media components, or transport containers greatly reduces the number of variables that need to be tested in product prototyping and preclinical testing phases. Additionally, historical data available for any technology can help develop testing strategies for a final prototype and even establish early clinical-phase designs.

### Production of TERMPs in GMP Facilities

With established product characteristics, standard operating procedures, and clinical production processes, a GMP-qualified facility can be deployed to manufacture the first clinical prototype. GMP facilities not only meet GMP guidelines, but they have specialized facility designs and highly trained personnel to produce faithfully the first clinical prototypes in a controlled and reproducible fashion. Considerations for GMP facilities include capacity limitations, availability restrictions, and costs to build, operate, and maintain. Furthermore, utilization of a particular GMP facility may be constrained by the controls needed to generate a particular product. Deploying contract manufacturing is a strategy that can hasten product-prototype production in a manner that complies with regulatory guidelines.

TERMP technologies can vary substantially, so it is not uncommon for small GMP facilities to be custom-built to meet the needs of a particular technology platform. Facility design considerations are outside of the scope of this chapter, but a GMP-qualified facility that can provide the required clean-room processing, shipping, and receiving procedures and HVAC systems for airflow maintenance should be identified before any consideration can be given to initiating clinical testing. This should be done as early as possible, but no later than the final stages of the prototyping phase, to ensure that the necessary facility design capable of producing products in compliance with GTPs and GMPs is available.

Contract manufacturing operations (CMOs) have emerged that produce scaffold-only and scaffold-plus-cell products. These operations have staff skilled in various aspects of product manufacturing and generic facilities that can accommodate a variety of cellular methods and biomaterial-handling needs. A technology transfer plan (Bergmann, 2004) should be established before engaging a CMO, to ensure optimal product generation and the success of the first clinical trial.

## Medical and Market Considerations

Entering clinical testing of TERMPs will not achieve the promise of impacting major unmet medical needs without consideration of market demands. These demands include third-party payers' willingness to support costs, follow-up care, and subsequent patient morbidity. The availability of lower-cost alternatives may be the most significant and practical barrier to clinical testing of a TERMP. Tissue-engineering technologies address medical needs unmet by pharmaceutical agents or devices, but these needs may be met by the modification of medical practices, lower-cost alternatives (e.g., cadaveric skin), or currently accepted medical procedures (e.g., tissue transplantation). Exploratory clinical testing strategies can incorporate these alternative approaches to establish the comparative clinical benefit of a prototype product.

As the science and technology of tissue engineering become more established and regulatory pathways are clarified, products will become more broadly applied. Strengths and limitations of TERMP technologies will determine market size and application to unmet medical needs. At present, products have few competitors in the marketplace, and the opportunities are driven largely by reducing a particular technology to practice.

## Regulatory Considerations and Governance Bodies

Multiple FDA review organizations oversee TERMPs, depending on their characteristics. For devicelike products,

CDRH is the regulatory center, for biological products it is CBER, and for combination products, the Office of Combination coordinates a time-bound process that begins with a Request for Designation to assign the combination product to the appropriate center. For example, scaffold and cell TERMPs having a cell-based primary mode of action would most likely be regulated by CBER's Office of Cellular, Tissue and Gene Therapies, with varying involvement from CDRH.

These regulatory organizations conduct evaluations under different regulatory authorities, depending on the designation of the product and the extent of clinical testing required. Lower-risk products that are minimally manipulated and intended for homologous use are considered under Section 361 of the Public Health Services Act and must comply with current good tissue practices (Table 3.6). Higher-risk products (e.g., cartilage that is implanted to provide bladder support) that are modified through tissue culture or genetic manipulation and not intended for homologous use are regulated under Section 351 of the Public Health Service Act and must comply with both the current good tissue practices and good manufacturing practices (Table 3.6) and go through a premarket approval review through an IND/BLA under 21 CFR 312/601 or IDE/PMA 21 CFR 812/814.

In moving toward clinical testing, nongovernmental groups guided by governmental regulations provide oversight of studies conducted in animals and humans. Animal care, use, and housing are governed by an institutional animal care and use committee (IACUC) whose operations are defined and established in 9 CFR 1–3. Although not a direct part of regulatory requirements to engage in clinical testing, institutions conducting animal studies in support of human trial testing are regulated by good laboratory practices (GLPs) and comply with United States Department of Agriculture (USDA) guidelines. Human subject testing is also governed by an institutional review board (IRB). IRB conduct and necessity are controlled by 21 CFR 56 whenever an application is submitted for a research and marketing permit. Specific IRB conduct may vary somewhat between institutions, but the IRB is consulted about necessary preclinical data prior to consideration of human subject testing in that institution.

**Table 3.6.** Regulated practices for consideration when taking a TERMP to clinical testing

Good tissue practices — 21 CFR 1271
Good manufacturing practices — 21 CFR 210 and 211
Good laboratory practices — 21 CFR 58
Good clinical practices — 21 CFR 50
Quality systems regulations — 21 CFR 820*

*Replaced cGMPs for TERMPs regulated as devices.

## VI. ESTABLISHING A REGULATORY PATHWAY

Substantial clarification about appropriate regulatory pathways for evaluating TERMPs has occurred in recent years and is currently most advanced in the United States. Since several regulatory pathways exist for these products and most of the product characteristics consist of a scaffold and cells isolated from a specified source, the Office of Combination Products (OCP) serves as the most common entry point for establishing regulatory authority (21 CFR 3) (FDA, 2003). Notably, some regenerative products may fit into existing regulatory pathways for drugs, devices, or biologics. Since regulatory pathways for individual products are well established, consideration here will focus on TERMPs composed of a combination of materials (biologics, drugs, and/or devices).

A sponsor seeking to obtain regulatory guidance for a combination product prepares a Request for Designation (RFD) document laying out key information requested by the FDA (Table 3.7). This document presents the sponsor's recommendation and rationale for how the combination product should be regulated. The FDA's decision on how to regulate a combination product is based on the primary mode of action, a judgment that focuses on the scaffold and cellular components of the TERMP. If the product is sufficiently close to a product already regulated by a particular center and pathway, the FDA's decision about requirements for clinical trials may mirror that product's regulatory pathway. The FDA has 60 days from the time of RFD submission to render a decision. Once a pathway has been identified, the sponsor can engage that particular reviewing authority for the optimal study plan to support their first clinical trials.

Specific guidance on engaging the Office of Combination Products and establishing communications with the FDA can be found on the FDA website (FDA, updated regularly). Interacting with this office prior to clinical testing can assist in linking to the proper regulatory authority and necessary regulatory guidelines. For some products the primary mode of action is not readily apparent, and the primary mode of action assignment may be based on the most relevant therapeutic activity, intended therapeutic use, similarity of the product to an existing product, or the most relevant safety and efficacy questions. This designation is then used to establish the most relevant regulatory center and potential regulatory pathway for entry into the clinics and ultimate product registration. A current assignment algorithm and flowchart can be obtained on the FDA website.

**Table 3.7.** Request for designation — information requested by the FDA

Name of product
Composition of product
Primary mode of action
Method of manufacture
Related products currently regulated by the FDA
Duration of product use by the patient
Science supporting product development
Primary route of administration

## VII. CONCLUSIONS

It is inevitable that regenerative medicine–based products will represent an important class of treatments for future patients. These products have the potential to satisfy significant unmet medical needs with an almost unimaginable benefit — a cure, not just a treatment. Regenerative medicine products can be customized to heal the specific needs of the patient in need. Currently, many regenerative medical products have little downside risk, since they eliminate rejection — autologous products representing the clearest example. These products may offer unmatched benefit:risk profiles with the potential to be rapidly approved for introduction into the appropriate patient populations and bring reductions in health care costs and substantial patient benefits, particularly when there are no medically acceptable alternatives.

The path to clinical entry has already been paved for these breakthroughs, which emerge from applying established processes — in cell biology and scaffold engineering — in a knowledgeable way. It is possible that regenerative medical products can be brought to market more rapidly and efficiently than traditional medical products (e.g., pharmaceuticals). The logistical advantages include development that can occur quickly with patient studies (rather than time-consuming and costly large-scale preclinical studies to define unknown risks), smaller trial sizes (customized nature of the products), and long-term follow-up that occurs postregistration (these products, once implanted, become part of the patient). One could easily envisage that once there is a dramatic success that combines effective therapy with compelling clinical data, industrial-scale efforts will open the floodgates to developing treatments for diseases that today fill patients with fear and little hope. The responsibility of tissue engineers for today will be to deliver on the promise of the hope and bring forward the promise of their scientific endeavors.

## VIII. ACKNOWLEDGMENTS

The authors thank Randall McKenzie (rmac3@att.net) for his remarkable work to illustrate this chapter. AJR also thanks the Department of Defense for its support of the National Tissue Engineering Center through a series of grants.

# IX. REFERENCES

Bergmann, D. (2004). Successful biopharmaceutical technology transfer. *Contract Pharma.* (June), 52–59.

Cruise, G. M., Hegre, O. D., Lamberti, F. V., Hager, S. R., Hill, R., Scharp, D. S., and Hubbell, J. A. (1999). *In vitro* and *in vivo* performance of porcine islets encapsulated in interfacially photopolymerized (ethylene glycol) diacrylate membranes. *Cell Transplant.* **8**(3), 293–306.

FDA. (1991). Intercenter agreement between the Center for Biologics Evaluation and Research and the Center for Devices and Radiological Health. http://www.fda.gov/oc/ombudsman/bio-dev.htm.

FDA. (1995). Required biocompatibility training and toxicology profiles for evaluation of medical devices. http://www.fda.gov/cdrh/g951.html.

FDA. (1997). Proposed approach to regulation of cellular and tissue-based products. http://www.fda.gov/cber/gdlns/celltissue.txt.

FDA. (1999). Medical device quality systems manual: a small entity compliance guide. http://www.fda.gov/cdrh/dsma/gmpman.html.

FDA. (2001). Human cells, tissues, and cellular and tissue-based products, establishment registration and listing, final rule, Vol. 66, 5459. Federal Register.

FDA. (2003). 21 CFR chapter I subchapter A — general part 3 — product jurisdiction. http://www.fda.gov/oc/ombudsman/part3&5.htm.

FDA. (2005a). 21 CFR part 1271. http://www.accessdata.fda.gov/scripts/cdrh/cfdocs/cfcfr/CFRSearch.cfm?CFRPart=1271.

FDA. (2005b). 21 CFR part 820. http://www.accessdata.fda.gov/scripts/cdrh/cfdocs/cfCFR/CFRSearch.cfm?CFRPart=820.

FDA. (2006). Office of Combination Products. http://www.fda.gov/oc/combination/.

Linberg, C., and Carrel, A. (1938). "The Culture of Organs." Paul B. Hober, New York.

Nishida, K., Yamato, M., Hayashida, Y., Watanabe, K., Yamamoto, K., Adachi, E., Nagai, S., Kikuchi, A., Maeda, N., Watanabe, H., Okano, T., and Tano, Y. (2004). Corneal reconstruction with tissue-engineered cell sheets composed of autologous oral mucosal epithelium. *N. Engl. J. Med.* **351**(12), 1187–1196.

Panza, J. L., Wagner, W. R., Rilo, H. L., Rao, R. H., Beckman, E. J., and Russell, A. J. (2000). Treatment of rat pancreatic islets with reactive PEG. *Biomaterials.* **21**(11), 1155–1164.

Preti, R. A. (2005). Bringing safe and effective cell therapies to the bedside. *Nat. Biotechnol.* **23**(7), 801–804.

Rice, L. L., and Lowery, A. (1995). Premarket notification 510(k): regulatory requirements for medical devices. http://www.fda.gov/cdrh/manual/510kprt1.html.

Shinoka, T., Matsumura, K., Hibino, N., Naito, Y., Murata, A., Kosaka, Y., and Kurosawa, H. (2003). Clinical practice of transplantation of regenerated blood vessels using bone marrow cells. *Nippon Naika Gakkai Zasshi.* **92**(9), 1776–1780.

U.S. Dept. of Health and Human Services. (2005). "2020 A New Vision: A Future for Regenerative Medicine." Washington, DC: U.S. Government Printing Office.

Weber, D. J. (2004). Navigating FDA regulations for human cells and tissues. *BioProcess Internat.* **2**(8), 22–27.

# Chapter Four

# Future Perspectives

*Mark E. Furth and Anthony Atala*

I. Clinical Need
II. Current State of the Field
III. Current Challenges

IV. Future Directions
V. Future Challenges
VI. References

## I. CLINICAL NEED

Tissue engineering combines principles of materials and cell transplantation to develop substitute tissues and/or promote endogenous regeneration. The approach initially was conceived to address the critical gap between the growing number of patients on the waiting list for organ transplantation due to end-stage failure and the limited number of donated organs available for such procedures (Lavik and Langer, 2004; Nerem, 2000). Increasingly, tissue engineering and, more broadly, regenerative medicine will focus on even more prevalent conditions in which the restoration of functional tissue would answer a currently unmet medical need. The development of therapies for patients with severe chronic disease affecting major organs such as the heart, kidney, and liver but not yet on transplantation waiting lists would vastly expand the potential impact of tissue-engineering technologies. A notable example is congestive heart failure, with nearly 5 million patients in the United States alone who might benefit from successful engineering of cardiac tissue (Murray-Thomas and Cowie, 2003). Similarly, diabetes mellitus is now recognized as an exploding epidemic, with approximately 16 million patients in the United States and over 217 million worldwide (Smyth and Heron, 2006). Patients with both type 1 and type 2 disease have insufficient pancreatic β-cell mass and potentially could be treated by transplantation of surrogate β-cells or neo-islets (Weir, 2004). A recent report from the U.S. National Academy of Sciences on *Stem Cells and the Future of Regenerative Medicine* highlighted these and other condi-

tions, including osteoporosis (10 million U.S. patients), Alzheimer's and Parkinson's diseases (5.5 million patients), severe burns (0.3 million), spinal cord injuries (0.25 million), and birth defects (0.15 million), as targets of regenerative medicine (Research, 2002).

## II. CURRENT STATE OF THE FIELD

Significant progress has been realized in tissue engineering since its principles were defined (Langer and Vacanti, 1993) and its broad medical and socioeconomic promise were recognized (Lysaght and O'Loughlin, 2000; Vacanti and Langer, 1999). However, to date only a handful of products incorporating cells together with scaffolds, notably bioartificial skin grafts and replacement cartilage, have gained regulatory approval, and these have achieved limited market penetration (Lysaght and Hazlehurst, 2004). Nonetheless, recent clinical reports with multiple years of patient follow-up document the maturation of the field and validate the significance of creating living replacement structures.

In one study, vascular grafts utilizing autologous bone marrow cells seeded onto biodegradable synthetic conduits or patches were implanted into 42 pediatric patients with congenital heart defects (Mastumura *et al.*, 2003; Shin'oka *et al.*, 2005). Safety data were encouraging; there was no evidence of aneurysms or other adverse events after a mean follow-up of 490 days (maximum 32 months) postsurgery. The grafted engineered vessels remained patent and functional and, most importantly, increased in diameter as the patients grew.

Encouraging clinical data also have been reported from work on tissue-engineered bladder constructs. Grafts comprising autologous urothelium and smooth muscle cells expanded *ex vivo* and seeded onto a biodegradable collagen or collagen-PLGA composite scaffold were implanted into seven pediatric patients with high-pressure or poorly compliant bladders in need of cystoplasty (Atala *et al.*, 2006). Serial follow-up data obtained over 22–61 months (mean 46 months) postsurgery provide evidence for the safety and efficacy of the procedure and highlight advantages over previous surgical approaches.

## III. CURRENT CHALLENGES

Technical as well as economic hurdles must be overcome before therapies based on tissue engineering will be able to reach the millions of patients who might benefit from them. One long-recognized challenge is the development of methods to enable engineering of tissues with complex three-dimensional architecture. A particular aspect of this problem is to overcome the mass transport limit by enabling provision of sufficient oxygen and nutrients to engineered tissue prior to vascularization and enhancing the formation of new blood vessels after implantation. The use of angiogenic factors, improved scaffold materials, printing technologies, and accelerated *in vitro* maturation of engineered tissues in bioreactors may help to address this problem. Of particular interest is the invention of novel scaffold materials designed to serve an instructive role in the development of engineered tissues. Methods to prepare improved cell–scaffold constructs by growth in bioreactors before implantation will serve a complementary role in generating more robust clinical products.

A second key challenge centers on a fundamental dichotomy in strategies for sourcing of cells for engineered tissues — the use of autologous cells versus allogeneic or even xenogeneic cells. On the one hand, it appears most cost effective and efficient for manufacturing, regulatory approval, and wide delivery to end users to employ a minimal number of cell donors, unrelated to recipient patients, to generate an off-the-shelf product. On the other hand, grafts can be generated from autologous cells obtained from a biopsy of each individual patient. Such grafts present no risk of immune rejection because of genetic mismatches, thereby avoiding the need for immunosuppressive drug therapy. Thus, the autologous approach, though likely more laborious and costly, appears to have a major advantage. Nonetheless, there are many tissue-engineering applications for which appropriate autologous donor cells may not be available. Therefore, new sources of cells for regenerative medicine are being sought and assessed, mainly from among progenitor and stem cell populations.

## IV. FUTURE DIRECTIONS

### Smarter Biomaterials

Scaffolds provide mechanical support and shape for neotissue construction *in vitro* and/or through the initial period after implantation as cells expand, differentiate, and organize (Stock and Vacanti, 2001). Materials that mainly have been used to date to formulate degradable scaffolds include synthetic polymers, such as poly(L-lactic acid) (PLLA) and poly(glycolic acid) (PLGA), and polymeric biomaterials, such as alginate, chitosan, collagen, and fibrin (Langer and Tirrell, 2004). Composites of these synthetic or natural polymers with bioactive ceramics such as hydroxyapatite or certain glasses can be designed to yield materials with a range of strengths and porosities, particularly for the engineering of hard tissues (Boccaccini and Blaker, 2005).

### Extracellular Matrix

A scaffold used for tissue engineering can be considered an artificial extracellular matrix (ECM) (Rosso *et al.*, 2005). It has long been appreciated that the normal biological ECM, in addition to contributing to mechanical integrity, has important signaling and regulatory functions in the development, maintenance, and regeneration of tissues. ECM components, in synergy with soluble signals provided by growth factors and hormones, participate in the tissue-specific control of gene expression through a variety of transduction mechanisms (Blum *et al.*, 1989; Jones *et al.*, 1993; Juliano and Haskill, 1993; Reid *et al.*, 1981). Furthermore, the ECM is itself a dynamic structure that is actively remodeled by the cells with which it interacts (Behonick and Werb, 2003; Birkedal-Hansen, 1995). An important future area of tissue engineering will be to develop improved scaffolds that more nearly recapitulate the biological properties of authentic ECM (Lutolf and Hubbell, 2005).

Decellularized tissues or organs can serve as sources of biological ECM for tissue engineering. The relatively high degree of evolutionary conservation of many ECM components allows the use of xenogeneic materials (often porcine). Various extracellular matrices have been utilized successfully for tissue engineering in animal models, and products incorporating decellularized heart valves, small intestinal submucosa (SIS), and urinary bladder have received regulatory approval for use in human patients (Gilbert *et al.*, 2006). The use of decellularized matrices is likely to expand, because they retain the complex set of molecules and three-dimensional structure of authentic ECM. Despite many advantages, there are also concerns about the use of decellularized materials. These include the potential for immunogenicity, the possible presence of infectious agents, variability among preparations, and the inability to completely specify and characterize the bioactive components of the material.

### Electrospinning

Current developments foreshadow the development of a new generation of biomaterials that use defined, purified components to mimic key features of the ECM. Electrospinning allows the production of highly biocompatible micro- and nano-fibrous scaffolds from synthetic materials, such as poly(epsilon-caprolactone), and from diverse matrix proteins, such as collagen, elastin, fibrinogen, and silk fibroin (Boland *et al.*, 2004; M. Li *et al.*, 2005; W. Li *et al.*, 2003; Matthews *et al.*, 2002; McManus *et al.*, 2006; Pham *et al.*, 2006; Shields *et al.*, 2004). Electrospun protein materials have fiber diameters in the range of those found in native ECM and display improved mechanical properties over hydrogels. The electrospun scaffolds may incorporate additional important ECM components, such as particular subtypes of collagen, glycosaminoglycans, and laminin, either in the spun fibers or as coatings, to promote cell adhesion, growth, and differentiation (Ma *et al.*, 2005; Rho *et al.*, 2006; Zhong *et al.*, 2005). The use of specialized proteins such as silk fibroin offers the opportunities to design scaffolds with enhanced strength or other favorable features (Ayutsede *et al.*, 2006; Jin *et al.*, 2004; Kim *et al.*, 2005; Min *et al.*, 2004), while the use of inexpensive materials such as wheat gluten may enable the production of lower-cost electrospun biomaterials (Woerdeman *et al.*, 2005).

Electrospinning technology also facilitates the production of scaffolds blending proteins with synthetic polymers to confer desired properties. Blending of collagen type I with biodegradable, elastomeric poly(ester urethane)urea generated strong, elastic matrices with improved capacity to promote cell binding and expression of specialized phenotypes as compared to the synthetic polymer alone (W. He *et al.*, 2005; Kwon and Matsuda, 2005; Stankus *et al.*, 2004). Novel properties not normally associated with the ECM may be introduced. For example, nanofibers coelectrospun from polyaniline and gelatin yielded an electrically conductive scaffold with good biocompatibility (M. Li *et al.*, 2006).

One demanding application of scaffold technology is in the production of a biological vascular substitute (Niklason *et al.*, 1999). Electrospun combinations of collagen and elastin or collagen and synthetic polymers have been considered for the development of vascular scaffolds (Boland *et al.*, 2004; W. He *et al.*, 2005; Kwon and Matsuda, 2005; Ma *et al.*, 2005). Recently, electrospinning was utilized to fabricate scaffolds blending collagen type I and elastin with PLGA for use in neo–blood vessels (Stitzel *et al.*, 2006). These scaffolds showed compliance, burst pressure, and mechanics comparable to native vessels and displayed good biocompatibility both *in vitro* and after implantation *in vivo*. When seeded with endothelial and smooth muscle cells, such scaffolds may provide a basis to produce functional vascular grafts suitable for clinical applications such as cardiac bypass procedures.

It may be problematic to introduce cells into a nanofibrillar structure in which pore spaces are considerably smaller than the diameter of a cell (Lutolf and Hubbell, 2005). However, remarkably, it is possible to utilize electrospinning to incorporate living cells into a fibrous matrix. A recent proof-of-concept study documented that smooth muscle cells could be concurrently electrospun with an elastomeric poly(ester urethane)urea, leading to "microintegration" of the cells in strong, flexible fibers with mechanical properties not greatly inferior to those of the synthetic polymer alone (Stankus *et al.*, 2006). The cell population retained high viability, and, when maintained in a perfusion bioreactor, the cellular density in the electrospun fibers doubled over four days in culture. In a similar vein, it has been found that cells can survive inkjet printing (Nakamura *et al.*, 2005; Roth *et al.*, 2004; Xu *et al.*, 2005). Printing of cells together with matrix biomaterials will allow the production of three-dimensional structures that mimic the architectural complexity and cellular distribution of complex tissues. The technology can be applied even to highly specialized, fragile cells, such as neurons. After inkjet printing of hippocampal and cortical neurons, the cells retained their specialized phenotype, as judged by both immunohistochemical staining and whole-cell patch-clamping, a stringent functional test of electrical excitability (Xu *et al.*, 2006). Incorporation of cells by electrospinning or printing generates, in a sense, the ultimate smart biomaterials.

### Smart Polymers

At the chemical level, a number of groups have begun to explore the production of biomaterials that unite the advantages of smart synthetic polymers with the biological activities of proteins. The notion of smart polymers initially described materials that show large conformational changes in response to small environmental stimuli, such as temperature, ionic strength, pH, and light (Galaev and Mattiasson, 1999; Williams, 2005). The responses of the polymer may include precipitation or gelation, reversible adsorption on a surface, collapse of a hydrogel or surface graft, and alternation between hydrophilic and hydrophobic states (A. S. Hoffman *et al.*, 2000). In many cases the change in the state of the polymer is reversible. Biological applications of this technology currently under development span diverse areas, including bioseparation, drug delivery, reusable enzymatic catalysts, molecular switches, biosensors, regulated protein folding, microfluidics, and gene therapy (Roy and Gupta, 2003). In tissue engineering, smart polymers offer promise for revolutionary improvements in scaffolds. Beyond the physical properties of polymers, a major goal is to invest smart biomaterials with specific properties of signaling proteins, such as ECM components and growth factors.

One approach is to link smart polymers to proteins (A. S. Hoffman, 2000; A. S. Hoffman *et al.*, 2000). The proteins can be conjugated either randomly or in a site-specific manner, through engineering of the protein to introduce a reactive amino acid at a particular position. If a conjugation site is introduced near the ligand-binding domain of a protein, induction of a change in conformational state of the smart polymer can serve to regulate the protein's activity (Stayton *et al.*, 1995). This may allow selective capture and recovery of specific cells, delivery of cells to a desired location, and modulation of enzymes, such as matrix metalloproteases, that influence tissue remodeling.

## Proteins and Mimetics

More broadly, the design of genetically modified proteins or of hybrid polymers incorporating peptides and protein domains will enable the creation of a wealth of novel biomaterials that also can be designated smart (Anderson et al., 2004a). These include engineered mutant variants of existing proteins, semisynthetic scaffold materials incorporating protein domains, scaffold materials linked to synthetic peptides, and engineered peptides capable of self-assembly into nanofibers.

Genetic engineering may improve on natural proteins for applications in tissue engineering (van Hest and Tirrell, 2001). For example, a collagen-like protein was generated by using recombinant DNA technology to introduce tandem repeats of the domain of human collagen II most critically associated with the migration of chondrocytes (Ito et al., 2006). When coated onto a PLGA scaffold and seeded with chondrocytes, the engineered collagen was superior to wild-type collagen II in promoting artificial cartilage formation. Similarly, recombinant technology has been employed to generate a series of elastin-mimetic protein triblock copolymers (Nagapudi et al., 2005). These varied broadly in their mechanical and viscoelastic properties, offering substantial choices for the production of novel materials for tissue engineering.

The incorporation of bioactive signals into scaffold materials of the types just described can be accomplished by the chemical linkage of synthetic peptides as tethered ligands. Numerous studies have confirmed that incorporation of the integrin-binding motif arginine-glycine-aspartic acid (RGD), first identified in fibronectin (Ruoslahti and Pierschbacher, 1987), enhances the binding of many types of cells to a variety of synthetic scaffolds and surfaces (Alsberg et al., 2002; Hersel et al., 2003; Liu et al., 2004). The CS5 cell–binding domain of fibronectin (Mould et al., 1991) also has been incorporated into scaffolds and its activity shown to be subject to regulation by sequence context (Heilshorn et al., 2005). It is likely that greater selectivity and potency in cellular binding and enhancement of growth and function will be achieved in the future by taking advantage of the growing understanding of the role of additional binding motifs in addition to and/or in concert with RGD (Salsmann et al., 2006; Takagi, 2004). The integrin family comprises two dozen heterodimeric proteins, so there is great opportunity to expand the set of peptide-binding motifs that could be utilized on tissue-engineering scaffolds, with the hope of achieving greater selectivity and control.

The modification of matrices with bioactive peptides and proteins can extend well beyond binding motifs to promote cell adhesion (Boontheekul and Mooney, 2003). Cells also need to migrate in order to form remodeled tissues. Thus, the rate of degradation of scaffolds used for tissue engineering is a crucial parameter affecting successful regeneration (Alsberg et al., 2003). Regulation of the degradation rate can be achieved by varying physical parameters of the scaffold. Alternatively, target sites for proteolytic degradation can be built into the scaffold (Halstenberg et al., 2002; S. H. Lee et al., 2005; Mann et al., 2001). For example, the incorporation into a cross-linked synthetic hydrogel of target sequences for matrix metalloproteases known to play an important role in cell invasion was shown to enhance the migration of fibroblasts in vitro and the healing of bony defects in vivo (Lutolf et al., 2003). Biodegradation of the synthetic matrix was efficiently coupled to tissue regeneration.

## Growth and Angiogenic Factors

Growth factors that drive cell growth and differentiation can be added to the matrix in the form of recombinant proteins or, alternatively, expressed by regenerative cells via gene therapy. Factors of potential importance in tissue engineering and methods to deliver them have been reviewed recently (Vasita and Katti, 2006). Ideally, for optimized tissue formation without risk of hyperplasia, the growth factors should be presented to cells for a limited period of time and in the correct local environment. Biodegradable electrospun scaffolds are capable of releasing growth factors at low rates over periods of weeks to months (Chew et al., 2005; W. He et al., 2005; C. Li et al., 2006). Biologically regulated release of growth factors from scaffolds appears particularly promising as a means to ensure that cells in regenerating neotissues receive these signals when and in the amounts required. For example, by physically entrapping recombinant bone morphogenetic protein-2 (BMP-2) in a hydrogel so that it would be released by matrix metalloproteases, Lutolf et al. (2003) achieved excellent bone healing in a critical-size rat calvarial defect model. Similarly, incorporation of a neurotrophic factor in a degradable hydrogel was shown to promote local extension of neurites from explanted retina, and gels were designed to release multiple neurotrophin family members at different rates (Burdick et al., 2006).

Controlled presentation of angiogenic factors such as vascular endothelial growth factor (VEGF) should promote the well-regulated neovascularization of engineered regenerating tissue (Lei et al., 2004; Nomi et al., 2002). Again, it is possible to covalently couple an angiogenic factor to a matrix (Zisch et al., 2001) and to regulate its release based on cellular activity and demand (Zisch et al., 2003). The selection of a sulfated tetrapeptide that mimics the VEGF-binding capability of heparin, a sulfated glycosaminoglycan, provides another potential tool for the construction of scaffolds able to deliver an angiogenic factor to cells in a regulated manner (Maynard and Hubbell, 2005).

Spatial gradients can be generated in the presentation of growth factors within scaffold constructs. This may help to guide the formation of complex tissues and, in particular, to direct migration of cells within developing neotissues (Campbell et al., 2005; DeLong et al., 2005). The introduction of more sophisticated manufacturing technologies, such as solid free-form fabrication, will allow the production of tissue-engineering constructs comprising scaffolds, incorporated cells, and growth factors in precise, complex three-dimensional structures (Hutmacher *et al.*, 2004).

## Discovery of New Materials

A next stage of smart biomaterials development extends to the design or discovery of bioactive materials not necessarily based directly on naturally occurring carbohydrate or protein structures. At one level this may entail the relatively straightforward chemical synthesis of new materials, coupled with a search for novel activities. By adapting the combinatorial library approach already well established for synthetic peptides and druglike structures, together with even moderately high-throughput assays, thousands of candidate scaffold materials can be generated and tested. Thus, screening of a combinatorial library derived from commercially available monomers in the acrylate family revealed novel synthetic polymers that influenced the attachment, growth, and differentiation of human embryonic stem cells in unexpected ways (Anderson *et al.*, 2004b).

Potentially more revolutionary developments in biomaterials will continue to arise at the interface of tissue engineering with nanotechnology. Basic understanding of the three-dimensional structure of existing biological molecules is being applied to a "bottom-up" approach to generate new, self-assembling supramolecular architectures (Zhang, 2003; Zhao and Zhang, 2004). In particular, self-assembling peptides offer promise because of the large variety of sequences that can be made easily by automated chemical synthesis, the potential for bioactivity, the ability to form nanofibers, and responsiveness to environmental cues (Fairman and Akerfeldt, 2005). Recent advances include the design of short peptides (e.g., heptamers) based on coiled-coil motifs that reversibly assemble into nanofilaments and nanoropes, without excessive aggregation (Wagner *et al.*, 2005). These smart peptide amphiphiles can be induced to self-assemble by changes in concentration, pH, or level of divalent cations (Hartgerink *et al.*, 2001, 2002). Branched structures can be designed to present bioactive sequences such as RGD to cells via nanofiber gels or as coatings on conventional tissue-engineering scaffolds (Guler *et al.*, 2006; Harrington *et al.*, 2006). In addition, assembly can occur under conditions that permit the entrapment of viable cells in the resulting nanofiber matrix (Beniash *et al.*, 2005). The entrapped cells retain motility and the ability to proliferate.

Further opportunities exist to expand the range of peptidic biomaterials by utilizing additional chemical components, such as porphyrins, which can bind to peptides and induce folding (Kovaric *et al.*, 2006). Porphyrins and similar structures also may add functionality, such as oxygen storage, catalysis or photosensitization of chemical reactions, or transfer of charge or molecular excitation energy.

Peptide-based nanofibers may be designed to present bioactive sequences to cells at very high density, substantially exceeding that of corresponding peptide epitopes in biological ECM. For example, a pentapeptide epitope of laminin, isoleucine-lysine-valine-alanine-valine (IKVAV), known to promote neurite extension from neurons, was incorporated into peptide amphiphiles (PA) capable of self-assembly into nanofibers that form highly hydrated (>99.5 weight % water) gels (G. A. Silva *et al.*, 2004). When neural progenitor cells capable of differentiating into neurons or glia were encapsulated during assembly of the nanofibers, they survived over several weeks in culture. Moreover, even without the addition of neurotrophic growth factors, they displayed neuronal differentiation, as exemplified by the extension of large neurites, already obvious after one day, and by expression of βIII-tubulin. The production of neuron-like cells from the neural progenitors, whether dissociated or grown as clustered "neurospheres," was more rapid and robust in the IKVAV-PA gels than on laminin-coated substrates or with soluble IKVAV. By contrast, the production of cells expressing glial fibrillary acidic protein (GFAP), a marker of astrocytic differentiation, was suppressed significantly in the IKVAV-PA gels, even when compared to growth on laminin, which favors neuronal differentiation. The ability to direct stem or progenitor cell differentiation via a chemically synthesized biomaterial, without the need to incorporate growth factors, offers many potential advantages in regenerative medicine.

## Bioreactors

After seeding of cells onto scaffolds, a period of growth *in vitro* is often required prior to implantation. Static cell culture conditions generally have proven suboptimal for the development of engineered neotissues because of limitations on seeding efficiency and transport of nutrients, oxygen, and wastes. Bioreactor systems have been designed to overcome these difficulties and to facilitate the reproducible production of tissue-engineered constructs under tightly controlled conditions. The rapidly developing field of reactors for regenerative medicine applications has been reviewed recently (I. Martin *et al.*, 2004; Portner *et al.*, 2005; Visconti *et al.*, 2006; Wendt *et al.*, 2005). Future advances will likely come through improved understanding of the requirements for tissue development, coupled with increasingly sophisticated reactor engineering.

One area in which basic knowledge must increase is the level of oxygen most appropriate for formation of particular tissues. Contrary to conventional wisdom, for some tissues or cell types it appears that low oxygen tension is important for optimal growth and specialized function. For example, in tissue engineering of cartilage, whereas aerobic conditions are essential for adequate tissue production (Obradovic *et al.*, 1999), cultivation in bioreactors at reduced oxygen tension (e.g., 5% instead of the 20% found in room air) improves the production of glycoasminoglycans and the expression of additional characteristic phenotypic markers and functions (Kurz *et al.*, 2004; Mizuno and Glowacki, 2005; Saini and Wick, 2004). Growth of stem and progenitor cells at reduced oxygen tension also may enhance the production of differentiated derivatives (Betre *et al.*, 2006; Grayson *et al.*, 2006; D. W. Wang *et al.*, 2005).

It has become increasingly clear that, in addition to regulating mass transport, bioreactors may be used to enhance tissue formation through mechanical stimulation. For example, pulsatile flow helps the maturation of blood vessels (Niklason *et al.*, 1999), while mechanical stretch improves engineered muscle (Barron *et al.*, 2003). Engineering of bone, cartilage, blood vessels, and both skeletal and cardiac muscle all are likely to continue to advance, in part through more sophisticated mechanical conditioning of developing neotissues.

A third area of great importance will be the use of bioreactors to improve the manufacture of engineered grafts for clinical use (I. Martin *et al.*, 2004; Naughton, 2002; Wendt *et al.*, 2005). Key goals will be to standardize production in order to eliminate wasted units, to control costs, and to meet regulatory constraints, including good manufacturing practice (GMP) regulations.

The direct interface between man and bioreactors represents another significant challenge in the bioreactor field. On one hand, the patient is increasingly viewed as a potential *in vivo* bioreactor, providing an optimal environment for cell growth and differentiation to yield neotissues (Warnke *et al.*, 2006). There also are circumstances in which a bioreactor may serve as a bioartificial organ, attached directly to a patient's circulation. The most significant case is the effort to develop a bioartificial liver that can be used to sustain life during acute liver failure, until the patient's endogenous organ regenerates or can be replaced by orthotopic transplantation (Jasmund and Bader, 2002; Sauer *et al.*, 2001, 2003). Most designs to date have focused on the use of hollow-fiber bioreactors seeded either with human hepatic lineage cell lines or xenogeneic (e.g., porcine) hepatocytes. Despite intensive efforts, leading to at least nine clinical trials, no bioartificial liver assist device has yet achieved full regulatory approval (Park and Lee, 2005). However, improved bioreactor systems and the use of primary human hepatocytes show promise for enhanced functionality that may lead to clinical success (Gerlach, 2005; Guthke *et al.*, 2006; Zeilinger *et al.*, 2004).

The creation of a robust bioartificial pancreas to provide a physiologically responsive supply of insulin to diabetes patients represents a comparable major challenge for bioreactor development. Despite three decades of effort, no design has yet proved entirely successful (Kizilel *et al.*, 2005; A. I. Silva *et al.*, 2006), but recent reports offer encouragement (Ikeda *et al.*, 2006; Pileggi *et al.*, 2006). If bioartificial organ technology continues to advance, the demand for new sources of functional human cells such as hepatocytes and pancreatic β-cells will expand dramatically.

## Cell Sources

Both allogeneic and autologous cell sourcing have proven successful in certain tissue-engineering applications. Clinical trials have led to regulatory approval of products based on both types of sources.

Among the approved living, engineered skin products, Dermagraft (Smith & Nephew) and Apligraf (Organogenesis) both utilize allogeneic cells expanded greatly from donated human foreskins to treat many unrelated patients. Despite the genetic mismatch between donor and recipient, the skin cells in Dermagraft and Apligraf do not induce acute immune rejection, possibly because of the absence of antigen-presenting cells in the grafts (Briscoe *et al.*, 1999; Curran and Plosker, 2002; Eaglstein *et al.*, 1999; Horch *et al.*, 2005; Mansbridge, 1998). Thus, these products can be utilized without immunosuppressive drug therapy, which is essential for almost all organ transplantation and would be required for most regenerative-medicine applications using allogeneic cells (Moller *et al.*, 1999). Eventually, the donated skin cells may be rejected, but after sufficient time has passed for the patient's endogenous cells to take their place.

Tissue-engineered products based on autologous cells also have achieved regulatory approval and reached the market. Epicel (Genzyme Biosurgery), a permanent skin replacement product for patients with life-threatening burns, and Carticel (Genzyme Biosurgery), a chondrocyte-based treatment for large articular cartilage lesions, are examples of products based on harvesting and expanding autologous cells.

For some tissue-engineering applications currently under development, such as bladder augmentation, the ability to obtain a tissue biopsy and expand a sufficient number of autologous cells is well established. In other circumstances it is not clear how a patient's own cells could be harvested and/or expanded to yield enough material for production of the needed neotissue or organ. Cardiomyocytes, neurons of the central nervous system, hepatocytes and other liver cells, kidney cells, osteoblasts, and insulin-producing pancreatic beta-cells are examples of differentiated cell types for which new sources could enable novel therapies to address significant unmet medical needs.

Immature precursor cells present within tissue samples are essential for the expansion of cells from biopsies of skin, bladder, or cartilage that enables the engineering of the corresponding neotissues (Bianco and Robey, 2001). The ability to extend tissue engineering to other tissue and organ systems will depend greatly on finding sources of appropriate stem and progenitor cells. Three major sources currently are under intensive investigation by many laboratories: (1) embryonic stem (ES) and embryonic germ (EG) cells derived from discarded human embryos and germ line stem cells, respectively; (2) ES cells created by somatic cell nuclear transfer (therapeutic cloning); and (3) "adult" stem cells from fetal, neonatal, or adult tissue, either autologous or allogeneic. It appears likely that multiple tissue-engineered products based on each of these sources will be tested in the clinic in the coming years. They pose certain common challenges, and each also has specific drawbacks that must be overcome if clinical use is to be achieved.

## Embryonic Stem Cells

ES cells and EG cells appear very similar and will likely have comparable applications in tissue engineering. In fact, recent evidence suggests that the most closely related *in vivo* cell type to the ES cell is an early germ cell (Zwaka and Thomson, 2005). The ES cells can self-renew, apparently without limit, in culture and are pluripotent — that is, they can give rise to any cell type in the body (Amit *et al.*, 2000; Evans and Kaufman, 1981; G. R. Martin, 1981; Shamblott *et al.*, 1998; Thomson *et al.*, 1998). This great degree of plasticity represents both the strongest attraction and a significant potential limitation to the use of ES cells for regenerative medicine. A major remaining challenge is to direct the efficient production of pure populations of specific desired cell types from human ES cells (Odorico *et al.*, 2001).

ES cells appear unique among normal stem cells in being tumorigenic. Undifferentiated ES cells of murine, nonhuman primate, and human origin form teratomas *in vivo* containing an array of cell types, including representatives of all three embryonic germ layers (Cowan *et al.*, 2004; G. R. Martin, 1981; Thomson *et al.*, 1995, 1998; Vrana *et al.*, 2003). Therefore, it will be important to document rigorously the exclusion of undifferentiated stem cells from any tissue-engineered products derived from ES cells (Lawrenz *et al.*, 2004; Odorico *et al.*, 2001). Strategies have been envisaged to increase safety by introducing into ES cells a suicide gene, for example, that encoding the thymidine kinase of *Herpes simplex* virus, which would render any escaping tumor cells sensitive to the drug ganciclovir (Odorico *et al.*, 2001; Schuldiner *et al.*, 2003). However, the genetic manipulation is itself not without risk, and the need to validate the engineered cell system would likely extend and complicate regulatory review of therapeutic products.

A central issue that must be addressed for tissue-engineered products derived from ES cells, and also from any nonautologous adult stem cells, is immune rejection based on mismatches at genetic histocompatibility loci (Lysaght, 2003). It generally has been assumed that, because human ES cells and their differentiated derivatives can be induced to express high levels of MHC Class I antigens (e.g., HLA-A and HLA-B), any ES cell–based product will be subject to graft rejection (Drukker *et al.*, 2002).

Therapeutic cloning offers a potential means to generate cells with the exact genetic constitution of each individual patient so that immune rejection of grafts based on mismatched histocompatibility antigens should not occur. The approach entails transferring the nucleus of a somatic cell into an enucleated oocyte (SCNT), generating a blastocyst, and then culturing the inner cell mass to obtain an ES cell line (Colman and Kind, 2000). If required, genetic manipulation of the cells may be carried out to correct an inherited defect prior to production of the therapeutic graft (Rideout *et al.*, 2002). Despite a published claim later withdrawn (Hwang *et al.*, 2005), the generation of human ES cells by SCNT has not yet been achieved. However, the concept of therapeutic cloning to provide cells for tissue-engineering applications has been clearly validated in a large-animal model. Adult bovine fibroblasts were used as nuclear donors and bioengineered tissues were generated from cloned cardiac, skeletal muscle, and kidney cells (Lanza *et al.*, 2002). The grafts, including functioning renal units capable of urine production, were successfully transplanted into the corresponding donor animals long term, with no evidence of rejection. Although SCNT is the subject of political, ethical, and scientific debate, intense efforts in both the private sector and academic institutions are likely to yield cloned human lines in the near future (Hall *et al.*, 2006; Lysaght and Hazlehurst, 2003).

The properties and differentiation potential of a number of human ES cell lines currently used for research were reviewed recently (L. M. Hoffman and Carpenter, 2005). The clinical application of ES cells for tissue engineering will depend on the development of robust methods to isolate and grow them under conditions consistent with good manufacturing practice and regulatory review for safety. In particular, it is important to eliminate the requirement for murine feeder cells by using human feeders or, better, feeder-free conditions. In addition, development of culture conditions without the requirement for nonhuman serum would be advantageous. Progress has been made in the derivation and expansion of human ES cells with human feeder cells (Amit *et al.*, 2003; Hovatta *et al.*, 2003; J. B. Lee *et al.*, 2004, 2005; Miyamoto *et al.*, 2004; Stacey *et al.*, 2006; Stojkovic *et al.*, 2005; Yoo *et al.*, 2005) or entirely without feeders (Amit *et al.*, 2004; Beattie *et al.*, 2005; Carpenter *et al.*, 2004; Cheon *et al.*, 2006; Choo *et al.*, 2006; Darr *et al.*, 2006; Hovatta and Skottman, 2005; Klimanskaya *et al.*, 2005; Rosler *et al.*, 2004; Sjogren-Jansson *et al.*, 2005; G. Wang *et al.*, 2005).

Perhaps the greater challenge remains in directing the differentiation of human ES cells to a given desired lineage with high efficiency. The underlying difficulty is that ES cells are developmentally many steps removed from adult, differentiated cells, and to date we have no general way to deterministically control the key steps in lineage restriction. To induce differentiation *in vitro*, ES cells are allowed to attach to plastic in monolayer culture or, more frequently, to form aggregates called embryoid bodies (Itskovitz-Eldor *et al.*, 2000). Over time within these aggregates cell types of many lineages are generated, including representatives of the three germ layers. The production of embryoid bodies can be enhanced and made more consistent by incubation in bioreactors (Gerecht-Nir *et al.*, 2004). Further selection of specific lineages generally requires sequential exposure to a series of inducing conditions, either based on known signaling pathways or identified by trial and error. In most cases lineage-specific markers are expressed by the differentiated cells, but cells often do not progress to a full terminally differentiated phenotype. As summarized in recent reviews, the cell lineages that have been generated *in vitro* include, among others, several classes of neurons, astrocytes, oligodendrocytes, multipotent mesenchymal precursor cells, osteoblasts, cardiomyocytes, keratinocytes, pneumocytes, hematopoietic cells, hepatocytes, and pancreatic beta-cells (Caspi and Gepstein, 2006; S. G. Nir *et al.*, 2003; Passier *et al.*, 2006; Priddle *et al.*, 2006; Raikwar *et al.*, 2006; Tian and Kaufman, 2005; Trounson, 2006).

In general, it appears easier to obtain adult cells derived from ectoderm, including neurons, and mesoderm, including cardiomyocytes, than cells derived from endoderm (Trounson, 2006). This may help determine the earliest areas in which ES-derived cells enter clinical translation, once the barriers just discussed are surmounted. Dopaminergic neurons generated from primate and human ES cells already have been tested in animal models of Parkinson's disease, with encouraging results (Perrier *et al.*, 2004; Sanchez-Pernaute *et al.*, 2005; Tabar *et al.*, 2005). Promising data also have been obtained with ES-derived oligodendrocytes in spinal cord injury models (Enzmann *et al.*, 2006; Faulkner and Keirstead, 2005; Keirstead *et al.*, 2005; Mueller *et al.*, 2005; Nistor *et al.*, 2005; Vogel, 2005). Cardiomyocytes derived from human ES cells, similarly, are candidates for future clinical use, although the functional criteria that must be met to ensure physiological competence will be stringent because of the risk of inducing arrhythmias (Caspi and Gepstein, 2004, 2006; Gerecht-Nir and Itskovitz-Eldor, 2004; G. Goh *et al.*, 2005; J. Q. He *et al.*, 2003; Heng *et al.*, 2005; Lev *et al.*, 2005; Liew *et al.*, 2005; Moore *et al.*, 2005; Mummery *et al.*, 2002; S. G. Nir *et al.*, 2003; Passier *et al.*, 2006).

The robust generation of pancreatic β-cells and bioengineered islets from human ES cells or other stem cells would represent a particularly important achievement, with potential to treat diabetes (T. Nir and Dor, 2005; Weir, 2004). Clusters of insulin-positive cells, resembling pancreatic islets and expressing various additional markers of the endocrine pancreatic lineage, have been produced from mouse (Lumelsky *et al.*, 2001; Morioh *et al.*, 2003) and from nonhuman primate and human ES cell lines (Assady *et al.*, 2001; Baharvand *et al.*, 2006; Brolen *et al.*, 2005; Lester *et al.*, 2004). The production of β-like cells can be enhanced by expression of pancreatic transcription factors (Miyazaki *et al.*, 2004; Shiroi *et al.*, 2005). However, the assessment of differentiation must take into account the uptake of insulin from the growth medium, in addition to *de novo* synthesis (Hansson *et al.*, 2004; Paek *et al.*, 2005). It seems fair to conclude that the efficient production of functional β-cells from ES cells remains a difficult objective to achieve. As in other bioengineering applications with ES-derived cells, efforts to reverse diabetes also will depend on the complete removal of nondifferentiated cells, to avoid the formation of teratoma tumors, which were observed after implantation of ES-derived β-cells in an animal model (Fujikawa *et al.*, 2005).

## Adult Stem Cells

Despite the acknowledged promise of ES cells, the challenges of controlling lineage-specific differentiation and eliminating residual stem cells are likely to extend the timeline for a number of tissue-engineering applications. In many cases adult stem cells may provide a more direct route to clinical translation.

Lineage-restricted stem cells have been isolated from both fetal and postnatal tissues based on selective outgrowth in culture and/or immunoselection for surface markers. Examples with significant potential for new applications in regenerative medicine include neural (Baizabal *et al.*, 2003; E. L. Goh *et al.*, 2003; Leker and McKay, 2004; Rothstein and Snyder, 2004), cardiac (Boltrami *et al.*, 2003; Oh *et al.*, 2003, 2004), muscle-derived (Cao *et al.*, 2005; Deasy *et al.*, 2005; Kuroda *et al.*, 2006; Payne *et al.*, 2005), and hepatic stem cells (Dabeva and Shafritz, 2003; Kamiya *et al.*, 2006; Kubota and Reid, 2000; Schmelzer *et al.*, 2006; Sicklick *et al.*, 2006; Walkup and Gerber, 2006; Zheng and Taniguchi, 2003). A significant feature of each of these populations is a high capacity for self-renewal in culture. Their ability to expand may be less than that for ES cells, but in some cases the cells have been shown to express telomerase and may not be subject to replicative senescence. These adult stem cells are multipotent. Neural stem cells can yield neurons, astrocytes, and oligodendrocytes. Cardiac stem cells are reported to yield cardiomyocytes, smooth muscle, and endothelial cells. Muscle-derived stem cells yield skeletal muscle and can be induced to produce chondrocytes. Hepatic stem cells yield hepatocytes and bile duct epithelial cells. The lineage-restricted adult stem cells all appear nontumorigenic. Thus, unlike ES cells, it is likely that they could be used safely for bioengineered products with or without prior differentiation.

It is possible that some lineage-specific adult stem cells are capable of greater plasticity than might be supposed based solely on their tissue of origin. For example, there is evidence that hepatic stem cells may be induced to generate cells of additional endodermal lineages, such as the endocrine pancreas (Nakajima-Nagata *et al.*, 2004; Yamada *et al.*, 2005; Yang *et al.*, 2002; Zalzman *et al.*, 2005). This type of switching of fates among related cell lineages may prove easier than inducing a full developmental program from a primitive precursor such as an ES cell.

Another class of adult cells with enormous potential value for regenerative medicine is the mesenchymal stem cells (MSC), initially described in bone marrow (Barry and Murphy, 2004; Bruder *et al.*, 1994; Pittenger *et al.*, 1999). These multipotent cells are able to give rise to differentiated cells of connective tissues, including bone, cartilage, muscle, tendon, and fat. The MSC have, therefore, generated considerable interest for musculoskeletal and vascular tissue engineering (Barry and Murphy, 2004; Gao and Caplan, 2003; Guilak *et al.*, 2004; Pelled *et al.*, 2002; Raghunath *et al.*, 2005; Riha *et al.*, 2005; Risbud and Shapiro, 2005; Tuan *et al.*, 2003). Cells with similar differentiation potential and marker profiles have been isolated from a number of tissues in addition to the bone marrow. A notable source is adipose tissue, in which the cells are abundant and easily obtained by processing of suction-assisted lipectomy (liposuction) specimens (Gimble and Guilak, 2003; Gimble, 2003; Zuk *et al.*, 2001).

In general it seems better to view MSC as mixed populations of progenitor cells with varying degrees of replicative potential, rather than homogeneous stem cells. However, some classes of MSC, including lines cloned from single cells in skin (Bartsch *et al.*, 2005), have been maintained in culture for extended periods. A very small subset of mesenchymal cells from bone marrow, termed MAPC, reportedly are capable of extensive self-renewal and of differentiation into cell lineages not observed with typical MSC, including examples from each embryonic germ layer (Jiang *et al.*, 2002).

Cells originating in a developing fetus and isolated from amniotic fluid or chorionic villi are a new source of stem cells of great potential interest for regenerative medicine (De Coppi *et al.*, 2001; De Coppi *et al.*, 2007; Siddiqui and Atala, 2004; Tsai *et al.*, 2006). Fetal-derived cells with apparently similar properties also have been described in the amnion of term placenta (Miki *ei al.*, 2005). Amniotic fluid stem (AFS) cells and amniotic epithelial cells can give rise to differentiated cell types representing the three embryonic germ layers (De Coppi *et al.*, 2007; Miki *et al.*, 2005; Siddiqui and Atala, 2004). Formal proof that single AFS cells can yield this full range of progeny cells was obtained using clones marked by retroviral insertion. The cells can be expanded for well over 200 population doublings, with no sign of telomere shortening or replicative senescence, and retain a normal diploid karyotype. They are readily cultured without need for feeder cells. The AFS cells express some markers in common with embryonic stem cells, such as the surface antigen SSEA4 and the transcription factor Oct3/4, while other markers are shared with mesenchymal and neural stem cells (De Coppi *et al.*, 2007). A broadly multipotent cell population obtained from umbilical cord blood may have certain key properties in common with AFS cells, and it was termed *unrestricted somatic stem cells* (USSCs) (Kogler *et al.*, 2004).

The full developmental potential of the various stem cell populations obtained from fetal and adult sources remains to be determined. It is possible that virtually all of the cell types that might be desired for tissue engineering could be obtained from AFS cells, equivalent stem cells from placenta, USSCs, or comparable populations. Similar approaches to those being taken with ES cells, such as genetic modification with expression vectors for lineage-specific transcription factors, may help in the generation of differentiated cell types for which it proves difficult to develop a straightforward induction protocol using external signals. However, it will remain necessary to show, beyond induction of a set of characteristic markers, that fully functional mature cells can be generated for any given lineage.

## Immune Compatibility

The growing number of choices of cell sources for bioengineered tissues opens up a range of strategies to obtain the desired cell populations. The issue of immune compatibility remains central. Although lifelong immunosuppression can be successful, as documented by its use in conjunction with orthotopic transplantation to treat terminal organ failure, it would be preferable to design bioengineering-based products that will be tolerated by recipients even without immunosuppressive drugs. The only cell-based therapies guaranteed to be histocompatible would contain autologous cells or those derived by therapeutic cloning (assuming mitochondrial differences are not critical). When a perfectly matched, personalized therapeutic product is not available, there still should be ways to limit the requirement for immunosuppression. First, there may be a strong intrinsic advantage to developing cell-based products from certain stem cells because there is evidence that they, and possibly differentiated cells derived from them, are immune privileged. Second, it may be possible to develop banks of cells that can be used to permit histocompatibility matching with recipient patients.

Human ES cells express low levels of Class I major histocompatibility complex (MHC) antigens (HLA-A, HLA-B) and are negative for MHC Class II (Drukker *et al.*, 2002). Differentiated derivatives of the ES cells remain negative for MHC II but show some increase in MHC Class I that is up-regulated by exposure to interferon. These observations gave rise to the natural assumption that ES cells and their differentiated progeny would be subject to rejection based on MHC mismatches and led to a search for strategies to induce immunological tolerance in recipients of transplanted cells derived from ES lines (Drukker, 2004; Drukker and Benvenisty, 2004). However, it was observed that ES cells in the mouse and similar stage stem cells in the rat could be transplanted successfully in immune-competent animals despite mismatches at the major histocompatibility loci. Furthermore, rodent ES cells may be able to induce immune tolerance in the recipient animals (Fandrich *et al.*, 2002a, 2002b). Even more remarkably, human ES cells and differentiated derivatives were not rejected by immune-competent mice *in vivo*, nor did they stimulate an immune response *in vitro* by human T lymphoctyes specific for mismatched MHC. Rather, the human cells appeared to inhibit the T-cell response (L. Li *et al.*, 2004). An independent study using mice with a "humanized" immune system confirmed a very low T-cell response to human ES cells and differentiated derivatives (Drukker *et al.*, 2006).

MSC from bone marrow and their differentiated derivatives also have been shown both to escape an allogeneic immune response and to possess immunomodulatory activity to block such a response (Aggarwal and Pittenger, 2005; Barry, 2003; Bartholomew *et al.*, 2002; Le Blanc, 2003; Potian *et al.*, 2003). The effect also is observed with MSC isolated from adipose tissue (Puissant *et al.*, 2005). The successful therapeutic use of allogeneic MSC has been confirmed in animal models (Arinzeh *et al.*, 2003; De Kok *et al.*, 2003). Beyond the use of MSC as regenerative cells, it is possible that they could be employed to induce immune tolerance to grafts of other cell types. The mechanisms underlying the immunodulatory properties of MSC are under active investigation, and understanding them may have profound impact on regenerative medicine (Krampera *et al.*, 2006a, 2006b; Plumas *et al.*, 2005; Sotiropoulou *et al.*, 2006).

Other stem cell populations should be examined for their ability to escape and/or modulate an allogeneic immune response. While it is important to exercise caution in interpreting the laboratory results and in designing clinical trials, there is some reason to hope that the use of allogeneic stem cell–based bioengineered products will not necessarily imply the need for lifelong use of immunosuppressive drugs. In the first FDA-approved clinical trial of allogeneic human neural stem cells, in children with a neural ceroid lipofuscinosis disorder known as Batten disease (Taupin, 2006), immunosuppressive therapy will be utilized for the initial year after cell implantation and then reevaluated.

Banking of stem cells for future therapeutic use extends possibilities both for autologous and allogeneic therapy paradigms, even if it turns out that histocompatibility matching is important for stem cell–based therapies. Amniocentesis specimens, placenta, and cord blood represent sources from which highly multipotent adult stem cells can be obtained and typed with minimal invasiveness. Prospective parents could opt for collection and cryopreservation of such cells for future use by their children in the event of medical need. Furthermore, collection and typing of a sufficient number of samples (ca. 100,000 for the U.S. population) to permit nearly perfect histocompatibility matching between unrelated donors and recipients would be readily achieved. Similarly, collection and banking of cells from adult adipose tissue appears straightforward. Although it would entail a greater level of effort and could be politically controversial, it also might be feasible to prepare and bank a relatively large set of human ES lines to facilitate histocompatibility matching. One recent study suggests that a surprisingly modest number of banked lines or specimens could provide substantial ability to match donor cells to recipients (Taylor *et al.*, 2005). Based on patients registered on a kidney transplant waiting list in the United Kingdom, the authors concluded that "Approximately 150 consecutive blood group–compatible donors, 100 consecutive blood group O donors, or 10 highly selected homozygous donors could provide the maximum practical benefit for HLA matching." The main criterion in this analysis was achieving at least an HLA-DR match. However, the possibility to select a small number of donors (ca. 10) homozygous for common HLA types from a pool of approximately 10,000 potential donors would allow complete matching for over one-third of patients and beneficial matching (one HLA-A or one HLA-B mismatch only) for two-thirds, at least for a relatively genetically homogeneous population. Taken together with the low immunogenicity of certain stem cells, these results support the concept that the use of allogeneic bioengineered products may not demand concomitant intensive immunosuppressive treatment.

## V. FUTURE CHALLENGES

The clinical application of tissue engineering lies largely ahead of us. Although a handful of products have achieved regulatory approval and entered the marketplace, many more are in the planning or proof-of-concept stage. In order to reach the large number of patients who might potentially benefit from bioengineered therapeutics, advances will be required in manufacturing and distributing complex products. This will be a fruitful area for engineers to address.

It also will be critical to develop a close partnership among academic and industrial scientists and the regulatory agencies (e.g., the U.S. Food and Drug Administration) that must assess new therapies for safety and efficacy. Products that may contain novel cellular components, biomaterials, and active growth or angiogenic factors will demand sophisticated, multifaceted review. Historically, regulatory agencies have had far greater experience with single drug entities or devices than with combination products. However, there is reason for optimism that the FDA's experiences to date with successful applications will pave the way to effective review of future bioengineered products.

## VI. REFERENCES

Aggarwal, S., and Pittenger, M. F. (2005). Human mesenchymal stem cells modulate allogeneic immune cell responses. *Blood* **105**, 1815–1822.

Alsberg, E., Anderson, K. W., Albeiruti, A., Rowley, J. A., and Mooney, D. J. (2002). Engineering growing tissues. *Proc. Natl. Acad. Sci. U.S.A.* **99**, 12025–12030.

Alsberg, E., Kong, H. J., Hirano, Y., Smith, M. K., Albeiruti, A., and Mooney, D. J. (2003). Regulating bone formation via controlled scaffold degradation. *J. Dent. Res.* **82**, 903–908.

Amit, M., Carpenter, M. K., Inokuma, M. S., Chiu, C. P., Harris, C. P., Waknitz, M. A., Itskovitz-Eldor, J., and Thomson, J. A. (2000). Clonally derived human embryonic stem cell lines maintain pluripotency and proliferative potential for prolonged periods of culture. *Dev. Biol.* **227**, 271–278.

Amit, M., Margulets, V., Segev, H., Shariki, K., Laevsky, I., Coleman, R., and Itskovitz-Eldor, J. (2003). Human feeder layers for human embryonic stem cells. *Biol. Reprod.* **68**, 2150–2156.

Amit, M., Shariki, C., Margulets, V., and Itskovitz-Eldor, J. (2004). Feeder layer– and serum-free culture of human embryonic stem cells. *Biol. Reprod.* **70**, 837–845.

Anderson, D. G., Burdick, J. A., and Langer, R. (2004a). Materials science. Smart biomaterials. *Science* **305**, 1923–1924.

Anderson, D. G., Levenberg, S., and Langer, R. (2004b). Nanoliter-scale synthesis of arrayed biomaterials and application to human embryonic stem cells. *Nat. Biotechnol.* **22**, 863–866.

Arinzeh, T. L., Peter, S. J., Archambault, M. P., van den Bos, C., Gordon, S., Kraus, K., Smith, A., and Kadiyala, S. (2003). Allogeneic mesenchymal stem cells regenerate bone in a critical-sized canine segmental defect. *J. Bone Joint. Surg. Am.* **85-A**, 1927–1935.

Assady, S., Maor, G., Amit, M., Itskovitz-Eldor, J., Skorecki, K. L., and Tzukerman, M. (2001). Insulin production by human embryonic stem cells. *Diabetes* **50**, 1691–1697.

Atala, A., Bauer, S. B., Soker, S., Yoo, J. J., and Retik, A. B. (2006). Tissue-engineered autologous bladders for patients needing cystoplasty. *Lancet* **367**, 1241–1246.

Ayutsede, J., Gandhi, M., Sukigara, S., Ye, H., Hsu, C. M., Gogotsi, Y., and Ko, F. (2006). Carbon nanotube–reinforced *Bombyx mori* silk nanofibers by the electrospinning process. *Biomacromolecules* **7**, 208–214.

Baharvand, H., Jafary, H., Massumi, M., and Ashtiani, S. K. (2006). Generation of insulin-secreting cells from human embryonic stem cells. *Dev. Growth Differ.* **48**, 323–332.

Baizabal, J. M., Furlan-Magaril, M., Santa-Olalla, J., and Covarrubias, L. (2003). Neural stem cells in development and regenerative medicine. *Arch. Med. Res.* **34**, 572–588.

Barron, V., Lyons, E., Stenson-Cox, C., McHugh, P. E., and Pandit, A. (2003). Bioreactors for cardiovascular cell and tissue growth: a review. *Ann. Biomed. Eng.* **31**, 1017–1030.

Barry, F. P. (2003). Biology and clinical application of mesenchymal stem cells. *Birth Defects Res. C. Embryo Today* **69**, 250–256.

Barry, F. P., and Murphy, J. M. (2004). Mesenchymal stem cells: clinical applications and biological characterization. *Int. J. Biochem. Cell. Biol.* **36**, 568–584.

Bartholomew, A., Sturgeon, C., Siatskas, M., Ferrer, K., McIntosh, K., Patil, S., Hardy, W., Devine, S., Ucker, D., Deans, R., *et al.* (2002). Mesenchymal stem cells suppress lymphocyte proliferation *in vitro* and prolong skin graft survival *in vivo*. *Exp. Hematol.* **30**, 42–48.

Bartsch, G., Yoo, J. J., De Coppi, P., Siddiqui, M. M., Schuch, G., Pohl, H. G., Fuhr, J., Perin, L., Soker, S., and Atala, A. (2005). Propagation, expansion, and multilineage differentiation of human somatic stem cells from dermal progenitors. *Stem Cells Dev.* **14**, 337–348.

Beattie, G. M., Lopez, A. D., Bucay, N., Hinton, A., Firpo, M. T., King, C. C., and Hayek, A. (2005). Activin A maintains pluripotency of human embryonic stem cells in the absence of feeder layers. *Stem Cells* **23**, 489–495.

Behonick, D. J., and Werb, Z. (2003). A bit of give and take: the relationship between the extracellular matrix and the developing chondrocyte. *Mech. Dev.* **120**, 1327–1336.

Beltrami, A. P., Barlucchi, L., Torella, D., Baker, M., Limana, F., Chimenti, S., Kasahara, H., Rota, M., Musso, E., Urbanek, K., *et al.* (2003). Adult cardiac stem cells are multipotent and support myocardial regeneration. *Cell* **114**, 763–776.

Beniash, E., Hartgerink, J. D., Storrie, H., Stendahl, J. C., and Stupp, S. I. (2005). Self-assembling peptide amphiphile nanofiber matrices for cell entrapment. *Acta Biomater.* **1**, 387–397.

Betre, H., Ong, S. R., Guilar, F., Chilkoti, A., Fermor, B., and Setton, L. A. (2006). Chondrocytic differentiation of human adipose-derived adult stem cells in elastin-like polypeptide. *Biomaterials* **27**, 91–99.

Bianco. P., and Robey, P. G. (2001). Stem cells in tissue engineering. *Nature* **414**, 118–121.

Birkedal-Hansen, H. (1995). Proteolytic remodeling of extracellular matrix. *Curr. Opin. Cell. Biol.* **7**, 728–735.

Blum, J. L., Zeigler, M. E., and Wicha, M. S. (1989). Regulation of mammary differentiation by the extracellular matrix. *Environ. Health Perspect.* **80**, 71–83.

Boccaccini, A. R., and Blaker, J. J. (2005). Bioactive composite materials for tissue engineering scaffolds. *Expert Rev. Med. Devices* **2**, 303–317.

Boland, E. D., Matthews, J. A., Pawlowski, K. J., Simpson, D. G., Wnek, G. E., and Bowlin, G. L. (2004). Electrospinning collagen and elastin: preliminary vascular tissue engineering. *Front. Biosci.* **9**, 1422–1432.

Boontheekul, T., and Mooney, D. J. (2003). Protein-based signaling systems in tissue engineering. *Curr. Opin. Biotechnol.* **14**, 559–565.

Briscoe, D. M., Dharnidharka, V. R., Isaacs, C., Downing, G., Prosky, S., Shaw, P., Parenteau, N. L., and Hardin-Young, J. (1999). The allogeneic response to cultured human skin equivalent in the hu-PBL-SCID mouse model of skin rejection. *Transplantation* **67**, 1590–1599.

Brolen, G. K., Heins, N., Edsbagge, J., and Semb, H. (2005). Signals from the embryonic mouse pancreas induce differentiation of human embryonic stem cells into insulin-producing beta-cell-like cells. *Diabetes* **54**, 2867–2874.

Bruder, S. P., Fink, D. J., and Caplan, A. I. (1994). Mesenchymal stem cells in bone development, bone repair, and skeletal regeneration therapy. *J. Cell. Biochem.* **56**, 283–294.

Burdick, J. A., Ward, M., Liang, E., Young, M. J., and Langer, R. (2006). Stimulation of neurite outgrowth by neurotrophins delivered from degradable hydrogels. *Biomaterials* **27**, 452–459.

Campbell, P. G., Miller, E. D., Fisher, G. W., Walker, L. M., and Weiss, L. E. (2005). Engineered spatial patterns of FGF-2 immobilized on fibrin direct cell organization. *Biomaterials* **26**, 6762–6770.

Cao, B., Deasy, B. M., Pollett, J., and Huard, J. (2005). Cell therapy for muscle regeneration and repair. *Phys. Med. Rehabil. Clin. N. Am.* **16**, 889–907, viii.

Carpenter, M. K., Rosler, E. S., Fisk, G. J., Brandenberger, R., Ares, X., Miura, T., Lucero, M., and Rao, M. S. (2004). Properties of four human embryonic stem cell lines maintained in a feeder-free culture system. *Dev. Dyn.* **229**, 243–258.

Caspi, O., and Gepstein, L. (2004). Potential applications of human embryonic stem cell-derived cardiomyocytes. *Ann. N.Y. Acad. Sci.* **1015**, 285–298.

Caspi, O., and Gepstein, L. (2006). Regenerating the heart using human embryonic stem cells — from cell to bedside. *Isr. Med. Assoc. J.* **8**, 208–214.

Cheon, S. H., Kim, S. J., Jo, J. Y., Ryu, W. J., Rhee, K., and Roh, S., 2nd (2006). Defined feeder-free culture system of human embryonic stem cells. *Biol. Reprod.* **74**, 611.

Chew, S. Y., Wen, J., Yim, E. K., and Leong, K. W. (2005). Sustained release of proteins from electrospun biodegradable fibers. *Biomacromolecules* **6**, 2017–2024.

Choo, A., Padmanabhan, J., Chin, A., Fong, W. J., and Oh, S. K. (2006). Immortalized feeders for scale-up of human embryonic stem cells in feeder and feeder-free conditions. *J. Biotechnol.* **122**, 130–141.

Colman, A., and Kind, A. (2000). Therapeutic cloning: concepts and practicalities. *Trends Biotechnol.* **18**, 192–196.

Committee on the Biological and Biomedical Applications of Stem Cell Research. (2002). "Stem Cells and the Future of Regenerative Medicine." National Academy Press, Washington, DC.

Cowan, C. A., Klimanskaya, I., McMahon, J., Atienza, J., Witmyer, J., Zucker, J. P., Wang, S., Morton, C. C., McMahon, A. P., Powers, D., and Melton, D. A. (2004). Derivation of embryonic stem-cell lines from human blastocysts. *N. Engl. J. Med.* **350**, 1353–1356.

Curran, M. P., and Plosker, G. L. (2002). Bilayered bioengineered skin substitute (Apligraf): a review of its use in the treatment of venous leg ulcers and diabetic foot ulcers. *BioDrugs* **16**, 439–455.

Debeva, M. D., and Shafritz, D. A. (2003). Hepatic stem cells and liver repopulation. *Semin. Liver Dis.* **23**, 349–362.

Darr, H., Mayshar, Y., and Benvenisty, N. (2006). Overexpression of NANOG in human ES cells enables feeder-free growth while inducing primitive ectoderm features. *Development* **133**, 1193–1201.

Deasy, B. M., Gharaibeh, B. M., Pollett, J. B., Jones, M. M., Lucas, M. A., Kanda, Y., and Huard, J. (2005). Long-term self-renewal of postnatal muscle-derived stem cells. *Mol. Biol. Cell* **16**, 3323–3333.

De Coppi, P., Bartsch, G., Dal Cin, P., Yoo, J. J., Soker, S., and Atala, A. (2001). Human fetal stem cell isolation from amniotic fluid. Paper presented at: American Academy of Pediatrics National Conference (San Francisco), 210–211.

De Coppi, P., Bartsch, G., Siddiqui, M. M., Xu, T., Santos, C. C., Perin, L., Mostoslavsky, G., Serre, A. C., Snyder, E. Y., Yoo, J. J., Furth, M. E., Soker, S., and Atala, A. (2007). Isolation of amniotic stem cell lines with potential for therapy. *Nat. Biotechnol.* **25**, 100–106.

De Kok, I. J., Peter, S. J., Archambault, M., van den Bos, C., Kadiyala, S., Aukhil, I., and Cooper, L. F. (2003). Investigation of allogeneic mesenchymal stem cell–based alveolar bone formation: preliminary findings. *Clin. Oral Implants Res.* **14**, 481–489.

DeLong, S. A., Moon, J. J., and West, J. L. (2005). Covalently immobilized gradients of bFGF on hydrogel scaffolds for directed cell migration. *Biomaterials* **26**, 3227–3234.

Drukker, M. (2004). Immunogenicity of human embryonic stem cells: can we achieve tolerance? *Springer Semin. Immunopathol.* **26**, 201–213.

Drukker, M., and Benvenisty, N. (2004). The immunogenicity of human embryonic stem-derived cells. *Trends Biotechnol.* **22**, 136–141.

Drukker, M., Katz, G., Urbach, A., Schuldiner, M., Markel, G., Itskovitz-Eldor, J., Reubinoff, B., Mandelboim, O., and Benvenisty, N. (2002). Characterization of the expression of MHC proteins in human embryonic stem cells. *Proc. Natl. Acad. Sci. U.S.A.* **99**, 9864–9869.

Drukker, M., Katchman, H., Katz, G., Even-Tov Friedman, S., Shezen, E., Hornstein, E., Mandelboim, O., Reisner, Y., and Benvenisty, N. (2006). Human embryonic stem cells and their differentiated derivatives are less susceptible to immune rejection than adult cells. *Stem Cells* **24**, 221–229.

Eaglstein, W. H., Alvarez, O. M., Auletta, M., Leffel, D., Rogers, G. S., Zitelli, J. A., Norris, J. E., Thomas, I., Irondo, M., Fewkes, J., et al. (1999). Acute excisional wounds treated with a tissue-engineered skin (Apligraf). *Dermatol. Surg.* **25**, 195–201.

Enzmann, G. U., Benton, R. L., Talbott, J. F., Cao, Q., and Whittemore, S. R. (2006). Functional considerations of stem cell transplantation therapy for spinal cord repair. *J. Neurotrauma* **23**, 479–495.

Evans, M. J., and Kaufman, M. H. (1981). Establishment in culture of pluripotential cells from mouse embryos. *Nature* **292**, 154–156.

Fairman, R., and Akerfeldt, K. S. (2005). Peptides as novel smart materials. *Curr. Opin. Struct. Biol.* **15**, 453–463.

Fandrich, F., Dresske, B., Bader, M., and Schulze, M. (2002a). Embryonic stem cells share immune-privileged features relevant for tolerance induction. *J. Mol. Med.* **80**, 343–350.

Fandrich, F., Lin, X., Chai, G. X., Schulze, M., Ganten, D., Bader, M., Holle, J., Huang, D. S., Parwaresch, R., Zavazava, N., and Binas, B. (2002b). Preimplantation-stage stem cells induce long-term allogeneic graft acceptance without supplementary host conditioning. *Nat. Med.* **8**, 171–178.

Faulkner, J., and Keirstead, H. S. (2005). Human embryonic stem cell–derived oligodendrocyte progenitors for the treatment of spinal cord injury. *Transpl. Immunol.* **15**, 131–142.

Fujikawa, T., Oh, S. H., Pi, L., Hatch, H. M., Shupe, T., and Petersen, B. E. (2005). Teratoma formation leads to failure of treatment for type I diabetes using embryonic stem cell–derived insulin-producing cells. *Am. J. Pathol.* **166**, 1781–1791.

Galaev, I. Y., and Mattiasson, B. (1999). "Smart" polymers and what they could do in biotechnology and medicine. *Trends Biotechnol.* **17**, 335–340.

Gao, J., and Caplan, A. I. (2003). Mesenchymal stem cells and tissue engineering for orthopaedic surgery. *Chir. Organi. Mov.* **88**, 305–316.

Gerecht-Nir, S., and Itskovitz-Eldor, J. (2004). Human embryonic stem cells: a potential source for cellular therapy. *Am. J. Transplant.* **4**(Suppl 6), 51–57.

Gerecht-Nir, S., Cohen, S., and Itskovitz-Eldor, J. (2004). Bioreactor cultivation enhances the efficiency of human embryoid body (hEB) formation and differentiation. *Biotechnol. Bioeng.* **86**, 493–502.

Gerlach, J. C. (2005). Prospects of the use of hepatic cells for extracorporeal liver support. *Acta Gastroenterol. Belg.* **68**, 358–368.

Gilbert, T. W., Sellaro, T. L., and Badylak, S. F. (2006). Decellularization of tissues and organs. *Biomaterials* **27**, 3675–3683.

Gimble, J. M. (2003). Adipose tissue-derived therapeutics. *Expert Opin. Biol. Ther.* **3**, 705–713.

Gimble, J., and Guilak, F. (2003). Adipose-derived adult stem cells: isolation, characterization, and differentiation potential. *Cytotherapy* **5**, 362–369.

Goh, E. L., Ma, D., Ming, G. L., and Song, H. (2003). Adult neural stem cells and repair of the adult central nervous system. *J. Hematother. Stem Cell Res.* **12**, 671–679.

Goh, G., Self, T., Barbadillo Munoz, M. D., Hall, I. P., Young, L., and Denning, C. (2005). Molecular and phenotypic analyses of human embryonic stem cell–derived cardiomyocytes: opportunities and challenges for clinical translation. *Thromb. Haemost.* **94**, 728–737.

Grayson, W. L., Zhao, F., Izadpanah, R., Bunnell, B., and Ma, T. (2006). Effects of hypoxia on human mesenchymal stem cell expansion and plasticity in 3D constructs. *J. Cell. Physiol.* **207**, 331–339.

Guilak, F., Awad, H. A., Fermor, B., Leddy, H. A., and Gimble, J. M. (2004). Adipose-derived adult stem cells for cartilage tissue engineering. *Biorheology* **41**, 389–399.

Guler, M. O., Hsu, L., Soukasene, S., Harrington, D. A., Hulvat, J. F., and Stupp, S. I. (2006). Presentation of RGDS epitopes on self-assembled nanofibers of branched peptide amphiphiles. *Biomacromolecules* **7**, 1855–1863.

Guthke, R., Zeilinger, K., Sickinger, S., Schmidt-Heck, W., Buentemeyer, H., Iding, K., Lehmann, J., Pfaff, M., Pless, G., and Gerlach, J. C. (2006). Dynamics of amino acid metabolism of primary human liver cells in 3D bioreactors. *Bioprocess. Biosyst. Eng.* **28**, 331–340.

Hall, V. J., Stojkovic, P., and Stojkovic, M. (2006). Using therapeutic cloning to fight human disease: a conundrum or reality? *Stem Cells* **24**, 1628–1637.

Halstenberg, S., Panitch, A., Rizzi, S., Hall, H., and Hubbell, J. A. (2002). Biologically engineered protein-graft-poly(ethylene glycol) hydrogels: a cell adhesive and plasmin-degradable biosynthetic material for tissue repair. *Biomacromolecules* **3**, 710–723.

Hansson, M., Tonning, A., Frandsen, U., Petri, A., Rajagopal, J., Englund, M. C., Heller, R. S., Hakansson, J., Fleckner, J., Skold, H. N., *et al.* (2004). Artifactual insulin release from differentiated embryonic stem cells. *Diabetes* **53**, 2603–2609.

Harrington, D. A., Cheng, E. Y., Guler, M. O., Lee, L. K., Donovan, J. L., Claussen, R. C., and Stupp, S. I. (2006). Branched peptide-amphiphiles as self-assembling coatings for tissue engineering scaffolds. *J. Biomed. Mater. Res.* **78**, 157–167.

Hartgerink, J. D., Beniash, E., and Stupp, S. I. (2001). Self-assembly and mineralization of peptide-amphiphile nanofibers. *Science* **294**, 1684–1688.

Hartgerink, J. D., Beniash, E., and Stupp, S. I. (2002). Peptide-amphiphile nanofibers: a versatile scaffold for the preparation of self-assembling materials. *Proc. Natl. Acad. Sci. U.S.A.* **99**, 5133–5138.

He, J. Q., Ma, Y., Lee, Y., Thomson, J. A., and Kamp, T. J. (2003). Human embryonic stem cells develop into multiple types of cardiac myocytes: action potential characterization. *Circ. Res.* **93**, 32–39.

He, W., Yong, T., Teo, W. E., Ma, Z., and Ramakrishna, S. (2005). Fabrication and endothelialization of collagen-blended biodegradable polymer nanofibers: potential vascular graft for blood vessel tissue engineering. *Tissue Eng.* **11**, 1574–1588.

Heilshorn, S. C., Liu, J. C., and Tirrell, D. A. (2005). Cell-binding domain context affects cell behavior on engineered proteins. *Biomacromolecules* **6**, 318–323.

Heng, B. C., Haider, H. K., Sim, E. K., Cao, T., Tong, G. Q., and Ng, S. C. (2005). Comments about possible use of human embryonic stem cell-derived cardiomyocytes to direct autologous adult stem cells into the cardiomyogenic lineage. *Acta Cardiol.* **60**, 7–12.

Hersel, U., Dahmen, C., and Kessler, H. (2003). RGD-modified polymers: biomaterials for stimulated cell adhesion and beyond. *Biomaterials* **24**, 4385–4415.

Hoffman, A. S. (2000). Bioconjugates of intelligent polymers and recognition proteins for use in diagnostics and affinity separations. *Clin. Chem.* **46**, 1478–1486.

Hoffman, A. S., Stayton, P. S., Bulmus, V., Chen, G., Chen, J., Cheung, C., Chilkoti, A., Ding, Z., Dong, L., Fong, R., *et al.* (2000). Really smart bioconjugates of smart polymers and receptor proteins. *J. Biomed. Mater. Res.* **52**, 577–586.

Hoffman, L. M., and Carpenter, M. K. (2005). Characterization and culture of human embryonic stem cells. *Nat. Biotechnol.* **23**, 699–708.

Horch, R. E., Kopp, J., Kneser, U., Beier, J., and Bach, A. D. (2005). Tissue engineering of cultured skin substitutes. *J. Cell. Mol. Med.* **9**, 592–608.

Hovatta, O., and Skottman, H. (2005). Feeder-free derivation of human embryonic stem-cell lines. *Lancet* **365**, 1601–1603.

Hovatta, O., Mikkola, M., Gertow, K., Stromberg, A. M., Inzunza, J., Hreinsson, J., Rozell, B., Blennow, E., Andang, M., and Ahrlund-Richter, L. (2003). A culture system using human foreskin fibroblasts as feeder cells allows production of human embryonic stem cells. *Hum. Reprod.* **18**, 1404–1409.

Hutmacher, D. W., Sittinger, M., and Risbud, M. V. (2004). Scaffold-based tissue engineering: rationale for computer-aided design and solid free-form fabrication systems. *Trends Biotechnol.* **22**, 354–362.

Hwang, W. S., Roh, S. I., Lee, B. C., Kang, S. K., Kwon, D. K., Kim, S., Kim, S. J., Park, S. W., Kwon, H. S., Lee, C. K., *et al.* (2005). Patient-specific embryonic stem cells derived from human SCNT blastocysts. *Science* **308**, 1777–1783.

Ikeda, H., Kobayashi, N., Tanaka, Y., Nakaji, S., Yong, C., Okitsu, T., Oshita, M., Matsumoto, S., Noguchi, H., Narushima, M., *et al.* (2006). A newly developed bioartificial pancreas successfully controls blood glucose in totally pancreatectomized diabetic pigs. *Tissue Eng.* **12**, 1799–1809.

Ito, H., Steplewski, A., Alabyeva, T., and Fertala, A. (2006). Testing the utility of rationally engineered recombinant collagen-like proteins for applications in tissue engineering. *J. Biomed. Mater. Res. A* **76**, 551–560.

Itskovitz-Eldor, J., Schuldiner, M., Karsenti, D., Eden, A., Yanuka, O., Amit, M., Soreq, H., and Benvenisty, N. (2000). Differentiation of human embryonic stem cells into embryoid bodies compromising the three embryonic germ layers. *Mol. Med.* **6**, 88–95.

Jasmund, I., and Bader, A. (2002). Bioreactor developments for tissue engineering applications by the example of the bioartificial liver. *Adv. Biochem. Eng. Biotechnol.* **74**, 99–109.

Jiang, Y., Jahagirdar, B. N., Reinhardt, R. L., Schwartz, R. E., Keene, C. D., Ortiz-Gonzalez, X. R., Reyes, M., Lenvik, T., Lund, T., Blackstad, M., *et al.* (2002). Pluripotency of mesenchymal stem cells derived from adult marrow. *Nature* **418**, 41–49.

Jin, H. J., Chen, J., Karageorgiou, V., Altman, G. H., and Kaplan, D. L. (2004). Human bone marrow stromal cell responses on electrospun silk fibroin mats. *Biomaterials* **25**, 1039–1047.

Jones, P. L., Schmidhauser, C., and Bissell, M. J. (1993). Regulation of gene expression and cell function by extracellular matrix. *Crit. Rev. Eukaryot. Gene Expr.* **3**, 137–154.

Juliano, R. L., and Haskill, S. (1993). Signal transduction from the extracellular matrix. *J. Cell. Biol.* **120**, 577–585.

Kamiya, A., Gonzalez, F. J., and Nakauchi, H. (2006). Identification and differentiation of hepatic stem cells during liver development. *Front. Biosci.* **11**, 1302–1310.

Keirstead, H. S., Nistor, G., Bernal, G., Totoiu, M., Cloutier, F., Sharp, K., and Steward, O. (2005). Human embryonic stem cell-derived oligodendrocyte progenitor cell transplants remyelinate and restore locomotion after spinal cord injury. *J. Neurosci.* **25**, 4694–4705.

Kim, K. H., Jeong, L., Park, H. N., Shin, S. Y., Park, W. H., Lee, S. C., Kim, T. I., Park, Y. J., Seol, Y. J., Lee, Y. M., *et al.* (2005). Biological efficacy of silk fibroin nanofiber membranes for guided bone regeneration. *J. Biotechnol.* **120**, 327–339.

Kizilel, S., Garfinkel, M., and Opara, E. (2005). The bioartificial pancreas: progress and challenges. *Diabetes Technol. Ther.* **7**, 968–985.

Klimanskaya, I., Chung, Y., Meisner, L., Johnson, J., West, M. D., and Lanza, R. (2005). Human embryonic stem cells derived without feeder cells. *Lancet* **365**, 1636–1641.

Kogler, G., Sensken, S., Airey, J. A., Trapp, T., Muschen, M., Feldhahn, N., Liedtke, S., Sorg, R. V., Fischer, J., Rosenbaum, C., *et al.* (2004). A new human somatic stem cell from placental cord blood with intrinsic pluripotent differentiation potential. *J. Exp. Med.* **200**, 123–135.

Kovaric, B. C., Kokona, B., Schwab, A. D., Twomey, M. A., de Paula, J. C., and Fairman, R. (2006). Self-assembly of peptide porphyrin complexes: toward the development of smart biomaterials. *J. Am. Chem. Soc.* **128**, 4166–4167.

Krampera, M., Cosmi, L., Angeli, R., Pasini, A., Liotta, F., Andreini, A., Santarlasci, V., Mazzinghi, B., Pizzolo, G., Vinante, F., *et al.* (2006a). Role for interferon-gamma in the immunomodulatory activity of human bone marrow mesenchymal stem cells. *Stem Cells* **24**, 386–398.

Krampera, M., Pasini, A., Pizzolo, G., Cosmi, L., Romagnani, S., and Annunziato, F. (2006b). Regenerative and immunomodulatory potential of mesenchymal stem cells. *Curr. Opin. Pharmacol.* **6**, 435–41.

Kubota, H., and Reid, L. M. (2000). Clonogenic hepatoblasts, common precursors for hepatocytic and biliary lineages, are lacking classical major histocompatibility complex class I antigen. *Proc. Natl. Acad. Sci. U.S.A.* **97**, 12132–12137.

Kuroda, R., Usas, A., Kubo, S., Corsi, K., Peng, H., Rose, T., Cummins, J., Fu, F. H., and Huard, J. (2006). Cartilage repair using bone morphogenetic protein 4 and muscle-derived stem cells. *Arthritis Rheum.* **54**, 433–442.

Kurz, B., Domm, C., Jin, M., Sellckau, R., and Schunke, M. (2004). Tissue engineering of articular cartilage under the influence of collagen I/III membranes and low oxygen tension. *Tissue Eng.* **10**, 1277–1286.

Kwon, I. K., and Matsuda, T. (2005). Co-electrospun nanofiber fabrics of poly(L-lactide-co-epsilon-caprolactone) with type I collagen or heparin. *Biomacromolecules* **6**, 2096–2105.

Langer, R., and Tirrell, D. A. (2004). Designing materials for biology and medicine. *Nature* **428**, 487–492.

Langer, R., and Vacanti, J. P. (1993). Tissue engineering. *Science* **260**, 920–926.

Lanza, R. P., Chung, H. Y., Yoo, J. J., Wettstein, P. J., Blackwell, C., Borson, N., Hofmeister, E., Schuch, G., Soker, S., Moraes, C. T., *et al.* (2002). Generation of histocompatible tissues using nuclear transplantation. *Nat. Biotechnol.* **20**, 689–696.

Lavik, E., and Langer, R. (2004). Tissue engineering: current state and perspectives. *Appl. Microbiol. Biotechnol.* **65**, 1–8.

Lawrenz, B., Schiller, H., Willbold, E., Ruediger, M., Muhs, A., and Esser, S. (2004). Highly sensitive biosafety model for stem-cell-derived grafts. *Cytotherapy* **6**, 212–222.

Le Blanc, K. (2003). Immunomodulatory effects of fetal and adult mesenchymal stem cells. *Cytotherapy* **5**, 485–489.

Lee, J. B., Song, J. M., Lee, J. E., Park, J. H., Kim, S. J., Kang, S. M., Kwon, J. N., Kim, M. K., Roh, S. I., and Yoon, H. S. (2004). Available human feeder cells for the maintenance of human embryonic stem cells. *Reproduction* **128**, 727–735.

Lee, J. B., Lee, J. E., Park, J. H., Kim, S. J., Kim, M. K., Roh, S. I., and Yoon, H. S. (2005). Establishment and maintenance of human embryonic stem cell lines on human feeder cells derived from uterine endometrium under serum-free condition. *Biol. Reprod.* **72**, 42–49.

Lee, S. H., Miller, J. S., Moon, J. J., and West, J. L. (2005). Proteolytically degradable hydrogels with a fluorogenic substrate for studies of cellular proteolytic activity and migration. *Biotechnol. Prog.* **21**, 1736–1741.

Lei, Y., Haider, H., Shujia, J., and Sim, E. S. (2004). Therapeutic angiogenesis. Devising new strategies based on past experiences. *Basic Res. Cardiol.* **99**, 121–132.

Leker, R. R., and McKay, R. D. (2004). Using endogenous neural stem cells to enhance recovery from ischemic brain injury. *Curr. Neurovasc. Res.* **1**, 421–427.

Lester, L. B., Kuo, H. C., Andrews, L., Nauert, B., and Wolf, D. P. (2004). Directed differentiation of rhesus monkey ES cells into pancreatic cell phenotypes. *Reprod. Biol. Endocrinol.* **2**, 42.

Lev, S., Kehat, I., and Gepstein, L. (2005). Differentiation pathways in human embryonic stem cell-derived cardiomyocytes. *Ann. N.Y. Acad. Sci.* **1047**, 50–65.

Li, C., Vepari, C., Jin, H. J., Kim, H. J., and Kaplan, D. L. (2006). Electrospun silk-BMP-2 scaffolds for bone tissue engineering. *Biomaterials* **27**, 3115–3124.

Li, L., Baroja, M. L., Majumdar, A., Chadwick, K., Rouleau, A., Gallacher, L., Ferber, I., Lebkowski, J., Martin, T., Madrenas, J., and Bhatia, M. (2004). Human embryonic stem cells possess immune-privileged properties. *Stem Cells* **22**, 448–456.

Li, M., Mondrinos, M. J., Gandhi, M. R., Ko, F. K., Weiss, A. S., and Lelkes, P. I. (2005). Electrospun protein fibers as matrices for tissue engineering. *Biomaterials* **26**, 5999–6008.

Li, M., Guo, Y., Wei, Y., MacDiarmid, A. G., and Lelkes, P. I. (2006). Electrospinning polyaniline-contained gelatin nanofibers for tissue engineering applications. *Biomaterials* **27**, 2705–2715.

Li, W. J., Danielson, K. G., Alexander, P. G., and Tuan, R. S. (2003). Biological response of chondrocytes cultured in three-dimensional nanofibrous poly(epsilon-caprolactone) scaffold. *J. Biomed. Mater. Res.* **67**, 1105–1114.

Liew, C. G., Moore, H., Ruban, L., Shah, N., Cosgrove, K., Dunne, M., and Andrews, P. (2005). Human embryonic stem cells: possibilities for human cell transplantation. *Ann. Med.* **37**, 521–532.

Liu, J. C., Heilshorn, S. C., and Tirrell, D. A. (2004). Comparative cell response to artificial extracellular matrix proteins containing the RGD and CS5 cell-binding domains. *Biomacromolecules* **5**, 497–504.

Lumelsky, N., Blondel, O., Laeng, P., Velasco, I., Ravin, R., and McKay, R. (2001). Differentiation of embryonic stem cells to insulin-secreting structures similar to pancreatic islets. *Science* **292**, 1389–1394.

Lutolf, M. P., and Hubbell, J. A. (2005). Synthetic biomaterials as instructive extracellular microenvironments for morphogenesis in tissue engineering. *Nat. Biotechnol.* **23**, 47–55.

Lutolf, M. P., Weber, F. E., Schmoekel, H. G., Schense, J. C., Kohler, T., Muller, R., and Hubbell, J. A. (2003). Repair of bone defects using synthetic mimetics of collagenous extracellular matrices. *Nat. Biotechnol.* **21**, 513–518.

Lysaght, M. J. (2003). Immunosuppression, immunoisolation and cell therapy. *Mol. Ther.* **7**, 432.

Lysaght, M. J., and Hazlehurst, A. L. (2003). Private sector development of stem cell technology and therapeutic cloning. *Tissue Eng.* **9**, 555–561.

Lysaght, M. J., and Hazlehurst, A. L. (2004). Tissue engineering: the end of the beginning. *Tissue Eng.* **10**, 309–320.

Lysaght, M. J., and O'Loughlin, J. A. (2000). Demographic scope and economic magnitude of contemporary organ replacement therapies. *Asaio J.* **46**, 515–521.

Ma, Z., He, W., Yong, T., and Ramakrishna, S. (2005). Grafting of gelatin on electrospun poly(caprolactone) nanofibers to improve endothelial cell spreading and proliferation and to control cell orientation. *Tissue Eng.* **11**, 1149–1158.

Mann, B. K., Gobin, A. S., Tsai, A. T., Schmedlen, R. H., and West, J. L. (2001). Smooth muscle cell growth in photopolymerized hydrogels with cell adhesive and proteolytically degradable domains: synthetic ECM analogs for tissue engineering. *Biomaterials* **22**, 3045–3051.

Mansbridge, J. (1998). Skin substitutes to enhance wound healing. *Expert Opin. Investig. Drugs* **7**, 803–809.

Martin, G. R. (1981). Isolation of a pluripotent cell line from early mouse embryos cultured in medium conditioned by teratocarcinoma stem cells. *Proc. Natl. Acad. Sci. U.S.A.* **78**, 7634–7638.

Martin, I., Wendt, D., and Heberer, M. (2004). The role of bioreactors in tissue engineering. *Trends Biotechnol.* **22**, 80–86.

Matsumura, G., Hibino, N., Ikada, Y., Kurosawa, H., and Shin'oka, T. (2003). Successful application of tissue-engineered vascular autografts: clinical experience. *Biomaterials* **24**, 2303–2308.

Matthews, J. A., Wnek, G. E., Simpson, D. G., and Bowlin, G. L. (2002). Electrospinning of collagen nanofibers. *Biomacromolecules* **3**, 232–238.

Maynard, H. D., and Hubbell, J. A. (2005). Discovery of a sulfated tetrapeptide that binds to vascular endothelial growth factor. *Acta Biomater.* **1**, 451–459.

McManus, M. C., Boland, E. D., Koo, H. P., Barnes, C. P., Pawlowski, K. J., Wnek, G. E., Simpson, D. G., and Bowlin, G. L. (2006). Mechanical properties of electrospun fibrinogen structures. *Acta Biomater.* **2**, 19–28.

Miki, T., Lehmann, T., Cai, H., Stolz, D. B., and Strom, S. C. (2005). Stem cell characteristics of amniotic epithelial cells. *Stem Cells* **23**, 1549–1559.

Min, B. M., Lee, G., Kim, S. H., Nam, Y. S., Lee, T. S., and Park, W. H. (2004). Electrospinning of silk fibroin nanofibers and its effect on the adhesion and spreading of normal human keratinocytes and fibroblasts *in vitro*. *Biomaterials* **25**, 1289–1297.

Miyamoto, K., Hayashi, K., Suzuki, T., Ichihara, S., Yamada, T., Kano, Y., Yamabe, T., and Ito, Y. (2004). Human placenta feeder layers support undifferentiated growth of primate embryonic stem cells. *Stem Cells* **22**, 433–440.

Miyazaki, S., Yamato, E., and Miyazaki, J. (2004). Regulated expression of pdx-1 promotes in vitro differentiation of insulin-producing cells from embryonic stem cells. *Diabetes* **53**, 1030–1037.

Mizuno, S., and Glowacki, J. (2005). Low oxygen tension enhances chondroinduction by demineralized bone matrix in human dermal fibroblasts in vitro. *Cells Tissues Organs* **180**, 151–158.

Moller, E., Soderberg-Naucler, C., and Sumitran-Karuppan, S. (1999). Role of alloimmunity in clinical transplantation. *Rev. Immunogenet.* **1**, 309–322.

Moore, J. C., van Laake, L. W., Braam, S. R., Xue, T., Tsang, S. Y., Ward, D., Passier, R., Tertoolen, L. L., Li, R. A., and Mummery, C. L. (2005). Human embryonic stem cells: genetic manipulation on the way to cardiac cell therapies. *Reprod. Toxicol.* **20**, 377–391.

Moritoh, Y., Yamato, E., Yasui, Y., Miyazaki, S., and Miyazaki, J. (2003). Analysis of insulin-producing cells during *in vitro* differentiation from feeder-free embryonic stem cells. *Diabetes* **52**, 1163–1168.

Mould, A. P., Komoriya, A., Yamada, K. M., and Humphries, M. J. (1991). The CS5 peptide is a second site in the IIICS region of fibronectin recognized by the integrin alpha 4 beta 1. Inhibition of alpha 4 beta 1 function by RGD peptide homologues. *J. Biol. Chem.* **266**, 3579–3585.

Mueller, D., Shamblott, M. J., Fox, H. E., Gearhart, J. D., and Martin, L. J. (2005). Transplanted human embryonic germ cell–derived neural stem cells replace neurons and oligodendrocytes in the forebrain of neonatal mice with excitotoxic brain damage. *J. Neurosci. Res.* **82**, 592–608.

Mummery, C., Ward, D., van den Brink, C. E., Bird, S. D., Doevendans, P. A., Opthof, T., Brutel de la Riviere, A., Tertoolem, L., van der Heyden, M., and Pera, M. (2002). Cardiomyocyte differentiation of mouse and human embryonic stem cells. *J. Anat.* **200**, 233–242.

Murray-Thomas, T., and Cowie, M. R. (2003). Epidemiology and clinical aspects of congestive heart failure. *J. Renin Angiotensin Aldosterone Syst.* **4**, 131–136.

Nagapudi, K., Brinkman, W. T., Thomas, B. S., Park, J. O., Srinivasarao, M., Wright, E., Conticello, V. P., and Chaikof, E. L. (2005). Viscoelastic and mechanical behavior of recombinant protein elastomers. *Biomaterials* **26**, 4695–4706.

Nakajima-Nagata, N., Sakurai, T., Mitaka, T., Katakai, T., Yamato, E., Miyazaki, J., Tabata, Y., Sugai, M., and Shimizu, A. (2004). *In vitro* induction of adult hepatic progenitor cells into insulin-producing cells. *Biochem. Biophys. Res. Commun.* **318**, 625–630.

Nakamura, M., Kobayashi, A., Takagi, F., Watanabe, A., Hiruma, Y., Ohuchi, K., Iwasaki, Y., Horie, M., Morita, I., and Takatani, S. (2005). Biocompatible inkjet printing technique for designed seeding of individual living cells. *Tissue Eng.* **11**, 1658–1666.

Naughton, G. K. (2002). From lab bench to market: critical issues in tissue engineering. *Ann. N.Y. Acad. Sci.* **961**, 372–385.

Nerem, R. M. (2000). Tissue engineering: confronting the transplantation crisis. *Proc. Inst. Mech. Eng. [H]* **214**, 95–99.

Niklason, L. E., Gao, J., Abbott, W. M., Hirschi, K. K., Houser, S., Marini, R., and Langer, R. (1999). Functional arteries grown *in vitro*. *Science* **284**, 489–493.

Nir, S. G., David, R., Zaruba, M., Franz, W. M., and Itskovitz-Eldor, J. (2003). Human embryonic stem cells for cardiovascular repair. *Cardiovasc. Res.* **58**, 313–323.

Nir, T., and Dor, Y. (2005). How to make pancreatic beta cells — prospects for cell therapy in diabetes. *Curr. Opin. Biotechnol.* **16**, 524–529.

Nistor, G. I., Totoiu, M. O., Haque, N., Carpenter, M. K., and Keirstead, H. S. (2005). Human embryonic stem cells differentiate into oligodendrocytes in high purity and myelinate after spinal cord transplantation. *Glia* **49**, 385–396.

Nomi, M., Atala, A., Coppi, P. D., and Soker, S. (2002). Principals of neovascularization for tissue engineering. *Mol. Aspects. Med.* **23**, 463–483.

Obradovic, B., Carrier, R. L., Vunjak-Novakovic, G., and Freed, L. E. (1999). Gas exchange is essential for bioreactor cultivation of tissue engineered cartilage. *Biotechnol. Bioeng.* **63**, 197–205.

Odorico, J. S., Kaufman, D. S., and Thomson, J. A. (2001). Multilineage differentiation from human embryonic stem cell lines. *Stem Cells* **19**, 193–204.

Oh, H., Bradfute, S. B., Gallardo, T. D., Nakamura, T., Gaussin, V., Mishina, Y., Pocius, J., Michael, L. H., Behringer, R. R., Garry, D. J., *et al.* (2003). Cardiac progenitor cells from adult myocardium: homing, differentiation, and fusion after infarction. *Proc. Natl. Acad. Sci. U.S.A.* **100**, 12313–12318.

Oh, H., Chi, X., Bradfute, S. B., Mishina, Y., Pocius, J., Michael, L. H., Behringer, R. R., Schwartz, R. J., Entman, M. L., and Schneider, M. D. (2004). Cardiac muscle plasticity in adult and embryo by heart-derived progenitor cells. *Ann. N.Y. Acad. Sci.* **1015**, 182–189.

Paek, H. J., Morgan, J. R., and Lysaght, M. J. (2005). Sequestration and synthesis: the source of insulin in cell clusters differentiated from murine embryonic stem cells. *Stem Cells* **23**, 862–867.

Park, J. K., and Lee, D. H. (2005). Bioartificial liver systems: current status and future perspective. *J. Biosci. Bioeng.* **99**, 311–319.

Passier, R., Denning, C., and Mummery, C. (2006). Cardiomyocytes from human embryonic stem cells. *Handb. Exp. Pharmacol.* **174**, 101–122.

Payne, T. R., Oshima, H., Sakai, T., Ling, Y., Gharaibeh, B., Cummins, J., and Huard, J. (2005). Regeneration of dystrophin-expressing myocytes in the mdx heart by skeletal muscle stem cells. *Gene Ther.* **12**, 1264–1274.

Pelled, G., G, T., Aslan, H., Gazit, Z., and Gazit, D. (2002). Mesenchymal stem cells for bone gene therapy and tissue engineering. *Curr. Pharm. Des.* **8**, 1917–1928.

Perrier, A. L., Tabar, V., Barberi, T., Rubio, M. E., Bruses, J., Topf, N., Harrison, N. L., and Studer, L. (2004). Derivation of midbrain dopamine neurons from human embryonic stem cells. *Proc. Natl. Acad. Sci. U.S.A.* **101**, 12543–12548.

Pham, Q. P., Sharma, U., and Mikos, A. G. (2006). Electrospinning of polymeric nanofibers for tissue engineering applications: a review. *Tissue Eng.* **12**, 1197–1211.

Pileggi, A., Molano, R. D., Ricordi, C., Zahr, E., Collins, J., Valdes, R., and Inverardi, L. (2006). Reversal of diabetes by pancreatic islet transplantation into a subcutaneous, neovascularized device. *Transplantation* **81**, 1318–1324.

Pittenger, M. F., Mackay, A. M., Beck, S. C., Jaiswal, R. K., Douglas, R., Mosca, J. D., Moorman, M. A., Simonetti, D. W., Craig, S., and Marshak, D. R. (1999). Multilineage potential of adult human mesenchymal stem cells. *Science* **284**, 143–147.

Plumas, J., Chaperot, L., Richard, M. J., Molens, J. P., Bensa, J. C., and Favrot, M. C. (2005). Mesenchymal stem cells induce apoptosis of activated T cells. *Leukemia* **19**, 1597–1604.

Portner, R., Nagel-Heyer, S., Goepfert, C., Adamietz, P., and Meenen, N. M. (2005). Bioreactor design for tissue engineering. *J. Biosci. Bioeng.* **100**, 235–245.

Potian, J. A., Aviv, H., Ponzio, N. M., Harrison, J. S., and Rameshwar, P. (2003). Veto-like activity of mesenchymal stem cells: functional discrimination between cellular responses to alloantigens and recall antigens. *J. Immunol.* **171**, 3426–3434.

Priddle, H., Jones, D. R., Burridge, P. W., and Patient, R. (2006). Hematopoiesis from human embryonic stem cells: overcoming the immune barrier in stem cell therapies. *Stem Cells* **24**, 815–824.

Puissant, B., Barreau, C., Bourin, P., Clavel, C., Corre, J., Bousquet, C., Taureau, C., Cousin, B., Abbal, M., Laharrague, P., *et al.* (2005). Immunomodulatory effect of human adipose tissue-derived adult stem cells: comparison with bone marrow mesenchymal stem cells. *Br. J. Haematol.* **129**, 118–129.

Raghunath, J., Salacinski, H. J., Sales, K. M., Butler, P. E., and Seifalian, A. M. (2005). Advancing cartilage tissue engineering: the application of stem cell technology. *Curr. Opin. Biotechnol.* **16**, 503–509.

Raikwar, S. P., Mueller, T., and Zavazava, N. (2006). Strategies for developing therapeutic application of human embryonic stem cells. *Physiology (Bethesda)* **21**, 19–28.

Reid, L., Morrow, B., Jubinsky, P., Schwartz, E., and Gatmaitan, Z. (1981). Regulation of growth and differentiation of epithelial cells by hormones, growth factors, and substrates of extracellular matrix. *Ann. N.Y. Acad. Sci.* **372**, 354–370.

Rho, K. S., Jeong, L., Lee, G., Seo, B. M., Park, Y. J., Hong, S. D., Roh, S., Cho, J. J., Park, W. H., and Min, B. M. (2006). Electrospinning of collagen nanofibers: effects on the behavior of normal human keratinocytes and early-stage wound healing. *Biomaterials* **27**, 1452–1461.

Rideout, W. M., 3rd, Hochedlinger, K., Kyba, M., Daley, G. Q., and Jaenisch, R. (2002). Correction of a genetic defect by nuclear transplantation and combined cell and gene therapy. *Cell* **109**, 17–27.

Riha, G. M., Lin, P. H., Lumsden, A. B., Yao, Q., and Chen, C. (2005). Review: application of stem cells for vascular tissue engineering. *Tissue Eng.* **11**, 1535–1552.

Risbud, M. V., and Shapiro, I. M. (2005). Stem cells in craniofacial and dental tissue engineering. *Orthod. Craniofac. Res.* **8**, 54–59.

Rosler, E. S., Fisk, G. J., Ares, X., Irving, J., Miura, T., Rao, M. S., and Carpenter, M. K. (2004). Long-term culture of human embryonic stem cells in feeder-free conditions. *Dey. Dyn.* **229**, 259–274.

Rosso, F., Marino, G., Giordano, A., Barbarisi, M., Parmeggiani, D., and Barbarisi, A. (2005). Smart materials as scaffolds for tissue engineering. *J. Cell. Physiol.* **203**, 465–470.

Roth, E. A., Xu, T., Das, M., Gregory, C., Hickman, J. J., and Boland, T. (2004). Inkjet printing for high-throughput cell patterning. *Biomaterials* **25**, 3707–3715.

Rothstein, J. D., and Snyder, E. Y. (2004). Reality and immortality — neural stem cells for therapies. *Nat. Biotechnol.* **22**, 283–285.

Roy, I., and Gupta, M. N. (2003). Smart polymeric materials: emerging biochemical applications. *Chem. Biol.* **10**, 1161–1171.

Ruoslahti, E., and Pierschbacher, M. D. (1987). New perspectives in cell adhesion: RGD and integrins. *Science* **238**, 491–497.

Saini, S., and Wick, T. M. (2004). Effect of low oxygen tension on tissue-engineered cartilage construct development in the concentric cylinder bioreactor. *Tissue Eng.* **10**, 825–832.

Salsmann, A., Schaffner-Reckinger, E., and Kieffer, N. (2006). RGD, the Rho'd to cell spreading. *Eur. J. Cell. Biol.* **85**, 249–254.

Sanchez-Pernaute, R., Studer, L., Ferrari, D., Perrier, A., Lee, H., Vinuela, A., and Isacson, O. (2005). Long-term survival of dopamine neurons derived from parthenogenetic primate embryonic stem cells (cyno-1) after transplantation. *Stem Cells* **23**, 914–922.

Sauer, I. M., Obermeyer, N., Kardassis, D., Theruvath, T., and Gerlach, J. C. (2001). Development of a hybrid liver support system. *Ann. N.Y. Acad. Sci.* **944**, 308–319.

Sauer, I. M., Kardassis, D., Zeillinger, K., Pascher, A., Gruenwald, A., Pless, G., Irgang, M., Kraemer, M., Puhl, G., Frank, J., *et al.* (2003). Clinical extracorporeal hybrid liver support — phase I study with primary porcine liver cells. *Xenotransplantation* **10**, 460–469.

Schmelzer, E., Wauthier, E., and Reid, L. M. (2006). The phenotypes of pluripotent human hepatic progenitors. *Stem Cells* **24**, 1852–1858.

Schuldiner, M., Itskovitz-Eldor, J., and Benvenisty, N. (2003). Selective ablation of human embryonic stem cells expressing a "suicide" gene. *Stem Cells* **21**, 257–265.

Shamblott, M. J., Axelman, J., Wang, S., Bugg, E. M., Littlefield, J. W., Donovan, P. J., Blumenthal, P. D., Huggins, G. R., and Gearhart, J. D. (1998). Derivation of pluripotent stem cells from cultured human primordial germ cells. *Proc. Natl. Acad. Sci. U.S.A.* **95**, 13726–13731.

Shields, K. J., Beckman, M. J., Bowlin, G. L., and Wayne, J. S. (2004). Mechanical properties and cellular proliferation of electrospun collagen type II. *Tissue Eng.* **10**, 1510–1517.

Shin'oka, T., Matsumura, G., Hibino, N., Naito, Y., Watanabe, M., Konuma, T., Sakamoto, T., Nagatsu, M., and Kurosawa, H. (2005). Midterm clinical result of tissue-engineered vascular autografts seeded with autologous bone marrow cells. *J. Thorac. Cardiovasc. Surg.* **129**, 1330–1338.

Shiroi, A., Ueda, S., Ouji, Y., Saito, K., Moriya, K., Sugie, Y., Fukui, H., Ishizaka, S., and Yoshikawa, M. (2005). Differentiation of embryonic stem cells into insulin-producing cells promoted by Nkx2.2 gene transfer. *World J. Gastroenterol.* **11**, 4161–4166.

Sicklick, J. K., Li, Y. X., Melhem, A., Schmelzer, E., Zdanowicz, M., Huang, J., Caballero, M., Fair, J. H., Ludlow, J. W., McClelland, R. E., *et al.* (2006). Hedgehog signaling maintains resident hepatic progenitors throughout life. *Am. J. Physio. Gastrointest. Liver Physiol.* **290**, G859–870.

Siddiqui, M. M., and Atala, A. (2004). Amniotic fluid-derived pluripotential cells: adult and fetal. *In* "Handbook of Stem Cells" (R. Lanza, H. Blau, D. Melton, M. Moore, E. D. Thomas, C. Verfaillie, I. Weissman, and M. West, eds.), pp. 175–180. Elsevier Academic Press, Amsterdam.

Silva, A. I., de Matos, A. N., Brons, I. G., and Mateus, M. (2006). An overview on the development of a bio-artificial pancreas as a treatment of insulin-dependent diabetes mellitus. *Med. Res. Rev.* **26**, 181–222.

Silva, G. A., Czeisler, C., Niece, K. L., Beniash, E., Harrington, D. A., Kessler, J. A., and Stupp, S. I. (2004). Selective differentiation of neural progenitor cells by high-epitope density nanofibers. *Science* **303**, 1352–1355.

Sjogren-Jansson, E., Zetterstrom, M., Moya, K., Lindqvist, J., Strehl, R., and Eriksson, P. S. (2005). Large-scale propagation of four undifferentiated human embryonic stem cell lines in a feeder-free culture system. *Dev. Dyn.* **233**, 1304–1314.

Smyth, S., and Heron, A. (2006). Diabetes and obesity: the twin epidemics. *Nat. Med.* **12**, 75–80.

Sotiropoulou, P. A., Perez, S. A., Gritzapis, A. D., Baxevanis, C. N., and Papamichail, M. (2006). Interactions between human mesenchymal stem cells and natural killer cells. *Stem Cells* **24**, 74–85.

Stacey, G. N., Cobo, F., Nieto, A., Talavera, P., Healy, L., and Concha, A. (2006). The development of "feeder" cells for the preparation of clinical grade hES cell lines: challenges and solutions. *J. Biotechnol.*

Stankus, J. J., Guan, J., and Wagner, W. R. (2004). Fabrication of biodegradable elastomeric scaffolds with sub-micron morphologies. *J. Biomed. Mater. Res. A* **70**, 603–614.

Stankus, J. J., Guan, J., Fujimoto, K., and Wagner, W. R. (2006). Microintegrating smooth muscle cells into a biodegradable, elastomeric fiber matrix. *Biomaterials* **27**, 735–744.

Stayton, P. S., Shimoboji, T., Long, C., Chilkoti, A., Chen, G., Harris, J. M., and Hoffman, A. S. (1995). Control of protein-ligand recognition using a stimuli-responsive polymer. *Nature* **378**, 472–474.

Stitzel, J., Liu, J., Lee, S. J., Komura, M., Berry, J., Soker, S., Lim, G., Van Dyke, M., Czerw, R., Yoo, J. J., and Atala, A. (2006). Controlled fabrication of a biological vascular substitute. *Biomaterials* **27**, 1088–1094.

Stock, U. A., and Vacanti, J. P. (2001). Tissue engineering: current state and prospects. *Annu. Rev. Med.* **52**, 443–451.

Stojkovic, P., Lako, M., Stewart, R., Przyborski, S., Armstrong, L., Evans, J., Murdoch, A., Strachan, T., and Stojkovic, M. (2005). An autogeneic feeder cell system that efficiently supports growth of undifferentiated human embryonic stem cells. *Stem Cells* **23**, 306–314.

Tabar, V., Panagiotakos, G., Greenberg, E. D., Chan, B. K., Sadelain, M., Gutin, P. H., and Studer, L. (2005). Migration and differentiation of neural precursors derived from human embryonic stem cells in the rat brain. *Nat. Biotechnol.* **23**, 601–606.

Takagi, J. (2004). Structural basis for ligand recognition by RGD (Arg-Gly-Asp)-dependent integrins. *Biochem. Soc. Trans.* **32**, 403–406.

Taupin, P. (2006). HuCNS-SC (stem cells). *Curr. Opin. Mol. Ther.* **8**, 156–163.

Taylor, C. J., Bolton, E. M., Pocock, S., Sharples, L. D., Pedersen, R. A., and Bradley, J. A. (2005). Banking on human embryonic stem cells: estimating the number of donor cell lines needed for HLA matching. *Lancet* **366**, 2019–2025.

Thomson, J. A., Kalishman, J., Golos, T. G., Durning, M., Harris, C. P., Becker, R. A., and Hearn, J. P. (1995). Isolation of a primate embryonic stem cells line. *Proc. Natl. Acad. Sci. U.S.A.* **92**, 7844–7848.

Thomson, J. A., Itskovitz-Eldor, J., Shapiro, S. S., Waknitz, M. A., Swiergiel, J. J., Marshall, V. S., and Jones, J. M. (1998). Embryonic stem cell lines derived from human blastocysts. *Science* **282**, 1145–1147.

Tian, X., and Kaufman, D. S. (2005). Hematopoietic development of human embryonic stem cells in culture. *Methods Mol. Med.* **105**, 425–436.

Trounson, A. (2006). The production and directed differentiation of human embryonic stem cells. *Endocr. Rev.* **27**, 208–219.

Tsai, M. S., Hwang, S. M., Tsai, Y. L., Cheng, F. C., Lee, J. L., and Chang, Y. J. (2006). Clonal amniotic fluid-derived stem cells express characteristics of both mesenchymal and neural stem cells. *Biol. Reprod.* **74**, 545–551.

Tuan, R. S., Boland, G., and Tuli, R. (2003). Adult mesenchymal stem cells and cell-based tissue engineering. *Arthritis Res. Ther.* **5**, 32–45.

Vacanti, J. P., and Langer, R. (1999). Tissue engineering: the design and fabrication of living replacement devices for surgical reconstruction and transplantation. *Lancet* **354**(Suppl. 1), S132–134.

van Hest, J. C., and Tirrell, D. A. (2001). Protein-based materials, toward a new level of structural control. *Chem. Commun. (Camb.),* **19**, 1879–1904.

Vasita, R., and Katti, D. S. (2006). Growth factor–delivery systems for tissue engineering: a materials perspective. *Expert Rev. Med. Devices* **3**, 29–47.

Visconti, R., Mironov, V., Kasyanov, V. A., Yost, M. J., Twal, W., Trusk, T., Wen, X., Ozolanta, I., Kadishs, A., Prestwich, G. D., *et al.* (2006). Cardiovascular tissue engineering I. Perfusion bioreactors: a review. *J. Long-Term Eff. Med. Implants* **16**, 111–130.

Vogel, G. (2005). Cell biology. Ready or not? Human ES cells head toward the clinic. *Science* **308**, 1534–1538.

Vrana, K. E., Hipp, J. D., Goss, A. M., McCool, B. A., Riddle, D. R., Walker, S. J., Wettstein, P. J., Studer, L. P., Tabar, V., Cunniff, K., *et al.* (2003). Nonhuman primate parthenogenetic stem cells. *Proc. Natl. Acad. Sci. U.S.A.* **100**(Suppl. 1), 11911–11916.

Wagner, D. E., Phillips, C. L., Ali, W. M., Nybakken, G. E., Crawford, E. D., Schwab, A. D., Smith, W. F., and Fairman, R. (2005). Toward the development of peptide nanofilaments and nanoropes as smart materials. *Proc. Natl. Acad. Sci. U.S.A.* **102**, 12656–12661.

Walkup, M. H., and Gerber, D. A. (2006). Hepatic stem cells: in search of. *Stem Cells* **24**, 1833–1840.

Wang, D. W., Fermor, B., Gimble, J. M., Awad, H. A., and Guilak, F. (2005). Influence of oxygen on the proliferation and metabolism of adipose-derived adult stem cells. *J. Cell. Physiol.* **204**, 184–191.

Wang, G., Zhang, H., Zhao, Y., Li, J., Cai, J., Wang, P., Meng, S., Feng, J., Miao, C., Ding, M., *et al.* (2005). Noggin and bFGF cooperate to maintain the pluripotency of human embryonic stem cells in the absence of feeder layers. *Biochem. Biophys. Res. Commun.* **330**, 934–942.

Warnke, P. H., Wiltfang, J., Springer, I., Acil, Y., Bolte, H., Kosmahl, M., Russo, P. A., Sherry, E., Lutzen, U., Wolfart, S., and Terheyden, H. (2006). Man as living bioreactor: fate of an exogenously prepared customized tissue-engineered mandible. *Biomaterials* **27**, 3163–3167.

Weir, G. C. (2004). Can we make surrogate beta-cells better than the original? *Semin. Cell Dev. Biol.* **15**, 347–357.

Wendt, D., Jakob, M., and Martin, I. (2005). Bioreactor-based engineering of osteochondral grafts: from model system to tissue manufacturing. *J Biosci. Bioeng.* **100**, 489–494.

Willams, D. (2005). Environmentally smart polymers. *Med. Device Technol.* **16**, 9–10, 13.

Woerdeman, D. L., Ye, P., Shenoy, S., Parnas, R. S., Wnek, G. E., and Trofimova, O. (2005). Electrospun fibers from wheat protein: investigation of the interplay between molecular structure and the fluid dynamics of the electrospinning process. *Biomacromolecules* **6**, 707–712.

Xu, T., Jin, J., Gregory, C., Hickman, J. J., and Boland, T. (2005). Inkjet printing of viable mammalian cells. *Biomaterials* **26**, 93–99.

Xu, T., Gregory, C. A., Molnar, P., Cui, X., Jalota, S., Bhaduri, S. B., and Boland, T. (2006). Viability and electrophysiology of neural cell structures generated by the inkjet printing method. *Biomaterials* **27**, 3580–3588.

Yamada, S., Terada, K., Ueno, Y., Sugiyama, T., Seno, M., and Kojima, I. (2005). Differentiation of adult hepatic stem — like cells into pancreatic endocrine cells. *Cell. Transplant.* **14**, 647–653.

Yang, L., Li, S., Hatch, H., Ahrens, K., Cornelius, J. G., Petersen, B. E., and Peck, A. B. (2002). In vitro trans-differentiation of adult hepatic stem cells into pancreatic endocrine hormone-producing cells. *Proc. Natl. Acad. Sci. U.S.A.* **99**, 8078–8083.

Yoo, S. J., Yoon, B. S., Kim, J. M., Song, J. M., Roh, S., You, S., and Yoon, H. S. (2005). Efficient culture system for human embryonic stem cells using autologous human embryonic stem cell-derived feeder cells. *Exp. Mol. Med.* **37**, 399–407.

Zalzman, M., Anker-Kitai, L., and Efrat, S. (2005). Differentiation of human liver-derived, insulin-producing cells toward the beta-cell phenotype. *Diabetes* **54**, 2568–2575.

Zeilinger, K., Holland, G., Sauer, I. M., Efimova, E., Kardassis, D., Obermayer, N., Liu, M., Neuhaus, P., and Gerlach, J. C. (2004). Time course of primary liver cell reorganization in three-dimensional high-density bioreactors for extracorporeal liver support: an immunohistochemical and ultrastructural study. *Tissue Eng.* **10**, 1113–1124.

Zhang, S. (2003). Fabrication of novel biomaterials through molecular self-assembly. *Nat. Biotechol.* **21**, 1171–1178.

Zhao, X., and Zhang, S. (2004). Fabrication of molecular materials using peptide construction motifs. *Trends Biotechnol.* **22**, 470–476.

Zheng, Y. W., and Taniguchi, H. (2003). Diversity of hepatic stem cells in the fetal and adult liver. *Semin. Liver Dis.* **23**, 337–348.

Zhong, S., Teo, W. E., Zhu, X., Beuerman, R., Ramakrishna, S., and Yung, L. Y. (2005). Formation of collagen-glycosaminoglycan blended nanofibrous scaffolds and their biological properties. *Biomacromolecules* **6**, 2998–3004.

Zisch, A. H., Schenk, U., Schense, J. C., Sakiyama-Elbert, S. E., and Hubbell, J. A. (2001). Covalently conjugated VEGF — fibrin matrices for endothelialization. *J. Control Release* **72**, 101–113.

Zisch, A. H., Lutolf, M. P., Ehrbar, M., Raeber, G. P., Rizzi, S. C., Davies, N., Schmokel, H., Bezuidenhout, D., Djonov, V., Zilla, P., and Hubbell, J. A. (2003). Cell-demanded release of VEGF from synthetic, biointeractive cell ingrowth matrices for vascularized tissue growth. *Faseb J.* **17**, 2260–2262.

Zuk, P. A., Zhu, M., Mizuno, H., Huang, J., Futrell, J. W., Katz, A. J., Benhaim, P., Lorenz, H. P., and Hedrick, M. H. (2001). Multilineage cells from human adipose tissue: implications for cell-based therapies. *Tissue Eng.* **7**, 211–228.

Zwaka, T. P., and Thomson, J. A. (2005). A germ cell origin of embryonic stem cells? *Development* **132**, 227–233.

# The Basis of Growth and Differentiation

# Chapter Five

# Molecular Biology of the Cell

*Jonathan Slack*

## I. INTRODUCTION

This chapter is a general introduction to the properties of animal or human cells. It deals with gene expression, metabolism, protein synthesis and secretion, membrane properties, response to extracellular factors, cell division, properties of the cytoskeleton, cell adhesion, and the extracellular matrix. It shows how these cellular properties underlie the specific conditions required for successful tissue culture. In particular cells require effective access to nutrients, removal of waste products, and their growth and behavior are controlled by a variety of extracellular hormones and growth factors present in the medium. The properties of individual cells are also the basis for understanding how cells can become organized into tissues, which are normally composed of more than one cell type and have a specific microarchitecture appropriate to their function.

To a naive observer the term *tissue engineering* might seem a contradiction. The word *engineering* conjures up a vision of making objects from hard components, such as metals, plastics, concrete, and silicon, that are mechanically robust and will withstand a range of environmental conditions. The components themselves are often relatively simple, and the complexity of a system emerges from the number and connectivity of the parts. By contrast, the cells of living organisms are themselves highly delicate and highly complex. Despite our knowledge of a vast amount of molecular biological detail concerning cell structure and function, their properties are still understood only in qualitative

terms, and so any application using cells involves a lot of craft skill as well as rational design. What follows is a very brief account of cell properties intended for newcomers to tissue engineering who have an engineering or physical science background. It is intended to alert readers to some of the issues involved in working with cells and to pave the way for understanding how cells form tissues and organs, topics dealt with in more detail in the later chapters. Because it comprises very general material, it is not specifically referenced, although some further reading is provided at the end.

Cells are the basic building blocks of living organisms, in the sense that they can survive in isolation. Some organisms, such as bacteria, protozoa, and many algae, actually consist of single free-living cells. But most cells are constituents of multicellular organisms, which, though they can survive in isolation, need very carefully controlled conditions to do so. A typical animal cell suspended in liquid will be a sphere of the order of about 20 microns in diameter (Fig. 5.1).

Most cells will not grow well in suspension, and so they are usually grown attached to a substrate, where they flatten and may be quite large in horizontal dimensions but only a few microns in vertical dimension. All eukaryotic cells contain *a nucleus*, in which is located the genetic material that ultimately controls everything the cell is composed of and all the activities it carries out. This is surrounded by *cytoplasm*, which has a very complex structure and contain

*Principles of Tissue Engineering, 3rd Edition*
*ed. by Lanza, Langer, and Vacanti*

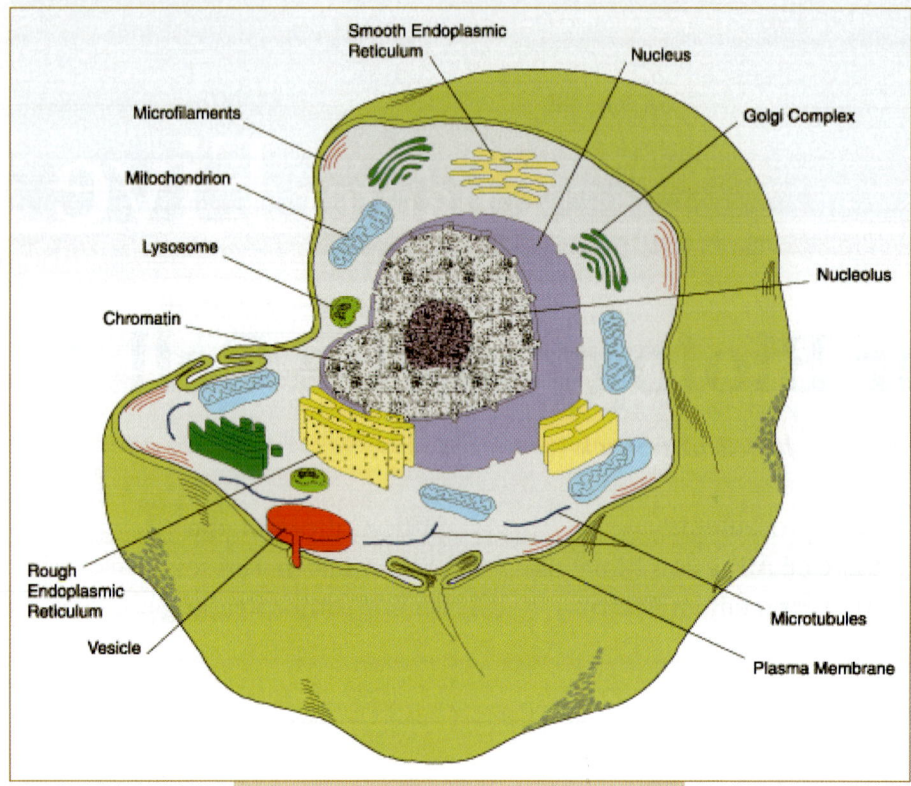

**FIG. 5.1.** Structure of a generalized animal cell. (From paternityexperts.com website.)

substructures called *organelles* that are devoted to specific biochemical functions. The outer surface of the cell is the *plasma membrane*, which is of crucial importance as the frontier across which all materials must pass on their way in or out. The complexity of a single cell is awesome, since it will contain thousands of different types of protein molecules, arranged in many very complex, multimolecular aggregates comprising both hydrophobic and aqueous phases, and also many thousands of low-molecular-weight metabolites, including sugars, amino acids, nucleotides, fatty acids, and phospholipids, among many others. Although some individual steps of metabolism may be near to thermodynamic equilibrium, the cell as a whole is very far from equilibrium and is maintained in this condition by a continuous interchange of substances with the environment. Nutrients are chemically transformed, with release of energy that is used to maintain the structure of the cell and to synthesize the tens of thousands of different macromolecules on which its continued existence depends. Maintaining cells in a healthy state means to provide them continuously with all the substances they need, in the right overall environment of substrate, temperature, and osmolarity, and also continuously to remove all potentially toxic waste products.

## II. THE CELL NUCLEUS

The nucleus contains the genes that control the life of the cell. A *gene* is a sequence of DNA that codes for a protein, or for a nontranslated RNA, and it is usually considered also to include the associated regulatory sequences as well as the coding region itself. The vast majority of eukaryotic genes are located in the nuclear chromosomes, although a few genes are also carried in the DNA of mitochondria and chloroplasts. The genes encoding nontranslated RNAs include those for ribosomal (transfer) RNAs and also a large number of microRNAs that are probably involved in controlling expression of protein-coding genes. The total number of protein-coding genes in vertebrate animals is about 30,000, and every nucleus contains all the genes, irreversible DNA modifications being confined to cells of the immune system in respect of the genes encoding antibodies and T-cell receptors.

The DNA is complexed into a higher-order structure called *chromatin* by the binding of basic proteins called *histones*. Protein-coding genes are transcribed into messenger RNA (mRNA) by the enzyme RNA polymerase II. Transcription commences at a transcription start sequence and finishes at a transcription termination sequence. Genes are usually divided into several *exons*, each of which codes for a part of the mature mRNA. The primary RNA transcript is extensively processed before it moves from the nucleus to the cytoplasm. It acquires a "cap" of methyl guanosine at the 5′ end and a polyA tail at the 3′ end both of which stabilize the message by protecting it from attack by exonucleases. The DNA sequences in between the exons are called *introns*, and the portions of the initial transcript complementary to

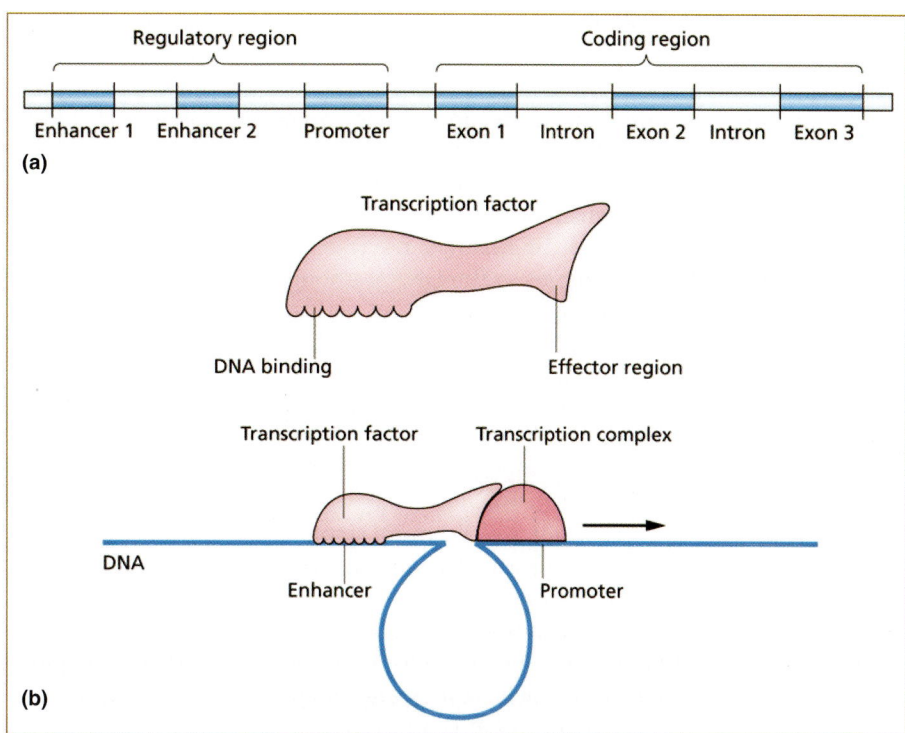

**FIG. 5.2.** **(a)** Structure of a typical gene. **(b)** Operation of a transcription factor. (From Slack, 2005.)

the introns are removed by splicing reactions catalyzed by *snRNPs* (small nuclear ribonucleoprotein particles). It is possible for the same gene to produce several different mRNAs as a result of alternative splicing, whereby different combinations of exons are spliced together from the primary transcript. In the cytoplasm the mature mRNA is translated into a polypeptide by the ribosomes. The mRNA still contains a 5′ leader sequence and a 3′ untranslated sequence flanking the protein-coding region, and these untranslated regions may contain specific sequences responsible for translational control or intracellular localization.

### Control of Gene Expression

There are many genes whose products are required in all tissues at all times, for example, those concerned with basic cell structure, protein synthesis, or metabolism. These are referred to as *housekeeping* genes. But there are many others whose products are specific to particular cell types, and indeed the various cell types differ from each other *because* they contain different repertoires of proteins. This means that the control of gene expression is central to tissue engineering. Control may be exerted at several points. Most common is control of *transcription*, and we often speak of genes being "on" or "off" in particular situations, meaning that they are or are not being transcribed. There are also many examples of *translational regulation*, where the mRNA exists in the cytoplasm but is not translated into protein until some condition is satisfied. Control may also be exerted at the stage of nuclear RNA processing or indirectly via the stability of individual mRNAs or proteins.

Control of transcription depends on regulatory sequences within the DNA and on proteins called *transcription factors* that interact with these sequences. The promoter region of a gene is the region just upstream from the transcription start site to which the RNA polymerase binds. The RNA polymerase is accompanied by a set of general transcription factors, which together make up a *transcription complex*. In addition to the general factors required for the assembly of the complex, there are numerous specific transcription factors that bind to specific regulatory sequences that may be either adjacent to or at some distance from the promoter (Fig. 5.2).

### Transcription Factors

Transcription factors are the proteins that regulate transcription. They usually contain a DNA-binding domain and a regulatory domain, which will either activate or repress transcription. Looping of the DNA may bring these regulatory domains into contact with the transcription complex and either promotes or inhibits its activity. There are many families of transcription factors, classified by the type of DNA-binding domain they contain, such as the homeodomain and the zinc-finger domain. Most are nuclear proteins, although some exist in the cytoplasm until they are activated and then enter the nucleus. Activation often occurs in response to intercellular signaling (see later). One type of transcription factor, the nuclear receptor family, is directly activated by lipid-soluble signaling molecules, such as retinoic acid and glucocorticoids.

Each type of DNA-binding domain in a protein has a corresponding type of target sequence in the DNA, usually 20 nucleotides or less. The activation domains of transcription factors often contain many acidic amino acids making up an *acid blob*, which accelerates the formation of the general transcription complex. Some transcription factors recruit histone acetylases, which open up the chromatin by neutralizing amino groups on the histones by acetylation and allow access of other proteins to the DNA. Although it is normal to classify transcription factors as activators or repressors of transcription, their action is also sensitive to context, and the presence of other factors may on occasion cause an activator to function as a repressor, or vice versa.

### Other Controls of Gene Activity

Some aspects of gene control are of a more stable and longer-term character than that exerted by combinations of positive and negative transcription factors. To some extent this depends on the remodeling of the chromatin structure, which is still poorly understood. The chromosomal DNA is complexed with histones into nucleosomes and is coiled into a 30-nm-diameter filament, which is in turn arranged into higher-order structures. In much of the genome the nucleosomes are to some extent mobile, allowing access of transcription factors to the DNA. This type of chromatin is called *euchromatin*. In other regions the chromatin is highly condensed and inactive, then being called *heterochromatin*. In the extreme case of the nucleated red blood cells of nonmammalian vertebrates, the entire genome is heterochromatic and inactive. Chromatin structure is regulated to some degree by protein complexes (such as the well-known polycomb and trithorax groups), which affect the expression of many genes but are not themselves transcription factors.

An important element of the chromatin remodeling is the control through acetylation of lysines on the exposed N-termini of histones. This partially neutralizes the binding of the histones to the negatively charged phosphodiester chains of DNA and thus opens up the chromatin structure and enables transcription complexes to assemble on the DNA. The degree of histone acetylation is controlled, at least partly, by DNA methylation, because histone deacetylases are recruited to methylated regions and will tend to inhibit gene activity in these regions. DNA methylation occurs on cytosine residues in CG sequences of DNA. Because CG on one strand will pair with GC on the other, antiparallel, strand, potential methylation sites always lie opposite one another on the two strands. There are several DNA methyl transferase enzymes, including *de novo* methylases, which methylate previously unmethylated CGs, and maintenance methylases, which methylate the other CG of sites bearing a methyl group on only one strand. Once a site is methylated, it will be preserved through subsequent rounds of DNA replication, because the hemimethylated site resulting from replication will be a substrate for the maintenance methylase.

There are many other chemical modifications of the histones in addition to acetylation, and it is probable that these too can be retained on chromosomes when the DNA is replicated. So both DNA methylation and histone modifications provide means for maintaining the state of activity of genes in differentiated cells, even after the original signals for activation or repression have disappeared.

## III. THE CYTOPLASM

The cytoplasm consists partly of proteins in free solution, although it also possesses a good deal of structure, which can be visualized as the cytoskeleton (see later). Generally considered to be in free solution, although probably in macromolecular aggregates, are the enzymes that carry out the central metabolic pathways. In particular, the pathway called *glycolysis* leads to the degradation of glucose to pyruvate. Glucose is an important metabolic fuel for most cells. Mammalian blood glucose is tightly regulated around 5–6 mm, and glucose is a component of most tissue culture media. Glycolysis leads to the production of two molecules of ATP per molecule of glucose, with a further 36 molecules of ATP produced by oxidative phosphorylation, which is needed for a very wide variety of synthetic and maintenance activities.

The cytoplasm contains many types of *organelles*, which are structures composed of phospholipid bilayers. *Phospholipids* are molecules with a polar head group and a hydrophobic tail. They tend to aggregate to form sheets in which all the head groups are exposed on the surface and the hydrophobic tails associate with each other to form a hydrophobic phase. Most cell organelles are composed of membranes comprising two sheets of phospholipid molecules with their hydrophobic faces joined. The mitochondria are the organelles responsible for oxidative metabolism as well as for other metabolic processes, such as the synthesis of urea. They are composed of an outer and an inner phospholipid bilayer. The oxidative degradation of sugars, amino acids, and fatty acids is accompanied by the production of ATP. Pyruvate produced by glycolysis is converted to acetyl CoA, and this is oxidized to two molecules of $CO_2$ by the citric acid cycle, with associated production of 12 molecules of ATP in the electron transport chain of the mitochondria. Because of the importance of oxidative metabolism for ATP generation, cells need oxygen to support themselves.

Tissue culture cells are usually grown in atmospheric oxygen concentration (about 20% by volume), although the optimum concentration may be somewhat lower than this since the oxygen level within an animal body is often lower than in the external atmosphere. Too much oxygen can be deleterious because it leads to the formation of free radicals, which cause damage to cells. Tissue culture systems may therefore be run at lower oxygen levels, such as 5%. The oxidation of pyruvate and acetyl CoA also results in the production of $CO_2$, which needs to be removed continuously to avoid acidification.

Apart from the central metabolic pathways, the cell is also engaged in the continuous synthesis and degradation of a wide variety of lipids, amino acids, and nucleotides.

The cytoplasm contains the *endoplasmic reticulum*, which is a ramifying system of phospholipid membranes. The interior of the endoplasmic reticulum can communicate with the exterior medium through the exchange of membrane vesicles with the plasma membrane. Proteins that are secreted from cells or that come to lie within the plasma membrane are synthesized by ribosomes that lie on the cytoplasmic surface of the endoplasmic reticulum, and the products are passed through pores into the endoplasmic reticulum lumen. From here they move to the *Golgi apparatus*, which is another collection of internal membranes, in which carbohydrate chains are often added. From there they move to the cell surface or the exterior medium. Secretion of materials is a very important function of all cells, and it needs to be remembered that their environment in tissue culture depends not only on the composition of the medium provided but also on what the cells themselves have been making and secreting.

The intracellular proteins are synthesized by ribosomes in the soluble cytoplasm. There is a continuous production of new protein molecules, the composition depending on the repertoire of gene expression of the cell. There is also a continuous degradation of old protein molecules, mostly in a specialized structure called the *proteosome*. This continuous turnover of protein requires a lot of ATP.

## The Cell Surface

The plasma membrane is the frontier between the cell and its surroundings. It is a phospholipid bilayer incorporating many specialized proteins. Very few substances are able to enter and leave cells by simple diffusion, in fact this method is really only available to low-molecular-weight hydrophobic molecules such as retinoic acid, steroids, and thyroid hormones. The movement of inorganic ions across the membrane is very tightly controlled. The main control is exerted by a sodium–potassium exchanger, which expels sodium and concentrates potassium. Differential back-diffusion of these ions then generates an electric potential difference across the membrane that ranges from about 10 mV in red blood cells to 80–90 mV (negative inside) in excitable cells such as neurons. Calcium ions are very biologically active within the cell and are normally kept at a very low intracellular concentration, about $10^{-7}$ M. This is about $10^4$ times lower than the typical exterior concentration, which means that any damage to the plasma membrane is likely to let in a large amount of calcium, which will damage the cell beyond repair. The proteins of the plasma membrane may be very hydrophobic molecules entirely contained within the lipid phase, but more usually they

have hydrophilic regions projecting to the cell exterior or to the interior cytoplasm or both. These proteins have a huge range of essential functions. Some are responsible for anchoring cells to the substrate or to other cells through adhesion molecules and junctional complexes. Others are responsible for transporting molecules across the plasma membrane. These include ion transporters and carriers for a large range of nutrients. Then there are the receptors for extracellular signaling molecules, which are critical for controlling cellular properties and behavior. These include hormones, neurotransmitters, and growth factors. Some receptors serve as ion channels, for example, admitting a small amount of calcium when stimulated by their specific ligand. Other receptors are enzymes and initiate a metabolic cascade of intracellular reactions when stimulated. These reaction pathways often involve protein phosphorylation and frequently result in the activation of a transcription factor and thereby the activation of specific target genes. The repertoire of responses that a cell can show depends on which receptors it possesses, how these are coupled to signal transduction pathways, and how these pathways are coupled to gene regulation. The serum that is usually included in tissue culture media contains a wide range of hormones and growth factors and is likely to stimulate many of the cell surface receptors.

## Signal Transduction

Lipid-soluble molecules, such as steroid hormones, can enter cells by simple diffusion. Their receptors are multi-domain molecules that also function as transcription factors. Binding of the ligand causes translocation to the nucleus, where the receptor complex can activate its target genes (Fig. 5.3a).

Most signalling molecules are proteins, which cannot diffuse across the plasma membrane and so work by binding to specific cell surface receptors. There are three main classes of these: enzyme-linked receptors, G-protein-linked receptors, and ion channel receptors. Enzyme-linked receptors are often tyrosine kinases or Ser/Thr kinases (Fig. 5.3b). All have a ligand-binding domain on the exterior of the cell, a single transmembrane domain, and the enzyme active site on the cytoplasmic domain. For receptor tyrosine kinases, the ligand binding brings about dimerization of the receptor, which results in an autophosphorylation whereby each receptor molecule phosphorylates and activates the other. The phosphorylated receptors can then activate a variety of targets. Many of these are transcription factors that are activated by phosphorylation and move to the nucleus, where they activate their target genes. In other cases, a cascade of kinases activate each other down the chain, culminating in the activation of a transcription factor. Roughly speaking, each class of factors has its own associated receptors and a specific signal transduction pathway; however, different receptors may be linked to the same signal transduction pathway, or one receptor may feed into more than one

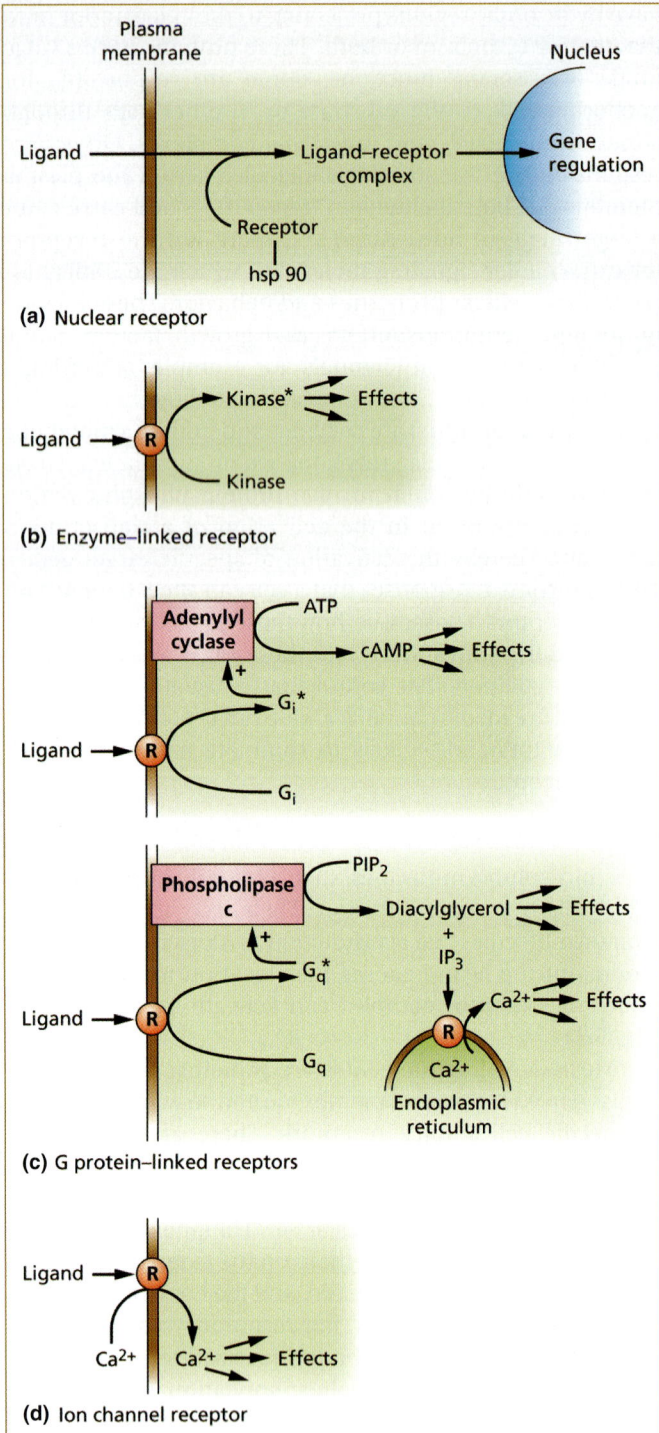

**FIG. 5.3.** Different types of signal transduction. (From Slack, 2005.)

There are several classes of G-protein-linked receptors (Fig. 5.3c). The best known are seven-pass membrane proteins, meaning that they are composed of a single polypeptide chain crossing the membrane seven times. These are associated with trimeric G proteins composed of α, β, and γ subunits. When the ligand binds, the activated receptor causes exchange of guanosine diphosphate (GDP) bound to the α subunit for guanosine triphosphate (GTP); the activated α subunit is released and can interact with other membrane components. The most common target is adenylyl cyclase, which converts adenosine triphosphate (ATP) to cyclic adenosine monophosphate (AMP). Cyclic AMP activates protein kinase A (PKA), which phosphorylates various further target molecules affecting both intracellular metabolism and gene expression.

Another large group of G-protein-linked receptors uses a different trimeric G protein to activate the inositol phospholipid pathway (Fig. 5.3c). Here the G protein activates phospholipase C β, which breaks down phosphatidylinositol bisphosphate ($PIP_2$) to diacylglycerol (DAG) and inositol trisphosphate ($IP_3$). The DAG activates an important membrane-bound kinase, protein kinase C. Like protein kinase A, this has a large variety of possible targets in different contexts and can cause both metabolic responses and changes in gene expression. The IP binds to an $IP_3$ receptor ($IP_3R$) in the endoplasmic reticulum and opens calcium channels, which admit calcium ions into the cytoplasm. Normally cytoplasmic calcium is kept at a very low concentration of around $10^{-7}$ m. An increase caused either by opening of an ion channel in the plasma membrane or as a result of $IP_3$ action can again have a wide range of effects on diverse target molecules.

Ion channel receptors (Fig. 5.3d) are also very important. They open on stimulation to allow passage of Na, K, Cl, or Ca ions. Na and K ions are critical to the electrical excitability of nerve or muscle. As mentioned earlier, Ca ions are very potent and can have a variety of effects on cell structure at low concentration.

## IV. GROWTH AND DEATH

Tissue engineering inevitably involves the growth of cells in culture, so the essentials of cell multiplication need to be understood. A typical animal cell cycle is shown in Fig. 5.4, and some typical patterns of cell division are shown in Fig. 5.5. The cell cycle is conventionally described as consisting of four phases. M indicates the phase of mitosis, S indicates the phase of DNA replication, and G1 and G2 are the intervening phases. For growing cells, the increase in mass is continuous around the cycle, and so is the synthesis of most of the cell's proteins. Normally the cell cycle is coordinated with the growth of mass. If it were not, cells would increase or decrease in size with each division. There are various internal controls built into the cycle, for example, to ensure that mitosis does not start before DNA replication is completed. These controls operate at checkpoints around

pathway. The effect of one pathway on the others is often called *cross-talk*. The significance of cross-talk can be hard to assess from biochemical analysis alone, but is much easier to assess using genetic experiments in which individual components are mutated to inactivity and the overall effect on the cellular behavior can be assessed.

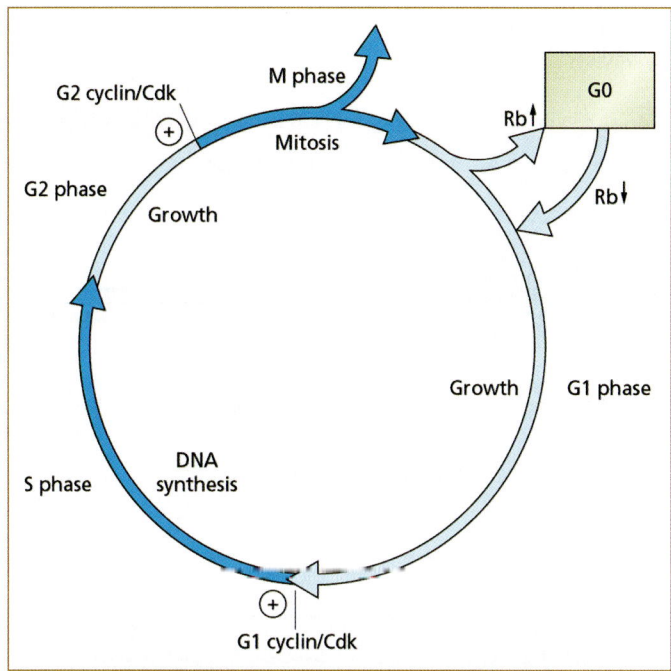

**FIG. 5.4.** The cell cycle, with phases of growth, DNA replication, and division. (From Slack, 2005.)

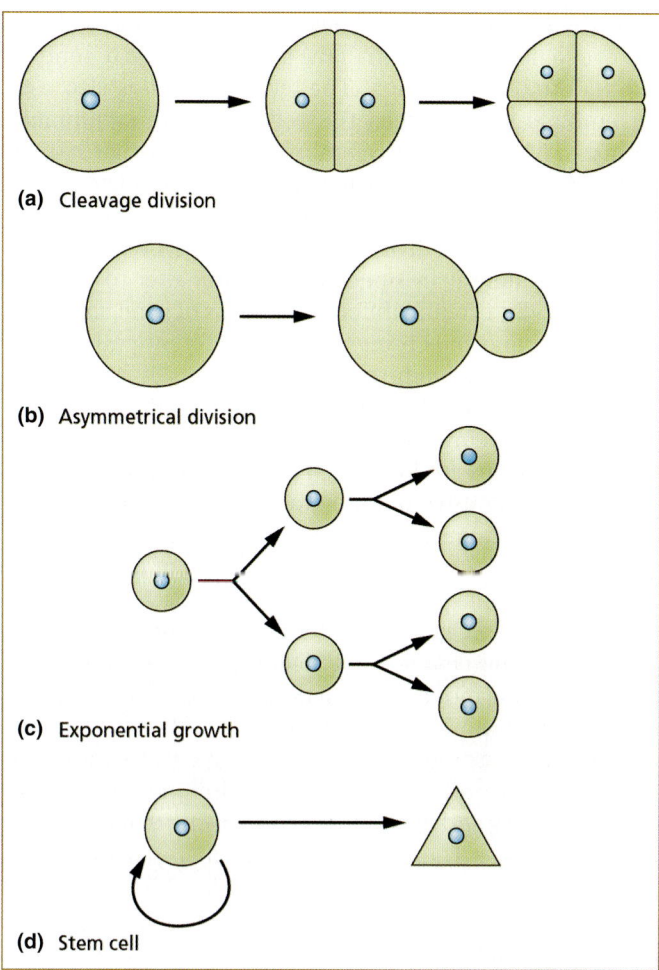

**FIG. 5.5.** Types of cell division. **(a)** Cleavage as found in early embryos. **(b)** Asymmetrical division, also found in early embryos. **(c)** Exponential growth found in tissue culture. **(d)** Stem cell division, found in renewal tissues in animals. (From Slack, 2005.)

the cycle at which the process stops unless the appropriate conditions are fulfilled.

Control of the cell cycle depends on a metabolic oscillator comprising a number of proteins called *cyclins* and a number of cyclin-dependent protein kinases (Cdks). In order to pass the M checkpoint and enter mitosis, a complex of cyclin and Cdk (called M-phase promoting factor, MPF) has to be activated. This phosphorylates and thereby activates the various components required for mitosis (nuclear breakdown, spindle formation, chromosome condensation). Exit from M phase requires the inactivation of MPF, via the destruction of cyclin, so by the end of the M phase it has disappeared. Passage of the G1 checkpoint depends on a similar process operated by a different set of cyclins and Cdks, whose active complexes phosphorylate and activate the enzymes of DNA replication. This is also the point at which the cell size is assessed. The cell cycle of the G1, S, G2, and M phases is universal, although there are some modifications in special circumstances. The rapid-cleavage cycles of early development have short or absent G1 and G2 phases, and there is no size check, the cells halving in volume with each division. The meiotic cycles require the same active MPF complex to get through the two nuclear divisions, but there is no S phase in between.

In the mature organism most cells are quiescent unless they are stimulated by growth factors. In the absence of growth factors, cells enter a state called G0, in which the Cdks and cyclins are absent. Restitution of growth factors induces the resynthesis of these proteins and the resump-

tion of the cycle, starting from the G1 checkpoint. One factor maintaining the G0 state is a protein called Rb (retinoblastoma protein). This becomes phosphorylated, and hence deactivated, in the presence of growth factors. In the absence of Rb, a transcription factor called E2F becomes active and initiates a cascade of gene expression culminating in the resynthesis of cyclins, Cdks, and other components needed to initiate the S phase.

Cells often have the capability for exponential growth in tissue culture (Fig. 5.5c), but this is very rarely found in animals. Although some differentiated cell types can go on dividing, there is a general tendency for differentiation to be accompanied by a slowdown or cessation of division. In postembryonic life, most cell division is found among stem cells and their immediate progeny, called *transit amplifying cells*. Stem cells are cells that can both reproduce themselves and generate differentiated progeny for their particular tissue type (Fig. 5.5d). This does not necessarily mean that every division of a stem cell has to be an asymmetrical one, but over a period of time half the progeny will go to renewal and half to differentiation. The term *stem cell* is also used for embryonic stem cells (ES cells) of early mammalian embryos. These are early embryo-type cells that can be grown in culture and are capable of repopulating embryos and contributing to all tissue types.

Asymmetric cell divisions necessarily involve the segregation of different cytoplasmic determinants to the two daughter cells, evoking different patterns of gene activity in their nuclei and thus bringing about different pathways of development. The nature of determinants is still poorly understood but often involves autosegregation of a self-organizing protein complex called the PAR complex.

Some tissues are formed by growth and differentiation of cells in the embryo but are quiescent in the adult organism. These include neurons and muscle. In fact there is now known to be some limited production of new neurons in the brain from stem cells and of new muscle fibers from muscle satellite cells. Some tissues are capable of expansion but remain quiescent most of the time unless stimulated by damage or hormonal stimulation. These would include most of the glandular-type tissues, such as the liver, kidney, and pancreas. Some tissues are in a state of continuous renewal, with a proliferative zone containing stem cells constantly dividing and generating new progeny that differentiate and then die. These include the haematopoietic system in the bone marrow, which forms all cells of the blood and immune system. It also includes the epithelial lining of the gut and the epidermis of the skin.

## V. CYTOSKELETON

The cytoskeleton is important for three distinct reasons. First, the orientation of cell division may be important. Second, animal cells move around a lot, either as individuals or as part of moving cell sheets. Third, the shape of cells is an essential part of their ability to carry out their functions. All of these activities are functions of the cytoskeleton.

The three main components of the cytoskeleton are:
* microfilaments, made of actin
* microtubules, made of tubulin
* intermediate filaments, made of cytokeratins in epithelial cells, vimentin in mesenchymal cells, neurofilament proteins in neurons, and glial fibrilliary acidic protein (GFAP) in glial cells

Microtubules and microfilaments are universal constituents of eukaryotic cells, while intermediate filaments are found only in animals.

### Microtubules

Microtubules (Fig. 5.6) are hollow tubes of 25-nm diameter composed of tubulin. Tubulin is a generic name for a family of globular proteins that exist in solution as heterodimers of α- and β-type subunits, and they are one of the more

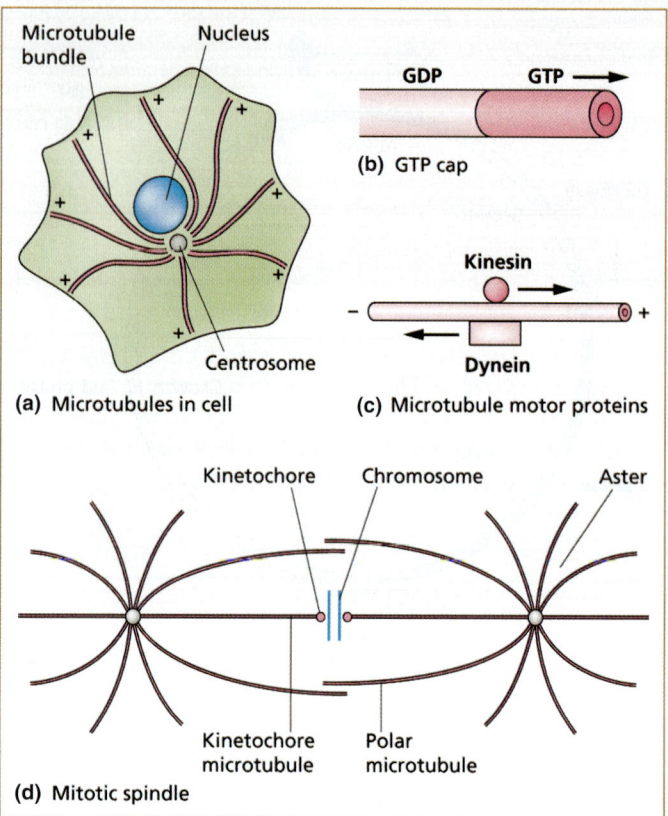

**FIG. 5.6.** Microtubules. **(a)** Arrangement in cell. **(b)** The GTP cap. **(c)** Motor proteins move along the tubules. **(d)** Structure of the cell division spindle. (From Slack, 2005.)

abundant cytoplasmic proteins. The microtubules are polarized structures, with a minus end anchored to the centrosome and a free plus end, at which tubulin monomers are added or removed. Microtubules are not contractile but exert their effects through length changes based on polymerization and depolymerization. They are very dynamic, either growing by addition of tubulin monomers or retracting by loss of monomers, and individual tubules can grow and shrink over a few minutes. The monomers contain GTP bound to the β subunit, and in a growing plus end this stabilizes the tubule. But if the rate of growth slows down, hydrolysis of GTP to GDP will catch up with the addition of monomers. The conversion of bound GTP to GDP renders the plus end of the tubule unstable, and it will then start to depolymerize. The drugs colchicine and colcemid bind to monomeric tubulin and prevent polymerization. Among other effects this causes the disassembly of the mitotic spindle. These drugs cause cells to become arrested in mitosis and are often used in studies of cell kinetics.

The shape and polarity of cells can be controlled by locating capping proteins in particular parts of the cell cortex that bind the free plus ends of the microtubules and stabilize them. The positioning of structures within the cell also depends largely on microtubules. There exist special

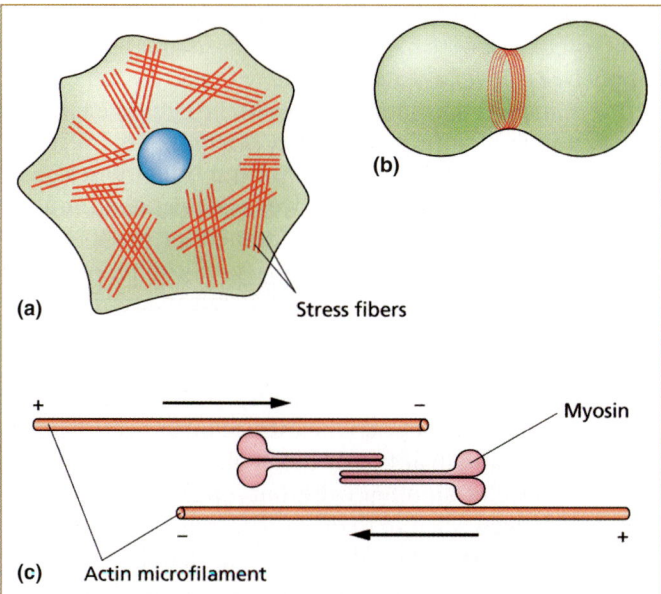

**FIG. 5.7.** Microfilaments. **(a)** Arrangement in cell. **(b)** Role in cell division. **(c)** Contraction achieved by movement of myosin along microfilament. (From Slack, 2005.)

motor proteins that can move along the tubules, powered by hydrolysis of ATP, and thereby can transport other molecules to particular locations within the cell. The kinesins move toward the plus ends of the tubules, while the dyneins move toward the minus ends.

Microtubules are prominent during cell division. The minus ends of the tubules originate in the centrosome, which is a microtubule-organizing center able to initiate the assembly of new tubules. In mitotic prophase the centrosome divides, and each of the radiating sets of microtubules becomes known as an *aster*. The two asters move to the opposite sides of the nucleus to become the two poles of the *mitotic spindle*. The spindle contains two types of microtubules. The *polar* microtubules meet each other near the center and become linked by plus-directed motor proteins. These tend to drive the poles apart. Each chromosome has a special site, called a *kinetochore*, that binds another group of microtubules, called *kinetochore* microtubules. At anaphase the kinetochores of homologous chromosomes separate. The polar microtubules continue to elongate, while the kinetochore microtubules shorten by loss of tubulin from both ends and draw the chromosome sets into the opposite poles of the spindle.

## Microfilaments

Microfilaments (Fig. 5.7) are polymers of actin, which is the most abundant protein in most animal cells. In vertebrates there are several different gene products, of which α actin is found in muscle and β/γ actins in the cytoskeleton of nonmuscle cells. For all actin types the monomeric soluble form is called G-actin. Actin filaments have an inert

minus end and a growing plus end to which new monomers are added. G-actin contains ATP, and this becomes hydrolyzed to ADP shortly after addition to the filament. As with tubules, a rapidly growing filament will bear an ATP cap that stabilizes the plus end. Microfilaments are often found to undergo *treadmilling*, such that monomers are continuously added to the plus end and removed from the minus end while leaving the filament at the same overall length. Microfilament polymerization is prevented by a group of drugs called *cytochalasins*, and existing filaments are stabilized by another group, called *phalloidins*. Like microtubules, microfilaments have associated motor proteins that will actively migrate along the fiber. The most abundant of these is myosin II, which moves toward the plus end of microfilaments, the process being driven by the hydrolysis of ATP. To bring about contraction of a filament bundle, the myosin is assembled as short bipolar filaments with motile centers at both ends. If neighboring actin filaments are arranged with opposite orientation, then the motor activity of the myosin will draw the filaments past each other, leading to a contraction of the filament bundle.

Microfilaments can be arranged in various different ways, depending on the nature of the accessory proteins with which they are associated. Contractile assemblies contain microfilaments in antiparallel orientation associated with myosin. These are found in the *contractile ring*, which is responsible for cell division, and in the *stress fibers*, by which fibroblasts exert traction on their substratum. Parallel bundles are found in filopodia and other projections from the cell. Gels composed of short, randomly oriented filaments are found in the cortical region of the cell.

## Small GTPases

There are three well-known GTPases, which activate cell movement in response to extracellular signals: Rho, Rac, and cdc42. They are activated by numerous tyrosine kinase-, G-coupled-, and cytokine-type receptors. Activation involves exchange of GDP for GTP, and many downstream proteins can interact with the activated forms. Rho normally activates the assembly of stress fibers. Rac activates the formation of lamellipodia and ruffles. Cdc42 activates formation of filopodia. In addition, all three promote the formation of focal adhesions, which are integrin-containing junctions to the extracellular matrix. These proteins can also affect gene activity through the kinase cascade signal transduction pathways.

## VI. CELL ADHESION MOLECULES

Organisms are not just bags of cells; rather, each tissue has a definite cellular composition and microarchitecture. This is determined partly by the cell surface molecules, by which cells interact with each other, and partly by the components of the extracellular matrix (ECM). Virtually all proteins on the cell surface or in the ECM are glycoproteins,

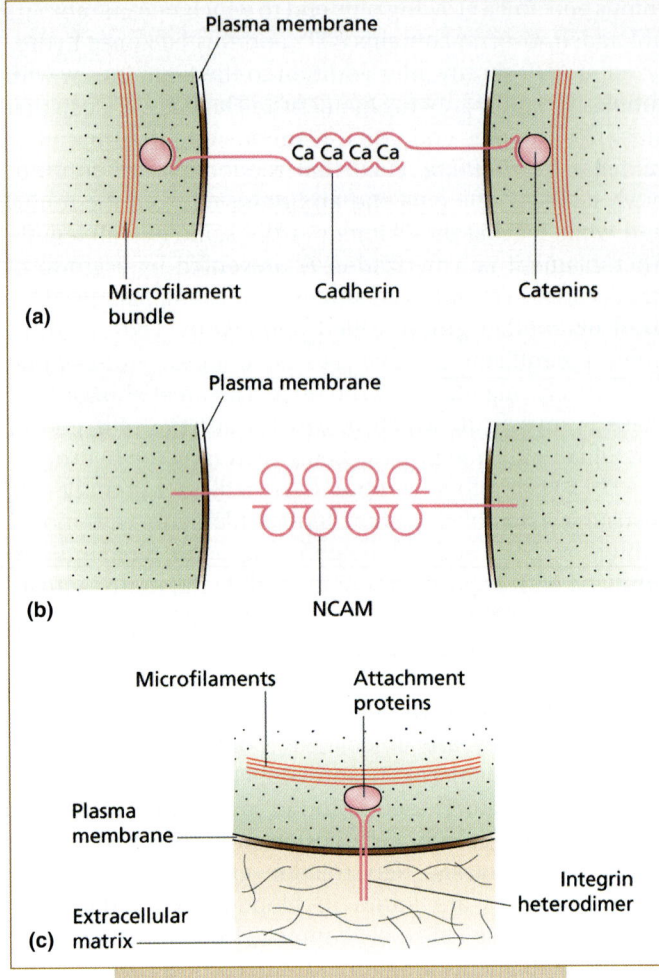

**FIG. 5.8.** Cell adhesion molecules. **(a)** Calcium-dependent system. **(b)** Calcium-independent system. **(c)** Adhesion to the extracellular matrix. (From Slack, 2005.)

containing oligosaccharide groups added in the endoplasmic reticulum or Golgi apparatus after translation and before secretion from the cell. These carbohydrate groups often have rather little effect on the biological activity of the protein, but they may affect its physical properties and stability.

Cells are attached to each other by *adhesion molecules* (Fig. 5.8). Among these are the *cadherins*, which stick cells together in the presence of Ca, the *cell adhesion molecules* (CAMs), which do not require Ca, and the *integrins*, which attach cells to the extracellular matrix. When cells come together they often form *gap junctions* at the region of contact. These consist of small pores joining the cytosol of the two cells. The pores, or *connexons*, are assembled from proteins called *connexins*. They can pass molecules up to about 1000 molecular weight by passive diffusion.

Cadherins are a family of single-pass transmembrane glycoproteins that can adhere tightly to similar molecules on other cells in the presence of calcium. They are the main factors attaching embryonic cells together, which is why embryonic tissues can often be caused to disaggregate simply by removal of calcium. The cytoplasmic tail of cadherins is anchored to actin bundles in the cytoskeleton by a complex including proteins called *catenins*. One of these, β-catenin, is also a component of the important Wnt signaling pathway, providing a link between cell signaling and cell association. Cadherins were first named for the tissues in which they were originally found, so E-cadherin occurs mainly in epithelia and N-cadherin occurs mainly in neural tissue.

The immunoglobulin superfamily is made from single-pass transmembrane glycoproteins, with a number of disulphide-bonded loops on the extracellular region, similar to the loops found in antibody molecules. They also bind to similar molecules on other cells; but, unlike the cadherins, they do not need calcium to do so. The neural cell adhesion molecule (N-CAM) is composed of a large family of different proteins formed by alternative splicing. It is most prevalent in the nervous system but also occurs elsewhere. It may carry a large amount of polysialic acid on the extracellular domain, and this can inhibit cell attachment because of the repulsion between the concentrations of negative charge on the two cells. Related molecules include L1 and ICAM (intercellular cell adhesion molecule).

The integrins are cell-surface glycoproteins that interact mainly with components of the extracellular matrix. They are heterodimers of α and β subunits and require either magnesium or calcium for binding. There are numerous different α and β chain types, and so there is a very large number of potential heterodimers. Integrins are attached by their cytoplasmic domains to microfilament bundles, so, like cadherins, they provide a link between the outside world and the cytoskeleton. They are also thought on occasion to be responsible for the activation of signal transduction pathways and new gene transcription following exposure to particular extracellular matrix components.

## VII. EXTRACELLULAR MATRIX

*Glycosaminoglycans (GAGs)* are unbranched polysaccharides composed of repeating disaccharides of an amino sugar and a uronic acid, usually substituted with some sulphate groups. GAGs are constituents of proteoglycans, which have a protein core to which the GAG chains are added in the Golgi apparatus before secretion. One molecule of a proteoglycan may carry more than one type of GAG chain. GAGs have a high negative charge, and a small amount can immobilize a large amount of water into a gel. Important GAGs, each of which has different component disaccharides, are heparan sulphate, chondroitin sulphate, and keratan sulphate. Heparan sulphate, closely related to the anticoagulant heparin, is particularly important for cell signaling, because it is required to present various growth factors, such as the fibroblast growth factors (FGFs), to their receptors. Hyaluronic acid differs from other GAGs because it occurs free and not as a constituent of a proteoglycan. It consists of repeating disaccharides of glucuronic acid and *N*-acetyl glucosamine, and it is not sulphated. It is synthesized by enzymes at the cell surface and is abundant in early embryos.

*Collagens* are the most abundant proteins by weight in most animals. The polypeptides, called $\alpha$ chains, are rich in proline and glycine. Before secretion, three $\alpha$ chains become twisted around each other to form a stiff triple helical structure. In the extracellular matrix, the triple helices become aggregated together to form the collagen fibrils visible in the electron microscope. There are many types of collagen, which may be composed of similar or of different $\alpha$ chains in the triple helix. Type I collagen is the most abundant and is a major constituent of most extracellular material. Type II collagen is found in cartilage and in the notochord of vertebrate embryos. Type IV collagen is a major constituent of the basal lamina underlying epithelial tissues. Collagen helices may become covalently cross-linked through their lysine residues, and this contributes to the changing mechanical properties of tissues with age.

*Elastin* is another extracellular protein with extensive intermolecular cross-linking. It confers the elasticity on tissues in which it is abundant, and it also has some cell signaling functions.

*Fibronectin* is composed of a large disulphide-bonded dimer. The polypeptides contain regions responsible for binding to collagen, to heparan sulphate, and to integrins on the cell surface. These latter, cell-binding domains are characterized by the presence of the amino acid sequence Arg-Gly-Asp (= RGD). There are many different forms of fibronectin produced by alternative splicing.

*Laminin* is a large extracellular glycoprotein, found particularly in basal laminae. It is composed of three disulphide-bonded polypeptides joined in a cross shape. It carries domains for binding to type IV collagen, heparan sulphate, and another matrix glycoprotein, entactin.

## VIII. CULTURE MEDIA

Mammalian cells will only remain in good condition very close to the normal body temperature, so good temperature control is essential. Because water can pass across the plasma membranes of animal cells, the medium must match the osmolarity of the cell interior, otherwise cells will swell or shrink due to osmotic pressure difference. Mammalian cell media generally have a total osmolarity about 350 mosm. The pH needs to be tightly controlled; usually 7.4 is normal. The pH control is typically achieved with bicarbonate-$CO_2$ buffers (2.2 gm/L bicarbonate and 5% $CO_2$ being a common combination). These give better results with most animal cells than other buffers, perhaps because bicarbonate is also a type of nutrient. The medium must contain a variety of components: salts, amino acids, and sugars plus low levels of specific hormones and growth factors required for the particular cells in question. Because

of the complexity of tissue culture media, they are rarely optimized for a given purpose by varying every one of the components. Usually changes are incremental and the result of a "gardening" approach rather than a systematic one.

The requirement for hormones and growth factors is usually met by including some animal serum, often 10% fetal calf serum. This is long-standing practice but has two substantial disadvantages. Serum can never be completely characterized, and there are often differences between batches of serum, which can be critical for experimental results. Also, there is currently a drive to remove serum from the preparation of cells intended for implantation into human patients. This is because of the perceived possibility, actually very remote, of transmitting animal diseases to patients. Assuming cells are kept in a near-optimal medium, they can, in principle, grow exponentially, with a constant doubling time. Indeed it is possible to grow many types of cells in exponential cultures, rather like microorganisms. In order to keep them growing at maximal rate, they need to have their medium renewed regularly and to be subcultured and replated at lower density whenever they approach confluence, which means covering all the available surface. Subculturing is usually carried out by treatment with the enzyme trypsin, which degrades much of the extracellular and cell surface protein and makes the cells drop off the substrate and become roughtly spherical bodies in suspension. Once the trypsin is diluted out, the cells can be transferred at lower density into new flasks. The cells take an hour or two to resynthesize their surface molecules, and they can then adhere to the new substrate and carry on growing.

Although exponential growth is often sought and encountered in tissue culture, it is important to bear certain things in mind. First, cells very rarely grow exponentially in the body. Most cells are quiescent, rarely undergoing any division at all, thus resembling static confluent tissue cultures more than growing ones. Some tissues undergo continual renewal, such that the production of new cells is balanced by the death and shedding of old ones, so the cell number remains constant even though proliferation is occurring. Growth also involves increase of cell size, which depends largely on the overall rate of protein synthesis relative to protein degradation. This needs to balance cell division such that the volume should exactly double in each cell cycle. If it did not, then the cells would get progressively bigger or smaller.

### Cell Types

On the basis of light microscopy it is estimated that there are about 210 different types of differentiated cells in the mammalian body. This number is certainly an underestimate, since many subdivisions of cells cannot be seen in the light microscope, particularly the different types of neuron in the nervous system and different types of T-lymphocyte in the immune system. Cells types are differ-

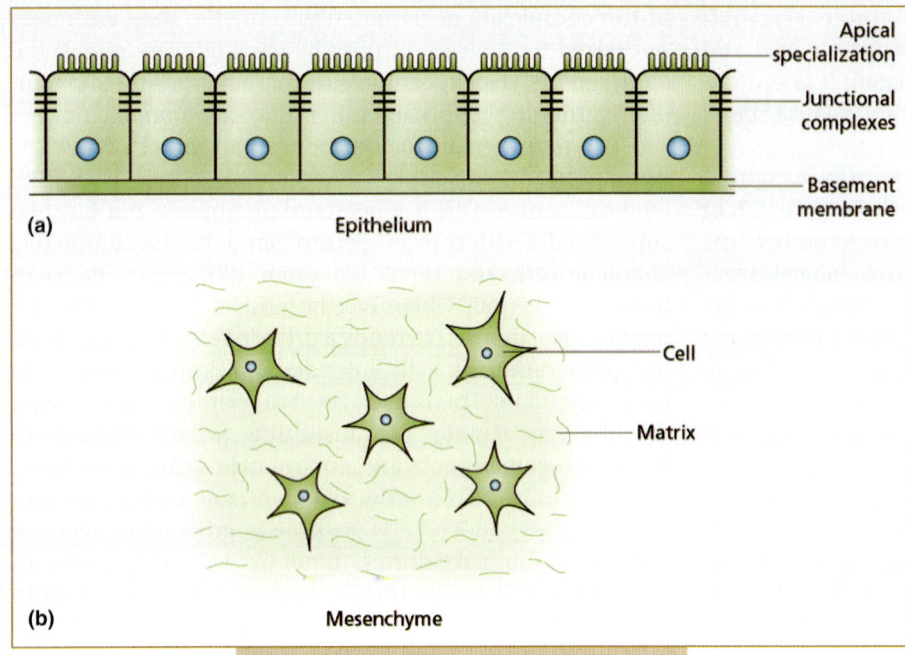

**FIG. 5.9.** Most tissues are composed of epithelial **(a)** and mesenchymal **(b)** components. (From Slack, 2005.)

ent from one another because they are expressing different subsets of genes and hence contain different proteins. The products of a relatively small number of genes may dominate the appearance of a differentiated cell, for example, the proteins of the contractile apparatus of skeletal muscle are very abundant. However, a typical cell will express many thousands of genes, and its character will also depend on the genes that are not expressed. It is possible to control cell differentiation to some extent. Certain special culture media are favorable for differentiation of particular cell types, such as adipocytes, muscle, or bone. Also, some regulatory genes are known that can force the differentiation of a particular cell type if they are overexpressed. For example, the MyoD gene, encoding a basic helix-loop-helix transcription factor, will force differentiation of muscle in a wide range of cultured cells. The runt domain factor Cbfa-1 plays a similar role for the differentiation of bone.

In some cases differentiated cells can continue to grow in pure culture. But in many cases differentiation causes slowing or cessation of cell division. Furthermore, sometimes differentiated cells are formed from stem cells that undergo unequal divisions, yielding one differentiated daughter and one stem cell. In such cases it will not be possible to obtain a pure culture of a single differentiated cell type.

## IX. CELLS IN TISSUES

For the purposes of tissue engineering it is useful to consider how tissues are structured in the normal body. There is very wide range of arrangements, but we can cite some general principles.

- All tissues contain more than one cell type.
- These are drawn from different embryological lineages.

- Even more cell types may be generated *in situ*.
- A vascular supply is essential for survival.

From a morphological point of view, most cells can be regarded as epithelial or mesenchymal (Fig. 5.9). These terms relate to cell shape and behavior rather than to embryonic origin. An *epithelium* is a sheet of cells, arranged on a basement membrane, each cell joined to its neighbors by specialized junctions and showing a distinct apical–basal polarity. *Mesenchyme* is a descriptive term for scattered stellate cells embedded in loose extracellular matrix. It fills up much of the embryo and later forms fibroblasts, adipose tissue, smooth muscle, and skeletal tissues. A tissue normally has both an epithelial and a mesenchymal component. Usually these depend on each other: Each secretes growth factors needed by the other for its survival and proliferation.

The epithelium is usually the functional part of the tissue, for example, the epithelial linings of the various segments of the gut have specific properties of protection, absorption, or secretion, while the underlying mesenchyme provides mechanical support, growth factors, and physiological response, in terms of muscular movements.

Vertebrate epithelial cells are bound together by tight junctions, adherens junctions, and desmosomes, the latter two types involving cadherins as major adhesion components. Mesenchymal cells may also adhere by means of cadherins, but usually more loosely. The adhesion of early embryo cells is usually dominated by the cadherins, and because of this most types of early embryo can be fully disaggregated into single cells by removal of calcium from the medium.

There is some qualitative specificity to cell adhesion. Cadherin-based adhesion is homophilic, and so cells carrying E-cadherin will stick more strongly to each other than to cells bearing N-cadherin. The calcium-independent immunoglobulin superfamily–based adhesion systems, such as N-CAM (neural cell adhesion molecule), particularly important on developing neurons and glia, are different again, and they also promote adhesion of similar cells. This qualitative specificity of adhesion systems provides a mechanism for the assembly of different types of cell aggregates in close proximity and also prevents individual cells from wandering off into neighboring domains. If cells with different adhesion systems are mixed, they will sort out into separate zones, eventually forming a dumbbell-like configuration or even separating altogether.

With the exception of the kidney, all other tissues draw their epithelium and mesenchyme from different germ layers of the embryo. The implication of this for tissue engineering is that it will probably be necessary to assemble tissues from separate epithelial and mesenchymal cells, designed such that each population can support the other.

Furthermore, the epithelium itself normally contains more than one cell type. Many tissues can be regarded as being organized into structural-proliferative units, of which the intestinal crypt serves as a good example. The small intestinal epithelium contains four cell types, all thought to be produced continually from a population of stem cells located near the base of the intestinal crypts. The four types are the absorptive cells, the goblet cells secreting mucus, the Paneth cells at the base of the crypts involved in defense against infection, and the endocrine cells, which themselves are of many subtypes, secreting a varitey of hormones controlling the physiology of the gut.

The intestine also provides an example of a renewal tissue, already referred to, which means that the epithelium is in a state of constant turnover, with cells being produced from the stem cells, dividing a few times, differentiating, and then dying and being shed from the tips of the villi. If tissues like this are going to be created by tissue engineers, they need to be organized into proliferative and differentiated zones, and this spatial organization needs to be stable, despite the flux of cells through the system.

The final consideration is that cells need a continuous supply of nutrients and oxygen and continuous removal of waste products. *In vivo* this is achieved by means of the blood vascular system, which culminates in capillary beds of enormous density such that all cells are within a few cell diameters of the blood. For tissue engineering the lesson is clear: It is possible to grow large avascular structures only so long as they are two-dimensional. For example, large sheets of epidermis a few cells thick can be grown *in vitro* and used successfully for skin grafting. But any tissue more than a fraction of a millimeter in thickness will need to be provided with some sort of vascular system.

Tissue engineering needs not attempt to copy everything found in the normal body. However, it is necessary to be aware of the constraints provided by the molecular biology of the cell. Factors to be considered include:

- How to keep cells in the desired state by providing the correct substrate and medium
- How to create an engineered tissue containing two or more cell types of different origins that can sustain one another
- How to provide a vascular system capable of delivering nutrients and removing waste products
- How to establish the structural-proliferative units of the tissue
- How to control cell division (renewal type with stem cells or quiescent type with regenerative growth)

# X. FURTHER READING

## General

Alberts, B., Johnson, A., Lewis, J., Raff, M., Roberts, K., and Walter, P. (2002). "Molecular Biology of the Cell," 4th ed. Garland Publishing, New York.

Darnell, J. E. (2003). "Molecular Cell Biology," 5th ed. W. H. Freeman, New York.

Slack, J. M. W. (2005). "Essential Developmental Biology," 2nd ed. Blackwell Science, Oxford, UK.

## Molecular and General Genetics

Brown, T. A. (2001). "Gene Cloning and DNA Analysis: An Introduction," 4th ed. Blackwell Science, Oxford, UK.

Hartl, D. L., and Jones, E. W. (2001). "Genetics: Analysis of Genes and Genomes," 5th ed. Jones and Bartlett, Sudbury, MA.

Hartwell, L. H., Hood, L., Goldberg, M. L., Reynolds, A. E., Silver, L. M., and Veres, R. C. (2004). "Genetics: From Genes to Genomes," 2nd ed. McGraw-Hill, New York.

Latchman, D. S. (2003). "Eukaryotic Transcription Factors." Academic Press, New York.

Primrose, S. B., Twyman, R. M., and Old, R. W. (2002). "Principles of Gene Manipulation," 6th ed. Blackwell Science, Oxford, UK.

Wolffe, A. (1998). "Chromatin: Structure and Function," 3rd ed. Academic Press, San Diego.

## Cell Signaling

Downward, J. (2001). The ins and outs of signaling. *Nature* **411**, 759–762.

Hancock, J. T. (1997). "Cell Signaling." Longman, Harrow, UK.

Heath, J. K. (2001). "Principles of Cell Proliferation." Blackwell Science, Oxford, UK.

Hunter, T. (2000). Signaling — 2000 and beyond. *Cell* **100**, 113–127.

## Cytoskeleton, Adhesion Molecules and Extracellular Matrix

Beckerle, M. C. (2002). "Cell Adhesion." Oxford University Press, Oxford, UK.

Kreis, T., and Vale, R. (1999a). "Guidebook to the Cytoskeletal and Motor Proteins," 2nd ed. Oxford University Press, Oxford, UK.

Kreis, T., and Vale, R. (1999b). "Guidebook to the Extracellular Matrix and Adhesion Proteins," 2nd ed. Oxford University Press, Oxford, UK.

## Cell Cycle and Apoptosis

Lawen, A. (2003). Apoptosis — an introduction. *Bioessays* **25**, 888–896.

Murray, A., and Hunt, T. (1994). "The Cell Cycle: An Introduction." Oxford University Press, Oxford, UK.

# Chapter Six

# Organization of Cells into Higher-Ordered Structures

*Jon D. Ahlstrom and Carol A. Erickson*

## I. INTRODUCTION

Multicellular tissues exist in one of two types of cellular arrangements, epithelial or mesenchymal. Epithelial cells adhere tightly to each other at their lateral surfaces and to an organized extracellular matrix (ECM) at their basal domain, thereby producing a sheet of cells with an apical, or adhesion-free, surface. Mesenchymal cells, in contrast, are individual cells with a bipolar morphology that are held together as a tissue within a three-dimensional ECM. The conversion of epithelial cells into mesenchymal cells, an *epithelial-to-mesenchymal transition* (EMT), plays an essential role in embryonic morphogenesis as well as a number of disease states. The reverse process, whereby mesenchymal cells coalesce into an epithelium, is a *mesenchymal-to-epithelial transition* (MET). Understanding the molecular mechanisms of EMTs and METs offers important insights into the basic mechanistic processes of embryonic morphogenesis and tissue organization in the adult.

The early embryo is structured as one or more epithelia. The emergence of the EMT during evolution has allowed rearrangements of cells and tissues to create novel morphological features (reviewed in Hay, 2005). There are several well-studied examples of EMTs during embryonic development. The migration of sea urchin primary mesenchyme cells (PMCs) from the vegetal plate of the blastocyst (epithelium) into the blastocoel cavity to initiate skeleton formation occurs by an EMT in the pregastrula embryo (reviewed in Shook and Keller, 2003). The process of gastrulation in amniotes (reptiles, birds, and mammals) occurs by an EMT as the epithelial epiblast at the primitive streak gives rise to mesenchymal cells — the precursors to mesoderm and endoderm. EMTs also occur later in vertebrate development, such as during the delamination of neural crest cells from the neural tube, the invasion of endothelial cells into the cardiac jelly to form the cardiac cushions, the formation of the sclerotome (connective tissue precursors) from epithelial somites, and the creation of mesenchymal cells in the palate from the epithelial seam where the palate shelves fuse (Hay, 2005; Shook and Keller, 2003). The reverse process of MET is likewise crucial to development, and examples include the condensation of mesenchymal cells to form the notochord and somites, kidney tubule formation from nephrogenic mesenchyme (Barasch, 2001), and the creation of heart valves from cardiac mesenchyme (Eisenberg and Markwald, 1995). In the adult organism, EMTs and METs occur during wound healing and tissue remodeling (Kalluri and Neilson, 2003). The conversion of neoplastic epithelial cells into invasive cancer cells is an EMT process (Thiery, 2002), as is the disintegration of epithelial kidney tissue into

*Principles of Tissue Engineering, 3rd Edition*
*ed. by Lanza, Langer, and Vacanti*

fibroblastic cells during end-stage renal disease (Iwano *et al.,* 2002).

The focus of this chapter is on the molecular agents that control the organization of tissues into epithelium or mesenchyme. We first discuss the cellular changes that occur during the EMT, including changes in cell–cell and cell–ECM adhesions, changes in cell polarity, the stimulation of cell motility, and the increased protease activity that accompanies invasion of the basal lamina. Then we consider the molecules and mechanisms that control the EMT or MET, including the transcription factors that initiate the changes in gene activity involved in the EMT and the upstream signal transduction pathways that regulate these transcription factors. We also identify gaps in our current understanding of these regulatory processes.

## II. MOLECULAR MECHANISMS OF THE EMT

The conversion of an epithelial sheet into individual migratory cells requires the coordinated changes of many distinct families of molecules. As an example of a typical EMT, we give a brief overview of the ingression of PMCs that occurs in sea urchin embryos just prior to gastrulation (for a recent review see Shook and Keller, 2003). The pregastrula sea urchin embryo is a hollow sphere of epithelial cells (blastula) in which the basal domain of the epithelium rests on a basal lamina and faces the inner surface of the sphere. The apical domain, with its microvilli, comprises the outer surface of the sphere. As the primary mesenchyme cells detach from the epithelium to enter the blastocoel, the apical adherens junctions that tether them in the epithelium are endocytosed, and the PMCs lose cell–cell adhesion, gain adhesion to the inner basal lamina, and migrate on the inner surface of the blastocoel cavity. The basal lamina is degraded at sites where PMCs enter the blastocoel. Similar events are observed in other EMTs. Thus, the basic steps of an EMT are: (1) the loss of cell–cell adhesion, (2) the gain of cell–ECM adhesion, (3) change in cell polarity and the stimulation of cell motility, and (4) invasion across the basal lamina. We now examine the components of an EMT in more detail.

### Changes in Cell–Cell Adhesion

Epithelial cells are held together by specialized cell–cell junctions, including adherens junctions, desmosomes, and tight junctions. These are localized in the lateral domain near the apical surface and establish the apical polarity of the epithelium. In order for an epithelial sheet to produce individual migrating cells, cell–cell adhesions must be disrupted. The transmembrane proteins of the adherens junctions and desmosomes that mediate cell–cell adhesions are members of the cadherin superfamily. During the ingression of PMCs in sea urchin embryos, cadherin protein is lost from the lateral membrane domain, and cadherins

are subsequently found in subcellular vesicles, suggesting that the cadherins are endocytosed (Miller and McClay, 1997). Cadherins are essential for establishing adherens junctions and desmosomes and in maintaining the epithelial phenotype. E-cadherin and N-cadherin (*E* for epithelial and *N* for neuronal) are classical cadherins that interact homotypically through their extracellular IgG domains with like cadherins on adjacent cells. Function-blocking antibody against E-cadherin causes the epithelial Madin–Darby canine kidney (MDCK) cell line to dissociate into individual migratory cells (reviewed in Thiery, 2002), and E-cadherin-mediated adhesion is necessary to maintain the epithelial integrity of embryonic epidermis (Levine *et al.,* 1994). Furthermore, E-cadherin is sufficient for promoting cell–cell adhesion and assembly of adherens junctions, since the overexpression of E-cadherin in fibroblasts results in the formation of cell–cell adhesions (Nagafuchi *et al.,* 1987). In epithelial cancers (carcinomas), E-cadherin acts as a tumor suppressor by inhibiting invasion and metastasis. Partial or complete loss of E-cadherin in carcinomas is associated with increased metastasis, and conversely, E-cadherin overexpression in cultured cancer cells reduces invasiveness *in vitro* and *in vivo* (Thiery, 2002). In a mouse model for β-cell pancreatic cancer, the loss of E-cadherin is the rate-limiting step for transformed epithelial cells to become invasive (Perl *et al.,* 1998). Although the loss of cadherin-mediated cell–cell adhesion is necessary for an EMT, the loss of cadherins is not always sufficient to generate a complete EMT *in vivo*. For example, the neural tube epithelium in mice expresses N-cadherin and not E-cadherin; and in the N-cadherin knockout mouse, the neural tube is ill formed (cell adhesion defect), but an EMT is not induced (Radice *et al.,* 1997). Hence, cadherins are essential for maintaining epithelial integrity, and the loss of cell–cell adhesion due to the reduction of cadherin function is an important step in an EMT.

Changes in cadherin expression, or *cadherin switching*, is characteristic of an EMT or an MET. For example, epithelia that express E-cadherin will down-regulate its expression at the time of the EMT and express a different cadherin, such as N-cadherin. When mesenchymal tissue becomes epithelial again (MET), N-cadherin is lost and E-cadherin is re-expressed. Cadherin switching occurs during the EMT that generates the neural crest. Just before neural crest cells detach from the neural tube, N-cadherin is down-regulated, and the mesenchymal cadherins, cadherin-11 and cadherin-7, are expressed. When neural crest cells cease migration and coalesce into ganglia, they express N-cadherin again (Pla *et al.,* 2001). Likewise, in various cultured mammary epithelial cell lines, TGF-β exposure results in the loss of E-cadherin, increased expression of N-cadherin, the loss of adherens junctions, and the induction of cell motility. N-cadherin misexpression in these cell lines is sufficient for increased cell motility in the absence of TGF-β. Conversely, when N-cadherin expression is knocked down by siRNA, the adherens junctions are still down-regulated in response to TGF-β, but the cells do not become motile. Hence, while cadherin switching is not sufficient to bring about a complete EMT, cadherin switching is necessary for cell motility (Maeda *et al.*, 2005).

There are several ways that cadherin expression and function can be regulated. The transcription factors that are central to an EMT, such as Snail-1, Snail-2 (previously Snail and Slug, respectively, see Barrallo-Gimeno and Nieto, 2005, for nomenclature), Sip1, δEF-1, Twist, and E2A, all bind to the *E-cadherin* promoter and repress the transcription of E-cadherin (reviewed in De Craene *et al.*, 2005). At the protein level, E-cadherin is regulated by trafficking and protein turnover pathways. The precise endocytic pathways for E-cadherin are still unclear, and there is evidence for both caveolae-dependent endocytosis and clathrin-dependent endocytosis of E-cadherin (for a recent review, see Bryant and Stow, 2004). E-cadherin can also be ubiquitinated in cultured cells by the E3-ligase Hakai, which targets E-cadherin to the proteasome (Y. Fujita *et al.*, 2002). Another mechanism by which E-cadherin function is disrupted is through extracellular proteases, such as matrix metalloproteases, which degrade the extracellular domain of E-cadherin and consequently reduce cadherin-mediated cell–cell adhesion (Egeblad and Werb, 2002). Some or all of these mechanisms may occur simultaneously during an EMT to disrupt cell–cell adhesion and promote motility.

In some cases, the delamination of cells from an epithelium occurs without the complete loss of cell–cell adhesion. In the sea urchin species *Mespilia*, the loss of cell–cell adhesions by ingressing PMCs is incomplete, and the PMCs tear themselves away from the epithelium, leaving behind a portion of their apical domain. However, this inefficient loss of cell–cell adhesion is not observed in other sea urchin species, such as *Arbacia* and *Lytechinus* (Shook and Keller, 2003). Similarly, in the delamination of the cranial neural crest and neuronal precursors from the trigeminal placodes in mice, apical adhesions are not completely down-regulated, but rather, the adherens junctions of departing mesenchymal cells remain intact and are pulled along the plane of the membranes of adjacent epithelial cells until they eventually rupture (Nichols, 1987). Therefore, the importance of a complete loss of cell–cell adhesion in EMTs is debatable.

Another potential mechanism of delamination from an epithelial sheet involves an asymmetric cell division, in which the basal parent cell retains adherens junctions (and therefore remains tethered to the epithelium) while the apical daughter cell is separated from the adherens junctions by the cleavage furrow and is released from the epithelium. An asymmetric mitosis has not yet been associated with well-studied EMTs, but it has been observed in the detachment of neurons from the ventricular zone of the ferret brain (Chenn and McConnell, 1995), neuroblast delamination in *Drosophila* (Urbach *et al.*, 2003), and the translocation of cells from the dermamyotome to the myotome (Gros *et al.*, 2005). These latter three events are not widely considered to be EMTs because the resultant cells do not exhibit complete mesenchymal behavior, such as active migration and invasiveness. At present, there is no direct evidence that an asymmetric cell division is involved in canonical EMTs, such as sea urchin PMC ingression, amniote primitive streak mesenchyme formation, neural crest delamination, and heart valve formation.

In summary, epithelial integrity is maintained principally by cadherins, and changes in cadherin expression are usually necessary for an EMT.

### Changes in Cell–ECM Adhesion

Altering the way that a cell interacts with the ECM is also important in EMTs and METs. For example, sea urchin PMCs lose cell–cell adhesions but simultaneously acquire adhesion to basal lamina components such as fibronectin and laminin during the EMT (Fink and McClay, 1985). Cell–ECM adhesion is mediated principally by integrins. Integrins are transmembrane proteins composed of two noncovalently linked subunits, α and β, and require $Ca^{2+}$ or $Mg^{2+}$ for binding to ECM components, such as fibronectin, laminin, and collagen. The cytoplasmic domain of integrins links to the cytoskeleton and interacts with signaling molecules. Changes in integrin function are required for many EMTs. During neural crest delamination, β1 integrin is necessary for neural crest adhesion to fibronectin, and it becomes functional just a few hours before the EMT occurs (Delannet and Duband, 1992). Likewise, as epiblast cells undergo an EMT to form mesoderm during mouse primitive streak formation, the cells exhibit increased adhesion to ECM molecules (for a review, see Hay, 2005). In both neural crest and primitive streak epiblast cells, inhibiting integrin function with function-blocking antibodies prevents cell migration. Various integrins are also markers for metastasis in certain cancers (reviewed in Hood and Cheresh, 2002). However, the misexpression of integrin subunits does not appear to be sufficient to bring about a full EMT *in vitro* (Valles *et al.*, 1996) or *in vivo* (Carroll *et al.*, 1998).

The presence and function of integrins can also be modulated in several ways. For example, the transcriptional activation of integrin β6 during colon carcinoma metastasis is mediated by the transcription factor Ets-1 (Bates *et al.*, 2005). Membrane trafficking and ubiquitination may also regulate the presence of integrin protein at the cell surface, but at present this process is poorly characterized. More importantly, most integrins can cycle between "On" (high-affinity) and "Off" (low-affinity) states. This inside-out regulation of integrin adhesion occurs at the integrin cytoplasmic tail (Hood and Cheresh, 2002). In addition to integrin activation, the spatial arrangement of integrins on the cell surface — or clustering — also affects the overall strength of integrin–ECM interactions. The increased adhesiveness of integrins due to clustering (known as *avidity*) can be activated by chemokines and is dependent on RhoA and phosphatidylinositol 3′ kinase (PI3K) activity.

In summary, adhesion to the ECM is required for the EMT. Cell–ECM adhesions are maintained by integrins, and integrins have varying degrees of adhesiveness, depending on the presence, activity, or avidity of the integrin subunits.

### Changes in Cell Polarity and Stimulation of Cell Motility

In order for mesenchymal cells to migrate away after detaching from the epithelium, they also must become motile. The asymmetric arrangement of the cytoskeleton and organelles in epithelial versus mesenchymal cells produces a distinct cellular polarity. Epithelial cell polarity is characterized by cell–cell junctions found at the apical-lateral domain and integrin-mediated adhesions at the basal side contacting basal lamina. Mesenchymal cells in contrast do not have apical/basal polarity, but, rather, front-end/back-end polarity, with actin-rich lamellipodia and Golgi localized at the leading edge (reviewed in Hay, 2005). Molecules that establish cell polarity include Cdc42, PAK1, PI3K, PTEN, Rac, and the PAR proteins. For example, the loss of cell polarity in the TGF-β-stimulated EMT of mammary epithelial cells in culture is mediated by the polarity protein Par6. The stimulated TGF-β receptor II causes Par6 to activate the E3 ubiquitin ligase Smurf1, and Smurf1 then targets GTPase RhoA to the proteasome. The loss of RhoA activity results in the loss of cell–cell adhesion and epithelial cell polarity (Ozdamar et al., 2005).

The cellular programs responsible for down-regulating cell–cell adhesion and stimulating cell motility are separable. For example, in EpH4 cells that undergo an EMT by activating the transcription factor Jun, there is a complete loss of epithelial polarity, but cell migration is not stimulated (Fialka et al., 1996). Similarly there are two steps during the EMT that generates the cardiac cushion cells. First, the cardiac endothelium is activated, whereby the cells lose their adhesions to each other, become hypertrophic and polarize the Golgi toward one end of the cell. Second, these activated cells become motile and invasive (Boyer et al., 1999). The process of mesenchymal motility begins with the polarization and elongation of the cell, followed by the extension of a lamellipodium in the direction of migration. The cell body is propelled forward by the contraction of actin-myosin cytoskeleton and traction provided by adhesion to the ECM. How cell motility is activated and the extent to which cell motility is required for an EMT must be the subject of further research.

### Invasion of the Basal Lamina

In most EMTs the emerging mesenchymal cells must penetrate a basal lamina. The basal lamina consists of ECM components such as collagen type IV, fibronectin, and laminin, and it functions to stabilize the epithelium and act as a barrier to migratory cells (Erickson, 1987). One mechanism that cells use to breach the basal lamina is to produce enzymes that degrade it, including plasminogen activator and matrix-metalloproteases (MMPs). Plasminogen activator is associated with a number of EMTs, including neural crest delamination (Erickson and Isseroff, 1989) and the formation of cardiac cushion cells during heart morphogenesis (McGuire and Alexander, 1993). Experimentally blocking plasminogen activity reduces the number of migratory cells in either model system. MMPs are also involved in a number of EMTs. MMP-2 is necessary for the EMT that generates neural crest cells, because when inhibitors of MMP-2 are added to chicken embryos *in vivo* or if MMP-2 translation is blocked with MMP-2 antisense oligonucleotides, neural crest delamination — but not neural crest motility — is inhibited (Duong and Erickson, 2004). In mouse mammary cells, MMP-3 misexpression is sufficient for an EMT *in vitro* and *in vivo* (Sternlicht *et al.*, 1999). Recently, the mechanism for MMP-3-induced EMT was elucidated. MMP-3 misexpression induces an alternatively spliced form of Rac1 (Rac1b), which then causes an increase in reactive oxygen species (ROS) intracellularly. Either Rac1b activity or ROS are necessary and sufficient for an MMP-3-induced EMT. Rac1b or ROS can also induce the expression of the transcription factor Snail-1 (Radisky *et al.*, 2005). The role of Rac1b or ROS in an EMT is unexpected, and it is not known if they control other EMT events during development or pathogenesis.

## III. THE EMT TRANSCRIPTIONAL PROGRAM

At the foundation of every EMT or MET program are the transcription factors that control the expression of genes required for this cellular transition. While many of the transcription factors that regulate EMTs have been identified, the complex transcriptional networks are still incomplete. Here we review the transcription factors that are known to promote the various phases of an EMT: loss of cell–cell adhesion, increase in cell–ECM adhesion, stimulation of cell motility, and invasion across the basal lamina. Then we examine how these EMT transcription factors themselves are regulated at the transcriptional and protein levels.

### Transcription Factors That Regulate EMTs

The Snail family of zinc-finger transcription factors, including Snail-1 and Snail-2 (formerly Snail and Slug, see Barrallo-Gimeno and Nieto, 2005), are emerging as direct regulators of cell–cell adhesion and motility during EMTs (Barrallo-Gimeno and Nieto, 2005; De Craene and Nieto, 2005). Snail-1 and Snail-2 are evolutionarily conserved in vertebrates and invertebrates, and, to our knowledge, Snail-1 or Snail-2 is expressed singly or in combination during every EMT yet examined. Blocking Snail-2 in the chicken primitive streak or during neural crest delamination with antisense oligos against *Snail-2* inhibits these EMTs (Nieto *et al.*, 1994). The disruption of *Snail-1* in mice is lethal early in gestation, and mutant embryos display defects in the primitive streak EMT required to generate mesoderm. While some mesodermal markers are expressed, these presumptive mesodermal cells still retain apical/basal polarity and adherens junctions, and express E-cadherin mRNA (Carver *et al.*, 2001). Snail-1 expression is sufficient for breast cancer recurrence in a mouse model *in vivo*, and high levels of Snail-1 expression predict the relapse of breast cancer in women (Moody *et al.*, 2005). One mode of Snail-1 or Snail-2 activity is to decrease cell–cell adhesion, particularly by repressing the *E-cadherin* promoter (reviewed in De Craene *et al.*, 2005). This repression requires the mSin3A corepressor complex and histone deacetylases. Snail-1 is also a transcriptional repressor of the tight junction proteins Claudin and Occludin (De Craene *et al.*, 2005). The misexpression of Snail-1 and Snail-2 leads to the transcription of genes important for cell motility. In MDCK cells, the misexpression of Snail-1 indirectly up-regulates fibronectin and vimentin, which are essential for mesenchymal cell adhesion (Cano *et al.*, 2000), and Snail-2 misexpression induces RhoB mRNA in avian neural crest cells (Del Barrio and Nieto, 2002). Snail-1 expression can also promote invasion across the basal lamina. In MDCK cells, the misexpression of Snail-1 indirectly up-regulates *mmp-9* transcription and subsequently increases basal lamina invasion (Jorda *et al.*, 2005). Hence, Snail-1 or Snail-2 is necessary and sufficient for bringing about many of the processes of an EMT, including loss of cell–cell adhesion, changes in cell polarity, gain of cell motility, and invasion of the basal lamina.

Snail-1 and Snail-2 have been well characterized as transcriptional *repressors*, and it is still mysterious how the expression of Snail-1 and Snail-2 results in the *activation* of genes important for an EMT. In the avian neural crest it was recently shown that the *Snail-2* promoter is activated by the binding of Snail-2 to an E-box motif, indicating that in this case Snail-2 can act as a transcriptional activator of itself (Sakai *et al.*, 2006). Hence, the role of Snail-2 (and also likely Snail-1) as a transcriptional repressor or activator may be context dependent. Much is still to be learned about the downstream roles of Snail-1 and Snail-2 in regulating genes critical to an EMT.

Two other zinc-finger transcription factors that regulate EMTs are delta-crystallin enhancer-binding factor 1 (δEF1; also known as ZEB1) and Smad-interacting protein-1 (Sip1, also known as ZEB2). δEF1 is necessary and sufficient for an EMT in mammary cells transformed by the transcription factor c-Fos in a process that is apparently independent of Snail-1 (Eger *et al.*, 2005). Sip1 is structurally similar to δEF1, and Sip1 overexpression is sufficient to down-regulate E-cadherin, dissociate adherens junctions, and increase motility in MDCK cells (Comijn *et al.*, 2001). The cranial neural crest cells of *Sip1* mutant mice do not undergo delamination properly (Wakamatsu *et al.*, 2001). Both δEF1 and Sip1 can bind to the *E-cadherin* promoter and repress transcription (De Craene *et al.*, 2005).

The lymphoid enhancer–binding factor/T-cell factor (LEF/TCF) transcription factors also play a role in EMTs. For example, the misexpression of Lef-1 in cultured colon cancer cells causes the down-regulation of E-cadherin and loss of cell–cell adhesion. Reversing Lef-1 misexpression (by removing Lef-1 retrovirus from the culture medium) causes cultured cells to revert back to an epithelium (Kim *et al.*, 2002). One important role for the LEF/TCF transcription factors in EMTs is to activate genes that regulate cell motility. The LEF/TCF pathway activates the promoter of the L1 adhesion molecule, a protein that is associated with increased motility and invasive behavior of colon cancer cells (Gavert *et al.*, 2005). β-catenin and LEF/TCF activate the *fibronectin* gene (Gradl *et al.*, 1999), and LEF/TCF transcription factors also activate genes that are required for basal lamina invasion, including *mmp-3* and *mmp-7* (Gustavson *et al.*, 2004).

## Regulation of the Snail and LEF/TCF Transcription Factors

Given the importance of the Snail and LEF/TCF transcription factors in orchestrating the various phases of an EMT, we need to understand how these EMT-inducing transcription factors are themselves regulated. Transcription factor activity can be controlled both at the transcriptional and at the protein level.

The activation of *Snail-1* transcription in *Drosophila* requires the transcription factors Dorsal (NF-κB) and Twist, and the *Snail-1* promoter includes both Dorsal and Twist binding sites (reviewed in De Craene *et al.*, 2005). The human *Snail-1* promoter also has functional NF-κB sites; and in cultured human cells transformed by Ras and induced by TGF-β, NF-κB is essential for EMT initiation and maintenance (Huber *et al.*, 2004). Also, a region of the *Snail-1* promoter is responsive to integrin-linked kinase (ILK) overexpression in cultured cells (reviewed in De Craene *et al.*, 2005), and preliminary results suggest that ILK can activate Snail-1 expression via poly-ADP-ribose polymerase (PARP, Lee *et al.*, 2006). There are also *Snail-1* transcriptional repressors. In breast cancer cell lines, metastasis-associated protein 3 (MTA3) binds directly to and represses the transcription of *Snail-1* in combination with the Mi-2/NuRD complex (N. Fujita *et al.*, 2003). MTA3 is induced by the estrogen receptor (ER, nuclear hormone) pathway, and the absence of ER signaling or MTA3 leads to the activation of *Snail-1* expression. This suggests a mechanism whereby loss of the estrogen receptor in breast cancer contributes to metastasis. The role of MTA3 in regulating the transcription of Snail-1 mRNA in other EMTs is not known.

*Snail-2* transcriptional regulators have also been identified. In *Xenopus*, the *Snail-2* promoter has functional LEF/TCF binding sites, and in the mouse, MyoD (transcription factor central to muscle cell development) binds to the

*Snail-2* promoter and activates *Snail-2* transcription. In humans, the oncogene E2A-HLF and the pigment cell regulator MITF also can bind to the *Snail-2* promoter and activate its transcription (De Craene *et al.*, 2005). As mentioned earlier, in the avian neural crest, Snail-2 is also able to activate its own promoter, either alone or in synergy with Sox9 (Sakai *et al.*, 2006).

LEF/TCF transcription factors can be activated by TGF-β signaling. For instance, exposure of the medial edge epithelium of the palatal shelves to TGF-β3 induces the binding of the phosphorylated Smads2/4 complex to the *Lef-1* promoter and activates *Lef-1* transcription (Nawshad and Hay, 2003). While β-catenin is necessary for the function of the LEF/TCF proteins, the presence of high levels of nuclear β-catenin is not necessary for the transcription of Lef-1 in the fusion of the mouse palate (Nawshad and Hay, 2003). The misexpression of Snail-1 also activates the transcription of δ*EF-1* and *Lef-1* through a yet unknown mechanism (see De Craene *et al.*, 2005).

The activity of EMT transcription factors is also regulated at the protein level, including protein stability (targeting to the proteasome) and nuclear localization. GSK-3β, the same protein kinase that phosphorylates β-catenin and targets it for destruction, also phosphorylates Snail-1. The human Snail-1 protein contains two GSK-3β phosphorylation consensus sites between amino acids 97–123. Inhibiting GSK-3β prevents Snail-1 degradation and results in the loss of E-cadherin in cultured epithelial cells (Zhou *et al.*, 2004). Therefore, the inhibition of GSK-3β activity by Wnt signaling may have multiple roles in an EMT, leading to the stabilization of both β-catenin and Snail-1. Two other proteins that play a role in preventing GSK-3β-mediated phosphorylation of Snail-1 are lysyl-oxidase-like proteins LOXL2 and LOXL3. LOXL2 and LOXL3 form a complex with the Snail-1 protein near the GSK-3β phosphorylation sites, thus preventing GSK-3β from interacting with Snail-1. The misexpression of LOXL2 or LOXL3 reduces Snail-1 protein degradation and induces an EMT in cultured epithelial cells (Peinado *et al.*, 2005). The importance of LOXL2 and LOXL3 in other EMTs is not yet known.

The function of Snail-1 also depends on its nuclear localization. Snail-1 has a nuclear localization sequence (NLS). The phosphorylation of human Snail-1 by p21-activated kinase 1 (Pak1) at Ser[246] promotes the nuclear localization of Snail-1 (and therefore Snail-1 activation) in breast cancer cells. Pak1 can be activated by RTK signaling, and knocking down Pak1 by siRNA blocks Pak1-mediated Snail-1 phosphorylation, increases the cytoplasmic accumulation of Snail-1, and reduces the invasive behavior of these breast cancer cells (Yang *et al.*, 2005). The protein that mediates the translocation of Snail-1 into the nucleus in human cells is not yet known, although a Snail-1 nuclear importer has been described in zebrafish. The zinc-finger transporting protein LIV1 is required for Snail-1 to localize to the nucleus during zebrafish gastrulation, and LIV1 is activated by STAT3 signaling (Yamashita *et al.*, 2004). In zebrafish, the protein kinase that phosphorylates the NLS sequence of Snail-1 to promote the translocation of Snail-1 to the nucleus has not yet been identified. Snail-1 also contains a nuclear export sequence (NES) at amino acids 132–143 that is necessary and sufficient for the export of Snail-1 from the nucleus to the cytoplasm and is dependent on the calreticulin nuclear export pathway (Dominguez *et al.*, 2003). This NES sequence is activated by the phosphorylation of the same lysine residues targeted by GSK-3β, which suggests a mechanism whereby phosphorylation of Snail-1 by GSK-3β results in the export of Snail-1 from the nucleus.

LEF/TCF transcriptional activity is also regulated by other proteins. β-catenin is required as a cofactor for the activation of LEF/TCF transcription factors, and Lef-1 can also associate with cofactor Smads to activate the transcription of additional EMT genes (Labbe *et al.*, 2000).

In summary, EMT transcription factors such as Snail-1 and Lef-1 are regulated by a variety of mechanisms, both at the transcriptional level and at the protein level by protein degradation, nuclear localization, and cofactors. Many questions remain. What activates *Snail-1* transcription? What promoters are targeted by the Snail and LEF/TCF transcription factors? And what other cofactors regulate Snail-1 and Snail-2?

## IV. MOLECULAR CONTROL OF THE EMT

The initiation of an EMT or an MET is a tightly regulated event during development and tissue repair, since the deregulation of either program is disastrous to the organism. A variety of external and internal signaling mechanisms coordinate the complex events of the EMT, and these same signaling pathways are often disrupted or reactivated during disease. Many of the molecules that trigger EMTs or METs have been identified, and in some cases the downstream effectors are known. EMTs or METs can be induced by either diffusible signaling molecules or ECM components, and these inductive signals act either directly on cell adhesion/structural molecules themselves or by regulating EMT transcriptional regulators. We first discuss the role of signaling molecules and ECM in triggering an EMT, and then we present a summary model for the induction of EMTs.

### Signaling Molecules

During development, five ligand–receptor signaling pathways are primarily employed: TGF-β, Wnt, RTK, Notch, and Hedgehog signaling pathways. These pathways all have a role in triggering EMTs. Although the activation of a single signaling pathway can be sufficient for an EMT, in most cases an EMT or MET is initiated by multiple signaling pathways acting in concert.

#### TGF-b Pathway

The transforming growth factor-beta (TGF-β) superfamily includes TGF-β, activin, and bone morphogenetic protein (BMP) families. These ligands signal through receptor serine/threonine kinases to activate a variety of signaling molecules, including Smads, MAPK, PI3K, and ILK. Most of the EMTs studied to date are induced, in part or solely, by TGF-β superfamily members (for a recent review, see Zavadil and Bottinger, 2005). During embryonic heart formation, an EMT occurs as the endocardium produces mesenchymal cells that invade the cardiac jelly to form the endocardial cushions (reviewed in Eisenberg and Markwald, 1995). In chicken embryos, TGF-β2 and TGF-β3 have sequential and necessary roles in activating the endocardium and in signaling mesenchymal invasion, respectively (Camenisch et al., 2002a). The TGF-β superfamily member BMP2 may play a similar role in the mouse, since in *BMP2* or *BMP receptor 1A* (*BMPR1A*) mouse mutants the EMT that generates endocardial cushion cells does not occur. Moreover, BMP2 can induce this EMT *in vitro* (Sugi et al., 2004). TGF-β3 also triggers the EMT that occurs in the fusing palate of mice (Nawshad et al., 2004). In the avian neural crest, BMP4 induces *Snail-2* expression, an important transcription factor in the neural crest EMT (Liem et al., 1995).

In the EMT that transforms epithelial tissue into metastatic cancer cells, it is generally accepted that TGF-β can act both as a tumor suppressor and as a tumor/EMT inducer. For example, transgenic mice expressing TGF-β1 in keratinocytes are more resistant to the development of chemically induced skin tumors than controls, suggesting a tumor suppressor effect of TGF-β1 on epithelial cells. However, a greater portion of the tumors that do form in the keratinocyte-TGF-β1 transgenic mice are highly invasive spindle-cell carcinomas, indicating that TGF-β1 can induce an EMT in later stages of skin cancer development (Cui et al., 1996). Similar effects of TGF-β are observed in breast cancer progression, where the TGF-β pathway initially inhibits tumor growth but later promotes metastasis to the lung (Zavadil and Bottinger, 2005). Expression of dominant-negative TGF-βR II in cancer cells transplanted into nude mice blocks TGF-β-induced metastasis (Portella et al., 1998). Multiple signaling pathways may be involved in TGF-β-induced EMT. For example, in cultured breast cancer cells, activated Ras and TGF-β induce an irreversible EMT (Janda et al., 2002); and in pig thyroid epithelial cells, TGF-β and epidermal growth factor (EGF) synergistically stimulate the EMT (Grande et al., 2002).

One outcome of TGF-β signaling is to immediately signal changes in cell polarity. As cited earlier, in TGF-β-induced EMTs of mammary epithelial cells, TGF-βR II phosphorylates the polarity protein, Par6, and phosphorylated Par6 causes the E3 ubiquitin ligase, Smurf1, to target the GTPase, RhoA, for degradation. RhoA is required for the stability of tight junctions, and loss of RhoA leads to their dissolution (Ozdamar et al., 2005). TGF-β signaling also regulates gene expression through the phosphorylation and activation of several Smads. Smad3 is necessary for a TGF-β-induced EMT, since the deletion of Smad3 in a mouse model leads to the inhibition of injury-induced lens and kidney tissue EMT (Roberts et al., 2006). The precise role of Smads in EMTs and the gene targets that Smads regulate will require further investigation. TGF-βR I can also bind to and activate PI3K (Yi et al., 2005), which in turn can activate ILK and downstream pathways.

ILK is emerging as an important positive regulator of EMTs (reviewed in Larue and Bellacosa, 2005). ILK has binding sites to allow interactions with integrins, the actin skeleton, focal adhesion complexes, PI3K, and growth factor receptors (TGF-β, Wnt, or RTK). ILK can directly phosphorylate and regulate either Akt or GSK-3β, and ILK activity indirectly results in the activation of downstream transcription factors such as AP-1, NF-κB, and Lef1. Overexpression of ILK in cultured cells causes the suppression of GSK-3β activity (Delcommenne et al., 1998), translocation of β-catenin to the nucleus, activation of Lef-1/β-catenin transcription factors, and the down-regulation of E-cadherin (Novak et al., 1998). Inhibition of ILK in cultured colon cancer cells leads to the stabilization of GSK-3β activity and decreased nuclear β-catenin localization and results in the suppression of *Lef-1* and *Snail-1* transcription and the reduced invasive behavior of these colon cancer cells (Tan et al., 2001). ILK activity also results in the expression of MMPs via Lef-1 transcriptional activity (Gustavson et al., 2004). Hence, ILK (inducible by TGF-β signaling) is capable of orchestrating major events in an EMT, including the loss of cell–cell adhesion and invasion across the basal lamina.

### Wnt Pathway

Many EMTs or METs are also regulated by Wnt signaling. Wnts signal through seven-pass transmembrane proteins of the Frizzled family and activate G-proteins, PI3K, and β-catenin nuclear signaling. During zebrafish gastrulation, Wnt11 activates the GTPase Rab5c, which results in the endocytosis of E-cadherin and subsequent loss of cell–cell adhesion (Ulrich et al., 2005). Wnt6 signaling is sufficient for the induction of *Snail-2* transcription in the neural crest in the chicken embryo, and perturbation of the Wnt pathway reduces neural crest induction (Garcia-Castro et al., 2002). Wnts can also signal METs. For instance, Wnt4 is required for the coalescence of nephrogenic mesenchyme into epithelial tubules during murine kidney formation (Stark et al., 1994), and Wnt6 is necessary and sufficient for the MET that forms somites (Schmidt et al., 2004).

One of the downstream signaling molecules activated by Wnt signaling is β-catenin. β-catenin is a structural component of adherens junctions, acting as a bridge between cadherins and the cytoskeleton. Nuclear β-catenin is also a limiting factor for the activation of LEF/TCF transcription factors. β-catenin is pivotal for regulating most EMTs. In the sea urchin embryo, β-catenin expression is observed in the nuclei of PMCs prior to ingression, and nuclear β-catenin expression is lost in PMCs after the EMT is complete. Misexpression of an intracellular cadherin domain in sea urchin embryos to interfere with nuclear β-catenin signaling blocks the ingression of PMCs (Logan et al., 1999). In mouse knockouts for β-catenin, the primitive streak EMT does not occur, and no mesoderm is formed (Huelsken *et al.*, 2000). β-catenin is also necessary for the EMT that occurs during cardiac cushion development (Liebner *et al.*, 2004). In breast cancer, β-catenin expression is highly correlated with metastasis and poor survival (Cowin *et al.*, 2005), and blocking β-catenin function in tumor cells inhibits their invasion *in vitro* (Wong and Gumbiner, 2003). It is unclear if β-catenin overexpression alone is sufficient for all EMTs. If β-catenin is misexpressed in cultured cells, it causes apoptosis (Kim *et al.*, 2000). However, the misexpression of a stabilized form of β-catenin in mouse epithelial cells *in vivo* results in metastatic skin tumors (Gat *et al.*, 1998).

## Signaling by RTK Ligands

The receptor tyrosine kinase (RTK) family of receptors and the growth factors that activate them also regulate EMTs or METs. Ligand binding promotes RTK dimerization and activation of their intracellular kinase domains by the auto-phosphorylation of tyrosine residues. These phospho-tyrosines act as docking sites for intracellular signaling molecules, which can activate signaling cascades such as Ras/MAPK, PI3K/Akt, JAK/STAT, and ILK. We now cite a few examples.

Hepatocyte growth factor (HGF, also known as scatter factor) acts through the RTK c-met. HGF is important for the MET in the developing kidney, since HGF/SF function-blocking antibodies inhibit the assembly of metanephric mesenchymal cells into kidney epithelium in organ culture (Woolf *et al.*, 1995). HGF signaling is required for the EMT that produces myoblasts (limb muscle precursors) from somite tissue in the mouse, because in knockout mice for c-met, myoblasts fail to detach from the myotome and migrate into the limb bud (reviewed in Thiery, 2002).

Fibroblast growth factor (FGF) signaling regulates mouse primitive streak formation. In FGFR1 mouse mutants, E-cadherin is not down-regulated, β-catenin does not relocate to the nucleus, *Snail-1* expression is down-regulated, and few FGFR1 −/− cells contribute to the mesoderm. Interestingly, if E-cadherin function is also inhibited in FGFR1 mutants by the addition of function-blocking E-cadherin antibodies, the primitive streak EMT proceeds normally. The suggested mechanism is that failure to remove E-cadherin (mediated by FGFR1 signaling) allows E-cadherin to sequester cytoplasmic β-catenin and therefore attenuate later Wnt signaling required to complete the primitive streak EMT (Ciruna and Rossant, 2001). FGF signaling also stimulates cell motility and activates MMPs. In studies with various epithelial cultured cancer cells, sustained FGF2 and N-cadherin signaling results in increased cell motility (increased invasion of uncoated filters), MMP-9 activation, and the ability to invade ECM (invasion of matrigel-coated filters) (Suyama *et al.*, 2002).

Insulin growth factor (IGF) signaling can also induce an EMT. In epithelial cell lines derived from breast tumors, IGF receptor I (IGFR I) hyperstimulation results in increased cell survival and motility, apparently through the activation of Akt2 and suppression of Akt1 (Irie *et al.*, 2005). In several cultured epithelial cell lines, IGFR1 is associated with the complex of E-cadherin and β-catenin, and the ligand IGF-II causes the redistribution of β-catenin from the membrane to the nucleus, activation of the transcription factor TCF-3, partial degradation of E-cadherin, and a subsequent EMT (Morali *et al.*, 2001).

Another RTK known for its role in EMTs is the ErbB2/HER-2/Neu receptor, whose ligand is heregulin/neuregulin. Overexpression of HER-2 occurs in 25% of human breast cancers, and the misexpression of HER-2 in mouse mammary tissue *in vivo* is sufficient to cause metastatic breast cancer (Muller *et al.*, 1988). Herceptin® (antibody against the HER-2 receptor) treatment is effective in reducing the recurrence of HER-2-positive metastatic breast cancers. HER-2 signaling activates *Snail-1* expression in breast cancer through an unknown mechanism (Moody *et al.*, 2005). Given these several examples, it appears that the RTK signaling pathway is important for the induction of EMTs.

## Notch Pathway

The Notch signaling family is well known for its role in cell specification, and it is now emerging as a regulator of EMTs. When the Notch receptor is activated by its ligand delta, an intracellular portion of the Notch receptor ligand is cleaved and transported to the nucleus, where it binds to the transcription factor Su(H) to regulate target genes. In zebrafish Notch1 mutants, cardiac endothelium expresses very little *Snail-1* and does not undergo the EMT required to make the cardiac cushions (Timmerman *et al.*, 2004). This mutation can be phenocopied by treating embryonic heart explants with inhibitors of the Notch pathway. Conversely, misexpression of activated Notch1 is sufficient to activate *Snail-1* expression and promote an EMT in cultured endothelial cells. In the heart, Notch functions via lateral induction to make cells competent to respond to TGF-β2, which we have previously discussed as a regulator of the cardiac cushion EMT (Timmerman *et al.*, 2004). In the avian neural crest EMT, Notch signaling is required for the induction and/or maintenance of *BMP4* expression and, hence, the EMT (Endo *et al.*, 2002). Similarly, Notch signaling is required for the TGF-β-induced EMT of epithelial cell lines. The use of antisense oligonucleotides against Hey1 mRNA, siRNA against Jagged1 mRNA (encodes a Notch-ligand), or γ-secretase inhibitor GSI treatment (to block Notch receptor activation) each can inhibit a TGF-β-induced EMT (Zavadil *et al.*, 2004). Therefore, the general role of Notch signaling in EMTs may be to induce competence to undergo an EMT in response to TGF-β signaling.

## Hedgehog Pathway

The hedgehog pathway also regulates EMTs. Metastatic prostate cancer cells express high levels of hedgehog and *Snail-1*. If prostate cancer cell lines are treated with the hedgehog-pathway inhibitor, cyclopamine, levels of *Snail-1* are decreased. Likewise, if the hedgehog-activated transcription factor, Gli, is misexpressed, Snail-1 mRNA expression increases and E-cadherin mRNA levels decrease (Karhadkar *et al.*, 2004).

## ECM Signaling

In addition to diffusible signaling molecules, the extracellular environment can regulate EMTs or METs. This was first dramatically demonstrated when lens or thyroid epithelium was embedded in collagen gels, and they promptly underwent an EMT (reviewed in Hay, 2005). Integrin signaling appears to be important in this process, because if function-blocking antibodies against integrins are present in the collagen gels, the EMT is inhibited (Zuk and Hay, 1994). Hyaluronan is another ECM component that may regulate EMTs. In the hyaluronan synthase-2 knockout mouse (*Has2 −/−*, which has defects in hyaluronan synthesis and secretion), the cardiac endothelium fails to undergo an EMT and produce the migratory mesenchymal cells to form the heart valve. The role of hyaluronan in this EMT may be to activate the RTK ErbB2/HER-2/Neu, because

treating cultured *Has2 −/−* heart explants with heregulin (ligand for ErbB2) rescues the EMT. Consistent with this hypothesis, treating cardiac explants with hyaluronan activates ErbB2, and blocking ErbB2 signaling with the drug herstatin reproduces the *Has2*-knockout phenotype (Camenisch *et al.*, 2002b). A third ECM component that can stimulate an EMT is the gamma-2 chain of laminin 5, which is cleaved from laminin 5 by MMP-2. The gamma-2 chain causes the scattering and migration of epithelial cancer cells (Koshikawa *et al.*, 2000) and may be a marker of epithelial tumor cell invasion (Katayama *et al.*, 2003).

During EMT, the loss of cadherin expression is associated with the gain of integrin function. One molecule that has been shown to coordinate the loss of cell–cell adhesion with the gain of cell–ECM adhesion during EMT is the GTPase Rap1. In several cultured cell lines, the endocytosis of E-cadherin activates the Ras family member Rap1. Activated Rap1 is required to form integrin-mediated adhesions, since the overexpression of the Rap1-inactivating enzyme, Rap1GAPv, blocks integrin-ECM adhesion formation (Balzac *et al.*, 2005). The molecules with which Rap1 interacts to activate integrin function are not yet known.

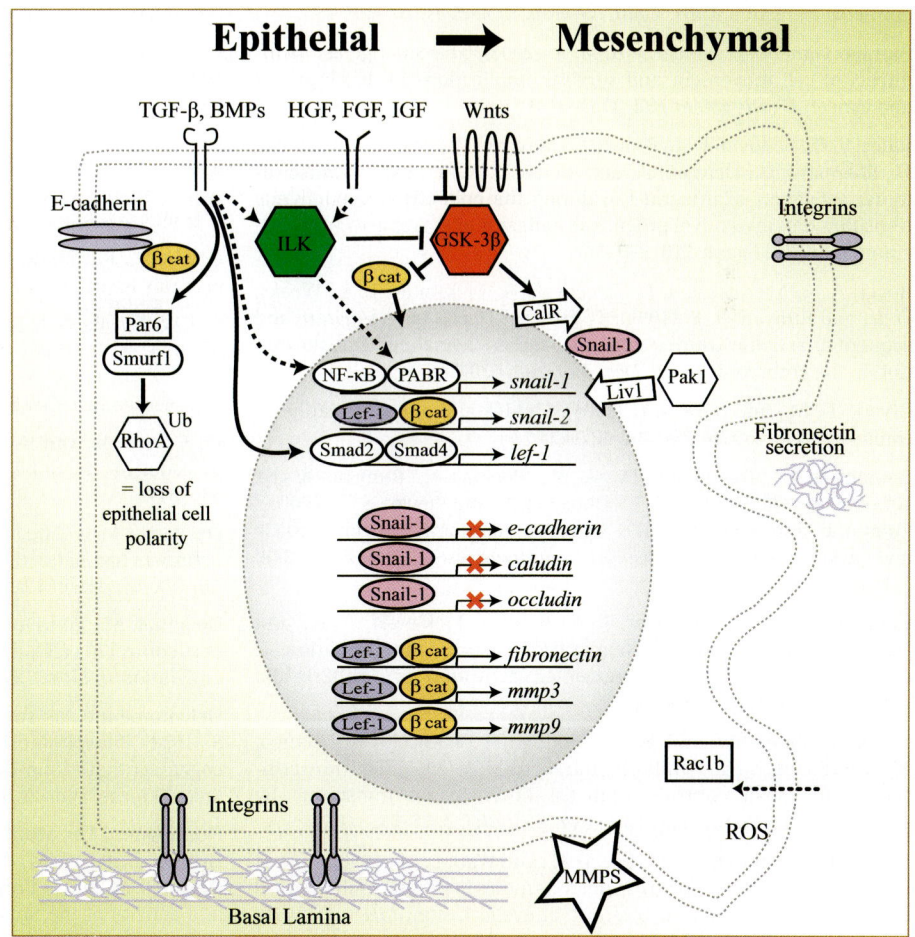

**FIG. 6.1.** Induction of an EMT. This summary figure emphasizes some of the important molecules that bring about an EMT. The direct action of proteins on downstream targets are indicated by solid arrows, whereas a dashed arrow represents signaling pathways that are not yet defined. Progression of the EMT proceeds from left to right.

## A Framework for EMT Induction

Much of the experimental work on EMT mechanisms is piecework, and in no system is the entire inductive pathway and downstream effectors for an EMT completely worked out. However, developing a framework in an attempt to define the EMT molecular pathways can lead to insights and testable hypotheses. Figure 6.1 summarizes many of the various signaling mechanisms, although in reality only a few of the inductive signals and pathways may be utilized in particular EMT events. From experimental evidence to date, it appears that many of the EMT signaling pathways converge on ILK and nuclear β-catenin signaling to activate Snail and LEF/TCF transcription factors. Snail and LEF/TCF transcription factors then act on a variety of targets to suppress cell–cell adhesion, induce changes in cell polarity, stimulate cell motility, and promote invasion of the basal lamina (see Fig. 6.1).

## V. CONCLUSION

Over the past 20+ years since the term *EMT* was coined (reviewed in Thiery, 2002), important insights have been made in this rapidly expanding field of research. EMT and MET events occur during development and disease, and many of the molecules that regulate the various EMTs or METs have been characterized, thanks in large part to the advent of cell culture models. Despite this progress, there are still major gaps in our understanding of the regulatory networks for any EMT or MET. Mounting evidence suggests that disease processes such as the metastasis of epithelial-derived cancers and kidney fibrosis are regulated by the same molecular mechanisms that allow an epithelium to produce migratory and invasive cells during development. A clearer understanding of EMT and MET pathways in the future will lead to more effective strategies for tissue engineering and novel therapeutic targets.

## VI. REFERENCES

Balzac, F., Avolio, M., Degani, S., Kaverina, I., Torti, M., Silengo, L., Small, J. V., and Retta, S. F. (2005). E-cadherin endocytosis regulates the activity of Rap1: a traffic light GTPase at the crossroads between cadherin and integrin function. *J. Cell Sci.* **118**, 4765–4783.

Barasch, J. (2001). Genes and proteins involved in mesenchymal to epithelial transition. *Curr. Opin. Nephrol. Hypertens.* **10**, 429–436.

Barrallo-Gimeno, A., and Nieto, M. A. (2005). The Snail genes as inducers of cell movement and survival: implications in development and cancer. *Development* **132**, 3151–3161.

Bates, R. C., Bellovin, D. I., Brown, C., Maynard, E., Wu, B., Kawakatsu, H., Sheppard, D., Oettgen, P., and Mercurio, A. M. (2005). Transcriptional activation of integrin ß-6 during the epithelial–mesenchymal transition defines a novel prognostic indicator of aggressive colon carcinoma. *J. Clin. Invest.* **115**, 339–347.

Boyer, A. S., Ayerinskas, I. I., Vincent, E. B., McKinney, L. A., Weeks, D. L., and Runyan, R. B. (1999). TGFß2 and TGFß3 have separate and sequential activities during epithelial–mesenchymal cell transformation in the embryonic heart. *Dev. Biol.* **208**, 530–545.

Bryant, D. M., and Stow, J. L. (2004). The ins and outs of E-cadherin trafficking. *Trends Cell Biol.* **14**, 427–434.

Camenisch, T. D., Molin, D. G. M., Person, A., Runyan, R. B., Gittenberger-de Groot, A. C., McDonald, J. A., and Klewer, S. E. (2002a). Temporal and distinct TGFß ligand requirements during mouse and avian endocardial cushion morphogenesis. *Dev. Biol.* **248**, 170–181.

Camenisch, T. D., Schroeder, J. A., Bradley, J., Klewer, S. E., and McDonald, J. A. (2002b). Heart-valve mesenchyme formation is dependent on hyaluronan-augmented activation of ErbB2-ErbB3 receptors. *Nat. Med.* **8**, 850–855.

Cano, A., Perez-Moreno, M. A., Rodrigo, I., Locascio, A., Blanco, M. J., del Barrio, M. G., Portillo, F., and Nieto, M. A. (2000). The transcription factor Snail controls epithelial–mesenchymal transitions by repressing E-cadherin expression. *Nat. Cell Biol.* **2**, 76–83.

Carroll, J. M., Luetteke, N. C., Lee, D. C., and Watt, F. M. (1998). Role of integrins in mouse eyelid development: studies in normal embryos and embryos in which there is a failure of eyelid fusion. *Mech. Dev.* **78**, 37–45.

Carver, E. A., Jiang, R., Lan, Y., Oram, K. F., and Gridley, T. (2001). The mouse snail gene encodes a key regulator of the epithelial–mesenchymal transition. *Mol. Cell. Biol.* **21**, 8184–8188.

Chenn, A., and McConnell, S. K. (1995). Cleavage orientation and the asymmetric inheritance of Notch1 immunoreactivity in mammalian neurogenesis. *Cell* **82**, 631–641.

Ciruna, B., and Rossant, J. (2001). FGF signaling regulates mesoderm cell fate specification and morphogenetic movement at the primitive streak. *Dev. Cell* **1**, 37–49.

Comijn, J., Berx, G., Vermassen, P., Verschueren, K., van Grunsven, L., Bruyneel, E., Mareel, M., Huylebroeck, D., and van Roy, F. (2001). The two-handed E box–binding zinc finger protein SIP1 down-regulates E-cadherin and induces invasion. *Mol. Cell* **7**, 1267–1278.

Cowin, P., Rowlands, T. M., and Hatsell, S. J. (2005). Cadherins and catenins in breast cancer. *Curr. Opin. Cell Biol.* **17**, 499–508.

Cui, W., Fowlis, D. J., Bryson, S., Duffie, E., Ireland, H., Balmain, A., and Akhurst, R. J. (1996). TGFß1 inhibits the formation of benign skin tumors, but enhances progression to invasive spindle carcinomas in transgenic mice. *Cell* **86**, 531–542.

De Craene, B., van Roy, F., and Berx, G. (2005). Unraveling signaling cascades for the Snail family of transcription factors. *Cell. Signal.* **17**, 535–547.

Del Barrio, M. G., and Nieto, M. A. (2002). Overexpression of Snail family members highlights their ability to promote chick neural crest formation. *Development* **129**, 1583–1593.

Delannet, M., and Duband, J. L. (1992). Transforming growth factor-beta control of cell-substratum adhesion during avian neural crest cell emigration *in vitro*. *Development* **116**, 275–287.

Delcommenne, M., Tan, C., Gray, V., Rue, L., Woodgett, J., and Dedhar, S. (1998). Phosphoinositide-3-OH kinase–dependent regulation of glycogen synthase kinase 3 and protein kinase B/AKT by the integrin-linked kinase. *Proc. Natl. Acad. Sci.* **95**, 11211–11216.

Dominguez, D., Montserrat-Sentis, B., Virgos-Soler, A., Guaita, S., Grueso, J., Porta, M., Puig, I., Baulida, J., Franci, C., and Garcia de Herreros, A. (2003). Phosphorylation regulates the subcellular location and activity of the Snail transcriptional repressor. *Mol. Cell. Biol.* **23**, 5078–5089.

Duong, T. D., and Erickson, C. A. (2004). MMP-2 plays an essential role in producing epithelial–mesenchymal transformations in the avian embryo. *Dev. Dyn.* **229**, 42–53.

Egeblad, M., and Werb, Z. (2002). New functions for the matrix metlloproteinases in cancer progression. *Nat. Rev. Cancer* **2**, 161–174.

Eger, A., Aigner, K., Sonderegger, S., Dampier, B., Oehler, S., Schreiber, M., Berx, G., Cano, A., Beug, H., and Foisner, R. (2005). DeltaEF1 is a transcriptional repressor of E-cadherin and regulates epithelial plasticity in breast cancer cells. *Oncogene* **24**, 2375–2385.

Eisenberg, L. M., and Markwald, R. R. (1995). Molecular regulation of atrioventricular valvuloseptal morphogenesis. *Circ. Res.* **77**, 1–6.

Endo, Y., Osumi, N., and Wakamatsu, Y. (2002). Bimodal functions of Notch-mediated signaling are involved in neural crest formation during avian ectoderm development. *Development* **129**, 863–873.

Erickson, C. A. (1987). Behavior of neural crest cells on embryonic basal laminae. *Dev. Biol.* **120**, 38–49.

Erickson, C. A., and Isseroff, R. R. (1989). Plasminogen activator activity is associated with neural crest cell motility in tissue culture. *J. Exp. Zool.* **251**, 123–133.

Fialka, I., Schwarz, H., Reichmann, E., Oft, M., Busslinger, M., and Beug, H. (1996). The estrogen-dependent c-JunER protein causes a reversible loss of mammary epithelial cell polarity involving a destabilization of adherens junctions. *J. Cell Biol.* **132**, 1115–1132.

Fink, R. D., and McClay, D. R. (1985). Three cell recognition changes accompany the ingression of sea urchin primary mesenchyme cells. *Dev. Biol.* **107**, 66–74.

Fujita, N., Jaye, D. L., Kajita, M., Geigerman, C., Moreno, C. S., and Wade, P. A. (2003). MTA3, a Mi-2/NuRD complex subunit, regulates an invasive growth pathway in breast cancer. *Cell* **113**, 207–219.

Fujita, Y., Krause, G., Scheffner, M., Zechner, D., Leddy, H. E. M., Behrens, J., Sommer, T., and Birchmeier, W. (2002). Hakai, a c-Cbl-like protein, ubiquitinates and induces endocytosis of the E-cadherin complex. *Nat. Cell Biol.* **4**, 222–231.

Garcia-Castro, M. I., Marcelle, C., and Bronner-Fraser, M. (2002). Ectodermal Wnt function as a neural crest inducer. *Science* **297**, 848–851.

Gat, U., DasGupta, R., Degenstein, L., and Fuchs, E. (1998). *De novo* hair follicle morphogenesis and hair tumors in mice expressing a truncated ß-catenin in skin. *Cell* **95**, 605–614.

Gavert, N., Conacci-Sorrell, M., Gast, D., Schneider, A., Altevogt, P., Brabletz, T., and Ben-Ze'ev, A. (2005). L1, a novel target of ß-catenin signaling, transforms cells and is expressed at the invasive front of colon cancers. *J. Cell Biol.* **168**, 633–642.

Gradl, D., Kuhl, M., and Wedlich, D. (1999). The Wnt/Wg signal transducer ß-catenin controls fibronectin expression. *Mol. Cell. Biol.* **19**, 5576–5587.

Grande, M., Franzen, A., Karlsson, J. O., Ericson, L. E., Heldin, N.-E., and Nilsson, M. (2002). Transforming growth factor-ß and epidermal growth factor synergistically stimulate epithelial-to-mesenchymal transition (EMT) through a MEK-dependent mechanism in primary cultured pig thyrocytes. *J. Cell Sci.* **115**, 4227–4236.

Gros, J., Manceau, M., Thome, V., and Marcelle, C. (2005). A common somitic origin for embryonic muscle progenitors and satellite cells. *Nature* **435**, 954–958.

Gustavson, M. D., Crawford, H. C., Fingleton, B., and Matrisian, L. M. (2004). Tcf binding sequence and position determines ß-catenin and Lef-1 responsiveness of MMP-7 promoters. *Mol. Carcinog.* **41**, 125–139.

Hay, E. D. (2005). The mesenchymal cell, its role in the embryo, and the remarkable signaling mechanisms that create it. *Dev. Dyn.* **233**, 706–720.

Hood, J. D., and Cheresh, D. A. (2002). Role of integrins in cell invasion and migration. *Nat. Rev. Cancer* **2**, 91–100.

Huber, M. A., Azoitei, N., Baumann, B., Grunert, S., Sommer, A., Pehamberger, H., Kraut, N., Beug, H., and Wirth, T. (2004). NF-kB is essential for epithelial–mesenchymal transition and metastasis in a model of breast cancer progression. *J. Clin. Invest.* **114**, 569–581.

Huelsken, J., Vogel, R., Brinkmann, V., Erdmann, B., Birchmeier, C., and Birchmeier, W. (2000). Requirement for beta-catenin in anterior-posterior axis formation in mice. *J. Cell Biol.* **148**, 567–578.

Irie, H. Y., Pearline, R. V., Grueneberg, D., Hsia, M., Ravichandran, P., Kothari, N., Natesan, S., and Brugge, J. S. (2005). Distinct roles of Akt1 and Akt2 in regulating cell migration and epithelial–mesenchymal transition. *J. Cell Biol.* **171**, 1023–1034.

Iwano, M., Plieth, D., Danoff, T. M., Xue, C., Okada, H., and Neilson, E. G. (2002). Evidence that fibroblasts derive from epithelium during tissue fibrosis. *J. Clin. Invest.* **110**, 341–350.

Janda, E., Lehmann, K., Killisch, I., Jechlinger, M., Herzig, M., Downward, J., Beug, H., and Grunert, S. (2002). Ras and TGFß cooperatively regulate epithelial cell plasticity and metastasis: dissection of Ras signaling pathways. *J. Cell Biol.* **156**, 299–314.

Jorda, M., Olmeda, D., Vinyals, A., Valero, E., Cubillo, E., Llorens, A., Cano, A., and Fabra, A. (2005). Upregulation of MMP-9 in MDCK epithelial cell line in response to expression of the Snail transcription factor. *J. Cell Sci.* **118**, 3371–3385.

Kalluri, R., and Neilson, E. G. (2003). Epithelial–mesenchymal transition and its implications for fibrosis. *J. Clin. Invest.* **112**, 1776–1784.

Karhadkar, S. S., Steven Bova, G., Abdallah, N., Dhara, S., Gardner, D., Maitra, A., Isaacs, J. T., Berman, D. M., and Beachy, P. A. (2004). Hedgehog signaling in prostate regeneration, neoplasia and metastasis. *Nature* **431**, 707–712.

Katayama, M., Sanzen, N., Funakoshi, A., and Sekiguchi, K. (2003). Laminin gamma 2-chain fragment in the circulation: a prognostic indicator of epithelial tumor invasion. *Cancer Res.* **63**, 222–229.

Kim, K., Lu, Z., and Hay, E. D. (2002). Direct evidence for a role of ß-Catenin/LEF-1 signaling pathway in induction of EMT. *Cell Biol. Int.* **26**, 463–476.

Kim, K., Pang, K. M., Evans, M., and Hay, E. D. (2000). Overexpression of ß-catenin induces apoptosis independent of its transactivation function with LEF-1 or the involvement of major G1 cell cycle regulators. *Mol. Biol. Cell* **11**, 3509–3523.

Koshikawa, N., Giannelli, G., Cirulli, V., Miyazaki, K., and Quaranta, V. (2000). Role of cell surface metalloprotease MT1-MMP in epithelial cell migration over laminin-5. *J. Cell Biol.* **148**, 615–624.

Labbe, E., Letamendia, A., and Attisano, L. (2000). Association of Smads with lymphoid enhancer binding factor 1/T cell-specific factor mediates cooperative signaling by the transforming growth factor-beta and Wnt pathways. *Proc. Natl. Acad. Sci.* **97**, 8358–8363.

Larue, L., and Bellacosa, A. (2005). Epithelial–mesenchymal transition in development and cancer: role of phosphatidylinositol 3' kinase//AKT pathways. *Oncogene* **24**, 7443–7454.

Lee, J. M., Dedhar, S., Kalluri, R., and Thompson, E. W. (2006). The epithelial–mesenchymal transition: new insights in signaling, development, and disease. *J. Cell Biol.* **172**, 973–981.

Levine, E., Lee, C. H., Kintner, C., and Gumbiner, B. M. (1994). Selective disruption of E-Cadherin function in early *Xenopus* embryos by a dominant negative mutant. *Development* **120**, 901–909.

Liebner, S., Cattelino, A., Gallini, R., Rudini, N., Iurlaro, M., Piccolo, S., and Dejana, E. (2004). ß-Catenin is required for endothelial–mesenchymal transformation during heart cushion development in the mouse. *J. Cell Biol.* **166**, 359–367.

Liem, J., Karel F., Tremml, G., Roelink, H., and Jessell, T. M. (1995). Dorsal differentiation of neural plate cells induced by BMP-mediated signals from epidermal ectoderm. *Cell* **82**, 969–979.

Logan, C., Miller, J., Ferkowicz, M., and McClay, D. (1999). Nuclear beta-catenin is required to specify vegetal cell fates in the sea urchin embryo. *Development* **126**, 345–357.

Maeda, M., Johnson, K. R., and Wheelock, M. J. (2005). Cadherin switching: essential for behavioral but not morphological changes during an epithelium-to-mesenchyme transition. *J. Cell Sci.* **118**, 873–887.

McGuire, P. G., and Alexander, S. M. (1993). Inhibition of urokinase synthesis and cell surface binding alters the motile behavior of embryonic endocardial-derived mesenchymal cells *in vitro*. *Development* **118**, 931–939.

Miller, J. R., and McClay, D. R. (1997). Characterization of the role of cadherin in regulating cell adhesion during sea urchin development. *Dev. Biol.* **192**, 323–339.

Moody, S. E., Perez, D., Pan, T.-C., Sarkisian, C. J., Portocarrero, C. P., Sterner, C. J., Notorfrancesco, K. L., Cardiff, R. D., and Chodosh, L. A. (2005). The transcriptional repressor Snail promotes mammary tumor recurrence. *Cancer Cell* **8**, 197–209.

Morali, O. G., Delmas, V., Moore, R., Jeanney, C., Thiery, J. P., and Larue, L. (2001). IGF-II induces rapid beta-catenin relocation to the nucleus during epithelium to mesenchyme transition. *Oncogene* **20**, 4942–4950.

Muller, W. J., Sinn, E., Pattengale, P. K., Wallace, R., and Leder, P. (1988). Single-step induction of mammary adenocarcinoma in transgenic mice bearing the activated c-neu oncogene. *Cell* **54**, 105–115.

Nagafuchi, A., Shirayoshi, Y., Okazaki, K., Yasuda, K., and Takeichi, M. (1987). Transformation of cell adhesion properties by exogenously introduced E-cadherin cDNA. *Nature* **329**, 341–343.

Nawshad, A., and Hay, E. D. (2003). TGFß3 signaling activates transcription of the LEF1 gene to induce epithelial–mesenchymal transformation during mouse palate development. *J. Cell Biol.* **163**, 1291–1301.

Nawshad, A., LaGamba, D., and Hay, E. D. (2004). Transforming growth factor ß (TGFß) signaling in palatal growth, apoptosis and epithelial–mesenchymal transformation (EMT). *Arch. Oral Biol.* **49**, 675–689.

Nichols, D. H. (1987). Ultrastructure of neural crest formation in the midbrain/rostral hindbrain and preotic hindbrain regions of the mouse embryo. *Am. J. Anat.* **179**, 143–154.

Nieto, M. A., Sargent, M. G., Wilkinson, D. G., and Cooke, J. (1994). Control of cell behavior during vertebrate development by Slug, a zinc finger gene. *Science* **264**, 835–839.

Novak, A., Hsu, S.-C., Leung-Hagesteijn, C., Radeva, G., Papkoff, J., Montesano, R., Roskelley, C., Grosschedl, R., and Dedhar, S. (1998). Cell adhesion and the integrin-linked kinase regulate the LEF-1 and ß-catenin signaling pathways. *Proc. Natl. Acad. Sci.* **95**, 4374–4379.

Ozdamar, B., Bose, R., Barrios-Rodiles, M., Wang, H.-R., Zhang, Y., and Wrana, J. L. (2005). Regulation of the polarity protein Par6 by TGFß receptors controls epithelial cell plasticity. *Science* **307**, 1603–1609.

Peinado, H., Del Carmen Iglesias-de la Cruz, M., Olmeda, D., Csiszar, K., Fong, K. S., Vega, S., Nieto, M. A., Cano, A., and Portillo, F. (2005). A molecular role for lysyl oxidase-like 2 enzyme in Snail regulation and tumor progression. *EMBO* **24**, 3446–3458.

Perl, A.-K., Wilgenbus, P., Dahl, U., Semb, H., and Christofori, G. (1998). A causal role for E-cadherin in the transition from adenoma to carcinoma. *Nature* **392**, 190–193.

Pla, P., Moore, R., Morali, O. G., Grille, S., Martinozzi, S., Delmas, V., and Larue, L. (2001). Cadherins in neural crest cell development and transformation. *J. Cell. Physiol.* **189**, 121–132.

Portella, G., Cumming, S., Liddell, J., Cui, W., Ireland, H., Akhurst, R., and Balmain, A. (1998). Transforming growth factor beta is essential for spindle cell conversion of mouse skin carcinoma *in vivo*: implications for tumor invasion. *Cell Growth Differ.* **9**, 393–404.

Radice, G. L., Rayburn, H., Matsunami, H., Knudsen, K. A., Takeichi, M., and Hynes, R. O. (1997). Developmental defects in mouse embryos lacking N-cadherin. *Dev. Biol.* **181**, 64–78.

Radisky, D. C., Levy, D. D., Littlepage, L. E., Liu, H., Nelson, C. M., Fata, J. E., Leake, D., Godden, E. L., Albertson, D. G., Angela Nieto, M., *et al.* (2005). Rac1b and reactive oxygen species mediate MMP-3-induced EMT and genomic instability. *Nature* **436**, 123–127.

Roberts, A. B., Tian, F., Byfield, S. D., Stuelten, C., Ooshima, A., Saika, S., and Flanders, K. C. (2006). Smad3 is key to TGF-ß-mediated epithelial-to-mesenchymal transition, fibrosis, tumor suppression and metastasis. *Cytokine Growth Factor Rev.* **17**, 19–27.

Sakai, D., Suzuki, T., Osumi, N., and Wakamatsu, Y. (2006). Co-operative action of Sox9, Snail2 and PKA signaling in early neural crest development. *Development* **133**, 1323–1333.

Schmidt, C., Stoeckelhuber, M., McKinnell, I., Putz, R., Christ, B., and Patel, K. (2004). Wnt 6 regulates the epithelialisation process of the segmental plate mesoderm leading to somite formation. *Dev. Biol.* **271**, 198–209.

Shook, D., and Keller, R. (2003). Mechanisms, mechanics and function of epithelial–mesenchymal transitions in early development. *Mech. Dev.* **120**, 1351–1383.

Stark, K., Vainio, S., Vassileva, G., and McMahon, A. P. (1994). Epithelial transformation of metanephric mesenchyme in the developing kidney regulated by Wnt-4. *Nature* **372**, 679–683.

Sternlicht, M. D., Lochter, A., Sympson, C. J., Huey, B., Rougier, J. P., Gray, J. W., Pinkel, D., Bissell, M. J., and Werb, Z. (1999). The stromal proteinase MMP3/stromelysin-1 promotes mammary carcinogenesis. *Cell* **98**, 137–146.

Sugi, Y., Yamamura, H., Okagawa, H., and Markwald, R. R. (2004). Bone morphogenetic protein-2 can mediate myocardial regulation of atrio-ventricular cushion mesenchymal cell formation in mice. *Dev. Biol.* **269**, 505–518.

Suyama, K., Shapiro, I., Guttman, M., and Hazan, R. B. (2002). A signaling pathway leading to metastasis is controlled by N-cadherin and the FGF receptor. *Cancer Cell* **2**, 301–314.

Tan, C., Costello, P., Sanghera, J., Dominguez, D., Baulida, J., de Herreros, A. G., and Dedhar, S. (2001). Inhibition of integrin linked kinase (ILK) suppresses beta-catenin-Lef/Tcf-dependent transcription and expression of the E-cadherin repressor, snail, in APC–/– human colon carcinoma cells. *Oncogene* **20**, 133–140.

Thiery, J. P. (2002). Epithelial–mesenchymal transitions in tumor progression. *Nat. Rev. Cancer* **2**, 442–454.

Timmerman, L. A., Grego-Bessa, J., Raya, A., Bertran, E., Perez-Pomares, J. M., Diez, J., Aranda, S., Palomo, S., McCormick, F., Izpisua-Belmonte,

J. C., *et al.* (2004). Notch promotes epithelial–mesenchymal transition during cardiac development and oncogenic transformation. *Genes Dev.* **18**, 99–115.

Ulrich, F., Krieg, M., Schotz, E.-M., Link, V., Castanon, I., Schnabel, V., Taubenberger, A., Mueller, D., Puech, P.-H., and Heisenberg, C.-P. (2005). Wnt11 functions in gastrulation by controlling cell cohesion through Rab5c and E-Cadherin. *Dev. Cell* **9**, 555–564.

Urbach, R., Schnabel, R., and Technau, G. M. (2003). The pattern of neuroblast formation, mitotic domains and proneural gene expression during early brain development in *Drosophila*. *Development* **130**, 3589–3606.

Valles, A., Boyer, B., Tarone, G., and Thiery, J. (1996). Alpha 2 beta 1 integrin is required for the collagen and FGF-1 induced cell dispersion in a rat bladder carcinoma cell line. *Cell Adhes. Commun.* **4**, 187–199.

Wakamatsu, N., Yamada, Y., Yamada, K., Ono, T., Nomura, N., Taniguchi, H., Kitoh, H., Mutoh, N., Yamanaka, T., Mushiake, K., *et al.* (2001). Mutations in SIP1, encoding Smad interacting protein-1, cause a form of Hirschsprung disease. *Nat. Genet.* **27**, 369–370.

Wong, A. S. T., and Gumbiner, B. M. (2003). Adhesion-independent mechanism for suppression of tumor cell invasion by E-cadherin. *J. Cell Biol.* **161**, 1191–1203.

Woolf, A. S., Kolatsi-Joannou, M., Hardman, P., Andermarcher, E., Moorby, C., Fine, L. G., Jat, P. S., Noble, M. D., and Gherardi, E. (1995). Roles of hepatocyte growth factor/scatter factor and the met receptor in the early development of the metanephros. *J. Cell Biol.* **128**, 171–184.

Yamashita, S., Miyagi, C., Fukada, T., Kagara, N., Che, Y.-S., and Hirano, T. (2004). Zinc transporter LIVI controls epithelial–mesenchymal transition in zebra fish gastrula organizer. *Nature* **429**, 298–302.

Yang, Z., Rayala, S., Nguyen, D., Vadlamudi, R. K., Chen, S., and Kumar, R. (2005). Pak1 phosphorylation of Snail, a master regulator of epithelial-to-mesenchyme transition, modulates Snail's subcellular localization and functions. *Cancer Res.* **65**, 3179–3184.

Yi, J. Y., Shin, I., and Arteaga, C. L. (2005). Type I transforming growth factor beta receptor binds to and activates phosphatidylinositol 3-kinase. *J. Biol. Chem.* **280**, 10870–10876.

Zavadil, J., and Bottinger, E. P. (2005). TGF-ß and epithelial-to-mesenchymal transitions. *Oncogene* **24**, 5764–5774.

Zavadil, J., Cermak, L., Soto-Nieves, N., and Bottinger, E. P. (2004). Integration of TGF-ß/Smad and Jagged1/Notch signaling in epithelial-to-mesenchymal transition. *EMBO J.* **23**, 1155–1165.

Zhou, B. P., Deng, J., Xia, W., Xu, J., Li, Y. M., Gunduz, M., and Hung, M.-C. (2004). Dual regulation of Snail by GSK-3ß-mediated phosphorylation in control of epithelial–mesenchymal transition. *Nat. Cell Biol.* **6**, 931–940.

Zuk, A., and Hay, E. D. (1994). Expression of ß1 integrins changes during transformation of avian lens epithelium to mesenchyme in collagen gels. *Dev. Dyn.* **201**, 378–393.

## Chapter Seven

# The Dynamics of Cell–ECM Interactions

*M. Petreaca and Manuela Martins-Green*

## I. INTRODUCTION

Most of the success in performing tissue and organ replacement that has led to improvement in patient length/quality of life and health care can be attributed to the inter-disciplinary approaches to tissue engineering. Today, scientists with diverse backgrounds, including molecular, cellular, and developmental biologists, collaborate with bioengineers to develop tissue analogues that allow physicians to improve, maintain, and restore tissue function. Several approaches have been taken to achieve these goals. One approach involves the use of matrices containing specific cells and growth factors. Recently, therapies based on stem cells are being implemented in conjunction with specific matrices and growth factors. Investigations of the basic cell and molecular mechanisms of the interactions between cells and extracelllar matrix (ECM) during development and development-like processes such as wound healing, have contributed to advancements in preparation of tissue substitutes. In this article, we provide an historical perspective on the importance of ECM in cell and tissue function, discuss some of the key findings that led to the understanding of how the dynamics of cell–ECM interactions contribute to cell migration, proliferation, differentiation, and programmed death, all of which are important parameters to consider when preparing and using tissue analogues.

### Historical Background

In the first part of the last century, the extracellular matrix (ECM) was thought to serve only as a structural support for tissues. However, in 1966 Hauschka and Konigsberg showed that interstitial collagen promotes the conversion of myoblasts to myotubes, and shortly thereafter it was shown that both collagen and glycosaminoglycans are crucial for salivary gland morphogenesis. Based on these and other findings, in 1977 Hay put forth the idea that the ECM is an important component in embryonic inductions, a concept that implicated the presence of binding sites (receptors) for specific matrix molecules on the surface of cells. The stage was then set for further investigations into the mechanisms by which ECM molecules influence cell behavior. Bissell and colleagues (1982) proposed the model of *dynamic reciprocity*. In this model, ECM molecules interact with receptors on the surface of cells, which then transmit signals across the cell membrane to molecules in the cytoplasm. These signals initiate a cascade of events through the cytoskeleton into the nucleus, resulting in the expression of specific genes, whose products, in turn, affect the ECM in various ways. Through the years, it has become clear that cell–ECM interactions participate directly in promoting cell adhesion, migration, growth, differentiation, and apoptosis (a form of programmed cell death) as well as in modu-

*Principles of Tissue Engineering, 3rd Edition*
*ed. by Lanza, Langer, and Vacanti*

lating the activities of cytokines and growth factors and in directly activating intracellular signaling.

## ECM Composition

The ECM is a molecular complex that consists of molecules such as collagens, other glycoproteins, hyaluronic acid, proteoglycans, glycosaminoglycans, and elastins that reside outside the cells, and that harbor proteins, including growth factors, cytokines, and matrix-degrading enzymes and their inhibitors. The distribution and organization of these molecules is not static but, rather, varies from tissue to tissue and during development from stage to stage, which has significant implications for tissue function. For example, mesenchymal cells are immersed in an interstitial matrix that confers specific biomechanical and functional properties to connective tissue (Suki et al., 2005), whereas epithelial and endothelial cells contact a specialized matrix, the basement membrane, via their basal surfaces only, conferring mechanical strength and specific physiological properties on the epithelia. This diversity of composition, organization, and distribution of ECM results not only from differential gene expression of the various molecules in specific tissues, but also from the existence of differential splicing and posttranslational modifications of those molecules. For example, alternative splicing may change the binding potential of proteins to other matrix molecules or to their receptors (Ghert et al., 2001; Mostafavi-Pour et al., 2001), and variations in glycosylation can lead to changes in cell adhesion (Anderson et al., 1994).

The local concentration and biological activity of growth factors and cytokines can be influenced by the ECM serving as a reservoir that binds these molecules and protects them from being degraded, by presenting them more efficiently to their receptors, or by affecting their synthesis (Nathan and Sporn, 1991; Sakakura et al., 1999; Miralem et al., 2001). Growth factor binding to ECM molecules may also exert an inhibitory effect (Kupprion et al., 1998; Francki et al., 2003), and, in some cases, only particular forms of these growth factors and cytokines bind to specific ECM molecules (Pollock and Richardson, 1992; Poltorak et al., 1997; Martins-Green et al., 1996). Importantly, binding of specific forms of these factors to specific ECM molecules can lead to their localization to particular regions within tissues and affect their biological activities.

ECM/growth factor interactions can also involve the ability of specific domains of ECM molecules (e.g., laminin-5, tenascin-C, and decorin) to bind and activate growth factor receptors (Tran et al., 2004); the EGF-like repeats of laminin and tenascin-C bind and activate the EGFR (Swindle et al., 2001; Schenk et al., 2003). In the case of laminin, the EGF-like repeats interact with EGFR following their release by MMP-mediated proteolysis (Schenk et al., 2003), whereas tenascin-C repeats are thought to bind EGFR in the context of the full-length protein (Swindle et al., 2001). Decorin also binds and activates EGFR, although this binding occurs via leucine-rich repeats rather than EGF-like repeats (Santra et al., 2002). This ability of ECM molecules to serve as ligands for growth factor receptors may facilitate a stable signaling environment for the associated cells due to the inability of the ligand either to diffuse or to be internalized, thus serving as a long-term pro-migratory and/or pro-proliferative signal (Tran et al., 2004).

## Receptors for ECM Molecules

Integrins, a family of heterodimeric transmembrane proteins composed of $\alpha$ and $\beta$ subunits, were the first ECM receptors to be identified. At least 18 $\alpha$ and 8 $\beta$ subunits have been identified so far; they pair with each other in a variety of combinations, giving rise to a large family that recognizes specific sequences on the ECM molecules (Fig. 7.1; Hynes, 2002). Some integrin receptors are very specific, whereas others bind several different epitopes, which may be on the same or different ECM molecules (Fig. 7.1), thus facilitating plasticity and redundancy in specific systems (Dedhar, 1999; Hynes, 2002). Although the $\alpha$ and $\beta$ subunits of integrins are unrelated, there is 40–50% homology within each subunit, with the highest divergence in the intracellular domain of the $\alpha$ subunit. All but one of these subunits ($\beta_4$) have large extracellular domains and very small intracellular domains. The extracellular domain of the $\alpha$ subunits contains four regions that serve as binding sites for divalent cations, which appear to augment ligand binding and increase the strength of the ligand–integrin interactions (Pujades et al., 1997; Leitinger et al., 2000).

Transmembrane proteoglycans are another class of proteins that can also serve as receptors for ECM molecules (Jalkehen, 1991; Couchman and Woods, 1996). Several proteoglycan receptors that bind to ECM molecules have been isolated and characterized. Syndecan, for example, binds cells to ECM via chondroitin- and heparan-sulfate glycosaminoglycans, whose composition varies based on the type of tissue in which syndecan is expressed. These differential glycosaminoglycan modifications alter the binding capacity of particular ligands (Salmivirta and Jalkanen, 1995). Furthermore, syndecan also associates with the cytoskeleton, promoting intracellular signaling events and cytoskeletal reorganization through activation of Rho GTPases (Bass and Humphries, 2002; Yoneda and Couchman, 2003). Another receptor, CD44, also carries chondroitin sulfate and heparan sulfate chains on its extracellular domain and undergoes tissue-specific splicing and glycosylation to yield multiple isoforms (Brown et al., 1991; Ehnis et al., 1996). One of the extracellular domains of CD44 is structurally similar to the hyaluronan-binding domain of the cartilage link protein and aggrecan, which suggested that CD44 also serves as a hyaluronan receptor. Using a variety of techniques involving antibody binding and mutagenesis, it has been shown that this domain of CD44, as well as an additional domain outside this region, can interact directly with hyaluronan. These regions can also mediate CD44 binding to other

**FIG. 7.1.** Members of the integrin family of ECM receptors and their respective ligands. These heterodimeric receptors are composed of one α and one β subunit and are capable of binding a variety of ligands, including Ig super-family cell adhesion molecules, complement and clotting factors, and ECM molecules. Cell–cell adhesion is largely mediated through integrins containing $\beta_2$ subunits, while cell–matrix adhesion is mediated primarily via integrins containing $\beta_1$ and $\beta_3$ subunits. In general, the $\beta_1$ integrins interact with ligands found in the connective tissue matrix, including laminin, fibronectin, and collagen, whereas the $\beta_3$ integrins interact with vascular ligands, including thrombospondin, vitronectin, fibrinogen, and von Willebrand factor. *Abbreviations*: CO, collagens; C3bi, complement component; FG, fibrinogen; FN, fibronectin; FX, Factor X; ICAM-1, intercellular adhesion molecule-1; ICAM-2, intercellular adhesion molecule-2; ICAM-3, intercellular adhesion molecule-3; LN, laminin; OSP, osteopontin; TN, tenascin; TSP, thrombospondin; VCAM-1, vascular cell adhesion molecule-1; VN, vitronectin; vWF, von Willebrand factor; ECADH, E-cadherin; LAPβ1, latent activating protein β1.

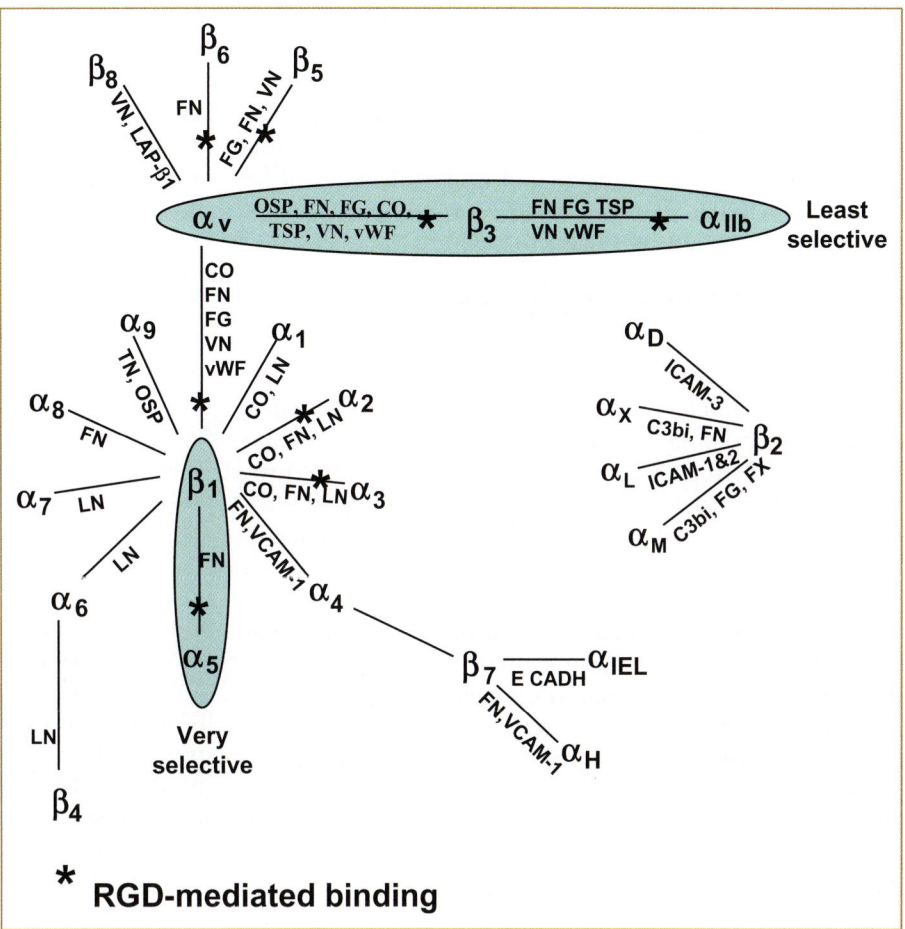

proteoglycans, although hyaluronic acid is its primary ligand (Marhaba and Zoller, 2004). CD44 can also interact with collagen, laminin, and fibronectin, although the exact binding sites of these molecules to CD44, as well as the functional significance of such interactions *in vivo*, are not well understood (Ehnis *et al.*, 1996; Ponta *et al.*, 2003; Marhaba and Zoller, 2004). RHAMM (receptor for hyaluronate-mediated motility) has been identified as an additional hyaluronic acid receptor (Hardwick *et al.*, 1992), which is responsible for hyaluronic-acid-mediated cell motility in a number of cell types and also appears to be important in trafficking of hematopoietic cells (Pilarski *et al.*, 1999; Savani *et al.*, 2001).

Other cell surface receptors for ECM have also been identified. A nonintegrin 67-kDa protein known as the *elastin-laminin receptor* (ELR) recognizes the YIGSR sequence of laminin and the VGVAPG sequence of elastin, neither of which recognizes integrins. The ELR colocalizes with cytoskeleton-associated and signaling proteins on laminin ligation, suggesting a role in laminin-mediated signaling (Massia *et al.*, 1993; Bushkin-Harav and Littauer, 1998), and has more recently been implicated in the signaling downstream of elastin and laminin during mechanotransduction (Spofford and Chilian, 2003). CD36, another

receptor for ECM, functions as a scavenger receptor for long-chain fatty acids and oxidized LDL, but also binds collagen I and IV, thrombospondin, and malaria-infected erythrocytes to endothelial cells and some types of epithelial cells. Each of these ligands has a separate binding site, but all are located in the same external loop of CD36, and the intracellular signals occurring after ligand binding lead to activation of a variety of signal transduction molecules (Febbraio *et al.*, 2001). For example, the antiangiogenic effects of thrombospondin are dependent on signaling downstream of CD36 (Jimenez *et al.*, 2000). Another alternative type of cell surface receptor, annexin II, is known to interact with alternative splice variants of tenascin-C, potentially mediating the cellular responses to these various forms of tenascin C (Chung and Erickson, 1994). In addition, ECM molecules have been shown to bind and activate tyrosine kinase receptors, including the EGFR via EGF-like domains (see earlier) as well as the discoidin domain receptors DDR1 and DDR2. DDR1 and DDR2 function as receptors for various collagens and mediate cell adhesion and signaling events (Vogel *et al.*, 1997). The DDR receptors have also been implicated in ECM remodeling because their overexpression decreases the expression of multiple matrix molecules and their receptors, including collagen, syndecan-1, and integrin $\alpha_3$, while simultaneously increasing MMP activity (Ferri *et al.*, 2004).

We first discuss selected examples that illustrate the dynamics of cell–ECM interactions during development and wound healing as well as the potential mechanisms involved in the signal transduction pathways initiated by these interactions. Finally, we discuss the implications of cell–ECM interactions in tissue engineering.

## II. CELL–ECM INTERACTIONS

Multiple biological processes, including those relevant to development and wound healing, require interactions between cells and their environment as well as modulation of such interactions. During development, the cellular cross-talk with the surrounding extracellular matrix promotes the formation of patterns, the development of form (morphogenesis), and the acquisition and maintenance of differentiated phenotypes during embryogenesis. Similarly, during wound healing these interactions contribute to the processes of clot formation, inflammation, granulation tissue development, and remodeling. As outlined earlier, the current body of research in the fields of both embryogenesis and wound healing implicates multiple cellular behaviors, including cell adhesion/deadhesion, migration, proliferation, differentiation, and apoptosis, in these critical events.

### Development

#### Adhesion and Migration

Today, there is a vast body of experimental evidence that demonstrates the direct participation of ECM in cell adhesion and migration. Some of the most compelling experiments come from studies in gastrulation, migration of neural crest cells (NCC), angiogenesis, and epithelial organ formation. Cell interactions with fibronectin are important during gastrulation. Microinjection of antibodies to fibronectin into the blastocoel cavity of *Xenopus* embryos causes disruption of normal cell movements and leads to abnormal development (Boucaut *et al.*, 1984a). Furthermore, injection of RGD-containing peptides (which compete with integrins for ECM binding) during this same stage of development induces randomization of the bilateral asymmetry of the heart and gut (Yost, 1992). Similarly, administration of RGD-containing peptides and/or antibodies to the β1 subunit of the integrin receptor for fibronectin perturbs gastrulation in salamander embryos (Boucaut *et al.*, 1984b; Yost, 1992). These effects are not unique to fibronectin. They can also be introduced by manipulation of other molecules; competition of heparan sulfate proteoglycans with heparin for target molecule binding perturbs gastrulation and neurulation (Erickson and Reedy, 1998).

The NCC develop in the dorsal portion of the neural tube just after closure of the tube, migrate extensively throughout the embryo in ECM-filled spaces, and give rise to a variety of phenotypes. The importance of cell–ECM interactions in the deadhesion process is supported by studies performed in the white mutant of Mexican axolotl embryos. The NCC that give rise to pigment cells fail to emigrate from the neural tube in these embryos. But when microcarriers containing subepidermal ECM from normal embryos are implanted into the appropriate area in these mutants, the NCC pigment cell precursors emigrate normally (Perris and Perissinotto, 2000). An RGD domain–carrying ECM molecule is known to promote the secretion of adhesion-degrading enzymes on integrin ligation, thereby facilitating emigration (Damsky and Werb, 1992). This may be due primarily to the RGD domain of fibronectin, for fibronectin appears between chick NCC just prior to their emigration from the neural tube (Martins-Green and Bissell, 1995). This fibronectin may consist predominantly or exclusively of the RGD domain–carrying segment that, when bound to its integrin receptor, promotes secretion of adhesion-degrading enzymes, thereby facilitating emigration (Damsky and Werb, 1992). Indeed, microinjection of antibodies to fibronectin (Poole and Thiery, 1986) or to the β1 subunit of the integrin receptor (Bronner-Fraser, 1985) into the crest pathways in chick embryos reduces the number of NCC that leave the tube and causes abnormal neural tube development. Other ECM molecules, such as laminin, also affect NCC adhesion and migration. For example, the YIGSR synthetic peptide known to inhibit laminin binding to cells inhibits NCC migration (Runyan *et al.*, 1986). NCC migration on laminin may also involve ligation of $\alpha_1\beta_1$ integrin, because function-blocking antibodies of this integrin largely prevent such migration *in vitro* (Desban and Duband, 1997).

Endothelial cell interactions with ECM molecules and the type and conformation of the matrix are also crucial during angiogenesis (the development of blood vessels from preexisting vessels; Li *et al.*, 2003). Early indications of the role of ECM in angiogenesis were observed when human umbilical vein endothelial cells (HUVEC) were cultured on matrigel, a matrix synthesized by Engelbreth-Holm-Swarm (EHS) tumors. This specialized matrix has many of the properties of basement membrane. It consists of large amounts of laminin as well as collagen IV, entactin/nidogen, and proteoglycans. When HUVEC are cultured on matrigel for 12 hours, they migrate and form tubelike structures. In contrast, when these cells are cultured with collagen I, they form tubelike structures only after they are maintained *inside* the gels for one week, at which time the cells have secreted their own basement membrane molecules (Kubota *et al.*, 1988; Grant *et al.*, 1989). The observation that tube formation occurs more rapidly on matrigel than within collagen gels strongly suggested an important role for one or more of the matrix molecules present within the basement membrane in the development of the capillary-like endothelial tubes. Indeed, laminin, the predominant matrix molecule of the basement membrane, was later shown to participate in endothelial tube formation and angiogenesis. *In vitro*, antibodies against laminin prevented the formation of endothelial tubes on matrigel, whereas treatment with synthetic peptides containing the YIGSR sequence derived from the B1 chain of laminin facilitates tube formation (Grant *et al.*, 1989). Another sequence, SIKVAV, found in the laminin A chain, promotes endothelial cell adhesion, elongation, and angiogenesis (Grant *et al.*, 1992). Interestingly, application of YIGSR-containing synthetic peptides prevent endothelial cell migration and angiogenesis (Sakamoto *et al.*, 1991). It is possible that the YIGSR peptides exert antiangiogenic properties due to competition for receptor binding with the intact laminin present *in vivo*. Indeed, if YIGSR peptides can successfully compete with laminin, it is possible that the displacement of YIGSR in the context of the whole molecule by the soluble YIGSR peptides will alter the presentation of the ligand to its receptor, resulting in changes in mechanical resistance that alter signaling events downstream of the receptor, ultimately resulting in different cellular responses. A similar hypothesis has been proposed for the interactions of integrins with soluble versus intact ligands (Stupack and Cheresh, 2002). Although the mechanisms generating different cellular outcomes are currently unknown, the mere fact that soluble and intact ECM receptor ligands may, at times, lead to alternative outcomes is likely of importance *in vivo* following matrix degradation. During angiogenesis, for example, endothelial cell migration and invasion into surrounding tissues is accompanied by the activation of matrix-degrading enzymes, which then cleave the matrix and release both matrix-bound growth factors as well as ECM fragments, providing additional angiogenic or antiangiogenic cues to influence the process further (Rundhaug, 2005). As such, matrix molecules that initially facilitate angiogenesis may be proteolytically cleaved at later angiogenic stages to create YIGSR peptides or some other antiangiogenic matrix fragment, preventing additional blood vessel formation and/or resulting in vessel maturation (Sakamoto *et al.*, 1991). Thus, the temporal and spatial production and cleavage of matrix molecules may have important consequences for tissue homeostasis.

### Proliferation

Some of the effects of cell–ECM interactions modulate cell proliferation. For example, a domain in the A chain of laminin that is rich in EGF-like repeats stimulates proliferation of a variety of different cell lines, and the entire molecule appears to promote proliferation of bone marrow–derived macrophages. These pro-proliferative effects are likely mediated, at least in part, by the activation of the EGFR by the EGF-like repeats (Schenk *et al.*, 2003). In contrast, there are also matrix molecules that inhibit cell proliferation. Heparin and heparin-like molecules are inhibitors of vascular smooth muscle cell (VSMC) proliferation. The conditioned medium of endothelial cells cultured from bovine aortas inhibits the proliferation of VSMC, and this inhibition is obliterated by treatment of the medium with heparinase but not with condroitinases or proteases. This suggests that heparan-type molecules have a direct antiproliferative effect on aortic VSMC. However, it is also possible that heparinase treatment may release proproliferative molecules that interact with heparin or heparan sulfate, thus allowing them to interact with their receptors and either promote proliferation or block any antiproliferative effects. One such mitogenic heparin-binding ECM molecule is thrombospondin, which is known to exert its mitogenic activities on VSMC via its amino terminal heparin-binding domain. Heparin has been shown to block thrombospondin binding to smooth muscle cells and also to block its mitogenic effects (Majack *et al.*, 1988). These results suggest that interactions between heparin and thrombospondin may interfere with thrombospondin-induced smooth muscle cell proliferation and that the observed increases in VSMC proliferation following heparinase treatment may result, at least in part, from the removal of such inhibitory interactions. It has also been proposed that the effects of heparin on VSMC may result from its regulation of TGF-$\beta$, an inhibitor of VSMC proliferation; heparin increases TGF-$\beta$ activation, and heparin-mediated antiproliferative effects are blocked by addition of a TGF-$\beta$ antibody. As such, heparinase treatment may prevent TGF-$\beta$ activation, abolishing the antiproliferative effects and explaining the conditioned media results. However, if heparin's effects are mediated by the inhibition of thrombospondin or activation of TGF-$\beta$, one would expect that treatment of the endothelial-cell-conditioned medium with proteases should also eliminate the antiproliferative effect. However, the protease treatment does not prevent these effects; it is likely that heparin-like molecules also have a direct antiproliferative effect.

The possibility that certain ECM molecules may exert antiproliferative effects is further supported by various studies performed in culture. For example, normal human breast cells do not growth arrest when cultured on plastic, but they do so if grown in a basement membrane matrix (Petersen *et al.*, 1992; Weaver *et al.*, 1997). Furthermore, growth of a mammary epithelial cell line is stimulated by overexpression of Id-1, a protein that binds to and inhibits the function of basic helix-loop-helix (HLH) transcription factors, which are important in cell differentiation. However, when these Id-1-overexpressing cells are cultured on EHS, they arrest growth and assume a normal 3D structure (Desprez *et al.*, 1995; Lin *et al.*, 1995). Similarly, EHS suppresses the growth of cultured hepatocytes, apparently due to the decreased expression of immediate-early growth response genes and the concomitant increased levels of C/EBP$\alpha$, which is necessary for the expression of hepatocyte-specific genes and also for growth arrest (Rana *et al.*, 1994).

Growth factors are critical in stimulation of cell proliferation. Indeed it has been found that some of the ECM effects on cell proliferation involve cooperation with growth factors. bFGF, IL-1, IL-2, IL-6, hepatocyte growth factor, PDGF-AA, and TGFβ are found in association with ECM at high concentrations and are released at specific times for interaction with their receptors (Schonherr and Hausser, 2000). In the case of TGFβ, cooperation with the ECM occurs during the early developmental stages of the mammary gland during puberty (virgin gland; Daniel *et al.*, 1996). During this period, inductive events take place between the epithelium and the surrounding mesenchyme that are mediated by the basement membrane (basal lamina and closely associated ECM molecules) and that play an important role in epithelial proliferation during branching of the gland. Endogenous TGFβ produced by the ductal epithelium and surrounding mesenchyme forms complexes with mature periductal ECM. This TGF-β may participate in stabilizing the epithelium by inhibiting both cell proliferation and the activity of matrix-degrading enzymes. However, TGFβ is absent from newly synthesized ECM deposited in the branching areas; thus its inhibitory effects on epithelial cell proliferation and on production of matrix-degrading enzymes do not occur, allowing the basement membrane to undergo remodeling. In these regions, proteases that are released locally partially degrade the matrix, thereby promoting cell proliferation and branching morphogenesis. An example of a protease important in this process is MMP-3/stromelysin-1, a protease important in basement membrane degradation and tissue remodeling; in mice transgenic for the autoactivated isoform of the MMP-3, the virgin glands are morphologically similar to the pregnant glands of normal mice (Sympson *et al.*, 1994). Furthermore, growth factor–induced branching morphogenesis in primary mammary organoids was shown to be MMP dependent, and application of recombinant MMP-3 to these organoids promoted morphogenesis in the absence of exogenous factors (Simian *et al.*, 2001). Taken together, these results suggest that MMP-3 stimulates the precocious proliferation of the epithelium and development of the alveoli due to the release of growth factors following matrix degradation.

## Differentiation

Processes leading to differentiation of keratinocyte, hepatocyte, and mammary gland epithelium illustrate well how ECM can affect cell behavior. Keratinocytes form the stratified epidermal layers of the skin. The basal layer is highly proliferative, does not express the markers for terminal differentiation, and is the only cell layer in contact with the basement membrane. As these cells divide, the daughter cells lose contact with the basement membrane, move up to the suprabasal layers, and begin to express differentiation markers, such as involucrin (Fuchs and Raghavan, 2002). This suggests that physical interaction with the basement membrane is responsible for the less differentiated basal keratinocytes. Indeed, it was first shown by Howard Green in 1977 that keratinocytes grown in suspension undergo

premature terminal differentiation. It has also been shown that human and mouse keratinocytes adhere to fibronectin via its $\alpha_5\beta_1$ integrin receptor and that the expression of both is inversely proportional to the expression of involucrin, a differentiation marker for these cells (Nicholson and Watt, 1991). However, the role of keratinocyte adhesion to the basement membrane in the regulation of differentiation status is unclear, for the keratinocytes of a conditional integrin β1 skin knockout mouse do not undergo premature terminal differentiation, suggesting that further studies are necessary to better understand the contribution of the basement membrane in differentiation. A major advance in such studies has been the ability to culture keratinocytes on feeder layers of 3T3 cells or collagen gels containing human dermal fibroblasts, which then form stratified sheets of cells that behave very much like epidermis does *in vivo* (Green, 1977; Schoop *et al.*, 1999). This latter development has had profound application in treating patients that have suffered extensive burns (Ehrenreich and Ruszczak, 2006).

Similarly, hepatocytes in culture remain differentiated and expressing liver-specific genes only when they are grown in the presence of extracellular matrix molecules, such as EHS, laminin, or collagen I. This process appears to involve $a_3$ integrin; down-regulation of this integrin using antisense RNA decreases hepatocyte adhesion to laminin and collagen I and prevents the differentiation-specific effects mediated by collagen-I (Lora, 1998). The specific mechanisms whereby these cell–ECM interactions regulate differentiation have not been fully elucidated. However, three liver-specific transcription factors, eE-TF, eG-TF/HNF3, and eH-TF, are activated when cells are cultured on or with matrix molecules, conditions that favor hepatocyte differentiation. In particular, the transcription factor eG-TF/HNF3 appears to be regulated by ECM (DiPersio *et al.*, 1991).

In the mouse mammary gland, the basement membrane and its individual components, in conjunction with lactogenic hormones, are responsible for the induction of the differentiated phenotype of the epithelial cells. When midpregnant mammary epithelial cells are cultured on plastic, they do not express mammary-specific genes. However, when the same cells are plated and maintained on EHS, they form alveolar-like structures and exhibit the fully differentiated phenotype with expression of the genes encoding milk proteins (e.g., Nelson and Bissell, 2005). Cultures of single mammary epithelial cells inside EHS showed that the molecules involved in induction of the differentiated phenotype act via transmembrane receptors rather than involving cell polarity or growth factors. It was later found that laminin is the ECM molecule present in EHS that is ultimately responsible for the observed differentiation and that the b1 integrin is critical in maintaining the differentiated state (Faraldo *et al.*, 2002; Nelson and Bissell, 2005). The impact of ECM molecules on the expression of milk proteins may be indirect, by altering the secretion of a growth factor that then affects milk protein, or more directly through signal transduction leading to changes in gene expression. An example of the former is seen in the stimulation of whey acidic protein (WAP) in mouse mammary gland epithelial cells. WAP expression is inhibited when cells are cultured on plastic; but if they are grown on ECM, expression is up-regulated. It has been found that cells cultured on plastic produce TGFα, which inhibits the expression of WAP, whereas the EHS matrix inhibits the production of TGFα, thus leading to the up-regulation of this milk protein (Lin *et al.*, 1995). An example that demonstrates the direct influence of ECM on the expression of milk protein genes comes from work performed on the expression of β-casein. It has been shown that there are two components to β-casein induction by ECM: One involves cell rounding (and therefore a change in the cytoskeleton) and the other a tyrosine kinase signal transduction pathway through integrin β$_1$ and potentially also integrin α$_6$β$_4$, leading to the activation of elements in the promoter region of the β-casein gene (Muschler *et al.*, 1999; Nelson and Bissell, 2005).

## Apoptosis

Programmed cell death occurs during embryogenesis of higher vertebrates in areas undergoing remodeling, such as in the development of the digits, palate, and nervous system, in the positive selection of thymocytes in the thymus, during mammary gland involution, and during angiogenesis. For example, basement membrane molecules appear to suppress apoptosis of the epithelial cells during the involution of the mammary gland (Strange *et al.*, 1992). The numerous alveoli that produce milk during lactation regress and are resorbed during involution due to enzymatic degradation of alveolar basement membrane and programmed cell death (Strange *et al.*, 1992; Talhouk *et al.*, 1992). During this involution, apoptosis appears to proceed in two distinct phases. An early phase characterized by increased expression of apoptosis-associated proteins, including interleukin-1β–converting enzyme (ICE), a protein known to be important in promoting mammary epithelial cell apoptosis (Boudreau *et al.*, 1995) is followed by a later apoptotic phase in which cell–ECM interactions are decreased due to both matrix degradation (Lund *et al.*, 1996) and reduced expression of integrin β1 and FAK (McMahon *et al.*, 2004). This disruption of cell–ECM binding is important for the apoptosis of mammary epithelial cells because ECM adhesion imparts critical survival signals. Indeed, these cells undergo apoptosis when an antibody is used to disrupt interactions between α$_1$ integrin and its ECM ligands (Boudreau *et al.*, 1995). Similarly, it has been found that α$_v$β$_3$ integrin interactions with ECM play a crucial role in endothelial cell survival during angiogenesis in embryogenesis. Disruption of these interactions with an antibody to α$_v$β$_3$ inhibits the development of new blood vessels in the chorioallantoic membrane (CAM) by causing the endothelial cells to undergo apoptosis

(Brooks *et al.*, 1994). In addition, tumstatin, a proteolytic fragment of collagen IV, induces endothelial cell apoptosis and thereby prevents angiogenesis via interaction with α$_v$β$_3$ (Maeshima *et al.*, 2001). This interaction may promote apoptosis by interfering with normal integrin–ECM binding, thus removing a critical survival signal. Tumstatin may also promote apoptosis through a separate mechanism, such as via the recruitment and activation of caspase 8, as has been suggested previously for such soluble ligands (Stupack *et al.*, 2001). All in all, these findings suggest that disruption of cell–ECM interactions may lead to an increase in the expression or activation of pro-apoptotic molecules, and may also lead to the removal of pro-survival signals, which then directly or indirectly cause apoptosis.

## Wound Healing

### Adhesion and Migration

Early in the wound-healing process, blood components and tissue factors are released into the wounded area in response to tissue damage, promoting both the activation and adhesion of platelets and the formation of a clot consisting of platelets, cross-linked fibrin, and plasma fibronectin as well as lesser amounts of SPARC (secreted protein acidic and rich in cysteine), tenascin, and thrombospondin. This is accompanied by the degranulation of mast cells, releasing factors important in vasodilation and in polymorphonuclear cell chemotaxis to the injured area, thereby initiating the inflammatory response. During these early stages of wound healing, a temporary extracellular matrix consisting of the fibrin–fibronectin meshwork facilitates the migration of keratinocytes to close the wound as well as the migration of leukocytes into the wounded area. Leukocyte adhesion, migration, and secretion of inflammatory mediators are further affected by their interactions with various ECM molecules (Vaday and Lider, 2000). Pro-inflammatory cytokine release from tissue macrophages, for example, occurs after CD44-mediated binding to low-molecular-weight hyaluronic acid (Hodge-Dufour *et al.*, 1997). As such, the types of ECM molecules present in the injured area may greatly affect the inflammatory phase of wound healing by influencing leukocyte behavior (Vaday and Lider, 2000). Furthermore, specific ECM molecules can bind chemokines, creating a stable gradient to promote leukocyte chemotaxis into the injured area. ECM–chemokine binding is critical for appropriate leukocyte recruitment, for mutant chemokines lacking the ability to bind glycosaminoglycans failed to induce chemotaxis *in vivo* (Handel *et al.*, 2005).

As mentioned earlier, keratinocytes participating in the re-epithelialization phase of cutaneous wound healing migrate on a provisional matrix composed of fibrin/fibrinogen, fibronectin, collagen type III, tenascin, and vitronectin. The keratinocytes express multiple receptors for these matrix molecules, including the integrins α$_2$β$_1$, α$_3$β$_1$, α$_5$β$_1$, α$_6$β$_1$, α$_5$β$_4$, and α$_v$; cell migration and the subsequent wound closure are facilitated by cell–ECM interactions via these receptors. The fibrin/fibrinogen meshwork appears to be of particular importance in re-epithelialization, as evidenced by the disordered re-epithelialization seen in fibrinogen-deficient mice (Drew *et al.*, 2001). This process also appears dependent on the synthesis and deposition of laminin, because keratinocyte migration on collagen and fibronectin was inhibited by an antilaminin antibody (Decline and Rousselle, 2001).

Interactions between epithelial cells and ECM are also critical in the wound closure of other types of epithelial wounds. After wounding, retinal pigment epithelial cells exhibit a sequential pattern of ECM molecule deposition that is critical in the epithelial cell adhesion and migration associated with wound closure. Within 24 hours of wounding, these epithelial cells secrete fibronectin, followed shortly by laminin and collagen IV. If the cell adhesion to these ECM molecules is blocked with either cyclic peptides or specific antibodies, the epithelial cells fail to migrate and close the wound, underscoring the importance of such interactions in wound closure (Hergott *et al.*, 1993; Hoffmann *et al.*, 2005). Similarly, the inhibition of various integrins or fibronectin in airway epithelial cells following mechanical injury largely prevented cell migration and wound healing.

During later stages of wound healing, macrophages and fibroblasts in the injured area deposit embryonic-type cellular fibronectin, which is important in the generation of the granulation tissue, a temporary connective tissue consisting of multiple types of ECM molecules and newly formed blood vessels (Li *et al.*, 2003). The cellular fibronectin provides a substrate for the migration of endothelial cells into the granulation tissue, thus forming the wound vasculature, and also facilitates the chemotaxis of myofibroblasts and lymphocytes stimulated by a variety of chemotactic cytokines (chemokines) that are produced by tissue fibroblasts and macrophages (Greiling and Clark, 1997; Feugate *et al.*, 2002). Many chemokines have been characterized in multiple species, including humans, other mammals, and birds, and have been grouped into a large superfamily that is further subdivided based on the position of the N-terminal cysteine residues (Gillitzer and Goebeler, 2001). These chemokines, along with cell–ECM interactions, are critical for the adhesion and chemotaxis/migration of the cells that ultimately enter the wounded area and generate the granulation tissue (Martins-Green and Feugate, 1998; Feugate *et al.*, 2002).

One prototypical chemokine, IL-8, has several functions important in wound healing. These functions have largely been elucidated in studies performed in the chick model system using chicken IL-8 (cIL-8/cCAF) (Martins-Green, 2001). After wounding, fibroblasts in the injured area produce large quantities of cIL-8, most likely resulting from their stimulation by thrombin, a coagulation enzyme activated on wounding that is known to induce fibroblasts to express and secrete cIL-8. The initial rapid increase in IL-8 generates a gradient that attracts neutrophils (Martins-Green, 2001). These cells, in turn, produce monocyte chemoattractant protein, a potent chemottractant for monocytes that differentiate into macrophages when in the wound environment. In addition, our *in vitro* studies using human THP-1-derived macrophages show that these cells can be stimulated to produce high levels of IL-8 (Zheng and Martins-Green, 2006), further increasing the levels of this chemokine in the wound tissue and potentially leading to angiogenesis. IL-8 is also secreted by the endothelial cells of the wound vasculature and is capable of binding to various matrix components of the granulation tissue, further increasing the presence of IL-8 in the granulation tissue. Therefore, IL-8 not only functions in the inflammatory phase of wound healing by serving as a leukocyte chemoattractant, but also plays an important role in granulation tissue formation by stimulating angiogenesis and matrix deposition (Martins-Green, 2001; Feugate *et al.*, 2002). Angiogenesis occurring during granulation tissue formation relies heavily on cell–ECM interactions, as mentioned earlier under "Development."

## Proliferation

After wounding, the keratinocytes alter their proliferation and migration in order to close the wound, a process known as *re-epithelialization*. As this process occurs, the cells at the edge of the wound migrate, whereas the cells around the wound proliferate in order to provide the additional cells needed to cover the wounded area. The proliferative state of these latter keratinocytes may be sustained by interactions with the ECM of the remaining basement membrane. Indeed, during the remodeling of normal skin, the proliferation of the basal layer of keratinocytes needed to replace the upper keratinocyte layers requires the presence of fibronectin in the epithelial basal lamina (see earlier). In addition, in a dermal wound model, ECM derived from the basement membrane can maintain the keratinocytes in a proliferative state for several days. It is likely that, in addition to fibronectin, laminin participates in keratinocyte proliferation, because previous work indicates that laminin can promote proliferation of these cells *in vitro* (Pouliot *et al.*, 2002). On the other hand, fibrin present in the provisional matrix may have an inhibitory effect on keratinocyte proliferation, as evidenced by the abnormal keratinocyte proliferation seen during the re-epithelialization of fibrinogen-deficient mice (Drew *et al.*, 2001).

The granulation tissue begins to form as re-epithelialization proceeds. This tissue is composed of ECM molecules, including embryonic fibronectin, type III collagen, type I collagen, and hyaluronic acid, along with multiple cell types, such as monocyte/macrophages, lymphocytes, fibroblasts, myofibroblasts, and the endothelial cells of the wound vasculature. Growth factors released by these cells and platelets cooperate with the aforementioned surrounding ECM molecules, to provide pro-proliferative signals to the granulation tissue fibroblasts and endothelial cells. In the case of endothelial cells, the increased proliferation can participate in the formation of the wound vasculature via angiogenesis. In this process, ECM molecules interact with VEGFs and FGFs, angiogenic factors that then stimulate endothelial cell proliferation and migration to form new blood vessels (Sottile, 2004). The importance of ECM–growth factor binding in blood vessel formation is underscored by recent studies suggesting that the antiangiogenic molecules thrombospondin and endostatin may exert their antiangiogenic effects by competing with pro-angiogenic growth factors for ECM binding (Gupta et al., 1999; Reis et al., 2005). Furthermore, some growth factors appear to promote proliferation only when specific ECM molecules are present, as is seen in the fibronectin requirement for TGF-b1-mediated fibroblast proliferation (Clark et al., 1997). In contrast, VEGF is unable to induce proliferation when bound to SPARC, indicating that interactions between growth factors and ECM can also be inhibitory (Kupprion et al., 1998). While ECM–growth factor interactions can significantly impact cell proliferation, specific ECM molecules also affect proliferation directly. Fibronectin, specific fragments of fibronectin, laminin, collagen VI, and SPARC/osteonectin can directly induce fibroblast and endothelial cell proliferation (e.g., Ruhl et al., 1999; Sage et al., 2003; Sottile, 2004). Previous studies suggest that the proliferative ability of laminin is mediated by its EGF-like domains, implicating EGFR activation in its pro-proliferative effects (Panayotou et al., 1989; Schenk et al., 2003). In addition, certain ECM molecules and/or proteolytic fragments can inhibit proliferation. SPARC and decorin as well as peptides derived from SPARC, decorin, collagen IV (tumstatin), and collagens XVIII and XV (endostatin) are antiangiogenic due to their inhibitory effects on endothelial cell proliferation (Sage et al., 2003; Sottile, 2004; Sulochana et al., 2005).

## Differentiation

As the granulation tissue forms, some of the fibroblasts within the wounded area differentiate into myofibroblasts, cells that express the protein a–smooth muscle actin (aSMA) and thus function similarly to smooth muscle cells (Desmouliere et al., 2005). This differentiation process is influenced by various matrix molecules, such as heparin, which decreases fibroblast proliferation while stimulating aSMA expression in vitro. Similarly, although the in vivo application of tumor necrosis factor a (TNFa) promotes granulation tissue formation, myofibroblasts were only detected when heparin was also added (Desmouliere et al., 1992). The effects of heparin on myofibroblast differentiation and aSMA expression are probably not due to its anticoagulant activity, but more likely result from the ability of heparin and heparan sulfate proteoglycans to interact with cytokines and/or growth factors such as TGF-b1, which then modulate myofibroblast differentiation (Li et al., 2004). This

TGF-b1-induced differentiation is prevented when $a_v$ or $b_1$ integrins or the ED-A-containing form of fibronectin are inhibited (Lygoe et al., 2004; Desmouliere et al., 2005). Furthermore, cardiac fibroblasts differentiate into myofibroblasts after plating on collagen VI (Naugle et al., 2006). Interstitial collagens, in conjunction with mechanical tension, also participate in the differentiation process. Fibroblasts cultured on collagen-coated plates or relaxed collagen gels fail to differentiate, whereas fibroblasts cultured under conditions that more closely mimic the granulation tissue, on anchored collagen gels with aligned collagen fibers, exhibit myofibroblast characteristics (Arora et al., 1999). In addition, more recent observations in vitro and during wound healing in vivo have further established a role for mechanical tension in myofibroblast differentiation (Wang et al., 2003).

## Apoptosis

Late in the wound-healing process, the granulation tissue undergoes remodeling to form scar tissue. This remodeling phase is characterized by decreased tissue cellularity due to the disappearance of multiple cell types, including fibroblasts, myofibroblasts, endothelial cells, and pericytes, and by the accumulation of ECM molecules, particularly interstitial collagens. The observed reduction in cell numbers during the remodeling phase occurs due to apoptosis. The number of apoptotic cells in the granulation tissue was shown to increase 20–25 days after wounding, with the significant decrease in cellularity apparent after 25 days (Desmouliere et al., 1995). Many of these apoptotic cells are endothelial cells and myofibroblasts, as shown by studies using in situ DNA fragment end-labeling in conjunction with transmission electron microscopy. Moreover, the release of mechanical tension in a system mimicking the formation of granulation tissue and its subsequent regression stimulates human fibroblast and myofibroblast apoptotic cell death. The apoptosis observed in this system was regulated by a combination of growth factors and the mechanical tension exerted by contractile collagens, underscoring the importance of such collagens in regulating apoptosis within the healing tissue. The fibroblast apoptosis regulated by mechanical tension also appears to involve interactions between thrombospondin-1 and the $a_vb_3$ integrin–CD47 complex (Graf et al., 2002). Apoptosis of fibroblasts and myofibroblasts may be important in preventing excessive scarring and facilitating the resolution of wound healing. Indeed, in keloids and hypertrophic scars there is a decrease in apoptosis of these cells, leading to increased matrix deposition and scarring. In keloids, the lack of apoptosis is thought to be caused by mutations in p53 or by growth factor receptor overexpression (e.g., Ladin et al., 1998; Ishihara et al., 2000; Moulin et al., 2004). In hypertrophic scars, however, the reduced apoptosis may result from increased expression of tissue transglutaminase, resulting in enhanced matrix degradation and diminished collagen contraction (Linge *et al.*, 2005). There is also evidence to suggest that alternative types of cell death may have roles in wound healing. For example, bronchoalveolar lavage fluid collected after lung injury during the remodeling phase stimulated fibroblast death in a manner that is not consistent with either apoptotic or necrotic cell death (Polunovsky *et al.*, 1993).

## III. SIGNAL TRANSDUCTION EVENTS DURING CELL–ECM INTERACTIONS

As discussed earlier, ECM molecules are capable of interacting with a variety of receptors. Such interactions activate signal transduction pathways within the cell, altering levels of both gene expression and protein activation, thus ultimately changing outcomes in cell adhesion, migration, proliferation, differentiation, and death. The signaling pathways linked to these specific outcomes have been studied for many of the ligand–receptor interactions, particularly those involving integrins. Based on these studies, we postulate the existence of three categories of cell–ECM interactions that lead to the aforementioned cellular events (Fig. 7.2).

### Type I Interactions

These are generally mediated by integrin and proteoglycan receptors and are important in the adhesion/deadhesion processes that accompany cell migration (Fig. 7.2A). These interactions are exemplified by fibronectin-mediated cell migration, which occurs when this matrix molecule simultaneously binds integrins and proteoglycans, the latter via its heparin-binding domain (Dedhar, 1999; Mercurius and Morla, 2001). These fibronectin receptors then colocalize and interact at cell adhesion sites, where the microfilaments interact with the cytoplasmic domain of integrin $\beta_1$ through the structural proteins talin and $\alpha$-actinin. The fact that integrins interact with the cytoskeleton suggests that the integrin-induced signaling involved in adhesion and migration may be mediated, in part, by the cytoskeleton itself. Additional integrin-mediated signaling occurs via the activation of the focal adhesion tyrosine kinase pp125[FAK], which also interacts with the cytoplasmic domain of integrin $\beta_1$. On activation, pp125[FAK] phosphorylates itself at tyrosine 397 (Hildebrand *et al.*, 1995), which then serves as the binding site for the SH2 domain of the c-Src tyrosine kinase. This kinase subsequently phosphorylates multiple proteins present in the focal adhesion plaques, including FAK itself, at position 925, as well as paxillin, tensin, vinculin, and p130[cas]. FAK phosphotyrosine 925 binds the Grb2/Sos complex, thus promoting the activation of Ras GTPase and the MAP kinase cascade, which may be involved in cell adhesion/deadhesion and migration events (Schlaepfer and Hunter, 1998; Dedhar, 1999). Paxillin may also participate in integrin-mediated signaling and motility, as evidenced by the reduced migration and decreased phosphorylation/activation of various signaling molecules observed in paxillin-deficient fibroblasts (Hagel *et al.*, 2002). The contribution of tensin to cell adhesion and motility is poorly understood, although it is known to interact with the cytoskeleton and various phosphorylated signaling molecules via its SH2 domain. Therefore, tensin may facilitate various signaling events downstream of integrin ligation (Lo, 2004). Active p130[cas] interacts with Crk and Nck, which function as adaptor molecules that appear to increase cell migration by promoting the localized activation of Rac-GTPase and the MAP/JNK kinase pathways (Chodniewicz and Klemke, 2004).

### Type II Interactions

These involve processes in which the matrix–receptor interactions, in conjunction with growth factor or cytokine receptors, affect proliferation, survival, differentiation, and/or maintenance of the differentiated phenotype (Fig. 7.2B). These cooperative effects may occur in a direct manner, for example, by the direct interaction of EGF-like repeats present in certain ECM molecules with the EGF receptor, thereby promoting cell proliferation (Swindle *et al.*, 2001; Tran *et al.*, 2004). Indirect cooperative effects are better understood at this time, particularly with regard to the

**FIG. 7.2.** Schematic diagrams illustrating the three categories of cell/ECM interactions proposed here. These categories are represented by sketches of the binding elements. **(A)** *Type I interactions* are generally mediated by integrin and proteoglycan receptors and are important in the adhesion/deadhesion processes that accompany cell migration. At focal adhesions, proteoglycan (treelike) and integrin (heterodimer) receptors on the plasma membrane (pm) bind to different epitopes on the same ECM molecule, leading to cytoskeletal reorganization. A variety of proteins become phosphorylated (e.g., pp125[FAK] and src), leading to activation of genes important for cell adhesion/deadhesion and for migration. **(B)** *Type II interactions* involve processes in which matrix–receptor interactions, in conjunction with growth factor or cytokine receptors, affect proliferation, survival, differentiation, and/or maintenance of the differentiated phenotype. Integrin receptors bind to their ligands, leading to activation of cytoskeletal elements as in Type I; but, also, growth factors bound to matrix molecules (triangle) bind to their receptors, which have kinase activity. This kinase activates phospholipase $C_\gamma$ which, in turn, cleaves $PIP_2$, leading to inisitoltrisphosphate ($IP_3$) and diacylglycerol (DAG); $IP_3$ binds to its receptor on the smooth endoplasmic reticulum, inducing the release of $Ca^{++}$, which can lead directly to activation of gene expression or indirectly by cooperation with DAG through protein kinase C (PKC). In this case, the genes activated are important in cell proliferation, differentiation and maintenance of the differentiated phenotype. **(C)** *Type III interactions* involve mostly processes leading to apoptosis and epithelial-to-mesenchymal transitions. Integrin receptors bind to fragments of ECM molecules containing specific domains. This leads to activation of matrix protease genes whose products (represented by purple ellipses) degrade the matrix and release peptides (squiggles) that can further interact with cell surface receptors and/or release growth factors (triangles and diamonds), which, in turn, bind to their own receptors, activating G proteins and kinases leading to expression of genes important in morphogenesis and cell death.

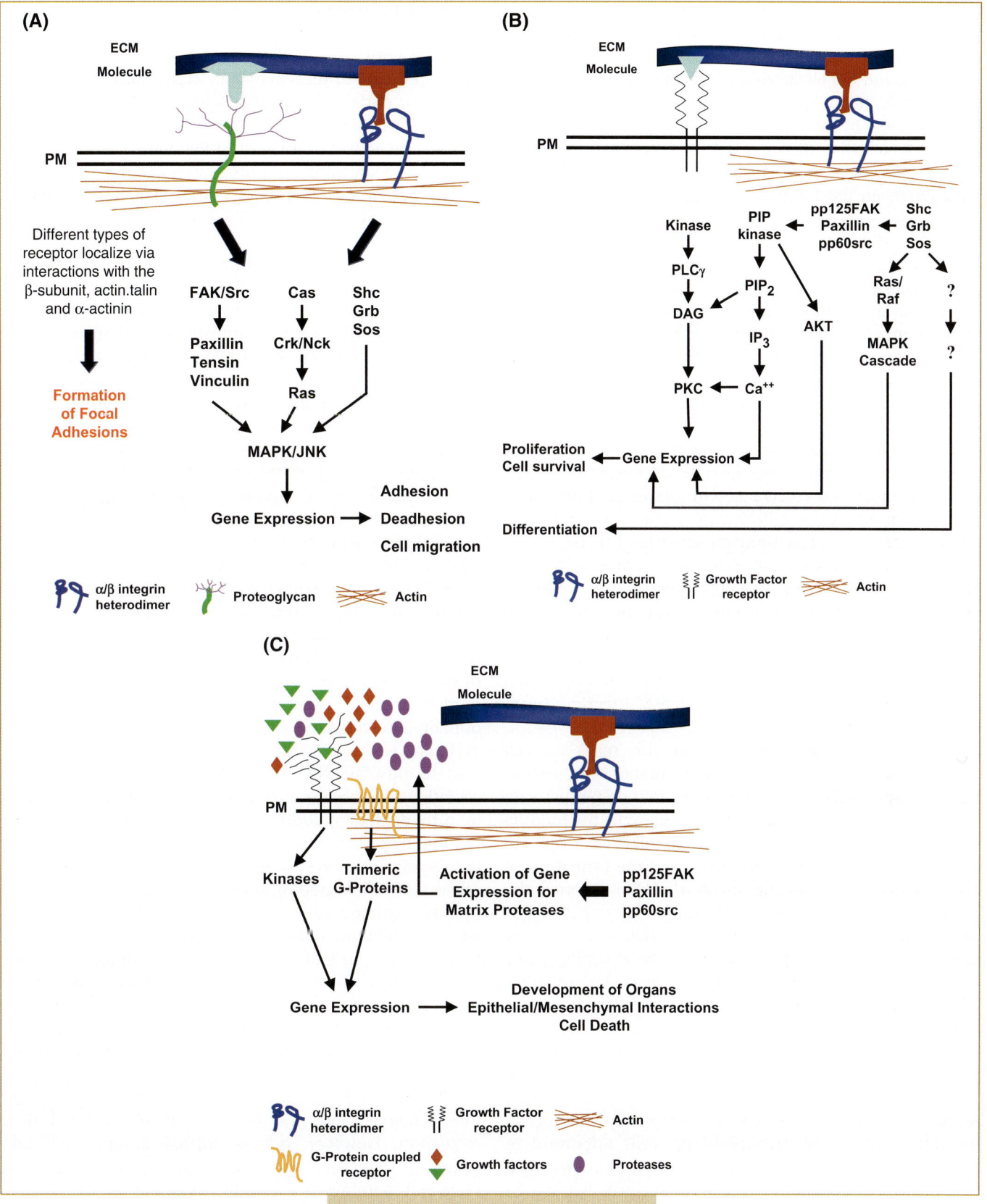

anchorage dependence of cell growth. S-phase entry, even when growth factors are present, requires the interaction of cells with a substrate, underscoring the critical role of cell–ECM adhesion in cell survival and proliferation (Giancotti, 1997; Hynes, 2002). Specifically, integrin ligation promotes the activation of Fyn and its binding to the Shc adaptor protein, which then recruits Grb2, thus activating the Ras/ERK pathway, resulting in the phosphorylation of the transcription factor Elk-1 and the activation of genes important in cell cycle progression. Furthermore, cell–ECM interactions are critical for the efficient and prolonged activation of MAPK by growth factors (Howe *et al.*, 2002). Ras-mediated signaling also leads to the activation of PI-3 kinase and thus of the Akt serine/threonine kinase; the activation of this pathway prevents the apoptosis of suspended cells. Integrin ligation also appears to promote cell proliferation through the degradation of cell cycle inhibitors, as is seen in the degradation of p21 downstream of fibronectin-mediated Cdc42 and Rac-1 activation. The critical role of the Rac/JNK pathway in this process is also seen in the $\beta_1$ integrin cytoplasmic domain mutant, in which the decreased activation of this pathway was correlated with diminished fibroblast proliferation and survival. Both of these effects were reversed on the expression of constitutively active Rac1 (Hirsch *et al.*, 2002). Negative affects on cell proliferation were also observed in other studies, in which integrins were inhibited or knocked out. For example, fibroblasts derived from mice lacking the $\alpha_1\beta_1$ integrin proliferated at a reduced rate, despite the fact that they were able to attach normally (Pozzi *et al.*, 1998). A similar result was seen in mammary epithelial cells overexpressing a dominant negative $\beta_1$ integrin subunit (Faraldo *et al.*, 2001).

Similarly, cellular differentiation also relies on cell interactions with ECM molecules, hormones, and growth factors, particularly those interactions that do not activate Shc and the MAP kinase cascade. For example, the binding of laminin to integrin $\alpha_2\beta_1$ in endothelial cells fails to activate the Shc pathway and promotes the formation of capillary-like structures (Kubota *et al.*, 1988), whereas the binding of fibronectin to integrin $\alpha_5\beta_1$ in these cells leads to cell proliferation (Wary *et al.*, 1998). Additional signaling molecules are required to generate these capillary-like tubes. One such molecule is integrin-linked kinase (ILK), which, when overexpressed, rescues capillary-like tube formation in the absence of ECM molecules (Cho *et al.*, 2005), while expression of a dominant negative version of ILK blocks tube formation even when ECM and VEGF are present (Watanabe *et al.*, 2005). Integrin-mediated signaling is also important in other differentiated phenotypes, e.g., in the differentiation of myofibroblasts, cells important in wound healing; the myofibroblast differentiation induced by TGF-$\beta1$ is dependent on specific integrin ligation as well as the activation of FAK and its associated signaling pathways (Thannickal *et al.*, 2003; Lygoe *et al.*, 2004).

## Type III Interactions

These primarily involve processes leading to apoptosis and epithelial-to-mesenchymal transitions (Fig. 7.2C). Apoptotic pathways have been identified for endothelial cells and leukocytes and appear to involve primarily tyrosine kinase activity (Ilan *et al.*, 1998; Avdi *et al.*, 2001). For example, neutrophil apoptosis stimulated by TNF-$\alpha$ is dependent on $\beta_2$ integrin–mediated signaling events involving the activation of the Pyk2 and Syk tyrosine kinases as well as JNK1 (Avdi *et al.*, 2001). In other cell types, alterations in the ligand presentation by ECM can also regulate apoptosis. Studies have suggested that integrin ligation by soluble, rather than intact, ligands can function as integrin antagonists and promote apoptosis rather than survival or proliferation (Stupack and Cheresh, 2002). Such soluble ligands may be created by matrix degradation during tissue remodeling. The apoptosis stimulated by soluble ligands or other antagonists appears to occur via the recruitment and activation of caspase 8 by clustered integrins, without any requirement for death receptors. In such cases, matrix remodeling is critical, because enzymatic degradation of the ECM causes the release of both soluble factors as well as ECM fragments that contain specific sequences that affect cell behavior and/or exhibit altered receptor interactions. For example, when fibronectin binds only through its cell-binding domain, the cells are stimulated to produce ECM-remodeling enzymes. There are at least three possible ways in which such a process could be initiated. (1) Changes in expression of fibronectin receptors would allow cells to bind fibronectin, predominantly through its cell-binding domain, and activate $\alpha_5\beta_1$ interactions with the actin cytoskeleton, with subsequent transduction of signals that lead to up-regulation of ECM-degrading enzymes. The secretion of these enzymes would start a positive-feedback loop by degrading additional fibronectin to produce cell-binding fragments that would bind to $\alpha_5\beta_1$, activate it, and in this way keep the specific event going. (2) Very localized release of ECM-degrading enzymes could degrade fibronectin into fragments containing only the cell-binding domain, which would bind to $\alpha_5\beta_1$ and initiate the positive-feedback loop. (3) At a particular time during development, specific cells would produce spliced forms of fibronectin that are only capable of interacting via their cell-binding domain. Binding of these fragments to $\alpha_5\beta_1$ would trigger the feedback loop. This positive-feedback loop and consequent runaway process of ECM degradation is advantageous locally for such events as cell growth, epithelial-to-mesenchymal transitions, or cell death, relieving their tight regulation. However, during normal development and wound healing, there must be a signal that can break this cycle and thereby bring it under control at the appropriate time and place. Without application of such a brake, these processes can lead to abnormal development or wound healing or to pathological situations, such as tumor growth and invasion.

Although these three categories may not be exhaustive of the general types of cell–ECM interactions that occur during development and wound healing, they encapsulate the major interactions documented to date. Each category has its place in many developmental and repair events, and they may operate in sequence. A compelling example of the latter is the epithelial-to-mesenchymal transition and morphogenesis of the neural crest cell system (Martins-Green and Bissell, 1995). These cells originate in the neural epithelium that occupies the crest of the neural folds. After the delamination event that separates the neural epithelium from the epidermal ectoderm (Martins-Green, 1988), the folds fuse to form the tube. At this time, the NCC occupy the dorsalmost portion of the tube, they are not covered by basal lamina, and the subepidermal space above them contains large amounts of fibronectin (Martins-Green and Erickson, 1987). Just before the NCC emigrate from the neural tube, fibronectin appears between them; they separate from each other and migrate away, carrying fibronectin on their surfaces (Martins-Green, 1987). During the period of emigration at any particular level of the neural tube, basal lamina is deposited progressively toward the crest from the sides of the tube (Martins-Green and Erickson, 1986, 1987). NCC emigration terminates as deposition reaches the crest of the tube. The NCC then follow specific migration pathways throughout the embryo, arriving at a wide variety of locations, where they differentiate into many different phenotypes in response to external cues (Perris and Perissinotto, 2000).

The appearance of fibronectin between the NCC just before emigration must be the result of secretion by the adjacent cells or introduction from the epithelial cells after loss of cell–cell adhesions. In keeping with the cell–ECM interaction mechanism of Type III, either alternative could initiate a positive-feedback loop and release the NCC, leading to emigration. Enzymatic degradation of the stabilizing domain of fibronectin above the tube could cause enhanced secretion of specific enzymes by the NCC in response to the effect of the cell-binding domain acting alone, thus severing the cell adhesions and producing additional fibronectin fragments containing the cell-binding domain. These fragments, in turn, would bind to adjacent cells and stimulate further enzymatic secretion that would be self-perpetuating. NCC emigration occurs in an anterior-to-posterior wave; thus, following initiation of enzymatic activity in the head of the embryo, it could propagate in a posterior direction, triggering NCC emigration in a wave from head to tail.

Clearly some controlling event(s) must terminate NCC emigration at each location along the neural tube. Such an event has already been identified. At the time of NCC emigration, the ventral and lateral surfaces of the neural tube are covered by an intact basal lamina, which stabilizes the epithelium and separates it from the fibronectin layer around the tube (Martins-Green and Erickson, 1986). During the few hours of emigration at any one site, as the NCC are leaving from the dorsalmost portion of the neural tube, basal lamina deposition progresses quickly up the sides of the tube and terminates local emigration when it becomes complete over the crest of the tube (Martins-Green and Erickson, 1986, 1987). After they have emigrated from the neural tube, the NCC find themselves in an extracellular space filled with intact fibronectin and other ECM molecules that stimulate the focal adhesions of cell–ECM interactions of Type I, thereby providing the substrate for migration. On arrival at their final destination, further interactions of Type II stimulate differentiation into a wide range of phenotypes (Perris and Perissinotto, 2000).

# IV. RELEVANCE FOR TISSUE ENGINEERING

Designing tissue and organ replacements that closely simulate nature is a challenging endeavor. One avenue to achieve this goal is to study how tissues and organs arise during embryogenesis and during normal processes of repair and how those functions are maintained. When developing tissue replacements, one needs to consider the following (Fig. 7.3).

1. *Avoiding an immune response that can cause inflammation and/or rejection.* Ideally, one would like to manipu-

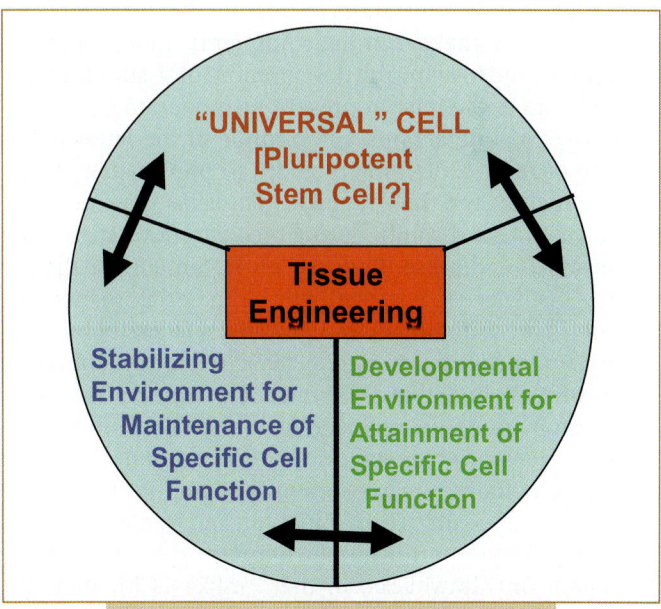

**FIG. 7.3.** Conceptualization of the interactions of a "universal" cell, i.e., a pluripotent stem cell, and environments in which it is conditioned to a particular function (developmental environment) and maintained in that function (stabilizing environment).

late cells *in vitro* to make them more universal and thereby decrease the possibility of immune responses. In theory, these cells could then differentiate in the presence of an environment conducive to expression of the appropriate phenotype. However, little progress has been made toward this elusive goal. Alternatively, engineered tissues could incorporate progenitor cells that may suppress host immune responses directly or indirectly through decreased expression of MHC; these cells could be induced at a later time to differentiate into various cell types (Barry and Murphy, 2004). One example of a progenitor cell that appears to decrease immune responses and also maintains a broad differentiation capacity is the mesenchymal stem cell, which is capable of differentiating into multiple cell types and may thus prove to be an invaluable asset in tissue engineering (Barry and Murphy, 2004).

2. *Creating the proper substrate for cell survival and differentiation.* One of the strategies to fulfill this goal is the use of biocompatible implants composed of extracellular matrix molecules seeded with autologous cells or with heterologous cells in conjunction with immunosuppressant drugs. Addition of growth and differentiation factors to these matrices as well as agonists or antagonists that favor cell–ECM interactions can potentially increase the rate of successful tissue replacement. One example in which the knowledge obtained in studies of cell–ECM interactions has proven useful in tissue engineering was the discovery that most integrins bind to their ECM ligands via the tripeptide RGD. This small sequence of amino acids has been used as an agonist to make synthetic implants more biocompatible and to allow the development of tissue structure or as an antagonist to prevent or moderate unwanted cell–ECM interactions. An example of the latter is the use of RGD-containing peptides to prevent fibrinogen interaction and thus modulate platelet aggregation and formation of thrombi during reconstructive surgery or in vascular disease (Bennett, 2001). Similarly, collagen has been used to coat synthetic biomaterials to increase their biocompatibility and promote successful biological interactions (Ma *et al.*, 2005). While the foregoing examples show that ECM molecules can be used successfully in tissue engineering, the use of natural ECM molecules in engineered tissue has several disadvantages, including the possibility of generating an immune response, possible contamination, and ease of degradation. Likewise, artificial biocompatible materials have significant drawbacks, in that, unlike ECM, they are generally incapable of transmitting growth and differentiation cues to cells (e.g., Rosso *et al.*, 2005). One future alternative to these approaches may be preparation of "semisynthetic biomaterials," in which func-

tional regions of ECM molecules, including those that interact with receptors or growth factors or those that are cleaved by proteases, are incorporated into artificial biomaterials to impart additional functionality (Lutolf and Hubbell, 2005; Rosso *et al.*, 2005). The inclusion of ECM-like cell-binding sites that promote cell adhesion, growth, and/or differentiation into such biomaterials may be critical in developing and maintaining functional engineered tissues by providing the appropriate cellular microenvironment. However, the use of either native ECM molecules or engineered ECM-like biomaterials in engineered tissues requires additional knowledge regarding the types of cell–ECM interactions that result in the desired cellular effects.

3. *Providing the appropriate environmental conditions for tissue maintenance.* To maintain tissue homeostasis, it is crucial to create a balanced environment with the appropriate cues for preservation of specific cell function(s). It is important to realize that such stasis on the level of a tissue is achieved via tissue remodeling — the *dynamic equilibrium* between cells and their environment. However, little is known about the crosstalk between cells and ECM under such "normal" conditions. As indicated earlier, the same ECM molecule may have multiple cellular effects. The ultimate cellular outcome likely depends on the combination of variables, such as the domain of the molecule involved in the cellular interactions, the receptor used for these interactions, and the cellular microenvironment. These variables can, in turn, be influenced by matrix remodeling, because enzymatic degradation of the ECM can release functional fragments of ECM that then alter cell–ECM interactions by removing certain binding sites while exposing others.

Because organ transplantation is one of the least cost-effective therapies and is not always available, tissue engineering offers hope for more consistent and rapid treatment of those in need of a body part replacement, and it therefore has greater potential to improve patient quality of life. The selected examples presented illustrate that further advances in tissue engineering require additional knowledge of the basic mechanisms of cell function and of the ways they interact with the environment. The recent surge in research on ECM molecules themselves and their interactions with particular cells and cell surface receptors has led to realization that these interactions are many and complex, allow the modulation of fundamental events during development and wound repair, and are crucial for the maintenance of the differentiated phenotype and tissue homeostasis. As such, the manipulation of specific cell–ECM interactions has the potential to modulate particular cellular functions and processes in order to maximize the effectiveness of engineered tissues.

# V. REFERENCES

Anderson, S. S., Kim, Y., and Tsilibary, E. C. (1994). Effects of matrix glycation on mesangial cell adhesion, spreading and proliferation. *Kidney Int.* **46**, 1359–1367.

Arora, P. D., Narani, N., and McCulloch, C. A. (1999). The compliance of collagen gels regulates transforming growth factor-beta induction of alpha-smooth muscle actin in fibroblasts. *Am. J. Pathol.* **154**, 871–882.

Avdi, N. J., Nick, J. A., Whitlock, B. B., Billstrom, M. A., Henson, P. M., Johnson, G. L., and Worthen, G. S. (2001). Tumor necrosis factor-alpha activation of the c-Jun N-terminal kinase pathway in human neutrophils. Integrin involvement in a pathway leading from cytoplasmic tyrosine kinases apoptosis. *J. Biol. Chem.* **276**, 2189–2199.

Barry, F. P., and Murphy, J. M. (2004). Mesenchymal stem cells: clinical applications and biological characterization. *Int. J. Biochem. Cell Biol.* **36**, 568–584.

Bass, M. D., and Humphries, M. J. (2002). Cytoplasmic interactions of syndecan-4 orchestrate adhesion receptor and growth factor receptor signalling. *Biochem. J.* **368**, 1–15.

Bennett, J. S. (2001). Novel platelet inhibitors. *Annu. Rev. Med.* **52**, 161–184.

Boucaut, J. C., Darribere, T., Boulekbache, H., and Thiery, J. P. (1984a). Prevention of gastrulation but not neurulation by antibodies to fibronectin in amphibian embryos. *Nature* **307**, 364–367.

Boucaut, J. C., Darribere, T., Poole, T. J., Aoyama, H., Yamada, K. M., and Thiery, J. P. (1984b). Biologically active synthetic peptides as probes of embryonic development: a competitive peptide inhibitor of fibronectin function inhibits gastrulation in amphibian embryos and neural crest cell migration in avian embryos. *J. Cell Biol.* **99**, 1822–1830.

Boudreau, N., Sympson, C. J., Werb, Z., and Bissell, M. J. (1995). Suppression of ICE and apoptosis in mammary epithelial cells by extracellular matrix. *Science* **267**, 891–893.

Bronner-Fraser, M. (1985). Alterations in neural crest migration by a monoclonal antibody that affects cell adhesion. *J. Cell Biol.* **101**, 610–617.

Brooks, P. C., Clark, R. A., and Cheresh, D. A. (1994). Requirement of vascular integrin alpha v beta 3 for angiogenesis. *Science* **264**, 569–571.

Brown, T. A., Bouchard, T., St John, T., Wayner, E., and Carter, W. G. (1991). Human keratinocytes express a new CD44 core protein (CD44E) as a heparan-sulfate intrinsic membrane proteoglycan with additional exons. *J. Cell Biol.* **113**, 207–221.

Bushkin-Harav, I., and Littauer, U. Z. (1998). Involvement of the YIGSR sequence of laminin in protein tyrosine phosphorylation. *FEBS Lett.* **424**, 243–247.

Cho, H. J., Youn, S. W., Cheon, S. I., Kim, T. Y., Hur, J., Zhang, S. Y., Lee, S. P., Park, K. W., Lee, M. M., Choi, Y. S., *et al.* (2005). Regulation of endothelial cell and endothelial progenitor cell survival and vasculogenesis by integrin-linked kinase. *Arterioscler Thromb. Vasc. Biol.* **25**, 1154–1160.

Chodniewicz, D., and Klemke, R. L. (2004). Regulation of integrin-mediated cellular responses through assembly of a CAS/Crk scaffold. *Biochim. Biophys. Acta.* **1692**, 63–76.

Chung, C. Y., and Erickson, H. P. (1994). Cell surface annexin II is a high affinity receptor for the alternatively spliced segment of tenascin-C. *J. Cell Biol.* **126**, 539–548.

Clark, R. A., McCoy, G. A., Folkvord, J. M., and McPherson, J. M. (1997). TGF-beta 1 stimulates cultured human fibroblasts to proliferate and produce tissue-like fibroplasia: a fibronectin matrix-dependent event. *J. Cell Physiol.* **170**, 69–80.

Couchman, J. R., and Woods, A. (1996). Syndecans, signaling, and cell adhesion. *J. Cell Biochem.* **61**, 578–584.

Damsky, C. H., and Werb, Z. (1992). Signal transduction by integrin receptors for extracellular matrix: cooperative processing of extracellular information. *Curr. Opin. Cell Biol.* **4**, 772–781.

Daniel, C. W., Robinson, S., and Silberstein, G. B. (1996). The role of TGF-beta in patterning and growth of the mammary ductal tree. *J. Mammary Gland Biol. Neoplasia* **1**, 331–341.

Decline, F., and Rousselle, P. (2001). Keratinocyte migration requires alpha2beta1 integrin-mediated interaction with the laminin 5 gamma2 chain. *J. Cell Sci.* **114**, 811–823.

Dedhar, S. (1999). Integrins and signal transduction. *Curr. Opin. Hematol.* **6**, 37–43.

Desban, N., and Duband, J. L. (1997). Avian neural crest cell migration on laminin: interaction of the alpha1beta1 integrin with distinct laminin-1 domains mediates different adhesive responses. *J. Cell Sci.* **110** (Pt 21), 2729–2744.

Desmouliere, A., Rubbia-Brandt, L., Grau, G., and Gabbiani, G. (1992). Heparin induces alpha-smooth muscle actin expression in cultured fibroblasts and in granulation tissue myofibroblasts. *Lab. Invest.* **67**, 716–726.

Desmouliere, A., Redard, M., Darby, I., and Gabbiani, G. (1995). Apoptosis mediates the decrease in cellularity during the transition between granulation tissue and scar. *Am. J. Pathol.* **146**, 56–66.

Desmouliere, A., Chaponnier, C., and Gabbiani, G. (2005). Tissue repair, contraction, and the myofibroblast. *Wound Repair Regen.* **13**, 7–12.

Desprez, P. Y., Hara, E., Bissell, M. J., and Campisi, J. (1995). Suppression of mammary epithelial cell differentiation by the helix-loop-helix protein Id-1. *Mol. Cell Biol.* **15**, 3398–3404.

DiPersio, C. M., Jackson, D. A., and Zaret, K. S. (1991). The extracellular matrix coordinately modulates liver transcription factors and hepatocyte morphology. *Mol. Cell Biol.* **11**, 4405–4414.

Drew, A. F., Liu, H., Davidson, J. M., Daugherty, C. C., and Degen, J. L. (2001). Wound-healing defects in mice lacking fibrinogen. *Blood* **97**, 3691–3698.

Ehnis, T., Dieterich, W., Bauer, M., Lampe, B., and Schuppan, D. (1996). A chondroitin/dermatan sulfate form of CD44 is a receptor for collagen XIV (undulin). *Exp. Cell Res.* **229**, 388–397.

Ehrenreich, M., and Ruszczak, Z. (2006). Update on Tissue-Engineered Biological Dressings. *Tissue Eng.*

Erickson, C. A., and Reedy, M. V. (1998). Neural crest development: the interplay between morphogenesis and cell differentiation. *Curr. Top. Dev. Biol.* **40**, 177–209.

Faraldo, M. M., Deugnier, M. A., Thiery, J. P., and Glukhova, M. A. (2001). Growth defects induced by perturbation of beta1-integrin function in the mammary gland epithelium result from a lack of MAPK activation via the Shc and Akt pathways. *EMBO Rep.* **2**, 431–437.

Faraldo, M. M., Deugnier, M. A., Tlouzeau, S., Thiery, J. P., and Glukhova, M. A. (2002). Perturbation of beta1-integrin function in involuting mammary gland results in premature dedifferentiation of secretory epithelial cells. *Mol. Biol. Cell* **13**, 3521–3531.

Febbraio, M., Hajjar, D. P., and Silverstein, R. L. (2001). CD36: a class B scavenger receptor involved in angiogenesis, atherosclerosis, inflammation, and lipid metabolism. *J. Clin. Invest.* 108, 785–791.

Ferri, N., Carragher, N. O., and Raines, E. W. (2004). Role of discoidin domain receptors 1 and 2 in human smooth muscle cell-mediated collagen remodeling: potential implications in atherosclerosis and lymphangioleiomyomatosis. *Am. J. Pathol.* 164, 1575–1585.

Feugate, J. E., Wong, L., Li, Q. J., and Martins-Green, M. (2002). The CXC chemokine cCAF stimulates precocious deposition of ECM molecules by wound fibroblasts, accelerating development of granulation tissue. *BMC Cell Biol.* 3, 13.

Francki, A., Motamed, K., McClure, T. D., Kaya, M., Murri, C., Blake, D. J., Carbon, J. G., and Sage, E. H. (2003). SPARC regulates cell cycle progression in mesangial cells via its inhibition of IGF-dependent signaling. *J. Cell Biochem.* 88, 802–811.

Fuchs, E., and Raghavan, S. (2002). Getting under the skin of epidermal morphogenesis. *Nat. Rev. Genet.* 3, 199–209.

Ghert, M. A., Qi, W. N., Erickson, H. P., Block, J. A., and Scully, S. P. (2001). Tenascin-C splice variant adhesive/anti-adhesive effects on chondrosarcoma cell attachment to fibronectin. *Cell Struct. Funct.* 26, 179–187.

Giancotti, F. G. (1997). Integrin signaling: specificity and control of cell survival and cell cycle progression. *Curr. Opin. Cell Biol.* 9, 691–700.

Gillitzer, R., and Goebeler, M. (2001). Chemokines in cutaneous wound healing. *J. Leukoc. Biol.* 69, 513–521.

Graf, R., Freyberg, M., Kaiser, D., and Friedl, P. (2002). Mechanosensitive induction of apoptosis in fibroblasts is regulated by thrombospondin-1 and integrin associated protein (CD47). *Apoptosis* 7, 493–498.

Grant, D. S., Tashiro, K., Segui-Real, B., Yamada, Y., Martin, G. R., and Kleinman, H. K. (1989). Two different laminin domains mediate the differentiation of human endothelial cells into capillary-like structures *in vitro. Cell* 58, 933–943.

Grant, D. S., Kinsella, J. L., Fridman, R., Auerbach, R., Piasecki, B. A., Yamada, Y., Zain, M., and Kleinman, H. K. (1992). Interaction of endothelial cells with a laminin A chain peptide (SIKVAV) *in vitro* and induction of angiogenic behavior *in vivo. J. Cell Physiol.* 153, 614–625.

Green, H. (1977). Terminal differentiation of cultured human epidermal cells. *Cell* 11, 405–416.

Greiling, D., and Clark, R. A. (1997). Fibronectin provides a conduit for fibroblast transmigration from collagenous stroma into fibrin clot provisional matrix. *J. Cell Sci.* 110 (Pt 7), 861–870.

Gupta, K., Gupta, P., Wild, R., Ramakrishnan, S., and Hebbel, R. P. (1999). Binding and displacement of vascular endothelial growth factor (VEGF) by thrombospondin: effect on human microvascular endothelial cell proliferation and angiogenesis. *Angiogenesis* 3, 147–158.

Hagel, M., George, E. L., Kim, A., Tamimi, R., Opitz, S. L., Turner, C. E., Imamoto, A., and Thomas, S. M. (2002). The adaptor protein paxillin is essential for normal development in the mouse and is a critical transducer of fibronectin signaling. *Mol. Cell Biol.* 22, 901–915.

Handel, T. M., Johnson, Z., Crown, S. E., Lau, E. K., and Proudfoot, A. E. (2005). Regulation of protein function by glycosaminoglycans — as exemplified by chemokines. *Annu. Rev. Biochem.* 74, 385–410.

Hardwick, C., Hoare, K., Owens, R., Hohn, H. P., Hook, M., Moore, D., Cripps, V., Austen, L., Nance, D. M., and Turley, E. A. (1992). Molecular cloning of a novel hyaluronan receptor that mediates tumor cell motility. *J. Cell Biol.* 117, 1343–1350.

Hergott, G. J., Nagai, H., and Kalnins, V. I. (1993). Inhibition of retinal pigment epithelial cell migration and proliferation with monoclonal antibodies against the beta 1 integrin subunit during wound healing in organ culture. *Invest. Ophthalmol. Vis. Sci.* 34, 2761–2768.

Hildebrand, J. D., Schaller, M. D., and Parsons, J. T. (1995). Paxillin, a tyrosine phosphorylated focal adhesion-associated protein binds to the carboxyl terminal domain of focal adhesion kinase. *Mol. Biol. Cell* 6, 637–647.

Hirsch, E., Barberis, L., Brancaccio, M., Azzolino, O., Xu, D., Kyriakis, J. M., Silengo, L., Giancotti, F. G., Tarone, G., Fassler, R., *et al.* (2002). Defective Rac-mediated proliferation and survival after targeted mutation of the beta1 integrin cytodomain. *J. Cell Biol.* 157, 481–492.

Hodge-Dufour, J., Noble, P. W., Horton, M. R., Bao, C., Wysoka, M., Burdick, M. D., Strieter, R. M., Trinchieri, G., and Pure, E. (1997). Induction of IL-12 and chemokines by hyaluronan requires adhesion-dependent priming of resident but not elicited macrophages. *J. Immunol.* 159, 2492–2500.

Hoffmann, S., He, S., Jin, M., Ehren, M., Wiedemann, P., Ryan, S. J., and Hinton, D. R. (2005). A selective cyclic integrin antagonist blocks the integrin receptors alphavbeta3 and alphavbeta5 and inhibits retinal pigment epithelium cell attachment, migration and invasion. *BMC Ophthalmol.* 5, 16.

Howe, A. K., Aplin, A. E., and Juliano, R. L. (2002). Anchorage-dependent ERK signaling — mechanisms and consequences. *Curr. Opin. Genet. Dev.* 12, 30–35.

Hynes, R. O. (2002). Integrins: bidirectional, allosteric signaling machines. *Cell* 110, 673–687.

Ilan, N., Mahooti, S., and Madri, J. A. (1998). Distinct signal transduction pathways are utilized during the tube formation and survival phases of *in vitro* angiogenesis. *J. Cell Sci.* 111 (Pt 24), 3621–3631.

Ishihara, H., Yoshimoto, H., Fujioka, M., Murakami, R., Hirano, A., Fujii, T., Ohtsuru, A., Namba, H., and Yamashita, S. (2000). Keloid fibroblasts resist ceramide-induced apoptosis by overexpression of insulin-like growth factor I receptor. *J. Invest. Dermatol.* 115, 1065–1071.

Jalkehen, M., Jalkanen, S., and Bernfield, M. (1991). Binding of extracellular effector molecules by cell surface proteoglycans. *In* "Receptors for Extracellular Matrix," (J. A. McDonald and R. P. Mecham, eds.), pp. 1–38. Academic Press, San Diego, CA.

Jimenez, B., Volpert, O. V., Crawford, S. E., Febbraio, M., Silverstein, R. L., and Bouck, N. (2000). Signals leading to apoptosis-dependent inhibition of neovascularization by thrombospondin-1. *Nat. Med.* 6, 41–48.

Kubota, Y., Kleinman, H. K., Martin, G. R., and Lawley, T. J. (1988). Role of laminin and basement membrane in the morphological differentiation of human endothelial cells into capillary-like structures. *J. Cell Biol.* 107, 1589–1598.

Kupprion, C., Motamed, K., and Sage, E. H. (1998). SPARC (BM-40, osteonectin) inhibits the mitogenic effect of vascular endothelial growth factor on microvascular endothelial cells. *J. Biol. Chem.* 273, 29635–29640.

Ladin, D. A., Hou, Z., Patel, D., McPhail, M., Olson, J. C., Saed, G. M., and Fivenson, D. P. (1998). p53 and apoptosis alterations in keloids and keloid fibroblasts. *Wound Repair Regen.* 6, 28–37.

Leitinger, B., McDowall, A., Stanley, P., and Hogg, N. (2000). The regulation of integrin function by $Ca^{(2+)}$. *Biochim. Biophys. Acta* 1498, 91–98.

Li, J., Zhang, Y. P., and Kirsner, R. S. (2003). Angiogenesis in wound repair: angiogenic growth factors and the extracellular matrix. *Microsc. Res. Tech.* 60, 107–114.

Li, J., Kleeff, J., Kayed, H., Felix, K., Penzel, R., Buchler, M. W., Korc, M., and Friess, H. (2004). Glypican-1 antisense transfection modulates TGF-beta-dependent signaling in Colo-357 pancreatic cancer cells. *Biochem. Biophys. Res. Commun.* **320**, 1148–1155.

Lin, C. Q., Dempsey, P. J., Coffey, R. J., and Bissell, M. J. (1995). Extracellular matrix regulates whey acidic protein gene expression by suppression of TGF-alpha in mouse mammary epithelial cells: studies in culture and in transgenic mice. *J. Cell Biol.* **129**, 1115–1126.

Linge, C., Richardson, J., Vigor, C., Clayton, E., Hardas, B., and Rolfe, K. (2005). Hypertrophic scar cells fail to undergo a form of apoptosis specific to contractile collagen — the role of tissue transglutaminase. *J. Invest. Dermatol.* **125**, 72–82.

Lo, S. H. (2004). Tensin. *Int. J. Biochem. Cell Biol.* **36**, 31–34.

Lora, J. M., R. K., Soares, L., Giancotti, F., and Zaret, K. S. (1998). Alpha-3beta1-integrin as a critical mediator of the hepatic differentiation response to the extracellular matrix. *Hepatology* **28**, 1095–1104.

Lund, L. R., Romer, J., Thomasset, N., Solberg, H., Pyke, C., Bissell, M. J., Dano, K., and Werb, Z. (1996). Two distinct phases of apoptosis in mammary gland involution: proteinase-independent and -dependent pathways. *Development* **122**, 181–193.

Lutolf, M. P., and Hubbell, J. A. (2005). Synthetic biomaterials as instructive extracellular microenvironments for morphogenesis in tissue engineering. *Nat. Biotechnol.* **23**, 47–55.

Lygoe, K. A., Norman, J. T., Marshall, J. F., and Lewis, M. P. (2004). AlphaV integrins play an important role in myofibroblast differentiation. *Wound Repair Regen.* **12**, 461–470.

Ma, Z., Gao, C., Gong, Y., and Shen, J. (2005). Cartilage tissue engineering PLLA scaffold with surface-immobilized collagen and basic fibroblast growth factor. *Biomaterials* **26**, 1253–1259.

Maeshima, Y., Yerramalla, U. L., Dhanabal, M., Holthaus, K. A., Barbashov, S., Kharbanda, S., Reimer, C., Manfredi, M., Dickerson, W. M., and Kalluri, R. (2001). Extracellular matrix–derived peptide binds to alpha(v)beta(3) integrin and inhibits angiogenesis. *J. Biol. Chem.* **276**, 31959–31968.

Majack, R. A., Goodman, L. V., and Dixit, V. M. (1988). Cell surface thrombospondin is functionally essential for vascular smooth muscle cell proliferation. *J. Cell Biol.* **106**, 415–422.

Marhaba, R., and Zoller, M. (2004). CD44 in cancer progression: adhesion, migration and growth regulation. *J. Mol. Histol.* **35**, 211–231.

Martins-Green, M. (1987). Ultrastructural and immunolabeling studies of the neural crest: processes leading to neural crest cell emigration. PhD Dissertation, University of California, Davis, p. 194.

Martins-Green, M. (1988). Origin of the dorsal surface of the neural tube by progressive delamination of epidermal ectoderm and neuroepithelium: implications for neurulation and neural tube defects. *Development* **103**, 687–706.

Martins-Green, M. (2001). The chicken chemotactic and angiogenic factor (cCAF), a CXC chemokine. *Int. J. Biochem. Cell Biol.* **33**, 427–432.

Martins-Green, M., and Bissell, M. J. (1995). Cell–extracellular matrix interactions in development. *Sems. Dev. Biol.* **6**, 149–159.

Martins-Green, M., and Erickson, C. A. (1986). Development of neural tube basal lamina during neurulation and neural crest cell emigration in the trunk of the mouse embryo. *J. Embryol. Exp. Morphol.* **98**, 219–236.

Martins-Green, M., and Erickson, C. A. (1987). Basal lamina is not a barrier to neural crest cell emigration: documentation by TEM and by immunofluorescent and immunogold labeling. *Development* **101**, 517–533.

Martins-Green, M., and Feugate, J. E. (1998). The 9E3/CEF4 gene product is a chemotactic and angiogenic factor that can initiate the wound-healing cascade *in vivo*. *Cytokine*. **10**, 522–535.

Massia, S. P., Rao, S. S., and Hubbell, J. A. (1993). Covalently immobilized laminin peptide Tyr-Ile-Gly-Ser-Arg (YIGSR) supports cell spreading and colocalization of the 67-kilodalton laminin receptor with alpha-actinin and vinculin. *J. Biol. Chem.* **268**, 8053–8059.

McMahon, C. D., Farr, V. C., Singh, K., Wheeler, T. T., and Davis, S. R. (2004). Decreased expression of beta1-integrin and focal adhesion kinase in epithelial cells may initiate involution of mammary glands. *J. Cell Physiol.* **200**, 318–325.

Mercurius, K. O., and Morla, A. O. (2001). Cell adhesion and signaling on the fibronectin 1st type III repeat: requisite roles for cell surface proteoglycans and integrins. *BMC Cell Biol.* **2**, 18.

Miralem, T., Steinberg, R., Price, D., and Avraham, H. (2001). VEGF(165) requires extracellular matrix components to induce mitogenic effects and migratory response in breast cancer cells. *Oncogene* **20**, 5511–5524.

Mostafavi-Pour, Z., Askari, J. A., Whittard, J. D., and Humphries, M. J. (2001). Identification of a novel heparin-binding site in the alternatively spliced IIICS region of fibronectin: roles of integrins and proteoglycans in cell adhesion to fibronectin splice variants. *Matrix Biol.* **20**, 63–73.

Moulin, V., Larochelle, S., Langlois, C., Thibault, I., Lopez-Valle, C. A., and Roy, M. (2004). Normal skin wound and hypertrophic scar myofibroblasts have differential responses to apoptotic inductors. *J. Cell Physiol.* **198**, 350–358.

Muschler, J., Lochter, A., Roskelley, C. D., Yurchenco, P., and Bissell, M. J. (1999). Division of labor among the alpha6beta4 integrin, beta1 integrins, and an E3 laminin receptor to signal morphogenesis and beta-casein expression in mammary epithelial cells. *Mol. Biol. Cell* **10**, 2817–2828.

Nathan, C., and Sporn, M. (1991). Cytokines in context. *J. Cell Biol.* **113**, 981–986.

Naugle, J. E., Olson, E. R., Zhang, X., Mase, S. E., Pilati, C. F., Maron, M. B., Folkesson, H. G., Horne, W. I., Doane, K. J., and Meszaros, J. G. (2006). Type VI collagen induces cardiac myofibroblast differentiation: implications for postinfarction remodeling. *Am. J. Physiol. Heart Circ. Physiol.* **290**, H323–330.

Nelson, C. M., and Bissell, M. J. (2005). Modeling dynamic reciprocity: engineering three-dimensional culture models of breast architecture, function, and neoplastic transformation. *Semin. Cancer Biol.* **15**, 342–352.

Nicholson, L. J., and Watt, F. M. (1991). Decreased expression of fibronectin and the alpha 5 beta 1 integrin during terminal differentiation of human keratinocytes. *J. Cell Sci.* **98** (Pt 2), 225–232.

Panayotou, G., End, P., Aumailley, M., Timpl, R., and Engel, J. (1989). Domains of laminin with growth-factor activity. *Cell* **56**, 93–101.

Perris, R., and Perissinotto, D. (2000). Role of the extracellular matrix during neural crest cell migration. *Mech. Dev.* **95**, 3–21.

Petersen, O. W., Ronnov-Jessen, L., Howlett, A. R., and Bissell, M. J. (1992). Interaction with basement membrane serves to rapidly distinguish growth and differentiation pattern of normal and malignant human breast epithelial cells. *Proc. Natl. Acad. Sci. U.S.A.* **89**, 9064–9068.

Pilarski, L. M., Pruski, E., Wizniak, J., Paine, D., Seeberger, K., Mant, M. J., Brown, C. B., and Belch, A. R. (1999). Potential role for hyaluronan and the hyaluronan receptor RHAMM in mobilization and trafficking of hematopoietic progenitor cells. *Blood* **93**, 2918–2927.

Pollock, R. A., and Richardson, W. D. (1992). The alternative-splice isoforms of the PDGF A-chain differ in their ability to associate with the extracellular matrix and to bind heparin in vitro. *Growth Factors* **7**, 267–277.

Polunovsky, V. A., Chen, B., Henke, C., Snover, D., Wendt, C., Ingbar, D. H., and Bitterman, P. B. (1993). Role of mesenchymal cell death in lung remodeling after injury. *J. Clin. Invest.* **92**, 388–397.

Ponta, H., Sherman, L., and Herrlich, P. A. (2003). CD44: from adhesion molecules to signalling regulators. *Nat. Rev. Mol. Cell Biol.* **4**, 33–45.

Poole, T. J., and Thiery, J. P. (1986). Antibodies and a synthetic peptide that block cell-fibronectin adhesion arrest neural crest cell migration in vivo. *Prog. Clin. Biol. Res.* **217B**, 235–238.

Pouliot, N., Saunders, N. A., and Kaur, P. (2002). Laminin 10/11: an alternative adhesive ligand for epidermal keratinocytes with a functional role in promoting proliferation and migration. *Exp. Dermatol.* **11**, 387–397.

Pozzi, A., Wary, K. K., Giancotti, F. G., and Gardner, H. A. (1998). Integrin alpha1beta1 mediates a unique collagen-dependent proliferation pathway in vivo. *J. Cell Biol.* **142**, 587–594.

Pujades, C., Alon, R., Yauch, R. L., Masumoto, A., Burkly, L. C., Chen, C., Springer, T. A., Lobb, R. R., and Hemler, M. E. (1997). Defining extracellular integrin alpha-chain sites that affect cell adhesion and adhesion strengthening without altering soluble ligand binding. *Mol. Biol. Cell* **8**, 2647–2657.

Rana, B., Mischoulon, D., Xie, Y., Bucher, N. L., and Farmer, S. R. (1994). Cell–extracellular matrix interactions can regulate the switch between growth and differentiation in rat hepatocytes: reciprocal expression of C/EBP alpha and immediate-early growth response transcription factors. *Mol. Cell Biol.* **14**, 5858–5869.

Reis, R. C., Schuppan, D., Barreto, A. C., Bauer, M., Bork, J. P., Hassler, G., and Coelho-Sampaio, T. (2005). Endostatin competes with bFGF for binding to heparin-like glycosaminoglycans. *Biochem. Biophys. Res. Commun.* **333**, 976–983.

Rosso, F., Marino, G., Giordano, A., Barbarisi, M., Parmeggiani, D., and Barbarisi, A. (2005). Smart materials as scaffolds for tissue engineering. *J. Cell Physiol.* **203**, 465–470.

Ruhl, M., Sahin, E., Johannsen, M., Somasundaram, R., Manski, D., Riecken, E. O., and Schuppan, D. (1999). Soluble collagen VI drives serum-starved fibroblasts through S phase and prevents apoptosis via down-regulation of Bax. *J. Biol. Chem.* **274**, 34361–34368.

Rundhaug, J. E. (2005). Matrix metalloproteinases and angiogenesis. *J. Cell Mol. Med.* **9**, 267–285.

Runyan, R. B., Maxwell, G. D., and Shur, B. D. (1986). Evidence for a novel enzymatic mechanism of neural crest cell migration on extracellular glycoconjugate matrices. *J. Cell Biol.* **102**, 432–441.

Sage, E. H., Reed, M., Funk, S. E., Truong, T., Steadele, M., Puolakkainen, P., Maurice, D. H., and Bassuk, J. A. (2003). Cleavage of the matricellular protein SPARC by matrix metalloproteinase 3 produces polypeptides that influence angiogenesis. *J. Biol. Chem.* **278**, 37849–37857.

Sakakura, S., Saito, S., and Morikawa, H. (1999). Stimulation of DNA synthesis in trophoblasts and human umbilical vein endothelial cells by hepatocyte growth factor bound to extracellular matrix. *Placenta* **20**, 683–693.

Sakamoto, N., Iwahana, M., Tanaka, N. G., and Osada, Y. (1991). Inhibition of angiogenesis and tumor growth by a synthetic laminin peptide, CDPGYIGSR-NH2. *Cancer Res.* **51**, 903–906.

Salmivirta, M., and Jalkanen, M. (1995). Syndecan family of cell surface proteoglycans: developmentally regulated receptors for extracellular effector molecules. *Experientia* **51**, 863–872.

Santra, M., Reed, C. C., and Iozzo, R. V. (2002). Decorin binds to a narrow region of the epidermal growth factor (EGF) receptor, partially overlapping but distinct from the EGF-binding epitope. *J. Biol. Chem.* **277**, 35671–35681.

Savani, R. C., Cao, G., Pooler, P. M., Zaman, A., Zhou, Z., and DeLisser, H. M. (2001). Differential involvement of the hyaluronan (HA) receptors CD44 and receptor for HA-mediated motility in endothelial cell function and angiogenesis. *J. Biol. Chem.* **276**, 36770–36778.

Schenk, S., Hintermann, E., Bilban, M., Koshikawa, N., Hojilla, C., Khokha, R., and Quaranta, V. (2003). Binding to EGF receptor of a laminin-5 EGF-like fragment liberated during MMP-dependent mammary gland involution. *J. Cell Biol.* **161**, 197–209.

Schlaepfer, D. D., and Hunter, T. (1998). Integrin signaling and tyrosine phosphorylation: just the FAKs? *Trends Cell Biol.* **8**, 151–157.

Schonherr, E., and Hausser, H. J. (2000). Extracellular matrix and cytokines: a functional unit. *Dev. Immunol.* **7**, 89–101.

Schoop, V. M., Mirancea, N., and Fusenig, N. E. (1999). Epidermal organization and differentiation of HaCaT keratinocytes in organotypic coculture with human dermal fibroblasts. *J. Invest. Dermatol.* **112**, 343–353.

Simian, M., Hirai, Y., Navre, M., Werb, Z., Lochter, A., and Bissell, M. J. (2001). The interplay of matrix metalloproteinases, morphogens and growth factors is necessary for branching of mammary epithelial cells. *Development* **128**, 3117–3131.

Sottile, J. (2004). Regulation of angiogenesis by extracellular matrix. *Biochim. Biophys. Acta* **1654**, 13–22.

Spofford, C. M., and Chilian, W. M. (2003). Mechanotransduction via the elastin-laminin receptor (ELR) in resistance arteries. *J. Biomech.* **36**, 645–652.

Strange, R., Li, F., Saurer, S., Burkhardt, A., and Friis, R. R. (1992). Apoptotic cell death and tissue remodelling during mouse mammary gland involution. *Development* **115**, 49–58.

Stupack, D. G., and Cheresh, D. A. (2002). Get a ligand, get a life: integrins, signaling and cell survival. *J. Cell Sci.* **115**, 3729–3738.

Stupack, D. G., Puente, X. S., Boutsaboualoy, S., Storgard, C. M., and Cheresh, D. A. (2001). Apoptosis of adherent cells by recruitment of caspase-8 to unligated integrins. *J. Cell Biol.* **155**, 459–470.

Suki, B., Ito, S., Stamenovic, D., Lutchen, K. R., and Ingenito, E. P. (2005). Biomechanics of the lung parenchyma: critical roles of collagen and mechanical forces. *J. Appl. Physiol.* **98**, 1892–1899.

Sulochana, K. N., Fan, H., Jois, S., Subramanian, V., Sun, F., Kini, R. M., and Ge, R. (2005). Peptides derived from human decorin leucine-rich repeat 5 inhibit angiogenesis. *J. Biol. Chem.* **280**, 27935–27948.

Swindle, C. S., Tran, K. T., Johnson, T. D., Banerjee, P., Mayes, A. M., Griffith, L., and Wells, A. (2001). Epidermal growth factor (EGF) like repeats of human tenascin-C as ligands for EGF receptor. *J. Cell Biol.* **154**, 459–468.

Sympson, C. J., Talhouk, R. S., Alexander, C. M., Chin, J. R., Clift, S. M., Bissell, M. J., and Werb, Z. (1994). Targeted expression of stromelysin-1 in mammary gland provides evidence for a role of proteinases in branching morphogenesis and the requirement for an intact basement membrane for tissue-specific gene expression. *J. Cell Biol.* **125**, 681–693.

Talhouk, R. S., Bissell, M. J., and Werb, Z. (1992). Coordinated expression of extracellular matrix–degrading proteinases and their inhibitors regulates mammary epithelial function during involution. *J. Cell Biol.* **118**, 1271–1282.

Thannickal, V. J., Lee, D. Y., White, E. S., Cui, Z., Larios, J. M., Chacon, R., Horowitz, J. C., Day, R. M., and Thomas, P. E. (2003). Myofibroblast differentiation by transforming growth factor-beta1 is dependent on cell adhesion and integrin signaling via focal adhesion kinase. *J. Biol. Chem.* **278**, 12384–12389.

Tran, K. T., Griffith, L., and Wells, A. (2004). Extracellular matrix signaling through growth factor receptors during wound healing. *Wound Repair Regen.* **12**, 262–268.

Vaday, G. G., and Lider, O. (2000). Extracellular matrix moieties, cytokines, and enzymes: dynamic effects on immune cell behavior and inflammation. *J. Leukoc. Biol.* **67**, 149–159.

Vogel, W., Gish, G. D., Alves, F., and Pawson, T. (1997). The discoidin domain receptor tyrosine kinases are activated by collagen. *Mol. Cell.* **1**, 13–23.

Wang, J., Chen, H., Seth, A., and McCulloch, C. A. (2003). Mechanical force regulation of myofibroblast differentiation in cardiac fibroblasts. *Am. J. Physiol. Heart Circ. Physiol.* **285**, H1871–1881.

Wary, K. K., Mariotti, A., Zurzolo, C., and Giancotti, F. G. (1998). A requirement for caveolin-1 and associated kinase Fyn in integrin signaling and anchorage-dependent cell growth. *Cell* **94**, 625–634.

Watanabe, M., Fujioka-Kaneko, Y., Kobayashi, H., Kiniwa, M., Kuwano, M., and Basaki, Y. (2005). Involvement of integrin-linked kinase in capillary/tube-like network formation of human vascular endothelial cells. *Biol. Proced. Online* **7**, 41–47.

Weaver, V. M., Petersen, O. W., Wang, F., Larabell, C. A., Briand, P., Damsky, C., and Bissell, M. J. (1997). Reversion of the malignant phenotype of human breast cells in three-dimensional culture and *in vivo* by integrin blocking antibodies. *J. Cell Biol.* **137**, 231–245.

Yoneda, A., and Couchman, J. R. (2003). Regulation of cytoskeletal organization by syndecan transmembrane proteoglycans. *Matrix Biol.* **22**, 25–33.

Yost, H. J. (1992). Regulation of vertebrate left–right asymmetries by extracellular matrix. *Nature* **357**, 158–161.

Zheng, L., and Martins-Green, M. Molecular mechanisms of thrombin-induced Interleukin-8 expression in macrophages. (in press).

# Chapter Eight

# Matrix Molecules and Their Ligands

*Bjorn Reino Olsen*

## I. INTRODUCTION

Successful repair, regeneration or replacement of tissues and organs by tissue engineering requires insights into the processes, tested and refined during a billion years of evolution, by which cells form, maintain, and repair tissues. It is based on understanding what goes on inside cells as well as knowledge about what goes on between them; how they generate their extracellular matrix (ECM) environment; how they fill it with molecules that allow them be buffered against mechanical and chemical stress; how they use it to communicate with each other and to proliferate, differentiate, migrate, and survive within it. This chapter describes some of the major classes of molecules that allow the ECM to meet the needs of the cells within it. It describes polymer-forming proteins such as collagen, elastin, and fibrillin that allow cells to be organized in space and provide the basis for spatially defined interactions between cells. It discusses adhesive glycoproteins that bind to integrins and other cell surface receptors regulating attachment, shape, proliferation, and differentiation of cells. It further describes large proteoglycans that generate hydrophilic tissue compartments for both facilitating and blocking of cell migration. Finally, it provides examples of how matrix molecules, in addition to serving in structural roles, can regulate cell

behavior by stimulating and inhibiting growth factor activities or by releasing peptide fragments that act directly on cells.

Cellular growth and differentiation, in two-dimensional cell culture as well as in the three-dimensional space of the developing organism, requires the presence of a structured environment with which the cells can interact. This extracellular matrix (ECM) is composed of polymeric networks of several types of macromolecules in which smaller molecules, ions, and water are bound. The major types of macromolecules are polymer-forming proteins, such as collagens, elastin, fibrillins, fibronectin, and laminins, and hydrophilic heteropolysaccharides, such as glycosaminoglycan chains in hyaluronan and proteoglycans. It is the combination of protein polymers and hydrated proteoglycans that gives extracellular matrices their resistance to tensile and compressive mechanical forces.

The macromolecular components of the polymeric assemblies of the ECM are in many cases secreted by cells as precursor molecules that are significantly modified (proteolytically processed, oxidized, and cross-linked) before they assemble with other components into functional polymers (Fig. 8.1). The formation of matrix assemblies *in vivo* is therefore in most instances a unidirectional, irreversible process, and the disassembly of the matrix is not a simple reversal of assembly, but involves multiple, highly regulated processes. One consequence of this is that polymers reconstituted in the laboratory with components extracted from extracellular matrices do not have all the properties they have when assembled by cells *in vivo*. The ECM *in vivo* is also modified by cells as they proliferate, differentiate, and migrate, and cells in turn continuously interact with the matrix and communicate with each other through it (Hay, 1991).

The ECM is therefore not an inert product of secretory activities, but influences cellular shape, fate, and metabolism in ways that are as important to tissue and organ structure and function as the effects of many cytoplasmic processes. This realization has led to a reassessment of the need for a detailed molecular understanding of ECM. In the past, the ECM was appreciated primarily for its challenge to biochemists interested in protein and complex carbohydrate structure; a detailed characterization of ECM constituents is now considered essential for understanding cell behavior in the context of tissue and organ development and function. Some of these constituents are obviously most important for their structural properties (collagens and elastin), while others (fibronectin, fibrillin, laminin, thrombospondin, tenascin, perlecan, and other proteoglycans) are multidomain molecules that are both structural constituents as well as regulators of cell behavior (Fig. 8.1). In a third category are matrix-bound signaling molecules (matrix-bound FGFs, TGF-β, and BMPs).

## II. COLLAGENS – MAJOR CONSTITUENTS OF ECM

### Fibrillar Collagens Are Major Tissue Scaffold Proteins

Collagens constitute a large family of proteins that represent the major proteins (about 25%) in mammalian tissues (Kielty and Grant, 2002). A subfamily of these proteins, the fibrillar collagens, contains rigid, rodlike molecules with three subunits, α-chains, folded into a right-handed collagen triple helix. Within a fibrillar collagen triple helical domain, each α-chain consists of about 1000 amino acid residues and is coiled into an extended, left-handed polyproline II helix; three α-chains are in turn twisted into a right-handed superhelix (Fig. 8.2). The extended conformation of each α-chain does not allow the formation of intrachain hydrogen bonds; the stability of the triple helix is instead due to interchain hydrogen bonds. Such interchain bonds can form only if every third residue of each α-chain does not have a side chain and is packed close to the triple helical axis. Only glycine residues can therefore be accomodated in this position. This explains why the amino acid sequence of each α-chain in fibrillar collagens consists of about 300 Gly-X-Y tripeptide repeats, where X and Y can be any residue but Y is frequently proline or hydroxyproline. It

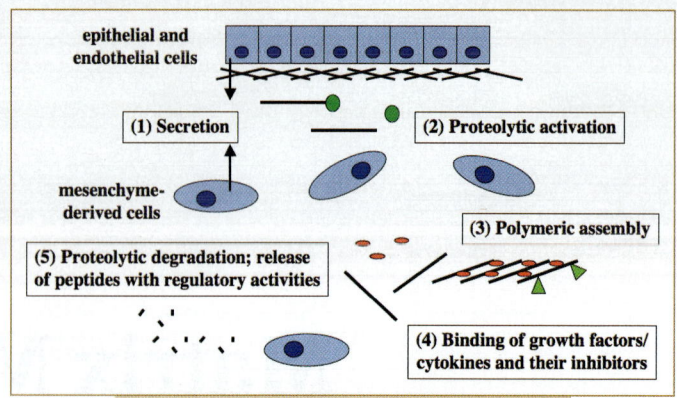

**FIG. 8.1.** The life cycle of extracellular matrix molecules. Soluble matrix molecules are secreted by cells, modified by proteolysis, and assembled into polymeric complexes. These complexes serve as scaffolds for cells and as binding sites for small molecules, such as growth and differentiation factors. Depending on the growth factor and cellular context, this may either inhibit or stimulate growth factor activity. Degradation of the scaffolds, during normal tissue turnover or during wound healing, may release bound growth factors and/or release peptide fragments from the larger scaffold proteins; such fragments may bind to cellular receptors and regulate cellular behavior.

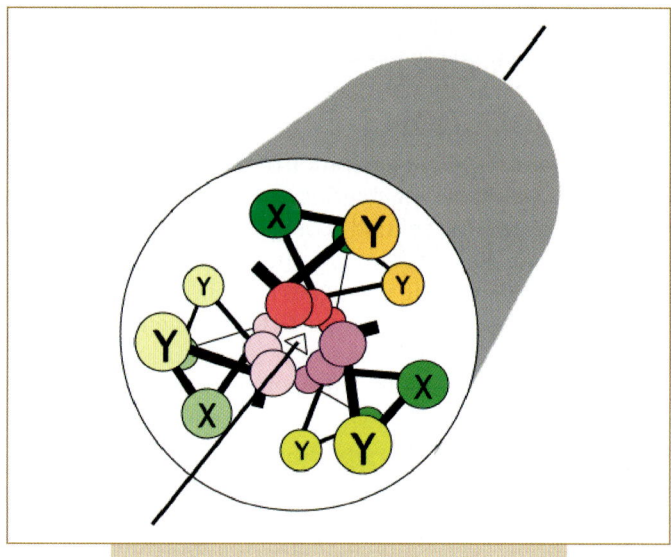

**FIG. 8.2.** Diagram showing a segment of a triple helical collagen molecule. The triple helix is composed of three left-handed helices (α-chains) that are twisted into a right-handed superhelix. The sequence of each α-chain is a repeat of the tripeptide Gly-X-Y. The Gly residues are packed close to the triple helical axis (indicated by a line through a triangle). Only glycine (without a side chain) can be accommodated in this position. Although any residue can fit into the X- and Y-positions, Pro is frequently found in the Y-position.

also provides an explanation for why mutations in collagens that lead to a replacement of triple helical glycine residues with more bulky residues can cause severe abnormalities.

Fibrillar collagen molecules are the major components of collagen fibrils. Their α-chains are synthesized as precur-

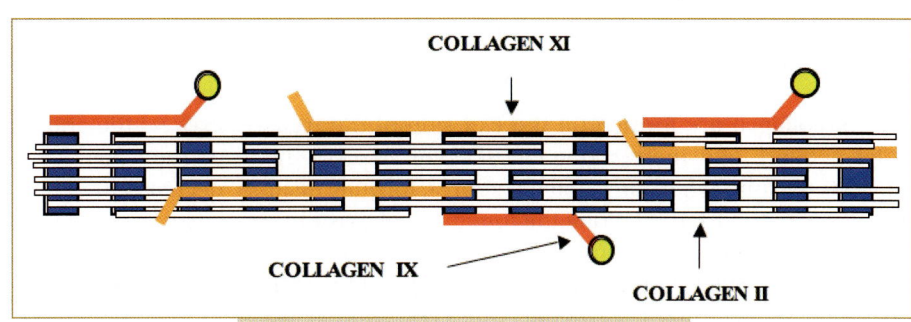

**FIG. 8.3.** Diagram of a cartilage collagen fibril. Collagen II molecules are the major components. Molecules of collagens XI and IX are located on the surface. Collagen XI molecules, heterotrimers of three different α-chains, have amino-terminal domains that are thought to sterically block the addition of collagen II molecules at the fibril surface.

sors, proα-chains, with large propeptide regions flanking the central triple helical domain. The carboxyl propeptide (C-propeptide) is important for the assembly of trimeric molecules in the RER. Formation of C-propeptide trimers, stabilized by intra- and interchain disulfide bonds, is the first step in the intracellular assembly and folding of trimeric procollagen molecules (McAlinden et al., 2003; Olsen, 1991). The folding of the triple helical domain at body temperature requires post translational hydroxylation of about 50% of the prolyl residues by prolyl hydroxylases and proceeds in a zipperlike fashion from the carboxyl toward the amino end of procollagen molecules. Mutations in fibrillar procollagens that affect the structure and folding of the C-propeptide domain are therefore likely to affect the participation of the mutated chains in triple helical assemblies. In contrast, mutations upstream of the C-propeptide, such as in-frame deletions or glycine substitutions in the triple helical domains, exert a dominant negative effect, in that the mutated chains will participate in trimer assembly but will interfere with subsequent folding of the triple helical domain.

Fibrillar procollagen chains are the products of 11 genes. The similarities between these genes suggest that they arose by multiple duplications from a single ancestral gene. Despite their similarities and the high degree of sequence identity between their protein products, they exhibit specificity in the interactions of their C-propeptides during intracellular trimeric assembly in the RER. Thus, a relatively small number of chain combinations are found among triple helical procollagens; these combinations represent fibrillar collagen types.

### Collagens V/XI — Regulators of Fibril Assembly, Spatial Organization, and Cell Differentiation

Some collagen types are heterotrimers (types I, V, and XI), while others are homotrimers (types II, III, XXIV, and XXVII). Some chains participate in more than one type: For example, the α1(II) chain (encoded by the COL2A1 gene) forms the homotrimeric collagen II but is also one of three different chains in collagen XI molecules. Between collagens V and XI there is extensive sharing of polypeptide subunits, and fibrillar collagen molecules previously described as belonging to either collagen V or XI are now referred to as

belonging to the V/XI type. Thus, fibrillar procollagen molecules secreted by cells are members of a group of homologous proteins. They all contain a C-propeptide that is completely removed by an endoproteinase after secretion, and their triple helical rodlike domains polymerize in a staggered fashion into fibrillar arrays (Fig. 8.3). They differ, however, in the structure of their amino propeptide (N-propeptide) domains and in the extent to which this domain is proteolytically removed. For some collagen types, such as collagens I and II, the N-propeptide processing is complete in molecules within mature fibrils. For other types, such as collagens V/XI, this is not the case, in that a large portion of the N-propeptides in these molecules remain attached to the triple helical domain (Fig. 8.3). The incomplete processing of type V/XI molecules allows them to serve as regulators of fibril assembly (Linsenmayer et al., 1993). Collagen fibrils are heterotypic, i.e., contain more than one collagen type, such that collagen I fibrils in skin, tendon, ligaments, and bone contain 5–10% collagen V, and collagen II fibrils in cartilage contain 5–10% collagen XI. The presence of N-propeptide domains on V/XI molecules represents a steric hindrance to addition of molecules at fibril surfaces. This heterotypic/steric hindrance model predicts that collagen fibril diameters in a tissue are determined by the ratio of the minor component (V/XI) to the major component (I or II). A high ratio results in thin fibrils; a low ratio results in thick fibrils. Direct support for this comes from studies of mutant and transgenic mice. For example, mice that are homozygous for a functional null mutation in α1(XI) collagen and transgenic mice overexpressing collagen II have cartilage collagen fibrils that are abnormally thick (Garofalo et al., 1993; Li et al., 1995).

A characteristic feature of collagen fibrillar scaffolds is their precise three-dimensional patterns. These patterns follow mechanical stress lines and ensure a maximum of tensile strength with a minimum of material. Examples are the crisscrossing lamellae of collagen fibers in lamellar bone or in cornea, the arcades of collagen fibrils under the surface of articular cartilage, and the parallel-fiber bundles in tendons and ligaments. Ultimately, cells are responsible for establishing these patterns, but the cellular mechanisms involved are only beginning to be understood. A study by Canty et al. (2004) suggests that the orientation of collagen fibrils in the extracellular space is linked to the cytoskeletal organization induced by cellular responses to mechanical stress. In tendon fibroblasts, Golgi-to-plasma transport carriers of collagen are formed on the exit side of the trans-Golgi network and move along cytoskeletal "tracks" into long cytoplasmic extensions. Collagen fibrils, forming inside the carriers, are oriented along the longitudinal axis of the carriers. When the membrane at the distal tip of the carrier fuses with the cell membrane covering the tip of the extension, the space within the carrier becomes continuous with the extracellular space and the fibrils are, in effect, moved from an intracellular to an extracellular compartment. Thus, the parallel orientation of collagen fibrils in a tendon is a consequence of the polarized structure, intracellular movement, and polarized exocytosis of fibril-containing Golgi-derived transport carriers. This cellular mechanism for orientation of collagen fibrils is consistent with data showing that the same kind of heterotypic fibril can be part of scaffolds with very different spatial organization. Transgenic mice with an alteration in the N-propeptide region of collagen V molecules show a disruption of the lamellar arrangement of fibrils in the cornea of the eye, suggesting a role for fibril surface domains in generating and/or stabilizing the spatial pattern (Andrikopoulos *et al.*, 1995). Finally, members of a unique subfamily of collagens, FACIT collagens (Olsen *et al.*, 1995), are good candidates for molecules that modulate the surface properties of fibrils and allow tissue-specific fibril patterns to be generated and stabilized by cells.

The phenotypic consequences of mutations in fibrillar collagen genes indicate that a major function of these proteins is to provide elements of high tensile strength at the tissue level. Thus, mutations in COL1A1 or COL1A2, the human genes encoding the α1 and α2 subunits of fibrillar collagen I (in bone, ligaments, tendons, and skin), cause osteogenesis imperfecta (brittle bone disease) or clinical forms of Ehlers–Danlos syndrome, characterized by skin hyperextensibility and fragility and joint hypermobility, with or without bone abnormalities (Byers and Cole, 2002; Steinmann *et al.*, 2002). Mutations in COL2A1, the gene encoding the α-chains of collagen II (in cartilage), cause a spectrum of human disorders, ranging from lethal deficiency in cartilage formation to relatively mild deficiencies in cartilage mechanical properties and function (Horton and Hecht, 2002). Fibrillar collagens also have regulatory functions. For example, mutations in collagen V/XI genes suggest that fibrillar collagen scaffolds are essential for normal cellular growth and differentiation. A functional null mutation in α1(XI) collagen resulting in complete lack of collagen XI in cartilage causes a severe disproportionate dwarfism in mice and perinatal death of homozygotes (Li *et al.*, 1995). Histology of mutant long-bone growth plates reveals a disorganized spatial distribution of cells and a defect in chondrocyte differentiation to hypertrophy. The explanation for this is likely related to the fact that proliferation and differentiation of chondrocytes in growth plates are regulated by locally produced growth factors and cytokines. Cells that produce these factors are localized close to cells that express the appropriate receptors. Lack of collagen XI may disrupt this relationship, since it results in a dramatic decrease in cohesive properties of the matrix and a loss of cellular organization. Transgenic mice with a mutation in α2(V) collagen have a large number of hair follicles of unusual localization in the hypodermis; this may be related to a defect in the mechanical properties of the fibrillar collagen scaffold but could also be mediated by an effect on extracellular signaling molecules (Andrikopoulos *et al.*, 1995).

## FACIT Collagens — Modulators of Collagen Fibril Surface Properties

Molecules that are associated with collagen fibrils, contain two or more triple helical domains, and share characteristic protein domains (modules) are classified as FACIT collagens (Olsen *et al.*, 1995; Shaw and Olsen, 1991). Of the eight known members in the group (collagens IX, XII, XIV, XVI, XIX, XX, XXI, and XXII), collagen IX is the best characterized both structurally and functionally. Collagen IX molecules are heterotrimers of three different gene products (van der Rest *et al.*, 1985). Each of the three α-chains contains three triple helical domains separated and flanked by non–triple helical sequence regions (Fig. 8.4). Between the amino-terminal and central triple helical domains, a flexible hinge gives the molecule a kinked structure with two arms. Type IX molecules are located on the surface of type II/XI containing fibrils with the long arm parallel to the fibril surface and the short arm projecting into the perifibrillar space (Vaughan *et al.*, 1988) (Fig. 8.3). Collagen IX functions as a bridging molecule between fibrils, between fibrils and other matrix constituents (Pihlajamaa *et al.*, 2004), and between fibrils and cells (Kapyla *et al.*, 2004). Transgenic mice with a dominant-negative mutation in the α1(IX)-chain (Nakata *et al.*, 1993), as well as mice that are homozygous for null alleles of the gene (Col9a1) coding for α1(IX) (Faessler *et al.*, 1994), exhibit osteoarthritis in knee joints and mild chondrodysplasia. In humans, mutations in the α1(IX), α2(IX), or α3(IX) collagen chains cause a form of multiple epiphyseal dysplasia, an autosomal dominant disorder characterized by early-onset osteoarthritis in large joints associated with short stature and stubby fingers (Jakkula *et al.*, 2005; Muragaki *et al.*, 1996).

Molecules of collagens XII and XIV are homotrimers of chains that are made up of several kinds of modules. Some modules are homologous to modules found in collagen IX, while others show homology to von Willebrand factor A domains and fibronectin type 3 repeats. Both types of molecules contain a central globule with three fingerlike extensions and a thin triple helical tail attached (Fig. 8.4). For collagen XII, two forms that differ greatly in the lengths of

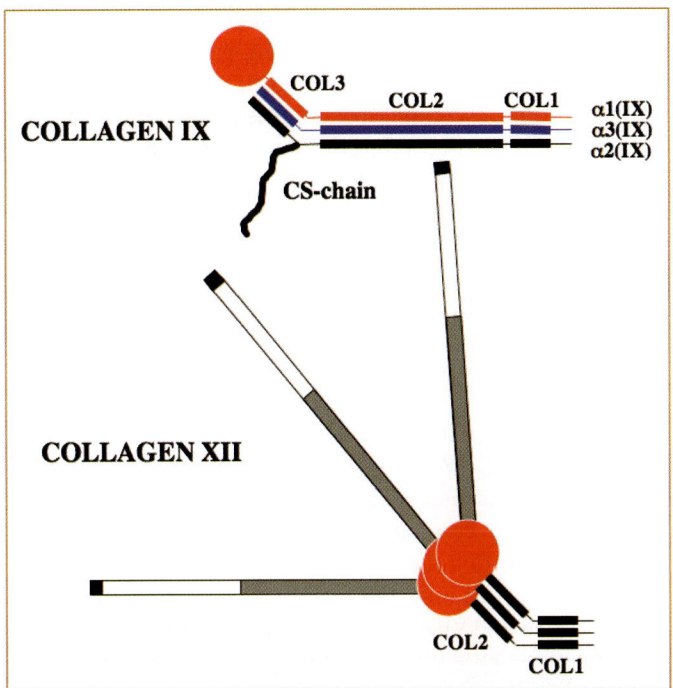

**FIG. 8.4.** Diagrams of collagen IX and XII (long-form) molecules. Collagen IX molecules contain the three chains α1(IX), α2(IX), and α3(IX). Each chain contains three triple helical domains (COL1, COL2, COL3), interrupted and flanked by non–triple helical sequences. In cartilage, the α1(IX)-chain contains a large globular amino-terminal domain. The α2(IX)-chain serves as a proteoglycan core protein, in that it contains a chondroitin sulfate (CS-) side chain attached to the non–triple helical region between the COL2 and COL3 domains. Collagen XII molecules are homotrimers of α1(XII)-chains. The three chains form two short triple helical domains separated by a flexible hinge region. A central globule is composed of three globular domains that are homologous to the amino-terminal globular domain of α1(IX) collagen chains. The amino-terminal region of the three α1(XII)-chains contain multiple fibronectin type 3 repeats and von Willebrand factor A–like domains. These regions form three "fingers" that extend from the central globule. Through alternative splicing a portion of the "fingers" (white region in the diagram) is spliced out in the short form of collagen XII. Hybrid molecules with both long and short "fingers" can be extracted from tissues.

the fingerlike extensions are generated by alternative splicing of RNA transcripts. Variations in the carboxyl regions also occur (Olsen *et al.*, 1995). Both collagens XII and XIV are found in connective tissues containing type I collagen fibrils, except mineralized bone matrix, and immunolabeling studies show a fibril-associated distribution. Type XIV collagen can bind to heparin sulfate and the small fibril-associated proteoglycan decorin (Brown *et al.*, 1993; Font *et al.*, 1993). This would suggest an indirect fibril association. A direct association is also possible, since collagen XII molecules form copolymers with collagen I even in the absence of proteoglycans. A functional interaction between fibrils and collagens XII and XIV is implied by studies showing that addition of the two collagens to type I collagen gels promote gel contraction mediated by fibroblasts

(Nishiyama *et al.*, 1994). The effect is dose dependent and can be prevented by denaturation or addition of specific antisera. The association of collagens XII and XIV with fibrils may therefore modulate the frictional properties of fibril surfaces. The synthesis of different isoforms could be important in this context, since they could bind to fibrils with different affinities. Also, since the long form of collagen XII is a proteoglycan, whereas the short form is not, variations in the relative proportion of the two splice variants may serve to modulate the hydrophilic properties of interfibrillar matrix compartments. Finally, the discovery that the collagen I N-propeptide–processing enzyme (see later) binds to collagen XIV and can be purified as part of a complex with antibodies against collagen XIV suggests that the FACIT collagens provide binding sites for fibril-modifying extracellular matrix enzymes (Colige *et al.*, 1995).

## Basement Membrane Collagens and Associated Collagen Molecules

At epithelial (and endothelial)–stromal boundaries, basement membranes serve as specialized areas of ECM for cell attachment. Collagen IV molecules form a networklike scaffold in basement membranes by end-to-end and lateral interactions (Yurchenco *et al.*, 2004). Six different collagen IV genes exist in mammals, and their products interact to form at least three different types of heterotrimeric collagen IV molecules. These different isoforms show characteristic tissue-specific expression patterns. The physiological importance of collagen IV isoforms is highlighted by Alport syndrome (Tryggvason and Martin, 2002). This disease, characterized by progressive hereditary nephritis associated with sensorineural hearing loss and ocular lesions, can be caused by mutations within α3(IV) and α4(IV) collagen genes (autosomal Alport syndrome) or mutations in α5(IV) collagen (X-linked Alport syndrome). In cases of large deletions including both the α5(IV) and the neighboring α6(IV) collagen genes, renal disease is associated with inherited smooth muscle tumors.

Within basement membranes, the collagen IV networks are associated with a large number of noncollagenous molecules, such as various isoforms of laminin, nidogen, and the heparin sulfate proteoglycan perlecan (Fig. 8.5). Additional collagens are also associated with basement membranes. These include the transmembrane collagen XVII in hemidesmosomes and collagen VII in anchoring fibrils. Collagens XVII and VII are important in regions of significant mechanical stress, such as skin, in that they anchor epithelial cells to the basement membrane (collagen XVII) and strap the basement membrane to the underlying stroma (collagen VII) (Fig. 8.6). In bullous pemphigoid, autoantibodies against collagen XVII cause blisters that separate epidermis from the basement membrane; dominant and recessive forms of epidermolysis bullosa can be caused by mutations in collagens VII and XVII (Franzke *et al.*, 2003; Uitto and Richard, 2005).

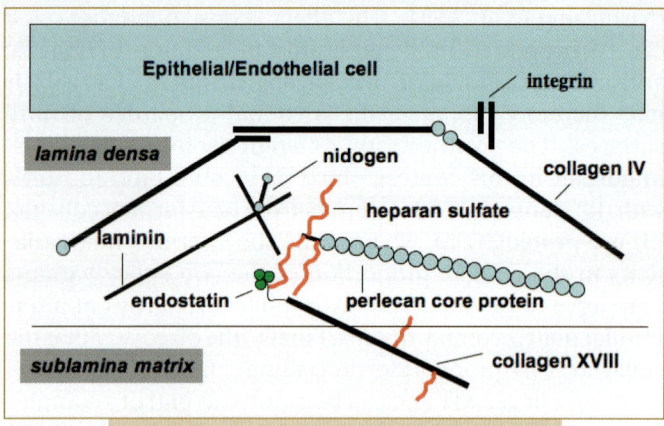

**FIG. 8.5.** Components of basement membranes. Basement membranes contain interconnected networks of collagen IV and laminin polymers, together with nidogen, perlecan, and collagen XVIII. Collagen XVIII molecules are located at the boundary between the lamina densa and the sublamina matrix, with their carboxyl endostatin domain within the lamina densa and the amino end projecting into the underlying matrix.

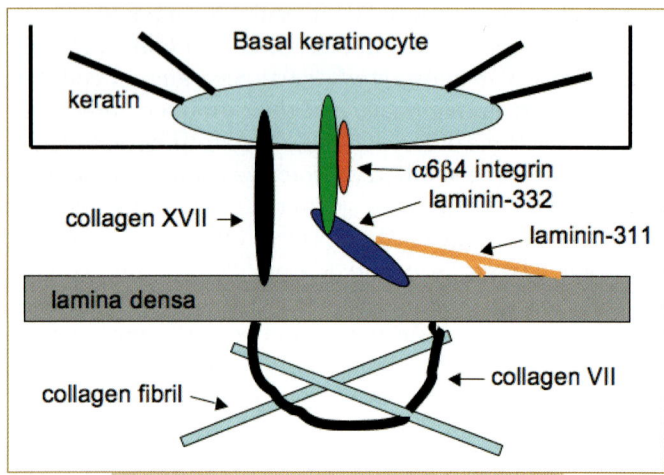

**FIG. 8.6.** Epidermal basement membrane and associated collagens and laminins. Basal portion of a keratinocyte with hemidesmosome, anchoring filaments of collagen XVII and anchoring fibril of collagen VII. A complex of laminin-332, laminin-311, and integrin α6β4 provides further strength to the cell–basement membrane junction.

Two additional basement membrane–associated collagens, collagen VIII and collagen XVIII, are of interest because of their function in vascular physiology and pathology. Collagen VIII is a short-chain, nonfibrillar collagen with significant homology to collagen X, a product of hypertrophic chondrocytes in long-bone growth plates and cartilage growth regions (synchondroses) at the skull base. Collagen VIII expression is up-regulated during heart development (Iruela-Arispe and Sage, 1991), in human atherosclerotic lesions (MacBeath *et al.*, 1996), and following experimental damage to the endothelium in large arteries (Bendeck *et al.*, 1996). Collagen VIII may be important in facilitating the

migration of smooth muscle cells from the medial layer into the intima during neointimal thickening following endothelial cell injury. Collagen VIII molecules are also major building blocks of Descemet's membrane, the thick basement membrane that bridges the corneal stroma with the corneal endothelium on the inside of the cornea (Hopfer *et al.*, 2005). Mutations in collagen VIII can cause clouding of the cornea and blurred vision (corneal dystrophy) in humans (Biswas *et al.*, 2001).

Collagen XVIII, together with collagen XV, belongs to a distinct subfamily of collagens called *multiplexins* because of their multiple triple-helix domains and interruptions (Oh *et al.*, 1994; Rehn and Pihlajaniemi, 1994). Because of the alternative utilization of two promoters and alternative splicing, the COL18A1 gene gives rise to three different transcripts that are translated into three protein variants. These are localized in various basement membranes (Fig. 8.5), including those that separate vascular endothelial cells from the underlying intima in blood vessels (Marneros and Olsen, 2001). Collagen XVIII α-chains contain several consensus sequences for attachment of heparan sulfate side chains, and studies have, in fact, confirmed that collagen XVIII forms the core protein of a basement membrane proteoglycan (Halfter *et al.*, 1998). Proteolytic processing of the carboxyl non–triple helical domain of collagen XVIII in tissues leads to the release of a heparin binding fragment with anti-angiogenic activity.

This fragment, named *endostatin*, represents the carboxyl-terminal 20-kDa portion of collagen XVIII chains (Fig. 8.5) (O'Reilly *et al.*, 1997). Endostatin has been shown to inhibit the proliferation and migration of vascular endothelial cells, inhibit the growth of tumors in mice and rats, and cause regression of tumors in mice (Marneros and Olsen, 2001). The antitumor effects are mediated by inhibition of tumor-induced angiogenesis. The x-ray crystallographic structure of mouse and human endostatin proteins (Ding *et al.*, 1998; Hohenester *et al.*, 1998) shows a compact structure consisting of two α-helices and a number of β-strands, stabilized by two intramolecular disulfide bonds. A coordinated zinc atom is part of the structure, and on the surface a patch of basic residues forms a binding site for heparin. Studies of mutant endostatins have shown that specific arginines within this patch are required for heparin binding (Yamaguchi *et al.*, 1999).

The physiological function of collagen XVIII is highlighted by the consequences of loss-of-function mutations in this basement membrane component (Marneros and Olsen, 2005). In humans, collagen XVIII mutations cause Knobloch syndrome, a recessive eye disorder in which affected individuals lose their eyesight at an early age because of degeneration of the retina and the vitreous. Mice with inactivated collagen XVIII genes exhibit age-dependent changes in the retina and the pigment epithelial layer behind the retina. These changes are similar to what is seen in age-dependent macular degeneration in humans and lead (as in humans) to gradual loss of eyesight (Marneros and Olsen, 2005).

Of considerable interest is the finding that proteolytic fragments of basement membrane components other than collagen XVIII also have antiangiogenic properties (Bix and Iozzo, 2005). Molecules that give rise to such fragments include collagen IV and perlecan, the major heparan sulfate proteoglycan in basement membranes. The fragments involved show no sequence homology with endostatin, so it is likely that the molecular mechanisms underlying the antiangiogenic effects are different. For example, endostatin is a heparin-binding molecule, and its ability to inhibit angiogenesis is in many contexts heparan sulfate dependent. In contrast, the fragment from perlecan, called *endorepellin*, does not bind to heparin, and its inhibitory effects on vascular endothelial cells is heparan sulfate independent. In any case, release of such fragments as vascular basement membranes are degraded at sites of sprouting angiogenesis are likely to provide a local mechanism of negative control to balance the effects of proangiogenic factors.

## III. ELASTIC FIBERS AND MICROFIBRILS

Collagen molecules and fibers evolved as structures of high tensile strength, equivalent to that of steel when compared on the basis of the same cross-sectional area but three times lighter on a per-unit weight basis. In contrast, elastic fibers, composed of molecules of elastin, provide tissues with elasticity so that they can recoil after transient stretch (Rosenbloom *et al.*, 1993; von der Mark and Sorokin, 2002). In organs such as the large arteries, skin, and lungs, elasticity is obviously crucial for normal functioning. Elastin fibers derive their impressive elastic properties, an extensibility that is about five times that of a rubber band with the same cross-sectional area, from the structure of elastin molecules. Each molecule is composed of alternating segments of hydrophobic and $\alpha$-helical Ala- and Lys-rich sequences. Oxidation of the lysine side chains by the enzyme lysyl oxidase leads to formation of reactive aldehydes and extensive covalent cross-links between neighboring molecules in the fiber. It is thought that the elasticity of the fiber is due to the tendency of the hydrophobic segments to adopt a random-coil configuration following stretch.

On the surface of elastic fibers one finds a cover of microfibrils, beaded filaments with molecules of fibrillin as their major components (Corson *et al.*, 2004; Sakai and Keene, 1994). The fibrillins, products of genes on chromosomes 5 (FIB5), 15 (FIB15), and 19 (FIB19) in humans, also form microfibrils that are found in almost all extracellular matrices in the absence of elastin. Fibrillin molecules are composed of multiple repeat domains, the most prominent being calcium-binding EGF-like repeats; similar repeats in latent TGF-$\beta$-binding proteins suggest that the fibrillins belong to a superfamily of proteins. The physiological importance of fibrillin is highlighted by mutations causing the Marfan syndrome and congenital contractural arachnodactyly in humans (Pyeritz and Dietz, 2002). The Marfan syndrome is caused by mutations in FIB15 and is characterized by dislocation of the eye lens due to weakening of the suspensory ligaments of the zonule, congestive heart failure, aortic aneurysms, and skeletal growth abnormalities resulting in a tall frame, scoliosis, chest deformities, arachnodactyly, and hypermobile joints. In patients with congenital contractural arachnodactyly, mutations in FIB5 lead to similar skeletal abnormalities and severe contractures but no ophthalmic and cardiovascular manifestations. The tall stature and arachnodactyly seen in patients with the Marfan syndrome suggest that FIB15 is a negative regulator of longitudinal bone growth. Since fibrillin microfibrils are found in growth plate cartilage, it is conceivable that they affect chondrocyte proliferation and/or maturation.

A regulatory role for fibrillin in growth plates would be consistent with the function of fibrillin in other tissues (Isogai *et al.*, 2003). In lung tissue and blood vessel walls, fibrillin functions as a regulator of TGF-$\beta$ activity. Fibrillin mutations in mouse models of Marfan syndrome are associated with increased TGF-$\beta$ activity in lungs and the aorta, causing impaired alveolar septation in lungs and widening and weakening (aneurysms) of the aorta. Inhibition of TGF-$\beta$ largely prevents these defects (Habashi *et al.*, 2006; Neptune *et al.*, 2003). Some of the major clinical abnormalities in patients with Marfan syndrome are therefore likely a consequence of altered fibrillin-mediated control of TGF-$\beta$ activity and not loss of fibrillin as a structural molecule. The current data suggest that drugs to inhibit TGF-$\beta$ activity may prevent early death caused by aortic aneurysms in Marfan syndrome patients. Clinical trials are under way to test this hypothesis. If successful, this would represent an exciting example of how a genetic disease may be effectively treated by pharmacological modulation of pathogenetic consequences of the mutation, without correcting the mutation.

## IV. OTHER MULTIFUNCTIONAL PROTEINS IN ECM

Several proteins in the extracellular matrix contain binding sites for structural macromolecules and for cells, thus contributing to both the structural organization of ECM and its interaction with cells (von der Mark and Sorokin, 2002). The prototype of these adhesive proteins is fibronectin.

### Fibronectin Is a Multidomain, Multifunctional Adhesive Glycoprotein

Fibronectin is a disulfide-bonded dimer of 220- to 250-kDa subunits (Hynes, 1990). Each subunit is folded into rodlike domains separated by flexible "joints." The domains are composed of three types of multiple repeats or modules, Fn1, Fn2, and Fn3. Fn1 modules are found in the fibrin-

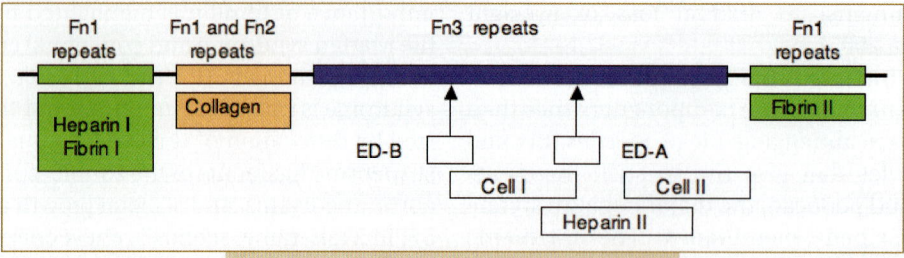

**FIG. 8.7.** Diagram of a fibronectin polypeptide chain. The polypeptide chain is composed of several repeats (Fn1, Fn2, and Fn3) and contains binding sites for several matrix molecules and cells. Two regions can bind heparin and fibrin, and two regions are involved in cell binding as well. By alternative splicing, isoforms are generated that may or may not contain certain Fn3 domains (labeled ED-A and ED-B in the diagram). Additional splice variations in the second cell-binding domain (Cell II) generate other isoforms.

binding amino- and carboxyl-terminal regions of fibronectin and in a collagen (gelatin)-binding region (Fig. 8.7). Single copies of Fn1 modules are also found in other proteins, such as tissue-type plasminogen activator (t-PA) and coagulation factor XII (Potts and Campbell, 1994).

NMR studies of Fn1 modules demonstrate the presence of two layers of antiparallel β-sheets (two strands in one layer and three strands in the other) held together by hydrophobic interactions. The structure is further stabilized by disulfide and salt bridges. Fn2 modules are found together with Fn1 modules in the collagen-binding region of fibronectin and in many other proteins. Their structure, two double-stranded antiparallel β-sheets connected by loops, suggests that a ligand such as collagen may bind to this module through interactions of hydrophobic amino acid side chains with its hydrophobic surface. Fn3 modules are the major structural units in fibronectin and are found in a large number of other proteins as well. Some of these proteins (for example, the long-splice variant of collagen XII) contain more Fn3 modules than fibronectin itself. The structure of Fn3 is that of a sandwich of antiparallel β-sheets (three strands in one layer and four strands in the other) with a hydrophobic core. The binding of fibronectin to some integrins involves the tripeptide sequence Arg-Gly-Asp in the 10th Fn3 module; these residues lie in an exposed loop between two of the strands in one of the β-sheets (Potts and Campbell, 1994).

Fibronectin can assemble into a fibrous network in the ECM through interactions involving cell surface receptors and the amino-terminal region of fibronectin (Mosher *et al.*, 1991). A fibrin-binding site is also contained in this region; a second site is in the carboxyl domain. The ability to bind to collagen ensures association between the fibronectin network and the scaffold of collagen fibrils. Binding sites for heparin and chondroitin sulfate further make fibronectin an important bridging molecule between collagens and other matrix molecules (Fig. 8.7).

Transcripts of the fibronectin gene are alternatively spliced in a cell- and developmental stage–dependent manner. As a result there are many different isoforms of fibronectin (Schwarzbauer, 1990). The main form produced by the liver and circulating in plasma lacks two of the Fn3 repeats found in cell- and matrix-associated fibronectin. One alternatively spliced domain is adjacent to the heparin-binding site, and this region binds to integrins α4β1 and α4β7. Thus, there is a mechanism for fine-tuning of fibronectin structure and interaction properties. Not surprisingly, mice that are homozygous for fibronectin null alleles die early in embryogenesis with multiple defects (George *et al.*, 1993).

The biologically most important activity of fibronectin is its interaction with cells. The ability of fibronectin to serve as a substrate for cell adhesion, spreading, and migration is based on the activities of several modules. The Arg-Gly-Asp sequence in the 10th Fn3 module plays a key role in the interaction with the integrin receptor α5β1, but synergistic interactions with other Fn3 modules are essential for high-affinity binding of cells to fibronectin (Aota *et al.*, 1991).

## Laminins Are Major Components of Basement Membranes

Laminins are trimeric basement membrane molecules of α-, β-, and γ-chains (Timpl and Brown, 1994; Yurchenco *et al.*, 2004). With a large number of genetically distinct chains, more than 15 different trimeric isoforms are known from mice and humans. A recently proposed nomenclature introduced a systematic approach to naming the different trimers; they are now named on the basis of their chain composition (i.e., α1β1γ1) or by numbers, only without the greek letters (i.e., 111 instead of α1β1γ1) (Aumailley *et al.*, 2005). Several forms have a cross-shaped structure as visualized by rotary shadowing electron microscopy; some forms contain T-shaped molecules (Fig. 8.8).

In basement membranes, laminins provide interaction sites for many other constituents, including cell surface receptors (Timpl, 1996). The functional and structural mapping of these sites and the complete sequencing of many laminin chains has provided detailed insights into the organization of laminin molecules. Within the cross-shaped laminin-111 molecule, three similar short arms are formed by the N-terminal regions of the α1-, β1-, and γ1-chains, whereas a long arm is composed of the carboxyl regions of

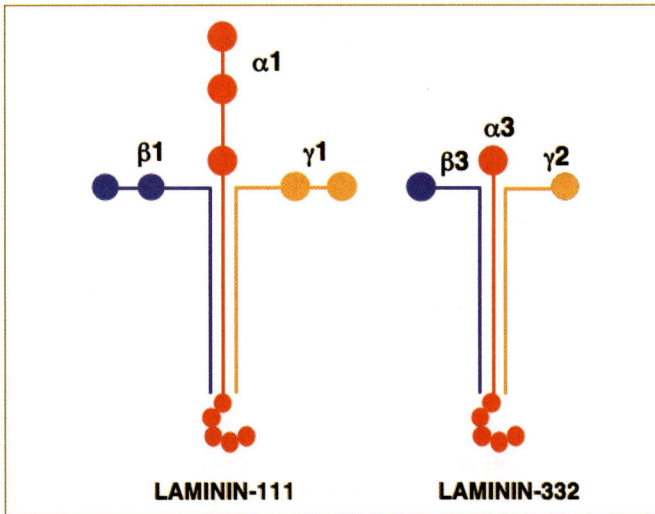

**FIG. 8.8.** Diagrams of two types of laminins. Laminin-111 has a cross-shaped structure; laminin-332 is T-shaped, due to a shorter α-chain.

all three chains (Fig. 8.8). The three chains are connected at the center of the cross by interchain disulfide bridges. The short arms contain multiple EGF-like repeats of about 60 amino acid residues, terminated and interrupted by globular domains. The long arm consists of heptad repeats covering about 600 residues in all three chains folded into a coiled-coil structure. The α1-chain is about 1000 amino acid residues longer than the β1- and γ1-chains and forms five homologous globular repeats at the base of the cross; these globular repeats are similar to repeats found in the proteoglycan molecule perlecan, also a component of basement membranes (Fig. 8.5) (Olsen, 1999).

Calcium-dependent polymerization of laminin is based on interactions between the globular domains at the N-termini and is thought to be important for the assembly and organization of basement membranes. Of significance for the assembly of basement membranes is also the high-affinity interaction with nidogen (Yurchenco et al., 2004). The binding site in laminin for nidogen is on the γ1-chain, close to the center of the cross (Fig. 8.5). On nidogen, a rodlike molecule with three globular domains, the binding site for laminin is in the carboxyl globular domain, while another globular domain binds to collagen IV. Thus, nidogen is a bridging molecule that connects the laminin and collagen IV networks and is important for the assembly of normal basement membranes.

Laminin does not bind directly to collagen IV, but has binding sites for several other molecules besides nidogen. These are heparin, perlecan, and fibulin-1, which bind to the end of the long arm of the laminin cross. However, the biologically most significant interactions of laminin involve a variety of both integrin and nonintegrin cell surface receptors. Several integrins are laminin receptors. They show distinct preferences for different laminins and recognize

binding sites on either the short or long arms of laminin molecules.

The different laminin genes likely arose through duplication of a single ancestral gene (Miner and Yurchenco, 2004). The laminins most closely related to this ancestral gene, laminin-111 and laminin-511, are crucial for early steps in development, including gastrulation, placentation, and neural tube closure. In contrast, laminins that have evolved more recently are adapted to more specialized functions in the development and function of specific organs. For example, laminin α2β1γ1 is a major laminin in the basement membranes surrounding skeletal muscle fibers, where it provides binding sites for the dystroglycan–dystrophin complex, linking the muscle cell cytoskeleton to the basement membrane. In skin, a disulfide-linked complex of laminins α3β3γ2 and α3β1γ1 is crucial for the firm attachment of keratinocytes to the basement membrane by its interaction with α6β4 integrins in hemidesmosomes (Fig. 8.6). Mutations in any one of three genes encoding the subunits of laminin-332 cause autosomal recessive junctional epidermolysis bullosa, a lethal skin-blistering disorder in which the epidermal cell layers are separated from the underlying epidermal basement membrane. Loss-of-function mutations in the laminin α2 gene cause congenital muscular dystrophy both in mice and in humans.

Mice with targeted disruption of the laminin α3 gene develop a blistering skin disease similar to the disorder in human patients. In addition, the kidneys of the mutant animals show arrested development of glomeruli, with a failure to develop glomerular capillaries with fenestrated endothelial cells and lack of migration of mesangial cells into the glomeruli (Abrass et al., 2006). In humans, mutations resulting in laminin β2 deficiency cause a syndrome of loss of albumin and other plasma proteins through the glomerular basement membrane (congenital nephrotic syndrome), combined with sclerosis of the glomerular mesangium and severe impairment of vision and neurodevelopment (Zenker et al., 2004).

## Other Modulators of Cell–Matrix Interactions

Whereas proteins such as fibronectin and laminin are important for adhesion of cells to extracellular matrices, other ECM molecules function as both positive and negative modulators of such adhesive interactions. Examples of such modulators are thrombospondin (Adams and Lawler, 2004) and tenascin (Chiquet-Ehrismann, 2004). Thrombospondins (TSPs) are a group of homologous trimeric (TSP-1 and TSP-2) and pentameric (TSP-3, TSP-4 and TSP-5/COMP) matrix proteins composed of several Ca++-binding (type 3) domains, EGF-like repeats (type 2), as well as other modules (Fig. 8.9). Different members of the group show differences in cellular expression and functional properties. The most highly conserved regions of the different thrombospondins are the carboxyl halves of the molecules, all consisting of a variable number of EGF-like domains, seven Ca++-binding

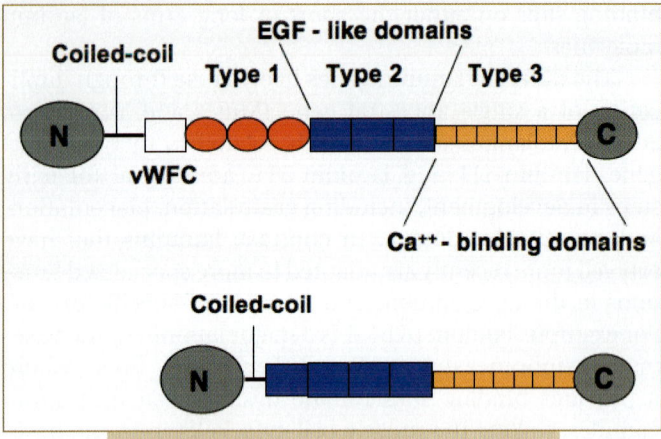

**FIG. 8.9.** Diagram of thrombospondins. Diagram of trimeric thrombospondins (TSP-1 and TSP-2) on top showing the multidomain structure and the location of the coiled-coil domain important for trimerization. Diagram of pentameric thrombospondins (TSP-3, TSP-4, and TSP-5/COMP) at bottom, showing the lack of von Willebrand factor C–like domain (vWC) and type 1 repeats in this group of thrombospondins.

TSP type 3 repeats, and a C-terminal globular domain. In contrast, the N-terminal regions are quite variable, but all members have a short coiled-coil domain of heptad repeats in this region that is crucial for oligomerization into trimers in TSP-1 and TSP-2 or pentamers in TSP-3, TSP-4, and TSP-5/COMP. These oligomerization domains are stabilized by interchain disulfide bonds, but they are quite stable even with the disulfides reduced (Engel, 2004). Since the subunits are held together at the coiled-coil domains, the assembled molecules have a flowerlike appearance, with three or five "petals" extending out from the center, available for binding to cell surface receptors and other ECM molecules. The crystal structure of the five-stranded coiled-coil domain of TSP-5 shows that it forms a hydrophobic channel with some similarity to ion channels and can bind vitamin D and all-trans retinoic acid. One function of pentameric thrombospondins may therefore be to store small hydrophobic signaling molecules in the ECM.

Interestingly, TSP type 1 repeats are found in many other proteins, including the large family of matrix metalloproteases called ADAMTS enzymes; in some members of this family, there are more copies of TSP type 1 domains than in TSP-1 or TSP-2 themselves (Tucker, 2004). Members of the ADAMTS family have important biological functions (Apte, 2004). For example, ADAMTS2, ADAMTS3, and ADAMTS14 are procollagen propeptidases, responsible for processing the amino propeptide in fibrillar procollagens (see earlier), and ADAMTS4 and ADAMTS5 are aggrecanases, able to degrade the major proteoglycan component of cartilage.

The carboxyl regions of thrombospondins can bind to a variety of ECM molecules, extracellular proteases, and cell surface components such as integrins (Adams, 2004).

Their oligomeric structure enables thrombospondins to be involved in multiple interactions and to modulate both cellular behavior and ECM assembly. All thrombospondins support attachment of cells in a Ca$^{++}$-dependent manner and stimulate cell migration, proliferation, chemotaxis, and phagocytosis. Trimeric thrombospondins (TSP-1 and TSP-2) have additional activities associated with the type 1 domains within their N-terminal regions (Bornstein et al., 2004). These activities include inhibition of angiogenesis through mechanisms by which thrombospondin induces endothelial cell apoptosis and inhibits mobilization of vascular endothelial growth factor (VEGF). The two trimeric thrombospondins have also recently been shown to promote the formation of synapses in the central nervous system (Christopherson et al., 2005).

Cartilage oligomeric matrix protein (TSP-5/COMP) is thought to have evolved from the TSP-3 and TSP-4 branches of the thrombospondin (Posey et al., 2004). COMP is a secretory product of chondrocytes and is localized in their territorial matrix in cartilage. Beyond the fact that COMP interacts with collagens II and IX, little is known about its normal function in cartilage. Mice lacking COMP develop a normal skeleton and have no significant abnormalities. However, mutations in COMP cause pseudoachondroplasia and multiple epiphyseal dysplasia in humans (Briggs et al., 1995; Horton and Hecht, 2002). At birth, affected individuals have normal weight and length but show reduced growth of long limb bones and striking defects in growth plate regions. These defects are caused by retention of mutant protein in the RER of chondrocytes, causing premature cell death. Mutations in COMP appear therefore to generate a mutant phenotype by a mechanism involving RER stress in chondrocytes.

The four members of the vertebrate tenascin family (C, R, W, and X) are large multimeric proteins with subunits composed of multiple protein modules (Chiquet-Ehrismann, 2004). The modules include heptad repeats, fibronectin type 3 repeats, EGF-like domains, and a carboxyl domain with homology to the carboxyl-terminal domains of β- and γ-fibrinogen chains. These modules form rodlike structures that interact with their amino-terminal domains to form oligomers. Alternative splicing of tenascin-C generates multiple isoforms. The tenascins are differentially expressed in different tissues and at different times during development and growth (Chiquet-Ehrismann and Tucker, 2004). For example, tenascin-R is expressed only in the central nervous system, in contrast to tenascin-C, which is found in both the central nervous system as well as peripheral nerves. Tenascin-C expression is high during development and inflammation and around tumors, but it is otherwise relatively low in postnatal tissues, with some interesting exceptions. In tissue regions of high mechanical stress, the levels of tenascin expression are high, suggesting a role for tenascin-C in the mechanisms used by cells to cope with mechanical stress. In fact, tenascin-C was first identified as a myotendinous antigen because of the high level of expression at tendon–muscle junctions (Chiquet and Fambrough, 1984). It is also expressed by other cells that are exposed to mechanical stress, including osteoblasts, perichondrial cells around cartilage, smooth muscle cells, and fibroblasts in healing wounds. This association between mechanical stress and expression is also seen for other tenascins. Thus, tenascin-W is expressed by both osteoblasts and smooth muscle cells, and tenascin-X is expressed at high level in the connective tissue "wraps" in skeletal muscle. Consistent with a mechanical role in connective tissues is also the finding that a form of Ehlers–Danlos syndrome, with hypermobile joints and hyperelastic skin, is caused by a deficiency in tenascin-X (Schalkwijk *et al.*, 2001).

The finding that collagen XII (see earlier) interacts with tenascin-X, combined with data showing that tenascin-X can bind to collagen I fibrils, possibly via interaction with the small proteoglycans decorin, suggests that a complex of collagen XII and tenascin-X serves as important interfibrillar bridges in skin (Veit *et al.*, 2006). This complex may also mediate attachment of collagen fibrils to cells, since tenascin can bind to integrin receptors. That a similar complex between collagen XII and tenascin-C may be present at myotendinous junctions is suggested by the high-level expression of both collagen XII and tenascin-C at such junctions (Böhme *et al.*, 1995).

The interactions between tenascins and cells are relatively weak compared to other proteins, such as fibronectin and thrombospondin. In certain experimental conditions, tenascin-C can be an adhesive molecule for cells; it can also, however, have antiadhesive effects (Chiquet-Ehrismann and Tucker, 2004; Orend and Chiquet-Ehrismann, 2006). The adhesive activity can be mediated by either cell surface proteoglycans or integrins, depending on cell type. Tenascin-C can bind heparin, and this may be responsible for interactions with cell surface proteoglycans such as glypican. Tenascin-C can also block adhesion by covering up adhesive sites in other matrix molecules, such as fibronectin, and sterically block their interactions with cells. Tenascin-C has therefore been characterized as a cell adhesion–modulating protein. Likewise, tenascin-R can both promote neuronal cell adhesion and act as a repellant for neurites.

# V. PROTEOGLYCANS – MULTIFUNCTIONAL MOLECULES IN THE EXTRACELLULAR MATRIX AND ON CELL SURFACES

A variety of proteoglycans play important roles in cellular growth and differentiation and in matrix structure. They range from the large hydrophilic space–filling complexes of aggrecan and versican with hyaluronan, to the cell surface syndecan receptors. In basement membranes the major heparan sulfate proteoglycan is perlecan (Timpl, 1994). With three heparan sulfate side chains attached to the amino-terminal region, its core protein is multimodular in structure, having borrowed structural motifs from a variety of other genes. These include an LDL receptor-like module, regions with extensive homology to laminin chains, a long stretch of N-CAM-like IgG repeats, and a carboxyl-terminal region with three globular and four EGF-like repeats similar to a region of laminin (Olsen, 1999). Alternative splicing can generate molecules of different lengths. Perlecan is present in a number of basement membranes, but it is also found in the pericellular matrix of fibroblasts and in cartilage ECM. In fact, fibroblasts, rather than epithelial cells, appear to be major producers of perlecan (for example, in skin). In liver, perlecan is expressed by sinusoidal endothelial cells and is localized in the perisinusoidal space. Mutations in the *unc-52* gene in *C. elegans*, encoding a short version of perlecan, cause disruptions of skeletal muscle (Rogalski *et al.*, 1993). This indicates that the molecule, as a component of skeletal muscle basement membranes, is important for assembly of myofilaments and their attachment to cell membranes. Binding of growth factors and cytokines to the heparan sulfate side chains also enables perlecan to serve as a storage vehicle for biologically active molecules such as bFGF. The critical role of perlecan is further highlighted by the dramatic effects of knocking out the perlecan gene in mice (Costell *et al.*, 1999). Most of the mutant embryos die halfway through pregnancy, and the few embryos that survive to birth have severe defects in the brain and the skeleton. The skeletal defects include severe shortening of axial and limb bones and disruption of normal growth plate structure.

Several small leucine-rich repeat proteins and proteoglycans with homologous core proteins are found in a variety of tissues, where they interact with other matrix macromolecules and regulate their functions (McEwan *et al.*, 2006). They include decorin, biglycan, lumican, and fibromodulin. Decorin binds along collagen fibrils and plays a role in regulating fibril assembly and mechanical properties (Reed and Iozzo, 2002; Robinson *et al.*, 2005). It also modulates the binding of cells to matrix constituents such as collagen, fibronectin, and tenascin (Ameye and Young, 2002). Through binding of TGF-β isoforms, the small proteoglycans help sequester growth factors within the ECM and thus regulate their activities (Hildebrand *et al.*, 1994).

A variety of proteoglycans also have important functions at cell surfaces. These include members of the syndecan family, transmembrane molecules with highly conserved cytoplasmic domains, and glypican-related molecules that are linked to the cell surface via glycosyl phosphatidylinositol. Through their heparan sulfate side chains these molecules can bind growth factors, protease inhibitors, enzymes, and matrix macromolecules. They are therefore important modulators of cell signaling pathways and cell–matrix

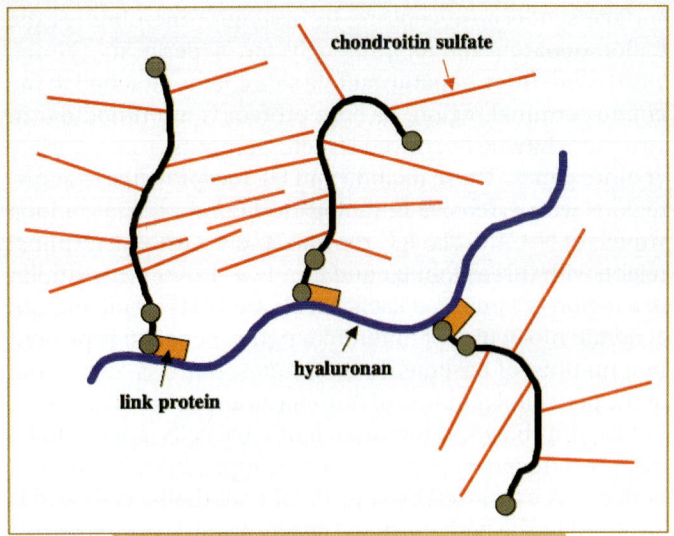

**FIG. 8.10.** Diagram of a portion of a large proteoglycan complex from cartilage. Monomers of aggrecan, composed of core proteins with glycosaminoglycan side chains (mostly chondroitin sulfate), are bound to hyaluronan. The binding is stabilized by link proteins. For clarity, only some of the glycosaminoglycan side chains are shown in the monomers.

contacts (De Cat and David, 2001; Tkachenko *et al.*, 2005; Zimmermann and David, 1999).

Hyaluronan is an important component of most extracellular matrices (Laurent and Fraser, 1992). It serves as a ligand for several proteins, including cartilage link protein and aggrecan and versican core proteins. In cartilage, based on such interactions, it is the backbone for the large proteoglycan complexes responsible for the compressive properties of cartilage (Morgelin *et al.*, 1994) (Fig. 8.10). It also is a ligand for cell surface receptors and regulates cell proliferation and migration (Tammi *et al.*, 2002; Turley *et al.*, 2002). One receptor for hyaluronan is the transmembrane molecule CD44. By alternative splicing and variations in post-translational modifications, a family of CD44 proteins is generated (Lesley *et al.*, 1993). These show cell- and tissue-specific expression patterns and are thought to have distinct functional roles. Hyaluronan-mediated cell motility is based on the interaction of hyaluronan with a cell surface–associated protein called RHAMM (receptor for hyaluronate-mediated motility) (Nedvetzki *et al.*, 2004). As a space-filling molecule and through its interaction with cell surface receptors, hyaluronan is important for several morphogenetic processes during development. It creates cell free spaces through which cells (for example, neural crest cells) can migrate, and its degradation by hyaluronidase is probably important for processes of cellular condensation. Since hyaluronan is not immunogenic, is readily available, and can easily be manipulated and chemically modified, it is receiving considerable attention as a tissue-engineering biopolymer (Allison and Grande-Allen, 2006).

## VI. CONCLUSION

Research efforts since the mid-1970s have led to significant insights into the composition of extracellular matrices and the structure and function of the major components. We now realize that the evolution of vertebrates and mammals was associated with an expansion of several families of matrix molecules, providing cells with an increasing repertoire of isoforms and homologs to build different tissues. It is also evident that the increasing number of different families of genes encoding matrix molecules during evolution of more complex organisms involved shuffling and recombination of genes encoding a relatively small number of structural and functional modules. Finally, the data suggest that cells are building matrices by adding layer upon layer of components that can interact with various affinities (but mostly on the low side) and in multiple ways with their neighbors. The result is an extracellular matrix that readily can be fine-tuned to meet the demands of the moment, but one that is relatively resistant to the effects of mutations that may cause dysfunction of specific components. As we learn to use these insights to identify the most critical matrix properties from a cellular point of view, exciting and rapid advances in tissue engineering should follow.

## VII. REFERENCES

Abrass, C. K., Berfield, A. K., Ryan, M. C., Carter, W. G., and Hansen, K. M. (2006). Abnormal development of glomerular endothelial and mesangial cells in mice with targeted disruption of the lama3 gene. *Kidney Int.* **70**, 1062–1071.

Adams, J. C. (2004). Functions of the conserved thrombospondin carboxy-terminal cassette in cell–extracellular matrix interactions and signaling. *Int. J. Biochem. Cell Biol.* **36**, 1102–1114.

Adams, J. C., and Lawler, J. (2004). The thrombospondins. *Int. J. Biochem. Cell Biol.* **36**, 961–968.

Allison, D. D., and Grande-Allen, K. J. (2006). Review. Hyaluronan: a powerful tissue-engineering tool. *Tissue Eng.* **12**, 2131–2140.

Ameye, L., and Young, M. F. (2002). Mice deficient in small leucine-rich proteoglycans: novel *in vivo* models for osteoporosis, osteoarthritis,

Ehlers–Danlos syndrome, muscular dystrophy, and corneal diseases. *Glycobiology* **12**, 107R–116R.

Andrikopoulos, K., Liu, X., Keene, D. R., Jaenisch, R., and Ramirez, F. (1995). Targeted mutation in the col5a2 gene reveals a regulatory role for type V collagen during matrix assembly. *Nat. Genet.* **9**, 31–36.

Aota, S., Nagai, T., and Yamada, K. M. (1991). Characterization of regions of fibronectin besides the arginine-glycine-aspartic acid sequence required for adhesive function of the cell-binding domain using site-directed mutagenesis. *J. Biol. Chem.* **266**, 15938–15943.

Apte, S. S. (2004). A disintegrin-like and metalloprotease (reprolysin type) with thrombospondin type 1 motifs: the ADAMTS family. *Int. J. Biochem. Cell Biol.* **36**, 981–985.

Aumailley, M., Bruckner-Tuderman, L., Carter, W. G., Deutzmann, R., Edgar, D., Ekblom, P., Engel, J., Engvall, E., Hohenester, E., Jones, J. C., *et al.* (2005). A simplified laminin nomenclature. *Matrix Biol.* **24**, 326–332.

Bendeck, M. P., Regenass, S., Tom, W. D., Giachelli, C. M., Schwartz, S. M., Hart, C., and Reidy, M. A. (1996). Differential expression of .a1 type VIII collagen in injured platelet-derived growth factor-BB-stimulated rat carotid arteries. *Circ. Res.* **79**, 524–531.

Biswas, S., Munier, F. L., Yardley, J., Hart-Holden, N., Perveen, R., Cousin, P., Sutphin, J. E., Noble, B., Batterbury, M., Kielty, C., *et al.* (2001). Missense mutations in COL8A2, the gene encoding the alpha2 chain of type VIII collagen, cause two forms of corneal endothelial dystrophy. *Hum. Mol. Genet.* **10**, 2415–2423.

Bix, G., and Iozzo, R. V. (2005). Matrix revolutions: "tails" of basement-membrane components with angiostatic functions. *Trends Cell Biol.* **15**, 52–60.

Böhme, K., Li, Y., Oh, S. P., and Olsen, B. R. (1995). Primary structure of the long- and short-splice variants of mouse collagen XII and their tissue-specific expression during embryonic development. *Dev. Dyn.* **204**, 432–445.

Bornstein, P., Agah, A., and Kyriakides, T. R. (2004). The role of thrombospondins 1 and 2 in the regulation of cell-matrix interactions, collagen fibril formation, and the response to injury. *Int. J. Biochem. Cell Biol.* **36**, 1115–1125.

Briggs, M. D., Hoffman, S. M. G., King, L. M., Olsen, A. S., Mohrenweiser, H., Leroy, J. G., Mortier, G. R., Rimoin, D. L., Lachman, R. S., Gaines, E. S., *et al.* (1995). Pseudoachondroplasia and multiple epiphyseal dysplasia due to mutations in the cartilage oligomeric matrix protein gene. *Nat. Genet.* **10**, 330–336.

Brown, J. C., Mann, K., Wiedemann, H., and Timpl, R. (1993). Structure and binding properties of collagen type XIV isolated from human placenta. *J. Cell Biol.* **120**, 557–567.

Byers, P., and Cole, W. G. (2002). Osteogenesis imperfecta. *In* "Connective Tissue and Its Heritable Disorders. Molecular, Genetic, and Medical Aspects" (P. Royce and B. Steinmann, eds.), pp. 385–430. Wiley-Liss, New York.

Canty, E. G., Lu, Y., Meadows, R. S., Shaw, M. K., Holmes, D. F., and Kadler, K. E. (2004). Coalignment of plasma membrane channels and protrusions (fibripositors) specifies the parallelism of tendon. *J. Cell Biol.* **165**, 553–563.

Chiquet, M., and Fambrough, D. M. (1984). Chick myotendinous antigen. I. A monoclonal antibody as a marker for tendon and muscle morphogenesis. *J. Cell Biol.* **98**, 1926–1936.

Chiquet-Ehrismann, R. (2004). Tenascins. *Int. J. Biochem. Cell Biol.* **36**, 986–990.

Chiquet-Ehrismann, R., and Tucker, R. P. (2004). Connective tissues: signalling by tenascins. *Int. J. Biochem. Cell Biol.* **36**, 1085–1089.

Christopherson, K. S., Ullian, E. M., Stokes, C. C., Mullowney, C. E., Hell, J. W., Agah, A., Lawler, J., Mosher, D. F., Bornstein, P., and Barres, B. A. (2005). Thrombospondins are astrocyte-secreted proteins that promote CNS synaptogenesis. *Cell* **120**, 421–433.

Colige, A., Beschin, A., Samyn, B., Goebels, Y., Van Beeumen, J., Nusgens, B. V., and Lapiere, C. M. (1995). Characterization and partial amino acid sequencing of a 107-kDa procollagen I N-proteinase purified by affinity chromatography on immobilized type XIV collagen. *J. Biol. Chem.* **270**, 16724–16730.

Corson, G. M., Charbonneau, N. L., Keene, D. R., and Sakai, L. Y. (2004). Differential expression of fibrillin-3 adds to microfibril variety in human

and avian, but not rodent, connective tissues. *Genomics* **83**, 461–472.

Costell, M., Gustafsson, E., Aszodi, A., Morgelin, M., Bloch, W., Hunziker, E., Addicks, K., Timpl, R., and Fassler, R. (1999). Perlecan maintains the integrity of cartilage and some basement membranes. *J. Cell Biol.* **147**, 1109–1122.

De Cat, B., and David, G. (2001). Developmental roles of the glypicans. *Semin. Cell Dev. Biol.* **12**, 117–125.

Ding, Y. H., Javaherian, K., Lo, K. M., Chopra, R., Boehm, T., Lanciotti, J., Harris, B. A., Li, Y., Shapiro, R., Hohenester, E., *et al.* (1998). Zinc-dependent dimers observed in crystals of human endostatin. *Proc. Natl. Acad. Sci. U.S.A.* **95**, 10443–10448.

Engel, J. (2004). Role of oligomerization domains in thrombospondins and other extracellular matrix proteins. *Int. J. Biochem. Cell Biol.* **36**, 997–1004.

Faessler, R., Schnegelsberg, P. N. J., Dausman, J., Muragaki, Y., Shinya, T., McCarthy, M. T., Olsen, B. R., and Jaenisch, R. (1994). Mice lacking a1(IX) collagen develop noninflammatory degenerative joint disease. *Proc. Natl. Acad. Sci. U.S.A.* **91**, 5070–5074.

Font, B., Aubert-Foucher, E., Goldschmidt, D., Eichenberger, D., and van der Rest, M. (1993). Binding of collagen XIV with the dermatan sulfate side chain of decorin. *J. Biol. Chem.* **268**, 25015–25018.

Franzke, C. W., Tasanen, K., Schumann, H., and Bruckner-Tuderman, L. (2003). Collagenous transmembrane proteins: collagen XVII as a prototype. *Matrix Biol.* **22**, 299–309.

Garofalo, S., Metsaranta, M., Ellard, J., Smith, C., Horton, W., Vuorio, E., and de Crombrugghe, B. (1993). Assembly of cartilage collagen fibrils is disrupted by overexpression of normal type II collagen in transgenic mice. *Proc. Natl. Acad. Sci. U.S.A.* **90**, 3825–3829.

George, E. L., Georges-Labouesse, E. N., Patel-King, R. S., Rayburn, H., and Hynes, R. O. (1993). Defects in mesoderm, neural tube and vascular development in mouse embryos lacking fibronectin. *Development* **119**, 1079–1091.

Habashi, J. P., Judge, D. P., Holm, T. M., Cohn, R. D., Loeys, B. L., Cooper, T. K., Myers, L., Klein, E. C., Liu, G., Calvi, C., *et al.* (2006). Losartan, an AT1 antagonist, prevents aortic aneurysm in a mouse model of Marfan syndrome. *Science* **312**, 117–121.

Halfter, W., Dong, S., Schurer, B., and Cole, G. J. (1998). Collagen XVIII is a basement membrane heparan sulfate proteoglycan. *J. Biol. Chem.* **273**, 25404–25412.

Hay, E. D. (1991). "Cell Biology of Extracellular Matrix," 2nd ed. Plenum Press, New York.

Hildebrand, A., Romaris, M., Rasmussen, L. M., Heinegard, D., Twardzik, D. R., Border, W. A., and Ruoslahti, E. (1994). Interaction of the small interstitial proteoglycans biglycan, decorin and fibromodulin with transforming growth factor beta. *Biochem. J.* **302**, 527–534.

Hohenester, E., Sasaki, T., Olsen, B. R., and Timpl, R. (1998). Crystal structure of the angiogenesis inhibitor endostatin at 1.5-Å resolution. *EMBO J.* **17**, 1656–1664.

Hopfer, U., Fukai, N., Hopfer, H., Wolf, G., Joyce, N., Li, E., and Olsen, B. R. (2005). Targeted disruption of Col8a1 and Col8a2 genes in mice leads to anterior segment abnormalities in the eye. *Faseb J.* **19**, 1232–1244.

Horton, W. A., and Hecht, J. T. (2002). Chondrodysplasias: general concepts and diagnostic and management considerations. *In* "Connective Tissue and Its Heritable Disorders: Molecular, Genetic, and Medical Aspects" (P. M. Royce and B. Steinmann, eds.), pp. 901–908. Wiley-Liss, New York.

Hynes, R. O. (1990). "Fibronectin." Springer-Verlag, New York.

Iruela-Arispe, M. L., and Sage, E. H. (1991). Expression of type VIII collagen during morphogenesis of the chicken and mouse heart. *Dev. Biol.* **144**, 107–118.

Isogai, Z., Ono, R. N., Ushiro, S., Keene, D. R., Chen, Y., Mazzieri, R., Charbonneau, N. L., Reinhardt, D. P., Rifkin, D. B., and Sakai, L. Y. (2003). Latent transforming growth factor beta–binding protein 1 interacts with fibrillin and is a microfibril-associated protein. *J. Biol. Chem.* **278**, 2750–2757.

Jakkula, E., Makitie, O., Czarny-Ratajczak, M., Jackson, G. C., Damignani, R., Susic, M., Briggs, M. D., Cole, W. G., and Ala-Kokko, L. (2005). Mutations in the known genes are not the major cause of MED: distinctive phenotypic entities among patients with no identified mutations. *Eur. J. Hum. Genet.* **13**, 292–301.

Kapyla, J., Jaalinoja, J., Tulla, M., Ylostalo, J., Nissinen, L., Viitasalo, T., Vehvilainen, P., Marjomaki, V., Nykvist, P., Saamanen, A. M., *et al.* (2004). The fibril-associated collagen IX provides a novel mechanism for cell adhesion to cartilaginous matrix. *J. Biol. Chem.* **279**, 51677–51687.

Kielty, C. M., and Grant, M. E. (2002). The collagen family: structure, assembly, and organization in the extracellular matrix. *In* "Connective Tissue and Its Heritable Disorders: Molecular, Genetic, and Medical Aspects" (P. M. Royce and B. Steinmann, eds.), pp. 159–221. Wiley-Liss, New York.

Laurent, T. C., and Fraser, J. R. (1992). Hyaluronan. *FASEB J.* **6**, 2397–2404.

Lesley, J., Hyman, R., and Kincade, P. W. (1993). CD44 and its interaction with extracellular matrix. *Adv. Immunol.* **54**, 271–335.

Li, Y., Lacerda, D. A., Warman, M. L., Beier, D. R., Yoshioka, H., Ninomiya, Y., Oxford, J. T., Morris, N. P., Andrikopoulos, K., Ramirez, F., *et al.* (1995). A fibrillar collagen gene, Col11a1, is essential for skeletal morphogenesis. *Cell* **80**, 423–430.

Linsenmayer, T. F., Gibney, E., Igoe, F., Gordon, M. K., Fitch, J. M., Fessler, L. I., and Birk, D. E. (1993). Type V collagen: molecular structure and fibrillar organization of the chicken alpha 1(V) NH2-terminal domain, a putative regulator of corneal fibrillogenesis. *J. Cell Biol.* **121**, 1181–1189.

MacBeath, J. R., Kielty, C. M., and Shuttleworth, C. A. (1996). Type VIII collagen is a product of vascular smooth-muscle cells in development and disease. *Biochem. J.* **319**, 993–998.

Marneros, A. G., and Olsen, B. R. (2001). The role of collagen-derived proteolytic fragments in angiogenesis. *Matrix Biol.* **20**, 337–345.

Marneros, A. G., and Olsen, B. R. (2005). Physiological role of collagen XVIII and endostatin. *FASEB J.* **19**, 716–728.

McAlinden, A., Smith, T. A., Sandell, L. J., Ficheux, D., Parry, D. A., and Hulmes, D. J. (2003). Alpha-helical coiled-coil oligomerization domains are almost ubiquitous in the collagen superfamily. *J. Biol. Chem.* **278**, 42200–42207.

McEwan, P. A., Scott, P. G., Bishop, P. N., and Bella, J. (2006). Structural correlations in the family of small leucine-rich repeat proteins and proteoglycans. *J. Struct. Biol.* **155**, 294–305.

Miner, J. H., and Yurchenco, P. D. (2004). Laminin functions in tissue morphogenesis. *Annu. Rev. Cell Dev. Biol.* **20**, 255–284.

Morgelin, M., Heinegard, D., Engel, J., and Paulsson, M. (1994). The cartilage proteoglycan aggregate: assembly through combined protein–carbohydrate and protein–protein interactions. *Biophys. Chem.* **50**, 113–128.

Mosher, D. F., Fogerty, F. J., Chernousov, M. A., and Barry, E. L. (1991). Assembly of fibronectin into extracellular matrix. *Ann. New York Acad. Sci.* **614**, 167–180.

Muragaki, Y., Mariman, E. C., van Beersum, S. E., Perala, M., van Mourik, J. B., Warman, M. L., Olsen, B. R., and Hamel, B. C. (1996). A mutation in the gene encoding the alpha 2 chain of the fibril-associated collagen IX, COL9A2, causes multiple epiphyseal dysplasia (EDM2). *Nat. Genet.* **12**, 103–105.

Nakata, K., Ono, K., Miyazaki, J., Olsen, B. R., Muragaki, Y., Adachi, E., Yamamura, K., and Kimura, T. (1993). Osteoarthritis associated with mild chondrodysplasia in transgenic mice expressing alpha 1(IX) collagen chains with a central deletion. *Proc. Natl. Acad. Sci. U.S.A.* **90**, 2870–2874.

Nedvetzki, S., Gonen, E., Assayag, N., Reich, R., Williams, R. O., Thurmond, R. L., Huang, J. F., Neudecker, B. A., Wang, F. S., Turley, E. A., *et al.* (2004). RHAMM, a receptor for hyaluronan-mediated motility, compensates for CD44 in inflamed CD44-knockout mice: a different interpretation of redundancy. *Proc. Natl. Acad. Sci. U.S.A.* **101**, 18081–18086.

Neptune, E. R., Frischmeyer, P. A., Arking, D. E., Myers, L., Bunton, T. E., Gayraud, B., Ramirez, F., Sakai, L. Y., and Dietz, H. C. (2003). Dysregulation of TGF-beta activation contributes to pathogenesis in Marfan syndrome. *Nat. Genet.* **33**, 407–411.

Nishiyama, T., McDonough, A. M., Bruns, R. R., and Burgeson, R. E. (1994). Type XII and XIV collagens mediate interactions between banded collagen fibers *in vitro* and may modulate extracellular matrix deformability. *J. Biol. Chem.* **269**, 28193–28199.

Oh, S. P., Kamagata, Y., Muragaki, Y., Timmons, S., Ooshima, A., and Olsen, B. R. (1994). Isolation and sequencing of cDNAs for proteins with multiple domains of Gly-X-Y repeats identify a novel family of collagenous proteins. *Proc. Natl. Acad. Sci. U.S.A.* **91**, 4229–4233.

Olsen, B. R. (1991). Collagen biosynthesis. *In* "Cell Biology of Extracellular Matrix" (E. D. Hay, ed.), pp. 177–220. Plenum Publishing, New York.

Olsen, B. R. (1999). Life without perlecan has its problems. *J. Cell Biol.* **147**, 909–911.

Olsen, B. R., Winterhalter, K. H., and Gordon, M. K. (1995). FACIT collagens and their biological roles. *Trends Glycosci. Glycotechnol.* **7**, 115–127.

Orend, G., and Chiquet-Ehrismann, R. (2006). Tenascin-C–induced signaling in cancer. *Cancer Lett.*

O'Reilly, M. S., Boehm, T., Shing, Y., Fukai, N., Vasios, G., Lane, W. S., Flynn, E., Birkhead, J. R., Olsen, B. R., and Folkman, J. (1997). Endostatin: an endogenous inhibitor of angiogenesis and tumor growth. *Cell* **88**, 277–285.

Pihlajamaa, T., Lankinen, H., Ylostalo, J., Valmu, L., Jaalinoja, J., Zaucke, F., Spitznagel, L., Gosling, S., Puustinen, A., Morgelin, M., *et al.* (2004). Characterization of recombinant amino-terminal NC4 domain of human collagen IX: interaction with glycosaminoglycans and cartilage oligomeric matrix protein. *J. Biol. Chem.* **279**, 24265–24273.

Posey, K. L., Hayes, E., Haynes, R., and Hecht, J. T. (2004). Role of TSP-5/COMP in pseudoachondroplasia. *Int. J. Biochem. Cell Biol.* **36**, 1005–1012.

Potts, J. R., and Campbell, I. D. (1994). Fibronectin structure and assembly. *Curr. Opin. Cell Biol.* **6**, 648–655.

Pyeritz, R. E., and Dietz, H. C. (2002). Marfan syndrome and other microfibrillar disorders. *In* "Connective Tissue and Its Heritable Disorders: Molecular, Genetic, and Medical aspects" (P. M. Royce and B. Steinmann, eds.), pp. 585–626. Wiley-Liss, New York.

Reed, C. C., and Iozzo, R. V. (2002). The role of decorin in collagen fibrillogenesis and skin homeostasis. *Glycoconj. J.* **19**, 249–255.

Rehn, M., and Pihlajaniemi, T. (1994). α1(XVIII), a collagen chain with frequent interruptions in the collagenous sequence, a distinct tissue distribution, and homology with type XV collagen. *Proc. Natl. Acad. Sci. U.S.A.* **91**, 4234–4238.

Robinson, P. S., Huang, T. F., Kazam, E., Iozzo, R. V., Birk, D. E., and Soslowsky, L. J. (2005). Influence of decorin and biglycan on mechanical properties of multiple tendons in knockout mice. *J. Biomech. Eng.* **127**, 181–185.

Rogalski, T. M., Williams, B. D., Mullen, G. P., and Moerman, D. G. (1993). Products of the unc-52 gene in *Caenorhabditis elegans* are homologous to the core protein of the mammalian basement membrane heparan sulfate proteoglycan. *Genes Dev.* **7**, 1471–1484.

Rosenbloom, J., Abrams, W. R., and Mecham, R. (1993). Extracellular matrix 4: the elastic fiber. *FASEB J.* **7**, 1208–1218.

Sakai, L. Y., and Keene, D. R. (1994). Fibrillin: monomers and microfibrils. *In* "Methods in Enzymology" (E. Ruoslahti and E. Engvall, eds.), pp. 29–52. Academic Press, San Diego.

Schalkwijk, J., Zweers, M. C., Steijlen, P. M., Dean, W. B., Taylor, G., van Vlijmen, I. M., van Haren, B., Miller, W. L., and Bristow, J. (2001). A recessive form of the Ehlers–Danlos syndrome caused by tenascin-X deficiency. *N. Engl. J. Med.* **345**, 1167–1175.

Schwarzbauer, J. (1990). The fibronectin gene. *In* "Extracellular Matrix Genes" (L. J. Sandell and C. D. Boyd, eds.), pp. 195–219. Academic Press, San Diego.

Shaw, L. M., and Olsen, B. R. (1991). Collagens in the FACIT group: diverse molecular bridges in extracellular matrices. *Trends Biochem. Sci.* **16**, 191–194.

Steinmann, B., Royce, P. M., and Superti-Furga, A. (2002). The Ehlers–Danlos syndrome. *In* "Connective Tissue and Its Heritable Disorder: Molecular, Genetic, and Medical Aspects" (P. M. Royce and B. Steinmann, eds.), pp. 431–523. Wiley-Liss, New York.

Tammi, M. I., Day, A. J., and Turley, E. A. (2002). Hyaluronan and homeostasis: a balancing act. *J. Biol. Chem.* **277**, 4581–4584.

Timpl, R. (1994). Proteoglycans of basement membranes. *Exs* **70**, 123–144.

Timpl, R. (1996). Macromolecular organization of basement membranes. *Curr. Opin. Cell Biol.* **8**, 618–624.

Timpl, R., and Brown, J. C. (1994). The laminins. *Matrix Biol.* **14**, 275–281.

Tkachenko, E., Rhodes, J. M., and Simons, M. (2005). Syndecans: new kids on the signaling block. *Circ. Res.* **96**, 488–500.

Tryggvason, K., and Martin, P. (2002). Alport syndrome. *In* "Connective Tissue and Its Heritable Disorders: Molecular, Genetic, and Medical Aspects" (P. M. Royce and B. Steinmann, eds.), pp. 1069–1102. Wiley-Liss, New York.

Tucker, R. P. (2004). The thrombospondin type 1 repeat superfamily. *Int. J. Biochem. Cell Biol.* **36**, 969–974.

Turley, E. A., Noble, P. W., and Bourguignon, L. Y. (2002). Signaling properties of hyaluronan receptors. *J. Biol. Chem.* **277**, 4589–4592.

Uitto, J., and Richard, G. (2005). Progress in epidermolysis bullosa: from eponyms to molecular genetic classification. *Clin. Dermatol.* **23**, 33–40.

van der Rest, M., Mayne, R., Ninomiya, Y., Seidah, N. G., Chretien, M., and Olsen, B. R. (1985). The structure of type IX collagen. *J. Biol. Chem.* **260**, 220–225.

Vaughan, L., Mendler, M., Huber, S., Bruckner, P., Winterhalter, K. H., Irwin, M. I., and Mayne, R. (1988). D-periodic distribution of collagen type IX along cartilage fibrils. *J. Cell Biol.* **106**, 991–997.

Veit, G., Hansen, U., Keene, D. R., Bruckner, P., Chiquet-Ehrismann, R., Chiquet, M., and Koch, M. (2006). Collagen XII interacts with avian tenascin-X through its NC3 domain. *J. Biol. Chem.* **281**, 27461–27470.

von der Mark, K., and Sorokin, L. (2002). Adhesive glycoproteins. *In* "Connective Tissue and Its Heritable Disorders: Molecular, Genetic, and Medical Aspects" (P. M. Royce and B. Steinmann, eds.), pp. 293–328. Wiley-Liss, New York.

Yamaguchi, N., Anand-Apte, B., Lee, M., Sasaki, T., Fukai, N., Shapiro, R., Que, I., Lowik, C., Timpl, R., and Olsen, B. R. (1999). Endostatin inhibits VEGF-induced endothelial cell migration and tumor growth independently of zinc binding. *EMBO J.* **18**, 4414–4423.

Yurchenco, P. D., Amenta, P. S., and Patton, B. L. (2004). Basement membrane assembly, stability and activities observed through a developmental lens. *Matrix Biol.* **22**, 521–538.

Zenker, M., Aigner, T., Wendler, O., Tralau, T., Muntefering, H., Fenski, R., Pitz, S., Schumacher, V., Royer-Pokora, B., Wuhl, E., *et al.* (2004). Human laminin beta2 deficiency causes congenital nephrosis with mesangial sclerosis and distinct eye abnormalities. *Hum. Mol. Genet.* **13**, 2625–2632.

Zimmermann, P., and David, G. (1999). The syndecans, tuners of transmembrane signaling. *Faseb J.* **13**(Suppl), S91–S100.

# Chapter Nine

# Morphogenesis and Tissue Engineering

*A. H. Reddi*

## I. INTRODUCTION

*Morphogenesis* is the developmental cascade of pattern formation, the establishment of the body plan and architecture of mirror-image bilateral symmetry of musculoskeletal structures, culminating in the adult form. *Tissue engineering* is the emerging discipline of fabrication of spare parts for the human body, including the skeleton, for functional restoration and aging of lost parts due to cancer, disease, and trauma. It is based on rational principles of molecular developmental biology and morphogenesis and is further governed by bioengineering. The three key ingredients for both morphogenesis and tissue engineering are inductive morphogenetic signals, responding stem cells, and the extracellular matrix scaffolding (Reddi, 1998) (Fig. 9.1). Recent advances in molecular cell biology of morphogenesis will aid in the design principles and architecture for tissue engineering and regeneration.

The long-term goal of tissue engineering is to engineer functional tissues *in vitro* for implantation *in vivo* to repair, enhance, and replace, to preserve physiological function. Tissue engineering is based on the principles of develop-mental biology, evolution and self-assembly of supramolecular assemblies and higher hierarchal tissues and even whole embryos and organisms (Figs. 9.2 and 9.3). Regeneration recapitulates embryonic development and morphogenesis. Among the many tissues in the human body, bone has considerable powers for regeneration and, therefore, is a prototype model for tissue engineering. On the other hand, articular cartilage, a tissue adjacent to bone, is recalcitrant to repair and regeneration. Implantation of demineralized bone matrix into subcutaneous sites results in local bone induction. The sequential cascade of bone morphogenesis mimics sequential skeletal morphogenesis in limbs and permits the isolation of bone morphogens. Although it is traditional to study morphogenetic signals in embryos, bone morphogenetic proteins (BMPs), the primordial inductive signals for bone were isolated from demineralized bone matrix from adults. BMPs initiate, promote, and maintain chondrogenesis and osteogenesis and have actions beyond bone. The recently identified cartilage-derived morphogenetic proteins (CDMPs) are critical for cartilage and joint morphogenesis. The symbiosis of bone inductive and con-

## KEY INGREDIENTS FOR TISSUE ENGINEERING

- Morphogenetic Signals
- Responding Stem Cells
- Extracellular Matrix Scaffolding

**FIG. 9.1.** The tissue-engineering triad consists of signals, stem cells, and scaffolding.

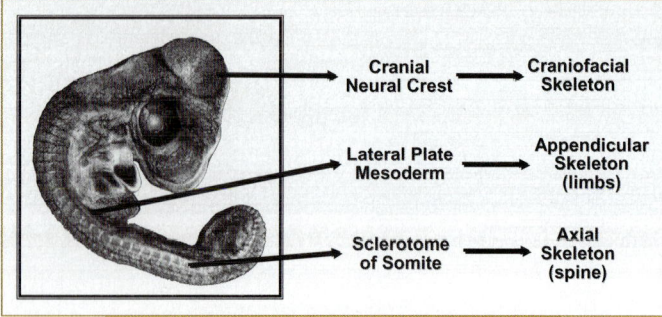

**FIG. 9.3.** Developmental origins of skeleton in the chick embryo. The cranial neural crest gives rise to craniofacial skeleton. The lateral plate mesoderm gives rise to the limbs of the appendicular skeleton. The sclerotome of the somite gives rise to spine and the axial skeleton.

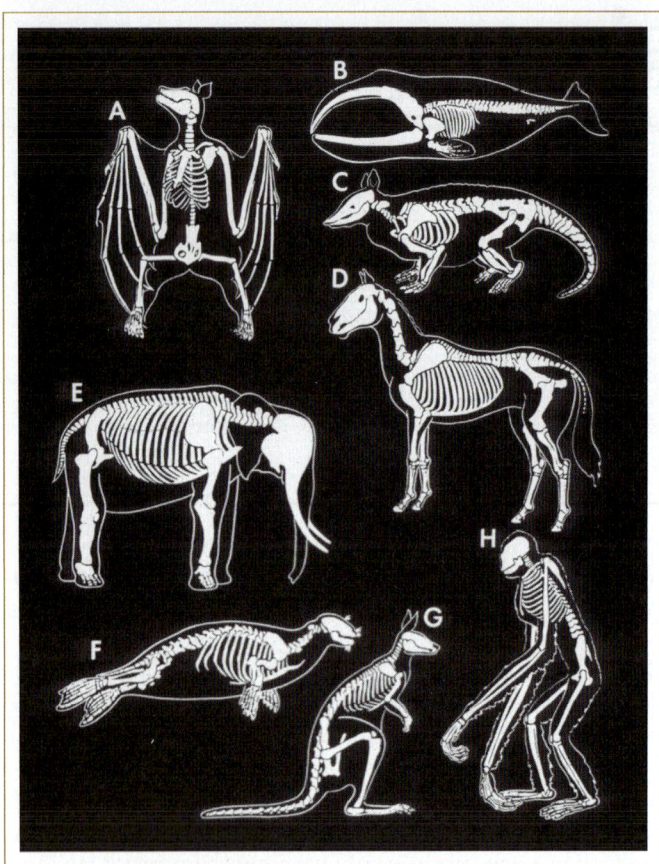

**FIG. 9.2.** Evolution of skeletal structures in a variety of mammals adapted for flight (bat, A) and aquatic life (whale, B) and the use of hands in humans (H).

ductive strategies is critical for tissue engineering and is in turn governed by the context and biomechanics. The context is the microenvironment, consisting of extracellular matrix scaffolding, and can be duplicated by biomimetic biomaterials, such as collagens, hydroxyapatite, proteoglycans, and cell adhesion proteins, including fibronectins and laminins. The rules of architecture for tissue engineering are an imitation and adoption of the laws and signals of developmental biology and morphogenesis, and thus they may be universal for all tissues, including bones and joints and associated musculoskeletal tissues in the limbs.

The traditional approach for identification and isolation of morphogens is first to identify genes in fly and frog embryos by means of genetic approaches, differential displays, substractive hybridization, and expression cloning. This information is subsequently extended to mice and men. An alternative approach is to isolate morphogens from bone, the premier tissue with the highest regenerative potential. Morphogenesis is the developmental cascade of pattern formation, the establishment of the body plan and architecture of mirror-image bilateral symmetry of musculoskeletal structures in the appendicular skeleton, culminating in the adult form. The expanding knowledge in bone and cartilage morphogenesis is a prototypical paradigm for all of tissue engineering. The principles gleaned from bone morphogenesis and BMPs can be extended to tissue engineering of bone and cartilage and other tissues.

## II. BONE MORPHOGENETIC PROTEINS (BMPs)

Bone grafts have been used by orthopedic surgeons for nearly a century to aid and abet recalcitrant bone repair. Decalcified bone implants have been used to treat patients with osteomyelitis (Senn, 1989). It was hypothesized that bone might contain a substance, osteogenin, that initiates bone growth (Lacroix, 1945). Urist (1965) made the key discovery that demineralized, lyophilized segments of rabbit bone, when implanted intramuscularly, induced new bone formation. The diaphysis (shafts) of long bones of rats were cleaned of marrow, pulverized, and sieved. The demineralization of matrix was accomplished by 0.5 M HCl (Fig. 9.4). Bone induction, a sequential multistep cascade, is depicted in Fig. 9.5 (Reddi and Huggins, 1972; Reddi and Anderson, 1976; Reddi, 1981). The key steps in this cascade are chemotaxis, mitosis, and differentiation. Chemotaxis is the directed migration of cells in response to a chemical gradient of signals released from the insoluble demineralized bone

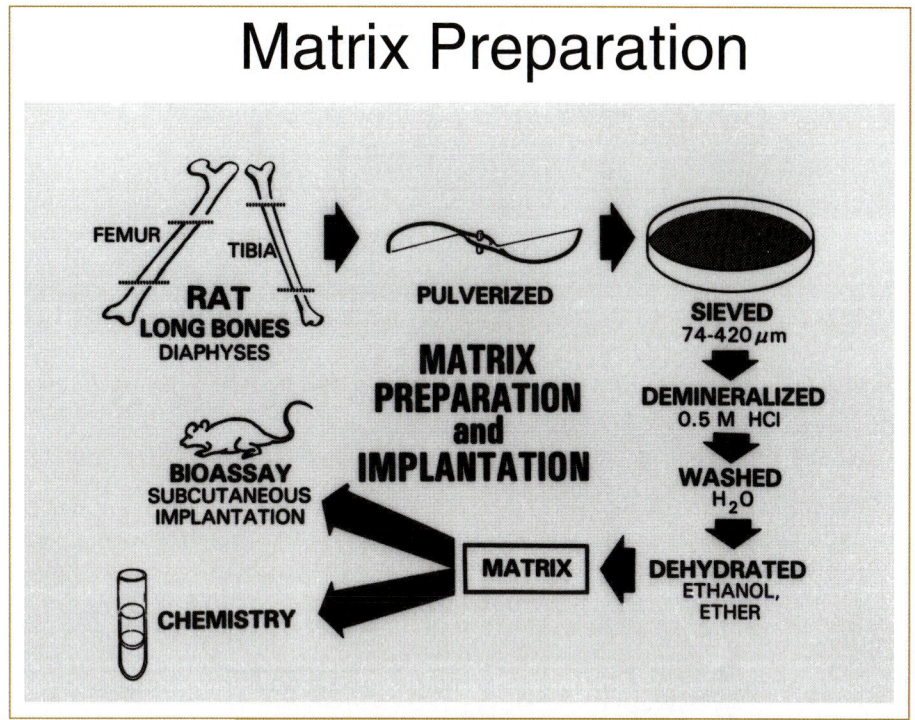

# Matrix Preparation

**FIG. 9.4.** Preparation of the demineralized bone matrix (DBM). The diaphysis (shafts) of femur and tibia are cleansed of marrow and dried prior to pulverization. The pulverized bone matrix is sieved to a particle size of 74–420 µm and demineralized by 0.5 M HCl, dehydrated in ethanol and diethyl ether. The resulting matrix is a potent inducer of cartilage and bone, and we isolate, by means of chemical techniques, the active osteoinductive agent BMP.

matrix. The demineralized bone matrix is composed predominantly of type I insoluble collagen, and it binds plasma fibronectin (Weiss and Reddi, 1980). Fibronectin has domains for binding to collagen, fibrin, and heparin. The responding mesenchymal cells attached to the collagenous matrix and proliferated as indicated by [³H]thymidine autoradiography and incorporation into acid-precipitable DNA on day 3 (Rath and Reddi, 1979). Chondroblast differentiation was evident on day 5, chondrocytes on days 7–8, and cartilage hypertrophy on day 9 (Fig. 9.5). Vascular invasion was concomitant on day 9 with osteoblast differentiation. On days 10–12 alkaline phosphatase was maximal. Osteocalcin, bone γ-carboxyglutamic acid containing gla protein (BGP), increased on day 28. Hematopoietic marrow differentiated in the ossicle and was maximal by day 21. This entire sequential bone development cascade is reminiscent of bone and cartilage morphogenesis in the limb bud (Reddi, 1981, 1984). Hence, it has immense implications for isolation of inductive signals initiating cartilage and bone morphogenesis. In fact, a systematic investigation of the chemical components responsible for bone induction was undertaken.

The foregoing account of the demineralized bone matrix–induced bone morphogenesis in extraskeletal sites demonstrated the potential role of morphogens tightly associated with the extracellular matrix. Next, we embarked on a systematic study of the isolation of putative morphogenetic proteins. A prerequisite for any quest for novel morphogens is the establishment of a battery of bioassays for new bone formation. A panel of *in vitro* assays were established for chemotaxis, mitogenesis, and chondrogenesis, and an *in vivo* bioassay was established for bone formation. Although the *in vitro* assays are expedient, we monitored routinely a labor-intensive *in vivo* bioassay, for it was the only bona fide bone induction assay.

A major stumbling block in the approach was that the demineralized bone matrix is insoluble and is in the solid state. In view of this, dissociative extractants such as 4 M guanidine HCl or 8 M urea as 1% sodium dodecyl sulfate (SDS) at pH 7.4 were used (Sampath and Reddi, 1981) to solubilize proteins. Approximately 3% of the proteins were solubilized from demineralized bone matrix, and the remaining residue was mainly insoluble type I bone collagen. The extract alone or the residue alone was incapable of new bone induction. However, addition of the extract to the residue (insoluble collagen) and then implantation in a subcutaneous site resulted in bone induction (Fig. 9.6). Thus, it would appear that for optimal osteogenic activity there was a collaboration between the soluble signal in the extract and insoluble substratum or scaffolding (Sampath and Reddi, 1981). Thus, an operational concept of tissue engineering was established that soluble signals bound to extracellular matrix scaffold act on responding stem/progenitor cells to induce tissue digestion. This bioassay was a useful advance in the final purification of bone morphogenetic proteins and led to the determination of limited tryptic peptide sequences leading to the eventual cloning of BMPs (Wozney *et al.*, 1988; Luyten *et al.*, 1989; Ozkaynak *et al.*, 1990).

In order to scale up the procedure, a switch was made to bovine bone. Demineralized bovine bone was not osteo-

# Sequential Cascade

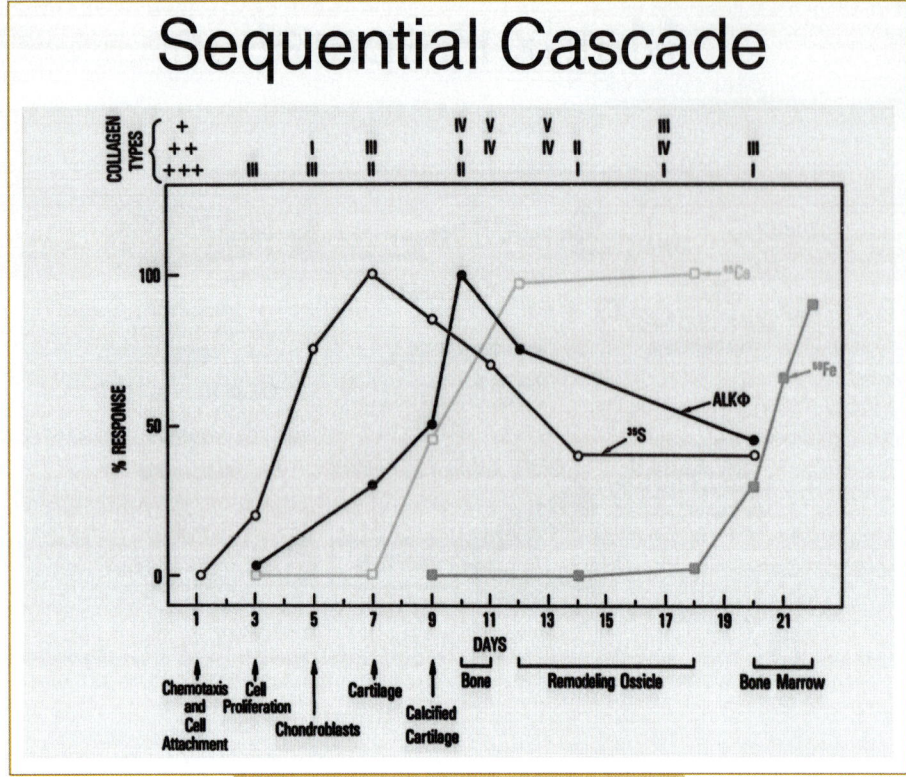

**FIG. 9.5.** Developmental sequence of extracellular matrix–induced cartilage, bone, and marrow formation. Changes in $^{35}SO_4$ incorporation into proteoglycans and $^{45}Ca$ incorporation into the mineral phase indicate peaks of cartilage and bone formation, respectively. The $^{59}Fe$ incorporation into heme is an index of erythropoiesis, as plotted from the data of Reddi and Anderson (1976). The values for alkaline phosphatase indicate early stages of bone formation (Reddi and Huggins 1972). The transitions in collagen types I to IV, summarized on top of the figure, are based on immunofluorescent localization = polymorphonuclear leukocytes. (*Source*: Reddi (1981), with permission.)

## Dissociative Extraction and Reconstitution

| DBM | Activity |
|---|---|
| | +++ |
| 4M ↓ Guanidine | |
| Collagen | — |
| Extract | — |
| + | +++ |

**FIG. 9.6.** Dissociative extraction by chaotropic reagents such as 4 M guanidine and reconstitution of osteoinductive activity with insoluble collagenous matrix. The results demonstrate a collaboration between a soluble signal and insoluble extracellular matrix. This experiment further established the basic tenets of tissue engineering in 1981 as signals, scaffolds, and responding stem cells.

inductive in rats, and the results were variable. However, when the guanidine extracts of demineralized bovine bone were fractionated on an S-200 molecular sieve column, fractions less than 50 kD were consistently osteogenic when bioassayed after reconstitution with allogeneic insoluble collagen (Sampath and Reddi, 1983; Reddi, 1994). Thus, protein fractions inducing bone were not species specific and appear to be homologous in several mammals. It is likely that larger molecular mass fractions and/or the insoluble xenogeneic (bovine and human) collagens were inhibitory or immunogenic. Initial estimates revealed 1 μg of active osteogenic fraction in a kilogram of bone. Hence, over a ton of bovine bone was processed to yield optimal amounts for animo acid sequence determination. The amino acid sequences revealed homology to TGF-$\beta_1$ (Reddi, 1994). The important work of Wozney and colleagues (1988) cloned BMP-2, BMP-2B (now called BMP-4), and BMP-3 (also called osteogenin). Osteogenic protein-1 and -2 (OP-1 and OP-2) were cloned by Ozkaynak and colleagues (1990). There are nearly 10 members of the BMP family (Table 9.1). The other members of the extended TGF-$\beta$/BMP superfamily include inhibins and activins (implicated in follicle-stimulating hormone release from pituitary), Müllerian duct inhibitory substance (MIS), growth/differentiation factors (GDFs), nodal, and lefty, a gene implicated in establishing right/left asymmetry (Reddi, 1997; Cunningham *et al.*, 1995). BMPs are also involved in embryonic induction (Lemaire and Gurdon, 1994; Melton 1991; Lyons *et al.*, 1995; Reddi, 1997).

| Table 9.1. | Bone morphogenetic proteins | |
|---|---|---|
| *BMP* | *Other names* | *Function* |
| BMP-2 | BMP-2A | Bone and cartilage morphogenesis |
| BMP-3 | Osteogenin | Bone morphogenesis |
| BMP-3B | GDF-10 | Intramembranous bone formation |
| BMP-4 | BMP-2B | Bone morphogenesis |
| BMP-5 | | Bone morphogenesis |
| BMP-6 | | Cartilage hypertrophy |
| BMP-7 | Osteogenic protein-1 | Bone formation |
| BMP-8 | Osteogenic Protein-2 | Bone formation |
| BMP-8B | | Spermatogenesis |
| BMP-9 | | Liver differentiation |
| BMP-10 | | ? |
| | | Tooth differentiation |
| BMP-11 | GDF-11 | Odontoblast regulation |
| BMP-15 | | ? |

| Table 9.2. | Cartilage-derived morphogenetic proteins | |
|---|---|---|
| *CDMP* | *Other names* | *Function* |
| CDMP 1 | GDF-5 | Cartilage condensation |
| CDMP 2 | GDF-6 | Cartilage formation, hypertrophy |
| CDMP 3 | GDF-7 | Tendon/ligament morphogenesis |

BMPs are dimeric molecules, and the conformation is critical for biological actions. Reduction of the single interchain disulfide bond resulted in the loss of biological activity. The mature monomer molecule consists of about 120 amino acids, with seven canonical cysteine residues. There are three intrachain disulfides per monomer and one interchain disulfide bond in the dimer. In the critical core of the BMP monomer is the cysteine knot. The crystal structure of BMP-7 has been determined (Griffith *et al.*, 1996). It is a good possibility that in the near future the crystal structure of BMP-receptor and receptor contact domains will be determined (Griffith *et al.*, 1996).

## III. CARTILAGE-DERIVED MORPHOGENETIC PROTEINS (CDMPs)

Morphogenesis of the cartilage is the key rate-limiting step in the dynamics of bone development. Cartilage is the initial model for the architecture of bones. Bone can form either directly from mesenchyme, as in intramembranous bone formation, or with an intervening cartilage stage, as in endochondral bone development (Reddi, 1981). All BMPs induce, first, the cascade of chondrogenesis, and therefore in this sense they are cartilage morphogenetic proteins. The hypertrophic chondrocytes in the epiphyseal growth plate mineralize and serve as a template for appositional bone morphogenesis. Cartilage morphogenesis is critical for both bone and joint morphogenesis. The two lineages of cartilage are clear-cut. The first, at the ends of bone, forms articulating articular cartilage. The second is the growth plate chondrocytes, which, through hypertrophy, synthesize cartilage

matrix destined to calcify prior to replacement by bone, and are the "organizer" centers of longitudinal and circumferental growth of cartilage, setting into motion the orderly program of endochondral bone formation. The phenotypic stability of the articular (permanent) cartilage is at the crux of the osteoarthritis problem. The maintenance factors for articular chondrocytes include TGF-β isoforms and the BMP isoforms (Luyten *et al.*, 1992).

An *in vivo* chondrogenic bioassay with soluble purified proteins and insoluble collagen scored for chondrogenesis. A concurrent reverse transcription–polymerase chain reaction (RT-PCR) approach was taken with degenerate oligonucleotide primers. Two novel genes for cartilage-derived morphogenetic proteins (CDMPs) 1 and 2 were identified and cloned (Chang *et al.*, 1994). CDMPs 1 and 2 are also called GDF-5 and GDF-6 (Storm *et al.*, 1994). CDMPs are related to bone morphogenetic proteins (Table 9.2). CDMPs are critical for cartilage and joint morphogenesis (Tsumaki *et al.*, 1999). CDMPs stimulate proteoglycan synthesis in cartilage. CDMP 3 (also known as GDF-7) initiates tendon and ligament morphogenesis (Wolfman *et al.*, 1998).

## IV. PLEIOTROPY AND THRESHOLDS

Morphogenesis is a sequential multistep cascade. BMPs regulate each of the key steps: chemotaxis, mitosis, and differentiation of cartilage and bone (Fig. 9.7). BMPs initiate chondrogenesis in the limb (S. Chen *et al.*, 1991; Duboule, 1994). The apical ectodermal ridge is the source of BMPs in the developing limb bud. The intricate dynamic, reciprocal interactions between the ectodermally derived epithelium and mesodermally derived mesenchyme sets into motion the train of events culminating in the pattern of phalanges, radius, ulna, and the humerus.

The chemotaxis of human monocytes is optimal at femtomolar concentration (Cunningham *et al.*, 1992). The apparent affinity was 100–200 pM. The mitogenic response was optimal at the 100-pM range. The initiation of differentiation was in the nanomolar range in solution. However, caution should be exercised, because BMPs may be sequestered by extracellular matrix components, and the local concentration may be higher when BMPs are bound on the extracellular matrix. A single recombinant BMP human 4 can govern chemotaxis and mitosis differentiation of cartilage and bone, maintain phenotype, stimulate extracellular

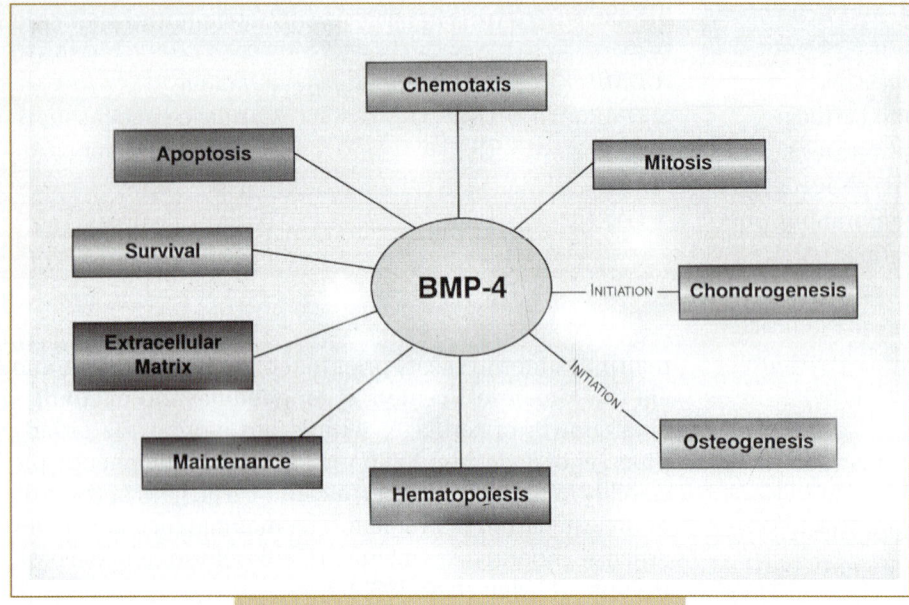

**FIG. 9.7.** BMPs are pleiotropic molecules. Pleiotropy is the property of a single gene or protein to act on a multiplicity of cellular phenomena and targets.

matrix, and promote survival of some cells but cause the death of others (Fig. 9.7). Thus BMPs are pleiotropic regulators that act in concentration-dependent thresholds.

## V. BMPs BIND TO EXTRACELLULAR MATRIX

It is well known that extracellular matrix components play a critical role in morphogenesis. The structural macromolecules and their supramolecular assembly in the matrix do not explain their role in epithelial–mesenchymal interaction and morphogenesis. This riddle can now be explained by the binding of bone morphogenetic proteins to heparan sulfate heparin and type IV collagen (Paralkar *et al.*, 1990, 1991, 1992) of the basement membranes. In fact, this might explain in part the necessity for angiogenesis prior to osteogenesis during development. In addition, the actions of activin in development of the frog, in terms of dorsal mesoderm induction, is modified to neuralization by follistatin (Hemmati-Brivanlou *et al.*, 1994). Similarly, Chordin and Noggin from the Spemann organizer induce neuralization via binding and inactivation of BMP-4 (Fig. 9.2). Thus neural induction is likely to be a default pathway when BMP-4 is nonfunctional (Piccolo *et al.*, 1996; Zimmerman *et al.*, 1996). Thus, an emerging principle in development and morphogenesis is that binding proteins can terminate a dominant morphogen's action and initiate a default pathway. Finally, the binding of a soluble morphogen to extracellular matrix converts it into an insoluble matrix–bound morphogen to act locally in the solid state (Paralkar *et al.*, 1990).

Although BMPs were isolated and cloned from bone, work with gene knockouts have revealed a plethora of actions beyond bone. Mice with targeted disruption of BMP-2 caused embryonic lethality. The heart development is abnormal, indicating a need for BMP-2 in heart development (Zhang and Bradley, 1996). BMP-4 knockouts exhibit no mesoderm induction, and gastrulation is impaired (Winnier *et al.*, 1996). Transgenic overexpression of BMPs under the control of keratin 10 promoter leads to psoriasis. The targeted deletion of BMP-7 revealed the critical role of this molecule in kidney and eye development (Luo *et al.*, 1995; Dudley *et al.*, 1995; Vukicevic *et al.*, 1996). Thus the BMPs are really true morphogens for such disparate tissues as skin, heart, kidney, and eye. In view of this, BMPs may also be called body morphogenetic proteins (Reddi, 2005).

## VI. BMP RECEPTORS

Recombinant human BMP-4 and BMP-7 bind to BMP receptor IA (BMPR-IA) and BMP receptor IB (BMPR-IB) (ten Dijke *et al.*, 1994). CDMP-1 also binds to both type I BMP receptors. There is a collaboration between type I and type II BMP receptors (Nishitoh *et al.*, 1996). The type I receptor serine/threonine kinase phosphorylates a signal-transducing protein substrate called Smad 1 or 5 (S. Chen *et al.*, 1996). *Smad* is a term derived from the fusion of the *Drosophila* Mad gene and the *Caenorhabtitis elegans* (nematode) Sma gene. Smads 1 and 5 signal in partnership with a common co-Smad, Smad 4 (Fig. 9.8). The transcription of BMP-response genes are initiated by Smad 1/Smad 4 heterodimers. Smads are trimeric molecules, as gleaned via x-ray crystallography. The phosphorylation of Smads 1 and 5 by type I BMP receptor kinase is inhibited by inhibitory Smads 6 and 7 (Hayashi *et al.*, 1997). Smad-interacting protein (SIP) may interact with Smad 1 and modulate BMP-response gene expression (Heldin *et al.*, 1997; Reddi, 1997; Miyazono *et al.*, 2005). The downstream targets of BMP signaling are likely to be homeobox genes, the cardinal genes for morphogenesis and transcription. BMPs in turn may be

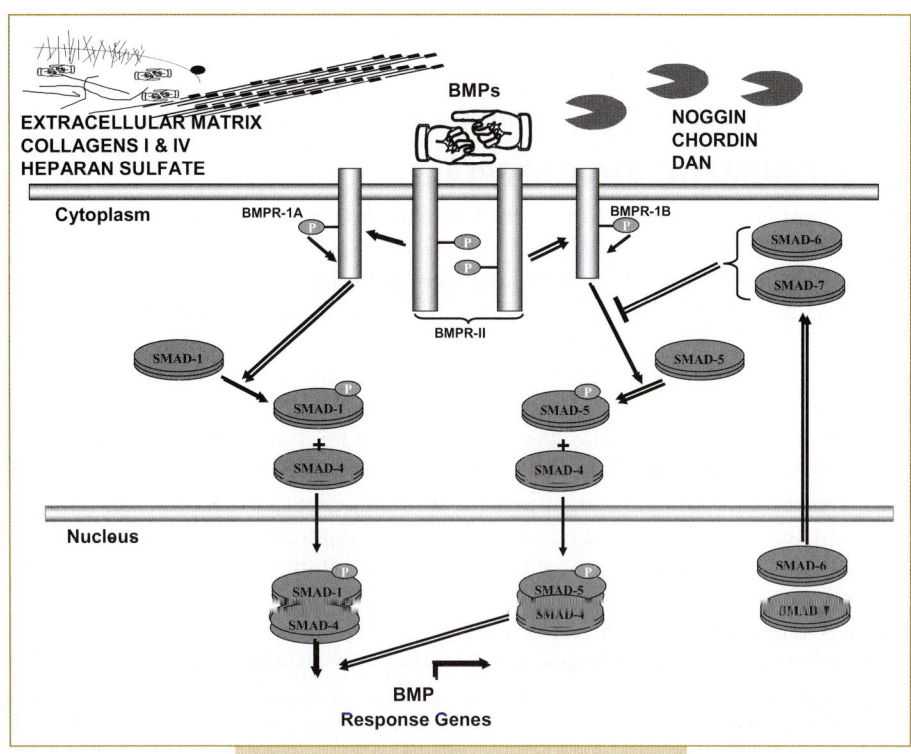

**FIG. 9.8.** BMP receptors and signaling cascades. BMPs are dimeric ligands with cysteine knot in each monomer fold. Each monomer has two β-sheets, represented as two pointed fingers. In the functional dimer, the fingers are oriented in opposite directions. BMPs interact with both type I and II BMP receptors. The exact stoichiometry of the receptor complex is currently being elucidated. BMPR-II phosphorylates the GS domain of BMPR-I. The collaboration between type I and II receptors forms the signal-transducing complex. BMP type I receptor kinase complex phosphorylates the trimeric signaling substrates Smad 1 and Smad 5. This phosphorylation is inhibited and modulated by inhibitory Smads 6 and 7. Phosphorylated Smad 1 or 5 interacts with Smad 4 (functional partner) and enters the nucleus to activate the transcriptional machinery for early BMP-response genes. A novel Smad-interacting protein (SIP) may interact and modulate the binding of heteromeric Smad 1/Smad 4 complexes to the DNA.

regulated by members of the hedgehog family of genes, such as Sonic and Indian hedgehog (Johnson and Tabin, 1997), including receptors patched and smoothened and transcription factors such as Gli 1, 2, and 3. The actions of BMPs can be terminated by specific binding proteins, such as noggin (Zimmerman *et al.*, 1996).

## VII. RESPONDING STEM CELLS

It is well known that the embryonic mesoderm-derived mesenchymal cells are progenitors for bone, cartilage, tendons, ligaments, and muscle. However, certain stem cells in adult bone marrow, muscle, and fascia can form bone and cartilage. The identification of stem cells readily sourced from bone marrow may lead to banks of stem cells for cell therapy and perhaps gene therapy with appropriate "homing" characteristics to bone marrow and hence to the skeleton. The pioneering work of Friedenstein *et al.* (1968) and Owen and Friedenstein (1988) identified bone marrow stromal stem cells. These stromal cells are distinct from the hematopoietic stem cell lineage. The bone marrow stromal stem cells consist of inducible and determined osteoprogenitors committed to osteogenesis. Determined osteogenic precursor cells have the propensity to form bone cells, without any external cues or signals. On the other hand, inducible osteogenic precursors require an inductive signal, such as BMP or demineralized bone matrix. It is noteworthy that operational distinctions between stromal stem cells and hematopoietic stem cells are getting more and more blurry! The stromal stem cells of Friedenstein and Owen are also called mesenchymal stem cells (Caplan, 1991), with

potential to form bone, cartilage, adipocytes, and myoblasts in response to cues from the environment and/or intrinsic factors. There is considerable hope and anticipation that these bone marrow stromal cells may be excellent vehicles for cell and gene therapy (Prockop, 1997).

From a practical standpoint, these stromal stem cells can be obtained by bone marrow biopsies and expanded rapidly for use in cell therapy after pretreatment with bone morphogenetic proteins. The potential uses in both cell and gene therapy is very promising. There are continuous improvements in the viral vectors and efficiency of gene therapy (Bank, 1996; Mulligan, 1993). For example, it is possible to use BMP genes transfected in stromal stem cells to target the bone marrow.

## VIII. MORPHOGENS AND GENE THERAPY

The recent advances in morphogens are ripe for techniques of regional gene therapy for orthopedic tissue engineering. The availability of cloned genes for BMPs and CDMPs and the requisite platform technology of gene therapy may have immediate applications. Whereas protein therapy provides an immediate bolus of morphogen, gene therapy achieves a sustained, prolonged secretion of gene products. Furthermore, recent improvements in regulated gene expression allows the turning on and off of gene expresssion. The progress in vectors for delivering genes also bodes well. The use of adenoviruses, adeno-associated viruses, and tetroviruses is poised for applications in bone and joint repair (Bank, 1996; Kozarsky and Wilson, 1993; Morsy *et al.*, 1993; Mulligan, 1993). Although gene therapy has some advantages for orthopedic tissue engineering, an optimal delivery system for protein and gene therapy is needed, especially in the replacement of large segmented defects and in fibrous nonunions and malunions.

## IX. BIOMIMETIC BIOMATERIALS

Our earlier discussions of inductive signals (BMPs) and responding stem cells (stromal cells) lead us to the scaffolding (the microenvironment/extracellular matrix) for optimal tissue engineering. The natural biomaterials in the composite tissue of bones and joints are collagens, proteoglycans, and glycoproteins of cell adhesion, such as fibronectin and the mineral phase. The mineral phase in bone is predominantly hydroxyapatite. In native state, the associated citrate, fluoride, carbonate, and trace elements constitute the physiological hydroxyapatite. The high protein-binding capacity makes hydroxyapatite a natural delivery system. Comparison of insoluble collagen, hydroxyapatite, tricalcium phosphate, glass beads, and polymethylmethacrylate as carriers revealed collagen to be an optimal delivery system for BMPs (Ma *et al.*, 1990). It is well known that collagen is an ideal delivery system for growth factors in soft and hard tissue wound repair (McPherson, 1992). Hydrogels may be of great utility in cartilage tissue engineering (Fisher *et al.*, 2004).

During the course of systematic work on hydroxyapatite of two pore sizes (200 or 500 μm) in two geometrical forms (beads or discs), an unexpected observation was made. The geometry of the delivery system is critical for optimal bone induction. The discs were consistently osteoinductive with BMPs in rats, but the beads were inactive (Ripamonti *et al.*, 1992). The chemical composition of the two hydroxyapatite configurations were identical. In certain species, the hydroxyapatite alone appears to be "osteoinductive" (Ripamonti, 1996). In subhuman primates, the hydroxyapatite induces bone, albeit at a much slower rate. One interpretation is that osteoinductive endogenous BMPs in circulation progressively bind to an implanted disc of hydroxyapatite. When an optimal threshold concentration of native BMPs is achieved, the hydroxyapatite becomes osteoinductive. Strictly speaking, most hydroxyapatite substrata are ideal osteoconductive materials. This example in certain species also serves to illustrate how an osteoconductive biomimetic biomaterial can progressively function as an osteoinductive substance by binding to endogenous BMPs. Thus, there is a physiological-physicochemical continuum between the hydroxyapatite alone and progressive composites with endogenous BMPs. Recognition of this experimental nuance will save unnecessary arguments among biomaterials scientists about the osteoinductive action of a conductive substratum such as hydroxyapatite.

Complete regeneration of baboon craniotomy defect was accomplished via recombinant human osteogenic protein (rhOP-1; human BMP-7) (Ripamonti *et al.*, 1996). Recombinant BMP-2 was delivered by poly.(-hydroxy acid) carrier for calvarial regeneration (Hollinger *et al.*, 1996).

Copolymer of polylactic acid and polyglycolic acid with recombinant BMP-2 were used in a nonunion model in rabbit ulna and complete unions were achieved in the bone (Bostrom *et al.*, 1996).

An important problem in the clinical application of biomimetic biomaterials with BMPs and/or other morphogens would be the sterilization. Although gas (ethylene oxide) is used, one should always be concerned about reactive free radicals. Using allogeneic demineralized bone matrix with endogenous native BMPs, as long as low temperature (4°C or less) is maintained, the samples tolerated up to 5–7 M rads of irradiation (Weintroub and Reddi, 1988; Weintroub *et al.*, 1990). The standard dose acceptable to the Food and Drug Administration is 2.5 M rads. This information would be useful to the biotechnology companies preparing to market recombinant BMP-based osteogenic devices. Perhaps, the tissue-banking industry, with its interest in bone grafts (Damien and Parson, 1991), could also use this critical information. The various freeze-dried and demineralized allogeneic bone may be used in the interim as satisfactory carriers for BMPs. The moral of this experiment is it is not the irradiation dose but the ambient sample temperature during irradiation that is absolutely critical.

## X. TISSUE ENGINEERING OF BONES AND JOINTS

Unlike bone, with its considerable prowess for repair and even regeneration, cartilage is recalcitrant. But why? In part this may be due to the relative avascularity of hyaline cartilage and the high concentration of protease inhibitors and perhaps even of growth inhibitors. The wound-debridement phase is not optimal for preparing the cartilage wound bed for the optimal *milieu interieur* for repair. Although cartilage has been successfully engineered to predetermined shapes (Kim *et al.*, 1994), true repair of the tissue continues to be a real challenge, in part due to hierarchical organization and geometry (Mow *et al.*, 1992). However, considerable excitement in the field has been generated by a group of Swedish workers in Gōteborg, using autologous culture-expanded human chondrocytes (Brittberg *et al.*, 1994). A continuous challenge in chondrocyte cell therapy is progressive dedifferentiation and loss of characteristic cartilage phenotype. The redifferentiation and maintenance of the chondrocytes for cell therapy can be aided by BMPs, CDMPs, TGF-β isoforms, and IGFs. It is also possible to repair cartilage using muscle-derived mesenchymal stem cells (Grande *et al.*, 1995). The potential possibility of the problems posed by cartilage proteoglycans in preventing cell immigration for repair was investigated by means of chondroitinase ABC and trypsin pretreatment in partial-thickness defects (Hunziker and Rosenberg, 1996), with and without TGF. Pretreatment with chondroitinase ABC followed by TGF revealed a contiguous layer of cells from the synovial membrane, hinting at the potential source of repair

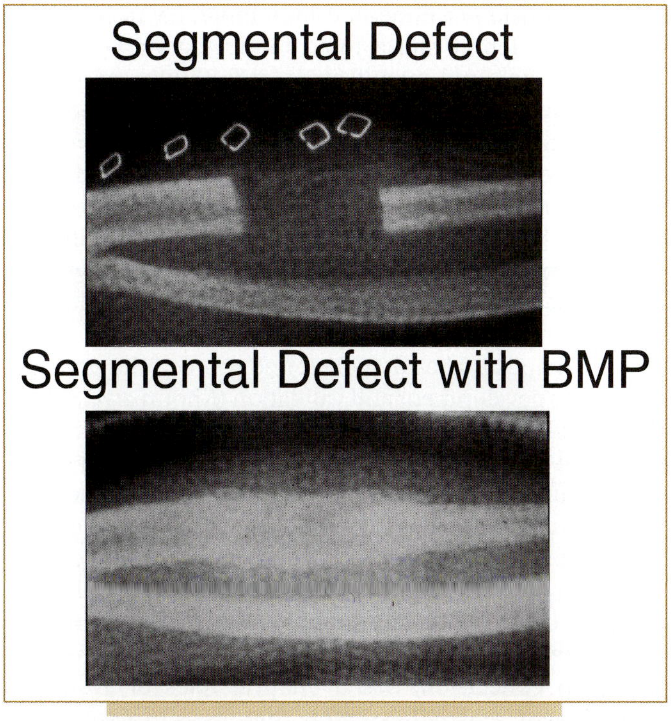

## Segmental Defect

## Segmental Defect with BMP

**FIG. 9.9.** Repair of segmental defects in a primate bone by means of recombinant human BMP-7 and collagenous matrix scaffold. (Photographs provided by Dr. T. K. Sampath.)

cells from synovium. Multiple avenues of cartilage morphogens, cell therapy with chondrocytes and stem cells from marrow and muscle, and a biomaterial scaffolding may lead to an optimal tissue-engineered articular cartilage.

Recombinant human and BMP-2 and BMP-7 were approved in 2002 by the Food and Drug Administration (FDA) for tibial nonunions and single-level spine fusion. BMPs have been used in healing segmental defects (Fig. 9.9). The proof of the principles of tissue engineering based on BMPs as signals, mold of a scaffold, and responding cells can be demonstrated (Fig. 9.10).

## XI. FUTURE CHALLENGES

It is inevitable during the aging of humans that one will confront impaired locomotion due to wear and tear in bones and joints. Therefore, the repair and possibly complete regeneration of the musculoskeletal system and other vital organs, such as skin, liver, and kidney, may potentially need optimal repair or a spare part for replacement. Can we create spare parts for the human body? There is much reason for optimism that tissue engineering can help patients. We are living in an extraordinary time with regard to biology, medicine, surgery, bioengineering, computer modeling of predictive tissue engineering, and technology. The confluence of advances in molecular developmental biology and attendant advances in inductive signals for morphogenesis, stem cells, biomimetic biomaterials, and extracellular matrix biology augers well for imminent breakthroughs.

The symbiosis of biotechnology and biomaterials has set the stage for systematic advances in tissue engineering

**FIG. 9.10.** Proof of the principles of tissue engineering was established *in vivo* by Khouri *et al.* (1991). A mold was used to contain the vascularized muscle flap and treated with purified BMPs and collagen scaffold. The newly formed bone faithfully reproduced the shape of the mold. In the future, one can use stem cells directed by recombinant BMPs to induce bone.

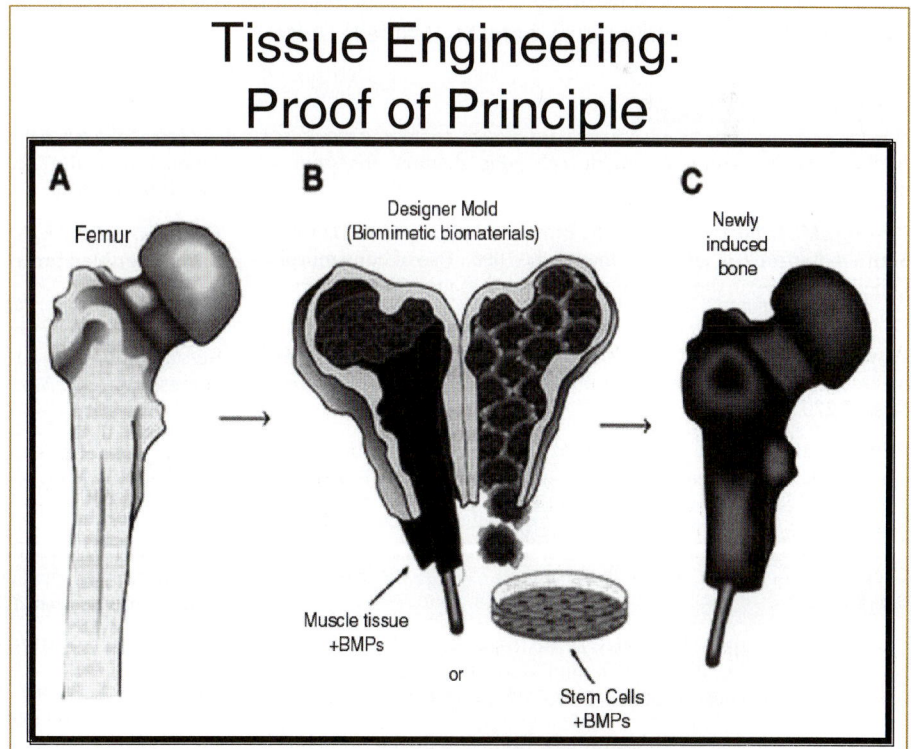

## Tissue Engineering: Proof of Principle

A — Femur

B — Designer Mold (Biomimetic biomaterials)

C — Newly induced bone

Muscle tissue +BMPs

or

Stem Cells +BMPs

(Reddi, 1994; Langer and Vacanti, 1993; Hubbel, 1995). The recent advances in enabling platform technology includes molecular imprinting (Mosbach and Ramstrom, 1996). In principle, specific recognition and catalytic sites are imprinted using templates. The applications include biosensors, catalytic applications to antibody, and receptor recognition sites. For example, the cell-binding RGD site in fibronectin (Ruoslahti and Pierschbacher, 1987) or the YIGSR domain in laminin can be imprinted in complementary sites (Vukicevic *et al.*, 1990).

The rapidly advancing frontiers in morphogenesis with BMPs, hedgehogs, homeobox genes, and a veritable cornucopia of general and specific transcription factors, coactivators, and repressors will lead to cocrystallization of ligand–receptor complexes, protein–DNA complexes, and other macromolecular interactions. This will lead to peptidomimetic agonists for large proteins, as exemplified by erythroprotein (Livnah *et al.*, 1996). To such advances one can add new developments in self-assembly of millimeter-scale structures floating at the interface of perfluorodecalin and water and interacting by means of capillary forces controlled by the pattern of wettablity (Bowden *et al.*, 1997). The final self-assembly is due to minimization of free energy in the interface. These are truly incredible advances that will lead to man-made materials that mimic extracellular matrix in tissues. Superimpose on such chemical progress a bio-logical platform in a bone-and-joint mold. Let us imagine a head of the femur and a mold fabricated via computer-assisted design and manufacture. It faithfully reproduces the structural features and may be imprinted with morphogens, inductive signals, and cell adhesion sites. This assembly can be loaded with stem cells and BMPs and other inductive signals, with a nutrient medium optimized for the number of cell cycles, and then it predictably exits into the differentiation phase to reproduce a totally new bone femoral head. In fact, such a biological approach with vascularized muscle flap and BMPs yielded new bone with a defined shape and has demonstrated proof of the principle for further development and validation (Khouri *et al.*, 1991). We indeed are entering a brave new world of prefabricated biological spare parts for the human body, based on sound architectural rules of inductive signals for morphogenesis, responding stem cells with lineage control, and with growth factors immobilized on a template of biomimetic biomaterial based on extracellular matrix. Like life itself, such technologies evolve with continuous refinements to benefit humankind by reducing the agony of human pain and suffering. In conclusion, based on principles of evolution, development, and self-assembly, the fields of tissue engineering and regenerative medicine are poised to make explosive advances with immense applications in the clinic.

## XII. ACKNOWLEDGMENTS

This work is supported by the Lawrence Ellison Chair in Musculoskeletal Molecular Biology of the Lawrence Ellison Center for Tissue Regeneration. I thank Ms. Danielle Neff for outstanding bibliographic assistance and help with the figures. Our research is supported by grants from the NIH.

## XIII. REFERENCES

Bank, A. (1996). Human somatic cell gene therapy. *Bioessays* **18**, 999–1007.

Bostrom, M., Lane, J. M., Tomin, E., Browne, M., Berbian, W., Turek, T., Smith, J., Wozney, J., and Schildhauer, T. (1996). Use of bone morphogenetic protein-2 in the rabbit ulnar nonunion model. *Clin. Orthop. Rel. Res.* **327**, 272–282.

Bowden, N., Terfort, A., Carbeck, J., and Whitesides, G. M. (1997). Self-assembly of mesoscale objects into ordered two-dimensional assays. *Science* **276**, 233–235.

Brittberg, M., Lindahl, A., Nilsson, A., Ohlsson, C., Isaksson, O., and Peterson, L. (1994). Treatment of deep cartilage defects in the knee with autologous chondrocyte transplantation. *New England J. Med.* **331**, 889–895.

Caplan, A. I. (1991). Mesenchymal stem cell. *J. Orthop. Res.* **9**, 641–650.

Chang, S. C., Hoang, B., Thomas, J. T., Vukicevic, S., Luyten, F. P., Ryban, N. J. P., Kozak, C. A., Reddi, A. H., and Moos, Jr., M. (1994). Cartilage-derived morphogenetic proteins. *J. Biol. Chem.* **269**, 28227–28234.

Chen, P., Carrington, J. L., Hammonds, R. G., and Reddi, A. H. (1991). Stimulation of chondrogenesis in limb bud mesodermal cells by recombinant human BMP-2B and modulation by TGF-$\beta_1$ and TGF-$\beta_2$. *Exp. Cell Res.* **195**, 509–515.

Chen, S., Rubbock, M. J., and Whitman, M. (1996). A transcriptional partner for Mad proteins in TGF-signaling. *Nature* **383**, 691–696.

Cunningham, N. S., Paralkar, V., and Reddi, A. H. (1992). Osteogenin and recombinant bone morphogenetic protein-2B are chemotactic for human monocytes and stimulate transforming growth factor-$\beta_1$ mRNA expression. *Proc. Natl. Acad. Sci. USA* **89**, 11740–11744.

Cunningham, N. S., Jenkins, N. A., Gilbert, D. J., Copeland, N. G., Reddi, A., and Hari, Lee Se-Jin. (1995). Growth/Differentiation Factor-10: a new member of the transforming growth factor–superfamily related to bone morphogenetic protein-3. *Growth Factors* **12**, 99–109.

Damien, C. J., and Parson, J. R. (1991). Bone graft and bone graft substitutes: a review of current technology and applications. *J. Applied Biomaterials* **2**, 187–208.

Duboule, D. (1994). How to make a limb? *Science* **266**, 575–576.

Dudley, A. T., Lyons, K. M., and Robertson, E. J. (1995). A requirement for bone morphogenetic protein-7 during development of the mammalian kidney and eye. *Genes and Development* **9**, 2795–2807.

Fisher, J. P., Jo, S., Mikos, A. G., and Reddi, A. H. (2004). Thermoreversible hydrogel scaffolds for articular cartilage engineering. *J. Biomed. Materials Res.* **71A**(2), 268–274.

Friedenstein, A. J., Petrakova, K. V., Kurolesova, A. I., and Frolora, G. P. (1968). Heterotopic transplants of bone marrow: analysis of precursor cell for osteogenic and hemopoietic tissues. *Transplantation* **6**, 230–247.

Grande, D. A., Southerland, S. S., Manji, R., Pate, D. W., Schwartz, R. E., and Lucas, P. A. (1995). Repair of articular cartilage defects using mesenchymal stem cells. *Tissue Engineering* **1**, 345–353.

Griffith, D. L., Keck, P. C., Sampath, T. K., Rueger, D. C., and Carlson, W. D. (1996). Three-dimensional structure of recombinant human osteogenic protein-1: structural paradigm for the transforming growth factor-$ superfamily. *Proc. Natl. Acad. Sci. USA* **93**, 878–883.

Hayashi, H., Abdollah, S., Qui, Y., Cai, J., Xu, Y. Y., Grinnell, B. W., *et al.* (1997). The MAD-related protein Smad 7 associates with the TGFβ receptor and functions as an antagonist of TGFβ signaling. *Cell* **89**, 1165–1173.

Heldin, C. H., Miyazono, K., and ten Dijke, P. (1997). TGFβ signaling from cell membrane to nucleus through Smad proteins. *Nature* **300**, 465–471.

Hemmati-Brivanlou, A., Kelly, O. G., and Melton, D. A. (1994). Follistatin an antagonist of activin is expressed in the Spemann organizer and displays direct neuralizing activity. *Cell* **77**, 283–295.

Hollinger, J., Mayer, M., Buck, D., Zegzula, H., Ron, E., Smith, J., Jin, L., and Wozney, J. (1996). Poly (α-hydroxy acid) carrier for delivering recombinant human bone morphogenetic protein-2 for bone regeneration. *J. Controlled Release* **39**, 287–304.

Hubbell, J. A. (1995). Biomaterials in tissue engineering. *Biotechnology* **13**, 565–575.

Hunzinker, E. B., and Rosenberg, L. C. (1996). Repair of partial-thickness defects in articular cartilage: cell recruitment from the synovial membrane. *J. Bone Jt. Surg.* **78-A**, 721–733.

Johnson, R. L., and Tabin, C. J. (1997). Molecular models for vertebrate limb development. *Cell* **90**, 979–990.

Khouri, R. K., Koudsi, B., and Reddi, A. H. (1991). Tissue transformation into bone *in vivo*. *JAMA* **266**, 1953–1955.

Kim, W. S., Vacanti, J. P., Cima, L., Mooney, D., Upton, J., Puelacher, W. C., and Vacanti, C. A. (1994). Cartilage engineered in predetermined shapes employing cell transplantation on synthetic biodegradable polymers. *Plast. Reconstruc. Surgery* **94**, 233–237.

Kozarsky, K. F., and Wilson, J. M. (1993). Gene therapy: adenovirus vectors, *Curr. Opinion Gen. Dev.* **3**, 499–503.

Lacroix, P. (1945). Recent investigations on the growth of bone. *Nature* **156**, 576.

Langer, R., and Vacanti, J. P. (1993). Tissue engineering. *Science* **260**, 930–932.

Lemaire, P., and Gurdon, J. B. (1994). Vertebrate embryonic inductions. *Bio Essays* **16**(9), 617–620.

Livnah, O., Stura, E. A., Johnson, D. L., Middleton, S. A., Mulcahy, L. S., Wrighton, N. D., Dower, W. J., Jolliffe, L. K., and Wilson, I. A. (1996). Functional mimicry of a protein hormone by a peptide agonist: the EPO receptor complex at 2.8 C. *Science* **273**, 464–471.

Luo, G., Hoffman, M., Bronckers, A. L. J., Sohuki, M., Bradley, A., and Karsenty, G. (1995). BMP-7 is an inducer of morphogens and is also required for eye development, and skeletal patterning. *Genes Dev.* **9**, 2808–2820.

Luyten, F., Cunningham. N. S., Ma, S., Muthukumaran, S., Hammonds, R. G., Nevins, W. B., Wood, W. I., and Reddi, A. H. (1989). Purification and partial amino acid sequence of osteogenin, a protein initiating bone differentiation. *J. Biol. Chem.* **265**, 13377–13380.

Luyten, F. P., Yu, Y. M., Yanagishita, M., Vukicevic, S., Hammonds, R. G., and Reddi, A. H. (1992). Natural bovine osteogenin and recombinant BMP-2B are equipotent in the maintenance of proteoglycans in bovine articular cartilage explant cultures. *J. Biol. Chem.* **267**, 3685–3691.

Lyons, K. M., Hogan, B. L. M., and Robertson, E. J. (1995). Colocalization of BMP-2 and BMP-7 RNA suggest that these factors cooperatively act in tissue interactions during murine development. *Mechanisms Dev.* **50**, 71–83.

Ma, S., Chen, G., and Reddi, A. H. (1990). Collaboration between collagenous matrix and osteogenin is required for bone induction. *Ann. N.Y. Acad. Sci.* **580**, 524–525.

McPherson, J. M. (1992). The utility of collagen-based vehicles in delivery of growth factors for hard and soft tissue wound repair. *Clinical Materials* **9**, 225–234.

Melton, D. A. (1991). Pattern formation during animal development. *Science* **252**, 234–241.

Miyazono, K., Maeda, S., and Imamura, T. (2005). BMP receptor signaling: transcriptional targets, regulation of signals and signaling cross-talk. *Cytokine Growth Factor Rev.* **16**, 251–263.

Morsy, M. A., Mitani, K, Clemens, P., and Caskey, T. (1993). Progress toward human gene therapy. *JAMA* **270**(19), 2338–2345.

Mosbach, K., and Ramstrom, O. (1996). The emerging technique of molecular imprinting and its future impact on biotechnology. *Biotechnology* **14**, 163–170.

Mow, V. C., Ratcliffe, A., and Poole, A. R. (1992). Cartilage and diarthrodial joints as paradigms for hierarchical materials and structures. *Biomaterials* **13**, 67–97.

Mulligan, R. C. (1993). The basic science of gene therapy. *Science* **260**, 926–932.

Nishitoh, H., Ichijo, H., Kimura, M., Matsumoto, T., Makishima, F., Yamaguchi, A., Yamashita, H., Enomoto, S., and Miyazono, K. (1996). Identification of type I and type II serine/threonine kinase receptors for growth and differentiation factor-5. *J. Biol. Chem.* **271**, 21345–21352.

Owen, M. E., and Friedenstein, A. J. (1988). Stromal stem cells: marrow-derived osteogenic precursors. *CIBA Foundation Symposium* **136**, 42–60.

Ozkaynak, E., Rueger, D. C., Drier, E. A., Corbett, C., Ridge, R. J., Sampath, T. K., and Opperman, H. (1990). OP-1 cDNA encodes an osteogenic protein in the TGF-$ family. *EMBO J.* **9**, 2085–2093.

Paralkar, V. M., Nandedkar, A. K. N., Pointers, R. H., Kleinman, H. K., and Reddi, A. H. (1990). Interaction of osteogenin, a heparin-binding bone morphogenetic protein, with type IV collagen. *J. Biol. Chem.* **265**, 17281–17284.

Paralkar, V. M., Vukicevic, S., and Reddi, A. H. (1991). Transforming growth factor-β type I binds to collagen IV of basement membrane matrix: implications for development. *Dev. Biol.* **143**, 303–308.

Paralkar, V. M., Weeks, B. S., Yu, Y. M., Kleinman, H. K., and Reddi, A. H. (1992). Recombinant human bone morphogenetic protein 2B stimulates PC12 cell differentiation: potentiation and binding to type IV collagen. *J. Cell Biol.* **119**, 1721–1728.

Piccolo, S., Sasai, Y., Lu, B., and De Robertis, E. M. (1996). Dorsoventral patterning in *Xenopus*: inhibition of ventral signals by direct binding of chordin to BMP-4. *Cell* **86**, 589–598.

Prockop, D. J. (1997). Marrow stromal cells and stem cells for non hematopoietic tissues. *Science* **276**, 71–74.

Rath, N. C., and Reddi, A. H. (1979). Collagenous bone matrix is a local mitogen. *Nature* **278**, 855–857.

Reddi, A. H. (1981). Cell biology and biochemistry of endochondral bone development. *Collagen Rel. Res.* **1**, 209–226.

Reddi, A. H. (1984). Extracellular matrix and development. *In* "Extracellular Matrix Biochemistry" (K. A. Piez and A. H. Reddi, eds.), pp. 375–412. Elsevier, New York.

Reddi, A. H. (1994). Bone and cartilage differentiation. *Curr. Opinion Gen. Dev.* **4**, 937–944.

Reddi, A. H. (1997). Bone morphogenetic proteins: an unconventional approach to isolation of first mammalian morphogens. *Cytokine Growth Factor Rev.* **8**, 11–20.

Reddi, A. H. (1998). Role of morphogenetic proteins in skeletal tissue engineering and regeneration. *Nature Biotechno.* **16**, 247–252.

Reddi, A. H. (2005). BMPs: from bone morphogenetic proteins to body morphogenetic proteins. *Cytokine Growth Factor Rev.* **16**, 249–250.

Reddi, A. H., and Anderson, W. A. (1976). Collagenous bone matrix-induced endochondral ossification and hemopoiesis. *J. Cell Biol.* **69**, 557–572.

Reddi, A. H., and Huggins, C. B. (1972). Biochemical sequences in the transformation of normal fibroblasts in adolescent rat. *Proc. Natl. Acad. Sci. USA* **69**, 1601–1605.

Ripamonti, U. (1996). Osteoinduction in porous hydroxyapatite implanted in heterotopic sites of different animal models. *Biomaterials* **17**, 31–35.

Ripamonti, U., Ma, S., and Reddi, A. H. (1992). The critical role of geometry of porus hydroxyapatite delivery system induction of bone by osteogenin, a bone morphogenetic protein. *Matrix* **12**, 202–212.

Ripamonti, U., Van den Heever, B., Sampath, T. K., Tucker, M. M., Rueger, D. C., and Reddi, A. H. (1996). Complete regeneration of bone in the baboon by recombinant human osteogenic protein-1 (hOP-1, bone morphogenetic protein-7). *Growth Factors* **123**, 273–289.

Ruoslahti, E., and Pierschbacher, M. D. (1987). New perspectives in cell adhesion: RGD and integrins. *Science* **238**, 491–497.

Sampath, T. K., and Reddi, A. H. (1981). Dissociative extraction and reconstitution of bone matrix components involved in local bone differentiation. *Proc. Natl. Acad. Sci. USA* **78**, 7599–7603.

Sampath, T. K., and Reddi, A. H. (1983). Homology of bone inductive proteins from human, monkey, bovine, and rat extracellular matrix. *Proc. Natl. Acad. Sci. USA* **80**, 6591–6595.

Senn, N. (1989). On the healing of aseptic bone cavities by implantation of antiseptic decalcified bone. *Am. J. Med. Sci.* **98**, 219–240.

Shubin, N., Tabin, C., and Carroll, S., (1997). Fossils, genes, and the evolution of animal limbs. *Nature* **388**, 639–648.

Storm, E. E., Huynh, T. V., Copeland, N. G., Jenkins, N. A., Kingsley, D. M., and Lee, S.-J. (1994). Limb alterations in brachypodism mice due to mutations in a new member of TGF-$ superfamily. *Nature* **368**, 639–642.

ten Dijke, P., Yamashita, H., Sampath, T. K., Reddi, A. H., Riddle, D., Helkin, C. H., and Miyazono, K. (1994). Identification of type I receptors for OP-1 and BMP-4. *J. Biol. Chem.* **269**, 16986–16988.

Tsumaki, N., Tanaka, K., Arikawa-Hirasawa, E., Nakase, T., Kimura, T., Thomas, J. T., Ochi, T., Luyten, F. P., and Yamada, Y. (1999). Role of CDMP-1 in skeletal morphogenesis: promotion of mesenchymal cell recruitment and chondrocyte differentiation. *J. Cell Bio.* **144**, 161–173.

Urist, M. R. (1965). Bone: formation by autoinduction. *Science* **150**, 893–899.

Vukicevic, S., Luyten, F. P., Kleinman, H. K., and Reddi, A. H. (1990). Differentiation of canalicular cell processes in bone cells by basement membrane matrix component: regulation by discrete domains of laminin. *Cell* **64**, 437–445.

Vukicevic, S., Luyten, F. P., Kleinman, H. K., and Reddi, A. H. (1990). Differentiation of canalicular cell processes in bone cells by basement membrane matrix components: regulation by discrete domains of laminin. *Cell* **63**, 437–445.

Vukicevic, S., Kopp, J. B., Luyten, F. P., and Sampath, K. (1996). Induction of nephrogenic mesenchyme by osteogenic protein-1 (bone morphogenetic protein-7). *Proc. Natl. Acad. Sci. USA* **92**, 9021–9026.

Weintroub, S., and Reddi, A. H. (1988). Influence of irradiation on the osteoinductive potential of demineralized bone matrix. *Calcif. Tissue Int.* **42**, 255–260.

Weintroub, S., Weiss, J. F., Catravas, G. N., and Reddi, A. H. (1990). Influence of whole-body irradiation and local shielding on matrix-induced endochondral bone differentiation. *Calcif. Tissue Int.* **46**, 38–45.

Weiss, R. E., and Reddi, A. H. (1980). Synthesis and localization of fibronectin during collagenous matrix mesenchymal cell interaction and differentiation of cartilage and bone *in vivo*. *Proc. Natl. Acad. Sci. USA* **77**, 2074–2078.

Winnier, G., Blessing, M., Labosky, P. A., and Hogan, B. L. M. (1996). Bone morphogenetic protein-4 is required for mesoderm formation and patterning in the mouse. *Genes Dev.* **9**, 2105–2116.

Wozney, J. M., Rosen, V., Celeste, A. J., Mitsock, L. M., Whittiers, M., Kriz, W. R., Heweick, R. M., and Wang, E. A. (1988). Novel regulators of bone formation: molecular clones and activities. *Science* **242**, 1528–1534.

Zhang, H., and Bradley, A. (1996). Mice deficient of BMP-2 are nonviable and have defects in amnion/chorion and cardiac development. *Development* **122**, 2977–2986.

Zimmerman, L. B., Jesus-Escobar, J. M., and Harland, R. M. (1996). The Spemann organizer signal Noggin binds and inactivates bone morphogenetic protein-4. *Cell* **86**, 599–606.

# Chapter Ten

# Gene Expression, Cell Determination, and Differentiation

*William Nikovits, Jr., and Frank E. Stockdale*

## I. INTRODUCTION

Studies of skeletal muscle development were the first to provide the principles for understanding the genetic and molecular bases of determination and differentiation. Molecular signals from adjacent embryonic structures activate specific genetic pathways within target cells. Important families of transcriptional regulators are expressed in response to these cues to initiate these important developmental processes in skeletal muscle as well as in other tissues and organs. Both activators and repressors are essential to control the time and location at which development occurs, and self-regulating, positive feedback loops assure that once begun development can proceed normally. An understanding of the mechanistic basis of embryonic commitment to a unique developmental pathway, and the subsequent realization of the adult phenotype, are essential for understanding stem cell behavior and how they might be manipulated for therapeutic goals.

This chapter focuses on determination and differentiation, classical embryological concepts that emerged from descriptive embryology. It has been through the study of muscle development that the genetic basis of these processes was first revealed, laying a mechanistic basis for understanding determination and differentiation. Following the success in studies of skeletal muscle (reviewed in Berkes and Tapscott, 2005; Brand-Saberi, 2005), genetic pathways involved in the determination of other systems, some of which are detailed in other chapters, have also been uncovered, largely because of the underlying conservation of structure among the various effector molecules and mechanisms.

## II. DETERMINATION AND DIFFERENTIATION

Determination describes the process whereby a cell becomes committed to a unique developmental pathway, which, under conditions of normal development, appears to be a stable state. In many cases cells become committed early in development yet remain highly proliferative, expanding exponentially for long periods of time before differentiation occurs. Until recently, determination could

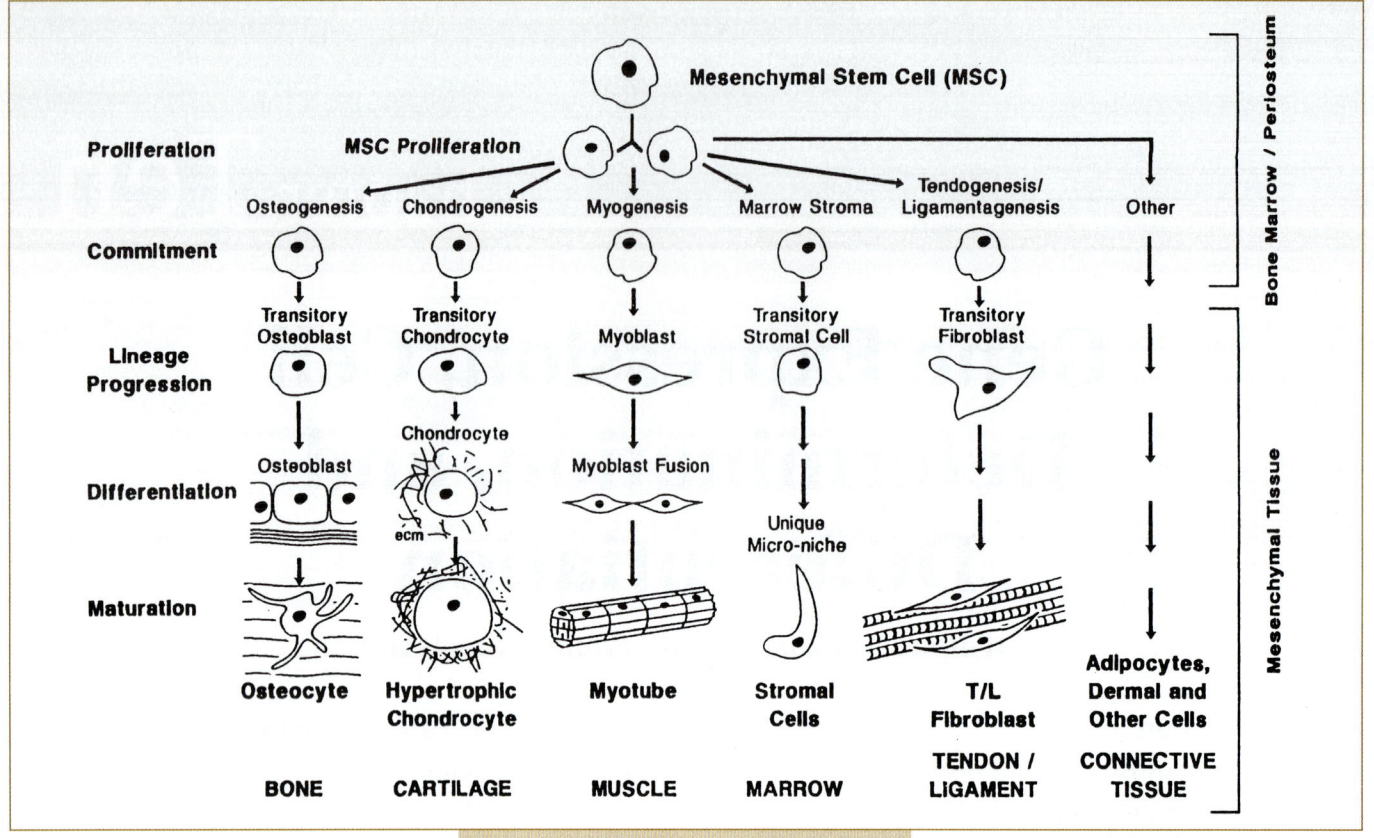

**FIG. 10.1.** The process of commitment and differentiation. Cells arise during gastrulation in the vertebrate embryo that subsequently produce all the different cell type of the body. Cells that can be designated as mesenchymal stem cells (MSC) proliferate and, in response to cues from the cellular environment, enter lineages that undergo differentiation and subsequent maturation into the mature cell types.

only be defined post hoc. Prior to the discovery of transcriptional regulators there were few markers to indicate whether or not a cell was committed to a unique phenotype. Thus *determination* was operationally defined as that state that existed immediately prior to differentiation, that is, before expression of a cell type–specific phenotype. The identification of transcription factors that control the differential expression of large families of genes changed this concept.

Determination and differentiation are processes that are coupled during embryogenesis, where a small number of pluripotent cells (stem cells), expand and enter pathways where they form the diverse cell types of the adult. The process of differentiation describes the acquisition of the phenotype of a cell, most often identified by the expression of specific proteins achieved as a result of differential gene expression. The differentiated state is easily determined by simple observation in most instances, because most differentiated cells display a unique phenotype as a result of the expression of specific structural proteins. Skeletal muscle cells are an extreme example of this, having a cytoplasmic matrix filled with highly ordered myosin, actin, and other contractile proteins within sarcomeres — the functional

units of contraction — giving the fibers their cross-striated pattern. As development proceeds, there is a gradual narrowing of the possible final cell phenotypes that individual cells can adopt, with the final cell fate (set of genes expressed) determined by factors both extrinsic and intrinsic to the cell (Fig. 10.1). Changes in gene expression responsible for directing cells to differentiate along particular developmental pathways result from a response to stimuli received from surrounding cells and the specific cellular phenotype of the cell itself at the time of interaction. For example, cells of recently formed somites have the potential to form either skeletal muscle or cartilage in response to adjacent tissues, and the fate adopted is a result of their location with respect to adjacent structures — the notochord, neural tube, and overlying ectoderm — that produce signaling molecules that determine phenotype (Borycki and Emerson, 2000). In addition to the activation of muscle-specific structural and enzyme-encoding genes, the differentiated state is maintained by the continued expression of specific regulatory transcription factors that can now be identified using modern tools of cellular and molecular biology, including monoclonal antibodies, antisense nucleic acid probes, and gene chip analyses.

Commitment and differentiation to a skeletal muscle fate begins in the somites of the early vertebrate embryo. Within the embryonic somites, two distinct anatomical regions contain muscle progenitors. Specified by signals from the adjacent structures, the dorsal portion of each somite forms an epithelial structure, the *dermomyotome*, which contains the precursor cells of all skeletal muscles that will form in the vertebrate body (with the exception of those found in the head). The medial portion of dermomyotome contains cells that form the axial musculature surrounding the vertebral column, while cells of the ventral-lateral portion of the dermomyotome undergo a process of delamination and migrate into the forming appendages to produce the appendicular musculature of the limbs and body wall. While the muscle fibers that form from the different regions of the somite are nearly indistinguishable, myogenesis in the axial and appendicular muscles is regulated by different effectors, demonstrating the complexity of determination and differentiation in the early embryo.

## III. MyoD AND THE bHLH FAMILY OF DEVELOPMENTAL REGULATORY FACTORS

It was not until late in the 20th century that experiments first demonstrated that cellular commitment to specific developmental fates could be determined by the expression of a single gene or a very small number of genes (O'Neill and Stockdale, 1974; Taylor and Jones, 1979; Konieczny and Emerson, 1984; Lassar *et al.*, 1986; Tapscott *et al.*, 1989). With improvements in tissue culture methods and rapid advances in molecular biology that permitted the introduction of foreign genes into mammalian cells, the first factor capable of specifying a cell to a particular cellular phenotype, MyoD, was isolated and characterized in the laboratory of Dr. Harold Weintraub (Davis *et al.*, 1987; Tapscott *et al.*, 1988). MyoD expression is specific to skeletal muscle, and introduction of MyoD cDNA into fibroblasts of the 10T1/2 cell line converts them at a high frequency into stable myoblasts, which in turn express skeletal muscle proteins. MyoD was only the first of a family of myogenic regulatory factors (MRFs) to be discovered; others include myf-5 (Arnold and Winter, 1998), myogenin (Wright *et al.*, 1989), and MRF4 (Rhodes and Konieczny, 1989; reviewed by Berkes and Tapscott, 2005). The importance of MyoD and myf-5 to the determination of skeletal muscle was demonstrated when double knockout of these two genes in transgenic mice resulted in a nearly complete absence of skeletal muscle (Rudnicki *et al.*, 1993).

MRF members share a common structure, a stretch of basic amino acids followed by a stretch of amino acids that form two amphipathic helices separated by an intervening loop (the helix-loop-helix (HLH) motif), and they are nuclear-located DNA-binding proteins that act as transcrip-

tional regulators (Berkes and Tapscott, 2005). Experiments have demonstrated that the basic amino acids are required for DNA binding and essential for the myogenic conversion of fibroblasts to muscle, while the HLH motif plays an essential role in the formation of heterodimers with other ubiquitously expressed HLH proteins (products of the E2a gene) as well as in DNA binding (Murre *et al.*, 1989).

Nature being conservative, it is not surprising that the bHLH motif was found in transcriptional factors regulating the determination of cell types other than muscle. Based on homology to MyoD, the transcription factor NeuroD was isolated by the Weintraub laboratory and shown to act as a neuronal determination factor (Lee *et al.*, 1995). Expression of NeuroD in presumptive epidermal cells of *Xenopus* embryos converted many into fully differentiated neurons. Interestingly, NeuroD also plays an important role in the differentiation of pancreatic endocrine cells (Naya *et al.*, 1997; Itkin-Ansari *et al.*, 2005). While NeuroD is involved primarily in neuronal differentiation and survival, neurogenin, whose expression precedes that of NeuroD in the embryo, functions more like a determination factor (Ma *et al.*, 1996). Overexpression of *Xenopus* neurogenin induces ectopic neurogenesis as well as ectopic expression of NeuroD. Additional bHLH family members, including HES, Math-5, and Mash-1, have been isolated and participate in the determination of neural cells as well.

Differences in the expression of various members of the neurogenic bHLH family help to explain the diversity of neuronal cell types. For example, genetic deletion of the Mash-1 gene eliminates sympathetic and parasympathetic neurons and enteric neurons of the foregut (Lo *et al.*, 1994), while knockout of NeuroD leads to a loss of pancreatic endocrine cells as well as cells of the central and peripheral nervous system (Naya *et al.*, 1997). In addition to the various bHLH activators, other homeodomain-containing transcription factors are required for specification of neuronal subtypes.

Because cardiac muscle has so much in common with skeletal muscle, including a large number of contractile proteins, an exhaustive search was made for MyoD family members in the heart. Surprisingly, MyoD family members were not found in the developing heart, and thus they play no part in the differentiation of cardiac muscle cells. However, a different family of bHLH-containing factors, including dHAND and eHAND, were found in the developing heart, autonomic nervous system, neural crest, and deciduum. In the heart these factors are important for cardiac morphogenesis and the specification of cardiac chambers (Srivastava *et al.*, 1995). Unlike their MyoD family cousins, neither of the HAND proteins plays a role in differentiation of cardiac muscle cells.

Acting as dominant negative regulators of the bHLH family of transcriptional regulators is a ubiquitously expressed family of proteins that contain the helix-loop-helix structure but lack the upstream run of basic amino acids essential for specific DNA binding by MyoD family members. Termed *inhibitors of differentiation* (Id), these proteins can associate specifically with MyoD or products of the E2A gene and attenuate their ability to bind DNA by forming nonfunctional heterodimeric complexes (Benezra *et al.*, 1990). Id in proliferating myoblasts inhibit the terminal differentiation program by complexing with the E12/E47 protein until the cell receives an appropriate stimulus. Id levels decrease on terminal differentiation.

Neuronal development is also regulated in part by repressors of the neurogenic family of bHLH activators. The HES family of HLH-containing proteins is expressed in neural stem cells, where they maintain proliferation of neuronal precursors and prevent premature differentiation in cells expressing NeuroD and neurogenin. The interaction of unique sets of positively acting bHLH activators and negatively acting members of the HES family helps explain how different subsets of neurons undergo differentiation at different times during development so that the complex structure of the brain can be achieved (Hatakeyama *et al.*, 2004; Kageyama *et al.*, 2005).

## IV. MEFs — COREGULATORS OF DEVELOPMENT

The myocyte enhancer factor 2 (MEF2) family of MADS-box regulatory factors, originally described by Olsen and colleagues (Gossett *et al.*, 1989), participate with MyoD family members to regulate skeletal muscle differentiation (Molkentin and Olson, 1996). By themselves, MEF2 factors do not specify the muscle fate, but they are present in early stages of development, where they interact with MyoD family members to initiate muscle cell specific–gene expression (Molkentin *et al.*, 1995).

MEF2 proteins are transcriptional activators that bind to A+T-rich DNA sequences found in many muscle-specific genes, including those encoding contractile proteins, muscle fiber enzymes, and the muscle differentiation factor myogenin. While some members of this MADS-box regulatory factor family show a nearly ubiquitous distribution among tissue types, a few show more restricted expression to striated muscle (Martin *et al.*, 1993).

Unable to act alone, MEF2 family members must physically interact with MyoD family members at their DNA-binding domains to positively regulate transcription of downstream muscle-specific differentiation genes (Yun and Wold, 1996). Additionally, the transcriptional activation of muscle-specific genes requires that either the MyoD or MEF2 protein provide a transcriptional activation domain. Interestingly, although the wide tissue distribution of some MEF2 family members suggested that they may act in combination with bHLH family members found in other cell types (such as neurogenin in neural precursors) to activate downstream genes (Molkentin and Olson, 1996), no evidence has been found to that effect.

## V. PAX IN DEVELOPMENT

Much of what has been learned about the role of MyoD and myf-5n in the determination of skeletal muscle has come from studies on transgenic mice in which various genetic loci have been deleted (Braun *et al.*, 1992; Rudnicki *et al.*, 1992; Hasty *et al.*, 1993; Nabeshima *et al.*, 1993; Rudnicki *et al.*, 1993; Tajbakhsh and Buckingham, 1994). This work suggested that MyoD and myf-5 acted as redundant activators of myogenesis, albeit with some slight distinctions. Subsequently, Pax3, a DNA-binding protein with both a paired box and a paired-type homeodomain, was identified as a key regulator of myogenesis (Goulding *et al.*, 1991; Relaix *et al.*, 2004). Double knockout of Pax3 and myf-5 leads to a complete absence of skeletal muscle and places Pax3 genetically upstream of MyoD (Tajbakhsh *et al.*, 1997). Using knock-in experiments where the lacZ marker gene replaced Pax3, Buckingham's group demonstrated that Pax3 and myf-5 are activated independent of one another. The implication is that axial muscle (myf-5-dependent) and appendicular muscle (MyoD-dependent) are specified separately in the embryonic somites by two different pathways (Hadchouel *et al.*, 2000, 2003) and that a Pax gene(s) is required for this specification (determination).

A second member of the paired box transcription factor family, Pax7, has also been implicated in myogenesis. Pax7 was isolated from satellite cells, a population of muscle-committed stem cells found in intimate association with mature muscle fibers and involved in muscle growth and repair in the adult. Pax7 is specifically expressed in proliferative myogenic precursors, both embryonic myoblasts as well as satellite cells, and is down-regulated at differentiation (Seale *et al.*, 2000). Transgenic mice lacking the Pax7 gene have normal musculature, albeit with reduced muscle mass, but a complete absence or markedly reduced numbers of satellite cells (Seale *et al.*, 2000; Relaix *et al.*, 2006). These investigators found that in these Pax7 mutants, satellite cells, cells responsible for postnatal growth of skeletal muscle, are progressively lost by cell death. These results suggest that specification of skeletal muscle satellite cells requires Pax7 expression or that Pax7 expression is responsible for survival of satellite cells. The interplay of the many factors that control the initiation and maintenance of myogenesis are diagrammed in Fig. 10.2.

## VI. CONCLUSIONS

Determination and differentiation are in large part controlled by the expression of transcriptional regulators. The processes begin early in development and involve the formation of stem cells that become committed to specific pathways of regulated gene expression. The regulators responsible were first characterized in studies examining commitment to, and differentiation of, skeletal muscle, and muscle development serves as a model for the mechanisms involved. Some other developing organs, such as the central

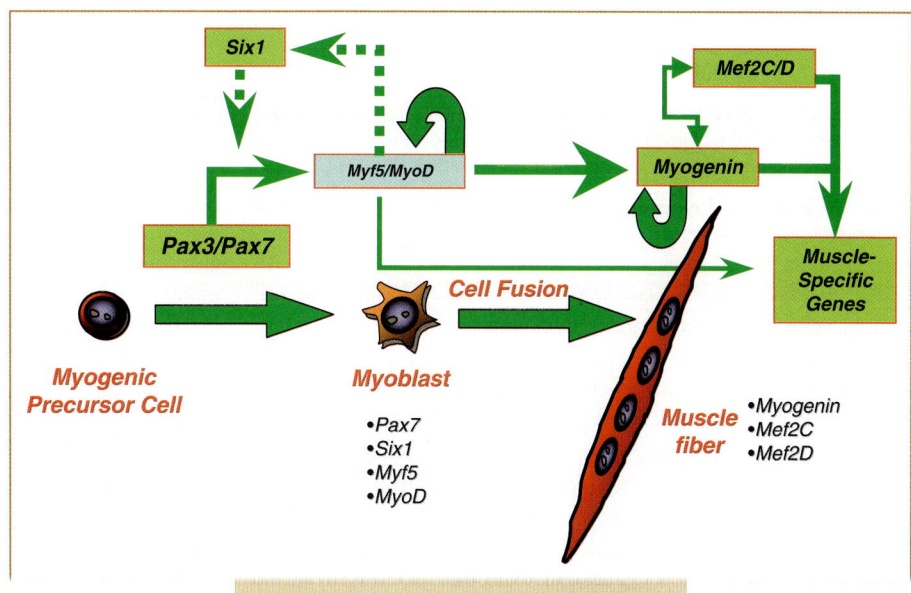

**FIG. 10.2.** A regulatory network controls muscle cell differentiation. (Provided by Dr. Michael Rudnicki and modified.)

and peripheral nervous systems and the pancreas, employ very similar mechanisms and closely related members of the HLH family of proteins to drive development. However, for reasons that are not clear, other organs, such as the heart, in which cardiac cells express many of the same contractile protein genes as their skeletal muscle cousins, use other mechanisms. As our understanding of the mechanisms and effectors of determination and differentiation during embryonic development increase, we will be better able to conceive and apply strategies to engineer stem cells, embryonic- or adult-derived, to address medical problems through transplantation.

## VII. REFERENCES

Arnold, H. H., and Winter, B. (1998). Muscle differentiation: more complexity to the network of myogenic regulators. *Curr. Opin. Genet. Dev.* **8**, 539–544.

Benezra, R., Davis, R. L., Lockshon, D., Turner, D. L., and Weintraub, H. (1990). The protein id: a negative regulator of helix-loop-helix DNA-binding proteins. *Cell* **61**, 49–59.

Berkes, C. A., and Tapscott, S. J. (2005). Myod and the transcriptional control of myogenesis. *Semin. Cell Dev. Biol.* **16**, 585–595.

Borycki, A. G., and Emerson, Jr., C. P. (2000). Multiple tissue interactions and signal transduction pathways control somite myogenesis. *Curr. Top. Dev. Biol.* **48**, 165–224.

Brand-Saberi, B. (2005). Genetic and epigenetic control of skeletal muscle development. *Ann. Anat.* **187**, 199–207.

Braun, T., Rudnicki, M. A., Arnold, H. H., and Jaenisch, R. (1992). Targeted inactivation of the muscle regulatory gene myf-5 results in abnormal rib development and perinatal death. *Cell* **71**, 369–382.

Davis, R. L., Weintraub, H., and Lassar, A. B. (1987). Expression of a single transfected cDNA converts fibroblasts to myoblasts. *Cell* **51**, 987–1000.

Gossett, L. A., Kelvin, D. J., Sternberg, E. A., and Olson, E. N. (1989). A new myocyte-specific enhancer-binding factor that recognizes a conserved element associated with multiple muscle-specific genes. *Mol. Cell Biol.* **9**, 5022–5033.

Goulding, M. D., Chalepakis, G., Deutsch, U., Erselius, J. R., and Gruss, P. (1991). Pax-3, a novel murine DNA-binding protein expressed during early neurogenesis. *EMBO J.* **10**, 1135–1147.

Hadchouel, J., Tajbakhsh, S., Primig, M., Chang, T.-T., Daubas, P., Rocancourt, D., and Buckingham, M. (2000). Modular long-range regulation of *myf5* reveals unexpected heterogeneity between skeletal muscles in the mouse embryo. *Development* **127**, 4455–4467.

Hadchouel, J., Carvajal, J. J., Daubas, P., Bajard, L., Chang, T., Rocancourt, D., Cox, D., Summerbell, D., Tajbakhsh, S., Rigby, P. W., and Buckingham, M. (2003). Analysis of a key regulatory region upstream of the *myf5* gene reveals multiple phases of myogenesis, orchestrated at each site by a combination of elements dispersed throughout the locus. *Development* **130**, 3415–3426.

Hasty, P., Bradley, A., Morris, J. H., Edmondson, D. G., Venuti, J. M., Olson, E. N., and Klein, W. H. (1993). Muscle deficiency and neonatal death in mice with a targeted mutation in the myogenin gene. *Nature* **364**, 501–506.

Hatakeyama, J., Bessho, Y., Katoh, K., Ookawara, S., Fujioka, M., Guillemot, F., and Kageyama, R. (2004). Hes genes regulate size, shape and histogenesis of the nervous system by control of the timing of neural stem cell differentiation. *Development* **131**, 5539–5550.

Itkin-Ansari, P., Marcora, E., Geron, I., Tyrberg, B., Demeterco, C., Hao, E., Padilla, C., Ratineau, C., Leiter, A., Lee, J. E., and Levine, F. (2005). NeuroD1 in the endocrine pancreas: localization and dual function as an activator and repressor. *Dev. Dyn.* **233**, 946–953.

Kageyama, R., Ohtsuka, T., Hatakeyama, J., and Ohsawa, R. (2005). Roles of *bhlh* genes in neural stem cell differentiation. *Exp. Cell Res.* **306**, 343–348.

Konieczny, S. F., and Emerson, Jr., C. P. (1984). 5-Azacytidine induction of stable mesodermal stem cell lineages from 10t1/2 cells: evidence for regulatory genes controlling determination. *Cell* **38**, 791–800.

Lassar, A. B., Paterson, B. M., and Weintraub, H. (1986). Transfection of a DNA locus that mediates the conversion of 10t1/2 fibroblasts to myoblasts. *Cell* **47**, 649–656.

Lee, J. E., Hollenberg, S. M., Snider, L., Turner, D. L., Lipnick, N., and Weintraub, H. (1995). Conversion of *Xenopus* ectoderm into neurons by NeuroD, a basic helix-loop-helix protein. *Science* **268**, 836–844.

Lo, L., Guillemot, F., Joyner, A. L., and Anderson, D. J. (1994). Mash-1: a marker and a mutation for mammalian neural crest development. *Perspect. Dev. Neurobiol.* **2**, 191–201.

Ma, Q., Kintner, C., and Anderson, D. J. (1996). Identification of neurogenin, a vertebrate neuronal determination gene. *Cell* **87**, 43–52.

Martin, J. F., Schwarz, J. J., and Olson, E. N. (1993). Myocyte enhancer factor (MEF) 2c: a tissue-restricted member of the mef-2 family of transcription factors. *Proc. Natl. Acad. Sci. USA* **90**, 5282–5286.

Molkentin, J. D., and Olson, E. N. (1996). Combinatorial control of muscle development by basic helix-loop-helix and mads-box transcription factors. *Proc. Natl. Acad. Sci. USA* **93**, 9366–9373.

Molkentin, J. D., Black, B. L., Martin, J. F., and Olson, E. N. (1995). Cooperative activation of muscle gene expression by mef2 and myogenic bhlh proteins. *Cell* **83**, 1125–1136.

Murre, C., McCaw, P. S., and Baltimore, D. (1989). A new DNA-binding and dimerization motif in immunoglobulin enhancer binding, daughterless, MyoD, and MYC proteins. *Cell* **56**, 777–783.

Nabeshima, Y., Hanaoka, K., Hayasaka, M., Esumi, E., Li, S., Nonaka, I., and Nabeshima, Y. (1993). Myogenin gene disruption results in perinatal lethality because of severe muscle defect. *Nature* **364**, 532–553.

Naya, F. J., Huang, H. P., Qiu, Y., Mutoh, H., DeMayo, F. J., Leiter, A. B., and Tsai, M. J. (1997). Diabetes, defective pancreatic morphogenesis, and abnormal enteroendocrine differentiation in beta2/NeuroD-deficient mice. *Genes Dev.* **11**, 2323–2334.

O'Neill, M. C., and Stockdale, F. E. (1974). 5-Bromodeoxyuridine inhibition of differentiation. Kinetics of inhibition and reversal in myoblasts. *Dev. Biol.* **37**, 117–132.

Relaix, F., Rocancourt, D., Mansouri, A., and Buckingham, M. (2004). Divergent functions of murine pax3 and pax7 in limb muscle development. *Genes Dev.* **18**, 1088–1105.

Relaix, F., Montarras, D., Zaffran, S., Gayraud-Morel, B., Rocancourt, D., Tajbakhsh, S., Mansouri, A., Cumano, A., and Buckingham, M. (2006).

Pax3 and pax7 have distinct and overlapping functions in adult muscle progenitor cells. *J. Cell Biol.* **172**, 91–102.

Rhodes, S. J., and Konieczny, S. F. (1989). Identification of mrf4: a new member of the muscle regulatory factor gene family. *Genes Dev.* **3**, 2050–2061.

Rudnicki, M. A., Braun, T., Hinuma, S., and Jaenisch, R. (1992). Inactivation of MyoD in mice leads to up-regulation of the myogenic hlh gene *myf-5* and results in apparently normal muscle development. *Cell* **71**, 383–390.

Rudnicki, M. A., Schnegelsberg, P. N. J., Stead, R. H., Braun, T., Arnold, H.-H., and Jaenisch, R. (1993). Myod or myf-5 is required for the formation of skeletal muscle. *Cell* **75**, 1351–1359.

Seale, P., Sabourin, L. A., Girgis-Gabardo, A., Mansouri, A., Gruss, P., and Rudnicki, M. A. (2000). Pax7 is required for the specification of myogenic satellite cells. *Cell* **102**, 777–786.

Srivastava, D., Cserjesi, P., and Olson, E. N. (1995). A subclass of bhlh proteins required for cardiac morphogenesis. *Science* **270**, 1995–1999.

Tajbakhsh, S., and Buckingham, M. E. (1994). Mouse limb muscle is determined in the absence of the earliest myogenic factor myf-5. *Proc. Natl. Acad. Sci. USA* **91**, 747–751.

Tajbakhsh, S., Rocancourt, D., Cossu, G., and Buckingham, M. (1997). Redefining the genetic hierarchies controlling skeletal myogenesis: *pax-3* and *myf-5* act upstream of *MyoD*. *Cell* **89**, 127–138.

Tapscott, S. J., Davis, R. L., Thayer, M. J., Cheng, P. F., Weintraub, H., and Lassar, A. B. (1988). Myod1: a nuclear phosphoprotein requiring a myc homology region to convert fibroblasts to myoblasts. *Science* **242**, 405–411.

Tapscott, S. J., Lassar, A. B., Davis, R. L., and Weintraub, H. (1989). 5-Bromo-2'-deoxyuridine blocks myogenesis by extinguishing expression of myod1. *Science* **245**, 532–536.

Taylor, S. M., and Jones, P. A. (1979). Multiple new phenotypes induced in 10t1/2 and 3t3 cells treated with 5-azacytidine. *Cell* **17**, 771–779.

Wright, W. E., Sassoon, D. A., and Lin, V. K. (1989). Myogenin, a factor regulating myogenesis, has a domain homologous to MyoD. *Cell* **56**, 607–617.

Yun, K., and Wold, B. (1996). Skeletal muscle determination and differentiation: story of a core regulatory network and its context. *Curr. Opin. Cell Biol.* **8**, 877–889.

# In Vitro Control of Tissue Development

## Chapter Eleven

# Engineering Functional Tissues

*Lisa E. Freed and Farshid Guilak*

## I. INTRODUCTION

Tissue engineering is a rapidly growing field that seeks to restore the function of diseased or damaged tissues through the use of implanted cells, biomaterials, and biologically active molecules. Within the context of many organ systems, such as the musculoskeletal and cardiovascular systems, tissue function has a large biomechanical component involving the transmission or generation of mechanical forces. For example, articular cartilage and cardiac tissue possess highly specialized structures and compositions that provide unique biomechanical properties required to move the limbs and circulate the blood. The loss of function of cartilage and cardiac tissue due to injury, disease, or aging accounts for a significant number of clinical disorders, at a tremendous social and economic cost (Praemer *et al.*, 1999; Thom *et al.*, 2006). Although different in many respects, cartilage and cardiac tissue share two features that are highly relevant to functional tissue engineering: (1) lack of intrinsic capacity for self-repair and (2) performance of critical biomechanical functions *in vivo*.

Despite many early successes, few engineered tissue products are available for clinical use, and significant challenges still remain in exploiting tissue-engineering technologies for the successful long-term repair of mechanically functional tissues. The precise reasons for graft failure in experimental animal studies and preclinical trials are not fully understood, but they include a combination of factors that lead to the breakdown of repair tissues under conditions of physiologic loading. The magnitudes of stresses and frequency of loading that tissues are subjected to *in vivo* can be quite large, and few engineered tissue constructs possess the biomechanical properties to withstand such stresses at the time of implantation. For many native tissues, however, the potential range of *in vivo* stresses and strains are not well characterized, thus making it difficult to incorporate a true "safety factor" into the design criteria for engineered tissues. Furthermore, the challenge is not as simple as matching a single mechanical parameter, such as modulus or strength; instead most tissues possess complex viscoelastic, nonlinear, and anisotropic mechanical and physicochemical properties that vary with age, site, and other host factors. Finally, a number of complex interactions must be considered, for the graft and surrounding host tissues are expected to grow and remodel in response to their changing environments postimplantation (Badylak *et al.*, 2002).

An evolving discipline referred to as ***functional tissue engineering*** has sought to address these challenges by developing guidelines for rationally investigating the role of biological and mechanical factors in tissue engineering. A series of formal goals and principles for functional tissue engineering have been proposed in a generalized format (Butler *et al.*, 2000; Guilak *et al.*, 2003). In brief, these guidelines include development of: (1) improved definitions of functional success for tissue engineering applications; (2) improved understanding of the *in vivo* mechanical requirements and intrinsic properties of native tissues; (3) improved understanding of the biophysical environment of cells within engineered constructs; (4) scaffold design criteria that aim to enhance cell survival, differentiation, and tissue mechanical function; (5) bioreactor design criteria that aim to enhance cell survival and the regeneration of functional 3D tissue constructs; (6) construct design criteria that aim to meet the metabolic and mechanical demands of specific tissue-engineering applications; and (7) improved understanding of biological and mechanical responses of an engineered tissue construct following implantation.

In this chapter, we focus on how *in vitro* culture parameters, including convective mixing, perfusion, culture duration, and mechanical conditioning (i.e., compression, tensile stretch, pressure, and shear), can affect the development and functional properties of engineered tissues. We further focus on engineered cartilage and cardiac tissues, although the concepts discussed are also of relevance to other tissues and organs that serve some mechanical function (e.g., muscle, tendon, ligament, bone, blood vessels, heart valves, bladder) and are the targets of tissue-engineering research efforts.

## II. KEY CONCEPTS

One approach to tissue engineering involves the *in vitro* culture of cells on biomaterial scaffolds to generate functional engineered tissues for *in vivo* applications, such as the repair of damaged articular cartilage or myocardium (Fig. 11.1). The **working hypothesis** is that *in vitro* culture

parameters determine the structural and mechanical properties of engineered tissues and, therefore, can be exploited to manipulate the growth and functionality of engineered tissues. ***In vitro culture parameters*** refer to tissue-engineering bioreactors, scaffolds, and mechanical conditioning that can mediate cell behavior and functional tissue assembly (Freed *et al.*, 2006).

***Bioreactors*** are defined as *in vitro* culture systems designed to perform some or all of the following functions: (1) provide control over the initial cell distribution on 3D scaffolds; (2) provide efficient mass transfer of gases, nutrients, and regulatory factors to tissue-engineered constructs during their *in vitro* cultivation; and (3) expose the developing constructs to convective mixing, perfusion, and/or mechanical conditioning. Tissue-engineering bioreactors are also discussed in Chapter 12 and were reviewed in Darling and Athanasiou (2003), I. Martin *et al.* (2004), and Martin and Vermette (2005). ***Scaffolds*** are defined as 3D porous solid biomaterials designed to perform some or all of the following functions: (1) promote cell–biomaterial interactions, cell adhesion, and extracellular matrix (ECM) deposition; (2) permit sufficient transport of gases, nutrients, and regulatory factors to allow cell survival, proliferation, and differentiation; (3) biodegrade at a controllable rate that approximates the rate of tissue regeneration under the culture conditions of interest; and (4) provoke a minimal degree of inflammation or toxicity *in vivo*. Tissue-engineering scaffolds are also discussed in Chapters 19, 20, and 22–25 and were reviewed in Langer and Tirrell (2004), Muschler *et al.* (2004), Hollister (2005), and Lutolf and Aubbell (2005).

The biological and mechanical requirements of an engineered tissue depend on the specific application (i.e., engineered cartilage should provide a low-friction, articulating

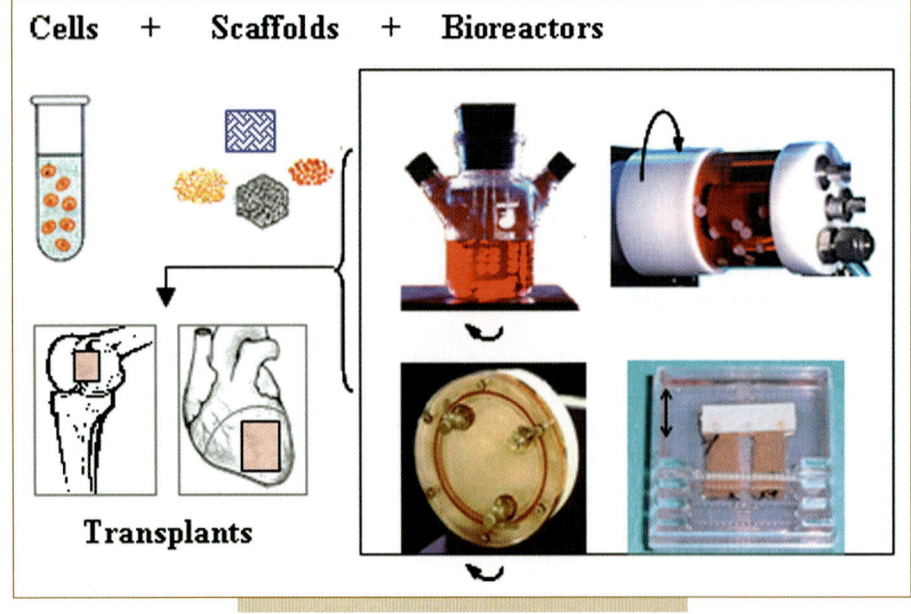

**FIG. 11.1.**   Model system. Cells are cultured on porous solid biomaterial scaffolds in representative bioreactors (clockwise from the upper left image are shown the spinner flask, slow-turning lateral vessel, high-aspect-ratio vessel, and Biostretch®). Functional engineered cartilage or cardiac tissue can potentially be used to repair damaged articular cartilage or myocardium.

surface and be able to withstand and transmit loading in compression, tension, and shear, whereas engineered cardiac tissue should propagate electrical signals, contract in a coordinated manner, and withstand dynamic changes in pressure, tension, and shear). Tissue-specific requirements translate to design principles for functional tissue engineering as follows: (1) At the time of implantation, an engineered tissue should possess sufficient size and mechanical integrity to allow for handling and permit survival under physiological conditions; (2) Once implanted, an engineered tissue should provide some minimal level of biomechanical function immediately postimplantation that should improve progressively until normal tissue function has been restored; and (3) Once implanted, an engineered tissue should mature and integrate with surrounding host tissues.

One of the key challenges in realizing the approach shown in Fig. 11.1 is to optimize the *in vitro* culture environment in order to achieve the best possible conditions for functional tissue engineering. In particular, the ability precisely to define and control *in vitro* culture parameters, such as mass transport and biophysical signaling, can potentially be exploited to improve and ultimately control the structure, composition, and functional properties of engineered tissues. The following sections of this chapter consider *in vitro* studies aimed at clinical translation; representative *in vitro* culture environments; and the effects of convective mixing, perfusion, culture duration, and mechanical conditioning on engineered tissue constructs. Illustrative examples and alternative approaches for engineering cartilage and cardiac tissue are provided.

## III. *IN VITRO* STUDIES AIMED AT CLINICAL TRANSLATION

Osteochondral-defect repair remains an important, unsolved clinical problem, and a number of tissue-engineering approaches have been attempted to promote the functional integration of an engineered cartilage implant with adjacent host tissues (Hunziker, 1999, 2001). For example, we implanted composites based on engineered cartilage into defects in adult rabbit knees and found that the six-month repair cartilage exhibited physiologic thickness and Young's modulus but integrated in a variable and incomplete manner with adjacent host cartilage (Fig. 11.2 A–B) (Schaefer *et al.*, 2002, 2004). Further studies aimed at clinical translation are clearly needed, but *in vivo* models (e.g., orthotopic implants in animal knee joints) are complicated by high variability and biological and mechanical environments very different from those existing in human joint lesions (Hunziker, 1999, 2001). Moreover, further studies are needed to design appropriate physical rehabilitation protocols aimed at preventing overt failure, dislodgment, or fatigue of a tissue-engineered construct postimplantation. For example, a rehabilitation period consisting

of joint immobilization was shown to hinder the delamination of periosteal grafts in the goat knee (Driesang and Hunziker, 2000).

*In vitro* studies have been suggested to: (1) address the challenges of *in vivo* complexity in controllable model systems, and (2) define how an *in vitro*–grown construct may behave when implanted *in vivo* (Guilak *et al.*, 2001; Tognana *et al.*, 2005b). For example, we used rotating bioreactors and composites consisting of engineered cartilage discs within rings of native cartilage, vital bone, or devitalized bone to demonstrate significant effects of chondrogenic potential of the cells, degradation rate of the scaffold, developmental stage of the construct, and architecture of the adjacent tissue on construct development and integration (Obradovic *et al.*, 2001; Tognana *et al.*, 2005a, 2005b). Engineered cartilage constructs interfaced with the solid matrix of adjacent cartilage without any gaps or intervening capsule (Fig. 11.2C–D) and focal intermingling between construct collagen fibers and native cartilage collagen fibers provided evidence of structural integration (Fig. 11.2E). Interestingly, the composition and mechanical properties (e.g., adhesive strength, Fig. 11.2F) were superior for constructs cultured adjacent to bone as compared to cartilage and best for constructs cultured adjacent to devitalized bone. These findings could be rationalized by considering the differences in adjacent-tissue architecture (histological features) and transport properties (diffusivity) (Tognana *et al.*, 2005a). Consistently, perfused bioreactors were used to demonstrate significantly higher amounts of GAG and total collagen in engineered cartilage cocultured adjacent to engineered bone than to either engineered cartilage, native cartilage, or native bone (Mahmoudifar and Doran, 2005b).

Hypothesis-driven experiments aimed at elucidating cell- and tissue-level responses to biological, hydrodynamic, and mechanical stimuli are expected to improve our understanding of complex *in vivo* phenomena and to promote clinical translation of tissue-engineering technologies. In this context, the *in vitro* culture environment plays a key role by allowing for reproducible test conditions, and tissue culture bioreactors represent a controllable model system for (1) studying the effects of biophysical stimuli on cells and developing tissues, (2) simulating responses of an *in vitro*–grown construct to *in vivo* implantation and thereby helping to define its potential for survival and functional integration, and (3) developing and testing physical therapy regimens for patients who have received engineered tissue implants.

## IV. REPRESENTATIVE *IN VITRO* CULTURE ENVIRONMENTS

### Spinner Flasks

A spinner flask system (Table 11.1 column 1; Fig. 11.1, upper left image) has been developed for cell seeding of

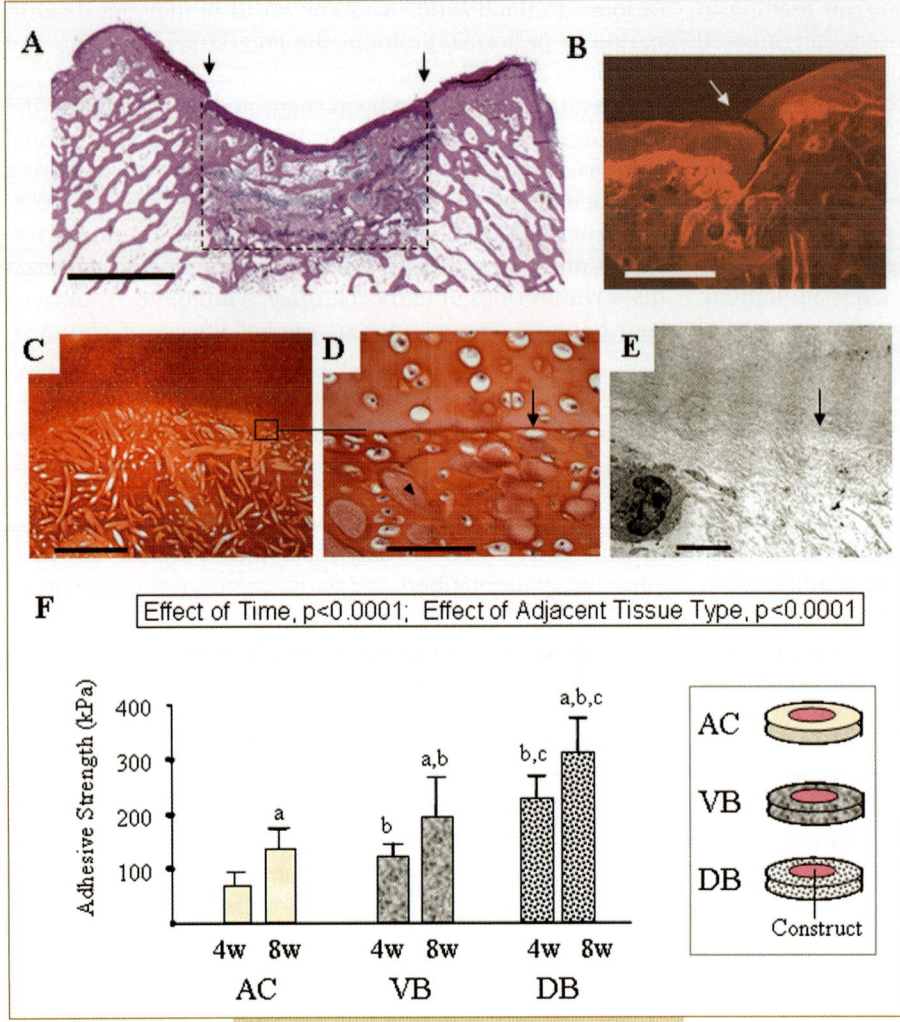

**FIG. 11.2.** Studies of the clinically relevant problem of engineered cartilage integration. **(A–B)** Integration between an engineered cartilage implant and adjacent host cartilage was variable and incomplete (arrows) six months after osteochondral-defect repair in adult rabbit knee joints. **(A)** Alcian blue stain; scale bar: 2.5 mm; dashed line shows borders of original defect. **(B)** Immunofluorescence microscopy; scale bar: 400 µm. **(C–F)** Engineered cartilage integration was studied by using rotating bioreactors to culture construct discs within rings of articular cartilage (AC), vital bone (VB), and devitalized bone (DB). **(C, D, E)** Histology of the construct–AC interface; arrows at interfaces point toward the construct; arrowheads indicate the scaffold. **(C and D)** Safranin-O stain; scale bars 500 µm and 50 µm, respectively. **(E)** TEM; scale bar 5 µm. **(F)** Adhesive strength for construct discs cultured in rings of AC (open bars), VB (grey bars), or DB (stipled bars) for four or eight weeks (4 w, 8 w). a: significant difference due to time; b: significantly different from the corresponding AC composite; c: significantly different from the corresponding VB composite. **(A–B)** Reproduced with kind permission of the publisher from Schaefer *et al.* (2004). **(C–D)** Reproduced with kind permission of the publisher from Tognana *et al.* (2005b). **(E–F)** Reproduced with kind permission of the publisher from Tognana *et al.* (2005a).

3D scaffolds and cultivation of 3D tissue constructs (e.g., Freed and Vunjak-Novakoric, 1995, 1997; Freed *et al.*, 1998; Martin *et al.*, 1998, 2001; Vunjak-Novakovic *et al.*, 1998; Bursac *et al.*, 1999; Gooch *et al.*, 2001a, 2001b; Papadaki *et al.*, 2001; Pei *et al.*, 2002b; Schaefer *et al.*, 2002; Mahmoudifar *et al.*, 2005a, 2005b). In brief, spinner flasks are 12-cm-high × 6.5-cm-diameter vessels that provide gas exchange via side arms with loose screw caps and mixing via a nonsuspended 4.5 × 0.8-cm magnetic stir bar. Scaffolds are fixed on two to four needles placed symmetrically in a stopper in the mouth of the flask. Each needle holds up to three scaffolds separated by silicone spacers. The flask is filled with 100–120 mL of media, inoculated with cells, and then stirred at ~50 rpm. Smaller spinner flasks with operating volumes of 60 mL can also be used.

## Rotating Bioreactors

Two rotating bioreactor systems were developed at NASA for *in vitro* tissue culture on earth: the slow-turning lateral vessel (STLV) (Table 11.1, column 2; Fig. 11.1, upper right image) and high-aspect-ratio vessel (HARV) (Table 11.1, column 3; Fig. 11.1, lower left image) (e.g., Schwarz *et al.*, 1992; Freed and Vunjak-Novakovic, 1995, 1997; Freed *et al.*, 1998; Riesle *et al.*, 1998; Unsworth and Lelkes, 1998; Carrier *et al.*, 1999; Vunjak-Novakovic *et al.*, 1999; Obradovic *et al.*, 2001; Madry *et al.*, 2002; Pei *et al.*, 2002a, 2002b; Bursac *et al.*, 2003; Yu *et al.*, 2004; Tognana *et al.*, 2005a, 2005b; Marolt *et al.*, 2006). Also, a rotating-wall perfused vessel (RWPV) was developed at NASA for tissue culture during spaceflight (Table 11.1, column 4) (Freed *et al.*, 1997; Jessup *et al.*, 2000; Freed and Vunjak-Novakovic, 2002).

In the STLV, tissues are housed in the annular space between two concentric cylinders (outer and inner diameters of 5.75 and 2 cm, respectively; 5 cm high), whereas the HARV is a discoid vessel (10 cm in diameter, 1.3 cm high). In the STLV and HARV, gas exchange is provided by an internal, fiber-reinforced, 175-µm-thick silicone membrane. A smaller STLV and HARV, with operating volumes of 50–60 mL, are also available. The STLV or HARV is completely filled with medium and then rotated around its central axis. Rotation suspends the constructs within the

**Table 11.1.** Representative culture environments

| Bioreactor vessel | Spinner flask (stirred flask, SF) | Rotating vessel (slow-turning lateral vessel, STLV) | Rotating vessel (high aspect-ratio vessel, HARV) | Rotating-wall perfused vessel (RWPV) |
|---|---|---|---|---|
| **Engineered constructs** | Discoid, 5–10 × 2–5 mm; fixed in place in vessel; up to 12 discs per vessel | Discoid, 5–10 × 2–5 mm; freely suspended in vessel; up to 12 discs per vessel | Discoid, 5–10 × 2–5 mm; freely suspended in vessel; up to 12 discs per vessel | Discoid, 5–10 × 2–5 mm; freely suspended in vessel; up to 10 discs per vessel |
| **Operating volume** | 60 or 120 mL | 55 or 110 mL | 50 or 110 mL | 125 mL |
| **Operating parameters:** | | | | |
| (i) *Medium exchange* | Batchwise (replace 50% every 2–3 days) | Batchwise (replace 50% every 2–3 days) | Batchwise replace 50% (every 2–3 days) | Batchwise or continuous |
| (ii) *Gas exchange* | Continuous, via surface aeration | Continuous, via an internal membrane | Continuous, via an internal membrane | Intermittent or continuous, via an external membrane |
| (iii) *Mixing mechanism* | Magnetic stirring (50–60 rpm) | Solid body rotation of vessel; Construct settling | Solid body rotation of vessel; Construct settling | Discoid centrifugal pump and differential rotation of cylinders |
| **Fluid flow pattern** | Turbulent | Laminar | Laminar | Laminar |
| **Mass transfer in bulk medium** | Convection | Convection | Convection | Convection |

culture medium due to the combined effects of gravitational (weight and buoyancy) and flow-induced (drag) forces (Freed and Vunjak-Novakovic, 1995). The rotation rate is adjusted as needed (e.g., 10–40 rpm) to maintain the constructs freely suspended within the vessel during *in vitro* culture.

In the RWPV, tissues are housed in the annular space between two concentric cylinders (outer and inner diameters of 5 and 1.5 cm, respectively; 6.3 cm high). In the microgravity environment of space, where gravitational settling of constructs does not occur, convective mixing is provided by a flat disc (3.8 cm in diameter) that is attached to one end of the inner cylinder and serves as a centrifugal pump and by differential rotation of the inner and outer cylinders (e.g., at 10 and 1 rpm) (Begley *et al.*, 2000). In the RWPV system, culture medium is periodically circulated through the vessel via an inlet at one end of the annulus and a spin-filter outlet at its inner cylinder, and gas exchange is provided by an external, fiber-reinforced, 175-μm-thick silicone membrane (Freed and *et al.*, 1997; Jessup et al., 2000; Freed and Vunjak-Novakovic, 2002).

## Mechanical Conditioning

A variety of devices have been custom designed and built to study effects of mechanical conditioning (i.e., compression, tension, pressure, or shear) on cells and tissues *in vitro* (reviewed in Brown, 2000; Darling and Athanasiou, 2003). For engineering cartilage, devices typically apply dynamic compression (e.g., Buschmann *et al.*, 1995; Mauck *et al.*, 2000), hydrostatic pressure (e.g., Mizuno *et al.*, 2002; Toyoda *et al.*, 2003), or mechanical shear (Waldman *et al.*, 2003). For engineering muscular and cardiovascular tissues, devices typically apply dynamic tensile strain (Vandenburgh and Karlisch, 1989; Vandenburgh *et al.*, 1991; Kim *et al.*, 1999; Niklason *et al.*, 1999; Fink *et al.*, 2000; Sodian *et al.*, 2001; Akhyari *et al.*, 2002; Powell *et al.*, 2002; Zimmermann *et al.*, 2002; Gonen-Wadmany *et al.*, 2004; Boublik *et al.*, 2005) or pulsatile hydrostatic pressure (Niklason *et al.*, 1999; Sodian *et al.*, 2001). In one representative example (Fig. 11.1, lower right image), a commercially available electromagnetic device (Biostretch®, ICCT, Ontario, Canada) (Liu *et al.*, 1999) was used to study effects of cyclic stretch on engineered cardiac constructs (Boublik *et al.*, 2005).

# V. CONVECTIVE MIXING, FLOW, AND MASS TRANSFER

## Cell Seeding of 3D Scaffolds

Cell seeding of a biomaterial scaffold is the first step in tissue engineering and plays a critical role in determining subsequent tissue formation (Freed *et al.*, 1994a; Kim *et al.*, 1998). We showed that high and spatially uniform initial cell densities were associated with increases in extracellular matrix (ECM) deposition and compressive modulus in engineered cartilage (Vunjak-Novakovic *et al.*, 1996, 1999; Freed

*et al.*, 1998) and with increases in contractile proteins and contractility in engineered cardiac tissue (Carrier *et al.*, 2002a; Radisic *et al.*, 2003). For example, the spinner flask system (Table 11.1, Fig. 11.1) improved the efficiency and spatial uniformity of cell seeding throughout porous solid 3D scaffolds (e.g., fiber-based textiles and porogen-leached sponges) as compared with controls seeded statically (Vunjak-Novakovic *et al.*, 1996, 1998). The probable mechanism of cell seeding in the spinner flask system is convection of suspended cells into the porous scaffold leading to inertial impacts between cells and scaffold and then to cell adhesion. However, cell seeding in spinner flasks is not perfectly uniform, and initial cell densities are highest at the construct surfaces (Freed *et al.*, 1998; Mahmoudifar and Doran, 2005a). Moreover, for cardiac tissue engineering, the cell seeding efficiency in spinner flasks is only ~60% (Carrier *et al.*, 1999).

Alternative systems for cell seeding of 3D scaffold systems are based on convective flow of a cell suspension directly through a porous solid 3D scaffold via filtration (Li *et al.*, 2001) or bi-directional perfusion (Radisic *et al.*, 2003; Wendt *et al.*, 2003). Engineered cartilage seeded in perfused bioreactors with alternating medium flow reportedly exhibited higher cell viability and uniformity than controls seeded statically and in spinner flasks (Wendt *et al.*, 2003). Likewise, engineered cardiac tissue seeded in perfused bioreactors with alternating medium flow exhibited higher cell viability and spatial uniformity than controls seeded in mixed petri dishes (Radisic *et al.*, 2003).

## Cultivation of 3D Tissue Constructs

Convective mixing, flow, and mass transport are required to supply the oxygen, nutrients, and regulatory factors that are in turn required for the *in vitro* cultivation of large tissue constructs (Karande *et al.*, 2004; Muschler *et al.*, 2004; Martin and Vermette, 2005). Oxygen is the factor that generally limits cell survival and tissue growth, due to its relatively low stability slow diffusion rate and high consumption rate (Martin and Vermette, 2005). Different tissue types have different mass transport requirements, depending on cell type(s), concentrations, and metabolic activities. For example, articular cartilage, an avascular tissue, has a lower requirement for oxygen than myocardium, a highly vascularized tissue. Convective mixing of the culture media in rotating bioreactors supported the growth of engineered cartilage constructs 5–8 mm thick (Freed *et al.*, 1997, 1998; Vunjak-Novakovic *et al.*, 1999). Chondrocyte metabolic function was in between aerobic and anaerobic (assessed by a ratio of lactate produced to glucose consumed, L/G ~ 1.5), and synthesis rates of glycosaminoglycans (GAG) and collagen were high in rotating bioreactors with gas exchange membranes, whereas anaerobic metabolism (L/G ~ 2.0) and significantly lower ECM synthesis rates were measured in control bioreactors without gas exchange membranes (Fig. 11.3A–C) (Obradovic *et al.*, 1999). In the case of engineered cardiac tissue, convective mixing in rotating bioreactors and spinner flasks supported the growth of a tissue-like surface layer ~100 μm thick (Carrier *et al.*, 1999, 2002b; Papadaki *et al.*, 2001). However, convective flow of culture medium directly through an engineered cardiac construct within a perfused bioreactor can significantly improve its thickness and spatial homogeneity (Carrier *et al.*, 2002a; Radisic *et al.*, 2003, 2004b, 2006) (also see Chapter 38). Experimental and modeling studies have correlated oxygen gradients within engineered tissues with morphology and composition (Obradovic *et al.*, 2000; Malda *et al.*, 2003, 2004; Martin and Vermette, 2005).

Regulatory factors can be used as culture medium supplements to selectively induce chondrogenic or osteogenic differentiation of progenitor cells harvested from the bone marrow (e.g., Martin *et al.*, 2001; Muschler *et al.*, 2004; Marott *et al.*, 2006) and to enhance the growth, composition, and mechanical function of engineered cartilage based on

chondrocytes (Gooch *et al.*, 2001a; Pei *et al.*, 2002a; Mauck *et al.*, 2003). In one study of engineered cartilage cultured in medium that was supplemented, or not, with insulin-like growth factor (IGF-I) using three different culture environments (rotating bioreactors, spinner flasks, and static), the growth and hydrodynamic factors (1) independently modulated construct structure, composition, and mechanical properties and (2) in combination produced effects superior to those that could be obtained by modifying the factors individually (Gooch *et al.*, 2001a). In particular, construct size was increased by IGF-I in all culture environments (Fig. 11.3D), whereas the fractional amount of GAG was significantly increased only if IGF-I was used in combination with rotating bioreactors (Fig. 11.3E). Consistently, others have demonstrated synergistic effects of growth factors and dynamic mechanical loading on the structure, composition, and mechanical function of engineered cartilage constructs (Mauck *et al.*, 2003).

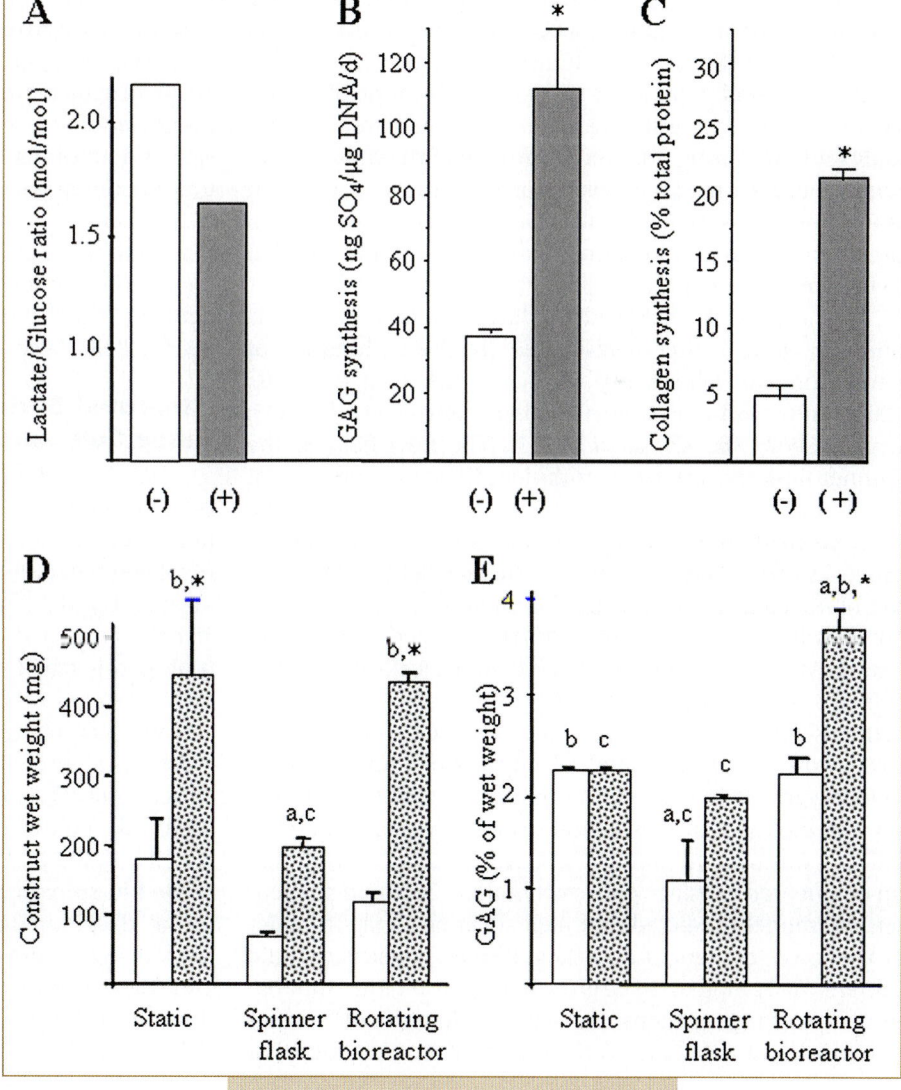

**FIG. 11.3.** Convective flow and mass transport affect cell function and the size and composition. **(A–C)** Engineered cartilage cultured for five weeks in rotating bioreactors with (+) or without (–) a gas exchange membrane. **(A)** Cell metabolism (ratio of lactate to glucose); **(B and C)** biosynthesis rates of GAG (normalized per cell) and collagen (as a percentage of the total protein). *: Significant effect of gas exchange. **(D–E)** Engineered cartilage constructs cultured for four weeks in a static flask, spinner flask, or rotating bioreactor, using basal media (white bars) or media supplemented, with insulin-like growth factor (IGF-I, 100 ng/mL, stipled bars). **(D and E)** Construct wet weight (mg) and GAG (% of wet weight). a: Significantly different from static flask; b: Significantly different from spinner flask; c: Signficantly different from rotating bioreactor; *: Significantly different from basal media. **(A–C)** Data from Obradovic *et al.* (1999). **(D–E)** Reproduced with kind permission of the publisher from Gooch *et al.* (2001).

Hydrodynamic forces associated with convective mixing affect the morphology of tissue-engineered cartilage. For example, engineered cartilage cultured in rotating bioreactors had thinner surface capsules and higher fractional amounts of GAG than constructs grown in spinner flasks (Freed and Vunjak-Novakovic, 1995, 1997; Vunjak-Novakovic et al., 1999; Martin et al., 2000). Moreover, engineered cartilage constructs cultured in rotating bioreactors in normal gravity (on earth) tended to maintain their initial discoid shape, whereas constructs cultured in microgravity (aboard the Mir space station) tended to become spherical, and constructs cultured in normal gravity had higher GAG contents and compressive moduli than constructs cultured in microgravity (Freed et al., 1997). Importantly, different flow fields were utilized in normal gravity and in microgravity, in order to ensure mass transfer in the two environments; i.e., on earth convective mixing was achieved by gravitational settling of constructs that were freely suspended by synchronous rotation of the inner and outer cylinders at 28 rpm, whereas aboard Mir convective mixing was induced by differential rotation of the inner and outer cylinders at 10 and 1 rpm, respectively. Therefore, on earth gravitational settling of initially discoid constructs tended to align their flat circular areas perpendicular to the direction of motion, increasing shear and mass transfer circumferentially and promoting preferential growth in the radial direction, whereas on Mir exposure of constructs to uniform shear and mass transfer at all surfaces promoted equal tissue growth in all directions such that constructs tended to become spherical.

The effects of hydrodynamic forces on construct morphology were further investigated by the techniques of particle-image velocimetry (Brown, 1998; Neitzel et al., 1998) and computational modeling (Neitzel et al., 1998; Lappa, 2003; Sucosky et al., 2004). The flow field in the spinner flask was unsteady, turbulent (Reynolds number of 1758), and characterized by large spatial variations in the velocity field and maximum shear stresses (Fig. 11.4A) (Sucosky et al., 2004). In contrast, the flow field in the rotating bioreactor (STLV) was predominately laminar (Brown, 1998), with shear stresses of ~1 dyn/cm² (Freed and Vunjak-Novakovic, 1995; Neitzel et al., 1998) and a well-mixed interior due to secondary flow patterns induced by the freely settling constructs (Fig. 11.4B and C, which show, respectively, flow-visualization and velocity-vector fields). A model of tissue growth in the rotating bioreactor, which accounted for the intensity of convection over six weeks of in vitro culture, was used to predict the morphological evolution of an engineered cartilage construct (Lappa, 2003). In particular, the model predicted that high shear and mass transfer at the lower corners of a settling, discoid construct would preferentially induce tissue growth in these regions and that temporal changes in construct size and shape would further enhance local variations in the flow field in a manner that accentuated localized tissue growth (Fig. 11.4D–F). The computed velocity fields and shear stress data corresponded well with the morphological evolution of engineered cartilage, as shown by superimposing a calculated flow field on a histological cross section of an actual 42-day construct (Fig. 11.4F). The foregoing examples suggest that combining experimental studies and modeling of hydrodynamic shear stresses and concentration gradients in bioreactors may be exploited to derive underlying mechanisms and potentially control the growth of engineered tissue constructs.

Alternative systems for cultivation of 3D tissue constructs are based on the convective flow of the culture medium directly through a porous construct via multipass perfusion. In particular, a single construct is placed within a bioreactor such that there is a tight fit between the circumference of the construct and the inner wall of the bioreactor, and then a pump is used to provide the flow of culture medium directly through the construct. Perfusion enhances mass transfer and generates compressive and shear forces, the magnitude of which can be controlled by varying the fluid flow rate. In the case of engineered cartilage, constructs cultured in perfused bioreactors contained higher amounts of ECM as compared with static controls (Dunkelman et al., 1995; Pazzano et al., 2000; Davisson et al., 2002b). Furthermore, periodic reversal of flow direction enhanced construct size and amounts of cartilaginous ECM as compared with unidirectional flow (Mahmoudifar and Doran, 2005a). In the case of engineered cardiac tissue, constructs cultured in perfused bioreactors exhibited enhanced cell survival and contractile function as compared with nonperfused controls (Carrier et al., 2002a, 2002b; Radisic et al., 2003, 2004b, 2006) (also see Chapter 38).

## Integrated Systems for Cell Seeding and Tissue Cultivation

We showed advantages of using spinner flasks for cell seeding and then rotating bioreactors for long-term cultivation of engineered tissue constructs in a systematic study involving two different scaffold materials (benzylated hyaluronan, Hyaff-11® (Fidia Advanced Biopolymers) and polyglycolic acid, PGA) and three different scaffold structures (porogen-leached sponge and nonwoven and woven textiles) (Pei et al., 2002b). The culture system was the parameter with the highest impact on the size, composition, and mechanical properties of three-day and four-week constructs (Fig. 11.5, Table 11.2). Importantly, findings attributed to the culture system represented the integrated effects of cell seeding and tissue culture. The three-day constructs seeded in spinner flasks had higher numbers of more uniformly distributed cells than statically seeded controls, and the four-week constructs cultured in bioreactors were larger and thicker and had higher amounts of cartilaginous ECM than controls seeded and cultured in petri dishes (Fig. 11.5, Table 11.2). Bioreactor-grown constructs had compressive

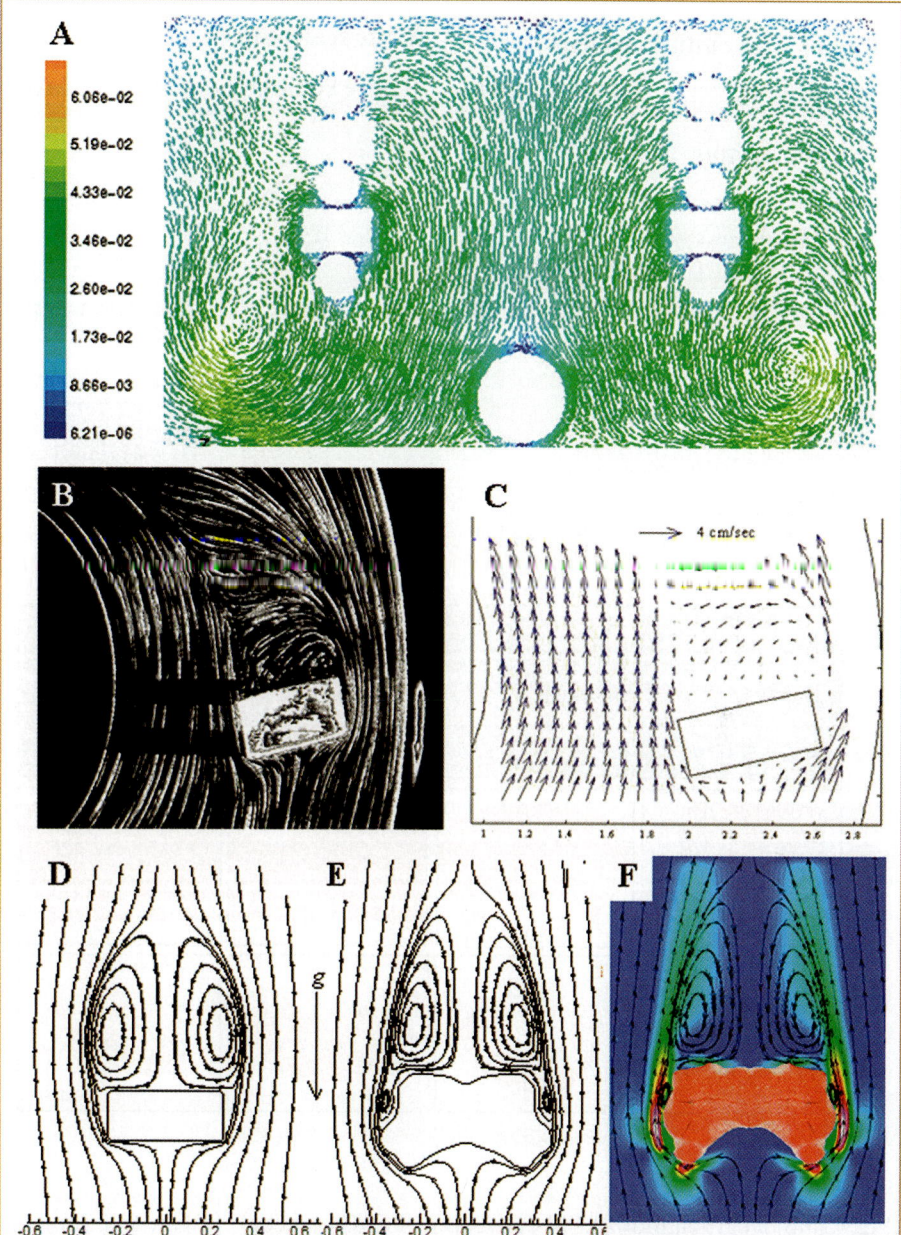

**FIG. 11.4.** Convective mixing and mass transfer in bioreactors; effects on construct morphology. Flow patterns in **(A)** a spinner flask containing two needles, each with three constructs and three spacers, and a nonsuspended stir bar and **(B–F)** a rotating vessel containing one construct. Flow-visualization **(B)** and velocity-vector **(C)** fields (the latter obtained by particle-image velocimetry) corresponded well with computed velocity fields on culture day 0 **(D)** and day 40 **(E–F)**. Computed velocity fields and shear stress data also corresponded well with the morphology of an actual cartilage construct (calculations and histological cross section of a 42-day construct are superimposed) **F**). **(A)** Reproduced with kind permission of the publisher from Sucosky *et al.* (2004). **(B, C)** Reproduced with kind permission of Academic Press, *Principles of Tissue Engineering,* 2000 (Lanza *et al.*, eds.), Fig. 13.4, p. 148. **(D–F)** Reproduced with kind permission of the publisher from Lappa (2003).

moduli in the range of native articular cartilage (0.13–0.54 MPa, Table 11.2), whereas mechanical properties of controls cultured in petri dishes could not be properly measured due to their poor structural integrity and spatial inhomogeneity. In an alternative approach, an integrated bioreactor system was custom built to provide rotational flow during cell seeding and then perfusion during construct cultivation (Sodian *et al.*, 2002). One advantage this approach may offer over the use of different bioreactors for cell seeding and long-term tissue cultivation is a lower risk of contamination.

## VI. CULTURE DURATION AND MECHANICAL CONDITIONING

### Effects of Culture Duration

With increasing time of *in vitro* culture, chondrocytes assemble a mechanically functional ECM (e.g., Buschmann *et al.*, 1995) and cardiomyocytes develop contractile responsiveness to electrical impulses (e.g., Radisic *et al.*, 2004a). For example, primary bovine calf chondrocytes seeded on nonwoven PGA mesh in spinner flasks and then cultured in rotating bioreactors synthesized (Fig. 11.6A) and deposited

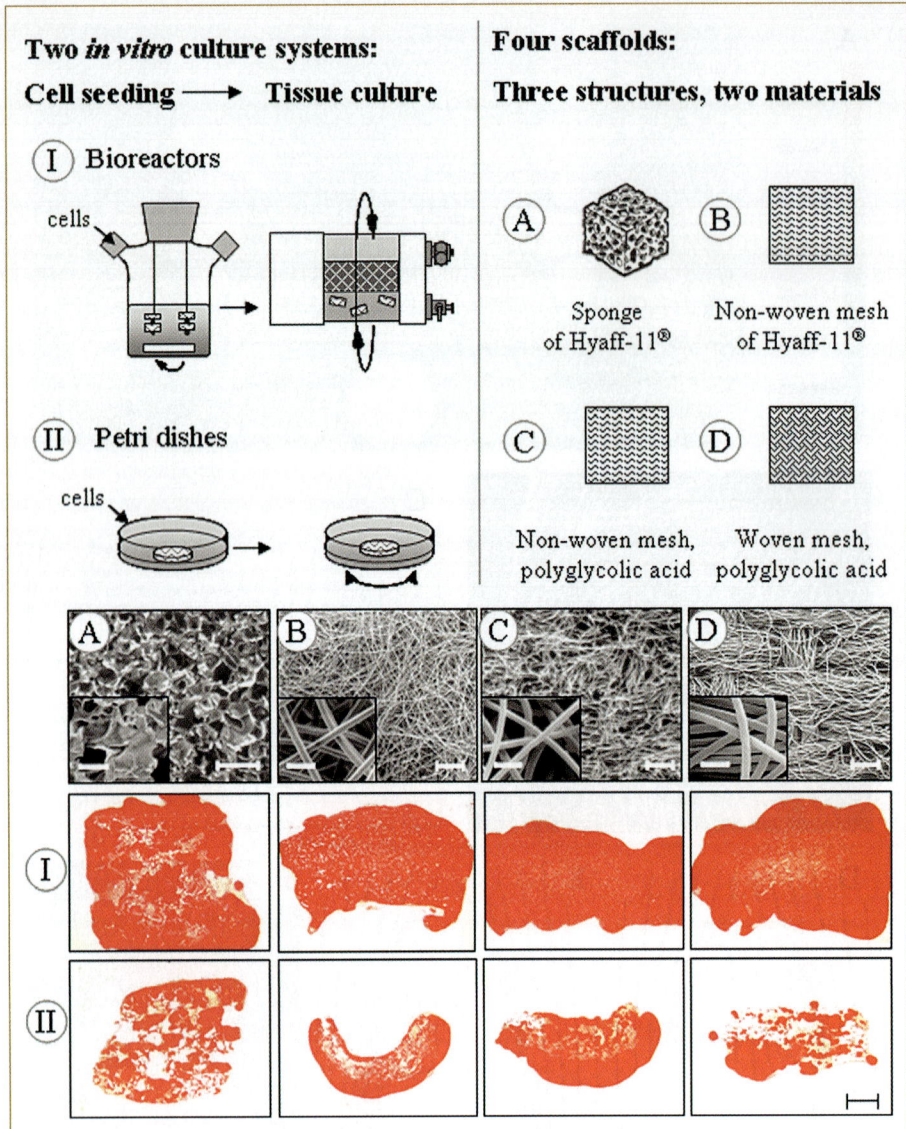

**Two *in vitro* culture systems:**

**Cell seeding ⟶ Tissue culture**

Ⓘ Bioreactors

cells

Ⓘ Petri dishes

cells

**Four scaffolds:**

**Three structures, two materials**

Ⓐ Sponge of Hyaff-11®

Ⓑ Non-woven mesh of Hyaff-11®

Ⓒ Non-woven mesh, polyglycolic acid

Ⓓ Woven mesh, polyglycolic acid

**FIG. 11.5.** Culture system and scaffold structure individually and interactively affect construct morphology. Two culture systems were used to study four different scaffolds. In culture system I, cells were seeded in spinner flasks and constructs were cultured for four weeks in rotating bioreactors. In culture system II, cells were seeded statically and constructs were cultured for four weeks in mixed petri dishes. **(A–D)** Schematics and SEM of the scaffolds. Scale bars: 500 μm; insets: 50 μm. (I, II) Histological cross sections of four-week constructs made with scaffolds **(A–D)** in culture systems I and II. Stain: safranin-O; scale bars: 1 mm. Reproduced with kind permission of the publisher from Pei *et al.* (2002b).

(Fig. 11.6B) a cartilaginous ECM consisting of GAG and collagen type II (Freed *et al.*, 1998). Importantly, the relatively high rates of ECM synthesis and deposition by the calf chondrocytes matched approximately the relatively high degradation rate of the PGA scaffold (Fig. 11.6B), a finding that did not hold true when the same scaffold was studied with other cell types (e.g., bone marrow stromal cells (Martin *et al.*, 1999)) or in other culture environments (e.g., mixed petri dishes (Freed *et al.*, 1994b)).

The structural and functional properties of tissue-engineered constructs can be improved to some degree by extending the duration of *in vitro* culture. For example, seven-month-long cultures carried out in rotating bioreactors operated on earth yielded engineered cartilage constructs with very high GAG fractions (~8% of wet weight) and compressive moduli (~0.9 MPa) that were comparable to normal articular cartilage (Fig. 11.6C–D), although the col-

lagen fraction and dynamic stiffness of the seven-month constructs remained subnormal (Freed *et al.*, 1997). Also, the adhesive strength of engineered cartilage to adjacent cartilage and bone improved between four and eight weeks of culture (Fig. 11.2F). Engineered cardiac tissue cultured for up to eight days exhibited a temporal increase in contractile amplitude (Radisic *et al.*, 2004a), but at present the maximal contractile force generation reported for an engineered cardiac construct (~4 mN/mm$^2$) remains more than an order of magnitude below that of normal heart muscle (Eschenhagen and Zimmermann, 2005).

## Effects of Mechanical Conditioning

It is well known that mechanical forces play a key role in determining the architecture of native tissues such as bone (Thompson, 1977), and a wide variety of laboratory devices have been developed for mechanical conditioning

**Table 11.2.** Culture system and scaffold structure individually and interactively affect the properties of engineered cartilage

| Culture system (CS): | Spinner flask → Rotating bioreactor | | | | Petri dishes | | | | CS | SM | SS | CS* SM | CS* SS |
|---|---|---|---|---|---|---|---|---|---|---|---|---|---|
| Scaffold material (SM): | Hyaff-11® | Hyaff-11® | PGA | PGA | Hyaff-11® | Hyaff-11® | PGA | PGA | | | | | |
| Scaffold structure (SS): | Sponge | NWM | NWM | WM | Sponge | NWM | NWM | LM | | | | | |
| **3-day constructs** | | | | | | | | | | | | | |
| Cells (millions/C, $n = 3$) | 3.04 ± 0.16 | 3.95 ± 0.74 | 5.29 ± 1.22 | 4.64 ± 1.30 | 1.99 ± 0.13* | 2.51 ± 0.34* | 1.88 ± 0.95* | 2.35 ± 0.44* | <.0001 | NS | NS | 0.046 | NS |
| Wet weight (mg, $n = 3$) | 48.4 ± 1.5 | 29.6 ± 3.7 | 33.7 ± 2.1 | 47.7 ± 2.0‡ | 44.6 ± 3.1 | 25.4 ± 3.2 | 30.7 ± 4.9 | 45.9 ± 2.9‡ | 0.0097 | 0.017 | <.0001 | NS | NS |
| **1-month constructs** | | | | | | | | | | | | | |
| Cells (millions/C, $n = 3$) | 9.38 ± 0.88 | 12.4 ± 1.5§ | 12.8 ± 0.6 | 15.6 ± 0.36‡ | 5.07 ± 0.53‡ | 6.00 ± 0.41* | 4.51 ± 0.82* | 5.91 ± 1.49* | <.0001 | NS | 0.0003 | NS | NS |
| Wet weight (mg, $n = 7$) | 144 ± 11 | 198 ± 29§ | 210 ± 8 | 231 ± 14‡ | 75.6 ± 4.9* | 66.2 ± 9.0* | 48.5 ± 12.9* | 70.5 ± 16.4* | <.0001 | NS | <.0001 | 0.033 | <.0001 |
| Glycosaminoglycans | | | | | | | | | | | | | |
| (mg/C, $n = 3$) | 4.82 ± 0.74 | 6.72 ± 1.27 | 8.63 ± 0.25 | 9.95 ± 0.79 | 1.26 ± 0.15* | 2.51 ± 0.18* | 1.24 ± 0.98* | 1.68 ± 0.63* | <.0001 | NS | 0.0022 | 0.0016 | NS |
| (% wet weight, $n = 3$) | 3.29 ± 0.14 | 3.70 ± 0.09§ | 4.10 ± 0.20† | 4.39 ± 0.22 | 1.58 ± 0.06* | 3.57 ± 0.13 | 2.20 ± 1.06* | 2.19 ± 0.45* | <.0001 | NS | 0.0007 | 0.0025 | 0.016 |
| Total collagen | | | | | | | | | | | | | |
| (mg/C, $n = 3$) | 4.90 ± 0.53 | 6.36 ± 0.94 | 7.29 ± 0.39 | 8.27 ± 0.72 | 2.37 ± 0.14* | 2.95 ± 0.32* | 1.77 ± 0.53* | 2.42 ± 0.58* | <.0001 | NS | 0.0040 | 0.0053 | NS |
| (% wet weight, $n = 3$) | 3.35 ± 0.07 | 3.52 ± 0.25 | 3.46 ± 0.15 | 3.65 ± 0.25 | 2.99 ± 0.15* | 4.20 ± 0.33* | 3.48 ± 0.13 | 3.20 ± 0.18 | NS | 0.0042 | 0.0001 | 0.014 | 0.0008 |
| Compressive moduli | | | | | | | | | | | | | |
| (MPa, $n = 4$) | 0.13 ± 0.06 | 0.52 ± 0.22§ | 0.40 ± 0.08 | 0.54 ± 0.08 | NM | NM | NM | NM | NM | NM | 0.0099 | NM | NM |

*Abbreviations:* CS = culture system; SM = scaffold material; SS = scaffold structure; PGA = polyglycolic acid; NWM = nonwoven mesh; WM = composite mesh; C = construct; NM = not measured; NS = not statistically significant. Data represent the mean ± SD of $n = 3$–7 independent samples.

*Significantly different ($p < 0.05$) from corresponding constructs cultured in the bioreactor culture system.

§Significantly different ($p < 0.05$) from corresponding constructs based on Hyaff-11® sponges.

†Significantly different ($p < 0.05$) from corresponding constructs based on Hyaff-11® NWM.

‡Significantly different ($p < 0.05$) from corresponding constructs based on PGA NWM.

Reproduced with kind permission of the publisher from Pei *et al.* (2002b).

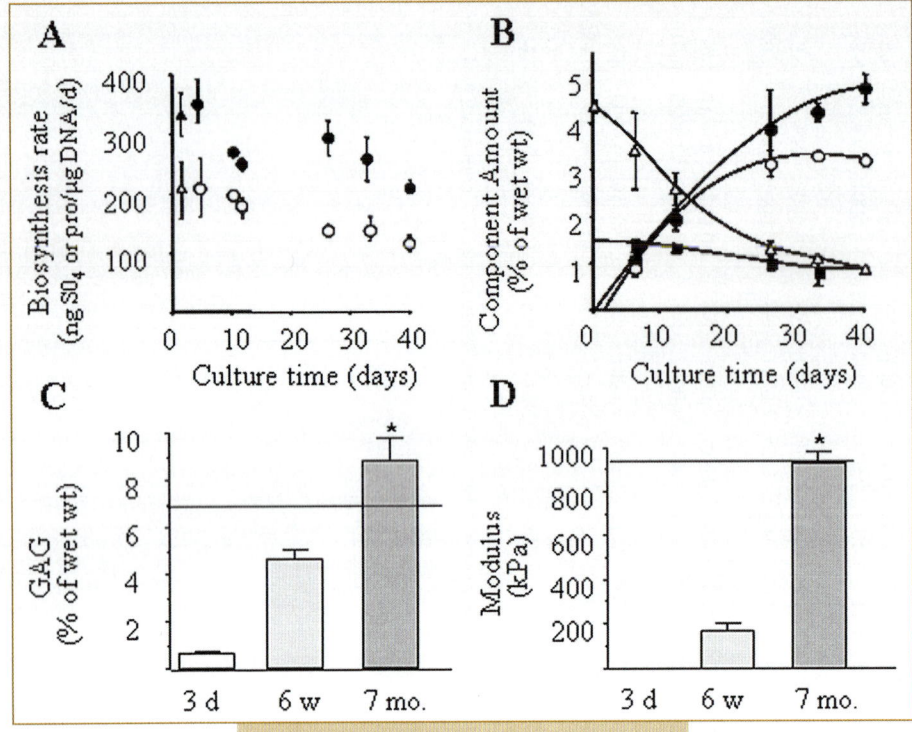

**FIG. 11.6.** Effects of culture duration on the composition and function of engineered cartilage. **(A)** Construct biosynthesis rates of GAG (closed circles) and collagen (open circles), are compared to corresponding synthesis rates obtained for native bovine calf cartilage (triangles). **(B)** Fractional construct amounts of GAG (closed circles), collagen (open circles), cells (squares), and PGA scaffold (triangles). **(C,D)** Construct amounts of GAG and compressive modulus measured at 3 days, 6 weeks, and 7 months (3 d, 6 w, and 7 mo). Lines indicate average values obtained for native bovine calf cartilage. Reproduced with kind permission of the publisher from Guilak *et al.* (2003), Figure 27.4, p. 365.

of cell and tissue cultures (reviewed in Brown, 2000). There is growing evidence that mechanical factors can play an important role in determining stem cell fate (Estes *et al.*, 2004). In particular, the role of *in vitro* mechanical stress in maintaining and promoting the chondrogenic phenotype has been the topic of several recent investigations, but the specific influences of different physical stimuli and their interactions with the biochemical environment are not fully understood. For example, dynamic compression caused an approximately two-fold increase in cartilage nodule density and a 2.5-fold increase in GAG synthesis in stage 23/24 chick limb-bud cells cultured in agarose gel (Elder *et al.*, 2000). Likewise, cyclic hydrostatic pressure significantly increased the amounts of proteoglycan and collagen in aggregates of human bone marrow stromal cells (Angele *et al.*, 2003). In another study, compression enhanced chondrogenic differentiation in mouse embryonic E10 stage cells embedded in collagen type I as compared to unloaded controls, as shown by up-regulating SOX-9 and down-regulating IL-1β expression (Takahashi *et al.*, 1998).

In the case of engineered cartilage, a number of studies have shown that mechanical conditioning can enhance chondrogenesis. Dynamic loading can increase GAG accumulation and ECM assembly and, hence improve the mechanical properties of constructs based on bovine calf articular chondrocytes and a variety of 3D scaffolds, including agarose gel (Buschmann *et al.*, 1995; Mauck *et al.*, 2000, 2002), PGA nonwoven mesh (Davisson *et al.*, 2002a), and

self-assembling peptide gel (Kisiday *et al.*, 2004). The response of chondrocytes to dynamic loading depended on the amount and composition of ECM in the developing construct (Demarteau *et al.*, 2003), and in some studies loading increased both synthesis of new GAG and its loss into the culture media (Kisiday *et al.*, 2004; Seidel *et al.*, 2004). Dynamic loading also increased the amount of ECM and the compressive modulus of engineered cartilage based on calf chondrocytes and a porous calcium polyphosphate scaffold, and, interestingly, application of shear stress yielded constructs with higher amounts of ECM and higher compressive moduli than compressive stress (Waldman *et al.*, 2003). Likewise, application of dynamic hydrostatic pressure affected chondrogenesis in 3D cultures of bovine (Mizuno *et al.*, 2002) and human (Toyoda *et al.*, 2003) chondrocytes.

In the case of engineered muscle (skeletal, smooth, and cardiac), dynamic tensile and pulsatile loading can affect construct composition, contractility, and pharmacological responsiveness (Vandenburgh *et al.*, 1991; Kim *et al.*, 1999; Niklason *et al.*, 1999; Fink *et al.*, 2000; Sodian *et al.*, 2001; Akhyari *et al.*, 2002; Powell *et al.*, 2002; Zimmermann *et al.*, 2002; Gonen-Wadmany *et al.*, 2004; Boublik *et al.*, 2005). In one representative study, cyclic stretch (applied continuously for a one-week period at a frequency of 1 Hz and strain amplitudes ranging from 1% to 10%) affected the composition and mechanical properties of engineered cardiac constructs based on a knitted, elastomeric fabric

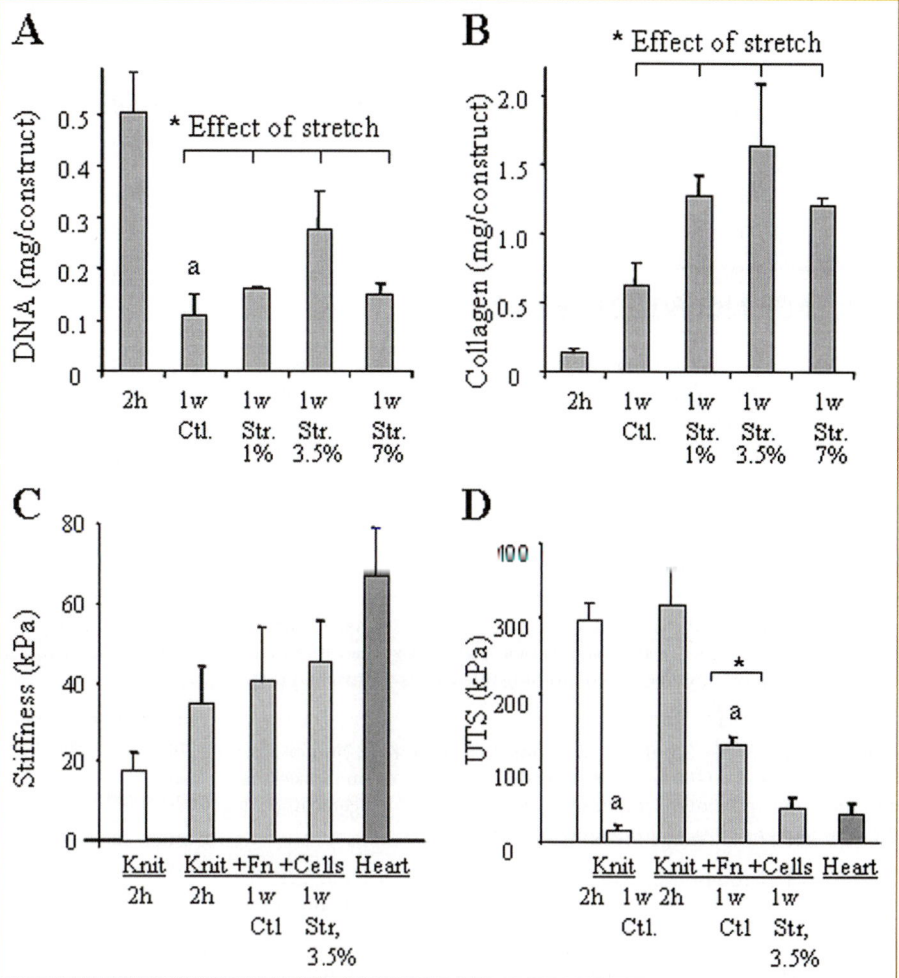

**FIG. 11.7.** Effects of culture duration and cyclic stretch on the composition and function of engineered cardiac tissue. **(A)** DNA content, **(B)** collagen content, **(C)** tensile stiffness, and **(D)** ultimate tensile strength (UTS) of constructs based on cardiomyocytes cultured for two hours (2 h) or one week (1 w) without (Ctl.) or with cyclic stretch (Str.) applied at different strain amplitudes (1%, 3.5%, and 7%). The knitted elastomeric scaffold and the fibrin hydrogel are respectively denoted by (Knit) and (Fn) respectively. a: Significant effect of culture time; *: Significant effect of cyclic stretch. Reproduced with kind permission of the publisher from Boublik et al. (2005).

(Hyalonect®, Fidia Advanced Biopolymers), fibrin gel, and cardiomyocytes (Fig. 11.7) (Boublik et al., 2005). As compared with unstretched controls, dynamic stretch increased construct amounts of DNA and collagen in a manner that depended on strain amplitude (Fig. 11.7A–B), did not affect tensile stiffness (Fig. 11.7C), and decreased ultimate tensile strength (Fig 11.7D). Likewise, in engineered cardiac constructs based on vascular cells and poly-4-hydroxy-butyrate scaffolds, pulsatile stimulation enhanced construct amounts of DNA and collagen but did not affect mechanical properties (Yang et al., 2006). Notably, in the two aforementioned studies, construct biomechanical properties were determined mainly by the scaffold. In contrast, for engineered cardiac tissues based on collagen scaffolds, construct biomechanical properties were cell and ECM derived, and cyclic stretch not only affected construct structure (e.g., orientation of cells and collagen) (Gonen-Wadmany et al., 2004) but also enhanced mechanical function (i.e., contractility) (Fink et al., 2000; Akhyari et al., 2002;

Zimmermann et al., 2002). Together, mechanical-conditioning studies demonstrate that specific loading protocols may be exploited to stimulate specific responses in engineered tissue constructs and emphasize the potential utility of bioreactors in studying and promoting ex vivo construct formation.

## VII. CONCLUSIONS

As the field of tissue engineering progresses, it is becoming apparent that the next generation of functional tissue replacements may require additional exogenous influences to achieve many of the important requirements for long-term success. Two such needs — physiologic mechanical properties and the ability to grow and remodel in a manner that allows for restoration of physiologic functions — can potentially be achieved by the judicious selection of in vitro culture parameters. For example, scaffolds can provide tissue-mimetic mechanical properties (Moutos et al., 2007), and bioreactors can be exploited to enhance in vitro generation of 3D engineered tissue constructs and to test their physiologic responses in environments mimicking those in which they will eventually be implanted (Freed *et al.*, 2006). In this chapter, we provided representative studies of tissue engineered cartilage and cardiac constructs in which *in vitro* culture parameters were manipulated to accelerate the development and integration of 3D engineered tissue constructs and to improve their biomechanical properties. Effects of *in vitro* convective flow (i.e., flow regimen and associated mixing, mass transfer and shear) and mechanical conditioning (i.e., compression, tensile stretch, pressure and shear) were discussed. Given the rapid evolution of the field of functional tissue engineering, it is important to consider *in vitro* culture parameters in conjunction with novel growth factors, biomaterials, gene therapies, and other emerging technologies.

## VIII. ACKNOWLEDGMENTS

We would like to thank J. Boublik, J. Brown, R. Carrier, B. Estes, J. Gimble, K. Gooch, M. Lappa, I. Martin, F. Moutos, B. Obradovic, M. Pei, D. Schaefer, L. Setton, P. Sucosky, E. Tognana, and G. Vunjak-Novakovic for contributing to the studies described in this chapter; R. Langer, M. Moretti, and G. P. Neitzel for advice and critical reviews; and S. Kangiser for help with manuscript preparation. This work was supported by National Aeronautics and Space Administration Grant NNJ04HC72G and National Institutes of Health Grants AR49294, AG15768, AR50245, and AR48182.

## IX. REFERENCES

Akhyari, P., Fedak, P. W. M., Weisel, R. D., Lee, T. Y. J., Verma, S., Mickle, D. A. G., and Li, R. K. (2002). Mechanical stretch regimen enhances the formation of bioengineered autologous cardiac muscle grafts. *Circulation* **106**, I137–I142.

Angele, P., Yoo, J. U., Smith, C., Mansour, J., Jepsen, K. J., Nerlich, M., and Johnstone, B. (2003). Cyclic hydrostatic pressure enhances the chondrogenic phenotype of human mesenchymal progenitor cells differentiated *in vitro*. *J. Orthop. Res.* **21**, 451–457.

Badylak, S. F., Grompe, M., Caplan, A. I., Greisler, H. P., Guldberg, R. E., and Taylor, D. A. (2002). *In vivo* remodeling: breakout session summary. *Ann. N. Y. Acad. Sci.* **961**, 319–322.

Begley, C. M., and Kleis, S. J. (2000). The fluid dynamic and shear environment in the NASA/JSC rotating-wall perfused-vessel bioreactor. *Biotechnol. Bioeng.* **70**, 32–40.

Boublik, J., Park, H., Radisic, M., Tognana, E., Chen, F., Pei, M., Vunjak-Novakovic, G., and Freed, L. E. (2005). Mechanical properties and remodeling of hybrid cardiac constructs made from heart cells, fibrin, and biodegradable, elastomeric knitted fabric. *Tissue Eng.* **11**, 1122–1132.

Brown, J. B. (1998). An experimental facility for the investigation of the flow in a circular-Couette flow bioreactor. MS thesis in mechanical engineering, Georgia Institute of Technology, Atlanta, GA.

Brown, T. D. (2000). Techniques for mechanical stimulation of cells *in vitro*: a review. *J. Biomech.* **33**, 3–14.

Bursac, N., Papadaki, M., Cohen, R. J., Schoen, F. J., Eisenberg, S. R., Carrier, R., Vunjak-Novakovic, G., and Freed, L. E. (1999). Cardiac muscle tissue engineering: toward an *in vitro* model for electrophysiological studies. *Am. J. Physiol. Heart Circ. Physiol.* **277**, H433–H444.

Bursac, N., Papadaki, M., White, J. A., Eisenberg, S. R., Vunjak-Novakovic, G., and Freed, L. E. (2003). Cultivation in rotating bioreactors promotes maintenance of cardiac myocyte electrophysiology and molecular properties. *Tissue Eng.* **9**, 1243–1253.

Buschmann, M. D., Gluzband, Y. A., Grodzinsky, A. J., and Hunziker, E. B. (1995). Mechanical compression modulates matrix biosynthesis in chondrocyte/agarose culture. *J. Cell Sci.* **108**, 1497–1508.

Butler, D. L., Goldstein, S. A., and Guilak, F. (2000). Functional tissue engineering: the role of biomechanics. *J. Biomech. Eng.* **122**, 570–575.

Carrier, R. L., Papadaki, M., Rupnick, M., Schoen, F. J., Bursac, N., Langer, R., Freed, L. E., and Vunjak-Novakovic, G. (1999). Cardiac tissue engineering: cell seeding, cultivation parameters and tissue construct characterization. *Biotechnol. Bioeng.* **64**, 580–589.

Carrier, R. L., Rupnick, M., Langer, R., Schoen, F. J., Freed, L. E., and Vunjak-Novakovic, G. (2002a). Perfusion improves tissue architecture of engineered cardiac muscle. *Tissue Eng.* **8**, 175–188.

Carrier, R. L., Rupnick, M., Langer, R., Schoen, F. J., Freed, L. E., and Vunjak-Novakovic, G. (2002b). Effects of oxygen on engineered cardiac muscle. *Biotechnol. Bioeng.* **78**, 617–625.

Darling, E. M., and Athanasiou, K. A. (2003). Articular cartilage bioreactors and bioprocesses. *Tissue Eng.* **9**, 9–26.

Davisson, T., Kunig, S., Chen, A., Sah, R., and Ratcliffe, A. (2002a). Static and dynamic compression modulate matrix metabolism in tissue-engineered cartilage. *J. Orthop. Res.* **20**, 842–848.

Davisson, T. H., Sah, R. L., and Ratcliffe, A. R. (2002b). Perfusion increases cell content and matrix synthesis in chondrocyte three-dimensional cultures. *Tissue Eng.* **8**, 807–816.

Demarteau, O., Wendt, D., Braccini, A., Jakob, M., Schafer, D., Heberer, M., and Martin, I. (2003). Dynamic compression of cartilage constructs engineered from expanded human articular chondrocytes. *Biochem. Bioph. Res. Co.* **310**, 580–588.

Driesang, I. M., and Hunziker, E. B. (2000). Delamination rates of tissue flaps used in articular cartilage repair. *J. Orthop. Res.* **18**, 909–911.

Dunkelman, N. S., Zimber, M. P., Lebaron, R. G., Pavelec, R., Kwan, M., and Purchio, A. F. (1995). Cartilage production by rabbit articular chondrocytes on polyglycolic acid scaffolds in a closed bioreactor system. *Biotechnol. Bioeng.* **46**, 299–305.

Elder, S. H., Kimura, J. H., Soslowsky, L. J., Lavagnino, M., and Goldstein, S. A. (2000). Effect of compressive loading on chondrocyte differentiation in agarose cultures of chick limb-bud cells. *J. Orthop. Res.* **18**, 78–86.

Eschenhagen, T., and Zimmermann, W. H. (2005). Engineering myocardial tissue. *Circ. Res.* **97**, 1220–1231.

Estes, B. T., Gimble, J. M., and Guilak, F. (2004). Mechanical signals as regulators of stem cell fate. *Curr. Top. Dev. Biol.* **60**, 91–126.

Fink, C., Ergun, S., Kralisch, D., Remmers, U., Weil, J., and Eschenhagen, T. (2000). Chronic stretch of engineered heart tissue induces hypertrophy and functional improvement. *FASEB J.* **14**, 669–679.

Freed, L. E., and Vunjak-Novakovic, G. (1995). Cultivation of cell-polymer constructs in simulated microgravity. *Biotechnol. Bioeng.* **46**, 306–313.

Freed, L. E., and Vunjak-Novakovic, G. (1997). Microgravity tissue engineering. *In Vitro Cell. Dev. Biol. — An.* **33**, 381–385.

Freed, L. E., and Vunjak-Novakovic, G. (2002). Spaceflight bioreactor studies of cells and tissues. *In* "Advances in Space Biology and Medicine" (A. Cogoli, ed.), pp. 177–195. Elsevier, Amsterdam.

Freed, L. E., Marquis, J. C., Vunjak-Novakovic, G., Emmanual, J., and Langer, R. (1994a). Composition of cell-polymer cartilage implants. *Biotechnol. Bioeng.* **43**, 605–614.

Freed, L. E., Vunjak-Novakovic, G., Biron, R., Eagles, D., Lesnoy, D., Barlow, S., and Langer, R. (1994b). Biodegradable polymer scaffolds for tissue engineering. *Nature Biotechnology* **12**, 609–693.

Freed, L. E., Langer, R., Martin, I., Pellis, N., and Vunjak-Novakovic, G. (1997). Tissue engineering of cartilage in space. *Proc. Natl. Acad. Sci. USA* **94**, 13885–13890.

Freed, L. E., Hollander, A. P., Martin, I., Barry, J. R., Langer, R., and Vunjak-Novakovic, G. (1998). Chondrogenesis in a cell-polymer-bioreactor system. *Exp. Cell Res.* **240**, 58–65.

Freed, L. E., Guilak, F., Guo, X. E., Gray, M. L., Tranquillo, R., Holmes, J. W., Radisic, M., Sefton, M. V., Kaplan, D., and Vunjak-Novakovic, G. (2006). Advanced tools for tissue engineering: scaffolds, bioreactors, and signaling. *Tissue Eng.* **12**, 3285–3305.

Gonen-Wadmany, M., Gepstein, L., and Seliktar, D. (2004). Controlling the cellular organization of tissue-engineered cardiac constructs. *Ann. N.Y. Acad. Sci.* **1015**, 299–311.

Gooch, K. J., Blunk, T., Courter, D. L., Sieminski, A. L., Bursac, P. M., Vunjak-Novakovic, G., and Freed, L. E. (2001a). IGF-I and mechanical environment interact to modulate engineered cartilage development. *Biochem. Bioph. Res. Co.* **286**, 909–915.

Gooch, K. J., Kwon, J. H., Blunk, T., Langer, R., Freed, L. E., and Vunjak-Novakovic, G. (2001b). Effects of mixing intensity on tissue-engineered cartilage. *Biotechnol. Bioeng.* **72**, 402–407.

Guilak, F., Butler, D. L., and Goldstein, S. A. (2001). Functional tissue engineering: the role of biomechanics in articular cartilage repair. *Clin. Orthop. Rel. Res.* **391** Suppl. S295–S305.

Guilak, F., Butler, D. L., Golstein, S. A., and Mooney, D. J. (2003). "Functional Tissue Engineering." Springer-Verlag, New York.

Hollister, S. J. (2005). Porous scaffold design for tissue engineering. *Nat. Mater.* **4**, 518–524.

Hunziker, E. B. (1999). Biologic repair of articular cartilage. Defect models in experimental animals and matrix requirements. *Clin. Orthop. Rel. Res.* **367S**, S135–S146.

Hunziker, E. B. (2001). Articular cartilage repair: basic science and clinical progress. A review of the current status and prospects. *Osteoarthr. Cartilage* **10**, 432–463.

Jessup, J. M., Frantz, M., Sonmez-Alpan, E., Locker, J., Skena, K., Waller, H., Battle, P., Nachman, A., Weber, M. E., Thomas, D. A., *et al.* (2000). Microgravity culture reduces apoptosis and increases the differentiation of a human colorectal carcinoma cell line. *In Vitro Cell. Dev. Biol. — An.* **36**, 367–373.

Karande, T. S., Ong, J. L., and Agrawal, C. M. (2004). Diffusion in musculoskeletal tissue engineering scaffolds: design issues related to porosity, permeability, architecture, and nutrient mixing. *Ann. Biomed. Eng.* **32**, 1728–1743.

Kim, B. S., Putnam, A. J., Kulik, T. J., and Mooney, D. J. (1998). Optimizing seeding and culture methods to engineer smooth muscle tissue on biodegradable polymer matrices. *Biotechnol. Bioeng.* **57**, 46–54.

Kim, B. S., Nikolovski, J., Bonadio, J., and Mooney, D. J. (1999). Cyclic mechanical strain regulates the development of engineered smooth muscle tissue. *Nat. Biotechnol.* **17**, 979–983.

Kisiday, J. D., Jin, M., DiMicco, M. A., Kurz, B., and Grodzinsky, A. J. (2004). Effects of dynamic compressive loading on chondrocyte biosynthesis in self-assembling peptide scaffolds. *J. Biomech.* **37**, 595–604.

Langer, R., and Tirrell, D. A. (2004). Designing materials for biology and medicine. *Nature* **428**, 487–492.

Lappa, M. (2003). Organic tissues in rotating bioreactors: fluid-mechanical aspects, dynamic growth models., and morphological evolution. *Biotechnol. Bioeng.* **84**, 518–532.

Li, Y., Ma, T., Kniss, D. A., Lasky, L. C., and Yang, S. T. (2001). Effects of filtration seeding on cell density, spatial distribution, and proliferation in nonwoven fibrous matrices. *Biotechnol. Prog.* **17**, 935–944.

Liu, M., Montazeri, S., Jedlovsky, T., Van Wert, R., Zhang, J., Li, R. K., and Yan, J. (1999). Bio-stretch, a computerized cell strain apparatus for three-dimensional organotypic cultures. *In Vitro Cell. Dev. Biol. — An.* **35**, 87–93.

Lutolf, M. P., and Hubbell, J. A. (2005). Synthetic biomaterials as instructive extracellular microenvironments for morphogenesis in tissue engineering. *Nat. Biotechnol.* **23**, 47–55.

Madry, H., Padera, R., Seidel, J., Langer, R., Freed, L. E., Trippel, S. B., and Vunjak-Novakovic, G. (2002). Gene transfer of a human insulin-like growth factor I cDNA enhances tissue engineering of cartilage. *Hum. Gene Ther.* **13**, 1621–1630.

Mahmoudifar, N., and Doran, P. M. (2005a). Tissue engineering of human cartilage in bioreactors using single and composite cell-seeded scaffolds. *Biotechnol. Bioeng.* **91**, 338–355.

Mahmoudifar, N., and Doran, P. M. (2005b). Tissue engineering of human cartilage and osteochondral composites using recirculation bioreactors. *Biomaterials* **26**, 7012–7024.

Malda, J., Martens, D. E., Tramper, J., van Blitterswijk, C. A., and Riesle, J. (2003). Cartilage tissue engineering: controversy in the effect of oxygen. *Crit. Rev. Biotechnol.* **23**, 175–194.

Malda, J., Rouwkema, J., Martens, D. E., Le Comte, E. P., Kooy, F. K., Tramper, J., van Blitterswijk, C. A., and Riesle, J. (2004). Oxygen gradients in tissue-engineered PEGT/PBT cartilaginous constructs: measurement and modeling. *Biotechnol. Bioeng.* **86**, 9–18.

Marolt, D., Augst, A., Freed, L. E., Vepari, C., Fajardo, R., Patel, N., Gray, M., Farley, M., Kaplan, D., and Vunjak-Novakovic, G. (2006). Bone and cartilage tissue constructs grown using human bone marrow stromal cells, silk scaffolds and rotating bioreactors. *Biomaterials* **27**, 6138–6149.

Martin, I., Padera, R. F., Vunjak-Novakovic, G., and Freed, L. E. (1998). *In vitro* differentiation of chick embryo bone marrow stromal cells into cartilaginous and bone-like tissues. *J. Orthop. Res.* **16**, 181–189.

Martin, I., Shastri, V., Padera, R. F., Langer, R., Vunjak-Novakovic, G., and Freed, L. E. (1999). Bone marrow stromal cell differentiation on porous polymer scaffolds. *Trans. Orthop. Res. Soc.* **24**, 57.

Martin, I., Obradovic, B., Treppo, S., Grodzinsky, A. J., Langer, R., Freed, L. E., and Vunjak-Novakovic, G. (2000). Modulation of the mechanical properties of tissue-engineered cartilage. *Biorheology* **37**, 141–147.

Martin, I., Shastri, V. P., Padera, R. F., Yang, J., Mackay, A. J., Langer, R., Vunjak-Novakovic, G., and Freed, L. E. (2001). Selective differentiation of mammalian bone marrow stromal cells cultured on three-dimensional polymer foams. *J. Biomed. Mater. Res.* **55**, 229–235.

Martin, I., Wendt, D., and Heberer, M. (2004). The role of bioreactors in tissue engineering. *Trends Biotechnol.* **22**, 80–86.

Martin, Y., and Vermette, P. (2005). Bioreactors for tissue mass culture: design, characterization, and recent advances. *Biomaterials* **26**, 7481–7503.

Mauck, R. L., Soltz, M. A., Wang, C. C. B., Wong, D. D., Chao, P. G., Valhmu, W. B., Hung, C. T., and Ateshian, G. A. (2000). Functional tissue engineering of articular cartilage through dynamic loading of chondrocyte-seeded agarose gels. *J. Biomech. Eng.* **122**, 252–260.

Mauck, R. L., Seyhan, S. L., Ateshian, G. A., and Hung, C. T. (2002). Influence of seeding density and dynamic deformational loading on the developing structure/function relationships of chon-drocyte-seeded agarose hydrogels. *Ann. Biomed. Eng.* **30**, 1046–1056.

Mauck, R. L., Nicoll, S. B., Seyhan, S. L., Ateshian, G. A., and Hung, C. T. (2003). Synergistic action of growth factors and dynamic loading for articular cartilage tissue engineering. *Tissue Eng.* **9**, 597–611.

Mizuno, S., Tateishi, T., Ushida, T., and Glowacki, J. (2002). Hydrostatic fluid pressure enhances matrix synthesis and accumulation by bovine chondrocytes in three-dimensional culture. *J. Cell. Physiol.* **193**, 319–327.

Moutos, F. T., Freed, L. E., and Guilak, F. (2007). A biomimetic three-dimensional woven composite scaffold for functional tissue engineering of cartilage. *Nat. Mater.* **6**, 162–167.

Muschler, G. F., Nakamoto, C., and Griffith, L. G. (2004). Engineering principles of clinical cell-based tissue engineering. *J. Bone Joint Surg. Am.* **86-A**, 1541–1558.

Neitzel, G. P., Nerem, R. M., Sambanis, A., Smith, M. K., Wick, T. M., Brown, J. B., Hunter, C., Jovanovic, I. P., Malaviya, P., Saini, S., *et al.* (1998). Cell function and tissue growth in bioreactors: fluid mechanical and chemical environments. *J. Japan. Soc. Microgr. Appl.* **15**, 602–607.

Niklason, L. E., Gao, J., Abbott, W. M., Hirschi, K. K., Houser, S., Marini, R., and Langer, R. (1999). Functional arteries grown *in vitro*. *Science* **284**, 489–493.

Obradovic, B., Carrier, R. L., Vunjak-Novakovic, G., and Freed, L. E. (1999). Gas exchange is essential for bioreactor cultivation of tissue-engineered cartilage. *Biotechnol. Bioeng.* **63**, 197–205.

Obradovic, B., Meldon, J. H., Freed, L. E., and Vunjak-Novakovic, G. (2000). Glycosaminoglycan deposition in engineered cartilage: Experiments and mathematical model. *AIChE J.* **46**, 1860–1871.

Obradovic, B., Martin, I., Padera, R. F., Treppo, S., Freed, L. E., and Vunjak-Novakovic, G. (2001). Integration of engineered cartilage. *J. Orthop. Res.* **19**, 1089–1097.

Papadaki, M., Bursac, N., Langer, R., Merok, J., Vunjak-Novakovic, G., and Freed, L. E. (2001). Tissue engineering of functional cardiac muscle: molecular, structural and electrophysiological studies. *Am. J. Physiol. Heart Circ. Physiol.* **280**, H168–H178.

Pazzano, D., Mercier, K. A., Moran, J. M., Fong, S. S., DiBiasio, D. D., Rulfs, J. X., Kohles, S. S., and Bonassar, L. J. (2000). Comparison of chondrogensis in static and perfused bioreactor culture. *Biotechnol. Prog.* **16**, 893–896.

Pei, M., Seidel, J., Vunjak-Novakovic, G., and Freed, L. E. (2002a). Growth factors for sequential cellular de- and redifferentiation in tissue engineering. *Biochem. Bioph. Res. Co.* **294**, 149–154.

Pei, M., Solchaga, L. A., Seidel, J., Zeng, L., Vunjak-Novakovic, G., Caplan, A. I., and Freed, L. E. (2002b). Bioreactors mediate the effectiveness of tissue-engineering scaffolds. *FASEB J.* **16**, 1691–1694.

Powell, C. A., Smiley, B. L., Mills, J., and Vandenburgh, H. H. (2002). Mechanical stimulation improves tissue-engineered human skeletal muscle. *Am. J. Physiol. Cell Physiol.* **283**, C1557–C1565.

Praemer, A., Furner, S., and Rice, D. P. (1999). "Musculoskeletal Conditions in the United States." American Academy of Orthopedic Surgeons, Rosemont, IL.

Radisic, M., Euloth, M., Yang, L., Langer, R., Freed, L. E., and Vunjak-Novakovic, G. (2003). High-density seeding of myocyte cells for tissue engineering. *Biotechnol. Bioeng.* **82**, 403–414.

Radisic, M., Park, H., Shing, H., Consi, T., Schoen, F. J., Langer, R., Freed, L. E., and Vunjak-Novakovic, G. (2004a). Functional assembly of engineered myocardium by electrical stimulation of cardiac myocytes cultured on scaffolds. *Proc. Nat. Acad. Sci. USA* **101**, 18129–18134.

Radisic, M., Yang, L., Boublik, J., Cohen, R. J., Langer, R., Freed, L. E., and Vunjak-Novakovic, G. (2004b). Medium perfusion enables engineering of compact and contractile cardiac tissue. *Am. J. Physiol. Heart Circ. Physiol.* **286**, H507–H516.

Radisic, M., Malda, J., Epping, E., Geng, W., Langer, R., and Vunjak-Novakovic, G. (2006). Oxygen gradients correlate with cell density and cell viability in engineered cardiac tissue. *Biotechnol. Bioeng.* **93**, 332–343.

Radisic, M., Park, H., Chen, F., Salazar-Lazzaro, J. E., Wang, Y., Dennis, R. G., Langer, R., Freed, L. E., and Vunjak-Novakovic, G. (2006). Biomimetic approach to cardiac tissue engineering: Oxygen carriers and channeled scaffolds. *Tissue Eng.* **12**, 1–15.

Riesle, J., Hollander, A. P., Langer, R., Freed, L. E., and Vunjak-Novakovic, G. (1998). Collagen in tissue-engineered cartilage: types, structure and crosslinks. *J. Cell. Biochem.* **71**, 313–327.

Schaefer, D., Martin, I., Jundt, G., Seidel, J., Heberer, M., Grodzinsky, A. J., Bergin, I., Vunjak-Novakovic, G., and Freed, L. E. (2002). Tissue-engineered composites for the repair of large osteochondral defects. *Arthritis Rheum.* **46**, 2524–2534.

Schaefer, D., Seidel, J., Martin, I., Jundt, G., Heberer, M., Grozinsky, A., Vunjak-Novakovic, G., and Freed, L. E. (2004). Engineering and characterization of functional osteochondral repair tissue. *Orthopade* **33**, 721–726.

Schwarz, R. P., Goodwin, T. J., and Wolf, D. A. (1992). Cell culture for three-dimensional modeling in rotating-wall vessels: an application of simulated microgravity. *J. Tiss. Cult. Meth.* **14**, 51–58.

Seidel, J. O., Pei, M., Gray, M. L., Langer, R., Freed, L. E., and Vunjak-Novakovic, G. (2004). Long-term culture of tissue-engineered cartilage in a perfused chamber with mechanical stimulation. *Biorheology* **41**, 445–458.

Sodian, R., Lemke, T., Loebe, M., Hoerstrup, S. P., Potapov, E. V., Hausmann, H., Meyer, R., and Hetzer, R. (2001). New pulsatile bioreactor for fabrication of tissue-engineered patches. *J. Biomed. Mater. Res.* **58**, 401–405.

Sodian, R., Lemke, T., Fritsche, C., Hoerstrup, S. P., Fu, P., Potapov, E. V., Hausmann, H., and Hetzer, R. (2002). Tissue-engineering bioreactors: a new combined cell-seeding and perfusion system for vascular tissue engineering. *Tissue Eng.* **8**, 863–870.

Sucosky, P., Osorio, D. F., Brown, J. B., and Neitzel, G. P. (2004). Fluid mechanics of a spinner-flask bioreactor. *Biotechnol. Bioeng.* **85**, 34–46.

Takahashi, I., Nuckolis, G. H., Takahashi, K., Tanaka, O., Semba, I., Dashner, R., Shum, L., and Slavkin, H. C. (1998). Compressive force promotes Sox9, type II collagen and aggrecan and inhibits IL-1b expression resulting in chondrogenesis in mouse embryonic limb bud mesenchymal cells. *J. Cell Sci.* **111**, 2067–2076.

Thom, T., Haase, N., Rosamond, W., Howard, V. J., Rumsfeld, J., Manolio, T., Zheng, Z. J., Flegal, K., O'Donnell, C., Kittner, S., et al. (2006). Heart disease and stroke statistics — 2006 update — A report from the American Heart Association Statistics Committee and Stroke Statistics Subcommittee. *Circulation* **113**, e85.

Thompson, D. W. (1977). "On Growth and Form." Cambridge University Press, New York.

Tognana, E., Chen, F., Padera, R. F., Leddy, H. A., Christensen, S. E., Guilak, F., Vunjak-Novakovic, G., and Freed, L. E. (2005a). Adjacent tissue (cartilage, bone) affects the functional integration of engineered calf cartilage *in vitro*. *Osteoarthr. Cartilage* **13**, 129–138.

Tognana, E., Padera, R. F., Chen, F., Vunjak-Novakovic, G., and Freed, L. E. (2005b). Development and remodeling of engineered cartilage-explant composites *in vitro* and *in vivo*. *Osteoarthr. Cartilage* **13**, 896–905.

Toyoda, T., Seedhom, B. B., Yao, J. Q., Kirkham, J., Brookes, S., and Bonass, W. A. (2003). Hydrostatic pressure modulates proteoglycan metabolism in chondrocytes seeded in agarose. *Arthritis Rheum.* **48**, 2865–2872.

Unsworth, B. R., and Lelkes, P. I. (1998). Growing tissues in microgravity. *Nat. Med.* **4**, 901–907.

Vandenburgh, H. H., and Karlisch, P. (1989). Longitudinal growth of skeletal myotubes *in vitro* in a new horizontal mechanical cell stimulator. *In Vitro Cell. Dev. B.* **25**, 607–616.

Vandenburgh, H. H., Swadison, S., and Karlisch, P. (1991). Computer-aided mechanogenesis of skeletal muscle organs from single cells in vitro. *FASEB J.* **5**, 2860–2867.

Vunjak-Novakovic, G., Freed, L. E., Biron, R. J., and Langer, R. (1996). Effects of mixing on the composition and morphology of tissue-engineered cartilage. *AIChE J.* **42**, 850–860.

Vunjak-Novakovic, G., Obradovic, B., Bursac, P., Martin, I., Langer, R., and Freed, L. E. (1998). Dynamic cell seeding of polymer scaffolds for cartilage tissue engineering. *Biotechnol. Prog.* **14**, 193–202.

Vunjak-Novakovic, G., Martin, I., Obradovic, B., Treppo, S., Grodzinsky, A. J., Langer, R., and Freed, L. E. (1999). Bioreactor cultivation conditions modulate the composition and mechanical properties of tissue-engineered cartilage. *J. Orthop. Res.* **17**, 130–138.

Waldman, S. D., Spiteri, C. G., Grynpas, M. D., Pilliar, R. M., Hong, J., and Kandel, R. A. (2003). Effect of biomechanical conditioning on cartilaginous tissue formation *in vitro*. *J. Bone Joint Surg. Am.* **85-A**(Suppl 2), 101–105.

Wendt, D., Marsano, A., Jakob, M., Heberer, M., and Martin, I. (2003). Oscillating perfusion of cell suspensions through three-dimensional scaffolds enhances cell seeding efficiency and uniformity. *Biotechnol. Bioeng.* **84**, 205–214.

Yang, C., Sodian, R., Fu, P., Luders, C., Lemke, T., Du, J., Hubler, M., Weng, Y., Meyer, R., and Hetzer, R. (2006). *In vitro* fabrication of a tissue-engineered human cardiovascular patch for future use in cardiovascular surgery. *Ann. Thorac. Surg.* **81**, 57–63.

Yu, X., Botchwey, E. A., Levine, E. M., Pollack, S. R., and Laurencin, C. T. (2004). Bioreactor-based bone tissue engineering: the influence of dynamic flow on osteoblast phenotypic expression and matrix mineralization. *Proc. Nat. Acad. Sci. USA* **101**, 11203–11208.

Zimmermann, W. H., Schneiderbanger, K., Schubert, P., Didie, M., Munzel, F., Heubach, J. F., Kostin, S., Nehuber, W. L., and Eschenhagen, T. (2002). Tissue engineering of a differentiated cardiac muscle construct. *Circ. Res.* **90**, 223–230.

# Chapter Twelve

# Principles of Tissue Culture and Bioreactor Design

R. I. Freshney, B. Obradovic, W. Grayson, C. Cannizzaro, and Gordana Vunjak Novakovic

I. Introduction
II. Basic Principles of Cell and Tissue Culture
III. Principles of Bioreactor Design

IV. Summary
 V. Acknowledgments
VI. References

## I. INTRODUCTION

In this chapter, we review the principles of cell culture and bioreactor design in the context of tissue engineering. In the first part, we describe the principles of cell and tissue culture, with emphasis on the initiation of culture, transformation, preservation, and validation of the cells and differentiation methods that utilize simple and well-established petri dish protocols. In the second part, we describe the design and operation of tissue engineering bioreactors, with focus on mass transport considerations associated with environmental control, and the provision of biophysical signals associated with the modulation of cell differentiation and functional tissue assembly. We provide eight examples of representative bioreactor designs that illustrate how some of the differentiation factors are provided in a physiologically relevant fashion.

## II. BASIC PRINCIPLES OF CELL AND TISSUE CULTURE

Historically, tissue culture arose from the concept that small fragments of tissue could somehow be maintained outside of the body in an appropriate environment such that the tissue cells survived. Harrison (1907) was the first to show that not only could cells survive but that they could perform functions, in his case neurite outgrowth, comparable to their normal *in vivo* activity. Some years later, tissue culture diverged into *cell culture*, where survival and propagation of the cells was paramount, and *organ culture*, where the retention of histological structure was seen as the major endpoint and indicative of the retention of *in vivo* function, so clearly lost in dispersed cell culture.

Cell culture has the advantage that it is more quantifiable and allows the development of optimized culture conditions, initially for cell proliferation but ultimately for functional expression as well. However, it was clear from early work that functional expression depended heavily on cell–cell interaction, both *homotypic* (among cells of the same lineage) and *heterotypic* (among cells of different lineages), and on the interaction of the cells with the extracellular matrix, itself a product of the interacting cells. For many years this was achievable only via organ culture. But recent advances in organotypic culture, including filter well inserts and engineered tissue constructs, mean that controlled characterized cell populations can be recombined and the nature of specific interactions defined. In some cases, e.g., with skin, it is possible to create a construct that allows complete terminal differentiation of keratinized squames by providing the correct cellular interactions, matrix, and oxygen tension (Maas-Szabowski *et al.*, 2002).

As a resource, cell culture has contributed historically not just to fundamental mechanistic studies of genotypic

**Table 12.1.** Selective pressure during isolation and culture

| Stage | Process | Selection |
|---|---|---|
| **Primary culture** | | Survival *in vitro* |
| | Explant | Adhesion; cell migration into outgrowth |
| | Trypsin digestion | Cells resistant to trypsin |
| | Collagenase digestion | Epithelial or endothelial cell clusters |
| | Mechanical disaggregation | Cells resistant to mechanical damage; eliminates fragile cells |
| **Subculture** | | Resistance to trypsin; ability to reattach; proliferation |
| | Explant transfer | Slowly migrating cells, e.g., epithelium |
| | Trypsinization | Cells resistant to trypsin |
| | Scraping | Cells resistant to mechanical damage |
| | Dispase | Epithelial cells detach preferentially |
| | Selective agents | Drug resistance, selective markers, serum-free selective media |

and phenotypic expression and other cellular processes but also to the production of biopharmaceuticals. Now, with the elaboration of the correct substrates or scaffolds and the ability to tune the microenvironment, it has become a reliable resource of material for tissue reconstruction and grafting. The various aspects of this are discussed in other chapters, and it is the role of this chapter to fill in some of the background detail on the basics of cell culture.

## Initiation of the Culture

There are several ways to isolate cells or tissue to initiate a culture (Table 12.1). The following are the commonest methods.

1. *Primary explantation.* A fragment of tissue is placed at a solid–liquid interface and allowed to adhere, and the outgrowth of cells from the explant is collected enzymatically and subcultured (Fig. 12.1). It is a very gentle method, suitable for small amounts of tissue, but has a low yield.

2. *Enzymatic disaggregation.* Finely chopped tissue is subjected to proteolysis, usually in low $Ca^{2+}$ and $Mg^{2+}$, to digest the extracellular matrix and dissociate junctional complexes. This method has a high yield, but some cells may be sensitive to the action of the protease. This can be reduced (a) by exposing to enzyme at 4°C for a prolonged period followed by a short incubation at 37°C, or (b) selecting enzymes that are less aggressive, e.g., collagenase or Dispase. These enzymes often give poorer cell dispersal, but that can be an advantage, particularly with epithelial and endothelial cells, where clusters of cells can be separated by sedimentation from dispersed stromal cells and often survive better than single cells.

3. *Mechanical disaggregation.* The tissue is dispersed by repeated pipetting or syringing or by forcing it through a plastic or metal mesh or screen. Mechanical disaggregation is useful for large amounts of tissue, particularly where the cells required are not strongly adherent,

e.g., hematopoietic cells from bone marrow or spleen or tumor cells from soft, highly anaplastic tumors.

### Primary Culture

The cells that grow out from an explant or attach to the substrate after enzymatic or mechanical disaggregation are referred to as the *primary culture*. In some cases, e.g., some hematopoietic cells, pleural effusions, or ascites, the cells may survive in suspension and not attach.

The primary culture will contain a selection of cells from the tissue, dependent on their ability to migrate out of the explant or adhere after disaggregation. At this stage the culture's cellular heterogeneity is closest to that of the tissue but still by no means totally representative, for many of the cells will not survive, either because they cannot attach or because they are unsuited to the isolation medium. The most important component of these is the differentiated cell pool, which may require highly specialized conditions to survive and is unlikely to proliferate. Those cells that survive best tend to be progenitor cells, either stem cells, present in very small numbers, or committed precursor cells. This is evident when embryonic tissues are compared with adult, where embryonic tissues will give a much higher survival, particularly from younger embryos.

Clearly there is considerable selection during the primary culture phase (see Table 12.1), and this selection is further enhanced when the cells start to proliferate. Cell proliferation will be restricted to precursor and stem cells, and, given that the culture medium has often been selected for its growth-promoting ability, this leads to the amplification of the precursor compartment. There is usually little cell differentiation, and the stem cells may or may not regenerate. Because the cycle time of the stem cells is usually quite long (~36 h), they will tend to be diluted out by the more rapidly dividing precursor cells. Some cells may enter the precursor pool from the differentiated cell pool, e.g., differentiated hepatocytes after partial hepatectomy (although their propagation *in vitro* is still difficult), fibro-

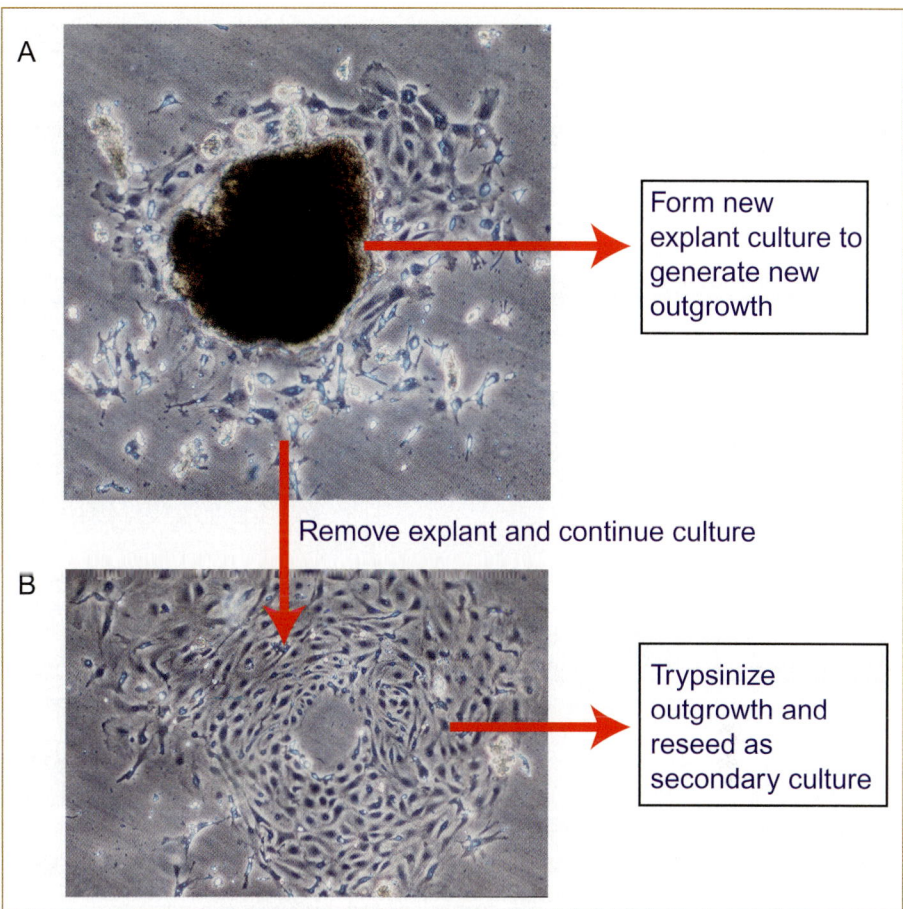

Form new explant culture to generate new outgrowth

Remove explant and continue culture

Trypsinize outgrowth and reseed as secondary culture

**FIG. 12.1.** Propagation of primary explant. Appearance of primary explant culture about one week after explantation. **(A)** Explant (dark area) with cells growing out from it. **(B)** Outgrowth with explant removed. There are two methods of subculture or passage: (1) The explant may be reseeded to generate a second outgrowth; (2) the outgrowth may be trypsinized and reseeded. Modified from Freshney (2005), Fig. 12.6.

cytes may convert to fibroblasts, and vascular endothelial cells may reenter the cell cycle.

While a high proportion of the cells in an embryo may survive and proliferate in culture, in the adult the tissue is dominated by the differentiated cell compartment, and the frequency of proliferative cells may be quite low and that of stem cells very low. Although vascular and connective tissue cells may enter the cycle quite readily and come to dominate the culture, specialized lineages of cells will often require very stringent conditions in order to survive, proliferate, and, ultimately, differentiate. The size of the proliferative (or potentially proliferative) compartment will vary from tissue to tissue, but in general it may be between $10^{-2}$ and $10^{-4}$ of the total cell mass, while the size of the stem cell compartment will be even smaller ($10^{-7}$–$10^{-9}$). This has necessitated the development of purification and selection techniques capable of amplifying the desired population of cells and minimizing stromal overgrowth. These procedures may involve selective isolation and physical cell separation applied at the primary culture stage and selective culture conditions applied at primary culture and at each successive subculture. Typically, isolation may involve collagenase digestion, which is less aggressive than trypsinization, followed by purification. Density-gradient separation allows for partial purification and has been used for the enrichment of mesenchymal stem cells (Lennon and Caplan, 2006; Gregory and Prockop, in press) but will often require being supplemented with, or replaced by, magnetic sorting (MACS) (Zborowski *et al.*, 1999) and/or flow cytometry (FACS) (Levenberg *et al.*, 2006) if antibodies to cell surface markers are available.

## Subculture and Propagation

When the entire growth surface is occupied (or when the concentration of the cells growing in suspension exhausts the medium), the culture may be transferred to a fresh culture vessel, or *subcultured*. If the cells grow in suspension, this simply means diluting the cells into fresh medium; but if the cells are adherent, then they must be dislodged, usually by trypsinization (Fig. 12.2). This process, also referred to as *passage*, exerts additional selection pressure (see Table 12.1), for only those cells that can survive the trypsinization process will be carried forward. Also, because an increase in cell number is usually the criterion for subculture, the process favors proliferating cells; and with each successive subculture, the growth fraction increases until by the third passage it usually exceeds 90%.

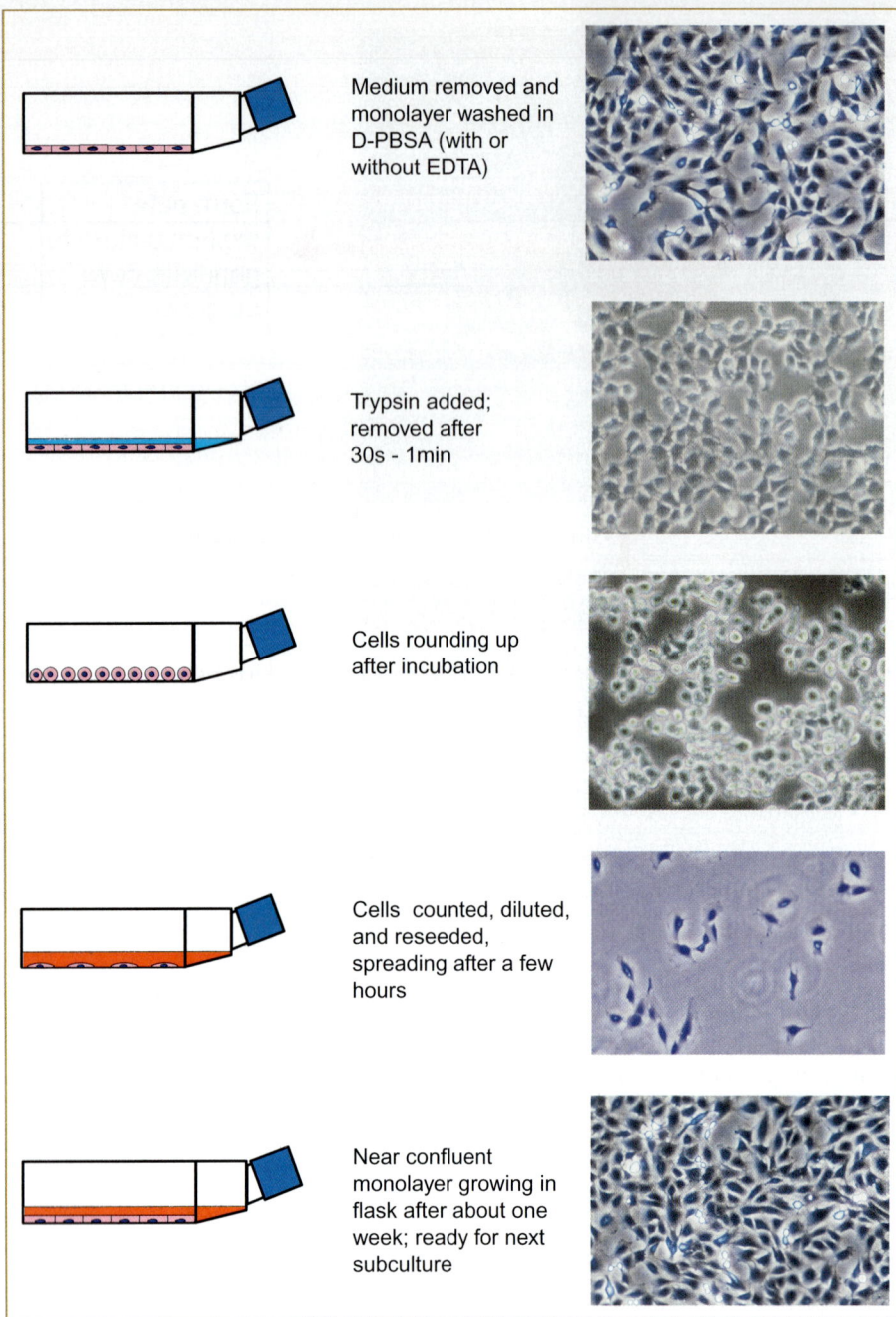

Medium removed and monolayer washed in D-PBSA (with or without EDTA)

Trypsin added; removed after 30s - 1min

Cells rounding up after incubation

Cells counted, diluted, and reseeded, spreading after a few hours

Near confluent monolayer growing in flask after about one week; ready for next subculture

**FIG. 12.2.** Trypsinization and subculture. Standard method for trypsinization and reseeding of attached monolayer cells. Modified from Freshney (2005), Fig. 13.3 and Plates 4 and 5.

Once a primary culture is subcultured it becomes a *cell line* (Table 12.2). If it dies out after a set number of population doublings (senesces), it is referred to as a *finite cell line*. If it survives indefinitely, either without change or following transformation or immortalization (spontaneous or induced), it is referred to as a *continuous cell line*. If it is selected, either by cloning or by physical cell sorting, to produce a purified cell line with defined characteristics, it is referred to as a *cell strain*.

Senescence is regulated by a number of dominantly acting genes (Sasaki *et al.*, 1996) some of which negatively regulate the enzyme telomerase. Telomerase maintains the length of the telomeres, the shortening of which accompa-

nies senescence, and is found to be active in stems cells, which often have a longer life span in culture than normal somatic cells (Flores *et al.*, 2006). However, MSCs have been shown to lose telomerase with extended culture (Izadpanah *et al.*, 2006), and this has led to immortalization by telomerase transfection (Serakinci, 2005), which may prove to be a valuable tool in the expansion of adult stem cell lines (Shay and Wright, 2005) (see next subsection, on transformation and immortalization).

## Sourcing Cell Lines and Cell Strains

Ethical and regulatory considerations have limited the number of embryonal stem cell lines generated in the

| Table 12.2. | Definitions and stages of culture | |
|---|---|---|
| *Term* | *Definition* | *Characteristics* |
| Primary culture | Culture period following explantation until first subculture | Heterogeneous; low growth fraction |
| Cell line | Culture following first subculture | More homogeneous than primary, but still multilineage; increased growth fraction |
| Finite cell line | Cell line with finite life span limited by entry into senescence | Mortal |
| Continuous cell line | Cell line with infinite life span achieved by spontaneous or induced transformation | Immortal; genetically unstable; heterogeneous |
| Cell strain | Purified cell line selected by cloning or physical cell separation and with defined characteristics | Homogeneous (but if continuous will become progressively heterogeneous, even if cloned) |
| Subculture | Transfer of culture to new vessel | Transit of growth cycle |
| Seeding | Inoculation of new culture vessel | Cells enter lag period |
| Seeding efficiency | Percentage of cells attaching and entering cell cycle after subculture | Low and variable for primary culture (0.1–50%); >90% for established* finite cell lines and continuous cell lines |
| Plating efficiency | Percentage of cells that will form clones in dilute culture | Index of survival; low in primary cultures; variable in finite cell lines, depending on medium; high in continuous cell lines |
| Saturation density | Highest density that the culture can achieve in nonlimiting medium supply | Low in normal cells; high in transformed cells |
| Cell density | Cells/cm² of available growth surface | Low at beginning of growth cycle, high at end; applied to adherent cultures only |
| Cell concentration | Cells/mL culture medium | Low at beginning of growth cycle, high at end; applied to adherent and suspension cultures; limited by nutrients and growth factors |

*Established no longer used for immortal cell lines; now called *continuous*. Now simply means that the culture grows and can be subcultured routinely.

United States, but some are undergoing characterization (Loring and Rao, 2006; Plaia *et al.*, 2006; Ludwig *et al.*, 2006) and are available through WiCell [www.wicell.org/], NIH [http://stemcells.nih.gov/research/registry/], ATCC [www.atcc.org], NIBSC [www.nibsc.ac.uk/spotlight/ukstemcell.html], and ECACC [www.ecacc.org.uk]. Cell lines and cell strains are frequently obtained from collaborating institutions, for they may be the only source, but, where possible, they should be obtained from a reliable cell bank, such as those just listed, which will have performed the necessary authentication (see upcoming subsection on validation). Continuous cell lines, in particular, are subject to problems of misidentification and cross-contamination (Freshney, 2005), so it is essential to confirm the identity of any cell line that you receive if it does not come from a reliable cell bank. Even obtaining cells from the originator is no guarantee, for a survey of misidentified cell lines (MacLeod *et al.*, 1999) showed that the majority of the problems stemmed from the originator. Samples of cells received from a cell bank will come with handling instructions (see upcoming subsection on cryopreservation). Cells should be kept in quarantine until they are shown to be free of contamination when cultured in the absence of antibiotics (Freshney, 2005).

## Growth Cycle

A growing culture is diluted when it is passaged. The degree of dilution is determined by the ability of the cells to reattach and start to proliferate (*seeding efficiency*) and is limited by the time taken to reach the density at which further subculture will be required. This not the maximum attainable density (or *saturation density*), for it is usually preferable to subculture before the cells stop growing at their saturation density.

It is important to determine both the specific cell concentration required for optimum seeding and the maximum cell concentration that would be expected before the next subculture. These data monitor the growth rate of the culture for consistency and ensure that the culture

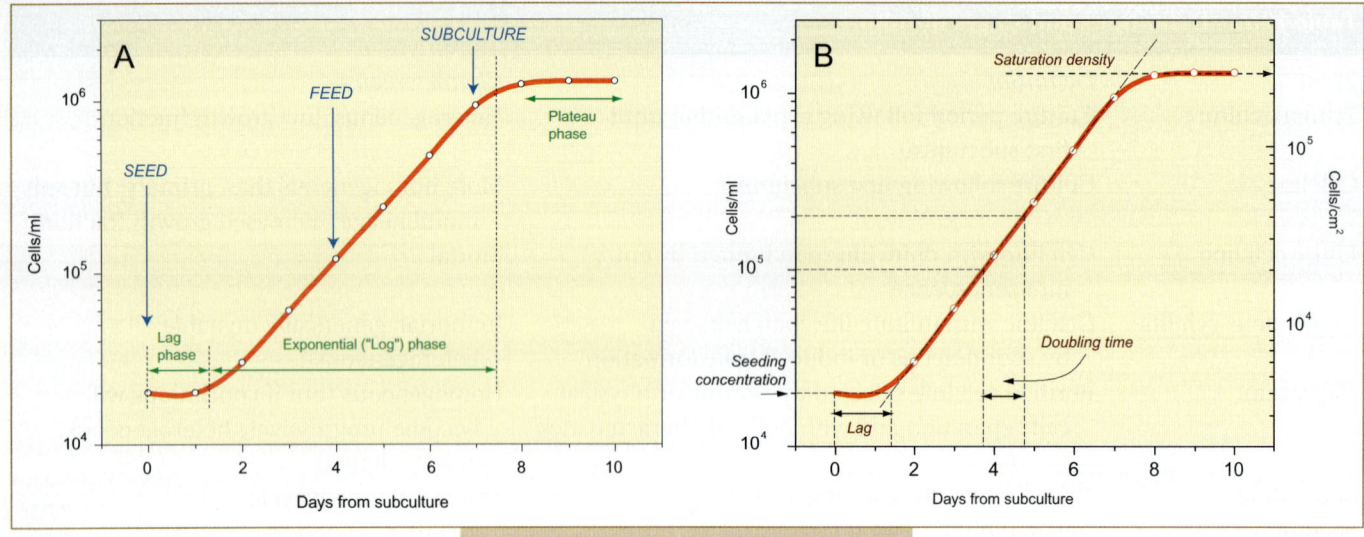

**FIG. 12.3.** Growth of cells following subculture. Semilog plot of cell number against time from seeding. **(A)** Phases of the growth curve so produced, and times when maintenance is required. **(B)** Analysis of the growth curve, showing the derivation of the lag period, the doubling time, and the saturation density (measured in cells/cm²). Reproduced from Freshney (2005), Figs. 13.2 and 21.6, by permission of the publisher.

is always seeded at the optimal cell concentration. These critical concentrations are best determined by generating a growth curve (Fig. 12.3), at three different seeding concentrations (Freshney, 2005), once the cell line has passed the third passage, and several passages later if the cell line is to be propagated for a long time. The *lag period* (time before growth commences), the *doubling time* (time taken for the cell population to double), and the time at which the culture enters *plateau* (the period when cell number remains stable with time) can all be determined from this curve and used to establish the optimal seeding concentration and subculture interval that both are convenient and give a long exponential phase and high viable cell yield.

### Passage Number and Generation Number

Every time a culture is subcultured it advances by one *passage number*. However, most normal cell lines will survive for only a fixed time in continuous culture, and this is determined by the number of population doublings, or generations. This can be calculated at each passage from the equation PD = ln($N_{finish}$/$N_{start}$)/ln 2, where PD is the number of population doublings, ln is the natural logarithm, $N_{start}$ is the number of cells originally seeded, and $N_{finish}$ is the total number of cells recovered at subculture (Serakinci, 2005).

However, the simplest way to deal with this is to ensure that the cells are diluted by a power of 2 at each subculture. The dilution factor is called the *split ratio*; a split ratio of 2 means that the culture will have to go through one population doubling to grow back to the same density, a split ratio of 4 ($2^2$) requires two doublings, 8 ($2^3$) three doublings, and so on. Finite cell lines will continue to proliferate for between 10 and 60 population doublings before they senesce, so it is important to be able to predict when the decline in growth

will occur so that fresh cultures can be initiated from primary culture or frozen stock (see upcoming subsection on cryopreservation).

### Transformation and Immortalization

Cultures of normal human cells rarely, if ever, transform spontaneously, although normal cells from many other species, particularly rodents, do so quite readily. Hence most normal human cell lines will have a finite life span unless immortalized with SV40, Ad5 E1a, HPV16 E6/E7, or EBV. Immortalization is also possible by means of transfection with hTRT, the catalytic subunit of the human telomerase gene (Serakinci, 2005; Shay and Wright, 2005), which blocks the shortening of the telomeres that accompanies senescence. Although immortal, these cell lines are not necessarily fully transformed, because their growth kinetics (anchorage dependence, growth at saturation density) are often similar to those of the finite cell line and they are non-tumorigenic. Further transformation of these lines with carcinogens or oncogene transfection can induce malignant transformation, but it appears not to happen spontaneously, although the risk has yet to be fully determined in stem cell lines.

Transformed or immortalized cell lines survive indefinitely, so recording the generation number becomes redundant, and only the passage number (usually the number of subcultures since thawing from storage) is normally recorded.

### Cryopreservation

Once a cell line has been shown to grow satisfactorily and to possess valuable characteristics, it is advisable to freeze samples in a cryostorage container, usually cooled to −196°C by liquid nitrogen. Initially, a token freeze of one to three ampoules is prepared; then, when characterization has been completed and the absence of contamination (see later) confirmed, a *seed stock* of 10–20 ampoules is stored. One of these ampoules is thawed, to confirm viability, identity, and freedom from contamination, and then grown up to generate a *distribution stock*, which can be issued to end users. End users may choose to prepare their own *using stock* for the lifetime of a particular project, based on serial replacement of culture stocks every three months, but these stocks should be discarded at the end of the project and not distributed to other users, because they lack proper validation.

To freeze cells for storage, they are cooled slowly (1°C/min) at a high cell concentration ($(1–10) \times 10^6$ cells/mL) in the presence of a preservative (usually 5–10% DMSO or glycerol) down to below −50°C (usually at least −70°C) and then transferred quickly to the cryofreezer at −196°C. It is preferable to store ampoules in the gas phase or in perfused jacket freezers rather than submerged in liquid nitrogen, for storing in the liquid phase increases the risk of contamination of the contents and of ampoules exploding (they must be thawed in a covered container). Where human material is used, this also presents a potential biological hazard.

When required, an ampoule is selected, thawed rapidly in warm water (37°C), diluted at least 10-fold, and seeded into a flask. For adherent cells, it is usually sufficient to remove the preservative the following day, once the cells have attached, by replacing the medium. However, for cells growing in suspension, it is preferable to centrifuge the cells to remove the preservative before seeding. For direct seeding and for centrifugation, it is important to dilute the cells slowly in medium (at least for DMSO), for rapid dilution can lead to osmotic damage.

It is vital to ensure that proper records are kept detailing (a) the origin, characteristics, and maintenance of the cells and (b) the position of the ampoules in the freezer, how many ampoules are left in any batch, and where free spaces are to be found. It is possible to devise a simple database, e.g., in Microsoft Access, to cover this, although commercial inventory control systems are available (e.g., CryoTrack [www.cryotrack.com] and Planer [www.planer.co.uk/]).

## Validation

The term *validation* is used to imply that the cell line is safe and reliable to use. There are four main elements to validation of a cell line: authentication, provenance, characterization, and the demonstration of freedom from contamination.

### Authentication

Authentication requires confirmation of the identity of the cell line. The generally accepted method of achieving this is DNA profiling or fingerprinting. This requires that DNA or tissue is available from the tissue donor or that an already authenticated stock exists elsewhere, e.g., in a cell bank. When deriving new cell lines, it is important to retain some tissue, even blood, from the donor, to use to confirm the origin of the eventual cell line.

### Causes of Misidentification of a Cell Line

1. Poor culture technique allows cross-contamination with another cell line to occur. Inadequate labeling allows one culture flask to be confused with another.
2. Ampoules in the cryofreezer may be inadequately or erroneously labeled or may lose their label.
3. An ampoule may be located in the wrong place or inadvertently thawed from the wrong place.

### Prevention of Misidentification and Cross-Contamination

1. Obtain cell lines from a reputable source.
2. Do not handle more than one cell line at a time.
3. Do not share media or other reagents or pipettes among cell lines or with other operators.
4. Label flasks and ampoules clearly and distinctly.
5. Do not put a pipette back into a bottle of medium or other reagent after it has been used to dispense cells or has been inside a culture flask containing cells.
6. Perform a routine visual check on all cell lines before any maintenance or experimental procedure. Keep an album, photos on the wall above the microscope, or readily accessible computerized digital images of each cell line with cells at different densities. Changes in culture morphology can often be the first indication of cross-contamination.
7. Double-check the identity of cryovials before thawing; use color coding of caps and rectangular arrays of vials to provide coordinates in the freezer, and ensure that labels are not erased or dislodged by freezing.
8. Confirm identity after thawing.

### Provenance

Each cell line should have details of how it was derived, what has been done to it (transformation, transfection, infection, decontamination, or any other manipulation, including feeding and subculture regimes and transfers to other laboratories). This makes up the *provenance* of the cell line. Although trivial at the time observed, some events may prove important in retrospect, if a cell line becomes an important resource, so careful record taking is essential. Daily written records should be summarized in a database.

### Contamination

Microbial contamination remains one of the major recurring problems in cell culture, although, given strict control of technique and environment, occurrences should be infrequent. Most media reagents and plastics are obtained

| Table 12.3. | Criteria for characterization and authentication | |
| --- | --- | --- |
| *Criterion* | *Method* | *Reference* |
| Karyotype | Chromosome spread with banding | Rothfels and Siminovitch (1958); Rooney and Czepulkowski (1986) |
| Isoenzyme analysis | Agar gel electrophoresis | Hay (2000) |
| Cell surface antigens | Immunohistochemistry, flow cytometry | Burchell *et al.* (1983, 1987); Hay (2000) |
| Cytoskeleton | Immunocytochemistry with antibodies to specific cytokeratins | Lane (1982); Moll *et al.* (1982) |
| DNA fingerprint | Restriction enzyme digest; PAGE; satellite DNA probes | Jeffreys *et al.* (1985); Hay (2000) |
| DNA Profile | PCR of microsatellite repeats | Masters *et al.* (2001) |

Modified from Freshney (2005), Table 16.1.

presterilized, and excellent handling protection is provided by a laminar flow hood. Even without antibiotics, the frequency of contamination should be low. When it does occur, it is caused most frequently by lapses in technique, overcrowding, or faulty or poorly maintained equipment. Laminar flow hoods must be serviced regularly (every six months), incubators must be cleaned out frequently (every one to two weeks for a humid incubator, monthly for a regular incubator), and newcomers to the laboratory must be given proper training (Freshney, 2005) by an experienced cell culturist. The use of antibiotics should be discouraged, except in high-risk situations such as primary culture, for they tend to mask contaminations and engender careless technique.

## Characterization

The characteristics of a cell line will help to confirm its origin, in terms of cell lineage, its status (stem cell, precursor cells, or differentiated cells), product formation, virus susceptibility, or any other specialized property. Characteristics may be of general interest (e.g., cell lineage and status) or special interest to the user (e.g., drug resistance or virus susceptibility) and may also contribute to the authentication and provenance by confirming species and tissue of origin and identifying a marker property, such as high expression of a particular protein or a chromosomal abnormality. The type of characterization that is done will depend on the available technology and the use to which the cell line is put, but it is also a useful check on consistency and alerts the user to any change in the cell line that might imply contamination or transformation. Features used in characterization are too numerous to deal with here in detail and are summarized in Table 12.3.

## Differentiation

Because the conditions employed to generate cell lines tend to favor cell proliferation, most cell lines do not express the differentiated properties of the tissues from which they were derived. Historically this was blamed on *dedifferentia-*

*tion*, implying that differentiated cells grown in culture lost their differentiated phenotype. However, it is now apparent that many of the earlier attempts to grow differentiated cells, e.g., with liver, selected stromal cells rather than the functional parenchyma. Now, with selective media and a better understanding of the signaling capacity of the microenvironment, it is clear that the bulk of the cells cultured are precursor cells; although of the correct lineage, they require a shift in the culture conditions away from mitogenesis and toward cytostasis and differentiation (Table 12.4).

The factors required for differentiation can be divided into six groups (although there is a degree of overlap and interaction among them).

## Cell–Cell Interaction

• Homotypic interactions between like cells via gap–junctional communication or contact-mediated cell adhesion molecules.
• Heterotypic interactions between dissimilar cell types, such as epidermal keratinocytes and dermal fibroblasts, usually mediated by soluble paracrine factors (see later) or interaction with the extracellular matrix separating the cells (see following section).

## Matrix Interactions

Extracellular matrix forms between cells, and its constitution will depend on the cells participating in its formation. Hence the matrix laid down between an epithelial layer and an underlying mesenchymal layer will be different from that laid down between pairs of epithelial cells or pairs of fibroblasts and from that laid down between endothelial cells and astrocytes (blood–brain barrier) or endothelial cells and smooth muscle cells. These matrices contain specific protein and proteoglycan constituents, which act as ligands to receptors on the cell surface, initiating a signaling cascade that can alter cell attachment, cell shape, cell motility, and a number of other attributes, many of which are mediated via changes in gene transcription.

| Table 12.4. | Conditions favoring mitogenesis or cytostasis and differentiation | |
| --- | --- | --- |
| *Condition* | *Mitogenesis* | *Cytostasis/differentiation* |
| Cell density | Low | High |
| Cell spreading | High | Low |
| Cell interaction | Low | High |
| Anchorage independence | In transformed cells | In normal cells |
| Serum | High (some exceptions, e.g., TGF-β will favor cytostasis) | Low |
| Growth factors | Mitogenic at low cell density | May induce differentiation; high cell density |
| Hydrocortisone | Mitogenic at low cell density | May induce differentiation; high cell density |

## Soluble Factors

These fit into four main classes:

- *Inorganic Ions*, particularly $Ca^{2+}$, which induces epithelial differentiation above 2 mM.
- *Hormones*, such as hydrocortisone (low concentrations often mitogenic, high concentrations cytostatic and differentiation inducing, particularly at high cell densities), triiodothyronine, epinephrine, melanocyte-stimulating hormone (MSH), estrogens, and androgens.
- *Vitamins*, such as all-*trans*-retinoic acid, vitamin D3, and vitamin E.
- *Growth factors and cytokines.* These act mainly as paracrine factors, where the factor originates in one cell type and binds to receptors in a different cell type; e.g., KGF is produced by mesenchymal cells in the developing prostate but binds to receptors in the prostatic epithelium to induce growth and differentiation. Many of these paracrine interactions are reciprocal; e.g., IL-1α is secreted by epidermal keratinocytes and stimulates dermal fibrocytes to make keratinocyte growth factor (KGF) and granulocyte-macrophage colony–stimulating factor (GM-CSF), which, in turn, induces proliferation and differentiation of the keratinocytes.

The extracelluar matrix plays a pivotal role in the action of many of these paracrine factors (Iozzo, 2005). Matrix proteoglycans, particularly heparan sulfate proteoglycans (HSPGs), bind growth factors, such as FGF, may stabilize them, translocate them to the interacting cell, where low-affinity transmembrane proteoglycan receptors, with no intracellular signaling domain, bind the growth factor and translocate it tangentially within the cell membrane to the high-affinity signaling receptor. These proteoglycans are themselves subject to regulation, e.g., by hydrocortisone (Yevdokimova and Freshney, 1997), adding further complexity to the interaction.

## Cell Shape and Polarity

Many cells need to adopt the correct shape and become polarized for complete differentiation to occur. The paradigm is well illustrated by thyroid epithelium. Polarity is hard to achieve by growing cells on a solid plastic substrate. But if the cells are plated onto collagen in a filter well, the cells become crowded, less well spread (inhibiting mitosis), and more cuboidal or columnar. In this orientation the cell surface in contact with the collagen or filter becomes the basal surface, and the opposite surface becomes the apical and the sides the lateral surfaces (Chambard *et al.*, 1987). This is enabled by the tagging of newly synthesized proteins, such that some, e.g., receptors for thyroid-stimulating hormone (TSH), are directed toward the basal surface, proteins that form the junctional complex are directed to the apico-lateral membrane, and secreted proteins, e.g., thyroglobulin, are directed toward the apical surface. This does not happen unless the cell is allowed access to nutrient from below; nutrient above is not required and may even be detrimental.

In some cases, hepatocytes plated on collagen gel will show complete differentiation only if the collagen gel is detached from the substrate (Sattler *et al.*, 1978), allowing it to shrink (although medium access from below is also enhanced) and permitting the required shape change in the hepatocyte from a flattened squamous-like morphology to a cuboidal one. Clearly, several processes are involved here: shape change, withdrawal from cycle, basal nutrient access, enhanced lateral cell adhesion and junction formation, enhanced/altered integrins on the basal surface, promoting matrix interaction and binding of paracrine factors, formation of signaling receptors on the basal surface, and simulation of the equivalent of a luminal space on the nutrient-free upper side, promoting exocytosis. All of these may have implications in the construction of scaffolds to host functional transplants.

## Dynamic Stress

It is now clear that different types of dynamic stress are required to simulate the correct microenvironment for certain cell types (see upcoming section on biophysical signaling considerations). Skeletal myocytes require tensile stress (Shansky, 2006), cardiac myocytes require pulsatile tensile stress (Eschenhagen, 2006), chondrocytes (Darling and Athanasiou, 2003) and osteocytes (Mullender *et al.*, 2004) require compressive stress, and endothelial cell/smooth muscle cell constructs require pulsatile flow (Niklason, 2006). This is not altogether surprising when one looks at the evidence *in vivo*, e.g., muscle wasting without exercise, but it is revealing to find that the physical forces can be shown to act at the cellular level rather than as a systemic effect as one might have supposed.

### Oxygen Tension

Epidermal keratinocytes will differentiate terminally only if the upper side of the filter well construct is not submerged (Maas-Szabowski *et al.*, 2002). Likewise, bronchial epithelium will differentiate into mucus-secreting cells if at the air–liquid interface (ALI), but it will undergo squamous differentiation at the solid substrate/liquid interface (Kaartinen *et al.*, 1993; Dobbs *et al.*, 1997). It is assumed that the higher oxygen tension achievable at the ALI is the main factor, but other influences, such as surface tension, may also play a part.

The processes of cell differentiation and functional assembly during embryogenesis, early development, and tissue remodeling in an adult organism are directed by these six groups of factors acting in concert. In most cases, multiple molecular and physical factors are presented to the cells in the form of rather complex spatial and temporal sequences. It is thought that the cell differentiation and the functional assembly of engineered tissues *in vitro* can be modulated by these same factors, by using bioreactors for the cultivation of cells on three-dimensional scaffolds. The following sections explain in more detail the principles of bioreactor design for tissue engineering and provide representative examples on how some of the differentiation factors are provided in a physiologically relevant fashion.

## III. PRINCIPLES OF BIOREACTOR DESIGN

The most fundamental, biologically inspired, approach to engineering tissues is to direct the organization of metabolically active cells into three-dimensional (3D) spatial arrangements (via hydrogels or porous scaffolds) and to establish the environmental conditions such that the cells are induced to reconstruct immature — but functional — tissues. In nature, cells are the key architects of developing tissues and organs. Once they are provided with suitable environments, they remodel their immediate microenvironments and also integrate into a functional whole via homotypic or heterotypic interactions with neighboring cells. In both *in vivo* (development/regeneration) and *in vitro* (tissue engineering) settings, the cues with which cells are presented are the principal determinants of the phenotypic nature of the resultant tissues. Hence, for tissue-engineering applications extraordinary efforts have been expended to character-

ize the native environments of these tissues and express them as a combination of parameters that may be recapitulated experimentally. Bioreactors are the primary tools for mimicking these environments and provide cell-based constructs with physiologically relevant stimuli that facilitate and orchestrate the conversion of a "collection of cells" into a specific tissue phenotype (Fig. 12.4).

Two general classes of bioreactors germane to our objectives are described. The first is the tissue-engineering bioreactor, which incorporates a quantitative understanding of both the native environments and structural characteristics of the tissue into its overall design. Thus, tissue-engineering bioreactors are designed to precisely regulate the cellular microenvironment in order to facilitate construct uniformity and overall cell viability. Key requirements of the reactors include (1) increased cell-seeding efficiency into the 3D scaffold, (2) improved mass transfer, (3) adequate gas exchange to the bulk culture media, (4) regular replenishment of spent medium, (5) temperature/pH control, and (6) physiological stimuli (Vunjak-Novakovic *et al.*, 1996, 1998, 1999, 2002; Obradovic *et al.*, 1999, 2003). Current designs incorporate biological, mechanical, or electrical stimuli in order to exert greater influence over cellular differentiation and development into functional tissue constructs. Contemporary bioreactors are therefore custom designed to account for specific structure–function relationships, mechanisms of nutrient transfer, and physiological forces inherent in the native tissue.

The second class of bioreactors is aimed at optimizing process variables prior to actual tissue-engineering applications. They are utilized primarily for screening purposes and are particularly advantageous for the economy of cells required as well as for investigating the physiological ranges of parameters and their synergistic effects. These bioreactors are typically designed to be modular, miniscaled, and multiparametric. The 3D spatial organization of cells in these reactors is not critical, and they typically are not cell or tissue specific. For example, systems may be used to investigate the effects of oxygen levels or shear stresses on the survival, proliferation, or interactions of cells on planar surfaces. As a consequence of the modular design, these systems may also test a range of these parameters to optimize conditions on small scales prior to use in larger tissue-engineering reactors, where cell numbers are generally orders of magnitude higher and culture times are longer, making them less pragmatic for screening and optimization studies.

Various examples of both types of reactors are provided in the subsequent sections of the chapter. They not only illustrate how the universal requirements are incorporated into the design, but emphasize how design specificity and relevance are influenced by tissue types. These reactors diverge considerably from "bioreactors" used in the pharmaceutical industries, where, for example, the kinetics of biological growth (unicellular organisms) or chemical reac-

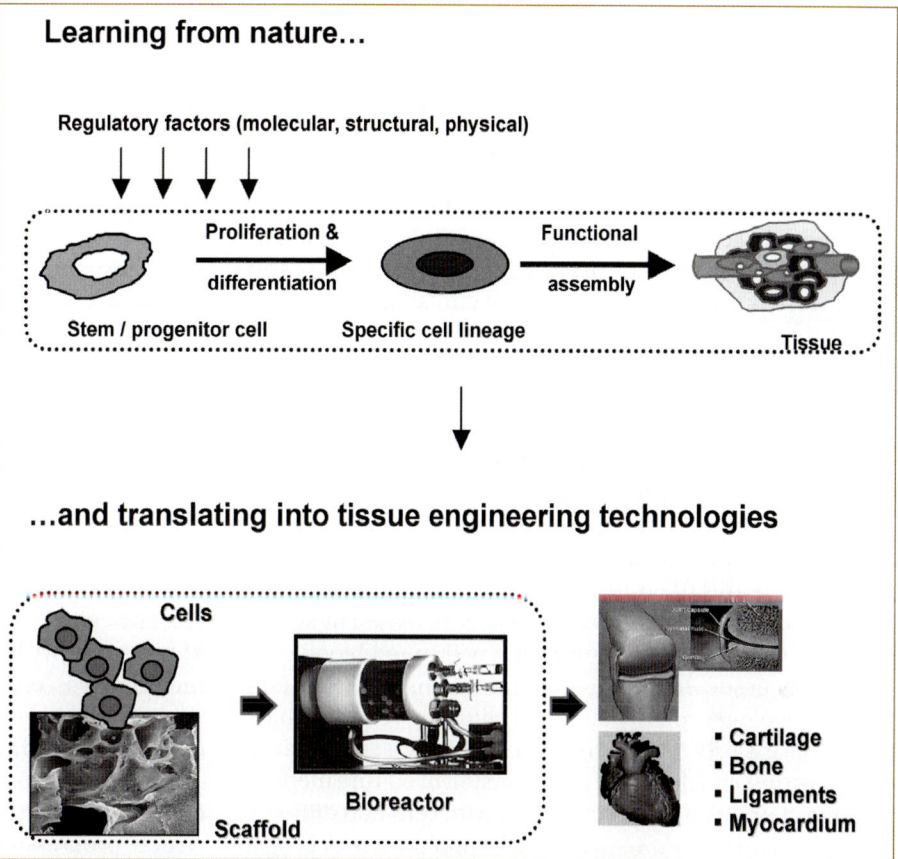

**FIG. 12.4.** Biomimetic approach to tissue engineering. **(Top panel)** Cell fate and tissue assembly during early development and tissue remodeling in an adult organism are regulated by multiple cues acting across different-length scales and time sequences. Please see Table 12.4 for a description of the differentiation factors. **(Bottom panel)** Tissue engineering attempts to mimic the cell context present *in vivo* by culturing cells on biomaterial scaffolds (providing a structural and logistic template for tissue development) in bioreactors (providing an environment and regulatory signals necessary for functional tissue assembly). Adapted from Gerecht-Nir *et al.* (2006), Figs. 1 and 2.

tions are well defined and the relationship between the yield or quality of a product is well correlated to changes in specific parameters. In the tissue-engineering system, the quality of the cell population, the spatial organization of the cells, and the effects of parametric variables are all components of the culture microenvironment that interact in complex ways to influence the biochemical, phenotypic, or mechanical characteristics of the resultant tissue. Thus, the ability to provide exquisite control of the cellular microenvironment by precise manipulation of process parameters is essential to successful tissue engineering.

## Mass Transport Considerations

Compared to two-dimensional (2D) monolayer cultures, 3D constructs impose more stringent requirements for efficient mass transport to cells in order to achieve tissue uniformity and avoid widespread cell death in the inner regions of scaffolds. Nutrients, oxygen, and regulatory molecules have to be efficiently transported from the bulk culture medium to the tissue surfaces (external mass transfer) and then through the tissue to the cells (internal mass transfer). Similarly, metabolites and $CO_2$ are removed from the cells through the tissue matrix to the surfaces and then to the bulk medium. External mass transfer rates depend primarily on hydrodynamic conditions in the bioreactor. In contrast, internal rates may depend on a combination of

diffusion and convection (induced by medium perfusion or scaffold deformation due to imposed mechanical loads) and may depend significantly on the scaffold/tissue structure. Integrated design of bioreactor configuration and biomaterial support is aimed at mimicking the *in vivo* mass transfer mechanisms and cellular microenvironment specific to the tissue structure.

Controlled delivery of nutrients and bioregulatory molecules in stem and primary cell cultures is essential to induce the expression of a desired cellular phenotype. However, in this section the primary focus is on the efficient mass transport of oxygen, because of its importance as a metabolic and signaling molecule and because it poses a challenging transport dilemma due to its extremely low solubility in aqueous media, with concentrations reaching up to 220 μm for fully oxygenated culture medium at 37°C. For comparison, oxygen concentration in blood plasma is 130 μm, but the total oxygen concentration in oxyhemoglobin can be as high as 8600 μm. Oxygen transfer is limiting even in conventional 2D tissue culture (Randers-Eichorn *et al.*, 1996) and becomes critically inadequate when cells adopt 3D structures (Martin *et al.*, 2004; Muschler *et al.*, 2004), and the diffusional penetration depth of oxygen within native tissues is in the range of 100–200 μm (Muschler *et al.*, 2004). Typically, tissue-engineering bioreactors are designed to improve external mass transfer of oxygen. However, internal mass transfer mechanisms should reflect the particular characteristics and functions of the tissue.

In this section we describe approaches to bioreactor design with appropriate mass transport for engineering of two tissue types (cartilage and myocardium) with distinctly different structures and requirements for oxygen supply. Native cartilage (an avascular tissue with low cell density and low consumption of oxygen per unit volume) is supplied with oxygen by a combination of diffusion and loading-induced convective flow through the tissue. Therefore, the systems for cartilage tissue engineering rely largely on convection around the constructs and molecular diffusion within constructs, with a few examples of loading-enhanced transport rates. In contrast, cardiac muscle consists of densely packed cells that are highly metabolically active and are supplied by oxygen via a dense capillary network that reduces the diffusional distances to only 10–20 $\mu$m and thereby increases the transport rates of oxygen. In addition, the oxygen-carrying capacity of blood is increased by several orders of magnitude by hemoglobin within red blood cells. Therefore, tissue engineering of cardiac muscle that contains physiologic cell density over clinically relevant thicknesses necessarily involves oxygen supply by a combination of convection and diffusion (by perfusion of culture medium through channeled scaffolds seeded with cells and diffusion through the tissue between channels) and the use of oxygen carriers that increase the capacity of culture medium for oxygen. For both cartilage and cardiac muscle, the bioreactor systems contain a gas exchanger that supplies oxygen to the culture medium and removes carbon dioxide. The selection of the exact conditions of bioreactor cultivation that provide the necessary level of oxygen transport is in both cases supported by the use of mathematical models of oxygen supply and consumption within the cultured tissue.

## Example 1. Engineered Cartilage: Diffusional Transport of Oxygen

Skeletally mature articular cartilage is an avascular tissue that is supplied by oxygen and nutrients from the synovial fluid by a combination of molecular diffusion through the tissue and "physiologic pump" transport associated with joint loading and cartilage deformation during normal activities (O'Hara et al., 1990). Concentrations of dissolved oxygen range from 45 to 57 mmHg at the cartilage surface to less than 7.6 mmHg in the deep zone (Brighton and Heppenstall, 1971; Urban, 1994). Chondrocytes are therefore well adapted to hypoxic conditions and provide normal turnover of the cartilaginous matrix in healthy tissue. However, adult cartilage under physiologic conditions has practically no capacity for self-repair following injury. In contrast, immature cartilage is vascularized and efficiently supplied with oxygen and nutrients and exhibits high biosynthetic activity.

The relatively low oxygen requirements of cartilage can be met using bioreactors that (1) control the level of oxygen in culture medium (via gas exchange with incubator air), (2) eliminate external gradients via convective transport around the tissue surfaces, and (3) allow diffusional transport between the construct surfaces and inner tissue phase. Rotating bioreactors are an example of a system that can provide oxygen transport in a manner that promoted chondrogenesis (Obradovic et al., 2000). Engineered tissue constructs formed by seeding cells into highly porous polymer meshes or sponges were freely suspended in the rotating flow of culture medium between two concentric cylinders, where the inner cylinder was covered by a silicone membrane providing continuous gas exchange with the incubator gas (Fig. 12.5A). The vessel rotation rate was adjusted to maintain each construct settling at a stationary point within the vessel. The flow conditions were dynamic and laminar, with tissue constructs settling in a tumble-slide regime associated with fluctuations in fluid pressure, velocity, and shear (Freed and Vunjak-Novakovic, 1995). Mass transport between the tissue constructs and culture medium was enhanced by dynamic laminar convection, whereas the transport within the tissue remained governed by molecular diffusion.

Under normal oxygen tensions (~80 mmHg) in the culture medium in rotating bioreactors, engineered constructs progressively deposited ECM components over the cultivation time (Fig. 12.5B). Chondrogenesis started at the construct periphery and, over time, progressed both inward toward the construct center and outward from its surface, resulting in significant increase in the construct size and weight. By six weeks of culture, self-regulated cell proliferation and deposition of cartilaginous matrix yielded constructs that had physiological cell density and spatially uniform distributions of matrix components comprising 75% as much GAG and 40% as much total collagen (hydroxyproline) per unit wet weight as normal cartilage (Fig. 12.5B, C) (Vunjak-Novakovic et al., 1999). Over prolonged cultivation for up to seven months, GAG contents exceeded physiologic levels, while collagen content remained constant (Fig. 12.5C). Biomechanical construct properties correlated with construct compositions (Fig. 12.5D) such that in seven-month constructs, both the equilibrium modulus and the hydraulic permeability became comparable with those measured for native cartilage (Freed et al., 1997; Vunjak-Novakovic et al., 1999), while dynamic stiffness averaged only 46% of that of native cartilage (Fig. 12.5D), suggesting the importance of a functional collagen network (Martin et al., 2000).

In contrast to the cultures at normal oxygen tensions, cultivation at relatively low oxygen tensions under otherwise identical conditions in rotating bioreactors resulted in significantly lower and nonunifom ECM deposition and yielded significantly smaller engineered constructs. After five weeks of cultivation at low ($42.7 \pm 4.5$ mmHg) and normal

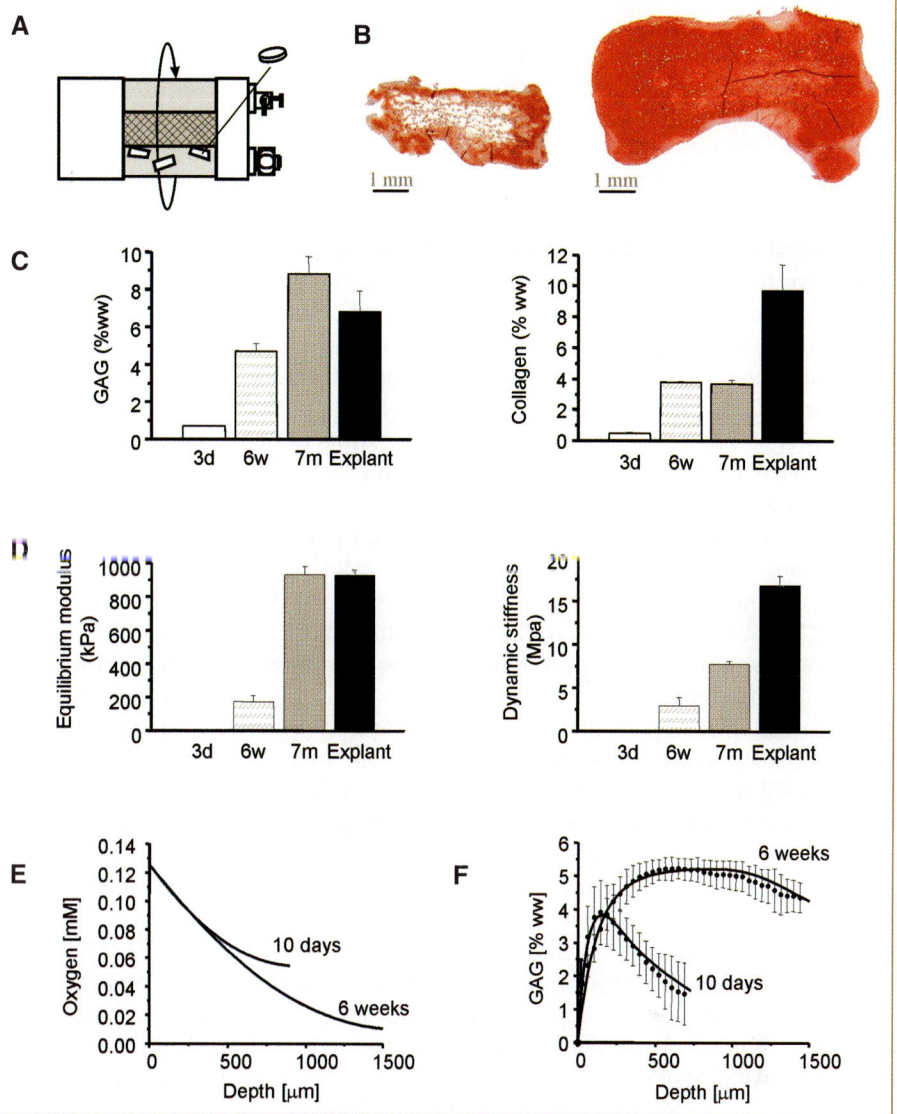

**FIG. 12.5.** Cultivation of engineered cartilage in rotating bioreactors. **(A)** Cell-polymer constructs are freely suspended in dynamic laminar flow in the annular reactor space; continuous gas exchange is provided by the silicone membrane covering the inner cylinder, through which sterile incubator gas is blown such that oxygen tension in the medium is maintained at a constant level (86.5 ± 7.3 mmHg). **(B)** *Progression of chondrogenesis:* histological cross sections of engineered constructs cultured for 10 days and 6 weeks stained with safranin-O (Obradovic *et al.*, 2000), chondrogenesis started at the constructs' periphery and progressed both inward and outward, resulting in uniformly cartilaginous tissue at the 6-week time point. **(C)** *Deposition of ECM over the cultivation time:* contents of GAG and collagen in engineered constructs as functions of the cultivation time; data for cartilage explants are shown for comparison (data from Freed *et al.*, 1997, 1998). **(D)** *Biomechanical properties of engineered cartilage:* equilibrium modulus and dynamic stiffness as functions of the cultivation time; data for cartilage explants are shown for comparison (data from Freed *et al.*, 1997, 1998). *Mathematical modeling of GAG deposition within the engineered cartilage over the cultivation time:* **(E)** model predictions of oxygen concentrations, **(F)** model predictions (lines) and experimental data (symbols) of GAG concentrations (Obradovic *et al.*, 2000).

(86.5 ± 7.3 mmHg) oxygen tensions, engineered constructs had 101 ± 8 and 139 ± 12 mg wet weight, 3.07 ± 0.28 and 4.18 ± 0.22%ww GAG, and 0.77 ± 0.03 and 2.76 ± 0.03%ww collagen, respectively (Obradovic *et al.*, 1999). Similarly, in cultures of cartilage explants, a decrease in medium oxygen tension from ~80 mmHg to ~40 mmHg resulted in a shift from aerobic to anaerobic cell metabolism and the suppression of tissue growth (Obradovic *et al.*, 1997).

In order to quantify the effects of oxygen on the deposition of cartilaginous matrix in engineered tissue constructs, a mathematical model was formulated that yielded concentrations of glycosaminoglycan (GAG) as a function of time and position within constructs cultivated in rotating bioreactors (Obradovic *et al.*, 2000). Production of GAG was taken as a marker of chondrogenesis, based on previous findings that GAG deposition in engineered constructs coincided with deposition of collagen type II, the other major component of cartilage tissue matrix (Freed *et al.*, 1998). The model

accounted for consumption of oxygen, synthesis of GAG as a function of the local oxygen concentration, and diffusion of both species. Due to the oxygen consumption by the cells, oxygen concentration decreased from the construct surface toward its center. The decrease in oxygen concentration was markedly higher in 6-week as compared to 10-day constructs, consistent with the increase in the total number of cells and diffusion distances with time in culture (Fig. 12.5E). The model was validated by comparing its predictions to the local GAG concentrations, measured by image processing of histological tissue sections (Martin *et al.*, 1999). Qualitative and quantitative agreements of the model predictions with experimental data for both time points were obtained (average SD = ±0.2% wet weight GAG) (Fig. 12.5F). In addition, the model adequately predicted GAG concentration profiles in constructs cultivated under low oxygen concentrations in culture medium (Obradovic *et al.*, 2000). These studies indicated the importance of oxygen concentration for GAG synthesis, in particular in the beginning of cultivation (about 10 days), and the suitability of hydrodynamically active rotating bioreactors for cartilage tissue engineering.

## Example 2. Engineered Cardiac Muscle: Convective-Diffusive Transport of Oxygen

The myocardium is a highly differentiated tissue that (via highly specialized structures) couples electrical and mechanical outputs to provide blood flow throughout the organism. It is composed of cardiac myocytes and fibroblasts with a dense supporting vasculature and collagen-based extracellular matrix. Cardiomyocytes comprise only 20–40% of the cells in the heart, but they occupy 80–90% of the heart volume. Their high metabolic activity is supported by the high density of mitochondria in the cells (MacKenna et al., 1994; Brilla et al., 1995) and by vast amounts of oxygen supplied by convection of blood through capillary networks and diffusion into the tissue space surrounding each capillary. Under physiological conditions, oxygen dissolved in blood plasma accounts for only ~1.5% of total oxygen content of the blood (Fournier, 1998), while the majority of oxygen-carrying capacity of the blood comes from hemoglobin, a natural oxygen carrier.

Conventional approaches to engineer cardiac tissues based on cell-polymer constructs immersed in medium flow providing convective transport of oxygen to the construct surfaces and molecular diffusion through the interior resulted in constructs with ~100-μm-thick outer layer of viable cells and acellular or necrotic core due to hypoxia (Bursac et al., 1999). In order to provide an in vivo–like oxygen supply to the cells within engineered cardiac constructs, a biomimetic approach based on highly porous scaffolds, culture medium supplemented with oxygen carriers, and perfusion bioreactors was developed (Radisic et al., 2006) (Fig. 12.6A).

To mimic the capillary network, rat cardiomyocytes alone or in a coculture with fibroblasts were cultivated on a highly porous elastomer (Wang et al., 2002), with a parallel array of channels that were perfused with culture medium at a flow velocity of 500 μm/s (comparable to that of blood flow in native heart). In this way, convective transport within the cultivated tissue was provided.

To mimic the role of hemoglobin, the culture medium was supplemented with a synthetic oxygen carrier (Oxygent™, perfluorocarbon emulsion). Oxygent™ is a 60% w/v (32% v/v) phospholipid-stabilized emulsion of perfluorooctyl bromide as a principal component and a small percentage of perfluorodecyl bromide (Kraft et al., 1998). Since perfluorocarbon (PFC) droplets are immiscible with the aqueous phase, they served as rechargeable oxygen reservoirs. As oxygen was depleted from the aqueous phase of culture medium, it was replenished by diffusion of dissolved oxygen from PFC particles (Fig. 12.6B). Overall, the oxygen partial pressures measured in the aqueous phase of PFC-supplemented and unsupplemented (control) media were

the same, and the PFC particles replenished oxygen consumed by the cells without increasing oxygen concentration in the culture medium.

A steady-state mathematical model of oxygen distribution in the cardiac tissue construct with a parallel channel array perfused with PFC-supplemented medium was developed and solved for a set of parameters (Radisic et al., 2005). The model accounted for oxygen transfer in the channels by axial convection and radial diffusion as well as for axial and radial diffusion and consumption in the tissue region surrounding the channels. Supplementation of culture medium by PFC emulsion was predicted to improve mass transport by increasing convective term and effective diffusivity of culture medium, resulting in increased total oxygen concentration (Fig. 12.6C). Model predictions were consistent with experimental results showing lower oxygen decrease in PFC-supplemented medium passing through the tissue construct (28 mmHg) as compared to the group perfused with pure culture medium (45 mmHg) (Radisic et al., 2006). Higher availability of oxygen resulted in markedly and significantly higher contents of DNA and cardiac proteins as well as enhanced contractile properties of tissue constructs cultured with the PFC-supplemented medium as compared to those cultured in control medium (Fig. 12.6D–F) (Radisic et al., 2006).

In sum, these examples demonstrate that the biomimetic approach to the design of 3D tissue culture systems can provide in vivo–like transport of oxygen, resulting in physiologic densities of viable, differentiated cells.

## Biophysical Signaling Considerations

Tissues and organs in the body are subjected to complex biomechanical environments with dynamic stresses and strains, fluid flow, and electrical signals. It is widely accepted that biophysical signals that play a role in cell physiology in vivo can also modulate the activity of cells within engineered tissues cultured in vitro. Innovative bioreactors have been developed to apply one or more regimes of controlled physical and/or electrical stimuli to 3D-engineered constructs in an attempt to improve or accelerate the generation of a functional tissue. In this section, we describe several examples of biomimetic approaches to engineer different tissue types under suitable biophysical regulation, with the aim to stimulate ECM synthesis, improve structural organization of the tissue, direct cell differentiation, and enhance a specific cell/tissue function.

## Example 3. Engineered Cartilage: Dynamic Compression

Articular cartilage lines bone ends in synovial joints and allows joint mobility while transferring compressive and shear forces. During joint movements, articular cartilage is dynamically exposed to compressive stresses, which can be as high as 6–18 MPa (Ateshian and Hung, 2003). Mechanical loading was shown to strongly affect metabolic activity of articular chondrocytes, consequently altering composition,

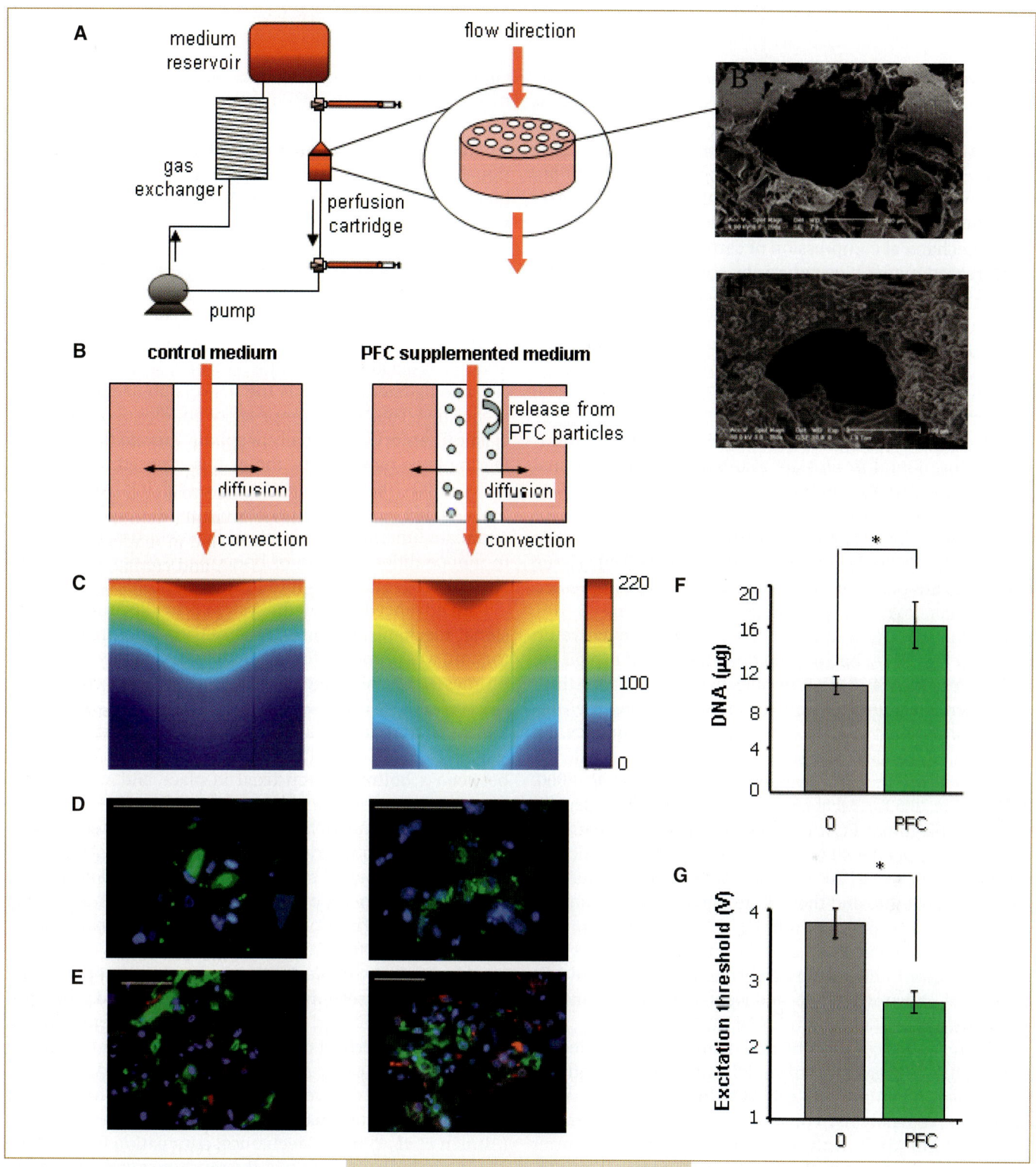

**FIG. 12.6.** Cultivation of engineered cardiac tissue in perfusion bioreactors. **(A)** *Perfusion loop:* Channeled elastomer scaffolds seeded with cells were placed in perfusion cartridges connected to a pump, a medium reservoir, and a gas exchanger. Cell-polymer constructs were perfused either with the control medium or medium supplemented with PFC. *Right:* SEMs of a channel in the initial Biorubber scaffold (upper) and the cardiac construct perfused for 7 days with PFC-supplemented media. **(B)** $O_2$ is transported by convection through the channel and diffusion to the tissue space. In PFC-supplemented media, $O_2$ is replenished from the PFC particles. **(C)** Predictions of $O_2$ profiles in the control (*left*) and PFC-supplemented media (*right*); vertical lines designate channel walls. (Adapted with modifications from Radisic *et al.*, 2005.) Immunostains of cross sections of constructs cultivated in the control (*left*) and PFC-supplemented media (*right*). Scale in C is in micromoles/L oxygen. **(D)** Presence of connexin-43 (green dots). **(E)** Presence of troponin I (green) and vimentin (red). Nuclei (blue) were stained with DAPI; scale bar = 50 $\mu$m. **(F)** DNA contents in constructs cultivated in the control and PFC-supplemented media. *: Significant difference between the groups by rank-sum test ($p < 0.05$), Ave $\pm$ SD ($n = 4$). **(G)** Contractile properties of constructs cultivated in the control and PFC-supplemented media: excitation thresholds (V), measured as the minimal amplitude of electrical stimuli (square pulses, 2-ms duration, 60 bpm) needed to induce synchronous macroscopic contractions of the constructs. *: Significant difference between the groups by *t*-test ($p < 0.05$), Ave $\pm$ SD ($n = 4$–7) (adapted with modifications from Radisic *et al.*, 2006).

structure, and biomechanical properties of articular cartilage *in vivo* and *in vitro*. *In vivo* studies clearly showed that moderate exercise leads to an increase in the proteoglycan content of articular cartilage (Saamanen *et al.*, 1987; Kiviranta *et al.*, 1988), whereas immobilization of joints leads to a loss of proteoglycans (Palmoski *et al.*, 1979; Vanwanseele *et al.*, 2002). *In vitro* studies indicated that static mechanical compression of cartilage explants results in decreased matrix biosynthesis by chondrocytes (Larsson *et al.*, 1991; Wong *et al.*, 1997), whereas cyclic loading may stimulate synthesis of proteoglycans and proteins, in a way that depends on the frequency, intensity, and duration of the stimulus (Gray *et al.*, 1989; Sah *et al.*, 1989; Parkkinen *et al.*, 1993; Steinmeyer and Knue, 1997; Wong *et al.*, 1999; Grodzinsky *et al.*, 2000).

It is believed that suitable biomechanical stimuli mimicking those found *in vivo* are also required for functional cartilaginous tissue assembly under *in vitro* conditions. Bioreactors with controlled dynamic loading were proposed for *in vitro* cultivation of engineered cartilage (Mauck *et al.*, 2000, 2002, 2003a, 2003b; Demarteau *et al.*, 2003a, 2003b). Tissue constructs are placed in wells and dynamically compressed by plungers to mimic physiological joint loading (5–10% strain, 0.1- to 1-Hz frequency). In the experimental setup shown in Fig. 12.7A, bovine chondrocytes seeded in alginate and agarose gels (4.76- to 6.76-mm-diameter, 2.2-mm-thick discs) were cultivated for up to two months under intermittent deformational loading (10% strain, 1-Hz frequency, 1 h on/1 h off, 3 h per day, 5 days per week) at different cell densities and culture medium compositions (Mauck *et al.*, 2000, 2002, 2003a, 2003b). Mechanical loading stimulated deposition of cartilaginous ECM components in cultures that were efficiently supplied with nutrients as compared to static free swelling (Mauck *et al.*, 2002, 2003b). At the seeding density of 10 million cells/mL and the culture medium supplemented with 10% fetal bovine serum (FBS), GAG and collagen contents continually increased over the cultivation time, resulting in significantly higher wet weight fractions in mechanically stimulated constructs at six- and eight-week time points as compared to the static controls (Fig. 12.7B).

Correspondingly, mechanical stimulation resulted in significantly higher values of equilibrium Young's modulus and dynamic modulus (Fig. 12.7C). In addition, dynamic loading of cell–polymer constructs based on dually transfected bovine MSCs with an aggrecan-luciferase reporter construct and a constituitive CMV-Renilla plasmid, modulated aggrecan promoter activity, such that it increased over time, resulting in increased proteoglycan content (Mauck *et al.*, 2006). It should be noted that intermittent dynamic compression enhanced biomechanical properties of engineered constructs (Mauck *et al.*, 2002, 2003b), while no significant differences in GAG and collagen contents were measured, which implies different structural organizations of the produced ECM (Mauck *et al.*, 2003b). Growth factors provided in conjunction with dynamic loading during culture had a synergystic effect on the compositions and mechanical properties of cultured constructs (Mauck *et al.*, 2003a).

Results of these studies suggest that dynamic deformational loading, with optimized parameters, and appropriate biochemical signals can facilitate the growth of a structurally organized cartilage tissue substitute.

### Example 4. Engineered Ligament: Dynamic Torsion and Tension

Ligaments consist of bundles of fibrous connective tissue that stabilize the joint and help hold it together. The anterior cruciate ligament (ACL) works with the posterior cruciate ligament (PCL) as the major stabilizing ligaments of the knee. The ACL is composed of bundles of collagen and elastic fibers oriented axially or spirally around the long axis. During knee movements, the ACL is exposed to axial tensions of 100–700 N and torsions up to 140°. It is hypothesized that suitable biomechanical environments could be used to stimulate functional assembly of ligament tissue as well as to induce differentiation of bone marrow–derived mesenchymal progenitor cells into ligament fibroblasts.

An advanced bioreactor system providing multidimensional mechanical strains (axial tension/compression and torsion) was proposed for ligament tissue engineering based on mesenchymal progenitor cells cultured on collagen or silk scaffolds. It consisted of individual cartridges containing tissue constructs, each within a separate perfusion loop (Fig. 12.8A). Cell–polymer constructs were anchored between a bottom platen fixed in place and a top platen connected to gear trains driven by motors. The bioreactor provided independent but concurrent control over axial and torsional strains imparted to the growing tissue. Axial strain (5–10%) and torsional strain (<90°) were applied via the top platen at frequencies of 0.01–1 Hz, with periods of loading and rest (e.g., 20 min of mechanical stimulation every 8 h) (Altman *et al.*, 2002a, 2002b, 2002c).

Mechanical stimulation *in vitro*, without supplementation of ligament-selective exogenous growth and differentiation factors, induced the preferential differentiation of bone marrow–derived mesenchymal progenitor cells into a ligament cell lineage. A total of 24 engineered ligaments based on bovine or human cells were cultured for up to 21 days with 90° rotational and 10% axial strain at 0.0167 Hz (Altman *et al.*, 2002b). Mechanical stimulation fostered cell alignment in the direction of the resulting force, resulted in the formation of oriented collagen fibers, and up-regulated collagen types I and III and tenascin-C, all features characteristic of ligament cells. On collagen matrices, helically organized collagen fibers (20-μm-diameter bundles) with characteristic collagen banding patterns were formed in the direction of the applied load at the periphery of the mechanically stimulated ligaments, a feature absent in the controls (Fig. 12.8B). On silk matrices, the expression of collagen types I and III and tenascin-C was observed; the ratio of

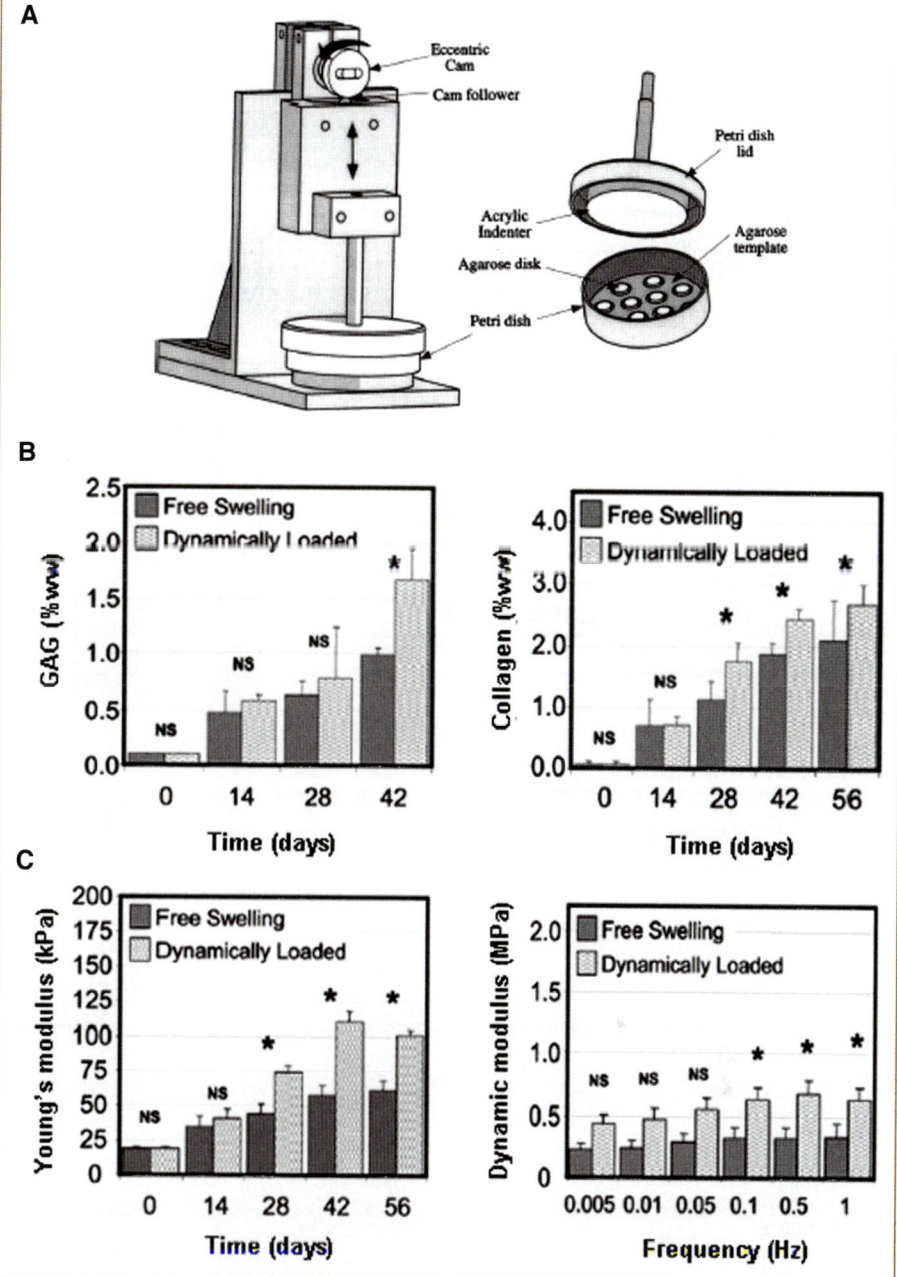

**FIG. 12.7.** Cultivation of engineered cartilage in a bioreactor with dynamic compression. **(A)** Chondrocyte-seeded agarose discs are placed in an agarose template and subjected to unconfined dynamic compression (Mauck *et al.*, 2000). **(B)** Deposition of ECM in constructs seeded with bovine chondrocytes at $10 \times 10^6$ cells/mL and cultured in medium supplemented with 10% FBS under free-swelling conditions or subjected to dynamic loading (10% strain, 1 Hz, 1 h on/1 h off, 3 h per day, 5 days per week): contents of GAG and collagen as functions of the cultivation time. **(C)** Biomechanical properties of engineered cartilage: equilibrium Young's modulus as a function of the cultivation time and dynamic modulus at frequencies ranging from 0.005 to 1 Hz for constructs on day 56. Dynamic compression stimulated functional ECM assembly such that dynamically loaded constructs exhibited significantly better biomechanical properties as compared to static cultures under free-swelling conditions (* indicates $p < 0.05$). Adapted from Mauck *et al.* (2003).

collagen type I expression to collagen type III was 8.9:1, consistent with that of native cruciate ligaments (Fig. 12.8C). At the same time, no up-regulation of bone or cartilage-specific cell markers was observed (Altman *et al.*, 2002b).

These studies support the notion that advanced bioreactors providing physiologically relevant mechanical loading are essential for meeting the complex requirements of *in vitro* engineering of functional skeletal tissues.

### Example 5. Engineered Bone: Hydrodynamic Shear

Native bone is a load-bearing tissue with a multiplicity of roles within the organism. Notably, the composition of bone varies only a little within the skeleton. The cells are embedded in a matrix composed of approximately 50% fibrous organic matrix (collagen, GAG, and glycoproteins) and 50% mineral (hydroxyapatite crystals and amorphous calcium phosphate). In contrast, the hierarchical organization of bone varies tremendously across distinct skeletal sites, such that the biomechanical properties also change considerably. Macroscopically, bone has a dense cortical compartment and a porous trabecular bone compartment. Microscopically, bone consists of discrete assemblies: osteons in the cortical bone and trabecular packets, with an arrangement of lamellae composed of cells embedded within the mineralized collagen type I–dominated matrix (Vunjak-Novakovic and Goldstein, 2005). To fully describe bone, it is

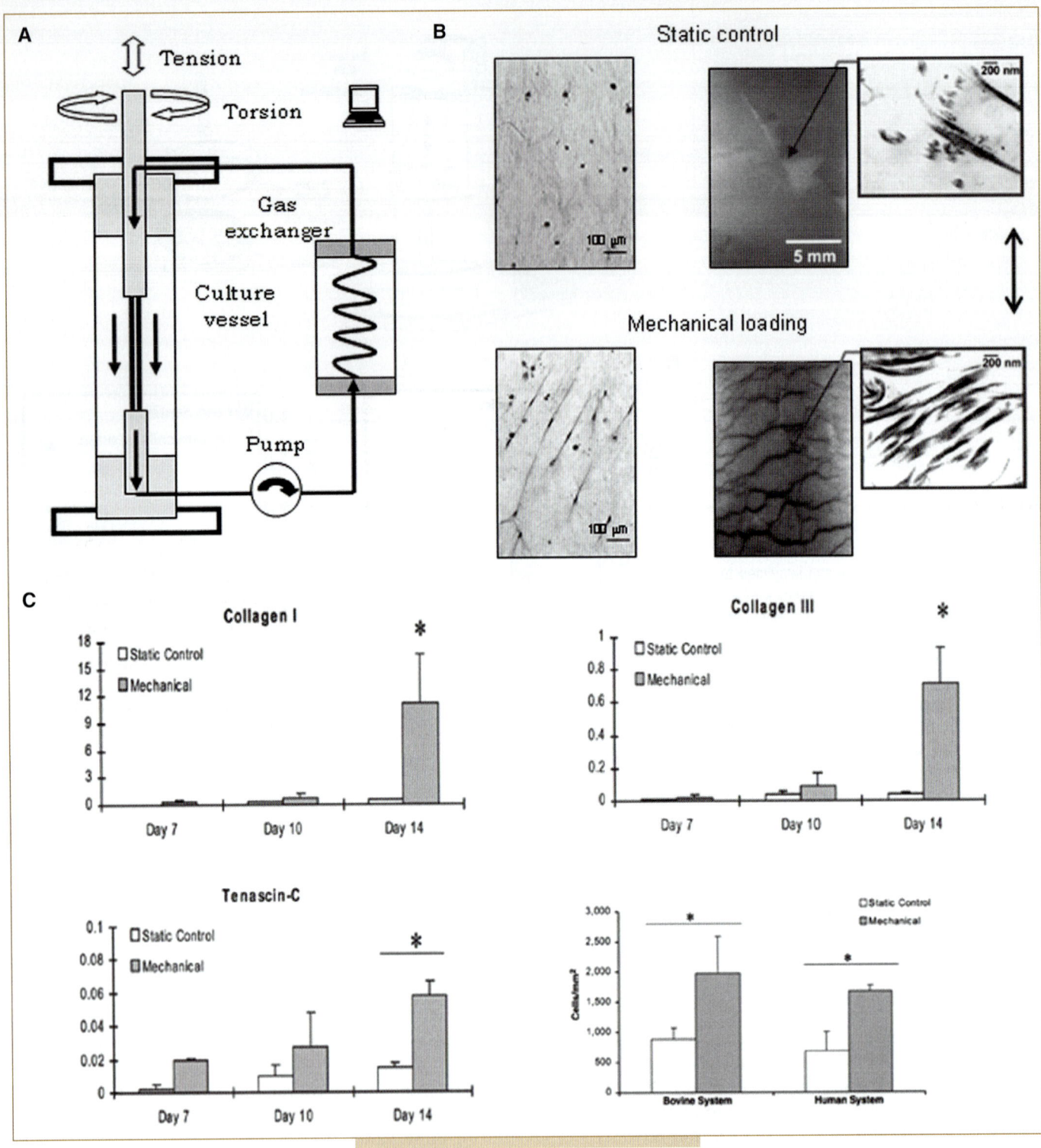

**FIG. 12.8.** Cultivation of engineered ligament in the bioreactor system with multidimensional mechanical strains. **(A)** Cell-polymer constructs are anchored in cartridges between two platens, the upper one of which is connected to computer-controlled gear trains to provide axial (5–10%) and torsional (<90°) strains at a specified frequency (0.01–1 Hz). Each cartridge is connected to a separate perfusion loop provided with a gas exchanger and connected to a pump. **(B)** Appearances of gels seeded with bovine bone marrow–derived cells and cultivated for 21 days in the bioreactor system under static conditions (*upper panel*) and under mechanical loading (10% axial strain, 90° torsion, 0.0167 Hz for 20 min every 8 h; *lower panel*); double-headed arrow indicates the longitudinal axis of the ligaments: longitudinal sections of gels stained with H&E (*left*) and gel images with TEM inserts (*right*). Mechanical stimulation induced cell alignment in the direction of the resulting force (45°) and collagen fiber bundle formation and organization in a helical manner at the tissue periphery. **(C)** Composition of engineered tissue over time: relative expressions of collagen type I and III and Tenascin-C in static and mechanically loaded constructs obtained by RT-PCR (* indicates $p < 0.05$) and cross-sectional cell densities of gels seeded with bovine cells after 21 days of culture and gels seeded with human cells after 14 days of culture (* indicates $p < 0.001$). Adapted from Altman et al. (2002a, 2002b, 2002c).

necessary to measure biomechanical properties of the tissue at multiple scales. Macroscopically, bone has a compressive modulus ranging from 10 to 50 MPa, and therefore it has the ability to withstand high levels of mechanical forces.

In the *in vivo* settings, compressive stresses have been shown to be potent anabolic stimuli during bone formation and remodeling. In contrast, a sustained absence of mechanical stimulation (e.g., microgravity) leads to significant reductions in mineral content and strength of bone. Likewise, bone cells *in vitro* respond to a variety of mechanical signals, including hydrostatic pressure and hydrodynamic shear. Paradoxically, compressive stimulation during culture has a relatively minor influence on *in vitro* bone growth and mineralization since the load is borne almost exclusively by the scaffold substrate and not efficiently transferred to the cells within the pore spaces (Owan *et al.*, 1997; Smalt *et al.*, 1997). Rather, tissue-engineered bone growth is considerably enhanced by hydrodynamic shear induced by perfusion of culture medium. It is thought that this mimics the interstitial flow in canalicular spaces during bone deformation associated with natural compressive loading. Contemporary bone bioreactors therefore are designed for direct medium flow through the construct in order to expose all cells to shear (Bancroft *et al.*, 2002; Cartmell *et al.*, 2003; Sikavitsas *et al.*, 2003; Braccini *et al.*, 2005; Holtorf *et al.*, 2005).

It has been widely demonstrated that incorporation of medium flow enhances tissue-engineered bone growth. In a well-designed series of experiments, the dose-dependent effects of perfusion/shear on osteoblasts was investigated using a system designed for direct perfusion of culture medium through six independent chambers (Bancroft *et al.*, 2002). Each chamber contained a titanium-fiber mesh scaffold sandwiched between two O-rings to ensure flow directly through the scaffold (Fig. 12.9A). Three flow rates, corresponding to superficial velocities ranging from 64 μm/s to 640 μm/s, were used, and the scaffolds were cultured for up to 16 days with continuous perfusion. In osteoblast culture, a direct correlation was observed between the perfusion rate and calcium deposition within the scaffolds on day 16 of culture (Fig. 12.9B). There were also increases in the alkaline phosphatase levels in perfused cultures relative to static cultures, although the enzyme activity was independent of perfusion rates. Osteopontin, a marker of early-stage osteogenesis, exhibited temporal increases in perfusion cultures. Peak osteopontin levels occurred earlier at higher perfusion rates, indicating acceleration of the osteogenic process that is flow-rate dependent.

In order to elucidate whether the effects were due to increased shear or enhanced nutrient transfer, dextran molecules were used to vary medium viscosities and obtain modified culture media with viscosities twice and three times that of the control (Sikavitsas *et al.*, 2003). Then the same flow rate (~64 μm/s) was used for three experimental groups to impart different shear forces (1×, 2×, and 3×)

throughout the constructs without increasing levels of nutrient transfer in the various groups. Enhanced calcium deposition/matrix distribution was observed that was directly related to shear forces through the scaffolds (Fig. 12.9C, D).

Using the identical system, marrow stromal cells were cultivated in the presence and absence of dexamethasone (a potent osteoinductive hormone) and under static and flow conditions (64 μm/s for 1 day, 640 μm/s thereafter) to investigate whether mechanostimulatory effects were sufficient to induce an osteoblastic phenotype (Holtorf *et al.*, 2005). Medium flow alone was sufficient to induce increased AP activity, although it appeared that peak levels were higher when both flow and dexamethasone were used (Fig. 12.9E). Similarly, flow with dexamethasone maintained greater calcium deposition relative to the other three groups, but calcium deposition also significantly increased using only flow stimulation. Osteopontin levels were only higher than the control (static, no dexamethasone) in perfused constructs (Fig. 12.9F). These results demonstrate that mechanostimulation of marrow stem cells, even in the absence of a biological stimulus, may be sufficient to induce some characteristics of a differentiated phenotype. Moreover, it emphasizes that the mechanical stimuli work synergistically with biological signals to enhance the expression of the osteoblastic phenotype and to improve the functionality of the tissue-engineered bone construct.

### Example 6. Engineered Arteries: Pulsatile Radial Strains

Arteries transport blood in the body under high pulsatile pressures (80–120 mmHg), inducing cyclic radial distensions ranging up to 10%. Arterial walls are composed of three layers: tunica interna, consisting of endothelial cells (EC); tunica media, consisting of smooth muscle cells (SMC) and connective tissue, comprising collagen and elastin fibers; and tunica externa, consisting of irregular connective tissue. Endothelium is exposed to blood flow and provides low thrombogenicity; tunica media provides strength, elasticity, and contractibility of blood vessels; while tunica externa provides support and attachment of the blood vessel to the surrounding tissue. Large-diameter arteries carrying high blood flow rates have been successfully replaced using synthetic grafts. However, synthetic grafts performed poorly when implanted to replace small-diameter arteries (<6 mm).

A biomimetic approach was developed to culture autologous tissue-engineered arteries using specialized scaffolds and bioreactors (Fig. 12.10) (Niklason, 1999; Niklason *et al.*, 1999; McKee *et al.*, 2003). Biodegradable PGA scaffolds (Prabhakar *et al.*, 2003) were chemically modified (Gao *et al.*, 1998) to enhance SMC attachment and sewn into tubes. Tubular scaffolds were placed over silicone tubing in bioreactors and seeded with SMC (Fig. 12.10A). After seeding, bioreactors were filled with culture medium that was supplemented by factors regulating vascular development, and a pulsatile flow was applied through the silicone tubing to

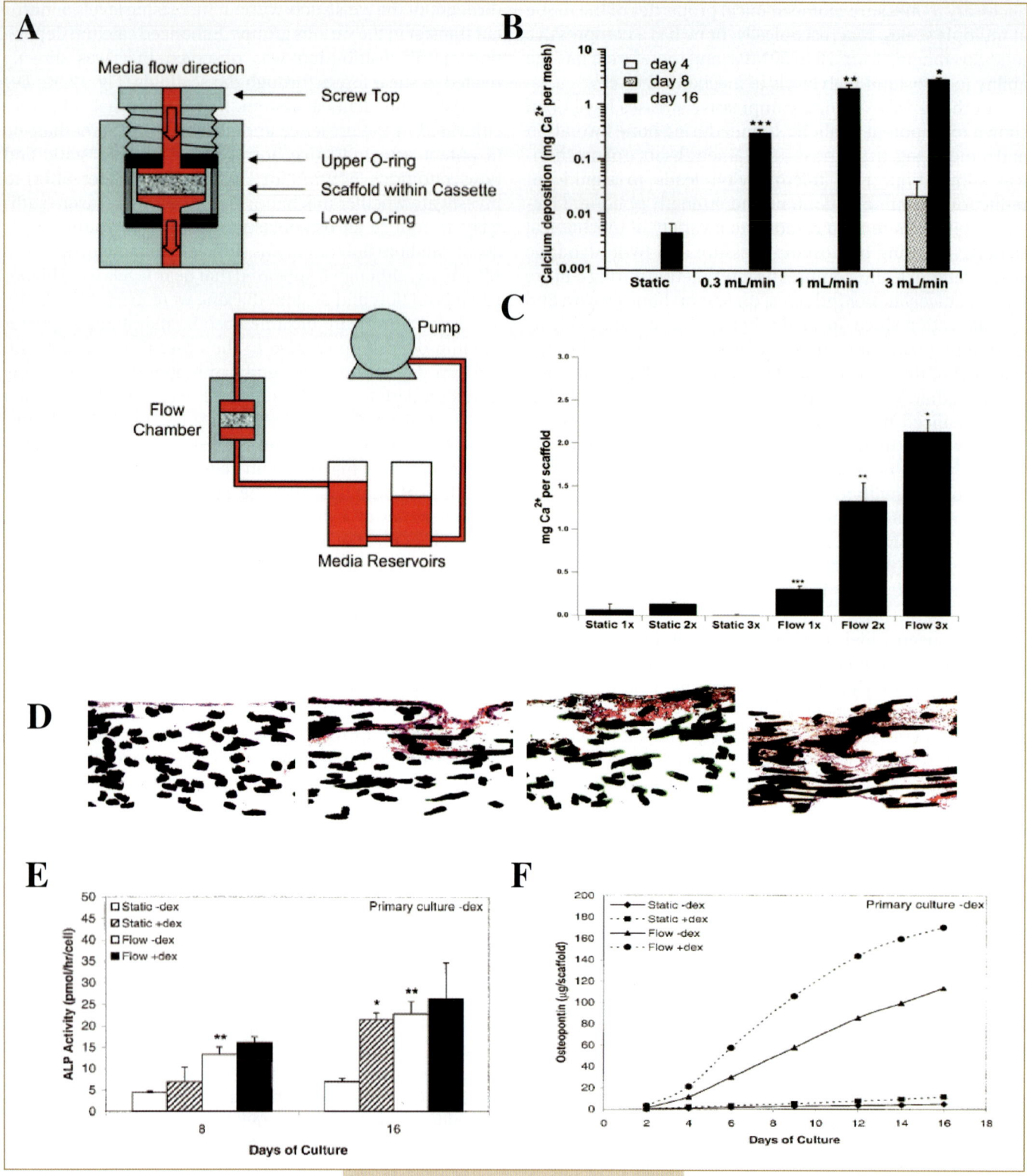

**FIG. 12.9.** Cultivation of tissue-engineered bone in perfusion bioreactors. **(A)** Schematic of single bioreactor compartment and perfusion flow diagram. Rat marrow stromal osteoblasts cultured in titanium meshes under static and flow perfused conditions. **(B)** Calcium deposition in scaffolds increases with increased rate of medium perfusion. Each group is statistically different from all other groups. (Data from Bancroft *et al.*, 2002.) **(C)** Calcium deposition in cultured scaffolds (mg Ca2+/scaffold). Static and flow-perfused cultures with control dextran free medium (1×), 3% dextran at twice the viscosity of control (2×) and 6% dextran at three times the viscosity of control (3×). Calcium deposition increases with increased shear in constructs. **(D)** Representative histological sections of scaffolds cultured for 16 days with static control medium (left), flow control medium (2nd from left), flow 2× medium (2nd from right) and flow 3× medium (right). (Adapted from Sikavitsas *et al.*, 2003.) Passage 2 rat marrow stromal cell-seeded scaffolds cultured in static and flow conditions with and without dexamethasone. **(E)** AP activity expressed in pmol/hr/cell. **(F)** Osotepontin secretion into culture medium expressed in µg per scaffold. (Data from Holtorf *et al.*, 2005.)

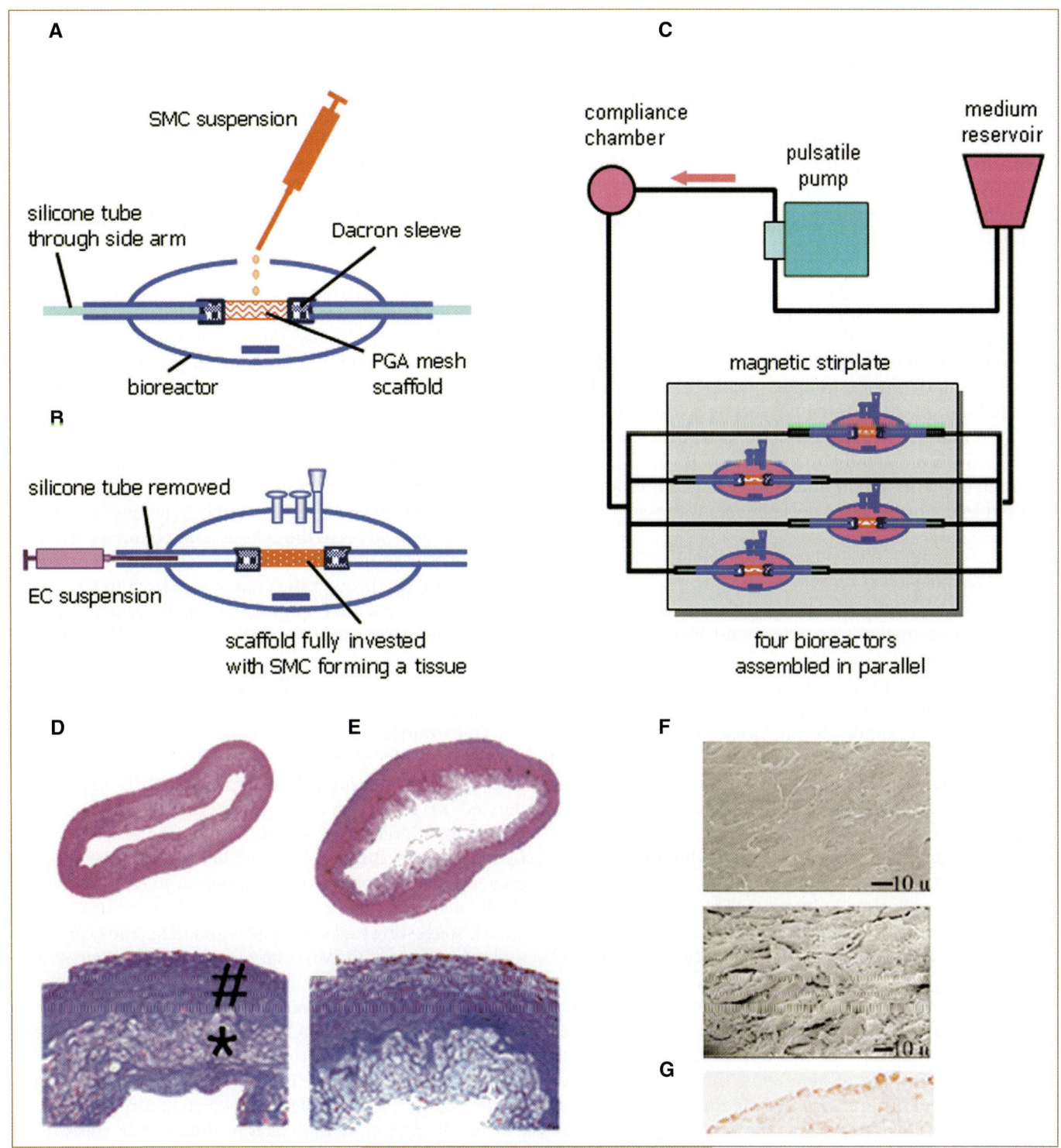

**FIG. 12.10.** Cultivation of engineered blood vessels in a bioreactor with pulsatile flow. *Cell seeding:* **(A)** Tubular PGA scaffolds were secured with polyester fiber (Dacron) sleeves to the bioreactor side openings, with silicone tubing protruding through the vessel lumen, and then seeded with SMCs by direct pipetting of concentrated cell suspension. **(B)** After cultivation period under pulsatile flow through the silicone tubing, the tubing was removed and EC suspension was injected into engineered vessel via side arm. **(C)** *Bioreactor system:* Each bioreactor was filled with culture medium that was supplemented by factors regulating vascular development and stirred with a magnetic stirrer; silicone tubing was connected to a perfusion system providing pulsatile radial stresses. Histological cross sections of engineered bovine vessels cultured for 8 weeks and stained with Verhoff's elastin stain (*upper panels*, original magnification ×20) and Masson's trichrome stain (*lower panels*, collagen stains blue; original magnification ×100): **(D)** engineered vessel cultured under pulsatile conditions, **(E)** control nonpulsed engineered vessel (# indicates the dense cellular region, * indicates the region with undegraded polymer fibers). (Adapted from Niklason *et al.*, 1999.) **(F)** Scanning electron micrographs of luminal surfaces of nonendothelialized (*upper*) and engineered vessel seeded with ECs and perfused for 3 days in which a confluent EC layer can be observed (*lower*). Scale bar = 10 μm. **(G)** Platelet-endothelial cell adhesion molecule immunoperoxidase staining confirms EC identity on luminal surface. Original magnification: ×200. (F and G adapted from Niklason *et al.*, 2001.)

mimic the flow of blood through developing blood vessels (Fig. 12.10B). In this way, cell–polymer constructs were subjected to pulsatile radial strains of 1.5–5% at a controlled frequency (165 bpm), for up to eight weeks. After eight weeks, the silicone tubing was removed, the EC suspension was injected into the lumen, and the flow rate was then gradually increased over three days of culture (Niklason *et al.*, 2001).

Using these approaches, arteries that are 3 mm in diameter and 8 cm in length have been cultured from bovine (Niklason *et al.*, 2001), porcine (Niklason *et al.*, 1999), and human cells (McKee *et al.*, 2003). Pulsatile flow enhanced the structure and mechanical integrity of engineered vessels as compared to nonpulsed vessels (Fig. 12.10C). Vessels engineered under pulsatile flow developed wall thickness of 200–400 μm and rupture strengths greater than 2000 mmHg. Endothelium in engineered vessels was confluent and displayed multiple EC-specific markers (Fig. 12.10D). Engineered vessels remained patent for more than three weeks following implantation in miniature swine, but only if precultured in the presence of pulsatile radial stress.

### Example 7. "Smart Well Plates" for Screening and Directed Differentiation of Cells

The clinical utility of cell-based therapies depends on simple, robust methods to select and expand the cells, to precisely regulate their differentiation into specific phenotype(s), and to fully characterize their potential for tissue repair. Ideally, the properties of cells should be tailored for a specific application and a specific patient. This is particularly true for human stem cells (embryonic and adult) and their use in clinical settings, due to the strong and complex dependence of cell responses on exogenous factors.

At this time, there are two distinctly different complementary approaches to cell cultivation in a clinical setting of stem cells: well-plate cultures and bioreactor cultures. The use of conventional *well plates* is still most widely spread (both in research labs and clinically) because of their simplicity and well-established experimental protocols. The key disadvantage of well plates is the inability to control some of the key regulatory factors of stem cell function, most notably (a) hydrodynamic shear, and (b) oxygen levels. The use of *bioreactors* involving convective mixing and/or perfusion of culture medium enables precise control of the cellular microenvironment, a feature that can substantially advance the repertoire of regulatory regimes we can use to modulate cell function. Unfortunately, bioreactors are in most cases larger in volume than well plates, and their use requires some technical skill. Most notably, the use of bioreactors involves the change from tissue culture plastics to other, less preferred substrates (such as glass and either natural and synthetic polymers). At present, a researcher can only choose between well plates and bioreactors, i.e., between routinely used protocols and precise

control of environmental factors. Clearly, this is a significant trade-off that limits the cell outcomes.

The importance of hydrodynamic shear for the maturation and differentiation of stem cells has raised interest in using microfluidic devices, which again involve the need to switch to a substrate different from the tissue culture plastics (Leclerc *et al.*, 2004; Shin *et al.*, 2004; B. L. Gray *et al.*, 2002). Also, there are various problems with cell seeding and attachment within microfluidic devices. In this respect, microfluidic devices are not necessarily user friendly.

One recently developed system, smart well plates, combines the advantages of well plates and bioreactors within a single, user-friendly device. The smart well plates (SWP) add functionality to the conventional well plate while retaining its inherent simplicity. The clinician preparing cells can simply replace the well plate cover with the SWP, to get the capability to apply medium flow and to control hydrodynamic shear, oxygen and pH, and medium supplements. The cover is designed and built to fit over a standard six-well plate (Fig. 12.11) and contains six plungers (Ø 35 mm) extended into each well and precisely machined to provide a defined shear stress to the cells. A multichannel peristaltic pump is used to drive medium across the lower surface of each piston, up through the center, and back to each well. A gas exchanger is not required if the device is in an incubator, but one can be used if studies under low and high oxygen levels are conducted. The plungers and cover are constructed of clear polycarbonate to allow direct microscope observation of the cells.

One example of application of this device is the modulation of vascular differentiation of human embryonic stem cells (hESC). Vascular smooth muscle cells (v-SMC) were derived in this device using our previously developed protocols (Gerecht-Nir *et al.*, 2003), under continuous hydrodynamic shear stress for 24 hours. Cells adopted characteristic perpendicular orientation to the direction of shear, and they had enhanced expression of αSMA (Fig. 12.11), suggesting that shear stress enhanced the expression and maturation of the vascular phenotype. Shear stress also affected the organization of stress fibers, thus enhancing the functionality of hES-derived v-SMCs.

Other applications include cell screening and directed differentiation for a wide variety of cell therapies. In all cases, the outcome of the clinical cell-based therapy depends on the quality of cell preparation, which in turn depends on control of environmental factors during cell cultivation. Some of the important factors are listed in Table 12.5, which explains briefly how they can be manipulated in SWPs.

### Example 8. Engineered Cardiac Muscle: Electrical Stimulation

The contractile apparatus of cardiac myocytes consists of sarcomeres, with high density of mitochondria supporting high metabolic activity and electrical signal propagation provided by specialized intercellular connections, gap

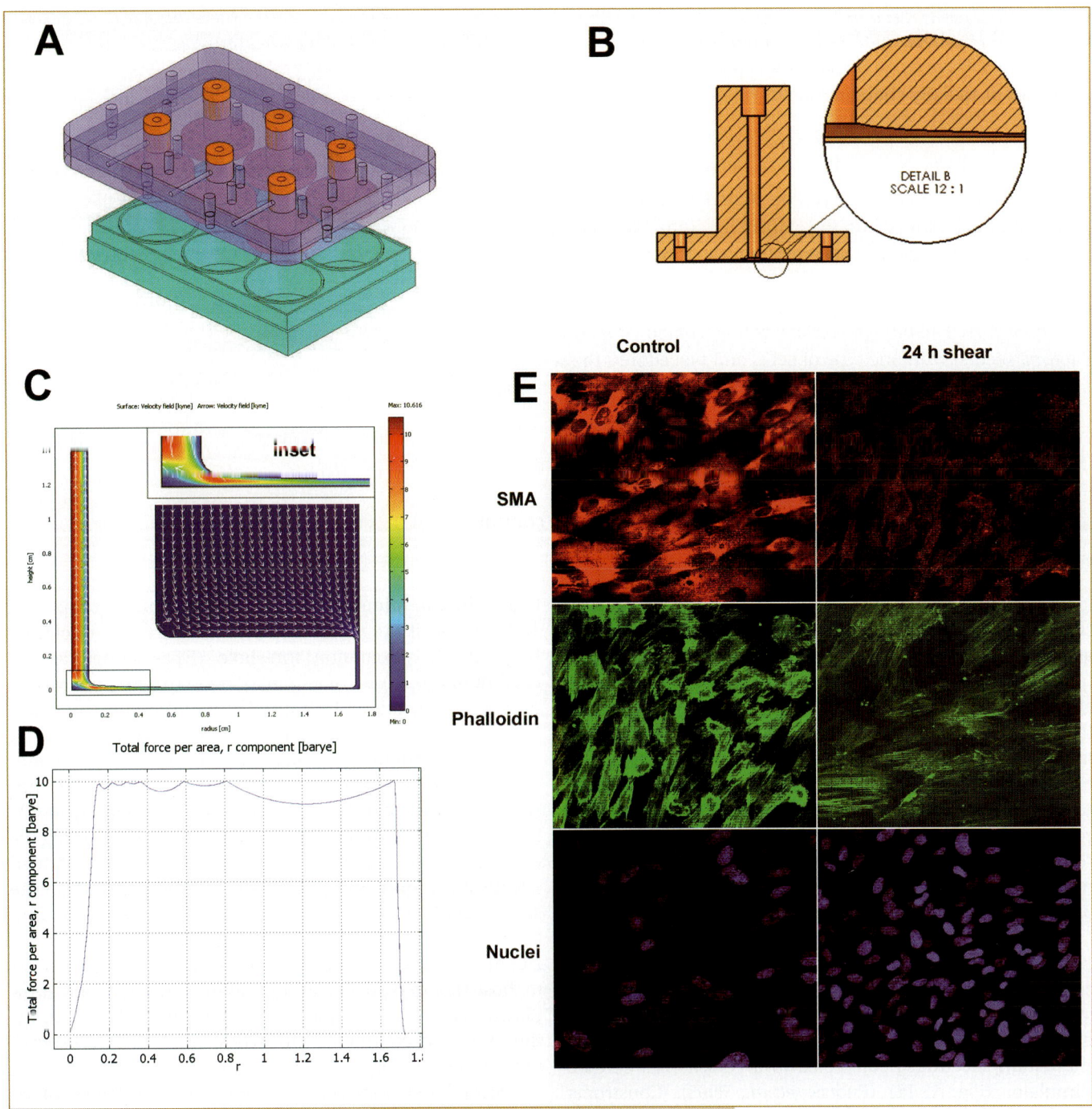

**FIG. 12.11.** Multiwell device with controlled application of hydrodynamic shear. **(A)** Custom cover placed over a standard six-well plate contains plungers designed to provide well-defined levels of hydrodynamic shear at cell monolayers in completely independent culture wells. **(B)** Section view of the plunger. Culture medium enters through the center of the plunger. **(C, D)** The curvature of the plunger is designed to maintain constant hydrodynamic shear at all radial positions in the well. **(E)** Modulation of vascular differentiation of human embryonic stem cells (hESC) by hydrodynamic shear. The application of shear stress during cultivation markedly enhanced the expression and organization of $\alpha$-SMA (red) and phalloidin (green). Nuclei are shown in blue.

| Table 12.5. | Examples of applications of "smart well plates" for optimizing cell culture parameters |
| --- | --- |

*Shear stress.* Device can be used for rapid screening under either constant or radially changing shear.

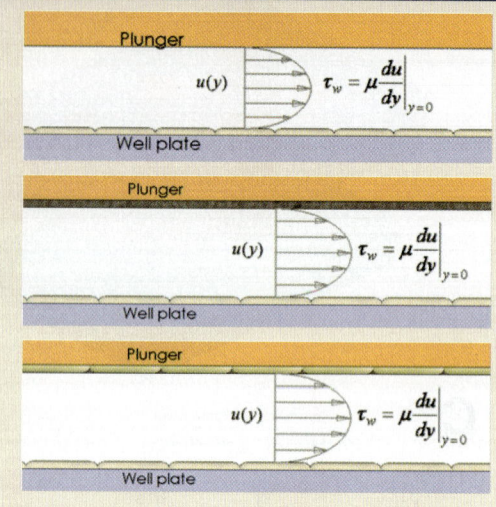

*Growth factors.* The plunger can be coated with growth factors that will diffuse slowly into the culture medium. Plungers can be exchanged throughout an experiment to direct cell development.

*Cell coculture.* The plunger surface can be coated with adhesion factors, seeded with one type of cells, and placed into the well plate in close proximity to a second type of cells already in the well plate.

*Hypoxia.* A controlled gas environment can be achieved by integrating a gas exchanger into the perfusion loop. Optical pH and $pO_2$ sensors can be used for on-line monitoring and control.

junctions (MacKenna *et al.*, 1994; Brilla *et al.*, 1995). The control of heart contractions is self-contained and can be attributed to the groups of specialized cardiac myocytes (pace makers), the fastest of which are located in the sinoatrial node. Contraction of the cardiac muscle is driven by the waves of electrical excitation generated by pacing cells that spread rapidly along the membranes of adjoining cardiac myocytes and trigger release of calcium, which in turn stimulates contraction of the myofibrils. Electromechanical coupling of the myocytes is crucial for their synchronous response to electrical pacing signals (Severs, 2000). Notably, the cardiovascular system is the first organ system that develops in the body.

To enhance the differentiation of cultured cardiac myocytes and their functional assembly into engineered cardiac constructs, cardiac-like electrical stimuli were applied to induce contractions of cultured constructs. The stimulation chamber was designed by fitting a petri dish with two stimulating electrodes connected to a commercial cardiac stimulator (Fig. 12.12) (Radisic *et al.*, 2003). Constructs were stimulated using suprathreshold square biphasic pulses (2-ms duration, 1 Hz, 5 V), and the stimulation was initiated after one to five days of scaffold seeding and applied for up to eight days. Immunohistochemical staining and contractile responses indicated that the stimulation should be initiated three days after seeding, a period that appeared to be sufficient for the cells to recover from isolation and attach to the scaffolds.

Cardiac constructs stimulated during cultivation had significantly better contractile responses to pacing as compared to unstimulated controls. The measured amplitude of synchronous contractions was on average seven times higher in the stimulated than nonstimulated group (Fig. 12.12C), and it progressively increased with time in culture (Fig. 12.12D). Excitation threshold (ET), defined as the minimum amplitude of stimulation at 60 bpm necessary to observe synchronous beating of the entire construct, was lower for stimulated than nonstimulated constructs (Fig. 12.12E). Consistently, the maximum capture rate was significantly higher for stimulated than for nonstimulated constructs (Fig. 12.12F). Excitation–contraction coupling of CM in stimulated constructs was also evidenced by recording the electrical activity using a platinum electrode positioned ~2 mm away from the pair of stimulating electrodes. The shape, amplitude (~100 mV), and duration (~200 ms) of electrical potentials recorded for cells in constructs that were electrically stimulated during culture (Fig. 12.12) were similar to those reported previously for cells from mechanically stimulated constructs (Zimmermann *et al.*, 2002). Importantly, the improved contractile properties were not the result of increased cellularity.

Stimulated constructs exhibited higher levels of α-myosin heavy-chain (α-MHC), connexin 43, creatin kinase-MM, and cardiac troponin I expression as assessed from Western blots and compared to unstimulated constructs. The ratio of α-MHC (adult) and β-MHC (neonatal) isoforms (~1.5 at day 3) decreased to 1.4 in the nonstimulated group and increased to 1.8 in the stimulated group by the end of the cultivation period, indicating cell maturation in the stimulated group. Stimulated constructs had thick, aligned myofibers that expressed sarcomeric α-actin, troponin I, α-MHC and β-MHC resembling myofibers in the native heart, and elongated nuclei. Stimulated constructs also exhibited stronger staining for connexin-43 than either early or

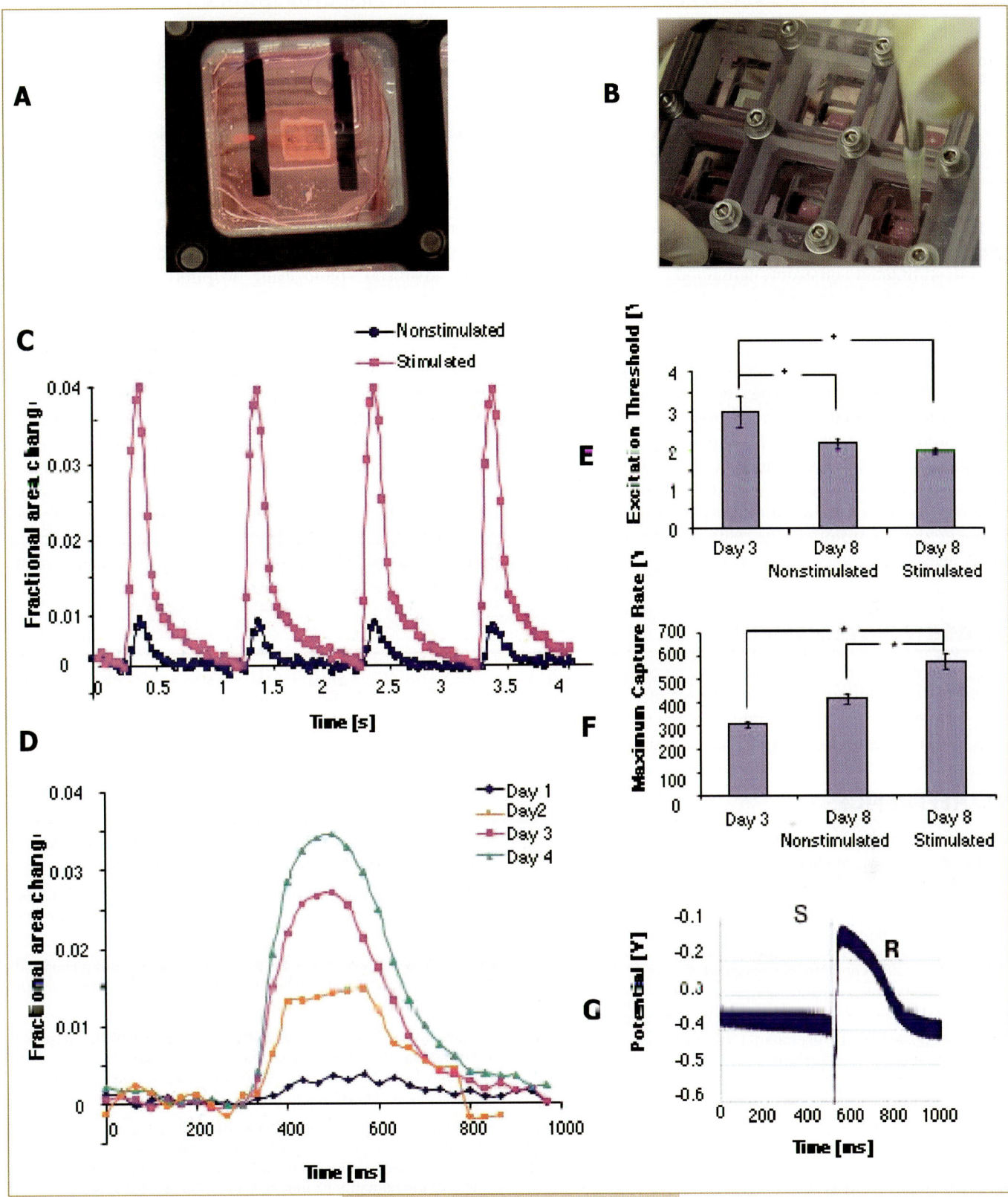

**FIG. 12.12.** Cardiac tissue engineering with electrical stimulation. **(A, B)** *Experimental setup:* Multiple culture wells are fitted with stimulating electrodes. Cardiac constructs prepared by seeding collagen sponges ($6 \times 8 \times 1.5$ mm) with neonatal rat ventricular cells ($5 \times 10^6$) were stimulated using suprathreshold square biphasic pulses (2 ms in duration, 1 Hz, 5 V) starting at day 3 of culture, up to day 8 of culture. **(C)** Contraction amplitude of constructs cultured for a total of 8 days, shown as a fractional change in the construct size. Electrical stimulation increased the amplitude of contractions by a factor of 7. **(D)** Amplitude of contractions progressively developed with time in culture and was associated with transmembrane potentials similar to action potential measured in native cardiac myocytes. **(E)** Excitation threshold (ET) decreased and **(F)** maximum capture rate (MCR) increased significantly both with time in culture and due to electrical stimulation. (*) denotes statistically significant differences ($p < 0.05$; $n = 5$–10). **(G)** Contractile activity was associated with the propagation of action potentials. (C–G reproduced with permission from Radisic *et al.*, 2004.)

nonstimulated constructs. In contrast, early and nonstimulated constructs exhibited a lower level of tissue organization and contained round nuclei.

Cells and nuclei in stimulated constructs were elongated and exhibited specialized cytoplasmic features characteristic of native myocardium. Gap junctions, intercalated discs, and microtubules were all more frequent in the stimulated than in the nonstimulated group and qualitatively more similar to those of neonatal rat heart. Cells in stimulated constructs contained aligned myofibrils and well-developed sarcomeres with clearly visible M, Z lines, H, I, and A bands. In most cells, Z lines were aligned, and the intercalated discs were positioned between two Z lines. Mitochondria (between myofibrils) and abundant glycogen were detected. In contrast, nonstimulated constructs had poorly developed cardiac-specific organelles and poor organization of ultrastructural features.

Overall, cardiac-like electrical stimulation during *in vitro* tissue culture progressively enhanced the excitation–contraction coupling and improved the properties of engineered myocardium at the cellular, ultrastructural, and tissue levels.

## IV. SUMMARY

Much progress has been made in the design of advanced bioreactors, based on our improved understanding of the conditions required for directing cell differentiation and assembly into functional tissue structures. Most of the published work has been based on the use of animal cells or nonstandardized human cell populations. One major challenge ahead of us is to extend the application of technologically advanced, biologically inspired tissue-engineering systems to the cultivation of functional human tissue grafts, using clinically suitable cell sources. Another challenge is to further advance the capabilities of our *in vitro* culture systems. The growing interactions between the fields of tissue engineering and developmental biology are expected to facilitate further the translation of biological principles into our engineering designs. The application of engineered tissues in regenerative medicine will also require the demonstration of their safety and efficacy in rigorous animal models. Many current efforts are directed at moving the science and practice of tissue engineering from observational to mechanistic and rational and from the limited current products to many more that will improve people's lives.

## V. ACKNOWLEDGMENTS

Bioreactor designs described in this chapter and their application to tissue engineering are a result of a meritorious, long-term effort of many different research groups, and the individual contributions are specifically credited. The senior author is grateful to the NIH (NIDCR, NHLBI, NIBIB) for supporting the tissue-engineering work in her laboratory. The authors thank Sue Kangiser for her help in the preparation of the manuscript.

## VI. REFERENCES

Altman, G. H., Horan, R. L., Lu, H. H., Moreau, J., Martin, I., Richmond, J. C., and Kaplan, D. L. (2002a). Silk matrix for tissue-engineered anterior cruciate ligaments. *Biomaterials* **23**, 4131–4141.

Altman, G. H., Horan, R. L., Martin, I., Farhadi, J., Stark, P. R., Volloch, V., Richmond, J. C., Vunjak-Novakovic, G., and Kaplan, D. L. (2002b). Cell differentiation by mechanical stress. *Faseb J.* **16**, 270–272.

Altman, G. H., Lu, H. H., Horan, R. L., Calabro, T., Ryder, D., Kaplan, D. L., Stark, P., Martin, I., Richmond, J. C., and Vunjak-Novakovic, G. (2002c). Advanced bioreactor with controlled application of multidimensional strain for tissue engineering. *J. Biomech. Eng.* **124**, 742–749.

Ateshian, G. A., and Hung, C. T. (2003). Functional properties of native articular cartilage. *In* "Functional Tissue Engineering" (F. Guilak, D. L. Butler, S. A. Goldstein, and D. J. Mooney, eds.), pp. 46–68. Springer-Verlag, New York.

Bancroft, G. N., Sikavitsas, V. I., van den Dolder, J., Sheffield, T. L., Ambrose, C. G., Jansen, J. A., and Mikos, A. G. (2002). Fluid flow increases mineralized matrix deposition in 3D perfusion culture of marrow stromal osteoblasts in a dose-dependent manner. *Proc. Natl. Acad. Sci. USA* **99**, 12600–12605.

Braccini, A., Wendt, D., Jaquiery, C., Jakob, M., Heberer, M., Kenins, L., Wodnar-Filipowicz, A., Quarto, R., and Martin, I. (2005). Three-dimensional perfusion culture of human bone marrow cells and generation of osteoinductive grafts. *Stem Cells* **23**, 1066–1072.

Brighton, C. T., and Heppenstall, R. B. (1971). Oxygen tension in zones of the epiphyseal plate, the metaphysis and diaphysis. *J. Bone Joint Surg. Am.* **53A**, 719–728.

Brilla, C. G., Maisch, B., Rupp, H., Sunck, R., Zhou, G., and Weber, K. T. (1995). Pharmacological modulation of cardiac fibroblast function. *Herz* **20**, 127–135.

Burchell, J., Durbin, H., and Taylor-Papadimitriou, J. (1983). Complexity of expression of antigenic determinants recognised by monoclonal antibodies HMFG 1 and HMFG 2 in normal and malignant human mammary epithelial cells. *J. Immunol.* **131**, 508–513.

Burchell, J., Gendler, S., Taylor-Papadimitriou, J., Girling, A., Lewis, A., Millis, R., and Lamport, D. (1987). Development and characterisation of breast cancer reactive monoclonal antibodies directed to the core protein of the human milk mucin. *Cancer Res.* **47**, 5476–5482.

Bursac, N., Papadaki, M., Cohen, R. J., Schoen, F. J., Eisenberg, S. R., Carrier, R., Vunjak-Novakovic, G., and Freed, L. E. (1999). Cardiac muscle tissue engineering: toward an *in vitro* model for electrophysiological studies. *Am. J. Physiol. Heart Circ. Physiol.* **277**, H433–H444.

Cartmell, S. H., Porter, B. D., Garcia, A. J., and Guldberg, R. E. (2003). Effects of medium perfusion rate on cell-seeded three-dimensional bone constructs *in vitro*. *Tissue Eng.* **9**, 1197–1203.

Chambard, M., Mauchamp, J., and Chaband, O. (1987). Synthesis and apical and basolateral secretion of thyroglobulin by thyroid cell monolayers on permeable substrate: modulation by thyrotropin. *J. Cell Physiol.* **133**, 37–45.

Darling, E. M., and Athanasiou, K. A. (2003). Articular cartilage bioreactors and bioprocesses. *Tissue Eng.* **9**, 9–26.

Demarteau, O., Jakob, M., Schafer, D., Heberer, M., and Martin, I. (2003a). Development and validation of a bioreactor for physical stimulation of engineered cartilage. *Biorheology* **40**, 331–336.

Demarteau, O., Wendt, D., Braccini, A., Jakob, M., Schafer, D., Heberer, M., and Martin, I. (2003b). Dynamic compression of cartilage constructs engineered from expanded human articular chondrocytes. *Biochem. Bioph. Res. Co.* **310**, 580–588.

Dobbs, L. G., Pian, M., Dumars, S., Maglio, M., and Allen, L. (1997). Maintenance of the differentiated type II cell phenotype by culture with an apical air surface. *Am. J. Physiol. Lung Cell. Physiol.* **273**, L347–L354.

Eschenhagen, T. (2006). Engineered heart tissue. *In* "Culture of Cells for Tissue Engineering" (G. Vunjak-Novakovic and R. I. Freshney, eds.), pp. 259–291. John Wiley & Sons, Hoboken, NJ.

Flores, I., Benetti, R., and Blasco, M. A. (2006). Telomerase regulation and stem cell behaviour. *Curr. Opin. Cell Biol.* **18**, 254–260.

Fournier, R. L. (1998). "Basic Transport Phenomena in Biomedical Engineering." Taylor & Francis, Philadelphia.

Freed, L. E., and Vunjak-Novakovic, G. (1995). Cultivation of cell-polymer constructs in simulated microgravity. *Biotechnol. Bioeng.* **46**, 306–313.

Freed, L. E., Langer, R., Martin, I., Pellis, N., and Vunjak-Novakovic, G. (1997). Tissue engineering of cartilage in space. *Proc. Natl. Acad. Sci. USA* **94**, 13885–13890.

Freed, L. E., Hollander, A. P., Martin, I., Barry, J. R., Langer, R., and Vunjak-Novakovic, G. (1998). Chondrogenesis in a cell-polymer-bioreactor system. *Exp. Cell Res.* **240**, 58–65.

Freshney, R. I. (2005). "Culture of Animal Cells, a Manual of Basic Technique." John Wiley & Sons, Hoboken, NJ.

Gao, J., Niklason, L., and Langer, R. (1998). Surface hydrolysis of poly(glycolic acid) meshes increases the seeding density of vascular smooth muscle cells. *J. Biomed. Mater. Res.* **42**, 417–424.

Gerecht-Nir, S., Ziskind, A., Cohen, S., and Itskovitz-Eldor, J. (2003). Human embryonic stem cells as an *in vitro* model for human vascular development and the induction of vascular differentiation. *Lab. Invest.* **83**, 1811–1820.

Gerecht-Nir, S., Radisic, M., Park, H., Boublik, J., Cannizzaro, C., Langer, R., and Vunjak-Novakovic, G. (2006). Biophysical regulation of cardiogenesis. *Int. J. Dev. Biology* **50**, 233–243.

Gray, B. L., Lieu, D. K., Collins, S. D., Smith, R. L., and Barakat, A. I. (2002). Microchannel platform for the study of endothelial cell shape and function. *Biomed. Microdevices* **4**, 9–16.

Gray, M. L., Pizzanelli, A. M., Lee, R. C., Grodzinsky, A. J., and Swann, D. A. (1989). Kinetics of the chondrocyte biosynthetic response to compressive load and release. *Biochim. Biophys. Acta* **991**, 415–425.

Gregory, C. A., and Prockop, D. J. (in press). Stem/progenitor cells (MSCs) from bone marrow stroma. *In* "Culture of Human Stem Cells" (R. I. Freshney, G. N. Stacey, and J. A. Auerbach, eds.), John Wiley & Sons, Hoboken, NJ.

Grodzinsky, A. J., Levenston, M. E., Jin, M., and Frank, E. H. (2000). Cartilage tissue remodeling in response to mechanical forces. *Annu. Rev. Biomed. Eng.* **2**, 691–713.

Harrison, R. G. (1907). Observations on the living developing nerve fiber. *Proc. Soc. Exp. Biol. Med.* **4**, 140–143.

Hay, R. J. (2000). Cell line preservation and characterization. *In* "Animal Cell Culture, a Practical Approach" (J. R. W. Masters, ed.), pp. 95–148. IRL Press at Oxford University Press, Oxford, UK.

Holtorf, H. L., Jansen, J. A., and Mikos, A. G. (2005). Flow perfusion culture induces the osteoblastic differentiation of marrow stromal cell-scaffold constructs in the absence of dexamethasone. *J. Biomed. Mater. Res.* **72A**, 326–334.

Iozzo, R. V. (2005). Basement membrane proteoglycans: from cellar to ceiling. *Nat. Rev. Mol. Cell Biol.* **6**, 646–656.

Izadpanah, R., Trygg, C., Patel, B., Kriedt, C., Dufour, J., Gimble, J. M., and Bunnell, B. A. (2006). Biologic properties of mesenchymal stem cells derived from bone marrow and adipose tissue. *J. Cell. Biochem.* Jun 22 (Epub ahead of print).

Jeffreys, A. J., Wilson, V., and Thein, S. L. (1985). Individual specific "fingerprints" of human DNA. *Nature* **316**, 76–79.

Kaartinen, L., Nettesheim, P., Adler, K. B., and Randell, S. H. (1993). Rat tracheal epithelial cell differentiation *in vitro*. *In Vitro Cell. Dev. Biol. — An.* **29A**, 481–492.

Kiviranta, I., Tammi, M., Jurvelin, J., Saamanen, A. M., and Helminen, H. J. (1988). Moderate running exercise augments glycosaminoglycans and thickness of articular cartilage in the knee joint of young beagle dogs. *J. Orthop. Res.* **6**, 188–195.

Kraft, M. P., Riess, J. G., and Weers, J. G. (1998). The design and engineering of oxygen-delivering fluorocarbon emulsions. *In* "Submicron Emulsions in Drug Targeting and Delivery" (S. Benita, ed.), pp. 235–333. Harwood Academic, Amsterdam.

Lane, E. B. (1982). Monoclonal antibodies provide specific intramolecular markers for the study of tonofilament organisation. *J. Cell Biol.* **92**, 665–673.

Larsson, T., Aspden, R. M., and Heinegard, D. (1991). Effects of mechanical load on cartilage matrix biosynthesis *in vitro*. *Matrix Biol.* **11**, 388–394.

Leclerc, E., Sakai, Y., and Fujii, T. (2004). Perfusion culture of fetal human hepatocytes in microfluidic environments. *Biochem. Eng. J.* **20**, 143–148.

Lennon, D. P., and Caplan, A. I. (2006). Mesenchymal stem cells for tissue engineering. *In* "Culture of Cells for Tissue Engineering" (G. Vunjak-Novakovic and R. I. Freshney, eds.), pp. 23–59. John Wiley & Sons, Hoboken, NJ.

Levenberg, S., Khademhosseini, A., Macdonald, M., Fuller, J., and Langer, R. (2006). Human embryonic stem cell culture for tissue engineering. *In* "Culture of Cells for Tissue Engineering" (G. Vunjak-Novakovic and R. I. Freshney, eds.), pp. 62–82. John Wiley & Sons, Hoboken, NJ.

Loring, J. F., and Rao, M. S. (2006). Establishing standards for the characterization of human embryonic stem cell lines. *Stem Cells* **24**, 145–150.

Ludwig, T. E., Levenstein, M. E., Jones, J. M., Berggen, W. T., Mitchen, E. R., Frane, J. L., Crandal, L. J., Daigh, C. A., Conrad, K. R., Piekarczyk, M. S., Llanas, R. A., and Thomson J. A. (2006). Derivation of human embryonic stem cells in defined conditions. *Nat Biotechnol.* **24**, 185–187.

Maas-Szabowski, N., Stark, H. J., and Fusenig, N. E. (2002). Cell interaction and epithelial differentiation. *In* "Culture of Epithelial Cells" (R. I. Freshney and M. G. Freshney, eds.), pp. 31–63. Wiley-Liss, Hoboken, NJ.

MacKenna, D. A., Omens, J. H., McCulloch, A. D., and Covell, J. W. (1994). Contribution of collagen matrix to passive left ventricular mechanics in isolated rat heart. *Am. J. Physiol.* **266**, H1007–H1018.

MacLeod, R. A., Dirks, W. G., Matsuo, Y., Kaufmann, M., Milch, H., and Drexler, H. G. (1999). Widespread intraspecies cross-contamination of human tumor cell lines arising at source. *Int. J. Cancer* **83**, 555–563.

Martin, I., Obradovic, B., Freed, L. E., and Vunjak-Novakovic, G. (1999). A method for quantitative analysis of glycosaminoglycan distribution in cultured natural and engineered cartilage. *Ann. Biomed. Eng.* **27**, 656–662.

Martin, I., Obradovic, B., Treppo, S., Grodzinsky, A. J., Langer, R., Freed, L. E., and Vunjak-Novakovic, G. (2000). Modulation of the mechanical properties of tissue-engineered cartilage. *Biorheology* **37**, 141–147.

Martin, I., Wendt, D., and Heberer, M. (2004). The role of bioreactors in tissue engineering. *Trends Biotechnol.* **22**, 80–86.

Masters, J. R. W., Thomson, J. A., Daly-Burns, B., Reid, Y. A., Dirks, W. G., Packer, P., Toji, L. H., Ohno, T., Tanabe, H., Arlett, C. F., *et al.* (2001). STR profiling provides an international reference standard for human cell lines. *Proc. Natl. Acad. Sci. USA* **98**, 8012–8017.

Mauck, R. L., Soltz, M. A., Wang, C. C. B., Wong, D. D., Chao, P. G., Valhmu, W. B., Hung, C. T., and Ateshian, G. A. (2000). Functional tissue engineering of articular cartilage through dynamic loading of chondrocyte-seeded agarose gels. *J. Biomech. Eng.* **122**, 252–260.

Mauck, R. L., Seyhan, S. L., Ateshian, G. A., and Hung, C. T. (2002). Influence of seeding density and dynamic deformational loading on the developing structure/function relationships of chondrocyte-seeded agarose hydrogels. *Ann. Biomed. Eng.* **30**, 1046–1056.

Mauck, R. L., Nicoll, S. B., Seyhan, S. L., Ateshian, G. A., and Hung, C. T. (2003a). Synergistic action of growth factors and dynamic loading for articular cartilage tissue engineering. *Tissue Eng.* **9**, 597–611.

Mauck, R. L., Wang, C. C.-B., Oswald, E. S., Ateshian, G. A., and Hung, C. T. (2003b). The role of cell seeding density and nutrient supply for articular cartilage tissue engineering with deformational loading. *Osteoarthr. Cartilage* **11**, 879–890.

Mauck, R. L., Byers, B. A., Yuan, X., Rackwitz, L., and Tuan, R. S. (2006). Cartilage tissue engineering with MSC-laden hydrogels: effect of seeding density, exposure to chondrognic medium and loading. 52nd Annual Meeting of the Orthopedic Research Society, Chicago, ILL, March 19–22, Paper No. 0336.

McKee, J. A., Banik, S. S., Boyer, M. J., Hamad, N. M., Lawson, J. H., Niklason, L. E., and Counter, C. M. (2003). Human arteries engineered *in vitro*. *EMBO Rep.* **4**, 633–638.

Moll, R., Franke, W. W., and Schiller, D. L. (1982). The catalog of human cytokeratins: patterns of expression in normal epithelia, tumors and cultured cells. *Cell* **31**, 11–24.

Mullender, M., El Haj, A. J., Yang, Y., van Duin, M. A., Burger, E. H., and Klein-Nulend, J. (2004). Mechanotransduction of bone cells *in vitro*: mechanobiology of bone tissue. *Med. Biol. Eng. Comput.* **42**, 14–21.

Muschler, G. F., Nakamoto, C., and Griffith, L. G. (2004). Engineering principles of clinical cell-based tissue engineering. *J. Bone Joint Surg. Am.* **86-A**, 1541–1558.

Niklason, L. (2006). Tissue-engineered blood vessels. *In* "Culture of Cells for Tissue Engineering" (G. Vunjak-Novakovic and R. I. Freshney, eds.), pp. 293–322. John Wiley & Sons, Hoboken, NJ.

Niklason, L. E. (1999). Replacement arteries made to order. *Science* **286**, 1493–1494.

Niklason, L. E., Gao, J., Abbott, W. M., Hirschi, K. K., Houser, S., Marini, R., and Langer, R. (1999). Functional arteries grown *in vitro*. *Science* **284**, 489–493.

Niklason, L. E., Abbott, W., Gao, J., Klagges, B., Conroy, N., Jones, R., Vasanawala, A., Sanzgiri, S., and Langer, R. (2001). Morphologic and mechanical characteristics of bovine engineered arteries. *J. Vasc. Surg.* **33**, 628–638.

Obradovic, B., Freed, L. E., Langer, R., and Vunjak-Novakovic, G. (1997). Bioreactor studies of natural and engineered cartilage metabolism. *In* "Proceedings of the Topical Conference on Biomaterials, Carriers for Drug Delivery, and Scaffolds for Tissue Engineering" (N. A. Peppas, D. J. Mooney, A. G. Mikos, and L. Brannon-Peppas, eds.), pp. 335–337. AIChE, New York.

Obradovic, B., Carrier, R. L., Vunjak-Novakovic, G., and Freed, L. E. (1999). Gas exchange is essential for bioreactor cultivation of tissue engineered cartilage. *Biotechnol. Bioeng.* **63**, 197–205.

Obradovic, B., Meldon, J. H., Freed, L. E., and Vunjak-Novakovic, G. (2000). Glycosaminoglycan deposition in engineered cartilage: experiments and mathematical model. *AIChE J.* **46**, 1860–1871.

Obradovic, B., Martin, I., Freed, L. E., and Vunjak-Novakovic, G. (2003). Towards functional cartilage equivalents: bioreactor cultivation of cell-polymer constructs. *In* "Contemporary Studies in Advanced Materials and Processes" (D. P. Uskokovic, G. A. Battiston, S. K. Milonjic, and D. I. Rakovic, eds.), pp. 251–256. Materials Science Forum, Vol. 413.

O'Hara, B. P., Urban, J. P. G., and Maroudas, A. (1990). Influence of cyclic loading on the nutrition of articular cartilage. *Ann. Rheum. Dis.* **49**, 536.

Owan, I., Burr, D. B., Turner, C. H., Qiu, J. Y., Tu, Y., Onyia, J. E., and Duncan, R. L. (1997). Mechanotransduction in bone: osteoblasts are more responsive to fluid forces than mechanical strain. *Am. J. Physiol. Cell Physiol.* **42**, C810–C815.

Palmoski, M., Perricone, E., and Brandt, K. D. (1979). Development and reversal of a proteoglycan aggregation defect in normal canine knee cartilage after immobilization. *Arthritis Rheum.* **22**, 508–517.

Parkkinen, J. J., Ikonen, J., Lammi, M. J., Laakkonen, J., Tammi, M., and Helminen, H. J. (1993). Effects of cyclic hydrostatic pressure on proteoglycan synthesis in cultured chondrocytes and articular cartilage explants. *Arch. Biochem. Biophys.* **300**, 458–465.

Plaia, T. W., Josephson, R., Liu, Y., Zeng, X., Ording, C., Toumadje, A., Brimble, S. N., Sherrer, E. S., Uhl, E. W., Freed, W. J., *et al.* (2006). Characterization of a new NIH-registered variant human embryonic stem cell line, BG01V: a tool for human embryonic stem cell research. *Stem Cells* **24**, 531–546.

Prabhakar, V., Grinstaff, M. W., Alarcon, J., Knors, C., Solan, A. K., and Niklason, L. E. (2003). Engineering porcine arteries: effects of scaffold modification. *J. Biomed. Mater. Res.* **67A**, 303–311.

Radisic, M., Euloth, M., Yang, L., Langer, R., Freed, L. E., and Vunjak-Novakovic, G. (2003). High-density seeding of myocyte cells for tissue engineering. *Biotechnol. Bioeng.* **82**, 403–414.

Radisic, M., Park, H., Shing, H., Consi, T., Schoen, F., Langer, R., Freed, L. E., and Vunjak-Novakovic, G. (2004). Functional assembly of engineered myocardium by electrical stimulation of cardiac myocytes cultured on scaffolds. *PNAS* **101**, 18129–18134.

Radisic, M., Deen, W., Langer, R., and Vunjak-Novakovic, G. (2005). Mathematical model of oxygen distribution in engineered cardiac tissue with parallel channel array perfused with culture medium containing oxygen carriers. *Am. J. Physiol. Heart Circ. Physiol.* **288**, H1278–H1289.

Radisic, M., Park, H., Chen, F., Salazar-Lazzaro, J. E., Wang, Y., Dennis, R. G., Langer, R., Freed, L. E., and Vunjak-Novakovic, G. (2006). Biomimetic approach to cardiac tissue engineering: oxygen carriers and channeled scaffolds. *Tissue Eng.* **12**, 1–15.

Randers-Eichorn, L., Bartlett, R., Frey, D., and Rao, G. (1996). Noninvasive oxygen measurements and mass transfer considerations in tissue culture flasks. *Biotechnol. Bioeng.* **51**, 466–478.

Rooney, D. E., and Czepulkowski, B. H., eds. (1986). Human cytogenetics, a practical approach. IRL Press at Oxford University Press, Oxford, UK.

Rothfels, K. H., and Siminovitch, L. (1958). An air-drying technique for flattening chromosomes in mammalian cells growth *in vitro. Stain Technol.* **33**, 73–77.

Saamanen, A. M., Tammi, M., Kiviranta, I., Jurvelin, J., and Helminen, H. J. (1987). Maturation of proteoglycan matrix in articular cartilage under increased and decreased joint loading. A study in young rabbits. *Connect. Tissue Res.* **16**, 163–175.

Sah, R. L. Y., Kim, Y. J., Doong, J. Y. H., Grodzinsky, A. J., Plaas, A. H. K., and Sandy, J. D. (1989). Biosynthetic response of cartilage explants to dynamic compression. *J. Orthop. Res.* **7**, 619–636.

Sasaki, M., Honda, T., Yamada, H., Wake, N., and Barrett, J. C. (1996). Evidence for multiple pathways to cellular senescence. *Cancer Res.* **54**, 6090–6093.

Sattler, G. A., Michalopoulos, G., Sattler, G. L., and Pitot, H. C. (1978). Ultrastructure of adult rat hepatocytes cultured on floating collagen membranes. *Cancer Res.* **38**, 1539–1549.

Serakinci, N. (2005). Telomerase-induced immortalization. *In* "Culture of Animal Cells" (R. I. Freshney, ed.), pp. 299–301. John Wiley & Sons, Hoboken, NJ.

Severs, N. J. (2000). The cardiac muscle cell. *Bioessays* **22**, 188–199.

Shansky, J., Ferland, P., McGuire, S., Powell, C., DelTatto, M., Nackman, M., Hennessey, J., and Vandenburgh, H. H. (2006). Tissue engineering human skeletal muscle for clinical applications. *In* "Culture of Cells for Tissue Engineering" (G. Vunjak-Novakovic and R. I. Freshney, eds.), pp. 239–257. John Wiley & Sons, Hoboken, NJ.

Shay, J. W., and Wright, W. E. (2005). Use of telomerase to create bioengineered tissues. *Ann. N.Y. Acad. Sci.* **1057**, 479–491.

Shin, M., Matsuda, K., Ishii, O., Terai, H., Kaazempur-Mofrad, M., Borenstein, J., Detmar, M., and Vacanti, J. P. (2004). Endothelialized networks with a vascular geometry in microfabricated poly(dimethyl siloxane). *Biomedi. Microdevices* **6**, 269–278.

Sikavitsas, V. I., Bancroft, G. N., Holtorf, H. L., Jansen, J. A., and Mikos, A. G. (2003). Mineralized matrix deposition by marrow stromal osteoblasts in 3D perfusion culture increases with increasing fluid shear forces. *Proc. Nat. Acad. Science USA* **100**, 14683–14688.

Smalt, R., Mitchell, F. T., Howard, R. L., and Chambers, T. J. (1997). Induction of NO and prostaglandin E-2 in osteoblasts by wall-shear stress but not mechanical strain. *Am. J. Physiol. Endocrinol. Metab.* **36**, E751–E758.

Steinmeyer, J., and Knue, S. (1997). The proteoglycan metabolism of mature bovine articular cartilage explants superimposed to continuously applied cyclic mechanical loading. *Biochem. Bioph. Res. Co.* **240**, 216–221.

Urban, J. P. G. (1994). The chondrocyte: a cell under pressure. *Br. J. Rheumatol.* **33**, 901–908.

Vanwanseele, B., Lucchinetti, E., and Stussi, E. (2002). The effects of immobilization on the characteristics of articular cartilage: current concepts and future directions. *Osteoarthr. Cartilage* **10**, 408–419.

Vunjak-Novakovic, G., and Goldstein, S. A. (2005). Biomechanical principles of cartilage and bone tissue engineering. *In* "Basic Orthopaedic Biomechanics and Mechanobiology" (V. C. Mow and R. Huiskes, eds.), pp. 343–408. Lippincott-Williams and Wilkens, Philadelphia.

Vunjak-Novakovic, G., Freed, L. E., Biron, R. J., and Langer, R. (1996). Effects of mixing on the composition and morphology of tissue-engineered cartilage. *AIChE J.* **42**, 850–860.

Vunjak-Novakovic, G., Obradovic, B., Bursac, P., Martin, I., Langer, R., and Freed, L. E. (1998). Dynamic cell seeding of polymer scaffolds for cartilage tissue engineering. *Biotechnol. Prog.* **14**, 193–202.

Vunjak-Novakovic, G., Martin, I., Obradovic, B., Treppo, S., Grodzinsky, A. J., Langer, R., and Freed, L. E. (1999). Bioreactor cultivation conditions modulate the composition and mechanical properties of tissue-engineered cartilage. *J. Orthop. Res.* **17**, 130–138.

Vunjak-Novakovic, G., Obradovic, B., Martin, I., and Freed, L. E. (2002). Bioreactor studies of native and tissue-engineered cartilage. *Biorheology* **39**, 259–268.

Wang, Y., Ameer, G. A., Sheppard, B. J., and Langer, R. (2002). A tough biodegradable elastomer. *Nat. Biotechnol.* **20**, 602–606.

Wong, M., Wuethrich, P., Buschmann, M. D., Eggli, P., and Hunziker, E. (1997). Chondrocyte biosynthesis correlates with local tissue strain in statically compressed adult articular cartilage. *J. Orthop. Res.* **15**, 189–196.

Wong, M., Siegrist, M., and Cao, X. (1999). Cyclic, compression of articular cartilage explants is associated with progressive consolidation and altered expression pattern of extracellular matrix proteins. *Matrix Biol.* **18**, 391–399.

Yevdokimova, N., and Freshney, R. I. (1997). Activation of paracrine growth factors by heparan sulphate induced by glucocorticoid in A549 lung carcinoma cells. *Br. J. Cancer* **76**, 261–289.

Zborowski, M., Sun, L., Moore, L. R., and Chalmers, J. J. (1999). Rapid cell isolation by magnetic cell sorting for applications in tissue engineering. *ASAIO J.* **45**, 127–130.

Zimmermann, W. H., Schneiderbanger, K., Schubert, P., Didie, M., Munzel, F., Heubach, J. F., Kostin, S., Nehuber, W. L., and Eschenhagen, T. (2002). Tissue engineering of a differentiated cardiac muscle construct. *Circ. Res.* **90**, 223–230.

# Chapter Thirteen

# Regulation of Cell Behavior by Extracellular Proteins

*Amy D. Bradshaw and E. Helene Sage*

## I. INTRODUCTION

The extracellular milieu is critical for the control of the behavior of every cell in all tissues. Many factors can contribute to the environment of a cell, for example, cell : cell contact, growth factors, extracellular matrix (ECM) proteins, and matricellular proteins. All of these components act together to regulate cell surface protein activity, intracellular signal transduction, and subsequent gene expression, which leads to proliferation, migration, differentiation, and ultimately the formation of complex tissues. This chapter focuses on matricellular proteins as modulators of extracellular signals. The matricellular proteins are characterized as secreted, modular proteins that are associated with the extracellular matrix but do not act as structural constituents (Bornstein and Sage, 2002). Presumably, the function of matricellular proteins is to provide a link between the extracellular matrix and cell surface receptors or between cytokines and proteases localized in the extracellular environment whose activity might be affected by this interaction (Bornstein and Sage, 2002). Thrombospondin 1 and 2, tenascin-C, osteopontin, and SPARC are representatives of this class of proteins. A growing body of evidence points to these proteins as important mediators of growth factor, ECM, and cell signaling pathways (Table 13.1). Consequently, *in vitro*

systems designed to mimic tissue conditions should consider the influence of matricellular proteins.

## II. THROMBOSPONDIN-1

Thrombospondin-1 is a 450,000-dalton glycoprotein with seven modular domains (Bornstein, 2001). To date, five different paralogs of thrombspondin have been identified, termed thrombospondins 1–5. This chapter reviews the two more characterized forms, thrombospondin-1 and thrombospondin-2. At least five different ECM-associated proteins are able to bind to thrombospondin-1: collagens I and V, fibronectin, laminin, fibrinogen, and SPARC (Bornstein, 2001). Likewise, cell surface receptors for thrombospondin are multiple and include the integrin family of extracellular matix receptors (Bornstein, 2001). Given the significant number and variety of thrombospondin-1–binding proteins, it is of little surprise that a wide variety of functions has been attributed to this protein, some of which appear to be contradictory. Many of these disparities may, however, actually reflect the dynamic interaction of thrombospondin-1 with other extracellular factors that can influence cells in different ways to give rise to distinct cellular outcomes — a common theme in matricellular protein biology.

*Principles of Tissue Engineering, 3rd Edition*
ed. by Lanza, Langer, and Vacanti

**Table 13.1.** Matricellular protein interactions and activities

| | Extracellular matrix interaction | Receptor | Adhesive (+) vs. counteradhesive (−) | Growth factor modulation | Extracellular matrix formation |
|---|---|---|---|---|---|
| **Thrombospondin-1** | Col I & V Fn, Ln, Fg | Integrin CD-36 LRP | (−) | HGF (−) TGF-β (+) | ? |
| **Thrombospondin-2** | MMP-2 | LRP | (−) | ? | + |
| **Tenascin-C** | Fn | Integrin Annexin II | (−) (+) | EGF (+) bFGF (+) PDGF (+) | + |
| **Osteopontin** | Fn, Col I, II, III, IV, & V | Integrin CD-44 | (+) | ? | + |
| **SPARC** | Col I, III, IV, & V | Stabilin-1 | (−) | bFGF (−) VEGF (−) PDGF (−) TGF-β (+) | + |
| **Hevin** | Col I | ? | (−) | ? | + |

A summary of the activities for the matricellular proteins described in this chapter. This table is not a complete list of all activities and receptors for the proteins included in the table nor are all the matricellular proteins included. For references refer to the text. *Abbreviations:* Col: collagen; Fn: fibronectin; Ln, laminin; Fg: fibrinogen. Growth factor abbreviations are defined in the text.

Several studies have established an antiangiogenic function of thrombospondin-1 (Armstrong and Bornstein, 2003). *In vitro*, aortic endothelial cells transfected with antisense thrombospondin-1 cDNA generate twice as many capillary-like structures on a gelled basement membrane than control cells that produce higher levels of thrombospondin-1 (DiPietro *et al.*, 1994). *In vivo*, increased expression of thrombospondin-1 by tumor cells results in decreased vascularization and an accompanied reduction in tumor progression by a number of different types of tumor-derived cells (de Fraipoint *et al.*, 2001).

One mechanism by which thrombospondin-1 might influence angiogenesis is by promoting cell death in microvascular endothelial cells (Jimenez *et al.*, 2000). CD-36, a thrombospondin-1 cell surface receptor, has been shown to be required for caspase-3-mediated induction of apoptosis by thrombopspondin-1. Antimigratory effects elicited by CD-36 interaction with thrombospondin-1 in endothelial cells has also been demonstrated. As the first natural inhibitor of angiogenesis, characterization of the mechanisms by which thrombospondin-1 functions has provided valuable insight into strategies to control blood vessel formation.

Thrombospondin-1, like many of the matricellular proteins, influences cytokine activity. For example, thrombospondin-1 has been shown to interact specifically with transforming growth factor (TGF-β) (Schultz-Cherry and Murphy-Ullrich, 1993). This interaction leads to activation of the latent form of TGF-β, presumably through a conformational change in the cytokine, which allows interaction with cell surface receptors (Schultz-Cherry and Murphy-

Ullrich, 1993). Consequently, the presence of thrombospondin-1 in the extracellular milieu might affect the activity of this potent, multifunctional cytokine. Generation of a thrombospondin-1-null mouse lends support to the importance of thrombospondin-1-mediated activation of TGF-β *in vivo*. The phenotype in the absence of thrombospondin-1 expression mimics to some degree the phenotype of the TGFβ-null mouse (Crawford *et al.*, 1998). Specifically, similar pathologies of the lung and pancreas are observed in both mice, and, importantly, thrombospondin-1-null mice treated with thrombospondin-1 peptides show a partial restoration of active TGF-β levels and a reversion of the lung and pancreatic abnormalities toward tissues of wild-type mice. Importantly, thrombospondin-1 is a potent chemoattractant for inflammatory cells (Bornstein, 2001). Therefore, a reduction in active TGF-β levels in thrombospondin-1-null mice might be influenced by inflammatory cell recruitment.

Thrombospondin-1 influences cell adhesion and cell shape. For example, it will diminish the number of focal adhesions of bovine aortic endothelial cells and thus will promote a migratory phenotype (Murphy-Ullrich, 2001). Thrombospondin-1, therefore, is proposed to modulate cell : matrix interaction to allow for cell migration when necessary. An intermediate stage of adhesion describes cell responses to counteradhesive matricellular proteins and, as such, are predicted to promote cell motility (Murphy-Ullrich, 2001). Accordingly, thrombospondin-1 expression is observed during events such as dermal wound repair, when cell movement is required. Thrombospondin-1 is also expressed by many tumor cells and might facilitate metastatic migration (Tusczynski and Nicosia, 1996). In fact, metastatic human carcinoma cells transfected with antisense thrombospondin-1 cDNA to reduce thrombospondin-1 protein expression lose their capacity to proliferate and metastasize in athymic mice (Castle *et al.*, 1991).

## III. THROMBOSPONDIN-2

Similar to thrombospondin-1, thrombospondin-2 inhibits angiogenesis *in vivo*. Recombinant thrombospondin-2 contained in pellets implanted in the rat cornea inhibits basic-fibroblast growth factor (bFGF)–induced vessel invasion to nearly the same degree as thrombospondin-1 (Volpert *et al.*, 1995). Thus, the observation that the thrombospondin-2-null mouse appears to have a greater degree of vascularization in the skin is in accordance with the hypothesis that thrombospondin-2 might serve as an endogenous regulator of angiogenesis in mice (Kyriakides *et al.*, 1998). In fact, the expression pattern of thrombospondin-2 is more consistent with the importance of this protein in vascular formation, because the mRNA for thrombospondin-2 is more closely associated with the vasculature in developing tissues than that of thrombospondin-1 (Armstrong and Bornstein, 2003).

How might thrombospondin-2 mediate inhibition of angiogenesis? Abrogation of thrombospondin-2 expression results in increased amounts of matrix metalloproteinase (MMP)-2 activity (Yang *et al.*, 2000). Thrombospondin-2 binds to MMP-2 and facilitates uptake by scavenger receptors on cell surfaces (Yang *et al.*, 2001). Hence without thrombospondin-2, MMP-2 levels accumulate in the extracellular milieu and result in significant decreases in cell adhesion. The increase in MMP-2 levels is correlated with decreased amounts of tissue transglutaminase on thrombospondin-2-null cell surfaces (Agah *et al.*, 2005). Because transglutaminase is known to enhance integrin-mediated cell attachment and serve as a substrate of MMP-2, the increase in MMP-2 activity is proposed to elicit decreased cell adhesion in the absence of thrombospondin-2 by diminishment of the amounts of transglutaminase on cell surfaces. Hence, blood vessel formation appears to be sensitive to levels of MMP-2 in tissues. Likewise, tissue transglutaminase on cell surfaces might also contribute to angiogenesis.

Another observation from the thrombospondin-2-null mouse is that of altered collagen fibrillogenesis in the skin, relative to that seen in wild-type mice. The collagen fibrils in the null mice are larger in diameter, have aberrant contours, and are disordered (Kyriakides *et al.*, 1998). Presumably this effect on collagen fibril formation contributes to the reduced tensile strength of the skin in thrombospondin-2-null versus wild-type mice. An intriguing relationship between aberrant collagen fibril morphology and decreased transglutaminase activity is suggested by the thrombospondin-2-null phenotype.

Thrombospondins-1 and -2 can act as negative regulators of cell growth. In particular, endothelial cells are susecptible to an inhibition of proliferation by both proteins, resulting in their classification as inhibitors of angiogenesis (Armstrong and Bornstein, 2003). However, the variety of cell surface receptors for thrombospondins allows for diverse signaling events in different cell types; consequently, there might be situations when thrombospondin appears to support angiogenesis as well (Nicosia and Tusczynski, 1994). Thrombospondin-1 has been shown to modulate the activity of at least two cytokines, TGF-β and hepatocyte growth factor (HGF): diminished activity, in the case of HGF, or enhanced activity, as seen for TGF-β. Finally, thrombospondin-1 can alter cell shape to promote a migratory phenotype but can also inhibit migration. Such conflicting conclusions illustrate the importance of contextual presentation of matricellular proteins in various assays. Further characterization of the thrombospondins, including closer examination of the remaining family members (thrombospondins 3–5), will no doubt yield fascinating insight into this multifunctional gene family.

## IV. TENASCIN-C

Tenascin-C is a matricellular protein with a widespread pattern of developmental expression, in comparison to a restricted pattern in adult tissues. In addition to tenascin-C, three other, less characterized forms of tenascin have been identified: tenascin-R, tenascin-X, and tenascin-W (Chiquet-Ehrismann, 2004). This review focuses on tenascin-C, as the best characterized of the tenascin gene family. Tenascin-C consists of six subunits (or arms) linked by disulfide bonds to form a 2000-kDa molecule that can associate with fibronectin in the extracellular matrix (Hsai and Schwarzbauer, 2005). Like thrombospondin-1, a number of different functions have been attributed to tenascin-C, and, accordingly, a number of cell surface receptors appear to mediate distinct properties of this matricellular protein. At least five different integrins are known receptors for tenascin-C, as well as annexin II (Crossin, 1996). Whereas the integrins appear to support cell adhesion to tenascin-C, annexin II is thought to mediate the counteradhesive function attributed to this protein. Hence, tenascin-C can act as either an adhesive or a counteradhesive substrate for different cell types, dependent on the profile of receptors expressed on the cell surface (Crossin, 1996).

Tenascin-C has also been shown to modulate the activity of growth factors. Specifically, it promotes epidermal growth factor (EGF)–dependent and bFGF-dependent cell growth (End *et al.*, 1992; Jones *et al.*, 1997). In fact, Jones *et al.* (1997) have shown that smooth muscle cells plated in a collagen gel secrete MMPs that degrade the collagen to expose integrin receptor–binding sites. Engagement of these receptors induces tenascin-C expression; tenascin-C is subsequently deposited into the extracellular matrix and can itself serve as an integrin ligand. The deposition of tenascin-

C leads to cell shape changes initiated by a redistribution of focal adhesion complexes concomitant with a clustering of EGF receptors on the cell surface. Presumably, clustering of the EGF receptors facilitates EGF signaling and thereby enhances the mitogenic effect of EGF. Conversely, when MMP activity is inhibited, tenascin-C expression is decreased and the cells become apoptotic (Jones *et al.*, 1997). Thus, tenascin-C is able to modulate EGF activity such that the presence of this matricellular protein supports cell growth and its absence induces programmed cell death.

Given the widespread expression of tenascin-C in the developing embryo, the lack of an overt phenotype in the tenascin-C-null mouse is surprising (Chiquet-Ehrismann and Tucker, 2004). In particular, the high level of tenascin-C expression in the central and peripheral nervous systems had indicated that the absence of this protein might lead to neuronal abnormalities. Although no histological differences could be detected in the brains of adult tenascin-C-null mice, they displayed behavioral aberrations, including reduced anxiety and enhanced novelty-induced activity (Steindler *et al.*, 1995; Morellini and Schachner, 2006). In addition, altered numbers of embryonic central nervous system stem cells were noted in the absence of tenascin-C expression, an observation confirming that the composition of the extracellular matrix is an important factor in cell differentiation (Garcion *et al.*, 2004).

The genetic background of the tenascin-C-null mouse is likely to be a major factor in the identification of tissues in which tenascin-C might be functionally important. For example, Nakao *et al.* (1998) used three different congenic mouse lines to study the effect of Habu-snake venom-induced glomerulonephritis in a tenascin-C-null background. Although the disease is worse in all tenascin-C-null mice than in wild-type controls, each line exhibits a different level of severity. Induction of the disease in one strain, GRS/A, results in death from irreversible renal failure (Nakao *et al.*, 1998). Moreover, mesangial cells cultured from tenascin-C-null animals do not respond to cytokines such as platelet-derived growth factor (PDGF) unless exogenous tenascin-C is included in the culture medium (Nakao *et al.*, 1998). Hence, tenascin-C can also modulate the activity of this growth factor, as observed previously for EGF. Tenascin-C provides another example of a matricellular protein able to affect growth factor efficacy.

Although tenascin-C shows a limited pattern of expression in the adult, an induction of tenascin-C is seen in many tissues undergoing wound repair or neoplasia (Chiquet-Ehrismann and Tucker, 2004). Thus tenascin-C, like the other matricellular proteins, is ideally suited to act as a modulator of cell shape, migration, and growth. One mechanism by which tenascin-C influences cell behavior is through the modulation of fibronectin interaction with cells. Tenascin-C decreases cell adhesion to fibronectin through competition with a heparan sulfate proteoglycan, syndecan 4 (Midwood *et al.*, 2004). Syndecan 4 is required for efficient cell attachment to fibronectin and for tenascin-C inhibition of adhesion to fibronectin. Hence, a scenario in which tenascin-C competes for syndecan 4 binding to fibronectin is consistent with these results.

The paucity of developmental abnormalities manifested in the tenascin-C-null mouse points to the greater importance of tenascin-C in remodeling events that take place in response to injury or transformation. For example, tenascin-C has been shown to regulate cardiac neovascularization by bone marrow–derived endothelial progenitor cells in response to angiogenic stimuli in adult mice (Ballard *et al.*, 2006). In addition, high levels of tenascin-C expression are frequently associated with several types of malignancies including tumors of the brain (Chiquet-Ehrismann and Tucker, 2004).

## V. OSTEOPONTIN

As the name implies, osteopontin was originally classified as a bone protein. A more thorough examination, however, reveals a widespread expression pattern for this protein, with multiple potential functions (Rangaswami *et al.*, 2006). Osteopontin associates with the extracellular matrix, for it binds to fibronectin and to collagens I, II, III, IV, and V. Osteopontin also affects cellular signaling pathways by virtue of its capacity to act as a ligand for multiple integrin receptors as well as cluster designation (CD)-44 (Rangaswami *et al.*, 2006). Thus osteopontin, like most of the matricellular proteins, is able to act as a bridge between the extracellular matrix and the cell surface. Since matricellular proteins might be synthesized, secreted, and incorporated into the extracellular matrix with greater ease than more complex secreted proteins, which must be incorporated into fibrils and assembled into a network, a bridging function might be useful during remodeling events in the organism when rapid conversion of the cellular substrata is required for cell movement.

In support of this concept, Weintraub *et al.* (1996) report that transfection of vascular smooth muscle cells with anti-osteopontin cDNA reduces adhesion, spreading, and invasion of three-dimensional collagen matrices. Addition of osteopontin to the collagen gel restores the capacity of these cells to invade the gel. Osteopontin also appears to be susceptible to modification by extracellular proteases, which have revealed cryptic integrin-binding sites within the sequence. The protease thrombin cleaves osteopontin at the N-terminal domain and exposes binding sites for integrin $\alpha9\beta1$ and $\alpha4\beta1$ (Smith *et al.*, 1996). In addition, Senger *et al.* (1996) report that endothelial cells treated with vascular endothelial growth factor (VEGF) increase their expression of the integrin $\alpha v\beta3$, another osteopontin cell surface receptor, concomitantly with an increase in osteopontin. These investigators also show an increase in the amount of thrombin-processed osteopontin in tissues injected with VEGF and radiolabeled osteopontin. The significance of this result lies in the enhanced support of endothelial cell migration *in vitro* by thrombin-cleaved versus full-length protein. Since the migratory activity of the endothelial cells is blocked partially by an anti-αvβ3 antibody, the remainder of the activity might be attributed to β1-integrins or to other cell surface receptors (Senger *et al.*, 1996).

The capacity of osteopontin to influence cell migration might be linked to matrix metalloproteinase activity. Osteopontin is a member of the recently classified SIBLING (small integrin-binding ligand N-linked glycoprotein) family of proteins (Fisher and Fedarko, 2003), which have been shown to bind to and regulate the activity of MMPs. Osteopontin binds to recombinant proMMP-3 and active MMP-3 (Fedarko *et al.*, 2004). In addition, a decrease in MMP-9 activity has been reported in osteopontin-null myofibroblasts and vascular smooth muscle cells (Lai *et al.*, 2006).

The promotion of cell survival is another property ascribed to osteopontin. Denhardt and Noda (1998) have reported that human umbilical vein endothelial cells plated in the absence of growth factors will undergo apoptosis. If these cells are plated on an osteopontin substrate, however, apoptosis is inhibited. Furthermore, rat aortic endothelial cells subjected to serum withdrawal undergo programmed cell death, a response inhibited by an osteopontin substratum. In fact, it is the ligation of integrin αvβ3 by osteopontin at the cell surface that induces nuclear factor-kappa B (NF-κB), a transcription factor that controls a variety of genes through direct binding to their promoters. Thus osteopontin and other αvβ3 ligands can protect cells from apoptosis (Scatena *et al.*, 1998).

Osteopontin is involved in inflammatory responses. Expression of osteopontin is found to increase during intradermal macrophage infiltration, and purified osteopontin injected into the rat dermis leads to an increase in the number of macrophages at the site of admininstration. Importantly, antiosteopontin antibodies inhibit macrophage accumulation in a rat intradermal model after a potent macrophage chemotactic peptide is used to induce an inflammatory response (Giachelli *et al.*, 1998). Osteopontin expression is required for development of an effective T<sub>H</sub>1 immune response (Ashkar *et al.*, 2000). Interestingly, a recently described intracellular form of osteopontin has been implicated as the mediator of interferon-α production in plasmacytoid dendritic cells (Shinohara *et al.*, 2006).

The phenotype of the osteopontin-null mouse supports the hypothesis that osteopontin affects macrophage activity. Although the number of macrophages does not appear to differ significantly in incisional wounds of wild-type and osteopontin-null mice, the amount of cell debris was higher in wounds of the latter animals. Because macrophages are thought to be primary mediators of wound debridement, osteopontin could be important in the regulation of macrophage function (Liaw *et al.*, 1995). Collagen fibril formation in the deeper dermal layers of the wounded osteopontin-null mice also appear to be affected. Osteopontin-null mice have smaller collagen fibrils than wild-type controls. Similar to thrombospondin-2, osteopontin might affect collagen fibrillogenesis, especially at wound sites, for no differences are seen in the size of collagen fibrils in unwounded skin (Liaw *et al.*, 1995).

An additional function of osteopontin as a negative regulator of calcification is supported by a number of different studies (Giachelli *et al.*, 2005). Osteopontin inhibits apatite crystal formation *in vitro* and facilitates resorption of cellular minerals. As shown in an *in vivo* model of ectopic calcification, the capacity of osteopontin to abrogate calcification is dependent on phosphorylation of osteopontin and the presence of the integrin-binding RGD sequence in osteopontin (Giachelli *et al.*, 2005).

Osteopontin and its repertoire of cell surface receptors represent components of a pathway used by cells in need of rapid movement or migration. In addition, the ligation of osteopontin by certain cell surface receptors supports cell survival and thus provides a mechanism for a given tissue to protect a subset of cells, expressing the appropriate receptor, from apoptosis. Hence, osteopontin is a useful protein in events that require cell movement and cell survival, such as wound healing and angiogenesis, and has the potential to contribute to cancer progression (Rangaswami *et al.*, 2006).

## VI. SPARC

SPARC (secreted protein acidic and rich in cysteine; BM-40; osteonectin) was first identified as a primary component of bone but has since been shown to have a wider distribution (Brekken and Sage, 2001). Increased expression of SPARC is observed in many tissues undergoing different types of remodeling. For example, SPARC is found in the gut epithelium, which normally exhibits rapid turnover, and in healing wounds. An increase in SPARC is observed during glomerulonephritis and liver fibrosis and in association with many different tumors. Like the other matricellular proteins, SPARC interacts with the extracellular matrix by binding to collagens I, III, IV, and V (Brekken and Sage, 2001).

SPARC has also been shown to bind to a variety of growth factors present in the extracellular space. For example, SPARC binds to PDGF-AB and -BB and prevents their interaction with PDGF cell surface receptors. PDGF-stimulated mitogenesis is inhibited by the addition of SPARC (Brekken and Sage, 2001). SPARC can also prevent VEGF-induced endothelial cell growth, for it binds directly to the growth factor and thereby prevents VEGF receptor stimulation of the mitogen-activated protein kinases Erk-1 and Erk-2 (Kupprion *et al.*, 1998). *In vivo*, regulation of VEGF activity by SPARC via VEGF receptor 1 occurs during choroidal neovascularization after injury (Nozaki *et al.*, 2006).

Interestingly, a variety of mitogenic stimulators are inhibited by SPARC in culture, some of which do not necessarily associate physically with this protein. For example, SPARC is thought not to bind to bFGF, but SPARC inhibits bFGF-stimulated endothelial cell cycle progression (Brekken and Sage, 2001). Apparently, the effect of SPARC on cell proliferation is complex and could occur through (1) a direct prevention of receptor activation and/or (2) a pathway mediated by a cell surface receptor recognizing SPARC. To date, cell surface receptors for SPARC include stabilin-1 on macrophages and VCAM-1 on leukocytes (Kzhyshkowska *et al.*, 2006; Kelly *et al.*, 2007). Whether there are other cell surface receptors for SPARC on different cell types remains to be determined.

Another significant effect of SPARC on cells in culture is its capacity to elicit changes in cell shape. Many cell types plated on various substrata retract their filopodia and lamellopodia and assume a rounded phenotype after exposure to SPARC. Bovine aortic endothelial cells, for example, are prevented from spreading in the presence of SPARC (Brekken and Sage, 2001). Clearly, cell rounding could contribute to the inhibition of cell cycle mentioned previously; however, these two effects of SPARC appear to be independent. Motamed and Sage (1998) have shown that inhibition of tyrosine kinases reverses the counteradhesive function of SPARC but has no effect on cell-cycle inhibition. Thus, at least in endothelial cells, SPARC appears to mediate two aspects of cell behavior through different mechanisms.

Recently the targeted inactivation of the SPARC gene in mice has allowed a more thorough examination of SPARC activity. In support of SPARC as a regulator of cell proliferation, primary mesenchymal cells isolated from SPARC-null mice proliferate faster than their wild-type counterparts (Bradshaw *et al.*, 1999). The majority of SPARC-null phenotypic abnormalities, however, are seen as aberrant extracellular matrix assembly. Early-onset cataractogenesis reported in two independently-generated SPARC-null mice appears to be caused by inappropriate basement membrane synthesis by lens epithelial cells (Brekken and Sage, 2001). In the absence of SPARC, collagen IV, a SPARC ligand, is not localized to the outer border of the lens capsule, in contrast to its distribution in wild-type capsules (Yan *et al.*, 2003). In skin, collagen fibrils are small and uniform in diameter in comparison to those of wild-type dermis (Bradshaw *et al.*, 2003). In fact, fibrosis in SPARC-null mice in response to a variety of stimuli is decreased in several tissues, including lung, kidney, and skin. Hence a growing body of evidence supports a function of SPARC in assembly and stability of fibrillar collagen and basement membranes.

SPARC interaction with collagen I was recently investigated by atomic force microscopy (Wang *et al.*, 2005). SPARC bound to collagen I is predicted to interfere with integrin engagement of collagen I. Similar to inhibition of fibronectin binding by tenascin-C and consistent with a counteradhesive activity, production of SPARC might be induced by cells to modulate interaction of integrin receptors with collagens.

In addition to its effect on cell cycle, recent evidence shows that SPARC influences TGF-β-dependent pathways in some cell types. Mesangial cells isolated from SPARC-null mice show a decrease in collagen I expression, accompanied by a decrease in the levels of the cytokine TGFβ-1, in comparison to wild-type cells (Francki *et al.*, 1999). TGF-β is a known positive regulator of extracellular matrix synthesis. Addition of recombinant SPARC to SPARC-null mesangial cultures restores the levels of collagen I and TGF-β nearly to those of wild-type cells (Francki *et al.*, 1999). SPARC also promotes the TGF-β signal transduction pathway in epithelial cells (Schiemann *et al.*, 2003). Thus it appears that SPARC can act as a regulator of TGF-β activity and, perhaps by extension, influences collagen I production.

Once again, we see SPARC as an example of a multifunctional protein able to regulate cell shape and modulate growth factor activity. In addition, SPARC appears to be a key factor in extracellular matrix assembly in both basal lamina and interstitium. Consistent with these functions, a homolog of SPARC termed *hevin* was recently shown to be counteradhesive and modulatory of extracellular matrix structure (Sullivan and Sage, 2004).

## VII. CONCLUSIONS

In addition to the proteins described here, others are potential candidates for inclusion in the family of matricellular proteins. These include small proteoglycans, such as the matrix-associated protein decorin, which has been shown to be an endogenous regulator of TGF-β activity (Border *et al.*, 1992), as well as certain members of the recently classified CCN gene family (Brigstock *et al.*, 2003). The CCN proteins are secreted and modular and exhibit functions based, at least in part, on integrin-mediated mechanisms. Whereas CCN1 (CYR 61) enhances apoptosis in fibroblasts, endothelial cell adhesion to CCN1 promotes cell survival (Todoravic *et al.*, 2005). Further analysis of these proteins and their actions will expand our comprehension of matricellular proteins and the functions they serve in regulating cell interaction with components of their immediate environment.

In any cellular environment, numerous extracellular signals are in place to control cell behavior. In adult tissues, injury and disease both lead to a wide-scale release of multiple factors, either secreted from cells or sequestered in the extracellular matrix, that are capable of eliciting potent cellular responses. Appropriately, an increase in the expression of many matricellular proteins is associated with pathological events. Hence, the matricellular proteins appear to be ideally suited to act as modulators of these extracellular signals. They are able to serve as bridges between the extracellular matrix and cell surface receptors, such that cell shape changes or cell movements can be initiated prior to matrix breakdown or synthesis. At least three matricellular proteins, thrombospondin-2, osteopontin, and SPARC, appear to participate in matrix synthesis, through either the promotion of collagen fibrillogenesis or the enhancement of collagen production. Matricellular proteins can either inhibit or potentiate specific growth factor signal transduction pathways. Thus, different growth factor effects may be amplified or subdued by the presence or absence of these proteins. The fact that multiple receptors exist for most of the matricellular proteins allows for diverse functional consequences for different cell types in a complex tissue in response to a matricellular ligand. A given repertoire of cell surface receptors can stimulate (or inhibit) a particular pathway in a cell-type-dependent manner. Apparently, evolution has fine-tuned these proteins to serve as specialized mediators of extracellular signals that provide a coordinated, efficient resolution of tissue injury.

## VIII. REFERENCES

Agah, A., Kyriakides, T. R., and Bornstein, P. (2005). Proteolysis of cell-surface tissue transglutaminase by matrix metalloproteinase-2 contributes to the adhesive defect and matrix abnormalities in thrombospondin-2-null fibroblasts and mice. *Am. J. Pathol.* **167**, 81–88.

Armstrong, L. C., and Bornstein P. (2003). Thrombospondins 1 and 2 function as inhibitors of angiogenesis. *Matrix Biol.* **22**, 63–71.

Ashkar, S., Weber, G. F., Panoutsakopoulou, V., Sanchirico, M. E., Jansson, M., Zawaideh, S., Rittling, S. R., Denhardt, D. T., Glimcher, M. J., and Cantor, H. (2000). Eta-1 (osteopontin): an early component of type-1 (cell-mediated) immunity. *Science* **287**, 860–864.

Ballard, V. L. T., Sharma, A., Duignan, I., Holm, J. M., Chin, A., Choi, R., Hajjar, K. A., Wong, S. -C., and Edelberg, J. M. (2006). Vascular tenascin-C regulates cardiac endothelial phenotype and neovascularization. *FASEB J.* **20**, 717–719.

Border, W. A., Noble, N. A., Yamamoto, T., Harper, J. R., Yamaguchi, Y., Pierschbacher, M. D., and Ruoslahti, E. (1992). Natural inhibitor of transforming growth factor-β protects against scarring in experimental kidney disease. *Nature* **360**, 361–364.

Bornstein, P. (2001). Thrombospondins as matricellular modulators of cell function. *J. Clin. Invest.* **107**, 929–934.

Bornstein, P., and Sage, E. H. (2002). Matricellular proteins: extracellular modulators of cell function. *Curr. Opin. Cell Biol.* **14**, 608–616.

Bradshaw, A. D., Francki, A., Motamed, K., Howe, C., and Sage, E. H. (1999). Primary mesenchymal cells isolated from SPARC-null mice exhibit altered morphology and rates of proliferation. *Mol. Biol. Cell* **10**, 1569–1579.

Bradshaw, A. D., Puolakkainen, P., Dasgupta, J., Davidson, J. M., Wight, T. N., and Sage, E. H. (2003). SPARC-null mice display abnormalities in the dermis characterized by decreased collagen fibril diameter and reduced tensile strength. *J. Invest. Dermatol.* **120**, 949–955.

Brekken, R. A., and Sage, E. H. (2001). SPARC, a matricellular protein: at the crossroads of cell-matrix communication. *Matrix Biol.* **19**, 816–827.

Brigstock, D. R., Goldschmeding, R., Katsube, K. I., Lam, S. C., Lau, L. F., Lyons, K., Naus, C., Perhal, B., Riser, B., Takigawa, M., and Yeger, H. (2003). Proposal for a unified CCN nomenclature. *Mol. Pathol.* **56**, 127–128.

Castle, V., Varani, J., Fligiel, S., Prochownik, E. V., and Dixit, V. (1991). Anti-sense-mediated reduction in thrombospondin reverses the malignant phenotype of a human squamous carcinoma. *J. Clin. Invest.* **6**, 1883–1888.

Chiquet-Ehrismann, R. (2004). Molecules in focus: tenascins. *Int. J. Biochem. Cell Biol.* **36**, 986–990.

Chiquet-Ehrismann, R., and Tucker, R. (2004). Connective tissues: signaling by tenascins. *Int. J. Biochem. Cell Biol.* **36**, 1085–1089.

Crawford, S. E., Stellmach, V., Murphy-Ullrich, J. E., Ribeiro, S. M. F., Lawler, J., Hynes, R. O., Boivin, G. P., and Bouck, N. (1998). Thrombospondin-1 is a major activator of TGF-β *in vivo*. *Cell* **93**, 1159–1170.

Crossin, K. L. (1996). Tenascin: a multifunctional extracellular matrix protein with a restricted distribution in development and disease. *J. Cell. Biochem.* **61**, 592–598.

de Fraipont, F., Nicholson, A. C., Feige, J. J., and Van Meir, E. G. (2001). Thrombospondins and tumor angiogenesis. *Trends Mol. Med.* **7**, 401–407.

Denhardt, D. T., and Noda, M. (1998). Osteopontin expression and function: role in bone remodeling. *J. Cell. Bioch.* **30/31**(Suppl.), 92–102.

DiPietro, L. A., Negben, D. R., and Polverini, P. J. (1994). Down-regulation of endothelial cell thrombospondin 1 enhances *in vitro* angiogenesis. *J. Vasc. Res.* **31**, 178–185.

End, P., Panayotou, G., Entwhistle, A., Waterfield, M. D., and Chiquet, M. (1992). Tenascin: a modulator of cell growth. *Eur. J. Biochem.* **209**, 1041–1051.

Fedarko, N. S., Jain, A., Karadag, A., and Fisher, L. W. (2004). Three small integrin binding ligand N-linked glycoproteins (SIBLINGs) bind and activate specific matrix metalloproteinases. *FASEB J.* **18**, 734–736.

Fisher, L. W., and Fedarko, N. S. (2003). Six genes expressed in bones and teeth encode the current members of the SIBLING family of proteins. *Connect. Tissue Res.* **44**, 33–40.

Francki, A., Bradshaw, A. D., Bassuk, J. A., Howe, C. C., Couser, W. G., and Sage, E. H. (1999). SPARC regulates the expression of collagen type I and transforming growth factor-beta1 in mesangial cells. *J. Biol. Chem.* **274**(45), 32145–32152.

Garcion, E., Halilagic, A., Faissner, A., and ffrench-Constant, C. (2004). Generation of an environmental niche for neural stem cell development by the extracellular matrix molecule tenascin C. *Development* **131**, 3423–3432.

Giachelli, C. M., Lombardi, D., Johnson, R. J., Murry, C. E., and Almeida, M. (1998). Evidence for a role of osteopontin in macrophage infiltration in response to pathological stimuli *in vivo*. *Am. J. Pathol.* **152**, 353–358.

Giachelli, C. M., Speer, M. Y., Li, X., Rajachar, R. M., and Yang, H. (2005). Regulation of vascular calcification: roles of phosphate and osteopontin. *Circ. Res.* **96**, 717–722.

Hsai, H. C., and Schwarzbauer, J. E. (2005). Meet the tenascins: multifunctional and mysterious. *J. Biol. Chem.* **280**, 26641–26644.

Jones, P. L., Crack, J., and Rabinovitch, M. (1997). Regulation of tenascin-C, a vascular smooth muscle cells survival factor that interacts with αvβ3 integrin to promote epidermal growth factor receptor phosphorylation and growth. *J. Cell Biol.* **139**, 279–293.

Kelly, K. A., Allport, J. R., Yu, A. M., Sinh, S., Sage, E. H., Gerszten, R. E., and Weissleder, R. (2007). SPARC is a VCAM-1 counter-ligand that mediates leukocyte transmigration. *J. Leukoc. Biol.* **81**, 748–756.

Kupprion C., Motamed, K., and Sage, E. H. (1998). SPARC (BM-40, osteonectin) inhibits the mitogenic effect of vascular endothelial growth factor on microvascular endothelial cells. *J. Biol. Chem.* **273**, 29635–29640.

Kyriakides, T. R., Zhu, Y-H., Smith, L. T., Bain, S. D., Yang, Z., Lin, M. T., Danielson, K. G., Iozzo, R. V., LaMarca, M., McKinney, C. E., Ginns, E. I., and Bornstein, P. (1998). Mice that lack thrombospondin 2 display connective tissue abnormalities that are associated with disordered collagen fibrillogenesis, an increased vascular density, and a bleeding diathesis. *J. Cell Biol.* **140**, 419–430.

Kzhyshkowska, J., Workman, G., Cardo-Vila, M., Arap, W., Pasqualini, R., Gratchev, A., Krusell, L., Goerdt, S., and Sage, E. H. (2006). Novel function of alternatively activated macrophages: stabilin-1-mediated clearance of SPARC. *J. Immunol.* **176**, 5825–5832.

Lai, C. F., Seshadri, V., Huang, K., Shao, J. S., Cai, J., Vattikuti, R., Schumacher, A., Loewy, A. P., Denhardt, D. T., Rittling, S. R., and Towler, D. A. (2006). An Osteopontin-NADPH oxidase signaling cascade promotes pro-matrix metalloproteinase 9 activation in aortic mesenchymal cells. *Circ Res.* **98**, 1479–1489.

Liaw, L., Skinner, M. P., Raines, E. W., Ross, R., Cheresh, D. A., and Schwartz, S. M. (1995). The adhesive and migratory effects of osteopontin are mediated via distinct cell surface integrins. *J. Clin. Invest.* **95**, 713–724.

Midwood, K., Valenick, L. V., Hsia, H. C., and Schwarzbauer, J. E. (2004). Coregulation of fibronectin signaling and matrix contraction by tenascin-C and syndecan-4. *Mol. Biol. Cell* **15**, 5670–5677.

Morellini, F., and Schachner, M. (2006). Enhanced novelty-induced activity, reduced anxiety, delayed resynchronization to daylight reversal and weaker muscle strength in teanscin-C-deficient mice. *Eur. J. Neurosci.* **23**, 1255–1268.

Motamed, K., and Sage, E. H. (1998). SPARC inhibits endothelial cell adhesion but not proliferation through a tyrosine phosphorylation-dependent pathway. *J. Cell. Bioch.* **70**, 543–552.

Murphy-Ullrich, J. E. (2001). The de-adhesive activity of matricellular proteins: is intermediate cell adhesion an adaptive state? *J. Clin. Invest.* **107**, 785–790.

Nakao, N., Hiraiwa, N., Yoshiki, A., Ike, F., and Kusakabe, M. (1998). Tenascin-C promotes healing of habu-snake-venom-induced glomerulonephritis. *Am. J. Pathol.* **152**, 1237–1245.

Nozaki, M., Sakurai, E., Raisler, B. J., Baffi, J. Z., Witta, J., Ogura, Y., Brekken, R. A., Sage, E. H., Ambati, B. K., and Ambati, J. (2006). Loss of SPARC-mediated VEGFR-1 suppression after injury reveals a novel anti-angiogenic activity of VEGF-A. *J. Clin. Invest.* **116**, 422–429.

Rangaswami, H., Bulbule, A., and Kundu, G. C. (2006). Osteopontin: role in cell signaling and cancer progression. *Trends Cell Biol.* **16**, 79–87.

Scatena, M., Almeida, M., Chaisson, M. L., Fausto, N., Nicosia, R. F., and Giachelli, C. M. (1998). NF-κB mediates αvβ3 integrin-induced endothelial cell survival. *J. Cell Biol.* **141**, 1083–1093.

Schiemann, B. J., Neil, J. R., and Schiemann, W. P. (2003). SPARC inhibits epithelial cell proliferation in part through stimulation of the transforming growth factor-beta-signaling system. *Mol. Biol. Cell* **14**, 3977–3988.

Schultz-Cherry, S., and Murphy-Ullrich, J. E. (1993). Thrombospondin causes activation of latent transforming growth factor-beta secreted by endothelial cells by a novel mechanism. *J. Cell Biol.* **122**, 923–932.

Senger, D. R., Ledbetter, S. R., Claffey, K. P., Papdopoulos-Serfgiou, A., Perruzzi, C. A., and Detmar, M. (1996). Stimulation of endothelial cell migration by vascular permeability factor/vascular endothelial growth factor through the cooperative mechanisms involving the αvβ3 integrin, osteopontin and thrombin. *Am. J. Pathol.* **149**, 293–305.

Shinohara, M. L., Lu, L., Bu, J., Werneck, M. B., Kobayashi, K. S., Glimcher, L. H., and Cantor, H. (2006). Osteopontin expression is essential for interferon-alpha production by plasmacytoid dendritic cells. *Nat. Immunol.* **7**, 498–506.

Smith, L. L., Cheung, H.-K., Ling, L. E., Chen, J., Sheppard, D., Pytela, R., and Giachelli, C. M. (1996). Osteopontin N-terminal domain contains a cryptic adhesive sequence recognized by α9β1 integrin. *J. Biol. Chem.* **45**, 28485–28491.

Sullivan, M. M., and Sage, E. H. (2004). Hevin/SC1, a matricellular glycoprotein and potential tumor suppressor of the SPARC/BM-40/Osteonectin family. *Int. J. Biochem. Cell Biol.* **36**, 991–996.

Steindler, D. A., Settles, D., Erickson, H. P., Laywell, E. D., Yoshiki, A., Faissner, A., and Kusakabe, M. (1995). Tenascin knockout mice: barrels, boundary, molecules and glial scars. *J. Neurosci.* **15**, 1971–1983.

Todorovic, V., Chen, C. C., Hay, N., and Lau, L. F. (2005). The matrix protein CCN1 (CYR61) induces apoptosis in fibroblasts. *J. Cell Biol.* **171**, 559–568.

Tusczynski, G. P., and Nicosia, R. F. (1996). The role of thrombospondin-1 in tumor progression and angiogenesis. *Bioessays* **18**, 71–76.

Volpert, O. V., Tolsma, S. S., Pellerin, S., Feige, J.-J., Chen, H., Mosher, D. F., and Bouck, N. (1995). Inhibition of angiogenesis by thrombospondin-2. *Biochem. Biophys. Res. Comm.* **217**, 326–332.

Wang, H., Fertala, A., Ratner, B. D., Sage, E. H., and Jiang, S. (2005). Identifying the SPARC binding sites on collagen I and procollagen I by atomic force microscopy. *Anal. Chem.* **77**, 6765–6771.

Weintraub, A. S., Giachelli, C. M., Krauss, R. S., Almeida, M., and Taubman, M. B. (1996). Autocrine secretion of osteopontin by vascular smooth muscle cells regulates their adhesion to collagen gels. *Am. J. Pathol.* **149**, 259–272.

Yan, Q., Blake, D., Clark, J. I., and Sage, E. H. (2003). Expression of the matricellular protein SPARC in murine lens: SPARC is necessary for the structural integrity of the capsular basement membrane. *J. Histochem. Cytochem.* **51**, 503–511.

Yang, A., Kyriakides, T. R., and Bornstein, P. (2000). Matricelluar proteins as modulators of cell-matrix interactions: the adhesive defect in thrombospondin 2–null fibroblasts is a consequence of increased levels of matrix metalloproteinase-2. *Mol. Biol. Cell* **11**, 3353–3364.

Yang, Z., Strickland, D. K., and Bornstein P. (2001). Extracellular MMP2 levels are regulated by the LRP scavenger receptor and thrombospondin 2. *J. Biol. Chem.* **276**, 8403–8408.

# Chapter Fourteen

# Growth Factors

*Thomas F. Deuel and Yunchao Chang*

## I. INTRODUCTION

Tissue remodeling is an essential component in the process of wound healing and thus in the normal maintenance and survival of all organisms. Tissue remodeling occurs throughout the entire temporal span of injury and repair. However, although abundant evidence using histologic methods has accumulated to detail the sequence of appearance of cell types that characterize wound healing, the mechanisms that initiate and sustain the process remain to be established (Ross, 1968; Deuel et al., 1991; Pierce and Mustoe, 1995). What has become increasingly clear from analysis of properties of growth factors and other cytokines is their importance in the initiation and propagation of the processes that begin with injury and end with a healed wound. These factors and their receptors dictate cell type and tissue specificity of response, and the regulation of their genes at the level of transcription has established the time course of their expression. Their properties are those needed in the wound-healing process.

Because growth factors applied directly to wounds accelerate the normal process of wound healing *in vivo*, it seems likely that the properties described for the growth factors *in vitro* are recapitulated *in vivo* (Grotendorst et al., 1985; Paulsson et al., 1987; Sprugel et al., 1987; Pierce et al., 1988, 1989, 1992; Terracio et al., 1988; Greenhalgh et al., 1990; Hart et al., 1990; Quaglino et al., 1990; Shaw et al., 1990; Antoniades et al., 1991; Deuel et al., 1991; Mustoe et al., 1991; Nagaoka et al., 1991; Pepper et al., 1991; Soma et al., 1992; Reuterdahl et al., 1993; Sundberg et al., 1993; Pierce and Mustoe, 1995). It is also important to recognize that properties of growth factors may be important in the development of abnormalities in the vascular wall, such as atherosclerosis, and in tumor growth, suggesting that deregulation of growth factors and other cytokines may have serious consequences for the host and that tight regulation of their activities is needed for normal function (Broadley, 1989; Golden, 1991; D'Amore, 1993; Forsberg, 1993). Compelling evidence is therefore available to indicate

*Principles of Tissue Engineering, 3rd Edition*
*ed. by Lanza, Langer, and Vacanti*

the importance of growth factors and other cytokines in the wound-healing process. However, the specific roles of these endogenous factors remain to be established in normal wound healing, and the signals that regulate the sites and time of their expression are little understood.

This chapter provides an overview of roles of the platelet-derived growth factor (PDGF) and other cytokines as models of factors that initiate and propagate a number of the normal processes required for the orderly progression of inflammation, tissue remodeling, and repair.

PDGF was initially described as a potent mitogen. However, PDGF directs cell migration, activates diverse cell types, and up-regulates expression of genes that otherwise are quiescent and not expressed at significant levels. This last property of PDGF appears to establish gene activation pathways that are unique to different cell types. It provides a temporal sequence of expression of different pathways that may be of central importance in tissue remodeling and repair, a property required of a signaling molecule in the wound-healing process. Experiments in which PDGF has been directly applied to wounds have shown that PDGF accelerates the normal healing process; it does not distort it, indicating that PDGF can up-regulate genes needed for wound healing in a precise order. However, there is little if any information to establish the relative contributions of PDGF and different growth factors to the healing of normal wounds, and thus experimental strategies continue to focus on *in vitro* properties of growth factors. Additional models and reagents are needed to test these results directly *in vivo*.

## II. WOUND HEALING

Normal wound healing may be separated operationally into different stages that overlap and proceed in a clear sequence to a healed wound. Initially, platelets release granule products, and products of the coagulation process are deposited locally. The sequential migration of neutrophils, monocytes, and fibroblasts into wounds begins immediately and continues over the first several days. Wound macrophages (derived from circulating monocytes) and fibroblasts become activated, to initiate a cascade of new gene expression that results in *de novo* synthesis of growth factors and other cytokines, synthesis of extracellular matrix proteins (including collagen), and proliferation of fibroblasts. Later, tissue remodeling results in active collagen turnover and cross-linking that lasts from two weeks to one year postwounding.

Factors that arise locally account for the cell migration and activation of fibroblasts in wounds (Arey, 1936; Dunphy, 1963; Liebovich, 1975). However, circulating platelets are invariably associated with wounds and are now known to release factors originally stored within intracellular compartments, such as the platelet granules, into wounds. These factors attract neutrophils, monocytes, fibroblasts, and other (tissue-specific) cells, such as the smooth muscle cells. Platelets are now considered to be important mediators in the initiation of inflammation and subsequent tissue remodeling and repair. Platelet release is initiated by products of coagulation, such as thrombin, and by platelet exposure to foreign surfaces, such as collagen. Platelet release occurs within seconds of platelet activation. Synthesis of other potent molecules, such as the prostaglandins and leukotrienes, also influences intracellular responses and extracellular activities that contribute significantly to early events within wounds. These factors also initiate the transcriptional activation of genes that encode proteins with other signaling roles, serving to attract inflammatory cells locally. Because inflammatory cells are in close proximity to platelets in a number of models of inflammation (immune complex disease and atherosclerosis), it is likely that platelets are important in the earliest stages of different inflammatory states. Platelets and inflammatory cells appear to "talk" to each other in essential ways, through the release of cytokines and growth factors to attract additional inflammatory cells and to activate transcription of factors that regulate or encode additional mediators of inflammation and wound repair. Studies with PDGF in the *in vivo* context support these functional roles for platelet secretory proteins.

PDGF was originally identified as a potent mitogen in serum for mesenchymally derived cells. Purified human PDGF is predominantly a heterodimeric molecule consisting of two separate but highly related polypeptide chains (A and B chains) (Hart *et al.*, 1990). The B chain of PDGF is ~92% identical with the protein product of the v-sis gene, the oncogene of simian sarcoma virus, an acute transforming retrovirus, and thus is a protooncogene. Both BB and AA homodimers of PDGF also have been isolated from natural sources and have been shown to be expressed in many cell types of diverse origins, including macrophages, endothelial cells, smooth and skeletal muscle cells, fibroblasts, glial cells, and neurons. These cells secrete molecules with PDGF-like activity. The detection of these molecules has been associated with the expression of the mRNAs for A and B chains that can be specifically associated with chain isoforms of PDGF. However, identification of the specific isoform(s) at the level of the protein has been difficult because reagents to detect each isoform specifically are not readily available. Multiple PDGF-AA transcripts have been detected in cells that appear to be alternative splice variants of a single seven-exon gene (Collins *et al.*, 1987; Tong *et al.*, 1987). Three splice variants result in short and long processed proteins, 110 and 125 amino acids (Collins *et al.*, 1987; Tong, Auer *et al.*, 1987; Matoskova *et al.*, 1989) in length, and may be important because of differential regulation of the transcripts and because they appear to bind with different affinities to the extracellular matrix (Ostman *et al.*, 1989; Nagaoka *et al.*, 1991). The transcripts for both proteins are widely expressed in multiple cell types.

Each of the three isoforms (BB, AB, and AA) binds to the PDGF-$\alpha$ receptor, whereas only the BB isoform binds with significant affinity to the type of PDGF receptor-β (Hart *et al.*, 1988; Heldin *et al.*, 1988; Matsui *et al.*, 1989). Both receptor types are highly homologous transmembrane tyrosine kinases and are expressed in a cell type–specific manner, not only governing cell target specificity of the PDGF response (Eriksson *et al.*, 1992), but also adding both complexity and thus diversity to the potential roles of PDGF in inflammation and wound healing. The functions of PDGF *in vivo* thus are regulated by the need for specific ligand–receptor interactions to transduce the signals required for various functions of PDGF and the up-regulation of other genes that propagate the cascade of events initiated by PDGF.

## III. GROWTH FACTORS AND CYTOKINES ACTIVE AS EARLY MEDIATORS OF THE INFLAMMATORY PROCESS

In order to investigate and thus to understand the roles of growth factors and other cytokine mediators of inflammation, tissue remodeling, and wound healing, experimental strategies were designed first to isolate and characterize these factors, to analyze their properties *in vitro*, to determine the sites and levels of expression achievable *in vivo*, and to develop reagents and experimental models to analyze normal functions of growth factors and other cytokines in the intact animal. Early work focused on the roles of PDGF as a mitogen, largely because it was the first mitogen to be purified. However, in addition to the striking potency of PDGF as a mitogen, PDGF is strongly chemotactic for human monocytes and neutrophils, a property of PDGF that may be more important in early phases of inflammation. Optimal concentrations of PDGF that are required for maximum chemotactic and mitogenic activity are 20 and 1 ng/mL, respectively (Deuel *et al.*, 1982), concentrations of PDGF that are well below those measured in human serum (~50 ng/mL) (Deuel *et al.*, 1982). Because platelets aggregate and release at wound sites, the concentrations of PDGF at these sites may be substantially higher than those in serum. Furthermore, PDGF is a chemoattractant for fibroblasts and smooth muscle cells but requires a somewhat higher concentration (Seppa *et al.*, 1981; Senior *et al.*, 1983), suggesting that PDGF may also be involved later in wound healing through its influence on fibroblast and smooth muscle cell movement (Senior *et al.*, 1983).

In contrast to the PDGF AB isoform, the chemotactic potential of PDGF AA has been controversial. This controversy was important to resolve because the A chain homodimers of PDGF (PDGF AA) are widely expressed in normal and transformed cells. However, PDGF AA has been independently established as a strong chemoattractant for human monocytes, granulocytes, and fetal bovine ligament fibroblasts. Relevant to the roles of monocytes/tissue macrophages in the wound-healing process, it was shown that highly purified (>98%) monocytes require the addition of

lymphocytes or interleukin 1 (IL-1) for chemotactic responsiveness to PDGF AA but not for the chemotactic activity of formyl-methionyl-leucyl-phenylalanine (fMLP) or C5a. Monocytes therefore require activation before becoming responsive to PDGF as a chemoattractant. Activation of lymphocytes or exogenous cytokines is required for monocytes to respond hemotactically to PDGFAA but not to fMLP or C5a, suggesting that regulation of the chemotactic response of the monocyte to PDGF AA by the lymphocyte or cytokines may be another regulatory level of importance in the ultimate pathway to tissue remodeling and wound healing *in vivo* (Shure *et al.*, 1992).

The effect of PDGF on cell migration was extended to other platelet granule proteins when it was shown that β-thromboglobulin (TG-β) is a highly active chemotactic factor for fibroblasts (Senior *et al.*, 1983). Platelet factor 4 also is a potent chemoattractant. Lymphokines, a peptide derived from the fifth component of complement, collagen-derived peptides, fibronectin, tropoelastin, and elastin-derived peptides are wound-associated factors that also are potent chemoattractants with differing cell type specificity (Postlethwaite *et al.*, 1976, 1978, 1979; Senior *et al.*, 1982).

In addition to its roles in cell migration, PDGF activates the polymorphonuclear leukocyte (Tzeng *et al.*, 1984) and is capable of inducing monocyte activation responses, as evaluated by generation of superoxide anion ($O_2$) from the membrane-associated oxidase system, release of granule enzymes, and enhanced cell adherence and cell aggregation. Superoxide anion release in monocytes is observed at 10 ng/mL, and the levels of PDGF-dependent release are comparable to the release induced by $10^{-7}$ mL/L fMLP. The potency of PDGF to induce this response in monocytes is of the same magnitude as that observed in polymorphonuclear leukocytes (PMNs). Similarly, lysozyme release and monocyte adherence are increased in a dose-dependent manner in response to PDGF and achieve maximal responses at 40 ng/mL PDGF. The PDGF concentration required to achieve maximal monocyte aggregation was twofold (60 ng/mL) that found for PMNs, and, at this concentration of PDGF, it is likely to occur only in the immediate vicinity of platelet aggregates. PDGF thus induces the full sequence of cell activation events in human monocytes, similar to human PMNs, and this property may well be of major significance in the release of enzymes and other factors required to begin the process of tissue breakdown required for remodeling.

Other early cellular effects important to inflammation and repair also may be initiated by the release of PDGF from platelets and cells within the wound. For example, collagenase expression was studied in normal human skin fibroblasts that were cultured for 24 h in the presence of PDGF. Collagenase release is required for the breakdown of collagen necessary for the later remodeling of collagen and for tissue repair (Bauer *et al.*, 1985; Pierce *et al.*, 1989). A dose-dependent, saturable increase in collagenase activity in culture media was observed with paralleled increases in immunoreactive collagenase protein, suggesting the enhanced synthesis of an immunologically unaltered enzyme. The specificity of this effect was demonstrated by collagenase stimulatory effect with that on total protein synthesis and DNA synthesis. Under *in vitro* conditions that produced a 2.5-fold increase in collagenase synthesis, there was an ~20% increase in total protein synthesis and no detectable change in DNA synthesis. In addition, as a second control, platelet factor 4, another platelet-derived protein, caused a <20% increase in collagenase expression. In time-course studies, stimulation of collagenase synthesis was first observed 8–10 h after exposure to the growth factor. Furthermore, when cells were exposed to PDGF for ~24 h, an increase in the rate of collagenase synthesis was seen for ~6 h after PDGF was removed, after which the rate reverted to control levels. Growth factor–regulated expression of collagenase may be of unique importance to tissue remodeling, because it occurs somewhat later than the most immediate responses to PDGF at the time collagen degradation may be essential for progression of remodeling and tissue repair (Bauer *et al.*, 1985).

Patients with Werner's syndrome, an autosomal recessive disorder, undergo an accelerated aging process and premature death. Fibroblasts from such patients typically grow poorly in culture but have a markedly attenuated mitogenic response to platelet-derived growth factor and fibroblast growth factor (FGF). In contrast, these cells have a full mitogenic response to fetal bovine serum. The Werner's syndrome cells express high constitutive levels of collagenase *in vitro*. However, no induction of collagenase occurs in the Werner's syndrome fibroblasts in response to PDGF. The coupling of the failure of one or more PDGF-mediated pathways in Werner's syndrome cells with the phenotype of this disorder is perhaps important support for roles of PDGF and other cytokines in tissue remodeling (Bauer *et al.*, 1986).

## IV. INFLAMMATORY RESPONSE MEDIATORS ACTIVATE TRANSCRIPTION OF QUIESCENT GENES

Another potentially highly important role of PDGF and other growth factors and cytokines in wound healing is the up-regulation of early-intermediate response genes in cells activated by PDGF. In response to PDGF and other cytokines, fibroblasts and other cells initiate transcription and expression of genes that are dormant in the absence of stimulation. The products of some of these genes are important in the intracellular propagation of the mitogenic signal, but others are important for intercellular communication, direct cell migration, and activation of cells of different types. The products of these genes (Cochran *et al.*, 1983; Kohase *et al.*, 1987; Rollins *et al.*, 1988; Pierce *et al.*, 1989a, 1989b) may

serve to coordinate a second wave of diverse responses designated to combat wound infection and to initiate and propagate the healing process.

The PDGF early/early-intermediate response genes *JE* and *KC* were first isolated by differential colony hybridization (Cochran *et al.*, 1983) and only later were identified as cytokines belonging to a newly described superfamily of small inducible genes (SIGs) (Kawahara and Deuel, 1989) (now known as the chemokine family). This family surprisingly includes platelet granule proteins, such as the platelet basic protein and platelet factor 4. These PDGF-induced cytokines suggest pathways of gene activation that may account in part for cell type specificity and the normal temporal responses of wound healing. The *JE* gene encodes a basic (PI = 10.4) 148-amino acid polypeptide (Rollins *et al.*, 1988; Kawahara and Deuel, 1989) that shows 82% amino acid identity with the murine *JE* gene (Timmers *et al.*, 1990). Both rat and murine *JE* gene products contain N-terminal hydrophobic leader sequences, virtually identical alternating hydrophobic and hydrophilic domains, a single N-linked glycosylation site at position 126 that is conserved in both the murine and rat *JE* genes, and identical intron–exon splice junctions.

The importance of the *JE* gene product was suggested when the human homolog of the rodent *JE* gene was identified as monocyte chemotactic protein-l (MCP-1) (Yoshimura *et al.*, 1989b). MCP-l is identical with monocyte chemotactic and activating factor (MCAF) (Furutani *et al.*, 1989), MCP (Decock *et al.*, 1990), HC11 (H. C. Chang *et al.*, 1989), and the smooth muscle cell chemotactic factor (SMC-CF) (Graves *et al.*, 1989). Both the MCP-l and rodent *JE* genes are induced in fibroblasts by PDGF [also serum, phorbol myristate acetate (PMA), double-stranded RNA, and interleukin-l], and each cross-hybridizes to the same bands on Southern blots (Rollins *et al.*, 1989). Antibodies raised to the murine *JE* cross-react with MCP-l. MCP-l is a potent factor chemotactic for monocytes but not for neutrophils (Yoshimura *et al.*, 1989), a specificity of cell type recognition that is not seen with many of the other chemotactic factors that have been characterized (Senior *et al.*, 1989), suggesting a unique role in the inflammatory process. Other cytokines with sharply defined functional activity have also been identified in the chemokine family. MCP-l/SMC-CF or related molecules appear to be the major chemotactic factor(s) released by a number of tumor cell lines as well (Graves *et al.*, 1989). This relationship calls to attention the surprising parallels that have been found between tumor cells and activated cells such as the fibroblast.

The genes of this family are normally not expressed at detectable levels but can be induced in the presence of the appropriate stimuli. Remarkably, each of the inducers has an important role in cell growth, inflammation, or immune responses. Increasing knowledge of this family of genes and their products has suggested an enlarging role for PDGF and other cytokines in the sequential and cell type–specific development of inflammation and the evolution of the wound-healing process (Kawahara *et al.*, 1991).

## V. BIOLOGIC PROPERTIES OF SIG (CHEMOKINE) FAMILY MEMBERS

Neutrophil activation protein/IL-8 (NAP-l/IL-8) is a highly cell type–specific activator and chemotactic factor for neutrophils but not for monocytes (Yoshimura *et al.*, 1987). NAP-l/IL-8 was identified in conditioned medium of lipopolysaccharide-stimulated monocytes (Walz *et al.*, 1987; Yoshimura *et al.*, 1987), and its gene (*3-10C*) was cloned independently as a staphylococcal enterotoxin A–induced gene from peripheral blood leukocytes (Schmid and Weissmann, 1987). NAP-l is also chemotactic for T-lymphocytes and basophils (Larsen *et al.*, 1989; Leonard *et al.*, 1990) and is the N-terminal processed form of endothelial-derived leukocyte adhesion inhibitor (LAI), an inhibitor of neutrophil adhesion that protects endothelial cells from neutrophil-mediated damage (Gimbrone *et al.*, 1989; Kawahara *et al.*, 1991).

MCP-l is chemotactic for monocytes. It was purified from the culture media of human glioma cell line U-105MG (Yoshimura *et al.*, 1989), THP-l cells (Matsushima *et al.*, 1989), phytohemagglutinin-stimulated mononuclear cells (Robinson *et al.*, 1989), and double-stranded RNA [poly(rI). poly(rC)]–stimulated fibroblasts (Van Damme *et al.*, 1989; Kawahara *et al.*, 1991).

Macrophage inflammatory protein-l (MIP-l) was purified in two forms (MIP-l, MIP-2) from lipopolysaccharide-stimulated RAW264.7 cells and is chemotactic for polymorphonuclear cells (Wolpe *et al.*, 1988). MIP-l induces oxidative burst and degranulation in neutrophils and promotes inflammatory reactions. MIP-l is also a prostaglandin-independent pyrogen (Davatelis *et al.*, 1989). MIP-2 is a potent chemotactic factor for neutrophils and causes neutrophil degranulation, but it does not induce an oxidative burst (Kawahara *et al.*, 1991).

Platelet factor 4 (PF4) is a platelet granule protein that is chemotactic for both neutrophils and monocytes (Deuel *et al.*, 1981). It has immunoregulatory activity (Katz *et al.*, 1986) and can inhibit angiogenesis (Maione *et al.*, 1990). Platelet basic protein, connective tissue–activating peptide III (CTAP-III), β-thromboglobulin, and neutrophil-activating peptide-2 (NAP-2) are N-terminal proteolytic processed forms of a single gene product, CTAP-III is mitogenic for human dermal fibroblasts (Castor *et al.*, 1983), and NAP-2 is induced in monocytes when they are exposed to lipopolysaccharides and can stimulate the release of elastase from human neutrophils (Walz *et al.*, 1987; Kawahara *et al.*, 1991).

Melanoma growth stimulatory activity (MGSA) is a mitogen for cultured melanoma and pigmented (nevus) cells (Richmond *et al.*, 1988). Primary melanoma cells secrete MGSA, and for this reason it has been suggested

to be an autocrine regulator of growth. The protein product of MGSA is identical to the *gro-α* gene isolated by subtractive hybridization of mRNA from transformed hamster CHEF/16 cells. The rat analog, cytokine-induced neutrophil chemoattractant (CINC) is chemotactic for neutrophils (Watanabe *et al.*, 1989). The murine analog (*KC*) is a PDGF-inducible gene (Cochran *et al.*, 1983; Kawahara *et al.*, 1991).

## VI. REGULATION OF *JE* GENE EXPRESSION BY GLUCOCORTICOIDS

The role of PDGF and other cytokines in the induction of otherwise unexpressed genes has resulted in additional insights into the wound-healing process. The addition of dexamethasone and other steroid drugs to PDGF- and serum-stimulated BALB/c 3T3 fibroblasts prevents the PDGF and serum-dependent induction of the *JE* gene in a dose-dependent manner. Furthermore, induction follows the rank order of potency expected for glucocorticoid receptor–mediated antiinflammatory activity. The inhibition appears to be highly specific for inflammatory mediators, suggesting that the role of PDGF in the induction of selective SIG family members may be a highly important step at which glucocorticoids negatively influence both inflammation and repair (Kawahara *et al.*, 1991).

## VII. GROWTH FACTORS AND ACCELERATED HEALING

Much new information about the roles of growth factors and other cytokines has resulted from the direct application of PDGF to experimental wounds in animals. Morphometric methods and histologic techniques have been used to compare wounds treated with supraphysiological concentrations of growth factors to untreated wounds. This approach, coupled with the *in vitro* approaches cited earlier, has greatly clarified roles of PDGF-enhanced tissue repair and remodeling and has stimulated similar experiments with other cytokines as well.

In one set of experiments, exogenous PDGF was applied locally, and once only, to incisional wounds in rats and rabbits. PDGF accelerated tissue repair in a dose-dependent manner (Pierce *et al.*, 1988, 1991). The single application of PDGF to rat incisions enhanced by 150–170% over a three-week time course the strength required to "break" the healing wound (Pierce *et al.*, 1988). PDGF accelerated healing by four to six days over the first two weeks and by five to ten days between two and seven weeks postwounding (Pierce *et al.*, 1989). At 89 days postwounding, both PDGF-treated and control wounds had achieved similar wound strength (~90% of unwounded dermis), and doses therefore induced a long-lasting enhancement rate of healing, supporting the view that PDGF and other growth factors initiate events required for normal healing of wounds and enhance the normal rate of wound healing and tissue remodeling.

In these wounds, PDGF significantly increased the rate of increase of cellularity and granulation tissue formation. The early appearance of wounds treated with PDGF is that of a highly exaggerated inflammatory response. An increased neutrophil influx was seen at 12 h. A marked increase in influx of macrophages and, later, fibroblasts was found from day 2, persisting through 21 days postwounding (Pierce et al., 1988). Furthermore, enhanced cellular function of fibroblasts could be established in treated wounds. Increased numbers of procollagen-containing fibroblasts were strikingly apparent as early as two days postwounding. At 28 days, it was not possible to distinguish activated from nonactivated fibroblasts in treated and untreated wounds. These results indicate that PDGF accelerates the healing response.

Furthermore, within the limits of detection, PDGF also results in a healed wound that cannot be distinguished from its untreated counterpart (Pierce et al., 1989). PDGF also appears to be multifunctional in diverse ways, perhaps through the recruitment and activation of macrophages and fibroblasts and through up-regulation of expression at different cytokine genes whose functions and cell type specificity provide additional diversity to the normal healing process.

Another polypeptide growth factor that is stored in the granules of platelets, transforming growth factor β (TGF-β), may also be very important in tissue repair (Sporn et al., 1986). Both PDGF and TGF-β increase collagen formation, DNA content, and protein levels in wound chambers implanted in rats (Sporn et al., 1983; Grotendorst et al., 1985). TGF-β enhances the reversible formation of granulation tissue when injected subcutaneously into newborn mice (Roberts et al., 1986). A single application of human TGF-β at the time of wounding advanced by two to three days the breaking strength required to rupture the incision margins when assayed at one week. The marked augmentation of wound healing using both human PDGF (hPDGF) and recombinant c-sis homodimers (rPDGF-B) revealed a unique pattern of response when compared with the results obtained with TGF-β (Mustoe et al., 1987; Pierce et al., 1988).

Circular excisional wounds also were placed through the dermis of the rabbit ear to the level of underlying cartilage, a model that does not allow contraction from the wound margins (Mustoe et al., 1991). Ingrowth of extracellular matrix and granulation tissue was measured by histomorphometric techniques (Mustoe et al., 1991). A single application of PDGF increased granulation tissue ~200% at seven days in association with a predominance of fibroblasts and collagen-containing extracellular matrix (Mustoe et al., 1991). The healed wound again was indistinguishable from the control, non-PDGF-treated healed wound. Interestingly, PDGF also nearly doubled the rate at which re-

epithelialization occurred, and neovascularization was prominent in granulation tissue. These unexpected findings, in view of the known cell type specificity of PDGF, suggested up-regulation either of PDGF receprors or of other cytokine genes, with resultant cell type–specific epithelial and/or endothelial cell functions. Whereas again the mechanisms of the PDGF effect are not established, the influence of a single application of a growth factor remarkably accelerated the inflammatory response and the subsequent tissue remodeling to result in a healed wound of normal appearance.

In this model, PDGF accelerated deposition of matrix largely composed of glycosaminoglycans and later of collagen (Pierce et al., 1991). The normal sequence in repair and the acceleration of wound healing, to all appearances, are identical to those in an untreated wound, supporting the view that the role of PDGF is to accelerate normal wound healing in human wounds. It has been demonstrated in human wounds that PDGF AA expression is increased within healing pressure ulcers. The accumulation of PDGF AA is accompanied by activated fibroblasts, extracellular matrix deposition, and active neovessel formation. However, far less PDGF AA is present in chronic nonhealing wounds. Thus, up-regulation of PDGF AA may be important in the normal repair process as well (Robson et al., 1992; Mustoe et al., 1994; Pierce et al., 1994).

PDGF BB, bFGF, and epidermal growth factor (EGF), applied locally at the time of wounding, cause a twofold increase in complete reepithelialization of treated wounds, whereas TGF-β significantly inhibits reepithelializarion. Both PDGF BB and TGF-β increased the depth and area of new granulation tissue, the influx of fibroblasts, and the deposition of new matrix into wounds. Explants from treated wounds remained metabolically more active than controls, incorporating 473% more [3H]-thymidine into DNA and significantly more [3H]-leucine and [3H]-proline into collagenase-sensitive protein (Mustoe et al., 1991). The different response to growth factors underscores the importance of tissue and cell type specificity of the growth factors to tissue repair and perhaps also the need for exquisite timing of induction of different factors in the orderly progression of the healing process.

The use of these wound-healing models provided the basis to suggest at least one mechanism whereby PDGF may function in vivo (Mustoe et al., 1989; Pierce et al., 1989). Animals that were pretreated with glucocorticoids or total-body irradiation to reduce circulating monocytes to nondetectable levels had a sharply reduced influx of wound macrophages. PDGF was tested in these animals. It was shown that the PDGF-induced acceleration of incisional repair was abrogated in glucocorticoid-treated animals (Pierce et al., 1989). However, PDGF attracted a significant influx of fibroblasts into the compromised wounds of animals. The fibroblasts lacked procollagen type I, suggesting the requirement of the wound macrophage to support stimulation of fibroblast procollagen type I synthesis. PDGF alone does not stimulate procollagen type I synthesis when added directly to fibroblasts *in vitro* (Blatti *et al.*, 1988; Rossi *et al.*, 1988). However, PDGF-activated macrophages synthesize TGF-β (Pierce *et al.*, 1989), a potent activator of the fibroblast procollagen type I gene (Ignotz and Massague, 1986; Rossi *et al.*, 1988). It is suggested that TGF-β may be a second signal messenger to induce collagen formation in PDGF-treated wounds. This suggestion is supported directly in experiments in which macrophages and fibroblasts in PDGF-treated incisional wounds were shown to contain increased levels of intracellular TGF-β (Pierce *et al.*, 1989).

The results with wound-healing models and PDGF have illustrated much about inflammation, tissue remodeling, and the process of wound healing. Other *in vivo* models of PDGF-treated wounds also indicate that PDGF stimulates normal and reverses deficient dermal repair (Grotendorst *et al.*, 1985; Lynch *et al.*, 1987; Sprugel *et al.*, 1987; Greenhalgh *et al.*, 1990) and have suggested roles for this growth factor in inflammation (Circolo *et al.*, 1990), uterine smooth muscle hypertrophy (Mendoza *et al.*, 1990), lens growth and transparency (Brewitt and Clark, 1988), and central nervous system gliogenesis (Richardson *et al.*, 1988). What may be the most important conclusions are that PDGF has the potential to initiate and accelerate the process and to sustain the process to an endpoint that is effectively identical to untreated wounds. The most important mechanisms of its activity may be the consequences of its potent chemotactic activity and its ability to induce multiple autocrine loops that lead to an endogenous cascade of new gene activation and cell-specific cytokine synthesis within the healing wound. Clinical trials have now established the efficacy of PDGF in advancing the healing of diabetic ulcers. PDGF is now "in the clinic."

## VIII. ROLE OF BASIC FIBROBLAST GROWTH FACTOR AND ANGIOGENESIS

Normal wound healing can be divided into three phases: inflammation, fibroplasia, and maturation provoked by liberated angiogenic factors. Vessel-dense granulation tissue is central to the process of tissue repair. The formation of new blood vessels provides a route for oxygen and nutrient delivery as well as a conduit for components of the inflammatory response. Endogenous bFGF, like PDGF, is also found in the wound site and is presumed to be a necessary part of natural wound healing. Mechanically damaged endothelial cells and burned wound tissue release significant amounts of bFGF into the wound area, suggesting that damaged endothelial cells at the cut edges of blood vessels, as well as other damaged cells at the site of a wound, may provide an early source of bFGF (McNeil *et al.*, 1989; Muthukrishnan *et al.*, 1991; Gibran *et al.*, 1994). A slightly later and more sustained supply of bFGF at the wound site is delivered by the invading activated macrophages (Baird *et al.*, 1985; Rappolee

*et al.*, 1988; Fiddes *et al.*, 1991). Moreover, recombinant bFGF has been shown to enhance healing, if added exogenously to a wound site (Brew *et al.*, 1995). It is now clear that bFGF can act as an angiogenic factor *in vitro* (Montesano *et al.*, 1986) and *in vivo* (Folkman and Klagsbrun, 1987), directing endothelial cell migration and proliferation, and that bFGF is a mitogen to endothelial cells. It also up-regulates expression of plasminogen activator and the αvβ3 integrin that promotes both migration of endothelial cells and luminal formation (Pepper *et al.*, 1991; Friedlander *et al.*, 1995). The effects of stimulating endothelial cell growth, capillary differentiation, and connective tissue cell growth by bFGF contribute to wound healing and tissue regeneration (Gibran *et al.*, 1994; Steenfos, 1994).

Wound healing involves the interactions of many cell types and is controlled in part by growth factors. Intercellular communication mediated by gap junctions is considered to play an important role in the coordination of cellular metabolism during the growth and development of tissues and organs. Basic fibroblast growth factor, known to be important in wound healing, has been found to increase gap junctional protein annexin 43 expression and intercellular communication in endothelial cells and cardiac fibroblasts (Abdullah *et al.*, 1999). It has been proposed that an increased coupling is necessary for the coordination of these cells in wound healing and angiogenesis and that one of the actions of bFGF is to modulate intercellular communication. In a variety of animal models, exogenous bFGF has been shown to accelerate wound healing by speeding granulation tissue formation and increasing fibroblast proliferation and collagen accumulation at the site of a subcutaneously implanted sponge in rats (Davidson *et al.*, 1985; Buckley-Sturrock *et al.*, 1989). Fresh wound tensile strength has also been increased (McGee *et al.*, 1988). However, the greatest benefit from the application of exogenous bFGF can probably be obtained in cases in which wound healing is impaired. Research on skin flaps indicates that bFGF has the potential to increase viability by accelerating flap revascularization when administered in a sustained-release manner (Hom and Winters, 1998). This may have applications to open or nonunion fractures with impaired wound healing. Despite encouraging experimental results to date coming from studies showing that treatment of genetically diabetic mice with recombinant bFGF not only increased fibroblast and capillary density in the wound but also accelerated wound closure, clinical application of recombinant human bFGF has fallen short of expectations. In a randomized double-blind placebo-controlled study, direct application of bFGF to nonhealing ulcers of diabetic patients did not accelerate wound healing (Richard *et al.*, 1995). In a subsequent randomized placebo controlled trial to study the effect of recombinant bovine bFGF on burn healing, all patients treated with bFGF had faster granulation tissue formation and epidermal regeneration, compared with those in the placebo group. Superficial and deep second-degree burns treated with bFGF healed in a mean of 9.9 days (SD 2.5) and 17.0 days (SD 4.6), respectively, compared with 12.4 (2.7) and 21.2 (4.9) days. No adverse effects were seen locally or systematically with bFGF. It is indicated that bFGF effectively decreased healing time and improved healing quality. Clinical benefits for use of bFGF in burn wounds would be shorter hospital stays and the patients' skin quickly becoming available for harvesting and grafting (Fu *et al.*, 1998).

## IX. PLEIOTROPHIN REMODELS TUMOR MICROENVIRONMENT AND STIMULATES ANGIOGENESIS

Pleiotrophin (PTN, *Ptn*) is a PDGF-inducible, heparin-binding cytokine and thus a downstream effector of PDGF. Pleiotrophin was first purified in this laboratory as an 18-kD heparin-binding, mitogenic polypeptide from bovine uterus (Milner *et al.*, 1989) and as neurite-outgrowth-promoting activity from neonatal rat brain (Rauvala, 1989); it is a lysine-rich, highly basic protein of 168 amino acids with a 32-amino acid signal peptide that is nearly 50% identical in amino acid sequence with Midkine (MK, *Mk*) (Li *et al.*, 1990), a gene expressed in early stages of retinoic acid–induced differentiation in mouse embryonal carcinoma cells (Kadomatsu *et al.*, 1988). Pleiotrophin and *Mk* are the only known members of this family. Pleiotrophin lacks homology with members of the heparin-binding growth factor (HBGF) family of structurally related polypeptides (Goldfarb, 1990).

The *Ptn* gene is expressed during neuroectodermal and mesodermal development but not in endoderm, ectoderm, or trophoblasts (Silos-Santiago *et al.*, 1996). Expression levels of *Ptn* peak in late embryonic development and the immediate postnatal period. They correlate with peaks of growth and differentiation in both neoectoderm and mesoderm (Rauvala, 1989; Li *et al.*, 1990). In adults, PTN expression is limited to selected cell populations and is constitutive; however, the *Ptn* gene expression is up-regulated after injury and during repair. In cultured cells, its expression is modulated by growth factors, including PDGF, bFGF, and steroid hormones (Li *et al.*, 1992a, 1992b; Vacherot *et al.*, 1995).

Pleiotrophin normally signals a diversity of responses that suggest its potential importance in dysregulated growth and angiogenesis when it is inappropriately expressed in the cancer cell. PTN has mitogenic activity for fibroblasts (Milner *et al.*, 1989), for brain capillary endothelial cells (Courty *et al.*, 1991), and for SW-13 (adrenal carcinoma) cells in soft-agar culture (Fang *et al.*, 1992). Pleiotrophin also stimulates angiogenesis *in vivo* and *in vitro* by different criteria (Courty *et al.*, 1991). Furthermore, the *Ptn* gene is a proto-oncogene; *Ptn*-transformed NIH 3T3 cells develop into rapidly growing and highly vascular tumors in flanks of nude mice (Chauhan *et al.*, 1993; Deuel *et al.*, 2002). Importantly, it is highly expressed in many human breast cancers

(Fang *et al.*, 1992; Tsutsui *et al.*, 1993). Introduction of the exogenous *Ptn* gene into premalignant SW13 (adrenal carcinoma) cells and into MCF-7 (human premalignant breast carcinoma) cells stimulates highly malignant subcutaneous tumor growth at sites of implantation in flanks of nude mice (Zhang and Deuel, 1999; Chang and Zuka, in press), and the highly malignant phenotype of human malignant breast cancer cells and other cells with constitutive activation of the *Ptn* gene is reversed when a dominant negative inhibitor of PTN signaling or a *Ptn* gene–targeted ribozyme is introduced (Czubayko *et al.*, 1996; Zhang and Deuel, 1999; Aigner *et al.*, 2002; Chang and Deuel, Submitted), thus establishing that a single gene mutation that initiates activation of the *Ptn* gene alone is sufficient to switch the premalignant phenotype of premalignant cells to that of a highly malignant phenotype. These findings may be highly significant, since many different human malignancies of different origins have activated endogenous *Ptn* (Mailleux *et al.*, 1992; Nakagawara *et al.*, 1995; Czubayko *et al.*, 1996; Zhang *et al.*, 1997; Weber *et al.*, 2000; Mentlein and Held-Feindt, 2002). Importantly, in those cell lines studied, expression of the endogenous *Ptn* gene is constitutive, and PTN signaling has been demonstrated to be a major contributor to the highly malignant phenotype of these cells (Mailleux *et al.*, 1992; Nakagawara *et al.*, 1995; Czubayko *et al.*, 1996; Zhang *et al.*, 1997; Weber *et al.*, 2000).

Pleiotrophin signals by inactivating the tyrosine phosphatase activity of the transmembrane receptor protein tyrosine phosphatase (RPTP)β/ζ (Meng *et al.*, 2000). Inactivation of RPTPβ/ζ by the interaction of PTN and RPTPβ/ζ leaves unchecked the persistent activity of a tyrosine kinase that phosphorylates the identical tyrosine that is dephosphorylated by RPTPβ/ζ (Meng *et al.*, 2000). Thus, the tyrosine phosphorylation levels of the substrates of RPTPβ/ζ are sharply increased. Pleiotrophin is the first natural ligand to be discovered for any of the RPTPs (Meng *et al.*, 2000). Anaplastic Lymphoma Kinase (ALK) also has been proposed as a receptor for PTN (Stoica *et al.*, 2001), as has syndecan 3 (Rauvala *et al.*, 2000), leaving open the place of these different proteins in PTN signaling. The downstream targets of PTN/RPTPβ/ζ signaling pathway are β-catenin (Meng *et al.*, 2000), β-adducin (Pariser, Perez-Pinera, *et al.*, 2005), and Fyn (Pariser, Ezquerra *et al.*, 2005). β-catenin, β-adducin, and Fyn are substrates of RPTPβ/ζ; the steady-state levels of tyrosine phosphorylation of each of these proteins is sharply increased in PTN-stimulated cells. However, each of these proteins is phosphorylated in non-PTN-stimulated cells as well, suggesting their steady-state levels of tyrosine phosphorylation are maintained by the endogenous PTN/RPTPβ/ζ signaling pathway. Each of the downstream targets of PTN signaling is an important regulator of cytoskeletal structure (Meng *et al.*, 2000; Pariser and Ezquerra *et al.*, 2005; Pariser and Perez-Pinera *et al.*, 2005), and destabilization of cytoskeleton and loss of cell–cell adhesion is the immediate response of cells stimulated by PTN (Meng *et al.*,

2000), suggesting that a major target of exogenous PTN is desestabilization of cytoskeletal elements and cell–cell adhesion through the coordinated increase in tyrosine phosphorylation of these substrates of RPTPβ/ζ.

Collagens and elastin are the major fibrillar elements in the cardiovascular system and thus are major determinants in vascular tone, distensibility, and other properties of aorta and other vessels. Analysis of aorta of *Ptn* –/– and *Mk* –/– mice uncovered the previously unknown but remarkable findings that PTN and MK are major regulators of the transcripts of the catecholamine biosynthesis pathway (Ezquerra *et al.*, 2004, 2006) and the transcripts of the renin–angiotensin pathway (Herradon *et al.*, 2004, 2005) in mouse. Since procollagen and elastin synthesis is known to be regulated through both the catecholamine and the angiotensin II signaling pathways, the striking down-regulation of the elastin precursor gene in aorta of *Ptn* –/– and *Mk* –/– mice define PTN and MK as critical regulators of elastin expression (Ezquerra *et al.*, Submitted). Furthermore, tumors of MMTV-*PyMT* transgenic mice with very high levels of the MMTV-*Ptn* transgene have strikingly increased synthesis of collagen and elastin interspersed with activated stroma and increased tumor angiogenesis which is also seen in the wound-healing process (Chang and Deuel, Submitted). These data, collectively, suggest that pleiotrophin plays an important role in the remodeling of the microenvironment of cells in both normal and disordered conditions.

## X. OTHER ROLES OF GROWTH FACTORS AND CYTOKINES

A growing appreciation for roles of growth factors in many normal and abnormal processes is emerging from numerous other investigations of a diverse nature. For example, TGF-α and TGF-β are expressed with high specificity in developing mouse embryo (Twardzik, 1985; Heine *et al.*, 1987); PDGF may mediate normal gliogenesis (Richmond *et al.*, 1988); maternally encoded FGF, TGF-β, and PDGF have been implicated as important in the developing *Xenopus* embryo (Kimelman and Kirschner, 1987;

Weeks and Melton, 1987; Mercola *et al.*, 1988); and bFGF is a potential neurotrophin during development (Anderson *et al.*, 1988). PDGF also has been identified within plaques and is a potent vasoconstrictor and thus has been implicated in the genesis of atherosclerosis (Berk *et al.*, 1986; Libby *et al.*, 1988). Furthermore, it is secreted from endothelial cells and arterial smooth muscle cells (Jaye *et al.*, 1985; Sejersen *et al.*, 1986; Martinet *et al.*, 1987; Tong *et al.*, 1987; Majesky *et al.*, 1988; Rubin *et al.*, 1988) and activated monocytes/macrophages (Martinet *et al.*, 1987). Elevated levels of PDGF receptors are found on synovial cells in patients with rheumatoid arthritis (Rubin *et al.*, 1988). In each instance, however, when the *in vitro* studies are considered in the context of roles of the growth factors in inflammation and tissue repair, it seems likely that the common roles of growth factors are associated with normal development and that the abnormal remodeling is associated with disease states, indicating the importance of attention to mechanisms that regulate cell type and temporal levels of expression of the growth factors and their cognate receptors.

## XI. CONCLUSIONS

Much needs to be learned concerning the roles of growth factors and cytokines if we are to establish fully how they function in both normal and dysregulated tissue remodeling. Remarkable progress over the past several years has resulted from the identification, isolation, cloning, and characterization of the properties of these molecules and their use in wound-healing models. The most important lesson learned from their use appears to be that the growth factors and cytokines are important to initiate and accelerate the normal processes involved in going from injury to repair. Questions for future avenues of investigation should address mechanisms by which these factors initiate and propagate the processes and the regulatory signals that govern cell type specificity, differentiation responses in cells migrating into and dividing within wounds, and the temporal sequences needed for orderly progression to a healed and ultimately functional tissue.

## XII. REFERENCES

Abdullah, K. M., Luthra, G., *et al.* (1999). Cell-to-cell communication and expression of gap junctional proteins in human diabetic and nondiabetic skin fibroblasts: effects of basic fibroblast growth factor. *Endocrine* 10(1), 35–41.

Aigner, A., Fischer, D., *et al.* (2002). Delivery of unmodified bioactive ribozymes by an RNA-stabilizing polyethylenimine (LMW-PEI) efficiently down-regulates gene expression. *Gene Ther.* 9(24), 1700–1707.

Anderson, K. J., Dam, D., *et al.* (1988). Basic fibroblast growth factor prevents death of lesioned cholinergic neurons *in vivo*. *Nature* 332(6162), 360–361.

Antoniades, H. N., Galanopoulos, T., *et al.* (1991). Injury induces *in vivo* expression of platelet-derived growth factor (PDGF) and PDGF receptor mRNAs in skin epithelial cells and PDGF mRNA in

connective tissue fibroblasts. *Proc. Natl. Acad. Sci. USA* 88(2), 565–569.

Arey, l. (1936). Wound healing. *Physiol. Rev.* 16, 327–406.

Baird, A., Culler, F., *et al.* (1985). Angiogenic factor in human ocular fluid. *Lancet* 2(8454), 563.

Bauer, E. A., Cooper, T. W., *et al.* (1985). Stimulation of *in vitro* human skin collagenase expression by platelet-derived growth factor. *Proc. Natl. Acad. Sci. USA* 82(12), 4132–4136.

Bauer, E. A., Silverman, N., *et al.* (1986). Diminished response of Werner's syndrome fibroblasts to growth factors PDGF and FGF. *Science* 234(4781), 1240–1243.

Berk, B. C., Alexander, R. W., *et al.* (1986). Vasoconstriction: a new activity for platelet-derived growth factor. *Science* 232(4746), 87–90.

Blatti, S. P., Foster, D. N., *et al.* (1988). Induction of fibronectin gene transcription and mRNA is a primary response to growth-factor stimulation of AKR-2B cells. *Proc. Natl. Acad. Sci. USA* 85(4), 1119–1123.

Brew, E. C., Mitchell, M. B., *et al.* (1995). Fibroblast growth factors in operative wound healing. *J. Am. Coll. Surg.* 180(4), 499–504.

Brewitt, B., and Clark, J. I. (1988). Growth and transparency in the lens, an epithelial tissue, stimulated by pulses of PDGF. *Science* 242(4879), 777–779.

Buckley-Sturrock, A., Woodward, S. C., *et al.* (1989). Differential stimulation of collagenase and chemotactic activity in fibroblasts derived from rat wound repair tissue and human skin by growth factors. *J. Cell Physiol.* 138(1), 70–78.

Castor, C. W., Miller, J. W., *et al.* (1983). Structural and biological characteristics of connective tissue activating peptide (CTAP-III), a major human platelet-derived growth factor. *Proc. Natl. Acad. Sci. USA* 80(3), 765–769.

Chang, H. C., Hsu, F., *et al.* (1989). Cloning and expression of a gamma-interferon-inducible gene in monocytes: a new member of a cytokine gene family. *Int. Immunol.* 1(4), 388–397.

Chang, Y., and Deuel, T. F. (Submitted). Dominant negative mutant of pleiotrophin reverses malignant phenotype of U87MG glioblastoma cells and is an angiostatic factor *in vivo*. *J. Biol. Chem.*

Chang, Y., Zuka, M., *et al.* Inappropriate expression of pleiotrophin (PTN) stimulates breast cancer progression through secretion of PTN and PTN-dependent remodelling of the tumor microenvironment. Proceedings of National Academy of Sciences U.S.A. (in press).

Chauhan, A. K., Li, Y. S., *et al.* (1993). Pleiotrophin transforms NIH 3T3 cells and induces tumors in nude mice. *Proc. Natl. Acad. Sci. USA* 90(2), 679–682.

Circolo, A., Pierce, G. F., *et al.* (1990). Antiinflammatory effects of polypeptide growth factors. Platelet-derived growth factor, epidermal growth factor, and fibroblast growth factor inhibit the cytokine-induced expression of the alternative complement pathway activator factor B in human fibroblasts. *J. Biol. Chem.* 265(9), 5066–5071.

Cochran, B. H., Reffel, A. C., *et al.* (1983). Molecular cloning of gene sequences regulated by platelet-derived growth factor. *Cell* 33(3), 939–947.

Collins, T., Bonthron, D. T., *et al.* (1987). Alternative RNA splicing affects function of encoded platelet-derived growth factor A chain. *Nature* 328(6131), 621–624.

Courty, J., Dauchel, M. C., *et al.* (1991). Mitogenic properties of a new endothelial cell growth factor related to pleiotrophin. *Biochem. Biophys. Res. Commun.* 180(1), 145–151.

Czubayko, F., Schulte, A. M., *et al.* (1996). Melanoma angiogenesis and metastasis modulated by ribozyme targeting of the secreted growth factor pleiotrophin. *Proc. Natl. Acad. Sci. USA* 93(25), 14753–14758.

Davatelis, G., Wolpe, S. D., *et al.* (1989). Macrophage inflammatory protein-1: a prostaglandin-independent endogenous pyrogen. *Science* 243(4894 Pt 1), 1066–1068.

Davidson, J. M., Klagsbrun, M., *et al.* (1985). Accelerated wound repair, cell proliferation, and collagen accumulation are produced by a cartilage-derived growth factor. *J. Cell Biol.* 100(4), 1219–1227.

Decock, B., Conings, R., *et al.* (1990). Identification of the monocyte chemotactic protein from human osteosarcoma cells and monocytes: detection of a novel N-terminally processed form. *Biochem. Biophys. Res. Commun.* 167(3), 904–909.

Deuel, T. F., Senior, R. M., *et al.* (1981). Platelet factor 4 is chemotactic for neutrophils and monocytes. *Proc. Natl. Acad. Sci. USA* 78(7), 4584–4587.

Deuel, T. F., Senior, R. M., *et al.* (1982). Chemotaxis of monocytes and neutrophils to platelet-derived growth factor. *J. Clin. Invest.* 69(4), 1046–1049.

Deuel, T. F., Kawahara, R. S., *et al.* (1991). Growth factors and wound healing: platelet-derived growth factor as a model cytokine. *Annu. Rev. Med.* 42, 567–584.

Deuel, T. F., Zhang, N., *et al.* (2002). Pleiotrophin: a cytokine with diverse functions and a novel signaling pathway. *Arch. Biochem. Biophys.* 397(2), 162–171.

Dunphy, J. (1963). The fibroblast-A ubiquitous ally for the surgeon. *N. Engl. J. Med.* 268, 1367–1377.

Eriksson, A., Siegbahn, A., *et al.* (1992). PDGF alpha- and beta-receptors activate unique and common signal transduction pathways. *Embo. J.* 11(2), 543–550.

Ezquerra, L., Herradon, G., *et al.* (2004). Pleiotrophin is a major regulator of the catecholamine biosynthesis pathway in mouse aorta. *Biochem. Biophys. Res. Commun.* 323(2), 512–517.

Ezquerra, L., Herradon, G., *et al.* (2005). Midkine, a newly discovered regulator of the renin-angiotensin pathway in mouse aorta: significance of the pleiotrophin/midkine developmental gene family in angiotensin II signaling. *Biochem. Biophys. Res. Commun.* 333(2), 636–643.

Ezquerra, L., Herradon, G., *et al.* (2006). Midkine is a potent regulator of the catecholamine biosynthesis pathway in mouse aorta. *Life Sci.* 79, 1049–1055.

Ezquerra, L., Herradon, G., and Deuel, T. F. (Submitted). Pleiotrophin and midkine are potent regulators of collagen and elastin synthesis. *Biochem. Biophys. Res. Commun.*

Fang, W., Hartmann, N., *et al.* (1992). Pleiotrophin stimulates fibroblasts and endothelial and epithelial cells and is expressed in human cancer. *J. Biol. Chem.* 267(36), 25889–25897.

Fiddes, J. C., Hebda, P. A., *et al.* (1991). Preclinical wound-healing studies with recombinant human basic fibroblast growth factor. *Ann. NY Acad. Sci.* 638, 316–328.

Folkman, J., and Klagsbrun, M. (1987). Angiogenic factors. *Science* 235(4787), 442–447.

Friedlander, M., Brooks, P. C., *et al.* (1995). Definition of two angiogenic pathways by distinct alpha v integrins. *Science* 270(5241), 1500–1502.

Fu, X., Yang, Y., *et al.* (1998). Ischemia and reperfusion impair the gene expression of endogenous basic fibroblast growth factor (bFGF) in rat skeletal muscles. *J. Surg. Res.* 80(1), 88–93.

Furutani, Y., Nomura, H., *et al.* (1989). Cloning and sequencing of the cDNA for human monocyte chemotactic and activating factor (MCAF). *Biochem. Biophys. Res. Commun.* 159(1), 249–255.

Gibran, N. S., Isik, F. F., *et al.* (1994). Basic fibroblast growth factor in the early human burn wound. *J. Surg. Res.* 56(3), 226–234.

Gimbrone, M. A., Jr., Obin, M. S., *et al.* (1989). Endothelial interleukin-8: a novel inhibitor of leukocyte–endothelial interactions. *Science* 246(4937), 1601–1603.

Goldfarb, M. (1990). The fibroblast growth factor family. *Cell Growth Differ.* 1(9), 439–445.

Graves, D. T., Jiang, Y. L., *et al.* (1989). Identification of monocyte chemotactic activity produced by malignant cells. *Science* **245**(4925), 1490–1493.

Greenhalgh, D. G., Sprugel, K. H., *et al.* (1990). PDGF and FGF stimulate wound healing in the genetically diabetic mouse. *Am. J. Pathol.* **136**(6), 1235–1246.

Grotendorst, G. R., Martin, G. R., *et al.* (1985). Stimulation of granulation tissue formation by platelet-derived growth factor in normal and diabetic rats. *J. Clin. Invest.* **76**(6), 2323–2329.

Hart, C. E., Forstrom, J. W., *et al.* (1988). Two classes of PDGF receptor recognize different isoforms of PDGF. *Science* **240**(4858), 1529–1531.

Hart, C. E., Bailey, M., *et al.* (1990). Purification of PDGF-AB and PDGF-BB from human platelet extracts and identification of all three PDGF dimers in human platelets. *Biochemistry* **29**(1), 166–172.

Heine, U., Munoz, E. F., *et al.* (1987). Role of transforming growth factor-beta in the development of the mouse embryo. *J. Cell Biol.* **105**(6 Pt 2), 2861–2876.

Heldin, C. H., Backstrom, G., *et al.* (1988). Binding of different dimeric forms of PDGF to human fibroblasts: evidence for two separate receptor types. *Embo. J.* **7**(5), 1387–1393.

Herradon, G., Ezquerra, L., *et al.* (2004). Pleiotrophin is an important regulator of the renin-angiotensin system in mouse aorta. *Biochem. Biophys. Res. Commun.* **324**(3), 1041–1047.

Hom, D. B., and Winters, M. (1998). Effects of angiogenic growth factors and a penetrance enhancer on composite grafts. *Ann. Otol. Rhinol. Laryngol.* **107**(9 Pt 1), 769–774.

Ignotz, R. A., and Massague, J. (1986). Transforming growth factor-beta stimulates the expression of fibronectin and collagen and their incorporation into the extracellular matrix. *J. Biol. Chem.* **261**(9), 4337–4345.

Jaye, M., McConathy, E., *et al.* (1985). Modulation of the sis gene transcript during endothelial cell differentiation *in vitro*. *Science* **228**(4701), 882–825.

Kadomatsu, K., Tomomura, M., *et al.* (1988). cDNA cloning and sequencing of a new gene intensely expressed in early differentiation stages of embryonal carcinoma cells and in mid-gestation period of mouse embryogenesis. *Biochem. Biophys. Res. Commun.* **151**(3), 1312–1318.

Katz, I. R., Thorbecke, G. J., *et al.* (1986). Protease-induced immunoregulatory activity of platelet factor 4. *Proc. Natl. Acad. Sci. USA* **83**(10), 3491–3495.

Kawahara, R. S., and Deuel, T. F. (1989). Platelet-derived growth factor-inducible gene JE is a member of a family of small inducible genes related to platelet factor 4. *J. Biol. Chem.* **264**(2), 679–682.

Kawahara, R. S., Deng, Z. W., *et al.* (1991). PDGF and the small inducible gene (SIG) family: roles in the inflammatory response. *Adv. Exp. Med. Biol.* **305**, 79–87.

Kimelman, D., and Kirschner, M. (1987). Synergistic induction of mesoderm by FGF and TGF-beta and the identification of an mRNA coding for FGF in the early Xenopus embryo. *Cell* **51**(5), 869–877.

Kohase, M., May, L. T., *et al.* (1987). A cytokine network in human diploid fibroblasts: interactions of beta-interferons, tumor necrosis factor, platelet-derived growth factor, and interleukin-1. *Mol. Cell Biol.* **7**(1), 273–280.

Larsen, C. G., Anderson, A. O., *et al.* (1989). The neutrophil-activating protein (NAP-1) is also chemotactic for T lymphocytes. *Science* **243**(4897), 1464–1466.

Leonard, E. J., Skeel, A., *et al.* (1990). Leukocyte specificity and binding of human neutrophil attractant/activation protein-1. *J. Immunol.* **144**(4), 1323–1330.

Li, Y. S., Milner, P. G., *et al.* (1990). Cloning and expression of a developmentally regulated protein that induces mitogenic and neurite outgrowth activity. *Science* **250**(4988), 1690–1694.

Li, Y. S., Gurrieri, M., *et al.* (1992a). Pleiotrophin gene expression is highly restricted and is regulated by platelet-derived growth factor. *Biochem. Biophys. Res. Commun.* **184**(1), 427–432.

Li, Y. S., Hoffman, R. M., *et al.* (1992). Characterization of the human pleiotrophin gene. Promoter region and chromosomal localization. *J. Biol. Chem.* **267**(36), 26011–26016.

Libby, P., Warner, S. J., *et al.* (1988). Production of platelet-derived growth factor–like mitogen by smooth-muscle cells from human atheroma. *N. Engl. J. Med.* **318**(23), 1493–1498.

Liebovich, S. (1975). The role of the macrophage in wound repair. *Am. J. Pathol.* **78**, 71–100.

Lynch, S. E., Nixon, J. C., *et al.* (1987). Role of platelet-derived growth factor in wound healing: synergistic effects with other growth factors. *Proc. Natl. Acad. Sci. USA* **84**(21), 7696–7700.

Mailleux, P., Vanderwinden, J. M., *et al.* (1992). The new growth factor pleiotrophin (HB-GAM) mRNA is selectively present in the meningothelial cells of human meningiomas. *Neurosci. Lett.* **142**(1), 31–35.

Maione, T. E., Gray, G. S., *et al.* (1990). Inhibition of angiogenesis by recombinant human platelet factor-4 and related peptides. *Science* **247**(4938), 77–79.

Majesky, M. W., Benditt, E. P., *et al.* (1988). Expression and developmental control of platelet-derived growth factor A-chain and B-chain/Sis genes in rat aortic smooth muscle cells. *Proc. Natl. Acad. Sci. USA* **85**(5), 1524–1528.

Martinet, Y., Rom, W. N., *et al.* (1987). Exaggerated spontaneous release of platelet-derived growth factor by alveolar macrophages from patients with idiopathic pulmonary fibrosis. *N. Engl. J. Med.* **317**(4), 202–209.

Matoskova, B., Rorsman, F., *et al.* (1989). Alternative splicing of the platelet-derived growth factor A-chain transcript occurs in normal as well as tumor cells and is conserved among mammalian species. *Mol. Cell Biol.* **9**(7), 3148–3150.

Matsui, T., Pierce, J. H., *et al.* (1989). Independent expression of human alpha or beta platelet-derived growth factor receptor cDNAs in a naive hematopoietic cell leads to functional coupling with mitogenic and chemotactic signaling pathways. *Proc. Natl. Acad. Sci. USA* **86**(21), 8314–8318.

Matsushima, K., Larsen, C. G., *et al.* (1989). Purification and characterization of a novel monocyte chemotactic and activating factor produced by a human myelomonocytic cell line. *J. Exp. Med.* **169**(4), 1485–1490.

McGee, G. S., Davidson, J. M., *et al.* (1988). Recombinant basic fibroblast growth factor accelerates wound healing. *J. Surg. Res.* **45**(1), 145–153.

McNeil, P. L., Muthukrishnan, L., *et al.* (1989). Growth factors are released by mechanically wounded endothelial cells. *J. Cell Biol.* **109**(2), 811–822.

Mendoza, A. E., Young, R., *et al.* (1990). Increased platelet-derived growth factor A chain expression in human uterine smooth muscle cells during the physiologic hypertrophy of pregnancy. *Proc. Natl. Acad. Sci. USA* **87**(6), 2177–2181.

Meng, K., Rodriguez-Pena, A., *et al.* (2000). Pleiotrophin signals increased tyrosine phosphorylation of beta-catenin through inactivation of the intrinsic catalytic activity of the receptor-type protein tyrosine phosphatase beta/zeta. *Proc. Natl. Acad. Sci. USA* **97**(6), 2603–2608.

Mentlein, R., and Held-Feindt, J. (2002). Pleiotrophin, an angiogenic and mitogenic growth factor, is expressed in human gliomas. *J. Neurochem.* **83**(4), 747–753.

Mercola, M., Melton, D. A., *et al.* (1988). Platelet-derived growth factor-A chain is maternally encoded in *Xenopus* embryos. *Science* **241**(4870), 1223–1225.

Milner, P. G., Li, Y. S., *et al.* (1989). A novel 17-kD heparin-binding growth factor (HBGF-8) in bovine uterus: purification and N-terminal amino acid sequence. *Biochem. Biophys. Res. Commun.* **165**(3), 1096–1103.

Montesano, R., Vassalli, J. D., *et al.* (1986). Basic fibroblast growth factor induces angiogenesis *in vitro*. *Proc. Natl. Acad. Sci. USA* **83**(19), 7297–7301.

Mustoe, T. A., Pierce, G. F., *et al.* (1987). Accelerated healing of incisional wounds in rats induced by transforming growth factor-beta. *Science* **237**(4820), 1333–1336.

Mustoe, T. A., Purdy, J., *et al.* (1989). Reversal of impaired wound healing in irradiated rats by platelet-derived growth factor-BB. *Am. J. Surg.* **158**(4), 345–350.

Mustoe, T. A., Pierce, G. F., *et al.* (1991). Growth factor–induced acceleration of tissue repair through direct and inductive activities in a rabbit dermal ulcer model. *J. Clin. Invest.* **87**(2), 694–703.

Mustoe, T. A., Cutler, N. R., *et al.* (1994). A phase II study to evaluate recombinant platelet-derived growth factor-BB in the treatment of stage 3 and 4 pressure ulcers. *Arch. Surg.* **129**(2), 213–219.

Muthukrishnan, L., Warder, E., *et al.* (1991). Basic fibroblast growth factor is efficiently released from a cytolsolic storage site through plasma membrane disruptions of endothelial cells. *J. Cell Physiol.* **148**(1), 1–16.

Nagaoka, I., Someya, A., *et al.* (1991). Comparative studies on the platelet-derived growth factor-A and -B gene expression in human monocytes. *Comp. Biochem. Physiol. B* **100**(2), 313–319.

Nakagawara, A., Milbrandt, J., *et al.* (1995). Differential expression of pleiotrophin and midkine in advanced neuroblastomas. *Cancer Res.* **55**(8), 1792–1797.

Ostman, A., Backstrom, G., *et al.* (1989). Expression of three recombinant homodimeric isoforms of PDGF in *Saccharomyces cerevisiae*: evidence for difference in receptor binding and functional activities. *Growth Factors* **1**(3), 271–281.

Pariser, H., Ezquerra, L., *et al.* (2005). Fyn is a downstream target of the pleiotrophin/receptor protein tyrosine phosphatase beta/zeta-signaling pathway: regulation of tyrosine phosphorylation of Fyn by pleiotrophin. *Biochem. Biophys. Res. Commun.* **332**(3), 664–669.

Pariser, H., Perez-Pinera, P., *et al.* (2005). Pleiotrophin stimulates tyrosine phosphorylation of beta-adducin through inactivation of the transmembrane receptor protein tyrosine phosphatase beta/zeta. *Biochem. Biophys. Res. Commun.* **335**(1), 232–239.

Paulsson, Y., Hammacher, A., *et al.* (1987). Possible positive autocrine feedback in the prereplicative phase of human fibroblasts. *Nature* **328**(6132), 715–717.

Pepper, M. S., Ferrara, N., *et al.* (1991). Vascular endothelial growth factor (VEGF) induces plasminogen activators and plasminogen activator inhibitor-1 in microvascular endothelial cells. *Biochem. Biophys. Res. Commun.* **181**(2), 902–906.

Pierce, G. F., and Mustoe, T. A. (1995). Pharmacologic enhancement of wound healing. *Annu. Rev. Med.* **46**, 467–481.

Pierce, G. F., Mustoe, T. A., *et al.* (1988). *In vivo* incisional wound healing augmented by platelet-derived growth factor and recombinant c-sis gene homodimeric proteins. *J. Exp. Med.* **167**(3), 974–987.

Pierce, G. F., Mustoe, T. A., *et al.* (1989a). Transforming growth factor beta reverses the glucocorticoid-induced wound-healing deficit in rats: possible regulation in macrophages by platelet-derived growth factor. *Proc. Natl. Acad. Sci. USA* **86**(7), 2229–2233.

Pierce, G. F., Mustoe, T. A., *et al.* (1989b). Platelet-derived growth factor and transforming growth factor-beta enhance tissue repair activities by unique mechanisms. *J. Cell Biol.* **109**(1), 429–440.

Pierce, G. F., Vande Berg, J., *et al.* (1991). Platelet-derived growth factor-BB and transforming growth factor beta 1 selectively modulate glycosaminoglycans, collagen and myofibroblasts in excisional wounds. *Am. J. Pathol.* **138**(3), 629–646.

Pierce, G. F., Tarpley, J. E., *et al.* (1992). Platelet-derived growth factor (BB homodimer), transforming growth factor-beta 1, and basic fibroblast growth factor in dermal wound healing. Neovessel and matrix formation and cessation of repair. *Am. J. Pathol.* **140**(6), 1375–1388.

Pierce, G. F., Tarpley, J. E., *et al.* (1994). Tissue repair processes in healing chronic pressure ulcers treated with recombinant platelet-derived growth factor BB. *Am. J. Pathol.* **145**(6), 1399–1410.

Postlethwaite, A. E., Snyderman, R., *et al.* (1976). The chemotactic attraction of human fibroblasts to a lymphocyte-derived factor. *J. Exp. Med.* **144**(5), 1188–1203.

Postlethwaite, A. E., Seyer, J. M., *et al.* (1978). Chemotactic attraction of human fibroblasts to type I, II, and III collagens and collagen-derived peptides. *Proc. Natl. Acad. Sci. USA* **75**(2), 871–875.

Postlethwaite, A. E., Snyderman, R., *et al.* (1979). Generation of a fibroblast chemotactic factor in serum by activation of complement. *J. Clin. Invest.* **64**(5), 1379–1385.

Quaglino, D., Jr., Nanney, L. B., *et al.* (1990). Transforming growth factor-beta stimulates wound healing and modulates extracellular matrix gene expression in pig skin. I. Excisional wound model. *Lab. Invest.* **63**(3), 307–319.

Rappolee, D. A., Mark, D., *et al.* (1988). Wound macrophages express TGF-alpha and other growth factors *in vivo*: analysis by mRNA phenotyping. *Science* **241**(4866), 708–712.

Rauvala, H. (1989). An 18-kD heparin-binding protein of developing brain that is distinct from fibroblast growth factors. *Embo. J.* **8**(10), 2933–2941.

Reuterdahl, C., Sundberg, C., *et al.* (1993). Tissue localization of beta receptors for platelet-derived growth factor and platelet-derived growth factor B chain during wound repair in humans. *J. Clin. Invest.* **91**(5), 2065–2075.

Richard, J. L., Parer-Richard, C., *et al.* (1995). Effect of topical basic fibroblast growth factor on the healing of chronic diabetic neuropathic ulcer of the foot. A pilot, randomized, double-blind, placebo-controlled study. *Diabetes Care* **18**(1), 64–69.

Richardson, W. D., Pringle, N., *et al.* (1988). A role for platelet-derived growth factor in normal gliogenesis in the central nervous system. *Cell* **53**(2), 309–319.

Richmond, A., Balentien, E., *et al.* (1988). Molecular characterization and chromosomal mapping of melanoma growth stimulatory activity, a growth factor structurally related to beta-thromboglobulin. *Embo. J.* **7**(7), 2025–2033.

Roberts, A. B., Sporn, M. B., *et al.* (1986). Transforming growth factor type beta: rapid induction of fibrosis and angiogenesis *in vivo* and stimulation of collagen formation *in vitro*. *Proc. Natl. Acad. Sci. USA* **83**(12), 4167–4171.

Robinson, E. A., Yoshimura, T., *et al.* (1989). Complete amino acid sequence of a human monocyte chemoattractant, a putative mediator of cellular immune reactions. *Proc. Natl. Acad. Sci. USA* **86**(6), 1850–1854.

Robson, M. C., Phillips, L. G., *et al.* (1992). Platelet-derived growth factor BB for the treatment of chronic pressure ulcers. *Lancet* **339**(8784), 23–35.

Rollins, B. J., Morrison, E. D., *et al.* (1988). Cloning and expression of JE, a gene inducible by platelet-derived growth factor and whose product has cytokine-like properties. *Proc. Natl. Acad. Sci. USA* **85**(11), 3738–3742.

Rollins, B. J., Stier, P., *et al.* (1989). The human homolog of the JE gene encodes a monocyte secretory protein. *Mol. Cell Biol.* **9**(11), 4687–4695.

Ross, R. (1968). The fibroblast and wound repair. *Biol. Rev. Camb. Philos. Soc.* **43**(1), 51–96.

Rossi, P., Karsenty, G., *et al.* (1988). A nuclear factor 1 binding site mediates the transcriptional activation of a type I collagen promoter by transforming growth factor-beta. *Cell* **52**(3), 405–414.

Rubin, K., Tingstrom, A., *et al.* (1988). Induction of B-type receptors for platelet-derived growth factor in vascular inflammation: possible implications for development of vascular proliferative lesions. *Lancet* **1**(8599), 1353–1356.

Schmid, J., and Weissmann, C. (1987). Induction of mRNA for a serine protease and a beta-thromboglobulin-like protein in mitogen-stimulated human leukocytes. *J. Immunol.* **139**(1), 250–256.

Sejersen, T., Betsholtz, C., *et al.* (1986). Rat skeletal myoblasts and arterial smooth muscle cells express the gene for the A chain but not the gene for the B chain (c-sis) of platelet-derived growth factor (PDGF) and produce a PDGF-like protein. *Proc. Natl. Acad. Sci. USA* **83**(18), 6844–6848.

Senior, R. M., Griffin, G. L., *et al.* (1982). Chemotactic responses of fibroblasts to tropoelastin and elastin-derived peptides. *J. Clin. Invest.* **70**(3), 614–618.

Senior, R. M., Griffin, G. L., *et al.* (1983). Chemotactic activity of platelet alpha granule proteins for fibroblasts. *J. Cell Biol.* **96**(2), 382–385.

Senior, R. M., Griffin, G. L., *et al.* (1989). Platelet alpha-granule protein-induced chemotaxis of inflammatory cells and fibroblasts. *Methods Enzymol.* **169**, 233–244.

Seppa, H. E., Yamada, K. M., *et al.* (1981). The cell binding fragment of fibronectin is chemotactic for fibroblasts. *Cell Biol. Int. Rep.* **5**(8), 813–819.

Shaw, R. J., Doherty, D. E., *et al.* (1990). Adherence-dependent increase in human monocyte PDGF(B) mRNA is associated with increases in c-fos, c-jun, and EGR2 mRNA. *J. Cell Biol.* **111**(5 Pt 1), 2139–2148.

Shure, D., Senior, R. M., *et al.* (1992). PDGF AA homodimers are potent chemoattractants for fibroblasts and neutrophils, and for monocytes activated by lymphocytes or cytokines. *Biochem. Biophys. Res. Commun.* **186**(3), 1510–1514.

Silos-Santiago, I., Yeh, H. J., *et al.* (1996). Localization of pleiotrophin and its mRNA in subpopulations of neurons and their corresponding axonal tracts suggests important roles in neural–glial interactions during development and in maturity. *J. Neurobiol.* **31**(3), 283–296.

Soma, Y., Dvonch, V., *et al.* (1992). Platelet-derived growth factor AA homodimer is the predominant isoform in human platelets and acute human wound fluid. *Faseb. J.* **6**(11), 2996–3001.

Sporn, M. B., Roberts, A. B., *et al.* (1983). Polypeptide transforming growth factors isolated from bovine sources and used for wound healing *in vivo*. *Science* **219**(4590), 1329–1331.

Sporn, M. B., Roberts, A. B., *et al.* (1986). Transforming growth factor-beta: biological function and chemical structure. *Science* **233**(4763), 532–534.

Sprugel, K. H., McPherson, J. M., *et al.* (1987). Effects of growth factors *in vivo*. I. Cell ingrowth into porous subcutaneous chambers. *Am. J. Pathol.* **129**(3), 601–613.

Steenfos, H. H. (1994). Growth factors and wound healing. *Scand. J. Plast. Reconstr. Surg. Hand Surg.* **28**(2), 95–105.

Stoica, G. E., Kuo, A., *et al.* (2001). Identification of anaplastic lymphoma kinase as a receptor for the growth factor pleiotrophin. *J. Biol. Chem.* **276**(20), 16772–16779.

Sundberg, C., Ljungstrom, M., *et al.* (1993). Microvascular pericytes express platelet-derived growth factor-beta receptors in human healing wounds and colorectal adenocarcinoma. *Am. J. Pathol.* **143**(5), 1377–1388.

Terracio, L., Ronnstrand, L., *et al.* (1988). Induction of platelet-derived growth factor receptor expression in smooth muscle cells and fibroblasts upon tissue culturing. *J. Cell Biol.* **107**(5), 1947–1957.

Timmers, H. T., Pronk, G. J., *et al.* (1990). Analysis of the rat JE gene promoter identifies an AP-1 binding site essential for basal expression but not for TPA induction. *Nucleic Acids Res.* **18**(1), 23–34.

Tong, B. D., Auer, D. E., *et al.* (1987). cDNA clones reveal differences between human glial and endothelial cell platelet-derived growth factor A-chains. *Nature* **328**(6131), 619–621.

Twardzik, D. R. (1985). Differential expression of transforming growth factor-alpha during prenatal development of the mouse. *Cancer Res.* **45**(11 Pt 1), 5413–5416.

Tzeng, D. Y., Deuel, T. F., *et al.* (1984). Platelet-derived growth factor promotes polymorphonuclear leukocyte activation. *Blood* **64**(5), 1123–1128.

Vacherot, F., Laaroubi, K., *et al.* (1995). Up-regulation of heparin-affin regulatory peptide by androgen. *In Vitro Cell Dev. Biol. Anim.* **31**(9), 647–648.

Van Damme, J., Decock, B., *et al.* (1989). Identification by sequence analysis of chemotactic factors for monocytes produced by normal and transformed cells stimulated with virus, double stranded RNA or cytokine. *Eur. J. Immunol.* **19**(12), 2367–2373.

Walz, A., Peveri, P., *et al.* (1987). Purification and amino acid sequencing of NAF, a novel neutrophil-activating factor produced by monocytes. *Biochem. Biophys. Res. Commun.* **149**(2), 755–761.

Watanabe, K., Konishi, K., *et al.* (1989). The neutrophil chemoattractant produced by the rat kidney epithelioid cell line NRK-52E is a protein related to the KC/gro protein. *J. Biol. Chem.* **264**(33), 19559–19563.

Weber, D., Klomp, H. J., *et al.* (2000). Pleiotrophin can be rate-limiting for pancreatic cancer cell growth. *Cancer Res.* **60**(18), 5284–5288.

Weeks, D. L., and Melton, D. A. (1987). A maternal mRNA localized to the vegetal hemisphere in *Xenopus* eggs codes for a growth factor related to TGF-beta. *Cell* **51**(5), 861–867.

Wolpe, S. D., Davatelis, G., *et al.* (1988). Macrophages secrete a novel heparin-binding protein with inflammatory and neutrophil chemokinetic properties. *J. Exp. Med.* **167**(2), 570–581.

Yoshimura, T., Matsushima, K., *et al.* (1987). Purification of a human monocyte-derived neutrophil chemotactic factor that has peptide sequence similarity to other host defense cytokines. *Proc. Natl. Acad. Sci. USA* **84**(24), 9233–9237.

Yoshimura, T., Robinson, E. A., *et al.* (1989a). Purification and amino acid analysis of two human glioma-derived monocyte chemoattractants. *J. Exp. Med.* **169**(4), 1449–1459.

Yoshimura, T., Yuhki, N., *et al.* (1989b). Human monocyte chemoattractant protein-1 (MCP-1). Full-length cDNA cloning, expression in mitogen-stimulated blood mononuclear leukocytes, and sequence similarity to mouse competence gene JE. *FEBS Lett.* **244**(2), 487–493.

Zhang, N., Zhong, R., *et al.* (1997). Human breast cancer growth inhibited *in vivo* by a dominant negative pleiotrophin mutant. *J. Biol. Chem.* **272**(27), 16733–16736.

Zhang, N., and Deuel, T. F. (1999). Pleiotrophin and midkine, a family of mitogenic and angiogenic heparin-binding growth and differentiation factors. *Curr. Opin. Hematol.* **6**(1), 44–50.

# Chapter Fifteen

# Mechanochemical Control of Cell Fate Switching

*Donald E. Ingber*

## I. INTRODUCTION

Tissue engineering has as its main goal the fabrication of artificial tissues for use as replacements for damaged body structures. Great advances have been made in terms of developing prosthetic devices that can repair structural defects (e.g., vascular grafts) and even replace complex mechanical behaviors (e.g., artificial joints). However, the challenge for the future is to develop tissue substitutes that restore the normal biochemical functions of living tissues in addition to their structural features. To accomplish this feat, we must first establish precise design criteria for tissue fabrication. These design features should be based on a thorough understanding of the molecular and cellular basis of tissue regulation. They also must take into account the important role that insoluble extracellular matrix (ECM) scaffolds and mechanical stresses play in control of cell phenotypes or "fates" (e.g., growth, differentiation, motility, apoptosis, different stem cell lineages) during tissue formation and repair. This latter point is critical since the spatial organization of cells and the mechanical constraints imposed on them as they grow actively regulate tissue development (Ingber, 2006a) as well as tissue and organ function throughout adult life (Ingber, 2006b).

The goal of this chapter is not to provide an extensive review of literature in the field of ECM biology or tissue development. Rather, I will summarize the known functions of ECM and describe some insights we have made relating to how ECM regulates cell growth and function as well as tissue morphogenesis. Our analysis of this regulatory mechanism should be of particular interest to the tissue engineer, because it has led to the identification of critical chemical and mechanical features of ECM that are responsible for control of cell fate switching in developing tissues. In addition, I hope to introduce the reader to some unanswered puzzles in developmental biology that, if deciphered, could provide powerful new approaches to tissue regeneration and repair.

## II. EXTRACELLULAR MATRIX STRUCTURE AND FUNCTION

### Composition and Organization

One of the most critical elements of tissue engineering is the ability to mimic the ECM scaffolds that normally serve to organize cells into tissues. ECMs are composed of different collagen types, large glycoproteins (e.g., fibronectin, laminin, entactin, osteopontin), and proteoglycans that contain large glycosaminoglycan side chains (e.g., heparan sulfate, chondroitin sulfate, dermatan suflate, keratan sulfate, hyaluronic acid) (see Chapter 8, by Olsen, for more details). While all ECMs share these components, the organization, form, and mechanical properties of ECMs can vary widely in different tissues, depending on the chemical composition and three-dimensional (3D) organization of the specific ECM components that are present. For example, interstitial collagens (e.g., types I, III) self-assemble into 3D lattices, which, in turn, bind fibronectin and proteoglycans. This type of native ECM hydrogel forms the backbone of loose connective tissues, such as dermis. In contrast, basement membrane collagens (types IV, V) assemble into planar arrays; when these collagenous sheets interact with fibronectin, laminin, and heparan sulfate proteoglycan, a planar ECM results (i.e., the basement membrane). The ability of tendons to resist tension and of cartilage and bone to resist compression similarly result from local differences in the organization and composition of the ECM.

### In Vivo Foundation for Cell Anchorage

The first and foremost function of ECM in tissue development is its role as a physiological substratum for cell attachment. This feature is easily visualized by treating whole tissues with ECM-degrading enzymes (collagenase, proteases); cell detachment and loss of cell and tissue morphology rapidly result. Cells that are dissociated in this manner can reattach to an artificial culture substrate (e.g., plastic, glass). However, adhesion is again mediated by cell binding to ECM components that are either experimentally immobilized on the culture surface, deposited *de novo* by the adhering cells, or spontaneously adsorbed from serum (e.g., fibronectin, vitronectin) (Kleinman *et al.*, 1981; Madri and Stenn, 1982; Stenn *et al.*,1979; Wicha *et al.*, 1979; Salomon *et al.*, 1981; Bissell *et al.*, 1986; Ingber, 1990). In fact, standard tissue culture plates are actually bacteriological (nonadhesive) plastic dishes that have been chemically treated using proprietary methods to enhance adsorption of serum- and cell-derived ECM proteins. To summarize, cells do not attach directly to the culture substrate (i.e., plastic or glass); rather, they bind to intervening ECM components that are adsorbed (or derivatized) to these substrates. For this reason, cell adhesion can be prevented by coating normally adhesive culture surfaces with polymers that prevent protein adsorption, such as poly(hydroxyethylmethacrylate) (Folkman and Moscona, 1978; Ingber, 1990).

### Spatial Organizer of Polarized Epithelium

Living cells exhibit polarized form as well as function (e.g., basal nuclei, supranuclear Golgi complex, apical secretory granules in secretory epithelia). Dissociated cells fail to orient in a consistent manner when cultured on standard tissue culture substrata or on interstitial connective tissue. In the case of epithelial cells, normal polarized form is often restored if the cells synthesize and accumulate their own ECM or if they are cultured on exogenous basement membrane (i.e., the specialized epithelial ECM) (Emerman and Pitelka, 1977; Chambard *et al.*, 1981; Ingber *et al.*, 1986). These types of studies suggest that basement membrane normally serves to integrate and maintain individual cells within a polarized epithelium. Clearly, there are many intracellular and intercellular determinants of polarized cell form and function (e.g., cytoskeletal organization, organelle movement, junctional complex formation). However, anchorage to ECM appears to provide an initial point of orientation and stability on which additional steps in the epithelial organization cascade can build. ECM may regulate the orientation of other cell types (e.g., chondrocytes, osteoclasts) as well.

### Scaffolding for Orderly Tissue Renewal

All tissues are dynamic structures that exhibit continual turnover of all molecular and cellular components. Thus, it is the maintenance of tissue pattern integrity that is most critical to the survival of the organism, and regeneration of normal tissue architecture is one of the main goals of tissue engineering. Maintenance of specialized tissue form requires that cells lost due to injury or aging must be replaced in an organized fashion. Importantly, orderly tissue renewal depends on the continued presence of insoluble ECM scaffoldings, which act as templates that maintain the original architectural form and ensure accurate regeneration of pre-existing structures (Vracko, 1974). For example, when cells within a tissue are killed by freezing or treatment with toxic chemicals, all of the cellular components die and are removed; however, the basement membranes often remain intact. These residual ECM scaffoldings ensure correct repositioning of cells (e.g., cell polarity) and restore different cell types to their correct locations (e.g., muscle cells within muscle basement membrane, nerve cells in nerve sheaths, endothelium within vessels) in addition to promoting the cell migration and growth that are required for repair of all the component tissues. Conversely, loss of ECM integrity during wound healing results in disorganization of tissue pattern and, thus, scar formation. Uncoupling between basement membrane extension and cell doubling also leads to disorganization of tissue morphology during neoplastic transformation (Ingber *et al.*, 1981).

### Establishment of Tissue Microenvironments

ECMs often establish a physical boundary between neighboring tissues. For example, the basement membrane normally restricts mixing between the epithelium and underlying connective tissue, and compromise of basement membrane integrity is indicative of the onset of malignant invasion when seen in the context of tumor formation (Ingber *et al.*, 1981). The ECM boundary also may regulate macromolecular transport between adjacent tissues, given that the basement membrane forms the semipermeable filtration barrier in the kidney glomerulus (Farquhar, 1978).

However, little is known about this potential function of the ECM in other local tissue microenvironments.

## Sequestration, Storage, and Presentation of Soluble Regulatory Molecules

ECMs may modulate tissue growth and morphogenesis through their ability to bind, store, and eventually release soluble regulators of morphogenesis. For example, the soluble mitogen, basic fibroblast growth factor (FGF) exists in an immobilized form in ECMs deposited by cells cultured *in vitro* (Vlodavsky *et al.*, 1987) as well as in normal tissues (e.g., corneal basement membranes) *in vivo* (Folkman *et al.*, 1988). In fact, many cytokines (e.g., FGF, VEGF, HGF, insulin-like growth factor, hematopoietic growth factors) are normally stored bound to heparin, heparan sulfate, or some other specific binding proteins within natural ECMs (Gordon, 1991; Folkman and Shing, 1992). The low growth rate observed in most normal tissues may result from sequestration of mitogens by ECM, whereas release of these stored factors (*stormones*) due to injury or hormonally-induced changes in ECM turnover may help to switch growth on locally. Binding of other types of regulatory molecules to the endothelial basement membrane (e.g., plasminogen activator inhibitor; Pollanen *et al.*, 1987) also may play a role in tissue physiology (e.g., blood coagulation, cell migration).

## Regulator of Cell Growth, Differentiation, and Apoptosis

Most normal (nontransformed) cells grow only when attached and spread on a solid substrate (Folkman and Moscona, 1978). Cells attach and spread *in vitro* either by depositing new ECM components or by binding to exogenous ECM (Kleinman *et al.*, 1981; Madri and Stenn, 1982; Stenn *et al.*, 1979; Wicha *et al.*, 1979; Salomon *et al.*, 1981; Bissell *et al.*, 1986; Ingber, 1990). In fact, cell spreading and growth can be suppressed by inhibiting ECM deposition *in vitro* using drugs (Madri and Stenn, 1982; Stenn *et al.*, 1979; Wicha *et al.*, 1979 ). Cell growth stimulated by soluble mitogens also has been shown to vary, depending on the type of ECM component used for cell attachment (e.g., collagen versus fibronectin; Kleinman *et al.*, 1981; Bissell *et al.*, 1986; Ingber *et al.*, 1987) as well as on the mechanical properties of the ECM (e.g., malleable gel versus rigid ECM-coated dish; Li *et al.*, 1987; Kubota *et al.*, 1987; Ben Ze'ev *et al.*, 1988; Opas, 1989; Vernon *et al.*, 1992).

Interestingly, ECM substrates that promote growth tend to suppress differentiation, and vice versa. For example, many cells proliferate and lose differentiated features when cultured on attached type I collagen gels that can resist cell tension and promote cell spreading. In contrast, the same cells cease growing and increase expression of tissue-specific functions (e.g., albumin secretion in hepatocytes, milk secretion and acinus formation by mammary cells, capillary tube formation by endothelial cells) if cultured on the same gels

that are made flexible by floating them free in medium or on attached ECM gels that exhibit high malleability (e.g., basement membrane gels, such as Matrigel). Under these conditions, the cells exert tension across their adhesions, resulting in contraction of the ECM gel and cell rounding, which, in turn, shut off growth and turn on differentiation-specific gene functions. The differentiation-inducing effects of these malleable ECM substrates also can be suppressed by making the gels rigid through chemical fixation (Li *et al.*, 1987; Opas, 1989), thus confirming the critical role of cell-generated mechanical forces in this response.

While local changes in ECM turnover may promote tissue remodeling, large-scale breakdown of the ECM may force the same growing tissues to undergo involution. Many cultured cells rapidly lose viability and undergo programmed cell death (i.e., apoptosis) when detached from ECM and maintained in a round form (Meredith *et al.*, 1993; Chen *et al.*, 1997). Loss of basement membrane integrity is also observed in regions of tissues that are actively regressing (Trelstad *et al.*, 1982; Ingber *et al.*, 1986), and growing tissues (e.g., capillaries, mammary gland) can be induced to involute using pharmacological agents (e.g., proline analogs) that inhibit ECM deposition and lead to basement membrane dissolution *in vivo* (Ingber *et al.*, 1986; Wicha *et al.*, 1980; Ingber and Folkman, 1988). Recent transgenic mice studies confirm that growing tissues can be made to involute by shifting the endogenous proteolytic balance such that total ECM breakdown results (Talhouk *et al.*, 1992). These findings suggest that local changes in ECM composition and flexibility may regulate cell sensitivity to soluble mitogens and thereby control cell growth, viability, and function in the local tissue microenvironment.

## III. PATTERN FORMATION THROUGH ECM REMODELING

### Mesenchymal Control of Epithelial Pattern

Probably the greatest insight into the role of ECM in tissue development comes from analysis of embryogenesis. In the embryo, genesis of a tissue's characteristic form (e.g., tubular versus acinar) and deposition of ECM are both controlled by complex interactions between adjacent epithelial and mesenchymal cell societies. The epithelial cell is genetically programmed to express tissue-specific (differentiated) functions and to deposit the insoluble basement membrane, which functions as a common attachment foundation that both separates adjacent tissues and stabilizes tissue form (Banerjee *et al.*, 1977; Dodson and Hay, 1971). However, while production of tissue-specific cell products (cytodifferentiation) is determined by the epithelium, tissue pattern (histodifferentiation) is usually directed by the surrounding mesenchyme. For example, when embryonic mammary epithelium is isolated and combined with salivary mesenchyme, the mammary epithelial cells take on the form of the salivary gland, although they still produce milk proteins

(Sakakura *et al.*, 1976). However, the specificity of these epithelial–mesenchymal interactions can vary widely from organ to organ. For instance, embryonic salivary epithelia specifically require salivary mesenchyme for successful development, while pancreatic epithelia will undergo normal cytodifferentiation and histodifferentiation in response to mesenchyme isolated from a variety of embryonic tissues (Golosow and Grobstein, 1962).

### Tissue Patterning Through Localized ECM Remodeling

The complex tissue patterns generated through epithelial–mesenchymal interactions result from the establishment of local differentials in tissue growth and expansion in a microenvironment that is likely saturated with soluble mitogens. The classic work on salivary gland development by Bernfield and coworkers revealed that the epithelium imposes morphological stability through production of its basement membrane, whereas the mesenchyme produces local changes in tissue form, specifically by degrading basement membrane at selective sites (Banerjee *et al.*, 1977; Bernfield and Banerjee, 1978; Smith and Bernfield, 1982; Bernfield *et al.*, 1972; David and Bernfield, 1979). An increased rate of cell division is observed in the tips of growing lobules, where the highest rate of ECM breakdown and resynthesis (i.e., highest turnover rate) is also observed. At the same time, the mesenchyme slows basement membrane turnover and suppresses epithelial cell growth in the clefts of the glands. This is accomplished by secretion of fibrillar collagens that slow ECM degradation locally and thereby promote basement membrane accumulation in these regions. Similar local coupling between ECM turnover, cell growth rates, and tissue expansion is observed in many other developing tissues, including growing capillary blood vessels (Folkman, 1982).

It is important to note that increased ECM turnover involves enhanced rates of matrix synthesis as well as degradation. In fact, net basement membrane accumulation (i.e., increased area available for cell attachment) must result for epithelial tissues to grow and expand laterally; thus, the local rate of ECM synthesis must be greater than that of degradation in these high-turnover regions. If the rate of ECM degradation is significantly greater than the rate of synthesis, then net basement membrane dissolution along with cell retraction and rounding result. As described earlier, this would lead to cell death and tissue regression rather than expansion.

### Role of Mechanical Stresses

Before ending discussion of the role of ECM in pattern formation, it is critical to emphasize that while chemical regulators mediate tissue morphogenesis, the signals that are actually responsible for dictating tissue pattern are often mechanical in nature (Ingber, 2006b). The pattern-generating effects of compression on bone, shear on blood vessels, and tension on muscle are just a few examples. Mechanical stresses are also important for embryological development; however, internal cell-generated forces appear to play a more critical role (Ingber, 2006a). For example, mechanical tension that is generated via actomyosin filament sliding within the cytoskeleton of the cells that compose the embryo play a key role during gastrulation (Beloussov *et al.*, 1975) as well as during tissue morphogenesis (Ash *et al.*, 1973). In fact, the pattern of development can be experimentally altered by applying external stresses to the embryo using micropipettes (Beloussov *et al.*, 1988). Tensile forces that are generated internally within mesenchyme and transmitted across ECM are also likely required for the condensation of mesenchyme that commonly precedes formation of new organ rudiments. The pattern-generating capabilities of mesenchyme isolated from different developing tissues also correlates with differences in their ability to exert mechanical tension on external substrates (e.g., Nogawa and Nakanishi, 1987). Local changes in ECM turnover may drive morphogenetic patterning of tissues, in part by altering the mechanical compliance of the ECM and thereby changing cell shape or cytoskeletal tension (Ingber and Jamieson, 1985; Huang and Ingber, 1999), as is described next.

## IV. MECHANOCHEMICAL SWITCHING BETWEEN CELL FATES

Given the pivotal role that ECM plays in tissue development, many studies have been carried out to analyze how changes in cell–ECM interactions act locally to regulate cell sensitivity to soluble mitogens and thereby establish the differentials of growth, differentiation, apoptosis, and motility that are required for tissue morphogenesis. To accomplish this, simplified *in vitro* model systems have been developed that retain the minimal determinants necessary for maintenance of the physiological functions of interest (i.e., cell growth and differentiation) (Ingber, 1990; Ingber and Folkman, 1989; Mooney *et al.*, 1992). For example, to determine the effects of varying cell–ECM contacts directly, we precoated bacteriological petri dishes that were otherwise nonadhesive with different densities of purified ECM molecules, such as fibronectin, laminin, or different collagen types. Quiescent, serum-deprived cells were plated on these dishes in chemically defined medium that contained a constant and saturating amount of soluble growth factor.

When capillary endothelial cells were studied, DNA synthesis and cell-doubling rates increased in an exponential fashion as the density of immobilized ECM ligand was raised and cell spreading was promoted (Ingber and Folkman, 1989; Ingber, 1990). When higher cell-plating numbers were used to promote cell–cell interactions as well as cell–ECM contact formation, the capillary cells could be switched between growth and differentiation (capillary tube formation) in the presence of saturating amounts of soluble mitogen (FGF) simply by varying the ECM coating density (Ingber and Folkman, 1989). Specifically, when plated on a high ECM density (e.g., >500 ng/cm$^2$ fibronectin), the cells attached, spread extensively, formed many cell–cell contacts, and organized into a planar cell monolayer. When the same cells were plated on a low ECM density (<100 ng/cm$^2$), the cells attached but could not spread, and thus only cell clumps or aggregates were observed. When the same capillary cells were plated on a moderate density, cells first attached, spread, and formed cell–cell contacts, as they did on the higher ECM densities. However, the tensile forces generated by the cells appeared to overcome the resistance provided by their relatively weak ECM adhesions, and thus the cell aggregates began to retract over a period of hours until a mechanical equilibrium was attained. Under these conditions, formation of an extensive network comprising interconnected capillary tubes resulted. Many of these capillary tubes became elevated above the culture surface, although the network remained adherent to the culture dish at discrete points through interconnected multicellular aggregates. The importance of mechanical forces for this switching between growth and differentiation was confirmed by demonstrating that similar capillary tube formation could be induced on the high ECM density that normally promoted spreading and growth simply by increasing the cell-plating numbers and thereby amplifying the level of cell tension.

The same system was used to demonstrate similar shape- (stretch-) dependent switching between growth and differentiation in other cell types. For example, the growth and differentiation of primary rat hepatocytes could be controlled independent of cell–cell contact formation by varying cell–ECM contacts and cell spreading using the method described earlier (Mooney et al., 1992). Additional studies confirmed that ECM exerts its regulatory effects at the level of gene expression (Mooney et al., 1992; Hansen et al., 1994) and that these effects are mediated at least in part through modulation of the cytoskeleton (Mooney et al., 1995). These results are consistent with those from other laboratories, which demonstrate that malleable ECM gels (e.g., Matrigel, native collagen gels) that promote cell rounding also induce differentiation and suppress growth, whereas the opposite effects are produced when these gels are fixed and made rigid (Li et al., 1987; Kubota et al., 1987, Ben Ze'ev et al., 1988; Opas, 1989).

Altering cell–ECM contacts by varying ECM coating densities influences cell function via two distinct, but integrated, mechanisms. First, increasing the local density of immobilized ECM ligand promotes clustering of transmembrane ECM receptors on the cell surface, which are known as *integrins* (Hynes, 1987). Integrin clustering, in turn, activates a number of different chemical signaling pathways (e.g., tyrosine phosphorylation, inositol lipid turnover, Na$^+$/H$^+$ exchange, MAP kinase) that are also utilized by growth factor receptors to alter cellular biochemistry and gene expression (Ingber et al., 1990; Schwartz et al., 1991; McNamee et al., 1993; Plopper et al., 1995; Geiger et al., 2001). Activation of these signaling pathways likely plays an important role in control of cell differentiation and survival; however, integrin-dependent chemical signaling alone is not sufficient to explain how cells are induced to enter S phase and proliferate (Ingber, 1990; Hansen et al., 1994; Chen et al., 1997; Huang et al., 1998; Mammoto et al., 2004). A second mechanism that involves tension-dependent changes in cell shape and cytoskeletal organization or changes in the level prestress (isometric tension) within the cell and linked ECM also comes into play.

The importance of tension-dependent changes of cell shape and cytoskeletal organization was demonstrated directly by developing a model system in which cell distortion was varied independent of the local density of immobilized ECM molecule by controlling the spatial distribution of ECM anchors that can resist cell-generated traction forces. This was made possible by adapting a technique for forming spontaneously assembled monolayers (SAMs) of alkanethiols (Prime and Whitesides, 1991) to create micropatterned surfaces containing adhesive ECM islands with defined surface chemistry, shape, and position on the cell (micrometer) scale, separated by nonadhesive regions (Singhvi et al., 1994; Chen et al., 1997). The method involves fabrication of a flexible elastomeric stamp that exhibits the particular surface features of interest using photolithographic techniques. The topographic high points on the stamp (e.g., 40 × 40-μm square plateaus raised above recessed intervening regions) are coated with an alkanethiol ink, and the stamp is then apposed to a gold-coated surface. The alkanethiol forms SAMs covering only the regions where the stamp contacts the surface (i.e., 40 × 40-μm squares). Then the surrounding uncoated regions are filled with a SAM composed of similar alkanethiols that are conjugated to poly(ethylene glycol) (PEG), which prevents protein adsorption. The result is a chemically defined culture surface that is completely covered with a continuous SAM of alkanethiols; however, the local adhesive islands of defined geometry support protein adsorption, whereas the surrounding boundary regions coated with PEG do not. Thus, when these substrates are coated with a high density of purified ECM protein, such as laminin or fibronectin, adhesive islands of defined shape and position coated with a saturating density of matrix molecule result. Using this technique, cell position and shape can be precisely controlled because the cells only attach to the ECM-coated adhesive islands. In fact, even square and rectangular-shaped cells exhibiting 90° corners can be engineered using this approach (Singhvi et al., 1994; Chen et al., 1997).

This micropatterning method was first used to ask the following question: If cells are restricted to a small size similar to that produced on a low ECM coating concentration but the local density of immobilized integrin ligand is increased 1000-fold, which is the critical determinant of cell function — the ECM density or cell shape? The answer was cell shape. Primary hepatocytes remained quiescent on small adhesive islands coated with a high ECM density, even though the cells were stimulated with high concentrations of soluble growth factors, and cell growth (DNA synthesis) increased in parallel as the size of the adhesive island was increased (Singhvi *et al.*, 1994). Similar results were obtained with capillary endothelial cells (Chen *et al.*, 1997). Moreover, inhibition of hepatocyte growth on the small islands was accompanied by a concomitant increase in albumin secretion. In the case of endothelial cells, the cells were similarly induced to differentiate into capillary tubes when cultured on linear patterns that supported cell–cell contact as well as a moderate degree of cell distension, whereas the cells were induced to undergo apoptosis (cellular suicide) when cultured on the smallest islands, which fully prevented cell extension (Chen *et al.*, 1997; Dike *et al.*, 1999). Thus, cell shape and function can be engineered simply by altering the geometry of the cell's adhesive substrate.

Intracellular signals elicited by integrin receptor clustering due to ECM binding have been shown to be critical for control of cell growth and function (Hynes, 1987; Ingber *et al.*, 1990; Schwartz *et al.*, 1991; Geiger *et al.*, 2001). Thus, one could argue that cell shape and mechanical distortion of the CSK are not important. Instead, it might be the increase in the total area of cell–ECM contacts and the associated enhancement of integrin binding that dictate whether cells will grow, differentiate, or die on large versus small adhesive islands. To explore this further, substrates were designed in which a single small adhesive island (which would not support spreading or growth) was effectively broken up into many smaller islands (3–5 μm in diameter) that were separated by nonadhesive barrier regions (Chen *et al.*, 1997). When capillary cells were plated on these substrates, their processes stretched from island to island and the cells exhibited an overall extended form similar to cells on large islands. However, the total area of cell–ECM contact was almost identical to that exhibited by nongrowing cells on the smaller islands. These studies revealed that in the presence of optimal growth factors and high ECM binding, DNA synthesis was high in the cells that spread over multiple small islands, whereas apoptosis was completely shut off (Chen *et al.*, 1997), thus confirming that cell shape distortion is the critical governor of this response.

Importantly, cell distortion also impacts cell migration as well as stem cell lineage switching. For instance, when cells on square islands are stimulated with motility factors, the cells preferentially extend new motile processes (e.g., lamellipodia, filopodia, microspikes) from their corners, whereas they extend in all directions along the edge of round cells on circular islands (Parker *et al.*, 2002). Cells on polygonal ECM islands also prefer to form new lamellipodia from corners with acute angles rather than obtuse angles (Brock *et al.*, 2003). In contrast, when mesenchymal stem cells are cultured on different-size ECM islands, those on small

islands switch on the fat cell lineage, whereas the spread cells on larger ECM islands become bone cells (McBeath *et al.*, 2004). Thus, taken together, these results suggest that microfabrication methods might lead to new approaches to tissue engineering using microfabricated substrates. With this approach, it may be possible to direct cell migration, growth, and differentiation of stem cells in specific locations by modifying the surface chemistry and topography of artificial materials, instead of creating gradients of soluble chemokines.

The effects of shape distortion on cell fate switching may, in part, be mediated by changes in the level of isometric tension or prestress within the cytoskeleton. For example, both flexible and poorly adhesive culture dishes that inhibit growth and induce differentiation dissipate prestress, whereas substrates that stimulate cell spreading and growth (e.g., rigid dishes) support high levels of cytoskeletal tension. Small ECM islands, which inhibit cell spreading, also prevent pulmonary vascular smooth muscle cells from responding to vasconstrictors, such as endothelin-1 (as measured by increases of myosin light-chain phosphorylation), and similar effects are produced by culturing cells on flexible ECM substrates that dissipate tensional prestress in the ECM and in the interconnect cytoskeleton (i.e., linked through transmembrane integrins) (Polte *et al.*, 2004; Kumar *et al.*, 2006). In addition, cells preferentially differentiate on ECM substrates that are compliance matched with their cytoskeleton and that have a mechanical stiffness similar to that of their natural tissues (Griffen *et al.*, 2004; Engler *et al.*, 2004; Georges and Janmey, 2005). Cells also migrate up gradients of ECM stiffness, a process known as *durotaxis* (Lo *et al.*, 2000). These observations suggest that tissue formation may be controlled by changing the physical properties of either the ECM or the cell or by altering tension generation in the cytoskeleton.

Regional variations of cell shape distortion or of cytoskeletal prestress may similarly drive tissue patterning in the embryo. For example, during tissue morphogenesis, only a subset of cells must respond to soluble growth factors by proliferating locally and sprouting or budding outward relative to neighboring nongrowing cells. This process is repeated along the sides of the newly formed sprouts and buds, and the whole process reiterates over time; this is how the fractal-like patterns of tissues develop. This process is mediated by regional changes in ECM structure: The ECM thins in regions where new buds or sprouts will form due to local enzymatic degradation (Bernfield and Banerjee, 1978). Because tissues are prestressed (due to the action of cytoskeletal contractile forces), a local region of the tensed ECM may thin more than the rest, like a run in a woman's stocking. Cells anchored to this region will also stretch or spread, whereas the shape of neighboring cells on intact ECM would remain unchanged. If cell spreading promotes growth, then local cell growth differentials would result.

This possibility, that tissue morphogenesis may be controlled through changes in the mechanical force balance between the cytoskeleton and the ECM, is supported by recent experimental studies in embryonic systems. Analysis of the effects of cell shape on cell cycle progression *in vitro* have revealed that these effects are mediated by altering signal transduction through the small GTPase Rho and its downstream target, Rho-associated kinase (ROCK), which controls cytoskeletal tension generation (Huang *et al.*, 1998; Mammoto *et al.*, 2004). Importantly, when cytoskeletal contractility was suppressed in whole embryonic lung rudiments by inhibiting Rho or ROCK, both epithelial branching morphogenesis and angiogenesis were inhibited, and this was accompanied by decreased basement thinning (as if there was no tension in the stocking) (Moore *et al.*, 2005). In contrast, stimulating the Rho pathway had the opposite effect (i.e., it increased morphogenetic branching) at moderate levels of activation and caused complete organ contraction and inhibition of tissue growth at very high levels of stimulation (Moore *et al.*, 2005). Thus, local changes in ECM structure and cell shape caused by altering cytoskeletal prestress may govern how individual cells respond to chemical signals in their microenvironment *in vivo*, just as experimental studies using microfabricated ECM islands have demonstrated *in vitro*. This mechanism for establishing local growth differentials may play a critical role in morphogenesis in all developing systems (Ingber and Jamieson, 1985; Huang and Ingber, 1999).

## V. SUMMARY

In summary, our work has shown that the development of functional tissues, such as branching capillary networks, requires both soluble growth factors and insoluble ECM molecules. The ECM appears to be the dominant regulator, however, since it dictates whether individual cells will either proliferate, differentiate, undergo directional motility, or die locally in response to soluble stimuli. This local control mechanism is likely critical for the establishment of the local differentials of cell growth, motility, differentiation, apoptosis, and stem cell commitment that mediate pattern formation in all developing tissues.

Analysis of the molecular basis of these effects revealed that ECM molecules alter cell growth via both biochemical and biomechanical signaling mechanisms. ECM molecules cluster specific integrin receptors on the cell surface and thereby activate intracellular chemical signaling pathways (Ingber *et al.*, 1990; Schwartz *et al.*, 1991, McNamee *et al.*, 1993; Plopper *et al.*, 1995), stimulate expression of early growth response genes (e.g., c-fos, jun-B) (Hansen *et al.*, 1994), and induce quiescent cells to pass through the $G_0/G_1$ transition. However, in addition, the immobilized ECM components must physically resist cell tension and promote changes of cell shape (Ingber, 1990; Ingber and Folkman, 1989; Mooney *et al.*, 1992; Singhvi *et al.*, 1994; Chen *et al.*, 1997) and cytoskeletal tension (Mooney *et al.*,

1995; Huang *et al.*, 1998; Ingber *et al.*, 1995; Mammoto *et al.*, 2004) in order to promote full progression through $G_1$ and entry into S phase. In fact, studies with living and membrane-permeabilized cells confirm that changes in cell shape result from the action of mechanical tension generated within microfilaments and balanced by resistance sites within the underlying ECM (Sims *et al.*, 1992; Kumar *et al.*, 2006). Local changes of ECM structure or mechanics will therefore alter the ECM's ability to resist cell tractional forces exerted on integrins and thereby produce changes in cytoskeletal organization. This will, in turn, alter the level of isometric tension (prestress) in the cytoskeleton as well as modulate the activity of various signal transduction pathways inside the cell (Ingber, 2006). Taken together, this work suggests that the pattern-regulating information ECM conveys to cells is both chemical and mechanical in nature. Thus, design of future artificial ECMs for tissue-engineering applications must take into account both of these features.

## VI. THE FUTURE

Early tissue-engineering efforts by reconstructive surgeons and material scientists started with knowledge of the clinical need and of the mechanical behavior of connective tissues on the macroscopic scale and worked backwards. The long-term goal for the field is to design and fabricate tissue substitutes starting from first principles. This includes incorporating biologically inspired mechanical and architectural principles as well as an in-depth understanding of the molecular and biophysical basis of tissue regulation. Clearly, given the potent and varied functions of the ECM and fabrication of artificial ECMs will play a central role in all of these future efforts. We and others have already begun to explore the utility of synthetic bioerodible polymers as cell attachment substrates (Cima *et al.*, 1991; Mikos *et al.*, 1993) and the potential usefulness of immobilized synthetic ECM peptides for controlling cell growth and function (Hansen *et al.*, 1994; Mooney *et al.*, 1992; Roberts *et al.*, 1998). These materials offer a major advantage in terms of biocompatibility, since the artificial substrates completely disappear over time, and thus the implanted donor cells become fully incorporated into the host. They also provide great chemical versatility as well as the potential for large-scale production at relatively low cost. In addition, use of synthetic chemistry reduces the likelihood of batch-to-batch variation during large-scale production, a problem that can potentially complicate use of purified ECM components. For these reasons, polymer chemistry and novel fabrication techniques will likely lead to development of more effective tissue substitutes.

However, if we understood the fundamental principles that guide ECM remodeling and pattern formation in tissues, perhaps tissue engineering might take a different approach in the future. For example, one could envision entirely new methods of clinical intervention if we understood how tissue-specific mesenchyme generates tissue pattern; how ECM turnover is controlled *locally*; how cell-generated contractile forces contribute to tissue repair and remodeling; or how compressing or pulling tissues alter their growth and form. This knowledge could lead to methods for identifying and isolating relevant pattern-generating cells; for developing pharmacological modifiers of ECM remodeling that may be incorporated in local regions of implants to promote or suppress tissue expansion locally; and for fabricating biomimetic scaffolds that mimic the mechanical and architectural features of natural ECMs necessary to switch on or off the function of interest (e.g., growth vs. differentiation or apoptosis) at a particular time or place. These are just a few of the challenges for the future.

# VII. ACKNOWLEDGMENTS

I would like to acknowledge that the success of many of the studies summarized would not have been possible without the assistance of my major collaborators, including Drs. Judah Folkman, Robert Langer, Martin Schwartz, Jay Vacanti, and George Whitesides, and the multiple students and post-doctoral fellows who executed these experiments in our laboratories. This work was supported by grants from NIH, NASA, DARPA, ARO, and NSF.

# VIII. REFERENCES

Ash, J. F., Spooner, B. S., and Wessells, N. K. (1973). Effects of papaverine and calcium-free medium on salivary gland morphogenesis. *Dev. Biol.* **33**, 463–469.

Banerjee, S. D., Cohn, R. H., and Bernfield, M. R. (1977). Basal lamina of embryonic salivary epithelia. Production by the epithelium and role in maintaining lobular morphology. *J. Cell Biol.* **73**, 445–463.

Beloussov, L. V., Dorfman, J. G., and Cherdantzev, V. G. (1975). Mechanical stresses and morphological patterns in amphibian embryos. *J. Embryol. Exp. Morph.* **34**, 559–574.

Beloussov, L. V., Lakirev, A. V., and Naumidi, I. I. (1988). The role of external tensions in differentiation of Xenopus laevis embryonic tissues. *Cell Differ. Dev.* **25**, 165–176.

Ben Ze'ev, A. G., Robinson, S., Bucher, N. L., *et al.* (1988). Cell–cell and cell–matrix interactions differentially regulate the expression of hepatic and cytoskeletal genes in primary cultures of rat hepatocytes. *Proc. Natl. Acad. Sci. USA* **85**, 1–6.

Bernfield, M. R., and Banerjee, S. D. (1978). The basal lamina in epithelial–mesenchymal interactions. *In* "Biology and Chemistry of Basement Membranes" (N. Kefalides, ed.), pp. 137–148. Academic Press, New York.

Bernfield, M. R., Banerjee, S. D., and Cohn, R. H. (1972). Dependence of salivary epithelial morphology and branching morphogenesis upon acid mucopolysaccharide-protein proteoglycan at the epithelial cell surface. *J. Cell Biol.* **52**, 674–689.

Bissell, D. M., Stamatoglou, S. C., Nermut, M. V., *et al.* (1986). Interactions of rat hepatocytes with type IV collagen, fibronectin, and laminin matrices. Distinct matrix-controlled modes of attachment and spreading. *Eur. J. Cell Biol.* **40**, 72–78.

Brock, A., Chang, E., Ho, C.-C., *et al.* (2003). Geometric determinants of directional cell motility revealed using microcontact printing. *Langmuir.* **19**, 1611–1617.

Chambard, M., Gabrion, J., and Mauchamp, J. (1981). Influence of collagen gel on the orientation of epithelial cell polarity: follicle formation from isolated thyroid cells and from preformed monolayers. *J. Cell Biol.* **91**, 157–166.

Chen, C. S., Mrksich, M., Huang, S., Whitesides, G., and Ingber, D. E. (1997). Geometric control of cell life and death. *Science* **276**, 1425–1428.

Cima, L., Vacanti, J., Vacanti, C., *et al.* (1991). Tissue engineering by cell transplantation using degradable polymer substrates. *J. Biomech. Eng.* **113**, 143–151.

David, G., and Bernfield, M. R. (1979). Collagen reduces glycosaminoglycan degradation by cultured mammary epithelial cells: possible mechanism for basal lamina formation. *Proc. Natl. Acad. Sci. USA* **76**, 786–790.

Dike, L., Chen, C. S., Mrkisch, M., *et al.* (1999). Geometric control of switching between growth, apoptosis, and differentiation during angiogenesis using micropatternd substrates. *In Vitro Cell Dev. Biol.* **35**, 441–448.

Dodson, J. W., and Hay, E. D. (1971). Control of corneal differentiation by extracellular materials. Collagen as promotor and stabilizer of epithelial stroma production. *Dev. Biol.* **38**, 249–270.

Engler, A. J., Griffin, M. A., Sen, S., Bonnemann, C. G., Sweeney, H. L., and Discher, D. E. (2004). Myotubes differentiate optimally on substrates with tissue-like stiffness: pathological implications for soft or stiff microenvironments. *J. Cell Biol.* **166**, 877–887.

Emerman, J. T., and Pitelka, D. R. (1977). Maintenance and induction of morphological differentiation in dissociated mammary epithelium on floating collagen membranes. *In Vitro* **13**, 316–328.

Farquhar, M. G. (1978). Structure and function in glomerular capillaries: role of the basement membrane in glomerular filtration. *In* "Biology and Chemistry of Basement Membranes" (N. Kefalides, ed.), pp. 137–148. Academic Press, New York.

Folkman, J. (1982). Angiogenesis: initiation and control. *Ann. N.Y. Acad. Sci.* **401**, 212–227.

Folkman, J., and Moscona, A. (1978). Role of cell shape in growth control. *Nature* **273**, 345–349.

Folkman, J., and Shing, Y. (1992). Control of angiogenesis by heparin and other sulfated polysaccharides. *Adv. Exp. Med. Biol.* **313**, 355–364.

Folkman, J., Klagsbrun, M. K., Sasse, J., *et al.* (1988). A heparin-binding angiogenic protein — basic fibroblast growth factor — is stored within basement membrane. *Am. J. Pathol.* **130**, 393–400.

Geiger, B., Bershadsky, A., Pankov, R., and Yamada, K. M. (2001). Transmembrane cross-talk between the extracellular matrix–cytoskeleton cross-talk. *Nat. Rev. Mol. Cell Biol.* **2**, 793–805.

Georges, P. C., and Janmey, P. A. (2005). Cell type–specific response to growth on soft materials. *J. Appl. Physiol.* **98**, 1547–1553.

Golosow, N., and Grobstein, C. (1962). Epitheliomesenchymal interaction in pancreatic morphogenesis. *Dev. Biol.* **4**, 242–255.

Gordon, M. Y. (1991). Hemopoietic growth factors and receptors: bound and free. *Cancer Cells* **3**, 127–133.

Griffin, M. A., Sen, S., Sweeney, L., and Discher, D. E. (2004). Adhesion-contractile balance in myotube differentiation. *J. Cell Sci.* **117**, 5855–5863.

Hansen, L., Mooney, D., Vacanti, J. P., *et al.* (1994). Integrin binding and cell spreading on extracellular matrix act at different points in the cell cycle to promote hepatocyte growth. *Mol. Biol. Cell* **5**, 967–975.

Huang, S., and Ingber, D.E. (1999). Cell growth control in context. *Nature Cell Biol.* **1**, E131–E138.

Huang, S., Chen, C. S., and Ingber, D. E. (1998). Control of cyclin D1, p27$^{Kip1}$ and cell cycle progression in human capillary and endothelial cells by cell shape and cytoskeletal tension. *Mol. Biol. Cell* **9**, 3179–3193.

Hynes, R. O. (1987). Integrins: a family of cell surface receptors. *Cell* **48**, 549–554.

Ingber, D. E. (1990). Fibronectin controls capillary endothelial cell growth by modulating cell shape. *Proc. Natl. Acad. Sci. USA* **87**, 3579–3583.

Ingber, D. E. (2006a). Mechanical control of tissue morphogenesis during embryological development. *Intl. J. Dev. Biol.* **50**, 255–266.

Ingber, D. E. (2006b). Cellular mechanotransduction: putting all the pieces together again. *FASEB J.* **20**, 811–827.

Ingber, D. E., and Folkman, J. (1988). Inhibition of angiogenesis through inhibition of collagen metabolism. *Lab. Invest.* **59**, 44–51.

Ingber, D. E., and Folkman, J. (1989). Mechanochemical switching between growth and differentiation during fibroblast growth factor–stimulated angiogenesis *in vitro*: role of extracellular matrix. *J. Cell Biol.* **109**, 317–330.

Ingber, D. E., Madri, J. A., and Jamieson, J. D. (1981). Role of basal lamina in the neoplastic disorganization of tissue architecture. *Proc. Natl. Acad. Sci. USA* **78**, 3901–3905.

Ingber, D. E., Madri, J. A., and Folkman, J. (1986a). A possible mechanism for inhibition of angiogenesis by angiostatic steroids: induction of capillary basement membrane dissolution. *Endocrinology* **119**, 1768–1775.

Ingber, D. E., Madri, J. A., and Jamieson, J. D. (1986b). Basement membrane as a spatial organizer of polarized epithelia: exogenous basement membrane reorients pancreatic epithelial tumor cells *in vitro*. *Am. J. Pathol.* **122**, 129–139.

Ingber, D. E., Madri, J. A., and Folkman, J. (1987). Extracellular matrix regulates endothelial growth factor action through modulation of cell and nuclear expansion. *In Vitro Cell Dev. Biol.* **23**, 387–394.

Ingber, D. E., Prusty, D., Frangione, J., *et al.* (1990). Control of intracellular pH and growth by fibronectin in capillary endothelial cells. *J. Cell Biol.* **110**, 1803–1012.

Ingber, D. E., Prusty, D., Sun, Z., *et al.* (1995). Cell shape, cytoskeletal mechanics and cell cycle control in angiogenesis. *J. Biomechan.* **28**, 1471–1484.

Kleinman, H. K., Klebe, R. J., and Martin, G. R. (1981). Role of collagenous matrices in the adhesion and growth of cells. *J. Cell Biol.* **88**, 473–485.

Kubota, Y., Kleinman, H. K., Martin, G. R., *et al.* (1987). Role of laminin and basement membrane in the morphological differentiation of human endothelial cells into capillary-like structures. *J. Cell Biol.* **107**, 1589–1598.

Kumar, S., Maxwell, I. Z., Heisterkamp, A., *et al.* (2006). Viscoelastic retraction of single living stress fibers and its impact on cell shape, cytoskeletal organization and extracellular matrix mechanics. *Biophys. J.* **90**, 1–12.

Li, M. L., Aggeler, J., Farson, D. A., *et al.* (1987). Influence of a reconstituted basement membrane and its components on casein gene expression and secretion in mouse mammary epithelial cells. *Proc. Natl. Acad. Sci. USA* **84**, 136–140.

Lo, C. M., Wang, H. B., Dembo, M., and Wang, Y. L. (2000). Cell movement is guided by the rigidity of the substrate. *Biophys. J.* **79**, 144–152.

Madri, J. A., and Stenn, K. S. (1982). Aortic endothelial cell migration: I. Matrix requirements and composition. *Am. J. Pathol.* **106**, 180–186.

Mammoto, A., Huang, S., Moore, K., *et al.* (2004). Role of RhoA, mDia, and ROCK in cell shape–dependent control of the Skp2–p27$^{kip1}$ pathway and the $G_1/S$ transition. *J. Biol. Chem.* **279**, 26323–25330.

McNamee, H., Ingber, D., and Schwartz, M. (1993). Adhesion to fibronectin stimulates inositol lipid synthesis and enhances PDGF-induced inositol lipid breakdown. *J. Cell Biol.* **121**, 673–678.

McBeath, R., Pirone, D. M., Nelson, C. M., *et al.* (2004). Cell shape, cytoskeletal tension, and RhoA regulate stem cell lineage commitment. *Dev. Cell* **6**, 483–495.

Meredith, J. E., Jr., Fazeli, B., and Schwartz, M. A. (1993). The extracellular matrix as a cell survival factor. *Mol. Biol. Cell* **4**, 953–961.

Mikos, A. G., Bao, Y., Cima, L. G., *et al.* (1993). Preparation of poly(glycolic acid) bonded fiber structures for cell attachment and transplantation. *J. Biomed. Mater. Res.* **27**, 183–189.

Mooney, D. J., Langer, R., Hansen, L. K., *et al.* (1992). Induction of hepatocyte differentiation by the extracellular matrix and an RGD-containing synthetic peptide. *Proc. Mat. Res. Soc. Symp.* **252**, 199–204.

Mooney, D., Hansen, L., Farmer, S., *et al.* (1992). Switching from differentiation to growth in hepatocytes: control by extracellular matrix. *J. Cell Physiol.* **151**, 497–505.

Mooney, D., Langer, R., Ingber, D. E. (1995). Cytoskeletal filament assembly and the control of cell shape and function by extracellular matrix. *J. Cell Sci.* **108**, 2311–2320.

Moore, K. A., Polte, T., Huang, S., Shi, B., Alsberg, E., Sunday, M. E., and Ingber, D. E. (2005). Control of basement membrane remodeling and epithelial branching morphogenesis in embryonic lung by Rho and cytoskeletal tension. *Dev. Dynamics* **232**, 268–281.

Nogawa, H., and Nakanishi, Y. (1987). Mechanical aspects of the mesenchymal influence on epithelial branching morphogenesis of mouse salivary gland. *Development.* **101**, 491–500.

Opas, M. (1989). Expression of the differentiated phenotype by epithelial cells *in vitro* is regulated by both biochemistry and mechanics of the substratum. *Dev. Biol.* **131**, 281–293.

Parker, K. K., Brock, A. L., Brangwynne, *et al.* (2002). Directional control of lamellipodia extension by constraining cell shape and orienting cell tractional forces. *Faseb. J.* **16**, 1195–1204.

Plopper, G., McNamee, H., Dike, L., *et al.* (1995). Convergence of integrin and growth factor receptor signaling pathways within the focal adhesion complex. *Mol. Biol. Cell* **6**, 1349–1365.

Prime, K. L., and Whitesides, G. M. (1991). *Science* **252**, 1164–1167.

Pollanen, J., Saksela, O., Salonen, E. M., *et al.* (1987). Distinct localizations of urokinase-type plasminogen activator and its type 1 inhibitor under cultured human fibroblasts and sarcoma cells. *J. Cell Biol.* **104**, 1085–1096.

Roberts, C., Chen, C. S., and Mrksich, M., *et al.* (1998). Using mixed self-assembled monolayers presenting GRGD and EG3OH groups to characterize long-term attachment of bovine capillary endothelial cells to surfaces. *J. Am. Chem. Soc.* **120**, 6548–6555.

Sakakura, T., Nishizura, Y., and Dawe, C. (1976). Mesenchyme-dependent morphogenesis and epithelium-specific cytodifferentiation in mouse mammary gland. *Science* **194**, 1439–1441.

Salomon, D. S., Liotta, L. A., and Kidwell, W. R. (1981). Differential response to growth factor by rat mammary epithelium plated on different collagen substrates in serum-free medium. *Proc. Natl. Acad. Sci. USA* **78**, 382–386.

Schwartz, M. A., Lechene, C., and Ingber, D. E. (1991). Insoluble fibronectin activates the $Na^+/H^+$ antiporter by inducing clustering and immobilization of its receptor, independent of cell shape. *Proc. Natl. Acad. Sci. USA* **88**, 7849–7853.

Sims, J., Karp, S., and Ingber, D. E. (1992). Altering the cellular mechanical force balance results in integrated changes in cell, cytoskeletal, and nuclear shape. *J. Cell Sci.* **103**, 1215–1222.

Singhvi, R., Kumar, A., Lopez, G., *et al.* (1994). Engineering cell shape and function. *Science* **264**, 696–698.

Smith, R. L. and Bernfield, M. R. (1982). Mesenchyme cells degrade epithelial basal lamina glycosaminoglycan. *Dev. Biol.* **94**, 378–390.

Stenn, K. S., Madri, J. A., and Roll, F. J. (1979). Migrating epidermis produces AB2 collagen and requires continual collagen synthesis for movement. *Nature* **277**, 229–232.

Talhouk, R. S., Bissell, M. J., and Werb, Z. (1992). Coordinate expression of extracellular matrix–degrading proteinases and their inhibitors regulate mammary epithelial function during involution. *J. Cell Biol.* **118**, 1271–1282.

Trelstad, R. L., Hayashi, A., Hayashi, K., *et al.* (1982). The epithelial–mesencymal interface of the male rat Mullerian duct: loss of basement membrane integrity and ductal regression. *Dev. Biol.* **92**, 27–40.

Vernon, R. B., Angello, J. C., Iruela-Arispe, L., *et al.* (1992). Reorganization of basement membrane matrices by cellular traction promotes the formation of cellular networks in vitro. *Lab. Invest.* **66**, 536.

Vlodavsky, I., Folkman, J., Sullivan, R., *et al.* (1987). Endothelial cell–derived basic fibroblast growth factor: synthesis and deposition into subendothelial extracellular matrix. *Proc. Natl. Acad. Sci. USA* **84**, 2292–2296.

Vracko, R. (1974). Basal lamina scaffold anatomy and significance for maintenance of orderly tissue structures. *Am. J. Pathol.* **77**, 314–346.

Wicha, M. S., Liotta, L. A., Garbisa, G., *et al.* (1979). Basement membrane collagen requirements for attachment and growth of mammary epithelium. *Exp. Cell. Res.* **124**, 181–190.

Wicha, M. S., Liotta, L. A., Vonderhaar, B. K., *et al.* (1980). Effects of inhibition of basement membrane collagen deposition on rat mammary gland development. *Dev. Biol.* **80**, 253–261.

# *In Vivo* Synthesis of Tissues and Organs

Part Three

# Chapter Sixteen

# *In Vivo* Synthesis of Tissues and Organs

Brendan A. Harley and Ioannis V. Yannas

## I. INTRODUCTION

With increasing fetal development, the mechanism of mammalian wound healing transitions from regeneration to a repair process characterized by organized wound contraction and scar synthesis. Recently, a variety of tissue engineering constructs have been developed to block the contraction and scar formation mechanisms of repair and induce regeneration following injury. Such constructs, mostly scaffolds that are analogs of extracellular matrix and possess specific biological activity, have become the basis of studies of *in vivo* synthesis of tissues and organs. In this article, we provide an overview of mammalian wound healing processes following severe injury as well as a description of the tissue triad and the regenerative capacity of the three distinct tissue types that comprise the triad. We also discuss the critical structural elements of an active extracellular matrix analog that induces regeneration, and describe the use of standardized wound models for study of *in vivo* regeneration processes. We conclude this review describing recent data from studies utilizing active extracellular matrix analogs (scaffolds) that have shown regenerative activity.

The goal of achieving *in vivo* induced regeneration for a variety of tissue and organs following severe injury remains at the forefront of current tissue engineering investigations. Typically, an analog of the extracellular matrix is utilized as a template that, when properly formulated, induces regeneration of lost or damaged tissue. Currently, successful regeneration has been induced in the skin and peripheral nerves (Yannas, 2001), while progress has been made in developing appropriate extracellular matrix analogs to alter the typical organismic response to injury in a variety of tissues, including kidney, cartilage, bone, central nervous system, and brain dura. These investigations, active since the mid-1970s, have focused primarily on identifying the optimal extracellular matrix analog components to block organized wound contraction and scar tissue formation while inducing regeneration of physiological tissue (Yannas, 2001). Although many tissue engineering investigations currently focus on developing techniques appropriate for synthesis of tissues and organs *in vitro*, such products must eventually be implanted in the appropriate anatomical site of the host. Since implantation of an organ construct is almost always preceded by a surgical procedure that generates a severe wound, it is essential to master the evolving methodology of *in vivo* wound healing in order to synthesize appropriate neo-organ constructs *in vitro* for eventual implantation. The goal of this chapter is to inform the reader

*Principles of Tissue Engineering, 3rd Edition*
ed. by Lanza, Langer, and Vacanti

about the salient features of the organismic response to injury, the historical underpinnings of current studies of induced tissue regeneration, induced tissue regeneration as a technique to treat severe injuries to a variety of tissue and organs, and the future directions of research in this field.

## II. MAMMALIAN RESPONSE TO INJURY

### Defect Scale

Treatment options for organ injury depend significantly on the scale of the defect. Microscopic defects can be treated using a wide variety of soluble factors (i.e., herbs, potions, pharmaceuticals, vitamins, hormones, and antibiotics). However, organ-scale defects present a significantly larger wound site, require considerably different treatment practices, and constitute the focus of this chapter. These defects, created primarily by disease or by an acute or chronic insult that results in millimeter- or centimeter-scale wounds, cannot be treated with drugs because the problem is the failure of a mass of tissue, including cells, soluble proteins and cytokines, and insoluble extracellular matrix (ECM). Significant loss of function in the affected tissue or organ, termed the *missing organ*, leads to consequences such as lack of social acceptance in cases of severe burns and facial scars, loss of mobility and sensory function in the case of neuroma, and life-threatening symptoms in cases such as cirrhotic liver, large-scale severe burns, and ischemic heart muscle.

### Regeneration versus Repair

Certain organisms have the ability to regenerate significant portions of damaged tissue. An example is the amphibian newt, which regenerates functional limbs following amputation. The mammalian fetus has displayed the ability to regenerate damaged organs and tissue spontaneously up to the third trimester of gestation; however, adult mammals do not typically exhibit spontaneous regeneration following severe organ injuries (Mast *et al.*, 1992; Yannas, 2001). Instead, the adult mammal response to severe injury is closure of the wound by contraction and scar tissue formation, a process termed *repair*. Cell-mediated contraction of the wound site is observed in many different species, to varying degrees, at many organ sites (Yannas, 2001). Distinct from repair, *regeneration* is characterized by synthesis of a physiological (normal, functional) replacement tissue in the wound site that is structurally and functionally similar to the original tissue. Based on these observations, for the remainder of this chapter, *early fetal* refers to the fetal response to injury that leads to regeneration, while *adult* refers to all mammals (adult as well as late fetal) that respond to injury via repair.

The contractile fibroblast phenotype, termed the *myofibroblast*, plays a critical role in determining the nature (repair or regeneration) of mammalian wound healing. During adult repair, myofibroblast-mediated organized wound contraction and scar tissue synthesis is observed (Yannas, 2001). During early fetal healing, differentiation of myofibroblasts has not occurred, and regeneration of damaged tissue and organs occurs in the absence of contraction. The data suggest that induced organ regeneration in the adult may be encouraged by developing techniques to stimulate partial reversion to early fetal healing. Additionally, the transforming growth factor-β (TGF-β) family of molecules has been implicated in this ontogenetic transition between fetal regeneration and adult repair response to injury. TGF-βs are multifunctional cytokines with widespread effects on cell growth and differentiation, embryogenesis, immune regulation, inflammation, and wound healing (Border *et al.*, 1995). In terms of their relationship with repair processes, TGF-β1 and TGF-β2 are known to promote scar, while TGF-β3 may reduce scar (Lin *et al.*, 1995; Shah *et al.*, 1995). As such, deficient levels of TGF-β1 and -β2 and increased levels of TGF-β3 are observed in early-gestational ("fetal" regenerative healing response) compared to late-gestational ("adult" repair healing response) mice. These results implicate increased TGF-β1, -β2, and decreased TGF-β3 expression along with myofibroblast activity in late gestation and postpartum fetal scar formation (Soo *et al.*, 2003). The available evidence suggests future experiments utilizing procedures for control of TGF-β in conjunction with other tissue-engineering constructs that modify myofibroblast behavior, such as bioactive scaffolds, to induce regeneration of tissues that are known to be nonregenerative.

### Tissue Triad

There are three distinct tissue types, termed the *tissue triad*, which together define the structure of most organs: the epithelial layer, the basement membrane layer, and the stroma (Fig. 16.1) (Martinez-Hernandez, 1988; Yannas, 2001). Developmental and functional similarities between this triad in a variety of tissues and organs, such as skin and peripheral nerves, have been observed, suggesting that it can be used as an illustrative device to understand injury response in other organs as well (Yannas, 2001). A layer of epithelial cells (the *epithelial layer*) covers all surfaces, tubes, and cavities of the body. This layer is cell continuous and avascular. Unlike the basement membrane and stroma, the epithelial layer does not comprise a significant amount of extracellular matrix (ECM). The *basement membrane* is an acellular, avascular, continuous layer of ECM components separating the epithelial layer and the stroma. The *stroma* is cellular, contains ECM and connective tissue components, is heavily vascularized, and provides a reservoir for nutrient uptake to and waste removal from the basement membrane and epithelia.

Following injury to a variety of tissues, such as the skin, peripheral nerves, blood vessels, lung, kidney, and pancreas, the epithelial and basement membrane layers regenerate spontaneously when the stroma remains intact, while the damaged stroma heals through repair-mediated contraction and scar formation processes (Fu and Gordon, 1997;

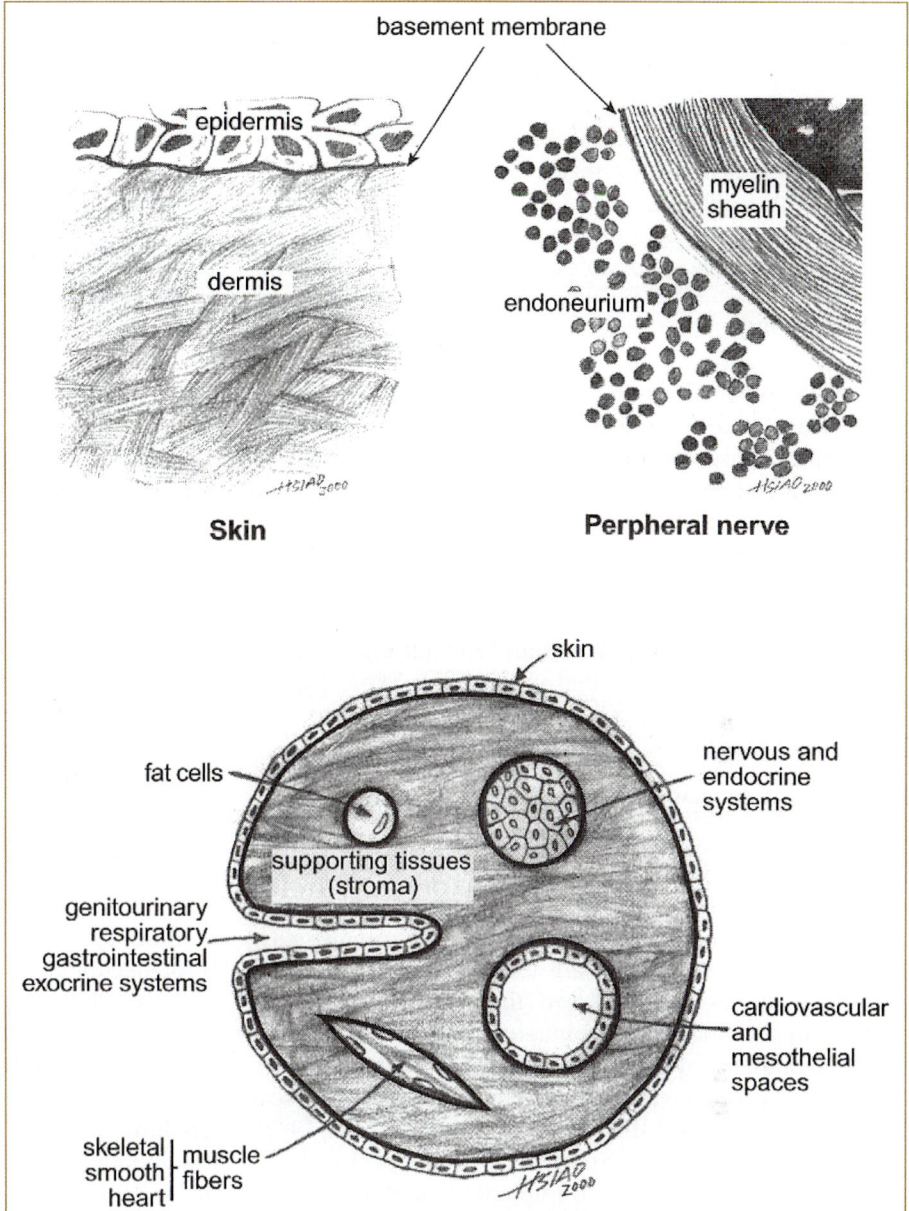

**FIG. 16.1.** Schematics of the tissue triad structure observed in mammalian tissue. *Top*: Tissue triad of skin and peripheral nerves. The basement membrane is a thin extracellular matrix tissue located between the cellular and nonvascular epithelia (epidermis, myelin sheath) and the cellular, vascularized stroma (dermis, endoneurium). *Bottom*: Diagram of the distribution of epithelial, basement membrane, and stromal tissues in the mammalian system. Examples of stromal tissues are bone, cartilage, and their associated cell types as well as elastin and collagen. Examples of epithelial tissues are those covering the surface of the genitourinary, respiratory, and gastrointestinal tracts as well as surfaces of the mesothelial cells in body cavities, muscle fibers, fat cells, and endothelial cells in the cardiovascular system (Yannas, 2001).

Yannas, 2001; Yannas *et al.*, 1989). Understanding the injury response of the tissue triad has suggested a paradigm for inducing regeneration in nonregenerative tissues: repair mechanisms appear to be activated by disruption of the stromal architecture, and proper replacement of the stromal layer is critical for any regeneration to occur. As such, development of materials to act as analogs of the ECM to replace the lost stromal architecture has been a primary focus of studies of regenerative medicine since the mid-1980s.

## III. METHODS TO TREAT LOSS OF ORGAN FUNCTION

Six approaches have been used to restore some level of functionality to a damaged tissue or organ: transplantation, autografting, implantation of a permanent prosthetic device,

use of stem cells, *in vitro* synthesis of organs, and induced regeneration. The last three techniques are often collectively referred to as *tissue engineering* (Lanza *et al.*, 1997). *In vivo* induced regeneration is the only methodology to date that has modified the adult mammalian wound-healing response to induce regeneration. All six techniques are briefly discussed in the following sections; a more detailed description has been previously published (Yannas, 2001).

### Transplantation

Transplantation is widely used to replace complex tissues and organs, but it is limited by two significant factors. While transplantation of a select few tissues, such as the eye and testis, occur without rejection, a significant challenge facing transplantation is the immunological barrier between the donor and the host. After transplantation, the donor organ is attacked and rejected by the host's immune system. The primary clinical method for preventing such rejection is the use of immunosuppressive drugs for the remainder of the host's life to suppress the immune system. However, immunosuppression also makes the host vulnerable to infections (Wickelgren, 1996). A second major obstacle is the difficulty in finding immunocompatible donors and the shortness of supply of suitable organs (Lanza *et al.*, 1997).

## Autografting

With autografting, the donor and the recipient are the same individual. A fraction of the tissue or organ is harvested from an uninjured site and grafted at the nonfunctional site (Medawar, 1944). Autografting removes issues related to immune response, but it necessitates the creation of a second wound site (donor site), subjecting the patient to a second severe trauma and additional loss of functionality. Therefore, autografting is utilized only when sufficient autograft tissue is available and when the loss of functionality or morbidity at the primary wound site outweighs that at the harvest site, giving it limited applicability. Typical applications of autografting are following severe burns and peripheral nerve injuries in the hand.

## Permanent Prosthetic Device

Permanent prosthetic devices are typically fabricated from biologically inert materials, such as metals, ceramics, and synthetic polymers, that do not provoke the immune response problems inherent to many transplanted tissues. Even though these devices are fabricated from bio-inert materials, interactions between the prosthesis and the surrounding biological environment still lead to a number of unfavorable physical and biological manifestations. Specific examples are the formation of a thick, fibrous scar tissue capsule around the implant (Rudolph *et al.*, 1992), stress-shielding of the surrounding tissue (Spector *et al.*, 1993), platelet aggregation to implanted surfaces (Snyder *et al.*, 2002), and accumulation of wear particles both at the site of implantation and in the lymphatic system (Urban *et al.*, 2000). The spontaneous remodeling process of the tissues surrounding the implant can also be significantly altered, leading to further tissue degradation (Yannas, 2001). These often-serious side effects illustrate the difficulty of replacing bioactive tissues with bio-inert implants fabricated from materials possessing drastically different material and mechanical properties.

## Stem Cells

The pluripotent nature of stem cells offers a multitude of therapeutic possibilities (Kondo *et al.*, 2003; Lerou and Daley, 2005). Current efforts in stem cell research have focused on understanding stem cell plasticity and ways of controlling stem cell differentiation (Wagers and Weissman, 2004). Previously, mesenchymal (Pittenger *et al.*, 1999), epi-

thelial (Slack, 2000), and neural stem cells (Gage, 2000) have been grown *in vitro* and studied. More recently, experimental investigations of stem cells have also focused on utilizing hematopoietic and embryonic stem cells. In particular, techniques to harvest and identify them (Surdez *et al.*, 2005), expand and differentiate them in culture (Daley, 2005; Lengerke and Daley, 2005; Passegue *et al.*, 2005; Udani *et al.*, 2005; Wang *et al.*, 2005), and reimplant them at an injury site (Kunisaki *et al.*, 2006; Udani *et al.*, 2005) have been at the forefront of stem cell research. While few significant advances have been made to this point, stem cell technologies present a great deal of promise. However, improved understanding of stem cell behavior and development of stem cell–based technologies also raise a number of important ethical questions (Daley *et al.*, 2005), consideration of which will play a significant role in the development of stem cell–based tissue-engineering solutions.

## *In Vitro* Synthesis

*In vitro* synthesis requires the growth of a functional volume of tissue *in vitro*. *In vitro* synthesis allows for total control over the culture environment, such as soluble regulator content (i.e., growth factors, cytokines), insoluble regulator content (i.e., ECM proteins), and a variety of cell culture media and loading conditions. In order to develop large (critical dimension >1 cm), bioactive scaffolds, it is important to support the cells metabolically within these constructs. There are two mechanisms available for transport of metabolites to and waste products from cells in a scaffold: diffusion and, with *in vivo* applications, transport through capillary networks formed in the scaffold via angiogenesis. While angiogenesis becomes the limiting factor *in vivo*, significant angiogenesis is not observed for the first few days after implantation, and it is not present at all *in vitro*. As a result, current tissue-engineering constructs are size limited (<1 cm) due to diffusion constraints. Improving metabolite influx is critical for larger, more complex scaffolds. Additionally, the complexity of biological systems, specifically cytokine, growth factor, and intercellular signaling needs, throughout the volume of developing tissue has to date precluded, with few exceptions (such as *in vitro* culture of replacement heart valves (Rabkin-Aikawa *et al.*, 2005) and epithelial sheets for severe burn patients (Woodley *et al.*, 1988)), the formation of complex tissues *in vitro*.

## Induced *In Vivo* Organ Synthesis (Induced Regeneration)

Induced organ synthesis *in vivo* relies on the processes inherently active in the wound site to regenerate lost or damaged tissue. A highly porous analog of the ECM, also termed a *scaffold*, is utilized to induce regeneration at a wound site where the organism would normally respond via repair processes. Induced organ synthesis was made possible by the development of fabrication techniques to produce ECM analogs with well-defined pore microstructure, specific surface area, chemical composition, and degradation rate (Yannas, 2001; Yannas *et al.*, 1989). Its first application was the use of a collagen–glycosaminoglycan (CG) scaffold (termed *dermal regeneration template* — DRT) that induced skin regeneration following severe injury. The DRT displayed high biological activity when implanted into a full-thickness skin wound and was capable of inducing regeneration of the underlying dermal layer of skin as well as the epidermal and basement membrane layers (Yannas, 2001; Yannas *et al.*, 1989).

For the remainder of this review, we focus on studies of the structure and function of ECM analogs used in tissue-engineering applications as well as an in-depth discussion of the tissue-engineering approaches utilized to induce regeneration of a number of tissues, namely skin, peripheral nerves, and orthopedic tissues.

# IV. ACTIVE EXTRACELLULAR MATRIX ANALOGS

## Fundamental Design Principles for Tissue Regeneration Scaffolds

Porous, three-dimensional scaffolds have been used extensively as biomaterials in the field of tissue engineering for *in vitro* study of cell–scaffold interactions and tissue synthesis and *in vivo* induced tissue regeneration studies. These scaffolds, analogs of the ECM, act as a physical support and, more importantly, as an insoluble regulator of biological activity that affects cell processes such as migration, contraction, and division. For the remainder of this review, the term *active* or *bioactive ECM analog* will refer to scaffolds that induce regeneration of normally nonregenerative tissues following severe injury. It has been hypothesized that these active ECM analogs induce regeneration by establishing an environment that selectively inhibits wound contraction by preventing the organized contractile response and blocking scar synthesis, the two processes normally responsible for closing a wound following severe injury (Yannas, 2001).

Four physical and structural properties must be controlled to critical levels for the fabrication of an active ECM analog: the degradation rate, which defines the template residence time, the chemical composition, the pore microstructure (mean pore size, shape, and orientation), and the scale of the scaffold, which is defined by the critical cell path length (Yannas, 2001). These characteristics and any governing models that help to describe cell behavior in active ECM analogs are discussed briefly in the following sections.

### Template Residence Time

The length of time the scaffold remains insoluble in the wound site (*residence time*) is critical in defining its bioactivity. For physiologic tissue to be synthesized at a wound site, the scaffold must initially support cell migration, prolifera-

tion, and organization, but the scaffold must then degrade in such a manner that it does not interfere with the native tissue synthesis and remodeling processes. These considerations require a scaffold residence time with an upper and lower bound, a concept that has been formalized as the isomorphous tissue replacement model: scaffold residence time must be approximately equal to the time required to synthesize a mature tissue via regeneration at the specific tissue site under study (Yannas, 2005). In the case of a full-thickness skin wound, for example, this healing time is approximately 25 days, and the degradation kinetics of the active ECM analog that induces skin regeneration was optimized for that time period (Yannas *et al.*, 1989). Alternatively, the healing time of a peripheral nerve injury is dependent on the gap length: Peripheral nerve trunks grow unidirectionally, from the proximal toward the distal stump, at a rate of approximately 1 mm/day (Williams *et al.*, 1983). Accordingly, during induced regeneration of peripheral nerves across a gap, a scaffold in contact with the nerve stumps (either the tube into which the stumps are inserted or the scaffold structure in the tube lumen) must remain in an insoluble state over a period approximately equal to the time, of order 10–20 days, during which axon elongation proceeds from the proximal to the distal stump.

An intact scaffold cannot diffuse away from the wound bed. Therefore the simplest method for achieving isomorphous tissue replacement requires the insoluble scaffold structure to be degraded by enzymes in the wound bed into low-molecular-weight fragments. The lifetime of the scaffold is defined by the degradation time constant ($t_d$) and can be compared to the normal healing-process time constant for a wound at the anatomic site of interest ($t_h$) (Yannas, 2001). When the scaffold remained in the wound bed as a nondegradable implant ($t_d/t_h > 1$), dense fibrous tissue similar to scar was synthesized underneath the scaffold (Yannas, 2001; Yannas *et al.*, 1989). When the scaffold degraded rapidly ($t_d/t_h < 1$), wound healing was marked by wound contraction and scar synthesis, similar to the reparative healing process observed in an ungrafted wound, and no regeneration was observed (Yannas, 2001). Because different wound sites and even the same wound site in different species may have different time constants for healing ($t_h$), it is necessary to adjust the degradation rate of the ECM analog for each wound site and species in order to satisfy the isomorphous tissue replacement requirement and to induce regeneration.

Scaffold resistance to degradation can be increased by increasing the relative density (solid content) of the scaffold. However, a more elegant option that does not involve changing the structural characteristics of the scaffold is also available: Scaffold degradation resistance increases with increasing cross-link density between the fibers that make up the scaffold structure (Harley *et al.*, 2004; Yannas *et al.*, 1975, 1989). In the case of collagen-based scaffolds, degradation in the wound bed is accomplished primarily by native collagenases. Decreased degradation rates for collagen-based scaffolds have been achieved by introducing glycosaminoglycans (GAGs) into the collagen mixture, which results in the formation of additional cross-links, and by further cross-linking the resultant scaffold using a multitude of physical and chemical cross-linking processes. Dehydrothermal (DHT) cross-linking is a physical technique where the scaffold is exposed to a high temperature under vacuum, leading to the removal of water from the scaffold. Drastic dehydration (<1% water content) of the scaffold leads to the formation of interchain amide bonds through condensation (Yannas and Tobolsky, 1967). DHT cross-linking is adjustable; exposure to higher temperatures or longer lengths of time produces a higher cross-link density and a slower degradation rate (Harley *et al.*, 2004; Yannas, 2001; Yannas *et al.*, 1989). UV treatment is a second physical cross-linking technique that can create cross-links between collagen fibers due to the effects of radiation (J. E. Lee *et al.*, 2001). Chemical cross-linking treatments, such as glutaraldehyde or carbodiimide (J. E. Lee *et al.*, 2001) exposure can also be utilized to induce covalent bonds between collagen fibers. These cross-linking techniques are considerably more powerful, resulting in significantly higher cross-link densities and slower degradation rates. However, chemically cross-linked scaffolds must be extensively washed to remove all traces of the typically cytotoxic chemicals. Additionally, some chemical cross-linkers integrate a portion of the chemical compound into the cross-link. Degradation of these scaffolds releases cytotoxic agents into the wound site, so these techniques must be used with care (J. E. Lee *et al.*, 2001; Olde Damink *et al.*, 1996). While described in detail for collagen-based scaffolds, these cross-linking tools and techniques can be applied to a multitude of scaffolds, fabricated from both natural and synthetic materials.

## Chemical Composition

The chemical composition defines the ligands displayed on the scaffold surface. Cell behaviors such as attachment, migration, proliferation, and contraction are all mediated by interactions between the focal adhesions and integrins expressed on the cell surface and the ligands available on the scaffold surface. An active ECM analog must be fabricated in a manner, and from specific materials, that leads to expression of a chemical environment conducive to cell–scaffold interactions that prevent organized wound contraction and scar synthesis and that instead induce regeneration.

A multitude of different materials has been used to fabricate scaffolds for many tissue-engineering applications, including studies of induced *in vivo* regeneration. Several synthetic, nondegradable polymers such as (poly)dimethyl siloxane have occasionally been utilized. These polymers, which parenthetically violate the principle of isomorphous tissue replacement, do not express ligands on their surface and have not induced regeneration. Degradable synthetic polymers, such as poly(L-lactide) and poly(lactide-co-glycolide) variants, have been fabricated to satisfy isomorphous tissue replacement and with surfaces that have been modified chemically to display appropriate ligands. However, these materials have not been observed to prevent contraction and scar formation or to induce regeneration of stromal tissue. A chemical composition that has been used successfully to induce regeneration is a graft copolymer of type I collagen and a sulfated glycosaminoglycan (GAG) (Yannas, 2001; Yannas *et al.*, 1989). Collagen is a significant constituent of the natural ECM, and collagen-based scaffolds have been used in a variety of applications due to many useful properties: low antigenic response and a high density of ligands that interact specifically with integrin receptors in fibroblasts (the cell type predominantly responsible for cell-mediated contractions processes during repair). Particular collagen scaffolds have been observed to promote cell and tissue attachment and growth as well as to induce regeneration of tissue following injury (O'Brien *et al.*, 2005; van Tienen *et al.*, 2002; Yannas, 2001; Yannas *et al.*, 1989). However, in order to induce regeneration, the periodic banding (~64 nm) of the collagen fiber structure must be selectively abolished to prevent platelet aggregation that leads to repair (Yannas, 2001). A number of other natural protein-based scaffolds (i.e., hyaluronate-based, fibrin-based, and chitosan-based scaffolds) have also shown great promise in the field of tissue engineering. These natural polymers are also capable of facilitating cell binding similar to that observed with collagen scaffolds, in part due to their expression of natural ligand-binding sites, but at this point these materials have not been developed to the point of inducing regeneration.

## Template Pore Microstructure

The biological activity of any ECM analog also depends significantly on its three-dimensional pore microstructure. Having migrated into the scaffold, the cell interacts with the surrounding scaffold environment, making use of its cell surface receptors to bind to specific ligands on the scaffold surface. The first critical component of ECM analog microstructure to consider is the open- or closed-cell nature of the scaffold. An *open-cell* pore microstructure exhibits pore interconnectivity, whereas a *closed-cell* microstructure exhibits membranelike faces between adjacent pores, effectively sealing the environment of one pore from its neighbors. Pore interconnectivity is critical for scaffold bioactivity in order for cells to be able to migrate through the construct and to interact with other cells in a manner similar to that observed *in vivo*. A second critical structural feature is the relative density ($R_d$) or porosity of the ECM analog. The *relative density* of the scaffold is calculated as the ratio of the scaffold density to the density of the solid from which the scaffold is made. *Porosity*, a measure of how porous the scaffold is, is the pore volume fraction of the scaffold. Additionally, the relative density and porosity are inversely related: Scaffold porosity can also be defined as $(1 - R_d)$. Both of these variables define the amount of solid material in the scaffold. When the pores are closed or too small or when the relative density is too large, cells are not able to migrate through the scaffold, a significant impediment for a tissue-engineering scaffold. An active ECM analog must possess an open-cell pore structure with a relative density below a critical value that is characteristic of each application but typically significantly less than 10% (bioactive porosity typically greater than 90%). This structural criteria, determined from the results of a number of experiments studying cell–scaffold interactions, suggest that a critical number of cells are required within a bioactive scaffold (Yannas, 2001).

The mean pore size of the ECM analog significantly influences its bioactivity. The effect of mean pore size will now be discussed in conjunction with the application of cellular solids modeling and a discussion of scaffold specific surface area. To describe even a simple cell–scaffold interaction, a highly detailed model describing the number of receptors utilized per bound cell and the nature of the binding and receptor sites needs to be developed. However, a more generic explanation can be used to describe the complexity of the cell–scaffold interaction and the significant influence the pore size has on its bioactivity. At first pass, it is apparent that there is a minimum pore size requirement for a bioactive ECM analog: The pores must be large enough to allow cells to fit through the pore structure and populate the analog. There is also an observed upper bound to pore size (Yannas, 2001; Yannas *et al.*, 1989). It has been hypothesized that this upper bound in mean pore size is due to the effects of scaffold-specific surface area. To further test this hypothesis, cellular solids modeling techniques have been integrated to better describe scaffold microstructure using a quantitative framework. The complex geometry of foams (and of scaffolds) is difficult to model exactly; instead, dimensional arguments that rely on modeling the mechanisms of deformation and failure in the foam, but not the exact cell geometry, are used (Gibson and Ashby, 1997). Scaffold relative density ($R_d$) and mean pore size ($d$) together define the scaffold specific surface area ($SA/V$), the total surface area of pore walls available for cell attachment. A cellular solids model [Eq. (16.1)] has been developed and experimentally validated to accurately predict $SA/V$ of collagen-based scaffolds (O'Brien *et al.*, 2005):

$$\frac{SA}{V} = \frac{10.17}{d}(R_d)^{1/2} \qquad (16.1)$$

Increasing mean pore size in a series of constructs and keeping $R_d$ constant decreases the overall $SA/V$ while increasing $R_d$, and keeping the mean pore size constant increases $SA/V$. The primary feature of this structural analysis to consider is that a change in the construct $SA/V$ indicates a change in the available area to which cells bind. This

calculation and experimental result suggests the significance of the scaffold $SA/V$ in defining its bioactivity. If the $SA/V$ is too small (i.e., due to a relatively large mean pore size), an insufficient number of cells will be able to bind to the scaffold and the cells that remain free will contribute to the spontaneous repair mechanism.

As previously discussed, there is also a minimum mean pore size, defined by the characteristic dimension of the cell: approximately 10–50 µm for most cells. When the scaffold pore size is smaller than this critical dimension, cells will be unable to migrate through the scaffold; whereas when the pore size is too large, an insufficient specific surface area will be available. These upper and lower bounds of the scaffold mean pore size, mediated by cell size and specific surface requirements, have been determined experimentally for each cell type for tissues where regenerative templates have been used (Yannas, 2001).

The shape of the pores that make up the porous scaffold must also be considered. Cells have been observed to be exquisitely sensitive to the mechanical properties of the underlying substrate (Lo *et al.*, 2000), and slight changes in the mean shape of the pores can result in significant variations in the extracellular mechanical properties (Gibson and Ashby, 1997) and overall construct bioactivity (Chang and Yannas, 1992). Changes in mean pore shape may also play a role in defining the areas of the scaffold available or unavailable for binding and the predominant direction of cell migration as well as in the geometrical organization of cells within the scaffold.

## Critical Cell Path Length

Successful migration of cells into an active ECM analog and their initial survival is critically important for successful regeneration. While the effect of the structural characteristics (i.e., pore size, shape, relative density) on the bioactivity of an ECM analog has been discussed, there is another important characteristic to consider: an adequate source of metabolites (i.e., oxygen, nutrients). There are two mechanisms available for transport of metabolites to and waste products from the cells: diffusion to and from the surrounding wound bed or transport along capillaries that have sprouted into the scaffold as a result of angiogenesis. While angiogenesis becomes the limiting factor for long-term cell survival and growth, significant angiogenesis is not observed for the first few days after implantation. Therefore, early cell survival inside the scaffold is defined solely by diffusion. The critical scaffold thickness, the maximum scaffold thickness than can be supported by metabolite diffusion, has been observed empirically to be on the order of a few millimeters (Yannas, 2001).

A quantitative model of cell metabolic requirements and nutrient diffusion characteristics that defines the critical cell path length for cell migration into a scaffold has been developed to describe the salient features of this process (Harley *et al.*, 2006; Yannas, 2000, 2001). Here, the complexity of the nutritional requirements of the cell is simplified by generically considering a critical nutrient that is required for normal cell function; such a nutrient is assumed to be metabolized by the cell at a rate $R$ in moles/mm$^3$/s. The nutrient is pictured being transported from the wound bed, where the concentration of nutrient is assumed to be a constant, $C_o$, over a distance $L$ via the exudate until it reaches the cell within the scaffold. Immediately following implantation of the scaffold, nutrient transport is performed exclusively via diffusion that can be modeled using the scaffold diffusivity $D$ in mm$^2$/sec. Dimensional analysis [Eq. (16.2)] readily yields the cell lifeline number ($S$):

$$S = RL^2/DC_o \qquad (16.2)$$

The cell lifeline number characterizes the relative ratio of the rate of nutrient supply to nutrient consumption by the cell. If the rate of consumption of the critical nutrient exceeds greatly the rate of supply, $S \gg 1$, the cell will soon die. At steady state ($S = 1$), the rate of consumption of nutrient by the cell equals the rate of transport via diffusion over a distance $L$; at steady state, the value of $L$ is the critical cell path length, $L_c$, the longest distance away from the wound bed that the cell can migrate without requiring nutrient in excess of that supplied by diffusion. For many cell nutrients of low molecular weight $L_c$ is of the order of a few hundred micrometers to a few millimeters, a distance short enough to suggest the need for very close proximity between wound bed and implant and to indicate that *in vivo* regeneration of large tissue or organs requires incorporation of special promoters to angiogenesis (Yannas, 2001).

As a result, recent experimental work has focused on understanding the relationship between scaffold pore microstructure and permeability. Scaffold permeability controls diffusion-based metabolite and waste transport to and from the scaffold and influences the final hydrostatic pressure distribution in the scaffold. Both of these parameters can significantly influence cell behavior and overall scaffold bioactivity. Cellular solids modeling tools that quantitatively describe scaffold permeability in terms of salient microstructural features have been utilized in this analysis. Both scaffold pore size and compressive strain can vary significantly with different applications, making them the primary features to characterize in terms of scaffold permeability. An open-cell foam, cellular solids model has been developed to accurately model scaffold permeability ($K$) in terms of scaffold mean pore size ($d$), percent compression ($\varepsilon$), a system constant ($A$), and scaffold relative density ($\rho^*/\rho_s$) (Harley *et al.*, 2006) (O'Brien, F. J., Harley, B. A., Waller, M. A., Yannas, I. V., Gibson, L. J., and Prendergast, P. J., 2007):

$$K = A \cdot d^2 \cdot (1-\varepsilon)^2 \cdot \left(1 - \frac{\rho^*}{\rho_s}\right)^{3/2} \qquad (16.3)$$

The cellular solids model [Eq. (16.3)] of scaffold permeability suggests that scaffold permeability increases with

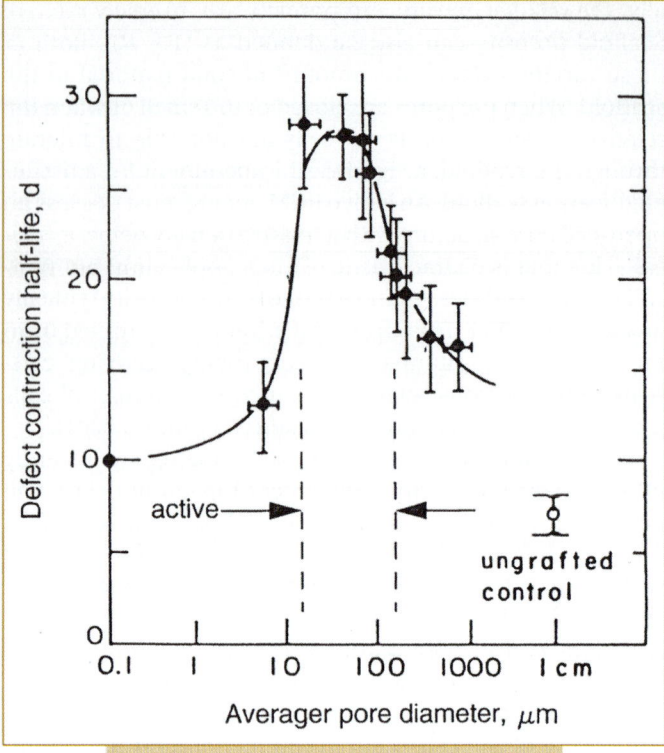

**FIG. 16.2.** Bioactivity of ECM analog variants for skin regeneration versus scaffold mean pore diameter (Yannas *et al.*, 1989).

increasing pore size and decreases with increasing compressive strain, a result that has also been observed experimentally for a series of collagen-based scaffolds. The excellent comparison between experimentally measured permeability and what is predicted from the cellular solids model suggests that such predictive modeling tools can be used to describe scaffold permeability for many different scaffold architectures under a variety of physiological loading conditions.

## Active Collagen–Glycosaminoglycan Scaffolds

Collagen–glycosaminoglycan scaffolds have been observed to have a high degree of bioactivity and to be able to induce regeneration of nonregenerative tissues at a variety of anatomical sites. Three specific wounds and the appropriate ECM analogs have been studied in our laboratory: skin regenerated via the dermal regeneration template (DRT) (Yannas, 2001; Yannas *et al.*, 1989), peripheral nerves regenerated by the nerve regeneration template (NRT) (Chamberlain *et al.*, 1998; Chang and Yannas, 1992; Harley *et al.*, 2004), and the conjunctiva regenerated by a modified DRT (Hsu *et al.*, 2000). As predicted, the bioactivity of the DRT and NRT has been found to be closely related to specific physical parameters of the scaffold. Scaffold mean pore size (Fig. 16.2) (Yannas *et al.*, 1989), degradation rate (Figs. 16.3 and 16.4) (Harley *et al.*, 2004; Yannas *et al.*, 1989), chemical composition (Yannas *et al.*, 1989), pore orientation

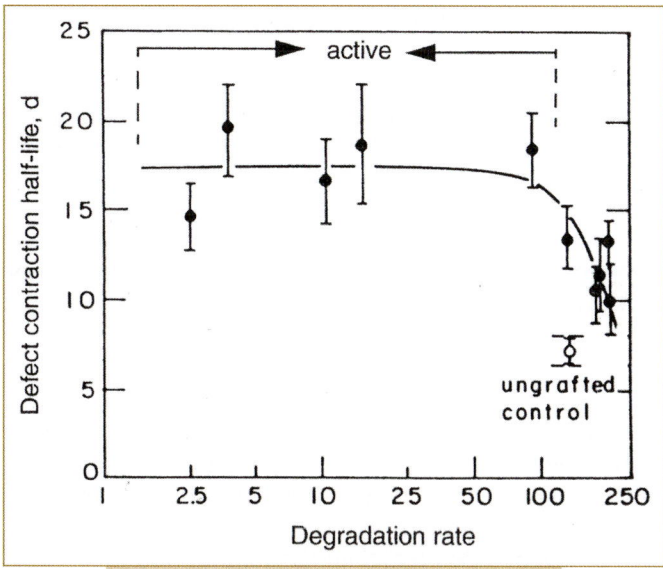

**FIG. 16.3.** Bioactivity of ECM analog variants for skin regeneration versus scaffold degradation rate (Yannas *et al.*, 1989).

(Chang and Yannas, 1992), and the pore volume fraction have all been shown to have a significant effect on the quality of regeneration, and it appears that only a narrow range of structural features satisfies the criteria for bioactivity (Yannas, 2001).

## V. BASIC PARAMETERS FOR *IN VIVO* REGENERATION STUDIES: REPRODUCIBLE, NONREGENERATIVE WOUNDS

*In vivo* regeneration of injured or excised tissue can be modeled as a process taking place within a bioreactor surrounded by a reservoir with constant properties; the bioreactor symbolizes the entire organism with its complex homeostatic mechanisms. From an engineering standpoint, the bioreactor must have a defined and consistent anatomical and physicochemical environment for quantitative study of any biological process. In order to study induced *in vivo* regeneration, it is therefore critical to standardize the wound site where studies are performed. Without a standardized,

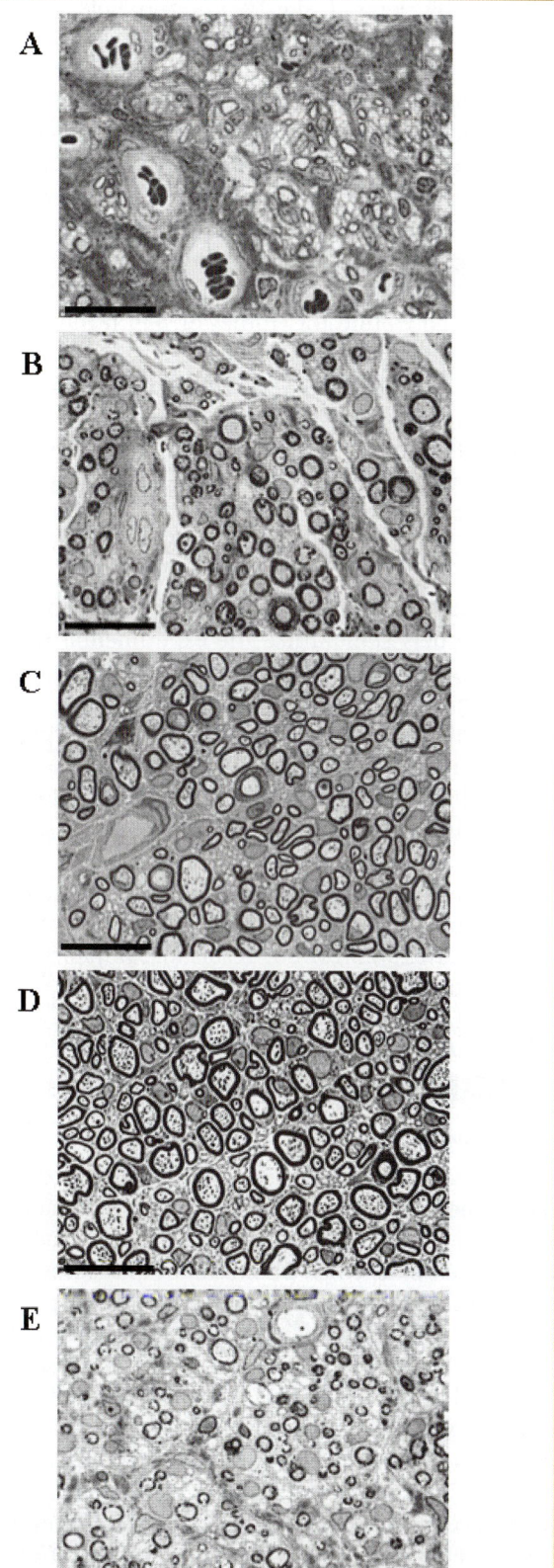

**FIG. 16.4.** Histomorphometic, cross-sectional images of the regenerated nerve trunk from the five collagen tubes with distinct degradation rates, devices A to E. The devices are arranged in order of lowest to highest cross-link density, or most rapid to slowest degradation rate, from A to E. Nerve trunks regenerated in devices C and D, characterized by intermediate levels of the cross-link density and degradation rate (device half-life: 2–3 weeks), showed superior morphology, with significantly larger axons, a more well-defined myelin sheath, and a significantly larger N-ratio. Light-micrograph images of toluidine blue–stained cross sections of regenerated axons taken from the midpoint of the 15-mm gap. Scale bar = 25 μm (Harley *et al.*, 2004).

reproducible wound site, it is impossible to assess differences accurately and statistically between ECM analogs within a single laboratory or to compare results from different, independent laboratories. Billingham and Medawar (1955) introduced the concept of an anatomically constant wound for the study of massive skin injuries. This concept has been amplified and standardized. It is referred to as the *anatomically well-defined wound* (Yannas, 2001) and has been applied to the study of injuries in skin, peripheral nerves, and a variety of other wound sites. The anatomically well-defined wound for studies of skin regeneration is the full-thickness skin wound where all tissue (epithelia, basement membrane, stroma) is removed down to the underlying fascia (Yannas, 2001). For peripheral nerve regeneration studies, the standardized wound is complete transection (total or full axotomy) of the nerve trunk, typically at the midpoint of the sciatic nerve (Yannas, 2001). For studies of cartilage repair following injury, the standardized wound model is complete removal of the cartilage down to the underlying subchondral bone (Capito and Spector, 2003). These models standardize the experimental wound environment, making it possible to obtain statistically significant results that can be compared meaningfully. However, in clinical cases, the nature of the wounds typically varies on a case-by-case basis, and it is important to understand the applicability of each treatment methodology to the range of injuries encountered. For example, while a tubular scaffold can be used to study peripheral nerve regeneration using the full axotomy model, a tubular implant would be difficult to implement clinically in the case of severe crush injuries that do not result in complete nerve trunk transection.

## VI. EXAMPLES OF *IN VIVO* ORGAN REGENERATION

### Skin Regeneration

Patients exhibiting skin wounds with loss of a substantial fraction of total body surface area (TBSA) face an immediate threat to their survival, originating primarily from the loss of their epidermis. One result of this loss is an increase by an order of magnitude of the moisture evaporation rate, which, if left uncorrected, leads to excessive dehydration and shock. Another result is a sharp increase in risk for massive bacterial infection, which, if allowed to progress, frequently resists treatment and leads to sepsis. Even when patients manage to survive these immediate threats, there is a residual serious problem of quality of life, originating in the occurrence of crippling contractures and disfiguring scars due to the physiological repair process. Conventional treatment is based on use of autografting, which yields excellent results at the treatment site but is burdened by the trauma caused at the donor site as well as by the unavailability of donor sites when the TBSA exceeds about 40–50%.

Of the three major tissue types that together comprise skin, the epidermis and basement membrane layers regenerate spontaneously following injury, provided there is a dermal substrate (stromal layer) underneath. Complete regeneration of the damaged epithelia and basement membrane is typically observed following first- and second-degree burns, small cuts and scrapes, and blisters. However, complete skin regeneration is not observed in the case where substantial damage to the underlying stroma occurs. Such cases include third-degree burns, deep cuts, and scrapes. In these cases, similar to the full-thickness, anatomically well-defined wounds created for studies of skin regeneration, organized wound contraction and scar synthesis is observed. A more in-depth description of the skin tissue triad, salient anatomical features, and its regenerative capacity has been previously published (Yannas, 2001).

In addition to the clinical success of the collagen-based dermal regeneration template (DRT), four other technologies developed to induce skin regeneration have achieved variable levels of success and are briefly discussed here to motivate further thought about scaffold-based options for treating severe injuries. In the following sections, the DRT, cultured epithelial autograft (CEA), living dermal replacement (LDR), living skin equivalent (LSE), and the naturally derived collagen matrix (NDCM, Alloderm) are discussed.

### Dermal Regeneration Template (DRT)

The dermal regeneration template (DRT) is a collagen–glycosaminoglycan scaffold whose microstructural and materials properties have been optimized to produce a bioactive ECM analog that, when implanted, induces sequential regeneration of the underlying dermis and resultant regeneration of the basement membrane and epithelial layers. The effectiveness of the DRT has been demonstrated clinically with a population of massively burned patients (Yannas, 2001) and in animal experiments utilizing a full-thickness (anatomically well-defined skin wound) skin wound (Yannas, 2001; Yannas *et al.*, 1989).

The DRT is typically used as an acellular implant that induces regeneration of the dermis, thereby providing the essential substrate for spontaneous regeneration of the epidermis and basement membrane layers. Following dermal regeneration by the unseeded DRT, epidermal cells from the wound edges migrate into the center of the wound and form a mature epidermis and basement membrane in a process termed *sequential regeneration* (Yannas, 2001). The resultant regeneration of appropriate tissue layers (tissue triad), along with associated structures (i.e., rete ridges), has indicated that the DRT is capable of inducing regeneration of mature skin in a full-thickness skin wound model (Fig. 16.5) (Compton *et al.*, 1998).

Clinically, the DRT is used as an acellular, bilayer device consisting of an inner layer of the active ECM analog and an outer layer of elastomeric poly(dimethyl siloxane). The layer of silicone is removed about two weeks after grafting the

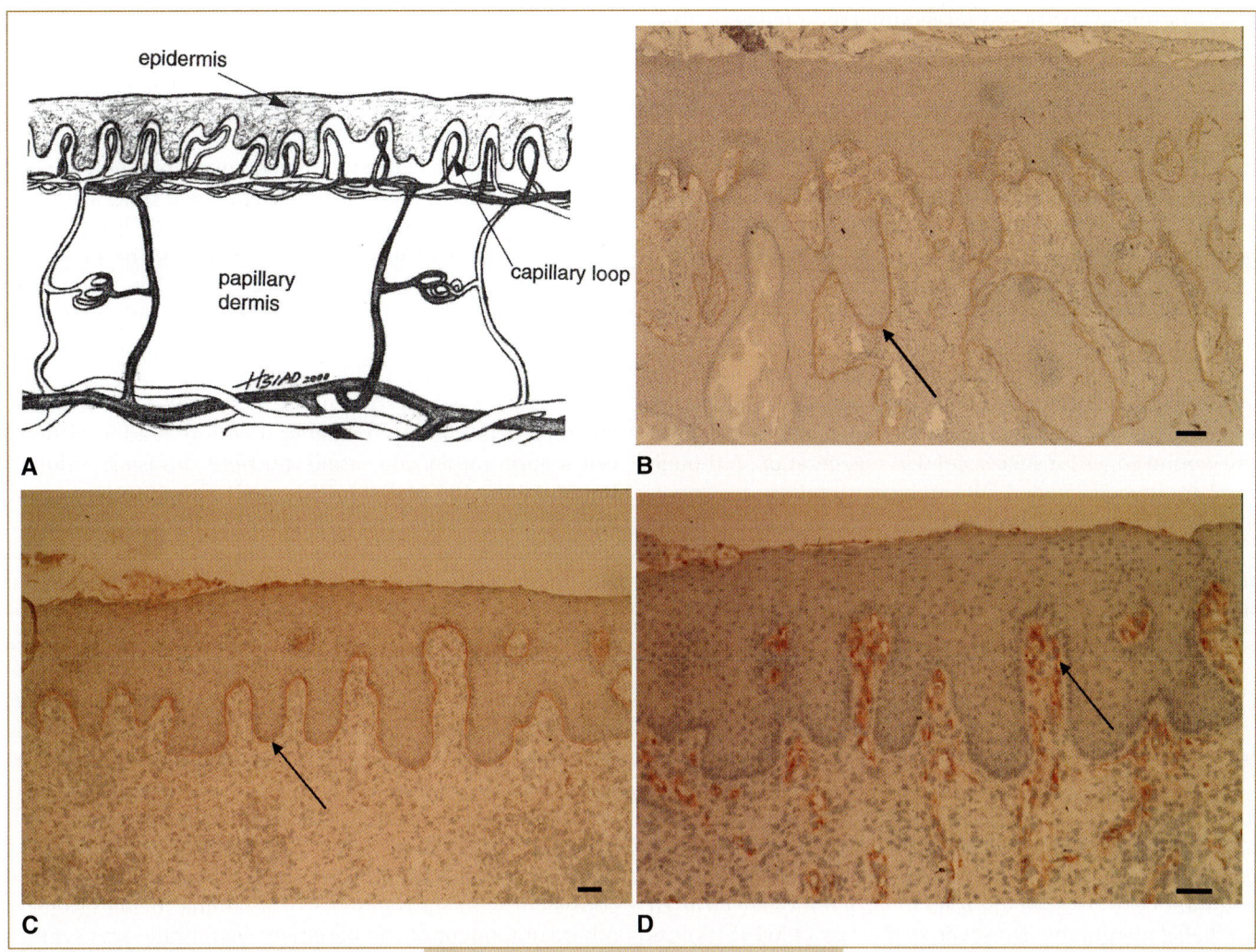

**FIG. 16.5.** **(A)** Diagram of normal skin showing the characteristic rete ridges at the dermo–epidermal junction and the vascular network (capillary loops) that populates the subepidermal region (Yannas, 2001). **(B)** As early as 12 days after grafting a full-thickness skin wound with a keratinocyte-seeded DRT, anchoring fibrils were observed in the regenerating basement membrane (arrow). The basal surface epithelium and the periphery of the epithelial cords are labeled with type VII collagen immunostaining, identifying the anchorage structures at the dermo–epidermal interface. Bar: 150 μm. **(C)** As early as 35 days after grafting a full-thickness skin wound with a keratinocyte-seeded DRT, a confluent hemidesmosomal staining pattern is observed at the dermo–epidermal junction (arrow) by immunostaining for the $\alpha_6\beta_4$ integrin. The pattern observed in the regenerating skin is identical to that observed in physiologic skin. Bar: 100 μm. **(D)** A full-thickness skin wound grafted with a keratinocyte-seeded DRT was observed to regenerate many of the structures observed in normal skin. Immunostaining for factor VIII 35 days after grafting revealed that capillary loops had formed in the rete ridges of the regenerated dermis (arrow), similar to those observed in physiologic skin. Bar: 75 μm (Compton *et al.*, 1998)

device, having served the important temporary role of controlling moisture flux and bacterial invasion at the site of organ synthesis. The DRT is fabricated from collagen–glycosaminoglycan copolymer, with a 98 : 2 ratio of microfibrillar, type I collagen to chondroitin 6-sulfate. The microstructure of the DRT has been optimized with both a lower and an upper pore size bound of $20 \pm 4$ μm and $125 \pm 35$ μm, respectively. Additionally, the biodegradation rate of the DRT has been optimized with lower and upper bounds of residence time in the wound bed of 5 and 15 days, respectively (Yannas, 2001; Yannas *et al.*, 1989). Clinical use of the DRT has emphasized treatment of patients with massive

burns as well as those who require resurfacing of large or small scars from burns. The DRT is responsible for regenerating the underlying dermal architecture, providing the appropriate stromal layer on which native regenerative processes of the basement membrane and epithelial layer can take place. For wounds of relatively small characteristic dimension, e.g., 1 cm, epithelial cells migrating at speeds on the order of 0.5 mm/day from the wound edges can provide a confluent epidermis within 10 days. In such cases, the unseeded DRT fulfills all the design specifications set for successful *in vivo* skin regeneration. However, the wounds incurred by a massively burned patient are typically of characteristic dimension of several centimeters, often more than 20–30 cm. These wounds are large enough to preclude formation of a new epidermis by cell migration alone within a clinically acceptable time frame. Current clinical protocol favors harvesting and use of a very thin autoepidermal graft from another part of the patient's body, thin enough not to leave behind a scarred donor site, to cover the neodermis (Yannas, 2001), or application of a cultured epithelial autograft (described later) on top of the DRT following removal of the poly(dimethyl siloxane) membrane.

A keratinocyte-seeded DRT has been used in guinea pig and swine animal models. The keratinocyte-seeded DRT induces simultaneous formation of a dermis as well as the epidermal and basement membrane layers and removes the requirement for epithelial cell–mediated migration from the wound edges for successful skin regeneration. Although the keratinocyte-seeded DRT induces *simultaneous regeneration* of skin in the guinea pig model, an animal model characterized by extensive organized wound contraction that is significantly more severe than that observed with humans, the sum total of the results of studies using the DRT suggest that *in vivo* synthesis of skin in the clinical setting does not require anything more than the DRT (Yannas, 2001).

## Cultured Epithelial Autograft (CEA)

Cultured epithelial autografts (CEAs) are an epidermal graft formed from a sheet of keratinocytes grown *in vitro* and then implanted into the wound site. A small epidermal sample is removed from the patient and dissociated, and the resultant cells are then cultured *in vitro* until they form an epithelial membrane. However, in the case of full-thickness skin wounds where there is no underlying dermis to support the epidermal sheet, rapid CEA degradation is typically observed and long-term clinical applicability has not been observed. In particular, one persistent problem is the formation of blisters under large areas of the graft (avulsion). However, in the case of partial-thickness wounds where a significant portion of the stroma remains, the "take" of the CEA graft has been very good. As such, the CEA has been used to cover significant areas, as large as 50% of the TBSA, for partial-thickness wounds and as a temporary solution for full-thickness wounds to prevent immediate wound dehydration and infection, making the CEA a life-saving, although temporary, graft (Yannas, 2001). Additionally, regardless of whether the graft was placed on a partial- or full-thickness skin wound, the resulting CEA graft exhibited mechanical fragility due to a lack of the 7-S domain of type IV collagen, anchoring fibrils, and rete ridges (Woodley *et al.*, 1988). These structures, required for the formation of a physiological dermo–epidermal junction as well as for the formation of a physiological collagen and elastin fiber architecture observed in the normal, adult dermis, are not formed after grafting of the wound site with the CEA (Yannas, 2001). Without these structures, the CEA cannot be used as a per-

manent skin replacement; instead, the CEA is often used as a temporary coverage as part of a more substantial treatment regimen, such as autografting or DRT implantation.

## Living Dermal Replacement (LDR)

The LDR (living dermal replacement) is a polyglactin-910 nylon surgical mesh scaffold fabricated from a copolymer of 90 wt% glycolic acid and 10 wt% lactic acid termed *PGL* that was developed to be an analog of the ECM and induce skin regeneration following injury. The PGL fibers, approximately 100 µm in thickness, were knitted into a mesh that exhibits a pore microstructure with a characteristic dimension of 280–400 µm. This mesh structure presented a large-weave structure relative to the characteristic cell dimension of approximately 10 µm that allowed rapid cell incorporation and ample nutrient diffusion. Prior to implantation, the acellular PGL mesh was cultured *in vitro* with fibroblasts until the cells were observed to synthesize several important ECM components, such as collagen and elastin, *in vitro*. Immediately prior to implantation, the upper surface of the PGL scaffold was seeded with keratinocytes to produce a bilayer graft. Once the keratinocytes reached confluence on the surface of the PGL mesh, the entire structure was grafted into the wound site (Hansbrough *et al.*, 1993). A thin, fragile epidermal layer typically developed by 10 days postgrafting and became cornified as early as 20 days postgrafting. Additionally, by 20 days postgrafting the LDR scaffold had degraded completely, with minimal inflammatory response. While the interface between the graft and the wound bed stained positive for laminin, consistent with the synthesis of a lamina lucida layer, no other component of the basement membrane was synthesized. In addition, rete ridges were not synthesized, and a thick layer of fibrotic tissue with a large fibroblast population and vascular ingrowth, characteristic of scar tissue formation, was observed below the newly synthesized epidermal layer. In summary, while the LDR showed the ability to induce regeneration of a neoepidermis, it did not exhibit the ability to induce regeneration of a complete basement membrane or dermal layer (Hansbrough *et al.*, 1993).

## Living-Skin Equivalent (LSE)

The LSE (living-skin equivalent) was formed by populating a collagen lattice with dermal fibroblasts that *in vitro* contracted the lattice and synthesized additional ECM proteins, forming a neo-dermal layer (Bell *et al.*, 1983). Similar to the LDR, after the initial culture period, the upper surface of the neo-dermal layer was seeded with a suspension of keratinocytes. Once seeded, the keratinocytes attached to the collagen construct, proliferated, and differentiated to form a multilayered, epidermal structure within one to two weeks, all *in vitro*. Although short segments of the lamina densa were observed along the dermo–epidermal convergence *in vitro*, the cultured LSE did not exhibit a complete basement membrane layer, rete ridges, or any other skin appendages at the end of the *in vitro* culture period (Nolte *et al.*, 1994). At this point, the construct was then implanted into the wound site. Continued structural and biological changes were observed in the LSE following grafting, indicating that remodeling was taking place (Yannas, 2001). A functional, fully differentiated epidermis was observed as early as seven days following grafting, and by 14 days a vascularized subepidermal layer with many of the structural characteristics of normal dermis, such as a "basketweave" collagen fiber pattern, was present. However, continued maturation into a formalized dermal region was not observed. Experimental results across multiple animal models and experimental trials were consistent in indicating that the LSE was able to induce regeneration of a mature epidermis and basement membrane, but a mature dermal layer was not observed and the dermo–epidermal junction remained flat, without any sign of rete ridges (Yannas, 2001).

## Naturally Derived Collagen Matrix (NDCM)

Instead of relying on technologies to fabricate a three-dimensional scaffold structure from either synthetic or natural materials (i.e., LSE, LDR, DRT), the naturally derived collagen matrix (NDCM) technology uses decellularized dermal tissue as a scaffold to induce skin regeneration. The NDCM was designed to act as an appropriate ECM analog, because it is the dermal ECM that supports normal tissue remodeling following severe injury, thereby inducing regeneration. The main advantage of the NDCM over a homograft (cellularized dermal tissue from a human donor) and xenograft (cellularized dermal tissue from an animal donor) is that owing to decellularization, the antigenicity of the scaffold is significantly reduced, thereby eliminating the incidence of rejection. The NDCM has been used primarily to treat full-thickness burns and burns to areas of the skin where contraction and scar formation would inhibit functionality (i.e., feet and hands). The NDCM is typically implanted into full-thickness skin wounds and is often covered by a thin autograft of the patient's own epidermis to speed the healing process. This treatment, similar to that observed with the use of the DRT, results in a high percentage of graft take. Additionally, the thin autograft of epidermal tissue significantly decreases the time for complete graft reepithelialization. Patients showed normal range of motion, grip strength, motor control, and functionality along with formation of appropriate anatomical structures (i.e., rete ridges) following treatment (Tsai *et al.*, 1999), demonstrating the utility of a naturally derived material in tissue-engineering applications.

## Peripheral Nerve Regeneration

The mammalian peripheral nervous system also consists of a distinct tissue triad. Thousands of axons make up a nerve trunk, where each axon is surrounded sequentially by an epithelial, basement membrane, and stromal layer. Schwann cells wrap around the individual axon, forming the myelin sheath that constitutes the cell-continuous epithelia (Bunge and Bunge, 1983). Surrounding the epithelia is the acellular ECM layer (basement membrane) connecting the myelin sheath and the stromal layer (Fig. 16.1). The endoneurium is the stromal layer of cells, blood vessels, and ECM that surrounds, insulates, and protects all nerve fibers (Yannas, 2001). Similar to that observed with skin, the myelin sheath (epithelia) and basement membrane regenerate spontaneously following injury (Fig. 16.6), while the endoneurium (stromal layer) does not (Chamberlain *et al.*, 1998; Yannas, 2001).

## General Findings

While there have been a large number of studies investigating many different device designs, few have been shown to perform as well as the peripheral nerve autograft (typically the sural nerve from the leg), and none appears to have been able to improve over the autograft for gaps larger than 10 mm (Chamberlain *et al.*, 1998). For studies of peripheral nerve regeneration, complete axotomy (total severance of the nerve trunk) was utilized as the standardized wound model. An analysis of efficacy of the various devices is presented here using the critical axon elongation method (Yannas, 2001) in an attempt to identify optimal device parameters for successful and optimal peripheral nerve regeneration. This method identifies the relative regenerative capacity of different devices by comparing their critical axon elongation, the gap length at which 50% of the implanted devices induce reinnervation between the proximal and distal nerve stumps via elongating axons. Using this technique, increased critical axon elongation length ($L_c$) indicates superior regenerative capacity.

## Effect of Tubulation

Peripheral nerve regeneration is not observed in the absence of a tubular device connecting the two ends of a transected nerve stump if the gap between the two transected stumps is greater than a few millimeters (Yannas, 2001). It has also been observed that insertion of the distal stump into the tube is required for successful regeneration (Chamberlain *et al.*, 1998; Yannas, 2001). This result indicates that even though axon elongation takes place from the proximal to the distal stump, the distal stump appears to provide a critical cytokine field responsible for guiding axon elongation (neurotrophic effect). However, while it has been observed that a tube is all that is required to induce regeneration across a gap of modest length following complete transection, the physical parameters of the tube and any material in the tube lumen significantly affect the kinetics and quality of regeneration; these are discussed in the remaining sections of this chapter.

## Effect of Tube Chemical Composition

A variety of natural and synthetic polymers have been used in peripheral nerve regeneration conduits, and the

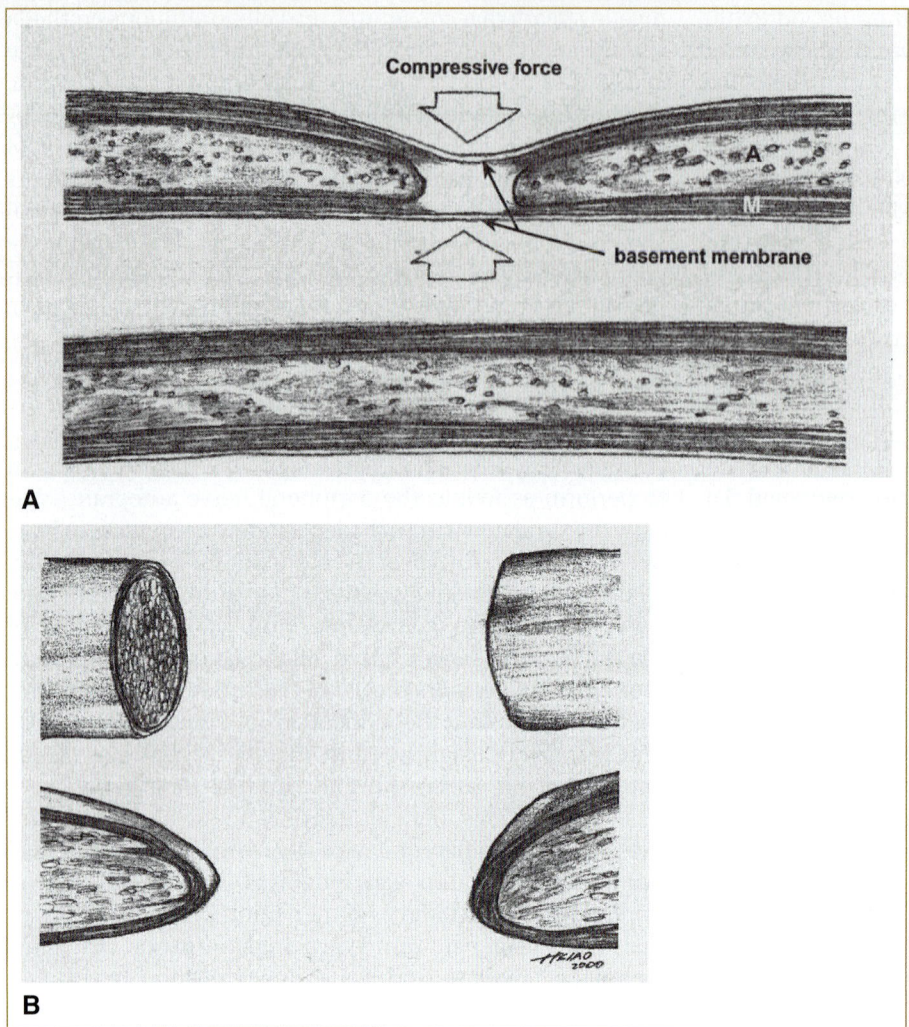

**FIG. 16.6.** **(A)** Axons and myelin sheath inside a nerve fiber are regenerative. Following mild crush injuries, the axoplasm (A) and myelin sheath (M) degenerate along the length of the crushed nerve fiber, but the basement membrane remains intact throughout. Spontaneous regeneration of the nerve fiber occurs within a few weeks of injury. **(B)** Most of the supporting tissues (stroma) surrounding nerve fibers are not regenerative. Although axons are regenerative following transection, the remainder of the nerve trunk is not. Following transection, each stump is closed via cell-mediated contraction and neuroma synthesis (Yannas, 2001).

choice of tube (conduit) composition has been found to significantly affect the quality of regeneration. For a complete review of the literature, please consult previous publications by the authors (Yannas, 2001; Harley and Yannas, 2006). Early conduits were produced from naturally occurring materials that were easily harvested and implanted, such as bone, dura, perineurium, and blood vessels (Yannas, 2001). More recent devices have utilized nondegradable, synthetic materials, such as stainless steel, rayon, and silicone, as well as degradable, synthetic polymers, such as polyester, polyglactin, and polylactate, and natural polymers, such as collagen, laminin, and fibronectin (Yannas, 2001).

Nondegradable tubes have typically resulted in a poor quality of regeneration marked by significant formation of a neural scar (neuroma) (Chamberlain *et al.*, 1998; Yannas, 2001). Conversely, conduits fabricated from ECM components, specifically collagen, fibronectin, and laminin, have been generally shown to enhance the quality of peripheral nerve regeneration (Bailey *et al.*, 1993; Chamberlain *et al.*, 1998; Ohbayashi *et al.*, 1996; Yannas, 2001). Collagen tubes in particular have been observed to induce the highest

quality of regeneration, as measured by both morphological and electrophysiological methods (Archibald *et al.*, 1995; Chamberlain *et al.*, 1998; Yannas, 2001). It has been hypothesized that this optimal chemical composition was observed because collagen tubes contain inherently bioactive binding sites (ligands) for attachment and migration of various cell types and can be manufactured in such a way that the tube wall pore structure can display a range of cell and protein permeability. Additionally, the degradation rate of the collagen tubes can be tailored to degrade with half-lives that vary over a very wide range to meet the requirements of the isomorphous tissue replacement model (Chang and Yannas, 1992; Harley *et al.*, 2004).

### Effect of Tube Permeability

Conduit permeability significantly affects the mechanism and quality of peripheral nerve regeneration. *Tube permeability* can be defined by its initial structural features (i.e., scaffold porosity, pore size) or its degradation characteristics (i.e., rapid degradation quickly permeabilizes the tube). Additionally, the conduits may be protein permeable but cell impermeable or cell and protein permeable. No appreciable increase in the quality of peripheral nerve regeneration was observed for the protein-permeable and cell-impermeable conduit compared to the impermeable conduit (Jenq and Cosseshall, 1987). However, cell- (and therefore protein-) permeable tubes exhibited significantly superior regenerative capacity as compared to impermeable as well as protein-permeable and cell-impermeable conduits (Jenq and Coggeshall, 1987; Jenq *et al.*, 1987; Li *et al.*, 1992; Rodriguez *et al.*, 1999; Yannas, 2001). Device permeability also significantly influences the contractile response following peripheral nerve injury; permeable collagen tubes display a significantly thinner contractile capsule surrounding the regenerating nerve trunk than biodurable silicone tubes (Chamberlain *et al.*, 1998). It is hypothesized that device permeability reduces myofibroblast-mediated contraction of the wound site by permitting migration of the contractile cells away from the wound site through the tube wall (Yannas, 2001) and by allowing connective tissue cells from the surrounding environment access to the tube lumen (Jenq and Coggeshall, 1987).

## Effect of Tube Degradation Rate

A biodurable conduit, such as a silicone tube, initially induces partial reinnervation between proximal and distal stumps following implantation (Chamberlain *et al.*, 1998). However, as the initially regenerated nerve trunk remodels and matures, the silicone tube constricts this process, resulting in pain and eventual degeneration of the regenerated nerve; prevention of the ultimate degeneration of the initially regenerated nerve trunk requires a difficult second surgical procedure that can further harm the nerve trunk. For this reason, one of the historical goals in studies of peripheral nerve regeneration studies is identification of a suitable degradable conduit. Superior performance of biodegradable tubes compared to tubes made from materials that were either biodurable or which had a very low degradation rate has since been reported in studies by several investigators (Harley *et al.*, 2004; Itoh *et al.*, 2001; Navarro *et al.*, 1996; Rodriguez *et al.*, 1999; Valero-Cabre *et al.*, 2001; Yannas, 2001). Examples include comparison of tubes made of two biodurable polymers, silicone and poly(tetrafluorethylene), and two degradable polymers, a copolymer of poly(lactic acid) and ε-caprolactone (PLA/PCL) as well as collagen (Navarro *et al.*, 1996). This study indicated that the degradable tubes induced a higher quality of regeneration as compared to the biodurable tubes over a 6-mm gap in the mouse sciatic nerve. Additionally, a number of degradable collagen tubes have been shown to perform as well as the autograft, considered to be clearly superior to regeneration via a silicone tube and commonly thought to fall short functionally only of physiological (preinjury) nerve (Archibald *et al.*, 1995; Chamberlain *et al.*, 1998; den Dunnen *et al.*, 1996; Molander *et al.*, 1983; Robinson *et al.*, 1991; Yannas, 2001).

While the regenerative advantage of a degradable tube versus a biodurable tube is supported by the evidence, there have not been many extensive studies on the effect of the magnitude of the degradation rate and whether an optimal degradation rate exists, as has been found in the study of ECM analogs (the DRT) to induce skin regeneration (Yannas *et al.*, 1989). Recently, one of the first comprehensive studies of the effect of the degradation rate of collagen tubes on the quality of peripheral nerve regeneration has been published by these authors. This study utilized a 15-mm gap in the rat sciatic nerve and evaluated the regenerative capacity of a homologous series of porous, collagen tubes, showing a significant effect of degradation rate on the quality of peripheral nerve regeneration (Harley *et al.*, 2004). The chemical composition, pore structure, and permeability of the conduits in this study were kept constant, while the cross-link density was steadily increased to create a series of five devices with *in vivo* degradation half-lives varying between <1 week and >100 weeks. The quality of peripheral nerve regeneration was observed to vary significantly with tube degradation rate; the highest quality of peripheral nerve regeneration was observed for tubes with a degradation half-life of two to three weeks (Fig. 16.4) (Harley *et al.*, 2004). These data suggest that the positive effects of tubulation in treating peripheral nerve injuries are due to the presence of the tube immediately following injury; tubulation appears to significantly affect the early mechanisms of peripheral nerve regeneration. A speculative view of the maximum regenerative activity observed in the present study can be based on the putative existence of a lower and higher limit in tube degradation rate, similar to that observed for the dermal regeneration template (Yannas *et al.*, 1989). When the degradation rate of the collagen tubes is excessively slow, the tubes behave as if they were biodurable, remaining intact long enough to interfere with tissue remodeling. In contrast, tubes that degrade very rapidly fail to maintain a protected environment for regeneration, which is the basis for the use of tubulation to promote nerve regeneration.

## Effect of Tube Filling

A wide variety of solutions, ECM analogs, and cell suspensions have been introduced into the tube lumen in an effort to improve the quality of peripheral nerve regeneration. Use of ECM macromolecules, such as collagen, laminin, and fibronectin, in both solution and gel form, has been observed to have no significant effect on peripheral nerve regeneration. Furthermore, when gel concentrations exceeded certain critical levels, a negative effect on the quality of regeneration is observed (Bailey *et al.*, 1993). However, a laminin-coated collagen-based scaffold has been observed to improve the quality of regeneration, implying the requirement of an insoluble structure within the tube lumen in addition to any soluble regulators to improve the quality of regeneration (Ohbayashi *et al.*, 1996).

Several insoluble substrates, ECM analogs inserted into the empty lumen of the tube prior to implantation, have shown significant regenerative activity. Examples include highly oriented fibrin fibers and axially oriented polyamide filaments that significantly improved the quality of regeneration (Lundborg *et al.*, 1997; Williams *et al.*, 1987). Specific ECM analogs have been observed to significantly increase the maximal gap length that can be bridged by axonal tissue (Madison *et al.*, 1988; Yannas, 2001), the speed of axonal bridging (Madison *et al.*, 1988; Ohbayashi *et al.*, 1996), and the quality of regeneration (Chamberlain *et al.*, 1998; Chang and Yannas, 1992; Yannas, 2001).

The ECM analog that has been found to induce the highest quality of regeneration is a collagen-based scaffold termed the *nerve regeneration template* (NRT) (Chang and Yannas, 1992; Yannas, 2001). The NRT has induced regeneration of a functional peripheral nerve across gaps varying from 10 to 15 mm in the rat sciatic nerve (Chamberlain *et al.*, 1998; Yannas, 2001). Like the DRT, the microstructural and material properties of the NRT have been optimized. The highly bioactive NRT is characterized by axially (extending between the proximal and distal stumps) elongated pore tracks defined by axially oriented ellipsoidal pores with a mean pore size of approximately 35 µm (Chang and Yannas, 1992). This pore structure is hypothesized to improve the quality of peripheral nerve regeneration by providing directional guidance to the formation of linear Schwann cell columns, which act as tracks for axon elongation (microtube hypothesis) (Yannas, 2001; Zhao *et al.*, 1997). The positive effect of an axially oriented fiber structure has also been observed using a fibrin fiber–based ECM analog in the tube lumen (Williams *et al.*, 1987). The degradation rate of the NRT has also been found to significantly affect the quality of regeneration, with an *in vivo* degradation half-life on the order of six weeks found to be optimal; NRT variants that degraded too rapidly or too slowly led to significantly poorer functional recovery (Chang and Yannas, 1992; Yannas, 2001). The long-term morphological structure and electrophysiological function of nerves regenerated using the NRT have been observed to be at the level of an autografted nerve, the current gold standard for peripheral nerve injury treatment (Chamberlain *et al.*, 1998).

Soluble regulators and cell suspensions have also been introduced into the tube lumen in an effort to improve the quality of regeneration. Suspensions of Schwann cells showed very significant regenerative activity (Ansselin *et al.*, 1997), further supporting the microtube hypothesis, which describes nerve regeneration as dependent on early formation of linear columns of Schwann cells extending from the proximal toward the distal nerve trunk. In addition to the use of a cell suspension, solutions of acidic (aFGF) and basic fibroblast growth factor (bFGF) were also observed to improve the quality of regeneration (Walter *et al.*, 1993). However, the use of nerve growth factor (NGF) was not found to improve the quality of regeneration (Hollowell *et al.*, 1990).

A variety of conclusions have been drawn about the relative efficacy of the various biomaterials and devices employed in the study of peripheral nerve regeneration. Nerve chamber configurations that had the highest regenerative activity were those in which the tube wall comprised certain synthetic biodegradable polymers, such as collagen, was cell permeable rather than protein permeable or impermeable, and had an *in vivo* degradative half-life on the order of two to three weeks. Introduction of an insoluble regulator (ECM analog in the form of a scaffold) into the tube lumen, but not of ECM components in a gel or solution form, also significantly improves the quality of regeneration. The optimal ECM structures were found to be highly porous, with controlled degradation rates and an axially aligned micro structure. In addition, suspensions of Schwann cells as well as solutions of either acidic or basic fibroblast growth factor placed within the tube lumen with or without an insoluble ECM analog have been shown to improve the quality of regeneration.

## Cartilage Regeneration

Articular cartilage contains an avascular, non-neural ECM composed primarily of type II collagen and glycosaminoglycans. Compared to other tissues, cartilage possesses a very low cell density and a very high ECM density. The chondrocytes populating the cartilaginous ECM display low proliferative activity and, due to the high ECM protein density, are unable to migrate through the tissue. Due to the low proliferative activity, avascularity, and high ECM protein density, injuries to cartilage display an injury response distinct from traditional injury responses, characterized by inflammatory processes, cell-mediated contraction, and scar synthesis. No repair or regeneration processes are observed *in vivo* following cartilage injuries; instead, the scope of injury increases as the wound edges become increasingly degraded, eventually compromising the joint (Capito and Spector, 2003). Severe cartilage injuries are extremely prevalent in today's active society, resulting in pain, decreased patient activity, and eventually disability, profoundly impacting quality of life. The current methods for treating such focal cartilage defects include microfracture, autologous chondrocyte implantation, and osteochondral autografting (Capito and Spector, 2003). In the case of microfracture and autologous chondrocyte implantation, a flap of periosteum is sewn over the cartilage defect area and is sealed with fibrin glue. With microfracture, immediately prior to application of the periosteal flap, microfractures in the subchondral bone are created to allow bone marrow cells access to the then-sealed cartilage defect in an attempt to utilize the stem cell population in the bone marrow to regenerate the damaged cartilage. With autologous chondrocyte implantation, a biopsy of cartilage tissue is removed from the patient prior to surgery and cultured *in vitro* to obtain a large chondrocyte population that is then injected back into the periosteal flap–sealed cartilage defect. In the case of osteochondral autografting, a series of osteochondral plugs consisting of cartilage, the underlying subchondral bone, and the tidemark region separating them are removed from a nonloading region at the edge of the damage joint. These plugs are then implanted into the primary cartilage defect using an approach termed *mosaicplasty*, named after the mosaic pattern of the implanted osteochondral plugs (Capito and Spector, 2003). These procedures, however, exhibit limited long-term success in treating the cartilage injury. Preclinical studies implementing tissue-engineering approaches have yielded results that represent improvements over the currently employed cartilage repair procedures.

Various synthetic and natural materials have been employed to fabricate porous, bioresorbable scaffolds for articular cartilage tissue engineering. Among the list of bioresorbable or partially resorbable materials used for cartilage repair are collagen, hyaluronan, fibrin, polylactic acid (PLA) and polyglycolic acid (PGA), chitosan scaffolds and gels, devitalized cartilage, hydroxyapatite, demineralized bone matrix, and bioactive glass (Capito and Spector, 2003). Natural polymers, such as collagen, provide a more native surface to cells and have been the primary focus of tissue-engineering studies due to previous successes in peripheral nerve and skin regeneration. Studies have confirmed that the addition of cells seeded within these 3D scaffolds enhance matrix synthesis and increase type II collagen production *in vivo* and *in vitro* (Capito and Spector, 2003; Kinner *et al.*, 2005; Lee *et al.*, 2003). Various cell types may be used to enhance cartilage synthesis when seeded into matrices, including articular chondrocytes and chondroprogenitor cells derived from bone marrow, periosteum, or perichondrium (Capito and Spector, 2003; C. R. Lee *et al.*, 2003).

Preliminary investigations have been focused on the bioactivity of chondrocytes in a series of ECM analogs. Promising results have been observed for adult articular chondrocytes cultured *in vitro* in type II collagen–glycosaminoglycan (CG) scaffolds, where the chondrocytes have retained high biosynthetic capacity for producing type II collagen, the predominant ECM component of cartilage (Capito and Spector, 2003; C. R. Lee *et al.*, 2001). Type II CG scaffolds seeded with autologous, articular chondrocytes have also been evaluated for their *in vivo* regenerative capacity in a full-thickness cartilage injury model; the full-thickness cartilage injury, with cartilage removed down to the subchondral bone, is the standardized, anatomically well-defined wound utilized for *in vivo* studies of cartilage regeneration. The greatest total amount of reparative tissue was found in the cell-seeded type II CG scaffolds, as opposed to unseeded type II CG scaffolds and seeded or unseeded type I CG scaffolds. Moreover, examination of the reparative tissue formed in the subchondral region of defects treated with the chondrocyte-seeded type II collagen scaffolds indicated that the majority of the tissue was positive for type II collagen and that good integration was observed between the implant and the surrounding cartilage. These results indicate an influence of the exogenous chondrocytes on the process of chondrogenesis (Nehrer *et al.*, 1998). Such studies of the healing of chondral defects in animal models have revealed that there is some potential for regeneration of this connective tissue. The introduction of certain biomaterial scaffolds along with selected surgical procedures and cell therapies has been demonstrated in animal studies to facilitate the cartilage reparative process and now offers the promise of extending the longevity of clinical treatments of cartilage defects (Kinner *et al.*, 2005; C. R. Lee *et al.*, 2003; Samuel *et al.*, 2002).

In a separate series of studies, the regenerative potential of chondrocytes encapsulated in photopolymerized poly(ethylene oxide) hydrogels and self-assembling peptide hydrogels has been tested using an *in vitro* culture model followed by *in vivo* implantation. Preliminary results suggest that photo cross linked (Bryant and Anseth, 2001) and self assembling hydrogels (Kisiday *et al.*, 2002; Zhang *et al.*, 1993) are promising scaffolds for tissue-engineering cartilage, for cell viability was maintained, uniform cell seeding was achieved, and the biochemical content of the ECM proteins synthesized within the construct were similar to those found in native cartilage (Bryant and Anseth, 2001; Kisiday *et al.*, 2002). Additionally, the importance of the biomechanical environment during *in vitro* culture of chondrocytes within a three-dimensional construct has been observed. Improved biosynthesis of ECM components, notably native proteoglycans, is observed when a cyclic loading environment is applied to *in vitro* chondrocyte cultures within a series of hydrogel and scaffold constructs (Kisiday *et al.*, 2004).

Collectively these findings provide the basis for the rational development of approaches for the more complete regeneration of articular cartilage and demonstrate that meaningful clinical outcomes can be achieved even if complete regeneration is not achieved (Kinner *et al.*, 2005; C. R. Lee *et al.*, 2003; Samuel *et al.*, 2002).

## VII. CONCLUSIONS

Following severe injury, the typical physiological response is characterized by a complex inflammatory response, cell-mediated wound contraction, and scar tissue synthesis characterized as repair. However, introduction of a suitable analog of the extracellular matrix into the wound site has been observed to block cell-mediated contraction of the wound site and to induce regeneration of physiological tissue. Although several ECM analogs have been studied, only those with a narrowly defined structure have been shown to be capable of regeneration. The microstructural, chemical compositional, and biodegradation rate specificity of these templates appears to be related to the requirement for inhibition of wound contraction prior to the incidence of regeneration. A number of appropriate materials have been developed and tested for use in particular *in vivo* wound sites, notably the skin, peripheral nerve, conjunctiva, and cartilage. The results of *in vitro* and *in vivo* investigations of many biomaterial constructs suggest that the properties of an ECM analog must be tailored to the specific wound site and species, requiring further development of appropriate *in vivo* and *in vitro* models to further understand the complexity of cell–scaffold interactions and the process of induced regeneration.

## VIII. ACKNOWLEDGMENTS

We are grateful for the funding provided by the Cambridge-MIT Institute (BAH, IVY) as well as the financial contributions of the Whitaker-MIT Health Science Fund Fellowship (BAH).

## IX. REFERENCES

Ansselin, A. D., Fink, T., and Davey, D. F. (1997). Peripheral nerve regeneration through nerve guides seeded with adult Schwann cells. *Neuropathol. Appl. Neurobiol.* **23**, 387–398.

Archibald, S. J., Shefner, J., Krarup, C., and Madison, R. D. (1995). Monkey median nerve repaired by nerve graft or collagen nerve guide tube. *J. Neurosci.* **15**, 4109–4123.

Bailey, S. B., Eichler, M. E., Villadiego, A., and Rich, K. M. (1993). The influence of fibronectin and laminin during Schwann cell migration and peripheral nerve regeneration through silicon chambers. *J. Neurocytol.* **22**, 176–184.

Bell, E., Sher, S., Hull, B., Merrill, C., Rosen, S., Chamson, A., Asselinea, D., Dubertret, L., Coulomb, B., and Lapiere, C. (1983). The reconstitution of living skin. *J. Invest. Dermatol.* **81**, 2s–10s.

Billingham, R. E., and Medawar, P. B. (1955). Contracture and intussusceptive growth in the healing of expensive wounds in mammalian skin. *J. Anat.* **89**, 114–123.

Border, W. A., Noble, N. A., and Ketteler, M. (1995). TGF-B: a cytokine mediator of glomerulosclerosis and a target for therapeutic intervention. *Kidney Int. Suppl.* **49**, S59–S61.

Bryant, S. J., and Anseth, K. S. (2001). The effects of scaffold thickness on tissue engineered cartilage in photocrosslinked poly(ethylene oxide) hydrogels. *Biomaterials* **22**, 619–626.

Bunge, R. P., and Bunge, M. B. (1983). Interrelationship between Schwann cell function and extracellular matrix production. *Trends Neurosci.* **6**, 499–505.

Capito, R. M., and Spector, M. (2003). Scaffold-based articular cartilage repair. *IEEE Eng. Med. Biol. Mag.* **22**, 42–50.

Chamberlain, L. J., Yannas, I. V., Hsu, H. -P., Strichartz, G., and Spector, M. (1998). Collagen-GAG substrate enhances the quality of nerve regeneration through collagen tubes up to level of autograft. *Exper. Neurol.* **154**, 315–329.

Chang, A. S., and Yannas, I. V. (1992). Peripheral nerve regeneration. *In* "Encyclopedia of Neuroscience" (G. Adelman, ed.), pp. 125–126. Birkhauser, Boston.

Compton, C. C., Butler, C. E., Yannas, I. V., Warland, G., and Orgill, D. P. (1998). Organized skin structure is regenerated *in vivo* from collagen-GAG matrices seeded with autologous keratinocytes. *J. Invest. Dermatol.* **110**, 908–916.

Daley, G. Q. (2005). Customized human embryonic stem cells. *Nat. Biotechnol.* **23**, 826–828.

Daley, G. Q., Sandel, M. J., and Moreno, J. D. (2005). Stem cell research: science, ethics and policy. *Med. Ethics.* **12**, 5.

den Dunnen, W. F. A., Stokroos, I., Blaauw, E. H., Holwerda, A., Pennings, A. J., Robinson, P. H., and Schakenraad, J. M. (1996). Light-microscopic and electron-microscopic evaluation of short-term nerve regeneration using a biodegradable poly(DL-lactide-0-caprolacton) nerve guide. *J. Biomed. Mater. Res.* **31**, 105–115.

Fu, S. Y., and Gordon, T. (1997). The cellular and molecular basis of peripheral nerve regeneration. *Mol. Neurobiol.* **14**, 67–116.

Gage, F. H. (2000). Mammalian neural stem cells. *Science* **287**, 1433–1438.

Gibson, L. J., and Ashby, M. F. (1997). "Cellular Solids: Structure and Properties," Cambridge University Press, Cambridge, U.K.

Hansbrough, J. F., Morgan, J. L., Greenleaf, G. E., and Bartel, R. (1993). Composite grafts of human keratinocytes grown on a polyglactin mesh-cultured fibroblast dermal substitute function as a bilayer skin replacement in full-thickness wounds on athymic mice. *J. Burn Care Rehab.* **14**, 485–494.

Harley, B. A., and Yannas, I. V. (2006). Skin, tissue engineering for regeneration. *In* "Wiley Encyclopedia of Medical Devices and Instrumentation" (J. G. Webster, ed.), pp. 179–202. John Wiley & Sons, New York.

Harley, B. A., Spilker, M. H., Wu, J. W., Asano, K., Hsu, H.-P., Spector, M., and Yannas, I. V. (2004). Optimal degradation rate for collagen chambers used for regeneration of peripheral nerves over long gaps. *Cells Tissues Organs.* **176**, 153–165.

Harley, B. A., O'Brien, F. J., Waller, M. A., Prendergast, P. J., Yannas, I. V., and Gibson, L. J. (2006). Mean pore size and compressive strain effects on the permeability of collagen-GAG scaffolds: cellular solids modeling and experimental results. *Trans. Soc. Biomater.* **32**, 206.

Hollowell, J. P., Villadiego, A., and Rich, K. M. (1990). Sciatic nerve regeneration across gaps within silicone chambers: long-term effects of NGF and consideration of axonal branching. *Exp. Neurol.* **110**, 45–51.

Hsu, W. C., Spilker, M. H., Yannas, I. V., and Rubin, P. A. (2000). Inhibition of conjunctival scarring and contraction by a porous collagen-glycosaminoglycan implant. *Invest. Ophthalmol. Vis. Sci.* **41**, 2404–2411.

Itoh, S., Takakuda, K., Ichinose, S., Kikuchi, M., and Schinomiya, K. (2001). A study of induction of nerve regeneration using bioabsorbable tubes. *J. Reconstr. Microsurg.* **17**, 115–123.

Jenq, C. B., and Coggeshall, R. E. (1987). Permeable tubes increase the length of the gap that regenerating axons can span. *Brain Res.* **408**, 239–242.

Jenq, C. B., Jenq, L. L., and Coggeshall, R. E. (1987). Nerve regeneration changes with filters of different pore sizes. *Exp. Neurol.* **97**, 662–671.

Kinner, B., Capito, R. M., and Spector, M. (2005). Regeneration of articular cartilage. *Adv. Biochem. Eng./Biotechnol.* **94**, 91–123.

Kisiday, J., Jin, M., Kurz, B., Hung, H., Semino, C., Zhang, S., and Grodzinsky, A. J. (2002). Self-assembling peptide hydrogel fosters chondrocyte extracellular matrix production and cell division: implications for cartilage tissue repair. *Proc. Natl. Acad. Sci. USA* **99**, 9996–10001.

Kisiday, J. D., Jin, M., DiMicco, M. A., Kurz, B., and Grodzinsky, A. J. (2004). Effects of dynamic compressive loading on chondrocyte biosynthesis in self-assembling peptide scaffolds. *J. Biomech.* **37**, 595–604.

Kondo, M., Wagers, A. J., Manz, M. G., Prohaska, S. S., Scherer, D. C., Beilhack, G. F., Shizuru, J. A., and Weissman, I. L. (2003). Biology of hematopoietic stem cells and progenitors: implications for clinical application. *Annu. Rev. Immunol.* **21**, 759–806.

Kunisaki, S. M., Fuchs, J. R., Kaviani, A., Oh, J. T., LaVan, D. A., Vacanti, J. P., Wilson, J. M., and Fauza, D. O. (2006). Diaphragmatic repair through fetal tissue engineering: a comparison between mesenchymal amniocyte- and myoblast-based constructs. *J. Pediatr. Surg.* **41**, 34–39.

Lanza, R. P., Cooper, D. K. C., and Chick, W. L. (1997). Xenotransplantation. *Sci. Am.* July, 54–59.

Lee, C. R., Grodzinsky, A. J., and Spector, M. (2001). The effects of crosslinking of collagen-glycosaminoglycan scaffolds on compressive stiffness, chondrocyte-mediated contraction, proliferation, and biosynthesis. *Biomaterials* **22**, 3145–3154.

Lee, C. R., Grodzinsky, A. J., Hsu, H. -P., and Spector, M. (2003). Effects of a cultured autologous chondrocyte-seeded type II collagen scaffold on the healing of a chondral defect in a canine model. *J. Orthop. Res.* **21**, 272–281.

Lee, J. E., Park, J. C., Hwang, Y. S., Kim, J. K., Kim, J. G., and Sub, H. (2001). Characterization of UV-irradiated dense/porous collagen membranes: morphology, enzymatic degradation, and mechanical properties. *Yonsei Med. J.* **42**, 172–179.

Lengerke, C., and Daley, G. Q. (2005). Patterning definitive hematopoietic stem cells from embryonic stem cells. *Exp. Hematol.* **33**, 971–979.

Lerou, P. H., and Daley, G. Q. (2005). Therapeutic potential of embryonic stem cells. *Blood Rev.* **19**, 321–331.

Li, S. T., Archibald, S. J., Krarup, C., and Madison, R. D. (1992). Peripheral nerve repair with collagen conduits. *Clin. Mater.* **9**, 195–200.

Lin, R. Y., Sullivan, K. M., Argenta, P. A., Meuli, M., Lorenz, H. P., and Adzick, N. S. (1995). Exogenous transforming growth factor-B amplifies its own expression and induces scar formation in a model of human fetal skin repair. *Ann. Surg.* **222**, 146–154.

Lo, C.-M., Wang, H.-B., Dembo, M., and Wang, Y.-L. (2000). Cell movement is guided by the rigidity of the substrate. *Biophys. J.* **79**, 144–152.

Lundborg, G., Dahlin, L., Dohi, D., Kanje, M., and Terada, N. (1997). A new type of "bioartificial" nerve graft for bridging extended defects in nerves. *J. Hand Surg. (Br. Eur.).* **22B**, 299–303.

Madison, R. D., da Silva, C. F., and Dikkes, P. (1988). Entubulation repair with protein additives increases the maximum nerve gap distance successfully bridged with tubular prostheses. *Brain Res.* **447**, 325–334.

Martinez-Hernandez, A. (1988). Repair, regeneration, and fibrosis. *In* "Pathology" (E. Rubin and J. L. Farber, eds.), pp. 66–95. J. B. Lippincott, Philadelphia.

Mast, B. A., Nelson, J. M., and Krummel, T. M. (1992). Tissue repair in the mammalian fetus. *In* "Wound Healing" (I. K. Cohen, R. F. Diegelmann, and W. J. Lindblad, eds.), pp. 326–343. W.B. Saunders, Philadelphia.

Medawar, P. B. (1944). The behavior and fate of skin autografts and skin homografts in rabbits. *J. Anat.* **78**, 176–199.

Molander, H., Engkvist, O., Hagglund, J., Olsson, Y., and Torebjork, E. (1983). Nerve repair using a polyglactin tube and nerve graft: an experimental study in the rabbit. *Biomaterials* **4**, 276–280.

Navarro, X., Rodriguez, F. J., Labrador, R. O., Buti, M., Ceballos, D., Gomez, N., Cuadras, J., and Perego, G. (1996). Peripheral nerve regeneration through bioresorbable and durable nerve guides. *J. Peripher. Nerv. Syst.* **1**, 53–64.

Nehrer, S., Breinan, H. A., Ramappa, A., Hsu, H. -P., Minas, T., Shortkroff, S., Sledge, C. B., Yannas, I. V., and Spector, M. (1998). Chondrocyte-seeded collagen matrices implanted in a chondral defect in a canine model. *Biomaterials* **19**, 2313–2328.

Nolte, C. J. M., Oleson, M. A., Hansbrough, J. F., Morgan, J., Greenleaf, G., and Wilkins, L. (1994). Ultrastructural features of composite skin cultures grafted onto athymic mice. *J. Anat.* **185**, 325–333.

O'Brien, F. J., Harley, B. A., Yannas, I. V., and Gibson, L. J. (2005). The effect of pore size on cell adhesion in collagen-GAG scaffolds. *Biomaterials* **26**, 433–441.

O'Brien, F. J., Harley, B. A., Waller, M. A., Yannas, I. V., Gibson, L. J., and Prendergast, P. J. (2007). The effect of pore size on permeability and cell attachment in collagen scaffolds for tissue engineering. *Technol. Health Care* **15**, 3–17.

Ohbayashi, K., Inoue, H. K., Awaya, A., Kobayashi, S., Kohga, H., Nakamura, M., and Ohye, C. (1996). Peripheral nerve regeneration in a silicone tube: effect of collagen sponge prosthesis, laminin, and pyrimidine compound administration. *Neurol. Med. Chir.* **36**, 428–433.

Olde Damink, L. H. H., Dijkstra, P. J., van Luyn, M. J. A., Van Wachem, P. B., Nieuwenhuis, P., and Feijen, J. (1996). Cross-linking of dermal sheep collagen using a water-soluble carbodiimide. *Biomaterials* **17**, 765–773.

Passegue, E., Wagers, A. J., Giuriato, S., Anderson, W. C., and Weissman, I. L. (2005). Global analysis of proliferation and cell cycle gene expression in the regulation of hematopoietic stem and progenitor cell fates. *J. Exp. Med.* **202**, 1599–1611.

Pittenger, M. F., Mackay, A. M., Berk, S. C., Jaiswal, R. K., Douglas, R., Mosca, J. D., Morman, M. A., Simonetti, D. W., Craig, S., and Marshak, D. R. (1999). Multilineage potential of adult human mesenchymal stem cells. *Science* **284**, 143–147.

Rabkin-Aikawa, E., Mayer, J. E., and Schoen, E. J. (2005). Heart valve regeneration. *In* "Regenerative Medicine II" (I. V. Yannas, ed.), pp. 141–179. Springer, Berlin.

Robinson, P. H., van der Lei, B., Hoppen, H. J., Leenslag, J. W., Pennings, A. J., and Nieuwenhuis, P. (1991). Nerve regeneration through a two-ply biodegradable nerve guide in the rat and the influence of ACTH4-9 nerve growth factor. *Microsurgery* **12**, 412–419.

Rodriguez, F. J., Gomez, N., Perego, G., and Navarro, X. (1999). Highly permeable polylactide-caprolactone nerve guides enhance peripheral nerve regeneration through long gaps. *Biomaterials* **20**, 1489–1500.

Rudolph, R., Van de Berg, J., and Ehrlich, P. (1992). Wound contraction and scar contracture. *In* "Wound Healing" (I. K. Cohen, R. F. Diegelmann, and W. J. Lindblad, eds.), pp. 96–114. W.B. Saunders, Philadelphia.

Samuel, R. E., Lee, C. R., Ghivizzani, S. C., Evans, C. H., Yannas, I. V., Olsen, B. R., and Spector, M. (2002). Delivery of plasmid DNA to articular chondrocytes via novel collagen-glycosaminoglycan matrices. *Human Gene Therapy* **13**, 791–802.

Shah, M., Foreman, D. M., and Ferguson, M. W. (1995). Neutralization of TGF-B1 and TGF-B2 or exogenous addition of TGF-B3 to cutaneous rat wounds reduces scarring. *J. Cell Sci.* **108**, 985–1002.

Slack, J. M. W. (2000). Stem cells in epithelial tissues. *Science* **287**, 1431–1433.

Snyder, T. A., Watach, M. J., Litwak, K. N., and Wagner, W. R. (2002). Platelet activation, aggregation, and life span in calves implanted with axial flow ventricular assist devices. *Ann. Thorac. Surg.* **73**, 1933–1938.

Soo, C., Beanes, S. R., Hu, F. -Y., Zhang, X., Catherine Dang, C., Chang, G., Wang, Y., Nishimura, I., Freymiller, E., Longaker, M. T., *et al.* (2003). Ontogenetic transition in fetal wound transforming growth factor-B regulation correlates with collagen organization. *Am. J. Pathol.* **163**, 2459–2476.

Spector, M., Heyligers, I., and Roberson, J. R. (1993). Porous polymers for biological fixation. *Clin. Orth.* **235**, 207–219.

Surdez, D., Kunz, B., Wagers, A. J., Weissman, I. L., and Terskikh, A. V. (2005). Simple and efficient isolation of hematopoietic stem cells from H2K-zFP transgenic mice. *Stem Cells* **23**, 1617–1625.

Tsai, C. C., Lin, S. D., Lai, C. S., and Lin, T. M. (1999). The use of composite acellular allodermis-ultrathin autograft on joint area in major burn patients — one-year follow-up. *Kaohsiung J. Med. Sci.* **15**, 651–658.

Udani, V. M., Santarelli, J. G., Yung, Y. C., Wagers, A. J., Cheshier, S. H., Weissman, I. L., and Tse, V. (2005). Hematopoietic stem cells give rise to perivascular endothelial-like cells during brain tumor angiogenesis. *Stem Cells Dev.* **14**, 478–486.

Urban, R. M., Jacobs, J. J., Tomlinson, M. J., Gavrilovic, J., Black, J., and Peoc'h, M. (2000). Dissemination of wear particles to the liver, spleen, and abdominal lymph nodes of patients with hip and knee replacement. *J. Bone Joint Surg. Am.* **82**, 457–476.

Valero-Cabre, A., Tsironis, K., Skouras, E., Perego, G., Navarro, X., and Neiss, W. F. (2001). Superior muscle reinnervation after autologous nerve graft or poly-l-lactide-epsilon-caprolactone (PLC) tube implantation in comparison to silicone tube repair. *J. Neurosci. Res.* **63**, 214–223.

van Tienen, T. G., Heijkants, R. G. J. C., Buma, P., de Groot, J. H., Pennings, A. J., and Veth, R. P. H. (2002). Tissue ingrowth and degradation of two biodegradable porous polymers with different porosities and pore sizes. *Biomaterials* **23**, 1731–1738.

Wagers, A. J., and Weissman, I. L. (2004). Plasticity of adult stem cells. *Cell* **116**, 639–648.

Walter, M. A., Kurouglu, R., Caulfield, J. B., Vasconez, L. O., and Thompson, J. A. (1993). Enhanced peripheral nerve regeneration by acidic fibroblast growth factor. *Lymphokine Cytokine Res.* **12**, 35–41.

Wang, Y., Yates, F., Naveiras, O., Ernst, P., and Daley, G. Q. (2005). Embryonic stem cell–derived hematopoietic stem cells. *Proc. Natl. Acad. Sci. USA* **102**, 19081–19086.

Wickelgren, I. (1996). Muscling transplants in mice. *Science* **273**, 33.

Williams, L. R., Longo, F. M., Powell, H. C., Lundborg, G., and Varon, S. (1983). Spatial-temporal progress of peripheral nerve regeneration within a silicone chamber: parameters for a bioassay. *J. Comp. Neurol.* **218**, 460–470.

Williams, L. R., Danielsen, N., Muller, H., and Varon, S. (1987). Exogenous matrix precursors promote functional nerve regeneration across a 15-mm gap within a silicone chamber in the rat. *J. Comp. Neurol.* **264**, 284–290.

Woodley, D. T., Peterson, H. D., Herzog, S. R., Stricklin, G. P., Burgeson, R. E., Briggaman, R. A., Cronce, D. J., and O'Keefe, E. J. (1988). Burn wounds resurfaced by cultured epidermal autografts show abnormal reconstitution of anchoring fibrils. *JAMA* **259**, 2566–2571.

Yannas, I. V. (2000). *In vitro* synthesis of tissues and organs. *In* "Principles of Tissue Engineering" (R. P. Lanza, R. Langer, and J. Vacanti, eds.), pp. 167–178. Academic Press, San Diego.

Yannas, I. V. (2001). "Tissue and Organ Regeneration in Adults." Springer-Verlag, New York.

Yannas, I. V. (2005). Facts and theories of induced organ regeneration. *Adv. Biochem. Eng. Biotechnol.* **93**, 1–31.

Yannas, I. V., and Tobolsky, A. V. (1967). Cross-linking of gelatine by dehydration. *Nature* **215**, 509–510.

Yannas, I. V., Burke, J. F., Huang, C., and Gordon, P. L. (1975). Correlation of *in vivo* collagen degradation rate with *in vitro* measurements. *J. Biomed. Mater. Res.* **9**, 623–628.

Yannas, I. V., Lee, E., Orgill, D. P., Skrabut, E. M., and Murphy, G. F. (1989). Synthesis and characterization of a model extracellular matrix that induces partial regeneration of adult mammalian skin. *Proc. Natl. Acad. Sci. USA* **86**, 933–937.

Zhang, S., Holmes, T., Lockshin, C., and Rich, A. (1993). Spontaneous assembly of a self-complementary oligopeptide to form a stable macroscopic membrane. *Proc. Natl. Acad. Sci. USA* **90**, 3334–3338.

Zhao, Q., Lundborg, G., Danielsen, N., Bjursten, L. M., and Dahlin, L. B. (1997). Nerve regeneration in a "pseudo-nerve" graft created in a silicone tube. *Brain Res.* **769**, 125–134.

Part Four

# Models for Tissue Engineering

# Chapter Seventeen

# Models as Precursors for Prosthetic Devices

*Eugene Bell*

## I. INTRODUCTION

Work *in vitro* with intact tissues or tissue fragments in the form of organ cultures has had a long history in developmental biology in studies devoted to causal relationships between tissues on which the emergence of differentiated structures depend. Before the discovery of pluripotent embryonic stem cells and adult mesenchymal stem cells it was believed that, while monolayer cell cultures usually lost the characteristic histological organization and most of the physiological functions of the original tissue, organ cultures retained both and permitted the process of development to unfold. But organ culturing, unlike cell culturing, consists of explanting organ anlage or pieces of intact tissues under conditions that preserve tissue integrity and discourage cells from moving out of the explant and from engaging in DNA synthesis that leads to cell division. The disadvantages of organ cultures include their short life, particularly when adult tissues are explanted, their tendency to undergo central necrosis, the occurrence of vascular occlusion, and the difficulty of preparing reproducibly standard models. Nonetheless, in retrospect, stability of the organ cultured explant was attributed, in large measure, to the matrix scaffolds with which the cells in tissues were associated.

As long ago as 1963, Dodson, in Honor Fell's laboratory in Cambridge, England, was concerned with the role of scaffolding in studies of tissue interactions *in vitro*. He isolated dermis from the metatarsus of 12-day chick embryos and subjected it to freeze–thaw cycles before combining it with epidermis that alone was unable to differentiate (Dodson, 1963). Although it lacked living cells, the frozen–thawed dermis, constituting a structurally intact collagen scaffold, and probably rich in growth and differentiation factors, was sufficient to support keratinocyte differentiation *in vitro*. A control, consisting of heat-killed dermis, failed to promote keratinocyte differentiation. Similar combinations using adult acellular dermis and keratinocytes made many years later (Heck *et al.*, 1985; Cuono *et al.*, 1987) were used successfully as transplants to patients requiring skin.

Now we believe that what cells did in organotypic explants, as *in vivo*, was regulated by instructive scaffolds and cell signaling. The use of media unsupplemented with growth factors and containing high concentrations of serum may have had a detrimental effect on signal-induced differentiation. The development of low-serum or serum-free media has significantly improved the value of *in vitro* models as well as the effectiveness of their use as remedial implants *in vivo* (Boyce and Ham, 1983).

An alternative approach to the study of tissue interactions *in vitro* consists of reconstituting each of the tissues with scaffolds into which cells of one or more interrelated tissues of an organ are seeded (Bell *et al.*, 1979). The field of tissue engineering had its beginning in our lab at MIT in the late 1970s. We realized nearly immediately that a source of stemlike cells was required for tissue propagation, both for research and particularly for products designed for clinical use. For that reason we elected to focus first on skin, a self-renewing tissue whose principal cells, dermal fibroblasts and keratinocytes, were capable of dividing and of expressing their phenotype for at least the life span of the human organism. To maximize their functional life span it made sense to choose newborn foreskin as a starting source for cells. Foreskin is often a discard tissue, coming mainly from newborns. The focus now on the use of stem or "stemlike" cells to develop tissue or organ equivalents for *in vitro* analysis or for subsequent transplantation has sharpened with the discovery of multipotent mesenchymal stem cells in the postnatal human organism (Pittenger *et al.*, 1999) and the discovery of pluripotent embryonic stem cells (Thompson *et al.*, 1998).

For tissue engineering the second key requirement was that of selecting an appropriate matrix scaffold recognizable and remodelable by the cells that would be seeded into it. The requirements for the original scaffold selected were based on the bias that what has evolved in nature might be imitated *in vitro*. We are still not sure that conditions exactly similar to those exhibited by a collagen gel prevail *in vivo* in early tissue formation in the embryo. References and a discussion of the role of the extracellular matrix appear in a paper on strategy for the selection of scaffolds for tissue engineering (Bell, 1995). The subject is touched on in a book on the organization of the extracellular matrix (ECM) (Modis, 1990). The author uses a polarization microscope to describe the ECM (p. 178), seemingly comparing a collagen gel to what he sees *in vivo*. He states: "Collagen gel shows parameters that are more similar to those observed *in vivo*." Unfortunately, he elaborates no further. Since the assembly of collagen fibrils and fibers can be visualized with the polarization microscope, it would be possible to provide a detailed description of the cell-directed organization of the early dermis compared with what cells do in a contracting collagen gel. We considered ideal properties for the scaffold to be the following: (1) that it have binding sites normally used by the cells that populate it, and (2) that it have, or come to have, the structural organization cells need for their orientation. These are cues that dictate tissue design, on which function depends. We found, for example, that if two opposite edges of a forming dermis were constrained, the force field created by cells as they collected collagen fibrils in the gel, causing fluid to be expressed from the gel, resulted in cell polarization. The long axes of cells and the forming collagen fibers became oriented in parallel with the lines of force between the constrained edges of the gel. We underscore that *in vitro* models might well be designed to incorporate physical forces that act on scaffolds and cells, contributing to their full developmental potential. An early relevant paper appeared shortly after we reported on formation of a dermal equivalent. The authors developed a model system using distortable sheets of silicone rubber on which they plated various types of cells to measure the relative strengths of traction exhibited (Harris *et al.*, 1981).

Coculturing interactive tissue-specific subsets of cells, such as the two principal phenotypes that make up skin, in an appropriate scaffold is reviewed later as *in vitro* models for studying disease states, bioprocesses such as melanogenesis, the immunogenicity of engineered tissues, and wound healing. Other factors may be implicated in extracellular matrix remodeling. Platelets and platelet lysate up-regulate collagen gel contraction *in vitro* as a function of time and concentration. Release of PDGF-AA/AB and TGF-$\beta$ from platelets partially mediates the increased gel contraction (Zagai *et al.*, 2003). We keep in mind that living-skin equivalent (LSE) was the first tissue-engineered product to make its way into the medical market and should be considered an early product subject to improvement. We also attempt to cover several other models by tissue type and to touch on some of the biomaterials and techniques used for engineering scaffolds.

The first LSE consisting of "living" dermis and epidermis was made up of fibroblasts and keratinocytes derived from foreskin. The dermal fibroblasts contracted the collagen gel scaffolding (Bell *et al.*, 1979; Bell *et al.*, 1991) making it tissue like. The time required to construct a skin model was 14–21 days, in the course of which at least three different media were used. First, an initial medium supports the contractile activity of dermal fibroblasts. The dermal cells are responsible for compacting the starting collagen lattice by attaching podial processes, to collagen fibrils and withdrawing the processes, with the attached fibrils toward the cell body. As a result of these local contractions, fluid is expressed from the collagen lattice, and the volume of the lattice diminishes by a factor of 30–40 until an equilibrium size is attained and no further contraction occurs. The volume reduction is related to the number of cells used and the concentration of collagen (Bell *et al.*, 1979). Second, another medium is applied at the time of keratinocyte plating when skin is being fabricated, since the first phase of epidermal development, keratinocyte proliferation, requires low $Ca^{++}$, EGF, and an up-regulator of cyclic AMP. A third medium is used during the second phase of epidermal development, calling for high $Ca^{++}$ and air-lifting to expose the epidermis to the atmosphere to promote keratinization and differentiation of a stratum cornium (Bell *et al.*, 1991; Parenteau *et al.*, 1991).

The LSE was shown to undergo virtually complete differentiation *in vitro*, lacking, however, pigment, sweat glands, neurogenic elements, a microcirculation, and hair follicles. The model was reproduced faithfully and kept alive *in vitro* for months, at least. Although collagenolytic activity is high in young dermal equivalents, it has been observed that the resistance of dermal equivalents to breakdown by collagenase is greatly enhanced by 30 days of cultivation *in vitro*, suggesting that extensive cross-linking of the collagen fibrils by cell-secreted lysyl oxidase has occurred, as shown by Rowling *et al.*, 1990. Continued differentiation of the skin equivalent model *in vitro* and the resemblance of differentiated cells in the matrix to their *in vivo* counterparts, rather than to cells grown on plastic in two dimensions, are distinguishing features (Coulomb *et al.*, 1983). Collagen processing by cells grown on 2D substrates is known to be deficient, in that the exported molecule retains the pro collagen C and N-terminal telopeptides normally cleaved by peptidases (Freiberger *et al.*, 1980). In the collagen lattice scaffold, processing and subsequent polymerization, for which it is a requirement, proceed normally (Nusgens *et al.*, 1984).

Aware of the potential of using embryonic stem cells (ESC) (Thompson *et al.*, 1998) as a source of cells for the tissue engineering of skin, we have looked for publications announcing successful induction of either the dermal fibroblast or keratinocyte phenotype or both in ESC. Some early efforts may need further trials. For example, a paper dealing with the subject reports reconstituted skin from murine ESC. Keratinocyte differentiation was prompted by use of ECM secreted by primary cultures (Soraux *et al.*, 2003). It was not stated that ESC were harvested from the blastula's inner cell mass, nor are the putative ESC shown to proliferate for many doublings without differentiating, nor was it shown that the genotypes of the differentiated skin cells were the same as that of the ESC. Rigorous criteria laid down by Tompson need to be observed. There appear to be no reports that postnatal mesenchymal stem cells (MSC) can be induced to express either skin phenotype. The seminal paper by Pittenger *et al.*, 1999 lists the lineages into which adult human MSC can be induced to differentiate, consisting exclusively of adipocytes, chondrocytes, osteocytes, tendon, muscle cells, and marrow stroma. Subsequent reports suggest that the repertoire of MSC can be expanded to include neural cells [in the mouse, for example (Croft and Przyborski, 2006; Hong *et al.*, 2006)]. Croft and Przyborski suggest that trans-differentiation may occur as a result of culture conditions for cells under test *in vitro*. The cells spontaneously express neuronal phenotypic markers in culture, and, when implanted into the lateral ventricles of the neonatal mouse brain, they differentiate into olfactory bulb granule cells and periventricular astrocytes. The foregoing work and related papers suggest that the full range of potential phenotypes of MSC has yet to be realized.

In the light of what appear to be expanding cell sources in adult organisms hopefully available for use in reconstitut-

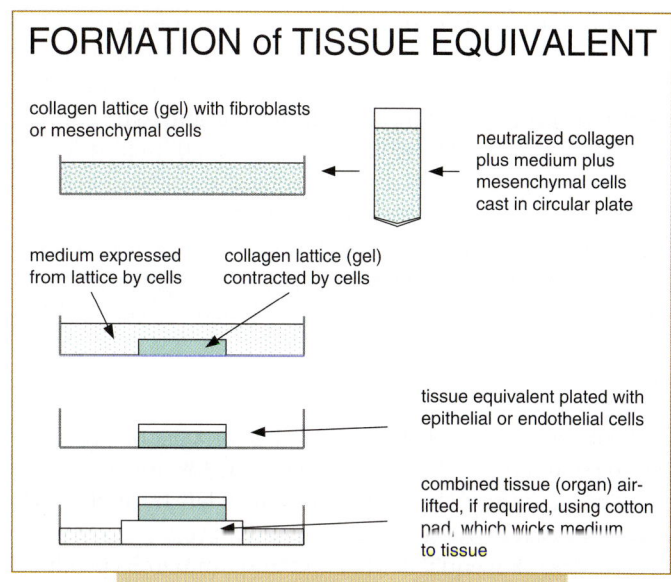

**FORMATION of TISSUE EQUIVALENT**

collagen lattice (gel) with fibroblasts or mesenchymal cells

neutralized collagen plus medium plus mesenchymal cells cast in circular plate

medium expressed from lattice by cells

collagen lattice (gel) contracted by cells

tissue equivalent plated with epithelial or endothelial cells

combined tissue (organ) air-lifted, if required, using cotton pad, which wicks medium to tissue

**FIG. 17.1.**

ing needed prosthetic devices, there is optimism that adult organisms have reservoirs waiting to be tapped. Reported sources of progenitor endothelial cells in bone marrow and blood (Asahara *et al.*, 1997; Shi *et al.*, 1998; Lin *et al.*, 2000; Asahara *et al.*, 1999; Kalka *et al.*, 2000; and Peichev *et al.*, 2000) and of smooth muscle from the same sources (Shimizu *et al.*, 2001) bring the prospect of populating arterial scaffolds closer to clinical reality.

Reports of other sources of multipotent stem cells have been surfacing. Of interest have been preliminary discoveries that adult stem cells from rodent dermis have the capacity to proliferate and differentiate *in vitro*. Evidence was gathered (Toma *et al.*, 2001) that neural, smooth muscle, and adipocyte phenotypes arose in clonal cultures *in vitro*. A similar finding in glabrous fetal human skin, that dermal cells can give rise to a broad range of phenotypes each having been induced by undefined tissue-specific signals, was published several months later (Dai *et al.*, 2002). The phenotypes induced included the following: cartilage, liver, lung, heart, endocrine pancreas, smooth muscle, and bone.

However, we point out that the ease with which strains of skin cells can be established from foreskins has already been shown by the company Organogenesis to be ample for the commercialization of skin products for medical use. Much emphasis has been placed on markers that identify skin stem cells, particularly keratinocytes. Keratin-19 (K-19) as a marker of stem cells was found to be expressed at glabrous sites, such as foreskin in deep epidermal rete ridges (Michel *et al.*, 1996). Cells expressing K-19 were also rich in alpha-3 beta-1 integrin. K-19 positive cells were more numerous in newborns as compared with older foreskins. The subject of keratinocyte stem cell markers has been reviewed and covered in a number of key papers (Watt, 1998; Lavker and Sun, 2000; Pellegrini *et al.*, 1999; Li *et al.*, 1998).

Populating matrix scaffolds, other than collagen lattices (gels), with dermal fibroblasts to produce dermal equivalents followed the introduction of the original work that reported the fabrication of both dermal and skin equivalents (Bell *et al.*, 1979, 1981). Materials used include Vicryl (Hansbrough *et al.*, 1993) and Biobrane (Hansborough *et al.*, 1994). For example, there is the commercial product Dermagraft, no longer being manufactured by Advanced Tissue Sciences but adopted by Smith and Nephew. The product was made using the technique of seeding the Vicryl scaffold with allogeneic dermal fibroblasts, first shown to be possible in fibroblast-contracted collagen gel scaffolds covered by differentiating epidermis in animal experiments by Bell and colleagues at MIT. See Table 17.1 for a list of collagen-based products for skin replacement and/or repair.

The company Organogenesis, founded by Bell *et al.*, conducted extensive animal trials and human clinical trials of the living-skin equivalent before the commercial product, called Apligraft™, was launched, and they confirmed the early rat trials by Bell and his group carried out at MIT (Sher *et al.*, 1983). In a collaboration with Dr. Jack Gaisford at the Western Pennsylvania Hospital in Pittsburgh it was found that the living-skin equivalent performed poorly when applied to burned skin. The prosthesis began to degrade in a matter of days. This response did not discourage us, because of the known difficulties of replacing badly burned skin. Dr. Gaisford's continuing interest was unflagging. He had raised capital for support of the work at the Western Pennsylvania Hospital and continued to do so for the newly formed Organogenesis. Organogenesis continued the MIT experiments with animals and, with Dr. Gaisford, carried out the first trials in humans. These took place before Apligraft was approved by the FDA and launched commercially. In the course of extensive patient use for curing ulcers and for use on donor sites it was established that the living-skin equivalent had significant remedial value and that, being built with both allogeneic fibroblasts and keratinocytes, from which donor immune cells are deleted, it did not provoke a rejection response, confirming the early rat trials carried out by Bell and his group at MIT.

## II. PIGMENTATION OF THE LIVING-SKIN EQUIVALENT (LSE) *IN VITRO*

Neonatal melanocytes cultured from foreskin have been incorporated into the living-skin equivalent (Topol *et al.*, 1986) by plating them on the cell-populated contracting dermal equivalent before the application of keratinocytes. Punch biopsies (2 mm) of unpigmented foreskin were overlaid on the dermal equivalents after third-passage melanocytes were added to it. By 14 days the dermal lattice was covered by keratinocytes. By that time a basal lamina developed at the dermal–epidermal junction *in vitro*. The procedure led to skin pigmentation *in vitro*, as pigment was donated by melanocytes to keratinocytes. Exposure of some

living-skin equivalents to UVB irradiation for 14 days stimulated and enhanced pigment transfer. Pigment transfer was documented by light and electron microscopic studies. Donated melanosomes, identified by their pigment as well as by dopa-oxidase staining, were found dispersed throughout the cytoplasm of the keratinocytes.

## III. THE LIVING-SKIN EQUIVALENT AS AN IMMUNOLOGICAL MODEL

The LSE is constituted with cultivated parenchymal cells free of any subsets of immune cells normally found in the dermis and epidermis. Using the X chromosome as a genetic marker, LSEs made up with female cultivated rat cells were transplanted to male rat hosts across a major histocompatibility barrier (Sher *et al.*, 1983). Allografted fibroblasts were accepted by recipient rats after a transient mononuclear cell response, as shown by cell karyotyping. LSEs containing xenogeneic fibroblasts were rejected. A second LSE allograft from the same donor strain did not provoke rejection, either of the original allograft or of the challenge allograft. In a further experiment a secondary graft of allogeneic skin did not provoke rejection of the original skin, while an actual skin graft was rejected. In an additional control, grafting the recipient first with actual allogeneic skin and then with the LSE allograft led to rejection of the actual skin but not of the LSE graft, ruling out the possibility that suppressor T-cells were responsible for LSE allograft acceptance. It can be said, then, that allografted fibroblasts do not provoke a rejection response, even in presensitized animals, that they do not render the recipient tolerant to allogeneic skin, and that they do not act as targets when active rejection is taking place. We propose that cells bearing class I antigens generally may be acceptable as graft constituents if incorporated in tissue equivalents excluding cells bearing class II antigens. Similar results in humans have been reported by Organogenesis Inc. using *in vitro* cultivated cells. Clinical trials of living-skin equivalents made up with human allogeneic keratinocytes do not provoke an immune reaction in recipients (Parenteau *et al.*, 1994). The model should be a valuable tool for determining the roles played by cells of the immune system and microcirculation in allograft rejection of actual skin. It should allow use of cells of any genotype and of human origin to study genetic abnormalities as well as the contribution of specific genes to skin development by transplanting skin equivalents to immunodeficient organisms.

## IV. THE LIVING-SKIN EQUIVALENT AS A DISEASE MODEL

Since either normal or aberrant skin cells can be incorporated into living-skin equivalents, models of disease states can be fabricated. A psoriasis model was fabricated to test the contribution of psoriatic dermal fibroblasts to the expression of features of the disease *in vitro* (Saiag *et al.*,

**Table 17.1. Collagen-based products for skin replacement or repair**

| Name | Manufacturer | Composition | Indications | Incorporated into wound bed? |
|---|---|---|---|---|
| Promogran® | Johnson & Johnson | Animal collagen (55%) and oxidized regenerated cellulose | Most partial- and full-thickness wounds | No data |
| PuraPly® | Royce Medical | Porcine type I collagen | Most partial- and full-thickness wounds, including chronic wounds, donor cites, and MOHS surgery | No data |
| Fibrocol Plus® | Johnson & Johnson | Denatured animal collagen (90%) with alginate (10%) | Partial-thickness wounds, venous ulcers, acute traumatic wounds, second-degree burns | No data |
| Integra® | Integra Life Sciences | Collagen-glycosaminoglycan copolymer matrix coated with polysiloxane elastomer | Grade I–III noninfected ulcers that involve tendon, bone, joint. Full-thickness burns | Incorporated within 2–3 weeks |
| OrCel® | Ortec International | Collagen sponge that contains allogeneic living skin cells | Epidermolysis bullosa, skin graft donor sites, venous and diabetic foot ulcers | No data |
| EZ Derm® | Brennan | Porcine chemically modified xenografts | Partial-thickness wounds, donor sites, ulcers, and meshed autografts | Not incorporated and must be removed |
| OASIS® | Cook | Porcine xenograft (small intestinal submucosa) | Partial- and full-thickness wounds, ulcers, trauma, and surgical wounds | Incorporated within few weeks |
| APLIGRAF® | Organogenesis | Living bilayered skin substitute on type I bovine collagen matrix | Venous leg ulcers, diabetic foot ulcers | Incorporated within few weeks |
| AlloDerm | Life Cell | Human decellularized dermis | Soft tissue repair and reconstruction, abdominal wall repair, burns, and wound coverage | Incorporated within 2–3 weeks |
| PriMatrix™ | TEI Biosciences | Fetal bovine acellular dermal matrix | Partial- and full-thickness wounds, ulcers, trauma and surgical wounds, general orthopedics, second-degree burns | Incorporated within 2–3 weeks |
| GRAFTJACKET® | Wright Medical | Human decellularized dermis | Soft tissue repair and reconstruction, abdominal wall repair, burns, and wound coverage | Incorporated within 2–3 weeks |
| GammaGraft® | Promethean LifeSciences | Irradiated human skin allograft | Partial- and full-thickness wounds, ulcers, trauma and surgical wounds, burns | Incorporated within few weeks |
| Mediskin | Novamedical GmbH | Fresh sterile porcine skin | Burns | Temporary, not incorporated, and must be removed |

*From: burnsurgery.org; Podiatry Management, June/July 2006.*

1985). A button of normal keratinocytes was plated on the centers of dermal equivalent discs constituted with human normal or psoriatic dermal fibroblasts, and the rate of growth of the sheet of keratinocytes over the dermal substrate was measured. It was observed that the psoriatic fibroblasts induced hyperproliferation of keratinocytes as compared with the growth induced by control fibroblasts. In addition to the study of psoriasis and other epidermal diseases, such as epidermolysis bullosa, the model should provide an *in vitro* basis for studying disorders in dermal connective tissues, including dermatosparaxis and sclerosis. It is obvious that any pair of mesenchymal cells and epithelial cells of which one or both is diseased or aberrant can be used in the three-dimensional coculturing system for studying the expression of features of the disease.

## V. WOUND-HEALING MODEL

LSE models can be used *in vitro* to analyze the role of dermis in epidermal wound healing (Bell and Scott, 1992). After making a differentiated skin equivalent in a 6- or 12-well multiwell plate insert (24 mm, e.g.), a central disc of the skin is removed with a punch (Fig. 17.2). The acellular pad in contact with the membrane of the insert is replaced, and the remainder of the gap is filled with a collagen scaffold to the level of the interface between the dermis and epidermis. As the collagen scaffold is being populated by fibroblasts migrating into it from the surrounds, keratinocytes from the edges of the epidermis migrate over the neodermis and undergo differentiation. We suppose that the migrating epidermal cells are stem cells whose properties are waiting to be assessed. The rate of overgrowth of the neodermis by keratinocytes and the development of the epidermis can be

taken as measures of the effectiveness of neodermis as an interacting substrate. The model lends itself as an assay system for determining the effect of growth factors and other signals on each of the events of wound healing. The new scaffold developed at TEI Biosciences as an acellular skin replacement, called EBM after its discoverer, (commercially, it is called PriMatrix, which is a freeze-dried product) and discussed in detail later in the context of vascular prostheses, supports the overgrowth of epidermis from adjacent host tissue in a rat model. Note in Fig. 17.3 that in-growth of blood vessels is extensive, as is the growth and differentiation of overlying epidermis, which in the example shown has covered a circle 8.0 mm in diameter in one week. The neodermis is populated by dermal fibroblasts and other cells. The EBM material is now in the market.

Generic models for many types of epithelial–mesenchymal combinations were reported in the early years of tissue engineering. Some of the combinations for organotypic coculturing are the following: for skin references (Bell *et al.*, 1979, 1981; Topol *et al.*, 1986); for cornea reference (Minami *et al.*, 1993); for blood–brain barrier references (Bell *et al.*, 1990; Beigel and Pachter, 1994); for gut reference (Montgomery, 1986); for vessel references (Weinberg and Bell, 1986); for nerve tissue references (Guthrie and Pini, 1995; Bergsteinsdottir *et al.*, 1993; Tomooka *et al.*, 1993); for capillary network references (Mori *et al.*, 1989); for pancreatic islet references (Lanza *et al.*, 1995; Soon-Shiong *et al.*, 1993); for hair follicle references (Watt, 1998); for thyroid gland references (Bell *et al.*, 1984). Many others are of course possible. We are not suggesting that combinations need be restricted to epithelial–mesenchymal combinations. For example, interactions between subpopulations that are coderivatives embryologically can also be modeled in a 3D system *in vitro*. In the development of the nervous system, both attractive and repellent signals guide the course of axons en route to endpoint targets (Guthrie and Pini, 1995). The authors of the foregoing citation seeded a collagen gel

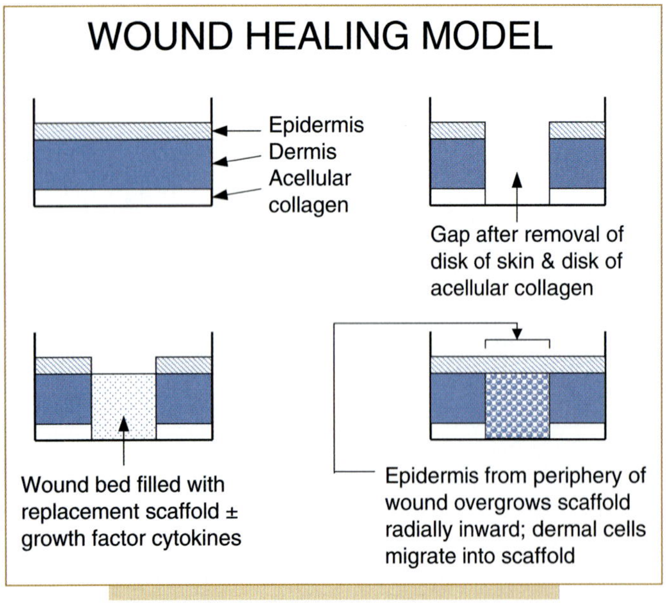

## WOUND HEALING MODEL

Epidermis
Dermis
Acellular collagen

Gap after removal of disk of skin & disk of acellular collagen

Wound bed filled with replacement scaffold ± growth factor cytokines

Epidermis from periphery of wound overgrows scaffold radially inward; dermal cells migrate into scaffold

**FIG. 17.2.**

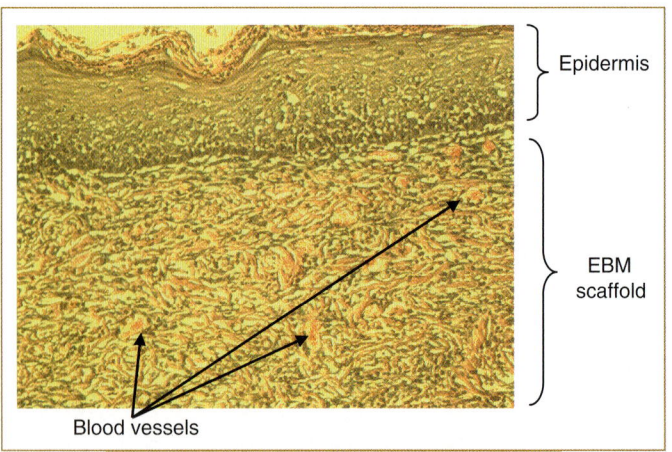

Epidermis

EBM scaffold

Blood vessels

**FIG. 17.3.**

scaffold with two components, either ventral explants of rat hind brain or spinal cord and floor plate explants from which the former tissues were separated in the scaffold gel. It is shown that outgrowth of motor axons from aspects of hind brain or cord that faced the plate were greatly reduced, suggesting that diffusible factors from the floor plate can exclude motor axons from certain regions of the developing nervous system. The existence of chemorepellents as well as attractants has been discussed recently (Marx, 1995). They divide into at least two families of proteins, nestins and semaphorins, responsible for growth cone guidance. It is suggested that a single factor can induce attraction or repulsion, depending on the cells involved.

## VI. VASCULAR MODELS WITHOUT CELLS

Vascular prostheses constructed from Dacron and other synthetics have been in use for many years but are known to elicit persistent inflammatory reactions and to become occluded. The thermosetting polymers are not biodegradable and do not integrate with host tissues, but some successes have been reported under limited conditions in experimental animals (Sandusky et al., 1992; Leenslag et al., 1988; Stronck et al., 1992). For the foregoing reasons and because the synthetics do not come to life, other materials have been proposed and tried as arterial substitutes. Animal tissues that resemble arteries have been used with some success, in particular the porcine small intestine (Sandusky et al., 1992). The mucosal cells are scraped off the luminal side, and the muscular layers are removed from the abluminal side, leaving the stratum compactum, a dense, highly organized fibrillar collagen matrix, and the looser connective tissue of the mucosa. The material can be used as a scaffold for cells in vitro and has been used in animal experiments. Implanted in vivo without cells it is invaded by capillaries, which, it has been claimed, contribute cells that provide an intima, and, similarly, smooth muscle cells migrate from the anastomoses to provide a media.

Using engineered vessels made up of air-dried EBM wrapped onto a sutured tube-shaped vessel, we began a program of testing it in a closed circulatory loop through which saline and/or human blood was pumped under physiological conditions at a pressure of 90 mmHg. While the air-dried EBM itself was impermeable, the tube leaked at the suture holes and between the layers of the doubly wrapped tube (experiments were conducted by V. Russakovsky,* T. Fofonoff,* and E. Bell at TEI Biosciences Inc.).

## VII. VASCULAR MODELS WITH CELLS ADDED

Vascular models that examine the effect of shear and other forces on monolayers of endothelial cells in vitro have been developed by Nerem et al., 1993, who have shown that the rate of endothelial cell proliferation is decreased by flow and that entry of cells into a cycling state is inhibited. They point out the limitations of the 2D system, since the important component of cyclic stretch of basement membrane, which the endothelium normally synthesizes and on which it rests, is absent. They suggest that a coculture system in which the endothelium is supported on smooth muscle tissue would be superior for providing a more physiological environment. Such a system was developed by Weinberg and Bell, 1986, who showed that a basal lamina was laid down between the endothelium and the contiguous smooth muscle tissue cast in the form of a small-caliber tube in vitro. Fabricating the vessel was a three-step procedure.

The first tissue layer cast around a small-caliber mandrel was a smooth muscle cell media whose ends were anchored in a Velcro cuff or held fast by ridges and grooves in the mandrel until radial contraction made space for bands that were secured around the ends. Hence the mechanical restraint imposed on the contracting tube allowed contraction to occur radially, but not longitudinally, since each end of the vessel equivalent was held fast. The second tissue layer cast was the adventitia surrounding the smooth muscle media. To make room for it, the fluid expressed from the collagen gel scaffold was drawn out of the casting tube, and the mixture of adventitial fibroblasts suspended in medium containing neutralized collagen was introduced into the space between the media and the wall of the cylindrical casting chamber. After the adventitia had contracted radially, but not longitudinally, since the media provided a frictional surface, which prevented it, the mandrel was extracted, leaving the lumen of the tissue tube empty. The lumen was then filled with a suspension of endothelial cells. The cells came to rest on the inner surface of the media as the tube was rotated.

Histological analysis showed that the smooth muscle cells of the media were oriented in parallel with the long axis of the tissue tube that had formed, rather than circumferentially as they dispose themselves in vivo. By modifying the foregoing procedure the vessel was allowed to contract freely in its length (L'Heureux et al., 1993; Hirai et al., 1994). That is by eliminating the restraining ties at the ends of the mandrel, the smooth muscle cells oriented circumferentially as they do in vivo. Similar results are obtained by applying pulsitile pressures, causing the media of the vessel to expand radially (Kanda et al., 1993). Another method (Tranquililo et al., 1996) is based on the observation that polymerizing collagen fibrils align in magnetic fields (Torbert and Ronziere, 1984). By orienting the model devised by Weinberg and Bell, in which the ends of the developing vessel are tethered, with its long axis parallel to a strong magnetic field during fibrilogenesis, it has been shown that collagen fiber orientation is circumferential. The fiber orientation provides contact guidance for smooth muscle cells that encircle the vessel in a direction perpendicular to its length. We found that the more normal organization of the cells of the media reduces the excessive compliance of the structure that resulted from the longitudinal orientation of the smooth muscle cells.

The cell-seeded arterial blood vessel equivalent has been steadily evolving. In the course of early animal trials, we concluded that the collagen gel lattice does not provide the structural resistance typical of an actual artery, and we resorted to the use of synthetics to provide added strength. We think now that the stratagem was not useful, because the synthetic structure, designed to provide strength, does not satisfy the binding- and substrate-dictated cell orientation needs of medial or adventitial cells (Weinberg *et al.*, 1993). We were then led to try a new tissue-engineering approach that incorporates aspects of issued patents and patents pending. We have been developing and using products in the market, various versions of EBM (U.S. Patent #6,696,074 — issued 2004), made up from bovine fetal embryonic skin harvested before hair follicle differentiation. In addition, for the new version of the vessel we have been using various types of spun collagen threads (U.S. Patents: #5,562,946, #5,851,290, #5,911,942, #6,705,850 B1). To prepare a vascular prosthesis, a layer of hydrated collagen threads is spun spirally around a mandrel of the diameter of the vessel to be produced. The spirals are then freeze-dried. When populated by cells, the freeze-dried threads provide the orienting substrate for smooth muscle cells, which dispose themselves perpendicularly to the long axis of the vessel. We refer to some of the history of work concerned with the organization of smooth muscle cells in the media of arterial vessels. In order for the cells to contract the vessel radially and then to relax to allow for vessel expansion, they must be oriented circumferentially. What we are describing now is yet another method for directing the desired disposition of cells in a prosthesis.

Back to the incorporation of processed sheet EBM into the vessel being formed. The EBM is designed to endow the vessel with the strength it requires. The tensile-breaking strength of unmeshed EBM is ~25 MPa, while the breaking strength of 1/1.5 meshed EBM using a Zimmer Skin Graft Mesher machine is ~16 MPa. The meshed EBM is wrapped around the spiral thread layer so that the edges of the sheet meet exactly. It is then sewn longitudinally, making an EBM cylinder adhering tightly around the subjacent spiral thread layer, the layer to be seeded with smooth muscle cells. After sterilization, a solution of the desired concentration of cells, medium, and collagen, of a concentration able to form a gel, is prepared, and the mandrel with its acellular wraps is inserted and held centrally in a cylindrical tube containing a total volume that will form a thin gel when contracted by medial cells after the tube is transferred to a bath sterilely, where the sealed tube is rotated for a period of 1 h at a temperature too low to allow gelling of the collagen but long enough to be populated by the cells that will make up the media of the vessel. After allowing cells to populate the collagen fibers, the tube is transferred to a 37° incubator, where it is again rotated while the collagen gel forms. After gel formation, the tube is oriented vertically and the stopper of the tube is exchanged for one that admits $CO_2$ sterilely. The new vessel, built by methods described earlier, is currently under test. Results are expected to be published shortly.

In using multipotent stem cells in *in vitro* models, the requirements for inducing expression of a specific phenotype include a sequence of molecular signals, a matrix of extracellular design populated by the cells, and physical forces to which cells and matrix are subjected that modulate tissue development. Needed as well are the medium or media, growth factors, and other conditions of culture to promote cell growth and differentiation and tissue building, *in vitro*. The foregoing should lead to the fabrication of standardized models that closely resemble actual tissues in many respects. A generic model of two different tissues, that is, an organotypic system, is shown in Fig. 17.1. A collagen scaffold is seeded with mesenchymal cells, which condense it, together providing a substrate on which epithelial cells, such as keratinocytes, are plated.

## VIII. CONCLUSIONS

One of the major goals of model tissue building is the reconstitution of tissue or organ equivalents that faithfully resemble actual body parts in order to better understand cell and tissue function under controlled and reproducible conditions. From this point of view the model offers an opportunity to study disease states and to seek remedies for them. The model can also serve as a test system for toxicity testing, instead of depending on the use of animal cells. It has the advantage, when constituted with human cells, of giving responses to toxins that may be different from those of animal cells. Cells in organized tissue scaffolds exhibit reduced susceptibility to damage as compared with cells grown on a monolayer (Bell *et al.*, 1991). Other uses of model tissues, in addition to those detailed in this chapter, include diagnostic systems and cultivation of viruses that depend on cycling and differentiating cells.

Not least of all, as model building *in vitro* improves, so do prostheses for implantation in people. Model making and *in vitro* testing go hand in hand with testing in experimental animals. Model tissues can take many forms. (1) They can be seeded with parenchymal cells that normally populate the tissue being replaced; (2) they can be acellular but designed to mobilize cells from contiguous tissues after implantation *in vivo*; (3) they can be constituted with degradable biopolymer scaffolds that are permissive, with the expectation that the cells with which they are seeded or which populate them will create a tissue-specific extracellular matrix and associated scaffolding; or (4) they can be instructive, being provided with signals able to influence the activities of cells. The ultimate goal of tissue engineering is that of building replacement parts for people who need them.

# IX. REFERENCES

Asahara, T., Murohara, T., Sullivan, A., Silver, M., van der Zee, R., Li, T., Witzenbichler, B., Schatteman, G., and Isner, J. M. (1997). Isolation of putative progenitor endothelial cells for angiogenesis. *Science* **275**, 964.

Asahara, T., Masuda, H., Takahashi, T., Kalka, C., Pastore, C., Silver, ••., Kearne, M., Magner, M., and Isner, J. M. (1999). Bone marrow origin of endothelial progenitor cells responsible for postnatal vasculogenesis in physiological and pathological neovascularization. *Circ. Res.* **85**, 221.

Beigel, C., and Pachter, J. S. (1994). Growth of brain microvessel endothelial cells on collagen gels: applications to the study of blood–brain barrier physiology and CNS inflammation. *In Vitro Cell Dev. Biol. Anim.* **30A**, 581.

Bell, E. (1995). Strategy for the selection of scaffolds for tissue engineering. *Tissue Eng.* **1**, 163.

Bell, E., and Scott, S. (1992). Tissue fabrication: reconstitution and remodeling *in vitro*. *Mat. Res. Soc. Symp. Proc.* **252**, 141.

Bell, E., Iverson, B., and Merrill, C. (1979). Production of a tissue-like structure by contraction of collagen lattices by human fibroblasts of different proliferative potential *in vitro*. *Proc. Natl. Acad. Sci. USA* **76**, 1274.

Bell, E., Ehrlich, H. P., Buttle, D. J., and Nakatsuji, T. (1981). Living tissue formed *in vitro* and accepted as a skin-equivalent tissue of full thickness. *Science* **211**, 1052.

Bell, E., Moore, H., Mitchie, C., Sher, S., and Coon, H. (1984). Reconstitution of thyroid gland equivalent from cells and matrix materials. *J. Exp. Zool.* **232**, 277.

Bell, E., Kagan, D., Nolte, C., Mason, V., and Hastings, C. (1990). Demonstration of barrier properties exhibited by brain capillary endothelium cultured as a monolayer on a tissue equivalent. *J. Cell Biol.* **5**, 185a.

Bell, E., Rosenberg, M., Kemp, P., Gay, R., Green, G., Muthukumaran, N., and Nolte, C. (1991). Recipes for reconstituting skin. *J. Biomech. Eng.* **113**, 113.

Bergsteinsdottir, K., Hashimoto, Y., Brennan, A., Mirsky, R., and Jessen, K. R. (1993). The effect of three-dimensional collagen type I preparation on the structural organization of guinea pig enteric ganglia in culture. *Exp. Cell Res.* **209**, 64.

Boyce, S. T., and Ham, R. G. (1983). Calcium-regulated differentiation of normal human epidermal keratinocytes in chemically defined clonal culture and serum-free serial culture. *J. Invest. Dermatol.* **81**(Suppl.1), 33s.

Coraux, C., Hilmi, C., Rouleau, M., Spadafora, A., Hinnrasky, J., Ortonne, J. P., Dani, C., and Aberdam, D. (2003). Reconstituted skin from murine embryonic stem cells. *Curr. Biol.* **13**, 849.

Coulomb, B., Dubertret, L., Bell, E., Merrill, C., Fosse, M., Breton-Gorius, J., Prost, C., and Touraine, R. (1983). Endogenous peroxidases in normal human dermis: a marker of fibroblast differentiation. *J. Invest. Derm.* **8**, 75.

Croft, A., and Przyborski, S. (2006). Formation of neurons by non-neural adult stem cells: potential mechanism implicates an artifact of growth in culture. *Stem. Cell* **24**, 1841.

Cuono, C. B., Langdon, R., Birdcall, N., Barttelbort, S., and McGuire, J. (1987). Composite autologous-allogeneic skin replacement: development and clinical application. *Plast. Reconstr. Surg.* **80**, 626.

Dai, J., Kumar, J., Feng, Y, Asrican, R., Kim, J., Fofonoff, T., Russakovsky, V., Churchill, R., Roy, N., and Bell, E. (2002). The specificity of phenotypic induction of mouse and human stem cells by signaling complexes. *In Vitro Cell Dev. Biol. Anim.* **38**, 198.

Dodson, J. W. (1963). On the nature of tissue interactions in embryonic skin. *Exp. Cell Res.* **31**, 233.

Freiberger, H., Grove, D., Sivarajah, A., and Pinnell, S. R. (1980). Procollagen I synthesis in human skin fibroblasts: effect of culture conditions on biosynthesis. *J. Invest. Dermatol.* **75**, 425.

Guthrie, S., and Pini, A. (1995). Chemorepulsion of developing motor axons by the floor plate. *Neuron* **14**, 1117.

Hansbrough, J. F., Morgan, J. L., Greenleaf, G. E., and Bartel, R. (1993). Composite grafts of human keratinocytes grown on a polyglactin mesh-cultured fibroblast dermal substitute function as a bilayer skin replacement in full-thickness wounds on athymic mice. *J. Burn Care Rehab.* **14**, 485.

Hansbrough, J. F., Morgan, J. L., Greenleaf, G., and Underwood, J. (1994). Development of a temporary living skin replacement composed of human neonatal fibroblasts cultured in Biobrane, a synthetic dressing material. *Surgery* **115**, 633.

Harley, B. A., and Yannas, I. V. (2006). Induced peripheral nerve regeneration using scaffolds. *Minerva Biotechnologica*. **18**, 97–120.

Harris, A. K., Stopak, D., and Wild, P. (1981). Fibroblast traction as a mechanism for collagen morphogenesis. *Nature* **290**, 249.

Heck, E. L., Bergstresser, P. R., and Baxter, C. R. (1985). Composite skin graft: frozen dermal allografts support the engraftment and expansion of autologous epidermis. *J. Trauma* **25**, 106.

Hirai, J., Kanda, K., Oka, T., and Matsuda, T. (1994). Highly oriented, tubular hybrid vascular tissue for a low-pressure circulatory system. *Asaio J.* **40**, M383.

Hong, Y., Dong, F., Suresh, M. K., Ling, L., Thiennga, K. N., Geza, A., Meenhard, H., and Xiaowei, X. (2006). Isolation of a novel population of multipotent adult stem cells from human hair follicles. *Am. J. Pathol.* **168**, 1879.

Kalka, C., Masuda, H., Takahashi, T., Kalka-Moll, W. M., Silver, M., Kearney, M., Li, T., Isner, J. M., and Asahara, T. (2000). Transplantation of *ex vivo* expanded endothelial progenitor cells for therapeutic neovascularization. *Proc. Natl. Acad. Sci. USA* **97**, 3422.

Kanda, K., Matsuda, T., and Oka, T. (1993). Mechanical stress–induced cellular orientation and phenotypic modulation of 3-D cultured smooth muscle cells. *Asaio J.* **39**, M686.

Lanza, R. P., Ecker, D., Kuhtreiber, W. M., Staruk, J. E., Marsh, J., and Chick, W. L. (1995). A simple method for transplanting discordant islets into rats using alginate gel spheres. *Transplantation* **59**, 1485.

Lavker, S. M., and Sun, T. T. (2000). Epidermal stem cells: properties, markers and location. *Proc. Natl. Acad. Sci. USA* **97**, 13473.

Leenslag, J. W., Kroes, M. T., Pennings, A. J., and van der Lei, B. (1988). A compliant, biodegradable, vascular graft: basic aspects of its construction and biological performance. *New Polymeric Mater* **1**, 111.

L'Heureux, N., Germain, L., Labbe, R., and Auger, F. A. (1993). *In vitro* construction of a human blood vessel from cultured vascular cells: a morphologic study. *J. Vasc. Surg.* **17**, 499.

Li, A., Simmons, P. J., and Kaur, P. (1998). Identification and isolation of candidate human keratinocyte stem cells based on cell surface phenotype. *Proc. Natl. Acad. Sci. USA* **95**, 3902.

Lin, Y., Weisdorf, D. J., Solovey, A., and Hebbel, R. P. (2000). Origins of circulating endothelial cells and endothelial outgrowth from blood. *J. Clin. Invest.* **105**, 71.

Marx, J. (1995). Research news: helping neurons find their way. *Science* **268**, 971.

Michel, M., Torok, N., Godbout, M. J., Lussier, M., Gaudreau, P., Royal, A., and Germain, L. (1996). Keratin 19 as a biochemical marker of skin stem cells *in vivo* and *in vitro*: keratin 19 expressing cells are differentially localized in function of anatomic sites, and their number varies with donor age and culture stage. *J. Cell Sci.* **109**, 1017.

Minami, Y., Sugihara, H., and Oono, S. (1993). Reconstruction of cornea in three-dimensional collagen gel matrix culture. *Invest. Ophthalmol. Vis. Sci.* **34**, 2316.

Modis, L. (1990). "Organization of the Extracellular Matrix: A Polarization Microscopic Approach." CRC Press, Boca Raton, FL.

Montgomery, R. K. (1986). Morphogenesis *in vitro* of dissociated fetal rat small intestine cells upon an open surface and subsequent to collagen gel overlay. *Dev. Biol.* **117**, 64.

Mori, M., Sadahira, V., and Kawasaki, S. (1989). Formation of capillary networks from bone marrow cultured in collagen gel. *Cell Struct. Funct.* **14**, 393.

Nerem, R. M., Mitsumata, M., Ziegler, T., Berk, B. C., and Alexander, R. W. (1993). Mechanical stress effects on vascular endothelial cell growth. *In* "Tissue Engineering: Current Perspectives" (E. Bell, ed.), p. 120. Birkhäuser, Boston.

Nusgens, B., Merrill, C., Lapiere, C., and Bell, E. (1984). Collagen biosynthesis by cells in a tissue equivalent matrix *in vitro*. *Collagen Rel. Res.* **4**, 351.

Parenteau, N., Nolte, C., Bilbo, P., Rosenberg, M., Wilkins, L., Johnson, E., Watson, S., Mason, V., and Bell, E. (1991). Epidermis generated *in vitro*: practical considerations and applications. *J. Cell Biochem.* **45**, 245.

Parenteau, N. L., Sabolinski, M. L., Wilkins, L. M., and Rovee, D. T. (1994). Development of a bilayered "skin equivalent": from basic science to clinical use. *J. Cell Biochem.* **18C**(Suppl.), 273.

Peichev, M., Naiyer, A. J., Pereira, D., Zhu, Z., Lane, W. J., Williams, M., Oz, M. C., Hicklin, D. J., Witte, L., A. S. Moore, M., and Rafii, S. (2000). Expression of VEGFR-2 and AC133 by circulating human CD34$^+$ cells identifies a population of functional endothelial precursors. *Blood* **95**, 952.

Pellegrini, G., Golisano, O., Paterna, P., Lambiase, A., Bonini, S., Rama, P., *et al.* (1999). Location and clonal analysis of stem cells and their differentiated progeny in the human ocular surface. *J. Cell Biol.* **145**, 769.

Pittenger, M. F., Mackay, A. M., Beck, S. C., Douglas, R., Mosca, J. D., Mormon, M. A., Simonetti, D. W., Craig, S., and Marshak, D. R. (1999). Multilineage potential of adult human mesenchymal stem cells. *Science* **284**, 143.

Rowling, R. J. E., Raxworthy, M. J., Wood, E. J., and Kearney, J. N. (1990). Fabrication and reorganization of dermal equivalents suitable for skin grafting after major cutaneous injury. *Biomaterials* **11**, 181.

Saiag, P., Coulomb, B., Lebreton, C., Bell, E., and Dubertret, L. (1985). Psoriatic fibroblasts induce hyperproliferation of normal keratinocytes in a skin equivalent model *in vitro*. *Science* **230**, 669.

Sandusky G. E., Jr., Badylak, S. F., Morff, R. J., Johnson, W. D., and Lantz, G. (1992). Histologic findings after *in vivo* placement of small intestine submucosal vascular grafts and saphenous vein grafts in the carotid artery in dogs. *Am. J. Pathhol.* **140**, 317.

Sher, S., Hull, B., Rosen, S., Church, D., Friedman, L., and Bell, E. (1983). Acceptance of allogeneic fibroblasts in skin-equivalent transplants. *Transplant* **36**, 552.

Shi, Q., Rafii, S., Hong-De Wu, M., Wijelath, E. S., Yu, C., Ishida, A., Fujita, Y., Kothari, S., Mohle, R., Sauvage, L. R., A. S. Moore, M., Storb, R. F., and Hammond, W. P. (1998). Evidence for circulating bone marrow–derived endothelial cells. *Blood* **92**, 362.

Shimizu, K., Sugiyama, S., Aikawa, M., Fukumoto, Y., Rabkin, E., Libby, P., and Mitchell, R. N. (2001). Host bone-marrow cells are a source of donor intimal smooth muscle–like cells in murine aortic transplant arteriopathy. *Nature Med.* **7**, 738.

Soon-Shiong, P., Feldman, E., Nelson, R., Heintz, R., Yao, Q., Yao, Z., Zheng, T., Merideth, N., Skjak-Braek, G., Espevik, T., Smidsrod, O., and Sandford, P. (1993). Long-term reversal of diabetes by the injection of immunoprotected islets. *Proc. Natl. Acad. Sci. USA* **90**, 5843.

Stronck, J. W. S., van der Lei, B., and Wildevuur, C. R. H. (1992). Improved healing of small-caliber polytetrafluoroethylene vascular prostheses by increased hydrophilicity and by enlarged fibril length. *J. Thoracic Cardovasc. Surg.* **103**, 146.

Thompson, J. A., Itskovitz-Eldor, J., Shapiro, S. S., Waknitz, M. A., Swiergiel, J. J., Marshall, V. S., and Jonesw, J. M. (1998). Embryonic stem cell lines derived from human blastocysts. *Science* **282**, 1145.

Toma, J. G., Akhavan, M., J. L. Fernandes, K., Barnabe-Heider, F., Sadikot, A., Kaplan, D. R., and Miller, M. D. (2001). Isolation of multipotent adult stem cells from the dermis of mammalian skin. *Nature Cell Biol.* **3**, 778.

Tomooka, Y., Kitani, H., Jing, N., Matsushima, M., and Sakakura, T. (1993). Reconstruction of neural tube–like structures *in vitro* from primary neural precursor cells. *Proc. Natl. Acad. Sci. USA* **90**, 9683.

Topol, B. M., Haimes, H. B., Dubertret, L., and Bell, E. (1986). Transfer of melanosomes in a skin equivalent model *in vitro*. *J. Invest. Derm.* **87**, 642.

Torbert, J., and Ronziere, M. C. (1984). Magnetic alignment of collagen during self-assembly. *Biochem. J.* **219**, 1057.

Tranqulilo, R. T., Girton, T. S., Bromberek, B. A., Triebes, T. G., and Mooradian, D. L. (1996). Magnetically oriented tissue-equivalent tubes: application to a circumferentially oriented media-equivalent. *Biomaterials* **17**, 349.

Watt, F. M. (1998). Epidermal stem cells: markers, patterning and the control of stem cell fate. *Philos. Trans. R. Soc. Lond. B Biol. Sci.* **353**, 831.

Weinberg, C., and Bell, E. (1986). A blood vessel model constructed from collagen and cultured vascular cells. *Science* **231**, 397.

Weinberg, C. B., O'Neil, K. D., Carr, R. M., Cavallaro, J. F., Ekstein, B. A., Kemp, P. D., Rosenberg, M., Garcia, J. P., Tantillo, M., and Khuri, S. F. (1993). Matrix engineering: remodeling of dense fibrillar collagen vascular grafts *in vivo*. *In* "Tissue Engineering: Current Perspectives" (E. Bell, ed.) p. 141. Birkhäuser, Boston.

Zagai, U., Fredriksson, K., Rennard, S. I., Lundahl, J., and Sköld, C. M. (2003). Platelets stimulate fibroblast-mediated contraction of collagen gels. *Resp. Res.* **4**, 13.

# Chapter Eighteen

# Quantitative Aspects

*Alan J. Grodzinsky, Roger D. Kamm, and Douglas A. Lauffenburger*

## I. INTRODUCTION

A chapter on quantitative aspects of tissue engineering remains difficult to write at this point in the development of tissue regeneration and cell therapy processes. Although cell functional behavior and underlying molecular mechanisms are becoming increasingly amenable to quantitative approaches, incorporating these into the complex interactions among multiple cell types in organized tissue is currently beyond the scope of feasibility. A historical parallel may be drawn to the decades-long lag between the advent of the petrochemical industry and the useful introduction of rigorous analysis, in terms of fundamental physicochemical theories, decades later. Thus, it is clear that today's beginning attempts in tissue engineering must be highly empirical, with design based largely on intuition and experience, much like the petrochemical industry in the 1930s.

Following this parallel, however, it can be recalled that in the 1940s the concept of *unit operations* appeared, in which each particular chemical production plant could be broken down into some component processes that had some similarities to processes in different types of production plants. In this way, basic principles of chemical reactors, heat exchangers, material separation equipment, and so forth were elucidated in very simple quantitative terms. It was not that mathematical models for these unit operations were immediately combined into a comprehensive quantitative description of the entire plant — which is, of course, common practice today — but instead merely that

important design parameters of the components could be identified, and in some cases the direction of their manipulation for process improvement could be indicated.

This, then, appears to be the state of affairs for tissue engineering at the beginning of the 21st century: that key parameters governing components of the overall device or procedure might be identifiable, and some design principles for how they could be altered toward an improved device or procedure might be developed. In a manner analogous to the unit operations, simple mathematical descriptions of cell and tissue processes can be constructed for purposes of elucidating what properties matter and how they can be manipulated. This chapter therefore provides a brief overview of how major cell and tissue properties can be described and quantified in most basic form. These properties include molecular and cell transport through tissue, molecular interactions with cells, and tissue mechanics.

## II. MOLECULAR INTERACTIONS WITH CELLS

There are three main classes of molecules that must be dealt with in the context of cell interactions in tissue: soluble nutrients, soluble signaling molecules, and molecules associated with the extracellular matrix. The latter two classes interact with cells primarily via cell surface receptors, whereas the former class can either bind to cell surface receptors or pass directly across the cell membrane.

*Principles of Tissue Engineering, 3rd Edition*
*ed. by Lanza, Langer, and Vacanti*

## Nutrients

For nutrient molecules that pass directly across the cell membrane — either by passive diffusion or via carrier proteins — such as oxygen, glucose, and amino acids, the kinetics of uptake and metabolism generally follow a Michaelis–Menten type of dependence. That is, at low concentrations the rate is first order in concentration, but as concentration increases the rate asymptotically approaches a constant plateau. In combination with transport rates through tissue, as described later, the consumption of nutrients can be analyzed in a fairly straightforward manner. If transport is modeled as simple diffusion through a medium in which cells are embedded as point sinks, the nutrient concentration distribution, $L(x)$, in the tissue is governed by the combined diffusion/reaction equation:

$$\frac{\partial L}{\partial t} = D\frac{\partial^2 L}{\partial x^2} - \frac{\rho k L}{K_M + L} \tag{18.1}$$

where $D$ is the diffusion coefficient, $k$ is the maximal uptake rate constant per cell, $K_M$ is the saturation constant, $\rho$ is the cell density, and $x$ is the spatial distance from the source. For simple nutrients, the source is typically the bloodstream. Significant depletion of nutrient, and hence possible nutrient deprivation, will occur when the ratio of uptake to diffusion becomes small. This ratio can be usefully expressed in terms of the Thiele modulus: $\phi = (\rho k X^2/D)^{1/2}$, where $X$ is the overall distance from the source — for simple nutrients, this can be considered the mean distance between microcirculatory blood vessels (this is an approximate expression, assuming that the uptake rate is constant, with value $k$, throughout the tissue.) When $\phi < 1$, the steady-state nutrient concentration becomes significantly less than the exogenous level. Transplantation of cells within a polymeric matrix can often give rise to depletion of important nutrients from levels required for sustained cell viability, when $\rho$ is great and $X$ is large, because the implant is not adequately vascularized. Nutrient transport limitations have been examined both theoretically and experimentally for the important case of encapsulated cells (Colton, 1995).

## Growth Factors and Other Regulatory Molecules

Many molecules important in tissue engineering, however, cannot be dealt with quite so simply. Regulatory molecules, such as growth factors, along with nutrient carriers, such as iron-bearing transferrin, bind reversibly to plasma membrane receptors and are then internalized by the cell by means of invaginating membrane structures in a process known as *endocytosis*. The receptor/ligand complexes are carried to intracellular organelles termed *endosomes*, from which they are sorted to a variety of fates, including lysosomal degradation and recycling to the cell surface (see Fig. 18.1). This entire process, known as *trafficking*, is quite complicated, and the distribution of fates for a

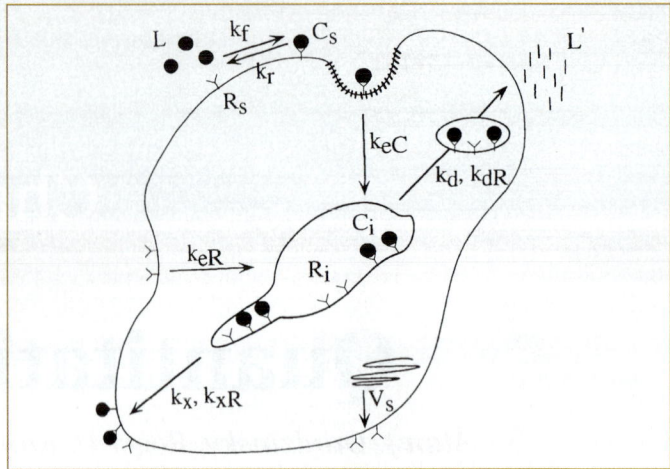

**FIG. 18.1.** Schematic illustration of receptor/ligand trafficking. $L$ is intact ligand, $L'$ is undegraded ligand; $R_s$, $C_s$, $R_i$, and $C_i$ are cell surface and intracellular free and bound receptors; $k_f$ and $k_r$ are receptor/ligand association and dissociation rate constants; $k_{eR}$ and $k_{eC}$ are internalization rate constants for free and bound receptors; $k_d$ and $k_{dR}$ are degradation rate constants for ligand and receptor; $k_x$ and $k_{xR}$ are recycling rate constants for ligand and receptor; $V_s$ is the receptor synthesis rate.

particular ligand can vary quantitatively as a function of its concentration as well as some of its biochemical and biophysical properties.

In the absence of trafficking, and when there is only a simple one-step reversible binding process, kinetic mass action equations can be written for the number of bound and free surface receptors and their ligands ($L$). At equilibrium, the number ($C$) of receptors ($R$) found in the bound state is given by the expression

$$C = \frac{LR_\tau}{K_D + L} \tag{18.2}$$

$R_\tau$ is the total number of surface receptors, constant only when trafficking processes are eliminated. $K_D$ is the dissociation equilibrium constant, equal to the ratio of the receptor/ligand dissociation rate constant to the association rate constant: $k_r/k_f$. It is essentially the reciprocal of the receptor/ligand-binding affinity. Commonly, however, ligands can bind to multiple receptors on the surface of a given cell, or receptors can be found in various states possessing different ligand-binding affinities. The interested reader is referred elsewhere for a detailed discussion of how to treat these more complicated situations (Lauffenburger and Linderman, 1996).

However, trafficking is almost universally present under physiological conditions, and the overall dynamics of uptake and metabolism of ligands and their receptors are strongly influenced by internalization, degradation, and recycling. Kinetic mass action equations can be written for the set of events shown in Fig. 18.1, and the dynamics of ligands and receptors in the various cell and tissue compartments can be thereby analyzed for purposes of understanding key design parameters (Lauffenburger and Linderman, 1996).

For instance, consider a molecular ligand delivered to a tissue at rate $V_L$. For ligands such as growth factors, the source may be other cells in the tissue — perhaps implanted cells expressing a desired growth factor — or a polymeric controlled-release device or the bloodstream. Assuming a spatially homogeneous source, the steady-state ligand concentration will be given approximately by (assuming, for simplicity, that receptor and ligand recycling are negligible) the expression

$$L = K_D \left( \frac{k_{eR}}{k_{eC}} \right) \left[ \left( \frac{\rho V_R' + k_{\deg} k_{eR}/k_f}{V_L} - 1 \right) \right]^{-1} \quad (18.3)$$

where $V_R'$ is the receptor synthesis rate per cell, $k_{eR}$ and $k_{eC}$ are the free receptor and bound receptor internalization rate constants, respectively, and $k_{\deg}$ is a rate constant for extracellular proteolytic ligand degradation.

Following Wiley (1985), the corresponding steady-state number of receptor/ligand complexes per cell, which typically governs the functional response, is then given approximately by

$$C = \frac{L R_{\max}}{K_{SS} + L} \quad (18.4)$$

$R_{\max}$ is equal to $V_R'/k_{eR}$, and $K_{ss}$ is equal to $k_{eC} k_f / [k_{eR}(k_r + k_{eC})]$. Note the similarity in form to Eq. (18.2), which governs the number of complexes in the absence of trafficking. Thus, trafficking determines the number of available surface receptors and the effective ligand-binding affinity.

When ligand concentrations are not spatially homogeneous, then the local value of $L$ at any spatial position within the tissue must be determined by including the effects of diffusion, as in Eq. (18.1). Here $x$ would be the thickness of tissue away from the source. Moreover, in place of the simple metabolic uptake rate constant, $k$, the net dynamics of trafficking must be incorporated to account for ligand consumption.

Binding and trafficking rate constants can be measured in cell culture assays, using ligands labeled with fluorescent or radioactive tags. Descriptions of basic experimental procedures and corresponding methods of analysis can be found elsewhere (Lauffenburger and Linderman, 1996).

Extracellular matrix components can compete with cells for binding ligand, and in some circumstances they may serve as a reservoir due to the reversible nature of protein/protein association. At equilibrium, this binding can be characterized using an expression analogous to Eq. (18.2). Extracellular matrix components may exhibit binding interactions with cell surface receptors; when they are sufficiently immobilized by their linkage within the matrix, Eq. (18.2) again can represent these interactions in the simplest case. As with soluble ligands, though, more complicated interactions can arise with multiple receptors for different domains of a particular matrix component.

Expressions such as Eqs. (18.1)–(18.4) can be used to estimate the rate of ligand delivery required to maintain a desired level of cell surface complexes and/or free ligand concentration. The number of surface complexes typically is important for determining the cell functional response. We emphasize, though, that it is likely that the rate of ligand binding to receptors to form signaling complexes may govern cell responses as much as, or more than, the static level of complexes (Wiley, 1992). Quantification of the dependence of cell behavior on complex levels and dynamics, however, remains in its infancy at present. A small number of particular examples can be offered as guidelines, including proliferation responses of fibroblasts to epidermal growth factor (Knauer et al., 1984), leading to improvements in cell-level principles for growth factor design and delivery (Reddy et al., 1996a, 1996b). A further level of complication in analysis arises from the presence of autocrine ligands in a wide spectrum of tissue physiological processes, though quantitative modeling and experimental studies are now being pursued on this aspect as well (Forsten and Lauffenburger, 1992; Lauffenburger et al., 1998).

## III. MOLECULAR AND CELL TRANSPORT THROUGH TISSUE

Movements of molecular as well as cellular species through tissue is a vital aspect of most tissue-engineering processes, because the whole point of the intervention is to alter the composition of a local region from what it had been prior to the therapy. As mentioned earlier, transport of nutrients and regulatory factors into a cellular implant is generally required. Also, the goal of certain implanted devices is to release regulatory factors into the surrounding tissue. The synthesis of neotissues involves placing cells within a matrix composed of natural materials (e.g., collagen) or synthetic polymers (e.g., biodegradable templates). Within this environment, the cells also secrete and regulate the assembly of their own extracellular matrix, composed of collagens, proteoglycans, glycoproteins, and a variety of macromolecules important in cell–matrix and cell–cell interactions. Solute transport through this tortuous and often dense matrix plays a critical role in the maintenance of cell viability and in the successful growth and development of the intact tissue. Moreover, colonization of implants or grafts by cultured or host cells is often desirable. And, in some cell therapies, mainly those involving immune white blood cells, the movement of cells from an injection site into certain tissue regions is essential for them to carry out their intended functions.

### Molecular Diffusion and Convection

Cell biosynthesis requires transport of low-molecular-weight metabolites and waste products, which occurs primarily by diffusion. Many other solutes play key roles as mediators of cell behavior and in cell production and turnover of extracellular matrix, including growth factors, hormones, cytokines, and endogenous proteinases and their natural inhibitors. Transport of regulatory factors can be crucial in tissue regeneration (Perez *et al.*, 1995) and can be reduced severely by interactions with matrix (Dowd *et al.*, 1999). In addition, the assembly of a physicochemically and mechanically functional matrix requires transport of newly synthesized matrix macromolecules to appropriate locations within the extracellular space. These matrix proteins and proteoglycans (as large as several megadaltons) have such low intratissue diffusivities that even slight fluid convection within the matrix can significantly enhance macromolecular transport (Garcia *et al.*, 1996).

In general, transport depends on the size and charge of the solute and on the density (tortuosity), water content, and charge of the matrix (Garcia *et al.*, 1998) (see Fig. 18.2). Although the principal modes of transport are molecular diffusion and convection (Deen, 1987), electrostatic interactions can also affect the partitioning and transport of charged solutes into and within a charged matrix (Grodzinsky, 1983). In this regard, it is useful to note that matrix charge density at the macroscale is often associated with the content of glycosaminoglycan chains attached to proteoglycans. At the scale of the cell, the dense, negatively charged glycocalyx at the surface of many cell types can also affect transport across the cell membrane.

Each cell and tissue type presents unique challenges for providing adequate transport in tissue-engineering design. For example, the density of cells in liver and kidney is typically much higher than that in connective tissues such as ligament, tendon, and cartilage. On the other hand, connective tissue cells grow within a very dense, hydrated, negatively charged matrix that may be several millimeters thick but with little or no vascular or lymph supply. Some cells are best grown on surfaces, whereas others require three-dimensional matrix encapsulation. The mechanical environment of cells also varies greatly: Vascular cells are subjected to fluid shear and to pulsatile stresses and strains within vessel walls; cartilage, bone, and other soft connective tissue cells must sustain peak mechanical stresses as high as 10–20 MPa (100–200 atm). These dynamic stresses produce fluid flows within the matrix that can affect intratissue solute transport and the concentration of solute species at the boundaries of tissues *in vivo* (O'Hare *et al.*, 1990). Such effects must therefore be taken into account in tissue-engineering design *in vitro*.

Given these diverse tissue types and geometric constructs, mathematical modeling of transport processes can be critically important in the design of tissue reactors, the basic understanding of the spatial and temporal distribution of small and large solutes in newly developing tissues, and the understanding of failure modes that might result from inadequate transport. We briefly summarize next the governing transport laws for solutes within tissues and the boundary conditions at the tissue/media interface. These general laws may then be adapted to the particular cell and tissue types of interest.

First, the one-dimensional flux $N_i$ of the $i$th species within the tissue due to diffusion, convection, and electrical migration (ohmic conduction) is described by the flux equation:

$$N_i = \phi\left(-D_i \frac{\partial c_i}{\partial x} + \frac{z_i}{|z_i|}\mu_i c_i E\right) + W_i c_i U \qquad (18.5)$$

where $c_i$ is the solute concentration at the position $x$, $\phi$ is the matrix (tissue) porosity, $z_i$ is the valence of the solute (if charged), $U$ is the fluid velocity relative to the matrix (averaged over the total matrix area), and $E$ is the electric field within the matrix (e.g., induced by diffusion of ionic species or by electrokinetic effects, such as flow-induced streaming potentials through a charged matrix). The intrinsic transport parameters $D_i$, $\mu_i$, and $W_i$ represent, respectively, the intratissue diffusivity of the solute, the electrical mobility, and a hindrance factor for convection that incorporates hydrodynamic and steric interactions between solute molecules and the matrix molecules forming the pore walls. Even in the absence of net fluid convection, frictional hydrodynamic and steric interactions between solute, matrix, and solvent can play a critical role in transport and may cause significant differences between the intratissue values of $D_i$, $\mu_i$, and their corresponding values in free solution (Deen, 1987). In general, the flux Eq. (18.5) is general, in that it applies to ionic solutes in the medium, small neutral and charged solutes (e.g., amino acids, glucose), and large neutral and charged macromolecular solutes. For small solutes, $W \to 1$, and the values of $D_i$ and $\mu_i$ may be close to their free solution values; for larger solutes, $W < 1$, and $D_i$ and $\mu_i$ as written in Eq. (18.5) implicitly incorporate associated hindrance factors (Deen, 1987).

It is therefore necessary to measure or estimate values for the intrinsic transport parameters $D_i$, $\mu_i$, and $W_i$ in order to predict solute distribution profiles within the tissue construct. To predict the spatial and temporal distribution of

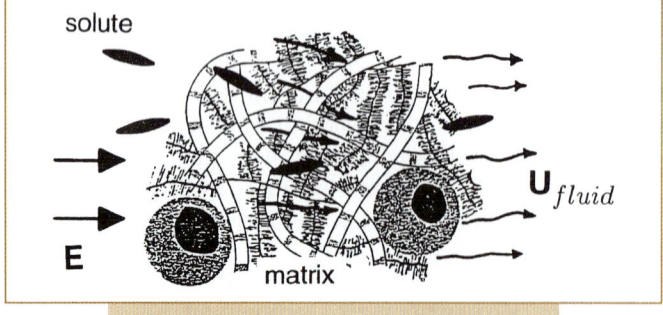

**FIG. 18.2.** The flux of macromolecular and small solutes within a cell-seeded matrix results from diffusion, convection (fluid velocity $U$), and electrical migration if the solute is charged; an electric field $E$ exists within the tissue (caused by fluid flow or by ion concentration gradients).

solutes in the neotissue, the flux equation is incorporated into the continuity law:

$$\frac{\partial c_i}{\partial t} = \frac{\partial N_i}{\partial x} + \sum_i (G_i - R_i) \qquad (18.6)$$

where $G_i$ and $R_i$ represent chemical reaction rates associated with binding of solutes to cells or matrix molecules. In the absence of binding, electric migration, and convection effects, the continuity law, Eq. (18.6), reduces to Fick's second law for molecular diffusion. For the case of purely diffusive nutrient transport within a matrix containing cells acting as point sinks, the continuity law, Eq. (18.6), would reduce to the diffusion/reaction Eq. (18.1).

Electroneutrality within the tissue, including charges on all solute molecules ($i$) and matrix molecules ($m$), requires that

$$z_m c_m + \sum_i (z_i c_i) = 0 \qquad (18.7)$$

Although tissues may have a high water content, the concentration of solutes within tissue water may be very different from that in the bathing medium because of size and charge effects. In general, the solute partition coefficient $\kappa$ is defined as the ratio of the solute concentration within the tissue fluid to that in the bathing medium, $\kappa = c^{\text{tiss}}/c^{\text{med}}$. This partition coefficient acts as a boundary condition for the flux and continuity equations, which are written specifically in terms of solute concentrations inside the tissue. In addition, for charged solutes in the presence of a charged matrix, this partitioning is further modified due to Donnan equilibrium considerations for positively and negatively charged species, giving the Donnan partition boundary conditions (Maroudas, 1979):

$$(c_{i+}^{\text{med}}/c_{i+}^{\text{tiss}})^{1/|z+|} = (c_j^{\text{tiss}}/c_{j-}^{\text{med}})^{1/|z-|} \qquad (18.8)$$

Finally, it is useful to recast the continuity law in terms of the dimensionless Peclet number, Pe, which is defined as the ratio of solute convection plus migration fluxes to the diffusive flux (Grimshaw *et al.*, 1989):

$$\frac{\partial c_i}{\partial t} = D_i \left[ \frac{\partial^2 c_i}{\partial x^2} + \frac{\text{Pe}}{X} \left( \frac{\partial c_i}{\partial x} \right) \right] \qquad (18.9)$$

$$\text{Pe} = X \left( \frac{W_i U + \phi \mu_i E}{\phi D_i} \right) \qquad (18.10)$$

where $X$ is the characteristic tissue thickness over which transport must occur. Thus, Eq. (18.9) takes the form of a modified Pick's second law and puts into perspective the relative importance of the convective and electrical migration terms, given values for $X$, $D_i$, $W_i$, and $\mu_i$. It is apparent that for solutes having small enough $D_i$ and/or large enough fluid flow, convection may have a significant effect in determining the solute distribution within the tissue. Said another way, convection and intratissue electrokinetic (e.g., electrophoretic) effects could significantly augment the transport of nutrients and macromolecules, as compared to diffusion alone, over tissue dimensions $X$.

## Cell Migration

Polymeric scaffolds are often introduced, into which it is hoped that certain cell types will migrate while others do not. For instance, regeneration of bone tissue for enhanced healing of full-thickness injuries requires that osteoprogenitor cells from surrounding tissue colonize an implanted scaffold and that endothelial cells migrate in, to achieve neovascularization, whereas connective tissue fibroblast influx is undesirable. There are two approaches for controlling migration: (1) surface chemistry of the implanted material for selective cell adhesion interactions with the surface, influencing migratory behavior in a differential manner among various cell types, and (2) release of chemotactic attractants from within the implanted material to induce enhanced migratory responses of particular cell types.

The rate of cell population movement into a tissue can be quantified using a cell transport flux expression analogous to that for molecular diffusion and convection (Tranquillo *et al.*, 1988):

$$J_{\text{cells}} = -\mu \frac{\partial N}{\partial x} + \chi \left( \frac{\partial A}{\partial x} \right) N \qquad (18.11)$$

where $\mu$ is the cell random motility coefficient, analogous to a molecular diffusion coefficient, and $\chi$ is the chemotaxis coefficient; $\chi$ is a biphasic function of the attractant concentration, $A$, because chemotaxis arises from a spatial difference in the number of receptor/attractant complexes across a cell length. The first term in Eq. (18.11) represents cell dispersion down the spatial gradient of cell density, whereas the second term yields a directed, convective velocity for cell migration in the direction of an attractant concentration gradient. The values $\mu$ and $\chi$ can be measured in a variety of cell population assays, such as movement through a porous filter or under an agarose gel. They can alternatively be calculated from individual cell-tracking assays in which cell speed, directional persistence time, and directional orientation bias are determined, because the cell population parameters are related by theory to the individual cell parameters. Details on these approaches can be found elsewhere (Lauffenburger and Linderman, 1996).

A fair amount of effort has been devoted to quantitative understanding of how substrata properties of materials affect cell migration. A central concept that is beginning to show general verification is that the speed of migration depends in a biphasic manner on the strength of cell/substratum adhesiveness (see Fig. 18.3) (DiMilla *et al.*, 1991). This principle can account for a wide range of experimental observations of how migration speed varies with the density of extracellular matrix ligands immobilized on biomaterial surfaces, and for the effects of ligand affinity and competing soluble ligand (Palecek *et al.*, 1997; Wu *et al.*, 1994). This lack of monotonicity is unfortunate, of course, because it makes

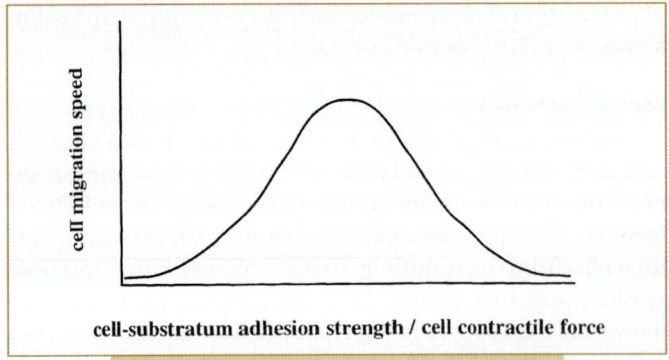

**FIG. 18.3.** Qualitative representation of typical dependence of cell migration speed on the ratio of cell-substratum adhesion strength to cell-generated contractile force.

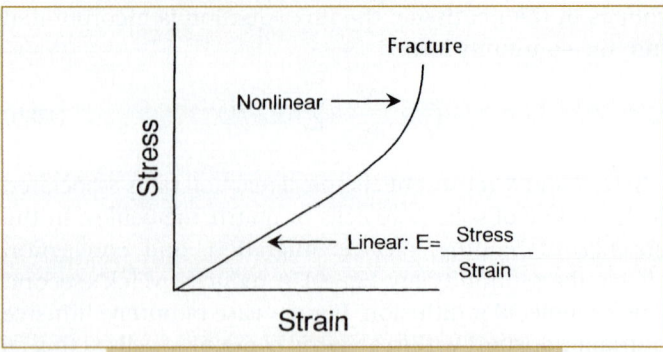

**FIG. 18.4.** Plot of stress (σ) vs. strain (ε), showing an initial linear region, in which the Young's modulus (E) is constant, and a nonlinear region, with increasing E, leading ultimately to fracture of the specimen.

design of a material quite problematic; unless the full quantitative picture is well characterized, one may produce a substratum either too strongly adhesive or too weakly adhesive for effective cell locomotion. Additional modulating approaches may be possible, though, by means of other parameters influencing migration speed. For instance, the dependence illustrated in Fig. 18.3 actually entails the ratio of cell/substratum adhesiveness to effective cell contractile force, so on a given substratum migration can be enhanced or inhibited by factors altering cell force generation or the substratum compliance.

## IV. CELL AND TISSUE MECHANICS

The mechanics of a tissue or tissue substrate plays a central role in many situations, ranging from the fabrication of cell-based vascular grafts to wound healing. The transmission of stresses between the extracellular matrix and the cell or between the cell membrane and its nucleus is an important factor in controlling the biological response of the cell to its environment. Even cell motility and adhesion are influenced by the elastic characteristics of the substrate on which they are grown. To appreciate these factors requires an understanding of mechanics, both at the microscale (e.g., a single protein and its interaction with an actin filament) and at the macroscale (e.g., the elastic properties of a cell-based tissue implant). Some of the basic concepts are outlined here, along with some examples of how these concepts are applied in the field of tissue engineering.

### Elasticity

Before dealing with some of the complexities of real biologic materials, consider first the case of a simple uniform and isotropic elastic medium. The elasticity of a material is typically characterized by its stress–strain relationship, where the stress is the force acting per unit area and the strain is the fractional change in length of the specimen, both of which are tensorial quantities. In the case of uniaxial stress, a material might exhibit a stress–strain relation of the

type shown in Fig. 18.4. In this instance, the relationship is linear (or nearly so) for small strains but becomes nonlinear as strains increase above a certain level. The elastic, or Young's, modulus (E) is defined as the ratio of stress (σ) to strain (ε); in this example, E would be constant for small strains, but eventually it would increase as the material experiences increasing strain.

The elastic modulus of most biologic tissues is highly nonlinear, exhibiting an increasing elastic modulus for higher strains. The fibrous elements that comprise the tissue matrix, however, are more likely to be linearly elastic over an appreciable range of strain. Elastin, for example, which can be stretched up to 300% of its initial length, has a nearly constant modulus. Table 18.1 shows the elastic moduli for several tissues and tissue components. Another important property of the material is the maximum amount of strain it can experience before it fractures, some values of which are also given in the table.

In contrast to elastic deformation, in which there exists a unique relation between the applied stress and material strain, in plastic deformation the material experiences an irreversible deformation, usually at high levels of stress, and fails to return to its original length when the stress is removed.

All the foregoing applies to uniaxial loading. If a material is isotropic, then its elasticity is the same, regardless of the direction in which it is tested, and its mechanical properties can be completely characterized by two parameters, e.g., its elastic modulus and Poisson ratio, $v$. Anisotropic materials, however, exhibit different stress–strain characteristics along the different coordinate directions. The wall of an artery, for example, is extremely stiff in the circumferential and axial directions but relatively compliant in the radial direction. The description of such materials can lead to considerable complexity, as evidenced by the fact that in a truly anisotropic material, *21 constants* analogous to the Young's modulus are required to characterize completely a material's elasticity. Intermediate between these two extremes can

| Table 18.1. | Elastic constants for a variety of biologic materials[a] | | | |
|---|---|---|---|---|
| *Material* | *Elastic modulus* | *Yield stress* | *Max. strain* | *Ref.* |
| Cortical bone | 6–30 GPa | 50–200 MPa | — | Cowin *et al.* (1987) |
| Collagen fibers | 500 MPa | 50 MPa | 0.1 | Kato *et al.* (1989) |
| Elastin | 100 kPa | 300 kPa | 3.0 | Mulcherjee *et al.* (1976) |
| Cartilage | 10 MPa | 8–20 MPa | 0.7–1.2 | Woo *et al.* (1987) |
| Skin | 35 MPa | 15 MPa | 1.1 | Yamada (1970) |
| Muscle facia | 340 MPa | 15 MPa | 1.17 | Yamada (1970) |
| Tendon | 700 MPa | 60 MPa | 0.10 | Yamada (1970) |

[a]Note that in many instances, the stress–strain relationship is highly nonlinear; the elastic moduli in those cases represent an approximate, characteristic value.

be found materials that are transversely isotropic, that is, materials that exhibit isotropic behavior in two dimensions but different properties in the third. Many biological materials (blood vessels, ocular sclera) can be well characterized as transversely isotropic because the primary force-supporting structures (collagen and elastin fibers) are oriented primarily in one plane.

The existence of contractile elements, such as smooth muscle cells in vessel walls or the myocardium or actomyosin filaments in the cytoskeleton, lends further complexity to material properties. As far as the elastic properties of the material are concerned, active constriction can be thought of as influencing both the elastic modulus of the material and the length of the sample under zero stress.

## Viscoelasticity and Pseudoelasticity

Although some materials (e.g., elastin, collagen) show an immediate elastic response such that they exhibit an immediate and constant deformation following the application of stress, tissues more often respond in a time-dependent fashion (Fig. 18.5). The initial elastic response of such materials is followed by a period of additional deformation, or creep, asymptotically approaching an equilibrium state at long times. Alternatively, when subjected to a periodic load, these materials display a degree of hysteresis when stress is plotted against strain. This is termed a *viscoelastic* response and is a result of several processes internal to the tissue. One is the transient movement of liquid through the tissue due to the nonuniform distribution of interstitial pressure that results from the application of stress. For example, if a disk of cartilage is subjected to a load as in Fig. 18.5, a gradient in fluid pressure is set up between the central regions and the edge of the specimen, which causes fluid to be expelled. The rate of deformation during this phase is clearly dependent on the permeability of the tissue.

Biological tissues are often referred to as *pseudoelastic* (Fung, 1981), in that, with periodic loading, they evolve to a stress–strain pattern that is repeatable from one cycle to the next and relatively insensitive to the strain rate. The resulting curve typically differs on extension from relaxation,

exhibiting some degree of hysteresis (Fig. 18.6). By treating the loading and unloading maneuvers separately, the viscoelastic characteristics of the material can be incorporated into an otherwise elastic description of the material.

It is often necessary to simulate the true viscoelastic response of a tissue specimen, as, for example, when subjected to an abrupt increase in load (Fig. 18.5). Several simple models have been devised for this purpose. Although these make no attempt to mimic the mechanism responsible for the viscoelasticity, they are useful descriptors that can be used to represent certain viscoelastic characteristics. The most common among these models are the Maxwell, Voigt, and Kelvin models. Each is composed of one or more springs and dashpots, as shown in Fig. 18.7. These simple models are sometimes used as a basis for constructing more complex descriptions of the material. For example, a continuum description has been used to study the mechanical characteristics of leukocytes (Schmid-Schönbein *et al.*, 1981), based on the Kelvin model of Fig. 18.7, by arranging them in various series and parallel networks. More recent models for the cytoskeleton take explicit account of the fibrous microstructure, treating it alternatively by a structure of interconnected beams undergoing deformation by bending (Satcher and Dewey, 1996), by an interconnected structure with some elements under tension and some under compression (Stamenovic *et al.*, 1996), or as a cross-linked network of long filaments exhibiting bending stiffness and experiencing thermal excitation (Storm *et al.*, 2005).

## Measurement of Mechanical Properties

A variety of methods have been employed to ascertain the mechanical characteristics of biological specimens, the method being determined by the elastic characteristics of the material, its size, and the particular property of interest.

For large-scale samples such as tissues and scaffolds used in tissue engineering, the simplest measurement that yields the elastic modulus of a specimen is the *uniaxial strain test*, in which the sample is grasped at two ends and pulled while axial strain and stress are simultaneously

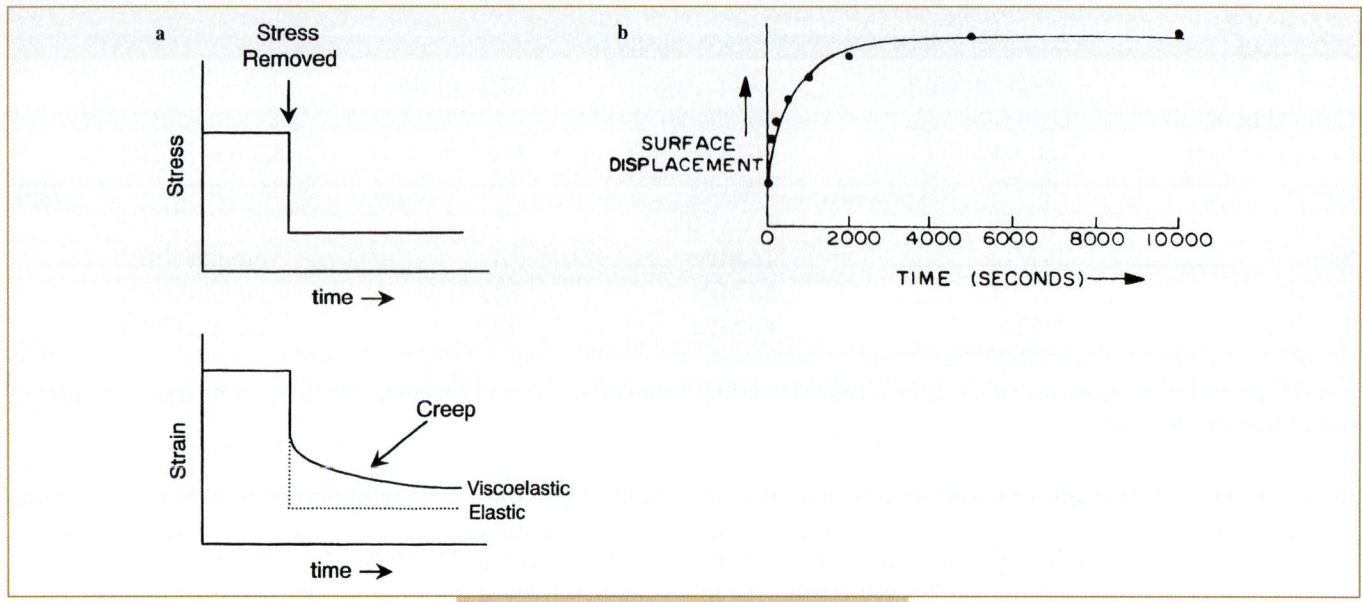

**FIG. 18.5.** **(a)** Stress ($\sigma$; top) and strain ($\varepsilon$; bottom) plotted against time in a maneuver in which the stress is abruptly reduced. In a purely elastic material (dotted line), the strain immediately adjusts to a new level consistent with the new applied stress. A viscoelastic material (solid line) exhibits a further reduction in strain (creep), indicative of continued deformation following an initial elastic response. **(b)** The response of a cylindrical sample of cartilage subjected to a sudden compressive stress. The sample continues to compress with time as a result of water being expelled from the tissue through a porous platen. Reproduced with permission from Mow and Mak (1987).

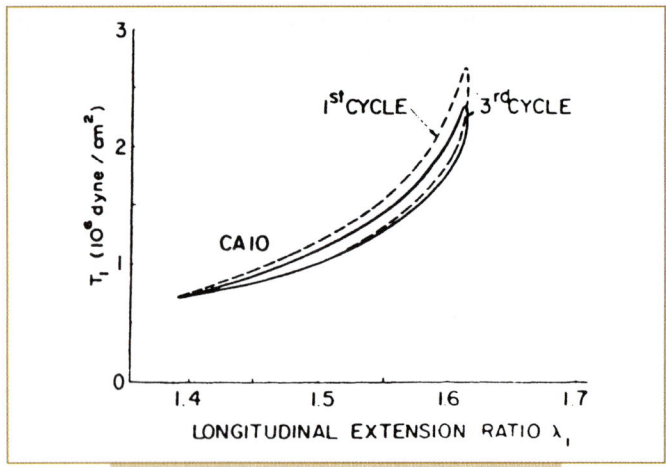

**FIG. 18.6.** Pseudoelastic behavior of biological tissue. A stress–strain plot for a segment of a canine carotid artery undergoing cyclic loading; first and third cycles shown. $T_1$ is the axial tension, $\lambda_1$ is the axial stretch ratio, and $L(t)/L_0$ with $L_0$ is the relaxed length. The sample exhibits different curves for loading and unloading, which are highly reproducible. Reproduced with permission from Lee *et al.* (1967).

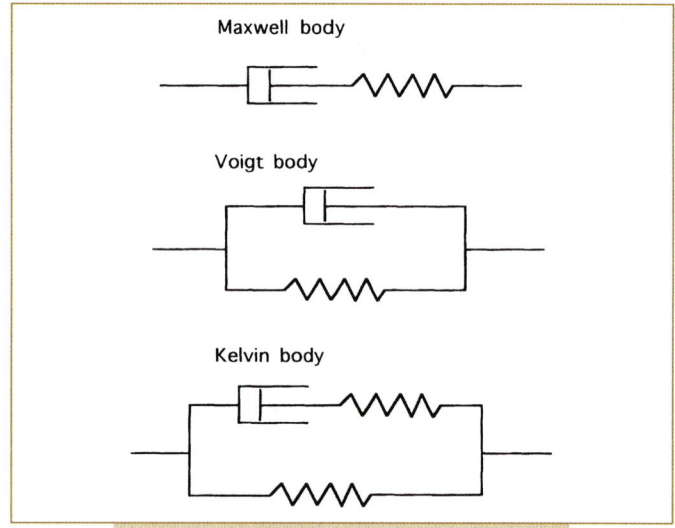

**FIG. 18.7.** Three commonly used models for viscoelastic behavior composed of viscous elements (dashpots) and elastic elements (springs).

measured. In order to minimize end effects, the sample is often necked down to a lateral dimension in the central section, smaller than the ends, and the strain is measured directly in the necked region. Because stresses in both directions perpendicular to the axis of the specimen are zero, the elastic modulus is determined by the ratio of stress to strain,

$E = \sigma_x/\varepsilon_x$. Poisson's ratio can be determined from the change of thickness of the specimen in the direction perpendicular to the applied stress. If the sample is anisotropic, additional uniaxial tests in the other two coordinate directions can be used. Alternatively, stresses can be applied in two (biaxial) or three (triaxial) dimensions simultaneously. These tests yield more information about the material, but they do so at the cost of greater complexity.

These tests are most often used to measure the properties of a material in tension. The compressive characteristics are also often of interest, especially in materials such as bone and cartilage, which are often subjected to compressive loads *in vivo*. For this purpose, specimens are generally cut into the shape of a short cylinder and compressed between two platens. In the case of cartilage, the platens may be permeable to allow water to escape as the sample is compressed, thereby obtaining information on the permeability of the sample from the time-dependent compression following the application of a load (Grodzinsky, 1983). Both confined and unconfined compression tests provide useful information. The advantage of confined compression, in which the sample is placed in a rigid cylindrical chamber, typically with nonpermeable side walls, is that the stress, strain, and flow of water are purely axial and the results can be more easily interpreted.

A third type of test measures the shear modulus, the stiffness of the sample to shear strain. Specimens are placed between two platens, one of which can be rotated. Shear strains with parallel flat platens vary as a function of radius, complicating the situation in nonlinear materials, so the upper platen is sometimes replaced by a cone with very shallow angle, in contact at the center of rotation (a cone-and-plate rheometer). This has the advantage that the entire sample is subjected to the same level of shear strain.

Any of the tests just mentioned are capable of measuring nonlinearities in the stress–strain behavior of a material, although the calculation of nonlinear properties is much easier with those methods that subject the sample to uniform strains. As mentioned earlier, biological materials typically become stiffer as the sample is stretched. Therefore, it is important to test the material with loads in the range of those the specimen is likely to experience *in vivo*. It is equally important to know the zero-stress state of a tissue, which, as has been demonstrated for a variety of vessels, is not necessarily achieved simply by reducing the transmural pressure to zero (Liu and Fung, 1988).

All these tests require excised specimens to be cut or machined into a particular shape. Often in the case of biological tissues, it is desirable to obtain measurements from the *in vivo* state so as to avoid the artifacts that are necessarily introduced by specimen removal. Although advantageous in many respects, *in vivo* testing also limits the choice of testing methods available. Fung and Liu (1995) argued for the importance of making measurements *in vivo*, and they have demonstrated how, in the case of a blood vessel, a battery of tests may be used to measure the mechanical properties of the arterial wall. Other methods — for example, the simple indentation of the material by a probe (Lai-Fook *et al.*, 1976) — can also be used.

Methods for testing the elastic and viscoelastic properties of cells and even individual molecules are rapidly becoming available. One approach for whole-cell measurements, micropipette aspiration, has been used in numerous studies (e.g., Evans and Yeung, 1989; Guilak *et al.*, 2000) on various cell types. In this test, introduced over 20 years ago to study red blood cell mechanics and later for leukocytes, the cell is partially drawn into a micropipette by means of a negative pressure. Knowing the pressures acting and monitoring the extent of cell deformation by microscopy allow for estimation of cell elasticity. Cell poking, a method analogous to the indentation studies just discussed, can also be applied to individual cells (Zahalak *et al.*, 1990; Peeters *et al.*, 2003), as well as compression or extension between two surfaces (Caille *et al.*, 2002; Broers *et al.*, 2004).

In recent years, methods that probe the cell or cellular constituents on a smaller length scale have become more common, in part for their ability to measure subcellular structures, but also because they are less likely to interfere with normal biological function, leaving the cell in a more natural state. These methods might employ 1 to 5 μm diameter microparticles or an atomic force microscopy (AFM) probe. In AFM measurements, the probe is typically used to indent the cell locally, and the force displacement curve is used to infer the local modulus. This method has the capability of providing detailed maps of cell elasticity (Yamane *et al.*, 2000). Microparticle tracking methods can be either active or passive. In passive methods the thermal motion of either one or multiple particles is monitored to determine mean squared displacement over time, from which the shear modulus of the cell can be computed (Mason, 2000). Microparticles used for this purpose can either be membrane tethered or, if sufficiently small (~1 μm or less), internalized by the cell. Specificity for particular structures can be conferred by coating with specific ligands. Active methods use particles of similar size, but in conjunction with a magnetic trap (Glogauer and Ferrier, 1998) or optical tweezers (Block *et al.*, 1989; Kuo and Sheetz, 1993; Kuo, 1995) to actively oscillate or displace the particle and monitor particle displacements with high-speed video (Savin *et al.*, 2005), a quadrant detector (Yamada *et al.*, 2000), or laser interferometry (Pralle *et al.*, 1999). One advantage of all of these methods is their ability to provide measures of the complex shear modulus over decades in frequency (see e.g., Fabry *et al.*, 2001). Methods of analyzing the data assume the structures to be linearly viscoelastic, but that appears to be a reasonably good approximation for typical cell deformations. While any of these methods could, in principle, be applied to cells in a three-dimensional matrix, practical limitations either in introducing and attaching the microparticles or in observing their motion with sufficient resolution have proved difficult to overcome, although some success has recently been achieved (Panorchan *et al.*, 2006). Also of interest in tissue constructs is the level of force exerted by the cell on its surroundings. This can be done by imaging deformations of the matrix using, for example, either second harmonic imaging of collagen (Williams *et al.*, 2005) or displacements of microparticles embedded in the matrix (Dembo and Wang, 1999).

## V. REFERENCES

Block, S. M., Blair, D. F., and Berg, H. C. (1989). Compliance of bacterial flagella measured with optical tweezers. *Nature (Lon.)* **338**, 514–518.

Colton, C. K. (1995). Implantable biohybrid artificial organs. *Cell Transplant* **4**, 415–436.

Cowin, S. C., Van Buskirk, W. C., and Ashman, R. B. (1987). Properties of bone. *In* "Handbook of Bioengineering" (R. Skalak and S. Chien, eds.). McGraw-Hill, New York.

Deen, W. M. (1987). Hindered transport of large macromolecules in liquid-filled pores. *AIChE J.* **33**, 1409–1425.

Dembo, M., and Wang, Y. L. (1999). Stresses at the cell-to-substrate interface during locomotion of fibroblasts. *Biophys. J.* **76**, 2307–2316.

DiMilla, P. A., Barbee, K., and Lauffenburger, D. A. (1991). Mathematical model for the effects of adhesion and mechanics on cell migration speed. *Biophys. J.* **60**, 15–37.

Dowd, C. J., Cooney, C. L., and Nugent, M. A. (1999). Heparan sulfate mediates bFGF transport through basement membrane by diffusion with rapid reversible binding. *J. Biol. Chem.* **274**, 5236–5244.

Evans, E., and Yeung, A. (1989). Apparent viscosity and cortical tension of blood granulocytes determined by micropipette aspiration. *Biophys. J.* **56**, 151–160.

Forsten, K. E., and Lauffenburger, D. A. (1992). Autocrine ligand binding to cell receptors. *Biophys. J.* **61**, 518.

Fung, Y. C. (1981). "Biomechanics: Mechanical Properties of Living Tissues." Springer-Verlag, New York.

Fung, Y. C., and Liu, S. Q. (1995). Determination of the mechanical properties of the different layers of blood vessels *in vivo*. *Proc. Natl. Acad. Sci. USA.* **92**, 2169–2173.

Garcia, A. M., Frank, E. H., Grimshaw, P. E., and Grodzinsky, A. J. (1996). Contribution of fluid convection and electrical migration to molecular transport: relevance to loading. *Arch. Biochem. Biophys.* **333**, 317–325.

Garcia, A. M., Lark, M. W., Trippel, S. B., and Grodzinsky, A. J. (1998). Transport of TIMP-1 through cartilage. *J Orthop. Res.* **16**, 734–742.

Glogauer, M., and Ferrier, J. A. (1998). A new method for application offeree to cells via ferric oxide beads. *Pfluegers Arch.* **435**(2), 320–327.

Grimshaw, P. E., Grodzinsky, A, J., Yarmush, M. L., and Yarmush, D. M. (1989). Dynamic membranes for protein transport: chemical and electrical control. *Chem. Eng. Sci.* **44**, 827–840.

Grodzinsky, A. J. (1983). Electromechanical and physicochemical properties of connective tissues. *CRC Crit. Rev. Biomed. Eng.* **9**, 133–199.

Kato, Y. P., Christiansen, D. L., Hahn, R A., Shieh, J. J., Goldstein, J. D., and Silver, F. H. (1989). Mechanical properties of collagen fibers: a comparison of reconstituted and rat tail tendon fibers. *Biomaterials* **10**, 38–42.

Knauer, D. J., Wiley, H. S., and Cunningham, D. D. (1984). Relationship between epidermal growth factor receptor occupancy and mitogenic response: quantitative analysis using a steady-state model system. *J. Biol. Chem.* **259**, 5623–5631.

Kuo, S. C. (1995). Optical tweezers: a practical guide. *JMSA* **1**(2), 65–74.

Kuo, S. C., and Sheetz, M. P. (1993). Force of single kinesin molecules measured with optical tweezers. *Science* **260**, 232–234.

Lai-Fook, S. J., Wilson, T. A., Hyatt, R. E., and Rodarte, J. R. (1976). Elastic constants of inflated lobes of dog lungs. *J. Appl. Physiol.* **40**(4), 508–513.

Lauffenburger, D. A., and Linderman, J. J. (1996). "Receptors: Models for Binding, Trafficking, and Signaling." Oxford University Press, New York.

Lauffenburger, D. A., Oehrtman, G. T., Walker, L., and Wiley, H. S. (1998). Real-time quantitative measurement of autocrine ligand binding indicates that autocrine loops are spatially localized. *Proc. Natl. Acad. Sci. USA* **95**, 15368.

Lee, J. S., Frasher, W. G., and Fung, Y. C. (1967). "Two-Dimensional Finite-Deformation on Experiments on Dog's Arteries and Veins," Tech. Rep. No. AFOSR 67–1980. University of California at San Diego.

Liu, S. Q., and Fung, Y. C. (1988). Zero-stress states of arteries. *ASME J. Biomech. Eng.* **110**, 82–84.

MacKintosh, F. C., and Janmey, P. A. (1997). Actin gels. *Curr. Opin. Solid State Mater. Sci.* **2**, 350–357.

Maroudas, A. (1979). "Adult Articular Cartilage," pp. 215–290. Pitman Medical, London.

Mow, V. C., and Mak, A. F. (1987). Lubrication of diarthroidal joints. *In* "Handbook of Bioengineering" (R. Skalak and S. Chien, eds.), Chapter 5. McGraw-Hill, New York.

Mukherjee, D. P., Kagan, H. M., Jordan, R. E., and Franzblau, C. (1976). Effect of hydrophobic elastin ligands on the stress–strain properties of elastin fibers. *Connect. Tissue Res.* **4**, 177.

O'Hare, B. P., Urban, J. P. G., and Maroudas, A. (1990). Influence of cyclic loading on the nutrition of articular cartilage. *Ann. Rheum. Dis.* **49**, 536–539.

Palacek, S. P., Loftus, J. C., Ginsberg, M. H., Lauffenburger, D. A., and Horwitz, A. F. (1997). Integrin/ligand-binding properties govern cell migration speed through cell/substratum adhesiveness. *Nature (Lond.)* **385**, 537.

Perez, E. P., Merrill, E. W., Miller, D., and Griffith Cima, L. (1995). Corneal epithelial wound healing on bilayer composite hydrogels. *Tissue Eng.* **1**, 263–277.

Reddy, C. C., Niyogi, S. K., Wells, A., Wiley, H. S., and Lauffenburger, D. A. (1996a). Engineering epidermal growth factor for enhanced mitogenic potency. *Nat. Biotechnol.* **14**, 1696.

Reddy, C. C., Wells, A., and Lauffenburger, D. A. (1996b). Receptor-mediated effects on ligand availability influence relative mitogenic potencies of epidermal growth factor and transforming growth factor alpha. *J. Cell. Physiol.* **166**, 512.

Satcher, R. L. Jr., and Dewey, C. F. Jr. (1996). Theoretical estimates of mechanical properties of the endothelial cell cytoskeleton. *Biophys. J.* **71**, 109–118.

Schmid-Schönbein, G. W., Sung, K. L. P., Tözeren, H., Skalak, R., and Chien, S. (1981). Passive mechanical properties of human leukocytes. *Biophys. J.* **36**, 243–256.

Stamenovic, D., Fredberg, J. J., Wang, N., Butler, J. P., and Ingber, D. E. (1996). A microstructural approach to cytoskeletal mechanics based on tensegrity. *J. Theor. Biol.* **181**, 125–136.

Tranquillo, R. T., Zigmond, S. H., and Lauffenburger, D. A. (1988). Measurement of the chemotaxis coefficient for human neutrophils in the under-agarose migration assay. *Cell Motil. Cytoskel.* **11**, 1–15.

Wang, N., and Ingber, D. (1994). Control of cytoskeletal mechanics by extracellular matrix, cell shape, and mechanical tension. *Biophys. J.* **66**, 2181–2189.

Wiley, H. S. (1985). Receptors as models for the mechanisms of membrane protein turnover and dynamics. *Curr. Top. Membr. Transp.* **24**, 369–412.

Wiley, H. S. (1992). Receptors: topology, dynamics, and regulation. *Fundam. Med. Cell Biol.* **5A**, 113–142.

Williams, R. M., Zipfel, W. R., and Webb, W. W. (2005). Interpreting second-harmonic generation images of collagen I fibrils. *Biophys. J.* **88**, 1377–1386.

Woo, S. L.-Y., Mow, V. C., and Lai, W. M. (1987). Biomechanical properties of articular cartilage. *In* "Handbook of Bioengineering" (R. Skalak and S. Chien, eds.). McGraw-Hill, New York.

Wu, P., Hoying, J. B., Williams, S. K., Kozikowski, B. A., and Lauffenburger, D. A. (1994). Integrin-binding peptide in solution inhibits or enhances endothelial cell migration, predictably from cell adhesion. *Ann. Biomed. Eng.* **22**, 144.

Yamada, H. (1970). *In* "Strength of Biological Materials" (F. G. Evans, ed.). Williams & Wilkins, Baltimore, MD.

Zahalak, G. I., McConnaughey, W. B., and Elson, E. L. (1990). Determination of cellular mechanical properties by cell poking, with an application to leukocytes. *ASME J. Biomech. Eng.* **112**, 283–294.

Part Five

# Biomaterials in Tissue Engineering

# Chapter Nineteen

# Micro-Scale Patterning of Cells and Their Environment

*Xingyu Jiang, Shuichi Takayama, Robert G. Chapman,*
*Ravi S. Kane, and George M. Whitesides*

## I. INTRODUCTION

Control of the cellular environment is crucial for understanding the behavior of cells and for engineering cellular function (Jiang and Whitesides, 2003; Whitesides *et al.*, 2001; Xia and Whitesides, 1998). This chapter describes the use of a set of tools in microfabrication, called *soft lithography*, for patterning the substrate to which cells attach, the location and shape of the areas to which cells are confined, and the fluid environment surrounding the cells, all with micrometer precision. We summarize examples where these tools have helped to control the microenvironment of cells and have been useful in solving problems in fundamental cell biology. The methods described here are experimentally simple, inexpensive, and well suited for patterning biological materials.

How do tissues assemble *in vivo*? How do cells interact with each other in tissues? How do cells respond to stimuli? How do abnormal stimuli give rise to pathological conditions? Answering these fundamental biological questions and using the information thus obtained for medical applications requires understanding the behavior of cells in well-controlled microenvironments. Many of the challenges in

trying to control the environment experienced by individual cells lie in the relevant scales of size as well as the character of the stimuli. These scales of size range from angstroms (for molecular detail), through micrometers (for an individual cell), to millimeters and centimeters (for groups of cells); the types of stimuli that must be addressed include the molecular composition of the liquid in which the cell is immersed, the topographical and chemical composition of the surface to which the cells attach, the nature of neighboring cells, and the temperature.

Microfabrication and micropatterning using stamps or molds fabricated from elastomeric polymers (*soft lithography*) provide versatile methods for generating patterns of proteins and ligands on surfaces, microscale chambers for culturing cells, and laminar flows of media in capillaries, all in the size range of 0.1–100 micrometers (Gates *et al.*, 2005; Jiang and Whitesides, 2003; Whitesides *et al.*, 2001; Xia and Whitesides, 1998). Soft lithographic methods are relatively simple and inexpensive. The elastomeric polymer most often used in these procedures — polydimethylsiloxane (PDMS) — has several characteristics (optical transparency, ease of manipulation, and low cost) that make it attractive for biological applications (McDonald and Whitesides, 2002). Although a new technology when compared to molecular biology, soft lithography is being increasingly used in cell biology, due to its biocompatibility, simplicity, and adaptability to biological and biochemical problems. This chapter gives an overview of the application of soft lithography to the patterning of cells and their fluidic environment, using micro-scale features and laminar flows.

Researchers have used a number of techniques to pattern cells and their environment (Letourneau, 1975). Before the 1990s, the most common was photolithography. This technique has been highly developed for the micro-electronics industry; it has also been adapted, with varying degrees of success, for biological studies (Kleinfeld et al., 1988; Letourneau, 1975; Ravenscroft et al., 1998). Examples have included topographical features that confine the growth of snail neurons to silicon chips, as first demonstrations of interfacing natural computation with artificial ones (Merz and Fromherz, 2005; Zeck and Fromherz, 2001). As useful and powerful as photolithography is (it is capable of mass production at 70-nm resolution of multilevel, registered structures), it is not always the technique best suited for biological studies. It is an expensive and time-consuming technology; it is poorly suited for patterning nonplanar surfaces; it provides too little control over surface chemistry to pattern sufficiently diverse types of biomolecules on surfaces; it is poorly suited for patterning materials such as hydrogels; the equipment required to use it is rarely routinely accessible to biologists; and it is directly applicable to patterning only a limited set of photosensitive materials (e.g., photoresists).

## II. SOFT LITHOGRAPHY

Soft lithography solves many of the problems that required the application of microfabrication to biological problems (Chen et al., 2005; Jiang and Whitesides, 2003; Whitesides et al., 2001). Soft lithographic techniques are inexpensive, are relatively procedurally simple, are applicable to the complex and delicate molecules often dealt with in biochemistry and biology, can be used to pattern a variety of different materials, are applicable to both planar and curved substrates (Jackman et al., 1995), and do not require stringent control (such as a clean-room environment) over the environment in which they are fabricated beyond that required for routine experiments with cultured cells (Whitesides et al., 2001; Xia and Whitesides, 1998). Access to photolithographic technology is required only to create a master for casting the elastomeric stamps or membranes; even then, the requirement for chrome masks — the preparation of which is one of the slowest and most expensive steps in conventional photolithography — can often be bypassed in favor of high-resolution printing (Deng et al., 2000; Linder et al., 2003). Soft lithography offers special advantages for biological applications, in that the elastomer most often used (PDMS) is compatible with most types of

optical microscopy commonly used in cell biology, is permeable to gases such as $O_2$ and $CO_2$, is mechanically flexible, seals conformally to a variety of surfaces (including most types of petri dishes), is generally biocompatible (Lee et al., 2004), and can be implantable in vivo. The soft lithographic techniques that we discuss include microcontact printing, micromolding, patterning with microfluidic channels, and laminar flow patterning.

## III. SELF-ASSEMBLED MONOLAYERS (SAMs)

### Introduction to SAMs

Since many of the studies involving the patterning of proteins and cells using soft lithography have been carried out on self-assembled monolayers (SAMs) of alkanethiolates on gold, we give a brief introduction to SAMs (Allara and Nuzzo, 1985; Bain and Whitesides, 1988; Jiang et al., 2004b; Prime and Whitesides, 1991; Ulman, 1996). SAMs are organized organic monolayer films normally formed by exposing a surface of a gold film to a solution containing an alkanethiol (RSH). SAMs allow control at the molecular level by chemical synthesis of derivatized alkanethiol(s); this molecular control, in turn, gives control over the properties of the interface. The properties of surfaces covered with SAMs are often largely or entirely determined by the nature of the terminal groups of these alkanethiols. The ease of formation of SAMs and their ability to present a range of chemical functionality at their interface with aqueous solution make them particularly useful as model surfaces in studies involving biological components. Furthermore, SAMs can be easily patterned by simple methods such as microcontact printing (μCP) with features down to 100 nanometers in size (Love et al., 2005; Xia and Whitesides, 1998). These features of SAMs make them structurally the best-defined substrates for use in patterning proteins and cells. SAMs on gold are used for many experiments requiring the patterning of proteins and cells, because they are biocompatible, easily handled, and chemically stable. SAMs on silver, although better defined structurally than those on gold, cannot be used in most experiments with cultured cells, due to the toxicity of silver (Ostuni et al., 1999). SAMs on palladium and platinum are just starting to be explored (Jiang et al., 2004a; Petrovykh et al., 2006).

The substrates for SAMs are easy to prepare; once formed, SAMs are stable for weeks under conditions typical for culturing cells. Gold substrates are prepared on glass coverslips or silicon wafers by evaporating a thin layer of titanium or chromium (1–5 nm) to promote the adhesion of gold to the support, followed by a thin layer of gold (10–200 nm) (Lopez et al., 1993). SAMs formed on these gold substrates are stable to the conditions used for cell culture, but care should be taken to avoid strong light and temperatures above ~70°C since both can result in degradation of the SAM (J. Huang and Hemminger, 1993; Love et al., 2005).

## Preventing Protein Adsorption: "Inert Surfaces"

Proteins play an integral part in the adhesion of cells with surfaces: Cells require adsorbed proteins (or peptides that mimic parts of a protein) to adhere to the surface (Ruoslahti and Pierschbacher, 1987). Control of the interaction of proteins with a surface, therefore, enables the control of the interactions of cells with that surface. Most solid surfaces — especially hydrophobic surfaces — adsorb proteins. Thus, the main challenge in controlling the interactions of proteins and cells with surfaces lies in finding surfaces that *resist* nonspecific adsorption of proteins (surfaces that we call *inert*, for brevity). Inert surfaces provide the background necessary for spatially restricting protein adsorption or for preparing surfaces that bind only specific proteins and are used in patterning proteins and cells, as biomaterials (Andrade *et al.*, 1996; Han *et al.*, 1998), and in the construction of biosensors (Mrksich and Whitesides, 1995).

SAMs terminated in oligo(ethylene glycol) (EG$_n$, $n > 2$) resist the adsorption from solution of all known proteins and their mixtures (Prime and Whitesides, 1991, 1993; Whitesides *et al.*, 2001). We know that EG$_n$ groups are not unique in their ability to make inert SAMs. For example, several polar functional groups that do not contain H-bond donors often make good components of inert surfaces (Chapman *et al.*, 2000; Kane *et al.*, 2003). The combination of inert and adsorptive surfaces with soft lithographic techniques enables the facile patterning of proteins and cells. Patterns of hydrophobic regions (for example, SAMs terminated in methyl groups) and regions that are "inert" provide the basis for most work using patterned cells.

A number of other substances also make the surface more or less inert. Many of them are used in connection with soft lithography, for example, bovine serum albumin (BSA) and related proteins (Sweryda-Krawiec *et al.*, 2004), man-made polymeric materials (e.g., polyethyleneglycol, or PEG) (Jeon *et al.*, 1991), and dextran (Frazier *et al.*, 2000).

# IV. MICROCONTACT PRINTING AND MICROFEATURES USED IN CELL BIOLOGY

## Patterning Ligands, Proteins, and Cells Using Microcontact Printing on SAMs

Microcontact printing (µCP) is a technique that uses topographic patterns on the surface of an elastomeric PDMS stamp to form patterns on the surfaces of various substrates (Fig. 19.1) (Xia and Whitesides, 1998). The stamp is first "inked" with a solution containing the patterning component, the solvent is allowed to evaporate under a stream of air, and the stamp is brought into conformal contact with the surface of the gold film for intervals ranging from a few seconds to minutes. The thiol transfers to the gold film in the regions of contacts. Other components used as ink for µCP include activated silanes that react with the SiOH groups (RSiCl$_3$ or RSi(OCH$_3$)$_3$) present on the surface of silicon (with a native film of SiO$_2$) and various ligands (such

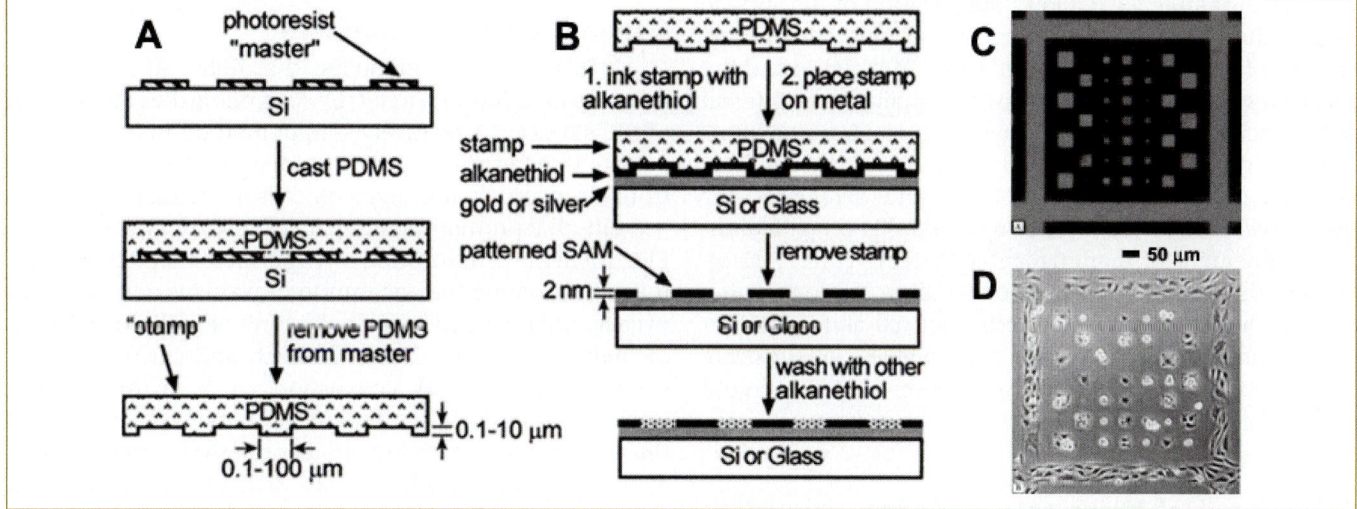

**FIG. 19.1.** Molding from a master, microcontact printing (µCP), and patterning of proteins and cells. **(A)** A method for generating stamps (also applicable for channels and other molds) of PDMS for µCP. A PDMS stamp is prepared by pouring PDMS liquid prepolymer on a "master" (generally generated by photolithography), followed by curing the PDMS and removing it (as an elastomeric solid) from the master. **(B)** A typical procedure used for µCP. A solution containing the patterning component of interest (ink) is applied to this stamp and the solution allowed to dry. This inked stamp is placed on a substrate to allow the ink to transfer to the substrate. A substrate patterned with SAMs remains after removal of the stamp. **(C)** Selective adsorption of fibronectin onto a surface patterned by SAMs into areas that either promote or resist the adsorption of proteins, by µCP, visualized by immunostaining. **(D)** Patterned attachment and spreading of cells on the protein-patterned substrate in C.

as amine-containing compounds) that react with activated SAMs (usually resulting in the formation of a peptide bond that tethers ligands with surfaces) (Lahiri *et al.*, 1999; Yan *et al.*, 1998).

The most general and reliable method for patterning proteins is accomplished by preparing areas of SAMs that promote the adsorption of proteins, surrounded by regions that resist adsorption of proteins (regions that we call *inert background*), and allowing proteins to adsorb onto the adsorbing regions from solutions. For example, we used microcontact printing to pattern gold surfaces into regions terminated in methyl groups and then filled the rest of the surface of gold with an oligo(ethylene glycol)-terminated thiol to form inert regions (Lopez *et al.*, 1993). Immersion of the patterned SAMs in solutions of proteins such as fibronectin, fibrinogen, pyruvate kinase, streptavidin, and immunoglobulins resulted in adsorption of the proteins exclusively on the methyl-terminated regions (Lopez *et al.*, 1993). Characterization of patterns of adsorbed proteins with electron and optical microscopy shows that the layers of adsorbed protein appeared to be homogeneous. Alternatively, proteins can be anchored to ligands patterned onto surfaces by μCP; for example, μCP of biotin onto activated SAMs allows the biospecific immobilization of avidin on the surface (Lahiri *et al.*, 1999).

The ability of μCP to create patterns of ligands and proteins allows the patterning of many anchorage-dependent cells (most normal cells in multicellular organisms are anchorage dependent) (Alberts *et al.*, 2002); this patterning confines them to specific regions of a substrate and allows the precise control of the size and shape of the cells (Fig. 19.1). For example, μCP allows the partition of the surface of gold into regions presenting $EG_n$ groups and methyl groups (Mrksich and Whitesides, 1996). After we coated the substrates with fibronectin, bovine capillary endothelial cells attached only to the methyl-terminated, fibronectin-coated regions of the patterned SAMs. The cells remained attached in the patterns defined by the underlying SAMs for five to seven days. We have also used SAMs on palladium for confinement of mammalian cells (Jiang *et al.*, 2004a). EG-terminated SAMs on palladium allow the patterning of individual cells, groups of cells, as well as focal adhesions (subcellular complexes that enable cell-substrate attachment, FA) for over four weeks; similar SAMs on gold confined cells to patterns for one to two weeks.

## Other Types of Microcontact Printing

It is possible to pattern certain proteins (ones that can withstand drying onto the surface of the stamp and stamping) directly onto surfaces (Bernard *et al.*, 1998, 2000, 2001; Mayer *et al.*, 2004; St. John *et al.*, 1998). Direct patterning of proteins, however, is typically applicable only to structurally stable proteins and is usually more demanding experimentally than patterning via SAMs (Kam and Boxer, 2001). The surface of the PDMS stamp used in this type of procedure must be rendered hydrophilic by exposure to a plasma before use (Bernard *et al.*, 2000).

Patterning cells directly with μCP was thought to be unfeasible, because most cells are too delicate to be dried or stamped. Recently, however, we have demonstrated the stamping of proteins and cells directly with a soft hydrogel stamp (agarose) that contains large amounts of water (Mayer *et al.*, 2004; Stevens *et al.*, 2005). The resolution of this technique (tens of micrometers) is not comparable to μCP with PDMS and SAMs, but it makes patterning cells at this large size range easier than patterning with SAMs.

Other workers have used μCP for different types of cells on other types of surfaces. Craighead and coworkers patterned polylysine on surfaces of electrodes to confine the growth of neurons (James *et al.*, 2000). Zhang *et al.* have synthesized oligopeptides containing a cell-adhesion motif at the N-terminus connected by an oligo (alanine) linker to a cysteine residue at the C-terminus (Zhang *et al.*, 1999). The thiol group of cysteine allowed the oligopeptides to form monolayers on gold-coated surfaces. They used a combination of microcontact printing and these self-assembling oligopeptide monolayers to pattern gold surfaces into regions presenting cell-adhesion motifs and oligo(ethylene glycol) groups that resist protein adsorption. Wheeler *et al.* created patterns of covalently bound ligands and proteins on glass coverslips and used these patterns to control nerve cell growth (Branch *et al.*, 1998). In addition, polymers of EG and supported phospholipids have been used on a series of different substrates for pattering cells (Amirpour *et al.*, 2001; Kam and Boxer, 2001; Michel *et al.*, 2002; Tourovskaia *et al.*, 2003).

## Dynamic Control of Surfaces

It is possible to modulate the ability of surfaces to promote the adhesion of cells by controlling the composition of the surfaces. A relatively simple method to achieve this control is to desorb EG-terminated SAMs electrochemically from a substrate patterned with cells in a buffer containing proteins that promote attachment of cells (Jiang *et al.*, 2003). Electrochemical desorption converts inert areas into regions that can promote the adsorption of proteins and adhesion of cells, and thus it allows initially confined cells to move out of their patterns (Fig. 19.2). Mrksich and coworkers have used electrochemical conversion of a hydroquinone-terminated SAM into a quinone-terminated SAM to allow the attachment of a cyclopentadiene-tether peptide (which allows the immobilization of cells via the binding of integrin receptors on cell surface) to "turn on" an otherwise-inert SAM for adhesion of cells (Yousaf *et al.*, 2001a). Mrksich has used this technique for the patterning of multiple types of cells on surfaces (Yousaf *et al.*, 2001b). A newer technique from the same group involves first desorption of an immobilized ligand (patterned in certain areas on the surface) and detachment of cells adhered to the surface via this ligand, on application of an electrical reduction, and a subsequent electrical oxidation of the substrate for immobilizing another

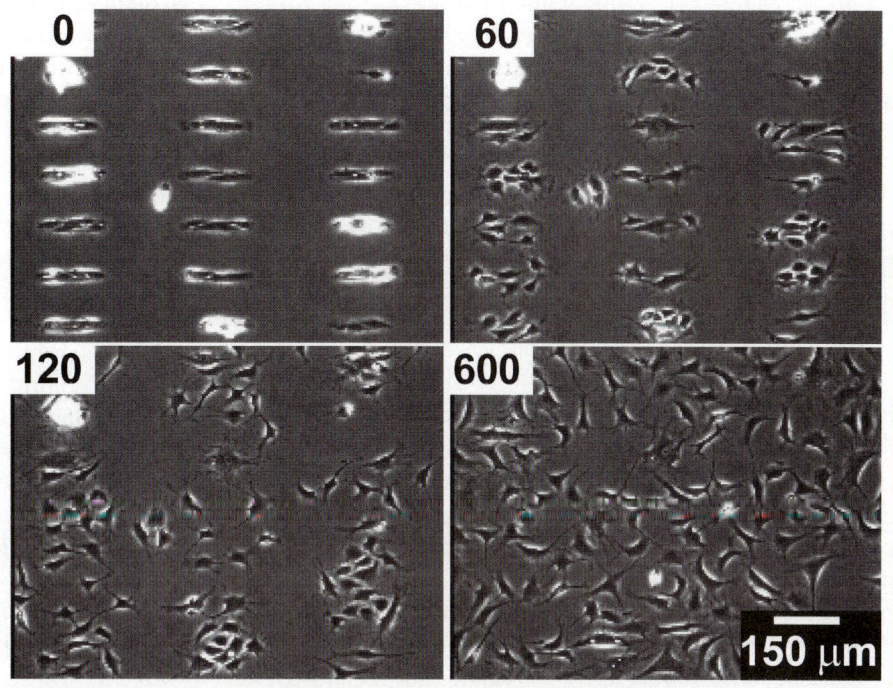

**FIG. 19.2.** Using electrochemical desorption of EG-terminated SAMs to release confined BCE cells and allow their free movement. Initially, bovine capillary endothelial cells are confined to a patterned array of hydrophobic regions (SAMs terminated in methyl groups) and surrounded by regions that are "inert" (SAMs terminated in EG-terminated groups). Application of a cathodic voltage desorbs the EG-terminated SAMs to allow cells initially confined to move freely on the surface. The numbers indicate time after application of electrochemical treatment (in minutes).

ligand that allows the migration of adherent cells (that were initially confined to certain areas on the surface) (Yeo *et al.*, 2003). Some of the methods for preparing the appropriate thiols employed by Mrksich *et al.* may be too complicated for routine use in a regular cell biology laboratory, but the demonstration of such sophisticated control of the interactions between the cell and solid substrates is unprecedented. Another set of electrochemical methods can be used to activate BSA-rendered inert surfaces for cell adhesion (Kaji *et al.*, 2004, 2005).

There is a set of photochemical methods that allow the tuning of the inertness of SAMs. Mrksich and coworkers have devised a SAM that is initially inert but has in it a nitro-veratryloxycarbonyl-protected hydroquinone, which can be oxidized photochemically to generate a benzoquinone group, which, in turn, allows the attachment of a ligand for immobilization of cells on surfaces (Dillmore *et al.*, 2004). Another method for photoactivation of the surface utilized a photochemical process that desorbed a ligand from SAMs on surfaces. BSA, initially physically adsorbed on the surface via a photocleavable 2-nitrobenzyl group–terminated SAM, made this surface inert; photochemistry-mediated desorption of the 2-nitrobenzyl group, and therefore BSA, allowed cells to adhere to the surface (Nakanishi *et al.*, 2004). Both techniques allow the use of the mercury lamp on a fluorescent microscope of the kind typically used for experiments with cultured cells to carry out the required electrochemistry.

Takayama and his coworkers fabricated substrates that allow the dynamic control of focal adhesions (FAs). They first generated a slab of PDMS with a brittle surface by means of oxygen plasma treatment and then made the surface inert by means of physical adsorption of a polymer containing moieties of PEG (Zhu *et al.*, 2005). By stretching the slab, they created cracks on the surface, which are not covered with the polymer-containing PEG; these cracks could thus promote the formation of FAs and adhesion of cells. Releasing the stress on the slab PDMS closed these cracks and again prevented the adhesion of cells. This stretch-and-release process could be recycled multiple times.

A number of other techniques also allow the patterning and dynamic patterning of proteins and cells in time and space. These techniques are related to soft lithography in one way or another (Co *et al.*, 2005; Kumar *et al.*, 2005; Ryan *et al.*, 2004).

## Patterning with Microtopographies

Microtopographies include membranes with microholes, microwells, microneedles, and grooves or steps on surfaces. It is possible to confine cells to micropatterns using either elastic membranes that carry holes or microwells (Folch *et al.*, 2000; Ostuni *et al.*, 2000, 2001). Some of these techniques also allowed the initial confinement, then release, of groups of cells (Folch *et al.*, 2000; Ostuni *et al.*, 2000).

Chen and his coworkers fabricated stamps with multiple levels that allowed the patterning of several different types of proteins and cells at once (Tien *et al.*, 2002). Chen *et al.* also used bowtie-shaped microwells of agarose gel both to confine individual cells to particular shapes and to allow cells to be close to each other without mutual contact (Nelson and Chen, 2002). Positioning cells next to each other while preventing their direct contact is difficult to achieve with μCP alone. They also fabricated arrays of micropillars (in sizes much smaller than a single cell) to probe forces that cells apply to the substrate as they adhere to and migrate on solid surfaces (Tan *et al.*, 2003). Tien and coworkers succeeded in molding microstructures in hydrogels of resolution larger than 5 micrometers and used these structures to generate arrays of cells in three dimensions (Tang *et al.*, 2003).

We also studied the issue of topographical contact guidance — how cells interact with chemically homogeneous surfaces that have topographical features. These studies provide simple methods for further studies of this interesting and complex type of interactions (Jiang *et al.*, 2002; Lam *et al.*, 2006; Takayama *et al.*, 2001b).

## Fundamental Studies in Cell Biology Using Patterned Substrates

The ability to pattern proteins, groups of cells, single cells, and their FAs has led to new studies on the effect of patterned surface environments and cell shape on cell behavior.

Our first attempt in these studies was to prepare (via µCP) substrates consisting of square and rectangular islands of laminin surrounded by inert regions and to study the behavior of rat hepatocytes on them (Singhvi *et al.*, 1994). The cells conformed to the shape of the laminin patterns; the patterning allowed the control of cell shape independent of the density of ligands in the extracellular matrix (ECM). We observed that cell size, regardless of ECM ligand density, was the major determinant of cell growth and differentiation. We then used µCP to prepare substrates that presented circular cell-adhesive islands of various diameters and interisland spacings (Chen *et al.*, 1997; Dike *et al.*, 1999). Such patterns allowed the control of the extent of cell spreading without varying the total cell-matrix contact area. We found that the extent of spreading (the projected surface area of the cell), rather than the area of the adhesive contact, controlled whether the cell divided, remained in stationary phase, or entered apoptosis.

Dike *et al.* (1999) used µCP to prepare substrates with cell-adhesive lines of varying widths. They found that bovine capillary endothelial cells (BCE) cultured on 10-µm-wide lines underwent differentiation to capillary tube–like structures containing a central lumen. Cells cultured on wider (30 µm) lines formed cell–cell contacts, but these cells continued to proliferate and did not form tubes.

Recent progress in understanding how cell adhesion regulates cell physiology has used methods related to micropatterning. In several model types of cells, the strength of cell–substrate adhesion increased as the allowed area of cell adhesion increased, for small areas (typically less than 300 µm²); but the strength of adhesion remained constant for larger areas (Gallant *et al.*, 2005; Tan *et al.*, 2003). These results related cell spreading and the strength of cell attachment empirically. Within the focal adhesion (FA), integrin receptors need to aggregate in order to activate the appropriate biochemical pathways for adhesion of cells

(Assoian and Zhu, 1997; Hotchin and Hall, 1995; Schwartz *et al.*, 1991). It has not been clear, however, at what maximum separation integrin receptors can still perform their normal functions. Bastmeyer has used different combinations of micropatterns to determine that cells could adhere to surfaces with arrays of circles with area of 0.1 µm², when the spacing between these circles is less than 5 µm; but when the separation between these circles is larger than 30 µm (and when the circles are larger than 1 µm²), cells fail to adhere and spread on these surfaces (Lehnert *et al.*, 2004). This result gives a semiquantitative description of the geometrical requirements for clustered integrins.

Using a combination of self-assembly of nanoparticles and micropatterns, Spatz and colleagues definitively determined the maximum distances (73 nm) between individual integrin receptors in order for normal cell adhesion to occur (Arnold *et al.*, 2004). Another recent report shows that when FAs mature into larger-than-normal sizes, they appear to exert four times as much stress on the surface than normal FAs; this result might be important for myogenesis, the process of the formation of muscle fibers (Goffin *et al.*, 2006).

Micro-scale features enable the studies of the movement of mammalian cells. The extension of lamellipodia is an important process in the movement of cells. Bailly *et al.* (1998) used micropatterned substrates to study the regulation of lamellipodia during chemotactic responses of mammalian carcinoma cells to growth factors. On stimulation with epidermal growth factor, the cells extended their lamellipodia laterally out of their patterns of confinement, over the inert part of the substrate. This result showed that the extension of lamellipodia could occur independent of any contact with the substrate. Contact formation was, however, necessary for stabilizing the protrusion. We further observed that when endothelial cells were confined to patterns with corners (such as triangles and squares), their lamellipodia tended to spread most actively from the corners of these shapes (Brock *et al.*, 2003; Parker *et al.*, 2002).

Most moving mammalian cells adopt asymmetric shapes. We have examined whether the asymmetry in the shape of a cell is connected to the direction of its movement (Jiang *et al.*, 2005a) (Fig. 19.3A). Since moving cells often appear to have a teardrop shape, we confined cells to teardrop-shaped patterns and then used electrochemistry to allow free cell movement. It is tempting to assume that the released cells would move toward the sharp end of the teardrop pattern, considering that lamellipodia tend to extend from sharp corners; but a teardrop-shaped cell resembles a naturally moving cell (with the *blunt* end being the front) (Fig. 19.3). The conflicting observations left unclear the direction in which a teardrop-shaped cell would move once released. We released teardrop-shaped cells electrochemically and determined that these preshaped cells appear to prefer moving toward their blunt ends.

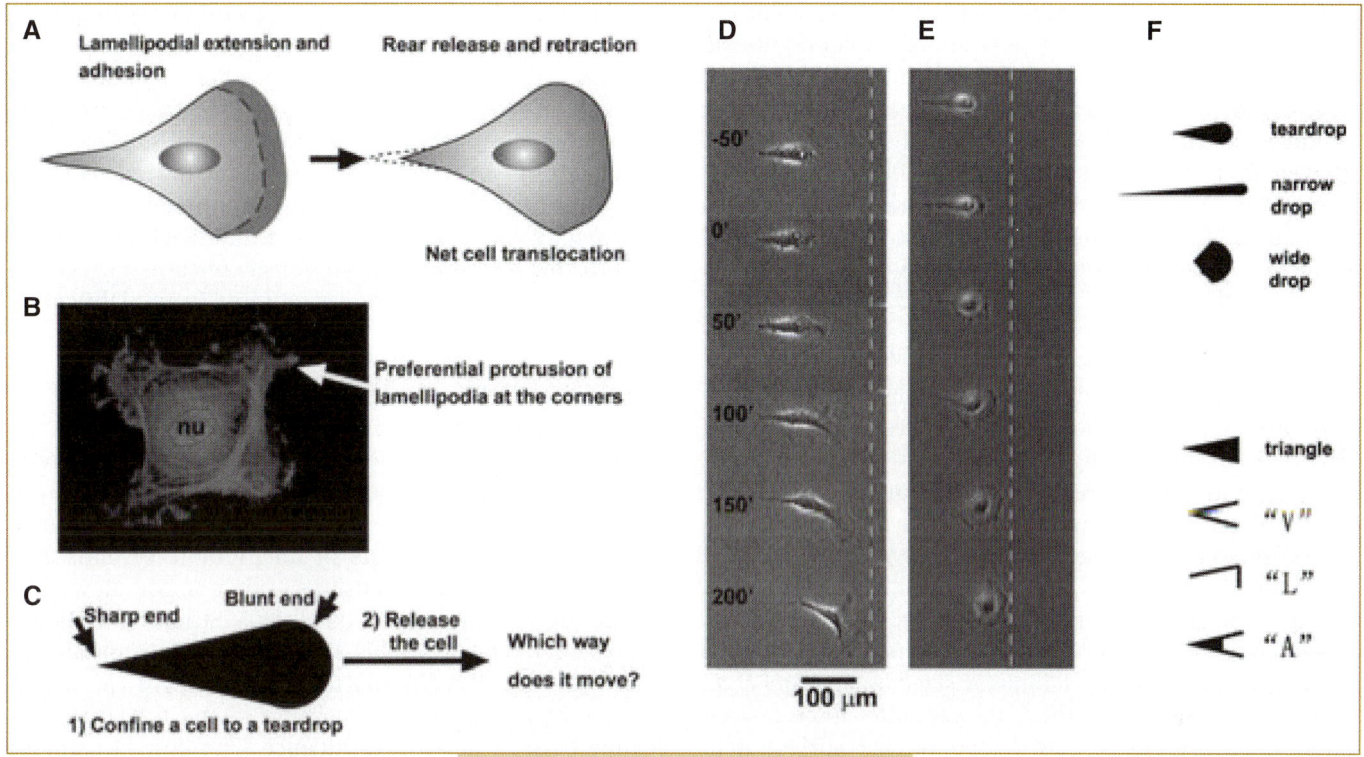

**FIG. 19.3.** Understanding the relationship between cell shape and the direction of cell migration. **(A)** A typical migrating cell has an asymmetric, teardrop-like shape and moves toward the blunt end. **(B)** When confined to shapes that have sharp corners, lamellipodia tend to extend out of the sharp corners. **(C)** A cell initially confined to a teardrop shape and then subsequently released may choose which way it moves: toward its blunt or sharp end. **(D) (E)** Individual 3T3 cells and COS cells, respectively, were initially confined to teardrop shapes and then released to move freely across the surface. The movement was predominantly toward the blunt end. Numbers indicate time after the application of electrochemical potential (minutes). **(F)** Different initial patterns used to confine cells.

To understand the issue in detail, we varied the initial shapes. It appeared that "narrow drops" (regardless of details of the shape) had a similar capability to direct cell migration, while "wide drops" failed to do so. Triangular patterns (having a similar aspect ratio to the teardrop) also direct cell migration, to the same extent as the teardrop, further confirming that the asymmetry of the initial shape alone could direct cell movement. In contrast, symmetric patterns such as rectangles, squares, and circles failed to direct cell migration. To determine whether it was the shape of the spread cell or the uneven distribution of FAs underneath the cell that was responsible for directed cell motion, we started cells on "L"-, "V"-, and "A"-shaped patterns. Because individual cells could span these patterns, the immobilized cells resembled each other in overall geometry, and their geometry was similar to that of cells confined to triangular patterns. But these patterns allowed different distributions of FAs. Triangles and "L"-shaped cells allowed more FAs in the blunt end, "V"-shaped patterns allowed the same amount of the FAs in the blunt end as in the sharp end, while "A"-shaped patterns allowed more FAs in the sharp end. All these patterns appeared to direct cell migration to the same extent, thus confirming that the overall shape of cells was the determining factor in directing cell motion.

Micron-scale tools based on soft lithography have also been used in studies of cell division. Even though most mammalian cells round up and almost completely detach from the substrate when they divide, Bornens' group has used µCP to show that the shape of the ECM to which cells initially attach determines the direction of cell division (Thery *et al.*, 2005).

Micropatterns could also bias the differentiation of human mesenchymal stem cells: When allowed to adhere and spread, the stem cells became osteoblasts; when spreading was prohibited by small patterns of confinement, stem cells became adipocytes (McBeath *et al.*, 2004).

These tools offer opportunities to study not just single cells, but groups of cells. For example, Chen and coworkers devised experiments to control the size and the contact between a pair of two cells, thus definitively proving that cell–cell contact, not soluble factors alone, enable cell contact–mediated proliferation of cells in culture (Nelson and Chen, 2002). Toner and colleagues showed, by patterning of hepatocytes and nonparenchymal cells with precise geometrical parameters, that the interface between the two types of cells is critical for the function of hepatocytes (Bhatia *et al.*, 1999). Ingber and his coworkers discovered, using patterned groups of two or more endothelial cells, that spontaneous ordering arises and patterns that resemble the Chinese Yin-Yang ideograph would emerge while endothelial cells migrate on the patterns (Brangwynne *et al.*, 2000; S. Huang *et al.*, 2005). Studying groups of tens to hundreds of cells patterned into defined geometries, Chen and coworkers realized that the shapes of sheets of cells influence the mechanical forces each cell within experiences, and these forces affect the physiology of individual cells differently, as a function of where in the sheets these cells are located (Nelson *et al.*, 2005).

## V. MICROFLUIDIC PATTERNING

The use of microfluidic channels allows patterning surfaces by restricting the flow of fluids to desired regions of a substrate. The patterning components — such as ligands, proteins, and cells — are deposited from the solution to create a pattern on the substrate.

Delamarche *et al.* used microfluidic patterning (µFP) to pattern immunoglobins with submicron resolution on a variety of substrates including gold, glass, and polystyrene (Delamarche *et al.*, 1997). Only microliters of reagent were required to cover square millimeter-sized areas. Patel *et al.* (1998) developed a method to generate micron-scale patterns of any biotinylated ligand on the surface of a biodegradable polymer. These investigators prepared biotin-presenting polymer films and patterned the films by allowing solutions of avidin to flow over them through 50-µm channels fabricated in PDMS. The avidin moieties bound to the biotin groups on the surface and served as a bridge between the biotinylated polymer and biotinylated ligands. Patterns created with biotinylated ligands containing the RGD or IKVAV oligopeptide sequences determined the adhesion and spreading of bovine aortic endothelial cells and PC12 nerve cells.

Both our group and Toner's group used µFP to produce patterns of adsorbed proteins and adherent cells on biocompatible substrates (Chiu *et al.*, 2000; Folch *et al.*, 1999; Folch and Toner, 1998). We formed micropatterns of proteins deposited from fluids in separately addressable capillaries.

By allowing different cell suspensions to flow through different channels, we could pattern two types of cell on surfaces with high spatial precision. After the adhesion of two types of cells in different areas on the surface, we could remove the elastomeric stamps to allow the studies of the movement of two types of cells. By filling individual channels with different fluids, multiple components could be patterned at the same time without the need for multiple steps or the accompanying technical concerns of registration (although registration was required in the fabrication of the stamp itself).

## VI. LAMINAR FLOW PATTERNING

Laminar flow patterning (LFP) is a technique that can pattern surfaces and the positions of cells on them in useful ways (Takayama *et al.*, 2001b). It can also pattern fluids themselves (Takayama *et al.*, 1999). This technique utilizes a phenomenon that occurs in microfluidic systems as a result of their small dimensions — that is, low-Reynolds-number flow (Squires and Quake, 2005; Stroock and Whitesides, 2003). The Reynolds number (Re) is a parameter describing the ratio of inertial to viscous forces in a particular flow configuration; it is a measure of the tendency of a flowing fluid to develop turbulence. The flow of aqueous fluids in capillaries usually has a low Re and is laminar. Laminar flow allows two or more streams of fluid to flow next to each other without any mixing other than by diffusion of their constituent molecules across the boundary between them (which is usually fairly slow). Diffusional motion of particulate components (e.g., cells) is even slower.

In a typical setup for LFP experiments, a network of capillaries is made by sealing a patterned PDMS slab with a glass slide or the surface of a petri dish (Fig. 19.4) (Takayama *et al.*, 1999). By passing streams of fluid with different compositions from the inlets, patterns of parallel stripes of flowing fluid are created in the main channel. It is possible, therefore, to treat different parts of a single cell with different reagents if a cell happens to span these stripes. Fig. 19.4

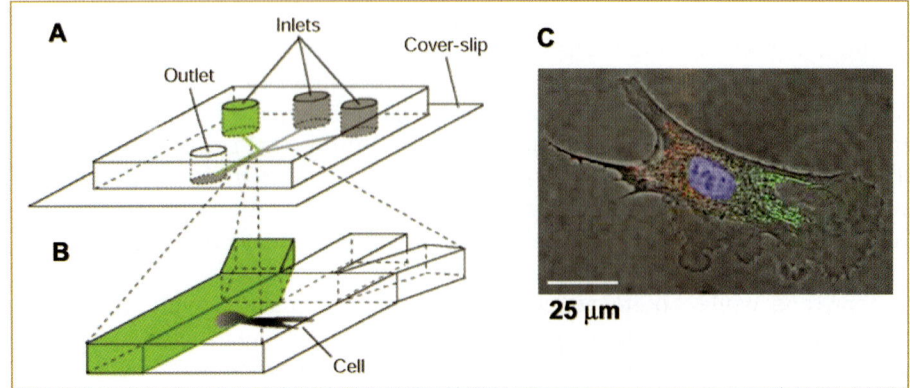

**FIG. 19.4.** Manipulation of two regions of a single bovine capillary endothelial cell using multiple laminar flows. **(A)** Experimental setup; **(B)** shows a close-up of the point at which the inlet channels combine into one main channel. **(C)** Fluorescence images of a single cell after treatment of its right pole with Mitotracker Green FM and its left pole with Mitotracker Red CM-H2XRos. The entire cell is treated with the DNA-binding dye Hoechst 33342.

illustrates the painting of a single cell with dyes that stained mitochondria located in different parts of the same cell (Takayama *et al.*, 2001a, 2003).

Using this technique, the positions and micro-environments of cells can be controlled simultaneously in several stripes in the same channel (Gu *et al.*, 2004; Sawano *et al.*, 2002). Using a similar approach, we could pattern the substrate with different proteins and cells (Takayama *et al.*, 1999). We can pattern the culture media over an individual cell by delivering chemicals selectively to cells. Since no physical barriers are required to separate the different liquid streams, different liquids can flow over different portions of a single cell.

Ismagilov and colleagues have used LFP to generate a step gradient in temperature to study the development of the embryos of the fruit fly *Drosophila melanogaster* (Lucchetta *et al.*, 2005). They treated the anterior (front) and posterior (back) of the embryo with media of different temperatures and observed that the fly embryos developed normally under such a condition. They concluded that in complex biochemical systems, there exist mechanisms for compensation. They further showed that if they reverse the temperature gradient within a certain time, embryos failed to develop normally. This observation shows there is a limitation of time for the mechanisms for compensation.

Another type of LFP generates gradients with parallel streams of flow of increasing or decreasing concentrations. We have generated gradients of biomolecules both in solution and on surfaces (Fig. 19.5) (Dertinger *et al.*, 2001; Jeon *et al.*, 2000). Because we can control the input concentration and the width of the microfluidic channel, it is possible to generate gradients of virtually any characteristics (e.g., the length and the slope), both in solution and on surfaces. We studied the generation of neuronal

polarity — the process of the selective formation of one axon and several dendrites from a number of initially equivalent neurites projecting from a single neuron — and found that a surface gradient of laminin was sufficient to guide the orientation of this process (Fig. 19.5) (Dertinger *et al.*, 2002). We further quantified the slope of the gradient and determined the minimum slope of the gradient required for this process to take place. We have also studied the chemotaxis of neutrophils in a solution of gradient of interleukin-8 (IL8) (Jeon *et al.*, 2002). The neutrophils migrate directionally toward increasing concentrations of IL8 in linear gradients. Neutrophils halt abruptly when encountering a sudden drop in the chemoattractant concentration (from maximum directly to zero). When neutrophils encounter a gradual increase and decrease in chemoattractant (from maximum gradually to zero), however, the cells initially cross the crest of maximum concentration but then head back toward the maximum. It would be very difficult to carry out experiments to answer questions of these types — questions covering the response of cells to the details of gradients over scales of microns — without the ability to form precisely controlled gradients.

LFP has some features that make it complementary to other patterning techniques used for biological applications. It takes advantage of easily generated multiphase laminar flows to pattern fluids and to deliver components for patterning. The ability to pattern the growth medium itself is a special feature that cannot be achieved by other processes. This method can pattern even delicate structures, such as portions of a mammalian cell. This type of patterning is difficult by other techniques. LFP can also give simultaneous control over the surface patterns, cell positioning, and the fluid environment in the same channel.

One may ask if the fluid flow required in the generation of laminar flows would cause problems for certain

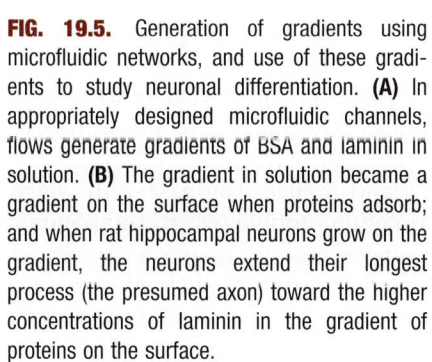

**FIG. 19.5.** Generation of gradients using microfluidic networks, and use of these gradients to study neuronal differentiation. **(A)** In appropriately designed microfluidic channels, flows generate gradients of BSA and laminin in solution. **(B)** The gradient in solution became a gradient on the surface when proteins adsorb; and when rat hippocampal neurons grow on the gradient, the neurons extend their longest process (the presumed axon) toward the higher concentrations of laminin in the gradient of proteins on the surface.

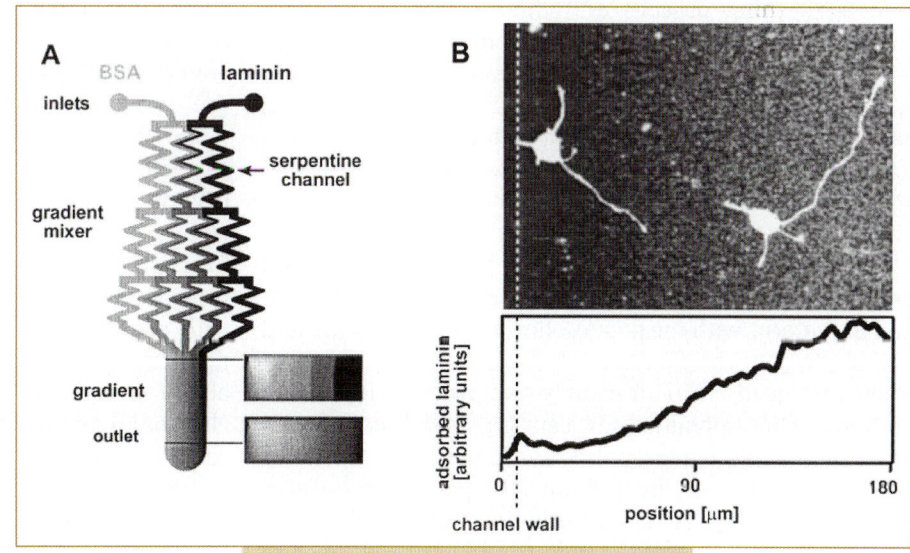

experiments, such as the measurement of chemotaxis. Wikswo and coworkers addressed this issue by measuring the motility of HL60 leukemia cells (which express CXCR2 receptors) in a gradient of CXCL8 (Walker *et al.*, 2005). They found that high rates of flow can affect the motility of cells. Reasonably low rates of flow, however, do not affect measurements on motile cells.

A few recent examples have combined patterned substrates with patterned flows. We have fabricated gradients of proteins on surfaces in microchannels whose floors carry patterns generated by μCP (Jiang *et al.*, 2005b). Jeon and his coworkers have combined substrates patterned in topography and patterned flows to form a model system that conveniently isolates axons of rat hippocampal neurons from the rest of the cell for studies of their molecular biology (Taylor *et al.*, 2005). Langer and coworkers have fabricated microchannels that have micropatterns within them to immobilize proteins and cells (Khademhosseini *et al.*, 2004). Folch and coworkers have used micropatterns to form myotubes from myoblasts and then used laminar flows to deliver agrin, a proteoglycan found in the neuromuscular junction, precisely to these myotubes. In these experiments, he monitored the clustering of acetylcholine receptor (AChR), and his results corroborated the hypothesis that focalized release of agrin causes the clustering the AChR (Tourovskaia *et al.*, 2006).

## VII. CONCLUSION AND FUTURE PROSPECTS

Soft lithography brings to microfabrication low cost, simple procedures, rapid prototyping of custom-designed devices, three-dimensional capability, easy integration with existing instruments such as optical microscopy, molecular level control of surfaces, and biocompatibility (Jiang and Whitesides, 2003; Whitesides *et al.*, 2001). These techniques allow patterning of cells and their environments with convenience and flexibility at dimensions smaller than micrometers. We have described several complementary soft lithographic techniques — microcontact printing, patterning with microtopography, patterning using fluids in microfluidic channels, and laminar flow patterning — that are useful in their ability to pattern the microenvironment of cultured cells.

Microcontact printing is perhaps the simplest method for patterning surfaces. It also provides the highest resolution in patterns with the greatest flexibility in the shape and size of the patterns generated. It provides the best control when one needs to pattern only two types of ligands or proteins. Microtopographies can be useful for certain experiments where micropatterning alone is not sufficient. Microfluidic channels are well suited for patterning surfaces using delicate objects such as proteins and cells. They are also useful when multiple ligands, proteins, or cells need to be patterned. Laminar flow patterning is similar to pattern-ing with individual microfluidic channels, except the indi-vidual flows are kept from mixing with each other by laminar flow, not by physical walls. The ability to pattern the fluid environment is the distinguishing feature of this method, and it enables laminar flow to be used to pattern the distri-bution of different fluids over the surface of a single mam-malian cell. This capability allows patterning of portions of a single cell and remodeling of the cell culture environment, both in the presence of living cells. The combination of two or more of these techniques is starting to become useful for more sophisticated experiments.

Soft lithography is still practiced by a relatively small number of biologists, but its use is growing rapidly. There are many cell culture environments and related technologies that we have not discussed in this chapter; many of them relate to soft lithographic methods. For example, there are a number of methods of manipulation of chemistry on SAMs and tools of micropatterns to allow for molecular level control at the cell–substrate interface (Chapman *et al.*, 2000; Kandere-Grzybowska *et al.*, 2005; Kato and Mrksich, 2004).

Although we are starting to have more techniques for fabrication in three dimensions, patterning of cells in three dimensions is still difficult (Gates *et al.*, 2004, 2005; Shin *et al.*, 2004). We are making rapid progress, however, in the fourth dimension, i.e., time. Since we can turn the surface on and off for adhesion of cells and change the media at will in laminar flows, real-time monitoring of temporal changes in individual cells is possible (Jiang *et al.*, 2003; Takayama *et al.*, 1999; Yousaf *et al.*, 2001a). The optical trans-parency of PDMS makes it straightforward to pattern the intensity of light in cell cultures (Whitesides *et al.*, 2001). The gas permeability of PDMS may be useful in patterning the gas surrounding cells. PDMS is electrically insulating, and molding or fabricating electrically conducting wires in it should allow patterning of electric fields around cells (Kenis *et al.*, 1999; Takayama *et al.*, 1999). Gravitational fields can also be affected: Microfluidic culture chambers with adherent cells can be turned upside down without loss of the culture media. Temperature, fluid shear, and other factors may also be accurately patterned.

The functional potential of a cell is determined by its genetics. Realization of that potential depends, *inter alia*, on whether the cell is exposed to the appropriate environment for expression of particular sets of genes. Soft lithography provides tools for patterning cells and their environment with precise spatial control. This capability aids efforts to understand fundamental cell biology and advances our ability to engineer cells and tissues. The ease with which electronic components or other "nonbiological" components can be fabricated with soft lithography also paves the way for the engineering of cells and tissues for use in biosensors and hybrid systems (e.g., interfaces between semiconductor-based computation and biological computation) that combine living and nonliving components.

## VIII. ACKNOWLEDGMENTS

Supported by NIH GM065364. The content of the information does not necessarily reflect the position or the policy of the government, and no official endorsement should be inferred. X. J. thanks the National Center for Nanoscience and Technology (China) and the Chinese Academy of Sciences for a startup fund. S. T. is a Leukemia Society of America Fellow and thanks the society for a fellowship. R. G. C. thanks the Natural Sciences and Engineering Research Council of Canada for a fellowship.

## IX. REFERENCES

Alberts, B., Johnson, A., Lewis, J., Raff, M., Keith, R., and Walter, P. (2002). "Molecular Biology of the Cell," 4th ed. Garland Science, Taylor & Francis Group, New York.

Allara, D. L., and Nuzzo, R. G. (1985). Spontaneously organized molecular assemblies. 2. Quantitative infrared spectroscopic determination of equilibrium structures of solution-adsorbed N-alkanoic acids on an oxidized aluminum surface. *Langmuir* **1**, 52–66.

Amirpour, M. L., Ghosh, P., Lackowski, W. M., Crooks, R. M., and Pishko, M. V. (2001). Mammalian cell cultures on micropatterned surfaces of weak-acid, polyelectrolyte hyperbranched thin films on gold. *Anal. Chem.* **73**, 1560–1566.

Andrade, J. D., Hlady, V., and Jeon, S. I. (1996). Poly(ethylene oxide) and protein resistance. Principles, problems, and possibilities. *Adv. Chem. Series* **248**, 51–59.

Arnold, M., Cavalcanti-Adam, E. A., Glass, R., Blummel, J., Eck, W., Kantlehner, M., Kessler, H., and Spatz, J. P. (2004). Activation of integrin function by nanopatterned adhesive interfaces. *Chemphyschem.* **5**, 383–388.

Assoian, R. K., and Zhu, X. (1997). Cell anchorage and the cytoskeleton as partners in growth factor–dependent cell cycle progression. *Curr. Opin. Cell. Biol.* **9**, 93–98.

Bailly, M., Yan, L., Whitesides, G. M., Condeelis, J. S., and Segall, J. E. (1998). Regulation of protrusion shape and adhesion to the substratum during chemotactic responses of mammalian carcinoma cells. *Exp. Cell Res.* **241**, 285–299.

Bain, C. D., and Whitesides, G. M. (1988). Depth sensitivity of wetting — monolayers of omega-mercapto ethers on gold. *J. Am. Chem. Soc.* **110**, 5897–5898.

Bernard, A., Delamarche, E., Schmid, H., Michel, B., Bosshard, H. R., and Biebuyck, H. (1998). Printing patterns of proteins. *Langmuir* **14**, 2225–2229.

Bernard, A., Renault, J. P., Michel, B., Bosshard, H. R., and Delamarche, E. (2000). Microcontact printing of proteins. *Adv. Mater.* **12**, 1067–1070.

Bernard, A., Fitzli, D., Sonderegger, P., Delamarche, E., Michel, B., Bosshard, H. R., and Biebuyck, H. (2001). Affinity capture of proteins from solution and their dissociation by contact printing. *Nat. Biotechnol.* **19**, 866–869.

Bhatia, S. N., Balis, U. J., Yarmush, M. L., and Toner, M. (1999). Effect of cell–cell interactions in preservation of cellular phenotype: cocultivation of hepatocytes and nonparenchymal cells. *FASEB J.* **13**, 1883–1900.

Branch, D. W., Corey, J. M., Weyhenmeyer, J. A., Brewer, G. J., and Wheeler, B. C. (1998). Microstamp patterns of biomolecules for high-resolution neuronal networks. *Med. Biol. Eng. Comput.* **36**, 135–141.

Brangwynne, C., Huang, S., Parker, K. K., and Ingber, D. E. (2000). Symmetry breaking in cultured mammalian cells. *In Vitro Cell Dev. Biol. Anim.* **36**, 563–565.

Brock, A., Chang, E., Ho, C.-C., LeDuc, P., Jiang, X., Whitesides, G. M., and Ingber, D. E. (2003). Geometric determinants of directional cell motility revealed using microcontact printing. *Langmuir* **19**, 1611–1617.

Chapman, R. G., Ostuni, E., Takayama, S., Holmlin, R. E., Yan, L., and Whitesides, G. M. (2000). Surveying for surfaces that resist the adsorption of proteins. *J. Am. Chem. Soc.* **122**, 8303–8304.

Chen, C. S., Mrksich, M., Huang, S., Whitesides, G. M., and Ingber, D. E. (1997). Geometric control of cell life and death. *Science* **276**, 1425–1428.

Chen, C. S., Jiang, X. Y., and Whitesides, G. M. (2005). Microengineering the environment of mammalian cells in culture. *MRS Bull.* **30**, 194–201.

Chiu, D. T., Jeon, N. L., Huang, S., Kane, R. S., Wargo, C. J., Choi, I. S., Ingber, D. E., and Whitesides, G. M. (2000). Patterned deposition of cells and proteins onto surfaces by using three-dimensional microfluidic systems. *Proc. Natl. Acad. Sci. USA* **97**, 2408–2413.

Co, C. C., Wang, Y. C., and Ho, C. C. (2005). Biocompatible micropatterning of two different cell types. *J. Am. Chem. Soc.* **127**, 1598–1599.

Delamarche, E., Bernard, A., Schmid, H., Michel, B., and Biebuyck, H. (1997). Patterned delivery of immunoglobulins to surfaces using microfluidic networks. *Science* **276**, 779–781.

Deng, T., Wu, H., Brittain, S. T., and Whitesides, G. M. (2000). Prototyping of masks, masters, and stamps/molds for soft lithography using an office printer and photographic reduction. *Anal. Chem.* **72**, 3176–3180.

Dertinger, S. K. W., Chiu, D. T., Jeon, N. L., and Whitesides, G. M. (2001). Generation of gradients having complex shapes using microfluidic networks. *Anal. Chem.* **73**, 1240–1246.

Dertinger, S. K. W., Jiang, X., Li, Z., Murthy, V. N., and Whitesides, G. M. (2002). Gradients of substrate-bound laminin orient axonal specification of neurons. *Proc. Natl. Acad. Sci. USA* **99**, 12542–12547.

Dike, L. E., Chen, C. S., Mrksich, M., Tien, J., Whitesides, G. M., and Ingber, D. E. (1999). Geometric control of switching between growth, apoptosis, and differentiation during angiogenesis using micropatterned substrates. *In Vitro Cell Dev. Biol. Anim.* **35**, 441–448.

Dillmore, W. S., Yousaf, M. N., and Mrksich, M. (2004). A photochemical method for patterning the immobilization of ligands and cells to self-assembled monolayers. *Langmuir* **20**, 7223–7231.

Folch, A., and Toner, M. (1998). Cellular micropatterns on biocompatible materials. *Biotechnol. Prog.* **14**, 388–392.

Folch, A., Ayon, A., Hurtado, O., Schmidt, M. A., and Toner, M. (1999). Molding of deep polydimethylsiloxane microstructures for microfluidics and biological applications. *J. Biomech. Eng.* **121**, 28–34.

Folch, A., Jo, B. H., Hurtado, O., Beebe, D. J., and Toner, M. (2000). Microfabricated elastomeric stencils for micropatterning cell cultures. *J. Biomed. Mater. Res.* **52**, 346–353.

Frazier, R. A., Matthijs, G., Davies, M. C., Roberts, C. J., Schacht, E., and Tendler, S. J. B. (2000). Characterization of protein-resistant dextran monolayers. *Biomaterials* **21**, 957–966.

Gallant, N. D., Michael, K. E., and Garcia, A. J. (2005). Cell adhesion strengthening: contributions of adhesive area, integrin binding, and focal adhesion assembly. *Mol. Biol. Cell.* **16**, 4329–4340.

Gates, B. D., Xu, Q. B., Love, J. C., Wolfe, D. B., and Whitesides, G. M. (2004). Unconventional nanofabrication. *Ann. Rev. Mater. Res.* **34**, 339–372.

Gates, B. D., Xu, Q. B., Stewart, M., Ryan, D., Willson, C. G., and Whitesides, G. M. (2005). New approaches to nanofabrication: molding, printing, and other techniques. *Chem. Rev.* **105**, 1171–1196.

Goffin, J. M., Pittet, P., Csucs, G., Lussi, J. W., Meister, J. J., and Hinz, B. (2006). Focal adhesion size controls tension-dependent recruitment of alpha-smooth muscle actin to stress fibers. *J. Cell. Biol.* **172**, 259–268.

Gu, W., Zhu, X., Futai, N., Cho, B. S., and Takayama, S. (2004). Computerized microfluidic cell culture using elastomeric channels and Braille displays. *Proc. Natl. Acad. Sci. USA* **101**, 15861–15866.

Han, D. K., Park, K. D., Hubbell, J. A., and Kin, Y. H. (1998). Surface characteristics and biocompatibility of lactide-based poly(ethylene glycol) scaffolds for tissue engineering. *J. Biomater. Sci. Polymer. Edn.* **9**, 667–680.

Hotchin, N. A., and Hall, A. (1995). The assembly of integrin adhesion complexes requires both extracellular matrix and intracellular rho/rac GTPases. *J. Cell Biol.* **131**, 1857–1865.

Huang, J., and Hemminger, J. C. (1993). Photooxidation of thiols in self-assembled monolayers on gold. *J. Am. Chem. Soc.* **115**, 3342–3343.

Huang, S., Brangwynne, C. P., Parker, K. K., and Ingber, D. E. (2005). Symmetry-breaking in mammalian cell cohort migration during tissue pattern formation: role of random-walk persistence. *Cell Motil. Cytosk.* **61**, 201–213.

Jackman, R. J., Wilbur, J. L., and Whitesides, G. M. (1995). Fabrication of submicrometer features on curved substrates by microcontact printing. *Science* **269**, 664–666.

James, C. D., Davis, R., Meyer, M., Turner, A., Turner, S., Withers, G., Kam, L., Banker, G., Craighead, H., Isaacson, M., *et al.* (2000). Aligned microcontact printing of micrometer-scale poly-L-lysine structures for controlled growth of cultured neurons on planar microelectrode arrays. *IEEE Trans. Biomed. Eng.* **47**, 17–21.

Jeon, S. I., Lee, J. H., Andrade, J. D., and De Gennes, P. G. (1991). Protein–surface interactions in the presence of polyethylene oxide. I. Simplified theory. *J. Colloid. Interf. Sci.* **142**, 149–158.

Jeon, N. L., Dertinger, S. K. W., Chiu, D. T., Choi, I. S., Stroock, A. D., and Whitesides, G. M. (2000). Generation of solution and surface gradients using microfluidic systems. *Langmuir* **16**, 8311–8316.

Jeon, N. L., Baskaran, H., Dertinger, S. K. W., Whitesides, G. M., Van De Water, L., and Toner, M. (2002). Neutrophil chemotaxis in linear and complex gradients of interleukin-8 formed in a microfabricated device. *Nat. Biotechnol.* **20**, 826–830.

Jiang, X., and Whitesides, G. M. (2003). Engineering microtools in polymers to study cell biology. *Eng. Life Sci.* **3**, 475–480.

Jiang, X., Takayama, S., Qian, X., Ostuni, E., Wu, H., Bowden, N., LeDuc, P., Ingber, D. E., and Whitesides, G. M. (2002). Controlling mammalian cell spreading and cytoskeletal arrangement with conveniently fabricated continuous wavy features on poly (dimethylsiloxane). *Langmuir* **18**, 3273–3280.

Jiang, X., Ferrigno, R., Mrksich, M., and Whitesides, G. M. (2003). Electrochemical desorption of self-assembled monolayers noninvasively releases patterned cells from geometrical confinements. *J. Am. Chem. Soc.* **125**, 2366–2367.

Jiang, X., Bruzewicz, D. A., Thant, M. M., and Whitesides, G. M. (2004a). Palladium as a substrate for self-assembled monolayers used in biotechnology. *Anal. Chem.* **76**, 6116–6121.

Jiang, X., Lee, J. N., and Whitesides, G. M. (2004b). Self-assembled monolayers in mammalian cell cultures. In "Tissue Scaffolding" (P. Ma, ed.), Marcel Dekker, New York.

Jiang, X., Bruzewicz, D. A., Wong, A. P., Piel, M., and Whitesides, G. M. (2005a). Directing cell migration with asymmetric micropatterns. *Proc. Natl. Acad. Sci. USA* **102**, 975–978.

Jiang, X., Xu, Q., Dertinger, S. K., Stroock, A. D., Fu, T. M., and Whitesides, G. M. (2005b). A general method for patterning gradients of biomolecules on surfaces using microfluidic networks. *Anal. Chem.* **77**, 2338–2347.

Kaji, H., Kanada, M., Oyamatsu, D., Matsue, T., and Nishizawa, M. (2004). Microelectrochemical approach to induce local cell adhesion and growth on substrates. *Langmuir* **20**, 16–19.

Kaji, H., Tsukidate, K., Hashimoto, M., Matsue, T., and Nishizawa, M. (2005). Patterning the surface cytophobicity of an albumin-physisorbed substrate by electrochemical means. *Langmuir* **21**, 6966–6969.

Kam, L., and Boxer, S. G. (2001). Cell adhesion to protein-micro-patterned-supported lipid bilayer membranes. *J. Biomed. Mater. Res.* **55**, 487–495.

Kandere-Grzybowska, K., Campbell, C., Komarova, Y., Grzybowski, B. A., and Borisy, G. G. (2005). Molecular dynamics imaging in micro-patterned living cells. *Nat. Methods* **2**, 739–741.

Kane, R. S., Deschatelets, P., and Whitesides, G. M. (2003). Kosmotropes form the basis of protein-resistant surfaces. *Langmuir* **19**, 2388–2391.

Kato, M., and Mrksich, M. (2004). Rewiring cell adhesion. *J. Am. Chem. Soc.* **126**, 6504–6505.

Kenis, P. J., Ismagilov, R. F., and Whitesides, G. M. (1999). Microfabrication inside capillaries using multiphase laminar flow patterning. *Science* **285**, 83–85.

Khademhosseini, A., Suh, K. Y., Jon, S., Eng, G., Yeh, J., Chen, G.-J., and Langer, R. (2004). A soft lithographic approach to fabricate patterned microfluidic channels. *Anal. Chem.* **76**, 3675–3681.

Kleinfeld, D., Kahler, K. H., and Hockberger, P. E. (1988). Controlled outgrowth of dissociated neurons on patterned substrates. *J. Neurosci.* **8**, 4098–4120.

Kumar, G., Meng, J. J., Ip, W., Co, C. C., and Ho, C. C. (2005). Cell motility assays on tissue culture dishes via noninvasive confinement and release of cells. *Langmuir* **21**, 9267–9273.

Lahiri, J., Isaacs, L., Tien, J., and Whitesides, G. M. (1999). A strategy for the generation of surfaces presenting ligands for studies of binding based on an active ester as a common reactive intermediate: a surface plasmon resonance study. *Anal. Chem.* **71**, 777–790.

Lam, M. T., Sim, S., Zhu, X., and Takayama, S. (2006). The effect of continuous wavy micropatterns on silicone substrates on the alignment of skeletal muscle myoblasts and myotubes. *Biomaterials* **27**, 4340–4347.

Lee, J. N., Jiang, X., Ryan, D., and Whitesides, G. M. (2004). Compatibility of mammalian cells on surfaces of poly(dimethylsiloxane). *Langmuir* **20**, 11684–11691.

Lehnert, D., Wehrle-Haller, B., David, C., Weiland, U., Ballestrem, C., Imhof, B. A., and Bastmeyer, M. (2004). Cell behavior on micropatterned substrata: limits of extracellular matrix geometry for spreading and adhesion. *J. Cell Sci.* **117**, 41–52.

Letourneau, P. C. (1975). Cell-to-substratum adhesion and guidance of axonal elongation. *Dev. Biol.* **44**, 92–101.

Linder, V., Wu, H., Jiang, X., and Whitesides, G. M. (2003). Rapid prototyping of 2D structures with feature sizes larger than 8 microns. *Anal. Chem.* **75**, 2522–2527.

Lopez, G. P., Biebuyck, H. A., Harter, R., Kumar, A., and Whitesides, G. M. (1993). Fabrication and imaging of 2-dimensional patterns of proteins adsorbed on self-assembled monolayers by scanning electron microscopy. *J. Am. Chem. Soc.* **115**, 10774–10781.

Love, J. C., Estroff, L. A., Kriebel, J. K., Nuzzo, R. G., and Whitesides, G. M. (2005). Self-assembled monolayers of thiolates on metals as a form of nanotechnology. *Chem. Rev.* **105**, 1103–1169.

Lucchetta, E. M., Lee, J. H., Fu, L. A., Patel, N. H., and Ismagilov, R. F. (2005). Dynamics of *Drosophila* embryonic patterning network perturbed in space and time using microfluidics. *Nature* **434**, 1134–1138.

Mayer, M., Yang, J., Gitlin, I., Gracias, D. H., and Whitesides, G. M. (2004). Micropatterned agarose gels for stamping arrays of proteins and gradients of proteins. *Proteomics* **4**, 2366–2376.

McBeath, R., Pirone, D. M., Nelson, C. M., Bhadriraju, K., and Chen, C. S. (2004). Cell shape, cytoskeletal tension, and RhoA regulate stem cell lineage commitment. *Dev. Cell* **6**, 483–495.

McDonald, J. C., and Whitesides, G. M. (2002). Poly(dimethylsiloxane) as a material for fabricating microfluidic devices. *Acc. Chem. Res.* **35**, 491.

Merz, M., and Fromherz, P. (2005). Silicon chip interfaced with a geometrically defined net of snail neurons. *Adv. Func. Mater.* **15**, 739–744.

Michel, R., Lussi, J. W., Csucs, G., Reviakine, I., Danuser, G., Ketterer, B., Hubbell, J. A., Textor, M., and Spencer, N. D. (2002). Selective molecular assembly patterning: a new approach to micro- and nanochemical patterning of surfaces for biological applications. *Langmuir* **18**, 3281–3287.

Mrksich, M., and Whitesides, G. M. (1995). Patterning self-assembled monolayers using microcontact printing — a new technology for biosensors. *Trends Biotechnol.* **13**, 228–235.

Mrksich, M., and Whitesides, G. M. (1996). Using self-assembled monolayers to understand the interactions of man-made surfaces with proteins and cells. *Annu. Rev. Biophys. Biomol. Struct.* **25**, 55–78.

Nakanishi, J., Kikuchi, Y., Takarada, T., Nakayama, H., Yamaguchi, K., and Maeda, M. (2004). Photoactivation of a substrate for cell adhesion under standard fluorescence microscopes. *J. Am. Chem. Soc.* **126**, 16314–16315.

Nelson, C. M., and Chen, C. S. (2002). Cell-cell signaling by direct contact increases cell proliferation via a PI3K-dependent signal. *FEBS Lett.* **514**, 238–242.

Nelson, C. M., Jean, R. P., Tan, J. L., Liu, W. F., Sniadecki, N. J., Spector, A. A., and Chen, C. S. (2005). Emergent patterns of growth controlled by multicellular form and mechanics. *Proc. Natl. Acad. Sci. USA* **102**, 11594–11599.

Ostuni, E., Yan, L., and Whitesides, G. M. (1999). The interaction of proteins and cells with self-assembled monolayers of alkanethiols on gold and silver. *Colloids Surface B* **15**, 3–30.

Ostuni, E., Kane, R., Chen, C. S., Ingber, D. E., and Whitesides, G. M. (2000). Patterning mammalian cells using elastomeric membranes. *Langmuir* **16**, 7811–7819.

Ostuni, E., Chen, C. S., Ingber, D. E., and Whitesides, G. M. (2001). Selective deposition of proteins and cells in arrays of microwells. *Langmuir* **17**, 2828–2834.

Parker, K. K., Brock, A. L., Brangwynne, C., Mannix, R. J., Wang, N., Ostuni, E., Geisse, N. A., Adams, J. C., Whitesides, G. M., and Ingber, D. E. (2002). Directional control of lamellipodia extension by constraining cell shape and orienting cell tractional forces. *FASEB J.* **16**, 1195–1204.

Patel, N., Padera, R., Sanders, G. H., Cannizzaro, S. M., Davies, M. C., Langer, R., Roberts, C. J., Tendler, S. J., Williams, P. M., and Shakesheff, K. M. (1998). Spatially controlled cell engineering on biodegradable polymer surfaces. *FASEB J.* **12**, 1447–1454.

Petrovykh, D. Y., Kimura-Suda, H., Opdahl, A., Richter, L. J., Tarlov, M. J., and Whitman, L. J. (2006). Alkanethiols on platinum: multicomponent self-assembled monolayers. *Langmuir* **22**, 2578–2587.

Prime, K. L., and Whitesides, G. M. (1991). Self-assembled organic monolayers: model systems for studying adsorption of proteins at surfaces. *Science* **252**, 1164–1167.

Prime, K. L., and Whitesides, G. M. (1993). Adsorption of proteins onto surfaces containing end-attached oligo(ethylene oxide) — a model system using self-assembled monolayers. *J. Am. Chem. Soc.* **115**, 10714–10721.

Ravenscroft, M. S., Bateman, K. E., Shaffer, K. M., Schessler, H. M., Jung, D. R., Schneider, T. W., Montgomery, C. B., Custer, T. L., Schaffner, A. E., Liu, Q. Y., *et al.* (1998). Developmental neurobiology implications from fabrication and analysis of hippocampal neuronal networks on patterned silane-modified surfaces. *J. Am. Chem. Soc.* **120**, 12169–12177.

Ruoslahti, E., and Pierschbacher, M. D. (1987). New perspectives in cell adhesion: RGD and integrins. *Science* **238**, 491–497.

Ryan, D., Parviz, B. A., Linder, V., Semetey, V., Sia, S. K., Su, J., Mrksich, M., and Whitesides, G. M. (2004). Patterning multiple aligned self-assembled monolayers using light. *Langmuir* **20**, 9080–9088.

Sawano, A., Takayama, S., Matsuda, M., and Miyawaki, A. (2002). Lateral propagation of EGF signaling after local stimulation is dependent on receptor density. *Dev. Cell* **3**, 245–257.

Schwartz, M. A., Lechene, C., and Ingber, D. E. (1991). Insoluble fibronectin activates the Na/H antiporter by clustering and immobilizing integrin alpha-5-beta-1, independent of cell shape. *Proc. Natl. Acad. Sci. USA* **88**, 7849–7853.

Shin, M., Matsuda, K., Ishii, O., Terai, H., Kaazempur-Mofrad, M., Borenstein, J., Detmar, M., and Vacanti, J. P. (2004). Endothelialized networks with a vascular geometry in microfabricated poly-(dimethyl siloxane). *Biomed. Microd.* **6**, 269–278.

Singhvi, R., Kumar, A., Lopez, G. P., Stephanopoulos, G. N., Wang, D. I. C., Whitesides, G. M., and Ingber, D. E. (1994). Engineering cell shape and function. *Science* **264**, 696–698.

Squires, T. M., and Quake, S. R. (2005). Microfluidics: fluid physics at the nanoliter scale. *Rev. Mod. Phys.* **77**, 977–1026.

St. John, P. M., Davis, R., Cady, N., Czajka, J., Batt, C. A., and Craighead, H. G. (1998). Diffraction-based cell detection using a microcontact printed antibody grating. *Anal. Chem.* **70**, 1108–1111.

Stevens, M. M., Mayer, M., Anderson, D. G., Weibel, D. B., Whitesides, G. M., and Langer, R. (2005). Direct patterning of mammalian cells onto porous tissue-engineering substrates using agarose stamps. *Biomaterials* **26**, 7636–7641.

Stroock, A. D., and Whitesides, G. M. (2003). Controlling flows in microchannels with patterned surface charge and topography. *Accounts Chem. Res.* **36**, 597–604.

Sweryda-Krawiec, B., Devaraj, H., Jacob, G., and Hickman, J. J. (2004). A new interpretation of serum albumin surface passivation. *Langmuir* **20**, 2054–2056.

Takayama, S., McDonald, J. C., Ostuni, E., Liang, M. N., Kenis, P. J., Ismagilov, R. F., and Whitesides, G. M. (1999). Patterning cells and their environments using multiple laminar fluid flows in capillary networks. *Proc. Natl. Acad. Sci. USA* **96**, 5545–5548.

Takayama, S., Ostuni, E., LeDuc, P., Naruse, K., Ingber, D. E., and Whitesides, G. M. (2001a). Subcellular positioning of small molecules. *Nature* **411**, 1016.

Takayama, S., Ostuni, E., Qian, X., McDonald, J. C., Jiang, X., LeDuc, P., Wu, M.-H., Ingber, D. E., and Whitesides, G. M. (2001b). Topographical micropatterning of poly(dimethylsiloxane) using laminar flows of liquids in capillaries. *Adv. Mater.* **13**, 570–574.

Takayama, S., Ostuni, E., LeDuc, P., Naruse, K., Ingber, D. E., and Whitesides, G. M. (2003). Selective chemical treatment of cellular microdomains using multiple laminar streams. *Chem. Biol.* **10**, 123–130.

Tan, J. L., Tien, J., Pirone, D. M., Gray, D. S., Bhadriraju, K., and Chen, C. S. (2003). Cells lying on a bed of microneedles: an approach to isolate mechanical force. *Proc. Natl. Acad. Sci. USA* **100**, 1484–1489.

Tang, M. D., Golden, A. P., and Tien, J. (2003). Molding of three-dimensional microstructures of gels. *J. Am. Chem. Soc.* **125**, 12988–12989.

Taylor, A. M., Blurton-Jones, M., Rhee, S. W., Cribbs, D. H., Cotman, C. W., and Jeon, N. L. (2005). A microfluidic culture platform for CNS axonal injury, regeneration and transport. *Nat. Methods* **2**, 599–605.

Thery, M., Racine, V., Pepin, A., Piel, M., Chen, Y., Sibarita, J. B., and Bornens, M. (2005). The extracellular matrix guides the orientation of the cell division axis. *Nat. Cell Biol.* **7**, 947–953.

Tien, J., Nelson, C. M., and Chen, C. S. (2002). Fabrication of aligned microstructures with a single elastomeric stamp. *Proc. Natl. Acad. Sci. USA* **99**, 1758–1762.

Tourovskaia, A., Barber, T., Wickes, B. T., Hirdes, D., Grin, B., Castner, D. G., Healy, K. E., and Folch, A. (2003). Micropatterns of chemisorbed cell adhesion–repellent films using oxygen plasma etching and elastomeric masks. *Langmuir* **19**, 4754–4764.

Tourovskaia, A., Kosar, T. F., and Folch, A. (2006). Local induction of acetylcholine receptor clustering in myotube cultures using microfluidic application of agrin. *Biophys. J.* **90**, 2192–2198.

Ulman, A. (1996). Formation and structure of self-assembled monolayers. *Chem. Rev.* **96**, 1533–1554.

Walker, G. M., Sai, J., Richmond, A., Stremler, M., Chung, C. Y., and Wikswo, J. P. (2005). Effects of flow and diffusion on chemotaxis studies in a microfabricated gradient generator. *Lab. Chip* **5**, 611–618.

Whitesides, G. M., Ostuni, E., Takayama, S., Jiang, X., and Ingber, D. E. (2001). Soft lithography in biology and biochemistry. *Annu. Rev. Biomed. Eng.* **3**, 335–373.

Xia, Y., and Whitesides, G. M. (1998). Soft lithography. *Angew. Chem.* **37**, 550–575.

Yan, L., Zhao, X.-M., and Whitesides, G. M. (1998). Patterning a preformed, reactive SAM using microcontact printing. *J. Am. Chem. Soc.* **120**, 6179–6180.

Yeo, W.-S., Yousaf, M. N., and Mrksich, M. (2003). Dynamic interfaces between cells and surfaces: electroactive substrates that sequentially release and attach cells. *J. Am. Chem. Soc.* **125**, 14994–14995.

Yousaf, M. N., Houseman, B. T., and Mrksich, M. (2001a). Turning on cell migration with electroactive substrates. *Angew. Chem. Int. Edit.* **40**, 1093–1096.

Yousaf, M. N., Houseman, B. T., and Mrksich, M. (2001b). Using electroactive substrates to pattern the attachment of two different cell populations. *Proc. Natl. Acad. Sci. USA* **98**, 5992–5996.

Zeck, G., and Fromherz, P. (2001). Noninvasive neuroelectronic interfacing with synaptically connected snail neurons immobilized on a semiconductor chip. *Proc. Natl. Acad. Sci. USA* **98**, 10457–10462.

Zhang, S., Yan, L., Altman, M., Lassle, M., Nugent, H., Frankel, F., Lauffenburger, D. A., Whitesides, G. M., and Rich, A. (1999). Biological surface engineering: a simple system for cell pattern formation. *Biomaterials* **20**, 1213–1220.

Zhu, X., Mills, K. L., Peters, P. R., Bahng, J. H., Liu, E. H., Shim, J., Naruse, K., Csete, M. E., Thouless, M. D., and Takayama, S. (2005). Fabrication of reconfigurable protein matrices by cracking. *Nat. Mater.* **4**, 403–406.

# Chapter Twenty

# Cell Interactions with Polymers

*W. Mark Saltzman and Themis R. Kyriakides*

## I. INTRODUCTION

Scaffolds composed of synthetic and natural polymers have been essential components of tissue engineering since its inception (Vacanti, 1988). Polymers are currently used in a wide range of biomedical applications, including applications in which the polymer remains in intimate contact with cells and tissues for prolonged periods (Table 20.1). Many of these polymer materials have been tested for tissue-engineering applications as well. To select appropriate polymers for tissue engineering, it is helpful to understand the influence of these polymeric materials on viability, growth, and function of attached or adjacent cells. In addition, it is now possible to synthesize polymers that interact in predictable ways with cells and tissues; understanding the nature of interactions between cells and polymers provides the foundation to this approach.

This chapter reviews previous work on the interactions of tissue-derived cells with polymers, particularly the types of synthetic polymers that have been employed as biomaterials. The focus of the chapter is on interactions of cells with polymers that might be used in tissue engineering; therefore, both *in vitro* and *in vivo* methods for measuring cell–polymer interactions are described. The interactions of cells in flowing blood, particularly platelets, with synthetic polymer surfaces are also an important aspect of biomaterials design, but they are not considered here.

## II. METHODS FOR CHARACTERIZING CELL INTERACTIONS WITH POLYMERS

### In Vitro Cell Culture Methods

Cell interactions with polymers are often studied *in vitro*, using cell culture techniques. While *in vitro* studies do not reproduce the wide range of cellular responses observed after implantation of materials, the culture environment provides a level of control and quantification that cannot be obtained *in vivo*. Cells in culture are generally plated over a polymer surface, and interaction is allowed to proceed for several hours; the extent of cell adhesion and spreading on the surface is then measured. By maintaining the culture for longer periods, perhaps for many days, the influence of the substrate on cell viability, function, and motility can also be determined. Since investigators use different techniques to assess cell interactions with polymers and the differences between techniques are critically important for interpretation of interactions, some of the most frequently used *in vitro* methods are reviewed in this section.

To perform any measurement of cell interaction with a polymer substrate, the polymeric material and the cells must come into contact. Preferably, the experimenter should control (or at least understand) the nature of the contact; this is a critical, and often overlooked, aspect of all of these measurements. Some materials are easily

**Table 20.1.** Some of the polymers that might be useful in tissue engineering, based on past use in biomedical devices

| Polymer | Medical applications |
| --- | --- |
| Polydimethylsiloxane, silicone elastomers (PDMS) | Breast, penile, and testicular prostheses |
| | Catheters |
| | Drug delivery devices |
| | Heart valves |
| | Hydrocephalus shunts |
| | Membrane oxygenators |
| Polyurethanes (PEU) | Artificial hearts and ventricular assist devices |
| | Catheters |
| | Pacemaker leads |
| Poly(tetrafluoroethylene) (PTFE) | Heart valves |
| | Vascular grafts |
| | Facial prostheses |
| | Hydrocephalus shunts |
| | Membrane oxygenators |
| | Catheters |
| | Sutures |
| Polyethylene (PE) | Hip prostheses |
| | Catheters |
| Polysulphone (PSu) | Heart valves |
| | Penile prostheses |
| Poly(ethylene terephthalate) (PET) | Vascular grafts |
| | Surgical grafts and sutures |
| Poly(methyl methacrylate) (pMMA) | Fracture fixation |
| | Intraocular lenses |
| | Dentures |
| Poly(2-hydroxyethylmethacrylate) (pHEMA) | Contact lenses |
| | Catheters |
| Polyacrylonitrile (PAN) | Dialysis membranes |
| Polyamides | Dialysis membranes |
| | Sutures |
| Polypropylene (PP) | Plasmapheresis membranes |
| | Sutures |
| Poly(vinyl chloride) (PVC) | Plasmapheresis membranes |
| | Blood bags |
| Poly(ethylene-*co*-vinyl acetate) | Drug delivery devices |
| Poly(L-lactic acid), Poly(glycolic acid), and | Drug delivery devices |
| Poly(lactide-*co*-glycolide) (PLA, PGA, and PLGA) | Sutures |
| Polystyrene (PS) | Tissue culture flasks |
| Poly(vinyl pyrrolidone) (PVP) | Blood subsitutes |

Collected from Marchant and Wang (1994); Peppas and Langer (1994).

fabricated in a format suitable for study; polystyrene films, for example, are transparent, durable, and strong. Other materials must be coated onto a rigid substrate (such as a glass coverslip) prior to study. But since cell function is sensitive to chemical, morphological, and mechanical properties of the surface, almost every aspect of material preparation can introduce variables that are known to influence cell interactions.

## Adhesion and Spreading

Most tissue-derived cells require attachment to a solid surface for viability and growth. For this reason, the initial events that occur when a cell approaches a surface are of fundamental interest. In tissue engineering, cell adhesion to a surface is critical because adhesion precedes other cell behaviors, such as cell spreading, cell migration, and, often, differentiated cell function.

A number of different techniques for quantifying the extent and strength of cell adhesion have been developed. In fact, so many different techniques are used that it is usually difficult to compare studies performed by different investigators. This situation is further complicated by the fact that cell adhesion depends on a large number of experimental parameters (Lauffenburger and Linderman, 1993), many of which are difficult to control. The simplest methods for quantifying the extent of cell adhesion to a surface involve three steps: (1) suspension of cells over a surface, (2) incubation of the sedimented cells in culture medium for some period of time, and (3) detachment of loosely adherent cells under controlled conditions. The extent of cell adhesion, which is a function of the conditions of the experiment, is determined by quantifying either the number of cells that remain associated with the surface (the adherent cells) or the number of cells that were extracted with the washes. Radiolabeled or fluorescently labeled cells can be used to permit measurement of the number of attached cells. Alternatively, the number of attached cells can be determined by direct visualization, by measurement of concentration of an intracellular enzyme, or by binding of a dye to an intracellular component such as DNA. In many cases, the adherent cells are further categorized based on morphological differences (e.g., extent of spreading, formation of actin filament bundles, presence of focal contacts). This technique is simple, rapid, and, since it requires simple equipment, common. Unfortunately, it is often difficult to control the force that is provided to dislodge the nonadherent cells, making it difficult to compare results obtained from different laboratories, even when they are using the same technique.

This disadvantage can be overcome by using a centrifuge or a flowing fluid to provide a reproducible detachment force. In centrifugal detachment assays, the technique just described is modified slightly: Following the incubation period, the plate is inverted and subjected to a controlled detachment force by centrifugation (McClay et al., 1981). In most flow chambers, the fluid is forced between two parallel plates (Lawrence et al., 1987). Prior to applying the flow field, a cell suspension is injected into the chamber, and the cells are permitted to settle onto the surface of interest and adhere. After some period of incubation, flow is initiated between the plates. These chambers can be used to measure the kinetics of cell attachment, detachment, and rolling on surfaces under conditions of flow. Usually, the overall flow rate is adjusted so that the flow is laminar, and the shear stresses at the wall approximate those found in the circulatory system; however, these chambers can be used to characterize cell detachment under a wide range of conditions.

Radial-flow detachment chambers have also been used to measure forces of cell detachment (Cozens-Roberts et al., 1990). Because of the geometry of the radial-flow chamber, where cells are attached uniformly to a circular plate and fluid is circulated from the center to the periphery of the chamber along radial paths, the fluid shear force experienced by the attached cells decreases with radial position from the center to the periphery. Therefore, in a single experiment, the influence of a range of forces on cell adhesion can be determined. A spinning disc apparatus can be used in a similar fashion (F. Lee et al., 2004). Finally, micropipette techniques can be used to measure cell membrane deformability or forces of cell–cell or cell–surface adhesion (Qin et al., 2005).

## Migration

The migration of individual cells within a tissue is a critical element in the formation of the architecture of organs and organisms (Trinkhaus, 1984). Similarly, cell migration is likely to be an important phenomenon in tissue engineering, since the ability of cells to move, either in association with the surface of a material or through an ensemble of other cells, will be an essential part of new tissue formation or regeneration. Cell migration is also difficult to measure, particularly in complex environments. Fortunately, a number of useful techniques for quantifying cell migration in certain situations have been developed. As in cell adhesion, however, no technique has gained general acceptance, so it is difficult to correlate results obtained by different techniques or different investigators.

Experimental methods for characterizing cell motility can be divided into two general categories. In *visual assays*, the movements of a small number of cells (usually ~100) are observed individually (Tan, 2001). *Population techniques*, on the other hand, allow the observation of the collective movements of groups of cells: In filter chamber assays the number of cells migrating through a membrane or filter is measured (Boyden, 1962), while in under-agarose assays the leading front of cell movement on a surface under a block of agarose is monitored (Nelson et al., 1975). Both visual and population assays can be quantitatively analyzed, enabling the estimation of intrinsic cell motility parameters, such as the random motility coefficient and the persistence time (see review in Saltzman, 2004).

## Aggregation

Cell aggregates are important tools in the study of tissue development, permitting correlation of cell–cell interactions with cell differentiation, viability, and migration as well as subsequent tissue formation. The aggregate morphology permits reestablishment of cell–cell contacts normally present in tissues; therefore, cell function and survival are often enhanced in aggregate culture. Because of this, cell aggregates may also be useful in tissue engineering, enhancing the function of cell-based hybrid artificial organs or reconstituted tissue transplants (Mahoney and Saltzman, 2001).

Gentle rotational stirring of suspensions of dispersed cells is the most common method for making cell aggregates (Moscona, 1961). While this method is suitable for aggregation of many cells, serum or serum proteins must be added to promote cell aggregation in many cases, thus making it difficult to characterize the aggregation process and to control the size and composition of the aggregate. Specialized techniques can be used to produce aggregates in certain cases, principally by controlling cell detachment from a solid substratum. For example, stationary culture of hepatocytes above a nonadherent surface (Parsons-Wingerter and Saltzman, 1993) or attached to a temperature-sensitive polymer substratum (Takezawa et al., 1993) have been used to form aggregates. Synthetic polymers produced by linking cell-binding peptides (such as RGD and YIGSR) to both ends of poly(ethylene glycol) (PEG) have been used to promote aggregation of cells in suspension (Dai, 1994).

The kinetics and extent of aggregation can be measured by a variety of techniques. Often, direct visualization of aggregate size is used to determine the extent of aggregation, following the pioneering work of Moscona (1961). The kinetics of aggregation can be monitored in this manner as well, by measuring aggregate size distributions over time. This procedure is facilitated by the use of computer image analysis techniques or electronic particle counters, where sometimes the disappearance of single cells (instead of the growth of aggregates) is followed. Specialized aggregometers have been constructed to provide reproducible and rapid measurements of the rate of aggregation. In one such device, small-angle light scattering through rotating sample cuvettes is used to produce continuous records of aggregate growth (Thomas, 1980).

### Cell Phenotype

In tissue-engineering applications, particularly those in which cell–polymer hybrid materials are prepared, one is usually interested in promotion of some cell-specific function. For example, protein secretion and detoxification are essential functions for hepatocytes used for transplantation or liver support devices; therefore, measurements of protein secretion and intracellular enzyme activity (particularly the hepatic $P_{450}$ enzyme system) are frequently used to assess hepatocyte function. Similarly, the expression and activity of enzymes involved in neurotransmitter metabolism (such as choline acetyltransferase and tyrosine hydroxylase) are often used to assess the function of neurons. Production of extracellular matrix (ECM) proteins is important in the physiology of many cells; collagen and glycosaminoglycan production has been used as an indicator of cell function in chondrocytes, osteoblasts, and fibroblasts. In some cases, the important cell function involves the coordinated activity of groups of cells, such as the formation of myotubules in embryonic muscle cell cultures, the contraction of the matrix surrounding the cells, or the coordinated contraction of cardiac muscle cells. In these cases, cell function is measured by observation of changes in morphology of cultured cells or cell communities.

### In Vivo Methods

The context of cell–polymer interactions *in vivo* is inherently complex, due to the presence of blood, interstitial fluids, and multiple cell types in various activation states. Proteins from the blood and interstitial fluid adsorb to polymer surfaces in a nonspecific manner and provide a substrate for cell adhesion (Ratner, 2004). Thus, it is believed that cells interact directly with this proteinaceous layer instead of the polymer surface, an assumption that explains the similar *in vivo* reactions elicited by a diverse array of polymers. Almost all implanted polymers induce a unique inflammatory response termed the *foreign-body response* (FBR) (Ratner, 2004).

The FBR can be divided into several overlapping phases that include nonspecific protein adsorption, inflammatory cell recruitment — predominantly of neutrophils and macrophages — macrophage fusion to form foreign-body giant cells (FBGC), and involvement of fibroblasts and endothelial cells (Fig. 20.1). The end result of the FBR is the formation of foreign-body giant cells directly on the polymer surface and the subsequent encapsulation of the implant by a fibrous capsule that is largely avascular. To date, the molecular and cellular determinants of the FBR remain largely unknown. However, recent studies have suggested specific roles for several proteins and a key role for macrophages in the FBR.

A number of implantation techniques in rodents and larger animals (typically rabbits, pigs, or sheep) have been adopted for the investigation of cell–polymer interactions. Most notably, short-term studies for the analysis of protein adsorption, inflammatory cell recruitment and adhesion, and macrophage fusion most often employ either intraperitoneal (IP) implantation or the subcutaneous (SC) cage implant, also known as the wound-chamber model. In the IP implant model, a sterilized piece of polymer is placed in the peritoneal cavity through a surgical incision (Tang et al., 1998). This model has been used extensively in the investigation of protein adsorption onto polymer surfaces and its impact during the early inflammatory response (0–72 h). In general, due to the short duration of such studies, it is difficult to extrapolate the information obtained from such studies to the fate of the FBR. Implant studies of intermediate duration (days to weeks) have been executed using an SC cage implant; in this system, a polymer is placed within a cage made of stainless steel wire mesh and then implanted through a surgical incision (Kao et al., 1995). The SC cage implant is useful because biologically active soluble agents

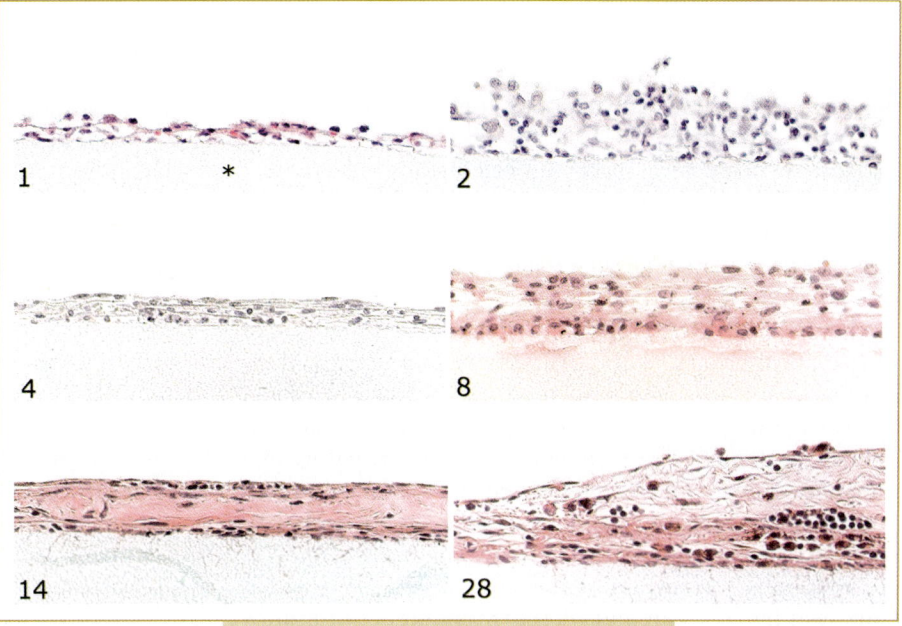

**FIG. 20.1.** Development of the foreign-body response. Representative images of hematoxylin and eosin-stained Millipore filter disks (mixed cellulose ester) implanted in the peritoneal cavity of wild-type mice. The number on each panel indicates the number of days after implantation. In the first two days following implantation, the discs display loose and expanding layers of attached inflammatory cells. Between four and eight days, the cell layers show increased organization that appears to coincide with the deposition of collagen fibers. Continuous deposition of collagenous matrix and an overall reduction in cellular content are characteristic of the two- and four-week time points. * indicates the disc. Original magnification 400×.

can be injected through the wire mesh in the immediate vicinity of the polymer.

Both IP and SC implantations allow for the analysis of recruited, nonadherent cell populations, which are collected by lavage of the peritoneal cavity or aspiration of host-derived exudates from the cage. In addition, implant materials can be recovered and analyzed for the presence of adherent cells by various techniques, immunohistochemistry being the most reliable. Thus, both implantation models can provide information regarding the migration of cells to the implantation site and the interaction of cells (adhesion, survival, fusion) with the polymer surface. Other commonly used *in vivo* models for the analysis of cell–polymer interactions involve the implantation of test polymers directly into host tissues such as dermis (subcutaneous), muscle, bone, and brain. Generally, such studies have been carried for longer incubation periods (weeks to months). Furthermore, polymer-based constructs have been surgically implanted at sites where they are expected to perform biological functions, for example, vascular grafts (L'Heureux *et al.*, 2006), heart valves (Vesely, 2005), and cardiac patches (Robinson *et al.*, 2005). Thus, in addition to biocompatibility and toxicity, polymers can be evaluated for their role in the function of a tissue or an organ. The direct implant model can be considered the most representative for most polymer applications and can provide information regarding the FBR, such as inflammation (FBGC formation), fibrosis (collagen content), encapsulation (capsule thickness), and angiogenesis (vascular density). Finally, in cases where the polymer is susceptible to degradation, additional analyses can be performed to evaluate the role of chemical and physical changes of the material in interactions with the tissue.

## III. CELL INTERACTIONS WITH POLYMERS

### Protein Adsorption to Polymers

A polymeric material that is placed in solution or implanted in the body becomes coated with proteins quickly, usually within minutes. Many of the subsequent interactions of cells with the material depend on, or derive from, the composition of the protein layer that forms on the surface. Polymers have been shown to adsorb a large number of proteins *in vitro* (Andrade and Hlady, 1987). For example, proteomic analysis of PP,* PET, and PDMS incubated with serum identified immunoglobulins, transferrin, albumin, serum amyloid P, and complement C4, among other proteins, on the polymer surface (Kim *et al.*, 2005). The C3 component of complement has also been shown to adsorb to medical-grade PEU, and adhesion of inflammatory cells to the polymer was reduced in the presence of C3-depleted serum but not fibronectin-depleted serum (Kao, 1999). In contrast, *in vivo* studies employing the IP implantation method suggest that adsorption of fibrinogen is the critical determinant of the FBR (Tang, 1993). Specifically, adsorption and denaturation of fibrinogen on polymer surfaces might lead to the exposure of cryptic cell adhesion motifs (integrin-binding sites) that influence subsequent cellular interactions.

### Effect of Polymer Chemistry on Cell Behavior

#### Synthetic Polymers

For cells attached to a solid substrate, cell behavior and function depend on the characteristics of the substrate.

* See Table 20.1 for abbreviations of polymers.

Consider, for example, experiments described by Folkman and Moscona, in which cells were allowed to settle onto surfaces formed by coating conventional tissue-culture polystyrene (TCPS) with various dilutions of pHEMA (Folkman and Moscona, 1978). As the amount of pHEMA added to the surface was increased, cell spreading decreased, as refected by the average cell height on the surface. The degree of spreading, or average height, correlated with the rate of cell growth (Fig. 20.2), suggesting that cell shape, which was determined by the adhesiveness of the surface, modulated cell proliferation. In these experiments, two simple polymers (TCPS and pHEMA) were used to produce a series of surfaces with graded adhesivity, permitting the identification of an important aspect of cell physiology. These experiments clearly demonstrate that the nature of a polymer surface will have important conquences for cell function, an observation of considerable significance with regard to the use of polymers in tissue engineering.

Following an experimental design similar to that employed by Folkman and Moscona, a number of groups have examined the relationship between chemical or physical characteristics of the substrate and the behavior or function of attached cells. For example, in a study of cell adhesion, growth, and collagen synthesis on synthetic polymers, fetal fibroblasts from rat skin were seeded onto surfaces of 13 different polymeric materials (Tamada and Ikada, 1994). The polymer surfaces had a range of surface energies, as determined by static water contact angles, from very hydrophilic to very hydrophobic (Table 20.2). On a few of the surfaces (PVA and cellulose), little cell adhesion and no cell growth was observed. On most of the remaining surfaces,

however, a moderate fraction of the cells adhered to the surface and proliferated. The rate of proliferation was relatively insensitive to surface chemistry: The cell doubling time is ~24 hr, with slightly slower growth observed for two very hydrophobic surfaces (PTFE and PP; see Table 20.2). Collagen biosynthesis was also correlated with contact angle, with higher rates of collagen synthesis per cell for the most hydrophobic surfaces.

Results from a number of similar studies are summarized in Table 20.2. Cell adhesion appears to be maximized on surfaces with intermediate wettability (Fig. 20.3). For most surfaces, adhesion requires the presence of serum; therefore, this optimum is probably related to the ability of proteins to adsorb to the surface. In the absence of serum, adhesion is enhanced on positively charged surfaces. This general principle was further confirmed for cell spreading on copolymers of HEMA (hydrophilic) and EMA (hydrophobic), which was highest at an intermediate HEMA content, again corresponding to intermediate wettability (Horbett *et al.*, 1988); in this case, spreading correlated with fibronectin adsorption (Horbett and Schway, 1988). Similarly, the rate of fibroblast growth on polymer surfaces appears to be relatively independent of surface chemistry. Cell viability may also be related to interactions with the surface, although this is not yet predictable. The migration of surface-attached fibroblasts, endothelial cells, and corneal epithelial cells is also a function of polymer surface chemistry (Table 20.2).

## Surface Modification

Polymers can frequently be made more suitable for cell attachment and growth by surface modification. In fact, polystyrene (PS) substrates used for tissue culture are usually treated by glow discharge or exposure to chemicals, such as sulfuric acid, to increase the number of charged groups at the surface, which improves attachment and growth of many types of cells. Other polymers can also be modified in this manner. Treatment of pHEMA with sulfuric acid, for example, improves adhesion of endothelial cells and permits cell proliferation on the surface (Hannan and McAuslan, 1987). Modification of PS or PET by radio-frequency plasma deposition enhances attachment and spreading of fibroblasts and myoblasts (Chinn *et al.*, 1989). Again, the effects of these surface modifications appear to be secondary to increased adsorption of cell attachment proteins, such as fibronectin and vitronectin, to the surface. On the other hand, some reports have identified specific chemical groups at the polymer surface — such as hydroxyl (–OH) (Curtis, 1983) or surface C–O functionalities (Chinn *et al.*, 1989) — as important factors in modulating the fate of surface-attached cells.

So far, no general principles that would allow prediction of the extent of attachment, spreading, or growth of cultured cells on different polymer surfaces have been identified. Correlations have been made with parameters such as the density of surface hydroxyl groups (Curtis *et al.*, 1983),

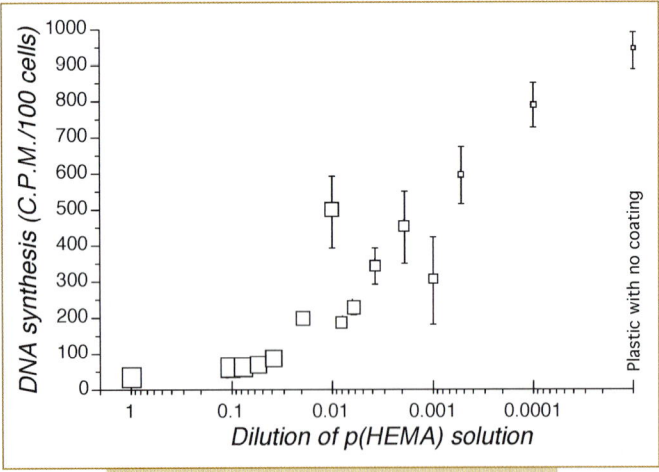

**FIG. 20.2.** Cell shape and growth are modulated by properties of a polymer surface. Cell culture surfaces were produced by evaporating diluted solutions of pHEMA onto TCPS. The uptake of [³H]thymidine was used as a measure of proliferation. The size of the symbol represents the relative cell height. Small symbols represent cells with small heights and therefore significant spreading; large symbols represent cells with large heights and therefore negligible spreading. Data replotted from Folkman and Moscona (1978).

| Table 20.2. | Cell attachment, growth, and motility on commonly used biomedical materials | | | |
|---|---|---|---|---|
| Polymer | Water contact angle (°) | Cell adhesion (% control) | Fibroblast doubling time (h) | Cell migration (mm) |
| PTFE | 105–116[a,l,m] | 28[f] 50[a] 70[m] | 60[a] | nm[f] 0.5[j] |
| PE | 94–97[a,k,l,m,n] | 83[a] 78[m] | 25[a] | 1.0[j] |
| PP | 92–97[a,m] | 67[a] 75[m] | 36[a] | |
| PS | 75–90[a,k] | 100[a] 85[m] | 20[a] | 1.2[j] |
| PET | 61–65[a,k] | 53[f] 20[g] 100[a] 77[m] | 24[a] | nm[f] 1.1[j] |
| Nylon | 61[a] | 92[a] 78[m] | 27[a] | |
| TCPS | 35–68[a,b,k,l,n] | | 22[a] | 1.8[j] |
| Poly(vinyl alcohol) (PVA) | 42[a] | | ng[a] | |
| Glass | 26–45[a,l,n] | 75[a] 100[m] | 19[a] 21[e] | 1.2[j] |
| Cellulose | 18[a] | 4[a] 27[m] | ng[a] | |
| PEU[h] | 57–86[h,k,l] | | | |
| PMMA | 65–80[k,l,n] | 85[m] | | 1.0[j] |
| PVF | 75[l] | | | 0.6[j] |
| PVDF | | | | 0.5[j] |
| FEP | 105–111[l,m,n] | 62[m] | | 0.4[j] |
| PDMS | 100–104[k,l] | 67[m] | | |
| pHEMA | 60–80[k] | | | |

[a]Fetal rat skin fibroblasts with 10% fetal calf serum; static water contact angles were measured; n.g. = no growth (Tamada and Ikada, 1994).

[b]Human umbilical vein endothelial cells with 20% human serum: receding water contact angles (van Wachem *et al.*, 1987).

[c]Bovine aortic endothelial cell with 10% fetal calf serum (McAuslan and Johnson, 1987).

[d]Mouse 3T3 fibroblasts with no serum (Horbett *et al.*, 1988).

[e]Mouse 3T3 fibroblasts with 10% calf serum (Horbett and Schway, 1988).

[f]Human adult endothelial cells with 20% fetal calf serum; n.m. = no motility (Hasson *et al.*, 1987).

[g]Human adult endothelial cells with 20% bovine serum, polymers pretreated with fibronectin (Pratt *et al.*, 1989).

[h]Polyurethane (PEU) based on poly(tetramethylene oxide)methylene diphenylene diisocyanate and 1,4-butanediol; receding water contact angles (Lin *et al.*, 1994).

[i]Dissociated fetal neurons with 20% serum; cell adhesion relative to poly-L-lysine control; the HEMA preparations contain trace MAA (Woerly *et al.*, 1993).

[j]Cell motility determined as outgrowth area in $cm^2$, corneal epithelial cells (Pettit *et al.*, 1994).

[k]Water-in-air contact angles (Ertel *et al.*, 1994).

[l]PBS-in-air contact angles (Schakenraad *et al.*, 1986).

[m]L cells with 10% serum (Ikada, 1994).

[n]Water contact angles (van der Valk *et al.*, 1983).

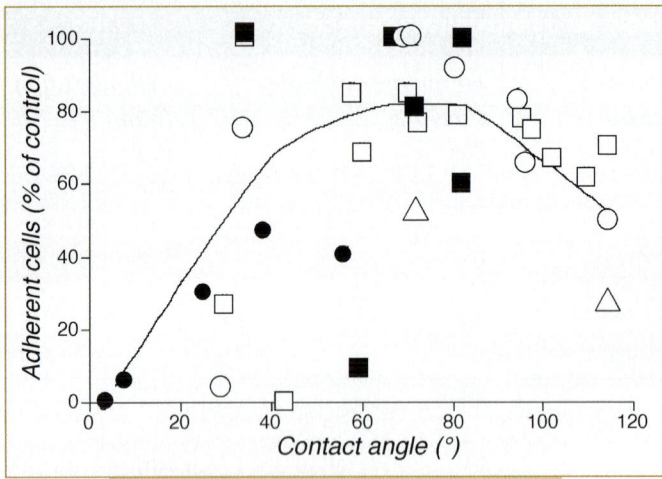

**FIG. 20.3.** Relationship between cell adhesion and water-in-air contact angle. Data replotted from Tamada and Ikada for fibroblasts (open squares) (Tamada and Ikada, 1994), Ikada for L cells (open circles) (Ikada, 1994), Hasson *et al.* for endothelial cells (open triangles) (Hasson *et al.*, 1987), van Wachem *et al.* for endothelial cells (filled circles) (van Wachem *et al.*, 1987).

density of surface sulfonic groups (Kowalczynska and Kaminski, 1991), surface free energy (Schakenraad *et al.*, 1986; van der Valk *et al.*, 1983), fibronectin adsorption (Chinn *et al.*, 1989), and equilibrium water content (Lydon *et al.*, 1985) for specific cells, but exceptions to these correlations are always found. Perhaps general predictive correlations will emerge, as complete characterization of polymer systems — including bulk properties, surface chemical properties, and nanoscale topography (as described later) — are collected. But many commonly used polymer materials are complicated mixtures, containing components that are added to enhance polymerization or to impart desired physical properties, often in trace quantities. Lot-to-lot variations in the properties of commercially available polymers can be significant.

The surface chemistry of polymers appears to influence cell interactions *in vivo*. For example, the ability of macrophages to form multinucleated giant cells at the material surface correlates with the presence of certain chemical groups at the surface of hydrogels: Macrophage fusion decreases in the order $(CH_3)_2N- > -OH = -CO-NH- > -SO_3H > -COOH$ $(-COONa)$ (Smetana *et al.*, 1990). A similar hierarchy has been observed for CHO (Chinese hamster ovary) cell adhesion and growth on surfaces with grafted functional groups: CHO cell attachment and growth decreased in the order $-CH_2NH_2 > -CH_2OH > -CONH_2 > -COOH$ (J. H. Lee *et al.*, 1994).

## Biodegradable Polymers

Biodegradable polymers slowly degrade and then dissolve following implantation. This feature may be important for many tissue-engineering applications, since the polymer will disappear as functional tissue regenerates. For this reason, interactions of cells with a variety of biodegradable polymers have been studied. Biodegradable polymers may provide an additional level of control over cell interactions: During polymer degradation, the surface of the polymer is constantly renewed, providing a dynamic substrate for cell attachment and growth.

Homopolymers and copolymers of lactic and glycolic acid (PLA, PGA, PLGA) have been frequently examined as cell culture substrates, since they have been used as implanted sutures for several decades. Many types of cells will attach and grow on these materials. For example, chondrocytes proliferate and secrete glycosaminoglycans within porous meshes of PGA and foams of PLA (Freed *et al.*, 1993). Similarly, rat hepatocytes attach to blends of biodegradable PLGA polymers and secrete albumin when maintained in culture (Cima *et al.*, 1991). Neonatal rat osteoblasts also attach to PLA, PGA, and PLGA substrates and synthesize collagen in culture (Ishaug *et al.*, 1994).

Cell adhesion and function have been examined on materials made from other biodegradable polymers. As an example, cells from an osteogenic cell line attach onto polyphosphazenes produced with a variety of side groups; the rate of cell growth as well as the rate of polymer degradation depend on side-group chemistry (Laurencin *et al.*, 1993).

Use of degradable polymers increases the complexity analysis in the *in vivo* setting because their degradation products can cause excessive and prolonged biological responses. Depending on the material used, degradation products can be released by hydrolysis, enzymatic digestion, degradative activity of macrophages and FBGC, or combinations of these mechanisms. The effects of polymer-derived products that are released by hydrolysis or enzymatic digestion depend on the chemical properties of the polymer and are not difficult to determine. On the other hand, the effects of products released due to cellular activities are difficult to predict, because the chemical and biological mechanisms responsible for their generation and release have not been established. Furthermore, the activity of cells, and thus the rate of release, can be influenced by several parameters, such as local concentrations of growth factors. It is expected that biodegradable polymers should allow for the favorable resolution of the FBR following their complete removal from the host tissue. The disappearance of FBGC from implantation sites following degradation of polymers to microparticles smaller than 10 μm in size has been observed (Kyriakides *et al.*, unpublished data). Furthermore, macrophages were observed to persist until the degradation of the polymer was complete. In SC implantation sites the formation of a vascularized collagenous work as a replacement for the polymer was observed. Presumably, the physiology of the implantation site, the acuteness of the inflammatory response, and the rate of polymer degradation all influence the nature of the FBR. Favorable *in vivo* results have been obtained with biodegradable elastomeric polyesters composed of poly(diol citrates) (J. Yang *et al.*, 2006). However, the duration of the FBR was shown to exceed 12 mo in implanted compressed naltrexone-poly[trans-3, 6-dimethyl-1, 4-dioxane-2, 5-dione] (DL-lactide) loaded microspheres (Hulse *et al.*, 2005).

## Synthetic Polymers with Adsorbed Proteins

As mentioned earlier, cell interactions with polymer surfaces appear to be mediated by proteins adsorbed from the local environment. Since it is difficult to study these effects *in situ* during cell culture, often the polymer surfaces are pretreated with purified protein solutions. In this way, the investigators hope that subsequent cell behavior on the surface will represent cell behavior in the presence of a stable layer of surface-bound protein. A major problem with this approach is the difficulty in determining whether surface conditions, i.e., the density of protein on the surface, change during the period of the experiment.

As described earlier, cell spreading, but not attachment, correlates with fibronectin adsorption to a variety of surfaces. Rates of cell migration on a polymer surface are usually sensitive to the concentration of preadsorbed adhesive proteins (Calof, 1991), and migration can be modified by addition of soluble inhibitors to cell adhesion (Wu *et al.*, 1994). It appears that the rate of migration is optimal at intermediate substrate adhesiveness, as one would expect from mathematical models of cell migration (Lauffenburger and Linderman, 1993). In fact, a recent study shows a clear correlation between adhesiveness and migration for CHO cells (Palecek *et al.*, 1997).

The outgrowth of corneal epithelial cells from explanted rabbit corneal tissue has been used as an indicator of cell attachment and migration on biomaterial surfaces (Pettit *et al.*, 1990). When corneal cell outgrowth was measured on 10 different materials that were preadsorbed with fibronectin, outgrowth generally increased with the ability of fibronectin to adsorb to the material (Table 20.2, footnote j). Exceptions to this general trend could be found, suggesting that other factors (perhaps stability of the adsorbed protein layer) are also important.

## Hybrid Polymers with Immobilized Functional Groups

Surface modification techniques have been used to produce polymers with surface properties that are more suitable for cell attachment (Ikada, 1994). For example, chemical groups can be added to change the wettability of the surface, which often influences cell adhesion (Fig. 20.3), as described earlier. Alternatively, whole proteins, such as collagen, can be immobilized to the surface, providing the cell with a substrate that more closely resembles the ECM found in tissues. Collagen and other ECM molecules have also been incorporated into hydrogels either by adding the protein to a reaction mixture containing monomers and initiating the radical polymerization or by mixing the protein with polymerized polymer, such as pHEMA, in appropriate

solvents. To isolate certain features of ECM molecules and to produce surfaces that are simpler and easier to characterize, smaller biologically active functional groups have been used to modify surfaces. These biologically active groups can be oligopeptides, saccharides, or glycolipids.

Certain short amino acid sequences, identified by analysis of active fragments of ECM molecules, appear to bind to receptors on cell surfaces and mediate cell adhesion. For example, the cell-binding domain of fibronectin contains the tripeptide RGD (Arg-Gly-Asp) (Pierschbacher and Ruoslahti, 1984).* Cells attach to surfaces containing adsorbed oligopeptides with the RGD sequence, and soluble, synthetic peptides containing the RGD sequence reduce the cell-binding activity of fibronectin, demonstrating the importance of this sequence in the adhesion of cultured cells. A large number of ECM proteins (fibronectin, collagen, vitronectin, thrombospondin, tenascin, laminin, and entactin) contain the RGD sequence. The sequences YIGSR and IKVAV on the A chain of laminin also have cell-binding activity and appear to mediate adhesion in certain cells.

Because RGD appears to be critical in cell adhesion to ECM, many investigators have examined the addition of this sequence to synthetic polymer substrates. The addition of cell-binding peptides to a polymer can induce cell adhesion to otherwise nonadhesive or weakly adhesive surfaces. Cell spreading and focal contact formation are also modulated by the addition of peptide (Fig. 20.4). Since cells contain cell adhesion receptors that recognize only certain ECM molecules, use of an appropriate cell-binding sequence can lead to cell-selective surfaces, where the population of the cells that adhere to the polymer is determined by the peptide (Hubbell *et al.*, 1991).

The presence of serum proteins attenuates the adhesion activity of peptide-grafted PEU surfaces (Lin *et al.*, 1994), highlighting a difficulty in using these peptide-grafted materials *in vivo*. This problem may be overcome, however, through the development of base materials that are biocompatible yet resistant to protein adsorption. One of the most successful approachs for reducing protein adsorption or cell adhesion is to produce a surface rich in PEG. A variety of techniques have been used, including surface grafting of PEG, adsorption of PEG-containing copolymers, semi-interpenetrating networks, and immobilization of PEG-star polymers to increase the density of PEG chains at the surface.

Alternatively, matrices formed directly from synthetic polypeptides may also be useful for cell adhesion. Genes coding the β-sheet of silkworm silk have been combined with genes coding fragments of fibronectin to produce

---

* Amino acids are identified by their one-letter abbreviations: A = alanine, R = arginine, N = asparagine, D = aspartic acid, B = asparagine or aspartic acid, C = cysteine, Q = glutamine, E = glutamic acid, Z = glutamine or glutamic acid, G = glycine, H = histidine, I = isoleucine, L = leucine, K = lysine, M = methionine, F = phenylalanine, P = proline, S = serine, T = threonine, W = trypotophan, Y = tyrosine, V = valine.

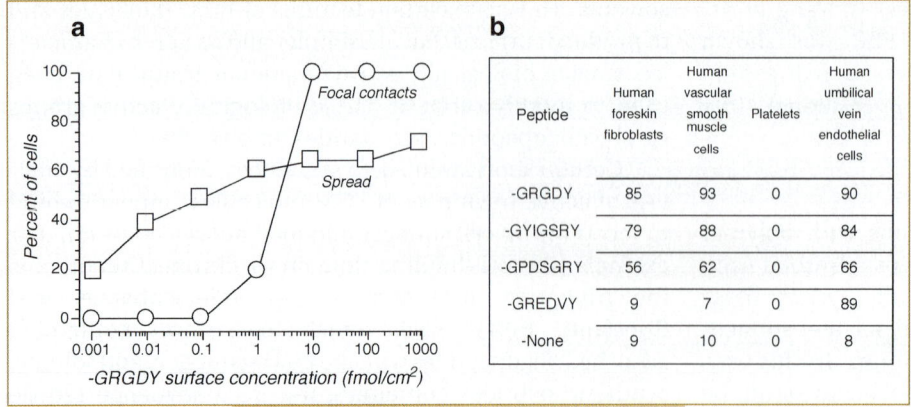

**a**

**b**

| Peptide | Human foreskin fibroblasts | Human vascular smooth muscle cells | Platelets | Human umbilical vein endothelial cells |
|---|---|---|---|---|
| -GRGDY | 85 | 93 | 0 | 90 |
| -GYIGSRY | 79 | 88 | 0 | 84 |
| -GPDSGRY | 56 | 62 | 0 | 66 |
| -GREDVY | 9 | 7 | 0 | 89 |
| -None | 9 | 10 | 0 | 8 |

**FIG. 20.4.** Cell adhesion to surfaces with immobilized peptides. Data from Hubbell *et al.* (1991); Massia and Hubbell (1991). **(a)** Fibroblast spreading on surfaces with immobilized –GRGDY. **(b)** Cell-selective surfaces: fraction of cells spread for several immobilized peptides.

proteins that form very stable matrices with cell adhesion domains (Pronectin F, Sanyo Chemical, Inc.). Synthetic proteins based on peptide sequences from elastin have been used as cell culture substrates: In the presence of serum, fibroblasts and endothelial cells adhered to the surfaces of matrices formed by γ-irradiation cross-linking of polypeptides containing repeated sequences GGAP, GGVP, GGIP, and GVGVP (Nicol *et al.*, 1993).

Surface adsorption of homopolymers of basic amino acids, such as polylysine and polyornithine, are frequently used to enhance cell adhesion and growth on polymer surfaces. Similarly, covalently bound amine groups can also influence cell attachment and growth. Polymerization of styrene with monamine- or diamine-containing monomers produced copolymers with ~8% mono- or diamine side chains, which enhanced spreading and growth: diamine-PS > monoamine-PS > PS (Kikuchi *et al.*, 1992).

The immobilization of saccharide units to polymers can also influence cell attachment and function. As an example, *N-p*-vinylbenzyl-*o*-β-D-galactopyranosyl-(1-4)-D-gluconamide has been polymerized to form a polymer with a polystyrene backbone and pendant lactose functionalities (Kobayashi *et al.*, 1992). Rat hepatocytes adhere to surfaces formed from this polymer, via asialoglycoprotein receptors on the cell surface, and remain in a rounded morphology consistent with enhanced function in culture. In the absence of serum, rat heptocytes will adhere to similar polymers with pendant glucose, maltose, or maltotriose. Similar results have been obtained with polymer surfaces derivatized with N-acetyl glucosamine, which is recognized by a surface lectin on chicken hepatocytes (Gutsche *et al.*, 1994).

### Electrically Charged or Electrically Conducting Polymers

A few studies have examined cell growth and function on polymers that are electrically charged. Piezoelectric polymer films, which were produced by high-intensity corona poling of poly(vinylidene fluoride) or poly(vinylidene fluoride-*co*-trifluoroethylene) and should generate transient surface charge in response to mechanical forces,

enhanced the attachment and differentiation of mouse neuroblastoma cells (Nb2a), as determined by neurite number and mean neurite length (Valentini *et al.*, 1992). These observations may be important *in vivo*, as well. For example, positively poled poly(vinylidene fluoride-*co*-trifluoroethylene) nerve guidance channels produced greater numbers of myelinated axons than either negatively poled or unpoled channels (Fine *et al.*, 1991). Electrically conducting polymers might be useful for tissue-engineering applications, because their surface properties can be changed by application of an applied potential. For example, endothelial cells attached and spread on fibronectin-coated polypyrrole films in the oxidized state, but they became rounded and ceased DNA synthesis when the surface was electrically reduced (Wong *et al.*, 1994).

### Influence of Surface Morphology on Cell Behavior

The microscale texture of an implanted material can have a significant effect on the behavior of cells in the region of the implant. This has long been observed *in vivo*. For example, fibrosarcomas developed with high frequency, approaching 50% in certain situations, around implanted Millipore filters; the tumor incidence increased with decreasing pore size in the range of 450–50 µm (Goldhaber, 1961).

The behavior of cultured cells on surfaces with edges, grooves, or other textures is different than behavior on smooth surfaces. In many cases, cells oriented and migrated along fibers or ridges in the surface, a phenomenon called *contact guidance*, from early studies on neuronal cell cultures (Weiss, 1934). Fibroblasts orient on grooved surfaces (Brunette, 1986), particularly when the texture dimensions are 1–8 µm (Dunn and Brown, 1986). The degree of cell orientation depends on both the depth and the pitch of the grooves. Not all cells exhibit the same degree of contact guidance when cultured on identical surfaces: BHK and MDCK cells orient on 100-nm-scale grooves in fused quartz, while cerebral neurons do not (Clark *et al.*, 1991). Fibroblasts, monocytes and macrophages, but not keratinocytes or neutrophils, spread when cultured on silicon oxide with grooves with a 1.2-µm depth and a 0.9-µm pitch (Meyle *et al.*, 1995). The variation in responses to surfaces with grooves and edges is shown in Table 20.3.

Substrates with peaks and valleys also influence the function of attached cells (Schmidt and von Recum, 1992). PDMS surfaces with 2- to 5-μm texture maximized macrophage spreading. Similarly, PDMS surfaces with 4- or 25-μm$^2$ peaks uniformly distributed on the surface provided better fibroblast growth than 100-μm$^2$ peaks or 4-, 25-, or 100-μm$^2$ valleys.

The microscale structure of a surface has a significant effect on cell migration, at least for the migration of human neutrophils. In one study, microfabrication technology was used to create regular arrays of micron-size holes (2 μm × 2 μm × 210 nm) on fused quartz and photosensitive polyimide surfaces (Tan *et al.*, 2001). The patterned surfaces, which possessed a basic structural element of a three-dimensional network (i.e., spatially separated mechanical edges), were used as a model system for studying the effect of substrate microgeometry on neutrophil migration. The edge-to-edge spacing between features was systematically varied from 6 μm to 14 μm with an increment of 2 μm. The presence of evenly distributed holes at the optimal spacing of 10 μm enhanced μ by a factor of 2 on polyimide, a factor of 2.5 on collagen-coated quartz, and a factor of 10 on uncoated quartz. The biphasic dependence on the mechanical edges of neutrophil migration on two-dimensional patterned substrate was strikingly similar to that previously observed during neutrophil migration within three-dimensional networks, suggesting that microfabricated materials provide relevant models of three-dimensional structures with precisely defined physical characteristics. Perhaps more importantly, these results illustrate that the microgeometry of a substrate, when considered separately from adhesion, can play a significant role in cell migration.

### Use of Patterned Surfaces to Control Cell Behavior

A variety of techniques have been used to create patterned surfaces containing cell adhesive and nonadhesive regions. Patterned surfaces are useful for examining fundamental determinants of cell adhesion, growth, and function. For example, individual fibroblasts were attached to adhesive microislands of palladium that were patterned onto a nonadhesive pHEMA substrate using microlithographic techniques (O'Neill *et al.*, 1986). By varying the size of the microisland, the extent of spreading and hence the surface area of the cell was controlled. On small islands (~500 μm$^2$), cells attached but did not spread. On larger islands (4000 μm$^2$), cells spread to the same extent as in unconfined monolayer culture. Cells on large islands proliferate at the same rate as cells in conventional culture, and most cells attached to small islands proliferate at the same rate as suspended cells. For 3T3 cells, however, contact with the surface enhanced proliferation, suggesting that anchorage can stimulate cell division by simple contact with the substrate as well as by increases in spreading.

A number of other studies have employed patterned surfaces in cell culture. Micrometer-scale adhesive islands of self-assembled alkanethiols were created on gold surfaces using a simple stamping procedure (Singhvi *et al.*, 1994), which served to confine cell-spreading islands. When hepatocytes were attached to these surfaces, larger islands (10,000 μm$^2$) promoted growth, while smaller islands (1600 μm$^2$) promoted albumin secretion. Stripes of a monoamine-derivatized surface were produced on fluorinated ethylene propylene films by radio-frequency glow discharge (Ranieri *et al.*, 1993). Since proteins adsorbed differently to the monamine-derivatized and the untreated stripes, striped patterns of cell attachment were produced. A similar approach, using photolithography to produce hydrophilic patterns on a hydrophobic surface, produced complex patterns of neuroblastoma attachment and neurite extension (Matsuda *et al.*, 1992). A variety of substrate microgeometries were created by photochemical fixation of hydrophilic polymers onto TCPS or hydrophobic polymers onto PVA through patterned photomasks: Bovine endothelial cells attached and proliferated preferentially on either the TCPS surface (on TCPS/hydrophilic patterns) or the hydrophobic surface (on PVA/hydrophobic patterns) (Matsuda and Sugawara, 1995). When chemically patterned substrates were produced on self-assembled monolayer films using microlithographic techniques, neuroblastoma cells attached to and remained confined within amine-rich patterns on these substrates (Matsuzawa *et al.*, 1993).

## IV. CELL INTERACTIONS WITH POLYMERS IN SUSPENSION

Most of the studies reviewed in the preceding section concerned the growth, migration, and function of cells attached to a solid polymer surface. This is a relevant paradigm for a variety of tissue-engineering applications where polymers will be used as substrates for the transplantation of cells or as scaffolds to guide tissue regeneration *in situ*. Polymers may be important in other aspects of tissue engineering, as well. For example, polymer microcarriers can serve as substrates for the suspension culture of anchorage-dependent cells, and therefore they might be valuable for the *in vitro* expansion of cells or cell transplantation (Demetriou *et al.*, 1986). In addition, immunoprotection of cells suspended within semipermeable polymer membranes is another important approach in tissue engineering, since these encapsulated cells may secrete locally active proteins or function as small endocrine organs within the body.

The idea of using polymer microspheres as particulate carriers for the suspension culture of anchorage-dependent cells was introduced by van Wezel (1967). As already described for planar polymer surfaces, the surface characteristics of microcarriers influence cell attachment, growth, and function. In the earliest studies, microspheres composed of diethylaminoethyl (DEAE)-dextran were used; these spheres have a positively charged surface and are

**Table 20.3.** Summary of the effect of parallel ridges/grooves on cell behavior

| Cells | Material | h/d (μm) | w (μm) | s (μm) | Result |
|---|---|---|---|---|---|
| Chick heart fibroblasts (Dunn, 1982) | Glass | * | 2 | 2 | Not aligned |
| | | * | 4 | 4–12 | Aligned |
| Human gingival fibroblasts (Brunette, 1986) | Epon | 5 | 36–78 | 36–78 | Aligned |
| | | 92 | 100–162 | 101–162 | Not aligned |
| Teleost fin mesenchymal cells (Wagers et al., 2002) | Quartz | 0.8–1.1 | 1–4 | 1–4 | Aligned with increasing width |
| BHK, MDCK, and Chick embryo neurites (Clark et al., 1990) | Silicon or ECM protein-coated silicon | 0.2–1.9 | 2–12 | 2–12 | Aligned with increasing depth, and alignment depended on depth |
| Hippocampal neurons (Rajnicek et al., 1997) | Quartz | 0.014 | 1 | 1 | Not aligned |
| | | 1.1 | 4 | 4 | Aligned |
| Xenopus spinal cord neurons (Rajnicek et al., 1997) | Quartz | 0.014–1.1 | 1–4 | 1–4 | Aligned |
| Epithelial tissue and cells (Dalton et al., 2001) | Polystyrene | 1 or 5 | 1–10 | 1–10 | Migration was enhanced along the grooves; more significant effect on deeper grooves |
| Osteoblasts (Perizzolo et al., 2001) | Ti- or Ca-P-coated silicon | 3, 10, or 30 | 5 | 42 | Aligned and increased bonelike nodule formation |
| Murine macrophage P388D1 (Wojciak-Stothard et al., 1996) | Fused silica or ECM protein-coated silica | 0.03–0.282 | 2 or 10 | * | Aligned with increasing depth or decreasing width |
| Human neutrophils (Tan and Saltzman, 2002) | Silicon and silicon coated with metals | 3, 5 | 2 | 6–14 | Rate of migration depended on s |
| Bovine pulmonary artery smooth muscle cells (Hu et al., 2005) | Polystyrene | 0.2–0.9 | 0.1–10 | 0.1–10 | Alignment of attached cells, which was enhanced on the nanoscale features |

Adapted from Tan and Saltzman (2002) and Saltzman (2004). *Abbreviations:* BHK, baby hamster kidney; MDCK, Madin Darby canine kidney; ECM, extracellular matrix; *h*, height of ridges; *d*, depth of grooves; *w*, width of ridges; *s*, spacing between ridges; *, data not specified.

routinely used as anion-exchange resins. DEAE-dextran microcarriers support the attachment and growth of both primary cells and cell lines, particularly when the surface charge is optimized. In addition to dextran-based microcarriers, microspheres that support cell attachment can be produced from PS, gelatin, and many of the synthetic and naturally occuring polymers described in the preceding sections. The surface of the microcarrier can be modified chemically or by immobilization of proteins, peptides, or carbohydrates.

Suspension culture techniques can be used to permit cell interactions with complex three-dimensional polymer formulations, as well. For example, cells seeding onto polymer fiber meshes during suspension culture often result in more uniform cell distribution within the mesh than can be obtained by inoculation in static culture (Freed and Vunjak-Novakovic, 1995).

In cell encapsulation techniques, cells are suspended within thin-walled capsules or solid matrices of polymer. Alginate forms a gel with the addition of divalent cations under very gentle conditions and, therefore, has been frequently used for cell encapsulation. Certain synthetic polymers, such as polyphosphazenes, can also be used to encapsulate cells by cation-induced gelation. Low-melting-temperature agarose has also been studied extensively for cell encapsulation. Methods for the microencapsulation of cells within hydrophilic or hydrophobic polyacrylates by interfacial precipitation have been described (Dawson et al., 1987), although the thickness of the capsule can limit the permeation of compounds, including oxygen, through the semipermeable membrane shell. Interfacial polymerization can be used to produce conformal membranes on cells or cell clusters (Sawhney et al., 1994), thereby providing immunoprotection while reducing diffusional distances.

Hollow fibers are frequently used for macroencapsulation; cells and cell aggregates are suspended within thin fibers composed of a porous, semipermeable polymer. Chromaffin cells suspended within hollow fibers formed from copolymer of vinyl chloride and acrylonitrile, which are commonly used as ultrafiltration membranes, have been studied as potential treatments for cancer patients with pain (Joseph et al., 1994), Alzheimer's disease (Emerich et al., 1994), and retinitis pigmentosa (Tao et al., 2002). Other polymer materials — such as chitosan, alginate, and agar — have been added to the interior of the hollow fibers to provide an internal matrix that enhances cell function or growth.

# V. CELL INTERACTIONS WITH THREE-DIMENSIONAL POLYMER SCAFFOLDS AND GELS

Cells within tissues encounter a complex chemical and physical environment that is quite different from commonly used cell culture conditions. Three-dimensional cell culture methods are frequently used to simulate the chemical and physical environment of tissues. Often, tissue-derived cells cultured in ECM gels will reform multicellular structures that are reminiscent of tissue architecture.

Gels of agarose have also been used for three-dimensional cell culture. Chondrocytes dedifferentiate when cultured as monolayers, but they reexpress a differentiated phenotype when cultured in agarose gels (Benya and Shaffer, 1982). When fetal striatal cells are suspended in three-dimensional gels of hydroxylated agarose, ~50% of the cells extended neurites in gels containing between 0.5% and 1.25% agarose, but no cells extended neurites at concentrations above 1.5%. This inhibition of neurite outgrowth correlates with an average pore radius of greater than 150 nm (Bellamkonda et al., 1995). Neurites produced by PC12 cells within agarose gels, even under optimal conditions, are much shorter and fewer in number than neurites produced in gels composed of ECM molecules (Krewson et al., 1994).

Macroporous hydrogels can also be produced from pHEMA-based materials, using either freeze–thaw or porosigen techniques. These materials, when seeded with chondrocytes, may be useful for cartilage replacement (Corkhill et al., 1993). Similar structures may be produced from PVA by freeze–thaw cross-linking. Recently, a PEG-based macroporous gel was used as a scaffold for endothelial cells to form microvessel networks in vivo (Ford et al., 2006). Although cells adhere poorly to pHEMA, PVA, and PEG materials, adhesion proteins or charged polymers can be added during the formation to encourage cell attachment and growth. Alternatively, water-soluble, nonadhesive polymers containing adhesive peptides, such as RGDS, can be photopolymerized to form a gel matrix around cells (see Moghaddam and Matsuda, 1991, for example).

Fiber meshes and foams of PLGA, PLA, and PGA have been used to create three-dimensional environments for cell proliferation and function and to provide structural scaffolds for tissue regeneration. When cultured on three-dimensional PGA fiber meshes, chondrocytes proliferate, produce both glycosaminoglycans and collagen, and form structures that are histologically similar to cartilage (Puelacher et al., 1994). The internal structure of the material as well as the physical dimensions of the polymer fiber mesh influence cell growth rate, with slower growth in thicker meshes. Changing the fluid mechanical forces on the cells during the tissue formation also appears to influence the development of tissue structure.

In addition to fiber meshes, porosity can be introduced into polymer films by phase separation, freeze-drying, salt leaching, and a variety of other methods (reviewed in S. Yang et al., 2001). It is now possible to make porous, degradable scaffolds with controlled pore architectures and oriented pores (Ma and Choi, 2001; Tong et al., 2002). Fabrication methods that provide control over the structure at different length scales may be useful in the production of three-dimensional tissue-like structures.

Most methods for producing fiber meshes are limited to producing fibers ~10 microns in diameter, which is much larger than the diameter of natural fibers that occur in the extracellular matrix and also larger than many of the features that are known to be important in orienting or guiding cell activity (Table 20.3). Electrospinning techniques can be used to make small-diameter fibers and nonwoven meshes of a variety of materials, including poly(caprolactone), PLA, collagen, and elastin-mimetic polymers.

## VI. CELL INTERACTIONS UNIQUE TO THE *IN VIVO* SETTING

While cell interactions with polymers *in vitro* can be described by examination of cell behaviors — such as adhesion, migration, and gene expression — or the coordinated behavior of cell groups — such as aggregation, cell interactions with polymers *in vivo* can lead to other responses, involving cells that are recruited to the implantation site and remodeling of the tissue space surrounding, or even within, the polymeric material. Inflammation, the foreign-body response, and angiogenesis are three examples of these more global responses to an implanted material.

There is much still to learn in this area, but it is clear that both the implant material and the physiology of the implant site are important variables. A recent paper describing a relatively simple experiment, in which ePTFE implants were placed in adipose tissue, in subcutaneous tissue, or epicardially, illustrates the variability of these responses (Kellar *et al.*, 2002). This short section introduces these physiological responses to implanted materials.

### Inflammation

The implantation of polymers through surgical incisions means that an initial component of the FBR involves a wound healing–like response, and it is reasonable to assume that the early inflammatory response is mediated, at least in part, by wound-derived factors. Analysis of inflammatory cells has been pursued in several implantation models and was shown to involve predominantly neutrophils (early) and monocyte/macrophages (late). Subsequent to their recruitment, these cells are believed to utilize adhesion receptors to interact with adsorbed proteins. Studies in mice that lack specific integrins or fibrinogen have provided supporting evidence for this hypothesis (Busuttil *et al.*, 2004; Lu *et al.*, 1997). Specifically, short-term (18-h) IP implantation of polyethylene terephthalate (PET) discs in mice that lack fibrinogen indicated normal recruitment but reduced adhesion of macrophages and neutrophils to the polymer. In the same study, analysis of the response in mice that lack plasminogen indicated no changes in cell adhesion to the polymer, despite a reduction in the recruitment of both cell types in the peritoneal cavity. Thus, in addition to fibrinogen for adhesion, inflammatory cells can utilize plasminogen for migration/recruitment. Surprisingly, the chemokine CCL2 (also known as monocyte chemoattractant protein, or MCP-1) was shown not to be important in monocyte/macrophage recruitment in long-term implants (Kyriakides *et al.*, 2004). However, CCL2 was shown to be important for macrophage fusion leading to FBGC formation. FBGC can cause damage to polymer surfaces through their degradative and phagocytic activities and, thus, pose a significant obstacle to the successful application of polymer-based biomaterials and devices. *In vivo* studies have identified a critical role for interleukin (IL)-4 in the formation of FBGC (Kao *et al.*, 1995), but little is known about the regulation of macrophage fusion. On the other hand, several studies have focused on the role of polymer surface chemistry on macrophage function and FBGC formation. For example, analysis of macrophage adhesion, apoptosis, and fusion on hydrophobic (PET- and BDEDTC-coated), hydrophilic (PAAm), anionic (PAANa), and cationic (DMAPAAmMel) surfaces implanted in the rat cage implant model revealed that PAAm and PAANa induced more apoptosis and reduced adhesion and fusion (Christenson *et al.*, 2005).

### Fibrosis and Angiogenesis

Unlike wound healing, the resolution of the polymer-associated inflammatory response is characterized by the excessive deposition of a highly organized collagenous matrix and a striking paucity of blood vessels (Kyriakides and Bornstein, 2003; Mikos *et al.*, 1998). The collagenous capsule can vary in thickness but usually exceeds 100 μm, presumably to limit diffusion of small molecules to and from the polymer. The dense and organized nature of the collagen fibers in the capsule could play a role in limiting blood vessel formation. Implantation studies in mice that lack the angiogenesis inhibitor TSP2 indicated that an increase in vascular density in capsules surrounding PDMS discs was associated with significant loosening of the collagenous matrix (Kyriakides *et al.*, 1999). However, a direct link between the arrangement of collagen fibers in the capsule and blood vessel formation has not been established. Interestingly, the modification of the PDMS surface from a hydrophobic to a hydrophilic state altered its cell adhesive properties *in vitro* but did not cause a change in the FBR *in vivo* (Kyriakides *et al.*, 1999). Such observations underscore the significance of *in vivo* evaluation of cell- and tissue–polymer interactions. Reduced encapsulation of polydimethylsiloxane (silicone rubber) discs and cellulose Millipore filters implanted SC has been reported in mice that lack SPARC (secreted protein, acidic and rich in cysteine), a matricellular glycoprotein that modulates the interactions of cells with the extracellular matrix. Interestingly, mice that lack SPARC and its close homolog, hevin, display diminished vascular density in encapsulated Millipore filters (type HA, mixed cellulose ester) (Barker *et al.*, 2005). Taken together, implantation studies in genetically modified mice suggest that members of the matricellular protein group play critical roles in the FBR (Kyriakides and Bornstein, 2003). The process however, can also be influenced by parameters such as polymer special geometry and porosity. Comparison of the FBR elicited by expanded and condensed PTFE showed similar encapsulation but more mature fibrous capsule formation in the latter (Voskerician *et al.*, 2006). In addition, the effect of polymer porosity in the FBR was examined in SC-implanted PTFE membranes in rats, where it was shown that the vascular density could be increased in capsules surrounding polymers with a pore size in the range of 5 μm (Brauker *et al.*, 1995). However, it is unclear whether the same porosity would enhance the vascular density of capsules surrounding other polymers. Finally, an additional concern with polymer encapsulation is the presence of contractile cells, myofibroblasts, which can cause contraction of the capsule and misshape or damage polymer implants. For example, silicone-based breast implants have been shown to be susceptible to this phenomenon (Granchi *et al.*, 1995).

# VII. REFERENCES

Andrade, J. D., and Hlady, V. (1987). Plasma protein adsorption: the big twelve. *Ann. NYork Acad. Sci.* **516**, 158–172.

Barker, T. H., Framson, P., Puolakkainen, P. A., Reed, M., Funk, S. E., and Sage, E. H. (2005). Matricellular homologs in the foreign body response: hevin suppresses inflammation, but hevin and SPARC together diminish angiogenesis. *Am. J. Pathol.* **166**, 923–933.

Bellamkonda, R., Ranieri, J. P., Bouche, N., and Aebischer, P. (1995). Hydrogel-based three-dimensional matrix for neural cells. *J. Biomed. Mater. Res.* **29**, 663–671.

Benya, P. D., and Shaffer, J. D. (1982). Dedifferentiated chondrocytes reexpress the differentiated collagen phenotype when cultured in agarose gels. *Cell* **30**, 215–224.

Boyden, S. V. (1962). The chemotactic effect of mixtures of antibody and antigen on polymorphonuclear leukocytes. *J. Exp. Med.* **115**, 453–466.

Brauker, J. H., Carr-Brendel, V. E., Martinson, L. A., Crudele, J., Johnston, W. D., and Johnson, R. C. (1995). Neovascularization of synthetic membranes directed by membrane microarchitecture. *J. Biomed. Mater. Res.* **29**, 1517–1524.

Brunette, D. (1986). Fibroblasts on micromachined substrata orient hierarchically to grooves of different dimensions. *Exp. Cell Res.* **164**, 11–26.

Busuttil, S. J., Ploplis, V. A., Castellino, F. J., Tang, L., Eaton, J. W., and Plow, E. F. (2004). A central role for plasminogen in the inflammatory response to biomaterials. *J. Thromb. Haemost.* **2**, 1798–1805.

Chinn, J., Horbett, T., Ratner, B., Schway, M., Haque, Y., and Hauschka, S. (1989). Enhancement of serum fibronectin adsorption and the clonal plating efficiencies of swiss mouse 3T3 fibroblast and MM14 mouse myoblast cells on polymer substrates modified by radiofrequency plasma deposition. *J. Colloid Interface Sci.* **127**, 67–87.

Christenson, E. M., Dadsetan, M., and Hiltner, A. (2005). Biostability and macrophage-mediated foreign body reaction of silicone-modified polyurethanes. *J. Biomed. Mater. Res. A* **74**, 141–155.

Cima, L., Ingber, D., Vacanti, J., and Langer, R. (1991). Hepatocyte culture on biodegradable polymeric substrates. *Biotechnol. Bioengin.* **38**, 145–158.

Clark, P., Connolly, P., Curtis, A. S., Dow, J. A., and Wilkinson, C. D. (1990). Topographical control of cell behavior II. Multiple grooved substrata. *Development* **108**, 635–644.

Clark, P., Connolly, P., Curtis, A. S. G., Dow, J. A. T., and Wilkinson, C. D. W. (1991). Cell guidance by ultrafine topography in vitro. *J. Cell Sci.* **99**, 73–77.

Corkhill, P. H., Fitton, J. H., and Tighe, B. J. (1993). Towards a synthetic articular cartilage. *J. Biomater. Sci. Polymer Edn.* **4**, 6150–6630.

Cozens-Roberts, C., Quinn, J., and Lauffenburger, D. (1990). Receptor-mediated adhesion phenomena: model studies with the radial-flow detachment assay. *Biophys. J.* **58**, 107–125.

Curtis, A., Forrester, J., McInnes, C., and Lawrie, F. (1983). Adhesion of cells to polystyrene surfaces. *J. Cell Biol.* **97**, 1500–1506.

Dalton, B. A., Walboomers, X. F., Dziegielewski, M., Evans, M. D. M., Taylor, S., Jansen, J. A., and Steele, J. G. (2001). Modulation of epithelial tissue and cell migration by microgrooves. *J. Biomed. Mater. Res.* **56**, 195–207.

Dawson, R. M., Broughton, R. L., Stevenson, W. T. K., and Sefton, M. V. (1987). Microencapsulation of CHO cells in a hydroxyethyl methacrylate-methyl methacrylate copolymer. *Biomaterials* **8**, 360–366.

Demetriou, A., Whiting, J., Feldman, D., Levenson, S., Chowdhury, N., Moscioni, A., Kram, M., and Chowdhury, J. (1986). Replacement of liver function in rats by transplantation of microcarrier-attached hepatocytes. *Science* **23**, 1190–1192.

Dunn, G. A. (1982). Contact guidance of cultured tisse cells: a survey of potentially relevant properties of the substratum. *In* "Cell Behavior" (R. Bellairs, A. Curtis, and G. Dunn, eds.). Cambridge University Press, Cambridge, UK.

Dunn, G. A., and Brown, A. F. (1986). Alignment of fibroblasts on grooved surfaces described by a simple geometric transformation. *J. Cell Sci.* **83**, 313–340.

Emerich, D. F., Hammang, J. P., Baetge, E. E., and Winn, S. R. (1994). Implantation of polymer-encapsulated human nerve growth factor–secreting fibroblasts attenuates the behavioral and neuropathological consequences of quinolinic acid injections into rodent striatum. *Exp. Neurol.* **130**, 141–150.

Ertel, S. I., Ratner, B. D., Kaul, A., Schway, M. B., and Horbett, T. A. (1994). *In vitro* study of the intrinsic toxicity of synthetic surfaces to cells. *J. Biomed. Mater. Res.* **28**, 667–675.

Fine, E. G., Valentini, R. F., Bellamkonda, R., and Aebischer, P. (1991). Improved nerve regeneration through piezoelectric vinylidenefluordie-trifluoroethylene copolymer guidance channels. *Biomaterials* **12**, 775–780.

Folkman, J., and Moscona, A. (1978). Role of cell shape in growth control. *Nature* **273**, 345–349.

Ford, M. C., Bertram, J. P., Hynes, S. R., Michaud, M., Li, Q., Young, M., Segal, S. S., Madri, J. A., and Lavik, E. B. (2006). A macroporous hydrogel for the coculture of neural progenitor and endothelial cells to form vascular networks in vivo. *Proc. Natl. Acad. Sci. USA* **103**, 2512–2517.

Freed, L. E., and Vunjak-Novakovic, G. (1995). Cultivation of cell-polymer tissue constructs in simulated microgravity. *Biotechnol. Bioeng.* **46**, 306–313.

Freed, L., Marquis, J., Nohria, A., Emmanual, J., Mikos, A., and Langer, R. (1993). Neocartilage formation *in vitro* and *in vivo* using cells cultured on synthetic biodegradable polymers. *J. Biomed. Mater. Res.* **27**, 11–23.

Granchi, D., Cavedagna, D., Ciapetti, G., Stea, S., Schiavon, P., Giuliani, R., and Pizzoferrato, A. (1995). Silicone breast implants: the role of immune system on capsular contracture formation. *J. Biomed. Mater. Res.* **29**, 197–202.

Gutsche, A. T., Parsons-Wingerter, P., Chand, D., Saltzman, W. M., and Leong, K. W. (1994). N-acetylglucosamine and adenosine derivatized surfaces for cell culture: 3T3 fibroblast and chicken hepatocyte response. *Biotechnol. Bioeng.* **43**, 801–809.

Hannan, G., and McAuslan, B. (1987). Immobilized serotonin: a novel substrate for cell culture. *Exp. Cell Res.* **171**, 153–163.

Hasson, J., Wiebe, D., and Abbott, W. (1987). Adult human vascular endothelial cell attachment and migration on novel bioabsorbable polymers. *Arch. Surg.* **122**, 428–430.

Horbett, T., and Schway, M. (1988). Correlations between mouse 3T3 cell spreading and serum fibronectin adsorption on glass and hydroxyethylmethacrylate-ethylmethacrylate copolymers. *J. Biomed. Mater. Res.* **22**, 763–793.

Horbett, T. A., Waldburger, J. J., Ratner, B. D., and Hoffman, A. S. (1988). Cell adhesion to a series of hydrophilic-hydrophobic copolymers studied with a spinning disc apparatus. *J. Biomed. Mater. Res.* **22**, 383–404.

Hu, W., Yim, E. K. F., Reano, R. M., Leong, K. W., and Pang, S. W. (2005). Effects of nanoimprinted patterns in tissue-culture polystyrene on cell behavior. *J. Vac. Sci. Technol. B* **23**, 2984–2989.

Hubbell, J. A., Massia, S. P., Desai, N. P., and Drumheller, P. D. (1991). Endothelial cell-selective materials for tissue engineering in the vascular graft via a new receptor. *Bio/Technology* **9**, 568–572.

Hulse, G. K., Stalenberg, V., McCallum, D., Smit, W., O'Neil, G., Morris, N., and Tait, R. J. (2005). Histological changes over time around the site of sustained release naltrexone-poly(DL-lactide) implants in humans. *J. Control. Release* **108**, 43–55.

Ikada, Y. (1994). Surface modification of polymers for medical applications. *Biomaterials* **15**, 725–736.

Ishaug, S. L., Yaszemski, M. J., Bizios, R., and Mikos, A. G. (1994). Osteoblast function on synthetic biodegradable polymers. *J. Biomed. Mater. Res.* **28**, 1445–1453.

Joseph, J. M., Goddard, M. B., Mills, J., Padrun, V., Zurn, A., Zielinski, B., Favre, J., Gardaz, J. P., Mosimann, F., Sagen, J., Christenson, L., and Aebischer, P. (1994). Transplantation of encapsulated bovine chrommafin cells in the sheep subarachnoid space: a preclinical study for the treatment of cancer pain. *Cell Transpl.* **3**, 355–364.

Kao, W. J. (1999). Evaluation of protein-modulated macrophage behavior on biomaterials: designing biomimetic materials for cellular engineering. *Biomaterials* **20**, 2213–2221.

Kao, W. J., McNally, A. K., Hiltner, A., and Anderson, J. M. (1995). Role for interleukin-4 in foreign-body giant cell formation on a poly(etherurethane urea) in vivo. *J. Biomed. Mater. Res.* **29**, 1267–1275.

Kellar, R. S., Kleinert, L. B., and Williams, S. K. (2002). Characterization of angiogenesis and inflammation surrounding ePTFE implanted on the epicardium. *J. Biomed. Mater. Res.* **61**, 226–233.

Kikuchi, A., Kataoka, K., and Tsuruta, T. (1992). Adhesion and proliferation of bovine aortic endothelial cells on monoamine- and diamine-containing polystyrene derivatives. *J. Biomater. Sci. Polymer Ed.* **3**, 253–260.

Kim, J. K., Scott, E. A., and Elbert, D. L. (2005). Proteomic analysis of protein adsorption: serum amyloid P adsorbs to materials and promotes leukocyte adhesion. *J. Biomed. Mater. Res. A* **75**, 199–209.

Kobayashi, A., Kobayashi, K., and Akaike, T. (1992). Control of adhesion and detachment of parenchymal liver cells using lactose-carrying polystyrene as substratum. *J. Biomater. Sci. Polymer ed.* **3**, 499–508.

Kowalczynska, H. M., and Kaminski, J. (1991). Adhesion of L1210 cells to modified styrene copolymer surfaces in the presence of serum. *J. Cell Sci.* **99**, 587–593.

Krewson, C. E., Chung, S. W., Dai, W., and Saltzman, W. M. (1994). Cell aggregation and neurite growth in gels of extracellular matrix molecules. *Biotechnol. Bioengin.* **43**, 555–562.

Kyriakides, T. R., and Bornstein, P. (2003). Matricellular proteins as modulators of wound healing and the foreign body response. *Thromb. Haemost.* **90**, 986–992.

Kyriakides, T. R., Leach, K. J., Hoffman, A. S., Ratner, B. D., and Bornstein, P. (1999). Mice that lack the angiogenesis inhibitor, thrombospondin 2, mount an altered foreign body reaction characterized by increased vascularity. *Proc. Natl. Acad. Sci. USA* **96**, 4449–4454.

Kyriakides, T. R., Foster, M. J., Keeney, G. E., Tsai, A., Giachelli, C. M., Clark-Lewis, I., Rollins, B. J., and Bornstein, P. (2004). The CC chemokine ligand, CCL2/MCP1, participates in macrophage fusion and foreign body giant cell formation. *Am. J. Pathol.* **165,** 2157–2166.

Lauffenburger, D. A., and Linderman, J. J. (1993). "Receptors: Models for Binding, Trafficking, and Signaling." Oxford University Press, New York.

Laurencin, C. T., Norman, M. E., Elgendy, H. M., El-Amin, S. F., Allcock, H. R., Pucher, S. R., and Ambrosio, A. A. (1993). Use of polyphosphazenes for skeletal tissue regeneration. *J. Biomed. Mater. Res.* **27**, 963–973.

Lawrence, M., McIntire, L., and Eskin, S. (1987). Effect of flow on polymorphonuclear leukocyte/endothelial cell adhesion. *Blood* **70**, 1284–1290.

Lee, F., Haskell, C., Charo, I., and Boettiget, D. (2004). Receptor-ligand binding in the cell-substrate contact zone: a quantitative analysis using CX3CR1 and CXCR1 chemokine receptors. *Biochemistry* **43**, 7179–7186.

Lee, J. H., Jung, H. W., Kang, I.-K., and Lee, H. B. (1994). Cell behavior on polymer surfaces with different functional groups. *Biomaterials* **15**, 705–711.

L'Heureux, N., Dusserre, N., Konig, G., Victor, B., Keire, P., Wight, T. N., Chronos, N. A., Kyles, A. E., Gregory, C. R., Hoyt, G., Robbins, R. C., and McAllister, T. N. (2006). Human tissue-engineered blood vessels for adult arterial revascularization. *Nat. Med.* **12**, 361–365.

Lin, H., Sun, W., Mosher, D. F., Garcia-Echeverria, C., Schaufelberger, K., Lelkes, P. I., and Cooper, S. L. (1994). Synthesis, surface, and cell-adhesion properties of polyurethanes containing covalently grafted RGD-peptides. *J. Biomed. Mater. Res.* **28**, 329–342.

Lu, H., Smith, C. W., Perrard, J., Bullard, D., Tang, L., Shappell, S. B., Entman, M. L., Beaudet, A. L., and Ballantyne, C. M. (1997). LFA-1 is sufficient in mediating neutrophil emigration in Mac-1-deficient mice. *J. Clin. Invest.* **99**, 1340–1350.

Lydon, M., Minett, T., and Tighe, B. (1985). Cellular interactions with synthetic polymer surfaces in culture. *Biomaterials* **6**, 396–402.

Ma, P. X., and Choi, J. W. (2001). Biodegradable polymer scaffolds with well-defined interconnected spherical pore networks. *Tissue Eng.* **7**, 23–33.

Mahoney, M. J., and Saltzman, W. M. (2001). Transplantation of brain cells assembled around a programmable synthetic microenvironment. *Nat. Biotechnol.* **19**, 934–939.

Marchant, R. E., and Wang, I. (1994). Physical and chemical aspects of biomaterials used in humans. *In* "Implantation Biology" (R. S. Greco, ed.), pp. 13–53. CRC Press, Boca Raton, FL.

Massia, S. P., and Hubbell, J. A. (1991). An RGD spacing of 440 nm is sufficient for integrin αVß3-mediated fibroblast spreading and 140 nm for focal contact and stress fiber formation. *J. Cell Biol.* **114**, 1089–1100.

Matsuda, T., and Sugawara, T. (1995). Development of surface photochemical modification method for micropatterning of cultured cells. *J. Biomed. Mater. Res.* **29**, 749–756.

Matsuda, T., Sugawara, T., and Inoue, K. (1992). Two-dimensional cell manipulation technology: an artificial neural circuit based on surface microprocessing. *ASAIO Journal* **38**, M243–M247.

Matsuzawa, M., Potember, R. S., Stenger, D. A., and Krauthamer, V. (1993). Containment and growth of neuroblastoma cells on chemically patterned substrates. *J. Neurosci. Meth.* **50**, 253–260.

McAuslan, B., and Johnson, G. (1987). Cell responses to biomaterials I: adhesion and growth of vascular endothelial cells on poly(hydroxyethyl methacrylate) following surface modification by hydrolytic etching. *J. Biomed. Mater. Res.* **21**, 921–935.

McClay, D. R., Wessel, G. M., and Marchase, R. B. (1981). Intercellular recognition: quantitation of initial binding events. *Proc. Nat. Acad. Sci.* **78**, 4975–4979.

Meyle, J., Gultig, K., and Nisch, W. (1995). Variation in contact guidance by human cells on a microstructured surface. *J. Biomed. Mater. Res.* **29**, 81–88.

Mikos, A. G., McIntire, L. V., Anderson, J. M., and Babensee, J. E. (1998). Host response to tissue-engineered devices. *Adv. Drug Deliv. Rev.* **33**, 111–139.

Moghaddam, M. J., and Matsuda, T. (1991). Development of a 3D artificial extracellular matrix. *Trans. Am. Soc. Artif. Internal Organs* **37**, M437–M438.

Moscona, A. A. (1961). Rotation-mediated histogenic aggregation of dissociated cells. A quantifiable approach to cell interactions *in vitro. Exp. Cell Res.* **22**, 455–475.

Nelson, R. D., Quie, P. G., and Simmons, R. L. (1975). Chemotaxis under agarose: a new and simple method for measuring chemotaxis and spontaneous migration of human polymorphonuclear leukocytes and monocytes. *J. Immunol.* **115**, 1650–1656.

Nicol, A., Gowda, D. C., Parker, T. M., and Urry, D. W. (1993). Elastomeric polytetrapeptide matrices: hydrophobicity dependence of cell attachment from adhesive (GGIP)n to nonadhesive (GGAP)n even in serum. *J. Biomed. Mater. Res.* **27**, 801–810.

O'Neill, C., Jordan, P., and Ireland, G. (1986). Evidence for two distinct mechanisms of anchorage stimulation in freshly explanted and 3T3 swiss mouse fibroblasts. *Cell* **44**, 489–496.

Palecek, S. P., Loftus, J. C., Ginsberg, M. H., Lauffenburger, D. A., and Horwitz, A. F. (1997). Integrin-ligand binding properties govern cell migration speed through cell-substratum adhesiveness. *Nature* **385**, 537–540.

Parsons-Wingerter, P., and Saltzman, W. M. (1993). Growth versus function in three-dimensional culture of single and aggregated hepatocytes within collagen gels. *Biotechnol. Progr.* **9**, 600–607.

Peppas, N. A., and Langer, R. (1994). New challenges in biomaterials. *Science* **263**, 1715–1720.

Perizzolo, D., Lacefield, W. R., and Brunette, D. M. (2001). Interaction between topography and coating in the formation of bone nodules in culture for hydroxyapatite- and titanium-coated micromachined surfaces. *J. Biomed. Mater. Res.* **56**, 494–503.

Pettit, D. K., Horbett, T. A., Hoffman, A. S., and Chan, K. Y. (1990). Quantitation of rabbit corneal epithelial-cell outgrowth on polymeric substrates *in vitro. Invest. Ophthalmol. Visual Sci.* **31**, 2269–2277.

Pettit, D. K., Hoffman, A. S., and Horbett, T. A. (1994). Correlation between corneal epithelial cell outgrowth and monoclonal antibody binding to the cell binding domain of adsorbed fibronectin. *J. Biomed. Mater. Sci.* **28**, 685–691.

Pierschbacher, M. D., and Ruoslahti, E. (1984). Cell attachment activity of fibronectin can be duplicated by small synthetic fragments of the molecule. *Nature* **309**, 30–33.

Pratt, K., Williams, S., and Jarrell, B. (1989). Enhanced adherence of human adult endothelial cells to plasma discharge modified polyethylene terephthalate. *J. Biomed. Mater. Res.* **23**, 1131–1147.

Puelacher, W. C., Mooney, D., Langer, R., Upton, J., Vacanti, J. P., and Vacanti, C. A. (1994). Design of nasoseptal cartilage replacements synthesized from biodegradable polymers and chondrocytes. *Biomaterials* **15**, 774–778.

Qin, T., Yang, Z., Wu, Z., Xie, H., Qin, H., and Cai, S. (2005). Adhesion strength of human tenocytes to extracellular matrix component-modified poly(DL-lactide-co-glycolide) substrates. *Biomaterials* **26**, 6635–6642.

Rajnicek, A. M., Britland, S., and McCaig, C. D. (1997). Contact guidance of CNS neurites on grooved quartz: influence of groove dimensions, neuronal age, and cell type. *J. Cell Sci.* **110**, 2905–2913.

Ranieri, J. P., Bellamkonda, R., Jacob, J., Vargo, T. G., Gardella, J. A., and Aebischer, P. (1993). Selective neuronal cell adhesion to a covalently patterned monoamine on fluorinated ethylene propylene films. *J. Biomed. Mater. Res.* **27**, 917–925.

Robinson, K. A., Li, J., Mathison, M., Redkar, A., Cui, J., Chronos, N. A., Matheny, R. G., and Badylak, S. F. (2005). Extracellular matrix scaffold for cardiac repair. *Circulation* **112**, 1135–1143.

Saltzman, W. M. (2004). "Tissue Engineering: Engineering Principles for the Design of Replacement Organs and Tissues." Oxford University Press, New York.

Sawhney, A. S., Pathak, C. P., and Hubbell, J. A. (1994). Modification of islet of Langerhans surfaces with immunoprotective poly(ethylene glycol) coatings via interfacial polymerization. *Biotechnol. Bioeng.* **44**, 383–386.

Schakenraad, J. M., Busscher, H. J., Wildevuur, C. R. H., and Arends, J. (1986). The influence of substratum surface free energy on growth and spreading of human fibroblasts in the presence and absence of serum proteins. *J. Biomed. Mater. Res.* **20**, 773–784.

Schmidt, J. A., and von Recum, A. F. (1992). Macrophage response to microtextured silicone. *Biomaterials* **12**, 385–389.

Singhvi, R., Kumar, A., Lopez, G. P., Stephanopoulos, G. N., Wang, I. C., Whitesides, G. M., and Ingber, D. E. (1994). Engineering cell shape and function. *Science* **264**, 696–698.

Smetana, K., Vacik, J., Souckova, D., Krcova, Z., and Sulc, J. (1990). The influence of hydrogel functional groups on cell behavior. *J. Biomed. Mater. Res.* **24**, 463–470.

Takezawa, T., Mori, Y., Yonaha, T., and Yoshizato, K. (1993). Characterization of morphology and cellular metabolism during the spheroid formation by fibroblasts. *Exp. Cell Res.* **208**, 430–441.

Tamada, Y., and Ikada, Y. (1994). Fibroblast growth on polymer surfaces and biosynthesis of collagen. *J. Biomed. Mater. Res.* **28**, 783–789.

Tan, J., and Saltzman, W. M. (2002). Topographical control of human neutrophil motility on micropatterned materials with various surface chemistry. *Biomaterials* **23**, 3215–3225.

Tan, J., Shen, H., and Saltzman, W. M. (2001). Micron-scale positioning of features influences the rate of polymorphonuclear leukocyte migration. *Biophys. J.* **81**, 2569–2579.

Tang, L., Jennings, T. A., and Eaton, J. W. (1998). Mast cells mediate acute inflammatory responses to implanted biomaterials. *Proc. Natl. Acad. Sci. USA* **95**, 8841–8846.

Tao, W., Wen, R., Goddard, M. B., Sherman, S. D., O'Rouke, P. J., Stabila, P. F., Bell, W. J., Dean, B. J., Kauper, K. A., Budz, V. A., Tsiaras, W. G., Acland, G. M., Pearce-Kelling, S., Laties, A. M., and Aguirre, G. D. (2002). Encapsulated cell-based delivery of CNTF reduces photoreceptor degeneration in animal models of retinitis pigmentosa. *Invest. Ophthalmol. Visual Sci.* **43**, 3292–3298.

Teng, Y. D., Lavik, E. B., Qu, X., Park, K. I., Ourednik, J., Zurakowski, D., Langer, R., and Snyder, E. Y. (2002). Functional recovery following traumatic spinal cord injury mediated by a unique polymer scaffold seeded with neural stem cells. *Proc. Natl. Acad. Sci.* **99**, 3024–3029.

Trinkhaus, J. P. (1984). "Cells into Organs: The Forces That Shape the Embryo." Prentice-Hall, Englewood Cliffs, NJ.

Valentini, R. F., Vargo, T. G., Gardella, J. A., and Aebischer, P. (1992). Electrically charged polymeric substrates enhance nerve fiber outgrowth *in vitro. Biomaterials* **13**, 183–190.

van der Valk, P., van Pelt, A., Busscher, H., de Jong, H., Wildevuur, R., and Arends, J. (1983). Interaction of fibroblasts and polymer surfaces: relationship between surface free energy and fibroblast spreading. *J. Biomed. Mater. Res.* **17**, 807–817.

van Wachem, P. B., Hogt, A. H., Beugeling, T., Feijen, J., Bantjes, A., Detmers, J. P., and van Aken, W. G. (1987). Adhesion of cultured human endothelial cells onto methacrylate polymers with varying surface wettability and charge. *Biomaterials* **8**, 323–328.

van Wezel, A. L. (1967). Growth of cell strains and primary cells on microcarriers in homogeneous culture. *Nature* **216**, 64–65.

Vesely, I. (2005). Heart valve tissue engineering. *Circ. Res.* **97**, 743–755.

Voskerician, G., Gingras, P. H., and Anderson, J. M. (2006). Macroporous condensed poly(tetrafluoroethylene). I. *In vivo* inflammatory response and healing characteristics. *J. Biomed. Mater. Res. A* **76**, 234–242.

Wagers, A. J., Sherwood, R. I., Christensen, J. L., and Weissmann, I. L. (2002). Little evidence for developmental plasticity of adult hematopoietic stem cells. *Science* **297**, 2256–2259.

Woerly, S., Maghami, G., Duncan, R., Subr, V., and Ulbrich, K. (1993). Synthetic polymer derivatives as substrata for neuronal cell adhesion and growth. *Brain Res. Bull.* **30**, 423–432.

Wojciak-Stothard, B., Curtis, A., Monaghan, W., Macdonald, K., and Wilkinson, C. (1996). Guidance and activation of murine macrophages by nanometric scale topography. *Exp. Cell Res.* **223**, 426–435.

Wong, J. Y., Langer, R., and Ingber, D. E. (1994). Electrically conducting polymers can noninvasively control the shape and growth of mammalian cells. *Proc. Natl. Acad. Sci. USA* **91**, 3201–3204.

Wu, P., Hoying, J. B., Williams, S. K., Kozikowski, B. A., and Lauffenburger, D. A. (1994). Integrin-binding peptide in solution inhibits or enhances endothelial cell migration, predictably from cell adhesion. *Ann. Biomed. Eng.* **22**, 144–152.

Yang, J., Webb, A. R., Pickerill, S. J., Hageman, G., and Ameer, G. A. (2006). Synthesis and evaluation of poly(diol citrate) biodegradable elastomers. *Biomaterials* **27**, 1889–1898.

Yang, S., Leong, K. F., Du, Z., and Chua, C. K. (2001). The design of scaffolds for use in tissue engineering Part 1. Traditional factors. *Tissue Eng.* **7**, 679–689.

# Chapter Twenty-One

# Matrix Effects

*Jeffrey A. Hubbell*

## I. INTRODUCTION

The extracellular matrix is a complex chemically and physically cross-linked network of proteins and glycosaminoglycans. The matrix serves to organize cells in space, to provide them with environmental signals to direct site-specific cellular regulation, and to separate one tissue space from another. The interaction between cells and the extracellular matrix is bidirectional and dynamic: Cells are constantly accepting information on their environment from cues in the extracellular matrix, and cells are frequently remodeling their extracellular matrix. In this chapter, the proteins in the extracellular matrix and their cell surface receptors are introduced, and mechanisms by which cells transduce chemical information in their extracellular matrix are discussed. Methods for spatially displaying matrix recognition factors on and in biomaterials are described, both in the context of model systems for investigation of cellular behavior and from the perspective of the creation of bioactive biomaterials for tissue-engineering therapies.

The extracellular matrix serves at least three functions in its role of controlling cell behavior: It provides adhesion signals, it provides growth factor–binding sites, and it provides degradation sites to give way to the enzymatic activity of cells as they migrate. An understanding of these interactions is important in tissue engineering, where one may desire to mimic the biological recognition molecules that control the relationships between cells and their natural bio-

material interface, namely the extracellular matrix. The components of the extracellular matrix are on one level immobilized, but not necessarily irreversibly. Cell-derived enzymes, such as tissue transglutaminase and lysyl oxidase, serve chemically to cross-link certain components of the extracellular matrix, such as fibronectin chains to other fibronectin chains and to fibrillar collagen chains. Other components are more transiently immobilized, such as growth factors within the extracellular matrix proteoglycan network. This network can be partially degraded, and the growth factors themselves can be proteolytically cleaved, to mobilize the growth factors under cellular control. Not all of the signals of the extracellular matrix are biochemical in nature. A biomechanical interplay between cells and their extracellular matrix also plays an important role in the functional regulation in many tissues, particular in load-bearing tissues. This chapter considers only the biochemical aspects of biological recognition; the reader is referred elsewhere for treatments of the role of the extracellular matrix as a biomechanical regulator of cell behavior (Grinnell, 2003; Discher *et al.*, 2005).

## II. EXTRACELLULAR MATRIX PROTEINS AND THEIR RECEPTORS

Interactions between cells and the extracellular matrix are mediated by cell surface glycoprotein and proteoglycan receptors interacting with proteins bound within the extracellular matrix. This section begins with an introduction of the glycoprotein receptors on cell surfaces involved in cell adhesion. The discussion then turns to the proteins in the extracellular matrix to which those receptors bind, including the active domains of those proteins that bind to the cell surface receptors. Finally, the roles of cell surface–associated enzymes in the processing and remodeling of the extracellular matrix are addressed.

Four major classes of glycoprotein adhesion receptors are present on cell surfaces, three of which are involved primarily in cell–cell adhesion and one of which is involved in both cell adhesion to other cells and cell adhesion to the extracellular matrix. The first three are introduced briefly and the fourth more extensively.

The cadherins are a family of cell surface receptors that participate in homophilic binding (i.e., the binding of a cadherin on one cell with an identical cadherin on another cell) (Gumbiner, 2005; Leckband and Prakasam, 2006). These molecules allow a cell of one type (e.g., endothelial cells) to recognize other cells of the same type and are important in the early stages of organogenesis. These interactions depend on the presence of extracellular $Ca^{++}$ and may be dissociated by calcium ion chelation. Since all cadherins are present on cell surfaces, cadherins are not involved directly in cell interactions with the extracellular matrix. They may be involved indirectly, in that they may organize cell–cell contacts in concert with another receptor system that is involved in regulation of cell–extracellular matrix binding.

A second class of receptors is the selectin family (Vestweber and Blanks, 1999). These membrane-bound proteins are involved in heterophilic binding between cells, such as blood cells and endothelial cells, in a manner that depends, as with the cadherins, on extracellular $Ca^{++}$. These proteins contain lectin-like features and recognize branched oligosaccharide structures in their ligands, namely the sialyl Lewis X and the sialyl Lewis A structures. As with the cadherins, these receptor–ligand interactions are important primarily in cell–cell interactions, and they are particularly important in the context of inflammation.

A third class of receptors represents members of the immunoglobulin superfamily, cell adhesion molecule proteins that are denoted as Ig-CAMs or simply CAMs (Walsh and Doherty, 1997). These proteins bind their protein ligands in a manner that is independent of extracellular $Ca^{++}$, and they participate in both homophilic and heterophilic interactions. As for the cadherins and the selectins, they bind to other cell surface proteins and are thus primarily involved in cell–cell interactions. One class of ligand for these receptors includes selected members of the integrin class of adhesion receptors, discussed later.

The fourth class of adhesion receptors is the integrin family (van der Flier and Sonnenberg, 2001; Bokel and Brown, 2002; Arnaout et al., 2005; Ginsberg et al., 2005; Wiesner et al., 2005). While the other three classes of receptors just described briefly are involved primarily in cell–cell recognition, the integrins are involved in both cell–cell and cell–

extracellular matrix binding. The integrins are dimeric proteins, consisting of an $\alpha$ and a $\beta$ subunit assembled noncovalently into an active dimer. There are many known $\alpha$ and $\beta$ subunits, with at least 18 such $\alpha$ subunits and eight such $\beta$ subunits that are capable of assembly into at least 24 $\alpha\beta$ combinations. Some of the $\alpha\beta$ combinations present in the $\beta_1$, $\beta_2$, $\beta_3$, and $\beta_4$ subclasses are shown in Table 21.1; these are the most commonly expressed integrins and are thus arguably the most generally important. The $\beta_2$ integrins are involved primarily in cell–cell recognition; for example, the integrin $\alpha_L\beta_2$ binds to ICAM-1 and ICAM-2, both members of the immunoglobulin superfamily subclass of cell adhesion molecules described in the preceding paragraph. By contrast, the $\beta_1$, $\beta_3$, and $\beta_4$ integrins are involved primarily in cell–extracellular matrix interactions.

The $\beta_1$ and $\beta_3$ integrins bind to numerous proteins present in the extracellular matrix, as illustrated in Table 21.1. These proteins include collagen, fibronectin, vitronectin, von Willebrand factor, and laminin.

Collagen is the primary structural protein of the tissues; the reader is referred elsewhere for a focused review on this extensive topic (Fratzl et al., 1998). Many forms of collagen exist, several of which are multimeric and fibrillar. To these collagens many other adhesion proteins bind, thus putting collagen in the role of organizing many other proteins that interact with and organize cells. Collagen also interacts directly with integrins, primarily $\alpha_1\beta_1$ and $\alpha_2\beta_1$.

Fibronectin is a globular protein present in nearly all tissues; fibronectin has been extensively reviewed elsewhere (Magnusson and Mosher, 1998). Fibronectin also exists in many forms, depending on the site in the tissues and the regulatory state of the cell that synthesized the fibronectin. Almost all cells interact with fibronectin, primarily through the so-called fibronectin receptor $\alpha_5\beta_1$, and to a lesser extent through the $\beta_3$ integrin $\alpha_v\beta_3$ as well as other integrins, as is described later.

Vitronectin is a multifunctional adhesion protein found in the circulation and in many tissues (Preissner and Seiffert, 1998). The protein is active in promoting the adhesion of nu-merous cell types and binds primarily to the so-called vitronectin receptor, $\alpha_v\beta_3$, as well as $\alpha_v\beta_1$ and to the platelet receptor $\alpha_{IIb}\beta_3$.

The von Willebrand factor is an adhesion protein that is involved primarily in the adhesion of vascular cells; the reader is referred elsewhere for a detailed review (Sadler, 1998). It is synthesized by the megakaryocyte, the platelet-generating cells of the bone marrow, and is stored in the $\alpha$-granules of circulating platelets. Activation of the platelet leads to the release of the granule contents, including the von Willebrand factor. The von Willebrand factor is also synthesized by and stored within the endothelial cell. A multimeric form of the protein, where tens of copies of the protein may be linked together into insoluble form, is found in the subendothelium and is involved in blood platelet adhesion to the subendothelial tissues on vascular injury.

| Table 21.1. | Selected members of the integrin receptor class and their ligands |
|---|---|
| *Integrin heterodimer* | *Ligands* |
| $\alpha_1\beta_1$ | Collagen, laminin |
| $\alpha_2\beta_1$ | Collagen, laminin |
| $\alpha_3\beta_1$ | Collagen, fibronectin, laminin, thrombospondin-1 |
| $\alpha_4\beta_1$ | Fibronectin, osteopontin, vascular cell adhesion molecule-1 |
| $\alpha_5\beta_1$ | RGD, fibronectin, L1 |
| $\alpha_6\beta_1$ | Laminin |
| $\alpha_7\beta_1$ | Laminin |
| $\alpha_8\beta_1$ | RGD, fibronectin, tenascin |
| $\alpha_9\beta_1$ | Collagen, laminin, osteopontin, tenascin, vascular cell adhesion molecule-1 |
| $\alpha_{10}\beta_1$ | Collagen |
| $\alpha_{11}\beta_1$ | Collagen |
| $\alpha_V\beta_1$ | RGD, collagen, fibrinogen, fibronectin, vitronectin, von Willebrand factor |
| $\alpha_X\beta_2$ | Complement protein C3bi, fibrinogen |
| $\alpha_M\beta_2$ | Complement protein C3bi, fibrinogen, intercellular adhesion molecule-1, vascular cell adhesion molecule-1 |
| $\alpha_L\beta_2$ | Intercellular adhesion molecule-1 — intercellular adhesion molecule-5 |
| $\alpha_D\beta_2$ | Intercellular adhesion molecule-3, vascular cell adhesion molecule-1 |
| $\alpha_V\beta_3$ | RGD, bone sialoprotein, fibrinogen, fibronectin, thrombospondin, vitronectin, von Willebrand factor |
| $\alpha_{IIb}\beta_3$ | Fibrinogen, fibronectin, thrombospondin, vitronectin, von Willebrand factor |
| $\alpha_6\beta_4$ | Laminin, hemidesmosomes |

Van der Flier and Sonnenberg (2001).

Laminin is a very complex adhesion protein generally present in the basement membrane, the proteins immediately beneath epithelia and endothelia, as well as in many other tissues, as reviewed in detail elsewhere (Miner and Yurchenko, 2004; Sasaki *et al.*, 2004). Laminin is present in a family of forms (Aumailley *et al.*, 2005). The classic form was purified from the extracellular matrix of Engelbreth–Holm–Swarm tumor cells and consists of a disulfide cross-linked trimer of one $\alpha 1$ (400,000 Da), one $\beta 1$ (210,000 Da), and one $\gamma 1$ (200,000 Da) polypeptide chains. This form binds to the $\beta_1$ integrins $\alpha_1\beta_1$, $\alpha_2\beta_1$, $\alpha_3\beta_1$, $\alpha_6\beta_1$, and $\alpha_7\beta_1$ and to the $\beta_3$ integrins $\alpha_V\beta_3$ and $\alpha_{IIb}\beta_3$, as well as to other integrins. A number of other laminin forms exist, composed of $\alpha\beta\gamma$ combinations of $\alpha 1$, $\alpha 2$, $\alpha 3$, $\alpha 4$, or $\alpha 5$, $\beta 1$, $\beta 2$, or $\beta 3$, and $\gamma 1$ or $\gamma 2$ chains. The details of the differences in function of all of the various laminin forms remain only partially elucidated, but it is clear that several of them do stimulate very different behaviors in a variety of cell types. Laminin is a particularly important component of the basal lamina, i.e., the extracellular matrix beneath monolayer structures such as epithelia, mesothelia, and endothelia (Schwarzbauer, 1999). For example, laminin contains numerous domains that bind to endothelial cells (Ponce *et al.*, 1999), and these are undoubtedly important in regulating a variety of cell-type-specific functions, as discussed later.

There are several cell morphological hallmarks of integrin binding to adhesion proteins in the extracellular matrix. These include spreading of the cells, extension of cellular membrane processes called *focal contacts* to within approximately 20 nm of the extracellular matrix surface (Shemesh *et al.*, 2005), clustering of integrin receptors at the sites of focal contacts, and assembly of intracellular accessory proteins at the site of clustered integrins to assist in the attachment of the integrin complex to the f-actin cytoskeleton (Ward and Hammer, 1993). These sites of clustered receptors and interaction of the transmembrane receptors with the intracellular cytoskeleton carry most of the stress between the cells and the extracellular matrix or artificial surfaces; indeed, both theoretical analysis and experimental results demonstrate that without the formation of focal contacts and without the connection of numerous transmembrane integrin $\alpha\beta$ heterodimer complexes into much larger multi-integrin complexes by intracellular proteins such as talin, vinculin, and $\alpha$-actinin, cell adhesion would be very much weaker than in reality (Ward and Hammer, 1993). The focal contact serves as an important center for regulation of cell signaling, both mechanically and chemically, as discussed later.

The extracellular matrix proteins just described are very complex. They contain sites responsible for binding to collagen, for binding to glycosaminoglycans (as described later), for cross-linking to other extracellular matrix proteins via transglutaminase activity, for degradation by proteases (as described briefly later), and for binding to integrin and other adhesion receptors (as described in detail next). Since the proteins must be so multifunctional, the sites that serve the singular function of binding to integrins comprise a small fraction of the protein mass. In most cases, the receptor-binding domain can be localized to an oligopeptide sequence less than 10 amino acid residues in length, and this site can be mimicked by linear or cyclic oligopeptides of identical or similar sequence as that found in the protein (Ruoslahti, 1991). The first such minimal sequence to be identified was the tripeptide RGD (Ruoslahti and Piersch-bacher, 1987) (using the single-letter amino acid code, shown in a footnote to Table 21.2). Synthetic RGD-containing peptide, when appropriately coupled to a surface or a carrier molecule (see later), is capable of recapitulating much of the adhesive interactions of the RGD site in the protein fibronectin, including integrin binding. At least for the case of integrin binding via $\alpha_v\beta_3$, the RGD ligand alone is capable of also inducing integrin clustering and, when the signal is presented at sufficient surface concentration, focal

contact formation, and cytoskeletal organization (Massia and Hubbell, 1991). Many receptor-binding sequences other than the RGD tripeptide have been identified by a variety of methods, and a few of these sequences are shown in Table 21.2. In these receptor-binding sequences, the affinity is highly specific to the particular ordering of the amino acids in the peptide; for example, the peptide RDG, containing the same amino acids but in a different sequence, is completely inactive in binding to integrins.

One class of the adhesion peptides contains the central RGD sequence. These are modified by their flanking residues, which modify the receptor specificity of the receptor-binding sequence. For example, the sequence found in fibronectin is RGDS, in vitronectin is RGDV, in laminin is RGDN, and in collagen is RGDT. Other adhesion peptides maintain the central D residue, such as the REDV and LDV sequences of fibronectin. The REDV and LDV sequences are relatively specific in their binding and interact with the integrin $\alpha_4\beta_1$; RGD peptides also bind to $\alpha_4\beta_1$, but the REDV and LDV sequences bind essentially only to $\alpha_4\beta_1$.

In addition to peptides that bind to the integrin adhesion receptors, there are other peptide sequences that bind to other, nonintegrin receptors. As an example, laminin bears several such sequences, such as the YIGSR, SIKVAV, and RNIAEIIKDI peptides. The YIGSR sequence binds to a 67-kDa monomeric nonintegrin laminin receptor (Meecham, 1991). As do the integrin receptors, this laminin receptor also interacts via its cytoplasmic domain with intracellular proteins involved with linkage to the f-actin cytoskeleton. The YIGSR sequence is involved in the adhesion and spreading of numerous cell types (see later). The SIKVAV sequence in laminin binds to a neuronal cell surface receptor and stimulates the extension of neurites (Tashiro *et al.*, 1989).

In addition to the highly sequence-specific binding of adhesion peptides to cell surface receptors, most of the adhesion proteins also bind to cell surface components by less specific mechanisms. These proteins contain a heparin-binding domain (so called because of the use of heparin affinity chromatography in purification of the protein) that binds to cell surface proteoglycans that contain heparan sulfate or chondroitin sulfate glycosaminoglycans (Lyon and Gallagher, 1998). The peptide sequences that bind to cell surface proteoglycans are rich in cationic residues, such as arginine (R) and lysine (K), relative to their content in the anionic residues aspartic acid (D) and glutamic acid (E), and they also contain hydrophobic amino acids, such as alanine (A), isoleucine (I), leucine (L), proline (P), and valine (V). Several of these sequences are shown in Table 21.3. For example, the heparin-binding sequence in fibronectin bears a sequence of PRRARV, having a motif of XBBXBX, which is observed in several cell adhesion proteins, X being a hydrophobic residue and B being a basic residue, either K or R. These sites within adhesion proteins such as fibronectin and laminin bind to cell surface proteoglycan in parallel

**Table 21.2.** Selected cell-binding domain sequences of extracellular matrix proteins

| Protein | Sequence[a] | Role |
|---|---|---|
| Fibronectin | RGDS | Adhesion of most cells, via $\alpha_5\beta_1$ |
| | LDV | Adhesion |
| | REDV | Adhesion |
| Vitronectin | RGDV | Adhesion of most cells, via $\alpha_v\beta_3$ |
| Laminin A | LRGDN | Adhesion |
| | SIKVAV | Neurite extension |
| Laminin B1 | YIGSR | Adhesion of many cells, via 67-kDa laminin receptor |
| | PDSGR | Adhesion |
| Laminin B2 | RNIAEIIKDI | Neurite extension |
| Collagen I | RGDT | Adhesion of most cells |
| | DGEA | Adhesion of platelets, other cells |
| Thrombospondin | RGD | Adhesion of most cells |
| | VTXG | Adhesion of platelets |

From Hubbell (1995), after Y. Yamada and Kleinman, (1992).
[a]Single-letter amino acid code: A, alanine; C, cysteine; D, aspartic acid; E, glutamic acid; F, phenylalanine; G, glycine; H, histidine; I, isoleucine; K, lysine; L, leucine; M, methionine; N, asparagine; P, proline; Q, glutamine; R, arginine; S, serine; T, threonine; V, valine; W, tryptophan; Y, tyrosine.

**Table 21.3.** Proteoglycan-binding domain sequences of extracellular matrix proteins

| Sequence[a] | Protein |
|---|---|
| X*BB*X*B*X | Consensus sequence |
| P*RRAR*V | Fibronectin |
| YE*K*PGSPP*R*EVVP*R*P*R*PGV | Fibronectin |
| *RP*SLA*KKQR*F*RH*R*N*RKG*Y*RSRQ*GHS*RGR* | Vitronectin |
| *R*IQNLL*K*ITNL*R*I*K*FV*K* | Laminin |

After Hubbell (1995); references contained in Massia and Hubbell (1992).
[a]X indicates a hydrophobic amino acid. Basic amino acids are shown in italics.

with interaction by the integrin-binding sites to stabilize the adhesion complex (Lebaron *et al.*, 1988). Cell–cell adhesion molecules also employ cell surface proteoglycan-binding affinity, e.g., N-CAM bears the domain KHKGRDVILKKDVR, which binds to heparan sulfate and chondroitin sulfate proteoglycans (Kallapur and Akeson, 1992). The interactions with cell surface proteoglycans are much less specific than those with integrins; the binding is not as sensitive to the order of the oligopeptide sequence, and, moreover, the effect can be mimicked simply by R or K residues immobilized on a surface, albeit certainly with a great loss in specificity (Massia and Hubbell, 1992).

The extracellular matrix is subject to dynamic remodeling under the influence of cells in contact with it. Cells seeded *in vitro* on an extracellular matrix of one composition may adhere, spread, form focal contacts, remove the initial protein, secrete a new extracellular matrix of different protein composition, and form new focal contacts. Cell surface–bound and cell-derived free enzymes play an important role in this remodeling of the extracellular matrix (Kleinman *et al.*, 2003). For example, cell-released protein disulfide isomerases are released from cells to a covalently cross-linked protein in the extracellular matrix by disulfide bridging. Cell-derived transglutaminases also form an amide linkage between the ε-amino group of lysine and the side group amide of glutamine to chemically cross-linked proteins in the extracellular matrix. These processes are responsible, for example, for the assembly of the globular adhesion protein fibronectin into fibrils within the extracellular matrix beneath cells.

Membrane-bound and cell-released enzymes are also involved in degradation of the extracellular matrix to permit matrix remodeling and cell migration (Shapiro, 1998). Cell-released matrix metalloproteinases such as collagenase and gelatinase, serine proteases such as urokinase plasminogen activator and plasmin, and cathepsins are each involved in both remodeling and degradation during cell migration. Accordingly, the matrix–cell interaction

should be understood to be bidirectional: the cell accepting information from the matrix, and the matrix being tailored by the cell.

One of the important roles of cell-associated enzymatic degradation of the extracellular matrix is in mobilization of growth factor activity. Because growth factors are such powerful regulators of biological function, their activity must be highly spatially regulated. One means by which this occurs in nature is by high-affinity-binding interactions between growth factors and the three-dimensional extracellular matrix in which they exist. Many such growth factors bind heparin, meaning that they, like adhesion proteins, bear domains that bind extracellular matrix heparan sulfate and chondroitin sulfate proteoglycans. For example, basic fibroblast growth factor binds heparin with high affinity (Faham *et al.*, 1998). Vascular endothelial growth factor is another example of a heparin-binding growth factor (Fairbrother *et al.*, 1998). These growth factors are strongly immobilized by binding to extracellular matrix proteoglycans, and they can be mobilized under local cellular activity, e.g., by degradation of these proteoglycans or, in the case of vascular endothelial cell growth factor, by cleavage of the main chain of the growth factor away from the heparin-binding domain by plasmin activated at the surface of a nearby cell. Fun-damental studies have demonstrated that the interaction between growth factors and the extracellular matrix can dramatically alter their local behavior, where the length scale of the local response is measured in single cell diameters. Specifically, very low interstitial flows can convect cell-derived proteases directionally downstream of a cell, and these proteases can liberate matrix-bound growth factors that were previously homogeneously distributed throughout the matrix. Since the protease activity is preferentially downstream of the cell, growth factor liberation is also pre-ferentially downstream. This can create gradients of growth factor, allowing the cell to sense the directionality of flow, even when the flows are extremely subtle (Helm *et al.*, 2005; Fleury *et al.*, 2006).

## III. MODEL SYSTEMS FOR STUDY OF MATRIX INTERACTIONS

Since the extracellular matrix adhesion proteins may be mimicked, at least to some degree, by small synthetic peptides, it is possible to investigate cell–substrate interactions with well-defined systems. Foundational to them all are the interactions of cells with the surface, other than with adhesion peptides intentionally endowed on the surface, that would produce cell adhesion. These so-called nonspecific interactions are between cell surface receptors and proteins that have adsorbed to the surface. Due to this role played by adsorbing proteins, some introduction to the protein and surface interactions leading to adsorption is warranted. The thermodynamic and kinetic aspects of protein adsorption have been reviewed and the reader is referred elsewhere for a more detailed treatment (Andrade and Hlady, 1986). The primary driving force for protein adsorption is the hydrophobic effect: Water near a hydrophobic material surface fails to hydrogen bond with that surface and thus assumes a more highly ordered structure in which the water is more thoroughly hydrogen bonded to itself than is the case in water far away from the surface. A protein can adsorb to this surface, acting like a surfactant, and thus replace the hydrophobic material surface with a polar surface capable of hydrogen bonding with water. This releases the order in the water, with a net result of a large entropic gain. Electrostatic interactions, e.g., between charges on D, E, K, or R residues on the protein with cationic or anionic functions on the material surface, play a lesser but important role as well; since proteins generally bear a net negative charge, anionic surfaces typically adsorb less protein than do cationic surfaces. These observations, overly generalized in the preceding sentences, guide one to examine model surfaces that are hydrophilic and nonionic as well as being derivatizable to permit coupling of the adhesion peptide under study.

The tendency for proteins to adsorb to material surfaces has been exploited as a method by which to immobilize peptides onto substrates for study. Pierschbacher *et al.* have described the peptide Ac–GRGDSPASSKGGGGSRLLLLLLR–NH$_2$ (where the Ac indicates that the N-terminus is acetylated and the –NH$_2$ indicates that the C-terminus is amidated to block the terminal charges) for this purpose (Ruoslahti and Pierschbacher, 1986). The LLLLLL stretch is hydrophobic and adsorbs avidly to hydrophobic polymer surfaces and effectively immobilizes the cell-binding RGDS sequence from fibronectin. Nonadhesive proteins such as albumin have also been grafted with RGD peptide, e.g., by binding to amine groups on lysine residues on the albumin; adsorption of the albumin conjugate thus immobilizes the attached RGD peptide (Danilov and Juliano, 1989).

Surfaces coated with hydrophilic polymers have also been employed to graft adhesion peptides. One simple system that has been useful is glass modified with a silane, 3-glycidoxypropyl triethoxysilane; once the silane is grafted to the surface, the epoxide group is hydrolyzed to produce –CH$_2$CH(OH)CH$_2$OH groups pendant from the surface (glycophase glass). The hydroxyl groups serve as sites for covalent immobilization of adhesion peptide, e.g., via the N-terminal primary amine (Massia and Hubbell, 1990). Titration of the surface density of grafted RGD peptides versus cell response using this system revealed quantitative information on the number density of interactions required to establish morphologically complete cell spreading (Massia and Hubbell, 1991). A surface density of approximately 10 fmol/cm$^2$ of RGD was required to induce spreading, focal contact formation, integrin $\alpha_v\beta_3$ clustering, $\alpha$-actinin and vinculin colocalization with $\alpha_v\beta_3$, and f-actin cytoskeletal assembly in human fibroblasts cultured on this synthetic extracellular matrix. This surface density corresponds to a spacing of roughly 140 nm between immobilized RGD sites, demonstrating that far less than monolayer coverage is sufficient to promote cell responses. Silane-modified quartz has been employed as a surface to study the role of RGD sites and heparin-binding domains of adhesion proteins in osteoblast adhesion and mineralization, demonstrating a strong benefit for the involvement of both modes of adhesion (Rezania and Healy, 1999). The base material, glycophase glass, is only modestly resistant to protein adsorption and thus to nonspecific cell adhesion; accordingly, the investigation of long-term interactions, during which the adherent cells may be synthesizing and secreting their own extracellular matrix to adsorb to the synthetic one experimentally provided, are difficult to investigate. This has motivated exploration with substrates that are more resistant to protein adsorption.

An enormous amount of research has been expended into grafting material surfaces with water-soluble, nonionic polymers such as polyethylene glycol, HO(CH$_2$CH$_2$O)$_n$H, abbreviated PEG. This vast body of research has been extensively reviewed elsewhere (Otsuka *et al.*, 2003). Polyethylene glycol has been immobilized on surfaces by numerous means; three particularly effective means are addressed in the following paragraphs.

Thiol compounds bind by chemisorption avidly to gold surfaces (Love *et al.*, 2005; Whitesides *et al.*, 2005). When those thiols are terminal to an alkane group, R–(CH$_2$)$_n$–SH, the thiol adsorbs in perfect self-assembling monolayers; the thiol–gold interaction contributes about half of the energy of interaction, and the alkane–alkane van der Waals interaction contributes the other half. Accordingly, it is easy to employ alkanethiols to display, in very regular fashion, some functionality R on a gold-coated substrate (so long as the R group is not so large as to sterically inhibit monolayer packing, in which case it can be diluted with a nonfunctional alkanethiol). Using this approach, Prime and Whitesides (1993) immobilized oligoethylene glycol–containing alkanethiol, HS–(CH$_2$)$_{11}$(OCH$_2$CH$_2$)$_n$OH, on gold surfaces. Protein adsorption was investigated on surfaces formed with this alkanethiol and a hydrophobic coreactant, HS–(CH$_2$)$_{10}$CH$_3$. Degrees of polymerization (*n*) as low as 4 were observed to dramatically limit the adsorption of even very large proteins, such as fibronectin. When the oligoethylene glycol monolayer was incomplete, i.e., when the monolayer was mixed with the hydrophobic alkanethiol, longer oligoethylene glycol functions were able to preserve the protein repulsiveness of the surface. Since the background amount of protein adsorption on these materials is so low, one would expect them to be very useful as substrates for peptide attachment for studies with model synthetic extracellular matrices, e.g., with HS–(CH$_2$)$_{11}$(OCH$_2$CH$_2$)$_n$–NH–RGDS.

Drumheller and Hubbell have developed a polymeric material that was highly resistant to cell adhesion for use in peptide grafting (Drumheller and Hubbell, 1994;

Drumheller *et al.*, 1994). Materials that contain large amounts of polyethylene glycol generally swell extensively, rendering material properties sometimes unsuitable for cell culture or medical devices. To circumvent this, the polyethylene glycol swelling was constrained by distributing it as a network throughout a densely cross-linked network of a hydrophobic monomer, trimethylolpropane triacrylate. This yielded a material with the surface hydrophilicity of a hydrogel but with mechanical and optical properties of a glass. These materials were highly resistant to protein adsorption, even after an adsorptive challenge to the material with a very large adhesion protein, laminin, and even over multiweek durations. When the polymer network was formed with small amounts of acrylic acid as a comonomer, the polymer still remained cell nonadhesive. The carboxyl groups near the polymer surface were useful, however, as sites for derivatization with adhesion peptides such as the RGD and YIGSR sequences. Since the adsorption of proteins to those surfaces was so low, materials endowed with inactive peptides such as the RDG supported no cell adhesion.

Numerous other approaches are possible. One of particular interest, because of its ease of use, is physisorption. Block copolymers, consisting of adsorbing domains and nonadsorbing domains, can be adsorbed to material surfaces and can be used to regulate biological interactions. For example, when the nonadsorbing domains are polyethylene glycol, surfaces can be generated that display very low levels of nonspecific adhesion (Amiji and Park, 1992). One convenient class of polymers are ABA block copolymers of polyethylene glycol (the A blocks) and polypropylene glycol (B), in which the hydrophilic and cell-repelling polyethylene glycol domains flank the central hydrophobic and adsorbing polypropylene glycol block. The central hydrophobic block adsorbs well to hydrophobic surfaces, thus immobilizing the hydrophilic polymer and thereby resisting cell adhesion (Amiji and Park, 1992). Cell adhesion peptides can be displayed at the tips of these hydrophilic chain termini, and a very effective and simple model surface can be obtained (Neff *et al.*, 1998, 1999). Similar constructions can be designed for anionic surfaces, e.g., by using a polycationic block as a binding domain, with polyethylene glycol chains attached thereto (Elbert and Hubbell, 1998; Kenausis *et al.*, 2000; Huang *et al.*, 2001). Adhesion-promoting peptides may be grafted to the termini of the dangling polyethylene glycol chains, to permit cell attraction to these ligands on an otherwise remarkably nonadhesive background (VandeVondele *et al.*, 2003).

Model systems have also been employed for ligand discovery, i.e., to determine which parts of an adhesion protein are responsible for binding to an adhesion receptor. This has been most convincingly implemented using peptide arrays, i.e., surfaces in which peptides have been immobilized or even more powerfully synthesized in small domains on an otherwise-passive substrate (Mrksich, 2002; Min and Mrksich, 2004). Arrays of peptides that constitute overlap-

ping sequences order of 10 amino acids long spanning the entire length of a candidate adhesion protein can be constructed. If a receptor for the putative binding site is already known, it can be labeled, e.g., with a fluor, and the identity of peptides that bind can be determined by the location of the spots that bind the fluorescent protein. If a candidate receptor is not known, then larger spots can be synthesized (which somewhat limits the size of the peptide library that is formed), and the identity of adhesion-promoting peptides can be determined by the location of spots that promote cell adhesion and spreading, for example. Once the identity of a binding peptide is determined, the binding receptor can be determined by affinity chromatography of cell-derived proteins on columns containing bound peptide, with proteomic analysis of the protein that bind.

The aforementioned model systems for the study of cell–matrix interactions are all two-dimensional systems. Three-dimensional systems have also been developed in an effort to mimic the spatial complexity of the natural extracellular matrix (Lutolf and Hubbell, 2005; Pedersen and Swartz, 2005; Griffith and Swartz, 2006). Some of these utilize natural proteins, such as cell-derived extracellular matrix, fibrin, and collagen (Helm *et al.*, 2005; Mao and Schwarzbauer, 2005; Ng and Swartz, 2006). Some systems enable the identity and amounts of adhesion ligands, and potentially other ligands, to be precisely controlled. Two approaches are presented in the following paragraphs by way of example.

When fibrin forms spontaneously, nonfibrin proteins, such as fibronectin, are grafted into the nascent fibrin network by the enzymatic activity of the coagulation transglutaminase factor XIIIa. This feature of coagulation has been exploited to engineer fibrin matrices, by coagulating fibrinogen in the presence of exogenous and even synthetic factor XIIIa substrates (Schense and Hubbell, 1999). For example, if an adhesion ligand is synthesized as a fusion with a factor XIIIa substrate peptide, the adhesion ligand will be immobilized within the fibrin network. This approach has been carried out with small synthetic adhesion peptides (Schense *et al.*, 2000), with recombinant proteins that are fusions with a factor XIIIa substrate domain (Hall and Hubbell, 2004), and with peptides that bind glycosaminoglycans, which can in turn bind to growth factors (Sakiyama-Elbert and Hubbell, 2000). One can incorporate other bioactive molecules, such as growth factors, directly within the fibrin matrices, also by expressing them as recombinant fusion proteins with factor XIIIa substrate domains (Ehrbar *et al.*, 2004). Using these approaches, three-dimensional matrices for cell culture investigations of basic cellular processes can be constructed.

Synthetic three-dimensional matrices that allow precise control of cell adhesion ligand display have also been developed. In one system, reactively functionalized branched polyethylene glycol is cross-linked by a counterreactive peptide, the peptide being designed with a sequence that is a substrate for proteases that cells use when they migrate, such as matrix metalloproteinases or plasmin (Lutolf *et al.*, 2003). To the ends of some of the polyethylene glycol arms are also grafted adhesion ligands, such as a reactive RGD peptide. Although the porosity of these hydrogels is very small compared to the length scale of the smallest of processes that cells extend when they migrate, local activation of proteases at the cell membrane surface enables the cells to proteolytically tunnel their way through the materials (Raeber *et al.*, 2005). Like the fibrin materials described earlier, these materials are useful both as model systems for study of cell biology as well as for therapeutic ends.

## IV. CELL PATTERN FORMATION BY SUBSTRATE PATTERNING

The ability to control material surface properties precisely enables the formation of designed architectures of multiple cells in culture and potentially *in vivo* as well. Large-scale architectures have been formed by patterning adhesive surfaces. Four methods for patterning have been particularly powerful: photolithography, mechanical stamping, microfluidics, and lift-off.

Photolithographic methods have been employed to impart patterns on cell adhesion surfaces. Alkoxysilanes have been chemisorbed to glass surfaces (using the same grafting chemistry as with the glycophase glass described earlier), and ultraviolet light was employed to selectively degrade the alkoxy group to yield patterns of surface hydroxyl groups (Healy *et al.*, 1994). These hydroxyl groups were used as sites for reaction with a second layer of an amine-containing alkoxysilane. These aminated regions supported cell adhesion and thus formed the cell-binding regions on the patterned substrate (Kleinfeld *et al.*, 1988). Patterned amines on polymer surfaces have also been employed to induce cell patterning via adhesive domains patterned on a nonadhesive background (Ranieri *et al.*, 1993). These approaches have been combined with the bioactive peptide technology described earlier. For example, patterned amines have been used as grafting sites for the adhesive peptide YIGSR to pattern neurite extension on material surfaces (Ranieri *et al.*, 1994). One of the goals of this work was to create neuronal networks as a simple system in which to study communication among networks of neurons. A powerful system for such work has been provided by using adhesive aminoalkylsilanes patterned on a nonadhesive perfluoroalkylsilane background (Stenger *et al.*, 1992). Polymers have been synthesized explicitly for the purpose of attaching adhesive peptide sequences such as RGD, and these will be very useful in future studies of cell–cell interactions in neuronal and other cell systems (Herbert *et al.*, 1997). Such patterned surfaces have been formed to control cell shape and size, in order to gain deeper insight into the interplay between cell biomechanics and cell function (Thomas *et al.*, 1999). It is particularly convenient in photopatterning studies to develop photo-chemistries on materials specifically for the purpose of immobilizing polymers and polymer-peptide adducts (Moghaddam and Matsuda, 1993); this has been addressed, e.g., with phenylazido-derivatized surfaces (Matsuda and Sugawara, 1996; Sugawara and Matsuda, 1996).

Alkanethiols on gold have also been patterned using simple methods. Contact printing has been employed for this purpose (Love *et al.*, 2005; Whitesides *et al.*, 2005). Conventional photolithographic etching of silicon was employed to make a master printing stamp, a negative of which was then formed in silicone rubber. Structures as small as 200 nm were preserved in the silicone rubber stamp. The stamp was then wetted with a cell adhesion–promoting alkanethiol HS–$(CH_2)_{15}$–$CH_3$. Stamping a gold substrate resulted in creation of a pattern of the hydrophobic alkane group. The stamped gold substrate was then treated with the cell-resistant alkanethiol HS–$(CH_2)_{11}(OCH_2CH_2)_6OH$. Using this system it was possible to create adhesive patches of defined size on a very cell nonadhesive substrate (Singhvi *et al.*, 1994). Microcontact printing can also be employed with binding approaches other than alkane thiols binding to gold. For example, adhesion proteins such as laminin have been stamped onto reactive silane-modified surfaces to produce patterns to guide neurite outgrowth in culture (Wheeler *et al.*, 1999). This very flexible and powerful system will be useful in a wide variety of cell biological and tissue-engineering applications.

A third powerful method is based on microfluidic systems, in which silicone rubber stamps are formed with silicon masters; the stamps are pressed to a surface, and the thin spaces patterned thereby are employed as capillaries to draw up fluid, containing a treatment compound, onto desired regions of the surface. The fluid can contain a soluble, adsorbing polymer with an attached adhesion peptide (Neff *et al.*, 1998), or it can contain a peptide with some affinity linker for the surface. In the practice of the latter, it is powerful to employ the very high-affinity streptavidin–biotin pair, e.g., by biotinylating the polymer at the surface and exposing, with the aid of the microfluidics channels, to peptide conjugated to streptavidin (Patel *et al.*, 1998).

In a fourth method, also involving silicone layers on material surfaces, silicone layers can be used to pattern directly the locations in which cells adhere to surfaces, and the silicone layers can be lifted off the substrate, if desired, after such cell attachment (Sniadecki *et al.*, 2006). Using such approaches, it is possible to pattern two-dimensional surfaces as well as three-dimensional microwells atop such two-dimensional surfaces.

## V. CONCLUSIONS

While it is tempting to think of the matrix to which cells attach as providing primarily a mechanical support, it is clear from the preceding text that this is only a small part of the picture. Cell interaction with adhesive substrates is known to provide signaling information to the cells via numerous means, both biochemical and biomechanical, and this topic has been extensively reviewed (Roskelley *et al.*, 1995; Katz and Yamada, 1997; Otoole, 1997; K. M. Yamada, 1997; Danen *et al.*, 1998; Discher *et al.*, 2005; Pedersen and Swartz, 2005; Griffith and Swartz, 2006). The biochemical mechanisms underlying these interactions are likely numerous and have not yet been fully elucidated. One key mechanism involves the focal contact as a site for catalysis. Integrin clustering induces tyrosine phosphorylation of several proteins, many of which still have unknown function (Cohen and Guan, 2005). One of these proteins is a 125-kDa tyrosine kinase that localizes, after it is tyrosine phosphorylated, at the sites of focal contacts; this protein has been accordingly named pp125 focal adhesion kinase, or pp125$^{fak}$. Thus, although the cytoplasmic domain of integrins bears no direct catalytic activity, clustering of integrins is known to stimulate tyrosine phosphorylation, and further specific kinases are known to assemble at the sites of clustered integrins. Interestingly, when cells were permitted to spread via a non-integrin-mediated mechanism, specifically by interaction of cell surface proteoglycans with surface-adsorbed polycations, phosphorylation of intracellular proteins did not occur (Cohen and Guan, 2005). The phosphorylation of these proteins, associated with focal contact formation, is known to be an important signal for survival of a variety of cell types (Frisch *et al.*, 1996). Thus, the matrix plays not only a mechanical role as a support for cell adhesion and migration, but also a key signaling role in determining the details of cell behavior, ranging from survival to differentiation.

Engineered biomaterials will play an increasingly important role in deciphering the language of the interaction between cells and their extracellular matrix (Lutolf and Hubbell, 2005). Indeed, this represents one of the key challenges for researchers as the field moves forward, to represent more faithfully the complexity of the natural extracellular matrix in synthetic analogs. It is clear that the complex cellular interactions that exist with the three-dimensional milieu *in vivo* cannot be represented well by culture of cells on simple two-dimensional substrates like cell culture flasks (Griffith and Swartz, 2006), and it falls to the tissue engineer to develop more physiologically representative models.

While one goal of biomaterials and tissue-engineering research is certainly to develop systems for the quantitative study of biological interactions, another is to develop practical novel therapeutics. Many of the concepts described herein, both in terms of development of model surfaces and especially three-dimensional matrices and with regard to manipulating cellular behavior, are directly transferable; however, some cautionary comments should be made. It is not only the chemical identity of an adhesion peptide that determines its biological activity, but also its amount and distribution. This was very clearly demonstrated by Palecek *et al.* (1997), who showed that small amounts of an adhesion molecule could enhance cell migration, whereas larger amounts could inhibit it. They further demonstrated that this effect depended on, among other features, the affinity of the receptor–ligand pair, the number of receptors, and the polarization of receptors from the leading to the trailing edge of the cell. Given that many of these features depend on on the state of the cell and can be modulated by the cell's biological environment, e.g., by the growth factors to which the cell is exposed (Maheshwari *et al.*, 1999), many confounding features must be considered in translation from model to practical application (Lutolf and Hubbell, 2005).

# VI. REFERENCES

Amiji, M., and Park, K. (1992). Prevention of protein adsorption and platelet adhesion on surfaces by PEO PPO PEO triblock copolymers. *Biomaterials* **13**, 682–692.

Andrade, J. D., and Hlady, V. (1986). Protein adsorption and materials biocompatibility: a tutorial review and suggested hypotheses. *Adv. Polym. Sci.* **79**, 1–63.

Arnaout, M. A., Mahalingam, B., and Xiong, J. P. (2005). Integrin structure, allostery, and bidirectional signaling. *Ann. Rev. Cell Dev. Biol.* **21**, 381–410.

Aumailley, M., Bruckner-Tuderman, L., Carter, W. G., Deutzmann, R., Edgar, D., Ekblom, P., Engel, J., Engvall, E., Hohenester, E., Jones, J. C. R., *et al.* (2005). A simplified laminin nomenclature. *Matrix Biol.* **24**, 326–332.

Bokel, C., and Brown, N. H. (2002). Integrins in development: moving on, responding to, and sticking to the extracellular matrix. *Dev. Cell* **3**, 311–321.

Cohen, L. A., and Guan, J. L. (2005). Mechanisms of focal adhesion kinase regulation. *Curr. Cancer Drug Targets* **5**, 629–643.

Danen, E. H. J., Lafrenie, R. M., Miyamoto, S., and Yamada, K. M. (1998). Integrin signaling: cytoskeletal complexes, map kinase activation, and regulation of gene expression. *Cell Adh. Commun.* **6**, 217–224.

Danilov, Y. N., and Juliano, R. L. (1989). (ARG-GLY-ASP)N-albumin conjugates as a model substratum for integrin-mediated cell adhesion. *Exp. Cell Res.* **182**, 186–196.

Discher, D. E., Janmey, P., and Wang, Y.-L. (2005). Tissue cells feel and respond to the stiffness of their substrate. *Science* **310**, 1139–1143.

Drumheller, P. D., and Hubbell, J. A. (1994). Polymer networks with grafted cell-adhesion peptides for highly biospecific cell adhesive substrates. *Anal. Biochem.* **222**, 380–388.

Drumheller, P. D., Elbert, D. L., and Hubbell, J. A. (1994). Multifunctional poly(ethylene glycol) semiinterpenetrating polymer networks as highly selective adhesive substrates for bioadhesive peptide grafting. *Biotechnol. Bioeng.* **43**, 772–780.

Ehrbar, M., Djonov, V. G., Schnell, C., Tschanz, S. A., Martiny-Baron, G., Schenk, U., Wood, J., Burri, P. H., Hubbell, J. A., and Zisch, A. H. (2004). Cell-demanded liberation of vegf121 from fibrin implants induces local and controlled blood vessel growth. *Circ. Res.* **94**, 1124–1132.

Elbert, D. L., and Hubbell, J. A. (1998). Self-assembly and steric stabilization at heterogeneous, biological surfaces using adsorbing block copolymers. *Chem. Biol.* **5**, 177–183.

Faham, S., Linhardt, R. J., and Rees, D. C. (1998). Diversity does make a difference: fibroblast growth factor–heparin interactions. *Curr. Opin. Str. Biol.* **8**, 578–586.

Fairbrother, W. J., Champe, M. A., Christinger, H. W., Keyt, B. A., and Starovasnik, M. A. (1998). Solution structure of the heparin-binding domain of vascular endothelial growth factor. *Structure* **6**, 637–648.

Fleury, M. E., Boardman, K. C., and Swartz, M. A. (2006). Autologous morphogen gradients by subtle interstitial flow and matrix interactions. *Biophys. J.* **91**, 113–121.

Fratzl, P., Misof, K., Zizak, I., Rapp, G., Amenitsch, H., and Bernstorff, S. (1998). Fibrillar structure and mechanical properties of collagen. *J. Str. Biol.* **122**, 119–122.

Frisch, S. M., Vuori, K., Ruoslahti, E., and ChanHui, P. Y. (1996). Control of adhesion-dependent cell survival by focal adhesion kinase. *J. Cell Biol.* **134**, 793–799.

Ginsberg, M. H., Partridge, A., and Shattil, S. J. (2005). Integrin regulation. *Curr. Opin. Cell Biol.* **17**, 509–516.

Griffith, L. G., and Swartz, M. A. (2006). Capturing complex 3D tissue physiology *in vitro. Nat. Rev. Mol. Cell Biol.* **7**, 211–224.

Grinnell, F. (2003). Fibroblast biology in three-dimensional collagen matrices. *Trends Cell Biol.* **13**, 264–269.

Gumbiner, B. M. (2005). Regulation of cadherin-mediated adhesion in morphogenesis. *Nat. Rev. Mol. Cell Biol.* **6**, 622–634.

Hall, H., and Hubbell, J. A. (2004). Matnix-bound sixth Ig-like domain of cell adhesion molecule l1 acts as an angiogenic factor by ligating alpha-V-beta-3-integrin and activating vegf-r2. *Microvasc. Res.* **68**, 169–178.

Healy, K. E., Lom, B., and Hockberger, P. E. (1994). Spatial distribution of mammalian cells dictated by material surface chemistry. *Biotechnol. Bioeng.* **43**, 792–800.

Helm, C. L. E., Fleury, M. E., Zisch, A. H., Boschetti, F., and Swartz, M. A. (2005). Synergy between interstitial flow and vegf directs capillary morphogenesis in vitro through a gradient amplification mechanism. *Proc. Nat. Acad. Sci. USA* **102**, 15779–15784.

Herbert, C. B., McLernon, T. L., Hypolite, C. L., Adams, D. N., Pikus, L., Huang, C. C., Fields, G. B., Letourneau, P. C., Distefano, M. D., and Hu, W. S. (1997). Micropatterning gradients and controlling surface densities of photoactivatable biomolecules on self-assembled monolayers of oligo(ethylene glycol) alkanethiolates. *Chem. Biol.* **4**, 731–737.

Huang, N. P., Michel, R., Voros, J., Textor, M., Hofer, R., Rossi, A., Elbert, D. L., Hubbell, J. A., and Spencer, N. D. (2001). Poly(L-lysine)-G-poly(ethylene glycol) layers on metal oxide surfaces: surface-analytical characterization and resistance to serum and fibrinogen adsorption. *Langmuir* **17**, 489–498.

Hubbell, J. A. (1995). Biomaterials in tissue engineering. *Biotechnology* **13**, 565–576.

Kallapur, S. G., and Akeson, R. A. (1992). The neural cell-adhesion molecule (nCAM) heparin-binding domain binds to cell-surface heparan-sulfate proteoglycans. *J. Neurosci. Res.* **33**, 538–548.

Katz, B. Z., and Yamada, K. M. (1997). Integrins in morphogenesis and signaling. *Biochimie* **79**, 467–476.

Kenausis, G. L., Voros, J., Elbert, D. L., Huang, N. P., Hofer, R., Ruiz-Taylor, L., Textor, M., Hubbell, J. A., and Spencer, N. D. (2000). Poly(L-lysine)-G-poly(ethylene glycol) layers on metal oxide surfaces: attachmnet mechanism and effects of polymer architecture on resistance to protein adsorption. *J. Phys. Chem. B* **104**, 3298–3309.

Kleinfeld, D., Kahler, K. H., and Hockberger, P. E. (1988). Controlled outgrowth of dissociated neurons on patterned substrates. *J. Neurosci.* **8**, 4098–4120.

Kleinman, H. K., Philp, D., and Hoffman, M. P. (2003). Role of the extracellular matrix in morphogenesis. *Curr. Opin. Biotechnol.* **14**, 526–532.

Lebaron, R. G., Esko, J. D., Woods, A., Johansson, S., and Hook, M. (1988). Adhesion of glycosaminoglycan-deficient Chinese hamster ovary cell mutants to fibronectin substrata. *J. Cell Biol.* **106**, 945–952.

Leckband, D., and Prakasam, A. (2006). Mechanism and dynamics of cadherin adhesion. *Annu. Rev. Biomed. Eng.* Epub ahead of print.

Love, J. C., Estroff, L. A., Kriebel, J. K., Nuzzo, R. G., and Whitesides, G. M. (2005). Self-assembled monolayers of thiolates on metals as a form of nanotechnology. *Chem. Rev.* **105**, 1103–1169.

Lutolf, M. P., and Hubbell, J. A. (2005). Synthetic biomaterials as instructive extracellular microenvironments for morphogenesis in tissue engineering. *Nature Biotechnol.* **23**, 47–55.

Lutolf, M. P., Raeber, G. P., Zisch, A. H., Tirelli, N., and Hubbell, J. A. (2003). Cell-responsive synthetic hydrogels. *Adv. Mater.* **15**, 888+.

Lyon, M., and Gallagher, J. T. (1998). Bio-specific sequences and domains in heparan sulphate and the regulation of cell growth and adhesion. *Matrix Biol.* **17**, 485–493.

Magnusson, M. K., and Mosher, D. F. (1998). Fibronectin — structure, assembly, and cardiovascular implications. *Arterioscl. Thromb. Vasc. Biol.* **18**, 1363–1370.

Maheshwari, G., Wells, A., Griffith, L. G., and Lauffenburger, D. A. (1999). Biophysical integration of effects of epidermal growth factor and fibronectin on fibroblast migration. *Biophys. J.* **76**, 2814–2823.

Mao, Y., and Schwarzbauer, J. E. (2005). Stimulatory effects of a three-dimensional microenvironment on cell-mediated fibronectin fibrillogenesis. *J. Cell Sci.* **118**, 4427–4436.

Massia, S. P., and Hubbell, J. A. (1990). Covalent surface immobilization of ARG-GLY-ASP-containing and TYR-ILE-GLY-SER-ARG-containing peptides to obtain well-defined cell-adhesive substrates. *Anal. Biochem.* **187**, 292–301.

Massia, S. P., and Hubbell, J. A. (1991). An RGD spacing of 440 nm is sufficient for integrin alpha-V-beta-3-mediated fibroblast spreading and 140 nm for focal contact and stress fiber formation. *J. Cell Biol.* **114**, 1089–1100.

Massia, S. P., and Hubbell, J. A. (1992). Immobilized amines and basic amino acids as mimetic heparin-binding domains for cell-surface proteoglycan-mediated adhesion. *J. Biol. Chem.* **267**, 10133–10141.

Matsuda, T., and Sugawara, T. (1996). Control of cell adhesion, migration, and orientation on photochemically microprocessed surfaces. *J. Biomed. Mater. Res.* **32**, 165–173.

Meecham, R. P. (1991). Laminin receptors. *Annu. Rev. Cell Biol.* **7**, 71–91.

Min, D. H., and Mrksich, M. (2004). Peptide arrays: towards routine implementation. *Curr. Opin. Chem. Biol.* **8**, 554–558.

Miner, J. H., and Yurchenco, P. D. (2004). Laminin functions in tissue morphogenesis. *Annu. Rev. Cell Dev. Biol.* **20**, 255–284.

Moghaddam, M. J., and Matsuda, T. (1993). Molecular design of 3-dimensional artificial extracellular matrix — photosensitive polymers containing cell adhesive peptide. *J. Polym. Sci. A Polym. Chem.* **31**, 1589–1597.

Mrksich, M. (2002). What can surface chemistry do for cell biology? *Curr. Opin. Chem. Biol.* **6**, 794–797.

Neff, J. A., Caldwell, K. D., and Tresco, P. A. (1998). A novel method for surface modification to promote cell attachment to hydrophobic substrates. *J. Biomed. Mater. Res.* **40**, 511–519.

Neff, J. A., Tresco, P. A., and Caldwell, K. D. (1999). Surface modification for controlled studies of cell–ligand interactions. *Biomaterials* **20**, 2377–2393.

Ng, C. P., and Swartz, M. A. (2006). Mechanisms of interstitial flow-induced remodeling of fibroblast–collagen cultures. *Ann. Biomed. Eng.* **34**, 446–454.

Otoole, T. E. (1997). Integrin signaling: building connections beyond the focal contact? *Matrix Biol.* **16**, 165–171.

Otsuka, H., Nagasaki, Y., and Kataoka, K. (2003). Pegylated nanoparticles for biological and pharmaceutical applications. *Adv. Drug Deliv. Rev.* **55**, 403–419.

Palecek, S. P., Loftus, J. C., Ginsberg, M. H., Lauffenburger, D. A., and Horwitz, A. F. (1997). Integrin-ligand-binding properties govern cell migration speed through cell–substratum adhesiveness. *Nature* **385**, 537–540.

Patel, N., Padera, R., Sanders, G. H. W., Cannizzaro, S. M., Davies, M. C., Langer, R., Roberts, C. J., Tendler, S. J. B., Williams, P. M., and Shakesheff, K. M. (1998). Spatially controlled cell engineering on biodegradable polymer surfaces. *FASEB J.* **12**, 1447–1454.

Pedersen, J. A., and Swartz, M. A. (2005). Mechanobiology in the third dimension. *Ann. Biomed. Eng.* **33**, 1469–1490.

Ponce, M. L., Nomizu, M., Delgado, M. C., Kuratomi, Y., Hoffman, M. P., Powell, S., Yamada, Y., Kleinman, H. K., and Malinda, K. M. (1999). Identification of endothelial cell binding sites on the laminin gamma 1 chain. *Circ. Res.* **84**, 688–694.

Preissner, K. T., and Seiffert, D. (1998). Role of vitronectin and its receptors in haemostasis and vascular remodeling. *Thromb. Res.* **89**, 1–21.

Prime, K. L., and Whitesides, G. M. (1993). Adsorption of proteins onto surfaces containing end-attached oligo(ethylene oxide) — a model system using self-assembled monolayers. *J. Am. Chem. Soc.* **115**, 10714–10721.

Raeber, G. P., Lutolf, M. P., and Hubbell, J. A. (2005). Molecularly engineered PEG hydrogels: a novel model system for proteolytically mediated cell migration. *Biophys. J.* **89**, 1374–1388.

Ranieri, J. P., Bellamkonda, R., Jacob, J., Vargo, T. G., Gardella, J. A., and Aebischer, P. (1993). Selective neuronal cell attachment to a covalently patterned monoamine on fluorinated ethylene-propylene films. *J. Biomed. Mater. Res.* **27**, 917–925.

Ranieri, J. P., Bellamkonda, R., Bekos, E. J., Gardella, J. A., Mathieu, H. J., Ruiz, L., and Aebischer, P. (1994). Spatial control of neuronal cell attachment and differentiation on covalently patterned laminin oligopeptide substrates. *Int. J. Dev. Neurosci.* **12**, 725–735.

Rezania, A., and Healy, K. E. (1999). Biomimetic peptide surfaces that regulate adhesion, spreading, cytoskeletal organization, and mineralization of the matrix deposited by osteoblast-like cells. *Biotechnol. Prog.* **15**, 10–32.

Roskelley, C. D., Srebrow, A., and Bissell, M. J. (1995). A hierarchy of ECM-mediated signaling regulates tissue-specific gene expression. *Curr. Opin. Cell Biol.* **7**, 736–747.

Ruoslahti, E. (1991). Integrins. *J. Clin. Invest.* **87**, 1–5.

Ruoslahti, E., and Pierschbacher, M. (1986). Tetrapeptide. U.S. Pat. 4,578,079.

Ruoslahti, E., and Pierschbacher, M. (1987). New perspectives in cell adhesion: RGD and integrins. *Science* **238**, 491–497.

Sadler, J. E. (1998). Biochemistry and genetics of von willebrand factor. *Annu. Rev. Biochem.* **67**, 395–424.

Sakiyama-Elbert, S. E., and Hubbell, J. A. (2000). Development of fibrin derivatives for controlled release of heparin-binding growth factors. *J. Control. Rel.* **65**, 389–402.

Sasaki, T., Fassler, R., and Hohenester, E. (2004). Laminin: the crux of basement membrane assembly. *J. Cell Biol.* **164**, 959–963.

Schense, J. C., and Hubbell, J. A. (1999). Cross-linking exogenous bifunctional peptides into fibrin gels with factor XIIIa. *Bioconj. Chem.* **10**, 75–81.

Schense, J. C., Bloch, J., Aebischer, P., and Hubbell, J. A. (2000). Enzymatic incorporation of bioactive peptides into fibrin matrices enhances neurite extension. *Nature Biotechnol.* **18**, 415–419.

Schwarzbauer, J. L. (1999). Basement membranes: putting up the barriers. *Curr. Biol.* **9**, R242–R244.

Shapiro, S. D. (1998). Matrix metalloproteinase degradation of extracellular matrix: biological consequences. *Curr. Opin. Cell Biol.* **10**, 602–608.

Shemesh, T., Geiger, B., Bershadsky, A. D., and Kozlov, M. M. (2005). Focal adhesions as mechanosensors: a physical mechanism. *Proc. Nat. Acad. Sci. USA* **102**, 12383–12388.

Singhvi, R., Kumar, A., Lopez, G. P., Stephanopoulos, G. N., Wang, D. I. C., Whitesides, G. M., and Ingber, D. E. (1994). Engineering cell shape and function. *Science* **264**, 696–698.

Sniadecki, N., Desai, R. A., Ruiz, S. A., and Chen, C. S. (2006). Nanotechnology for cell–substrate interactions. *Ann. Biomed. Eng.* **34**, 59–74.

Stenger, D. A., Georger, J. H., Dulcey, C. S., Hickman, J. J., Rudolph, A. S., Nielsen, T. B., McCort, S. M., and Calvert, J. M. (1992). Coplanar molecular assemblies of aminoalkylsilane and perfluorinated alkylsilane — characterization and geometric definition of mammalian-cell adhesion and growth. *J. Am. Chem. Soc.* **114**, 8435–8442.

Sugawara, T., and Matsuda, T. (1996). Synthesis of phenylazido-derivatized substances and photochemical surface modification to immobilize functional groups. *J. Biomed. Mater. Res.* **32**, 157–164.

Tashiro, K., Sephel, G. C., Weeks, B., Sasaki, M., Martin, G. R., Kleinman, H. K., and Yamada, Y. (1989). A synthetic peptide containing the ikvav sequence from the A-chain of laminin mediates cell attachment, migration, and neurite outgrowth. *J. Biol. Chem.* **264**, 16174–16182.

Thomas, C. H., Lhoest, J. B., Castner, D. G., McFarland, C. D., and Healy, K. E. (1999). Surfaces designed to control the projected area and shape of individual cells. *J. Biomech. Eng.* **121**, 40–48.

van der Flier, A., and Sonnenberg, A. (2001). Function and interactions of integrins. *Cell Tissue Res.* **305**, 285–298.

VandeVondele, S., Voros, J., and Hubbell, J. A. (2003). RGD-grafted poly-L-lysine-graft-(polyethylene glycol) copolymers block nonspecific protein adsorption while promoting cell adhesion. *Biotechnol. Bioeng.* **82**, 784–790.

Vestweber, D., and Blanks, J. E. (1999). Mechanisms that regulate the function of the selectins and their lignads. *Physiol. Rev.* **79**, 181–213.

Walsh, F. S., and Doherty, P. (1997). Neural cell adhesion molecules of the immunoglobulin superfamily: role in axon growth and guidance. *Annu. Rev. Cell Dev. Biol.* **13**, 1997.

Ward, M. D., and Hammer, D. A. (1993). A theoretical analysis for the effect of focal contact formation on cell–substrate attachment strength. *Biophys. J.* **64**, 936–959.

Wheeler, B. C., Corey, J. M., Brewer, G. J., and Branch, D. W. (1999). Microcontact printing for precise control of nerve cell growth in culture. *J. Biomech. Eng.* **121**, 73–78.

Whitesides, G. M., Kriebel, J. K., and Love, J. C. (2005). Molecular engineering of surfaces using self-assembled monolayers. *Sci. Prog.* **88**, 17–48.

Wiesner, S., Legate, K. R., and Fassler, R. (2005). Integrin–actin interactions. *Cell. Molec. Life Sci.* **62**, 1081–1099.

Yamada, K. M. (1997). Integrin signaling. *Matrix Biol.* **16**, 137–141.

Yamada, Y., and Kleinman, H. K. (1992). Functional domains of cell adhesion molecules. *Curr. Opin. Cell Biol.* **4**, 819–823.

# Chapter Twenty-Two

# Polymer Scaffold Fabrication

*Matthew B. Murphy and Antonios G. Mikos*

## I. INTRODUCTION

In the modern age of medicine, tissue engineering has become a viable option for the replacement of tissue and organ function. The creation of such substitutes requires a three-dimensional, porous, biocompatible, and preferably biodegradable scaffold. Tissue-engineering scaffolds should have geometries that direct new tissue formation and mass transport properties sufficient for the exchange of biological nutrients and waste. The scaffolds also provide temporary mechanical support to the regenerating tissue. They must degrade into biocompatible byproducts, ideally on a time scale comparable to that of new tissue development. Such scaffolds are typically fabricated with biocompatible polymers, proteins, peptides, and inorganic materials. Aside from the properties of the raw material, the major factor determining the final scaffold characteristics is the fabrication technique utilized to produce the scaffold. Mechanical strength, porosity, degradation rates, surface chemistry, and the ability to incorporate biologically active molecules are all aspects affected by the manner of fabrication. This chapter discusses many established processing and fabrication methods using various polymeric components, including fiber bonding, electrostatic fiber spinning, solvent casting and particulate leaching, melt molding, membrane lamination, extrusion, freeze-drying, phase separation, high-internal-phase emulsion, gas foaming, polymer/ceramic composite fabrication, rapid prototyping, peptide self-assembly, and *in situ* polymerization.

In an era of decreasing availability of organs for transplantation and a growing need for suitable replacements, the emerging field of tissue engineering gives hope to patients who desperately require tissue and organ substitutes. Scaffolding is essential in this endeavor to act as a three-dimensional template for tissue ingrowth by mimicking the extracellular matrix (ECM) for cell adhesion and proliferation (Freed *et al.*, 1994). Since the mid-1980s, researchers have developed many novel techniques to shape polymers into complex architectures that exhibit the desired properties for specific tissue-engineering applications. These fabrication techniques result in reproducible scaffolds for the regeneration of specific tissues. Polymer scaffolds can provide mechanical strength, interconnected porosity and surface area, varying surface chemistry, and unique geometries to direct tissue regeneration (Hutmacher, 2001). These key scaffold characteristics can be tailored to

*Principles of Tissue Engineering, 3rd Edition*
ed. by Lanza, Langer, and Vacanti

the application by careful selection of the polymers, additional scaffold components, and the fabrication technique.

Patient safety is the paramount concern for any tissue-engineering application. The bulk material and degradation products of the scaffold must be biocompatible and clearable by the body. It is equally critical that the elected processing strategy not affect the biocompatibility and biodegradability of the scaffolding materials. Restoring the function of a tissue or replacing an organ entirely requires a porous scaffold that degrades on an appropriate time scale so that the new tissue replaces the resorbing scaffold. The primary function of the scaffold is to direct the growth and migration of cells from surrounding tissues into the defect or to facilitate the growth of cells seeded into the scaffold prior to implantation. Surface chemistry favorable to cell attachment and proliferation is desirable. Large pore diameters and high pore interconnectivity are essential for confluent tissue formation, transport of nutrients and metabolic wastes, and sufficient vascularization of the new tissue. Increased porosity and pore diameter can result in increased surface-area-to-volume ratios within the scaffold or more surfaces for cell adhesion. Control over the scaffold's size and shape provides increased utility for differing tissue-engineering applications.

The mechanical properties of a scaffold arise from a combination of the properties of the bulk polymer, the geometry of the scaffold, incorporation of strength-enhancing materials, and the scaffold fabrication technique. For example, polymers with higher crystallinity exhibit increased tensile strength at the expense of slower degradation rates. Processing methods that reduce crystallinity or the molecular weight of polymer chains diminish the strength of the scaffold and reduce the scaffold's lifetime. Elevated mechanical strength is preferable in the regeneration of load-bearing tissues such as bone and cartilage. Mechanical stimulation via force transduction can be beneficial in the differentiation of many cell types (Tan *et al.*, 1996). While hydrophobic polymers typically offer greater mechanical properties, adsorbing proteins may become denatured through interaction with the surface (Gray, 2004). Typical materials utilized in tissue engineering scaffolds include synthetic polymers [e.g., poly(glycolic acid) (PGA), poly(L-lactic acid) (PLLA), poly(D,L-lactic-*co*-glycolic acid) (PLGA) copolymers, poly(ε-caprolactone) (PCL), and ethylene glycol–based copolymers], natural polymers (e.g., collagens, gelatins, fibrin, carbohydrates, peptides, and nucleic acids), and inorganic materials (e.g., hydroxyapatite, tricalcium phosphate, and titanium).

The inclusion of bioactive molecules is another major consideration in the design of porous scaffolds. Bioactive molecules include proteins, ECM-like peptides, and DNA. Because the bioactive molecules are incorporated for cell adhesion, cell signaling, or drug/gene delivery, fabrication techniques that do not inactivate the molecules must be utilized. Local drug and gene delivery to promote cell migration, proliferation, and differentiation is an enormous tool to improve the required time and quality of tissue regeneration (Jang *et al.*, 2004).

The fabrication technique for tissue-engineering scaffolds depends almost entirely on the bulk and surface properties of the material and the proposed function of the scaffold. However, the cost and time of manufacturing scaffolds must be considered for the viability of patient treatment. Most techniques involve the application of heat and/or pressure to the polymer or dissolving it in an organic solvent to mold the material into its desired shape. Evolving techniques have been studied that reduce potentially harsh conditions of older scaffold fabrication schemes to protect incorporated cells and bioactive molecules. While each method presents distinct advantages and disadvantages, the appropriate technique must be selected to meet the requirements for the specific type of tissue.

## II. FIBER BONDING

Polymer fibers exhibit an excellent surface-area-to-volume ratio for enhanced cell attachment, making them a viable option as a scaffold material. The earliest tissue-engineering scaffolds were fiber mesh, nonbonded PGA tassels or felts that lacked the mechanical integrity to be used for *in vivo* organ regeneration (Cima *et al.*, 1991). To overcome this problem, fiber-bonding techniques were developed to bind the fibers together at points of intersection. The original examples of fiber-bonded scaffolds used PGA and PLLA polymers (Mikos *et al.*, 1993a). Briefly, PGA fibers are arranged in a nonwoven mesh. At temperatures above the melting point of the polymer, the fibers will bond at their contact points. To prevent a structural collapse of the melting polymer, PGA fibers are encapsulated prior to heat treatment. PLLA, dissolved in methylene chloride (not a solvent for PGA), is cast over the meshed fibers and dried, resulting in a PGA–PLLA composite matrix. After heat treatment and fiber bonding, the PLLA is dissolved in methylene chloride and the solvent is removed from the scaffold by vacuum drying. Another method involves rotating a nonwoven PGA fiber mesh while spraying it with an atomized PLLA or PLGA solution (Mooney *et al.*, 1996a). The polymer solution builds up on the PGA fibers and bonds them at contact points. This method provides the mechanical properties of PGA while exposing cells to the surface properties of PLLA or PLGA. This method is excellent for producing tubular structures, but it lacks the ability to create complex three-dimensional structures and increases the original fiber diameter. Similar methods exist for other biocompatible polymer fibers.

The fiber-bonding scaffold fabrication technique is desirable for its simplicity, the retention of the PGA fibers' original properties, the use of only biocompatible materials, and the structural advantages over tassel or felt arrangements. The drawbacks of fiber bonding are the lack of control over porosity and pore size, the availability of suitable solvents, immiscibility of the two polymers in the melt state, and the required relative melting temperatures of the polymers.

## III. ELECTROSPINNING

A modern method for creating porous scaffolds composed of nano- and microscale biodegradable fibers employs electrostatic fiber spinning, or electrospinning, a technology derived from the electrostatic spraying of polymer coatings. Electrospinning fabricates highly porous scaffolds of nonwoven and ultrafine fibers. Many biocompatible polymers, including PGA, PLGA, and PCL, can be electrospun into scaffolds of nanofibers with porosities greater than 90% (Yoshimoto *et al.*, 2003). Scaffolds are prepared by dissolving the selected polymer in an appropriate solvent (e.g., PCL in chloroform). The polymer solution is loaded into a syringe and then expelled through a metal capillary at a constant rate via syringe pump. A high voltage (10–15 kV) is applied to the capillary, charging the polymer and ejecting it toward a grounded collecting surface. As the thin fibers assemble on the plate, the solvent evaporates, leaving a nonwoven porous scaffold. Fiber thickness, scaffold diameter, and average pore diameter are adjusted by factors including polymer concentration, choice of solvent, ejection rate, applied voltage, capillary diameter, collecting plate material, and the distance between the capillary and the collecting plate. Examples of electrospun P(LLA-CL) fiber meshes are shown in Fig. 22.1. Electrospun scaffolds exhibit promise in mesenchymal stem cell culture for bone and cartilage tissue engineering (Pham *et al.*, 2006).

## IV. SOLVENT CASTING AND PARTICULATE LEACHING

For enhanced control over porosity and pore diameter as compared to most fabrication methods, a solvent-casting and particulate-leaching technique was developed. With careful system selection, porous scaffolds can be manufactured with specific pore size, porosity, surface-area-to-volume ratio, and crystallinity. This technique involves casting a dissolved polymer around a suitable porogen, drying and solidifying the polymer, and leaching out the porogen to yield a polymer scaffold with an interconnected porous network. Early systems utilized PLLA and PLGA polymers with sieved salt particles as a porogen (Mikos *et al.*, 1994). To adjust the crystallinity, the composite material is heated above the polymer melting temperature and annealed at the appropriate rate prior to porogen leaching. Afterwards, the composite is immersed in water to remove the salt particles, leaving a porous PLLA membrane. Similar techniques have utilized alternative biocompatible porogens, such as sugars (Holy *et al.*, 1999) and lipids (Hacker *et al.*, 2003). A solvent exchange system, where the second organic phase dissolves the porogen but is a nonsolvent for the polymer, eliminates the traditional leaching step and presents an advantage in the total leaching time required.

**FIG. 22.1.** Scanning electron micrographs of P(LLA-CL) fibers electrospun at an applied voltage of 12 kV from different polymer concentration solutions: **(A)** 3 wt.%; **(B)** 5 wt.%; **(C)** 7 wt.%; **(D)** 9 wt.%. Reprinted from X. M. Mo, C. Y. Xu, M. Kotaki, and S. Ramakrishna (2004), Electrospun P(LLA-CL) nanofiber: a biomimetic extracellular matrix for smooth muscle cell and endothelial cell proliferation, *Biomaterials* **25**, pp. 1883–1890. Copyright 2003, with permission of Elsevier Science.

For polymers preloaded with bioactive molecules, salt leaching can remove molecules or decrease their bioactivity during the leaching process.

This technique can produce scaffolds with controlled porosity (up to 93%), pore size (up to 500 µm), and crystallinity. By adjusting the fabrication parameters and the type, amount, and size of porogen, porous scaffolds can be tailored to the tissue-engineering application of interest. The primary advantage to this technique is the relatively small amount of polymer required to create a scaffold. PLGA and poly(ethylene glycol) (PEG) blends have been utilized to produce porous foams with the solvent casting and particulate leaching technique that are less brittle and more suitable for soft-tissue regeneration (Wake *et al.*, 1996). To overcome problems with cell seeding due to the polymer's hydrophobicity, scaffolds can be prewetted using ethanol (Mikos *et al.*, 1994). The scaffolds are submerged first in ethanol, followed by water. Prewet scaffolds show higher cell attachment for chondrocytes and hepatocytes. As an alternative, PLGA scaffolds have been soaked and coated with more hydrophilic polymers, such as poly(vinyl alcohol) (PVA) (Mooney *et al.*, 1994). The attachment of hepatocytes was greatly increased for PVA-coated scaffolds as compared to untreated PLGA scaffolds.

## V. MELT MOLDING

An alternative method for the production of three-dimensional scaffolds is melt molding. This technique calls for polymer and porogen particles to be combined in a mold and heated above the polymer's glass transition temperature (for amorphous polymers) or melting temperature (for semicrystalline polymers). After the reorganization of the polymer, the composite material is removed from the mold, cooled, and soaked in an appropriate liquid to leach out the porogen. The resulting porous scaffold has the exact external shape as the mold. PLGA/gelatin microparticle composites have been formed in this fashion with gelatin leaching in distilled-deionized water (Thomson *et al.*, 1995a). Melt molding allows for the formation of scaffolds of any desired geometry by altering the size and shape of the mold. Adjusting the amount and size of porogen used, respectively, can control the porosity and pore size of the scaffold. The melt-molding protocol can be adapted to incorporate materials such as hydroxyapatite fibers (Thomson *et al.*, 1995b). Such fibers provide additional mechanical support and a bioactive surface for cells when uniformly distributed throughout the polymer prior to melting. Melt molding is advantageous for the inclusion and delivery of bioactive molecules because the materials are not exposed to harsh organic solvents, although excessively high molding temperatures can degrade and inactivate the molecules.

## VI. MEMBRANE LAMINATION

Tissue engineering often requires precise three-dimensional anatomical geometries for hard tissues with shape-dependent function like bone and cartilage. Thin layers of porous polymer produced in the previously mentioned manners can be cut, stacked, and bonded by means of membrane lamination (Mikos *et al.*, 1993b). The layers are chemically joined, but there is no distinguishable boundary at the interface of two adjacent membranes. The key to this method is the creation of a three-dimensional contour plot of the desired scaffold shape. Each layer of the scaffold is cut from a highly porous membrane into its corresponding shape for that level. A small amount of solvent, such as chloroform, is coated on the interfacial surface, and a bond is formed between membranes. This process is repeated for all subsequent layers until the completion of the final three-dimensional structure.

Porous polymers used in membrane lamination include PLLA and PLGA membranes formed by solvent casting and particulate leaching. As previously mentioned, there is no detectable boundary between layers in the finished scaffold. Membrane lamination provides a method for fabricating three-dimensional anatomical shapes with identical bulk properties to the individual membranes. Membrane lamination has also been utilized in the preparation of degradable tubular stents (Mooney *et al.*, 1994). Porous membranes of PLGA are produced by solvent casting and particulate leaching and wrapped around a Teflon cylinder. The overlapped edges are bonded with a small volume of solvent, and the Teflon is removed, yielding a hollow cylinder of porous PLGA for applications such as intestinal and vascular regeneration.

## VII. EXTRUSION

While extrusion is a well-documented processing method for industrial polymers such as polyethylene, this method is relatively new for biocompatible porous scaffold production. The first extrusion of polymers for tissue engineering utilized PLGA and PLLA to form tubular scaffolds for peripheral nerve regeneration (Widmer *et al.*, 1998). Extruded tubular PLGA scaffolds are illustrated in Fig. 22.2. The polymers were fabricated into membranes using solvent casting, with sodium chloride as a porogen. The membranes were cut to appropriate sizes and loaded into a customized extrusion tool. The extruder applies heat and pressure to the composite material and forces it through a die and out the nozzle to form cylindrical conduits. After the conduits are cooled, they are soaked in water, to leach the salt, and vacuum dried. Higher temperatures require less pressure, and vice versa. While high pressures may require a powerful hydraulic press, high temperatures can adversely affect the crystallinity and porosity of the scaffold and the activity of incorporated biomolecules. As with other methods, porogen content and size are the most important parameters of porosity and average pore diameter. Extruded polymer scaffolds can be fabricated to support the loading of cells or growth factors for tissue engineering. PLGA, PCL, and most biocompatible polymers can be extruded at appropriate temperatures and pressures.

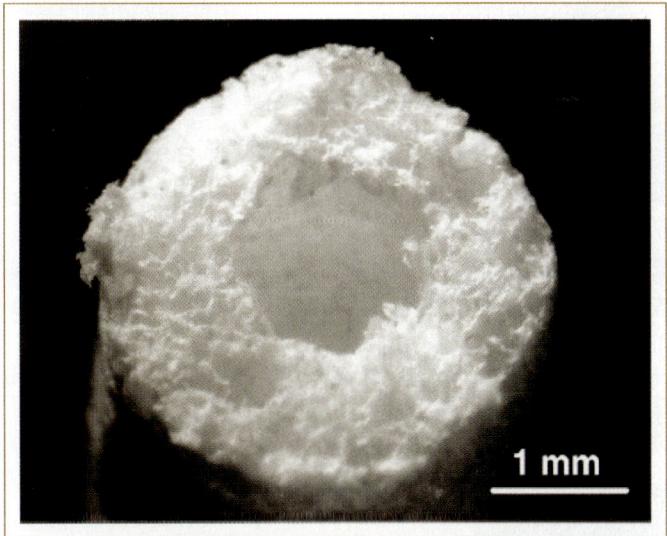

**FIG. 22.2.** Optical micrograph of a conduit fabricated by extrusion from PLGA and salt crystals, a salt weight fraction of 90%, and an extrusion temperature of 250°C. Reprinted from Widmer *et al.* (1998). Copyright 1998, with permission of Elsevier Science.

## VIII. FREEZE-DRYING

Another method for rapid fabrication of scaffolds with controllable porosity and average pore diameter employs emulsion and freeze-drying. An organic solution containing dissolved polymer is combined with a suitable amount of water and emulsified until homogeneity is achieved (Whang *et al.*, 1995). The resulting emulsion is poured into a metal mold of specified dimensions and frozen with liquid nitrogen. Freeze-drying removes the water and solvent to yield scaffolds of highly interconnected pores, porosities up to 90%, and median pore diameters from 15 to 35 μm. This technique has been utilized with many biocompatible polymers, including PGA, PLLA, PLGA, and poly(propylene fumarate) (PPF) blends. Inclusion of polymers like PPF in composite scaffolds is beneficial for adjustment of compressive strength and properties related to hydrophobicity (e.g., water penetration, scaffold degradation rates, and drug diffusion) (Hsu *et al.*, 1997). PLGA/PPF foam scaffolds exhibit a closed-pore morphology, however, an unattractive quality for most tissue-engineering applications. PLGA and PLGA-blend polymer scaffolds of greater than 1-cm thickness can be manufactured by emulsion and freeze-drying.

Non-emulsion-based freeze-drying is also capable of producing porous polymer scaffolds. Synthetic polymers dissolved in glacial organic solvents are frozen, and then the solvent is removed by freeze-drying (Hsu *et al.*, 1997). Similar techniques were utilized to create collagen scaffolds by dispersing the protein in water and freeze-drying the suspension (Yannas *et al.*, 1980). Sublimed ice crystals generate pores, with pore size being controlled by solution parameters such as freezing rate, temperature, ionic concentrations, and pH.

## IX. PHASE SEPARATION

The potential to deliver drugs and other bioactive molecules from a degradable tissue-engineering scaffold is advantageous for modulating cell differentiation and guiding tissue regeneration. Such scaffolds can be produced by a phase separation technique that does not expose the bioactive molecules to harsh organic chemicals or temperatures (Lo *et al.*, 1995). Briefly, a biocompatible polymer such as PLGA or a poly(phosphoester) is dissolved in an appropriate solvent (e.g., phenol at 552°C, dioxane at 63°C, or naphthalene at 85°C). While stirring, the bioactive molecules are added and dispersed into a homogeneous mixture and cooled below the solvent melting point until the liquid phases separate (Hua *et al.*, 2002). The polymer and solvent are quenched with liquid nitrogen, resulting in a two-phase solid. The solvent is removed by sublimation, which yields a porous scaffold with bioactive molecules embedded inside the polymer. Porosity and architecture are affected by the cooling rate and the melting temperature of the solvent relative to the polymer. Tailoring specific drug-release rates and incorporating large proteins are the major obstacles with phase separation methods of polymer scaffold fabrication.

## X. HIGH-INTERNAL-PHASE EMULSION

Porous scaffolds are typically prepared by bulk polymerization or condensation with the use of porogenic materials. An alternative method of fabrication is the polymerization of the continuous phase around aqueous droplets in an emulsion (Busby *et al.*, 2001). The setup involves a water-in-oil emulsion system with an organic phase containing the specified monomers. When the internal (droplet) phase volume fraction exceeds 74%, the emulsion is defined as a high-internal-phase emulsion (HIPE) (Lissant, 1974). Under desired HIPE conditions, polymers are synthesized and/or cross-linked to yield a solid network with interconnected pores. Polymers derived from HIPEs are dubbed PolyHIPEs. PolyHIPE foams resemble the structure of emulsion-formed scaffolds at the gel point. The morphology of the structure depends primarily on the volume fraction and the droplet radius, which can be controlled by the physical conditions of the emulsion. Total porosity is based on phase volume fraction, and scaffolds of more than 90% porosity have been produced from PolyHIPE systems. Porogens can also be incorporated into PolyHIPEs for additional porosity. Early research with PolyHIPE scaffolds used nondegradable polymers like poly(styrene), but recent work has utilized biodegradable polymers such as PLLA and PCL (Busby *et al.*, 2002). Images of PLA-MMA PolyHIPEs are shown in Fig. 22.3.

## XI. GAS FOAMING

A major concern with typical solvent-casting and particulate-leaching strategies is the use of organic solvents, remnants of which might lead to an inflammatory response after implantation. A method that avoids any organic sol-

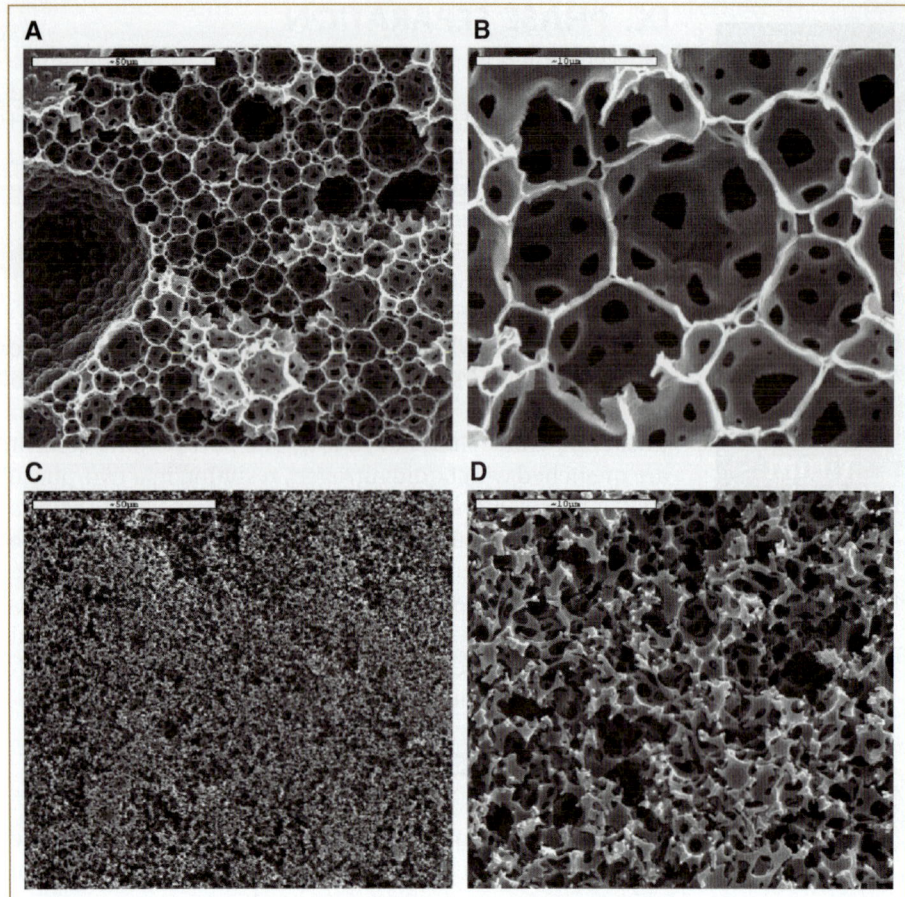

**FIG. 22.3.** Scanning electron micrographs of PLA-MMA PolyHIPEs: **(A)** 0.2-M at low magnification; **(B)** 0.2-M at high magnification; **(C)** 0.4-M at low magnification; **(D)** 0.4-M at high magnification. Reprinted from Busby *et al.* (2002). Copyright 2002, with permission of John Wiley & Sons, Ltd.

vents is gas-foaming scaffold fabrication (Mooney *et al.*, 1996b). Compressed polymer disks (e.g., PLGA) are treated with high-pressure $CO_2$. As the pressure is decreased, nucleation and pore formation occur in the polymer matrix based on the amount and reduction rate of pressure. The average pore size ranges from 100 to 500 μm; however, a drawback of this method remains its closed-pore morphology. Incorporation of a particle-leaching technique has been shown to create an open-pore network in scaffolds produced by gas foaming (Harris *et al.*, 1998). Smooth muscle cells have exhibited enhanced adhesion and proliferation to scaffolds fabricated in this manner.

## XII. POLYMER/CERAMIC COMPOSITE FABRICATION

Tissue-engineering strategies for bone replacement are unique, in that they must account for the irregular shape of most bone defects and the required mechanical strength of the scaffold. While scaffolds of polymers such as the poly($\alpha$-hydroxyester) family provide sufficient support in orthopedic applications, increasing the scaffold porosity drastically reduces the compressive strength (Thomson *et al.*, 1995). PLGA scaffolds containing hydroxyapatite (HA, the mineral component of bone) fibers have been assembled using melt-molding and solvent-casting techniques (Thomson *et al.*, 1998). The greatest effects of HA on the scaffold's mechanical properties are observed when the fibers are fully dispersed throughout the polymer to maximize polymer–HA contact.

More recent methods incorporate microparticles of HA, rather than fibers, into the scaffold network. One technique uses an emulsion of PLGA and HA dissolved in chloroform with an aqueous PVA solution (Devin *et al.*, 1996). After the mixture is emulsified, it is cast into molds and vacuum dried to yield a porous PLGA/HA composite foam. The compressive strength of the scaffold was found to be proportional to its HA content. Such scaffolds exhibited compressive strengths on the same order of magnitude as cancellous bone (10–1000 MPa) (Hollister, 2005). Another method that integrates HA powder into PLGA scaffolds employs phase separation (R. Zhang and Ma, 1999). HA is dispersed in a PLGA/dioxane solution; then the blend is injected into molds and frozen. Following phase separation, the material is freeze-dried to remove the solvent. The resulting composite scaffolds contain an interconnected-pore network, with pore sizes from 30 to 100 μm and porosity up to 95%. PLGA/HA composite scaffolds produced by solvent casting or gas

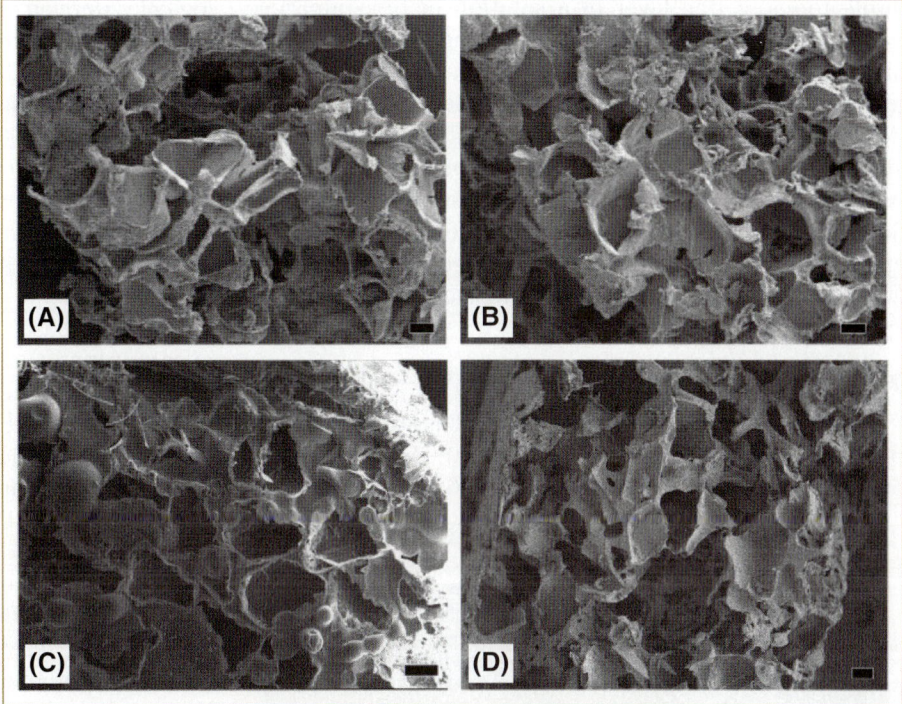

**FIG. 22.4.** Scanning electron micrographs of **(A,C)** surfaces and **(B,D)** cross sections of the PLGA/HA composite scaffolds fabricated by **(A,B)** the solvent-casting/particulate-leaching method and **(C,D)** the gas-foaming/particulate-leaching method. Reprinted from Kim *et al.* (2005). Copyright 2005, with permission of Elsevier Science.

foaming, followed by particulate leaching, are pictured in Fig. 22.4.

## XIII. RAPID PROTOTYPING OF SOLID FREE FORMS

Another technique for the creation of scaffolds with specific three-dimensional structures is the *rapid prototyping* of solid free-form structures, which includes three-dimensional printing, laser sintering, and stereolithography. These methods require a computer model of the desired scaffold architecture from computer-assisted design (CAD) or computed tomography (CT). Although there are several approaches to this family of scaffold production, the result is a three-dimensionally accurate structure with a fully interconnected network of pores (Lam *et al.*, 2002). These methods have an advantage over conventional fabrication techniques due to their ability to create geometries with complex architectures on the micron scale.

Three-dimensional printing utilizes a simple inkjet printing system directed by the CAD program. Briefly, a thin layer of polymer powder, such as PLGA, is spread over a piston surface. The inkjet dispenses a binding liquid, which is a solvent for the polymer, in the desired pattern of the scaffold layer. After a short bonding time, the piston is lowered by the thickness of a single layer and the subsequent layers of powder and binding liquid are applied. Unbound polymer remains in the network during the fabrication process to support disconnected sections in the layer. PLLA and PLGA scaffolds produced in this manner have properties similar to those made via compression molding

(Giordano *et al.*, 1996) and show great promise in cell transplantation and vascular penetration into the implanted structure (Kim *et al.*, 1998).

Fused-deposition modeling (FDM) combines the elements of extrusion and melt molding with free-form scaffold fabrication (Leong *et al.*, 2003). Polymer stock is heated and extruded through a computer-controlled nozzle. With each layer deposited and cooled, the nozzle changes the direction of deposition to yield a porous, honeycomb-type structure. Scaffolds produced via FDM have controlled pore size, porosity, and total pore interconnectivity. FDM is used with many synthetic polymers, including PCL, PLGA, and high-density polyethylene. An FDM-fabricated scaffold with three-dimensional pore interconnectivity is shown in Fig. 22.5.

Laser sintering is similar to three-dimensional printing, but it uses a high-powered laser to sinter the polymer instead of dispensing a binding liquid. The laser selectively scans the powder polymer surface, directed by the CAD or CT computer program (K. H. Tan *et al.*, 2003). The laser beam heats the polymer above its melt temperature and fuses particles into a solid structure. Additional layers of polymer are added to the top surface and sintered accordingly. This technique has been used with biocompatible materials such as PLLA, PCL, PVA, and hydroxyapatite (K. H. Tan *et al.*, 2005). Such scaffolds were shown to be biocompatible, highly porous, and accurate to design specifications.

A popular method of fabrication by rapid prototyping is stereolithography. Stereolithography uses light to polymerize, cross-link, or harden a photosensitive material (Dhariwala *et al.*, 2004). Typically for tissue-engineering applications, a fine layer of a solution of biocompatible polymer, photo-cross-linking initiating agent, porogen, and an appropriate solvent is placed beneath the laser. Like previous methods, the CAD software guides the laser in the desired pattern for the designed scaffold. The laser's ultraviolet light reacts with the photo-initiator to form chemical bonds between polymer chains in the specified locations. Subsequent layers of polymer solution are added and photo-cross-linked. The final product is washed to remove unreacted polymer and yield a three-dimensional structure with specific microarchitectures.

## XIV. PEPTIDE SELF-ASSEMBLY

Since the mid-1990s, new research has studied the use of peptide nanofibers as a synthetic ECM in a tissue-engineering scaffold (S. Zhang *et al.*, 2006). While other bio-

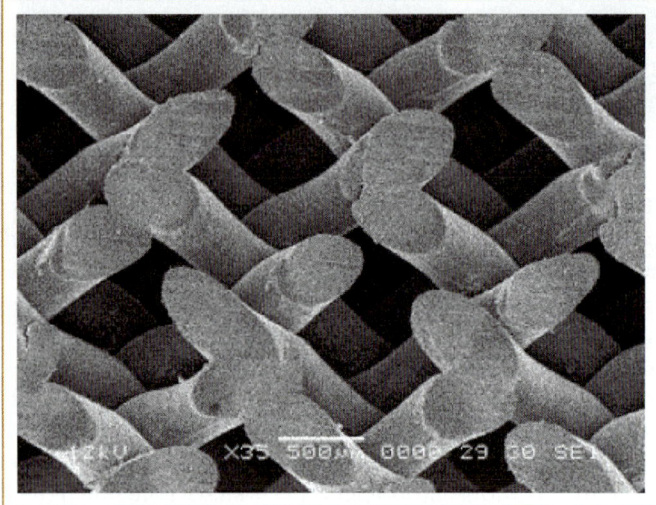

**FIG. 22.5.** Scanning electron micrograph of scaffold with three-dimensional pore interconnectivity fabricated by means of FDM. Reprinted from Leong *et al.* (2003). Copyright 2003, with permission of Elsevier Science.

logically derived materials, such as collagen, gelatin, and fibrin, can interact favorably with cells as compared to synthetic polymers, designer peptide fibers can self-assemble to form stable, highly ordered scaffolds on the nanoscale (Yokoi *et al.*, 2005). Self-assembling peptides typically consist of ionic, self-complementary sequences with alternating hydrophobic and hydrophilic domains (S. Zhang *et al.*, 1995). They can also include motifs favorable to cell attachment, such as the popular arginine-glycine-aspartate (RGD) peptide.

Peptide-based scaffolds have shown promise in the *in vitro* culture of osteoblasts, chondrocytes, and hepatocytes. Self-assembling peptide structures form on the nanoscale, allowing attached cells to remain in their native three-dimensional shape and not flattened like cells attached to some microscale surfaces. While the individual fibers can be as small as 5 nm, the aggregate scaffolds can reach sizes in the centimeters (Hartgerink *et al.*, 2002). A scanning electron micrograph of self-assembling peptide nanofibers is seen in Fig. 22.6. By controlling the spacing of charged and hydrophobic residues in the amino acid sequence, the geometries of the forming scaffold can be manipulated. Noncovalent bonds and ionic interactions within and between peptide molecules create functional and dynamic structures in these synthetic biological systems. Adjacent fibers can be permanently cross-linked with disulfide bonds by the strategic placement of cysteine residues. Self-assembling peptides typically form stable β-sheets in water or physiological solutions. Peptide amphiphiles have also been shown to form more complex architectures, such as sheets, rods, spheres, and discs. Scaffold assembly and size can be controlled by pH, peptide concentration, and divalent ion induction.

## XV. *IN SITU* POLYMERIZATION

The previous scaffold fabrication techniques discuss the production of prefabricated scaffolds for surgical implantation within a defect. Although these scaffolds are

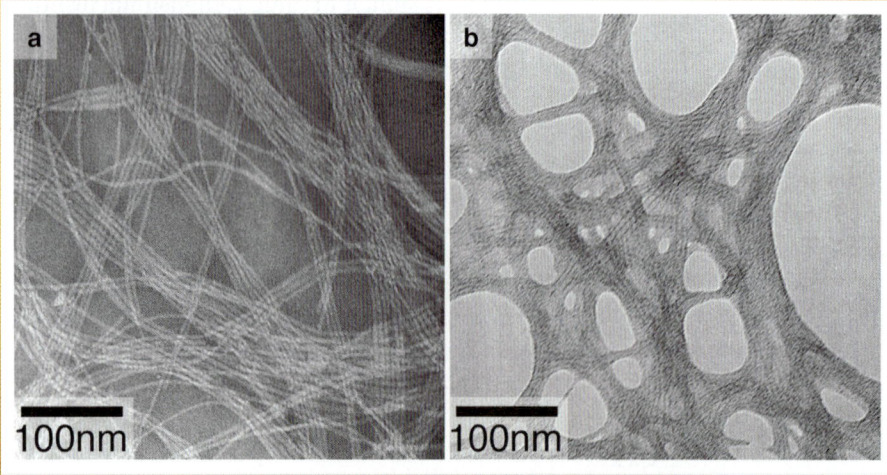

**FIG. 22.6.** Transmission electron microscopy images of peptide nanofibers. **(a)** Self-assembled by drying without adjusted pH; **(b)** self-assembled by mixing with $CaCl_2$. Reprinted from Hartgerink *et al.* (2002). Copyright 2002, with permission of the National Academy of Sciences.

**FIG. 22.7.** Scanning electron micrograph of the cross section of a PCLF scaffold thermally cross-linked with 75 vol% salt content. Reprinted from Jabbari *et al.* (2005). Copyright 2005, with permission of the American Chemical Society.

useful in most tissue-engineering applications, many orthopedic procedures require immediate treatment in defects of irregular or unpredictable shape. In such situations, an injectable, *in situ* polymerizing or hardening polymer is advantageous. Early bone cements composed of PMMA were injected into the bone fracture space (Yaszemski *et al.*, 1996). A degradable alternative for cementing bone defects is PPF, which can be thermally cross-linked with the addition of N-vinyl pyrrolidone. Unlike PMMA, injected PPF may not result in necrosis of local tissues from the elevated temperatures of polymerization or any residual toxic monomer. Incorporation of mineral into the polymer mixture can provide added mechanical properties to the scaffold. PPF with β-tricalcium phosphate has shown strength similar to that of human trabecular bone (Peter *et al.*, 1997). More self-cross-linkable macromers, such as poly(ε-caprolactone-fumarate) (PCLF), have been developed to harden *in situ* without the aid of low-molecular-weight cross-linking agents, but with the addition of an initiator and accelerator, to form porous, biodegradable scaffolds (Jabbari *et al.*, 2005). A cross section of a thermally cross-linked PCLF scaffold is presented in Fig. 22.7.

For cartilage and most soft tissues, less compressive strength is required during tissue repair. Water-based polymer gels, or *hydrogels*, are often favorable for promoting cell migration, angiogenesis, high water content, and rapid nutrient diffusion (Bryant and Anseth, 2001). Most hydrogels are formed by the aqueous cross-linking of poly(ethylene glycol) (PEG)–based synthetic polymers or biologically derived molecules such as gelatin and fibrin. Prior to injection, cells cultured *in vitro* can be loaded into the polymer solution and encapsulated within the cross-linked hydrogel to accelerate tissue regeneration. Like PPF, modified PEG and oligo(poly(ethylene glycol) fumarate) (OPF) can be *in situ* cross-linked with inclusion of a thermal initiator (Temenoff *et al.*, 2004).

There are a variety of strategies to create porosity within *in situ* cross-linked scaffolds. Salts or other small biocompatible molecules included in the polymer solution are able to leach out *in vivo* to create a pore network over time (Peter *et al.*, 1997). Gelatin microparticles incorporated into hydrogels are enzymatically degraded to leave pores for tissue penetration (Kasper *et al.*, 2005). Hydrogels can utilize gas bubbling to form pores during cross-linking (Behravesh *et al.*, 2002). Carbon dioxide produced from the reaction of L-ascorbic acid with sodium bicarbonate, both mixed into the polymer solution prior to injection, has been used in the synthesis of poly(propylene fumarate-*co*-ethylene glycol) hydrogels with greater than 80% porosity.

## XVI. CONCLUSIONS

To meet the diverse needs of tissue reconstruction and replacement, tissue-engineering strategies attempt to provide artificial, yet permanent, biological solutions. As a key component of any tissue-engineering application, scaffolds require a high porosity, adequate pore size for cell migration and nutrient diffusion, biocompatibility, biodegradability, and mechanical integrity. The selected scaffold processing technique can have a profound effect on the final properties and geometry of the scaffold. The fabrication schemes in this chapter offer a practical and promising solution for scaffolds to repair and regenerate different tissues. Each method presents distinctive advantages (e.g., the ease of processing, the ability to incorporate bioactive molecules, or increased structural properties) and limitations (e.g., applicable polymers, cost of materials or equipment). Thus there is no universal scaffold fabrication technique for all tissue-engineering applications (see Table 22.1). Depending on the tissue type and extent of regeneration, scaffold properties must be prioritized in order to select the most appropriate manufacturing method. At present, tissue engineers are working to incorporate bioactive molecules into the scaffolds, develop new scaffold materials, produce constructs with mechanical properties that match those of the specific tissue, and improve the time and costs of scaffold production.

## XVII. ACKNOWLEDGMENTS

We acknowledge financial support by the National Institutes of Health for development of tissue-engineering scaffolds (R01-AR42639, R01-AR48756, and R01-DE15164). MBM also acknowledges financial support by the National Science Foundation Integrative Graduate Education and Research Training Grant (NSF-IGERT, DGE 0114264).

**Table 22.1.  Summary of scaffold fabrication techniques**

| Technique | Description | Advantages | Disadvantages |
|---|---|---|---|
| Fiber bonding | Individual polymer fibers bonded at intersection points | Simple procedure; high porosity and surface-area-to-volume ratio | Poor mechanical properties; limited polymer types |
| Electrospinning | Fibers are electrostatically spun into a nonwoven scaffold | Control over pore sizes, porosity, and fiber thickness | Pore size decreases with fiber thickness; limited mechanical properties |
| Solvent casting/ particulate leaching | Polymer scaffold containing porogen is solidified; then the porogen is leached out | Control over porosity, pore sizes, and crystallinity; high porosity | Limited mechanical properties; residual solvents and porogen material |
| Melt molding | Polymer and porogen are heated, homogenized, and cooled in a mold | Control over macrogeometry, porosity, and pore size | Temperatures unsuitable for biomolecules |
| Membrane lamination | Thin layers of porous scaffolds are surface bonded to yield complex architectures | Control over macrogeometry, porosity, and pore size | Limited mechanical properties; inadequate pore interconnectivity; residual solvents |
| Extrusion | Prefabricated membranes are extruded through nozzle; then the porogen is leached out | Control over porosity and pore size; unique macrogeometry (e.g., tubular shapes) | Limited mechanical properties; temperatures unsuitable for biomolecules |
| Freeze-drying | Dissolved polymer and water are emulsified and freeze-dried to remove water and solvent | Good porosity and pore interconnectivity | Limited pore sizes |
| Phase separation | Polymer is dissolved in melted organic solvent and then solidified with liquid nitrogen | High porosity; ability to incorporate biomolecules | Limited pore sizes; residual solvents; no control over microgeometry |

| Technique | Description | Advantages | Disadvantages |
|---|---|---|---|
| High-internal-phase emulsion | Polymers are synthesized and/or cross-linked as the organic phase of a HIPE | Control over porosity, pore size, and interconnectivity | Limited mechanical properties; limited polymer types |
| Gas foaming | Compressed polymer is treated with high-pressure gas, leading to pore formation on depressurization | Free of harsh organic solvents; control over porosity; ability to incorporate biomolecules | Limited mechanical properties; inadequate pore interconnectivity |
| Polymer/ceramic composite fabrication | Scaffolds of polymers and inorganic molecules or fibers are formed by solvent-casting or melt-molding techniques | Control over porosity and pore size; enhanced mechanical properties | Residual solvents |
| Rapid prototyping | CAD-controlled fabrication using solvent dispensing, fused deposition, laser sintering, or stereolithography | Excellent control over geometry (macro and micro) and porosity | Limited polymer types; high equipment cost |
| Peptide self-assembly | Designer peptide sequences are self-assembled into spheres, fibers, or complex scaffolds | Control over porosity, pore size, and fiber diameter; bioactive degradation products | Expensive materials; complex design parameters; limited macrosizes and mechanical properties |
| In situ polymerization | Polymers are polymerized or cross-linked; scaffolds are formed postimplantation | Injectable; control over mechanical properties; ability to incorporate biomolecules | Limited porosity; residual monomers and cross-linking agents |

# XVIII. REFERENCES

Behravesh, E., Jo, S., Zygourakis, K., and Mikos, A. G. (2002). Synthesis of *in situ* cross-linkable macroporous biodegradable poly(propylene fumarate-*co*-ethylene glycol) hydrogels. *Biomacromolecules* **3**, 374–381.

Bryant, S. J., and Anseth, K. S. (2001). The effects of scaffold thickness on tissue-engineered cartilage in photocrosslinked poly(ethylene oxide) hydrogels. *Biomaterials* **22**, 619–626.

Busby, W., Cameron, N. R., and Jahoda, C. A. B. (2001). Emulsion-derived foams (PolyHIPEs) containing poly(ε-caprolactone) as matrixes for tissue engineering. *Biomacromolecules* **2**, 154–164.

Busby, W., Cameron, N. R., and Jahoda, C. A. B. (2002). Tissue-engineering matrixes by emulsion templating. *Polym. Int.* **51**, 871–881.

Cima, L. J., Vacanti, J. P., Vacanti, C., Ingber, D., Mooney, D., and Langer, R. (1991). Tissue engineering by cell transplantation using degradable polymer substrates. *J. Biomech. Eng.* **113**, 143–151.

Devin, J. E., Attawia, M. A., and Laurencin, C. T. (1996). Three-dimensional degradable porous polymer–ceramic matrices for use in bone repair. *J. Biomater. Sci. Polym. Ed.* **7**, 661–669.

Dhariwala, B., Hunt, E., and Boland, T. (2004). Rapid prototyping of tissue-engineering constructs, using photopolymerizable hydrogels and stereolithography. *Tissue Eng.* **10**, 1316–1322.

Freed, L. E., Vunjak-Novakovic, G., Biron, R. J., Eagles, D. B., Lesnoy, D. C., Barlow, S. K., and Langer, R. (1994). Biodegradable polymer scaffolds for tissue engineering. *Biotechnology* **12**, 689–693.

Giordano, R. A., Wu, B. M., Borland, S. W., Cima, L. G., Sachs, E. M., and Cima, M. J. (1996). Mechanical properties of dense polylactic acid structures fabricated by three-dimensional printing. *J. Biomater. Sci. Polym. Ed.* **8**, 63–75.

Gray, J. J. (2004). The interaction of proteins with solid surfaces. *Curr. Opin. Struct. Biol.* **14**, 110–115.

Hacker, M., Tessmar, J., Neubauer, M., Blaimer, A., Blunk, T., Gopferich, A., and Schulz, M. B. (2003). Towards biomimetic scaffolds: anhydrous scaffold fabrication from biodegradable amine-reactive diblock copolymers. *Biomaterials* **24**, 4459–4473.

Harris, L. D., Kim, B. S., and Mooney, D. J. (1998). Open-pore biodegradable matrices formed with gas foaming. *J. Biomed. Mater. Res.* **42**, 396–402.

Hartgerink, J. D., Beniash, E., and Stupp, S. I. (2002). Peptide–amphiphile nanofibers: a versatile scaffold for the preparation of self-assembling materials. *Proc. Natl. Acad. Sci. USA* **99**, 5133–5138.

Hollister, S. J. (2005). Porous scaffold design for tissue engineering. *Nat. Mater.* **4**, 518–524.

Holy, C. E., Dang, S. M., Davies, J. E., and Shoichet, M. S. (1999). *In vitro* degradation of a novel poly(lactide-*co*-glycolide) 75/25 foam. *Biomaterials* **20**, 1177–1185.

Hsu, Y. Y., Gresser, J. D., Trantolo, D. J., Lyons, C. M., Gangadharam, P. R., and Wise, D. L. (1997). Effect of polymer foam morphology and density on kinetics of *in vitro* controlled release of isoniazid from compressed foam matrices. *J. Biomed. Mater. Res.* **35**, 107–116.

Hua, F. J., Kim, G. E., Lee, J. D., Son, Y. K., and Lee, D. S. (2002). Macroporous poly(L-lactide) scaffold 1. Preparation of a macroporous scaffold by liquid–liquid phase separation of a PLLA–dioxane–water system. *J. Biomed. Mater. Res.* **63**, 61–167.

Hutmacher, D. W. (2001). Scaffold design and fabrication technologies for engineering tissues — state of the art and future perspectives. *J. Biomat Sci. Polym. Ed.* **12**, 107–124.

Jabbari, E., Wang, S., Lu, L., Gruetzmacher, J. A., Ameenuddin, S., Hefferan, T. E., Currier, B. L., Windebank, A. J., and Yaszemski, M. J. (2005). Synthesis, material properties, and biocompatibility of a novel self-cross-linkable poly(caprolactone fumarate) as an injectable tissue-engineering scaffold. *Biomacromolecules* **6**, 2503–2511.

Jang, J. H., Houchin, T. L., and Shea, L. D. (2004). Gene delivery from polymer scaffolds for tissue engineering. *Exp. Rev. Med. Dev.* **1**, 127–38.

Kasper, F. K., Kushibiki, T., Kimura, Y., Mikos, A. G., and Tabata, Y. (2005). *In vivo* release of plasmid DNA from composites of oligo(poly(ethylene glycol)fumarate) and cationized gelatin microspheres. *J. Control. Release* **107**, 547–561.

Kim, S. S., Utsunomiya, H., Koski, J. A., Wu, B. M., Cima, M. J., Sohn, J., Mukai, K., Griffith, L. G., and Vacanti, J. P. (1998). Survival and function of hepatocytes on a novel three-dimensional synthetic biodegradable polymer scaffold with an intrinsic network of channels. *Ann. Surg.* **228**, 8–13.

Kim, S. S., Sun Park, M., Jeon, O., Yong Choi, C., and Kim, B. S. (2005). Poly(lactide-*co*-glycolide)/hydroxyapatite composite scaffolds for bone tissue engineering. *Biomaterials* **27**, 871–881.

Lam, C. X. F., Mo, X. M., Teoh, S. H., and Hutmacher, D. W. (2002). Scaffold development using 3D printing with a starch-based polymer. *Mater. Sci. Eng. C Biol. Sci.* **20**, 49–56.

Leong, K. F., Cheah, C. M., and Chua, C. K. (2003). Solid free-form fabrication of three-dimensional scaffolds for engineering replacement tissues and organs. *Biomaterials* **24**, 2363–2378.

Lissant, K. J. (1974). "Emulsions and Emulsion Technology." Marcel Dekker, New York.

Lo, H., Ponticiello, M. S., and Leong, K. W. (1995). Fabrication of controlled-release biodegradable foams by phase separation. *Tissue Eng.* **1**, 15–27.

Mikos, A. G., Bao, Y., Cima, L. G., Ingber, D. E., Vacanti, J. P., and Langer, R. (1993a). Preparation of poly(glycolic acid) bonded fiber structures for cell attachment and transplantation. *J. Biomed. Mater. Res.* **27**, 183–189.

Mikos, A. G., Sarakinos, G., Leite, S. M., Vacanti, J. P., and Langer, R. (1993b). Laminated three-dimensional biodegradable foams for use in tissue engineering. *Biomaterials* **14**, 323–330.

Mikos, A. G., Lyman, M. D., Freed, L. E., and Langer, R. (1994). Wetting of poly(L-lactic acid) and poly(DL-lactic-*co*-glycolic acid) foams for tissue engineering. *Biomaterials* **15**, 55–58.

Mooney, D. J., Kaufmann, P. M., Sano, K., McNamara, K. M., Vacanti, J. P., and Langer, R. (1994). Transplantation of hepatocytes using porous, biodegradable sponges. *Transplant. Proc.* **26**, 3425–3426.

Mooney, D. J., Mazzoni, C. L., Breuer, C., McNamara, K., Hern, D., Vacanti, J. P., and Langer, R. (1996a). Stabilized polyglycolic acid fibre-based tubes for tissue engineering. *Biomaterials* **17**, 115–124.

Mooney, D. J., Baldwin, D. F., Suh, N. P., Vacanti, J. P., and Langer, R. (1996b). Novel approach to fabricate porous sponges of poly(D,L-lactic-*co*-glycolic acid) without the use of organic solvents. *Biomaterials* **17**, 1417–1422.

Peter, S. J., Nolley, J. A., Widmer, M. S., Merwin, J. E., Yaszemski, M. J., Yasko, A, W., Engel, P. S., and Mikos, A. G. (1997). *In vitro* degradation of a poly(propylene fumarate)/B-tricalcium phosphate injectible composite scaffold. *Tissue Eng.* **3**, 207–215.

Pham, Q. P., Sharma, U., and Mikos, A. G. (2006). Electrospinning of polymeric nanofibers for tissue-engineering applications. *Tissue Eng.* **12**, 1197–1211.

Tan, E. P. S., and Lim, C. T. (2006). Characterization of bulk properties of nanofibrous scaffolds from nanomechanical properties of single nanofibers. *J. Biomed. Mater. Res. Part A.* **7**, 526–533.

Tan, K. H., Chua, C. K., Leong, K. F., Cheah, C. M., Cheang, P., Abu Bakar, M. S., and Cha, S. W. (2003). Scaffold development using selective laser sintering of polyetheretherketone–hydroxyapatite biocomposite blends. *Biomaterials* **24**, 3115–3123.

Tan, K. H., Chua, C. K., Leong, K. F., Cheah, C. M., Gui, W. S., Tan, W. S., and Wiria, F. E. (2005). Selective laser sintering of biocompatible polymers for applications in tissue engineering. *Biomed. Mater. Eng.* **15**, 113–124.

Temenoff, J. S., Park, H., Jabbari, E., Conway, D. E., Sheffield, T. L., Ambrose, C. G., and Mikos, A. G. (2004). Thermally cross-linked oligo(poly(ethylene glycol) fumarate) hydrogels support osteogenic differentiation of encapsulated marrow stromal cells *in vitro. Biomacromolecules* **5**, 5–10.

Thomson, R. C., Yaszemski, M. J., Powers, J. M., and Mikos, A. G. (1995a). Fabrication of biodegradable polymer scaffolds to engineer trabecular bone. *J. Biomater. Sci., Polym. Ed.* **7**, 23–28.

Thomson, R. C., Yaszemski, M. J., Powers, J. M., and Mikos, A. G. (1995b). Poly(alpha-hydroxy ester)/short-fiber hydroxyapatite composite foams for orthopedic applications. *Polym. Med. Pharm.* **394**, 25–30.

Thomson, R. C., Yaszemski, M. J., Powers, J. M., and Mikos, A. G. (1998). Hydroxyapatite fiber–reinforced poly(alpha-hydroxy ester) foams for bone regeneration. *Biomaterials* **19**, 1935–1943.

Wake, M. C., Gupta, P. K., and Mikos, A. G. (1996). Fabrication of pliable biodegradable polymer foams to engineer soft tissues. *Cell Transplant.* **5**, 465–473.

Whang, K., Thomas, H., and Healy, K. E. (1995). A novel method to fabricate bioabsorbable scaffolds. *Polymer* **36**, 837–841.

Widmer, M. S., Gupta, P. K., Lu, L., Meszlenyi, R. K., Evans, G. R., Brandt, K., Savel, T., Gurlek, A., Patrick, C. W. Jr., and Mikos, A. G. (1998). Manufacture of porous biodegradable polymer conduits by an extrusion process for guided tissue regeneration. *Biomaterials* **19**, 1945–1955.

Yannas, I. V., Burke, J. F., Gordon, P. L., Huang, C., and Rubenstein, R. H. (1980). Design of an artificial skin. Part II. Control of chemical composition. *Biomaterials* **14**, 107–131.

Yaszemski, M. J., Payne, R. G., Hayes, W. C., Langer, R., and Mikos, A. G. (1996). *In vitro* degradation of a poly(propylene fumarate)-based composite material. *Biomaterials* **17**, 2127–2130.

Yokoi, H., Kinoshita, T., and Zhang, S. (2005). Dynamic reassembly of peptide RADA16 nanofiber scaffold. *Proc. Natl. Acad. Sci. USA* **102**, 8414–8419.

Yoshimoto, H., Shin, Y. M., Terai, H., and Vacanti, J. P. (2003). A biodegradable nanofiber scaffold by electrospinning and its potential for bone tissue engineering. *Biomaterials* **24**, 2077–2082.

Zhang, R., and Ma, P. X. (1999). Poly(alpha-hydroxyl acids)/hydroxyapatite porous composites for bone-tissue engineering. I. Preparation and morphology. *J. Biomed. Mater. Res.* **44**, 446–455.

Zhang, S., Holmes, T., DiPersio, M., Hynes, R. O., Su, X., and Rich, A. (1995). Self-complementary oligopeptide matrices support mammalian cell attachment. *Biomaterials* **16**, 1385–1393.

Zhang, S., Zhao, X., and Spirio, L. (2006). PuraMatrix: self-assembling peptide nanofiber scaffolds. In "Scaffolding in Tissue Engineering" (P. X. Ma and J. Elisseeff, eds.), pp. 217–236. CRC Press, Boca Raton, FL.

Chapter **Twenty-Three**

# Biodegradable Polymers

*James M. Pachence, Michael P. Bohrer, and Joachim Kohn*

## I. INTRODUCTION

The design and development of tissue-engineered products has benefited from many years of clinical utilization of a wide range of biodegradable polymers. Newly developed biodegradable polymers and novel modifications of previously developed biodegradable polymers have enhanced the tools available to create clinically important tissue-engineering applications. Insights gained from studies of cell–matrix interactions, cell–cell signaling, and organization of cellular components are placing increased demands on biomaterials for novel sophisticated medical implants, such as tissue engineering constructs, and continue to fuel the interest in improving the performance of existing medical-grade polymers and developing new synthetic polymers. This chapter surveys those biologically derived and synthetic biodegradable polymers that have been used or are under consideration for use in tissue-engineering applications. The polymers are described in terms of their chemical composition, breakdown products, mechanism of breakdown, mechanical properties, and clinical limitations. Also discussed are product design considerations in processing of biomaterials into a final form (e.g., gel, membrane, matrix) that will effect the desired tissue response.

## II. BIODEGRADABLE POLYMER SELECTION CRITERIA

The selection of biomaterials plays a key role in the design and development of tissue-engineering product development. While the classical selection criterion for a safe, stable implant dictated choosing a passive, inert material, it is now understood that any such device will elicit a cellular response (Peppas and Langer, 1994; Langer and Tirrell, 2004). Therefore, it is now widely accepted that a biomaterial must interact with tissue to repair, rather than act simply as a static replacement. Furthermore, biomaterials used directly in tissue repair or replacement applications (e.g., artificial skin) must be more than biocompatible; they must elicit a desirable cellular response. Consequently, a major focus of biomaterials for tissue-engineering applications centers around harnessing control over cellular interactions with biomaterials, often including components to manipulate cellular response within the supporting biomaterial as a key design component. Specific examples include protein growth factors, anti-inflammatory drugs, gene delivery vectors, and other bioactive factors to elicit the desired cellular response (see recent reviews by Murphy and Mooney, 1999, and Davies, 2004).

*Principles of Tissue Engineering, 3rd Edition*
*ed. by Lanza, Langer, and Vacanti*

It is important for the tissue-engineering product developer to have several biomaterials options available, for each application calls for a unique environment for cell–cell interactions. Such applications include (1) support for new tissue growth (wherein cell–cell communication and cell availability to nutrients, growth factors, and pharmaceutically active agents must be maximized); (2) prevention of cellular activity (where tissue growth, such as in surgically induced adhesions, is undesirable); (3) guided tissue response (enhancing a particular cellular response while inhibiting others); (4) enhancement of cell attachment and subsequent cellular activation (e.g., fibroblast attachment, proliferation, and production of extracellular matrix for dermis repair); (5) inhibition of cellular attachment and/or activation (e.g., platelet attachment to a vascular graft); and (6) prevention of a biological response (e.g., blocking antibodies against homograft or xenograft cells used in organ replacement therapies).

Biodegradable polymers are applicable to those tissue-engineering products in which tissue repair or remodeling is the goal, but not where long-term materials stability is required. Biodegradable polymers must also possess (1) manufacturing feasibility, including sufficient commercial quantities of the bulk polymer; (2) the capability to form the polymer into the final product design; (3) mechanical properties that adequately address short-term function and do not interfere with long-term function; (4) low or negligible toxicity of degradation products, in terms of both local tissue response and systemic response; and (5) drug delivery compatibility in applications that call for release or attachment of active compounds.

## III. BIOLOGICALLY DERIVED BIORESORBABLES

### Type I Collagen

Collagen is the major component of mammalian connective tissue, animal protein, accounting for approximately 30% of all protein in the human body. It is found in every major tissue that requires strength and flexibility (e.g., skin, bone). Fourteen types of collagens have been identified, the most abundant being type I (van der Rest *et al.*, 1990). Because of its abundance (it makes up more than 90% of all fibrous proteins) and its unique physical and biological properties, type I collagen has been used extensively in the formulation of biomedical materials (Pachence *et al.*, 1987; Pachence, 1996). Type I collagen is found in high concentrations in tendon, skin, bone, and fascia, which are consequently convenient and abundant sources for isolation of this natural polymer.

The structure, function, and synthesis of type I collagen has been thoroughly investigated (Piez, 1984; Tanzer and Kimura, 1988). Collagen proteins, by definition, are characterized by a unique triple helix formation extending over a large portion of the molecule. The three peptide subunits that make up the triple helix of collagen have similar amino acid composition, each chain comprising approximately 1050 amino acid residues. The length of each subunit is ~300 nm, and the diameter of the triple helix is ~1.5 nm. The primary structure of collagen (with its high content of proline and hydroxyproline and with every third amino acid being glycine) shows a strong sequence homology across genus and adjacent family line. Because of its phylogenetically well-conserved primary sequence and its helical structure, collagen is only mildly immunoreactive (De Lustro *et al.*, 1987; Anselme *et al.*, 1990).

The individual collagen molecules will spontaneously polymerize *in vitro* into strong fibers that can be subsequently formed into larger organized structures (Piez, 1984). The collagen may be further modified to form intra- and intermolecular cross-links, which aid in the formation of collagen fibers, fibrils, and then macroscopic bundles that are used to form tissue (Nimni and Harkness, 1988). For example, tendon and ligaments comprise mainly oriented type I collagen fibrils, which are extensively cross-linked in the extracellular space. Added strength via *in vivo* cross-linking is imparted to the collagen fibers by several enzymatic (such as lysyl oxidase) and nonenzymatic reactions. The most extensive cross-linking occurs at the telopeptide portion of the molecule.

Collagen cross-linking can be enhanced after isolation through a number of well-described physical or chemical techniques (Pachence *et al.*, 1987). Increasing the intermolecular cross-links (1) increases biodegradation time, by making collagen less susceptible to enzymatic degradation; (2) decreases the capacity of collagen to absorb water; (3) decreases its solubility; and (4) increases the tensile strength of collagen fibers. The free ε-amines on lysine residues on collagen can be utilized for cross-linking or can similarly be modified to link or sequester active agents. These simple chemical modifications provide a variety of processing possibilities and, consequently, the potential for a wide range of tissue-engineering applications using type I collagen.

It has long been recognized that substrate attachment sites are necessary for growth, differentiation, replication, and metabolic activity of most cell types in culture. Collagen and its integrin-binding domains (e.g., RGD sequences) assist in the maintenance of such attachment-dependent cell types in culture. For example, fibroblasts grown on collagen matrices appear to differentiate in ways that mimic *in vivo* cellular activity and to exhibit nearly identical morphology and metabolism (Silver and Pins, 1992). Chondrocytes can also retain their phenotype and cellular activity when cultured on collagen (Toolan *et al.*, 1996). Such results suggest that type I collagen can serve as tissue regeneration scaffold for any number of cellular constructs.

The recognition that collagen matrices could support new tissue growth was exploited to develop the original formulations of artificial extracellular matrices for dermal replacements (Yannas and Burke, 1980; Yannas *et al.*, 1980; Burke *et al.*, 1981). Yannas and Burke were the first to show that the rational design and construction of an artificial dermis could lead to the synthesis of a dermislike structure whose physical properties "would resemble dermis more than they resembled scar" (Burke *et al.*, 1981). They created a collagen–chondroitin sulfate composite matrix with a well-described pore structure and cross-linking density that optimizes regrowth while minimizing scar formation (Dagalakis *et al.*, 1980). The reported clinical evidence and its simplicity of concept make this device an important potential tool for the treatment of severely burned patients (Heimbach *et al.*, 1988).

Collagen gels were used by Eugene Bell at the Massachusetts Institute of Technology to create a cell-based system for dermal replacement (Bell *et al.*, 1991; Parenteau, 1999). This living-skin equivalent (commercially known as Alpligraf™) is composed of a mixture of live human fibroblasts and soluble collagen in the form of a contracted gel, which is then seeded with keratinocytes. A number of clinical investigators have tested such cell-based collagen dressings for use as a skin graft substitute for chronic wounds and burn patients.

The advantageous properties of collagen for supporting tissue growth have been used in conjunction with the superior mechanical properties of synthetic biodegradable polymer systems to make hybrid tissue scaffolds for bone and cartilage (Hsu *et al.*, 2006; Chen *et al.*, 2006, 2004; Sato *et al.*, 2004). These hybrid systems show superior cell adhesion, interaction, and proliferation as compared to the synthetic polymer system alone. Collagen has also been used to improve cell interactions with electrospun nanofibers of poly(hydroxy acids), such as poly(lactic acid), poly(glycolic acid), poly(ε-caprolactone), and their copolymers (Venugopal *et al.*, 2005; He *et al.*, 2005a, 2005b).

## Glycosaminoglycans

Glycosaminoglycans (GAGs), which consist of repeating disaccharide units in linear arrangement, usually include a uronic acid component (such as glucuronic acid) and a hexosamine component (such as N-acetyl-D-glucosamine). The predominant types of GAGs attached to naturally occurring core proteins of proteoglycans include chondroitin sulfate, dermatan sulfate, keratan sulfate, and heparan sulfate (Heinegard and Paulson, 1980; Naeme and Barry, 1993). The GAGs are attached to the core protein by specific carbohydrate sequences containing three or four monosaccharides. The largest GAG, hyaluronic acid (hyaluronan), is an anionic polysaccharide with repeating disaccharide units of N-acetylglucosamine and glucuronic acid, with unbranched units ranging from 500 to several thousand. Hyaluronic acid can be isolated from natural sources (e.g., rooster combs) or via microbial fermentation (Balazs, 1983). Because of its water-binding capacity, dilute solutions of hyaluronic acid form viscous solutions.

Like collagen, hyaluronic acid can be easily chemically modified, as by esterification of the carboxyl moieties, which reduces its water solubility and increases its viscosity (Balazs, 1983; Sung *et al.* and Topp, 1994). Hyaluronic acid can be cross-linked to form molecular weight complexes in the range 8 to $24 \times 10^6$ or to form an infinite molecular network (gels). In one method, hyaluronic acid is cross-linked using aldehydes and small proteins to form bonds between the C—OH groups of the polysaccharide and the amino or imino groups of the protein, thus yielding high-molecular-weight complexes (Balazs and Leshchiner, 1986). Other cross-linking techniques include the use of vinyl sulfone, which reacts to form an infinite network through sulfonyl-bis-ethyl cross-links (Balazs and Leshchiner, 1985). The resultant infinite network gels can be formed into sheaths, membranes, tubes, sleeves, and particles of various shapes and sizes. No species variations have been found in the chemical and physical structure of hyaluronic acid. The fact that it is not antigenic, eliciting no inflammatory or foreign-body reaction, make it desirable as a biomaterial. Its main drawbacks in this respect are its residence time and the limited range of its mechanical properties.

Because of its relative ease of isolation and modification and its superior ability in forming solid structures, hyaluronic acid has become the preferred GAG in medical device development. It has been used as a viscoelastic during eye surgery since 1976 and has undergone clinical testing as a means of relieving arthritic joints (Weiss and Balazs, 1987). In addition, gels and films made from hyaluronic acid have shown clinical utility to prevent postsurgical adhesion formation (Urmann *et al.*, 1991; Holzman *et al.*, 1994; Medina *et al.*, 1995).

The benzyl ester of hyaluronic acid, sold under the trade name HYAFF-11, has been studied for use in vascular grafts (Lepidi *et al.*, 2006; Turner *et al.*, 2004), to support chondrocyte growth (Grigolo *et al.*, 2002; Solchaga *et al.*, 2000) and for bone tissue engineering (Giordano *et al.*, 2006; Sanginario *et al.*, 2006).

## Chitosan

Chitosan is a biosynthetic polysaccharide that is the deacylated derivative of chitin. Chitin is a naturally occurring polysaccharide that can be extracted from crustacean exoskeletons or generated via fungal fermentation processes. Chitosan is a β-1,4-linked polymer of 2-amino-2-deoxy-*d*-glucose; it thus carries a positive charge from amine groups (Kaplan *et al.*, 1994). It is hypothesized that the major path for chitin and chitosan breakdown *in vivo* is through lysozyme, which acts slowly to depolymerize the polysaccharide (Taravel and Domard, 1993). The biodegradation rate of the polymer is determined by the amount of residual acetyl content, a parameter that can easily be varied. Chemical modification of chitosan produces materials with a variety of physical and mechanical properties (Muzzarelli *et al.*, 1988; Wang *et al.*, 1988; Laleg and Pikulik,

1991). For example, chitosan films and fibers can be formed utilizing cross-linking chemistries, adapted techniques for altering from other polysaccharides, such as treatment of amylose with epichlorohydrin (Wei *et al.*, 1977). Like hyaluronic acid, chitosan is not antigenic and is a well-tolerated implanted material (Malette *et al.*, 1986).

Chitosan has been formed into membranes and matrices suitable for several tissue-engineering applications (Hirano, 1989; Sandford, 1989; Byrom, 1991; Madihally and Matthew, 1999; Shalaby *et al.*, 2004) as well as conduits for guided nerve regeneration (Huang *et al.*, 2005; Bini *et al.*, 2005). Chitosan matrix manipulation can be accomplished using the inherent electrostatic properties of the molecule. At low ionic strength, the chitosan chains are extended via the electrostatic interaction between amine groups, whereupon orientation occurs. As ionic strength is increased, and chain–chain spacing diminished, the consequent increase in the junction zone and stiffness of the matrix result in increased average pore size. Chitosan gels, powders, films, and fibers have been formed and tested for such applications as encapsulation, membrane barriers, contact lens materials, cell culture, and inhibitors of blood coagulations (East *et al.*, 1989).

## Polyhydroxyalkanoates

Polyhydroxyalkanoate (PHA) polyesters are degradable, biocompatible, thermoplastic materials made by several microorganisms (Miller and Williams, 1987; Gogolewski *et al.*, 1993). They are intracellular storage polymers whose function is to provide a reserve of carbon and energy (Dawes and Senior, 1973). Depending on growth conditions, bacterial strain, and carbon source, the molecular weights of these polyesters can range from tens into the hundreds of thousands. Although the structures of PHA can contain a variety of *n*-alkyl side-chain substituents (see Structure 23.1), the most extensively studied PHA is the simplest: poly(3-hydroxyburtyrate) (PHB).

ICI developed a biosynthetic process for the manufacture of PHB, based on the fermentation of sugars by the bacterium *Alcaligenes eutrophus*. PHB homopolymer, like all other PHA homopolymers, is highly crystalline, extremely brittle, and relatively hydrophobic. Consequently, the PHA homopolymers have degradation times *in vivo* on the order of years (Holland *et al.*, 1987; Miller and Williams, 1987). The copolymers of PHB with hydroxyvaleric acid are less crystalline, more flexible, and more readily processible, but they suffer from the same disadvantage of being too hydrolytically stable to be useful in short-term applications when resorption of the degradable polymer within less than one year is desirable.

PHB and its copolymers with up to 30% of 3-hydroxyvaleric acid are now commercially available under the trade name Biopol. It was found previously that a PHA copolymer of 3-hydroxybutyrate and 3-hydroxyvalerate, with a 3-hydroxyvalerate content of about 11%, may have an optimum balance of strength and toughness for a wide range of possible applications. PHB has been found to have low toxicity, in part due to the fact that it degrades *in vivo* to D-3-hydroxybutyric acid, a normal constituent of human blood. Applications of these polymers previously tested or now under development include controlled drug release, artificial skin, and heart valves as well as such industrial applications as paramedical disposables (Yasin *et al.*, 1989; Doi *et al.*, 1990; Sodian *et al.*, 2000). Sutures are the main usage for polyhydroxyalkanonates, although a number of clinical applications and trials are ongoing (Ueda and Tabata, 2003).

## Experimental Biologically Derived Bioresorbables

Synthetic biomolecules are beginning to find a place in the repertoire of biomaterials for medical applications. Model synthetic proteins structurally similar to elastin have been formulated by Urry and coworkers (Nicol *et al.*, 1992; Urry, 1995). Using a combination of solid-phase peptide chemistry and genetically engineered bacteria, they synthesized several polymers having homologies to the elastin repeat sequences of valine-proline-glycine-valine-glycine repeat (VPGVG). The constructed amino acid polymers were formed into films and then cross-linked. The resultant films have intriguing mechanical responses, such as a reverse phase transition. When a film is heated, its internal order increases, translating into substantial contraction with increasing temperature (Urry, 1995). The films can be mechanically cycled many times, and the phase transition of the polymers can be varied by amino acid substitution. Copolymers of VPGVG and VPG*X*G have been constructed (where *X* is the substitution) that show a wide range of transition temperatures (Urry, 1995). Several medical applications are under consideration for this system, including musculoskeletal repair mechanisms, ophthalmic devices, and mechanical and/or electrically stimulated drug delivery.

Other investigators, notably Tirrell and Cappello, have combined techniques from molecular and fermentation biology to create novel protein-based biomaterials

**STRUCTURE 23.1.** Poly(*b*-hyroxybutyrate) and copolymers with hydroxyvaleric acid. For a homopolymer of HB, Y = 0; commonly used copolymer ratios are 7, 11, or 22 mole percent of hydroxyvaleric acid.

(Cappello, 1992; J. P. Anderson *et al.*, 1994; Tirrell *et al.*, 1994; van Hest and Tirrell, 2001). These protein polymers are based on repeat oligomeric peptide units, which can be controlled via the genetic information inserted into the producing bacteria. It has been shown that the mechanical properties and the biological activities of these protein polymers can be programmed, suggesting a large number of potential biomedical applications (Krejchi *et al.*, 1994).

Another approach to elicite an appropriate cellular response to a biomaterial is to graft active peptides to the surface of a biodegradable polymer. For example, peptides containing the RGD sequence have been grafted to various biodegradable polymers to provide active cell-binding surfaces (Hubbell, 1995). Similarly, Panitch *et al.* (1999) incorporated oligopeptides containing the REDV sequence to stimulate endothelial cell binding for vascular grafts.

## IV. SYNTHETIC POLYMERS

From the beginnings of the material sciences, the development of highly stable materials has been a major research challenge. Today, many polymers are available that are virtually indestructible in biological systems, e.g., Teflon, Kevlar, and poly(ether-ether-ketone). On the other hand, the development of degradable biomaterials is a relatively new area of research. The variety of available degradable biomaterials is still too limited to cover a wide enough range of diverse material properties. Thus, the design and synthesis of new degradable biomaterials is currently an important research challenge.

Due to the efforts of a wide range of research groups, a large number of different polymeric compositions and structures have been suggested as degradable biomaterials. However, in most cases no attempts have been made to develop these new materials for specific medical applications. Thus, detailed toxicological studies *in vivo*, investigations of degradation rate and mechanism, and careful evaluations of the physicomechanical properties have so far been published for only a very small fraction of those polymers. This leaves the tissue engineer with only a relatively limited number of promising polymeric compositions to choose from. The following section is limited to a review of the most commonly investigated classes of biodegradable synthetic polymers.

### Poly(α-hydroxy acids)

Naturally occurring hydroxy acids, such as glycolic, lactic, and ε-caproic acids, have been utilized to synthesize an array of useful biodegradable polymers for a variety of medical product applications. As an example, bioresorbable surgical sutures made from poly(α-hydroxy acids) have been in clinical use since 1970; other implantable devices made from these versatile polymers (e.g., internal fixation devices for orthopedic repair) are becoming part of standard surgical protocol (Helmus and Hubbell, 1993; Shalaby and Johnson, 1994; Hubbell, 1995).

The ester bond of the poly(hydroxy acids) are cleaved by hydrolysis, which results in a decrease in the polymer molecular weight (but not mass) of the implant (Vert and Li, 1992). This initial degradation occurs until the molecular weight is less than 5000 Da, at which point cellular degradation takes over. The final degradation and resorption of the poly(hydroxy acid) implants involves inflammatory cells (such as macrophages, lymphocytes, and neutrophils). Although this late-stage inflammatory response can have a deleterious effect on some healing events, these polymers have been successfully employed as matrices for cell transplantation and tissue regeneration (Freed *et al.*, 1994a, 1994b). The degradation rate of these polymers is determined by initial molecular weight, exposed surface area, crystallinity, and (in the case of copolymers) the ratio of the hydroxy acid monomers.

The poly(hydroxy acid) polymers have a modest range of mechanical properties and a correspondingly modest range of processing conditions. Nevertheless, these thermoplastics can generally be formed into films, tubes, and matrices using such standard processing techniques as molding, extrusion, solvent casting, and spin casting. Ordered fibers, meshes, and open-cell foams have been formed to fulfill the surface area and cellular requirements of a variety of tissue-engineering constructs (Helmus and Hubbell, 1993; Freed *et al.*, 1994; Hubbell, 1995; Wintermantel *et al.*, 1996). The poly(hydroxy acid) polymers have also been combined with other materials, e.g., poly(ethylene glycol), to modify the cellular response elicited by the implant and its degradation products (Sawhney *et al.*, 1993).

### Poly(glycolic acid), Poly(lactic acid), and Their Copolymers

Poly(glycolic acid) (PGA), poly(lactic acid) (PLA), and their copolymers are the most widely used synthetic degradable polymers in medicine. Of this family of linear aliphatic polyesters, PGA has the simplest structure (see Structure 23.2) and consequently enjoys the largest associated literature base. Since PGA is highly crystalline, it has a high melting point and low solubility in organic solvents. PGA was used in the development of the first totally synthetic absorbable suture (Frazza and Schmitt, 1971). The crystallinity of PGA in surgical sutures is typically in the range 46–52% (Gilding and Reed, 1979). Due to its hydrophilic nature, surgical sutures made of PGA tend to lose their mechanical strength rapidly, typically over a period of two to four weeks post-implantation (Reed and Gilding, 1981).

**STRUCTURE 23.2.** Poly(glycolic acid) (PGA).

$$\left[\!\!\begin{array}{c} \text{CH}_3 \quad \text{O} \\ | \quad \quad \| \\ \text{O}\!-\!\text{CH}\!-\!\text{C} \end{array}\!\!\right]_n$$

**STRUCTURE 23.3.** Poly(lactic acid) (PLA).

In order to adapt the materials properties of PGA to a wider range of possible applications, researchers undertook an intensive investigation of copolymers of PGA with the more hydrophobic poly(lactic acid) (PLA). Alternative sutures composed of copolymers of glycolic acid and lactic acid are currently marketed under the trade names Vicryl and Polyglactin 910.

Due to the presence of an extra methyl group in lactic acid, PLA (Structure 23.3) is more hydrophobic than PGA. The hydrophobicity of high-molecular-weight PLA limits the water uptake of thin films to about 2% (Gilding and Reed, 1979) and results in a rate of backbone hydrolysis lower than that of PGA (Reed and Gilding, 1981). In addition, PLA is more soluble in organic solvents than is PGA.

It is noteworthy that there is no linear relationship between the ratio of glycolic acid to lactic acid and the physicomechanical properties of their copolymers. Whereas PGA is highly crystalline, crystallinity is rapidly lost in PGA–PLA copolymers. These morphological changes lead to an increase in the rates of hydration and hydrolysis. Thus, copolymers tend to degrade more rapidly than either PGA or PLA (Gilding and Reed, 1979; Reed and Gilding, 1981).

Since lactic acid is a chiral molecule, it exists in two stereoisomeric forms that give rise to four morphologically distinct polymers. D-PLA and L-PLA are the two stereoregular polymers, D,L-PLA is the racemic polymer obtained from a mixture of D- and L-lactic acid, and meso-PLA can be obtained from D,L-lactide. The polymers derived from the optically active D and L monomers are semicrystalline materials, while the optically inactive D,L-PLA is always amorphous. Generally, L-PLA is more frequently employed than D-PLA, since the hydrolysis of L-PLA yields L(+)-lactic acid, which is the naturally occurring stereoisomer of lactic acid.

The differences in the crystallinity of D,L-PLA and L-PLA have important practical ramifications: Since D,L-PLA is an amorphous polymer, it is usually considered for applications such as drug delivery, where it is important to have a homogeneous dispersion of the active species within a monophasic matrix. On the other hand, the semicrystalline L-PLA is preferred in applications where high mechanical strength and toughness are required — for example, sutures and orthopedic devices (Christel et al., 1982; Leenstag et al., 1987; Vainionpaa et al., 1987).

Recently, PLA, PGA, and their copolymers have been combined with bioactive ceramics such as Bioglass particles and hydroxyapatite that stimulate bone regeneration while greatly improving the mechanical strength of the composite material (Rezwan et al., 2006). Bioglass particles combined with D,L-PLA-co-PGA have also been shown to be angiogenic, suggesting a novel approach for providing a vascular supply to implanted devices (Day et al., 2005). Chu has recently reviewed the most significant and successful biomedical applications of the poly(hydroxy acids) (Chu, 2003). PLA, PGA, and their copolymers are also being intensively investigated for a large number of drug delivery applications. This research effort has been comprehensively reviewed by Lewis (1990).

Some controversy surrounds the use of these materials for orthopedic applications. According to one review of short- and long-term response to resorbable pins made from either PGA or PGA:PLA copolymer in over 500 patients, 1.2% required reoperation due to device failure, 1.7% suffered from bacterial infection of the operative wound, and 7.9% developed a late noninfectious inflammatory response that warranted operative drainage (Böstman, 1991). Subsequently it has become evident that the delayed inflammatory reaction represents the most serious complication of the use of the currently available degradable fixation devices. The mean interval between fixation and the clinical manifestation of this reaction is 12 weeks for PGA and can be as long as three years for the more slowly degrading PLA (Böstman, 1991). Whether avoiding reoperation to remove a metal implant outweighs an approximately 8% risk of severe inflammatory reaction is a difficult question; in any event, an increasing number of trauma centers have suspended the use of these degradable fixation devices. It has been suggested that the release of acidic degradation products (glycolic acid for PGA, lactic acid for PLA, and glyoxylic acid for polydioxanone) contributes to the observed inflammatory reaction. Thus, the late inflammatory response appears to be a direct consequence of the chemical composition of the polymer degradation products, for which there is currently no prophylactic measure (Böstman, 1991). In vitro and animal experiments indicate that incorporation of alkaline salts or antibodies to inflammatory mediators may diminish the risk of a late inflammatory response (Böstman and Pihlajamaki, 2000). A more desirable solution to these problems for orthopedic (and perhaps other) applications requires the development of a polymer that is more hydrophobic than PGA or PLA, degrades somewhat more slowly, and does not release acidic degradation products on hydrolysis.

## Polydioxanone (PDS)

This poly(ether-ester) is prepared by a ring-opening polymerization of p-dioxanone. PDS has gained increasing interest in the medical field and pharmaceutical field due to its degradation to low toxicity monomers in vivo. PDS has a lower modulus than PLA or PGA; thus it became the first degradable polymer to be used to make a monofilament suture. PDS has also been introduced to the market as a suture clip as well as more recently as a bone pin marketed under the name ORTHOSRB in the United States and Ethipin in Europe (Ray *et al.*, 1981; Greisler *et al.*, 1987; Mäkelä *et al.*, 1989).

## Poly(ε-caprolactone)

Poly(ε-caprolactone) (PCL) (Structure 23.4) is an aliphatic polyester that has been intensively investigated as a biomaterial (Pitt, 1990). The discovery that PCL can be degraded by microorganisms led to evaluation of PCL as a biodegradable packaging material (Pitt, 1990). Later, it was discovered that PCL can also be degraded by a hydrolytic mechanism under physiological conditions (Pitt *et al.*, 1981). Under certain circumstances, cross-linked PCL can be degraded enzymatically, leading to "enzymatic surface erosion" (Pitt *et al.*, 1981). Low-molecular-weight fragments of PCL are reportedly taken up by macrophages and degraded intracellularly, with a tissue reaction similar to that of other poly(hydroxy acids) (Pitt *et al.*, 1984). Compared with PGA or PLA, the degradation of PCL is significantly slower. PCL is therefore most suitable for the design of long-term, implantable systems such as Capronor, a one-year implantable contraceptive device (Pitt, 1990).

Poly(ε-caprolactone) exhibits several unusual properties not found among the other aliphatic polyesters. Most noteworthy are its exceptionally low glass transition temperature of −62°C and its low melting temperature of 57°C. Another unusual property of poly(ε-caprolactone) is its high thermal stability. Whereas other tested aliphatic polyesters had decomposition temperatures ($T_d$) between 235 and 255°C, poly(ε-caprolactone) has a $T_d$ of 350°C, which is more typical of poly(ortho esters) than of aliphatic polyesters (Engelberg and Kohn, 1991). PCL is a semicrystalline polymer with a low glass transition temperature of about −60°C. Thus, PCL is always in a rubbery state at room temperature. Among the more common aliphatic polyesters,

this is an unusual property, which undoubtedly contributes to the very high permeability of PCL for many therapeutic drugs (Pitt *et al.*, 1987).

Another interesting property of PCL is its propensity to form compatible blends with a wide range of other polymers (Koleske, 1978). In addition, ε-caprolactone can be copolymerized with numerous other monomers (e.g., ethylene oxide, chloroprene, THF, δ-valerolactone, 4-vinylanisole, styrene, methyl methacrylate, vinylacetate). Particularly noteworthy are copolymers of ε-caprolactone and lactic acid that have been studied extensively (Pitt *et al.*, 1981; Feng *et al.*, 1983). PCL and copolymers with PLA have been electron-spun to create nanofibrous tissue-engineered scaffolds that show promise for vascular applications (Venugopal *et al.*, 2005; He *et al.*, 2005a, 2005b; Xu *et al.*, 2004). The toxicology of PCL has been extensively studied as part of the evaluation of Capronor. Based on a large number of tests, the monomer, ε-caprolactone, and the polymer, PCL, are currently regarded as nontoxic and tissue-compatible materials. Consequently, clinical studies of the Capronor system are currently in progress (Kovalevsky and Barnhart, 2001).

It is interesting to note that in spite of its versatility, PCL has so far been predominantly considered for controlled-release drug delivery applications. In Europe, PCL is being used as a biodegradable staple, and it stands to reason that PCL (or blends and copolymers with PCL) will find additional medical applications in the future. The most recent, comprehensive review of the status of PCL has been by Pitt (1990).

## Poly(orthoesters)

Poly(ortho esters) are a family of synthetic degradable polymers that have been under development for several years (Heller *et al.*, 1990) (Structure 23.5). Devices made of poly(ortho esters) can be formulated in such a way that the device undergoes "surface erosion" — that is, the polymeric device degrades at its surface only and will thus tend to become thinner over time rather than crumbling into pieces. Since surface-eroding, slablike devices tend to release drugs embedded within the polymer at a constant rate, poly(ortho esters) appear to be particularly useful for controlled-release drug delivery (Heller, 1988); this interest is reflected by the many descriptions of these applications in the literature (Heller and Daniels, 1994).

**STRUCTURE 23.4.** Poly(*e*-caprolactone).

**STRUCTURE 23.5.** Poly(orthoesters). The specific composition shown here is a terpolymer of hexadecanol (1,6-HD), *trans*-cyclohexyldimethanol (*t*-CDM), and DETOSU.

There are two major types of poly(ortho esters). Originally, poly(ortho esters) were prepared by the condensation of 2,2-diethoxytetrahydrofuran and a dialcohol (Cho and Heller, 1978) and marketed under the trade names Chronomer and Alzamer. Upon hydrolysis, these polymers release acidic by-products that autocatalyze the degradation process, resulting in degradation rates that increase with time. Later, Heller *et al.* (1980) synthesized a new type of poly(ortho ester) based on the reaction of 3,9-bis(ethylidene 2,4,8,10-tetraoxaspiro {5,5} undecane) (DETOSU) with various dialcohols. These poly(ortho esters) do not release acidic by-products upon hydrolysis and thus do not exhibit autocatalytically increasing degradation rates.

## Polyurethanes

Polyurethanes, polymers in which the repeating unit contains a urethane moiety, were first produced by Bayer in 1937 (Structure 23.6). These polymers are typically produced through the reaction of a diisocyanate with a polyol. Conventional polyols are polyethers or polyesters. The resulting polymers are segmented block copolymers, with the polyol segment providing a low-glass-transition-temperature (i.e., <25°C) soft segment and the diisocyanate component, often combined with a hydrocarbon chain extender, providing the hard segment. A wide range of physical and mechanical properties have been realized with commercial polyurethanes.

Polyurethanes have been used for nearly 50 years in biomedical applications, particularly as the blood-contacting material in cardiovascular devices. Intended as nonbiodegradable coatings, polyurethanes fell out of favor with the failure of pacemaker leads and breast implant coatings. Subsequent studies, as reviewed by Santerre *et al.* (2005), have elucidated much about the behavior of polyurethanes in biological systems. Elucidation of the biodegradation mechanism and its dependence on the polyurethane structure and composition have led to the development of biodegradable polyurethanes for a variety of tissue-engineering applications, such as myocardial repair (McDevitt *et al.*, 2003) and vascular tissues (Stankus *et al.*, 2005). Design of biodegradable polyurethanes has required alternative diisocyanate compounds. The traditional aromatic diisocyanates are putative carcinogenic compounds. Biodegradable polymers are made from diisocyanates, such as lysine-diisocyanate or hexamethylene diisocyanate, that release nontoxic degradation products such as lysine.

The poyol or soft-segment portion of biodegradable polyurethanes is used to modify the degradation rate (Santerre *et al.*, 2005). Poly(α-hydroxy acids), including PLA, PGA, and PCL, have been used as soft segments for biodegradable polyurethanes (Gorna and Gogolewski, 2002; Cohn *et al.*, 2002).

An interesting applicaton of polyurethanes was developed by Santerre *et al.* where fluoroquinolone antimicrobial drugs were incorporated into the polymer as hard-segment monomers (Woo *et al.*, 2000; Santerre *et al.*, 2005). This led to the design of drug polymers (trade name, Epidel) that release the drug when degraded by enzymes generated by an inflammatory response. This is an example of a smart system, in that antibacterial agents are released only while inflammation is present. Once healing occurs, the enzyme level drops and the release of drug diminishes.

## Poly(anhydrides)

Polyanhydrides (Structure 23.7) were first investigated in detail by Hill and Carothers (1932) and were considered in the 1950s for possible applications as textile fibers (Conix, 1958). Their low hydrolytic stability, their major limitation for industrial applications, was later recognized as a potential advantage by Langer *et al.* (Rosen *et al.*, 1983), who suggested the use of polyanhydrides as degradable biomaterials. A study of the synthesis of high-molecular-weight polyanhydrides has been published by Domb *et al.* (1989, 1994).

A comprehensive evaluation of the toxicity of the polyanhydrides showed that, in general, the polyanhydrides possessed excellent *in vivo* biocompatibility (Laurencin *et al.*, 1990). The most immediate applications for polyanhydrides are in the field of drug delivery, although tissue-engineering applications are also being developed (Katti and Laurencin, 2003). Drug-loaded devices are best prepared by compression molding or microencapsulation (Mathiowitz *et al.*, 1988). A wide variety of drugs and proteins including insulin, bovine growth factors, angiogenesis inhibitors (e.g., heparin and cortisone), enzymes (e.g., alkaline phosphatase and *b*-galactosidase), and anesthetics have been incorporated into polyanhydride matrices, and

PEO soft segment       lysine diisocyanate hard segment       PCL soft segment

**STRUCTURE 23.6.**   Polyurethane.

**STRUCTURE 23.7.** Poly(SA-HDA anhydride). This composition represents one of many polyanhydrides that were explored. The clinically relevant polyanhydrides are copolymers of sebacic acid and *p*-carboxy-phenoxypropane.

**STRUCTURE 23.8.** Polyphosphazene. Shown here is a polymer containing an amino acid ester attached to the phosphazene backbone.

their *in vitro* and *in vivo* release characteristics have been evaluated (Chasin *et al.*, 1990). One of the most aggresively investigated uses of the polyanhydrides is for the delivery of chemotherapeutic agents. One particular example of this application is the delivery of BCNU (bis-chloroethylnitrosourea) to the brain for the treatment of glioblastoma multiformae, a universally fatal brain cancer (Langer, 1990). For this application, polyanhydrides derived from bis-*p*-(carboxyphenoxy propane) and sebacic acid received FDA regulatory clearance in the fall of 1996 and are currently being marketed under the name Gliadel®.

## Polyphosphazenes

Polyphosphazenes (Structure 23.8) consist of an inorganic phosphorous-nitrogen backbone, in contrast to the commonly employed hydrocarbon-based polymers (Scopelianos, 1994; Heyde and Schacht, 2004). Consequently, the phosphazene backbone undergoes hydrolysis to phosphate and ammonium salts, with the concomitant release of the side group. Of the numerous polyphosphazenes that have been synthesized, those that have some potential use for medical products are substituted with amines of low p$K_a$ and those with activated alcohol moieties (Allcock, 1990; Crommen *et al.*, 1992; Laurencin *et al.*, 1993). Singh *et al.* (2006) have modified the side groups to tune polyphophazine properties, such as glass transition temperature, degradation rate, surface wettability, tensile strength, and elastic modulus, enabling these polymers to be considered for a wider range of biomedical applications. The most extensively studied polyphosphazenes are hydrophobic, having fluoroalkoxy side groups. In part, these materials are of interest because of their expected minimal tissue interaction, which is similar to Teflon.

Aryloxyphosphazenes and closely related derivatives have also been extensively studied. One such polymer can be cross-linked with dissolved cations such as calcium to form a hydrogel matrix because of its polyelectrolytic nature (Allcock and Kwon, 1989). Using methods similar to alginate encapsulation, microspheres of aryloxyphosphazene have been used to encapsulate hybridoma cells without affecting their viability or their capacity to produce antibodies. Interaction with poly(L-lysine) produced a semipermeable membrane. Similar materials have been synthesized that show promise in blood contacting and with novel drug delivery applications.

## Poly(amino acids) and Pseudo-Poly(amino acids)

Since proteins are composed of amino acids, many researchers have tried to develop synthetic polymers derived from amino acids to serve as models for structural, biological, and immunological studies. In addition, many different types of poly(amino acids) have been investigated for use in biomedical applications (J. M. Anderson *et al.*, 1985). Poly(amino acids) are usually prepared by the ring-opening polymerization of the corresponding *N*-carboxy anhydrides that are obtained by reaction of the amino acid with phosgene (Bamford *et al.*, 1956).

Poly(amino acids) have several potential advantages as biomaterials. A large number of polymers and copolymers can be prepared from a variety of amino acids. The side chains offer sites for the attachment of small peptides, drugs, cross-linking agents, or pendent groups that can be used to modify the physicomechanical properties of the polymer. Since these polymers release naturally occurring amino acids as the primary products of polymer backbone cleavage, their degradation products may be expected to show a low level of systemic toxicity.

Poly(amino acids) have been investigated as suture materials (Spira *et al.*, 1969), as artificial skin substitutes (Aiba *et al.*, 1985), and as drug delivery systems (Mitra *et al.*, 1979; McCormick-Thomson and Duncan, 1989). Various drugs have been attached to the side chains of poly(amino acids), usually via a spacer unit that distances the drug from the backbone. Poly(amino acid)–drug combinations investigated include poly(L-lysine) with methotrexate and pepstatin (Campbell *et al.*, 1980) and poly(glutamic acid) with adriamycin and norethindrone (van Heeswijk *et al.*, 1985). Short amino acid sequences such as RGD and RGDS, strong promotors of specific cell adhesion, have been coupled to other biodegradable polymers to promote cell growth in tissue-engineering applications (Masuko *et al.*, 2005; Yang *et al.*, 2005).

Despite their apparent potential as biomaterials, poly(amino acids) have actually found few practical applications. *N*-carboxy anhydrides, the starting materials, are expensive to make and difficult to handle because of their high reactivity and moisture sensitivity. Most poly(amino acids) are highly insoluble and nonprocessible materials. Since poly(amino acids) degrade via enzymatic hydrolysis of the amide bond, it is difficult to reproduce and control their degradation *in vivo* because the level of enzymatic activity varies from person to person. Furthermore, the antigenicity of polymers containing three or more amino acids excludes their use in biomedical applications (J. M. Anderson *et al.*, 1985). Because of these difficulties, only a few poly(amino acids), usually derivatives of poly(glutamic acid) carrying various pendent chains at the γ-carboxylic acid group, are currently being investigated as implant materials (Lescure *et al.*, 1989).

As an alternative approach, Kohn *et al.* have replaced the peptide bonds in the backbone of synthetic poly(amino acids) by a variety of such "nonamide" linkages as ester, iminocarbonate, urethane, and carbonate bonds (James and Kohn, 1997; Kemnitzer and Kohn, 1997). The term *pseudo-poly(amino acid)* is used to denote this new family of polymers, in which naturally occurring amino acids are linked together by nonamide bonds.

The use of such backbone-modified pseudo-poly(amino acids) as biomaterials was first suggested in 1984 (Kohn and Langer, 1984). The first pseudo-poly(amino acids) investigated were a polyester from *N*-protected *trans*-4-hydroxy-L-proline and a polyiminocarbonate from tyrosine dipeptide (Kohn and Langer, 1985, 1987). Several studies indicate that the backbone modification of conventional

poly(amino acids) in general improves their physicomechanical properties (Ertel and Kohn, 1994; Fiordeliso, *et al.*, 1994; James and Kohn, 1997). This approach is applicable to, among other materials, serine, hydroxyproline, threonine, tyrosine, cysteine, glutamic acid, and lysine. It is limited only by the requirement that the nonamide backbone linkages give rise to polymers with desirable material properties. Additional pseudo-poly(amino acids) can be obtained by considering dipeptides as monomeric starting materials. Hydroxyproline-derived polyesters (Kohn and Langer, 1987; Yu *et al.*, 1987; Yu-Kwon and Langer, 1989), serine-derived polyesters (Zhou and Kohn, 1990), and tyrosine-derived polyiminocarbonates (Kohn and Langer, 1987; Pulapura *et al.*, 1990) and polycarbonates (Pulapura and Kohn, 1992) represent specific embodiments of these synthetic concepts.

# V. CREATING MATERIALS FOR TISSUE-ENGINEERED PRODUCTS

As described in this chapter, the bulk polymer properties and the cellular response to biomaterials are important selection criteria for the design of a tissue-engineered product. In addition, the ability to mold the biomaterial into the appropriate cellular-level architecture must be considered, and such architecture must be compatible with the desired tissue response. Details such as pore size, pore structure (isotropic vs. anisotropic), and oriented topography, such as aligned grooves, are often critical for forming tissue with the proper cell morphology, orientation, arrangement of intercellular material, and the relationship between different cell types (Curtis and Riehle, 2001).

In recent years, the selection of appropriate biomaterials has been aided by the development of combinatorial methods and sophisticated modeling techniques that allow prediction of polymer properties and celluar response to the material (Smith *et al.*, 2005, 2004; Kholodovych *et al.*, 2004; Thorstenson and Narasimhan, 2006). Such techniques promise to greatly expand the universe of biodegradable polymers for tissue-engineering applications.

Hubbell classified approaches to choosing biomaterials for various tissue-engineering applications according to type of tissue response sought: (1) conducting tissue responses and architectures; (2) inducing tissue responses and architectures; and (3) blocking tissue responses (Hubbell, 1995; Wintermantel *et al.*, 1996). Implicit in this consideration is that the specific material architecture (e.g., membrane, gel, matrix, tube) is critical to the tissue-engineering product design and thus influences the choice of biomaterials.

**STRUCTURE 23.9.** A poly(amide carbonate) derived from desaminotyrosyl tyrosine alkyl esters. This is an example for a group of new, amino acid–derived polymers.

## Barriers: Membranes and Tubes

Design formats requiring cell activity on one surface of a device while precluding transverse movement of surrounding cells onto that surface call for a barrier material.

For example, peripheral nerve regeneration must allow for axonal growth and at the same time preclude fibroblast activity that could produce neural-inhibiting connective tissue. Structures such as collagen tubes can be fabricated to yield a structure dense enough to inhibit connective tissue formation along the path of repair while allowing axonal growth through the lumen (S. Li *et al.*, 1992). Similarly, collagen membranes for periodontal repair provide an environment for periodontal ligament regrowth and attachment while preventing epithelial ingrowth into the healing site (van Swol *et al.*, 1993). Antiadhesion formulations using hyaluronic acid, which prevent ingrowth of connective tissue at a surgically repaired site, also work on this concept (Urmann *et al.*, 1991).

## Gels

Gels are used to provide a hydrogel scaffold, encapsulate, or provide a specialized environment for isolated cells. For example, collagen gels for tissue engineering were first used to maintain fibroblasts, which were the basis of a living-skin equivalent (Parenteau, 1999). Gels have also been used for the maintenance and immunoprotection of xenograft and homograft cells, such as hepatocytes, chondrocytes, and islets of Langerhans, used for transplantation (Sullivan *et al.*, 1991; Chang, 1992; Lacy, 1995). Semipermeable gels have been created to limit cell–cell communication and interaction with surrounding tissue and to minimize movement of peptide factors and nutrients through the implant. Injectable biodegradable gel materials that form through cross-linking *in situ* show promise for regeneration of bone and cartilage. Temenoff and Mikos (2000) review a number of injectable systems that demonstrate appropriate properties for these applications. Lee *et al.* (2005) discuss the use of biodegradable polyester dendrimers, highly branched synthetic polymers with layered architectures, to form hydrogels for tissue-engineering applications such as corneal wound sealants. In general, nondegradable materials are used for cell encapsulation to maximize long-term stability of the implant. In the future, however, it may be possible to formulate novel smart gels, in which biodegradation is triggered by a specific cellular response instead of simple hydrolysis.

## Matrices

It has been recognized since the mid-1970s that three-dimensional structures are an important component of engineered tissue development (Yannas and Burke, 1980; Yannas *et al.*, 1980). Yannas and his coworkers were the first to show that pore size, pore orientation, and fiber structure are important characteristics in the design of cell scaffolds. Several techniques have subsequently been developed to form well-defined matrices from synthetic and biologically derived polymers, and the physical characteristics of these matrices are routinely varied to maximize cellular and tissue responses (Fenkel *et al.*, 1997; Langer and Vacanti, 1993; Salem *et al.*, 2002; Hodde, 2002). Examples of engineered matrices that have led to several resorbable templates are oriented pore structures designed for regeneration of trabecular bone (Borden *et al.*, 2003; Lin *et al.*, 2003).

## VI. CONCLUSION

Research in the use of currently available biomaterials and in developing novel bioresorbable polymers has helped to drive the establishment of the field of tissue engineering. Despite a wide range of possible choices, there is a tendency to choose those bioresorbable polymers that have a history of regulatory approval, instead of letting the application guide the choice of material. The latter approach, moreover, may require lengthy and costly polymer development work. Nonetheless, in order to gain the sort of precise control over cell response and cell interactions with surrounding tissues that is expected of tissue-engineering applications, continued research on new bioresorbable polymers will be necessary.

## VII. REFERENCES

Aiba, S., Minoura, N., Fujiwara, Y., Yamada, S., and Nakagawa, T. (1985). Laminates composed of polypeptides and elastomers as a burn wound covering. Physicochemical properties. *Biomaterials* **6**, 290–296.

Allcock, H. R. (1990). Polyphosphazenes as new biomedical and bioactive materials. *In* "Biodegradable Polymers as Drug Delivery Systems" (M. Chasin and R. Langer, eds.), pp. 163–193. Marcel Dekker, New York.

Allcock, H. R., and Kwon, S. (1978). *Macromolecules* **22**, 75.

Allcock, H. R., and Kwon, S. (1989). An ionically crosslinkable polyphosphazene: poly[bis(carboxylatophenoxy)phosphazene] and its hydrogels and membranes. *Macromolecules* **22**, 75–79.

Anderson, J. M., Spilizewski, K. L., and Hiltner, A. (1985). Poly-a-amino acids as biomedical polymers. *In* "Biocompatibility of Tissue Analogs" (D. F. Williams, ed.), pp. 67–88. Press, Boca Raton, FL.

Anderson, J. P., Cappello, J., and Martin, D. C. (1994). Morphology and primary crystal structure of a silk-like protein polymer synthesized by genetically engineered *E. coli* bacteria. *Biopolymers* **34**, 1049–1058.

Anselme, K., Bacques, C., Charriere, G., Hartmann, D. J., Herbage, D., and Garrone, R. (1990). Tissue reaction to subcutaneous implantation of a collagen sponge. *J. Biomed. Mat. Res.* **24**, 689–703.

Balazs, E. A. (1983). Sodium hyaluronate and viscosurgery. *In* "Healon (Sodium Hyaluronate): A Guide to Its Use in Ophthalmic Surgery" (D. Miller and R. Stegmann, eds.), pp. 5–28. John Wiley & Sons, New York.

Balazs, E. A., and Denlinger, J. L. (1984). Sodium hyaluronate and joint function. *J. Equine Vet. Sci.* **5**, 217–288.

Balazs, E. A., and Leshchiner, A. (1985). Hyaluronate Modified Polymeric Articles. U.S. Patent 4,500,676.

Balazs, E. A., and Leshchiner, A. (1986). Cross-Linked Gels of Hyaluronic Acid and Products Containing such Gels. U.S. Patent 4,582,865.

Balazs, E. A., and Leshchiner, E. A. (1988). Hyaluronan, its cross-linked derivative — Hylan — and their medical applications. *Appl. Sco. Publ., Tokyo* 1–10.

Bamford, C. H., Elliot, A., and Hanby, W. E. (1956). Synthetic polypeptides — preparation, structure and properties. *In* "Physical Chemistry — A Series of Monographs," Vol. 5. (E. Hutchinson, ed.). Academic Press, New York.

Bell, E., Parenteau, R., Gay, R., Rosenberg, M., Kemp, P., Green, G. D., Muthukumaran, N., and Nolte, C. (1991). The living-skin equivalent: its manufacture, its organotypic properties, and its responses to irritants. *Toxic. in Vitro* **5**, 591–596.

Bini, T. B., Gao, S., Wang, S., and Ramakrishna, S. (2005). Development of fibrous biodegradable polymer conduits for guided nerve regeneration. *J. Mater. Sci. Mater. Med.* **16**, 367–375.

Borden, M., El-Amin, S. F., Attawia, M., and Laurencin, C. T. (2003). Structural and human cellular assessment of a novel microsphere-based tissue-engineered scaffold for bone repair. *Biomaterials* **24**, 597–609.

Borten, M., and Friedman, E. A. (1983). Translaparoscopic hemostasis with microfibrillar collagen in lieu of laparotomy. *J. Reprod. Med.* **28**, 804–806.

Böstman, O. M. (1991). Absorbable implants for the fixation of fractures. *J. Bone Joint Surg.* **73**, 148–153.

Böstman, O., and Pihlajamaki, H. (2000). Clinical biocompatibility of biodegradable orthopedic implants for internal fixation: a review. *Biomaterials* **21**, 2615–2621.

Burd, D. A. R., Greco, R. M., Regauer, S., Longaker, M. T., Siebert, J. W., and Garg, H. G. (1991). Hyaluronan and wound healing: a new perspective. *Brit. J. Plastic Surg.* **44**, 579–584.

Burke, J. F., Yannas, I. V., Quinby Jr., W. C., Bondoc, C. C., and Jung, W. K. (1981). Successful use of a physiologically acceptable artificial skin in the treatment of extensive burn injury. *Ann. Surg.* **194**, 413.

Byrom, D. (1991). Chitosan and chitosan derivatives. *In* "Biomaterials: Novel Materials from Biological Sources." (D. Byron, ed.), pp. 333–359. Stockton Press, New York.

Campbell, P., Glover, G. I., and Gunn, J. M. (1980). Inhibition of intracellular protein degradation by pepstatin, poly(L-lysine) and pepstatinyl-poly(L-lysine). *Arch. Biochem. Biophys.* **203**, 676–680.

Cappello, J. (1992). Genetic production of synthetic protein polymers. *MRS Bull.* **17**, 48–53.

Chang, T. M. S. (1992). Artificial liver support based on artificial cells with emphasis on encapsulated hepatocytes. *Artif. Org.* **16**, 71–74.

Chasin, M., Domb, A., Ron, E., Mathiowitz, E., Langer, R., Leone, K., Laurencin, C., Brem, H., and Grossman (1990). Polyanhydrides as drug delivery systems. *In* "Biodegradable Polymers as Drug Delivery Systems" (C. G. Pitt, ed.), pp. 43–69. Marcel Dekker, New York.

Chen, G., Sato, T., Ushida, T., Ochiai, N., and Tateishi, T. (2004). Tissue engineering of cartilage using a hybrid scaffold of synthetic polymer and collagen. *Tissue Eng.* **10**, 323–330.

Chen, G., Sato, T., Tanaka, J., and Tateishi, T. (2006). Preparation of a biphasic scaffold for osteochondral tissue engineering. *Mater. Sci. Eng. C Biomimetic Supramolec. Sys.* **26**, 118–123.

Cho, N. J., and Heller, J. (1978). Drug Delivery Devices Manufactured from Polyorthoesters and Polyorthocarbonates. U.S. Patent 4,078,038.

Christel, P., Chabot, F., Leray, J. L., Morin, C., and Vert, M. (1982). Biodegradable composites for internal fixation. *In* "Biomaterials" (G. O. Winter, D. F. Gibbons, and H. Pienkj, eds.), pp. 271–280. John Wiley & Sons, New York.

Chu, C. (2003). Biodegradable polymeric biomaterials: an updated overview. *In* "Biomaterials" (J. B. Park and J. D. Bronzino, eds.), pp. 95–115. CRC Press, Boca Raton, FL.

Cima, L. G., Ingber, D. E., Vacanti, J. P., and Langer, R. (1991). Hepatocyte culture on biodegradable polymeric substrates. *Biotechnol. Bioeng.* **38**, 145–158.

Cohn, D., Stern, T., Gonzalez, M. F., and Epstein, J. (2002). Biodegradable poly(ethylene oxide)/poly(μ-caprolactone) multiblock copolymers. *J. Biomed. Mater. Res.* **59**, 273–281.

Conix, A. (1958). Aromatic polyanhydrides, a new class of high-melting fiber-forming polymers. *J. Polym. Sci.* **29**, 343–353.

Crommen, J. H. L., Schacht, E. H., and Mense, E. H. G. (1992). Biodegradable polymers. II. Degradation characteristics of hydrolysis-sensitive poly[(organo)phosphazenes]. *Biomaterials* **13**, 601–611.

Curtis, A., and Riehle, M. (2001). Tissue engineering: the biophysical background. *Physics Med. Biol.* **46**, R47–R65.

Dagalakis, N., Flink, J., Stasikelis, P., Burke, J. F., and Yannas, I. V. (1980). Design of an artificial skin: control of pore structure. *J. Biomed. Mater. Res.* **14**, 511–528.

Davies, N. (2004). Gene-activated matrix. *In* "Encyclopedia of Biomaterials and Biomedical Engineering" (G. E. B. Wnek and L. Gary eds.), pp. 662–669. Marcel Dekker, New York.

Dawes, E. A., and Senior, P. J. (1973). The role and regulation of energy reserve polymers in microorganisms. *Adv. Microb. Physiol.* **10**, 135–266.

Day, R. M., Maquet, V., Boccaccini, A. R., Jerome, R., and Forbes, A. (2005). *In vitro* and *in vivo* analysis of macroporous biodegradable poly(D,L-lactide-*co*-glycolide) scaffolds containing bioactive glass. *J. Biomed. Mater. Res. A* **75A**, 778–787.

De Lustro, F., Condell, R. A., Nguyen, M., and McPherson, J. (1987). A comparative study of the biologic and immunologic response to medical devices derived from dermal collagen. *J. Biomed. Mater. Res.* **20**, 109–120.

De Lustro, F., Keefe, J., Fong, A. T., and Jolivette, D. M. (1991). Biochemistry, biology, and immunology of injectable collagen implant in the treatment of urinary incontinence. *Pediat. Surg. Int.* **6**, 245–251.

Dessipri, E., and Tirrell, D. A. (1994). Trifluoroalanine *N*-carboxyanhydride: a reactive intermediate for the synthesis of low surface energy polypeptides. *Macromolecules* **27**, 5463–5470.

Doi, Y., Kanesawa, Y., Kunioka, M., and Saito, T. (1990). Biodegradation of microbial copolyesters: poly(3-hydroxy-butyrate-*co*-3-hydroxyvalerate) and poly(3-hydroxybutyrate-*co*-4-hydroxyvalerate). *Macromolecules* **23**, 26–31.

Doillon, C. J., and Silver, F. H. (1986). Collagen-based wound dressing: effects of hyaluronic acid and firbonectin on wound healing. *Biomaterials* **7**, 3–8.

Domb, A. J., Gallardo, C. F., and Langer, R. (1989). Poly(anhydrides). 3. Poly(anhydrides) based on aliphatic–aromatic diacids. *Macromolecules* 22, 3200–3204.

Domb, A. J., Amselem, S., Langer, R. and Maniar, M. (1994). Polyanhydrides as carriers of drugs. *In* "Biomedical Polymers" (S. Shalaby, ed.), pp. 69–96. Hanser, New York.

East, G. C., McIntyre, J. E. and Qin, Y. (1989). Medical use of chitosan. *In* "Chitin and Chitosan" (G. Skjak-Braek, T. Anthonsen, and P. Sandford, eds.), pp. 757–764. Elsevier, London.

Ellingsworth, L. R., DeLustro, F., Brennan, J. E., Sawamura, S., and McPherson, J. (1986). The human immune response to reconstituted bovine collagen. *J. Immunol.* **136**, 877–882.

Eloy, R., Baguet, J., Christe, G., Rissoan, M. C., Paul, J., and Belleville, J. (1988). An *in vitro* evaluation of the hemostatic activity of topical agents. *J. Biomed. Mat. Res.* **22**, 149–157.

El-Samaligy, M. S., and Rohdewald, P. (1983). Reconstituted collagen nanoparticles, a novel drug carrier delivery system. *J. Pharm. Pharmacol.* **35**, 537–539.

Engelberg, I., and Kohn, J. (1991). Physico-mechanical properties of degradable polymers used in medical applications: a comparative study. *Biomaterials* **12**, 292–304.

Ertel, S. I., and Kohn, J. (1994). Evaluation of a series of tyrosine-derived polycarbonates as degradable biomaterials. *J. Biomed. Mater. Res.* **28**, 919–930.

Feng, X. D., Song, C. X., and Chen, W. Y. (1983). Synthesis and evaluation of biodegradable block copolymers of E-caprolactone and D,L-lactide. *J. Polym. Sci. (Polym. Lett. Edn.)* **21**, 593–600.

Fenkel, S. R., Toolan, B., Menche, D., Pitman, M. I., and Pachence, J. M. (1997). Chondrocyte transplantation using a collagen bilayer matrix for cartilage repair. *J. Bone Joint Surg. Br.* **79**, 831–836.

Fiordeliso, J., Bron, S., and Kohn, J. (1994). Design, synthesis, and preliminary characterization of tyrosine-containing polyarylates: new biomaterials for medical applications. *J. Biomater. Sci. Polymer Ed.* **5**, 497–510.

Foster, A. B., and Webber, J. M. (1960). Chitin. *Adv. Carbohyd. Chem.* **15**, 371–393.

Frazza, E. J., and Schmitt, E. E. (1971). A new absorbable suture. *J. Biomed. Mater. Res. Symp.* **1**, 43–58.

Freed, L. E., Grande, D. A., Lingbin, Z., Emmanual, J., Marquis, J. C., and Langer, R. (1994a). Joint resurfacing using allograft chondrocytes and synthetic biodegradable polymer scaffords. *J. Biomed. Mater. Res.* **28**, 891–899.

Freed, L. E., Vunjak-Novakvic, G., Biron, R. J., Eagles, D. B., Lesnoy, D. C., Barlow, S. K., and Langer, R. (1994c). Biodegradable polymer scaffolds for tissue engineering. *Bio/Technology* **12**, 689–693.

Fuller, R. A., and Rosen, J. A. (1986). Materials for medicine. *Scientific Am.* **260**, 119–125.

Gay, S., and Miller, E. J. (1978). The biochemistry and metabolism of collagen. *In* "Collagen in the Physiology and Pathology of Connective Tissue" (E. J. Miller, ed.), Ch. 1. Gustav Fischer Verlag, New York.

Gilding, D. K., and Reed, A. M. (1979). Biodegradable polymers for use in surgery — poly(glycolic)/poly(lactic acid) homo and copolymers. *Polymer* **20**, 1459–1464.

Giordano, C., Sanginario, V., Ambrosio, L., Di Silvio, L., and Santin, M. (2006). Chemical-physical characterization and *in vitro* preliminary biological assessment of hyaluronic acid benzyl ester–hydroxyapatite composite. *J. Biomater. Appl.* **20**, 237–252.

Gogolewski, S., Jovanovic, M., Perren, S. M., Dillo, J. G., and Hughers, M. K. (1993). Tissue response and *in vivo* degradation of selected poly-hydroxyacids: polylactides (PLA), poly(3-hydroxybutyrate) (PHB), and poly(3-hydroxybutyrate-*co*-3-hydroxyvalerate) (PHB/VA). *J. Biomed. Mater. Res.* **27**, 1135–1148.

Gorna, K., and Gogolewski, S. (2002). Biodegradable polyurethanes for implants. II. *In vitro* degradation and calcification of materials from poly(epsilon-caprolactone)-poly(ethylene oxide) diols and various chain extenders. *J. Biomed. Mate. Res.* **60**, 592–606.

Greisler, H. P., Ellinger, J., Schwarcz, T. H., Golan, J., Raymond, R. M., and Kim, D. U. (1987). Arterial regeneration over polydioxanone prostheses in the rabbit. *Arch. Surg.* **122**, 715–721.

Grigolo, B., Lisignoli, G., Piacentini, A., Fiorini, M., Gobbi, P., Mazzotti, G., Duca, M., Pavesio, A., and Facchini, A. (2002). Evidence for redifferentiation of human chondrocytes grown on a hyaluronan-based biomaterial (HYAff 11): molecular, immunohistochemical and ultrastructural analysis. *Biomaterials* **23**, 1187–1195.

Hansbrough, J., Boyce, S., Cooper, T., and Foreman, T. (1989). Burn wound closure with cultured autologous keratinocytes and fibroblasts attached to a collagen–glycosaminoglycan substrate. *J. Am. Med. Assoc.* **262**, 2125–2130.

Hav, D. L., von Fraunhofe, J. A., Chegini, N., and Masterson, B. J. (1988). Locking mechanism strength of absorbable ligating devices. *J. Biomed. Mater. Res.* **22**, 179–190.

He, W., Ma, Z., Yong, T., Teo, W. E., and Ramakrishna, S. (2005a). Fabrication of collagen-coated biodegradable polymer nanofiber mesh and its potential for endothelial cells growth. *Biomaterials* **26**, 7606–7615.

He, W., Yong, T., Teo, W. E., Ma, Z., and Ramakrishna, S. (2005b). Fabrication and endothelialization of collagen-blended biodegradable polymer nanofibers: potential vascular graft for blood vessel tissue engineering. *Tissue Eng.* **11**, 1574–1588.

Heimbach, D., Luterman, A., Burke, J., Cram, A., Herndon, D., Hunt, J., and Jordan, M. (1988). Artificial dermis for major burns. *Ann. Surg.* **208**, 313–320.

Heinegard, D., and Paulsson, M. (1980). Proteoglycans and matrix proteins in cartilage. *In* "The Biochemistry of Glycoproteins and Proteoglycans" (W. J. Lennarz, ed.), pp. 297–328. Plenum Press, New York.

Heller, J. (1988). Synthesis and use of poly(ortho esters) for the controlled delivery of therapeutic agents. *J. Bioact. Compat. Polym.* **3**, 97–105.

Heller, J. and Daniels, A. U. (1994). Poly(ortho esters). *In* "Biomedical Polymers" (S. Shalaby, ed.), pp. 35–67. Hanser New York.

Heller, J., Penhale, D. W. H., and Helwing, R. F. (1980). Preparation of poly(ortho esters) by the reaction of diketen acetals and polyolis. *J. Polym. Sci. (Polym. Lett. Ed.)* **18**, 619–624.

Heller, J., Sparer, R. V., and Zentner, G. M. (1990). Poly(ortho esters) for the controlled delivery of therapeutic agents. *J. Bioact. Comper. Polym.* **3**, 97–105.

Helmus, M. N., and Hubbell, J. A. (1993). Materials selection. *Cardiovasc. Pathol.* **2**, 53S–71S.

Heyde, M., and Schacht, E. (2004). Biodegradable polyphosphazenes for biomedical applications. *In* "Phosphazenes" (M. D. J. Gleria, ed.), pp. 367–398. Nova Science, Hauppauge, N Y.

Hill, J. W., and Carothers, W. H. (1932). Studies of polymerizations and ring formations (XIV): a linear superpolyanhydride and a cyclic dimeric anhydride from sebacic acid. *J. Am. Chem. Soc.* **54**, 1569–1579.

Hirano, S. (1989). Chitosan wound dressings. *In* "Chitin and Chitosan" (G. Skjak-Braek, T. Anthonsen, and P. Sandford, eds.), pp 1–835. Elsevier, London.

Hodde, J. (2002). Naturally occurring scaffolds for soft tissue repair and regeneration. *Tissue Eng.* **8**, 295–308.

Holland, S. J., Jolly, A. M., Yasin, M., and Tighe, B. J. (1987). Polymers for biodegradable medical devices II. Hydroxybutyrate-hydroxyvalerate copolymers: hydrolytic degradation studies. *Biomaterials* **8**, 289–295.

Holmes, P. A. (1985). Applications of PHB — a microbially produced biodegradable thermoplastic. *Phys. Technol.* **16**, 32–36.

Holzman, S., Connolly, R. J., and Schwaitzberg, S. D. (1994). Effect of hyaluronic acid solution on healing of bowel anastomoses. *J. Invest. Surg.* **7**, 431–437.

Hsu, S., Chang, S., Yen, H., Whu, S. W., Tsai, C., and Chen, D. C. (2006). Evaluation of biodegradable polyesters modified by type II collagen and Arg-Gly-Asp as tissue-engineering scaffolding materials for cartilage regeneration. *Artificial Organs* **30**, 42–55.

Huang, Y., Huang, Y., Huang, C., and Liu, H. (2005). Manufacture of porous polymer nerve conduits through a lyophilizing and wire-heating process. *J. Biomed. Mater. Res.* B *Appl. Biomat.* **74B**, 659–664.

Huang-lee, L. L. H., Wu, J. H., and Nimni, M. E. (1994). Effects of hyaluronan on collagen fibrillar matrix contraction by fibroblasts. *J. Biomed. Mater. Res.* **28**, 123–132.

Hubbell, J. A. (1995). Biomaterials in tissue engineering. *Bio/Technology* **13**, 565–576.

Hubel, A., Toner, M., Cravalho, E. G., Yarmush, M. L., and Tompkins, R. G. (1991). Intracellular ice formation during the freezing of hepatocytes cultured in a double collagen gel. *Biotechnol. Prog* **7**, 554–559.

James, K., and Kohn, J. (1997). Pseudo-poly(amino acid)s: examples for synthetic materials derived from natural metabolites. In "Controlled Drug Delivery: Challenges and Strategies" (K. Park, ed.), pp. 389–403. American Chemical Society, Washington, DC.

Kaplan, D. L., Wiley, B. J., Mayer, J. M., Arcidiacono, S., Keith, J., Lombardi, S. J., Ball, D. and Allen, A. L. (1994). Biosynthetic polysaccharides. In "Biomedical Polymers" (S. Shalaby, ed.), pp. 189–212. Hanser, New York.

Katti, D. S., and Laurencin, C. T. (2003). Synthetic biomedical polymers for tissue engineering and drug delivery. In "Advanced Polymeric Materials" (G. O. Shonaike and S. G. Advani, eds.), pp. 479–525. CRC Press, Boca Raton, FL.

Kemnitzer, J., and Kohn, J. (1997). Degradable polymers derived from the amino acid L-tyrosine. In "Handbook of Biodegradable Polymers" (A. J. Domb, J. Kost, and D. M. Wiseman, eds.), pp. 251–272. Harwood Academic, Amsterdam.

Kholodovych, V., Smith, J. R., Knight, D., Abramson, S., Kohn, J., and Welsh, W. J. (2004). Accurate predictions of cellular response using QSPR: a feasibility test of rational design of polymeric biomaterials. *Polymer* **45**, 7367–7379.

Kohn, J., and Langer, R. (1984). A new approach to the development of bioerodible polymers for controlled-release applications employing naturally occurring amino acids. *Polymeric Materials, Science and Engineering*. Washington, DC, American Chemical Society, **51**, 119–121.

Kohn, J., and Langer, R. (1985). Nonpeptide poly(amino acids) for biodegradable drug delivery systems. In "12th International Symposium on Controlled Release of Bioactive Materials Geneva, Switzerland" (N. A. Peppas and R. J. Haluska, eds.), pp. 51–52. Controlled Release Society, Lincolnshire, IL.

Kohn, J., and Langer, R. (1986). Poly(iminocarbonates) as potential biomaterials. *Biomaterials* **7**, 176–181.

Kohn, J., and Langer, R. (1987). Polymerization reactions involving the side chains of a-L-amino acids. *J. Am. Chem. Soc.* **109**, 817–820.

Kohn, J., Niemi, S. M., Albert, E. C., Murphy, J. C., Langer, R., and Fox, J. G. (1986). Single-step immunization using a controlled-release, biodegradable polymer with sustained adjuvant activity. *J. Immunol. Meth.* **95**, 31–38.

Koleske, J. V. (1978). Blends containing poly-e-caprolactone and related polymers. In "Polymer Blends" (O. R. Paul and S. Newman, eds.), pp. 369–389. Academic Press, New York.

Kovalevsky, G., and Barnhart, K. (2001). Norplant and other implantable contraceptives. *Clini. Obstet. Gynecol.* **44**, 92–100.

Krejchi, M. T., Atkins, E. D., Waddon, A. J., Fournier, M. J., Mason, T. L., and Tirrell, D. A. (1994). Chemical sequence control of beta-sheet assembly in macromolecular crystals of periodic polypeptides. *Science* **265**, 1427–1432.

Kuroyanagi, Y., Kenmouchi, M., and Ishihara, S. (1993). A cultured skin substitute of fibroblasts and keratinocytes with a collagen matrix: preliminary results of clinical trials. *Ann. Plas. Surg.* **31**, 340–351.

Kvam, B. J., Atzori, M., Toffanin, R., Paoletti, S., and Biviano, F. (1992). $^{1H}$-NMR and $^{13C}$-NMR studies of solutions of hyaluronic acid esters and salts in methyl sufoxide: comparison of hydrogen bond patterns and conformational behavior. *Carbohyd. Res.* **130**, 1–13.

Lacy, P. E. (1995). Treating diabetes with transplanted cells. *Scientific Am.* 51–58.

Laleg, M., and Pikulik, I. (1991). Wet-web strength increase by chitosan. *Nordic Pulp Paper Res. J.* **9**, 99–103.

Langer, R. (1990). Novel drug delivery systems. *Chemistry Britain* **26**, 232–236.

Langer, R., and Tirrell, D. A. (2004). Designing materials for biology and medicine. *Nature* **428**, 487–492.

Langer, R., and Vacanti, J. (1993). Tissue engineering. *Science* **260**, 920–926.

Larson, P. O. (1988). Topical hemostatic agents for dermatologic surgery. *J. Dermatol. Surg. Oncol* **14**, 623–632.

Laurencin, C., Domb, A., Morris, C., Brown, V., Chasin, M., McConnel, R., Lange, N., and Langer, R. (1990). Poly(anhydride) administration in high doses *in vivo*: studies of biocompatibility and toxicology. *J. Biomed. Mater. Res.* **24**, 1463–1481.

Laurencin, C. T., Norman, M. E., Elgendy, H. M., El-Amin, S. F., Allcock, H. R., Pucher, S. R., and Ambrosio, A. A. (1993). Use of polyphosphazenes for skeletal tissue regeneration. *J. Biomed. Mat. Res.* **27**, 963–973.

Lee, C. C., MacKay, J. A., Frechet, J. M. J., and Szoka, F. C. (2005). Designing dendrimers for biological applications. *Nat. Biotechnol.* **23**, 1517–1526.

Leenstag, J. W., Pennings, A. J., Bos, R. R. M., Roxema, F. R., and Boenng, G. (1987). Resorbable materials of polyl-lactides VI. Plates and screws for internal fracture fixation. *Biomaterials* **8**, 70–73.

Leong, K. W., D'Amore, P. D., Marletta, M., and Langer, R. (1986). Bioerodible polyanhydrides as drug-carrier matrices. II: Biocompatibility and chemical reactivity. *J. Biomed. Mater. Res.* **20**, 51–64.

Lepidi, S., Abatangelo, G., Vindigni, V., Paolo Deriu, G., Zavan, B., Tonello, C., and Cortivo, R. (2006). *In vivo* regeneration of small-diameter (2-mm) articles using a polymer scaffold. *FASEB J.* **20**, 103–105.

Lescure, F., Gurny, R., Doelker, E., Pelaprat, M. L., Bichon, D., and Anderson, J. M. (1989). Acute histopathological response to a new biodegradable, polypeptidic polymer for implantable drug delivery system. *J. Biomed. Mater. Res.* **23**, 1299–1313.

Lewis, D. H. (1990). Controlled release of bioactive agents from lactide/glycolide polymers. *In* "Biodegradable Polymers as Drug Delivery Systems" (M. Chasin and R. Langer, eds.), pp. 1–41. Marcel Dekker, New York.

Li, C., and Kohn, J. (1989). Synthesis of poly(iminocarbonates): degradable polymers with potential applications as disposable plastics and as biomaterials. *Macromolecules* **22**, 2029–2036.

Li, S. T., Archibald, S. J., Krarup, C., and Madison, R. D. (1992). Peripheral nerve repair with collagen conduits. *Clin. Mater.* **9**, 195–200.

Lin, A. S. P., Barrows, T. H., Cartmell, S. H., and Guldberg, R. E. (2003). Microarchitectural and mechanical characterization of oriented porous polymer scaffolds. *Biomaterials* **24**, 481–489.

Lohmann, D. (1990). *In* "Novel Biodegradable Microbial Polymers" (E. A. Dawes, ed.), pp. 333–348. Kluwer Academic, London.

Madihally, S. V., and Matthew, H. W. T. (1999). Porous chitosan scaffolds for tissue engineering. *Biomaterials* **20**, 1133–1142.

Mäkelä, E. A., Vainionpää, S., Vihtonen, K., Mero, M., Helevirta, P., Törmälä, P., and Rokkanen, P. (1989). The effect of a penetrating biodegradable implant on the growth plate: an experimental study on growing rabbits, with special reference to polydioxanone. Clin. Orthopaed. (US) **241**, 300–308.

Maldonado, B. A., and Oegema Jr., T. R. (1992). Initial characterization of the metabolism of intervertebral disc cells encapsulated in microspheres. *J. Orthopaed. Res.* **10**, 677–690.

Malette, W., Quigley, M., and Adicks, E. (1986). Chitosan effect in vascular surgery, tissue culture and tissue regeneration. *In* "Chitin in Nature and Technology" (R. Muzzarelli, C. Jeuniaux, and G. Gooday, eds.), pp. 435–442. Plenum Press, New York.

Mast, B. A., Haynes, J. H., Krummel, T. M., Diegelmann, R. F., and Cohen, I. K. (1992). *In vivo* degradation of fetal wound hyaluronic acid results in increased fibroplasia, collagen deposition, and neovascularization. *Plast. Reconstruct. Surg.* **53**, 503–509.

Masuko, T., Iwasaki, N., Yamane, S., Funakoshi, T., Majima, T., Minami, A., Ohsuga, N., Ohta, T., and Nishimura, S. (2005). Chitosan–RGDSGGC conjugate as a scaffold material for musculoskeletal tissue engineering. *Biomaterials* **26**, 5339–5347.

Mathiowitz, E., Saltzman, W. M., Domb, A., Dor, P., and Langer, R. (1988). Polyanhydride microspheres as drug carriers. II. Microencapsulation by solvent removal. *J. Appl. Polym. Sci.* **35**, 755–774.

McCormick-Thomson, L. A., and Duncan, R. (1989). Poly(amino acid) copolymers as a potential soluble drug delivery system. 1. Pinocytic uptake and lysosomal degradation measured in vitro. *J. Bioact. Biocompat. Polym.* **4**, 242–251.

McDevitt, T. C., Woodhouse, K. A., Hauschka, S. D., Murry, C. E., and Stayton, P. S. (2003). Spatially organized layers of cardiomyocytes on biodegradable polyurethane films for myocardial repair. *J. Biomed. Mater. Res. A* **66A**, 586–595.

McPherson, D. T., Morrow, C., Minehan, D. S., Wu, J. G., Hunter, E., and Urry, D. W. (1992). Production and purification of a recombinant elastomeric polypeptide G (VPGVG)19VPGV from *Escherichia coli*. *Biotechnol. Progress* **8**, 347–352.

McPherson, J. M. (1992). The utility of collagen-based vehicles in delivery of growth factors for hard and soft tissue wound repair. *Clin. Mater.* **9**, 225–234.

Medina, M., Paddock, H. N., Connolly, R. J., and Schwaitzberg, S. D. (1995). Novel anti-adhesion barrier does not prevent anastomotic healing in a rabbit model. *J Invest Surg* **8**, 179–186.

Miller, N. D., and Williams, D. F. (1987). On the biodegradation of poly-*b*-hydroxybutyrate (PHB) homopolymer and poly-*b*-hydroxybutyrate-hydroxyvalerate copolymers. *Biomaterials* **8**, 129–137.

Mitra, S., van Dress, M., Anderson, J. M., Peterson, R. V., Gregonis, D., and Feijen, J. (1979). Pro-drug controlled release from poly(glutamic acid). *Polym. Preprints* **20**, 32–35.

Murphy, W. L., and Mooney, D. J. (1999). Controlled delivery of inductive proteins, plasmid DNA and cells from tissue-engineering matrices. *J. Periodont. Res.* **34**, 413–419.

Muzzarelli, R., Baldassara, V., Conti, F., Ferrara, P., Biagini, G., Gazzarelli, G., and Vasi, V. (1988). Biological activity of chitosan: ultrastructural study. *Biomaterials* **9**, 247–252.

Naeme, P. J., and Barry, F. P. (1993). The link proteans. *Experimentia* **49**, 393–402.

Nicol, A., Gowda, D. C., and Urry, D. W. (1992). Cell adhesion and growth on synthetic elastomeric matrices containing Arg-Gly-Asp-Ser$_3$. *J. Biomed. Mater. Res.* **26**, 393–413.

Nimni, M. E., and Harkness, R. D. (1988). Molecular structure and functions of collagen. *In* "Collagen" (M. E. Nimni, ed.), pp. 10–48. CRC Press, Boca Raton, FL.

Pachence, J. M. (1996). Collagen-based devices for soft tissue repair. *J. Appl. Biomat.* **33**, 35–40.

Pachence, J. M., Berg, R. A., and Silver, F. H. (1987). Collagen: its place in the medical device industry. *MD&DI* 49–55.

Pachence, J. M., Frenkel, S. R., and Lin, H. (1992). Development of a tissue analog for cartilage repair. *In* "Tissue-Inducing Biomaterials" (L. G. Cima and E. S. Ron, eds.), pp. 125–131. Materials Research Society, Warrendale, PA.

Panitch, A., Yamaoka, T., Fournier, M. J., Mason, T. L., and Tirrell, D. A. (1999). Design and biosynthesis of elastin-like artificial extracellular matrix proteins containing periodically spaced fibronectin CS5 domains. *Macromolecules* **32**, 1701–1703.

Parenteau, N. (1999). The first tissue-engineered products. *Scientific Am.* **280**, 83–84.

Peacock, E. E., Seigler, H. F., and Biggers, P. W. (1965). Use of tanned collagen sponges in the treatment of liver injuries. *Ann. Surg.* **161**, 238–247.

Peppas, N. A., and Langer, R. L. (1994). New challenges in biomaterials. *Science* **263**, 1715–1720.

Piez, K. A. (1982). Structure and assembly of the native collagen fibril. *Connect. Tissue Res.* **10**, 25.

Piez, K. A. (1984). Molecular and aggregate structures of the collagens. *In* "Extracellular Matrix Biochemistry" (K. A. Piez and A. H. Reddi, eds.), Ch. 1. Elsevier, New York.

Pitt, C. G. (1990). Poly-*e*-caprolactone and its copolymers. *In* "Biodegradable Polymers as Drug Delivery Systems" (M. Chasin and R. Langer, eds.), pp. 71–119. Marcel Dekker, New York.

Pitt, C. G., Chasalow, F. I., Hibionada, Y. M., Klimas, D. M., and Schindler, A. (1981a). Aliphatic polyesters 1. The degradation of poly-*e*-caprolactone *in vivo*. *J. Appl. Polym. Sci.* **26**, 3779–3787.

Pitt, C. G., Gratzl, M. M., Kimmel, G. L., Surles, J., and Schindler, A. (1981b). Aliphatic polyesters II. The degradation of poly-D,L-lactide, poly-*e*-caprolactone, and their copolymer *in vivo*. *Biomaterials* **2**, 215–220.

Pitt, C. G., Hendren, R. W., Schindler, A., and Woodward, S. C. (1984). The enzymatic surface erosion of aliphatic polyesters. *J. Control. Rel.* **1**, 3–14.

Pitt, C. G., Andrady, A. L., Bao, Y. T., and Sarnuei, N. K. P. (1987). Estimation of the rate of drug diffusion in polymers. In "Controlled-Release Technology, Pharmaceutical Applications" (P. I. Lee and W. R. Good, eds.), pp 49–77. American Chemical Society, Washington, DC.

Pons, J. E., Clandinning, R. A., and Cohen, S. (1975). Biodegradable plastic containers for seedling transplants. *Soc. Plastr. Eng. Tech. Pap.* **21**, 567–569.

Pulapura, S., and Kohn, J. (1992). Tyrosine-derived polycarbonates: backbone-modified, "pseudo"-poly(amino acids) designed for biomedical applications. *Biopolymers* **32**, 411–417.

Pulapura, S., Li, C., and Kohn, J. (1990). Structure–property relationships for the design of polyiminocarbonates. *Biomaterials* **11**, 666–678.

Putnam, D., and Cappello, J. (1993). Improving the growth of anchorage-dependent cells upon abrupt passage to serum-free media. *Am. Biotechnol. Lab.* **11**, 14.

Ray, J. A., Doddi, N., Regula, D., Williams, J. A., and Melveger, A. (1981). Polydioxanone (PDS), a novel monofilament synthetic absorbable suture. *Surg. Gynecol. Obstet.* **153**, 497–507.

Reed, A. M., and Gilding, D. K. (1981). Biodegradable polymers for use in surgery — poly(glycolic)/poly(lactic acid) homo and copolymers: 2. *In vitro* degradation. *Polymer* **22**, 494–498.

Rezwan, K., Chen, Q. Z., Blaker, J. J., and Boccaccini, A. R. (2006). Biodegradable and bioactive porous polymer/inorganic composite scaffolds for bone tissue engineering. *Biomaterials* **27**, 3413–3431.

Rosen, H. B., Chang, J., Wnek, G. E., Linhardt, R. J., and Langer, R. (1983). Bioerodible polyanhydrides for controlled drug delivery. *Biomaterials* **4**, 131–133.

Salem, A. K., Stevens, R., Pearson, R. G., Davies, M. C., Tendler, S. J. B., Roberts, C. J., Williams, P. M., and Shakesheff, K. M. (2002). Interactions of 3T3 fibroblasts and endothelial cells with defined pore features. *J. Biomed. Mater. Res.* **61**, 212–217.

Sams, A. E., and Nixon, A. J. (1995). Chondrocyte-laden collagen scaffolds for resurfacing extensive articular cartilage defects. *Osteoarthritis Cartilage* **3**, 47–59.

Sandford, P. A. (1989). Chitosan chemistry. In "Chitin and Chitosan" (G. Skjak-Braek, T. Anthonsen, and P. Sandford, eds.), pp. 51–69. Elsevier, London.

Sanginario, V., Ginebra, M. P., Tanner, K. E., Planell, J. A., and Ambrosio, L. (2006). Biodegradable and semibiodegradable composite hydrogels as bone substitutes: morphology and mechanical characterization. *J. Mater. Sci. Mater. Med.* **17**, 447–454.

Santerre, J. P., Woodhouse, K., Laroche, G., and Labow, R. S. (2005). Understanding the biodegradation of polyurethanes: from classical implants to tissue engineering materials. *Biomaterials* **26**, 7457–7470.

Sato, T., Chen, G., Ushida, T., Ishii, T., Ochiai, N., Tateishi, T., and Tanaka, J. (2004). Evaluation of PLLA-collagen hybrid sponge as a scaffold for cartilage tissue engineering. *Mater. Sci. Eng. C Biomimetic Supramolec. Sys.* **C24**, 365–372.

Sawhney, A. S., Pathak, C. P., and Hubbell, J. A. (1993). Bioerodible hydrogels based on photopolymerized poly(ethylene glycol)-*co*-poly(*a*-hydroxy acid) diacrylate macromers. *Macromolecules* **26**, 581–587.

Scopelianos, A. G. (1994). Polyphosphazenes as new biomaterials. In "Biomedical Polymers" (S. Shalaby, ed.), pp. 153–171. Hanser, New York.

Shalaby, S. W., and Johnson, R. A. (1994). Synthetic absorbable polyesters. *Biomed. Polym.* 2–34.

Shalaby, S. W., DuBose, J. A., and Shalaby, M. (2004). Chitosan-based systems. In "Absorbable and Biodegradable Polymers" (S. W. B. Shalaby and J. L. Karen, eds.), pp. 77–89. CRC Press, Boca Raton, FL.

Silver, F. H., and Pins, G. (1992). Cell growth on collagen: a review of tissue engineering using scaffolds containing extracellular matrix. *J. Long-Term Eff. Med. Implants* **2**, 67–80.

Singh, A., Krogman, N. R., Sethuraman, S., Nair, L. S., Sturgeon, J. L., Brown, P. W., Laurencin, C. T., and Allcock, H. R. (2006). Effect of side-group chemistry on the properties of biodegradable L-alanine cosubstituted polyphosphazenes. *Biomacromolecules* **7**, 914–918.

Smith, J. R., Knight, D., Kohn, J., Rasheed, K., Weber, N., Kholodovych, V., and Welsh, W. J. (2004). Using surrogate modeling in the prediction of fibrinogen adsorption onto polymer surfaces. *J. Chem. Infor. Computer Sci.* **44**, 1088–1097.

Smith, J., Kholodovych, V., Knight, D., Welsh, W. J., and Kohn, J. (2005). QSAR models for the analysis of bioresponse data from combinatorial libraries of biomaterials. *QSAR Combinator. Sci.* **24**, 99–113.

Sodian, R., Hoerstrup, S. P., Sperling, J. S., Martin, D. P., Daebritz, S., Mayer, J. E., Jr., and Vacanti, J. P. (2000). Evaluation of biodegradable, three-dimensional matrices for tissue engineering of heart valves. *ASAIO J.* **46**, 107–110.

Solchaga, L. A., Yoo, J. U., Lundberg, M., Dennis, J. E., A., H. B., Goldberg, V. M., and I., C. A. (2000). Hyaluronan-based polymers in the treatment of osteochondral defects. *J. Orthopaed. Res.* **18**, 773–780.

Spira, M., Fissette, J., Hall, C. W., Hardy, S. B., and Gerow, F. J. (1969). Evaluation of synthetic fabrics as artificial skin grafts to experimental burn wounds. *J. Biomed. Mater. Res.* **3**, 213–234.

Stankus, J. J., Guan, J., Fujimoto, K., and Wagner, W. R. (2005). Microintegrating smooth muscle cells into a biodegradable, elastomeric fiber matrix. *Biomaterials* **27**, 735–744.

Sullivan, S. J., Maki, T., Borland, K. M., Mahoney, M. D., Solomon, B. A., Muller, T. E., Monaco, A. P., and Chick, W. L. (1991). Biohybrid artificial pancrease: long-term implantation studies in diabetic, pancreatectomized dogs. *Science* **252**, 718–721.

Sung, K. C., and Topp, E. M. (1994). Swelling properties of hyaluronic acid ester membranes. *J. Membrane Sci.* **92**, 157–167.

Tanzer, M. L., and Kimura, S. (1988). Phylogenetic aspects of collagen structure and function. In "Collagen" (M. E. Nimni, ed.), pp. 55–98. CRC Press, Boca Raton, FL.

Taravel, M. N., and Domard, A. (1993). Relation between the physicochemical characteristics of collagen and its interactions with chitosan: I. *Biomaterials* **14**, 930–938.

Temenoff, J. S., and Mikos, A. G. (2000). Injectable biodegradable materials for orthopedic tissue engineering. *Biomaterials* **21**, 2405–2412.

Thorstenson, J. B., and Narasimhan, B. (2006). Combinatorial methods for the high-throughput characterization and screening of biodegradable polymers. In "Handbook of Biodegradable Polymeric Materials and Their Applications" (S. K. N. Mallapragada, ed.), pp. 1–11. American Scientific, Stevenson Ranch, CA.

Tirrell, J. G., Fournier, M. J., Mason, T. L., and Tirrell, D. A. (1994). Biomolecular materials. *Chem. Eng. News* 40–51.

Toolan, B. C., Frenkel, S. R., Yalowitz, B. S., Pachence, J. M., and Alexander, H. (1996). An analysis of a collagen–chondrocyte composite for cartilage repair. *J. Biomed. Mat. Res.* **31**, 273–280.

Turner, N. J., Kielty, C. M., Walker, M. G., and Canfield, A. E. (2004). A novel hyaluronan-based biomaterial (Hyaff-11) as a scaffold for endothelial cells in tissue engineered vascular grafts. *Biomaterials* **25**, 5955–5964.

Ueda, H., and Tabata, Y. (2003). Polyhydroxyalkanonate derivatives in current clinical applications and trials. *Adv. Drug Delivery Rev.* **55**, 501–518.

Urmann, B., Gomel, V., and Jetha, N. (1991). Effect of hyaluronic acid on postoperative intraperitoneal adhesion prevention in the rat model. *Fertil. Steril.* **56**, 563–567.

Urry, D. (1995). Elastic biomolecular machines. *Scientific Am.* **272**, 64–69.

Urry, D. W. (1993). Molecular machines: how motion and other functions of living organisms can result from reversible chemical changes. *Angew. Chem. Int. Ed. Engl.* **32**, 819–841.

Vainionpaa, S., Kilpukart, J., Latho, J., Heleverta, P., Rokkanen, P., and Tormala, P. (1987). Strength and strength retention *in vitro*, of absorbable, self-reinforced polyglycolide (PGA)rodes for fracture fixation. *Biomaterials* **8**, 45–48.

van der Rest, W. J., Dublet, B., and Champliaud, M. (1990). Fibril-associated collagens. *Biomaterials* **11**, 28.

van Heeswijk, W. A. R., Hoes, C. J. T., Stoffer, T., Eenink, M. J. D., Potman, W., and Feijen, J. (1985). The synthesis and characterization of polypeptide-adriamycin conjugates and its complexes with adriamycin. Part 1. *J. Control. Rel.* **1**, 301–315.

van Hest, J. C. M., and Tirrell, D. A. (2001). Protein-based materials, toward a new level of structural control. *Chem. Comm.* **19**, 1897–1904.

van Swol, R. L., Ellinger, R., Pfeifer, J., Barton, N. E., and Blumenthal, N. (1993). Collagen membrane barrier therapy to guide regeneration in class II furcations in humans. *J. Periondontol.* **64**, 622–629.

Veis, A., and Payne, K. (1988). Collagen fibrillogenesis. *In* "Collagen" (M. E. Nimni, ed.), Ch. 4. CRC Press, Boca Raton, FL.

Venugopal, J., Zhang, Y. Z., and Ramakrishna, S. (2005). Fabrication of modified and functionalized polycaprolactone nanofiber scaffolds for vascular tissue engineering. *Nanotechnology* **16**, 2138–2142.

Vert, M., and Li, S. M. (1992). Bioresorbability and biocompatibility of aliphatic polyesters. *J. Mater. Sci. Mater. Med.* **3**, 432–446.

Wade, C. W. R., Gourlay, S., Rice, R., Hegyeli, A., Singler, R., and White, J. (1978). *In* "Organometallic Polymers" (C. E. Carraher, J. E. Sheats, and C. V. Pittman, eds.), p. 289. Academic Press, New York.

Wang, E., Overgaard, S. E., Scharer, J. M., Bols, N. C., and Moo-Young, M. (1988). Occlusion immobilization of hybridoma cells. *In* "Chitosan: Biotechnology Techniques" (ed.), pp. 133–136.

Wasserman, D., and Versfeit, C. C. (1975). Use of Stannous Octoate Catalyst in the Manufacture of L-Lactide-Glycolide Copolymer Sutures. U.S. Patent 3,839,297.

Wei, J. C., Hudson, S. M., Mayer, J. M., and Kaplan, D. L. (1977). A novel method for crosslinking carbohydrates. *J. Polymer Sci.* **30**, 2187–2193.

Weiss, C., and Balazs, E. A. (1987). Arthroscopic viscosurgery. *Arthroscopy* **3**, 138.

Weiss, C., Levy, H. J., Denlinger, J. L., Suros, J. M., and Weiss, H. E. (1986). The role of Nahylan in reducing postsurgical tendon adhesions. *Bull. Hosp. J. Dis. Ortho. Inst.* **46**, 9–15.

Wintermantel, E., Mayer, J., Blum, J., Eckert, K. L., and Luscher, P. (1996). Tissue-engineering scaffolds using superstructures. *Biomaterials* **17**, 83–91.

Woo, G. L. Y., Mittelman, M. W., and Santerre, J. P. (2000). Synthesis and characterization of a novel biodegradable antimicrobial polymer. *Biomaterials* **21**, 1235–1246.

Woodward, S. C., Brewer, P. S., Montarned, F., Schindler, A., and Pitt, C. G. (1985). The intracellular degradation of poly-*e*-caprolactone. *J. Biomed Mater. Res.* **19**, 437–444.

Xu, C. Y., Inai, R., Kotaki, M., and Ramakrishna, S. (2004). Aligned biodegradable nanofibrous structure: a potential scaffold for blood vessel engineering. *Biomaterials* **25**, 877–886.

Yamauchi, M., and Mechanic, G. (1988). Crosslinking of collagen. *In* "Collagen" (M. E. Nimni, ed.), Ch. 6. CRC Press, Boca Raton, FL.

Yang, F., Williams, C. G., Wang, D., Lee, H., Manson, P. N., and Elisseeff, J. (2005). The effect of incorporating RGD adhesive peptide in polyethylene glycol diacrylate hydrogel on osteogenesis of bone marrow stromal cells. *Biomaterials* **26**, 5991–5998.

Yannas, I. V., and Burke, J. F. (1980). Design of an artificial skin: basic design principles. *J. Biomed. Mater. Res.* **14**, 65–81.

Yannas, I. V., Burke, J. F., Gordon, P. L., Huang, C., and Rubenstein, R. H. (1980). Design of an artificial skin: control of chemical composition. *J. Biomed. Mater. Res.* **14**, 107–131.

Yasin, M., Holland, S. J., Jolly, A. M., and Tighe, B. J. (1989). Polymers for biodegradable medical devices VI. Hydroxybutyrate-hydroxyvalerate copolymers: accelerated degradation of blends with polysaccharides. *Biomaterials* **10**, 400–412.

Yoshikawa, E., Fourneir, M. J., Mason, T. L., and Tirrell, D. A. (1994). Genetically engineered fluoropolymers: synthesis of repetive polypeptides containing *p*-fluorophenylalanine residues. *Macromolecules* **27**, 5471–5475.

Yu, H., Lin, J., and Langer, R. (1987). Preparation of hydroxyproline polyesters. *In* "14th International Symposium on Controlled Release of Bioactive Materials, Toronto, Canada" (P. I. Lee, and B. A. Leonhardt, eds.), pp. 109–110. Controlled Release Society, Lincolnshire, IL.

Yu-Kwon, H., and Langer, R. (1989). Pseudopoly(amino acids): a study of the synthesis and characterization of poly(trans-4-hydroxy-*N*-acyl-L-proline esters). *Macromolecules* **22**, 3250–3255.

Zhou, Q. X., and Kohn, J. (1990). Preparation of poly(L-serine ester): a structural analogue of conventional poly(L-serine). *Macromolecules* **23**, 3399–3406.

# Chapter Twenty-Four

# Micro- and Nanofabricated Scaffolds

*Christopher J. Bettinger, Jeffrey T. Borenstein, and Robert Langer*

## I. INTRODUCTION

The design and fabrication of biodegradable scaffolds are keystones to advancing the field of tissue engineering and organ regeneration. Similarly, the widespread application of microfabrication strategies has proven to be beneficial both in elucidating complex biological processes and improving cell function through a variety of avenues. Although a wide number of techniques and approaches have been developed to study the behavior of cells and their components *in vitro*, expanding micro-scale functionalities to tissue engineering scaffolds could prove beneficial in controlling cell function *in vivo*. Incorporating micro-scale systems and strategies for tissue and organ regeneration into scaffolds requires the ability to develop advanced microfabrication techniques tailored specifically for biomaterials. This chapter is dedicated to describing current strategies for the microfabrication of biomaterials within the context of realizing an ultimate goal of fabricating biodegradable scaffolds with micron- and nanometer-scale features. In general, these processes and approaches implement modifications to traditional micro-scale fabrication techniques, such as replica molding, soft lithography, and electrospinning, thereby expanding processing capabilities to include either natural or synthetic biomaterials. More advanced techniques, such as solid free-form fabrication and the production of *in situ* cell-seeded scaffolds, are also reviewed. The next generation of scaffold fabrication will benefit from adapting nascent generalized materials-processing strategies to expand functionality and match corresponding advances in novel biomaterial development.

The design and engineering of suitable biodegradable scaffolds are central to the field of tissue engineering and organ regeneration. Traditional advancements in scaffold fabrication have focused on developing new types of biomaterial systems with more desirable characteristics, such as reduced toxicity or immune response, increased strength, and elastomeric properties. Parallel fabrication strategies have also been improved and developed to accommodate novel biopolymeric systems and to some extent have also been modified to improve and control a similarly defined parameter space, including, for example, biocompatibility, pore size, porosity, and pore connectivity (Murphy *et al.*, 2002). Integrating drug delivery techniques to administer appropriate growth factors or growth factor–encoding plasmids can lead to improved cell and tissue function and in some cases can promote improved tissue function and vascularization of the construct (Lee *et al.*, 2000; Shea *et al.*, 1999). This general approach has been shown to be useful in the application of designing systems to support the growth of small volumes of simple tissues of primarily one cell type, such as the epidermis, cartilage, and the bladder (Oberpenning *et al.*, 1999). Complex or highly vascularized organs and tissues such as the muscle, liver, and kidney require the integration of many cell phenotypes, where the functionality of the organ is highly dependent on spatially defined microenvironments and subsequent heterotypic cell–cell interactions. While there have been substantial advancements in producing vascularized scaffolds (Levenberg *et al.*, 2005), the inability to produce large volumes of organs is still problematic. Controlling the mechanical, chemical, and spatial cellular microenvironments within a scaffold is essential to designing tissue engineering systems.

The integration of micro- and nanoscale technologies with biology and bioengineering has led to significant advancements in the field of tissue engineering. Probing cells and biological systems with tools that operate at the micron- and submicron-length scales have led to the elucidation of some of the fundamental parameters of the cellular microenvironment that influence cell processes and phenotype. Studying and controlling cell–matrix interactions is also of extreme importance. While the chemistry and biology of cell–matrix interactions have been studied extensively, the topography of this interface also plays an important role in regulating cell function. The extracellular matrix is known to contain nanometer-scale features, which provide cues that influence essential cell functions such as proliferation, migration, and spreading. Numerous synthetic systems with a variety of submicron-scale feature sizes and geometries have been used to study the behavior of cells in response to substrates rich in nanometer-scale topographical cues (Flemming *et al.*, 1999). Cells have also been known to respond to randomly oriented topography such as nano-scale roughness in addition to well-defined substrates with submicron-scale fabricated features. Topographic features on the order of 1 micron or smaller can influence a number of cell functions, including cell attachment, morphology, and directed migration, which are important cellular processes to control in fabricating cell–scaffold constructs. Cell alignment, for example, has been shown to play an important role in developing stronger tissues in the cases of smooth muscle cells, skeletal muscle, and fibroblasts. Topography has also been shown to influence the gene profile (Dalby *et al.*, 2003b), including the up-regulation of fibronectin mRNA levels in fibroblasts (Chou *et al.*, 1995). The generalized reaction of cell to topography has been extensively reviewed elsewhere (Curtis and Wilkinson, 1997, 1998; Flemming *et al.*, 1999).

An understanding of the interactions between cells and chemical, topographical, and spatial microenvironments is an important aspect for the rational design of tissue engineering systems. The corpus of work performed in the field of microfabrication for tissue engineering has focused primarily on studying cellular interactions in two dimensions.

Translating the systems and techniques developed to control cell function in two dimensions must be expanded and applied to demonstrate similar control in three-dimensional scaffolds in order to utilize tissue engineering as a viable therapeutic option. The current paradigm of biomimicry in the fabrication of tissue engineering scaffolds requires the ability to control the cellular environment on a micron and submicron level. A wide range of top-down and bottom-up processes have been developed to meet the corresponding increase in demand of micro- and nanometer-scale precision in developing tissue and organ regeneration systems. This chapter focuses on the design and fabrication of tissue-engineering scaffolds with micron- and submicron-scale features by surveying the current state of the art in a wide range of systems and approaches.

## II. ADAPTATION OF TRADITIONAL MICRO-SCALE TECHNIQUES FOR SCAFFOLD FABRICATION

Photolithographic-based processes originally employed in the microfabrication of integrated circuits, used in combination with techniques such as replica molding of polymers and patterning of biomolecules, function as the primary method for studying the effect of microenvironmental parameters on cell function. Manipulating cell geometries on the micron scale can lead to precise control over cell functions such as differentiation, migration (Jiang *et al.*, 2005), proliferation, and cell fate (Chen *et al.*, 1997). In addition to spatially dependent signals in homogeneous cell populations, heterotypic cell–cell interactions have been proven critical in controlling cell function (Bhatia *et al.*, 1999). Because nearly all functional tissues are heterotypic, defining the spatial organization of multiple cell types precisely has been shown to lead to improved function of said tissues. As the complexity of the organ increases, so does the importance of defining appropriate cell–cell interactions precisely. The following section is dedicated to describing the application of traditional microfabrication techniques to the fabrication of tissue-engineering scaffolds.

### Photolithographic-Dependent Processes

Photolithography for use in silicon micromachining and soft lithography is a keystone for developing microfabricated systems for tissue engineering with precise spatial control of structures. The inherent two-dimensional nature of photolithographic-based systems led to the development of microdevices that interface directly with biomolecules and cells, which has proven useful in the study of cell–matrix and heterotypic cell–cell interactions. Microfabrication of inorganic materials such as silicon and quartz for etching and replica molding of poly(di-methyl-siloxane) (PDMS) has been used extensively for biomedical applications, including biosensors and microfluidic networks (Borenstein *et al.*, 2002; Duffy *et al.*, 1998). While they are biocompatible, microscale systems constructed using inorganic materials found in traditional microfabrication techniques are inherently limited to *in vitro* applications such as diagnostic systems. Nevertheless, microfabricated silicon or replica-molded PDMS molds produced from soft lithography can also be interfaced directly with natural or synthetic polymers to create microsystems for biomedical applications. Biodegradable polymers can be cast onto microfabricated molds to produce structures on substrates with feature resolutions as small as 20 nm. Thermoplastic biopolymers such as poly($\varepsilon$-caprolactone) (PCL), poly(L-lactic acid) (PLA), and poly(L-lactic-*co*-glycolic acid) (PLGA) have been processed in this manner for various biomedical applications, including tissue engineering scaffolds and devices for controlled drug release (Armani and Liu, 2000; Richards-Grayson *et al.*, 2003). Melt-casting, solvent-casting, and hot-embossing can all be employed to attain well-defined feature geometries in polycrystalline or amorphous synthetic biomaterials. Composite materials featuring various synthetic biodegradable materials or inert metals such as gold and silicon have been fabricated for applications in drug delivery (Santini *et al.*, 1999). Similar processes can be adapted for fabricating tissue-engineering scaffolds with conductive polymers for potential in nerve regeneration applications (Schmidt *et al.*, 1997). Thermoset biomaterials, including cross-linked elastomeric networks such as poly(glycerol-sebacate) (PGS) (Wang *et al.*, 2002) and poly(1,8-octanediol-co-citric acid) (POC) (Yang *et al.*, 2004), require that the material be set into a given shape when initially molded, followed by a chemical or physical cross-linking process. Therefore, processing of such materials may require the use of a sacrificial mold release layer consisting of a biologically benign material to prevent adhesion of these aggressive materials to the mold used during the final polymerization. Dilute sucrose solutions, which are typically used to prevent flocculation and coagulation in microparticle formulations, can be used to create thin films for aid in mold release of films while maintaining the fidelity of submicron-sized features (Bettinger *et al.*, 2006a).

Fabricating tissue engineering scaffolds with therapeutic potential requires the expansion of two-dimensional microfabrication techniques to enable the production of three-dimensional systems. Lamination techniques are suitable for the integration of multiple micromolded biopolymer layers into a complex three-dimensional structure. Soft-lithography techniques have been used to fabricate molds with PDMS for use in the solvent casting of concentrated PLGA solutions or embossing of solid PLGA to produce microfabricated layers (King *et al.*, 2004; Vozzi *et al.*, 2003). Multiple thin layers of PLGA on the order of hundreds of microns in thickness can be laminated together to form substantially thick structures. Bonding of micromolded PLGA has been accomplished by applying pressure and elevated temperatures to produce scaffolds in a variety of hierarchical geometries. The processing window to bond layers and maintain microfluidic structures is heavily dependent on the temperature and the duration of bonding. The combination of heat and pressure results in a fusion of PLGA molecules at the points of contact, leading to a nearly indistinguishable interface. An alternative method of lamination of synthetic biopolymers can be achieved by solvent fusion at the interface. These processes, while originally demonstrated in PLGA systems, can be applied in principle to any thermoplastic polymer with similar physical properties. Chemical bonding between interfaces can also be employed to laminate multiple layers together. Cross-linking of thermoset polymers involving an additional polymerization step has been used to produce three-dimensional networks by laminating multiple microfabricated layers (Bettinger *et al.*, 2006b) (see Fig. 24.1). Altering the processing conditions of the thermoset layers used in lamination can vary material properties, such as cross-linking density, which directly influences subsequent properties, such as mechanical moduli, ultimate tensile strength, adhesiveness, and permeability. Fine-tuning the adhesiveness is especially critical in developing lamination processes able to preserve the microstructures within the layers. The selective implementation of chemical reagents such as (1-ethyl-3-[3-dimethylaminopropyl]carbodiimide hydrochloride) (EDC) and *N*-hydroxysuccinimide (NHS), an established chemical route for bioconjugation of amines to carboxylates, could also theoretically be employed to laminate polymeric sheets. One potential challenge for expanding the use of biodegradable microfluidic networks is the incorporation of proteins such as growth factors. Microfabrication processes often generate nonphysiological conditions through the use of high temperatures or the inclusion of organic solvents, both of which would denature and eliminate the bioactivity of most proteins or lead to low cell viabilities.

One typical application of microfabricated biopolymers is the development of biodegradable microfluidic devices (Bettinger *et al.*, 2006b; King *et al.*, 2004). Polymeric microfluidic systems provide a platform technology for the development of microfluidic scaffolds, which are advantageous for traditional porous scaffolds, due to the potential for rapid perfusion and improved control over the micro-environment. These systems have been seeded with a variety of cell types, which maintain their long-term function and viability via perfusion. Furthermore, after fabrication and *in vitro* cell seeding, these devices have the potential for host implantation. One could also envision the expansion of biodegradable polymeric microfabrication to produce implantable diagnostic systems, such as temporary, resorbable biosensors. Microfluidic devices have also been fabricated from calcium alginate hydrogels, another versatile material used extensively in drug delivery and tissue-engineering applications (Cabodi *et al.*, 2005). In this reported method, calcium alginate gels are patterned using an adapted soft-lithography technique, which are able to

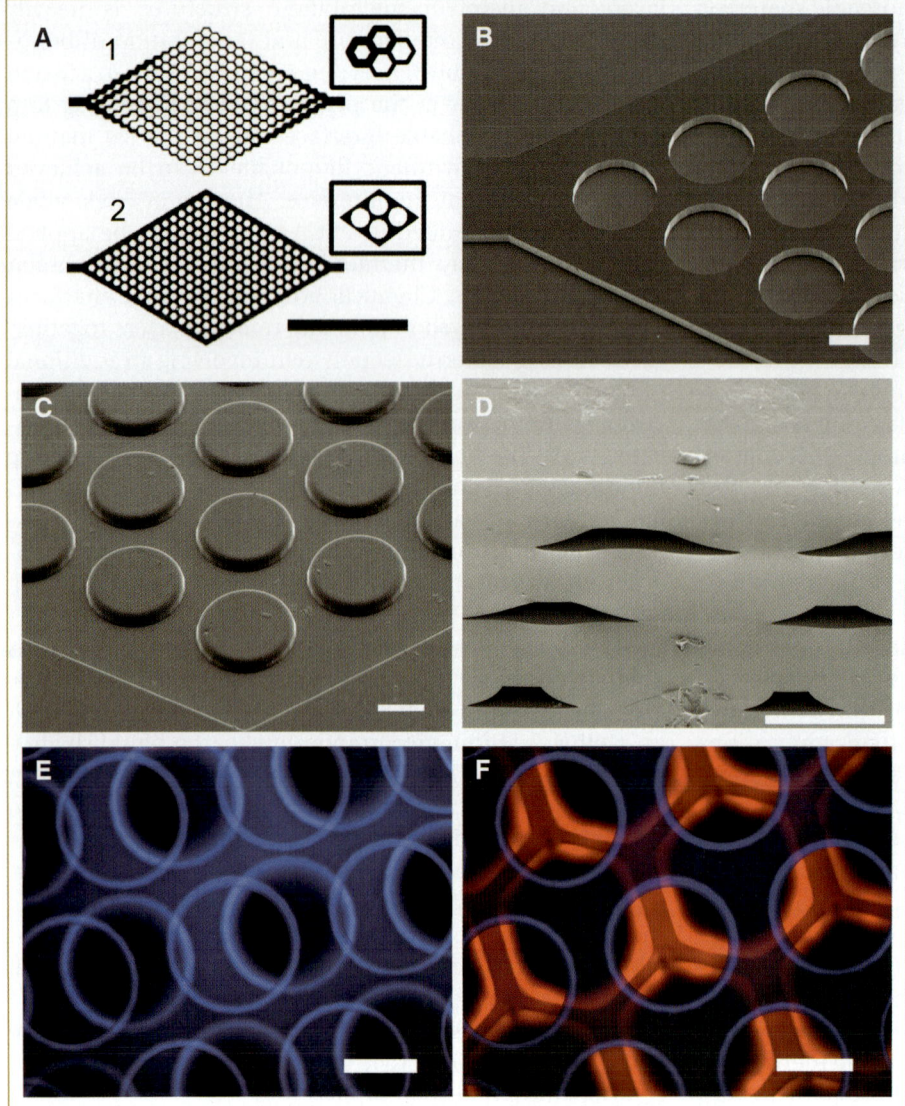

**FIG. 24.1.** Application of traditional replica-molding techniques to microfluidic scaffolds. **(A)** Two-dimensional layouts can be designed in arbitrary patterns using graphics programs such as AutoCAD 2000. Two example layouts are shown (scale bar is 5 mm). Insets show detailed schematics of the representative microstructures used in each two-dimensional layout. **(B)** SEM of micromachined silicon "negative" masters designed for polymer replica molding of micro-fluidic networks. **(C)** Replica-molded poly(glycerol sebacate) layer of silicon master in B. Replica molding of microstructures has also been developed for numerous other materials, including PLA, PLGA, alginate hydrogels, and photo-cross-linkable biomaterials (see text). **(D)** SEM cross sections of three-dimensional PGS microfluidic device. **(E)**–**(F)** Composite fluorescent micrographs of devices after flowing rhodamine and DAPI solutions to demonstrate the patency of multilayer microfluidic devices (scale bars in B–F are all 200 microns). Multiple fluids can be perfused into various microfluidic layers without mixing. Reproduced from Bettinger *et al.* (2006b) with kind permission of Wiley Interscience and John Wiley & Sons, Inc. Copyright 2006. All rights reserved.

produce microchannels with cross-sectional areas between $100 \times 200$ microns and $25 \times 25$ microns. The channels are sealed via chemical cross-linking by first chelating calcium from each of the faces to remove cross-links, laminating the hydrogel slabs together, and then inducing chemical linkages via the addition of calcium chloride. Flowing solutions of both small molecules and high-molecular-weight compounds results in a variety of possibilities of transient concentration profiles throughout the hydrogel network. Parameters such as flow geometry, volumetric flow rate, concentration of solute, molecular weight of solute, and cross-linking density of hydrogel can be adjusted to control the spatial and temporal coordinates of concentration. The utility of such a system lies in the ability to potentially perfuse the ambient hydrogel network, which would presumably be housing seeded cells, to enhance nutrient supply and waste removal. This application would be especially useful for highly metabolically active cells, such as hepato-

cytes. Microfluidic systems have also been fabricated via photo cross linking of acrylated PCL and PLA hybrid polymers (Leclerc *et al.*, 2004). These systems have been fabricated using an adapted soft-lithography technique as well, in which a PDMS mold is used to form microchannels during the UV-initiated photopolymerization process. The mold is removed immediately after cross-linking, and the microfabricated material remains. Human umbilical endothelial cells (HUVECs), fibroblasts, and HepG2, a human hepatocarcinoma cell line, have been cultured on microstructures fabricated in this manner.

Modification to traditional photolithographic techniques can be used to expand the set of potential fabrication methods to produce more complicated systems. The creation of three-dimensional structures is a desirable thrust for use as masters in the fabrication of advanced microscale systems, such as microfluidic systems with complex layouts or biomimetic structures. The formation of three-

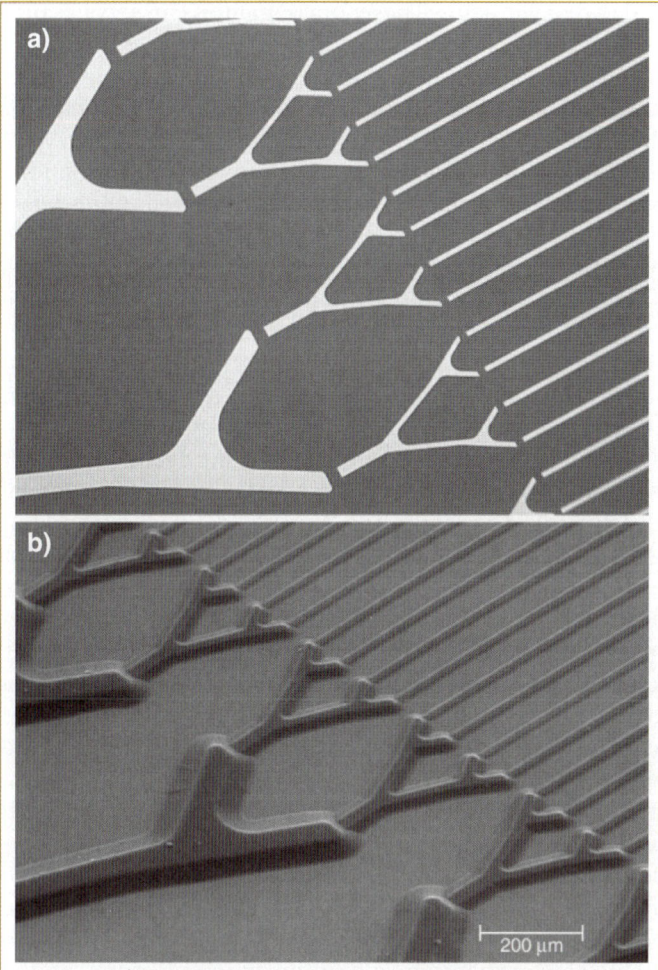

**FIG. 24.2.** A 3D master pattern to cast a microfluidic mimic of a vascular network. The gaps in the original pattern determine the height of each section. **(a)** The original two-dimensional conductive pattern — a gold film patterned onto a silicon nitride covered wafer; **(b)** SEM image of the resulting 3D structure; the smallest lines are 10 μm high, the tallest are 80 μm. The deposition originated from the left side of the image. Reproduced from Lavan *et al.* (2003) with kind permission of Wiley Interscience and John Wiley & Sons, Inc. Copyright 2003. All rights reserved.

dimensional structures via traditional two-dimensional photolithographic techniques requires numerous masks, multiple photolithographic steps, and precise alignment between each cycle of photolithography and etching. These complications lead to increased cost and a reduction in speed. A convenient method for rapidly producing three-dimensional structures using a modified electroplating technique can circumvent many of the previously defined issues (Lavan *et al.*, 2003) (see Fig. 24.2). First, a patterned substrate consisting of an array of conductive islands with a precise, predefined lateral arrangement is patterned on an insulating substrate through a single lithographic metal patterning step, such as gold liftoff. A wire lead is connected to a single conductive island and a corresponding counter-

electrode, respectively, and placed in a plating solution of a conductive material such as nickel or poly(pyrrole), an electrically conducting biocompatible polymer (George *et al.*, 2005). At the start of electrodeposition, material is plated only on the island initially in contact with the lead. As electroplating continues, the conductive material grows laterally, spanning the insulating regions, and vertically, increasing the feature height, at rates that are governed by the plating material and processing parameters. When the conductive material bridges an insulating region and comes in contact with a nearby conductive island, material is then plated on both patterns. Due to the lag in electroplating imparted by the insulating region, the structures plated earlier in the first structure will be taller and wider than features plated on subsequent conductive regions. This process of bridging insulating gaps is repeated as electroplating continues, where the resulting differences in feature height and width of subsequent features is governed by the lateral spacing of the conducting islands and the intrinsic anisotropy of plating velocities of the electrodeposited material. In addition to being helpful for creating a wide variety of MEMS components and devices, this technique can be used to fabricate masters for the production of microfluidic vascular scaffolds with channels that possess aspect ratios of the order of unity. This technique may serve to overcome the issue of connecting macroscopic fluid-handling systems with microscale fluidics by eliminating the presence of high-aspect-ratio microchannels at the inlet and outlet, which create difficulty in establishing patency. Creating a microfluidic master with similar flow characteristics using a traditional process consisting of a single photolithographic step would lead to undesirable feature geometries with extremely high aspect ratios that can exceed $1:500$.

## Electrospinning

Photolithography is a time-consuming and expensive method to produce well-defined microfabricated features with accurate spatial characteristics. Alternative methods to photolithographic-based processes allow the fabrication of biomaterial surfaces with nanoscale topography, which can improve cell functions such as adhesion, migration, and proliferation. Electrospinning is a convenient method for producing arrays of loose fibrous networks containing rich nanometer-scale texture with individual fiber diameters on the order of hundreds of nanometers. The electrospinning process draws a continuous narrow stream of material from a reservoir of polymer melt or solution to a collecting plate, where the material accumulates, producing the fibrous mat. This is accomplished by inducing charge buildup on the surface of the solution through the application of strong voltages. When the voltage is sufficiently strong, the electrostatic potential overcomes the energy associated with surface tension of the bulk material at the orifice, and the solution is accelerated toward the grounded collector. As the polymer solution is propelled to the collector, the solvent evaporates, resulting in a continuous stream of ultrathin fibers. Electrospinning has been used to fabricate fibrous scaffolds using a variety of natural materials, such as silk fibroin (Jin *et al.*, 2002), collagen, polypeptides, and synthetic polymers such as PLGA, PCL, poly(vinyl alcohol) (PVA), and poly(ethylene oxide) (PEO). Ceramics and composite fiber networks have also been produced using electrospinning techniques (Li *et al.*, 2003). The diameters of individual fibers can range from approximately 30 nm to 1 micron, and they are varied by controlling the properties of the solution, such as polymer concentration, viscosity, and conductivity, while also carefully defining electrospinning parameters, such as the applied voltage. Nanofibrous scaffolds produced by electrospinning typically result in a thin three-dimensional film of nonwoven mesh consisting of randomly oriented fibers. These mesh constructs can then be laminated together by heating the scaffold to temperatures slightly above the melting point of the polymer or by solvent dissolution and melding. These processes produce fiber bonds at the points of contact between the fibers and result in a cross-linked network. Although random-fiber mats are useful in producing surfaces with rich texture and topography, improving the utility for electrospun materials in tissue engineering is dependent on developing processes to better control the large-scale fiber alignment. Nanofibrous systems with aligned fibers have also been synthesized by using a rotating drum as a collector (Xu *et al.*, 2004) or, in the case of PLGA, by annealing the fibrous network after applying mechanical forces (Zhong *et al.*, 2003). Fabricating topographically rich scaffolds with aligned nanofibers can exploit the contact guidance response in cells and lead to engineered cell functions, such as directed motility. Detailed processes and applications concerning electrospinning nanofibrous scaffolds have been reviewed extensively elsewhere (Ma *et al.*, 2005).

### Laser Patterning of Biodegradable Polymers

Laser micromachining presents a method for rapid production of micro-scale features without the use of expensive photomasks. Microfabrication of polymeric substrates using ablation is typically performed with UV laser types such as excimer, argon–ion, fluorine, helium–cadmium, metal vapor, and nitrogen. Polymers are etched when the energy of the incident photoelectron is large enough to dissociate chemical bonds directly while imparting little thermal damage to the nonmachined regions. Geometries such as holes and trenches can be produced in biodegradable substrates with minimum feature sizes of approximately 10 microns in polymers such as PVA, PCL, and PLA (Kancharla and Chen, 2002). Laser ablation, while proven directly useful for micromachining, can simultaneously functionalize surfaces that have undergone laser treatment. Laser irradiation can lead to the incorporation of nitrogen or oxygen molecules, thereby creating functional groups such as amines

and carboxylic acids. These functionalities can serve as precursors to surface modifications such as covalent linkages of peptides or non-biofouling agents. Laser ablating techniques can be used in combination with nanoparticles for the rapid fabrication of polymer substrates with arrays of nanoscale features with dimensions much smaller than the wavelength of light. One approach, termed *nanosphere lithography*, uses an ordered template of nanometer-scale particles as a method to focus laser irradiation that is directed at the surface of the polymer substrate to create arrays of nano-scale pits (Lu *et al.*, 2003b). A solution of silica nanoparticles is deposited onto the surface of the substrate. Upon evaporation of the solvent, surface tension effects largely outweigh the thermal energy ($kT$), which produces in an energetically favored packing event. The result is spontaneous ordering of nanoparticles into a hexagonal closely packed configuration. A nearly identical method can serve to create ordered micro-scale sacrificial templates for use in porogen-leaching scaffold fabrication, which is described in the upcoming section on micro-scale templates and self-assembly. The ordered silica nanoparticles focus the UV irradiation on illumination, causing localized ablation, which can produce features, such as pits, as small as 30 nanometers in diameter. The size and geometry of the features can be varied by adjusting processing parameters, including the diameter of the nanoparticles, the wavelength of the laser, the energy of radiation, and the angle of incidence of the irradiation (Kosiorek *et al.*, 2005). Nanosphere lithography provides a convenient method for creating large ordered arrays of feature dimensions that would otherwise be too slow, expensive, and, in the case of features on the order of 10 nanometers, impossible using traditional photolithographic methods. While the set of potential feature geometries and sizes is inherently limited by the top-down nature of nanosphere lithography processes, the arrays of materials that can be used are fundamentally similar to those involved in laser ablation. Nonetheless, this method could be useful in engineering polymeric substrates to study or enhance cell–matrix interactions by providing nanotopographic signals.

## III. THREE-DIMENSIONAL SCAFFOLDS WITH MICRO-SCALE ARCHITECTURE

The behavior of cells in three-dimensional microenvironments is well known to differ vastly from cells grown on two-dimensional surfaces (Cukierman *et al.*, 2002). This fact has prompted the drive to design and fabricate tissue engineering scaffolds that provide a suitable three-dimensional structure to support native cell function. The gravitation toward three-dimensional scaffolds has presented significant challenges in integrating control over spatial distribution and geometries of micron-scale structures. However, average scaffold geometry characteristics can be designed by controlling parameters such as polymer composition in conjunction with physical processing parameters. This technique has been accomplished using a variety of material processing approaches, some of which are presented in the following section.

## Polymer Demixing and Microphase Separation

Capitalizing on characteristic phase separation behavior of polymer solutions and polymer blends is another route toward developing systems for use as scaffolds with ordered features on the micron- and sub micron-length scale. The microphase morphology and characteristic length scale of features in polymer blends can be tuned by adjusting thermodynamic-dependent parameters such as the composition of the blend and kinetic parameters such as temperature, which govern the potential for spinodal decomposition and subsequent polymer demixing. The most commonly observed microphase morphologies are spherical, cylindrical, and lamellar, although numerous other classifications of morphologies have been observed in a variety of systems including block copolymer blends (Tang *et al.*, 2004). Modification of these bulk polymer blends via selective polymer etching or dissolution could lead to the potential for development of ordered tissue engineering scaffolds with micron-scale geometries that are physiologically relevant for numerous applications. The phenomenon of polymer demixing and phase separation can also be adapted to the development of thin films as a method for chemical-induced topological modification. For example, partially compatible polymer blends spin-coated into a thin film can produce topographically rich surfaces with structures of rounded pillar morphologies on the order of 100 nm in height and width (Dalby *et al.*, 2003a). Surfaces with polymer demixed nanotopography have been shown to affect the growth and spreading of human fibroblasts, for example. Further refinement of a similar process could lead to the development of integrating topographically rich surfaces with complex three-dimensional porous structures that enable the uniform presentation of topographical cues throughout the spatial extent of the scaffold. Electrospinning polymer blends provides another route for fabrication of nanometer-scale fibers with cocontinuous phase separation by using ternary mixtures of PLA and poly(pyrrolidone) (Bognitzki *et al.*, 2001).

Freeze-drying presents another possible route to producing scaffolds with micron-scale features (Whang *et al.*, 1995), although the control of spatial locations and distributions is reduced. A typical freeze-drying procedure begins with the addition of water to a solution of PLGA in methylene chloride. The solution is homogenized, poured into the appropriate mold, and then quenched rapidly at approximately –180°C using liquid nitrogen. This rapid-quenching step leads to immediate phase separation within the scaffold, which is then freeze-dried to remove the water and residual solvent. This process can produce scaffolds with pore sizes in the range of 15–35 microns, which can be as

small as one order of magnitude smaller than the standard PLGA scaffolds fabricated using the traditional porogen-leaching process. A similar method has been demonstrated in the production of porous PLA scaffolds with incorporated bioactive molecules (Lo *et al.*, 1995). The potential advantages of this and similar systems for tissue-engineering applications are numerous and somewhat obvious. Drug release experiments suggest that slightly elevated temperatures in the range of 55°C to 85°C and the presence of organic solvents during preparation of these scaffolds has shown to impact slightly negatively the activity of the released molecules. However, in terms of engineering the microscale structure of these scaffolds, precise control of pore size, geometry, and interconnectivity is severely limited by the deferral to equilibrium processes in the preparation of such scaffolds. This leads to control of the averaged scaffold features, which can be manipulated through adjusting processing parameters.

## Micro-Scale Templates and Self-Assembly

Porogen leaching is a common approach to developing large, three-dimensional, porous scaffolds. In this established technique, the scaffold material is incorporated with a chemically or physically incompatible porogen. Upon scaffold fabrication via a process such as solvent evaporation or chemical and physical cross-linking, the porogen is selectively removed. One consequence of this method of fabrication technique is the high degree of porosity (typically in the range of 90% or higher), which is necessary to achieve an interconnected network for random porogen organization. Scaffolds with extremely high porosities are advantageous for tissue engineering metabolically active tissues by allowing for more rapid diffusion of nutrients and removal of waste products. High porosities also provide large surface areas per volume to allow cell attachment and proliferation. However, the mechanical properties of highly porous networks are severely compromised, including a dramatic reduction in the relative stiffness of the porous and solid scaffolds, $E_{\text{cellular}}$ and $E_{\text{solid}}$, respectively, with increasing porosity, $P$. This relationship is proportional to a geometrically defined constant through the following equation:

$$\frac{E_{\text{Cellular}}}{E_{\text{Solid}}} = C(1-P)^2 \qquad (24.1)$$

Reducing the porosity required for interconnection can be obtained through ordered assembly of uniform, micron-scale, spherical, colloidal particles (Stachowiak *et al.*, 2005). Physical templating through geometric constraints can produce scaffolds that exhibit dramatically increased stiffness and improved toughness, which has been shown to affect cell behavior, when compared to randomly assembled porogen structures. This fabrication technique has been employed for creating hydrogel networks with increased moduli using a porogen network consisting of ordered microspheres. Poly(methyl-methacrylate) (PMMA) microspheres of diameters ranging from 20 to 60 microns in a 30 : 70 water/ethanol mixture were deposited under physical perturbation to enhance characteristic hexagonal close-packed ordering. A solution of low-molecular-weight PEG dimethacrylate (PEG-DMA) modified with adhesion-promoting peptides was administered to the template and irradiated with UV to induce cross-linking of the scaffold. The PMMA porogen network was etched with acetic acid, leaving a porous hydrogel network that was shown to support the attachment and migration of NR6 fibroblasts. Using an ordered sacrificial pore structure has demonstrated an increase in stiffness by up to several orders of magnitude, to the range of 10 kilopascals, which is comparable to that of soft tissues. In principle, this process could be applied to any number of polymer–porogen systems that exhibit physical properties that permit removal of the sacrificial phase, such as relative selectively in dissolution, degradation, or differential melting points.

Ordered template assembly can also be achieved through self-assembly of biomolecules such as proteins and DNA. Self-assembly of peptide systems into supramolecular structures requires the rational incorporation of specific and selective interactions among the amino acid building blocks. The resulting structures can be tuned by manipulating amino acid sequences that provide the appropriate combinations of properties, such as charge and residue hydrophilicity. Beta-sheet structures can be produced by incorporating peptides with one hydrophobic surface and one hydrophilic surface in an aqueous solution. The hydrophobic surfaces become shielded while the charged hydrophilic residues remain in contact with the aqueous environment in an organization event that is similar in nature to *in vivo* protein folding. The hierarchical structure of this assembly process is evident by the ability to synthesize structures on the order of 30–100 nanometers by virtue of single-peptide building blocks with lengths on the order of 1 nanometer. This technique has been utilized with natural amino acid sequences to produce a wide range of structures with nanometer-scale features, including modified surfaces, filaments, and fibrils and peptide nanotubes for a variety of applications, including tissue engineering and drug delivery (Zhang, 2002). Macromolecular structures can also assemble to produce stable networks of nanometer-scale fibers. Self-assembling peptide networks have been used to support three-dimensional cell culture systems and feature fibers diameters on the order of 10 nm, with pore sizes between 200 and 500 nm (Zhang, 2003). Reversible hydrogel structures have also been synthesized by engineering synthetic tri-block peptide structures (Petka *et al.*, 1998). Peptide constructs have been used to produce ordered peptide nanotubes, which may include other materials for various applications. In one approach, amyloid fiber peptides, the precursors that form prions, which eventually produce the onset of Alzheimer's disease, can self-assemble to produce silver nanowires on the order of 100 nm in diam-

eter (Reches and Gazit, 2003). Although the original intention of studying these processes was to aid in the assembly of inorganic materials such as metallic nanowires for applications in electronics (Scheibel *et al.*, 2003), similar approaches could be applied to develop advanced tissue engineering constructs. One can envision the development of a two-phase scaffold that may incorporate a functional inorganic material or organic polymer in conjunction with a peptide shell that could introduce additional desirable properties. One potential limitation of this class of materials for tissue engineering applications is the extremely small characteristic length scale of peptide self-assembly. It may be necessary to fabricate tissue engineering constructs with feature definition that ranges over a wide range of length scales. Although nano-scale features can modulate biological function on the cellular level, controlling the structures of scaffolds on the length scale of hundreds of microns or larger may be required for some tissues to create biomimetic macroscopic geometries, improve mechanical strength, or optimize porosity connections to enhance nutrient transport and waste removal. Expanding the operable length scale of this technique could lead to the advancement in meso-scale assembly of larger structures with feature sizes on the order of tens of microns, which may result in a tissue engineering microenvironment that better supports spatially dependent cell processes such as migration, proliferation, and matrix remodeling. Combining template or self-assembly techniques with other scaffold fabrication techniques with much higher minimum resolutions may be a near-term solution to controlling scaffold features and geometry across several orders of magnitude.

# IV. THREE-DIMENSIONAL SCAFFOLD ASSEMBLY

The motivation for scaffold design must be focused on the ability to create scaffolds with predefined and precisely controlled unit cell geometries to enable optimization of pore structure that can accommodate constraining parameters such as mechanical properties and nutrient transport. Controlling pore structure on a micron scale requires the recruitment of computational topology-design tools. The general field of rapid prototyping forms three-dimensional objects from computer-generated solid or surface models. Programs such as computer-aided design (CAD) coupled with the ability to fabricate arbitrary and complex three-dimensional structures through the use of rapid prototyping and solid free-form fabrication (SFF) techniques have allowed the production of designer scaffolds with predefined microarchitecture. One of the major limitations in microfabricated scaffolds has been the inherent limitation of photolithography to two-dimensional micropatterning. Although some photolithographic processes can be stepped to create three-dimensional structures, SFF techniques, through the implementation of controlled deposition using computer programs, do not require the use of multiple photomasks or an alignment step between layers. Systems that can freely incorporate controlled microfabrication processes in the $z$-direction have complete control over the microscale geometries and the relevant resulting macroscopic properties. A number of fabrication routes that control the fabrication in three dimensions have been devised to fabricate many classes of materials, including polymers, composites, and ceramics (see Fig. 24.3). Of particular interest is controlling the mechanical moduli of the scaffold to match the range of moduli found in either soft tissues (0.5–350 MPa) or hard tissues (10–1500 MPa). The following section outlines the current state of the art of SFF and outlines potential advancements and applications for use in scaffold fabrication.

## Stereolithography, Selective Laser Sintering, and Three-Dimensional Printing

Stereolithography (SLA) and selective laser sintering (SLS) are SFF techniques that use a laser-based curing device to photopolymerize liquid monomers or to sinter powered materials, respectively. Both processes begin by designing a three-dimensional structure via CAD, which is then converted into geometrically patterned two-dimensional slices of approximately 100 microns in thickness. A computer-controlled servo mechanism transmits information to the scanning laser and guides the location within the $x$-$y$ plane, selectively bonding the material to form the structure. The build platform is then stepped down the same distance as the slice thickness and a new slice is scanned. SLA selectively polymerizes material for a vat of photopolymeric solution in a layer-by-layer process. SLA has been used to fabricate scaffolds from poly(propylene fumarate) and diethyl fumarate mixtures for hard tissues, including the repair of bone defects (Cooke *et al.*, 2002). PEG-DMA scaffolds incorporated with PLGA microparticles have also been fabricated for use in the tissue engineering of soft tissues (Lu *et al.*, 2003a). One previous limitation of SLA-based processes had been the inability to fabricate structures, such as scaffolds, that were large enough for clinical applications. This limitation has been overcome with the work of Cooke *et al.* because relevant length scales for repairing critical-sized bone defects can be fabricated using SLA. SLS, an SFF method that is operationally similar to SLA, selectively bonds material from a powder bed using a $CO_2$ laser that is directed on a powder bed of material. The laser is scanned across the surface and sinters polymers or composites in a preprogrammed pattern for each layer. Low compaction forces during fabrication in SLS lead to highly porous structures. SLS has been employed in the fabrication of scaffolds using calcium phosphate, PLA, PCL, and poly(ether ether ketone).

The spatial resolution and minimum feature size of SLA and SLS are governed by the laser spot size in the $x$-$y$ plane and by the step size in the $z$-dimension. Laser spot sizes on the order of 250 microns are commonly attainable, while spot sizes as low as 70 microns have been achieved for specialty systems. Feature resolution in the $z$-axis are theoretically limited by the precision of the mechanical stepping system that governs the slice thickness and is typically no smaller than 100 microns. In general, reducing feature resolution and slice thickness in SLA and SLS processes results in a dramatic decrease in the production speed, which prompts the need to solve the engineering optimization problem to balance the benefits of these two fabrication characteristics. Variations of SLA processes could overcome the limitations due to slice thickness. Cross-linking of biomaterials via multiphoton excitation could also provide an efficient method of three-dimensional scaffold microfabrication. Multiphoton excitation cross-linking using UV has been applied to forming three-dimensional matrices of a wide range of polymers and bulk protein formulations, such as collagen, laminin, fibronectin, polyacrylamide, bovine serum albumin, alkaline phosphatase, and various blends of these (Basu and Campagnola, 2004b). The distribution of intensity of irradiation in the $z$-direction theoretically could reduce the minimum feature sizes in this axis to approximately 20 microns or smaller. Remarkably, protein matrices fabricated via this method have been shown to remain active in this polymer matrix environment (Basu and Campagnola, 2004a). Further pursuit of this technology could lead to the rapid and efficient fabrication of three-dimensional structures using a wide variety of biomaterials while simultaneously maintaining delicate conditions that can preserve biological activity.

Three-dimensional printing (3DP) is similar in practice to that of SLA and SLS, in that a three-dimensional object is fabricated one two-dimensional slice at a time using a stepping system. One exception is that 3DP utilizes inkjet printer technology to control the deposition of a chemical binder and to selectively fuse material in a powder bed in order to create the object. Drug delivery and tissue-engineering devices have been fabricated from PEO, PCL, and PLGA using 3DP. However, any biological material could be fabricated in principle using 3DP given the selection of an appropriate chemical binder. The resolution of objects created using 3DP varies with the complexity of the object. Lines 200 and 500 microns in width have been produced using 3DP of polymer solutions and pure solvent, respectively. Feature resolutions of approximately 1 millimeter are more likely for complex geometries. One significant limitation in the application of 3DP for scaffolds is the addition of the chemical binder. Even if the composition of the chemical binder itself is found to be nontoxic and biocompatible, the introduction of cytotoxic organic solvents such as chloroform and methylene chloride is undesirable. Post-fabrication efforts to remove residual solvent, such as vacuum drying, are not completely effective; therefore the issue of cytotoxicity in 3DP-fabricated scaffolds remains.

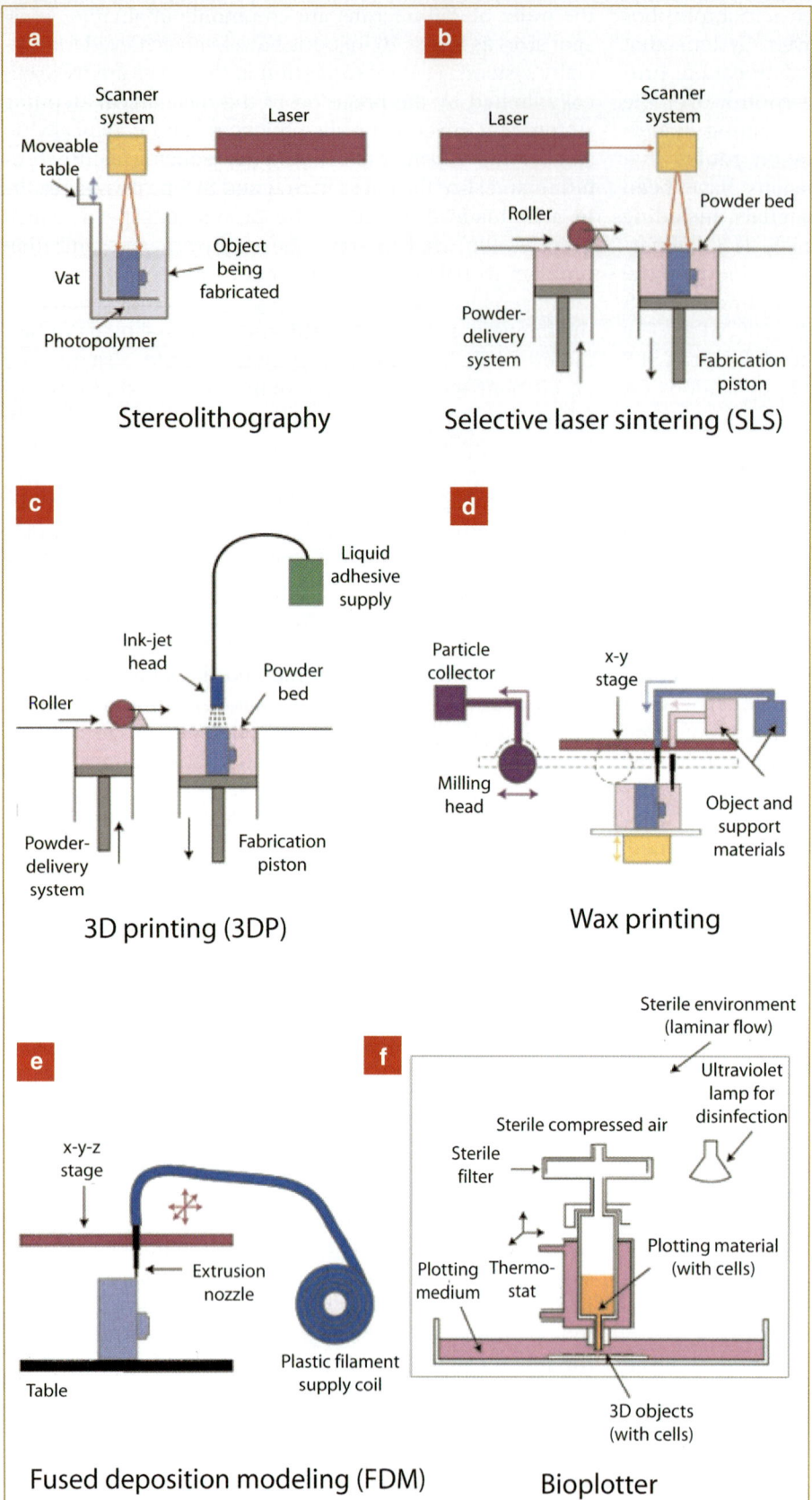

**FIG. 24.3.** Schematics of SFF systems categorized by the processing technique. **(a, b)** Laser-based processing systems include the stereolithography system, which photopolymerizes a liquid (a) and the SLS systems, which sinter powdered material (b). In each system, material is swept over a build platform that is lowered for each layer. **(c, d)** Printing-based systems, including 3D printing (c) and a wax-printing machine (d). 3DP prints a chemical binder onto a powder bed. The wax-based system prints two types of wax material in sequence. **(e, f)** Nozzle-based systems. The fused-deposition modeler prints a thin filament of material that is heated through a nozzle (e). The Bioplotter prints material that is processed either thermally or chemically (f). Reprinted by permission of Macmillan Publishers Ltd. from Hollister (2005). Copyright 2005 and Worldwide Guide to Rapid Prototyping website. Copyright Castle Island Co. [http://home.att.net/~castleisland/]. All rights reserved.

## Fused-Deposition Modeling

Fused-deposition modeling (FDM) is another method for SFF that has been employed in the fabrication of three-dimensional scaffolds. FDM uses a layer-by-layer deposition technique, in which molten polymers or ceramics are extruded through a nozzle with a small orifice, which merges with the material on the previous layer. The pattern for each layer is controlled by mechanical manipulation of the $x$-$y$ position of the nozzle and can be different or arbitrary for each deposited layer. This technique has been applied to the production of three-dimensional scaffolds using polymers (PCL, high-density polyethylene) and composites (PCL/ hydroxyapatite). A modification of FDM, termed *fused deposition of ceramics*, has also been developed for fabrication of scaffolds from ß-tricalcium phosphate. FDM processes can achieve pore sizes ranging from 160 to 700 microns, with porosities ranging from 48% to 77%. PCL scaffolds fabricated via FDM have a compressive stiffness ranging from 4 to 77 MPa, which spans the mechanical stiffness for both soft and hard tissues (Hollister, 2005). One of the primary advantages of FDM and related processes is the high degree of precision that can be achieved in the $x$-$y$ plane. However, control in the $z$-direction is limited and governed by the diameter of the material extruded through the nozzle. Additional limitations, including pore anisotropy and the geometry of pore connectivity, are substantially limited due to the continuous deposition process. The extrusion process limits the types of materials that can be processed and therefore has limited the wide applicability of FDM to scaffold fabrication. FDM is typically limited to synthetic thermoplastic polymers, thereby eliminating many natural biomaterials and thermoset synthetic polymers.

## Microsyringe Deposition

Microsyringe deposition is similar in practice to FDM, in that a polymer is patterned onto a precision-controlled surface using a continuous stream of material via layer-by-layer deposition in order to create three-dimensional structures (Vozzi *et al.*, 2003). Pressure-assisted microsyringe-based deposition (PAM) uses compressed air to eject a solution of polymer in volatile solvent through a narrow capillary needle, which has a diameter of between 10 and 20 microns. Control of the placement of solution in the $x$-$y$ plane is controlled by a micropositioning system, which can achieve lateral precisions of 0.1 microns, while the physical dimensions of the structures can range from 5 to 600 microns, depending on various processing parameters. PAM has been used to fabricate three-dimensional microfabricated scaffolds from PLGA, but it could theoretically be expanded to any polymer, synthetic or natural, that is soluble in a volatile solvent. PAM presents several distinct advantages over traditional SFF methods, including dramatically improved spatial resolution and feature dimensions. Traditional SFF processes are known to have minimum resolutions on the order of hundreds of microns or larger and have principal limitations in minimum feature size. PAM also offers a convenient method for fabrication of multiphase scaffolds with micron-scale precision placement of multiple polymers. Low temperatures also allow potential integration systems for controlled release of proteins and other biomolecules to create favorable microenvironments for tissue regeneration.

## Future of Solid Free-Form Fabrication for Tissue Engineering

SFF has been established as a convenient set of methods for use in three-dimensional scaffold fabrication of various materials, including polymers, hydrogels, ceramics, and metallic biomaterials. Rational engineering design coupled with computer-controlled materials-processing techniques has allowed precision fabrication techniques that enable the control over the microarchitecture of three-dimensional porous scaffolds. In the near term, designing scaffolds using SFF techniques allows the possibility of indirectly controlling macroscopic properties, such as the bulk mechanical properties and the permeability for nutrient transport considerations. Controlling the characteristics of pores on the micron scale, such as pore geometry, connectivity, and porosity, can lead to tissue-specific scaffold fabrication. SFF also allows the possibility of fabricating designer scaffolds for patient-specific organ regeneration applications. In general, the concept of made-to-order scaffolds is an appealing endeavor, which could serve currently unmet clinical needs, such as scaffolds for reconstructive surgery. However, the eventual use of tissue engineering scaffolds in general relies on the ability to induce predetermined cellular responses. Although SFF has proven to be useful in engineering bulk properties through hierarchical design, future tissue engineering systems may require scaffolds with features of resolution sizes that are smaller than what is achievable with current SFF methods. Current design and fabrication strategies operate on a length scale of hundreds of microns, which may be too large to control cell-specific function. Consider the example of bone tissue engineering using a porous scaffold. Although optimal pore sizes between 200 and 600 microns have been suggested, empirical data for a number of systems suggest otherwise. For example, no significant differences in bone growth of osteocytes on 3DP PLGA scaffolds with pores of 500 microns and 1600 microns was detected (Roy and Al, 2003; Simon and Al, 2003). Hydroxyapatite scaffolds with pore sizes ranging from 400 microns to 1200 microns were used in a mandibular defect model. No significant differences in bone growth were detected across the range of pore sizes used in this study. These findings are consistent with the experimentally verified notion that the length scales of structures that govern many important cell functions, such as adhesion, migration, proliferation, and differentiation, are on the order of tens of microns or smaller. While the value of controlling macroscopic properties via hierarchical design should not be fully discounted, current SFF techniques as a stand-alone for scaffold fabrication may not be sufficient for inducing appropriate biological responses. One option for fabricating scaffolds with nanoscale features using SFF is simply to improve the current resolution of the processes, which may prove to be too technically demanding or costly to justify future pursuit. Improving resolution to submicron-scale precision would likely slow fabrication speed and would reduce the upper limit of scaffold volume. Therefore, more near-term advances in this field may stem from the integration of micron and submicron structures as a postfabrication modification for scaffolds produced by SFF. One can envision the possibility of using SFF to fabricate a scaffold with the appropriate macroscopic properties, such as scaffold stiffness and geometry. Next, other techniques for producing smaller features, such as the possible integration of another polymer system, could be applied, such as polymer demixing or freeze-drying. The final result would essentially be a two-phase scaffold, where one phase with features on the order of hundreds of microns governs macroscopic properties that would support tissue-level function, while another phase would feature micron-scale geometries to enhance cell–matrix and cell–cell interactions on the cellular level.

# V. MICROFABRICATION OF CELL-SEEDED SCAFFOLDS

Understanding cell–matrix as well as heterotypic cell–cell interactions, both *in vitro* and *in vivo*, is a critical component in designing tissue engineering systems. Coordinating the placement of cells on scaffolds is a promising method of capitalizing on specific cell–cell interactions that may lead to favorable conditions for tissue formation. However, three-dimensional scaffolds that are postseeded with cells cannot typically retain spatial segregation or control specific cell–cell interactions across multiple cell types. While photolithography-based cell-patterning techniques are a powerful method for controlling such interactions in a two-dimensional microenvironment, these methods translate to three-dimensional systems with much difficulty. Advances in microfabrication of biomaterials and SFF have led to the potential to control the placement of polymers and cells with micron-scale precision. Combining aspects of these existing technologies with a layer-by-layer approach is a promising way to fabricate scaffolds that are preloaded with cells that are spatially controlled in arbitrary geometries. Typical scaffold fabrication techniques often employ harsh conditions, such as cytotoxic solvents or elevated temperatures. Hence, developing novel scaffold fabrication processes that are capable of maintaining viable cells is essential. This limitation also serves to reduce drastically the number of biomaterials that are suitable for such processes. Despite these challenges, a number of possibilities have been demonstrated using this approach to scaffold fabrication.

## Layer-by-Layer Deposition of Cell–Matrix Components

Microfluidic platforms provide a convenient method for controlling many aspects of the cellular microenvironment, including spatial orientation. One-step cell patterning in microfluidic devices has been used to create cellular arrays for diagnostic purposes. When used in combination with a layer-by-layer assembly strategy, micropatterning and microfluidics have proven useful for the controlled assembly of hierarchical constructs of multiple extracellular matrices and cell types. A microfluidic patterned co-culture system with three cell types has been demonstrated for potential use in vascular tissue engineering applications (Tan and Desai, 2004) (see Fig. 24.4). Endothelial cells, smooth muscle cells, and fibroblasts were each mixed with ECM protein solutions such as collagen, one-to-one mixture of collagen–chitosan, and matrigel and deposited sequentially. The result is a three-dimensional layer of multiple cell types that are each contained in user-defined ECM proteins. First, silicon oxide surfaces were modified with 3-aminopropyltriethoxysilane using vapor deposition to enhance the adhesion of ECM proteins. PDMS-on-glass microfluidic systems were fabricated using the modified silicon oxide surfaces. The two-dimensional layout of the microfluidic network can be designed to dictate the eventual location of the nascent tissue structures. One selected cell–matrix solution is then flowed into the microfluidic network at approximately 2°C, at which point the system is heated to 37°C to induce polymerization of the ECM solution. Subsequent layers of cell–matrix systems can be added in a similar manner. After formation of the co-culture structures, the PDMS layer is removed to allow for further characterization of cell function or matrix properties. The obvious advantage of this process is the use of cell-friendly conditions during fabrication, such as the absence of toxic chemicals such as solvents and photoinitiators and UV irradiation. Designing a process that is free of these potentially cytotoxic components of fabrication can produce higher cell viabilities and improve cell function. Tailoring specific cell–matrix combinations can be engineered to produce optimal conditions that improve the overall performance of cell-seeded scaffolds. The use of one microfluidic system limits the flexibility in patterning multiple cell types in an independent manner. This limitation could be potentially overcome by the use of multiple microfluidic patterning techniques, which has been demonstrated in other systems (Khademhosseini *et al.*, 2005).

## Photoencapsulation of Cell–Scaffold Constructs

PEG hydrogels modified with photoactive acrylate groups (PEG-DMA) have been used extensively for photo-

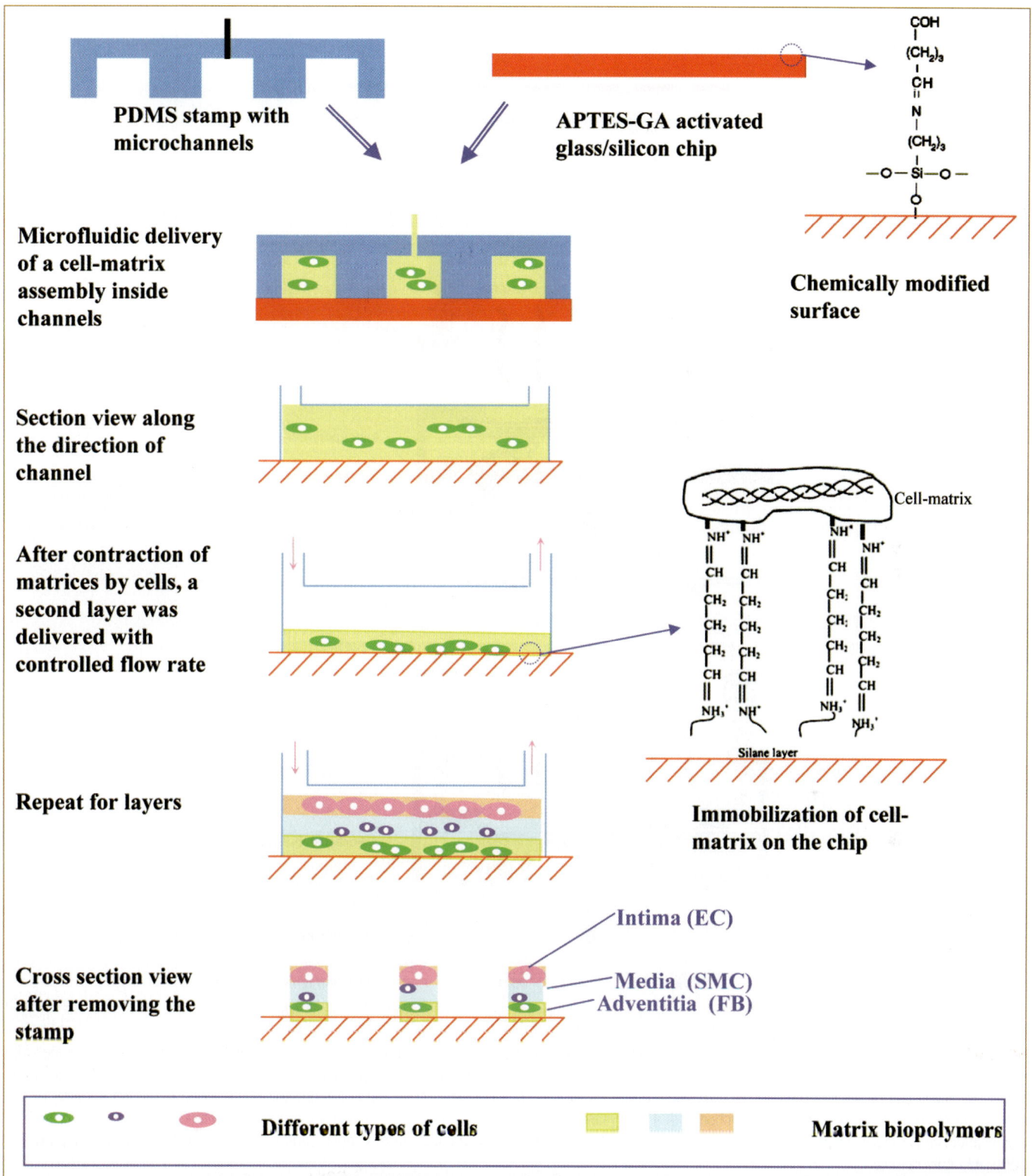

**FIG. 24.4.** Schematic illustration of our approach using microfluidics to create a 3D hierarchical system for three layers of cells within biopolymer matrices. Reproduced from Tan and Desai (2005) with kind permission of Wiley Interscience and John Wiley & Sons, Inc. Copyright 2005. All rights reserved.

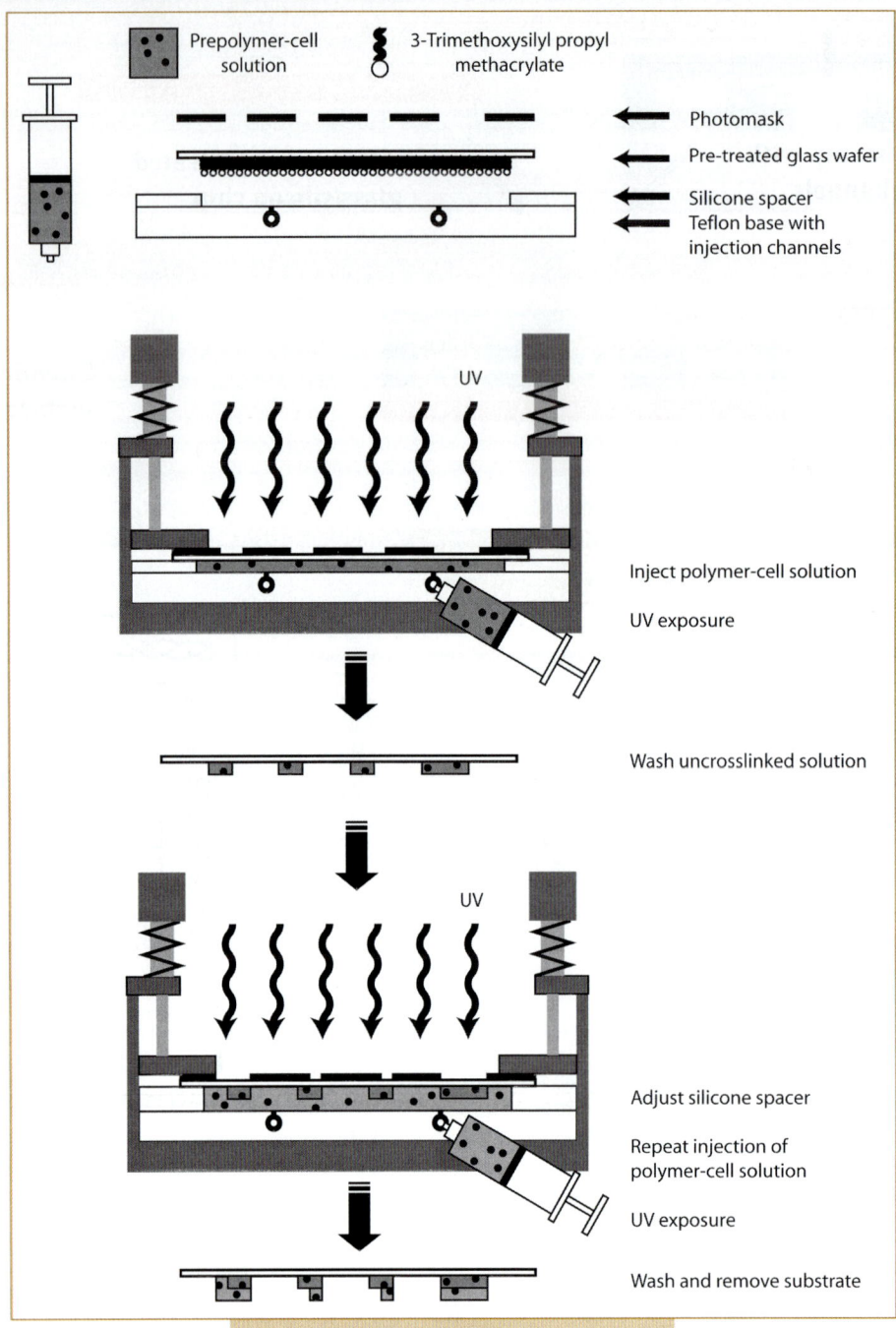

**FIG. 24.5.** Process for formation of hydrogel microstructures containing living cells. The apparatus is assembled, including a pretreated glass wafer with reactive methacrylate groups on its surface and a Teflon base with an inlet and outlet. Once the cells and prepolymer solution are injected, the inlet and outlet are closed and the unit is irradiated with UV light. The resulting patterned hydrogels containing cells are covalently bound to the glass wafer. At this time, a thicker spacer can be used in conjunction with a new mask to add another layer of cells. This process can be repeated multiple times to produce cell–scaffold constructs in a hierarchical manner. Adapted from Liu and Bhatia (2002) with kind permission of Springer Science and Business Media. Copyright 2002. All rights reserved.

cross-linkable systems for drug delivery and tissue engineering. Masks typically used for photolithography in traditional microfabrication processes can be used in combination with PEG-DMA solutions to pattern hydrogel structures with micron-scale resolution. In one reported method, multiple steps of micropatterned photopolymerization processes can be coupled to produce three-dimensional cell matrix structures with micron-scale resolution (Liu and Bhatia, 2002). Silicon dioxide substrates were first functionalized with 3-(trimethoxysilyl)-propyl methacrylate to allow for covalent bonding of hydrogels. Thin layers of solutions of PEG-DMA hydrogel with HepG2 cells were applied to the functionalized surface and exposed to UV irradiation through a photomask. The substrate is washed to remove the un-cross-linked regions and is stepped in the z-direction using a thicker silicone spacer (see Fig. 24.5). This process can be repeated using arbitrary suspensions of cells in PEG-DMA solutions to create additional micropatterned layers. Photopolymerization of cell-encapsulated hydrogels in a layer-by-layer technique provides the flexibility of patterning cell layers in arbitrary geometries. PEG-DMA molecules modified with specific peptides that promote cell functions such as adhesion and migration can be integrated into the scaffold with ease. Furthermore, variables of peptide-modified PEG hydrogels, such as peptide sequences and concentration, can be optimized and incorporated given the tissue of interest. The result is a flexible hydrogel scaffold fabrication strategy that can be used to control the microenvironment of multiple cell types while simultaneously governing the spatial configuration with micron-scale precision. One drawback of such a system is the use of potentially cytotoxic photoinitiators in combination with UV irradiation as a means of creating microstructures. In the reported method, the viability of cells, as measured by MTT assay, is reduced in a dose-dependent manner as the time of UV exposure and concentration of acetophenone photoinitiator was varied between 0 and 60 seconds and 0 and 3 mg/mL, respectively. Another factor that may limit the utility of microfabricated cell-seeded scaffolds using photo cross linkable hydrogels is the maintenance of pattern fidelity of small features with a high degree of precision. While the resolution of features greater than 200 microns is maintained at approximately 10% of the feature size, the resolution is greatly increased as the feature size is reduced below 100 microns. The actual dimensions of features can be increased up to 200% larger than the feature sizes on the photomask, where the ratio is dependent primarily on the exposure time. One mechanism that might explain the increase in feature size is the inherent swelling nature of PEG hydrogels. Other possibilities may be the rapid migration of photoreactive species within the hydrogel during polymerization, which would lead to an overall deficiency in producing features smaller than 50 microns with tight resolution. Although this process has been adapted for the fabrication of PEG hydrogels, it could be expanded to other types of polymeric systems. The design criteria for such future systems are strict and would require a biocompatible, biodegradable, photo cross linkable, water-soluble polymer.

SFF techniques have demonstrated great promise in the design and fabrication of customizable tissue engineering scaffolds. Postseeded cell scaffolds, fabricated via SFF or other methods, limit the attachment of cells to the surface of the scaffold alone, which results in relatively low cell densities. Combining aspects of stereolithography (SLA)

with cell–polymer solutions allows for the possibility of incorporating cells directly in the scaffold to achieve high cell densities. Such a system has been reported where Chinese hamster ovary (CHO) cells were photoencapsulated in PEG-DMA hydrogels using a modified SLA process (Dhariwala et al., 2004). The technical issues associated with this fabrication technique are simply the union of those with standard SLA processes and photoencapsulation of living cells. Such challenges include attaining a reasonably small minimum feature size, producing tight resolutions, and maintaining high cell viability of encapsulated cells. In the process developed by Dhariwala et al., minimum feature sizes of 150 microns in the z-direction and 250 microns in the x-y plane were achieved, which is similar to traditional SLA processes. As reported in previous photoencapsulation studies, the cell viability was highly dependent on the concentration of photoinitiator used during processing.

### Direct Deposition of Cells

The concept of direct deposition of cells is another promising prospective approach to obtaining patterns of cells in arbitrary geometries. The direct deposition of cells using a spraying technique in combination with a patterning mask is a convenient method for creating two-dimensional patterning of cells. In this relatively straightforward method, a solution of cells is aerosolized using an off-the-shelf airbrush apparatus, creating a suspension of liquid droplets containing cells, which is directed toward a surface of collagen gel through a mask. Another collagen gel, approximately 400 microns in thickness, can be adsorbed to the surface, followed by additional repeated sequences of cell and collagen deposition. Increased cytochrome P450 activity, an assay for liver-specific cell function, was reported when patterning NIH 3T3 fibroblasts in conjunction with plated primary hepatocytes in a short-term culture. The ability to create arbitrary geometries of patterned cells in a collagen sandwich is an appealing technology to preserve polarity and function of hepatocytes cultured *in vitro* (Moghe et al., 1996). This technique has been shown to produce features 120 microns in thickness from a mask with 100-micron features, and, though a collagen gel layer of 400 microns was used in the reported work, this thickness could be reduced significantly. The spraying process was shown not to dramatically affect cell viability or function of cultured human umbilical endothelial cells or primary hepatocytes. Although this approach is similar in principle to other techniques for cell patterning, including the use of micropatterned stamps or stencils (Folch et al., 2000), the ability to combine patterned cell deposition with a layer-by-layer method can control the location of cells in the z-direction for newfound control. However, this approach requires the use of multiple masks for patterning cells in different geometries for each corresponding slice in the x-y plane.

Another method of direct cell deposition employs an off-the-shelf inkjet printer that uses a thermal ink deposition method, which is loaded with a suspension of CHO cells in a buffer solution (see Fig. 24.3f). Cells are then deposited in a precise manner onto hydrogel-based substrates, using a simple graphics program to define the location of the deposited cells in the x-y plane. This process requires little modification to the commercially available printer, and, despite the extreme stresses on cells that are associated with the printing process, viabilities of approximately 90% were maintained in the deposition of the robust CHO cell line as a proof of concept. The resolution of inkjet printing of cell suspensions was shown to be on the order of hundreds of microns, which is typical of most previously described SFF methods. Although this approach has demonstrated only two-dimensional patterned deposition of one cell type, this technology represents a flexible platform for future development of three-dimensional cell-seeded scaffolds.

The noncontact manipulation of cells through the use of single-beam gradient optical traps is an efficient method of controlling the spatial coordinates of a variety of cell types in suspension. This technology, also termed *optical tweezers*, has been demonstrated in the manipulation of a variety of biological components, including cells, subcellular components, and microparticles coated with functional biomolecules. The use of high-numerical-aperture lenses leads to tight confinement of the micro- and nanoscale components along the beam axis, limiting the range of mobility in this direction, which ultimately leads to the drastic reduction in efficiency for depositing cells in this manner. By simply reducing the numerical aperture of the lens, micron- and submicron-sized particles can be directed along the beam path in a continuous stream and absorbed onto a target of interest. This general technique, termed *laser-guided direct writing*, has been applied to a variety or organic and inorganic particles as well as mammalian cells (Odde and Renn, 2000). During laser-guided deposition of cells, a laser emitting a near-infrared radiation is configured near a lens that is controlled using a three-axis micromanipulator and directed into a chamber containing a suspension of embryonic chick spinal cord cells with a target substrate of untreated glass. As cells randomly drift into the path of the laser, a gradient force dominates and brings the cell to the center of the beam. The laser beam then directs the cell along the beam path until it reaches the substrate. The position of the single cells or streams of cells can be directed by manipulating the position of the lens. Cells have been deposited in multiple layers with resolutions of 1 micrometer, which offers a dramatic improvement over other methods of cell patterning. Furthermore, multiple sequences of cell deposition followed by the adsorption of a thin biomaterial layer could be cycled to create three-dimensional cell-seeded constructs. One potential drawback to large-scale implementation of this technology could be the speed of deposition. In the reported system, cells were deposited onto the surface at a rate of 2.5 cells per minute. Although this rate could theoretically be raised by simply increasing the concentration of cells within the suspension, the patterned deposition of 100–1000 cells would take on the order of hours to complete.

## VI. SUMMARY AND FUTURE DIRECTION

Engineering the spatial and temporal microenvironments on a precise level is critical for inducing the desired cellular response, which is especially important in the field of tissue engineering and organ regeneration. Consequently, designing the microarchitecture of scaffolds used for said applications is an important first step in inducing favorable cell–matrix and cell–cell interactions. The need for microscale control has led to the development of a number of fabrication routes that are able to control the spatial location of scaffold features. One viable model that has produced the current state of the art and may continue to direct progress in the field of advanced scaffold manufacturing is the adaptation of traditional materials-processing techniques. Replica molding, electrodeposition, and solid freeform fabrication all represent examples of general materials-processing techniques that have been applied to develop advanced processing platforms for scaffold fabrication. This trend will likely continue as future technologies begin to mature, such as self-assembly and three-dimensional photolithography. The adaptation of nascent materials-processing technologies in the development of scaffold systems, where appropriate, could lead to advances in organ regeneration, with the potential for clinical therapy.

Often, the driving force for scaffold design and fabrication is the desire to create a biomimetic system. This manifests itself as designing materials and processes that try to mimic tissue properties such as microarchitecture, mechanical properties, surface topography, or chemical cues. The notion of the pursuit of biomimicry as effective design criteria for organ regeneration may or may not prove to be the most efficient means. However, for the time being, biomimetic strategies for scaffold fabrication provide challenges that drive the overall improvement of tissue engineering systems and technology. Advancing engineering systems at the present will therefore mature the technology in anticipation of the application of knowledge to be gained by future advancements in unraveling the complexity of biological systems, such as the further elucidation of the proteome and the mapping of cell signaling networks. Hence, the full potential of organ and tissue regeneration therapy in the future may only be realized on combining advanced scaffold fabrication techniques with design criteria outlined by information on fundamental biological processes derived from integrated systems biology.

# VII. REFERENCES

Armani, D. K., and Liu, C. (2000). Microfabrication technology for poly-caprolactone, a biodegradable polymer. *J. Micromech. Microeng.* **10**, 80–84.

Basu, S., and Campagnola, P. J. (2004a). Enzymatic activity of alkaline phosphatase inside protein and polymer structures fabricated via multiphoton excitation. *Biomacromolecules* **5**, 572–579.

Basu, S., and Campagnola, P. J. (2004b). Properties of crosslinked protein matrices for tissue engineering applications synthesized by multiphoton excitation. *J. Biomed. Mater. Res. A* **71A**, 359–368.

Bettinger, C. J., Orrick, B., Misra, A., Langer, R., and Borenstein, J. T. (2006a). Microfabrication of poly(glycerol-sebacate) for contact guidance applications. *Biomaterials* **27**, 2558–2565.

Bettinger, C. J., Weinberg, E. J., Kulig, K. M., Vacanti, J. P., Wang, Y., Borenstein, J. T., and Langer, R. (2006b). Three-dimensional microfluidic tissue engineering scaffolds using a flexible biodegradable polymer. *Adv. Mater.* **18**, 165–169.

Bhatia, S. N., Balis, U. J., Yarmush, M. L., and Toner, M. (1999). Effect of cell–cell interactions in preservation of cellular phenotype: cocultivation of hepatocytes and nonparenchymal cells. *FASEB J.* **13**, 1883–1900.

Bognitzki, M., Frese, T., Steinhart, M., Greiner, A., and Wendorff, J. H. (2001). Preparation of fibers with nanoscaled morphologies: electrospinning of polymer blends. *Polym. Eng. Sci.* **41**, 982–989.

Borenstein, J. T., Terai, H., King, K. R., Weinberg, E. J., Kaazempur-Mofrad, M. R., and Vacanti, J. P. (2002). Microfabrication technology for vascularized tissue engineering. *Biomed. Microdevices* **4**, 167–175.

Cabodi, M., Choi, N. W., Gleghorn, J. P., Lee, C. S. D., Bonassar, L. J., and Stroock, A. D. (2005). A microfluidic biomaterial. *J. Amer. Chem. Soc.* **127**, 13788–13789.

Chen, C. S., Mrksich, M., Huang, S., Whitesides, G. M., and Ingber, D. E. (1997). Geometric control of cell life and death. *Science* **276**, 1425–1428.

Chou, L., Firth, J. D., Uitto, V. J., and Brunette, D. M. (1995). Substratum surface topography alters cell shape and regulates fibronectin mRNA level, mRNA stability, secretion and assembly in human fibroblasts. *J. Cell Sci.* **108**, 1563–1573.

Cooke, M. N., Fisher, J. P., Dean, D., Rimnac, C., and Mikos, A. G. (2002). Use of stereolithography to manufacture critical-sized 3D biodegradable scaffolds for bone ingrowth. *J. Biomed. Mater. Res.* **64B**, 65–69.

Cukierman, E., Pankov, R., and Yamada, K. M. (2002). Cell interactions with three-dimensional matrices. *Curr. Opin. Cell Biol.* **14**, 633–639.

Curtis, A., and Wilkinson, C. (1997). Topographical control of cells. *Biomaterials* **18**, 1573–1581.

Curtis, A. S. G., and Wilkinson, C. D. W. (1998). Reactions of cells to topography. *J. Biomater. Sci. Polym. Ed.* **9**, 1313–1329.

Dalby, M. J., Childs, S., Riehle, M. O., Johnstone, H., Attrossman, S., and Curtis, A. S. G. (2003a). Fibroblast reaction to island topography: changes in cytoskeleton and morphology with time. *Biomaterials* **24**, 927–935.

Dalby, M. J., Riehle, M. O., Yarwood, S. J., Wilkinson, C. D. W., and Curtis, A. S. G. (2003b). Nucleus alignment and cell signaling in fibroblasts: response to a micro-grooved topography. *Exp. Cell Res.* **284**, 274–282.

Dhariwala, B., Hunt, E., and Boland, T. (2004). Rapid prototyping of tissue-engineering constructs using photopolymerizable hydrogels and stereolithography. *Tissue Eng.* **10**, 1316–1322.

Duffy, D. C., McDonald, J. C., Schueller, J. A., and Whitesides, G. M. (1998). Rapid prototyping of microfluidic systems in poly (dimethylsiloxane). *Anal. Chem.* **70**, 4974–4984.

Flemming, R. G., Murphy, C. J., Abrams, G. A., Goodman, S. L., and Nealey, P. F. (1999). Effects of synthetic micro- and nanostructured surfaces on cell behavior. *Biomaterials* **20**, 573–588.

Folch, A., Jo, B. H., Hurtado, O., Beebe, D. J., and Toner, M. (2000). Microfabricated elastomeric stencils for micropatterning cell cultures. *J. Biomed. Mater. Res.* **52**, 346–353.

George, P. M., Lyckman, A. W., LaVan, D. A., Hegde, A., Leung, Y., Avasare, R., Testa, C., Alexander, P. M., Langer, R., and Sur, M. (2005). Fabrication and biocompatibility of polypyrrole implants suitable for neural prosthetics. *Biomaterials* **26**, 3511–3519.

Hollister, S. J. (2005). Porous scaffold design for tissue engineering. *Nat. Mater.* **4**, 518–524.

Jiang, X., Bruzewicz, D. A., Wong, A. P., Piel, M., and Whitesides, G. M. (2005). Directing cell migration with asymmetric micropatterns. *Proc. Natl. Acad. Sci. U.S.A.* **102**, 975–978.

Jin, H. J., Fridrikh, S. V., Rutledge, G. C., and Kaplan, D. L. (2002). Electrospinning *Bombyx mori* silk with poly(ethylene oxide). *Biomacromolecules* **3**, 1233–1239.

Kancharla, V. V., and Chen, S. (2002). Fabrication of biodegradable polymeric microdevices using laser micromachining. *Biomed. Microdevices* **4**, 105–109.

Khademhosseini, A., Yeh, J., Eng, G., Karp, J., Kaji, H., Borenstein, J., Farokhzad, O. C., and Langer, R. (2005). Cell docking inside microwells within reversibly sealed microfluidic channels for fabricating multiphenotype cell arrays. *Lab Chip* **5**, 1380–1386.

King, K. R., Wang, C. C. J., Kaazempur-Mofrad, M. R., Vacanti, J. P., and Borenstein, J. T. (2004). Biodegradable microfluidics. *Adv. Mater.* **16**, 2007–2009.

Kosiorek, A., Kandulski, W., Glaczynska, H., and Giersig, M. (2005). Fabrication of nanoscale rings, dots, and rods by combining shadow nanosphere lithography and annealed polystyrene nanosphere masks. *Small* **1**, 439–444.

Lavan, D. A., George, P. M., and Langer, R. (2003). Simple, three-dimensional microfabrication of electrodeposited structures. *Angew. Chem.* **42**, 1262–1265.

Leclerc, E., Furukawa, K. S., Miyata, F., Sakai, Y., Ushida, T., and Fujii, T. (2004). Fabrication of microstructures in photosensitive biodegradable polymers for tissue engineering applications. *Biomaterials* **25**, 4683–4690.

Lee, K. Y., Peters, M. C., Anderson, K. W., and Mooney, D. J. (2000). Controlled growth factor release from synthetic extracellular matrices. *Nature* **408**, 998–1000.

Levenberg, S., Rouwkema, J., Macdonald, M., Garfein, E. S., Kohane, D. S., Darland, D. C., Marini, R., Blitterwijk, C. A. V., Mulligan, R. C., D'Amore, P. A., et al. (2005). Engineering vascularized skeletal muscle tissue. *Nat. Biotechnol.* **23**, 879–884.

Li, D., Wang, Y., and Xia, Y. (2003). Electrospinning of polymeric and ceramic nanofibers as uniaxially aligned arrays. *Nano Lett.* **3**, 1167–1171.

Liu, V. A., and Bhatia, S. N. (2002). Three-dimensional patterning of hydrogels containing living cells. *Biomed. Microdevices* **4**, 257–266.

Lo, H., Ponticiello, M. S., and Leong, K. W. (1995). Fabrication of controlled-release biodegradable foams by phase separation. *Tissue Eng.* **1**, 15–27.

Lu, Y., Mapili, G., Chen, S. C., and Roy, K. (2003a). *In* "Proceedings of the 30th Annual Meeting and Exposition." Glasgow, Scotland.

Lu, Y., Theppakuttai, S., and Chen, S. C. (2003b). Marangoni effect in nanosphere-enhanced laser nanopatterning of silicon. *Appl. Phys. Lett.* **82**, 4143–4145.

Ma, Z., Kotaki, M., Inai, R., and Ramakrishna, S. (2005). Potential of nanofiber matrix as tissue engineering scaffolds. *Tissue Eng.* **11**, 101–109.

Moghe, P. V., Berthiaume, F., Ezzell, R. M., Toner, M., Tompkins, R. G., and Yarmush, M. L. (1996). Culture matrix configuration and composition in the maintenance of hepatocyte polarity and function. *Biomaterials* **17**, 373–385.

Murphy, W. L., Dennis, R. G., Kileny, J. L., and Mooney, D. J. (2002). Salt fusion: an approach to improve pore interconnectivity within tissue engineering scaffolds. *Tissue Eng.* **8**, 43–52.

Oberpenning, F., Meng, J., Yoo, J. J., and Atala, A. (1999). De novo reconstitution of a functional mammalian urinary bladder by tissue engineering. *Nature* **17**, 149–155.

Odde, D. J., and Renn, M. J. (2000). Laser-guided direct writing of living cells. *Biotechnol. Bioeng.* **67**, 312–318.

Petka, W. A., Harden, J. L., McGrath, K. P., Wirtz, D., and Tirrell, D. A. (1998). Reversible hydrogels from self-assembling artificial proteins. *Science* **281**, 389–392.

Reches, M., and Gazit, E. (2003). Casting metal nanowires within discrete self-assembled peptide nanotubes. *Science* **300**, 625–627.

Richards-Grayson, A. C., Choi, I. S., Tyler, B. M., Wang, P. P., Brem, H., Cima, M. J., and Langer, R. (2003). Multi-pulse drug delivery from a resorbable polymeric microchip device. *Nat. Mater.* **2**, 767–772.

Roy, T. D., and Al, E. (2003). Performance of degradable composite bone repair products made via three-dimensional fabrication techniques. *J. Biomed. Mater. Res.* **66B**, 574–580.

Santini, J. T., Cima, M. J., and Langer, R. (1999). A controlled-release microchip. *Nature* **397**, 335–338.

Scheibel, T., Parthasarathy, R., Sawicki, G., Lin, X. M., Jaeger, H., and Lindquist, S. L. (2003). Conducting nanowires built by controlled self-assembly of amyloid fibers and selective metal deposition. *Proc. Natl. Acad. Sci. U.S.A.* **100**, 4527–4532.

Schmidt, C. E., Shastri, V. R., Vacanti, J. P., and Langer, R. (1997). Stimulation of neurite outgrowth using an electrically conducting polymer. *Proc. Natl. Acad. Sci. U.S.A.* **94**, 8948–8953.

Shea, L. D., Smiley, E., Bonadio, J., and Mooney, D. J. (1999). DNA delivery from polymer matrices for tissue engineering. *Nature* **17**, 551–554.

Simon, J. L., and Al, E. (2003). Engineered cellular response to scaffold architecture in a rabbit trephine defect. *J. Biomed. Mater. Res.* **66A**, 275–282.

Stachowiak, A. N., Bershteyn, A., Tzatzalos, E., and Irvine, D. J. (2005). Bioactive hydrogels with an ordered cellular structure combine interconnected macroporosity and robust mechanical properties. *Adv. Mater.* **14**, 399–403.

Tan, W., and Desai, T. A. (2004). Layer-by-layer microfluidics for biomimetic three-dimensional structures. *Biomaterials* **24**, 1355–1364.

Tan, W., and Desai, T. A. (2005). Microscale multilayer cocultures for biomimetic blood vessels. *J. Biomed. Mater. Res.* **72A**, 146–160.

Tang, P., Qui, F., Zhang, H., and Yang, Y. (2004). Morphology and phase diagram of complex block copolymers: ABC linear triblock copolymers. *Phys. Rev. E* **69**, 1–8.

Vozzi, G., Flaim, C., Ahluwalia, A., and Bhatia, S. (2003). Fabrication of PLGA scaffolds using soft lithography and microsyringe deposition. *Biomaterials* **24**, 2533–2540.

Wang, Y., Ameer, G. A., Sheppard, B. J., and Langer, R. (2002). A tough biodegradable elastomer. *Nat. Biotechnol.* **20**, 602–606.

Whang, K., Thomas, H., and Healy, K. E. (1995). A novel method to fabricate bioabsorbable scaffolds. *Polymer* **36**, 837–841.

Xu, C. Y., Inai, R., Kotaki, M., and Ramakrishna, S. (2004). Aligned biodegradable nanofibrous structure: a potential scaffold for blood vessel engineering. *Biomaterials* **25**, 877–886.

Yang, J., Webb, A. R., and Ameer, G. A. (2004). Novel citric acid–based biodegradable elastomers for tissue engineering. *Adv. Mater.* **16**, 511–515.

Zhang, S. (2002). Emerging biological materials through molecular self-assembly. *Biotechnol. Adv.* **20**, 321–339.

Zhang, S. (2003). Fabrication of novel biomaterials through molecular self-assembly. *Nat. Biotechnol.* **21**, 1171–1178.

Zhong, X. II., Ran, S. F., Fang, D. F., Hsiao, B. S., and Chu, B. (2003). Control of structure, morphology, and property in electrospun poly(glycolide-*co*-lactide) nonwoven membranes via postdraw treatments. *Polymer* **44**, 4959.

# Chapter Twenty-Five

# Three-Dimensional Scaffolds

*Ying Luo, George Engelmayr, Debra T. Auguste, Lino da Silva Ferreira, Jeffrey M. Karp,*
*Rajiv Saigal, and Robert Langer*

## I. INTRODUCTION

Three-dimensional (3D) scaffolds serve as temporary substrates for supporting and guiding tissue formation in various *in vitro* and *in vivo* tissue-regeneration settings. Recent years have seen the emergence of scaffolds designed explicitly to control various aspects of the tissue-formation process. Spatiotemporal control of scaffold biochemical and physical properties has been investigated via incorporating fundamental molecular and cellular biomechanisms into scaffolds. Because living cells and tissues are highly complex systems, a central theme in designing tissue-engineering scaffolds is to understand how scaffold structures, properties, and functions are correlated. In this chapter, fundamental scaffold design variables (i.e., mass transport, mechanics, electrical conductivity, surface chemistry, and topology) are decoupled and analyzed in terms of how they have been engineered to control cell behavior through different constituent biomaterials and architectural structures. Approaches for temporal and spatial control are further described to illustrate principles that may guide integrated rational design of 3D scaffolds.

A key concept in tissue engineering (TE) is using material-based porous 3D scaffolds to provide physical support and a local environment for cells to enable and facilitate tissue development (Langer and Vacanti, 1993). Since the mid-1980s, 3D scaffolds have evolved to serve complex functions of guiding cellular behavior in different TE applications. Scaffolds can be seeded with embryonic or adult stem cells, progenitor cells, mature differentiated cells, or cocultures of cells to induce tissue formation *in vitro* and *in vivo*; scaffolds can also be directly implanted *in vivo* with capabilities to deliver soluble/insoluble biomolecules and temporal/spatial cues to guide the function regeneration in defected tissues and organs. While the specific functions vary with tissue type and clinical need, scaffolds may potentially coordinate biological events at the molecular, cellular, and tissue levels on time and length scales ranging from seconds to weeks and nanometers to centimeters, respectively (Griffith, 2002). A central theme in designing tissue-engineering scaffolds is to understand the correlations between scaffold properties and biological functions.

During tissue development, cells constantly decode and release different morphogenetic factors in their surroundings. In response, cells make decisions on how to divide, differentiate, migrate, degrade/produce extracellular matrix (ECM), and orient themselves. Based on our understanding of cell–ECM interactions in native tissues, it should theoretically be possible to design scaffolds capabilities to promote tissue formation via logical, systematic variations in one or more chemical and/or physical properties. Living cells and tissues, however, are not simple, linear systems. The complex roles that ECM structures play in orchestrating cellular and tissue processes can be thought of as a multiband signal to which different cell types may exhibit radically different sensitivities. Moreover, the performance of any scaffold property must be evaluated not only in the relatively controlled *in vitro* environment, but also in the context of the tissue's physiological function (e.g., under mechanical or electrical stimulation) and ultimately *in vivo*. The dynamic superposition of multiple scaffold properties with multiple environmental factors can profoundly impact the outcome of the tissue formation process. Correlations between individual scaffold properties and certain aspects of tissue formation (e.g., cellular phenotype and concomitant proliferation or ECM production) can be derived, and they have even been applied to the optimization of scaffold designs (Hollister, 2005). From an engineering standpoint, it is desirable that scaffold properties and their regulatory functions be reproducibly integrated within a scaffold design such that tissue formation can be induced and controlled in a stepwise manner *in vitro* and/or at the site of implantation *in vivo* (Lutolf and Hubbell, 2005; Stevens and George, 2005) (Fig. 25.1).

Owing to the innate complexity of biological systems, reproducing natural processes and engineering tissues with the aid of scaffolds presents a formidable task. Complications can arise on many fronts, such as from our incomplete understanding of native cell–ECM interactions, the difficulty of analyzing and standardizing biological systems, and even from our increasing technical capabilities to synthesize scaffolds, which may outstrip our understanding of how best to apply them. More questions can be raised, such as to what degree of complexity a tissue-engineering scaffold should be designed, how to coordinate different types of cellular and tissue events, and how scaffolds can be designed to function robustly in response to patient-specific variations and different clinical situations. In this chapter, we summarize current scaffold design approaches and essential principles that can guide the rational multifunctional design of scaffolds for tissue engineering. Design variables are decoupled and analyzed in terms of how they have been controlled using different materials and fabrication techniques and what implications they have in controlling the tissue formation process.

## II. THREE-DIMENSIONAL SCAFFOLD DESIGN AND ENGINEERING

From the perspective of material properties, mass transport, mechanics, electrical conductivity, surface chemistry, and topology all mediate cell behavior at the nano-, micro-, and macroscopic scales. These properties are often correlated with scaffold chemical compositions and structures in an interdependent way. For example, while a higher porosity may result in a beneficial increase in the mass transport capabilities of a scaffold, it may also compromise the scaffold's mechanical strength. Achieving optimal scaffold functional performance for a particular tissue-engineering application, therefore, requires balancing different factors.

An understanding of the fundamental scaffold properties and functions must form the foundation of any higher-order hierarchical scaffold design involving temporal and/or spatial controls. Temporally, the scaffold properties at the onset of *in vitro* seeding (i.e., time zero) or during the acute phase of the host response following direct implantation *in vivo* will provide an initial set of environmental conditions for cells and tissues. These initial scaffold properties and associated functions will then evolve over time in response to passive and active interactions with the culture environment and cells, respectively. Temporal features, enabled mainly by scaffold degradation and the release of bioactive factors, are obligatory for presenting time-appropriate signals corresponding with each particular stage of tissue development. Spatially, tissues and organs are 3D structures having a characteristic organization consisting of patterns of different types of cells and ECM constituents. Anisotropic biochemical and structural properties are designed in scaffolds to induce cell and ECM orientations and to guide multiscale organizations from molecular and cellular level up to the gross tissue or organ level.

### Mass Transport and Pore Architectures

The mass transport characteristics of a scaffold affect how fluid, solutes, and cells move in and out of a tissue con-

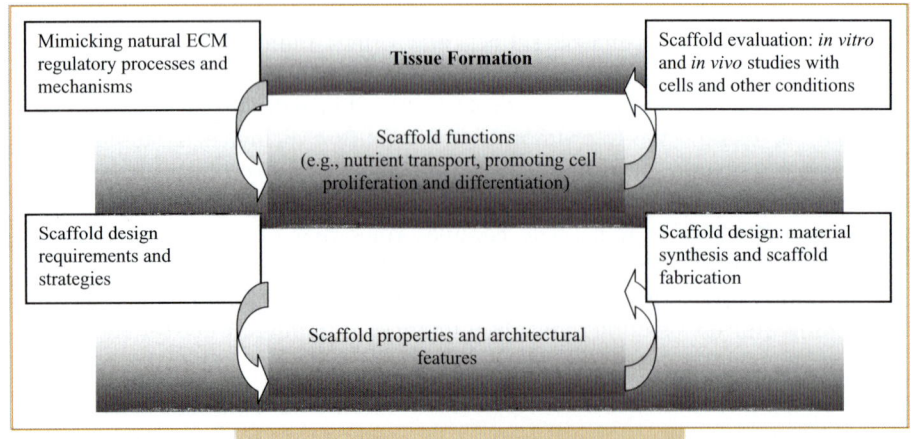

**FIG. 25.1.** Scaffold synthesis consists of processes to understand the design principles and develop enabling methodologies. On one hand, scaffold design is a deductive process entailing the proposition and testing of candidate scaffold properties and the cumulative elucidation of their independent and coupled reactions with cells and tissues. On the other hand, scaffold design is an inductive process, an exercise in recapitulating the wealth of tissue functions and cellular mechanisms observed throughout the body.

struct. As a porous 3D substrate, the mass transport in a scaffold at the initial stage is correlated with the scaffolds chemical compositions and the structural parameters of its pores or voids, characterized mainly by pore size, porosity, pore interconnectivity/tortuosity, and surface area. These scaffold characteristics interdependently affect the diffusive or convective behavior of soluble nutrients, growth factors, and cytokines. In addition to structural parameters, the local biochemical environment cellular activities can also influence the scaffold behavior (e.g., pH dependence of the hydrolysis of poly(lactic-*co*-glycolic acid) (PLGA) scaffolds). In scaffolds designed to deliver bioactive soluble factors, the release profile and distribution pattern of these factors will be influenced by the mass transport characteristics of the scaffold. Different from the vascularized tissues containing convective flows, diffusion is often the primary mass transport mechanism in engineered tissue constructs cultivated *in vitro* and prior to inosculation with the host vasculature following implantation *in vivo*. Theoretical models to simulate and predict the nutrient, protein, and cell diffusion processes in scaffolds have been developed (Botchwey *et al.*, 2003; Malda *et al.*, 2004).

Once a scaffold is impregnated with cells, the solute diffusion and distribution profiles are influenced mainly by the cell organization. In metabolically active tissues *in vivo*, most cells reside within 100 μm of a capillary. In engineered tissue constructs cultivated in static culture without medium perfusion, neotissue formation is generally limited to the peripheral 100–200 μm of the scaffold, due to diffusion limitations (Colton, 1995; Ishaug *et al.*, 1997; Freed *et al.*, 1998). Subsequently, the organization and density of cells in this peripheral region can further influence the distribution and availability of nutrients to cells within the scaffold interior (Malda *et al.*, 2004). Although certain types of cells may tolerate nutrient deficiency to some extent (such as chondrocytes under hypoxia), the induction of rapid vascularization following implantation is considered essential to establishing nutrient exchange and to the retention of a viable cell population within an engineered tissue construct. The structural characteristics of the pores and void spaces in scaffolds are the primary variables governing the initial cell distribution and organization in scaffolds.

One main parameter that affects the efficiency of initial cell impregnation is the pore size of the scaffold. The resulting geometries and spatial characteristics of individual cells or cell aggregates within the void space of scaffolds direct the subsequent cell proliferation, differentiation, and tissue formation events. As observed in nonwoven fibrous scaffolds made of poly(ethylene terephthalate) (PET) and seeded with human placenta trophoblast cells *in vitro*, the size of cell aggregates increased with larger pore volumes between fibers (T. Ma *et al.*, 1999). Variations in cell behavior were also observed, with attenuated differentiation and suppressed proliferation activities in larger cell aggregates compared to smaller ones (T. Ma *et al.*, 1999, 2000). Optimal

scaffold pore sizes have been suggested for regenerating various types of tissues *in vivo*. The critical pore size was found to be above 500 μm for rapid fibrovascularization in poly(L-lactic acid) (PLLA) scaffolds with cylindrical pores (Wake *et al.*, 1994). In ECM analog scaffolds (made of cross-linked collagen and glycosaminoglycans [GAGs]), the average pore diameters required to induce dermis and peripheral nerve regeneration were within 20–120 μm and 5–10 μm, respectively (Yannas, 2005). In bone engineering, early studies by Hulbert *et al.* (1970) showed that a minimum pore size of 100 μm was required to allow bone tissue ingrowth in ceramic scaffolds. Further investigations were carried out to understand the pore size requirement for bone tissue engineering. Although the optimal pore size varied with scaffold material and other parameters, such as tortuosity, the general consensus is that larger pore sizes (e.g., >100 μm) may favor higher alkaline phosphatase activity and more bone formation (Tsuruga *et al.*, 1997; Karageorgiou and Kaplan, 2005).

Cell transport and vascularization as a result of scaffold pore size can also affect the tissue types and tissue formation process in scaffolds. When bone morphogenetic proteins were loaded into honeycomb-shaped hydroxyapatite scaffolds to induce osteogenesis, it was found that smaller diameters (90–120 μm) induced cartilage formation followed by bone formation, whereas those with larger diameters (350 μm) induced bone formation directly (Kuboki *et al.*, 2001). The difference was likely caused by the different onset time of vascularization and cell differentiation.

In addition to pore size, cell transport within a scaffold such as diffusion, attachment, and migration are controlled by porosity (the fraction of pore volume), pore interconnectivity, and available surface area in scaffolds. While a high porosity is often desired, it is inversely related to the surface area available for cell attachment in 3D scaffolds. Achieving an optimal cell density in scaffolds therefore necessitates a high surface-area-to-volume ratio. In order to facilitate the transport of cells and bioactive chemicals, scaffolds may also need to have pores at both macro and micro scales, features that may be difficult to obtain via traditional scaffold fabrication techniques, such as particle leaching, gas foaming, and phase separation. Rapid-prototyping techniques such as solid free-form fabrication (SFF) are emerging as important methods to generate highly controlled scaffold structures. Compared to scaffolds fabricated with traditional methods, the pore size and tortuosity in rapid-prototyped scaffolds have much narrower variations in structural distribution. Local topologies can also potentially be optimized by computational algorithms to control the permeability and mechanical properties (Hollister, 2005). Studies have been carried out to compare scaffolds fabricated by controlled processes with those containing irregular structures fabricated by conventional methods. In one such study, scaffolds with similar porosities were prepared by particle leaching or by 3D fiber deposition and compared

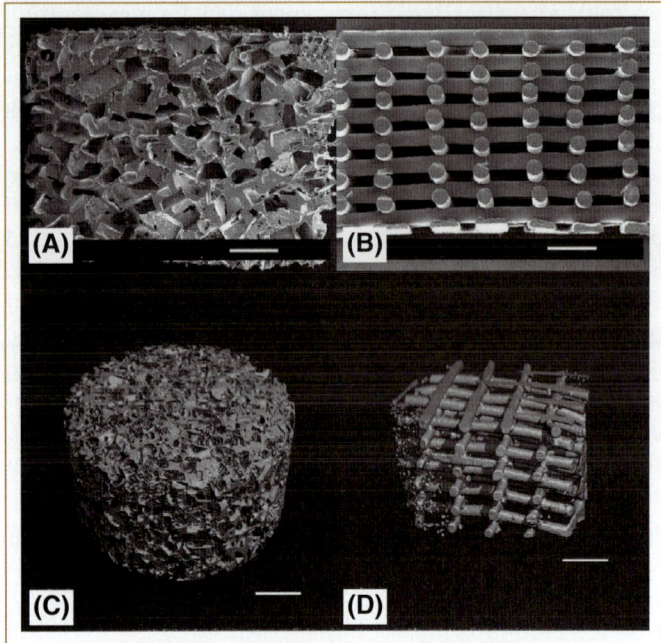

**FIG. 25.2.** Designer scaffolds offer opportunities to control scaffold properties such as pore structures, surface area, and mechanical strength. Electron micrographs of scaffolds fabricated by the conventional methods of compression molding and particle leaching **(A)** and 3D fiber deposition **(B)**. **(C)** and **(D)** 3D images of scaffolds in A and B, respectively. Reprinted from Malda *et al.* (2005), with permission from Elsevier.

for their ability to support cartilage tissue growth (Malda *et al.*, 2005) (Fig. 25.2). A significantly higher glycosaminoglycan (GAG) content was observed in the scaffolds fabricated by the controlled 3D fiber deposition process. Besides the fibrous morphology and highly accessible pores, more uniform and efficacious cell diffusion/attachment may also have contributed to the observed up-regulation of GAG production and the better scaffold function. Advanced scaffolds fabricated by tightly controlled methods, with more uniform and controllable structures and properties, therefore hold promise in promoting the reproducible formation of functional engineered tissues.

## Mechanics

The mechanical properties of the natural ECM are of paramount importance in dictating macroscopic tissue functions (e.g., bearing load) and regulating cellular behavior via mechanotransduction signaling. In designing tissue constructs, scaffold mechanical properties are often sought that resemble native tissue properties. Foremost, in the acute phase following implantation the scaffold must fulfill the key mechanical functions of the tissue being replaced. For example, the earliest TE blood vessels based on cell-contracted collagen gels were not strong enough to withstand physiologic blood pressures, and thus they had to be reinforced by a tubular synthetic polymer mesh to ensure structural integrity (Weinberg and Bell, 1986). More recent

TE blood vessels based on relatively strong nonwoven poly(glycolic acid) (PGA) scaffolds exhibited burst pressures exceeding physiological requirements upon implantation (>2000 mmHg) (Niklason *et al.*, 1999). In addition to appropriately matching gross tissue mechanical properties, the scaffold must also provide an internal micromechanical environment conducive of the *de novo* synthesis and organization of ECM. For example, while nonwoven PGA scaffolds have been successfully employed in blood vessel (Niklason *et al.*, 1999) and heart valve TE (Sutherland *et al.*, 2005), their application to myocardial tissue has been comparatively challenging (Papadaki *et al.*, 2001). In contrast to the predominant load-bearing functions of vessels and valves, the primary function of myocardial tissue is cyclic contraction. While the out-of-plane compressive modulus of a typical nonwoven PGA scaffold is relatively low (~6.7 ± 0.5 kPa) (Kim and Mooney, 1998), the in-plane tensile and compressive moduli resisting cardiomyocyte-mediated contraction are comparable, at ~284 ± 34 kPa (Engelmayr and Sacks, 2006). To understand the role of material elasticity on cell behavior, myoblasts were cultured on collagen strips attached to glass or polymer gels of varied elasticity (Engler *et al.*, 2004). Cells were found to differentiate into a striated, contractile phenotype only on substrates within a very narrow range of musclelike stiffnesses (i.e., 8–11 kPa) (Fig. 25.3). To optimize a scaffold design for a particular application requires consideration of the gross organ and tissue-level functional requirements as well as the micromechanical requirements for appropriate tissue formation at the cellular level.

The mechanical properties of TE scaffolds are determined in part by the bulk properties of their constituent materials (e.g., modulus of elasticity, degradation rate). For example, most hydrogel materials exhibit a much lower strength and stiffness than hydrophobic polyester materials. Because traditional PLGA-based scaffolds have a limited subset of mechanical properties, new biodegradable materials have been developed, such as poly(hydroxyalkanoates) and poly(glycerol sebacates), to improve scaffold toughness and elasticity (Zinn *et al.*, 2001; Wang *et al.*, 2003). Because of the high porosity and concomitant low material content, the mechanical properties of TE scaffolds are very often dictated primarily by the structural arrangement of their constituent materials (e.g., pore size, fiber diameter, and orientation; Table 25.1) and associated modes of structural degeneration (e.g., fiber fragmentation, bond disruption). For example, in a recent study the effective stiffness (**E**) (equivalent to initial tensile modulus) of nonwoven PGA scaffolds was predictably modulated by tuning the fiber diameter via NaOH-mediated hydrolysis (Engelmayr and Sacks, in press).

In addition to the initial structure imparted during the fabrication of the TE scaffold, the modes of structural degeneration manifested by the scaffold need to be considered. For example, while 50 : 50 blend PGA/PLLA scaffolds do

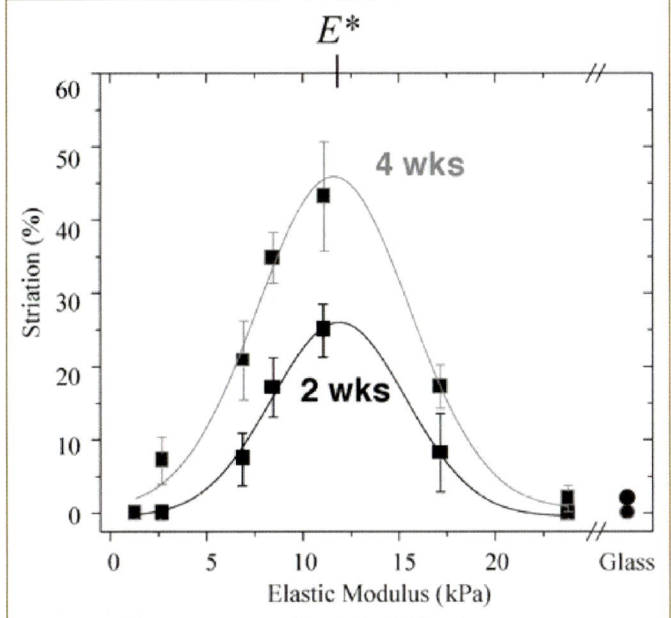

**FIG. 25.3.** To understand the role of material elasticity on cell behavior, myoblasts were cultured on collagen strips attached to glass or poly(acrylamide) gels of varied elasticity. Substrate stiffness was found to have a profound influence on myocyte differentiation, with optimal differentiation (as assessed by striation) occurring within a very narrow range of musclelike stiffnesses (i.e., 8–11 kPa) (Engler *et al.*, 2004). These results suggest that cell differentiation within 3D tissue-engineering scaffolds may exhibit a similar sensitivity to the local micromechanical properties. Figure reproduced with permission of the Rockefeller University Press.

not undergo significant mechanical degeneration over a period of three weeks (Engelmayr *et al.*, 2005), scaffolds dip-coated with the biologically derived thermoplastic poly(4-hydroxybutyrate) (P4HB) incur a rapid loss of rigidity with cyclic flexural mechanical loading as the P4HB bonds between fibers are disrupted (Engelmayr *et al.*, 2003). Depending on the kinetics of scaffold hydrolysis, structural degeneration may be more pronounced and thus such kinetics represent a more important consideration in scaffold design.

Because the foremost role of the scaffold following implantation is temporarily to fulfill the key mechanical functions of the replaced tissue, it is essential to consider the physiological loading state of the native tissue. While the physiological loading state may be highly complex and virtually impossible to reproduce *ex vivo*, certain mechanical testing configurations are generally more relevant than others. For example, the physiological loading state of a semilunar heart valve leaflet depends on time-varying solid–fluid coupling (i.e., leaflet tissue–blood) and includes multiaxial flexural, tensile, and fluid shear stress components. In light of the strong planar anisotropy and trilayered structures exhibited by native leaflet tissues, biaxial tensile testing (Grashow *et al.*, 2006) and flexural testing (Mirnajafi *et al.*, 2005) have been employed to characterize their behavior.

Because engineered tissues based on synthetic polymer scaffolds inherently begin development as composite materials, their effective mechanical properties will be determined by the combined effects of the cells, ECM, and scaffold and their unique micromechanical interactions. Thus, the appropriate formulation and validation of a mathematical model to simulate and/or predict the mechanical properties of a scaffold are critical prerequisites for developing a mathematical model to simulate and/or predict the

**Table 25.1.** Dependence of a scaffold mechanical property (initial tensile modulus) on the bulk material mechanical property and scaffold structure[a]

| Material | Initial tensile modulus (MPa) | | |
| --- | --- | --- | --- |
| | Fibrous scaffold | Foam scaffold | Bulk material |
| Poly(glycolic acid) (PGA) | 0.284 ± 0.034 (nonwoven) (Engelmayr and Sacks, in press) | 0.919 ± 0.067[b] (salt leach) (Beatty *et al.*, 2002) | 18,780 ± 3430 (fiber) (Engelmayr and Sacks, in press) |
| Poly(ester urethane)urea (PEUU) | 8 ± 2 (electrospun) (Stankus *et al.*, 2004) | ~1.4[c] (TIPS[d]) (Guan *et al.*, 2005) | 60 ± 10 (film) (Stankus *et al.*, 2004) |
| Poly(glycerol sebacate) (PGS) | N/A | 0.004052 ± 0.0013 (salt leach) (Gao *et al.*, in press) | 0.282 ± 0.0250 (film) (Wang *et al.*, 2002) |

[a]Several order-of-magnitude differences in modulus can be realized by starting with different bulk materials and/or by converting the bulk material into different porous scaffold structures (e.g., foam or fibrous). For comparison, the initial tensile modulus of a typical passive muscle tissue was reported to be 0.012 ± 0.004 MPa (Engler *et al.*, 2004).
[b]Aggregate modulus obtained from creep indentation testing of PGA-PLLA scaffold.
[c]Estimated from PEUU1020 stress–strain curve (Fig. 3, Guan *et al.*, 2005).
[d]Thermally induced phase separation.

mechanical properties of an engineered tissue. While standard phenomenological models may be useful in characterizing the gross mechanical behavior of a scaffold or engineered tissue construct (i.e., for meeting organ- and tissue-level functional requirements), only an appropriately formulated structural-based model can be used to design the micromechanical environment presented at the cellular level. Structural-based models can be either computationally driven, as in the work of Hollister et al. (Hollister *et al.*, 2002; Hollister, 2005) or purely analytical, as in the case of a model for nonwoven scaffolds by Engelmayr and Sacks (in press). Irrespective of the solution method, the goals of structure-based modeling are both to bridge the gap between the disparate length scales of cells, tissues, and scaffolds and, in particular, to simulate accurately the micromechanical environment presented at the cellular level. For example, while traditional rule-of-mixtures theories accounting for the volume fractions and orientations of individual composite constituents are often invoked in describing native and engineered tissues (Gibson, 1994), higher-order reinforcement effects observed via the structural-based modeling of commercially available nonwoven PGA and PLLA scaffolds preclude the use of rule-of-mixtures approaches. These higher-order reinforcement effects, which yield proportional increases in fiber effective stiffness with increased ECM stiffness, predict a very different micromechanical environment than that predicted by a rule of mixtures, highlighting the importance of an accurate micromechanical representation of the TE scaffold.

## Electrical Conductivity

Electrical conduction is an important mechanism that enables cellular signaling and function in many types of tissues. The cardiac electrical conduction system is essential to maintaining synchronous beats that pump blood in an ordered fashion. In the process of bone regeneration, naturally occurring piezoelectric properties of the apatite crystal are hypothesized to generate electric fields involved in bone remodeling. The nervous system possesses the well-known system of electrochemical signaling. Much research has been carried out using materials to record from and influence bioelectric fields. In making tissue scaffolds, electrically conductive biomaterials have been studied to understand their abilities to interface with bioelectrical fields in cells and tissues to replicate normal electrophysiology.

A wide variety of electrically conductive polymers is available to the tissue engineer, each with differing characteristics that may direct the choice for a given application. These organic compounds include poly(pyrrole), poly(vinylidene fluoride), poly(tetrafluoroethylene), poly(aniline), poly(thiophene), and poly(acetylene). Such polymers generally contain delocalized pi bonds and can be considered semiconductors, with conductivity determined by degree of doping. Polypyrrole, an aromatic heterocycle, is perhaps the most widely studied conductive biomaterial,

due to both ease of fabrication and demonstrated biocompatibility. In contrast to static charge seen in conductive materials like poly(pyrrole), piezoelectric materials such as poly(vinylidene fluoride) display transient charge in response to mechanical deformation. For transplantation purposes, it is often desirable to have the material biodegrade after cell delivery and incorporation into the host environment have been accomplished. Degradable conductive polymers have been synthesized using several different approaches, including incorporation of three-substituted hydrolyzable side groups (Zelikin *et al.*, 2002), degradable ester linkages connecting oligomers of pyrrole and thiophene (Rivers *et al.*, 2002), and emulsion/precipitation of poly(pyrrole) in poly(D,L-lactide) (Shi *et al.*, 2004).

In order to tailor conductive biomaterials to a given application, scalable technologies have been developed to fabricate conductive structures of arbitrary geometry. Due to horizontal and vertical growth of poly(pyrrole) during electropolymerization, two-dimensional photolithographically defined gold layers can be used to pattern three-dimensional structures (LaVan *et al.*, 2003; George *et al.*, 2005). Hollow polymer tubes ranging from the nano to the micro scale have been fabricated using a range of techniques, including polymerization via electrodes on opposite ends of a silicone tube (Chen *et al.*, 2000) or via oxidation on platinum wire molds followed by reduction (Saigal *et al.*, 2005).

A growing number of reports are demonstrating synergistic effects of conductive biomaterials and electrical stimulation. Neonatal cardiomyocytes cultured on collagen sponges and matrigel show synchronized contraction in response to applied electric fields (Radisic *et al.*, 2004; Gerecht-Nir *et al.*, 2006). Applied potentials can also be used to change surface properties of conductive polymers, altering cell shape and function, including DNA synthesis and extension (Wong *et al.*, 1994). Others have shown that electrical stimulation promotes neurite outgrowth on conductive polymers beyond that seen on indium-tin oxide, an inorganic conductive substrate (Schmidt *et al.*, 1997). Composite conductive polymer films have been manufactured to include substrates that direct cell function, such as hyaluronic acid to promote angiogenesis (Collier *et al.*, 2000).

The mechanistic basis for each of these electrical–material–tissue interactions is not fully understood and will be an important area for future study. However, some leading hypotheses have emerged. This enhanced function of engineered cardiac tissues may be due to greater ultrastructural organization in response to electric fields (Radisic *et al.*, 2004). Increased neurite outgrowth with electrical stimulation may be caused by better ECM protein adsorption (Kotwal and Schmidt, 2001) rather than direct effects on the cell itself, although there is ample evidence of the latter. Electrophoretic redistribution of cell surface receptors likely governs the galvanotropic response of neurons to a horizontally oriented two-dimensional applied field

(Patel and Poo, 1982). Such mechanisms do not fully explain altered cell function in response to stimulation applied to the substrate or material relative to the medium or a distant ground. For such depolarization, signaling through voltage-gated calcium channels can activate ubiquitous second messengers such as cAMP and alter gene transcription, affecting learning, memory, survival, and growth (West *et al.*, 2001). Such secreted gene products might be used to enhance the survival of host cells surrounding an implanted scaffold. Given that applied electric fields can so profoundly affect cell function, conductive three-dimensional scaffolds will be important tools for harnessing this interaction to create functional tissues.

## Surface Properties

Cells interact with scaffolds primarily through the material surface, which can be dominated by the surface chemical and topological features. The surface chemistry here refers to the insoluble chemical environment that the scaffold surface presents to cells. Such environment is dictated by the biochemical compositions of the bulk and/or the substances resulting from the surface adsorption or chemical reactions. Besides mediating cell behavior and functions inside scaffolds, controlled surface properties are of central importance in directing the inflammatory and immunological responses. Controlled surface properties may be useful for ameliorating the foreign-body reaction at the host–scaffold interface *in vivo* (Mikos *et al.*, 1998; Hu *et al.*, 2001).

### Surface Chemistry

Each type of synthetic, natural, or composite scaffold gives rise to a set of distinct surface chemical characteristics governed by the material chemistry and its physical form (such as crystallinity, charge, and topology). Although numerous efforts have been made to tailor the scaffold surface, the chemical environment can exhibit extremely complicated patterns within the biological milieu. Complex processes, such as the spontaneous adsorption of a diversity of proteins from biological fluids to the scaffold surface and the protein surface conformation, are difficult to analyze, though they exert profound effects on the scaffold performance. To tailor the scaffold chemical properties, the interactions of scaffolds with different environmental factors need to be considered.

Scaffolds derived from natural ECM materials, such as collagen, fibrin, hyaluronic acid (HA), proteoglycans, or their composites, have the advantages of directly containing innate biological ligands that cells can recognize and provide natural mechanisms for tissue remodeling. ECM analogs have been created to emulate an appropriate tissue-regeneration environment. For example, as an essential ECM component in natural cartilage tissue, collagen type II scaffolds may have better biochemical properties to maintain chondrocyte phenotype and enhance the biosynthesis of glycosaminoglycans compared to collagen type I (Nehrer

*et al.*, 1997). Fibrin, the native provisional matrix of blood clots, can provide ligands to initiate cell attachment and ECM remodeling (Hubbell, 2003). HA plays a role in morphogenesis, inflammation, and wound repair, and the cell–ECM interactions mediated through receptors such as CD44 and RHAMM can be activated in scaffolds with HA constituents (Hubbell, 2003). The natural cell–ECM interactions, however, can also alter after purification, manufacturing, and scaffold fabrication processes. For example, acid-treated type I collagen polymers contained only 5% residual crystallinity as compared to native collagen. The loss of crystalline structure led to platelets binding with diminished degranulation, which in turn limited the myofibroblast numbers and the related inflammatory response at the injured site (Yannas, 2005).

Synthetic scaffolds offer a variety of mechanisms to modulate cell behavior. Chemical reactions with biological fluid remodel the scaffold surface and affect tissue growth through both reaction dynamics and kinetics. Studies have shown that bioactive glasses (Class A bioglass composed of 45–52% $SiO_2$, 20–24% CaO, 20–24% $Na_2O$, and 6% $P_2O_5$) have osteoproductive properties superior to those of either bioactive hydroxyapatite or bioinert metals and plastics. The difference was found due to the surface reaction kinetics in physiological fluid. The rapid reaction rate that converts amorphous silicate to polycrystalline hydroxyl-carbonate apatite on the bioglass surface is the key to positively regulating the cell cycle and bone formation (Hench *et al.*, 2000) (Fig. 25.4A). Physical processes also play active roles in controlling the material–cell interface. Because of the adsorbed protein moieties from serum or body fluid, many polyester-based scaffolds, such as those made from poly(α-hydoxy esters), exhibit adequate adhesion to support cell attachment and tissue growth in some *in vitro* and *in vivo* applications. Methods that alter the surface hydrophobicity, e.g., by changing monomer compositions or by chemical surface treatment, can potentially improve the scaffold performance (Mikos *et al.*, 1994; Gao *et al.*, 1998; Harrison *et al.*, 2004). For example, biodegradable foams of hydrophobic polymers (e.g., PLLA and PLGA) can be efficiently wet by two-step immersion in ethanol and water. This surface treatment could overcome the hindered entry of water into air-filled pores to facilitate cell seeding (Mikos *et al.*, 1994).

Surface modifications of scaffolds have been developed to generate surface chemical specificity and recognition. The surface chemistry can be created by either incorporating bioactive moieties directly in the scaffold bulk or modifying the surface. These moieties bound to scaffolds trigger desired specific intracellular signaling. In particular, many synthetic and natural hydrogel materials (e.g., poly[ethylene glycol], poly[vinyl alcohol], alginate, and dextran) are protein repellent, and immobilizing biomolecules to such hydrogel scaffolds may be especially useful in tailoring the surface chemistry for cell–material interactions at the molecular level.

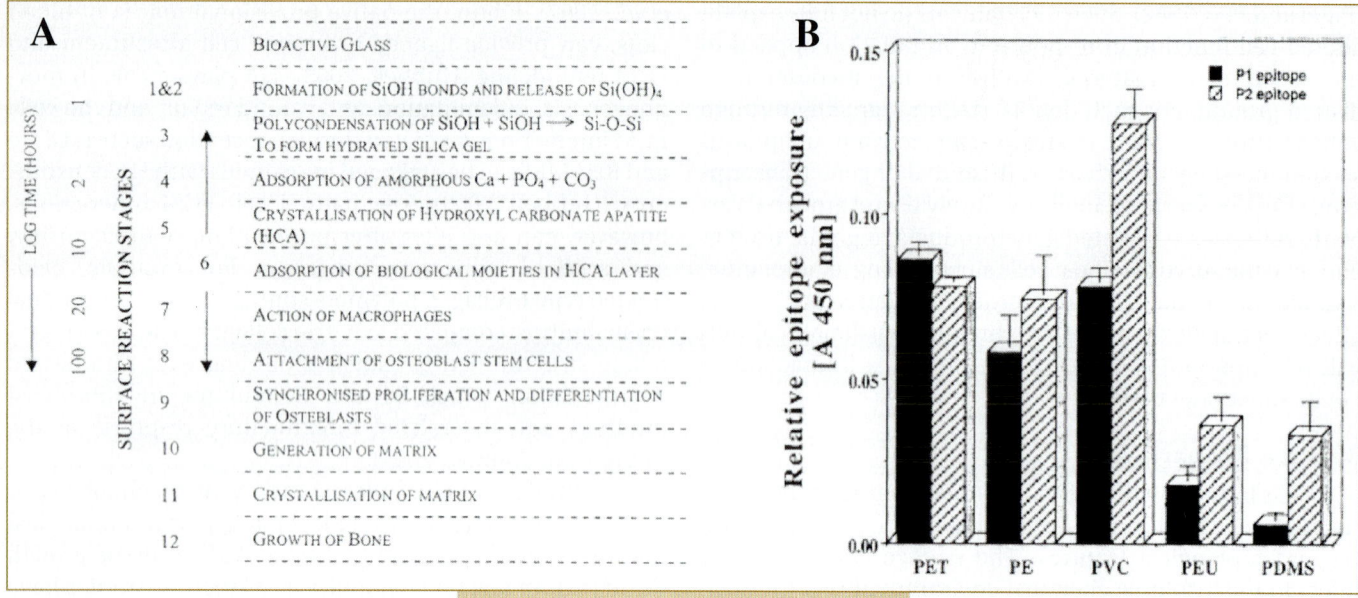

**FIG. 25.4.** Understanding the molecular basis of surface interactions is essential to controlling surface functions. **(A)** A series of surface reactions on Class A bioglass (Hench *et al.*, 2004). Figure reproduced with permission from Springer. **(B)** Dynamic and kinetic mechanisms of material interactions with proteins can lead to different cell response at the scaffold surface. The example shows the exposure of two epitopes (P1 and P2) in fibrinogen on a surface varied with the substrate materials (PET, PE, PVC, PEU, and PDMS disks coated with human fibrinogen) (Hu *et al.*, 2001). Figure reproduced with permission from the American Society of Hematology.

Cell adhesion mediated through extracellular adhesive proteins is involved in many intracellular signaling pathways that regulate most fundamental cell behaviors, including differentiation, proliferation, and migration. Enriching scaffold surfaces with specific ECM-derived adhesion proteins has been widely applied to scaffold modification. PLGA-based scaffolds have been coated with fibronectin by physical adsorption for supporting growth and differentiation of human embryonic stem cells in 3D (Levenberg *et al.*, 2003). Fibronectin was covalently attached to PVA hydrogels for improved cell adhesion, proliferation, and migration (Nuttelman *et al.*, 2001). Fibrinogen was also denatured and fused into the backbone of a PEG hydrogel material (Seliktar, 2005) to elicit cellular responses. Elucidating the underlying molecular mechanisms on scaffold surface, however, is not a trivial task. Fundamental studies have been carried out to understand how the adsorption and denaturing of proteins can lead to different cellular responses at the material surface and may provide a molecular basis to control cell–material interaction and specificity for rational scaffold design (Hu *et al.*, 2001) (Fig. 25.4B).

Immobilizing peptide ligands derived from the active domains of ECM adhesion proteins to scaffolds is another major approach to generate specific surface-bound biological signals. For example, integrins, the principal adhesion receptor mediating cell–ECM attachment, comprise a family of more than 20 subtypes of heterodimeric transmembrane proteins. Each of them recognizes and interacts with certain types of ECM adhesion proteins to activate a cascade of signaling pathways to regulate essential cell activities and functions (Plow *et al.*, 2000). Integrins can be activated by short peptides in similar ways, e.g., Arg-Gly-Asp (RGD) from fibronectin and Tyr-Ile-Gly-Ser-Arg (YIGSR) from laminin. Provided that peptides are relatively stable and economical to use, incorporating them into scaffolds has become an important way to generate surface biomimicry and enhance tissue regeneration (Hersel *et al.*, 2003; H. Shin *et al.*, 2003). To introduce peptide moieties, the scaffold materials need to contain appropriate functional groups, which may not be available in most hydrophobic polyesters. Methods have been developed to functionalize polyesters. For example, poly(lactic acid-*co*-lysine) has been synthesized, and the RGD peptide can be immobilized through the side-chain amine groups of the lysine residues (Barrera *et al.*, 1993; Cook *et al.*, 1997).

Like the biomolecules in natural ECM, the functions of immobilized bioactive ligands in modulating membrane receptors and intracellular signaling are influenced by their spatial characteristics. For example, integrin affinity to ECM affects cell attachment and migration. As a result, 3D neurite migration demonstrated a biphasic dependence on RGD concentration, with intermediate adhesion site densities (between 0.2 and 1.7 mol of peptide/mol of fibrinogen) yielding maximal neurite extension as compared with higher densities, which inhibited outgrowth (Schense and Hubbell, 2000). In another study, integrin clustering, a prerequisite to many integrin-mediated signaling pathways, was recapitulated by RGD nanoclusters immobilized on a comb–polymer substrate (Maheshwari *et al.*, 2000).

## Surface Topography

The material–cell interactions mediated through topographical features have traditionally been studied through planar substrates. Surface modification techniques, including photolithography, contact printing, and chemical treatments, have been developed to generate micro- and nano-scale surface topographical features. Surface topographical features such as ridges, steps, and grooves were found to guide cytoskeletal assembly and cell orientation. Surfaces with textures such as nodes, pores, or random patterns are often associated with marked changes of cell morphology, cell activities, and the production of autocrine/paracrine regulatory factors as compared to smooth surfaces (Flemming *et al.*, 1999). In general, surface roughness increases cell adhesion, migration, and the production of ECM. Cells sense and respond to topographical features in a dimension-dependent way. As demonstrated on titanium surfaces, whereas microtextures increased osteoblast attachment and growth, only the presence of nanoscale roughness led to enhanced cell differentiation in connection with elevated growth-factor production (Zinger *et al.*, 2005).

Current fabrication techniques can be used to generate a wide variety of topographical features in scaffolds. Scaffolds can be randomly packed with regular or irregular geometries and shapes (e.g., particles, pellets, and fibers) or condensed with amorphous structures (e.g., foam and sponge) or fabricated with specifically designed architectures. Based on the cell–material interactions, scaffolds provide different topographical properties correlated with the dimensions of scaffold geometries and shapes. When the feature size is larger than or comparable to that of cells, e.g., the fiber diameter in a nonwoven mesh and pores and walls in a foam, the scaffold may provide curved surfaces for cell attachment. Pore size and surface area constitute the major topological features in an extracellular environment. As demonstrated in a study of mesenchymal stem cells seeded onto polymeric fibrous fabrics, increased fiber diameter favored cell attachment and proliferation by providing more surface area (Takahashi and Tabata, 2004). Surface treatment techniques such as sodium hydroxide etching have been used to generate nanoscale roughness to increase cell adhesion, growth, and ECM production (Pattison *et al.*, 2005).

The size scale of most natural ECM components, e.g., fibrous elastin and collagen, fall into the range of several to tens of nanometers. The extracellular environment is dominated by nanoscale topographical features, such as nanopores, ridges, fibers, ligand clusters, and high surface-area-to-volume ratios in 3D. Such native topographies can be recapitulated to a degree in scaffolds made of natural ECM polymers, such as collagen and elastin. Because synthetic materials have the advantage of greater control over scaffold properties, interest is growing in developing synthetic nanofibrous scaffolds. Three-dimensional PLLA-based scaffolds containing nanofibers have been produced by thermally induced phase separation processes (Woo *et al.*, 2003; F. Yang *et al.*, 2004), and selective surface adsorption of adhesion proteins was observed. In a more versatile method, electrospinning techniques have been used to fabricate a variety of synthetic and natural materials with different hydrophobicities into fibers with diameters ranging from a few to hundreds of nanometers (Z. Ma *et al.*, 2005). Another important approach involves building scaffolds from the bottom up. Polypeptides made of 12–16 amino acids have been designed to form hydrogel scaffolds through β-sheet assembling (Zhang, 2002). Amphiphilic molecules consisting of a hydrophilic peptide head and a hydrophobic alkyl tail self-assemble into nanocylinders to form interwoven scaffolds (Hartgerink *et al.*, 2002; Silva *et al.*, 2004).

Nanofibrous scaffolds have demonstrated abilities to support cell and tissue growth. For example, when cardiomyocytes were cultured in meshes made of electrospun poly(ε-caprolactone) nanofibers, they expressed cardiac-specific markers and were contractile in 3D scaffolds (M. Shin *et al.*, 2004). In a scaffold based on self-assembled peptide amphiphilic molecules containing the laminin epitope IKVAV, neural progenitor cells selectively differentiated into neurons (Silva *et al.*, 2004). The potential of designing nanoscale topographies in scaffolds remains largely unknown with regard to exactly how cell cycle, gene expression patterns, and other cell activities are regulated. Some possible mechanisms may be related to cell receptor regulation (clustering, density, and ligand-binding affinity) on nanofibers, nutrient gradients in nanoporous matrices, mechanotransduction induced by the unique matrix mechanics, and the conformation of adhered proteins for cellular recognition sites.

## Temporal Control

### Scaffold Degradation

Unlike permanent or slowly degrading implants, which may serve to augment or replace organ function (e.g., hip implants, artificial hearts, or craniofacial plates), tissue-engineering scaffolds serve as temporary devices to facilitate the tissue healing and regeneration process. The regeneration of a fully functional tissue ideally coincides in time with complete scaffold degradation and resorption. Controlling degradation mechanisms allows scaffolds to cooperate temporally with cell and tissue events via changes in scaffold properties and functions. Tuning the scaffold degradation rate to make it kinetically match with the evolving environment during tissue healing and regeneration is an important design criterion. Due to the multiple roles of scaffolds, the interrelations between scaffold property variables, and the different wound conditions in individual patients, it remains a challenge to design degradation properties that can be tailored to meet various clinical tissue regeneration requirements.

During scaffold degradation, some scaffold properties and functions may weaken or diminish with time. In general, there exist lower and upper limits on the optimal degradation rate, which may vary with different cellular or tissue processes, scaffold chemical compositions, and scaffold functions. For example, scaffolds often need to serve mechanical functions, such as in bone implants to support compressive loading while maintaining an environment permissive to new bone formation. If a material degrades prior to transferring mechanical load to the new tissue, the therapy would fail (K. Y. Lee *et al.*, 2001). Alternatively, materials in bone implants that degrade too slowly may cause stress shielding, thereby impeding the regeneration process and potentially endangering surrounding tissues (Cristofolini, 1997). In skin wound models, healing can be compromised when the scaffold degradation occurs too quickly, whereas scar tissue occurs when the degradation is too slow (Yannas, 2005). The optimal skin synthesis and prevention of scar formation could be achieved when the template was replaced by new tissue in a synchronous way; i.e., the time constant for scaffold degradation ($t_d$) and the time constant for new tissue synthesis during wound healing ($t_h$) were approximately equal (Yannas, 2005). Matching tissue formation with material degradation thus requires coupling of specific temporal aspects of tissue formation processes with chemical properties of the scaffold.

Scaffold degradation can occur through mechanisms that involve physical or chemical processes and/or biological processes that are mediated by biological agents, such as enzymes in tissue remodeling. Degradation results in scaffold dismantling and material dissolution/resorption through the scaffold bulk and/or surface. In the passive degradation mode, the degradation is often triggered by reactions that cleave the polymer backbone or cross-links within the polymer network. Many polyester scaffolds made of lactic acid and glycolic acid, e.g., PLLA and PLGA, undergo bulk backbone degradation due to their wettability and water penetration through the surface. Hydrophilic scaffolds such as hydrogels made of natural or synthetic materials cross-linked by hydrolyzable bonds (e.g., ester, carbonate, or hydrazone bonds) also convert to soluble degradation products, predominantly through the bulk (K. Y. Lee *et al.*, 2000; Ferreira *et al.*, 2005). Chemical degradation can be conveniently varied through scaffold physical and chemical properties, such as the backbone hydrophobicity, crystallinity, glass transition temperature, and cross-link density. Because of this flexibility, the degradation rate can be engineered principally for optimal tissue regeneration (K. Y. Lee *et al.*, 2001; Tognana *et al.*, 2005).

Scaffolds degrading through passive mechanisms exhibit limited capabilities to match with tissue growth and wound healing. In bulk degradation, the accumulation of degradation products may exert adverse effects on tissue, e.g., acidic products from PLGA degradation. It is also difficult to tailor the degradation to match the healing rate, which may vary with wound conditions, such as age of the patient, severity of the defect, and presence of other diseases. In order to exert more control over the degradation properties of a scaffold and to attempt to tailor the degradation of the scaffolds, consideration of pertinent wound-healing and tissue-regeneration mechanisms is required. For example, would healing is a highly complex yet orchestrated cascade of events, controlled by a vast array of cytokines and growth factors, that generally involves three phases, including inflammation, granulation tissue formation, and remodeling of the ECM. Scaffolds should be designed to degrade *in vivo* during the formation of granulation tissue and/or during the remodeling process. Ideally these materials should withstand uncontrolled dissolution or degradation at physiologic conditions while being resorbed by natural cell-mediated processes. Many inorganic scaffolds for bone tissue engineering demonstrate biodegradable and bioresorbable characteristics to facilitate new tissue formation (Pietrzak and Ronk, 2000; Yuan *et al.*, 2000; Hench *et al.*, 2004).

To integrate natural biological mechanisms of ECM remodeling, a new class of hydrogels that degrade in response to proteases have been developed (Gobin and West, 2002; Lutolf *et al.*, 2003a). In these scaffolds, degradation occurs through cellular proteolytic activities mediated by enzymes such as collagenase and plasmin. In one of these studies, a poly(ethylene glycol) (PEG) hydrogel modified with adhesion ligands was cross-linked with molecules containing matrix metalloproteinase (MMP) peptide substrates. Migration of human primary fibroblasts inside the gel was observed and found to be dependent on the substrate sensitivity to the enzyme (Lutolf *et al.*, 2003b). When used for delivering recombinant human bone morphogenetic protein-2 (rhBMP-2) into critical-sized defects in rat crania, the PEG hydrogel matrix was remodeled through the MMP-mediated mechanism and supported bone regeneration within five weeks (Lutolf *et al.*, 2003c). The approach demonstrated a paradigm of how scaffold degradation and intervention can be engineered to synchronize with wound healing and new tissue synthesis via natural mechanisms.

## Delivery of Soluble Bioactive Factors

The incorporation of delivery systems in 3D scaffolds offers an indispensable platform for enabling temporal and spatial control in tissue constructs. Compared to systemic administration, using a local controlled-release system to deliver soluble inductive and therapeutic factors has the advantages of preventing rapid factor clearance, metering factors in a desired pharmacokinetic manner, and allowing therapeutic doses for an appropriate duration while limiting side effects at unwanted body locations. Although numerous material-based controlled-release systems are at the disposal of tissue engineers, most of them need to be adapted when applied to tissue scaffolds. The release of soluble factors from scaffolds can be mediated through single or multiple mechanisms, e.g., by diffusion, dissolution, scaffold or carrier degradation, or external stimuli. In particular, delivery of growth factors has been studied for various tissues, due to their important roles in instructing cell behavior.

Built on established particle-based delivery systems, one common method for controlling the release from a scaffold involves prefabricating biofactor-loaded particles and embedding them into a scaffold matrix (Hedberg *et al.*, 2002; J. E. Lee *et al.*, 2004). The release of biofactors in these systems can be delayed with a minimized burst effect, compared to the particles used alone. Typical particle carriers include PLGA and hydrogel microspheres. This method takes advantage of established systems but involves double matrices, which influence the release profile. Alternatively, soluble factors may be incorporated directly into the scaffold itself without a secondary carrier/matrix. This often requires the scaffold to be fabricated under mild physiological conditions to preserve the bioactivity of proteins or other biofactors. Growth factors and proteins have been incorporated in scaffolds through surface coating (Park *et al.*, 1998), emulsion freezing-drying (Whang *et al.*, 2000), gas-foaming/particulate leaching (Murphy *et al.*, 2000; Jang and Shea, 2003), and nanofiber electrospinning (Luu *et al.*, 2003; Z. Ma *et al.*, 2005). Different delivery profiles of growth factors or DNA plasmids were achieved. Due to their hydrophilic and biocompatible nature, hydrogel scaffolds are amenable to incorporating proteins and plasmid DNA, yielding both higher loading efficiencies and bioactivity as compared to PLGA-based materials. Biofactors have been immobilized in hydrogel matrices via physical interactions and/or covalent chemical bonds for prolonging retention time and controlling release via designed mechanisms (Sakiyama-Elbert and Hubbell, 2000; Tabata, 2003).

Scaffolds that integrate controlled-release methods have been used in conjunction with scaffolds for a variety of purposes, including enhancing tissue formation, stimulating angiogenesis, guiding cell differentiation, and facilitating wound healing (Babensee *et al.*, 2000; Tabata, 2003). Delivering growth factors from scaffolds has demonstrated advantages over using the free form directly (Yamamoto *et al.*, 2000). Synergistic effects on accelerating tissue regeneration have been observed when scaffolds, cells, and growth factors are combined. For example, autologous bone marrow–derived cells transplanted with scaffolds containing bone morphogenetic protein-7 resulted in the greatest bone formation as compared to constructs without either growth factor or cells (Borden *et al.*, 2004). A major challenge in delivering biofactors involves achieving meaningful

pharmacokinetic delivery. The dosage, release kinetics, and duration time should be optimized and tailored to tissue growth/healing mechanisms. For example, VEGF acts as an initiator of angiogenesis, while PDGF provides essential stimuli for blood vessel maturation. To simulate the process, PLGA-based scaffolds have been developed to deliver these two angiogenic factors with distinctive kinetics for rapid formation of a mature vascular network (Richardson *et al.*, 2001).

Instead of releasing soluble growth factors directly into the environment, they can also be initiated by cellular activities. Many growth factors are tightly sequestered in the ECM as inactive precursors and released through their interaction with cells via specific protease-mediated mechanisms. To simulate the biological process, growth factors were covalently conjugated to hydrogel matrices via proteolytically cleavable linkages. The immobilized growth factors remained active, and their release was elicited by cells through plasmin activity (Sakiyama-Elbert *et al.*, 2001; Zisch *et al.*, 2003). In these systems, the mode of growth factor release could be varied and controlled by cellular activities to achieve precise temporal control in different clinical situations.

## Spatial Control

Tissues consist of hierarchically ordered structures of cells and ECM; an important tissue-engineering design principle is incorporating spatial cues into 3D scaffolds to guide structural tissue formation. Such guidance involves designing anisotropic scaffold properties. By generating directional variations in cell–cell and cell–ECM communications, various cell behaviors and new ECM depositions can potentially be guided.

At the macroscopic or tissue level, scaffolds are configured to have appropriate geometries that correspond to tissue/organ anatomical features. The scaffolds are seeded with cells and/or direct the ingrowth of cells from host tissues to promote spatially compartmentalized new tissue growth and wound healing. One of the first methods introduced to generate macroscopic, anatomical shapes utilized highly interconnected pore structures from laminating porous membranes of PLLA and PLGA (Mikos *et al.*, 1993). In another example, polymeric conduits were constructed for growing blood vessels and guiding nerve tissue regeneration. Smooth muscle cells and endothelial cells have been seeded onto tubular biodegradable PGA scaffolds as an approach for engineering vascular grafts (Niklason *et al.*, 1999). Nerve-guidance channels are used for connecting damaged nerve stumps. The entubulation strategy has demonstrated abilities to guide axonal spouting, directing growth-factor diffusion, and blocking undesired fibrous-tissue ingrowth. To fabricate scaffolds with more complex shapes, rapid-prototyping techniques use tissue or defect images recorded by medical imaging modalities, such as computed tomography (CT) and magnetic resonance imaging (MRI) (Hutmacher, 2001; S. Yang *et al.*, 2002; Hollister, 2005). Another approach to shape scaffold materials to fit into individual tissue defects relies on the liquid–solid transformation process. *In situ* forming hydrogels solidify on external stimuli (e.g., chemicals, light, and pH) and have been designed as injectable scaffolds for minimally invasive cell and biomolecule transplantation. For example, methacrylated PEG and HA polymers were able to photopolymerize *in situ* with chondrocytes and support neocartilage formation *in vivo* (Elisseeff *et al.*, 1999; Burdick *et al.*, 2005).

At the microscopic or cellular level, various techniques have been developed to couple cell-/tissue-guidance mechanisms to regenerate tissues that require directional cell growth (e.g., nerve). Anisotropic characteristics in pore architectures, mechanics, surface properties, degradation, and delivery can potentially generate signals recapitulating haptotaxis and chemotaxis for guiding cell behavior and ECM deposition in 3D. Various scaffolds have been developed containing aligned longitudinal regions. For example, PLGA conduit devices containing physical channels have been created by low-pressure injection molding; oriented lumen surfaces facilitated and guided Schwann cell attachment for peripheral nerve regeneration (Hadlock *et al.*, 2000). Guidance can also be generated through fibrous topographical features. Aligned nanofibers made of poly(L-lactide-*co*-ε-caprolactone) were fabricated by an electrospinning technique, and the oriented fiber structure elicited directional growth of smooth muscle cells. Fibrils in natural scaffold materials, such as collagen and fibrin, have been aligned using magnetic fields (Dubey *et al.*, 1999, 2001). The resulting scaffolds increased the rate and depth of axonal elongation *in vitro* and improved sciatic nerve regeneration *in vivo* as compared to scaffolds with random fiber orientations (Dubey *et al.*, 1999).

Creating heterogeneous chemistry in scaffolds is another approach that has been explored to achieve spatial control for tissue guidance. Adhesive RGD peptides have been photoimmobilized in selected regions of agarose hydrogel matrices (Luo and Shoichet, 2004). The patterns of adhesive and nonadhesive regions induced oriented axonal elongation and migration from dorsal root ganglion cell aggregates *in vitro*. Studies have also been carried out to incorporate chemical gradients in scaffolds. Gradients of proteins play important roles in tissue formation/remodeling during embryogenesis and wound healing. Combining fluidic systems and *in situ* forming hydrogel materials, concentration gradients of peptides and proteins have been generated in 3D matrices and exhibited abilities to modu-

late cell functions. For example, entrapment of nerve growth factor in a poly(2-hydroxyethyl methacrylate) hydrogel induced directional axonal growth from PC12 cells *in vitro* (Kapur and Shoichet, 2004). A microfluidic device was used to create gradients of immobilized molecules and cross-linking densities in photo-cross-linked hydrogels; the gradients of immobilized RGD adhesion ligands modulated the spatial distribution of attached endothelial cells (Burdick *et al.*, 2004).

Creating spatial features involves processing of scaffolds to integrate different control mechanisms. To this end, biomaterials with improved processing may need to be combined with different fabrication techniques so that architectural structures, biomolecules, and cells can all be combined in a desired manner. For example, to generate complex tissue patterns, it is desirable to position cells with defined microstructures. Encapsulation of cells using photopolymerizable hydrogel materials has been combined with stereolithography in rapid prototyping to create programmed cell organization in 3D (Tsang and Bhatia, 2004). In designing devices for spinal cord injury repair, as molecular-, cellular-, and tissue-level treatments are discovered, the combinations of such treatments will be necessary to synergistically promote tissue regeneration. Multiple-channel, biodegradable scaffolds have been fabricated with capabilities locally to deliver molecular agents and control cell spatial distribution for transplantation (Moore *et al.*, 2006).

## III. CONCLUSIONS

Tissue-engineering scaffolds need to be built with functions to interact with cells at different spatial and temporal scales to invoke complex, tissuelike patterns. Since the mid-1980s, scaffold design criteria have evolved from simply inducing tissue formation to explicitly controlling tissue formation. Tissue engineers have at their disposal an ever-broadening array of techniques to fabricate scaffolds incorporating spatially and temporally varying biochemical and physical cues. Based on our collective understanding of natural cellular and tissue processes, optimal integration of these scaffold structures and properties should in principle allow us to explicitly control the tissue formation process. The challenge, therefore, is to develop a system-level understanding of how fundamental scaffold properties (e.g., mass transport, mechanics, electrical conductivity, and surface properties) are interrelated in affecting cell behavior and how they can be rationally programmed — spatially and temporally — to provide the necessary signals at the right time and place to aid tissue formation/regeneration.

# IV. REFERENCES

Babensee, J. E., McIntire, L. V., et al. (2000). Growth factor delivery for tissue engineering. *Pharm. Res.* **17**(5), 497–504.

Barrera, D. A., Zylstra, E., et al. (1993). Synthesis and RGD peptide modification of a new biodegradable copolymer-poly(lactic acid-*co*-lysine). *J. Am. Chem. Soc.* **115**(23), 11010–11011.

Borden, M., Attawia, M., et al. (2004). Tissue-engineered bone formation *in vivo* using a novel sintered polymeric microsphere matrix. *J. Bone Joint. Surg. Br.* **86**(8), 1200–1208.

Botchwey, E. A., Dupree, M. A., et al. (2003). Tissue-engineered bone: measurement of nutrient transport in three-dimensional matrices. *J. Biomed. Mater. Res. A* **67**(1), 357–367.

Burdick, J. A., Khademhosseini, A., et al. (2004). Fabrication of gradient hydrogels using a microfluidics/photopolymerization process. *Langmuir* **20**(13), 5153–5156.

Burdick, J. A., Chung, C., et al. (2005). Controlled degradation and mechanical behavior of photopolymerized hyaluronic acid networks. *Biomacromolecules* **6**(1), 386–391.

Chen, S. J., Wang, D. Y., et al. (2000). Template synthesis of the polypyrrole tube and its bridging *in vivo* sciatic nerve regeneration. *J. Mater. Sci. Lett.* **19**, 2157–2159.

Collier, J. H., Camp, J. P., et al. (2000). Synthesis and characterization of polypyrrole-hyaluronic acid composite biomaterials for tissue-engineering applications. *J. Biomed. Mater. Res.* **50**(4), 574–584.

Colton, C. K. (1995). Implantable biohybrid artificial organs. *Cell Transplant* **4**(4), 415–436.

Cook, A. D., Hrkach, J. S., et al. (1997). Characterization and development of RGD-peptide-modified poly(lactic acid-*co*-lysine) as an interactive, resorbable biomaterial. *J. Biomed. Mater. Res.* **35**(4), 513–523.

Cristofolini, L. (1997). A critical analysis of stress shielding evaluation of hip prostheses. *Crit. Rev. Biomed. Eng.* **25**(4–5), 409–483.

Dubey, N., Letourneau, P. C., et al. (1999). Guided neurite elongation and Schwann cell invasion into magnetically aligned collagen in simulated peripheral nerve regeneration. *Exp. Neurol.* **158**(2), 338–350.

Dubey, N., Letourneau, P. C., et al. (2001). Neuronal contact guidance in magnetically aligned fibrin gels: effect of variation in gel mechano-structural properties. *Biomaterials* **22**(10), 1065–1075.

Elisseeff, J., Anseth, K., et al. (1999). Transdermal photopolymerization for minimally invasive implantation. *Proc. Natl. Acad. Sci. U.S.A.* **96**(6), 3104–3107.

Engelmayr, G. C., Jr., and Sacks, M. S. (2006). A structural model for the flexural mechanics of nonwoven tissue-engineering scaffolds. *J. Biomech. Eng.* **128**(4), 610–622.

Engelmayr, G. C., Jr., Hildebrand, D. K., et al. (2003). A novel bioreactor for the dynamic flexural stimulation of tissue-engineered heart valve biomaterials. *Biomaterials* **24**(14), 2523–2532.

Engelmayr, G. C., Jr., Rabkin, E., et al. (2005). The independent role of cyclic flexure in the early *in vitro* development of an engineered heart valve tissue. *Biomaterials* **26**(2), 175–187.

Engler, A. J., Griffin, M. A., et al. (2004). Myotubes differentiate optimally on substrates with tissue-like stiffness: pathological implications for soft or stiff microenvironments. *J. Cell Biol.* **166**(6), 877–887.

Ferreira, L., Gil, M. H., et al. (2005). Biocatalytic synthesis of highly ordered degradable dextran-based hydrogels. *Biomaterials* **26**(23), 4707–4716.

Flemming, R. G., Murphy, C. J., et al. (1999). Effects of synthetic micro- and nano-structured surfaces on cell behavior. *Biomaterials* **20**(6), 573–588.

Freed, L. E., Hollander, A. P., et al. (1998). Chondrogenesis in a cell–polymer–bioreactor system. *Exp. Cell Res.* **240**(1), 58–65.

Gao, J., Niklason, L., et al. (1998). Surface hydrolysis of poly(glycolic acid) meshes increases the seeding density of vascular smooth muscle cells. *J. Biomed. Mater. Res.* **42**(3), 417–424.

George, P. M., Lyckman, A. W., et al. (2005). Fabrication and biocompatibility of polypyrrole implants suitable for neural prosthetics. *Biomaterials* **26**(17), 3511–3519.

Gerecht-Nir, S., Radisic, M., et al. (2006). Biophysical regulation during cardiac development and application to tissue engineering. *Int. J. Dev. Biol.* **50**(2–3), 233–243.

Gibson, R. F. (1994). "Principles of Composite Material Mechanics." McGraw-Hill, New York.

Gobin, A. S., and West, J. L. (2002). Cell migration through defined, synthetic extracellular matrix analogues. *FASEB J.* **26**, 26.

Grashow, J. S., Yoganathan, A. P., et al. (2006). Biaxial stress–stretch behavior of the mitral valve anterior leaflet at physiologic strain rates. *Ann. Biomed. Eng.* 1–11.

Griffith, L. G. (2002). Emerging design principles in biomaterials and scaffolds for tissue engineering. *Ann. N.Y. Acad. Sci.* **961**, 83–95.

Hadlock, T., Sundback, C., et al. (2000). A polymer foam conduit seeded with Schwann cells promotes guided peripheral nerve regeneration. *Tissue Eng.* **6**(2), 119–127.

Harrison, J., Pattanawong, S., et al. (2004). Colonization and maintenance of murine embryonic stem cells on poly(alpha-hydroxy esters). *Biomaterials* **25**(20), 4963–4970.

Hartgerink, J. D., Beniash, E., et al. (2002). Peptide–amphiphile nanofibers: a versatile scaffold for the preparation of self-assembling materials. *Proc. Natl. Acad. Sci. U.S.A.* **99**(8), 5133–5138.

Hedberg, E. L., Tang, A., et al. (2002). Controlled release of an osteogenic peptide from injectable biodegradable polymeric composites. *J. Controlled Release* **84**(3), 137–150.

Hench, L. L., Polak, J. M., et al. (2000). Bioactive materials to control cell cycle. *Mat. Res. Innovat.* **3**, 313–323.

Hench, L. L., Xynos, I. D., et al. (2004). Bioactive glasses for *in situ* tissue regeneration. *J. Biomater. Sci. Polym. Ed.* **15**(4), 543–562.

Hersel, U., Dahmen, C., et al. (2003). RGD-modified polymers: biomaterials for stimulated cell adhesion and beyond. *Biomaterials* **24**(24), 4385–4415.

Hollister, S. J. (2005). Porous scaffold design for tissue engineering. *Nat. Mater.* **4**(7), 518–524.

Hollister, S. J., Maddox, R. D., et al. (2002). Optimal design and fabrication of scaffolds to mimic tissue properties and satisfy biological constraints. *Biomaterials* **23**(20), 4095–4103.

Hu, W. J., Eaton, J. W., et al. (2001). Molecular basis of biomaterial-mediated foreign body reactions. *Blood* **98**(4), 1231–1238.

Hubbell, J. A. (2003). Materials as morphogenetic guides in tissue engineering. *Curr. Opin. Biotechnol.* **14**(5), 551–558.

Hulbert, S. F., Young, F. A., et al. (1970). Potential of ceramic materials as permanently implantable skeletal prostheses. *J. Biomed. Mater. Res.* **4**(3), 433–456.

Hutmacher, D. W. (2001). Scaffold design and fabrication technologies for engineering tissues — state of the art and future perspectives. *J. Biomater. Sci. Polym. Ed.* **12**(1), 107–124.

Ishaug, S. L., Crane, G. M., et al. (1997). Bone formation by three-dimensional stromal osteoblast culture in biodegradable polymer scaffolds. *J. Biomed. Mater. Res.* **36**(1), 17–28.

Jang, J. H., and Shea, L. D. (2003). Controllable delivery of nonviral DNA from porous scaffolds. *J. Controlled Release* **86**(1), 157–168.

Kapur, T. A., and Shoichet, M. S. (2004). Immobilized concentration gradients of nerve growth factor guide neurite outgrowth. *J. Biomed. Mater. Res.* **68**(2), 235–243.

Karageorgiou, V., and Kaplan, D. (2005). Porosity of 3D biomaterial scaffolds and osteogenesis. *Biomaterials* **26**(27), 5474–5491.

Kim, B. S., and Mooney, D. J. (1998). Engineering smooth muscle tissue with a predefined structure. *J. Biomed. Mater. Res.* **41**(2), 322–332.

Kotwal, A., and Schmidt, C. E. (2001). Electrical stimulation alters protein adsorption and nerve cell interactions with electrically conducting biomaterials. *Biomaterials* **22**(10), 1055–1064.

Kuboki, Y., Jin, Q., et al. (2001). Geometry of carriers controlling phenotypic expression in BMP-induced osteogenesis and chondrogenesis. *J. Bone Joint Surg. Am.* **83-A**(Suppl 1, Pt 2), S105–S115.

Langer, R., and Vacanti, J. P. (1993). Tissue engineering. *Science* **260**(5110), 920–926.

LaVan, D. A., George, P. M., et al. (2003). Simple, three-dimensional microfabrication of electrodeposited structures. *Angew. Chem., Int. Ed. Engl.* **42**(11), 1262–1265.

Lee, J. E., Kim, K. E., et al. (2004). Effects of the controlled-released TGF-beta 1 from chitosan microspheres on chondrocytes cultured in a collagen/chitosan/glycosaminoglycan scaffold. *Biomaterials* **25**(18), 4163–4173.

Lee, K. Y., Bouhadir, K. H., et al. (2000). Degradation behavior of covalently cross-linked poly(aldehyde guluronate) hydrogels. *Macromolecules* **33**, 97–101.

Lee, K. Y., Alsberg, E., et al. (2001). Degradable and injectable poly(aldehyde guluronate) hydrogels for bone tissue engineering. *J. Biomed. Mater. Res.* **56**(2), 228–233.

Levenberg, S., Huang, N. F., et al. (2003). Differentiation of human embryonic stem cells on three-dimensional polymer scaffolds. *Proc. Natl. Acad. Sci. U.S.A.* **100**(22), 12741–12746.

Luo, Y., and Shoichet, M. S. (2004). A photolabile hydrogel for guided three-dimensional cell growth and migration. *Nat. Mater.* **3**(4), 249–253.

Lutolf, M. P., and Hubbell, J. A. (2005). Synthetic biomaterials as instructive extracellular microenvironments for morphogenesis in tissue engineering. *Nat. Biotechnol.* **23**(1), 47–55.

Lutolf, M. P., Lauer-Fields, J. L., et al. (2003a). Synthetic matrix metalloproteinase-sensitive hydrogels for the conduction of tissue regeneration: engineering cell-invasion characteristics. *Proc. Natl. Acad. Sci. U.S.A.* **100**(9), 5413–5418.

Lutolf, M. P., Raeber, G. P., et al. (2003b). Cell-responsive synthetic hydrogels. *Adv. Mater.* **15**, 888–892.

Lutolf, M. P., Weber, F. E., et al. (2003c). Repair of bone defects using synthetic mimetics of collagenous extracellular matrices. *Nat. Biotechnol.* **21**(5), 513–518.

Luu, Y. K., Kim, K., et al. (2003). Development of a nanostructured DNA delivery scaffold via electrospinning of PLGA and PLA-PEG block copolymers. *J. Control Release* **89**(2), 341–353.

Ma, T., Li, Y., et al. (1999). Tissue engineering human placenta trophoblast cells in 3D fibrous matrix: spatial effects on cell proliferation and function. *Biotechnol. Prog.* **15**(4), 715–724.

Ma, T., Li, Y., et al. (2000). Effects of pore size in 3D fibrous matrix on human trophoblast tissue development. *Biotechnol. Bioeng.* **70**(6), 606–618.

Ma, Z., Kotaki, M., et al. (2005). Potential of nanofiber matrix as tissue-engineering scaffolds. *Tissue Eng.* **11**(1–2), 101–109.

Maheshwari, G., Brown, G., et al. (2000). Cell adhesion and motility depend on nanoscale RGD clustering. *J. Cell Sci.* **113**(Pt 10), 1677–1686.

Malda, J., Woodfield, T. B., et al. (2004). The effect of PEGT/PBT scaffold architecture on oxygen gradients in tissue-engineered cartilaginous constructs. *Biomaterials* **25**(26), 5773–5780.

Malda, J., Woodfield, T. B., et al. (2005). The effect of PEGT/PBT scaffold architecture on the composition of tissue-engineered cartilage. *Biomaterials* **26**(1), 63–72.

Mikos, A. G., Sarakinos, G., et al. (1993). Laminated three-dimensional biodegradable foams for use in tissue engineering. *Biomaterials* **14**(5), 323–330.

Mikos, A. G., Lyman, M. D., et al. (1994). Wetting of poly(L-lactic acid) and poly(DL-lactic-co-glycolic acid) foams for tissue culture. *Biomaterials* **15**(1), 55–58.

Mikos, A. G., McIntire, L. V., et al. (1998). Host response to tissue-engineered devices. *Adv. Drug Deliv. Rev.* **33**(1–2), 111–139.

Mirnajafi, A., Raymer, J., et al. (2005). The effects of collagen fiber orientation on the flexural properties of pericardial heterograft biomaterials. *Biomaterials* **26**(7), 795–804.

Moore, M. J., Friedman, J. A., et al. (2006). Multiple-channel scaffolds to promote spinal cord axon regeneration. *Biomaterials* **27**(3), 419–429.

Murphy, W. L., Peters, M. C., et al. (2000). Sustained release of vascular endothelial growth factor from mineralized poly(lactide-co-glycolide) scaffolds for tissue engineering. *Biomaterials* **21**(24), 2521–2527.

Nehrer, S., Breinan, H. A., et al. (1997). Canine chondrocytes seeded in type I and type II collagen implants investigated *in vitro*. *J. Biomed. Mater. Res.* **38**(2), 95–104.

Niklason, L. E., Gao, J., et al. (1999). Functional arteries grown *in vitro*. *Science* **284**(5413), 489–493.

Nuttelman, C. R., Mortisen, D. J., et al. (2001). Attachment of fibronectin to poly(vinyl alcohol) hydrogels promotes NIH3T3 cell adhesion, proliferation, and migration. *J. Biomed. Mater. Res.* **57**(2), 217–223.

Papadaki, M., Bursac, N., et al. (2001). Tissue engineering of functional cardiac muscle: molecular, structural, and electrophysiological studies. *Am. J. Physiol. Heart Circ. Physiol.* **280**(1), H168–H178.

Park, Y. J., Ku, Y., et al. (1998). Controlled release of platelet-derived growth factor from porous poly(L-lactide) membranes for guided tissue regeneration. *J. Controlled Release* **51**(2–3), 201–211.

Patel, N., and Poo, M. M. (1982). Orientation of neurite growth by extracellular electric fields. *J. Neurosci.* **2**(4), 483–496.

Pattison, M. A., Wurster, S., et al. (2005). Three-dimensional, nanostructured PLGA scaffolds for bladder tissue replacement applications. *Biomaterials* **26**(15), 2491–2500.

Pietrzak, W. S., and Ronk, R. (2000). Calcium sulfate bone void filler: a review and a look ahead. *J. Craniofac. Surg.* **11**(4), 327–333; discussion 334.

Plow, E. F., Haas, T. A., et al. (2000). Ligand binding to integrins. *J. Biol. Chem.* **275**(29), 21785–21788.

Radisic, M., Park, H., et al. (2004). Functional assembly of engineered myocardium by electrical stimulation of cardiac myocytes cultured on scaffolds. *Proc. Natl. Acad. Sci. U.S.A.* **101**(52), 18129–18134.

Richardson, T. P., Peters, M. C., et al. (2001). Polymeric system for dual growth-factor delivery. *Nat. Biotechnol.* **19**, 1029–1034.

Rivers, T. J., Hudson, T. W., et al. (2002). Synthesis of a novel, biodegradable electrically conducting polymer for biomedical applications. *Adv. Funct. Mater.* **12**(1), 33–37.

Saigal, R., George, P. M., et al. (2005). "Conductive Polymer Scaffolds for Delivery of Human Neural Stem Cells." Society for Neuroscience, Washington, DC.

Sakiyama-Elbert, S. E., and Hubbell, J. A. (2000). Controlled release of nerve growth factor from a heparin-containing fibrin-based cell ingrowth matrix. *J. Controlled Release* **69**(1), 149–158.

Sakiyama-Elbert, S. E., Panitch, A., et al. (2001). Development of growth factor fusion proteins for cell-triggered drug delivery. *FASEB J.* **15**(7), 1300–1302.

Schense, J. C., and Hubbell, J. A., (2000). Three-dimensional migration of neurites is mediated by adhesion site density and affinity. *J. Biol. Chem.* **275**(10), 6813–6818.

Schmidt, C. E., Shastri, V. R., et al. (1997). Stimulation of neurite outgrowth using an electrically conducting polymer. *Proc. Natl. Acad. Sci. U.S.A.* **94**(17), 8948–8953.

Seliktar, D. (2005). Extracellular stimulation in tissue engineering. *Ann. N.Y. Acad. Sci.* **1047**, 386–394.

Shi, G., Rouabhia, M., et al. (2004). A novel electrically conductive and biodegradable composite made of polypyrrole nanoparticles and polylactide. *Biomaterials* **25**(13), 2477–2488.

Shin, H., Jo, S., et al. (2003). Biomimetic materials for tissue engineering. *Biomaterials* **24**(24), 4353–4364.

Shin, M., Ishii, O., et al. (2004). Contractile cardiac grafts using a novel nanofibrous mesh. *Biomaterials* **25**(17), 3717–3723.

Silva, G. A., Czeisler, C., et al. (2004). Selective differentiation of neural progenitor cells by high-epitope-density nanofibers. *Science* **303**(5662), 1352–1355.

Stevens, M. M., and George, J. H. (2005). Exploring and engineering the cell surface interface. *Science* **310**(5751), 1135–1138.

Sutherland, F. W., Perry, T. E., et al. (2005). From stem cells to viable autologous semilunar heart valve. *Circulation* **111**(21), 2783–2791.

Tabata, Y. (2003). Tissue regeneration based on growth factor release. *Tissue Eng.* **9**(Suppl 1), S5–S15.

Takahashi, Y., and Tabata, Y. (2004). Effect of the fiber diameter and porosity of nonwoven PET fabrics on the osteogenic differentiation of mesenchymal stem cells. *J. Biomater. Sci. Polym. Ed.* **15**(1), 41–57.

Tognana, E., Padera, R. F., et al. (2005). Development and remodeling of engineered cartilage-explant composites *in vitro* and *in vivo*. *Osteoarthritis Cartilage* **13**(10), 896–905.

Tsang, V. L., and Bhatia, S. N. (2004). Three-dimensional tissue fabrication. *Adv. Drug Deliv. Rev.* **56**(11), 1635–1647.

Tsuruga, E., Takita, H., et al. (1997). Pore size of porous hydroxyapatite as the cell-substratum controls BMP-induced osteogenesis. *J. Biochem. (Tokyo)* **121**(2), 317–324.

Wake, M. C., Patrick, C. W., Jr., et al. (1994). Pore morphology effects on the fibrovascular tissue growth in porous polymer substrates. *Cell Transplant.* **3**(4), 339–343.

Wang, Y., Kim, Y. M., et al. (2003). *In vivo* degradation characteristics of poly(glycerol sebacate). *J. Biomed. Mater. Res. A* **66**(1), 192–197.

Weinberg, C. B., and Bell, E. (1986). A blood vessel model constructed from collagen and cultured vascular cells. *Science* **231**(4736), 397–400.

West, A. E., Chen, W. G., et al. (2001). Calcium regulation of neuronal gene expression. *Proc. Natl. Acad. Sci. U.S.A.* **98**(20), 11024–11031.

Whang, K., Goldstick, T. K., et al. (2000). A biodegradable polymer scaffold for delivery of osteotropic factors. *Biomaterials* **21**(24), 2545–2551.

Wong, J. Y., Langer, R., et al. (1994). Electrically conducting polymers can noninvasively control the shape and growth of mammalian cells. *Proc. Natl. Acad. Sci. U.S.A.* **91**(8), 3201–3204.

Woo, K. M., Chen, V. J., et al. (2003). Nanofibrous scaffolding architecture selectively enhances protein adsorption contributing to cell attachment. *J. Biomed. Mater. Res. A* **67**(2), 531–537.

Yamamoto, M., Tabata, Y., et al. (2000). Bone regeneration by transforming growth factor beta-1 released from a biodegradable hydrogel. *J. Controlled Release* **64**(1–3), 133–142.

Yang, F., Murugan, R., et al. (2004). Fabrication of nanostructured porous PLLA scaffold intended for nerve tissue engineering. *Biomaterials* **25**(10), 1891–1900.

Yang, S., Leong, K. F., et al. (2002). The design of scaffolds for use in tissue engineering. Part II. Rapid prototyping techniques. *Tissue Eng.* **8**(1), 1–11.

Yannas, I. V. (2005). Facts and theories of induced organ regeneration. *Adv. Biochem. Eng. Biotechnol.* **93**, 1–38.

Yuan, H., Li, Y., et al. (2000). Tissue responses of calcium phosphate cement: a study in dogs. *Biomaterials* **21**(12), 1283–1290.

Zelikin, A. N., Lynn, D. M., et al. (2002). Erodible conducting polymers for potential biomedical applications. *Angew. Chem. Int. Ed. Engl.* **41**(1), 141–144.

Zhang, S. (2002). Emerging biological materials through molecular self-assembly. *Biotechnol. Adv.* **20**(5–6), 321–339.

Zinger, O., Zhao, G., et al. (2005). Differential regulation of osteoblasts by substrate microstructural features. *Biomaterials* **26**(14), 1837–1847.

Zinn, M., Witholt, B., et al. (2001). Occurrence, synthesis and medical application of bacterial polyhydroxyalkanoate. *Adv. Drug Deliv. Rev.* **53**(1), 5–21.

Zisch, A. H., Lutolf, M. P., et al. (2003). Cell-demanded release of VEGF from synthetic, biointeractive cell-ingrowth matrices for vascularized tissue growth. *FASEB J.* **17**(15), 2260–2262.

Part Six

# Transplantation of Engineered Cells and Tissues

# Chapter Twenty-Six

# Tissue Engineering and Transplantation in the Fetus

*Dario O. Fauza*

## I. INTRODUCTION

The fetus is, arguably, the quintessential subject for tissue engineering, both as a host and as a donor. The developmental and long-term impacts of tissue implantations into a fetus, along with the many unique characteristics of fetal cells, add new dimensions that greatly expand the reach of tissue engineering, to extents unmatched by most other age groups. Indeed, perhaps not surprisingly, attempts at harnessing these prospective benefits started long before the modern era of transplantation. The first reported transplantation of human fetal tissue took place in 1922, when a fetal adrenal graft was transplanted into a patient with Addison's disease (Hurst *et al.*, 1922). A few years later, in 1928, fetal pancreatic cells were transplanted in an effort to treat diabetes mellitus. In 1957, a fetal bone marrow transplantation program was first undertaken. All those initial experiments involving human fetal tissue transplantation failed. It was only since around 1980 that fetal tissue transplantation in humans started to yield favorable outcomes. A number of therapeutic applications of fetal tissue have already been explored, with variable results. Although the majority of studies to date have simply involved fetal cell, tissue, or organ transplantation, a number of engineered open systems

using fetal cells have been explored in animal models, and their first clinical applications are seemingly imminent.

Fetal tissue has also been utilized as a valuable investigational tool in biomedical science since the 1930s. Embryologists, anatomists, and physiologists have long studied fetal metabolism, feto-placental unit function, premature life support, and brain activity in previable fetuses. *In vitro* applications of fetal tissue are well established and somewhat common. Cultures of different fetal cell lines, as well as commercial preparations of human fetal tissue, have been routinely used in the study of normal human development and neoplasias, in genetic diagnosis, in viral isolation and culture, and to produce vaccines. Biotechnology, pharmaceutical, and cosmetic companies have employed fetal cells and extraembryonic structures, such as placenta, amnion, and the umbilical cord, to develop new products and to screen them for toxicity, teratogenicity, and carcinogenicity. Fetal tissue banks have been operating in the United States and abroad for many years as a source of various fetal cell lines for research.

Considering that a large body of data has come out of research involving fetal cells or tissues and that attempts at engineering virtually every mammalian tissue have already

*Principles of Tissue Engineering, 3rd Edition*
*ed. by Lanza, Langer, and Vacanti*

taken place, comparatively little has been done on the true engineering of fetal tissue, through culture and placement of fetal cells into matrices or membranes or through other *in vitro* manipulations prior to implantation. Human trials of open-system tissue engineering have yet to be performed, and a relatively small number of animal experiments have been reported thus far. Fetal cells were first used experimentally in engineered constructs by Vacanti *et al.* (1988). Interestingly, this investigation was part of the introductory study on selective cell transplantation using bioabsorbable, synthetic polymers as matrices. This same group performed another study involving fetal cells in 1995 (Cusick *et al.*, 1995). Both experiments, in rodents, did not include structural replacement or functional studies. The use of fetal constructs as a means of structural and functional replacement in large animal models was first reported experimentally only in 1997 (Fauza *et al.*, 1998a, 1998b).

This chapter offers a look at the still-infantile field of fetal tissue engineering along with a general overview of fetal cell and tissue transplantation.

## II. GENERAL CHARACTERISTICS OF FETAL CELLS

Immunological rejection (in nonautologous applications), growth limitations, differentiation and function restraints, incorporation barriers, and cell/tissue delivery difficulties are all well-known complications of tissue engineering. Many of those problems can be better managed, if not totally prevented, when fetal cells are used. Due to their properties both *in vitro* and *in vivo*, fetal cells are an excellent raw material for tissue engineering.

### In Vitro

Compared with cells harvested postnatally, most fetal cells multiply more rapidly and more often in culture. Depending on the cell line considered, however, this increased proliferation is more or less pronounced or, in a few cases, not evident at all. Due, at least in part, to their proliferation and differentiation capacities, fetal cells have long been recognized as ideal targets for gene transfers.

Because they are very plastic in their differentiation potential, fetal cells respond better than mature cells to environmental cues. Data from fetal myoblasts and osteoblasts and mesenchymal amniocytes suggest that purposeful manipulations in culture or in a bioreactor can be designed to steer fetal cells to produce improved constructs. Younger mesenchymal stem cells (MSC) from midgestational fetal tissues are more plastic and grow faster than adult, bone marrow–derived MSC. Mesenchymal stem cells have also been isolated earlier in fetal development, from first-trimester blood, liver, and bone marrow. These cells are biologically closer to embryonic stem cells and have unique markers and characteristics not found in adult bone marrow MSC, which are potentially advantageous for cell

therapy. Fetal MSC typically express HLA class I but not HLA class II. The presence of interferon gamma (IFN-gamma) in the growth medium could initiate the intracellular synthesis and cell surface expression of HLA class II, but neither undifferentiated nor differentiated fetal MSC induced proliferation of allogeneic lymphocytes in mixed cultures. Actually, fetal MSC treated with IFN-gamma suppressed alloreactive lymphocytes in this setting. These data indicate that both undifferentiated and differentiated fetal MSC may not elicit much alloreactive lymphocyte proliferation, thus potentially rendering these cells particularly suitable for heterologous transplantation.

Fetal cells can survive at lower oxygen tensions than those tolerated by mature cells and are therefore more resistant to ischemia during *in vitro* manipulations. They also commonly lack long extensions and strong intercellular adhesions. Probably because of those characteristics, fetal cells display better survival after refrigeration and cryopreservation protocols when compared with adult cells. This enhanced endurance during cryopreservation, however, seems to be tissue specific. For instance, data from primates and humans have shown that fetal hematopoietic stem cells, as well as fetal lung, kidney, intestine, thyroid, and brain tissues, can be well preserved at low temperatures, whereas nonhematopoietic liver and spleen tissues can also be cryopreserved, but not as easily.

### In Vivo

The expression of major histocompatibility complex (H-2) antigens in the fetus and, hence, fetal allograft survival in immunocompetent recipients is age and tissue specific. The same applies to fetal allograft growth, maturation, and function. At least in fetal mice, the precise gestational time of detection of H-2 antigen expression and the proportion of cells expressing these determinants depend on inbred strain, specific haplotype, tissue of origin, and antiserum batch employed. Nevertheless, the precise factors governing the timing and tissue specificity of H-2 antigen expression are yet to be determined in most species, including humans.

Other mechanisms, in addition to H-2 antigen expression, also seem to govern fetal immunogenicity. It has been suggested that, by catabolizing tryptophan, the mammalian conceptus suppresses T-cell activity and defends itself against rejection by the mother. Fetal cells can be found in the maternal circulation in most human pregnancies, and fetal progenitor cells have been found to persist in the circulation of women decades after childbirth (Bianchi *et al.*, 2002). Interestingly, a novel population of fetal cells, the so-called pregnancy-associated progenitor cells (PAPC), appears to differentiate in diseased/injured maternal tissue. The precise original phenotypical identity of these cells remains unknown. They are thought to be a hematopoietic stem cell, a mesenchymal stem cell, or possibly a novel cell type. What is known is that pregnancy results in the acquisition of cells with stem-cell-like properties that may influence maternal health postpartum, by triggering disease and/or avoiding/combating it.

These data allow us to suppose that engineered constructs made with fetal cells should be less susceptible to rejection in allologous applications. Xenograft implantations may also become viable, for studies suggest that fetal cells are also better tolerated in cross-species transplantations, including in humans (Liechty *et al.*, 2000).

On the other hand, while less immunogenic, some fetal cells may be too immature and functionally limited if harvested too early. Yet experimental models of fetal islet pancreatic cell transplantation have shown that, with time, the initially immature and functionally limited cells will grow, develop, and eventually function normally. Conversely, however, certain cells, such as those from the rat's striatum, actually function better after implantation if harvested early, as opposed to late, in gestation.

Fetal cells may produce high levels of angiogenic and trophic factors, which enhance their ability to grow once grafted. By the same token, those factors may also facilitate regeneration of surrounding host tissues. Interestingly, significant clinical and hematological improvement has been described following fetal liver stem cell transplantation in humans, even when there is no evidence of engraftment. These improvements have been attributed to regeneration of autologous hematopoiesis and inhibition of tumor cell growth promoted by the infused cells, through mechanisms yet to be determined. The underdifferentiated state of fetal cells also optimizes engraftment, by allowing them to grow, elongate, migrate, and establish functional connections with other cells.

## Applications

Because of all the general benefits derived from the use of fetal cells, along with others specific to each cell line, several types of fetal cellular transplantation have been investigated experimentally or employed in humans for decades now. Clinically, fetal cells have been (mostly anecdotally) useful in a number of different conditions, including: Parkinson's disease and Huntington's disease; diabetes mellitus; aplastic anemia; Wiskott–Aldrich syndrome; thymic aplasia (DiGeorge syndrome) and thymic hypoplasia with abnormal immunoglobulin syndrome (Nezelof syndrome); thalassemia; Fanconi anemia; acute myelogenous and lymphoblastic leukemia; Philadelphia chromosome-positive chronic myeloid leukemia; X-linked lymphoproliferative syndrome; neuroblastoma; severe combined immunodeficiency disease; hemophilia; osteogenesis imperfecta; skin reconstruction; acute fatty liver of pregnancy; and neurosensory hypoacusis. They have also been applied to treat inborn errors of metabolism, including Gaucher's disease, Fabry's disease, fucosidosis, Hurler's syndrome, metachromatic leukodystrophy, Hunter's syndrome, glycogenosis, Sanfilippo's syndrome, Morquio syndrome type B, and Niemann–Pick disease. Experimentally, fetal cell and organ transplantation has been studied in an ever-expanding array of diseases. *In utero* hematopoietic stem cell transplantation is a promising approach for the treatment of a variety of genetic disorders. The rationale is to take advantage of prenatal hematopoietic and immunological ontogeny to facilitate allogeneic hematopoietic engraftment (Flake, 2004). It is an entirely nonmyeloablative approach to achieve mixed hematopoietic chimerism and associated donor-specific tolerance. Nonetheless, actual fetal tissue engineering as a therapeutic means has barely started to be explored, with comparatively few studies undertaken thus far.

## III. FETAL TISSUE ENGINEERING

Vacanti *et al.* (1988) were the first to make use of fetal cells in engineered constructs. The experiment, in rats, used fetal cells from the liver, intestine, and pancreas, which were cultured, seeded on bioabsorbable matrices, and later implanted. The fetal constructs were implanted in heterologous fashion and heterotopically, namely in the interscapular fat, omentum, and mesentery, with no structural replacement. They were removed for histological analysis no later than two weeks after implantation. Successful engraftment was observed in some animals that received hepatic and intestinal constructs but in none that received pancreatic ones.

Only in 1995 was a second study performed, by the same group, involving fetal liver constructs, also implanted in heterologous and heterotopic fashion in rats (Cusick *et al.*, 1995). Then fetal hepatocytes were shown to proliferate to a greater extent than adult ones in culture and to yield higher cross-sectional cell area at the implant. As in the first experiment, neither structural replacement studies nor functional studies were included.

Fetal constructs as a means of structural and functional replacement, in autologous fashion, in large animal models, were first reported experimentally in 1997 (Fauza *et al.*, 1998a, 1998b). Those studies introduced a novel concept in perinatal surgery, involving the minimally invasive harvest of fetal cells, which are then used to engineer tissue *in vitro* in parallel to the remainder of gestation, so an infant or a fetus with a prenatally diagnosed birth defect can benefit from having autologous, expanded tissue readily available for surgical implantation in the neonatal period or before birth.

## Congenital Anomalies

Major congenital anomalies are present in approximately 3% of all newborns. Those diseases are responsible for nearly 20% of deaths occurring in the neonatal period and even higher morbidity rates during childhood. By definition, birth defects entail loss and/or malformation of tissues or organs. Treatment of many of those congenital anomalies is often hindered by the scarce availability of normal tissues or organs, either in autologous or allologous fashion, mainly at birth. Autologous grafting is frequently not an option in newborns, due to donor site size limitations, and the well-known severe donor shortage observed in practically all areas of transplantation is even more critical during the neonatal period.

Although yet to be fully explored, several studies utilizing fetal tissue engineering as a means to treat congenital anomalies have already been reported in large-animal models (Fauza *et al.*, 1998a, 1998b, 2001; Kaviani *et al.*, 2001, 2002b, 2003; Fuchs *et al.*, 2002, 2003a, 2003b, 2004, 2005, 2006; Krupnick *et al.*, 2004; Kunisaki *et al.*, 2006a, 2006, 2006d). So far, these studies have involved different models of congenital anomalies/structural replacements, involving the skin, bladder, trachea, diaphragm, myocardium, blood vessels, and chest wall. However, many other anomalies are likely to benefit from this therapeutic principle. The first clinical application of fetal tissue engineering is thought to be imminent.

## Alternative Sources of Fetal Cells

In the experiments reported thus far, fetal cells amenable to processing for tissue engineering have been obtained from a variety of sources besides the fetus. These have included the amniotic fluid, the placenta, and the umbilical cord blood. Of all these sources, the amniotic fluid and the placenta are the least invasive ones for both the mother and the fetus, also because amniocentesis and chorionic villus sampling are widely accepted forms of prenatal diagnostic screening (Fauza, 2004). This is particularly true for the amniotic fluid, in that a diagnostic amniocentesis is routinely offered to any mother with a fetus in whom a structural anomaly has been diagnosed by prenatal imaging (Fig. 26.1).

Many amniotic and placental cells share a common origin: the inner cell mass of the morula, which gives rise to the embryo itself, the yolk sac, the mesenchymal core of the chorionic villi, the chorion, and the amnion. Most, if not all, types of progenitor cells that can be isolated from the amniotic fluid and the placenta seem to share many characteristics. However, a common origin of amniotic and placental progenitor cells remains to be unequivocally demonstrated. The full spectrum of cell types that can be obtained from amniotic and placental progenitor cells remains to be determined. In addition, certain fetal pathologic states, such as neural tube and body wall defects, may lead to the availability of cells not normally found in healthy pregnancies, which could be clinically useful.

### Amniotic Fluid

Embryonic and fetal cells from all three germ layers have long been identified in the amniotic fluid (Milunsky, 1979; Hoehn and Salk, 1982; Gosden, 1983; Prusa *et al.*, 2003). However, the specific origins of many subsets of these cell populations remain to be determined. The cellular profile of the amniotic fluid varies with gestational age. In

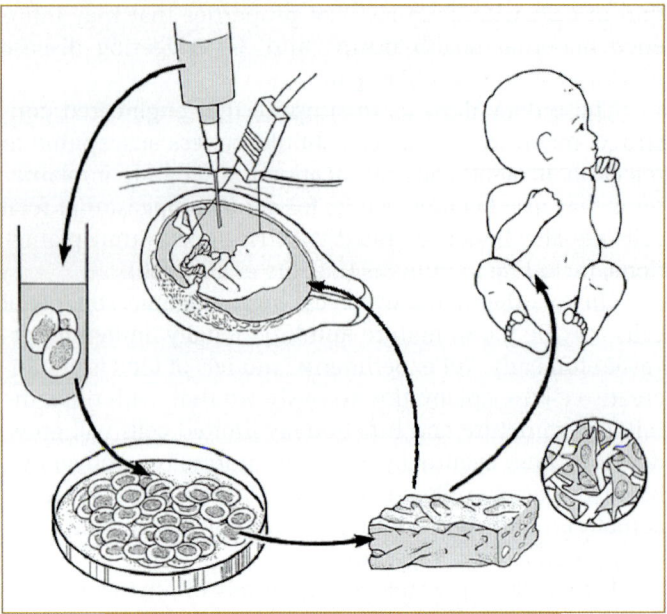

**FIG. 26.1.** Diagram representing the concept of fetal tissue engineering from amniotic fluid cells for the treatment of birth defects. A small aliquot of amniotic fluid is obtained from a routine amniocentesis, typically performed when a structural anomaly is diagnosed by routine prenatal imaging screening. Fetal tissue is then engineered *in vitro* from amniotic progenitor cells while pregnancy continues, so the newborn or a fetus can benefit from having autologous, expanded tissue promptly available for surgical reconstruction at birth or *in utero*.

addition to the common origin with the mesenchymal portion of the placenta, as mentioned earlier, the amniotic cavity/fluid receives cells shed from the fetus and, quite possibly, from the placenta as well (the latter is yet to be definitely confirmed).

The mechanisms responsible for the production and turnover of the amniotic fluid are thought also to determine the cell types present in the amniotic cavity. In the first half of gestation, most of the amniotic fluid derives from active sodium and chloride transport across the amniotic membrane and fetal skin, with concomitant passive movement of water. In the second half, most of the fluid comes from fetal micturition. An additional substantial source of amniotic fluid is secretion from the respiratory tract. Fetal swallowing and gastrointestinal tract excretions, while not voluminous, also play a role in the composition of the amniotic fluid. As a result of such fluid dynamics, cells present in the urinary, respiratory, and gastrointestinal tracts are shed into the amniotic cavity.

Overall amniotic fluid composition changes predictably throughout gestation. In humans, it is isotonic with fetal plasma in early pregnancy, due to transudation of fetal plasma through the maternal decidua or through the fetal skin prior to keratinization, which occurs at approximately 24 weeks. Afterwards and until term, it becomes increasingly hypotonic relative to maternal or fetal plasma.

All these variables that play a role in amniotic fluid composition seem to contribute to the changeable profile of the cellular component of the amniotic fluid. Still, much remains to be clarified about the ontogeny of many subsets of amniocytes at any gestational age. This is particularly true before the 12th week of gestation, due, to a large extent, to the limitations of performing amniocentesis before that time.

The fact that certain progenitor cells can be found in the amniotic fluid was apparently first reported in 1993, when small, nucleated, round cells identified as hematopoietic progenitor cells were found therein only before the 12th week of gestation, possibly coming from the yolk sac (Torricelli et al., 1993). A study from 1996 was the first to suggest the possibility of multilineage potential of nonhematopoietic cells present in the amniotic fluid, by demonstrating myogenic conversion of amniocytes (Streubel et al., 1996). That study did not specify the identity of the cells that responded to the myogenic culture conditions, in that case the supernatant of a rhabdomyosarcoma cell line. The presence of mesenchymal cells in the amniotic fluid has been proposed for decades. However, the differentiation potential of mesenchymal amniocytes started to be determined only very recently (Kaviani et al., 2002a; In't Anker et al., 2003; Kunisaki et al., 2006). Likewise, the presence of embryonic-like stem cells in the amniotic fluid was suggested only in the last few years (Coppi et al., 2002; Prusa and Hengstschlager, 2002; Prusa et al., 2003).

Human amniotic epithelial cells have shown pluripotency, being able to differentiate at least into neural and glial cells and into hepatocyte precursors. These cells have been employed therapeutically in animal models of cerebral ischemia and spinal cord injury and as experimental transgene carriers into the liver. However, they are not yet universally considered amniotic fluid cells and will not be discussed further here. Also, given the other, more practical sources of fetal hematopoietic stem cells, other than the difficult pre-12th week amniocentesis, it is unlikely that the amniotic fluid will be a useful option for a source of these cells in clinical practice. Hence, this overview focuses on the mesenchymal and embryonic-like stem cells.

## Mesenchymal Stem Cells

The amniotic fluid is rich in MSC. We have described a very simple protocol for isolation of these mesenchymal amniocytes, based on mechanical separation and natural selection by the culture medium (Kaviani et al., 2001, 2003). Other protocols for isolation of mesenchymal amniocytes have also been described (Hurych et al., 1976; In't Anker et al., 2003). Previous data on amniotic cell culture, without description of the specific nature of the cells grown (possibly predominantly mesenchymal), show that low oxygen tension in the gas phase can be an effective means of enhancing clonal cell expansion (Held and Sonnichsen, 1984). Although not routinely employed, if necessary, molecular HLA typing can be performed on DNA obtained

from expanded MSC, as well as from fetal and maternal blood cells, by polymerase chain reaction/sequence-specific oligonucleotide using a reverse dot blot method, in order to confirm the fetal origin of the cultured cells.

The precise origins of the MSC found in the amniotic fluid remain to be determined. At first, these cells were thought simply to be shed by the fetus at the end of their life cycle. However, they may actually come from the fetus itself and/or the placenta and/or the inner cell mass of the morula, staying viable within the fluid. In no way are mesenchymal amniocytes at the end of their life cycle. Ovine data have shown that amniotic fluid–derived MSC proliferate significantly faster in culture than immunocytochemically comparable cells derived from fetal or adult subcutaneous connective tissue, neonatal bone marrow, and umbilical cord blood (Kaviani et al., 2001; Kunisaki et al., 2006). In humans, the expansion potential of mesenchymal amniocytes exceeds that of bone marrow MSC (In't Anker et al., 2003; Kaviani et al., 2003). The phenotype of human mesenchymal amniocytes expanded in vitro is similar to that reported for MSC derived from second-trimester fetal tissue and adult bone marrow (Pittenger et al., 1999; Noort et al., 2002; In't Anker et al., 2003).

Human mesenchymal amniocytes have shown potential for mesodermal differentiation into fibroblasts, adipocytes, chondrocytes, osteocytes, and myogenic lineages after exposure to previously described specific culture conditions (Noort et al., 2002; In't Anker et al., 2003; Kaviani et al., 2003; Zhao et al., 2005; Kunisaki et al., 2006d). The progenitor nature of these cells impart the possibility that they could be used to engineer constructs to correct a wide variety of defects. For example, recent large-animal studies have shown that the repair of congenital diaphragmatic hernia and tracheal defects can be enhanced by the respective use of tendon and cartilaginous grafts engineered from mesenchymal amniocytes during surgical reconstruction (Fig. 26.2 and 26.3) (Fuchs et al., 2004; Kunisaki et al., 2006a, 2006c). Another recently proposed use of an engineered construct based on amniotic mesenchymal (and epithelial) cells is for the repair of premature rupture of membranes during pregnancy, but this has yet to be done in vivo.

An intriguing characteristic of mesenchymal amniocytes, at this time studied only in animals, is what seems to be a unique immunological profile, manifest (at least) when they are isolated and expanded in vitro. In culture, these cells down-regulate the expression of immune-associated antigens, including MHC-1, when compared to mesenchymal cells obtained from fetal or adult tissue. Hence, mesenchymal amniocytes may be immunologically privileged, when compared to mesenchymal cells derived from fetal or adult tissue.

It has been shown that MSC enhance the engraftment of umbilical cord blood–derived CD34+ hematopoietic cells. The amniotic fluid has recently been proposed as a useful

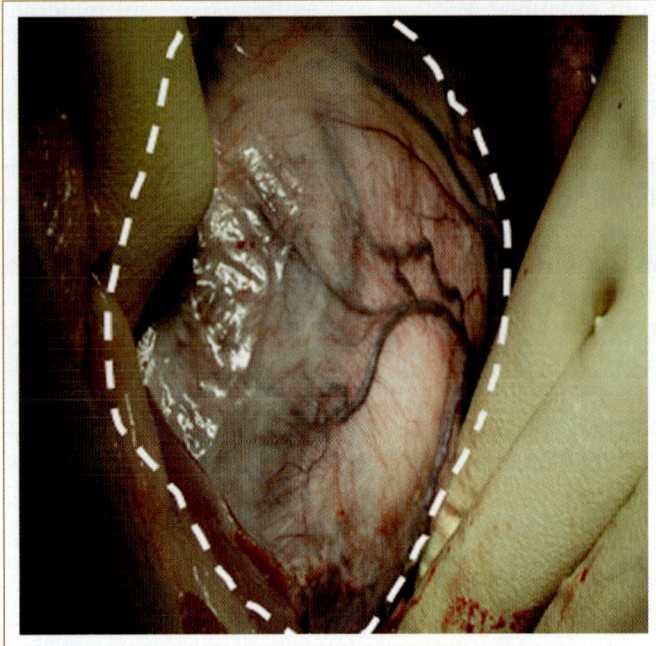

**FIG. 26.2.** An intact ovine diaphragmatic tendon seen from the chest, 12 months after autologous repair with an engineered, mesenchymal amniocyte-based construct. The dotted line encircles the area of the graft. Reproduced, with permission, from Kunisaki *et al.* (2006c).

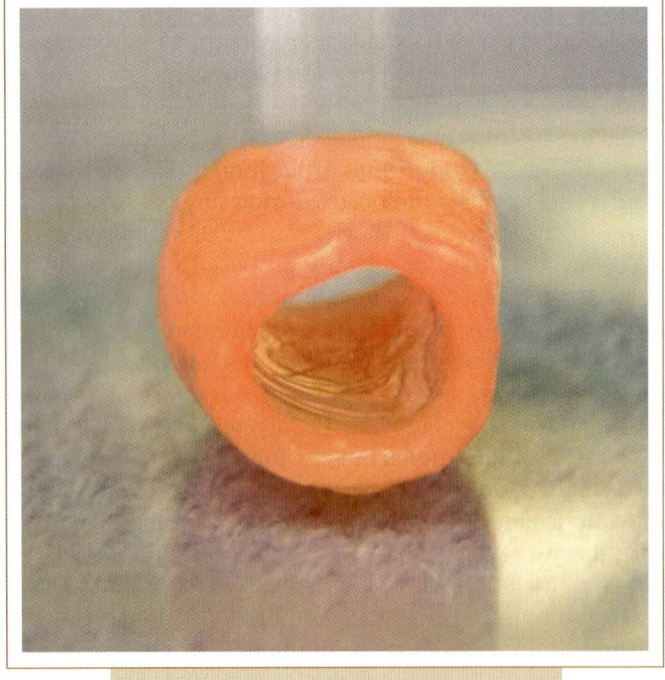

**FIG. 26.3.** A representative, cross-sectional view of a 3D cartilaginous tube engineered from mesenchymal amniocytes seeded onto a polyglycolic acid matrix, previously maintained in a bioreactor under chondrogenic conditions.

source of MSC to be cotransplanted with hematopoietic stem cells from the same donor (In't Anker *et al.*, 2003). This could be particularly useful in the setting of umbilical cord blood transplantation between siblings.

*In utero*, transamniotic gene transfer has shown some promise in animal models. Young, highly proliferative, less differentiated fetal cells are natural targets for the optimization of gene therapy. Thus, amniotic and placental stem cells could certainly make useful targets for genetic manipulation. This is a potentially exciting development for the foreseeable future.

### Embryonic-like Stem Cells

From the very little currently known about what could be more primitive, embryonic-like stem cells (ESC) present in the human amniotic fluid, unlike their mesenchymal counterparts, these cells are very scarce, representing 0% (i.e., they cannot always be isolated) to less than 1% of the cells present in amniocentesis samples (Coppi *et al.*, 2002; Prusa and Hengstschlager, 2002; Prusa *et al.*, 2003). In addition, these cells have been identified mostly through markers commonly present in ESC, such as nuclear Oct-4, stem cell factor, vimentin, alkaline phosphatase, CD34, CD105, and cKit, but that were not necessarily concomitantly expressed, nor are they exclusive of ESC proper. Specifically, these markers can also be expressed, alone or in various combinations, in embryonic germ cells; embryonic fibroblasts; embryonal carcinoma cells; mesenchymal stem cells; hematopoietic stem cells; ectodermal, neural, and pancreatic progenitor cells; and fetal and adult nerve tissue; among others. Another marker for human stem cell pluripotency, telomerase activity, has been detected in amniotic fluid cell samples, albeit in a study unrelated to the ones devoted to stem cell isolation (Mosquera *et al.*, 1999).

It is well known that markers alone are not considered enough to characterize human ESC. A uniform and universal differentiation potential needs to be demonstrated, which remains to be verified in these so-called amniotic ESC. So far, some of the amniotic cells that express the markers just mentioned have been shown to differentiate into muscle, adipogenic, osteogenic, nephrogenic, neural, and endothelial cells but not necessarily from a uniform population of undifferentiated cells (Coppi *et al.*, 2002). On the other hand, in accordance with an ESC profile, these pluripotent cells seem to be clonogenic, for at least the Oc-4 positive ones express cyclin A, a cell cycle regulator (Hengstschlager *et al.*, 1999; Prusa *et al.*, 2003). Therefore, although the data currently available can indeed be considered promising, final proof that true ESC can be consistently isolated from the amniotic fluid, and at which gestational ages, remains to be established.

### Placenta

Different cell types are found in the placenta at different gestational ages, as a result of the mechanisms behind placental development. In humans, placental villous development starts between 12 and 18 days postconception (p.c.), when the trophoblastic trabeculae of the placental anlage proliferate and form trophoblastic protrusions, the primary villi, into the maternal blood surrounding the trabeculae. Two days later, embryonic connective tissue from the extra-embryonic mesenchyme invades these villi, which then become secondary villi. Between days 18 and 20 p.c., the first fetal capillaries begin to appear in the now-abundant mesenchyme of the villous stroma, marking the development of the tertiary villi, the first generation of which are the mesenchymal villi. Mesenchymal villi are the first structures providing the morphological requisites for materno–fetal exchange of gases, nutrients, and waste. They are also the precursors of all other villous types, namely immature intermediate villi, stem villi, mature intermediate villi, and terminal villi. At approximately the fifth week p.c., all placental villi are of the mesenchymal type. From about the 23rd week p.c. until term, the pattern of villous growth changes, with the mesenchymal villi transforming into mature intermediate villi, rather than into immature ones. During that phase, few mesenchymal and immature intermediate villi remain, in the centers of the villous trees, where they comprise a poorly differentiated growth reserve.

As shown earlier, placental villous sprouting entails active growth of both trophoblastic and mesenchymal tissue components in a coordinated fashion. Further differentiation of the mesenchymal villi into immature or mature intermediate villi is a determining factor in the balance between growth and maturation of the placenta, which, in turn, has a direct impact on the cell types that can be isolated from the placenta at different gestational ages.

The genetic and molecular mechanisms behind placental development have hardly begun to be clarified and should also have a bearing on the pluripotency of placental cells. Interestingly, against the conventional wisdom of searching for placental-specific genes that would control this process, most of the genes that have been shown to be essential for placental development are also involved in the development of other organs. No more than a very limited set of genes is expressed exclusively in the placenta.

Since much of the placenta comes from the inner cell mass of the morula, the presence of embryonic progenitor cells in the placenta has long been proposed (Crane and Cheung, 1988). More specific types of stem cells, such as trophoblastic, hematopoietic, and mesenchymal stem cells, have also been identified in the placenta for many years. Trophoblast stem cells, also known as trophoendoderm stem cells, are defined as cells with the potential to give rise to all differentiated trophoblast cell subtypes as well as to yolk sac phenotypes. Interestingly, one trophoblast sub-population, the cytotrophoblasts of the basal decidua, may undergo a full transition to mesenchymal phenotype when they infiltrate maternal mesenchymally derived uterine stroma and arterial walls. At present, trophoblast stem cells

have limited, if any, foreseeable potential for clinical application and will not be reviewed here. Placental/umbilical hematopoietic stem cells have already shown high clinical relevance, but they can also be isolated from umbilical cord blood and are reviewed in another chapter. Placental mesenchymal and embryonic stem cells are further discussed next.

### Mesnchymal Stem Cells

As described earlier, the placenta has a large mesenchymal component. While its role in placental development remains to be better understood, we already know that it relates, in part, to the pluripotent potential of placental mesenchymal cells. For example, recruitment of these cells supports the so-called vasculogenesis that occurs during vascularization of the villous sprouts, in addition to the angiogenesis based on the proliferation of endothelial precursors. These mesenchymal cells also play other roles in placental development, such as the paracrine signals that they send to control the stability of the cytotrophoblast column, which in turn determines the degree of trophoblast invasiveness. Yet, much about the mesenchymal core of the placenta remains to be elucidated.

Placental mesenchymal cells can be isolated by a number of different protocols (Haigh et al., 1999; Kaviani et al., 2002b; Romanov et al., 2003). We have described an easy method, analog to the one described earlier for the separation of mesenchymal amniocytes (i.e., based on mechanical separation and natural selection by the culture medium), which can be employed in both "full-thickness" and chorionic villus sampling placental specimens (Kaviani et al., 2002b). These cells have unique characteristics, when compared to other mesenchymal cells. They proliferate more quickly in culture than comparable cells harvested from fetal or adult tissue, at a similar rate to that of mesenchymal amniocytes. Also like mesenchymal amniocytes, they are often stained by monoclonal cytokeratin antibodies, which is a rare finding in mature mesenchymal cells but common in fetal and umbilical stroma, smooth muscle tumors, and stromal cells associated with reactive processes. It is as yet unclear whether this immunoreactivity is a result of cross-reactivity between a common epitope found in smooth muscle cells, myofibroblasts, and intermediate-sized filaments of cytokeratin or whether it actually indicates cytokeratin expression in nonepithelial cells.

In addition to their natural differentiation into smooth muscle cells during placental development, placental mesenchymal cells have been shown to be able to differentiate into all mesodermal cell lineages, as well as some neural cells lineages, in vitro (Portmann-Lanz et al., 2006; Zhang et al., 2006). Given the many similarities between placental mesenchymal cells and mesenchymal amniocytes, as shown earlier, it is reasonable to speculate that these two cell subsets are actually the same. This, however, remains to be unequivocally demonstrated.

*Embryonic-like Stem Cells*

The embryonary origins of the placenta render it a natural candidate for a reservoir of ESC (Crane and Cheung, 1988). From the few reported attempts to isolate ESC from the placenta, like their amniotic counterparts, what could be ESC are present in 0% to less than 1% of the cells present in placental samples (Coppi *et al.*, 2002). These cells have also been identified mostly through certain markers, including CD34, CD105, and cKit, that, again, were not necessarily concomitantly expressed and are not exclusive of ESC.

These pluripotent cells have shown potential for self-renewal, but a uniform and universal differentiation potential of these cells also remains to be verified. Like their amniotic counterparts, thus far some of these cells have been shown to differentiate into a variety of cell types, but not necessarily from a uniform population of undifferentiated cells. Another limitation of the existing data is the fact that the placenta is rich in hematopoietic stem cells, known not to be committed solely to the hematopoietic lineage. In addition to blood cells, they can give rise, at least, to neurons, hepatocytes, and muscle cells and might have contributed to the differentiation findings reported to date. Again, although the existing data are promising and in accordance with the origins of the placenta, final proof that true ESC can be consistently isolated from that organ, and at which gestational ages, remains to be verified.

Another interesting potential development of the study of placental cells is the establishment of 3D models of the placenta and of maternal–fetal circulation through the maintenance of live engineered placental models in bioreactors (Ma *et al.*, 1999). This could lead to a better understanding of normal and pathological placental physiology, possibly with indirect therapeutic benefits.

## IV. ETHICAL CONSIDERATIONS

The use of fetal tissue has always been object of intense ethical debate. The main reason for the ethical controversies comes from the fact that the primary source of fetal tissue is induced abortion. Spontaneous abortion usually does not raise many moral issues. The National Institutes of Health, the American Obstetrical and Gynecological Society, and the American Fertility Society, in accordance with the provisions that control the use of adult human tissue, have long regulated the use of fetal specimens from this latter source. However, spontaneous abortion generally yields unsuitable fetal tissue, for it is frequently compromised by pathology such as chromosomal abnormalities, infections, and/or anoxia. Arguments on the use of fetal tissue from induced abortion are based largely on somewhat limited scientific evidence, along with clashing religious and customary beliefs about the beginning of life. Not surprisingly, despite the efforts of numerous national and international ethical committees and governmental bodies, a consensus has not yet been reached. In the United States, in spite (or perhaps because) of an ongoing moratorium on federal funding for fetal tissue transplantation research, an agreement on this issue may be forming slowly, though a stable solution could still be years away.

This polemic notwithstanding, tissue engineering, as a novel development in fetal tissue processing, adds a new dimension to the discussion concerning the use of fetal tissue for therapeutic or research purposes. If specimens from a live, diseased fetus or cells from a routine prenatal diagnostic procedure such as an amniocentesis or chorionic villus sampling are to be used for the engineering of tissue, which, in turn, is to be implanted in autologous fashion, then no ethical objections should be anticipated, as long as the procedure is a valid therapeutic choice for a given perinatal condition. In that scenario, ethical considerations are the very same that apply to any fetal intervention. On the other hand, if fetal engineered tissue is to be implanted in heterologous fashion, ethical issues are analog to the ones involving fetal tissue/organ transplantation, regardless of whether the original specimen comes from a live or deceased fetus or from banked fetal cells obtained from the amniotic, placenta, or umbilical cord blood.

The distinction between autologous and heterologous implantation of engineered fetal tissue is a critical one, in that, again, no condemnation of autologous use could be ethically justified. At the same time, regardless of whether an autologous or heterologous application is being considered, should the amniotic fluid or the placenta be definitely confirmed as dependable sources of ESC, then the ethical objections to embryo disposal now plaguing the progress of ESC research could possibly be avoided.

## V. THE FETUS AS A TRANSPLANTATION HOST

One could envision a number of advantages of a fetus receiving an engineered construct *in utero*, not only from a theoretical perspective but also from clinical and experimental evidence derived from intrauterine cellular transplantation studies already reported. Those potential advantages encompass induction of graft tolerance in the fetus, due to its immunologic immaturity; induction of donor-specific tolerance in the fetus by concurrent or previous intrauterine transplantation of hematopoietic progenitor cells; a completely sterile environment; the presence of hormones, cytokines, and other intercellular signaling factors that may enhance graft survival and development; the unique wound-healing properties of the fetus; and early prevention of clinical manifestations of disease, before they can cause irreversible damage. Most of those advantages should be more or less evident, depending on the gestational timing of transplantation.

## Fetal Immune Development

Among the potential benefits of *in utero* transplantation, the singularity of the fetal immune system deserves special attention. In that respect, basic research on fetal development as well as studies involving pre- and postnatal transplantation of lymphohematopoietic fetal cells have contributed to a better understanding of the fetal immune response.

For almost a century now, fetal tolerance resulting in permanent chimerism has been known to occur in nature in nonidentical twins with shared placental circulation. Little is known, however, about precisely when and by what mechanism this tolerance is lost. During fetal development, the precursors of the hematopoietic stem cells arise in the yolk sac and migrate to the fetal liver and then to the thymus, spleen, and bone marrow. The fetal liver has its highest concentration of hematopoietic stem cells between the 4th and the 20th week of gestation. Because of their cellular immunologic "immaturity," the fetal liver and, to a lesser extent, the fetal thymus have been studied as potential sources of hematopoietic stem cells for major histocompatibility complex–incompatible bone marrow transplantation for almost four decades now. Umbilical cord blood has been increasingly employed as a source of hematopoietic stem cells, in both autologous and heterologous applications (Benito *et al.*, 2004; Rocha *et al.*, 2005).

Lymphocytes capable of eliciting graft-versus-host disease (GVHD) are found in the thymus by the 14th week of gestation but are not detectable in the liver until the 18th week. Despite considerable numbers of granulocyte-macrophage colony–forming cells, there is an almost complete absence of mature T-cells up to the 14th week in human fetal livers. During gestation, while B-cell development takes place mostly in the liver, T-cell development occurs predominantly in the thymus. This fact is probably the reason why fetal liver cells are immunoincompetent for cell-mediated and T-cell-supported humoral reactions, such as graft rejections and GVHD. Thus, in principle, tissue matching is not necessary in fetal liver transplantation if this is harvested up to a certain point in gestation. In a number of animal models and small clinical series, fetal liver cells have induced no or merely moderate GVHD in histoincompatible donor/recipient pairs.

Umbilical cord blood and placental blood, on the other hand, while rich in hematopoietic progenitor cells, contain alloreactive lymphocytes. Although those lymphocytes are also immature, it is unclear, however, whether they are more or less reactive than adult ones. Compared with those from adult blood, the proportions of activated T-cells and helper-inducer subsets (CD4/29) are significantly reduced, while the helper-suppressor (CD4/45A) subset is significantly increased. Cord blood natural killer cell activity is low or similar to that in adult blood, but lymphokine-activated killer cell activity may be higher.

Although fetal liver stem cells should not cause GVHD, they could still be subject to rejection. Because of that, fetal liver stem cell transplantation has been attempted in the clinical setting preferably in patients with depressed immune function, such as in immunodeficiencies, replacement therapy during bone marrow compromise, and during fetal life (*in utero* transplantation). The same principle applies to the use of fetal thymus. Fatal cases of GVHD are considered much less likely in patients who receive fetal liver stem cells harvested before the 14th week of gestation. This complication, however, has been reported in a patient who received liver cells from a 16-week-old fetus. With umbilical cord blood stem cell transplantation, the incidence of GVHD has been minimal (Benito *et al.*, 2004; Rocha *et al.*, 2005).

## Applications of *In Utero* Transplantation

Cellular intrauterine transplantation has been employed clinically to treat a variety of diseases, including lymphohematopoietic diseases, beta-thalassemia, inborn errors of metabolism, and genetic disorders, with some success. The optimal gestational age for transplantation along with cell selection, route of cell administration, and postinterventional tocolysis is still evolving.

*In utero* hematopoietic stem cell transplantation is a nonmyeloablative approach to achieve mixed hematopoietic chimerism and associated donor-specific tolerance, improving survival of other grafts later in life. Through prenatal transplantation of hematopoietic progenitor cells, both allogeneic and xenogeneic chimerisms have been induced in animal models, and allogeneic chimerism has been achieved in humans. Tolerance of allogeneic intrauterine implantation of an engineered construct has recently been first demonstrated in an ovine model of fetal tracheal reconstruction with heterologous cartilage engineered from amniotic fluid–derived MSC (Kunisaki *et al.*, 2006a). Other potential therapeutical benefits of prenatal implantation of engineered tissue, as well as likely advantages stemming from the commonly scarless fetal wound healing, are yet to be fully explored.

## VI. CONCLUSIONS

Fetal tissue engineering may become a preferred perinatal alternative for the treatment of a number of birth defects. Given the recently proven viability of minimally invasive fetal cell sources, such as amniotic fluid, placenta, and umbilical cord blood, the promise of fetal tissue engineering should apply to both life-threatening and non-life-threatening anomalies. Fetal progenitor cells from various sources are becoming progressively relevant, if not indispensable, tools in research related to stem cells, tissue engineering, gene therapy, and maternal–fetal medicine.

Still, much remains to be learned, and a variety of evolutionary paths, including unsuspected ones, have yet to be pursued in this relatively new branch of tissue engineering. The whole subfield of *in utero* implantation of engineered tissue also waits to be fully explored. Fetal tissue engineering shall also benefit from the progress expected for the different aspects of tissue engineering in general. Fertile experimental work from an increasing number of groups has introduced promising novel therapeutic concepts utilizing fetal cells. As long as progress is made in the ethical debate over the use of these cells and their banking, the reach of fetal tissue engineering, nonetheless, will likely go beyond the perinatal period, offering unique therapeutic perspectives for different age groups.

## VII. REFERENCES

Benito, A. I., Diaz, M. A., Gonzalez-Vicent, M., Sevilla, J., and Madero, L. (2004). Hematopoietic stem cell transplantation using umbilical cord blood progenitors: review of current clinical results. *Bone Marrow Transplant.* **33**(7), 675–690.

Bianchi, D. W., Johnson, K. L., and Salem, D. (2002). Chimerism of the transplanted heart. *N. Engl. J. Med.* **346**(18), 1410–1412; author reply 1410–1412.

Coppi, P. D., Filippo, R. D., Soker, S., Bartsch Jr., G., Yoo, J. J., Cin, P. D., and Atala, A. (2002). Human embryonic and fetal stem-cell isolation from amniotic fluid and placenta for tissue reconstruction. *J. Am. Coll. Surg.* **195**(Suppl.), S93 [abstract].

Crane, J. P., and Cheung, S. W. (1988). An embryogenic model to explain cytogenetic inconsistencies observed in chorionic villus versus fetal tissue. *Prenat. Diagn.* **8**, 119–129.

Cusick, R. A., Sano, K., Lee, H., and Al, E. (1995). Heterotopic fetal rat hepatocyte transplantation on biodegradable polymers. *Surg. Forum* **XLVI**, 658–661.

Fauza, D. O. (2004). Amniotic fluid and placental stem cells. *Best Pract. Res. Clin. Obstet. Gynaecol.* **18**(6), 877–891.

Fauza, D. O., Fishman, S. J., Mehegan, K., and Atala, A. (1998a). Video-fetoscopically assisted fetal tissue engineering: bladder augmentation. *J. Pediatr Surg.* **33**(1), 7–12.

Fauza, D. O., Fishman, S. J., Mehegan, K., and Atala, A. (1998b). Video-fetoscopically assisted fetal tissue engineering: skin replacement. *J. Pediatr. Surg.* **33**(2), 357–361.

Fauza, D. O., Marler, J. J., Koka, R., Forse, R. A., Mayer, J. E., and Vacanti, J. P. (2001). Fetal tissue engineering: diaphragmatic replacement. *J. Pediatr. Surg.* **36**(1), 146–151.

Flake, A. W. (2004). *In utero* stem cell transplantation. *Best Pract. Res. Clin. Obstet. Gynaecol.* **18**(6), 941–958.

Fuchs, J. R., Terada, S., Ochoa, E. R., Vacanti, J. P., and Fauza, D. O. (2002). Fetal tissue engineering: *in utero* tracheal augmentation in an ovine model. *J. Pediatr. Surg.* **37**(7), 1000–1006.

Fuchs, J. R., Hannouche, D., Terada, S., Vacanti, J. P., and Fauza, D. O. (2003a). Fetal tracheal augmentation with cartilage engineered from bone marrow–derived mesenchymal progenitor cells. *J. Pediatr. Surg.* **38**(6), 984–987.

Fuchs, J. R., Terada, S., Hannouche, D., Ochoa, E. R., Vacanti, J. P., and Fauza, D. O. (2003b). Fetal tissue engineering: chest wall reconstruction. *J Pediatr. Surg.* **38**(8), 1188–1193.

Fuchs, J. R., Kaviani, A., Oh, J. T., LaVan, D., Udagawa, T., Jennings, R. W., Wilson, J. M., and Fauza, D. O. (2004). Diaphragmatic reconstruction with autologous tendon engineered from mesenchymal amniocytes. *J. Pediatr. Surg.* **39**(6), 834–8; discussion 834–838.

Fuchs, J. R., Hannouche, D., Terada, S., Zand, S., Vacanti, J. P., and Fauza, D. O. (2005). Cartilage engineering from ovine umbilical cord blood mesenchymal progenitor cells. *Stem Cells* **23**(7), 958–964.

Fuchs, J. R., Nasseri, B. A., Vacanti, J. P., and Fauza, D. O. (2006). Postnatal myocardial replacement through fetal tissue engineering. *Surgery,* **140**, 100–107.

Gosden, C. M. (1983). Amniotic fluid cell types and culture. *Br. Med. Bull.* **39**, 348–354.

Haigh, T., Chen, C., Jones, C. J., and Aplin, J. D. (1999). Studies of mesenchymal cells from 1st-trimester human placenta: expression of cytokeratin outside the trophoblast lineage. *Placenta* **20**(8), 615–625.

Held, K. R., and Sonnichsen, S. (1984). The effect of oxygen tension on colony formation and cell proliferation of amniotic fluid cells *in vitro*. *Prenat. Diagn.* **4**(3), 171–179.

Hengstschlager, M., Braun, K., Soucek, T., Miloloza, A., and Hengstschlager-Ottnad, E. (1999). Cyclin-dependent kinases at the G1–S transition of the mammalian cell cycle. *Mutat. Res.* **436**, 1–9.

Hoehn, H., and Salk, D. (1982). Morphological and biochemical heterogeneity of amniotic fluid cells in culture. *Methods Cell Biol.* **26**, 11–34.

Hurst, A. F., Tanner, W. E., and Osman, A. A. (1922). Addison's disease with severe anemia treated by suprarenal grafting. *Proc. R. Soc. Med.* **15**, 19.

Hurych, J., Macek, M., Beniac, F., and Rezacova, D. (1976). Biochemical characteristics of collagen produced by long-term cultivated amniotic fluid cells. *Hum. Genet.* **31**(3), 335–340.

In't Anker, P. S., Scherjon, S. A., Kleijburg-van der Keur, C., Noort, W. A., Claas, F. H., Willemze, R., Fibbe, W. E., and Kanhai, H. H. (2003). Amniotic fluid as a novel source of mesenchymal stem cells for therapeutic transplantation. *Blood* **102**(4), 1548–1549.

Kaviani, A., Perry, T. E., Dzakovic, A., Jennings, R. W., Ziegler, M. M., and Fauza, D. O. (2001). The amniotic fluid as a source of cells for fetal tissue engineering. *J. Pediatr. Surg.* **36**(11), 1662–1665.

Kaviani, A., Jennings, R. W., and Fauza, D. O. (2002a). Amniotic fluid–derived fetal mesenchymal cells differentiate into myogenic precursors *in vitro*. *J. Am. Coll. Surg.* **195**(Suppl.), S29 [abstract].

Kaviani, A., Perry, T. E., Barnes, C. M., Oh, J. T., Ziegler, M. M., Fishman, S. J., and Fauza, D. O. (2002b). The placenta as a cell source in fetal tissue engineering. *J. Pediatr. Surg.* **37**(7), 995–999.

Kaviani, A., Guleserian, K., Perry, T. E., Jennings, R. W., Ziegler, M. M., and Fauza, D. O. (2003). Fetal tissue engineering from amniotic fluid. *J. Am. Coll. Surg.* **196**(4), 592–597.

Krupnick, A. S., Balsara, K. R., Kreisel, D., Riha, M., Gelman, A. E., Estives, M. S., Amin, K. M., Rosengard, B. R., and Flake, A. W. (2004). Fetal liver as a source of autologous progenitor cells for perinatal tissue engineering. *Tissue Eng.* **10**(5–6), 723–735.

Kunisaki, S. M., Freedman, D. A., and Fauza, D. O. (2006a). Fetal tracheal reconstruction with cartilaginous grafts engineered from mesenchymal amniocytes. *J. Pediatr. Surg.* **41**(4), 675–682.

Kunisaki, S. M., Fuchs, J. R., Azpurua, H., Zurakowski, D., and Fauza, D. O. (2006b). A comparison of different perinatal sources of mesenchymal progenitor cells: implications for tissue engineering. Thirty-Seventh Annual Meeting of the American Pediatric Surgical Association, Hilton Head, SC.

Kunisaki, S. M., Fuchs, J. R., Kaviani, A., Oh, J. T., LaVan, D. A., Vacanti, J. P., Wilson, J. M., and Fauza, D. O. (2006c). Diaphragmatic repair through fetal tissue engineering: a comparison between mesenchymal amniocyte- and myoblast-based constructs. *J. Pediatr. Surg.* **41**(1), 34–39; discussion 34–39.

Kunisaki, S. M., Jennings, R. W., and Fauza, D. O. (2006d). Fetal cartilage engineering from amniotic mesenchymal progenitor cells. *Stem Cells Dev.* **15**(2), 245–253.

Liechty, K. W., MacKenzie, T. C., Shaaban, A. F., Radu, A., Moseley, A. M., Deans, R., Marshak, D. R., and Flake, A. W. (2000). Human mesenchymal stem cells engraft and demonstrate site-specific differentiation after *in utero* transplantation in sheep. *Nat. Med.* **6**(11), 1282–1286.

Ma, T., Yang, S. T., and Kniss, D. A. (1999). Development of an *in vitro* human placenta model by the cultivation of human trophoblasts in a fiber-based bioreactor system. *Tissue Eng.* **5**(2), 91–102.

Milunsky, A. (1979). Amniotic fluid cell culture. *In* "Genetic Disorder of the Fetus" (A. Milunsky, ed.), pp. 75–84. Plenum Press, New York.

Mosquera, A., Fernandez, J. L., Campos, A., Goyanes, V. J., Ramiro-Dias, J. R., and Gosalvez, J. (1999). Simultaneous decrease of telomerase length and telomerase activity with aging of human amniotic fluid cells. *J. Med. Genet.* **36**, 494–496.

Noort, W. A., Kruisselbrink, A. B., In't Anker, P. S., Kruger, M., van Bezooijen, R. L., de Paus, R. A., Heemskerk, M. H., Lowik, C. W., Falkenburg, J. H., Willemze, R., *et al.* (2002). Mesenchymal stem cells promote engraftment of human umbilical cord blood–derived CD34(+) cells in NOD/SCID mice. *Exp. Hematol.* **30**(8), 870–878.

Pittenger, M. F., Mackay, A. M., Beck, S. C., Jaiswal, R. K., Douglas, R., Mosca, J. D., Moorman, M. A., Simonetti, D. W., Craig, S., and Marshak, D. R. (1999). Multilineage potential of adult human mesenchymal stem cells. *Science* **284**(5411), 143–147.

Portmann-Lanz, C. B., Schoeberlein, A., Huber, A., Sager, R., Malek, A., Holzgreve, W., and Surbek, D. V. (2006). Placental mesenchymal stem cells as potential autologous graft for pre- and perinatal neuroregeneration. *Am. J. Obstet. Gynecol.* **194**(3), 664–673.

Prusa, A. R., and Hengstschlager, M. (2002). Amniotic fluid cells and human stem cell research: a new connection. *Med. Sci. Monit.* **8**(11), RA253–RA257.

Prusa, A. R., Marton, E., Rosner, M., Bernaschek, G., and Hengstschlager, M. (2003a). Oct-4-expressing cells in human amniotic fluid: a new source for stem cell research? *Hum. Reprod.* **18**(7), 1489–1493.

Prusa, A. R., Marton, E., Rosner, M., Freilinger, A., Bernaschek, G., and Hengstschlager, M. (2003b). Stem cell marker expression in human trisomy 21 amniotic fluid cells and trophoblasts. *J. Neural. Transm. Suppl* **67**, 235–242.

Rocha, V., Garnier, F., Ionescu, I., and Gluckman, E. (2005). Hematopoietic stem-cell transplantation using umbilical-cord blood cells. *Rev. Invest. Clin.* **57**(2), 314–323.

Romanov, Y. A., Svintsitskaya, V. A., and Smirnov, V. N. (2003). Searching for alternative sources of postnatal human mesenchymal stem cells: candidate MSC-like cells from umbilical cord. *Stem Cells* **21**(1), 105–110.

Streubel, B., Martucci-Ivessa, G., Fleck, T., and Bittner, R. E. (1996). [*In vitro* transformation of amniotic cells to muscle cells — background and outlook]. *Wien Med. Wochenschr.* **146**(9–10), 216–217.

Torricelli, F., Brizzi, L., Bernabei, P. A., Gheri, G., Di Lollo, S., Nutini, L., Lisi, E., Di Tommaso, M., and Cariati, E. (1993). Identification of hematopoietic progenitor cells in human amniotic fluid before the 12th week of gestation. *Ital. J. Anat. Embryol.* **98**(2), 119–126.

Vacanti, J. P., Morse, M. A., Saltzman, W. M., Domb, A. J., Perez-Atayde, A., and Langer, R. (1988). Selective cell transplantation using bioabsorbable artificial polymers as matrices. *J. Pediatr. Surg.* **23**(1 Pt. 2), 3–9.

Zhang, X., Mitsuru, A., Igura, K., Takahashi, K., Ichinose, S., Yamaguchi, S., and Takahashi, T. A. (2006). Mesenchymal progenitor cells derived from chorionic villi of human placenta for cartilage tissue engineering. *Biochem. Biophys. Res. Commun.* **340**(3), 944–952.

Zhao, P., Ise, H., Hongo, M., Ota, M., Konishi, I., and Nikaido, T. (2005). Human amniotic mesenchymal cells have some characteristics of cardiomyocytes. *Transplantation* **79**(5), 528–535.

# Chapter Twenty-Seven

# Immunomodulation

*Denise L. Faustman*

## I. INTRODUCTION

A long-standing goal of the transplant community has been to overcome immunologic rejection of transplanted tissues and organs. Although much of the research has been devoted to identifying new immunosuppressive agents and new combinations of existing agents for allotransplantation, immunosuppression still carries significant long-term side effects, especially enhanced susceptibility to infection. For xenografts, even stronger immunosuppression is necessary. One solution to allograft or xenograft acceptance may reside in novel therapeutic strategies that strive to prevent the need for long-term and high-dose immunosuppression.

Our laboratory has attempted to avert transplant rejection by immunologically modifying the foreign proteins on cells and tissues from the donor, instead of treating the host. Treated tissues and cells before transplantation allow the concealment or elimination of the antigens that summon immune rejection. This technology is sometimes referred to as *donor antigen modification* or, more stylistically, *designer tissue and organs* (Faustman and Coe, 1991). The goal is to avoid or reduce the need for immunosuppression by modi-

fying directly the graft at the molecular or protein level. Modification at the *cellular* level, by eliminating donor lymphoid cell populations, had been previously successful in murine allografts. These cellular modifications were effective for whole organs and cell transplants in lower species, even for the avoidance of xenograft rejection. Designer tissues and organs modified at the *molecular* or *protein* level offer an alternative approach with greater selectivity and latitude in tailoring the graft to escape immune detection by the host.

Without the risks of massive doses of systemic immunosuppression, designer tissues and organs could be therapeutic for a broad range of chronic conditions that are not life threatening. They also could be offered to conventional organ transplant patients at earlier stages of their disease, when patients are healthier and better able to withstand surgical intervention. The main risk with designer tissues and organs is that the modified donor tissue could be rejected, a risk commonly encountered with any transplant. Because designer tissue can potentially be rendered safer and may ultimately avoid host intervention with toxic drugs, a broader spectrum of diseases could be treated and at earlier time points.

*Principles of Tissue Engineering, 3rd Edition*
*ed. by Lanza, Langer, and Vacanti*

The purpose of this chapter is to describe the evolution of research on designer tissues and to trace its contribution to a growing body of transplant research. The first success was with murine hosts given transplants of xenogeneic cells and tissues that had been modified at the *protein* level. The success allowed extension of the concept to the molecular level, in the form of DNA modifications. The ability to modify proteins at the DNA, RNA, or protein level offers tremendous versatility in developing new cells or organs as biological therapies. Variations on the designer tissue concept are being probed in animal models of diabetes, solid organ failure, and neurological diseases. A concept launched in the laboratory in the early 1990s has already culminated in landmark human clinical trials for Parkinson's disease and Huntington's disease using donor-modified pig neurons in humans.

The wide range of applications of designer tissues and organs for allo- and xenotransplantation is readily apparent; modified donor cells, tissues, and organs can be considered potential therapies for an almost limitless number of conditions as long as the dominant antigens are identified and then effectively shielded or eliminated prior to implantation.

## II. ORIGIN OF THE DESIGNER TISSUE CONCEPT

The idea of targeting donor antigens instead of modifying the host immune system was stimulated by research on the molecular events surrounding the killing of cancer cells by cytotoxic T-lymphocytes (CTLs; hereinafter referred to as T-cells). In an elegant study, Spits and co-workers (1986) identified the sequential stages of T-cell destruction of the cancer cell. In contrast to earlier research, which focused primarily on the T-cell, they examined the roles of cell surface markers on both the T-cell and the target cell. By using antibodies against different cell surface markers on the cancer cells, they were able to block distinct obligatory stages of T-cell activation and destruction. They were able to tease apart *in vitro* the interactions between molecules on the T-cell and the target cell. Spits and colleagues were among the first to identify three stages in T-cell cytotoxicity: (1) adhesion between T-cell and target cell through two adhesion proteins, (2) T-cell receptor activation through class I, and (3) T-cell lysis of the tumor target cell through persistent class I and T-cell receptor binding.

A critical finding was that one of the major classes of histocompatibility antigens on the surface of the cancer cell — the major histocompatibility complex (MHC) class I molecules — was involved in both adhesion to and activation of the T-cell. Class I antigens and other classes of antigens encoded by the MHC complex serve to distinguish "self" from "nonself" because they differ across species and between members of the same species. Class I antigens had long been suspected of eliciting T-cell cytotoxicity, but this study offered more detailed insight into their pivotal role. What this study also demonstrated was that antibodies to class I antigens on the cancer cell could block T-cell adhesion and activation, thereby preventing lysis of the target cell.

Spits and coworkers crystallized the importance of class I antigens in immune rejection of cancer cells. While this study's goal was to dissect and promote CTL killing of the target, our goal was the opposite: to prevent tissue rejection at the same T-cell–target interface. We chose to "mask" class I antigens on donor cells by using antibody fragments to class I, a system that prevented death of the target, and then to transplant the modified donor cells into nonimmunosuppressed hosts.

## III. FIRST DEMONSTRATION OF THE CONCEPT

The first successful demonstration of the designer tissue concept used a xenogeneic model (Faustman and Coe, 1991). The cellular graft, in the form of insulin-secreting islets, was coated with antibody fragments to conceal class I antigens; the grafted cells (of human origin) indefinitely eluded the immune system of the murine host. The grafted cells also functioned normally. The host even developed tolerance to the treated graft because secondary transplants of untreated tissue were later accepted. The mechanism of donor-specific tolerance is still not fully defined, but it may involve induction of T-cell anergy through altered donor class I density. Altered class I density may be pivotal in T-cell shaping in both the periphery as well as the thymus, as defined by Pestano *et al.* (1999).

The graft consisted of purified human cadaveric islets that had been incubated with antibody fragments before being implanted into nonimmunosuppressed mice. Pure antibody fragments that lack the portion of the antibody molecule that binds complement, the Fc fragment, were obligatory to prevent lysis of the target. When the Fc fragment is enzymatically cleaved from the F(ab')$_2$ fragment, the purified F(ab')$_2$ fragment binds to the graft for several days without fixing complement; this prevents the graft from being destroyed. Grafts survived for 200 days and functioned appropriately, as determined by assays for human C' peptide, a proinsulin-processing product. Finally, human liver cells similarly treated with antibody fragments also survived for an extended period.

Treatment with whole antibodies to class I antigens failed to prolong graft survival; whole antibody class I proteins coated the donor cell but also, upon transplantation, killed the cell, due to host-derived complement. Treatments with antibodies and antibody fragments to CD29 was effective in an allogeneic transplant barrier using cells but not in a xenogeneic barrier. CD29 is an antigen with restricted expression on the passenger lymphocytes that accompany the graft. Passenger lymphocyte elimination can be important for allografts but less important for cross-species transplants. Treatment of cellular grafts with polyclonal antibody fragments to all antigenic determinants prior to transplantation did prolong allograft and xenograft survival. Polyclonal antibody fragments with class I antibody fragment removal had minimal effects at facilitating cellular transplants.

For the sake of rigor, it is generally accepted that xenografts represent a more challenging transplant barrier than allografts. Furthermore, if experimental test concepts are applied to xenografts with success, similarly enhanced allograft survival is likely to ensue. If designer tissues could succeed in a difficult xenogeneic case, then these same procedures could be considered for simpler allogeneic transplant models. Additionally, cellular transplant models represent the intermediate model of simplicity. In the setting of cells, the graft tissue contains one dominant antigen that needs to be masked. For instance, human islet tissue highly expresses class I antigens while displaying only minimal expression of two adhesion molecules, intercellular adhesion molecule-I (ICAM-I) and lymphocyte function–associated antigen-3 (LFA-3) (which in other tissues are thought to stabilize binding and to contribute to T-cell activation). Second, xenogeneic cellular transplants lack the vasculature that can be separated from the tissue to avoid hyperacute rejection, the earliest and most formidable barrier to discordant xenograft acceptance. Third, targeted antigens at the surface protein level rather than at the genetic level allow greater flexibility and less expense than the creation of genetically engineered pigs. Our goal was to conceal protein antigens that already appeared on the surface of graft cells. Other approaches, discussed later in this chapter, targeted antigens, not at the protein level, but at the DNA and RNA level. The use of more sophisticated transgenic and antisense technology, respectively, can similarly prevent antigen expression. It also can validate the donor antigens as a target for the immune response.

In summary, this tough xenogeneic model establishes the paramount role of class I antigens on cellular transplants. By providing a challenging test of the designer tissue concept, it also helped to launch a novel therapeutic strategy.

## IV. EXPANSION OF RESEARCH ON DESIGNER TISSUES

The potential of designer tissues to produce tissues and organs with reduced surface proteins can use a variety of donor antigen modification techniques in various transplantation settings. This body of early research is, in part, summarized in Table 27.1 and can be classified by the method that interferes with the expression of the surface protein of interest. The methods to remove or disguise surface proteins in the donor cells can include antibody "masking," gene ablation, and antisense and enzymatic treatments and are the topic of this chapter.

## V. ANTIBODY MASKING

Antibody masking of xenogeneic tissue is the first of the donor antigen modification techniques to progress to primate and human trials. Pancreatic islet cells and neurons are the most common types of donor cells to be camouflaged with antibody masking, although any type of tissue can theoretically be treated once the dominant antigens

| Table 27.1. | Designer donor tissues | | | |
|---|---|---|---|---|
| Technique/tissues | Allo/xeno | Donor/recipient | Target | Reference |
| **Masking antibody** | | | | |
| Islets | Xeno | Human/mouse | Class I | Faustman and Coe (1991) |
| Islets | Xeno | Human/mouse | All surface antigens | Faustman and Coe (1991) |
| Islets | Allo/xeno | Human or monkey/monkey | Class I | Steele et al. (1994) |
| Islets | Allo | Mouse/mouse | Class I | Osorio et al. (1994) |
| Neurons | Xeno | Pig/rat | Class I | Osorio et al. (1994), Pakzaban et al. (1995) |
| Neurons | Xeno | Pig/monkey | Class I | Burns et al. (1994) |
| Liver cells | Xeno | Human/mouse | Class I | Faustman and Coe (1991) |
| **Gene ablation** | | | | |
| Islets | Allo | Mouse/mouse | Class I | Markmann et al. (1992) |
| Islets | Allo | Mouse/mouse | Class I | Osorio et al. (1993) |
| Islets | Allo | Mouse/mouse | Class I | Osorio et al. (1994a, 1994b, 1994c, 1994d) |
| Liver cells | Allo/xeno | Mouse/mouse/guinea pig/frog | Class I | Li and Faustman (1993) |
| Kidneys | Allo | Mouse/mouse | Class I | Coffman et al. (1993) |

have been identified. All of the studies cited here targeted class I antigens using antibody fragments and relate to cellular transplants.

Islet cell masking for the treatment of diabetes was early on pursued, with mixed results. Osorio and colleagues (1994a, 1994b, 1994c, 1994d) targeted class I antigens in a mouse allograft model. To approximate a diabetic state, the mice first were treated with a drug that chemically induced hyperglycemia. Then they received islet allografts that had been pretreated with antibody fragments. Graft survival was prolonged relative to controls, but within one month the grafts were eventually rejected. Investigators attributed eventual allograft rejection to a variety of possibilities, including the absence of sufficient quantities of $F(ab')_2$ fragments, antiidiotypic antibodies against the $F(ab')_2$ fragment, and an immune pathway independent of class I activation of T-cells. Steele and colleagues (1994) investigated islet cell transplants in a primate model. Cynomolgus monkeys received either allogeneic or xenogeneic (human) islets. The grafts were pretreated with antibody fragments to class I antigens. Histologic evidence revealed that donor islets were present months after transplantation into nonimmunosuppressed monkeys.

Neuronal xenografts with antibody masking have been investigated for Huntington's disease (Pakzaban *et al.*, 1995) and Parkinson's disease (Burns *et al.*, 1994). In the first study, fetal pig striatal cells were implanted into rats whose striatum had been lesioned one week earlier with injections of quinolinic acid. These injections destroy striatal neurons in an attempt to simulate the dysfunction present in Huntington's disease. Rats received either untreated tissue or tissue pretreated with $F(ab')_2$ fragments against porcine class I antigens. Control rats receiving untreated tissue were either immunosuppressed with immunosuppressant cyclosporin A (CsA) or left untreated. Three to four months later, graft survival was found to be prolonged in animals given $F(ab')_2$-treated grafts and in the CsA-treated control animals given unaltered grafts. Grafts did not survive in nonimmunosuppressed controls. Graft volume, determined histologically with the aid of computer image analysis, was significantly larger in the CsA group as compared with the $F(ab')_2$ group. Yet in both of these groups, immunohistochemistry revealed graft cytoarchitecture to be well organized and graft axons to have grown correctly in the direction of their target nuclei. It was encouraging that pig neurons appeared to be capable of locating their murine target.

In a similar study design, Burns and coworkers (1994) applied antibody masking to a primate model of Parkinson's disease. Porcine mesencephalic neuroblasts were implanted into monkeys with 1-methyl-4-phenyl-1,2,3,6-tetrahydropyridine (MPTP)-induced Parkinsonism. Measures of locomotor activity and dopamine fiber density in the host striatum confirmed that pretreatment of donor tissue with $F(ab')_2$ fragments succeeded in restoring motor function and replenishing dopamine fibers at the site of implantation. A control animal, maintained on CsA after receiving untreated cells, showed similar improvement. A nonimmunosuppressed control showed no improvement after transplantation, suggesting graft rejection.

The studies described in this section highlight the range of potential applications of antibody masking in both allo- and xenotransplants. Neuronal cell masking has been sufficiently effective to usher in human clinical trials.

## VI. GENE ABLATION

Gene ablation, or gene "knockout" technology, offers another vehicle to modify donor tissue and organs. With gene ablation, the gene or genes encoding protein antigens can be permanently deleted, thereby eliminating antigen expression in all cells. In one common method of gene ablation, the target gene is inactivated in cultured embryonic stem cells via homologous recombination; a target vector containing an inactive version of the gene recombines with, and thereby replaces, the wild-type gene. Through reintroduction of the embryonic cells into a foster mother and through selective breeding, progeny can be produced that are homozygous for the mutation. Rejection by the host immune system is expected to be avoided when the targeted gene encodes a protein antigen slated for expression on the surface of donor cells.

The major advantage of gene ablation over protein modification of donor tissue is that the antigen is permanently eliminated in all cells, tissues, and organs in which it is expressed. The major limitation of this technology is that it is restricted to potential xenografts from pigs and requires pig "farming," an expensive, albeit potentially abundant, source of donor tissue. The permanent nature of the modification also can sometimes be a limitation, especially when the protein encoded by the gene has additional functions. Inactivation of the gene might lead to physiological changes that compromise the utility of the donor tissue. Gene ablation can also target genes that encode proteins essential for the target protein expression.

Several transplantation laboratories have exploited the availability of knockout mice deficient in $\beta_2$-microglobulin. $\beta_2$-microglobulin is a peptide that performs a chaperone function as part of the class I molecule and is necessary for its assembly and expression. Mice homozygous for the $\beta_2$-microglobulin mutation fail to display class I antigens on the cell surface, a feature that makes them highly desirable for transplantation studies. Class I ablation through the mutation are not necessarily a permanent depletion of class I, due to host $\beta_2$-microglobulin reconstituting the graft. Although much of the research described later has focused on islet cell and liver cell xenotransplants, one project examined whole-organ allografts. Coffman and coworkers (1993) found that kidneys from $\beta_2$-microglobulin-deficient mice functioned significantly better in allogeneic recipients than did kidneys from normal mice.

The fate of pancreatic islet transplants from $\beta_2$-microglobulin-deficient mice has been explored in several studies. The islets showed prolonged survival when implanted into a normal mouse strain (Markmann *et al.*, 1992; Osorio *et al.*, 1993). Most grafts survived indefinitely (>80% beyond 100 days) and were capable of reversing hyperglycemia that had been chemically induced prior to transplantation. Investigators attributed the few instances of graft rejection to three possibilities. First, some surface expression of class I antigens can occur in the absence of $\beta_2$-microglobulin. The presence of even a fraction of the total number of class I antigens may be sufficient for T-cell recognition and lysis. Second, $\beta_2$-microglobulin circulating in the serum of the recipient can be used to reconstitute class I antigens on the donor tissue (because $\beta_2$-microglobulin is a highly conserved protein not encoded by genes at the MHC locus). Third, rejection of the tissue from $\beta_2$-microglobulin-deficient donors may be mediated by other immune pathways, such as by natural killer cells. Support for the second and third possibilities was presented by Li and Faustman (1993) in their study of liver cell allo- and xenografts.

For whole-organ xenografts, it has long been recognized that certain sugars on donor tissues, especially the vascular endothelium, such as gal $\alpha(1,3)$galactosyltransferase, have a central role in hyperacute rejection. This hyperacute rejection is mediated by preformed antibodies in disparate species, often referred to as *natural* antibodies (Sandrin and McKenzie, 1994; Sandrin *et al.*, 1993). If hyperacute rejection of whole-organ xenografts can be prevented, then other cellular barriers, such as disparate class I expression, can be addressed.

Given the important role of donor sugar epitopes eliciting rapid organ rejection, knockout pigs were produced, with the first strategy applied to donor antigen modification for this epitope. The pigs had complete elimination of the gene that encoded $\alpha(1,3)$galactosyltransferase. To the surprise of the scientific community, the knockout pigs continued to express considerable levels of the carbohydrate sugars.

Therefore, in the pig, more than one genetically encoded enzyme is able successfully to synthesize Gal $\alpha$ (1,3) gal surface sugars (Milland *et al.*, 2006). These studies demonstrate that although three prior methods can change donor antigen expression in cells, i.e., gene ablation, antisense for RNA, and "masking" antibodies, some methods may have unexpected shortcomings specific to the interruption method.

## VII. RNA ABLATION

RNA ablation is another strategy that strives to prevent antigen expression by blocking gene transcription or translation. RNA ablation can be achieved through the creation of oligodeoxynucleotides that are complementary to, and hybridize with, DNA or RNA sequences to inhibit transcription or translation, respectively.

To date RNA ablation in the transplant field has been less frequently pursued. Ramanathan and coworkers (1994) identified an oligodeoxynucleotide that inhibited induction of class I and ICAM expression by interferon-$\gamma$. The studies were performed *in vitro* in a cell line, K562, that normally has low-level expression of class I antigens. They first postulated that the oligodeoxynucleotide acted in the early stages of interferon-$\gamma$ induction rather than posttranslationally. In their follow-up study, they showed that the oligodeoxynucleotide acted even earlier via a novel mechanism: It inhibited binding of interferon-$\gamma$ to the cell surface (Ramakrishnan and Houston, 1984; Ramanathan *et al.*, 1994). This may be an unusual mechanism for an oligodeoxynucleotide, but it only enhances the possibilities for xenotransplant research. RNA ablation, unlike protein modifications with masking or DNA modifications in transgenic donors, offers an additional challenge. For RNA ablation to work, all cells must be treated equally with the interfering RNA. Antisense technology or even newer methods, such as siRNA, often result in uneven distributions of interruptions in the cells or organs. What these studies provide is yet another means of blocking expression of class I antigens or other transplantation antigens.

## VIII. ENZYME ABLATION

Although MHC class I surface structures are extremely polymorphic, the polymorphisms of MHC class I are confined predominantly to the exterior region of the protein and the regions of the protein that are exterior facing. This feature allows the host's immune system, i.e., T-cells, the early opportunity to reject tissues transplanted from unrelated individuals.

As the genome effort has advanced and hundreds of MHC class I alleles of the gene have been sequenced, it has also become apparent that certain regions of the MHC structure are highly polymorphic, in contrast to other regions of the protein that are highly conserved. Even across species, the conserved regions of the MHC allele compose the protein portions that are near and within the cell surface membrane. This conservation of structure is maintained across species as diverse as mice and humans.

The conserved regions of MHC class I within and across species have afforded, in solubilized cells, the opportunity specifically to purify these proteins by enzymatic cleavage. Most typically, cell lysates have been treated with papain to solubilize all MHC structures from the cell surface, usually at very acid pH (Ezquerra *et al.*, 1985). Papain has the remarkable ability to cleave the conserved hydrophilic regions of the class I protein without cleaving other cellular proteins (Springer and Strominger, 1976). Papain's conserved-cleavage feature is preserved for mouse, human, and primate class I alleles.

We have gradually worked with this system in a mouse transplant model to allow papain to cleave class I from all cells and have the cells remain fully viable at the end of the brief treatment. This MHC class I modification method with pretreatment with papain, with modified buffers at physiological pH, now allows complete class I depletion and subsequent transplantation of cell transplants, with prolonged survival. Survival of the transplanted cells, in the form of insulin-secreting islets or liver cells, is comparable to either methods of genetically ablated class I–deficient tissues or "masked" class I–deficient tissues.

Each method of modifying donor cell surface proteins, such as MHC class I, prior to transplantation has features that are unique to the translation of these methods to clinical trials (Table 27.2). The production of genetically engineered donor pigs permits the generation of abundant tissue, but the donor tissue will be xenografts and thus encounter tough transplant barriers and xenosis risk. Also, the cost of developing the herds of animals is very expensive. The generation of masking antibodies allows the donor tissue modifications to be applied to allografts or xenografts. The masking antibodies are protein fragments and can be designed to different donor HLA types. The production of masking antibodies is easier than the production of genetically engineered pigs but requires manufacturing of a novel

protein fragment and standardization for FDA approval. Also, the preclinical efficacy studies using masking antibodies would be specific for the species chosen, so the translation to human studies would not use the identical antibody fragment. Finally, the use of specific enzymes to cleave off donor antigens, such as MHC class I, is inexpensive to develop, and the same enzyme product could be applied to preclinical murine studies, baboon studies, and human indications. All three donor tissue modifications could be combined with reduced or eliminated immunosuppression, thus increasing the safety margins for cellular transplants for nonlethal diseases.

## IX. MECHANISMS OF GRAFT SURVIVAL AFTER CLASS I DONOR ABLATION OR ANTIBODY MASKING

Research on the ability of class I modified tissues or cells to survive long term without host immunosuppression has been closely studied. Indeed it was appreciated that although the method for altering donor class I expression prior to transplantation could be diverse, i.e., masking antibodies or gene ablation of chaperone proteins or

**Table 27.2.** Positive and negative features of different methods of donor antigen modification compared to systemic immunosuppressant

| | Host treatment | Donor tissue modification[a] | | |
| | Immunosuppression | Donor Modification | | |
| | Chemicals/Biologics | Transgenic pigs | Masking Abs | Enzymatic |
|---|---|---|---|---|
| Experimental time for efficacy testing | | | | |
|   Animals | +++ | +++ | ++ | ++ (limited) |
|   Humans | +++ | N/A | +/− | N/A |
| Safety | | | | |
|   Side effects profiles | +++ | None | None +/− | None |
| Applicability/market size | | | | |
|   Allografts | +++ | 0 | +++ | +++ |
|   Xenografts | 0 | +++ | +++ | +++ |
| Development costs and time | | | | |
|   Basic research | Large | Very large | Average | Inexpensive |
|   Clinical testing | Large | Very large | Average | Moderate |
| Manufacturing | | | | |
|   Feasibility | Moderate | Difficult | Moderate | Easy |
|   Costs | Moderate | High | Moderate | Low |
| Regulatory/FDA | | | | |
|   Frequency | Proven | No prior approvals; xenosis | No prior approvals | +/− |
|   Time line | Average | Very long | Long | Short |

[a]Antisense or siRNA technology has had limited success in settings of cell or tissue transplantation and thus is not represented in the table as a method for donor antigen modification.

enzymes, a common theme in these models was long-term stable survival of the functioning tissue. The transplants were viable and certainly reexpressed donor class I with the passage of time. Insight into the role of systemic tolerance in these transplants is important for future refinements of cellular transplant methods to optimize success.

Systemic tolerance to the graft was initially addressed for donor tissues with modified class I due to ablation of the class I chaperone protein, i.e., $\beta_2$-microglobulin. First, as the donor and recipient diverge phylogenetically, so does the homology of the $\beta_2$-microglobulin proteins. Donor grafts from highly divergent species have slower reconstitution of surface class I with $\beta_2$-microglobulin from the host serum and shorter survival times (Hyafil and Strominger, 1979). The shortened survival times as compared to less divergent cross-species transplants suggests that some degree of donor class I antigen expression is beneficial after the transplant is established. Indeed, many have proposed that the intact natural killer cells of the host, a natural defense against class I–deficient tumors, may be the reason for more brisk rejection with more permanent class I depletion.

The reconstitution of donor surface class I with host $\beta_2$-microglobulin in the gene ablation model also explains why transient ways of interruption class I allows systemic tolerance. The primary cellular grafts with transient class I interruption permit secondary nongenetically manipulated or non-antibody-masked grafts to survive — if from the same donor (Faustman and Coe, 1991). The data using the masking method to conceal class I, taken together, suggest donor-specific tolerance occurs in these transplant models.

Insights into the mechanisms of T-cell tolerance in hosts receiving transient class I–ablated grafts may have been clarified. In a publication by the Cantor laboratory, peripheral CD8 T-cell tolerance was mechanistically characterized in terms of maintenance of peripheral tolerance (Pestano et al., 1999). It has been recognized in some experimental settings that the persistence of peripheral class I–expressing cells is necessary for peripheral CD8+ T-cell tolerance (Vidal-Puig and Faustman, 1994). Using $\beta_2$-microglobulin-deficient mice, transferred and potentially cytotoxic and mature CD8 T-cells from a normal donor were transferred into the mutant host. The transferred T-cells then failed to engage their T-cell receptors. The CD8 cells down-regulated their CD8 gene expression and underwent apoptosis. Thus, inhibition or interference of the T-cell receptor and CD8 binding to host class I triggered these CD8 cells into a pathway of cell death. This is an important observation, and lack of class I expression immediately after transplantation eliminated potential direct and immediate T-cell killing. As the genetically modified graft gradually reexpresses class I, it is protected from the next layer of the immune response, natural killer cell lysis of totally class I–negative transplants, and active host tolerance is additionally achieved.

Elimination of class I antigens also may be insufficient for select tissues and select combinations of donor and host species. Liver cells from $\beta_2$-microglobulin-deficient donors were implanted into allogeneic recipients and two different xenogeneic recipients (Li and Faustman, 1993). The allotransplants and the xenotransplants into guinea pigs were not effective. In contrast, xenotransplants of mouse cells into frog recipients survived; when liver cells were transplanted from humans to mice in antibody-masking experiments, they were accepted (Faustman and Coe, 1991). These studies show that results of class I antigen removal vary according to the species combination of donor and host.

Gene ablation experiments have demonstrated the advantages and limitations of eliminating class I antigens. They are clearly dominant antigens on islet cells and neurons in some allogeneic and xenogeneic combinations. However, class I antigens may play a secondary role, depending on the type of donor tissue, the species combination, and/or the disease state of the recipient. In most cases, reexpression of class I antigens on the donor cells was beneficial for long-term survival. Indeed, with antibody "masking" to class I, secondary transplants from the same donor were possible if the primary transplant was still in the host (Faustman and Coe, 1991).

Rare cases of stable allograft acceptance in humans who discontinue their immunosuppressive regimens have been documented in the setting of whole organs, such as kidney transplants. A study of a mouse model found that preengraftment of donor cells bearing single low-dose foreign MHC class I allele resulted in lifelong donor cell acceptance and an immune system that both *in vivo* and *in vitro* was unresponsive. Similar to the foregoing data presented with temporary class I ablation, removal of the primary transplant reversed, in a time-dependent fashion, the systemic tolerance. This suggests that the transplants can actively maintain host unresponsiveness toward a single MHC class I allele by continuously inactivating a reactive T-cell (Bonilla et al., 2006).

# X. ROLE OF CLASS I MODIFICATIONS IN RESISTANCE TO RECURRENT AUTOIMMUNITY

The fate of pancreatic islet transplants from $\beta_2$-microglobulin-deficient mice has been explored in several studies using chemically induced mice as well as spontaneously diabetic mice. NOD mice are a well-recognized model of spontaneous type 1 diabetes and have a long prodrome of prediabetes from six to eight weeks of age prior to spontaneous diabetes at approximately 18–24 weeks of age, the stage where islet destruction sufficient for hyperglycemia occurs. Although clinically asymptomatic, this stage of prediabetes is the most active phase of the disease.

Class I–depleted islets, either with masking antibodies or from donors with ablation of the $\beta_2$-microglobulin gene, show prolonged survival when implanted into allogeneic mouse strains without autoimmunity (Markmann *et al.*, 1992; Osorio *et al.*, 1993). Additional experiments also tested the transplantation of $\beta_2$-microglobulin hosts into the autoimmune-prone NOD mouse strain, both as prediabetic mice as well as fully diabetic mice. When genetically modified islets were transplanted into young NOD, not yet diabetic, graft rejection due to ongoing autoimmunity occurred almost invariably (Markmann *et al.*, 1992). Although NOD mice at this stage are clinically asymptomatic due to some islets still surviving, it is the most active phase of disease. Soon after implantation to prediabetic NOD mice, but at a slightly delayed rate, islet allografts from $\beta_2$-microglobulin-deficient mice or class I–masked xenogeneic islets are rejected after a twofold increased survival beyond control islets. This modest prolongation is in contrast to the almost complete success with recipients whose diabetes is chemically induced or when the NOD mouse are already hyperglycemic. Fully diabetic NOD mice, receiving MHC class I–deficient islets, demonstrate indefinite survival of the transplants (Young *et al.*, 2004). In conclusion, class I–deficient islets, rendered deficient by a number of methods, can show modest to dramatic prolongations in murine hosts with different stages of diabetic autoimmunity.

## XI. LAUNCHING OF XENOGENEIC HUMAN CLINICAL TRIALS IN THE UNITED STATES USING IMMUNOMODULATION

Few cross-species transplantation technologies have progressed to human clinical trials. In part this has been due to primate models showing minimal efficiency. Also, early whole-organ human clinical trials in the 1970s demonstrated minimal success, even with massive dosages of immunosuppressives. Finally, some segments of the medical community have been concerned about cross-species infections, therefore necessitating new technologies to avoid the use of donor species closely related to humans, e.g., baboons. The testing of novel transplantation approaches needs to avoid the limitation of a severely compromised host immune system with decreased resistance to fight infections.

Clinically close primate and rat models are available for neurological diseases such as Parkinson's disease. The approach of masking class I antibody fragments shows promise for porcine neurons for spinal cord injuries in rat models and porcine liver cells for transient liver failure from hypotension or infection. Fetal neuronal xenografts with antibody masking have been investigated for Huntington's disease and Parkinson's disease (Deacon *et al.*, 1997). In the first study, fetal pig striatal cells were implanted into rats whose striatum had been lesioned one week earlier with injections of quinolinic acid. These injections destroy striatal neurons and attempt to simulate the dysfunction present in Huntington's disease. Rats received either untreated

tissue or tissue pretreated with $F(ab')_2$ fragments against porcine class I antigens. Control rats received untreated tissue and were immunosuppressed with CSA, and others were nonimmunosuppressed. Graft volume, determined histologically with the aid of computer image analysis, was significantly larger and well organized, and graft axons had grown correctly in the direction of their target nuclei. It was encouraging that pig neurons appeared to be capable of locating their target. Also, because of the use of fetal tissue, the transplanted mass had become significantly enlarged at the time of autopsy, demonstrating posttransplantation survival as well as growth of the transplant. Six Parkinson's patients have now been treated with fetal pig neurons masked with class I antibody fragments and have reported long-term survival (>2 years), with patients demonstrating mild to marked functional improvements. Six more patients were similarly treated with fetal pig neurons and CSA. These patients showed less clinical improvement but still had function exceeding baseline. At eight months, one of these patients died of a thromboembolic event and an autopsy was performed. As reported by Deacon *et al.* (1997), similar to the primate studies performed before clinical trials, the fetal pig neurons survived and correctly sent out axons over long distances in the brain toward their target nuclei. This confirmed the optimism of the suitability of using pig tissue in this transplant setting and the masking approach to decrease tissue immunogenicity.

Additional clinical trials of masked neuron transplants have continued in the United States. To date, an additional 11 patients in a Food and Drug Administration (FDA)–scrutinized clinical trial have been enrolled in a phase II/III trial using pig cells for Parkinson's disease. The FDA has approved 36 patients for 18 months in this blinded human study for safety and efficacy. The advantages of the donor antigen modification methods include the lack of host interventions, thus allowing a broader audience for applications of cellular transplants for disease treatments.

## XII. COMMENT

Designer tissues and organs, achieved through donor antigen modification, hold tremendous promise for xenotransplantation and allotransplantation. Research in animal models has already demonstrated that long-term xenograft survival can be achieved without immunosuppression. This achievement has galvanized the transplantation community, for it shows that an overwhelming obstacle to graft acceptance can be alleviated in select settings. Immune rejection need not occur if graft antigens can be immunologically masked, enzymatically cleaved, or genetically eliminated. Researchers now have at their disposal a battery of techniques that operate at the DNA, RNA, or protein level to remove or conceal antigens. Improved xeno- or allogeneic transplantation is within reach, not just for patients with life-threatening conditions, but also for patients with chronic conditions.

The successes with cellular and tissue grafts still are not the final solution for the far more difficult task of whole-organ xenotransplantation or recurrent autoimmune disease. Solid organs have a multiplicity of antigens, particularly those that elicit hyperacute rejection from differences in the expression of sugars between species. Once all domi-nant antigens are identified, the therapeutic strategy for whole organs and tissues is conceptually identical: Modify the donor, not the host. The barriers for recurrent autoimmune disease still stand, but it is hoped that donor antigen modification may also be beneficial in this setting during select times in the disease process.

## XIII. REFERENCES

Bonilla, W. V., Geuking, M. B., Aichele, P., Ludewig, B., Hengartner, H., and Zinkernagel, R. M. (2006). Microchimerism maintains deletion of the donor cell–specific CD8+ T-cell repertoire. *J. Clin. Invest.* **116**, 156–162.

Burns, L. H., Pakzaban, P., Deacon, T. W., Dinsmore, J., and Isaacson, O. (1994). Xenotransplantation of porcine ventral mesencephalic neuroblasts restores function in primates with chronic MPTP-induced parkinsonism. *Soc. Neurosci.* **19**, 1330.

Coffman, T., Geier, S., Ibrahim, S., Griffiths, R., Spurney, R., Smithies, O., Koller, B., and Sanfilippo, F. (1993). Improved renal function in mouse kidney allografts lacking MHC class I antigens. *J. Immunol.* **151**, 425–435.

Deacon, T., Schumacher, J., Dinsmore, J., Thomas, C., Palmer, P., Kott, S., Edge, A., Penney, D., Kassissieh, S., Dempsey, P., and Isacson, O. (1997). Histological evidence of fetal pig neural cell survival after transplantation into a patient with Parkinson's disease. *Nat. Med.* **3**, 35–353.

Ezquerra, A., Bragado, R., Vega, M. A., Strominger, J. L., Woody, J., and Lopez de Castro, J. A. (1985). Primary structure of papain-solubilized human histocompatibility antigen HLA-B27. *Biochemistry* **24**, 1733–1741.

Faustman, D., and Coe, C. (1991). Prevention of xenograft rejection by masking donor HLA class I antigens. *Science* **252**, 1700–1702.

Hyafil, F., and Strominger, J. L. (1979). Dissociation and exchange of the $\beta_2$-microglobulin subunit of HLA-A and HLA-B antigens. *Proc. Natl. Acad. Sci., U.S.A.* **76**, 5834–5838.

Li, X., and Faustman, D. (1993). Use of donor $\beta_2$-microglobulin-deficient transgenic mouse liver cells for isografts, allografts, and xenografts. *Transplantation* **55**, 940–946.

Markmann, J. F., Bassiri, H., Desai, N. M., Odorico, J. S., Kim, J. I., Koller, B. H., Smithies, O., and Barker, C. F. (1992). Indefinite survival of MHC class I–deficient murine pancreatic islet allografts. *Transplantation* **54**, 1085–1089.

Milland, J., Christiansen, D., Lazarus, B. D., Taylor, S. G., Xing, P. X., and Sandrin, M. S. (2006). The molecular basis for Gal{alpha}(1,3)Gal expression in animals with a deletion of the {alpha}1,3galactosyltransferase gene. *J. Immunol.* **176**, 2448–2454.

Osorio, R. W., Ascher, N. L., Jaenisch, R., Freise, C. E., Roberts, J. P., and Stock, P. G. (1993). Major histocompatibility complex class I deficiency prolongs islet allograft survival. *Diabetes* **42**, 1520–1527.

Osorio, R. W., Ascher, N. L., Melzer, J. S., and Stock, P. G. (1994a). Beta-2-microglobulin gene disruption prolongs murine islet allograft survival in NOD mice. *Transplant. Proc.* **26**, 752.

Osorio, R. W., Ascher, N. L., Melzer, J. S., and Stock, P. G. (1994b). Enhancement of islet allograft survival in mice treated with MHC class I specific F(ab')2 alloantibody. *Transplant. Proc.* **26**, 749.

Osorio, R. W., Ascher, N. L., and Stock, P. G. (1994c). Prolongation of *in vivo* mouse islet allograft survival by modulation of MHC class I antigen. *Transplantation* **57**, 783–788.

Osorio, R. W., Asher, N. L., Melzer, J. S., and Stock, P. G. (1994d). Beta-2-microglobulin gene disruption prolongs murine islet allograft survival in NOD mice. *Transplant. Proc.* **26**, 752.

Pakzaban, P., Deacon, T. W., Burns, L. H., Dinsmore, J., and Isaacson, O. (1995). Enhanced survival of neural xenografts after masking of donor major histocompatibility complex class I. *Soc. Neurosci.* **16**, 1708.

Pestano, G. A., Zhou, Y., Trimble, L. A., Daley, J., Weber, G. F., and Cantor, H. (1999). Inactivation of misselected CD8 T-cells by CD8 gene methylation and cell death. *Science* **284**, 1187–1191.

Ramakrishnan, S., and Houston, L. L. (1984). Inhibition of human acute lymphoblastic leukemia cells by immunotoxins: potentiation by chloriquine. *Science* **233**, 58–61.

Ramanathan, M., Lantz, M., MacGregor, R. D., Garovoy, M. R., and Hunt, C. A. (1994). Characterization of the oligodeoxynucleotide-mediated inhibition of interferon-gamma-induced major histocompatibility complex class I and intercellular adhesion molecule-1. *J. Biol. Chem.* **269**, 24564–24574.

Sandrin, M. S., and McKenzie, I. F. (1994). Gal alpha (1,3)Gal, the major xenoantigen(s) recognized in pigs by human natural antibodies. *Immunol. Rev.* **141**, 169–190.

Sandrin, M. S., Vaughan, H. A., Dabkowski, P. L., and McKenzie, I. F. (1993). Anti-pig IgM antibodies in human serum react predominantly with Gal(alpha 1-3)Gal epitopes. *Proc. Natl. Acad. Sci. U.S.A.* **90**, 11391–11395.

Spits, H., van Schooten, W., Keizer, H., van Seventer, G., van de Rijn, M., Terhorst, C., and de Vries, J. E. (1986). Alloantigen recognition is preceded by nonspecific adhesion of cytotoxic T-cells and target cells. *Science* **232**, 403–405.

Springer, T. A., and Strominger, J. L. (1976). Detergent-soluble HLA antigens contain a hydrophilic region at the COOH-terminus and a penultimate hydrophobic region. *Proc. Natl. Acad. Sci. U.S.A.* **73**, 2481–2485.

Steele, D. J. R., Hertel-Wulff, B., Chappel, S., Wallstrom, A., Bleier, K., Tsang, W. G., Austen, J., and Auchincloss, H. (1994). Long-term survival of pancreatic islets in diabetic monkeys. *Cell Transplant.* **3**, 216.

Vidal-Puig, A., and Faustman, D. L. (1994). Tolerance to peripheral tissue is transient and maintained by tissue specific class I expression. *Transplant. Proc.* **26**, 3314–3316.

Young, H. Y., Zucker, P., Flavell, R. A., Jevnikar, A. M., and Singh, B. (2004). Characterization of the role of major histocompatibility complex in type 1 diabetes recurrence after islet transplantation. *Transplantation* **78**, 509–515.

# Chapter Twenty-Eight

# Immunoisolation

*Beth A. Zielinski and Michael J. Lysaght*

## I. INTRODUCTION

Replacement of the vital functional physiology of deteriorated or irreparably damaged native organs has been the goal of both transplantation medicine and immunoisolatory medicine. Although transplants of allogeneic tissue carry the promise of complete metabolic restoration, supply is constrained and the side effects of an effective immunosuppressive regimen are far from benign. Direct administration of therapeutic proteins designed to replace metabolic deficiencies are limited by administration, enzymatic degradation, inability to maintain therapeutic levels of drug, bioavailability, and, ultimately, cost. Immunoisolation and transplantation of protein-secreting cells is a viable option for the replacement of protein-secreting tissues and potentially of the function of whole organs.

Immunoisolation, or encapsulated cell therapy, is the process of encapsulating or sequestering metabolically active cells within a selective membrane barrier. This membrane allows for bidirectional diffusion of nutrients, oxygen, secretogues, and bioactive cell secreting while limiting the entry of host immune molecules and cells that could potentially destroy the cellular implant. In addition to delivering potentially inexhaustible supplies of therapeutic protein, implants such as these may also be responsive to metabolic changes in the patient, which results in feedback-mediated modification of cellular secretions.

The first serious investigative efforts in the field of immunoisolation began with the implantation of encapsulated islets for the treatment of diabetes. Chick *et al.* (1977)

and Lim and Sun (1980) successfully maintained glucose homeostasis in chemically induced diabetic rats using encapsulated allogeneic or syngeneic islets. Since that time, strides in the field of encapsulated cell therapy have been made not only in the area of diabetes but also in the areas of chronic pain (Sagen *et al.*, 1993; Joseph *et al.*, 1994; Decostard *et al.*, 1988), neurodegenerative diseases, including Parkinson's disease (Tresco *et al.*, 1992; Aebischer *et al.*, 1994a; Tseng *et al.*, 1997; Sautter *et al.*, 1998), amyotrophic lateral sclerosis (ALS) (Sagot *et al.*, 1995; Tan *et al.*, 1996), dwarfism (Chang *et al.*, 1993), anemia (Koo and Chang, 1993; Rinsch *et al.*, 1996), hemophilia (Colton, 1996; Brauker *et al.*, 1992), and cancer (Geller *et al.*, 1997b). Human trials have been initiated for diabetes (Soon-Shiong *et al.*, 1994; Scharp *et al.*, 1994), chronic pain (Aebischer *et al.*, 1994b; Buscher *et al.*, 1996), ALS (Ezzell, 1995; Aebischer *et al.*, 1996), and macular degeneration (Tao *et al.*, 2002).

This chapter addresses the principles of immunoisolation and the technological developments underlying progress in this field. The theory of immunoisolation, capsule format criteria, cell sourcing, and the issues of immunological recognition and rejection are discussed. Finally, modifications to the design of the immunoisolatory system are proposed for future study and review.

## II. THEORY AND CAPSULE FORMAT

Immunoisolation is based on the premise that allogeneic and xenogeneic cells, once sequestered within a selectively permeable membrane, are protected, completely or in part, from host immune destruction and are able to deliver specific therapeutic proteins to the host over an extended period of time. In addition to minimizing the vulnerability of transplanted cells to immune-mediated destruction, the semipermeable membrane also prevents outgrowth of the encapsulated cells into host parenchyma. This facilitates the use of mitotically active cells. Diffusion of oxygen, carbon dioxide, soluble nutrients, signaling molecules, and bioactive cellular secretions, including therapeutic proteins, allows for both sustained viability of transplanted cells and delivery to the host of the therapeutic molecules of interest. Membranes can be fabricated from either natural or synthetic material and can be designed to have various pore sizes, depending on the intended application. Membranes are classified according to their nominal molecular weight cutoffs (MWCOs). The MWCO determines the size of molecules that are able to diffuse into and out of the capsule. The molecular specificities of the membranes are approximate, since pore sizes are not uniform but, rather, vary widely around a mean. Although the MWCOs govern the movement of most molecules larger than the selective range, outlier pores can result in "leakage" of small quantities of larger molecules. This can lead to unintended sensitization of the host and immune vulnerability of transplanted cells. On the other hand, molecules within the selective range may be restricted from entering or leaving the capsule due to steric hindrance, charge, and hydrophobicity. This could lead to restricted delivery of therapeutic protein. Available barriers range from those considered semipermeable, with an MWCO of 30 kDa, to those considered microporous, having pore sizes up to 0.6 microns. Immunogenicity of the encapsulated cell types plays a key role in governing the selection of a membrane with the appropriate MWCO (Colton, 1996). Semipermeable membranes are usually considered for those applications requiring xenogeneic and highly immunogene allogeneic cells. Microporous membranes are more appropriate for those applications where larger amounts of soluble proteins are to be delivered and long-term viability of encapsulated cells is not a primary requirement. For example, this strategy could be employed for tumor cell encapsulation, where transient release of tumor antigen from encapsulated tumor cells results in host sensitization, generation of antitumor immune responses, and *in situ* as well as encapsulated tumor cell destruction (Geller *et al.*, 1997b).

The types of immunoisolatory systems used since the early 1980s can be categorized into two main groups: vascular perfusion devices that are implanted directly in contact with the host's circulatory system (Sullivan *et al.*, 1991; Maki *et al.*, 1993) and nonvascular devices that are implanted subcutaneously, intramuscularly, or intraperitoneally. Vascular perfusion devices have waned in popularity as a result of their intrinsic complexity and the need for long-term anticoagulation in order to prevent thrombosis. Recently, however, resurgence in the development of the artificial liver and artificial kidney has led to increased preclinical and clinical activity in this format (Stange *et al.*, 2002; Humes *et al.*, 2004).

Nonvascular devices can be subdivided into two classifications: spherical microcapsules and larger, polymer-based macrocapsules (Geller *et al.*, 1997a). Macrocapsules can be designed as sealed cylindrical hollow fibers, flat sheet, and planar devices (Ezzell, 1995). Although all of these devices are very different in configuration, they are engineered with one dimension below 1 mm in order to maximize bidirectional diffusion of nutrients and cellular secretogues. Conformal coatings of cells have also been investigated as a way to reduce diffusion distances between the encapsulated cells and the host interstitum (Hill *et al.*, 1997).

Spherical microcapsules can be been formed from organic polymers such as sodium alginate and poly-L-lysine (Lim and Sun, 1980), agarose (Iwata *et al.*, 1999; Kobayashi *et al.*, 2003), polyethylene glycol (Chen *et al.*, 1998), glycol chitosan, and multilayered glycol chitosan–alginate complexes (Sakai *et al.*, 2000). The method most commonly used to form microcapsules from organic polymers is interfacial precipitation (Chaikof, 1999), usually gelation of a polyanionic polymer-cell suspension, such as alginate and islets, in a bath containing a divalent cation such as calcium chloride. Once formed, the capsules can be laminated with alternating coats of polylysine and alginate. The alginate core may then be liquefied by chelating the calcium with sodium citrate. Liquefaction of the sphere's core allows for additional space within the capsule for cellular movement and growth. In the case of alginate microcapsules, permeability of the membrane and membrane strength are controlled by the concentration of alginate in the original suspension, the molecular weight of the poly-L-lysine, and the number of additional polylysine–alginate layers applied (Thu *et al.*, 1996a, 1996b). Modifications to enhance membrane strength by reducing polylysine molecular weight or altering alginate concentration usually have collateral effects that negatively impact membrane permeability and diffusive properties. Attempts have been made to increase the strength of microcapsule membranes by combining cells with water-insoluble polyacrylates and precipitating the selective membrane in an aqueous bath (Boag and Sefton, 1987; Brauker *et al.*, 1992, 1998; Broughton and Sefton, 1990). With this approach, viability of the encapsulated cells appears to be marginal due to contact between the encapsulated cells and organic solvent and inadequate diffusive properties of the encapsulating membrane.

Polyelectrolyte coacervation is another method used to construct hydrogel microcapsules with binary polymer blends. A hydrogel membrane is formed by complexation of oppositely charged polymers, resulting in the formation of an interpenetrating hydrogel network. Examples of binary polymer blends are alginate with protamine and carboxymethyl cellulose and chitosan (Chaikof, 1999). These complexes also exhibit an inverse relationship between permeability and molecular strength. A limiting factor in all of the methods discussed thus far is the inability to achieve independent control of both permeability and mechanical strength.

Macrocapsules are fabricated from preformed hollow-fiber or planar membranes prepared by phase inversion of solution of water-insoluble polymers quenched in aqueous bath, high-humidity atmosphere. Capsules consist of an asymmetric porous structure containing a selective skin that determines molecular weight cutoff and a more open structure responsible for mechanical support. Polymer precipitation time, polymer–solvent compatibility, and solvent concentration can influence phase separation, resulting in the formation of a wide range of selective membranes having vastly different molecular weight cutoffs (Chaikof, 1999). Membrane strength is strongly influenced by wall thickness and is often coupled to decreased diffusive capacity and lower permeability (Chaikof, 1999). Unlike cell-loaded microcapsules, prefabricated polymer macrocapsules can be analyzed and tested for specific characteristics, such as molecular weight cutoff, prior to cell loading. Once characterized, macrocapsules are loaded and sealed at the ends. In some cases the integrity of the seal may be verified before the capsule is depicted. Due to their inherently larger size, macrocapsules are able to sequester larger cell volumes (up to tens of millions of cells per vehicle) than microcapsules and can be scaled more easily to clinical applications. Furthermore, these larger capsules can accommodate the addition of a variety of luminal matrices, such as alginate, chitosan, and cross-linked collagen, in order to optimize cell viability and function (Lanza *et al.*, 1992; Zielinski and Aebischer, 1994). Several Phase I and Phase II clinical trials have been conducted using cell-loaded prefabricated polymer hollow fibers (Scharp *et al.*, 1994; Aebischer *et al.*, 1994b, 1996).

## III. CELL SOURCING

Encapsulations of primary cell types such as islets of Langerhans dominated the early literature (Chick *et al.*, 1997; Lim and Sun, 1980; Maki *et al.*, 1993; Lacy *et al.*, 1991; Lanza *et al.*, 1993; Soon-Shiong *et al.*, 1993). Primary cells are isolated from excised glands of donor animals. Following excision, the glands are enzymatically treated and then mechanically digested. Isolated cells are subsequently adapted to *in vitro* cell culture conditions and then finally encapsulated in the preferred system. Primary cells isolated in this manner offer advantages over cell lines, including the potential to provide regulated release of cellular products.

Alternatives to primary cell sources include mitotically active cells and genetically engineered cell lines. Some of these cell sources are immortalized and thus have the ability to proliferate indefinitely. Capsules can be seeded with a priming dose of cells and allowed to support continued growth until the cells reach the carrying capacity of the capsule. Within the encapsulated environment, however, cell proliferation may be constrained by contact inhibition

and metabolic factors (Lysaght *et al.*, 1994). Dividing cell lines can also be engineered to secrete desired gene products (Sautter *et al.*, 1998; Tan *et al.*, 1996; Rinsch *et al.*, 1996; Chang *et al.*, 1993). This allows investigators to use easily accessible cell types such as fibroblasts and tailor these cells and the encapsulated product for a very specific application. Xenogeneic as well as allogeneic cells can be used. The disparity between the host and cell source determines the selectivity and thus the molecular weight cutoff of the isolating membrane used. Although cell lines appear to be generally advantageous over primary cells, genetically engineered cell lines usually only have the capability of constitutive release of one bioactive cellular product at a time. Dividing cells that escaped through a capsule defect are likely to be destroyed by the host immune system, though risk of immune evasion is a concern when these cells are implanted into so-called immunoprivileged sites, such as the central nervous system (Morris, 1996).

## IV. HOST IMMUNE RESPONSES TO ENCAPSULATED CELLS

The initial premise of immunoisolation held that physical separation of implanted cells and the cellular components of the host's immune system was sufficient to preserve viability of grafted cells. Although this premise remains as an extremely important consideration, it is now known that host immune systems are modulated by myriad soluble molecules, including cytokines, and chemotactic factors that not only affect host immune cell reactions but also have the potential to affect the viability of the encapsulated cells directly. In order ultimately to achieve the goal of successful, long-term immunoisolation, issues such as capsule biocompatibility, the innate immune response, and direct and indirect pathways of antigen recognition need to be addressed.

The sine qua non characteristic of any implanted material to be used for cell encapsulation is biocompatibility. Biomaterials must be immunologically inert and therefore must not cause the development of chronic inflammatory responses and foreign-body reactions. Shortly following implantation, encapsulating polymer membranes become encased in a layer of host protein. Adsorption of protein onto capsule surfaces is a dynamic process that may lead to the accumulation and activation of local cell populations, including resident macrophages and fibroblasts (see Fig. 28.1). Activation of this innate immune response can continue for approximately seven days, at which time the response either diminishes and is replaced by fibrotic growth or continues to proliferate into chronic inflammation and rejection of the capsule (de Vos *et al.*, 2002; Grey, 2001). Either condition can result in impedance of bidirectional diffusion, resulting in chronic nutritional deprivation that eventually leads to partial or total necrosis of the encapsulated cells. By this pathway, cells can be damaged or destroyed without any direct interaction with the host

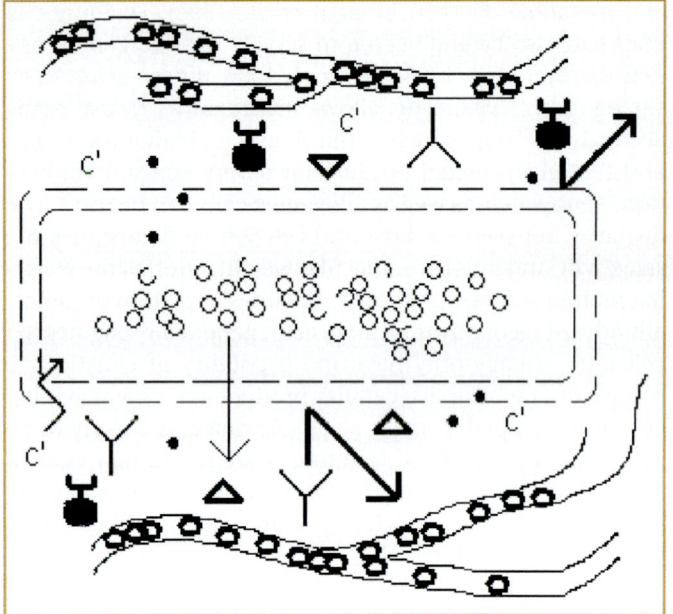

**FIG. 28.1.** Schematic depiction of encapsulated cells implanted into a nonautologous host. Bidirectional diffusion of nutrients and cellular secretogues into and out of the capsule occurs, along with movement of immune molecules that may have the potential to traverse the capsule wall. Movement into the capsule is influenced by the structural integrity of the membrane, the MWCO of the membrane, and the intensity of the host reaction that is elicited.

immune system. Immunoisolation is thus not synonymous with immunoprotection! The initial responses of protein adsorption and acute inflammation may be influenced by both the selection of foreign biomaterials and the surgical procedure of implantation.

Although the preliminary response of the host is generated toward the biomaterial component of the implant, the ultimate success or failure of the device may also be influenced by the type of cell encapsulated. Early models of encapsulation were based on the theory of direct antigen presentation (see Fig. 28.1). Graft-derived antigen presenting cells (APC) complex soluble antigen with major histocompatibility molecules (MHC II) and present this complex to host T-cells. APC–T-cell contact interactions, along with costimulation, lead to graft recognition and rejection. According to this theory, physical separation of graft and host by a semipermeable membrane is adequate for maintaining graft function and viability. Immune recognition of foreign tissue is much more complex, however, and involves capture of soluble antigen and the production of chemotactic factors and cytotoxic molecules such as cytokines, antibodies, and complement proteins by the host (Tao *et al.*, 2002). If the membrane is permissive, graft-derived antigens can potentially diffuse across the semipermeable membrane and be processed and presented by host APCs. This leads to the generation of antibodies with specificity for the grafted cells. In conjunction with complement proteins and soluble

immune factors, antibodies may be able to traverse the membrane and destroy encapsulated cells. The indirect pathway just described is responsible for the failures of many allogeneic and most xenogeneic implants (Gill, 2001). Lastly, in addition to the generation of both nonspecific and specific immune responses, animal models themselves can differ in the immune reactions they initiate toward implantations of the same encapsulated cells (Gill, 2001). Larger-animal models tend to be more sensitive to the presence of foreign molecules. Scale-up from traditional rodent and other small-animal models to preclinical and clinical trials has met with much failure and disappointment (Gill, 2001).

It is important to note that pathways to biocompatibility do not equate to failure of this approach. Such pathways can be contravened by implantation of the capsule in relatively nonimmuogenic sites (vitreous fluid or SF) rather than directly into solid tissue, by the use of relatively small numbers of graft cells and by deploying allogeneic rather than xenogeneic cells (because of the known tendency of the latter to shell soluble antigens).

## V. CONCLUSION

Continuous delivery of therapeutic proteins by encapsulated cells is a promising alternative to traditional modes of treatment for many conditions, such as neurodegenerative disorders, diabetes, chronic pain, and cancer. Immunoisolation provides a method for achieving sustained release of specific proteins that target host tissue directly. Issues of toxicity and unwanted side effects that plague traditional therapeutic approaches can be circumvented by using this unique delivery system. Immune recognition of encapsulated cells is evaded by incorporating a selectively permeable membrane that acts as an immunological sieve. Semipermeable membranes are designed to block the diffusion of soluble immune molecules as well as activated immune cells. Simultaneously, bioactive molecules produced by the encapsulated cells and essential nutrients are allowed access to the host and encapsulated cells, respectively. Balance between these components results in a therapeutic system that not only can maintain delivery of bioactive molecules within the therapeutic window but may also be reactive to metabolic changes in the host. This dynamic equilibrium is the goal of immunoisolatory technology. Genetic engineering of implanted cells has led to more targeted and effective delivery. Optimization of the selective membrane remains an engineering challenge. Investigators have focused their efforts on designing hydrogel composite membranes, uniform nanoporous and micromachined capsules, and vascularizing membranes to attain optimal cell viability and protein delivery (Risbud *et al.*, 2001; Leoni and Desai, 2001; Tao and Desai, 2003). With the advancement of genetic engineering and capsule design, immunoisolation offers the promise of biologically smart therapeutic systems that can be applied to virtually any physiological disorder.

# VI. REFERENCES

Aebischer, P., Scluep, M., Deglon, N., Joseph, J. M., Hirt, L., Heyd, B., Goddard, M., Hammang, J. P., Zurn, A. D., Kato, A. C., Regli, R., and Baetge, E. E. (1996). Intrathecal delivery of CNTF using encapsulated genetically modified xenogeneic cells in amyotrophic lateral sclerosis patients. *Nat. Med.* **2**, 1041.

Aebischer, P., Goddard, M., Signore, A. P., and Timpson, R. L. (1994a). Functional recovery in hemiparkinsonian primates transplanted with polymer-encapsulated PC12 cells. *Exp. Neurol.* **126**, 151–158.

Aebischer, P., Buscher, E., Joseph, J. M., *et al.* (1994b). Transplantation in humans of encapsulated xenogeneic cells without immunosuppression. *Transplantation* **58**, 1275–1277.

Boag, A. H., and Sefton, M. F. (1987). Microencapsulation of human fibroblasts in a water-soluble polyacrylate. *Biotechnol. Bioeng.* **30**, 954–962.

Brauker, J. H., Martinson, L. A., Hill, R. S., Young, S. K., Carr-Brendel, V. E., and Johnson, R. C. (1992). Neovascularization of immunoisolation membranes: the effect of membrane architecture and encapsulated tissue. *Transplant Proc.* **24**, 2924.

Brauker, J. H., Frost, G. H., Dwarki, V., Nijjar, T., Chin, R., Carr-Brendel, V., Jasunas, C., Hodgett, D., Stone, W., Cohen, L. K., and Johnson, R. C. (1998). Sustained expression of high levels of human factor IX from human cells implanted within an immunoisolation device into athymic rodents. *Hum. Gene Ther.* **9**, 879–888.

Broughton, R. L., and Sefton, M. V. (1990). Effect of capsule permeability on growth of CHO cells in Eudragit RL microcapsules: use of FITC-dextran as a marker of capsule quality. *Biomaterials* **10**, 462–465.

Buscher, E., Goddard, M., Heyd, B., *et al.* (1996). Immunoisolated xenogeneic chromaffin cell therapy for chronic pain. *Anesthesiology* **85**, 1005–1012.

Chaikof, E. L. (1999). Engineering and material considerations in islet cell transplantation. *Annu. Rev. Biomed. Eng.* **1**, 103–127.

Chang, P. L., Sheng, N., and Westcott, A. J. (1993). Delivery of recombinant gene products with microencapsulated cells in vivo. *Hum. Gene Ther.* **4**, 433–440.

Chen, J. P., Chu, I. M., and Shiao, M. Y. (1998). Microencapsulation of islets in PEG-amine modified alginate–poly(L-lysine)–alginate microcapsules for constructing bioartificial pancreas. *J. Ferment. Bioeng.* **86**, 185–190.

Chick, W. L., Perna, J., Lauras, V., Law, D., Galetti, P. M., Panol, G., Whittemore, A. D., Like, A. A., Colton, C. K., and Lysaght, M. J. (1997). Artificial pancreas using live beat cells: effects on glucose homeostasis in diabetic rats. *Science* **197**, 780–782.

Colton, C. K. (1996). Engineering challenges in cell-encapsulation technology. *Trends Biotechnol.* **14**, 158–162.

Decostard, I., Buscher, E., Gilliard, N., *et al.* (1988). Intrathecal implants of bovine chromaffin cells alleviate mechanical allodynia in a rat model of neuropathic pain. *Pain* **76**, 159–166.

de Vos, P., van Hoogmoed, C. G., de Haan, B. J., and Busscher, H. J. (2002). Tissue responses against immunoisolating alginate-PLL capsules in the immediate posttransplant period. *J. Biomed. Mater. Res.* **62**, 430–437.

Ezzell, C. (1995). Tissue engineering and the human body shop: encapsulated-cell transplants enter the clinic. *J. NIH Res.* **7**, 47–51.

Geller, R. L., Loudovaris, T., Neunfeldt, S., Johnson, R. C., and Brauker, J. H. (1997a). Use of an immunoisolation device for cell transplantation and tumor immunotherapy. *Ann. N.Y. Acad. Sci.* **831**, 438–451.

Geller, R. L., Neuenfeldt, S., Levon, S. A., Maryanov, D. A., Thomas, T. J., and Brauker, J. H. (1997b). Immunoisolation of tumor cells: generation of antitumor immunity through indirect presentation of antigen. *J. Immunother.* **20**, 131–137.

Gill, R. G. (2001). Use of small animal models for screening immunoisolation approaches to cellular transplantation. *Ann. N.Y. Acad. Sci.* **944**, 35–46.

Grey, D. W. R. (2001). An overview of the immune system with specific reference to membrane encapsulation and islet transplantation. *Ann. N.Y. Acad. Sci.* **944**, 226–239.

Hill, R. S., Cruise, G. M., Lamberti, F. V., Yu, X., Garufis, C. L., Yu, Y., Mundwiler, K. E., Cole, J. F., Hubbll, J. A., Hegre, O. D., and Scharp, D. W. (1997) Immunoisolation of adult porcine islets for the treatment of diabetes mellitus: the use of photopolymerizable polyethylene glycol in the conformal coating of mass-isolated porcine islets. *Ann. N.Y. Acad. Sci.* **831**, 332–343.

Humes, H. D., Weitzel, W. F., and Fissell, W. H. (2004). Renal cell therapy in the treatment of patients with acute and chronic renal failure. *Blood Purif.* **22**, 60–72.

Iwata, H. Y., Murakami, Y., and Ikada, Y. (1999). Control of complement activities for immunoisolation. *Ann. N.Y. Acad. Sci.* **875**, 7–23.

Joseph, J. M., Goddard, M. B., Mills, J., Padrun, V., Zurn, A., Zielinski, B., Favre, J., Gardaz, J. P., Mosimann, F., Sagen, J., *et al.* (1994). Transplantation of encapsulated bovine chromaffin cells in the sheep subarachnoid space: a preclinical study for the treatment of cancer pain. *Cell Transplant.* **3**, 355–364.

Kobayashi, T., Aomatsu, Y., Iwata, H., Kin, T., Kanehiro, H., Hisanaga, M., Ko, S., Nagao, M., and Nakajima, Y. (2003). Indefinite islet protection from autoimmune destruction in nonobese diabetic mice by agarose microencapsulation without immunosuppression. *Transplantation* **75**, 619–625.

Koo, J., and Chang, T. M. S. (1993). Secretion of erythropoietin from microencapsulated rat kidney cells: preliminary results. *Int. J. Artif. Organs* **16**, 557–560.

Lacy, P. E., Hegre, O. D., Gerasimidi-Vazeou, A., Gentile, F. T., and Dionne, K. E. (1991). Maintenance of normoglycemia in diabetic mice by subcutaneous xenografts of encapsulated islets. *Science* **254**, 1782–1784.

Lanza, R. P., Butler, D. H., Borland, K. M., Harvey, J. M., Fanstman, D. L., Soloman, B. A., Muller, T. E., Rupp, R. G., Maki, T., Monaco, A. P., *et al.* (1992). Successful xenotransplantation of a diffusion-based biohybrid artificial pancreas: a study using canine, bovine, and porcine islets. *Transplant. Proc.* **24**, 669–671.

Lanza, R. P., Lodge, P., Borland, K. M., Carretta, M., Sullivan, S. J., Beyer, A. M., Muller, T. E., Soloman, B. A., Maki, T., Monaco, A. P., *et al.* (1993). Transplantation of islet allografts using a diffusion-based biohybrid artificial pancreas: long-term studies in diabetic, pancreatectomized dogs. *Transplant. Proc.* **25**, 978–980.

Leoni, L., and Desai, T. A. (2001). Nanoporous biocapsules for the encapsulation of insulinoma cells: biotransport and biocompatibility considerations. *IEEE Trans. Biomed. Eng.* **48**, 1335–1341.

Lim, F., and Sun, A. M. (1980). Microencapsulated islets as bioartificial endocrine pancreas. *Science* **210**, 908–910.

Lysaght, M. J., Frydel, B., Gentile, F., Emerich, D., and Winn, S. (1994). Recent progress in immunoisolated cell therapy. *J. Cell Biochem.* **56**, 196–203.

Maki, T., Lodge, J. P. A., Carretta, M., Ohzato, H., Barland, K. M., Sullivan, S. J., Staruk, J., Muller, T. E., Soloman, B. A., Chick, W. L., et al. (1993). Treatment of severe diabetes mellitus for more than one year using a vascularized hybrid artificial pancreas. *Transplantation* **55**, 713–718.

Morris, P. J. (1996). Immunoprotection of therapeutic cell transplants by encapsulation. *Trends Biotechnol.* **14**, 163–167.

Rinsch, C., Regulier, E., Deglon, N., Dalle, B., Beuzard, Y., and Aebischer, P. (1996). A gene approach to regulated delivery of erythropoietin as a function of oxygen retention. *Hum. Gene Ther.* **8**, 1881–1889.

Risbud, M. V., Bhonde, M. R., and Bhonde, R. R. (2001). Effect of chitosan-polyvinyl pyrrolidone hydrogel on proliferation and cytokine expression of endothelial cells: implications in islet immunoisolation. *J. Biomed. Mater. Res.* **57**, 300–305.

Sagen, J., Hama, A. T., Winn, S. R., et al. (1993). Pain reduction by spinal implantation of xenogeneic chromaffin cells immunologically isolated in polymer capsules. *Neurosci. Abstr.* **19**, 234.

Sagot, Y., Tan, S. A., Baetge, E., Schmalbruch, H., Kato, A. C., and Aebischer, P. (1995). Polymer-encapsulated cell lines genetically modified to release ciliary neurotrophic factor can slow down progressive motor neuropathy in the mouse. *Eur. J. Neurosci.* **7**, 1313–1322.

Sakai, S., Ono, T., Ijima, H., and Kawakami, K. (2000). Control of molecular weight cutoff for immunoisolation by multilayering glycol chitosan–alginate polyion complex on alginate-based microcapsule. *J. Microencapsul.* **17**, 691–699.

Sautter, J., Tseng, J. L., Braguglia, D., Aebischer, P., Spenger, C., Seiler, R. W., Widmer, H. R., and Zurn, A. D. (1998). Implants of polymer-encapsulated genetically modified cells releasing glial cell line–derived neurotrophic factor improve survival, growth, and function of fetal dopaminergic grafts. *Exp. Neurol.* **149**, 230–236.

Scharp, D. W., Swanson, C. J., Olack, B. J., Latta, P. P., Hegre, O. D., Doherty, E. J., Gentile, F. T., Flavin, K. S., Ansara, M. F., and Lacy, P. E. (1994). Protection of encapsulated human islets implanted without immunosuppression in patients with type I or type II diabetes and in nondiabetic control subjects. *Diabetes* **43**, 1167–1170.

Soon-Shiong, P., Feldman, E., Nelson, R., Heintz, R., Yao, O., Zheng, T., Merideth, N., Skjak-Braek, G., Espevik, T., et al. (1994). Long-term reversal of diabetes by the injection of immunoprotected islets. *Proc. Natl. Acad. Sci. U.S.A.* **90**, 5843–5847.

Stange, J., Hassanein, T. I., Mehta, R., Mitzer, S. R., and Bartlett, R. H. (2002). The molecular adsorbents recycling system as a liver support system based on albumin dialysis: a summary of preclinical investigations, prospective, randomized, controlled clinical trial and clinical experience from 19 centers. *Artif. Organs* **26**, 103–110.

Sullivan, J., Maki, T., Borland, K. M., Mahoney, M. D., Soloman, B. A., Muller, T. E., Monaco, A. P., and Chick, W. L. (1991). Biohybrid artificial pancreas: long-term implantation studies in diabetic, pancreatectomized dogs. *Science* **252**, 718–721.

Tan, S. A., Deglon, N., Zurn, A. D., Baetge, E. E., Bamber, B., Kato, A. C., and Aebischer, P. (1996). Rescue of motor neurons from axotomy-induced cell death by polymer encapsulated cells genetically engineered to release CNTF. *Cell Transplant.* **5**, 577–587.

Tao, S. L., and Desai, T. A. (2003). Microfabricated drug delivery systems: from particles to pores. *Adv. Drug Deliv. Rev.* **24**, 315–328.

Tao, W., Wen, R., Goddard, M. B., Sherman, S. D., O'Rourke, P. J., Stabila, P. F., Bell, W. J., Dean, B. J., Kauper, K. A., Budz, V. A., Tsiaras, W. G., Acland, G. M., Pearce-Kelling, S., Laties, A. M., and Aguirre, G. D. (2002). Encapsulated cell–based delivery of CNTF reduces photoreceptor degeneration in animal models of retinitis pigmentosa. *Invest. Ophthalmol. Vis. Sci.* **43**, 3292–3298.

Thu, B., Bruheim, P., Espevik, T., Smidsrod, O., Soon-Shiong, P., and Skjak-Braek, G. (1996a). Alginate polycation microcapsules. I. Interaction between alginate and polycation. *Biomaterials* **17**, 1031–1040.

Thu, B., Bruheim, P., Espevik, T., Smidsrod, O., Soon-Shiong, P., and Skjak-Braek, G. (1996b). Alginate polycation microcapsules. II. Some functional properties. *Biomaterials* **17**, 1069–1079.

Tresco, P. A., Winn, S. R., Tan, S., Jaeger, C. B., Greene, L. A., and Aebischer, P. (1992). Polymer-encapsulated PC12 cells: long-term survival and associated reduction in lesion-induced rotational behaviour. *Cell Transplant.* **1**, 255–264.

Tseng, J. L., Baetge, E. E., Zurn, A. D., and Aebischer, P. (1997). GDNF reduces drug-induced rotational behavior after medial forebrain bundle transaction by a mechanism not involving striatal dopamine. *J. Neurosci.* **17**, 325–333.

Zielinski, B. A., and Aebischer, P. (1994). Chitosan as a matrix for mammalian cell encapsulation. *Biomaterials* **15**, 1049–1056.

# Chapter Twenty-Nine

# Engineering Challenges in Immunobarrier Device Development

*Amy S. Lewis and Clark K. Colton*

## I. INTRODUCTION

Immunobarrier devices can be used to transplant cells to treat a variety of human diseases without requiring immunosuppressive drugs. Transplanted cells are enclosed within a material that provides protection from the immune system while allowing adequate transport of nutrients, oxygen, waste products, and therapeutic products. There are three types of immunobarrier devices: (1) intravascular, (2) extravascular macrocapsules, and (3) microcapsules. Challenges exist that prevent the widespread applications of cell encapsulation therapies: tissue supply, effective immune protection of encapsulated cells, and maintenance of cell viability posttransplantation. This chapter discusses approaches that can be used to overcome these problems and focuses specifically on the maintenance of cell viability in the treatment of type 1 diabetes. One method to enhance cell viability by increasing oxygen permeability of the encapsulating material with perfluorocarbons (PFCs) is discussed in detail, and a mathematical model is used to predict the extent of enhancement of islet viability and function.

Cell therapies have the potential to treat a large range of diseases by providing *in vivo* delivery of a required protein. A problem that complicates and restricts the use of cell therapies is that transplanted cells need to be protected from the immune system by immunosuppressive drugs. An engineering approach that addresses this problem is the use of immunobarrier devices that provide a physical barrier to protect transplanted cells from the recipient's immune system through a semipermeable membrane. Some diseases that have been investigated for treatment with immunobarrier devices are diabetes (Omer *et al.*, 2005), hemophilia (Brauker *et al.*, 1998), anemia (Schwenter *et al.*, 2004), parathyroid disease (Hasse *et al.*, 1997), chronic pain (Buchser *et al.*, 1996), Parkinson's disease (Kishima *et al.*, 2004), Huntington's disease (Bloch *et al.*, 2004), and amyotrophic lateral sclerosis (Aebischer *et al.*, 1996). An immunobarrier device can contain cells that constantly produce a therapeutic protein, as is required for the treatment of most of the diseases just listed, or for the case of diabetes treatment the cells secrete insulin in response to

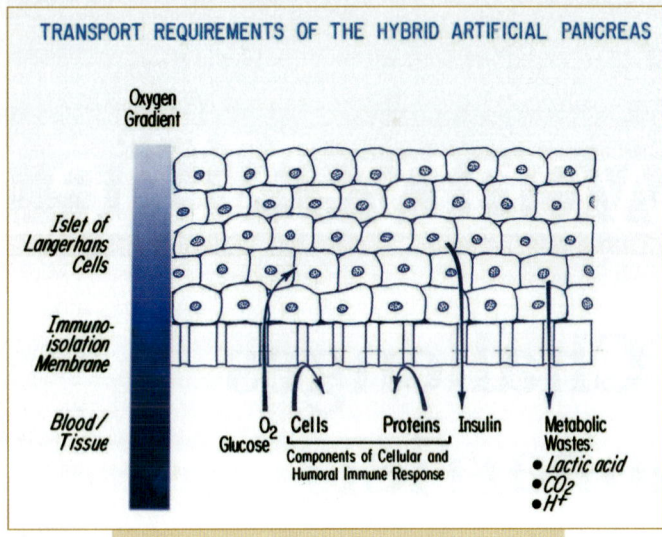

**FIG. 29.1.** Essential elements of an implanted device that incorporates encapsulated cells, illustrating the case of a biohybrid artificial pancreas. Reproduced, with permission, from Colton (1995).

changes in the blood glucose level in a feedback-controlled mechanism.

The essential elements of an implanted device that incorporates encapsulated cells are shown in Fig. 29.1, which is a conceptual illustration of a biohybrid artificial pancreas. The implanted tissues are separated from the host by an immunobarrier membrane. Cells can be encapsulated at a high, tissuelike density, as illustrated in Fig. 29.1, or dispersed in an extracellular gel matrix, such as agar, alginate, or chitosan. The membrane prevents access of immune cells and prevents or minimizes access of humoral immune components but permits passage of the secreted product (insulin). At the same time there must be sufficient access to nutrients, such as glucose and oxygen, and removal of secreted metabolic waste products, such as lactic acid, carbon dioxide, and hydrogen ions. Transplanted cells must be supplied with nutrients by diffusion from the nearest blood supply, through surrounding tissue, the immunobarrier membrane, and the graft tissue itself.

This chapter focuses on the engineering challenges associated with developing immunobarrier devices and methods that are under study to circumvent the problems: (1) supply of tissue, (2) protection from immune rejection, and (3) maintenance of cell viability and function. The emphasis of the chapter is on immunobarrier devices used in transplanting islets for treatment of type 1 diabetes and specifically focuses on the challenge of maintaining cell viability and function. After describing the challenges, approaches are discussed to improve immunobarrier device designs to overcome some of these difficulties. One particular approach is described in detail: use of encapsulating materials that contain perfluorocarbons to enhance oxygen delivery and increase islet survival and function. A mathe-

matical analysis is presented to demonstrate the benefits of this approach.

The material presented here builds on previous reviews of the field since the early 1990s (Colton and Avgoustiniatos, 1991; Colton, 1995; Avgoustiniatos and Colton, 1997a; Avgoustiniatos et al., 2000; Lewis and Colton, 2005). Consequently, in this chapter we focus on recently reported work.

## II. ENGINEERING CHALLENGES

### Device Designs

Immunobarrier device designs typically fall into three categories: intravascular devices, extravascular macrocapsule devices, and microcapsules (Colton and Avgoustiniatos, 1991; Colton, 1995). Intravascular devices create a vascular shunt between an artery and a vein. The device is typically made of a hollow fiber, where blood flows through the lumen and the transplanted cells are separated from the blood by an immunoisolating membrane. This device design is appealing because it brings blood into very close proximity with the transplanted tissue, which will aid in nutrient, waste, and therapeutic protein transport. However, the implantation of this device is associated with greater risks because you are disrupting the patient's vascular system, leading to a greater risk of complications. Human clinical trials of this type of device for islet transplantation were being planned when they were stopped by the FDA because of a mechanical failure of the cannula, in dogs, and have never been resumed. At this point in time this type of device is not under study.

The second device design type is the extravascular macrocapsule, which is typically a planar diffusion chamber or a hollow fiber. This type of device is typically implanted within a cavity of the body containing tissue within the device lumen surrounded by an immunoisolating membrane. Delivery of oxygen and other nutrients requires diffusion from the surrounding tissue to the device, across the device membrane, and then through the interior of the device itself to the tissue. This type of device can be limited in the amount of tissue that can be included within the device due to oxygen supply limitations. Oxygen supply limitations can be even further aggravated by overgrowth of fibrotic tissue, which is another transport barrier and which also consumes some of the oxygen that would normally be delivered to transplanted tissue (Avgoustiniatos and Colton, 1997b). Benefits of using macroencapsulation devices are that many tend to be made of materials that are very stable on implantation in the body, and if needed these types of devices can easily be retrieved because of their larger size. This type of device is under study for several applications, including diabetes (Desai et al., 2004) and diseases of the central nervous system (Hauser et al., 2004). Examples of materials used for macroencapsulation are polytetrafluoroethylene (Brauker et al., 1995), which can be used to form an immunoisolating membrane or an exterior vascularizing membrane, poly-ether-sulfone (Aebischer *et al.*, 1996), which is used for devices that have entered into clinical trials, alumina (La Flamme *et al.*, 2005), and nanoporous micromachined silicon-based membranes (Desai *et al.*, 2004), which can result in a more stringent control over the exact pore size of the material.

The final type of immunobarrier device is the microcapsule. Microcapsules are small spherical gels ranging in size from 200 μm for islet conformal coatings to 2 mm for macrobeads. Microcapsules are currently being studied the most extensively for islet transplantation to treat diabetes. Each microcapsule typically contains one to two islets, and the microcapsules are most commonly transplanted into the peritoneal cavity. The peritoneal cavity is the implantation site of choice, because there is ample space for the implant and immune responses are not as high as at other implantation sites, such as the subcutaneous space. Capsules in this location may be located far away from the blood supply and therefore have very limited oxygen, which can have detrimental effects on tissue survival. The feasibility of microcapsule implantation in the liver via intraportal injection (the location and method for naked human islet transplantations) is being examined in order to enhance microcapsule proximity to the blood supply (Toso *et al.*, 2005). A drawback of the liver as a transplantation site is that there is an increased immune response, which was found to be reduced by short-term immunosuppression with gadolinium chloride, rapamycin, or tacrolimus (Toso *et al.*, 2005).

The most common choice of materials for microcapsules is alginate, a polysaccharide derived from seaweed. Alginate can be dissolved in water to form a viscous solution that, on exposure to a divalent ion (e.g., calcium or barium) or a trivalent ion (e.g., gadolinium), is transformed into a hydrogel. This very gentle gelation process is the reason for alginate's widespread use, because tissue can be encapsulated without causing damage to the cells. However, because it is a naturally derived product, its properties are batch and source dependent, and the impurities in the alginate itself can have detrimental effects on the success of immunobarrier devices involving alginate. Alginate microcapsules usually come in one of two forms: alginate alone cross-linked with barium ions (Omer *et al.*, 2005; Schneider *et al.*, 2005) or alginate cross-linked with calcium and then coated with poly-L-lysine (De Vos *et al.*, 2004) or poly-L-ornithine (Calafiore *et al.*, 2004) to form a perm-selective barrier and enhance capsule stability. Alternative materials to alginate are being developed. One promising example is a synthetic Tetronic polymer that thermally gels and chemically cross-links to form more stable gels while still maintaining gentle processing steps (Cellesi *et al.*, 2004). The capsule formation process for the Tetronic polymers is adaptable to the machinery developed for making alginate capsules (Cellesi *et al.*, 2004). An informational review on the field of microencapsulation has been published (Orive *et al.*, 2004).

## Supply of Tissue

Tissue for cell therapy applications can be derived from three main types of sources: (1) human primary tissue, (2) primary xenogeneic tissue, and (3) cell lines. Each type of tissue has reasons why it is an attractive and an unattractive source. Allotransplantation, or use of primary human tissue, is desirable because this tissue, were it to be transplanted alone, would illicit less of an immune response than tissue derived from another animal (xenotransplantation). Therefore it should be easier to provide immunoprotection to allotransplanted tissue as compared to xenotransplanted tissue. However, the main drawback of primary human tissue is that it requires a cadavaric donor, and therefore the supply is extremely limited. Additionally, a procedure is required to isolate the tissue prior to transplantation, which can be costly and time consuming.

Some researchers are looking to use tissue derived from animal sources in immunobarrier devices. The main advantage of xenogeneic tissue is that its supply is not as limited. However, there is a risk of retroviral disease transmission, immunobarrier requirements are more stringent, and tissue isolation procedures are still required.

The final type of tissue that is under study for transplantation is cell lines, of which the supply of a particular cell type is infinite and there are no isolation requirements. Cell lines can be allogeneic or xenogeneic in nature. When immunobarrier devices are used to deliver a desired product at a continuous rate, as is typically desired for gene therapy strategies, the use of cell lines has proven to be an adequate cell source for this application, as is discussed in a review of encapsulation of genetically modified cells and their clinical applications (Hauser *et al.*, 2004). The main drawback of using cell lines for diabetes treatment is that the cell line must secrete insulin under the same feedback-control mechanisms as the islet and at the same rate. Also, in order to prevent hypoglycemia in diabetes treatments, the growth of the cell line should be arrested prior to transplantation. To date, this type of cell line has yet to be developed, although work is being done to address these issues in the hopes that a fully functional beta cell line will solve the tissue-shortage problem associated with islet transplantation and make the therapy available to all type 1 diabetic patients (Aoki *et al.*, 2005; Simpson *et al.*, 2005). Stem cells also offer great promise as an unlimited source of cells for immunobarrier devices, but a greater understanding and control of the differentiation process is required in order to generate cells with the desired phenotype for a particular application.

## Preventing Immune Rejection

Possible rejection pathways elicited by tissue in immunobarrier devices are shown in Fig. 29.2. The process may begin with diffusion across the immunobarrier of immuno-

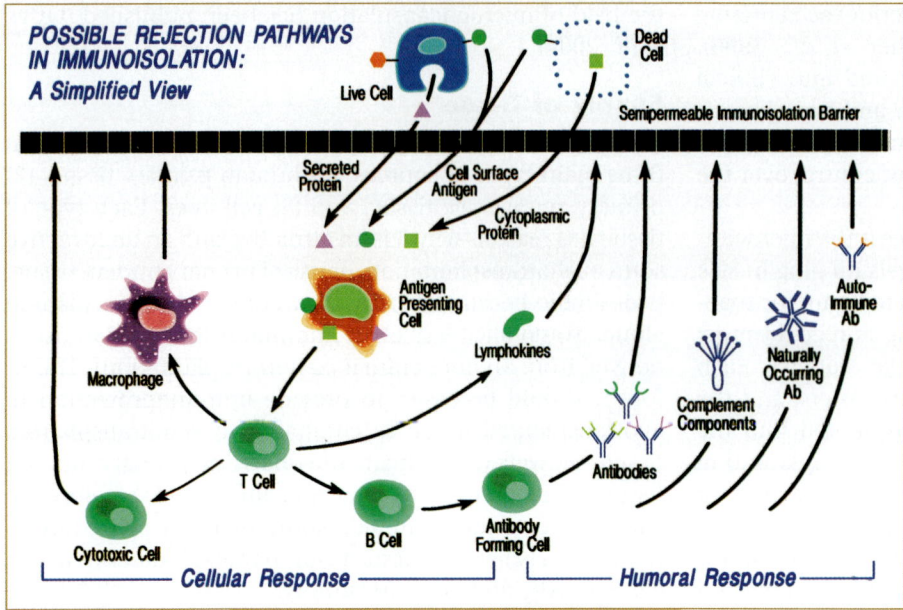

**FIG. 29.2.** Possible pathways for immune rejection mechanisms with encapsulated cells. Reproduced, with permission, from Colton (1995).

geneic tissue antigens that have been shed from the cell surface, secreted by live cells, or liberated from dead cells. Recognition and display of these antigens by host antigen-presenting cells initiate the cellular and humoral immune responses. The former response leads to the activation of cytotoxic T-cells, macrophages, and other immune cells. Preventing cells from entering the tissue compartment is easily achieved using microporous membranes. This may be the only requirement for immunoisolation of allogeneic tissue, at least for periods up to several months (Korsgren and Jansson, 1994). With xenografts it is also necessary to keep out components of the humoral immune response, which is more difficult to accomplish. Such components include cytokines and lymphokines (e.g., interleukin-1), which can have deleterious effects on β-cells, as well as newly formed antibodies to immunogenic antigens that have leaked across the barrier. In addition, there may be naturally occurring antibodies, most likely IgM, to cell surface antigens on xenografts. Antibodies produced during preexisting autoimmune disease, such as type 1 diabetes, might also bind to cell surface antigens. Last, macrophages and certain other immune cells can secrete low-molecular-weight reactive metabolites of oxygen and nitrogen, including free radicals, hydrogen peroxide, superoxide, and nitric oxide, which are toxic to cells in a nonspecific fashion. The extent to which these agents may play a role in causing rejection of immunoisolated tissue depends on how far they can diffuse before they are inactivated by chemical reactions. Based on diffusion and reaction parameters for nitric oxide and superoxide, the reaction-length scales for nitric oxide and superoxide in water are estimated to be on the order of 1 mm and 1 μm, respectively (Colton, 1995). Therefore a macrophage that is at the surface of an immunoisolating device secreting these molecules will be capable of delivering nitric oxide to the encapsulated tissue but not

superoxide, for it will react away before it can diffuse far enough into the capsule to damage the tissue.

Cytotoxic events occur if antibodies and complement components pass through the membrane. Binding of the first component (C1q) to IgM or two or more IgG molecules initiates a cascade that culminates in the formation of the membrane attack complex, which can lyse a single cell. IgM (910 kDa) and C1q (410 kDa) are both larger than IgG (see Fig. 29.3), so if host IgM and C1q can be prevented from crossing the barrier, then a specific, antibody-mediated attack on the islet cell should be averted. If the alternative complement pathway is activated and not inhibited by the implanted tissue, then passage of C3 (200 kDa) across the membrane must be prevented.

The precise mechanism(s) that may play a role in rejection of encapsulated tissue are, in general, incompletely understood, and they probably depend on the specific types of cells present, the species and its phlyogenetic distance from humans, and the concentration of humoral immune molecules to which the implanted tissue is exposed. The latter depends on the transport properties of the immuno-barrier membrane as well as on the total mass of cells implanted (which affects the magnitude of the immune response), the tissue density, and the diffusion distances between tissue and immunobarrier membrane (which affect access of the generated humoral components to the graft tissue). If complete retention of IgG (or even C1q and IgM) coupled with passage of albumin and iron-carrying transferrin (81 kDa) is required, there is a problem because of the size discrimination properties of membranes, in which there is invariably a wide distribution of pore sizes. If cytokine transport must also be prevented, the problem is even more difficult. A comparison of molecule sizes required for cell survival, the insulin that is required to be secreted so that the device is functional, and the immune system com-

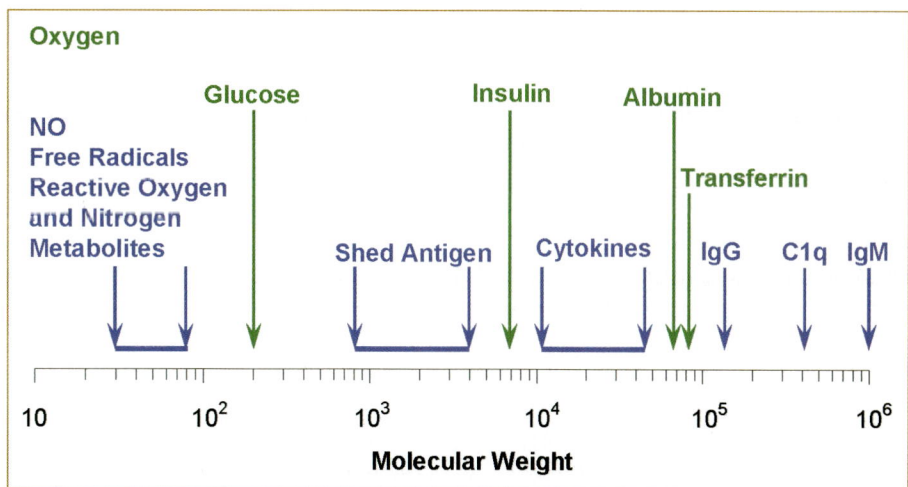

**FIG. 29.3.** Comparison of molecular weights of immune system molecules and molecules required by cells for survival. Modified from Colton (1995, Fig. 19).

ponents are shown in Fig. 29.3. This figure demonstrates that if the cells require albumin and transferrin, then the transport of cytokines to the tissue cannot be blocked by an immunobarrier alone. The membrane would have to allow for cytokine transportation but could potentially contain components that could deactivate the cytokines before they reach the encapsulated tissue. If free radicals and other reactive oxygen and nitrogen species pose a significant problem, no passive membrane barrier will be able to provide immunoprotection, and some other approach (e.g., scavenging of free radicals, immunomodulation of transplanted tissue, local suppression of host immune response) will be necessary. The portion of overgrown allotransplanted alginate-PLL microcapsules containing rat islets explanted four weeks posttransplantation were shown to have mostly macrophages, with some fibroblasts on the capsule surface, without any NK cells, granulocytes, T-cells, or B-cells (De Vos *et al.*, 2004). These results indicate that in an allotransplantation model it may be most important to block macrophage activation and neutralize the effects of nitric oxide and superoxide to prevent immune rejection.

## Maintenance of Cell Viability

Maintenance of cell viability and function is essential and is limited by the supply of nutrients and oxygen. Diffusion limitations of oxygen in tissue *in vivo* are far more severe than those of glucose because the concentration of glucose in tissue is manyfold higher (Tannock, 1972). The requirements of specific tissues for other small molecules and for large macromolecules are poorly understood or have not yet been quantified; transport limitations for large molecules are highly dependent on immunobarrier membrane properties, whereas oxygen limitations are always serious. Hypoxia at levels high enough to keep cells alive can nonetheless have deleterious effects on cell functions that require high cellular ATP concentrations — for example, ATP-dependent insulin secretion (Dionne *et al.*, 1993). Supply of oxygen to immunoisolated tissue is therefore the critical component that determines tissue survival.

Oxygen supply to encapsulated cells depends in a complicated way on a variety of factors, including (1) the site of implantation and the local oxygen partial pressure ($P_{O2}$) in the blood, (2) the spatial distribution of host blood vessels in the vicinity of the implant surface, (3) the oxygen permeability of the membrane or encapsulant, (4) the oxygen consumption rate of the encapsulated tissue, (5) the geometric characteristics of the implant device, and (6) the tissue density and spatial arrangement of the encapsulated cells or tissues. Despite encouraging results with various tissues and applications (Hauser *et al.*, 2004), the problem of oxygen transport limitations is one of the major hurdles that remain. The maximum $P_{O2}$ (about 40 mmHg for the microvasculature) available for extravascular devices limits the steady-state thickness of viable tissue that can be supported.

Islets are particularly prone to oxygen supply limitations because they have a relatively high oxygen consumption rate (Dionne *et al.*, 1993). In the normal physiologic state they are highly vascularized and are supplied with blood at arterial $P_{O2}$. When cultured *in vitro* under ambient normoxic conditions, islets develop a necrotic core, the size of which increases with increasing islet size, as is to be expected as a result of oxygen diffusion and consumption within the islet (Dionne *et al.*, 1993). Central necrosis of the encapsulated islet occurs after islet microcapsule transplantation, resulting in reduction in transplant volume when only a small fraction (~10%) of the capsules have fibrotic overgrowth (De Vos *et al.*, 1999). The fact that necrotic tissue is at the center instead of the periphery of the islet indicates that it is likely a nutrient or oxygen supply limitation causing the necrosis and not a mechanism of the immune system. Hypoxia causes encapsulated islets in culture to become necrotic, to up-regulate iNOS (inducible nitric oxide synthase), which indicates that islets are producing NO, which can cause damage to themselves, and to up-regulate MCP-1 (monocyte chemoattractant protein 1), which can attract macrophages and hence also induce islet damage postimplantation (De Groot *et al.*, 2003). The results of all of these studies indicate that oxygen transport limitations exist within transplanted islet microcapsules and can have serious effects on islet survival.

Previous modeling studies have investigated how device designs affect oxygen transport limitations. When comparing devices that are shaped like slabs, cylinders, or spheres, the spherical geometry is the most beneficial when comparing the volume fraction of nonanoxic tissue in each geometry (Avgoustiniatos and Colton, 1997a). When small amounts of tissue are required for transplantation, the size of the device that is required is feasible in all geometries; but as the amount of tissue increases, especially for the large amount required for islet transplantation, the surface area of the slab, the length of the cylinder, or the number of microcapsules increases significantly (Avgoustiniatos and Colton, 1997a). The size of a planar diffusion device and the volume of microcapsules needed was estimated by assuming that 500,000 islets may be required to treat type 1 diabetes (Ryan *et al.*, 2005). In a planar diffusion device with 120-$\mu$m islets sandwiched between 100 $\mu$m thick membranes and device surface oxygen partial pressure of 40 mmHg, the maximum islet density is 1100 islets/cm$^2$ for fully functional tissue, which corresponds to a total device surface area of 450 cm$^2$ (Avgoustiniatos, 2001). If it is assumed that there is 1 islet/500 $\mu$m diameter microcapsule, then 500,000 capsules would need to be transplanted, equaling a total volume of 33 mL.

## III. STRATEGIES FOR IMPROVING IMMUNOBARRIER DEVICES

### Enhancement of Immunoprotection Capacity of Devices

It will never be possible to create a device that provides complete immunoprotection. Therefore it is most likely necessary to augment the immunoprotection by short-term administration of immunosuppressive drugs, trapping or neutralizing small toxic molecules released during an immune response, or enhancing the resistance of the islet to the stresses experienced during an immune response. Administration of immunosuppressive drugs defeats the purpose of an immunobarrier device, but tissue survival can be enhanced if immunosuppressive drugs are administered to inhibit the immune response, when it is most severe, for a short time period posttransplantation. Administration of antibody molecules that block T-cell costimulation pathways (CTLA4-Ig, anti-CD154, and anti-LFA-1) enhances the length of graft function for neonatal porcine cell clusters or adult porcine islets in alginate microcapsules (Kobayashi *et al.*, 2005; Safley *et al.*, 2005). Alternatively, in order to decrease the immune response posttransplantation, macrophage activation can be prevented by systemic administration of gadolinium chloride (Toso *et al.*, 2005) or local administration of clodronate liposomes to reduce cell overgrowth of alginate microcapsules (Omer *et al.*, 2003). These studies demonstrate that short-term immunosuppression, which is far less risky than lifetime immunosuppression, can be beneficial in promoting graft survival and enhancing the efficacy of immunobarrier devices.

An alternative approach to suppressing the immune system to enhance graft survival is to neutralize or trap the toxic molecules of the immune system released to destroy the transplanted tissue. Macrophages play a role in immune responses to immunoisolated tissues, and it has been demonstrated that the release of nitric oxide from activated macrophages, and not cytokines, is responsible for islet destruction, in a dose-dependent manner (Wiegand *et al.*, 1993). Therefore, including a nitric oxide scavenger such as hemoglobin within the microcapsules can inhibit islet cell death through nitric oxide–mediated mechanisms. Results have shown that the inclusion of live erythrocytes, fixed erythrocytes, or cross-linked hemoglobin in alginate microcapsules enhances islet survival with exposure to activated macrophages or nitric oxide (Wiegand *et al.*, 1993; Chae *et al.*, 2004b).

### Enhancement of Oxygen Transport to Encapsulated Tissue

In order for immunobarrier devices to be successful, the transplanted cells must remain viable and functional posttransplantation. Methods have been developed to enhance mass transfer to immunoisolated tissue, most specifically transport of oxygen, by using vascularizing membranes, *in situ* oxygen generation, and thinner microcapsules and by enhancing the oxygen-carrying capacity of immunoisolating materials. Mass transfer of oxygen to an immunoisolated device can be enhanced if the vasculature is brought in very close contact with the device. The vasculature cannot come into direct contact with the tissue itself, which would be a breach of the immunobarrier, but vessels in close proximity to the device are beneficial. The Theracyte™ (Baxter Healthcare) planar diffusion chamber has two membranes: (1) an exterior vascularizing membrane that has an optimal pore size (5 $\mu$m) so that cells can penetrate the layer and vascularization is induced; and (2) an immunoisolating membrane, pore size 0.45 $\mu$m, both made from PTFE (Brauker *et al.*, 1995). The Theracyte™ device can be preimplanted in order to induce vascularization of the device prior to transplantation, which aids in islet survival posttransplantation (Rafael *et al.*, 2003). In addition, infusion of the device with VEGF (vascular endothelial growth factor) improves the density of blood vessels that form around the device (Trivedi *et al.*, 2000).

An alternative approach to overcome oxygen limitations is to supply implanted tissue with oxygen generated *in situ* adjacent to one side of the immunobarrier device (Wu *et al.*, 1999). On the other side, the exterior of the device is exposed to either culture medium for *in vitro* studies or the host tissue for *in vivo* conditions. *In situ* oxygen generation can occur by the electrolytic decomposition of water in an electrolyzer (Wu *et al.*, 1999). The electrolyzer is in the form of a thin, multilayer sheet, within which electrolysis reactions take place on the anode and the cathode to form oxygen and hydrogen, respectively (Wu *et al.*, 1999). *In vitro* studies with βTC3 cells in the *in situ* oxygen generation device show that the thickness of viable tissue increases with oxygen generation (Wu *et al.*, 1999).

Oxygen transport to encapsulated islets can be enhanced by reducing the diffusion distance through the use of smaller capsules or thinner membranes. There are drawbacks to reducing the diffusion distance, because free radicals released from immune system effector cells may not become inactivated prior to reaching the encapsulated tissue, and the amount of shed antigens from the encapsulated tissue can be increased, thereby enhancing the recipient's immune response to the transplant. Alginate capsules made using an electronic droplet generator, as opposed to an air-driven droplet generator, can be made to be <500 μm instead of ~800 μm. There are also techniques in which only a thin coherent membrane is used to coat the islets. Examples are the use of an emulsion procedure to form calcium alginate PLO microcapsules (Leung *et al.*, 2005) or by centrifuging an islet alginate cell suspension in a discontinuous gradient that contains a barium chloride layer (Zekorn *et al.*, 1992).

A final approach that can be employed to enhance oxygen delivery to encapsulated tissue is to enhance the oxygen-carrying capacity of the material and thus increase the rate at which oxygen can be delivered to the tissue. Components that can be used for this purpose are organic compounds with high oxygen solubility, such as perfluorocarbons or silicone or soybean oils. Perfluorocarbon emulsions have been developed as blood substitutes and could be incorporated into the encapsulation material to increase its oxygen permeability. Perfluorocarbons and silicone oils have been used to enhance oxygen transfer in bioreactors (McMillan and Wang, 1990; Poncelet *et al.*, 1993). There are reports in the literature that including a perfluorocarbon emulsion in islet culture medium enhances islet function (Zekorn *et al.*, 1991). Perfluorocarbons have also been used in the storage of the pancreas prior to islet isolation, to increase islet yield and storage time (Matsumoto and Kuroda, 2002). Inclusion of hemoglobin in alginate microcapsules increases the length of islet survival after transplantation (Chae *et al.*, 2004a), but this enhancement likely results from trapping of nitric oxide.

We envision that the type of microcapsule that results in successful islet transplantation will provide some immunoprotection by preventing contact between islets and activated immune cells, complement components, and antibodies. However, this is likely not enough to provide complete immune protection, and inclusion of a costimulatory blocker such as CTLA4-Ig and an NO scavenger such as red blood cells will greatly enhance the immunoprotective properties of the microcapsules. These modifications should help in preventing immune destruction of the islets, but increased delivery of oxygen is still required to maintain islet survival and function. Increased oxygen delivery can be accomplished by including a permeability enhancer, such as a PFC emulsion, in the alginate matrix. A picture of this ideal microcapsule that is likely to overcome many of the challenges that hamper immunobarrier devices is depicted in Fig. 29.4. The following section examines the oxygen levels within microcapsules that contain islets and a PFC emulsion for the purpose of improving islet survival and function.

# IV. THEORETICAL ANALYSIS OF PFC-CONTAINING MICROCAPSULES

One way to enhance oxygen delivery to immunoisolated tissue is to increase the permeability of the immunobarrier material. One method to enhance the permeability of alginate microcapsules for islet transplantation is to make the capsules from an alginate solution that contains a perfluorocarbon (PFC) emulsion. Perfluorocarbons are highly desirable materials for enhancing oxygen delivery, due to their very high oxygen solubility, approximately 25 times that of water, on a volumetric basis. Enhanced solubility will lead to enhanced permeability because the permeability is the product of the gas solubility and diffusivity in the material of interest. We have currently been studying a perfluorocarbon emulsion made from perfluorodecalin and 20 wt% (w/v) Intralipid® (Baxter), a soybean oil emulsion (Schweighardt and Kayhart, 1990). To assess the benefits of PFC-containing microcapsules, a theoretical model has been developed to predict the local partial pressure of oxygen, which, in turn, is used to assess tissue viability and the fraction of normal insulin secretion for encapsulated islets and single cells. Predictions of the model will be presented for a 500-μm capsule that contains a 150 μm diameter islet or single cells that have a total tissue volume equal to that of the 150 μm diameter islet. A diagram of the two geometries is presented in Fig. 29.5. The islets are assumed to have the properties of rodent islets, in which the beta cells (Layer 1), comprising about 75% of the islet volume, are at the center of the islet and the non-beta cells form an outer shell (Layer 2) that is then surrounded by the alginate microcapsule (Layer 3). It should be noted that human islet beta cells are distributed throughout the entire islet; consequently, predicted beta cell viability and insulin secretion predictions will be slightly higher than for rodent islets.

## Problem Formulation

We used the one-dimensional species conservation equation for reaction and diffusion in spherical coordinates to predict the oxygen profile within the microcapsule:

$$D_i \frac{1}{r^2} \frac{d}{dr}\left(r^2 \frac{dC_i}{dr}\right) = V_i \qquad \text{Eq. (29.1)}$$

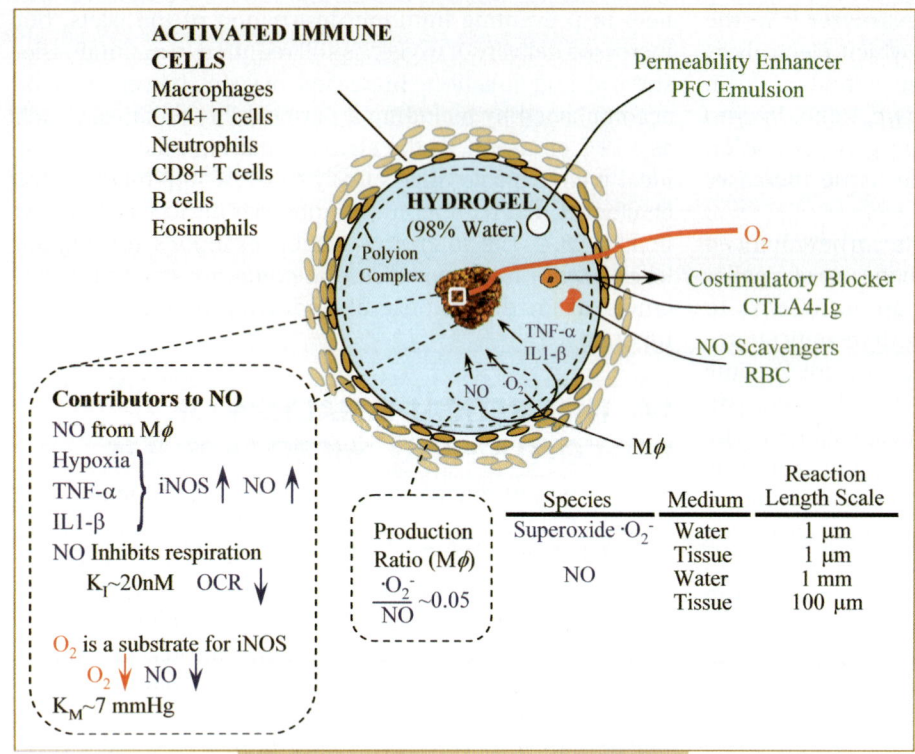

**FIG. 29.4.** Activated immune cells surround the microcapsule, causing islet damage. Immune cell types are listed roughly in decreasing order of involvement in inflammatory and immune reactions to implanted immunobarrier devices. Macrophages damage islets through the release of nitric oxide, super oxide, and cytokines (labeled in blue). Oxygen supply (labeled in red) to encapsulated islets is limited, causing hypoxic conditions that are detrimental to islet survival. Potential methods to protect tissue from immune reactions or enhance oxygen delivery to improve encapsulated islet survival are labeled in green.

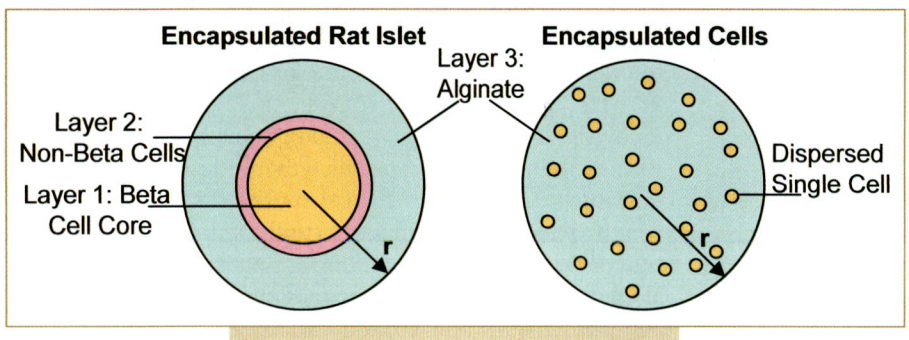

**FIG. 29.5.** Capsule geometry (drawing not to scale).

where $D_i$ (cm²/s) is the effective diffusivity of oxygen in layer $i$, $C_i$ (mol/cm³) is the concentration of oxygen in layer $i$, and $V_i$ (mol/cm³/s) is the local oxygen consumption rate per unit volume in layer $i$. For convenience, since partial pressures are equal across interfaces of different materials, we make use of oxygen partial pressure instead of concentration, which are related by

$$C = \alpha P \qquad \text{Eq. (29.2)}$$

where $\alpha$ (mol/cm³/mmHg) is the effective Bunsen solubility coefficient in layer $i$, and $P$ is the partial pressure of oxygen. Combining Eqs. (29.1) and (29.2) gives

$$(\alpha D)_i \frac{1}{r^2} \frac{d}{dr}\left(r^2 \frac{dP_i}{dr}\right) = V_i \qquad \text{Eq. (29.3)}$$

The oxygen consumption rate is assumed to follow Michaelis–Menten kinetics for all tissue,

$$V_i = \frac{V_{\max}(1-\varepsilon_i)P_i}{K_m + P_i} \qquad \text{Eq. (29.4)}$$

where $V_{\max}$ is the maximum oxygen consumption rate for the tissue, $\varepsilon_i$ is the tissue volume fraction in layer $i$, and $K_m$ is the Michaelis–Menten constant. For the capsule that contains an islet, no oxygen is consumed in the alginate layer; therefore $V_3$ is equal to zero, and the tissue volume fraction in the islet (Layers 1 and 2) is equal to 1.

Equation (29.3) is solved simultaneously for the three layers of the islet microcapsule, subject to the following boundary conditions. The islet is assumed to be centrally located. At the capsule center a symmetry boundary condition is used:

$$\left.\frac{dP_1}{dr}\right|_{r=0} = 0 \qquad \text{Eq. (29.5)}$$

Transport and reaction parameters in Layers 1 and 2 are identical. There is no need for a separate solution in regions 1 and 2 or for the associated boundary conditions. At the interface between Layers 2 and 3 (the interface between the non-beta-cell portion of the islet and alginate), where $r = R_2$,

$$r = R_2 \qquad P_2 = P_3 \qquad \text{Eq. (29.6)}$$

$$r = R_2 \qquad (\alpha D)_2 \frac{dP_2}{dr} = (\alpha D)_3 \frac{dP_3}{dr} \qquad \text{Eq. (29.7)}$$

The final boundary condition required to solve the equations is the assumption that the external partial pressure of oxygen is specified at the capsule surface ($P_S$):

$$r = R_3 \qquad P_3 = P_S \qquad \text{Eq. (29.8)}$$

For the case of a capsule containing dispersed single cells that have a total volume equal to a 150-μm islet, the model consists only of one layer, ε = 0.027, the boundary conditions are represented by Eqs. (29.5) and (29.8), and all subscripts referring to layers can be dropped.

## Material and Tissue Properties

Several material properties for each layer were determined from theoretical relationships. The ratio of the effective permeability $(\alpha D)_{\text{eff},i}$ for layer $i$, consisting of a dispersed ($d$) phase and a continuous ($c$) phase (as occurs in the alginate layers of the model system), to the permeability of the continuous phase $(\alpha D)_c$ was calculated from Maxwell's relationship:

$$\frac{(\alpha D)_{\text{eff},i}}{(\alpha D)_c} = \frac{2 - 2\phi + \rho(1 + 2\phi)}{2 + \phi + \rho(1 - \phi)} \qquad \text{Eq. (29.9)}$$

where $\alpha D$ is the effective permeability of the material (mol/cm/mmHg/s), $\rho = (\alpha D)_d / (\alpha D)_c$ is the ratio of the permeability of the dispersed phase to the permeability of the continuous phase, and $\phi$ is the volume fraction of the dispersed phase. For the multiple dispersed phases employed in our model system, Maxwell's relationship was used sequentially, starting with the phase with the smallest particle size and ending with the phase with the largest particle size. For particle types of the same size, Maxwell's relationship was used for one particle and then the other (and in the reverse order), and the two results were averaged.

In the final PFC alginate composite phase, there were perfluorodecalin and soybean oil droplets of approximately the same size; the alginate polymer itself was treated as an impermeable phase dispersed in water (Fig. 29.6). The effective permeability of PFC emulsion was estimated by first using Maxwell's relationship for PFC droplets in water and then for soybean oil droplets in the PFC and water emulsion. Secondly, Maxwell's relationship was used for soybean oil droplets in water and then for PFC droplets in the soybean oil and water emulsion. The two effective permeabilites as calculated were averaged (Fig. 29.7). Finally, the effective

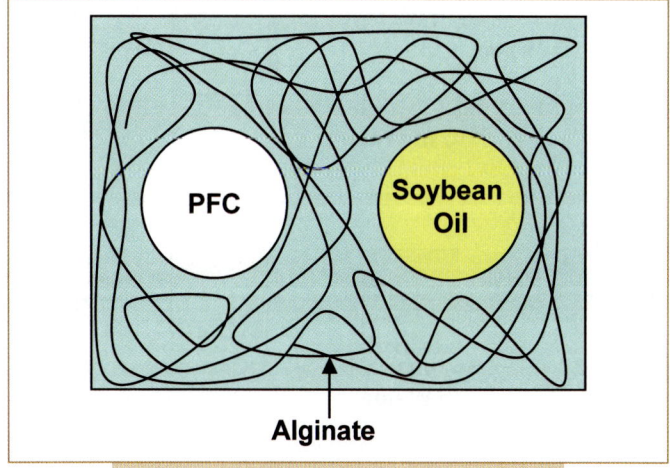

**FIG. 29.6.** PFC alginate consists of PFC and soybean oil droplets surrounded by an alginate matrix with a continuous water phase.

permeability of PFC alginate was estimated with alginate as the dispersed phase and PFC emulsion as the continuous phase. The pure component material transport properties are given in Table 29.1, and the effective properties for different alginate types are given in Table 29.2.

The enhancement of PFC emulsion permeability relative to that of water or to the Intralipid® emulsion used as the starting material is shown in Fig. 29.8. A 70 wt% PFC emulsion has approximately 2.8 times the permeability of water and 2.1 times the permeability of Intralipid®. There is a smaller relative enhancement in the permeability of the emulsion when compared to Intralipid® because Intralipid® contains soybean oil, which has an enhanced permeability to oxygen as compared to water. In comparison, a 90 wt% PFC emulsion (maximum PFC loading in our experimental system) has 3.6 times the permeability of water, and a 110 wt% PFC emulsion (maximum reported loading for a PFC emulsion in the literature) has 4.6 times the permeability of water.

The relationship used to predict the local fraction ($F_S$) of normal insulin secretion rate as a function of local oxygen partial pressure within the islet was developed from data on the effect of hypoxia on islet insulin secretion, estimating the oxygen partial pressure profile within the islet in the experimental system, and then determining the simplest model type and parameters(s) that best predict the insulin secretion level (Dionne *et al.*, 1993; Avgoustiniatos, 2001). The result for a value of $V_{\text{max}} = 4 \times 10^{-8}$ mol/cm³/s is

$$P < 5.1 \text{mmHg} \qquad F_S = \frac{P}{5.1}$$
$$P \geq \text{mmHg} \qquad F_S = 1.0 \qquad \text{Eq. (29.10)}$$

The fraction of normal insulin secretion averaged over the whole islet was determined by evaluating the volume integral of $F_s$ in Layer 1. The fractional viability of the beta cell

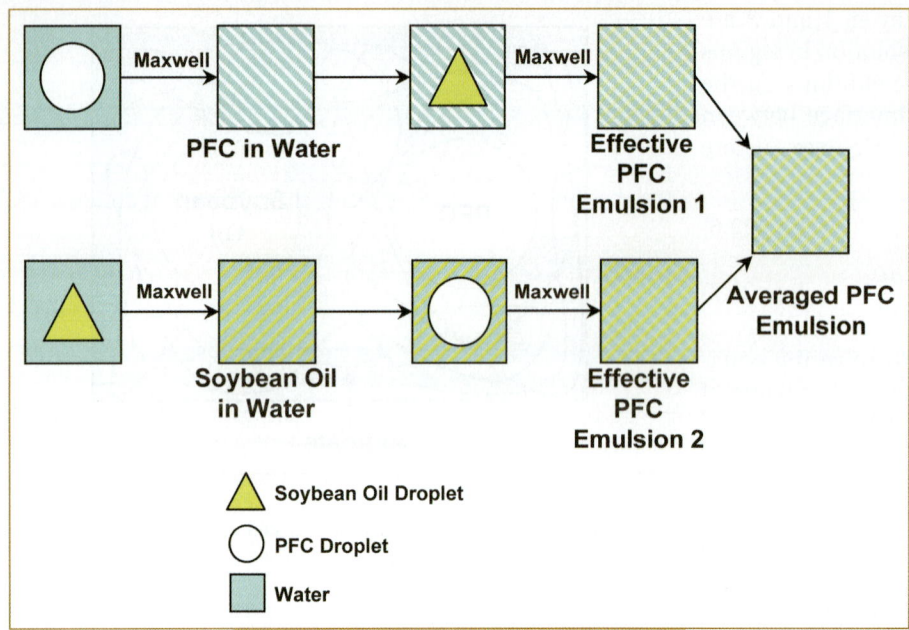

**FIG. 29.7.** Sequence of calculations performed using Maxwell's relationship to calculate the effective permeability of PFC emulsion that consists of three phases: PFC droplets (○), soybean oil droplets (△), and water (▣) as the continuous phase.

**Table 29.1.** Transport properties of materials in system

|  | $D$ | $\alpha$ | $\alpha D$ | Source |
|---|---|---|---|---|
|  | $(cm^2/s)$ | $(mol/mmHg/mL)$ | $(mol/cm/mmHg/s)$ |  |
| $H_2O$ | $2.78 \times 10^{-5}$ | $1.27 \times 10^{-9}$ | $3.53 \times 10^{-14}$ | Avgoustiniatos and Colton (1997b) |
| PFC | $5.61 \times 10^{-5}$ | $2.54 \times 10^{-8}$ | $1.42 \times 10^{-12}$ | Tham *et al.* (1973) |
| Soybean Oil | $2.13 \times 10^{-5}$ | $6.84 \times 10^{-9}$ | $1.46 \times 10^{-13}$ | Bailey (1979) Fillion and Morsi (2000) |
| Tissue | $1.24 \times 10^{-5}$ | $1.00 \times 10^{-9}$ | $1.24 \times 10^{-14}$ | Avgoustiniatos and Colton (1997b) |
| Alginate | $0$ | $0$ | $0$ | Avgoustiniatos and Colton (1997b) |

core (Layer 1) was estimated by determining the volume of tissue where $P_1 > P_C$. $P_C$ is the critical oxygen partial pressure below which tissue dies, and the value of $P_C$ is assumed to be 0.1 mmHg.

## Comparison of $O_2$ Profiles in Microcapsules Containing Single Cells or an Islet

The model equations were solved using the parameters given in Tables 29.1, 29.2, and 29.3. The resulting oxygen profiles for a surface oxygen partial pressure ($P_S$) of 36 mmHg are given in Fig. 29.9. The results demonstrate that for the same amount of total tissue, its distribution within the capsule can greatly affect tissue survival and the minimum oxygen level that the tissue experiences. In a pure-alginate capsule, the oxygen partial pressure drop from the capsule surface to its center is only 3 mmHg for dispersed cells, whereas $P$ decreases to values equal to $P_C$ for a capsule containing one islet in the center, thereby indicating that there will be some loss of islet tissue in the pure-alginate capsule. When the alginate microcapsule is made from a perfluoro-

**Table 29.2.** Effective permeability calculated using Maxwell's relationship

|  | $\alpha D$ |
|---|---|
|  | $(mol/cm/mmHg/s)$ |
| 70 wt% PFC emulsion | $9.92 \times 10^{-14}$ |
| 70 wt% PFC alginate | $9.63 \times 10^{-14}$ |
| Dispersed cells, 70 wt% PFC alginate | $9.31 \times 10^{-14}$ |
| Pure alginate | $3.43 \times 10^{-14}$ |
| Dispersed cells, pure alginate | $3.35 \times 10^{-14}$ |

All capsules are composed of 2 vol% alginate.

carbon emulsion, the drop in the partial pressure of oxygen is smaller in both examples. The enhanced permeability of the PFC–alginate composite prevents oxygen partial pressure from decreasing to $P_C$ in the case of an encapsulated islet.

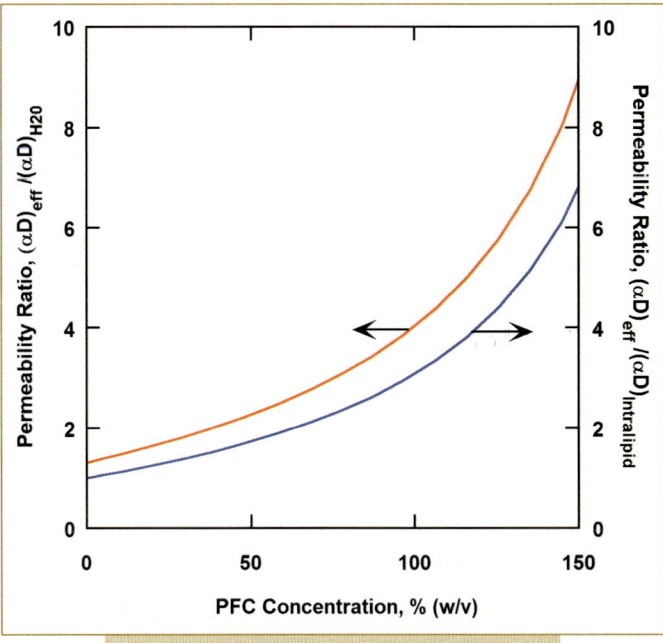

**FIG. 29.8.** Permeability enhancement of PFC emulsion relative to water (left axis) or Intralipid® (right axis). PFC wt% = $\rho_{PFC}$ × PFC vol%.

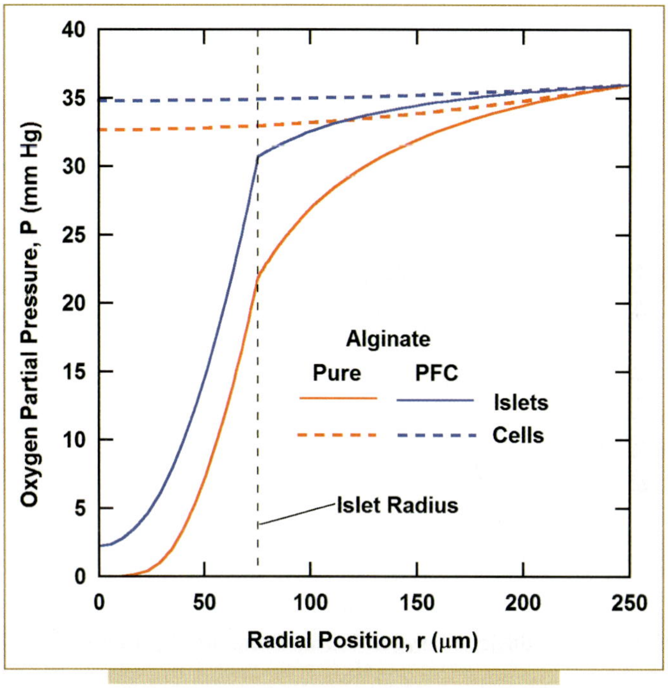

**FIG. 29.9.** Oxygen profiles for microencapsulated islets and single cells in pure alginate and alginate containing 70 wt% PFC emulsion. PFC wt% = $\rho_{PFC}$ × PFC vol%.

| Table 29.3. | Other model parameters | |
|---|---|---|
| | *Parameter value* | *Units* |
| $R_1$ | 68 | μm |
| $R_2$ | 75 | μm |
| $R_3$ | 250 | μm |
| $K_m$ | 0.44 | mmHg |
| $V_{max}$ | $4 \times 10^{-8}$ | mol/cm³/s |
| $P_C$ | 0.1 | mmHg |
| $\rho_{PFC}$ | 1.93 | g/mL |
| $\rho_{soybean\ oil}$ | 0.92 | g/mL |

## Predicted Fractional Viability and Insulin Secretion Rate

By using the oxygen profiles shown in Fig. 29.9, the overall tissue fractional viability and insulin secretion rate were calculated for the capsules containing an islet. Comparison of beta cell fractional viability and relative insulin secretion for an islet centrally located in microcapsules containing 0, 30, 70, and 110 wt% PFC–alginate microcapsules are shown in Figs. 29.10 and 29.11, respectively. The range in capsule surface oxygen partial pressure ($P_S$) studied corresponds to the ranges that could occur in the peritoneal cavity, the most common implantation site of microcapsules. Experimental measurements of $P_{O2}$ in empty perfluorocarbon-loaded capsules have resulted in measurements ranging from 38 to

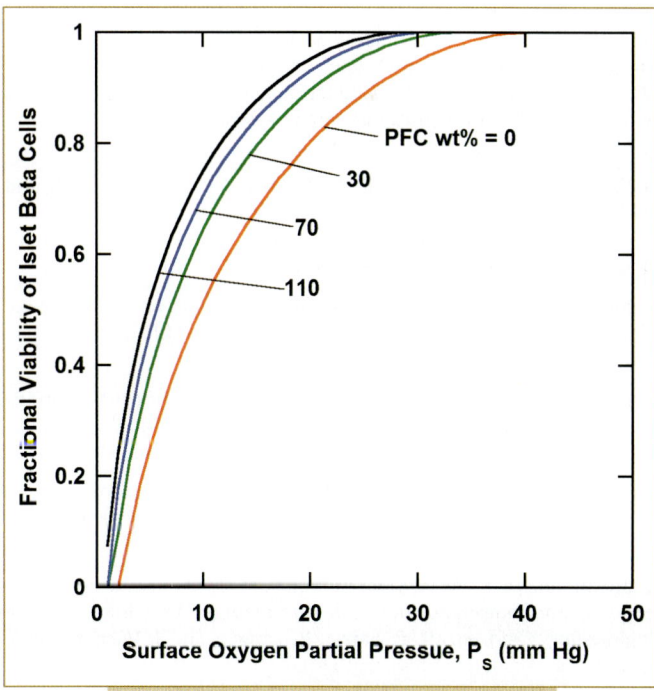

**FIG. 29.10.** Predicted beta cell fractional viability for a centrally located islet in pure- and PFC-containing alginate microcapsules with variable PFC loading. PFC wt% = $\rho_{PFC}$ × PFC vol%.

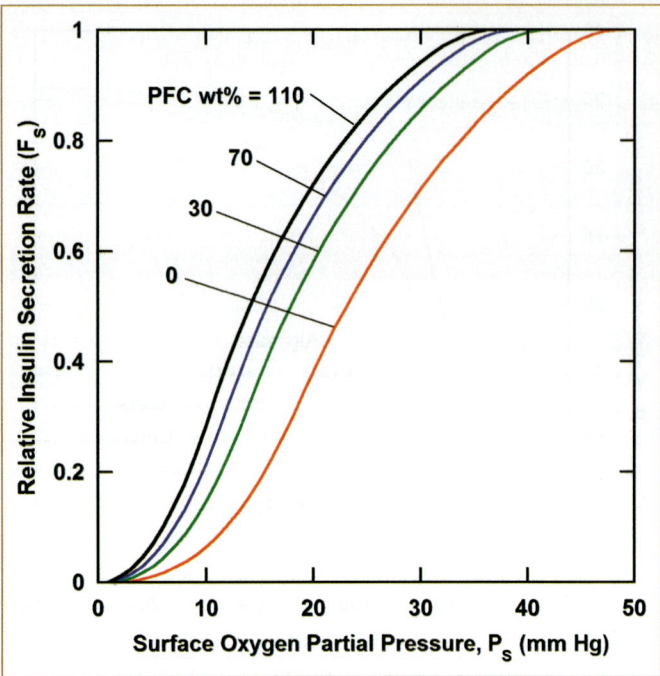

**FIG. 29.11.** Predicted fraction of normal insulin secretion for a centrally located islet in pure- and PFC-containing alginate microcapsules with variable PFC loading. PFC wt% = $\rho_{PFC} \times$ PFC vol%.

0 mmHg (Zimmermann *et al.*, 2000). Over a range of $P_S$ there is an enhancement of beta cell survival for islets encapsulated in PFC-containing microcapsules, and the degree of enhancement is a function of PFC loading. Maintenance of normal insulin secretion of islet beta cells is substantially enhanced by including PFC in the microcapsules in a dose-dependent fashion (Fig. 29.11). The results in Figs. 29.10 and 29.11 demonstrate that including PFCs in alginate microcapsules can increase islet survival and even further enhance islet function after capsule transplantation.

## V. FUTURE DIRECTIONS

Immunobarrier devices have the potential to provide treatments for many human diseases. Immunobarrier devices have entered into clinical trials in which transplantation of smaller numbers of cells or placement in less immune responsive sites is possible or even desirable, and recently a clinical trial with islet containing alginate microcapsules has begun (Hauser *et al.*, 2004; Calafiore *et al.*, 2006). No encapsulation method has yet come into widespread clinical use. Through studies of new immunobarrier materials, methods to enhance the immunoprotection properties of the materials used as immunobarriers, and methods to increase the oxygen supply to the encapsulated tissue, many human diseases, including type 1 diabetes, may one day be treated by cell therapies using immunobarrier devices.

## VI. REFERENCES

Aebischer, P., Schluep, M., Deglon, N., Joseph, J.-M., Hirt, L., Heyd, B., Goddard, M., Hammang, J. P., Zurn, A. D., Kato, A. C., *et al.* (1996). Intrathecal delivery of CNTF using encapsulated genetically modified xenogeneic cells in amyotrophic lateral sclerosis patients. *Nat. Med.* **2**(6), 696–699.

Aoki, T., Hui, H., Umehara, Y., LiCalzi, S., Demetriou, A. A., Rozga, J., and Perfetti, R. (2005). Intrasplenic transplantation of encapsulated genetically engineered mouse insulinoma cells reverses streptozotocin-induced diabetes in rats. *Cell Transplant.* **14**, 411–421.

Avgoustiniatos, E. S. (2001). Oxygen Diffusion Limitations in Pancreatic Islet Culture and Immunoisolation. PhD Thesis, M.I.T., Cambridge, MA.

Avgoustiniatos, E. S., and Colton, C. K. (1997a). Design considerations in immunoisolation. *In* "Principles of Tissue Engineering" (R. Lanza, R. Langer, and W. Chick, eds.), pp. 333–346. R. G. Landes, Austin, TX.

Avgoustiniatos, E. S., and Colton, C. K. (1997b). Effect of external oxygen mass transfer resistance on viability of immunoisolated tissue. *Ann. N.Y. Acad. Sci.* **831**, 145–167.

Avgoustiniatos, E. S., Wu, H., and Colton, C. K. (2000). Engineering challenges in immunoisolation device development. *In* "Principles of Tissue Engineering," 2nd ed. (R. P. Lanza, R. Langer, and J. Vacanti, eds.), pp. 331–350. Academic Press, San Diego.

Bailey, A. E. (1979). "Bailey's Industrial Oil and Fat Products." Wiley, New York.

Bloch, J., Bachoud-Levi, A. C., Deglon, N., Lefaucheur, J. P., Winkel, L., Palfi, S., Nguyen, J. P., Bourdet, C., Gaura, V., Remy, P., *et al.* (2004).

Neuroprotective gene therapy for Huntington's disease, using polymer-encapsulated cells engineered to secrete ciliary neurotrophic factor: results of a phase I study. *Hum. Gene Ther.* **15**, 968–975.

Brauker, J., Carr-Brendel, V., Martinson, L., Crudele, J., Johnston, W., and Johnson, R. C. (1995). Neovascularization of synthetic membranes directed by membrane microarchitecture. *J. Biomed. Mater. Res.* **29**(12), 1517–1524.

Brauker, J., Frost, G. H., Dwarki, V., Nijjar, T., Chin, R., Carr-Brendel, V., Jasunas, C., Hodgett, D., Stone, W., Cohen, L. K., *et al.* (1998). Sustained expression of high levels of human factor IX from human cells implanted within an immunoisolation device into athymic rodents. *Hum. Gene Ther.* **9**, 879–888.

Buchser, E., Goddard, M., Heyd, B., Joseph, J. M., Favre, J., de Tribolet, N., Lysaght, M., and Aebischer, P. (1996). Immunoisolated xenogeneic chromaffin cell therapy for chronic pain: initial clinical experience. *Anesthesiology* **85**(5), 1005–1012.

Calafiore, R., Basta, G., Luca, G., Calvitti, M., Calabrese, G., Racanicchi, L., Macchiarulo, G., Mancusso, F., Guido, L., and Brunetti, P. (2004). Grafts of microencapsulated pancreatic islet cells for the therapy of diabetes mellitus in nonimmunosuppressed animals. *Biotechnol. Appl. Biochem.* **39**, 159–164.

Calafiore, R., Basta, G., Luca, G., Lemmi, A., Racanicchi, L., Mancuso, F., Montanucci, M. P., and Brunetti, P. (2006). Standard technical procedures for microencapsulation of human islets for graft into nonimmunosuppressed patients with type 1 diabetes mellitus. *Transplant. Proc.* **38**, 1156–1157.

Cellesi, F., Weber, W., Fussenegger, M., Hubbell, J. A., and Tirelli, N. (2004). Towards a fully synthetic substitute of alginate: optimization of a thermal gelation/chemical cross-linking scheme ("tandem" gelation) for the production of beads and liquid-core capsules. *Biotechnol. Bioeng.* **88**(6), 740–749.

Chae, S. Y., Kim, Y. Y., Kim, S. W., and Bae, Y. H. (2004a). Prolonged glucose normalization of streptozotocin-induced diabetic mice by transplantation of rat islets coencapsulated with crosslinked hemoglobin. *Transplantation* **78**(3), 392–397.

Chae, S. Y., Lee, M., Kim, S. W., and Bae, Y. H. (2004b). Protection of insulin-secreting cells from nitric oxide–induced cellular damage by cross-linked hemoglobin. *Biomaterials* **25**, 843–850.

Colton, C. K. (1995). Implantable biohybrid artificial organs. *Cell Transplant.* **4**(4), 415–436.

Colton, C. K., and Avgoustiniatos, E. S. (1991). Bioengineering in development of the hybrid artificial pancreas. *J. Biomech. Eng.* **113**, 152–170.

De Groot, M., Schuurs, T. A., Keizer, P. P. M., Fekken, S., Leuvenink, H. G. D., and Van Schilfgaarde, R. (2003). Response of encapsulated rat pancreatic islets to hypoxia. *Cell Transplant.* **12**, 867–875.

Desai, T. A., West, T., Cohen, M., Boiarski, T., and Rampersaud, A. (2004). Nanoporous microsystems for islet cell replacement. *Adv. Drug Del. Rev.* **56**, 1661–1673.

De Vos, P., Van Straaten, J. F. M., Nieuwenhuizen, A. G., de Groot, M., Ploeg, R. J., De Haan, B. J., and Van Schilfgaarde, R. (1999). Why do microencapsulated islet grafts fail in the absence of fibrotic overgrowth? *Diabetes* **48**(7), 1381–1388.

De Vos, P., De Haan, B. J., De Haan, A., van Zanten, J., and Faas, M. M. (2004). Factors influencing functional survival of microencapsulated islet grafts. *Cell Transplant.* **13**, 515–524.

Dionne, K. E., Colton, C. K., and Yarmush, M. L. (1993). Effect of hypoxia on insulin secretion by isolated rat and canine islets of Langerhans. *Diabetes* **42**, 12–21.

Fillion, B., and Morsi, B. I. (2000). Gas–liquid mass-transfer and hydrodynamic parameters in a soybean oil hydrogenation process under industrial conditions. *Ind. Eng. Chem. Res.* **39**, 2157–2168.

Hasse, C., Klock, G., Schlosser, A., Zimmermann, U., and Rothmund, M. (1997). Parathyroid allotransplantation without immunosuppression. *Lancet North Am. Ed.* **350**, 1296–1297.

Hauser, O., Prieschl-Grassauer, E., and Salmons, B. (2004). Encapsulated, genetically modified cell producing *in vivo* therapeutics. *Curr. Opin. Mol. Ther.* **6**(4), 412–420.

Kishima, H., Poyot, T., Bloch, J., Dauguet, J., Conde, F., Dolle, F., Hinnen, F., Pralong, W., Palfi, S., Deglon, N., *et al.* (2004). Encapsulated GDNF-producing C2C12 cells for Parkinson's disease: a preclinical study in chronic MPTP-treated baboons. *Neurobiol. Dis.* **16**, 428–439.

Kobayashi, T., Harb, G., and Rayat, G. R. (2005). Prolonged survival of microencapsulated neonatal porcine islets in mice treated with a combination of anti-CD154 and anti-LFA-1 monoclonal antibodies. *Transplantation* **80**(6), 821–827.

Korsgren, O., and Jansson, J. (1994). Porcine islet-like cell clusters cure diabetic nude rats when transplanted under the kidney capsule, but not when implanted into the liver or spleen. *Cell Transplant.* **3**, 49–54.

La Flamme, K. E., Mor, G., Gong, D., La Tempa, T., Fusaro, V. A., Grimes, C. A., and Desai, T. A. (2005). Nanoporous alumina capsules for cellular macroencapsulation: transport and biocompatibility. *Diabetes Technol. Ther.* **7**(5), 684–694.

Leung, A., Ramaswamy, Y., Munro, P., Lawrie, G., Nielsen, L., and Trau, M. (2005). Emulsion strategies in the microencapsulation of cells: pathways to thin coherent membranes. *Biotechnol. Bioeng.* **92**(1), 45–53.

Lewis, A. S., and Colton, C. K. (2005). Tissue engineering for insulin replacement in diabetes. *In* "Scaffolding in Tissue Engineering" (P. X. Ma and J. Elisseeff, eds.), pp. 575–598. Taylor & Francis, Boca Raton, FL.

Matsumoto, S., and Kuroda, Y. (2002). Perfluorocarbon for organ preservation before transplantation. *Transplantation* **74**(12), 1804–1809.

McMillan, J. D., and Wang, D. I. C. (1990). Mechanisms of oxygen transfer enhancement during submerged cultivation in perfluorochemical-in-water dispersions. *Ann. N.Y. Acad. Sci.* **589**, 283–300.

Omer, A., Keegan, M., Czismadia, E., De Vos, P., Van Rooijen, N., Bonner-Weir, S., and Weir, G. C. (2003). Macrophage depletion improves survival of porcine neonatal pancreatic cell clusters contained in alginate macrocapsules transplanted into rats. *Xenotransplantation* **10**, 240–251.

Omer, A., Duvivier-Kali, V., Fernandes, J., Tchipashvili, V., Colton, C. K., and Weir, G. C. (2005). Long-term normoglycemia in rats receiving transplants with encapsulated islets. *Transplantation* **79**(1), 52–58.

Orive, G., Hernandez, R. M., Gascon, A. R., Calafiore, R., Chang, T. M. S., de Vos, P., Hortelano, G., Hunkeler, D., Lacik, I., and Pedraz, J. L. (2004). History, challenges and perspectives of cell microencapsulation. *Trends Biotechnol.* **22**(2), 87–92.

Poncelet, D., Leung, R., Centomo, L., and Neufeld, R. J. (1993). Microencapsulation of silicone oils within polyacrylamide-polyethylene membranes as oxygen carriers for bioreactor oxygenation. *J. Chem. Technol. Biot.* **57**, 253–263.

Rafael, E., Wu, G. S., Hultenby, K., Tibell, A., and Wernerson, A. (2003). Improved survival of macroencapsulated islets of Langerhans by pre-implantation of the immunoisolating device: a morphometric study. *Cell Transplant.* **12**, 407–412.

Ryan, E. A., Paty, B. W., Senior, P. A., Bigam, D., Alfadhli, E., Kneteman, N. M., Lakey, J. R. T., and Shapiro, A. M. J. (2005). Five-year follow-up after clinical islet transplantation. *Diabetes* **54**, 2060–2069.

Safley, S. A., Kapp, L. M., Tucker-Burden, C., Hering, B., Kapp, J. A., and Weber, C. J. (2005). Inhibition of cellular immune responses to encapsulated porcine islet xenografts by simultaneous blockade of two different costimulatory pathways. *Transplantation* **79**(4), 409–417.

Schneider, S., Feilen, P. J., Brunnenmeier, F., Minnemann, T., Zimmermann, H., Zimmermann, U., and Weber, M. M. (2005). Long-term graft function of adult rat and human islets encapsulated in novel alginate-based microcapsules after transplantation in immunocompetent diabetic mice. *Diabetes* **54**, 687–693.

Schweighardt, F. K., and Kayhart, C. R. (1990). "Concentrated Stable Fluorochemical Aqueous Emulsions Containing Triglycerides." United States, Patent 4, 895, 876.

Schwenter, F., Schneider, B. L., Pralong, W. F., Deglon, N., and Aebischer, P. (2004). Survival of encapsulated human primary fibroblasts and erythropoietin expression under xenogeneic conditions. *Hum. Gene Ther.* **15**, 669–689.

Simpson, N. E., Khokhlova, N., Oca-Cossio, J. A., McMarlane, S. S., Simpson, C. P., and Constantinidis, I. (2005). Effects of growth regulation on conditionally transformed alginate-entrapped insulin-secreting cell lines *in vitro*. *Biomaterials* **26**, 4633–4641.

Tannock, I. F. (1972). Oxygen diffusion and the distribution of cellular radiosensitivity in tumors. *Br. J. Radiol.* **45**, 515–524.

Tham, M. K., Walker, R. D., and Modell, J. H. (1973). Diffusion coefficients of $O_2$, $N_2$ and $CO_2$ in fluorinated ethers. *J. Chem. Eng. Data* **18**, 411–412.

Toso, C., Mathe, Z., Morel, P., Oberholzer, J., Bosco, D., Sainz-Vidal, D., Hunkeler, D., Buhler, L. H., Wandrey, C., and Berney, T. (2005). Effect of microcapsule composition and short-term immunosuppression on intraportal biocompatibility. *Cell Transplant.* **14**, 159–167.

Trivedi, N., Steil, G. M., Colton, C. K., Bonner-Weir, S., and Weir, G. C. (2000). Improved vascularization of planar membrane diffusion devices following continuous infusion of vascular endothelial growth factor. *Cell Transplant.* **9**, 115–124.

Wiegand, F., Kroncke, K.-D., and Kolb-Bachofen, V. (1993). Macrophage-generated nitric oxide as cytotoxic factor in destruction of alginate-encapsulated islets. *Transplantation* **56**(5), 1206–1212.

Wu, H., Avgoustiniatos, E. S., Swette, L., Bonner-Weir, S., Weir, G. C., and Colton, C. K. (1999). *In situ* electrochemical oxygen generation with an immunoisolated device. *Ann. N.Y. Acad. Sci.* **875**, 105–125.

Zekorn, T., Siebers, U., Bretzel, R. G., Heller, S., Meder, U., Ruttkay, H., Zimmermann, U., and Federlin, K. (1991). Impact of the perfluorochemical FC43 on function of isolated islets. *Horm. Metab. Res.* **23**, 302–303.

Zekorn, T., Siebers, U., Horcher, A., Schnettler, R., Zimmermann, U., Bretzel, R. G., and Federlin, K. (1992). Alginate coating of islets of Langerhans: *in vitro* studies on a new method for microencapsulation for immuno-isolated transplantation. *Acta Diabetol.* **29**, 41–45.

Zimmermann, U., Noth, U., Grohn, P., Jork, A., Ulrichs, K., Lutz, J., and Haase, A. (2000). Noninvasive evaluation of the location, the functional integrity, and the oxygen supply of implants: $^{19}$F nuclear magnetic resonance imaging of perfluorocarbon-loaded $Ba^{2+}$-alginate beads. *Artif. Cells Blood Substit. Immobil. Biotechnol.* **28**(2), 129–146.

Part Seven

# Stem Cells

# Chapter Thirty

# Embryonic Stem Cells

### *Alan Trounson*

## I. INTRODUCTION

Embryonic stem cells (ESC) are derived from human preimplantation embryos formed during infertility treatment of couples for *in vitro* fertilization (IVF). They are donated by the couples, with informed consent, for research on embryonic stem cells. In addition, couples having treatment for preimplantation genetic diagnosis (PGD) of their embryos, to avoid the birth of babies with severe genetic disease, may consent to provide their discarded early-cleavage-stage embryos for derivation of disease-specific ESC, such as those with Huntington's disease, thalassaemia, or cystic fibrosis. These ESC may be renewed *in vitro* and expanded indefinitely by regular passage in coculture with a wide range of feeder cells (e.g., mouse or human embryonic fibroblasts) and, more recently, in feeder- and serum-free culture conditions in the presence of fibroblast growth factor (FGF) and neurotrophic growth factors (NGFs). It is now strongly recommended that regular assays of karyotypic normality and the absence of genetic deletions be undertaken prior to experiments, to confirm the genomic type and normality of the ESC being studied. It is also possible to introduce reporter genes in a targeted manner into regulatory sequences of genes of interest for ESC renewal and differentiation. ESC are pluripotential and immortal, so they can potentially form very large numbers of cells of all the tissues of the body. They may be directed into a wide range of specific mature cell types and may be selected for specific lineages under differentiating conditions by immunohistochemistry, expression of marker genes, and cell morphology. These cell types can be purified by fluorescence-activated cell sorting (FACS) and shown to be functional by biological assays *in vitro* and by transplantation into preclinical animal models of human diseases. There is considerable optimism that ESC will be used for cell therapy, as vehicles for gene therapy, targets for drug discovery, and cells for tissue engineering, in the wide spectrum of human regenerative medicine.

Embryonic stem cells (ESC) are derived from the inner cell mass (ICM) of the developing blastocyst stage embryos five to eight days after fertilization. ESC are capable of unlimited expansion *in vitro* and are considered an immortal epiblast derivative that can be maintained at a natural checkpoint in differentiation and expanded as undifferentiated colonies in the presence of appropriate feeder cells.

*Principles of Tissue Engineering, 3rd Edition*
ed. by Lanza, Langer, and Vacanti

They can be maintained in this pluripotential phenotype indefinitely. When allowed to overgrow in culture or when grown in the absence of feeder cells, ESC will spontaneously differentiate as flat colonies in adherence cultures or as embryoid bodies, under nonadherent conditions, to produce a wide range of cell types representing ectoderm, mesoderm, and endoderm derivatives (Trounson, 2006). It has been difficult genetically to manipulate ESC, particularly for targeted mutagenesis because of the need to maintain minimal cell colonies of >10 cells. This interferes with the ability for selection to targeted transfection events, such as the introduction of antibiotic resistance, and with the clonal expansion of cells for identification of homologous recombination events. However, rapid progress has been made recently for targeting genes of interest with fluorescent marker constructs and the long-term inhibition of gene action using short hairpin RNAs (shRTNAs).

## II. DERIVATION OF HUMAN EMBRYONIC STEM CELLS (hESC)

Human embryos are usually eight cells surrounded by an acellular glycoprotein shell (zona pellucida) at three days in culture *in vitro* after insemination with sperm. They form a compacted grapelike cluster of 16–32 cells on the fourth day known as a *morula*. By the fifth to sixth day the embryo is a hollow ball of >64 cells, known as a *blastocyst*, that has an outer layer of trophectoderm cells and a cluster of ~10–30 internalized cells, termed the *inner cell mass* (ICM), within a fluid-filled cavity (blastocoele) (Jones, 2000; Jones *et al.*, 2002). Embryos are donated by patients treated by IVF who have consented to donate their spare embryos for research. The embryos are usually frozen and need to be thawed and grown to the blastocyst stage *in vitro*.

ESC are formed by the culture of ICM cell colonies on confluent cultures of embryonic fibroblasts inactivated by irradiation or mitomycin C (Reubinoff *et al.*, 2000; Thomson *et al.*, 1998). They may also be derived from the earlier-morula-stage embryos (Verlinsky *et al.*, 2005). There appears to be very little difference in the efficiency of producing hESC from these different stages of development. Mouse ESC may be maintained in the presence of leukaemia-inhibiting factor (LIF) in culture *in vitro* without feeder cell support. Unlike mouse ESC renewal, which is dependent on activation of the JAK-STAT pathway, human ESC cannot be maintained undifferentiated in the presence of LIF without feeders (Daheron *et al.*, 2004; Trounson, 2001b). The conventional methods used for hESC derivation were described by Thomson *et al.* (1998), Reubinoff *et al.* (2000), and others (Cowan *et al.*, 2004; Hovatta *et al.*, 2003; Mitalipova *et al.*, 2003; J. H. Park *et al.*, 2003; Stojkovic *et al.*, 2004a). It involves the isolation of ICM colonies from human blastocysts by immunosurgery or dissection and their coculture with mitotically inactivated murine or embryonic fibroblasts (MEFs) or selected human cell lines. They form typical colonies of undifferentiated cells that need to be passaged weekly, or more often, as mechanically dissected colonies of >10 cells. These methods were based on those used to derive rhesus monkey and marmoset ESC (Thomson *et al.*, 1995, 1996). Many hundreds of human ESC have now been derived [*Nature News* (2006) **442**, 336–337] using a variety of methods, which include the preference for microdissection of ICMs, use of entirely humanized reagents, selected human feeder cells, and derivation in specialized facilities for "Good Manufacturing or Laboratory Practice."

It is generally considered that ESC are an epiblast derivative or even a type of germ stem cell (Zwaka and Thomson, 2005) that can be maintained in the presence of basic fibroblast growth factor (bFGF) as an immortal and pluripotential cell type under strict culture conditions. Self-renewal of ESC involves the Wnt family signaling pathway (Sato *et al.*, 2004). Recently it has been shown that ESC have tyrosine receptor kinase (TRK) family receptors, which are responsive to brain-derived neurotrophic factor (BDNF), neurotrophin 3 (NT3), and neurotrophin 4 (NT4) (Pyle *et al.*, 2006). The presence of these neurotrophins in culture medium for ESC reduces apoptosis, stabilizes chromosome ploidy during high-throughput passaging, and enables single-cell clonal derivation of ESC in coculture and feeder-free conditions.

## III. SELECTING EMBRYOS FOR PRODUCING EMBRYONIC STEM CELLS

The criteria for choosing human embryos for deriving hESCs will determine the eventual success rates for their production. In parallel to their developmental potential, it is more efficient to derive ESC from fresh, nonfrozen, rather than frozen–thawed, embryos. Twelve selected fresh blastocyst-stage embryos, grown in coculture with human fallopian tube epithelial cells, were used by us (Reubinoff *et al.*, 2000; Trounson, 2001a) to produce six ESC lines, after preliminary experiments involving around 30 embryos. This was a very high success rate when compared with much larger numbers of frozen embryos (blastocysts) commonly used to derive ESC, donated by patients who have finished their IVF treatments (Hoffman and Carpenter, 2005). Mosaic human blastocysts have been constructed by aggregating uninuclear cells of poor-quality embryos that would normally be discarded because they lack developmental potential (Alikani and Willadsen, 2002). However, the genomic variation of such mosaic embryos would limit their usefulness as ESC. There are new ESC lines now available from a variety of mutant and chromosomally abnormal embryos that are identified by PGD (Pickering *et al.*, 2005; Verlinsky *et al.*, 2005). This includes ESC for Huntington's disease, cystic fibrosis, thalassaemia, Fanconi anaemia A, fragile X (FRMI gene), and Duchenne muscular dystrophy. These ESC are being studied for their differentiation *in vitro* and function *in vitro* and *in vivo* and for the screening of molecular libraries for candidate drugs that may interfere with expression of the disease phenotype.

## IV. MAINTAINING EMBRYONIC STEM CELLS

The long–term stability of ESC is an important issue. While normal karyotypes can be maintained for extended culture times *in vitro* (Buzzard *et al.*, 2004; Rosler *et al.*, 2004), others have reported instability of chromosomes 12 and 17 in conditions that are known to stress ESC (Brimble *et al.*, 2004; Draper *et al.*, 2004). Interestingly, Pyle *et al.* (2006) showed that NT3 and the neurotrophin receptor p75[NGFR] are encoded on chromosomes 12 and 17, and chromosomal amplification may enable a selective advantage of these cells in culture. The addition of neurotrophins to culture media will apparently minimize these selective pressures. It is important to reassess karyotypes regularly for ESC, particularly those passaged by enzymatic digestion into single-cell suspensions because they may continue to express pluripotent markers even when they have become aneuploid. It is also possible that microdeletions will appear after long-term culture, and it is recommended that regular screening by techniques such as comparative genome hybridization (CGH) be introduced for quality control. Similarly, there may be a need to monitor for mutations and epigenetic changes (Rugg-Gunn *et al.*, 2005) that might appear in critical genes, such as the oncogene family, which may influence differentiation and tumor formation.

Optimizing culture conditions for ESC is extremely important and is discussed in considerable detail by Hoffman and Carpenter (2005). It is likely that there will be a change to the production of hESC under the rigid regulations involved in good manufacturing practice (GMP). Completely humanized culture and preparation methods will have to be developed that involve no animal products. Bulk cultures are still in their infancy, and much work is needed to improve the systems for the optimal production of ESC and their derivatives.

Normally, blastocyst-stage embryos (Jones, 2000) are chosen for derivation of hESC, having well-developed ICMs that are isolated mechanically or by immunosurgery (Hovatta *et al.*, 2003; Mitalipova *et al.*, 2003; J. H. Park *et al.*, 2003; Reubinoff *et al.*, 2000; Richards *et al.*, 2002; Stojkovic *et al.*, 2004a; Thomson *et al.*, 1998). ICMs will form rounded cell colonies of small, tightly packed cells with a large nucleus-to-cytoplasmic ratio (Sathananthan, 2003). Serum-free culture systems containing serum substitutes and bFGF-2 may be used for propagation of hESC and reduced spontaneous differentiation. Amit *et al.* (2000) showed that hESC can be maintained in medium containing serum replacement, bFGF-2, transforming growth factor-β (TGF-β), LIF, and fibronectin extracellular matrix. Pebay *et al.* (2005) have demonstrated that sphingosine-1-phosphate and platelet-derived growth factor (PDGF) are active serum components that can replace the need to use serum for hESC culture. These data show that signaling pathways for hESC renewal are activated by tyrosine kinases synergistically with those downstream from lysophospholipid (LPL) receptors.

## V. PLURIPOTENTIAL MARKERS OF EMBRYONIC STEM CELLS

Using microarrays, Sperger *et al.* (2003) showed the genes *POU5F1* (*Oct4*) and *FLJ10713* [a homolog highly expressed in mouse ESC (Ramalho-Santos *et al.*, 2002)] are highly expressed in both the pluripotential human ESC and embryonal carcinoma cells (ECC) and in seminomas. Others that have been identified include a DNA methylase (*DNMT3B*), which functions in early embryogenesis (Watanabe *et al.*, 2002), and *FOXD3*, a forkhead family transcription factor that interacts with *Oct4*, which is essential for the maintenance of mouse primitive ectoderm (Hanna *et al.*, 2002). *Sox2* is also highly expressed in ESC and is known to be important in pluripotentiality (Avilion *et al.*, 2003). Serial analysis of gene expression (SAGE) has been reported by Richards *et al.* (2004) and compared with some cancer SAGE libraries. As expected, *Oct4*, *Nanog*, and *Sox2* transcripts appear abundantly, but there were differences between hESC in some other transcript abundance (e.g., *Rex-1*).

Markers that are now recognized as important for pluripotentiality of ECSs include *Oct4*, *Nanog*, *Sox2*, *Foxd3*, *Rex1*, and *UTF1* transcription factors; *TERF1*, *CHK2*, and *DNMT3* DNA modifiers; *GFA1* surface marker; *GDF3* growth factor; *TDGF1* receptor; and *Stella* and *FLJ10713* (Pera and Trounson, 2004a).

For characterization of hESC it is common to report one or more of the following: *Oct4* expression, alkaline phosphatase, and telomerase activities; stage-specific embryonic antigens (SSEA)-3 and -4; hESC antigens TRA-1-60, TRA-1-81, GCTM-2, TG-30, and TG-343; and CD9, Thy1, and major histocompatibility complex class 1 (Stojkovic *et al.*, 2004b). It is important to recognize the heterogeneity of cells positive for different markers in ESC colonies. For example, only a minority of cells that are positive for GCTM-2 and negative for TG-30 express *Oct4* protein, whereas the majority of GCTM-2- and TG-30-positive cells do.

Other stem cell antigens are also sometimes reported, e.g., AC133, c-kit (CD117), and flt3 (CD135), but these are frequently only expressed in a proportion of the hESC population, making them potential derivatives of interest in the heterogeneous hESC cell population [see the discussion in the review by Hoffman and Carpenter (2005)]. The presence of *Oct4* expression alone may be misleading, for this transcription factor takes some time to shut down RNA transcription in differentiating hESC, and it is also found in other pluripotent cell populations (e.g., embryonic germ stem cells) as well as in some adult and fetal multipotential stem cells. Target genes of *Oct4* include *Rex-1, Lefy-1, PDG-FalfaR,* and *Utf-1,* and those cooperating with *Oct4* include *Sox2* (Hoffman and Carpenter, 2005).

# VI. GENETIC MANIPULATION OF EMBRYONIC STEM CELLS

The therapeutic potential of ESC for regenerative medicine will depend on their ability to direct their differentiation into cell lineages and pure homogeneous progenitor cell types that can be screened for drug discovery, their reliance that they will remain of the tissue type required and not form undesirable teratomas or other tissue derivatives (Brustle *et al.,* 1997; Deacon *et al.,* 1998). It is often difficult to identify and monitor the appearance of specific cell types in culture, because the changes in gene expression may not be visibly identifiable or the cell morphology may not be easily recognizable, especially during the early stages of differentiation. Current methods of identification rely on tedious and time-consuming immunological methods and reverse transcriptase polymerase chain reaction (rtPCR) to detect expression of a specific marker. If the ESC derivatives of interest are to be used for transplantation studies, it is desirable to be able to separate them from the mixed population of cells.

Clonal derivation of ESC is difficult, and hence the efficiency of identifying homologous recombination for gene "knock-in" or "knockout" has been extremely low. Recently, these methods have been improved so that new ESC with fluorescent gene tags are becoming available for genes of interest in differentiation (A. Giudice and A. Trounson, submitted for publication). It is possible to transfect ESC by random integration with suitably designed DNA constructs. This can enable the determination of the role of up-regulation of transcription factors for the renewal and differentiation of ESC, identification of specific gene expression by reporter genes that enable purification of cells of interest in differentiation, and the tracking of hESC derivatives in mixed cultures or when transplanted into animal models of human disease and injury. Conventional transfection methods have been successful (Eiges *et al.,* 2001), as have lentiviral methods (Gropp *et al.,* 2003; Ma *et al.,* 2003). Zwaka and Thomson (2005) have shown that it is possible to electroporate ESC to achieve homologous recombination of ESC colony fragments. Gene function may be more appropriately determined in ESC by using small inhibitory RNAs (siRNA) (Vallier *et al.,* 2004) to explore renewal, differentiation, apoptosis, oncogenesis, etc. (Trounson, 2006).

# VII. DIFFERENTIATION OF EMBRYONIC STEM CELLS

Early morphological differentiation events may be observed in the cells within ESC colonies within a week of passage (Sathananthan *et al.,* 2002) and, together with heterogeneity in pluripotential markers (e.g., *Oct4* expression), can be observed in early differentiating hESC colonies. When hESC are permitted to overgrow in two-dimensional culture, cells begin to pile up, and differentiation begins at the leading-edge borders of the colony and also in the central piled-up areas of cells (Sathananthan *et al.,* 2002). A wide range of differentiating cell types can be observed in these flat cultures, including ectodermal neuroectoderm, mesodermal muscle, and endodermal organ tissue types (Conley *et al.,* 2004; Reubinoff *et al.,* 2000; Sathananthan and Trounson, 2005).

ESC may also be differentiated in three-dimensional cultures of "hanging drops" of culture medium or in plastic culture dishes, which do not favor attachment. The ESC ball up into embryoid bodies, with differentiation occurring within five to seven days into the primary embryonic germ lineages (Itskovitz-Eldor *et al.,* 2000). Human embryoid bodies have a consistent vesicular appearance and structure (Conley *et al.,* 2004; Gertow *et al.,* 2004; Sathananthan, 2003), with a variety of cell types that appear to develop in a more random organization than mouse embryoid bodies. Visceral endoderm is consistently identified in the outer layers of human embryoid bodies. With a wide variety of cell types produced in embryoid bodies, it is possible to select specific cell populations of interest in single cells using cell surface markers and cell separation techniques, including FACS of cell suspensions (Levenberg *et al.,* 2002). It is also possible to use lineage-specific promoters driving reporter genes [(e.g., transfection of the Tet-off system for driving the pdx-1 gene in mouse ESCs (Miyazaki *et al.,* 2004)] or selective cell morphology (Reubinoff *et al.,* 2001). Cultures have also been significantly enriched for ESC-derived cardiomyocytes, using buoyant density gradient separation methods and marker selection (Kehat *et al.,* 2001; Xu *et al.,* 2002).

When ESC (>1000 cells) are transplanted into animal tissues, they form solid teratomas of advanced development and mixed tissue lineage. The tissues that are recognized are often primitive but well-organized examples of embryonic or fetal organs. Normally ESC are transplanted under the kidney or testis capsule and recovered within one to six weeks for histological examination. The teratomas contain mouse microstructures with differentiating histotypic human tissue (Gertow *et al.,* 2004).

Blood islands appear in spontaneously differentiating ESC (Sathananthan and Trounson, 2005) and hematopoietic cells formed from ESCs *in vitro* (Kaufman *et al.,* 2001). By exposing embryoid bodies to a cocktail of hematopoietic cytokines and BMP-4, Chadwick *et al.* (2003) induced the formation of hematopoietic progenitors that could produce both erythroid and myeloid derivatives. The progenitors had an immunophenotype similar to hematopoietic progenitors of the dorsal aorta. The growth factors used were stem cell factor (SCF), interleukins-3 and -6 (IL-3, IL-6), granulocyte colony-stimulating factor (GCSF), and Flt-3 ligand. Enhancement of erythroid colony formation can be obtained with the addition of vascular endothelial growth factor-A (VEGF-A) to cultures (Cerdan *et al.*, 2004; Ng *et al.*, 2005b). A novel aggregation method has been described by Ng *et al.* (2005a) that results in the sequential expression of *Mixl1* and *Brachyury* (primitive streak markers) and *Flk1/KDR* (mesoderm marker), maturing to hematopoietic precursors at a rate of 1 : 500 ESC.

Selection of differentiating cells of the endodermal lineage has been difficult because of the lack of suitable selection markers of early endoderm progenitors. There is much interest in the production of pancreatic β-islet cells because of the potential to treat diabetes. Some cells of embryoid bodies will stain positive to insulin antibodies (Assady *et al.*, 2001; Segev *et al.*, 2004). But while they weakly express insulin-2, they do not express insulin-1 and do not stain for C-peptide, and it is probable that the insulin-positive cells are a result of the uptake up of exogenous insulin from the culture medium (Rajagopal *et al.*, 2003).

Rambhatla *et al.* (2003) reported differentiation of hESC into cells expressing markers of hepatocytes (albumin, alpha-1-antitrypsin, cytokeratin 8 and 18), which accumulate glycogen, by treatment of differentiating embryoid bodies with sodium butyrate or adherent hESC cultures with dimethyl sulfoxide followed by sodium butyrate. Others have reported hepatic-like endodermal cells in embryoid bodies (Lavon *et al.*, 2004).

## VIII. DIRECTING DIFFERENTIATION OF EMBRYONIC STEM CELLS

The enhancement of differentiation toward a specific lineage can be achieved by activating endogenous transcription factors, by transfection of ESC with ubiquitously expressing transcription factors, by exposure of ESC to selected growth factors, or by coculture of ESC with cell types capable of lineage induction (Lavon and Benvenisty, 2003; Pera and Trounson, 2004; Trounson, 2005a, 2005b). ESC may be induced to form cell types of interest by a combination of growth factors and/or their antagonists (Loebel *et al.*, 2003).

The formation of ectodermal derivatives is very common in spontaneously differentiating hESC (Reubinoff *et al.*, 2001, 2000; Sathananthan and Trounson, 2005), and this is commonly considered a developmental default pathway. The neural differentiating pathway can be enhanced in cultures (Carpenter *et al.*, 2001) and the neural progenitors transplanted into the ventricular spaces in the brain of newborn mice, resulting in diffuse migration of human neurones and astrocytes into the parenchyma and the olfactory rostral migratory pathway (Reubinoff *et al.*, 2001; Zhang *et al.*, 2001). ESC-derived neurones respond to neurotransmitters, generate action potentials, and make functional synapses (Carpenter *et al.*, 2001). Oligodendrocytes derived from ESC are able to remyelinate neurones of the shiver mouse model (Nistor *et al.*, 2005). Doperminergic neurones

can be formed from ESC (S. Park *et al.*, 2004; Perrier *et al.*, 2004), and these are of interest in preclinical transplantation experiments, where they are shown to be capable of reversing some motor behavioral abnormalities after transplantation into the brain of a parkinsonian rat model (Ben-Hur *et al.*, 2004).

ESC may be efficiently directed into neuroectoderm by culture in the presence of an antagonist to BMP signaling. Noggin blocks the BMP2 paracrine loop that drives hESC into flattened epithelial-expressing genes characteristic of extraembryonic endoderm. The "noggin colonies" are capable of renewal in culture as relatively homogeneous colonies of neuroectoderm and show facile conversion to neurones or glia in the appropriate culture systems (Pera *et al.*, 2004).

## IX. COCULTURE SYSTEMS FOR DIRECTED DIFFERENTIATION OF EMBRYONIC STEM CELLS

Coculture of monkey ESC with the rodent PA6 stromal cell line that produces stromal cell–derived inducing activity (SDIA) will result in midbrain neuronal cells that are tyrosine hydrolase positive (TH$^+$) and express *nurr1* and *LMX1b* genes (Kawasaki *et al.*, 2002). Manipulation of culture conditions with BMP4 induces epidermogenesis or neural crest and dorsal most central nervous system cells, and suppression of *sonic hedgehog* promotes motor neurone formation (Mizuseki *et al.*, 2003; Trounson, 2004).

Mummery *et al.* (2003, 2002) have shown that cultures of ESC cultured together with mouse visceral endoderm cell type END-2 will preferentially form beating heart muscle cell colonies. These cells express the cardiomyocyte markers, including alpha-myosin heavy chain, cardiac troponins, and atrial natriuretic factor, as well as transcription factors typical of cardiomyocytes (e.g., Nkx2.5, GATA4, and MEF3) (Kehat *et al.*, 2001; Mummery *et al.*, 2002; Xu *et al.*, 2002). These cells respond to pharmacological stimuli, and the action potentials of the muscle cells produced in this coculture system most commonly resemble that for human fetal left ventricular cardiomyocytes (He *et al.*, 2003; Mummery *et al.*, 2003). Atrial- and pacemaker-like cells may also be formed in the differentiating ESC cultures. These differentiated cardiomyocytes are capable of functionally integrating into rodent heart muscle when transplanted. The human cardiomyocytes form gap junctions with mouse adult cardiomyocytes (Hassink *et al.*, 2003; Kehat *et al.*, 2004; Xue *et al.*, 2005).

Differentiation strategies similar to those developed by Lumelsky *et al.* (2001), Rajagopal *et al.* (2003), and Sipione *et al.* (2004) in mouse ESC to produce insulin and C-peptide-expressing cells have differentiated human ESC into insulin-producing cells that coexpress insulin and C-peptide and glucagon or somatostatin (Segev *et al.*, 2004). They also expressed a number of pancreatic genes that appear to be similar to immature pancreatic beta-islet cells. In another study, reported by Brolen *et al.* (2005), human ESC allowed to differentiate spontaneously in prolonged (34-day) adherent two-dimensional cultures on mitotically inactivated MEFs, produced a heterogeneous population of cells at 14–19 days, some of which expressed Pdx-1, Foxa2, and Isl1. These cells did not produce insulin. These regions of the cultured ESC were mechanically dissected and transplanted together with dorsal pancreases of E11.5 or E13.5 mouse embryos beneath the kidney capsule of severe combined immunodeficient (SCID) mice for eight weeks. Insulin-positive human cells were organized into isletlike 5- to 25-cell clusters. These insulin-positive cells coexpressed pro-insulin and C-peptide and the key beta-cell transcription factors Foxa2, Pdx-1, and Isl1. No insulin-positive cells were found if the mouse embryonic tissue was not cotransplanted, but glucagon-positive and amylase-positive cells (but not insulin-positive cells) of ESC origin were found localized in ducts when ESC were transferred alone. These data indicate that instruction on the maturation of pancreatic islet cells may need to include components of the embryonic foregut mesenchyme in a manner analogous to that observed for ESC differentiated into mature functional human prostate tissue (Taylor *et al.*, 2006).

## X. CONCLUDING COMMENTS

Pluripotential embryonic stem cells provide a rich source of research opportunities to determine the primary nature of stemness characteristics, such as renewal and differentiation. It is clear that they are a source of a very wide range of progenitor and mature cell types, which can be used to explore differentiation pathways and what may be wrong. They will be used to study the cancer stem cell phenotype, and disease specific ESC will become the basic resource for determining early events that may indicate when disease phenotype will occur. This may enable early diagnostics to be developed for very serious genetic diseases and other complex diseases for which the causes remain obscure. The possible derivation of ESC from cancer cells and those from patients with degenerative tissue pathologies such as motor neurone disease, Alzheimer's disease, and muscular dystrophies may require nuclear transfer to merge diseased cells with enucleated human oocytes. While we await the developments necessary for this method in the human, it is already apparent that ESC can be derived from parthenogenetic activation of oocytes from patients with such disorders.

The rich research resource offered by the range of ESC developing will enable scientists to dissect the developmental events underpinning normal organ formation and the complex interaction of the transcriptome with the cell and matrix compartments to enable tissue formation and function. The control of cell and tissue maturation involves epigenetic regulation of gene expression, microRNAs that regulate translation and the proteomic profile. We have hardly scratched the surface of these dimensions of cell regeneration or of the obvious instruction of cells by factors in their microenvironment (niche). There are physical forces that provide regulation and education of cells necessary for deriving the components required for tissue construction, integration, and function. The possibilities for using ESC for tissue engineering are phenomenal, and the potential clinical applications are immense. However, there is much to learn about the cell and molecular biology of development and tissue regeneration, and we have only just begun to be informed.

## XI. REFERENCES

Alikani, M., and Willadsen, S. M. (2002). Human blastocysts from aggregated mononucleated cells of two or more nonviable zygote-derived embryos. *Reprod. Biomed. Online* **5**(1), 56–58.

Amit, M., Carpenter, M. K., Inokuma, M. S., Chiu, C. P., Harris, C. P., Waknitz, M. A., Itskovitz-Eldor, J., and Thomson, J. A. (2000). Clonally derived human embryonic stem cell lines maintain pluripotency and proliferative potential for prolonged periods of culture. *Dev. Biol.* **227**(2), 271–278.

Assady, S., Maor, G., Amit, M., Itskovitz-Eldor, J., Skorecki, K. L., and Tzukerman, M. (2001). Insulin production by human embryonic stem cells. *Diabetes* **50**(8), 1691–1697.

Avilion, A. A., Nicolis, S. K., Pevny, L. H., Perez, L., Vivian, N., and Lovell-Badge, R. (2003). Multipotent cell lineages in early mouse development depend on SOX2 function. *Genes Dev.* **17**(1), 126–140.

Ben-Hur, T., Idelson, M., Khaner, H., Pera, M., Reinhartz, E., Itzik, A., and Reubinoff, B. E. (2004). Transplantation of human embryonic stem cell–derived neural progenitors improves behavioral deficit in parkinsonian rats. *Stem Cells* **22**(7), 1246–1255.

Brimble, S. N., Zeng, X., Weiler. D. A., Luo, Y., Liu, Y., Lyons, I. G., Freed, W. J., Robins, A. J., Rao, M. S., and Schulz, T. C. (2004). Karyotypic stability, genotyping, differentiation, feeder-free maintenance, and gene expression sampling in three human embryonic stem cell lines derived prior to August 9, 2001. *Stem Cells Dev.* **13**(6), 585–597.

Brolen, G. K., Heins, N., Edsbagge, J., and Semb, H. (2005). Signals from the embryonic mouse pancreas induce differentiation of human embryonic stem cells into insulin-producing beta-cell-like cells. *Diabetes* **54**(10), 2867–2874.

Brustle, O., Spiro, A. C., Karram, K., Choudhary, K., Okabe, S., and McKay, R. D. (1997). *In vitro*–generated neural precursors participate in mammalian brain development. *Proc. Natl. Acad. Sci. U.S.A.* **94**(26), 14809–14814.

Buzzard, J. J., Gough, N. M., Crook, J. M., and Colman, A. (2004). Karyotype of human ES cells during extended culture. *Nat. Biotechnol.* **22**(4), 381–382; author reply 382.

Carpenter, M. K., Inokuma, M. S., Denham, J., Mujtaba, T., Chiu, C. P., and Rao, M. S. (2001). Enrichment of neurons and neural precursors from human embryonic stem cells. *Exp. Neurol.* **172**(2), 383–397.

Cerdan, C., Rouleau, A., and Bhatia, M. (2004). VEGF-A165 augments erythropoietic development from human embryonic stem cells. *Blood* **103**(7), 2504–2512.

Chadwick, K., Wang, L., Li, L., Menendez, P., Murdoch, B., Rouleau, A., and Bhatia, M. (2003). Cytokines and BMP-4 promote hematopoietic differentiation of human embryonic stem cells. *Blood* **102**(3), 906–915.

Conley, B. J., Trounson, A. O., and Mollard, R. (2004). Human embryonic stem cells form embryoid bodies containing visceral endoderm-like derivatives. *Fetal Diagn. Ther.* **19**(3), 218–223.

Cowan, C. A., Klimanskaya, I., McMahon, J., Atienza, J., Witmyer, J., Zucker, J. P., Wang, S., Morton, C. C., McMahon, A. P., Powers, D., and Melton, D. A. (2004). Derivation of embryonic stem-cell lines from human blastocysts. *N. Engl. J. Med.* **350**(13), 1353–1356.

Daheron, L., Opitz, S. L., Zaehres, H., Lensch, W. M., Andrews, P. W., Itskovitz-Eldor, J., and Daley, G. Q. (2004). LIF/STAT3 signaling fails to maintain self-renewal of human embryonic stem cells. *Stem Cells* **22**(5), 770–778.

Deacon, T., Dinsmore, J., Costantini. L. C., Ratliff, J., and Isacson, O. (1998). Blastula-stage stem cells can differentiate into dopaminergic and serotonergic neurons after transplantation. *Exp. Neurol.* **149**(1), 28–41.

Draper, J. S., Smith, K., Gokhale, P., Moore, H. D., Maltby, E., Johnson, J., Meisner, L., Zwaka, T. P., Thomson, J. A., and Andrews, P. W. (2004). Recurrent gain of chromosomes 17q and 12 in cultured human embryonic stem cells. *Nat. Biotechnol.* **22**(1), 53–54.

Eiges, R., Schuldiner, M., Drukker, M., Yanuka, O., Itskovitz-Eldor, J., and Benvenisty, N. (2001). Establishment of human embryonic stem cell–transfected clones carrying a marker for undifferentiated cells. *Curr. Biol.* **11**(7), 514–518.

Gertow, K., Wolbank, S., Rozell, B., Sugars, R., Andang, M., Parish, C. L., Imreh, M. P., Wendel, M., and Ahrlund-Richter, L. (2004). Organized development from human embryonic stem cells after injection into immunodeficient mice. *Stem Cells Dev.* **13**(4), 421–435.

Gropp, M., Itsykson, P., Singer, O., Ben-Hur, T., Reinhartz, E., Galun, E., and Reubinoff, B. E. (2003). Stable genetic modification of human embryonic stem cells by lentiviral vectors. *Mol. Ther.* **7**(2), 281–287.

Hanna, L. A., Foreman, R. K., Tarasenko, I. A., Kessler, D. S., and Labosky, P. A. (2002). Requirement for Foxd3 in maintaining pluripotent cells of the early mouse embryo. *Genes Dev.* **16**(20), 2650–2661.

Hassink, R. J., Brutel de la Riviere, A., Mummery, C. L., and Doevendans, P. A. (2003). Transplantation of cells for cardiac repair. *J. Am. Coll. Cardiol.* **41**(5), 711–717.

He, J. Q., Ma, Y., Lee, Y., Thomson, J. A., and Kamp, T. J. (2003). Human embryonic stem cells develop into multiple types of cardiac myocytes: action potential characterization. *Circ. Res.* **93**(1), 32–39.

Hoffman, L. M., and Carpenter, M. K. (2005). Characterization and culture of human embryonic stem cells. *Nat. Biotechnol.* **23**(6), 699–708.

Hovatta, O., Mikkola, M., Gertow, K., Stromberg, A. M., Inzunza, J., Hreinsson, J., Rozell, B., Blennow, E., Andang, M., and Ahrlund-Richter, L. (2003). A culture system using human foreskin fibroblasts as feeder cells allows production of human embryonic stem cells. *Hum. Reprod.* **18**(7), 1404–1409.

Itskovitz-Eldor, J., Schuldiner, M., Karsenti, D., Eden, A., Yanuka, O., Amit, M., Soreq, H., and Benvenisty, N. (2000). Differentiation of human embryonic stem cells into embryoid bodies compromising the three embryonic germ layers. *Mol. Med.* **6**(2), 88–95.

Jones, G. (2000). Growth and viability of human blastocysts *in vitro*. *Reprod. Biomed. Online* **8**, 241–287.

Jones, G., Figueiredo, F., Osianlis, T., Pope, A., Rombauts, L., Steeves, T., Thouas, G., and Trounson, A. (2002). Embryo culture, assessment, selection and transfer. *In* "Current Practices and Controversies in Assisted Reproduction" (E. Vayena, P. Rowe, and P. Griffin, eds.), pp. 177–209, World Health Organization, Geneva, Switzerland.

Kaufman, D. S., Hanson, E. T., Lewis, R. L., Auerbach, R., and Thomson, J. A. (2001). Hematopoietic colony-forming cells derived from human embryonic stem cells. *Proc. Natl. Acad. Sci. U.S.A.* **98**(19), 10716–10721.

Kawasaki, H., Suemori, H., Mizuseki, K., Watanabe, K., Urano, F., Ichinose, H., Haruta, M., Takahashi, M., Yoshikawa, K., Nishikawa, S., Nakatsuji, N., and Sasai, Y. (2002). Generation of dopaminergic neurons and pigmented epithelia from primate ES cells by stromal cell–derived inducing activity. *Proc. Natl. Acad. Sci. U.S.A.* **99**(3), 1580–1585.

Kehat, I., Kenyagin-Karsenti, D., Snir, M., Segev, H., Amit, M., Gepstein, A., Livne, E., Binah, O., Itskovitz-Eldor, J., and Gepstein, L. (2001). Human embryonic stem cells can differentiate into myocytes with structural and functional properties of cardiomyocytes. *J. Clin. Invest.* **108**(3), 407–414.

Kehat, I., Khimovich, L., Caspi, O., Gepstein. A., Shofti. R., Arbel, G., Huber, I., Satin, J., Itskovitz-Eldor, J., and Gepstein, L. (2004). Electromechanical integration of cardiomyocytes derived from human embryonic stem cells. *Nat. Biotechnol.* **22**(10), 1282–1289.

Lavon, N., and Benvenisty, N. (2003). Differentiation and genetic manipulation of human embryonic stem cells and the analysis of the cardiovascular system. *Trends Cardiovasc. Med.* **13**(2), 47–52.

Lavon, N., Yanuka, O., and Benvenisty, N. (2004). Differentiation and isolation of hepatic-like cells from human embryonic stem cells. *Differentiation* **72**(5), 230–238.

Levenberg, S., Golub, J. S., Amit, M., Itskovitz-Eldor, J., and Langer, R. (2002). Endothelial cells derived from human embryonic stem cells. *Proc. Natl. Acad. Sci. U.S.A.* **99**(7), 4391–4396.

Loebel, D. A., Watson, C. M., De Young, R. A., and Tam, P. P. (2003). Lineage choice and differentiation in mouse embryos and embryonic stem cells. *Dev. Biol.* **264**(1), 1–14.

Lumelsky, N., Blondel, O., Laeng, P., Velasco, I., Ravin, R., and McKay, R. (2001). Differentiation of embryonic stem cells to insulin-secreting structures similar to pancreatic islets. *Science* **292**(5520), 1389–1394.

Ma, Y., Ramezani, A., Lewis, R., Hawley, R. G., and Thomson, J. A. (2003). High-level sustained transgene expression in human embryonic stem cells using lentiviral vectors. *Stem Cells* **21**(1), 111–117.

Mitalipova, M., Calhoun, J., Shin, S., Wininger, D., Schulz, T., Noggle, S., Venable, A., Lyons, I., Robins, A., and Stice, S. (2003). Human embryonic stem cell lines derived from discarded embryos. *Stem Cells* **21**(5), 521–526.

Miyazaki, S., Yamato, E., and Miyazaki, J. (2004). Regulated expression of pdx-1 promotes *in vitro* differentiation of insulin-producing cells from embryonic stem cells. *Diabetes* **53**(4), 1030–1037.

Mizuseki, K., Sakamoto, T., Watanabe, K., Muguruma, K., Ikeya, M., Nishiyama, A., Arakawa, A., Suemori, H., Nakatsuji, N., Kawasaki, H., Murakami, F., and Sasai, Y. (2003). Generation of neural crest–derived peripheral neurons and floor plate cells from mouse and primate embryonic stem cells. *Proc. Natl. Acad. Sci. U.S.A.* **100**(10), 5828–5833.

Mummery, C., Ward, D., van den Brink, C. E., Bird, S. D., Doevendans, P. A., Opthof, T., Brutel de la Riviere, A., Tertoolen, L., van der Heyden, M., and Pera, M. (2002). Cardiomyocyte differentiation of mouse and human embryonic stem cells. *J. Anat.* **200**(Pt. 3), 233–242.

Mummery, C., Ward-van Oostwaard, D., Doevendans, P., Spijker, R., van den Brink, S., Hassink, R., van der Heyden, M., Opthof, T., Pera, M., de la Riviere, A. B., Passier, R., and Tertoolen, L. (2003). Differentiation of human embryonic stem cells to cardiomyocytes: role of coculture with visceral endoderm-like cells. *Circulation* **107**(21), 2733–2740.

Ng, E. S., Azzola, L., Sourris, K., Robb, L., Stanley, E. G., and Elefanty, A. G. (2005a). The primitive streak gene Mixl1 is required for efficient haematopoiesis and BMP4-induced ventral mesoderm patterning in differentiating ES cells. *Development* **132**(5), 873–884.

Ng, E. S., Davis, R. P., Azzola, L., Stanley, E. G., and Elefanty, A. G. (2005b). Forced aggregation of defined numbers of human embryonic stem cells into embryoid bodies fosters robust, reproducible hematopoietic differentiation. *Blood* **106**, 1601–1603.

Nistor, G. I., Totoiu, M. O., Haque, N., Carpenter, M. K., and Keirstead, H. S. (2005). Human embryonic stem cells differentiate into oligodendrocytes in high purity and myelinate after spinal cord transplantation. *Glia* **49**(3), 385–396.

Park, J. H., Kim, S. J., Oh, E. J., Moon, S. Y., Roh, S. I., Kim, C. G., and Yoon, H. S. (2003). Establishment and maintenance of human embryonic stem cells on STO, a permanently growing cell line. *Biol. Reprod.* **69**(6), 2007–2014.

Park, S., Lee. K. S., Lee, Y. J., Shin, H. A., Cho, H. Y., Wang, K. C., Kim, Y. S., Lee, H. T., Chung, K. S., Kim, E. Y., and Lim, J. (2004). Generation of dopaminergic neurons *in vitro* from human embryonic stem cells treated with neurotrophic factors. *Neurosci. Lett.* **359**(1–2), 99–103.

Pebay, A., Wong, R., Pitson, S., Wolvetang, E., Peh, G., Filipczyk, A., Koh, K., Tellis, I., Nguyen, L., and Pera, M. (2005). Essential roles of sphingosine-1-phosphate and platelet-derived growth factor in the maintenance of human embryonic stem cells. *Stem Cells* **23**, 1541–1548.

Pera, M. F., and Trounson, A. O. (2004). Human embryonic stem cells: prospects for development. *Development* **131**(22), 5515–5525.

Pera, M. F., Andrade, J., Houssami, S., Reubinoff, B., Trounson, A., Stanley, E. G., Ward-van Oostwaard, D., and Mummery, C. (2004). Regulation of human embryonic stem cell differentiation by BMP-2 and its antagonist noggin. *J. Cell. Sci.* **117**(Pt. 7), 1269–1280.

Perrier, A. L., Tabar, V., Barberi, T., Rubio, M. E., Bruses, J., Topf, N., Harrison, N. L., and Studer, L. (2004). Derivation of midbrain dopamine neurons from human embryonic stem cells. *Proc. Natl. Acad. Sci. U.S.A.* **101**(34), 12543–12548.

Pickering, S. J., Minger, S. L., Patel, M., Taylor, H., Black, C., Burns, C. J., Ekonomou, A., and Braude, P. R. (2005). Generation of a human embryonic stem cell line encoding the cystic fibrosis mutation deltaF508, using preimplantation genetic diagnosis. *Reprod. Biomed. Online* **10**(3), 390–397.

Pyle, A. D., Lock, L. F., and Donovan, P. J. (2006). Neurotrophins mediate human embryonic stem cell survival. *Nat. Biotechnol.* **24**(3), 344–350.

Rajagopal, J., Anderson, W. J., Kume, S., Martinez, O. I., and Melton, D. A. (2003). Insulin staining of ES cell progeny from insulin uptake. *Science* **299**(5605), 363.

Ramalho-Santos, M., Yoon, S., Matsuzaki, Y., Mulligan, R. C., and Melton, D. A. (2002). "Stemness": transcriptional profiling of embryonic and adult stem cells. *Science* **298**(5593), 597–600.

Rambhatla, L., Chiu, C. P., Kundu, P., Peng, Y., and Carpenter, M. K. (2003). Generation of hepatocyte-like cells from human embryonic stem cells. *Cell Transplant* **12**(1), 1–11.

Reubinoff, B. E., Pera, M. F., Fong, C. Y., Trounson, A., and Bongso, A. (2000). Embryonic stem cell lines from human blastocysts: somatic differentiation *in vitro*. *Nat. Biotechnol.* **18**(4), 399–404.

Reubinoff, B. E., Itsykson, P., Turetsky, T., Pera, M. F., Reinhartz, E., Itzik, A., and Ben-Hur, T. (2001). Neural progenitors from human embryonic stem cells. *Nat. Biotechnol.* **19**(12), 1134–1140.

Richards, M., Fong, C. Y., Chan, W. K., Wong, P. C., and Bongso, A. (2002). Human feeders support prolonged undifferentiated growth of human inner cell masses and embryonic stem cells. *Nat. Biotechnol.* **20**(9), 933–936.

Richards, M., Tan, S. P., Tan, J. H., Chan, W. K., and Bongso, A. (2004). The transcriptome profile of human embryonic stem cells as defined by SAGE. *Stem Cells* **22**(1), 51–64.

Rosler, E. S., Fisk, G. J., Ares, X., Irving, J., Miura, T., Rao, M. S., and Carpenter, M. K. (2004). Long-term culture of human embryonic stem cells in feeder-free conditions. *Dev. Dyn.* **229**(2), 259–274.

Rugg-Gunn, P. J., Ferguson-Smith, A. C., and Pedersen, R. A. (2005). Epigenetic status of human embryonic stem cells. *Nat. Genet.* **37**(6), 585–587.

Sathananthan, A. H. (2003). Origins of human embryonic stem cells and their spontaneous differentiation. Proceedings of First National Stem Cell Centre Scientific Conference, p. 225. Melbourne, Australia.

Sathananthan, A. H., and Trounson, A. (2005). Human embryonic stem cells and their spontaneous differentiation. *Ital. J. Anat. Embryol.* **110**(2 Suppl. 1), 151–157.

Sathananthan, H., Pera, M., and Trounson, A. (2002). The fine structure of human embryonic stem cells. *Reprod. Biomed. Online* **4**, 56–61.

Sato, N., Meijer, L., Skaltsounis, L., Greengard, P., and Brivanlou, A. H. (2004). Maintenance of pluripotency in human and mouse embryonic stem cells through activation of Wnt signaling by a pharmacological GSK-3-specific inhibitor. *Nat. Med.* **10**(1), 55–63.

Segev, H., Fishman, B., Ziskind, A., Shulman, M., and Itskovitz-Eldor, J. (2004). Differentiation of human embryonic stem cells into insulin-producing clusters. *Stem Cells* **22**(3), 265–274.

Sipione, S., Eshpeter, A., Lyon, J. G., Korbutt, G. S.,and Bleackley, R. C. (2004). Insulin-expressing cells from differentiated embryonic stem cells are not beta cells. *Diabetologia* **47**(3), 499–508.

Sperger, J. M., Chen, X., Draper, J. S., Antosiewicz, J. E., Chon, C. H., Jones, S. B., Brooks, J. D., Andrews, P. W., Brown, P. O., and Thomson, J. A. (2003). Gene expression patterns in human embryonic stem cells and human pluripotent germ cell tumors. *Proc. Natl. Acad. Sci. U.S.A.* **100**(23), 13350–13355.

Stojkovic, M., Lako, M., Stojkovic, P., Stewart, R., Przyborski, S., Armstrong, L., Evans, J., Herbert, M., Hyslop, L., Ahmad, S., Murdoch, A., and Strachan, T. (2004a). Derivation of human embryonic stem cells from day-8 blastocysts recovered after three-step *in vitro* culture. *Stem Cells* **22**(5), 790–797.

Stojkovic, M., Lako, M., Strachan, T., and Murdoch, A. (2004b). Derivation, growth and applications of human embryonic stem cells. *Reproduction* **128**(3), 259–267.

Taylor, R. A., Cowin, P. A., Cunha, G. R., Pera, M., Trounson, A. O., Pedersen, J., and Risbridger, G. P. (2006). Formation of human prostate tissue from embryonic stem cells. *Nat. Methods* **3**(3), 179–181.

Thomson, J. A., Kalishman, J., Golos, T. G., Durning, M., Harris, C. P., Becker, R. A., and Hearn, J. P. (1995). Isolation of a primate embryonic stem cell line. *Proc. Natl. Acad. Sci. U.S.A.* **92**(17), 7844–7848.

Thomson, J. A., Kalishman, J., Golos, T. G., Durning, M., Harris, C. P., and Hearn, J. P. (1996). Pluripotent cell lines derived from common marmoset (*Callithrix jacchus*) blastocysts. *Biol. Reprod.* **55**(2), 254–259.

Thomson, J. A., Itskovitz-Eldor, J., Shapiro, S. S., Waknitz, M. A., Swiergiel, J. J., Marshall, V. S., and Jones, J. M. (1998). Embryonic stem cell lines derived from human blastocysts. *Science* **282**(5391), 1145–1147.

Trounson, A. (2001a). Human embryonic stem cells: mother of all cell and tissues. *Reprod. Biomed. Online* **4**, 58–63.

Trounson, A. (2001b). The derivation and potential use of human embryonic stem cells. *Reprod. Fertil. Dev.* **13**(7–8), 523–532.

Trounson, A. (2004). Stem cells, plasticity and cancer — uncomfortable bed fellows. *Development* **131**(12), 2763–2768.

Trounson, A. (2005a). Derivation characteristics and perspectives for mammalian pluripotential stem cells. *Reprod. Fertil. Dev.* **17**(2), 135–141.

Trounson, A. (2005b). Human embryonic stem cell derivation and directed differentiation. *In* "The Promises and Challenges of Regenerative Medicine" (J. Morser and S. Nishikawa, eds.), pp. 27–44. Springer, Kobe, Japan.

Trounson, A. (2006). The production and directed differentiation of human embryonic stem cells. *Endocr. Rev.* **27**(2), 208–219.

Vallier, L., Rugg-Gunn, P. J., Bouhon, I. A., Andersson, F. K., Sadler, A. J., and Pedersen, R. A. (2004). Enhancing and diminishing gene function in human embryonic stem cells. *Stem Cells* **22**(1), 2–11.

Verlinsky, Y., Strelchenko, N., Kukharenko, V., Rechitsky, S., Verlinsky, O., Galat, V., and Kuliev, A. (2005). Human embryonic stem cell lines with genetic disorders. *Reprod. Biomed. Online* **10**(1), 105–110.

Watanabe, D., Suetake, I., Tada, T., and Tajima, S. (2002). Stage- and cell-specific expression of Dnmt3a and Dnmt3b during embryogenesis. *Mech. Dev.* **118**(1–2), 187–190.

Xu, C., Police, S., Rao, N., and Carpenter, M. K. (2002). Characterization and enrichment of cardiomyocytes derived from human embryonic stem cells. *Circ. Res.* **91**(6), 501–508.

Xue, T., Cho, H. C., Akar, F. G., Tsang, S. Y., Jones, S. P., Marban, E., Tomaselli, G. F., and Li, R. A. (2005). Functional integration of electrically active cardiac derivatives from genetically engineered human embryonic stem cells with quiescent recipient ventricular cardiomyocytes: insights into the development of cell-based pacemakers. *Circulation* **111**(1), 11–20.

Zhang, S. C., Wernig, M., Duncan, I. D., Brustle, O., and Thomson, J. A. (2001). *In vitro* differentiation of transplantable neural precursors from human embryonic stem cells. *Nat. Biotechnol.* **19**(12), 1129–1133.

Zwaka, T. P., and Thomson, J. A. (2005). A germ cell origin of embryonic stem cells? *Development* **132**(2), 227–233.

# Chapter Thirty-One

# Adult Epithelial Tissue Stem Cells

*Christopher S. Potten and James W. Wilson*

## I. INTRODUCTION

Stem cell concepts have evolved dramatically over the last few years from the simple ideas in the literature in the mid-20th century. This has culminated in a rapid expansion of interest in both embryonic and adult tissue stem cells in the last five years with the development of interest in gene therapy, tissue engineering, and stem cell therapies. This chapter explores the evolution of stem cell concepts as applied to adult epithelial tissues. These tissues are characterized by a high degree of polarization and very distinct cell maturation and migration pathways, which permit the identification of specific locations in the tissues that represent the origins of all this cell movement. Cells at the origin of the migratory pathways must represent the cells on which the tissue is ultimately dependent and the cells that have a long-term (permanent) residence in the tissue, i.e., the stem cells. A variety of cell kinetic studies, together with lineage-tracking experiments, have indicated that in the intestine, the dorsal surface of the tongue, and interfollicular epidermis, the proliferative compartment of the tissue is divided into discrete units of proliferation. Each unit has its own stem cell compartment and adjacent family of stem cell progeny, including dividing transit cells and differentiated functional cells. In the skin the evolving stem cell studies suggest at least three distinct stem cell populations, providing a source of cells for the epidermis, for the growing hair follicle, and a reserve regenerative, highly potent population in the upper follicle region. In the small intestine there are indications that the stem cell compartment itself is hierarchical, with a commitment to differentiation occurring two to three generations down the lineage, resulting in a population of actual stem cells that perform their function in steady state and a population of potential stem cells that can be called into action if the actual stem cells are killed. Until recently there have been no reliable markers for adult intestinal stem cells. However, recent developments have indicated ways in which these cells may be identified. Cancer is rare in the small intestinal epithelium, which is surprising, since this tissue represents a large mass, with many stem cells dividing many times. This suggests that effective genome protective mechanisms have evolved, and some aspects of these have now been identified.

In the 1950s and '60s, all proliferating cells in the renewing tissues of the body were regarded as a having an equal potential to self-maintain, with one daughter cell (on average) from each division of a proliferative cell being retained within the proliferative compartment. Thus all proliferating cells were regarded as stem cells. It proved somewhat difficult to displace this concept. However, a groundbreaking paper by Till and McCulloch (1961) provided the first clear evidence for one of the replacing tissues of the body, the bone marrow, that not all proliferative cells are identical. The approach they employed was to study the cells that were capable of repopulating hemopoietic tissues,

*Principles of Tissue Engineering, 3rd Edition*
*ed. by Lanza, Langer, and Vacanti*

following cellular depletion of the tissue by exposure to a cytotoxic agent, i.e., radiation. Specifically, mice were irradiated to deplete their bone marrow of endogenous, functional hematopoietic precursors; then they were injected with bone marrow–derived precursors obtained from another animal. The exogenous cells were subject to a variety of treatments prior to transplant, including irradiation. It was found that the hemopoietic precursors circulated in the host and seeded cells into various hemopoietic tissues, including the spleen. Those cells that seeded into the spleen and possessed extensive regenerative and differentiative potential grew by a process of clonal expansion to form macroscopically visible nodules of haemopoietic tissue 10–14 days after transplant. By appropriate genetic or chromosome tracking (marking), it could be shown that these nodules were derived from single cells, i.e., were clones, and that further clonogenic cells were produced within the clones. The colonies were referred to as *spleen colonies*, and the cells that formed the colonies were called *colony-forming units (spleen)* (CFU-s). These experiments provided the theoretical basis for subsequent human bone marrow transplant studies. By a variety of preirradiation manipulations and pre- and posttransplantation variables, this technique led to our current understanding of the bone marrow hierarchies, or cell lineages, and their stem cells. These studies showed that this tissue contained undifferentiated self-maintaining precursor cells that generated dependent lineages able to differentiate down a range of different pathways, generating a variety of cell types. Recent studies have suggested that these CFU-s are not the ultimate hemopoietic stem cells but are part of a stem cell hierarchy in the bone marrow.

Such clonal regeneration approaches have subsequently been developed for a variety of other tissues, notably by the imaginative approaches adopted by Rod Withers for epidermis, intestine, kidney, and testis. These clonal regeneration approaches were summarized and collected in a book produced in 1985 (Potten and Hendry, 1985), but readers are also specifically referred to Withers and Elkind (1970), which deals with the gut, and Withers (1967), for studies on the epidermis. These approaches implicated hierarchical organizations within the proliferative compartments of many tissues (Potten and Hendry, 1985). The stringency of the criteria defining a clone varied enormously, depending as it did on the number of cell divisions required to produce the detectable clones. For epidermis and intestine, the stringency was high, since the clones could be large and macroscopic, containing many cells resulting from many cell divisions, and in fact were very similar in appearance to the spleen colony nodules.

One difficulty with the interpretation of clonal regeneration studies and the generality of their application to stem cell populations is that in order to see the regenerating clones, the tissue has to be disturbed, generally by exposure to a dose of radiation, and this disturbance may alter the cellular hierarchies that one wishes to study and will certainly alter the nature (e.g., cell cycle status, responsiveness to signals, susceptibility to subsequent treatment) of the stem cell compartment. This has been referred to as the biological equivalent of the Heisenberg uncertainty principle in quantum physics. However, these clonal regeneration assays do still provide a valuable and, in some places, unique opportunity to study some aspects of stem cell biology *in vivo*, i.e., by using this approach to look at stem cell survival and functional competence under a variety of conditions.

## II. A DEFINITION OF STEM CELLS

There have been relatively few attempts to define what is meant by the term *stem cells*, which has resulted in some confusion in the literature and the use of a variety of terms, the relationship between which sometimes remain obscure and confusing. These include *precursors*, *progenitors*, and *founder cells*. The concept is further complicated by the use of terms such as *committed precursors* and *progenitors* and the sometimes confusing use, or implication, of the term *differentiation*. One of the difficulties in defining stem cells is that the definitions are often very context dependent; hence, different criteria are brought into the definition by embryologists, haematologists, dermatologists, gastroenterologists, etc. In 1990, in a paper in *Development* (Potten and Loeffler, 1990), we attempted to define a stem cell. This definition was, admittedly, formulated within the context of the gastrointestinal epithelium, but we felt it had a broader application. The definition still largely holds and can be summarized as follows. Within adult replacing tissues of the body, the stem cells can be defined as a small subpopulation of the proliferating compartment, consisting of relatively undifferentiated proliferative cells that maintain their population size when they divide while at the same time producing progeny that enter a dividing transit population, within which further rounds of cell division occur together with differentiation events, resulting in the production of the various differentiated functional cells required of the tissue. The stem cells persist in the tissue throughout the animal's lifetime, dividing a large number of times, and, as a probable consequence of this large division potential, these cells are the most efficient repopulators of the tissue following injury. If this repopulation requires a reestablishment of the full stem cell compartment, the *self-maintenance probability* of the stem cells at division will be raised from the steady-state value of 0.5 to a value between 0.5 and 1, which enables the stem cell population to be reestablished while at the same time maintaining the production of differentiated cells to ensure the functional integrity of the tissue.

The consequences of this definition are obvious, namely that stem cells are:

- Rare cells in the tissue, vastly outnumbered by the dividing transit population, and are the cells on which the entire lineage and ultimately the tissue are dependent.

- The only permanent long-term residents of the tissue.
- Cells at the origin of any cell lineages or migratory pathways that can be identified in the tissue, producing progeny that differentiate.
- Cells that in steady state have a self-maintenance probability of .5, but this can be changed according to requirements.

The concept of *differentiation* enters into the definition of stem cells, and this also often leads to confusion. In our view, differentiation is a qualitative and relative phenomenon. Cells tend to be differentiated relative to other cells, hence, adult tissue stem cells may, or may not, be differentiated relative to embryonic stem cells (a point of current debate, bearing in mind the controversy in the literature concerning bone marrow stem cell *plasticity*). Stem cells produce progeny that may differentiate down a variety of pathways, leading to the concept of *totipotency* and *pluripotency* or *multipotency* of stem cells, in terms of their differentiation. This is actually a strange concept to apply to a stem cell since it is their progeny that differentiate and not the stem cells themselves. The fact is that the progeny can differentiate down more than one differentiated lineage, as is very obviously the case in the bone marrow, resulting in bone marrow stem cells being referred to as pluripotent; the initial dividing transit cells that initiate a lineage, which ultimately leads to specific differentiated cells, can be thought of as committed precursors for that lineage.

Some of the instructive signals for differentiation in the hemopoietic cell lineage are now well understood, but such signals for other tissues organized on a cell lineage basis have yet to be determined. There is much debate in the literature at present concerning the extent to which stem cells may be instructed to produce progeny of specific differentiated types and whether this is limited or unlimited. This topic is referred to as the *degree of plasticity* for stem cells. There are two very distinct issues here.

- The first is whether a stem cell, like a bone marrow stem cell, is ever instructed by its environment or niche naturally, or in laboratory or clinical situations, to make an apparently unrelated tissue cell type, such as a liver, intestinal, or skin cell, and whether it can regenerate these tissues if they are injured. A subsidiary question is not whether this ever happens normally in nature, but whether we, as experimentalists or clinicians, can provide the necessary instructions or environment for this to happen in a controlled situation.
- The second issue relates not only to the stem cells but also to the early progeny of stem cells from, for example, the bone marrow and whether these cells that circulate round the body and may end up in a distant tissue can end up expressing differentiation markers unrelated to the bone marrow cell lineages but specific to the tissue in which the cell then resides.

The former is an issue of plasticity of the bone marrow stem cells, and the latter may be more an issue of the plasticity of the bone marrow–derived cell lineages. If a bone marrow stem cell can ever be instructed to be a gastrointestinal stem cell, it should be capable of undertaking all the functional duties of a gastrointestinal stem cell, including the regeneration of the gastrointestinal epithelium if it is subsequently injured. The cloning of animals by nuclear transfer technology into egg cytoplasm clearly demonstrates that all nuclei of the body contain a full complement of DNA and that, under the right environmental conditions, this can be reprogrammed (or unmasked) by environmental signals to make all the tissues of the body. It must be remembered that such cloning experiments, e.g., Dolly the sheep, are rare and inefficiently produced events. They do, however, clearly indicate the enormous potential that can be achieved if we provide the necessary instructive reprogramming signals. This should enable us in the future to reproducibly instruct any adult tissue stem cell to make any tissue of the body. If and when this becomes the case, the distinction between embryonic stem cells and adult tissue stem cells may disappear.

## III. HIERARCHICALLY ORGANIZED STEM CELL POPULATIONS

The issue here is what determines the difference between a dividing transit cell and a stem cell and whether that transition is an abrupt one or a gradual one. One can think of this transition as being a differentiation event that distinguishes a dividing transit cell from a stem cell. This is an old argument. Do differentiation signals act on preexisting stem cells, removing on average half the cells produced by previous symmetric divisions, or do the stem cells divide asymmetrically to produce a differentiated progeny at division and a stem cell? One possibility is that this distinction is made at the time that a stem cell divides. Indeed, does it need to divide to differentiate? In this case, such divisions must be regarded as asymmetric, with the dividing stem cell producing one stem cell (i.e., for self-maintenance) and one dividing transit cell. This type of asymmetric division may occur in some tissues, such as the epidermis. However, if this is the case, the stem cell must also retain the potential to alter its self-maintenance probability, which for an asymmetric division is 0.5 in steady state, and adopt a value somewhat higher than this if stem cells are killed and need to be repopulated. The current view regarding the bone marrow stem cells is that the transition between a stem cell and a dividing transit cell is a gradual one that occurs over a series of divisions within a cell lineage, which inevitably implies that one has a population of stem cells with a varying degree of stemness or, conversely, a varying degree of differentiation. For the bone marrow, one issue is whether experimentalists have ever identified the presence of the truly ancestral ultimate bone marrow stem cell. The difficulty here may be one of identifying and extracting such cells, the location of which is probably in the bone where they will be present in increasingly diminishing numbers, as one looks for the increasingly primitive cells.

Our current model for the gastrointestinal cellular organization, which is based on an attempt to accommodate as much experimental data as possible, is that the commitment to differentiation producing dividing transit cells does not occur at the level of the ultimate stem cell in the lineage but at a position two or three generations along the cell lineage. If such a concept is drawn as a cell lineage diagram, the proliferative units in the intestine, the crypts, each contain four to six cell lineages and, hence, four to six lineage ancestor stem cells but up to 30 second- and third-tier stem cells, which under steady-state circumstances are inevitably displaced and moved toward the dividing transit compartment. But if damage occurs in one or more of the ultimate stem cells, they can assume the mantle of the ultimate stem cell and repopulate the lineage (Potten *et al.*, 1997; Potten, 1998; Marshman *et al.*, 2002). This gives rise to the concept of *actual* vs. *potential stem cells* (see Fig. 31.1), which is discussed later in this chapter.

An analogy can be drawn here with the hierarchical structure within an organization such as the army, a concept that was discussed at the time that we were formulating the text for the *Development* paper in which we defined stem cells (Potten and Loeffler, 1990). In a military battlefield environment, the hierarchically organized army is under the control of and ultimately dependent on the highly trained (or so one hopes) general. In the event that the general is killed in the battlefield, there may be a reasonably well-

trained colonel or major who can take over command and assume the insignia and uniform as well as the function of the general. In the event the colonel also should be killed, there may be lesser trained officers who will attempt to assume the mantle of command (e.g., a captain). Ultimately, the vast majority of the troops, the privates, would be insufficiently trained or experienced to be able to adopt the functional role of the commander. However, the Dolly the sheep scenario suggests that occasionally a private given a crash course in military strategy might function as the officer in command. The analogy could be taken even further to relate to the apoptosis sensitivity that is seen in the gastrointestinal ultimate stem cells. These cells appear to adopt a strategy with complete intolerance to any genetic damage and a reluctance to undertake repair, since this may be associated with inherent genetic risk they commit an altruistic suicide: the general who undergoes a nervous breakdown or suffers serious injury and has to be removed from command.

In the small intestinal crypts there have been in the past no useful markers that permit the stem cells to be identified and, hence, studied. However, such markers are now being identified. In the absence of markers, the small intestine proved to be an invaluable biological model system to study stem cells because the cells of the intestinal cell lineage are arranged spatially along the long axis of the crypt. This can be demonstrated by cell migration tracking and mutational marker studies. As a consequence, the stem cells are known to be located at very specific positions in the tissue (crypts):

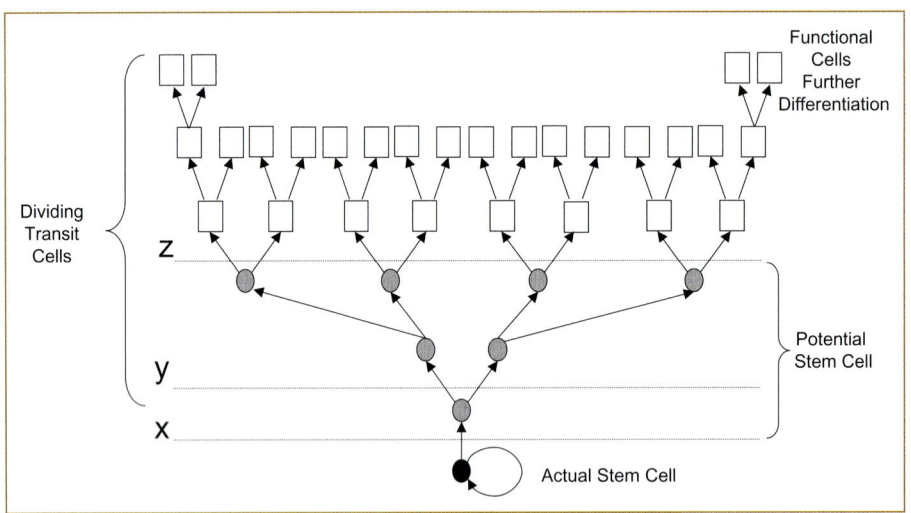

**FIG. 31.1.** A typical stem cell–derived cell lineage, which may be applicable to most epithelial tissues of the body. The lineage is characterized by a self-maintaining lineage ancestor actual stem cell (black), which divides and produces a progeny that enters a dividing transit population. The number of cell generations in the dividing transit population varies from tissue to tissue. The commitment to differentiation that separates the stem cell from the dividing transit population can occur at the point of the actual stem cell division (x), in which case the stem cells are dividing asymmetrically on average. However, this commitment may be delayed to point y or z, generating a population of potential stem cells that can replace the actual stem cell if it is killed. Under normal steady-state circumstances, the potential stem cells form part of the dividing transit population and are gradually displaced down the lineage, undergoing further differentiation events if required to produce the functional mature cells of the tissue.

the fourth to fifth cell position from the crypt base in the small intestine and at the very base of the crypt in the mid-colon of the large intestine (see Fig. 31.2) (Potten *et al.*, 1997; Potten, 1998; Marshman *et al.*, 2002).

## IV. SKIN STEM CELLS

The first suggestion that the proliferative compartment of the epidermis, the basal layer, was heterogeneous and contained only a small subpopulation of stem cells came with the development of the skin macrocolony (and subsequently microcolony) clonal regeneration assays developed by Withers (Withers, 1967; Withers and Elkind, 1970). This was soon combined with other cell kinetic and tissue organization data to formulate the epidermal proliferative unit (EPU) concept (Potten, 1974) (see Fig. 31.3). This suggested that the basal layer consisted of a series of small, functionally and cell lineage-related cells, with a spatial organization that related directly to the superficial functional cells of the epidermis, the stratum corneum. The concept indicated that the epidermis should be regarded as being made up of a series of functional proliferative units, each unit having a centrally placed self-maintaining stem cell and a short stem cell–derived cell lineage (with three generations). The differentiated cells, produced at the end of the lineage, migrated out of the basal layer into the suprabasal layers in an ordered fashion, where further maturation events occurred, eventually producing the thin, flattened, cornified cells at the skin surface, which were stacked into columns (like a pile of plates), with cell loss occurring at a constant rate from the surface of the column (Fig. 31.3). Such an organization is clearly evident in the body skin epidermis of the mouse, its ears, and a modified version of the proliferative unit can clearly be identified in the dorsal surface of the tongue (Hume and Potten, 1976). There has been, and continues to be, some debate as to whether this concept applies to human epidermis. It is clear that in many sites of the body in man

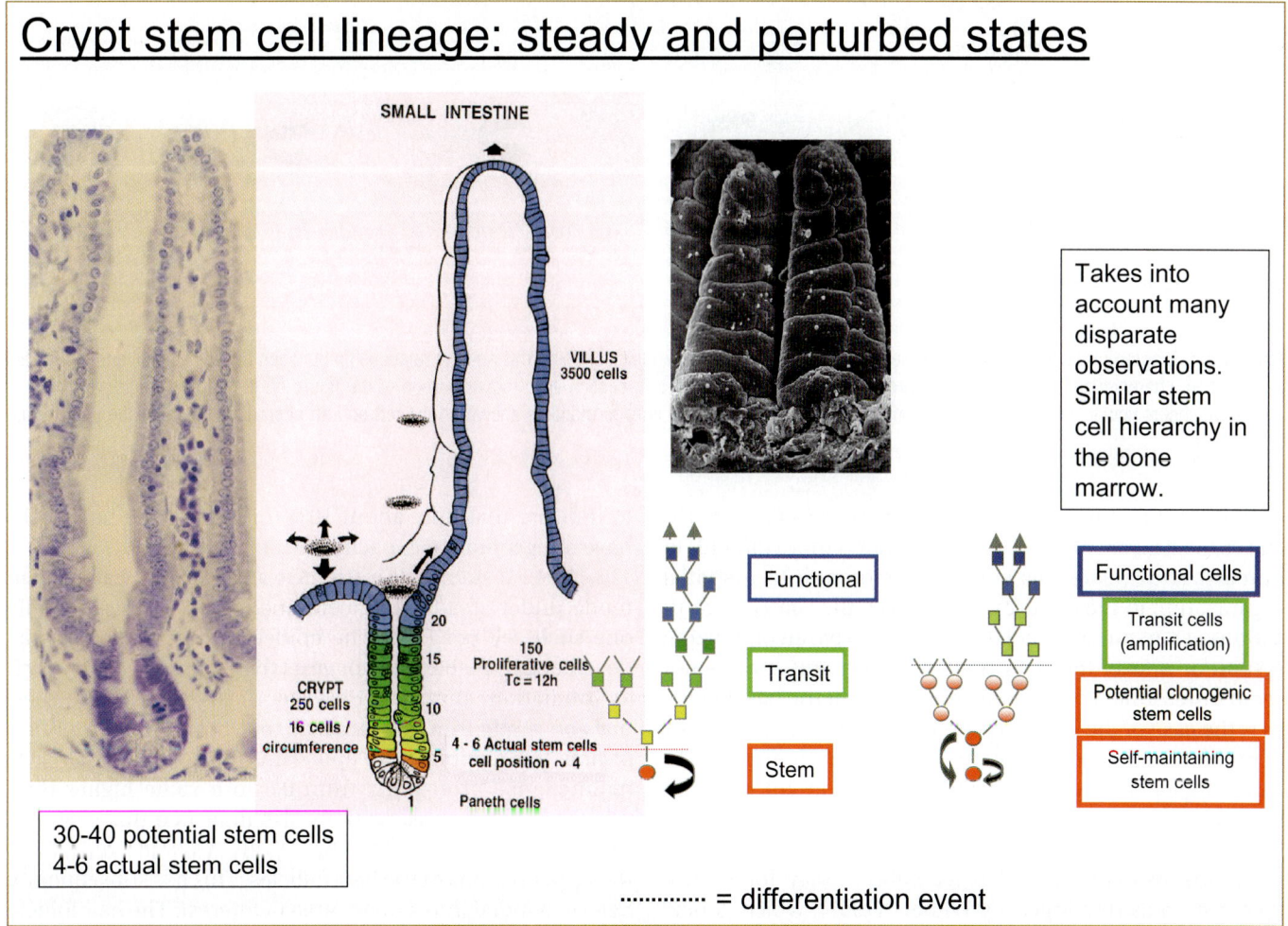

**FIG. 31.2.** The cell lineage for the small intestinal crypts. It is postulated that each crypt contains four to six such lineages and, hence, four to six lineage ancestor actual stem cells, and there are about six cell generations in each lineage with at least four distinct differentiated cell types being produced. The attractive feature of this as a cell biological model system is that the position of a cell in a lineage can be related to its topographical position in a longitudinal section through the crypt.

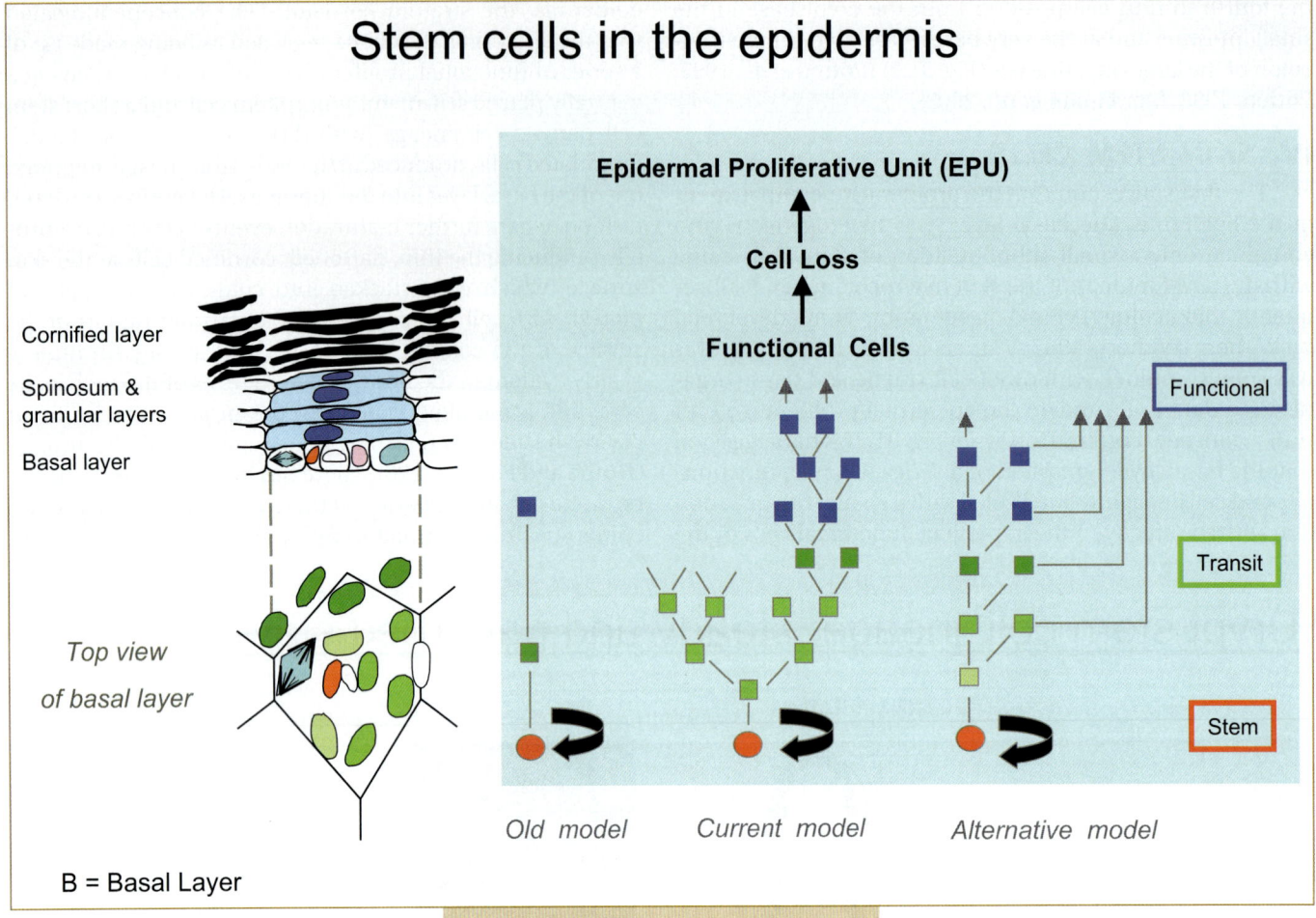

## Stem cells in the epidermis

**FIG. 31.3.** Diagrammatic representation of the cell lineage seen in the interfollicular epidermis and the relationship between the cell lineage and the spatial organization, characterized as the epidermal proliferative unit (EPU), as seen in section view (upper portion of the figure on the left) and in surface view in epidermal sheets (lower portion of the figure on the left). The old historical lineage model is shown on the left and an alternative model to the current one on the right.

a similar columnar organization can be seen in the superficial corneal layers of the epidermis. What is more difficult in humans is to relate this superficial structure to a spatial organization in the basal layer. However, the spatial organization seen in the superficial layers must have an organizing system at a level lower in the epidermis, and it does not seem unreasonable to assume that this is in the basal layer, as is the case of the mouse epidermis. The presence of units of proliferation in human epidermis has been demonstrated using β-galacosidase/GFP staining of skin in cultures and xenografts (MacKenzie, 1997; Ghazizadeh and Taichman, 2005).

A macroscopic, clonal regeneration assay for mouse epidermis was developed by Withers (1967), which generates nodules very similar in appearance to spleen colonies. Subsequently, Al-Barwari developed a microscopic clonal assay that required a shorter time interval between irradiation and tissue sampling (Al-Barwari and Potten, 1976). Together, these clonal regeneration assays were interpreted

to indicate that only about 10% (or less) of the basal cells have a regenerative capacity, i.e., are stem cells.

The EPU stem cells must have an asymmetrical division mode under steady-state cell kinetics, since there is only one such cell per EPU. The epidermal microcolony assay developed by Al-Barwari suggests that, following injury such as irradiation, surviving EPU stem cells can change their division mode from asymmetric to symmetric for a period of time to repopulate the epidermis (i.e., change their self-maintenance probability from 0.5 to a value higher than 0.5). Al-Barwari's observations also indicated that a significant contribution to reepithelialization could come from the upper regions of the hair follicles. This has subsequently been expanded into a major area of interest: The hair follicle lineage stem cells have been shown to possess a wide range of regenerative activities, including production of epidermis. It was also clear from studies on the structural organization of the epidermis following injury that in order to reestablish the spatial distribution of stem cells, the epidermis undergoes a reorganization involving hyperplasia, during which stem cells are redistributed and eventually establish their EPU spatial configurations.

The skin clearly contains another important stem cell population, namely that associated with the growing hair follicles. Hair is produced over a protracted period of time by rapid divisions in the germinal region of the growing hair follicle (termed an *anagen follicle*). This hair growth may be maintained for long periods of time, three weeks in a mouse (where the average cell cycle time may be 12 h), months to years in man, and for more indefinite periods for some animal species, such as Angora rabbits, merino sheep, and some strains of dog. This high level of cell division in the germinal matrix of the follicle, which has a considerable spatial polarity like the intestinal crypt, must have a fixed stem cell population residing in the lowest regions of the germinal matrix that can maintain the cell production for the required period of time. Very little is known about these stem cells. The complication with hair follicles is that in mouse and man, the growing follicles eventually contain a mature hair and cell proliferation activity ceases. The follicle shrinks and becomes quiescent (a *telogen follicle*). The simplest explanation here is that the telogen follicle, which consists of far fewer cells in total than in a growing follicle, contains a few quiescent hair follicle stem cells that can be triggered back into proliferation at the onset of a new hair growth cycle. However, as discussed later, there is some controversy concerning this concept.

It is now very clear that the skin contains a third stem cell compartment, which is located in the upper outer sheath of the hair follicle, below the sebaceous glands. This is sometimes identifiable by virtue of a small bulge in the outer root sheath, and so this population of cells has been referred to as the *bulge* cells. What is very evident from a whole series of extremely elegant but complicated experiments is that these bulge cells possess the ability, under specialized conditions, to reform the hair follicle if it is damaged and also to contribute to the reepithelialization of the epidermis. It is cells from this region of the follicle that were probably responsible for the epidermal reepithelialization from follicles seen by Al-Barwari. Cells from the bulge can make follicles during development of the skin and also reestablish the follicles if they are injured. The controversy concerns the issue of whether bulge stem cells, which are predominantly quiescent cells, ever contribute to the reestablishment of an anagen follicle under normal undamaged situations. The simplest interpretation is that these cells are not required for this process, since in order for this to happen some very complex cell division and cell migratory and homing pathways have to be inferred. This goes somewhat against the concept of stem cells being fixed or anchored and also against the concept of keratinizing epithelia being a tightly bound strong and impervious barrier. What seems likely for the skin is that the EPU stem cell and the hair follicle stem cell have a common origin

during the development of the skin to the bulge stem cells, which then become quiescent and are present as a versatile "reserve" stem cell population that can be called into action if the skin is injured and requires reepithelialization (see Figs. 31.4 and 31.5) (Potten and Booth, 2002).

Another issue relating to the stem cells of the bulge region of the follicle is that they are defined, usually by immunohistochemistry, as being present as a "cluster" of largely quiescent cells. If these "reserve" cells are called into action, which cells divide? If the cells toward the center of the cluster divide, this will inevitably push the peripheral cells out of the cluster — so they were not stem cells in the first place. If the cells at the periphery divide, there is no need for the cells toward the center — so they do not satisfy stem cell criteria. The most likely explanation is that the cluster contains only a few (one or two?) *real* stem cells at the center and that the rest are quiescent early lineage cells.

Quiescence is a property often stated as applicable to stem cells. It is certainly true that in many, but not all, replacing tissues the stem cells are cycling more slowly than the transit cells (longer G1 phase). The *Oxford English Dictionary* defines *quiescence* as "a state of motionless, inertness, silence, dormancy, or inactivity." Stem cells are usually defined by virtue of what they can do. If they are doing nothing, can they be stem cells?

## V. INTESTINAL STEM CELL SYSTEM

The intestinal epithelium, like all epithelia, is highly polarized and divided into discrete units of proliferation and differentiation. In the small intestine the differentiated units are the fingerlike villi protruding into the lumen of the intestine. These structures are covered by a simple columnar epithelium consisting of several thousand cells, which perform their specific function, become worn out, and are shed, predominantly from the tip of the villus. There is no proliferation anywhere on the villus. The cell loss from the villus tip is precisely balanced in steady state by cell proliferation at the base of the villi in units of proliferation called *crypts*. Each villus is served by about six crypts, and each crypt can produce cells that migrate onto more than one villus. The crypts in the mouse contain about 250 cells in total, 150 of which are proliferating rapidly and have an average cell cycle time of 12 hours. The cells move from the mouth of the crypt at a velocity of about one cell diameter per hour, and all this movement can be traced, in the small intestine, back to a cell position about four cell diameters from the base of the crypt. The very base of the crypt, in mice and humans, is occupied by a small population of functional differentiated cells called *Paneth* cells. Cell migration tracking and innumerable cell kinetic experiments all suggest that the stem cells that represent the origin of all this cell movement are located at the fourth position from the base of the crypt in the small intestine and right at the base of the crypt in some regions of the large bowel. The crypt is

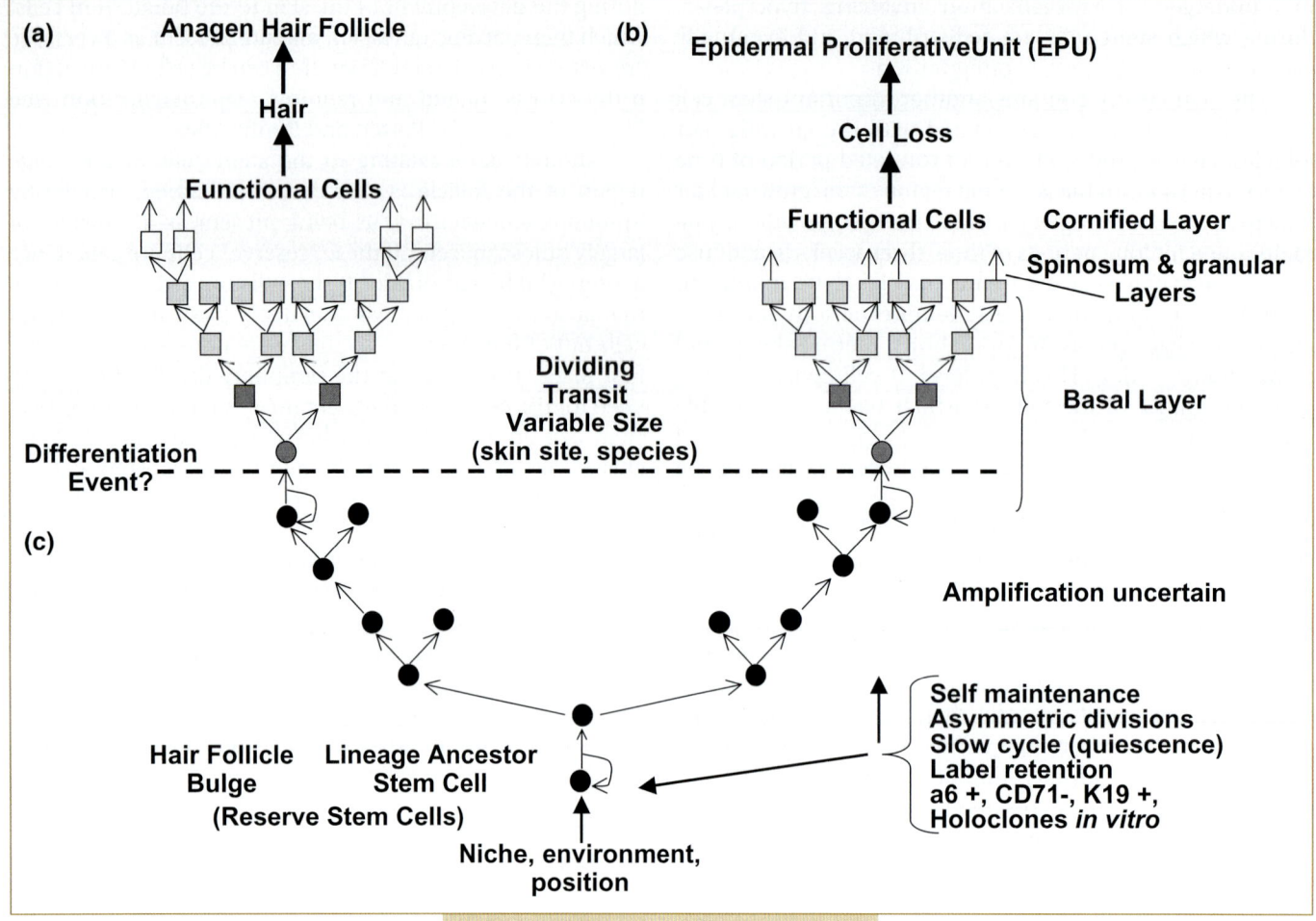

**FIG. 31.4.** The complexity of the stem cell populations in mammalian skin as characterized in the mouse. **(a)** A distinct cell lineage is proposed for the interfollicular epidermis (EPU), **(b)** another is proposed for the matrix region of the growing hair follicle (anagen follicle), and **(c)** a potent reserve regenerative stem cell compartment, which resides in the upper/outer root sheath or bulge region of the hair follicle, is proposed. The stem cells in the bulge region can clearly regenerate the epidermis, the hair follicle, and probably other structures, such as the sebaceous glands.

a flask-shaped structure, about 16 cells in circumference. Mathematical modeling suggests that each crypt contains about five cell lineages and, hence, five cell lineage ancestor stem cells. Under steady-state kinetics these cells are responsible for all the cell production, producing daughters that enter a dividing transit lineage of between six and eight generations in the small and large bowel, respectively (see Figs. 31.1 and 31.2). The stem cells in the small intestine divide with a cycle time of approximately 24 hours; hence, in the lifetime of a laboratory mouse they divide about 1000 times. It is assumed that these cells are anchored or fixed in a micro-environmental niche that helps determine their function and behavior. The uniquely attractive feature of this model system, from a cell biological point of view, is that in the absence of stem cell–specific markers, the behavior and characteristics and response to treatment of these crucial lineage ancestor cells can be assessed by studying the behavior of cells at the fourth position from the bottom of the crypt in the small intestine. When this is done, one of

the features that seems to characterize a small population of cells at this position (about five cells) is that they express an exquisite sensitivity to genotoxic damage, such as delivered by small doses of radiation (Potten, 1977; Hendry *et al.*, 1982; Potten and Grant, 1998). They appear to tolerate no DNA damage and activate a p53-dependent altruistic suicide (apoptosis). It is believed that this is part of the genome protection mechanisms that operate in the small intestine and account for the very low incidence of cancer in this large mass of rapidly proliferating tissue.

The clonal regeneration techniques for the intestine developed by Withers (Withers and Elkind, 1970) have been used extensively. These techniques suggest the presence of a second compartment of clonogenic or potential stem cells (about 30 per crypt) that possess a higher radioresistance and a good ability to repair DNA damage. These observations, together with others, suggest a stem cell hierarchy of the sort illustrated in Figs. 31.1 and 31.2, with the commitment to differentiation that distinguishes dividing transit

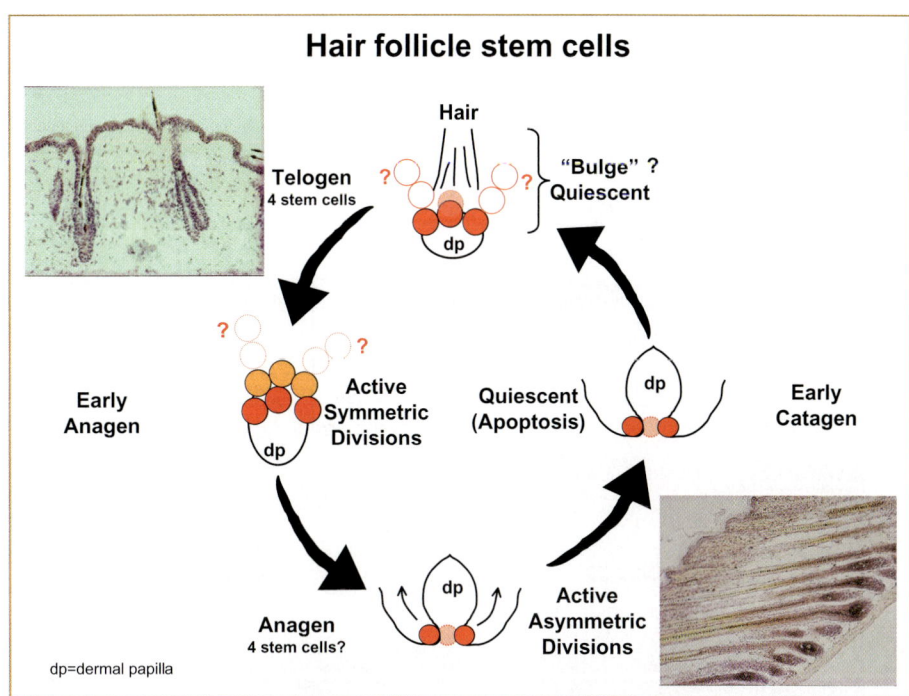

**Hair follicle stem cells**

**FIG. 31.5.** Diagrammatic representation of a growing anagen hair follicle and a resting or quiescent telogen follicle. Photomicrographs of representative anagen and telogen follicles are inset.

cells from stem cells, occurring about three generations along the lineage. Virtually identical lineage structures can be inferred for the colonic crypts (Cai *et al.*, 1997).

There has been an absence of stem cell–specific markers in the past. However, current work suggests that some may now be available. Antibodies to Musashi-1, an RNA-binding protein identified as playing a role in asymmetric division control in neural stem cells, appears to be expressed in very early lineage cells in the small intestine (see Fig. 31.6) (Potten *et al.*, 2003).

Recent studies have indicated that the ultimate stem cells in the crypt possess the ability selectively to segregate old and new strands of DNA at division and to retain the old template strands in the daughter cell destined to remain a stem cell while passing the newly synthesized strands, which may contain any replication-induced errors, to the daughter cell destined to enter the dividing transit population and to be shed from the tip of the villus five to seven days after birth from division (Potten *et al.*, 2002). This selective DNA segregation process provides a second level of genome protection for the stem cells in the small intestine (Cairns, 1975), protecting them totally from the risk of replication-induced errors, thus providing further protection against carcinogenic risk and an explanation for the very low cancer incidence in this tissue (see Table 31.1). This mechanism of selective DNA segregation allows the template strands to be labeled with DNA synthesis markers at times of stem cell expansion, i.e., during late tissue development and during tissue regeneration after injury. The incorporation of label into the template strands persists (label-retaining cells),

**Table 31.1.**

Why do small intestinal stem cells not develop more cancers?

When one considers that:

- the tissue is 3–4 times greater in mass (length)
- the cells are proliferating 1.5 times more rapidly
- there are 2–3 times more total stem cells
- there are 3–4 times more stem cell divisions in a lifetime

Compared with the large intestine, the small intestine has 70 times fewer cancers.

thus providing a truly specific marker for the lineage ancestor cells (see Fig. 31.6). This figure also illustrates some other ways in which intestinal stem cells may be distinguished from their rapidly dividing progeny.

The selective segregation of "old" and "new" DNA strands at mitosis is appearing to be a universal feature of the rare ultimate adult tissue stem cell. In addition to the small intestinal epithelium, this process has been detected in breast epithelial stem cells, both *in vivo* and *in vitro*, in neural stem cells *in vitro*, and in some muscle satellite cells (Merok *et al.*, 2002; Clarke *et al.*, 2005; Karpowicz *et al.*, 2005; Pare and Shirley *et al.*, 2006; Shinin *et al.*, 2006). It can also be inferred from work on the dorsal surface of the tongue and epidermis.

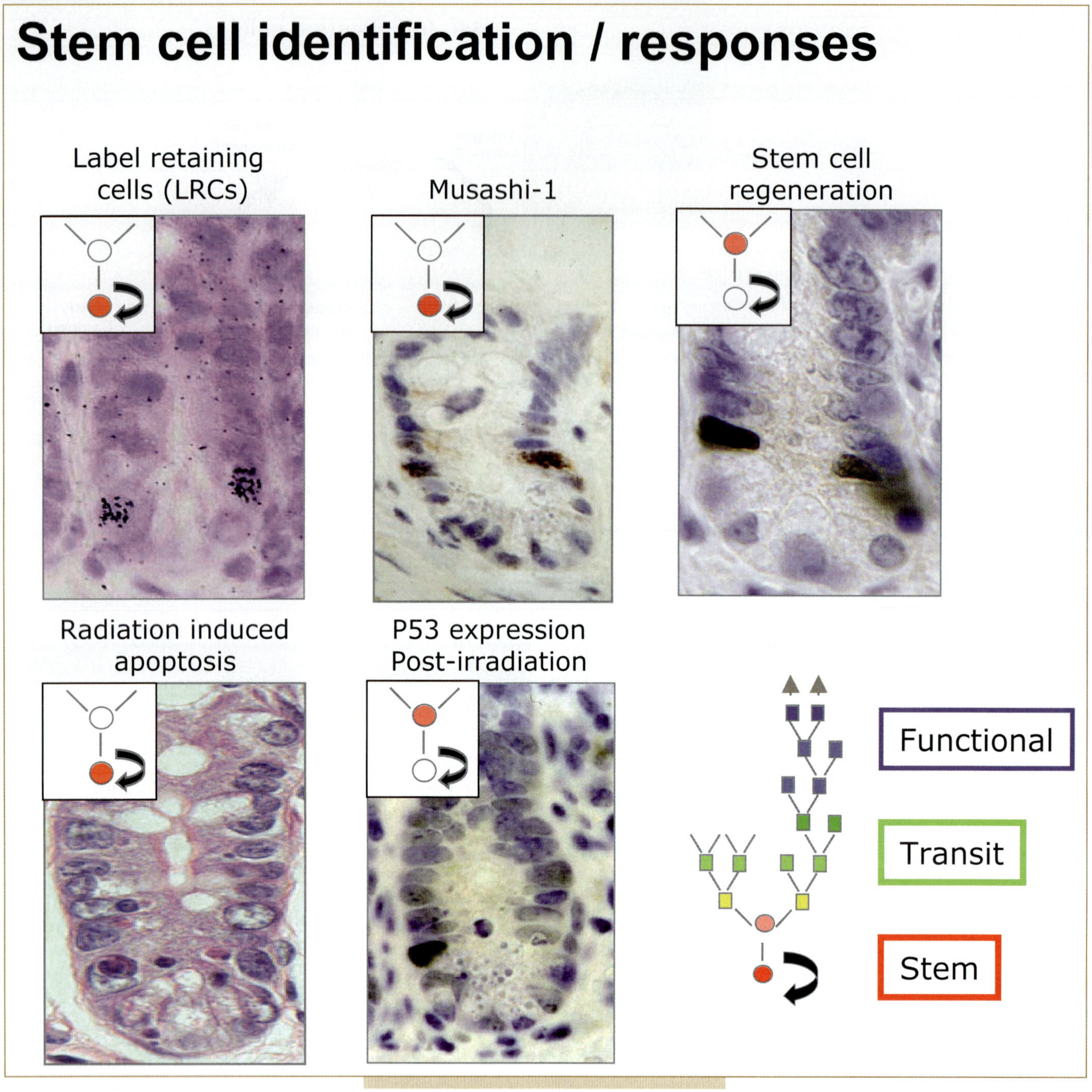

# Stem cell identification / responses

**FIG. 31.6.** Photomicrographs of longitudinal sections of the small intestinal crypts from the mouse, illustrating a range of possible ways of identifying the stem cell compartment. Making use of the selective strand segregation hypothesis, template strands of DNA can be labeled, generating label-retaining cells at the fourth position from the bottom of crypts. Musashi-1, an RNA-binding protein, is expressed in early lineage cells and under some labeling conditions can show specificity for individual cells at around cell position 4. Part of the regenerative or potential stem cell compartment can be seen by S-phase labeling (Bromodeoyuridine labeling) at critical phases following cytotoxic injury, when these cells are called into regenerative mode. The example shown here is a labeling pattern at 24 hours after two doses of 5-fluorouracil, when the only cells in S phase are a few cells scattered around the fourth position from the base of the crypt. As part of the genome protective mechanism it is postulated that the ultimate lineage ancestor stem cells have an exquisite sensitivity to radiation and the induction of genome damage. When this happens, the cells commit suicide via apoptosis, which can easily be recognized and occurs at about the fourth position from the base of the crypt. These cells do not express p53 protein, at least at the times studied and as detectable by immunohistochemistry. However, some cells do express p53 protein at high levels following radiation exposure, and it is postulated that these are the surviving potential stem cells in cell cycle arrest, to allow for repair prior to entering rapid regenerative cell cycles. Under appropriate immunohistochemical preparative procedures, individual wild-type P53 protein–expressing cells can be seen at around cell position 4. The cartoons represent the cell within the lineage, which is thought to be labeled by each marker (stem cell = red; stem cell daughter/early progenitor = pink).

## VI. STEM CELL ORGANIZATION IN FILIFORM PAPILLAE ON THE DORSAL SURFACE OF THE TONGUE

Oral mucosae are keratinizing, stratified epithelia, similar to epidermis in their structural organization. The dorsal surface of the tongue is composed of many small, filiform papillae that have a very uniform shape and size. Detailed histological investigations, together with cell kinetic studies performed by Hume (Hume and Potten, 1976), showed that each papilla is composed of four columns of cells: two dominant and two buttressing columns. The dominant anterior and posterior columns represent modified versions of the epidermal proliferative units and were called *tongue proliferative units*. The cell migratory pathways were mapped (like the studies in the intestinal crypts), which enabled the position in the tissue from which all migration originated to be identified, this being the presumed location of the stem cell compartment. The lineage characterizing this epithelium is similar to that seen in the dorsal epidermis of the mouse, i.e., self-replacing asymmetrically dividing stem cells, occurring at a specific position in the tissue, producing a cell lineage that has approximately three generations (Fig. 31.7). The stem cells here have a particularly pronounced circadian rhythm (Potten *et al.*, 1977a).

## VII. GENERALIZED SCHEME

It thus appears that for the major replacing tissues of the body, hierarchical or cell lineage schemes appear to explain the cell replacement processes. These schemes may involve isolated single stem cells that, under steady-state circumstances, must be presumed to divide asymmetrically, producing a dividing transit population. The size of the dividing transit population differs dramatically from tissue to tissue. The number of generations defining the degree of amplification that the transit population provides for each stem cell division is related inversely to the frequency that stem cells will be found within the proliferating compartment (see Fig. 31.8). For some systems, such as the bone marrow and the intestine, the commitment to differentiation that separates the dividing transit compartment from the stem cell compartment appears to be delayed until a few generations along the lineage. This generates a stem cell hierarchy, with cells of changing (decreasing) stemness or, conversely, increasing commitment, leading to the concept of committed precursor cells. In the small intestine this delay in the commitment to differentiation to a dividing transit

## Tongue proliferative units (TPUs) - Filiform papillae

 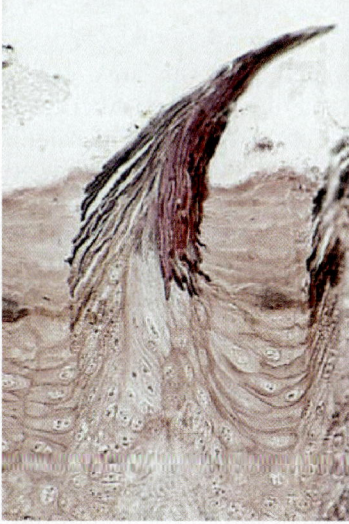 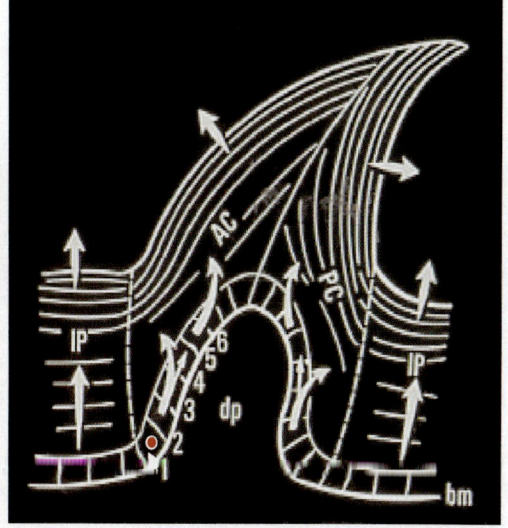

**FIG. 31.7.** A scanning electron micrograph of (left) and histological section through the dorsal surface of the tongue (center), and a diagrammatic representation of this tissue (right), showing the tongue proliferative units (the dominant anterior column, AC, and posterior column, PC). Cell migratory pathways have been identified based on cell positional analyses and cell marking and on the location of the stem cells identified in the basal layer (red diamond — cell position 1). The stem cells in this tissue express one of the strongest circadian rhythms in proliferation seen anywhere in the body.

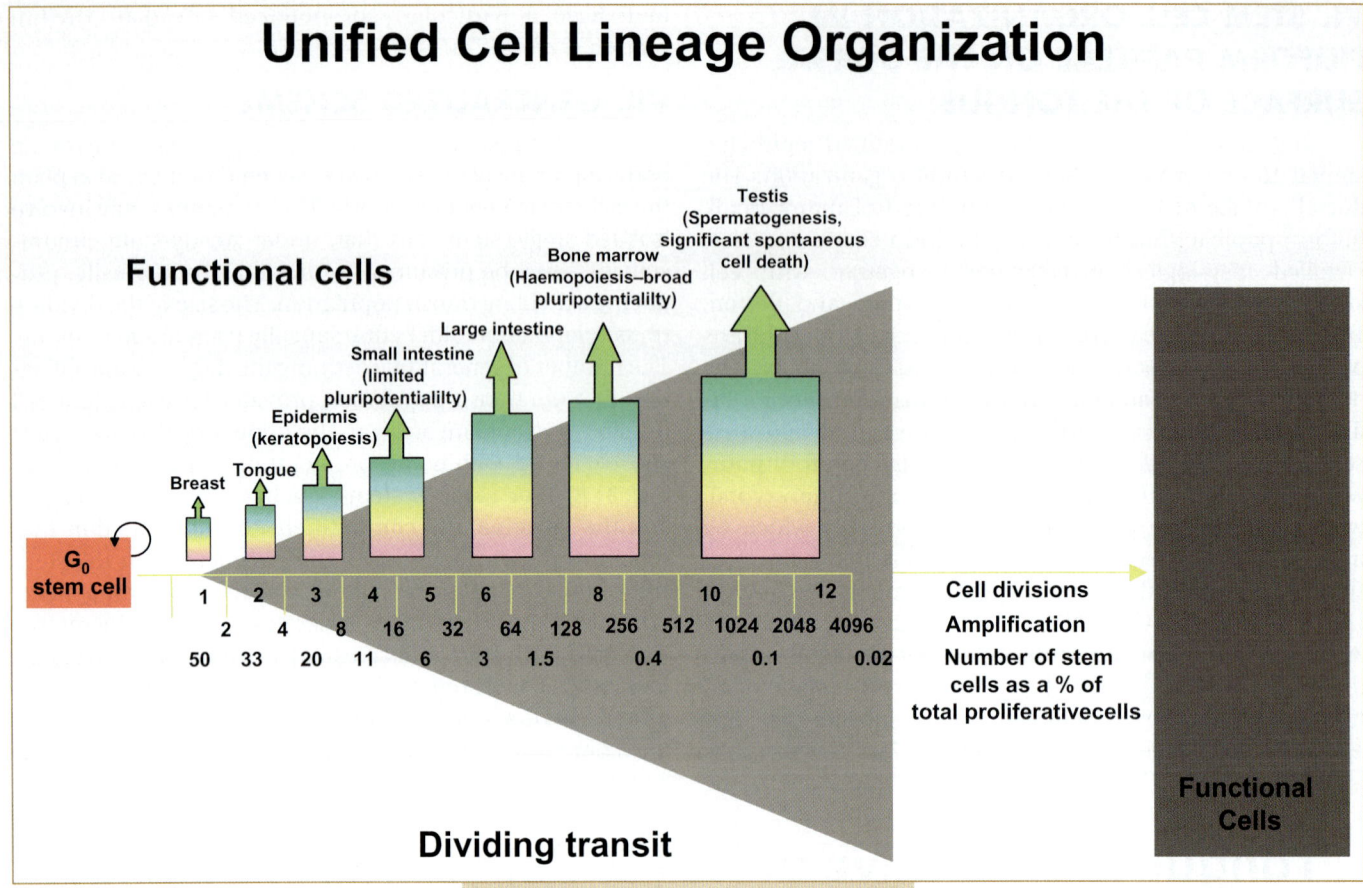

**FIG. 31.8.** A diagrammatic representation of a stem cell–derived cell lineage, showing the approximate positions for the number of cell generations in the dividing transit population for a range of murine tissues. Stratified keratinizing epithelia, such as the breast, tongue, and epidermis, tend to have the shortest lineages; the bone marrow and the testis tend to have the longest lineages. Also shown is the degree of theoretical amplification that the dividing transit lineage provides for each stem cell division and the inverse relationship between the degree of amplification and the proportion of the proliferative compartment that the stem cells occupy.

population provides the tissue with a reserve population of potential stem cells that can repopulate the tissue if the lineage ancestor cells are destroyed, an added level of tissue protection in this extremely well-protected tissue. With regard to the bone marrow, committed precursors (or even earlier cells) appear to circulate in the blood and may lodge in various tissues. Given appropriate microenvironments and local signals, some of these lodged cells may be instructed to differentiate down unusual pathways. This has prompted research into using such cells to correct genetic deficiencies. Although the transdifferention theory is attractive, recent research indicates that the apparent plasticity of stem cells may be less clear-cut. Transplantation experiments in mice with specific gene disorders suggest that transplanted bone marrow cells may "fuse" with cells in the "target" organ and, hence, complement any gene deficiency.

## VIII. ADULT STEM CELL PLASTICITY

Recently, a number of significant papers on the plasticity of bone marrow progenitor cells have been published that focus on their ability to contribute to the different pro-

liferative cell compartments of the intestinal epithelium (and other cell types of the intestinal mucosa). In an elegant series of experiments, bone marrow progenitor cells from a male donor were injected into female recipient mice, which were subsequently subject to induction of mucosal injury (Brittan *et al.*, 2005). Alternatively, bone marrow progenitors were injected into lethally irradiated female mice (Brittan *et al.*, 2002; Brittan and Wright, 2004). *In situ* hybridization was able to demonstrate cells with a male Y chromosome within the regenerating endothelium and within the pericryptal fibroblast cell population (the cells that surround the crypt epithelial cells on the basolateral side), but not within the colonic epithelium. The fibroblasts with male chromosomes appeared as columns of cells along the crypt axis within the tissue sections examined, suggesting that bone marrow derived cells had established a clonal lineage within the proliferative fibroblast cell population.

Other studies have suggested that bone marrow progenitors can contribute to all the different cell lineages within the intestinal epithelium (Rizvi *et al.*, 2006). In mice, injection of labeled female bone marrow progenitor cells into lethally irradiated male mice resulted in cells with female-specific label being found within the proliferative cells of the small intestinal crypts and within the various differentiated cell populations of the villi (columnar, goblet, enteroendocrine, Paneth). The conclusion was that female bone marrow progenitors "fused" with stem or early progenitor cells of the intestinal epithelium of the male recipient. Fusion rather than transdifferentiation was suggested as a mechanism in this study, based on the high frequency of cells that were positive for the female-specific marker and the male Y chromosome. In humans, cells from male donors have been found throughout the gastrointestinal tract epithelium of female recipients (Okamoto et al., 2002). It was suggested that the donor cells were able to incorporate into the epithelium and differentiate into mature epithelial cells but did not integrate into the stem cell compartment.

In addition to the regenerating epithelium postinjury, high levels of intestinal epithelial cell proliferation are to be found within tumors of epithelial origin. Further experiments examining the incorporation of female bone marrow progenitors into intestinal tumors of male Min/+ mice revealed a large number of epithelial cells within the tumors, which were double positive for male- and female-specific markers (Rizvi et al., 2006). These results paralleled those from the lethal-irradiation experiments and suggest that cell fusion between bone marrow progenitors and epithelial cells is common (within the intestinal epithelium at least) in situations when cell proliferation rates are elevated. It also raises the possibility that bone marrow progenitors can contribute to tumor progression.

This last point is rather controversial, and, to date, only one study has suggested that bone marrow progenitors are directly responsible for tumor development within an epithelial tissue (Houghton et al., 2004). Mice had their bone marrow ablated by irradiation and received a bone marrow transplant from ROSA26 transgenic mice. They were then infected with *Helicobacter felis*, in order to induce gastric carcinogenesis. The mice showed classic progression from metaplasia to dysplasia to carcinoma, over the period of one year, with most dysplastic gastric glands and all neoplasia being derived from the bone marrow progenitors, as assessed by their expression of the ROSA26 transgene. Experiments examining male donor/female recipient combinations failed to show colocalization of male- and female-specific markers in isolated gastric epithelial cells, which suggests that stable fusion events between the bone marrow progenitor cells and gastric epithelial cells do not take place in this experimental system. As evidence for transdifferentiation of bone marrow cells, the authors were able to present evidence of epithelial cell–specific gene expression in bone marrow–derived mesenchymal stem cell cultures exposed to conditioned medium from gastric epithelial cell primary cultures.

It can be clearly seen that different experimental models offer different perspectives on the relative merits of stem cell transdifferentiation and fusion and their ability to contribute to different cell populations within epithelial tissues. There is the possibility that in a highly proliferative tissue like the gastrointestinal epithelium, bone marrow transplantation may introduce into the tissue a population of cells that are a "low-threshold" target for induction of tumorigenesis; however, this still requires further evidence to show whether this is a real phenomenon or just a peculiarity of the experimental model. This aside, adult stem cell transplantation clearly has exciting potential for promoting tissue regeneration following injury.

## IX. REFERENCES

Al Barwari, S. E., and Potten, C. S. (1976). Regeneration and dose–response characteristics of irradiated mouse dorsal epidermal cells. *Int. J. Radiat. Biol. Relat. Stud. Phys. Chem. Med.* **30**, 201–216.

Brittan, M., and Wright, N. A. (2004). Stem cells in gastrointestinal structure and neoplastic development. *Gut* **53**, 899–910.

Brittan, M., Hunt, T., Jeffery, R., Poulsom, R., Forbes, S. J., Hodivala-Dilke, K., Goldman, J., Alison, M. R., and Wright, N. A. (2002). Bone marrow derivation of pericryptal myofibroblasts in the mouse and human small intestine and colon. *Gut* **50**, 752–757.

Brittan, M., Chance, V., Elia, G., Poulsom, R., Alison, M. R., MacDonald, T. T., and Wright, N. A. (2005). A regenerative role for bone marrow following experimental colitis: contribution to neovasculogenesis and myofibroblasts. *Gastroenterology* **128**, 1984–1995.

Cai, W. B., Roberts, S. A., Bowley, E., Hendry, J. H., and Potten, C. S. (1997). Differential survival of murine small and large intestinal crypts following ionizing radiation. *Int. J. Radiat. Biol.* **71**, 145–155.

Cairns, J. (1975). Mutation selection and the natural history of cancer. *Nature* **255**, 197–200.

Clarke, R. B., Spence, K., Anderson, E., Howell, A., Okano, H., and Potten, C. S. (2005). A putative human breast stem cell population is enriched for steroid receptor-positive cells. *Dev. Biol.* **277**, 443–456.

Ghazizadeh, S., and Taichman, L. B. (2005). Organisation of stem cells and their progeny in human epidermis. *J. Invest. Dermatol.* **124**, 367–372.

Hendry, J. H., Potten, C. S., Chadwick, C., and Bianchi, M. (1982) Cell death (apoptosis) in the mouse small intestine after low doses: effects of dose rate, 14.7-MeV neutrons, and 600-MeV (maximum energy) neutrons. *Int. J. Radiat. Biol. Relat. Stud. Phys. Chem. Med.* **42**, 611–620.

Houghton, J., Stoicov, C., Nomura, S., Rogers, A. B., Carlson, J., Li, H., Cai, X., Fox, J. G., Goldenring, J. R., and Wang, T. C. (2004). Gastric cancer originating from bone marrow–derived cells. *Science* **306**, 1568–1571.

Hume, W. J., and Potten, C. S. (1976). The ordered columnar structure of mouse filiform papillae. *J. Cell Sci.* **22**, 149–160.

Karpowicz, P., Morshead, C., Kam, A., Jervis, E., Ramunas, J., Cheng, V., and van der Kooy, D. (2005). Support for the immortal strand hypothesis: neural stem cells partition DNA asymmetrically *in vitro*. *J. Cell Biol.* **170**, 721–732.

MacKenzie, I. C. (1997). Retroviral transduction of murine epidermal stem cells demonstrates clonal units of epidermal structure. *J. Invest. Dermatol.* **109**, 377–383.

Marshman, E., Booth, C., and Potten, C. S. (2002). The intestinal epithelial stem cell. *Bioessays* **24**, 91–98.

Merok, J. R., Lansita, J. A., Tunstead, J. R., and Sherley, J. L. (2002). Cosegregation of chromosomes containing immortal DNA strands in cells that cycle with asymmetric stem cell kinetics. *Cancer Res.* **62**, 6791–6795.

Okamoto, R., Yajima, T., Yamazaki, M., Kanai, T., Mukai, M., Okamoto, S., Ikeda, Y., Hibi, T., Inazawa, J., and Watanabe, M. (2002). Damaged epithelia regenerated by bone marrow–derived cells in the human gastrointestinal tract. *Nat. Med.* **8**, 1011–1017.

Pare, J. F., and Sherley, J. L. (2006). Biological principles for *ex vivo* adult stem cell expansion. *Curr. Top. Dev. Biol.* **73**, 141–171.

Potten, C. S. (1974). The epidermal proliferative unit: the possible role of the central basal cell. *Cell Tissue Kinet.* **7**, 77–88.

Potten, C. S. (1977). Extreme sensitivity of some intestinal crypt cells to X and gamma irradiation. *Nature* **269**, 518–521.

Potten, C. S. (1998). Stem cells in gastrointestinal epithelium: numbers, characteristics and death. *Philos. Trans. R. Soc. London, Sev. B, Biol. Sci.* **353**, 821–830.

Potten, C. S., and Hendry, J. H. (1985). The micro-colony assay in mouse small intestine. *In* "Cell Clones: Manual of Mammalian Cell Techniques." (C. S. Potten and J. H. Hendry, eds.), pp. 50–60. Edinburgh, Churchill Livingstone.

Potten, C. S., and Loeffler, M. (1990). Stem cells: attributes, cycles, spirals, pitfalls and uncertainties. Lessons for and from the crypt. *Development* **110**, 1001–1020.

Potten, C. S., and Grant, H. K. (1998). The relationship between ionizing radiation-induced apoptosis and stem cells in the small and large intestine. *Br. J. Cancer* **78**, 993–1003.

Potten, C. S., and Booth, C. (2002). Keratinocyte stem cells: a commentary. *J. Invest. Dermatol.* **119**, 888–899.

Potten, C. S., Al Barwari, S. E., Hume, W. J., and Searle, J. (1977a). Circadian rhythms of presumptive stem cells in three different epithelia of the mouse. *Cell Tissue Kinet.* **10**, 557–568.

Potten, C. S., Booth, C., and Pritchard, D. M. (1997b). The intestinal epithelial stem cell: the mucosal governor. *Int. J. Exp. Pathol.* **78**, 219–243.

Potten, C. S., Owen, G., and Booth, D. (2002). Intestinal stem cells protect their genome by selective segregation of template DNA strands. *J. Cell Sci.* **115**, 2381–2388.

Potten, C. S., Booth, C., Tudor, G. L., Booth, D., Brady, G., Hurley, P., Ashton, G., Clarke, R., Sakakibara, S., and Okano, H. (2003). Identification of a putative intestinal stem cell and early lineage marker; musashi-1. *Differentiation* **71**, 28–41.

Rizvi, A. Z., Swain, J. R., Davies, P. S., Bailey, A. S., Decker, A. D., Willenbring, H., Grompe, M., Fleming, W. H., and Wong, M. H. (2006). Bone marrow–derived cells fuse with normal and transformed intestinal stem cells. *Proc. Natl. Acad. Sci. U.S.A.* **103**, 6321–6325.

Shinin, V., Gayraud-Morel, B., Gomes, D., and Tajbakhsh, S. (2006). Asymmetric division and cosegregation of template DNA strands in adult muscle satellite cells. *Nat. Cell Biol.* **8**, 677–687.

Till, J. E., and McCulloch, E. A. (1961). A direct measurement of the radiation sensitivity of normal bone marrow cells. *Rad. Res.* **14**, 213.

Withers, H. R. (1967). The dose–survival relationship for irradiation of epithelial cells of mouse skin. *Brit. J. Radiolo.* **40**, 187.

Withers, H. R., and Elkind, M. M. (1970). Micro-colony survival assay for cells of mouse intestinal mucosa exposed to radiation. *Int. J. Rad. Biol.* **17**, 261.

# Chapter Thirty-Two

# Embryonic Stem Cells as a Cell Source for Tissue Engineering

*Ali Khademhosseini, Jeffrey M. Karp, Sharon Gerecht, Lino Ferreira, Gordana Vunjak-Novakovic, and Robert Langer*

## I. INTRODUCTION

Tissue engineering may one day generate off-the-shelf organs for transplantation that may be able to treat a variety of debilitating ailments, such as diabetes and Parkinson's disease. One of the major barriers to realization of this enormous potential is the lack of a renewable source of cells for transplantation. Embryonic stem cells (ESCs) have the potential to provide such a source of cells because of their ability to differentiate to all somatic cells and their seemingly unlimited proliferative capability. In this chapter we address the potential and the challenges associated with the use of ESC in tissue engineering. In particular we address methods to proliferate and direct ESC differentiation, to isolate and transplant ESCs, and to incorporate these cells into existing tissue-engineering approaches. In addition, we address issues associated with the host's immune rejection, ESC-derived tumor formation, and scale-up processes.

It has been estimated that approximately 3000 people die every day in the United States from diseases that could have been treated with stem cell–derived tissues (Lanza et al., 2001). Given the therapeutic potential and growing public awareness of stem cells to treat disease, it is not surprising that embryonic stem cell (ESC) research has been rapidly expanding since mouse ESC (mESC) were first

isolated in 1981 (Evans and Kaufman, 1981; Martin, 1981) followed by the isolation of human embryonic stem cells (hESCs) in 1998 (Shamblott *et al.*, 1998; Thomson *et al.*, 1998) from the inner cell mass (ICM) of human blastocysts (Fig. 32.1). Adult stem cells have been used clinically for therapies such as bone marrow transplantation since the 1960s. Although adult stem cells hold great therapeutic promise, ESCs represent an alternative source of cells, with benefits including ease of isolation, ability to propagate rapidly without differentiation, and potential to form all cell types in the body. Additionally, ESCs represent an attractive cell source for the study of developmental biology, for drug/toxin screening studies, and for the development of therapeutic agents to aid in tissue or organ replacement therapies. Regarding the latter, which is the focus of this chapter, ESCs have the potential to exhibit a considerable impact on the field of tissue engineering, where current treatments for large tissue defects involve graft procedures having severe limitations. Specifically, many patients with end-stage organ disease are unable to yield sufficient cells for expansion and transplantation. In addition, there exists an inadequate supply of harvestable tissues, for grafting has associated risks, such as donor site morbidity, infection, disease transmission, and immune rejection. Tissue engineering–based

*Principles of Tissue Engineering, 3rd Edition*
*ed. by Lanza, Langer, and Vacanti*

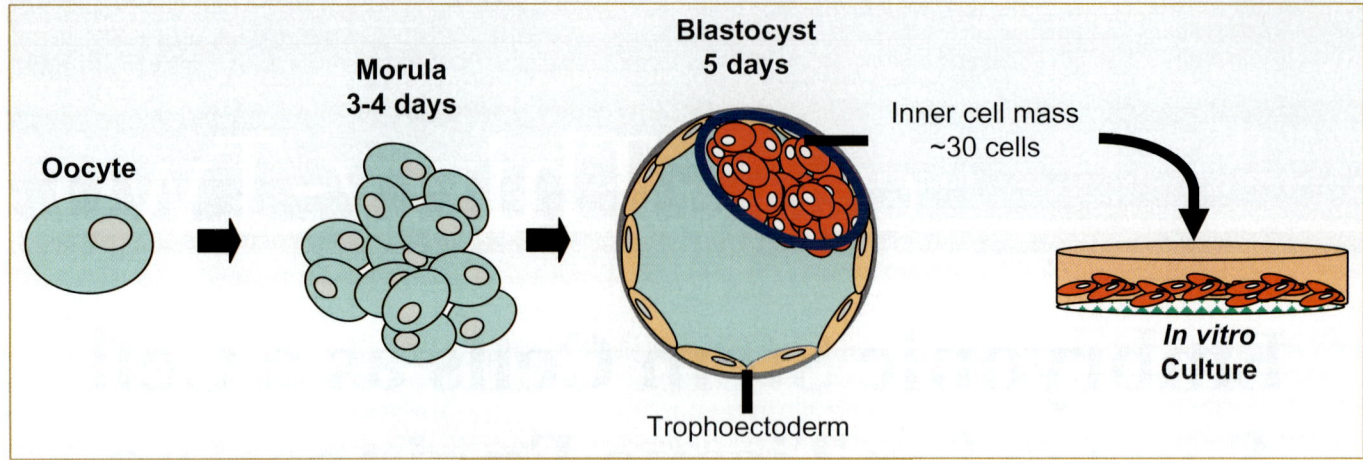

**FIG. 32.1.** Schematic diagram of the derivation of embryonic stem cells.

therapies may provide a possible solution to alleviate the current shortage of organs.

Tissue engineering has been defined as an interdisciplinary field that applies the principles of engineering, materials science, and life sciences toward the development of biologic substitutes that restore, maintain, or improve tissue function (Langer and Vacanti, 1993). Thus, tissue engineering may provide therapeutic alternatives for organ or tissue defects that are acquired congenitally or by cancer, trauma, infection, or inflammation. Tissue-engineered products would provide a life-long therapy and may greatly reduce the hospitalization and health care costs associated with drug therapy while simultaneously enhancing the patients' quality of life.

A central part of such promising strategies is the cell source to be used and the methods whereby sufficient numbers of viable differentiated cells can be obtained. ESCs represent a powerful source of cells capable of multilineage differentiation because they can potentially provide a renewable source of cells for transplantation. ES-derived cells can be used directly as cellular replacement parts or in combination with materials (typically in the form of scaffolds, Fig. 32.2). Despite this promise, the application of ESC to tissue engineering faces numerous challenges, including appropriately differentiating the cells to the desired lineage in a controlled and homogenous fashion and avoiding implantation of undifferentiated ESCs, which can potentially form teratocarcinomas. Currently, ESC-based tissue-engineering research is focused on elucidating soluble and immobilized cues and respective signaling mechanisms that direct cell fate, on characterization and isolation of differentiated progeny, and on establishing protocols to improve the expansion and homogeneity of differentiated cells.

This chapter discusses key concepts and approaches for (1) propagation of undifferentiated ESCs, (2) directed differentiation into tissue-specific cells, (3) isolation of pro-

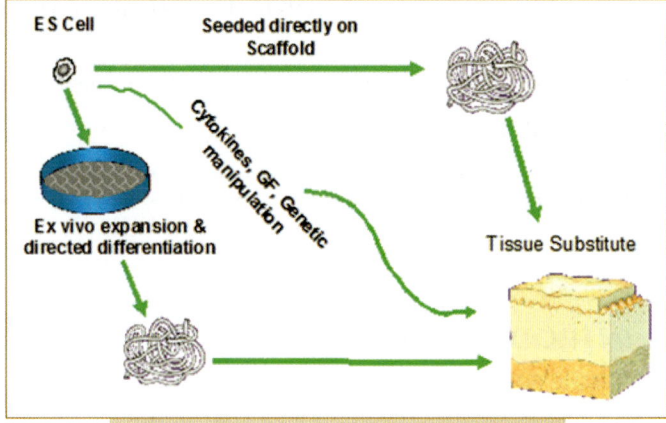

**FIG. 32.2.** Approaches for using ES cells for scaffold-based tissue-engineering applications. ES cells can be used in tissue-engineering constructs in a variety of methods. ES cells can be expanded in culture and then seeded directly onto scaffold, where they are allowed to differentiate. Alternatively, stem cells can be directed to differentiate into various tissues and enriched for desired cells prior to seeding the cells onto scaffolds.

genitor and differentiated phenotypes, (4) transplantation of progenitor and differentiated cells, and (5) remaining challenges for translating ESC-based tissue-engineering research into clinical therapies. Whenever possible, approaches using hESCs will be preferentially reported.

## II. MAINTENANCE OF ESCS

The self-renewal of ESCs is a prerequisite for generating a therapeutically viable amount of cells. Over the past few years, much insight has been gained into self-renewal of ESCs. Both mouse and human ESCs were first derived and routinely maintained in culture at an undifferentiated state on mouse embryonic fibroblast feeder (MEF) layers with medium containing serum. Through this research considerable behavioral, morphological, and biochemical differences have been observed between mESCs and hESC, and research on nonhuman embryonic cells is often not easily translated to humans (Ginis *et al.*, 2004; Park *et al.*, 2004; Thomson *et al.*, 1998). For example, mESCs form tight, rounded clumps, whereas hESCs form flatter, looser colonies, grow more slowly, and demand strict culture conditions to maintain their normal morphology and genetic integrity. Unlike mESC, which can be maintained in an undifferentiated state in the presence of leukemia-inhibitory factor (LIF), LIF alone is not sufficient to maintain hESC cultures, which require supplementation with basic fibroblast growth factor (bFGF) or the presence of a feeder layer. Although both mouse and human ESC express common transcription factors of stemness, such as Nanog, Oct4, and alkaline phosphatase, in the human system undifferentiated ESC express stage-specific embryonic antigen-3 (SSEA-3) and SSEA-4, while SSEA-1 is expressed only on differentiation; the opposite expression is observed in the mouse system. Due to these differences, efforts in hESC research focus on improving culture conditions to allow better expansion in undifferentiated state as well as on understanding the mechanisms of hESC self-renewal.

Since the primary therapeutic aim of the research on hESCs is to derive cells for the replacement of diseased or damaged tissues, difficulties concerning xenograft transplantation had arisen. Moreover, all potential applications depend largely on the routine availability of moderate to large numbers of cells requiring methods amenable to scale-up. Addressing the xenograft obstacle, Richards *et al.* examined the culture of hESCs cells on human feeders and found that human fetal muscle fibroblasts, human fetal skin fibroblasts, and adult fallopian tubal epithelial cells supported the pluripotency of hESCs culture *in vitro* (Richards *et al.*, 2002). Richards *et al.* further derived and established an hESC line on human fetal muscle fibroblasts in entirely animal-free conditions. Since then, different fetal and adult cells were examined and shown to support the continuous growth of hESC (Amit *et al.*, 2003; Cheng *et al.*, 2003; Hovatta *et al.*, 2003; Richards *et al.*, 2003). However, the use of hESCs for therapeutic applications requires defined culture medium and controlled cell derivation, maintenance, and scale-up. To overcome these obstacles, C. Xu *et al.* (2001) showed that hESCs can be maintained on Matrigel or laminin in MEF-conditioned medium. Cells grown in these conditions meet all the criteria for pluripotent cells: They maintain normal karyotypes, exhibit a stable proliferation rate and high telomerase activity, and differentiate into derivatives of all three germ layers, both *in vitro* and *in vivo*. Recent studies further defined culture conditions and show that hESCs can be expanded on human fibronectin using medium supplemented with bFGF and tumor growth factor β1 (TGFβ1) (Amit *et al.*, 2004). Noggin, an antagonist of bone morphogenetic protein (BMP), was found critical in preventing differentiation of hESC in culture. The combina-

tion of Noggin and bFGF is sufficient to maintain the proliferation of undifferentiated hESCs (R. H. Xu *et al.*, 2005). Furthermore, hESCs maintained in medium containing high concentrations of bFGF (24–36 ng/mL), alone or in combination with other factors, show characteristics similar to cultures maintained with feeder cell–conditioned medium (Wang *et al.*, 2005; C. Xu *et al.*, 2005). The derivation of hESCs has also been achieved with minimal exposure to animal-derived material, using serum replacement (SR) and human foreskin fibroblasts as feeder cells (Inzunza *et al.*, 2005), instead of the murine feeder layer (Klimanskaya *et al.*, 2005), which may provide well-defined culture conditions (Ludwig *et al.*, 2006). Research is currently under way to determine how these conditions maintain cell integrity over long-term culture.

Although laboratory-scale ESC cultures have been shown to produce differentiated progeny for both rodent and human ESCs, it is generally acceptable that these culturing methods are not feasible for large-scale production of ESCs for therapeutic applications. Although two-dimensional methods, such as the high-density cultures of undifferentiated mESCs, have been developed by combining automated feeding and culture of mESCs on petriperm dishes (Oh *et al.*, 2005), the three-dimensional culture of ESC may be a more suitable technology for large-scale expansion of ESCs.

At the present time, the aggregation of multiple ESCs is an obligatory process to initiate EB formation. A few methods have been developed for the differentiation of mESC in controlled cultures. Hanging drops and methylcellulose cultures were shown to be somewhat efficient in preventing the agglomeration of EBs, but their complex nature makes their up-scaling a rather difficult task. A much simpler process, however, involving spinner flasks, resulted in the formation of large cell clumps within a few days, indicative of significant cell aggregation in the cultures (Wartenberg *et al.*, 2001). An increase of the culture medium's stirring rate, to avoid agglomeration, resulted in massive hydrodynamic damage to the cells, due to the extensive mixing in the vessels. Therefore, to establish a scalable process for the development of EBs, there is a need for dynamic cultivation under controlled mixing conditions. One approach employed a static system for an initial aggregation period of four days, followed by a period in dynamic culture in spinner flasks, to successfully achieve the bulk production of cardiomyocytes from differentiating mESCs (Zandstra *et al.*, 2003).

Another dynamic approach found to be highly effective for hESCs is to generate and culture EBs within rotating cell-culture systems (Gerecht-Nir *et al.*, 2004a). These bioreactors provide exceptionally supportive environments for the cultivation of hESCs: minimal hydrodynamic damage to incipient EBs; reduced opportunity for EB fusion and agglomeration (the speed of rotation may be increased according to EB size); and the uniform growth and

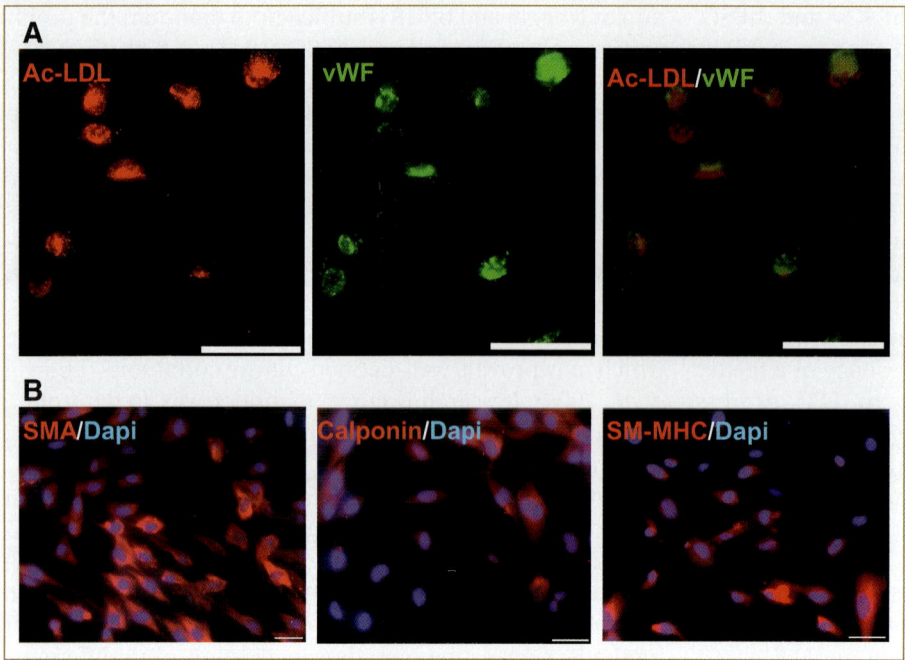

**FIG. 32.3.** Vascular differentiation of hES cells. Immunofluorescence analysis of mesodermal cells cultured in the presence of **(A)** hVEGF165 reveals expression of EC markers such as DiI-Ac-LDL and von Willebrand factor. **(B)** Similar analysis of cells cultured in the presence of hPDGF-BB reveals expression of VSMC markers such as SMA, calponin, and SM-MHC. Scale bars = 100 μm. Figure adapted from Gerecht-Nir *et al.*, 2003.

differentiation of EBs in three dimensions, as they oscillate and rotate evenly (Fig. 32.3A). hESCs cultured within these systems formed aggregates after 12 h and were smaller and more uniform in size and evenly rounded, due to minimal agglomeration; the yield of EBs was threefold higher than that measured for static cultures. Also, dynamically formed EBs exhibit steady and progressive differentiation, with cyst formation and elaboration of complex structures such as neuro-epithelial tubes, blood vessels, and glands, as observed in statically formed EBs (Gerecht-Nir *et al.*, 2004a) (Fig. 32.3B). Although still an area of active research, these technologies have demonstrated the potential of engineering for the development of scalable technologies to expand ESCs for research and therapies.

## III. DIRECTED DIFFERENTIATION

Perhaps the biggest challenge in using ESCs in clinical applications is the lack of knowledge for directing their differentiation ability. For example, although ESCs have been shown to generate cells of hematopoietic, endothelial, cardiac, neural, osteogenic, hepatic, and pancreatic tissues, it has been very difficult to achieve uniform and predictable differentiation into these tissues. This lack of homogeneous differentiation may be attributed to the intrinsic property of ESCs to differentiate stochastically in the absence of proper temporal and spatial signals from the surrounding microenvironment. In this section we describe some of the current approaches used to direct the differentiation of ESC and give examples of their use.

### Genetic Reprogramming

This approach includes the introduction of specific gene(s) into hESCs, which enable the production (by enhancement or selection) and propagation of specific cell type populations. Different techniques for knock-in and knockout genes into hESCs were already established. Transfection of undifferentiated hESCs with specific plasmid was established using either chemical reagents or electroporation. The latter was further shown to be useful for the generation of homologous recombination events (Zwaka and Thomson, 2003). Another technique is the introduction of transgenes into hESCs with self-inactivated lentiviruses. This transduction technique was shown to be efficient, with sustained expression in undifferentiated hESCs as well as in hESCs that undergo differentiation (Gropp *et al.*, 2003; Ma *et al.*, 2003). However, using adenoviral and adeno-associated viral vectors, both undifferentiated and differentiated hESCs were successfully infected (Smith-Arica *et al.*, 2003). Another approach, which uses genetic manipulation, is the introduction of suicidal genes, permitting the ablation of the cells if necessary (Fareed and Moolten, 2002). Using this approach, hESCs were transfected to express the *Herpes simplex* virus thymidine kinase gene, which renders them to ganciclovir (Schuldiner *et al.*, 2003).

Genetic techniques can be used to generate positive or negative regulators. The positive regulators include the constitutive or controlled expression of transcription factors that have been shown to derive the differentiation into particular tissues. For example, the overexpression of Nurr transcription factor has been shown to increase the frequency of ESCs that differentiate into functional neural cells (Kim *et al.*, 2002). Alternatively, the negative regulators could be incorporated to induce the apoptosis of cells that differentiate to varying pathways. For example, neomycin selection and suicide genes that are activated by certain transcription factors can be used (Soria *et al.*, 2000). Clearly, these techniques will benefit from further understanding of the inner workings of transient cells and a knowledge of the differentiation pathways and lineages. Further analysis into stem cell and progenitor hierarchy through high-throughput analysis of gene and protein profiles should accelerate this process. Despite the power of these approaches, one potential concern is that the genetic modifications may make the cells unsuitable for transplantation.

## Microenvironmental Cues

Another approach to direct ESC differentiation is through the use of microenvironmental cues that have been shown to be important in regulating adult and ESC fate decisions. During development, inner cell mass cells are exposed to a series of tightly regulated microenvironmental signals. However, in tissue culture much of the complex expression patterns and spatial orientation of this signaling can be lost. Currently, ESCs are grown in their primitive state as aggregated colonies of cells. To stimulate differentiation, two main methods have been examined. In one method, differentiated cells are derived from EBs. EBs can be formed from either single-cell suspensions of ESCs or from aggregates of cells. EBs mimic the structure of the developing embryo and recapitulate many of the stages involved during its differentiation, and clonally derived EBs can be used to locate and isolate tissue-specific progenitors. EBs initiate many developmental processes and create suitable conditions for differentiation of cells into all three germ layers and are generally formed through suspension or hanging drop methods. Alternatively, ESCs may be cultured in 2D monolayers and induced to differentiate. In general, differentiation of ESCs in EBs produces a wider spectrum of cell types, due to the EBs' ability to better mimic the temporal pattern of cell differentiation as seen in the embryo. In some applications the combined use of EBs and adherent cultures has resulted in better cell yields. For example, to induce ESCs to differentiate to cardiomyocytes, an EB formation in suspension cultures was followed by a differentiation in adhesion cultures. This was shown to optimize the percentage of cells that give rise to cardiomyocytes (Guan *et al.*, 1999; Klinz *et al.*, 1999). Similarly, the production of hepatocytes has been shown to be induced by first culturing the cell in EBs and then in 2D cultures (Hamazaki *et al.*, 2001).

One of the tissues that has been the focus of much study regarding ESC differentiation is neural tissue. Neural progenitor cells were isolated from hESCs that showed positive immunoreactivity to neuron-specific antigens, responded to neurotransmitter application, and presented voltage-dependent channels in the cell membrane (Carpenter *et al.*, 2001; Schulz *et al.*, 2003; Zeng *et al.*, 2004; Zhang *et al.*, 2001). Highly enriched cultures of neural progenitor cells were isolated from hESCs and grafted into the stratum of parkinsonian rats (Ben-Hur *et al.*, 2004). The grafted cells differentiated *in vivo* into dopaminergic neurons and partially corrected behavioral deficits in the transplanted animals. A subsequent study showed that hESCs implanted in the brain ventricles of embryonic mice can differentiate into functional neural lineages and generate mature, active human neurons that successfully integrate into the adult mouse forebrain (Muotri *et al.*, 2005). Oligodendrocytes and their progenitors were also isolated in high yield from hESCs (Nistor *et al.*, 2005). Transplantation of these cells into animal models of dysmyelination resulted in integration, differentiation into oligodendrocytes, and compact myelin formation, demonstrating that these cells displayed a functional phenotype.

ESCs have also been shown to give rise to functional vascular tissue. Early vascular progenitor cells isolated from differentiating mESCs were shown to give rise to all three blood vessel cell components: hematopoietic, endothelial, and smooth muscle cells (Yamashita *et al.*, 2000). Once injected into chick embryos, these vascular progenitors differentiated into endothelial and mural cells and contributed to the vascular development. Using similar protocol, differentiation into endothelial and smooth muscle cells was achieved with hESCs (Gerecht-Nir *et al.*, 2003, Fig. 32.4). Our group has shown that hESCs can differentiate into endothelial cells, and isolated these cells using platelet endothelial cell adhesion molecule-1 antibodies (Levenberg *et al.*, 2002). *In vivo*, when transplanted into immunodeficient mice, the cells appeared to form microvessels. Furthermore, it has been shown that monkey ESCs can give rise to endothelial cells when the embryonic cells were exposed to a medium containing different growth factors. The isolated cells were able to form vascular-like networks when implanted *in vivo* (Kaufman *et al.*, 2004). Endohemato progenitor cells have been isolated from hESCs that presented hematopoietic (L. Wang et al., 2004) or endothelial cells competency. hESCs have been reported to differentiate into hematopoietic precursor cells when cocultured with bone marrow and endothelial cell lines (Kaufman *et al.*, 2001). When these precursor cells are cultured on semisolid media with hematopoietic growth factors, they form characteristic myeloid, erythroid, and megakaryocyte colonies.

Cardiomyocytes have been also isolated from hES cells for the treatment of cardiac diseases. Beating cells were observed after one week of culture under differentiation conditions, increased in numbers with time, and could retain contractility for over 70 days (C. Xu *et al.*, 2002). The beating cells expressed markers characteristic of cardiomyocytes, such as cardiac α-myosin heavy chain, cardiac troponin I and T, atrial natriuretic factor, and cardiac transcription factors GATA 4, Nkx2.5, and MEF 2. In addition, cardiomyocytes isolated from hES cells expressed sarcomeric marker proteins, chronotropic responses, and ion channel expression (Mummery *et al.*, 2003). Electrophysiology demonstrated that most cells resembled human fetal ventricular cells.

Insulin-producing β-cells were also generated from hESC. These cells were observed through the spontaneous

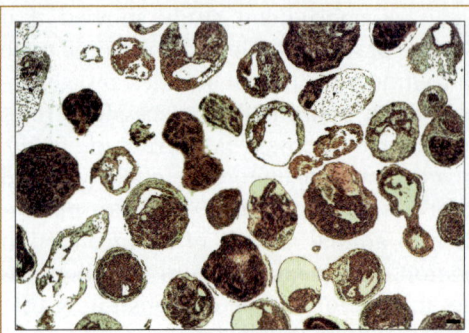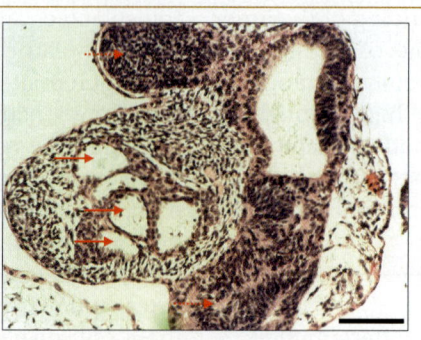

**FIG. 32.4.** Human EB formation in a rotating cell-culture system (Adapted from Gerecht-Nir *et al.*, 2004a). **(A–B)** Haematoxilin- and eosin-stained sections of EBs generated after one month in rotation culture, showing the formation of **(A)** a small and relatively homogenous population of human EBs and **(B)** a variety of cell types, such as epithelial neuronal tubes (dashed arrows) and blood vessels (solid arrows). Scale bars = 100 μm.

differentiation of hESC in adherent or suspension culture conditions (Assady *et al.*, 2001) and in medium containing growth factors (Segev *et al.*, 2004). Reverse transcription-polymerase chain reaction detected an enhanced expression of pancreatic genes in the different cells (Segev *et al.*, 2004). Immunofluorescence and *in situ* hybridization revealed high percentages of insulin-expressing cells (Segev *et al.*, 2004).

Recently there has been great interest in examining the osteogenic potential of ESCs, derived from both mice and humans. Human hESCs can differentiate into osteogenic cells with the same media supplements that are used to differentiate adult mesenchymal stem cells. Our group showed that culturing hESC without EBs leads to over a sevenfold increase in the number of osteogenic cells and to spontaneous bone nodule formation after 10–12 days (Karp *et al.*, 2005). In contrast, when hESC were differentiated as EBs for five days, followed by plating of single cells, bone nodules formed after four weeks only in the presence of dexamethasone.

## 3D Versus 2D Cell Culture Systems

ESCs can differentiate into complex 3D tissue structures. To study this phenomenon *in vitro*, 3D culture systems are necessary. The scaffold may act as a temporary extracellular matrix (ECM) providing physical cues for cell orientation, spreading, differentiation, and the remodeling of tissue structures (Ferreira *et al.*, 2007). For example, our group showed that the culture of hESCs in poly(lactic acid-*co*-glycolic acid) (PLGA) scaffolds in specific media containing transforming growth factor β, activin-A, or insulin-like growth factor induced the differentiation of the cells into 3D structures with characteristics of developing neural tissues, cartilage, or liver, respectively (Levenberg *et al.*, 2003). It was also demonstrated that the 3D environment created by cell encapsulation in Matrigel failed to support hESC growth and 3D organization, likely due to the fact that this gel was unable to resist the force of cell contraction. Furthermore, when these cells were cultured in PLGA and poly(lactic acid) (PLA) scaffolds in the presence of medium containing nerve growth factor and neurotrophin 3, enhanced numbers of neural structures were observed (Levenberg *et al.*, 2005a). In another study, the culture of ESCs in a 3D collagen scaffold,

stimulated with exogenous growth factors and hormones, led to the differentiation of the cells into hepatocyte-like cells. These cells were characterized by the expression of liver-specific genes and synthesis of albumin, and the differentiation pattern observed compared favorably to cells differentiated in a 2D system. It was also reported that the differentiation of rhesus monkey ESCs in 3D collagen matrixes was different from that in monolayers (S. S. Chen *et al.*, 2003). Alginate scaffolds were also used for the differentiation of hESCs (Gerecht-Nir *et al.*, 2004b). These scaffolds induced vasculogenesis in encapsulated cells to a larger extent than cells grown in bioreactors. Tantalum scaffolds also increased the differentiation of mESCs into hematopoietic cells as compared to traditional 2D cultures (Liu and Roy, 2005).

## High-Throughput Assays for Directing Stem Cell Differentiation

Today, chemists and engineers are equipped with tools to be able to synthesize molecules and test their effects on cells in a high-throughput manner. For example, libraries of small molecules, polymers, and genes have been generated and used to screen candidate molecules to induce osteogenesis (Wu *et al.*, 2002) and cardiomyogenesis (Wu *et al.*, 2004) in ESCs as well as the dedifferentiation of committed cells (S. B. Chen *et al.*, 2004). The use of chemical compound libraries may provide a method of addressing the complexities associated with native microenvironments by directing cell behavior through interacting with transcription factors and cell fate regulators.

Micro-scale technologies can miniaturize assays and facilitate high-throughput experimentation and therefore provide a potential tool for screening libraries. Recently, robotic spotters capable of dispensing and immobilizing nanoliters of material have been used to fabricate microarrays, where cell–matrix interactions can be tested and optimized in a high-throughput manner. For example, synthetic biomaterial arrays have been fabricated to test the interaction of stem cells with various extracellular signals. In this approach thousands of polymeric materials were synthesized and their effect on differentiation of hESCs (Anderson *et al.*, 2004) and human mesenchymal stem cells (hMSCs) (Anderson *et al.*, 2005) evaluated. These interactions have led to unexpected and novel cell–material interactions. Although the molecular mechanisms associated with the biological responses have yet to be clarified, such technology may be widely applicable in cell-microenvironment studies and in the identification of cues that induce desired cell responses. Also, the materials that yield desired responses could be used as templates for tissue-engineering scaffolds. Such an approach is a radical change from traditional methods of developing new biomaterials, where polymers have been individually developed and tested for their effect on cells.

In addition to analyzing synthetic-material libraries, the effect of natural ECM molecules on cell fate can be evaluated in a high-throughput manner. In one example, combinatorial matrices of various natural extracellular matrix proteins were tested for their ability to maintain the function of differentiated hepatocytes and to induce hepatic differentiation from murine ESCs (Flaim *et al.*, 2005).

Cell arrays have been used to pattern stem cells on 2D substrates. Arrays of cells can be used to localize and track individual cells, enabling clonal analysis of stem cell fates. For example, clonal populations of neural stem cells were immobilized within microfabricated structures, and their progeny were tracked using real-time microscopy, yielding information about cellular kinetics and cell fate decisions in a high-throughput manner (Chin *et al.*, 2004). Using this approach, it is possible to study the response of individual stem cells to various microenvironmental signals.

Cell patterning on geometrically defined shapes has been used to study the effects of cell shape on various cell fate decisions. As cells adhere onto micropatterned substrates, they align themselves to the shape of the underlying adhesive region. This shape change induces changes in the cytoskeletal features and has been shown to influence apoptosis, proliferation (Chen *et al.*, 1997), and stem cell differentiation. When hMSCs were patterned on fibronectin islands of various sizes, cells on large islands adhered and flattened, while cells on small islands generated spherically shaped cells. As these cells were stained for differentiation markers, it was observed that the spread cells generated osteoblasts and spherical cells gave rise to adipocytes (McBeath *et al.*, 2004). Further elucidation of the molecular mechanisms indicated that cell shape regulated the activation of the RhoA pathway, demonstrating that mechanical stresses can be crucial for directing stem cell differentiation. Therefore, controlling the cellular microenvironment using micropatterning may be used for directing cell fate for tissue-engineering applications.

It is known that mechanical forces affect the differentiation and functional properties of many cell types. Thus, mechanical stimuli may be required to direct the differentiation of ESCs. Understanding the effects of mechanical stimuli on ESC differentiation is still in its infancy, but tissue-engineering systems are already being developed that incorporate the effects of mechanical forces. For example, functional autologous arteries have been cultured using pulsatile perfusion bioreactors (Niklason *et al.*, 1999). Thus the use of mechanical stimuli may further enhance the ability of these cells to respond to exogenous signals. Other environmental factors that may be required include electrical signals. Hopefully, with time such techniques will allow for the development of ESC-based tissue-engineering applications. The design of bioreactors that control the spatial and temporal signaling that induces ESC differentiation requires further collaborative efforts between engineers and biologists.

# IV. ISOLATION OF SPECIFIC PROGENITOR CELLS FROM ESCs

Although hESCs can generate specific functional cell types from all three germ layers, it is typically not possible to differentiate the cells directly in culture and obtain pure cell populations. Isolation of a specific differentiated population of cells for transplantation will eliminate the presence of undifferentiated hESC, which have tumorigenic potential, and allow for efficient use of the various cell populations for therapeutic purposes. With the exception of a few cases where the enrichment of cells of interest was almost fully achieved (Ben-Hur *et al.*, 2004; Nistor *et al.*, 2005; Zeng *et al.*, 2004), the protocols adopted for the differentiation of hESCs do not yield pure cell populations. Therefore, there is a need for suitable techniques to isolate desired cells from a heterogeneous population of cells (Table 32.1). One approach is to isolate specific cells by using cell surface markers by fluorescence-activated cell sorting (FACS). In this case, the initial population of cells is immunostained by a single marker or a combination of different markers and the desired cell type isolated by FACS. Part of the initial population of cells is also labeled with isotype controls to gate the populations. The use of FACS yields pure populations of cells and allows one to select cells using different markers (L. Wang *et al.*, 2004). However, the intense flows in this technique may hamper the final cell survival.

Magnetic immunoselection has been used very often to isolate specific differentiated cells (Carpenter *et al.*, 2001; Kaufman *et al.*, 2001; Vodyanik *et al.*, 2005). Initially, the cells are labeled with relevant cell surface antibodies conjugated with magnetic beads. The magnetically labeled cells are then separated from the other ones by means of a magnetic column. The purity of isolated cells is generally higher than 80% (Vodyanik *et al.*, 2005). Although the purity achieved is slightly lower than the one obtained by FACS, the magnetic selection is generally less harmful to the cells than FACS. Another potential method for cell isolation is through reporter gene knock-in modifications (Lavon *et al.*, 2004). For example, to trace hepatic-like cells during differentiation of hESCs in culture, a reporter gene expressed under the control of a liver-specific promoter was used (Lavon *et al.*, 2004). For that purpose, hESCs underwent stable trans-

| Table 32.1. | Summary of methodologies to isolate specific lineages from hES cells | | |
|---|---|---|---|
| *Cell type* | *Methodology followed to isolate specific lineages* | *Cell lines* | *Reference* |
| **Mesoderm** | | | |
| Cardiomyocytes | Discontinuous percoll gradient | H1, H7, H9 | C. Xu *et al.* (2002) |
| Cardiomyocytes | Enzymatic and mechanical dissociation | H9.2 | Kehat *et al.* (2003) |
| Cardiomyocytes | Enzymatic dissociation | HES2 | Mummery *et al.* (2003) |
| Hematopoietic progenitor cells | Magnetic immunoselection | H1, H1.1, H9.2 | Kaufman *et al.* (2001) |
| Hematopoietic progenitor cells | Flow-activated cell sorting | H1, H9 | Chadwick *et al.* (2003) |
| Hematopoietic progenitor cells | Magnetic immunoselection | H1, H9 | Vodyanik *et al.* (2005) |
| Leucocytes | Selective adhesion of cells | H1 | Zhan *et al.* (2004) |
| Endothelial cells | Flow-activated cell sorting | H9 | Levenberg *et al.* (2002) |
| Endothelial and smooth muscle cells | Selective adhesion and size | H9, H9.2, H13,16 | Gerecht-Nir *et al.* (2003) |
| Endothelial-like cells | Flow-activated cell sorting | H1, H9 | L. Wang *et al.* (2004) |
| **Ectoderm** | | | |
| Neurons and glia | Magnetic immunoselection | H1, H7, H9 | Carpenter *et al.* (2001) |
| Neurons and glia | Enzymatic dissociation and selective adhesion of cells | H1, H9, H9.2 | Zhang *et al.* (2001) |
| Oligodendrocytes | Selective adhesion of cells | H7 | Keirstead *et al.* (2005) |
| **Endoderm** | | | |
| Hepatocyte-like cells | Introduction of a reporter gene and cell selection by flow-activated cell sorting | H9 | Lavon *et al.* (2004) |

fection with eGFP fused to the albumin minimal promoter sequence. This methodology allowed one to follow the differentiation pattern of hESCs into hepatic-like cells and to isolate those cells by FACS using the fluorescence of eGFP.

Isolation of a specific differentiated population of cells may also be accomplished by means of mechanical/enzymatic separation of cells exhibiting specific morphology, functional activity, or adhesion to a substrate. For example, cardiomyocytes have been isolated by dissecting contracting areas in embryoid bodies and dissociating those areas using collagenase (Kehat *et al.*, 2003). Oligodendroglial cells were isolated from stem cell aggregates that adhered to a specific substrate (Nistor *et al.*, 2005). In addition, neuroepithelial cells were isolated from embryoid bodies attached to a tissue culture–treated flask by using dispase (Zhang *et al.*, 2001). This enzyme selectively detached neuroepithelial islands from the embryoid bodies, leaving the surrounding cells adherent.

## V. TRANSPLANTATION

The first application of stem cells as a cellular replacement therapy is associated with bone marrow transplantation and blood transfusion, in which donor hematopoietic stem cells repopulated the host's blood cells (Till and McCulloch, 1980). Specific cell types from ESC can be an important cell-based therapy for numerous diseases, includ-

ing diabetes mellitus, Parkinson's disease, traumatic spinal cord injury, liver failure, muscular dystrophy, bone, vascular, and heart diseases, among others. Despite the significant advances in the use of mESC to treat several disease models (Verfaillie *et al.*, 2002), in the case of hESC, few studies have reported the *in vivo* functionality of differentiated cells. In most cases the cells are injected in a disease area and the functionality of those cells evaluated by immunohistochemistry and behavior tests. For instance, partial functional recovery of a mouse model of Parkinson's disease after injection of hESC-derived neural progenitor cells has been reported (Ben-Hur *et al.*, 2004). Moreover, the transplantation of oligodendroglial progenitor cells obtained from hESC in the *shiverer* model of dysmyelination resulted in integration, differentiation into oligodendrocytes, and compact myelin formation (McDonald and Howard, 2002).

Despite the ability of stem cells to differentiate to cells with the phenotypic and morphological structure of desired cell types, there has been very few scaffold-based tissue-engineering studies that use ESCs. ESCs may be differentiated in culture and desired cell types selected and subsequently seeded onto scaffolds. Ideally, this scaffold provides the cells with a suitable growth environment, optimum oxygen and nutrient transport properties, good mechanical integrity, and a suitable degradation rate. The scaffold brings the cells in close proximity and thereby enhances self-assembly and formation of tissue structures. Ultimately, *in vitro* differentiated constructs can potentially be used directly for transplantation.

For a tissue-engineering approach, the scaffold could comprise either synthetic or natural material or a composite of the two. Common scaffold materials include synthetic materials, such as hydroxyapatite, calcium carbonate, PLA, and poly(glycolic acid) (PGA), PLGA, and poly(propylene fumarate), or natural materials, such as collagen, Matrigel, and alginate. Natural materials are typically more favorable to cell adherence, while the properties of synthetic materials, such as degradation rate, mechanical properties, structure, and porosity, can be better controlled (Langer and Vacanti, 1993).

We have used a tissue-engineering approach to study the behavior of hESC-derived endothelial cells under *in vivo* conditions (Levenberg *et al.*, 2002). Human ESC–derived endothelial cells that were seeded onto highly porous PLGA biodegradable polymer scaffolds formed blood vessels that appeared to merge with the host vasculature when implanted into immunodeficient mice. In addition, those cells were able to vascularize constructs containing myoblasts to yield vascularized skeletal muscle (Levenberg *et al.*, 2005b). Osteoblast-like cells derived from hESC were also transplanted into an animal model by using a poly(D,L-lactide) scaffold. After 35 days of implantation, regions of mineralized tissue could be identified within the scaffold by Von Kossa staining and expression of human osteocalcin (Bielby *et al.*, 2004).

## Transplantation and Immune Response

One of the major obstacles for the successful transplantation of differentiated cells from hESCs is the potential immunogeneity of these cells. Because long-term immunosuppressive therapy would limit successful clinical applications, the creation of immunologic tolerance would enable stem cell–derived therapy. Methods currently under development and examination include (a) the establishment of hESC-line banks large enough to represent the majority of tissue types; (b) nuclear reprogramming the cells to carry patient-specific nuclear genome (therapeutic cloning); (c) creation of a "universal cell" suitable for all patients by manipulating the major histocompatibility complex (MHC) (Drukker *et al.*, 2002); (d) deletion of genes for immune-response proteins using homologous recombination (as mentioned earlier); and (e) the generation of hematopoietic chimerism, to create the required tolerance for tissues or cells derived from it (Bradley *et al.*, 2002). The latter was shown to be specifically useful with rat embryonic-like stem cells, where their injection into full MHC-mismatched rats resulted in permanent engraftment (Fandrich *et al.*, 2002).

Therapeutic cloning, or somatic cell nuclear transfer (SCNT), the process through which Dolly the sheep was cloned in 1997, might be an important tool to create hESCs from patient-specific genome, thus preventing immunore-

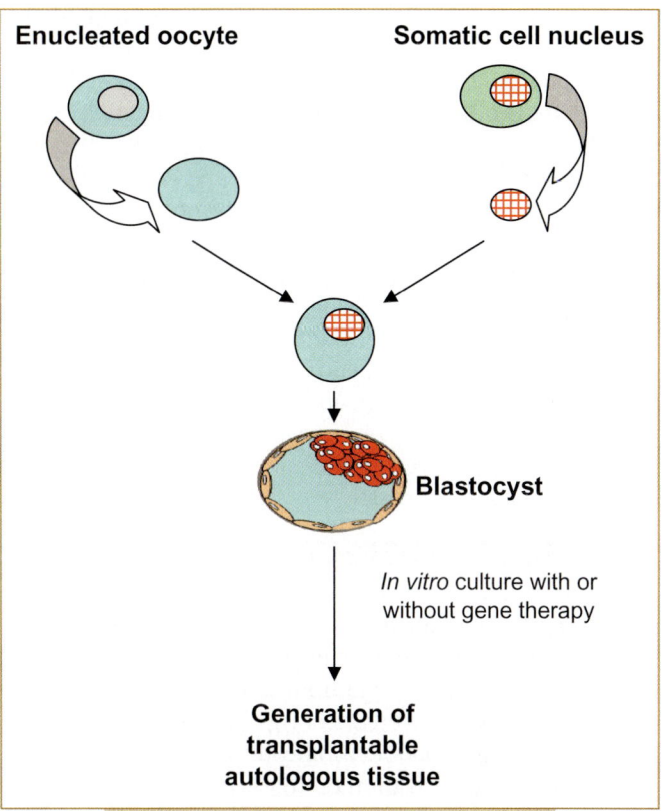

**FIG. 32.5.** Schematic diagram of the therapeutic cloning process to generate immunologically compatible tissues.

jection issues (Hwang *et al.*, 2004) (Fig. 32.5). This is very important for the application of hES cells in the area of tissue engineering where a transplantable population of cells can be generated with genes that are derived only from the original donor nucleus (i.e., from the patient). Studies to date have demonstrated that cells derived from SCNT can be expanded in culture and can organize into tissue structures after transplantation *in vivo* in combination with biodegradable scaffolds. However, before SCNT research can be translated into human therapies, the reliability and efficiency of the overall process need to be improved. Additional challenges include preventing alterations in gene expression and finding a sufficient source of oocytes.

Closed tissue-engineering systems, such as the use of immunoisolation systems, may overcome the immunological incompatibility of the tissue. Thus, immunoisolation of cells may prove to be particularly useful in conjunction with ESCs to overcome the immunological barrier associated with the ESC-based therapies. In such systems, cells may be immobilized within semipermeable polymeric matrices that provide a barrier for the immunological components of the host. Such semipermeable membranes are permeable to nutrients and oxygen while providing a barrier to immune cells, antibodies, and other components of the immune system (Lim and Sun, 1980; Uludag *et al.*, 2000). Within these systems, the implants can be either implanted into the patient or used as extracorpical devices. Closed tissue-engineering systems have been used particularly for the treatment of diabetes (Chicheportiche and Reach, 1988; Sefton *et al.*, 2000; Zekorn *et al.*, 1992), liver failure (Chandy *et al.*, 1999; Uludag and Sefton, 1993; L. Wang *et al.*, 2000), and Parkinson's disease (Aebischer *et al.*, 1991; Esposito *et al.*, 1996; Vallbacka *et al.*, 2001; Y. Wang *et al.*, 1997). For example, ESC-derived β-cells that can respond to insulin or dopamine-producing neurons can be used in clinics without fear of rejection. In addition, closed systems protect the host against potentially tumorigenic cells because it limits the cells within the polymeric barrier. Currently, engineering and biological limitations, such as material biocompatibility, molecular weight cutoff, and the immune system's reaction to shed antigens by the transplanted cells, are some of the challenges that prevent these systems from widespread clinical applications.

## VI. FUTURE PROSPECTS

Despite significant progress in the field of tissue engineering and ESC biology, there are number of challenges that provide a barrier to the use of ESCs for tissue engineering. These challenges range from understanding cues that direct stem cell fate, to engineering challenges on scale-up, to business questions of feasibility and pricing.

Although the derivation of hESCs from the ICM of pre-implantation blastocysts has become a standard procedure and has been performed in a variety of laboratories, live human embryos must be destroyed in the process, which is ethically problematic and has led to a variety of responses from around the world. However, recent reports show that it is possible to isolate embryonic stem cells without destroying the embryo (Chung *et al.*, 2006).

In August 2001, President Bush announced a policy in the United States to allow federal funding for research using an existing 60 lines of embryonic stem cells but prohibited funding for the creation of new lines or for using cells obtained from surplus embryos produced in fertility clinics. Aside from this decision to eliminate unregulated private creation of ESC lines, the existing federally approved lines are neither sufficiently available nor adequate for human therapies. Specifically, all of the approved stem cell lines were prepared using mouse cells and thus pose a risk of contaminating human subjects. However, advancements since 2001 have established methods to culture hESCs without mouse feeder layers. Interestingly, it has recently been proposed that a common ground for pursuing hESC research may exist through assessing the death of a human embryo in the ethical context surrounding organ donation. Specifically, Landry and Zucker (2004) argue that a significant fraction of embryos generated for *in vitro* fertilization undergo irreversible arrest of cell division and thus can be considered as organismically dead yet can be used as normal donors of blastomers. Thus donation of these embryos could ethically be considered analogous to the donation of essential organs from cadavers. Although criteria for determining the irreversible arrest of cell division have yet to be defined, it will certainly be interesting to see if these theories can be experimentally established and how these arguments will fare with those who currently oppose hESC research.

Synthetic scaffolds that support tissue growth by serving as the extracellular matrix for the cells do not represent the natural matrix molecules associated with each cell type and tissue. ESCs and their progeny during development reside in a dynamic environment; thus a scaffold should be designed to mimic the signaling and structural elements in the developing embryo. The use of "smart" scaffolds that release particular factors and/or control the temporal expression of various molecules released from the polymer can help induce differentiation of ESCs within the scaffolds (Richardson *et al.*, 2001). For example, dual delivery of vascular endothelial growth factor (VEGF)-165 and platelet-derived growth factor (PDGF), each with distinct kinetics, from a single, structural polymer scaffold resulted in a mature vascular network (Richardson *et al.*, 2001). An alternative approach to modify the surface exposed to the cells is to immobilize desired ligands onto the scaffold. For example, RGD peptides, the adherent domain of fibronectin, can be incorporated into polymers to provide anchorage for adherent cells.

Another difficulty with the current materials is their lack of control over the spatial organization within the scaffold. Spatial patterning is necessary to create tissues that resemble the natural structure of biological tissues. In the direct cell-patterning system, cells can be seeded into the scaffold at particular regions within the cells. For example, the direct attachment of two different cell types on the different sides of the scaffold has been used to generate cells of the bladder. Cell-patterning techniques such as those that have been developed for soft lithography for controlled coculture of hepatocytes and fibroblasts could be scaled up to tissue-engineering scaffolds to allow for more controlled and complex direct patterning.

## VII. CONCLUSIONS

A number of challenges remain in making ESC-based therapy clinically viable. These include directing the differentiation of ESCs (i.e., using controlled microenvironments or genetic engineering) to ensure their safety (i.e., by eliminating tumorogenicity), to ensure that the differentiated cells integrate functionally into the body, to ensure their immune compatibility with the patient, and to improve the cost and feasibility of cell-based therapies. Each of these challenges is an area of active research that must be validated and optimized. In particular, since ESCs can give rise to many different cell types, solving these challenges for the various possible tissue types will be a major undertaking. Further research is required to control and direct the differentiation of ESCs. If done in parallel with developing methods to generate tissues of various organs, this may lead to realizing the ultimate goal of tissue engineering. We are getting close to the day when ESCs can be manipulated in culture to produce fully differentiated cells that can be used to create and repair specific organs. Clearly, significant challenges remain, and the ability to overcome these diffi-culties is not confined to any single scientific discipline but, rather, involves an interdisciplinary approach. Innovative approaches to solve these challenges could lead to improved quality of life for a variety of patients who could benefit from tissue-engineering approaches.

## VIII. ACKNOWLEDGMENTS

The authors would like to acknowledge funding from NIH (HL60435, HL076485) Juvenile Diabetes Research Foundation (G V-N, Fellowship to SG), NSF, the Draper Laboratories, the Centre for Integration of Medicine and Innovative Technology, the Coulter Foundation, the Institute for Soldier Nanotechnology (DAAD-19-02-D-002), and Fundação para a Ciência e Tecnologia (grant SFRH/BPD/14502/2003 to L.F).

## IX. REFERENCES

Aebischer, P., Wahlberg, L., Tresco, P. A., and Winn, S. R. (1991). Macro-encapsulation of dopamine-secreting cells by coextrusion with an organic polymer solution. *Biomaterials* **12**(1), 50–56.

Amit, M., Margulets, V., Segev, H., Shariki, K., Laevsky, I., Coleman, R., and Itskovitz-Eldor, J. (2003). Human feeder layers for human embryonic stem cells. *Biol. Reprod.* **68**(6), 2150–2156.

Amit, M., Shariki, C., Margulets, V., and Itskovitz-Eldor, J. (2004). Feeder layer- and serum-free culture of human embryonic stem cells. *Biol. Reprod.* **70**(3), 837–845.

Anderson, D. G., Levenberg, S., and Langer, R. (2004). Nanoliter-scale synthesis of arrayed biomaterials and application to human embryonic stem cells. *Nat. Biotechnol.* **22**(7), 863–866.

Anderson, D. G., Putnam, D., Lavik, E. B., Mahmood, T. A., and Langer, R. (2005). Biomaterial microarrays: rapid, micro-scale screening of polymer–cell interaction. *Biomaterials* **26**(23), 4892–4897.

Assady, S., Maor, G., Amit, M., Itskovitz-Eldor, J., Skorecki, K. L., and Tzukerman, M. (2001). Insulin production by human embryonic stem cells. *Diabetes* **50**(8), 1691–1697.

Ben-Hur, T., Idelson, M., Khaner, H., Pera, M., Reinhartz, E., Itzik, A., and Reubinoff, B. E. (2004). Transplantation of human embryonic stem cell-derived neural progenitors improves behavioral deficit in parkinsonian rats. *Stem Cells* **22**(7), 1246–1255.

Bielby, R. C., Boccaccini, A. R., Polak, J. M., and Buttery, L. D. (2004). *In vitro* differentiation and *in vivo* mineralization of osteogenic cells derived from human embryonic stem cells. *Tissue Eng.* **10**(9–10), 1518–1525.

Bradley, J. A., Bolton, E. M., and Pedersen, R. A. (2002). Stem cell medicine encounters the immune system. *Nat. Rev. Immunol.* **2**(11), 859–871.

Carpenter, M. K., Inokuma, M. S., Denham, J., Mujtaba, T., Chiu, C. P., and Rao, M. S. (2001). Enrichment of neurons and neural precursors from human embryonic stem cells. *Exp. Neurol.* **172**(2), 383–397.

Chadwick, K., Wang, L., Li, L., Menendez, P., Murdoch, B., Rouleau, A., and Bhatia, M. (2003). Cytokines and BMP-4 promote hematopoietic differentiation of human embryonic stem cells. *Blood* **102**(3), 906–915.

Chandy, T., Mooradian, D. L., and Rao, G. H. (1999). Evaluation of modified alginate-chitosan-polyethylene glycol microcapsules for cell encapsulation. *Artif. Organs* **23**(10), 894–903.

Chen, C. S., Mrksich, M., Huang, S., Whitesides, G. M., and Ingber, D. E. (1997). Geometric control of cell life and death. *Science* **276**(5317), 1425–1428.

Chen, S. B., Zhang, Q. S., Wu, X., Schultz, P. G., and Ding, S. (2004). Dedifferentiation of lineage-committed cells by a small molecule. *J. Am. Chem. Soc.* **126**(2), 410–411.

Chen, S. S., Revoltella, R. P., Papini, S., Michelini, M., Fitzgerald, W., Zimmerberg, J., and Margolis, L. (2003). Multilineage differentiation of rhesus monkey embryonic stem cells in three-dimensional culture systems. *Stem Cells* **21**(3), 281–295.

Cheng, L., Hammond, H., Ye, Z., Zhan, X., and Dravid, G. (2003). Human adult marrow cells support prolonged expansion of human embryonic stem cells in culture. *Stem Cells* **21**(2), 131–142.

Chicheportiche, D., and Reach, G. (1988). *In vitro* kinetics of insulin release by microencapsulated rat islets: effect of the size of the microcapsules. *Diabetologia* **31**(1), 54–57.

Chin, V. I., Taupin, P., Sanga, S., Scheel, J., Gage, F. H., and Bhatia, S. N. (2004). Microfabricated platform for studying stem cell fates. *Biotechnol. Bioeng.* **88**(3), 399–415.

Chung, Y., Klimanskaya, I., Becker, S., Marh, J., Lu, S. J., Johnson, J., Meisner, L., and Lanza, R. (2006). Embryonic and extraembryonic stem cell lines derived from single mouse blastomeres. *Nature* **439**(7073), 216–219.

Drukker, M., Katz, G., Urbach, A., Schuldiner, M., Markel, G., Itskovitz-Eldor, J., Reubinoff, B., Mandelboim, O., and Benvenisty, N. (2002). Characterization of the expression of MHC proteins in human embryonic stem cells. *Proc. Natl. Acad. Sci. U.S.A.* **99**(15), 9864–9869.

Esposito, E., Cortesi, R., and Nastruzzi, C. (1996). Gelatin microspheres: influence of preparation parameters and thermal treatment on chemico-physical and biopharmaceutical properties. *Biomaterials* **17**(20), 2009–2020.

Evans, M. J., and Kaufman, M. H. (1981). Establishment in culture of pluripotential cells from mouse embryos. *Nature* **292**(5819), 154–156.

Fandrich, F., Lin, X., Chai, G. X., Schulze, M., Ganten, D., Bader, M., Holle, J., Huang, D. S., Parwaresch, R., Zavazava, N., and Binas, B. (2002). Preimplantation-stage stem cells induce long-term allogeneic graft acceptance without supplementary host conditioning. *Nat. Med.* **8**(2), 171–178.

Fareed, M. U., and Moolten, F. L. (2002). Suicide gene transduction sensitizes murine embryonic and human mesenchymal stem cells to ablation on demand — a fail-safe protection against cellular misbehavior. *Gene Ther.* **9**(14), 955–962.

Ferreira, L. S., Gerecht, S., Fuller, J., Shieh, H. F., Vunjak-Novakovic, G., Langer, R. (2007). Bioactive hydrogel scaffolds for controllable vascular differentiation of human embryonic stem cells. *Biomaterials* **28**, 2706–2717.

Flaim, C. J., Chien, S., and Bhatia, S. N. (2005). An extracellular matrix microarray for probing cellular differentiation. *Nat. Methods* **2**(2), 119–125.

Gerecht-Nir, S., Ziskind, A., Cohen, S., and Itskovitz-Eldor, J. (2003). Human embryonic stem cells as an *in vitro* model for human vascular development and the induction of vascular differentiation. *Lab. Invest.* **83**, 1811–1820.

Gerecht-Nir, S., Cohen, S., and Itskovitz-Eldor, J. (2004a). Bioreactor cultivation enhances the efficiency of human embryoid body (hEB) formation and differentiation. *Biotechnol. Bioeng.* **86**(5), 493–502.

Gerecht-Nir, S., Cohen, S., Ziskind, A., and Itskovitz-Eldor, J. (2004b). Three-dimensional porous alginate scaffolds provide a conducive environment for generation of well-vascularized embryoid bodies from human embryonic stem cells. *Biotechnol. Bioeng.* **88**(3), 313–320.

Ginis, I., Luo, Y., Miura, T., Thies, S., Brandenberger, R., Gerecht-Nir, S., Amit, M., Hoke, A., Carpenter, M. K., Itskovitz-Eldor, J., and Rao, M. S. (2004). Differences between human and mouse embryonic stem cells. *Dev. Biol.* **269**(2), 360–380.

Gropp, M., Itsykson, P., Singer, O., Ben-Hur, T., Reinhartz, E., Galun, E., and Reubinoff, B. E. (2003). Stable genetic modification of human embryonic stem cells by lentiviral vectors. *Mol. Ther.* **7**(2), 281–287.

Guan, K., Furst, D. O., and Wobus, A. M. (1999). Modulation of sarcomere organization during embryonic stem cell–derived cardiomyocyte differentiation. *Eur. J. Cell Biol.* **78**(11), 813–823.

Hamazaki, T., Iiboshi, Y., Oka, M., Papst, P. J., Meacham, A. M., Zon, L. I., and Terada, N. (2001). Hepatic maturation in differentiating embryonic stem cells in vitro. *FEBS Lett.* **497**(1), 15–19.

Hovatta, O., Mikkola, M., Gertow, K., Stromberg, A. M., Inzunza, J., Hreinsson, J., Rozell, B., Blennow, E., Andang, M., and Ahrlund-Richter, L. (2003). A culture system using human foreskin fibroblasts as feeder cells allows production of human embryonic stem cells. *Hum. Reprod.* **18**(7), 1404–1409.

Hwang, W. S., Ryu, Y. J., Park, J. H., Park, E. S., Lee, E. G., Koo, J. M., Jeon, H. Y., Lee, B. C., Kang, S. K., Kim, S. J., Ahn, C., Hwang, J. H., Park, K. Y., Cibelli, J. B., and Moon, S. Y. (2004). Evidence of a pluripotent human embryonic stem cell line derived from a cloned blastocyst. *Science* **303**(5664), 1669–1674.

Inzunza, J., Gertow, K., Stromberg, M. A., Matilainen, E., Blennow, E., Skottman, H., Wolbank, S., Ahrlund-Richter, L., and Hovatta, O. (2005). Derivation of human embryonic stem cell lines in serum replacement medium using postnatal human fibroblasts as feeder cells. *Stem Cells* **23**(4), 544–549.

Karp, J. M., Ferreira, L. S., Khademhosseini, A., Kwon, A. H., Yeh, J., and Langer, R. (2006). Cultivation of human embryonic stem cells without the embryoid body step enhances osteogenesis *in vitro*. *Stem Cells* **24**(4), 835–843.

Kaufman, D. S., Hanson, E. T., Lewis, R. L., Auerbach, R., and Thomson, J. A. (2001). Hematopoietic colony-forming cells derived from human embryonic stem cells. *Proc. Natl. Acad. Sci. U.S.A.* **98**(19), 10716–10721.

Kaufman, D. S., Lewis, R. L., Hanson, E. T., Auerbach, R., Plendl, J., and Thomson, J. A. (2004). Functional endothelial cells derived from rhesus monkey embryonic stem cells. *Blood* **103**(4), 1325–1332.

Kehat, I., Amit, M., Gepstein, A., Huber, I., Itskovitz-Eldor, J., and Gepstein, L. (2003). Development of cardiomyocytes from human ES cells. *Methods Enzymol.* **365**, 461–473.

Keirstead, H. S., Nistor, G., Bernal, G., Totoiu, M., Cloutier, F., Sharp, K., and Steward, O. (2005). Human embryonic stem cell–derived oligodendrocyte progenitor cell transplants remyelinate and restore locomotion after spinal cord injury. *J. Neurosci.* **25**(19), 4694–4705.

Kim, J. H., Auerbach, J. M., Rodriguez-Gomez, J. A., Velasco, I., Gavin, D., Lumelsky, N., Lee, S. H., Nguyen, J., Sanchez-Pernaute, R., Bankiewicz, K., and McKay, R. (2002). Dopamine neurons derived from embryonic stem cells function in an animal model of Parkinson's disease. *Nature* **418**(6893), 50–56.

Klimanskaya, I., Chung, Y., Meisner, L., Johnson, J., West, M. D., and Lanza, R. (2005). Human embryonic stem cells derived without feeder cells. *Lancet* **365**(9471), 1636–1641.

Klinz, F., Bloch, W., Addicks, K., and Hescheler, J. (1999). Inhibition of phosphatidylinositol-3-kinase blocks development of functional embryonic cardiomyocytes. *Exp. Cell Res.* **247**(1), 79–83.

Landry, D. W., and Zucker, H. A. (2004). Embryonic death and the creation of human embryonic stem cells. *J. Clin. Invest.* **114**(9), 1184–1186.

Langer, R., and Vacanti, J. P. (1993). Tissue engineering. *Science* **260**(5110), 920–926.

Lanza, R. P., Cibelli, J. B., West, M. D., Dorff, E., Tauer, C., and Green, R. M. (2001). The ethical reasons for stem cell research. *Science* **292**(5520), 1299.

Lavon, N., Yanuka, O., and Benvenisty, N. (2004). Differentiation and isolation of hepatic-like cells from human embryonic stem cells. *Differentiation* **72**(5), 230–238.

Levenberg, S., Burdick, J. A., Kraehenbuehl, T., and Langer, R. (2005a). Neurotrophin-induced differentiation of human embryonic stem cells on three-dimensional polymeric scaffolds. *Tissue Eng.* **11**(3–4), 506–512.

Levenberg, S., Golub, J. S., Amit, M., Itskovitz-Eldor, J., and Langer, R. (2002). Endothelial cells derived from human embryonic stem cells. *Proc. Natl. Acad. Sci. U.S.A.* **99**(7), 4391–4396.

Levenberg, S., Huang, N. F., Lavik, E., Rogers, A. B., Itskovitz-Eldor, J., and Langer, R. (2003). Differentiation of human embryonic stem cells on three-dimensional polymer scaffolds. *Proc. Natl. Acad. Sci. U.S.A.* **100**(22), 12741–12746.

Levenberg, S., Rouwkema, J., Macdonald, M., Garfein, E. S., Kohane, D. S., Darland, D. C., Marini, R., van Blitterswijk, C. A., Mulligan, R. C., D'Amore, P. A., and Langer, R. (2005b). Engineering vascularized skeletal muscle tissue. *Nat. Biotechnol.* **23**(7), 879–884.

Lim, F., and Sun, A. M. (1980). Microencapsulated islets as a bioartificial endocrine pancreas. *Science* **210**, 980–910.

Liu, H., and Roy, K. (2005). Biomimetic three-dimensional cultures significantly increase hematopoietic differentiation efficacy of embryonic stem cells. *Tissue Eng.* **11**(1–2), 319–330.

Ludwig, T. E., Levenstein, M. E., Jones, J. M., Berggren, W. T., Mitchen, E. R., Frane, J. L., Crandall, L. J., Daigh, C. A., Conard, K. R., Piekarczyk, M. S., Llanas, R. A., and Thomson, J. A. (2006). Derivation of human embryonic stem cells in defined conditions. *Nat. Biotechnol.* **24**(2), 185–187.

Ma, Y., Ramezani, A., Lewis, R., Hawley, R. G., and Thomson, J. A. (2003). High-level sustained transgene expression in human embryonic stem cells using lentiviral vectors. *Stem Cells* **21**(1), 111–117.

Martin, G. R. (1981). Isolation of a pluripotent cell line from early mouse embryos cultured in medium conditioned by teratocarcinoma stem cells. *Proc. Natl. Acad. Sci. U.S.A.* **78**(12), 7634–7638.

McBeath, R., Pirone, D. M., Nelson, C. M., Bhadriraju, K., and Chen, C. S. (2004). Cell shape, cytoskeletal tension, and RhoA regulate stem cell lineage commitment. *Dev. Cell* **6**(4), 483–495.

McDonald, J. W., and Howard, M. J. (2002). Repairing the damaged spinal cord: a summary of our early success with embryonic stem cell transplantation and remyelination. *Prog. Brain Res.* **137**, 299–309.

Mummery, C., Ward-van Oostwaard, D., Doevendans, P., Spijker, R., van den Brink, S., Hassink, R., van der Heyden, M., Opthof, T., Pera, M., de la Riviere, A. B., Passier, R., and Tertoolen, L. (2003). Differentiation of human embryonic stem cells to cardiomyocytes: role of coculture with visceral endoderm-like cells. *Circulation* **107**(21), 2733–2740.

Muotri, A. R., Nakashima, K., Toni, N., Sandler, V. M., and Gage, F. H. (2005). Development of functional human embryonic stem cell–derived neurons in mouse brain. *Proc. Natl. Acad. Sci. U.S.A.* **102**(51), 18644–18648.

Niklason, L. E., Gao, J., Abbott, W. M., Hirschi, K. K., Houser, S., Marini, R., and Langer, R. (1999). Functional arteries grown in vitro. *Science* **284**(5413), 489–493.

Nistor, G. I., Totoiu, M. O., Haque, N., Carpenter, M. K., and Keirstead, H. S. (2005). Human embryonic stem cells differentiate into oligodendrocytes in high purity and myelinate after spinal cord transplantation. *Glia* **49**(3), 385–396.

Oh, S. K., Fong, W. J., Teo, Y., Tan, H. L., Padmanabhan, J., Chin, A. C., and Choo, A. B. (2005). High-density cultures of embryonic stem cells. *Biotechnol. Bioeng.* **91**(5), 523–533.

Park, J. H., Kim, S. J., Lee, J. B., Song, J. M., Kim, C. G., Roh, S., 2nd, and Yoon, H. S. (2004). Establishment of a human embryonic germ cell line and comparison with mouse and human embryonic stem cells. *Mol. Cells* **17**(2), 309–315.

Richards, M., Fong, C. Y., Chan, W. K., Wong, P. C., and Bongso, A. (2002). Human feeders support prolonged undifferentiated growth of human inner cell masses and embryonic stem cells. *Nat. Biotechnol.* **20**(9), 933–936.

Richards, M., Tan, S., Fong, C. Y., Biswas, A., Chan, W. K., and Bongso, A. (2003). Comparative evaluation of various human feeders for prolonged undifferentiated growth of human embryonic stem cells. *Stem Cells* **21**(5), 546–556.

Richardson, T. P., Peters, M. C., Ennett, A. B., and Mooney, D. J. (2001). Polymeric system for dual growth factor delivery. *Nat. Biotechnol.* **19**(11), 1029–1034.

Schuldiner, M., Itskovitz-Eldor, J., and Benvenisty, N. (2003). Selective ablation of human embryonic stem cells expressing a "suicide" gene. *Stem Cells* **21**(3), 257–265.

Schulz, T. C., Palmarini, G. M., Noggle, S. A., Weiler, D. A., Mitalipova, M. M., and Condie, B. G. (2003). Directed neuronal differentiation of human embryonic stem cells. *BMC Neurosci.* **4**, 27.

Sefton, M. V., May, M. H., Lahooti, S., and Babensee, J. E. (2000). Making microencapsulation work: conformal coating, immobilization gels and in vivo performance. *J. Controlled Release* **65**(1–2), 173–186.

Segev, H., Fishman, B., Ziskind, A., Shulman, M., and Itskovitz-Eldor, J. (2004). Differentiation of human embryonic stem cells into insulin-producing clusters. *Stem Cells* **22**(3), 265–274.

Shamblott, M. J., Axelman, J., Wang, S., Bugg, E. M., Littlefield, J. W., Donovan, P. J., Blumenthal, P. D., Huggins, G. R., and Gearhart, J. D. (1998). Derivation of pluripotent stem cells from cultured human primordial germ cells. *Proc. Natl. Acad. Sci. U.S.A.* **95**(23), 13726–13731.

Smith-Arica, J. R., Thomson, A. J., Ansell, R., Chiorini, J., Davidson, B., and McWhir, J. (2003). Infection efficiency of human and mouse embryonic stem cells using adenoviral and adeno-associated viral vectors. *Cloning Stem Cells* **5**(1), 51–62.

Soria, B., Roche, E., Berna, G., Leon-Quinto, T., Reig, J. A., and Martin, F. (2000). Insulin-secreting cells derived from embryonic stem cells normalize glycemia in streptozotocin-induced diabetic mice. *Diabetes* **49**(2), 157–162.

Thomson, J. A., Itskovitz-Eldor, J., Shapiro, S. S., Waknitz, M. A., Swiergiel, J. J., Marshall, V. S., and Jones, J. M. (1998). Embryonic stem cell lines derived from human blastocysts. *Science* **282**(5391), 1145–1147.

Till, J. E., and McCulloch, E. A. (1980). Hemopoietic stem cell differentiation. *Biochim. Biophys. Acta* **605**(4), 431–459.

Uludag, H., and Sefton, M. V. (1993). Microencapsulated human hepatoma (HepG2) cells: in vitro growth and protein release. *J. Biomed. Mater. Res.* **27**(10), 1213–1224.

Uludag, H., De Vos, P., and Tresco, P. A. (2000). Technology of mammalian cell encapsulation. *Adv. Drug Deliv. Rev.* **42**(1–2), 29–64.

Vallbacka, J. J., Nobrega, J. N., and Sefton, M. V. (2001). Tissue engineering as a platform for controlled release of therapeutic agents: implantation of microencapsulated dopamine producing cells in the brains of rats. *J. Controlled Release* **72**(1–3), 93–100.

Verfaillie, C. M., Pera, M. F., and Lansdorp, P. M. (2002). Stem cells: hype and reality. *Hematology (Am. Soc. Hematol. Educ. Program)*, **1**, 369–391.

Vodyanik, M. A., Bork, J. A., Thomson, J. A., and Slukvin, II. (2005). Human embryonic stem cell–derived CD34+ cells: efficient production in the coculture with OP9 stromal cells and analysis of lymphohematopoietic potential. *Blood* **105**(2), 617–626.

Wang, L., Sun, J., Li, L., Harbour, C., Mears, D., Koutalistras, N., and Sheil, A. G. (2000). Factors affecting hepatocyte viability and CYPIA1 activity during encapsulation. *Artif. Cells Blood Substit. Immobil. Biotechnol.* **28**(3), 215–227.

Wang, L., Li, L., Shojaei, F., Levac, K., Cerdan, C., Menendez, P., Martin, T., Rouleau, A., and Bhatia, M. (2004). Endothelial and hematopoietic cell fate of human embryonic stem cells originates from primitive endothelium with hemangioblastic properties. *Immunity* **21**(1), 31–41.

Wang, L., Li, L., Menendez, P., Cerdan, C., and Bhatia, M. (2005). Human embryonic stem cells maintained in the absence of mouse embryonic fibroblasts or conditioned media are capable of hematopoietic development. *Blood* **105**(12), 4598–4603.

Wang, Y., Wang, S. D., Lin, S. Z., Chiou, A. L., Chen, L. K., Chen, J. F., and Zhou, F. C. (1997). Transplantation of microencapsulated PC12 cells provides long-term improvement of dopaminergic functions. *Chin. J. Physiol.* **40**(3), 121–129.

Wartenberg, M., Donmez, F., Ling, F. C., Acker, H., Hescheler, J., and Sauer, H. (2001). Tumor-induced angiogenesis studied in confrontation cultures of multicellular tumor spheroids and embryoid bodies grown from pluripotent embryonic stem cells. *FASEB J.* **15**(6), 995–1005.

Wu, X., Ding, S., Ding, Q., Gray, N. S., and Schultz, P. G. (2002). A small molecule with osteogenesis-inducing activity in multipotent mesenchymal progenitor cells. *J. Am. Chem. Soc.* **124**(49), 14520–14521.

Wu, X., Ding, S., Ding, G., Gray, N. S., and Schultz, P. G. (2004). Small molecules that induce cardiomyogenesis in embryonic stem cells. *J. Am. Chem. Soc.* **126**(6), 1590–1591.

Xu, C., Inokuma, M. S., Denham, J., Golds, K., Kundu, P., Gold, J. D., and Carpenter, M. K. (2001). Feeder-free growth of undifferentiated human embryonic stem cells. *Nat. Biotechnol.* **19**(10), 971–974.

Xu, C., Police, S., Rao, N., and Carpenter, M. K. (2002). Characterization and enrichment of cardiomyocytes derived from human embryonic stem cells. *Circ. Res.* **91**(6), 501–508.

Xu, C., Rosler, E., Jiang, J., Lebkowski, J. S., Gold, J. D., O'Sullivan, C., Delavan-Boorsma, K., Mok, M., Bronstein, A., and Carpenter, M. K. (2005). Basic fibroblast growth factor supports undifferentiated human embryonic stem cell growth without conditioned medium. *Stem Cells* **23**(3), 315–323.

Xu, R. H., Peck, R. M., Li, D. S., Feng, X., Ludwig, T., and Thomson, J. A. (2005). Basic FGF and suppression of BMP signaling sustain undifferentiated proliferation of human ES cells. *Nat. Methods* **2**(3), 185–190.

Yamashita, J., Itoh, H., Hirashima, M., Ogawa, M., Nishikawa, S., Yurugi, T., Naito, M., and Nakao, K. (2000). Flk1-positive cells derived from embryonic stem cells serve as vascular progenitors. *Nature* **408**(6808), 92–96.

Zandstra, P. W., Bauwens, C., Yin, T., Liu, Q., Schiller, H., Zweigerdt, R., Pasumarthi, K. B., and Field, L. J. (2003). Scalable production of embryonic stem cell-derived cardiomyocytes. *Tissue Eng.* **9**(4), 767–778.

Zekorn, T., Siebers, U., Horcher, A., Schnettler, R., Zimmermann, U., Bretzel, R. G., and Federlin, K. (1992). Alginate coating of islets of Langerhans: *in vitro* studies on a new method for microencapsulation for immuno-isolated transplantation. *Acta Diabetol.* **29**(1), 41–45.

Zeng, X., Cai, J., Chen, J., Luo, Y., You, Z. B., Fotter, E., Wang, Y., Harvey, B., Miura, T., Backman, C., Chen, G. J., Rao, M. S., and Freed, W. J. (2004). Dopaminergic differentiation of human embryonic stem cells. *Stem Cells* **22**(6), 925–940.

Zhan, X., Dravid, G., Ye, Z., Hammond, H., Shamblott, M., Gearhart, J., and Cheng, L. (2004). Functional antigen-presenting leucocytes derived from human embryonic stem cells *in vitro*. *Lancet* **364**(9429), 163–171.

Zhang, S. C., Wernig, M., Duncan, I. D., Brustle, O., and Thomson, J. A. (2001). *In vitro* differentiation of transplantable neural precursors from human embryonic stem cells. *Nat. Biotechnol.* **19**(12), 1129–1133.

Zwaka, T. P., and Thomson, J. A. (2003). Homologous recombination in human embryonic stem cells. *Nat. Biotechnol.* **21**(3), 319–321.

# Chapter Thirty-Three

# Postnatal Stem Cells

*Pamela Gehron Robey and Paolo Bianco*

## I. INTRODUCTION

The years since the mid-1990s have witnessed the emergence of stem cell–based tissue engineering as a means to restore normal structure and function to tissues lost to trauma or disease. Successful application will rely on (1) further characterization of the biological properties of postnatal stem cells (what they can and cannot do); (2) optimization of their isolation and *ex vivo* expansion; (3) formulation of appropriate scaffolds and carriers, which may also include growth factors; (4) development of more advanced bioreactors to create tissues with appropriate biomechanical properties; and (5) determination of mechanisms by which to mobilize and activate endogenous stem cells. This chapter highlights these aspects of stem cell–based tissue engineering and presents examples of current approaches and applications.

At first inception, tissue engineering was based on the use of natural or synthetic scaffolds seeded with organ-specific cells *ex vivo*. This approach was somewhat distinct from guided tissue regeneration, which utilized scaffolds and/or bioactive factors to encourage local cells to repair a defect *in situ*. These two approaches are now merged in the current field of tissue engineering, which encompasses multiple and diverse disciplines to use cells, materials, and bioactive factors in various combinations to restore and even improve tissue structure and function. Stem cell–based tissue engineering represents a major turn in the conceptual approach to reconstruction of tissues. By expanding the repertoire of available cells, of potential targets, and of technological means of generating functional tissues *ex vivo* and above all by making it possible (or promising) to engineer tissues that otherwise would not be amenable to recreation, advances in stem cell biology are having a profound impact on tissue engineering in general. In many cases, tissue engineering seems to have more use for stem cells than does nature itself. This is especially apparent in the case of stem cells derived from tissues with low turnover or no apparent turnover at all. If neural stem cells were able to repair the loss of dopaminergic neurons in the intact brain *in vivo*, there would be no Parkinson's disease for which to envision stem cell–based therapies. Dental pulp stem cells do not regenerate primary dentin *in vivo*, but vast numbers of cells can be made from the pulp of a single extracted tooth to generate copious amounts of new dentin. Current definitions of stem cells in fact include a technological dimension in many cases (a cell that can be expanded *ex vivo* and bent to the generation of differentiating cells). Importantly, the two neighboring fields of cell therapy (reconstruction of functional tissues *in vivo* using cells) and tissue engineering proper (reconstruction of functional tissues using cells and something else) merge significantly once a stem cell angle is adopted for either, not only with respect to the ultimate goals, but also to several biotechnological aspects.

## II. RESERVOIRS OF POSTNATAL STEM CELLS

Because of ethical constraints and for ease of harvest and control, current approaches to stem cell–based tissue engineering rely largely on the use of postnatal stem cells. Once restricted to a handful of constantly (and rapidly) self-renewing tissues, such as blood, skin, and the gastrointestinal tract, the repertoire of postnatal stem cells has expanded to include perhaps every single tissue in the body, regardless of the rate of tissue turnover or ability to regenerate (reviewed in Preston *et al.*, 2003; Wagers and Weissman, 2004). Not all of these tissue-specific stem cells, however, are equally accessible for safe harvest or available in sufficient quantity (or amenable for *ex vivo* expansion) to generate the number of cells needed for tissue regeneration. However, lessons on the dynamics of tissue homeostasis (growth and turnover) that can be learned from these cells have an obvious impact on the design of future tissue-engineering strategies nonetheless. In addition, mechanisms whereby postnatal cells maintain differentiated functions in tissues and organs are relevant to future embryonic stem cell–based approaches and can only be learned from postnatal cells (Wagers and Weissman, 2004).

## Plasticity and Transdifferentiation — Implications for Tissue Engineering

The recent clamor about the potential *plasticity* (the ability of a cell to convert from one type to another) and *transdifferentiation* (conversion of a cell from one lineage to another) of certain classes of postnatal stem cells unquestionably adds further questions (Moore and Quesenberry, 2003). Rigorous proof of true plasticity and transdifferentiation have been lacking in many reports and can only be demonstrated by the ability of a single (clonogenic) cell to form cells of multiple different phenotypes and which can be shown to function as those different cell types. Some would also add that conversion must be more than an isolated phenomenon; it must occur at a high frequency and be persistent or stable (reviewed in Lakshmipathy and Verfaillie, 2005; Pauwelyn and Verfaillie, 2006). Nonetheless, postnatal stem cell plasticity provides a new twist, once the perspective of usage for tissue reconstruction is directly addressed. It would be impractical, to say the least, to use neural stem cells for bone marrow transplantation; and a tissue engineer's enthusiasm for this unexpected finding may remain lukewarm. In contrast, deciding whether liver regeneration is successfully accomplished using previously unknown "hepatogenic" stem cells or transdifferentiation of hematopoietic stem cells (Lagasse *et al.*, 2000) or by reprogramming of a donor lymphocyte nucleus following cell fusion (Wang *et al.*, 2003) may not disrupt a tissue engineer's sleep, as long as the goal is met. Unquestionably, some examples of "unorthodox" differentiation of postnatal stem cells have been given (reviewed in Camargo *et al.*, 2004; Goodell, 2003; Wagers and Weissman, 2004). However, further investigation is needed to confirm not only the concept of postnatal stem cell plasticity, but specifically the technological aspects of its effective translation into clinical application, once proof of principle has been given. Even if true plasticity is a rare event, analyzing the way it occurs may provide clues as to how better to manipulate cell populations to increase its frequency.

## Bone Marrow — A Source of Multiple Stem Cells

It has long been known that bone marrow is the home of at least two different types of stem cells, the hematopoietic stem cell (HSC, reviewed in Bryder *et al.*, 2006) and the skeletal stem cell (SSC, also known as the bone marrow stromal stem cell or "mesenchymal" stem cell; reviewed in Bianco and Robey, 2004), each able to reconstitute the hematopoietic and skeletal systems, respectively. Both systems are thought to be able to contribute differentiated cell types outside of their physiological progeny. Highly purified HSCs were reported to give rise to cardiomyocytes (Orlic *et al.*, 2001), as well as hepatocytes (Lagasse *et al.*, 2000) and a host of epithelial tissues (Krause *et al.*, 2001). Bone marrow stromal cells have been reported to generate functional cardiomyocytes *in vitro* (murine) (reviewed in Minguell and Erices, 2006) and to be capable of neural differentiation (reviewed in Phinney and Isakova, 2005). A rare subset of murine bone marrow cells (multipotential adult progenitor cells, MAPCs) have been reported to be almost as multipotent as ES cells (Jiang *et al.*, 2002). AC133 (CD133) positive endothelial progenitors are found in the marrow (Gehling *et al.*, 2000), and endothelial cells themselves may generate cardiomyocytes *in vitro* (Badorff *et al.*, 2003). Circulating, marrow-derived cells contribute to regeneration of skeletal muscle in response to injury (G. Ferrari *et al.*, 1998) and, in mouse models, of muscular dystrophy (Bittner *et al.*, 1999; Gussoni *et al.*, 1999). Donor-derived cells have also been detected in neuronal tissue, in newly formed vasculature, in the kidney, and even in the oral cavity following bone marrow or mobilized peripheral blood transplantation (reviewed in Poulsom *et al.*, 2002). Ideally, the identification of a single accessible site that would contain reservoirs of cells with pluri- and multipotentiality that are easily harvested in large quantities would mark a major advantage in tissue engineering. If substantiated, this wealth of observations would make bone marrow the central organ of tissue engineering and stem cell therapy (Fig. 33.1). However, much more investigation is needed in order to substantiate these preliminary findings to determine the true efficacy of bone marrow for treatment of such a broad spectrum of abnormalities (reviewed in Pauwelyn and Verfaillie, 2006).

## III. CURRENT APPROACHES TO TISSUE ENGINEERING

Tissue-engineering approaches to the regeneration of functional tissue using postnatal stem cells can be envisioned by three different scenarios: (1) expansion of a population *ex vivo* prior to transplantation into the host, (2) re-creation of a tissue or organ *ex vivo* for transplantation, and (3) design of substances and/or devices for *in vivo* acti-

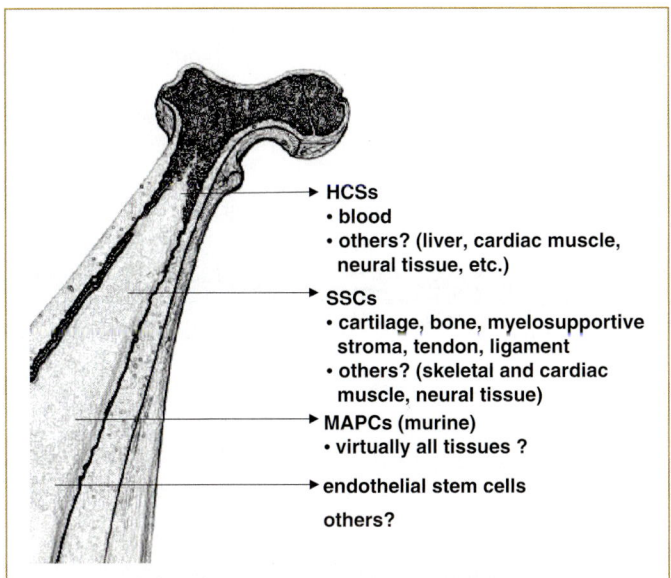

HCSs
- blood
- others? (liver, cardiac muscle, neural tissue, etc.)

SSCs
- cartilage, bone, myelosupportive stroma, tendon, ligament
- others? (skeletal and cardiac muscle, neural tissue)

MAPCs (murine)
- virtually all tissues ?

endothelial stem cells

others?

**FIG. 33.1.** Bone marrow as a central source of postnatal stem cells. Bone marrow consists of at least two well-defined populations of postnatal stem cells, the hematopoietic stem cell (HSC) and the skeletal stem cell (SSC), both of which form numerous phenotypes within their cellular system but may also form cells outside of them. Multipotential adult progenitor cells (MAPCs) are a subset of bone marrow stromal cells and have been reported to form virtually all cell types in mouse. Endothelial precursors (AC133+) have recently been identified, and other types, such as a hepatocyte-like stem cell, may also exist. These remarkable findings, if verified, place bone marrow high on the list of tissues that are easily accessible in sufficient quantity for use in tissue engineering.

vation of stem cells, either local or distant, to induce appropriate tissue repair (Fig. 33.2). In all of these cases, considerable knowledge of the stem cell population's dynamics is required in order to predict and control their activity under a variety of different circumstances.

## Ex Vivo Culture of Postnatal Stem Cells

*Ex vivo* expansion of tissue- or organ-specific cells, either used alone or added to carriers and scaffoldings or with growth factors at the time of transplantation, has been the primary approach in tissue engineering to date. However, *ex vivo* expansion in a fashion that maintains an appropriate proportion of stem cells within the population is a significant hurdle that must be overcome. For example, in spite of enormous effort, the culture conditions for maintaining HSCs (let alone expanding their number) are as yet undefined. It is perhaps for this very reason that currently there are only a handful of examples in which *ex vivo* expanded postnatal stem cells are used successfully to restore structure and function. The key to successful expansion will lie in understanding cell proliferation kinetics (asymmetric versus symmetric division) (Morrison and Kimble, 2006; Sherley, 2002). The efficacy achieved by the use of *ex vivo* expanded populations, whether stem or more

committed in character, may also depend, at least in part, on the nature of the tissue under reconstruction. Within this context, the rate of tissue turnover most likely defines the rate of success. In tissues with a high rate of turnover, such as blood and skin, it is clear that long-term success depends on the persistence of a stem cell within the transplanted population. More committed progenitors may provide some short-term advantage; but without a self-renewing population, failure is ultimate. However, in tissues that turn over more slowly (e.g., bone), long-lasting benefit may be achieved with more committed populations of cells.

While optimizing culture conditions represents one hurdle, the issues of time and quantities represent others. In most cases, the amount of time required to generate the number of cells sufficient to repair defects induced by trauma or disease is on the order of weeks. In cases of trauma, this poses a large problem if the use of autologous stem cells is considered. For that reason, the use of allogeneic populations that could be used "off the shelf" from unrelated donors would be preferable, but this raises the issue of rejection, as in organ transplantation. Although it is suggested that some postnatal stem cells appear to escape from immune surveillance in allogeneic settings and are immunomodulatory (Le Blanc and Pittenger, 2005), definitive proof is lacking. Furthermore, differentiation of stem cells would imply the expression of a mature tissue-specific phenotype, including a complete histocompatibility profile (Fang *et al.*, 2004). Use of allogeneic cells would most likely require concomitant immunosuppressive therapy, which has its own list of side effects. Cotransplantation of allogeneic bone marrow to reconstitute the hematopoietic system has been proposed as a way of inducing tolerance (Weissman, 2005). Recent studies in organ transplantation in conjunction with bone marrow or mobilized blood transplantation after immune ablation indicate a substantial improvement in long-term survival. In this type of approach, allogeneic cells could be better envisioned, provided that a single donor is the source of cells for the tissue to be restored along with the immune system.

## Delivery of Stem Cells

Delivery of bone marrow or mobilized blood by systemic infusion for restoration of the hematopoietic system by HSCs is clearly an example of the efficacy of this method of delivery, but it is perhaps the only example. The success of systemic infusion is based primarily on the fact that in terms of physical structure, blood is a simple, fluid entity. Although bone marrow transplantation has been attempted for the delivery of nonhematopoietic cells to sites of injury, it is not yet clear that there is long-term survival and engraftment of such cells (reviewed in Pauwelyn and Verfaillie, 2006). Reconstruction of two- or three-dimensional structures requires different approaches. Currently, there are a number of clinical applications, or ones soon to emerge, for the delivery of cell populations orthotopically into a site for

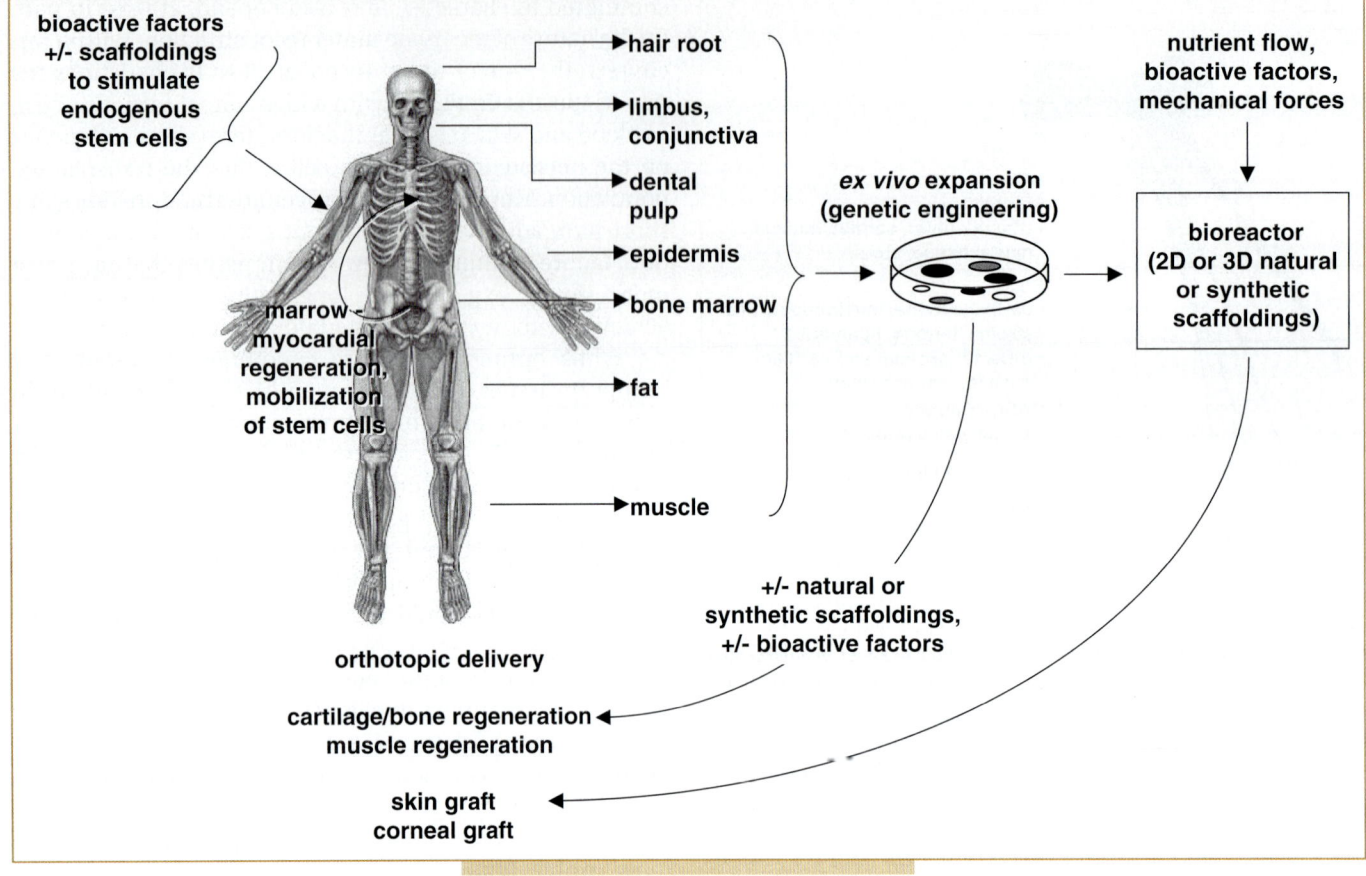

**FIG. 33.2.** Current applications of postnatal stem cells in tissue engineering. While virtually all tissues in the body have cells with some regenerative capabilities, current postnatal stem cell–based strategies, or ones in the foreseeable future, rely on a relatively limited number of source tissues. Autologous bone marrow is currently in trial for myocardial regeneration. Most approaches utilize *ex vivo* expanded cell populations that are then delivered orthotopically in various combinations with bioactive factors and scaffolds, with skeletal and muscle regeneration. Generation of 2D and 3D structures *ex vivo* requires the use of bioreactors, in which cells are seeded onto scaffoldings and subjected to nutrient flow, bioactive factors, and mechanical forces to induce formation of functional tissue for transplantation.

tissue regeneration. In some cases, specific populations of cells are expanded *ex vivo* and delivered either alone or in conjunction with a natural or synthetic material. As important as it is to control the *ex vivo* behavior of the cell population, it is also important to take into consideration what the cells' response will be to the host environment. It is the host that must provide the signals that will dictate the differentiation if uncommitted populations are utilized. Furthermore, the recipient host tissue must also support the maintenance of the stem cell's niche, if not the creation of it *de novo*.

### Reconstruction of the Skeletal Tissues — Bone, Cartilage, and Teeth

After numerous preclinical studies in a variety of animal models, it was demonstrated that bone marrow stromal cells, when used in conjunction with appropriate carriers, could regenerate new bone in critical-sized defects that would never heal without intervention (reviewed in Bianco and Robey, 2001; Cowan *et al.*, 2005). A wide variety of carriers were tested, ranging from collagen sponges and syn-

thetic biodegradable polymers to synthetic hydroxyapatite derivatives, although, to date, none appear to be optimal (reviewed in El-Ghannam, 2005). A carrier that provides immediate stability, especially in the case of weight-bearing bones, and yet can be completely resorbed as new bone is formed and remodeled has not yet been fabricated. Nonetheless, a small number of patients were treated with *ex vivo* expanded bone marrow stromal cells (BMSCs) in conjunction with hydroxyapatite/tricalcium phosphate ceramic particles, with good outcome (Quarto *et al.*, 2001), and other clinical studies are in progress. In addition to healing segmental defects, BMSCs were also used to construct vascularized bone flaps. In cases where morbidity of the recipient site is an issue, it can be envisioned that a new bone rudiment could be grown elsewhere in the body and then transferred, with vasculature intact (Mankani *et al.*, 2001). More recently, adipose tissue was identified as a source of stem cells that also have the potential to differentiate into bone and cartilage, at least, and may be equally as multipotential as BMSCs (reviewed in Gimble and Guilak, 2003). Naturally, if these studies are further substantiated, fat would become another prime source of easily accessible stem cells for tissue engineering (Parker and Katz, 2006).

Current cell-based therapy for the treatment of cartilage defects relies on growth of chondrocytes from a biopsy, followed by transplantation onto articular surfaces in conjunction with a periosteal flap (Brittberg *et al.*, 1994). While this procedure is practiced worldwide, long-term efficacy is questionable, and there is a clear need for better procedures. BMSCs and stem cells from adipose tissue are capable of forming cartilage in a test tube (Guilak *et al.*, 2004; Song *et al.*, 2004), but their use in the reconstruction of cartilage on articular surfaces will rely on inhibiting their further differentiation into hypertrophic chondrocytes. In addition, further development of suitable carriers that provide the appropriate microenvironment to maintain the cartilage phenotype and prevent hypertrophy is also required. Natural polymeric gels, such as hyaluronic acid, collagen, alginate, and chitosan, provide an adequate three-dimensional structure to maintain the chondrocyte phenotype, but these are not well modeled into specific shapes and have very poor biomechanical properties. For these reasons, synthetic biodegradable polymers, such as polylactic acid and polyglycolic acid, and mixtures of both that can be cross-linked and molded to form porous three-dimensional structures are thought to provide more adequate scaffolding and have been used to construct cartilage with predefined shapes (reviewed in Frenkel and Di Cesare, 2004; Kuo *et al.*, 2006). In designing constructs of stem cells and scaffolds, the differences between the elastic cartilages of the ear and the nose and articular cartilage must also be considered to ensure that tissues form with the appropriate biomechanical characteristics.

Using techniques developed for the isolation and characterization of bone marrow stromal cells, dental pulp from the permanent dentition was found to contain a population of stem cells (dental pulp stem cells, DPSCs) that has the ability to form copious amounts of primary dentin and a pulplike complex on *in vivo* transplantation with hydroxyapatite/tricalcium phosphate ceramic particles (Gronthos *et al.*, 2000). Stem cells are also present in the pulp of deciduous teeth (SHED, stem cells from human, exfoliated, deciduous teeth) that have a high proliferative capacity, and, in addition to forming dentin, they may be able to form other cell types (Miura *et al.*, 2003). Further characterization of their differentiation potential is under way. Cells isolated from the periodontal ligament re-create a PDL-like structure and form cementum when transplanted in conjunction with hydroxyapatite/tricalcium phosphate (Seo *et al.*, 2004). Based on these findings, reconstruction of structures in the craniofacial complex can be envisioned (reviewed in Robey, 2005; Robey and Bianco, 2006).

### Regeneration of Skeletal and Cardiac Muscle

Muscle contains a population of muscle-specific stem cells, the satellite cell (reviewed in Zammit *et al.*, 2006). Early attempts to use *ex vivo* expanded populations of allogeneic myoblastic cells derived from satellite cells to correct muscular dystrophies were all uniformly unsuccessful, due to immunological responses of the host to allogeneic cells (and the expression of a foreign protein by the donor cells) and the inability of donor cells to repopulate extensively (Cossu and Mavilio, 2000). In other studies, bone marrow transplantation provided a cell that could participate in the formation of new myotubes in a muscle injury model (G. Ferrari *et al.*, 1998) and in a model of muscular dystrophy (Gussoni *et al.*, 1999); however, the level of incorporation of donor cells was very low. Recently, "mesenchymal" stem cells isolated from the synovial membrane and sharing resemblances with bone marrow–derived stromal cells have been shown to repair dystrophic muscle in the mdx mouse (De Bari *et al.*, 2003). Mesoangioblasts, which are blood vessel–associated stem cells, may also be useful in skeletal muscle regeneration (Sampaolesi *et al.*, 2003).

Although skeletal muscle cells are functionally quite distinct from cardiomyocytes, autologous myoblastic cells have been viewed as a potential therapy for the treatment of myocardial infarct. Initial results suggest that there is functional improvement (Menasche *et al.*, 2003), even though the mechanisms are unclear, due to recent evidence that engrafted cells are not electromechanically linked with recipient cells (Leobon *et al.*, 2003). The utility of autologous bone marrow for myocardial regeneration is also currently receiving a great deal of attention, based on studies that attribute the ability of HSCs and BMSCs to form cardiomyocytes and the ability of marrow-derived endothelial precursors to participate in the development of new blood vessels in a number of animal models (reviewed in Itescu *et al.*, 2003). Based on these encouraging findings, transendocardial injection for the treatment of ischemic heart failure using marrow-derived cells is being tested in a number of small clinical trials. Patients receiving an autologous bone marrow transplant in conjunction with cardiac bypass surgery (Perin *et al.*, 2003) and those receiving AC133+ endothelial precursor cells alone (Stamm *et al.*, 2003) are reported to have increased cardiac perfusion and enhanced ventricular function. Interestingly, it appears that the use of many cell types results in a substantial improvement of cardiac function (Murry *et al.*, 2006). Whether they engraft at a substantial level and in fact differentiate into cardiomyocytes remains to be proven (Orlic, 2005; Stamm *et al.*, 2006). It should be considered that within this context, clinical benefit may be conveyed by a "helper" effect on putative local progenitors or by improvement of perfusion following angiogenesis induced by transplanted endothelial progenitors. Considering the high prevalence of ischemic heart disease, further testing of this application is warranted.

### Ex Vivo Reconstructions — Cells, Scaffolds, and Bioreactors

The creation of tissues and even organs prior to transplantation is a rapidly expanding area of research. This approach relies on expanding cell populations on or in either natural or synthetic scaffolds in a bioreactor. Whether natural or synthetic, scaffolds must be biocompatable, bioresorbable, and nonimmunogenic. Furthermore, they must also be instructive and dictate appropriate cell growth and differentiation, and they must support the recapitulation of a stem cell niche, which is essential for tissue renewal on transplantation. The most commonly used natural scaffolds include either individual purified extracellular matrix (ECM) proteins, such as collagen, fibronectin, and laminin (or peptides derived from them) or devitalized ECM from skin, submucosa of the small intestine, urinary bladder, and others, and they can be autologous, allogeneic, or even xenogeneic. Such devitalized ECMs contain not only the structural proteins (which also have other important biological activities) that define the three-dimensional organization of the tissue, but also the local repertoire of growth factors that are stored within them (reviewed in Gilbert *et al.*, 2006). Various chemical and nonchemical treatments have been applied to ECMs and their components in attempts to modify their biomechanical and immunological properties on transplantation, but these result in less-than-desirable effects due to inactivation of their conducive and inductive properties and inhibition of their resorption and replacement. If left unmodified, ECMs generally promote infiltration and proliferation of cells and differentiation *in vitro* and *in vivo* and are ultimately turned over on transplantation (reviewed in Badylak, 2002). Synthetic scaffolds are designed not only to mimic the biological properties of ECMs, but also to have enhanced material properties appropriate for a particular tissue. Polyglycolic acids, polyhyroxyalkonates, and hydrogels are the most common examples, and all can be manipulated to form a broad range of structures with varying degrees of porosity and rigidity (Hollister, 2005). Scaffolds can also be formulated to include morphogenetic and growth factors or even naked plasmid DNA to transduce ingrowing cells to produce these important factors, such that they can become as instructive as ECMs (reviewed in Hench and Polak, 2002).

In addition to having appropriate scaffolds, the ability to construct tissues and organs *ex vivo* depends on the design of appropriate bioreactors. The initial designs provided primarily nutrient flow through the developing tissue. However, many tissues grown under these circumstances fail to achieve the biomechanical properties required for tissue function *in vivo*. For example, the construction of blood vessels using a variety of biomaterials and cell populations using culture perfusion yields structures that histologically resemble native blood vessels. Yet when transplanted, these constructs failed, due to their inability to withstand changes in pressure. A substantial improvement in such constructs was achieved by subjecting the developing tissues to pulsatile flow conditions (reviewed in Levenberg, 2005). Perfusion-type bioreactors support the growth of tissue with only a nominal number of cell layers and are amenable to construction of relatively simple structures. Construction of larger, more complex three-dimensional structures requires the development of even better bioreactors that will support cell survival to achieve a significant cell mass (reviewed in Martin *et al.*, 2004). Another critical aspect of using large constructs is the development of supporting vasculature. While constructs can be placed in vascular beds in some cases, it will be essential to design constructs that rapidly induce vascularization, perhaps by including the angiogenic factor, VEGF, or cells that produce it or even by including endothelial cells and their precursors, which tend to assemble themselves into primitive tubular structures that may allow for more rapid establishment of organ perfusion (reviewed in Griffith and Naughton, 2002).

Next to bone marrow transplantation, skin grafts perhaps provide the best example of the impact of postnatal stem cell biology on the success of cell-based tissue reconstruction. Although the use of *ex vivo* expanded keratinocytes for the generation of skin grafts was first brought to a clinical setting by the pioneering work of Green and coworkers (Green, 1989), persistent engraftment has not been routine. It is now recognized that both the expansion culture conditions in which the cells are grown and the scaffolding on which they are placed must support self-renewal of epidermal stem cells (cells that give rise to holoclones, as opposed to transiently amplifying meroclones or more differentiated paraclones) (Barrandon and Green, 1987). Expansion of keratinocyte populations on fibrin substrates was found to maintain the stem cell population and to improve long-term maintenance of skin grafts (Pellegrini *et al.*, 1999). A number of studies are now under way to molecularly engineer autologous epidermal stem cells to correct a gene defect causing severe blistering prior to generation of skin grafts (reviewed in S. Ferrari *et al.*, 2006). Currently, a number of commercial products are available (both autologous and allogeneic), some of which incorporate dermal fibroblasts in a collagen gel, which is then layered with *ex vivo* expanded epidermal cells (reviewed in Supp and Boyce, 2005). They provide coverage of severe wounds and are maintained while the recipient generates new skin. The standard starting material for establishing epidermal cultures is skin. However, recent evidence indicates that a stem cell located in the bulge region of the root sheet of hair follicles has the ability to form not only new follicles and sebaceous glands, but also interfollicular epidermis under certain circumstances (reviewed in Blanpain and Fuchs, 2005). These cells, which can be obtained noninvasively, are highly proliferative and have recently been used successfully to treat leg ulcers (Limat and Hunziker, 2002). Based on lessons learned from skin regeneration, stem cells in the limbus of the sclera and conjunctiva have been expanded in culture on fibrin substrates to generate sheets of epithelial cells able to reconstitute a damaged corneal surface, with dramatic beneficial clinical results (reviewed in Vascotto and Griffith, 2006).

## Activation of Local and Distant Endogenous Stem Cells

While the *ex vivo* reconstruction of entire organs that are functional is the goal of the bioengineering world, what perhaps is even more challenging is the goal of inducing endogenous stem cells to become activated to reconstruct a tissue. Most tissues in the body display at least some sort of regenerative capability (Preston *et al.*, 2003; Tsonis, 2002) but in many cases remain insufficient to mount a spontaneous and successful repair response *in vivo*. The application or induction of growth factors and morphogens, perhaps in combination with appropriate scaffolds, might be envisioned to "encourage" local or distant progenitors to regenerate a functional tissue. This *in situ* process could occur through several different pathways, including activation of local or distant stem cells, transdifferentiation, or reprogramming, to generate adequate numbers of committed precursors. In all cases, a regulated morphogenetic process is needed to establish normal structure and function. Definition of the intrinsic properties of regenerative cells and extrinsic signals that may trigger a recapitulation of developmental processes is critical.

### Local Cells

Based on reasoning that, following an injury, signals that normally activate a local cell to initiate repair would be either obliterated or not of a high enough magnitude, numerous studies attempted to provide appropriate morphogenetic and growth factors. However, this approach has been uniformly disappointing, most likely due to the short half-life of such soluble factors and the lack of an appropriate scaffolding. Not only does scaffolding provide a template for the organized outgrowth of local cells, but it also provides a substrate on which to stabilize and/or orient factors (or parts thereof) for appropriate presentation to a cell (reviewed in Lutolf and Hubbell, 2005; Rosso *et al.*, 2005). However, in using "smart" constructs, tissue regeneration of large defects may not be complete, due to exhaustion of the local stem cell population; hence the need to maintain a stem cell niche. Furthermore, some clues as to the negative impact of local constraints on the native, albeit incomplete, repair capacity of cells and tissues may be found in studies of spinal cord regeneration. Following injury, there is a phase characterized by neurite outgrowth from the severed end. However, it is short-lived due to the production of inhibitory factors by cells within the myelin sheath, and gliosis impedes neurite outgrowth (reviewed in Di Giovanni, 2006; Schwab, 2002). Thus, regeneration by endogenous cells must take into consideration not only the bioactive factors and scaffolding that are necessary to maintain an ingrowing population with the appropriate balance of stem cells, transiently amplifying cells, and differentiating cells, but also the inactivation of local inhibitory factors that work against the process.

### Mobilization and Recruitment of Distant Stem Cells

Numerous studies have suggested that once introduced into the circulation, many types of cells have the propensity to end up at a diseased or injured site and to participate in local regeneration (Stocum, 2001), and it has been argued that HSCs might generate a broad variety of nonhematopoietic cell types. Ongoing tissue damage and repair, at the time when systemically infused stem cells reach a target organ, may be the critical determinant for triggering their homing, engraftment, and differentiation to local cell types.

Cytokine administration has become a well-established procedure for mobilization of HSCs into peripheral blood (reviewed in Winkler and Levesque, 2006), but it is not clear that other stem cell populations present in marrow would be equally as amenable to liberation by current procedures or whether different ones would have to be developed. Consequently, mobilization of marrow stem cells might provide a mechanism to enhance a local population and improve tissue regeneration, a hypothesis that might be tested in the context of extant local, tissue-specific damage and repair.

## IV. CONCLUSIONS

The recent advances in embryonic and postnatal stem cell biology have most certainly captured the public's attention and have raised expectations for miraculous cures in the near future. While certain applications that utilize postnatal stem cells are in practice or will be shortly, the majority are in their infancy and will take much more effort. However, the current sense of urgency in translating recent findings into clinical applications should not lead this field to repeat the mistakes experienced in the field of gene therapy. The development of tissue engineering as a viable medical practice must proceed in an evidence-based fashion in all of the associated disciplines that are involved, and several hurdles are yet to be overcome. First, our current understanding of postnatal stem cell biology is rudimentary, at best. The manner in which we isolate stem cells and manipulate them and their progeny appropriately relies heavily on a clear understanding of cell population kinetics. This also requires a complete understanding of their response not only to bioactive factors, scaffoldings, and delivery systems, but also to the host microenvironment in which they must survive and function. Second, more efficient bioreactors that adequately model the microenvironment and also allow for scale-up of the tissue-engineering process must be developed. Third, in order to translate what we learn into clinical application, development of appropriate preclinical models to prove the principle that stem cells do indeed have a positive biological impact is absolutely essential. In analyzing these models, stringent criteria must be defined to determine efficacy, and we must remain principled in assessing them. Exciting times, yes, but also a time for due diligence to bring what started off as a scientific curiosity into medical reality.

# V. ACKNOWLEDGMENTS

The support of Telethon Fondazione Onlus Grant E1029 (to P.B.) and of the DIR, NIDCR of the IRP, NIH (P.G.R.) is gratefully acknowledged.

# VI. REFERENCES

Badorff, C., Brandes, R. P., Popp, R., Rupp, S., Urbich, C., Aicher, A., Fleming, I., Busse, R., Zeiher, A. M., and Dimmeler, S. (2003). Transdifferentiation of blood-derived human adult endothelial progenitor cells into functionally active cardiomyocytes. *Circulation* **107**, 1024–1032.

Badylak, S. F. (2002). The extracellular matrix as a scaffold for tissue reconstruction. *Semin. Cell Dev. Biol.* **13**, 377–383.

Barrandon, Y., and Green, H. (1987). Three clonal types of keratinocyte with different capacities for multiplication. *Proc. Natl. Acad. Sci. U.S.A.* **84**, 2302–2306.

Bianco, P., and Robey, P. G. (2001). Stem cells in tissue engineering. *Nature* **414**, 118–121.

Bianco, P., and Robey, P. G. (2004). Skeletal stem cells. *In* "Handbook of Adult and Fetal Stem Cells" (R. P. Lanza, ed.), pp. 415–424. Academic Press, San Diego.

Bittner, R. E., Schofer, C., Weipoltshammer, K., Ivanova, S., Streubel, B., Hauser, E., Freilinger, M., Hoger, H., Elbe-Burger, A., and Wachtler, F. (1999). Recruitment of bone-marrow-derived cells by skeletal and cardiac muscle in adult dystrophic mdx mice. *Anat. Embryol. (Berl.)* **199**, 391–396.

Blanpain, C., and Fuchs, E. (2006). Epidermal stem cells of the skin. *Annu. Rev. Cell Dev. Biol.* **22**, 339–373.

Brittberg, M., Lindahl, A., Nilsson, A., Ohlsson, C., Isaksson, O., and Peterson, L. (1994). Treatment of deep cartilage defects in the knee with autologous chondrocyte transplantation. *N. Engl. J. Med.* **331**, 889–895.

Bryder, D., Rossi, D. J., and Weissman, I. L. (2006). Hematopoietic stem cells: the paradigmatic tissue-specific stem cell. *Am. J. Pathol.* **169**, 338–346.

Camargo, F. D., Chambers, S. M., and Goodell, M. A. (2004). Stem cell plasticity: from transdifferentiation to macrophage fusion. *Cell Prolif.* **37**, 55–65.

Cossu, G., and Mavilio, F. (2000). Myogenic stem cells for the therapy of primary myopathies: wishful thinking or therapeutic perspective? *J. Clin. Invest.* **105**, 1669–1674.

Cowan, C. M., Soo, C., Ting, K., and Wu, B. (2005). Evolving concepts in bone tissue engineering. *Curr. Top. Dev. Biol.* **66**, 239–285.

De Bari, C., Dell'Accio, F., Vandenabeele, F., Vermeesch, J. R., Raymackers, J. M., and Luyten, F. P. (2003). Skeletal muscle repair by adult human mesenchymal stem cells from synovial membrane. *J. Cell Biol.* **160**, 909–918.

Di Giovanni, S. (2006). Regeneration following spinal cord injury, from experimental models to humans: where are we? *Expert Opin. Ther. Targets* **10**, 363–376.

El-Ghannam, A. (2005). Bone reconstruction: from bioceramics to tissue engineering. *Expert Rev. Med. Devices* **2**, 87–101.

Fang, T. C., Alison, M. R., Wright, N. A., and Poulsom, R. (2004). Adult stem cell plasticity: will engineered tissues be rejected? *Int. J. Exp. Pathol.* **85**, 115–124.

Ferrari, G., Cusella-De Angelis, G., Coletta, M., Paolucci, E., Stornaiuolo, A., Cossu, G., and Mavilio, F. (1998). Muscle regeneration by bone marrow–derived myogenic progenitors. *Science* **279**, 1528–1530.

Ferrari, S., Pellegrini, G., Matsui, T., Mavilio, F., and De Luca, M. (2006). Gene therapy in combination with tissue engineering to treat epidermolysis bullosa. *Expert Opin. Biol. Ther.* **6**, 367–378.

Frenkel, S. R., and Di Cesare, P. E. (2004). Scaffolds for articular cartilage repair. *Ann. Biomed. Eng.* **32**, 26–34.

Gehling, U. M., Ergun, S., Schumacher, U., Wagener, C., Pantel, K., Otte, M., Schuch, G., Schafhausen, P., Mende, T., Kilic, N., *et al.* (2000). *In vitro* differentiation of endothelial cells from AC133-positive progenitor cells. *Blood* **95**, 3106–3112.

Gilbert, T. W., Sellaro, T. L., and Badylak, S. F. (2006). Decellularization of tissues and organs. *Biomaterials* **27**, 3675–3683.

Gimble, J. M., and Guilak, F. (2003). Differentiation potential of adipose-derived adult stem (ADAS) cells. *Curr. Top. Dev. Biol.* **58**, 137–160.

Goodell, M. A. (2003). Stem-cell "plasticity": befuddled by the muddle. *Curr. Opin. Hematol.* **10**, 208–213.

Green, H. (1989). Regeneration of the skin after grafting of epidermal cultures. *Lab. Invest.* **60**, 583–584.

Griffith, L. G., and Naughton, G. (2002). Tissue engineering — current challenges and expanding opportunities. *Science* **295**, 1009–1014.

Gronthos, S., Mankani, M., Brahim, J., Robey, P. G., and Shi, S. (2000). Postnatal human dental pulp stem cells (DPSCs) *in vitro* and *in vivo*. *Proc. Natl. Acad. Sci. U.S.A.* **97**, 13625–13630.

Guilak, F., Awad, H. A., Fermor, B., Leddy, H. A., and Gimble, J. M. (2004). Adipose-derived adult stem cells for cartilage tissue engineering. *Biorheology* **41**, 389–399.

Gussoni, E., Soneoka, Y., Strickland, C. D., Buzney, E. A., Khan, M. K., Flint, A. F., Kunkel, L. M., and Mulligan, R. C. (1999). Dystrophin expression in the mdx mouse restored by stem cell transplantation. *Nature* **401**, 390–394.

Hench, L. L., and Polak, J. M. (2002). Third-generation biomedical materials. *Science* **295**, 1014–1017.

Hollister, S. J. (2005). Porous scaffold design for tissue engineering. *Nat. Mater.* **4**, 518–524.

Itescu, S., Schuster, M. D., and Kocher, A. A. (2003). New directions in strategies using cell therapy for heart disease. *J. Mol. Med.* **81**, 288–296.

Jiang, Q. J., Izakovic, J., Zenker, M., Fartasch, M., Meneguzzi, G., Rascher, W., and Schneider, H. (2002). Treatment of two patients with Herlitz junctional epidermolysis bullosa with artificial skin bioequivalents. *J. Pediatr.* **141**, 553–559.

Krause, D. S., Theise, N. D., Collector, M. I., Henegariu, O., Hwang, S., Gardner, R., Neutzel, S., and Sharkis, S. J. (2001). Multi-organ, multilineage engraftment by a single bone marrow–derived stem cell. *Cell* **105**, 369–377.

Kuo, C. K., Li, W. J., Mauck, R. L., and Tuan, R. S. (2006). Cartilage tissue engineering: its potential and uses. *Curr. Opin. Rheumatol.* **18**, 64–73.

Lagasse, E., Connors, H., Al-Dhalimy, M., Reitsma, M., Dohse, M., Osborne, L., Wang, X., Finegold, M., Weissman, I. L., and Grompe, M. (2000). Purified hematopoietic stem cells can differentiate into hepatocytes *in vivo*. *Nat. Med.* **6**, 1229–1234.

Lakshmipathy, U., and Verfaillie, C. (2005). Stem cell plasticity. *Blood Rev.* **19**, 29–38.

Le Blanc, K., and Pittenger, M. (2005). Mesenchymal stem cells: progress toward promise. *Cytotherapy* **7**, 36–45.

Leobon, B., Garcin, I., Menasche, P., Vilquin, J. T., Audinat, E., and Charpak, S. (2003). Myoblasts transplanted into rat infarcted myocardium are functionally isolated from their host. *Proc. Natl. Acad. Sci. U.S.A.* **100**, 7808–7811.

Levenberg, S. (2005). Engineering blood vessels from stem cells: recent advances and applications. *Curr. Opin. Biotechnol.* **16**, 516–523.

Limat, A., and Hunziker, T. (2002). Use of epidermal equivalents generated from follicular outer root sheath cells *in vitro* and for autologous grafting of chronic wounds. *Cells Tissues Organs* **172**, 79–85.

Lutolf, M. P., and Hubbell, J. A. (2005). Synthetic biomaterials as instructive extracellular microenvironments for morphogenesis in tissue engineering. *Nat. Biotechnol.* **23**, 47–55.

Mankani, M. H., Krebsbach, P. H., Satomura, K., Kuznetsov, S. A., Hoyt, R., and Robey, P. G. (2001). Pedicled bone flap formation using transplanted bone marrow stromal cells. *Arch. Surg.* **136**, 263–270.

Martin, I., Wendt, D., and Heberer, M. (2004). The role of bioreactors in tissue engineering. *Trends Biotechnol.* **22**, 80–86.

Menasche, P., Hagege, A. A., Vilquin, J. T., Desnos, M., Abergel, E., Pouzet, B., Bel, A., Sarateanu, S., Scorsin, M., Schwartz, K., *et al.* (2003). Autologous skeletal myoblast transplantation for severe postinfarction left ventricular dysfunction. *J. Am. Coll. Cardiol.* **41**, 1078–1083.

Minguell, J. J., and Erices, A. (2006). Mesenchymal stem cells and the treatment of cardiac disease. *Exp. Biol. Med. (Maywood)* **231**, 39–49.

Miura, M., Gronthos, S., Zhao, M., Lu, B., Fisher, L. W., Robey, P. G., and Shi, S. (2003). SHED: stem cells from human exfoliated deciduous teeth. *Proc. Natl. Acad. Sci. U.S.A.* **100**, 5807–5812.

Moore, B. E., and Quesenberry, P. J. (2003). The adult hemopoietic stem cell plasticity debate: idols vs. new paradigms. *Leukemia* **17**, 1205–1210.

Morrison, S. J., and Kimble, J. (2006). Asymmetric and symmetric stem-cell divisions in development and cancer. *Nature* **441**, 1068–1074.

Murry, C. E., Reinecke, H., and Pabon, L. M. (2006). Regeneration gaps: observations on stem cells and cardiac repair. *J. Am. Coll. Cardiol.* **47**, 1777–1785.

Orlic, D. (2005). BM stem cells and cardiac repair: where do we stand in 2004? *Cytotherapy* **7**, 3–15.

Orlic, D., Kajstura, J., Chimenti, S., Limana, F., Jakoniuk, I., Quaini, F., Nadal-Ginard, B., Bodine, D. M., Leri, A., and Anversa, P. (2001). Mobilized bone marrow cells repair the infarcted heart, improving function and survival. *Proc. Natl. Acad. Sci. U.S.A.* **98**, 10344–10349.

Parker, A. M., and Katz, A. J. (2006). Adipose-derived stem cells for the regeneration of damaged tissues. *Expert Opin. Biol. Ther.* **6**, 567–578.

Pauwelyn, K. A., and Verfaillie, C. M. (2006). 7. Transplantation of undifferentiated, bone marrow–derived stem cells. *Curr. Top. Dev. Biol.* **74**, 201–251.

Pellegrini, G., Ranno, R., Stracuzzi, G., Bondanza, S., Guerra, L., Zambruno, G., Micali, G., and De Luca, M. (1999). The control of epidermal stem cells (holoclones) in the treatment of massive full-thickness burns with autologous keratinocytes cultured on fibrin. *Transplantation* **68**, 868–879.

Perin, E. C., Dohmann, H. F., Borojevic, R., Silva, S. A., Sousa, A. L., Mesquita, C. T., Rossi, M. I., Carvalho, A. C., Dutra, H. S., Dohmann, H. J., *et al.* (2003). Transendocardial, autologous bone marrow cell transplantation for severe, chronic ischemic heart failure. *Circulation* **107**, 2294–2302.

Phinney, D. G., and Isakova, I. (2005). Plasticity and therapeutic potential of mesenchymal stem cells in the nervous system. *Curr. Pharm. Des.* **11**, 1255–1265.

Poulsom, R., Alison, M. R., Forbes, S. J., and Wright, N. A. (2002). Adult stem cell plasticity. *J. Pathol.* **197**, 441–456.

Preston, S. L., Alison, M. R., Forbes, S. J., Direkze, N. C., Poulsom, R., and Wright, N. A. (2003). The new stem cell biology: something for everyone. *Mol. Pathol.* **56**, 86–96.

Quarto, R., Mastrogiacomo, M., Cancedda, R., Kutepov, S. M., Mukhachev, V., Lavroukov, A., Kon, E., and Marcacci, M. (2001). Repair of large bone defects with the use of autologous bone marrow stromal cells. *N. Engl. J. Med.* **344**, 385–386.

Robey, P. G. (2005). Postnatal stem cells for dental and craniofacial repair. *Oral Biosci. Med.* **2**, 83–90.

Robey, P. G., and Bianco, P. (2006). The use of adult stem cells in rebuilding the human face. *J. Am. Dent. Assoc.* **137**, 961–972.

Rosso, F., Marino, G., Giordano, A., Barbarisi, M., Parmeggiani, D., and Barbarisi, A. (2005). Smart materials as scaffolds for tissue engineering. *J. Cell Physiol.* **203**, 465–470.

Sampaolesi, M., Torrente, Y., Innocenzi, A., Tonlorenzi, R., D'Antona, G., Pellegrino, M. A., Barresi, R., Bresolin, N., De Angelis, M. G., Campbell, K. P., *et al.* (2003). Cell therapy of alpha-sarcoglycan null dystrophic mice through intra-arterial delivery of mesoangioblasts. *Science* **301**, 487–492.

Schwab, M. E. (2002). Repairing the injured spinal cord. *Science* **295**, 1029–1031.

Seo, B. M., Miura, M., Gronthos, S., Bartold, P. M., Batouli, S., Brahim, J., Young, M., Robey, P. G., Wang, C. Y., and Shi, S. (2004). Investigation of multipotent postnatal stem cells from human periodontal ligament. *Lancet* **364**, 149–155.

Sherley, J. L. (2002). Asymmetric cell kinetics genes: the key to expansion of adult stem cells in culture. *Stem Cells* **20**, 561–572.

Song, L., Baksh, D., and Tuan, R. S. (2004). Mesenchymal stem cell–based cartilage tissue engineering: cells, scaffold and biology. *Cytotherapy* **6**, 596–601.

Stamm, C., Westphal, B., Kleine, H. D., Petzsch, M., Kittner, C., Klinge, H., Schumichen, C., Nienaber, C. A., Freund, M., and Steinhoff, G. (2003). Autologous bone-marrow stem-cell transplantation for myocardial regeneration. *Lancet* **361**, 45–46.

Stamm, C., Liebold, A., Steinhoff, G., and Strunk, D. (2006). Stem cell therapy for ischemic heart disease: beginning or end of the road? *Cell Transplant* **15**(Suppl. 1), S47–S56.

Stocum, D. L. (2001). Stem cells in regenerative biology and medicine. *Wound Repair Regen.* **9**, 429–442.

Supp, D. M., and Boyce, S. T. (2005). Engineered skin substitutes: practices and potentials. *Clin. Dermatol.* **23**, 403–412.

Tsonis, P. A. (2002). Regenerative biology: the emerging field of tissue repair and restoration. *Differentiation* **70**, 397–409.

Vascotto, S. G., and Griffith, M. (2006). Localization of candidate stem and progenitor cell markers within the human cornea, limbus, and bulbar conjunctiva *in vivo* and in cell culture. *Anat. Rec. A Discov. Mol. Cell. Evol. Biol.* **288**, 921–931.

Wagers, A. J., and Weissman, I. L. (2004). Plasticity of adult stem cells. *Cell* **116**, 639–648.

Wang, X., Willenbring, H., Akkari, Y., Torimaru, Y., Foster, M., Al-Dhalimy, M., Lagasse, E., Finegold, M., Olson, S., and Grompe, M. (2003). Cell fusion is the principal source of bone-marrow-derived hepatocytes. *Nature* **422**, 897–901.

Weissman, I. (2005). Stem cell research: paths to cancer therapies and regenerative medicine. *JAMA* **294**, 1359–1366.

Winkler, I. G., and Levesque, J. P. (2006). Mechanisms of hematopoietic stem cell mobilization: when innate immunity assails the cells that make blood and bone. *Exp. Hematol.* **34**, 996–1009.

Zammit, P. S., Partridge, T. A., and Yablonka-Reuveni, Z. (2006). The skeletal muscle satellite cell: the stem cell that came in from the cold. *J. Histochem. Cytochem.* **54**, 1177–1191.

Part Eight

# Gene Therapy

# Chapter Thirty-Four

# Gene Therapy

*Stefan Worgall and Ronald G. Crystal*

## I. INTRODUCTION

Gene transfer is a therapeutic strategy using genetic information, usually in the form of DNA, to modify the phenotype of cells. Gene therapy strategies can be useful for tissue engineering by modifying cells directly or by providing a favorable growth environment for the engineered tissue. To accomplish this, cells are modified genetically *ex vivo* or *in vivo* using gene transfer vectors that mediate the transfer of therapeutic DNA into the nucleus, where it is transcribed in parallel with genomic DNA. A variety of non-viral and viral gene therapy vectors have been developed, including plasmids, plasmids combined with liposomes, adenovirus, adeno-associated virus, retrovirus, and lentivirus. A number of strategies have evolved to enhance the targeting of gene transfer vectors by genetic or chemical modification of the surface of the vector. Gene expression directed by the transferred gene can be regulated by including inducible promoters, tissue-specific promoters, and *trans*-splicing. Although still in the early stage of development, there is significant potential to combine gene therapy with stem cell strategies to aid in controlling cell growth and to circumvent immune rejection. There are still challenges to using gene transfer for tissue engineering, but the tech-

nology of gene transfer is sufficiently advanced that rapid progress should be made in the near future.

Gene therapy uses the transfer of genetic information to modify a phenotype for therapeutic purposes (Verma and Weitzman, 2005). The application of gene transfer to tissue engineering has myriad possibilities, including the transient or permanent genetic modification of the engineered tissue to produce proteins for internal, local, or systemic use, helping to protect the engineered tissue, and providing stimuli for the engineered tissue to grow and/or differentiate. To provide a background for the application of gene transfer to tissue engineering, this chapter reviews the general strategies of gene therapy, details the gene transfer vectors used to achieve these goals, and discusses the strategies being used to improve gene transfer by modifying the vectors to provide cell-specific targeting and by regulating the expression of the targeted gene. The applications combining gene therapy with stem cell therapy are reviewed. Our overall goal is to provide a state-of-the-art review of the technology of gene therapy, including the challenges to making gene therapy for tissue engineering a reality. For details regarding the applications of gene therapy to specific organs and clinical disorders, several reviews are available (Anderson, 1998; Crystal, 1995; Kaji and Leiden, 2001; Lundstrom, 2003; O'Connor and Crystal, 2006; Thomas *et al.*, 2003; Verma and Weitzman, 2005).

## II. STRATEGIES OF GENE THERAPY

The basic concept of gene therapy is to transfer nucleic acid, usually in the form of DNA (or RNA in retrovirus and lentivirus vectors), to target cells. The vector, with its gene cargo, can be administered *ex vivo*, where the gene is transferred to the cells of interest in the laboratory and the genetically modified cells restored to the patient, or *in vivo*, when the nucleic acid is administered directly to the individual (Fig. 34.1). Independent of the overall strategy, an expression cassette containing the genetic sequences to be delivered, typically a cDNA along with regulatory sequences to control expression, is inserted into a "vector," a nonviral or viral package used to improve efficiency and specificity of the gene transfer. Together, the choice of vector, the design of the expression cassette, and the coding sequences of the gene determine the pharmacokinetics of the resulting gene expression.

Simple in concept, gene transfer is complex in execution. In addition to choosing an *ex vivo* vs. *in vivo* strategy, the major issues relating to successful gene therapy are the design of vector, how the vector is delivered to the cell population of interest, translocating the vector/expression cassette from outside the cell to the nucleus, and the expression of the gene to obtain the desired therapeutic effect (Fig. 34.2). Independent of the choice and design of the gene transfer vector, successful gene therapy requires decisions regarding the quantity of vector required to modify the numbers of target cells necessary to obtain the desired therapeutic effect and whether or not the vector will evoke a host immune response and/or cause unacceptable toxicity. The

gene therapist must also decide how to get the vector to the target cells, including targeting specificity and avidity of the vector for its relevant receptor. Once the vector reaches the cells of interest, the gene cargo within the vector must be translocated from outside of the cell to within the nucleus. In designing the gene transfer strategy, it is critical to decide whether or not the gene is to be inserted into the chromosomal DNA of the target cells. Finally, the transferred gene must be expressed, with the concomitant issues of amount and control of expression and host immunity and toxicity that may be evoked by the expression of the gene. In the sections that follow, we discuss all of these issues.

## III. *EX VIVO* VS. *IN VIVO* GENE THERAPY

There are generally two strategies by which gene therapy technology can be used for genetic engineering: *ex vivo* transfer of genetic material, with subsequent transfer of the modified cells or tissue to the host, and *in vivo* transfer, with direct administration of the gene therapy vectors to the patient (Fig. 34.1).

### Ex Vivo

The *ex vivo* strategy has the advantage that the cell population can be purified and carefully defined, and the transfer of the gene is limited to that cell population and not to other cells or tissues. The challenge for this approach is that, for most applications, once returned to the patient, the genetically modified cells need to have a selective advantage to modulate the therapeutic goal. For applications where long-term expression of the gene is required and the transferred cells subsequently replicate within the patient, the vectors used to transfer the gene must mediate integration of the gene so that it persists when the cell divides (Verma and Weitzman, 2005).

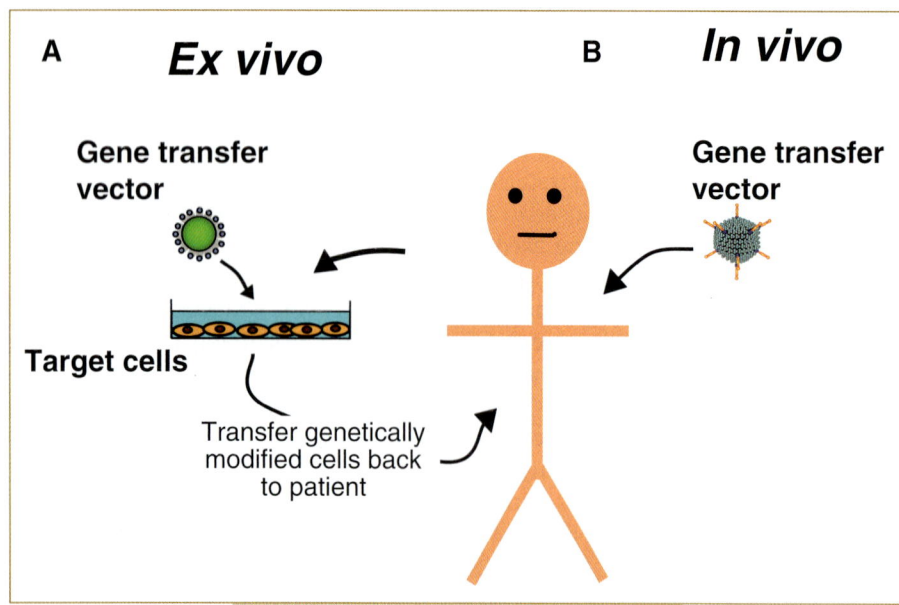

**FIG. 34.1.** General strategies for gene therapy for tissue engineering. **(A)** *Ex vivo* strategies use a gene transfer vector to genetically modify autologous target cells (e.g., skin fibroblasts) *in vitro*, followed by transfer of the genetically modified cells back to the patient. **(B)** *In vivo* strategies transfer the therapeutic gene by direct administration of a gene transfer vector to the patient.

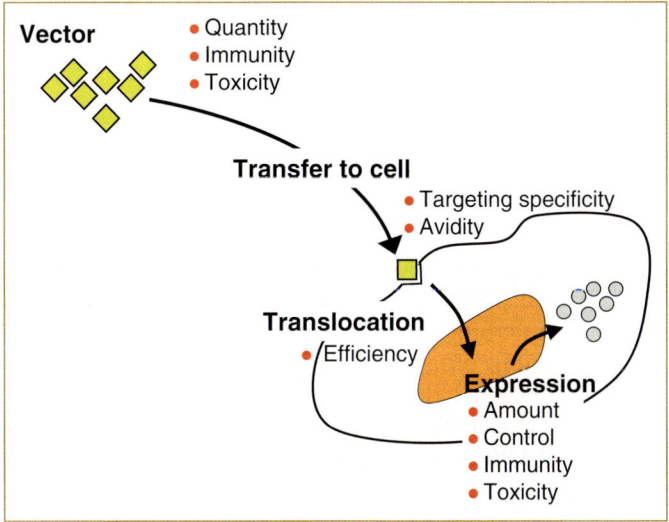

**FIG. 34.2.** Issues relating to successful gene transfer. In addition to the choice of vector, successful gene transfer requires decisions regarding the quantity of vector to be used and that the vector does not induce immunity and/or toxicity that will limit its use. The vector has to be transferred to the cell, a relatively easy task for *ex vivo* strategies, but *in vivo* may require enhancing targeting specificity and vector avidity to its receptor. Once reaching the target cells, the vector must ensure its gene cargo is efficiently translocated to the nucleus. In the nucleus, the gene must be expressed in appropriate amounts, with control if necessary and without appreciable immunity and/or toxicity induced by the gene product.

The *ex vivo* strategy typically is used to genetically modify hematopoietic cells, such as CD34+ cells derived from bone marrow, or skin fibroblasts (Kaji and Leiden, 2001). Examples of *ex vivo* gene transfer strategies include the correction of hereditary immunodeficiencies with retroviral vectors, transfer of the factor VIII gene to autologous fibroblasts to treat hemophilia A, and transfer of suicide genes to T cells to control graft vs. host disease (Cavazzana-Calvo *et al.*, 2005; Qasim *et al.*, 2005; Roth *et al.*, 2001). For tissue-engineering applications, the *ex vivo* strategy is applicable to providing genes to enhance and/or modulate the growth of the engineered tissue as well as to protect the engineered tissue from host responses or disease processes. The *ex vivo* strategy is also the most suitable strategy to use gene transfer to genetically modify stem cells.

Although *ex vivo* gene therapy can be carried out using cells derived from a nonautologous source, potential immune rejection of nonmatched cells generally requires that autologous, or closely matched donor, cells be used for the tissue-engineering strategy. However, the immune system can potentially recognize components of the vector and/or transferred gene product. For *ex vivo* strategies in general, immune recognition of the gene therapy vector is minimal, because the immune system will be in contact not with the total dose of gene therapy vector used to transfer the gene *in vitro*, but only the residue of the vector within the cells to be transferred. There is the possibility of immune recognition via MHC presentation of viral antigens inducing antivector immunity, and this may be responsible for the shutdown of gene expression over time.

### In Vivo

*In vivo* gene transfer strategies administer the gene therapy vector either directly to the target organ or via the vascular system into vessels feeding that organ. *In vivo* gene transfer has an advantage over *ex vivo* strategies in that it avoids the cumbersome (and costly) process of removing cells from the patient, manipulating the cells *in vitro*, and returning the genetically modified cells to the patient. Challenges that need to be overcome for *in vivo* gene transfer strategies include the induction of immunity by the gene transfer vector, transport of the gene therapy vector to the targeted cells/organ, efficient binding of the vector to the cell, translocation of the genetic material to the nucleus, and toxicity and immunity induced by expression of virus and/or transgene peptides.

## IV. CHROMOSOMAL VS. EXTRACHROMOSOMAL PLACEMENT OF THE TRANSFERRED GENE

One of the critical decisions in strategizing gene therapy is whether the transferred gene is to be inserted into the chromosomal DNA of the targeted cells or is designed to function in an extrachromosomal location within the nucleus. There are advantages and disadvantages to both strategies, and the choice of the strategy is determined by the specific application of the gene transfer.

As described later, some gene transfer vectors (e.g., retrovirus, lentivirus) insert their genome, and hence the transferred gene, directly into the chromosomal DNA of the target cells. This has the advantage that it is permanent; and when the genetically modified cells divide, both daughter cells have the newly transferred sequences, for they are now part of the genome of the genetically modified cell population. This is a desirable feature for applications where persistent gene expression is required, such as for the correction of a hereditary disorder. The disadvantage is that, once inserted, the gene cannot be removed, and thus unless controls are designed into the transferred sequences, cannot be shut down. Equally important is the issue of randomness of where the gene is inserted. If the gene is inserted into a relatively "silent" region of the genome, the resulting gene expression will be low, while gene expression from other regions will be high (Bushman *et al.*, 2005). This genome regional modulation of the level of gene expression will be different for each targeted cell. While there are some more favored regions of gene insertion, depending on the vector characteristics, the population of genetically modified cells essentially becomes a mixed population in terms of where the gene has been inserted. Thus, expression may be low for some cells, while other cells may be average or high expressors. More troubling is that if the gene is inserted into a region influencing cell proliferation, the result may be uncontrolled cell growth, e.g., malignancy. This phenomenon, referred to as *insertional mutagenesis*, has been observed in experimental animal and human applications of gene transfer (Bushman *et al.*, 2005).

Some vectors (e.g., nonviral, adenovirus, adeno-associated virus) transfer the gene into the nucleus, but mostly into an extrachromosomal location, where the gene is transcribed using the same transcriptional machinery as for genomic DNA (Verma and Weitzman, 2005). The consequences of this strategy is that as long as the cells do not proliferate and as long as host defenses do not recognize the genetically modified cell as foreign, expression of the transferred gene will persist. However, if the cell divides, the transferred gene will not be replicated, and gene expression in the daughter cells will eventually wane as proliferation continues. The consequences are transient expression of the gene, a "pharmacokinetic" result that is ideal for some applications (e.g., angiogenesis, most cancer therapies) but not desirable for applications requiring persistent expression (e.g., correction of a genetic disease). Independent of the issue of persistence of gene expression, extrachromosomal placement of the gene has the advantage that insertional mutagenesis and variability of gene expression secondary to variability of chromosomal insertion are of no concern.

## V. GENE TRANSFER VECTORS

There are two general classes of commonly used vectors for gene therapy: nonviral and viral (Fig. 34.3, Table 34.1). Although a variety of vectors have been developed in both classes, the most commonly used nonviral vectors are naked plasmids and plasmids combined with liposomes, and the commonly used viral vectors are adenovirus, adeno-associated virus, retrovirus, and lentivirus (Lundstrom, 2003; Thomas *et al.*, 2003; Verma and Weitzman, 2005).

Independent of the vector used, all carry an expression cassette, which includes the gene to be transferred together with the relevant regulatory sequences to control the expression of the gene once it has been transferred to the target cells (Fig. 34.3A). The typical expression cassette includes (5′ to 3′) a promoter, an intron (this is not critical, but it usually enhances gene expression and enables specific PCR identification of the cytoplasmic mature mRNA from the pre-mRNA and transferred expression cassette), the transgene itself (usually in the form of a cDNA, but it can contain one or more introns or can be the generic form of the gene), and finally the polyA/stop and other 3′ regulatory sequences, if desired. In special cases, such as *trans*-splicing (discussed in the section on regulation of expression), the expression cassette may contain only a fragment of the gene together with sequences to direct splicing into an endogenous nuclear pre-mRNA.

### Nonviral Vectors

The most simple gene transfer vectors are plasmids. To achieve relevant transduction of cells *in vivo*, the plasmids are usually combined with liposomes to facilitate attachment and entry into target cells (Lechardeur *et al.*, 2005; Miller, 1999). A variety of physical methods have also been developed to promote entry of plasmids into target cells, including microinjection, hydrodynamic administration, electroporation, ultrasound, and ballistic delivery (the so-called "gene gun") (Miller, 1999). Most of these physical methods of gene delivery are not applicable for *in vivo* gene transfer due to the inaccessibility of the target cells to direct manipulation.

Plasmids contain a relatively simple expression cassette, with the transgene driven by a promoter and flanked by an intron and the polyadenylation/stop site (Fig. 34.3A). Plasmid DNA has an unlimited size capacity; however, because plasmids larger than 10 kb are potentially unstable, during production most gene transfer strategies being considered for human applications use plasmids under 10 kb. Although plasmids can efficiently transduce cells *in vitro*, their efficiency *in vivo* is limited. There have been many attempts to correct hereditary disorders with plasmid gene transfer alone, but with little evidence of expression of the plasmid-directed mRNA in those trials (Montier *et al.*, 2004). Part of the reason for the inefficiency of plasmid-mediated gene transfer is that plasmids have no means to direct their traffic to the nucleus (Lechardeur *et al.*, 2005). For gene transfer applications for tissue engineering, the most that can be envisioned for the use of plasmid-based systems is to use them in an *ex vivo* approach, with the possible need for selection of the transduced cells.

### Adenovirus

Adenoviruses are nonenveloped viruses containing a linear double-stranded DNA genome of 36 kb (Shenk, 2001). Among the 49 different Ad strains that infect humans, serotype 5 and serotype 2 of subgroup C are widely used in gene transfer studies and are the only serotypes used in humans to date. The Ad5 genome is composed of early and late genes (Shenk, 2001). The E1 region controls the replication of the virus. Conventionally, the Ad gene transfer vectors have a deletion in E1 and E3 (a nonessential region). The expression cassette containing the promoter and the gene to be transferred are usually inserted into the E1 region. The vectors are produced in the 293 embryonic kidney cell line that provides the E1 information in *trans*, enabling replication of the recombinant vector.

Ad vectors can hold 7–7.5 kb of exogenous sequences (Shenk, 2001). If more space is needed, the E2 or E4 region can also be deleted and the vector made in cell lines providing these deleted sequences. The Ad capsid is an icosahe-

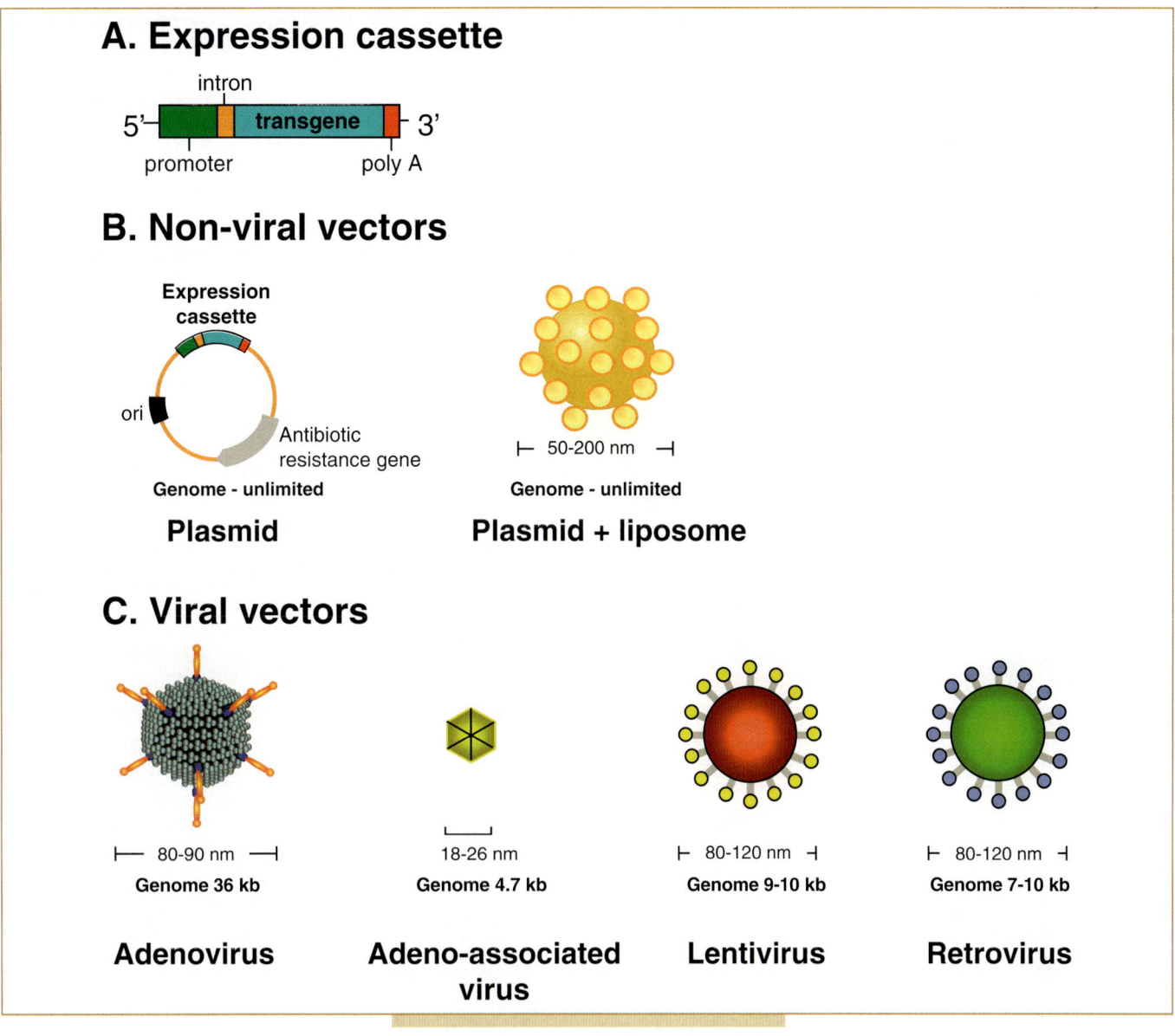

**FIG. 34.3.** Commonly used gene therapy vectors. **(A)** All gene transfer vectors contain an expression cassette with (5′ to 3′) a promoter, usually an intron, the transgene, and a polyA site/stop signal. **(B)** Commonly used nonviral vectors, including naked plasmids (typically comprising an origin of replication, the expression cassette, and an antibiotic-resistance gene) or plasmid combined with a liposome. The plasmid genome can be unlimited in size, but usually the expression cassette is under 10 kb; the liposome/plasmid combination ranges from 50 to 200 nm in diameter. **(C)** Commonly used viral vectors, including adenovirus, adeno-associate virus, lentivirus, and retrovirus. Shown is the size of the genome of each viral vector as well as the relative size of each vector. The size of the expression cassette depends on how much of the viral genome is included.

dral structure composed of 252 subunits, of which 240 are hexons and 12 are pentons. Hexon is the major structural component of the Ad capsid, forming 20 facets of the icosahedron, and is composed of three tightly associated molecules of polypeptide II forming a trimer. Polypeptide IX is associated with the hexon protein and serves to stabilize the structure. Each penton contains a base and a noncovalently projecting fiber. Sequences within the fiber are the primary means by which Ad interact with cells, with the penton base providing secondary attachment sequences.

Several cellular receptors have been identified for Ad vectors, and they differ for various serotypes (Table 34.2). The primary Ad receptor for the subgroup C Ad is the coxsackie adenovirus receptor (CAR) (Bergelson *et al.*, 1997). CAR is expressed on most cell types, and thus Ad group C vectors are capable of transferring genes to most organs. Besides the primary CAR receptor, epitopes in the penton base of the group C Ad vectors use $\alpha_v\beta_3$ (or $\alpha_v\beta_5$) surface integrins as coreceptors for virus internalization (Wickham *et al.*, 1993). Heparan sulfate has also been identified as a

**Table 34.1.** Characteristics of most commonly used gene transfer vectors

| Vector type | Maximum expression cassette capacity (kb) | Transfers genes to nondividing cells | Antivector immunogenicity | Chromosomal integration | Expression | Other characteristics |
|---|---|---|---|---|---|---|
| Plasmids/liposomes | <10[1] | Yes | None | No | Transient | Poor transduction efficiency |
| Adenovirus | 7–8[2] | Yes | High | No | Transient | Typically, mediates expression for 1–3 wk |
| Adeno-associated virus | 4.5 | Yes | Low | No[3] | Persistent in nondividing cell populations | Expression usually takes 1–3 wk to be initiated |
| Retrovirus | 8 | No | None | Yes | Persistent | Risk for insertional mutagenesis; difficult to produce in high titer |
| Lentivirus | 8 | Yes | None | Yes | Persistent | Theoretical safety concerns regarding HIV components; risk for insertional mutagenesis; difficult to produce in high titer |

[1]Plasmid size is generally not limited, but plasmids with >10-kb expression cassette capacity are more difficult to produce with consistent fidelity and thus are generally not suitable for gene human transfer.
[2]Typically, Ad vectors have a 7 to 8 kb capacity for the expression cassette; the capacity can be increased by removing additional viral genes.
[3]Wild-type adeno-associated virus (AAV) integrates in a site-specific fashion into chromosome 19, a process mediated by the *rev* gene; *rev* is deleted in the AAV gene transfer vectors; whether or not there is minimal integration of the genome AAV gene transfer vectors is debated (McCarty *et al.*, 2004; Qing *et al.*, 1999; Summerford *et al.*, 1999; Summerford and Samulski, 1998).

receptor for Ad2 and Ad5, and Ad5 has been shown to bind to vascular cell adhesion molecule 1 (VCAM 1) on endothelial cells and possibly also to MHC class I on the cell surface (Chu *et al.*, 2001; Hong *et al.*, 1997).

The group B serotypes Ad 11, 14, 16, 21, 35, and 50 utilize CD46 instead of CAR, enabling more efficient gene transfer into hematopoietic cells, cells of the urinary tract epithelium, and salivary glands (Sirena *et al.*, 2004). For Ad3 of the subgroup B Ad vectors, CD80 and CD86, which are usually expressed on antigen-presenting cells, have been identified as receptors for viral entry (Short *et al.*, 2006).

In addition to Ad vectors being effective in delivering genes to a wide variety of cell types for therapeutic purposes, Ad vectors interact rapidly with antigen-presenting cells such as dendritic cells, leading to the induction of immunity against the vector and potentially also against the transgene if it is foreign to the host (Hackett *et al.*, 2000; Thomas *et al.*, 2003). When Ad are administered directly in large doses to animals and humans, there is an innate and acquired immune response against the vector, resulting in local inflammation and infiltration of CD4, CD8, and dendritic cells (Hackett *et al.*, 2000). The immune response is

| Table 34.2. | Cell receptors of commonly used viral-based gene therapy vectors | | |
|---|---|---|---|
| *Vector* | *Virus group/serotype*[1] | *Receptor* | *References* |
| Adenovirus | Group C, serotypes 2, 5 | CAR[2]; coreceptors — $\alpha_V$*$b_3$, $\alpha_V$*$b_5$ integrins | Bergelson *et al.* (1997); Wickham *et al.* (1997) |
| | Group C, serotypes 2, 5 | Heparan sulfate | Dechecchi *et al.* (2000, 2001) |
| | Group C, serotype 5 | Vascular cell adhesion molecule I | Chu *et al.* (2001) |
| | Group C, serotype 5 | MHC class I $\alpha 2$ | Hong *et al.* (1997) |
| | Group D, serotype 37 | Sialic acid | Arnberg *et al.* (2000) |
| | Group B, serotypes 3, 11, 14, 16, 21, 35, 50 | CD46 | Segerman *et al.* (2003); Sirena *et al.* (2004) |
| | Group B, serotype 3 | CD 80, CD 86 | Short *et al.* (2006) |
| Adeno-associated virus | AAV 2 | Heparan sulfate proteoglycan; coreceptors — fibroblast growth factor receptor 1, $\alpha_V\beta_5$ integrin | Handa *et al.* (2000); Qing *et al.* (1999); Summerford *et al.* (1999); Summerford and Samulski (1998) |
| | AAV 3 | Heparan sulfate proteoglycan | Dechecchi *et al.* (2001); Handa *et al.* (2000) |
| | AAV4 | Sialic acid (O-linked) | Kaludov *et al.* (2001) |
| | AAV5 | Sialic acid (N-linked); coreceptor — platelet-derived growth factor receptor | Di *et al.* (2003); Walters *et al.* (2001) |
| Retrovirus | MoMLV[3] | Ecotropic | Buchholz *et al.* (1999); Markowitz *et al.* (1988) |
| | 470A MLV, VSV[4] | Amphotropic | Buchholz *et al.* (1999); Markowitz *et al.* (1988) |
| Lentivirus | HIV-1 (envelope protein) | CD4 — coreceptors — CCR5, CXCR4 | Pohlmann and Reeves (2006) |

[1]Adenoviruses are categorized into groups and serotypes; AAV is categorized into clades and serotypes, only the serotypes are listed.
[2]CAR — coxsackie adenovirus receptor.
[3]MoMLV — Moloney murine leukemia virus.
[4]VSV — vesicular stomatitis virus.

multifaceted, consisting of humoral and cellular immunity against both the capsid proteins and the transgene expressed by the vector if it is foreign to the host.

Several strategies have been investigated to circumvent the problem of host responses evoked against Ad gene transfer vectors, including the use of immunosuppressants administered together with the vector or including transgenes expressing immunomodulatory factors to suppress the immune responses against the vector (Hackett *et al.*, 2000). Considerable effort has also been placed on circumventing the host response by designing vectors with larger genomic deletions (e.g., of E1 plus E2 and/or E4 with complementing cell lines) that evoke a milder immune reaction (Hackett *et al.*, 2000; Lundstrom, 2003).

One challenge to the use of Ad vectors is preexisting immunity against the vector resulting from previous infection with a wild-type Ad virus from the same serotype (Hackett *et al.*, 2000). The acquired host responses to Ad

vector administration generally result in the inability to readminister a vector of the same serotype. To circumvent this problem, alternative serotypes can be administered, thus circumventing immunity against the first vector (Mastrangeli *et al.*, 1996). Also, different serotypes to which humans are usually not exposed have been developed as gene transfer vectors. One example is Ad serotype 48, to which humans rarely have preexisting immunity. A chimeric Ad vector containing the hexon loops of Ad5 replaced by those of Ad48 have been shown to be a possible strategy for an Ad-based genetic vaccine approach (Roberts *et al.*, 2006) and could also be envisioned to be useful for tissue-engineering applications.

Another strategy to circumvent preexisting anti-Ad immunity is the use of nonhuman Ad serotypes (Basnight *et al.*, 1971; Farina *et al.*, 2001; Hashimoto *et al.*, 2005). Non-human primate-derived Ad vectors were developed to overcome preexisting immunity to common human Ad serotypes and to broaden the repertoire of Ad when used as vaccines. For example, Ad vectors based on nonhuman primate serotypes C68, C6, and C7 do not circulate in the human population and are therefore not affected by preexisting immunity (Basnight *et al.*, 1971; Farina *et al.*, 2001; Hashimoto *et al.*, 2005).

## Adeno-Associated Virus (AAV)

AAV is a single-stranded DNA virus that belongs to the *Dependovirus* genus of the Parvoviridae family. AAVs were originally isolated as contaminants in laboratory stocks of adenoviruses (Muyzyczka and Berns, 2001). Over recent decades, six different isolates of AAV were characterized, most derived from laboratory stocks of adenoviruses and one isolated from a condylomatous wart. Nine subtypes of AAV have been described for which humans are the primary host (Muyzyczka and Berns, 2001). These isolates were found to be different, based on the antibody response generated against them, and were thus categorized as AAV serotypes 1–9. A substantial portion of humans have detectable antibodies against these serotypes. The exact nature and sequelae of the natural infection with AAV in humans are not known, and there is possibly no human disease associated with AAV.

AAV consists of a single-stranded 4.7-kb genome with characteristic termini of palindromic repeats that fold into a hairpin shape known as *inverted terminal repeats* (ITR) (Muyzyczka and Berns, 2001). During replication into a double-stranded form it expresses genes involved in replication (*rep*) and genes that code for the capsid proteins (*cap*). In the absence of a helper virus such as adenovirus or herpes simplex, wild-type AAV is capable of infecting nondividing cells and integrating its genome into chromosomal DNA at a specific region on chromosome 19, persisting in a latent form. In the gene vectors, *rep* and *cap* are deleted, and most of the vector genome resides and functions in an extrachromosomal location (Muyzyczka and Berns, 2001).

In the AAV vectors, the *rep* and *cap* genes are replaced with an expression cassette. During vector production, the *rep* and *cap* gene products as well as the necessary helpervirus elements (usually Ad-derived) are supplied *in trans*. AAV vectors are commonly produced by transfecting two plasmids into the 293 human embryonic kidney packaging cell line. DNA coding for the therapeutic gene is provided by one plasmid, and the AAV *rep* and *cap* functions plus the Ad helper functions are provided by the second plasmid. Titers are generally significantly lower than those obtained for Ad vectors, but they are sufficient to produce enough vectors for clinical trials.

Numerous studies in animals have been performed to assess the safety and efficacy of AAV-based vectors, and AAV2 serotype–based vectors have been assessed in humans (Carter, 2005). AAV vectors are capable of transducing nondividing cells *in vitro* and *in vivo*. The exact molecular intracellular state of the vector genome has not been completely elucidated, but there is little evidence that the vectors integrate when used in gene therapy *in vivo*. Most transgene expression is thought to be derived from extrachromosomal viral genomes that persist as double-stranded circular or linear episomes (Nakai *et al.*, 2000). This limits the usefulness of AAV for applications involving dividing cells, such as stem cells, since only one daughter cell will receive the vector genome (Nakai *et al.*, 2001).

One disadvantage of AAV vectors is their packaging capacity, limited to expression cassettes of about 4.5 kb. This size limitation is a challenge for the use of AAV vectors for clinically relevant large transgenes, such as dystrophin (11 kb) for muscular dystrophy or factor VIII (7–9 kb) for hemophilia A. Various efforts have been undertaken to deliver larger transgenes using AAV vectors, including co-administration of two vectors, each carrying one-half of the transgene, leading to intermolecular recombination during concatamerization, an intermediate state of the vector genome. *Trans*-splicing may also be employed as a strategy to circumvent the need to package the full-length cDNA into the AAV vector. Instead, delivering the full coding sequences, the gene is split into two pieces that then combine at the pre-mRNA level (Pergolizzi and Crystal, 2004).

The host immune response against AAV vectors is not as strong as that observed with Ad vectors (Hackett *et al.*, 2000; Jooss and Chirmule, 2003). AAV vectors generate humoral immunity against the capsid proteins, which impairs readministration of a vector of the identical serotype. Cellular immunity against AAV has been detected following administration to experimental animals and humans. But, unless high doses are used, there usually are no destructive cytotoxic T-cell responses generated against the vector or the transgene (Jooss and Chirmule, 2003).

There is usually persistent expression of the transgene directed by AAV vectors, particularly in organs with nondividing (or slowly dividing) cells, such as liver, muscle, heart, retina, and brain. The overall expression levels of the transgene appear to be lower with AAV than with Ad and are dependent on the target organ. Recently, novel AAVs have been identified from tissues of nonhuman primates and humans (Gao *et al.*, 2004, 2002). These AAVs have substantial heterogeneity in the capsid genes and are useful as chimeric capsids combined with the AAV2 genome, leading to improved infection of organs or tissues previously not considered to be valuable targets for AAV gene transfer.

AAV interacts with its target cells by binding to cell surface receptors (Table 34.2). For AAV2, the primary attachment receptor is heparan sulfate proteoglycan (Summerford and Samulski, 1998). AAV3 may share heparan sulfate proteoglycan as the primary attachment receptor. However, AAV3 may use other receptors, for AAV3 has been shown to infect hematopoietic cells, which were not effectively infected by AAV2 (Handa *et al.*, 2000). AAV4 and AAV5 use sialic acid as the primary attachment receptor. AAV4 uses O-linked sialic acid, whereas AAV5 uses N-linked sialic acid (Kaludov *et al.*, 2001; Walters *et al.*, 2001). For serotypes AAV1 and AAV6, no receptors have been identified (Halbert *et al.*, 2001), nor have the receptors for the nonhuman primate AAVs (Gao *et al.*, 2002).

AAV2 also uses coreceptors for efficient infection, including $\alpha_V\beta_5$ integrin (Summerford *et al.*, 1999) and fibroblast growth factor-1 (Qing *et al.*, 1999). Efficient infection with AAV5 appears to require a coreceptor; platelet-derived growth factor has been identified as a possible coreceptor for AAV5 but may also be able to act as the primary receptor (Di *et al.*, 2003).

## Retrovirus

Retrovirus and lentivirus vectors are both RNA-based vectors belonging to the family of retroviridae. Because lentiviruses are capable of infecting nondividing cells, they are discussed seperately herein. The original retroviral vectors used for gene therapy were based on endogenous murine viruses. Of these, the Moloney murine leukemia retrovirus (MMLV) was the first widely used gene transfer vector and was the first to be used to treat a hereditary disorder using an *ex vivo* strategy (Blaese *et al.*, 1995).

The genome of the retrovirus vector is a 7- to 10-kb single-stranded RNA containing long terminal repeats (LTRs) on both ends that flank *rev, gag, pol,* and other regulatory genes that are required for viral function. The RNA genome of the replication-deficient retroviral vectors contain an expression cassette up to 8 kb that replaces all viral protein–coding sequences. The LTRs flank the expression cassette and allow transcription initiation by host cell factors. The vectors are rendered self-inactivating by deletion of the promoter and enhancer in the 3' LTR, to prevent LTR-driven transcription. The packaging of the genomic RNA is controlled in *cis* by the packaging signal Ψ. The production of the retrovirus vectors requires a packaging or producer cell line in which the viral *gag, pol,* and *env* proteins are expressed in *trans* from separate helper products. Recombination between the helper constructs and the vector can be minimized by using nonretroviral regulatory sequences to control expression (Cosset *et al.*, 1995). Enhancer and promoter sequences can be deleted from the 3' LTR to create a transcriptionally silent 5' LTR during infection of the target cells. This strategy provides the basis for self-inactivating vectors and can also be used for the substitution with tissue-specific promoters (Yu *et al.*, 1986). The main reason why MMLV viruses can infect only nondividing cells is that they are unable to cross the nuclear membrane and can only achieve completion of the infection with provirus integration during cell division.

Retroviruses enter the cells via cell fusion of the envelope protein with the cell membrane. The murine viruses are able to infect only murine cells (ecotropic), whereas derivatives of MMLV or human retroviruses, such as vesicular stomatiitis virus (VSV), can infect both human and murine cells (amphotropic; Table 34.2). Providing the virus with a different coat, a process referred to as *pseudotyping,* can change the specificity of binding and entry into target cells. Over the past decade, there has been considerable effort in the development of pseudotyping strategies of retroviral vectors (Cosset *et al.*, 1995; Markowitz *et al.*, 1988; Miller *et al.*, 1991).

Most of the gene transfer strategies for retroviral vectors have been *ex vivo* approaches, due to the difficulty in producing high-titer concentrated preparations and the rapid inactivation of the retroviral vectors *in vivo* by complement. Due to their ability to infect rapidly dividing cells, retroviral vectors have been used extensively to develop gene transfer strategies to hematopoietic cells. The first clinical trial to treat a hereditary disorder used a retrovirus to transduce T-cells of patients with adenosine deaminase — severe combined immunodeficiency (SCID) (Blaese *et al.*, 1995).

One advantage of using retroviral vectors is the permanent integration of the vector genome into the host genome, providing long-term and stable expression of the transgene. This, however, also carries the greatest risk of retroviral vectors, in that they may induce insertional mutagenesis, with the subsequent development of malignancies. This has been observed in a clinical trial using an MMLV-based retroviral vector to correct hematopoietic stem cells to treat X-linked SCID (Cavazzana-Calvo *et al.*, 2005). Although long-term correction of the immunodeficiency has been observed in the study subjects, three children developed a clonal lymphoproliferative syndrome similar to acute lymphoblastic leukemia. In two of these cases, there was integration of the retroviral vector near the LIM-domain-only-2 (LMO2) promoter.

## Lentivirus

Based on the knowledge that the human retrovirus HIV-1 is able to infect nondividing cells such as macrophages and neurons, replication-defective versions of HIV were developed that were capable of infecting nondividing cells and achieving stable long-term expression through integration of the provirus into chromosomal DNA (Buchschacher and Wong-Staal, 2000). The genome of a lentivirus vector is a 9- to 10-kb single-stranded RNA genome, containing components of HIV-1 but otherwise similar to that of retroviral vectors. All viral protein–coding sequences are deleted and are replaced with an expression cassette that can be up to 8 kb.

Unique to lentiviruses are the central polypurine tract (cPPT) and central termination sequences (CTS), *cis*-acting sequences that coordinate the formation of a central DNA flap, which improves nuclear import (Sirven *et al.*, 2000). This may explain the increased transduction efficiency of lentiviral as compared to retroviral vectors. A *rev*-responsive element can be incorporated into the vector to facilitate the nuclear export of unspliced RNA. The packaging of the genomic RNA is controlled in *cis* by the packaging signal Θ.

Despite the greater complexity of the lentivirus genome, the basic principles of generating vectors free of a replication-competent virus are similar to those for retroviral vectors. Packaging systems use the HIV *gag* and *pol* genes, with or without the *rev* gene (Klages *et al.*, 2000; Kotsopoulou *et al.*, 2000). The HIV virulence genes *tat*, *vif*, *vpr*, *vpu*, and *nef* are completely absent from lentiviral vectors, thus making it theoretically impossible that a virus similar to HIV can be inadvertently produced (Kim *et al.*, 1998). The production of lentiviral vectors in high concentrated titers is still a challenge, but the production issues are slowly being solved (Buchschacher and Wong-Staal, 2000).

The cellular receptors for the lentivirus vectors are the same as for HIV1-1, including CD4, together with the chemokine coreceptors CCR5 and CXCR4 (Table 34.2). However, the lentiviral gene transfer vectors are usually pseudotyped with envelopes from other viruses, such as VSV-G, mediating the binding and entry into target cells similar to retroviral vectors. Efficient gene transfer of lentiviral vectors has been reported for a variety of dividing and nondividing cell types, including muscle, neurons, glia, and hematopoietic cells. Lentivirus vectors have been used successfully to correct disease phenotypes in experimental animals for CNS disorders such as metachromatic leukodystrophy using *in vivo* administration of the lentiviral vector to the brain (Consiglio *et al.*, 2001) or, using an *ex vivo* strategy, by infecting hematopoietic stem cells with the β-globin gene to correct β-thalassemia (Sadelain *et al.*, 2005).

Like retroviral vectors, insertional mutagenesis is a concern for the clinical use of lentiviral vectors, and liver cancers have been observed in mice infected *in utero* or neonatally with lentiviral vectors (Themis *et al.*, 2005). One hypothesis for this phenomenon is based on potentially oncogenic sequences present in the woodchuck hepatitis virus–derived posttranscriptional regulatory element (WPRE) that was included in the vector to increase mRNA stability (Themis *et al.*, 2005). The only clinical trial using lentiviral vectors that has been initiated is in patients infected with HIV receiving *ex vivo* lentivirus-modified autologous CD4 T-cells to treat the HIV-1 infection (MacGregor, 2001).

## VI. CELL-SPECIFIC TARGETING STRATEGIES

Viral gene transfer vectors use the receptors of their wild-type versions to enter into cells (Table 34.2). Modification of gene transfer vectors to target specific cell types or tissues is an attractive means to increase the specificity of the vectors to the target cell and may also enable the vectors to infect cells that are usually not infected by unmodified vectors. In general, targeting modifications can be accomplished either by genetic modification of the vector genome to change the properties of the outer surface of the vector or by chemical modification of the vectors via the addition of ligands (Table 34.3). Based on the targeting strategy used, the range of tropism can be widened or narrowed (Fig. 34.4).

## Targeting of Ad Vectors

Strategies have been developed to redirect Ad tropism and to enhance Ad tropism for cells difficult to transfer genes to because of lack of Ad receptors. In general, Ad vectors exhibit a broad tropism due to the widespread expression of the primary Ad receptor CAR and the secondary integrin receptors $\alpha_V\beta_3$ and $\alpha_V\beta_5$. While the widespread expression of the Ad receptors enables the efficient infection of a wide range of target cells, it poses the problem of unwanted uptake and gene expression in nontarget tissue when the vectors are administered *in vivo*. Some tissues have low expression of CAR (e.g., endothelial cells, antigen-presenting cells, and some tumor cells), which limits the use of Ad vectors for these targets.

Most genetic targeting strategies for Ad vectors have been focused on ablating CAR binding and have introduced new peptides or other ligands to the fiber knob domain, the primary site for the interaction of group C Ad vectors with CAR (Nicklin *et al.*, 2005). Fiber modifications to modify target–cell binding of Ad vectors include the introduction of poly(L) lysine to allow binding to heparan-containing receptors and the integrin-binding motif RGD, which is essential for penton-mediated internalization to allow integrin-mediated binding and uptake (Wickham *et al.*, 1997). Peptides can also be incorporated into other sites of the Ad capsid to achieve retargeting, such as incorporation of RGD into the hexon (Vigne *et al.*, 1999) and poly(L) lysine into polypeptide IX (Dmitriev *et al.*, 2002).

Another approach of genetic retargeting of virus vectors is to create chimeras of different serotypes, which are known to use different cellular receptors. Replacing the fiber or fiber knob domain of the Ad5-based vector with that of Ad3 or Ad7 has been shown to achieve CAR-independent infectivity (Gall *et al.*, 1996; Krasnykh *et al.*, 1996; Stevenson *et al.*, 1995). Replacement of the fiber of Ad2 with that of Ad17 has led to improved infectivity of airway epithelial cells (Zabner *et al.*, 1999). Replacement with the fiber of Ad35 achieved improved infectivity of hematopoietic cells (Saban *et al.*, 2005; Shayakhmetov *et al.*, 2002), and Ad16 enhanced infectivity of cardiovascular tissue (Havenga *et al.*, 2001). Ad5 vectors pseudotyped with fibers of the subgroup D Ad 19 and 37 increased the infectivity of endothelial and smooth muscle cells (Denby *et al.*, 2004). Recognition peptides for fiber modification have been identified by phage display (Nicklin *et al.*, 2000; Pereboev *et al.*, 2001; Xia *et al.*, 2000), and other complex genetic modifications of the fiber to retarget the Ad vector have been reported (Hong *et al.*, 2003; Krasnykh *et al.*, 2001). These cell-targeting modifications can also be combined with the use of tissue-specific promoters to achieve selective infection and transcription in the targeted cell type. For example, inserting the RGD motif into the fiber has been combined with the use of the endothelial cell–specific Flt-1 promoter and resulted in more specific infection and gene expression in endothelial cells

**Table 34.3.  Alteration of viral gene transfer vector cell targeting by modifying capsid/envelope structure**

| Modification | Viral vector | Examples | Purpose/target | References |
|---|---|---|---|---|
| Genetic alteration of capsid/envelope proteins | Adenovirus | Delete CAR[1] | CAR-independent infection | Kirby et al. (1999) |
| | | Add RGD[2] to fiber | $\alpha_v\beta_3$, $\alpha_v\beta_5$ integrins | Wickham et al. (1997) |
| | | Add polylysine to fiber or protein IX | Broadening infectivity by targeting heparan-containing molecules | Dmitriev et al. (2002); Wickham et al. (1997) |
| | | Replace Ad5 fiber with Ad19 or Ad37 fiber | Improved infection of smooth muscle and endothelial cells | Denby et al. (2004) |
| | | Replace Ad5 fiber with Ad37 fiber | Infection of hematopoietic cells | Shayakhmetov et al. (2000) |
| | | Replace Ad5 fiber knob with Ad3 fiber knob | CAR-independent infection | Krasnykh et al. (1996); Stevenson et al. (1997) |
| | | Replace Ad5 fiber with Ad7 fiber | CAR-independent infection | Gall et al. (1996) |
| | | Replace Ad2 fiber with Ad17 fiber | Improved infection of airway epithelial cells | Zabner et al. (1999) |
| | | Replace Ad5 fiber with Ad16 fiber | Improved infection of smooth muscle and endothelia cells | Havenga (2001) |
| | | Replace Ad5 fiber with trimerization motif of phage T4 fibritin | Chimeric fiber phage fibritin molecules targeted to artificial receptors | Krasnykh et al. (2001) |
| | | Replace Ad5 fiber knob with CAV2[3] fiber knob | Target to CAR-deficient cells | Glasgow et al. (2004) |
| | | Replace Ad5 fiber with fiber of ovine AdV7 | Targeting to kidney and "detargeting" of CAR | Nakayama et al. (2006) |
| | | Add VSV[4]-G epitope to Ad5 fiber knob | Targeting tropism to CAR-deficient cell expressing phosphatidylserine | Yun et al. (2003) |
| | | Replace 7 hypervariable regions (HVR) of Ad5 hexon with HVR of Ad48 | Circumvention of anti-Ad5 hexon immunity for Ad-based HIV vaccine | Roberts et al. (2006) |
| | Adeno-associated virus | Pseudotyping entire capsid with different serotype | Targeting tropism based on capsid serotype | Rabinowitz et al. (2002) |
| | | Pseudotyping with capsid of multiple serotypes | Generate mosaic AAV to target different serotype tropisms | Bowles et al. (2003); Hauck et al. (2003) |
| | | Addition of L14[5] to AAV2 capsid | Targeting to L14-binding integrins | Girod et al. (1999) |
| | | Incorporation of tumor-targeting peptide to AAV2 capsid | Targeting to CD13 | Grifman et al. (2001) |
| | | Insertion of serpin receptor ligand to AAV2 capsid | Targeting to serpin receptor | Wu et al. (2000) |
| | | Ad5 capsid with Ad37 fiber and partial AAV genome | Vector with increased capacity and tropism for hematopoietic cells | Shayakhmetov et al. (2002) |

**Table 34.3.** Alteration of viral gene transfer vector cell targeting by modifying capsid/envelope structure—cont'd

| Modification | Viral vector | Examples | Purpose/target | References |
|---|---|---|---|---|
| | Retrovirus | Pseudotyping entire envelope with envelope of vesicular stomatitis virus, Gibbon ape leukemia virus, murine leukemia virus | Targeting tropism of new envelope (from ecotropic to amphotropic) | Cosset et al. (1995); Markowitz et al. (1988); Miller et al. (1991) |
| | | Addition of peptides:<br>— Erythropoietin<br>— Heregulin<br>— Epidermal growth factor | Erythropoietin receptor<br>Heregulin receptor<br>Epidermal growth factor receptor | Han et al. (1995); Kasahara et al. (1994); Tai et al. (2003) |
| | Lentivirus | Pseudotyping entire envelope with envelope of vesicular stomatitis virus G | Improves infectivity to muscle | Gregory et al. (2004) |
| | | Rabies 6 envelope<br>Ebola | Improves infectivity to motor neurons<br>Improves infectivity for airway epithelial cells | Mazarakis et al. (2001); Kobinger et al. (2001) |
| | | Hantavirus<br>Fowl plague hemagglutinin (with expression of influenza M2 protein) | Improves infectivity to endothelium<br>Improves infection in airway epithelial cells | Qian et al. (2006); McKay et al. (2006) |
| Chimerical modifications of capsid/envelope | Adenovirus | Cationic lipids, cholesterol, polycationic polymers<br>Polyethylene glycol<br>Coupling of U7 peptide to Ad5 | Modify tropism<br>Decrease immunity<br>Targeting of urokinase plasminogen activator receptor on airway cells | Fasbender et al. (1997); Worgall et al. (2000)<br>Drapkin et al. (2000) |
| | Adeno-associated virus | Addition of single-chain antibodies<br>Chemical conjugation of avidin-linked ligands to AAV2<br>Incorporation of bispecific antibody to AAV2 | Target antibody ligand–expressing cells<br>Targeting to ligand receptors<br>Targeting to human megakaryocytes | Hedley et al. (2006)<br>Ponnazhagan et al. (2002)<br>Bartlett et al. (1999) |
| | Lentivirus | Polyethylene glycol | Prolonging half-life in serum by preventing inactivation | Croyle et al. (2004) |
| | Retrovirus | Addition of single-chain antibodies | Target antibody ligands | Marin et al. (1996) |

[1] CAR — coxsackie adenovirus receptor.
[2] RGD — integrin-binding peptide arginine (R), glycine (G), aspartate (D).
[3] CAV2 — canine adenovirus 2 (McCarty et al., 2004; Qing et al., 1999; Summerford et al., 1999; Summerford and Samulski, 1998).
[4] VSV — vesicular stomatitis virus (McCarty et al., 2004; Qing et al., 1999; Summerford et al., 1999; Summerford and Samulski, 1998).
[5] L14 — integrin-binding motif.

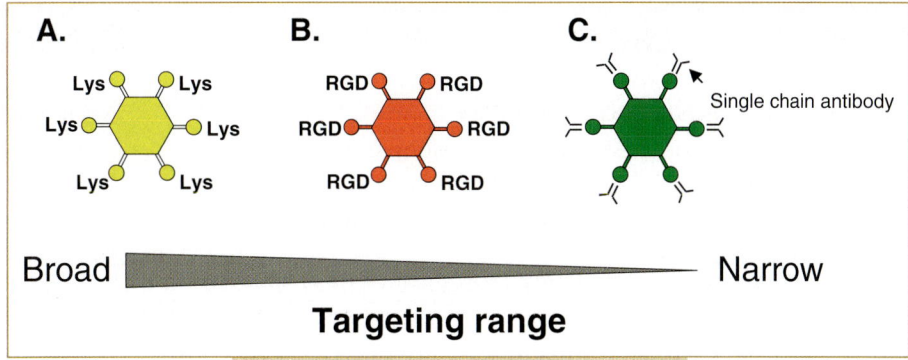

**FIG. 34.4.** Examples of modifications of tropism of gene transfer vectors. Shown are three examples of modification of adenovirus vector tropism. **(A)** The addition of polylysine to the Ad fiber protein provides broad tropism. **(B)** The addition of the integrin-binding motif RGD enhances targeting of cells expressing $\alpha_v,\beta_{3,5}$ integrins. **(C)** The addition of single-chain antibodies to the Ad fiber targets cells expressing the antibody ligand. Strategies A ≡ C provide broad to narrow cell-specific targeting.

(Work *et al.*, 2004). Finally, a capsid-modified Ad/AAV hybrid vector was able to achieve long-term expression in human hematopoietic cells (Shayakhmetov *et al.*, 2002).

Besides genetic modification of gene therapy vectors to modify tropism and target vector uptake to a particular cell type, chemical modifications of Ad vectors have been utilized for targeting. Ad vectors have been complexed to cationic lipids, polycationic polymers, or cholesterol to increase the efficiency of gene transfer *in vitro* and *in vivo* (Fasbender *et al.*, 1997; Worgall *et al.*, 2000). Bispecific monoclonal antibodies have been added to the fiber to target specific cell types expressing the ligand for that antibody.

### Targeting of AAV Vectors

The strategy of creating AAV chimeras by pseudotyping with capsids of different AAV serotypes has been used to broaden the tropism of AAV vectors (Bowles *et al.*, 2003; Hauck *et al.*, 2003; Rabinowitz *et al.*, 2002). Most of these vectors used the vector genome derived from AAV2 with the capsid from another serotype. For example, this strategy has enabled enhanced transduction in lung-directed gene transfer using an AAV5 pseudotype (Zabner *et al.*, 2000).

### Targeting of Retroviral and Lentiviral Vectors

The classic method to broaden the tropism of retroviral and lentiviral vectors is by pseudotyping, creating chimeras using envelope glycoproteins from other viruses. Most retrovirus vectors are based on MuMoLV, an ecotropic virus, which infects only murine cells. To achieve infection of human cells, the vectors are propagated in packaging cells that express the envelope of the amphotropic or nonmurine viruses, such as 4070A murine leukemia virus, gibbon ape leukemia virus, vesicular stomatitis virus (VSV), or the feline endogenous virus RD114 (Markowitz *et al.*, 1988; Miller *et al.*, 1991). Various envelopes have also been used for lentiviral vectors to increase infectivity of specific cell types, including VSV-G for muscle (Gregory *et al.*, 2004), Ebola for

airway epithelial cells (Kobinger *et al.*, 2001; Medina *et al.*, 2003), rabies-G for motor neurons (Mazarakis *et al.*, 2001), and hantavirus for endothelium (Qian *et al.*, 2006). Hybrid proteins of the murine amphotropic envelope have been combined with the extracellular domains of GALV or RD114 envelope to enhance infection of CD34+ cells (Sandrin *et al.*, 2002). As with Ad vectors, strategies have been developed to target retro- and lentiviral vectors by addition of ligands to the envelope glycoprotein. Examples include peptide sequences from erythropoietin (Kasahara *et al.*, 1994), heregulin (Han *et al.*, 1995), epithelial growth factor and ligands for the Ram-1 phosphate transporter (Cosset *et al.*, 1995), as well as the addition of single-chain antibodies (Marin *et al.*, 1996; Tai *et al.*, 2003). The efficiency of these modifications, however, has not been very high, and the vector production yield is significantly impaired.

Another strategy has been to utilize the membrane proteins that are incorporated during the budding process for targeting. For example, incorporation of the membrane-bound stem cell factor not only provided a growth signal for the CD34+ cells expressing c-kit, the receptor for stem cell factor, but also led to increased infection efficiency of the CD34 cells (Chandrashekran *et al.*, 2004). Monomethoxy poly(ethylene) glycol conjugated to VSV-G protects the vector from inactivation in the serum, leading to a prolonged half-life and increased transduction of bone marrow following intravenous administration in mice (Croyle *et al.*, 2004). Another strategy for targeting of retro- and lentiviruses *in vivo* has been to target retrovirus producer cells (Crittenden *et al.*, 2003).

## VII. REGULATED EXPRESSION OF THE TRANSFERRED GENE

A variety of strategies have been developed to regulate expression of the genes transferred by gene therapy vectors (Table 34.4). The ability to regulate gene expression is

| Table 34.4. | Regulation of expression of the transferred gene | | |

| Category | Strategy | Note | References |
|---|---|---|---|
| Inducible promoter | Glucocorticoid — responsive | Multiple response elements | Narumi *et al.* (1998) |
| | cGMP[1] — responsive | Multiple response elements | Suzuki *et al.* (1996, 1997) |
| | Heat shock protein — inducible | Hyperthermia and cellular stress induce gene expression | Isomoto *et al.* (2006); Luo *et al.* (2004) |
| | Radiation — inducible | | Kufe and Weichselbaum (2003) |
| | Insulin — responsive | | X. P. Wang *et al.* (2004) |
| | Tetracycline — responsive | Repressible-$\text{TET}_{\text{off}}$ or inducible-$\text{TET}_{\text{on}}$ systems | Aurisicchio *et al.* (2001); Fitzsimons *et al.* (2001); Haberman *et al.* (1998); Hofmann *et al.* (1996); Kafrie *et al.* (2000); Reiser *et al.* (2000); Rendahl *et al.* (2002) |
| | Antibiotic resistance | Streptogramin-class antibiotic (e.g., pristinamycin) induces pristinamycin-induced protein, preventing expression | Fussenegger *et al.* (2000) |
| | | Erythromycin binds to prokaryotic DNA–binding protein MphR (A)[2] | Weber *et al.* (2002) |
| | Chemical-induced dimerization (FKBP/FRAP[3]) | Rapamycin induces heterodimerization of FKBP and FRAP | Auricchio *et al.* (2002a, 2002b); Pollock and Clackson (2002) |
| | Steroid receptor | Transactivator (GLVP[4] or Glp65) targeting genes with GAL-4-binding site in the presence of mifepristone (RU486) | Ngan *et al.* (2002); Y. Wang *et al.* (1994) |
| | Insect ecdysone receptor | Ectodysone receptor ligand induces transactivation of transgene | Hoppe *et al.* (2000); Karzenowski *et al.* (2005) |
| Tissue-specific promoter | Liver | Albumin, α1-antitrypsin, liver-activated protein (LAP), transthyretin promoters | Costa and Grayson (1991); Pastore *et al.* (1999); Talbot *et al.* (1994) |
| | Smooth muscle | Smooth muscle actin, SM-22, smooth muscle myosin heavy-chain promoter | Dean (2005); Ribault *et al.* (2001) |
| | Prostate | Prostate-specific antigen promoter | Chapel-Fernandes *et al.* (2006); Latham *et al.* (2000) |
| *Trans*-splicing | Therapeutic *trans*-splicing | Target pre-mRNA is *trans*-spliced into independent pre-mRNA | Liu *et al.* (2005); Tahara *et al.* (2004) |

[1]cGMP — cyclic guanosine monophosphate.
[2]MphR/A — a prokaryotic DNA–binding protein that binds to a 35-bp operon sequence.
[3]FKBP (McCarty *et al.*, 2004) — FK506-binding protein; FRAP — FK506-binding protein rapamycin-binding).
[4]GLVP — a mifepristone-activated chimeric nuclear receptor.

particularly important for application where too much expression of the gene transfer product could lead to unwanted effects, e.g., sustained expression of a growth factor, which could potentially be tumorigenic. For *in vivo* strategies of gene transfer, tissue-specific regulation of gene expression may be warranted to avoid expression of genes in undesired cells or tissues. In the context of the use of gene transfer vectors to genetically modify stem cells, regulation of gene expression may be critical to avoid differentiation into an unwanted tissue or cell type. The regulation of the gene expression of the transgene could also be combined with inducible systems of suicide genes or factors that could destroy the genetically modified cells, should unwanted differentiation occur.

To turn gene expression on and off at will, a number of inducible promoters and inducible regulated systems have been developed that are applicable to be used in gene transfer vectors (Table 34.4). For example, inducible gene expression in gene transfer vectors can be achieved using inducible promoters, such as promoters responsive to glucocorticoids, cGMP, heat shock protein, radiation, and insulin. Inducible regulated systems also include systems based on response to tetracycline, antibiotic resistance, chemical-induced dimerization, steroid receptors, and insect ecdysone receptors. The basic mechanism for these systems is a combination of ligand-binding synthetic inducer or repressor proteins and a promoter control system that regulates transgene expression.

The tetracycline-responsive system has been widely used to study gene function and to generate conditional mutants in cell lines and transgenic animals (Gossen *et al.*, 1995). The transgene is placed behind a promoter that also contains binding sites for the tetracycline response element (TRE), which can act as a repressor or inducer of transgene expression. A fusion protein that binds to the inducer doxycycline needs to be present on a separate gene construct. The tetracycline transactivator binds to the TRE and activates the transcription in the absence of doxycycline. Upon addition of doxycycline, the expression is turned off (Tet-off). Another fusion protein that can be used is the reverse tetracycline transactivator, which only binds to TRE in the presence of doxycycline, causing induction of expression upon addition of doxycycline (Tet-on). Expression can be controlled in a graded manner; the more doxycycline is added, the greater the level of suppression or induction. The disadvantage of this system is the potential side effects of doxycycline. The tetracycline system has been used for regulated gene transfer with gene therapy vectors, including Ad (Aurisicchio *et al.*, 2001; Molin *et al.*, 1998), AAV (Fitzsimons *et al.*, 2001; Haberman *et al.*, 1998; Rendahl *et al.*, 2002), retroviral (Paulus, 1996; Hofmann *et al.*, 1996), and lentiviral vectors (Kafri *et al.*, 2000; Reiser *et al.*, 2000).

The steroid receptor systems use a mifepristone-binding progesterone receptor fused to the DNA-binding domain of the yeast GAL4 protein and the transactivation domain from the NfkB p65 subunit (Ngan *et al.*, 2002; Y. Wang *et al.*, 1994). The fusion protein binds to a GAL4-activating sequence to regulate gene expression. Upon addition to mifepristone, gene expression is induced; upon removal of the drug, gene expression returns to baseline within five days.

The insect ecdysone receptor system consists of a fusion protein of the transactivation domain of the glucocorticoid receptor fused to the ecdysone-binding nuclear receptor and an ecdysone-response element placed upstream of the promoter driving the transgene expression (Hoppe *et al.*, 2000; Karzenowski *et al.*, 2005). Upon addition of ecdysone, an insect hormone with no mammalian homologs, the fusion protein dimerizes and induces expression. Because there are no known mammalian factors binding to the insect protein, there is very low background expression in the absence of the drug, and expression can be very tightly controlled.

Another strategy to control gene expression at the desired location is the use of tissue-specific promoters (Table 34.4). The majority of tissue-specific promoters have been used to target expression to liver and the cardiovascular system by targeting muscle cells. The challenge in the use of tissue-specific promoters is that the level of expression is usually lower than for the commonly used strong viral promoters. However, viral promoters such as CMV have been shown to be subject to silencing after several weeks *in vivo*. This has been seen in airways, cardiomyocytes, and smooth muscle cells, in particular with nonviral gene transfer (Dean, 2005).

Another modality to regulate gene expression is by *trans*-splicing at the pre-mRNA level. In therapeutic *trans*-splicing, the sequence of the target pre-mRNA is modified by being *trans*-spliced to an independent pre-mRNA, the sequences for which are delivered exogenously by a gene transfer vector (Pergolizzi and Crystal, 2004; Puttaraju *et al.*, 1999). Therapeutic *trans*-splicing can be used to alter coding domains, to create novel fusion proteins, to direct gene products to various cellular compartments, and to enable gene therapy with large genes or genes coding for toxic products. *Trans*-splicing gene transfer strategies also offer the advantage that the expression of the *trans*-spliced sequence is controlled by endogenous regulation of the target pre-mRNA. *Trans*-splicing strategies have been used to correct animal models of hemophilia, X-linked immunodeficiency with hyper IgM, and cystic fibrosis (Chao *et al.*, 2003; Liu *et al.*, 2002; Tahara *et al.*, 2004).

## VIII. COMBINING GENE TRANSFER WITH STEM CELL STRATEGIES

In general, stem cells are categorized into embryonal stem cells, the bone marrow–derived stem/progenitor cells, including mesenchymal stem cells (MSC), endothelial progenitor cells (EPC), and the tissue-derived stem cell populations (Bryder *et al.*, 2006). Stem cells offer the potential for tissue regeneration, and combining gene therapy and stem cell approaches is a promising strategy to direct the differentiation of stem cells in the desired cell type and to regulate and control growth and differentiation of the stem cell. Stem cells have the potential for both self-renewal and differentiation and are dependent on signals from their microenvironment that direct stem cell maintenance or differentiation. The precise spatial and temporal presentation of these signals is critically important in development and will most likely be the decisive factor in the success of stem cell therapies (Alsberg *et al.*, 2006). As the potential of stem cells to be used clinically becomes more realistic, gene transfer strategies may play a pivotal role in controlling stem cell growth, by either preventing uncontrolled growth or directing differentiation into specific cell types, and regulating gene expression. In addition, as more of the microenvironmental cues that control stem cell differentiation into the desired phenotypes become known, the use of gene transfer strategies to express or inhibit these signals becomes a viable strategy to support stem cell therapy. Three general areas can be currently envisioned for gene transfer to be helpful for stem cell therapies: (1) to control unwanted stem cell growth or differentiation, (2) to provide environmental cues for stem cell differentiation, and (3) to regulate gene expression. In addition gene transfer may also be useful in marking stem cells as a modality to track the cells and their progeny *in vivo* (Yoneyama *et al.*, 2006).

## Gene Transfer to Stem Cells

The challenges in developing gene transfer strategies for stem cells are: (1) how most efficiently to infect the cells; (2) gene silencing during differentiation; (3) whether gene expression should be transient or persistent; and (4) if gene expression is persistent, whether it should terminate once the cell has differentiated into the desired phenotype. Gene transfer has been accomplished for a variety of stem cell types. Human and murine embryonal stem cells have been successfully infected, with subsequent gene expression, with adenovirus, AAV, and lentiviral vectors (Wobus and Boheler, 2005). The reported efficiency for each of these vector systems is limited, however, requiring selection to obtain pure populations of genetically modified cells. Gene expression can be affected by subsequent differentiation of the cells, and it is not clear how the stem cell properties are affected by the various gene therapy modalities themselves without transgene expression. Bone marrow–derived endothelial progenitor and mesenchymal stem cells have been successfully transduced with the most commonly used gene therapy vectors (Kaji and Leiden, 2001).

## Gene Transfer to Control Uncontrolled Stem Cell Growth

Gene transfer strategies developed to control the growth of cells for cancer gene therapy applications can be applied to controlling unwanted growth of stem cells, in particular the risk of the development of teratomas with embryonic stem cells. Some of the suicide gene transfer strategies, including the prodrug strategies using herpes simplex thymidine kinase (HSV-TK) (ganciclovir) and cytosine-deaminase (5-fluorocytosine) that lead to activation of cell death following administration of the prodrug may prove useful. Other suicide genes may be used with regulatable gene expression systems (as outlined earlier, Table 34.4) to control potential malignant stem cell growth, as long as there is no "leak" of baseline gene expression of the suicide gene.

## Gene Transfer to Instruct Stem Cell Differentiation

Gene transfer strategies may prove useful in providing the environmental factors necessary to develop specific phenotypes from stem cells. For example, expression of growth factors or other known differentiation signals secreted by the stem cell should aid in mediating differentiation; i.e., the genetically modified stem cell would create its own favorable microenvironment for differentiation. Expression of these factors could be regulated by inducible gene expression systems. For example, for muscle-derived mesenchymal stem cells, with their potential for the treatment of skeletal, cardiac, and smooth muscle injuries and disease, the cytokines required for the differentiation in the respective lineage have been identified (Deasy *et al.*, 2002).

## Gene Transfer to Regulate Gene Expression

Regulation of gene expression is critically important in the maintenance and differentiation of stem cells, and the transfer of genes coding for factors that regulate endogenous gene expression in response to specific stimuli could prove very helpful for stem cell therapies. This concept has been used in gene transfer strategy to regulate angiogenesis in the ischemic myocardium. AAV vector–mediated transfer of a hypoxia, a responsive element to ischemic myocardium, resulted in endogenous expression of the vascular endothelial growth factor (VEGF) (Su *et al.*, 2002). Similarly engineered transcription factors capable of activating endogenous VEGF expression have been successfully transferred with adenoviral vectors (Rebar *et al.*, 2002). As more of the gene expression and regulation patterns critical for stem cell differentiation and maintenance are known, the possibilities to direct or aid in gene expression using gene transfer are significant.

One potential barrier to using embryonic stem cells in humans is rejection of the transplanted cell by the immune system (Fairchild *et al.*, 2004). This could theoretically be circumvented by using gene transfer with the relevant gene to autologous stem cells. For example, transfer of a genetically corrected nucleus to an enucleated egg from an unrelated donor would result in the generation of genetically modified embryonic stem cells, which could be then differentiated and the corrected, differentiated cells transplanted to the same patient (Fig. 34.5).

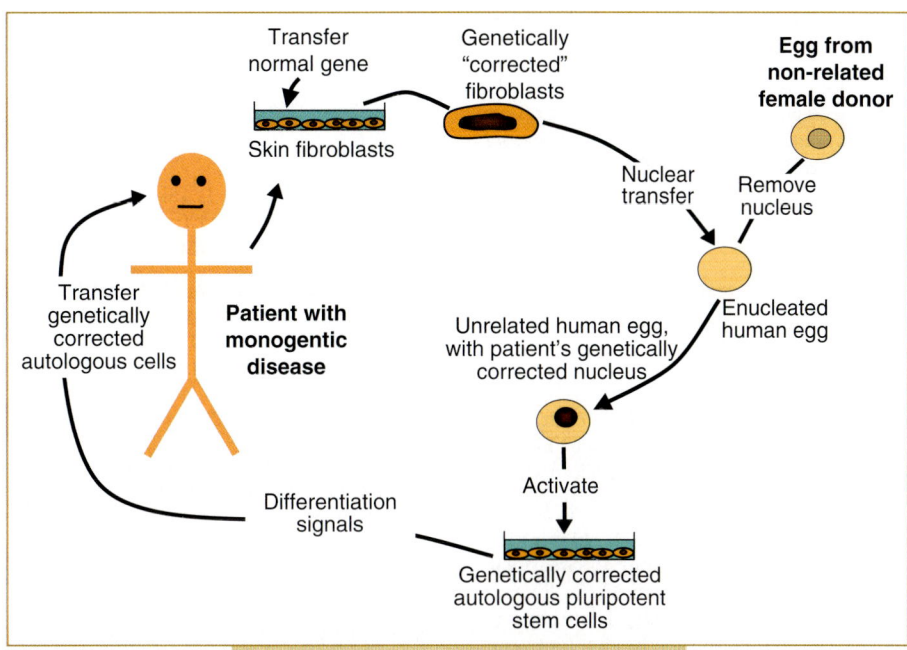

**FIG. 34.5.** Strategy to combine gene therapy with nuclear transfer and stem cell therapy. Shown is an example to genetically modify skin fibroblast of an individual with a monogenetic disease, to correct the abnormality. The nucleus of the genetically corrected fibroblast is then transferred to an enucleated egg of an unrelated donor to generate corrected autologous pluripotent stem cells that can be differentiated and then transferred back to the patient.

## IX. CHALLENGES TO GENE THERAPY FOR TISSUE ENGINEERING

Although gene therapy has been proven very effective in a variety of model systems, the major challenges for gene therapy to cure human diseases include circumvention of immune responses against viral vectors, transferring the genes to a sufficient number of cells to change the phenotype, and controlling the expression of the gene. The main hurdle for successful gene therapy to compensate for a missing or defective protein has been the host response to the gene therapy vector, the lack of long-term gene expression, and problems related to integration into the host genome. However, short-term expression of transgenes has been feasible in humans and shown to be efficient in a variety of cell types and tissues. Immunogenicity against a nonself transgene, as well as vector-derived proteins, may be an issue if gene transfer is being used for permanent expression of a transgene. Controlling the gene expression is a challenge that needs to be addressed, especially if gene transfer strategies are combined with stem cell strategies. Most regulatable gene expression systems show some baseline expression, which may be critical to eliminate, in particular to fine-tune the microenvironment for temporal expression of genetically transferred signals for stem cell strategies.

Over the past decade, progress has been made in addressing many of these challenges. Based on the continued focus on solving these issues, together with gene therapy, the knowledge gained from the successes and the setbacks will prove beneficial in their use for tissue-engineering and stem cell applications.

## X. ACKNOWLEDGMENTS

We thank N. Mohamed for help in preparing this manuscript. These studies were supported, in part, by U01 HL66952 and the Will Rogers Memorial Fund, Los Angeles, CA.

## XI. REFERENCES

Alsberg, E., von Recum, H. A., and Mahoney, M. J. (2006). Environmental cues to guide stem cell fate decision for tissue engineering applications. *Expert. Opin. Biol. Ther.* **6**, 847–866.

Anderson, W. F. (1998). Human gene therapy. *Nature* **392**, 25–30.

Arnberg, N., Edlund, K., Kidd, A. H., and Wadell, G. (2000). Adenovirus type 37 uses sialic acid as a cellular receptor. *J. Virol.* **74**, 42–48.

Auricchio, A., Gao, G. P., Yu, Q. C., Raper, S., Rivera, V. M., Clackson, T., and Wilson, J. M. (2002a). Constitutive and regulated expression of processed insulin following *in vivo* hepatic gene transfer. *Gene Ther.* **9**, 963–971.

Auricchio, A., Rivera, V. M., Clackson, T., O'Connor, E. E., Maguire, A. M., Tolentino, M. J., Bennett, J., and Wilson, J. M. (2002b). Pharmacological regulation of protein expression from adeno-associated viral vectors in the eye. *Mol. Ther.* **6**, 238–242.

Aurisicchio, L., Bujard, H., Hillen, W., Cortese, R., Ciliberto, G., La, M. N., and Palombo, F. (2001). Regulated and prolonged expression of mIFN(alpha) in immunocompetent mice mediated by a helper-dependent adenovirus vector. *Gene Ther.* **8**, 1817–1825.

Bartlett, J. S., Kleinschmidt, J., Boucher, R. C., and Samulski, R. J. (1999). Targeted adeno-associated virus vector transduction of nonpermissive cells mediated by a bispecific F(ab'gamma)2 antibody. *Nat. Biotechnol.* **17**, 181–186.

Basnight, M., Rogers, N. G., Gibbs, C. J., and Gajdusek, D. C. (1971). Characterization of four new adenovirus serotypes isolated from chimpanzee tissue explants. *Am. J. Epidemiol.* **94**, 166–171.

Bergelson, J. M., Cunningham, J. A., Droguett, G., Kurt-Jones, E. A., Krithivas, A., Hong, J. S., Horwitz, M. S., Crowell, R. L., and Finberg, R. W. (1997). Isolation of a common receptor for coxsackie b viruses and adenoviruses 2 and 5. *Science* **275**, 1320–1323.

Blaese, R. M., Culver, K. W., Miller, A. D., Carter, C. S., Fleisher, T., Clerici, M., Shearer, G., Chang, L., Chiang, Y., Tolstoshev, P., Greenblatt, J. J., Rosenberg, S. A., Klein, H., Berger, M., Mullen, C. A., Ramsey, W. J., Muul, L., Morgan, R. A., and Anderson, W. F. (1995). T-lymphocyte-directed gene therapy for ADA-SCID: initial trial results after four years. *Science* **270**, 475–480.

Bowles, D. E., Rabinowitz, J. E., and Samulski, R. J. (2003). Marker rescue of adeno-associated virus (AAV) capsid mutants: a novel approach for chimeric AAV production. *J. Virol.* **77**, 423–432.

Bryder, D., Rossi, D. J., and Weissman, I. L. (2006). Hematopoietic stem cells: the paradigmatic tissue-specific stem cell. *Am. J. Pathol.* **169**, 338–346.

Buchholz, C. J., Stitz, J., and Cichutek, K. (1999). Retroviral cell targeting vectors. *Curr. Opin. Mol. Ther.* **1**, 613–621.

Buchschacher, G. L., and Wong-Staal, F. (2000). Development of lentiviral vectors for gene therapy for human diseases. *Blood* **95**, 2499–2504.

Bushman, F., Lewinski, M., Ciuffi, A., Barr, S., Leipzig, J., Hannenhalli, S., and Hoffmann, C. (2005). Genome-wide analysis of retroviral DNA integration. *Nat. Rev. Microbiol.* **3**, 848–858.

Carter, B. J. (2005). Adeno-associated virus vectors in clinical trials. *Hum. Gene Ther.* **16**, 541–550.

Cavazzana-Calvo, M., Lagresle, C., Hacein-Bey-Abina, S., and Fischer, A. (2005). Gene therapy for severe combined immunodeficiency. *Annu. Rev. Med.* **56**, 585–602.

Chandrashekran, A., Gordon, M. Y., Darling, D., Farzaneh, F., and Casimir, C. (2004). Growth factor displayed on the surface of retroviral particles without manipulation of envelope proteins is biologically active and can enhance transduction. *J. Gene Med.* **6**, 1189–1196.

Chao, H., Mansfield, S. G., Bartel, R. C., Hiriyanna, S., Mitchell, L. G., Garcia-Blanco, M. A., and Walsh, C. E. (2003). Phenotype correction of hemophilia A mice by spliceosome-mediated RNA trans-splicing. *Nat. Med.* **9**, 1015–1019.

Chapel-Fernandes, S., Jordier, F., Lauro, F., Maitland, N., Chiaroni, J., de Micco, P., and Bagins, C. (2006). Use of the PSA enhancer core element to modulate the expression of prostate- and non-prostate-specific basal promoters in a lentiviral vector context. *Cancer Gene Ther.*

Chu, Y., Heistad, D., Cybulsky, M. I., and Davidson, B. L. (2001). Vascular cell adhesion molecule-1 augments adenovirus-mediated gene transfer. *Arterioscler. Thromb. Vasc. Biol.* **21**, 238–242.

Consiglio, A., Quattrini, A., Martino, S., Bensadoun, J. C., Dolcetta, D., Trojani, A., Benaglia, G., Marchesini, S., Cestari, V., Oliverio, A., Bordignon, C., and Naldini, L. (2001). *In vivo* gene therapy of metachromatic leukodystrophy by lentiviral vectors: correction of neuropathology and protection against learning impairments in affected mice. *Nat. Med.* **7**, 310–316.

Cosset, F. L., Morling, F. J., Takeuchi, Y., Weiss, R. A., Collins, M. K., and Russell, S. J. (1995). Retroviral retargeting by envelopes expressing an N-terminal binding domain. *J. Virol.* **69**, 6314–6322.

Costa, R. H., and Grayson, D. R. (1991). Site-directed mutagenesis of hepatocyte nuclear factor (HNF) binding sites in the mouse transthyretin (TTR) promoter reveal synergistic interactions with its enhancer region. *Nucleic Acids Res.* **19**, 4139–4145.

Crittenden, M., Gough, M., Chester, J., Kottke, T., Thompson, J., Ruchatz, A., Clackson, T., Cosset, F. L., Chong, H., Diaz, R. M., Harrington, K., Alvarez, V. L., and Vile, R. (2003). Pharmacologically regulated production of targeted retrovirus from T-cells for systemic antitumor gene therapy. *Cancer Res.* **63**, 3173–3180.

Croyle, M. A., Callahan, S. M., Auricchio, A., Schumer, G., Linse, K. D., Wilson, J. M., Brunner, L. J., and Kobinger, G. P. (2004). PEGylation of a vesicular stomatitis virus G pseudotyped lentivirus vector prevents inactivation in serum. *J. Virol.* **78**, 912–921.

Crystal, R. G. (1995). Transfer of genes to humans: early lessons and obstacles to success. *Science* **270**, 404–410.

Dean, D. A. (2005). Nonviral gene transfer to skeletal, smooth, and cardiac muscle in living animals. *Am. J. Physiol. Cell Physiol.* **289**, C233–C245.

Deasy, B. M., Qu-Peterson, Z., Greenberger, J. S., and Huard, J. (2002). Mechanisms of muscle stem cell expansion with cytokines. *Stem Cells* **20**, 50–60.

Dechecchi, M. C., Tamanini, A., Bonizzato, A., and Cabrini, G. (2000). Heparan sulfate glycosaminoglycans are involved in adenovirus type 5 and 2 host–cell interactions. *Virology* **268**, 382–390.

Dechecchi, M. C., Melotti, P., Bonizzato, A., Santacatterina, M., Chilosi, M., and Cabrini, G. (2001). Heparan sulfate glycosaminoglycans are receptors sufficient to mediate the initial binding of adenovirus types 2 and 5. *J. Virol.* **75**, 8772–8780.

Denby, L., Work, L. M., Graham, D., Hsu, C., von Seggern, D. J., Nicklin, S. A., and Baker, A. H. (2004). Adenoviral serotype 5 vectors pseudotyped with fibers from subgroup D show modified tropism *in vitro* and *in vivo*. *Hum. Gene Ther.* **15**, 1054–1064.

Di, P. G., Davidson, B. L., Stein, C. S., Martins, I., Scudiero, D., Monks, A., and Chiorini, J. A. (2003). Identification of PDGFR as a receptor for AAV-5 transduction. *Nat. Med.* **9**, 1306–1312.

Dmitriev, I. P., Kashentseva, E. A., and Curiel, D. T. (2002). Engineering of adenovirus vectors containing heterologous peptide sequences in the C terminus of capsid protein IX. *J. Virol.* **76**, 6893–6899.

Fairchild, P. J., Cartland, S., Nolan, K. F., and Waldmann, H. (2004). Embryonic stem cells and the challenge of transplantation tolerance. *Trends Immunol.* **25**, 465–470.

Farina, S. F., Gao, G. P., Xiang, Z. Q., Rux, J. J., Burnett, R. M., Alvira, M. R., Marsh, J., Ertl, H. C., and Wilson, J. M. (2001). Replication-defective vector based on a chimpanzee adenovirus. *J. Virol.* **75**, 11603–11613.

Fasbender, A., Zabner, J., Chillon, M., Moninger, T. O., Puga, A. P., Davidson, B. L., and Welsh, M. J. (1997). Complexes of adenovirus with polycationic polymers and cationic lipids increase the efficiency of gene transfer *in vitro* and *in vivo*. *J. Biol. Chem.* **272**, 6479–6489.

Fitzsimons, H. L., Mckenzie, J. M., and During, M. J. (2001). Insulators coupled to a minimal bidirectional tet cassette for tight regulation of rAAV-mediated gene transfer in the mammalian brain. *Gene Ther.* **8**, 1675–1681.

Fussenegger, M., Morris, R. P., Fux, C., Rimann, M., von, S. B., Thompson, C. J., and Bailey, J. E. (2000). Streptogramin-based gene regulation systems for mammalian cells. *Nat. Biotechnol.* **18**, 1203–1208.

Gall, J., Kass-Eisler, A., Leinwand, L., and Falck-Pedersen, E. (1996). Adenovirus type 5 and 7 capsid chimera: fiber replacement alters receptor tropism without affecting primary immune neutralization epitopes. *J. Virol.* **70**, 2116–2123.

Gao, G. P., Alvira, M. R., Wang, L., Calcedo, R., Johnston, J., and Wilson, J. M. (2002). Novel adeno-associated viruses from rhesus monkeys as vectors for human gene therapy. *Proc. Natl. Acad Sci. U.S.A.* **99**, 11854–11859.

Gao, G., Vandenberghe, L. H., Alvira, M. R., Lu, Y., Calcedo, R., Zhou, X., and Wilson, J. M. (2004). Clades of adeno-associated viruses are widely disseminated in human tissues. *J. Virol.* **78**, 6381–6388.

Girod, A., Ried, M., Wobus, C., Lahm, H., Leike, K., Kleinschmidt, J., Deleage, G., and Hallek, M. (1999). Genetic capsid modifications allow efficient retargeting of adeno-associated virus type 2. *Nat. Med.* **5**, 1052–1056.

Glasgow, J. N., Kremer, E. J., Hemminki, A., Siegal, G. P., Douglas, J. T., and Curiel, D. T. (2004). An adenovirus vector with a chimeric fiber derived from canine adenovirus type 2 displays novel tropism. *Virology* **324**, 103–116.

Gossen, M., Freundlieb, S., Bender, G., Muller, G., Hillen, W., and Bujard, H. (1995). Transcriptional activation by tetracyclines in mammalian cells. *Science* **268**, 1766–1769.

Gregory, L. G., Waddington, S. N., Holder, M. V., Mitrophanous, K. A., Buckley, S. M., Mosley, K. L., Bigger, B. W., Ellard, F. M., Walmsley, L. E., Lawrence, L., Al-Allaf, F., Kingsman, S., Coutelle, C., and Themis, M. (2004). Highly efficient EIAV-mediated *in utero* gene transfer and expression in the major muscle groups affected by Duchenne muscular dystrophy. *Gene Ther.* **11**, 1117–1125.

Grifman, M., Trepel, M., Speece, P., Gilbert, L. B., Arap, W., Pasqualini, R., and Weitzman, M. D. (2001). Incorporation of tumor-targeting peptides into recombinant adeno-associated virus capsids. *Mol. Ther.* **3**, 964–975.

Haberman, R. P., McCown, T. J., and Samulski, R. J. (1998). Inducible long-term gene expression in brain with adeno-associated virus gene transfer. *Gene Ther.* **5**, 1604–1611.

Hackett, N. R., Kaminsky, S. M., Sondhi, D., and Crystal, R. G. (2000). Antivector and antitransgene host responses in gene therapy. *Curr. Opin. Mol. Ther.* **2**, 376–382.

Halbert, C. L., Allen, J. M., and Miller, A. D. (2001). Adeno-associated virus type 6 (AAV6) vectors mediate efficient transduction of airway epithelial cells in mouse lungs compared to that of AAV2 vectors. *J. Virol.* **75**, 6615–6624.

Han, X., Kasahara, N., and Kan, Y. W. (1995). Ligand-directed retroviral targeting of human breast cancer cells *Proc Natl Acad Sci U S A* **92** 9747–9751.

Handa, A., Muramatsu, S., Qiu, J., Mizukami, H., and Brown, K. E. (2000). Adeno-associated virus (AAV)-3-based vectors transduce haematopoietic cells not susceptible to transduction with AAV-2-based vectors. *J. Gene Virol.* **81**, 2077–2084.

Hashimoto, M., Boyer, J. L., Hackett, N. R., Wilson, J. M., and Crystal, R. G. (2005). Induction of protective immunity to anthrax lethal toxin with a nonhuman primate adenovirus-based vaccine in the presence of preexisting anti-human adenovirus immunity. *Infect. Immun.* **73**, 6885–6891.

Hauck, B., Chen, L., and Xiao, W. (2003). Generation and characterization of chimeric recombinant AAV vectors. *Mol. Ther.* **7**, 419–425.

Havenga, M. J., Lemckert, A. A., Grimbergen, J. M., Vogels, R., Huisman, L. G., Valerio, D., Bout, A., and Quax, P. H. (2001). Improved adenovirus vectors for infection of cardiovascular tissues. *J. Virol.* **75**, 3335–3342.

Hedley, S. J., uf der, M. A., Hohn, S., Escher, D., Barberis, A., Glasgow, J. N., Douglas, J. T., Korokhov, N., and Curiel, D. T. (2006). An adenovirus vector with a chimeric fiber incorporating stabilized single-chain antibody achieves targeted gene delivery. *Gene Ther.* **13**, 88–94.

Hofmann, A., Nolan, G. P., and Blau, H. M. (1996). Rapid retroviral delivery of tetracycline-inducible genes in a single autoregulatory cassette. *Proc. Natl. Acad. Sci. U.S.A.* **93**, 5185–5190.

Hong, S. S., Karayan, L., Tournier, J., Curiel, D. T., and Boulanger, P. A. (1997). Adenovirus type 5 fiber knob binds to MHC class I alpha2 domain at the surface of human epithelial and B lymphoblastoid cells. *EMBO J.* **16**, 2294–2306.

Hong, S. S., Magnusson, M. K., Henning, P., Lindholm, L., and Boulanger, P. A. (2003). Adenovirus stripping: a versatile method to generate adenovirus vectors with new cell target specificity. *Mol. Ther.* **7**, 692–699.

Hoppe, U. C., Marban, E., and Johns, D. C. (2000). Adenovirus-mediated inducible gene expression *in vivo* by a hybrid ecdysone receptor. *Mol. Ther.* **1**, 159–164.

Isomoto, H., Ohtsuru, A., Braiden, V., Iwamatsu, M., Miki, F., Kawashita, Y., Mizuta, Y., Kaneda, Y., Kohno, S., and Yamashita, S. (2006). Heat-directed suicide gene therapy mediated by heat shock protein promoter for gastric cancer. *Oncol. Rep.* **15**, 629–635.

Jooss, K., and Chirmule, N. (2003). Immunity to adenovirus and adeno-associated viral vectors: implications for gene therapy. *Gene Ther.* **10**, 955–963.

Kafri, T., van, P. H., Gage, F. H., and Verma, I. M. (2000). Lentiviral vectors: regulated gene expression. *Mol. Ther.* **1**, 516–521.

Kaji, E. H., and Leiden, J. M. (2001). Gene and stem cell therapies. *JAMA* **285**, 545–550.

Kaludov, N., Brown, K. E., Walters, R. W., Zabner, J., and Chiorini, J. A. (2001). Adeno-associated virus serotype 4 (AAV4) and AAV5 both require sialic acid binding for hemagglutination and efficient transduction but differ in sialic acid linkage specificity. *J. Virol.* **75**, 6884–6893.

Karzenowski, D., Potter, D. W., and Padidam, M. (2005). Inducible control of transgene expression with ecdysone receptor: gene switches with high sensitivity, robust expression, and reduced size. *Biotechniques* **39**, 191–192.

Kasahara, N., Dozy, A. M., and Kan, Y. W. (1994). Tissue-specific targeting of retroviral vectors through ligand-receptor interactions. *Science* **266**, 1373–1376.

Kim, V. N., Mitrophanous, K., Kingsman, S. M., and Kingsman, A. J. (1998). Minimal requirement for a lentivirus vector based on human immunodeficiency virus type 1. *J. Virol.* **72**, 811–816.

Kirby, I., Davison, E., Beavil, A. J., Soh, C. P., Wickham, T. J., Roelvink, P. W., Kovesdi, I., Sutton, B. J., and Santis, G. (1999). Mutations in the DG loop of adenovirus type 5 fiber knob protein abolish high-affinity binding to its cellular receptor CAR. *J. Virol.* **73**, 9508–9514.

Klages, N., Zufferey, R., and Trono, D. (2000). A stable system for the high-titer production of multiply attenuated lentiviral vectors. *Mol. Ther.* **2**, 170–176.

Kobinger, G. P., Weiner, D. J., Yu, Q. C., and Wilson, J. M. (2001). Filovirus-pseudotyped lentiviral vector can efficiently and stably transduce airway epithelia in vivo. *Nat. Biotechnol.* **19**, 225–230.

Kotsopoulou, E., Kim, V. N., Kingsman, A. J., Kingsman, S. M., and Mitrophanous, K. A. (2000). A Rev-independent human immunodeficiency virus type 1 (HIV-1)-based vector that exploits a codon-optimized HIV-1 gag-pol gene. *J. Virol.* **74**, 4839–4852.

Krasnykh, V. N., Mikheeva, G. V., Douglas, J. T., and Curiel, D. T. (1996). Generation of recombinant adenovirus vectors with modified fibers for altering viral tropism. *J. Virol.* **70**, 6839–6846.

Krasnykh, V., Belousova, N., Korokhov, N., Mikheeva, G., and Curiel, D. T. (2001). Genetic targeting of an adenovirus vector via replacement of the fiber protein with the phage T4 fibritin. *J. Virol.* **75**, 4176–4183.

Kufe, D., and Weichselbaum, R. (2003). Radiation therapy: activation for gene transcription and the development of genetic radiotherapy-therapeutic strategies in oncology. *Cancer Biol. Ther.* **2**, 326–329.

Latham, J. P., Searle, P. F., Mautner, V., and James, N. D. (2000). Prostate-specific antigen promoter/enhancer driven gene therapy for prostate cancer: construction and testing of a tissue-specific adenovirus vector. *Cancer Res.* **60**, 334–341.

Lechardeur, D., Verkman, A. S., and Lukacs, G. L. (2005). Intracellular routing of plasmid DNA during non-viral gene transfer. *Adv. Drug Deliv. Rev.* **57**, 755–767.

Liu, X., Jiang, Q., Mansfield, S. G., Puttaraju, M., Zhang, Y., Zhou, W., Cohn, J. A., Garcia-Blanco, M. A., Mitchell, L. G., and Engelhardt, J. F. (2002). Partial correction of endogenous DeltaF508 CFTR in human cystic fibrosis airway epithelia by spliceosome-mediated RNA trans-splicing. *Nat. Biotechnol.* **20**, 47–52.

Liu, X., Luo, M., Zhang, L. N., Yan, Z., Zak, R., Ding, W., Mansfield, S. G., Mitchell, L. G., and Engelhardt, J. F. (2005). Spliceosome-mediated RNA trans-splicing with recombinant adeno-associated virus partially restores cystic fibrosis transmembrane conductance regulator function to polarized human cystic fibrosis airway epithelial cells [epub ahead of print] [Record Supplied by Publisher]. *Hum. Gene Ther.* 2005 Aug 5.

Lundstrom, K. (2003). Latest development in viral vectors for gene therapy. *Trends Biotechnol.* **21**, 117–122.

Luo, P., He, X., Tsang, T. C., and Harris, D. T. (2004). A novel inducible amplifier expression vector for high and controlled gene expression. *Int. J. Mol. Med.* **13**, 319–325.

MacGregor, R. R. (2001). Clinical protocol. A phase 1 open-label clinical trial of the safety and tolerability of single escalating doses of autologous CD4 T cells transduced with VRX496 in HIV-positive subjects. *Hum. Gene Ther.* **12**, 2028–2029.

Marin, M., Noel, D., Valsesia-Wittman, S., Brockly, F., Etienne-Julan, M., Russell, S., Cosset, F. L., and Piechaczyk, M. (1996). Targeted infection of human cells via major histocompatibility complex class I molecules by Moloney murine leukemia virus-derived viruses displaying single-chain antibody fragment-envelope fusion proteins. *J. Virol.* **70**, 2957–2962.

Markowitz, D., Goff, S., and Bank, A. (1988). Construction and use of a safe and efficient amphotropic packaging cell line. *Virology* **167**, 400–406.

Mastrangeli, A., Harvey, B. G., Yao, J., Wolff, G., Kovesdi, I., Crystal, R. G., and Falck-Pedersen, E. (1996). "Sero-switch" adenovirus-mediated *in vivo* gene transfer: circumvention of anti-adenovirus humoral immune defenses against repeat adenovirus vector administration by changing the adenovirus serotype. *Hum. Gene Ther.* **7**, 79–87.

Mazarakis, N. D., Azzouz, M., Rohll, J. B., Ellard, F. M., Wilkes, F. J., Olsen, A. L., Carter, E. E., Barber, R. D., Baban, D. F., Kingsman, S. M., Kingsman, A. J., O'Malley, K., and Mitrophanous, K. A. (2001). Rabies virus glycoprotein pseudotyping of lentiviral vectors enables retrograde axonal transport and access to the nervous system after peripheral delivery. *Hum. Mol. Genet.* **10**, 2109–2121.

McCarty, D. M., Young, S. M., and Samulski, R. J. (2004). Integration of adeno-associated virus (AAV) and recombinant AAV vectors. *Annu. Rev. Genet.* **38**, 819–845.

McKay, T., Patel, M., Pickles, R. J., Johnson, L. G., and Olsen, J. C. (2006). Influenza M2 envelope protein augments avian influenza hemagglutinin pseudotyping of lentiviral vectors. *Gene Ther.* **13**, 715–724.

Medina, M. F., Kobinger, G. P., Rux, J., Gasmi, M., Looney, D. J., Bates, P., and Wilson, J. M. (2003). Lentiviral vectors pseudotyped with minimal filovirus envelopes increased gene transfer in murine lung. *Mol. Ther.* **8**, 777–789.

Miller, A. D. (1999). Nonviral deliveery systems for gene therapy. *In* "Understanding Gene Therpy" (N. R. Lemoine, ed.). Springer-Verlag, New York.

Miller, A. D., Garcia, J. V., von, S. N., Lynch, C. M., Wilson, C., and Eiden, M. V. (1991). Construction and properties of retrovirus packaging cells based on gibbon ape leukemia virus. *J. Virol.* **65**, 2220–2224.

Molin, M., Shoshan, M. C., Ohman-Forslund, K., Linder, S., and Akusjarvi, G. (1998). Two novel adenovirus vector systems permitting regulated protein expression in gene transfer experiments. *J. Virol.* **72**, 8358–8361.

Montier, T., Delepine, P., Pichon, C., Ferec, C., Porteous, D. J., and Midoux, P. (2004). Non-viral vectors in cystic fibrosis gene therapy: progress and challenges. *Trends Biotechnol.* **22**, 586–592.

Muyzyczka, N., and Berns, K. I. (2001). Parvoviridae: the viruses and their replication. *In* "Fields Virology" (D. M. Knipe and P. M. Howley, eds.), pp. 2327–2359. Lippincott Williams & Wilkins, Philadelphia.

Nakai, H., Storm, T. A., and Kay, M. A. (2000). Recruitment of single-stranded recombinant adeno-associated virus vector genomes and intermolecular recombination are responsible for stable transduction of liver *in vivo*. *J. Virol.* **74**, 9451–9463.

Nakai, H., Yant, S. R., Storm, T. A., Fuess, S., Meuse, L., and Kay, M. A. (2001). Extrachromosomal recombinant adeno-associated virus vector genomes are primarily responsible for stable liver transduction *in vivo*. *J. Virol.* **75**, 6969–6976.

Nakayama, M., Both, G. W., Banizs, B., Tsuruta, Y., Yamamoto, S., Kawakami, Y., Douglas, J. T., Tani, K., Curiel, D. T., and Glasgow, J. N. (2006). An adenovirus serotype 5 vector with fibers derived from ovine atadenovirus demonstrates CAR-independent tropism and unique biodistribution in mice. *Virology* **350**, 103–115.

Narumi, K., Suzuki, M., Song, W., Moore, M. A. S., and Crystal, R. G. (1998). Intermittent, repetitive corticosteroid-induced up-regulation of platelet levels after adenovirus-mediated transfer to the liver of a chimeric glucocorticoid-responsive promoter controlling the thrombopoietin cDNA. *Blood* **92**, 822–833.

Ngan, E. S., Schillinger, K., De, M. F., and Tsai, S. Y. (2002). The mifepristone-inducible gene regulatory system in mouse models of disease and gene therapy. *Semin. Cell Dev. Biol.* **13**, 143–149.

Nicklin, S. A., White, S. J., Watkins, S. J., Hawkins, R. E., and Baker, A. H. (2000). Selective targeting of gene transfer to vascular endothelial cells by use of peptides isolated by phage display. *Circulation* **102**, 231–237.

Nicklin, S. A., Wu, E., Nemerow, G. R., and Baker, A. H. (2005). The influence of adenovirus fiber structure and function on vector development for gene therapy. *Mole. Ther.* **12**, 384–393.

O'Connor, T. P., and Crystal, R. G. (2006). Genetic medicines: treatment strategies for hereditary disorders. *Nat. Rev. Genet.* **7**, 261–276.

Pastore, L., Morral, N., Zhou, H., Garcia, R., Parks, R. J., Kochanek, S., Graham, F. L., Lee, B., and Beaudet, A. L. (1999). Use of a liver-specific promoter reduces immune response to the transgene in adenoviral vectors. *Hum. Gene Ther.* **10**, 1773–1781.

Pereboev, A., Pereboeva, L., and Curiel, D. T. (2001). Phage display of adenovirus type 5 fiber knob as a tool for specific ligand selection and validation. *J. Virol.* **75**, 7107–7113.

Pergolizzi, R. G., and Crystal, R. G. (2004). Genetic medicine at the RNA level: modifications of the genetic repertoire for therapeutic purposes by pre-mRNA trans-splicing. *C. R. Biol.* **327**, 695–709.

Pohlmann, S., and Reeves, J. D. (2006). Cellular entry of HIV: evaluation of therapeutic targets. *Curr. Pharm. Des.* **12**, 1963–1973.

Pollock, R., and Clackson, T. (2002). Dimerizer-regulated gene expression. *Curr. Opin. Biotechnol.* **13**, 459–467.

Ponnazhagan, S., Mahendra, G., Kumar, S., Thompson, J. A., and Castillas, M. (2002). Conjugate-based targeting of recombinant adeno-associated virus type 2 vectors by using avidin-linked ligands. *J. Virol.* **76**, 12900–12907.

Puttaraju, M., Jamison, S. F., Mansfield, S. G., Garcia-Blanco, M. A., and Mitchell, L. G. (1999). Spliceosome-mediated RNA trans-splicing as a tool for gene therapy. *Nat. Biotechnol.* **17**, 246–252.

Qasim, W., Gaspar, H. B., and Thrasher, A. J. (2005). T-cell suicide gene therapy to aid haematopoietic stem cell transplantation. *Curr. Gene Ther.* **5**, 121–132.

Qian, Z., Haessler, M., Lemos, J. A., Arsenault, J. R., Aguirre, J. E., Gilbert, J. R., Bowler, R. P., and Park, F. (2006). Targeting vascular injury using hantavirus-pseudotyped lentiviral vectors. *Mol. Ther.* **13**, 694–704.

Qing, K., Mah, C., Hansen, J., Zhou, S., Dwarki, V., and Srivastava, A. (1999). Human fibroblast growth factor receptor 1 is a coreceptor for infection by adeno-associated virus 2. *Nat. Med.* **5**, 71–77.

Rabinowitz, J. E., Rolling, F., Li, C., Conrath, H., Xiao, W., Xiao, X., and Samulski, R. J. (2002). Cross-packaging of a single adeno-associated virus (AAV) type 2 vector genome into multiple AAV serotypes enables transduction with broad specificity. *J. Virol.* **76**, 791–801.

Rebar, E. J., Huang, Y., Hickey, R., Nath, A. K., Meoli, D., Nath, S., Chen, B., Xu, L., Liang, Y., Jamieson, A. C., Zhang, L., Spratt, S. K., Case, C. C., Wolffe, A., and Giordano, F. J. (2002). Induction of angiogenesis in a mouse model using engineered transcription factors. *Nat. Med.* **8**, 1427–1432.

Reiser, J., Lai, Z., Zhang, X. Y., and Brady, R. O. (2000). Development of multigene and regulated lentivirus vectors. *J. Virol.* **74**, 10589–10599.

Rendahl, K. G., Quiroz, D., Ladner, M., Coyne, M., Seltzer, J., Manning, W. C., and Escobedo, J. A. (2002). Tightly regulated long-term erythropoietin expression *in vivo* using tet-inducible recombinant adeno-associated viral vectors. *Hum. Gene Ther.* **13**, 335–342.

Ribault, S., Neuville, P., Mechine-Neuville, A., Auge, F., Parlakian, A., Gabbiani, G., Paulin, D., and Calenda, V. (2001). Chimeric smooth muscle-specific enhancer/promoters: valuable tools for adenovirus-mediated cardiovascular gene therapy. *Circ. Res.* **88**, 468–475.

Roberts, D. M., Nanda, A., Havenga, M. J., Abbink, P., Lynch, D. M., Ewald, B. A., Liu, J., Thorner, A. R., Swanson, P. E., Gorgone, D. A., Lifton, M. A., Lemckert, A. A., Holterman, L., Chen, B., Dilraj, A., Carville, A., Mansfield, K. G., Goudsmit, J., and Barouch, D. H. (2006). Hexon-chimaeric adenovirus serotype 5 vectors circumvent preexisting antivector immunity. *Nature* **441**, 239–243.

Roth, D. A., Tawa, N. E., O'Brien, J. M., Treco, D. A., and Selden, R. F. (2001). Nonviral transfer of the gene encoding coagulation factor VIII in patients with severe hemophilia A. *N. Engl. J. Med.* **344**, 1735–1742.

Saban, S. D., Nepomuceno, R. R., Gritton, L. D., Nemerow, G. R., and Stewart, P. L. (2005). CryoEM structure at 9A resolution of an adenovirus vector targeted to hematopoietic cells. *J. Mol. Biol.* **349**, 526–537.

Sadelain, M., Lisowski, L., Samakoglu, S., Rivella, S., May, C., and Riviere, I. (2005). Progress toward the genetic treatment of the beta-thalassemias. *Ann. N.Y. Acad. Sci.* **1054**, 78–91.

Sandrin, V., Boson, B., Salmon, P., Gay, W., Negre, D., Le, G. R., Trono, D., and Cosset, F. L. (2002). Lentiviral vectors pseudotyped with a modified RD114 envelope glycoprotein show increased stability in sera and augmented transduction of primary lymphocytes and CD34+ cells derived from human and nonhuman primates. *Blood* **100**, 823–832.

Segerman, A., Atkinson, J. P., Marttila, M., Dennerquist, V., Wadell, G., and Arnberg, N. (2003). Adenovirus type 11 uses CD46 as a cellular receptor. *J. Virol.* **77**, 9183–9191.

Shayakhmetov, D. M., Papayannopoulou, T., Stamatoyannopoulos, G., and Lieber, A. (2000). Efficient gene transfer into human CD34(+) cells by a retargeted adenovirus vector. *J. Virol.* **74**, 2567–2583.

Shayakhmetov, D. M., Carlson, C. A., Stecher, H., Li, Q., Stamatoyannopoulos, G., and Lieber, A. (2002). A high-capacity, capsid-modified hybrid adenovirus/adeno-associated virus vector for stable transduction of human hematopoietic cells. *J. Virol.* **76**, 1135–1143.

Shenk, T. (2001). Adenoviridae: the viruses and their replication. *In* "Fields Virology" (B. Fields, D. Knipe, and P. Howley, eds.), pp. 2111–2148. Lippincott-Raven, Philadelphia.

Short, J. J., Vasu, C., Holterman, M. J., Curiel, D. T., and Pereboev, A. (2006). Members of adenovirus species B utilize CD80 and CD86 as cellular attachments receptors. *Virus Res.*

Sirena, D., Lilienfeld, B., Eisenhut, M., Kalin, S., Boucke, K., Beerli, R. R., Vogt, L., Ruedl, C., Bachmann, M. F., Greber, U. F., and Hemmi, S. (2004). The human membrane cofactor CD46 is a receptor for species B adenovirus serotype 3. *J. Virol.* **78**, 4454–4462.

Sirven, A., Pflumio, F., Zennou, V., Titeux, M., Vainchenker, W., Coulombel, L., Dubart-Kupperschmitt, A., and Charneau, P. (2000). The human immunodeficiency virus type-1 central DNA flap is a crucial determinant for lentiviral vector nuclear import and gene transduction of human hematopoietic stem cells. *Blood* **96**, 4103–4110.

Stevenson, S. C., Rollence, M., White, B., Weaver, L., and McClelland, A. (1995). Human adenovirus serotypes 3 and 5 bind to two different cellular receptors via the fiber head domain. *J. Virol.* **69**, 2850–2857.

Stevenson, S. C., Rollence, M., Marshall-Neff, J., and McClelland, A. (1997). Selective targeting of human cells by a chimeric adenovirus vector containing a modified fiber protein. *J. Virol.* **71**, 4782–4790.

Su, H., rakawa-Hoyt, J., and Kan, Y. W. (2002). Adeno-associated viral vector-mediated hypoxia response element-regulated gene expression in mouse ischemic heart model. *Proc. Natl. Acad. Sci. U.S.A.* **99**, 9480–9485.

Summerford, C., and Samulski, R. J. (1998). Membrane-associated heparan sulfate proteoglycan is a receptor for adeno-associated virus type 2 virions. *J. Virol.* **72**, 1438–1445.

Summerford, C., Bartlett, J. S., and Samulski, R. J. (1999). AlphaVbeta5 integrin: a coreceptor for adeno-associated virus type 2 infection. *Nat. Med.* **5**, 78–82.

Suzuki, M., Singh, R. N., and Crystal, R. G. (1996). Regulatable promoters for use in gene therapy applications: modification of the 5′-

flanking region of the CFTR gene with multiple cAMP response elements to support basal, low-level gene expression that can be up-regulated by exogenous agents that raise intracellular levels of cAMP. *Hum. Gene Ther.* **7**, 1883–1893.

Suzuki, M., Singh, R. N., and Crystal, R. G. (1997). Ability of a chimeric cAMP-responsive promoter to confer pharmacologic control of CFTR cDNA expression and cAMP-mediated Cl-secretion. *Gene Ther.* **4**, 1195–1201.

Tahara, M., Pergolizzi, R. G., Kobayashi, H., Krause, A., Luettich, K., Lesser, M. L., and Crystal, R. G. (2004). Trans-splicing repair of CD40 ligand deficiency results in naturally regulated correction of a mouse model of hyper-IgM X-linked immunodeficiency. *Nat. Med.* **10**, 835–841.

Tai, C. K., Logg, C. R., Park, J. M., Anderson, W. F., Press, M. F., and Kasahara, N. (2003). Antibody-mediated targeting of replication-competent retroviral vectors. *Hum. Gene Ther.* **14**, 789–802.

Talbot, D., Descombes, P., and Schibler, U. (1994). The 5′ flanking region of the rat LAP (C/EBP beta) gene can direct high-level, position-independent, copy number-dependent expression in multiple tissues in transgenic mice. *Nucleic Acids Res.* **22**, 756–766.

Themis, M., Waddington, S. N., Schmidt, M., von, K. C., Wang, Y., Al-Allaf, F., Gregory, L. G., Nivsarkar, M., Themis, M., Holder, M. V., Buckley, S. M., Dighe, N., Ruthe, A. T., Mistry, A., Bigger, B., Rahim, A., Nguyen, T. H., Trono, D., Thrasher, A. J., and Coutelle, C. (2005). Onco-genesis following delivery of a nonprimate lentiviral gene therapy vector to fetal and neonatal mice. *Mol. Ther.* **12**, 763–771.

Thomas, C. E., Ehrhardt, A., and Kay, M. A. (2003). Progress and problems with the use of viral vectors for gene therapy. *Nat. Rev. Genet.* **4**, 346–358.

Verma, I. M., and Weitzman, M. D. (2005). Gene therapy: twenty-first century medicine. *Annu. Rev. Biochem.* **74**, 711–738.

Vigne, E., Mahfouz, I., Dedieu, J. F., Brie, A., Perricaudet, M., and Yeh, P. (1999). RGD inclusion in the hexon monomer provides adenovirus type 5-based vectors with a fiber knob-independent pathway for infection. *J. Virol.* **73**, 5156–5161.

Walters, R. W., Yi, S. M., Keshavjee, S., Brown, K. E., Welsh, M. J., Chiorini, J. A., and Zabner, J. (2001). Binding of adeno-associated virus type 5 to 2,3-linked sialic acid is required for gene transfer. *J. Biol. Chem.* **276**, 20610–20616.

Wang, X. P., Yazawa, K., Yang, J., Kohn, D., Fisher, W. E., and Brunicardi, F. C. (2004). Specific gene expression and therapy for pancreatic cancer using the cytosine deaminase gene directed by the rat insulin promoter. *J. Gastrointest. Surg.* **8**, 98–108.

Wang, Y., O'Malley, B. W., Tsai, S. Y., and O'Malley, B. W. (1994). A regulatory system for use in gene transfer. *Proc. Natl. Acad. Sci. U.S.A.* **91**, 8180–8184.

Weber, W., Fux, C., oud-el, B. M., Keller, B., Weber, C. C., Kramer, B. P., Heinzen, C., Aubel, D., Bailey, J. E., and Fussenegger, M. (2002). Macrolide-based transgene control in mammalian cells and mice. *Nat. Biotechnol.* **20**, 901–907.

Wickham, T. J., Mathias, P., Cheresh, D. A., and Nemerow, G. R. (1993). Integrins alpha v beta 3 and alpha v beta 5 promote adenovirus internalization but not virus attachment. *Cell* **73**, 309–319.

Wickham, T. J., Tzeng, E., Shears, L. L., Roelvink, P. W., Li, Y., Lee, G. M., Brough, D. E., Lizonova, A., and Kovesdi, I. (1997). Increased *in vitro* and *in vivo* gene transfer by adenovirus vectors containing chimeric fiber proteins. *J. Virol.* **71**, 8221–8229.

Wobus, A. M., and Boheler, K. R. (2005). Embryonic stem cells: prospects for developmental biology and cell therapy. *Physiol. Rev.* **85**, 635–678.

Worgall, S., Worgall, T. S., Kostarelos, K., Singh, R., Leopold, P. L., Hackett, N. R., and Crystal, R. G. (2000). Free cholesterol enhances adenoviral vector gene transfer and expression in CAR-deficient cells. *Mol. Ther.* **1**, 39–48.

Work, L. M., Ritchie, N., Nicklin, S. A., Reynolds, P. N., and Baker, A. H. (2004). Dual targeting of gene delivery by genetic modification of adenovirus serotype 5 fibers and cell-selective transcriptional control. *Gene Ther.* **11**, 1296–1300.

Wu, P., Xiao, W., Conlon, T., Hughes, J., gbandje-McKenna, M., Ferkol, T., Flotte, T., and Muzyczka, N. (2000). Mutational analysis of the adeno-associated virus type 2 (AAV2) capsid gene and construction of AAV2 vectors with altered tropism. *J. Virol.* **74**, 8635–8647.

Xia, H., Anderson, B., Mao, Q., and Davidson, B. L. (2000). Recombinant human adenovirus: targeting to the human transferrin receptor improves gene transfer to brain microcapillary endothelium. *J. Virol.* **74**, 11359–11366.

Yoneyama, R., Chemaly, E. R., and Hajjar, R. J. (2006). Tracking stem cells *in vivo*. *Ernst. Schering. Res. Found. Workshop* 2006, 99–109.

Yu, S. F., von, R. T., Kantoff, P. W., Garber, C., Seiberg, M., Ruther, U., Anderson, W. F., Wagner, E. F., and Gilboa, E. (1986). Self-inactivating retroviral vectors designed for transfer of whole genes into mammalian cells. *Proc. Natl. Acad. Sci. U.S.A.* **83**, 3194–3198.

Yun, C. O., Cho, E. A., Song, J. J., Kang, D. B., Kim, E., Sohn, J. H., and Kim, J. H. (2003). DL-VSVG-LacZ, a vesicular stomatitis virus glycoprotein epitope-incorporated adenovirus, exhibits marked enhancement in gene transduction efficiency. *Hum. Gene Ther.* **14**, 1643–1652.

Zabner, J., Chillon, M., Grunst, T., Moninger, T. O., Davidson, B. L., Gregory, R., and Armentano, D. (1999). A chimeric type 2 adenovirus vector with a type 17 fiber enhances gene transfer to human airway epithelia. *J. Virol.* **73**, 8689–8695.

Zabner, J., Seiler, M., Walters, R., Kotin, R. M., Fulgeras, W., Davidson, B. L., and Chiorini, J. A. (2000). Adeno-associated virus type 5 (AAV5) but not AAV2 binds to the apical surfaces of airway epithelia and facilitates gene transfer. *J. Virol.* **74**, 3852–3858.

# Chapter Thirty-Five

# Gene Delivery into Cells and Tissues

*Ales Prokop and Jeffrey M. Davidson*

## I. INTRODUCTION

This chapter provides an overview of principles and barriers relevant to intracellular gene delivery by means of nanovehicles. The aim is to deliver a cargo to a particular intracellular site, if possible, to exert a local action. *Nanovehicles* are defined as a wide range of nanosized particles leading to colloidal objects capable of entering cells and tissues and delivering a cargo intracellularly. Different localization and targeting means are discussed. Limited discussion on pharmacokinetics and pharmacodynamics is also presented. Current nanotechnologies for gene delivery that are already in commercial use are also discussed. Newer developments in nanovector technologies are outlined, and future applications are stressed. Two issues are of prominent importance: lysosomal escape and subsequent cytoplasmic/nuclear delivery and intracellular polyplex stability. More complete treatments may be found in the references cited and in a more extensive review article (Prokop and Davidson, 2007) dealing with nanotechnologies relevant to both drug and gene delivery.

## II. GENE DELIVERY SYSTEMS

The design of delivery systems is an increasingly valuable discipline in pharmaceutical development, allowing rational manipulation of the pharmacological profiles of drugs and genes and their concomitant therapeutic indices. Delivery systems are now used to modify potentially therapeutic agents toward: (a) creation of new pharmaceutical moieties (e.g., liposomal gene delivery); (b) improvement in the effectiveness or reduction of the side effects of an existing therapeutic; (c) tissue-selective targeting (Mrsny, 2004); and (d) extension of the patent lifetime for an already marketed drug (Kostarelos, 2003). This is of advantage for drugs and gene products, which often exhibit a narrow therapeutic index, a short half-time in the bloodstream, and a high overall clearance rate. The *therapeutic index* is defined as the ratio of the toxic to the therapeutic dose of a drug.

### Definitions

*Therapeutic agent* is defined as a chemical, biological, genetic, or radiological entity to be delivered to a disease site for the purpose of treatment or detection (imaging). For the purpose of this chapter it is limited to genes and similar modalities.

The development of delivery systems able to alter the biological profiles (biodistribution, tissue uptake, pharmacokinetics, and pharmacodynamics) of therapeutic agents is considered of utmost importance in biomedical research and the pharmaceutical industry (Moses *et al.*, 2003). Definitions of biodistribution, the role of the reticuloendothelial system (RES), pharmacokinetics, pharmacodynamics, and bioavailability are expanded in a recent review (Prokop and Davidson, 2007). *Biodistribution* is the percentage of an administered dose that is deposited into specific organs throughout the body at specific time points. The tissue distribution of a particulate/macromolecular drug (gene) among different body locations is highly dependent on the nonspecific RES effects. The RES is the cellular system responsible for protection and clearance of "foreign" material. The RES consists primarily of phagocytic cells (e.g., midzonal and periportal Kupffer cells–liver macrophages) resident in the blood, liver, spleen, and lymph nodes as well as circulating monocytes. The RES is also called the mononuclear phagocytic system (MPS). The RES represents a preferential drug distribution site, following the first pass (Kostarelos, 2003). *Pharmacokinetics* is a study of the time course in the distribution of an administered drug concentration in the different organs throughout the body. *Pharmacodynamics* is a quantitative evaluation of any clinical parameter, such as body temperature, decrease in drug load, or therapeutic index, temporally fluctuating relative to the administered agent. *Bioavailability* is used to describe the fraction of an administered dose of medication that reaches the systemic circulation. Bioavailability is a measure of the rate and extent of a therapeutically active drug that reaches the systemic circulation after the first pass through the liver and is available at the site of action (Pond and Tozer, 1984). By definition, when a medication is administered intravenously, its bioavailability is 100%. However, when a medication is administered via other routes (such as orally), its bioavailability decreases (due to incomplete absorption and first-pass metabolism). Bioavailability is one of the essential considerations in pharmacokinetics, for bioavailability must be considered when calculating dosages for nonintravenous routes of administration. Even intravenous administration does not guarantee that a drug is freely available, because of plasma-protein binding. Drugs can bind to a variety of particles in the blood, including red blood cells, leukocytes, and platelets, in addition to proteins such as albumin (HSA), glycoproteins, basic drugs, including gene delivery vehicles, lipoproteins (neutral and basic drugs), and globulins (van de Waterbeemd and Gifford, 2003).

*Nanovehicles* (1–1000 nm) are typically composed of biodegradable and biocompatible polymers. Their degradation *in vivo* can occur over months to years (Okada and Toguchi, 1995) via enzymatic/hydrolytic scission mechanisms. For example, TCA cycle metabolism of PLGA and related polymers can result in the biotolerable metabolites of lactic acid and glycolic acid. Nanovehicles often bind very efficiently to cells, internalized, and sorted into different organelles or cytoplasm, where they exert their function. Since gene products require an intracellular uptake for their function, such nanovehicles are typically used as means of their delivery.

Nanovehicles are thus defined as quasi-soluble, nanoparticulate delivery systems for intracellular delivery. Many different nanoparticulate technology platforms have been employed, each with different properties, strengths, and weaknesses. Most frequently discussed among these are polymer-based nanoparticulate platforms (Vicent and Duncan, 2006), dendrimers (Lee *et al.*, 2005), liposomes (Sato and Sunamoto, 1992; Bendas, 2001), gold nano-shells (Paciotti *et al.*, 2004), semiconductor nanocrystals (Q-dots; Akerman *et al.*, 2002), silicon- and silica-based nanosystems (Chen *et al.*, 2004), and superparamagnetic nanoparticulates (Ito *et al.*, 2005), among others. They can all be employed for gene delivery.

The fundamental opportunities for nanoparticulate delivery are summarized in three closely interrelated aspects: (1) the *recognition of target cells and tissues*, (2) the ability *to reach the diseased sites* where the target cells and tissues are located, and (3) the ability to *deliver multiple therapeutic agents*. The first two aspects comprise the notion of achieving preferred, substantially higher concentration of a therapeutic agent at site, a phenomenon that will be called *localization*, as opposed to the term *targeting*, which is often used to identify drugs that provide specific action against a target biological pathway (Ferrari, 2005a).

The nanovehicular systems offer certain distinct advantages for drug and gene delivery. Due to their subcellular and submicron size, nanovehicles can penetrate deep into tissues through fine capillaries, cross the fenestrations present in the epithelial lining (e.g., liver), and generally be taken up efficiently in cells. This allows efficient delivery of therapeutic agents to target sites in the body. Also, by modulating polymer characteristics, one can control the release of a therapeutic agent from nanovehicles to achieve the desired therapeutic level in the target tissue. Nanovehicles can be delivered to specific sites by means of conjugation or adsorption of a biospecific ligand. Localized delivery can improve the therapeutic index of drugs by minimizing the toxic effects to healthy (nondiseased) tissues/cells, largely depending on the extent of circulation time of nanovehicles in the central compartment (see below), the release rate, and the rate of uptake of nanovehicles (their internalization; see below).

## Intracellular Delivery: Pharmacokinetics

### Defining the Elementary Steps

1. *Removal from the circulation.* It is essential that the nanovehicle, loaded with a drug/gene, not be cleared too quickly from the circulation. Rapid clearance may

prevent the vehicle from reaching the required concentration at the site of localization. Many vehicles/drugs will bind to plasma components (principally HSA) or within other compartments of the tissue. Binding can greatly influence the transport and elimination in individual organs and can influence the overall pharmacokinetics.

2. *Release of free payload at nontargeted sites.* Depending on the amount of the gene/vector, the release of the gene/vector away from the target site could nullify any benefits that might potentially come from delivering the gene/vector to the target site.

3. *Delivery of the gene vehicle to the target site.* If the nanovehicle reaches the target site too slowly, the supply of contents might never be sufficient to generate the concentration required to elicit the desired therapeutic effect at the site of action (delivery window).

4. *Release of free payload at the target site.* The capacity of the system selected for the release of payload from the nanovehicle should be considered at a rate that also ensures drug accumulation at the target site.

5. *Removal of free payload from the target site.* Agents that benefit most from localized delivery are those that are retained at the site while acting on their target of action.

6. *Elimination of the vehicle from the body.* For optimal localization, elimination of the agent vehicle should be minimal. Nanovehicles and their payloads are typically eliminated via the kidneys, liver, and spleen, depending on their size and other properties.

Some useful details on these elementary steps and on mathematical modeling tools that encompass the foregoing considerations can be found in references by Petrak (2005) and Boddy *et al.* (1989).

## Pharmacokinetics

Pharmacokinetics and access to different compartments are of crucial importance for gene delivery.

### Fundamentals

Some fundamentals of pharmacokinetic (PK) modeling were summarized by Saltzman (2001). The goal of pharmacokinetics is synthesis into a coherent model of physical and biological phenomena involved in drug distribution in the body, although development of a comprehensive (elementary) model may not be practical. The four components of PK are commonly referred to as absorption, distribution, metabolism, and excretion (ADME). Being empirical, the utility of compartmental models is limited, because they are not valid beyond their experimental domain. In contrast, detailed physiological models often contain parameters that are difficult to measure. A physiological pharmacokinetic model may include a multiple-compartment approach, whereas each organ is composed of multiple subcompartments, reflecting the anatomy/morphology of the organ. In a typical pharmacological model, the target is accessible directly from the central compartment (whereas delivery to a peripheral compartment is required in another case; e.g., in delivery to skin, muscle, peritoneum). The central compartment is defined as a blood/lymphatic circulation system. In the most favorable case, traditionally measured plasma drug concentration is connected with tissue distribution and pharmacokinetic measurements. The pharmacokinetic modeling may provide insights into appropriate dosing regimens (amount of drug, capacity, and dosing frequency), optimal binding affinity, and how receptor-mediated effects may be anticipated from natural or mimetic drug ligands (Mager *et al.*, 2006).

### Accessing the Tumor Compartment

The tumor compartment can be presented as a single entity composed of a vascular subcompartment with cell surface receptors and an intracellular tumor space (the target tissue). Nanovehicles accumulate in tumor tissue because of their extended circulation time in conjunction with preferential extravasation from leaky tumor vessels (EPR effect). The EPR effect features tumor hypervasculature, defective vascular architecture, and a deficient lymphatic drainage system. The attachment of the nanovehicle to the endothelium is either nonspecific or facilitated by a specific targeting (active targeting) motif directing nanovehicles to the tumor endothelium. This passive targeting process facilitates tumor tissue binding, followed by uptake (internalization). This phenomenon results in intracellular release of antitumor activity. Nanovehicles that fail to bind to tumor cells will reside in the extracellular (interstitial) space, where they eventually become destabilized because of enzymatic and phagocytic attack. This results in extracellular payload release, for eventual diffusion to nearby tumor cells and possible bystander effects. Targeted drug delivery systems may substantially affect both drug disposition and pharmacological properties.

### Lymphatic Administration

Lymphatic administration is a means of minimizing general systemic drug exposure to modify biodistribution. The primary function of the lymphatics is to drain the capillary beds and return extracellular fluid to the systemic circulation. Unlike the blood flow, lymph flow is single-pass, recovering fluid from the periphery and returning it to the vasculature. Drug transport through the lymph may be utilized to prolong the time course of drug delivery to the systemic circulation, bypassing liver and avoiding hepatic first-pass metabolism.

## III. EXPLORING THE ROLE OF RECEPTOR LIGAND SIGNALING AND RECEPTOR CLUSTERING IN AGENT DELIVERY

Cell surface receptors are complex transmembrane proteins that mediate highly specific interactions between cells and their extracellular milieu. The type, number, and functions of receptors are defined by cellular lineage, genetics, and a number of extracellular factors. Since receptors are differentially expressed in various cells types and tissues, they provide potential targets for gene delivery. Receptors, however, not only are cellular markers, but play an integral role in the regulation of virtually all cellular functions, including growth, differentiation, metabolism, secretion, contraction, and migration.

Receptor binding is an area of potential importance to localized gene delivery, including endocytosis, transcytosis, ligand–receptor interactions, and receptor regulation. Since biochemical and physiological properties of receptors vary depending on both receptor type and cellular background, it is likely that some receptor systems may be more suitable than others for receptor-mediated drug delivery. Next, we review some basics of the cellular entry of macromolecules and particles (King and Feener, 1998).

## Taking Advantage of Receptor-Mediated Endocytosis and Signaling

An important function for receptors is to facilitate ligand internalization via receptor-mediated endocytosis. In this process, a ligand bound to its receptor is endocytosed from clathrin-coated pits on the plasma membrane, forming an endocytotic vesicle.

For localization of gene delivery vehicles, various endogenous ligands, such as peptides, proteins, lipoproteins, growth factors, vitamins, and carbohydrates, can be used. Folate, transferrin, and EGF receptors are good examples of such ligands.

## Employment of a Proper Receptor Repertoire for Delivery

While the initial events in the endocytosis of the receptor–ligand complex are similar for most systems, the processing of the ligand can differ, depending on both receptor and cell type. In most instances, the role of receptor-mediated endocytosis is to internalize ligands for their subsequent proteolytic degradation in the acidic pH of the endosomal compartment. The receptor is then recycled back to the plasma membrane, whereas the ligand is destined for lysosomal degradation. This proteolysis of ligand can serve to remove a signal or to clear an undesirable protein or delivery vehicle. Cells can also reduce the level of ligand stimulation both by degrading the ligand and by reducing the number of receptors expressed on the plasma membrane (down-regulation of receptors).

A more detailed understanding of the biochemical and physiological characteristics of receptors is necessary to avoid the undesirable effects of selected binding to these active cell surface proteins and to realize their full potential for receptor-mediated gene delivery. Receptors' primary functions are to mediate protein trafficking and to transduce signals across the cellular membrane. Studies of receptor structure have revealed specific domains and motifs that regulate receptor trafficking, kinase activity, and coupling with intracellular signaling networks. These findings can be utilized at the design of gene delivery systems.

## Delivering Across the Cellular Barriers

Receptor-mediated transcytosis could be utilized as a method for the delivery of drugs and genes across cellular layers, including the blood–brain barrier (Olivier, 2005). In cells that are polarized into apical and basolateral surfaces, such as endothelial and epithelial cells, certain receptors can mediate transcytosis. This involves the vectorial trafficking of the ligand, via endocytic vesicles, across the cellular layer. A number of ligands have been reported to be transcytosed in these cell types, including insulin and EGF (Kurihara and Patridge, 1999; Brandli et al., 1991; Shah and Shen, 1996). The role of transcytosis is to selectively facilitate the transport of ligands across diffusional barriers, such as the vascular endothelium.

## Employing Receptor Clustering

Transmembrane receptors are increasingly being found to be organized into higher-order structures in the cytoplasmic membrane (Bollinger et al., 2005). Clustering may be an evolved strategy to achieve the greater effectiveness and/or higher strengthening of adhesion required by many biological functions other than uniform molecular distribution. Clustering renders high local densities of receptors and/or ligands, which may be independent of global or average densities, resulting in multiple bonds that do not obey existing criteria for receptor–ligand interaction. One desirable mechanism is thus multiple receptor interactions of ligands with multiple receptors on the cell surface.

One example of clustering effects is the family of transmembrane proteoglycans, the syndecans, which have important roles in cell adhesion. They participate in biological processes through binding of ligands to their glycosaminoglycan chains, clustering, and the induction of signaling cascades to modify the internal microfilament organization. Syndecans can modulate the type of adhesive responses induced by other matrix ligand–receptor interactions, such as those involving the integrins, and thus contribute to the control of cell morphology, adhesion, and migration. These proteoglycans have been used for localization of drug delivery vehicles.

Another example comes from a carbohydrate–carbohydrate and protein–glycan interaction. Characteristic features of these interactions are their specificity, their strong dependence on divalent cations, and their extreme low affinity, which has to be compensated by multivalent presentation of the ligands. Multivalent presentation (display) has been proved to be important in the study of carbohydrate–protein interaction (de la Fuente and Penades, 2004).

# IV. OVERVIEW OF NANOVEHICLE UPTAKE AND TRAFFICKING

## Background Information

The successful design of a delivery system requires a thorough understanding of the mechanisms involved in the interaction of the delivery systems with the target cells. In particular, an understanding of intracellular trafficking and targeting would help to design better and more efficient gene delivery systems. Intracellular targeting refers to the delivery of therapeutic agents to specific compartments or organelles within the cell. Endocytosis (and some other forms of uptake, phagocytosis) is considered the most useful mechanism to describe multiple forms of internalization. Because of many commonalities between drug and gene delivery systems, many references are cited on the behavior of drug delivery systems.

## Nanovehicle Size

Our studies and those of others show that particle size significantly affects cellular and tissue uptake; in some cell lines, only the submicron-size particles are taken up efficiently but not the larger microparticles (e.g., Win and Feng, 2005; Desai *et al.*, 1996; Zauner *et al.*, 2001; Rejman *et al.*, 2004). In Caco-2 cells the uptake of nanoparticles (PLGA with PVA coating) 100 nm in size was 2.3-fold greater as compared to that of 50-nm particles, 1.3-fold to that of 500-nm particles, and about 1.8-fold to that of 1000-nm particles. Thus, it has been demonstrated that nanoparticles 100–200 nm in size possess the best properties for cellular uptake (Win and Feng, 2005; Moghimi *et al.*, 2001).

## Spatial and Temporal Effects

Nanovehicle internalization is concentration and time dependent as well as cell-type dependent. Endocytosis results in internalization of the cell's plasma membrane, to form vesicles that capture macromolecules and particles present in the extracellular fluid and/or bound to membrane-associated receptors. These vesicles then undergo a complex series of fusion events directing the internalized cargo to an appropriate intracellular compartment (uptake of fluids, macromolecules, particles, and other ligands that sort to the cell's processing pathways). Uptake of particulate systems can occur through various processes, such as by phagocytosis, fluid-phase pinocytosis, transport via clathrin-coated pits, caveolae-mediated transport, and nonendocytic pathways. Transcytosis and exocytosis are pathways communicating with the cells' external environment (Panyam and Labhasetwar, 2003).

## Trafficking Phenomena

Following their uptake, nanovehicles have been shown to be transported to primary endosomes and then to sorting endosomes. From sorting endosomes, a fraction of nanovehicles are sorted back to the cell exterior through recycling endosomes (efflux, exocytosis), while the remaining fraction is transported to secondary endosomes, which then fuse with lysosomes. The nanovehicles then escape the endo-lysosomes and enter the cytosolic compartment. Time-dependent uptake studies showed that nanovehicles escaped the endo-lysosomes within 10 min of incubation and entered the cytoplasmic compartment. Selective surface-charge reversal of nanovehicles in the acidic pH of endo-lysosomes is proposed as the mechanism responsible for the endo-lysosomal escape of the nanoparticles. Surface-charge reversal results from the transfer of protons/hydronium ions from bulk solution to the nanovehicle surface under acidic conditions. This is often due to the typical cationic charge of the nanovehicle periphery, leading to localized destabilization of the membrane and the escape of nanoparticles into the cytoplasmic compartment. Polystyrene nanoparticles, which do not exhibit surface-charge reversal with a change in pH, were not seen escaping the endo-lysosomal compartment (Panyam and Labhasetwar, 2004).

## Exocytosis

Nanovehicle amounts inside the cell are maintained as long as nanovehicles are present in the outside medium. Once the external concentration gradient is removed, exocytosis of intact nanovehicles begins and results in a substantial drop in a matter of minutes. However, at least some initial nanovehicle levels are maintained over several hours of incubation *in vitro*. Some researchers have observed that a nanovehicle exocytosis was inhibited in serum-free medium. Protein (albumin) present in the serum was found to be responsible for inducing nanovehicle exocytosis (Panyam *et al.*, 2003). Very little information is available on nanovehicle exocytosis.

## Are *In Vitro* Data Translatable to *In Vivo*?

The translation of the *in vitro* data to the *in vivo* situation is not always straightforward. One can infer the internalization of nanovehicles from experiments on gene delivery as well as from those that follow the localization of nanovehicles in tissues and cells. This detection is possible due to recent developments in labeling and imaging. The intracellular localization of nanovehicles in tissues is not easily obtained.

## Mechanisms of Nanovehicle Uptake and Efflux

Endocytosis occurs by multiple mechanisms that fall into two broad categories: *phagocytosis*, or cell eating (the uptake of large particles), and *pinocytosis*, or cell drinking (the uptake of fluid and solutes) (Conner and Schmid, 2002). Specialized efflux mechanisms serve to dispose of drugs and the nanovehicles. A summary of different mechanisms is presented in Table 35.1.

**Table 35.1.** Efficiency of nanovehicle entry via multiple portals as differentiated by cargo chemistry/size[a] and cell type

| | Phagocytosis | Pinocytosis | | | | |
|---|---|---|---|---|---|---|
| | | Macropinocytosis (fluid phase) | Clathrin-mediated (CME) | Caveolae-mediated rafts | Lipid rafts | Clathrin- and Caveolin-independent |
| Vehicle size | 1–10 μm[b] | 1–5 μm[b] | <150 nm | <60 nm | 40–60 nm | 200–300 nm |
| Cell types | Dendritic, macrophages, monocytes | Many cells | Many cells | Differentiated endothelial cells, adipocytes, epithelial, and muscle cells | Lymphocytes, cancer cells, rodent macrophages | Specialized cells, endothelial cells, and others |
| Efficiency of uptake[c] | +++++ for specialized cells | +++ | +++++ | + | + | ++++ |

[a]Typically, cargo chemistry determines a size range due to technological limitations (e.g., dendrimer-based and micellar-based nanovehicles are limited to rather small sizes).
[b]Phagocytosis and macropinocytosis most likely do not contribute to the nanovehicular uptake.
[c]The efficiency of uptake may depend on NV type and chemistry. The proportion between different uptake mechanisms may vary. Estimated on scale of 1–5, 5 being the highest.

## Phagocytosis (PC)

Phagocytosis is typically restricted to specialized mammalian cells (dendritic, macrophages, monocytes, and neutrophils), whereas pinocytosis occurs in all cells by at least five basic morphological distinct mechanisms: macropinocytosis (MPC), clathrin-mediated endocytosis (CME), caveolae-mediated endocytosis (CavME), lipid raft–mediated endocytosis (LRME), and clathrin- and caveolae-independent endocytosis (CCLIE).

## Macropinocytosis (MPC), or Fluid-Phase Endocytosis

Macropinocytosis can be measured by the intracellular accumulation of a tracer molecule present in the medium. The degree of internalization of fluid-phase markers is directly proportional to their concentration in the medium and the volume encased by the transport vesicles. Greater efficiency of endocytosis is achieved by nonspecific binding of solutes to the cell membrane (adsorptive pinocytosis), but the most efficient uptake occurs when dilute solutes are captured by specific high-affinity receptors (receptor-mediated endocytosis), which are themselves concentrated into specialized endocytic transport vesicles. Macropinocytosis is accompanied by membrane ruffling and protrusion formation, which then collapse onto and fuse with the plasma membrane to generate large endocytic vesicles called *macropinosomes* (>1 μm in size, up to 5 μm), which sample large volumes of the extracellular milieu.

Macropinocytosis is generally considered a non-receptor-mediated process (nonspecific), where cells take up large volumes of extracellular fluids and solutes. It is constitutive in highly specialized cells (macrophages and dendritic cells) and in some tumors, and it can be induced by growth factors in epithelial cells. It is the principal endocytic pathway in endothelial cells (EC). Although the pH of macropinosomes decreases, they do not fuse into lysosomes. This pathway provides some advantageous aspects, such as the increased uptake of particles and macromolecules, the avoidance of lysosomal degradation, and the ease of escape from macropinosomes because of their relatively leaky nature.

## Clathrin-Mediated Endocytosis (CME)

CME occurs constitutively in all mammalian cells, and it carries out the continuous uptake of essential nutrients, antigens, growth factors, and pathogens. Clathrin-coated vesicles of the CME pathway carry concentrated receptor–ligand complexes into the cells. They range in size from 100 to 150 nm in diameter. Molecules entering via this pathway will rapidly experience a drop in pH from neutral to pH 5.9–6.0 in the lumen of early endosomes, with a further reduction to pH 5 during progression from late endosomes to lysosomes (Maxfield and McGraw, 2004). Ligands and receptors are then sorted to their appropriate cellular destinations, such as lysosomes, the Golgi apparatus, the nucleus, or the cell surface membrane. Some portion of the receptor–ligand complex also recycles back, via what is called *exocytosis* (Goldstein *et al.*, 1985).

## Caveolar-Mediated Endocytosis (CavME)

Caveolae are flask-shaped invaginations (50–60 nm in size) of the plasma membrane that are typically seen on the surface of endothelial cells, where they are extremely abundant. Caveolae are derived from the Golgi complex and are now known to be present on many specialized cells and to demarcate cholesterol- and sphingolipid-rich microdomains of the plasma membrane, in which many diverse signaling molecules and membrane transporters are concentrated (Sangiorgio *et al.*, 2004). A major difference from CME is that the caveolar uptake is a nonacidic and nondigestive route of internalization, bypassing lysosomes, potentially a route of advantage for drug delivery. However, caveolae are internalized slowly and are small in size, and their fluid-phase volume is small; this results in *low-capacity uptake*. Thus, it is unlikely that they contribute significantly to constitutive endocytosis, although the situation is different in endothelial cells, in which caveolae constitute 10–20% of the cell surface (Conner and Schmid, 2002).

## Lipid Rafts–Mediated Uptake (LRME)

For anionic and neutral-lipid liposomes, solid-lipid nanoparticles, and hydrophobic nanoparticles (e.g., synthetic polystyrene nanoparticles) there is another potential pathway, endocytosis-independent uptake. Lipid rafts are small (40–60 nm in diameter) and free cholesterol- and sphingolipid-rich planar microdomains characteristic of cell surfaces that lack caveolin and caveolae, such as lymphocytes, many human cancer cells, and rodent macrophages. A hydrophobic uptake via the lipid membrane fusion is likely to be involved. The likely mechanism is that of the scavenger receptor, involved in the uptake of both lipophilic and anionic groups (Fabriek *et al.*, 2005). Scavenger receptor class B (SRB1; CD36 superfamily of proteins), expressed on mature macrophages, is particularly promiscuous, and ligands include proteins, polyribonucleotides, polysaccharides, and lipids, for which the main common feature is that they are "polyanionic." The involvement of SRB1 in liposome uptake was demonstrated on the basis of polyinosinic acid (a strong polyanion) competitive abolishment of the association of neutral phospholipids/cholesterol liposomes with hepatic cells (Adrian *et al.*, 2006). As expected, anionically charged liposomes (heparin-functionalized) exhibit an extended circulation time and provide better pharmacokinetics for encapsulated doxorubicin (Dox), following their intravenous injection in mice (Han *et al.*, 2006). It appears that the lipid raft–mediated fusion is facilitated by a transbilayer membrane potential that exists across cell membranes. This constitutes a major difference between the caveolae- and lipid raft–mediated endocytosis: They lack an energy input requirement for lipid raft–mediated uptake, a strong argument for this distinction, although

not universal. In addition, this type of uptake features bypassing of the lysosomal compartment, of significance for gene delivery.

## Clathrin- and Caveolae-Independent Endocytosis (Nonendocytic Pathway) (CCLIE)

The recently discovered clathrin- and caveoli-independent pathway is of significance for ICAM-1-positive cells (CAM-mediated endocytosis; Muro *et al.*, 2004). Nanovehicles coated with anti-ICAM antibody induce ICAM-1 clustering by multivalent nanovehicle presentation. CAM-mediated endocytosis delivers materials to lysosomes (Muro *et al.*, 2006a). CAM-1, a transmembrane glycoprotein from the Ig-like superfamily of adhesion molecules, which is up-regulated and functionally involved in inflammation and thrombosis, may provide a unique target for lysosomal disease (LSD). It is expressed by cell types affected in the diverse forms of LSDs, including endothelial (EC), epithelial, glial, and Schwann cells; leukocytes; myocytes; and other cells. However, ICAM-1 and another Ig superfamily cell adhesion molecule, platelet endothelial cell adhesion molecule-1 (PECAM-1), are not readily internalized by endothelial cells. Little is known about the intracellular trafficking and fate of anti-CAM nanoparticles. Muro and others have also reported on a "monensin-switch" modulation (via $Na^+/H^+$ exchanger), allowing the bypassing of lysosomes, with the potential of improving drug cytoplasmic efficacy (Muro *et al.*, 2006b).

## Exocytosis (EC)

Exocytosis is a constitutive mechanism of metabolite elimination for every cell. It may be also operational for nanovehicles. Limited studies are available at this point, however. The underlying observation is that once the cells are devoid of any external reservoir of nanovehicles for uptake, once-internalized nanovehicles leave the intracellular sites and are expelled outside (Panyam *et al.*, 2003). This is probably because of the high local concentration of nanovehicles in the vicinity of cell membranes, resulting in their internalization, along with extracellular liquid and solutes, into the cell through endocytosis (pinocytosis). Recent observations testify to the role of targeting ligands associated with nanovehicles in exocytosis. Sahoo and Labhasetwar (2003) observed decreased exocytosis for Trf-conjugated PLGA nanoparticles, as compared to nonfunctionalized ones. Salaün *et al.* (2004) argue that there is a constitutive exocytosis pathway operating in all cells, in addition to a regulated one, facilitated by lipid rafts and cholesterol- and sphingolipid-rich microdomains, enriched in the plasma membrane. This constitutive pathway may be of significance for liposomal gene delivery.

## Transcytosis (TC)

Transcytosis reflects transport across the endothelial (and epithelial) cells to the subendothelial space. For transcellular transport, drugs need to cross two different membranes on the basal and apical sides. In the intestine, drugs are absorbed from the luminal side (brush border membrane) and excreted into the portal blood across the basolateral membrane. The endocytosis on the apical (luminal) surfaces is followed by transport through a series of intercellular compartments and delivery to the basolateral (ablumenal) plasma surface, where the substances are secreted or diffused via intercellular junctions (paracellular transport). Transporters are thus important as a determinant of the clearance in the body (Shitara *et al.*, 2006). The process of caveolar-mediated endocytosis is highly inefficient, as observed on passage through the gastrointestinal tract and brain–blood barrier (BBB), amounting to less than 1–2% of the luminal input (Florence and Hussain, 2001). Transcytosis of nanovehicles has been also documented (Oberdoster *et al.*, 2005; Florence and Hussain, 2001).

## Conclusions About Nanovehicular Uptake

Concluding on the foregoing portals of entry of nanovehicular drugs/vehicles, there is accumulating evidence that different endocytic pathways merge and are connected to the same early endosome that serves as a distribution site inside the cells. The cargo may be delivered to internal organelles (caveosomes) from noncaveolar uptake mechanisms (Sharma *et al.*, 2002). There are no definitive answers to all questions in terms of involvement of a certain trafficking pathway in different cells. Considerable confusion exists because many endocytosis studies relied on the use of pharmacological agents whose specificity remains a matter of debate. In addition, many artifacts have been generated and reported because of improper cell fixation prior to microscopic observation of the cells. The prevalent view is that the cargo (payload) also influences the type of uptake mechanism and trafficking (Johannes and Lamaze, 2002). A schematic illustration of different uptake/exocytosis mechanisms is depicted in Fig. 35.1. Figure 35.2 presents possible routes of the intracellular fate of nanovehicles.

Kinetically, three modes of endocytosis can be defined: fluid-phase, adsorptive, and receptor-mediated endocytosis (Amyere *et al.*, 2002). *Fluid-phase endocytosis* refers to the bulk uptake of solutes in exact proportion to their concentration in the extracellular fluid. This is a low-efficiency and nonspecific process. In contrast, in *adsorptive* and *receptor-mediated endocytosis*, macromolecules are bound to the cell surface and concentrated before internalization. In adsorptive endocytosis, molecules preferentially interact with generic complementary binding sites (e.g., by HSPG, lectin, or charged interaction) in a nonspecific manner. Such a distinction is of practical interest because of the uncertainties of the intimate uptake mechanism, as discussed earlier. Fluid-phase endocytosis has a lower internalization capability than adsorptive endocytosis.

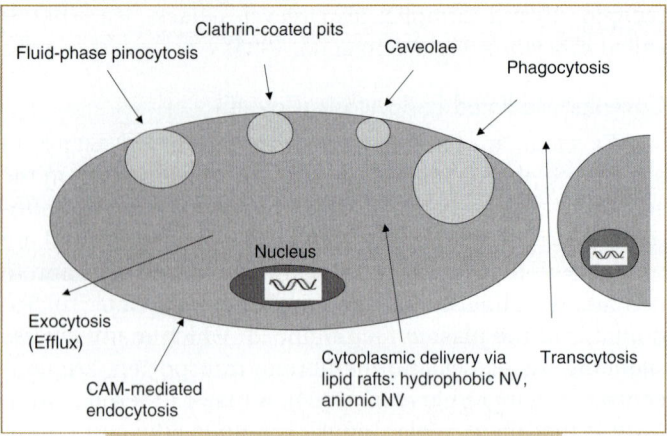

**FIG. 35.1.** Schematic representation of the uptake and exocytosis routes of nanovehicles. Adapted from Panyam and Labhasetwar (2003). Not drawn to scale. Courtesy of Springer Verlag.

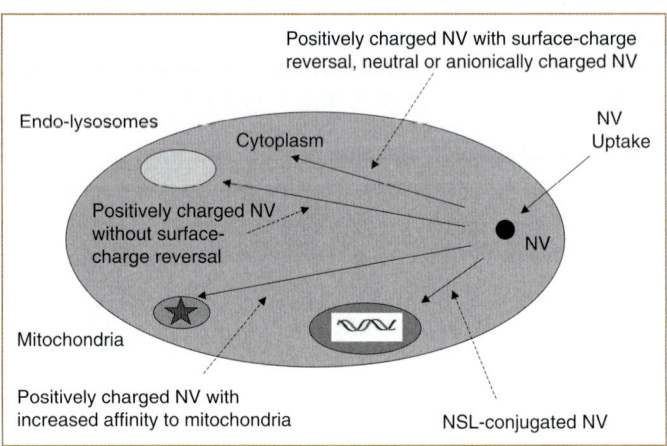

**FIG. 35.2.** Schematic representation of intracellular trafficking of nanovehicles. Although charge-based phenomena are shown to have a controlling role, a targeting ligand, attached to the nanovehicle periphery, may also contribute to the organelle distribution. Note that the endoplasmic reticulum (ER) has been eliminated from this drawing for the sake of simplicity. NSL — nuclear signaling ligand. Adapted from Panyam and Labhasetwar (2003). Not drawn to scale. Courtesy of Springer Verlag.

## V. STABILITY OF NANOVECTORS IN BUFFERS AND BIOLOGICAL FLUIDS: STERIC VS. ELECTROSTATIC STABILIZATION

### Steric Stabilization

To be effective for delivery, nanovehicles must remain in a colloidal form. This requires the minimization of interparticle interactions. To avoid nonspecific aggregation under physiological conditions, the long-term stability of the nanovehicles represents a serious problem (Moghimi, 2002). Steric stabilization is defined as reduction in particle interactions by means of a surface steric barrier. The prevailing view is that repulsive forces result from the "compression" of tethered layers presented on the nanovehicle surface. In addition, the hydrophilic side chains extending outward from the particle surface provide stability to the particle suspension via a repulsion effect through a steric mechanism of stabilization involving both enthalpic and entropic contributions, while the dimension of the stabilizing chains exceeds the range of Van der Waals force of attraction. Macrophages express surface receptors that do not recognize or internalize sterically protected nanovehicles.

## Electrostatic and Combined Stabilization

In contrast to the kinetic stability achieved by electrically neutral polymers, electrostatic stabilization employs repulsive, coulombic interactions between the charged surfaces (Pincus, 1991; Einarson and Berg, 1993). Note that most nanovehicles are endowed with a charge. Any ionizable groups in the hydrophilic domains presented at the nanovehicle periphery act as electrostatic stabilizers. This method is sensitive to dissolved salts, which suppress the screening effect. A coupled mechanism involving both electrostatic and steric mechanisms is denoted as *electrosteric stabilization*. A good example is the employment of anionic copolymers with suitably sized PEG molecules. Attached to gene delivery vectors via a coating, such complexes readily prevent opsonization by serum proteins, maintain their size, and do not aggregate at near electroneutrality (low zeta potential) (Finsinger *et al.*, 2000).

The stability of the electrostatic complex (the most frequently employed form of delivery for DNA; see before) *in vivo* is another issue. In general, PEC stability in plasma could present a problem. PLL/DNA complexes (poly L-lactate)/DNA can undergo rapid clearance from the plasma following intravenous administration. The plasma proteins, particularly serum albumin, can loosen the complexes and enlarge them. Dash *et al.* (1999) reported on the possibility of ternary complex between PLL/DNA and albumin that can undergo a disruption by positively charged serum proteins at physiological pH. All forms of PEC complexes may require stabilization to improve their circulation times *in vivo*. Cationic polymers with short side chains were shown to form smaller complexes, resistant to destabilization by polyanions. The destabilization of cationic lipid/DNA complexes *in vitro* and in plasma by HSPG polymers and heparin facilitates DNA release (Wiethoff *et al.*, 2001; Chittimalla *et al.*, 2005). The multicomponent PEC-based nanoparticles (Prokop *et al.*, 2002a, 2000b) are more resistant to the inter-polyelectrolyte exchange mechanism (competitive displacement of the components of the polyelectrolyte complex with other charged species, external polyelectrolytes, and small ions). The inter-polyelectrolyte exchange often leads to unpacking of the assembled particle (Zelikin *et al.*, 2003).

The PEC complex is considered a result of a reversible monomolecular reaction, while there is a competition between complexation and counterion binding. The multicomponent system apparently results from an interaction of two pairs of polyelectrolytes. Oupicky *et al.* (2002) demonstrated an increased resistance to polyelectrolyte exchange and prolonged plasma circulation time for polycation/DNA polyplexes modified with multivalent PHPMA [*N*-(2-hydroxypropyl)methacrylamide] laterally cross-linked polymer (of the surface of polyplex), for the presence of PEG molecules only has no beneficial effect on susceptibility to the polyelectrolyte exchange reaction. The stability of the intracellular complex should also be checked (Breunig *et al.*, 2006).

# VI. LOCALIZATION AND TARGETING

Ferrari (2005b) introduced a concept of localization and employment of additive, multimodal localization strategies, because they are noninterfering and each contributes favorably to the achievement of preferred spatial concentrations.

## Localization

### Localization by Size and Shape

This is applicable to liposomes, where it has been demonstrated that tailoring nanovector size to match the fenestrations of the cancer neovasculature yields preferential concentration at tumor sites — a phenomenon termed *enhanced penetration and retention* (EPR).

### Localization by Physical Properties

The surface charge of nanovectors influences tumor uptake. Differences between positively vs. negatively charged nanovectors have been noted. Several investigators demonstrated that negatively charged nanovehicles were distributed away from the RES system (Roser *et al.*, 1998; Yang and Zhu, 2002; Furumoto *et al.*, 2004), as opposed to the characteristic behavior of cationically charged nanovectors, which are typical in gene delivery.

### Localization by Remote or Environmental Activation

Regardless of the details of the distribution of injected nanovectors in the body, exquisite localization of the effect may be attained if the agent is activated only at the intended target sites by irradiation with an exogenous and locally focused source of energy. An excellent example of remote activation is a TNFerade Biologic™, an adenovirus carrying the TNFα gene under the control of a radiation-inducible promoter (McLoughlin *et al.*, 2005). It has entered phase II trials, where an intratumoral injection is used in combination with radiotherapy. In this way, control of different spatial and temporal levels is achieved.

## Targeting

Selective delivery and prolonged retention is desired via the control of the pharmacological parameters characteristic of the administered moieties. This can be achieved by designing delivery systems that either (a) enhance the deposition or accumulation of the therapeutic in a particular tissue; (b) associate the therapeutic with a particular cell population; (c) associate the therapeutic with a specific intracellular component; or (d) prolong the association of the therapeutic within a specific organ (e.g., brain, blood); or (e) any combination of these (Kostarelos, 2003).

### Active Targeting

This targeting refers to delivery systems designed to associate or interact with specific biological moieties, most commonly by attachment on their outer surface of ligands (peptides, antibodies, antibody fragments, and proteins) with an enhanced binding affinity for cellular receptors or surface matrix molecules. Besides a higher binding affinity, one should distinguish between the binding, enhanced vehicle internalization, and subsequent downstream biological activity.

### Passive Targeting

This targeting refers to all strategies attempting to achieve defined delivery without utilizing specific biological (ligand–receptor) interactions accompanied by correlating the physicochemical and surface characteristics of the delivery systems with the pathophysiology and anatomy of the target sites. Illustrative examples of passive targeting include the extravasation of sterically stabilized nanoparticles from leaky tumor capillaries into the interstitium, and the extended blood circulation half-lives of polymer-coated moieties (proteins, drugs, and liposomes). Passive targeting of macromolecular drugs, including gene delivery vectors, to tumors can be achieved by the EPR effect (Gunther et al., 2005).

### Externally Stimulated Targeting

The third type of targeting, gradually becoming more commonly studied, can be termed *externally stimulated targeting*, whereby localization of the therapeutic agent at a specific site is achieved by an externally applied stimulus, such as temperature, light, ultrasound, ionizing radiation, or magnetic force (Minko et al., 2004).

## Intracellular Targeting

Nanovectors can be used as efficient delivery vehicles for intracellular targeting. Types of intracellular targeting include (a) endo-lysosomal, (b) cytoplasmic, (c) cytoskeletal and caveolar, (d) mitochondrial, and (e) nuclear. An important feature of nanoparticles is that physical properties such as size, surface charge, hydrophobicity, and release characteristics can easily be varied by altering the composition of the formulation and/or the formulation method.

### Lysosomal Targeting

Lysosomal targeting, occurring in many forms of endocytic uptake, has shown an application in therapy of lysosomal storage diseases, including exogenous delivery of enzymes (or genes), such as glucocerebrosidase, fucosidases, phenylalanine ammonia lyases, and others (Sands and Davidson, 2006).

### Cytoplasmic Targeting

Nanoparticles are able to escape the endolysosomes depends on the surface charge of the nanoparticles (Panyam and Labhasetwar, 2004). Nanoparticles that show transition in their surface charge from anionic (at pH 7) to cationic in the acidic endosomal pH (pH 4–5) were found to escape the endosomal compartment, whereas the nanoparticles that remain negatively charged at pH 4 were retained, mostly in the endosomal compartment (Panyam et al., 2003). Thus, by varying the surface charge, one could potentially direct the nanoparticles either to lysosomes or to cytoplasm. Delivery to cytoplasmic targets is of special significance for siRNA delivery.

### Targeting to Cytoskeleton and Caveloae

A special targeting to subsurface structures is that of targeting to cytoskeleton and caveolae. Many diseases have now been associated with abnormalities in cytoskeletal and nucleoskeletal proteins (Ramaekers and Bosman, 2004). Very little has been proposed for such targeting. Targeting drugs to caveolae structures (a subcompartment of the plasma membrane) could potentially interfere with diseases such as muscular dystrophy and cancer, among other diseases (Campbell et al., 2001). Gene delivery to such a compartment is quite open to further exploration.

### Mitochondrial Targeting

It seems possible to localize nanovehicles to mitochondria by modifying the nanoparticle surface to obtain a high net positive charge (Panyam et al., 2002). Delivery to mitochondria has been reviewed (Murphy and Smith, 2000; Holcik, 2004). Permanently positive motifs with delocalized cationic charge, like oligoguaninidines, can target molecules and nanovehicles to mitochondria in response to the highly negative membrane potential (Rothbard et al., 2005). The correction of a mitochondrial deficiency has been demonstrated to be facilitated by this mechanism (Holt et al., 1988).

### Nuclear Targeting

Organelle-specific localization signals, such as fusogenic endosomal membrane peptide or nuclear localization signal (NLS), can be attached to the nanoparticle surface, enabling nanoparticles to target the nucleus (Simoes *et al.*, 1998). Nuclear targeting is of eminent significance for traditional gene delivery.

## Tissue Localization

Targeted delivery of therapeutic agents to specific tissues has been made feasible due to a number of developments, such as monoclonal antibodies, discovery of specific receptors that are either overexpressed or expressed only in specific tissues, and development of conjugation techniques to attach antibodies or ligands to drug delivery systems. Targeted delivery results in higher bioavailability of the therapeutic agent at its site of action and at the same time results in reduced side effects. Several methodologies have been developed to enable tumor-specific drug delivery.

### Specificity of Targeting

The development of nanovehicular delivery systems for targeted delivery has been reviewed recently by Moghimi *et al.* (2005). Active targeting of a therapeutic agent or a carrier is achieved by conjugating the therapeutic agent or the carrier to a tissue- or cell-specific ligand (Cegnar *et al.*, 2005). Passive targeting is achieved by coupling the therapeutic agent to a macromolecule or carrier that passively reaches the target organ (Nori and Kopecek, 2005). Nanoparticles can be passively targeted to the RES system and circulating macrophages. According to Senger's findings (Senger *et al.*, 1983), polymeric drugs (and gene vectors) accumulate in tumor tissue at 5–10 times greater concentrations in the 24 h after intravenous injection. Ideally, the nanovector size should not exceed 300 nm for optimum transport to tumor sites (Green *et al.*, 1998). Such sizes above a critical 300 nm are also most vulnerable to macrophage phagocytosis.

Only a few successful examples of targeting are accompanied by details on their mechanism. With regulatory approvals of drugs such as Herceptin (trastuzumab), for the treatment of metastatic breast cancer, and Gleevec (imatinib mesylate), for the treatment of chronic myelogenous leukemia, a new class of therapeutics, based on selective drug (gene) targeting to pathologic components, has been clinically validated. Within this growing class of targeted pharmaceuticals, folate–drug conjugates constitute a well-studied example of a distinct subclass of receptor-targeted therapeutics. Folic acid selectively binds to and delivers attached drugs and genes into any cell that expresses a cell surface folate receptor (FR). Because FR is expressed and accessible primarily on pathologic cells, folate conjugation allows delivery of nonspecific drugs selectively into pathologic cells. As a consequence, normal tissues lacking FR are spared the toxicity that commonly limits nontargeted therapies (Hilgenbrink and Low, 2005).

Targeting with help of transferrin represents another example. Iron-transporting protein (Trf) is up-regulated in tumor cells. DNA/lipid-based liposomes, coated with anti-transferrin receptor single-chain antibody fragment and PEG, lead to an efficient delivery to tumors, inhibiting the first-pass clearance observed with non-PEG-containing liposomes (Yu *et al.*, 2004). Specific targeting of genes by means of many different ligands has been readily used to guide gene delivery.

### Tumor Targeting

Gene delivery to cancer represents about two-thirds of present clinical trials. Targeting to the tumor vascular compartment provides improved cancer therapy. The advantages of targeting tumor vasculature are numerous (Alessi *et al.*, 2004):

- Better selectivity of the treatment against proliferative tumor-derived endothelial cells and minimal toxicity because angiogenesis in the adult is limited to wound healing, ovulation, and pregnancy.
- Easy access for drugs from the blood to tumor vascular endothelial cells (compared with drugs that have to penetrate large, bulky tumor masses).
- Low mutation rate (high, genetic stability of host endothelial and stromal cells) within the vasculature, and thus treatment is independent of tumor-cell resistance mechanisms.
- Broad applicability (both solid tumors and leukemia are dependent on angiogenesis for their survival) and high efficacy because each tumor capillary supplies hundreds of tumor cells.

There are some concerns on the development of resistance. To overcome different modes of resistance, broad-spectrum antiangiogenic modalities should be used or, alternatively, a combination of antiangiogenic agents or a combination of an antiangiogenic agent with drugs (or genes) that decrease the survival threshold of the tumor vascular compartment could be considered.

While considering endothelium as a target, we should differentiate between the endothelium as a portal for delivery vs. destruction of tumor vasculature. While attempting the latter, one should point out that avascularity of tumor (or of its regions) often translates into limited drug delivery to tumors (Baish *et al.*, 1996) because poorly vascularized portions of tumor will accumulate fewer drugs and genes.

## VII. DRUG LOADING

Nanovector cores serve as a reservoir for releasing and delivering drugs and genes. However, the small internal cavity or extravehicular surface does not permit high drug loading. For example, dendrimer-based nanovehicles, due to their small size, have a poor capacity to incorporate active compounds. Thus the delivery application is limited to very potent drugs (biological modifiers), such as peptides, gene vectors, oligonucleotides, antibody fragments, and cytokines. The capacity of a nanovehicle is determined primarily by the chemistry of the vehicle, of both the core and the surface. Polymeric vehicles are promising drug carriers, of which the critical parameters, such as size and drug loading and release, can be controlled by engineering the constituent polymers. As a rule, the core chemistry controls the drug loading (and release). A hydrophilic core is amenable to charged drugs and noncharged entities. Unfortunately, no systematic research is available to assess the drug loading as related to polymeric carrier chemistry. Most gene agents are hydrophilic and are typically associated with hydrophilic core polymers of nanovehicles or the inner hydrophilic cavity of certain liposomes.

## VIII. GENE DELIVERY SYSTEMS (GDS)

The ideal nanoparticle vector would achieve a long circulation time, a low immunogenicity, a good biocompatibility, selective localization, efficient penetration of physiological barriers such as vascular endothelium and the blood–brain barrier, external activation or self-regulating drug release, and have no clinical side effects. Universal tools, such as PEGylation and "antibody decoration," become requisite features to improve circulation time and enable selective targeting. In many delivery systems the gene vector or viral particle is entrapped within the core of the generated nanovehicle. The one exception is gene delivery based on condensed gene vectors with the help of cationically charged polymers.

Because there are many types and approaches, with many modifications and modalities, we will be able to list only a few examples of the concrete chemistry for gene delivery. In particular, we may describe some surface modifications that enable one to engineer a stealth character or to provide a targeting capability. A convenient summary is reported in Tables 35.2 and 35.3, listing examples of applications and of marketed gene nanoviclular products. Figure 35.3 demonstrates typical size ranges of different nanovehicular types. Sizes may be outside of the optimal range for given application. However, the present available data do not allow a solid generalization. The same applies to a selection/choice of a particular nanovehicular carrier and its specific application, while some guidance is provided with cited examples.

## Nanoparticles

### PLGA Nanoparticles

Aliphatic poly(α-hydroxy ester)s, such as poly(L-lactic acid) [PLA], poly(glycolic acid) [PGA], and their copolymers [PLGA], belong to the most widely utilized class of polymers approved for human clinical use. They were among the first

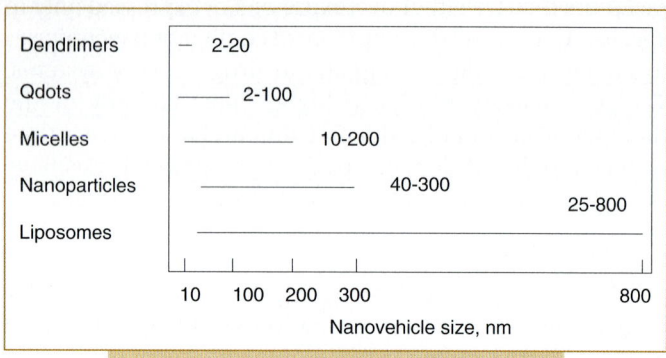

**FIG. 35.3.** Typical range of sizes of several representative classes of nanovehicles. Sizes beyond the 200-nm limit may be too large to enable efficient uptake, both *in vitro* and *in vivo*, and to minimize the RES system.

**Table 35.2.** Examples of gene delivery nanovehicle applications/products

| Type of product/generic vehicle | Reference (veh. descrip.) |
|---|---|
| Polyplex (PEI)/gene | Boussif *et al.* (1995) |
| Gold nanoparticle biolistic gene delivery | O'Brien and Lummis (2002) |
| Antisense nucleotide | Stephens and Rivers (2003) |
| Ribozyme | Ala-aho *et al.* (2004) |
| DNAzyme | Silverman (2005) |
| Aptamer | Fang *et al.* (2001) |
| siRNA | Sumimoto *et al.* (2004) |

**Table 35.3.** Examples of marketed gene nanovehicular products

| Product designation | Application | Company/reference (veh. descrip.) |
|---|---|---|
| PAMAM-dendrimer | PAMAM-based topical vaginal microbicide, VivaGel | Starpharma (Australia) hopes to bring VivaGel to market by 2008 |
| Vitravene | Antisense oligonucleotide for AIDS-related CMV retinitis | ISIS Pharmaceuticals Inc., Carlsbad, CA |
| Gendicine | Injectable AdV particles carrying the p53 gene | Approved in China |
| Rexin-G | Injectable retroviral vector carrying a mutant form of the cyclin G1 gene | Epeius Biotechnologies Corp., Glendale, CA |
| Accell® | Gene gun/generic | PowderJet Vaccines, Inc., Middleton, WI |
| Hellios™ | Gene gun/generic | Bio-Rad Laboratories, Hercules, CA |

synthetic, biodegradable polymers to be used clinically (Frazza and Schmitt, 1971) and show compatibility with living tissue. These polymers are commercially available with different molecular weights and copolymer compositions. The degradation time of these polymers in tissue can span from several months to several years, depending on the molecular weight and copolymer ratio (Gilding, 1981; Vert et al., 1994). The first PLA nanoparticles were reported by Bazile et al. (1992) and the first PLGA nanoparticles by Venier-Julienne and Benoit (1996). A recent review on PLGA nanoparticles (Bala et al., 2004) covers their potential use as carriers for several classes of drugs and genes. PLGA nanoparticles also has been designed for the site-specific delivery of genes (targeting) and evasion of the RES system.

## Polyalkylcyanoacrylate PACA Nanoparticles

Polyalkylcyanoacrylate nanoparticles seem to be an interesting drug carrier, owing to their size, structure, degradability, and drug sorptive properties. They were first prepared by Couvreur et al. (1979). Nanosphere preparation methods based on the polymerization of monomers generally consist in introducing a monomer into an aqueous phase or in dissolving the monomer in a nonsolvent of the polymer. The polymerization reaction in these systems generally occurs in two steps: a nucleation phase followed by a growth phase.

Polyalkylcyanoacrylates have been used for several years as surgical glues and are bioerodible (Lenaerts et al., 1984), which is the most significant advantage of alkylcyanoacrylates over other acrylic derivatives previously used. In contrast to other acrylic derivatives, requiring energy input for the polymerization, alkylcyanoacrylates can be polymerized easily without such a contribution, which is another advantage regarding the stability of the associated drug. These nanospheres are prepared by emulsion polymerization of cyanoacrylic monomers dispersed in an acidic, aqueous phase. The size of the nanospheres obtained is approximately 200 nm, but it can be reduced to 30–40 nm using a nonionic surfactant in the polymerization medium. A drug can be incorporated by adsorption or added to the reaction mixture before the polymerization. Lambert et al. (2005) reported on oligonucleotide (ODN) delivery with PACA nanoparticles. ODNs entrapped within the PACA nanoparticles were shown to be protected against degradation and to penetrate more easily into different types of cells. In vivo, PACA nanoparticles were able to efficiently distribute the ODNs to the liver and improve, in mice, the treatment of RAS cells expressing a mutant Ha-ras gene. Pakunlu et al. (2006) demonstrated that PEGylated liposomes are capable of penetrating directly into tumor cells after systemic administration in vivo and do successfully provide cytoplasmic and nuclear delivery of an encapsulated anticancer drug, doxorubicin (DOX) and antisense oligonucleotides (ASO). Encapsulation of DOX and ASO into liposomes substantially increased their specific activity.

Simultaneous suppression of pump and nonpump resistance dramatically enhanced the ability of DOX to induce apoptosis, leading to higher in vitro cytotoxicity and in vivo antitumor activity.

## Polyelectrolyte Complex Nanoparticles (PEC)

Nanoparticles can be made by means of a nonstoichiometric mixing of polyelectrolytes of opposite charge (via ionotropic gelation) by spontaneous association. Prokop et al. (2002a, 2002b) reported on multicomponent polymeric complexes that are candidates for delivery vehicles of biological molecules such as proteins and drugs. Biocompatible and largely natural polymers can be fabricated into thermodynamically stable nanoparticles insoluble in water and buffered media, in the absence of organic solvents, featuring core-shell morphology. A careful choice of construction materials and the superposition of several interacting principles during PEC production allows for the customization of the physicochemical properties of the structures. A recent development involves low-molecular-weight polymers. Typically, nanoparticles are made by mixing polyanionic core polymers, such as alginate, and chondroitin sulfate, and the corona polycations, such as spermine hydrochloride, and poly(methylene-co-guanidine) hydrochloride (PMCG). PMCG is a synthetic oligomer that mimics peptide structure. It contains highly charged cationic guanidine groups. This multipolymeric nanoparticulate system was shown to be very effective for in vitro gene transfer (Carlesso et al., 2003), particularly in cell systems that are normally refractory to gene transfer, such as pancreatic islets and antigen-presenting cells (Fig. 35.4). Recent data revealed that such nanoparticles are taken up by means of a nonspecific adsorptive endocytosis and/or via fluid-phase endocytosis.

Alonso et al. (Fernandez-Urrusuno, 1999; Calvo et al., 1997) introduced chitosan-based nanoparticles. Chitosan (CS) nanoparticles are obtained by the process of ionotropic gelation based on the interaction between the negative groups of the pentasodium tripolyphosphate (TPP) and the positively charged amino groups of CS. This process has been used to prepare nanoparticles for the delivery of peptides and proteins, insulin, and cyclosporine. A wider application of this product has been prevented by an instability resulting from the ion exchange (followed by dissolution) between the polyions incorporated into the nanoparticle and small ions present in the foregoing environments. However, this system offers an interesting potential for transmucosal delivery. Howard et al. (2006) employed chitosan nanoparticles to mediate knockdown of endogenous enhanced green fluorescent protein (eGFP) in both H1299 human lung carcinoma cells and murine peritoneal macrophages (77.9% and 89.3% reduction in eGFP fluorescence, respectively). Effective in vivo RNA interference was achieved in bronchiole epithelial cells of transgenic eGFP mice after nasal administration of chitosan/siRNA

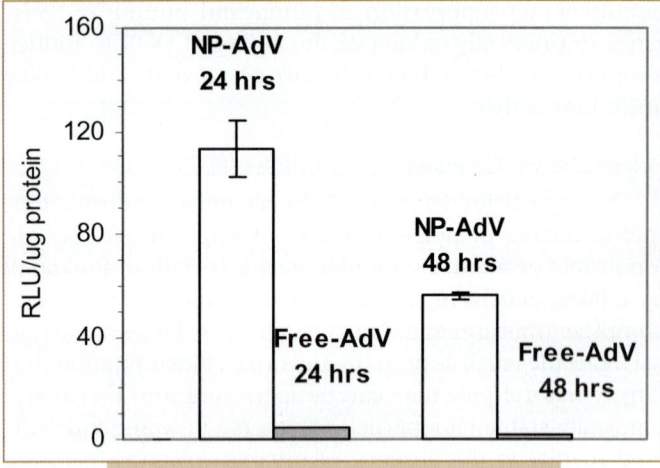

**FIG. 35.4.** Nanoparticle-mediated gene transfer into islets. Freshly isolated murine islet cells were transduced with either adenovirus containing NP (NP-AdV, MOI = 10) or free adenovirus (AdV, MOI = 10) in complete growth media. After 16 h, cells were washed twice with PBS and cultured with growth medium. At 24 and 48 h, cells were harvested and assayed for luciferase expression. Luciferase values were normalized on the basis of protein content. Courtesy of ACS.

formulations (37% and 43% reduction as compared to mismatch and untreated control, respectively).

Other powerful PEC gene delivery systems include polyethyleneimine (PEI) (Demeneix and Behr, 2005) and poly-L-lysine (PLL). Many other cationic polymers have been employed and tested.

## Polymeric Micelles

This technology has been pioneered and developed by K. Kataoka in Japan (Kataoka *et al.*, 2001). Polymeric micelles, core-shell type colloidal carriers for drug and gene targeting, prepared from the self-assemblies of block copolymers, are one of the most refined and promising modalities of drug delivery, since the critical parameters, such as size and drug loading and release, can be controlled by engineering the constituent block copolymers. It has been demonstrated that polymeric micelles show unique disposition characteristics in the body, suitable for drug targeting (e.g., prolonged blood circulation and significant tumor accumulation). Novel approaches for the formation of functionalized poly(ethylene glycol) (PEG) layers as an hydrophilic outer shell were developed to attain receptor-mediated drug and gene delivery through PEG-conjugated ligands with minimal nonspecific interaction with other proteins. Oishi *et al.* (2005) demonstrated efficient intracellular *in vitro* siRNA delivery.

## Dendrimers

Dendrimers are attractive macromolecular systems for drug delivery because of their nanometer size range (5–20 nm), ease of preparation and functionalization, and their ability to display multiple copies of surface groups for biological recognition processes (Morgan and Cloninger, 2002). Dendrimer molecules are monodisperse symmetric macromolecules built around a small molecule or in a linear polymer core using connectors and branching units. Interaction of dendrimer macromolecules with the molecular environment is controlled predominantly by their terminal groups. By modifying their termini, the interior of a dendrimer can be made hydrophilic while its exterior surface is hydrophobic, or vice versa. Dendrimers can be synthesized starting from the central core and working out toward the periphery (divergent synthesis) or in a top-down approach starting from the outermost residues (convergent synthesis). Commonly used and commercially available polyamido-amine dendrimers (PAMAM) form spherical polymers with good aqueous solubility, because of their highly charged exposed surface groups, with abundant primary amines for convenient functionalization (Majoros *et al.*, 2005). PEGylation of amino-terminated PAMAM dendrimers has reduced their immunogenicity, improving lifetime circulation (Bhadra *et al.*, 2003). A high *in vivo* activity, using a luciferase reporter, has been demonstrated with dendrimers (Banerjee *et al.*, 2004).

## Liposomes

Liposomes are small artificial vesicles of spherical shape that can be produced from natural nontoxic phospholipids and cholesterol. Because of their size, hydrophobic and hydrophilic compartments, as well as biocompatibility, liposomes are promising systems for drug delivery. Liposome properties vary substantially with lipid composition, size, surface charge, and the method of preparation. They are therefore grouped into three classes based on their size and the number of bilayers. Small unilamellar vesicles (SUV) are surrounded by a single lipid layer and are 25–50 nm in diameter. Large unilamellar vesicles (LUV) are a heterogeneous group of vesicles similar to SUVs. Multilamellar vesicles (MLV) consist of several lipid layers separated from one another by a layer of aqueous solution. Although SUVs can carry soluble substances in their lipid bilayer structure, MLVs can also carry membrane-bound lipophilic drugs for sustained release. A combination with a cationic lipid results in a cationically charged liposome (lipoplex). Liposomes are of significance for gene delivery because DNA can be incorporated within the aqueous-phase cavity.

Lipofection has been used with a variety of cell types *in vitro* as well as tissues *in vivo*. (Felgner and Ringold, 1989; Li and Hoffman, 1995). The main advantages of liposomes are that there are few constraints to the size of the gene which can be delivered, and liposomes are relatively nontoxic and can be applied repeatedly. A drawback of lipofection is the low frequency of stable transfection. The ability to target these liposome vectors to specific cells is also limited. Recent efforts to target liposomes to specific cell types have focused on the incorporation of a variety of ligands into the liposome-DNA complex (Trubetskoy, 1992). Several lipid-based nonviral vectors are currently under investigation for gene delivery (Rosenberg *et al.*, 2000).

Liposome shielding from the RES has been achieved by PEGylation (addition of PEG) (Blume and Cevc, 1990; Crosasso *et al.*, 2000; Cattel *et al.*, 2004). Targeting of liposomes has been readily demonstrated (e.g., antibody-conjugated liposomes) (Ishida *et al.*, 1999). Major progress is expected from developments of many different modalities of liposome preparation, including those of pH-sensitive status, cytosolic delivery of siRNA, delivery of cytotoxic agents and toxins, antigen delivery and viral gene delivery systems, and from functionalized liposomes (Drummond *et al.*, 2000). There are numerous commercial formulations for gene delivery.

### Cell-Penetrating Peptides (CPPs)

CPPs are up to 30 amino acid amphiphilic peptides, which can be internalized by cells via mechanisms that require no energy and can be receptor mediated. Protein-derived CPPs usually consist of the minimal effective partial sequence of the parent translocation protein and are also known as protein transduction domains or membrane translocation sequences (Zorko and Langel, 2005). CPPs from different classes do not share common amino acid sequence motifs. The only two common features of all CPPs appear to be a positive charge and amphipathicity.

Protein transduction occurs by a rapid, lipid raft–mediated macropinocytosis mechanism that is independent of receptors and transporters that all cells perform. Typically, an efficient cytoplasmic delivery occurs (Oehlke *et al.*, 2005). This is based on observations with an inhibition of this uptake by negatively charged GAG molecules (glycosaminoglycans, HSPG), dextran sulfate and heparin. This is supported by observations on inhibition of CPP uptake in GAG-deficient cells (Marty *et al.*, 2004).

CPPs do not belong to nanovehicles. These are listed here because such motifs are often used to functionalize nanovehicles in order to enable their uptake and quick penetration. It is implied that the mechanism of uptake of such vehicles may be similar to those of plain CPPs. Such nanovehicle functionalization is expected to play a significant role in future in gene delivery (Turner *et al.*, 2006).

### Special Generic Systems

Several inorganic systems have been reported for drug and gene delivery. These include calcium phosphate, nanogold, carbon material, silicon oxide, and iron oxide (Martin and Kohli, 2003; Kam *et al.*, 2005; Jain *et al.*, 2005; Maitra, 2005; Gemeinhart *et al.*, 2005). Carbon nanotubes are one-dimensional tubular objects of carbon that can be either multiwalled or single-walled. In addition, they can be functionalized, permitting an attachment of ligands and drugs (Klumpp *et al.*, 2005).

## IX. SPECIAL CONSIDERATIONS FOR GENE DELIVERY SYSTEMS

These vehicles comprise nonviral and viral systems. Nonviral gene delivery refers to the use of naked DNA, biolistic delivery, cationic lipids formulated into liposomes and complexed with DNA (lipoplexes), cationic polymers complexed with DNA (polyplexes), polymeric vesicles complexed with DNA (cholesterol vesicles), polymeric nanoparticles, or a combination of both cationic lipids and cationic polymers complexed with DNA (lipoplexes, Lipofectamine™). Many different chemistries and modifications were introduced earlier; some are special to gene delivery. Nonviral systems are treated first. This distinction has been introduced because differences in the outcome of gene expression are still quite great: Nonviral systems are still quite inefficient. A common feature is the nature of forces involved in the gene agent entrapment: In many of the formulations certain electrostatic forces predominate, as opposed to the many different mechanisms involved in generic nanovehicle assembly. For the former systems, the stability of such complexes is of crucial importance for *in vitro* and *in vivo* applications, as reviewed earlier. Surveys of the distinct advantages and disadvantages of gene delivery systems are available (Eming *et al.*, 2004).

### Nonviral Systems

Rapid advances have been made in the development of more physiological means (nonviral) to introduce genes into target cells. Nonviral gene transfer systems have the advantage of delivering genes to target cells without the inherent disadvantages of viral-based systems, such as antigenicity, potential for recombination with wild-type viruses, and possible cellular damage due to persistence or repeated exposure to the viral vectors (Felgner and Rhodes, 1991). These synthetic systems are also easier to manufacture on a large scale because they typically use plasmid constructs that can be grown with existing fermentation technology. Direct plasmid application, lipofection, and receptor-mediated delivery vectors are the most promising nonviral systems. Other nonviral transfection techniques are too inefficient for clinical use (i.e., electroporation, microinjection of DNA).

Typical nonviral gene delivery payload agents include gene vectors and therapeutic oligonucleotides, antisense, ribozymes, DNAzymes, aptamers, and small interfering RNAs — siRNAs. Selective silencing of genes was initially accomplished with short antisense oligonucleotides, pieces of DNA that hybridize with the target mRNA and inhibit their translation into the protein. A clinical trial is ongoing targeting of certain genes involved in oncogenesis (Stephens and Rivers, 2003). Specifically, in this case, oligonucleotides, sensitive to enzymatic breakdown, were replaced by derivatives, with the phosphorothioate backbone in place of original phosphoester backbone. Ribozymes are naturally occurring enzymes consisting of RNA, resulting in cleaving of RNA. A down-regulation of oncogenes resulting in tumor suppression has been reported (Ala-aho *et al.*, 2004). Deoxyribozymes or DNAzymes (enzymes made of DNA) are artificial molecules and are not found in nature. Silverman (2005) has isolated a multitude of DNAzymes that catalyze RNA ligation. RNA-cleaving DNAzymes, DNAzymes able to catalyze RNA hydrolysis in the absence of divalent metal ions, have been discovered. Aptamers are synthetic RNA molecules capable of capturing ligands with high binding selectivities and affinities for a specific target. They have proven to be valuable therapeutic agents with enhanced properties relative to antibodies (Nimjee *et al.*, 2005). Short interfering RNAs serve for a specific knockdown of genes. Sumimoto *et al.* (2004) demonstrated an *in vivo* melanoma inhibition with lentiviral vector encoding for siRNA directed against the mutated oncogene BRAF. While plasmid DNA needs to enter the nucleus, siRNA and other carriers mentioned earlier need only to reach the cytoplasm, since they act by mRNA degradation.

One of the most practical methods for delivering the DNA to the target tissue is the *direct injection* of DNA into the tissue. This approach was first investigated by Hengge and coworkers, who demonstrated that DNA injected directly into the skin can be taken up by keratinocytes, and its transient expression can induce a biological response (Hengge *et al.*, 1995). The efficacy of this approach has now been supported by reports of other groups (Sun *et al.*, 1997; Meuli *et al.*, 2001; Sawamura *et al.*, 2002). A more sophisticated technique for the delivery of naked DNA has been described by Ciernik *et al.* (1996) and Eriksson *et al.* (1998) and is named *puncture-mediated gene transfer* or *microseeding*, respectively. These techniques deliver the DNA directly to target cells by multiple perforations using oscillating solid microneedles. The authors state that transgene expression by microseeding is more efficient than delivering DNA by a single dermal injection.

Commonly used polymers include PEI, PLL, chitosans, and dendrimers. Nonviral vectors are receiving increasing attention due to compound stability, potential ease of chemical modification, low cost and consistent standard of production, and higher biosafety and flexibility (Dyer and Herrling, 2000). The cationic polymers usually bear protonable amines, enabling their complexation with nucleic acids and securing their transfection efficiency. Due to electrostatic interactions between the positively charged amino groups and negatively charged phosphate groups of plasmid DNA, polycations spontaneously form complexes with a display of rapid condensation. The ratio of polycation/DNA then allows for a charge control. Aggregation of originally colloidal structures (although of different submicroscopic shape) is one of the problems. Similarly, *in vivo*, biological fluids often cause an aggregation when complexes are introduced. Recent data suggest that coating the polycation/DNA complexes with polyanions can diminish particle aggregation (increasing colloidal stability) (Trubetskoy *et al.*, 2003).

Some targeting of gene complexes with suitable ligands has been achieved (e.g., Sagara and Kim, 2002).

Besides polycations, other polymers were suggested to condense large genes into smaller structures and to mask negative DNA charges. They include neutral polyvinyl alcohol, polyvinylpyrrolidone, and PEG copolymers. Many different polymeric approaches were summarized by Piskin *et al.* (2004).

Different nanoparticulate formulations have been shown to transfer genes in a significantly higher portion as compared to naked DNA (e.g., Hass *et al.*, 2005) but much less than the viral systems. Carlesso *et al.* (2003) have demonstrated highly efficient, nanoparticle (PEC)-based gene delivery *in vitro* and *in vivo*.

Biolistic method uses the handheld gene gun with a pulse of helium to fire small gold particles coated with DNA at target cells. This method of biolistic transfection is becoming increasingly popular as an effective means of rapid gene delivery into mammalian tissue. It is restricted to local expression in the dermis, muscle, or mucosal tissue. Current methods of microcarrier preparation, however, are slow (up to two days) and can result in variations in transfection efficiency due to a number of problems, including shearing of DNA, agglomeration, and adhesion of gold particles. One of the major applications of the gene gun is genetic vaccination in the form of intramuscular or cutaneous injection of DNA (Nanney *et al.*, 2000).

Nonviral gene transfer systems have the advantage of delivering genes to target cells without the inherent disadvantages possessed by viral-based systems (see later), such as antigenicity, potential for recombination with wild-type viruses, and possible cellular damage due to persistence or repeated exposure to the viral vectors (Eming *et al.*, 2004). Lipofection, and receptor-mediated delivery vectors (e.g., ligand-modified PEI–DNA complex) are the most promising nonviral systems. Ogris *et al.* (2003) demonstrated 10–100 times higher reporter gene expression (pCMVLuc) by employing Trf or an EGF-decorated PEI–PEG–liposomal delivery system in mice tumors. For RGD-decorated liposome-protamine DNA formulation, Harvie *et al.* (2003) showed 100-fold enhancement of transfection activity in tumor cells overexpressing appropriate integrin receptor *in vitro*, while Hofland *et al.* (2002) achieved a 50- to 100-fold reduction in RES tissue (lungs) while using folate-targeted liposomes as compared to nontargeted liposomes; however, the accumulation of complexes and gene transfer activity was not altered.

## Viral Systems

Having evolved to deliver their genes to target cells, viruses are currently the most effective means of gene delivery and can be manipulated to express therapeutic genes or to replicate specifically in certain cells. Gene therapy with attenuated viruses is being developed for a range of diseases, including inherited monogenic disorders and cardiovascular disease. But it is in the treatment of cancer that this approach has been most evident, resulting in the recent licensing in China of a gene therapy for the routine treatment of head and neck cancer. A variety of virus vectors have been employed to deliver genes to cells to provide either transient (e.g., adenovirus, vaccinia virus) or permanent (e.g., retrovirus, adeno-associated virus) transgene expression, and each approach has its own advantages and disadvantages (Eming *et al.*, 2004).

Adenoviruses (AdV) are the most commonly used therapeutic models, where mechanisms for initial fiber protein binding with CAR cell surface receptors (followed by endocytosis) have been actively investigated for tissue and organ targeting in human gene therapy (Roelvink *et al.*, 1999). The AdV vectors remain in an extrachromosomal location and are replication-defective, resulting in transient expression. AdV vector has an outstanding capability of escaping from endosome to cytoplasm and passing through the nuclear pore membrane. Despite extensive investigation of dozens of viral vectors for cancer treatment and promising results in clinical trials, the FDA has not approved any virus-based therapeutics because of toxicities and immunogenicities widely reported in gene therapy (Lin and Nemumaitis, 2004).

To date, adenoviral vectors have demonstrated the highest gene transfer efficiency *in vivo* (Wilson, 1996). Unlike retroviruses, adenoviruses can easily be concentrated to high viral titers and can infect dividing as well as nondividing cells with high efficiency. This is a major advantage for some applications, including gene transfer to the lung, where cell division is infrequent. A major drawback in the use of adenoviral vectors is the cytotoxicity of viral proteins and the host cellular immune response to the adenoviral proteins of the virus particle, which results in local inflammation and destruction of transduced cells.

Adeno-associated viruses are small nonenveloped viruses, which are not known to cause any human disease (Levine and Friedmann, 1993). The adeno-associated virus can transduce a wide range of dividing and nondividing cells, but its main advantage is that it integrates reliably into a specific position on chromosome 19, which reduces the probability of inadvertent activation of a protooncogene (Levine and Friedmann, 1993; Monahan and Samulski, 2000). The disadvantages of adeno-associated viruses are that the vector packages only small transgenes (up to 4.7 kbp), and the preparation of the recombinant vector is frequently contaminated with wild-type adenovirus, which must be separated or inactivated.

The major advantages of retroviral-mediated gene transfer are that recombinant retroviruses are capable of transferring genes to a wide range of different cell types, including normal diploid cells, the genes are stably integrated into the chromosomal DNA at relatively high frequency, without rearrangements, and retroviruses are capable of transferring genes simultaneously to large numbers of cells at high effi-

ciencies. The disadvantage of retroviral-mediated gene transfer is that a gene of only limited size (<6 kbp) can be packaged and transferred. Integration of the transferred gene is dependent on cell division, which limits the types of cells and tissues that can be modified. Although long-term gene expression has been reported in some systems, including fibroblasts and muscle cells, transient expression in retrovirally modified cells has also been described (Palmer *et al.*, 1991; Roemer and Friedmann, 1992).

Herpes simplex virus type-1 (HSV-1) is a human neurotropic virus used primarily as a vector for gene transfer to the nervous system, although the wild-type HSV-1 can infect and lyse other nonneuronal epithelial cells. Because of its large genome size, up to a 30–50-kbp transgene can be packaged into recombinant HSV-1 vectors. At present, two major classes of HSV-1 vectors have been developed: replication-defective viruses and replication-conditional mutants (Roizman, 1996; Glorioso *et al.*, 1997). However, the ability of HSV and these mutant recombinants to establish life-long latent infections raises concerns about the use of these vectors in humans. Efforts have been made to create HSV-1 amplicon vectors essentially free of HSV-1 helper virus, which might be a promising genetic vehicle for *in vivo* gene delivery (Costantini *et al.*, 2000).

## Targeting and Retargeting Modalities

Recently, Bazan-Peregrino *et al.* (2006) reviewed current therapeutic modalities and transductional and transcriptional detargeting and retargeting strategies for gene delivery of both nonviral and viral systems to tumor vasculature. The conclusions are of general significance. Transductional targeting is achieved by employing small peptides or antibodies attached to the viral particles genetically or chemically (by means of reactive bifunctional reagens) in order to eliminate their normal tropism and potentially introducing new ones. Transcriptional targeting can be achieved by placing the transgenes under the control of tumor vasculature–associated promoters (e.g., hypoxia-responsive, vascular endothelial growth factor, adhesion molecule, and other endothelial cell-specific promoters). A combination of transductional and transcriptional targeting is also possible. Specific examples are listed in the original review as well as those of antiangiogenic gene therapy targeted to nonendothelial or stem cells. The identification of promising new targets (e.g., pericytes, smooth muscle cells, tumor stroma) for molecular interventions using gene therapy will eventually lead to a combination therapy of these new treatment modalities.

## X. OUTLOOK

Nanovehicles loaded with an appropriate drug or gene offer improved spatiotemporal control (window) over drug kinetics and distribution. It should be restated that the ratio of the amount of drug released from nanovehicles within the bloodstream to the amount taken up intracellularly is governed by the circulatory properties of vehicles in the central compartment, by the release rate from nanovehicles in this compartment, and by the efficiency of intracellular uptake, no matter what the mechanism of uptake. This is the nanovehicular ratio of systemic to intracellular localization (S/I ratio). In order to optimize this ratio, one should maximize the circulation time and release and minimize the uptake. In other instances, intracellular delivery is desirable. Thus, there is a prospect for safer and mode-specific treatment therapies.

In the coming years, we expect vigorous progress on the cell and molecular biology sides, which will further enhance our understanding of the basic phenomena involved in drug/gene/vehicle uptake. Internalization of the delivery systems results in a poorly defined trafficking process. Elucidation of the nonendocytic pathway(s) mechanism may help the creation of rationally designed vehicles. The uptake of nanovehicles is a rather understudied area, even though a lot of similarities and ideas come from the area of the uptake of macromolecules. However, there may not be a single specific uptake pathway responsible for intracellular delivery. A series of inhibiting agents with preferably different modes of inhibition should be applied and combined with colocalization studies. Besides, it is not clear how the macromolecular properties of nanovector, particle size, sterical stabilization, and the specific design of homing devices as well as the cargo chemistry influence the mechanism of uptake. Such diversity in the nanovehicle uptake mechanisms employed by each cell type then offers the possibility of manipulating and enhancing the uptake via the desired route (Huth *et al.*, 2006).

The adjustable size of nanovehicles to a rather narrow-sized window determined by the size of blood vessels (both by the upper and lower limits of normal and pathological vasculature) and epithelial barrier openings in the case of oral applications and the opportunity of targeting preclude other delivery vehicles from gaining a strong foothold in future applications. In all cases, safety and economic considerations will control the ultimate application and its success.

A considerable research dilemma is the selection/choice of a suitable nanovehicular delivery vehicle for a specific application. Individual authors tend to prefer their own vehicle, and no generalizations or rules of thumb are available to guide the selection. The desirable delivery route and localization and biocompatibility of nanovehicles will, in the end, dictate the utility of individual constructs.

Because of the rather steep cost for preclinical toxicology testing of novel delivery vehicles, the pace of implementation of new delivery modes will not be high. Other considerations, such as patent expiration, may help drive the introduction of new or reformulated drug entities and place them on the market. Besides, large-scale manufacturing is most critical. Residual solvents where solvent-based methods are used to manufacture nanovehicles must be minimized via a complete evaporation in the final preparation. The nanoscale-based delivery strategies are beginning to make significant inroads on the global pharmaceutical industry.

## XI. ACKNOWLEDGMENT

We acknowledge the support of the National Institutes of Health Grant 1R01EB002825-01 (J.M.D. and A.P) and support from the Department of Veterans Affairs (J.M.D.).

## XII. REFERENCES

Adrian, J. E., Poelstra, K., Scherphof, G. L., Molema, G., Meijer, D. K., Reker-Smit, C., Morselt, H. W., and Kamps, J. A. (2006). Interaction of targeted liposomes with primary cultured hepatic stellate cells: involvement of multiple receptor systems. *J. Hepatol.* **44**(3), 560–567.

Akerman, M. E., Chan, W. C., Laakkonen, P., Bhatia, S. N., and Ruoslahti, E. (2002). Nanocrystal targeting *in vivo. Proc. Natl. Acad. Sci. U.S.A.* **99**(20), 12617–12621.

Ala-aho, R., Ahonen, M., George, S. J., Heikkila, J., Grenman, R., Kallajoki, M., and Kahari, V. M. (2004). Targeted inhibition of human collagenase-3 (MMP-13) expression inhibits squamous cell carcinoma growth *in vivo. Oncogene* **23**(30), 5111–5123.

Alessi, P., Ebbinghaus, C., and Neri, D. (2004). Molecular targeting of angiogenesis. *Biochim. Biophys. Acta* **1654**(1), 39–49.

Amyere, M., Mettlen, M., Van Der Smissen, P., Platek, A., Payrastre, B., Veithen, A., and Courtoy, P. J. (2002). Origin, originality, functions, subversions and molecular signalling of macropinocytosis. *Int. J. Med. Microbiol.* **291**, 487–494.

Baish, J. W., Gazit, Y., Berk, D. A., Nozue, M., Baxter, L. T., and Jain, R. K. (1996). Role of tumor vascular architecture in nutrient and drug delivery: an invasion percolation-based network model. *Microvasc. Res.* **51**(3), 327–346.

Bala, I., Hariharan, S., and Kumar, M. N. (2004). PLGA nanoparticles in drug delivery: the state of the art. *Crit. Rev. Ther. Drug Carrier. Syst.* **21**(5), 387–422.

Banerjee, P., Reichardt, W., Weissleder, R., and Bogdanov, A. Jr. (2004). Novel hyperbranched dendron for gene transfer *in vitro* and *in vivo. Bioconjug. Chem.* **15**(5), 960–968.

Bazan-Peregrino, M., Seymour, L. W., and Harris, A. L. (2006). Gene therapy targeting to tumor endothelium. *Cancer Gene Ther.* **14**(2), 117–127.

Bazile, D. V., Ropert, C., Huve, P., Verrecchia, T., Marlard, M., Frydman, A., Veillard, M., and Spenlehauer, G. (1992). Body distribution of fully biodegradable [14C]-poly(lactic acid) nanoparticles coated with albumin after parenteral administration to rats. *Biomaterials* **13**(15), 1093–1102.

Bendas, G. (2001). Immunoliposomes: a promising approach to targeting cancer therapy. *BioDrugs* **15**(4), 215–224.

Bhadra, D., Bhadra, S., Jain, S., and Jain, N. K. (2003). A PEGylated dendritic nanoparticulate carrier of fluorouracil. *Int. J. Pharm.* **257**(1–2), 111–124.

Blume, G., and Cevc, G. (1990). Liposomes for the sustained drug release *in vivo. Biochim. Biophys. Acta* **1029**(1), 91–97.

Boddy, A., Aarons, L., and Petrak, K. (1989). Efficiency of drug targeting: steady-state considerations using a three-compartment model. *Pharm. Res.* **6**(5), 367–372.

Bollinger, C. R., Teichgraber, V., and Gulbins, E. (2005). Ceramide-enriched membrane domains. *Biochim. Biophys. Acta* **1746**(3), 284–294.

Boussif, O., Lezoualc'h, F., Zanta, M. A., Mergny, M. D., Scherman, D., Demeneix, B., and Behr, J. P. (1995). A versatile vector for gene and oligonucleotide transfer into cells in culture and *in vivo*: polyethylenimine. *Proc. Natl. Acad. Sci. U.S.A.* **92**(16), 7297–7301.

Brandli, A. W., Adamson, E. D., and Simons, K. (1991). Transcytosis of epidermal growth factor. The epidermal growth factor receptor mediates uptake but not transcytosis. *J. Biol. Chem.* **266**(13), 8560–8566.

Breunig, M., Lungwitz, U., Liebl, R., Klar, J., Obermayer, B., Blunk, T., and Goepferich, A. (2006). Mechanistic insights into linear polyethylenimine-mediated gene transfer. *Biochim. Biophys. Acta* **1170**(2), 196–205.

Calvo, P., Remunan-Lopez, C., Vila-Jato, J. L., and Alonso, M. J. (1997). Novel hydrophilic chitosan-polyethylene oxide nanoparticles as protein carriers. *J. Appl. Polym. Sci.* **63**, 125–132.

Campbell, L., Gumbleton, M., and Ritchie, K. (2001). Caveolae and the caveolins in human disease. *Adv. Drug Deliv. Rev.* **49**(3), 325–335.

Carlesso, G., Kozlov, E., Prokop, A., Unutmaz, D., and Davidson, J. M. (2003). Nanoparticulate system for efficient gene transfer into refractory cell targets. *Biomacromolecules* **6**(3), 1185–1192.

Cattel, L., Ceruti, M., and Dosio, F. (2004). From conventional to stealth liposomes: a new frontier in cancer chemotherapy. *J. Chemother.* **16**(Suppl. 4), 94–97.

Cegnar, M., Kristl, J., and Kos, J. (2005). Nanoscale polymer carriers to deliver chemotherapeutic agents to tumours. *Expert Opin. Biol. Ther.* **5**(12), 1557–1569.

Chen, J. F., Ding, H. M., Wang, J. X., and Shao, L. (2004). Preparation and characterization of porous hollow silica nanoparticles for drug delivery application. *Biomaterials* **25**(4), 723–727.

Chittimalla, C., Zammut-Italiano, L., Zuber, G., and Behr, J. P. (2005). Monomolecular DNA nanoparticles for intravenous delivery of genes. *J. Am. Chem. Soc.* **127**(32), 11436–11441.

Ciernik, I. F., Krayenbuhl, B. H., and Carbone, D. P. (1996). Puncture-mediated gene transfer to the skin. *Hum. Gene Ther.* **7**, 893–899.

Conner, S. D., and Schmid, S. L. (2002). Regulated portals of entry into the cell. *Nature* **422**(6927), 37–44.

Costantini, L. C., Bakowska, J. C., Breakefield, X. O., Isacson, O. (2000). Gene therapy in the CNS. *Gene Ther.* **7**, 93–109.

Couvreur, P., Kante, B., Roland, M., Guiot, P., Bauduin, P., and Speiser, P. (1979). Polycyanoacrylate nanocapsules as potential lysosomotropic carriers: preparation, morphological and sorptive properties. *J. Pharm. Pharmacol.* **31**(5), 331–332.

Crosasso, P., Ceruti, M., Brusa, P., Arpicco, S., Dosio, F., and Cattel, L. (2000). Preparation, characterization and properties of sterically stabilized paclitaxel-containing liposomes. *J. Controlled Release* **63**(1–2), 19–30.

Dash, P. R., Read, M. L., Barrett, L. B., Wolfert, M. A., and Seymour, L. W. (1999). Factors affecting blood clearance and *in vivo* distribution of polyelectrolyte complexes for gene delivery. *Gene Ther.* **6**(4), 643–650.

de la Fuente, J. M., and Penades, S. (2004). Understanding carbohydrate–carbohydrate interactions by means of glyconanotechnology. *Glycoconj. J.* **21**(3–4), 149–163.

Demeneix, B., and Behr, J. P. (2005). Polyethylenimine (PEI). *Adv. Genet.* **53**, 217–230.

Desai, M. P., Labhasetwar, V., Amidon, G. L., and Levy, R. J. (1996). Gastrointestinal uptake of biodegradable microparticles: effect of particle size. *Pharm. Res.* **13**(12), 1838–1845.

Drummond, D. C., Zignani, M., and Leroux, J. (2000). Current status of pH-sensitive liposomes in drug delivery. *Prog. Lipid Res.* **39**(5), 409–460.

Dyer, M. R., and Herrling, P. L. (2000). Progress and potential for gene-based medicines. *Mol. Ther.* **1**(3), 213–224.

Einarson, M. B., and Berg, J. (1993). Electrosteric stabilization of colloidal latex dispersions. *J. Colloid. Interface Sci.* **155**, 165–172.

Eming, S. A., Krieg, T., and Davidson, J. M. (2004). Gene transfer in tissue repair: status, challenges and future directions. *Expert Opin. Biol. Ther.* **4**(9), 1373–1386.

Eriksson, E., YAO, F., Svensjo, T., *et al.* (1998). *In vivo* gene transfer to skin and wound by microseeding. *J. Surg. Res.* **78**, 85–91.

Fabriek, B. O., Dijkstra, C. D., and van den Berg, T. K. (2005). The macrophage scavenger receptor CD163. *Immunobiology* **210**(2–4), 153–160.

Fang, X., Cao, Z., Beck, T., and Tan, W. (2001). Molecular aptamer for real-time oncoprotein platelet-derived growth factor monitoring by fluorescence anisotropy. *Anal. Chem.* **73**(23), 5752–5757.

Felgner, P. L., and Rhodes, G. (1991). Gene therapeutics. *Nature* **349**, 351–352.

Felgner, P. L., and Ringold, G. M. (1989). Cationic liposome-mediated transfection. *Nature* **337**, 387–388.

Fernandez-Urrusuno, R., Calvo, P., Remunan-Lopez, C., Vila-Jato, J. L., and Alonso, M. J. (1999). Enhancement of nasal absorption of insulin using chitosan nanoparticles. *Pharm. Res.* **16**(10), 1576–1581.

Ferrari, M. (2005a). Cancer nanotechnology: opportunities and challenges. *Nat. Rev. Cancer* **5**(3), 161–171.

Ferrari, M. (2005b). Nanovector therapeutics. *Curr. Opin. Chem. Biol.* **9**(4), 343–346.

Finsinger, D., Remy, J. S., Erbacher, P., Koch, C., and Plank, C. (2000). Protective copolymers for nonviral gene vectors: synthesis, vector characterization and application in gene delivery. *Gene Ther.* **7**(14), 1183–1192.

Florence, A. T., and Hussain, N. (2001). Transcytosis of nanoparticle and dendrimer delivery systems: evolving vistas. *Adv. Drug Deliv. Rev.* **50**(Suppl. 1), 480–480.

Frazza, E. J., and Schmitt, E. E. (1971). A new absorbable suture. *J. Biomed. Mater. Res.* **1**, 43.

Friedmann, T. (1992). A brief history of gene therapy. *Nat. Genet.* **2**, 93–98.

Furumoto, K., Nagayama, S., Ogawara, K., Takakura, Y., Hashida, M., Higaki, K., and Kimura, T. (2004). Hepatic uptake of negatively charged particles in rats: possible involvement of serum proteins in recognition by scavenger receptor. *J. Controlled Release* **97**(1), 133–141.

Gemeinhart, R. A., Luo, D., and Saltzman, W. M. (2005). Cellular fate of a modular DNA delivery system mediated by silica nanoparticles. *Biotechnol. Prog.* **21**(2), 532–537.

Gilding, D. K. (1981). Biodegradable polymers. *In* "Biocompatibility of Clinical Implant Materials" (D. F. Williams, ed.), pp. 209–232. CRC Press, Boca Raton, FL.

Glorioso, J. C., Goins, W. F., and Schmidt, M. C. (1997). Engineering HSV vectors for human gene therapy. *Adv. Pharmacol.* **40**, 103–136.

Goldstein, J. L., Brown, M. S., Anderson, R. G., Russell, D. W., and Schneider, W. J. (1985). Receptor-mediated endocytosis: concepts emerging from the LDL receptor system. *Annu. Rev. Cell Biol.* **1**, 1–39.

Green, T. R., Fisher, J., Stone, M., Wroblewski, B. M., and Ingham, E. (1998). Polyethylene particles of a "critical size" are necessary for the induction of cytokines by macrophages *in vitro. Biomaterials* **19**(24), 2297–2302.

Gunther, M., Wagner, E., and Ogris, M. (2005). Specific targets in tumor tissue for the delivery of therapeutic genes. *Curr. Med. Chem. Anticancer Agents* **5**(2), 157–171.

Haas, J., Ravi Kumar, M. N., Borchard, G., Bakowsky, U., and Lehr, C. M. (2005). Preparation and characterization of chitosan and trimethyl-chitosan-modified poly(epsilon-caprolactone) nanoparticles as DNA carriers. *AAPS PharmSciTech* **6**(1), E22–E30.

Han, H. D., Lee, A., Song, C. K., Hwang, T., Seong, H., Lee, C. O., and Shin, B. C. (2006). *In vivo* distribution and antitumor activity of heparin-stabilized doxorubicin-loaded liposomes. *Int. J. Pharm.* **313**(1–2), 181–188.

Harvie, P., Dutzar, B., Galbraith, T., Cudmore, S., O'Mahony, D., Anklesaria, P., and Paul, R. (2003). Targeting of lipid-protamine-DNA (LPD) lipopolyplexes using RGD motifs. *J. Liposome Res.* **13**(3–4), 231–247.

Hengge, U. R., Chan, E. F., Foster, R. A., Walker, P. S., and Vogel, J. C. (1995). Cytokine gene expression in epidermis with both biological effects following injection of naked DNA. *Nat. Genet.* **10**, 161–166.

Hilgenbrink, A. R., and Low, P. S. (2005). Folate receptor-mediated drug targeting: from therapeutics to diagnostics. *J. Pharm. Sci.* **94**(10), 2135–2146.

Hofland, H. E., Masson, C., Iginla, S., Osetinsky, I., Reddy, J. A., Leamon, C. P., Scherman, D., Bessodes, M., and Wils, P. (2002). Folate-targeted gene transfer *in vivo. Mol. Ther.* **5**(6), 739–744.

Holcik, M. (2004). Targeting endogenous inhibitors of apoptosis for treatment of cancer, stroke and multiple sclerosis. *Expert Opin. Ther. Targets* **8**(3), 241–253.

Holt, I. J., Harding, A. E., and Morgan-Hughes, J. A. (1988). Deletions of muscle mitochondrial DNA in patients with mitochondrial myopathies. *Nature* **331**(6158), 717–719.

Howard, K. A., Rahbek, U. L., Liu, X., Damgaard, C. K., Glud, S. Z., Andersen, M. O., Hovgaard, M. B., Schmitz, A., Nyengaard, J. R., Besenbacher, F., and Kjems, J. (2006). RNA interference *in vitro* and *in vivo* using a novel chitosan/siRNA nanoparticle system. *Mol. Ther.* **14**(4), 476–484.

Huth, U. S., Schubert, R., and Peschka-Suss, R. (2006). Investigating the uptake and intracellular fate of pH-sensitive liposomes by flow cytometry and spectral bio-imaging. *J. Controlled Release* **110**(3), 490–504.

Ishida, O., Maruyama, K., Sasaki, K., and Iwatsuru, M. (1999). Size-dependent extravasation and interstitial localization of polyethyleneglycol liposomes in solid tumor-bearing mice. *Int. J. Pharm.* **190**(1), 49–56.

Ito, A., Shinkai, M., Honda, H., and Kobayashi, T. (2005). Medical application of functionalized magnetic nanoparticles. *J. Biosci. Bioeng.* **100**(1), 1–11.

Jain, T. K., Morales, M. A., Sahoo, S. K., Leslie-Pelecky, D. L., and Labhasetwar, V. (2005). Iron oxide nanoparticles for sustained delivery of anticancer agents. *Mol. Pharm.* **2**(3), 194–205.

Johannes, L., and Lamaze, C. (2002). Clathrin-dependent or not: is it still the question? *Traffic* **3**(7), 443–451.

Kam, N. W., O'Connell, M., Wisdom, J. A., and Dai, H. (2005). Carbon nanotubes as multifunctional biological transporters and near-infrared agents for selective cancer cell destruction. *Proc. Natl. Acad. Sci. U.S.A.* **102**(33), 11600–11605.

Kataoka, K., Harada, A., and Nagasaki, Y. (2001). Block copolymer micelles for drug delivery: design, characterization and biological significance. *Adv. Drug Deliv. Rev.* **47**(1), 113–131.

King, G. L., and Feener, E. P. (1998). The biochemical and physiological characteristics of receptors. *Adv. Drug Deliv. Rev.* **29**(3), 197–213.

Klumpp, C., Kostarelos, K., Prato, M., and Bianco, A. (2005). Functionalized carbon nanotubes as emerging nanovectors for the delivery of therapeutics. *Biochim. Biophys. Acta* **1758**(3), 404–412.

Kostarelos, K. (2003). Rational design and engineering of delivery systems for therapeutics: biomedical exercises in colloid and surface science. *Adv. Colloid Interface Sci.* **106**, 147–168.

Kurihara, A., and Pardridge, W. M. (1999). Imaging brain tumors by targeting peptide radiopharmaceuticals through the blood–brain barrier. *Cancer Res.* **59**(24), 6159–6163.

Lambert, G., Fattal, E., and Couvreur, P. (2005). Nanoparticulate systems for the delivery of antisense oligonucleotides. *Adv. Drug Deliv. Rev.* **47**(1), 99–112.

Lee, C. C., MacKay, J. A., Frechet, J. M., and Szoka, F. C. (2005). Designing dendrimers for biological applications. *Nat. Biotechnol.* **23**(12), 1517–1526.

Lenaerts, V., Couvreur, P., Christiaens-Leyh, D., Joiris, E., Roland, M., Rollman, B., and Speiser, P. (1984). Degradation of poly(isobutyl cyanoacrylate) nanoparticles. *Biomaterials* **5**, 65–68.

Levine, F., and Friedmann, T. (1993), Gene therapy. *Am. J. Dis. Child* **147**, 1167–1174.

Li, L., and Hoffman, R. M. (1995). The feasibility of targeted selective gene therapy of the hair follicle. *Nat. Genet.* **1**, 705–706.

Lin, E., and Nemunaitis, J. (2004). Oncolytic viral therapies. *Cancer Gene Ther.* **11**(10), 643–664.

Mager, D. E., Neuteboom, B., and Jusko, W. J. (2005). Pharmacokinetics and pharmacodynamics of PEGylated IFN-beta 1a following subcutaneous administration in monkeys. *Pharm. Res.* **22**(1), 58–61.

Maitra, A. (2005). Calcium phosphate nanoparticles: second-generation nonviral vectors in gene therapy. *Expert Rev. Mol. Diagn* **5**(6), 893–905.

Majoros, I. J., Thomas, T. P., Mehta, C. B., and Baker, J. R., Jr. (2005). Poly(amidoamine) dendrimer-based multifunctional engineered nanodevice for cancer therapy. *J. Med. Chem.* **48**(19), 5892–5899.

Martin, C. R., and Kohli, P. (2003). The emerging field of nanotube biotechnology. *Nat. Rev. Drug Discov.* **2**(1), 29–37.

Marty, C., Meylan, C., Schott, H., Ballmer-Hofer, K., and Schwendener, R. A. (2004). Enhanced heparan sulfate proteoglycan-mediated uptake of cell-penetrating peptide-modified liposomes. *Cell. Mol. Life Sci.* **61**(14), 1785–1794.

Maxfield, F. R., and McGraw, T. E. (2004). Endocytic recycling. *Nat. Rev. Mol. Cell Biol.* **5**, 121–132.

McLoughlin, J. M., McCarty, T. M., Cunningham, C., Clark, V., Senzer, N., Nemunaitis, J., and Kuhn, J. A. (2005). TNFerade, an adenovector carrying the transgene for human tumor necrosis factor alpha, for patients with advanced solid tumors: surgical experience and long-term follow-up. *Ann. Surg. Oncol.* **12**(10), 825–830.

Meuli, M., Liu, Y., Liggitt, D., *et al.* (2001). Efficient gene expression in skin wound sites following local plasmid injection. *J. Invest. Dermatol.* **131**, 131–135.

Minko, T., Dharap, S. S., Pakunlu, R. I., and Wang. Y. (2004). Molecular targeting of drug delivery systems to cancer. *Curr. Drug Targets* **5**(4), 389–406.

Moghimi, S. M. (2002). Liposome recognition by resident and newly recruited murine liver macrophages. *J. Liposome Res.* **12**(1–2), 67–70.

Moghimi, S. M., Hunter, A. C., and Murray, J. C. (2001). Long-circulating and target-specific nanoparticles: theory to practice. *Pharmacol. Rev.* **53**(2), 283–318.

Moghimi, S. M., Hunter, A. C., and Murray, J. C. (2005). Nanomedicine: current status and future prospects. *FASEB J.* **19**(3), 311–330.

Monahan, P. E., and Samulski, R. J. (2000). AAV vectors: is clinical success on the horizon? *Gene Ther.* **7**(1), 24–30.

Morgan, J. R., and Cloninger, M. J. (2002). Heterogeneously functionalized dendrimers. *Curr. Opin. Drug Discov. Devel.* **5**(6), 966–973.

Moses, M. A., Brem, H., and Langer, R. (2003). Advancing the field of drug delivery: taking aim at cancer. *Cancer Cell* **4**(5), 337–341.

Mrsny, R. J. (2004). Strategies for targeting protein therapeutics to selected tissues and cells. *Expert Opin. Biol. Ther.* **4**(1), 65–73.

Muro, S., Koval, M., and Muzykantov, V. (2004). Endothelial endocytic pathways: gates for vascular drug delivery. *Curr. Vasc. Pharmacol.* **2**(3), 281–299.

Muro, S., Schuchman, E. H., and Muzykantov, V. R. (2006a). Lysosomal enzyme delivery by ICAM-1-targeted nanocarriers bypassing glycosylation- and clathrin-dependent endocytosis. *Mol. Ther.* **3**(1), 135–141.

Muro, S., Mateescu, M., Gajewski, C., Robinson, M., Muzykantov, V. R., and Koval, M. (2006b). Control of intracellular trafficking of ICAM-1-targeted nanocarriers by endothelial Na$^+$/H$^+$ exchanger proteins. *Am. J. Physiol. Lung Cell. Mol. Physiol.* **290**(5), L809–L817.

Murphy, M. P., and Smith, R. A. (2000). Drug delivery to mitochondria: the key to mitochondrial medicine. *Adv. Drug. Deliv. Rev.* **41**(2), 235–250.

Nanney, L. B., Paulsen, S., Davidson, M. K., Cardwell, N. L., Whitsitt, J. S., and Davidson, J. M. (2000). Boosting epidermal growth factor receptor expression by gene gun transfection stimulates epidermal growth *in vivo*. *Wound Repair Regen.* **8**(2), 117–127.

Nimjee, S. M., Rusconi, C. P., and Sullenger, B. A. (2005). Aptamers: an emerging class of therapeutics. *Annu. Rev. Med.* **56**, 555–583.

Nori, A., and Kopecek, J. (2005). Intracellular targeting of polymer-bound drugs for cancer chemotherapy. *Adv. Drug Deliv. Rev.* **57**(4), 609–636.

Oberdorster, G., Oberdorster, E., and Oberdorster, J. (2005). Nanotoxicology: an emerging discipline evolving from studies of ultrafine particles. *Environ. Health Perspect.* **113**(7), 823–839.

O'Brien, J., and Lummis, S. C. (2002). An improved method of preparing microcarriers for biolistic transfection. *Brain Res. Protoc.* **10**(1), 12–15.

Oehlke, J., Lorenz, D., Wiesner, B., and Bienert, M. (2005). Studies on the cellular uptake of substance P and lysine-rich, KLA-derived model peptides. *J. Mol. Recognit.* **18**(1), 50–59.

Ogris, M., Walker, G., Blessing, T., Kircheis, R., Wolschek, M., and Wagner, E. (2003). Tumor-targeted gene therapy: strategies for the preparation of ligand–polyethylene glycol–polyethylenimine/DNA complexes. *J. Controlled Release* **91**(1–2), 173–181.

Oishi, M., Nagatsugi, F., Sasaki, S., Nagasaki, Y., and Kataoka, K. (2005). Smart polyion complex micelles for targeted intracellular delivery of PEGylated antisense oligonucleotides containing acid-labile linkages. *Chembiochem.* **6**(4), 718–725.

Okada, H., and Toguchi, H. (1995). Biodegradable microspheres in drug delivery. *Crit. Rev. Ther. Drug Carrier Syst.* **12**(1), 1–99.

Olivier, J. C. (2005). Drug transport to brain with targeted nanoparticles. *NeuroRx* **2**(1), 108–119.

Oupicky, D., Ogris, M., Howard, K. A., Dash, P. R., Ulbrich, K., and Seymour, L. W. (2002). Importance of lateral and steric stabilization of polyelectrolyte gene delivery vectors for extended systemic circulation. *Mol. Ther.* **5**(4), 463–472.

Paciotti, G. F., Myer, L., Weinreich, D., Goia, D., Pavel, N., McLaughlin, R. E., and Tamarkin, L. (2004). Colloidal gold: a novel nanoparticle vector for tumor directed drug delivery. *Drug Deliv.* **11**(3), 169–183.

Pakunlu, R. I., Wang, Y., Saad, M., Khandare, J. J., Starovoytov, V., and Minko, T. (2006). *In vitro* and *in vivo* intracellular liposomal delivery of antisense oligonucleotides and anticancer drug. *J. Controlled Release* **114**(2), 153–162.

Palmer, T. D., Rosman, G. J., Osborne, W. R. A., and Miller, A. D. (1991). Genetically modified skin fibroblasts persist long after transplantation but gradually inactivate introduced genes. *Proc. Natl. Acad. Sci. U.S.A.* **88**, 1330–1334.

Panyam, J., and Labhasetwar, V. (2003). Dynamics of endocytosis and exocytosis of poly(D,L-lactide-*co*-glycolide) nanoparticles in vascular smooth muscle cells. *Pharm. Res.* **20**(2), 212–220.

Panyam, J., and Labhasetwar, V. (2004). Targeting intracellular targets. *Curr. Drug Deliv.* **1**(3), 235–247.

Panyam, J., Zhou, W. Z., Prabha, S., Sahoo, S. K., and Labhasetwar, V. (2002). Rapid endo-lysosomal escape of poly(D,L-lactide-*co*-glycolide) nanoparticles: implications for drug and gene delivery. *FASEB J.* **16**(10), 1217–1226.

Panyam, J., Sahoo, S. K., Prabha, S., Bargar, T., and Labhasetwar, V. (2003). Fluorescence and electron microscopy probes for cellular and tissue uptake of poly(D,L-lactide-*co*-glycolide) nanoparticles. *Int. J. Pharm.* **262**(1–2), 1–11.

Petrak, K. (2005). Essential properties of drug-targeting delivery systems. *Drug Discov. Today* **10**(23–24), 1667–1673.

Pincus, P. (1991). Colloid stabilization with grafted polyelectrolytes. *Macromolecules* **24**, 2912–2919.

Piskin, E., Dincer, S., and Turk, M. (2004). Gene delivery: intelligent but just at the beginning. *J. Biomater. Sci. Polym. Ed.* **15**(9), 1181–1202.

Pond, S. M., and Tozer, T. N. (1984). First-pass elimination. Basic concepts and clinical consequences. *Clin. Pharmacokinet.* **9**(1), 1–25.

Prokop, A., and Davidson, J. M. (2007). Intracellular delivery to cells and tissues: drugs and genes, *J. Pharmaceut. Res.*, submitted.

Prokop, A., Holland, C. A., Kozlov, E., Moore, B., and Tanner, R. D. (2002a). Water-based nanoparticulate polymeric system for protein delivery. *Biotechnol. Bioeng.* **75**(2), 228–232.

Prokop, A., Kozlov, E, Carlesso, G., and Davidson, J. M. (2002b). Hydrogel-based colloidal polymeric system for protein and drug delivery: physical and chemical characterization, permeability control and applications. *Advan. Polymer Sci.* **160**, 119–173.

Ramaekers, F. C., and Bosman, F. T. (2004). The cytoskeleton and disease. *J. Pathol.* **204**(4), 351–354.

Rejman, J., Oberle, V., Zuhorn, I. S., and Hoekstra, D. (2004). Size-dependent internalization of particles via the pathways of clathrin- and caveolae-mediated endocytosis. *Biochem. J.* **377**(Pt. 1), 159–169.

Roelvink, P. W., Mi Lee, G., Einfeld, D. A., Kovesdi, I., and Wickham, T. J. (1999). Identification of a conserved receptor-binding site on the fiber proteins of CAR-recognizing adenoviridae. *Science* **286**(5444), 1568–1571.

Roemer, K., and Friedmann, T. (1992). Concepts and strategies for human gene therapy. *Eur. J. Biochem.* **208**, 211–225.

Roizman, B. (1996). The function of HSV genes: a primer for genetic engineering of novel vectors. *Proc. Natl. Acad. Sci. U.S.A.* **93**, 11307–11312.

Rosenberg, S. A., Blaese, R. M., Brenner, M. K., Deisseroth, A. B., Ledley, F. D., Lotze, M. T., Wilson, J. M., Nabel, G. J., Cornetta, K., Economou, J. S., *et al.* (2000). Human gene marker/therapy clinical protocols. *Hum. Gene Ther.* **11**(6), 919–979.

Roser, M., Fischer, D., and Kissel, T. (1998). Surface-modified biodegradable albumin nano- and microspheres. II: effect of surface charges on *in vitro* phagocytosis and biodistribution in rats. *Eur. J. Pharm. Biopharm.* **46**(3), 255–263.

Rothbard, J. B., Jessop, T. C., and Wender, P. A. (2005). Adaptive translocation: the role of hydrogen bonding and membrane potential in the uptake of guanidinium-rich transporters into cells. *Adv. Drug Deliv. Rev.* **57**(4), 495–504.

Sagara, K., and Kim, S. W. (2002). A new synthesis of galactose-poly(ethylene glycol)-polyethylenimine for gene delivery to hepatocytes. *J. Controlled Release* **79**(1–3), 271–281.

Sahoo, S. K., and Labhasetwar, V. (2003). Nanotech approaches to drug delivery and imaging. *Drug Discov. Today* **8**(24), 1112–1120.

Salaun, C., James, D. J., and Chamberlain, L. H. (2004). Lipid rafts and the regulation of exocytosis. *Traffic* **5**(4), 255–264.

Saltzman, W. M. (2001). "Drug Delivery: Engineering Principles for Drug Therapy." Oxford University Press, Oxford, UK.

Sands, M. S., and Davidson, B. L. (2006). Gene therapy for lysosomal storage diseases. *Mol. Ther.* **13**(5), 839–849.

Sangiorgio, V., Pitto, M., Palestini, P., and Masserini, M. (2004). GPI-anchored proteins and lipid rafts. *Ital. J. Biochem.* **53**(2), 98–111.

Sato, T., and Sunamoto, J. (1992). Recent aspects in the use of liposomes in biotechnology and medicine. *Prog. Lipid Res.* **31**(4), 345–372.

Sawamura, D., Yasukawa, K., Kodama, K., *et al.* (2002). The majority of keratinocytes incorporate intradermally injected plasmid DNA regardless of size, but only a small proportion of cells can express the gene product. *J. Invest. Dermatol.* **118**, 967–971.

Senger, D. R., Galli, S. J., Dvorak, A. M., Perruzzi, C. A., Harvey, V. S., and Dvorak, H. F. (1983). Tumor cells secrete a vascular permeability factor that promotes accumulation of ascites fluid. *Science* **219**(4587), 983–985.

Shah, D., and Shen, W. C. (1996). Transcellular delivery of an insulin–transferrin conjugate in enterocyte-like Caco-2 cells. *J. Pharm. Sci.* **85**(12), 1306–1311.

Sharma, P., Sabharanjak, S., and Mayor, S. (2002). Endocytosis of lipid rafts: an identity crisis. *Semin. Cell Dev. Biol.* **13**(3), 205–214.

Shitara, Y., Horie, T., and Sugiyama, Y. (2006). Transporters as a determinant of drug clearance and tissue distribution. *Europ. J. Pharmaceut. Sci.* **27**, 425–446.

Silverman, S. K. (2005). *In vitro* selection, characterization, and application of deoxyribozymes that cleave RNA. *Nucleic Acids Res.* **33**(19), 6151–6163.

Simoes, S., Slepushkin, V., Gaspar, R., de Lima, M. C., and Duzgunes, N. (1998). Gene delivery by negatively charged ternary complexes of DNA, cationic liposomes and transferrin or fusigenic peptides. *Gene Ther.* **5**(7), 955–964.

Stephens, A. C., and Rivers, R. P. (2003). Antisense oligonucleotide therapy in cancer. *Curr. Opin. Mol. Ther.* **5**(2), 118–122.

Sumimoto, H., Miyagishi, M., Miyoshi, H., Yamagata, S., Shimizu, A., Taira, K., and Kawakami, Y. (2004). Inhibition of growth and invasive ability of melanoma by inactivation of mutated BRAF with lentivirus-mediated RNA interference. *Oncogene* **23**(36), 6031–6039.

Sun, L., Xu, L., Chang, H., *et al.* (1997). Transfection with aFGF cDNA improves wound healing. *J. Invest. Dermatol.* **108**, 313–318.

Trubetskoy, V. S., Torchilin, V. P., Kennel, S., and Huang, L. (1992). Cationic liposomes enhance targeted delivery and expression of exogenous DNA mediated by N-terminal modified poly(L-lysine)-antibody conjugate in mouse lung endothelial cells. *Biochem. Biophys. Acta* **1131**, 311–313.

Trubetskoy, V. S., Wong, S. C., Subbotin, V., Budker, V. G., Loomis, A., Hagstrom, J. E., and Wolff, J. A. (2003). Recharging cationic DNA complexes with highly charged polyanions for *in vitro* and *in vivo* gene delivery. *Gene Ther.* **10**, 261–271.

Turner, J. J., Jones, S., Fabani, M. M., Ivanova, G., Arzumanov, A. A., and Gait, M. J. (2006). RNA targeting with peptide conjugates of oligonucleotides, siRNA and PNA. *Blood Cells Mol. Dis.* **38**(1), 1–7.

van de Waterbeemd, H., and Gifford, E. (2003). ADMET in silico modelling: towards prediction paradise? *Nat. Rev. Drug Discov.* **2**(3), 192–204.

Venier-Julienne, M. C., and Benoit, J. P. (1996). Preparation, purification and morphology of polymeric nanoparticles as drug carriers. *Pharm. Acta Helv.* **71**(2), 121–128.

Vert, M., Mauduit, J., and Li, S. (1994). Biodegradation of PLA/GA polymers: increasing complexity. *Biomaterials* **15**(15), 1209–1213.

Vicent, M. J., and Duncan, R. (2006). Polymer conjugates: nanosized medicines for treating cancer. *Trends Biotechnol.* **24**(1), 39–47.

Wiethoff, C. M., Smith, J. G., Koe, G. S., and Middaugh, C. R. (2001). The potential role of proteoglycans in cationic lipid-mediated gene delivery. Studies of the interaction of cationic lipid-DNA complexes with model glycosaminoglycans. *J. Biol. Chem.* **276**(35), 32806–32813.

Wilson, J. M. (1996). Adenovirus as gene-delivery vehicles. *N. Engl. J. Med.* **334**, 1185–1187.

Win, K. Y., and Feng, S. S. (2005). Effects of particle size and surface coating on cellular uptake of polymeric nanoparticles for oral delivery of anticancer drugs. *Biomaterials* **26**(15), 2713–2722.

Yang, S. C., and Zhu, J. B. (2002). Preparation and characterization of camptothecin solid lipid nanoparticles. *Drug Dev. Ind. Pharm.* **28**(3), 265–274.

Yu, W., Pirollo, K. F., Rait, A., Yu, B., Xiang, L. M., Huang, W. Q., Zhou, Q., Ertem, G., and Chang, E. H. (2004). A sterically stabilized immunolipoplex for systemic administration of a therapeutic gene. *Gene Ther.* **11**(19), 1434–1440.

Zauner, W., Farrow, N. A., and Haines, A. M. (2001). *In vitro* uptake of polystyrene microspheres: effect of particle size, cell line and cell density. *J. Controlled Release* **71**(1), 39–51.

Zelikin, A. N., Trukhanova, E. S., Putnam, D., Izumrudov, V. A., and Litmanovich, A. A. (2003). Competitive reactions in solutions of poly-L-histidine, calf thymus DNA, and synthetic polyanions: determining the binding constants of polyelectrolytes. *J. Am. Chem. Soc.* **125**(45), 13693–13699.

Zorko, M., and Langel, U. (2005). Cell-penetrating peptides: mechanism and kinetics of cargo delivery. *Adv. Drug. Deliv. Rev.* **57**(4), 529–545.

# Chapter Thirty-Six

# Breast Reconstruction

*Lamont Cathey, Kuen Yong Lee, Walter D. Holder, David J. Mooney, and Craig R. Halberstadt*

## I. INTRODUCTION

Reconstructive surgery after mastectomy or lumpectomy due to breast cancer may benefit greatly from tissue engineering for the creation of new soft-tissue replacements. Various cell types can be used for breast reconstruction, including preadipocytes, fibroblasts, smooth muscle cells, muscle myocytes, and chondrocytes. To introduce a cell–polymer construct into the body, proper selection of the polymeric materials should provide three-dimensional support for engineered new tissue. There are two strategies for introducing the cell–polymer constructs into the body. The first is to implant surgically, while the second approach is to deliver in a minimally invasive manner using injectable forms of materials. A critical challenge to engineer large tissue masses is vascularization, and approaches to optimize this process include delivery of angiogenic molecules or endothelial cells.

According to the American Cancer Society, approximately 182,000 new cases of breast cancer are detected each year that can potentially require breast reconstruction. Breast cancer represents the second most common cancer in the female population. In the United States approximately one out of every eight women will develop the disease (American Cancer Society, 2006), and it is the second leading cause of cancer-related death in women. Approximately $8.1 billion is spent each year in the United States for breast cancer treatment. Historically, breast cancer was treated with radical mastectomy. Radical breast surgery was developed at a time when breast cancer was typically very large and was often unresectable by then-current standards. This procedure removed the breast, underlying pectoral muscles, and axillary lymph nodes. The surgical treatment of breast cancer has expanded from the traditional radical mastectomy. Radical mastectomy gave way to modified radical mastectomy. Advances in cancer detection and increased awareness through self-examination have allowed cancers to be discovered at early stages. Small lesions can now be treated with lumpectomy (removal of the part of the breast containing the tumor plus some surrounding nontumor-bearing tissue) along with radiation. Axillary lymph node metastasis can be evaluated through sentinel lymph node biopsy, avoiding the morbidity of a full axillary dissection. Randomized prospective clinical trials conducted during the past 30 years comparing these less deformative techniques have demonstrated survival rates equivalent to those for modified radical mastectomy (Lichter, 1997). These studies have demonstrated unequivocally that for most women with small breast cancers, simple excision (lumpectomy) of the breast cancer with sampling of the axillary lymph nodes followed by radiation provides an outcome similar to that for radical mastectomy. However, in some women mastectomy remains the best treatment, due to extensive tumor involvement within the breast, which requires removal of all or most of the breast tissue.

*Principles of Tissue Engineering, 3rd Edition*
*ed. by Lanza, Langer, and Vacanti*

The current approaches to reconstructing breast tissue following a mastectomy include reconstructive surgery utilizing autologous tissue flaps or implants of synthetic materials. Complex flap reconstruction using transversus rectus abdominal myocutaneous (TRAM) flaps, latissimus dorsi flaps, or muscle-free flaps from the buttock or thigh are used to reconstruct large volumes of lost breast tissue. These are very extensive procedures with long recovery periods, and they require the harvest of tissue from another site in the body, with its attendant side effects and complications. The advantage of this approach is that a patient's own tissues are utilized, resulting in a more natural tactile tissue quality. Grafting of autologous fat is another technique that has been utilized for reconstruction. The procedure produces poor results, for several reasons. The cellular material used for transplantation is obtained from lipoaspiration, a process that can damage and lyse the cells. Second, the transplanted cells consist mainly of adipocytes that are terminally differentiated and incapable of proliferation (Patrick, 2004). A more common procedure involves the use of synthetic materials. In this process a tissue expander is placed, usually beneath the pectoral muscles, and gradually inflated over several weeks to the desired size. It is subsequently removed by a relatively simple procedure, and a saline-filled permanent silastic implant is placed into the cavity. Because the implant is a foreign body, there may be a substantial inflammatory response producing fibrosis, thickening, and an unnatural shape and tactile quality. Also, implants may leak and require replacement. Silicone-filled implants are now rarely used, due to fears of a possible induction of autoimmune disease caused by the leakage of silicone. These reconstructive approaches have been designed for the mastectomy procedure, and currently there is no reconstructive option for patients who are candidates for a lumpectomy procedure.

Tissue engineering is potentially a potent approach to treating the loss of tissues or organs (Langer and Vacanti, 1993; Mooney and Mikos, 1999), and it could find wide utility in breast-tissue reconstruction. There are myriad potential indications to apply tissue engineering for tissue reconstruction, including postoncologic resection (i.e., mastectomy, lumpectomy, and parotidectomy), soft-tissue defect from trauma, and congenital defects. Engineered breast reconstruction can be approached through one of two basic pathways: either *in situ* adipogenesis or *de novo* adipogenesis. The first technique employs the transplantation of a supportive matrix to encourage the development of adipose tissue *in situ*. The matrix is supplemented with specific growth factors that attract native cells, such as fibroblasts and preadipocytes, to migrate into the matrix (Patel and Patrick, 2004).

*De novo* adipogenesis takes on a slightly different method. In this approach, tissue-specific cells are isolated from a small tissue biopsy and expanded *in vitro*. The cells are subsequently placed onto polymeric scaffolds that act as synthetic extracellular matrices. These scaffolds deliver the cells to the desired site in the body, define a space for tissue formation, and potentially control the structure and function of the engineered tissue (Eiselt *et al.*, 1998; Kim and Mooney, 1998a; Putnam and Mooney, 1996). Utilization of biodegradable polymer scaffolds leads ultimately to a new tissue mass with no permanent synthetic element. New soft tissues, which can be transplanted into a patient to reconstruct lost breast tissue, can potentially be created with this process (Fig. 36.1). This approach may circumvent the main limitations of reconstruction of breast tissue using tissue transplantation while eliminating the need for the permanent implantation of synthetic prosthetic materials.

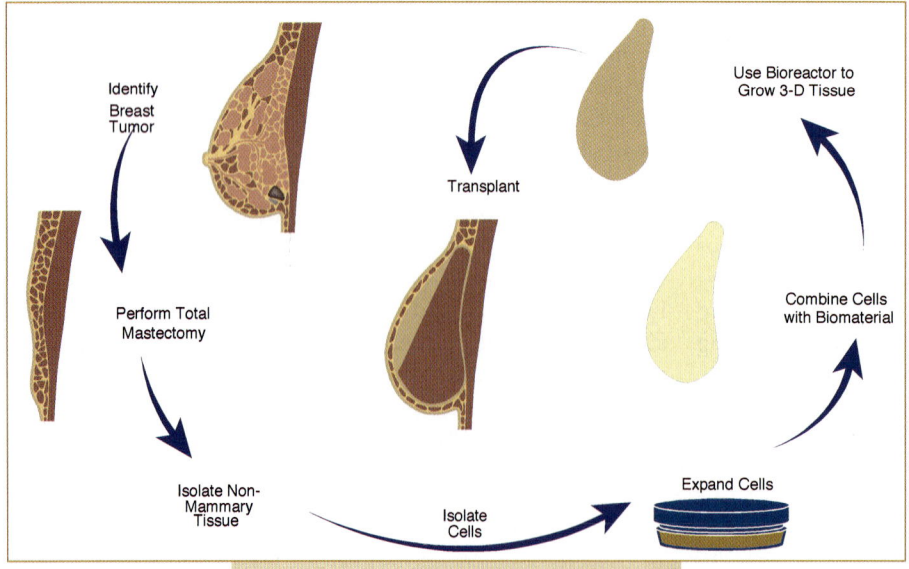

**FIG. 36.1.** Schematic of the approach to engineer a soft-tissue replacement for women with breast cancer. The first steps include identification and removal of the breast tissue and then isolation and multiplication *in vitro* of appropriate cells derived from another site in the body. These cells are then combined with a biodegradable polymer matrix and implanted in the body to re-create a soft tissue that emulates the lost breast tissue.

In this chapter, we summarize the various cell types that may be useful for breast reconstruction, the polymers that can potentially be utilized in this tissue-engineering approach, and the development of suitable animal models that allow one to test this concept for breast reconstruction. In addition, we review approaches being developed to promote vascularization of the engineered tissue, for this will be a critical challenge to engineering a large mass of soft tissue using a cell transplantation approach.

## II. CELL TYPES FOR SOFT-TISSUE ENGINEERING

Many considerations must be addressed when choosing which cell type(s) should be used in human-breast-tissue engineering. The first consideration is that there be substantial variability in the size, shape, and consistency of the breast. All three factors change over time, with a tendency for breast parenchyma (glands and ducts) to involute or regress as a woman ages, particularly after menopause, and to be replaced by fat. Also, comparing breast tissue among women of any given age, we find considerable variability in size and shape, and tactile, elastic, and tensile characteristics of the tissue. The tensile and elastic characteristics of the breast are produced by three major factors:

1. The amount and quality of fat within the breast.
2. The amount and quality of glandular and duct tissue in the breast.
3. The mechanical characteristics of the fibrous support structures of the breast (Cooper's ligaments).

In breast reconstruction, the creation of a functional breast with lactational ability is not needed and, in fact, may add to a woman's breast cancer risk by introducing mammary epithelial cells that may be predisposed to cancer development. The major goal of breast reconstruction is to produce a breast mound with all of the aesthetic properties of a normal breast. A major concern for a woman undergoing breast reconstruction for cancer or augmentation is that the right side match the left side and that the reconstruction be aesthetically correct. Also, it is important to understand that there is considerable variability in the expectations of women who undergo reconstructive or augmentation breast surgery in terms of the desired size and shape and with the elastic and tactile qualities of the breast reconstruction. With these observations, a projection can be made that fat cells would need to be a major part of the tissue-engineered construct and that other cell types (e.g., muscle) would be needed to contribute appropriate conformational, tensile, elastic, and tactile characteristics of the construct. Also, due to the variability in normal breast, some individual manipulation of these constituents may be required.

A major cell type comprising normal breast tissue, and likely the engineered replacement, is the adipocyte. Lipid-laden fat cells (adipocytes) are terminally differentiated and will not divide further *in vivo* or under cell culture conditions. If these cells were to be used, they would have to be harvested from fat in an equivalent volume required for the construct (Hang-Fu *et al.*, 1995; Kononas *et al.*, 1993; Nguyen *et al.*, 1990). Alternatively, preadipocytes could be used for engineering soft tissue (Coleman *et al.*, 2000; Patrick *et al.*, 1999) because these cells potentially can be expanded in culture. Preadipocytes are similar to fibroblasts in structure and possess the ability to expand in culture. Several studies have shown ways to induce differentiation into adipocytes. These cells can be successfully harvested and isolated from such sites as the subcutaneous tissues and the omentum. Investigations of autologous preadipocyte implantation with a sheep animal model have been promising (Halberstadt *et al.*, 2002). Technologies are being developed that may allow for simple harvesting and fast isolation of preadipocyte cells for immediate reimplantation into the patient to serve as stem cells (Fraser *et al.*, 2006).

Since breast tissue is more than fat alone in most cases, additional cell types must be considered to contribute to the shape and the tactile, elastic, and tensile characteristics of a breast reconstructed via tissue-engineering methods. Possible other cell types include smooth muscle cells, fibroblasts, skeletal muscle, and elastic cartilage. Fibroblasts contribute greatly to the support structure of the breast by laying down bands of collagen that connect the breast tissue to the skin and to the pectoral muscle as well as helping maintain the overall shape. Fibroblasts are an integral component of tissue-engineered skin products, and transplantation of fibroblast-containing tissues has been demonstrated to replace lost skin tissue successfully in several situations (Eaglstein and Falanga, 1998; Landeen *et al.*, 1993, 1992; Naughton *et al.*, 1997; Naughton, 2002). The density and firmness of the breast are primary effects of the glandular epithelium and ductal structure. Tissues that have similar tactile and elastic properties are almost exclusively muscle. Smooth muscle cells can readily be isolated from a number of organs and expanded in culture. Techniques have been developed to seed these cells efficiently onto three-dimensional scaffolds fabricated from various polymers (Kim *et al.*, 1998) and to grow new tissues from these cell–polymer constructs with a defined size and shape (Kim and Mooney, 1998b). Implantation of smooth muscle containing polymers can lead to the reformation of significant tissue masses, with reorganization of the smooth muscle tissue into appropriate three-dimensional structures (Oberpenning *et al.*, 1999). Muscle myocytes can also be greatly expanded *in vitro* and have been demonstrated to reform functional tissue masses under appropriate conditions (Vandenburgh and Kaufman, 1979; Vandenburgh *et al.*, 1991). However, it remains to be demonstrated that smooth or skeletal muscle myocytes will maintain a tissue mass over long periods of time without neural stimulation. An alternative approach is to utilize chondrocytes as a component of an engineered breast tissue. Elastic cartilage has many of the mechanical properties of glandular breast tissue that are potentially important for tissue engineering of breast (e.g., elasticity). Chondrocytes can be expanded in culture, and they are able to survive in low oxygen tensions. Chondrocytes have been used extensively in tissue engineering to engineer a variety of tissue constructs both *in vitro* and *in vivo* (Atala *et al.*, 1994; Breinan *et al.*, 1998; Wakitani *et al.*, 1998).

Other cell types, such as mesenchymal stem cells and embryonic stem cells, may be a viable source for breast-tissue engineering. However, in the case of embryonic stem cells, a great deal of work will need to be done to create terminally differentiated cells that, once transplanted, do not have the potential to form a tumor mass. The implantation of embryonic stem cells has demonstrated the potential of forming teratomas *in vivo* (J. A. Thomson *et al.*, 1998; Wobus and Boheler, 2005).

A number of issues must be ultimately addressed in relation to cell expansion for breast-tissue reconstruction. In considering the number of cell types that may potentially be used, it is very likely that a variety of growth factors will be needed for the *in vitro* expansion of cells. Implantable cell-bearing polymers will also need to be created to provide the short- or long-term stimulation of growth factors and other substances to enhance vascularization or to provide or sustain a local milieu that will maintain a stable, healthy tissue mass in an area that may be foreign to the tissue (Mooney *et al.*, 1996b; Peters *et al.*, 1999b; Sheridan *et al.*, 2000). Standard isolation and expansion protocols must also be developed, and critical factors for successful outcomes will include aseptic technique in the harvest, routine quality control testing of all cultures, and long-term cell storage. It may also be possible to isolate all the cell types required for breast-tissue engineering from a single tissue source. For example, fat, in addition to adipocytes, contains a large vascular network, composed primarily of capillary endothelial cells and some vascular smooth muscle cells as well as a collagen stromal structure produced by fibroblasts (Williams *et al.*, 1994). Hence, multiple cell types can potentially be obtained from this tissue. In attempting to produce fat tissue *in vivo*, it may be beneficial to purify and expand the cellular components of fat without isolating each component. This may be particularly important if, as suggested by several authors, vascular endothelial cells and others exhibit organ specificity (Bassenge, 1996; Craig *et al.*, 1998; Murphy *et al.*, 1998).

## III. MATERIALS

Several key characteristics make a biomaterial suitable for tissue engineering. First, favorable biomaterials must be biodegradable and resorb over time. The shape and texture of the material must resemble that of natural tissue. For the purpose of re-creating adipose tissue, the material must have a soft, pliable texture. Cellular adherence is another important aspect of any potential material. The biomaterial must interact favorably with its cellular components without negatively impacting them (the material cannot be tumori-

genic). A porous surface is another key feature that potentially allows for cellular ingrowth into the material, be it native or implanted cells. In addition, a porous surface allows for the diffusion of nutrients and the establishment of a vascular network into the biomaterial construct (Beahm *et al.*, 2003).

There are two general strategies for introducing the cell–polymer constructs back into the body in order to engineer tissues. The constructs can be implanted using an open surgical procedure (implantable material) (Fig. 36.1) or introduced in a minimally invasive manner utilizing syringes or endoscopic delivery (injectable materials) (Fig. 36.2). Implantable forms of materials are typically foams, sponges, films, and other solid devices. The typical injectable forms of materials include hydrogels and microbeads. Most synthetic polymers, such as aliphatic polyesters, polyanhydrides, poly(amino acid)s, and poly(ortho ester)s, are suitable for the fabrication of implantable devices. On the other hand, natural polymers, including alginate, chitosan, hyaluronic acid, and collagen, are good examples of injectable materials. Hydrogels are highly attractive due to the potential to implant into the body in a minimally invasive manner via injection (Lee and Mooney, 2001). Their properties can be engineered for biocompatibility, selective permeability, mechanical and chemical stability, and other requirements as specified by the application (Jen *et al.*, 1996). Certain polymers, such as the polyanhydrides and collagen, can be used in both types of applications (Anseth *et al.*, 1999; Doillon *et al.*, 1994).

### Implantable Materials

A variety of polymers can be utilized to form solid or macroporous scaffolds for tissue-engineering applications (Fig. 36.3). Aliphatic polyesters of poly(glycolic acid) (PGA) and poly(lactic acid) (PLA) are well-characterized synthetic biodegradable polymers that have been widely applied to the biomedical field generally and to the tissue-engineering arena specifically (Wong and Mooney, 1997). PGA has a high crystallinity, a high melting temperature, and low solubility in organic solvents. PLA has a much more hydrophobic character than PGA due to the introduction of the methyl group. PLA has low water uptake, and its ester bond is less labile to hydrolysis, owing to steric hindrance of the methyl group. Therefore, PLA degrades more slowly and has a higher solubility in organic solvents than PGA. Copolymers of PLA and PGA can be readily synthesized, and their physical properties are regulated by the ratio of glycolic acid to lactic acid. This enables these copolymers to be used in various applications as biodegradable matrices in tissue engineering (Mooney *et al.*, 1995).

These aliphatic polyesters can readily be processed into various physical forms appropriate for tissue-engineering applications. A number of techniques have been proposed to generate a highly porous structure of scaffolds, including solvent casting/particulate leaching (Mikos *et al.*, 1994),

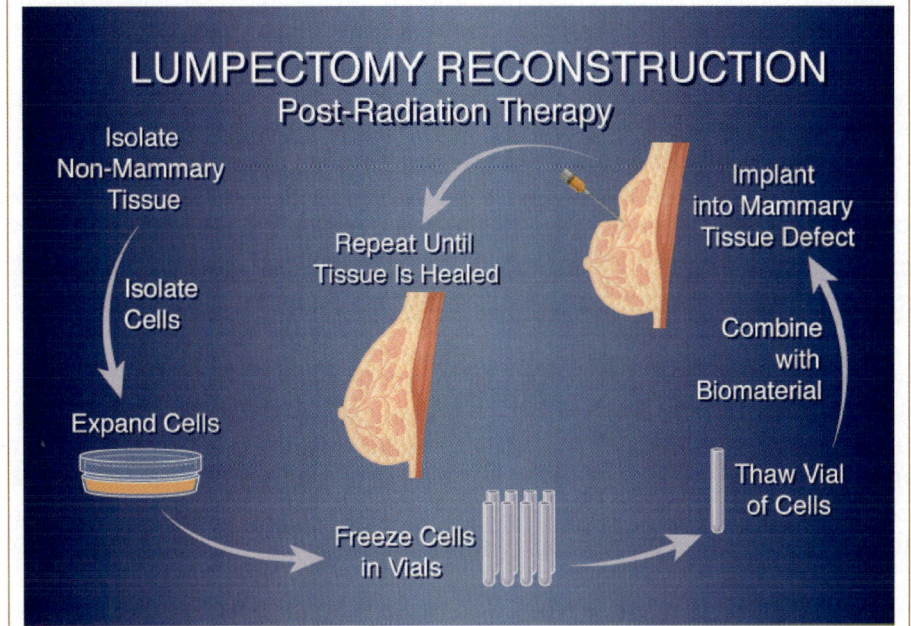

**FIG. 36.2.** Schematic of the approach to engineer a soft-tissue replacement for women with breast cancer following a lumpectomy procedure. The approach would combine nonmammary autologous cells seeded onto small biomaterial carriers that would provide the ability to be injectable. Multiple procedures could be necessary in order to reconstruct the appropriate cellular volume for space filling after tumor removal.

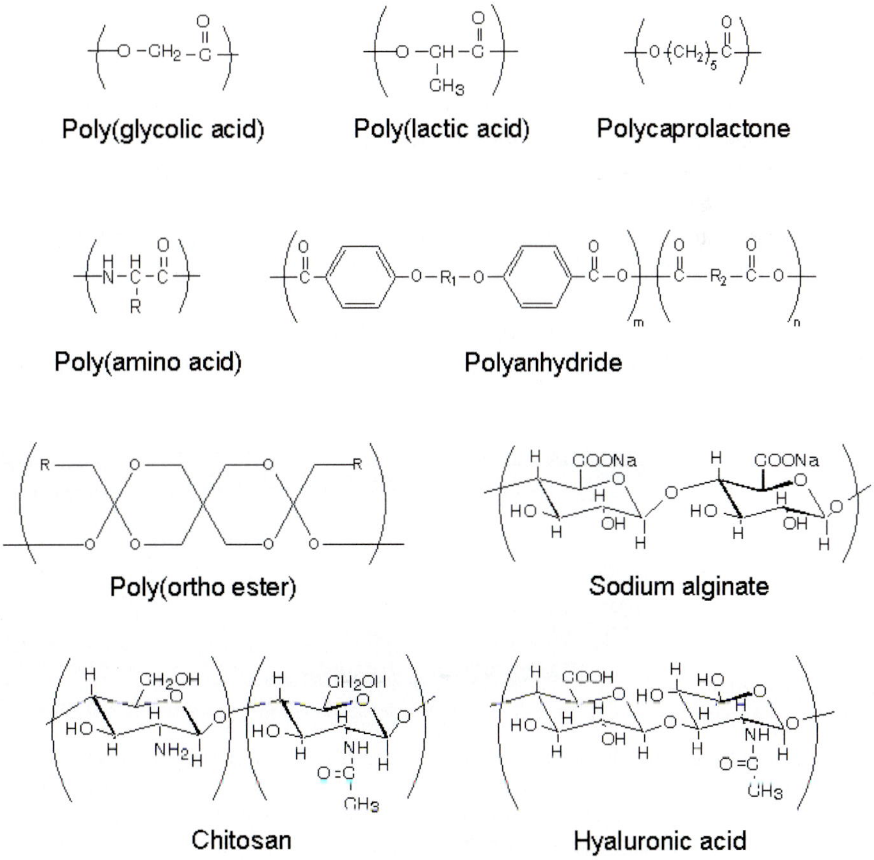

Poly(glycolic acid)   Poly(lactic acid)   Polycaprolactone

Poly(amino acid)   Polyanhydride

Poly(ortho ester)   Sodium alginate

Chitosan   Hyaluronic acid

**FIG. 36.3.** Chemical structure of implantable and injectable materials for tissue-engineering scaffold fabrication.

phase separation (Lo *et al.*, 1995), emulsion freeze-drying (Whang *et al.*, 1995), fiber extrusion and fabric formation (Freed *et al.*, 1994), and gas foaming (Harris *et al.*, 1998; Mooney *et al.*, 1996a). Nonwoven fabrics of PGA were stabilized by spraying with PLA or PLGA solution to resist significant compressional forces and used successfully to culture smooth muscle cells (Kim and Mooney, 1998b), even with a tubular structure (Mooney *et al.*, 1996c). The gas foaming and particulate leaching provided efficient techniques for generating open pore structures, and since this gas-foaming process requires no organic solvents or high temperature, biologically active molecules can be incorporated into these matrices without denaturation (Shea *et al.*, 1999; Sheridan *et al.*, 2000). Implantation of porous polymer scaffolds leads to host tissue ingrowth and formation of granulation tissue throughout the scaffold (Fig. 36.4). Investigations have been conducted using Matrigel (reconstituted basement membrane of mouse tumor) and fibroblast growth factor 2 (FGF-2) to induce *in situ* adipogenesis. Matrigel consists largely of type IV collagen, laminin, and perclan. Preliminary results have demonstrated the migration of native preadipocytes as well as endothelial cells into the gel matrix when injected into the subcutaneous tissue (Beahm *et al.*, 2003; Patrick, 2004; Walton *et al.*, 2004). Another approach, using the theme of *in situ* adipogenesis, involves the direct association of an implant with a preexisting blood supply. Experiments with nude mice have been performed in which silicone molds packed with poly(glycolic acid) (PGA) fibers were sewn to the inferior epigastric blood vessels. These silicone molds were injected with a combination of Matrigel and FGF-2. *In situ* adipogenesis was demonstrated over a 4- to 20-week time course. Additional studies have been conducted with microspheres composed of poly(L-lactic-*co*-glycolic acid)/polyethylene glycol (PLGA/PEG) that were implanted into the abdominal fascia of rats. These microspheres were injected with insulin, insulinlike growth factor, and FGF-2. Adipose tissue was evidenced in the microspheres four weeks following implantation (Patrick, 2004; Yuksel *et al.*, 2000).

Patel *et al.* (2005) conducted studies with four variations of polyethylene glycol (PEG) scaffolds. These groups consisted of unmodified PEG, PEG modified with a degradation peptide (LGPA) that allows enzymatic hydrolysis of the material, PEG modified with a laminin adhesion peptide (YIGSR), and PEG with both modifications. The individual groups of scaffolds were seeded with rat preadipocytes and evaluated for cell viability and proliferation. Out of the four groups, the PEG scaffolds that were modified with YIGSR and LGPA allowed for cellular adherence and proliferation. The seeded preadipocytes survived longer and proliferated to a greater extent than the other samples. The biodegradability of the scaffold was thought to allow the cells more physical space to proliferate, and thus they were not limited by contact inhibition (Patel *et al.*, 2005).

Nondegradable polymers have also been examined for tissue reconstruction. For example, *in vitro* studies have been conducted with fibronectin-coated polytetrafluoroethylene scaffolds (PTFE). Human preadipocytes have been shown successfully to attach, proliferate, and differentiate into adipocytes on the PTFE scaffolds (Kral and Crandall, 1999).

A number of other synthetic polymers could also be utilized to fabricate scaffolds for breast-tissue reconstruction. These include polycaprolactone, polyanhydrides, poly(amino acid)s, and poly(ortho ester)s. Polycaprolactone (PCL) is also one of the aliphatic polyesters. PCL is a semicrystalline polymer with high solubility in organic solvents and low melting temperature. Since the degradation rate of PCL is much slower than that of PGA or PLA, it has been previously used as a long-term drug delivery carrier (Pitt, 1990). The biocompatibility of PCL has been investigated in a rat model (Yamada *et al.*, 1997) and a rabbit model (Lowry *et al.*, 1997). Polyanhydrides are usually copolymers of aromatic diacids and aliphatic diacids. They usually degrade by surface erosion, and their degradation rate can be controlled, depending on the choice of diacids (Domb *et al.*, 1989). The degradation rate of polyanhydrides is much faster than those of poly(ortho ester)s in the absence of any

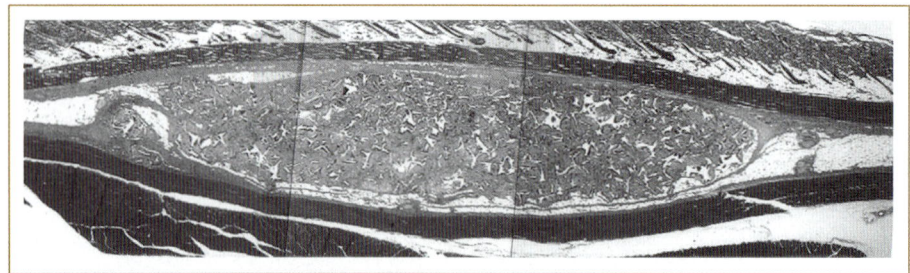

**FIG. 36.4.** Formation of new three-dimensional tissue following implantation of a highly porous biodegradable polymer matrix. Rat adipocytes were seeded for 24 h onto a 13-mm-diameter, 3-mm-thick PLA matrix and then implanted into the subcutaneous space of a female Lewis rat. The implants were harvested four weeks postimplantation, fixed in formalin, and embedded, and 4-μm sections were stained with hematoxylin and eosin. A 25× magnification of the graft is shown. A well-defined tissue is present throughout the implant site, with the establishment of a vascular bed, and the matrix size and shape predetermine the size and shape of the engineered tissue mass. Unpublished data from W. D. Holder and C. R. Halberstadt.

additives. Therefore, polyanhydrides have been widely used as biodegradable implants for local drug delivery of cefazolin (Park *et al.*, 1998), bupivacaine (Park *et al.*, 1998), methotrexate (Dang *et al.*, 1994), and taxol (Fung *et al.*, 1998; Park *et al.*, 1998). Poly[bis(*p*-carboxyphenoxy)propane-sebacic acid] was approved by the FDA for clinical trials (Engelberg and Kohn, 1991). It is biocompatible and nontoxic, even in the rat brain, when compared to standard neurosurgical implants (Tamargo *et al.*, 1989). Poly(amino acid)s have been studied due to their similarity to proteins, and they have been widely investigated for biomedical applications such as sutures and artificial skin (Anderson *et al.*, 1985). Poly(amino acid)s are usually polymerized by ring-opening of *N*-carboxyanhydrides, and versatile copolymers can be prepared from various combinations of amino acids. However, due to their low solubility and limited processability, pseudo-poly(amino acid)s have also been developed, by Kohn and Langer (1987). It was also reported that poly(amino acid)s containing L-arginine, L-lysine, or L-ornithine caused endothelium-dependent relaxation of bovine intrapulmonary artery and vein and stimulated the formation and/or release of an endothelium-derived relaxing factor identified as nitric oxide (Ignarro *et al.*, 1989). Poly(ortho ester)s are biodegradable polymers, which degrade by gradual surface erosion and have been known as useful materials for controlled drug delivery. Poly(ortho ester) membranes containing indomethacin (Solheim *et al.*, 1992), gentamycin (Pinholt *et al.*, 1992), or insulinlike growth factor (Busch *et al.*, 1996) have been prepared and implanted into rats without any noticeable inflammation after degradation.

Hyaluronic acid (HA) is a natural component of the extracellular matrix of many tissues. HA comprises repeated sequences of glucuronic acid and acetylglucosamine to form a polymer. This native form is susceptible to enzymatic degradation via hyaluronidase, making the molecule unstable for biomaterial purposes. Simple modifications have been made, such as cross-linking the chains to form insoluble hydrogels (Baier Leach *et al.*, 2003). In its natural form HA plays a role in enriching wound healing by promoting early inflammation and stimulating angiogenesis (Chen and Abatangelo, 1999). Hyaluronan benzyl ester (HYAFF® 11) scaffolds are derived from hyaluronic acid that is esterified with benzyl groups at the glucoronic acid monomer, and one group has experimented with these sponges by seeding them with human preadipocytes and surgically implanting them into subcutaneous tissue of athymic nude mice. They demonstrated good cellular penetration as well as development of new vascular networks within the sponges. However, adipose-tissue development remained sparse (Hemmrich *et al.*, 2005). This same group also compared collagen scaffolds to HYAFF® 11 scaffolds *in vivo* and found better results in terms of implant weight, adipose-tissue formation, and distribution of cells using the HYAFF® 11 scaffolds (von Heimburg *et al.*, 2001b).

Collagen is the best-known tissue-derived natural polymer. It is the main component of all mammalian tissues, including skin, bone, cartilage, tendon, and ligament. Collagen has been used as a tissue culture scaffold or artificial skin due to the easy attachment of many different cell types. However, collagen offers a limited range of physical properties, and it can be expensive (Pulapura and Kohn, 1992). Chemical modification and incorporation of fibronectin, chondroitin sulfate, or low levels of hyaluronic acid into the collagen matrix can affect cell adhesion (Srivastava *et al.*, 1990). Freeze-dried collagen scaffolds were compared to sponges and nonwoven mesh composed of hyaluronic acid. *In vivo* experiments were carried out in a murine animal model for eight months. The hyaluronic-based products were found to be superior to the freeze-dried collagen scaffold. Specifically, hyaluronic sponges displayed a greater number of adipocytes than the nonwoven mesh. This was attributed mainly to the porous nature of the sponge, which allowed for greater surface area for adipocyte cell distribution and growth (von Heimburg *et al.*, 2001a, 2001b).

### Injectable Materials

A number of hydrogels, mainly naturally derived, can be utilized as injectable materials. Interest in hydrogels as biomaterials has increased. These polymers are elastic in nature and generally contain over 90% water per volumetric weight. Hydrogel biomaterials possess many suitable features for tissue-engineering applications. First, they can be delivered via injection. Second, if necessary, some hydrogels can easily be modified with the addition of bioactive groups to enhance cellular interactions. These modifications include the addition of proteolytically degradable peptides to enhance the biodegradability. In addition, the adhesion peptides tyrosine, isoleucine, glycine, serine, and arginine (YIGSR) have been coupled to PEG to enhance cellular attachment (Patel and Patrick, 2004).

Alginate is a natural hydrogel that has found its widest use to date as an injectable scaffold material for tissue engineering (Atala *et al.*, 1994) due to its simple gelation properties when ionically cross-linked with divalent cations such as $Ca^{2+}$, $Mg^{2+}$, $Ba^{2+}$, and $Sr^{2+}$. Because of its favorable properties, including biocompatibility, nonimmunogenicity, hydrophilicity, and relatively low cost, there have been many attempts to utilize this material as a wound dressing, dental impression, immobilization matrix, and scaffolds for cultured and transplanted cells (Gombotz and Wee, 1998; Klock *et al.*, 1997; Shapiro and Cohen, 1997). Alginate can be used in an injectable form either by being preformed into small beads or by injection before or during the solidifying process. Due to the ease of implantation in a minimally invasive manner, alginate beads have been prepared and used for the transplantation of chondrocytes (Gregory *et al.*, 1999; Lemare *et al.*, 1998) or hepatocytes (Joly *et al.*, 1997).

Typical commercially available alginates are not ideal materials for tissue engineering because they do not mediate cell adhesion, and they lose their divalent ions into the surrounding tissue and then dissolve in a poorly controlled manner. In addition, the molecular weight of many alginates is typically above the renal clearance threshold of the kidney (Al-Shamkhani and Duncan, 1995). Alginate is also known to discourage protein adsorption due to its hydrophilic character, and this may decrease the survival of many cell types in alginate hydrogels (Smetana, 1993). Bouhadir *et al.* (1999) reported the development of hydrolytically degradable, covalently cross-linked hydrogels derived from alginate. To avoid the degradation problem of alginate, polyguluronate blocks with molecular weight of 6000 Da were isolated from alginate, oxidized, and covalently cross-linked with adipic dihydrazide. The gelling of these polymers could be readily controlled, and their mechanical properties depended on the cross-linking density. It has also been demonstrated that alginate gel degradation can readily be regulated by controlling the molecular weight distribution of the polymer chains in the gels and their susceptibility to hydrolytic scission by partial oxidation (Boontheekul *et al.*, 2005).

Cellular attachment to unmodified alginate is very poor. This limitation has been overcome with molecular modification. An RGD- (arginine-glycine-aspartic acid) containing cell adhesion ligand was covalently coupled to enhance the cellular interaction with alginate hydrogels (Rowley *et al.*, 1999). These modified alginate hydrogels provided for adhesion, proliferation, and expression of differentiated phenotype of mouse skeletal myoblasts. Our group developed a method for processing the alginate-coupled RGD hydrogel into a porous bead structure that supported cell attachment and ingrowth (Fig. 36.5a) (Halberstadt *et al.*, 2002; Loebsack *et al.*, 2001). These beads were developed to be small enough to be injected, making this material a potential candidate for cellular delivery (Halberstadt *et al.*, 2002). In addition, because of the cell adhesion peptides, these beads supported preadipocyte adhesion for transplantation (Fig. 36.5b) (Halberstadt *et al.*, 2002).

Other naturally occurring polymers can also be utilized as injectable materials for breast-tissue reconstruction. These include chitosan, hyaluronic acid, and collagen. Chitosan is the second-most-plentiful biomass, and coupling a variety of molecules to the reactive amino groups in the backbone (Lee *et al.*, 1995; Muzzarelli, 1983) has formed various derivatives of chitosan. Chitosan has been known to be biocompatible and biodegradable and has a low toxicity (Chandy and Sharma, 1990; Tomihata and Ikada, 1997). This enables chitosan to be useful for versatile biomedical applications such as plastic surgery (Biagini *et al.*, 1991) and wound healing (Muzzarelli *et al.*, 1993). Chitosan matrices were fabricated and proposed as cell substrates for bovine chromaffin cells and porcine hepatocytes due to its structural similarity to glycosaminoglycans (Elcin *et al.*, 1999; Eser Elcin *et al.*, 1998). Hyaluronic acid and its derivatives have been used as a drug delivery system (Larsen and Balazs,

1991). Hydrogels of hyaluronic acid have also been prepared by covalent cross-linking with various kinds of hydrazides (Vercruysse *et al.*, 1997). Human fibroblasts and keratinocytes were cultured on hyaluronic acid–derived biomaterials (Zacchi *et al.*, 1998), and cross-linked non-animal hyaluronic acid gel was used as facial intradermal implants (Duranti *et al.*, 1998).

The longest successful adipocyte construct utilized fibrin glue as a carrier for preadipocytes. The construct was implanted into a rat animal model, and adipose tissue was observed for one year after implantation (Wechselberger *et al.*, 2002).

## IV. ANIMAL MODELS

A primary problem with the development of tissue-engineered human-breast reconstruction or augmentation is that no specific animal models mimic the native human breast tissue. Human beings are the only animals to have well-developed breasts. Other primates may have small breast mounds, particularly when lactating, but these are inadequate for developing large tissue constructs for partial or complete mastectomy equivalent to what occurs in patients. Cattle and goats have udders that have no structural or functional similarity to humans. Rats and mice have multiple teats along mammary lines extending from the neck to the inguinal region, with small mammary glands beneath each teat. Pigs, dogs, and sheep similarly have subcutaneous mammary glands that are inconspicuous unless the animal is lactating. Despite the lack of an ideal model, major development of concepts can be achieved and important research questions answered using several animal models.

There has been excellent success using inbred female Lewis rats as a small-animal model for the development of transplantable tissues on absorbable polymers such as PGA, PLA, and PGLA and hydrogels such as alginate (Holder *et al.*, 1998). This model allows for transplantation of cells between individuals without concern for immunologic rejection, which parallels the likely autologous nature of cell transplantation for breast engineering. In addition, the Lewis rat is larger than many other strains and allows for the testing of larger (1–2 cm) or multiple constructs in the same animal. In order to prevent migration of implanted materials and to control the local formation of the engineered tissue, a "purse-string" technique has been developed to secure implants at specific sites (Loebsack *et al.*, 2001; Roland *et al.*, 1998). Briefly, a single nonabsorbable suture is passed in and out of the skin to form a circle. The skin is lifted with forceps, forming a pocket, and the implant is placed into the subcutaneous tissue in the center of the circle through a small incision that is then sutured together. The purse-string suture is then tightened to hold the implant in place. The suture may be removed two weeks later and the implant remains in place.

Another animal model used for the transplantation of human-derived cells is either the Nude (nu/nu) or SCID

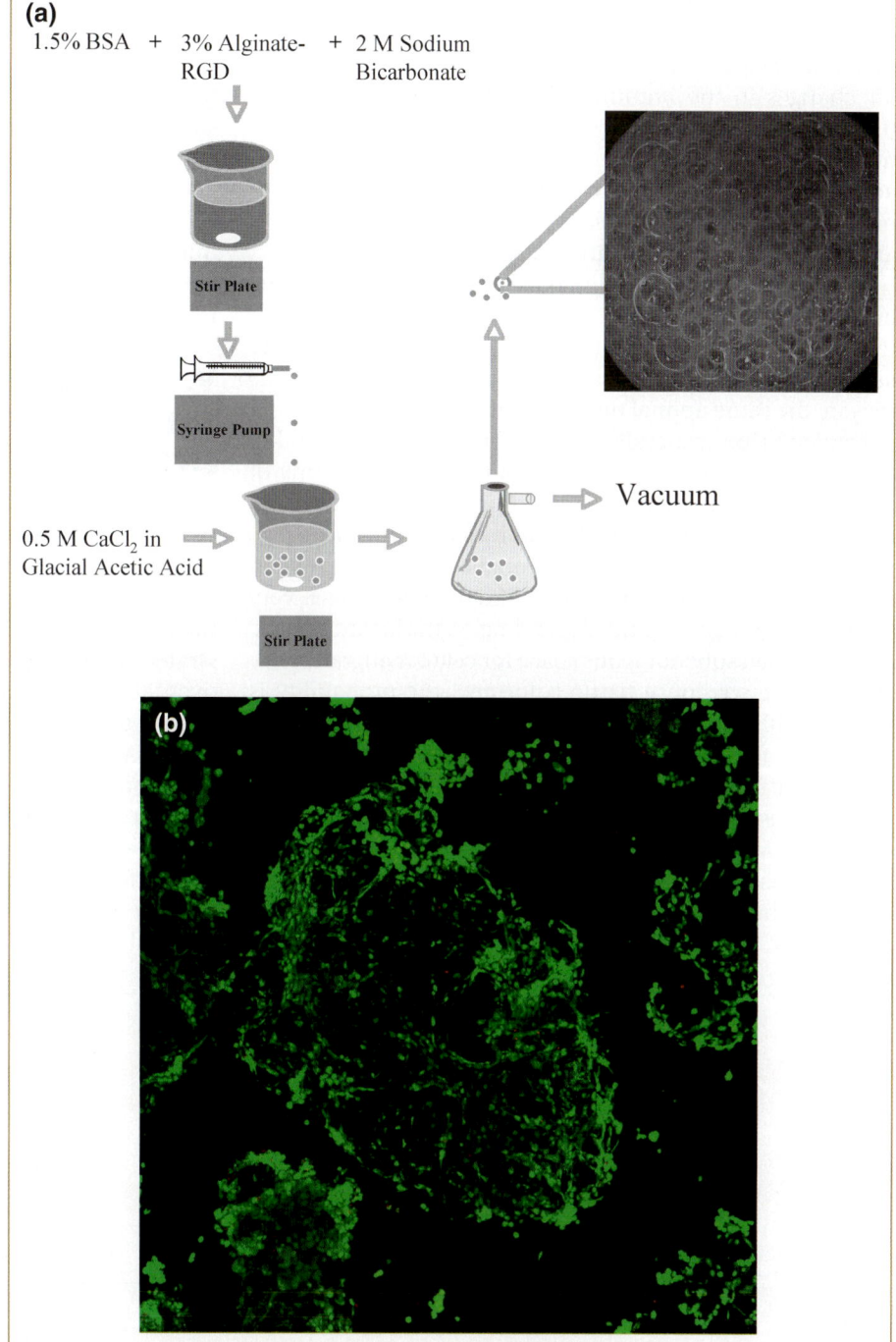

**FIG. 36.5.** **(a)** Cartoon of the generation of porous alginate beads. Briefly, a 3% alginate solution was mixed with 1.5% BSA and 2 M sodium bicarbonate. The solution was mixed rapidly and then drawn up into a syringe. The syringe was connected to a syringe pump, and individual drops were formed and added to a 0.5-M $CaCl_2$ solution made in glacial acetic acid. Upon formation of the beads, they were placed into a vacuum cylinder, and the vacuum was placed in order to remove the carbon dioxide gas and form the porous beads. The insert is a laser confocal microscopic image of the porous structure of a bead (taken with permission from Loebsack *et al.*, 2001). **(b)** Confocal image of sheep preadipocytes isolated from the omentum seeded onto porous alginate-RGD fragments. The cells were stained with a live/dead fluorescent probe (calcein AM/ethidium homodimer-1 — molecular probes) (5× magnification) (taken with permission from Halberstadt *et al.*, 2002).

mouse. These animals have compromised immune systems to the extent that they will often accept xenogeneic transplanted tissues, and these mice have been particularly useful for transplantation and immunologic studies of human tumors, bone marrow, skin, and other tissues (Bach-Mortensen *et al.*, 1976; Lopez Valle *et al.*, 1996; Ullmann *et al.*, 1998). They are potential models for evaluating various polymer constructs with and without human cells *in vivo* without the adverse effects of major immunologic reaction.

Further, basic questions about human cells and polymers *in vivo* may be answered without going to human trials prematurely. Development of these models may not be straightforward, since there are subtle differences between strains regarding the acceptance of various tissues and the growth of the implanted tissues in different sites (e.g., a tissue may grow in one mouse strain but not another or grow in a subcutaneous site but not in an internal location). Nude mice lack a T-lymphocyte response, while SCID mice lack both a

T- and a B-lymphocyte response. However, both types have natural killer cells that may interact with some transplanted materials. These mouse colonies must be monitored closely for changes in the immune status of the mice, which at times occur spontaneously. In addition, the transplanted human cells must be screened routinely for HIV, hepatitis, and mycoplasma before transplanting into immunosuppressed mice to ensure the safety of the animals and workers and the validity of the experiments.

In evaluating larger animals as models for breast-tissue engineering, animals with skin and subcutaneous tissues that are similar to humans are required in order to evaluate larger constructs in the subcutaneous position. For this reason, the same animal must be used as a tissue donor and a recipient. Dogs generally have little fat and very loose skin and present many of the problems of the rat in terms of migration of implanted materials. Porcine skin and subcutaneous tissues are very similar to humans, but most pigs continue to gain weight rapidly throughout their lives, which makes monitoring implants very difficult. Sheep have very little subcutaneous fat and, depending on the location, have a well-defined subcutaneous space for cellular engraftment. Hence, if new adipose tissue is formed, the probability is high that this would be developed from the implanted cells. One of the few locations in the sheep where there are significant fat deposits is the omentum. Halberstadt *et al.* (2002) demonstrated that preadipocytes isolated from the omentum could be expanded in culture and then seeded onto porous alginate-RGD fragments. The sheep preadipocytes attached, proliferated, and spread onto the biomaterial surface (Fig. 36.5b). This material was subsequently injected into the nape of the neck of some of the sheep to determine if new adipose tissue would form. Although the cells were autologous and not labeled with a tracking marker, there appeared to be *de novo* adipose tissue formation in the cell implant sites as compared to the biomaterial-only control sites (Fig. 36.6a and b).

Further research and development will be required to establish a technology for biopsies and cellular construct implantation that is efficient, technically straightforward, and minimally invasive in any of these large-animal models.

## V. STRATEGIES TO ENHANCE THE VASCULARIZATION OF ENGINEERED TISSUE

A critical challenge to engineer breast tissue, or any tissue of significant thickness, is to develop a vascular network that can support the metabolic needs of the engineered tissue and integrate it with the rest of the body. The need for a vascular network has been strikingly demonstrated in studies of hepatocyte transplantation, in which over 90% of the cells transplanted, even on relatively thin (3-mm-thick) scaffolds, died within days following implantation (Mooney *et al.*, 1997). This finding of significant cell death prior to tissue vascularization is not unique to hepatocytes; it has also been noted in smooth muscle cell transplantation (Cohn *et al.*, 1997) and is likely a general finding in all efforts to engineer vascularized tissues (Colton, 1995). Adipose tissue is highly vascularized, possessing a resting blood flow two to three times higher than that of skeletal

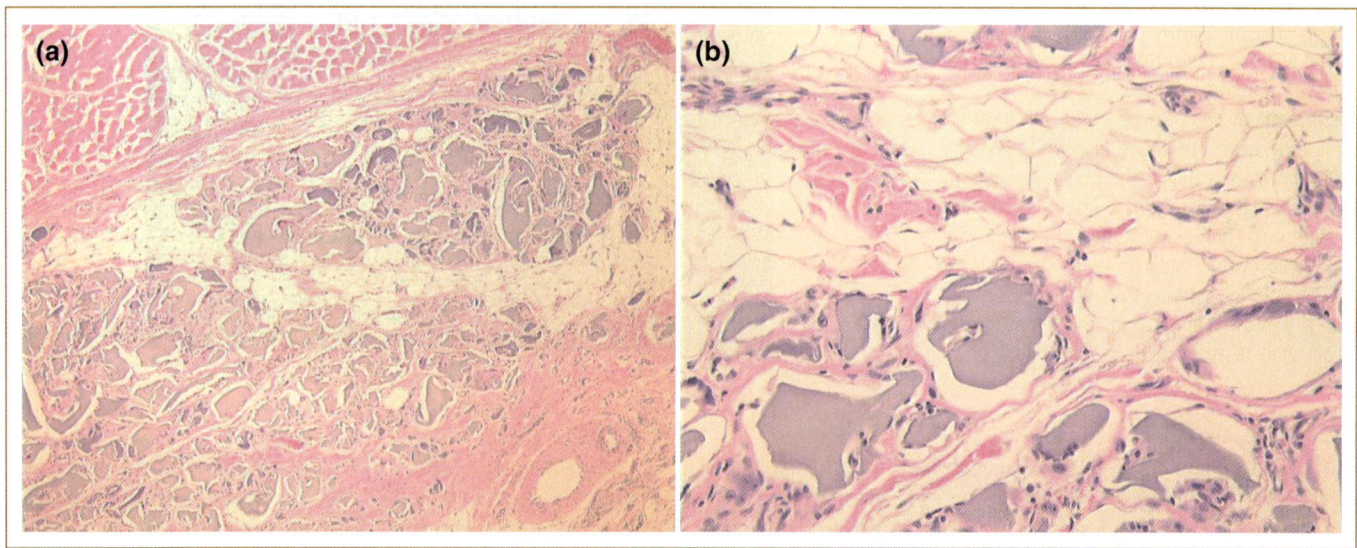

**FIG. 36.6.** Autologous sheep preadipocytes harvested from the omentum were seeded onto alginate-RGD fragments, grown for three days in culture, and 2-mL fractions were injected using an 18-gauge needle into the nape of the neck. **(a)** Histological section (40× magnification) of a three-month implant. Note the pocket of adipose tissue closely associated with the alginate matrix. **(b)** Higher magnification (200×) of Fig. 36.6a demonstrating a minimal inflammatory response to the alginate material and the development of adipose tissue. No adipose-tissue formation was observed in the acellular biomaterial-only controls (taken with permission from Halberstadt *et al.*, 2002).

muscle. The presence of vascularized networks in natural metabolic organs results in a short diffusion distance between the nutrient source and the cells (R. C. Thomson *et al.*, 1995), and these vascular networks must be created in engineered tissues as well. Nutrient diffusion *in vivo* is constrained to a distance of approximately 150 µm. Most metabolically active cells that are located farther than this distance from a nearby capillary are subject to hypoxia. It can be seen that the success of any large engineered tissue hinges on its blood supply. The important interrelationship between preadipocytes and endothelial cells was demonstrated in hypoxic culture experiments. Frye *et al.* (2005) exposed cell cultures of pure preadipocytes and mixed cultures of preadipocytes along with microvascular endothelial cells to hypoxic conditions (5–2% $O_2$) in specially designed chambers. Preadipocytes co-cultured with microvascular endothelial cells showed better viability than cultures consisting of preadipocytes alone. These results suggest the presence of an underlying humoral interaction between the two cell types that is stimulated by hypoxia (Frye *et al.*, 2005).

There are three general approaches taken to date to promote angiogenesis in engineered tissues. The first relies on the vascularization that accompanies the inflammatory response to any implanted foreign material. Optimization of the porosity and pore size of tissue-engineering scaffolds can increase the rate of granulation-tissue formation in engineered tissues (Mikos *et al.*, 1993; Mooney *et al.*, 1994). However, the blood vessel ingrowth occurs slowly with this approach and will likely not be sufficient to engineer large tissue volumes. The second two approaches attempt to actively modulate the vascularization process by either delivering angiogenic molecules or blood-vessel-forming cells (e.g., endothelial cells) to the site at which the tissue is being engineered.

One attractive approach to engineering tissues in these models is to deliver cellular constructs that are approximately 500 µm or less in thickness. This may optimize diffusional transport of nutrients to the cells while each small cell–polymer unit becomes vascularized. We have taken this approach by designing porous alginate-RGD beads (500 µm to 3 mm in diameter) for cell transplantation (Burg *et al.*, 2000; Eiselt *et al.*, 2000; Halberstadt *et al.*, 2000; Loebsack *et al.*, 2001). Because the beads are macroporous, the vascular endothelial cells can migrate into the beads and establish a three-dimensional vascular bed in a short time (Fig. 36.7a and b). This approach potentially can be combined with angiogenic factors and cells to enhance the timing of endothelialization to promote the survival of the transplanted cells.

Site-specific delivery of angiogenic molecules may provide an efficient means of stimulating localized vessel

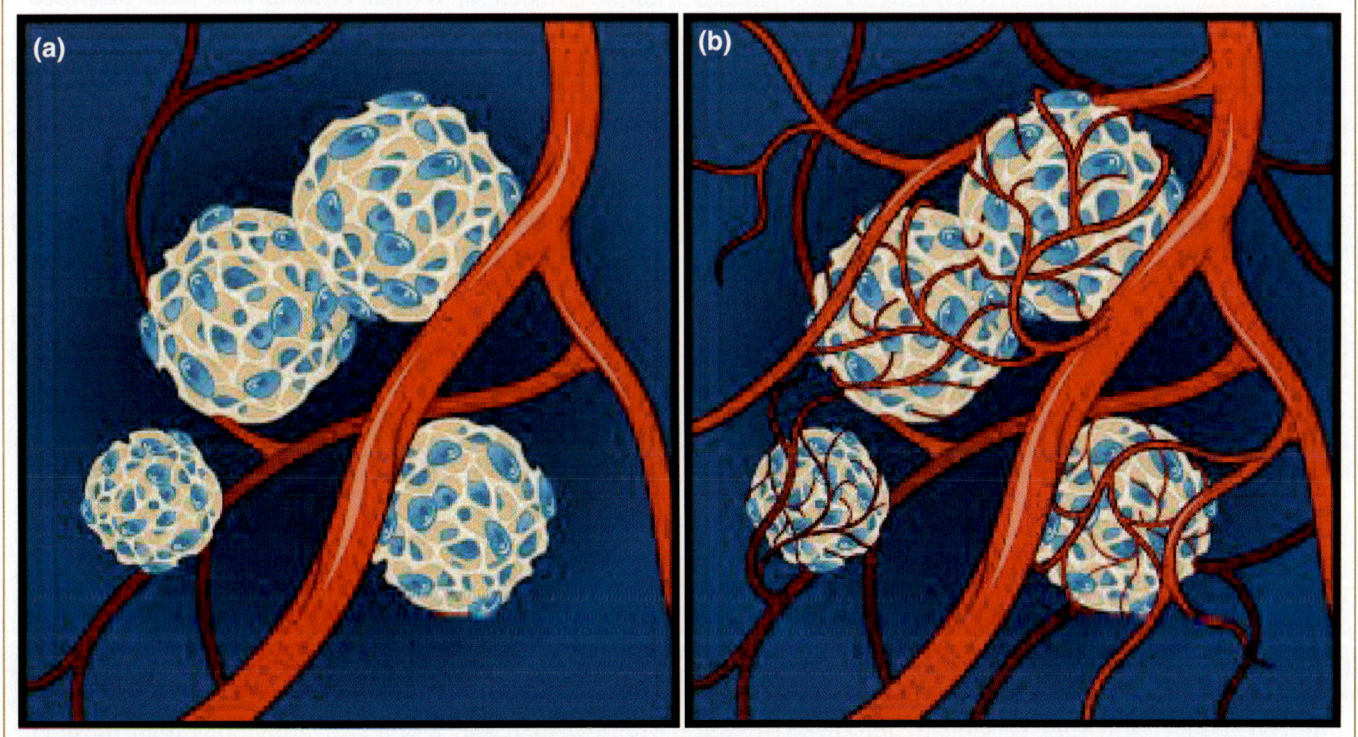

**FIG. 36.7.** Cartoon drawing of an approach to create three-dimensional tissue structures. **(a)** Small macroporous beads seeded with cells are injected into a tissue. **(b)** The foreign material, coupled with an inflammatory response, initiates a vascular ingrowth into the porous beads. The use of angiogenic factors and/or endothelial cells may further enhance the complete revascularization of this injectable construct over the course of a short period of time.

formation. Controlled and sustained release of angiogenic molecules from the tissue engineering scaffolds may allow optimization of this process (Eiselt *et al.*, 1998; Lee *et al.*, 2000; Peters *et al.*, 1999a; Sheridan *et al.*, 2000) and can significantly enhance perfusion of ischemic tissues (Sun *et al.*, 2005). Various growth factors, including vascular endothelial growth factor (VEGF), basic fibroblast growth factor (FGF-2), and epidermal growth factor (EGF), could be incorporated into a polymeric matrix together and released at a controlled and sustained rate in combinations or sequences (Richardson *et al.*, 2001) for an extended period of time. Alternatively, delivery of plasmid DNA encoding the angiogenic proteins may be another approach to generate the vascular networks in engineered tissues (Shea *et al.*, 1999). PLGA scaffolds containing a plasmid encoding for PDGF, a potent angiogenesis promoter, greatly increased the number of blood vessels and granulation tissue that formed as compared to scaffolds that did not deliver the plasmid DNA (Shea *et al.*, 1999).

Another potential approach to enhance angiogenesis in engineered tissues is to cotransplant endothelial cells or progenitors along with the primary cell type of interest. This approach is suggested by the observation that endothelial cells will spontaneously form capillarylike structures *in vitro* if cultured in an appropriate environment (Ingber and Folkman, 1989; Pepper *et al.*, 1992). Endothelial cells seeded into a tissue-engineering scaffold may be able to form capillaries or capillarylike structures, which then can merge with capillaries growing into the scaffold from the host tissue. This possibility is supported by our previous finding that transplantation of syngeneic endothelial cells on polymer scaffolds led to a statistically significant enhancement in the density of blood vessels in the scaffolds following two weeks implantation into rats (Holder *et al.*, 1997). In addition, the new capillaries that form in these scaffolds comprise both the transplanted endothelial cells and ingrowing cells from the host (Nor *et al.*, 1999), and these transplanted endothelial cells enhance the survival (Smith *et al.*, 2004) and tissue formation by transplanted cells (Kaigler *et al.*, 2005). The utility of embryonic stem cell–derived endothelial cells in enhancing skeletal muscle engineering has also been demonstrated (Levenberg *et al.*, 2005).

Recently, Khademhosseini *et al.* (2006) proposed a method of using microstamping and nanofluid chambers to create a three-dimensional vascular bed for liver-tissue engineering. The long-term goal for this group is to develop a three-dimensional vascular bed *ex vivo* that could be anastamosed to the host vasculature to support cellular engraftment. Depending on the materials used, this approach could also be developed for breast-tissue engineering.

Direct application of a vascular pedicle to the construct is another promising approach. The idea of using vascular pedicles is not a novel concept. Pre-fabricated flaps have been created with vascular pedicles since the 1960s. It stands to reason that preformed vascular pedicles will enhance the growth of tissue-engineered constructs. Vascular pedicles can be supplied in many different configurations. In general, conduits composed of an intact artery and vein together fare better than those composed of a single vein alone (Erol and Spira, 1990; Walton and Brown, 1993).

## VI. SPECIAL CONSIDERATIONS

Nipple reconstruction is an additional problem that is created after mastectomy. Traditional approaches have consisted of free composite grafts, local-tissue transfer, and prosthetic devices. Free composite grafts were used early on and were created from autologous tissues, such as the labia, inner thigh, cartilage (auricular or costal), the contralateral nipple, and the toe. Complications at the graft site have since made this technique less desirable. Local-tissue flaps are the most popular routes for nipple reconstruction. Commonly used techniques include the bell flap, the modified star flap, and the skate flap (Bernard and Beran, 2003; Eng, 1996; Guerra *et al.*, 2003; Little, 1984). Unfortunately these techniques can be hindered by flap necrosis and poor aesthetic results, including loss of nipple projection. Holton *et al.* (2005) have proposed a solution to this problem using tissue engineering. Their experiments utilized human acellular matrix implants in athymic mice. Three groups were created, consisting of traditional bell flap alone, bell flap plus a cylinder composed of an acellular dermal matrix, and bell flap plus intraflap injections of micro-sized acellular dermal matrix. Bell flaps alone maintained 44% of initial projection, while flaps with acellular dermal matrix cylinders maintained 70%. However, the acellular dermal matrix cylinders were extruded in 30% of the samples. The flaps that were microinjected with acellular dermal matrix retained 49% of their original projection.

## VII. CONCLUDING REMARKS

Tissue engineering may provide a new therapy to reconstruct breast tissue for women undergoing lumpectomy or mastectomy. A number of cell types and polymeric materials may be useful to reconstruct breast tissue. In addition, it is clear that optimization of vascularization is critical to engineering large tissue constructs. This approach may ultimately allow one to engineer and reconstruct breast tissue with less pain and less invasive methods than those currently utilized clinically.

# VIII. REFERENCES

Al-Shamkhani, A., and Duncan, R. (1995). Radioiodination of alginate via covalently bound tyrosinamide allows for monitoring of its fate *in vivo. J. Bioact. Compat. Polym.* **10**, 4–13.

American Cancer Society. (2006). Detailed Guide: Breast Cancer What Are the Key Statistics for Breast Cancer? The Society, www.cancer.org.

Anderson, J. M., Spilizewski, K. L., and Hiltner, A. (1985). Poly-α-amino acids as biomedical polymers. *In* "Biocompatibility of Tissue Analogues" (D. F. Williams, ed.), pp. 67–88. CRC Press, Boca Raton, FL.

Anseth, K. S., Shastri, V. R., and Langer, R. (1999). Photopolymerizable degradable polyanhydrides with osteocompatibility. *Nat. Biotechnol.* **17**, 156–159.

Atala, A., Kim, W., Paige, K. T., Vacanti, C. A., and Retik, A. B. (1994). Endoscopic treatment of vesicoureteral reflux with a chondrocyte–alginate suspension. *J. Urol.* **152**, 641–643; discussion 644.

Bach-Mortensen, N., Romert, P., and Ballegaard, S. (1976). Transplantation of human adipose tissue to nude mice. *Acta Pathol. Microbiol. Scand. [C]* **84**, 283–289.

Baier Leach, J., Bivens, K. A., Patrick, C. W., Jr., and Schmidt, C. E. (2003). Photo-cross-linked hyaluronic acid hydrogels: natural, biodegradable tissue-engineering scaffolds. *Biotechnol. Bioeng.* **82**, 578–589.

Bassenge, E. (1996). Endothelial function in different organs. *Prog. Cardiovasc. Dis.* **39**, 209–228.

Beahm, E. K., Walton, R. L., and Patrick, C. W., Jr. (2003). Progress in adipose tissue construct development. *Clin. Plast. Surg.* **30**, 547–558, viii.

Bernard, R. W., and Beran, S. J. (2003). Autologous fat graft in nipple reconstruction. *Plast. Reconstr. Surg.* **112**, 964–968.

Biagini, G., Pugnaloni, A., Damadei, A., Bertani, A., Belligolli, A., Bicchiega, V., and Muzzarelli, R. (1991). Morphological study of the capsular organization around tissue expanders coated with *N*-carboxybutyl chitosan. *Biomaterials* **12**, 287–291.

Boontheekul, T., Kong, H. J., and Mooney, D. J. (2005). Controlling alginate gel degradation utilizing partial oxidation and bimodal molecular weight distribution. *Biomaterials* **26**, 2455–2465.

Bouhadir, K. H., Hausman, D. S., and Mooney, D. J. (1999). Synthesis of cross-linked poly(aldehyde guluronate) hydrogels. *Polymer* **40**, 3575–3584.

Breinan, H. A., Minas, T., Barone, L., Tubo, R., Hsu, H. P., Shortkroff, S., Nehrer, S., Sledge, C. B., and Spector, M. (1998). Histological evaluation of the course of healing of canine articular cartilage defects treated with cultured autologous chondrocytes. *Tissue Eng.* **4**, 101–114.

Burg, K. J. L., Austin, C. E., Mooney, D. J., Eiselt, P., Yeh, J., Rowley, J. A., Culberson, C. R., Greene, K. G., Holder, W. D., Loebsack, A. B., Wyatt, S., and Halberstadt, C. R. (2000). Optimizing microstructure of porous alginate-RGD beads for tissue-engineering applications. *In* "Sixth World Biomaterials Congress", Hawaii, Hawaii.

Busch, O., Solheim, E., Bang, G., and Tornes, K. (1996). Guided tissue regeneration and local delivery of insulinlike growth factor I by bioerodible polyorthoester membranes in rat calvarial defects. *Int. J. Oral. Maxillofac. Implants.* **11**, 498–505.

Chandy, T., and Sharma, C. P. (1990). Chitosan — as a biomaterial. *Biomater. Artif. Cells Artif. Organs* **18**, 1–24.

Chen, W. Y., and Abatangelo, G. (1999). Functions of hyaluronan in wound repair. *Wound Repair Regen.* **7**, 79–89.

Cohn, N. A., Kim, B. S., Mooney, D. J., Emelianov, S. Y., and O'Donnell, M. (1997). Layer imaging in tissue engineering using an elasticity microscope. *In* "*Proc. IEEE Ultrasonics Symp.*, IEEE 97CH36118," pp. 1431–1434.

Coleman, S., Austin, C., Culberson, C., Loebsack, A., Morton, D., Holder, W., and Halberstadt, C. (2000). Isolation and characterization of pre-adipocytes obtained from liposuction for use in soft tissue engineering. *In* "Tissue Engineering Society Meeting," Orlando, FL.

Colton, C. K. (1995). Implantable biohybrid artificial organs. *Cell Transplant.* **4**, 415–436.

Craig, L. E., Spelman, J. P., Strandberg, J. D., and Zink, M. C. (1998). Endothelial cells from diverse tissues exhibit differences in growth and morphology. *Microvasc. Res.* **55**, 65–76.

Dang, W., Colvin, O. M., Brem, H., and Saltzman, W. M. (1994). Covalent coupling of methotrexate to dextran enhances the penetration of cytotoxicity into a tissue-like matrix. *Cancer Res.* **54**, 1729–1735.

Doillon, C. J., Deblois, C., Cote, M. F., and Fournier, N. (1994). Bioactive collagen sponges as connective-tissue substitute. *Mat. Sci. Eng. C Biomim.* **2**, 43–49.

Domb, A. J., Gallardo, C. F., and Langer, R. (1989). Polyanhydrides. 3. Polyanhydrides based on aliphatic-aromatic diacids. *Macromolecules* **22**, 3200–3204.

Duranti, F., Salti, G., Bovani, B., Calandra, M., and Rosati, M. L. (1998). Injectable hyaluronic acid gel for soft-tissue augmentation. A clinical and histological study. *Dermatol. Surg.* **24**, 1317–1325.

Eaglstein, W. H., and Falanga, V. (1998). Tissue engineering for skin: an update. *J. Am. Acad. Dermatol.* **39**, 1007–1010.

Eiselt, P., Kim, B. S., Chacko, B., Isenberg, B., Peters, M. C., Greene, K. G., Roland, W. D., Loebsack, A. B., Burg, K. J., Culberson, C., Halberstadt, C. R., Holder, W. D., and Mooney, D. J. (1998). Development of technologies aiding large-tissue engineering. *Biotechnol. Prog.* **14**, 134–140.

Eiselt, P., Yeh, J., Latvala, R. K., Shea, L. D., and Mooney, D. J. (2000). Porous carriers for biomedical applications based on alginate hydrogels. *Biomaterials* **21**, 1921–1927.

Elcin, Y. M., Dixit, V., Lewin, K., and Gitnick, G. (1999). Xenotransplantation of fetal porcine hepatocytes in rats using a tissue engineering approach. *Artif. Organs* **23**, 146–152.

Eng, J. S. (1996). Bell flap nipple reconstruction — a new wrinkle. *Ann. Plast. Surg.* **36**, 485–488.

Engelberg, I., and Kohn, J. (1991). Physico-mechanical properties of degradable polymers used in medical applications: a comparative study. *Biomaterials* **12**, 292–304.

Erol, O. O., and Spira, M. (1990). Reconstructing the breast mound employing a secondary island omental skin flap. *Plast. Reconstr. Surg.* **86**, 510–518.

Eser Elcin, A., Elcin, Y. M., and Pappas, G. D. (1998). Neural tissue engineering: adrenal chromaffin cell attachment and viability on chitosan scaffolds. *Neurol. Res.* **20**, 648–654.

Fraser, J. K., Wulur, I., Alfonso, Z., and Hedrick, M. H. (2006). Fat tissue: an underappreciated source of stem cells for biotechnology. *Trends Biotechnol.* **24**, 150–154.

Freed, L. E., Vunjak-Novakovic, G., Biron, R. J., Eagles, D. B., Lesnoy, D. C., Barlow, S. K., and Langer, R. (1994). Biodegradable polymer scaffolds for tissue engineering. *Biotechnology (NY)* **12**, 689–693.

Frye, C. A., Wu, X., and Patrick, C. W. (2005). Microvascular endothelial cells sustain preadipocyte viability under hypoxic conditions. *In Vitro Cell Dev. Biol. Anim.* **41**, 160–164.

Fung, L. K., Ewend, M. G., Sills, A., Sipos, E. P., Thompson, R., Watts, M., Colvin, O. M., Brem, H., and Saltzman, W. M. (1998). Pharmacokinetics of interstitial delivery of carmustine, 4-hydroperoxycyclophosphamide, and paclitaxel from a biodegradable polymer implant in the monkey brain. *Cancer Res.* **58**, 672–684.

Gombotz, W. R., and Wee, S. F. (1998). Protein release from alginate matrices. *Adv. Drug Deliv. Rev.* **31**, 267–285.

Gregory, K. E., Marsden, M. E., Anderson-MacKenzie, J., Bard, J. B., Bruckner, P., Farjanel, J., Robins, S. P., and Hulmes, D. J. (1999). Abnormal collagen assembly, though normal phenotype, in alginate bead cultures of chick embryo chondrocytes. *Exp. Cell Res.* **246**, 98–107.

Guerra, A. B., Khoobehi, K., Metzinger, S. E., and Allen, R. J. (2003). New technique for nipple areola reconstruction: arrow flap and rib cartilage graft for long-lasting nipple projection. *Ann. Plast. Surg.* **50**, 31–37.

Halberstadt, C. R., Mooney, D. J., Burg, K. J. L., Eiselt, P., Rowley, J., Beiler, R. J., Roland, W. D., Austin, C. E., Culberson, C. R., Greene, K. G., Wyatt, S., Loebsack, A. B., and Holder, W. D. (2000). The design and implementation of an alginate material for soft-tissue engineering. *In* "Sixth World Biomaterial Congress", Hawaii, Hawaii.

Halberstadt, C., Austin, C., Rowley, J., Culberson, C., Loebsack, A., Wyatt, S., Coleman, S., Blacksten, L., Burg, K., Mooney, D., and Holder, W. J. (2002). A hydrogel material for plastic and reconstructive applications injected into the subcutaneous space of a sheep. *Tissue Eng.* **8**, 309–319.

Hang-Fu, L., Marmolya, G., and Feiglin, D. H. (1995). Liposuction fat-fillant implant for breast augmentation and reconstruction. *Aesthetic Plast. Surg.* **19**, 427–437.

Harris, L. D., Kim, B. S., and Mooney, D. J. (1998). Open-pore biodegradable matrices formed with gas foaming. *J. Biomed. Mater. Res.* **42**, 396–402.

Hemmrich, K., von Heimburg, D., Rendchen, R., Di Bartolo, C., Milella, E., and Pallua, N. (2005). Implantation of preadipocyte-loaded hyaluronic acid–based scaffolds into nude mice to evaluate potential for soft-tissue engineering. *Biomaterials* **26**, 7025–7037.

Holder, W. D., Jr., Gruber, H. E., Roland, W. D., Moore, A. L., Culberson, C. R., Loebsack, A. B., Burg, K. J. L., and Mooney, D. J. (1997). Increased vascularization and heterogeneity of vascular structures occurring in polyglycolide matrices containing aortic endothelial cells implanted in the rat. *Tissue Eng.* **3**, 149.

Holder, W. D., Jr., Gruber, H. E., Moore, A. L., Culberson, C. R., Anderson, W., Burg, K. J., and Mooney, D. J. (1998). Cellular ingrowth and thickness changes in poly-L-lactide and polyglycolide matrices implanted subcutaneously in the rat. *J. Biomed. Mater. Res.* **41**, 412–421.

Holton, L. H., Haerian, H., Silverman, R. P., Chung, T., Elisseeff, J. H., Goldberg, N. H., and Slezak, S. (2005). Improving long-term projection in nipple reconstruction using human acellular dermal matrix: an animal model. *Ann. Plast. Surg.* **55**, 304–309.

Ignarro, L. J., Gold, M. E., Buga, G. M., Byrns, R. E., Wood, K. S., Chaudhuri, G., and Frank, G. (1989). Basic polyamino acids rich in arginine, lysine, or ornithine cause both enhancement of and refractoriness to formation of endothelium-derived nitric oxide in pulmonary artery and vein. *Circ. Res.* **64**, 315–329.

Ingber, D. E., and Folkman, J. (1989). Mechanochemical switching between growth and differentiation during fibroblast growth factor-stimulated angiogenesis *in vitro*: role of extracellular matrix. *J. Cell Biol.* **109**, 317–330.

Jen, A. C., Wake, M. C., and Mikos, A. G. (1996). Hydrogels for cell immobilization. *Biotech. Bioengin.* **50**, 357–364.

Joly, A., Desjardins, J. F., Fremond, B., Desille, M., Campion, J. P., Malledant, Y., Lebreton, Y., Semana, G., Edwards-Levy, F., Levy, M. C., and Clement, B. (1997). Survival, proliferation, and functions of porcine hepatocytes encapsulated in coated alginate beads: a step toward a reliable bioartificial liver. *Transplantation* **63**, 795–803.

Kaigler, D., Krebsbach, P. H., West, E. R., Horger, K., Huang, Y. C., and Mooney, D. J. (2005). Endothelial cell modulation of bone marrow stromal cell osteogenic potential. *FASEB J.* **19**, 665–667.

Khademhosseini, A., Langer, R., Borenstein, J., and Vacanti, J. P. (2006). Tissue engineering special feature: microscale technologies for tissue engineering and biology. *Proc. Natl. Acad. Sci. U.S.A.* **103**, 2480–2487.

Kim, B. S., and Mooney, D. J. (1998a). Development of biocompatible synthetic extracellular matrices for tissue engineering. *Trends Biotechnol.* **16**, 224–230.

Kim, B. S., and Mooney, D. J. (1998b). Engineering smooth muscle tissue with a predefined structure. *J. Biomed. Mater. Res.* **41**, 322–332.

Kim, B. S., Putnam, A. J., Kulik, T. J., and Mooney, D. J. (1998). Optimizing seeding and culture methods to engineer smooth muscle tissue on biodegradable polymer matrices. *Biotechnol. Bioeng.* **57**, 46–54.

Klock, G., Pfeffermann, A., Ryser, C., Grohn, P., Kuttler, B., Hahn, H. J., and Zimmermann, U. (1997). Biocompatibility of mannuronic acid–rich alginates. *Biomaterials* **18**, 707–713.

Kohn, J., and Langer, R. (1987). Polymerization reactions involving the side chains of α-L-amino acids. *J. Am. Chem. Soc.* **109**, 817–820.

Kononas, T. C., Bucky, L. P., Hurley, C., and May, J. W., Jr. (1993). The fate of suctioned and surgically removed fat after reimplantation for soft-tissue augmentation: a volumetric and histologic study in the rabbit. *Plast. Reconstr. Surg.* **91**, 763–768.

Kral, J. G., and Crandall, D. L. (1999). Development of a human adipocyte synthetic polymer scaffold. *Plast. Reconstr. Surg.* **104**, 1732–1738.

Landeen, L. K., Ziegler, F. C., Halberstadt, C. R., Cohen, C., and Slivka, S. R. (1992). Characterization of a human dermal replacement. *Wounds* **4**, 167.

Landeen, L. K., Halberstadt, C. R., King, B. D., and Naughton, G. K. (1993). A novel human dermal replacement: characterization and early clinical trials. Society for Investigative Dermatology, Washington, DC.

Langer, R., and Vacanti, J. P. (1993). Tissue engineering. *Science* **260**, 920–926.

Larsen, N. E., and Balazs, E. A. (1991). Drug delivery systems using hyaluronan and its derivatives. *Adv. Drug Deliv. Rev.* **7**, 279–308.

Lee, K. Y., and Mooney, D. J. (2001). Hydrogels for tissue engineering. *Chem. Rev.* **101**, 1869–1879.

Lee, K. Y., Ha, W. S., and Park, W. H. (1995). Blood compatibility and biodegradability of partially N-acylated chitosan derivatives. *Biomaterials* **16**, 1211–1216.

Lee, K. Y., Peters, M. C., Anderson, K. W., and Mooney, D. J. (2000). Controlled growth factor release from synthetic extracellular matrices. *Nature* **408**, 998–1000.

Lemare, F., Steimberg, N., Le Griel, C., Demignot, S., and Adolphe, M. (1998). Dedifferentiated chondrocytes cultured in alginate beads: restoration of the differentiated phenotype and of the metabolic responses to interleukin-1beta. *J. Cell Physiol.* **176**, 303–313.

Levenberg, S., Rouwkema, J., Macdonald, M., Garfein, E. S., Kohane, D. S., Darland, D. C., Marini, R., van Blitterswijk, C. A., Mulligan, R. C., D'Amore, P. A., and Langer, R. (2005). Engineering vascularized skeletal muscle tissue. *Nat. Biotechnol.* **23**, 879–884.

Lichter, A. S. (1997). The treatment of breast cancer without mastectomy. *Adv. Oncol.* **13**, 17–24.

Little, J. W., 3rd. (1984). Nipple-areola reconstruction. *Clin. Plast. Surg.* **11**, 351–364.

Lo, H., Ponticiello, M. S., and Leong, K. W. (1995). Fabrication of controlled-release biodegradable foams by phase separation. *Tissue Eng.* **1**, 15–28.

Loebsack, A., Greene, K., Wyatt, S., Culberson, C., Austin, C., Beiler, R., Roland, W., Eiselt, P., Rowley, J., Burg, K., Mooney, D., Holder, W., and Halberstadt, C. (2001). *In vivo* characterization of a porous hydrogel material for use as a tissue bulking agent. *J. Biomed. Mater. Res.* **57**, 575–581.

Lopez Valle, C. A., Germain, L., Rouabhia, M., Xu, W., Guignard, R., Goulet, F., and Auger, F. A. (1996). Grafting on nude mice of living skin equivalents produced using human collagens. *Transplantation* **62**, 317–323.

Lowry, K. J., Hamson, K. R., Bear, L., Peng, Y. B., Calaluce, R., Evans, M. L., Anglen, J. O., and Allen, W. C. (1997). Polycaprolactone/glass bioabsorbable implant in a rabbit humerus fracture model. *J. Biomed. Mater. Res.* **36**, 536–541.

Mikos, A. G., Sarakinos, G., Lyman, M. D., Ingber, D. E., Vacanti, J. P., and Langer, R. (1993). Prevascularization of porous biodegradable polymers. *Biotechnol. Bioeng.* **42**, 716.

Mikos, A. G., Thorsen, A. J., Czerwonka, L. A., Bao, Y., and Langer, R. (1994). Preparation and characterization of poly(L-lactic acid) foams. *Polymer* **35**, 1068–1077.

Mooney, D. J., and Mikos, A. G. (1999). Growing new organs. *Sci. Am.* **280**, 60–65.

Mooney, D. J., Kaufmann, P. M., Sano, K., McNamara, K. M., Vacanti, J. P., and Langer, R. (1994). Transplantation of hepatocytes using porous, biodegradable sponges. *Transplant. Proc.* **26**, 3425–3426.

Mooney, D. J., Breuer, M. D., McNamara, K., Vacanti, J. P., and Langer, R. (1995). Fabricating tubular devices from polymers of lactic and glycolic acid for tissue engineering. *Tissue Eng.* **1**, 107–118.

Mooney, D. J., Baldwin, D. F., Suh, N. P., Vacanti, J. P., and Langer, R. (1996a). Novel approach to fabricate porous sponges of poly(D,L-lactic-co-glycolic acid) without the use of organic solvents. *Biomaterials* **17**, 1417–1422.

Mooney, D. J., Kaufmann, P. M., Sano, K., Schwendeman, S. P., Majahod, K., Schloo, B., Vacanti, J. P., and Langer, R. (1996b). Localized delivery of epidermal growth factor improves the survival of transplanted hepatocytes. *Biotech. Bioengin.* **50**, 422–429.

Mooney, D. J., Mazzoni, C. L., Breuci, C., McNamara, K., Hern, D., Vacanti, J. P., and Langer, R. (1996c). Stabilized polyglycolic acid fiber-based tubes for tissue engineering. *Biomaterials* **17**, 115–124.

Mooney, D. J., Sano, K., Kaufmann, P. M., Majahod, K., Schloo, B., Vacanti, J. P., and Langer, R. (1997). Long-term engraftment of hepatocytes transplanted on biodegradable polymer sponges. *J. Biomed. Mater. Res.* **37**, 413–420.

Murphy, H. S., Bakopoulos, N., Dame, M. K., Varani, J., and Ward, P. A. (1998). Heterogeneity of vascular endothelial cells: differences in susceptibility to neutrophil-mediated injury. *Microvasc. Res.* **56**, 203–211.

Muzzarelli, R. A. A. (1983). Chitin and its derivatives: new trends of applied research. *Carbohydr. Polym.* **3**, 53–75.

Muzzarelli, R. A., Zucchini, C., Ilari, P., Pugnaloni, A., Mattioli Belmonte, M., Biagini, G., and Castaldini, C. (1993). Osteoconductive properties of methylpyrrolidinone chitosan in an animal model. *Biomaterials* **14**, 925–929.

Naughton, G. K. (2002). From lab bench to market: critical issues in tissue engineering. *Ann. N.Y. Acad. Sci.* **961**, 372–385.

Naughton, G., Mansbridge, J., and Gentzkow, G. (1997). A metabolically active human dermal replacement for the treatment of diabetic foot ulcers. *Artif. Organs* **21**, 1203–1210.

Nguyen, A., Pasyk, K. A., Bouvier, T. N., Hassett, C. A., and Argenta, L. C. (1990). Comparative study of survival of autologous adipose tissue taken and transplanted by different techniques. *Plast. Reconstr. Surg.* **85**, 378–386; discussion 387–389.

Nor, J. E., Christensen, J., Mooney, D. J., and Polverini, P. J. (1999). Vascular endothelial growth factor (VEGF)–mediated angiogenesis is associated with enhanced endothelial cell survival and induction of Bcl-2 expression. *Am. J. Pathol.* **154**, 375–384.

Oberpenning, F., Meng, J., Yoo, J. J., and Atala, A. (1999). De novo reconstitution of a functional mammalian urinary bladder by tissue engineering. *Nat. Biotechnol.* **17**, 149–155.

Park, E. S., Maniar, M., and Shah, J. C. (1998). Biodegradable polyanhydride devices of cefazolin sodium, bupivacaine, and taxol for local drug delivery: preparation, and kinetics and mechanism of *in vitro* release. *J. Controlled Release* **52**, 179–189.

Patel, P. N., and Patrick, C. W. (2004). Materials employed for breast augmentation and reconstruction. *In* "Scaffolding in Tissue Engineering" (P. X. Ma and J. Elisseeff, eds.), pp. 425–436. Marcel Dekker, New York.

Patel, P. N., Gobin, A. S., West, J. L., and Patrick, C. W., Jr. (2005). Poly(ethylene glycol) hydrogel system supports preadipocyte viability, adhesion, and proliferation. *Tissue Eng.* **11**, 1498–1505.

Patrick, C. W. (2004). Breast tissue engineering. *Annu. Rev. Biomed. Eng.* **6**, 109–130.

Patrick, C. W., Jr., Chauvin, P. B., Hobley, J., and Reece, G. P. (1999). Preadipocyte-seeded PLGA scaffolds for adipose tissue engineering. *Tissue Eng.* **5**, 139–151.

Pepper, M. S., Ferrara, N., Orci, L., and Montesano, R. (1992). Potent synergism between vascular endothelial growth factor and basic fibroblast growth factor in the induction of angiogenesis *in vitro. Biochem. Biophys. Res. Commun.* **189**, 824–831.

Peters, M. C., Isenberg, B. C., Rowley, J. A., and Mooney, D. J. (1999a). Release from alginate enhances the biological activity of vascular endothelial growth factor. *J. Biomat. Sci. Polym. Edn.* **9**, 1267–1278.

Peters, M. C., Shea, L. D., and Mooney, D. J. (1999b). Protein and plasmid DNA delivery from tissue engineering matrices. *Polym. Prep.* **40**, 372–374.

Pinholt, E. M., Solheim, E., Bang, G., and Sudmann, E. (1992). Bone induction by composites of bioresorbable carriers and demineralized bone in rats: a comparative study of fibrin–collagen paste, fibrin sealant, and polyorthoester with gentamicin. *J. Oral Maxillofac. Surg.* **50**, 1300–1304.

Pitt, C. G. (1990). Poly-ε-caprolactone and its copolymers. *In* "Biodegradable Polymers as Drug Delivery Systems" (M. Chasin and R. Langer, eds.), pp. 71–120. Marcel Dekker, New York.

Pulapura, S., and Kohn, J. (1992). Trends in the development of bioresorbable polymers for medical applications. *J. Biomater. Appl.* **6**, 216–250.

Putnam, A. J., and Mooney, D. J. (1996). Tissue engineering using synthetic extracellular matrices. *Nat. Med.* **2**, 824–826.

Richardson, T. P., Peters, M. C., Ennett, A. B., and Mooney, D. J. (2001). Polymeric system for dual growth factor delivery. *Nat. Biotechnol.* **19**, 1029–1034.

Roland, W. D., Holder, W. D., Culberson, C. R., Beiler, R. J., Burg, K. J. L., Greene, K. G., Loebsack, A. B., Wyatt, S., and Halberstadt, C. R. (1998). Optimizing cell culture time and seeding density on porous, absorbable constructs. Tissue Engineering Society, Orlando, FL.

Rowley, J. A., Madlambayan, G., and Mooney, D. J. (1999). Alginate hydrogels as synthetic extracellular matrix materials. *Biomaterials* **20**, 45–53.

Shapiro, L., and Cohen, S. (1997). Novel alginate sponges for cell culture and transplantation. *Biomaterials* **18**, 583–590.

Shea, L. D., Smiley, E., Bonadio, J., and Mooney, D. J. (1999). DNA delivery from polymer matrices for tissue engineering. *Nat. Biotechnol.* **17**, 551–554.

Sheridan, M. H., Shea, L. D., Peters, M. C., and Mooney, D. J. (2000). Bioabsorbable polymer scaffolds for tissue engineering capable of sustained growth factor delivery. *J. Controlled Release* **64**, 91–102.

Smetana, K., Jr. (1993). Cell biology of hydrogels. *Biomaterials* **14**, 1046–1050.

Smith, M. K., Peters, M. C., Richardson, T. P., Garbern, J. C., and Mooney, D. J. (2004). Locally enhanced angiogenesis promotes transplanted cell survival. *Tissue Eng.* **10**, 63–71.

Solheim, E., Pinholt, E. M., Bang, G., and Sudmann, E. (1992). Inhibition of heterotopic osteogenesis in rats by a new bioerodible system for local delivery of indomethacin. *J. Bone Joint Surg. Am.* **74**, 705–712.

Srivastava, S., Gorham, S. D., and Courtney, J. M. (1990). The attachment and growth of an established cell line on collagen, chemically modified collagen, and collagen composite surfaces. *Biomaterials* **11**, 162–168.

Sun, Q., Chen, R. R., Shen, Y., Mooney, D. J., Rajagopalan, S., and Grossman, P. M. (2005). Sustained vascular endothelial growth factor delivery enhances angiogenesis and perfusion in ischemic hind limb. *Pharm. Res.* **22**, 1110–1116.

Tamargo, R. J., Epstein, J. I., Reinhard, C. S., Chasin, M., and Brem, H. (1989). Brain biocompatibility of a biodegradable, controlled-release polymer in rats. *J. Biomed. Mater. Res.* **23**, 253–266.

Thomson, J. A., Itskovitz-Eldor, J., Shapiro, S. S., Waknitz, M. A., Swiergiel, J. J., Marshall, V. S., and Jones, J. M. (1998). Embryonic stem cell lines derived from human blastocysts. *Science* **282**, 1145–1147.

Thomson, R. C., Wake, M. C., Yaszemski, M. J., and Mikos, A. G. (1995). Biodegradable polymer scaffolds to regenerate organs. *Adv. Polym. Sci.* **122**, 245–274.

Tomihata, K., and Ikada, Y. (1997). *In vitro* and *in vivo* degradation of films of chitin and its deacetylated derivatives. *Biomaterials* **18**, 567–575.

Ullmann, Y., Hyams, M., Ramon, Y., Beach, D., Peled, I. J., and Lindenbaum, E. S. (1998). Enhancing the survival of aspirated human fat injected into nude mice. *Plast. Reconstr. Surg.* **101**, 1940–1944.

Vandenburgh, H., and Kaufman, S. (1979). *In vitro* model for stretch-induced hypertrophy of skeletal muscle. *Science* **203**, 265–268.

Vandenburgh, H. H., Swasdison, S., and Karlisch, P. (1991). Computer-aided mechanogenesis of skeletal muscle organs from single cells *in vitro*. *FASEB J.* **5**, 2860–2867.

Vercruysse, K. P., Marecak, D. M., Marecek, J. F., and Prestwich, G. D. (1997). Synthesis and *in vitro* degradation of new polyvalent hydrazide cross-linked hydrogels of hyaluronic acid. *Bioconjug. Chem.* **8**, 686–694.

von Heimburg, D., Zachariah, S., Heschel, I., Kuhling, H., Schoof, H., Hafemann, B., and Pallua, N. (2001a). Human preadipocytes seeded on freeze-dried collagen scaffolds investigated *in vitro* and *in vivo*. *Biomaterials* **22**, 429–438.

von Heimburg, D., Zachariah, S., Low, A., and Pallua, N. (2001b). Influence of different biodegradable carriers on the *in vivo* behavior of human adipose precursor cells. *Plast. Reconstr. Surg.* **108**, 411–420; discussion 421–422.

Wakitani, S., Goto, T., Young, R. G., Mansour, J. M., Goldberg, V. M., and Caplan, A. I. (1998). Repair of large full-thickness articular cartilage defects with allograft articular chondrocytes embedded in a collagen gel. *Tissue Eng.* **4**, 429–444.

Walton, R. L., and Brown, R. E. (1993). Tissue engineering of biomaterials for composite reconstruction: an experimental model. *Ann. Plast. Surg.* **30**, 105–110.

Walton, R. L., Beahm, E. K., and Wu, L. (2004). De novo adipose formation in a vascularized engineered construct. *Microsurgery* **24**, 378–384.

Wechselberger, G., Russell, R. C., Neumeister, M. W., Schoeller, T., Piza-Katzer, H., and Rainer, C. (2002). Successful transplantation of three tissue-engineered cell types using capsule induction technique and fibrin glue as a delivery vehicle. *Plast. Reconstr. Surg.* **110**, 123–129.

Whang, K., Thomas, C. H., Healy, K. E., and Nuber, G. (1995). A novel method to fabricate bioabsorbable scaffolds. *Polymer* **36**, 837–842.

Williams, S. K., Wang, T. F., Castrillo, R., and Jarrell, B. E. (1994). Liposuction-derived human fat used for vascular graft sodding contains endothelial cells and not mesothelial cells as the major cell type. *J. Vasc. Surg.* **19**, 916–923.

Wobus, A. M., and Boheler, K. R. (2005). Embryonic stem cells: prospects for developmental biology and cell therapy. *Physiol. Rev.* **85**, 635–678.

Wong, W. H., and Mooney, D. J. (1997). Synthesis and properties of biodegradable polymers used as synthetic matrices for tissue engineering. *In* "Synthetic Biodegradable Polymer Scaffolds" (A. Atala and D. J. Mooney, eds.), pp. 49–80. Birkhäuser Press, Boston.

Yamada, K., Miyamoto, S., Nagata, I., Kikuchi, H., Ikada, Y., Iwata, H., and Yamamoto, K. (1997). Development of a dural substitute from synthetic bioabsorbable polymers. *J. Neurosurg.* **86**, 1012–1017.

Yuksel, E., Weinfeld, A. B., Cleek, R., Wamsley, S., Jensen, J., Boutros, S., Waugh, J. M., Shenaq, S. M., and Spira, M. (2000). Increased free-fat-graft survival with the long-term, local delivery of insulin, insulin-like growth factor-I, and basic fibroblast growth factor by PLGA/PEG microspheres. *Plast. Reconstr. Surg.* **105**, 1712–1720.

Zacchi, V., Soranzo, C., Cortivo, R., Radice, M., Brun, P., and Abatangelo, G. (1998). *In vitro* engineering of human skin-like tissue. *J. Biomed. Mater. Res.* **40**, 187–194.

Part Ten

# Cardiovascular System

# Chapter Thirty-Seven

# Progenitor Cells and Cardiac Homeostasis

*Annarosa Leri, Toru Hosoda, Marcello Rota, Claudia Bearzi, Konrad Urbanek, Roberto Bolli, Jan Kajstura, and Piero Anversa*

## I. INTRODUCTION

This chapter discusses the role that endogenous and exogenous progenitor cells may play in the maintenance of cardiac homeostasis, organ repair after injury, and the treatment of the diseased heart. The recognition that the adult heart possesses a stem cell compartment that can regenerate myocytes and coronary vessels has raised the possibilities of rebuilding infarcted myocardium, replacing old hypertrophied and poorly contracting myocytes with new, better-functioning cells, and, perhaps, reversing ventricular dilation and wall thinning. Similarly, adult bone marrow cells are able to differentiate into cells beyond their own tissue boundary and to create cardiomyocytes and coronary vessels. This process has been termed *developmental plasticity* or *transdifferentiation*. These observations have challenged the view of the heart as a postmitotic organ and have proposed a new paradigm, in which cardiac cells are continuously replaced by newly formed myocytes and vascular cells.

## II. ORGAN HOMEOSTASIS

The balance between cell death and cell division is crucial for the preservation of cell number and organ mass in prenatal and postnatal life. The size of an organ is determined by the average volume of the parenchymal cells and the total number of these cells. These parameters are controlled by the rate of cell division, migration, growth, and death. The existence of an intrinsic mechanism that enables an organ to sense its own size and regulate its growth has been postulated, but the demonstration of this regulatory process remains elusive. When the human liver is surgically reduced, the parenchyma regenerates itself until the correct organ size is achieved. Similarly, the growth of fetal thymus glands transplanted into adult mice stops when the postnatal dimension is reached. The same pattern of growth takes place when a larval wing disc is transplanted into an adult fly. These observations support the notion that a conserved, organ-autonomous apparatus for the modulation of organ volume is present across species in prenatal life (Doseff *et al.*, 2004).

How organ growth in the embryo is regulated is largely unknown. However, it is reasonable to assume that the activation of a complex network of growth-inducing factors, negative feedback loops that decrease the rate of growth, together with the induction or inhibition of apoptosis, takes place when the organ either exceeds or does not reach the "organ-size checkpoint." Adult organ homeostasis may be modulated in a similar manner, but these processes are expected to become dysfunctional with aging and the appearance of diseases. Postnatally, decreases in the number of cells within an organ may be compensated by enhanced cell proliferation or by enlargement of the remaining parenchymal cells, which attempt to preserve organ dimension. In newts, *Drosophila*, and higher eukaryotes, the maintenance of the size of a given organ in the organism can be accomplished by a larger number of smaller cells or a smaller number of larger cells (Gomer *et al.*, 2001). These two distinct adaptations have different histological features but, most importantly, are not functionally equivalent. In this regard, the hypertrophic response of cells can become rapidly maladaptive because of the difficulty of the larger cells' performing efficiently a variety of specialized functions.

Although cellular hypertrophy can reconstitute a correct organ volume, actual repair is achieved through cell regeneration, which results in the replacement of the cells lost following injury by new, functionally competent cells. In physiological conditions, this process is highly efficient; cell dropout by normal wear and tear is counteracted by cell regeneration, and organ homeostasis is preserved. Conversely, in the presence of damage, restitutio ad integrum does not occur, and healing is associated with the formation of a scar. Scarring is crucial for rapid handling of the damage, to seclude the lesion from healthy tissue and to prevent a cascade of uncontrolled deleterious events (Mutsaers *et al.*, 1997). However, the fibrotic area does not possess the biochemical, physical, or functional properties of the uninjured tissue. By necessity, the reparative response, which occurs in damaged organs, leads to scarring and cellular hypertrophy. Both cellular phenomena affect the overall performance of the organ (Silver and Miller, 2004).

In spite of the presence of resident stem cells, spontaneous tissue regeneration is a rare event in adult self-renewing organs (Watanabe *et al.*, 2003). It is unknown why the evolution of a lesion is different in prenatal and postnatal life. In fact, tissue repair in early mammalian embryos is rapid, efficient, and scar free (Ferguson and O'Kane, 2004), and skin wounds heal with restitutio ad integrum, whereas wounds in adult mammals result in scarring. The difference in the inflammatory response may condition the distinct evolution of the healing process in embryos and adults. In the embryo, a lower number of less differentiated inflammatory cells accumulate in the region of damage, and the growth factors present at the site of healing are highly variable with respect to those in the adult tissue. In the latter case, the secretion of cytokines may influence negatively the regenerative ability of adult stem cells and might ultimately oppose their survival, interfering with the restoration of the damaged organ.

Interventions aiming at the recapitulation of the embryonic gene program could be fundamental for effective organ repair. Recent evidence points to the importance of morphogens for proper homeostasis. During normal development, cells have different fates, according to the distribution of morphogens, which are signaling proteins that set the positional value of a cell. They form concentration gradients across the developmental field in which the cells reside, determining their fate (Tabata and Takei, 2004). The effects of morphogens on the developing cells depend on the distance between the receiving cells and the morphogen-secreting cells. The *Drosophila* wing is considered a good model for the study of patterning; Decapentaplegic, Wingless, and Hedgehog proteins have been shown to act as morphogens in this system (Tabata and Takei, 2004).

In addition to the effect on cell fate, morphogens can function as chemoattractants or chemorepellents modulating the migration of cells. This phenomenon is apparent during neural crest development (Jones and Trainor, 2005). Morphogens regulate both induction and migration of this stem cell population that generates a variety of tissues, including Purkinje neurons, glia, heart, smooth muscle, connective tissue, melanocytes, craniofacial cartilage, and bone. Finally, the rates of cell division and cell apoptosis have to be tightly regulated to allow the survival of the correct cells in a system in which cells compete for the same trophic factors. Cell competition results in apoptosis of slower-growing cells located in proximity to faster-growing cells. C-myc up-regulation transforms cells in "supercompetitors," which become capable of clonal expansion (Secombe *et al.*, 2004). The cluster of supercompetitor cells influences the behavior of the surrounding cells, favoring the onset of apoptotic death in the more vulnerable subset. These cells are at a disadvantage, which is created by the low level of this oncogene. Although this cross-talk has been identified in *Drosophila* (Secombe *et al.*, 2004), it might be important in mammals as well. Cell competitions may lead to mechanical stresses within an organ or tissue, which, in turn, provide a feedback loop aiming at the stabilization and uniformity of cellular growth processes.

Based on this information, it might be reasonable to postulate that the objective of regenerative medicine is the polarization of cell-fate choices, pattern formation, and tissue reconstitution. To achieve this ambitious goal, cellular, genetic, biochemical, and molecular approaches need to be implemented concurrently to identify and characterize cellular mechanisms, modify their undesired evolution, reprogram their transcriptional targets, and force the acquisition of novel identities. The possibility of operating at these new levels of cell-fate determinants may alter the evolution of a damaging event and its impact on organ integrity, structurally and functionally. At present, our understanding of the fundamental biological principles of cell behavior is inadequate to address these issues, which are critical for the restoration of baseline physiological performance following injury.

## III. CARDIAC HOMEOSTASIS

According to the old view of cardiac biology, the heart survives and exerts its function until the death of the organism with the same number of cells as, or fewer cells than, are present at birth (Chien, 2004). Thus, the heart would respond to parenchymal losses only by increasing the size of its cells, without any form of regeneration. If this were the case, the protection of organ size in physiological situations would occur at the price of a progressive deterioration of cardiac function. In fact, normal wear and tear leads to continuous cell dropout, and this loss would be compensated by uniform enlargement of cardiomyocytes. Hypertrophied myocytes are poorly contracting cells characterized by profound alterations of intracellular calcium cycling, and this defect inevitably contributes over time to the onset of ventricular dysfunction and its progression to cardiac failure (Leri et al., 2005).

The novel view of the heart as an organ with regenerative ability implies that the reconstitution of muscle mass can be achieved not only by increasing myocyte volume but also by promoting myocyte proliferation. In embryonic and early postnatal life, the increase in cardiac mass reflects the delicate balance between the addition of new myocytes and the death of unnecessary cells (Olson et al., 1996). Shortly after birth, cardiac growth occurs by means of an increase in myocyte number and volume, which together participate in the development of the adult heart phenotype (Anversa and Olivetti, 2002). In adulthood, the heart is characterized by a surprisingly high and rapid turnover of its parenchymal cells that is regulated by a stem cell compartment (Anversa et al., 2006).

An in vivo functional assay that allows the evaluation of cell turnover in self-renewing organs consists of the pulse-chase labeling protocol. By this approach, the growth and phenotypic changes occurring in cardiac stem cells (CSCs) during their transition from a poorly differentiated to a specialized compartment can be identified.

To determine the activation of CSCs in the mouse heart, 4 and 12 injections of BrdU were performed at 12-hour intervals in two groups of mice over a period of two and six days, respectively. A third group of mice was treated in an identical manner for six days, and mice were sacrificed after a chasing period of 10 weeks. Bright and dim BrdU labeled Lin⁻CSCs were distinguished based on fluorescence intensity. The number of bright BrdU Lin⁻CSCs in the atria increased 140% from two to six days but decreased 91% after 10 weeks of chasing. Of 257 bright BrdU-Lin⁻CSCs at six days, only 23 cells retained this characteristic at 10 weeks, constituting the slow-cycling SC pool in this portion of the heart. A similar pattern of changes was seen in the base–midregion and apex. Dim BrdU-Lin⁻CSCs increased from to two to six days, but, in contrast to bright BrdU-Lin⁻CSCs, dim BrdU-Lin⁻CSCs increased dramatically after chasing, 16-fold in the atria, 13-fold at the base–midregion, and 16-fold at the apex. The aggregate number of Lin⁻CSCs did not change over a period of 10 weeks, indicating that the growth kinetics of Lin⁻CSCs tends to preserve the pool of primitive cells in the young, healthy heart (Urbanek et al., 2005).

The long-term label-retaining assay documented that ~10% of Lin⁻CSCs are slow cycling and conform to the paradigm of SCs. By employing the same approach, myocyte formation was assessed by measuring the fraction of BrdU$^{POS}$-cells. After 10 weeks of chasing, BrdU-bright myocytes were cells that experienced a limited number of divisions, while more rounds of doublings had to occur in myocytes with intermediate levels of labeling. These myocyte classes were assumed to correspond to amplifying myocytes, which incorporated BrdU at the time of injection and continued to divide and differentiate. Conversely, BrdU-dim myocytes were considered the progeny of cycling CSCs, which became BrdU$^{POS}$ at the time of injection and gave rise to a large number of committed cells. The percentage of BrdU-bright-myocytes detected at six days decreased markedly after 10 weeks, while BrdU-dim-myocytes increased. The apex had 19% BrdU-labeled myocytes, the atria 15%, and the base–midregion 10%. In two and a half months, ~10–19% of atrial and ventricular myocytes were formed. In spite of this large increase in the number of parenchymal cells, the total number of atrial and ventricular myocytes remained constant, suggesting a high turnover of this cell population.

However, myocyte renewal was not homogeneous, and the rate of cell regeneration was higher in the atria and apex than at the base–midregion of the left ventricle. Importantly, the heterogeneity of myocyte regeneration in these anatomical sites was paralleled by a similar heterogeneity in the number of CSCs, since the higher numerical density of CSCs per unit volume of myocardium was associated with the larger formation of cardiomyocytes. These data provide information on the magnitude of cell turnover in the myocardium. These results were complemented with the measurements of myocyte progenitors-precursors, which together with CSC number were utilized to evaluate the half-life of myocytes. Contrary to expectations, myocytes located at the base–midregion have a half-life twofold longer than atrial and apical myocytes, suggesting that myocyte turnover is high and heterogeneous. By this novel analysis, we found that 50% of ventricular myocytes are lost in approximately nine months, while this phenomenon is faster at the apex and atria, requiring less than four months. The degree of turnover is inversely related to the level of hemodynamic stress in the heart. It is unknown why higher workload is associated with a longer myocyte life span. The ventricle represents the region that mostly participates in the pump function, and the heart would fail if ventricular myocytes were to be engaged too often in mitotic division (Urbanek *et al.*, 2005). Together, these observations challenge the notion that the life span of myocytes coincides with the life span of the organ and the organism (MacLellan and Schneider, 2000) and the contention that myocyte regeneration is a slow, inconsequential process of heart homeostasis (Chien, 2004). Conversely, myocyte replication is constant and heterogeneous and involves a large number of cells.

Under physiologic conditions, the intrinsic growth reserve of the heart is sufficient to maintain cell homeostasis and adequate pump performance. However, when an increase in pressure and/or volume load is imposed on the heart, myocyte hypertrophy and proliferation, in combination with myocyte apoptosis and necrosis, constitute the elements of myocardial remodeling (Pfeffer and Braunwald, 1990). Novel methodological approaches applied to the analysis of the myocardium have defeated the paradigm, introduced more than 60 years ago, that the heart is a postmitotic organ and that myocytes are terminally differentiated cells that participate in cardiac function throughout life. A subset of parenchymal cells expresses the molecular components mediating the entry and progression of these cells through the cell cycle and karyokinesis and cytokinesis. The recognition that myocyte replication, hypertrophy, and death occur in the pathologic heart has significantly enhanced our understanding of the dynamic nature of the myocardium and the critical role these variables have in the preservation of cardiac performance or in the onset of ventricular dysfunction (Anversa and Kajstura, 1998).

Contrary to the general belief, the restricted regenerative capacity of the myocardium does not represent the initial causal event of impaired cardiac function. The nonischemic failing heart, in its early phases of decompensation, has a number of myocytes that often exceeds the number of cells in a normal heart. Alternatively, the modest reduction in myocyte number chronically is not consistent with the severe deterioration in ventricular performance. A typical example is found in hypertensive cardiomyopathy or chronic aortic stenosis in animals and humans (Olivetti *et al.*, 1987, 1994). The initiation and evolution of cardiac failure appears to depend mostly on two other crucial factors that are strictly interrelated in the determination of pump function: the accumulation of old, poorly contracting cells and the formation of multiple foci of myocardial scarring (Wei, 1992).

The formation of new myocytes defeats the notion that cardiac cells have the same age. During aging and in pathologic states, the continuous turnover of myocytes results in a heterogeneous cell population consisting of young, adult, old, and senescent myocytes. Data in humans suggest that this subdivision corresponds to cells of different sizes, raising the possibility that the life span of a cell is associated with a progressive increase in volume of the cell. Young,

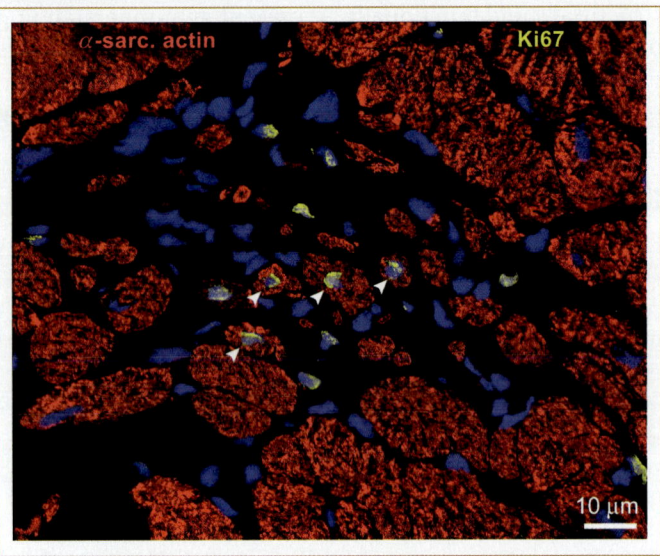

**FIG. 37.1.** Myocardial regeneration in the infarcted human heart. A cluster of highly proliferating, small, developing myocytes is visible. Myocytes are labeled by α-sarcomeric actin (red) and nuclei by DAPI (blue). Most of these cells are positive for the marker of the cell cycle, Ki67 (yellow, arrowheads). The proliferating myocytes have a small cross-sectional area.

amplifying, dividing myocytes are less than 180 μm$^2$ in cross-sectional area, adult nondividing cells are 200–250 μm$^2$ in cross-sectional area, old nondividing p16$^{INK4a}$-negative cells are 300–500 μm$^2$ in cross-sectional area, and senescent nondividing p16$^{INK4a}$-positive cells are 600–900 μm$^2$ in cross-sectional area (Urbanek *et al.*, 2003). Myocyte length, however, remains relatively constant in all these cell categories, varying at most from 90 to 120 μm. Importantly, young myocytes do not express inhibitors of the cell cycle, such as p53 and p16$^{INK4a}$ (Sherr and Roberts, 2004). In contrast, a large fraction of these cells, 10–15%, is cycling (Fig. 37.1), as demonstrated by BrdU, Ki67, MCM5, and Cdc6 labeling (Whitfield *et al.*, 2006). Adult myocytes are unable to reenter the cell cycle, since they are terminally differentiated. They express p21$^{Cip1}$ but not p53 and p16$^{INK4a}$. Old myocytes express p53 and p21$^{Cip1}$ in the absence of p16$^{INK4a}$, and senescent myocytes express p53 and p16$^{INK4a}$ (Urbanek *et al.*, 2003). p16$^{INK4a}$ and p53 are markers of cellular aging. Myocyte aging and the concomitant increase in myocyte volume typically result in a severe depression in cell function and in calcium metabolism.

Reduced regenerative responses dictated by deregulation of morphogens and/or limited cell proliferation has been associated with the onset of multiple cardiac diseases, and the possibility has been raised that the postnatal recapitulation of embryonic pathways promotes the regenerative response of the postnatal heart after injury. In this regard, intramyocardial gene transfer of naked DNA encoding human sonic hedgehog had a favorable impact on the recovery of the adult heart following acute and chronic ischemia (Kusano *et al.*, 2005). This positive response was mediated by enhanced vasculogenesis, reduced fibrosis and collagen deposition, and decreased myocyte apoptosis. This and similar observations strongly indicate that gene therapy may be beneficial in various diseases by triggering the expression of multiple trophic factors and promoting tissue repair in the adult heart.

This novel view of cardiac biology has introduced the concept of the heart as an organ permissive for myocardial regeneration mediated by the activation, proliferation, and differentiation of exogenous or endogenous primitive cells. Thus, tissue damage can be replaced by the formation of new, functionally competent myocytes and coronary vessels. Cardiac repair is conditioned by several factors, including (a) the number of cells to be administered, (b) cell death and survival in the hostile milieu of the acute and chronic injured areas, (c) cell engraftment, and (d) cell growth and differentiation. An additional critical variable when exogenous cells are involved in myocardial restoration is their level of plasticity, which is dictated by their ability to acquire the myocyte and vascular smooth muscle (SMC) and endothelial cell (EC) lineages. In an identical manner, resident cardiac progenitor cells (CPCs) have to promote the formation of a proportional number of parenchymal cells and coronary vessels. Moreover, progenitor cells can contribute indirectly to cardiac regeneration by releasing a variety of peptides that exert a paracrine action on the myocardium and its resident CPCs. These mechanisms are not mutually exclusive, and primitive cells may participate both directly and indirectly in the repair process (Gnecchi *et al.*, 2005; Yoon *et al.*, 2005). In all cases, progenitor cells have to engage themselves in homing into the myocardium to perform specific functions. These biological processes depend on a successful interaction between progenitor cell classes and tissue microenvironment and condition the efficiency of repair in multiple models of myocardial regeneration.

# IV. PROPERTIES OF EXOGENOUS AND ENDOGENOUS CELLS FOR CARDIAC REPAIR

Acutely after ischemic injury, the healing process inevitably results in scar formation, not only in the heart but also in other organs, whether their parenchymal cells are highly proliferating, slowly cycling, or terminally differentiated (Beltrami *et al.*, 1994; Silver and Miller, 2004; Leong and Freeman, 2005). The need to overcome this biological obstacle has favored the development of strategies aiming at the replacement of dead tissue with viable cells. Several interventions have been used in an attempt to promote cardiac regeneration experimentally. These protocols have employed different cell types, including fetal tissue and fetal and adult cardiomyocytes (Leor *et al.*, 1996), skeletal myoblasts (Menasche *et al.*, 2001), embryonic-derived myocytes and

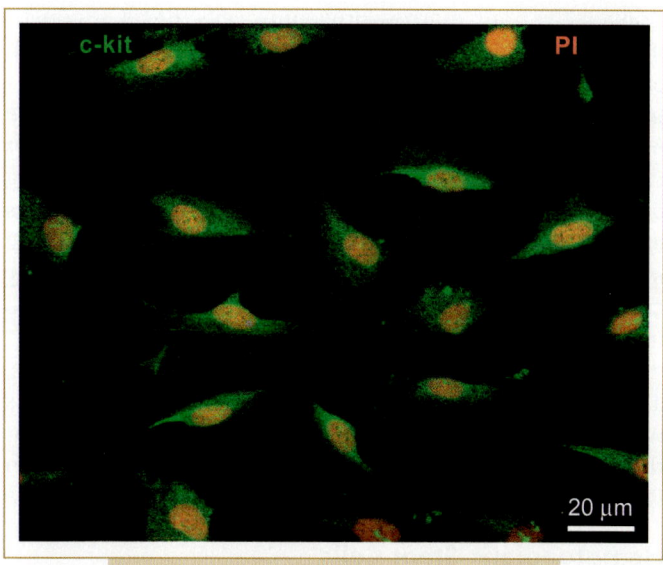

**FIG. 37.2.** Culture of cardiac stem cells. c-kit positive cardiac stem cells were isolated from a human heart by enzymatic dissociation and selection with immunomagnetic beads. Cells express c-kit in the plasma membrane (green), and nuclei are stained by propidium iodide (red).

endothelial cells (Etzion *et al.*, 2001), bone marrow–derived immature myocytes (Hattan *et al.*, 2005), fibroblasts, smooth muscle cells (Li *et al.*, 1999), and bone marrow c-kit positive and negative progenitor cells (Orlic *et al.*, 2001; Kawada *et al.*, 2004; Yoon *et al.*, 2005). In spite of these efforts, however, the most appropriate form of cell therapy for actual restitutio ad integrum of the damaged myocardium has not been identified yet.

The most logical and potentially powerful cell to be employed is the cardiac progenitor cell (Fig. 37.2). It is intuitively apparent that if the adult heart possesses a pool of primitive multipotent cells (Beltrami *et al.*, 2003; Oh *et al.*, 2003; Matsuura *et al.*, 2004; Martin *et al.*, 2004; Pfister *et al.*, 2005; Messina *et al.*, 2004; Rosenblatt-Velin *et al.*, 2005; Tomita *et al.*, 2005), these cells must be tested first, before more complex and unknown cells are explored. Cardiac regeneration would be accomplished by enhancing the normal turnover of myocardial cells. CPCs may be activated locally within the myocardium, and some of the growth factor–receptor systems that regulate CSC proliferation and differentiation have been recognized (Linke *et al.*, 2005; Urbanek *et al.*, 2005). Although the importance of resident CSCs in organ homeostasis is apparent and, in the future, CSCs may become the most appropriate form of cell therapy for the diseased heart, emphasis currently has been placed on the identification of exogenous sources of highly proliferating cells that can acquire the cardiac phenotypes.

Difficulties exist in the acquisition of myocardial samples in humans and in the isolation and expansion of CPCs in quantities that can be employed therapeutically.

Conversely, bone marrow progenitor cells (BMPCs) constitute an appealing form of cell intervention; BMPCs can easily be collected from bone marrow aspirates or the peripheral blood after their mobilization from the bone marrow with cytokines. At present, it is unknown whether CPCs and BMPCs are similarly effective in reconstituting dead myocardium after infarction or whether limitations exist in CPC growth and BMPC transdifferentiation resulting in inadequate restoration of lost tissue. Also, BMPCs may constitute a necessary initial form of intervention for the infarcted heart, whereas CPCs might be employed later during the chronic evolution of the cardiac myopathy. Thus, a fundamental question to be addressed in future studies is whether BMPCs are superior, equal, or inferior to CPCs for the regeneration of cardiomyocytes and coronary vessels in ischemic heart failure.

## Hematopoietic Stem Cells

The bone marrow can be subdivided into a hematopoietic cell compartment and a stroma composed of mesenchymal stem cells (MSCs), fibroblasts, adipocytes, and vascular and nervous structures. Hematopoietic stem cells (HSCs) are the first tissue-specific compartment of stem cells that have been described (Till *et al.*, 1964a). As a consequence, the features of all stem cells in adult organs are typically studied on the basis of the characteristics of these primitive blood-forming cells. It has to be emphasized, however, that these criteria cannot be translated without caveats to solid organs and their resident stem cells. HSCs were originally identified as clonogenic cells of bone marrow origin that gave rise to multilineage hematopoietic colonies in the spleen (Till *et al.*, 1964b). In addition to the three fundamental attributes of self-renewal, clonogenicity, and multipotentiality, HSCs were defined as units of a hierarchically structured system formed by three main cellular components: primitive cells, intermediate stages, and mature cells. On this basis, all stem cell–regulated organs were recognized to adhere to a model of progressive acquisition of differentiation by primitive stem cells, which undergo maturation into progenitors, precursors, amplifying cells, and eventually terminally differentiated cells.

This succession of events was considered irreversible. However, recent work has indicated that, at least at the stem/progenitor level, hematopoiesis may be a continuum of transcriptional opportunity. The most primitive HSCs are either continually cycling at a slow rate or entering and exiting cell cycle. The cell cycle passages are accompanied by changes in functional phenotype, including reversible modifications in homing and engraftment ability, adhesion protein expression, and cytokine receptor expression (Quesenberry *et al.*, 2002). Similarly, the differentiation potential of HSCs varies in the different phases of the cell cycle. Thus, the bone marrow corresponds to a system in which cells continually shift from the engraftable primitive state to the multifactor responsive progenitor cells. These

phenotypical changes occur in both directions and do not coincide with differentiation events. They are dictated by chromatin-remodeling and epigenetic mechanisms (Quesenberry *et al.*, 2004). These observations indicate that bone marrow stem cells and most likely all adult stem cells should be not be studied as discrete entities but defined as functional units (Blau *et al.*, 2001).

Cell epitopes are commonly employed to isolate HSCs from the whole bone marrow. Different antigens characterize HSCs in different species. Mouse HSCs correspond to a subpopulation of bone marrow cells (BMCs) that are lineage negative and c-kit$^+$Sca-1$^+$Thy-1$^{low}$ (Uchida and Weissman, 1992). This population is multipotent and repopulates the bone marrow of irradiated mice. However, the c-kit$^+$Sca-1$^+$Thy-1$^{low}$ cells are functionally heterogeneous. Based on the identification of additional surface markers and clonal analysis, the Mac-1$^-$CD4$^-$ cells within this category are enriched for long-term reconstituting cells (Morrison and Weissman, 1994). Conversely, the Mac-1$^{low}$CD4$^-$ negative pool contains short-term repopulating cells. Finally, the Mac-1$^{low}$CD4$^{low}$ cells correspond to transient multipotent progenitors together with B-lymphocyte progenitors (Morrison *et al.*, 1995). Even by using these complicated combinations of markers, only 20% of the intravenously injected cells give long-term multilineage reconstitution in most studies. The use of the Hoechst efflux dye protocol has led to the isolation of a more functionally homogeneous cell population (Goodell *et al.*, 1996). This approach, however, cannot be used for the recognition of stem and progenitor cells in tissue sections.

The expression of different members of the SLAM family of receptors has recently been correlated with the primitiveness of bone marrow cells. HSC activity was found to be present in the CD150$^+$/CD244$^-$/CD48$^-$ compartment but rarely in the CD150$^-$/CD244$^+$/CD48$^+$ fraction (Kiel *et al.*, 2005). The CD150$^+$/CD244$^-$/CD48$^-$ fraction of bone marrow cells represents only 0.008% of bone marrow cells and one in approximately five injected cells engrafts and yields long-term multilineage reconstitution. These results are similar to those obtained with lineage-negative c-kit$^+$Sca-1$^+$Thy-1$^{low}$ cells. Cells with SLAM receptors can easily be recognized *in vivo*, and their identification does not require the documentation of the lack of markers of commitment (Yilmaz *et al.*, 2006). The vast majority of CD150$^+$/CD244$^-$/CD48$^-$ cells have been found to be associated with both sinusoids and endosteum in close contact with endothelial cells and osteoblasts, respectively (Kiel *et al.*, 2005). This preferential localization within the osteoblastic and vascular niches represents an additional proof of the stemness of this cell subset.

The CD34 antigen has been used to sort human HSCs (Negrin *et al.*, 2000), which, however, are also present in the CD34 negative fraction. AC133 is another marker of human HSCs (Bhatia, 2001), but so far there is no specific epitope for the true stem cell in humans. A mixed population of bone marrow cells, rather than a highly purified stem cell pool, has commonly been used for the restoration of injured organs (Britten *et al.*, 2003; Poulsom *et al.*, 2003).

## Mesenchymal Stem Cells

Stromal cells or MSCs are easily defined *in vitro*, where they constitute the nonhematopoietic and adherent cell component in long-term cultures. *In vivo*, they contribute to the formation of the hematopoietic microenvironment. Over 40 years ago, Friedenstein (1961) described the isolation of stromal cells from the bone marrow. Once in culture dishes, these cells, later called MSCs, assume a spindle shape and proliferate to form colonies. These cells rapidly adhere to plastic and are able to differentiate under defined *in vitro* conditions into three lineages: osteoblasts, chondroblasts, and adipocytes. Recently, the differentiation or transdifferentiation potential of MSCs appears to be much broader. MSCs can give rise to multiple cell lineages of the three germ layers: mesodermal (skeletal myoblasts, cardiac myocytes, endothelial cells), endodermal (lung cells, hepatocytes, gut cells), and ectodermal (neural cells).

In a manner similar to other adult stem cell types, MSCs are clonogenic, self-renewing, and multipotent. However, not all reports agree that MSCs possess the self-renewal property (Dennis and Charbord, 2002). If self-renewal were a feature of MSCs, this property would not be crucial for MSC function. This is because of the high degree of plasticity of MSCs, which would confer on these cells phenotypic flexibility and the ability to switch from one cell type to another, at early as well as at late stages of differentiation (Bianchi *et al.*, 2001). In view of the great level of cellular heterogeneity and the absence of clear phenotype markers, it is, however, difficult to establish whether MSCs actually transdifferentiate or constitute progenitor cells with a broad differentiation potential into all the cell lineages of mesenchymal origin. In cases in which MSCs give rise to tissues of ectodermal or endodermal origin, like the brain and the pancreas, respectively, the occurrence of transdifferentiation can be claimed.

MSCs correspond to stromal cells, constituting also structural components of the bone marrow that support hematopoiesis *in vivo* and *in vitro* by providing extracellular matrix components, cytokines, and growth factors. One of the major difficulties encountered in the utilization of MSCs in the regeneration of the injured heart is that the nonhematopoietic compartment of bone marrow consists of multiple heterogeneous cell populations. In the last several years, efforts have been made to develop protocols for the separation and enrichment of the different cellular subsets. After density-gradient fractionation, the cells in the upper low-density portion of bone marrow aspirates are plated and the nonadherent hematopoietic cells are removed by medium replacement. Only 0.001–0.01% of the initial unfractionated bone marrow cell population consists of MSCs (Pittenger and Martin, 2004). Methods based on immunological properties have been used to enrich bone

marrow for the MSC subset. MSCs are negative for hematopoietic surface markers and positive for a wide array of surface proteins. The first antigens to be associated with MSCs were those reactive with the monoclonal antibodies SH2 and SH3, which recognize CD105 and CD73, respectively (Haynesworth *et al.*, 1992). Although CD105 is a relevant epitope of MSCs and angiogenesis, CD29, CD44, CD166, CD54, CD55, CD13, and CD90 have been proposed to be important determinants of the mesenchymal phenotype (Pittenger and Martin, 2004), while STRO-1 may identify an immature state (Gronthos *et al.*, 1994). Prockop and others (Smith *et al.*, 2004) support the notion that MSCs cannot be distinguished solely by antigen expression but necessitate functional assays demonstrating their multipotent growth and differentiation behavior. Thus, a definitive consensus on the properties of MSCs has not been reached yet, and discrepancies exist in the characteristics of the distinct MSC populations employed. The differences may be explained by dissimilarities in culture methods, donor age, and differentiation stage of the cells, but this phenomenon makes comparisons among studies extremely complex.

In adulthood, the bone marrow and the systemic circulation constitute the main sources of MSCs, although they have also been detected in tissues distant from the bone marrow. The oval cells of the liver, prostatic stem cells, metanephric mesenchymal cells, precursors of the Leydig cells in the testis, primitive osteoprogenitors, preadipocytes, and satellite cells of the skeletal muscle have been classified as MSCs. Because not all studies agree that MSCs reside in these mesenchymal tissues, the possibility has been raised that a long-distance traffic of MSCs occurs between the bone marrow and distant sites, through the bloodstream. Alternatively, embryonic primordia of MSCs may be stored in organs of mesodermal origin and, in response to injury, may participate in tissue repair (Pittinger and Martin, 2004).

During the phase of amplification, MSCs do not differentiate spontaneously but do so in the presence of growth factors and cytokines. They can acquire multiple phenotypes, including osteoblasts, chondrocytes, adipocytes, endothelial cells, and neuronal-like cells. Bone marrow MSCs differentiate into cardiomyocytes in the presence of 5-aza-cytidine (Makino *et al.*, 1999); the morphology of the cells changes from a spindle shape to a ball-like form and, subsequently, a rod-shaped configuration. The cells then fuse in a syncitiumlike structure that resembles a myotube. Although these characteristics mimic the organization of skeletal muscle cells, the committed progeny expresses markers commonly found in cardiomyocytes at different stages of fetal development (Makino *et al.*, 1999).

At the end of the 1990s, following transplantation in irradiated animals, MSCs durably engrafted in bone, cartilage, and lungs of mice (Pereira *et al.*, 1998), giving rise to fibroblast-like cells that could be reisolated and cultured from the lungs, calvaria, cartilage, long bones, tails, and skin. The presence of MSCs has been demonstrated in many tissues other than the bone marrow. The clear identification of MSCs is complicated by the fact that definitive *in vivo* markers are still lacking. Retrospective functional data strongly support the existence of common adult stem cells in mesenchymal tissues that have the capacity to differentiate along various specific differentiation lineages. Because MSCs are multipotent and can easily be expanded in culture, there has been much interest in their clinical potential for tissue repair and gene therapy. And numerous studies have been carried out demonstrating the migration and multiorgan engraftment potential of MSCs in animal models and in humans. However, understanding the mechanisms behind MSC cell fate determination is not easy, because the molecular processes that drive engraftment and differentiation are complex.

## Cardiac Stem Cells

As occurred for the central nervous system, an organ considered irreversibly terminally differentiated, CSCs have been isolated from the adult myocardium, and long-term cultures have been developed (Beltrami *et al.*, 2003). Subsequently, data collected in 10 distinct laboratories concur with the concept that primitive cells reside in the adult mammalian heart. However, disagreement exists concerning the surface phenotype of CPCs, their number, and their functional characteristics. A direct comparison of the *in vitro* and *in vivo* behavior of the distinct CPC subsets has not been performed yet, making difficult the interpretation of these recent findings. The different populations of CPCs may represent independent categories of cardiac primitive cells. If this were the case, the heart, in analogy with the bone marrow, could contain several compartments of stem cells. Alternatively, the same cellular compartment has been analyzed with noncomparable approaches and methodologies, and the apparent multiple CPC populations correspond to stages of maturation/differentiation of the same original cell. Thus, the postulated differences in CPC classes reflect stages of a common stem cell pool.

The first study focused on lineage-negative (Lin⁻) c-kit^POS cells, which were shown to have the components of stemness: self-renewal, clonogenicity, and multipotentiality (Beltrami *et al.*, 2003). Similar *in vitro* properties characterize cardiac cells that express MDR1 or Sca-1 only (Linke *et al.*, 2005). However, most primitive cardiac cells possess the three antigens together, and only a small percentage expresses only c-kit, MDR1, or Sca-1. *In vitro*, Lin⁻c-kit^POS cells grow as an adherent monolayer when seeded in dishes or form spheroids when in suspension, mimicking the biology of neural stem cells (Tropepe *et al.*, 1999). *In vivo*, CSCs implanted in the infarcted ventricle (Beltrami *et al.*, 2003) or mobilized to the dead myocardium (Urbanek *et al.*, 2005) migrate, engraft, undergo multilineage commitment, and promote cardiac regeneration (Fig. 37.3). Reports in mice, rats, dogs, and humans indicate that there is one

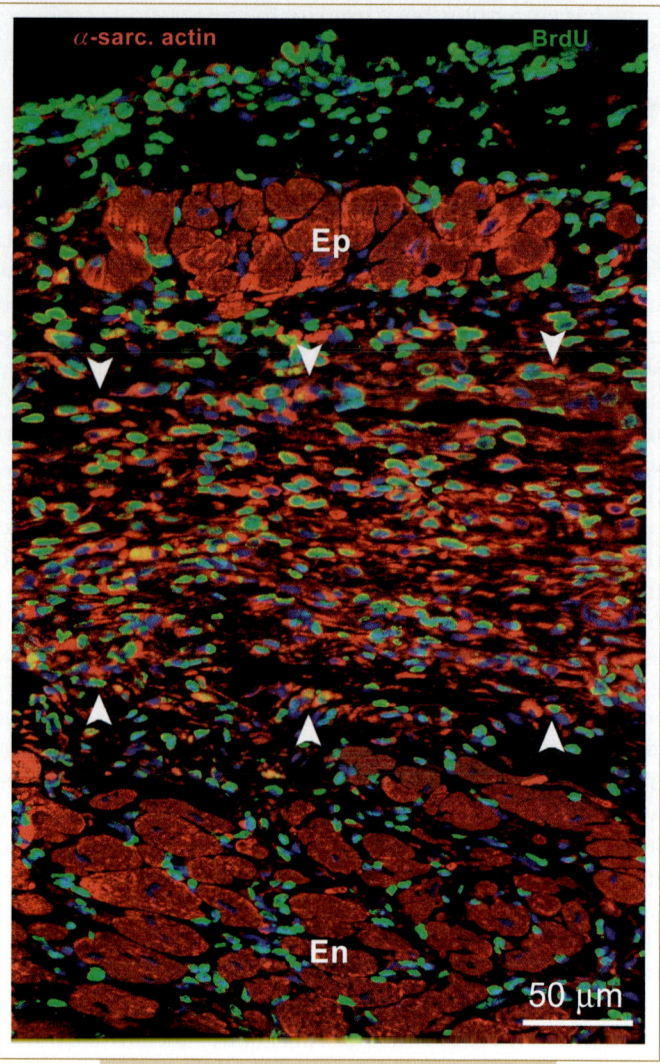

**FIG. 37.3.** Myocardial repair. Section of an infarcted rat myocardium at one month after coronary artery occlusion and implantation of clonogenic c-kit–positive cardiac stem cells. A band of regenerating myocardium is indicated by arrowheads. Newly formed myocytes are recognized by the red fluorescence of α-sarcomeric actin and green fluorescence of BrdU, which was administered after the injection of stem cells.

Lin⁻c-kit^POS stem cell per ~8,000–20,000 myocytes, or ~32,000–80,000 cardiac cells (Anversa *et al.*, 2006). This frequency is similar to that reported for HSCs in the bone marrow. However, some investigators have reported that primitive cells represent 2% of all cells, i.e., 2 per ~25 myocytes, or ~100 cardiac cells (Oh *et al.*, 2003). This high value may include some endothelial progenitors, which are lineage-committed Sca-1^POS cells.

The side population (SP) cells were first identified in the bone marrow (Goodell *et al.*, 1996). SP cells are characterized by their intrinsic capacity to efflux Hoechst dye through an ATP-binding cassette transporter. The bone marrow SP can be further subdivided in the upper side fraction that contains Lin⁻CD34^POS cells and the lower side fraction that is composed of KSL cells which are Lin⁻c-kit^POS-Sca-1^POS (Robinson *et al.*, 2005). SP cells have been shown to exist in multiple tissues and are capable of tissue-specific differentiation. A cardiac SP cell population has been found in the adult and developing heart. These putative CSCs are 93% Sca-1^POS and represent a subset, 1 per ~10,000 cells, of the 2% Sca-1–positive cells in the mouse heart (Pfister *et al.*, 2005). According to a different study, however, SP cells comprise 2% of all cells (2 per ~100 cardiac cells), express CD31, and form hematopoietic colonies *in vitro* (Martin *et al.*, 2004). The presence of CD31 and the peculiar growth behavior of these cells *in vitro* suggest that these cells could originate from the bone marrow and may include endothelial progenitors, raising doubts about the existence of a resident SP in the heart.

Recently, however, attention has been focused on a subset of the SP cells. This cell fraction retains the ability to exclude the Hoechst 33342 dye and express Sca-1 but is negative for CD31 (Pfister *et al.*, 2005). The modest expression of c-kit in these cells was attributed to methodological limitations inherent in the enzymatic cleavage of this receptor during digestion of the myocardium and cell isolation. The frequency of these cells in the myocardium is consistent with that found in other adult self-renewing organs. In fact, there is one SP cell per ~30,000 cardiac cells in the mouse heart. Importantly, CD31⁻ cardiac SP cells acquire the molecular and functional characteristics of adult myocytes. When cocultured with differentiated cardiomyocytes, SP-derived cells express cardiomyocyte-specific transcription factors and contractile proteins and, upon stimulation, exhibit calcium transients and contractile activity (Pfister *et al.*, 2005).

A distinct subset of the cardiac SP was found to form clonal spheroids in serum-free medium. In addition to the multidrug resistance transporter gene, these cells expressed nestin and Musashi-1, which are markers of undifferentiated neural precursor cells (Tomita *et al.*, 2005). After transplantation into chick embryos, nestin^POS-Musashi^POS cells undergo differentiation into cardiomyocytes. Importantly, undifferentiated and differentiated neural crest–derived cells are present in the fetal myocardium. In the adult heart, these cells have the features of stem cells: They are quiescent and multipotent, giving rise to myocytes and typical neural crest–derived cells, including neurons, glia, and smooth muscle.

Surprisingly, Isl1-positive cells have been alleged to correspond to the "true" cardiac stem cell, with important implications in myocardial regeneration of the failing human heart (Laugwitz *et al.*, 2005). Isl1-positive cells are neither self-renewing nor clonogenic or multipotent, lacking the indispensable elements of stem cells. Isl1-positive cells are undetectable in the atrial and ventricular myocardium of the adult mouse and rat heart or in the decompensated human heart. Even during development, Isl1 cells are not implicated in the formation of the left ventricle (Cai *et al.*,

2003), whose deterioration in the adult is the predominant cause of heart failure and death in the patient population. Whether direct activation, local implantation, or systemic mobilization of Isl1-positive cells can repair the infarcted heart remains to be shown. The lack of this information questions the therapeutic import, if any, of this immature myocyte.

## V. THE EMBRYO AS A MODEL OF PROGENITOR CELL ENGRAFTMENT AND PLASTICITY

Tissue injury is an essential aspect of successful cell engraftment, activation, and differentiation (Lapidot *et al.*, 2005). Therefore, the vast majority of the transplantation assays aiming at the documentation of the *in vivo* properties of progenitor cell populations has been performed in pathological models mimicking human cardiac diseases. However, the possibility has been raised that the capacity of stem cells to differentiate or transdifferentiate into parenchymal cells may be a nonphysiological artificial process. If this were the case, the behavior of stem cells would be entirely dictated by the unusual cytokine profile and extracellular matrix alterations present at the site of injury and would not reflect inherent features of the primitive cells. When stem cells fail to engraft and differentiate in intact organs and the results obtained in injured organs cannot be reproduced under physiological conditions, the artful nature of stem cell–mediated tissue repair seems to be confirmed. In these studies, the fact that the administered cells are at a growth disadvantage with respect to resident progenitors was not taken into account. Thus, it should not have been surprising that the implanted cells did not contribute significantly to physiological turnover and tissue homeostasis.

The embryo offers the unique opportunity to assess the behavior of progenitor cells in noninjured tissues and organs. Because of the multiple environmental cues that greatly favor cell growth and proliferation in embryos, the long-term engraftment of primitive cells is facilitated. Although the rapid pace at which cells renew themselves prenatally is not identical to the slow cell turnover in the postnatal heart, brain, or skeletal muscle, the intrinsic properties of progenitor cells may be better characterized when tissue damage is absent, such as during embryogenesis. Upon homing in embryonic and fetal tissues, adult progenitor cells can undergo growth and differentiation and actively participate in organ formation. Organogenesis involves the specification of primitive cell populations and the acquisition of specific fates. This process engages progenitor cells that reside at the embryonic site where the organ develops as well as migrating neighboring or circulating undifferentiated cells. The administered adult stem cells may colonize all organs or home preferentially to the organ from which they have been obtained or to a group of organs of the germ layer to which they belong. By this approach, the capacity of progenitor cells to form a progeny identical to or different from the organ of origin can be established. The generation of cells of unexpected lineage corresponds to transdifferentiation or stem cell plasticity, and, if a class of progenitor cells exhibits developmental plasticity in prenatal life, the possibility exists that this growth-and-differentiation characteristic is maintained in the adult.

An additional critical question is whether cardiac repair after injury recapitulates the developmental steps that occur in the embryonic heart. The unraveling of the molecular signaling pathways regulating the complex steps of embryonic heart formation is fundamental, in view of the option to manipulate progenitor cells genetically prior to their transplantation within the damaged myocardium. This information may help our understanding of the behavior of resident progenitor cells in response to pathologic states and their ability to initiate a repair process. The reactivation of the fetal gene program has been documented in the hypertrophic and failing heart. An extensively studied phenomenon is the switch of the myosin heavy-chain gene from the α-isoform, which is typical of the adult heart, to the β-isoform, which is highly expressed in the embryonic-fetal and neonatal heart (Tardiff *et al.*, 2000). Because the heart was viewed as a postmitotic organ, the possibility that activation of resident progenitor cells might lead to enhanced formation of amplifying myocytes has been ignored. This process may involve the identical spatial and temporal expression of the genes implicated in myocyte formation during embryonic maturation of the developing myocardium (Beltrami *et al.*, 2003). Thus, the expression of β-myosin may be part of the progressive acquisition of the adult phenotype by stem cell–derived myocytes.

Although primitive cells are present in the heart early in embryonic life and regulate heart morphogenesis and postnatal development, the properties of these primitive cells are still controversial. The hypothesis has been advanced that multiple unipotent progenitors give rise to distinct classes of myocytes that populate different regions of the heart: atria, outflow tract, and left and right ventricles (Mikawa and Fishman, 1996). Recently, the likelihood that a common precursor segregates into two lineages of myocytes has been raised (Meilhac *et al.*, 2004a). According to this theory, the primitive left ventricle and the outflow tract are derived from a single cell lineage, while all other regions of the heart, including the primitive atria and the right ventricle, are colonized by both lineages. This view is supported by the differential expression of the transcription factors Nkx2.5, Tbx5, Isl1, eHAND, and dHAND in distinct cardiac regions (Pandur, 2005).

The early common origin of cardiomyocytes has been recognized by utilizing a novel approach that involved the retrospective clonal analysis of the cells composing the mouse embryonic heart. Clones of developing myocytes derived from multiple precursors appear on different days of embryonic development and in distinct morphogenetic

regions but show a uniform genetic background. This characteristic documents their derivation from a shared pool of founder cells. These precursors account for ~140 cells at E8.5 and give rise to the whole heart tube in its entire length (Meilhac *et al.*, 2003, 2004a, 2004b). The initiation of separate transcriptional programs in the cells of the cardiac chambers is not the consequence of a diverse clonal origin from unipotent progenitors but a process that occurs later in development. This pattern of myocardial histogenesis has been confirmed by introducing the EGFP gene in the embryo at the very early stage of the morula–blastocyst transition. EGFP was placed under the control of the desmin or α-cardiac actin promoter, leading to the accumulation of EGFP in all cells that express these myocyte-specific proteins.

A similar strategy was employed to study the pattern of clonal myocyte distribution during heart development and to document the existence of an individual intracardiac common progenitor of cardiomyocytes in prenatal and postnatal life (Ebherard and Jockusch, 2005). Two eight-cell morulas were combined, giving rise to a chimeric morula that evolves into a chimeric blastocyst. Since the two morulas carried either the LacZ or EGFP transgene under the control of the desmin promoter, the different genetic origin of the developing myocytes could be studied by labeling techniques. Early chimeric blastocysts were implanted in pseudo-pregnant mice. This approach, termed the *aggregation method*, allows the determination of the number of progenitor cells that contributes to the development of the heart or other organs. The distribution of the progeny of the precursors may result in balanced or unbalanced mouse chimeras, which correspond to equal or unequal parental contribution, respectively.

Tissue sections obtained from the embryonic and adult heart of these chimeric mice were analyzed, and patchy areas of labeling were quantified. The presence of "holes," or nonlabeled areas, in the patches was also determined. The distribution of labeled areas did not change as a function of age, and it was dictated by a mechanism of coherent expansion of clones. In this case, patchy areas of the heart are β-gal-positive and others EGFP positive. However, limited intermingling of clonogenic cells occurs, and small foci of β-gal-positive cells are found in EGFP clonal patches, or vice versa. The possibility that this model of cardiac growth is mediated by a large number of progenitors would require a massive loss of clones early in development, a phenomenon that cannot be excluded. If thousands of progenitors were implicated in embryonic cardiac growth, a mechanism of coherent expansion together with extensive intermingling of clonogenic cells would lead to a dispersed pattern of cardiomyocyte formation. Importantly, the persistence of the embryonic clonal patches with small, intermingled foci in the adult myocardium excludes that new clonal regions contribute to the growth of the heart postnatally (Ebherard and Jockusch, 2005).

The view of the developing heart as an organ originated exclusively from the expansion of the cardiomyocyte precursors of the two paired mesodermal fields is a matter of debate (Eisenberg and Eisenberg, 2004). Evidence collected recently indicates that only a portion of the contractile muscle in the mature heart descends from this primordial myocardial tissue. Cells that reside in neighboring nonmuscle tissue or migrate to the myocardium from distant sites through the circulation participate in the growth of the cardiac muscle (Eisenberg et al., 2003). Whether this multiplicity of events is recapitulated in the postnatal heart is unknown, but it suggests the possibility that exogenous progenitor cells may maintain a certain degree of plasticity in adult life.

The integration of blood-borne cells in the developing heart has been documented in a model of parabiosis between quail and chicken (Sedmera et al., 2006). The formation of a common circulatory system occurs at embryonic day 15, and cells of quail origin are found in multiple tissues of the chick embryo. Circulating cells differentiate into vascular ECs and SMCs and cardiomyocytes. This process is the result of transdifferentiation; fusion of quail and chick cells is an extremely rare event. Similar results are obtained when adult bone marrow cells are coincubated with embryonic heart. In a manner comparable to the HSCs, the ability of human MSCs to contribute to organ growth and turnover has been investigated. The model of the fetal sheep has been employed because this animal is immunologically tolerant before 75 days of gestation and chimeric human–sheep organs can be formed without rejection and the need of immunosuppressive therapy. Moreover, the large difference between human and sheep antigens allows the unequivocal identification of the origin of cells and tissue structures. After transplantation in the fetal peritoneal cavity, MSCs traverse the endothelial barrier and engraft in nearly all tissues of the organism (Liechty et al., 2000). MSCs gave rise to multiple parenchymal cell types belonging to the ectodermal and mesodermal germ layers. Human cells have been found in the pulmonary epithelia, thymic epithelia, skin wounds, and perivascular regions of the central nervous system. Skeletal and cardiac muscle were among the nonlymphohemopoietic and mesodermal organs that were colonized by human MSCs. In the heart, cells positive for the Alu probe and SERCA-2 protein were identified, documenting the presence of human myocytes in the sheep heart. Cardiomyocytes derived predominantly from transdifferentiation of human MSCs, although scattered myocytes with human nucleus and sheep cytoplasm were seen. MSCs have long-term engraftment potential and maintain multipotentiality upon transplantation, representing good candidates for cellular and gene therapy applications.

## VI. CONCLUSIONS

This chapter has analyzed the different progenitor cell classes that are capable of differentiating into cardiomyocytes and coronary vessels in the adult heart. The recognition that HSCs and MSCs possess the ability to form myocardial tissue has significant clinical implications and supports the notion of stem cell plasticity in the adult organs and organism. The recent identification of CSCs, which are self-renewing, clonogenic, and multipotent *in vitro* and regenerate infarcted myocardium *in vivo* extends the possibility of implementing HSCs, MSCs, and CSCs in the treatment of cardiac failure of ischemic and nonischemic origin. However, the search for the most appropriate progenitor cell for cardiac repair continues.

## VII. REFERENCES

Anversa, P., and Kajstura, J. (1998). Ventricular myocytes are not terminally differentiated in the adult mammalian heart. *Circ. Res.* **83**, 1–14.

Anversa, P., and Olivetti, G. (2002). Cellular basis of physiological and pathological myocardial growth. *In* "Handbook of Physiology: the Cardiovascular System. The Heart" (Fozzard HA, Solaro RJ, eds.), pp. 75–144. New York, NY: Oxford University Press.

Anversa, P., Kajstura, J., Leri, A., and Bolli, R. (2006). Life and death of cardiac stem cells: a paradigm shift in cardiac biology. *Circulation.* **113**, 1451–1463.

Beltrami, C. A., Finato, N., Rocco, M., Feruglio, G. A., Puricelli, C., Cigola, E., Quaini, F., Sonnenblick, E. H., Olivetti, G., and Anversa, P. (1994). Structural basis of end-stage failure in ischemic cardiomyopathy in humans. *Circulation.* **89**, 151–163.

Beltrami, A. P., Urbanek, K., Kajstura, J., Yan, S. M., Finato, N., Bussani, R., Nadal-Ginard, B., Silvestri, F., Leri, A., Beltrami, C. A., and Anversa, P. (2001). Evidence that human cardiac myocytes divide after myocardial infarction. *N. Engl. J. Med.* **344**, 1750–1757.

Beltrami, A. P., Barlucchi, L., Torella, D., Baker, M., Limana, F., Chimenti, S., Kasahara, H., Rota, M., Musso, E., Urbanek, K., Leri, A., Kajstura, J.,

Nadal-Ginard, B., and Anversa, P. (2003). Adult cardiac stem cells are multipotent and support myocardial regeneration. *Cell.* **114**, 763–776.

Bhatia, M. (2001). AC133 expression in human stem cells. *Leukemia.* **15**, 1685–1688.

Bianchi, G., Muraglia, A., Daga, A., Corte, G., Cancedda, R., and Quarto, R. (2001). Microenvironment and stem properties of bone marrow-derived mesenchymal cells. *Wound Repair Regen.* **9**, 460–466.

Blau, H. M., Brazelton, T. R., and Weimann, J. M. (2001). The evolving concept of a stem cell: entity or function? *Cell.* **105**, 829–841.

Britten, M. B., Abolmaali, N. D., Assmus, B., Lehmann, R., Honold, J., Schmitt, J., Vogl, T. J., Martin, H., Schachinger, V., Dimmeler, S., and Zeiher, A. M. (2003). Infarct remodeling after intracoronary progenitor cell treatment in patients with acute myocardial infarction (TOPCARE-AMI): mechanistic insights from serial contrast-enhanced magnetic resonance imaging. *Circulation.* **108**, 2212–2218.

Cai, C. L., Liang, X., Shi, Y., Chu, P. H., Pfaff, S. L., Chen, J., and Evans, S. (2003). Isl1 identifies a cardiac progenitor population that proliferates prior to differentiation and contributes a majority of cells to the heart. *Dev. Cell.* **5**, 877–889.

Cerny, J., and Quesenberry, P. J. (2004). Chromatin remodeling and stem cell theory of relativity. *J. Cell. Physiol.* **201**, 1–16.

Chien, K. R. (2004). Stem cells: lost in translation. *Nature.* **428**, 607–608.

Dennis, J. E., and Charbord, P. (2002). Origin and differentiation of human and murine stroma. *Stem Cells.* **20**, 205–214.

Doseff, A. I. (2004). Apoptosis: the sculptor of development. *Stem Cells Dev.* **13**, 473–483.

Eberhard, D., and Jockusch, H. (2005). Patterns of myocardial histogenesis as revealed by mouse chimeras. *Dev. Biol.* **278**, 336–346.

Eisenberg, L. M., and Eisenberg, C. A. (2004). Adult stem cells and their cardiac potential. *Anat. Rec. A. Discov. Mol. Cell. Evol. Biol.* **276**, 103–112.

Etzion, S., Battler, A., Barbash, I. M., Cagnano, E., Zarin, P., Granot, Y., Kedes, L. H., Kloner, and R. A., and Leor, J. (2001) Influence of embryonic cardiomyocyte transplantation on the progression of heart failure in a rat model of extensive myocardial infarction. *J. Mol. Cell. Cardiol.* **33**, 1321–1330.

Ferguson, M. W., and O'Kane, S. (2004). Scar-free healing: from embryonic mechanisms to adult therapeutic intervention. *Philos. Trans. R. Soc. Lond. B. Biol. Sci.* **359**, 839–850.

Friedenstein, A. J. (1961). Osteogenetic activity of transplanted transitional epithelium. *Acta Anat. (Basel).* **45**, 31–59.

Gnecchi, M., He, H., Liang, O. D., Melo, L. G., Morello, F., Mu, H., Noiseux, N., Zhang, L., Pratt, R. E., Ingwall, J. S., and Dzau, V. J. (2005). Paracrine action accounts for marked protection of ischemic heart by Akt-modified mesenchymal stem cells. *Nat. Med.* **11**, 367–368.

Gomer, R. H. (2001). Not being the wrong size. *Nat. Rev. Mol. Cell. Biol.* **2**, 48–54.

Goodell, M. A., Brose, K., Paradis, G., Conner, A. S., and Mulligan, R.C. (1996). Isolation and functional properties of murine hematopoietic stem cells that are replicating in vivo. *J. Exp. Med.* **183**, 1797–1806.

Gronthos, S., Graves, S. E., Ohta, S., and Simmons, P. J. (1994). The STRO-1+ fraction of adult human bone marrow contains the osteogenic precursors. *Blood.* **84**, 4164–4173.

Hattan, N., Kawaguchi, H., Ando, K., Kuwabara, E., Fujita, J., Murata, M., Suematsu, M., Mori, H., and Fukuda. (2005). Purified cardiomyocytes from bone marrow mesenchymal stem cells produce stable intracardiac grafts in mice. *Cardiovasc. Res.* **65**, 334–344.

Haynesworth, S. E., Baber, M. A., and Caplan, A. I. (1992). Cell surface antigens on human marrow-derived mesenchymal cells are detected by monoclonal antibodies. *Bone.* **13**, 69–80.

Jones, N. C., and Trainor, P. A. (2005). Role of morphogens in neural crest cell determination. *J. Neurobiol.* **64**, 388–404.

Kawada, H., Fujita, J., Kinjo, K., Matsuzaki, Y., Tsuma, M., Miyatake, H., Muguruma, Y., Tsuboi, K., Itabashi, Y., Ikeda, Y., Ogawa, S., Okano, H., Hotta, T., Ando, K., and Fukuda, K. (2004). Nonhematopoietic mesenchymal stem cells can be mobilized and differentiate into cardiomyocytes after myocardial infarction. *Blood.* **104**, 3581–3587.

Kiel, M. J., Yilmaz, O. H., Iwashita, T., Terhorst, C., and Morrison, S. J. (2005). SLAM family receptors distinguished hematopoietic stem and progenitor cells and reveal endothelial niches for stem cells. *Cell.* **121**, 1109–1121.

Kim, I., He, S., Yilmaz, O. H., Kiel, M. J., and Morrison, S. J. (2006). Enhanced purification of fetal liver hematopoietic stem cells using SLAM family receptors. *Blood.* Mar 28, Epub ahead of print.

Kimmel, A. R., and Firtel, R. A. (2004). Beaking symmetrics: regulation of Dictyostlium development through chemoattractant and morphogen signal-response. *Curr. Opin. Genet. Dev.* **14**, 540–549.

Kusano, K. F., Pola, R., Murayama, T., Curry, C., Kawamoto, A., Iwakura, A., Shintani, S., Ii, M., Asai, J., Tkebuchava, T., et al. (2005). Sonic hedgehog myocardial gene therapy: tissue repair through transient reconstitution of embryonic signaling. *Nat. Med.* **11**, 1197–1204.

Lapidot, T., Dar, A., and Kollett, O. (2005). How do stem cells find their way home? *Blood.* **106**, 1901–1910.

Laugwitz, K. L., Moretti, A., Lam, J., Gruber, P., Chen, Y., Woodard, S., Lin, L. Z., Cai, C. L., Lu, M. M., Reth, M., Platoshyn, O., Yuan, J. X., Evans, S., and Chien, K. R. (2005). Postnatal isl1+ cardioblasts enter fully differentiated cardiomyocyte lineages. *Nature.* **433**, 647–653.

Le Douarin, N. M., Creuzet, S., Couly, G., and Dupin, E. (2004). Neural crest cell plasticity and its limits. *Development.* **131**, 4637–4650.

Leong, F. T., and Freeman, L. J. (2005). Acute renal infarction. *J. R. Soc. Med.* **98**, 121–122.

Leor, J., Patterson, M., Quinones, M. J., Kedes, L. H., and Kloner, R. A. (1996). Transplantation of fetal myocardial tissue into the infarcted myocardium of rat. A potential method for repair of infarcted myocardium? *Circulation.* **94**, 332–336.

Leri, A., Kajstura, J., and Anversa, P. (2005). Cardiac stem cells and mechanisms of myocardial regeneration. *Physiol. Rev.* **85**, 1373–1416.

Li, R. K., Jia, Z. Q., Weisel, R. D., Merante, F., and Mickle, D. A. (1999). Smooth muscle cell transplantation into myocardial scar tissue improves heart function. *J. Mol. Cell. Cardiol.* **31**, 53–522.

Liechty, K. W., MacKenzie, T. C., Shaaban, A. F., Radu, A., Moseley, A. M., Deans, R., Marshak, D. R., and Flake, A. W. (2000). Human mesenchymal stem cells engraft and demonstrate site-specific differentiation after in utero transplantation in sheep. *Nat. Med.* **6**, 1282–1286.

Linke, A., Muller, P., Nurzynska, D., Casarsa, C., Torella, D., Nascimbene, A., Castaldo, C., Cascapera, S., Bohm, M., Quaini, F., Urbanek, K., Leri, A., Hintze, T. H., Kajstura, J., and Anversa, P. (2005). Stem cells in the dog heart are self-renewing, clonogenic, and multipotent and regenerate infarcted myocardium, improving cardiac function. *Proc. Natl. Acad. Sci. U S A.* **102**, 8966–8971.

MacLellan, W. R., and Schneider, M. D. (2000).Genetic dissection of cardiac growth control pathways. *Annu. Rev. Physiol.* **62**, 289–319.

Makino, S., Fukuda, K., Miyoshi, S., Konishi, F., Kodama, H., Pan, J., Sano, M., Takahashi, T., Hori, S., Abe, H., Hata, J., Umezawa, A., and Ogawa, S. (1999). Cardiomyocytes can be generated from marrow stromal cells *in vitro. J. Clin. Invest.* **103**, 697–705.

Martin, C. M., Meeson, A. P., Robertson, S. M., Hawke, T. J., Richardson, J. A., Bates, S., Goetsch, S. C., Gallardo, T. D., and Garry, D. J. (2004). Persistent expression of the ATP-binding cassette transporter, Abcg2, identifies cardiac SP cells in the developing and adult heart. *Dev. Biol.* **265**, 262–275.

Matsuura, K., Nagai, T., Nishigaki, N., Oyama, T., Nishi, J., Wada, H., Sano, M., Toko, H., Akazawa, H., Sato, H., et al. (2004). Adult cardiac Sca-1-positive cells differentiate into beating cardiomyocytes. *J. Biol. Chem.* **279**, 11384–11391.

Meilhac, S. M., Kelly, R. G., Rocancourt, D., Eloy-Trinquet, S., Nicolas, J. F., and Buckingham, M. E. (2003). A retrospective clonal analysis of the myocardium reveals two phases of clonal growth in the developing mouse heart. *Development.* **130**, 3877–3889.

Meilhac, S. M., Esner, M., Kelly, R. G., Nicolas, J. F., and Buckingham, M. E. (2004a). The clonal origin of myocardial cells in different regions of the embryonic mouse heart. *Dev. Cell.* **6**, 685–698.

Meilhac, S. M., Esner, M., Kerszberg, M., Moss, J. E., and Buckingham, M. E. (2004b). Oriented clonal cell growth in the developing mouse myocardium underlies cardiac morphogenesis. *J. Cell Biol.* **164**, 97–109.

Menasche, P., Hagege, A.A., Scorsin, M., Pouzet, B., Desnos, M., Duboc, D., Schwartz, K., Vilquin, J-T., and Marolleau, J-P. (2001). Myoblast transplantation for heart failure. *Lancet.* **347**, 279–280.

Messina, E., De Angelis, L., Frati, G., Morrone, S., Chimenti, S., Fiordaliso, F., Salio, M., Battaglia, M., Latronico, M. V., Coletta, M., Vivarelli, E., Frati, L., Cossu, G., and Giacomello, A. (2004). Isolation and expansion of adult cardiac stem cells from human and murine heart. *Circ. Res.* **95**, 911–921.

Mikawa, T., and Fischman, D. A. (1996). The polyclonal origin of myocyte lineages. *Annu. Rev. Physiol.* **58**, 509–521.

Morrison, S. J., and Weissman, I. L. (1994). The long-term repopulating subset of hematopoietic stem cells is deterministic and isolatable by phenotype. *Immunity.* **1**, 661–673.

Morrison, S. J., Uchida, N., and Weissman, I. L. (1995). The biology of hematopoietic stem cells. *Annu. Rev. Cell. Dev. Biol.* **11**, 35–71.

Mutsaers, S. E., Bishop, J. E., McGrouther, G., and Laurent, G. J. (1997). Mechanisms of tissue repair: from wound healing to fibrosis. *Int. J. Biochem. Cell Biol.* **29**, 5–17.

Negrin, R. S., Atkinson, K., Leemhuis, T., Hanania, E., Juttner, C., Tierney, K., Hu, W. W., Johnston, L. J., Shizurn, J. A., Stockerl-Goldstein, K. E., Blume, K. G., Weissman, I. L., Bower, S., Baynes, R., Dansey, R., Karanes, C., Peters, W., and Klein, J. (2000). Transplantation of highly purified CD34+Thy-1+ hematopoietic stem cells in patients with metastatic breast cancer. *Biol. Blood Marrow Transplant.* **6**, 262–271.

Oh, H., Bradfute, S. B., Gallardo, T. D., Nakamura, T., Gaussin, V., Mishina, Y., Pocius, J., Michael, L. H., Behringer, R. R., Garry, D. J., and Schneider, M. D. (2003). Cardiac progenitor cells from adult myocardium: homing, differentiation, and fusion after infarction. *Proc. Natl. Acad. Sci. USA.* **100**, 12313–12318.

Olivetti, G., Melissari, M., Balbi, T., Quaini, F., Sonnenblick, E. H., and Anversa, P. (1994). Myocyte nuclear and possible cellular hyperplasia contribute to ventricular remodeling in the hypertrophic senescent heart in humans. *J. Am. Coll. Cardiol.* **24**, 140–149.

Olivetti, G., Ricci, R., and Anversa, P. (1987). Hyperplasia of myocyte nuclei in long-term cardiac hypertrophy in rats. *J. Clin. Invest.* **80**, 1818–1821.

Olson, E. N., and Srivastava, D. (1996). Molecular pathways controlling heart development. *Science.* **272**, 671–676.

Orlic, D., Kajstura, J., Chimenti, S., Jakoniuk, I., Anderson, S. M., Li, B., Pickel, J., McKay, R., Nadal-Ginard, B., Bodine, D. M., Leri, A. and Anversa, P. (2001). Bone marrow cells regenerate infarcted myocardium. *Nature.* **410**, 701–705.

Pandur, P. (2005). What does it take to make a heart? *Biol. Cell.* **95**, 197–210.

Pereira, R. F., O'Hara, M. D., Laptev, A. V., Halford, K. W., Pollard, M. D., Class, R., Livezey, K., and Prockop, D. J. (1998). Marrow stromal cells as a source of progenitor cells for nonhematopoietic tissues in transgenic mice with a phenotype of osteogenesis imperfecta. *Proc. Natl. Acad. Sci. USA.* **95**, 1142–1147.

Pfeffer, M. A., and Braunwald, E. (1990). Ventricular remodeling after myocardial infarction. Experimental observations and clinical implications. *Circulation.* **81**, 1161–1172.

Pfister, O., Mouquet, F., Jain, M., Summer, R., Helmes, M., Fine, A., Colucci, W. S., and Liao, R. (2005). CD31– but not CD31+ cardiac side population cells exhibit functional cardiomyogenic differentiation. *Circ. Res.* **97**, 52–61.

Pittenger, M. F., and Martin, B. J. (2004). Mesenchymal stem cells and their potential as cardiac therapeutics. *Circ. Res.* **95**, 9–20.

Poulsom, R., Alison, M. R., Cook, T., Jeffery, R., Ryan, E., Forbes, S. J., Hunt, T., Wyles, S., and Wright, N. A. (2003). Bone marrow stem cells contribute to healing of the kidney. *J. Am. Soc. Nephrol.* **Suppl 1**, S48–54.

Quesenberry, P. J., Colvin, G. A., and Lambert, J. F. (2002). The chiaroscuro stem cell: a unified stem cell theory. *Blood.* **100**, 4266–4271.

Quesenberry, P. J., Abedi, M., Aliotta, J., Colvin, G., Demers, D., Dooner, M., Greer, D., Hebert, H., Menon, M. K., Pimentel, J., and Paggioli, D. (2004). Stem cell plasticity: an overview. *Blood Cells Mol. Dis.* **32**, 1–4.

Robinson, S. N., Seina, S. M., Gohr, J. C., Kuszynski, C. A., and Sharp, J. G. (2005). Evidence for a qualitative hierarchy within the Hoechst-33342 "side population" (SP) of murine bone marrow cells. *Bone Marrow Transplant.* **35**, 807–818.

Rosenblatt-Velin, N., Lepore, M. G., Cartoni, C., Beermann, F., and Pedrazzini, T. (2005). FGF-2 controls the differentiation of resident cardiac precursors into functional cardiomyocytes. *J. Clin. Invest.* **115**, 1724–1733.

Secombe, J., Pierce, S. B., and Eisenman, R. N. (2004). Myc: a weapon of mass destruction. *Cell.* **117**, 153–156.

Sedmera, D., and Watanabe, M. (2006). Growing the coronary tree: the quail saga. *Anat. Rec. A Discov. Mol. Cell. Evol. Biol.* **288**, 933–935.

Sherr, C. J., and Roberts, J. M. (2004). Living with or without cyclins and cyclin-dependent kinases. *Genes Dev.* **15**, 2699–2711.

Silver, J., and Miller, J. H. (2004). Regeneration beyond the glial scar. *Nat. Rev. Neurosci.* **5**, 146–156.

Smith, J. R., Pochampally, R., Perry, A., Hsu, S. C., and Prockop, D. J. (2004). Isolation of a highly clonogenic and multipotential subfraction of adult stem cells from bone marrow stroma. *Stem Cells.* **22**, 823–831.

Tabata, T., and Takei, Y. (2004). Morphogens, their identification and regulation. *Development.* **131**, 703–712.

Tardiff, J. C., Hewett, T. E., Factor, S. M., Vikstrom, K. L., Robbins, J., and Leinwand, L.A. (2000). Expression of the beta (slow)-isoform of MHC in the adult mouse heart causes dominant-negative functional effects. *Am. J. Physiol. Heart Physiol.* **278**, H412–419.

Till, J. E., McCulloch, E. A., and Siminovitch, L. (1964a). A stochastic model of stem cell proliferation, based on the growth of spleen colony-forming cells. *Proc. Natl. Acad. Sci. U S A.* **51**, 29–36.

Till, J. E., McCulloch, E. A., and Siminovitch, L. (1964b). Isolation of variant cell lines during serial transplantation of hematopoietic cells derived from fetal liver. *J. Natl. Cancer Inst.* **33**, 707–720.

Till, J. E., and McCulloch, E. A. (1964c). Repair processes in irradiated mouse hematopoietic tissue. *Ann. N. Y. Acad. Sci.* **114**, 115–125.

Tomita, Y., Matsumura, K., Wakamatsu, T., Matsuzaki, Y., Shibuya, I., Kawaguchi, H., Ieda, M., Kanakubo, S., Shimazaki, T., Ogawa, S., et al. (2005). Cardiac neural crest cells contribute to the dormant multipotent stem cell in the mammalian heart. *J. Cell. Biol.* **170**, 1135–1146.

Tropepe, V., Sibilia, M., Ciruna, B. G., Rossant, J., Wagner, E. F., and van der Kooy, D. (1999). Distinct neural stem cells proliferate in response to EGF and FGF in the developing mouse telencephalon. *Dev. Biol.* **208**, 166–188.

Uchida, N., and Weissman, I. L. (1992). Searching for hematopoietic stem cells: evidence that Thy-1.1lo Lin- Sca-1+ cells are the only stem cells in C57BL/Ka-Thy-1.1 bone marrow. *J. Exp. Med.* **175**, 175–184.

Urbanek, K., Quaini, F., Tasca, G., Torella, D., Castaldo, C., Nadal-Ginard, B., Leri, A., Kajstura, J., Quaini, E., and Anversa, P. (2003). Intense myocyte formation from cardiac stem cells in human cardiac hypertrophy. *Proc. Natl. Acad. Sci. USA.* **100**, 10440–10445.

Urbanek, K., Rota, M., Cascapera, S., Bearzi, C., Nascimbene, A., De Angelis, A., Hosoda, T., Chimenti, S., Baker, M., Limana, F., Nurzynska, D., Torella, D., Rotatori, F., Rastaldo, R., Musso, E., Quaini, F., Leri, A., Kajstura, J., and Anversa, P. (2005). Cardiac stem cells possess growth factor-receptor systems that after activation regenerate the infarcted myocardium, improving ventricular function and long-term survival. *Circ. Res.* **97**, 663–673.

Urbanek, K., Cesselli, D., Rota, M., Nascimbene, A., Bearzi, C., Hosoda, T., Boni, A., Kajstura, J., Anversa, P., and Leri, Λ. (2006) Cardiac stem cell niches. *Proc. Natl. Acad. Sci. USA*, **103**, 9226–9231.

Watanabe, K., Abe, H., Mishima, T., Ogura, G., and Suzuki, T. (2003). Polyangitis overlap syndrome: a fatal case combined with adult Henoch-Schonlein purpura and polyarteritis nodosa. *Pathol. Inst.* **53**, 569–573.

Wei, J. Y. (1992). Age and the cardiovascular system. *N. Engl. J. Med.* **327**, 1735–1739.

Whitfield, M. L., George, L. K., Grant, G. D., and Perou, C.M. (2006). Common markers of proliferation. *Nat. Rev. Cancer.* **6**, 99–106.

Yilmaz, O. H., Kiel, M. J., and Morrison, S. J. (2006). SLAM family markers are conserved among hematopoietic stem cells from old and reconstituted mice and markedly increase their purity. *Blood.* **107**, 924–930.

Yoon, Y. S., Wecker, A., Heyd, L., Park, J. S., Tkebuchava, T., Kusano, K., Hanley, A., Scadova, H., Qin, G., Cha, D. H., et al. (2005). Clonally expanded novel multipotent stem cells from human bone marrow regenerate myocardium after myocardial infarction. *J. Clin. Invest.* **115**, 326–38.

Zhang, N., Mustin, D., Reardon, W., Almeida, A. D., Mozdziak, P., Mrug, M., Eisenberg, L. M., and Sedmera, D. (2006). Blood-borne stem cells differentiate into vascular and cardiac lineages during normal development. *Stem Cells Dev.* **15**, 17–28.

# Chapter Thirty-Eight

# Cardiac-Tissue Engineering

*M. Radisic, H. Park, and G. Vunjak-Novakovic*

## I. INTRODUCTION

Heart disease remains the leading cause of death in developed countries. Tissue engineering offers a potential to grow *in vitro* functional equivalents of native myocardium for use in tissue repair and to investigate new ways to treat or prevent the disease. In this chapter, we focus on a tissue-engineering approach to the generation of a functional cardiac patch. In the first part, we review the clinical problem and the requirements for the tissue engineering of a functional cardiac patch in a form suitable for surgical implantation. In the second part, we give an overview of the representative most significant research in cardiac tissue engineering:

1. The biomimetic approach through the integrated use of cells, scaffolds, and bioreactors.
2. The mechanical stimulation of cardiomyocytes embedded in the collagen gels.
3. The cell-sheet approach.

We then describe in more detail the biomimetic approach to tissue engineering, where the tissue-engineering approach is designed to mimic the main factors present in the native myocardium: high cell density with multiple cell types, convective-diffusive oxygen transport through a capillary network, and excitation–contraction coupling. Finally, we discuss some of the current challenges and research needs.

## II. CLINICAL PROBLEM

Cardiovascular disease is responsible for a preponderance of health problems in the developed countries as well as in many developing countries. Heart disease and stroke, the principal components of cardiovascular disease, are, respectively, the first and the third leading cause of death in the United States, accounting for nearly 40% of all deaths. Congenital heart defects, which occur in nearly 14 of every 1000 newborn children (Gillum, 1994), are the most common congenital defects and the leading cause of death in the first year of life. Cardiovascular diseases result in substantial disability and loss of productivity and contribute significantly to the escalating costs of health care. About 70 million Americans (almost one-fourth of the population) live with cardiovascular diseases, such as coronary heart disease, congenital cardiovascular defects, and congestive heart failure, and result in an economic burden of around $300 billion per year to treat these diseases (www.americanheart.org/presenter.jhtml?identifier=4478). The economic impact of cardiovascular disease is expected to grow further as the population ages.

Conventional therapies are limited by the inability of myocardium to regenerate after injury (Soonpaa and Field, 1998) and by the shortage of organs available for transplantation. For those reasons, cell-based therapies have been considered a novel and potentially curative treatment option (Reinlib and Field, 2000), either by utilization of cells alone (Koh *et al.*, 1993; Soonpaa *et al.*, 1994; Li *et al.*, 1996; Taylor *et al.*, 1998; Reinecke *et al.*, 1999; Sakai *et al.*, 1999; Condorelli *et al.*, 2001; Etzion *et al.*, 2001; Menasche *et al.*, 2001; Orlic *et al.*, 2001; Muller-Ehmsen *et al.*, 2002a, 2002b; Roell *et al.*, 2002; Laflamme and Murry, 2005) or by tissue engineering of cardiac grafts *in vitro* that can be surgically attached to the myocardium (Eschenhagen *et al.*, 1997; Akins *et al.*, 1999; Bursac *et al.*, 1999; Carrier *et al.*, 1999; Li *et al.*, 1999; Fink *et al.*, 2000; Leor *et al.*, 2000; Li *et al.*, 2000a; Papadaki *et al.*, 2001; Kofidis *et al.*, 2002; Shimizu *et al.*, 2002; Zimmermann *et al.*, 2002a, 2002b; Radisic *et al.*, 2003, 2004c). We believe that both treatment options are viable and that the appropriate choice will depend on time postinfarction.

Upon myocardial infarction, a vigorous inflammatory response is elicited, and dead cells are removed by marrow-derived macrophages. Over the subsequent weeks to months, fibroblasts and endothelial cells proliferate, forming granulation tissue that ultimately becomes a dense collagenous scar. Formation of scar tissue severely reduces contractile function of the myocardium and leads to ventricle wall thinning and dilatation, remodeling of the heart wall, and ultimately heart failure. The regeneration strategy thus depends on the time postinfarction; i.e., new and old infarcts most likely cannot be treated via the same approach.

Cell injection strategies will work best if applied shortly after myocardial infarction (MI). Application of cells and growth factors within hours and days after MI has a potential of directing the wound-repair process so that the minimum amount of scar tissue is formed, the contractile function is maintained in the border zone, and pathological remodeling is attenuated. Tissue-engineering strategies will work in the acute phase as well, but they may be more necessary after scar has formed. Then larger areas of heart must be replaced or augmented, and this is potentially where a scaffold-based approach may be most useful.

## III. PROBLEM DEFINITION

### The Myocardium

The myocardium is a highly differentiated tissue, 1.3–1.6 cm thick in humans, composed of cardiac myocytes and fibroblasts with a dense supporting vasculature and collagen-based extracellular matrix. Cardiac myocytes form a three-dimensional syncytium that enables propagation of electrical signals across specialized intracellular junctions to produce coordinated mechanical contractions that pump blood forward (Severs, 2000). Only 20–40% of the cells in the heart are cardiac myocytes, but they occupy 80–90% of the heart volume. The average cell density in the native rat myocardium is on the order of $5 \cdot 10^8$ cells/cm$^3$. Morphologically, intact cardiac myocytes have an elongated, rod-shaped appearance. The contractile apparatus of cardiac myocytes consists of sarcomeres arranged in parallel myofibrils. The

high metabolic activity of cardiac myocytes is supported by the high density of mitochondria. Electrical signal propagation is provided by specialized intercellular connections, gap junctions (Severs, 2000).

The control of heart contractions is almost entirely self-contained. Groups of specialized cardiac myocytes (pacemakers), the fastest of which are located in the sinoatrial node, drive periodic contractions of the heart. The majority of the cells in the myocardium are nonpacemaker cells, and they respond to the electrical stimuli generated by pacemaker cells. Excitation of each cardiac myocyte is followed by an increase in the amount of cytoplasmic calcium that triggers mechanical contraction. The propagation of the electrical excitation through the tissue by ion currents in the extracellular and in the intercellular space results in synchronous contractions that enable the pumping of blood from the heart.

### Requirements for a Functional Cardiac Patch

In order to provide a functional cardiac patch, tissue-engineering approaches must accurately mimic the structure of native myocardium over several different length scales. At the centimeter scale, tissue engineering should yield a mechanically stable construct of clinically relevant thickness, comparable to the thickness of the human myocardium (~1.5 cm) (Fig. 38.1A). This requirement is hampered by the diffusional limitations of the oxygen supply encountered in most tissue culture vessels, coupled with the high metabolic demand of cardiomyocytes for oxygen (27.6 nmol/mg protein/min; Yamada *et al.*, 1985).

At the millimeter scale, the tissue should consist of elongated myofibers aligned in parallel, with the angle changing along the thickness of the ventricle and capable of synchronous contractions, a requirement that is hampered by our ability to provide appropriate electromechanical stimulation during culture (Fig. 38.1B). At the micrometer scale, the tissue should consist of high-cell-density myocytes (~$10^8$ cells/cm$^3$) supported by a rich vasculature, with intercapillary distances of ~20 μm (Fig. 38.1C) (Rakusan and Korecky, 1982). At the nanometer scale, cells in the engineered cardiac tissue must be coupled by functional gap junctions and capable of propagating electrical impulses in order to prevent arrhythmia upon implantation. Finally, at the subcellular level, the excitation–contraction machinery of individual cardiomyocytes must be functional (Fig. 38.1D).

### Clinically Relevant Cell Sources

The major limitation in the progress of cardiac-tissue engineering toward clinical applications, via either the cardiac patch or drug testing, is the lack of an appropriate human cell source. Adult cardiac myocytes are terminally differentiated and have no ability to proliferate (Soonpaa and Field, 1998); thus they cannot be utilized as a source of autologous cells for tissue engineering. Cardiac myocytes

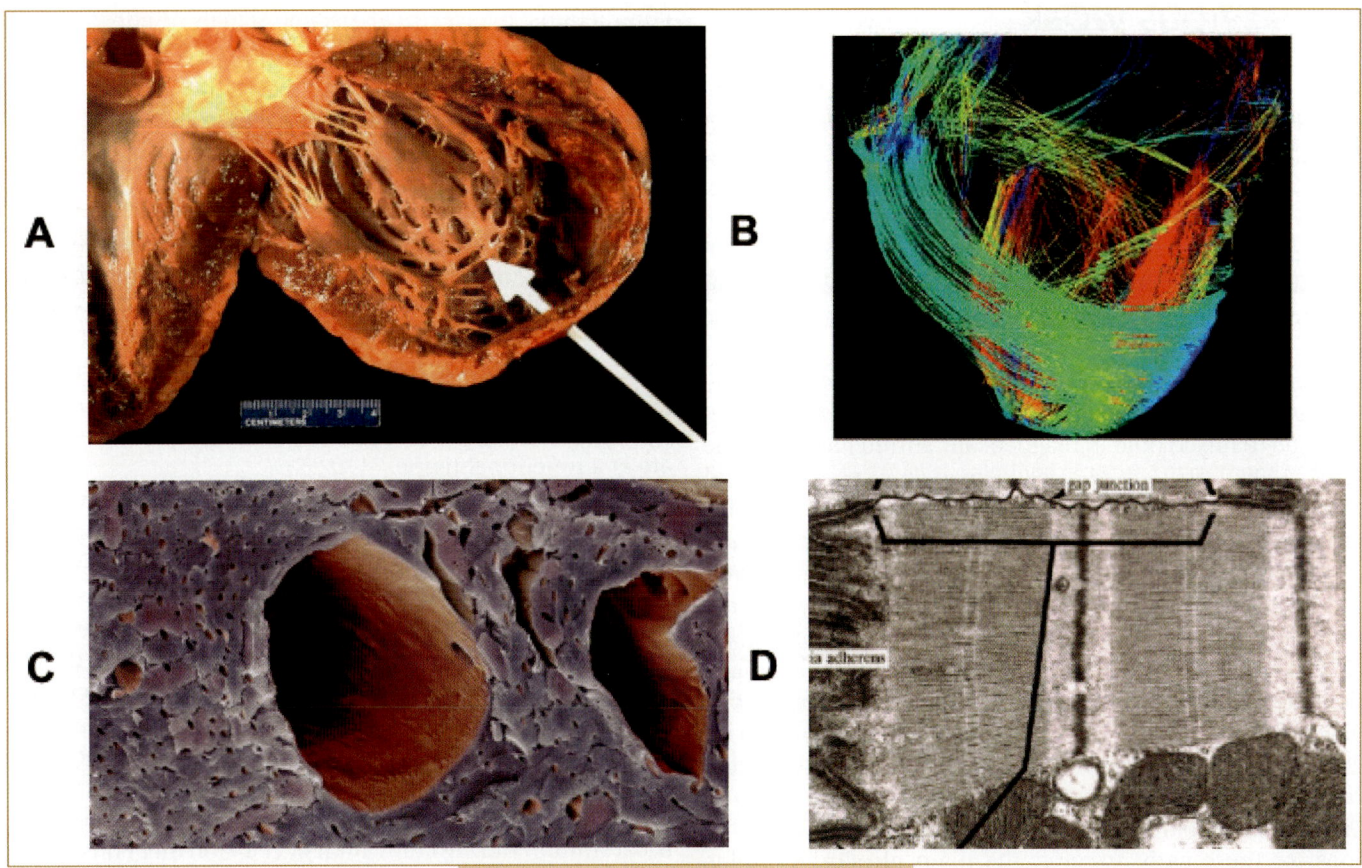

**FIG. 38.1.** Requirements for a functional cardiac patch. In order to provide appropriate function, the patch must satisfy structural requirements at different length scales: **(A)** thickness of 1.5 cm (Florida State University, The Internet Pathology Laboratory for Medical Education, http://www-medlib.med.utah.edu/WebPath/CVHTML/CV126.html); **(B)** appropriate myofiber orientation at the millimeter scale with permission from Helm *et al.* (2005); **(C)** high cell density and rich vasculature at the micrometer scale with permission from Zandonella (2003); and **(D)** functional excitation–contraction coupling pathway at the nanometer scale with permission from Severs (2000).

can be obtained in potentially unlimited quantities and at high purity from embryonic stem cells (Klug *et al.*, 1996; Zandstra *et al.*, 2003). However, nuclear transfer is required to make them autologous, and the presence of undifferentiated cells may lead to teratomas upon implantation (Laflamme and Murry, 2005). Recent studies report cultivation (Alperin *et al.*, 2005; Guo *et al.*, 2006) and implantation (Guo *et al.*, 2006) of cardiac grafts based on mouse embryonic stem cells.

Mesenchymal stem cells (MSC) from bone marrow (Makino *et al.*, 1999) or liposuction aspirates (Planat-Benard *et al.*, 2004) were reported to differentiate into cardiomyocytes *in vitro*. Transplantation of cell sheets based on adipose tissue–derived mesenchymal stem cells was demonstrated to improve contractile function in a rat model of coronary artery ligation (Miyahara *et al.*, 2006). It is thought that a small fraction of MSC differentiating into cardiomyocytes and newly formed blood vessels may contribute to the improvement in function.

Recent emerging work suggests that the heart may contain resident progenitor cells. This is an exciting possibility, for resident progenitor cells may be an ideal source of autologous cardiomyocytes. However, it appears that there is more than one heart cell subpopulation that fits the description of a cardiac progenitor. c-kit+ cells isolated from adult rat hearts and expanded under limited dilution gave rise to cardiomyocytes, smooth muscle, and endothelial cells when injected into ischemic myocardium (Beltrami *et al.*, 2003). Oh *et al.* (2003) reported Sca-1 as a marker of resident cardiac progenitors and expression of cardiac markers upon treatment with 5-azacytidine. LIM homeodomain islet 1 (isl1+) was also identified as a marker of resident cardiac progenitor cells (Laugwitz *et al.*, 2005). The isl1+ cells from mouse hearts were propagated in culture, and they differentiated into functional cardiac myocytes when in contact with terminally differentiated cardiomyocytes.

It remains to be seen if the progenitors, regardless of their marker, can be isolated from adult human biopsies and if sufficient numbers of cardiomyocytes (>10^8 cells/patient/patch) can be generated *in vitro*. Most of the current work in cardiac-tissue engineering is still performed with neonatal rat cardiomyocytes.

## IV. PREVIOUS WORK

Cardiac-tissue constructs that express structural and physiological features characteristic of native cardiac muscle have been engineered using rat cardiac myocytes cultured in collagen gels with mechanical stimulation (Eschenhagen *et al.*, 1997; Fink *et al.*, 2000; Zimmermann *et al.*, 2000, 2002b), on collagen fibers (Akins *et al.*, 1999), on synthetic fibrous scaffolds (Bursac *et al.*, 1999; Carrier *et al.*, 1999; Papadaki *et al.*, 2001; Carrier *et al.*, 2002a, 2002b), and on porous collagen scaffolds (Li *et al.*, 1999, 2000b; Radisic *et al.*, 2003). Culture systems included constructs cultured statically in well plates (Carrier *et al.*, 1999; Li *et al.*, 1999; Leor *et al.*, 2000; Papadaki *et al.*, 2001; Radisic *et al.*, 2003), exposed to mixed fluid in magnetically stirred flasks (Bursac *et al.*, 1999; Carrier *et al.*, 1999; Papadaki *et al.*, 2001), and freely suspended in convective flow in rotating bioreactor vessels (Akins *et al.*, 1999; Carrier *et al.*, 1999; Papadaki *et al.*, 2001). Transport of oxygen, carbon dioxide, nutrients, and metabolites to and from the cells was enhanced by convection at construct surfaces, but it was governed by molecular diffusion within the tissue constructs. At physiologic cell densities, oxygen transport by diffusion can support cell viability in only an ~100-μm-thick layer in contact with culture medium, but not in the construct interior, which remains largely acellular (Bursac *et al.*, 1999; Carrier *et al.*, 1999; Zimmermann *et al.*, 2000; Papadaki *et al.*, 2001). In most of these early systems, the main physical stimulus was the hydrodynamic shear associated with fluid flow that enhanced the rate of mass transport but — if excessive — could adversely affect the cells (Bursac *et al.*, 1999; Carrier *et al.*, 1999).

### Application of Cyclic Mechanical Stretch to Neonatal Rat Heart Cells in Collagen

Matrigel significantly improved the level of cell differentiation and the force of contraction, but some hallmarks of cardiac differentiation were still missing (including M bands and IC discs) (Zimmermann *et al.*, 2002b), and the distribution and frequency of gap junctions, critical for electrical signal propagation in the tissue, remained unclear. It appeared that the excitation–contraction coupling has not been fully established. More recent studies by the same group, discussed later in detail, have shown that multiple constructs engineered using this method and overlaid into a star-shaped structure, can be implanted on infarcted rat hearts, where it integrates with the host tissue and results in substantial improvement of the structural and functional properties of the grafted area (Zimmermann *et al.*, 2006).

Clinically sized ($6 \times 8 \times 2$ mm thick), compact cardiac constructs with physiologic cell densities were engineered *in vitro* by mimicking various aspects of the *in vivo* environment in native myocardium, including the oxygen supply by perfusion of culture medium supplemented with oxygen carriers (Radisic *et al.*, 2005) and the induction of construct contractions by electrical pacing signals (Radisic *et al.*, 2004a).

## V. BIOMIMETIC APPROACH TO CARDIAC-TISSUE ENGINEERING

We describe here in detail the "biomimetic" approach we have developed for cardiac-tissue engineering. The approach involves the *in vitro* creation of immature but functional tissues by an integrated use of: (1) *differentiated cells*, (2) *biomaterial scaffolds* that serve as a structural and logistic template for tissue development and that biodegrade at a controlled rate, and (3) *bioreactors* that provide environmental conditions necessary for the cells to regenerate at a functional tissue.

In this approach, a bioreactor should ideally provide all necessary conditions in the *in vitro* environment for rapid and orderly tissue development by cells cultured on a scaffold. In general, a bioreactor is designed to perform one or more of the following functions: (1) establish a desired spatially uniform cell concentration within the scaffold during cell seeding, (2) maintain controlled conditions in culture medium (e.g., temperature, pH, osmolality, levels of oxygen, nutrients, metabolites, regulatory molecules), (3) facilitate mass transfer, and (4) provide physiologically relevant physical signals (e.g., interstitial fluid flow, shear, pressure, electrical stimuli) during cultivation of cell–polymer constructs. We used the tissue-engineering model system to mimic the major factors present in the native myocardium. We focused on the following key parameters in the cell microenvironment: high cell density with multiple cell types, convective-diffusive oxygen transport through a capillary network, and orderly excitation contraction coupling (Table 38.1).

### Perfusion During Seeding Enabled Physiologic Cell Density

To provide an oxygen supply to the cells at levels necessary to maintain their viability, we developed a technique of seeding that involves (1) rapid cell inoculation into collagen sponges using Matrigel® as a cell delivery vehicle, and (2) transfer of inoculated scaffolds into perfused cartridges, with immediate establishment of the interstitial flow of the culture medium (Fig. 38.2A). Forward–reverse flow was used for the initial period of 1.5–4.5 h in order to further increase the rate and spatial uniformity of cell seeding (Radisic *et al.*, 2003). Unidirectional flow of culture medium was maintained for the duration of cultivation. In this system, cells were "locked" into the scaffold during a short (10 min) gelation period and supplied with oxygen at all times during culture.

Cell distributions in the top, center, and bottom areas of a 0.65-μm-wide strip extending from one construct surface to the other are shown in Fig. 38.2B. Constructs seeded in dishes had most cells located in the ~100-μm-thick layer at the top surface, and only a small number of

**Table 38.1.** Factors governing cardiac-tissue development *in vitro* and *in vivo*

| | *In vivo* | *In vitro* |
|---|---|---|
| **Cells** | High-density $5 \times 10^8$ cells/cm$^3$ | $0.5–1 \times 10^8$ cells/cm$^3$ |
| | Multiple cell types (myocytes, endothelial cells, fibroblasts) | Multiple cell types (myocytes, fibroblasts) |
| **Oxygen and nutrient supply** | Convection and diffusion | Convection and diffusion in perfusion bioreactor |
| **Geometry** | Capillary network (diameter 10 μm, spacing 20 μm) | Parallel channel array in the scaffold |
| **Oxygen carrier** | Hemoglobin | PFC emulsion (Oxygent®) |
| **Excitation–contraction coupling** | Electrical signal propagation | Electrical stimulation |
| | Ventricle contraction | Construct contraction |

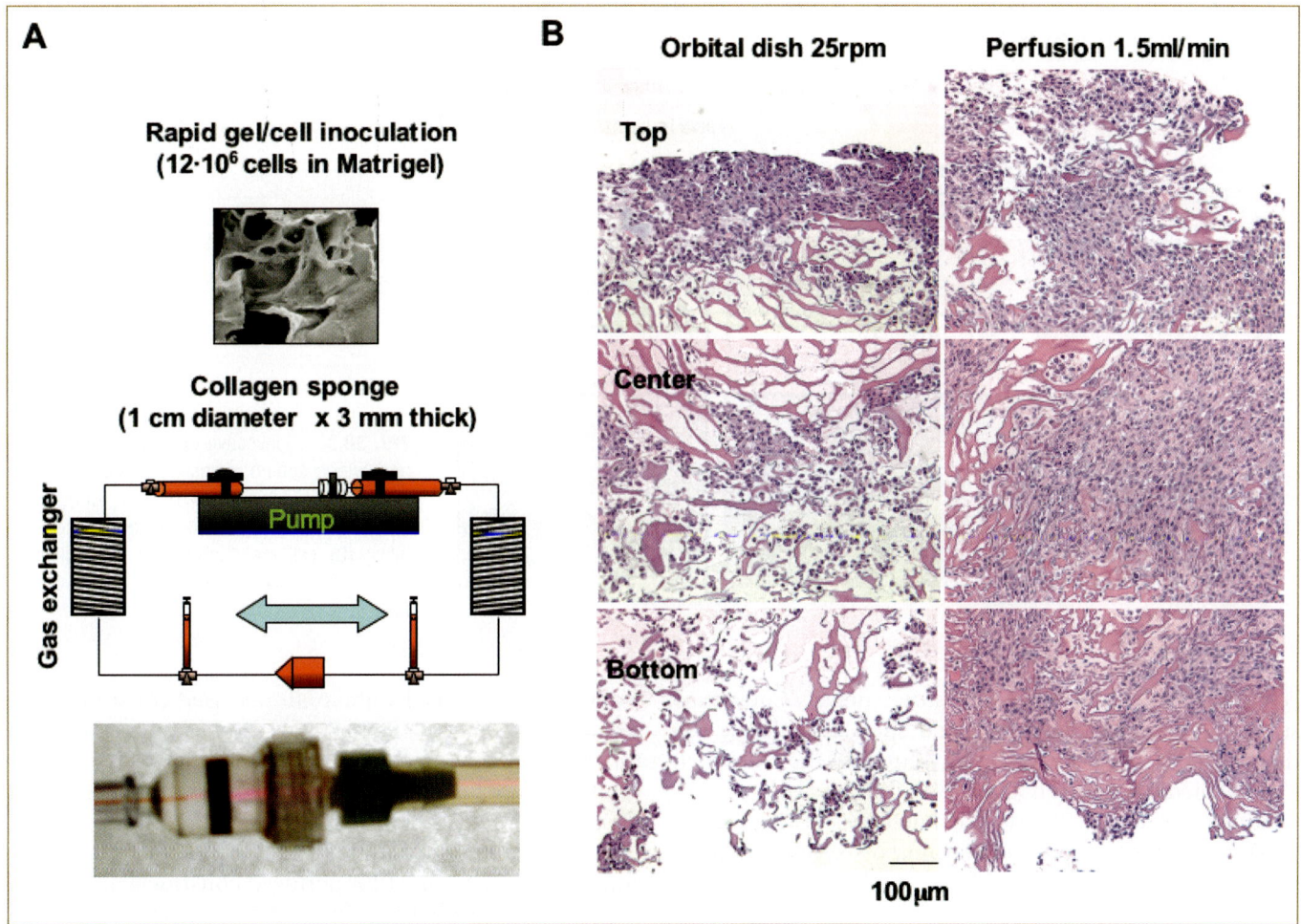

**FIG. 38.2.** High-density perfusion seeding. **(A)** Cells are rapidly inoculated into collagen sponge using Matrigel and seeded, with alternating flow direction **(B)** The method results in a spatially uniform distribution of cells at physiologically high concentrations and high cell viability. A full cross section through the center is shown for a construct inoculated with 12 million C2C12 cells and seeded for 4.5 h. Adapted from Radisic *et al.* (2003).

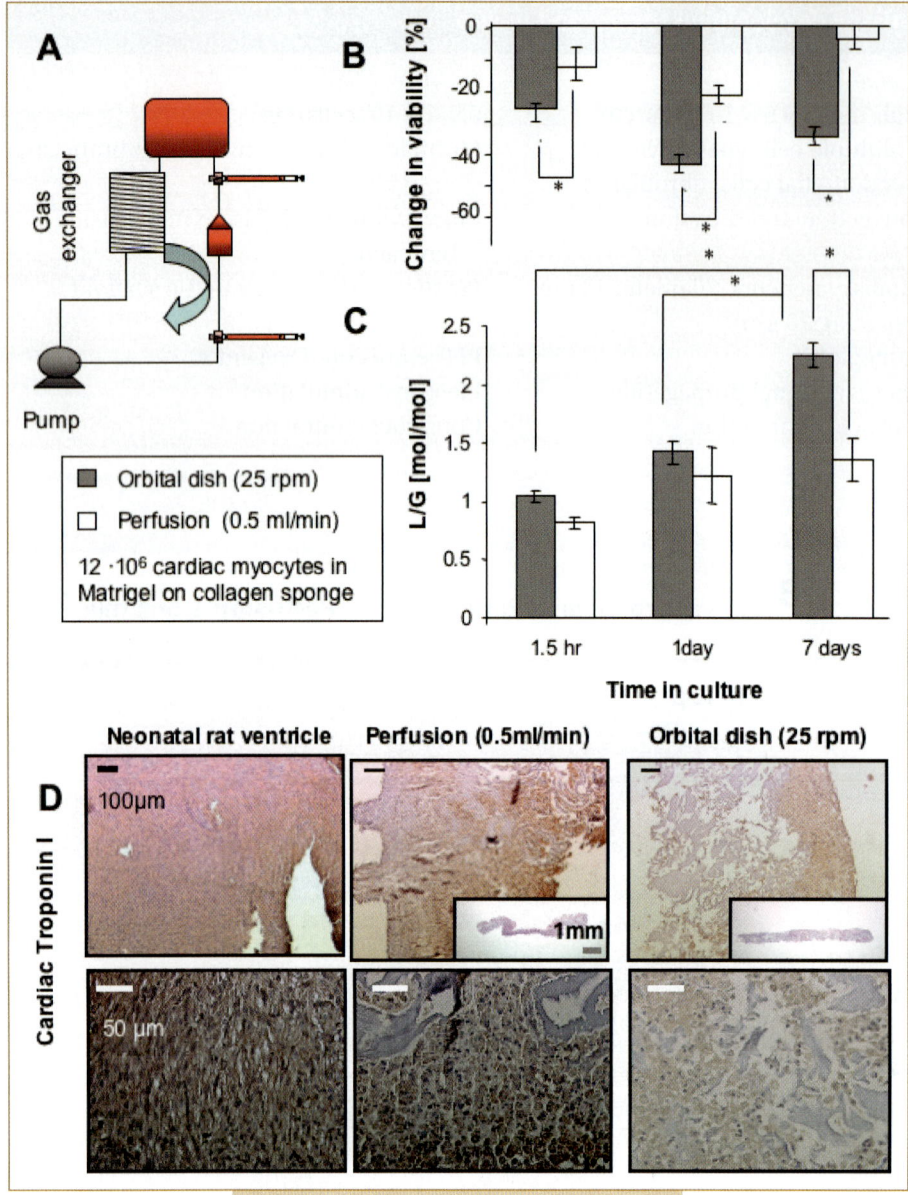

**FIG. 38.3.** Convective-diffusive oxygen transport during culture maintains a high density of viable cells, aerobic cell metabolism, and uniform cell distribution. **(A)** Perfusion loop; **(B)** cell viability; **(C)** cell metabolism; **(D)** distribution of cardiac Troponin I–positive cells. Adapted from Radisic *et al.* (2004c).

cells penetrated the entire construct depth. Constructs seeded in perfusion had high and spatially uniform cell density throughout the perfused volume of the construct. Clearly, medium perfusion during seeding was key for engineering thick constructs with high densities of viable cells, presumably due to enhanced transport of oxygen within the construct.

## Perfusion During Cultivation Enabled Cell Viability and Aerobic Metabolism

Throughout the cultivation, the number of live cells in perfused constructs was significantly higher than in dish-grown constructs, due to the perfusion of culture medium that was equilibrated in each pass with respect to oxygen and pH in an external-loop gas exchanger (Fig. 38.3A).

Notably, the final cell viability in perfused constructs cultured for eight days (81.6 ± 3.7%) was indistinguishable from the viability of the freshly isolated cells (83.8 ± 2.0%), and it was markedly higher than the cell viability in dish-grown constructs (47.4 ± 7.8%) (Radisic *et al.*, 2004c) (Fig. 38.3B). Consistently, the molar ratio of lactate produced to glucose consumed (L/G) was ~1 for perfused constructs, indicating aerobic cell metabolism. In contrast, in orbitally mixed dishes, with convective flow of medium around the constructs and molecular diffusion within constructs, L/G increased progressively from 1 to ~2, indicating a transition to anaerobic cell metabolism (Fig. 38.3C). Cell damage, assessed by monitoring the activity of lactate dehydrogenase (LDH) in culture medium, was at all time points significantly lower in perfusion than in dish cultures.

Perfused constructs and native ventricles had more cells in the S phase than in the G2/M phases, whereas the cells from dish-grown constructs appeared unable to complete the cell cycle and accumulated in the G2/M phase. Cells expressing cardiac-specific differentiation markers (sarcomeric α-actin, sarcomeric tropomyosin, cardiac troponin I) were present throughout the perfused constructs and only within approximately a 100-μm-thick surface layer in dish-grown constructs.

Spontaneous contractions were observed in some constructs early in culture and ceased after approximately five days of cultivation, indicating the maturation of engineered tissue. In response to electrical stimulation, perfused constructs contracted synchronously, had lower excitation thresholds, and recovered their baseline function levels following treatment with a gap junction blocker; dish-grown constructs exhibited arrhythmic contractile patterns and failed to recover their baseline levels. However, most cells were round and mononucleated, a situation likely due to the exposure of cardiac myocytes to hydrodynamic shear, in contrast to the native heart muscle, where blood is confined within the capillary bed and therefore not in direct contact with cardiac myocytes (Fig. 38.3D). This motivated the design of scaffolds with arrays of channels that provide a separate compartment for medium flow.

### In Vivo–Like Oxygen Supply: Medium Perfusion, Channeled Scaffolds, and Oxygen Carriers

To test the feasibility of using channeled scaffolds, cardiac constructs were first engineered using a channeled collagen scaffold (Ultrafoam™, 1 cm in diameter × 3 mm thick) seeded with neonatal cardiac myocytes and cultivated in perfusion at 0.5 mL/min for 10 days. The channel maintained its initial diameter and was surrounded with a 300-μm-thick tissue layer. However, collagen is not optimal for cardiac-tissue engineering due to its poor structural integrity. We thus explored the use of an elastomer, poly(glycerol sebacate), PGS (Wang et al., 2002), pretreated with neonatal rat cardiofibroblasts for three days in orbitally mixed dishes, followed by the addition of rat cardiomyocytes and perfusion cultivation (Radisic et al., 2006b) (Fig. 38.4A). After only three days of culture, cells on scaffolds formed constructs that contracted synchronously in response to electrical stimulation (Fig. 38.4C). The scaffold pores remained open, and the pressure drop measured across the construct was as low as 0.1 kPa/mm. To mimic the oxygen supply of hemoglobin, culture medium was supplemented by 10% v/v PFC emulsion [Oxygent™, kindly donated by Alliance Pharmaceuticals Corp. (San Diego, CA)]; constructs perfused with unsupplemented culture medium served as controls. Constructs were subjected to unidirectional medium flow at a flow rate of 0.1 mL/min provided by a multichannel peristaltic pump (IsmaTec).

As the medium flowed through the channel array, oxygen was depleted from the aqueous phase of the culture medium by diffusion into the construct space, where it was used for cell respiration. Depletion of oxygen in the aqueous phase acted as a driving force for the diffusion of dissolved oxygen from the PFC particles, thereby contributing to the maintenance of higher oxygen concentrations in the medium. Due to the small size of PFC particles, the passive diffusion of dissolved oxygen from the PFC phase into the aqueous phase was very fast, estimated not to be a rate-limiting step in this system. For comparison, in unsupplemented culture medium, oxygen was depleted faster, since there is no oxygen carrier phase that acts as a reservoir (Fig. 38.4B) (Radisic et al., 2005, 2006a).

In PFC-supplemented medium, the decrease in the partial pressure of oxygen in the aqueous phase was only 50% of that in control medium (28 mmHg vs. 45 mmHg between the construct inlet and outlet at the flow rate of 0.1 mL/min). Consistently, constructs cultivated in the presence of PFC had higher amounts of DNA, troponin I, and Cx-43 and significantly better contractile properties than control constructs. In both groups, cells were present at the channel surfaces as well as within constructs. Improved construct properties were correlated with the enhanced supply of oxygen to the cells within constructs.

In order to rationalize experimental data for oxygen transport and consumption in engineered cardiac constructs with an array of channels, we developed a mathematical model of oxygen distribution in cardiac constructs similar to the Krogh cylinder model. Concentration profiles of oxygen and cells within the constructs were obtained by numerical simulation of the diffusive-convective oxygen transport and its utilization by the cells. The model was used to evaluate the effects of the medium perfusion rate, oxygen carrier, and scaffold geometry on viable cell density (Radisic et al., 2005).

The model was used to define scaffold geometry and flow conditions necessary to cultivate cardiac constructs with clinically relevant thicknesses (5 mm). Oxygen profiles were modeled in a channel array consisting of channels 100 μm in diameter and 100 μm wall-to-wall spacing at physiologically high cell density $1 \cdot 10^8$ cells/cm³. At 0.049 cm/s, the oxygen concentration increased significantly in both the tissue space and the channel lumen with the increase in circulating PFC emulsion, from 0% to 6.4%. Although the oxygen concentration in the tissue space with physiological cell density increased considerably with the increase of circulating PFC concentration, from 0% to 6.4%, we had to increase the flow rate, keeping the shear stress in the physiological range 1 dyn/cm², in order to provide enough oxygen for the entire 0.5-cm-thick construct. At our best conditions (0.135 cm/s and 6.4% PFC) (Fig. 38.4D), oxygen is not depleted at any point within the tissue construct, and the minimum concentration of 33 μM is approximately five times above the $K_m$ (oxygen concentration at which the consumption rate in the tissue space is maintained at a maximum level) (Radisic et al., 2005).

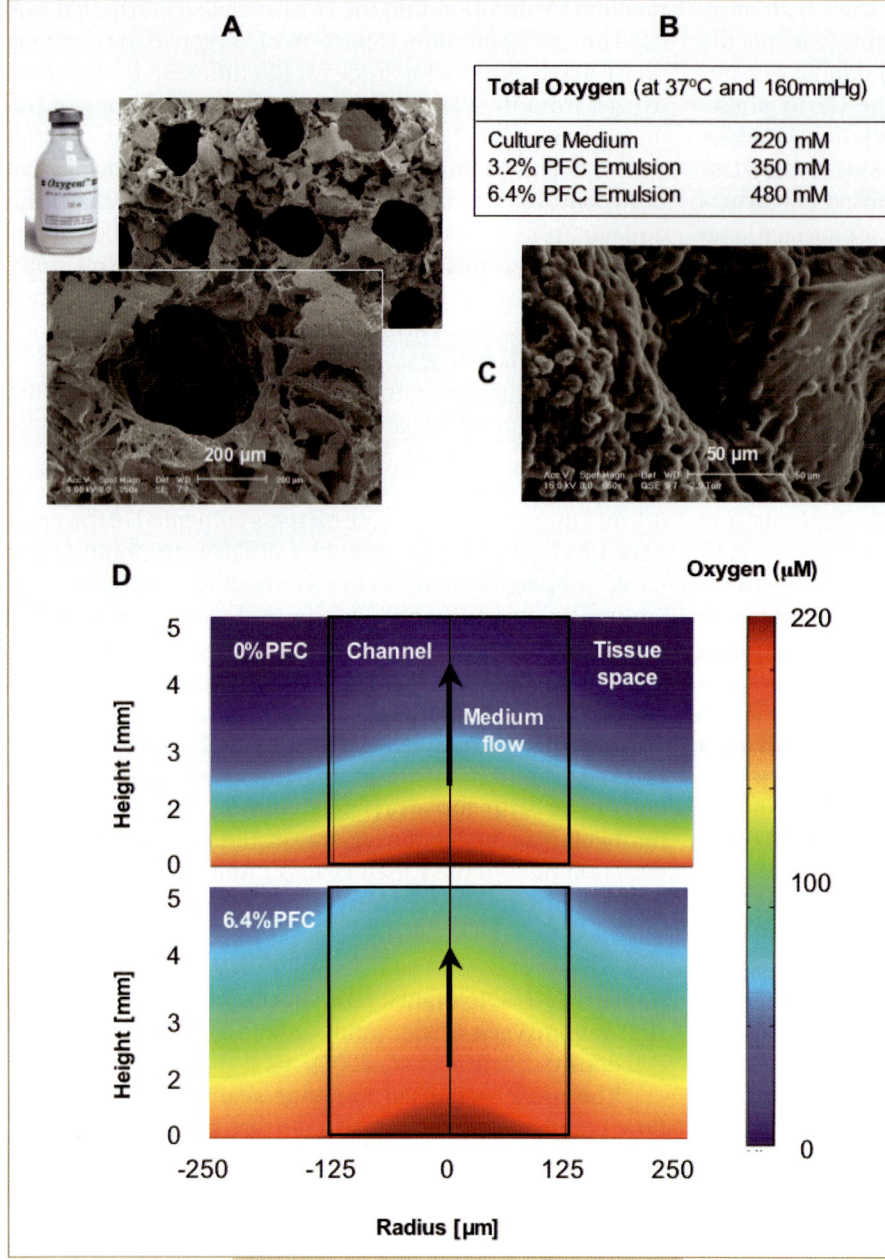

| Total Oxygen (at 37°C and 160mmHg) | |
|---|---|
| Culture Medium | 220 mM |
| 3.2% PFC Emulsion | 350 mM |
| 6.4% PFC Emulsion | 480 mM |

**FIG. 38.4.** PFC emulsion increases the average oxygen concentration in the channel lumen and tissue space by increasing the oxygen-carrying capacity of the culture medium. **(A)** Channeled scaffolds were seeded with cells and perfused with culture medium that was supplemented with a perfluorocarbon (PFC) emulsion. A scanning electron micrograph (SEM of a PGS scaffold with laser-bored channel array) is shown at two magnifications, before seeding with cells. **(B)** Supplementation of culture medium with PFC markedly increased the total oxygen content, as shown in table by data measured for two different concentrations of PFC. **(C)** SEM of cardiomyocytes cultured on channeled PGS scaffold perfused with PFC-supplemented culture medium. **(D)** Mathematical model of oxygen distribution in a perfused channeled scaffold with physiological cell density ($10^8$ cells/cm$^3$) in the tissue space. Data are shown for the superficial flow velocity of culture medium of 0.135 cm/s, for cultivations with unsupplemented culture medium (top image), and culture medium supplemented with 6.4% (vol/vol) of PFC (bottom image). Tissue constructs had channels that were 250 μm in diameter, spaced at 250 μm (wall-to-wall distance). Oxygen concentrations in the channel (outlined by a box in the center) and tissue space (on the sides) are shown by color code (on the right). Adapted from Radisic *et al.* (2005, 2006c).

## Excitation–Contraction Coupling: Electrical Stimulation

In native heart, mechanical stretch is induced by electrical signals, and the orderly coupling between electrical pacing signals and macroscopic contractions is crucial for the development and function of native myocardium (Severs, 2000). We therefore hypothesized that applying electrical signals designed to induce synchronous construct contractions would enhance cell differentiation and functional assembly of engineered tissue via physiologically relevant mechanisms. To test this hypothesis, cardiac constructs prepared by seeding collagen sponges ($6 \times 8 \times 1.5$ mm) with neonatal rat ventricular cells ($5 \cdot 10^6$) were stimulated using supra-threshold square biphasic pulses (2 ms duration, 1 Hz, 5 V). The stimulation was initiated after one to five days of scaffold seeding (three-day period was optimal) and applied for up to eight days.

The application of electrical stimulation during construct cultivation markedly enhanced the contractile behavior. After eight days of culture, the amplitude of contractions was sevenfold higher in stimulated than in nonstimulated constructs (Fig. 38.5A), a result of the progressive increase with the duration of culture. The excitation threshold (ET, the minimum voltage at which the entire construct was observed to beat) decreased (Fig. 38.5B) and the maximum

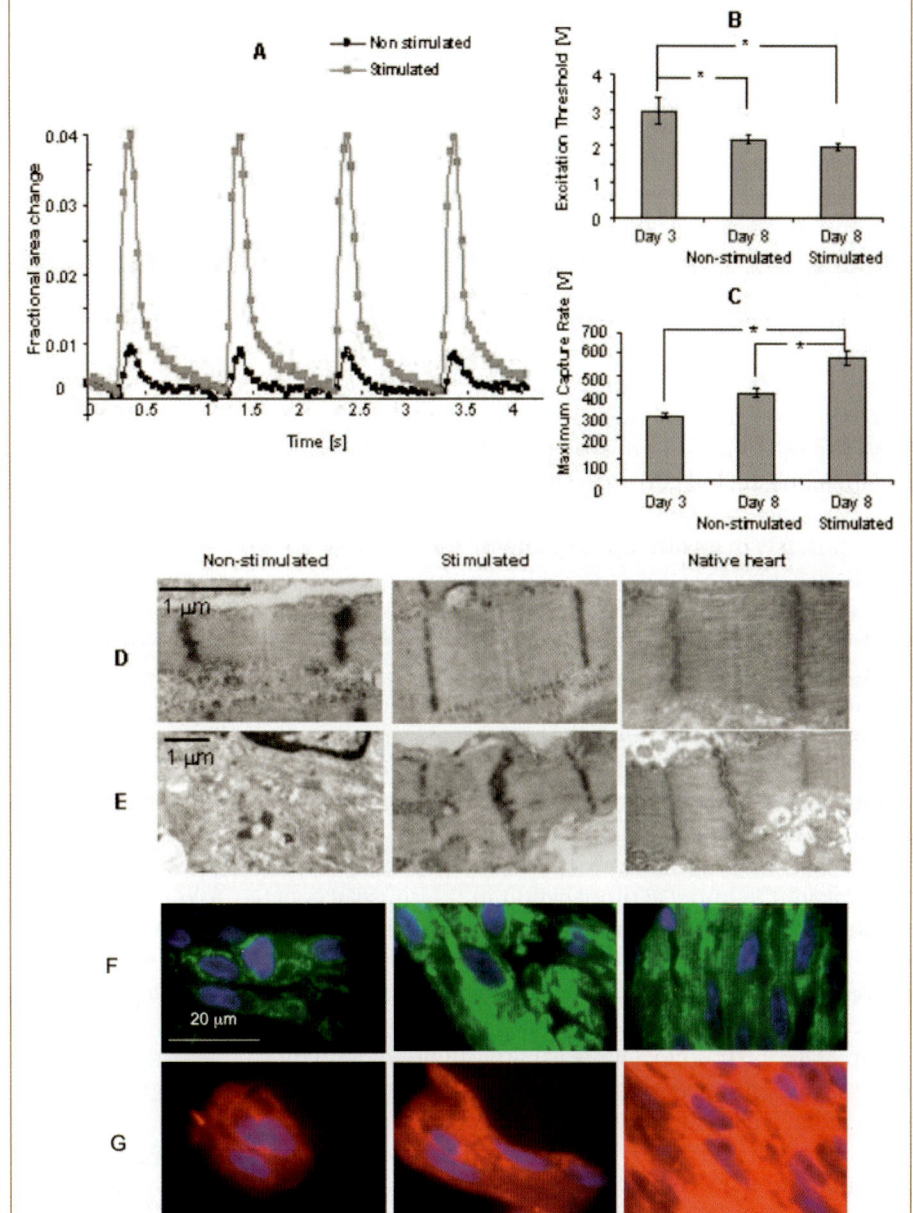

**FIG. 38.5.** Effects of electrical stimulation on functional assembly of engineered cardiac constructs. **(A)** Contraction amplitude of constructs cultured for a total of eight days, shown as a fractional change in the construct size. Electrical stimulation increased the amplitude of contractions by a factor of 7. **(B)** The excitation threshold (ET) decreased and **(C)** the maximum capture rate (MCR) increased significantly both with time in culture and due to electrical stimulation. (*) denotes statistically significant differences ($P < 0.05$; Tukey's post-hoc test with one-way ANOVA, $n = 5$–$10$ samples per group and time point). **(D)** The structure of sarcomeres and **(E)** gap junctions observed in micrographs of stimulated constructs after eight days of cultivation were remarkably similar to neonatal rat ventricles and markedly better developed than in control (nonstimulated) constructs. Representative sections of constructs stained for β-MHC (green, **E**) and α-MHC (red, **G**) cell nuclei are shown in blue. Adapted from Radisic *et al.* (2004b).

capture rate (MCR, the maximum pacing frequency for synchronous construct contractions) increased (Fig. 38.5C) both with time and due to electrical stimulation, suggesting functional coupling of the cells. The shape, amplitude (~100 mV), and duration (~200 ms) of the electrical activity recorded for cells in constructs stimulated during culture were similar to action potentials reported for constructs that were mechanically stimulated during culture (Zimmermann *et al.*, 2002b).

After eight days, stimulated constructs demonstrated a remarkable level of ultrastructural differentiation, comparable in several respects to that of native myocardium. Cells in stimulated constructs were aligned and elongated and contained centrally positioned elongated nuclei, in contrast

to round cells in nonstimulated constructs, which had a high nucleus-to-cytoplasm ratio. Stimulated constructs and neonatal ventricles contained abundant mitochondria positioned between myofibrils, in contrast to nonstimulated constructs, containing mitochondria scattered around the cytoplasm and substantially larger amounts of glycogen. Electrical stimulation induced the development of long, well-aligned registers of sarcomeres that closely resembled those in native myocardium (Fig. 38.5D), representing a hallmark of maturing cardiomyocytes (Severs, 2000). The volume fraction of sarcomeres in stimulated eight-day constructs was indistinguishable from that measured for neonatal ventricles (Olivetti *et al.*, 1980); in contrast, nonstimulated constructs contained only scattered and poorly organized sarcomeres. In stimulated constructs intercalated discs were positioned between aligned Z lines (Fig. 38.5E) and were as frequent as in neonatal ventricles; gap junctions were also substantially better developed and more frequent.

Myofibers aligned in the direction of the electrical field lines, possibly in an attempt to decrease the apparent ET in response to pacing (Tung *et al.*, 1991). In contrast, cells in nonstimulated constructs stayed round and expressed relatively low levels of cardiac markers. After eight days, stimulated constructs exhibited a markedly higher density of Cx-43 than either early (three-day) or nonstimulated constructs. Notably, the improved contractile properties of electrically stimulated constructs were not reflected in any apparent differences in construct cellularity, cell damage, or cell metabolism, but correlated instead with cell differentiation. Myofibers aligned in the direction of the electrical field lines, possibly in an attempt to decrease the apparent ET in response to pacing (Tung *et al.*, 1991). Stimulated constructs and neonatal ventricles expressed high levels of cardiac Tn-I, sarcomeric α-actin, Cx-43, α-MHC, and β-MHC and contained elongated cells aligned in parallel (Fig. 38.5F, G). In contrast, cells in nonstimulated constructs stayed round and expressed relatively low levels of cardiac markers. Cross-striations characteristic of mature cardiac myocytes were detected in stimulated constructs and native ventricles but not in nonstimulated constructs.

In our ongoing studies we performed optical mapping to measure the impulse propagation in the constructs cultivated in the presence or absence of electrical field stimulation. The constructs in the stimulated group had impulse propagation of $14.4 \pm 2.7$ V/cm, which was significantly higher than the impulse propagation in the nonstimulated group of $8.6 \pm 3.0$ V/cm. Our feasibility studies of implantation of the stimulated constructs in the rat model of myocardial infarction indicate that the constructs readily integrate with the host tissue and vascularize following implantation.

### Multiple Cell Types

Attempts to engineer cardiac tissue, including our previous studies, typically involve the use of cell populations enriched for cardiomyocytes (CM) by preplating and removing rapidly adhering cells (Bursac *et al.*, 1999; Carrier *et al.*, 1999; Fink *et al.*, 2000; Papadaki *et al.*, 2001; Zimmermann *et al.*, 2002b). Zimmermann *et al.* (2002b) were the first to recognize the importance of the presence of multiple cell types for the *in vitro* cultivation of heart tissue in their system, which involves the application of cyclic stretch. In separate studies, Akhyari *et al.* (2002) detected the presence of newly synthesized collagen network in the tissue constructs based on the passage of three human pediatric heart cells, suggesting that fibroblasts may play a role in the remodeling of engineered cardiac constructs *in vitro*.

We hypothesized that scaffold pretreatment with cardiac fibroblasts prior to the cultivation of cardiac myocytes can enhance functional assembly of the engineered cardiac constructs by creating an environment supportive of cardiomyocyte attachment, differentiation, and contractility. Neonatal rat heart cells were sorted and used to prepare three distinct cell populations: rapidly plating cells, identified as cardiac fibroblasts (CF); slowly plating cells, identified as cardiac myocytes (CM); and the unseparated initial population of cells (US). We investigated the structure and function of engineered constructs with respect to the use of the coculture (as compared to CM alone) and with respect to the regime of coculture (concurrent culture of CM and CF using US population; sequential coculture of CF and CM). Constructs formed with CF alone were used as an additional control. The cells were cultured for 11 days on porous scaffolds made of poly(glycerol sebacate) (PGS).

Constructs in the CF + CM and CM groups exhibited similar fractions of myocytes (~55%) and fibroblasts (~20%) and similar amounts of cardiac-specific proteins. However, the constructs in the CF + CM group had significantly higher contractile amplitude than the CM group ($p < 0.05$) and significantly lower excitation threshold than the US group ($p < 0.05$). Also, the CF + CM group exhibited a stratified, 100- to 200-μm-thick tissuelike structure that contained some elongated CM, in contrast to relatively less organized cells in the US and CM groups (Fig. 38.6). Thus, the sequential coculture of CF and CM on a synthetic elastomer scaffold created an environment supportive of cardiomyocyte attachment, differentiation, and contractile function. It is possible that CF played the role of a feeder layer for the cardiac cells and conditioned the scaffold by secreting collagen, which is a natural substrate for CM attachment. We believe that this approach can be used routinely to precondition the scaffolds for cardiac-tissue engineering.

## VI. ENGINEERED HEART TISSUE BY MECHANICAL STIMULATION OF CELLS IN COLLAGEN GELS

A significant approach to cardiac-tissue engineering, established by Eschenhagen and colleagues (Eschenhagen *et al.*, 1997; Fink *et al.*, 2000; Zimmermann *et al.*, 2000, 2002b), involves the cultivation of neonatal rat heart cells in collagen gel and Matrigel, in the presence of growth factors. This approach is also biomimetic in nature, because it involves the utilization of mechanical stretch as a physiologically relevant signal for cell differentiation. The cultured tissues are subjected to sustained mechanical strain. Under these conditions, cardiomyocytes and nonmyocytes form 3D cardiac organoids (Bursac *et al.*, 1999; Carrier *et al.*, 1999; Papadaki *et al.*, 2001), a well-organized and highly differentiated cardiac muscle syncytium, that exhibited the contractile and electrophysiological properties of working myocardium. The first implantation experi-

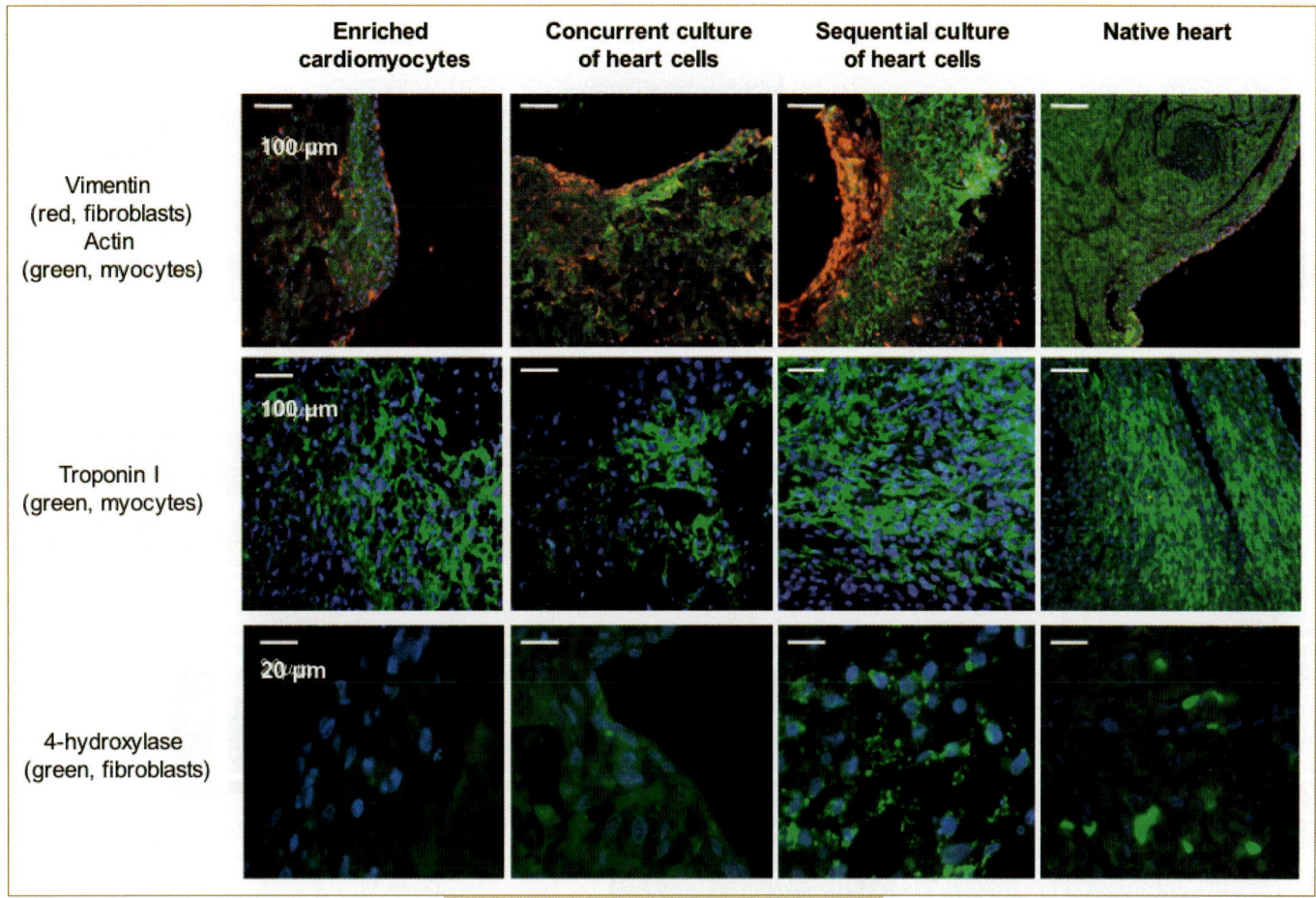

**FIG. 38.6.** Differential effects of cardiac cell subpopulations on construct appearance, fibroblast markers, and myocyte markers. Top row: vimentin/actin co-actin for fibroblasts (red) and myocytes (green). Middle row: cardiac myocytes are stained green for Tn-1. Bottom row: prolyl-4-hydroxylase, involved in collagen synthesis by fibroblasts stained green; nuclei are shown in blue (DAPI).

ments in healthy rats showed survival, strong vascularization, and signs of terminal differentiation of cardiac tissue grafts (Zimmermann *et al.*, 2002a). More recent work demonstrates rigorously, and using multiple molecular, structural, and functional assays, that these constructs can be engineered in forms suitable for implantation on infarcted rat hearts and that they delay the thinning of the heart wall and induce functional improvement (Zimmermann *et al.*, 2006).

In an early setup, neonatal rat ventricular myocytes were resuspended in a gel consisting of collagen I diluted in DMEM and Matrigel (Fink *et al.*, 2000). For each piece of tissue, 0.7 mL of the cell/gel mix was poured into a well (11 × 17 × 4 mm) made of silicone rubber containing one set of Velcro-coated silicone tubes (7-mm length, 3-mm OD, 2-mm ID). One tissue culture dish could hold six wells. The mixture was allowed to gel at 37°C for 60 min before culture medium was added. After four days in culture, a strip of biconcave tissue was formed, fixed at each side to a piece of silicone tubing. The tissues were transferred to a custom-made motorized stretching device that consisted of a spring attached between two stretching bars, one fixed, the other moving, with holders that attach to the pieces of silicone tubing at each end of the engineered cardiac tissue. A unidirectional and cyclic stretch at a frequency of 1.5 Hz and a strain rate of up to 20% was employed for six days. Stretched constructs exhibited an improved organization of cardiac myocytes into parallel arrays of rod-shaped cells. They had increased cell length and width and higher mitochondrial density and contained longer myofilaments. Compared with the unstretched controls, RNA/DNA increased by 100% and the protein/cell ratio increased by 50%. The force of contraction was higher in stretched constructs under basal conditions and after stimulation with isoprenaline (Fink *et al.*, 2000).

In an improved setup, neonatal rat cardiac cells were resuspended in the collagen/Matrigel mix and cast into circular molds (Zimmermann *et al.*, 2006). After seven days in

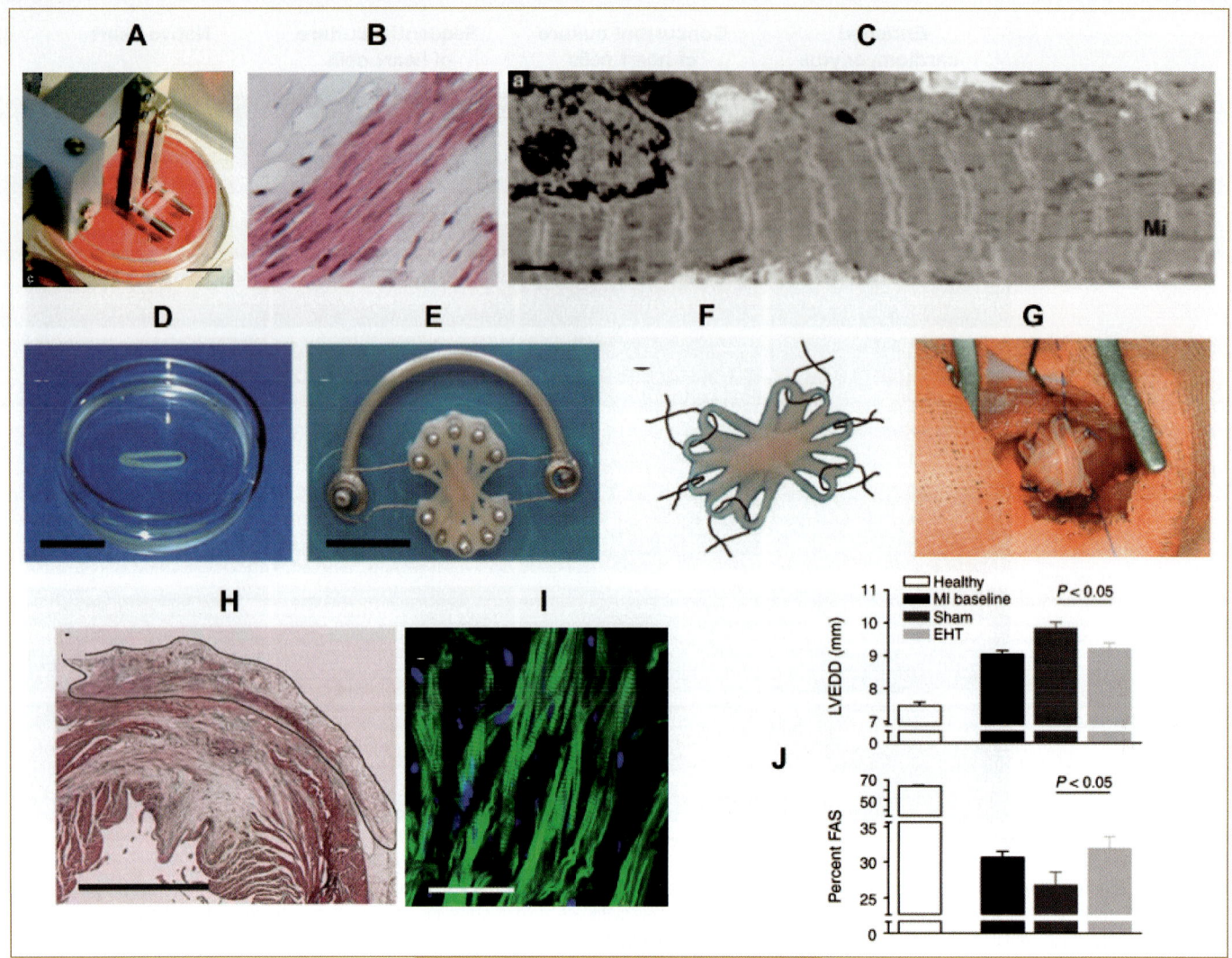

**FIG. 38.7.** Engineered heart tissue by mechanical stimulation. **(A)** Eschenhagen and Zimmermann designed a bioreactor that provides cyclic mechanical stretch to engineered heart tissue based on neonatal rat cardiomyocytes and collagen gel. Mechanical stimulation yielded elongated cardiomyocytes **(B)** with a remarkably well-developed contractile apparatus **(C)** (With permission from Zimmermann *et al.*, 2002b). **(D)–(F)** (With permission from Zimmermann *et al.*, 2006) A complex star-shaped engineered heart tissue was formed by stacking five ring-shaped constructs engineered using neonatal rat heart cells, collagen gel, and the bioreactor shown in A. The resulting synchronously contracting multiloop construct **(E)** was sutured **(F)** on top of an infarct scar caused by LAD ligation **(G)**. The patch integrated well with the host myocardium **(H)**. Implanted cells formed elongated myofibers with a well-developed contractile apparatus **(I,** green: actin; blue: nuclei). **(J)** Four weeks following implantation, left ventricular end-diastolic diameter (LVEDD, top graph) and fractional area shortening (FAS, bottom graph) measured by echocardiography showed that engineered heart tissue prevented further dilation and improved fractional area shortening of infarcted hearts as compared to controls. "Healthy" indicates untreated controls, "MI baseline" indicates values measured two weeks after myocardial infarction, "Sham" indicates four-week data for sham-operated rats, and "EHT" indicates four-week data for rats that received engineered constructs. Scale bars: A: 10 mm; B: 30 μm; C: 1 μm; D, E: 10 mm; H: 5 mm; I: 5 μm. Reproduced with permission from Zimmermann *et al.*, 2006.

culture, the strips of cardiac tissue were placed around two rods, each fixed to a stretching bar of a custom-made mechanical stretcher and subjected to unidirectional and cyclic stretch at a 10% strain rate and 2 Hz (Fig. 38.7A). The histology and immunohistochemistry revealed the formation of intensively interconnected, longitudinally oriented cardiac muscle bundles with morphological features resembling adult rather than immature native tissue (Fig. 38.7B). Fibroblasts and macrophages were found scattered through

the construct, and, at the high magnification, capillary structures positive for CD31 were noted. Cardiomyocytes exhibited well-developed ultrastructural features: sarcomeres arranged in myofibrils, with well-developed Z, I, A, H, and M bands, specialized cell–cell junctions (gap and adherence junctions), T tubules as well as well-developed basement membrane. The constructs exhibited contractile properties similar to those of the native tissue, with a high ratio of twitch to resting tension, a strong β-adrenegenic response, and action potentials characteristic of rat ventricular myocytes.

Zimmerman *et al.* (2002a) placed cardiac-tissue rings cultivated in the presence of mechanical stimulation onto uninjured hearts of Fisher 344 rats for 14 days. They noticed that although both cells and collagen were isolated from Fisher rats, immunosupporession was required for maintenance of heart tissue upon implantation. In the absence of immunosupression, even in the syngeneic approach, cardiac constructs completely degraded after only two weeks *in vivo*. It is unknown what exactly caused the response; it is possible that it was the remainder of serum or chick extract. Regardless, the finding has implications for the potential implantation of cardiac patches in clinical settings.

In recent studies (Zimmermann, 2006), the same group created large, 1- to 4-mm-thick tissue constructs from neonatal rat heart cells and demonstrated the capability of these constructs to support contractile function of infarcted hearts in a immunosupressed rat model. Four weeks following implantation, engineered heart-tissue grafts showed electrical coupling to the native myocardium. As compared to controls (untreated infarctions, sham operated rats, noncontractile constructs), engineered heart tissue prevented further dilation of the ventricle wall and improved fractional area shortening of infarcted hearts. Four weeks after grafting, engineered tissues had a high connective-tissue content and contained ~400-μm-thick strands of compact and well-differentiated cardiaclike muscles. The graft was electrically coupled to the host myocardium, as indicated by undelayed impulse propagation. The maximum conduction velocity in implanted grafts ($0.55 \pm 0.16$ m/s longitudinally, $0.16 \pm 0.07$ m/s transversally) was comparable to the conduction velocity in noninfarcted myocardium ($0.69 \pm 0.06$ m/s and $0.19 \pm 0.05$ m/s, respectively). Anisotropy, determined from the ratio of the longitudinal and transversal conduction velocities, in the grafts was similar to that in noninfarcted myocardium ($4.15 \pm 0.8$ versus $4.07 \pm 0.91$).

## VII. CELL-SHEET TISSUE ENGINEERING

Shimizu *et al.* (2002) created sheets of neonatal rat cardiomyocytes by plating isolated cells onto a temperature-sensitive ply(*N*-isopropylacrylamide) surface. The change in temperature from 37°C to 32°C causes the cell monolayer to detach. The initial surface area decreases during the detachment approximately fivefold, and the thickness of the cell sheet increases from approximately 20 μm to approximately 45 μm (Fig. 38.8A). When the cell sheets were layered over each other, the electrical communication was established, as evidenced by electrical signals measured by electrodes (Fig. 38.8B), and spontaneous beating was observed macroscopically. When implanted subcutaneously in nude rats, cell sheets survived for up to 12 weeks and vascularized. The outcomes were substantially better for implantation in three-week as compared to eight-week rats (Fig. 38.8C).

In separate recent studies (Furuta *et al.*, 2006), layered cell sheets formed using the neonatal rat cardiomyocytes and reinforced using an additional layer of collagen were studied in a nude rat model of myocardial infarction. The cell sheets were implanted immediately following ligation, and the outcomes were evaluated at one week and four weeks following implantation. Electrophysiological studies demonstrated functional integration of the grafted tissue and the propagation of APs without observable arrhythmias. The functional improvement was indicated by several measured parameters, including the AP amplitude, which was well maintained in the grafted group but not in untreated controls (Fig. 38.8D).

In another recent study, this method was extended to the formation of cell sheets using adipose-derived rat mesenchymal stem cells (MSC). The cell sheets were tested in a rat model of myocardial infarction. Four weeks following ligation, the cell sheets were implanted, and they were evaluated after an additional four weeks. The grafts formed a layer that contained some new blood vessels, viable undifferentiated cells, and a few cardiomyocytes (Fig. 38.8E). Notably, the implantation of cell sheets made using MSC reversed wall thinning in the scar area, in contrast to the control cell sheets made using dermal fibroblasts, which had no effect (Fig. 38.8F).

Overall, the method provides thin sheets of cells that express cardiac features *in vitro* and have the capability for cardiac repair *in vivo*. The studies involving rat MSC also represent an important step toward establishing clinically relevant cell sources.

## VIII. SUMMARY AND CURRENT RESEARCH NEEDS

### Summary

The overall goal of cardiac-tissue engineering is to direct the cells to establish the physiological structure and function of the heart tissue across different hierarchical scales. We described three significant and representative approaches to cardiac-tissue engineering that regulate cell differentiation and functional assembly of neonatal rat heart cells (the best-established cell model) by employing some of the signals that drive the development and function of normal heart. The differences in the conditions used to create cardiac tissues and the measured tissue outcomes between these approaches are instructive for the understanding of the state of the art of the field of cardiac-tissue engineering.

### Approaches to Cardiac-Tissue Engineering

The biomimetic approach involves an integrated use of cells, biomaterial scaffolds, and bioreactors to engineer compact, millimeters thick, synchronously contracting cardiac-tissue constructs. The system design is based on providing (1) convective-diffusive oxygen transport (critical

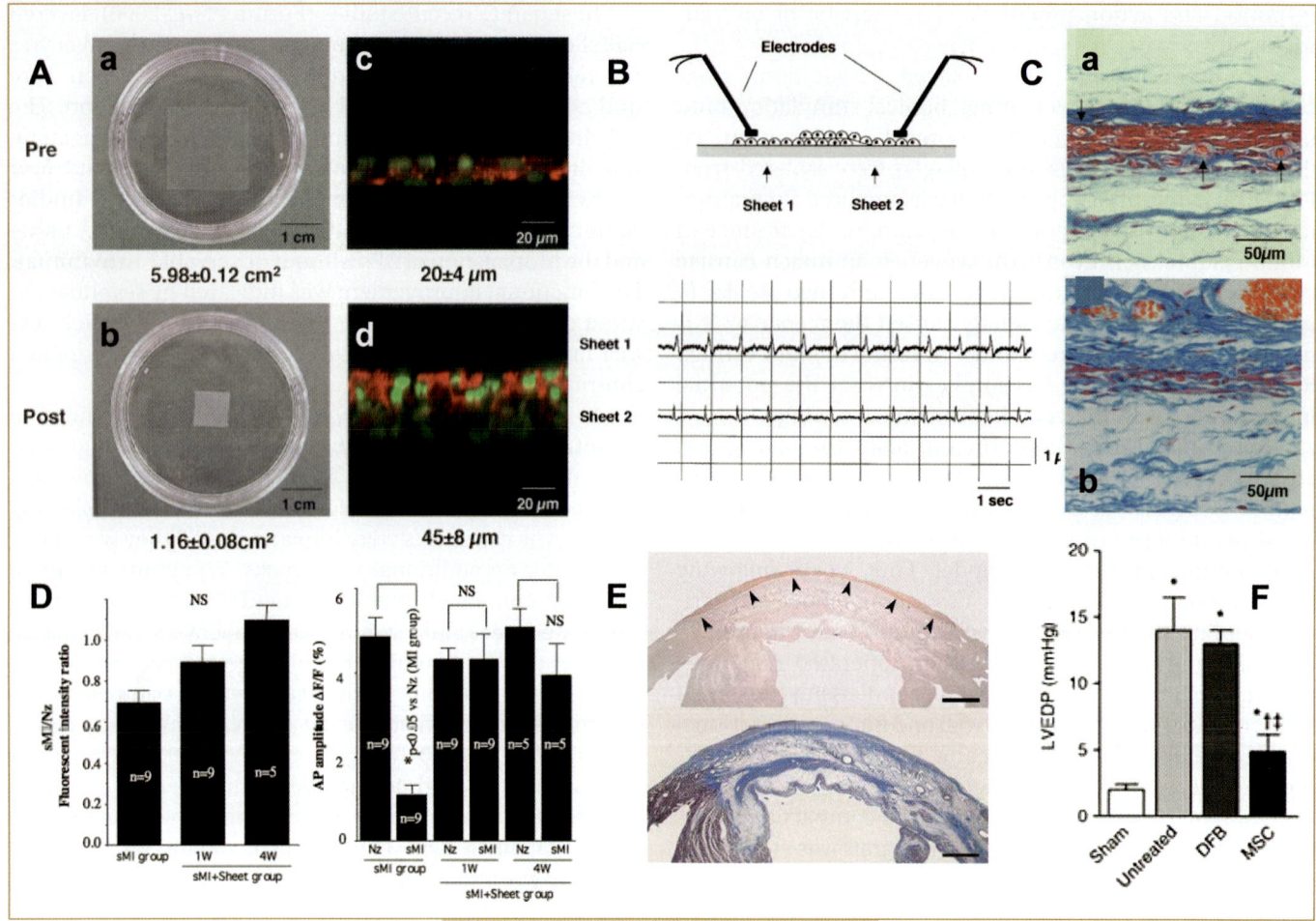

**FIG. 38.8.** Cell-sheet engineering. **(A)** Neonatal rat cardiomyocytes were cultured on thermoresponsive polymer surfaces for four days, and the formed cell sheets were detached by lowering the temperature from 37°C to 32°C. Cell sheets before **(a)** and after **(b)** detachment. The numbers refer to the surface area of the cell sheet. The respective sheet cross sections and thicknesses are shown on the right. Scale bar: 20 μm. **(B)** Two layered cardiomyocyte sheets established electrical communication, as evidenced by measured electrical potentials. **(C)** Histological analysis of cell sheets implanted subcutaneously in **(a)** three-week-old rats and **(b)** eight-week-old rats. Significant differences in tissue morphology were observed as a function of the age of the recipient. Scale bar: 50 μm. **(D)** Neonatal rat cardiomyocyte sheets described in images A–C were stabilized by a collagen layer and implanted, in separate studies, onto infarcted rat hearts immediately after injury. Action potentials (APs) were measured at explanted hearts one and four weeks following implantation, using an optical fluorescence–based method. The graph on the left shows the fluorescent intensity of the untreated (sMI) group and the treated (sMI + sheet) group. The graph on the right shows the preservation of the AP amplitude in the treated group and the loss of the AP in the sMI group. **(E)** Extent of mono-layered MSCs 48 h after transplantation (arrows). Scale bar: 100 μm. **(F)** Left ventricle end-diastolic pressure (LVEDP), measured by catheterization eight weeks after coronary ligation (four weeks after transplantation) in four groups of rats (sham; untreated; treated with a cell sheet made using dermal fibro-blasts, DFB; treated with a cell sheet made of rat adipose–derived mesenchymal stem cells). A, B, and C reproduced with permission from Shimizu *et al.* (2002); panels from Fig. 1, Fig. 2, and Fig. 3B, C; D was reproduced with permission from Furuta *et al.* (2006), Fig. 5C, D; E and F reproduced with permission from Miyahara *et al.* (2006), Fig. 2c and Fig. 5a.

for cell survival and function) and (2) excitation–contraction coupling (critical for cell differentiation and assembly). To mimic the capillary network, culture medium is perfused through a channeled scaffold seeded with cells at a physiologic density; to mimic the oxygen supply from hemoglobin, the culture medium is supplemented with an oxygen carrier (PFC emulsion). To promote cell differentiation and functional assembly, cultured constructs were subjected to electrical signals designed to mimic those in native heart and to induce synchronous construct contractions. Perfusion of culture medium containing PFC enabled the maintenance of a physiologic density of viable, differentiated cells in millimeters-thick constructs. Electrical field stimulation induced cell alignment and coupling, increased the amplitude of synchronous contractions, and resulted in a remarkable level of ultrastructural organization of engineered myocardium, over only eight days of cultivation (Radisic *et al.*, 2004b).

Eschenhagen and Zimmermann demonstrated that mechanical stimulation in a physiological regime yields functional and highly differentiated cardiac muscle, which was successfully utilized to improve contractile function of infarcted myocardium. Their approach has recently been validated by demonstration that it can be utilized to engineer cardiac tissue based on embryonic stem cell–derived cardiomyocytes (Guo *et al.*, 2006).

The cell-sheet approach pioneered by Okano and colleagues sets a novel paradigm for cardiac-tissue engineering that eliminated the need for a scaffold or a hydrogel for assembly of functional cardiac tissue. The method, based on layering confluent cell monolayers, has been validated by extensive *in vivo* and electrophysiological studies. This technique can render itself highly useful in co- and triculture studies.

## Challenge 1: Human Cell Source

One major challenge for cardiac-tissue engineering and its application in regenerative medicine is to extend the principles described earlier to the cultivation of functional cardiac grafts based on human cells. Practically all tissue-engineering work has been done using neonatal rat heart cells. Several different human cell sources are currently under consideration. Adult cardiac myocytes are not suitable because they are terminally differentiated and have no ability to proliferate (Soonpaa and Field, 1998). Transplantation of cell sheets based on adipose tissue–derived stem cells improved contractile function in a rat model of coronary artery ligation (Miyahara *et al.*, 2006). Emerging recent work suggests that the heart may contain resident progenitor cells. However, it appears that there is more than one heart cell subpopulation that fits the description of a cardiac progenitor. It remains to be determined if the progenitors, regardless of their marker, can be isolated from adult human biopsies and propagated to numbers sufficient to engineer a cardiac patch ($>10^8$ cells/patient). Also, for each of the candidate cell sources, the exact factors that need to be applied *in vitro* to induce and support cardiac cell differentiation and subsequent tissue development remain to be identified and optimized.

## Challenge 2: Improved Tissue-Engineering Systems

The second challenge is the need to advance the capabilities of our culture systems. Our most advanced existing bioreactors can provide *either* local microenvironmental control of oxygen and pH (via medium perfusion) *or* the application of physical stimuli (via electrical stimulation). A system providing *both factors simultaneously* was recently developed in our laboratory to further our biomimetic methods of cardiac-tissue engineering, by subjecting the cultured cells to *multiple signals* present in the native heart. The combined application of the oxygen supply and electrical stimulation designed to mimic those *in vivo* is expected to accelerate cell differentiation and functional assembly. This system could also enable controlled studies of cardiac development and function.

## Challenge 3: Vascularization

The third major challenge is the need for vascularization. In spite of many meritorious efforts in recent years (Ennett and Mooney, 2002; Jain, 2003; Levenberg *et al.*, 2005), vascularization of engineered tissues remains a fundamental and all-encompassing problem. Not surprisingly, tissue engineering has been most successful with tissues that either are thin (e.g., bladder) or have low oxygen requirements (e.g., cartilage). Our current inability to vascularize and perfuse thick cell masses has hindered efforts to build clinically useful tissues and, most critically, cardiac grafts. Vascular supply has multiple roles: (1) exchange of nutrients (most critically oxygen), metabolites, and regulatory factors between the cells and the environment; (2) paracrine signaling between the endothelial and cardiac cells; and (3) separation of the tissue phase with cardiac myocytes from the blood flow, and shielding of cardiomyocytes from hydrodynamic shear. Clearly, perfusion alone does not suffice, and the provision of the functions for the long term likely requires the formation of a stable and mature vascular network, functionally integrated with the macroscopic flow.

## Challenge 4: Interactions Between the Fields, Translational Research

The fourth and rather general challenge is for more effective interdisciplinary and translational research. The growing interactions between the fields of tissue engineering and developmental biology are expected to facilitate translation of biological principles into our engineering designs. The interactions between tissue engineers and clinicians are expected to help define the requirements, constraints, testing methods, and success criteria for translation into clinical practice. We thus need to continue to move the science and practice of tissue engineering from observational to mechanistic. It is the rational and interdisciplinary approach to tissue engineering that can bring us closer to producing biological grafts that can reestablish normal tissue structure and function across different-size scales, in the long term, and with the ability to remodel in response to environmental factors, growth, and aging.

# IX. ACKNOWLEDGMENTS

The cardiac-tissue-engineering work in our laboratories has been supported by the National Institutes of Health (NHLBI, NIBIB), the National Aeronautics and Space Administration, and the Juvenile Diabetes Research Foundation.

# X. REFERENCES

Akhyari, P., Fedak, P. W. M., Weisel, R. D., Lee, T. Y. J., Verma, S., Mickle, D. A. G., and Li, R. K. (2002). Mechanical stretch regimen enhances the formation of bioengineered autologous cardiac muscle grafts. *Circulation* **106**, I137–I142.

Akins, R. E., Boyce, R. A., Madonna, M. L., Schroedl, N. A., Gonda, S. R., McLaughlin, T. A., and Hartzell, C. R. (1999). Cardiac organogenesis *in vitro*: reestablishment of three-dimensional tissue architecture by dissociated neonatal rat ventricular cells. *Tissue Eng.* **5**, 103–118.

Alperin, C., Zandstra, P. W., and Woodhouse, K. A. (2005). Polyurethane films seeded with embryonic stem cell–derived cardiomyocytes for use in cardiac tissue engineering applications. *Biomaterials* **26**, 7377–7386.

Beltrami, A. P., Barlucchi, L., Torella, D., Baker, M., Limana, F., Chimenti, S., Kasahara, H., Rota, M., Musso, E., Urbanek, K., Leri, A., Kajstura, J., Nadal-Ginard, B., and Anversa, P. (2003). Adult cardiac stem cells are multipotent and support myocardial regeneration. *Cell* **114**, 763–776.

Bursac, N., Papadaki, M., Cohen, R. J., Schoen, F. J., Eisenberg, S. R., Carrier, R., Vunjak-Novakovic, G., and Freed, L. E. (1999). Cardiac muscle tissue engineering: toward an *in vitro* model for electrophysiological studies. *Am. J. Physiol. Heart Circ. Physiol.* **277**, H433–H444.

Carrier, R. L., Papadaki, M., Rupnick, M., Schoen, F. J., Bursac, N., Langer, R., Freed, L. E., and Vunjak-Novakovic, G. (1999). Cardiac tissue engineering: cell seeding, cultivation parameters and tissue construct characterization. *Biotechnol. Bioeng.* **64**, 580–589.

Carrier, R. L., Rupnick, M., Langer, R., Schoen, F. J., Freed, L. E., and Vunjak-Novakovic, G. (2002a). Effects of oxygen on engineered cardiac muscle. *Biotechnol. Bioeng.* **78**, 617–625.

Carrier, R. L., Rupnick, M., Langer, R., Schoen, F. J., Freed, L. E., and Vunjak-Novakovic, G. (2002b). Perfusion improves tissue architecture of engineered cardiac muscle. *Tissue Eng.* **8**, 175–188.

Condorelli, G., Borello, U., De Angelis, L., Latronico, M., Sirabella, D., Coletta, M., Galli, R., Balconi, G., Follenzi, A., Frati, G., Cusella De Angelis, M. G., Gioglio, L., Amuchastegui, S., Adorini, L., Naldini, L., Vescovi, A., Dejana, E., and Cossu, G. (2001). Cardiomyocytes induce endothelial cells to trans-differentiate into cardiac muscle: implications for myocardium regeneration. *Proc. Natl. Acad. Sci. U.S.A.* **98**, 10733–10738.

Ennett, A. B., and Mooney, D. J. (2002). Tissue engineering strategies for *in vivo* neovascularisation. *Expert. Opin. Biol. Ther.* **2**, 805–818.

Eschenhagen, T., Fink, C., Remmers, U., Scholz, H., Wattchow, J., Woil, J., Zimmermann, W., Dohmen, H. H., Schafer, H., Bishopric, N., Wakatsuki, T., and Elson, E. (1997). Three-dimensional reconstitution of embryonic cardiomyocytes in a collagen matrix: a new heart model system. *FASEB J.* **11**, 683–694.

Etzion, S., Battler, A., Barbash, I. M., Cagnano, E., Zarin, P., Granot, Y., Kedes, L. H., Kloner, R. A., and Leor, J. (2001). Influence of embryonic cardiomyocyte transplantation on the progression of heart failure in a rat model of extensive myocardial infarction. *J. Mol. Cell. Cardiol.* **33**, 1321–1330.

Fink, C., Ergun, S., Kralisch, D., Remmers, U., Weil, J., and Eschenhagen, T. (2000). Chronic stretch of engineered heart tissue induces hypertrophy and functional improvement. *FASEB J.* **14**, 669–679.

Furuta, A., Miyoshi, S., Itabashi, Y., Shimizu, T., Kira, S., Hayakawa, K., Nishiyama, N., Tanimoto, K., Hagiwara, Y., Satoh, T., Fukuda, K., Okano, T., and Ogawa, S. (2006). Pulsatile cardiac tissue grafts using a novel three-dimensional cell sheet manipulation technique functionally integrates with the host heart, *in vivo*. *Circ. Res.* **98**, 705–712.

Gillum, R. F. (1994). Epidemiology of congenital heart disease in the United States. *Am. Heart J.* **127**, 919–927.

Guo, X. M., Zhao, Y. S., Chang, H. X., Wang, C. Y., E, L. L., Zhang, X. A., Duan, C. M., Dong, L. Z., Jiang, H., Li, J., Song, Y., and Yang, X. J. (2006). Creation of engineered cardiac tissue *in vitro* from mouse embryonic stem cells. *Circulation* **113**, 2229–2237.

Helm, P., Beg, M. F., Miller, M. I., and Winslow, R. L. (2005). Measuring and mapping cardiac fiber and laminar architecture using diffusion tensor MR imaging. *Ann. N.Y. Acad. Sci.* **1047**, 296–307.

Jain, R. K. (2003). Molecular regulation of vessel maturation. *Nat. Med.* **9**, 685–693.

Klug, M. G., Soonpaa, M. H., Koh, G. Y., and Field, L. J. (1996). Genetically selected cardiomyocytes from differentiating embronic stem cells form stable intracardiac grafts. *J. Clin. Invest.* **98**, 216–224.

Kofidis, T., Akhyari, P., Boublik, J., Theodorou, P., Martin, U., Ruhparwar, A., Fischer, S., Eschenhagen, T., Kubis, H. P., Kraft, T., Leyh, R., Haverich, A. (2002). *In vitro* engineering of heart muscle: artificial myocardial tissue. *J. Thorac. Cardiovasc. Surg.* **124**, 63–69.

Koh, G. Y., Klug, M. G., Soonpaa, M. H., and Field, L. J. (1993). Differentiation and long-term survival od C2C12 myoblast grafts in heart. *J. Clin. Invest.* **92**, 1115–1116.

Laflamme, M. A., and Murry, C. E. (2005). Regenerating the heart. *Nat. Biotechnol.* **23**, 845–856.

Laugwitz, K. L., Moretti, A., Lam, J., Gruber, P., Chen, Y., Woodard, S., Lin, L. Z., Cai, C. L., Lu, M. M., Reth, M., Platoshyn, O., Yuan, J. X., Evans, S., and Chien, K. R. (2005). Postnatal isl1+ cardioblasts enter fully differentiated cardiomyocyte lineages. *Nature* **433**, 647–653.

Leor, J., Aboulafia-Etzion, S., Dar, A., Shapiro, L., Barbash, I. M., Battler, A., Granot, Y., and Cohen, S. (2000). Bioengineerred cardiac grafts: a new approach to repair the infarcted myocardium? *Circulation* **102**, III56–III61.

Levenberg, S., Rouwkema, J., Macdonald, M., Garfein, E. S., Kohane, D. S., Darland, D. C., Marini, R., van Blitterswijk, C. A., Mulligan, R. C., D'Amore, P. A., and Langer, R. (2005). Engineering vascularized skeletal muscle tissue. *Nat. Biotechnol.* **23**, 879–884.

Li, R.-K., Jia, Z. Q., Weisel, M. H., Mickel, D. A., Zhang, J., Mohabeer, M. K., Rao, V., and Ivanov, J. (1996). Cardiomyocyte transplantation improved heart function. *Ann. Thorac. Surg.* **62**, 654–660.

Li, R.-K., Jia, Z. Q., Weisel, R. D., Mickle, D. A. G., Choi, A., and Yau, T. M. (1999). Survival and function of bioengineered cardiac grafts. *Circulation* **100**, II63–II69.

Li, R.-K., Yau, T. M., Weisel, R. D., Mickle, D. A. G., Sakai, T., Choi, A., and Jia, Z. -Q. (2000a). Construction of a bioengineered cardiac graft. *J. Thorac. Cardiovasc. Surg.* **119**, 368–375.

Li, R.-K., Yau, T. M., Weisel, R. D., Mickle, D. A., Sakai, T., Choi, A., and Jia, Z. Q. (2000b). Construction of a bioengineered cardiac graft. *J. Thorac. Cardiovasc. Surg.* **119**, 368–375.

Makino, S., Fukuda, K., Miyoshi, S., Konishi, F., Kodama, H., Pan, J., Sano, M., Takahashi, T., Hori, S., Abe, H., Hata, J., Umezawa, A., and Ogawa, S. (1999). Cardiomyocytes can be generated from marrow stromal cells *in vitro*. *J. Clin. Invest.* **103**, 697–705.

Menasche, P., Hagege, A. A., Scorsin, M., Pouzet, B., Desnos, M., Duboc, D., Schwartz, K., Vilquin, J. T., and Marolleau, J. P. (2001). Myoblast transplantation for heart failure. *Lancet* **357**, 279–280.

Miyahara, Y., Nagaya, N., Kataoka, M., Yanagawa, B., Tanaka, K., Hao, H., Ishino, K., Ishida, H., Shimizu, T., Kangawa, K., Sano, S., Okano, T.,

Kitamura, S., and Mori, H. (2006). Monolayered mesenchymal stem cells repair scarred myocardium after myocardial infarction. *Nat. Med.* **12**, 459–465.

Muller-Ehmsen, J., Peterson, K. L., Kedes, L., Whittaker, P., Dow, J. S., Long, T. I., Laird, P. W., and Kloner, R. A. (2002a). Rebuilding a damaged heart: long-term survival of transplanted neonatal rat cardiomyocytes after myocardial infarction and effect on cardiac function. *Circulation* **105**, 1720–1726.

Muller-Ehmsen, J., Whittaker, P., Kloner, R. A., Dow, J. S., Sakoda, T., Long, T. I., Laird, P. W., and Kedes, L. (2002b). Survival and development of neonatal rat cardiomyocytes transplanted into adult myocardium. *J. Mol. Cell. Cardiol.* **34**, 107–116.

Oh, H., Bradfute, S. B., Gallardo, T. D., Nakamura, T., Gaussin, V., Mishina, Y., Pocius, J., Michael, L. H., Behringer, R. R., Garry, D. J., Entman, M. L., and Schneider, M. D. (2003). Cardiac progenitor cells from adult myocardium: homing, differentiation, and fusion after infarction. *Proc. Natl. Acad. Sci. U.S.A.* **100**, 12313–12318.

Olivetti, G., Anversa, P., and Loud, A. V. (1980). Morphometric study of early postnatal development in the left and right ventricular myocardium of the rat. II. Tissue composition, capillary growth, and sarcoplasmic alterations. *Circ. Res.* **46**, 503–512.

Orlic, D., Kajstura, J., Chimenti, S., Jakoniuk, I., Anderson, S. M., Li, B., Pickel, J., McKay, R., Nadal-Ginard, B., Bodine, D. M., Leri, A., and Anversa, P. (2001). Bone marrow cells regenerate infarcted myocardium. *Nature* **410**, 701–705.

Papadaki, M., Bursac, N., Langer, R., Merok, J., Vunjak-Novakovic, G., and Freed, L. E. (2001). Tissue engineering of functional cardiac muscle: molecular, structural and electrophysiological studies. *Am. J. Physiol. Heart Circ. Physiol.* **280**, H168–H178.

Planat-Benard, V., Menard, C., Andre, M., Puceat, M., Perez, A., Garcia-Verdugo, J. M., Penicaud, L., and Casteilla, L. (2004). Spontaneous cardiomyocyte differentiation from adipose tissue stroma cells. *Circ. Res.* **94**, 223–229.

Radisic, M., Euloth, M., Yang, L., Langer, R., Freed, L. E., and Vunjak-Novakovic, G. (2003). High-density seeding of myocyte cells for tissue engineering. *Biotechnol. Bioeng.* **82**, 403–414.

Radisic, M., Park, H., Shing, H., Consi, T., Schoen, F. J., Langer, R., Freed, L. E., and Vunjak-Novakovic, G. (2004a). Functional assembly of engineered myocardium by electrical stimulation of cardiac myocytes cultured on scaffolds. *Proc. Natl. Acad. Sci. U.S.A.* **101**, 18129–18134.

Radisic, M., Yang, L., Boublik, J., Cohen, R. J., Langer, R., Freed, L. E., and Vunjak-Novakovic, G. (2004b). Medium perfusion enables engineering of compact and contractile cardiac tissue. *Am. J. Physiol. Heart Circ. Physiol.* **286**, H507–H516.

Radisic, M., Deen, W., Langer, R., and Vunjak-Novakovic, G. (2005). Mathematical model of oxygen distribution in engineered cardiac tissue with parallel channel array perfused with culture medium containing oxygen carriers. *Am. J. Physiol. Heart Circ. Physiol.* **288**, H1278–H1289.

Radisic, M., Malda, J., Epping, E., Geng, W., Langer, R., and Vunjak-Novakovic, G. (2006a). Oxygen gradients correlate with cell density and cell viability in engineered cardiac tissue. *Biotechnol. Bioeng.* **93**, 332–343.

Radisic, M., Park, H., Chen, F., Salazar-Lazzaro, J. E., Wang, Y., Dennis, R., Langer, R., Freed, L. E., and Vunjak-Novakovic, G. (2006b). Biomimetic approach to cardiac tissue engineering: oxygen carriers and channeled scaffolds. *Tissue Eng.* **12**, 2077–2091.

Rakusan, K., and Korecky, B. (1982). The effect of growth and aging on functional capillary supply of the rat heart. *Growth* **46**, 275–281.

Reinecke, H., Zhang, M., Bartosek, T., and Murry, C. E. (1999). Survival, integration, and differentiation of cardiomyocyte grafts: a study in normal and injured rat hearts. *Circulation* **100**, 193–202.

Reinlib, L., and Field, L. (2000). Cell transplantation as future therapy for cardiovascular disease?: A workshop of the National Heart, Lung, and Blood Institute. *Circulation* **101**, e182–e187.

Roell, W., Lu, Z. J., Bloch, W., Siedner, S., Tiemann, K., Xia, Y., Stoecker, E., Fleischmann, M., Bohlen, H., Stehle, R., Kolossov, E., Brem, G., Addicks, K., Pfitzer, G., Welz, A., Hescheler, J., and Fleischmann, B. K. (2002). Cellular cardiomyoplasty improves survival after myocardial injury. *Circulation* **105**, 2435–2441.

Sakai, T., Li, R. K., Weisel, R. D., Mickle, D. A., Jia, Z. Q., Tomita, S., Kim, E. J., and Yau, T. M. (1999). Fetal cell transplantation: a comparison of three cell types. *J. Thorac. Cardiovasc. Surg.* **118**, 715–724.

Severs, N. J. (2000). The cardiac muscle cell. *Bioessays* **22**, 188–199.

Shimizu, T., Yamato, M., Isoi, Y., Akutsu, T., Setomaru, T., Abe, K., Kikuchi, A., Umezu, M., and Okano, T. (2002). Fabrication of pulsatile cardiac tissue grafts using a novel 3-dimensional cell sheet manipulation technique and temperature-responsive cell culture surfaces. *Cir. Res.* **90**, e40–e48.

Soonpaa, M. H., Koh, G. Y., Klug, M. G., and Field, L. J. (1994). Formation of nascent intercalated disks between grafted fetal cardiomyocytes and host myocardium. *Science* **264**, 98–101.

Soonpaa, M. H., and Field, L. J. (1998). Survey of studies examining mammalian cardiomyocyte DNA synthesis. *Circ. Res.* **83**, 15–26.

Taylor, D. A., Atkins, B. Z., Hungspreugs, P., Jones, T. R., Reedy, M. C., Hutcheson, K. A., Glower, D. D., and Kraus, W. E. (1998). Regenerating functional myocardium: improved performance after skeletal myoblast transplantation. *Nat. Med.* **4**, 929–933.

Tung, L., Sliz, N., and Mulligan, M. R. (1991). Influence of electrical axis of stimulation on excitation of cardiac muscle cells. *Circ. Res.* **69**, 722–730.

Wang, Y., Ameer, G. A., Sheppard, B. J., and Langer, R. (2002). A tough biodegradable elastomer. *Nat. Biotech.* **20**, 602–606.

Yamada, T., Yang, J. J., Ricchiuti, N. V., and Seraydarian, M. W. (1985). Oxygen consumption of mammalian myocardial cells in culture — measurements in beating cells attached to the substrate of the culture dish. *Anal. Biochem.* **145**, 302–307.

Zandonella, C. (2003). Tissue engineering: the beat goes on. *Nature* **421**, 884–886.

Zandstra, P. W., Bauwens, C., Yin, T., Liu, Q., Schiller, H., Zweigerdt, R., Pasumarthi, K. B., and Field, L. J. (2003). Scalable production of embryonic stem cell-derived cardiomyocytes. *Tissue Eng.* **9**, 767–778.

Zimmermann, W. H., Fink, C., Kralish, D., Remmers, U., Weil, J., and Eschenhagen, T. (2000). Three-dimensional engineered heart tissue from neonatal rat cardiac myocytes. *Biotechnol. Bioeng.* **68**, 106–114.

Zimmermann, W. H., Didie, M., Wasmeier, G. H., Nixdorff, U., Hess, A., Melnychenko, I., Boy, O., Neuhuber, W. L., Weyand, M., and Eschenhagen, T. (2002a). Cardiac grafting of engineered heart tissue in syngenic rats. *Circulation* **106**, I151–I157.

Zimmermann, W. H., Schneiderbanger, K., Schubert, P., Didie, M., Munzel, F., Heubach, J. F., Kostin, S., Nehuber, W. L., and Eschenhagen, T. (2002b). Tissue engineering of a differentiated cardiac muscle construct. *Circ. Res.* **90**, 223–230.

Zimmermann, W. H., Melnychenko, I., Wasmeier, G., Didie, M., Naito, H., Nixdorff, U., Hess, A., Budinsky, L., Brune, K., Michaelis, B., Dhein, S., Schwoerer, A., Ehmke, H., and Eschenhagen, T. (2006). Engineered heart tissue grafts improve systolic and diastolic function in infarcted rat hearts. *Nat. Med.* **12**, 452–458.

# Chapter Thirty-Nine

# Blood Vessels

*Luke Brewster, Eric M. Brey, and Howard P. Greisler*

## I. INTRODUCTION

Principles of tissue engineering are now being applied in the induction and development of microvascular networks as well as capacitance conduits, including the *in vitro* and *in vivo* biologic modification of synthetic vascular grafts and the generation of tissue-engineered blood vessels in bioreactors and *in vivo*. Yet 100 years since the development of vascular anastomoses, small-diameter vascular grafts continue to encounter significant translational barriers to widespread clinical application. This is probably because much of the reported progress in the field has been simplistic in design. Still these studies have led to a more complex and realistic understanding of the physiologic and pathologic processes occurring in a coordinated fashion both prior to and after implantation into the circulation. This understanding has allowed old ideas to become new again and may well provide the bridge necessary to cross this 100-year chasm that has separated patients from the ideal blood vessel substitute.

Alexis Carrel, the father of vascular surgery, was the first to describe the utility and shortfalls of autogenous and synthetic grafts. The main limitation discussed was the lack of durability for small-diameter synthetic grafts. For this reason he partnered with Charles Lindbergh to create new blood vessels *de novo* as a substitute for autogenous vessels, thus beginning the field of cardiovascular tissue engineering (Fig. 39.1). The first clinical successes of synthetic vascular grafts were subsequently delayed until the 1950s, when Voorhees developed the first fabric graft, termed the Vinyon N vascular prosthesis (Voorhees *et al.*, 1952); several ethylene-based synthetic grafts followed. Again, large-vessel grafts demonstrated excellent results, but small-diameter grafts did not. Now, more than 50 years after Voorhees and 100 years since the landmark discoveries of Carrel, small-diameter grafting continues to be associated with poor long-term patency rates as compared to autogenous vessels.

Unfortunately, cardiovascular disease has become more prevalent in the last 100 years, now being the leading cause of death in the world and the second leading cause of death for all Americans. Despite improvements in the medical therapy of cardiovascular disease, the number of vascular interventions (combining bypass grafting and angioplasty ± stenting) has increased in recent years. Currently there are more than 1.4 million bypass operations annually in the United States, and, although autogenous veins or arteries provide the best patency rates for cardiovascular bypass, many patients do not have suitable autogenous vessels available for use. For these reasons the creation of durable small-vessel vascular conduits is critical to the successful treatment of cardiovascular disease and injury.

Since the idea of a nonreactive vascular conduit as an ideal graft is no longer accepted, current initiatives emphasize its incorporation into the surrounding vessels and tissue. Such approaches include the development of non-

*Principles of Tissue Engineering, 3rd Edition*
*ed. by Lanza, Langer, and Vacanti*

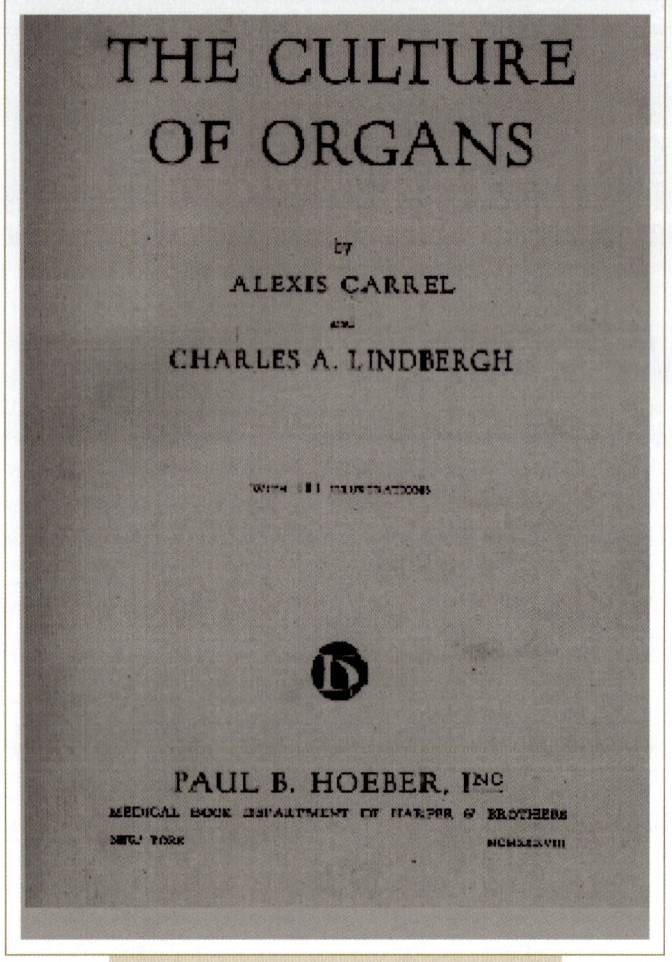

**FIG. 39.1.** "The culture of organs." Illustration of the classic 1938 text coauthored by Alexis Carrel and Charles Lindbergh outlining the potential uses of tissue engineering.

autologous biologics, synthetic grafts that promote or support tissue ingrowth, and tissue-engineered vascular grafts that mimic the functional properties of the blood vessel.

This chapter discusses the current status of vascular bypass grafts, the modifications of these grafts designed to facilitate healing, arteriogenesis, and therapeutic angiogenesis, and tissue-engineered vascular grafts. It concludes with tissue-engineering approaches to the rapidly advancing field of endovascular therapeutics.

## II. CURRENT STATUS OF VASCULAR CONDUITS

### Conduit Patency and Failure

Large-diameter grafts, such as those used for aortic reconstruction, have a superb five-year patency rate, approaching 95%. When synthetic grafts are used in the infrapopliteal region, the results are worse: One- and three-year patency rates for infrapopliteal bypass using ePTFE grafts was 43% and 30%, respectively (Veith *et al.*, 1986), and the outcomes are even more dire for those with comorbidities such as diabetes and renal insufficiency.

For smaller arteries the results with autogenous vein are also less than impressive, with the exception being the left internal mammary artery for coronary artery bypass grafting, which has a greater than 90% five-year patency. Currently, autogenous vein is the conduit of choice for concomitant coronary artery bypass grafting and infrainguinal vascular reconstruction. However, its primary and secondary patency rates at one year in a patient population with critical limb ischemia were recently reported to be 61% and 80%, respectively (Conte *et al.*, 2006); the results were not better in the coronary vasculature. These results are strikingly similar to those reported over 30 years ago.

A variety of mechanisms can lead to vascular graft occlusion and occur in a defined temporal sequence. Immediate graft failure is usually the result of technical error from the operation or from the patient's having a hypercoagulable status. Failure in the first month following graft placement is most likely the result of thrombosis secondary to distal flow resistance. Small-diameter grafts are prone to early thrombosis because of their lower flow rates and the higher resistance in their outflow vessels. Thus graft thrombogenicity is of primary concern early after graft placement.

Anastomotic pseudointimal hyperplasia[1] is the most common reason for graft failure from six months to three years after graft insertion, and later graft failure is frequently due to the progression of distal atherosclerotic disease. Small-diameter grafts are particularly susceptible to anastomotic myointimal hyperplasia (IH). Analogously, other vascular interventions, such as angioplasty and endarterectomy, fail because there is a limited regenerative endothelial cell (EC) capacity after injury in humans, leading directly to both the thrombotic and myointimal hyperplastic complications.

Pathophysiologically, denuded intima or an exposed luminal area of a graft may lead to thrombosis via platelet deposition and activation of the coagulation cascade, and over time it promotes pathologic smooth muscle cell (SMC) migration, proliferation, and extracellular matrix (ECM) deposition, leading to IH. IH in turn narrows the vessel lumen (restenosis), decreasing blood flow to the point where it may promote local thrombosis or lead to symptomatic ischemia in the relevant distal-end organs, such as the brain (cardiovascular accidents), heart (myocardial infarctions), and extremities (critical limb ischemia). Since thrombogenicity and intimal hyperplasia represent the most common causes of graft failure and both are mediated at the luminal interface of the vessel or graft, the inner lining of grafts has

---

[1] Although the terms *myointimal, neointimal,* and *pseudointimal hyperplasia* are often used interchangeably, here *neointima* is used to describe areas that include all normal intimal constituents, including ECs, whereas *pseudointima* is used to describe the relatively acellular tissue within clinically implanted vascular grafts; *myointima* is used more generically to include both types.

been the subject of much investigation. To date, however, we are still in the early stages of understanding and treating the complications of vascular interventions.

## Cellular and Molecular Mediators of Graft Outcome

### Normal and Pathologic Composition of the Vessel Wall

For arteries, the intima (*tunica intima*) is composed of a relatively quiescent EC monolayer and its surrounding basement membrane proteins (e.g., type IV collagen, perlecan). Together with the underlying internal elastic lamina, the intima maintains SMCs in their contractile state and inhibits pathologic SMC activity. Deep to the intima and separated by the internal elastic lamina is the medial layer (*tunica media*). It is the thickest arterial layer and in non-pathologic states is composed of SMCs and many ECM proteins (e.g., elastin and the fibrillar collagens, such as type I collagen). Medial SMCs here respond to intimal cues to dilate or contract the vessel. The medial layer in veins is difficult to define because it lacks the internal elastic lamina, but after exposure to arterial flow via vein graft bypass, they develop a defined media over time that has similarities to that of native artery. The external elastic lamina defines the abluminal edge of the media, and the vaso vasorum, which supplies most of the vessel wall's (outer two-thirds) metabolic needs, is prominent in the outer adventitial vessel layer (tunica adventitia). The adventitia is composed of loosely arranged connective tissue and fibroblasts, and it may play an important role in the progression of restenosis and late interventional failure after angioplasty due to vessel remodeling, since adventitial delivery of therapeutic treatments has reduced these complications (Scott *et al.*, 1996).

In the absence of disease or injury, native blood vessels possess an endothelial lining that constantly secretes bioactive substances inhibiting thrombosis, promoting fibrinolysis, and inhibiting SMCs from switching from a contractile to a synthetic phenotype. All the while the intima directly or indirectly (through cues to the vessel media) is maintaining blood flow to meet distal tissue demands. Thus, reestablishing this type of intimal lining quickly and completely in vascular grafts is vital to the normal function of small-diameter vascular grafts. Unfortunately, and unlike most animal models in use, which spontaneously endothelialize synthetic grafts, humans manifest only limited endothelial cell ingrowth, not extending beyond 1–2 cm of both anastomoses. However, endothelial islands have been described in the midportions of grafts at significant distances from the anastomosis, suggesting that other EC sources for graft endothelialization may exist. Interstitial tissue ingrowth accompanied by microvessels from the perigraft tissue is one potential source. There is also evidence that circulating endothelial cells, endothelial progenitor cells (EPC), or stem cells can be directed to these areas. Such homing can be promoted through affixation of EC-attractant antibodies to the grafts in a similar fashion as has been utilized in coronary artery stents (Aoki *et al.*, 2005).

However, ECs growing on prosthetic graft surfaces are not necessarily the same as their normal quiescent counterparts in uninjured vessels. These ECs are often activated, secreting bioactive substances (PDGF) that actually promote thrombogenesis and changes in SMC phenotype. This has been seen in the perianastomotic region, which is the most frequent site of interventional failure after operation. SMCs found within the myointima of prosthetic grafts are also functionally altered. They produce significantly higher amounts of PDGF, as well as various extracellular matrix proteins, than those of the adjacent vessel, which, along with the body's inflammatory reaction to prosthetic material, contribute to the development of intimal hyperplasia (Pitsch *et al.*, 1997).

### Inflammation

The inflammatory response to vascular interventions is complex. It is mediated by both inflammatory cells and proteins in a cooperative manner. Potent chemoattractants like complement 5a (C5a) and leukotriene $B_4$ recruit neutrophils to the graft surface, where they localize in the fibrin coagulum of the graft's inner and outer capsule via β2 integrins. Also, IgG binds to the neutrophils' Fcγ receptors, activating the neutrophils' pro-inflammatory response while inhibiting normal clearance of bacteria. Neutrophils also interact with various other deposited proteins, including C3bi and factor X, and they adhere to the endothelial cells in the perianastomotic region through selectin- and integrin-mediated mechanisms. L-Selectin is thought to modulate neutrophil/endothelial cell interactions by presenting neutrophil ligands to both the E- and P-selectins on the vascular endothelium. In addition selectin–carbohydrate bonds are important for the initial cellular contact, while the integrin–peptide bonds are responsible for strengthening of this adhesion as well as the transmigration of neutrophils. Both intercellular adhesion molecule-1 (ICAM-1) and vascular cell adhesion molecule-1 (VCAM-1) on the EC surface bind these integrins as well, and ECs up-regulate ICAM-1 and express VCAM-1 when stimulated by inflammatory agonists like IL-1, TNF, lipopolysaccharide, and thrombin. Further, activated neutrophils release oxygen-free radicals and various proteases, which result in matrix degradation and may inhibit both endothelialization and tissue incorporation of the vascular graft.

Circulating monocytes/macrophages are also attracted to areas of injured or regenerating endothelium, especially in response to IL-1 and TNF. There are many plasma monocyte–recruitment and –activating factors, including $LTB_4$, platelet factor 4, and PDGF. This process is propagated in the presence of these plasma-activating factors, driving monocytes to differentiate into macrophages that direct the host's chronic inflammatory response via the release of proteases and oxygen-free radicals.

A variety of cytokines are released from inflammatory cells activated by vascular grafts. Lactide/glycolide grafts are composed of bioresorbable materials that are phagocytosed by macrophages; in culture these materials stimulate macrophages to release mitogens that stimulate vascular cells. This mitogenic activity appears to be related to the secretion of FGF-2, since pretreatment of the culture media with a neutralizing anti-FGF-2 antibody significantly diminishes the stimulatory effect on smooth muscle cell growth (Greisler *et al.*, 1991). Cultured monocytes and macrophages incubated with Dacron and ePTFE have been demonstrated to produce different amounts of IL-1β, IL-6, and TNF-α that are biomaterial specific (Swartbol *et al.*, 1997). TNF-α is one of the factors that may contribute to the enhanced proliferation of SMCs caused by leukocyte–biomaterial interactions, while IL-1 may be partly responsible for the increased SMC proliferation caused by leukocyte–EC reaction. IL-1 also induces up-regulation of IGF-1 expression in ECs, and coculture of neutrophils with IL-1β-treated ECs dramatically increases PDGF release.

Since the inflammatory reaction elicits a cascade of growth processes, it has been proposed that approaches attenuating the initial inflammatory reaction may improve long-term graft patency. Alternatively, directing the inflammatory response to promote favorable cellular and protein responses may promote intimal generation and tissue incorporation.

## III. PHYSICAL OR CHEMICAL MODIFICATION OF CURRENT GRAFTS TO IMPROVE DURABILITY

The long-term patency of vascular grafts depends on the intrinsic properties of the graft itself and the hemodynamic environment in which the graft is placed, as well as patient variables (e.g., diabetes and renal failure) and may or may not be improved by prior or concomitant interventions such as proximal or distal angioplasty. Since it is now clear that tissue incorporation is important for graft function, grafts that have this ability are now desirable for medium- and small-caliber vessel replacement. Polyethylene terephthalate (Dacron) and expanded polytetrafluoroethylene (ePTFE) are the predominant materials currently used in prosthetic vascular grafts, but both Dacron and ePTFE react with blood components and perigraft tissues in concomitantly beneficial and detrimental fashions. All grafts, regardless of their composition and structure, evoke complex but predictable host responses that begin immediately upon restoration of perfusion, and this improved understanding of the cellular and molecular components of biomaterial/tissue interactions has led to more intelligent designs of grafts that maximize beneficial ingrowth while minimizing the chronic inflammatory changes that lead to graft dilation or occlusion. These approaches include protein adsorptive grafts (growth factors, anticoagulants, antibiotics, etc.) as well as improved graft skeletal construction via synthetic polymers or biologically derived structural proteins that can be bonded to various bioactive cytokines and growth factors to induce a favorable host response.

The humoral and cellular responses to synthetic materials include the deposition of plasma proteins and platelets; infiltration of neutrophils, macrophages, and circulating progenitor cells; and the migration and proliferation of adjacent vascular endothelial and smooth muscle cells. Together these complex interactions mediate the healing response at the graft/blood interface that is critical to future graft durability and patency. Thus, the chemical composition, construction parameters, and biomechanical characteristics of a vascular graft are the predominant mediators of host interaction and largely determine graft fate.

### Surface Characteristics

The thrombotic interaction at the graft's luminal interface is dependent on both the chemical and physical properties of the graft (e.g., surface charge, surface energy, and roughness). A negative surface charge attenuates platelet adhesion and a positive charge promotes it, while a heterogeneous charge density distribution is also thought to be thrombogenic. Myriad approaches have been designed to limit the thrombotic reaction, including modification of surface properties, incorporation of antiplatelet or anticoagulant substances onto the graft surface, and endothelialization of the luminal surface. In addition to thrombogenic reactions to the luminal surfaces, the rate and extent of endothelialization will vary, depending on the characteristics of the surface (Miller *et al.*, 2004). Therefore properties must be optimized for both reduced thrombogenic reaction and maximized endothelialization.

### Surface Modifications

The simplest modification of a graft surface is to coat it with an inert polymer. Carbon coating has been known since the 1960s to decrease surface thrombogenicity through its negative charge and hydrophobic nature. Experimentally, the carbon-impregnated prosthetic graft was found experimentally to reduce platelet deposition, but the advantage of these grafts was not confirmed in a prospective multicenter clinical study of 81 carbon-impregnated ePTFE and 79 standard ePTFE grafts for below-knee popliteal and distal bypass. Here the investigators failed to show a significant difference in patency rate between the two groups at up to 12 months after implantation (Bacourt, 1997).

Silicone polymer coating is another approach to altering the luminal surfaces of grafts; this process produces a smooth surface that is devoid of the usual ePTFE graft permeability and texture, and, when followed by plasma glow discharge polymerization, it effectively abolishes pannus tissue ingrowth as well as graft surface neointimal hyperplasia in a baboon arterial interposition graft model. Further, Nojiri *et al.* (1995) have developed a three-layered graft consisting of Dacron for the outer layer (to promote perigraft tissue incorporation), nonporous polyurethane in the middle layer (to obtain a smooth surface), and a 2-hydroxyethyl methacrylate and styrene (HEMA-st) copolymer coating for the inner layer (to establish a nonthrombogenic blood interface). These grafts with an ID of 3 mm were implanted in canine carotid arteries and remained patent for over one year. Only a monolayer of adsorbed proteins was described on the luminal surface of the grafts, with no pannus ingrowth from the adjacent artery, no thrombus, and no endothelial lining or neointimal formation.

## Protein Adsorption

Another approach is to cover vascular grafts' lumens with proteins. Protein coating has been used as an alternative to preclotting with blood to decrease the initial porosity of Dacron grafts in order to limit transmural blood loss. Knitted Dacron prostheses coated with albumin, gelatin, and collagen have all been available for clinical use. As the impregnated proteins are degraded, the graft undergoes tissue ingrowth.

The albumin coating created in the 1970s was found in animal models to diminish platelet and leukocyte adhesion. In the canine thoracoabdominal aortic model, knitted Dacron grafts impregnated with carbodiimide cross-linked human albumin were compared to identical Dacron grafts with the recipients' blood preclotting. Albumin-impregnated grafts displayed less transinterstitial blood loss at implantation and significantly thinner pseudointimas at 20 weeks (Kang *et al.*, 1997).

Collagen coating establishes a good matrix for cell ingrowth and induces the necessary myointimal formation, and, although native collagen is intrinsically thrombogenic, cross-linking of collagen limits this effect. Promising early results were reported with collagen coating, but a clinical study reported that there was no appreciable difference in graft patency between woven and collagen-impregnated knitted Dacron aorto–iliac grafts (Quarmby *et al.*, 1998). Gelatin is a derivative of collagen that is easily degraded when applied as a graft coating, and this degradation may be exploited via modifications of coating techniques to exploit this property.

The most abundant serum proteins are albumin, fibrinogen, and IgG. They adsorb to grafts almost instantaneously following exposure to the systemic circulation. Subsequently, there is a redistribution of proteins according to each protein's relative biochemical and electrical affinity for the graft surface and their relative abundance (Vroman, 1987). Since platelets and blood cells interact predominantly with the bound proteins and not with the prosthetic material itself, the constitution and concentration of bound protein has a profound influence over the type and degree of cellular interaction with the graft. Fibrinogen, laminin, fibronectin, and vitronectin all have an arginine-glycine-

aspartate (RGD) sequence that is recognized by platelets' glycoprotein (GPIIb/IIIa) receptor and initiate platelet activation. RGD sequences are also recognized by $\beta_2$ intergrin, which directs leukocyte adhesion to the graft. Additional plasma proteins, including complement components, can also be differentially activated directly by different synthetic surfaces. For example, the generation of the monocyte chemoattractant C5a is greater following implantation of Dacron as compared to ePTFE grafts in an animal model (Shepard *et al.*, 1984). In addition, the rapid accumulation of coagulant proteins such as thrombin and Factor Xa on the luminal surface after implantation contributes to the thrombogenicity of vascular grafts.

## Porosity

The prevalence of open spaces or pores determines the porosity of a scaffold or synthetic graft, while the permeability of a graft is defined by its ability to permit passage of a substance through itself. Since ePTFE is composed of a number of solid nodes interconnected by a matrix of thin fibrils with no uninterrupted transmural spaces, it is best categorized by the distance between these nodes, which is defined as the internodal distance (IND). This spacing allows for cellular ingrowth, but transinterstitial ingrowth is not strictly a function of porosity. We have shown that the extent of ingrowth varies greatly among different biomaterials (e.g., PGA and Dacron), despite these biomaterials' having similar porosity (Greisler *et al.*, 1985).

Still, the rate of tissue ingrowth can be improved by optimizing graft porosity or permeability. Clowes *et al.* (1986) have demonstrated enhanced tissue ingrowth and complete re-endothelialization of 60-μm or 90-μm-internodal-distance ePTFE grafts in a baboon model. However, transinterstitial capillary ingrowth was not seen with the more commonly used 30-μm internodal distance ePTFE. Human trials using ePTFE with these expanded internodal distances failed to show any advantage in platelet deposition as compared to standard 30-μm-internodal-distance ePTFE grafts (Clowes and Kohler, 1991).

## Compliance

The compliance mismatch between arteries and grafts causes flow disruption *in vivo*, which may contribute to anastomotic pseudointimal hyperplasia (Abbott and Cambria, 1982). It is for this reason that various surgeons have suggested interposing a segment of vein between the synthetic graft and artery, creating a composite graft at the distal anastomosis. This has led some investigators to design more compliant grafts using more flexible materials and/or changing the parameters of graft construction to improve graft compliance. Although animal experiments have suggested concept validity, the clinical benefit of this approach remains controversial. Many factors may contribute to this confusion, including longitudinal variability in the diameter and compliance of the arterial tree and the effect of activated endothelium on intimal hyperplasia. Further, there is a robust fibrotic response after implantation that leads compliant grafts to become incompliant after implantation; thus, even if a compliance match were attained initially, it would not likely persist. In the para-anastomotic region there are dynamic changes in compliance that vary over time. First a para-anastomotic hypercompliant zone exhibits a 50% gain in compliance; later its compliance is lessened 60% from baseline (Hasson *et al.*, 1985). It is likely that this bimodal effect limits the practical value of this approach.

## Thromboresistance

Early platelet deposition on vascular grafts is mediated by von Willebrand factor and platelet membrane glycoproteins. After adherence to a graft, platelets degranulate, releasing many bioactive substances, including serotonin, epinephrine, ADP, and thromboxane A2; these substances in turn activate additional platelets and promote a prothrombogenic reaction. Activated platelets also release growth factors, such as PDGF, EGF, and TGF-β, which promote SMC migration and proliferation as well as ECM degradation and ECM protein synthesis. In addition, platelets release monocyte chemoattractants such as platelet factor 4 and β-thromboglobulin, which mediate the recruitment of macrophages to the graft. Platelet deposition and activation continues chronically after graft implantation, as evidenced by increased thromboxane levels and decreased systemic platelet counts one year after Dacron graft implantation in a canine model (Ito *et al.*, 1990), and human studies have confirmed platelet adhesion to grafts up to one year after implantation.

Since the deposition and activation of platelets elicit various pathologic cascades, the thrombogenic nature of the synthetic graft surface can lead to both early and late graft failure. Myriad approaches have been studied to attenuate platelet deposition, aggregation, and degranulation. Antiplatelet agents directly targeting platelet/graft-binding molecules such as platelet surface GPIIb/IIIa and different functional domains of thrombin have been shown, at least transiently, to decrease the accumulation of platelets on Dacron grafts (Mazur *et al.*, 1994). Also, the surface thrombogenicity of grafts can be altered experimentally, as described earlier.

Similar approaches can be utilized to decrease thrombogenicity by disrupting the activation of the blood system's coagulation cascade on thrombogenic surfaces, such as cardiovascular stents and synthetic grafts. We have reported improved patency, thromboresistance, and the absence of intraluminal graft thrombus in heparin-analog-coated ePTFE grafts using a canine bilateral aorto–iliac artery model (Laredo *et al.*, 2004). Similarly, combined data tables of clinical data showed a decrease in subacute thrombosis with heparin-coated coronary artery stents (Gupta *et al.*, 2004). Genetic approaches to increase thromboresistance have been employed by multiple groups through the overexpression of thrombotic inhibitors. But since ECs themselves are antithrombotic, there may be limited benefit of this approach when compared to establishing a functioning endothelium.

## Resistance to Infection

Vascular graft infection is rare, but it is catastrophic when it occurs, as demonstrated by an amputation rate of approximately 50% and a reported mortality rate that ranges from 25% to 75%. In an attempt to limit this dreaded complication, penicillin and cephalosporins have been successfully bound to Dacron and ePTFE grafts and found to limit *Staphylococcus aureus* infection in animal models. Rifampin-bonded gelatin-sealed Dacron grafts have also been shown *in vitro* to lessen bacterial colonization (Vicaretti *et al.*, 1998). Intuitively, tissue ingrowth itself may also provide resistance to infection.

## Biological Modification Through Exogenous Sources

The delivery of potent angiogens or genes that promote EC-specific mitogenesis or chemotaxis on prosthetic surfaces may be used to regenerate a rapid and complete endothelium after vascular intervention. Such prosthetics could store these genes or proteins and provide a controlled expression or release of these genes or proteins locally to circulating or surrounding ECs in a cell-demanded fashion. Ideally, this kind of prosthetic could then be available as an off-the-shelf alternative to autogenous vein.

### Protein Therapy

Although tissue incorporation is a desirable process for implanted prostheses, excessive vascular cell proliferation as well as extracellular matrix (ECM) deposition can lead to intimal hyperplasia and ultimately graft failure. The ideal healing process in vascular grafts would be rapid endothelialization of blood-contacting surfaces, concomitant to a spatially and temporally limited subendothelial SMC growth and followed by phenotypic and functional differentiation of cellular components and the subsequent remodeling of a mature ECM. The recent expansion of knowledge concerning mechanisms responsible for the migration and proliferation of ECs and SMCs, angiogenesis, ECM deposition and remodeling, and physiologic parameters provides optimism for the possibility of manipulating the healing process through the directed manipulation of the microenvironment within the graft and perigraft tissue.

Since ECs have only limited capacity for regeneration, re-endothelialization of the relatively large surface areas encountered clinically exceeds the normal mitogenic capacity of surrounding ECs. Thus endothelialization requires the recruitment of ECs from sites beyond the anastomotic border via the circulation or through transinterstitial migration from the surrounding tissue and/or the vasa vasorum. This is possible under the direction of localized angiogenic stimuli, and to a limited degree this is what occurs *in vivo* as protease-driven ECM changes and local availability of

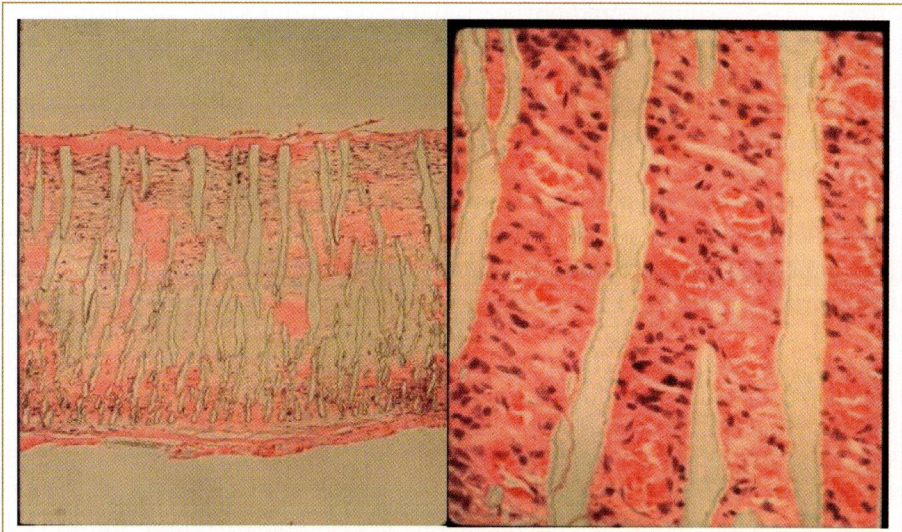

**FIG. 39.2.** Transmural induction of endothelialization by FGF-1 on prosthetic graft. Fibrin glue delivery of exogenous FGF-1 promotes ePTFE endothelialization. The untreated control is pictured on the left side at 117×, and the FGF-1 treated graft is seen on the right at 486×. The treated graft demonstrates robust capillary ingrowth and cellular coverage not seen in the control graft. Reprinted from Gray *et al.* (1994), pp. 600, 602, with permission from Elsevier.

growth factors stimulate ECs, SMCs, and fibroblasts to enter the cell cycle.

Platelet deposition and the perigraft inflammatory response lead to the release of many cytokines and proteolytic enzymes, which in turn incite secretion of a cascade of growth factors. Some of these growth factors are potent angiogens (e.g., VEGF and FGF), while others are angiogenic inhibitors (e.g., platelet factor 4, thrombospondin); still others (e.g., TNF-α and TGF-β) exhibit both activities.

The local delivery of growth factors or selected ECM may be utilized to promote desirable cellular events, such as endothelial ingrowth of synthetic grafts. For example, local delivery of exogenous angiogenic factors through a biologic delivery system (e.g., fibrin) may induce transmural capillary ingrowth *in vivo*, which can be the source of cells for an endothelial lining within synthetic grafts. Bioresorbable grafts may also facilitate local angiogenic responses by stimulating macrophage release of angiogenic proteins. Fibroblast growth factors, notably FGF-1 (acidic FGF) and FGF-2 (basic FGF), have potent mitogenic, chemotactic, and angiogenic activity on vascular cells. But, in order to deliver them locally to direct intimal regeneration, they require a delivery system that provides predictable local release, with bioactivities preserved over a specific interval of time. Their ability to promote endothelial ingrowth has been tested experimentally after application to grafts. Our lab has evaluated the affixation of FGF-1 to synthetic surfaces. In early attempts, FGF-1 was applied to various synthetic grafts via a fibrin glue delivery system that, due to its structural orientation and state of polymerization, had been found not to be thrombogenic (Zarge *et al.*, 1997). After delivering FGF-1 from fibrin glue to 60-μm IND ePTFE in both canine aorto–iliac and thoracoabdominal aortic models, there was a significant increase in luminal EC proliferation, as assayed by en face autoradiography, and a more rapid development of a confluent Factor VIII–positive endothelial blood-contacting surface (Greisler *et al.*, 1992;

Gray *et al.*, 1994) (Fig. 39.2). There was also extensive trans-interstitial capillary ingrowth observed throughout the graft wall. Cross-sectional autoradiography did find a significant increase in subendothelial myofibroblast proliferation in these treatment grafts at one month, but this returned to baseline at later time points. Still, treated grafts developed a significantly thicker pseudointima (139 μm ± 178 μm versus 93 μm ± 89 μm and 67 μm ± 151 μm) at 140 days. In order to limit this IH response, we have developed site-directed mutants based on the FGF-1 angiogen and bioactive chimeric proteins that promote favorable characteristics, including prolonged bioactivity, EC specificity, and increased potency, while removing unfavorable characteristics, such as heparin-dependent activity and susceptibility to thrombin-induced proteolysis (Erzurum *et al.*, 2003; Brewster *et al.*, 2004).

Coimmobilization of FGF-2 and heparin in a microporous polyurethane graft by cross-linked gelatin has also been demonstrated to accelerate tissue regeneration on synthetic grafts, associated with a greater extent of endothelialization via perianastomotic and transmural capillaries ingrowth, in a rat aortic grafting model (Doi and Matsuda, 1997). A consistent neointima of approximately 40-μm thickness with intermittent endothelialization as well as SMCs and fibroblasts underneath the luminal surface were observed in the middle portion of treatment grafts, compared to the control grafts, covered with only a fibrin layer. However, because of the cross-talk that exists in the vessel wall between ECs, SMCs, and fibroblasts, multimodal therapies that promote EC coverage while limiting activation of VSMCs or the delayed delivery of cell cycle inhibitors may be required to optimize graft healing.

### Gene Therapy

Gene therapy shares the same promise as proteins, but it may also allow sustained or controlled protein expression in a desired location that is not possible with protein formulations. These qualities could obviate some of the current limitations encountered with the direct application of growth factor proteins to tissue beds for the regeneration of the endothelium.

This approach shows much promise as a delivery system, but single gene therapy trials have not yielded straightforward results. For example, although VEGF does improve endothelialization consistently, it does not reliably limit IH (and sometimes even promotes IH) in the literature. Gene therapy still requires cellular transduction or infection, and this has a variable effect on the cellular behavior. Also, controversial results have been reported in the literature related to the proliferation, adhesion, and retention of genetically modified ECs on the surface of synthetic grafts. A major concern is that genetically modified ECs display poor retention on graft surfaces *in vivo*. This was demonstrated at six weeks in canine thorocoabdominal aortic ePTFE grafts seeded with lacZ-infected ECs as compared to noninfected control ECs (Baer *et al.*, 1996). Further, Dunn (Dunn *et al.*, 1996) reported only 6% retention of ECs that had been retrovirally infected with thromboplastin on Dacron grafts after two hours of exposure to flow *in vivo*. For these reasons, little success has been documented to date concerning the long-term benefit of genetically modified EC-seeded grafts *in vivo*.

### Cells

In addition to the emphasis on the endothelialization of the flow surface, the function of other cell types, particularly the SMCs, in the vascular wall have become better appreciated. It has been suggested that ECs by themselves cannot produce a stable intima without SMCs or fibroblasts underneath. In support of this contention, tissue fragments containing multiple cell types, including venous tissue, adipose tissue, and bone marrow, have been seeded onto grafts and found to accelerate graft-healing processes. Interestingly, bone marrow cell seeding was also reported to induce an abundant capillary ingrowth in the graft wall and a rapid, complete endothelialization of the inner surface without intimal hyperplasia. Since bone marrow stem cells have the ability to differentiate in response to their microenvironment and to proliferate as well as secrete cytokines critical to their survival, they may provide a useful cell source for blood-vessel-tissue engineering.

## IV. THERAPEUTIC ANGIOGENESIS AND ARTERIOGENESIS

### Therapeutic Angiogenesis

In addition to stimulating graft re-endothelialization via transmural angiogenesis, a number of cardiovascular pathologies and surgical interventions could benefit from the ability therapeutically to stimulate new blood vessel formation (neovascularization). Peripheral vascular disease, myocardial ischemia, wound healing, and tissue engineer-

ing are just a few of the many fields that may benefit from this approach.

### Growth Factor Therapy for Angiogenesis

The process of neovascularization is controlled by a complex spatial and temporal expression of proteins, and to date most strategies have attempted to stimulate angiogenesis by injecting growth factors (as proteins) systemically or directly into target tissues. VEGF, FGF-1, and FGF-2 have been utilized frequently, reaching the point of clinical trials for the treatment of myocardial and/or peripheral limb ischemia. Many other growth factors also play a role in neovascularization and are under investigation in animal and *in vitro* models. Angiopoietins (Ang-1 and Ang-2), placental growth factor (PlGF), FGF-4, hepatocyte growth factor (HGF), ephrin-B2, and platelet-derived growth factor BB (PDGF-BB) are a few of the many other proteins with the potential to be used for therapeutic neovascularization.

Due to short protein half-lives *in vivo* and their rapid diffusion out of target tissues, growth factor therapies require high initial concentrations, and repeat injections are required in order to achieve a noticeable response. Yet sustained protein levels are most likely needed in order to form a stable microvasculature. Since it is the local microenvironment concentration that determines the structure of the resultant microvasculature (Ozawa *et al.*, 2004), these initially high concentrations can lead to demonstrable side effects, such as hyperpermeable or malformed vessels. Even sustained local concentrations may not elicit a sufficient angiogenic response, because ischemic tissue may have an altered ability to respond to growth factors (Ruel *et al.*, 2003).

### Gene Therapy for Angiogenesis

The genes of these same proteins have also been investigated as neovascularization therapies. As with gene delivery to vascular grafts, the goal of these approaches is typically local overexpression of the protein. Delivery of a single gene or protein has shown promise in animal models, but clinical results have fallen short of expectations. Overexpression of multiple proteins may have a synergistic effect on neovascularization. Placental growth factor (PlGF) combined with VEGF has been shown to be significantly more potent than each factor alone in an animal model that was refractory to a single protein (Autiero *et al.*, 2003). Combined delivery of adenovirus-mediated VEGF and Ang-1 has also been shown to promote greater perfusion and vessel stability in muscle flaps than VEGF alone (Lubiatowski *et al.*, 2002). Combined therapies may also be more effective in treatment of diseased or elderly patients who are known to have an impaired response to a single protein.

### Cellular Therapy for Angiogenesis

Cellular therapies can also be used to enhance neovascularization. ECs injected into ischemic or engineered tissues are thought to mimic vasculogenesis and to assemble into capillary structures, but they may also act as sources of pro-angiogenic factors. Studies have shown that these cells can fuse with invading host vessels, recruit perivascular cells, and establish flow (Nor *et al.*, 2001) and that support cells can prolong the existence of transplanted vessels (Koike *et al.*, 2004). Here ECs seeded alone into fibronectin-collagen type I gels and implanted into mice showed little perfusion and regressed from the gels after 60 days, while ECs combined with mesenchymal precursor cells resulted in the formation of vessels that established flow through connections with the mouse circulatory system. Endothelial progenitor cells (EPCs) (S. Park *et al.*, 2004; Suh *et al.*, 2005) may also be used to increase neovascularization. These cells may have increased proliferative capacity relative to mature ECs and may be less sensitive to the short-term hypoxic conditions in tissues prior to establishing a blood flow. Injected EPCs selectively localize in ischemic tissues and increase vascular density (S. Park *et al.*, 2004). However, the mechanism for this increase is not clear. EPCs may develop into new vascular structures or may increase neovascularization indirectly by recruiting monocytes/macrophages that then secrete angiogenic factors (Suh *et al.*, 2005).

EPCs can also be genetically modified to enhance their therapeutic function. EPCs transfected to express VEGF stimulate a greater improvement in blood flow and angiogenesis in animal models of ischemia than EPCs alone (Iwaguro *et al.*, 2002). While they have a longer life than fully differentiated cells, EPCs isolated from adults have reduced telomerase activity and regenerative capacity relative to embryonic stem cells. EPCs isolated from bone marrow and transfected to express telomerase reverse transcriptase (TERT) are more resistant to apoptosis and increase neovascularization in an animal model of limb ischemia (Murasawa *et al.*, 2002).

### Arteriogenesis

Facilitating arterialization of capillary beds (arteriogenesis) rather than formation of new capillary beds (angiogenesis) may be critical to facilitating adequate perfusion needed to overcome tissue ischemia. Recent work has shown that nonviral monocyte chemoattractant protein-1 (MCP-1) gene delivery increases arteriogenesis in a rabbit ischemic limb (Muhs *et al.*, 2004), and results from clinical trials suggest that VEGF gene therapy may also stimulate arteriogenesis. Less specifically, cell delivery can also enhance arteriogenesis.

Monocytes injected intravenously into rabbits following femoral artery ligation have been shown to home to sites of ischemia and to stimulate arteriogenesis. By genetically modifying these monocytes with granulocyte-monocyte colony-stimulating factor, collateralization was increased significantly (Herold *et al.*, 2004). Although promising, monocytes are also active participants in atherosclerosis, and it may be difficult to direct arteriogenesis without concomitant atherogenic activity.

# V. TISSUE-ENGINEERED VASCULAR GRAFTS

Driven by the desire to develop an ideal vascular substitute, the construction of tissue-engineered arterial grafts has been attempted, with variable success. The potential benefits of tissue-engineered vascular grafts (TEVGs) include the creation of a responsive and self-renewing tissue graft with functional intimal, medial, and adventitial layers (including both cellular and ECM components) that can be remodeled by the body according to its needs. Such grafts may improve graft durability and reduce the potential for graft infection by lessening the foreign-body reaction and facilitating a more complete integration of the graft into the surrounding tissue.

## *In Vitro* Tissue-Engineered Vascular Grafts

Weinberg and Bell (1986) were the first to develop a TEVG *in vitro*. Using collagen and cultured bovine vascular cells, they demonstrated the feasibility of creating a TEVG, but their graft had prohibitively low burst pressures, requiring external Dacron for support. In the following decade L'Heureux *et al.* (1998) constructed a human blood vessel with an acceptable burst strength and a thromboresistant endothelium *in vitro* using cultured umbilical cord–derived human cells. But because of the immunogenic effects of the heterogeneic ECs *in vivo*, this graft, devoid of ECs, had only a 50% patency rate at eight weeks in a canine model. Since neonatal cells have a greater rejuvenative capacity, the foregoing TEVG was not considered applicable to the aged population that would most benefit from a TEVG. The prototype TEVG was created by seeding ECs onto ePTFE grafts *in vitro* and then implanting them clinically.

### Endothelial Cell Seeding

Ideally, EC-seeded grafts should have a confluent endothelial cell lining at the time of implantation, and the cells should be able to resist desquamation from sheer stress after the restoration of circulation, with the more desirable aspects of their physiologic function intact. Further, the void created after desquamation should be rapidly covered by proliferating ECs in response to autocrine/paracrine release of growth factors. Since ECs normally promote thrombo-resistance, EC-seeded grafts have a theoretical benefit in reducing graft thrombosis. In addition, a confluent EC monolayer may also prevent the development of myointimal hyperplasia by (1) preventing the deposition of platelets, which release bioactive factors responsible for SMC migration, proliferation, and production of ECM; (2) maintaining a mechanical barrier to VSMC invasion via intimal basement membrane and the internal elastic lamina; and (3) assuming a quiescent EC phenotype that does not stimulate SMC activity.

In 1978, Herring *et al.* (1978) first reported that endothelial cell seeding onto a graft surface enhanced graft survival. Since then, this subject has been intensively investigated. Considerable progress has been achieved, especially related to technical problems such as cell culture and retention. Initial difficulties in cell harvest, cell seeding and adhesion, and prevention of desquamation have been largely overcome. One of the early difficulties encountered was due to the relatively low cell density initially applied to the graft. Even though the cell density of the endothelial cell monolayer on a normal vein is approximately $10^3$ ECs/mm$^2$, an initial density of at least $5 \times 10^3$ ECs/mm$^2$ is required for immediate confluent EC coverage of a small-caliber vascular graft after exposure to the circulation.

Since ECs adhere very poorly to synthetic graft materials, many adhesive proteins, such as fibronectin, collagen, fibrin, laminin, cell-adhesive peptides (e.g., RGD, REDV, and YIGSR), and plasma, have been applied to the graft surface to improve the seeding efficiency. Studies on the kinetics of EC loss following seeding showed that between 20% and 70% of initially adherent cells are lost during the first hour and as few as 5% were retained after 24 hours (Rosenman *et al.*, 1985). Retained cells at least partially compensate for the cell loss by migration and proliferation. Preconditioning the seeded EC monolayer with graded shear stress promotes reorganization of the EC cytoskeleton and production of ECM, which in turn enhances the EC retention at flow exposure (Ott and Ballermann, 1995). The properties of the prostheses also affect EC attachment. Dacron yields higher cell-attachment rates as compared to ePTFE when both grafts are coated with the same matrix. Polyurethane also showed better cell attachment than ePTFE.

To maximize immediate cell inoculation density, a two-stage seeding procedure is often performed, in which endothelial cells are harvested, allowed to proliferate *in vitro*, and then seeded and grown to confluence on the vascular graft prior to implantation. The disadvantages of this technique include the increased potential for infection, the alterations of EC phenotype and function, the requirement of a waiting period of three to four weeks for expansion of the cell population, and the necessity for two operative procedures.

An alternative method involves the use of microvascular endothelial cells. Small-diameter Dacron grafts seeded with enzymatically harvested omental microvascular cells in a single-stage technique showed confluent endothelial linings, larger thrombus-free surface area, and improved patency rates at one year in a canine model (Pasic *et al.*, 1994). Zilla *et al.* demonstrated increased patency and decreased platelet deposition in clinically implanted endothelial cell–seeded ePTFE femoropopliteal bypass grafts over three years as compared to unseeded grafts. And this group more recently reported an overall seven-year primary patency rate of 62.8% for 153 endothelialized femoropopliteal ePTFE grafts (Meinhart *et al.*, 2001), which is comparable to the patency rate of saphenous vein grafts in this region. However, the seeded grafts have not been reproduc-

ibly shown significantly to reduce anastomotic pseudointimal hyperplasia.

In fact the potential for IH is a major concern with EC-seeded grafts, and this has been described in a couple of case reports. In one case, bilateral above-knee grafts seeded with cephalic vein ECs developed stenosis and had to be replaced 41 months after implantation. The central part of this graft was explanted and investigated. The graft had a confluent endothelial lining on a collagen IV positive basement membrane with subintimal tissue of $1.21 \pm 0.19$-mm thickness (Deutsch *et al.*, 1997). The unusually thick subendothelial layer was also found in another case, in which a microvascular EC-seeded Dacron graft was placed as a mesoatrial bypass and had to be resected because of external mechanical stricture nine months after implantation (P. K. Park *et al.*, 1990).

Besides technical problems, concern exists as to the ultimate function of those endothelial cells on the graft surface, the cells having been injured by the process of manipulation and/or chronic exposure to a nonphysiologic environment. Unlike their uninjured counterparts, injured endothelial cells produce a variety of procoagulants, such as von Willebrand Factor, plasminogen activator inhibitor, thrombospondin, and collagen. Higher levels of PDGF and bFGF have also been measured in EC-seeded grafts; this is particularly concerning, given their potential role in stimulating the migration and proliferation of SMCs, which can lead to IH.

### Arteries Engineered *In Vitro*

The two main components of arterial generation *de novo* are the cells and their scaffolds. These components are not separable in practice, but they are separated in this chapter to emphasize their respective contributions.

There is much discussion about the proper cells to use for cell seeding *ex vivo* or cell homing *in vivo*. The EC is the most fastidious vascular cell to grow, and heterogenous ECs are highly immunogenic. Therefore, in the absence of immunosuppression, autogenous ECs are currently a requirement for TEVG. TEVG media and adventitia can be created by using vascular smooth muscle cells or fibroblasts with or without exogenous matrix scaffolding. These cells can be harvested from the patient in need. But, since these patients are typically older, with significant comorbidities, their cells (particularly VSMC and ECs) may not retain sufficient doubling capacity necessary to generate these TEVGs. In part this may be due to the loss of telomere length that occurs with aging.

In support of this hypothesis, Niklason *et al.* have demonstrated an increase in the population doublings of adult VSMCs through retroviral infection with the telomerase reverse transcriptase subunit (hTERT), without evidence of inducing cellular transformation (Poh *et al.*, 2005). However these arteries have unacceptably low bursting strength, and this is probably because hTERT infection did not remediate the stunted production of essential ECM proteins (e.g., collagen) that confer vessel strength in these transformed aged cells. Using a clinically relevant aged fibroblast-derived media, L'Heureux demonstrated a TEVG that has suitable bursting strengths at insertion, has a functional endothelium, and demonstrates mechanical stability in a variety of animal models out to eight months (L'Heureux *et al.*, 2006). These functional characteristics of a TEVG provide proof of concept of the functional promise of this approach, begun approximately 100 years ago.

Clinical success has been seen when using a TEVG for vessel replacement in the venous system, which is potentially less tenuous hemodynamically. These TEVGs have been implanted in 42 pediatric patients with congenital heart defects. To date these grafts have been resistant to aneurysmal dilatation and have had superb patency (Shin'oka *et al.*, 2005).

Various cellular and acellular approaches to TEVG scaffolds have also been pursued, and the challenge of the scaffold begins with providing satisfactory retention of cells *in vivo* or *ex vivo* in clinically relevant bioreactors and promoting optimal phenotypic characteristics of those cells. In addition, the scaffold or resulting vessel must have sufficient bursting strength to withstand the physiologic stresses of the circulation. After implantation this scaffold will undergo remodeling, which can lead to aneurysmal dilatation and subsequent graft rupture. Niklason *et al.* (1999) reported encouraging results with a graft produced by seeding SMCs onto PGA scaffolds that were sodium hydroxide modified to promote cell attachment. This graft was cultured in an *in vitro* pulsatile radial stress environment for eight weeks prior to implantation. These grafts showed contractile responses to serotonin, endothelin-1, and prostaglandin $F_{2\alpha}$, and they expressed the SMC differentiation marker, myosin heavy chain. Further, the grafts cultured under pulsatile conditions produced more collagen than those grown without pulsatile stress and exhibited a mechanical strength comparable to native human saphenous veins; this graft ruptured at $2150 \pm 909$ mmHg versus $1680 + 307$ mmHg for saphenous vein. Autologous ECs were seeded onto the luminal surface and were cultured for three more days before implantation. Four of the grafts were then implanted into swine saphenous arteries, of which two were generated under pulsatile stress and two under static condition. Two pulsed grafts remained patent up to the fourth week without dilation or rupture, while two nonpulsed grafts thrombosed after three weeks; the polymer remnants were no longer visible at four weeks.

Decellularized tissue scaffolds are appealing because they are already composed of native vascular extracellular matrix proteins that exhibit reasonable structural characteristics as well as providing instructive cues for cellular ingrowth. Using bone marrow–derived cells incubated on decellularized canine carotid arteries, Byung-Soo Kim *et al.* demonstrated cellular incorporation into the scaffold and

subsequent differentiation of these cells into endothelial and vascular smooth muscle cells and subsequently into three distinct vessel layers (Cho *et al.*, 2005). Others have used a more focused approach and induced or applied endothelial progenitor cells (EPCs) into similar scaffolds, with promising results. The benefit of EPCs include a robust replication potential ideal for a TEVG, and these cells acquire mature endothelial cell markers and function upon seeding into TEVGs (Kaushal *et al.*, 2001). These attributes may be further augmented by gene therapy.

Using biologic gels, such as those composed of type I collagen or fibrin, one can promote tissue ingrowth and direct remodeling in a bioreactor, thereby promoting favorable characteristics, such as improved mechanical strength and vessel reactivity over time (Swartz *et al.*, 2005). Such approaches are easily modified by the addition of growth factors with refined delivery systems in order to enhance and sustain cellular ingrowth (Ehrbar *et al.*, 2004). Further refinement of these scaffolds can mimic the differential mechanical properties of the intimal and medial arterial layers.

### *In Vivo* Tissue-Engineered Vascular Grafts

Current clinically available synthetic vascular grafts, composed of ePTFE, Dacron, or polyurethane, are permanent prostheses within the host after implantation. Theoretically, it is possible with bioresorbable materials to stimulate a rapid and controlled ingrowth of tissue to assume the load bearing sufficient for resistance to dilation and to incorporate cellular and extracellular components with desirable physiologic characteristics to form a new artery *in vivo*, whereby the synthetic material itself would cease to be necessary following tissue ingrowth. Still, the limited regenerative capacity of aged or diseased cells discussed earlier may also compromise cellular ingrowth *in vivo*.

#### Bioresorbable Grafts

Aneurysmal dilatation of bioresorbable grafts can be expected to occur after sufficient degradation of the graft material and prior to adequate cell growth. The first published report of a fully bioresorbable graft was by Bowald in 1979 (Bowald *et al.*, 1979) and described the use of a rolled sheet of Vicryl (a copolymer of polyglycolide and polylactide). However, these early grafts were prone to aneurysmal dilation and rupture. We have determined that 10% of the PGA grafts have aneurysmal dilation within the first three months after implantation and that this does not increase over the next nine months, suggesting that the critical time for the development of aneurysms is during prosthetic resorption prior to the ingrowth of tissue with a sufficient strength to resist hemodynamic pressures. These studies also demonstrated the ability of bioresorbable grafts to support sufficient cellular ingrowth. Here, four weeks after implantation, these 24-mm by 4-mm grafts contained an inner capsule with a confluent layer of endothelial cells and myofibroblasts amidst dense collagen fibers (Greisler, 1982). Similarly constructed and implanted Dacron grafts demonstrated an inner capsule containing solely fibrin coagulum with minimal cellularity. Macrophage infiltration and phagocytosis paralleled the resorption of PGA, which was totally resorbed at three months.

In order to limit this catastrophic outcome, several approaches have been developed. One is to combine the bioresorbable material with a nonresorbable material in order to retain a mechanical strut. Another solution involves the combination of two or more bioresorbable materials with different resorption rates so that the more rapidly degraded material evokes a rapid tissue ingrowth while the second material provides temporary structural integrity to the graft. Thirdly, growth factors, chemoattractants, and/or cells can be applied to the graft to enhance tissue ingrowth, structure, and organization.

Using a more slowly resorbed compound, polydioxanone (PDS), we encouraged similar endothelialization of the regenerated luminal surface as earlier and prostacyclin and thromboxane production in similar concentrations as control rabbit aortas. PDS was retained up to six months, and only 1/28 PDS grafts exhibited aneurysmal dilation, with explant times as late as one year. The explanted specimens of these PDS grafts also demonstrated biomechanical characteristics similar to those of native arteries, being able to withstand static bursting pressures of 6000 mmHg and 2000 mmHg mean pulsatile pressure without fatigue (Greisler *et al.*, 1987).

Using the differentially resorbed approach with composite grafts woven from yarns of 74% PG910 and 26% PDS, we reported a one-year patency rate of 100% with no aneurysms in the rabbit aorta model. The PG910 was totally resorbed by two months and the PDS by six months. The regenerated arteries withstood 800 mmHg of pulsatile systolic pressure *ex vivo* without bursting, and a confluent, functional, von Willebrand factor–positive endothelial cell layer over circumferentially oriented smooth muscle–like myofibroblasts formed in the inner capsule of both these grafts. Additional experiments were done with a variety of resorbable materials in combination with Dacron (Greisler *et al.*, 1986). The results demonstrated that transanastomotic pannus ingrowth is not a critical source of cells replacing bioresorbable vascular prostheses. Rather, trans-interstitial ingrowth of myofibroblasts, capillaries, and endothelial cells is the principle mechanism. In addition, tissue ingrowth into all tested lactide/glycolide copolymeric grafts was observed to parallel the kinetics of macrophage phagocytosis and prosthetic resorption. *In vivo*, the rate of cell proliferation and collagen deposition in the inner capsule also paralleled the kinetics of macrophage-mediated prosthetic resorption (Greisler *et al.*, 1993). These studies also confirmed both the inhibitory effect of Dacron and the stimulatory effect of lactide/glycolides on tissue ingrowth and inner capsule cellularity. As described earlier in this chapter, activated macrophages release a variety of growth factors, including PDGF, interleukin-1, basic FGF, TGF-β, and tumor necrosis factor. Since the phagocytosis of various biomaterials can lead to macrophage activation, rabbit peritoneal macrophages were cultured in the presence of polyglactin-910, Dacron, or neither and the mitogenic activity in the resulting conditioned media compared using growth assays of quiescent BALB/c3T3 fibroblasts, rabbit aortic SMCs, and murine capillary ECs. The media of the macrophages grown in the presence of the bioresorbable polymer stimulated significantly more proliferation in all cell types than did the media of the macrophages grown in the absence of the material or in the presence of Dacron, with 60–80% of the mitogenic activity blocked by immunoprecipitation using an FGF-2 antibody (Greisler, 1991; Greisler *et al.*, 1996).

Partially resorbable grafts have also been investigated. Since Dacron was found to inhibit the macrophage-mediated arterial regeneration stimulated by the resorbable component PG910, polypropylene was evaluated as a nonresorbable component because of its high tensile strength, low fatigability, low degradation *in vivo*, and minimal inhibitory effect on cellular regeneration of grafts (Greisler *et al.*, 1992). Composite grafts constructed from yarns containing 69% PG 910 and 31% polypropylene implanted into rabbit and dog arteries again demonstrated superb results without aneurysmal dilation. Galletti (Galletti *et al.*, 1988) used Vicryl prostheses coated with retardant polyesters to protect the Vicryl temporarily from hydrolytic and cellular degradation. When implanted into the canine aorta, prosthetic resorption was noticed at four weeks and was complete by 24 weeks. No coated grafts developed aneurysmal dilatation, whereas one of the uncoated grafts ruptured. He theorized that the low pH generated in the microenvironment of the degrading bioresorbable polymers, such as polyglycolide and polylactide, stimulate macrophages to secrete growth factors that induce fibroblast proliferation. In separate experiments, another group evaluated grafts prepared from a mixture of 95% polyurethane and 5% poly-lactide (van der Lei *et al.*, 1987). They found that only relatively compliant grafts that induced circumferential smooth muscle development contained elastin and remained mechanically stable without dilating. They concluded that modifications of the graft preparation, including smooth muscle cell seeding, help enhance optimal orientation of the smooth muscle cells and prevent aneurysm formation.

### The Living Bioreactor

Cambell *et al.* have developed a modification of the Sparks' mandrel to create a TEVG. Here they utilize the abdominal peritoneum's reaction to foreign bodies as a living bioreactor (Hoenig *et al.*, 2005). After the graft has matured, it is removed from the mandrel and inverted; this creates a TEVG with a mesothelial inner lining. This graft's resistance to aneurismal dilatation and rupture has not been proven to date.

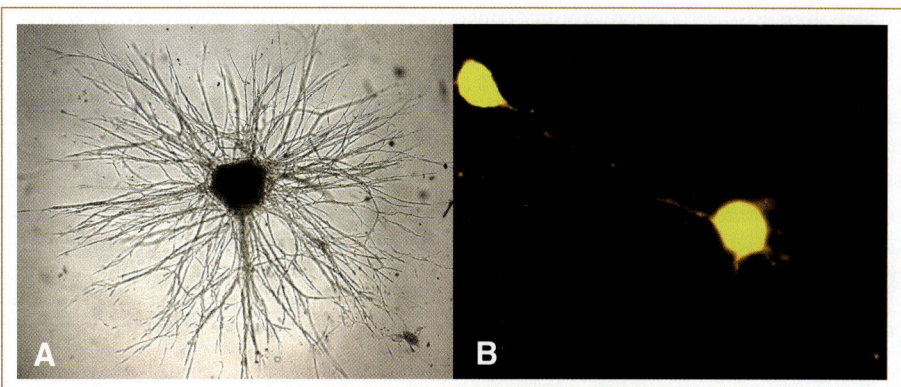

**FIG. 39.3.** Three-dimensional *in vitro* induction of a capillary network. Robust capillary formation can be induced in ECs stimulated with angiogenic growth factors in a three-dimensional fibrin matrix. **(A)** Spherical induction of a radially oriented capillary network from an EC aggregate induced by sustained low-dose FGF-1 (1 ng/mL). **(B)** When two EC aggregates are in proximity to each other, the EC sprouts are preferentially directed toward each other.

## Cellular Recruitment

Since autogenous cells cannot populate a scaffold by migration or proliferation alone, they must be recruited internally via circulating ECs or EPCs or externally from the surrounding tissue or exogenous source through angiogenic ingrowth during tissue incorporation of the graft. In addition to the benefit provided by the localization of cells, transmural ingrowth provides for cellular perfusion that extends beyond the distance supplied by simple diffusion (100 μm). We are currently designing three-dimensional capillary constructs (Fig. 39.3) that could provide grafts with the cellular and metabolic infrastructure requisite for the creation of a living TEVG through the induction of a vaso vasorum, which can be incorporated into preexisting capillary networks (inosculation). Proof of concept of this approach has been demonstrated in cardiac sheet grafts (Sekiya *et al.*, 2006). This induction can obviously be supplemented by the delivery of angiogenic proteins or genes to these constructs. With further research, a small-diameter, totally resorbable vascular graft may be able to improve the current dismal long-term patency rates of small-caliber grafts.

## VI. ENDOVASCULAR STENTS AND STENT GRAFTS

Endovascular stents were conceptualized prior to the introduction of angioplasty in 1969, and they have enjoyed increasing popularity in recent years. They have improved the durability of endovascular treatments and led a paradigm shift away from the traditional operative approach to vascular disease. Angioplasty, stents, and stent grafts are all endovascular interventions; they minimize the size of incisions, decrease lengths of hospital stays, and may confer some short-term survival benefit acutely after intervention. With over 1.5 million percutaneous coronary interventions occurring each year and a prevalence of in-stent stenosis ranging from 15% to 60% of patients, there is little wonder that much interest has been paid to the drug-eluting stent (DES). But DES release broadly suppressive drugs to the surrounding tissue in order to limit neointimal stenosis, and they likely retard myointimal thickening rather than

facilitate healing. This may be sufficient for short-term benefit but may compromise the long-term durability of these vascular interventions. Long-term durability is likely to become more critical in the coming years because of the improved mortality rates in patients with cardiovascular disease. Thus it is critical to avoid complacency in clinical thinking, for what comprises a good result today is dependent on mortality rates that may not be applicable to today's patients. We firmly believe that promoting healing and not limiting the adverse effects of healing will provide for durable results.

### Tissue-Engineered Cardiovascular Stents

There are three basic types of stents: balloon-expandable stents, which need balloon inflation to expand the stent into the arterial wall; self-expanding stents, allowing delivery in a collapsed form, with the stent expanding to its predetermined size after release from the delivery device; and thermal expanding stents, made by shape-memory metal alloys, which exist in an easily manipulated form and which restore their memorized shape at a certain transition temperature. All these stents have been successfully used in iliac arteries, with a two-year patency of approximately 84%. Improved endothelialization has also been reported using VEGF gene application to a modified metal stent (Walter *et al.*, 2004), and, although stenting itself has quickly become an important member of the endovascular armentareum, the development of biodegradable stents has progressed slowly, mostly due to difficulties in replicating the properties of stainless steel stents (Serruys *et al.*, 2006). Still, the promise of cell-demanded sustained drug delivery in biodegradable stents has led to a renewed interest in this approach. Currently preliminary evidence supports the short-term stability of these stents and the ability of these stents to deliver bioactive agents to the vessel lumen, providing proof of concept for this approach.

### Tissue-Engineered Stent Grafts

Stent grafts are a collapsible hybrid product composed of either Dacron or ePTFE, with stents providing radial support; they are delivered intravascularly to patients' large vessels like cardiovascular stents. These stent grafts maintain flow through their lumen and are commonly used to exclude flow into aneurysmal portions of arteries. Theoretically, the graft creates a barrier to exclude diseased arterial wall and provides a smooth flow conduit, while the stent support affixes the graft and may enhance luminal patency by resisting external compression. Stent grafts are delivered endovascularly, and since endovascular intervention reduces early operative morbidity and mortality, they have considerable consumer interest. However, the benefits of this approach over traditional operations is not clear, and it appears that outcomes are similar to those of the more traditional operative approach two years after intervention or operation. Further, a large study subgroup of nonoperative candidates failed to demonstrate a survival benefit from endovascular repair in these patients as compared to watchful waiting (EVAR Trial Participants, 2005).

Yet the rapid improvements in stent grafts will likely lead to improved durability and decreased frequency of reintervention. Therefore they will likely have a defined role in the treatment of patients with cardiovascular disease. For this reason it is important to define the known differences in healing between stent grafts and traditional bypass grafts.

Compliance changes between stent/unsupported graft/artery interfaces yield a remarkable hemodynamic disturbance. Also, delivery procedures, like balloon dilatation, may alter both graft intrinsic characteristics such as porosity and wall thickness as well as create mechanical injury to the surrounding artery. In addition, unlike conventional bypass, the endovascular graft is placed within the lumen of arteries with perigraft exposure to diseased arterial intima or to thrombus. All these factors change the healing characteristics of stent grafts as compared to synthetic grafts. An inflammatory reaction and progressive thickening of neointima have been observed in both ePTFE- and Dacron-based stent grafts. When comparing endovascular grafts to conventional bypass grafting using a canine iliac artery model, endovascular stent grafts composed of ePTFE grafts and balloon-expandable stents resulted in both a greater rate of endothelialization as well as an approximately five-times-thicker neointima in the midportion of the graft and a higher percentage of stenosis at the distal anastomosis when compared to conventional ePTFE grafts (Ohki *et al.*, 1997). Pathological remodeling has also been reported after anchoring to the arterial wall, and it has been reported that stent placement may cause a variety of flow disturbances. The stent components themselves also stimulate a nonspecific inflammatory reaction and induce neointimal formation, and tissue-engineered approaches, including cell seeding with genetically modified cells, may limit these complications (Panetta *et al.*, 2002; Eton *et al.*, 2004). The accumulation of experience, optimization of devices, and a better understanding of resultant pathological processes will likely allow the easy transfer of the recent advances in tissue-engineered grafts and vessels to stent grafts. This transfer is probably critical to enhancing their durability over time.

## VII. CONCLUSION

Much progress has been seen in recent years, and yet old things also become new again. As this publication comes to print there is renewed interest in the TEVG theorized by Carrel in the last century as well as in the use of biodegradable grafts. Currently, it is not correct to proclaim TEVGs either the promise of the future or a remnant of the past. Similarly, tissue-engineered approaches to promoting graft healing continue to provide scientific successes and clinical limitations. One can only hope that the successes seen with EC-seeded ePTFE can be translated to patients worldwide in the near future. As more patients live with cardiovascular disease than die from it, there is little doubt that promotion of graft healing will be more effective than inhibition of pathologic processes like IH. Still, for patients who are unlikely to live long enough to realize that benefit, inhibition of IH may be adequate. Thus both approaches may coexist for the near future.

## VIII. REFERENCES

Abbott, W., and Cambria, R. (1982). Control of physical characteristics elasticity and compliance of vascular grafts. *In* "Biological and Synthetic Vascular Prostheses" (J. Stanley, ed.), pp. 189ff. Grune and Stratton, New York.

Aoki, J., Serruys, P. W., *et al.* (2005). Endothelial progenitor cell capture by stents coated with antibody against CD34: the HEALING-FIM (Healthy Endothelial Accelerated Lining Inhibits Neointimal Growth-First In Man) Registry. *J. Am. Coll. Cardiol.* **45**(10), 1574–1579.

Autiero, M., Waltenberger, J., *et al.* (2003). Role of PlGF in the intra- and intermolecular cross-talk between the VEGF receptors Flt1 and Flk1. *Nat. Med.* **9**(7), 936–943.

Bacourt, F. (1997). Prospective randomized study of carbon-impregnated polytetrafluoroethylene grafts for below-knee popliteal and distal bypass: results at 2 years. The Association Universitaire de Recherche en Chirurgie. *Ann. Vasc. Surg.* **11**(6), 596–603.

Baer, R. P., Whitehill, T. E., *et al.* (1996). "Retroviral-mediated transduction of endothelial cells with the lac Z gene impairs cellular proliferation *in vitro* and graft endothelialization *in vivo*. *J. Vasc. Surg.* **24**(5), 892–899.

Bowald, S., Busch, C., *et al.* (1979). Arterial regeneration following polyglactin 910 suture mesh grafting. *Surgery* **86**(5), 722–729.

Brewster, L. P., Brey, E. M., *et al.* (2004). Heparin-independent mitogenicity in an endothelial and smooth muscle cell chimeric growth factor (S130K-HBGAM). *Am. J. Surg.* **188**(5), 575–579.

Cho, S. W., Lim, S. H., *et al.* (2005). Small-diameter blood vessels engineered with bone marrow–derived cells. *Ann. Surg.* **241**(3), 506–515.

Clowes, A. W., and Kohler, T. (1991). Graft endothelialization: the role of angiogenic mechanisms. *J. Vasc. Surg.* **13**(5), 734–736.

Clowes, A. W., Kirkman, T. R., *et al.* (1986). Mechanisms of arterial graft healing. Rapid transmural capillary ingrowth provides a source of intimal endothelium and smooth muscle in porous PTFE prostheses. *Am. J. Pathol.* **123**(2), 220–230.

Conte, M. S., Bandyk, D. F., *et al.* (2006). Results of PREVENT III: a multicenter, randomized trial of edifoligide for the prevention of vein graft failure in lower-extremity bypass surgery. *J. Vasc. Surg.* **43**, 742–751.

Deutsch, M., Meinhart, J., *et al.* (1997). *In vitro* endothelialization of expanded polytetrafluoroethylene grafts: a clinical case report after 41 months of implantation. *J. Vasc. Surg.* **25**(4), 757–763.

Doi, K., and Matsuda, T. (1997). Enhanced vascularization in a microporous polyurethane graft impregnated with basic fibroblast growth factor and heparin. *J. Biomed. Mater. Res.* **34**(3), 361–370.

Dunn, P. F., Newman, K. D., *et al.* (1996). Seeding of vascular grafts with genetically modified endothelial cells. Secretion of recombinant TPA results in decreased seeded cell retention *in vitro* and *in vivo*. *Circulation* **93**(7), 1439–1446.

Ehrbar, M., Djonov, V. G., *et al.* (2004). Cell-demanded liberation of VEGF121 from fibrin implants induces local and controlled blood vessel growth. *Circ. Res.* **94**(8), 1124–1132.

Erzurum, V. Z., Bian, J. F., *et al.* (2003). R136K fibroblast growth factor-1 mutant induces heparin-independent migration of endothelial cells through fibrin glue. *J. Vasc. Surg.* **37**(5), 1075–1081.

Eton, D., Yu, H., *et al.* (2004). Endograft technology: a delivery vehicle for intravascular gene therapy. *J. Vasc. Surg.* **39**(5), 1066–1073.

EVAR Trial Participants (2005). Endovascular aneurysm repair and outcome in patients unfit for open repair of abdominal aortic aneurysm (EVAR trial 2): randomized controlled trial. *Lancet* **365**(9478), 2187–2192.

Galletti, P. M., Aebischer, P., *et al.* (1988). Experience with fully bioresorbable aortic grafts in the dog. *Surgery* **103**(2), 231–241.

Gray, J. L., Kang, S. S., *et al.* (1994). FGF-1 affixation stimulates ePTFE endothelialization without intimal hyperplasia. *J. Surg. Res.* **57**(5), 596–612.

Greisler, H. P. (1982). Arterial regeneration over absorbable prostheses. *Arch. Surg.* **117**(11), 1425–1431.

Greisler, H. P. (1991). Bioresorbable materials and macrophage interactions. *J. Vasc. Surg.* **13**(5), 748–750.

Greisler, H. P., Kim, D. U., *et al.* (1985). Arterial regenerative activity after prosthetic implantation. *Arch. Surg.* **120**(3), 315–323.

Greisler, H. P., Schwarcz, T. H., *et al.* (1986). Dacron inhibition of arterial regenerative activities. *J. Vasc. Surg.* **3**(5), 747–756.

Greisler, H. P., Ellinger, J., *et al.* (1987). Arterial regeneration over polydioxanone prostheses in the rabbit. *Arch. Surg.* **122**(6), 715–721.

Greisler, H. P., Ellinger, J., *et al.* (1991). The effects of an atherogenic diet on macrophage/biomaterial interactions. *J. Vasc. Surg.* **14**(1), 10–23.

Greisler, H. P., Cziperle, D. J., *et al.* (1992a). Enhanced endothelialization of expanded polytetrafluoroethylene grafts by fibroblast growth factor type 1 pretreatment. *Surgery* **112**(2), 244–254; discussion 254–255.

Greisler, H. P., Tattersall, C. W., *et al.* (1992b). Polypropylene small-diameter vascular grafts. *J. Biomed. Mater. Res.* **26**(10), 1383–1394.

Greisler, H. P., Petsikas, D., *et al.* (1993). Kinetics of cell proliferation as a function of vascular graft material. *J. Biomed. Mater. Res.* **27**(7), 955–961.

Greisler, H. P., Petsikas, D., *et al.* (1996). Dacron stimulation of macrophage transforming growth factor-beta release. *Cardiovasc. Surg.* **4**(2), 169–173.

Gupta, V., Aravamuthan, B. R., *et al.* (2004). Reduction of subacute stent thrombosis (SAT) using heparin-coated stents in a large-scale, real-world registry. *J. Invasive. Cardiol.* **16**(6), 304–310.

Hasson, J. E., Megerman, J., *et al.* (1985). Increased compliance near vascular anastomoses. *J. Vasc. Surg.* **2**(3), 419–423.

Herold, J., Pipp, F., *et al.* (2004). Transplantation of monocytes: a novel strategy for *in vivo* augmentation of collateral vessel growth. *Hum. Gene Ther.* **15**(1), 1–12.

Herring, M., Gardner, A., *et al.* (1978). A single-staged technique for seeding vascular grafts with autogenous endothelium. *Surgery* **84**(4), 498–504.

Hoenig, M. R., Campbell, G. R., *et al.* (2005). Tissue-engineered blood vessels: alternative to autologous grafts? *Arterioscler. Thromb. Vasc. Biol.* **25**(6), 1128–1134.

Ito, R. K., Rosenblatt, M. S., *et al.* (1990). Monitoring platelet interactions with prosthetic graft implants in a canine model. *ASAIO Trans.* **36**(3), M175–M178.

Iwaguro, H., Yamaguchi, J., *et al.* (2002). Endothelial progenitor cell vascular endothelial growth factor gene transfer for vascular regeneration. *Circulation* **105**(6), 732–738.

Kang, S. S., Petsikas, D., *et al.* (1997). Effects of albumin coating of knitted Dacron grafts on transinterstitial blood loss and tissue ingrowth and incorporation. *Cardiovasc. Surg.* **5**(2), 184–189.

Kaushal, S., Amiel, G. E., *et al.* (2001). Functional small-diameter neovessels created using endothelial progenitor cells expanded *ex vivo*. *Nat. Med.* **7**(9), 1035–1040.

Koike, N., Fukumura, D., *et al.* (2004). Tissue engineering: creation of long-lasting blood vessels. *Nature* **428**(6979), 138–139.

L'Heureux, N., Paquet, S., *et al.* (1998). A completely biological tissue-engineered human blood vessel. *FASEB J.* **12**(1), 47–56.

L'Heureux, N., Dusserre, N., *et al.* (2006). Human tissue-engineered blood vessels for adult arterial revascularization. *Nat. Med.* **12**(3), 361–365.

Laredo, J., Xue, L., *et al.* (2004). Silyl–heparin bonding improves the patency and *in vivo* thromboresistance of carbon-coated polytetrafluoroethylene vascular grafts. *J. Vasc. Surg.* **39**(5), 1059–1065.

Lubiatowski, P., Gurunluoglu, R., *et al.* (2002). Gene therapy by adenovirus-mediated vascular endothelial growth factor and angiopoietin-1 promotes perfusion of muscle flaps. *Plast. Reconstr. Surg.* **110**(1), 149–159.

Mazur, C., Tschopp, J. F., *et al.* (1994). Selective alpha IIb beta 3 receptor blockage with peptide TP9201 prevents platelet uptake on Dacron vascular grafts without significant effect on bleeding time. *J. Lab. Clin. Med.* **124**(4), 589–599.

Meinhart, J. G., Deutsch, M., *et al.* (2001). Clinical autologous *in vitro* endothelialization of 153 infrainguinal ePTFE grafts. *Ann. Thorac. Surg.* **71**(5 Suppl.), S327–S331.

Miller, D. C., Thapa, A., *et al.* (2004). Endothelial and vascular smooth muscle cell function on poly(lactic-*co*-glycolic acid) with nanostructured surface features. *Biomaterials* **25**(1), 53–61.

Muhs, A., Lenter, M. C., *et al.* (2004). Nonviral monocyte chemoattractant protein-1 gene transfer improves arteriogenesis after femoral artery occlusion. *Gene. Ther.* **11**(23), 1685–1693.

Murasawa, S., Llevadot, J., *et al.* (2002). Constitutive human telomerase reverse transcriptase expression enhances regenerative properties of endothelial progenitor cells. *Circulation* **106**(9), 1133–1139.

Niklason, L. E., Gao, J., *et al.* (1999). Functional arteries grown *in vitro*. *Science* **284**(5413), 489–493.

Nojiri, C., Senshu, K., *et al.* (1995). Nonthrombogenic polymer vascular prosthesis. *Artif. Organs.* **19**(1), 32–38.

Nor, J. E., Peters, M. C., *et al.* (2001). Engineering and characterization of functional human microvessels in immunodeficient mice. *Lab. Invest.* **81**(4), 453–463.

Ohki, T., Marin, M. L., *et al.* (1997). Anastomotic intimal hyperplasia: a comparison between conventional and endovascular stent graft techniques. *J. Surg. Res.* **69**(2), 255–267.

Ott, M. J., and Ballermann, B. J. (1995). Shear stress-conditioned, endothelial cell–seeded vascular grafts: improved cell adherence in response to *in vitro* shear stress. *Surgery* **117**(3), 334–339.

Ozawa, C. R., Banfi, A., *et al.* (2004). Microenvironmental VEGF concentration, not total dose, determines a threshold between normal and aberrant angiogenesis. *J. Clin. Invest.* **113**(4), 516–527.

Panetta, C. J., Miyauchi, K., *et al.* (2002). A tissue-engineered stent for cell-based vascular gene transfer. *Hum. Gene. Ther.* **13**(3), 433–441.

Park, P. K., Jarrell, B. E., *et al.* (1990). Thrombus-free, human endothelial surface in the midregion of a Dacron vascular graft in the splanchnic venous circuit — observations after nine months of implantation. *J. Vasc. Surg.* **11**(3), 468–475.

Park, S., Tepper, O. M., *et al.* (2004). Selective recruitment of endothelial progenitor cells to ischemic tissues with increased neovascularization. *Plast. Reconstr. Surg.* **113**(1), 284–293.

Pasic, M., Muller-Glauser, W., *et al.* (1994). Superior late patency of small-diameter Dacron grafts seeded with omental microvascular cells: an experimental study. *Ann. Thorac. Surg.* **58**(3), 677–683; discussion 683–684.

Pitsch, R. J., Minion, D. J., *et al.* (1997). Platelet-derived growth factor production by cells from Dacron grafts implanted in a canine model. *J. Vasc. Surg.* **26**(1), 70–78.

Poh, M., Boyer, M., *et al.* (2005). Blood vessels engineered from human cells. *Lancet* **365**(9477), 2122–2124.

Quarmby, J. W., Burnand, K. G., *et al.* (1998). Prospective randomized trial of woven versus collagen-impregnated knitted prosthetic Dacron grafts in aortoiliac surgery. *Br. J. Surg.* **85**(6), 775–777.

Rosenman, J. E., Kempczinski, R. F., *et al.* (1985). Kinetics of endothelial cell seeding. *J. Vasc. Surg.* **2**(6), 778–784.

Ruel, M., Wu, G. F., *et al.* (2003). Inhibition of the cardiac angiogenic response to surgical FGF-2 therapy in a swine endothelial dysfunction model. *Circulation* **108** (Suppl. 1), II335–II340.

Scott, N. A., Cipolla, G. D., *et al.* (1996). Identification of a potential role for the adventitia in vascular lesion formation after balloon overstretch injury of porcine coronary arteries. *Circulation* **93**(12), 2178–2187.

Sekiya, S., Shimizu, T., *et al.* (2006). Bioengineered cardiac cell sheet grafts have intrinsic angiogenic potential. *Biochem. Biophys. Res. Commun.* **341**(2), 573–582.

Serruys, P. W., Kutryk, M. J., *et al.* (2006). Coronary-artery stents. *N. Engl. J. Med.* **354**(5), 483–495.

Shepard, A. D., Gelfand, J. A., *et al.* (1984). Complement activation by synthetic vascular prostheses. *J. Vasc. Surg.* **1**(6), 829–838.

Shin'oka, T., Matsumura, G., *et al.* (2005). Midterm clinical result of tissue-engineered vascular autografts seeded with autologous bone marrow cells. *J. Thorac. Cardiovasc. Surg.* **129**(6), 1330–1338.

Suh, W., Kim, K. L., *et al.* (2005). Transplantation of endothelial progenitor cells accelerates dermal wound healing with increased recruitment of monocytes/macrophages and neovascularization. *Stem. Cells.* **23**(10), 1571–1578.

Swartbol, P., Truedsson, L., *et al.* (1997). Tumor necrosis factor-alpha and interleukin-6 release from white blood cells induced by different graft materials *in vitro* are affected by pentoxifylline and iloprost. *J. Biomed .Mater. Res.* **36**(3), 400–406.

Swartz, D. D., Russell, J. A., *et al.* (2005). Engineering of fibrin-based functional and implantable small-diameter blood vessels. *Am. J. Physiol. Heart. Circ. Physiol.* **288**(3), H1451–H1460.

van der Lei, B., Nieuwenhuis, P., *et al.* (1987). Long-term biologic fate of neoarteries regenerated in microporous, compliant, biodegradable, small-caliber vascular grafts in rats. *Surgery* **101**(4), 459–467.

Veith, F. J., Gupta, S. K., *et al.* (1986). Six year prospective multicenter randomized comparison of autologous saphenous vein and expanded polytetrafluoroethylene grafts in infrainguinal arterial reconstructions. *J. Vasc. Surg.* **3**(1), 104–114.

Vicaretti, M., Hawthorne, W. J., *et al.* (1998). An increased concentration of rifampicin bonded to gelatin-sealed Dacron reduces the incidence of subsequent graft infections following a staphylococcal challenge. *Cardiovasc. Surg.* **6**(3), 268–273.

Voorhees, A. B., Jr., Jaretzki, 3rd, A., *et al.* (1952). The use of tubes constructed from vinyon "N" cloth in bridging arterial defects. *Ann. Surg.* **135**(3), 332–336.

Vroman, L. (1987). Methods of investigating protein interactions on artificial and natural surfaces. *Ann. N. Y. Acad. Sci.* **516**, 300–305.

Walter, D. H., Cejna, M., *et al.* (2004). Local gene transfer of phVEGF-2 plasmid by gene-eluting stents: an alternative strategy for inhibition of restenosis. *Circulation* **110**(1), 36–45.

Weinberg, C. B., and Bell, E. (1986). A blood vessel model constructed from collagen and cultured vascular cells. *Science* **231**(4736), 397–400.

Zarge, J. I., Gosselin, C., *et al.* (1997). Platelet deposition on ePTFE grafts coated with fibrin glue with or without FGF-1 and heparin. *J. Surg. Res.* **67**(1), 4–8.

# Chapter Forty

# Heart Valves

*Peter Marc Fong, Jason Park, and Christopher Kane Breuer*

## I. INTRODUCTION

Heart valve disease is a considerable medical problem around the world. The treatment for end-stage heart valve disease is valve replacement; however, the best current replacement heart valves suffer from significant shortcomings. Promising alternatives to current replacement heart valves are being developed in the field of tissue engineering. This multidisciplinary effort comprises the areas of tissue biomechanics, immunology, injury response, cellular and tissue development, and the chemical, physical, pharmacological, and genetic manipulation of both cells and biomaterials. Moreover, underpinning these advanced methodologies is a concerted effort focused on understanding the relationships of structure to function in normal and pathological tissues. In the development of a substitute tissue-engineered heart valve, significant advances have been made in several of these areas, including the development of different cell sources and cell-seeding techniques, advancements in polymeric and natural scaffold design, and the intoduction of bioreactors — biomimetic devices capable of biochemically and biomechanically modulating the *in vitro* development of tissue-engineered neotissue. This chapter highlights the need for a tissue-engineered option in heart valve replacement and reviews past and ongoing work in the field. The fundamental heart valve structure, function, and disease are described, including a review of the research leading to the development of a tissue-engineered heart valve and the challenges that remain.

## II. HEART VALVE FUNCTION AND STRUCTURE

The heart valve's physiological function is to maintain unidirectional, nonobstructed blood flow without damaging blood elements, causing thromboembolism, or placing excessive mechanical stress on the leaflets and cusps. The native heart valve is remarkably well designed to perform this function. This ability comes from an almost perfect correlation of structure with function, allowing the valve to reduce stress on the cusps and supporting tissues while enabling it to endure the wear and tear of billions of repetitive deformations (Schoen and Levy, 1999). There are four valves in the human heart, two semilunar and two atrioventricular. The semilunar heart valves include the structurally similar aortic and pulmonary valves; the atrioventricular heart valves consist of the tricuspid and mitral valves (Breuer et al., 2004). The following passages describe the relation of structure to function in the semilunar valve from a macroscopic (tissue) to a microscopic (cellular) scale.

Grossly, the semilunar heart valves comprise three thin cusps that open easily when exposed to the forward blood flow of ventricular systole and then rapidly close under the

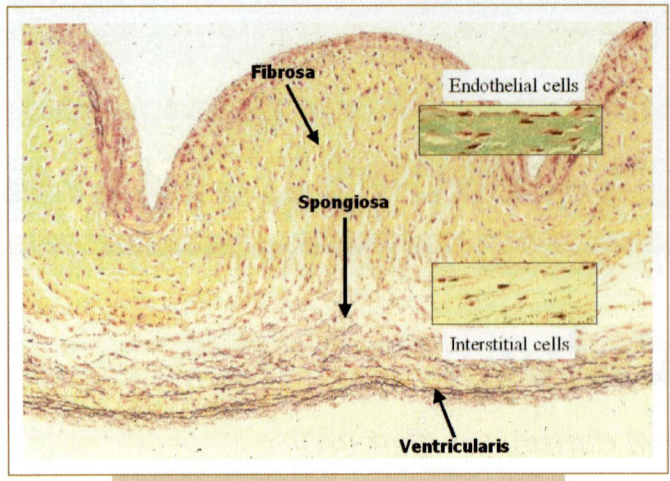

**FIG. 40.1.** Histological cross section of a heart valve, revealing the ventricularis, spongiosa, and fibrosa sections of the valve. Also shown are intersitial cells and endothelial cells.

minimal reverse flow of diastole (Schoen and Levy, 1999). The structure of the valve is such that these cusps are uniquely suited to deal with large forces and unidirectional flow. In the aortic valve, these cusps, simply called the *left, right,* and *noncoronary aortic valve cusps,* are attached to the aortic wall by a thick base known as the *commissures.* Despite the large forces applied, prolapse is prevented by substantial coaptation of the cusps (Rabkin-Aikawa *et al.,* 2005). In addition, these structural elements are anisotropically oriented in the tissue plane, resulting in disproportionate mechanical properties of the valve cusps, with greater compliance in the radial as opposed to the circumferential direction. This compliance allows the cusp thickness of the aortic valve to vary from 300 to 700 μm throughout the course of the cardiac cycle (Rabkin-Aikawa *et al.,* 2005) and decreases the impedance of flow. Further structural specializations that occur are lengthwise folding of collagen fibers and the orientation of collagen bundles in the fibrous layer toward the commissures. This reduces sagging in the cusp centers and conserves maximal coaptation, thereby preventing regurgitation from occurring. Thus, both the macroscopic valve geometry and the fibrous network inside the cusps work to transfer the stresses caused by the diastolic force to the aortic wall and annulus.

Microscopically, the semilunar heart valve comprises three layers: the fibrosa, ventricularis, and spongiosa, as shown in Fig. 40.1. The main constituents of the valvular extracellular matrix (ECM) of these layers are collagen (60%), proteoglycans (20%), elastin (10%), and some glycoprotein (Kunzelman *et al.,* 1993). It is the unique extracellular structural features within these layers that produce the specialized biomechanical profile necessary for proper function (see Fig. 40.2). Most of the mechanical strength of the valve is provided by the collagen, which is composed primarily of type I, type III, and some type V collagen (Cole *et al.,* 1984). The fibrosa, below the sinus surface, is composed predominantly of crimped, densely packed collagen fibers arranged parallel to the free edge of the cusp. The elastin of a heart valve forms an encompassing matrix that binds the collagen fibrous bundles throughout the heart valve. In this way, the elastin–collagen hybrid forms a larger network of interconnecting collagen fibers (Scott and Vesely, 1995). The ventricularis on the inflow surface is composed of collagenous tissue with radially aligned elastin fibers. Meanwhile, the centrally located spongiosa is composed of proteoglycans that have glycosaminoglycan (GAG) side chains. Matrix molecules can form covalent cross-links with the GAG side chains, thereby supporting other components of the ECM (Flanagan and Pandit, 2003).

## III. CELLULAR BIOLOGY OF THE HEART VALVE

### Valvular Endothelial Cells

Two main cell types are present in the heart valve: a covering layer of valvular endothelial cells (VECs) and inner valvular interstitial cells (VICs). Endothelial cells are critical to circulatory function and blood–tissue interaction. They also play a significant role in metabolism, synthesis, and the maintenance of homeostasis. Endothelial cells form a non-thrombogenic blood–tissue interface, have an interactive role in inflammatory and immune reactions, modulate the proliferation of other cell types, and make up the endothelium, a semipermeable membrane that regulates the transfer of large and small molecules through the vascular wall (Rabkin-Aikawa *et al.,* 2005).

VECs have an active role in the heart valve. They can respond to changing stimuli by altering their normal functions and by phenotypically expressing new, inducible properties through a process known as *endothelial activation.* Cytokines and hemodynamic forces are two examples of factors that induce endothelial activation. Once activated, the VECs produce other cytokines, chemokines, adhesion molecules, growth factors, and vasoactive molecules. This molecular cascade can result in vasodilation, vasoconstriction, or the production of procoagulant and anticoagulant moieties, major histocompatibility complex molecules, and a variety of other biologically active molecules. When the endothelial is functioning well, these factors are in equilibrium and the vessel can respond correctly to various stimuli. When the endothelium is not working properly, a form of endothelial activation known as *endothelial dysfunction* occurs, resulting in a surface that is adhesive to inflammatory cells or thrombogenic (Rabkin-Aikawa *et al.,* 2005). Also, a significant amount of endothelial cell phenotypic variation is associated with the different mechanical stresses of different anatomic locations. For example, endothelial cell populations developed from arterial sites, as opposed to venous sites, have significantly dissimilar characteristics (Shin *et al.,* 2001). This is also true in the heart valve, where researchers have found that the endothelium from aortic valves and human pulmonary valves

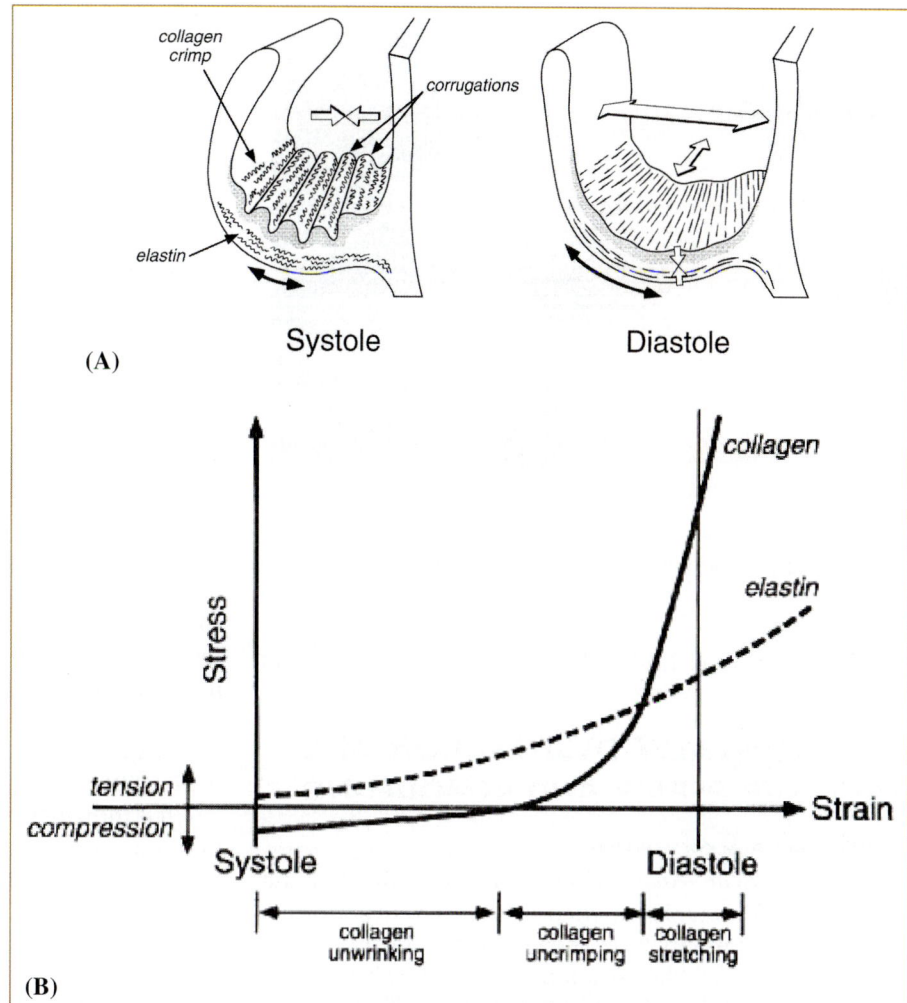

**FIG. 40.2.** **(A)** Schematic representation of the cuspal configuration and architecture of collagen and elastin in systole and diastole. **(B)** Stress–strain plot of biomechanical cooperativity between elastin and collagen during valve motion. During opening, elastin extends at minimal load during extension of collagen crimp and corrugations. Near full closure, when the collagen has fully unfolded, the load-bearing element shifts from elastin to collagen, and stress rises steadily while coaptation is maintained. In systole, elastin restores the contracted configuration of the cusp. A and B: Reproduced with kind permission from Schoen (1997).

have different phenotypic expression (Rabkin-Aikawa *et al.*, 2004).

Not surprisingly, there are substantial data suggesting that VECs differ from other types of endothelial cells in many ways. A recent report has suggested that the response of the endothelium to flow is different in the valves than in the aorta (Butcher *et al.*, 2004). Other investigators are beginning to propose the possibility of significant phenotypic diversity, even between the endothelium on the inflow and outflow aortic valve surfaces (Davies *et al.*, 2004).

## Valvular Interstitial Cells

While valvular endothelial cells' primary importance is their interaction with the blood and surrounding environment, valvular interstitial cells are believed to be the principal component in the maintenance of valvular structure (Flanagan and Pandit, 2003). VICs are elongated in shape and have many long, slender processes that extend throughout the valvular matrix (Filip *et al.*, 1986). By interconnecting, these processes create a three-dimensional network that permeates throughout the valve and is closely associated with the valve's ECM (Chester *et al.*, 2000, 2001).

It has been proposed that two morphologically and structurally distinct populations of VICs exist, the first having contractile characteristics of pronounced stress fibers, and the other having secretory characteristics, with pronounced rough Golgi apparatus and endoplasmic reticulum (Filip *et al.*, 1986; Zacks *et al.*, 1991; Flanagan and Pandit, 2003). In addition, VICs may have a functional capacity that extends even beyond the realm of fibroblast matrix secretion (Filip *et al.*, 1986). Specifically, it has been proposed that some VICs are capable of contracting, thereby sustaining a limited intrinsic valvular force while withstanding hemodynamic forces (Mulholland and Gotlieb, 1996). This is supported by data indicating that some VICs express both cardiac and skeletal contractile proteins (Roy *et al.*, 2000) and by observations that valve leaflets contract when exposed to a variety of vasoactive agents.

VICs are also of central importance in the valve's repair and maintenance operations (Flanagan and Pandit, 2003). The constant and rigorous mechanical movement of valves results in valvular damage and an ongoing regenerative process vital to sustained function (Schneider and Deck,

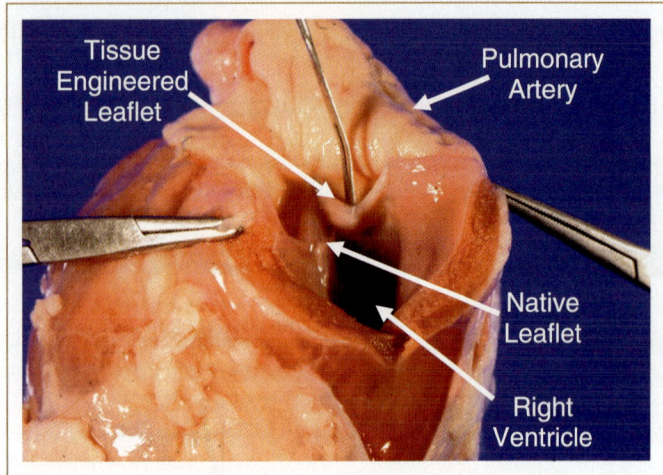

**FIG. 40.3.** Photograph of a tissue-engineered leaflet in lamb 11 weeks after implantation. Reproduced, with kind permission, from Shinoka (2002).

1981; Henney *et al.*, 1982). This is discussed further in the following sections.

## IV. HEART VALVE DYSFUNCTION AND VALVULAR REPAIR AND REMODELING

### Heart Valve Dysfunction

The American Heart Association has estimated that 87,000 heart valve replacement procedures were performed in the United States in 2000, and nearly 275,000 procedures are performed worldwide annually. Heart valve disease caused 19,737 deaths in the United States in 2000. The pathology of heart valve disease manifests itself in two ways: stenosis and insufficiency. Stenosis occurs when the valve does not completely open resulting in reduced forward flow, while insufficiency is the backflow of blood due to incomplete valvular closure (Rabkin-Aikawa *et al.*, 2005; Rabkin and Schoen, 2002; Flanagan and Pandit, 2003).

The causes and clinical frequency of heart valve dysfunction are not uniform. The most prevalent clinical problem is aortic stenosis, which usually occurs due to dystrophic calcification of the aortic valve cusps and annulus. In mitral valve stenosis the principal cause is deformity, which in most cases precipitates many years after an onset of rheumatic fever. Chronic aortic insufficiency is less common and is caused by aortic root dilation, resulting in distended and outwardly bowed commissures as well as impaired cuspal coaptation. In the tricuspid and pulmonary valves, the leading causes of dysfunction are congenital deformities. Of course, the physiologic function of any of the four cardiac valves can at any time be altered by endocarditis, an infection of the intercardiac structures. Endocarditis can destroy valve tissue in a matter of weeks and generally results in valvular insufficiency (Rabkin-Aikawa *et al.*, 2005).

### Valvular Repair and Remodeling

Valvular injury stimulates interstitial cell proliferation, migration, and apoptosis. The early events in valve repair are prominently characterized by the migration and proliferation of interstitial cells (Lester *et al.*, 1992; Lester and Gotlieb, 1988). This migration is likely governed by a series of events, including the activation of integrins and cell surface heterodimeric receptors that control cell–ECM and cell–cell adhesion. Furthermore, signaling mechanisms determine the allotment and activity of cytoskeleton proteins that orchestrate the cell spreading, contraction, and translocation typically involved in wound healing (Meredith and Schwartz, 1997; Woodard *et al.*, 1998; Rabkin-Aikawa *et al.*, 2005). Phenotypic changes of the interstitial cells themselves may also play a significant role. Recent investigations have suggested that VICs activate from a quiescent form in equilibrium to help determine the valve's biomechanical response when exposed to a new mechanical environment (Schoen, 1999; Rabkin-Aikawa *et al.*, 2005). In this manner, the VIC phenotype is determined by wall stresses in the valve leaflets. For example, in leaflets from patients with myxomatous mitral valve degeneration, interstitial cells express features of activated myofibroblasts (Rabkin *et al.*, 2001).

It is thought that VIC–matrix interactions play an important role in regulating interstitial cell-based remodeling, including the induction of cell migration, the secretion of ECM components, and the secretion of proteolytic enzymes — all critical functions in tissue repair. More specifically, enzymes such as interstitial collagenases, gelatinases, and other matrix metalloproteinase (MMP) enzymes have been localized in all four heart valves (Dreger *et al.*, 2002). In the heart valve's connective tissue, MMPs and tissue inhibitor metalloproteinases (TIMPs) count heavily in tissue remodeling and regeneration (Flanagan and Pandit, 2003), along with a number of other physiological and pathological processes (Nelson *et al.*, 2000; McCawley and Matrisian, 2001; Galis and Khatri, 2002). The interaction of TIMP, MMP, and their regulators are particularly important in cardiac and vascular remodeling (Rabkin-Aikawa *et al.*, 2005). The interstitial collagenases MMP-1 and MMP-13 mediate the preliminary phase of collagen degradation by disassembling the native helix of the fibrillar collagen. The resulting collagen fragments are then accessible to further proteases, such as gelatinases or elastases like Cathepsin K, for further catabolism (Krane *et al.*, 1996).

The pattern of TIMP and MMP expression in normal valves is highly specific (Dreger *et al.*, 2002). Excessive levels of MMP proteolytic enzymes, such as collagenases, gelatinases, and cysteine endoproteases (cathepsin S and K) overregulated by VIC, may cause collagen and elastin degradation, ultimately resulting in weakness of heart valve leaflets. Many investigations have hypothesized that the valve's own cells may cause ECM degradation in some degenerative diseases; in some heart valve pathologies an overexpression of MMPs can be detected (Rabkin *et al.*, 2001). This suggests that these cells are somehow being stimulated to produce and secrete soluble extracellular messengers that prompt indigenous cells to degrade the matrix. One such messenger is cardiac catabolic factor, derived from porcine heart valves and found to stimulate collagen and proteoglycan breakdown *in vitro* (Decker and Dingle, 1982). The mechanism of action and the importance of MMPs/TIMPs in valve tissue morphogenesis, repair, and remodeling remain to be investigated further. It is very probable that proteolytic enzymes play a significant role in matrix remodeling in tissue-engineered heart valves (Rabkin-Aikawa *et al.*, 2005).

### Heart Valve Replacement

Standard treatment for end-stage valvular dysfunction is heart valve replacement. The first successful implantation of a heart valve in a human was performed in 1952 (Hufnagel *et al.*, 1989). Throughout the following decades, surgeons saw more than 80 new designs of prosthetic heart valves (Vongpatanasin *et al.*, 1996). Like most medical devices, heart valve substitutes have experienced a progressive technological evolution; modifications have been made and new designs introduced to address specific deficiencies in earlier devices. The classification and fundamental technologies, however, remain the same: prosthetic heart valves are either mechanical, composed of purely synthetic components, or bioprosthetic, containing biological components. Slightly more than half of the world's implanted valves are mechanical; the remainder are bioprosthetic (Butany *et al.*, 2003). While each type of valve has (generally) improved the quality and length of life for its recipients, many problems persist.

Fifty to sixty percent of all patients will experience prosthesis-associated problems requiring reoperation within 10 years following initial heart valve replacement (Bloomfield *et al.*, 1991; Hammermeister *et al.*, 1993). The overall frequency of complication is similar for bioprostheses and mechanical prostheses. Four types of valve-related complications are most common: (1) thrombosis, thromboembolism, and secondary anticoagulation-related hemorrhage; (2) prosthetic valve endocarditis; (3) structural dysfunction, including failure or degeneration of the prosthetic biomaterials; and (4) nonstructural dysfunction, including complications arising from technical difficulties during surgical implantation, such as perivalvular leak and biological integration (Grunkemeier and Rahimtoola, 1990; Bloomfield *et al.*, 1991; Hammermeister *et al.*, 1993; Jamieson, 1993; Turina *et al.*, 1993; Schoen, 1995; Edmunds *et al.*, 1996; Vongpatanasin *et al.*, 1996; Schoen and Levy, 1999). Each type of valve has its own unique set of advantages and disadvantages.

### Mechanical Heart Valves

The mechanical heart valve has excellent durability, due to the mechanical properties of the synthetic materials used to construct them. Unfortunately, these synthetic materials also possess poor biocompatibility. Specifically, mechanical prosthetic valves are believed to increase the risk of thromboemboli and thrombotic occlusion. This is caused by the lack of an endothelial lining and the flow abnormalities that result from a rigid outflow structure (Grunkemeier and Rahimtoola, 1990; Jamieson, 1993; Turina *et al.*, 1993; Schoen, 1995; Vongpatanasin *et al.*, 1996; Cannegieter *et al.*, 1994). To reduce this risk, chronic anticoagulation therapy is necessary for all recipients of mechanical valves. Systemic anticoagulation, however, increases the potential for serious hemorrhagic complications (Cannegieter *et al.*, 1994). Thus, the combined risk of hemorrhage secondary to anticoagulation and thromboembolic complications constitutes the principal disadvantage of mechanical prosthetic heart valves. In the absence of antithrombotic therapy, meta-analysis found the occurrence of a major embolism to be four per 100 patient years. When antiplatelet therapy was administered this risk was 2.2 per 100 patient years, and with coumadin therapy the risk was reduced to one per 100 patient years (Cannegieter *et al.*, 1994). The incidence of major bleeding in coumadin-treated patients, however, was 1.4 per 100 patient years and increased significantly with the addition of antiplatelet therapy (Cannegieter *et al.*, 1994). An additional nontrivial issue is that mechanical replacement valves are more susceptible to serious infections, such as endocarditis (Mylonakis and Calderwood, 2001; Breuer *et al.*, 2004).

### Bioprosthetic Heart Valves

The bioprosthetic valve replacements, such as allografts and glutaraldehyde-fixed xenografts, are associated with a lower risk of hemolysis and thrombosis than the mechanical heart valve (Turina *et al.*, 1993; Vongpatanasin *et al.*, 1996). Patients that have glutaraldehyde-fixed xenograft valves do not require anticoagulation therapy and are thus not at risk of anticoagulation-associated bleeding (Cannegieter *et al.*, 1994). Due to their biomechanical characteristics, however, the durability of a glutaraldehyde-fixed valve is more limited than that of the mechanical valve. The predominant disadvantage of replacement valves made from tissue is a progressive structural deterioration that ultimately results in stenosis and/or regurgitation (Grunkemeier and Rahimtoola, 1990; Jamieson, 1993; Turina *et al.*, 1993; Schoen, 1991, 1995; Vongpatanasin *et al.*, 1996; Schoen and Levy, 1999). Structural valve failure nearly always requires reoperation. Because bioprosthetic valves degrade progressively, the rate of failure is time dependent. Fewer than 1% of porcine aortic valve bioprostheses in adults reveal structural dysfunction within five years of implantation (Bloomfield *et al.*, 1991; Hammermeister *et al.*, 1993; Schoen and Levy, 1999; Ferrans *et al.*, 1987; Cohn *et al.*, 1989; Grunkemeier *et al.*, 1994; Jamieson *et al.*, 1995). However, 20–30% become dysfunctional within 10 years, and more than half fail due to degeneration within 12–15 years postoperatively (Bloomfield *et al.*, 1991; Hammermeister *et al.*, 1993; Schoen and Levy, 1999; Ferrans *et al.*, 1987; Cohn *et al.*, 1989; Grunkemeier *et al.*, 1994; Jamieson *et al.*, 1995). The danger of structural breakdown is also age dependent, with individuals under the age of 35 — especially children and adolescents — having the highest failure rate. In those under the age of 35, near-uniform failure occurs in the first five years following implantation, but only 10% fail in 10 years for those older than 65 (Cohn *et al.*, 1989; Grunkemeier *et al.*, 1994; Jamieson *et al.*, 1995).

The primary source of bioprosthetic valve dysfunction is structural deterioration of the cuspal tissue (Schoen and Levy, 1984; Blackstone and Kirklin, 1985; Schoen and Hobson, 1985; Ferrans *et al.*, 1987; Schoen, 1987; Cohn *et al.*, 1989; Bloomfield *et al.*, 1991; Hammermeister *et al.*, 1993; Grunkemeier *et al.*, 1994; Jamieson *et al.*, 1995; Vongpatanasin *et al.*, 1996). Two distinct yet possibly synergistic mechanisms are implicated: (1) noncalcific degradation and (2) calcific degradation, which can be summarized as noncalcific mechanical fatigue and cuspal mineralization. Both will eventually cause failure of the connective-tissue matrix of the tissue valve (Schoen and Levy, 1984, 1994; Schoen and Hobson, 1985; Ferrans *et al.*, 1987; Schoen *et al.*, 1987; Turina *et al.*, 1993; Vesely *et al.*, 2001). Calcification occurs when blood plasma calcium binds with residual organic phosphorous of the nonviable, cross-linked cells of the preserved valve (Schoen *et al.*, 1985, 1986; Schoen, 1989). Additional weakening can be caused by proteolytic degradation of the collagenous ECM (Sacks and Schoen, 2002). Damage caused from only mechanical force has been observed in explanted porcine aortic valves and has been linked to elevated levels of ECM MMP enzymatic degrading activity in clinical prostheses (Simionescu *et al.*, 1993). It should be noted that the pattern of *in vivo* structural damage in explanted bioprosthetic valves resembles that found in mechanical *in vitro* models. This suggests that mechanical effects alone are probably highly significant in ECM degradation (Sacks and Schoen, 2002).

A different type of replacement bioprosthetic valve is the cryopreserved homograft valve. This has been shown to have advantages in select patient populations, in particular those that require aortic valve replacement or need congenital heart reconstruction with right-sided conduits (Kirklin *et al.*, 1989; O'Brien *et al.*, 1995). The utilization of the cryopreserved homograft, however, is a form of transplantation and therefore suffers from many transplant-associated problems. These include donor organ scarcity and transmission of infection — it has been documented that cryopreserved homografts can be transmitters of infectious disease (Kirklin *et al.*, 1989; O'Brien *et al.*, 1995; Fedalen *et al.*, 1999; Clark *et al.*, 1997). Despite these disadvantages, cryopreserved homografts are the most biocompatible heart valve replacements and used most often in pediatric cardiothoracic applications (Kirklin *et al.*, 1989; O'Brien *et al.*, 1995; McGiffin and Kirklin, 1995). When used in pediatric

cardiothoracic cases, however, they are significantly limited by their incapacity for growth and structural deterioration. This deterioration decreases their durability in a manner similar to that of their glutaraldehyde-fixed xenograft counterparts (Kirklin *et al.*, 1989, 1993; Cleveland *et al.*, 1992).

# V. APPLICATION OF TISSUE ENGINEERING TOWARD THE CONSTRUCTION OF A REPLACEMENT HEART VALVE

## Tissue-Engineering Theory

The ideal heart valve replacement would be readily available and perfectly biocompatible, have the potential for growth, and be durable (Mayer, 2001; Sapirstein and Smith, 2001). The manufacture of an autologous, tissue-engineered heart valve could potentially be a considerable improvement over the best current technology. An autologous tissue-engineered heart valve would be completely biocompatible, nonhemolytic, and nonthrombogenic. It would be capable of utilizing the natural mechanisms for repair, remodeling, and regeneration and thereby be highly durable. In addition it would possess the potential for growth. Since the cells could be harvested from the patient in need of a valve, immune-mediated rejection would be eliminated. In short, a tissue-engineered heart valve could be the ideal replacement heart valve (Shalak and Fox, 1988; Heineken and Skalak, 1991).

The central paradigm for many tissue-engineering projects is cells + matrix → neotissue (Vacanti *et al.*, 1988; Langer and Vacanti, 1993). Three-dimensional cell culture is the fundamental process underlying most tissue-engineering methodologies. The capability of creating an environment for three-dimensional cell growth and neotissue formation is the primary function of the matrix (Shalak and Fox, 1988; Heineken and Skalak, 1991; Vacanti *et al.*, 1988; Langer and Vacanti, 1993). According to this paradigm, cells are seeded onto a matrix composed of synthetic or natural material. Once seeded, the matrix acts as a three-dimensional scaffold on which proliferating cells assemble ECM *in vitro*. When this process yields a sufficiently stable construct, the construct can be implanted *in vivo* (Rabkin-Aikawa *et al.*, 2005) in the desired anatomic location, where further reorganization may happen, thus forming the normal architecture of the tissue. The entire sequence, from *in vitro* seeding to *in vivo* implantation, can be summarized as (1) cell proliferation, sorting, and differentiation; (2) ECM production; (3) degradation of the scaffold; (4) remodeling; and (5) growth of the tissue (Rabkin-Aikawa *et al.*, 2005).

In theory, the principles of tissue engineering can be used to create any tissue type. In actuality, there are several physiologic and biologic limitations when tissue engineering functional neotissue. Some of the more significant factors that currently limit the application of tissue engineering are the inability to control innervation of neotissue, the inability to construct a microvasculature *de novo*, and the challenges surrounding culturing certain cell types (Shalak and Fox, 1988; Heineken and Skalak, 1991; Vacanti *et al.*, 1988; Langer and Vacanti, 1993). Due to these limitations, the most successful tissue-engineering developments have been in neotissues unencumbered by these factors. These neotissues have come principally from tissues that are noninnervated and relatively avascular and whose function is determined primarily by the biomechanical properties of their three-dimensional structure and extracellular matrix (Shalak and Fox, 1988; Heineken and Skalak, 1991; Vacanti *et al.*, 1988; Langer and Vacanti, 1993). Accordingly, the semilunar heart valve is nearly an ideal candidate for tissue engineering (Mayer, 2001; Sapirstein and Smith, 2001).

## Tissue-Engineering Theory Applied to Heart Valves

The semilunar heart valve is not completely avascular, and nutrients and oxygen are supplied via two complementary pathways. The thickest section of the valve is vascularized for transport of oxygen and nutrients (Weind *et al.*, 2001). For the greater portion of the semilunar valve, however, vascularization is unnecessary because oxygen can simply diffuse through both the fibrosa and ventricularis surfaces, given that both surfaces are exposed to blood (Weind *et al.*, 2001). For this reason, it is conceivable that a tissue-engineered semilunar valve may not require development of microvascular circulation (Schoen and Levy, 1999; Mayer, 2001; Sapirstein and Smith, 2001; Weind *et al.*, 2001). Another advantage of tissue engineering a semilunar heart valve is its composition of easily cultured cells. This is a critical characteristic because it allows for the *in vitro* isolation and *in vitro* expansion of autologous cells. Another critical feature enabling the development of a tissue-engineered heart valve is the fact that valve function can be readily studied using well-established techniques, such as Doppler ultrasonography and arteriography (Mayer, 2001; Sapirstein and Smith, 2001). The pathophysiology of valvular dysfunction has also been expansively researched at a variety of levels using molecular, biochemical, and histological analysis (Schoen and Levy, 1999; Schoen, 1987). This analysis can then be used to evaluate the performance of the tissue-engineered heart valve, thus allowing for a prompt appraisal of the tissue-engineered valves in comparison with native valves (Shalak and Fox, 1988; Heineken and Skalak, 1991).

The blueprint for the tissue-engineered heart valve has evolved from a large body of research in bioprosthetic valves, diseased heart valves, and other tissue valve substitutes. Although these investigations were largely clinical, they have identified useful markers of cell function, matrix physiology, and matrix structure. This was summarized by Rabkin-Aikawa and others in five key concepts of functionally adaptive valvular remodeling/regeneration as follows (Rabkin-Aikawa *et al.*, 2005):

> *(1) The highly specialized arrangement of collagen and other ECM components (particularly elastin and proteoglycans) enables normal heart valve function and is the principal determinant of the durability of heart valves. (2) Structural deterioration of native and substitute valves is ultimately mediated by chemical and mechanical damage to collagen. (3) The quality of valvular ECM depends on valvular interstitial cell viability, function and ability to adapt to different environments. (4) Cell viability in nearly all current bioprosthetic tissue valve substitutes is compromised or completely eliminated during processing; thus, ECM damage, which occurs during valve function following implantation of current bioprosthetic valves, cannot be repaired. (5) The long-term success of a tissue-engineered valve replacement will, therefore, depend on the ability of its living cellular components to assume normal function, with the capacity to repair structural injury, remodel the ECM, and potentially grow.*

While there is substantial ongoing work to improve valve bioprostheses incrementally through anticalcification pretreatments and design optimization, the advantages of tissue engineering may result in drastically better treatments. This is most evident in pediatric applications, where there is still a large unmet need for very small-size replacement valves. Even with successful surgeries, repeat operations are often required as the patient grows. Tissue engineering with autologous living cells offers the potential to create nonthrombogenic, nonobstructive tissue valve substitutes capable of providing ongoing remodeling and repair and allowing growth in maturing recipients (Rabkin-Aikawa *et al.*, 2005).

## Biomaterials and Scaffolds

Tissue-engineering scaffolds must be biocompatible, biodegradable with safe by-products, and highly porous yet sufficiently mechanically stable to appropriate function (Rabkin-Aikawa *et al.*, 2005). The scaffold must possess properties that promote cell attachment and recapitulate complex tissues. For more complicated applications, the scaffolds may need to be additionally functionalized with bioactive molecules or ligands. Finally, one frequently overlooked requirement is that the scaffolds possess enough robustness to allow for handling by the end user, most usually a surgeon.

Scaffolds can be made from either synthetic or natural materials. Natural biomaterials are usually composed of ECM components: collagen, fibrin, elastin, glycosaminoglycans, and decellularized tissues. Some examples of decellularized tissues used for tissue engineering are decellularized heart valve, pericardium, and arterial wall (Schmidt and Baier, 2000; Hodde, 2002). Synthetic polymers have an advantage in that the raw materials are generally easily available, they can be made reproducibly, their chemistry is predictable, and their properties can be well controlled (Rabkin-Aikawa *et al.*, 2005). The most commonly utilized synthetic polymers in tissue engineering include the poly *a*-hydroxyesters poly(glycolic acid) (PGA), poly(L-lactic acid) (PLLA), and copolymers poly(lactic-*co*-glycolic acid) (PLGA), and poly(anhydrides) and poly(peptides). Another popular polymer is PHA, or polyhydroxyalkanoate (Rabkin-Aikawa *et al.*, 2005). PHA is a thermoplastic that is biocompatible, resorbable, and flexible and that causes minimal inflammatory response (Williams *et al.*, 1999). In addition, elastic biodegradable polymers are now available as well and may offer additional, unique mechanical advantages (Lendlein and Langer, 2002).

## Natural Matrices

Natural materials were originally picked for intrinsic properties that matched those of native tissue, most often the biomechanical profile. Furthermore, it was not unusual for availability and ease of use to dictate the choice of material. Decellularized small intestinal submucosal (SIS) matrix was one of the first natural materials isolated for use in tissue engineering (Badylak *et al.*, 1989, 1998). Investigators used porcine SIS scaffold as a resorbable matrix to make pulmonary valve leaflet replacements in porcine models (Matheny *et al.*, 2000). Interestingly, the implanted scaffolds seemed to undergo complete resorption (Flanagan and Pandit, 2003).

Another natural scaffold is biodegradable fibrin gels, which are easily made, using the patient's own blood as a source for raw material (Grassl *et al.*, 2002; Neidert *et al.*, 2002). In one experiment, cell–fibrin gel scaffolds were made from human aortic myofibroblasts suspended in a solution of thrombin, fibrinogen, and calcium chloride and allowed to polymerize at 37°C. In order to promote collagen synthesis, the growth media was supplemented with L-ascorbic acid 2-phosphate (Grinnell *et al.*, 1989). This technique was further advanced by developing a molding technique to form a trileaflet heart valve (Jockenhoevel *et al.*, 2001). An adjustable mold made from ventricular and aortic casts was constructed out of aluminum. Ultimately, however, even though the engineered tissue was robust enough for suture, it could not withstand surgical implantation (Flanagan and Pandit, 2003) and suffered from additional drawbacks, such as shrinkage with time. In a similar approach by Ye and colleagues, completely human autogeneic tissue was created without using a supporting scaffold. Briefly, myofibroblasts were cultured in a medium supplemented with L-ascorbic acid 2-phospate to promote ECM production, forming cell sheets that were then folded into quadrilaminar sheets and mounted on tailored culture frames. These constructs were then cultured for an additional month, after which a multilayer tissue pattern was observed, with active and viable cells surrounded by ECM (Ye *et al.*, 2000).

The majority of the mechanical and tensile strength of the heart valve comes from collagen (Flanagan and Pandit, 2003). For this reason, many investigations have attempted to use collagen as a matrix component. Collagen has been used as a material for tissue-engineered heart valve scaffolds in applications involving both human and porcine-derived cells (Rothenburger *et al.*, 2001; Taylor *et al.*, 2002). Rothenburger and colleagues utilized a collagen I derived from bovine skin tissue to make a scaffold that was 98% porous. The scaffold was sectioned into disc structures and serially seeded with either porcine or human aortic smooth muscle cells, followed by porcine aortic endothelial cells. After 28 days in culture, the tissue morphology had several layers of cells with newly synthesized ECM components, similar to native tissue. Although collagen scaffolds outperformed other natural materials, they still lacked the sophistication of native structures.

In 1995, a heart valve decellularization process was developed that utilized DNAse, RNAse, and the detergent Triton X-100 (Wilson *et al.*, 1995). Using this method, cellular components and debris such as nucleic acids, cell membranes, cytoplasmic structures, lipids, and soluble matrix molecules were removed while maintaining the elastin, collagen, and GAG components of the scaffold ECM (Zeltinger *et al.*, 2001). Thus, the sophistication of structure and desired biomechanical properties of the scaffold are theoretically preserved. When these scaffolds were used in a canine model, a four-week follow-up revealed no inflammatory process, partial endothelialization, or partial VIC infiltration at the valvular base (Wilson *et al.*, 1995). The same method for decellularization was used three years later with porcine matrices. These matrices were seeded with human endothelial cells. The resulting construct had a confluent and viable monolayer cell surface, an important feature in reducing thrombogenic risk (Bader *et al.*, 1998). The cross section of the leaflet was largely acellular, and the collagen framework was wavelike, similar to normal tissue. Unfortunately, there were significant interfibrillar spaces, an undesirable characteristic that can affect the biomechanical properties of the tissue. The investigators also had significant difficulties removing cellular remnants, which can act as a nidus for ectopic calcification (Schoen and Hobson, 1985; Valente *et al.*, 1985) or an immune response (Schmidt and Baier, 2000). This work continued in *in vivo* experimentation using a decellularized valve scaffold seeded with autogeneic cells (Steinhoff *et al.*, 2000). Endothelial cells and carotid artery myofibroblasts were added in succession to acellularized pulmonary valve conduits and implanted into ovine models. After one month, the valve leaflets were completely endothelialized. The endothelium remained confluent three months after implantation and had been densely infiltrated by myofibroblasts. There was, however, some indication that subvalvular calcification and inflammation had occurred, along with increased thickening of the valve leaflets (Flanagan and Pandit, 2003).

The commercially available SynerGraft™ valve, manufactured by CryoLife Inc., was based on the decellularized model for tissue engineering (O'Brien *et al.*, 1999). In the SynerGraft™ decellularization process, the cells on the matrix were lysed in sterile water, while remaining nucleic acid material was enzymatically digested. This was followed by an isotonic washout of several days. Histological evaluation 150 days postimplantation in porcine models revealed healthy leaflets, with local myofibroblasts in growth and no calcific mineralization (Flanagan and Pandit, 2003). This method was developed as an alternative to glutaraldehyde cross-linking to decrease xenograft antigenicity. Early failure of the valve, however, was reported in human trials, as described later in this chapter in the Clinical Applications section (Simon *et al.*, 2003).

As of yet, it is unclear as to what the ideal decellularizing agent for heart valves should be (Flanagan and Pandit, 2003). In a recent evaluation, Triton X-100, sodium dodecyl sulphate (SDS), sodium deoxycholate, MEGA 10, TnBP, CHAPS, and Tween 20 were compared for efficacy. The results revealed that only SDS (0.03–1%) or sodium deoxycholate (0.5–2%) resulted in complete decellularization after 24 hours (Booth *et al.*, 2002). It is possible, however, that any of these agents can have toxic side effects if even trace amounts are left in the scaffold.

### Synthetic Matrices

A number of biodegradable polymers have been explored for use as scaffolds. Biodegradable polymers have the advantage of offering a more biocompatible template on which cells can grow (Flanagan and Pandit, 2003). Proponents of the synthetic matrix believe that using biodegradable polymers can decrease biocompatibility/foreign-body complications. Additionally, they point out that customized polymers can be engineered to exact specifications in a reproducible manner.

Initial work focused on polymers composed of polyglycolic and polylactic acid (Shinoka *et al.*, 1995, 1996, 1997, 1998; Breuer *et al.*, 1996; Zund *et al.*, 1997; Bader *et al.*, 1998). These "off the shelf" polymers were chosen because they are biodegradable, biocompatible, well characterized, and already FDA approved for human use. Furthermore, cells readily attach to and grow on these polymers. The matrices for the heart valves were woven from PLGA microfibers and then layered between two nonwoven PGA mesh sheets. These scaffolds were then serially seeded with arterial myofibroblasts and arterial endothelial cells. Following the seeding, the constructs were transplanted into the pulmonary position as a single leaflet in sheep. The seeded cells were visible in the structure six weeks after implantation, and a postmortem evaluation revealed a native tissuelike architecture (Shinoka *et al.*, 1996). Furthermore, there was positive confirmation of elastin and collagen production in the leaflets (Shinoka *et al.*, 1996). The biomechanical profile of the construct, however, differs considerably from that of the native heart valve. Tissue-engineered heart valves using polyglycolic acid and polylactic acid copolymer–based matrices are thicker, stiffer, and less pliable than native valves.

Following this early work, Stock and collaborators evaluated the more flexible synthetic polymer polyhydroxyoctanoate (PHO) as a material for matrix construction (Stock *et al.*, 2000). The conduit design consisted of a layer of 240-μm-thick PHO film inserted between two layers of 1-mm-thick nonwoven PGA felt. Sutured to this conduit were leaflets composed of a monolayer of 120-μm-thick porous PHO. Postimplantation examination revealed a uniformly organized tissue with large amounts of collagen and proteoglycans but no elastin. In a different study, a porous PHO scaffold was molded into a trileaflet valved conduit by thermal processing. The resulting scaffold was subsequently seeded with ovine carotid jugular vein endothelial cells and arterial myofibroblasts (Sodian *et al.*, 2000). The constructs were surgically implanted into the pulmonary position of a sheep and harvested after 1–17 weeks. All of the valve constructs opened and closed in harmony and showed an increase in inner length and diameter. Whether this was caused by actual tissue growth or simple construct dilatation was indeterminable. Scanning electron microscopy revealed a smooth leaflet surface; however, histological evaluation did not confirm a confluent endothelium, necessary for long-term stability of the construct. There was some GAG and collagen formation, but no elastin was present.

It was concluded that the hydrolytical stability of PHO made it an undesirable choice of polymer. In addition to being insufficiently replaced by neotissue, the extended scaffold lifetime could increase the potential for unwanted host–tissue reactions, and, in fact, all valve constructs demonstrated some degree of mild stenosis and regurgitation as well as inflammatory responses (Stock *et al.*, 2000; Flanagan and Pandit, 2003).

Hoerstrup and colleagues introduced a unique composite matrix made from PGA coated with a thin layer of poly-4-hydroxybutyrate (P4HB). P4HB is a flexible, thermoplastic polymer that degrades more rapidly than PHO (Hoerstrup *et al.*, 2000; Martin and Williams, 2003). This polymer was used to weld trileaflet heart valve scaffolds under heat. Endothelial cells and myofibroblasts from lamb carotid artery were seeded onto the matrix and then placed in a bioreactor for two weeks (Hoerstrup *et al.*, 2000). The constructs were then implanted into the same animal of cell harvest, where they stayed for as long as five months. Exposure to the bioreactor was shown to induce a more structured internal architecture, increase ECM synthesis, and improve the construct's mechanical properties versus static controls. The leaflets revealed a layered architecture after four and five months, with a spongy layer comprising GAGs and elastin on the inflow surface and a largely fibrous layer composed mostly of collagen on the outflow surface. However, moderate regurgitation was noted at five months, with only partial endothelialization on the surface of the leaflet.

While it is unclear whether a synthetic or natural matrix type will be of most use in tissue engineering a heart valve, the scaffold design will be crucial for successful development of a tissue-engineered valve.

## The Search for Appropriate Cell Sources

The ideal cell source for tissue-engineered heart valves is a subject of intense investigation. The use of autologous cells as opposed to allograft or xenograft tissue has the advantage of avoiding an immunological response that could lead to rejection (Shinoka et al., 1995, 1996, 1997, 1998; Breuer et al., 1996; Zund et al., 1997; Bader et al., 1998; Stock and Mayer, 1999). In the earliest attempt to tissue engineer a heart valve, by Shinoka and others, valve leaflets were constructed and implanted in an ovine model by seeding cells from autogeneic and allogeneic sources onto a biodegradable polymeric scaffold (Shinoka et al., 1995). As expected, the autologous structures provoked less of an inflammatory response and rejection in the host, leading them to perform better and fail less often than the allogeneic structures.

Based on this finding, the logical choice of cell source for creating a tissue-engineered heart valve would be VICs and VECs harvested from the patient's own heart valve leaflets, eliminating the risk of rejection while theoretically capturing the requisite phenotypic profile. This proposal was investigated in an ovine model, using tissue from a valve biopsy as a VIC cell supply (Maish et al., 2003). An immediate concern in this process is the potential for wounding the valve. It did not appear, however, that the biopsy itself damaged leaflet function. Any negative long-term effects due to the biopsy, however, were indeterminable, for follow-up studies were never performed. One potential drawback of this approach is that for human patients, it is doubtful that a sufficient number of cells could be isolated and then cultured from the small portion of tissue resected in biopsy. Futhermore, the ease of culturing cells decreases with the patient's age, decreasing the utility of this approach in older patients. Additionally, patients requiring valve replacement often have diseased VECs and VICs, making them less than ideal as a basis for tissue engineering a replacement valve. Thus, the risks associated with valve biopsy are probably too great to make this a viable technique in human trials (Flanagan and Pandit, 2003).

Arterial cells were also briefly explored but ultimately rejected due to the requirement of sacrificing intact tissue and the potential for injury (Hoerstrup et al., 2000; Sodian et al., 2000; Stock et al., 2000; Flanagan and Pandit, 2003). In search of a more practical source, Shinoka and colleagues compared arterial myofibroblasts to dermal fibroblasts as cell sources for a tissue-engineered heart valve (Shinoka et al., 1997). The leaflets arising from dermal fibroblasts were more contracted, thicker, and less organized than leaflets arising from arterial myofibroblasts. Based on these findings it was hypothesized that mesodermal cells, such as arterial myofibroblasts, provide greater phenotypic specialization than ectodermally derived skin fibroblasts. In theory, this makes the mesodermal cell type more appropriate for the manufacture of a tissue-engineered heart valve.

Myofibroblasts derived from human saphenous vein represent a more realistic source for clinical applications of tissue-engineered heart valves because, unlike arterial cells, they can be harvested without the risk of limb ischemia (Schnell et al., 2001; Flanagan and Pandit, 2003). When cultured on polyurethane matrices, myofibroblasts were shown to be confluent and viable six weeks later. Moreover, in comparison with neotissue derived from aortic myofibroblasts, mechanical stability and collagen production were found to be higher in the saphenous vein–derived structure.

Less damaging cell sources from other tissues have also been evaluated. In 2002 a report demonstrated the feasability of using autologous umbilical cord cells (Kadner et al., 2002). The cells isolated were a mixed population derived from umbilical cord artery, vein, and the surrounding Wharton's jelly. These cells exhibited myofibroblast-like differentiation, including expression of vimentin, alpha smooth muscle actin, and deposition of collagen type I and collagen type III. Most importantly, they successfully adhered to matrices and formed layered tissuelike architecture similar to matrices seeded with vascular cells (Hoerstrup et al., 2000; Flanagan and Pandit, 2003). Unfortunately, elastin was not successfully produced and only low levels of GAG were found. More work is needed in characterization of the mixed cell population before the suitability of this source can be evaluated. Similarly, mesenchymal stem cells have shown promise for tissue-engineered heart valves, with their ability to differentiate. They can develop into a variety of connective tissues, including cartilage, bone, muscle, and fat, as well as easy collection via simple bone marrow puncture, making these cells an attractive possibility (Pittenger et al., 1999; Caplan and Bruder, 2001).

In a separate experiment, human bone marrow stromal cells were collected by bone marrow puncture and evaluated as a possible cell source (Hoerstrup et al., 2002; Kadner et al., 2002). Like umbilical cord cells, they were shown to undergo myofibroblast-like differentiation, expressing alpha smooth muscle actin and vimentin, and to produce collagen type I and collagen type III. When biodegradable polymeric scaffolds were seeded with these cells, the construct demonstrated advanced tissue development and an ordered internal structure. However, it is unknown if the bone marrow stromal cells differentiate into the desired cell types in the matrix.

The optimal replacement for both VICs and VECs is still being sought. However, given the remarkable differentiation potential of embryonic and adult stem cells, pluripotent cells may become an important feature of tissue-engineered heart valves.

## Cell-Seeding Techniques

The formation of neotissue has not been as simple as providing a scaffold and cells. While relying on cell growth into unseeded matrices has been undependable, active cell seeding by an assortment of techniques has yielded some success (Shinoka *et al.*, 1995, 1996, 1997, 1998; Breuer *et al.*, 1996; Zund *et al.*, 1997; Bader *et al.*, 1998; Gloeckner *et al.*, 1999; Mitka, 2000; Stock *et al.*, 2001; Rabkin *et al.*, 2002; Sutherland *et al.*, 2002; Engelmayr *et al.*, 2003; Nasseri *et al.*, 2003). Cell attachment is a crucial step in initiating cell growth and neotissue development. The majority of synthetic and natural scaffolds initially used for tissue engineering were selected for their biocompatibility, with little knowledge of how cells and tissues attached and grew on these materials. The scaffolds were made more porous in secondary matrix processing, such as salt leaching, to enhance the surface area for cell attachment, increase cell–cell interactions, and enable neotissue ingrowth. More advanced methods sought to augment cell attachment by coating the matrix prior to seeding with a variety of cell adhesion molecules, such as laminin (Shinoka *et al.*, 1995, 1996, 1997, 1998; Breuer *et al.*, 1996; Zund *et al.*, 1997; Bader *et al.*, 1998; Teebken *et al.*, 2000).

It also turns out that the actual approach to cell seeding plays a role in how the cells interact with the matrix — the difference between dynamic and static cell seeding is now well established (Sutherland *et al.*, 2002; Nasseri *et al.*, 2003). The traditional approach to polymer scaffold seeding employed established static cell culture techniques. Briefly, a concentrated suspension of cells is pipetted onto collagen-coated polymer matrices and allowed to incubate for a variable time period, during which cell adhesion occurs. In dynamic cell seeding, either the medium or both the medium and the scaffold are in continuous motion throughout the incubation period (Sutherland *et al.*, 2002; Nasseri *et al.*, 2003). Nasseri found that dynamic cell seeding onto tissue-engineering matrices increased cellular adhesion, seeding density alignment in the direction of flow, and cell infiltration (Sutherland *et al.*, 2002; Nasseri *et al.*, 2003). Obviously, many variables can be tuned in such a system, including the use of pure versus mixed cell populations, single-step versus sequential seeding of different cells, and the introduction of intervals (Zund *et al.*, 1999; Ye *et al.*, 2000). Various permutations of these methods are being systematically evaluated in an effort to optimize cell seeding (Gloeckner *et al.*, 1999; Pittenger *et al.*, 1999; Mitka, 2000; Guleserian *et al.*, 2001; Kaushal *et al.*, 2001; Stock *et al.*, 2001; Hoerstrup *et al.*, 2002; Kadner *et al.*, 2002; Rabkin *et al.*, 2002; Sutherland *et al.*, 2002; Engelmayr *et al.*, 2003; Nasseri *et al.*, 2003; Perry *et al.*, 2003).

Uniform, adequate, and reproducible cell seeding of both synthetic polymeric and natural scaffolds remains a challenge in tissue engineering. Optimization of rapid seeding techniques will be critical in the effort to tissue engineer a heart valve, for it hastens the proliferation and subsequent differentiation of cells, maximizes the use of donor cells, provides a uniform distribution of cells, and decreases the time in culture (Vunjak-Novakovic *et al.*,

1998). Technical advances in the alteration of matrix surfaces to enhance cell adhesion and function will be needed for short-term *ex vivo* culture of tissues prior to implantation (Flanagan and Pandit, 2003).

## Neotissue Development in the Tissue-Engineered Heart Valve

Upon formation of the cell–scaffold construct, neotissue begins its development. This process, however, is poorly understood and is controlled by many factors. In some characteristics it appears to be a recapitulation of ontogeny, while in others it appears to operate under the mechanisms of tissue repair. In fact, it is most likely a unique process governed by its own set of laws. The cell type implanted and their environment determine the type of tissue that ultimately develops from the cell–scaffold complex (Shalak and Fox, 1988; Heineken and Skalak, 1991; Langer and Vacanti, 1993).

The environment of the growing construct will influence the extracellular matrix and histological structure formed. Scientists have approached this phenomenon from two perspectives. The first is a "black box" approach, where the scaffold is used as a cell delivery device and implanted *in vivo* shortly after cell attachment has taken place (Vacanti *et al.*, 1988; Langer and Vacanti, 1993). The cell–matrix construct is surgically implanted *in vivo* into the environment in which it is expected to function, the justification for this method being that appropriate environmental signals for tissue repair and remodeling are intrinsically present in the *in vivo* milieu (Vacanti *et al.*, 1988; Langer and Vacanti, 1993). In addition, biomechanical forces of the environment provide important stimuli that influence the formation of extracellular matrix and determine the biomechanical properties of the growing neotissue. The assumption underlying this approach is that the cell–scaffold construct possesses the necessary mechanical physicality to provide temporary function until neotissue develops (Vacanti *et al.*, 1988; Langer and Vacanti, 1993).

In a second approach an instrument called a bioreactor is used. A bioreactor is a biomimetic system used for *in vitro* neotissue development. It is used to condition or stimulate the cell–scaffold construct *in vitro* to improve cell attachment, change cellular orientation, and increase the production of extracellular matrix proteins such as collagen and elastin. The bioreactor can also be utilized to adjust the biomechanical properties of the neotissue, which is of special interest in the development of a tissue-engineered heart valve (Gloeckner *et al.*, 1999; Niklason *et al.*, 1999; Hoerstrup *et al.*, 2000; Sodian *et al.*, 2001, 2002a; Jockenhoevel *et al.*, 2002; Engelmayr *et al.*, 2003; Mol *et al.*, 2003). One type of bioreactor used in tissue engineering a heart valve is the pulse duplicator. This pulsatile bioreactor supplies physiological flow and pressure to the developing tissue-engineered heart valve and promotes both the modulation of cellular function and the development of mechanical strength (Niklason *et al.*, 1999; Hoerstrup *et al.*, 2000; Sodian *et al.*, 2001, 2002a). A bioreactor capable of inducing laminar flow has also been employed and shown to modulate tissue development and extracellular matrix formation. Other versions of bioreactors for tissue engineering the heart valves include bioreactors that can provide increasing dynamic flexural strain or cyclic strain (Engelmayr *et al.*, 2003; Mol *et al.*, 2003). The mechanical straining can induce more pronounced and organized tissue formation. Neotissue made in this way possesses superior mechanical properties over unstrained controls and results in heart valve cusps that are considerably less stiff than static controls (Gloeckner *et al.*, 1999; Sodian *et al.*, 2002a; Engelmayr *et al.*, 2003; Mol *et al.*, 2003).

Tissue-engineered heart valves have been evaluated physiologically, biochemically, histologically, molecularly, biomechanically, and morphologically. In each analysis the tissue-engineered valve was compared with the native valve (Shinoka *et al.*, 1995, 1996, 1997, 1998; Breuer *et al.*, 1996; Zund *et al.*, 1997; Bader *et al.*, 1998). These data could then be examined to identify weaknesses in the tissue-engineered heart valve. This information could then be used for the rational design of an improved tissue-engineered heart valve. The use of feedback of this type is an essential strategy in any biomedical engineering project and will play an important part in tissue engineering a heart valve. For example, insufficiencies in the biomechanical profile of tissue-engineered cardiovascular tissues led to the incorporation of polyhydroxyalkanoate as a polymer for the matrix, and structural weaknesses of the cell–scaffold construct compared with the native valve led to the development of bioreactors (Gloeckner *et al.*, 1999; Niklason *et al.*, 1999; Hoerstrup *et al.*, 2000; Sodian *et al.*, 2001, 2002a; Jockenhoevel *et al.*, 2002; Engelmayr *et al.*, 2003; Mol *et al.*, 2003).

## Clinical Applications of the Tissue-Engineered Heart Valve

Clinical application of the tissue-engineered heart valve has yielded mixed results. Dohmen *et al.* (2002) reported the first success in using a tissue-engineered heart valve. A decellularized cryo-preserved pulmonary homograft was seeded with autologous vascular endothelial cells isolated from a segment of forearm vein. The tissue-engineered valve was used in a 43-year-old patient to reconstruct the right ventricular outflow tract. One year after implantation, the tissue-engineered heart valve demonstrated effectively normal function, with trivial central regurgitation, which had been present since the beginning, no visible calcifications, and smoothly moving leaflets (Dohmen *et al.*, 2002).

Unfortunately, there have been unsuccessful cases as well. Simon *et al.* (2003) reported the use of the Synergraft™ decellularized porcine valve in four children. These decellularized valves had not been seeded with cells or condi-tioned in a bioreactor before implantation as a simple decellularized scaffold. Three of the four children died from valvular complications, while the fourth had the valve removed. Pathological evaluation revealed a fibrous sheath covering the valves. In addition, a severe inflammatory response was visible, with no signs of cellular repopulation or of endothelialization (Simon *et al.*, 2003). These findings highlight the need for further rigorous laboratory investigation of new tissue-engineering technologies before any additional attempt at clinical application.

To date, there have been no clinical reports of synthetic scaffolds being used to fabricate replacement heart valves. Shinoka *et al.*, however, have applied heart valve tissue-engineering techniques to construct autologous vascular grafts and patches for use as venous conduits in over 40 children with varying forms of congenital heart disease. The synthetic matrix used to create the neotissue was composed of polyglycolic acid and ε-caprolactone copolymers reinforced with poly-L-lactide. The scaffolds were either seeded with cells expanded from the saphenous vein or directly seeded with bone marrow or mononuclear bone marrow. The first successful clinical demonstration of this method occurred in May 1999 (Shinoka *et al.*, 1997; Matsumura *et al.*, 2003a). Serial postoperative computerized tomography, angiographic, or magnetic resonance imaging examinations revealed no rupture or dilatation of grafts, and there have been no complications related to the tissue-engineered autografts (Isomatsu *et al.*, 2003; Matsumura *et al.*, 2003a, 2003b; Naito *et al.*, 2003). Although histological evaluation is not possible, calcification has been noted by current imaging studies in these patients (Isomatsu *et al.*, 2003; Matsumura *et al.*, 2003a, 2003b; Naito *et al.*, 2003). Shinoka's work proves that the principles of tissue engineering have the potential for a tremendous clinical contribution. The tissue-engineered vascular conduits appear to have minimal risk of infection, no risk of rejection (autologous cell source), reduced incidence of calcification, and potential for growth (Isomatsu *et al.*, 2003; Matsumura *et al.*, 2003a, 2003b; Naito *et al.*, 2003). All of these potential benefits would be possible in a tissue-engineered heart valve as well.

Despite the excellent results of Shinoka *et al.*, and Dohmen *et al.*, the clinical translation of cardiovascular tissue engineering is premature by U.S. standards. The preclinical experimentation needed to provide justification for clinical FDA investigations of tissue-engineered products is in its infancy. The rapid progress of cardiovascular tissue-engineering research has outpaced the regulatory agencies ability to develop governing policies of product development (Mayer *et al.*, 1997; Mayer, 2001). The completion of preclinical studies on which clinical trials can be based is essential for the rational and responsible development of this promising technology. It is time we move beyond feasibility studies and into randomized, controlled preclinical and clinical trials in order to evaluate the true benefits of tissue-engineered heart valves compared to currently used bioprosthetic and mechanical valves. Investigation of this type will allow us to identify the current limitations of this technology so that we can direct our research toward improving replacement heart valves.

In terms of basic research, recent studies have provided insight into the processes underlying tissue engineering (Stock *et al.*, 2001; Rabkin *et al.*, 2002). These investigations have begun to reveal the dynamics of the extracellular matrix in terms of production, remodeling, and degradation (Stock *et al.*, 2001; Rabkin *et al.*, 2002). These complex mechanisms require the deposition and accumulation of extracellular proteins as well as remodeling by matrix metalloproteinases and their endogenous tissue inhibitors of metalloproteinases (Stock *et al.*, 2001; Rabkin *et al.*, 2002). The cell phenotype and extracellular matrix in tissue-engineered heart valves developed *in vitro* and implanted *in vivo* are dynamic and reflect the capacity of vital tissue to remodel and potentially grow. As our understanding of this process expands, our capability to direct it will also evolve.

## VI. CONCLUSION

Successful development of a tissue-engineered replacement heart valve may hold the key to better treatment of end-stage valve disease. Although significant progress has been achieved since its inception in the early 1990s, the field is young and many key issues need to be resolved. As of yet, we are still limited by our knowledge of the cell biology and ECM production and maintenance of a normal valve. The identification of an ideal cell source and optimal matrix persists as the point of focus in tissue-engineering strategy. It is possible that a more complete understanding of embryonic and fetal heart valve development may provide insight that eventually enables tissue engineers to build consistently clinically acceptable replacement heart valves *ex vivo* (Flanagan and Pandit, 2003). Other important advances will undoubtedly also be needed to allow for the tissue-engineered heart valve to come to fruition and will require contributions from many different disciplines.

## VII. REFERENCES

Bader, A., Schilling, T., *et al.* (1998). Tissue engineering of heart valves — human endothelial cell seeding of detergent acellularized porcine valves. *Eur. J. Cardiothorac. Surg.* **14**(3), 279–284.

Badylak, S. F., Lantz, G. C., *et al.* (1989). Small intestinal submucosa as a large-diameter vascular graft in the dog. *J. Surg. Res.* **47**(1), 74–80.

Badylak, S. F., Record, R., *et al.* (1998). Small intestinal submucosa: a substrate for *in vitro* cell growth. *J. Biomater. Sci. Polym. Ed.* **9**(8), 863–878.

Bertipaglia, B., Ortolani, F., *et al.* (2003). Cell characterization of porcine aortic valve and decellularized leaflets repopulated with aortic valve interstitial cells: the VESALIO Project (Vitalitate Exornatum Succedaneum Aorticum Labore Ingenioso Obtenibitur). *Ann. Thorac. Surg.* **75**(4), 1274–1282.

Blackstone, E. H., and Kirklin, J. W. (1985). Death and other time-related events after valve-replacement. *Circulation* **72**(4), 753–767.

Bloomfield, P., Wheatley, D. J., *et al.* (1991). Twelve-year comparison of a Bjork–Shiley mechanical heart valve with porcine bioprostheses. *N. Engl. J. Med.* **324**(9), 573–579.

Booth, C., Korossis, S. A., *et al.* (2002). Tissue engineering of cardiac valve prostheses I: development and histological characterization of an acellular porcine scaffold. *J. Heart Valve Dis.* **11**(4), 457–462.

Breuer, C. K., Shinoka, T., *et al.* (1996). Tissue engineering lamb heart valve leaflets. *Biotechnol. Bioeng.* **50**(5), 562–567.

Breuer, C. K., Mettler, B. A., Anthony, T., Sales, V. L., Schoen, F. J., and Mayer, J. E. (2004). Application of tissue-engineering principles toward the development of a semilunar heart valve substitute. *Tissue. Eng.* **10**(11–12), 1725–1736.

Bursac, N., Papadaki, M., *et al.* (1999). Cardiac muscle tissue engineering: toward an *in vitro* model for electrophysiological studies. *Am. J. Physiol.* **277**(2 Pt. 2), H433–H444.

Butany, J., Ahluwalia, M. S., *et al.* (2003). Mechanical heart valve prostheses: identification and evaluation (erratum). *Cardiovasc. Pathol.* **12**(6), 322–344.

Butcher, J. T., Penrod, A. M., *et al.* (2004). Unique morphology and focal adhesion development of valvular endothelial cells in static and fluid flow environments. *Arteriosclerosis Thrombosis Vasc. Biol.* **24**(8), 1429–1434.

Cannegieter, S. C., Rosendaal, F. R., *et al.* (1994). Thromboembolic and bleeding complications in patients with mechanical heart-valve prostheses. *Circulation* **89**(2), 635–641.

Caplan, A. I., and Bruder, S. P. (2001). Mesenchymal stem cells: building blocks for molecular medicine in the 21st century. *Trends Mol. Med.* **7**(6), 259–264.

Chester, A. H., Misfeld, M., *et al.* (2000). Receptor-mediated contraction of aortic valve leaflets. *J. Heart Valve Dis.* **9**(2), 250–254; discussion 254–255.

Chester, A. H., Misfeld, M., *et al.* (2001). Influence of 5-hydroxytryptamine on aortic valve competence *in vitro*. *J. Heart Valve Dis.* **10**(6), 822–825; discussion 825–826.

Clark, E., Chia, J., and Waterman, S. (1997). *Candida albicans* endocarditis associated with a contaminated aortic valve allograft — California, 1996. *MMWR Morb. Mortal. Wkly. Rep.* **46**(12), 261–263.

Cleveland, D. C., Williams, W. G., *et al.* (1992). Failure of cryopreserved homograft valved conduits in the pulmonary circulation. *Circulation* **86**(5 Suppl.), II150–II153.

Cohn, L. H., Collins, J. J., *et al.* (1989). Fifteen-year experience with 1678 Hancock porcine bioprosthetic heart valve replacements. *Ann. Surg.* **210**(4), 435–443.

Cole, W. G., Chan, D., *et al.* (1984). Collagen composition of normal and myxomatous human mitral heart valves. *Biochem. J.* **219**(2), 451–460.

Davies, P. F., Passerini, A. G., *et al.* (2004). Aortic valve — turning over a new leaf(let) in endothelial phenotypic heterogeneity. *Arteriosclerosis Thrombosis Vasc. Biol.* **24**(8), 1331–1333.

Decker, R. S., and Dingle, J. T. (1982). Cardiac catabolic factors — the degradation of heart-valve inter-cellular matrix. *Science* **215**(4535), 987–989.

Dohmen, P. M., Lembcke, A., *et al.* (2002). Ross operation with a tissue-engineered heart valve. *Ann. Thorac. Surg.* **74**(5), 1438–1442.

Dreger, S. A., Taylor, P. M., *et al.* (2002). Profile and localization of matrix metalloproteinases (MMPs) and their tissue inhibitors (TIMPs) in human heart valves. *J. Heart Valve Dis.* **11**(6), 875–880; discussion 880.

Edmunds, L. H., Clark, R. E., *et al.* (1996). Guidelines for reporting morbidity and mortality after cardiac valvular operations. *Ann. Thorac. Surg.* **62**(3), 932–935.

Engelmayr, G. C., Hildebrand, D. K., *et al.* (2003). A novel bioreactor for the dynamic flexural stimulation of tissue engineered heart-valve biomaterials. *Biomaterials* **24**(14), 2523–2532.

Fedalen, P. A., Fisher, C. A., *et al.* (1999). Early fungal endocarditis in homograft recipients. *Ann. Thorac. Surg.* **68**(4), 1410–1411.

Ferrans, V. J., Tomita, Y., *et al.* (1987). Pathology of bioprosthetic cardiac valves. *Hum. Pathol.* **18**(6), 586–595.

Filip, D. A., Radu, A., *et al.* (1986). Interstitial cells of the heart valves possess characteristics similar to smooth muscle cells. *Circ. Res.* **59**(3), 310–320.

Flanagan, T. C., and Pandit, A. (2003). Living artificial heart-valve alternatives: a review. *Eur. Cell Mater.* **6**, 28–45; discussion 45.

Galis, Z. S., and Khatri, J. J. (2002). Matrix metalloproteinases in vascular remodeling and atherogenesis: the good, the bad, and the ugly. *Circ. Res.* **90**(3), 251–262.

Gloeckner, D. C., Billiar, K. L., *et al.* (1999). Effects of mechanical fatigue on the bending properties of the porcine bioprosthetic heart valve. *Asaio J.* **45**(1), 59–63.

Grassl, E. D., Oegema, T. R., *et al.* (2002). Fibrin as an alternative biopolymer to type-I collagen for the fabrication of a media equivalent. *J. Biomed. Mater. Res.* **60**(4), 607–612.

Grinnell, F., Fukamizu, H., *et al.* (1989). Collagen processing, cross-linking, and fibril bundle assembly in matrix produced by fibroblasts in long-term cultures supplemented with ascorbic acid. *Exp. Cell Res.* **181**(2), 483–491.

Grunkemeier, G. L., and Rahimtoola, S. H. (1990). Artificial-heart valves. *Annu. Rev. Med.* **41**, 251–263.

Grunkemeier, G. L., Jamieson, W. R. E., *et al.* (1994). Actuarial versus actual risk of porcine structural valve deterioration. *J. Thorac. Cardiovasc. Surg.* **108**(4), 709–718.

Guleserian, K. J., Sander, T. L., *et al.* (2001). Human umbilical cord blood–derived endothelial progenitor cells (HUCB-EPCs): a novel cell source for cardiovascular tissue engineering. *Circulation* **104**(17), 761–762.

Hammermeister, K. E., Sethi, G. K., *et al.* (1993). A comparison of outcomes in men 11 years after heart-valve replacement with a mechanical valve or bioprosthesis. Veterans Affairs Cooperative Study on Valvular Heart Disease. *N. Engl. J. Med.* **328**(18), 1289–1296.

Heineken, F. G., and Skalak, R. (1991). Tissue engineering — a brief overview. *J. Biomech. Eng. Trans. ASME* **113**(2), 111–112.

Henney, A. M., Parker, D. J., *et al.* (1982). Collagen biosynthesis in normal and abnormal human heart valves. *Cardiovasc. Res.* **16**(11), 624–630.

Hodde, J. (2002). Naturally occurring scaffolds for soft-tissue repair and regeneration. *Tissue. Eng.* **8**(2), 295–308.

Hoerstrup, S. P., Kadner, A., *et al.* (2002). Tissue engineering of functional trileaflet heart valves from human marrow stromal cells. *Circulation* **106**(12 Suppl. 1), I143–I150.

Hoerstrup, S. P., Sodian, R., *et al.* (2000a). Functional living trileaflet heart valves grown *in vitro*. *Circulation* **102**(19 Suppl. 3), III44–III49.

Hoerstrup, S. P., Sodian, R., *et al.* (2000b). New pulsatile bioreactor for *in vitro* formation of tissue-engineered heart valves. *Tissue. Eng.* **6**(1), 75–79.

Hufnagel, C. A., Harvey, W. P., *et al.* (1989). In the beginning. Surgical correction of aortic insufficiency. 1954. *Ann. Thorac. Surg.* **47**(3), 475–476.

Isomatsu, Y., Shin'oka, T., *et al.* (2003). Extracardiac total cavopulmonary connection using a tissue-engineered graft. *J. Thorac. Cardiovasc. Surg.* **126**(6), 1958–1962.

Jamieson, W. R. E. (1993). Modern cardiac-valve devices — bioprostheses and mechanical prostheses — state-of-the-art. *J. Card. Surg.* **8**(1), 89–98.

Jamieson, W. R. E., Munro, A. I., *et al.* (1995). Carpentier–Edwards standard porcine bioprosthesis — clinical-performance to 17 years. *Ann. Thorac. Surg.* **60**(4), 999–1007.

Jockenhoevel, S., Chalabi, K., *et al.* (2001). Tissue engineering: complete autologous valve conduit — a new moulding technique. *Thorac. Cardiovasc. Surg.* **49**(5), 287–290.

Jockenhoevel, S., Zund, G., *et al.* (2002). Cardiovascular tissue engineering: a new laminar flow chamber for *in vitro* improvement of mechanical tissue properties. *Asaio J.* **48**(1), 8–11.

Kadner, A., Hoerstrup, S. P., *et al.* (2002a). Human umbilical cord cells: a new cell source for cardiovascular tissue engineering. *Ann. Thorac. Surg.* **74**(4), S1422–S1428.

Kadner, A., Hoerstrup, S. P., *et al.* (2002b). A new source for cardiovascular tissue engineering: human bone marrow stromal cells. *Eur. J. Cardiothorac. Surg.* **21**(6), 1055–1060.

Kaushal, S., Amiel, G. E., *et al.* (2001). Functional small-diameter neovessels created using endothelial progenitor cells expanded *ex vivo*. *Nat. Med.* **7**(9), 1035–1040.

Kirklin, J. K., Pacifico, A. D., *et al.* (1989). Surgical treatment of prosthetic valve endocarditis with homograft aortic valve replacement. *J. Card. Surg.* **4**(4), 340–347.

Kirklin, J. K., Smith, D., *et al.* (1993). Long-term function of cryopreserved aortic homografts — a 10-year study. *J. Thorac. Cardiovasc. Surg.* **106**(1), 154–166.

Krane, S. M., Byrne, M. H., *et al.* (1996). Different collagenase gene products have different roles in degradation of type I collagen. *J. Biol. Chem.* **271**(45), 28509–28515.

Kunzelman, K. S., Cochran, R. P., *et al.* (1993). Differential collagen distribution in the mitral valve and its influence on biomechanical behaviour. *J. Heart Valve Dis.* **2**(2), 236–244.

Langer, R., and Vacanti, J. P. (1993). Tissue engineering. *Science* **260**(5110), 920–926.

Lendlein, A., and Langer, R. (2002). Biodegradable, elastic shape–memory polymers for potential biomedical applications. *Science* **296**(5573), 1673–1676.

Lester, W. M., and Gotlieb, A. I. (1988). *In vitro* repair of the wounded porcine mitral valve. *Circ. Res.* **62**(4), 833–845.

Lester, W. M., Damji, A. A., *et al.* (1992). Bovine mitral-valve organ culture — role of interstitial cells in repair of valvular injury. *J. Mol. Cell. Cardiol.* **24**(1), 43–53.

Maish, M. S., Hoffman-Kim, D., *et al.* (2003). Tricuspid-valve biopsy: a potential source of cardiac myofibroblast cells for tissue-engineered cardiac valves. *J. Heart Valve Dis.* **12**(2), 264–269.

Martin, D. P., and Williams, S. F. (2003). Medical applications of poly-4-hydroxybutyrate: a string flexible absorbable material. *Biochem. Eng. J.* **3738**, 1–9.

Matheny, R. G., Hutchison, M. L., *et al.* (2000). Porcine small intestine submucosa as a pulmonary-valve leaflet substitute. *J. Heart Valve Dis.* **9**(6), 769–774.

Matsumura, G., Hibino, N., *et al.* (2003a). Successful application of tissue-engineered vascular autografts: clinical experience. *Biomaterials* **24**(13), 2303–2308.

Matsumura, G., Miyagawa-Tomita, S., *et al.* (2003b). First evidence that bone marrow cells contribute to the construction of tissue-engineered vascular autografts *in vivo*. *Circulation* **108**(14), 1729–1734.

Mayer, J. E. (2001). In search of the ideal valve replacement device. *J. Thorac. Cardiovasc. Surg.* **122**(1), 8–9.

Mayer, J. E., Shinoka, T., *et al.* (1997). Tissue engineering of cardiovascular structures. *Curr. Opin. Cardiol.* **12**(6), 528–532.

McCawley, L. J., and Matrisian, L. M. (2001). Matrix metalloproteinases: they're not just for matrix anymore! *Curr. Opin. Cell Biol.* **13**(5), 534–540.

McGiffin, D. C., and Kirklin, J. K. (1995). The impact of aortic valve homografts on the treatment of aortic prosthetic valve endocarditis. *Semin. Thorac. Cardiovasc. Surg.* **7**(1), 25–31.

Meredith, J. E., and Schwartz, M. A. (1997). Integrins, adhesion and apoptosis. *Trends Cell Biol.* **7**(4), 146–150.

Mitka, M. (2000). Tissue engineering approaches utility. *J. Am. Med. Assoc.* **284**(20), 2582–2583.

Mol, A., Bouten, C. V., *et al.* (2003). The relevance of large strains in functional tissue engineering of heart valves. *Thorac. Cardiovasc. Surg.* **51**(2), 78–83.

Mulholland, D. L., and Gotlieb, A. I. (1996). Cell biology of valvular interstitial cells. *Can. J. Cardiol.* **12**(3), 231–236.

Mylonakis, E., and Calderwood, S. B. (2001). Medical progress: infective endocarditis in adults. *N. Engl. J. Med.* **345**(18), 1318–1330.

Naito, Y., Imai, Y., *et al.* (2003). Successful clinical application of tissue-engineered graft for extracardiac Fontan operation. *J. Thorac. Cardiovasc. Surg.* **125**(2), 419–420.

Nasseri, B. A., Pomerantseva, I., *et al.* (2003). Dynamic rotational seeding and cell culture system for vascular tube formation. *Tissue Eng.* **9**(2), 291–299.

Neidert, M. R., Lee, E. S., *et al.* (2002). Enhanced fibrin remodeling *in vitro* with TGF-beta1, insulin and plasmin for improved tissue equivalents. *Biomaterials* **23**(17), 3717–3731.

Nelson, A. R., Fingleton, B., *et al.* (2000). Matrix metalloproteinases: biologic activity and clinical implications. *J. Clin. Oncol.* **18**(5), 1135–1149.

Niklason, L. E., Gao, J., *et al.* (1999). Functional arteries grown *in vitro*. *Science* **204**(5413), 409–493.

O'Brien, M. F., Stafford, E. G., *et al.* (1995). Allograft aortic valve replacement: long-term follow-up. *Ann. Thorac. Surg.* **60**(2 Suppl.), S65–S70.

O'Brien, M. F., Goldstein, S., *et al.* (1999). The SynerGraft valve: a new acellular (nonglutaraldehyde-fixed) tissue heart valve for autologous recellularization first experimental studies before clinical implantation. *Semin. Thorac. Cardiovasc. Surg.* **11**(4 Suppl. 1), 194–200.

Perry, T. E., Kaushal, S., *et al.* (2003). Bone marrow as a cell source for tissue engineering heart valves. *Ann. Thorac. Surg.* **75**(3), 761–767.

Pittenger, M. F., Mackay, A. M., *et al.* (1999). Multilineage potential of adult human mesenchymal stem cells. *Science* **284**(5411), 143–147.

Rabkin, E., and Schoen, F. J. (2002). Cardiovascular tissue engineering. *Cardiovasc. Pathol.* **11**(6), 305–317.

Rabkin, E., Aikawa, M., *et al.* (2001). Activated interstitial myofibroblasts express catabolic enzymes and mediate matrix remodeling in myxomatous heart valves. *Circulation* **104**(21), 2525–2532.

Rabkin, E., Hoerstrup, S. P., *et al.* (2002). Evolution of cell phenotype and extracellular matrix in tissue-engineered heart valves during *in vitro* maturation and *in vivo* remodeling. *J. Heart Valve Dis.* **11**(3), 308–314.

Rabkin-Aikawa, E., Aikawa, M., *et al.* (2004). Clinical pulmonary autograft valves: pathologic evidence of adaptive remodeling in the aortic site. *J. Thorac. Cardiovasc. Surg.* **128**(4), 552–561.

Rabkin-Aikawa, E., Mayer, Jr., J. E., *et al.* (2005). Heart valve regeneration. *Adv. Biochem. Eng. Biotechnol.* **94**, 141–179.

Rafii, S. (2000). Circulating endothelial precursors: mystery, reality, and promise. *J. Clin. Invest.* **105**(1), 17–19.

Rothenburger, M., Vischer, P., *et al.* (2001). *In vitro* modeling of tissue using isolated vascular cells on a synthetic collagen matrix as a substitute for heart valves. *Thorac. Cardiovasc. Surg.* **49**(4), 204–209.

Roy, A., Brand, N. J., *et al.* (2000). Molecular characterization of interstitial cells isolated from human heart valves. *J. Heart Valve Dis.* **9**(3), 459–464; discussion 464–465.

Sacks, M. S., and Schoen, F. J. (2002). Collagen fiber disruption occurs independent of calcification in clinically explanted bioprosthetic heart valves. *J. Biomed. Mater. Res.* **62**(3), 359–371.

Sapirstein, J. S., and Smith, P. K. (2001). The "ideal" replacement heart valve. *Am. Heart J.* **141**(5), 856–860.

Schmidt, C. E., and Baier, J. M. (2000). Acellular vascular tissues: natural biomaterials for tissue repair and tissue engineering. *Biomaterials* **21**(22), 2215–2231.

Schneider, P. J., and Deck, J. D. (1981). Tissue and cell renewal in the natural aortic valve of rats: an autoradiographic study. *Cardiovasc. Res.* **15**(4), 181–189.

Schnell, A. M., Hoerstrup, S. P., *et al.* (2001). Optimal cell source for cardiovascular tissue engineering: venous vs. aortic human myofibroblasts. *Thorac. Cardiovasc. Surg.* **49**(4), 221–225.

Schoen, F. J. (1987). Surgical pathology of removed natural and prosthetic heart valves. *Hum. Pathol.* **18**(6), 558–567.

Schoen, F. J. (1989). "Interventional and Surgical Cardiovascular Pathology: Clinical Correlations and Basic Principles." W.B. Saunders Company, Philadelphia.

Schoen, F. J. (1991). Pathology of bioprostheses and other tissue heart valve replacements. *In* "Cardiovascular Pathology" (M. D. Silver, ed.), pp. 1547–1605. Churchill Livingstone, New York.

Schoen, F. J. (1995). Approach to the analysis of cardiac-valve prostheses as surgical pathology or autopsy specimens. *Cardiovasc. Pathol.* **4**(4), 241–255.

Schoen, F. J. (1997). Aortic valve structure-function correlations: role of elastic fibers no longer a stretch of the imagination. *J. Heart Valve Dis.* **6**, 1.

Schoen, F. J. (1999). Future directions in tissue heart valves: impact of recent insights from biology and pathology. *J. Heart Valve Dis.* **8**(4), 350.

Schoen, F. J., and Hobson, C. E. (1985). Anatomic analysis of removed prosthetic heart valves — causes of failure of 33 mechanical valves and 58 bioprostheses, 1980 to 1983. *Hum. Pathol.* **16**(6), 549–559.

Schoen, F. J., and Levy, R. J. (1984). Bioprosthetic heart valve failure: pathology and pathogenesis. *Cardiol. Clin.* **2**(4), 717–739.

Schoen, F. J., and Levy, R. J. (1994). Pathology of substitute heart valves — new concepts and developments. *J. Card. Surg.* **9**(2), 222–227.

Schoen, F. J., and Levy, R. J. (1999). Tissue heart valves: current challenges and future research perspectives. *J. Biomed. Mater. Res.* **47**(4), 439–465.

Schoen, F. J., Levy, R. J., et al. (1985). Onset and progression of experimental bioprosthetic heart-valve calcification. *Lab. Invest.* **52**(5), 523–532.

Schoen, F. J., Tsao, J. W., et al. (1986). Calcification of bovine pericardium used in cardiac-valve bioprostheses — implications for the mechanisms of bioprosthetic tissue mineralization. *Am. J. Pathol.* **123**(1), 134–145.

Schoen, F. J., Kujovich, J. L., et al. (1987). Chemically determined mineral content of explanted porcine aortic-valve bioprostheses — correlation with radiographic assessment of calcification and clinical data. *Circulation* **76**(5), 1061–1066.

Scott, M., and Vesely, I. (1995). Aortic valve cusp microstructure: the role of elastin. *Ann. Thorac. Surg.* **60**(2 Suppl.), S391–S394.

Shalak, R., and Fox, C. F. (1988). "Tissue Engineering: Proceedings for a Workshop Held at Granlibakken," Granlibakken, Lake Tahoe, CA.

Shin, D., Garcia-Cardena, G., et al. (2001). Expression of ephrinB2 identifies a stable genetic difference between arterial and venous vascular smooth muscle as well as endothelial cells, and marks subsets of microvessels at sites of adult neovascularization. *Dev. Biol.* **230**(2), 139–150.

Shinoka, T. (2002). Tissue engineered heart valves: autologous cell seeding on biodegradable polymer scaffold. *Artificial Organs* **26**(5), 402–406.

Shinoka, T., Breuer, C. K., et al. (1995). Tissue engineering heart valves: valve leaflet replacement study in a lamb model. *Ann. Thorac. Surg.* **60**(6 Suppl.), S513–S516.

Shinoka, T., Ma, P. X., et al. (1996). Tissue-engineered heart valves. Autologous valve leaflet replacement study in a lamb model. *Circulation* **94**(9 Suppl.), II164–II168.

Shinoka, T., ShumTim, D., et al. (1997). Tissue-engineered heart valve leaflets — does cell origin affect outcome? *Circulation.* **96**(9), 102–107.

Shinoka, T., Shum-Tim, D., et al. (1998). Creation of viable pulmonary artery autografts through tissue engineering. *J. Thorac. Cardiovasc. Surg.* **115**(3), 536–545.

Simionescu, D., Simionescu, A., et al. (1993). Detection of remnant proteolytic activities in unimplanted glutaraldehyde-treated bovine pericardium and explanted cardiac bioprostheses. *J. Biomed. Mater. Res.* **27**(6), 821–829.

Simon, P., Kasimir, M. T., et al. (2003). Early failure of the tissue-engineered porcine heart valve SYNERGRAFT in pediatric patients. *Eur. J. Cardiothorac. Surg.* **23**(6), 1002–1006; discussion 1006.

Simper, D., Stalboerger, P. G., et al. (2002). Smooth muscle progenitor cells in human blood. *Circulation* **106**(10), 1199–1204.

Sodian, R., Hoerstrup, S. P., et al. (2000a). Tissue engineering of heart valves: *in vitro* experiences. *Ann. Thorac. Surg.* **70**(1), 140–144.

Sodian, R., Loebe, M., et al. (2002b). Application of stereolithography for scaffold fabrication for tissue-engineered heart valves. *Asaio J.* **48**(1), 12–16.

Sodian, R., Lemke, T., et al. (2001). New pulsatile bioreactor for fabrication of tissue-engineered patches. *J. Biomed. Mater. Res.* **58**(4), 401–405.

Sodian, R., Lemke, T., et al. (2002a). Tissue-engineering bioreactors: a new combined cell-seeding and perfusion system for vascular tissue engineering. *Tissue. Eng.* **8**(5), 863–870.

Sodian, R., Sperling, J. S., et al. (2000b). Fabrication of a trileaflet heart-valve scaffold from a polyhydroxyalkanoate biopolyester for use in tissue engineering. *Tissue. Eng.* **6**(2), 183–188.

Steinhoff, G., Stock, U., et al. (2000). Tissue engineering of pulmonary heart valves on allogenic acellular matrix conduits — *in vivo* restoration of valve tissue. *Circulation* **102**(19), 50–55.

Stock, U. A., and Mayer, Jr., J. E. (1999). Valves in development for autogenous tissue valve replacement. *Semin. Thorac. Cardiovasc. Surg. Pediatr. Card. Surg. Annu.* **2**, 51–64.

Stock, U. A., Nagashima, M., et al. (2000). Tissue-engineered valved conduits in the pulmonary circulation. *J. Thorac. Cardiovasc. Surg.* **119**(4 Pt. 1), 732–740.

Stock, U. A., Wiederschain, D., et al. (2001). Dynamics of extracellular matrix production and turnover in tissue-engineered cardiovascular structures. *J. Cell. Biochem.* **81**(2), 220–228.

Sutherland, F. W. H., Perry, T. E., et al. (2002). Advances in the mechanisms of cell delivery to cardiovascular scaffolds: comparison of two rotating cell culture systems. *Asaio J.* **48**(4), 346–349.

Taylor, P. M., Allen, S. P., et al. (2002). Human cardiac valve interstitial cells in collagen sponge: a biological three-dimensional matrix for tissue engineering. *J. Heart Valve Dis.* **11**(3), 298–306; discussion 306–307.

Teebken, O. E., Bader, A., et al. (2000). Tissue engineering of vascular grafts: human cell seeding of decellularized porcine matrix. *Eur. J. Vasc. Endovasc. Surg.* **19**(4), 381–386.

Turina, J., Hess, O. M., et al. (1993). Cardiac bioprostheses in the 1990s. *Circulation* **88**(2), 775–781.

Vacanti, J. P., Morse, M. A., et al. (1988). Selective cell transplantation using bioabsorbable artificial polymers as matrices. *J. Ped. Surg.* **23**(1), 3–9.

Valente, M., Bortolotti, U., et al. (1985). Ultrastructural substrates of dystrophic calcification in porcine bioprosthetic valve failure. *Am. J. Pathol.* **119**(1), 12–21.

Vesely, I., Barber, J. E., et al. (2001). Tissue damage and calcification may be independent mechanisms of bioprosthetic heart valve failure. *J. Heart Valve Dis.* **10**(4), 471–477.

Vongpatanasin, W., Hillis, L. D., et al. (1996). Prosthetic heart valves. *N. Engl. J. Med.* **335**(6), 407–416.

Vunjak-Novakovic, G., Obradovic, B., et al. (1998). Dynamic cell seeding of polymer scaffolds for cartilage-tissue engineering. *Biotechnol. Prog.* **14**(2), 193–202.

Weind, K. L., Boughner, D. R., et al. (2001). Oxygen diffusion and consumption of aortic valve cusps. *Am. J. Physiol. Heart Circ. Physiol.* **281**(6), H2604–H2611.

Williams, S. F., Martin, D. P., et al. (1999). PHA applications: addressing the price performance issue I. Tissue engineering. *Int. J. Biol. Macromol.* **25**(1–3), 111–121.

Wilson, G. J., Courtman, D. W., *et al.* (1995). Acellular matrix: a bio-materials approach for coronary artery bypass and heart valve replacement. *Ann. Thorac. Surg.* **60**(2 Suppl.), S353–S358.

Woodard, A. S., Garcia-Cardena, G., *et al.* (1998). The synergistic activity of alpha(v)beta(3) integrin and PDGF receptor increases cell migration. *J. Cell Sci.* **111**, 469–478.

Ye, Q., Zund, G., *et al.* (2000a). Tissue engineering in cardiovascular surgery: new approach to develop completely human autologous tissue. *Eur. J. Cardiothorac. Surg.* **17**(4), 449–454.

Ye, Q., Zund, G., *et al.* (2000b). Scaffold precoating with human autologous extracellular matrix for improved cell attachment in cardiovascular tissue engineering. *Asaio J.* **46**(6), 730–733.

Zacks, S., Rosenthal, A., *et al.* (1991). Characterization of cobblestone mitral valve interstitial cells. *Arch. Pathol. Lab. Med.* **115**(8), 774–779.

Zeltinger, J., Landeen, L. K., *et al.* (2001). Development and characterization of tissue-engineered aortic valves. *Tissue Eng.* **7**(1), 9–22.

Zimmermann, W. H., Schneiderbanger, K., *et al.* (2002). Tissue engineering of a differentiated cardiac muscle construct. *Circ. Res.* **90**(2), 223–230.

Zund, G., Breuer, C. K., *et al.* (1997). The *in vitro* construction of a tissue-engineered bioprosthetic heart valve. *Eur. J. Cardiothorac. Surg.* **11**(3), 493–497.

Zund, G., Ye, Q., *et al.* (1999). Tissue engineering in cardiovascular surgery: MTT, a rapid and reliable quantitative method to assess the optimal human cell seeding on polymeric meshes. *Eur. J. Cardiothorac. Surg.* **15**(4), 519–524.

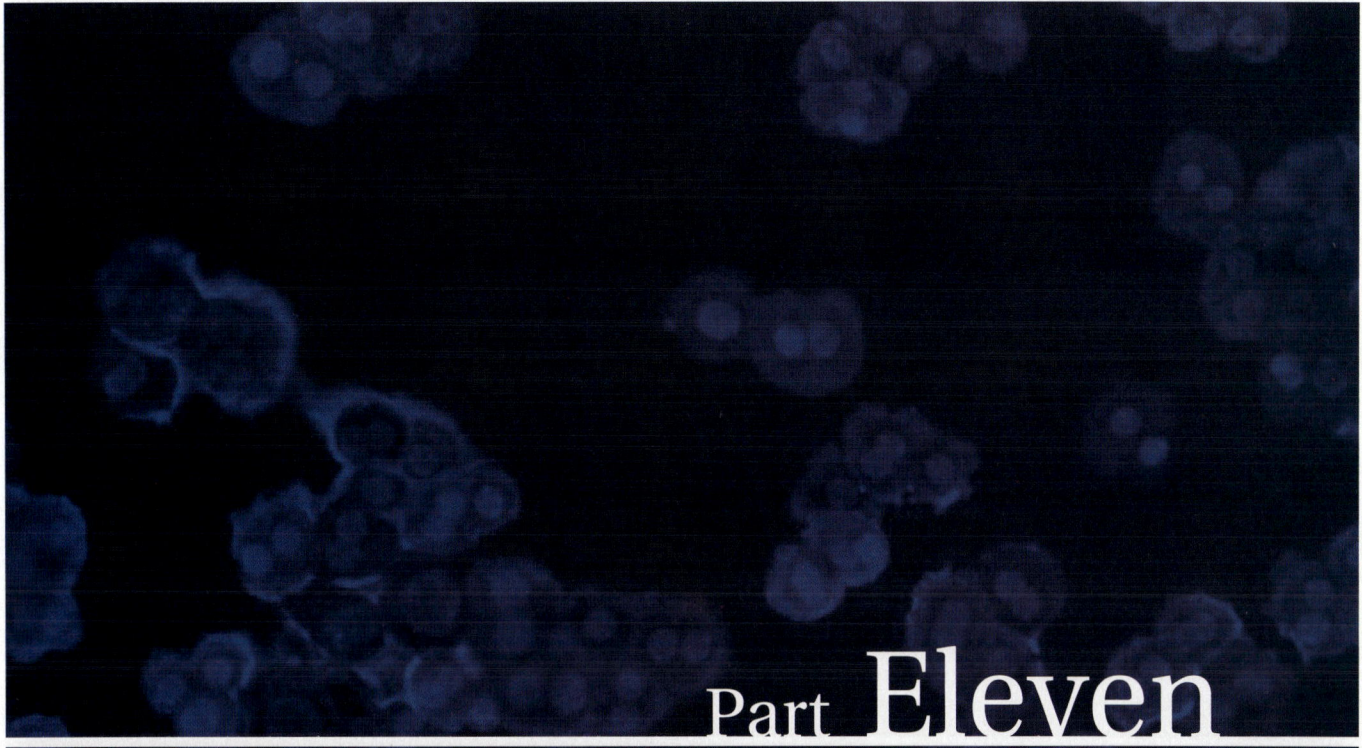

# Endocrinology and Metabolism

# Chapter Forty-One

# Generation of Islets from Stem Cells

*Bernat Soria, Abdelkrim Hmadcha, Francisco J. Bedoya, and Juan R. Tejedo*

## I. INTRODUCTION

Diabetes is a devastating disease affecting millions of people around the world. Islet transplantation has demonstrated that cell therapy works. Although this technique needs further improvements (amelioration islet isolation and survival, new immunosuppressive regimens, etc.), the "proof of principle" for future cell therapies has been established. However, the lack of sufficient donors, together with the need for immunosuppression, limits clinical applications of these techniques. New sources of insulin-producing cells are needed. The possibilities are xenogenic islets, surrogate beta-cells, adult stem cells from the pancreas and other tissues, and bioengineered embryonic stem cells. Macro- and microislet encapsulation may improve islet survival, avoid rejection, and permit immunosuppressive regimen-free transplantations. This chapter discusses these approaches.

Diabetes mellitus as a devastating metabolic disease affects around 2–5% of the world's population. While type 1 diabetes mellitus is characterized by autoimmune destruction of islets, it is now well recognized that reduced pancreatic beta-cell mass and insulin secretion failure play a pivotal role in the development and progression of type 2 diabetes mellitus (Roche *et al.*, 2005).

Daily insulin injections are necessary for patient survival, mainly in the case of type 1 diabetes. However, diabetic people very often develop late complications, such as neuropathy, nephropathy, retinopathy, and cardiovascular disorders, because the insulin injection does not mimic β-cell function exactly. The Diabetes Control and Complications Trial (DCCT, 1993) has demonstrated that intensive insulin therapy restores blood glucose homeostasis and subsequently reduces the appearance of diabetic complications. However, it requires educated and motivated patients; long-term follow-up of patients resulted in no clear differences. As a matter of fact, in both type 1 and type 2 diabetes mellitus, impaired blood glucose due to lack or death of β-cells is a common feature; thus, the best perspective for its treatment is beta-cell replacement.

Pancreas transplantation provides good glycemic control and insulin independence, together with an improvement in diabetic complications (Ryan *et al.*, 2006), but it has the associated morbidity of major surgery. Unfortunately, it requires long-term immunosuppression, with its attendant risks. Simultaneous pancreas and kidney transplantation is a worthwhile option to employ when renal transplantation is needed (Ryan *et al.*, 2006).

Clinical islet transplantation trials have been established as proof for beta-cell replacement therapy. Injection of islets isolated from cadaveric organ donor pancreata into the portal vein of type 1 diabetic patients results in insulin independence (Ryan *et al.*, 2005). However, this procedure most often requires the use of islets from more

*Principles of Tissue Engineering, 3rd Edition*
*ed. by Lanza, Langer, and Vacanti*

than one donor, and preserved graft function is based on the use of immunosuppressive medication, which has to counteract both immunorejection and recurrent beta-cell autoimmunity.

Lack of donors, the low yield of islet isolation procedures, and the need for immunosuppression are the most important caveats for future applications of cell therapy for diabetic patients.

## II. ISLET TRANSPLANTATION

Islet transplantation from cadaveric donors has been shown to improve the quality of life of severe diabetic patients. The Edmonton protocol (Shapiro et al., 2000) reported that insulin independence can be reached in type 1 diabetes subjects. The success was attributed mainly to two aspects: (1) the application of a steroid-free new immunosuppressive regime and (2) subsequent transplantation of freshly prepared islets from two or more donors (Shapiro et al., 2000). Five-year follow-up reported that 80% of transplanted patients are C-peptide positive, but only 5–10% maintained insulin independence (Ryan et al., 2005). The side effects of immunosuppressant drugs (mostly diabetogenic), together with recurrent autoimmune attacks, may explain the limited success. In order to achieve successful cell therapies, efficient islet isolation procedures, less toxic immunosuppressive regimes, and the exploration of new places for transplant have to be developed.

### Islet Isolation and Survival

One of the bottlenecks in pancreatic islet transplantation is reaching a high number of functional engrafted islets. It is estimated that only 15–30% of the allogeneic implanted islets are functional, thus forcing the use of two to three subsequent donors to achieve insulin independence. Islet isolation using the semiautomatic Ricordi method remains, with slight modifications, the best method for obtaining high amounts of viable islets.

The yield from islet isolation depends basically on the donor and processing parameters. Several studies have concluded that the islets from normal, overweight, and obese nondiabetic cadaver donors are also suitable (I. Matsumoto et al., 2004). In addition, islets from donors with a low body-mass index non-heart-beating were used successfully in an islet transplantation (Goto et al., 2005). Unfortunately, brain-dead donors promote a rapid anti-inflammatory reaction that predisposes the islets to a subsequent immunologic reaction in the recipient after transplantation (Takada et al., 2004). Limited experience with islets from living donors (S. Matsumoto et al., 2005) suggests better results. Islet transplant from a single donor is possible and results in more uniform islet preparation and makes possible a lower level immunosuppression therapy by searching for a compatible recipient (B. W. Lee et al., 2005).

The Ricordi method may be summarized as follows: organ extraction and transport, perfusion, digestion, and purification of islets. The maintenance of isolated islets prior to transplant and increasing islet engraftment may be instrumental to increasing the chance of success. *Organ extraction and transport* has requirements similar to those for pancreas transplantation. The use of a two-layer method (perfluorocarbon/University of Wisconsin solution) has been seen to increase the yield and quality of isolated islets (Hering et al., 2004; Papas et al., 2005). However, Papas et al. recently have shown that the two-layer method can ameliorate oxygenation in only around 15% of the pancreas (Papas et al., 2005). Preservation solutions, such as M-Kyoto solution, containing trehalose as a cytoprotector and ulinistatin as a trypsin inhibitor improve islet quality significantly as compared to University of Wisconsin solution in the two-layer method (Noguchi et al., 2006). In addition, intraductal glutamine administration improves islet yield, because it protects from lipoperoxidation and apoptosis (Avila et al., 2005).

During *perfusion and digestion* it is necessary to optimize enzymatic digestion by means of liberase H1 and to preserve islet integrity. Excess exposure to enzyme can produce islet fragmentation, decreased insulin secretory ability, and apoptosis (Balamurugan et al., 2005). On the other hand, the automatic method in the Ricordi chamber results in a strong mechanical shaking.

The difference in density between the islets and the rest of the pancreatic tissues allows for *islet purification* by means of separation on continuous density-gradient systems, using the COBE 2991 cell separator. The issues in the purification step are related to the purity and efficiency of the continuous gradient systems. Iodixanol gradients have clear advantages over other gradient systems. When purity is high, the islet fraction is small, whereas a great number of islets may be found in less pure fractionation methods. Recently the additional step named *rescue gradient purification* (RGP) has been described; islets recovered from these fractions were equivalent to islets obtained in a high-purity fraction (Ichii et al., 2005).

With regard to *islet maintenance*, the possibility of increasing viability via subsequent islet culture is being explored by many groups, for example, coculturing islets with Sertoli cells (Teng et al., 2005) as well as with small intestinal submucosa (Tian et al., 2005), the protective effect of L-glutamine and nitric oxide on the islet culture (Tejedo et al., 2004; Brandhorst et al., 2005) has also been described.

Furthermore, in recent years, several experimental strategies have been developed to enhance *islet engraftment*. For instance, antioxidant therapy with nicotinamide (Moberg et al., 2003) as well as sulforaphane (Solowiej et al., 2006), vitamin $D_3$ (Riachy et al., 2006), 17 beta-estradiol (Eckhoff et al., 2003), pentoxiphylline (Juang et al., 2000), caspase inhibitors (Brandhorst et al., 2003), or cholesterol-lowering agents such as simvastatin (Contreras et al., 2002) have all demonstrated a positive impact in the preclinical setting and suggest a potential role in future clinical trials designed to improve islet engraftment. Recently it has been shown that after intraportal islet transplant, a lipotoxic destruction of islets is induced by hepatic lipids, an effect that can be prevented when the islets have previously been covered with leptin (Unger, 2005).

Additionally, pancreatic islets express tissue factor, the major *in vivo* initiator of coagulation. During clinical islet transplantation, when islets come into direct contact with blood in the portal vein, islet-produced tissue factor triggers a detrimental clotting reaction, referred to as *instant blood-mediated inflammatory reaction* (IBMIR), which is characterized by activation of the coagulation and complement systems, rapid binding and activation of platelets, and infiltration of leukocytes to islets. Together, these effects cause a disruption of islet morphology, islet dysfunction, and death. At present, several forms to inhibit it or to counteract the effects have been proposed (Johansson *et al.*, 2006). Islet surface modification with poly(ethylene glycol) (PEG) was proposed as a strategy to prevent rejection (Contreras *et al.*, 2004; D. Y. Lee *et al.*, 2006). The rationale for this strategy is based on the concept that proteins and enzymes modified with PEG are nonimmunogenic. Incorporation of PEG into the islet surface has been proposed to "camouflage" surface antigens and prevent immunogenic reactions. In addition, PEG has been used successfully to reduce plasma protein adsorption and platelet adhesion to blood vessels and vascular devices because of its low interfacial free energy with water, high surface mobility, and steric stabilization effects. Contreras *et al.* (2004) using xenogenic porcine islets with their camouflaged surface with PEG and additional plus genetic modification to overexpress Bcl-2, and additional surface islet coverage by incorporation of albumin decreasing cytotoxicity mediated by XNA and complement. PEG incorporation onto the islet surface presumably will decrease the binding of platelets on its surface, thereby decreasing IBMIR. The combination of PEG camouflage with immunosuppressive medication would be highly effective in clinical islet transplantation (D. Y. Lee *et al.*, 2006).

## Immunosuppression

Most immunosuppressant agents are diabetogenic. Minimizing these effects while maintaining adequate potency to contend with both allograft rejection and autoimmune recurrence has been instrumental in making islet transplantation a clinical reality. A combination of sirolimus, a low dose of tacrolimus, and daclizumab avoided graft rejection and improved implant survival. Previous reports indicated that the combination of sirolimus and tacrolimus-based trials reported low rates of rejection in liver, kidney, and whole pancreas transplantation (McAlister *et al.*, 2000). Both compounds block T-cell activation. Daclizumab is a monoclonal antibody against IL-2 receptor, allowing the suppression of glucocorticoid administration in patients, which is very harmful to islet cells. The result of this immunosuppressor combination is the prevention of immune response activation and clonal expansion of T-lymphocytes. Sirolimus has

facilitated clinical islet transplant success by providing effective immunosuppression to contend with both auto- and alloimmunity. However, the agent is also responsible for many of the side effects encountered after islet transplantation, including nausea, vomiting, anemia, ovarian cysts, mouth ulcers, diarrhea, optic neuropathy, proteinuria, hypereosinophilic syndrome, parvovirus infection, aspiration pneumonia, severe depression, and hypertension (Hafiz *et al.*, 2005; Ryan *et al.*, 2005). It has also been suggested that sirolimus could still have a detrimental effect on islet engraftment and neovascularization as well as potential detrimental direct toxicity to islets (Hafiz *et al.*, 2005). On balance, however, this agent has proven to be advantageous as compared to former therapies based on steroid and high-dose calcineurin inhibitor. In contrast, combining glucocorticoid-free immunosuppressive strategy with low-dose FK506 tacrolimus and mycophenolate mofetil (MMF) could protect islet grafts in islet transplantation without diabetogenic side effects (Du and Xu, 2006). Pretransplant immunosuppression induction with sirolimus and humanized anti-CD-3 antibody, hOKT3γ1 (Ala-Ala), in recipients has resulted in engraftment and insulin independence after single-donor islet transplant (Hering *et al.*, 2004).

## Site of Transplantation

The liver is the most commonly used site; islet allografts are infused percutaneously into the portal vein (Shapiro *et al.*, 2000). Potential complications of an infusion into the liver include bleeding, portal venous thrombosis, and portal hypertension (Ryan *et al.*, 2005). Although portal blood pressure is monitored during the procedure and anticoagulant agents are used to prevent clotting, anticoagulation can promote hepatic bleeding at the sites of the percutaneous needle punctures. Furthermore, intrahepatic islets may be exposed to environmental toxins and potentially toxic prescribed medications, such as the immunosuppressive drugs, absorbed from the gastrointestinal tract and delivered into the portal vein. Thus islets transplanted are unable to release glucagon during hypoglycemia. In view of these problems, the use of nonhepatic sites for islet transplantation has been suggested.

The optimal site for transplantation of islets has not yet been defined. The implant site should provide an adequate microenvironment, vascularization, and nutritional support to maximize chances for the best engraftment of cells and to minimize morbidity. Several transplantation sites for islet engraftment have been reported in various experimental animal models, including intraperitoneal, intravenous, intrathecal, intrapancreatic, intrasalivary gland, intracerebral, muscle, spleen, liver via the portal vein, mammary fat pad, anterior eye chamber, omental pouch, testis, and renal capsule (Roche *et al.*, 2005). We used both spleen and kidney capsules as the recipient organ for the bioengineered insulin-positive aggregates (Soria *et al.*, 2000; Leon-Quinto *et al.*, 2004). Transplantation into the spleen is an experimental technique fast enough to minimize the risk of death in diabetic animals. Although the implanted cells could be traced by means of the expression of a reporter gene (i.e., *h*-galactosidase), graft monitoring could not be followed properly under these conditions because a portion of the cells migrated to the liver and the animal had to be sacrificed (Soria *et al.*, 2001). Kidney capsule appears to be a better place because the graft can easily be obtained via nephrectomy and then analyzed, which allows one to determine *in vivo* differentiation processes and structural changes in the implanted tissue

## III. ALTERNATIVE SOURCES OF ISLET CELLS

Organ procurement from cadaveric donors, even in countries like Spain, which ranks number one in the world in organ donation, will always be a limited source of islets. Search for alternative sources becomes a necessity. Xenogenic islets, bioengineered beta-cells, and directed differentiation of embryonic and adult stem cells could be considered.

### Xenogenic Islets

Xenogenic sources have been proposed, but they have problems related to immunological and physiological incompatibility, to the identity of insulin as compared to human insulin, and to the risk of zoonotic disease transmission, mainly retrovirus. Porcine islets are the most studied source. They have a control of glucose similar to that of humans, and the insulin has been used for more than 70 years in diabetic humans. Rejection of porcine islets is substantially greater than that of human islets. However, following the development of transgenic humanized pigs lacking xenoantigens, these are more resistant to immune attack (Phelps *et al.*, 2003), and the possibility exists of producing transgenic pigs with individualized matching for recipient HLA types. Zoonotic disease represents, however, a major drawback. Xenogenic islet encapsulation has been explored in detail (Lanza *et al.*, 1999; de Groot *et al.*, 2004; Qi *et al.*, 2004; Schaffellner *et al.*, 2005). *In vitro* experiments suggest that encapsulation is an effective barrier to diffusion of zoonotic disease (Petersen *et al.*, 2002).

### Bioengineered β-cells

Beta-cell surrogates have to achieve expression, processing, packaging, storing, and secretion of insulin in a glucose-dependent manner (Samson and Chan, 2006). Different cell types have been genetically modified to express GLUT 2, glucokinase, and insulin (Faradji *et al.*, 2001), although the results obtained so far are not satisfactory. Coexpression of GLUT 2 and glucokinase in intermediate lobe pituitary cells causes death by apoptosis due to an increment of cytotoxic effects in the presence of 3 mm of glucose (Faradji *et al.*, 2001). Nevertheless, the principal limitation to the use of transformed insulin-producing cell lines is the uncontrolled proliferation of these cells.

Hepatocytes are good candidates to be used as templates to obtain surrogate beta-cells. Hepatocytes possess similar glucose-sensing machinery to that of beta-cells and express glucokinase and glucose transporter 2 (GLUT2). Rat hepatocytes transfected with PDX1-VP16, a superactive version of PDX1, can be transdifferentiated into insulin-producing cells in the presence of high levels of glucose (Cao *et al.*, 2004). Human liver cells transfected with PDX-1 are able to transdifferentiate into insulin-producing cells in the presence of nicotinamide and epidermal growth factor (Sapir *et al.*, 2005). Hepatocytes transformed with human insulin cDNA express and secrete insulin that is regulated by low glucose and nitric oxide production (Qian *et al.*, 2005). These cells can modulate the hyperglycemia when they are transplanted in diabetic animal models. The expression of PDX1 human fetal liver cells induces the activation of beta-cell genes and has functional beta-cell characteristics. When these cells were cultured in starved-serum conditions in the presence of activin, they differentiated into insulin-producing cells that produce approximately 60% of the insulin content of normal beta-cells and regulate the hyperglycemia in diabetic animal models (Zalzman *et al.*, 2005). Taken together these results suggest that the liver is a potential autologous source of insulin-producing cells.

Another approach has exploited the characteristics of the gut-associated K-cell. This cell possesses a peptide secretory pathway, since one of its functions is to enhance insulin release through secretion of the glucose-dependent insulinotropic polypeptide (GIP). Intestinal mucosal K-cells transfected with the human preproinsulin gene have been injected into mouse embryos. Implanted mice not only produced human insulin within the gut, but also maintained sufficient synthesis and secretory capacity to protect the mice from diabetes following deliberate destruction of the endogenous beta-cell mass (Cheung *et al.*, 2000).

### Pancreatic Stem Cells

The pancreas is an endoderm-derived organ, consisting of exocrine and endocrine cells. Development of the endocrine cells begins with a common multipotent precursor that is directed along divergent pathways to form the different cell types contained in the islets of Langerhans: α-cells (glucagon), β-cells (insulin), δ-cells (somatostatin), and PP cells (pancreatic polypeptide). This process is dependent on a set of transcription factors (Samson and Chan, 2006).

A cascade of these factors coordinates the stepwise changes in gene expression that guide pluripotent pancreatic progenitor cells along the pathway to mature pancreatic cells (Fig. 41.1). During embryonic development these cells derived from a common stem cell, characterized by the expression of hepatocyte nuclear factor 6 (HFN-6) (Poll *et al.*, 2006) and Pdx-1 transcription factor (Wilding and

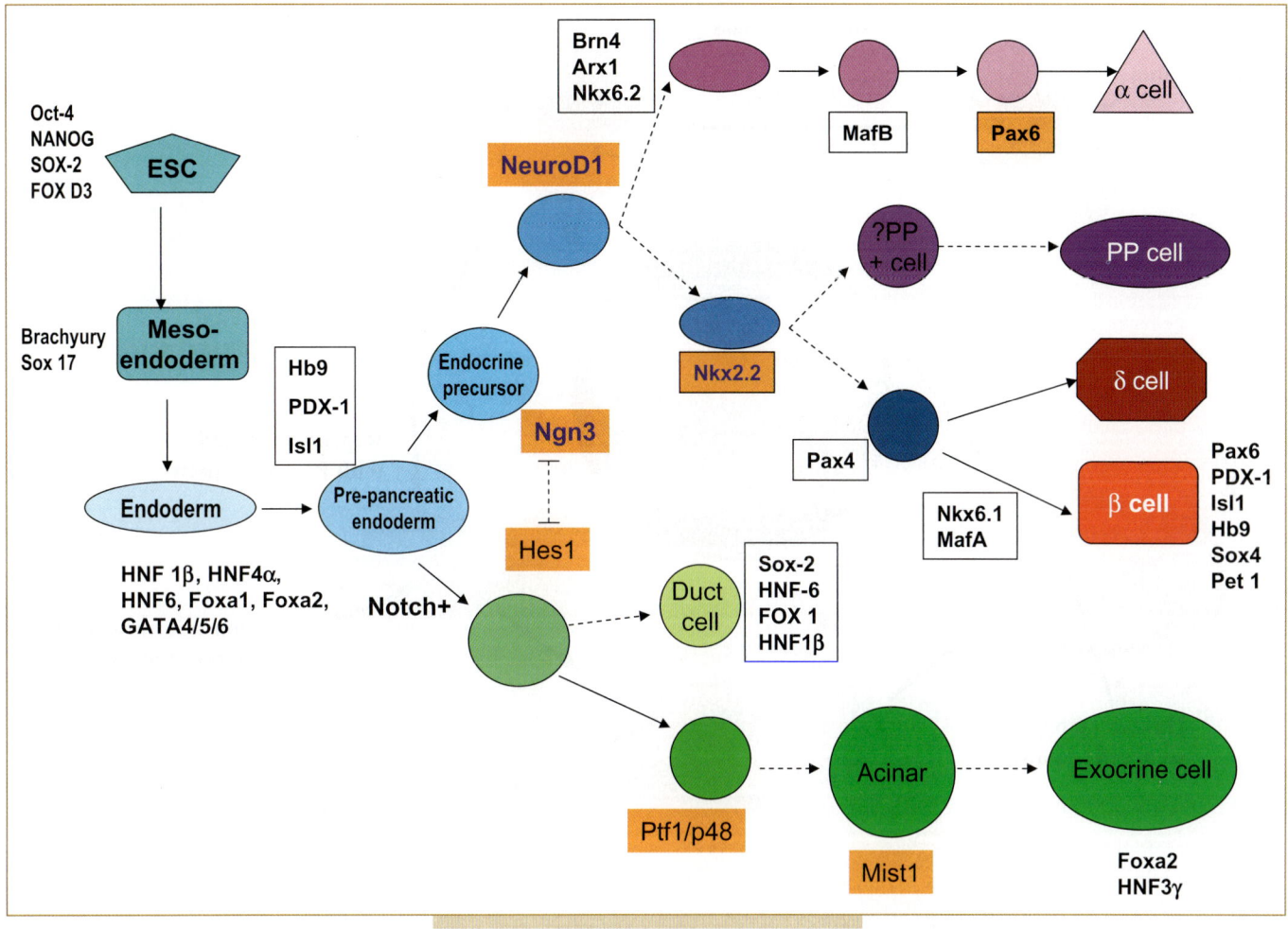

**FIG. 41.1.** Hierarchical cascade of transcription factor in the development of pancreatic tissue from embryonic stem cells. Adapted from Samson and Chan (2006).

Gannon, 2004). Pdx-1 leads downstream to Ngn3 factor expression, which is a member of the basic helix-loop-helix (bHLH) family. This factor specifies an endocrine-cell fate (Watada, 2004). Beta2/NeuroD, NeuroD/BETA2 is another basic helix-loop-helix pro-endocrine factor activated by Ngn3 downstream of PDX-1 in the endocrine developmental cascade. Beta2/NeuroD heterodimerizes with ubiquitous bHLH proteins of the E2A family to regulate transcription of the insulin gene and other β-cell-specific genes (Huang *et al.*, 2002). Isl 1 is a LIM homeodomain-family member that controls the differentiation of postmitotic endocrine progenitors. Isl 1 interacts with Beta2 to promote insulin gene transcriptional activation (Peng *et al.*, 2005). The activity of the NK-family member and homeodomain protein Nkx2.2 is necessary for the maturation of β-cells. Nkx2.2 expression is dependent on the NeuroD 1 only in β-cells (Sander *et al.*, 2000; Itkin-Ansari *et al.*, 2005), whereas its distant homolog Nkx6.1 controls their expansion (Sander *et al.*, 2000). The Pax-gene family encodes a group of transcription factors that are key regulators of vertebrate organogenesis, since

they play major roles in embryonic pattern formation, cell proliferation, and cell differentiation. Pax4 is a paired-box homeoprotein whose expression is restricted to the central nervous system and the developing pancreas (Sosa-Pineda, 2004). Thus, Pax4 functions early in the development of islet cells to promote the differentiation of β- and α-cells. Pax6, also important for β-cell development, regulates the promoters of insulin, glucagon, and somastotatin genes and molecules of adhesion (Sosa-Pineda, 2004).

## Embryonic Stem Cells

Embryonic stem cells (ESC) have self-renewal capacity in defined culture conditions and potential to differentiate into any cellular types present in the developed organism. ESC are derived from the inner mass of blastocyst (Fig. 41.2) and can be induced to differentiate into different lineages *in vitro* by means of the specific differentiation protocols to generate cardiomyocytes, endothelial cells, glial precursors, neurons cells, and oligodendrocytes (Wobus and Boheler, 2005).

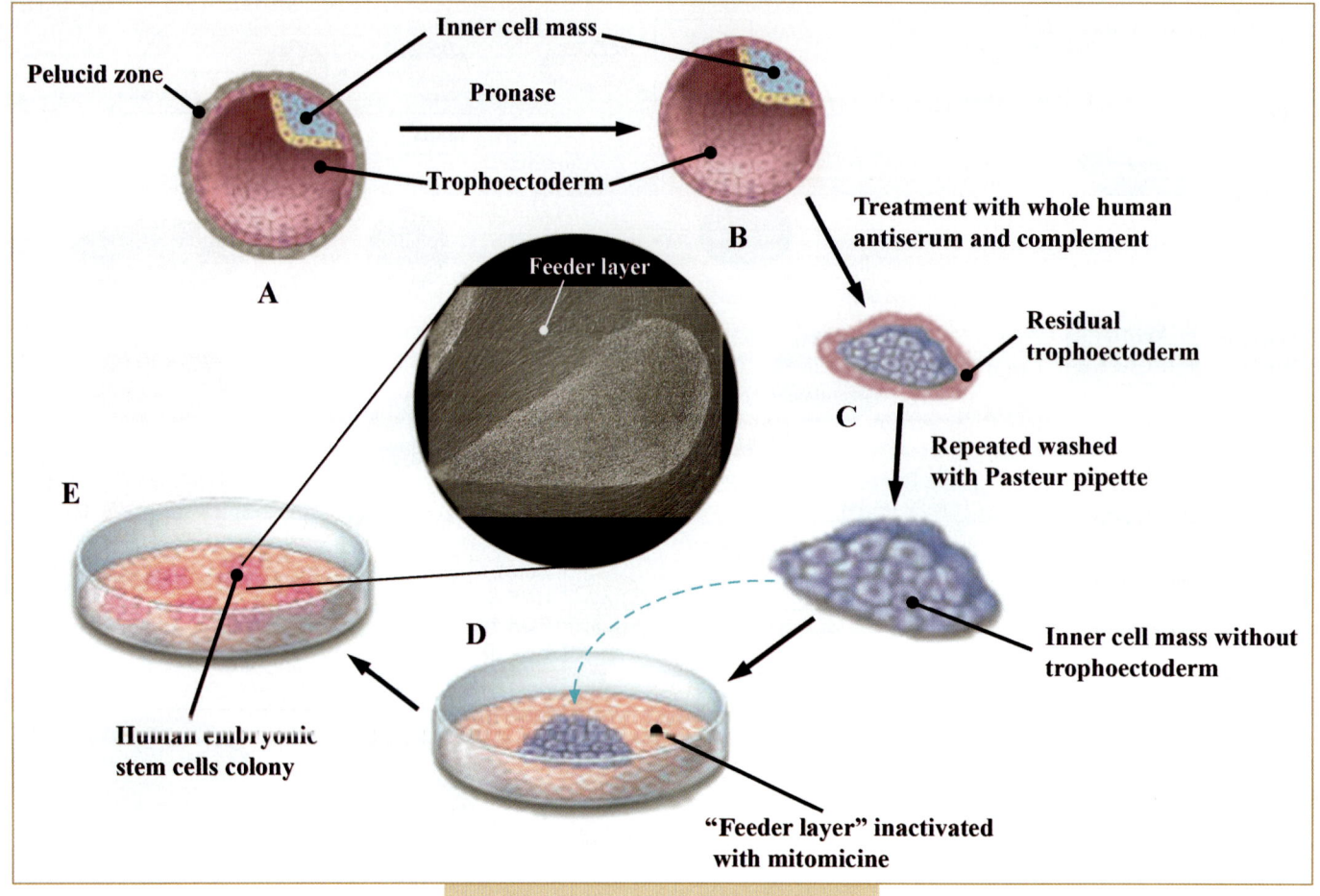

**FIG. 41.2.** Derivation of embryonic stem cells from the inner mass of the blastocyst.

Mouse ESC can be induced to differentiate into insulin-producing cells (Soria *et al.*, 2000; Soria *et al.*, 2001). Using a cell-trapping system we have been able to develop methods to obtain insulin-producing cells from embryonic stem cells. Insulin-producing cells correct hyperglycemia when transplanted into diabetic animal models (Soria *et al.*, 2000; Leon-Quinto *et al.*, 2004; Vaca *et al.*, 2006). ESC are transfected with a vector that contains two selection cassettes, one controlling the expression of antibiotic resistance that permits the selection of transfected cells, and another a chimeric gene containing a functional specific promoter driving the expression of a structural gene that codifies for antibiotic resistance allowing the selection of cells that express the selected gene (Fig. 41.3). This method has been shown to work with insulin and Nkx 6.1 promoters (Soria *et al.*, 2000; Leon-Quinto *et al.*, 2004; Vaca *et al.*, 2006).

Recently we published a protocol in which fetal soluble factor from pancreatic buds were used to direct ESC differentiation into insulin-producing cells further selected with a human insulin promoter—βgeo/PGK-hygro construction (Vaca *et al.*, 2006). Further improvements in the differentiation and selection procedure will result in better results.

Furthermore, expression of some β-cell development transcription factors may promote differentiation of ESC into betalike cells, i.e., Pax-4 constitutive expression (Sosa-Pineda, 2004), Pdx-1 overexpression regulated by a Tet-off regulation system (Miyazaki *et al.*, 2004), or Nkx2.2-transfected ESC (Shiroi *et al.*, 2005). Preliminary results suggest the potential use of human embryonic stem cells to obtain betalike-cells (Brolen *et al.*, 2005).

## Adult Stem Cells

Transdifferentiation of adult cells into insulin-producing cells also provides another exciting opportunity for β-cell expansion (Ruhnke *et al.*, 2005; Seeberger *et al.*, 2006). Recently our group has published the about obtaining betalike-cells from white blood cells by reprogramming blood monocytes in the presence of macrophage colony-stimulating factor and interleukin 3, followed by incubation with epidermal and hepatic growth factor and nicotinamide. These cells showed an *in vitro* glucose-dependent insulin secretion and normalized blood glucose levels in diabetic mice (Ruhnke *et al.*, 2005). While these strategies seem promising, more studies are needed.

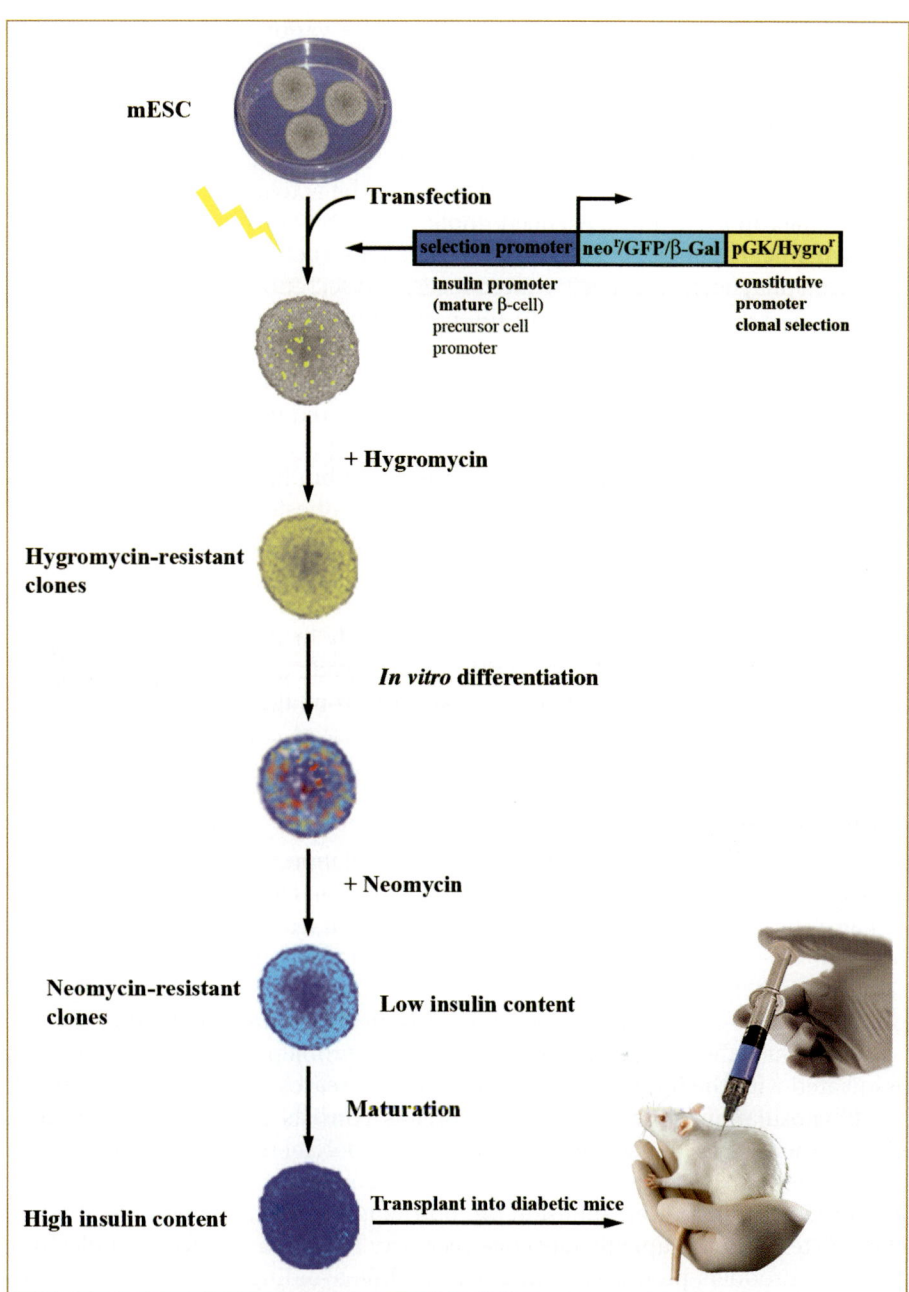

**FIG. 41.3.** Protocol for differentiation of ESC into insulin-producing cells. Adapted from Soria *et al.* (2001).

Generation of betalike cells from human adult pancreatic tissues has also been explored, such as development of differentiated islets from pancreatic duct cells and mesenchymal stem cells from human pancreatic ductal epithelium (Gershengorn *et al.*, 2004; Seeberger *et al.*, 2006), human adult pancreatic tissues (Lechner *et al.*, 2005), nonendocrine pancreatic epithelial cells (Hao *et al.*, 2006), and proliferating human islet–derived cells (Ouziel-Yahalom *et al.*, 2006). It is interesting that nonendocrine pancreatic epithelial cells were capable of endocrine differentiation when they were incubated with factors present in human fetal pancreas cells.

The strategy of redifferentiation with proliferating human islet–derived cells includes environmental activation of transcription factors such as Pdx-1, Neuro D, Nkx 2.2, and Nkx 6.1 by betacellulin, a member of the epidermal growth factor family (Ouziel-Yahalom *et al.*, 2006).

## IV. BIOMATERIALS

### Immunological Considerations

Transplanted islets are recognized as antigens by a host, triggering the process of recruiting and activating of immune cells, such as macrophages, fibroblasts, granulocytes, and lymphocytes. The activated immune cells secrete various cytokines and cytotoxic molecules, which can induce functional and structural damage to islets (Sigrist *et al.*, 2005). In addition, inflammatory cytokines such as interferon-γ stimulate the production and release of chemokines by transplanted islets. These chemokines promote the activation of macrophages and have been implicated in a series of biochemical, metabolic, and functional changes, such as increased phagocytosis capacity, chemotaxis, and secretion of chemoattractants, that result in the hampering of islet engraftment (Sigrist *et al.*, 2004). Interleukin (IL)-1 activates T-cells, resulting in the increased production and expression of IL-2 and IL-2 receptor (IL-2R), which are considered the major components of immune rejection. Selective prevention of the IL-2/IL-2R interaction using the IL-2R binding agent prolongs allograft survival. TNF-α increases adhesion of molecules on the grafted tissue, thereby stimulating graft rejection as well as inducing cell death by apoptosis. Recently it has been described that the granzyme B produced by allogenic cytotoxic T-lymphocytes participates actively in the rejection (Sutton *et al.*, 2006). In addition, nitric oxide and reactive oxygen species generated by activated macrophages act as a cytotoxic factor (Chae *et al.*, 2004; Sigrist *et al.*, 2005).

## Encapsulation

Encapsulation of pancreatic islets for transplantation into diabetic patients may solve two major obstacles in clinical applications. (1) Immunoisolation may allow for transplantation of islets in the absence of immunosuppression; (2) it may permit grafting of xenogenic islets, insulin-producing cells differentiated from stem cells, or surrogate beta-cell lines, thereby overcoming the logistical problems associated with the limited supply of human pancreases.

Encapsulation of cells or artificial organs consists of placing them inside bioactive materials (usually polymeric membranes, such as alginate), whose physicochemical properties facilitate free diffusion of nutrients, oxygen, electrolytes, and therapeutic bioactive secretory and cellular waste products produced in their inner while avoiding the intake of proteins of high molecular weight, such as the immunoglobulins, and cells of the immune system (Fig. 41.4). Nevertheless, cytokines and chemokines possess a molecular weight close to insulin, so a molecular filter cutoff will not exclude the cytokines and chemokines responsible for the immune response (Sigrist *et al.*, 2005). Encapsulation has not yet been applied in clinical practice, mainly because survival of encapsulated islet grafts is limited. The principal causes for the failure of microencapsulated islet grafts relate to lack of biocompatibility, limited immunoprotective properties, and hypoxia (de Groot *et al.*, 2004).

Encapsulated allogenic and xenogenic islets differ regarding the specific immunologic response when they are transplanted. Allograft rejection occurs as a result of the

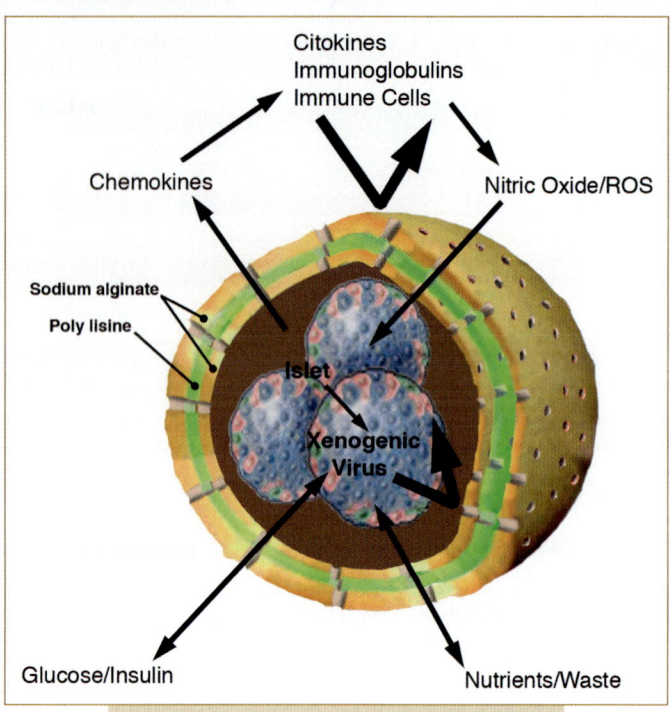

**FIG. 41.4.** Immunoisolation concept and microcapsule models. The islets are being enveloped in alginate-based microcapsules (alginate, poly-lysine, alginate), which permit the free diffusion of glucose, insulin, nutrients, wastes, and chemokines. The microcapsule protects from the immune system reactions and can protect from nitric oxide and reactive oxygen species. The size of the pore retains the xenogeneic virus.

activation of cellular immunity by interactions of host T-cells with the islet graft, as previously described, while humoral immunity, including antibodies and complement proteins, is mostly responsible for the rejection of xenografts. It is important to consider that xenografts are less capable of binding and responding to human cytokines. Immunoisolation may also protect against antigens of allogeneic or xenogeneic cells. Such antigens could be cell surface molecules and cell components, including those released on cell death. Shedding of antigens from encapsulated cells would initiate a molecular tissue response around the implant, which could affect the viability and function of the encapsulated cells. The recognition of antigens through this indirect pathway may lead to the activation of T helper cells, which then secrete cytokines and regulate the cell-mediated immune response and inflammation (Kizilel *et al.*, 2005). Other strategies that complement encapsulation should be explored to avoid the damage that free diffusion molecules such as nitric oxide and free-radical oxygen molecules can cause (Chae *et al.*, 2004).

Islets could be either macro- or microencapsulated. Macroencapsulation can use either extravascular or intravascular devices (Kim *et al.*, 2005; Kizilel *et al.*, 2005). Microencapsulation envelops islets in an alginate-based membrane and other biomaterials (Lanza *et al.*, 1999; Kim *et al.*, 2005).

**Table 41.1.** Advantages and disadvantages of transplantation of unencapsulated, microencapsulated, and macro-encapsulated islets (adapted from Emerich *et al.* 1992)

| *Unencapsulated* | *Microencapsulation* | *Macroencapsulation* |
|---|---|---|
| **Advantages** | | |
| Anatomical integration between host and transplanted islet | Use of allo- and xenoislets without immunosuppression | Use of allo- and xenoislets without immunosuppression |
| | Thin wall and spherical shape are optimal for cell viability and free diffusion of nutrients and insulin | Good mechanical stability |
| | | Good cell viability and free diffusion of nutrients and insulin |
| | | Retrievable |
| **Disadvantages** | | |
| Requires immunosuppression | Mechanically and chemically fragile | Internal characteristics (i.e., diameter) may potentially limit free diffusion of nutrients/insulin and cell viability |
| Tissue availability limited | Limited retrievability | Need for multiple implants may produce significant tissue displacement/damage |
| Tissue survival often poor | | |
| Limited retrievability | | |

The advantages and disadvantages of both are shown in Table 41.1.

## Macroencapsulation

Macroencapsulation refers to the enveloping of a large mass of islets in a chamber of a perm-selective membrane, forming either an intravascular macrocapsule (connected as a shunt to the systemic circulation) or an extravascular macrocapsule (transplanted subcutaneously or intraperitoneally) (Fig. 41.5). Intravascular macrocapsules are usually perfusion chambers of microporous or nanoporous material that are directly connected to the blood circulation. Extravascular macrocapsules are usually diffusion chambers in the shape of a tube or sphere. The typical dimension of macrocapsules is in the range of 0.5–1.5 mm inner diameter and 1–10 cm length (Kizilel *et al.*, 2005).

Polymers for macroencapsulation are mechanically more stable and the wall capsule generally thicker than those used in microencapsulation. So this technique gives greater long-term stability to the implant. However, the thicker wall and the larger diameter of the capsule can impair diffusion, threatening the viability of the islet and slowing the insulin kinetic release as a result of diffusion limitations of nutrients and oxygen. To guarantee adequate feeding of the cells, the islet density of the macrocapsules is kept quite low and never exceeds 5–10% of the volume fraction. Optimal results have been obtained by immobilization of the islets in a matrix before final macroencapsulation. As a consequence, large devices have to be implanted to provide sufficient masses of insulin-producing islets. These large graft volumes are impractical and cannot be implanted in conventional sites for transplantation of islets such as the liver, kidney capsule, or spleen. Even the relatively large space in the peritoneal cavity does not suffice for the large volume required for the long-term function of an islet graft in macrocapsules. Also, the relatively large surface-to-volume ratio of the macrocapsules interferes with adequate regulation of the glucose levels, because exchange of glucose and insulin occurs rather slowly. One advantage of the implantation of macrocapsules is the ease of retrieval in case of complications (Kizilel *et al.*, 2005). Macroencapsulated rat islets implanted in the peritoneal cavity of diabetic mice maintained an effective insulin secretory response (Qi *et al.*, 2004).

Bioartifical pancreases should avoid the possibility of contamination with virus from xenogenic islets while allowing insulin and nutrient exchanges. Hydroxymethylated polysulphone macrocapsules display insulin release kinetics very similar to those of free-floating islets and retain retrovirus release (Petersen *et al.*, 2002).

## Vascularization

Postimplantation conditions such as the design of the capsule, the environment at the implantation site, and the development of fibrosis around the construct can produce hypoxia-induced death of islet (Papas *et al.*, 1999). Enhancing the oxygenation of the capsule will increase the number of viable cells within and thus its overall secretory capacity. In this context, neovascularization around the capsule may be beneficial for the overall efficacy of such tissue substitutes *in vivo*. Encapsulated islets promote vascularization after transplantation due to VEGF release (Lembert *et al.*, 2005). A prevascularized site in the intermuscular space may be created by implanting a polyethylene

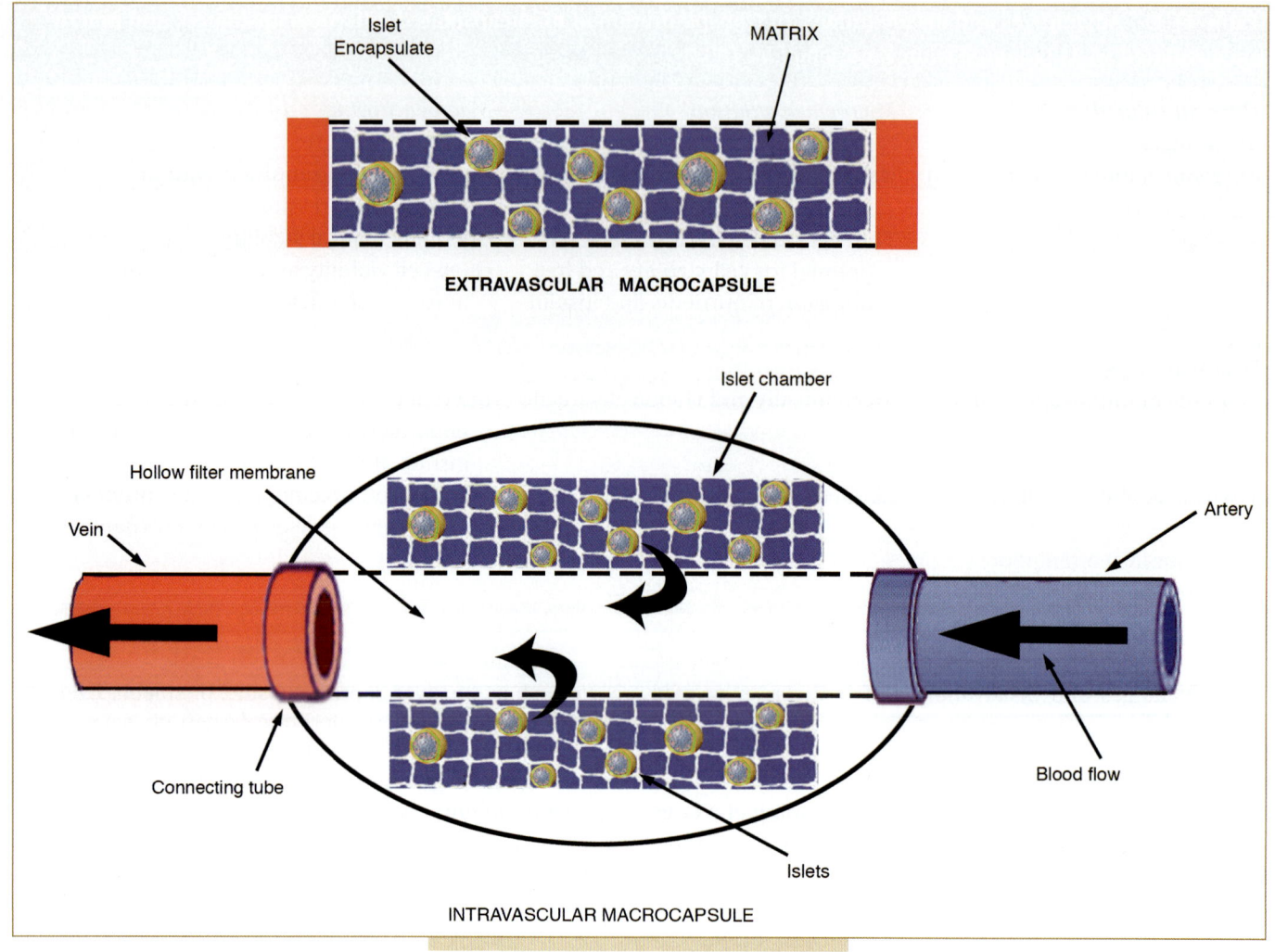

**FIG. 41.5.** Models of macrocapsules. Extravascular macrocapsule and intravascular macrocapsule.

terephthalate (PET) mesh bag with a collagen sponge and gelatin microspheres containing basic fibroblast growth factor (bFGF), before the implantation of islet macroencapsulates (Balamurugan *et al.*, 2003).

## Biocapsules

Nanoporous biocapsules are bulk and surface micromachined to present uniform and well-controlled pore sizes as small as 7 nm, tailored surface chemistries, and precise microarchitectures, in order to provide immunoisolating microenvironments for cells. Such a design may overcome some of the limitations associated with conventional encapsulation and delivery technologies, including chemical instabilities, material degradation or fracture, and broad membrane pore sizes (Desai *et al.*, 2004).

## Microencapsulation

Islet microencapsulation was first proposed in 1964 (Chang, 1964) and has been shown to be useful in maintain-

ing glucose homeostasis in rodents, dogs, and primates (Elliot *et al.*, 2005; Kizilel *et al.*, 2005; Calafiore *et al.*, 2006; Dufrane *et al.*, 2006).

Alginate–polylysine-alginate (Fig. 41.4) is most frequently used because it has been found not to interfere with cellular function and to be stable for years in small and large animals as well as in human beings (Lanza *et al.*, 1999; de Groot *et al.*, 2004; Elliot *et al.*, 2005; Calafiore *et al.*, 2006). Alginate is a polysaccharide extracted from seaweed. In solution it has a high viscosity, but in the presence of polyvalent cations, such as $CaCl_2$, it forms gels less insoluble in water. Impurities present in alginate polymers may contribute to the failure of the encapsulated islet implants. Alginates containing higher fractions of α-L-guluronic acid (G blocks) residues are more biocompatible (i.e., they do not induce a cytokine response from monocytes) than those containing a larger fraction of α-D-mannuronic acid (M blocks) residues. Omer *et al.* (2005) obtained smaller microcapsules with better stability and biocompatibility using highly purified alginate containing high G or M blocks with low endotoxin levels and $BaCl_2$ as cross-linked agent. In addition, simultaneous incorporation of alginate, $BaCl_2$ and human serum albumin improves the biocompatibility and protects adult rat and human islets against xenorejection for long periods after transplantation (Omer *et al.*, 2005). Several other strategies for increasing biocompatibility have been assayed, such as the use of alginate-poly-L-lysine-alginate modified with polyethylenglycol (Desai *et al.*, 2000), polyacrylates (Isayeva *et al.*, 2003), agarose (Kobayashi *et al.*, 2003), sodium cellulose sulfate (Schaffellner *et al.*, 2005), and the polymerization of acrylamide monomers on islet cells encapsulated in agarose microspheres (Dupuy *et al.*, 1988). Also, surface coating with polyethylene oxide improves the viability of microencapsulated islets by promoting oxygen supply and reduces the absorption of proteins associated with fibrotic reaction (Kim *et al.*, 2005). Despite all these efforts to increase biocompatibility, it has not been possible to eliminate the formation of fibrotic outgrowths.

An encapsulation system that improves strength and mechanical stability has been obtained using a double cross-linked simple physical mixture of sodium alginate (ionically cross-linked) and a polyethylene glycol-acrylate (PEGA) (covalently cross-linked) to form AP capsules, where the porosity and permeability depend on the ratio of alginate to PEGA. Tests *in vitro* have shown that the AP capsules provide a biocompatible and nontoxic environment for islets and are capable of retaining normal functions (Desai *et al.*, 2000).

The immunoprotective properties of the microcapsules are related to the physicochemical characteristics. Polymers containing sulfonic acid or sulfate groups, which have a strong affinity for complement proteins, improve these properties. With xenogenic islets, where the complement reaction is the principal humoral immune reaction, the poly(styrene sulfonic acid) mixed with agarose protects xenogenic islets in mice (Petersen *et al.*, 2002; Lembert *et al.*, 2005).

Since all polymers used for immunoisolation are not completely inert, several researchers are now studying the use of condrocytes and their matrix for islet encapsulation, to prevent immunorecognition and destruction of transplanted islets (Pollok *et al.*, 2001).

### Optimal Site for Transplantation

Selecting the optimal transplantation site for microencapsulated islets is an important consideration. Transplantation of microencapsulated xenogenic islets into intraperitoneal sites has been studied. Pig islets encapsulated in an alginate-based matrix survive for over 200 days in chemically diabetic animal models (Omer *et al.*, 2005), However, a high number of encapsulated pig islets and graft failures were also observed in most mice after 42 days of implantation. The variation observed is usually attributed to insufficient biocompatibility of the microcapsule, which causes an accumulation of macrophages and fibroblasts on the microcapsule, provoking necrosis of the islets. Dufrane *et al.* (2006) investigated the impact of implantation sites on the biocompatibility of alginate-encapsulated pig islets in nondiabetic rats. Thirty days after transplantation, explanted capsules from intraperitoneal sites demonstrated a higher degree of broken capsules and capsules with severe cellular overgrowth. On the other hand, capsules removed from subcutaneous and kidney subcapsule sites showed no such damage. They concluded that kidney subcapsular and subcutaneous spaces represent an interesting alternative.

Recently the Calafiore team (University of Perugia, Italy) published their results of pilot phase-1 clinical trials with the human allograft islet microencapsulated in 1.6% sodium alginate and sequentially double-coated with 0.12% and 0.06% poly-L-ornithine and finally with 0.04% sodium alginate. The islet microencapsulates were intraperitoneally (IP) implanted into two selected nonimmunosuppressed patients with type 1 diabetes. Both patients showed a rise in sCPR levels several weeks posttransplant, amelioration of their mean daily blood glucose levels, and a progressive decline in exogenous insulin consumption. GHb also decreased throughout months of posttransplant follow-up. At 60 days posttransplant, an OGTT in patient 1 showed a biphasic C-peptide response, compatible with the presence of differentiated islet β-cells. At 1 year (patient 1) and 6 months (patient 2) of posttransplant follow-up, sCPR was still being detected in these recipients (Calafiore *et al.*, 2006).

On the other hand, cytotoxic molecules produced by activated macrophages such as IL-1β, TNF-α, NO, and reactive oxygen species may also pass through the polymer membrane and damage the transplanted tissue (Chae *et al.*, 2004). The strategies used to protect the islet from these molecules are diverse. As an approach to preventing NO-induced damage, the rat islet and insulinoma cells (RINm5F) microencapsulated in alginate–poly-L-lysine-alginate were coencapsulated with cross-linked hemoglobin (Hb-C)–poly(ethylene glycol). These strategies protect from damage induced by nitric oxide and prevent islet death by hypoxia (Chae *et al.*, 2004).

## V. ACKNOWLEDGMENTS

We thank Giuseppe Pettinato and Sergio Mora for the artwork. This work has been partially supported by grants from Instituto de Salud Carlos III (PI0521/06), Ministerio de Educación y Ciencia (SAF 2003-367), and Fundación Progreso y Salud (Junta de Andalucía).

## VI. REFERENCES

Avila, J., Barbaro, B., *et al.* (2005). Intra-ductal glutamine administration reduces oxidative injury during human pancreatic islet isolation. *Am. J. Transplant.* **5**(12), 2830–2837.

Balamurugan, A. N., Gu, Y., *et al.* (2003). Bioartificial pancreas transplantation at prevascularized intermuscular space: effect of angiogenesis induction on islet survival. *Pancreas* **26**(3), 279–285.

Balamurugan, A. N., He, J., *et al.* (2005). Harmful delayed effects of exogenous isolation enzymes on isolated human islets: relevance to clinical transplantation. *Am. J. Transplant.* **5**(11), 2671–2681.

Brandhorst, H., Brandhorst, D., *et al.* (2003). Pretreatment of isolated islets with caspase-3 inhibitor DEVD increases graft survival after xenotransplantation. *Transplant. Proc.* **35**(6), 2142.

Brandhorst, H., Duan, Y., *et al.* (2005). Effect of stable glutamine compounds on porcine islet culture. *Transplant. Proc.* **37**(8), 3519–3520.

Brolen, G. K., Heins, N., *et al.* (2005). Signals from the embryonic mouse pancreas induce differentiation of human embryonic stem cells into insulin-producing beta-cell-like cells. *Diabetes* **54**(10), 2867–2874.

Calafiore, R., Basta, G., *et al.* (2006). Microencapsulated pancreatic islet allografts into nonimmunosuppressed patients with type 1 diabetes: first two cases. *Diabetes Care* **29**(1), 137–138.

Cao, L. Z., Tang, D. Q., *et al.* (2004). High glucose is necessary for complete maturation of Pdx1-VP16-expressing hepatic cells into functional insulin-producing cells. *Diabetes* **53**(12), 3168–3178.

Chae, S. Y., Lee, M., *et al.* (2004). Protection of insulin-secreting cells from nitric oxide–induced cellular damage by cross-linked hemoglobin. *Biomaterials* **25**(5), 843–850.

Chang, T. M. (1964). Semipermeable microcapsules. *Science* **146**, 524–525.

Cheung, A. T., Dayanandan, B., *et al.* (2000). Glucose-dependent insulin release from genetically engineered K cells. *Science* **290**(5498), 1959–1962.

Contreras, J. L., Smyth, C. A., *et al.* (2002). Simvastatin induces activation of the serine-threonine protein kinase AKT and increases survival of isolated human pancreatic islets. *Transplantation* **74**(8), 1063–1069.

Contreras, J. L., Xie, D., *et al.* (2004). A novel approach to xenotransplantation combining surface engineering and genetic modification of isolated adult porcine islets. *Surgery* **136**(3), 537–547.

DCCT — The Diabetes Control and Complications Trial Research group. (1993). The intensive treatment of diabetes in the development and progression of long-term complications in insulin-dependent diabetes mellitus. *N. Eng. J. Med.* **329**, 977–986.

de Groot, M., Schuurs, T. A., *et al.* (2004). Causes of limited survival of microencapsulated pancreatic islet grafts. *J. Surg. Res.* **121**(1), 141–150.

Desai, N. P., Sojomihardjo, A., *et al.* (2000). Interpenetrating polymer networks of alginate and polyethylene glycol for encapsulation of islets of Langerhans. *J. Microencapsul.* **17**(6), 677–690.

Desai, T. A., West, T., *et al.* (2004). Nanoporous microsystems for islet cell replacement. *Adv. Drug. Deliv. Rev.* **56**(11), 1661–1673.

Du, C. Y., and Xu, E. K. (2006). Protective effect of glucocorticoid-free immunosuppressive regimen in allogenic islet transplantation. *Hepatobiliary Pancreat. Dis. Int.* **5**(1), 43–47.

Dufrane, D., Steenberghe, M. V., *et al.* (2006). The influence of implantation site on the biocompatibility and survival of alginate-encapsulated pig islets in rats. *Biomaterials.* **27**(17), 3201–3208.

Dupuy, B., Gin, H., *et al.* (1988). *In situ* polymerization of a microencapsulating medium round living cells. *J. Biomed. Mater. Res.* **22**(11), 1061–1070.

Eckhoff, D. E., Smyth, C. A., *et al.* (2003). Suppression of the c-Jun N-terminal kinase pathway by 17beta-estradiol can preserve human islet functional mass from proinflammatory cytokine-induced destruction. *Surgery.* **134**(2), 169–179.

Elliot, R. B., Escobar, L., Calafiore, R., Basta, G., Garkavenko, O., Vasconcellos, A., and Bambra, C. (2005). Transplatation of micro- and macroencapsulated piglet islet into mice and monkeys. *Transplant. Proc.* **37**, 466–469.

Emerich, D. F., Winn, S. R., *et al.* (1992). A novel approach to neural transplantation in Parkinson's disease: use of polymer-encapsulated cell therapy. *Neurosci. Biobehav. Rev.* **16**(4), 437–447.

Faradji, R. N., Havari, E., *et al.* (2001). Glucose-induced toxicity in insulin-producing pituitary cells that coexpress GLUT2 and glucokinase. Implications for metabolic engineering. *J. Biol. Chem.* **276**(39), 36695–36702.

Gershengorn, M. C., Hardikar, A. A., *et al.* (2004). Epithelial-to-mesenchymal transition generates proliferative human islet precursor cells. *Science* **306**(5705), 2261–2264.

Goto, T., Tanioka, Y., *et al.* (2005). Successful islet transplantation from a single pancreas harvested from a young, low-BMI, non-heart-beating cadaver. *Transplant. Proc.* **37**(8), 3430–3432.

Hafiz, M. M., Faradji, R. N., *et al.* (2005). Immunosuppression and procedure-related complications in 26 patients with type 1 diabetes mellitus receiving allogeneic islet cell transplantation. *Transplantation* **80**(12), 1718–1728.

Hao, E., Tyrberg, B., *et al.* (2006). Beta-cell differentiation from non-endocrine epithelial cells of the adult human pancreas. *Nat. Med.* **12**(3), 310–316.

Hering, B. J., Kandaswamy, R., *et al.* (2004). Transplantation of cultured islets from two-layer preserved pancreases in type 1 diabetes with anti-CD3 antibody. *Am. J. Transplant.* **4**(3), 390–401.

Huang, H. P., Chu, K., *et al.* (2002). Neogenesis of beta-cells in adult BETA2/NeuroD-deficient mice. *Mol. Endocrinol.* **16**(3), 541–551.

Ichii, H., Pileggi, A., *et al.* (2005). Rescue purification maximizes the use of human islet preparations for transplantation. *Am. J. Transplant.* **5**(1), 21–30.

Isayeva, I. S., Kasibhatla, B. T., *et al.* (2003). Characterization and performance of membranes designed for macroencapsulation/implantation of pancreatic islet cells. *Biomaterials* **24**(20), 3483–3491.

Itkin-Ansari, P., Marcora, E., *et al.* (2005). NeuroD1 in the endocrine pancreas: localization and dual function as an activator and repressor. *Dev. Dyn.* **233**(3), 946–953.

Johansson, H., Goto, M., *et al.* (2006). Low-molecular-weight dextran sulfate: a strong candidate drug to block IBMIR in clinical islet transplantation. *Am. J. Transplant.* **6**(2), 305–312.

Juang, J. H., Kuo, C. H., *et al.* (2000). Beneficial effects of pentoxiphylline on islet transplantation. *Transplant. Proc.* **32**(5), 1073–1075.

Kim, Y. Y., Chae, S. Y., *et al.* (2005). Improved phenotype of rat islets in a macrocapsule by coencapsulation with cross-linked Hb. *J. Biomater. Sci. Polym. Ed.* **16**(12), 1521–1535.

Kizilel, S., Garfinkel, M., *et al.* (2005). The bioartificial pancreas: progress and challenges. *Diabetes Technol. Ther.* **7**(6), 968–985.

Kobayashi, T., Aomatsu, Y., *et al.* (2003). Indefinite islet protection from autoimmune destruction in nonobese diabetic mice by agarose microencapsulation without immunosuppression. *Transplantation* **75**(5), 619–625.

Lanza, R. P., Ecker, D. M., *et al.* (1999). Transplantation of islets using microencapsulation: studies in diabetic rodents and dogs. *J. Mol. Med.* **77**(1), 206–210.

Lechner, A., Nolan, A. L., *et al.* (2005). Redifferentiation of insulin-secreting cells after *in vitro* expansion of adult human pancreatic islet tissue. *Biochem. Biophys. Res. Commun.* **327**(2), 581–588.

Lee, B. W., Jee, J. H., *et al.* (2005). The favorable outcome of human islet transplantation in Korea: experiences of 10 autologous transplantations. *Transplantation* **79**(11), 1568–1574.

Lee, D. Y., Park, S. J., *et al.* (2006). A combination therapy of PEGylation and immunosuppressive agent for successful islet transplantation. *J. Controlled Release* **110**(2), 290–295.

Lembert, N., Wesche, J., *et al.* (2005). Encapsulation of islets in rough surface, hydroxymethylated polysulfone capillaries stimulates VEGF release and promotes vascularization after transplantation. *Cell Transplant.* **14**(2–3), 97–108.

Leon-Quinto, T., Jones, J., *et al.* (2004). *In vitro* directed differentiation of mouse embryonic stem cells into insulin-producing cells. *Diabetologia* **47**(8), 1442–1451.

Matsumoto, I., Sawada, T., *et al.* (2004). Improvement in islet yield from obese donors for human islet transplants. *Transplantation* **78**(6), 880–885.

Matsumoto, S., Okitsu, T., *et al.* (2005). Insulin independence of unstable diabetic patient after single living-donor islet transplantation. *Transplant. Proc.* **37**(8), 3427–3429.

McAlister, V. C., Gao, G. Z., Peltekian, K., Domingues, J., Mahalati, K., and MacDonald, A. S. (2000). Sirolimus–tacrolimus combination immunosuppression. *Lancet* **355**, 376–377.

Miyazaki, S., Yamato, E., *et al.* (2004). Regulated expression of pdx-1 promotes *in vitro* differentiation of insulin-producing cells from embryonic stem cells. *Diabetes* **53**(4), 1030–1037.

Moberg, L., Olsson, A., *et al.* (2003). Nicotinamide inhibits tissue factor expression in isolated human pancreatic islets: implications for clinical islet transplantation. *Transplantation* **76**(9), 1285–1288.

Noguchi, H., Ueda, M., *et al.* (2006). Modified two-layer preservation method (M-Kyoto/PFC) improves islet yields in islet isolation. *Am. J. Transplant.* **6**(3), 496–504.

Omer, A., Keegan, M., *et al.* (2003). Macrophage depletion improves survival of porcine neonatal pancreatic cell clusters contained in alginate macrocapsules transplanted into rats. *Xenotransplantation* **10**(3), 240–251.

Omer, A., Duvivier-Kali, V., *et al.* (2005). Long-term normoglycemia in rats receiving transplants with encapsulated islets. *Transplantation* **79**(1), 52–58.

Ouziel-Yahalom, L., Zalzman, M., *et al.* (2006). Expansion and redifferentiation of adult human pancreatic islet cells. *Biochem. Biophys. Res. Commun.* **341**(2), 291–298.

Papas, K. K., Long, Jr., R. C., *et al.* (1999). Development of a bioartificial pancreas: II. Effects of oxygen on long-term entrapped betaTC3 cell cultures. *Biotechnol. Bioeng.* **66**(4), 231–237.

Papas, K. K., Hering, B. J., *et al.* (2005). Pancreas oxygenation is limited during preservation with the two-layer method. *Transplant. Proc.* **37**(8), 3501–3504.

Peng, S. Y., Wang, W. P., *et al.* (2005). ISL1 physically interacts with BETA2 to promote insulin gene transcriptional synergy in non-beta-cells. *Biochim. Biophys. Acta* **1731**(3), 154–159.

Petersen, P., Lembert, N., *et al.* (2002). Hydroxymethylated polysulphone for islet macroencapsulation allows rapid diffusion of insulin but retains PERV. *Transplant. Proc.* **34**(1), 194–195.

Phelps, C. J., Koike, C., *et al.* (2003). Production of alpha 1,3-galactosyltransferase-deficient pigs. *Science* **299**(5605), 411–414.

Poll, A. V., Pierreux, C. E., *et al.* (2006). A vHNF1/TCF2-HNF6 cascade regulates the transcription factor network that controls generation of pancreatic precursor cells. *Diabetes* **55**(1), 61–69.

Pollok, J. M., Lorenzen, M., *et al.* (2001). *In vitro* function of islets of Langerhans encapsulated with a membrane of porcine chondrocytes for immunoisolation. *Dig. Surg.* **18**(3), 204–210.

Qi, M., Gu, Y., *et al.* (2004). PVA hydrogel sheet macroencapsulation for the bioartificial pancreas. *Biomaterials* **25**(27), 5885–5892.

Qian, Q., Williams, J. P., *et al.* (2005). Nitric oxide stimulates insulin release in liver cells expressing human insulin. *Biochem. Biophys. Res. Commun.* **329**(4), 1329–1333.

Riachy, R., Vandewalle, B., *et al.* (2006). 1,25-Dihydroxyvitamin D(3) protects human pancreatic islets against cytokine-induced apoptosis via down-regulation of the fas receptor. *Apoptosis* **11**(2), 151–159.

Roche, E., Reig, J. A., *et al.* (2005). Insulin-secreting cells derived from stem cells: clinical perspectives, hypes and hopes. *Transpl. Immunol.* **15**(2), 113–129.

Ruhnke, M., Ungefroren, H., *et al.* (2005). Differentiation of *in vitro*–modified human peripheral blood monocytes into hepatocyte-like and pancreatic islet–like cells. *Gastroenterology* **128**(7), 1774–1786.

Ryan, E. A., Paty, B. W., *et al.* (2005). Five-year follow-up after clinical islet transplantation. *Diabetes* **54**(7), 2060–2069.

Ryan, E. A., Bigam, D., *et al.* (2006). Current indications for pancreas or islet transplant. *Diabetes Obes. Metab.* **8**(1), 1–7.

Samson, S. L., and Chan, L. (2006). Gene therapy for diabetes: reinventing the islet. *Trends Endocrinol. Metab.* **17**(3), 92–100.

Sander, M., Sussel, L., *et al.* (2000). Homeobox gene Nkx6.1 lies downstream of Nkx2.2 in the major pathway of beta-cell formation in the pancreas. *Development* **127**(24), 5533–5540.

Sapir, T., Shternhall, K., *et al.* (2005). Cell-replacement therapy for diabetes: generating functional insulin-producing tissue from adult human liver cells. *Proc. Natl. Acad. Sci. U.S.A.* **102**(22), 7964–7969.

Schaffellner, S., Stadlbauer, V., *et al.* (2005). Porcine islet cells microencapsulated in sodium cellulose sulfate. *Transplant. Proc.* **37**(1), 248–252.

Seeberger, K. L., Dufour, J. M., *et al.* (2006). Expansion of mesenchymal stem cells from human pancreatic ductal epithelium. *Lab. Invest.* **86**(2), 141–153.

Shapiro, A. M., Lakey, J. R., *et al.* (2000). Islet transplantation in seven patients with type 1 diabetes mellitus using a glucocorticoid-free immunosuppressive regimen. *N. Engl. J. Med.* **343**(4), 230–238.

Shiroi, A., Ueda, S., *et al.* (2005). Differentiation of embryonic stem cells into insulin-producing cells promoted by Nkx2.2 gene transfer. *World J. Gastroenterol.* **11**(27), 4161–4166.

Sigrist, S., Oberholzer, J., *et al.* (2004). Activation of human macrophages by allogeneic islets preparations: inhibition by AOP-RANTES and heparinoids. *Immunology* **111**(4), 416–421.

Sigrist, S., Ebel, N., *et al.* (2005). Role of chemokine signaling pathways in pancreatic islet rejection during allo- and xenotransplantation. *Transplant. Proc.* **37**(8), 3516–3518.

Solowiej, E., Solowiej, J., *et al.* (2006). Application of sulforaphane: histopathological study of intraportal transplanted pancreatic islets into livers of diabetic rats. *Transplant. Proc.* **38**(1), 282–283.

Soria, B., Roche, E., *et al.* (2000). Insulin-secreting cells derived from embryonic stem cells normalize glycemia in streptozotocin-induced diabetic mice. *Diabetes* **49**(2), 157–162.

Soria, B., Skoudy, A., *et al.* (2001). From stem cells to beta cells: new strategies in cell therapy of diabetes mellitus. *Diabetologia* **44**(4), 407–415.

Sosa-Pineda, B. (2004). The gene Pax4 is an essential regulator of pancreatic beta-cell development. *Mol. Cells* **18**(3), 289–294.

Sutton, V. R., Estella, E., *et al.* (2006). A critical role for granzyme B, in addition to perforin and TNFalpha, in alloreactive CTL-induced mouse pancreatic beta cell death. *Transplantation* **81**(2), 146–154.

Takada, M., Toyama, H., *et al.* (2004). Augmentation of interleukin-10 in pancreatic islets after brain death. *Transplant. Proc.* **36**(5), 1534–1536.

Tejedo, J. R., Cahuana, G. M., *et al.* (2004). Nitric oxide triggers the phosphatidylinositol 3-kinase/Akt survival pathway in insulin-producing RINm5F cells by arousing Src to activate insulin receptor substrate-1. *Endocrinology* **145**(5), 2319–2327.

Teng, Y., Xue, W. J., *et al.* (2005). Isolation and culture of adult Sertoli cells and their effects on the function of cocultured allogeneic islets *in vitro. Chin. Med. J. (Engl.)* **118**(22), 1857–1862.

Thu, B., Bruheim, P., *et al.* (1996). Alginate polycation microcapsules. II. Some functional properties. *Biomaterials* **17**(11), 1069–1079.

Tian, X. H., Xue, W. J., *et al.* (2005). Small intestinal submucosa improves islet survival and function during *in vitro* culture. *World J. Gastroenterol.* **11**(46), 7378–7383.

Unger, R. H. (2005). Longevity, lipotoxicity and leptin: the adipocyte defense against feasting and famine. *Biochimie* **87**(1), 57–64.

Vaca, P., Martin, F., *et al.* (2006). Induction of differentiation of embryonic stem cells into insulin-secreting cells by fetal soluble factors. *Stem Cells* **24**(2), 258–265.

Watada, H. (2004). Neurogenin 3 is a key transcription factor for differentiation of the endocrine pancreas. *Endocr. J.* **51**(3), 255–264.

Wilding, L., and Gannon, M. (2004). The role of pdx1 and HNF6 in proliferation and differentiation of endocrine precursors. *Diabetes Metab. Res. Rev.* **20**(2), 114–123.

Wobus, A. M., and Boheler, K. R. (2005). Embryonic stem cells: prospects for developmental and cell therapy. *Physiol. Rev.* **85**, 635–678.

Zalzman, M., Anker-Kitai, L., *et al.* (2005). Differentiation of human liver-derived, insulin-producing cells toward the beta-cell phenotype. *Diabetes* **54**(9), 2568–2575.

# Chapter Forty-Two

# Bioartificial Pancreas

### *Athanassios Sambanis*

I. Introduction
II. Cell Types for Pancreatic Substitutes
III. Construct Technology
IV. *In Vivo* Implantation

V. Concluding Remarks
VI. Acknowledgments
VII. References

## I. INTRODUCTION

Diabetes is a significant health problem, affecting an estimated 20.8 million people in the United States alone, with nearly 1.8 million afflicted with type 1 diabetes [http://diabetes.niddk.nih.gov/dm/pubs/statistics/index.htm#7]. Type 1 diabetes results from the loss of insulin-producing cell mass (the β-cells of pancreatic islets) due to autoimmune attack. Type 2 diabetes has a more complicated disease etiology and can be the result of not producing enough insulin and/or the body's developing a resistance to insulin. Although initially controlled by diet, exercise, and oral medication, type 2 diabetes often progresses toward insulin dependence. It is estimated that insulin-dependent diabetics (both types 1 and 2) exceed 4 million people in the United States. Although insulin-dependent diabetes (IDD) is considered a chronic disease, even the most vigilant insulin therapy cannot reproduce the precise metabolic control present in the nondiseased state. The poor temporal match between glucose load and insulin activity leads to a number of complications, including increased risk of heart disease, kidney failure, blindness, and amputation due to peripheral nerve damage. Providing more physiological control would alleviate many of the diabetes-related health problems, as suggested by findings from the Diabetes Control and Complications Trial (The Diabetes Control and Complications Trial Research Group, 1997) and its continuation study (DCCT/EDIC NEJM 353(25):2643–53, 2005). Cell-based therapies, which provide continuous regulation of blood glucose through physiologic secretion of insulin, have the potential to revolutionize diabetes care.

Several directions are being considered for cell-based therapies of IDD, including implantation of immunoprotected allogeneic or xenogeneic islets, of continuous cell lines, or of engineered non-β-cells. For allogeneic islet transplantation, a protocol developed by physicians at the University of Edmonton (Shapiro *et al.*, 2001a, 2001b, 2001c, Bigam and Shapiro, 2004) has dramatically improved the survivability of grafts. The protocol uses human islets from cadaveric donors, which are implanted in the liver of carefully selected diabetic recipients via portal vein injection. The success of the Edmonton protocol is attributed to two modifications relative to earlier islet transplantation studies: the use of a higher number of islets and the implementation of a more benign, steroid-free immunosuppressive regimen. However, two barriers prevent the widespread application of this therapy. The first is the limited availability of human tissue, because generally more than one cadaveric donor pancreas is needed for the treatment of a single recipient. The second is the need for life-long immunosuppression, which, even with the more benign protocols, results in long-term side effects to the patients.

A tissue-engineered pancreatic substitute aims to address these limitations by using alternative cell sources, relaxing the cell availability limitation, and by reducing or eliminating the immunosuppressive regimen necessary for survival of the graft. A number of significant challenges are

*Principles of Tissue Engineering, 3ʳᵈ Edition*
*ed. by Lanza, Langer, and Vacanti*

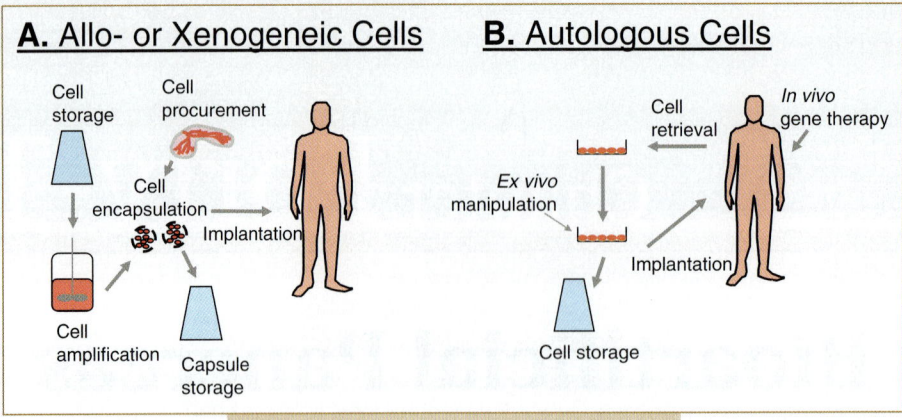

**FIG. 42.1.** Approaches for bioartificial pancreas development using allo- or xenogeneic cells (A) and autologous cells (B). In **(A)**, islets are procured from pancreatic tissue, or cell lines are thawed from cryostorage and expanded in culture; cells are encapsulated for immunoprotection before they are implanted to achieve a therapeutic effect; encapsulated cells may also be cryopreserved for inventory management and sterility testing. In **(B)**, cells are retrieved surgically from the patient; manipulated *ex vivo* phenotypically and/or genetically in order to express β-cell characteristics, and in particular physiologically responsive insulin secretion; the cells are implanted for a therapeutic effect either by themselves or, preferably, after incorporation in a three-dimensional substitute; some of the cells may be cryopreserved for later use by the same individual. In *in vivo* gene therapy approaches, a transgene for insulin expression is directly introduced into the host and expressed by cells in nonpancreatic tissues.

facing the development of such a substitute, however. These include procuring cells at clinically relevant quantities; the immune acceptance of the cells, which is exacerbated in type 1 diabetes by the resident autoimmunity in the patients; and the fact that diabetes is not an immediately life-threatening disease, so any other therapy will have to be more efficacious and/or less invasive than the current standard treatment of daily blood glucose monitoring and insulin injections.

In general, developing a functional living-tissue replacement requires advances and integration of several types of technology (Nerem and Sambanis, 1995). These are (1) *cell technology*, which addresses the procurement of functional cells at the levels needed for clinical applications; (2) *construct technology*, which involves combining the cells with biomaterials in functional three-dimensional configurations. Construct manufacturing at the appropriate scale, and preservation, as needed for off-the-shelf availability, also fall under this set of technologies; (3) technologies for *in vivo* integration, which address the issues of construct immune acceptance, *in vivo* safety and efficacy, and monitoring of construct integrity and function postimplantation. The same three types of technology need also be developed for a pancreatic substitute. It should be noted, however, that the critical technologies differ, depending on the type of cells used. With allogeneic or xenogeneic islets or beta-cells, the major challenge is the immune acceptance of the implant. In this case, encapsulation of the cells in permselective membranes, which allow passage of low-molecular-weight nutrients and metabolites but exclude larger antibodies and cytotoxic cells of the host, may assist the immune acceptance of the graft. With cell therapies based on potentially autologous nonpancreatic cells, targeted by gene expression vectors *in vivo*, or retrieved surgically,

engineered *ex vivo*, and returned to the host, the major challenge is engineering insulin secretion in precise response to physiologic stimuli. Lastly, with stem or progenitor cells, the primary hurdle is their reproducible differentiation into cells of the pancreatic β-phenotype. Figure 42.1 shows schematically the two general therapeutic approaches based on allo- or xenogeneic cells (Fig. 42.1A) or autologous cells (Fig. 42.1B).

This chapter is therefore organized as follows. We first describe the types of cells that have been used or are of potential use in engineering a pancreatic substitute. We then discuss issues of construct technology, specifically encapsulation methods and the relevant biomaterials, manufacturing issues, and preservation of the constructs. The challenges of *in vivo* integration and results from *in vivo* experiments with pancreatic substitutes are presented next. We conclude by offering a perspective on the current status and the future challenges in developing an efficacious, clinically applicable bioartificial pancreas.

## II. CELL TYPES FOR PANCREATIC SUBSTITUTES

### Islets

Despite several efforts, the *in vitro* expansion of primary human islets has met with limited success. Adult human islets are difficult to propagate in culture, and their expansion leads to dedifferentiation, generally manifested as loss of insulin secretory capacity. Although there exist reports on the redifferentiation of expanded islets (Lechner *et al.*, 2005; Ouziel-Yahalom *et al.*, 2006) and of nonislet pancreatic cells, which are discarded after islet isolation (Todorov *et al.*, 2006), the phenotypic stability and the *in vivo* efficacy of these cells remain unclear. Additionally, with expanded and redifferentiated islets, it remains unknown whether the insulin-secreting cells arose from the redifferentiation of mature endocrine cells or from an indigenous stem or progenitor cell population in the tissue isolate (Todorov *et al.*, 2006).

Animal, such as porcine, islets are amply available, and porcine insulin is very similar to human, differing by only one amino acid residue. However, the potential use of porcine tissue is hampered by the unlikely but possible transmission of porcine endogenous retroviruses (PERV) to human hosts as well as by the strong xenograft immunogenicity that they elicit. Use of closed, PERV-free herds is reasonably expected to alleviate the first problem. With regard to immunogenicity, a combination of less immunogenic islets, islet encapsulation in permselective barriers, and host immunosuppression may yield long-term survival of the implant. The use of transgenic pigs that do not express the α-Gal (α[1,3]-galactose) epitope is one possible approach for reducing the immunogenicity of the islets. Studies also indicate that neonatal pig islets induce a lower T-cell reactivity than adult islets (Bloch *et al.*, 1999), even though the α-Gal epitope is abundant in neonatal islets as well (Rayat *et al.*, 1998). Furthermore, it is possible that the primary antigenic components in islet tissue are the ductal epithelial and vascular endothelial cells, which express prominently the α-Gal epitope; on the other hand, β-cells express the epitope immediately after isolation but not after maintenance in culture (Heald *et al.*, 1999). It should also be noted that the large-scale isolation of porcine islets under conditions of purity and sterility that will be needed for eventual regulatory approval pose some major technical hurdles, which have not been addressed yet.

### β-Cell Lines

Recognizing the substantial difficulties involved with the procurement and amplification of pancreatic islets, several investigators have pursued the development of continuous cell lines, which can be amplified in culture yet retain key differentiated properties of normal β-cells. One of the first successful developments in this area was the generation of the βTC family of insulinomas, derived from transgenic mice carrying a hybrid insulin-promoted simian virus 40 tumor antigen gene; these cells retained their differentiated features for about 50 passages in culture (Efrat *et al.*, 1988). The hypersensitive glucose responsiveness of the initial βTC lines was reportedly corrected in subsequent lines by ensuring expression of glucokinase and of the high-$K_m$ glucose transporter Glut2, and with no or low expression of hexokinase and of the low-$K_m$ transporter Glut1 (Efrat *et al.*, 1993; Knaack *et al.*, 1994). A similar approach was used to develop the mouse MIN-6 cell line that exhibits glucose-responsive secretion of endogenous insulin (Miyazaki *et al.*, 1990). Subsequently, Efrat and coworkers developed the βTC-tet cell line, in which expression of the SV40 T antigen (Tag) oncoprotein is tightly and reversibly regulated by tetracycline. Thus, cells proliferate when Tag is expressed, and

shutting off Tag expression halts cell growth (Efrat *et al.*, 1995; Efrat, 1998). Such reversible transformation is an elegant approach in generating a supply of β-cells via proliferation of an inoculum, followed by arrest of the growth of cells when the desirable population size is reached. When retained in capsules, proliferating cells do not grow uncontrollably, since the dissolved-oxygen concentration in the surrounding milieu can support up to a certain number of viable, metabolically active cells in the capsule volume. This number of viable cells is maintained through equilibration of cell growth and death processes (Papas *et al.*, 1999a, 1999b). Thus, growth arrest is useful primarily in preventing the growth of cells that have escaped from broken capsules *in vivo* and in reducing the cellular turnover in the capsules. The latter reduces the number of accumulated dead cells in the implant and thus the antigenic load to the host affected by proteins from dead and lysed cells that pass through the capsule material.

In a different approach, Newgard and coworkers (Clark *et al.*, 1997) carried out a stepwise introduction of genes related to β-cell performance into a poorly secreting rat insulinoma (RIN) line. In particular, RIN cells were iteratively engineered to stably express multiple copies of the insulin gene, the glucose transporter Glut2, and the glucokinase gene, which are deemed essential for proper expression of β-cell function. Although this is an interesting methodology, it is doubtful that all genes necessary for reproducing β-cell function can be identified and stably expressed in a host cell. Recently, significant progress was made toward establishing a human pancreatic β-cell line that appears functionally equivalent to normal β-cells (Narushima *et al.*, 2005). This was accomplished through a complicated procedure involving retroviral transfection of primary β cells with the SV40 large T antigen and cDNAs of human telomerase reverse transcriptase. This resulted in a reversibly immortalized human β-cell clone, which secreted insulin in response to glucose, expressed β-cell transcriptional factors, prohormone convertases 1/3 and 2, which process proinsulin to mature insulin, and restored normoglycemia upon implantation in diabetic immunodeficient mice (Narushima *et al.*, 2005).

With regard to β-cell lines capable of proliferation under the appropriate conditions, key issues that remain to be addressed include (1) their long-term phenotypic stability, *in vitro* and *in vivo*; (2) their potential tumorigenicity, if cells escape from an encapsulation device, especially when these cells are allografts that may evade the hosts' immune defenses for a longer period of time than acutely rejected xenografts; and (3) their possible recognition by the autoimmune rejection mechanism in type 1 diabetic hosts.

### Engineered Non–β Pancreatic Cells

The use of non–β pancreatic cells from the same patient, engineered for insulin secretion, relaxes both the cell availability and immune acceptance limitations that exist with other types of cells. It has been shown that the A-chain/

C-peptide and B-chain/C-peptide cleavage sites on the proinsulin gene can be mutated so that the ubiquitous endopeptidase furin recognizes and completely processes proinsulin into mature insulin absent of any intermediates (Yanagita et al., 1992). Based on this concept, several nonendocrine cell lines have been successfully transfected to produce immunoreactive insulin, including hepatocytes, myoblasts, and fibroblasts (Yanagita et al., 1993). In a different approach, Lee and coworkers (2000) expressed a synthetic single-chain insulin analog, which does not require posttranslational processing, in hepatocytes. Although recombinant insulin expression is relatively straightforward, a key remaining challenge is achieving the tight regulation of insulin secretion in response to physiologic stimuli, which is needed for achieving normoglycemia in higher animals and, eventually, humans.

One approach for achieving regulation of insulin secretion is through regulation of biosynthesis at the gene transcription level, as realized in hepatocytes by Thule et al. (Thule et al., 2000; Thule and Liu, 2000) and Lee et al. (2000). Besides the ability to confer transcriptional-level regulation, hepatocytes are particularly attractive as producers of recombinant insulin due to their high synthetic and secretory capacity and their expression of glucokinase and Glut2 (Cha et al., 2000; Lannoy et al., 2002). Hepatic delivery by viral vectors and expression of the glucose-responsive insulin transgene in diabetic rats controlled the hyperglycemic state for extended periods of time (Lee et al., 2000; Thule and Liu, 2000; Olson et al., 2003). Nevertheless, transcriptional regulation is sluggish, involving long time lags between stimulation of cells with a secretagogue and induced insulin secretion as well as between removal of the secretagogue and down-regulation of the secretory response (Tang and Sambanis, 2003). The latter is physiologically more important, because it means that the cells continue to secrete insulin after glucose has been down-regulated, thus resulting in potentially serious hypoglycemic excursions. Increasing the number of stimulatory glucose elements in a promoter enhances the cellular metabolic responsiveness in vitro (Thule et al., 2000). With regard to secretion downregulation, Tang and Sambanis (2003) hypothesized that the slow kinetics of this process following removal of the transcriptional activator are due to the stability of the preproinsulin mRNA, which continues to become translated after transcription has been turned off. Using a modified preproinsulin cDNA that produced an mRNA with two more copies of the insulin gene downstream of the stop codon resulted in preproinsulin mRNA subjected to nonsense mediated decay and thus destabilized. This significantly expedited the kinetics of secretion down-regulation on turning off transcription (Tang and Sambanis, 2003). Thus, the combination of optimal transcriptional regulation with transgene message destabilization promises further improvements in insulin secretion dynamics from transcriptionally regulated hepatic cells. It should be noted, however, that despite the time delays inherent in transcriptional regulation, hepatic insulin gene therapy is sufficient to sustain vascular nitric oxide production and inhibit acute development of diabetes-associated endothelial dysfunction in diabetic rats (Thule et al., 2006). Hence, many aspects of the therapeutic potential of hepatic insulin expression remain to be explored.

Another appealing target cell type is endocrine cells, which possess a regulated secretory pathway and the enzymatic machinery needed to process authentic proinsulin into insulin. Early work in this area involved expression of recombinant insulin in the anterior pituitary mouse AtT-20 cell line (Moore et al., 1983), which can be subjected to repeated episodes of induced insulin secretion using nonmetabolic secretagogues (Sambanis et al., 1990). Cotransfection with genes encoding the glucose transporter Glut-2 and glucokinase resulted in glucose-responsive insulin secretion (Hughes et al., 1992, 1993). However, limitations of this approach include possible instabilities in the cellular phenotype and the continued secretion of endogenous hormones, such as adrenocorticotropic hormone from AtT-20 cells, which are not compatible with prandial metabolism.

In this regard, endocrine cells of the intestinal epithelium, or enteroendocrine cells, are especially promising. Enteroendocrine cells secrete their incretin products in a tightly controlled manner that closely parallels the secretion of insulin following oral glucose load in human subjects; incretin hormones are fully compatible with prandial metabolism and glucose regulation (Schirra et al., 1996; Kieffer and Habener, 1999). As with β-cells, enteroendocrine cells are polar, with sensing microvilli on their luminal side and secretory granules docked at the basolateral side, adjacent to capillaries. Released incretin hormones include the glucagon-like peptides (GLP-1 and GLP-2) from intestinal L-cells and glucose-dependent insulinotropic polypeptide (GIP) from K-cells, which potentiate insulin production from the pancreas after a meal (Drucker, 2002). The importance of enteroendocrine cells (and, in particular, L-cells) was first put forward by Creutzfeldt (1974), whose primary interest in these cells was for the prospect of using GLP-1 for the treatment of type 2 diabetes. Furthermore, groundbreaking work by Cheung et al. (2000) demonstrated that insulin produced and secreted by genetically modified intestinal K-cells of transgenic mice prevented the animals from becoming diabetic after injection with streptozotocin (STZ), which specifically kills the β-cells of the pancreas. This is an important proof-of-concept study, which showed that enteroendocrine cell–produced insulin can provide regulation of blood glucose levels. Subsequent work with a human intestinal L-cell line demonstrated that these cells can be effectively transduced to express recombinant human insulin, which colocalizes in secretory vesicles with endogenous GLP-1 and thus is secreted with identical kinetics to GLP-1 in response to stimuli (Tang and Sambanis, 2003,

2004). The intestinal tract could be considered an attractive target for gene therapy because of its large size, making it the largest endocrine organ in the body (Wang *et al.*, 2004); however, enteroendocrine cell gene therapy faces serious difficulties due to anatomic complexity, with the enteroendocrine cells being located at the base of invaginations of the gut mucosa called *crypts*, the very harsh conditions in the stomach and intestine, and the rapid turnover of the intestinal epithelium.

Contrary to direct *in vivo* gene delivery, *ex vivo* gene therapy involves retrieving the target cells surgically, culturing them and possibly expanding them *in vitro*, genetically engineering them to express the desired properties, and then returning them to the host, either as such or in a three-dimensional tissue substitute. It is generally thought that the *ex vivo* approach is advantageous, for it allows for the thorough characterization of the genetically engineered cells prior to implantation, possibly for the preservation of some of the cells for later use by the same individual and, importantly, for localization and retrievability of the implant. However, the challenges imposed by the *ex vivo* approach, including the surgical retrieval, culturing, and *in vitro* genetic engineering of the target cells are significant, so such methods are currently under development.

### Differentiated Stem or Progenitor Cells

Naturally, throughout life, islets turn over slowly, and new, small islets are continually generated from ductal progenitors (Finegood *et al.*, 1995; Bonner-Weir and Sharma, 2002). There is also considerable evidence that adult pluripotent stem cells may be a possible source of new islets (Bonner-Weir *et al.*, 2000; Ramiya *et al.*, 2000; Kojima *et al.*, 2003). However, efforts to regenerate β-cells *in vitro* or *in vivo* by differentiation of embryonic or adult stem or pancreatic progenitor cells have produced mixed results. Insulin-producing, glucose-responsive cells, as well as other pancreatic endocrine cells, have been generated from mouse embryonic stem cells (Lumelsky *et al.*, 2001). Insulin-secreting cells obtained from embryonic stem cells reversed hyperglycemia when implanted in mice rendered diabetic by STZ injection (Soria *et al.*, 2000). In another study, mouse embryonic stem cells transfected to express constitutively Pax4, a transcription factor essential for β-cell development, differentiated into insulin-producing cells and normalized blood glucose when implanted in STZ-diabetic mice (Blyszczuk *et al.*, 2003). On the other hand, other studies do not support differentiation of embryonic stem cells into the β-cell phenotype (Rajagopal *et al.*, 2003). Overall, the mixed and somewhat inconsistent results point to the considerable work that needs to be done before stem or progenitor cells can be reliably differentiated into β cells at a clinically relevant scale. Harnessing the *in vivo* regenerative capacity of the pancreatic endocrine system may present a promising alternative approach.

### Engineering of Cells for Enhanced Survival *In Vivo*

Because islets and other insulin-secreting cells experience stressful conditions during *in vitro* handling and *in vivo* postimplantation, several strategies have been implemented to enhance islet or nonislet cell survival in pancreatic substitutes. Strategies generally focus on improving the immune acceptance of the graft, enhancing its resistance to cytokines, and reducing its susceptibility to apoptosis. Phenotypic manipulations include extended culturing of neonatal and pig islets at 37°C, which apparently reduces their immunogenicity, possibly by down-regulating the major histocompatibility class 1 antigens on the islet surface; islet pretreatment with TGF-β1; and enzymatic treatment of pig islets with α-galactosidase to reduce the a-galactosyl epitope on islets (Prokop, 2001). However, the permanency of these modifications is unknown. For instance, a-galactosyl epitopes reappear on islets 48 hours after treatment with α-galactosidase. With proliferative cell lines destined for recombinant insulin expression, selection of clones resistant to cytokines appears feasible (Chen *et al.*, 2000). Gene chip analysis of resistant cells may then be used to identify the genes responsible for conferring cytokine resistance.

Genetic modifications for improving survival *in vivo* may offer prolonged expression of the desired properties relative to phenotypic manipulations, but they also present the possibility of modifying the islets in additional, undesirable ways. Notable among the various proposed approaches, reviewed in Jun and Yoon (2005), are the expression of the immunomodulating cytokines IL-4 or a combination of IL-10 and TGF-β, which promoted graft survival by preventing immune attack in mice; and the expression of the antiapoptotic bcl-2 gene using a replication defective herpes simplex virus, which resulted in protection of β-cells from a cytokine mixture of interleukin-1β, TNF-α, and IFN-γ *in vitro*.

## III. CONSTRUCT TECHNOLOGY

Construct technology focuses on associating cells with biocompatible materials in functional three-dimensional configurations. Depending on the type of cells used, the primary function of the construct can be one or more of the following: to immunoprotect the cells postimplantation, to enable cell function, to localize insulin delivery *in vivo*, or to provide retrievability of the implanted cells.

### Encapsulated Cell Systems

Encapsulation for immunoprotection involves surrounding the cells with a permselective barrier, in essence an ultrafiltration membrane, which allows passage of low-molecular-weight nutrients and metabolites, including insulin, but excludes larger antibodies and cytototoxic cells of the host. Figure 42.2 summarizes the common types of encapsulation devices, which include spherical microcapsules, tubular or planar diffusion chambers, thin sheets, and vascular devices.

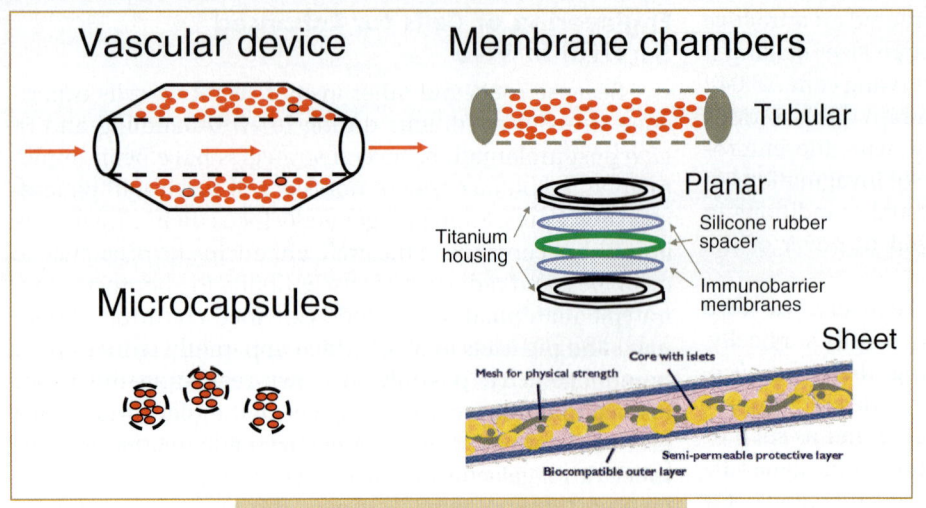

**FIG. 42.2.** Schematics of commonly used encapsulation devices for insulin-secreting cells. Vascular devices and membrane chambers of tubular, planar, or sheet architectures are generally referred to as macrocapsules, in distinction from the much smaller microcapsules.

Encapsulation can be pursued via one of two general approaches. With capsules fabricated using water-based chemistry, cells are first suspended in un-cross-linked polymer, which is then extruded as droplets into a solution of the cross-linking agent. A typical example here is the very commonly used alginate encapsulation. Alginate is a complex mixture of polysaccharides obtained from seaweeds, which forms a viscous solution in physiologic saline. Islets or other insulin-secreting cells are suspended in sodium alginate, and droplets are extruded into a solution of calcium chloride. Calcium cross-links alginate, instantaneously trapping the cells within the gel. The size of the droplets, hence also of the cross-linked beads, can be controlled by flowing air parallel to the extrusion needle so that droplets detach at a smaller size than if they were allowed to fall by gravity; or by using an electrostatic droplet generator, in which droplets are detached from the needle by adjusting the electrostatic potential between the needle and the calcium chloride bath. Capsules generated this way can have diameters from a few hundred micrometers to more than one millimeter. Alginate by itself is relatively permeable; to generate the permselective barrier, beads are treated with a polycationic solution, such as poly-L-lysine or poly-L-ornithine. The reaction time between alginate and the polycation determines the molecular weight cutoff of the resulting membrane. Poly-L-lysine is highly inflammatory *in vivo*, however, so beads are coated with a final layer of alginate to improve their biocompatibility. Hence, calcium alginate/poly-L-lysine/alginate (APA) beads are finally formed. Treating the beads with a calcium chelator, such as citrate, presumably liquefies the inner core, forming APA membranes. Other materials that have been used for cell microencapsulation include agarose, photo-cross-linked poly(ethylene glycol), and (ethyl methacrylate, methyl methacrylate, and dimethylaminoethyl methacrylate) copolymers (Mikos *et al.*, 1994; Sefton and Kharlip, 1994). Advantages of hydrogel microcapsules include a high surface-to-volume ratio, and thus good transport properties, as well as ease of handling and implantation. Small beads can be implanted in the peritoneal cavity of animals simply by injection, without the need for incision. Other common implantation sites include the subcutaneous space and the kidney capsules. Disadvantages include the fragility of the beads, especially if the cross-linking cation becomes chelated by compounds present in the surrounding milieu or released by lysed cells, and the lack of easy retrievability once the beads have been dispersed in the peritoneal cavity of a host. Earlier problems caused by the variable composition of alginates and the presence of endotoxins have been resolved through the development and commercial availability of ultrapure alginates of well-defined molecular weight and composition (Sambanis, 2000; Stabler *et al.*, 2001).

Hydrogels impose little diffusional resistance to solutes, and indeed effective diffusivities in calcium alginate and agarose hydrogels are in the range of 50–100% of the corresponding diffusivities in water (Tziampazis and Sambanis, 1995; Lundberg and Kuchel, 1997). However, with conventional microencapsulation, the volume of the hydrogel contributes significantly to the total volume of the implant. For example, with a 500-μm microcapsule containing a 300-μm islet, the polymer volume constitutes 78% of the total capsule volume. Additionally, conventional microcapsules may not be appropriate for hepatic portal vein implantation because, besides their higher implant volume relative to the same number of naked islets, they may result in higher portal vein pressure and more incidences of blood coagulation in the liver. To address this problem, methods have been developed for islet encapsulation in thin conformal polymeric coats. Materials that have been used for conformal coating include photopolymerized poly(ethylene glycol) diacrylate (Hill *et al.*, 1997; Cruise *et al.*, 1999) and hydroxyethyl methacrylate-methyl methacrylate (HEMA-MMA) (May and Sefton, 1999; Sefton *et al.*, 2000).

Encapsulated cell systems can also be fabricated by preforming the permselective membrane in a tubular or disc-shaped configuration, filling the construct with a suspension of islets or other insulin-secreting cells in an appropriate extracellular matrix, and then sealing the device. This approach is particularly useful when organic solvents or other chemicals harsh to the cells are needed for the fabrication of the membranes. Membrane chambers can be of tubular or planar geometry (Fig. 42.2). The cells are surrounded by the semipermeable membrane and can be implanted intraperitoneally, subcutaneously, or at other sites. Membrane materials used in fabricating these devices include polyacrylonitrile-polyvinyl chloride (PAN-PVC) copolymers, polypropylene, polycarbonate, cellulose nitrate, and polyacrylonitrile-sodium methallylsulfonate (AN69) (Lanza *et al.*, 1992; Mikos *et al.*, 1994; Prevost *et al.*, 1997; Delaunay *et al.*, 1998; Sambanis, 2000). Typical values of device thickness or fiber diameter are 0.5–1 mm. Advantages of membrane chambers are the relative ease of handling, the flexibility with regard to the matrix in which the cells are embedded, and retrievability after implantation. A major disadvantage is their inferior transport properties, since the surface-to-volume ratio is smaller than that of microcapsules and diffusional distances are longer.

Constructs connected to the vasculature via an arterio-venus shunt consist of a semipermeable tube surrounded by the cell compartment (Fig. 42.2). The tube is connected to the vasculature, and transport of solutes between the blood and the cell compartment occurs via the pores in the tube wall. A distinct advantage of the vascular device is the improved transport of nutrients and metabolites, which occurs by both diffusion and convection. However, the major surgery that is needed for implantation and problems of blood coagulation at the anastomosis sites have considerably reduced enthusiasm for these devices.

## Other Construct Systems

A common approach for improving the oxygenation of cells in diffusion chambers is to encourage the formation of neovasculature around the implant. This is discussed in the following "*In Vivo* Implantation" section. Other innovative approaches that have been proposed include the electro-chemical generation of oxygen in a device adjacent to a planar immunobarrier diffusion chamber containing the insulin-secreting cells (Wu *et al.*, 1999); and the coencapsulation of islets with algae, where the latter produce oxygen photosynthetically upon illumination (Bloch *et al.*, 2006). These were *in vitro* studies, however, and the ability to translate these approaches to effective *in vivo* configurations remains unknown.

In a different design, Cheng *et al.* (2004, 2006) combined constitutive insulin-secreting cells with a glucose-responsive material in a disc-shaped construct. As indicated earlier, it is straightforward to genetically engineer non-β-cells for constitutive insulin secretion; the challenge is in engineering appropriate cellular responsiveness to

physiologic stimuli. In this proposed device, a concanavalin A (con A)–glycogen material, sandwiched between two ultrafiltration membranes, acted as a control barrier to insulin release from an adjacent compartment containing the cells. Specifically, con A–glycogen formed a gel at a low concentration of glucose, which was reversibly converted to sol at a high glucose concentration, as glucose displaced glycogen from the gel network. Since insulin diffusivity is higher through the sol than through the gel, insulin released by the cells during low-glucose periods diffused slowly through the gel material; when switched to high glucose, the insulin accumulated in the cell compartment during the previous cycle was released at a faster rate through the sol-state polymer. Overall, this approach converted the constitutive secretion of insulin by the cells to a glucose-responsive insulin release by the device (Cheng *et al.*, 2006). Again, these were *in vitro* studies, and the *in vivo* efficacy of this approach remains to be evaluated.

## Construct Design and *In Vitro* Evaluation

Design of three-dimensional encapsulated systems can be significantly enabled using mathematical models of solute transport through the tissue and of nutrient consumption and metabolite production by the cells. Beyond the microvasculature surrounding the construct, transport of solutes occurs by diffusion, unless the construct is placed in a flow environment, in which case convective transport may also occur. Due to its low solubility, transport of oxygen to the cells is the critical issue. Models can be used to evaluate the dimensions and the cell density within the construct so that all cells are sufficiently nourished and the capsule as a whole is rapidly responsive to changes in the surrounding glucose concentration (Tziampazis and Sambanis, 1995). Experimental and modeling methods for determining transport properties and reaction kinetics have been described previously (Sambanis and Tan, 1999). Furthermore, models can be developed to account for the cellular reorganization that occurs in constructs with time as a result of cell growth, death, and possibly migration processes. Such reorganization is especially significant when proliferating insulin-secreting cell lines are encapsulated in hydrogel matrices (Stabler *et al.*, 2001; Simpson *et al.*, 2005; Gross *et al.*, 2007).

Pancreatic tissue substitutes should be evaluated *in vitro* prior to implantation, in terms of their ability to support the cells within over prolonged periods of time and to exhibit and maintain their overall secretory properties. Long-term cultures can be performed in perfusion bioreactors under conditions simulating aspects of the *in vivo* environment. In certain studies, the bioreactors and support perfusion circuits were made compatible with a nuclear magnetic resonance spectrometer. This allowed measuring intracellular metabolites, such as nucleotide triphosphates, as a function of culture conditions and time, without the need to extract the encapsulated cells (Papas *et al.*, 1999a, 1999b). Such studies produce a comprehensive understanding of the intrinsic tissue function in a well defined and controlled environment prior to introducing the additional complexity of host–implant interactions in *in vivo* experiments.

The secretory properties of tissue constructs can be evaluated with low time resolution in simple static culture experiments by changing the concentration of glucose in the medium and measuring the secreted insulin. In general, a square wave of insulin concentration is implemented, from basal to inducing basal conditions for insulin secretion. To evaluate the secretory response with a higher time resolution, perfusion experiments need to be performed, in which medium is flowed around the tissue and secreted insulin is assayed in the effluent. Again, a square wave of glucose concentration is generally implemented. By comparing the secretory dynamics of free and encapsulated cells, one can ensure that the encapsulation material does not introduce excessive time lags that might compromise the secretory properties of the construct. Indeed, properly designed hydrogel microcapsules introduce only minimal secretory time lags (Tziampazis and Sambanis, 1995; Sambanis *et al.*, 2002).

### Manufacturing Considerations

Fabrication of pancreatic substitutes of consistent quality requires the use of cells that are also of consistent quality. Although with clonal, expandable cells this is a rather straightforward issue, with islets isolated from human and animal tissues there can be significant variability in the quantity and quality of the cells in the preparations. With islets from cadaveric human donors, the quality of the isolates is assessed by microscopic observation, viability staining, and possibly a static insulin secretion test. It is generally recognized, however, that a quantitative, objective assessment of islet quality would help improve the consistency of the preparations and thus, possibly, the transplantation outcome.

It is conceivable that encapsulated cell systems could be fabricated at a central location from which they are distributed to clinical facilities for implantation. In this scheme, preservation of the constructs for long-term storage, inventory management, and, importantly, sterility control would be essential. Cryopreservation appears to be a promising method for maintaining fabricated constructs for prolonged time periods. Although there have been significant studies on the cryopreservation of single cells and some tissues, the problems pertaining to cryopreserving artificial tissues are only beginning to be addressed. Cryopreservation of macroencapsulated systems is expected to be particularly challenging and has not been reported in the literature. However, βTC-cells encapsulated in alginate beads have been preserved successfully (Mukherjee *et al.*, 2005; Song *et al.*, 2005). An especially promising approach involves using high concentrations of cryoprotective agents so that water is converted to a glassy, or vitrified, state at low temperatures; the absence of ice crystals in both the intracellular and extracellular domains appears helpful in maintaining not only cellular viability but also the structure and function of the surrounding matrix (Mukherjee *et al.*, 2005; Song *et al.*, 2005).

## IV. *IN VIVO* IMPLANTATION

This section highlights results from *in vivo* experiments using the different configurations outlined earlier. Results with encapsulated cell systems are presented first. Since *in vivo* experiments with non-β-cells engineered for insulin secretion are at present based mostly on *in vivo* gene therapy approaches, these are described next. Technologies for the *in vivo* monitoring of cells and constructs and the issue of implant retrievability are then discussed.

### Encapsulated Cell Systems

*In vivo* experiments with pancreatic substitutes are numerous in small animals, limited in large animals, and few in humans. Allogeneic and xenogeneic islets in hydrogel microcapsules implanted in diabetic mice and rats have generally restored normoglycemia for prolonged periods of time. In the early study of O'Shea *et al.* (1984), islet allografts encapsulated in APA membranes were implanted intraperitoneally in streptozotocin-induced diabetic rats. Of the five animals that received transplants, three remained normoglycemic for more than 100 days. One of these three animals remained normoglycemic 368 days postimplantation. In the later study of Lum *et al.* (1992), rat islets encapsulated in APA membranes and implanted in streptozotocin diabetic mice restored normoglycemia for up to 308 days, with a mean xenograft survival time of 220 days. With all recipients, normoglycemia was restored within two days postimplantation. Control animals receiving single injections of unencapsulated islets remained normoglycemic for less than two weeks (O'Shea *et al.*, 1984; Lum *et al.*, 1992). More recently, APA-encapsulated βTC6-F7 insulinomas restored normoglycemia in diabetic rats for up to 60 days (Mamujee *et al.*, 1997), and APA-encapsulated βTC-tet insulinomas in NOD mice for at least eight weeks (Black *et al.*, 2006). In the latter study, it was also observed that no host cell adherence occurred to microcapsules, and there were no significant immune responses to the implant, with cytokine levels being similar to those of sham-operated controls. These results are thus indicative of the potential use of an immunoisolated continuous β-cell line for the treatment of diabetes. With the recently developed human cell line (Narushima *et al.*, 2005), experiments were performed with unencapsulated cells transplanted into streptozotocin-induced diabetic severe combined immunodeficiency mice. Control of blood glucose levels started within two weeks postimplantation, and mice remained normoglycemic for longer than 30 weeks (Narushima *et al.*, 2005). Besides rodents, long-term restoration of normoglycemia with microencapsulated islets has been demonstrated in large animals, including spontaneously diabetic dogs, where normoglycemia was achieved with canine islet allografts for up to 172 days (Soon-Shiong *et al.*, 1992), and monkeys, where in one animal porcine islet xenografts normalized hyperglycemia for more than 150 days (Sun *et al.*, 1992). More recently, one of the companies working on islet encapsulation technology announced that primate subjects in ongoing studies have continued to exhibit improved glycemic regulation over a six-month period after receiving microencapsulated porcine islet transplants (MicroIslet Inc. Press Release, August 7, 2006).

*In vivo* results with vascular devices are reportedly mixed. Implantation of devices containing allogeneic islets as arteriovenous shunts in pancreatectomized dogs resulted in 20–50% of the dogs becoming normoglycemic up to 10 weeks postimplantation without exogenous insulin administration. When xenogeneic bovine or porcine islets were used, only 10% of the dogs remained normoglycemic 10 weeks postimplantation. All dogs were reported diabetic or dead after 15 weeks (Sullivan *et al.*, 1991). Recently, a hollow-fiber device composed of polyethylene-vinyl alcohol fibers and a poly-amino-urethane-coated, nonwoven polytetrafluoroethylene fabric seeded with porcine islets provided normalization of the blood glucose levels in totally pancreatectomized pigs when connected to the vasculature of the animals (Ikeda *et al.*, 2006). It should be noted, however, that the overall interest in vascular devices has faded, due to the surgical and blood coagulation challenges they pose.

Although several hypotheses exist, the precise cause of the eventual *in vivo* failure of encapsulated cell systems remains unclear. Encapsulation does not completely prevent the immune recognition of the implant. Although direct cellular recognition is prevented, antigens shed by the cells as a result of secretion and, more importantly, lysis in the capsules eventually pass through the permselective barrier and are recognized by the antigen-presenting cells of the host. For example, in one study, antibodies against islets in a tubular diffusion chamber were detected in plasma two to six weeks postimplantation, suggesting that islet antigens crossed the membrane and stimulated antibody formation in the host (Lanza *et al.*, 1994). In another study, alginate-encapsulated islets were lysed *in vitro* by nitric oxide produced by activated macrophages (Wiegand *et al.*, 1993). Passage of low-molecular-weight molecules cannot be prevented by immunoprotective membranes imposing a molecular weight cutoff on the order of 50 kDa. It should be noted that in one human study involving encapsulated allogeneic islets, the patient had to be provided with low levels of immunosuppression (Soon-Shiong *et al.*, 1994). In a more recent report, also with encapsulated allografts implanted peritoneally, type 1 diabetic patients remained nonimmunosuppressed but were unable to withdraw exogenous insulin (Calafiore *et al.*, 2006).

Nonspecific inflammation may also occur around the implant and develop into a fibrous capsule, reducing the oxygen available to the cells within. The fibrotic layer has been found to consist of several layers of fibroblasts and collagen with polymorphonuclear leukocytes, macrophages, and lymphocytes. The surface roughness of the membrane may also trigger inflammatory responses. In one study, membranes with smooth outside surfaces exhibited a minimal fibrotic reaction 10 weeks postimplantation, regardless of the type of encapsulated cells, whereas rough surfaces elicited a fibrotic response even one week postimplantation (Lanza *et al.*, 1991). Use of high-purity materials also helps to minimize inflammatory reactions. If a material is intrinsically inflammatory, such as poly-L-lysine, it can be coated with a layer of noninflammatory material, such as alginate, to minimize the host's reaction. Such coverage may not be sufficiently permanent, though, resulting in the eventual fibrosis of the implant. Indeed, several investigators report improved results with plain alginate beads without a poly-L-lysine layer, especially when allogeneic cells are used in the capsules.

Provision of nutrients to and removal of metabolites from encapsulated cells can be especially challenging *in vivo*. Normal pancreatic islets are highly vascularized and thus well oxygenated. There exists evidence that unencapsulated islets injected in the portal system of the liver become revascularized, which enhances their engraftment and function. On the other hand, encapsulation prevents revascularization, so the implanted tissue is nourished by diffusion alone. Promotion of vascularization around the immunoprotective membrane increases the oxygenation of the implanted islets (Prokop *et al.*, 2001). Interestingly, transformed cells, such as the βTC3 line of mouse insulinomas, are more tolerant of hypoxic conditions than intact islets; such cells may thus function better than islets in implanted capsules (Papas *et al.*, 1996). However, with transformed cells, too, enhanced oxygenation increases the density of functional cells that can be effectively maintained within the implant volume. Vascularization is dependent on the microarchitecture of the material, which should have pores 0.8–8.0 μm in size, allowing permeation of host endothelial cells (Brauker *et al.*, 1992, 1995). Vascularization is also enhanced by the delivery of angiogenic agents, such as FGF-2 and VEGF, possibly with controlled-release devices (Sakurai *et al.*, 2003). Although vascularization can be promoted around a cell-seeded device, improved success has been reported if a cell-free device is first implanted and vascularized and the cells are then introduced. One example of this procedure involved placing a cylindrical stainless steel mesh in the subcutaneous space of rats, with the islets introduced 40 days later (Pileggi *et al.*, 2006). Replacement of a vascularized implant is challenging, however, due to the bleeding that occurs. A solution to this problem may entail the design of a device that can be emptied and refilled with a suspension of cells in an extracellular matrix without disturbing the housing and the associated vascular network.

## Gene and Cell-Based Therapies

*In vivo* efficacy studies with gene therapy and non-β-cells genetically engineered for insulin secretion are generally limited to small animals. Intraportal injection of recombinant adenovirus expressing furin-compatible insulin under the control of a glucose-responsive promoter containing elements of the rat liver pyruvate kinase gene restored near-normal glycemia in streptozotocin diabetic rats for periods of 1–12 weeks (Thule and Liu, 2000). With hepatic delivery of a recombinant adeno-associated virus expressing a single-chain insulin analog under the control of an L-type pyruvate kinase promoter, Lee and coworkers (2000) controlled blood glucose levels in streptozotocin diabetic rats and NOD mice for periods longer than 20 weeks. However, transiently low blood glucose levels observed three to five hours after glucose loading indicated a drawback of the transcriptional regulation of insulin expression, which may result in hypoglycemic episodes (Lee *et al.*, 2000). Possible approaches toward ameliorating this problem include optimizing the number of glucose-regulatory and insulin-sensing elements in the promoter (Jun and Yoon, 2005) and destabilizing the preproinsulin mRNA; the latter has been shown to expedite significantly the down-regulation of secretion dynamics from transcriptionally controlled cells on removal of the secretory stimulus (Tang and Sambanis, 2003).

*In vivo* gene therapy with small animals has also shown success when the target cells for insulin expression were intestinal endocrine K- or gastric G-cells. Using a transgene expressing human insulin under the control of the glucose-dependent insulinotropic peptide (GIP) promoter, Cheung *et al.* (2000) expressed insulin specifically in gut K-cells of transgenic mice, which protected them from developing diabetes following STZ-mediated destruction of the native β-cells. Similarly, use of a tissue-specific promoter to express insulin in gastric G-cells of mice resulted in insulin release into circulation in response to meal-associated stimuli, suggesting that G-cell insulin expression is beneficial in the amelioration of diabetes (Lu *et al.*, 2005). Translation of these approaches to adult animals and, eventually, humans, requires the development of effective methods of gene delivery to intestinal endocrine or gastric cells *in vivo* or the development of effective *ex vivo* gene therapy approaches.

### In Vivo Monitoring

Monitoring of the number and function of insulin-secreting cells *in vivo* would provide valuable information directly on the implant and possibly offer early indications of implant failure. Additionally, in animal experiments, the ability to monitor an implant noninvasively reduces the number of animals that are needed in the experimental design and helps establish a critical link between implantation and endpoint physiologic effects, the latter commonly being blood glucose levels and animal weight.

Imaging techniques can provide unique insight into the structure/function relationship of a construct *in vitro* and *in vivo*. There are several imaging modalities that have been applied to monitor tissue-engineered constructs, including computed tomography (CT), positron emission tomography (PET), optical techniques, and nuclear magnetic resonance (NMR) imaging and spectroscopy. Among these, NMR offers the unique advantage of providing information on both construct integrity and function, without the need to modify the cells genetically (e.g., through the expression of green fluorescent protein, used in optical methods) or the introduction of radioactive labels (e.g., PET agents). Furthermore, since magnetic fields penetrate uniformly throughout the sample, NMR is ideally suited to monitor constructs implanted at deep-seated locations. Its disadvantage is its low sensitivity. Whereas optical and radionuclide techniques can detect tracer quantities, NMR detects metabolites that are available in the millimolar or, in some cases, submillimolar range.

The ability to monitor noninvasively *in vivo* a pancreatic substitute by NMR was reported recently (Stabler *et al.*, 2005). Agarose disc-shaped constructs containing mouse insulinoma βTC3-cells were implanted in the peritoneal cavity of mice. Construct integrity was visualized by MR imaging and the metabolic activity of the cells within by water-suppressed $^1$H NMR spectroscopy (Fig. 42.3). Control experiments established that the total choline (TCho) resonance at 3.2 ppm, which is attributed to three choline metabolites, correlated positively and linearly with the number of viable cells within the construct, measured with an independent assay. To obtain the TCho signal *in vivo* without interference from the surrounding host tissue, such as peritoneal fat, the central agarose disc containing the cells had to be surrounded by cell-free agarose buffer zones. This ensured that the MR signal arose only from the implanted cells, even as the construct moved due to animal breathing. A second problem that had to be resolved was that glucose diffusing into the construct produced a resonance that interfered with the TCho resonance at 3.2 ppm. For this, a unique glucose resonance at 3.85 ppm was used to correct for the interference at 3.2 ppm so that a corrected signal, uniquely attributed to TCho, could be obtained. The latter correlated positively and linearly with the number of viable cells, measured with an independent assay, on the constructs postexplantation (Fig. 42.3). Hence, with the appropriate implanted construct architecture and signal processing, the number of viable cells in an implant could be followed in the same animal as a function of time (Stabler *et al.*, 2005).

Labeling of cells with magnetic nanoparticles, which can be detected by magnetic resonance, and genetically engineering cells so that they express a fluorescent or luminescent marker that can be optically detected are other methods being pursued to track the location and possibly function of implanted cells *in vivo*. It is expected that development of robust monitoring methodologies will be helpful

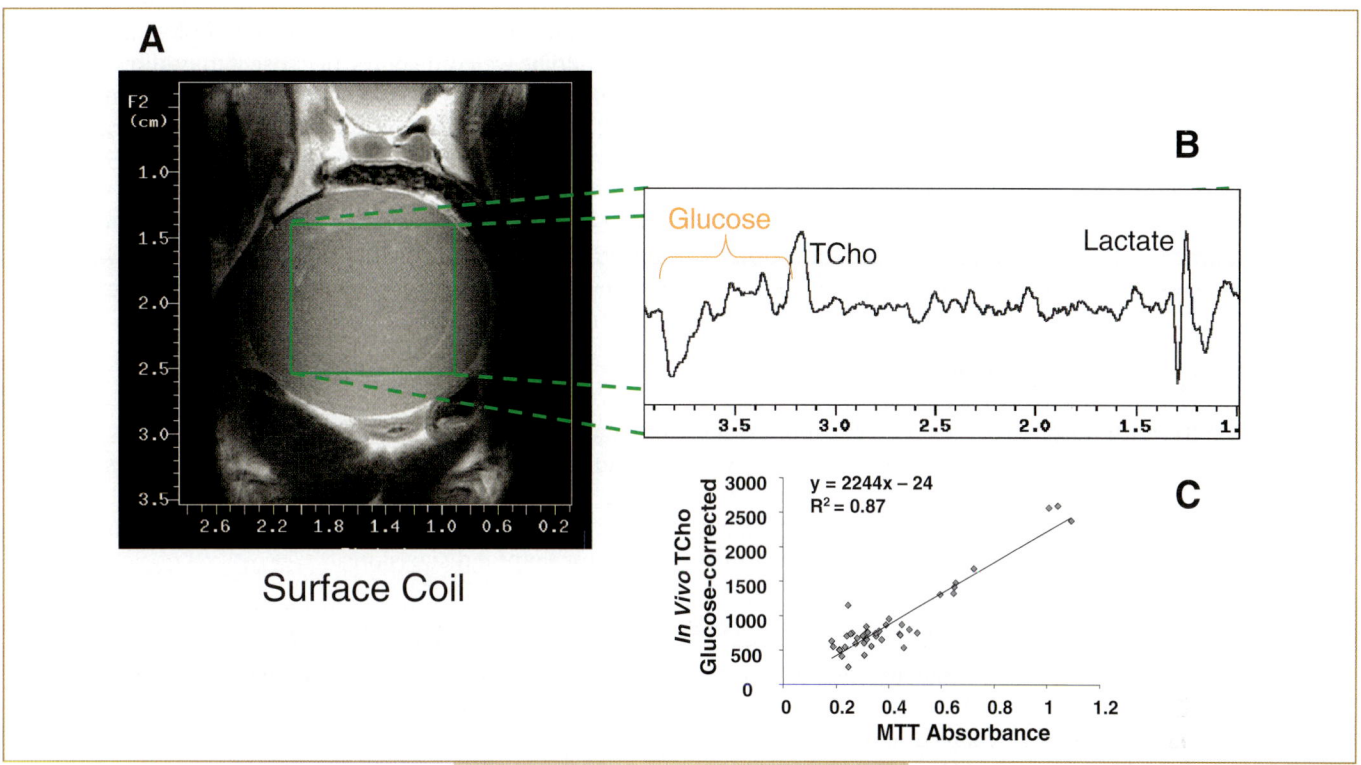

**FIG. 42.3.** Magnetic resonance imaging and localized spectroscopy of a disc-shaped agarose construct containing βTC3 mouse insulinoma cells at an initial density of $7 \times 10^7$ cells/mL agarose implanted in the peritoneal cavity of a mouse. **(A)** $^1$H NMR image obtained with a surface coil. The inner disc, containing the cells, is distinguishable from the surrounding cell-free buffer zone, implemented to exclude spectroscopic signal from the surrounding host tissue. **(B)** Localized, water-suppressed $^1$H NMR spectrum from the cells contained in the inner disc. Resonances due to total choline (TCho), glucose, and lactate are clearly visible. The time needed to collect the spectrum was 13 min. **(C)** Correlation between the glucose-corrected TCho resonance at 3.2 ppm and the viable cell number obtained postexplantation using the MTT [3-(4,5-dimethylthiazol-2-yl)-2,5-diphenyl-tetrazolium bromide] assay. Adapted from Stabler *et al.* (2005).

not only in experimental development studies but also in eventual clinical applications.

## Retrievability

The issue of construct retrievability needs to be considered for all pertinent applications. Useful lifetimes of constructs are limited, so repeated implantation of cells will be required. It is as yet unclear whether retrieval of constructs will be necessary at the end of their useful lives. Long-term studies on the safety challenges posed by accumulated implants in the host need to be carried out to address this question.

## V. CONCLUDING REMARKS

Tight glycemic regulation in insulin-dependent diabetics significantly improves their overall health and reduces the long-term complications of the disease. A pancreatic substitute holds significant promise at accomplishing this in a relatively noninvasive way. However, to justify the improved outcome, a substitute needs to be not only efficacious in terms of insulin secretion, but also immunologically acceptable. A number of approaches are being pursued

to address this obstacle and additionally develop constructs that can be manufactured at a clinically relevant scale. However, as the problems are being thoroughly investigated, their solutions become more challenging. Encapsulation in permselective barriers improves the immune acceptance of allo- and xenografts, but it is doubtful that encapsulation will, by itself, ensure the long-term survival and function of the implant in nonimmunosuppressed hosts. The development of specific, benign immune suppression protocols that work in concert with encapsulation appears necessary. Reducing the immunogenicity of the implanted cells and modifying them so that they better withstand the encapsulation and *in vivo* environment are appropriate strategies. To ensure that substitutes can be fabricated at the necessary scale, methods to expand pancreatic islets in culture, to produce β-cells from stem cells, or to generate expandable β-cell lines with appropriate phenotypic characteristics need to be pursued. In alternative approaches involving gene therapy of non-β-cells, or the *ex vivo* engineering of non-β-cells retrieved surgically from the host, the major problem is not that of cell procurement or immune acceptance but, rather, of ensuring precise regulation of insulin secretion by glucose or other physiologic stimuli. This poses a different set of problems, which, however, are equally challenging to those of β-cell procurement and immune acceptance. Methods for the preservation of substitutes and for the noninvasive monitoring of their integrity and functionality *in vivo* are integral parts of construct development and characterization with regard to construct manufacturing and assessment of *in vivo* efficacy, respectively.

As in many aspects of life, with challenges come opportunities. It is essential that multiple approaches be pursued in parallel, because it is currently unclear which ones will eventually develop into viable therapeutic procedures. If more than one method evolves into a clinical application, this would be welcome news, because it may allow flexibility in the personalization of therapy. For instance, in an adult type 2 insulin-dependent diabetic, use of an encapsulated allograft with low-level immunosuppression might constitute an appropriate therapeutic modality. In a juvenile type 1 diabetic with aggressive autoimmunity, however, use of autologous genetically engineered non-β-cells, which are not recognized by the resident autoimmunity, may constitute the therapeutic method of choice.

## VI. ACKNOWLEDGMENTS

The studies in the author's and coinvestigators' laboratories referenced in this chapter were supported by grants from the Georgia Tech/Emory Center for the Engineering of Living Tissues (GTEC), a National Science Foundation Engineering Research Center; and by grants from the National Institutes of Health, EmTech Bio, and the Juvenile Diabetes Research Foundation international. The author also wishes to thank Indra Neil Mukherjee and Heather Bara for critically reviewing the manuscript, as well as Drs. Constantinidis and Thulé for helpful discussions.

## VII. REFERENCES

Bigam, D. L., and Shapiro, A. J. (2004). Pancreatic transplantation: beta-cell replacement. *Curr. Treat. Options Gastroenterol.* **7**, 329–341.

Black, S. P., Constantinidis, I., Cui, H., Tucker-Burden, C., Weber, C. J., and Safley, S. A. (2006). Immune responses to an encapsulated allogeneic islet beta-cell line in diabetic NOD mice. *Biochem. Biophys. Res. Commun.* **340**, 236–243.

Bloch, K., Assa, S., Lazard, D., Abramov, N., Shalitin, S., Weintrob, N., Josefsberg, Z., Rapoport, M., and Vardi, P. (1999). Neonatal pig islets induce a lower T-cell response than adult pig islets in IDDM patients. *Transplantation* **67**, 748–752.

Bloch, K., Papismedov, E., Yavriyants, K., Vorobeychik, M., Beer, S., and Vardi, P. (2006). Photosynthetic oxygen generator for bioartificial pancreas. *Tissue Eng.* **12**, 337–344.

Blyszczuk, P., Czyz, J., Kania, G., Wagner, M., Roll, U., St-Onge, L., and Wobus, A. M. (2003). Expression of Pax4 in embryonic stem cells promotes differentiation of nestin-positive progenitor and insulin-producing cells. *Proc. Natl. Acad. Sci. U.S.A.* **100**, 998–1003.

Bonner-Weir, S., and Sharma, A. (2002). Pancreatic stem cells. *J. Pathol.* **197**, 519–526.

Bonner-Weir, S., Taneja, M., Weir, G. C., Tatarkiewicz, K., Song, K. H., Sharma, A., and O'Neil, J. J. (2000). *In vitro* cultivation of human islets from expanded ductal tissue. *Proc. Natl. Acad. Sci. U.S.A.* **97**, 7999–8004.

Brauker, J., Martinson, L. A., Hill, R. S., Young, S. K., Carr-Brendel, V. E., and Johnson, R. C. (1992). Neovascularization of immunoisolation membranes: the effect of membrane architecture and encapsulated tissue. *Transplant. Proc.* **24**, 2924.

Brauker, J. H., Carr-Brendel, V. E., Martinson, L. A., Crudele, J., Johnston, W. D., and Johnson, R. C. (1995). Neovascularization of synthetic membranes directed by membrane microarchitecture. *J. Biomed. Mater. Res.* **29**, 1517–1524.

Calafiore, R., Basta, G., Luca, G., Lemmi, A., Montanucci, M. P., Calabrese, G., Racanicchi, L., Mancuso, F., and Brunetti, P. (2006). Microencapsulated pancreatic islet allografts into nonimmunosuppressed patients with type 1 diabetes: first two cases. *Diabetes Care* **29**, 137–138.

Cha, J.-Y., Kim, H., Kim, K.-S., Hur, M.-W., and Ahn, Y. (2000). Identification of transacting factors responsible for the tissue-specific expression of human glucose transporter type 2 isoform gene. *J. Biol. Chem.* **275**, 18358–18365.

Chen, G., Hohmeier, H. E., Gasa, R., Tran, V. V., and Newgard, C. B. (2000). Selection of insulinoma cell lines with resistance to interleukin-1beta- and gamma-interferon-induced cytotoxicity. *Diabetes* **49**, 562–570.

Cheng, S. Y., Gross, J., and Sambanis, A. (2004). Hybrid pancreatic tissue substitute consisting of recombinant insulin-secreting cells and glucose-responsive material. *Biotechnol. Bioeng.* **87**, 863–873.

Cheng, S. Y., Constantinidis, I., and Sambanis, A. (2006). Use of glucose-responsive material to regulate insulin release from constitutively secreting cells. *Biotechnol. Bioeng.* **93**, 1079–1088.

Cheung, A. T., Dayanandan, B., Lewis, J. T., Korbutt, G. S., Rajotte, R. V., Bryer-Ash, M., Boylan, M. O., Wolfe, M. M., and Kieffer, T. J. (2000). Glucose-dependent insulin release from genetically engineered K cells. *Science* **290**, 1959–1962.

Clark, S. A., Quaade, C., Constandy, H., Hansen, P., Halban, P., Ferber, S., Newgard, C. B., and Normington, K. (1997). Novel insulinoma cell lines produced by iterative engineering of GLUT2, glucokinase, and human insulin expression. *Diabetes* **46**, 958–967.

Creutzfeldt, W. (1974). [Clinical significance of gastrointestinal hormones]. *Verh. Dtsch. Ges. Inn. Med.* **80**, 330–338.

Cruise, G. M., Hegre, O. D., Lamberti, F. V., Hager, S. R., Hill, R., Scharp, D. S., and Hubbell, J. A. (1999). *In vitro* and *in vivo* performance of porcine islets encapsulated in interfacially photopolymerized poly(ethylene glycol) diacrylate membranes. *Cell. Transplant.* **8**, 293–306.

Delaunay, C., Darquy, S., Honiger, J., Capron, F., Rouault, C., and Reach, G. (1998). Glucose-insulin kinetics of a bioartificial pancreas made of an AN69 hydrogel hollow fiber containing porcine islets and implanted in diabetic mice. *Artif. Organs* **22**, 291–299.

Drucker, D. J. (2002). Biological actions and therapeutic potential of the glucagon-like peptides. *Gastroenterology* **122**, 531–544.

Efrat, S. (1998). Cell-based therapy for insulin-dependent diabetes mellitus. *Eur. J. Endocrinol.* **138**, 129–133.

Efrat, S., Linde, S., Kofod, H., Spector, D., Delannoy, M., Grant, S., Hanahan, D., and Baekkeskov, S. (1988). Beta-cell lines derived from transgenic mice expressing a hybrid insulin gene-oncogene. *Proc. Natl. Acad. Sci. U.S.A.* **85**, 9037–9041.

Efrat, S., Leiser, M., Surana, M., Tal, M., Fusco-Demane, D., and Fleischer, N. (1993). Murine insulinoma cell line with normal glucose-regulated insulin secretion. *Diabetes* **42**, 901–907.

Efrat, S., Fusco-DeMane, D., Lemberg, H., al Emran, O., and Wang, X. (1995). Conditional transformation of a pancreatic beta-cell line derived from transgenic mice expressing a tetracycline-regulated oncogene. *Proc. Natl. Acad. Sci. U.S.A.* **92**, 3576–3580.

Finegood, D. T., Scaglia, L., and Bonner-Weir, S. (1995). Dynamics of beta-cell mass in the growing rat pancreas. Estimation with a simple mathematical model. *Diabetes* **44**, 249–256.

Gross, J. D., Constantinidis, I., and Sambanis, A. (2007). Modeling of encapsulated cell systems. *J. Theor. Biol.* **244**, 500–510.

Heald, K. A., Carless. N., Jay, T. R., Boucher, N., and Downing, R. (1999). Expression of the GALalpha(1–3)GAL epitope on pig islets. *J. Mol. Med.* **77**, 169–171.

Hill, R. S., Cruise, G. M., Hager, S. R., Lamberti, F. V., Yu, X., Garufis, C. L., Yu, Y., Mundwiler, K. E., Cole, J. F., Hubbell, J. A., Hegre, O. D., and Scharp, D. W. (1997). Immunoisolation of adult porcine islets for the treatment of diabetes mellitus. The use of photopolymerizable polyethylene glycol in the conformal coating of mass-isolated porcine islets. *Ann. N.Y. Acad. Sci.* **831**, 332–343.

Hughes, S. D., Johnson, J. H., Quaade, C., and Newgard, C. B. (1992). Engineering of glucose-stimulated insulin secretion and biosynthesis in non-islet cells. *Proc. Natl. Acad. Sci. U.S.A.* **89**, 688–692.

Hughes, S. D., Quaade, C., Johnson, J. H., Ferber, S., and Newgard, C. B. (1993). Transfection of AtT-20ins cells with GLUT-2 but not GLUT-1 confers glucose-stimulated insulin secretion. Relationship to glucose metabolism. *J. Biol. Chem.* **268**, 15205–15212.

Ikeda, H., Kobayashi, N., Tanaka, Y., Nakaji, S., Yong, C., Okitsu, T., Oshita, M., Matsumoto, S., Noguchi, H., Narushima, M., Tanaka, K., Miki, A., Rivas-Carrillo, J. D., Soto-Gutierrez, A., Navarro-Alvarez, N., Tanaka, K., Jun, H. S., Tanaka, N., and Yoon, J. W. (2006). A newly developed bioartificial pancreas successfully controls blood glucose in totally pancreatectomized diabetic pigs. *Tissue Eng.* **12**, 1799–1809.

Jun, H. S., and Yoon, J. W. (2005). Approaches for the cure of type 1 diabetes by cellular and gene therapy. *Curr. Gene Ther.* **5**, 249–262.

Kieffer, T. J., and Habener, J. F. (1999). The glucagon-like peptides. *Endocr. Rev.* **20**, 876–913.

Knaack, D., Fiore, D. M., Surana, M., Leiser, M., Laurance, M., Fusco-DeMane, D., Hegre, O. D., Fleischer, N., and Efrat, S. (1994). Clonal insulinoma cell line that stably maintains correct glucose responsiveness. *Diabetes* **43**, 1413–1417.

Kojima, H., Fujimiya, M., Matsumura, K., Younan, P., Imaeda, H., Maeda, M., and Chan, L. (2003). NeuroD-betacellulin gene therapy induces islet neogenesis in the liver and reverses diabetes in mice. *Nat. Med.* **9**, 596–603.

Lannoy, V. J., Decaux, J. F., Pierreux, C. E., Lemaigre, F. P., and Rousseau, G. G. (2002). Liver glucokinase gene expression is controlled by the onecut transcription factor hepatocyte nuclear factor-6. *Diabetologia* **45**, 1136–1141.

Lanza, R. P., Butler, D. H., Borland, K. M., Staruk, J. E., Faustman, D. L., Solomon, B. A., Muller, T. E., Rupp, R. G., Maki, T., Monaco, A. P., *et al.*
(1991). Xenotransplantation of canine, bovine, and porcine islets in diabetic rats without immunosuppression. *Proc. Natl. Acad. Sci. U.S.A.* **88**, 11100–11104.

Lanza, R. P., Sullivan, S. J., and Chick, W. L. (1992). Perspectives in diabetes. Islet transplantation with immunoisolation. *Diabetes* **41**, 1503–1510.

Lanza, R. P., Beyer, A. M., and Chick, W. L. (1994). Xenogenic humoral responses to islets transplanted in biohybrid diffusion chambers. *Transplantation* **57**, 1371–1375.

Lechner, A., Nolan, A. L., Blacken, R. A., and Habener, J. F. (2005). Redifferentiation of insulin-secreting cells after *in vitro* expansion of adult human pancreatic islet tissue. *Biochem. Biophys. Res. Commun.* **327**, 581–588.

Lee, H. C., Kim, S. J., Kim, K. S., Shin, H. C., and Yoon, J. W. (2000). Remission in models of type 1 diabetes by gene therapy using a single-chain insulin analogue. *Nature* **408**, 483–488.

Lu, Y.-C., Sternini, C., Rozengurt, E., and Zhokova, E. (2005). Release of transgenic human insulin from gastric G cells: a novel approach for the amelioration of diabetes. *Endocrinology* **146**, 2610–2619.

Lum, Z. P., Krestow, M., Tai, I. T., Vacek, I., and Sun, A. M. (1992). Xenografts of rat islets into diabetic mice. An evaluation of new smaller capsules. *Transplantation* **53**, 1180–1183.

Lumelsky, N., Blondel, O., Laeng, P., Velasco, I., Ravin, R., and McKay, R. (2001). Differentiation of embryonic stem cells to insulin-secreting structures similar to pancreatic islets. *Science* **292**, 1389–1394.

Lundberg, P., and Kuchel, P. W. (1997). Diffusion of solutes in agarose and alginate gels: 1H and 23Na PFGSE and 23Na TQF NMR studies. *Magn. Reson. Med.* **37**, 44–52.

Mamujee, S. N., Zhou, D., Wheeler, M. B., Vacek, I., and Sun, A. M. (1997). Evaluation of immunoisolated insulin-secreting beta TC6-F7 cells as a bioartificial pancreas. *Ann. Transplant.* **2**, 27–32.

May, M. H., and Sefton, M. V. (1999). Conformal coating of small particles and cell aggregates at a liquid–liquid interface. *Ann. N.Y. Acad. Sci.* **875**, 126–134.

Mikos, A., Papadaki, M., Kouvroukoglou, S., Ishaug, S., and Thomson, R. (1994). Mini-review: islet transplantation to create a bioartificial pancreas. *Biotechnol. Bioeng.* **43**, 673–677.

Miyazaki, J., Araki, K., Yamato, E., Ikegami, H., Asano, T., Shibasaki, Y., Oka, Y., and Yamamura, K. (1990). Establishment of a pancreatic beta cell line that retains glucose-inducible insulin secretion: special reference to expression of glucose transporter isoforms. *Endocrinology* **127**, 126–132.

Moore, H. P., Walker, M. D., Lee, F., and Kelly, R. B. (1983). Expressing a human proinsulin cDNA in a mouse ACTH-secreting cell. Intracellular storage, proteolytic processing, and secretion on stimulation. *Cell* **35**, 531–538.

Mukherjee, N., Chen, Z., Sambanis, A., and Song, Y. (2005). Effects of cryopreservation on cell viability and insulin secretion in a model tissue-engineered pancreatic substitute (TEPS). *Cell. Transplant.* **14**, 449–456.

Narushima, M., Kobayashi, N., Okitsu, T., Tanaka, Y., Li, S. A., Chen, Y., Miki, A., Tanaka, K., Nakaji, S., Takei, K., Gutierrez, A. S., Rivas-Carrillo, J. D., Navarro-Alvarez, N., Jun, H. S., Westerman, K. A., Noguchi, H., Lakey, J. R., Leboulch, P., Tanaka, N., and Yoon, J. W. (2005). A human beta-cell line for transplantation therapy to control type 1 diabetes. *Nat. Biotechnol.* **23**, 1274–1282.

Nerem, R. M., and Sambanis, A. (1995). Tissue engineering: from biology to biological substitutes. *Tissue Eng.* **1**, 3–13.

O'Shea, G. M., Goosen, M. F., and Sun, A. M. (1984). Prolonged survival of transplanted islets of Langerhans encapsulated in a biocompatible membrane. *Biochim. Biophys. Acta* **804**, 133–136.

Olson, D. E., Paveglio, S. A., Huey, P. U., Porter, M. H., and Thule, P. M. (2003). Glucose-responsive hepatic insulin gene therapy of spontaneously diabetic BB/Wor rats. *Hum. Gene Ther.* **14**, 1401–1413.

Ouziel-Yahalom, L., Zalzman, M., Anker-Kitai, L., Knoller, S., Bar, Y., Glandt, M., Herold, K., and Efrat, S. (2006). Expansion and redifferentiation of adult human pancreatic islet cells. *Biochem. Biophys. Res. Commun.* **341**, 291–298.

Papas, K. K., Long, R. C. Jr., Constantinidis, I., and Sambanis, A. (1996). Effects of oxygen on metabolic and secretory activities of beta TC3 cells. *Biochim. Biophys. Acta* **1291**, 163–166.

Papas, K. K., Long, R. C., Jr., Sambanis, A., and Constantinidis, I. (1999a). Development of a bioartificial pancreas: I. Long-term propagation and basal and induced secretion from entrapped betaTC3 cell cultures. *Biotechnol. Bioeng.* **66**, 219–230.

Papas, K. K., Long, R. C., Jr., Sambanis, A., and Constantinidis, I. (1999b). Development of a bioartificial pancreas: II. Effects of oxygen on long-term entrapped betaTC3 cell cultures. *Biotechnol. Bioeng.* **66**, 231–237.

Pileggi, A., Molano, R. D., Ricordi, C., Zahr, E., Collins, J., Valdes, R., and Inverardi, L. (2006). Reversal of diabetes by pancreatic islet transplantation into a subcutaneous, neovascularized device. *Transplantation* **81**, 1318–1324.

Prevost, P., Flori, S., Collier, C., Muscat, E., and Rolland, E. (1997). Application of AN69 hydrogel to islet encapsulation. Evaluation in streptozotocin-induced diabetic rat model. *Ann. N.Y. Acad. Sci.* **831**, 344–349.

Prokop, A. (2001). Bioartificial pancreas: materials, devices, function, and limitations. *Diabetes Technol. Ther.* **3**, 431–449.

Prokop, A., Kozlov, E., Nun Non, S., Dikov, M. M., Sephel, G. C., Whitsitt, J. S., and Davidson, J. M. (2001). Towards retrievable vascularized bioartificial pancreas: induction and long-lasting stability of polymeric mesh implant vascularized with the help of acidic and basic fibroblast growth factors and hydrogel coating. *Diabetes Technol. Ther.* **3**, 245–261.

Rajagopal, J., Anderson, W. J., Kume, S., Martinez, O. I., and Melton, D. A. (2003). Insulin staining of ES cell progeny from insulin uptake. *Science* **299**, 363.

Ramiya, V. K., Maraist, M., Arfors, K. E., Schatz, D. A., Peck, A. B., and Cornelius, J. G. (2000). Reversal of insulin-dependent diabetes using islets generated *in vitro* from pancreatic stem cells. *Nat. Med.* **6**, 278–282.

Rayat, G. R., Rajotte, R. V., Elliott, J. F., and Korbutt, G. S. (1998). Expression of Gal alpha(1,3)gal on neonatal porcine islet beta-cells and susceptibility to human antibody/complement lysis. *Diabetes* **47**, 1406–1411.

Sakurai, T., Satake, A., Nagata, N., Gu, Y., Hiura, A., Doo-Hoon, K., Hori, H., Tabata, Y., Sumi, S., and Inoue, K. (2003). The development of new immunoisolatory devices possessing the ability to induce neovascularization. *Cell Transplant.* **12**, 527–535.

Sambanis, A. (2000). Engineering challenges in the development of an encapsulated cell system for treatment of type 1 diabetes. *Diabetes Technol. Therap.* **2**, 81–89.

Sambanis, A., and Tan, S. A. (1999). Quantitative modeling of limitations caused by diffusion. *In* "Methods in Molecular Medicine, Vol. 18: Tissue Engineering Methods and Protocols" (J. R. Morgan and M. L. Yarmush, eds.). Humana Press, Totowa, NJ.

Sambanis, A., Stephanopoulos, G., Sinskey, A. J., and Lodish, H. F. (1990). Use of regulated secretion in protein production from animal cells: an evaluation with the AtT-20 model cell line. *Biotechnol. Bioeng.* **35**, 771–780.

Sambanis, A., Tang, S.-C., Cheng, S.-Y., Stabler, C. L., Long, R. C. J., and Constantinidis, I. (2002). Core technologies in tissue engineering and their application to the bioartificial pancreas. *In* "Tissue Engineering for Therapeutic Use" (Y. Ikada, Y. Umakoshi and T. Hotta, eds.), pp. 5–18. Elsevier, Boston.

Schirra, J., Katschinski, M., Weidmann, C., Schafer, T., Wank, U., Arnold, R., and Goke, B. (1996). Gastric emptying and release of incretin hormones after glucose ingestion in humans. *J. Clin. Invest.* **97**, 92–103.

Sefton, M., and Kharlip, L. (1994). Insulin release from rat pancreatic islets microencapsulated in a HEMA-MMA polyacrylate. *In* "Pancreatic Islet Transplantation. Volume III: Immunoisolation of Pancreatic Islets" (R. Lanza and W. Chick, eds.). RG Landes, Georgetown, TX.

Sefton, M. V., May, M. H., Lahooti, S., and Babensee, J. E. (2000). Making microencapsulation work: conformal coating, immobilization gels and *in vivo* performance. *J. Controlled Release* **65**, 173–186.

Shapiro, A. M., Ryan, E. A., and Lakey, J. R. (2001a). Clinical islet transplant—state of the art. *Transplant. Proc.* **33**, 3502–3503.

Shapiro, A. M., Ryan, E. A., and Lakey, J. R. (2001b). Diabetes. Islet cell transplantation. *Lancet* **358** (Suppl), S21.

Shapiro, A. M., Ryan, E. A., and Lakey, J. R. (2001c). Pancreatic islet transplantation in the treatment of diabetes mellitus. *Best Pract. Res. Clin. Endocrinol. Metab.* **15**, 241–264.

Simpson, N. E., Khokhlova, N., Oca-Cossio, J. A., McFarlane, S. S., Simpson, C. P., and Constantinidis, I. (2005). Effects of growth regulation on conditionally transformed alginate-entrapped insulin secreting cell lines *in vitro*. *Biomaterials* **26**, 4633–4641.

Song, Y. C., Chen, Z. Z., Mukherjee, N., Lightfoot, F. G., Taylor, M. J., Brockbank, K. G., and Sambanis, A. (2005). Vitrification of tissue-engineered pancreatic substitute. *Transplant. Proc.* **37**, 253–255.

Soon-Shiong, P., Feldman, E., Nelson, R., Heintz, R., Merideth, N., Sandford, P., Zheng, T., and Komtebedde, J. (1992). Long-term reversal of diabetes in the large animal model by encapsulated islet transplantation. *Transplant. Proc.* **24**, 2946–2947.

Soon-Shiong, P., Heintz, R. E., Merideth, N., Yao, Q. X., Yao, Z., Zheng, T., Murphy, M., Moloney, M. K., Schmehl, M., Harris, M., *et al.* (1994). Insulin independence in a type 1 diabetic patient after encapsulated islet transplantation. *Lancet* **343**, 950–951.

Soria, B., Roche, E., Berna, G., Leon-Quinto, T., Reig, J. A., and Martin, F. (2000). Insulin-secreting cells derived from embryonic stem cells normalize glycemia in streptozotocin-induced diabetic mice. *Diabetes* **49**, 157–162.

Stabler, C., Wilks, K., Sambanis, A., and Constantinidis, I. (2001). The effects of alginate composition on encapsulated betaTC3 cells. *Biomaterials* **22**, 1301–1310.

Stabler, C. L., Long, R. C., Jr., Constantinidis, I., and Sambanis, A. (2005). *In vivo* noninvasive monitoring of a tissue-engineered construct using 1H-NMR spectroscopy. *Cell. Transplant.* **14**, 139–149.

Sullivan, S. J., Maki, T., Borland, K. M., Mahoney, M. D., Solomon, B. A., Muller, T. E., Monaco, A. P., and Chick, W. L. (1991). Biohybrid artificial pancreas: long-term implantation studies in diabetic, pancreatectomized dogs. *Science* **252**, 718–721.

Sun, A. M., Vacek, I., Sun, Y. L., Ma, X., and Zhou, D. (1992). *In vitro* and *in vivo* evaluation of microencapsulated porcine islets. *Asaio J.* **38**, 125–127.

Tang, S.-C., and Sambanis, A. (2003a). Preproinsulin mRNA engineering and its application to the regulation of insulin secretion from human hepatomas. *FEBS Lett.* **537**, 193–197.

Tang, S. C., and Sambanis, A. (2003b). Development of genetically engineered human intestinal cells for regulated insulin secretion using rAAV-mediated gene transfer. *Biochem. Biophys. Res. Commun.* **303**, 645–652.

Tang, S. C., and Sambanis, A. (2004). Differential rAAV2 transduction efficiencies and insulin secretion profiles in pure and coculture models of human enteroendocrine L-cells and enterocytes. *J. Gene Med.* **6**, 1003–1013.

Thule, P. M., and Liu, J. M. (2000). Regulated hepatic insulin gene therapy of STZ-diabetic rats. *Gene Ther.* **7**, 1744–1752.

Thule, P. M., Liu, J., and Phillips, L. S. (2000). Glucose-regulated production of human insulin in rat hepatocytes. *Gene Ther.* **7**, 205–214.

Thule, P. M., Campbell, A. G., Kleinhenz, D. J., Olson, D. E., Boutwell, J. J., Sutliff, R. L., and Hart, C. M. (2006). Hepatic insulin gene therapy prevents deterioration of vascular function and improves adipocytokine profile in STZ-diabetic rats. *Am. J. Physiol. Endocrinol. Metab.* **290**, E114–E122.

Todorov, I., Omori, K., Pascual, M., Rawson, J., Nair, I., Valiente, L., Vuong, T., Matsuda, T., Orr, C., Ferreri, K., Smith, C. V., Kandeel, F., and Mullen, Y. (2006). Generation of human islets through expansion and differentiation of non-islet pancreatic cells discarded (pancreatic discard) after islet isolation. *Pancreas* **32**, 130–138.

Tziampazis, E., and Sambanis, A. (1995). Tissue engineering of a bioartificial pancreas: modeling the cell environment and device function. *Biotechnol. Prog.* **11**, 115–126.

Wang, S., Liu, J., Li, L., and Wice, B. M. (2004). Individual subtypes of enteroendocrine cells in the mouse small intestine exhibit unique patterns of inositol 1,4,5-trisphosphate receptor expression. *J. Histochem. Cytochem.* **52**, 53–63.

Wiegand, F., Kroncke, K. D., and Kolb-Bachofen, V. (1993). Macrophage-generated nitric oxide as cytotoxic factor in destruction of alginate-encapsulated islets. Protection by arginine analogs and/or coencapsulated erythrocytes. *Transplantation* **56**, 1206–1212.

Wu, H., Avgoustiniatos, E. S., Swette, L., Bonner-Weir, S., Weir, G. C., and Colton, C. K. (1999). *In situ* electrochemical oxygen generation with an immunoisolation device. *Ann. N.Y. Acad. Sci.* **875**, 105–125.

Yanagita, M., Nakayama, K., and Takeuchi, T. (1992). Processing of mutated proinsulin with tetrabasic cleavage sites to bioactive insulin in the nonendocrine cell line, COS-7. *FEBS Lett.* **311**, 55–59.

Yanagita, M., Hoshino, H., Nakayama, K., and Takeuchi, T. (1993). Processing of mutated proinsulin with tetrabasic cleavage sites to mature insulin reflects the expression of furin in nonendocrine cell lines. *Endocrinology* **133**, 639–644.

# Chapter Forty-Three

# Engineering Pancreatic Beta-Cells

*Hee-Sook Jun and Ji-Won Yoon*

## I. INTRODUCTION

The use of islet transplantation as a treatment for diabetes has been hampered by the limited availability of human islets; therefore, new sources of insulin-producing cells are needed. Expansion of beta-cells by the generation of reversibly immortalized beta-cells and creation of insulin-producing cells by exogenous expression of insulin in non-beta-cells have been investigated as new sources of beta-cells. Recently, embryonic and adult stem cells or pancreatic progenitor cells have been engineered to differentiate into insulin-producing cells, demonstrating the possible use of these cells for beta-cell replacement. Despite significant progress, further studies are needed to generate truly functional insulin-producing cells. In addition, the engineering of beta-cells to protect them from immune attack and to improve viability has been tried. Although the usefulness of engineered beta-cells has yet to be clinically proven, studies utilizing different engineering strategies and careful analysis of the resulting insulin-producing cells may offer potential methods to cure diabetes.

Diabetes mellitus is a metabolic disease characterized by uncontrolled hyperglycemia, which results in long-term clinical problems, including retinopathy, neuropathy, nephropathy, and heart disease. Diabetes affects over 150 million people worldwide and is considered an epidemic of the 21st century. Blood glucose homeostasis is controlled by endocrine beta-cells, located in the islets of Langerhans in the pancreas. When the concentration of blood glucose rises after a meal, insulin is produced and released from beta-cells. Insulin then induces glucose uptake by cells in the body and converts glucose to glycogen in the liver. When blood glucose concentration becomes low, glycogen is broken down to glucose in the liver and glucose is released into the blood.

There are two major forms of diabetes: type 1 diabetes, also known as insulin-dependent diabetes mellitus, and type 2 diabetes, also known as non-insulin-dependent diabetes mellitus. Both types are thought to result from a reduction in the number of insulin-producing beta-cells and deficits in beta-cell function. In type 1 diabetes, beta-cells are destroyed by autoimmune responses, resulting in a lack of insulin (reviewed in Adorini *et al.*, 2002; Yoon and Jun, 2005). In type 2 diabetes, both inadequate beta-cell function and insulin resistance of peripheral tissues contribute to the development of hyperglycemia, leading to eventual reduction in the number of beta-cells (reviewed in LeRoith, 2002). Intensive exogenous insulin therapy has been used for the treatment of type 1 diabetes, but it does not restore the tight control of blood glucose levels or completely prevent the development of complications. In addition, multiple daily injections are cumbersome and sometimes cause potentially life-threatening hypoglycemia. Islet transplantation has been considered an alternative and safe method for the treatment of diabetes (reviewed in Hatipoglu *et al.*, 2005).

With the improvement of islet isolation techniques, the success rate for independence from exogenous insulin is increasing. However, the lack of sufficient islets to meet the demands of patients and the side effects of immunosuppressive drugs that are required to prevent alloimmune and autoimmune attack against islet grafts are the major limitations of islet transplantation. Therefore, various alternative sources of insulin-producing cells are being investigated to provide a sufficient supply for the treatment of type 1 diabetes.

In this chapter, we discuss the use of cell engineering to produce and expand insulin-producing beta-cells; to create insulin-producing cells from non-beta-cells, embryonic stem cells, and adult stem cells; and to improve islet graft survival. Due to the publisher's restrictions, we are unable to cite all the references for primary data.

## II. ENGINEERING TO GENERATE INSULIN-PRODUCING CELLS

### Engineering Pancreatic Beta-Cells

The pancreas is composed of endocrine and exocrine tissues. The endocrine pancreas occupies less than 5% of the pancreatic tissue mass and is composed of cell clusters called the *islets of Langerhans*. The islets of Langerhans contain insulin-producing beta-cells (about 80% of cells in the islets), glucagon-producing alpha-cells, somatostatin-producing delta-cells, and pancreatic polypeptide-producing cells. The exocrine pancreas occupies more than 95% of the pancreas and is composed of ascinar and ductal cells, which produce digestive enzymes. The beta-cell mass is dynamic and increases in response to environmental changes such as pancreatic injury and physiological changes such as insulin resistance. In addition, mature beta-cells can replicate throughout life, although at a low level.

One approach to produce beta-cells for replacement therapy is to expand mature beta-cells *in vitro*. However, because mature beta-cells have limited proliferative capacity in culture, the expression of oncogenes has been tried as a method to establish beta-cell lines. The expression of simian virus (SV) 40 large T antigen in beta-cells under the control of the tet-on and tet-off regulatory system in transgenic mice resulted in a stable beta-cell line that could be expanded *in vitro*. These cells produced less insulin in the transformed state when T antigen was expressed, but insulin production increased after growth was arrested by cessation of T antigen expression, and insulin secretion was regulated as in normal mouse islets. When these cells were transplanted into streptozotocin-induced diabetic mice, the mice became normoglycemic, and normoglycemia was maintained for a prolonged time, without any treatment to prevent oncogene expression (Milo-Landesman *et al.*, 2001). In addition to beta-cell expansion, cell engineering has been used to improve beta-cell function. Rat insulinoma

cells that showed decreased glucose-responsive insulin secretion were transfected with a plasmid encoding a mutated form of GLP-1 that is resistant to the degrading enzyme dipeptidyl-peptidase IV. These engineered cells had increased insulin secretion in response to glucose, as compared with untransfected control cells (Islam *et al.*, 2005).

Expansion of human primary pancreatic islet cells has also been tried. Primary adult islet cells could be stimulated to divide when grown on an extracellular matrix in the presence of hepatocyte growth factor/scatter factor, but growth was arrested after 10–15 cell divisions, due to cellular senescence (Beattie *et al.*, 1999). Transformation of adult human pancreatic islets with a retroviral vector expressing SV40 large T antigen and H-ras$^{Val\ 12}$ oncogenes resulted in extended life span, but eventually the cells entered a crisis phase followed by altered morphology, lack of proliferation, and cell death, suggesting that immortalization of human beta-cells is more difficult than that of rodent beta-cells. However, introduction of human telomerase reverse transcriptase (hTERT) resulted in successful immortalization (Halvorsen *et al.*, 1999), because human cells do not express telomerase. This immortalized cell line, βlox5, initially expressed low levels of insulin, but insulin production subsequently fell to undetectable levels as a result of the loss of expression of key insulin gene transcription factors. A combination of the introduction of a beta-cell transcription factor (Pdx-1), treatment with exendin-4 (a glucagon-like peptide-1 [GLP-1] homolog), and cell–cell contact was required to recover beta-cell differentiated function and glucose-responsive insulin production (de la Tour *et al.*, 2001). However, Pdx-1 expression in this cell line resulted in a significant decrease in the growth rate of the cells. When streptozotocin-induced diabetic animals were transplanted with these Pdx-1-expressing cells, substantial levels of circulating human C-peptide were detected and diabetes was remitted. However, 10% of the animals developed tumors, even though the oncogenes and hTERT gene had been floxed by loxP sites so that they could be deleted by expression of Cre recombinase. This suggests that the Cre-expressing adenovirus and/or Cre-mediated deletion of the oncogenes was inefficient (de la Tour *et al.*, 2001).

The limitations of previously engineered beta-cell lines point to a need for a human beta-cell line that is functionally equivalent to primary beta-cells, can be expanded indefinitely, and can be rendered nontumorigenic. In another approach to establish a reversibly immortalized human beta-cell line, human islets were transduced with a combination of retroviral vectors expressing SV40 T antigen, hTERT, and enhanced green fluorescent protein to immortalize and mark terminally differentiated pancreatic beta-cells. These genes were floxed by loxP sites to allow excision of the immortalizing genes. Among 271 clones screened for tumorigenicity, 253 clones were selected for further study, and only one of these (NAKT-15) expressed insulin and the necessary beta-cell transcription factors, such as Isl-1, Pax-

6, Nkx6.1, Pdx-1, prohormone convertases, and secretory granule proteins. Addition of factors that enhance insulin expression and secretion during culture of the beta-cell line, such as troglitazone, a peroxisome proliferator-activated receptor-γ activator, and nicotinamide, helped to maintain the function of beta-cells, and culture of these cells on Matrigel matrix facilitated aggregate formation. Removal of the immortalizing genes by Cre recombinase expression stopped cell proliferation and increased the expression of beta-cell-specific transcription factors, resulting in reversion of the cells. These reverted NAKT-15 cells were functionally similar to normal human islets with respect to insulin secretion in response to glucose and nonglucose secretagogues, although the insulin content and amount of secreted insulin were lower than for human islets. However, NAKT-15 cells were able to remit diabetes and clear exogenous glucose when transplanted into diabetic severe combined immunodeficiency (SCID) mice. The insulin content of these cells was higher *in vivo* than *in vitro*, suggesting that the microenvironment may enhance cellular differentiation (Narushima *et al.*, 2005).

For clinical application of reversibly immortalized human beta-cells, safety issues, particularly tumorigenicity, should be considered. Reducing or eliminating tumorigenicity may be possible by using multiple selection procedures. In the case of NAKT-15 cells, nontumorigenic clones were first selected by screening for tumor formation in SCID mice. After infection of Cre-expressing adenovirus to remove the SV40 T antigen and hTERT, SV40T-negative cells were selected in the presence of a neomycin analog (the neomycin-resistance gene was positioned to be expressed after the loxP-flanked genes were deleted), and hTERT-negative cells were selected by purification of enhanced green fluorescent protein-negative cells. Finally, SV40T/hTERT-negative cells were selected by the addition of ganciclover, because the cells had been transduced with a suicide gene, herpes simplex thymidine kinase, which renders them susceptible to ganciclover. These multiple selection procedures resulted in no tumor development in SCID mice when reverted NAKT-15 cells were transplanted (Narushima *et al.*, 2005), although the possibility of tumorigenesis could not be completely eliminated. Nevertheless, there are advantages of reversibly immortalized human beta-cells as compared with primary beta-cells. They can be easily expanded to obtain sufficient cells for transplantation and genetically manipulated *in vitro* prior to transplantation, for example, to confer resistance to immune attack.

Establishment of insulin-producing beta-cell lines by reversible immortalization of primary islets is a promising approach for replacing insulin injections, for a beta-cell line can provide an abundant source of beta-cells for transplantation. In addition, beta-cell lines can be genetically manipulated to improve their function and survival. However, the functionality of the cell lines and safety issues remain to be further studied.

## Engineering Surrogate Beta-Cells

Non-beta-cells that are genetically engineered to produce insulin may have an advantage over intact islets or engineered beta-cells for transplantation therapy, because non-beta-cells should not be recognized by beta-cell-specific autoimmune responses. Pancreatic beta-cells have unique characteristics specific to the production of insulin, such as specific peptidases, glucose-sensing systems, and secretory granules that can release insulin promptly by exocytosis in response to extracellular glucose levels. Therefore, the ideal target cell to engineer for insulin production would be non-beta-cells possessing similar characteristics. A variety of cell types, including fibroblasts, hepatocytes, neuroendocrine cells, and muscle cells, have been engineered to produce insulin, with varying degrees of success (reviewed in Xu *et al.*, 2003; Yoon and Jun, 2002).

Neuroendocrine cells have received considerable attention because they have characteristics similar to those of beta-cells and contain components of the regulated secretory pathway, including prohormone convertases 2 and 3 and secretory granules. A mouse corticotrophic cell line derived from the anterior pituitary, AtT20, expressed active insulin after transfection with the insulin gene under the control of a viral or metallothionein promoter but lacked glucose responsiveness. Cotransfection of genes encoding glucose transporter (GLUT)2 and glucokinase conferred glucose-responsive insulin secretion in insulin-expressing AtT20 cells. Transgenic expression of insulin in the intermediate lobe of the pituitary of nonobese diabetic (NOD) mice under the control of the pro-opiomelanocortin promoter resulted in the production of biologically active insulin. Transplantation of this insulin-producing pituitary tissue into diabetic NOD mice restored normoglycemia, but insulin secretion was not properly regulated by glucose. Engineering primary rat pituitary cells to coexpress GLP-1 receptor and human insulin resulted in GLP-1-induced insulin secretion (Wu *et al.*, 2003).

Intestinal K-cells have been explored as possible surrogate beta-cells. K-cells are endocrine cells located in the gut that secrete the hormone glucose-dependent insulinotropic polypeptide (GIP), which facilitates insulin release after a meal. K-cells are also glucose responsive, have exocytotic mechanisms, and contain the necessary enzymes for processing proinsulin to insulin. A murine intestinal cell line containing K-cells transfected with human insulin DNA cloned under the control of the GIP promoter produced biologically active insulin in response to glucose, and transgenic mice expressing the human proinsulin gene under the GIP promoter were protected from diabetes after treatment with streptozotocin (Cheung *et al.*, 2000). These results suggest that K-cells may have great potential as surrogate beta-cells.

The strategy of engineering hepatocytes to produce insulin has been widely studied (Nett *et al.*, 2003). Hepatocytes have advantages for engineering as insulin-producing cells because they express components of a glucose-sensing system somewhat similar to that in pancreatic beta-cells, such as GLUT2 and glucokinase. In addition, there are several hepatocyte-specific gene promoters that respond to changes in glucose concentrations. The L-type pyruvate kinase promoter (LPK) and Spot14 promoter have been investigated as regulatory elements for glucose-responsive insulin production in liver. Using a chimeric promoter composed of three copies of the stimulatory glucose-responsive element from the LPK promoter and an inhibitory responsive element from the insulin-like growth factor–binding protein-1 basal promoter, the expression of a modified human proinsulin gene was stimulated by glucose and inhibited by insulin in hepatocytes. Engineering of rat hepatoma cells to express insulin under the control of the glucose-6-phosphatase promoter resulted in the stimulation of insulin production by glucose and self-limitation by insulin. However, insulin expression by the glucose-6-phosphatase promoter was low because of negative feedback by the produced insulin. It was recently reported that human hepatoma cells transduced with a furin-cleavable human preproinsulin gene under the control of the GLUT2 promoter expressed insulin in response to glucose (Burkhardt *et al.*, 2005).

A drawback for the regulation of insulin production by glucose-responsive promoters in hepatocytes is slow as compared with the rapid release by exocytosis from beta-cells. Because a longer period of time is required for transcriptional regulation to change the plasma levels of insulin in response to changes in blood glucose, hypoglycemia may occur. Therefore, the development of systems that mimic insulin secretory dynamics is required. Strategies that utilize synthetic promoters composed of multiple copies of glucose-responsive elements for the induction of high levels of insulin expression, insulin-sensitive elements for feedback regulation, and methods to control of the half-life of insulin mRNA so that it rapidly degrades may make it possible to mimic insulin production in a glucose-responsive fashion in non-beta-cells.

Another consideration is that most non-beta-cells do not have the appropriate endoproteases to convert proinsulin to insulin or secretory granules from which insulin can be rapidly released in response to physiological stimuli. One approach is the mutation of the proinsulin gene so that it can be cleaved and converted to insulin by the protease furin, which is expressed in a wide variety of cells. Another approach is the development of a single-chain insulin analog, which shows insulin activity without the requirement for processing. Artificially regulated insulin secretion in non-beta-cells has been tried by expressing insulin as a fusion protein containing an aggregation domain, which accumulates in the endoplasmic reticulum and is secreted when a drug that induces disaggregation is administered.

Although engineering somatic non-beta-cells to produce insulin is a very attractive method, no method has yet succeeded in imitating normal beta-cells regarding the rapid and tight regulation of glucose within a narrow physiological range. Improvements, including better control of glucose-responsive transcription of transgenic insulin mRNA and artificial secretory systems, provide hope for the potential use of insulin-producing non-beta-cells to cure diabetes.

## Engineering Stem and Progenitor Cells

An exciting advance in the last few years is the development of cell therapy strategies using stem cells. Stem cells are characterized by the ability to proliferate extensively and differentiate into one or more specialized cell types. Both embryonic and adult stem cells have been investigated as alternative sources for the generation of insulin-producing pancreatic islets. Although spontaneous differentiation of beta-cells from stem cells can be observed, engineering of stem cells for forced expression of key beta-cell or endocrine differentiation factors should be more efficient for driving beta-cell differentiation.

### Engineering Embryonic Stem Cells

In principle, embryonic stem (ES) cells have the potential to generate unlimited quantities of insulin-producing cells. ES cells can be expanded indefinitely in the undifferentiated state and differentiated into functional beta-cells. However, generation of fully differentiated beta-cells from ES cells has been difficult and controversial. Beta-cell differentiation from ES cells as determined on the basis of immunohistochemical evidence alone has been questioned, because insulin immunoreactivity can also result from insulin absorption from the medium as well as from genuine beta-cell differentiation. Therefore, these types of results should be interpreted with caution.

Some promising results have been reported for the differentiation of insulin-producing cells from mouse and human ES cells (reviewed in Bonner-Weir and Weir, 2005; Jun and Yoon, 2005; Montanya, 2004; Stoffel *et al.*, 2004). Pancreatic endocrine cells, including insulin-producing cells, could be generated from mouse embryonic stem cells by a five-step protocol, including the enrichment of nestin-positive cells from embryoid bodies, and these cells secreted insulin in response to glucose and other insulin secretagogues, such as tolbutamide and carbachol. However, these cells could not remit hyperglycemica when transplanted into diabetic mice. A modified protocol, in which a phosphoinositide kinase inhibitor was added to the medium to inhibit cell proliferation, resulted in improved insulin content and glucose-dependent insulin release. To enrich insulin-producing cells from mouse ES cells, a neomycin-resistance gene regulated by the insulin promoter was transferred to ES cells, which drove differentiation of insulin-secreting cells, and transplantation of these cells restored normoglycemia in streptozotocin-induced diabetic mice. In another report, mouse ES cells were transduced with a plasmid containing the Nkx6.1 promoter gene, followed by a neomycin-resistance gene to select the Nkx6.1-positive cells, and were differentiated in the presence of exogenous differentiating factors. The selected Nkx6.1-positive cells coexpressed insulin and Pdx-1, and transplantation of these cells into streptozotocin-induced diabetic mice resulted in normoglycemia.

Exogenous expression of beta-cell transcription factors has been used as a strategy to drive the differentiation of insulin-producing cells from ES cells. Overexpression of Pax4 in mouse ES cells promoted the differentiation of nestin-positive progenitor and insulin-producing cells, and these cells secreted insulin in response to glucose and normalized blood glucose when transplanted into diabetic mice (Blyszczuk et al., 2003). In the same study, the expression of Pdx-1 did not have a significant effect on the differentiation of insulin-producing cells from ES cells. However, another study demonstrated that the regulated expression of Pdx-1 in a murine ES cell line by the tet-off system enhanced the expression of insulin and other beta-cell transcription factors (Miyazaki et al., 2004).

It was shown that human ES cells can spontaneously differentiate in vitro into insulin-producing beta-cells, evidenced by the secretion of insulin and expression of other beta-cell markers (Assady et al., 2001). Differentiation of insulin-expressing cells from human ES cells was promoted when they were cultured in conditioned medium in the presence of low glucose and fibroblast growth factor, followed by nicotinamide (Segev et al., 2004). A recent report suggested that human ES cells differentiated into beta-cell-like clusters when cotransplanted with mouse dorsal pancreas (Brolen et al., 2005). Although several in vitro studies suggest the possibility of generating insulin-expressing cells from human ES cells, differentiation of truly functional beta-cells from human ES cells has not yet been reported.

Because of their proliferative ability and capacity to differentiate in culture, ES cells have received much attention as a potential source of unlimited quantities of beta-cells for transplantation therapy for diabetes. However, use of ES cells has ethical concerns, and the mechanisms by which ES cells differentiate to produce islets and beta-cells are not well understood. Therefore, further studies are needed to understand the details of the endoderm and beta-cell differentiation process so that an effective protocol for differentiating ES cells into insulin-producing cells can be developed.

## Engineering Adult Stem and Progenitor Cells

As with ES cells, adult stem cells have the potential to differentiate into other cell lineages, but they do not bring the ethical difficulties associated with ES cells. Beta-cell neogenesis in adults has been reported in animal models of experimentally induced pancreatic damage, suggesting the presence of adult stem/progenitor cells. These adult stem/progenitor cells could be potential sources for the production of new insulin-producing cells (reviewed in Jun and Yoon, 2005; Montanya, 2004; Nir and Dor, 2005). Bone marrow, mesenchymal splenocytes, neural stem cells, liver oval stem cells, and pancreatic stem cells have been investigated for their potential to differentiate into insulin-producing cells.

A large body of evidence suggests that the adult pancreatic ducts are the main site of beta-cell progenitors. Throughout life, the islets of Langerhans turn over slowly, and new small islets are continuously generated by differentiation of ductal progenitors (Finegood et al., 1995). It was found that isletlike clusters were generated in vitro from mouse pancreatic ducts and ductal tissue–enriched human pancreatic islets. In addition, multipotent precursor cells clonally identified from pancreatic islets and ductal populations could differentiate into cells with beta-cell function. The expression of the Pdx-1 gene or treatment of ductal cells with Pdx-1 protein increased the number of insulin-positive cells or induced insulin expression. Ectopic expression of neurogenin 3, a critical factor for the development of the endocrine pancreas in humans, in pancreatic ductal cells led to their conversion into insulin-expressing cells. In addition, treatment of human islets containing both ductal and ascinar cells with a combination of epidermal growth factor and gastrin induced neogenesis of islet beta-cells from the ducts and increased the functional beta-cell mass. In addition to ductal cells, exocrine acinar cells and other endocrine cells can generate beta-cells. An alpha-cell line transfected with Pdx-1 expressed insulin when treated with betacellulin. It was shown that treatment of rat exocrine pancreatic cells with epidermal growth factor and leukemia inhibitory factor could induce differentiation into insulin-producing beta-cells (Baeyens et al., 2005). Considerable evidence suggests that beta-cells in the pancreatic islets can be dedifferentiated, expanded, and redifferentiated into beta-cells by inducing the epithelial–mesenchymal transition process (Lechner et al., 2005). Nonendocrine pancreatic epithelial cells also have been reported to differentiate into beta-cells (Hao et al., 2006). These results suggest that pancreatic stem/progenitor cells are the source of new islets.

There is also the possibility of manipulating stem/progenitor cells from other organs to transform into the beta-cell phenotype (reviewed in Montanya, 2004; Nir and Dor, 2005). Although there are controversies regarding the differentiation of bone marrow–derived stem cells into insulin-producing cells, some successful studies have been reported. In vitro differentiation of mouse bone marrow cells resulted in the expression of genes related to pancreatic beta-cell development and function. These differentiated cells released insulin in response to glucose and reversed hyperglycemia when transplanted into diabetic mice. In addition, ectopic expression of key transcription factors of the endocrine pancreas developmental pathway, such as IPF1, HLXB9, and FOXA2, in combination with conditioned media in human bone marrow mesenchymal stem cells differentiated them into insulin-expressing cells (Moriscot *et al.*, 2005).

Because the liver and intestinal epithelium are derived from gut endoderm, as is the pancreas, the generation of islets from both developing and adult liver and intestinal cells has been tried. Rat hepatic oval stem cells could differentiate into insulin-producing isletlike cells when cultured in a high-glucose environment. Fetal human liver progenitor cells and mouse hepatocytes could differentiate into insulin-producing cells when engineered to produce Pdx-1, and transplantation of these cells reversed hyperglycemia in mice. It was reported that adult human liver cells engineered to express Pdx-1 produced insulin and secreted it in a glucose-regulated manner. Transplantation of these engineered cells under the renal capsule of diabetic mice resulted in prolonged reduction of hyperglycemia (Sapir *et al.*, 2005). As well, ectopic islet neogenesis in the liver could be induced by gene therapy with a combination of NeuroD, a transcription factor downstream of Pdx-1, and betacellulin, which reversed diabetes in streptozotocin-treated diabetic mice. Expression of Pdx-1 in a rat enterocyte cell line in combination with betacellulin treatment or coexpression of Isl-1 resulted in the expression of insulin. Treatment of developing as well as adult mouse intestinal cells with GLP-1 induced insulin production, and transplantation of these cells into streptozotocin-induced diabetic mice remitted diabetes. A recent study showed that neural progenitor cells could generate glucose-responsive, insulin-producing cells when exposed *in vitro* to a series of signals for pancreatic islet development (Hori *et al.*, 2005). These results suggest that the controlled differentiation of liver or intestinal cells into insulin-producing cells may provide an alternative source of beta-cells.

The use of adult stem/progenitor cells for generating beta-cells for transplantation therapy appears to be promising, although most of the studies have only been done in animal models. Further studies on the mechanisms for the differentiation of adult stem/progenitor cells into insulin-producing beta-cells and the characterization of the newly generated beta-cells are required before these cells can be considered for clinical application.

## III. ENGINEERING TO IMPROVE ISLET SURVIVAL

A hurdle to overcome for islet transplantation therapy is the rejection and autoimmune attack against the transplanted beta-cells. Immunosuppressive drugs have been used successfully, but they have many side effects. Therefore, it is desirable to develop drug-free strategies for the induction of tolerance to transplanted islet or beta-cells. A variety of approaches to protect islet grafts have been studied, such as bone marrow transplantation, treatment with anti-T-cell agents, and inhibition of activation of antigen-presenting cells. Another approach is engineering islets or beta-cells to express therapeutic genes to improve islet viability and function, such as genes for cytokines, antiapoptotic molecules, antioxidants, immunoregulatory molecules, and growth factors (reviewed in Giannoukakis and Trucco, 2005; Jun and Yoon, 2005; Van Linthout and Madeddu, 2005) (Table 43.1).

With regard to cytokines, introduction of genes for interleukin (IL)-4 or a combination of IL-10 and transforming growth factor-β improved islet graft survival by preventing immune attack in mice. In addition, islets expressing the p40 subunit of IL-12 could maintain normoglycemia when transplanted into diabetic NOD recipients by decreasing interferon-γ production and increasing transforming growth

| Table 43.1. Engineering islets for beta-cell survival | |
|---|---|
| *Strategy* | *Molecules used* |
| Cytokine expression | Erythropoietin |
| | Interleukin (IL)-1 receptor antagonist |
| | IL-4 |
| | IL-10 |
| | IL-12p40 |
| | Transforming growth factor-β |
| Antiapoptotic molecule expression | A20 |
| | Bcl-2 |
| | Bcl-xL |
| | Dominant-negative MyD88 |
| | Fas ligand |
| | Flice-like inhibitory protein |
| | IκB kinase inhibitor |
| | Tumor necrosis factor receptor-immunoglobulin (Ig) |
| Antioxidant molecule expression | Catalase |
| | c-Jun N-terminal kinase inhibitory peptide |
| | Glutathione peroxidase |
| | Heme oxygenase-1 |
| | Manganese superoxide dismutase |
| Immunoregulatory molecule expression | Adenoviral E3 genes |
| | CD40-Ig |
| | Cytotoxic T-lymphocyte antigen-4-Ig |
| | Dipeptide boronic acid |
| | Indoleamine 2,3-dioxygenase |
| Growth factor expression | Hepatocyte growth factor |
| | Insulinlike growth factor-1 |
| | Vascular endothelial growth factor |

factor-β at the transplantation site. Islets engineered to produce IL-1β receptor antagonist were also found to be more resistant to rejection. Adenoviral-mediated gene transfer of erythropoietin, a cytokine that promotes survival, in islets resulted in protection of islets from apoptosis in culture and destruction *in vivo*.

Expression of antiapoptotic molecules such as Bcl-2, Bcl-xL, and A20, which inhibit nuclear factor-κB activation, or an IκB kinase inhibitor was shown to protect from apoptosis. In addition, the expression of soluble human Fas ligand, dominant negative MyD88, flice-like inhibitory protein, or tumor necrosis factor receptor-immunoglobulin (Ig) improved allogeneic islet graft survival. A recent study demonstrated that silencing Fas expression with small interfering RNA in mouse insulinoma cells inhibited Fas-mediated beta-cell apoptosis (Burkhardt *et al.*, 2006).

Pancreatic islets are sensitive to oxidative stress because they produce relatively low amounts of antioxidant enzymes. Thus, expression of antioxidant molecules such as catalase, glutathione perioxidase, and manganese superoxide dismutase in islets or insulinoma cells could protect against oxidative stress and cytokine-induced damage. In addition, expression of heme oxygenase in pancreatic islets protected against IL-1β-induced islet damage. It was also found that delivery of a c-Jun-terminal kinase inhibitory peptide into isolated islets by the protein transduction system prevented apoptosis (Noguchi *et al.*, 2005).

Expression of immunoregulatory molecules that affect T-cell activation and proliferation have been tried. Expression in islets of cytotoxic T-lymphocyte antigen-4-immunoglobulin, which down-regulates T-cell activation, or CD40-Ig, which blocks CD40–CD40 ligand interactions, prolonged allogenic and xenogeneic graft survival. Transplantation of islets overexpressing indoleamine 2,3-dioxygenase prolonged survival in NOD/SCID mice after adoptive transfer of diabetogenic T-cells, probably by inhibiting T-cell proliferation by the depletion of tryptophan at the transplantation site. Similarly, a proteasome inhibitor, dipeptide boronic acid, was found to prevent islet allograft rejection by suppressing the proliferation of T-cells. Expression of adenoviral E3 transgenes in beta-cells was found to prevent islet destruction by autoimmune attack through the inhibition of major histocompatibility complex I expression.

With regard to growth factors, adenoviral-mediated transfer of hepatocyte growth factor resulted in an improved islet transplant outcome in animal models. As well, expression of insulinlike growth factor-1 in human islets prevented IL-1β-induced beta-cell dysfunction and apoptosis. Insufficient revascularization of transplanted islets can deprive them of oxygen and nutrients, contributing to graft failure. Therefore the expression of vascular endothelial growth factor, a key angiogenic molecule, enhanced islet revascularization and improved the long-term survival of murine islets after transplantation into the renal capsule of diabetic mice.

Another strategy to protect islets from immune attack is microencapsulation of islets within synthetic polymers (Kizilel *et al.*, 2005). Encapsulation of islets within a semipermeable membrane, such as alginate-poly-L-lysine-alginate, blocks the passage of larger cells but allows the passage of small molecules, thus conferring protection from autoimmune attack. However, this method has limitations for the long-term survival of islets within the microcapsules because of the lack of biocompatibility, ischemia, and limited protection from cytokine-induced damage. To overcome these limitations, a bioartificial pancreas has been developed, in which blood flows through artificial vessels in close proximity to insulin-producing cells.

A variety of approaches for engineering islets or beta-cells for improved islet graft survival and escape from immune rejection have been successful in animal models. However, the efficacy of these approaches in human diabetic patients remains to be determined.

## IV. VECTORS FOR ENGINEERING ISLETS AND BETA-CELLS

The cells of the pancreas divide very slowly; therefore gene transfer vehicles that can transduce quiescent cells have been used for the delivery of transgenes, such as nonviral plasmids and vectors based on lentivirus, adenovirus, helper-dependent adenovirus, adeno-associated virus (AAV), and herpes simplex virus. In addition, protein transduction using the cell penetrating peptide from HIV-1 transacting protein (reviewed in Becker-Hapak *et al.*, 2001) has been successfully used to engineer islets (Table 43.2). However, the choice of vector needs to be carefully made so that the vector itself does not affect islet function or viability.

Nonviral methods are considered safe, cost effective, and simple to use and do not induce an immune response, but they generally have a lower gene transfer efficacy as compared to viral-mediated gene transfer (reviewed in Nishikawa and Huang, 2001). Nonviral methods for transferring genetic material include the direct injection of DNA, either naked or enclosed in a liposome, electroporation, and the gene gun method. Cationic lipid and polymer-based plasmid delivery has been used to transduce islets, and the expression of cytotoxic T-lymphocyte antigen-4 by biolistically transfected islets improved graft survival.

Viral vectors (reviewed in Walther and Stein, 2000) have been widely used as a method of gene transfer to engineer islets and beta-cell surrogates. Retroviral vectors derived from Moloney murine leukemia virus can carry a gene efficiently and integrate it in a stable manner within the host chromosomal DNA, facilitating long-term expression of the gene. For immortalization of human islets, retroviral vectors expressing oncogenes or telomerase genes have been used (Narushima *et al.*, 2005). Although most retroviral vectors

| Table 43.2. | Vectors used for islet and beta-cell engineering | |
| --- | --- | --- |
| *Vector* | *Advantages* | *Disadvantages* |
| Nonviral plasmid vectors | Easy to produce | Low transduction efficiency |
| | Less toxic compared with viral vectors | Transient expression |
| Retrovirus | Long-term expression | Only infects dividing cells |
| | No expression of viral protein | Limited insertion capacity (8 Kb) |
| | | Random integration into host chromosomal DNA |
| Lentivirus | Stable, long-term expression | Produces low titers of virus |
| | Infects dividing and nondividing cells | Limited insertion capacity (8 Kb) |
| | Nonimmunogenic | Random integration into host chromosomal DNA |
| Adenovirus | Produces high titers of virus | Toxic |
| | High transduction efficiency | Host immune response to viral proteins |
| | Infects dividing and nondividing cells | Short-term expression |
| Gutless adenovirus | Reduced toxicity and prolonged expression compared with adenoviral vector | Helper-virus contamination |
| | Large insertion capacity | |
| Adeno-associated virus | Low immunogenicity, probably nonpathogenic | Produces low titers of virus |
| | | Low transduction efficiency |
| | Broad host range | Limited insertion capacity (4.8 Kb) |
| | Long-term expression | |
| | Infects dividing and nondividing cells | |
| Herpesvirus | Broad host range | Toxic |
| | High insertion capacity | Induces host immune response |
| Protein transduction | No immunogenicity | Short half-life |
| | Transduces many cell types | |

only infect proliferating cells, the lentivirus genus of retroviruses, which includes the human immunodeficiency virus, has all the advantages of Moloney murine leukemia virus–derived retroviral vectors and can infect nondividing as well as dividing cells. Lentiviral vectors have been successfully used to transduce islets with marker proteins (Okitsu *et al.*, 2003).

The adenoviral vector can harbor up to 30 Kb of foreign DNA and can transduce nondividing cells with high efficiency. In addition, a relatively high titer of virus, about $10^{12}$ plaque-forming units/mL, can be produced. The transferred genes are not integrated into the host genome, but remain as nonreplicating extrachromosomal DNA within the nucleus. Although there is no risk of alteration in cellular genotype by insertional mutation, the duration of gene expression may be short, and a strong cellular immune response to the viral proteins and, in some cases, to the transgene may be induced. Adenoviral vectors have been widely used to transduce islets for proof-of-concept experiments *in vitro* and *in vivo*. Although adenoviral vectors are toxic because of *de novo* synthesized viral proteins, islet viability and functional characteristics were not affected when transduced *in vitro*. However, transduction of islets with a high dose of recombinant adenovirus (500 MOI) markedly reduced glucose-stimulated insulin secretion, suggesting that an optimal dose is required to result in efficient transduction without compromising islet function. In general, adenoviral vectors result in transient transgene expression; however, long-term (20-week) expression of the transgene was observed in islets transduced with β-galactosidase and transplanted into syngeneic diabetic mice. It was reported that double-genetic modification of the adenovirus fiber with RGD polylysine motifs significantly reduced toxicity, inflammation, and immune responses (Contreras *et al.*, 2003).

A new generation of adenovirus vectors has been developed that are completely devoid of all viral protein–coding sequences and are therefore less immunogenic and less toxic. Although these gutless viruses require the presence of a helper virus for replication, contamination by the helper virus can be avoided by genetically engineering a conditional defect in the packaging domain of the helper virus or flanking the packing signal with loxP expression sites and encoding Cre recombinase in the supporting cell line. In addition, the gutless vectors are known to have a prolonged expression of the transgene. However, there is no report about islet engineering using these vectors.

AAVs are nonpathogenic, replication-defective parvoviruses that can infect both dividing and nondividing cells. AAVs generally have low immunogenicity; however, the generation of neutralizing antibodies may limit readministration. This problem can be overcome by selective capsid modification of AAV to evade recognition by preexisting antibodies or by direct administration of AAV to the target tissue. The recombinant AAV vector integrates randomly into the host chromosome or may stay in the episomal state. There is a limitation in the size of the DNA that can be inserted (a maximum of 4.8 Kb); however, larger inserts can be split over two vectors and delivered simultaneously, because AAVs tend to form concatemers, although the efficiency of transduction is often reduced (Young *et al.*, 2006). Efficient transduction of islets was achieved using a high dose of AAVs with an improved recombinant AAV purification method, which improved infectious titers and yield. Transduction of islets with AAV5 is more efficient than with AAV2, due to the low number of receptors for AAV2 on islet cells. AAV1 was found to be the most efficient serotype in transducing murine islets (Loiler *et al.*, 2003). However, it was recently demonstrated that intact human and murine islets could be efficiently transduced with a double-stranded AAV2-based vector, and the transduced murine islets showed normal glucose responsiveness and viability (Rehman *et al.*, 2005).

Herpes simplex virus type 1 (HSV-1) has also been used as a viral vector. Based on the persistence of latent herpes virus after infection, HSV-1 is attractive for its efficient infectivity in a wide range of target cells and its ability to infect both dividing and nondividing cells, including islets. Transfection of human islets with Bcl-2 protected beta-cells from cytokine-induced damage. However, the transduction may be unstable, and potential health risks of this vector remain to be determined.

Protein transduction is an emerging technology to deliver therapeutic proteins into cells as an alternative to gene therapy. This method uses peptides that can penetrate the cell membrane, such as antennapedia peptide, the HSV VP22 protein, and human immunodeficiency virus TAT protein transduction domain. The therapeutic molecule is linked to the penetrating peptide as a fusion protein, which is then used to transduce the cell (Becker-Hapak *et al.*, 2001). The protein transduction method is not immunogenic and can transduce a variety of cell types, but it has a short half-life. Delivery of antiapoptotic proteins such as Bcl-xL or anti-oxidant enzymes such as copper-zinc superoxide dismutase and heme oxygenase by protein transduction in human and rodent islets efficiently transduced the islets and improved their viability without affecting islet function (Embury *et al.*, 2001; Mendoza *et al.*, 2005).

## V. CONCLUSION

Engineering beta-cell lines and non-beta-cells, differentiating embryonic and adult stem cells, and transdifferentiating non-beta-cells have been studied as methods to provide new beta-cells for cell therapy for diabetes. Expansion of functional beta-cells by generation of reversibly immortalized human beta-cell lines has been reported, but the techniques have not been clinically proven. Generation of insulin-producing cells from non-beta-cells is an attractive method, but it has yet to achieve tight regulation of glucose-responsive insulin secretion. Differentiation of insulin-producing cells from ES cells and adult stem/progenitor cells is also a promising alternative to produce beta-cells; however, a better understanding of the mechanisms for the differentiation of beta-cells is needed to develop a successful strategy to engineer beta-cells from stem cells. Engineering of islets and beta-cells to improve the survival of islet transplants has also been investigated. Although much progress has been made, engineered beta-cells need to be carefully analyzed for true beta-cell function and possible tumorigenicity. It is hoped that continued research on beta-cell engineering will offer a potential cure for diabetes in the future.

## VI. REFERENCES

Adorini, L., Gregori, S., and Harrison, L. C. (2002). Understanding autoimmune diabetes: insights from mouse models. *Trends Mol. Med.* **8**, 31–38.

Assady, S., Maor, G., Amit, M., Itskovitz-Eldor, J., Skorecki, K. L., and Tzukerman, M. (2001). Insulin production by human embryonic stem cells. *Diabetes* **50**, 1691–1697.

Baeyens, L., De Breuck, S., Lardon, J., Mfopou, J. K., Rooman, I., and Bouwens, L. (2005). *In vitro* generation of insulin-producing beta cells from adult exocrine pancreatic cells. *Diabetologia* **48**, 49–57.

Beattie, G. M., Itkin-Ansari, P., Cirulli, V., Leibowitz, G., Lopez, A. D., Bossie, S., Mally, M. I., Levine, F., and Hayek, A. (1999). Sustained proliferation of PDX-1+ cells derived from human islets. *Diabetes* **48**, 1013–1019.

Becker-Hapak, M., McAllister, S. S., and Dowdy, S. F. (2001). TAT-mediated protein transduction into mammalian cells. *Methods* **24**, 247–256.

Blyszczuk, P., Czyz, J., Kania, G., Wagner, M., Roll, U., St-Onge, L., and Wobus, A. M. (2003). Expression of Pax4 in embryonic stem cells promotes differentiation of nestin-positive progenitor and insulin-producing cells. *Proc. Natl. Acad. Sci. U.S.A.* **100**, 998–1003.

Bonner-Weir, S., and Weir, G. C. (2005). New sources of pancreatic beta-cells. *Nat. Biotechnol.* **23**, 857–861.

Brolen, G. K., Heins, N., Edsbagge, J., and Semb, H. (2005). Signals from the embryonic mouse pancreas induce differentiation of human embryonic stem cells into insulin-producing beta-cell-like cells. *Diabetes* **54**, 2867–2874.

Burkhardt, B. R., Parker, M. J., Zhang, Y. C., Song, S., Wasserfall, C. H., and Atkinson, M. A. (2005). Glucose transporter-2 (GLUT2) promoter–mediated transgenic insulin production reduces hyperglycemia in diabetic mice. *FEBS Lett.* **579**, 5759–5764.

Burkhardt, B. R., Lyle, R., Qian, K., Arnold, A. S., Cheng, H., Atkinson, M. A., and Zhang, Y. C. (2006). Efficient delivery of siRNA into cytokine-stimulated insulinoma cells silences Fas expression and inhibits Fas-mediated apoptosis. *FEBS Lett.* **580**, 553–560.

Cheung, A. T., Dayanandan, B., Lewis, J. T., Korbutt, G. S., Rajotte, R. V., Bryer-Ash, M., Boylan, M. O., Wolfe, M. M., and Kieffer, T. J. (2000). Glucose-dependent insulin release from genetically engineered K cells. *Science* **290**, 1959–1962.

Contreras, J. L., Wu, H., Smyth, C. A., Eckstein, C. P., Young, C. J., Seki, T., Bilbao, G., Curiel, D. T., and Eckhoff, D. E. (2003). Double genetic modification of adenovirus fiber with RGD polylysine motifs significantly enhances gene transfer to isolated human pancreatic islets. *Transplantation* **76**, 252–261.

de la Tour, D., Halvorsen, T., Demeterco, C., Tyrberg, B., Itkin-Ansari, P., Loy, M., Yoo, S. J., Hao, E., Bossie, S., and Levine, F. (2001). Beta-cell differentiation from a human pancreatic cell line *in vitro* and *in vivo*. *Mol. Endocrinol.* **15**, 476–483.

Embury, J., Klein, D., Pileggi, A., Ribeiro, M., Jayaraman, S., Molano, R. D., Fraker, C., Kenyon, N., Ricordi, C., Inverardi, L., *et al.* (2001). Proteins linked to a protein transduction domain efficiently transduce pancreatic islets. *Diabetes* **50**, 1706–1713.

Finegood, D. T., Scaglia, L., and Bonner-Weir, S. (1995). Dynamics of beta-cell mass in the growing rat pancreas. Estimation with a simple mathematical model. *Diabetes* **44**, 249–256.

Giannoukakis, N., and Trucco, M. (2005). Gene therapy for type 1 diabetes. *Am. J. Ther.* **12**, 512–528.

Halvorsen, T. L., Leibowitz, G., and Levine, F. (1999). Telomerase activity is sufficient to allow transformed cells to escape from crisis. *Mol. Cell Biol.* **19**, 1864–1870.

Hao, E., Tyrberg, B., Itkin-Ansari, P., Lakey, J. R., Geron, I., Monosov, E. Z., Barcova, M., Mercola, M., and Levine, F. (2006). Beta-cell differentiation from nonendocrine epithelial cells of the adult human pancreas. *Nat. Med.* **12**, 310–316.

Hatipoglu, B., Benedetti, E., and Oberholzer, J. (2005). Islet transplantation: current status and future directions. *Curr. Diab. Rep.* **5**, 311–316.

Hori, Y., Gu, X., Xie, X., and Kim, S. K. (2005). Differentiation of insulin-producing cells from human neural progenitor cells. *PLoS Med.* **2**, e103.

Islam, M. S., Rahman, S. A., Mirzaei, Z., and Islam, K. B. (2005). Engineered beta-cells secreting dipeptidyl peptidase IV–resistant glucagon-like peptide-1 show enhanced glucose responsiveness. *Life Sci.* **76**, 1239–1248.

Jun, H. S., and Yoon, J. W. (2005). Approaches for the cure of type 1 diabetes by cellular and gene therapy. *Curr. Gene Ther.* **5**, 249–262.

Kizilel, S., Garfinkel, M., and Opara, E. (2005). The bioartificial pancreas: progress and challenges. *Diabetes Technol. Ther.* **7**, 968–985.

Lechner, A., Nolan, A. L., Blacken, R. A., and Habener, J. F. (2005). Redifferentiation of insulin-secreting cells after *in vitro* expansion of adult human pancreatic islet tissue. *Biochem. Biophys. Res. Commun.* **327**, 581–588.

LeRoith, D. (2002). Beta-cell dysfunction and insulin resistance in type 2 diabetes: role of metabolic and genetic abnormalities. *Am. J. Med.* **113**(Suppl. 6A), 3S–11S.

Loiler, S. A., Conlon, T. J., Song, S., Tang, Q., Warrington, K. H., Agarwal, A., Kapturczak, M., Li, C., Ricordi, C., Atkinson, M. A., *et al.* (2003). Targeting recombinant adeno-associated virus vectors to enhance gene transfer to pancreatic islets and liver. *Gene Ther.* **10**, 1551–1558.

Mendoza, V., Klein, D., Ichii, H., Ribeiro, M. M., Ricordi, C., Hankeln, T., Burmester, T., and Pastori, R. L. (2005). Protection of islets in culture by delivery of oxygen-binding neuroglobin via protein transduction. *Transplant. Proc.* **37**, 237–240.

Milo-Landesman, D., Surana, M., Berkovich, I., Compagni, A., Christofori, G., Fleischer, N., and Efrat, S. (2001). Correction of hyperglycemia in diabetic mice transplanted with reversibly immortalized pancreatic beta cells controlled by the tet-on regulatory system. *Cell Transplant.* **10**, 645–650.

Miyazaki, S., Yamato, E., and Miyazaki, J. (2004). Regulated expression of pdx-1 promotes *in vitro* differentiation of insulin-producing cells from embryonic stem cells. *Diabetes* **53**, 1030–1037.

Montanya, E. (2004). Islet- and stem-cell-based tissue engineering in diabetes. *Curr. Opin. Biotechnol.* **15**, 435–440.

Moriscot, C., de Fraipont, F., Richard, M. J., Marchand, M., Savatier, P., Bosco, D., Favrot, M., and Benhamou, P. Y. (2005). Human bone marrow mesenchymal stem cells can express insulin and key transcription factors of the endocrine pancreas developmental pathway upon genetic and/or microenvironmental manipulation *in vitro*. *Stem Cells* **23**, 594–603.

Narushima, M., Kobayashi, N., Okitsu, T., Tanaka, Y., Li, S. A., Chen, Y., Miki, A., Tanaka, K., Nakaji, S., Takei, K., *et al.* (2005). A human beta-cell line for transplantation therapy to control type 1 diabetes. *Nat. Biotechnol.* **23**, 1274–1282.

Nett, P. C., Sollinger, H. W., and Alam, T. (2003). Hepatic insulin gene therapy in insulin-dependent diabetes mellitus. *Am. J. Transplant.* **3**, 1197–1203.

Nir, T., and Dor, Y. (2005). How to make pancreatic beta cells — prospects for cell therapy in diabetes. *Curr. Opin. Biotechnol.* **16**, 524–529.

Nishikawa, M., and Huang, L. (2001). Nonviral vectors in the new millennium: delivery barriers in gene transfer. *Hum. Gene Ther.* **12**, 861–870.

Noguchi, H., Nakai, Y., Matsumoto, S., Kawaguchi, M., Ueda, M., Okitsu, T., Iwanaga, Y., Yonekawa, Y., Nagata, H., Minami, K., *et al.* (2005). Cell-permeable peptide of JNK inhibitor prevents islet apoptosis immediately after isolation and improves islet graft function. *Am. J. Transplant.* **5**, 1848–1855.

Okitsu, T., Kobayashi, N., Totsugawa, T., Maruyama, M., Noguchi, H., Watanabe, T., Matsumura, T., Fujiwara, T., and Tanaka, N. (2003). Lentiviral vector–mediated gene delivery into nondividing isolated islet cells. *Transplant. Proc.* **35**, 483.

Rehman, K. K., Wang, Z., Bottino, R., Balamurugan, A. N., Trucco, M., Li, J., Xiao, X., and Robbins, P. D. (2005). Efficient gene delivery to human and rodent islets with double-stranded (ds) AAV-based vectors. *Gene Ther.* **12**, 1313–1323.

Sapir, T., Shternhall, K., Meivar-Levy, I., Blumenfeld, T., Cohen, H., Skutelsky, E., Eventov-Friedman, S., Barshack, I., Goldberg, I., Pri-Chen, S., *et al.* (2005). Cell-replacement therapy for diabetes: generating functional insulin-producing tissue from adult human liver cells. *Proc. Natl. Acad. Sci. U.S.A.* **102**, 7964–7969.

Segev, H., Fishman, B., Ziskind, A., Shulman, M., and Itskovitz-Eldor, J. (2004). Differentiation of human embryonic stem cells into insulin-producing clusters. *Stem Cells* **22**, 265–274.

Stoffel, M., Vallier, L., and Pedersen, R. A. (2004). Navigating the pathway from embryonic stem cells to beta cells. *Semin. Cell Dev. Biol.* **15**, 327–336.

Van Linthout, S., and Madeddu, P. (2005). *Ex vivo* gene transfer for improvement of transplanted pancreatic islet viability and function. *Curr. Pharm. Des.* **11**, 2927–2940.

Walther, W., and Stein, U. (2000). Viral vectors for gene transfer: a review of their use in the treatment of human diseases. *Drugs* **60**, 249–271.

Wu, L., Nicholson, W., Wu, C. Y., Xu, M., McGaha, A., Shiota, M., and Powers, A. C. (2003). Engineering physiologically regulated insulin secretion in non-beta cells by expressing glucagon-like peptide 1 receptor. *Gene Ther.* **10**, 1712–1720.

Xu, R., Li, H., Tse, L. Y., Kung, H. F., Lu, H., and Lam, K. S. (2003). Diabetes gene therapy: potential and challenges. *Curr. Gene Ther.* **3**, 65–82.

Yoon, J. W., and Jun, H. S. (2002). Recent advances in insulin gene therapy for type 1 diabetes. *Trends Mol. Med.* **8**, 62–68.

Yoon, J. W., and Jun, H. S. (2005). Autoimmune destruction of pancreatic beta cells. *Am. J. Ther.* **12**, 580–591.

Young, L. S., Searle, P. F., Onion, D., and Mautner, V. (2006). Viral gene therapy strategies: from basic science to clinical application. *J. Pathol.* **208**, 299–318.

# Chapter Forty-Four

# Thymus and Parathyroid Organogenesis

*Craig Scott Nowell, Ellen Richie, Nancy Ruth Manley, and Catherine Clare Blackburn*

I. Introduction
II. Structure and Morphology of the Thymus
III. In Vitro T-Cell Differentiation
IV. Thymus Organogenesis

V. Summary
VI. Acknowledgments
VII. References

---

## I. INTRODUCTION

The thymus is the principal site of T-cell development and therefore is of central importance within the immune system: Congenital athymia results in profound immunodeficiency (Flanagan, 1966; Dodson *et al.*, 1969; J. Frank *et al.*, 1999), while perturbed thymic function can lead to autoimmunity. Although highly active in early life, the thymus undergoes premature involution, such that *de novo* T-cell development diminishes significantly with age. This has implications for immune function in the aging population and in clinical procedures such as bone marrow and solid organ transplantation, where thymic function is required for T-cell reconstitution and/or tolerance induction. Interest therefore exists in enhancing immune reconstitution through regenerative or cell therapies for boosting thymus activity *in vivo* or for providing customized *in vitro*–generated T-cell repertoires for adoptive transfer. The success of such strategies is likely to depend on a detailed knowledge of the mechanisms regulating thymus development and homeostasis. Here, we review current understanding of cellular and molecular regulation of thymus organogenesis, focusing on the epithelial component of the thymic stroma, which provides many of the specialist functions required to mediate T-cell differentiation and T-cell repertoire selection.

## II. STRUCTURE AND MORPHOLOGY OF THE THYMUS

The mature thymus is a highly dynamic cellular environment, comprising developing T-lymphocytes (thymocytes) — which make up over 95% of its cellularity — and a range of stromal elements that includes mesenchymal cells, bone marrow (BM)–derived cells, vasculature and the uniquely specialized thymic epithelium (Boyd *et al.*, 1993).

The mature thymus is encapsulated and lobulated and contains three principal histologically defined regions: the cortex, the medulla and the subcapsule (Fig. 44.1). The capsule and trabeculae consist of a thick layer of connective tissue and are separated from the cortex by a thin layer of simple epithelium, the subcapsule (Boyd *et al.*, 1993). The cortex and medulla each contain open networks of epithelial cells, which are densely packed with thymocytes (Van Vliet *et al.*, 1985; Boyd *et al.*, 1993; van Ewijk *et al.*, 1994), and each of these regions contains several different morphologically and phenotypically distinct epithelial subtypes (see later). The outer cortex also contains fibroblasts, and the organ as a whole is heavily vascularized. BM-derived stromal cells are found in both compartments, macrophages being distributed throughout the organ, while thymic dendritic

*Principles of Tissue Engineering, 3rd Edition*
ed. by Lanza, Langer, and Vacanti

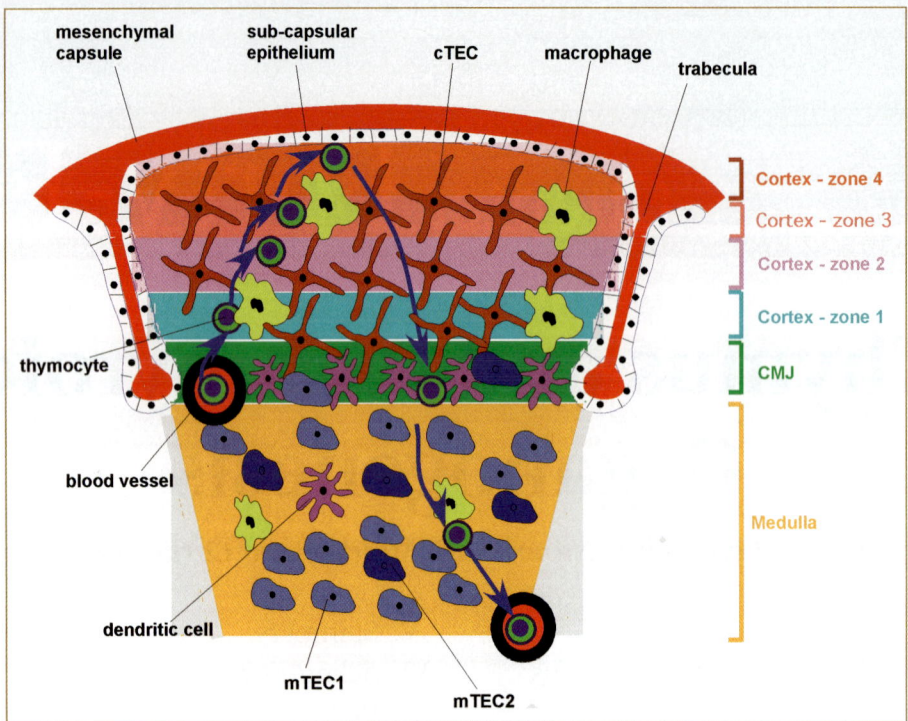

**FIG. 44.1.** Histology of the postnatal thymus. The postnatal thymus is surrounded by a capsule consisting of mesenchymal cells and connective tissue that penetrates into the thymus at regular intervals to form trabeculae. Underlying the capsule and trabeculae is the subcapsular epithelium, consisting of a layer of simple epithelium, which overlies the outer cortex. The cortex is populated with cortical thymic epithelial cells (cTEC), macrophages, and developing thymocytes at the triple negative (TN) and double positive (DP) stages of development. Thymocytes enter the thymus at the corticomedullary junction (CMJ) via the vasculature and migrate through the cortex to the subcapsule as they differentiate. The cortex can be divided into four zones based on the differentiation status of thymocytes that reside within it. Thus, zone 1 contains the most immature TN1 thymocytes and zone 4 contains thymocytes undergoing the TN4–DP transition. DP thymocytes are then screened for propensity to recognize self-MHC, a process termed *positive selection*, and those selected to mature into CD4⁺ or CD8⁺ single positive (SP) cells migrate into the medulla, where they undergo the final stages of maturation before being exported to the periphery. Central tolerance is established by deletion of self-reactive thymocytes in a process termed *negative selection*, and it is thought to occur principally at the corticomedullary junction and in the medulla: Negative selection is mediated by both thymic dendritic cells and medullary TECs, which supply self-peptides to medullary dendritic cells (DCs) in a process termed *cross-presentation*. Medullary TECs are also required for the generation of CD4⁺CD25⁺ T regulatory (Treg) cells and natural killer T-cells, both of which actively repress self-reactive T-cells.

cells — which are required for imposition of tolerance on the emerging T-cell repertoire — are found predominantly at the corticomedullary junction and in the medulla itself (Boyd *et al.*, 1993).

Thymus structure is intimately linked to its principal function, to support T-cell development. This encompasses the linked processes of T-cell differentiation and T-cell repertoire selection, which together ensure that the peripheral T-cell repertoire is populated predominantly by T-cells that have propensity to bind antigen in the context of self-major histocompatibility antigens (MHC) but that do not bind self-peptides. T-cell development has been extensively reviewed elsewhere (Zuniga-Pflucker and Lenardo, 1996; Petrie, 2002; Rothenberg and Dionne, 2002) and is not discussed in detail herein. In brief, hematopoietic progenitors enter the postnatal thymus at the corticomedullary junction, and subsequent T-cell development is then regulated such that thymocytes at different stages of development are found in different intrathymic locations. T-cell differentiation from the earliest postcolonization stages of thymocyte

maturation [termed *triple negative* (TN) cells because they do not express CD3 or the coreceptors CD4 and CD8] through to the CD4⁺CD8⁺ double positive (DP) stage occurs in the cortex, and the cortex itself can be subdivided into four regions based on the localization of different thymocyte populations. Thus, zone 1 contains the colonizing population of hematopoietic progenitor cells; these early thymocytes undergo proliferative expansion in zone 2; T-cell lineage commitment is completed in zone 3; and in zone 4, thymocytes differentiate to the DP stage of development, characterized by expression of both CD4 and CD8 coreceptors (Lind *et al.*, 2001; Porritt *et al.*, 2004). DP cells are then screened for their propensity to recognize self-MHC, a process termed *positive selection*, and those selected to mature into CD4⁺ or CD8⁺ single positive (SP) cells migrate into the medulla (Kurobe *et al.*, 2006). Central tolerance is established by deletion of self-reactive thymocytes at the DP–SP transition, in a process termed *negative selection* (Baldwin *et al.*, 2005), and is thought to occur principally at the corticomedullary junction and in the medulla. SP thymocytes proliferate and undergo the final stages of T-cell maturation in the medulla before exiting into the peripheral immune system. The outward migration of thymocytes, from the corticomedullary junction to the outer cortex, is regulated by chemokines (Plotkin *et al.*, 2003), as is the migration of positively selected cells from the cortex into the medulla (Ueno *et al.*, 2004; Kurobe *et al.*, 2006).

## Thymic Epithelial Cells

The thymic epithelium (TE) can be usefully classified into three broad subtypes: subcapsular/subtrabecular, cortical and medullary thymic epithelial cells. Within these subdivisions, ultrastructural and immunohistochemical analyses have revealed at least six different subsets (van de Wijngaert *et al.*, 1983), assigned as "clusters of thymic epithelial staining" (CTES) types I, II, III, IIIB, IIIC, and IV (Brekelmans and van Ewijk, 1990) based on different mAb staining profiles.

Thus, subcapsular/subtrabecular epithelium consists of type 1 epithelial cells in a simple epithelial layer. These cells are MHC Class II negative (Boyd *et al.*, 1992) and are reactive to CTES II mAbs (Godfrey *et al.*, 1990). The outermost cortical subpopulation comprises type II epithelial cells, characterized by their pale appearance in electronmicrographs (van de Wijngaert *et al.*, 1984). Immediately adjacent to the type II epithelia are type III thymic epithelial cells, which show intermediate electron lucency (van de Wijngaert *et al.*, 1984), while the innermost cTEC are type IV cells, which have high electron lucency and oval/spindle-shaped nuclei (van de Wijngaert *et al.*, 1984). Ultrastructural analysis has also revealed large complexes of type II and type III cells and developing thymocytes (van de Wijngaert *et al.*, 1984), termed *thymic nurse cells* (TNC) (Wekerle and Ketelsen, 1980a, 1980b). No mAbs are currently known to identify individual subpopulations corresponding to the types II–IV cells just described; however, all cTEC stain with CTES III mAbs (van de Wijngaert *et al.*, 1984), and types II and III cells are strongly MHC Class II positive (Boyd *et al.*, 1993). Medullary TEC (mTEC) predominantly express determinants reactive to CTES II and IV mAbs, with type III cells also identified by ultrastructural analysis (van de Wijngaert *et al.*, 1984). In addition, type V epithelia, classified as undifferentiated cells, exist in small isolated clusters at the corticomedullary junction (van de Wijngaert *et al.*, 1984), along with type VI cells, which have been proposed to be precursors of differentiated epithelial cells (von Gaudecker *et al.*, 1986). All mTEC express MHC Class I, while MHC Class II expression is variable (Jenkinson *et al.*, 1981; Farr and Nakane, 1983; Bofill *et al.*, 1985; Surh *et al.*, 1992).

The different thymic epithelial subpopulations are also defined by differential expression of cytokeratins (K). Two cortical populations have been identified: a predominant $K5^-K14^-K8^+K18^+$ subset and a minor subset also consisting of $K5^+K14^-K8^+K18^+$ cells (Klug *et al.*, 1998). Most mTEC display a $K5^+K14^+K8^-K18^-$ phenotype (Klug *et al.*, 1998) and also express the antigen reactive to mAb MTS10 (Godfrey *et al.*, 1990), and a minor $K5^-K14^-K8^+K18^+$ MTS10$^-$ medullary subset is also present (Klug *et al.*, 1998).

While precise functions are not ascribed to all of these TEC subpopulations, the clear functional dichotomy between the cortical and medullary compartments is reflected in functional differences between the cortical and medullary thymic epithelial cell types. Notably, cortical thymic epithelial cells (cTEC) are believed to express a ligand required for positive selection (Anderson *et al.*, 1994), while a subset of medullary thymic epithelial cells (mTEC) expresses AIRE1 (AIRE1 positively regulates expression of a cohort of tissue or developmentally restricted genes that play an essential role in the induction of central tolerance) (M. S. Anderson *et al.*, 2002). This high level of phenotypic and functional heterogeneity presents a significant challenge for attempts to support full T-cell development, including proper repertoire selection *in vitro*, and is also pertinent to cell replacement or regenerative strategies for enhancing thymus activity *in vivo*.

## III. *IN VITRO* T-CELL DIFFERENTIATION

The ability to generate T-cells in culture is widely used as a tool for investigating the regulation of T-cell differentiation (Hare *et al.*, 1999) and is also of interest for clinical and pharmaceutical purposes. Several methodologies exist that are based on the use of *ex vivo* thymic tissue. Thus, fetal thymic organ culture (FTOC) utilizes *ex vivo* thymic lobes usually derived from E15.5–E16.5 mouse embryos or second-trimester human fetuses to support the differentiation of T-cell progenitors from endogenous or exogenous sources (Jenkinson and Owen, 1990; Yeoman *et al.*, 1993; Barcena *et al.*, 1994; Plum *et al.*, 1994; Cumano *et al.*, 1996). The technique of reaggregate fetal thymic organ culture (RFTOC), in which defined TEC subpopulations are obtained by cell purification techniques, reaggregated with fibroblasts and defined lymphocyte populations, and then cultured further *in vitro*, was developed as an extension of FTOC and has proved invaluable for assessing the role of individual stromal components during specific stages of T-cell maturation (Jenkinson *et al.*, 1992; G. Anderson *et al.*, 1994). In addition, this approach has recently been adapted for testing the potency of different fetal and adult TEC subpopulations (Bennett *et al.*, 2002; Gill *et al.*, 2002; Rossi *et al.*, 2006).

Recently, it has been demonstrated that a tantalum-coated carbon matrix can be used to generate an *in vitro thymic organoid* when seeded with *ex vivo* murine thymic stromal cells (Poznansky *et al.*, 2000). When these structures were cocultured with human CD34$^+$ haematopoietic progenitors, efficient generation of mature CD4 and CD8 SP T-cells was observed after 14 days. The T-cells generated in this system were functional, as demonstrated by their proliferative response to mitogenic stimuli, and demonstrated a diverse TCR repertoire comparable to that of peripheral blood T-cells (Poznansky *et al.*, 2000). These findings demonstrate that the utilization of three-dimensional matrices in conjunction with thymic stromal cells can provide an efficient and reproducible method of *in vitro* T-cell generation. However, this approach currently relies on seeding with *ex vivo* thymus tissue and therefore is not highly scalable in its present form.

*In vitro* T-cell differentiation has also been investigated using a derivative of the BM stromal cell line OP-9, which expresses the Notch ligand Delta-like 1 (OP-9 DL-1). Recent studies demonstrate that OP-9 DL-1 monolayers can support the generation of CD4+CD8+ DP thymocytes from mouse fetal liver–, adult bone marrow–, or ES cell–derived hematopoietic progenitors (Schmitt and Zuniga-Pflucker, 2002; Schmitt *et al.*, 2004). Small numbers of CD8+ SP T-cells were also produced, although CD4+ SP T-cells were largely absent. This system was recently shown to support T-cell development from human cord blood– and human bone marrow– derived CD34+ cells (De Smedt *et al.*, 2002; La Motte-Mohs *et al.*, 2005). However, although it has been proposed that this system presents a scalable means of supporting *in vitro* T-cell differentiation (Lehar and Bevan, 2002), it remains unclear to what extent the T-cells generated on OP9-DL1 cells undergo positive and negative selection. In an interesting alternative to generation of mature T-cells, this system has recently been used as a means of expanding a CD4−CD8− DN precursor thymocyte population, which resulted in improved T-cell reconstitution after adoptive transfer of these DN cells in a mouse model of hematopoietic stem cell transplantation (Zakrzewski *et al.*, 2006). In addition, similar to RFTOC, the OP-9 DL-1 system is a powerful experimental tool for examining aspects of T-cell differentiation (Porritt *et al.*, 2004).

T-cell differentiation *in vitro* has also been achieved using preparations of cells derived from human skin in conjunction with a tantalum-coated carbon matrix (Clark *et al.*, 2005). In this system, cutaneous keratinocytes and fibroblasts were seeded onto the matrix and supplied with human CD34+ hematopoietic progenitor cells. After a period of three to four weeks, CD3+ T-cells were produced that exhibited functional maturity and were tolerant to self-MHC as assessed by the mixed lymphocyte reaction. Gene expression analysis demonstrated that the cells used to seed the matrix expressed transcription factors associated with TEC function, such as Foxn1, Aire, and Hoxa3, although at much lower levels than in the normal thymus. However, the efficiency of thymopoiesis in this system was low, with a relatively low number of mature T-cells generated despite the addition of a cocktail of prolymphopoietic cytokines.

## IV. THYMUS ORGANOGENESIS

### Cellular Regulation of Early Thymus Organogenesis

The thymus arises in the pharyngeal region of the developing embryo, in a common primordium with the parathyroid gland. This common primordium develops from the third pharyngeal pouch (3PP), one of a series of bilateral outpocketings of pharyngeal endoderm, termed the *pharyngeal pouches*, which form sequentially in a rostral-to-caudal manner.

In the mouse, outgrowth of the 3PP occurs from approximately E9.0 (Gordon *et al.*, 2004). At this stage, the epithelium of the 3PP consists of a single layer of columnar epithelium surrounded by a condensing population of neural crest cells (NCC) that will eventually form the capsule (Le Lievre and Le Douarin, 1975; Jiang *et al.*, 2000). Overt thymus organogenesis is evident from between E10.5 and E11.0, at which stage the epithelium begins to proliferate, assuming a stratified organization (Itoi *et al.*, 2001). Following this, at E12.5, the primordia separate from the pharynx and begin to resolve into discrete thymus and parathyroid organs. The thymus primordium subsequently migrates to its final anatomical location at the midline, while the parathyroid primodium associates with the lateral margins of the thyroid (Manley and Capecchi, 1995, 1998). In the case of the thymus at least, this migration is active and follows the path of the carotid artery and vagus nerve (Su *et al.*, 2001).

Within the common primordium, the prospective thymus is located in the ventral domain of the third pouch and the prospective parathyroid in the dorsal aspect. Patterning of these prospective organ domains appears to occur early in organogenesis, for the parathyroid domain is delineated by the transcription factor Gcm2 as early as E9.5. Possible mechanisms regulating the establishment/ maintenance of these domains are discussed later.

The mesenchymal capsule surrounding the thymus primordium is derived from the migratory neural crest, a transient population formed between the neural tube and the surface ectoderm. In the mouse, NCC migrate into the pharyngeal region from E9. Elegant chick-quail chimera studies provided the first evidence that NCC are the source of mesenchymal cells in the thymus (Le Lievre and Le Douarin, 1975), and this was recently confirmed in the mouse by heritable genetic labeling *in vivo* (Jiang *et al.*, 2000).

Colonization of the mouse thymus with hematopoietic progenitor cells occurs around E11.5 (Owen and Ritter, 1969; Cordier and Haumont, 1980; Jotereau *et al.*, 1987). Because vascularization has not occurred by this stage, the first colonizing cells migrate through the perithymic mesenchyme into the thymic epithelium. These cells have been reported to exhibit comparatively low T-cell progenitor activity, while a second colonizing wave, which arrives between E12 and E14, appears to display much higher levels of T-cell potential upon *in vivo* transfer (Douagi *et al.*, 2000).

The epithelial cells within the thymic primordium continue to proliferate strongly after E12.5, at least partly in response to factors supplied by the mesenchymal capsule. Concomitantly, TEC differentiation commences, with the first evidence of differentiation into cortical and medullary cell types appearing by E12.5 (Bennett *et al.*, 2002). Development of the two compartments then proceeds in a lymphocyte-independent manner until E15.5 (Klug *et al.*, 2002; Jenkinson *et al.*, 2005). The expression of MHC II and MHC I on the surface of thymic epithelial cells is first detected at E13.5 and ~E16, respectively (Jenkinson *et al.*, 1981; Van Vliet *et al.*, 1984), and is followed by the appearance of $CD4^+$ and $CD8^+$ SP thymocytes at E15.5 and E17.5 (Jenkinson *et al.*, 1981; Van Vliet *et al.*, 1984). Although a functional thymus is present in neonates, the full organization of the stroma is not achieved until two to three weeks postnatally in the mouse.

## Origin of Thymic Epithelial Cells

The precise embryonic origins of the thymic epithelium were until recently a matter of long-standing controversy; conflicting hypotheses suggested that the epithelium had a dual endodermal/ectodermal origin (Cordier and Heremans, 1975; Cordier and Haumont, 1980) or derived solely from the pharyngeal endoderm (Le Douarin and Jotereau, 1975; Manley and Blackburn, 2003; Blackburn and Manley, 2004; Gordon *et al.*, 2004). However, recent work from our laboratories has provided definitive evidence for a single endodermal origin in mice (Gordon *et al.*, 2004) through histological, fate, and potency analysis of the pharyngeal region. These data demonstrated that although the 3PP endoderm and third pharyngeal cleft ectoderm make contact at E10.5, as proposed in the dual-origin hypothesis, the germ layers subsequently separate, with apoptosis occurring in the contact region. Lineage tracing of pharyngeal surface ectoderm of E10.5 mouse embryos also failed to find evidence for an ectodermal contribution to the thymic primordium, and, finally, transplantation of pharyngeal endoderm isolated from E8.5–E9.0 embryos (i.e., prior to initiation of overt thymus organogenesis) indicated that the grafted endoderm was sufficient for complete thymus organogenesis, similar to previous results obtained using chick-quail chimeras (Le Douarin and Jotereau, 1975). Thus, pharyngeal endoderm alone is sufficient for the generation of both cortical and medullary thymic epithelial compartments.

## Thymic Epithelial Progenitor Cells (TEPC)

The phenotype of TEPC has been of considerable interest. Evidence suggestive of a progenitor/stem cell activity was initially provided by analysis of a subset of human thymic epithelial tumours that were found to contain cells that could generate both cortical and medullary subpopulations. This suggested that the tumourigenic targets were epithelial progenitor/stem cells (Schluep *et al.*, 1988). In addition, ontogenic studies suggested that the early thymus primordium in both mouse and human might be characterized by coexpression of markers that later segregated to either the cortical or medullary epithelium (Lampert and Ritter, 1988). However, the first genetic indication of a TEPC phenotype was provided by a study addressing the nature of the defect in *nude* mice, which fail to develop a functional thymus due to a single base deletion in the transcription factor Foxn1 (Blackburn *et al.*, 1996). Here, analysis of allophenic *nude*-wild-type aggregation chimeras demonstrated that cells homozygous for the *nude* mutation were unable to contribute to the major thymic epithelial subsets, establishing that the *nude* gene product (Foxn1) is required cell-autonomously for the development and/or maintenance of all mature TEC. However, a few *nude*-derived cells were present in the thymi of adult chimeras, and phenotypic analysis indicated that these cells expressed determinants reactive to mAbs MTS20 (Godfrey *et al.*, 1990) and MTS24 but did not express markers associated with mature TEC, including MHC Class II. Based on these findings, we suggested that in the absence of *Foxn1*, TEC lineage cells undergo maturational arrest and persist as $MTS20^+24^+$ progenitors (Blackburn *et al.*, 1996). This hypothesis was recently confirmed by an elegant study that demonstrated that functional thymus tissue containing organized cortical and medullary regions is generated on reactivation of a conditional null allele of Foxn1 in the postnatal thymus (Bleul *et al.*, 2006). Since clonal reactivation of Foxn1 was achieved in this study, it demonstrates unequivocally that in the absence of Foxn1, some persisting TECs have bipotent progenitor activity.

Further data regarding the phenotype of thymic epithelial progenitors came from analysis of mice, with a secondary block in thymus development resulting from a primary T-cell differentiation defect. The thymi of postnatal *CD3e26tg* mice, in which thymocyte development is blocked at the $CD44^+CD25^-$ TN1 stage (Hollander *et al.*, 1995), contain principally epithelial cells that coexpress K5 and K8 (Klug *et al.*, 1998) — which, in the normal postnatal thymus, are predominantly restricted to the medulla and cortex respectively and are coexpressed by only a small population of cells at the corticomedullary junction. In this study, Klug and colleagues (1998) demonstrated that transplantation of *CD3e26tg* thymi into $Rag1^{-/-}$ mice, which sustain a later block in T-cell differentiation, resulted in the development of $K5^-K8^+$ cells, suggesting that the $K5^+K8^+$ cells are progenitors of cTEC.

More recently, two studies have addressed the phenotypic and functional properties of $MTS20^+24^+$ cells within the fetal mouse thymus directly. Ontogenic analysis has demonstrated that the proportion of $MTS20^+24^+$ epithelial cells was highest in the early thymus primordium, decreasing to less than 1% in the postnatal thymus (Bennett *et al.*, 2002), consistent with the expression profile expected of markers of fetal tissue progenitor cells. Phenotypic analysis of the $MTS20^+24^+$ and $MTS20^-24^-$ populations of the E12.5 thymus indicated that all cells in the $MTS20^+24^+$ population coexpressed K5 and K8, while none expressed TEC differentiation markers, including MHC II (Bennett *et al.*, 2002). Importantly, the functional capacity of isolated $MTS20^+24^+$ cells and $MTS20^-24^-$ cells was then determined via ectopic transplantation. This analysis demonstrated that the MTS20⁺24⁺ population was sufficient for establishment of a functional thymus containing both cortical and medullary TEC populations, while, in this assay, the MTS20⁻24⁻ population fulfilled none of these functions (Bennett *et al.*, 2002; Gill *et al.*, 2002). The studies described earlier clearly demonstrated that progenitor activity resided in the MTS20⁺24⁺ population. However, although these data strongly suggested the existence of common thymic epithelial progenitor cells, this was not addressed at the clonal level in either study.

With regard to this issue, a recent study indicates that during initial organogenesis a bipotent progenitor exists that can form both cortical and medullary TECs and suggests that this activity may persist in the postnatal thymus (Bleul *et al.*, 2006). The approach used was to perform a lineage trace using hK14$^{CreERt2}$ transgenic mice crossed onto a ROSA26/silent eYFP background. In this system, spontaneous Cre recombinase activity in thymic epithelial cells activated the expression of eYFP in a small number of TECs at around 14 days postpartum. Analysis of the thymi by fluorescence microscopy revealed that the majority of eYFP⁺ cells were present in clusters and were not evenly distributed throughout the stroma, suggesting that fluorescent cells were derived from a single recombination event in cells with proliferative capacity. The location of the cell clusters was mixed; a small proportion of eYFP⁺ clusters were restricted to either the cortex or the medulla, but the majority (76%) appeared to span the corticomedullary junction. These results are consistent with the presence of a bipotent progenitor that may give rise to intermediate progenitors committed to either the cortical or medullary lineage. However, a major caveat for this interpretation is that stem cell activity per se was not assayed in these experiments and that the data would also be consistent with proliferation of differentiated epithelial cells. In an elegant extension of these experiments, the same hK14$^{CreERt2}$ deletor strain was used to reactivate a conditional null allele of Foxn1 in postnatal mice (discussed earlier). Here, clonal activation of Foxn1 resulted in the generation of small regions of thymus tissue that contained both cortical and medullary TEC, providing conclusive evidence for the existence of a common progenitor in initial organogenesis.

### Human Thymus Development

Early human thymus development closely parallels that of the mouse. Thus, the thymus forms from the third pharyngeal pouch in a common primordium with the parathyroid gland. The third pharyngeal pouch is evident from early in week 6 of human fetal development, and initially it develops as a tubelike lateral expansion from the pharynx, which makes contact with the ectoderm of the third pharyngeal cleft (Weller, 1933; Norris, 1938). A single endodermal origin has not been demonstrated directly for the human thymic epithelium. However, since the thymus has a single endodermal origin in mice and avians (Le Douarin and Jotereau,

1975), it is reasonable to assume that this is also the case in humans. Within the human common thymus/parathyroid primordia, the thymus and parathyroid domains are located ventrally and dorsally and are surrounded by condensing neural crest–derived mesenchyme from the onset of development. The thymus component of this primordium begins to migrate ventrally from week 7 to mid-week 8, forming a highly lobulated, elongated, cordlike structure. The upper part of this structure normally disappears at separation of the two organ rudiments, leaving the parathyroid in the approximate location in which it will remain throughout adulthood (Norris, 1938). The bilateral thymic primordia continue to migrate toward the midline, where they eventually meet and attach at the pericardium — the permanent location of the thymus into adulthood — by mid-week 8 (Norris, 1938). As in the mouse, the human early thymus primordium appears to contain undifferentiated epithelial cells (Bennett *et al.*, 2002), which express some markers that are later restricted to either cortical or medullary compartments (Lampert and Ritter, 1988; A. Farley and CCB, unpublished data). Nascent medullary development is evident from week 8, and by week 16 distinct cortical and medullary compartments are present. Other cell types penetrate the thymus from week 8, including mesenchymal, vascular and lymphoid cells, and mature lymphocytes begin to leave the thymus to seed the peripheral immune tissues between weeks 14 and 16 (van Dyke, 1941; Lobach and Haynes, 1986).

### Cervical Thymus in Mouse and Human

The presence of a cervical thymus in humans has been recorded for some time (van Dyke, 1941; Tovi and Mares, 1978; Ashour, 1995), and recent publications indicate that an ectopic cervical thymus is also a common occurrence in at least some mouse strains (Dooley *et al.*, 2006; Terszowski *et al.*, 2006). In terms of size and cellularity, the cervical thymus is much smaller than the thoracic thymus. However, the morphology of the two structures is very similar, with organized cortical and medullary regions and similar expression patterns of cytokeratin molecules (Dooley *et al.*, 2006; Terszowski *et al.*, 2006). Furthermore, the cervical thymus expresses the transcription factors Foxn1 and Aire and can produce functional T-cells that are tolerant to self-antigens (Dooley *et al.*, 2006; Terszowski *et al.*, 2006).

The origin of the cervical thymus is at present unclear. A plausible hypothesis is that it may arise from remnants of the thymus domain of the 3PP that become detached from the organ during separation of the thymus and parathyroid domains. Several alternative explanations exist, and the identification of cervical thymi in mice will allow the embryonic origins of these structures to be addressed experimentally. It appears that in mice the cervical thymus may mature postnatally (Terszowski *et al.*, 2006), while in humans it is clearly present in the second trimester of fetal development. Although the presence of Foxn1⁺ epithelial cells has not

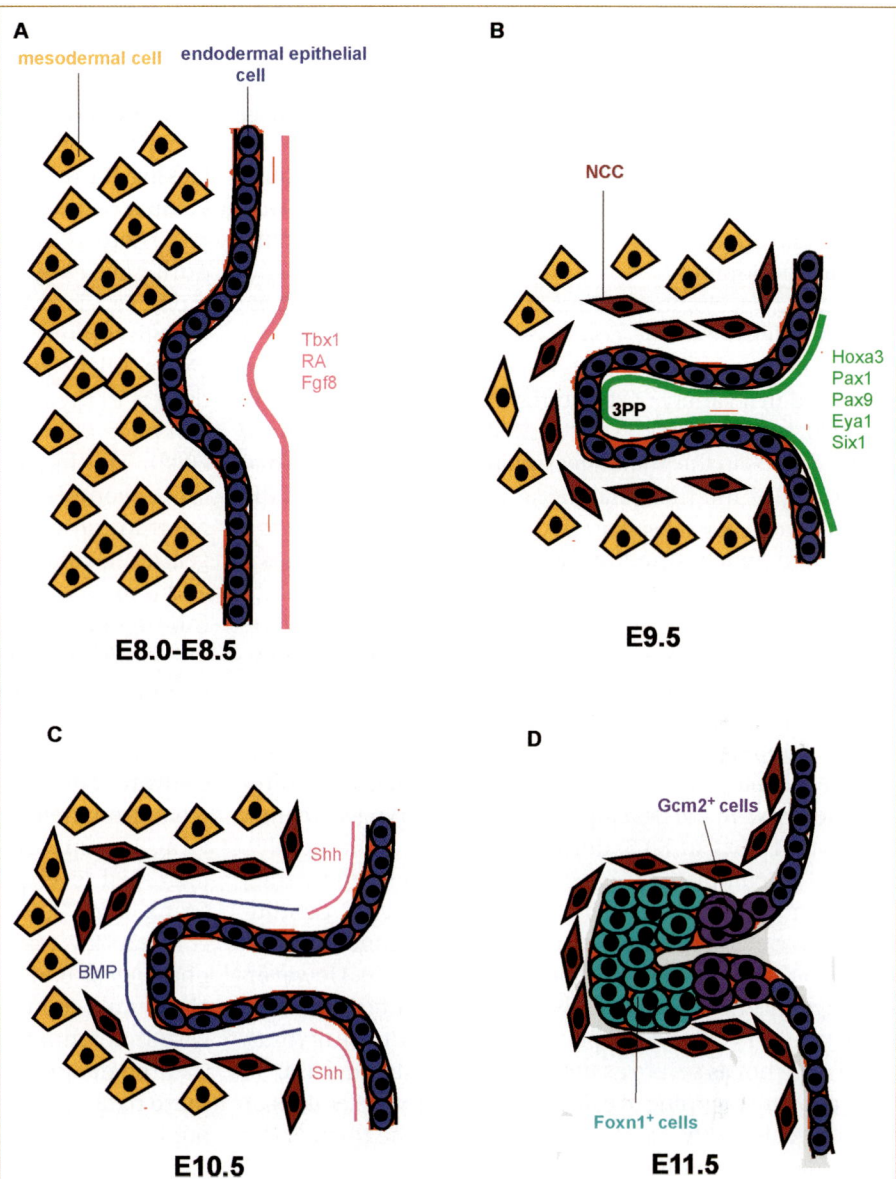

**FIG. 44.2.** Molecular regulation of early thymus organogenesis. **(A)** At approximately E8.0–E8.5 the formation of the third pharyngeal pouch (3PP) is initiated in the pharyngeal endoderm and is dependent on the expression of Tbx1 and retinoic acid (RA) and fibroblast growth factor 8 (Fgf8) signaling (pink). **(B)** At E9.5 the 3PP has formed and is surrounded by mesenchymal cells of mesodermal and neural crest cell (NCC) origin. Continued development is dependent on the expression of the transcription factors Hoxa3, Pax1, Pax9, Eya1, and Six1 (green). **(C)** Bone morphogenetic protein (BMP) (blue) and sonic hedgehog (Shh) (pink) signaling occur at E10.5 in the 3PP endoderm in the ventral and dorsal aspects, respectively. These factors may be involved in the specification of the 3PP into thymus- and parathyroid-specific domains. **(D)** At E11.5 epithelial cells in the ventral domain of the 3PP express the transcription factor Foxn1 (light blue) and will form the thymic epithelium. Epithelial cells in the dorsal domain express the transcription factor Gcm2 (purple) and will form the parathyroid gland. The differentiation and maintenance of both of these cell types are dependent on these factors.

been reported in the cervical regions of developing mouse embryos, because it is now clear that cells specified to the thymic epithelial lineage retain their identity in the absence of Foxn1 expression (Bleul *et al.*, 2006), Foxn1 may not be an appropriate lineage marker for tracing the origins of this tissue.

## Molecular Regulation of Thymus and Parathyroid Organogenesis

Although the regulation of thymus organogenesis is incompletely understood, studies of classical and genetically engineered mouse mutants have begun to reveal a network of transcription factors and signaling molecules that act in the pharyngeal endoderm and surrounding mesenchyme and mesoderm to regulate thymus and parathyroid organogenesis. The principal components of this network are discussed next and are summarized in Fig. 44.2.

## Molecular Control of Early Organogenesis

The earliest events in thymus organogenesis occur prior to overt organ development and relate to molecular control of 3PP formation. The T-box transcription factor Tbx1, retinoic acid (RA) signaling, and fibroblast growth factor 8 (Fgf8) signaling have been implicated as important regulators of this process.

Tbx1 was recently identified as the gene responsible for cardiovascular and glandular defects in Df1 mice, which carry a large deletion of chromosome 16 (Lindsay *et al.*, 1999). Df1 heterozygotes closely phenocopy a human condition known as 22q11.2 deletion syndrome (22q11.2DS, or DiGeorge Syndrome), in which a deletion in chromosome 22 covering an interval of approximately 30 genes (Scambler, 2000) results in a range of defects including thymus aplasia or, more frequently, hypoplasia (Paylor et al., 2001; Taddei, Morishima et al., 2001). Thus the Df1 mouse represents a useful model for 22q11.2DS and has allowed the identification of Tbx1 as a critical early regulator of pharyngeal development.

During development, Tbx1 is expressed in the pharyngeal endoderm and the core mesenchyme of the pharyngeal arches from approximately E7.5 and continues to be expressed in a variety of structures until E12.5 (Chapman et al., 1996; Hu et al., 2004; Xu et al., 2004). Tbx1 mutants have severe defects in the pharyngeal region, including abnormal patterning of the first pharyngeal arch; hypoplasia of the second arch; and absence of the third, fourth and sixth arches and pouches (Jerome and Papaioannou, 2001). As a result of this, Tbx1$^{-/-}$ mutants lack both thymus and parathyroid and display a spectrum of cardiovascular abnormalities and craniofacial defects (Jerome and Papaioannou, 2001). The phenotype of Tbx1$^{-/-}$ animals suggests an important role for this gene in the segmentation of the pharyngeal region. Supporting evidence for this hypothesis was provided by an elegant study addressing the temporal requirement for Tbx1 in the development of the pharyngeal region. Deletion of Tbx1 at E8.5, during the formation of the 3PP, resulted in complete absence of thymus and parathyroid, and complementary fate-mapping experiments demonstrated that cells that express Tbx1 at E8.5 contribute significantly to the thymic primordium (Xu et al., 2005). However, although deletion of Tbx1 at E9.5/E10.5 (after initial formation of the 3PP) caused morphological defects in the thymus, these were not as severe as the aplasia seen after deletion at E8.5, and fate mapping of cells expressing Tbx1 cells at E9.5/E10.5 revealed only a small contribution to the thymus (Xu et al., 2005). Taken together, these data suggest that Tbx1 is required for establishment of the 3PP but that it is not directly required for subsequent thymus development. Thus, Tbx1 may influence later stages of thymus organogenesis in a non-cell-autonomous manner. Furthermore, although Tbx1 is haplo-insufficient with respect to thymus development (Lindsay et al., 2001), the basis of this insufficiency remains to be determined and may result either from secondary effects resulting from mild defects in pouch formation or from dosage effects related to factor provision by non-NCC mesenchymal cells.

A role for RA in 3PP formation was suggested by experiments in which RA antagonist was administered to whole-embryo cultures. Here, blockade of RA signaling at E8.0 resulted in the absence of the growth factors Fgf8 and Fgf3 in the 3PP endoderm and impaired NCC migration to the third and fourth pharyngeal arches (Wendling et al., 2000). Expression of the transcription factor Pax9 (see later) was also absent in the third pouch, but was expanded in the second pouch endoderm. These data suggest that RA signaling is required for the specification of the third pouch endoderm, which confers subsequent competence to support NCC migration. In vivo evidence for a role for RA signaling was subsequently provided by the finding that fetal mice lacking RA receptors α and β display thymus agenesis and ectopia (Ghyselinck et al., 1997).

There is considerable evidence that Fgf8 is required during the early stages of thymus and parathyroid development. Fgf8 is expressed in the early gut endoderm and in the endoderm and ectoderm of the pharyngeal pouches and clefts. Mice carrying hypomorphic alleles of Fgf8 show defects in thymus development ranging from hypoplasia to complete aplasia (Abu-Issa et al., 2002; D. U. Frank et al., 2002): The initial impairment in thymus and parathyroid organogenesis is likely to occur at an early stage in development in these mice, because the third and fourth pharyngeal arches and pouches are usually hypoplastic/aplastic, in addition to other abnormalities in the pharyngeal region (Abu-Issa et al., 2002; D. U. Frank et al., 2002).

In terms of the cell types affected by impaired Fgf8 signaling, similarities between the phenotype of Fgf8 hypomorphs and the spectrum of abnormalities found during experimental NCC ablation (Bockman and Kirby, 1984; Conway et al., 1997) suggest that the glandular defects may result from defective NCC migration/differentiation or survival. In support of this, NCC of Fgf8 hypomorphs show increased levels of apoptosis (Abu-Issa et al., 2002; D. U. Frank et al., 2002) and reduced expression of Fgf10 (Frank et al., 2002), a factor that may mediate proliferation of the pharyngeal endoderm (D. U. Frank et al., 2002). There is also a mild reduction in the expression of genes associated with differentiated NCC (Abu-Issa et al., 2002), suggesting that the maintenance of NCC is perturbed. Taken together, these data implicate Fgf8 in maintaining a competent NCC population that can contribute to thymus organogenesis. However, Fgf8 may also act specifically on the 3PP endoderm, since ablation of Fgf8 in the endoderm and ectoderm or ectoderm alone results in different phenotypes; ablation in the ectoderm alone causes vascular and craniofacial defects seen in Fgf8 hypomorphs (Macatee et al., 2003), whereas when Fgf8 is also deleted in the endoderm, glandular defects are evident, including thymus hypoplasia and ectopia (Macatee et al., 2003). Since the NCC defects were the same in both cases, Fgf8 may directly influence development of the endoderm, although it remains possible that the differences observed result from a dosage effect.

### Transcription Factors and Regulation of 3PP Development

After the initial onset of 3PP formation, continued development is dependent on several transcription factors, most notably Hoxa3, Pax1, Pax 9, Eya1, Six1, and Six4. All of these transcription factors are expressed in the 3PP endoderm from approximately E9.5 to E10.5 and, with the exception of Pax1 and Pax9, are also expressed in associated NCC and the ectoderm.

Absence of functional Hoxa3 or Eya1 results in the failure to initiate overt thymus and parathyroid organogenesis once the 3PP has formed, revealing the essential roles of these factors (Manley and Capecchi, 1995, 1998; Xu et al., 2002; Zou et al., 2006). Tbx1 and Fgf8 are both down-regulated in the 3PP of Eya1$^{-/-}$ mice at E9.5 (Zou et al., 2006), indicating that Eya1 plays a role in the regulation of each of these factors. Lack of Six1, Six1 and 4, Pax1, or Pax9 causes much less severe phenotypes. The common primordium begins to develop in Six1$^{-/-}$ mice, and patterning into thymus and parathyroid domains (see the next section) is initiated. However, subsequent apoptosis of endodermally derived cells in the common primordium leads to complete disappearance of the organ rudiment by E12.5 (Zou et al., 2006). A similar phenotype is evident in Six1$^{-/-}$;Six4$^{-/-}$ embryos, though the size of the primordium is further diminished in the double mutants, indicating synergy between these gene products (Zou et al., 2006). Loss of function mutations in Pax9 results in failure of the primordia to migrate to the mediastinum and in severe hypoplasia from E14.5. However, the thymic lobes are vascularized and contain lymphocytes (Hetzer-Egger et al., 2002). Pax1$^{-/-}$ mutants show relatively mild thymus hypoplasia and aberrant TEC differentiation (Wallin et al., 1996; Su and Manley, 2000; Su et al., 2001). Since Pax1 and 9 are highly homologous, these phenotypes may reflect functional redundancy, as demonstrated in other tissues (Peters and Balling, 1999).

Expression of Hoxa3, Eya1 and Six1 in multiple germ layers complicates interpretation of the respective null phenotypes for these genes. However, Hoxa3 appears to regulate Pax1 and Pax9 either directly or indirectly, since Pax1 and Pax9 expression is initiated normally in Hoxa3$^{-/-}$ mutants but fails to be maintained at wild-type levels beyond E10.5 (Manley and Capecchi, 1995). Furthermore, Hoxa3$^{+/-}$;Pax1$^{-/-}$ compound mutants show delayed separation of the thymus/parathyroid primordium from the pharynx, resulting in thymic ectopia and a more severe hypoplasia than that seen in Pax1$^{-/-}$ single mutants (Su et al., 2001).

It has been suggested that Eya1 and Six1 act downstream of the Hox/Pax genes, for Eya1$^{-/-}$ embryos show normal expression of Hoxa3, Pax1, and Pax9 but reduced expression of Six1 in the endoderm of the third and fourth pouches and ectoderm of the second, third, and fourth pharyngeal arches (Xu et al., 2002). However, while it is likely that Six1 acts downstream of Eya1, recent evidence indicates that Eya1 and Six1 do not act downstream of the Pax genes. This has been shown by analysis of Pax9$^{-/-}$ and Pax1$^{-/-}$Pax9$^{-/-}$ mutants, which show normal expression of Eya1 and Six1 in the 3PP, and of Eya1$^{-/-}$;Six1$^{-/-}$ double mutants, which lack expression of Pax1 in the E10.5 3PP (Zou et al., 2006). Interestingly, Pax9 expression is unaffected in Eya1/Six1 double mutants (Zou et al., 2006). However, it remains possible that Eya1 and Six1 may be regulated by Hoxa3 independently of Pax1 and 9 function.

The data just reviewed indicate the power of compound-mutant analysis for unraveling genetic interactions. However, further work is required to elucidate exactly how these and other transcription regulators cooperate in the development of the 3PP endoderm and to determine the precise role of each of these factors at the cellular level.

## Specification of the Thymus and Parathyroid

Prior to the overt formation of the thymus and parathyroid, the 3PP is specified into organ-specific domains. At E9.5, epithelial cells within the anterior dorsal aspect initiate expression of Gcm2, a transcription factor required for the development of the parathyroid (Gordon et al., 2001). Current evidence suggests that Gcm2 acts downstream of Eya1 and Hoxa3, because Gcm2 is down-regulated in Hoxa3$^{-/-}$, Eya1$^{-/-}$ and Hoxa3$^{+/-}$;Pax1$^{-/-}$ compound mutant mice (Su et al., 2001; Xu et al., 2002; Blackburn and Manley, 2004). The first thymus specific transcription factor, Foxn1, is expressed at functionally relevant levels in the ventral domain of the 3PP from approximately E11.25, although low levels can be detected by PCR from E10.5 (Gordon et al., 2001; Balciunaite et al., 2002). Foxn1, a forkhead class transcription factor, is required for TEC differentiation and hair development, and is discussed in more detail in the next section. Upstream regulation of Foxn1 is currently not understood, and analysis of mutants in potential upstream regulators have not been informative in this regard. Hoxa3$^{-/-}$ and Eya1$^{-/-}$ embryos do not express Foxn1, but this is due to the block in primordium formation prior to the onset of Foxn1 expression. Although Foxn1 expression is unaltered in Pax9$^{-/-}$ and Hoxa3$^{+/-}$;Pax1$^{-/-}$ mutants (Hetzer-Egger et al., 2002; Su and Manley, 2000), the possibility remains of redundancy between Pax1 and Pax9. Although Six1$^{-/-}$ mutants display reduced Foxn1 expression, the 3PP exhibits increased cell death in the absence of Six1 (Zou et al., 2006), and therefore the reduced expression of Foxn1 may reflect poor survival of Foxn1$^{+}$ cells rather than a direct interaction between Foxn1 and Six1. Further studies are required to determine whether a regulatory relationship exists between these genes and Foxn1.

The expression patterns of Foxn1 and Gcm2 clearly define the thymus and parathyroid domains of the 3PP. However, these factors do not appear to be responsible for specification of their respective organs, indicating that specification must be mediated by an upstream factor or factors. With respect to Foxn1, this model is supported by the finding that epithelial cells in the ventral domain of the 3PP express the thymus specific cytokine IL-7, a factor required for thymocyte differentiation throughout development, at E11.5 (von Freeden-Jeffry et al., 1995; Zamisch et al., 2005), making IL-7 one of the earliest currently identified markers of thymus identity. In the thymus primordium, IL-7 is regulated independently of Foxn1, because Foxn1$^{-/-}$

embryos show normal IL-7 expression (Zamisch *et al.*, 2005), indicating that thymus identity is specified in the absence of Foxn1. Transplantation experiments also support this model, because E9.0 pharyngeal endoderm, which has not yet formed the 3PP, gives rise to a functional thymus when grafted ectopically (Gordon *et al.*, 2004), indicating that at this developmental stage some cells are already specified to the thymic epithelial lineage. Although Rhox4 has previously been identified as an early marker of thymus identity in the 3PP, the human orthologues of this gene are not expressed in human thymus organaogenesis, making it unlikely that it plays a critical role in lineage specification (Morris *et al.*, 2006).

Similarly, our recent studies conclude that Gcm2 is not required for specification of the parathyroid, for other parathyroid-specific markers, including CCL21 and CaSR, are initiated but not maintained in Gcm2$^{-/-}$ mice (Z. Liu, S. Yu, and N. R. M., unpublished). Thus, Gcm2 may play an analogous role in parathyroid development to that of Foxn1 in the thymus.

How, then, are the thymus and parathyroid domains within the 3PP established? Evidence suggests that opposing gradients of bone morphogenetic proteins (BMP) and sonic hedgehog (Shh) may play an important role in this process. During thymus development, BMP4 expression is first detected at E9.5, when it is expressed by a small number of mesenchymal cells in the third pharyngeal arch (Patel *et al.*, 2006). By E10.5, the BMP4 expression domain has expanded to include the ventral 3PP endoderm and the adjacent mesenchyme but remains absent from the dorsal 3PP. This expression pattern is maintained at E11.5, and by E12.5 BMP4 is expressed throughout the thymic primordium and the surrounding mesenchymal capsule (Patel *et al.*, 2006). The expression pattern of BMP4 suggests that it may be responsible for the initiation of Foxn1 expression, because BMP4 is restricted to the ventral domain of the 3PP immediately prior to the onset of Foxn1 expression. Furthermore, Noggin expression is restricted to the dorsal anterior region of the 3PP at E10.5 and E11.5 and there is some *in vitro* evidence that BMP4 can directly regulate Foxn1 expression (Tsai *et al.*, 2003; A. Farley and C. C. B., unpublished data).

*In vivo* evidence of a role for BMPs in thymus organogenesis has been provided by a transgenic approach in which the BMP inhibitor Noggin is driven by the Foxn1 promoter, thus impairing BMP signaling in the thymic stroma (Bleul and Boehm, 2005). These mice have hypoplastic and cystic thymi that fail to migrate to their normal position above the heart. It is highly likely that this is due to a direct effect on the thymic stroma, because impaired development is evident prior to the immigration of lymphocytes, which are able to differentiate into T-cells. Furthermore, mediators of BMP signaling, such as Msx1 and phosphorylated Smad proteins, are down-regulated in both the epithelium and surrounding mesenchyme, suggesting that impaired

communication between these cell types is the mechanism responsible for the phenotype observed. Interestingly, Foxn1 expression was not overtly affected in this transgenic model, arguing against a role for BMP in the regulation of this transcription factor. However, the inhibition of BMP signaling is driven by the Foxn1 promoter, and thus it may occur after the point at which it is required for initiation of Foxn1 expression. In addition, since BMP2 and BMP7 are also expressed in the 3PP and common primordium (C. C. B. and N. R. M., unpublished data), it is highly likely that redundancy operates between different BMP family members. It may thus be more pertinent to view BMPs in terms of the signaling they mediate during organogenesis rather than which specific member of the BMP family is involved.

During development of the 3PP, expression of the secreted glycoprotein Shh is restricted to cells of the pouch opening at E10.5 and E11.5, although its receptor, Patched1, is expressed by cells in close proximity to this region (Moore-Scott and Manley, 2005). Analysis of Shh$^{-/-}$ embryos revealed that both the BMP4 and Foxn1 expression domains are expanded in the 3PP, while the corresponding Gcm2$^+$ parathyroid domain is lost in these mutants (Moore-Scott and Manley, 2005). Thus, the role of Shh in thymus organogenesis may be to oppose the action of BMP4 to allow the specification and development of the parathyroid.

Wnt glycoproteins may also be important in regulating Foxn1 expression. Wnts are expressed by the thymic stroma and lymphoid cells, although Wnt receptors are expressed exclusively by TECs (Balciunaite *et al.*, 2002). The earliest reported expression of Wnt family members during thymus development is at E10.5, immediately prior to strong Foxn1 expression, when the epithelium of the 3PP and adjacent cells express Wnt4 (Balciunaite *et al.*, 2002). Given this expression pattern and the finding that TEC lines that overexpress Wnt4 display elevated levels of Foxn1 (Balciunaite *et al.*, 2002), it is possible that Wnt4 cooperates with BMP4 to regulate Foxn1. Wnt 1, Wnt 4, and Wnt1, 4-null mice all exhibit hypoplastic thymi characterized by reduced T-cell numbers but normal thymocyte developmental progression. However, because no histological analysis of the thymi in these mutants has been presented, it is not possible to evaluate whether the primary effect is on the thymic epithelium or on thymocytes (Mulroy *et al.*, 2002; Staal and Clevers, 2005). As with BMP family members, functional redundancy is again a possibility, because other Wnt family members are also expressed in the 3PP and surrounding mesenchyme (Mulroy *et al.*, 2002; Balciunaite *et al.*, 2002; C. C. B and N. R. M., unpublished data).

## Foxn1 and the Regulation of TEC Differentiation

As discussed earlier, the development of functionally mature TECs from the 3PP endoderm is cell autonomously dependent on Foxn1 (Blackburn *et al.*, 1996). Adult *nude* mice, which lack functional Foxn1, retain a cystic, alymphoid thymus consisting predominantly of apparently immature epithelial cells (Cordier, 1974; Cordier and Haumont, 1980; Gordon *et al.*, 2001). These and other data (see earlier) suggest that lack of Foxn1 results in developmental arrest of thymic-epithelial-lineage cells at the founder/progenitor-cell stage of development. Thus, whereas TEPCs form independent of Foxn1, their differentiation into mature TEC subtypes depends on it. In the thymus, Foxn1 is expressed by all thymic epithelial cells throughout development and is maintained postnatally (Nehls *et al.*, 1996). It is also expressed in the hair follicles and epidermis, where it is required for normal development of the skin (Flanagan, 1966).

The precise role of Foxn1 has not been completely elucidated in either the thymic or the cutaneous epithelial lineages. However, considerable evidence suggests that it regulates the balance between epithelial proliferation and differentiation. Keratinocytes derived from Foxn1$^{-/-}$ mice have reduced proliferative capacity *in vitro* and prematurely express markers associated with terminal differentiation (Brissette *et al.*, 1996). Furthermore, when Foxn1 is overexpressed, markers associated with earlier stages of differentiation are up-regulated and later markers are absent (Brissette *et al.*, 1996). More recently it was shown that primary human keratinocytes can be induced to initiate terminal differentiation by transient expression of Foxn1 but that completion of the differentiation program is dependent on the levels of activated Akt, which may in turn be regulated by Foxn1 (Janes *et al.*, 2004). Thus, in the epidermis Foxn1 may function to ensure that terminal differentiation proceeds in a temporally regulated manner. Whether this is the case in the thymus is unclear. The early thymic rudiment in *nude* mice also shows reduced proliferation (Itoi *et al.*, 2001). However, the epithelial cells resemble immature progenitors and do not express markers of terminal differentiation.

There is some evidence to suggest that in the thymus, Foxn1 is required for lymphoepithelial cross-talk. Comparative gene expression analysis between *nude* and wild-type thymus suggests that PD1 Ligand is a target of Foxn1, because it is down-regulated in *nude* embryos (Bleul and Boehm, 2001). The receptor for PD1 ligand is expressed by thymocytes and has been implicated in thymocyte survival, proliferation and positive selection (Nishimura *et al.*, 1996, 2000), lending weight to the hypothesis that Foxn1 may regulate genes required for lymphoepithelial cross-talk.

A separate publication suggested a role for Foxn1 in cross-talk by analyzing Foxn1$^{\Delta/\Delta}$ mice, which express a splice variant lacking exon 3 of the N-terminal domain (Su *et al.*, 2003). Skin and hair development is normal in these mice, but the differentiation of TECs is suspended at a time equivalent to E13.5. Unlike the *nude* thymus, the thymic stroma can attract lymphocytes and mediate T-cell differentiation, although there are postnatal abnormalities in thymocyte maturation and a severe reduction in thymocyte numbers. The thymus-specific defect in Foxn1$^{\Delta/\Delta}$ mice resembles to some degree the phenotype seen in hCD3ε26 mice, which have secondary blocks in TEC maturation due to suspended thymocyte development (Klug *et al.*, 1998, 2002). Based on this, we suggested that the N-terminal domain is required for lymphoepithelial cross-talk and as such has a thymus-specific function. However, the impairment of TEC differentiation in Foxn1$^{\Delta/\Delta}$ mice is more profound than in hCD3ε26 mutants and occurs prior to the stage when TEC development is dependant on lymphoepithelial cross-talk. It is therefore unclear if Foxn1 directly regulates genes involved in cross-talk or if its primary role is to regulate TEC differentiation, which has downstream effects on lymphoepithelial communication.

## Molecular Regulation of TEC Proliferation

Following the formation of the thymic primordium and the commitment of the epithelial cells to the TEC lineage, the thymus undergoes a period of expansion involving both the proliferation of stromal cells and an increase in thymocyte numbers. With regard to TECs, the growth factors Fgf7 and Fgf10 have been shown to be involved. *In vitro* experiments have demonstrated that both Fgf7 and Fgf10, which are expressed by the perithymic mesenchyme, can stimulate the proliferation of fetal TECs (Suniara *et al.*, 2000). Furthermore, mice lacking Fgfr2IIIb, the receptor for Fgf7 and Fgf10, have severely hypoplastic thymi, although they can support T-cell maturation, and Fgf10$^{-/-}$ mutants also develop hypoplastic thymi (Revest *et al.*, 2001).

## Noncanonical NF-kB Signaling Regulates mTEC Development

The development of mTECs depends on activation of the NF-kB signaling pathway, specifically the noncanonical pathway that culminates in RelB activation. Medullary TEC development is severely compromised in RelB-deficient mice (Burkly *et al.*, 1995; Weih *et al.*, 1995). In the noncanonical RelB activation pathway, ligand engagement of receptors in the TNFR family, such as lymphotoxinβ receptor (LTβR), activates NF-kB-inducing kinase (NIK), which phosphorylates homodimers of the downstream kinase, Ikkα. Activated Ikkα in turn phosphorylates the C-terminal region of NF-kB2 (p100), leading to ubiquitin-dependent degradation and release of the N-terminal polypeptide, p52. The formation of RelB/p52 heterodimers permits shuttling of RelB from the cytoplasm into the nucleus, where it functions as a transcriptional regulator (Bonizzi *et al.*, 2004). A range of medullary defects occurs in mice that are deficient in various components upstream of RelB in the alternative NF-kB activation pathway. Targeted disruption of the LTβR gene results in disorganized medullary regions that contain reduced numbers of both major mTECs subsets (Boehm *et al.*, 2003). Mice with a naturally occurring mutation in NIK (Barcena *et al.*, 1994) and Ikkα knockout mice have severe defects in medullary formation, including impairment of mTECs development and reduced expression of AIRE and tissue-restricted antigens (Kajiura *et al.*, 2004; Kinoshita *et al.*, 2006; E. Richie, unpublished observations). A similar medullary phenotype is found in TRAF6 knockout mice (Akiyama *et al.*, 2005). Interestingly, each of these mutant strains develops autoimmune manifestations, indicating a breakdown in the establishment of central tolerance.

### Role of Medullary Thymic Epithelial Cells in Establishing Central Tolerance

Medullary TECs have the unique ability to express genes encoding a wide array of antigens that initially were thought to be expressed only in peripheral tissues (Derbinski *et al.*, 2001). AIRE is expressed by mTECs and plays a prominent role in regulating tissue-restricted antigen expression. In humans, homozygous AIRE mutations result in a severe, multiorgan autoimmune disease termed *autoimmune polyglandular syndrome type 1* (Villasenor *et al.*, 2005). AIRE knockout mice develop a similar autoimmune phenotype due to defective clonal deletion, resulting in persistence of self-reactive thymocytes (Anderson *et al.*, 2002; Liston *et al.*, 2003). Although AIRE is clearly important for regulation of tissue-restricted antigen expression in mTECs, other as-yet-undefined factors are involved since certain tissue-restricted antigens are expressed even in the absence of AIRE (Derbinski *et al.*, 2005). Medullary TECs also affect central tolerance by supplying self-peptides to medullary dendritic cells (DCs) in a process termed *cross-presentation* (Gallegos and Bevan, 2004). Medullary TECs are also required for the generation of CD4$^+$CD25$^+$ T regulatory (Treg) cells and natural killer T-cells, both of which actively repress self-reactive T-cells (Kronenberg and Rudensky, 2005; Kim *et al.*, 2006). Given that AIRE and tissue-restricted antigens are highly expressed by mTECs, it is likely that the autoimmune phenotype in mice deficient in various components of the RelB activation pathway is a result of failed mTEC development. Taken together it is now clear that mTECs contribute directly and indirectly to establishing central tolerance and averting the development of autoimmunity.

## V. SUMMARY

Thymus organogenesis is a complex process in which a three-dimensional organ forms from the endoderm of the 3PP. The thymic epithelium, a critical regulator of thymopoiesis, comprises many subtypes of TEC, all of which arise from a common progenitor. Recent studies have begun to clarify the lineage relationships between these cell types and to identify the molecular factors that govern thymus organogenesis. However, several significant questions remain unresolved:

- What mechanisms regulate formation of the 3PP from the endoderm?
- How does the network of transcription factors and signaling molecules that are expressed in the endoderm and surrounding mesenchyme/mesoderm control subsequent 3PP development?
- What factor or factors specify thymus and parathyroid lineages within the 3PP?
- What are the factors responsible for the maintenance, proliferation and differentiation of TEPCs?

Additionally, important questions remain regarding homeostatic maintenance of the mature thymus, including elucidation of whether the postnatal organ is maintained by a stem cell mechanism, and the cellular and molecular mechanisms that operate to induce thymic involution. Resolution of these issues will permit the design of rational strategies for therapeutic reconstitution of the adaptive immune system. The development of such strategies should significantly impact the health of the aging population and other immunocompromised individuals.

## VI. ACKNOWLEDGMENTS

We wish to thank Lucy Morris (Carnegie Institution of Washington, D.C, and Howard Hughes Medical Institute, Baltimore), for critical reading of the manuscript, and our funding bodies, Leukaemia Research (CCB, CSN), the EU (CCB, CSN), and the NIH (NRM, ER), for support.

## VII. REFERENCES

Abu-Issa, R., Smyth, G., *et al.* (2002). Fgf8 is required for pharyngeal arch and cardiovascular development in the mouse. *Development* **129**(19), 4613–4625.

Akiyama, T., Maeda, S., *et al.* (2005). Dependence of self-tolerance on TRAF6-directed development of thymic stroma. *Science* **308**(5719), 248–251.

Anderson, G., Owen, J. J., *et al.* (1994). Thymic epithelial cells provide unique signals for positive selection of CD4+CD8+ thymocytes *in vitro*. *J. Exp. Med.* **179**(6), 2027–2031.

Anderson, M. S., Venanzi, E. S., *et al.* (2002). Projection of an immunological self-shadow within the thymus by the aire protein. *Science* **298**(5597), 1395–1401.

Ashour, M. (1995). Prevalence of ectopic thymic tissue in myasthenia gravis and its clinical significance. *J. Thorac. Cardiovasc. Surg.* **109**(4), 632–635.

Balciunaite, G., Keller, M. P., *et al.* (2002). Wnt glycoproteins regulate the expression of FoxN1, the gene defective in nude mice. *Nat. Immunol.* **3**(11), 1102–1108.

Baldwin, T. A., Sandau, M. M., *et al.* (2005). The timing of TCR alpha expression critically influences T-cell development and selection. *J. Exp. Med.* **202**(1), 111–121.

Barcena, A., Galy, A. H., *et al.* (1994). Lymphoid and myeloid differentiation of fetal liver CD34+-lineage cells in human thymic organ culture. *J. Exp. Med.* **180**(1), 123–132.

Bennett, A. R., Farley, A., *et al.* (2002). Identification and characterization of thymic epithelial progenitor cells. *Immunity* **16**(6), 803–814.

Blackburn, C. C., and Manley, N. R. (2004). Developing a new paradigm for thymus organogenesis. *Nat. Rev. Immunol.* **4**(4), 278–289.

Blackburn, C. C., Augustine, C. L., *et al.* (1996). The nu gene acts cell-autonomously and is required for differentiation of thymic epithelial progenitors. *Proc. Natl. Acad. Sci. U.S.A.* **93**(12), 5742–5746.

Bleul, C. C., and Boehm, T. (2001). Laser capture microdissection-based expression profiling identifies PD1-ligand as a target of the nude locus gene product. *Eur. J. Immunol.* **31**(8), 2497–2503.

Bleul, C. C., and Boehm, T. (2005). BMP signaling is required for normal thymus development. *J. Immunol.* **175**(8), 5213–5221.

Bleul, C. C., Corbeaux, T., *et al.* (2006). Formation of a functional thymus initiated by a postnatal epithelial progenitor cell. *Nature* **441**(7096), 992–996.

Bockman, D. E., and Kirby, M. L. (1984). Dependence of thymus development on derivatives of the neural crest. *Science* **223**(4635), 498–500.

Boehm, T., Scheu, S., *et al.* (2003). Thymic medullary epithelial cell differentiation, thymocyte emigration, and the control of autoimmunity require lympho-epithelial cross-talk via LTbetaR. *J. Exp. Med.* **198**(5), 757–769.

Bofill, M., Janossy, G., *et al.* (1985). Microenvironments in the normal thymus and the thymus in myasthenia gravis. *Am. J. Pathol.* **119**(3), 462–473.

Bonizzi, G., Bebien, M., *et al.* (2004). Activation of IKKalpha target genes depends on recognition of specific kappaB binding sites by RelB:p52 dimers. *Embo. J.* **23**(21), 4202–4210.

Boyd, R. L., Wilson, T. J., *et al.* (1992). Phenotypic characterization of chicken thymic stromal elements. *Dev. Immunol.* **2**(1), 51–66.

Boyd, R. L., Tucek, C. L., *et al.* (1993). The thymic microenvironment. *Immunol. Today* **14**(9), 445–459.

Brekelmans, P., and van Ewijk, W. (1990). Phenotypic characterization of murine thymic microenvironments. *Semin. Immunol.* **2**(1), 13–24.

Brissette, J. L., Li, J., *et al.* (1996). The product of the mouse nude locus, Whn, regulates the balance between epithelial cell growth and differentiation. *Genes Dev.* **10**(17), 2212–2221.

Burkly, L., Hession, C., *et al.* (1995). Expression of relB is required for the development of thymic medulla and dendritic cells. *Nature* **373**(6514), 531–536.

Chapman, D. L., Garvey, N., *et al.* (1996). Expression of the T-box family genes, Tbx1–Tbx5, during early mouse development. *Dev. Dyn.* **206**(4), 379–390.

Clark, R. A., Yamanaka, K., *et al.* (2005). Human skin cells support thymus-independent T cell development. *J. Clin. Invest.* **115**(11), 3239–3249.

Conway, S. J., Henderson, D. J., *et al.* (1997). Pax3 is required for cardiac neural crest migration in the mouse: evidence from the splotch (Sp2H) mutant. *Development* **124**(2), 505–514.

Cordier, A. C. (1974). Ultrastructure of the thymus in "Nude" mice. *J. Ultrastruct. Res.* **47**(20), 26–40.

Cordier, A. C., and Haumont, S. M. (1980). Development of thymus, parathyroids, and ultimo-branchial bodies in NMRI and nude mice. *Am. J. Anat.* **157**(3), 227–263.

Cordier, A. C., and Heremans, J. F. (1975). Nude mouse embryo: ectodermal nature of the primordial thymic defect. *Scand. J. Immunol.* **4**(2), 193–196.

Cumano, A., Dieterlen-Lievre, F., *et al.* (1996). Lymphoid potential, probed before circulation in mouse, is restricted to caudal intraembryonic splanchnopleura. *Cell* **86**(6), 907–916.

De Smedt, M., Reynvoet, K., *et al.* (2002). Active form of Notch imposes T cell fate in human progenitor cells. *J. Immunol.* **169**(6), 3021–3029.

Derbinski, J., Schulte, A., *et al.* (2001). Promiscuous gene expression in medullary thymic epithelial cells mirrors the peripheral self. *Nat. Immunol.* **2**(11), 1032–1039.

Derbinski, J., Gabler, J., *et al.* (2005). Promiscuous gene expression in thymic epithelial cells is regulated at multiple levels. *J. Exp. Med.* **202**(1), 33–45.

Dodson, W. E., Alexander, D., *et al.* (1969). The DiGeorge syndrome. *Lancet* **1**(7594), 574–575.

Dooley, J., Erickson, M., *et al.* (2006). Cervical thymus in the mouse. *J. Immunol.* **176**(11), 6484–6490.

Douagi, I., Andre, I., *et al.* (2000). Characterization of T cell precursor activity in the murine fetal thymus: evidence for an input of T cell precursors between days 12 and 14 of gestation. *Eur. J. Immunol.* **30**(8), 2201–2210.

Farr, A. G., and Nakane, P. K. (1983). Cells bearing Ia antigens in the murine thymus. An ultrastructural study. *Am. J. Pathol.* **111**(1), 88–97.

Flanagan, S. P. (1966). "Nude," a new hairless gene with pleiotropic effects in the mouse. *Genet. Res.* **8**(3), 295–309.

Frank, D. U., Fotheringham, L. K., *et al.* (2002). An Fgf8 mouse mutant phenocopies human 22q11 deletion syndrome. *Development* **129**(19), 4591–4603.

Frank, J., Pignata, C., *et al.* (1999). Exposing the human nude phenotype [letter]. *Nature* **398**(6727), 473–474.

Gallegos, A. M., and Bevan, M. J. (2004). Driven to autoimmunity: the nod mouse. *Cell* **117**(2), 149–151.

Ghyselinck, N. B., Dupe, V., *et al.* (1997). Role of the retinoic acid receptor beta (RARbeta) during mouse development. *Int. J. Dev. Biol.* **41**(3), 425–447.

Gill, J., Malin, M., *et al.* (2002). Generation of a complete thymic microenvironment by MTS24(+) thymic epithelial cells. *Nat. Immunol.* **3**(7), 635–642.

Godfrey, D. I., Izon, D. J., *et al.* (1990). The phenotypic heterogeneity of mouse thymic stromal cells. *Immunology* **70**(1), 66–74.

Gordon, J., Bennett, A. R., *et al.* (2001). Gcm2 and Foxn1 mark early parathyroid- and thymus-specific domains in the developing third pharyngeal pouch. *Mech. Dev.* **103**(1–2), 141–143.

Gordon, J., Wilson, V. A., *et al.* (2004). Functional evidence for a single endodermal origin for the thymic epithelium. *Nat. Immunol.* **5**(5), 546–553.

Hare, K. J., Jenkinson, E. J., *et al.* (1999). *In vitro* models of T cell development. *Semin. Immunol.* **11**(1), 3–12.

Hetzer-Egger, C., Schorpp, M., *et al.* (2002). Thymopoiesis requires Pax9 function in thymic epithelial cells. *Eur. J. Immunol.* **32**(4), 1175–1181.

Hollander, G. A., Wang, B., *et al.* (1995). Developmental control point in the induction of thymic cortex regulated by a subpopulation of prothymocytes. *Nature* **373**, 350–353.

Hu, T., Yamagishi, H., *et al.* (2004). Tbx1 regulates fibroblast growth factors in the anterior heart field through a reinforcing autoregulatory loop involving forkhead transcription factors. *Development* **131**(21), 5491–5502.

Itoi, M., Kawamoto, H., *et al.* (2001). Two distinct steps of immigration of haematopoietic progenitors into the early thymus anlage. *Int. Immunol.* **13**, 1203–1211.

Janes, S. M., Ofstad, T. A., *et al.* (2004). Transient activation of FOXN1 in keratinocytes induces a transcriptional program that promotes terminal differentiation: contrasting roles of FOXN1 and Akt. *J. Cell Sci.* **117**(Pt. 18), 4157–4168.

Jenkinson, E. J., and Owen, J. J. (1990). T-cell differentiation in thymus organ cultures. *Semin. Immunol.* **2**(1), 51–58.

Jenkinson, E. J., Van Ewijk, W., *et al.* (1981). Major histocompatibility complex antigen expression on the epithelium of the developing thymus in normal and nude mice. *J. Exp. Med.* **153**(2), 280–292.

Jenkinson, E. J., Anderson, G., *et al.* (1992). Studies on T cell maturation on defined thymic stromal cell populations *in vitro. J. Exp. Med.* **176**(3), 845–853.

Jenkinson, W. E., Rossi, S. W., *et al.* (2005). Development of functional thymic epithelial cells occurs independently of lymphostromal interactions. *Mech. Dev.* **122**(12), 1294–1299.

Jerome, L. A., and Papaioannou, V. E. (2001). DiGeorge syndrome phenotype in mice mutant for the T-box gene, Tbx1. *Nat. Genet.* **27**(3), 286–291.

Jiang, X., Rowitch, D. H., *et al.* (2000). Fate of the mammalian cardiac neural crest. *Development* **127**(8), 1607–1616.

Jotereau, F., Heuze, F., *et al.* (1987). Cell kinetics in the fetal mouse thymus: precursor cell input, proliferation, and emigration. *J. Immunol.* **138**(4), 1026–1030.

Kajiura, F., Sun, S., *et al.* (2004). NF-kappa B-inducing kinase establishes self-tolerance in a thymic stroma-dependent manner. *J. Immunol.* **172**(4), 2067–2075.

Kim, H. J., Hwang, S. J., *et al.* (2006). NKT cells play critical roles in the induction of oral tolerance by inducing regulatory T cells producing IL-10 and transforming growth factor beta, and by clonally deleting antigen-specific T cells. *Immunology* **118**(1), 101–111.

Kinoshita, D., Hirota, F., *et al.* (2006). Essential role of IkappaB kinase alpha in thymic organogenesis required for the establishment of self-tolerance. *J. Immunol.* **176**(7), 3995–4002.

Klug, D. B., Carter, C., *et al.* (1998). Interdependence of cortical thymic epithelial cell differentiation and T-lineage commitment. *Proc. Natl. Acad. Sci. U.S.A.* **95**(20), 11822–11827.

Klug, D. B., Carter, C., *et al.* (2002). Cutting edge: thymocyte-independent and thymocyte-dependent phases of epithelial patterning in the fetal thymus. *J. Immunol.* **169**(6), 2842–2845.

Kronenberg, M., and Rudensky, A. (2005). Regulation of immunity by self-reactive T cells. *Nature* **435**(7042), 598–604.

Kurobe, H., Liu, C., *et al.* (2006). CCR7-dependent cortex-to-medulla migration of positively selected thymocytes is essential for establishing central tolerance. *Immunity* **24**(2), 165–177.

La Motte-Mohs, R. N., Herer, E., *et al.* (2005). Induction of T-cell development from human cord blood hematopoietic stem cells by Delta-like 1 *in vitro. Blood* **105**(4), 1431–1439.

Lampert, I. A., and Ritter, M. A. (1988). The origin of the diverse epithelial cells of the thymus: is there a common stem cell? *In* "Thymus Update" (M. D. Kendall and M. A. Ritter, eds.), pp. 5–25. Harwood Academic, London.

Le Douarin, N. M., and Jotereau, F. V. (1975). Tracing of cells of the avian thymus through embryonic life in interspecific chimeras. *J. Exp. Med.* **142**(1), 17–40.

Le Lievre, C. S., and Le Douarin, N. M. (1975). Mesenchymal derivatives of the neural crest: analysis of chimaeric quail and chick embryos. *J. Embryol. Exp. Morphol.* **34**(1), 125–154.

Lehar, S. M., and Bevan, M. J. (2002). T cell development in culture. *Immunity* **17**(6), 689–692.

Lind, E. F., Prockop, S. E., *et al.* (2001). Mapping precursor movement through the postnatal thymus reveals specific microenvironments supporting defined stages of early lymphoid development. *J. Exp. Med.* **194**(2), 127–134.

Lindsay, E. A., Botta, A., *et al.* (1999). Congenital heart disease in mice deficient for the DiGeorge syndrome region. *Nature* **401**(6751), 379–383.

Lindsay, E. A., Vitelli, F., *et al.* (2001). Tbx1 haploinsufficieny in the DiGeorge syndrome region causes aortic arch defects in mice. *Nature* **410**(6824), 97–101.

Liston, A., Lesage, S., *et al.* (2003). Aire regulates negative selection of organ-specific T cells. *Nat. Immunol.* **4**(4), 350–354.

Lobach, D. F., and Haynes, B. F. (1986). Ontogeny of the human thymus during fetal development. *J. Clin. Immunol.* **7**, 81–97.

Macatee, T. L., Hammond, B. P., *et al.* (2003). Ablation of specific expression domains reveals discrete functions of ectoderm- and endoderm-derived FGF8 during cardiovascular and pharyngeal development. *Development* **130**(25), 6361–6374.

Manley, N. R., and Blackburn, C. C. (2003). A developmental look at thymus organogenesis: where do the non-hematopoietic cells in the thymus come from? *Curr. Opin. Immunol.* **15**(2), 225–232.

Manley, N. R., and Capecchi, M. R. (1995). The role of Hoxa-3 in mouse thymus and thyroid development. *Development* **121**(7), 1989–2003.

Manley, N. R., and Capecchi, M. R. (1998). Hox group 3 paralogs regulate the development and migration of the thymus, thyroid, and parathyroid glands. *Dev. Biol.* **195**(1), 1–15.

Moore-Scott, B. A., and Manley, N. R. (2005). Differential expression of Sonic hedgehog along the anterior–posterior axis regulates patterning of pharyngeal pouch endoderm and pharyngeal endoderm-derived organs. *Dev. Biol.* **278**(2), 323–335.

Morris, L., Gordon, J., *et al.* (2006). Identification of a tandem duplicated array in the Rhox locus on mouse chromosome X. *Mammalian Genome*, in press.

Mulroy, T., McMahon, J. A., *et al.* (2002). Wnt-1 and Wnt-4 regulate thymic cellularity. *Eur. J. Immunol.* **32**(4), 967–971.

Nehls, M., Kyewski, B., *et al.* (1996). Two genetically separable steps in the differentiation of thymic epithelium. *Science* **272**(5263), 886–889.

Nishimura, H., Agata, Y., *et al.* (1996). Developmentally regulated expression of the PD-1 protein on the surface of double-negative (CD4–CD8–) thymocytes. *Int. Immunol.* **8**(5), 773–780.

Nishimura, H., Honjo, T., *et al.* (2000). Facilitation of beta selection and modification of positive selection in the thymus of PD-1-deficient mice. *J. Exp. Med.* **191**(5), 891–898.

Norris, E. H. (1938). The morphogenesis and histogenesis of the thymus gland in man: in which the origin of the Hassall's corpuscle of the human thymus is discovered. *Contr. Embryol. Carnegie Instn.* **27**, 191–207.

Owen, J. J., and Ritter, M. A. (1969). Tissue interaction in the development of thymus lymphocytes. *J. Exp. Med.* **129**(2), 431–442.

Patel, S. R., Gordon, J., *et al.* (2006). Bmp4 and Noggin expression during early thymus and parathyroid organogenesis. *Gene Expr. Patterns* **6**(8), 794–799.

Paylor, R., McIlwain, K. L., *et al.* (2001). Mice deleted for the DiGeorge/velocardiofacial syndrome region show abnormal sensorimotor gating and learning and memory impairments. *Hum. Mol. Genet.* **10**(23), 2645–2650.

Peters, H., and Balling, R. (1999). Teeth. Where and how to make them. *Trends Genet.* **15**(2), 59–65.

Petrie, H. T. (2002). Role of thymic organ structure and stromal composition in steady-state postnatal T-cell production. *Immunol. Rev.* **189**(1), 8–20.

Plotkin, J., Prockop, S. E., *et al.* (2003). Critical role for CXCR4 signaling in progenitor localization and T cell differentiation in the postnatal thymus. *J. Immunol.* **171**(9), 4521–4527.

Plum, J., De Smedt, M., *et al.* (1994). Human CD34+ fetal liver stem cells differentiate to T cells in a mouse thymic microenvironment. *Blood* **84**(5), 1587–1593.

Porritt, H. E., Rumfelt, L. L., *et al.* (2004). Heterogeneity among DN1 prothymocytes reveals multiple progenitors with different capacities to generate T-cell and non–T-cell lineages. *Immunity* **20**(6), 735–745.

Poznansky, M. C., Evans, R. H., *et al.* (2000). Efficient generation of human T cells from a tissue-engineered thymic organoid. *Nat. Biotechnol.* **18**(7), 729–734.

Revest, J. M., Suniara, R. K., *et al.* (2001). Development of the thymus requires signaling through the fibroblast growth factor receptor R2-IIIb. *J. Immunol.* **167**(4), 1954–1961.

Rossi, S. W., Jenkinson, W. E., *et al.* (2006). Clonal analysis reveals a common progenitor for thymic cortical and medullary epithelium. *Nature* **441**(7096), 988–991.

Rothenberg, E. V., and Dionne, C. J. (2002). Lineage plasticity and commitment in T-cell development. *Immunol. Rev.* **187**, 96–115.

Scambler, P. J. (2000). The 22q11 deletion syndromes. *Hum. Mol. Genet.* **9**(16), 2421–2426.

Schluep, M., Willcox, N., *et al.* (1988). Myasthenia gravis thymus: clinical, histological and culture correlations. *J. Autoimmun.* **1**(5), 445–467.

Schmitt, T. M., and Zuniga-Pflucker, J. C. (2002). Induction of T-cell development from hematopoietic progenitor cells by delta-like-1 *in vitro. Immunity* **17**(6), 749–756.

Schmitt, T. M., de Pooter, R. F., *et al.* (2004). Induction of T-cell development and establishment of T-cell competence from embryonic stem cells differentiated *in vitro. Nat. Immunol.* **5**(4), 410–417.

Staal, F. J., and Clevers, H. C. (2005). WNT signaling and haematopoiesis: a WNT-WNT situation. *Nat. Rev. Immunol.* **5**(1), 21–30.

Su, D. M., and Manley, N. R. (2000). Hoxa3 and pax1 transcription factors regulate the ability of fetal thymic epithelial cells to promote thymocyte development. *J. Immunol.* **164**(11), 5753–5760.

Su, D., Ellis, S., *et al.* (2001). *Hoxa3* and *Pax1* regulate epithelial cell death and proliferation during thymus and parathyroid organogenesis. *Dev. Biol.* **236**, 316–329.

Su, D. M., Navarre, S., *et al.* (2003). A domain of Foxn1 required for cross-talk-dependent thymic epithelial cell differentiation. *Nat. Immunol.* **4**(11), 1128–1135.

Suniara, R. K., Jenkinson, E. J., *et al.* (2000). An essential role for thymic mesenchyme in early T-cell development. *J. Exp. Med.* **191**(6), 1051–1056.

Surh, C. D., Gao, E. K., *et al.* (1992). Two subsets of epithelial cells in the thymic medulla. *J. Exp. Med.* **176**(2), 495–505.

Taddei, I., Morishima, M., *et al.* (2001). Genetic factors are major determinants of phenotypic variability in a mouse model of the DiGeorge/del22q11 syndromes. *Proc. Natl. Acad. Sci. U.S.A.* **98**(20), 11428–11431.

Terszowski, G., Muller, S. M., *et al.* (2006). Evidence for a functional second thymus in mice. *Science* **312**(5771), 284–287.

Tovi, F., and Mares, A. J. (1978). The aberrant cervical thymus. Embryology, pathology, and clinical implications. *Am. J. Surg.* **136**(5), 631–637.

Tsai, P. T., Lee, R. A., *et al.* (2003). BMP4 acts upstream of FGF in modulating thymic stroma and regulating thymopoiesis. *Blood* **102**(12), 3947–3953.

Ueno, T., Saito, F., *et al.* (2004). CCR7 signals are essential for cortex–medulla migration of developing thymocytes. *J. Exp. Med.* **200**(4), 493–505.

van de Wijngaert, F. P., Rademakers, L. H., *et al.* (1983). Identification and *in situ* localization of the "thymic nurse cell" in man. *J. Immunol.* **130**(5), 2348–2351.

van de Wijngaert, F. P., Kendall, M. D., *et al.* (1984). Heterogeneity of epithelial cells in the human thymus. An ultrastructural study. *Cell Tissue Res.* **237**(2), 227–237.

van Dyke, J. H. (1941). On the origin of accessory thymus tissue. IV: The occurrence in man. *Anat. Rec.* **79**, 179–209.

van Ewijk, W., Shores, E. W., *et al.* (1994). Cross-talk in the mouse thymus. *Immunol. Today* **15**(5), 214–217.

Van Vliet, E., Melis, M., *et al.* (1984). Monoclonal antibodies to stromal cell types of the mouse thymus. *Eur. J. Immunol.* **14**(6), 524–529.

Van Vliet, E., Jenkinson, E. J., *et al.* (1985). Stromal cell types in the developing thymus of the normal and nude mouse embryo. *Eur. J. Immunol.* **15**(7), 675–681.

Villasenor, J., Benoist, C., *et al.* (2005). AIRE and APECED: molecular insights into an autoimmune disease. *Immunol. Rev.* **204**, 156–164.

von Freeden-Jeffry, U., Vieira, P., *et al.* (1995). Lymphopenia in interleukin (IL)-7 gene-deleted mice identifies IL-7 as a nonredundant cytokine. *J. Exp. Med.* **181**(4), 1519–1526.

von Gaudecker, B., Steinmann, G. G., *et al.* (1986). Immunohistochemical characterization of the thymic microenvironment. A light-microscopic and ultrastructural immunocytochemical study. *Cell Tissue Res.* **244**(2), 403–412.

Wallin, J., Eibel, H., *et al.* (1996). Pax1 is expressed during development of the thymus epithelium and is required for normal T-cell maturation. *Development* **122**(1), 23–30.

Weih, F., Carrasco, D., *et al.* (1995). Multiorgan inflammation and hematopoietic abnormalities in mice with a targeted disruption of RelB, a member of the NF-kappa B/Rel family. *Cell* **80**(2), 331–340.

Wekerle, H., and Ketelsen, U. P. (1980a). Thymic nurse cells — Ia-bearing epithelium involved in T-lymphocyte differentiation? *Nature* **283**(5745), 402–404.

Wekerle, H., Ketelsen, U. P., *et al.* (1980b). Thymic nurse cells. Lymphoepithelial cell complexes in murine thymuses: morphological and serological characterization. *J. Exp. Med.* **151**(4), 925–944.

Weller, G. L. (1933). Development of the thyroid, parathyroid and thymus glands in man. *Cont. Embryol.*

Wendling, O., Dennefeld, C., *et al.* (2000). Retinoid signaling is essential for patterning the endoderm of the third and fourth pharyngeal arches. *Development* **127**(8), 1553–1562.

Xu, P. X., Zheng, W., *et al.* (2002). Eya1 is required for the morphogenesis of mammalian thymus, parathyroid and thyroid. *Development* **129**(13), 3033–3044.

Xu, H., Morishima, M., *et al.* (2004). Tbx1 has a dual role in the morphogenesis of the cardiac outflow tract. *Development* **131**(13), 3217–3227.

Xu, H., Cerrato, F., *et al.* (2005). Timed mutation and cell-fate mapping reveal reiterated roles of Tbx1 during embryogenesis, and a crucial function during segmentation of the pharyngeal system via regulation of endoderm expansion. *Development* **132**(19), 4387–4395.

Yeoman, H., Gress, R. E., *et al.* (1993). Human bone marrow and umbilical cord blood cells generate CD4+ and CD8+ single-positive T-cells in murine fetal thymus organ culture. *Proc. Natl. Acad. Sci. U.S.A.* **90**(22), 10778–10782.

Zakrzewski, J. L., Kochman, A. A., *et al.* (2006). Adoptive transfer of T-cell precursors enhances T-cell reconstitution after allogeneic hematopoietic stem cell transplantation. *Nat. Med.*

Zamisch, M., Moore-Scott, B., *et al.* (2005). Ontogeny and regulation of IL-7-expressing thymic epithelial cells. *J. Immunol.* **174**(1), 60–67.

Zou, D., Silvius, D., *et al.* (2006). Patterning of the third pharyngeal pouch into thymus/parathyroid by Six and Eya1. *Dev. Biol.* **293**(2), 499–512.

Zuniga-Pflucker, J. C., and Lenardo, M. J. (1996). Regulation of thymocyte development from immature progenitors. *Curr. Opin. Immunol.* **8**(2), 215–224.

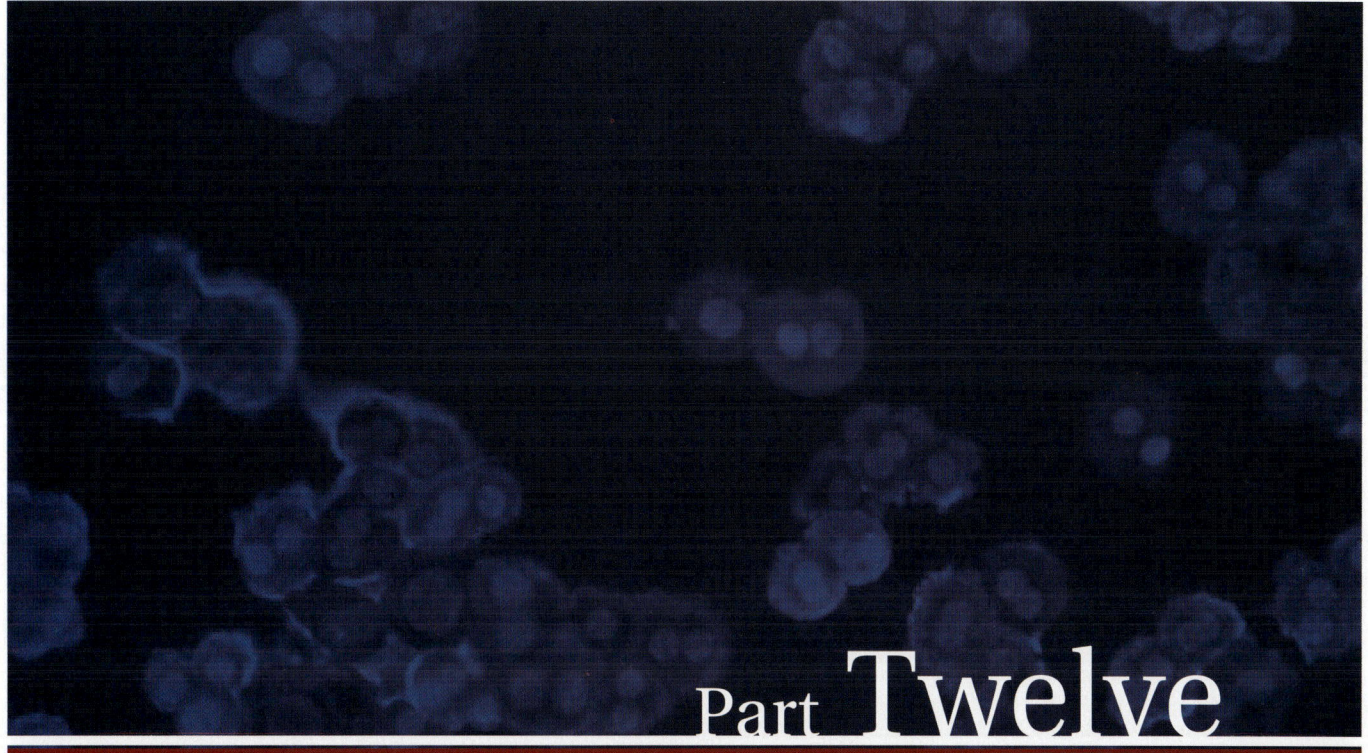

Part Twelve

# Gastrointestinal System

# Chapter Forty-Five

# Adult Stem Cells in Normal Gastrointestinal Function and Inflammatory Disease

*Mairi Brittan and Nicholas A. Wright*

## I. INTRODUCTION

This chapter defines the avenues via which stem cells contribute to the maintenance of the gastrointestinal tract in the normal state and in inflammatory disease. Much progress has recently been made to characterize and identify the intestinal epithelial stem cell. We have a strong interest in the intestinal stem cell niche, the key mediator of intestinal stem cell behavior. The stem cell niche is believed to be composed of the underlying mesenchymal cells, including the intestinal subepithelial myofibroblasts. The origins of the niche cells are not clear, although we have shown that a large proportion of intestinal myofibroblasts are derived from bone marrow cells and, moreover, that the contribution of bone marrow to intestinal myofibroblasts is significantly up-regulated in the inflamed colon. We have

shown that bone marrow cells also contribute to postnatal neovasculogenesis in the inflamed colon. We believe that this property of bone marrow stem cells, to redirect their lineage differentiation pathways and form mesenchymal and vascular lineages in response to specific regenerative signals, provides more clout than current evidence that bone marrow cells form epithelial cells in the gut as this is often not replicated, generally occurs to a very low degree, and the cellular mechanisms remain unclear with evidence that these cells may be a product of cell fusion rather than true differentiation. Indeed, bone marrow cells give rise to a far larger proportion of both myofibroblasts and vascular lineages in the gut, with a significant increase when placed under regenerative stress in the form of inflammatory disease. The role of bone marrow in tissue regeneration in mouse models of colitis will be considered, with increasing evidence that bone marrow transplantation can alleviate Crohn's disease in humans, highlighting a possible role of bone marrow cells in the treatment of inflammatory disease, and defining a possible mechanism by which these cells provide this disease alleviation.

## II. DEFINING PROPERTIES OF ADULT STEM CELLS

Stem cells are located in most adult tissues and display several collective criteria, allowing identification and classification of these otherwise-indefinable cells. It is generally accepted that adult stem cells are located within a *niche*, formed and maintained by the subjacent mesenchymal cells and their secreted basement membrane factors, which serve to provide a specific milieu conducive to stem cell longevity, which is epitomized by the capacity of these cells for prolonged self-renewal. Stem cells are also capable of multilineage differentiation and form a specific repertoire of adult lineages within their native tissue, which is initiated by asymmetrical stem cell division to produce a transit amplifying (TA) cell, which becomes committed to differentiation and the formation of a specific functional adult lineage. This latter characteristic has been somewhat revised in recent years, because we now know that stem cells within some tissues are not merely restricted to the formation of differentiated cells within their resident tissue, but can also form functional differentiated cell populations in tissues outside their own. This property, known as *stem cell plasticity*, has been confirmed in various disease and injury states in both humans and rodents, and we postulate that stem cells from one tissue can act as a *supplementary* stem cell population in response to specific stimuli released from a damaged or diseased tissue, possibly when regenerative pressure exceeds the capacity of the indigenous tissue stem cells. Stem cells appear to respond to these stimuli by homing to and engrafting within a damaged tissue, wherein they contribute to regeneration by differentiating to form a specific cell type. Stem cells in the bone marrow (BM) have frequently been shown to act in such a manner, i.e., as an auxiliary regenerative portion. This is fortuitous because BM provides a readily available source of well-characterized stem cells and has raised the potential of these cells as vectors for the delivery of therapeutic genes to a diseased tissue, although the extent of stem cell migration and differentiation between other tissues and the stimulatory signals and molecular mechanisms that initiate and regulate stem cell differentiation are unknown. Currently, the phenomenon of stem cell plasticity is not without stigma, for demonstrations of stem cell engraftment within a tissue often fail to be replicated, and the mechanism of stem cell differentiation is unclear, with evidence that in some tissues the cells merely fuse with an existing adult cell to form a heterokaryon.

In this chapter we discuss the role of the stem cells in the intestine in normal and inflamed states. We consider the recent growing evidence that BM transplantation appears to alleviate Crohn's disease, and we suggest a mechanism of action of the transplanted cells, i.e., via the formation of supporting regenerative cells in both the lamina propria and the blood vessels, which are important in tissue regeneration in inflammatory disease.

## III. CELLS OF THE INTESTINE

### Epithelial Cells

The epithelial lining of the intestine is folded into depressions known as *crypts*, which vary in size, depending on their location, and function to increase surface area to aid the absorption and digestion of nutrients from food. In the small intestine, in addition to the crypts, the surface area is further increased by the formation of luminal prominences known as *villi*. The fully differentiated cells of the crypts and villi are situated toward the luminal surface and are continually being shed into the lumen and replaced by a stream of proliferating progenitor cells. The columnar cells, termed *enterocytes* in the small intestine and *colonocytes* in the colon, are the most abundant epithelial cell type in the mucosa, responsible for secretion and absorption. Columnar cells are polarized cells with a basal nucleus and an apical brush border composed of microvilli, which increase the surface area further. The *goblet* cells, so-called because of their characteristic shape, are dispersed throughout the small intestinal and colonic epithelium and secrete mucus into the intestinal lumen to lubricate the mucosa and to trap and expel microorganisms. The gastrointestinal tract is the largest endocrine organ in the body; hence the *endocrine*, *neuroendocrine*, or *enteroendocrine* cells are abundant throughout the epithelium, although in the small intestine they are more common to the crypts than the villi. The *Paneth* cells are almost exclusive to the crypt base of the small intestine and ascending colon. These cells maintain a sterile environment in the crypt via phagocytosis and the secretion of various antibacterial substances.

### Lamina Propria

The lamina propria is a layer of loose connective tissue that lies immediately subjacent to the epithelial mucosa throughout the gastrointestinal tract. The lamina propria is composed of mesenchymal cells and their secreted basement membrane factors; it is highly vascular, with lymphatic capillaries and nervous tissue (reviewed in Powell *et al.*, 1999). The lamina propria contains two myofibroblast subpopulations: the interstitial cells of Cajal (ICC) and the intestinal subepithelial myofibroblasts (ISEMFs).

#### Intestinal Subepithelial Myofibroblasts

ISEMFs display cytologic characteristics of both fibroblasts and smooth muscle cells and are immunoreactive for α-smooth muscle actin (SMA), vimentin, and smooth muscle myosin antigens, though they do not express desmin under normal circumstances (reviewed in Powell *et al.*, 1999). This antigenic phenotype permits their distinction from the vimentin-negative fibroblasts and the strongly desmin-positive smooth muscle cells. ISEMFs exist as a cellular syncytium in the subepithelial lamina propria and are most prominently expressed around the lower two-thirds of the crypts. ISEMFs proliferate and migrate upward along the crypt–villus axis until they reach the tip of the crypt or villi, where they are then shed into the intestinal lumen. The pattern and time frame of this myofibroblast migration is similar to that of murine epithelial cells (Wright and Alison, 1984). ISEMFs are adjoined to epithelial cells via gap and adherens junctions, which permit epithelial:mesenchymal interaction, and these cells have a broad range of functions, including the regulation of epithelial cell homeostasis, mucosal protection and wound healing, contraction and motility of small intestinal villi, and water and electrolyte transport in the colon (reviewed in Powell *et al.*, 1999).

### Intestinal Epithelial Stem Cells

The epithelial cells of the intestine are a continuously and rapidly renewing population. For example, the lining of the gastrointestinal tract is replaced every two to three days in rodents and other mammals (Wright and Alison, 1984). To regulate homeostasis, a vital balance between cell apoptosis, senescence, and the proliferation and differentiation of new cells must be maintained. This role is attributed to the intestinal stem cell, although, despite its significance as the most important regulatory element of intestinal function, limited evidence exists to definitively substantiate the location, quantity, regulatory pathways, or function of this elusive cell.

### Location of the Intestinal Epithelial Stem Cell

Since the polarized orientation of the differentiated cells in the intestinal crypts and villi is well characterized, it is presumed that a pool of progenitor cells exists at the origin of cellular migration that is responsible for this continuous cellular flux. The epithelial cells of the small intestine originate in the lower regions of the crypt and undergo proliferation and differentiation as they travel upward to be shed into the intestinal lumen, with the exception of the Paneth cells, which migrate downward to the crypt base. It is therefore generally accepted that the epithelial stem cell compartment is in the crypt base of the colon and at approximately cell position 4–5 in the small intestinal crypts, just superior to the Paneth cells (Potten *et al.*, 1997).

### Intestinal Stem Cell Niche

As with stem cells of most adult tissues, it is thought that the intestinal epithelial stem cells are located and maintained within a mesenchymal niche, situated near the base of the intestinal crypts. However, because intestinal stem cells remain unidentified, it is not possible to localize the niche that underlies them; thus, our knowledge of the intestinal niche is based on circumstantial evidence and speculation. It is generally believed that epithelial stem cells are regulated by nonepithelial components, i.e., the mesenchymal cells and associated secreted factors in the lamina propria, which fits with the classical view of the composition of the stem cell niche (Spradling *et al.*, 2001) and is further supported by evidence of interactions between the ISEMFs and epithelial cells of the mucosa. For example, ISEMFs play vital roles in epithelial cell restitution, remodeling, fibrosis, and immunological and inflammatory responses via the secretion of specific growth factors and inflammatory cytokines (reviewed in Powell *et al.*, 1999). This epithelial:mesenchymal interaction highlights the ISEMF as a candidate component of the intestinal epithelial stem cell niche, although it is not clear if mesenchymal signals act directly on the stem cells or on the transiently differentiated daughter cells.

### Intestinal Stem Cell Number/Clonal Origins of Intestinal Crypts

#### Mouse

The number of stem cells within the gastrointestinal niche is a subject of ongoing debate. Investigations into the clonal origins of intestinal crypts have attempted to deduce whether they are monoclonal populations derived from a single stem cell or whether multiple stem cells proliferate to produce polyclonal crypts. The *Unitarian hypothesis* states that a single intestinal stem cell can clonally expand to produce the entire adult cell repertoire in the intestinal crypts (Cheng and Leblond, 1974). This is supported by evidence that a single surviving stem cell can recreate entire monoclonal crypts following irradiation damage (Ponder *et al.*, 1985), although these studies of damaged epithelium do not confirm that intestinal crypts are monoclonal under normal circumstances.

Mouse aggregation chimeras, wherein one parental strain bears a specific marker for the tissue of interest, have been used to investigate the Unitarian hypothesis. The binding capacity of the *Dolichos biflorus* agglutinin (DBA) lectin to the cells of the intestinal epithelium can be abolished by spontaneous mutation of the *Dlb-1* locus on chromosome 1 or by treatment with the chemical mutagen ethylnitrosourea (ENU). The SWR mouse has a carbohydrate polymorphism of *Dlb-1*; consequently, DBA binds to sites on the C57Bl/6J derived, but not SWR derived, cells in heterozygous C57Bl/6J(B6)↔SWR mouse embryo aggregation chimeras. Intestinal crypts in neonatal B6↔SWR mice were polyclonal for the first two weeks after birth, suggesting that multiple stem cells exist during development (Schmidt *et al.*, 1988). However, all crypts eventually become monoclonal, possibly due to the positive selection of a single dominant clone or by the segregation of lineages due to the division of crypts by *crypt fission*, which occurs frequently during this developmental period (Bjerknes and Cheng, 1999; Park *et al.*, 1995). The epithelial cells in the crypts remain monoclonal in the adult mouse intestine, confirming the existence of a single, sustainable stem cell (Bjerknes and Cheng, 1999; Ponder *et al.*, 1985; Winton and Ponder, 1990). Similarly, ENU treatment in B6↔SWR mice results in loss of DBA binding in mutated cells and revealed that small intestinal crypts are initially partially and then entirely negative for DBA staining (Winton *et al.*, 1988). It is proposed that ENU causes mutation of the *Dlb-1* locus in a stem cell, which then expands stochastically to produce a clone of cells that cannot bind DBA and remain unstained (Winton *et al.*, 1988). If this is the case, then a single stem cell can give rise to all the epithelial lineages within the small intestinal crypts.

In female mice with a heterozygous polymorphism of the X-linked gene glucose-6-phosphate dehydrogenase (*G6PD*), individual X chromosomes are distinguished by their natural "mosaic" pattern of *G6PD* immunohistological staining. Use of this model to investigate clonality in the intestinal crypts excludes the possibility that crypts derived from distinct strains in chimeric mice segregate differentially during organogenesis. The monophenotypic origin of murine intestinal crypts was confirmed following histochemical analyses in these mice, although the small intestinal villi showed a polyclonal derivation and are presumably formed by the upward migration of epithelial cells from multiple crypts (Thomas *et al.*, 1988). This is concordant with observations that crypts, although smaller than villi, are sevenfold more numerous in the mouse duodenum and fourfold more numerous in the ileum (Wright and Alison, 1984).

In both chimeric mice and the *G6PD* mouse model, the time taken for the decrease in partially mutated crypts and the emergence of entirely negative crypts to reach a plateau was approximately four weeks in the small intestine and 12 weeks in the large intestine, which was initially thought to be due to tissue differences in cell-cycle duration. However, an alternative explanation can be found in the *stem cell zone* hypothesis, which states that multiple stem cells occupy the crypt base at cell positions 1–4 and can undergo proliferation but do not differentiate until they have migrated to cell position 5. Paneth cells also appear initially at position 5, although these cells migrate downward to the crypt base (Bjerknes and Cheng, 1981). Based on this hypothesis, it was suggested that larger numbers of stem cells are present in the colon than in the small intestine, causing the difference in time taken for the mutant stem cells to expand stochastically and create monoclonal crypts (Williams *et al.*, 1992).

Alternatively, the temporal differences in the appearance of monophenotypical crypts in specific regions of the gastrointestinal tract may be due to differential rates of crypt fission at the time of mutagen administration (Park *et al.*, 1995). It was suggested that the stem cell number fluctuates throughout the crypt cycle until a threshold

number of cells is reached, thereby signaling crypt fission to occur (Loeffler *et al.*, 1997). Other reports state that between four and six stem cells located at cell position 4–5 from the base of the crypt, just superior to the Paneth cells, comprise the small intestinal stem cell population (Cai *et al.*, 1997; Potten *et al.*, 1997), and others claim that up to 16 or more stem cells can exist in a single intestinal crypt (Roberts *et al.*, 1995).

### Human

In humans, the intestinal crypts appear to be monoclonal and the small intestinal villi are polyclonal, analogous to the situation in the mouse. Perhaps the best evidence for the clonal origin of human intestinal crypts comes from a study of the colon of a rare XO/XY patient who had received a prophylactic colectomy for familial adenomatous polyposis (FAP). Nonisotopic *in situ* hybridization (NISH) using Y chromosome–specific probes showed that normal intestinal crypts were composed almost entirely of either Y chromosome–positive or Y chromosome–negative cells, with approximately 20% of crypts being XO. Immunostaining for neuroendocrine cell–specific markers was combined with NISH detection of the Y chromosome to show that these cells shared the same genotype as other crypt cells. The small intestinal villi epithelia were a mixture of XO and XY cells, in keeping with the notion that villi derive from stem cells of more than one crypt. Of the 12,614 crypts examined, only four crypts were composed of XO and XY cells, which was explained by nondisjunction with loss of the Y chromosome in a crypt stem cell (Novelli *et al.*, 1996). These observations agree with previous findings in chimeric mice that intestinal crypts are monoclonal and derive from a single multipotential stem cell.

Clonality studies are not strictly sustainable if they overlook the vital consideration of *patch size*, in which a *patch* is described as the number of cells of a single genotype within an area of tissue that is derived either from a single clone or from the coalescence of multiple clones of the same lineage (Schmidt *et al.*, 1985). Clonality must be determined at the patch edge, because it is not possible to decipher whether cells within the center of a patch are truly clonal or simply monophenotypic, formed from several stem cells of the same genotype. Heterozygosity for the *G6PD* Mediterranean mutation (563 C→T) is present in 17% of Sardinian females, permitting analyses of patch size by *G6PD* immunohistochemical staining. Of 10,538 colonic crypts analyzed from nine patients carrying the *G6PD* Mediterranean mutation, patch size in the colon was observed to be relatively large, containing up to 450 crypts. No evidence of any crypts with a mixed phenotype was observed in 2260 crypts located at the periphery of a patch, indicating that colonic crypts are indeed monoclonally derived (Novelli *et al.*, 2003). The Unitarian hypothesis, that all the differentiated epithelial lineages within the gastrointestinal tract share a common cell of origin, appears to apply to both mice and humans.

# IV. IDENTIFICATION OF INTESTINAL STEM CELLS

The maintenance of homeostasis of the intestinal epithelium involves an intricate and complex interplay of multiple regulatory mechanisms. It is therefore of great interest to uncover the molecular and cellular pathways that are required for normal intestinal function, starting, of course, with the ongoing quest to identify the intestinal epithelial stem cell and its niche.

## Crypt Fission and the Crypt Cycle

The intestinal crypts divide and replicate by a branching process known as *crypt fission*. In this process, the intestinal crypts undergo basal bifurcation and budding, leading to longitudinal division and the formation of identical daughter crypts. Elucidation of the process of crypt multiplication led to the notion that the intestinal crypts have a *crypt cycle* and finite life span, which was subsequently shown to be approximately two years in the mouse small intestine (Loeffler *et al.*, 1997). Because crypt fission begins at the bottom of the crypt, i.e., the location of the putative stem cells, it is possible that the stem cells initiate and regulate the formation of new crypts. Indeed, evidence is emerging that crypt fission may occur in the intestine when a maximal crypt size is reached (Park *et al.*, 1997).

A stochastic state–dependent model of stem cell growth proposes three possible outcomes when stem cells divide: *r*, in which a stem cell divides asymmetrically to produce one stem cell that is retained within the niche and one daughter cell that leaves the niche and becomes committed to differentiation; *p*, in which two stem cells are produced following symmetrical stem cell division; and *q*, in which stem cells divide symmetrically to produce two daughter cells that leave the niche and become transit amplifying cells (Loeffler *et al.*, 1993).

## DNA-Labeling Studies

A recent study of the mechanisms of stem cell division in the mouse small intestine has revealed that during asymmetrical division, stem cells retain an innate mechanism of genome protection. By labeling DNA template strands of the intestinal stem cells with tritiated thymidine ([3H]TdR) during development or in tissue regeneration and by bromodeoxyuridine (BrdUrd) labeling the newly synthesized daughter strands, both DNA strands can be visualized during cell division. Results showed that the original template DNA is retained within the stem cell and that the newly synthesized BrdUrd-labeled strands are passed on to the daughter cells, which leave the stem cell niche and become committed to differentiation. By discarding the newly transcribed DNA during asymmetrical cell division, the intestinal stem cell utilizes an inherent mechanism of genome protection, for this strand is more prone to replication-induced mutation (Potten *et al.*, 2002). This study supports the hypothesis of Cairns (1975), which suggests that selective retention of the template DNA strand during stem cell division provides a means of protection against DNA replication errors.

## DNA Mutation Studies

An insight into stem cell organization in the crypts was gained from studies of the methylation pattern of nonexpressed genes in the colon. Phylogenetic analyses of random sequence variations in methylation tags of three neutral loci in cells of human colonic crypts can predict stem cell division histories and map cell fate. Cells in separate, unrelated crypts show variations in these methylation patterns, as expected, and cells closely opposed within one crypt display identical or closely related sequential methylation tags. This indicates that human colonic crypts contain multiple stem cells, which are constantly lost and replaced, eventually leading to a "bottleneck" effect, in which all cells are related to the closest stem cell ascendant and are monoclonally derived (Yatabe *et al.*, 2001).

Both inherited and spontaneous mutations occur frequently in the human mitochondrial genome. The high frequency of mitochondrial DNA (mtDNA) mutations and their accumulation within individual cells are a result of clonal expansion by genetic drift (Chinnery and Samuels, 1999) and can produce a defect in oxidative phosphorylation (Sciacco *et al.*, 1994). Cells demonstrating >50% cytochrome-c oxidase enzyme function and normal succinate dehydrogenase (SDH) activity generally have increased numbers of mtDNA mutations (Johnson *et al.*, 1993). The biochemical function of samples of both normal and cancerous human colonic mucosa was deduced by sequencing their mtDNA and by histological analysis of cytochrome-c oxidase and SDH function. Cells with mutated mtDNA occupy part of the crypt and then clonally expand, resulting in entire colonization of a crypt by mutated cell, thus demonstrating the same process of *monoclonal conversion* as was observed in the *G6PD* mice, mentioned previously (Taylor *et al.*, 2003). Therefore, in humans, it appears that a single stem cell can form an entire crypt and then divide by crypt fission, demonstrated by the presence of stochastic mutations in mtDNA in both bifurcating arms of crypts undergoing fission (Taylor *et al.*, 2003). Moreover, mutations in crypts undergoing fission share the same genotype, which is different in cytochrome c–positive crypts; moreover, the genotype of neighboring mutated crypts is identical but differs from adjacent, nonmutated crypts. Patches of neighboring crypts deficient in cytochrome-c oxidase are clustered, and patches of mutated crypts increase in size with age, providing further evidence of the development of clonal patches in the human colon (Greaves *et al.*, 2006).

## Molecular Markers of Intestinal Stem Cells

### Musashi and Hairy and Enhancer-of-Split Proteins

Musashi-1 (Msi-1) is highly expressed in mammalian neural stem cells (Sakakibara *et al.*, 1996). The transcriptional repressor Hairy and Enhancer-of-split (Hes)-1 is essential for neural stem cell self-renewal and suppression of neural stem cell differentiation (Nakamura *et al.*, 2000). Msi-1 positively regulates transcription of Hes-1, suggesting a close interaction between the two proteins, which is supported by their coexpression in cells just superior to the Paneth cells in the small intestine, the postulated stem cell zone. Hes-1 expression in the mouse small intestine is more widespread than Msi-1 expression, because Hes-1 is also expressed, albeit in reduced levels, in epithelial cells migrating toward the villus tip. It is suggested that colocalization of Msi-1 and Hes-1 in cells just superior to the Paneth cells is indicative of the stem cell population in the mouse small intestine and that Hes-1 expression alone represents proliferating cells committed to differentiation that have migrated outside the stem cell niche (Kayahara, 2003). Musashi-1 mRNA and protein expression has also been confirmed in the putative stem cells in neonatal and adult intestinal crypts in mice (Potten *et al.*, 2003) and has recently been demonstrated in the human colon in epithelial cells located between positions 1 and 10 in the crypts (Nishimura *et al.*, 2003). These studies implicate Musashi-1 as a possible gastrointestinal epithelial stem cell marker.

### Repression of E-Cadherin

E-cadherin is a cell adhesion molecule with a well-established role in tumor suppression. E-cadherin is expressed in high levels in the intestinal epithelium and has been shown to regulate intestinal cell polarity and differentiation via the β-catenin/Tcf/LEF signaling pathway (Gottardi *et al.*, 2001), which is vital for normal intestinal epithelial cell function. A recent publication claims that differential E-cadherin expression in the epithelial cells along the human small intestine crypt–villus axis provides an insight into the nature of the hierarchical polar distribution of these cells and can be used to locate the intestinal epithelial stem cells. E-cadherin is absent from the cells in the base of the human small intestinal crypts, at approximately cell position 5–7. Forced expression of E-cadherin in those cells caused a decrease in proliferation and promoted cell junction formation, thereby inhibiting cell migration. It is therefore proposed that E-cadherin represents a marker of intestinal epithelial stem cells and plays an important role in regulating stem cell behavior. However, the inability of E-cadherin to induce terminal differentiation of enterocytes indicates that it acts in conjunction with other proteins (Escaffit *et al.*, 2005).

## V. PATHWAYS OF CELLULAR DIFFERENTIATION IN THE INTESTINE

An increasing number of genes and their ligands and receptors are being identified that are expressed by epithelial and mesenchymal cells in the gastrointestinal tract and are involved in the regulatory molecular pathways of epithelial cell function in both the normal and neoplastic gastrointestinal mucosa. The ongoing elucidation of these molecular pathways is key to providing an insight into the location and the behavior of the intestinal epithelial stem cell.

### Wnt/β-Catenin Signaling Pathway

The central player in the canonical Wnt signaling pathway is the cytoplasmic protein beta (β)-catenin, whose stability is regulated by the APC tumor suppressor complex. When Wnt receptors are not engaged, APC forms a subcellular, trimeric complex with axin and glycogen synthase kinase-3β (GSK3β), which triggers the phosphorylation, ubiquitination, and proteosomal degradation of β-catenin. Wnt stimulation activates the cytoplasmic phosphoprotein "disheveled" through its receptor, "frizzled," causing inhibition of GSK3β and a resultant accumulation of cytosolic β-catenin. Beta-catenin then translocates into the nucleus and interacts with members of the T-cell factor/lymphoid-enhancing factor (Tcf/LEF) family of DNA-binding proteins, transiently converting them from transcriptional repressors to activators, and signaling the activation of downstream target genes that increase cellular proliferation. When the Wnt signal is removed, APC extracts β-catenin from the nucleus, and the transcriptional repressor function of Tcf is restored (reviewed in Sancho *et al.*, 2004).

The Wnt/β-catenin pathway plays a role in malignant transformation. A mutation in APC renders the GSK3β/Axin/APC complex incapable of destabilizing β-catenin, leading to an accumulation of nuclear β-catenin/Tcf/LEF complexes and an increase in target gene transcription and cell proliferation, which can cause tumor formation (Wielenga *et al.*, 1999).

### Tcf/LEF DNA-Binding Protein Family

There are four known members of this family of transcription factors: Tcf-1, LEF1, Tcf-3, and Tcf-4. Tcf-4 is expressed in high levels in the developing intestine from embryonic day (E) 13.5 and in the epithelium of the adult small intestine and colon and in colon carcinomas (Barker *et al.*, 1999). In patients with a loss of function of the APC or β-catenin genes, nuclear accumulation of β-catenin/Tcf-4/LEF complexes in colonic epithelial cells implicates Tcf-4 in the ensuing uncontrolled target gene transcription and subsequent up-regulation of cell proliferation, which can often lead to tumorigenesis (Barker *et al.*, 1999). In the absence of β-catenin, the Tcf/LEF family recruits the corepressor proteins "Groucho" and CREB-binding protein (CBP) to the downstream Wnt target genes and inhibits their transcription (reviewed in Barker *et al.*, 2000). The Tcf-4 knockout mouse is devoid of proliferating cells in the small intestine and is presumed to lack a functional stem cell compartment. It is postulated that Tcf-4 is vital for the establishment of the intestinal stem cell population and is activated by a Wnt signal from the underlying stroma (Korinek *et al.*, 1998), further evidence that mesenchymal cells in the lamina propria comprise the intestinal stem cell niche.

## Cdx-1 and Cdx-2 Homeobox Genes

The mammalian homeobox proteins Cdx-1 and Cdx-2 also play an important role in intestinal epithelial stem cell transcriptional regulation, with a particular influence on gastrointestinal metaplasia. Cdx-1 is expressed throughout the proliferative compartment of the developing and adult mouse intestinal crypt epithelium (Subramanian et al., 1998), and both Cdx-1 and Cdx-2 mRNAs show restricted expression in the epithelial mucosa of the human small intestine and colon (Mizoshita et al., 2001). The Tcf-4 knockout mouse, mentioned earlier, does not express Cdx-1 in the small intestinal epithelium, implying that Cdx-1 is a direct downstream target of the Tcf-4/β-catenin complex in the Wnt signaling pathway and is employed in the development of the epithelial stem cell niche (Lickert et al., 2000). Expression of Cdx-1 is reduced in proliferating epithelial cell nuclei in colonic crypts concurrent with their progression to adenomas and adenocarcinomas (Lickert et al., 2000), although, because no colonic tumors develop in the Cdx-1 null mouse, this molecule does not appear to have direct tumor-suppressing properties (Subramanian et al., 1998). Cdx-2 is expressed in all epithelial cell nuclei in the upper regions of the crypts of the descending colon to the rectum, with decreasing expression parallel to an increasing degree of dysplasia in these cells (Ee et al., 1995). Region-specific genes such as Cdx-1, Cdx-2, and Tcf-4 appear to define the morphological features of the specialized regions of the intestinal epithelium and regulate stem cell function.

## Forkhead Family of Transcription Factors

The forkhead, or winged helix, family of transcription factors, of which there are nine members identified in mice, produce the Fox (forkhead box) proteins (reviewed in Kaestner et al., 2000). Mice with a heterozygous targeted mutation of the forkhead homolog 6 (fkh-6), or Fox1 gene, which is ordinarily expressed by gastrointestinal mesenchymal cells, display an atypical gastrointestinal epithelium with branched and elongated glands in the stomach, elongated villi, hyperproliferative crypts, and goblet cell hyperplasia due to increased epithelial cell proliferation. These Fox1 mutants have increased levels of heparin sulfate proteoglycans (HSPGs), causing overactivation of the Wnt/β-catenin pathway and an increase in target cell proliferation, demonstrating an indirect regulation of the Wnt/β-catenin pathway by Fox1-mediated HSPG production (Kaestner et al., 1997).

## TGF-β/Smad Signaling Pathway

The transforming growth factor TGF-β family inhibits gastrointestinal epithelial cell proliferation and is predominantly expressed in the differentiated compartments of the gut (Winesett et al., 1996). Under normal circumstances, TGF-β forms a multimeric complex with two serine-threonine kinase surface receptor molecules: TGF-β type I (TGFβRI) and type II (TGFβRII). The intracellular messengers of the TGF-β signaling pathways are the cytoplasmic Smad proteins, and Smad2 and Smad3 are phosphorylated upon activation of the TGF-β receptors and form a heteromeric complex with Smad4. This complex then translocates to the nucleus, where it interacts with transcriptional coactivators and corepressors to regulate TGF-β target gene transcription (Miyazono, 2000).

Disruption of TGF-β/Smad signaling causes intestinal epithelial cell hyperproliferation, and Smad2, Smad4, and TGFβRII are frequently inactivated in human cancers, confirming their function as tumor suppressor genes (Grady et al., 1999). Lack of TGF-β signaling in the intestine appears to contribute to the progression of early preexisting lesions (reviewed in Sancho et al., 2004). For example, mice with targeted heterozygous mutations in the Smad4 and APC genes develop adenomatous polyps in the small intestine and colon due to loss of heterozygosity (LOH) of the APC and Smad4 wild-type alleles. This implies a reciprocal interaction between the TGF-β and Wnt signaling pathways in the progression of intestinal carcinogenesis, in which LOH of genes from both pathways is required before malignant transformation can occur (Takaku et al., 1998).

## Notch Signaling Pathway

Notch encodes a large, single transmembrane receptor, of which there are four mammalian isoforms (Notch 1–4) and five corresponding ligands (Delta-like 1, 3, and 4; Jagged 1 and 2) (reviewed in Sancho et al., 2004). The Notch signaling pathway regulates gastrointestinal epithelial cell fate and the differentiation of the four specialized epithelial lineages of the gastrointestinal tract. This pathway supports the Unitarian hypothesis, that a single stem cell gives rise to all mature intestinal epithelial cell lineages (Cheng and Leblond, 1974). Increased levels of Notch protein negatively regulate transcription of Math1, via upregulation of the Hes1 transcriptional repressor. Mice with a targeted deletion of Math1 fail to develop goblet, Paneth, and enteroendocrine cell lineages in the small intestine, and these Math1-negative epithelial cell progenitors solely form enterocytes (Yang et al., 2001). Conversely, a reduced Notch expression and subsequent accumulation of its ligand, Delta, increases Math1 expression by blocking Hes1, causing cells to trans-differentiate to form goblet, Paneth, and enteroendocrine lineages in the small intestine (van Den Brink et al., 2001). Hes1 knockout mice display an elevated Math1 expression, with a concurrent increase in the numbers of goblet, Paneth, and enteroendocrine cells and a reduced enterocyte population (Jensen et al., 2000). These results suggest that the absorptive versus secretory cell fate decision is established through the Notch-Hes1 pathways, although aberrant Notch signaling in intestinal tumorigenesis has not been described to date (reviewed in Sancho et al., 2004).

## VI. ADULT STEM CELL PLASTICITY

An expanse of evidence regarding the surprisingly flexible potential of adult stem cells in their differentiation repertoires has recently challenged the classical belief that tissue stem cells are restricted to the production of adult lineages within their tissue of residence. Consequent to this discovery, stem cells now represent an entirely new field of regenerative medicine and may hold the key to the treatment of a number of diseases, such as cancer, cardiovascular disease, neurodegenerative disease, and diabetes. Recently, a vast number of reports have emerged showing that adult stem cells retain a large degree of plasticity and can differentiate to form many functional adult cells within extraneous tissues. Much of this research has focused on the adult BM as an easily accessible source of cells that have the potential to cross lineage boundaries, and transplanted adult BM cells have been shown to form adult cell types in many different tissues, including the liver, kidney, heart, CNS, gastrointestinal tract, lung, and skin (reviewed in Poulsom *et al.*, 2002).

It appears that selection pressure induced by target organ damage can intensify the efficacy of this process; e.g., increased numbers of cells of BM origin were observed in gastrointestinal epithelia of BM transplanted humans with graft-versus-host disease (GvHD) or intestinal ulceration (Okamoto *et al.*, 2002). It is proposed that BM cells respond to specific signals from damaged or diseased tissues, whereupon they migrate to and engraft within this tissue and aid regeneration and remodeling by contributing to differentiated adult cells.

It is important to demonstrate that BM-derived adult cells are functional, with a capacity to restore diseased or damaged tissues, and thus convey a possible clinical relevance. The fumarylaceto-acetate hydrolase-deficient (Fah$^{-/-}$) mouse model of the human metabolic liver disease, Type 1 tyrosinaemia, was used to confirm the functional significance of BM cell plasticity (Lagasse *et al.*, 2000). Fah$^{-/-}$ mice were transplanted with as few as 50 purified hematopoietic stem cells (HSCs), which traveled to the liver and restored enzyme levels to normal by forming Fah-expressing hepatocytes, thereby rescuing the mice from their fatal deficiency (Lagasse *et al.*, 2000).

## VII. BONE MARROW CONTRIBUTION TO THE CELLS IN THE ADULT INTESTINE

There is a growing number of reports that BM cells can repopulate both epithelial and mesenchymal lineages in the gastrointestinal tract of animals and man. A seminal paper by Krause and colleagues (2001) demonstrated the remarkable plasticity of a single purified BM cell, forming epithelial cells in the murine lung, skin, and gastrointestinal tract. A highly purified population of mesenchymal stem cells (MSCs), termed MAPCs, were isolated from ROSA26 mice and constitutively express β-galactosidase and were microinjected into mouse blastocysts following 55–65 population doublings, which were then transferred to foster mothers and allowed to develop and be born. Chimeric offspring were killed at 6–20 weeks, and BM-derived epithelial cells were observed in many somatic tissues, including the small intestine (Jiang, 2002).

A recent paper investigating long-term repopulation of murine intestinal epithelium by BM cells found that BM contributes to all intestinal lineages in both normal and neoplastic epithelium. However, donor-derived cells were present in only 0.05% of total crypts, and only a very small proportion of epithelial cells were BM derived. The authors report that of these cells, 51% were a product of cell fusion, presumably with an existing intestinal progenitor or stem cell (Rizvi *et al.*, 2006). However, it cannot be concluded that fusion is the sole means by which BM cells contribute to intestinal epithelial lineages in this model, for up to 40% of BM-derived cells remained unaccounted for, and other mechanisms, including true differentiation, cannot be discounted.

Perhaps the most striking demonstration of BM contribution to cells in the gastrointestinal tract was shown by Houghton and colleagues (2004), using a well-established model of gastric carcinoma. Donor BM from ROSA26 mice was transplanted into wild-type mice that were then infected with *Helicobacter felis*. Gastric mucosal apoptosis was present at six to eight weeks postinfection, donor-derived mucosal cells were observed in the gastric glands from 20 weeks posttransplant, and by 52 weeks up to 90% of the gastric mucosa was composed of BM-derived cells, which thus appear to home to and repopulate the gastric mucosa and contribute to metaplasia, dysplasia, and cancer. Three separate methods were utilized to investigate the fusion hypothesis, leading to the conclusion that stable fusion did not occur (Houghton *et al.*, 2004).

BM-derived epithelial cells have also been observed throughout the gastrointestinal tract in humans following sex-mismatched BM transplantation (Korbling *et al.*, 2002; Okamoto *et al.*, 2002; Spyridonidis *et al.*, 2004), and heterokaryon formation was ruled out as the mechanism by which BM cells form intestinal epithelial cells (Matsumoto *et al.*, 2005).

It is important to note that observed donor-derived epithelial cells in the mouse and human gastrointestinal tract are most frequently single cells randomly interspersed throughout the crypts and villi and that the incidence of these cells is low, ranging from 0.04% (Spyridonidis *et al.*, 2004) to approximately 10% (Jiang, 2002). However, the BM contribution to epithelial cells in human intestinal biopsies was increased between 5- and 50-fold in tissue damaged by GvHD, compared to intact epithelium (Okamoto *et al.*, 2002; Spyridonidis *et al.*, 2004). The low levels of engraftment and the lack of clusters of donor-derived cells indicate that it is unlikely that the BM cells form intestinal epithelial stem cells, supported recently by a lack of Musashi-1 expression by BM-derived intestinal epithelial cells in humans (Matsumoto *et al.*, 2005). However, a small proportion of BM-derived gastrointestinal epithelial cells expressed the proliferative marker Ki-67 and a larger proportion of expressed markers of terminal differentiation (interestingly, transplanted cells contributed to all four lineages present in the gastrointestinal epithelia), which indicates that BM cells contribute to tissue turnover (Matsumoto *et al.*, 2005).

## VIII. ORIGIN OF THE ISEMFS

The origin of the ISEMFs is unresolved, with reports of transdifferentiation between different mesenchymal cell types, including resident tissue fibroblasts (Gabbiani, 1996) or smooth muscle cells (Ronnov-Jessen *et al.*, 1995), or a putative stem cell population in the lamina propria near the crypt base (Bockman and Sohal, 1998; Marsh and Trier,

1974a, 1974b). However, more recent data from *in vitro* studies indicate that cells in the BM can differentiate to produce myofibroblast-like cells (Ball *et al.*, 2004; Emura *et al.*, 2000). We have recently confirmed this finding *in vivo* in the intestines of both mice and humans. Following lethal irradiation and transplantation of whole BM cells in mice, almost 60% of ISEMFs were derived from the donor cells six weeks posttransplant. These cells are present as interconnected rows, indicative of clonal expansion, and appear to be located predominantly around the crypt base, strong evidence that they contribute to the putative intestinal epithelial stem cell niche (Fig. 45.1A–F; Brittan *et al.*, 2002). Since this initial finding, that BM cells contribute to a large proportion of myofibroblasts in the gut, transplanted BM cells

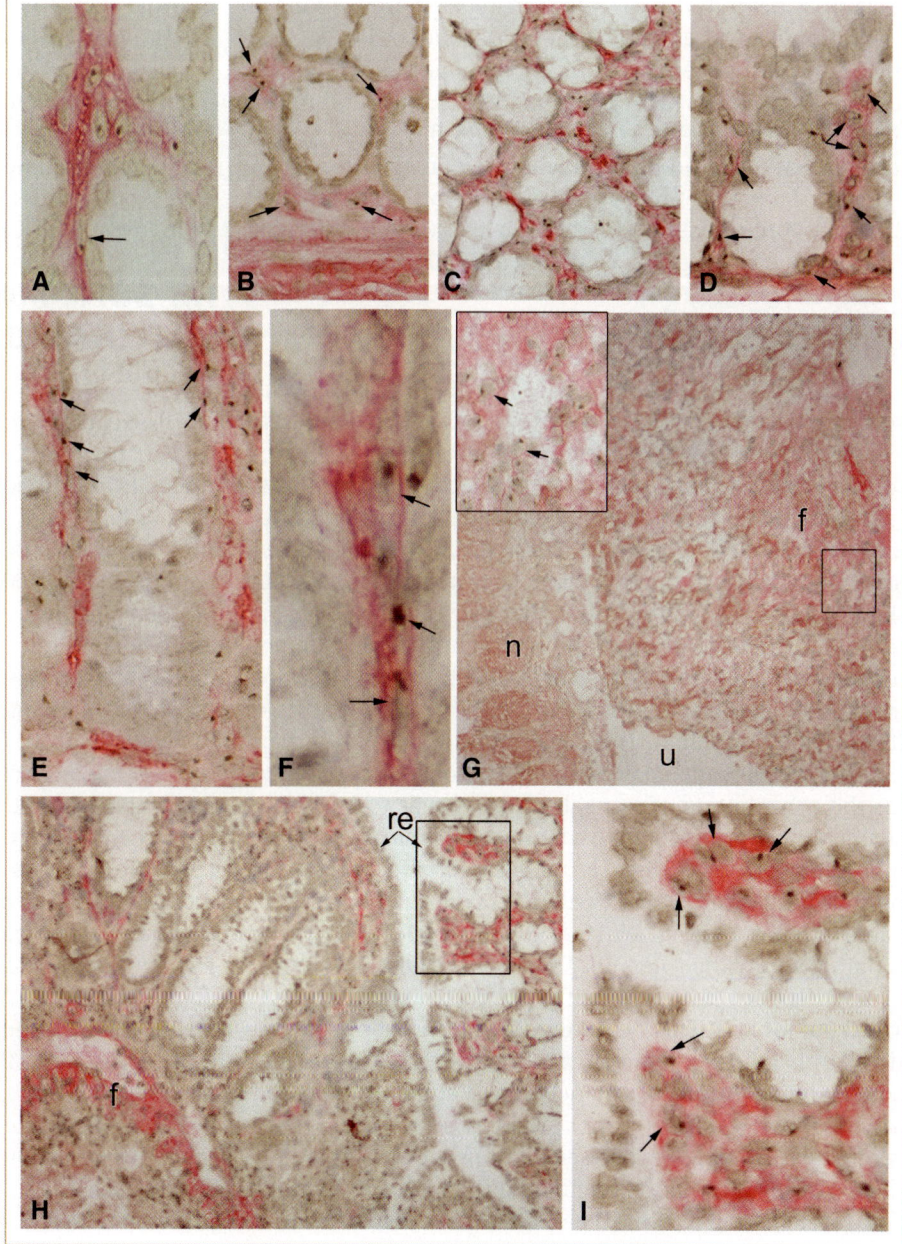

**FIG. 45.1.** BM-derived myofibroblasts in a model of chemically induced colitis. Female mice received a bone marrow transplant from male donor mice, and BM-derived myofibroblasts were identified by morphologic criteria, SMA immunoreactivity (red cytoplasm), and Y chromosome-expression (dark brown nuclear spot). These cells were frequently identified at all time points in both **(A)** ethanol-treated controls (four days after administration of ethanol, arrow) and **(B and C)** TNBS-treated mice (noninflamed mucosa adjacent to colitis eight days after administration of TNBS, arrows). BM-derived myofibroblasts were often present as cellular columns spanning from the epithelial crypt base to the intestinal lumen (**D**: six days after administration of ethanol; **E and F** [inset]: six days after administration of TNBS, arrows). **(G)** Activated myofibroblasts in colitis were identified by their location in regions of fibrosis and by their large, flat shape (six days after administration of TNBS, arrows). **(H and I** [inset]) BM-derived activated myofibroblasts were present in the submucosa in regions of regenerating epithelium (eight days after administration of TNBS, arrows). n, necrotic epithelial crypts; u, mucosal ulcer; f, fibrosis; re, epithelialization.

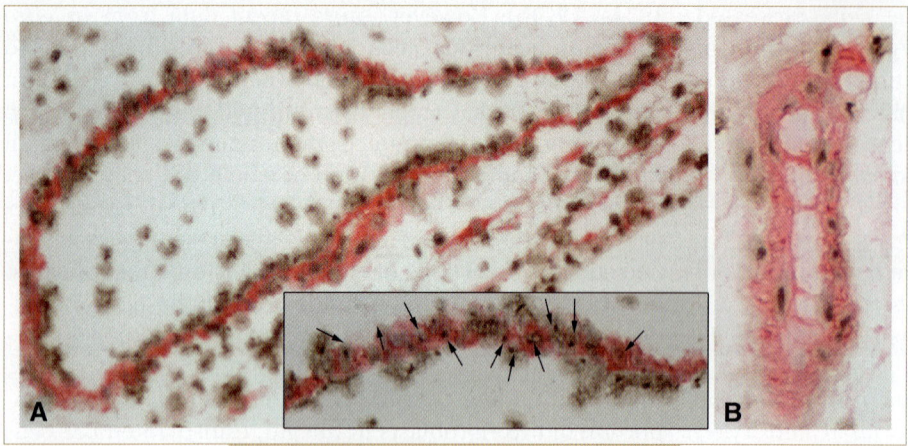

**FIG. 45.2.** BM contributions can form entire blood vessels in TNBS-induced colitis. **(A and B)** BM-derived vascular smooth muscle lining cells were identified as smooth muscle–like cells surrounding the blood vessels that were SMA immunoreactive (red cytoplasm) and possessed the Y chromosome (dark brown nuclear spot). The role of BM in postnatal vasculogenesis was confirmed by the presence of whole blood vessels composed entirely of BM-derived cells (arrows).

have since been shown to contribute to large proportions of myofibroblasts and fibroblasts in several other organs, including the mouse lung, stomach, esophagus, skin, kidney, and adrenal glands (Direkze *et al.*, 2003) and the human liver in fibrosis (Forbes *et al.*, 2004), and BM can maintain differentiated cell production within a damaged tissue for extensive lengths of time; e.g., in one patient, approximately 23% of myofibroblasts in the fibrotic liver were BM derived two years posttransplant (Forbes *et al.*, 2004).

### ISEMFs in Tissue Regeneration in Inflammatory Bowel Disease

In the normal, healthy gut, ISEMFs are present as spindle-shaped, transiently differentiated cells. However, in inflamed tissue, ISEMFs become activated and assume a round, flattened morphology, with an accelerated proliferation rate (McKaig *et al.*, 2002), increased expression of cytokines, chemokines, growth factors, and adhesion molecules, and enhanced secretion of soluble mediators of inflammation and extracellular matrix factors (reviewed in Powell *et al.*, 1999a, 1999b). Furthermore, overactivation and persistence of activated ISEMFs causes tissue fibrosis and scarring in Crohn's disease (Aigner *et al.*, 1997; Isaji *et al.*, 1994; Kinzler and Vogelstein, 1998). We showed that BM-derived ISEMFs appear capable of activated response in the mouse, for the contribution of transplanted BM cells to ISEMFs is significantly up-regulated in a model of chemically induced inflammatory bowel disease, similar to Crohn's disease in humans, and that these BM-derived cells display an activated phenotype (Fig. 45.1G–I; Brittan *et al.*, 2005). In a separate study, the role of BM was investigated in a model of spontaneous colonic inflammation, the interleukin-10 knockout mouse (IL10$^{-/-}$). Following whole BM transplantation from wild-type mice, inflammatory disease in the colons of IL10$^{-/-}$ mice was ameliorated, and approximately 30% of ISEMFs were of donor origin, whereas IL10$^{-/-}$ mice receiving IL10$^{-/-}$ BM progressed to extensive colitis, with up to 45% of myofibroblasts originating from the transplanted cells. Interestingly, wild-type mice that received bone

marrow from either IL10$^{-/-}$ mice or wild-type mice showed no progression to an inflammatory phenotype, despite replacement of the hematopoietic lineages by the transplanted cells (Bamba *et al.*, 2006). The implications of these findings, that the contribution of BM to ISEMFs is significantly up-regulated in the inflamed gut in response to regenerative pressure, is discussed later in relation to a possible therapy for Crohn's disease.

## IX. BONE MARROW–DERIVED VASCULAR LINEAGES CONTRIBUTE TO TISSUE REGENERATION IN IBD

The contribution of BM-derived EPCs to neovascularization in many inflammatory conditions is now well documented. In addition to the significant contribution of BM cells to ISEMFs in colitis, we also observed a large contribution of BM cells to multiple vascular lineages in these mice (Brittan *et al.*, 2005). In patches of severe inflammation, vessels were observed that appeared to be composed largely, and even entirely, of BM-derived cells, suggesting that BM is involved in the creation of entire new blood vessels via postnatal neovasculogenesis (Fig. 45.2). Quantification of the proportion of BM-derived mural cell–containing vessels revealed a statistically significant 6.4-fold increase in colitis as compared to ethanol-treated controls ($p = 0.013$); moreover, the total percentage of BM-derived mural cells within blood vessels was calculated, showing an almost twofold increase in the contribution of BM to mural cells in colitis, compared to ethanol-control tissue ($p = 0.017$). BM cells also formed endothelial cells within blood vessels in inflamed colons, ascertained by their morphology and their coexpression of ICAM-1 and the Y chromosome (Brittan *et al.*, 2005).

Expression of the intracellular adhesion molecule ICAM-3 (CD50) is up-regulated in Crohn's disease (Bretscher *et al.*, 2000; Vainer and Nielsen, 2000), and ICAM-3 has recently been shown to be expressed on BM-derived endothelial cells and to play a regulatory role in endothelial cell junction formation (van Buul *et al.*, 2004), possibly supporting our evidence for neovasculogenesis by BM-derived endothelial cells in colitis.

There are numerous reports of BM-derived endothelial cells, although few studies describe BM-derived mural cells. A recent report describes a BM origin of the periendothelial vascular mural cells, which are believed to be mural cell progenitors (Rajantie *et al.*, 2004). It has not yet been possible to purify EPCs from the mouse BM, although it is suggested that they derive from a population of HSCs that express Flk-1, c-Kit, Sca-1, CD34, and CD45 but lack markers of lineage differentiation (Lin-) (Asahara and Kawamoto, 2004). MSC-derived MAPCs, mentioned previously, isolated from both murine and human BM have also been shown to have endothelial differentiation potential, both *in vivo* and *in vitro* (Jiang, 2002; Reyes *et al.*, 2002).

Paris and colleagues (2001) presented data showing that the initiating factor in radiation-induced crypt death in mice is apoptosis of the endothelial cells in the tissue stroma, which precedes intestinal epithelial stem cell dysfunction and crypt damage. The authors proposed that the endothelial cells are the primary targets of gamma radiation and that the concurrent death of the epithelial stem cells is a secondary event, caused by a loss of endothelial cell support. However, these claims have never been corroborated and were refuted due to the lack of tissue ischemia and hypoxia that would be expected following such severe damage to the vasculature (Hendry *et al.*, 2001). With respect to our current findings of BM-derived endothelial cells in the mouse colon, it is important to note that if the endothelial cells in the intestine are the primary targets of radiation-induced intestinal mucosal damage, then this implicates the BM with a further role in intestinal epithelial stem cell regulation, via the formation of endothelial cells.

## X. STEM CELL PLASTICITY: *DE NOVO* CELL GENERATION OR HETEROKARYON FORMATION

It is becoming increasingly apparent that two separate mechanisms of stem cell plasticity exist: *de novo* cell generation through BM cell differentiation, and fusion of BM cells with preexisting cells to form a cell with multiple, genetically variant nuclei, a heterokaryon. The classic gold standard proof-of-principle model for adult stem cell plasticity was demonstrated by the rescue of the Fah$^{-/-}$ mouse from its fatal metabolic disease by BM stem cell transplantation (Lagasse *et al.*, 2000), mentioned previously, which definitively showed that BM-derived hepatocytes were functionally normal. However, cytogenetic analyses showed that almost all the newly formed, Fah-synthesizing hepatocytes in these transplanted mice expressed a karyotype indicative that fusion had occurred between a transplanted BM cell and a host hepatocyte (Vassilopoulos *et al.*, 2003; Wang *et al.*, 2003). It is important to consider that hepatocytes are frequently polyploid and express multiple sets of chromo-

somes, and it is thereby possible that the liver presents a unique environment for the formation of fused hybrid cells (reviewed in Alison *et al.*, 2004).

Transplanted BM stem cells have been shown to fuse with Purkinje cells in brains of both mice (Weimann *et al.*, 2003b) and humans (Weimann *et al.*, 2003a). In the SOD1 mouse, which develops a lethal form of human ALS, BM cells can form adult cells in the CNS, heart, and skeletal muscle, thereby prolonging animals' survival (Corti *et al.*, 2004). Examination of nuclear content by FISH analyses suggested that BM-derived neurons may be a result of cell fusion, although in the brains of some male mice that had received a female BM transplant, neurons were seen that did not express a Y chromosome, which could be explained by *de novo* cell formation by transplanted female BM cells or, alternatively, may simply be a result of incomplete detection by the FISH technique (Corti *et al.*, 2004). Conversely, in the human brain, transplanted BM cells appear capable of neurogenesis and contribute to approximately 1% of all neurons for up to six years posttransplant. Analyses of autopsy brain specimens from three female patients that received a male BM transplant showed no evidence of fusion, for only one X chromosome was present in all male-derived neurons (Cogle *et al.*, 2004).

Whereas studies of sex chromosomal content may prove ambiguous due to artifacts created by the *in situ* hybridization technique, the Cre/lox recombinase system provides an elegant means of studying the mechanism of adult stem cell plasticity, using mice of the Z/EG Cre-reporter strain. The Z/EG transgenic mouse expresses a transgene cassette consisting of the chicken β-actin promoter driving expression of a β-galactosidase/neomycin-STOP reporter gene. This is flanked by two loxP sites and followed downstream by enhanced green fluorescent protein (eGFP) expression. Expression of β-galactosidase alone is driven by the β-actin promoter until Cre-mediated recombination occurs, at which point the β-galactosidase-STOP DNA is excised and the downstream eGFP is expressed instead. BM from male Z/EG mice was transplanted into female mice that ubiquitously expressed Cre, and cells within multiple tissues were analyzed for eGFP or β-galactosidase expression, i.e., any cell resulting from the fusion of a BM cell with a host cell should express eGFP, and any cell that expresses β-galactosidase is the product of BM cell differentiation. However, recent reports using the Cre/lox recombinase system have also proved inconsistent, because BM-derived epithelial cells in the mouse liver, lung, and skin were shown by this method to form without cell fusion (Harris *et al.*, 2004), whereas results of Alvarez-Dolado and colleagues (2003), using a similar system, clearly demonstrated that BM cells spontaneously fuse with hepatocytes in the liver, Purkinje cells in the CNS, and cardiomyocytes in the heart, with no evidence of *de novo* generation.

Evidence of heterokaryon formation following stem cell transplantation in the gastrointestinal tract or skin has not been reported to date, and gastrointestinal epithelial cells derived from transplanted mobilized peripheral blood stem cells were reported to contain a normal ratio of sex chromosomes (Korbling *et al.*, 2002). An important observation was made, in that in some reported incidences of spontaneous cell fusion, the original stem cell markers are repressed and the transplanted cells appear to reprogram their new "host" nuclei to produce a sustainable, functional cell (Weimann *et al.*, 2003). Additionally, the incidence of heterokaryons was shown to increase as a tissue ages, suggesting that BM-derived cells can divide (Weimann *et al.*, 2003). The underlying principle, that transplanted BM cells have the capacity to rescue an otherwise-fatal metabolic disease, is maintained and should not be overlooked, although it is irrefutable that before we overlook, or indeed overstate, the capacity of adult stem cells to regenerate damaged or diseased tissue, further clarification of the mechanisms involved in this apparent stem cell plasticity is essential.

However, despite our lack of knowledge of the mechanisms involved, the desired endpoint of bone marrow transplantation as a cure for otherwise-lethal conditions has been demonstrated. Marcus Grompe and colleagues showed that hematopoietic bone marrow stem cell transplantation prevented lethality in the fumaryloacetate hydrolase (FAH)–deficient mouse, a model of fatal hereditary Type 1 tyrosinaemia (Lagasse *et al.*, 2000). Following karyotyping, this was shown to be due to the transplanted cells' fusing with the indigenous hepatocytes, forming cells with an abnormal number of chromosomes, and somehow stimulating their production of the defunct enzyme (Wang *et al.*, 2003). The authors have since reported that newly formed heterokaryons may undergo *reduction division*, in which nuclear chromosome content is returned to normal.

Therefore, it is apparent that much progress must be made before a clinical application of adult stem cells should be considered, although at this stage it would surely be injudicious to ignore the obvious potential of adult stem cells for tissue regeneration.

## XI. BONE MARROW TRANSPLANTATION AS A POTENTIAL THERAPY FOR CROHN'S DISEASE

As discussed previously, in mice with either chemically induced or spontaneous inflammatory bowel disease, the engraftment of transplanted bone marrow cells into the damaged gut and their formation of activated myofibroblasts, presumably in response to inflammation, is significantly enhanced.

Crohn's disease runs a chronic relapsing and remitting course, with a significant disease-associated mortality rate of between 6% and 24% of patients observed within a 13- to 20-year follow-up (Cooke *et al.*, 1980; Farmer *et al.*, 1985; Trnka *et al.*, 1982). Crohn's disease responds to immunosuppressive medications and antiinflammatory agents, although as yet no curative therapy has been found. Several case studies showing long-term remission from Crohn's disease following autologous (Kashyap and Forman, 1998; Soderholm *et al.*, 2002) or allogeneic BM transplantation (Lopez-Cubero *et al.*, 1998; Talbot *et al.*, 1998) have been documented. Prior to BM transplant, a patient's own BM is eradicated, and it is therefore possible that the observed remission of Crohn's disease following BM transplant is due to the elimination of aberrant BM cells that play a critical role in the pathogenesis of Crohn's disease. Furthermore, no immunosuppressive medication is required following BM transplantation in Crohn's disease patients, indicating that the observed long-term remission is a direct result of the transplantation procedure (reviewed in James, 1998). We now suggest a possible mechanism by which BM cells provide remission from Crohn's disease, by their formation of mesenchymal and vascular lineages in the inflamed colon, thereby contributing directly to tissue regeneration.

## XII. REFERENCES

Aigner, T., Neureiter, D., Muller, S., Kuspert, G., Belke, J., and Kirchner, T. (1997). Extracellular matrix composition and gene expression in collagenous colitis. *Gastroenterology* **113**, 136–143.

Alison, M. R., Poulsom, R., Otto, W. R., Vig, P., Brittan, M., Direkze, N. C., Lovell, M., Fang, T. C., Preston, S. L., and Wright, N. A. (2004). Recipes for adult stem cell plasticity: fusion cuisine or readymade? *J. Clin. Pathol.* **57**, 113–120.

Alvarez-Dolado, M., Pardal, R., Garcia-Verdugo, J. M., Fike, J. R., Lee, H. O., Pfeffer, K., Lois, C., Morrison, S. J., and Alvarez-Buylla, A. (2003). Fusion of bone-marrow-derived cells with Purkinje neurons, cardiomyocytes and hepatocytes. *Nature* **425**, 968–973.

Asahara, T., and Kawamoto, A. (2004). Endothelial progenitor cells for postnatal vasculogenesis. *Am. J. Physiol. Cell Physiol.* **287**, C572–C579.

Ball, S. G., Shuttleworth, A. C., and Kielty, C. M. (2004). Direct cell contact influences bone marrow mesenchymal stem cell fate. *Int. J. Biochem. Cell Biol.* **36**, 714–727.

Bamba, S., Lee, C. Y., Brittan, M., Preston, S., Direkze, N., Poulsom, R., Alison, M., Wright, N., and Otto, W. (2006). Bone marrow transplantation ameliorates pathology in interleukin-10 knockout colitic mice. *J. Pathol.* **209**(2), 265–273.

Barker, N., Huls, G., Korinek, V., and Clevers, H. (1999). Restricted high-level expression of Tcf-4 protein in intestinal and mammary gland epithelium. *Am. J. Pathol.* **154**, 29–35.

Barker, N., Morin, P. J., and Clevers, H. (2000). The Yin-Yang of TCF/beta-catenin signaling. *Adv. Cancer Res.* **77**, 1–24.

Bjerknes, M., and Cheng, H. (1981). The stem-cell zone of the small intestinal epithelium. I. Evidence from Paneth cells in the adult mouse. *Am. J. Anat.* **160**, 51–63.

Bjerknes, M., and Cheng, H. (1999). Clonal analysis of mouse intestinal epithelial progenitors. *Gastroenterology* **116**, 7–14.

Bockman, D. E., and Sohal, G. S. (1998). A new source of cells contributing to the developing gastrointestinal tract demonstrated in chick embryos. *Gastroenterology* **114**, 878–882.

Bretscher, A., Chambers, D., Nguyen, R., and Reczek, D. (2000). ERM-Merlin and EBP50 protein families in plasma membrane organization and function. *Annu. Rev. Cell Dev. Biol.* **16**, 113–143.

Brittan, M., Hunt, T., Jeffery, R., Poulsom, R., Forbes, S. J., Hodivala-Dilke, K., Goldman, J., Alison, M. R., and Wright, N. A. (2002). Bone marrow derivation of pericryptal myofibroblasts in the mouse and human small intestine and colon. *Gut* **50**, 752–757.

Brittan, M., Chance, V., Elia, G., Poulsom, R., Alison, M. R., Macdonald, T. T., and Wright, N. A. (2005). A regenerative role for bone marrow following experimental colitis: contribution to neovasculogenesis and myofibroblasts. *Gastroenterology* **128**, 1984–1995.

Cai, W. B., Roberts, S. A., and Potten, C. S. (1997). The number of clonogenic cells in crypts in three regions of murine large intestine. *Int. J. Radiat. Biol.* **71**, 573–579.

Cairns, J. (1975). Mutation selection and the natural history of cancer. *Nature* **255**, 197–200.

Cheng, H., and Leblond, C. P. (1974). Origin, differentiation and renewal of the four main epithelial cell types in the mouse small intestine. V. Unitarian theory of the origin of the four epithelial cell types. *Am. J. Anat.* **141**, 537–561.

Chinnery, P. F., and Samuels, D. C. (1999). Relaxed replication of mtDNA: a model with implications for the expression of disease. *Am. J. Hum. Genet.* **64**, 1158–1165.

Cogle, C. R., Yachnis, A. T., Laywell, E. D., Zander, D. S., Wingard, J. R., Steindler, D. A., and Scott, E. W. (2004). Bone marrow transdifferentiation in brain after transplantation: a retrospective study. *Lancet* **363**, 1432–1437.

Cooke, W. T., Mallas, E., Prior, P., and Allan, R. N. (1980). Crohn's disease: course, treatment and long term prognosis. *Q. J. Med.* **49**, 363–384.

Corti, S., Locatelli, F., Donadoni, C., Guglieri, M., Papadimitriou, D., Strazzer, S., Del Bo, R., and Comi, G. P. (2004). Wild-type bone marrow cells ameliorate the phenotype of SOD1-G93A ALS mice and contribute to CNS, heart and skeletal muscle tissues. *Brain* **127**, 2518–2532.

Direkze, N. C., Forbes, S. J., Brittan, M., Hunt, T., Jeffery, R., Preston, S. L., Poulsom, R., Hodivala-Dilke, K., Alison, M. R., and Wright, N. A. (2003). Multiple organ engraftment by bone-marrow-derived myofibroblasts and fibroblasts in bone-marrow-transplanted mice. *Stem Cells* **21**, 514–520.

Ee, H. C., Erler, T., Bhathal, P. S., Young, G. P., and James, R. J. (1995). Cdx-2 homeodomain protein expression in human and rat colorectal adenoma and carcinoma. *Am. J. Pathol.* **147**, 586–592.

Emura, M., Ochiai, A., Horino, M., Arndt, W., Kamino, K., and Hirohashi, S. (2000). Development of myofibroblasts from human bone marrow mesenchymal stem cells cocultured with human colon carcinoma cells and TGF beta 1. *In Vitro Cell Dev. Biol. Anim.* **36**, 77–80.

Escaffit, F., Perreault, N., Jean, D., Francoeur, C., Herring, E., Rancourt, C., Rivard, N., Vachon, P. H., Pare, F., Boucher, M. P., Auclair, J., and Beaulieu, J. F. (2005). Repressed E-cadherin expression in the lower crypt of human small intestine: a cell marker of functional relevance. *Exp. Cell Res.* **302**, 206–220.

Farmer, R. G., Whelan, G., and Fazio, V. W. (1985). Long-term follow-up of patients with Crohn's disease. Relationship between the clinical pattern and prognosis. *Gastroenterology* **88**, 1818–1825.

Forbes, S. J., Russo, F. P., Rey, V., Burra, P., Rugge, M., Wright, N. A., and Alison, M. R. (2004). A significant proportion of myofibroblasts are of bone marrow origin in human liver fibrosis. *Gastroenterology* **126**, 955–963.

Gabbiani, G. (1996). The cellular derivation and the life span of the myofibroblast. *Pathol. Res. Pract.* **192**, 708 711.

Gottardi, C. J., Wong, E., and Gumbiner, B. M. (2001). E-cadherin suppresses cellular transformation by inhibiting beta-catenin signaling in an adhesion-independent manner. *J. Cell Biol.* **153**, 1049–1060.

Grady, W. M., Myeroff, L. L., Swinler, S. E., Rajput, A., Thiagalingam, S., Lutterbaugh, J. D., Neumann, A., Brattain, M. G., Chang, J., Kim, S. J., Kinzler, K. W., Vogelstein, B., Willson, J. K., and Markowitz, S. (1999). Mutational inactivation of transforming growth factor beta receptor type II in microsatellite stable colon cancers. *Cancer Res.* **59**, 320–324.

Greaves, L. C., Preston, S. L., Tadrous, P. J., Taylor, R. W., Barron, M. J., Oukrif, D., Leedham, S. J., Deheragoda, M., Sasieni, P., Novelli, M. R., Jankowski, J. A., Turnbull, D. M., Wright, N. A., and McDonald, S. A. (2006). Mitochondrial DNA mutations are established in human colonic stem cells, and mutated clones expand by crypt fission. *Proc. Natl. Acad. Sci. U.S.A.* **103**, 714–719.

Harris, R. G., Herzog, E. L., Bruscia, E. M., Grove, J. E., Van Arnam, J. S., and Krause, D. S. (2004). Lack of a fusion requirement for development of bone marrow–derived epithelia. *Science* **305**, 90–93.

Hendry, J. H., Booth, C., and Potten, C. S. (2001). Endothelial cells and radiation gastrointestinal syndrome. *Science* **294**, 1411.

Houghton, J., Stoicov, C., Nomura, S., Rogers, A. B., Carlson, J., Li, H., Cai, X., Fox, J. G., Goldenring, J. R., and Wang, T. C. (2004). Gastric cancer originating from bone marrow–derived cells. *Science* **306**, 1568–1571.

Isaji, M., Momose, Y., Tatsuzawa, Y., and Naito, J. (1994). Modulation of morphology, proliferation and collagen synthesis in fibroblasts by the exudate from hypersensitive granulomatous inflammation in rats. *Int. Arch. Allergy Immunol.* **104**, 340–347.

James, S. P. (1998). Allogeneic bone marrow transplantation in Crohn's disease. *Gastroenterology* **114**, 596–598.

Jensen, J., Pedersen, E. E., Galante, P., Hald, J., Heller, R. S., Ishibashi, M., Kageyama, R., Guillemot, F., Serup, P., and Madsen, O. D. (2000). Control of endodermal endocrine development by Hes-1. *Nat. Genet.* **24**, 36–44.

Jiang, Q. (2002). Pluripotency of mesenchymal stem cells derived from adult marrow. *Nature* **418**, 41–49.

Johnson, M. A., Bindoff, L. A., and Turnbull, D. M. (1993). Cytochrome C oxidase activity in single muscle fibers: assay techniques and diagnostic applications. *Ann. Neurol.* **33**, 28–35.

Kaestner, K. H., Silberg, D. G., Traber, P. G., and Schutz, G. (1997). The mesenchymal winged helix transcription factor Fkh6 is required for the control of gastrointestinal proliferation and differentiation. *Genes Dev.* **11**, 1583–1595.

Kaestner, K. H., Knochel, W., and Martinez, D. E. (2000). Unified nomenclature for the winged helix/forkhead transcription factors. *Genes Dev.* **14**, 142–146.

Kashyap, A., and Forman, S. J. (1998). Autologous bone marrow transplantation for non-Hodgkin's lymphoma resulting in long-term remission of coincidental Crohn's disease. *Br. J. Haematol.* **103**, 651–652.

Kayahara, T. (2003). Musashi-1 and Hes-1 are expressed in crypt base columnar cells in mouse small intestine. *Gastroenterology* **124**(4) Supplement 1, A501.

Kinzler, K. W., and Vogelstein, B. (1998). Landscaping the cancer terrain. *Science* **280**, 1036–1037.

Korbling, M., Katz, R. L., Khanna, A., Ruifrok, A. C., Rondon, G., Albitar, M., Champlin, R. E., and Estrov, Z. (2002). Hepatocytes and epithelial cells of donor origin in recipients of peripheral-blood stem cells. *N. Engl. J. Med.* **346**, 738–746.

Korinek, V., Barker, N., Moerer, P., van Donselaar, E., Huls, G., Peters, P. J., and Clevers, H. (1998). Depletion of epithelial stem-cell compartments in the small intestine of mice lacking Tcf-4. *Nat. Genet.* **19**, 379–383.

Krause, D., Theise, N., Collector, M., Henegariu, O., Hwang, S., Gardner, R., Neutzel, S., and Sharkis, S. (2001). Multi-organ, multi-lineage engraftment by a single bone marrow–derived stem cell. *Cell* **105**, 369–377.

Lagasse, E., Connors, H., Al-Dhalimy, M., Reitsma, M., Dohse, M., Osborne, L., Wang, X., Finegold, M., Weissman, I. L., and Grompe, M. (2000). Purified hematopoietic stem cells can differentiate into hepatocytes *in vivo*. *Nat. Med.* **6**, 1229–1234.

Lickert, H., Domon, C., Huls, G., Wehrle, C., Duluc, I., Clevers, H., Meyer, B. I., Freund, J. N., and Kemler, R. (2000). Wnt/(beta)-catenin signaling regulates the expression of the homeobox gene Cdx1 in embryonic intestine. *Development* **127**, 3805–3813.

Loeffler, M., Birke, A., Winton, D., and Potten, C. (1993). Somatic mutation, monoclonality and stochastic models of stem cell organization in the intestinal crypt. *J. Theor. Biol.* **160**, 471–491.

Loeffler, M., Bratke, T., Paulus, U., Li, Y. Q., and Potten, C. S. (1997). Clonality and life cycles of intestinal crypts explained by a state-dependent stochastic model of epithelial stem cell organization. *J. Theor. Biol.* **186**, 41–54.

Lopez-Cubero, S. O., Sullivan, K. M., and McDonald, G. B. (1998). Course of Crohn's disease after allogeneic marrow transplantation. *Gastroenterology* **114**, 433–440.

Marsh, M. N., and Trier, J. S. (1974a). Morphology and cell proliferation of subepithelial fibroblasts in adult mouse jejunum. I. Structural features. *Gastroenterology* **67**, 622–635.

Marsh, M. N., and Trier, J. S. (1974b). Morphology and cell proliferation of subepithelial fibroblasts in adult mouse jejunum. II. Radioautographic studies. *Gastroenterology* **67**, 636–645.

Matsumoto, T., Okamoto, R., Yajima, T., Mori, T., Okamoto, S., Ikeda, Y., Mukai, M., Yamazaki, M., Oshima, S., Tsuchiya, K., Nakamura, T., Kanai, T., Okano, H., Inazawa, J., Hibi, T., and Watanabe, M. (2005). Increase of bone marrow–derived secretory lineage epithelial cells during regeneration in the human intestine. *Gastroenterology* **128**, 1851–1867.

McKaig, B. C., Hughes, K., Tighe, P. J., and Mahida, Y. R. (2002). Differential expression of TGF-beta isoforms by normal and inflammatory bowel disease intestinal myofibroblasts. *Am. J. Physiol. Cell Physiol.* **282**, C172–C182.

Miyazono, K. (2000). Transforming growth factor-beta signaling and cancer. *Hum. Cell* **13**, 97–101.

Mizoshita, T., Inada, K., Tsukamoto, T., Kodera, Y., Yamamura, Y., Hirai, T., Kato, T., Joh, T., Itoh, M., and Tatematsu, M. (2001). Expression of Cdx1 and Cdx2 mRNAs and relevance of this expression to differentiation in human gastrointestinal mucosa — with special emphasis on participation in intestinal metaplasia of the human stomach. *Gastric Cancer* **4**, 185–191.

Nakamura, Y., Sakakibara, S., Miyata, T., Ogawa, M., Shimazaki, T., Weiss, S., Kageyama, R., and Okano, H. (2000). The bHLH gene hes1 as a repressor of the neuronal commitment of CNS stem cells. *J. Neurosci.* **20**, 283–293.

Nishimura, S., Wakabayashi, N., Toyoda, K., Kashima, K., and Mitsufuji, S. (2003). Expression of Musashi-1 in human normal colon crypt cells: a possible stem cell marker of human colon epithelium. *Dig. Dis. Sci.* **48**, 1523–1529.

Novelli, M. R., Williamson, J. A., Tomlinson, I. P., Elia, G., Hodgson, S. V., Talbot, I. C., Bodmer, W. F., and Wright, N. A. (1996). Polyclonal origin of colonic adenomas in an XO/XY patient with FAP. *Science* **272**, 1187–1190.

Novelli, M., Cossu, A., Oukrif, D., Quaglia, A., Lakhani, S., Poulsom, R., Sasieni, P., Carta, P., Contini, M., Pasca, A., Palmieri, G., Bodmer, W., Tanda, F., and Wright, N. (2003). X-inactivation patch size in human female tissue confounds the assessment of tumor clonality. *Proc. Natl. Acad. Sci. U.S.A.* **100**, 3311–3314.

Okamoto, R., Yajima, T., Yamazaki, M., Kanai, T., Mukai, M., Okamoto, S., Ikeda, Y., Hibi, T., Inazawa, J., and Watanabe, M. (2002). Damaged epithelia regenerated by bone marrow–derived cells in the human gastrointestinal tract. *Nat. Med.* **8**, 1011–1017.

Paris, F., Fuks, Z., Kang, A., Capodieci, P., Juan, G., Ehleiter, D., Haimovitz-Friedman, A., Cordon-Cardo, C., and Kolesnick, R. (2001). Endothelial apoptosis as the primary lesion initiating intestinal radiation damage in mice. *Science* **293**, 293–297.

Park, H. S., Goodlad, R. A., and Wright, N. A. (1995). Crypt fission in the small intestine and colon. A mechanism for the emergence of G6PD locus-mutated crypts after treatment with mutagens. *Am. J. Pathol.* **147**, 1416–1427.

Park, H. S., Goodlad, R. A., Ahnen, D. J., Winnett, A., Sasieni, P., Lee, C. Y., and Wright, N. A. (1997). Effects of epidermal growth factor and dimethylhydrazine on crypt size, cell proliferation, and crypt fission in the rat colon. Cell proliferation and crypt fission are controlled independently. *Am. J. Pathol.* **151**, 843–852.

Ponder, B. A., Schmidt, G. H., Wilkinson, M. M., Wood, M. J., Monk, M., and Reid, A. (1985). Derivation of mouse intestinal crypts from single progenitor cells. *Nature* **313**, 689–691.

Potten, C. S., Booth, C., and Pritchard, D. M. (1997). The intestinal epithelial stem cell: the mucosal governor. *Int. J. Exp. Pathol.* **78**, 219–243.

Potten, C. S., Owen, G., and Booth, D. (2002). Intestinal stem cells protect their genome by selective segregation of template DNA strands. *J. Cell Sci.* **115**, 2381–2388.

Potten, C. S., Booth, C., and Hargreaves, D. (2003). The small intestine as a model for evaluating adult tissue stem cell drug targets. *Cell Prolif.* **36**, 115–129.

Poulsom, R., Alison, M. R., Forbes, S. J., and Wright, N. A. (2002). Adult stem cell plasticity. *J. Pathol.* **197**, 441–456.

Powell, D. W., Mifflin, R. C., Valentich, J. D., Crowe, S. E., Saada, J. I., and West, A. B. (1999a). Myofibroblasts. II. Intestinal subepithelial myofibroblasts. *Am. J. Physiol.* **277**, C183–C201.

Powell, D. W., Mifflin, R. C., Valentich, J. D., Crowe, S. E., Saada, J. I., and West, A. B. (1999b). Myofibroblasts. I. Paracrine cells important in health and disease. *Am. J. Physiol.* **277**, C1–C9.

Rajantie, I., Ilmonen, M., Alminaite, A., Ozerdem, U., Alitalo, K., and Salven, P. (2004). Adult bone marrow–derived cells recruited during angiogenesis comprise precursors for periendothelial vascular mural cells. *Blood* **104**, 2084–2086.

Reyes, M., Dudek, A., Jahagirdar, B., Koodie, L., Marker, P. H., and Verfaillie, C. M. (2002). Origin of endothelial progenitors in human postnatal bone marrow. *J. Clin. Invest.* **109**, 337–346.

Rizvi, A. Z., Swain, J. R., Davies, P. S., Bailey, A. S., Decker, A. D., Willenbring, H., Grompe, M., Fleming, W. H., and Wong, M. H. (2006). Bone marrow–derived cells fuse with normal and transformed intestinal stem cells. *Proc. Natl. Acad. Sci. U.S.A.* **103**, 6321–6325.

Roberts, S. A., Hendry, J. H., and Potten, C. S. (1995). Deduction of the clonogen content of intestinal crypts: a direct comparison of two-dose and multiple-dose methodologies. *Radiat. Res.* **141**, 303–308.

Ronnov-Jessen, L., Petersen, O. W., Koteliansky, V. E., and Bissell, M. J. (1995). The origin of the myofibroblasts in breast cancer. Recapitulation of tumor environment in culture unravels diversity and implicates converted fibroblasts and recruited smooth muscle cells. *J. Clin. Invest.* **95**, 859–873.

Sakakibara, S., Imai, T., Hamaguchi, K., Okabe, M., Aruga, J., Nakajima, K., Yasutomi, D., Nagata, T., Kurihara, Y., Uesugi, S., Miyata, T., Ogawa, M., Mikoshiba, K., and Okano, H. (1996). Mouse-Musashi-1, a neural RNA-binding protein highly enriched in the mammalian CNS stem cell. *Dev. Biol.* **176**, 230–242.

Sancho, E., Batlle, E., and Clevers, H. (2004). Signaling pathways in intestinal development and cancer. *Annu. Rev. Cell Dev. Biol.* **20**, 695–723.

Schmidt, G. H., Garbutt, D. J., Wilkinson, M. M., and Ponder, B. A. (1985). Clonal analysis of intestinal crypt populations in mouse aggregation chimaeras. *J. Embryol. Exp. Morphol.* **85**, 121–130.

Schmidt, G. H., Winton, D. J., and Ponder, B. A. (1988). Development of the pattern of cell renewal in the crypt-villus unit of chimaeric mouse small intestine. *Development* **103**, 785–790.

Sciacco, M., Bonilla, E., Schon, E. A., DiMauro, S., and Moraes, C. T. (1994). Distribution of wild-type and common deletion forms of mtDNA in normal and respiration-deficient muscle fibers from patients with mitochondrial myopathy. *Hum. Mol. Genet.* **3**, 13–19.

Soderholm, J. D., Malm, C., Juliusson, G., and Sjodahl, R. (2002). Long-term endoscopic remission of Crohn disease after autologous stem cell transplantation for acute myeloid leukaemia. *Scand. J. Gastroenterol.* **37**, 613–616.

Spradling, A., Drummond-Barbosa, D., and Kai, T. (2001). Stem cells find their niche. *Nature* **414**, 98–104.

Spyridonidis, A., Schmitt-Graff, A., Tomann, T., Dwenger, A., Follo, M., Behringer, D., and Finke, J. (2004). Epithelial tissue chimerism after human hematopoietic cell transplantation is a real phenomenon. *Am. J. Pathol.* **164**, 1147–1155.

Subramanian, V., Meyer, B., and Evans, G. S. (1998). The murine Cdx1 gene product localizes to the proliferative compartment in the developing and regenerating intestinal epithelium. *Differentiation* **64**, 11–18.

Takaku, K., Oshima, M., Miyoshi, H., Matsui, M., Seldin, M. F., and Taketo, M. M. (1998). Intestinal tumorigenesis in compound mutant mice of both Dpc4 (Smad4) and Apc genes. *Cell* **92**, 645–656.

Talbot, D. C., Montes, A., Teh, W. L., Nandi, A., and Powles, R. L. (1998). Remission of Crohn's disease following allogeneic bone marrow transplant for acute leukaemia. *Hosp. Med.* **59**, 580–581.

Taylor, R. W., Barron, M. J., Borthwick, G. M., Gospel, A., Chinnery, P. F., Samuels, D. C., Taylor, G. A., Plusa, S. M., Needham, S. J., Greaves, L. C., Kirkwood, T. B., and Turnbull, D. M. (2003). Mitochondrial DNA mutations in human colonic crypt stem cells. *J. Clin. Invest.* **112**, 1351–1360.

Thomas, G. A., Williams, D., and Williams, E. D. (1988). The demonstration of tissue clonality by X-linked enzyme histochemistry. *J. Pathol.* **155**, 101–106.

Trnka, Y. M., Glotzer, D. J., Kasdon, E. J., Goldman, H., Steer, M. L., and Goldman, L. D. (1982). The long-term outcome of restorative operation in Crohn's disease: influence of location, prognostic factors and surgical guidelines. *Ann. Surg.* **196**, 345–355.

Vainer, B., and Nielsen, O. H. (2000). Changed colonic profile of P-selectin, platelet-endothelial cell adhesion molecule-1 (PECAM-1), intercellular adhesion molecule-1 (ICAM-1), ICAM-2, and ICAM-3 in inflammatory bowel disease. *Clin. Exp. Immunol.* **121**, 242–247.

van Buul, J. D., Mul, F. P., van der Schoot, C. E., and Hordijk, P. L. (2004). ICAM-3 activation modulates cell–cell contacts of human bone marrow endothelial cells. *J. Vasc. Res.* **41**, 28–37.

van Den Brink, G. R., de Santa Barbara, P., and Roberts, D. J. (2001). Development. Epithelial cell differentiation — a Mather of choice. *Science* **294**, 2115–2116.

Vassilopoulos, G., Wang, P. R., and Russell, D. W. (2003). Transplanted bone marrow regenerates liver by cell fusion. *Nature* **422**, 901–904.

Wang, X., Willenbring, H., Akkari, Y., Torimaru, Y., Foster, M., Al-Dhalimy, M., Lagasse, E., Finegold, M., Olson, S., and Grompe, M. (2003). Cell fusion is the principal source of bone-marrow-derived hepatocytes. *Nature* **422**, 897–901.

Weimann, J. M., Charlton, C. A., Brazelton, T. R., Hackman, R. C., and Blau, H. M. (2003a). Contribution of transplanted bone marrow cells to Purkinje neurons in human adult brains. *Proc. Natl. Acad. Sci. U.S.A.* **100**, 2088–2093.

Weimann, J. M., Johansson, C. B., Trejo, A., and Blau, H. M. (2003b). Stable reprogrammed heterokaryons form spontaneously in Purkinje neurons after bone marrow transplant. *Nat. Cell Biol.* **5**, 959–966.

Wielenga, V. J., Smits, R., Korinek, V., Smit, L., Kielman, M., Fodde, R., Clevers, H., and Pals, S. T. (1999). Expression of CD44 in Apc and Tcf mutant mice implies regulation by the WNT pathway. *Am. J. Pathol.* **154**, 515–523.

Williams, E. D., Lowes, A. P., Williams, D., and Williams, G. T. (1992). A stem cell niche theory of intestinal crypt maintenance based on a study of somatic mutation in colonic mucosa. *Am. J. Pathol.* **141**, 773–776.

Winesett, M. P., Ramsey, G. W., and Barnard, J. A. (1996). Type II TGF(beta) receptor expression in intestinal cell lines and in the intestinal tract. *Carcinogenesis* **17**, 989–995.

Winton, D. J., and Ponder, B. A. (1990). Stem-cell organization in mouse small intestine. *Proc. R. Soc. Lond. B. Biol. Sci.* **241**, 13–18.

Winton, D. J., Flaks, B., and Flaks, A. (1988). Quantitative electron microscopy of carcinogen-induced alterations in hepatocyte rough endoplasmic reticulum. I. Chronic effect of 3'MeDAB and short-term effects of azo dyes of different carcinogenic potentials. *Carcinogenesis* **9**, 987–999.

Wright, N. A., and Alison, M. R. (1984). The Biology of Epithelial Populations. Oxford University Press.

Yang, Q., Bermingham, N. A., Finegold, M. J., and Zoghbi, H. Y. (2001). Requirement of Math1 for secretory cell lineage commitment in the mouse intestine. *Science* **294**, 2155–2158.

Yatabe, Y., Tavare, S., and Shibata, D. (2001). Investigating stem cells in human colon by using methylation patterns. *Proc. Natl. Acad. Sci. U.S.A.* **98**, 10839–10844.

# Chapter Forty-Six

# Alimentary Tract

*Shaun M. Kunisaki and Joseph Vacanti*

## I. INTRODUCTION

Currently, two conventional strategies are aimed at replacing absent or severely damaged portions of the alimentary tract. The first approach attempts to substitute native tissue with autogenous tissue derived from another anatomic location. For example, in cases of congenital long-gap esophageal atresia, the esophagus can be reconstructed using portions of stomach, colon, or jejunum. Similarly, the distal small intestine can be reconstructed into a colonlike pouch after removal of the large intestine. Unfortunately, these replacement tissues do not completely mimic the functions of the native tissue and therefore often result in poor function while incurring substantial donor site morbidity.

A second approach to replacing gastrointestinal tissue is through the transplantation of an entire organ from one person to another. This strategy is now a reasonable and potentially lifesaving option for carefully selected patients with intestinal failure. However, transplantation remains severely limited by the morbidity associated with immune rejection and chronic immunosuppression. Moreover, the continued shortage of available donor organs continues to be a major problem in the field of transplantation in general. As a result, transplantation can be justified only in selected clinical situations associated with high mortality, such as in cases of short bowel syndrome with recurrent sepsis.

Since the mid-1980s, a third approach to reconstructing the alimentary tract based on the principles of tissue engineering has emerged as another potential treatment option (Vacanti, 1988, 2003). Recently, tissue engineering has made significant progress in terms of successfully recapitulating all major structures within the alimentary tract. To date, esophagus, stomach, small intestine, and large intestine have all been created experimentally using the methods of tissue engineering. Despite significant challenges unique to the fabrication of the alimentary tract, ongoing work suggests enormous therapeutic potential in the years to come.

The purpose of this chapter is to provide an overview of the principles and current status of alimentary tract tissue engineering. We discuss the basic anatomy and physiology relevant to the field and review the rationale for tissue engineering as replacement therapy for severely damaged or absent alimentary tract tissue. We then evaluate the different state-of-the-art tissue-engineering strategies that have been employed in recent years using natural as well as synthetic scaffolds. Avenues for further exploration are also proposed.

## II. TISSUE-ENGINEERED SMALL INTESTINE

The native small intestine, which is composed of duodenum, jejunum, and ileum, is the most critical portion of the gastrointestinal tract. It is a remarkably complex and unique organ because of its digestion and vectorial transport (i.e., absorption and secretion) capabilities. From the pylorus to the ileocecal valve, the adult small intestine measures approximately five to six meters in length. The pancreas and liver secrete important digestive juices into the duodenum to help facilitate the breakdown of food (Brunicardi *et al.*, 2004).

The digestion and absorption capacity of the small intestine is facilitated by its highly specialized columnar epithelium, which features microscopic, fingerlike projections, known as *villi*, that help to increase total surface area. Brush border enzymes facilitate the digestion of carbohydrates as they are absorbed. In addition, each villus contains a lacteal and capillaries. The lacteal absorbs the digested fat into the lymphatic system, whereas the capillaries absorb all other digested nutrients. By the time the remaining products have reached the end of the small intestine, almost all nutrients and 90% of the water have been absorbed.

The smooth muscle layers of the small intestine are well organized and consist of an inner circular layer and an outer longitudinal layer. Intestinal smooth muscle helps to initiate antegrade propulsion of luminal contents (peristalsis). Both muscle layers are innervated intrinsically by autonomic nerve fibers and extrinsically by vagal and sympathetic nerve fibers. The celiac and superior mesenteric arteries provide essential arterial blood flow to the entire small intestine. The organ is a rich source of regulatory peptides, including secretin, cholecystokinin, and somatostatin, which are important for the control of various aspects of gut function. The mucosal immune system within the small intestine is critically important for defense against toxic and pathogenic threats from the luminal environment (Brunicardi *et al.*, 2004).

Major surgical resection of small intestine is required in a variety of pediatric and adult disorders, including necrotizing enterocolitis, gastroschisis, midgut volvulus, and Crohn's disease. When less than 25% of the native small intestine is left, the remaining intestine cannot absorb all of the necessary calories (energy) required for survival. This leads to a state of malabsorption, malnutrition, and dehydration known as the *short bowel syndrome* (SBS). Intestinal adaptation can occur within the remaining native intestine. Unfortunately, these adaptive changes can take up to two years to occur and may not be enough to help the most severe cases of SBS.

The current mainstay of therapy for severe SBS, which was uniformly lethal until the late 1960s, is total parenteral nutrition (TPN). TPN is delivered through a central intravenous catheter and provides essential calories, vitamins, and minerals. Although this therapy has markedly improved survival rates, the costs of TPN can be prohibitive, and complications resulting from long-term intravenous access, including catheter sepsis, venous thrombosis, and liver failure, are common (Gupte *et al.*, 2006).

Given the high morbidity and mortality associated with SBS, a variety of interesting and often-innovative intestinal

operations have been developed over the years. These procedures include the tapering enteroplasty, intestinal valve placement, Bianchi procedure, and, more recently, the serial transverse enteroplasty (STEP procedure). All of these operations are designed to slow intestinal transit and/or increase intestinal surface area. Unfortunately, most of these techniques have since been abandoned, and no single approach with demonstrable long-term results has been reported in a large series of patients (Gupte *et al.*, 2006). Small intestinal transplantation is becoming a more realistic option in carefully selected patients (Abu-Elmagd, 2006). However, one-year graft survival rates are still far below those with other organs, and problems associated with rejection, immunosuppressive therapy, and donor organ shortages are not likely to be solved, at least in the near future.

## Early Studies Using Synthetic Scaffold–Based Approaches

In the 1980s, early studies were established that laid the groundwork for modern tissue engineering of small intestine. Thompson (1987) published the first report on the use of an absorbable biomaterial, polyglycolic acid (PGA), as a scaffold for the ingrowth of intestinal neomucosa. In a leporine model, he repaired ileal defects with PGA, Dacron, or PTFE. At eight weeks, minimal (15%) neomucosal development was present over each of these materials. Interestingly, Thompson was so discouraged by his results that he concluded that "the use of prosthetic materials is not likely to be used in the clinical management of short bowel syndrome."

Shortly after publication of this study, our laboratory, in collaboration with Robert Langer's group, began initial efforts to create tissue-engineered small intestine using cells transplanted onto biodegradable prosthetic materials. In the seminal study, fetal rodent intestine was enzymatically digested in collagenase, and the resulting enterocytes were seeded onto biodegradable polymers of polyglactin 910, polyandhydrides, and polyorthoester (Vacanti *et al.*, 1988). After four days in culture, the cell–polymer constructs were implanted into the omentum, interscapular fat pad, or mesentery of syngeneic adult hosts in order to facilitate neovascularization. After seven days in this *in vivo* bioreactor, microscopic examination of these implants revealed well-differentiated epithelium contained with a cystic structure. The cells retained their polarity and appeared to be secreting mucus, suggesting ongoing specialized function. From these results, we concluded that tissue-engineered small intestine was potentially feasible and was therefore worthy of further study.

In an effort to increase the efficiency of producing small intestinal neomucosa, subsequent work by Organ in our laboratory was aimed at isolating pure populations of intestinal crypt stem cells (Organ *et al.*, 1992). This approach is based on the premise that enterocytes represent a rapidly renewing cell population characterized by continuous growth and differentiation. Using a method to isolate separate fractions of villus tip cells and crypt cells without the *in situ* leukocytes or fibroblasts, Organ showed that these intestinal crypt cells could be reliably isolated and used to reconstitute an epithelial layer on PGA scaffolds. However, the implanted constructs consistently demonstrated stratified epithelium five to seven layers thick. This mucosa was reminiscent of fetal gut development rather than that of normal columnar epithelium seen with mature small intestinal mucosa. In addition, no goblet cells or villus formation suggestive of advanced cellular or architectural differentiation was observed by 28 days *in vivo*. Based on these data, we concluded that this intestinal stem cell technique, as currently practiced, was limited by the absence of epithelial–mesenchymal cell–cell interactions, thought to be critical during embryonal organogenesis (Patel *et al.*, 1996).

## Synthetic Scaffold–Based Approaches Using Organoid Units

The use of organoid units, which preserve the important epithelial–mesenchymal interactions between cells as intestinal tissue is reconstituted, represents a major advance in small intestinal engineering. These unique cell clusters, termed *epithelial organoids*, were first described by Evans and Potten in the early 1990s in an attempt to establish primary cultures of intestinal cells (Evans *et al.*, 1992). In their initial study, the investigators found that a limited enzymatic digestion of neonatal rat intestine using a crude mixture of collagenase XI and dispase produced multicellular aggregates within a villus core composed of functional epithelial cells, mesenchymal tissue, and perhaps intestinal stem cells (Fig. 46.1). Approximately 50,000 organoid units, each approximately 100–200 microns in diameter, could be isolated per neonatal rat intestine. *In vitro* culture of these organoids has been demonstrated for up to 14 days.

In subsequent experiments, these small intestinal organoid units were used to regenerate small intestine–like structures *in vivo* in the absence of biodegradable scaffolds. This was first demonstrated when organoid units were implanted subcutaneously in a rodent model (Tait *et al.*, 1994a). Tait and Evans then showed that these organoid units could be used to regenerate a neomucosa with typical small bowel morphology and digestive enzyme activity when seeded onto the mucosectomized large intestine of adult rats (Tait *et al.*, 1994b, 1995). In contrast, control large intestine that underwent a mucosectomy without organoid implantation showed no epithelial regrowth. More recently, in an attempt to develop a treatment for bile acid malabsorption, Stelzner and his colleagues have transplanted organoid units derived from rat ileum onto denuded segments of jejunum and have shown increased bile acid uptake as well as increased ileal bile acid transporter protein expression within a neoileal segment (Avansino *et al.*, 2005).

The successful seeding of small intestinal organoid units on biodegradable scaffolds was first demonstrated by

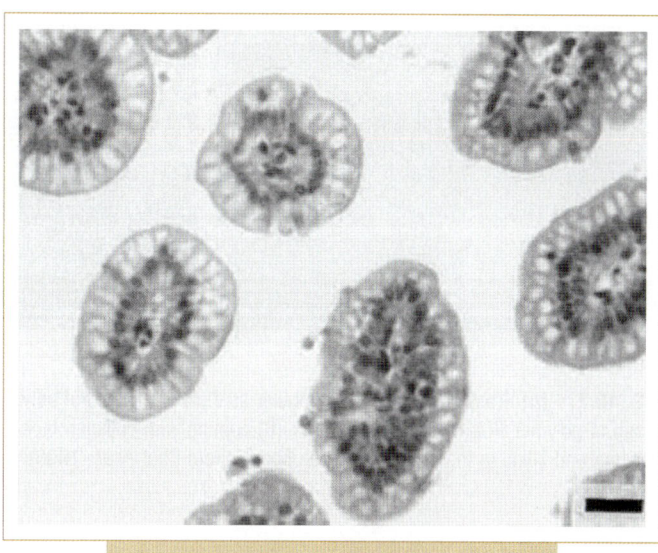

**FIG. 46.1.** Transverse section through isolated small intestinal epithelial organoid units derived from a neonatal rat after hematoxylin and eosin staining (bar, 35 µm). Integrity between the epithelial cell compartment and the underlying stromal core of lamina propria is maintained. From Tait *et al.* (1994a), p. 94.

Choi in our laboratory. In her initial work, organoid units derived from neonatal rat intestine were seeded on nonwoven PGA tubes. Each tube was highly porous (e.g., 95%) and had a mean pore size of 250 microns to accommodate the size of these cell clusters (Choi and Vacanti, 1997). The outer surface of the scaffold was sprayed with 5% poly-L-lactic acid to help increase the resistance to compressional forces that would otherwise lead to luminal collapse after implantation. During the final step in the scaffold fabrication process, the tubes were coated with collagen type I to enhance organoid unit attachment (Fig. 46.2). The seeded constructs were then implanted into the omentum of syngeneic adult rats to allow for vascular ingrowth. As early as two weeks after implantation, the harvested constructs resembled cysts lined by a neomucosa composed of columnar epithelium with goblet cells and Paneth cells (Fig. 46.3). In addition, immunohistochemical staining for smooth muscle alpha actin was positive in the stroma adjacent to the neomucosa, suggestive of the appropriate reconstitution of an intestinal smooth muscle layer (Fig. 46.4). Using an S100 stain, neuronal cells have also been detected in these constructs. Immunofluorescent staining of the neomucosa revealed the apical expression of brush border enzymes as well as basolateral staining for laminin (Choi *et al.*, 1998). Ussing chamber studies of the engineered constructs have shown a similar decreased active transepithelial resistance when compared to native small intestine, thus fulfilling major functional criteria for intestinal epithelia. Therefore, for the first time these studies established a reliable approach for generating tissue on biodegradable scaf-

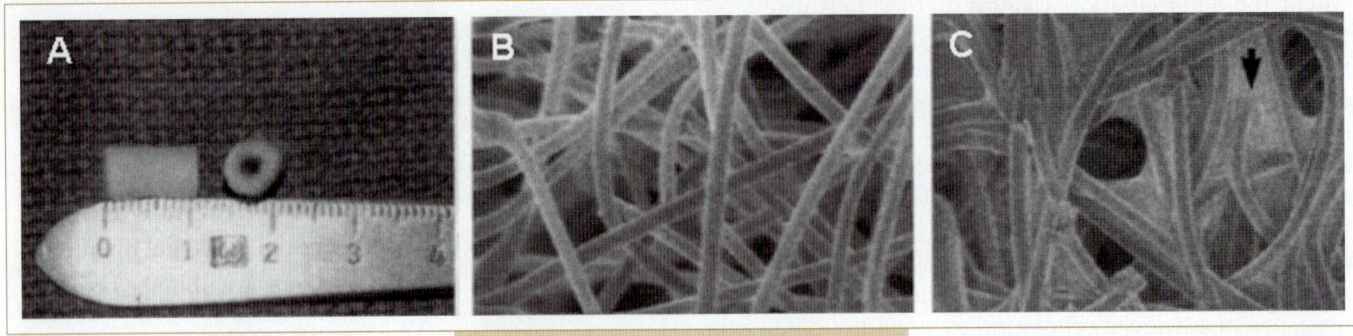

**FIG. 46.2.** **(A)** Gross appearance of a tubular scaffold composed of polyglycolic acid (PGA) and sprayed with 5% polylactic acid. Scanning electron micrograph of polymer-PGA before **(B)** and after **(C)** coating with collagen type I (original magnification, ×100). The black arrow marks the collagen coating the polymer and filling in the spaces between fibers. From Choi *et al.* (1998), p. 993.

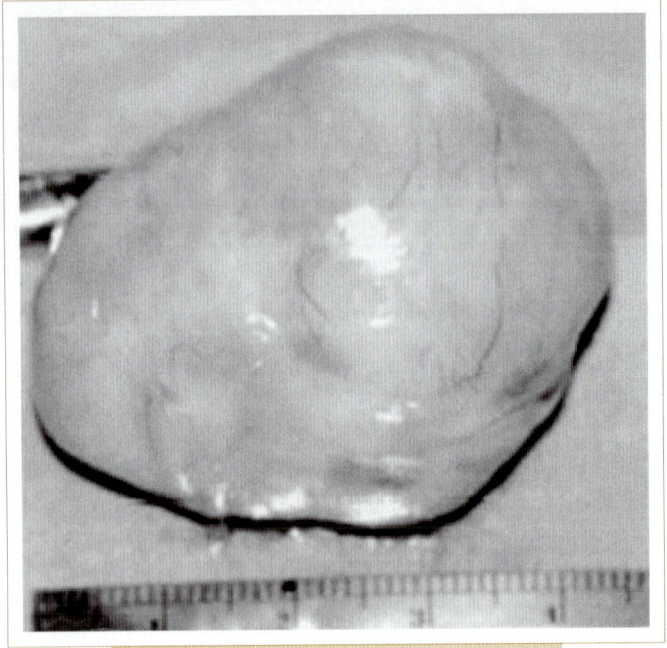

**FIG. 46.3.** Gross appearance of a representative tissue-engineered small intestinal cyst harvested from the omentum of an adult rat after four weeks. From Grikscheit *et al.* (2004), p. 751.

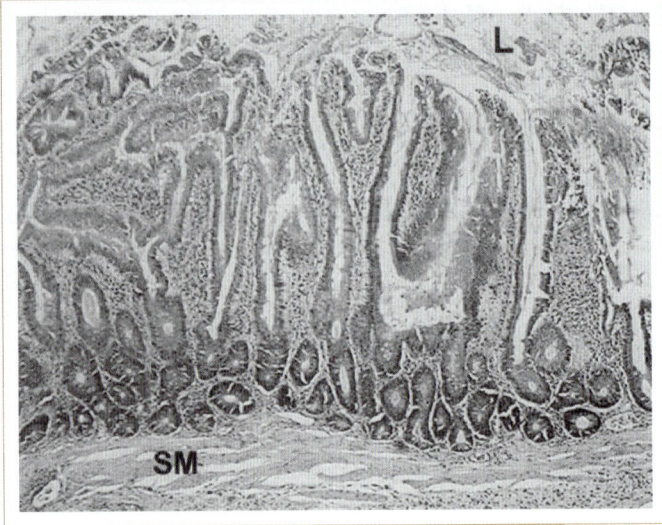

**FIG. 46.4.** Hematoxylin- and eosin-stained photomicrograph of rat tissue–engineered small intestine at 10 weeks after small bowel resection and side-to-side anastomosis (original magnification, ×80). L, cyst lumen; SM, smooth muscle. From Kim *et al.* (1999b), p. 11.

folds that had a striking resemblance to native, full-thickness small intestinal mucosa.

Over the ensuing years, our laboratory has explored the organoid approach to engineering small intestinal cysts in considerable detail. Prolonged *in vitro* culture of these cells for days to weeks has not led to better results when compared to implantation within hours of organoid unit isolation. However, other modifications in this protocol have proven to be successful in creating more robust intestinal neomucosa. For instance, both side-to-side and end-to-end anastomosis of engineered intestine with native jejunum improve the histology within these cysts when harvested in rodents up to 10 weeks postoperatively (Fig. 46.5) (Kaihara *et al.*, 1999a, 1999b; Kim *et al.*, 1999). Increased regenerative stimuli induced by small bowel resection, and to a lesser extent portacaval shunting, have also been shown to enhance the formation of tissue-engineered intestine in morphometric analyses. These results suggest a possible role for trophic factors such as hepatocyte growth factor (HGF) and epidermal growth factor (EGF) in neomucosal growth (Kim *et al.*, 1999). Patency of these engineered intestinal grafts has been maintained for up to 36 weeks (Kaihara *et al.*, 2000). Through ongoing collaboration with Whang and his colleagues, we have been able to further characterize the degree of angiogenesis and lymphangiogenesis within these tissue-engineered grafts (Gardner-Thorpe *et al.*, 2003; Duxbury *et al.*, 2004).

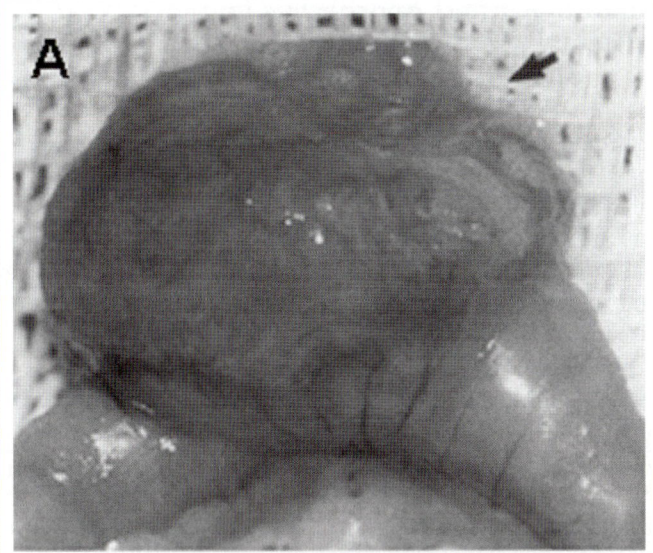

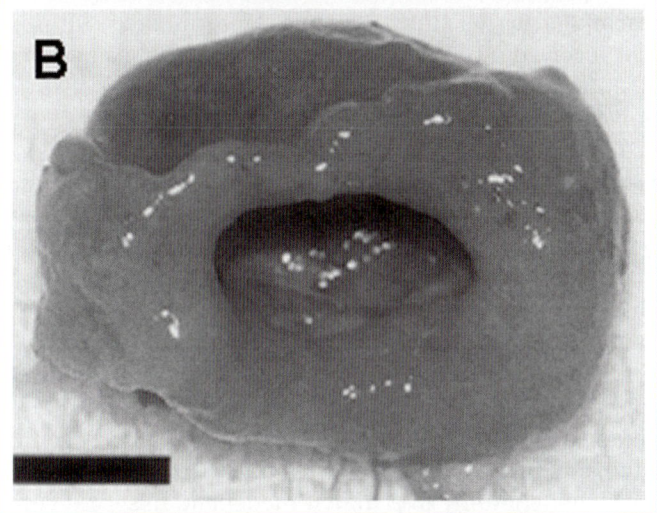

**FIG. 46.5.** **(A)** Gross appearance of rat tissue–engineered small intestine after side-to-side anastomosis to native small intestine at the time of harvest (black arrow denotes tissue-engineered small intestine). **(B)** Cross-sectional view of the patent anastomosis between the tissue-engineered construct and native small intestine (bar, 5 mm). From Kim *et al.* (1999b), p. 9.

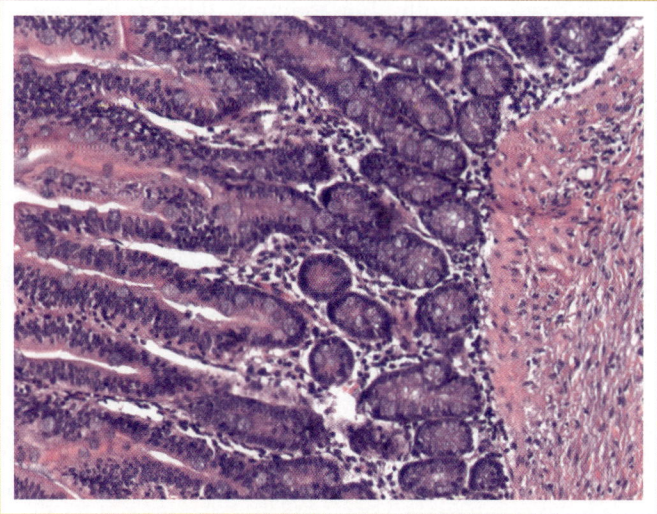

**FIG. 46.6.** Hematoxylin- and eosin-stained photomicrograph of tissue-engineered small intestine harvested from the omentum after seven weeks in a juvenile porcine model (original magnification, ×200). Unpublished data, courtesy of Tracy C. Grikscheit and Erin R. Ochoa.

Using a slight modification of the organoid unit isolation protocol, our laboratory has recently shown that these intestinal cysts, when anastomosed in a side-to-side fashion to native intestine after an 85% recipient enterectomy, can slow intestinal transit times and improve weight gain when compared to rats that had an enterectomy without tissue-engineered replacement (Grikscheit *et al.*, 2004). These data suggest that tissue-engineered intestinal cysts measuring as little as 10% of the length of native small intestine may be enough to provide adequate absorptive function in a rat model. In light of these promising results, we have initiated some large-animal studies using the organoid unit approach and have since demonstrated successful generation of tissue-engineered small intestine in a juvenile porcine model (Fig. 46.6, previously unpublished).

Engineered intestinal cysts have also provided a unique experimental model system for the study of intestinal morphogenesis. For example, in anastomosed tissue-engineered small intestinal cysts harvested at 20 weeks, the density and topographical distribution of immune cell subsets, namely T-cells, B-cells, and NK-cells, were found to be identical to those of normal jejunum (Perez *et al.*, 2002). Furthermore, these immunocyte population densities were only rudimentary in nonanastomosed cysts. These findings imply that exposure to luminal stimuli may be critical to the formation of a mucosal immune system.

It has also been shown that anastomosis of intestinal cysts to native intestine enhances expression of the sodium-dependent glucose transporter, SGLT1 (Tavakkolizadeh *et al.*, 2003). Mucosal growth and SGLT1 expression can also be augmented in this model if rats are given glucogon-like peptide 2 (GLP-2), an endogenous regulatory peptide that has potent trophic activity specific for intestinal mucosa (Ramsanahie *et al.*, 2003). Third-generation biomaterials, which can be designed for the controlled release of substances embedded with the biodegradable scaffold, have been used in other tissue-engineering applications and may be an approach to generate more robust intestinal neomucosal growth in our model (Kent Leach *et al.*, 2006).

Although organoid units have been highly successful in regenerating features of native small intestinal mucosa, it is important to realize the limitations of this strategy as it is performed currently. For example, despite the use of tubular scaffolds based on PGA, these intestinal constructs form not cylindrical tubes but, rather, oblate spheroids reminiscent of intestinal duplication cysts. This phenomenon occurs, at least in part, because of the increased intraluminal pressure as the intestinal neomucosa regenerates and produces secretions. Additionally, PGA clearly does not have adequate mechanical strength to withstand the increased intraluminal pressure, despite a degradation time of approximately six weeks. Regardless, the resultant oblate spheroid architecture suggests that meaningful and coordinated peristalsis within a sizable construct is unlikely, even in the presence of an innervated smooth muscle layer. Methods to decompress the lumen during the process of intestinal epithelial regeneration may be helpful in producing a more desirable architecture. In addition, the use of novel biomaterials that have more favorable mechanical and degradation characteristics may allow for better control of the three-dimensional geometry of neointestine. Studies utilizing alternative scaffolds, such as poly(epsilon-caprolactone) (PCL) and poly(ester urethane)urea (PEUU), remain ongoing in our laboratory, and the early data are promising.

Another limitation of the organoid unit approach to creating tissue-engineered small intestine is the inability consistently to expand these cell clusters *ex vivo*, as was originally described by Evans and colleagues. At present, a relatively high number of organoid units (approximately 15,000 cm$^2$) are required to create an intestinal construct using the omentum as an *in vivo* bioreactor. Unfortunately, we have been unable to culture normal organoid units for more than a week *in vitro*. Transformed epithelial cell lines could technically be used for small intestinal tissue engineering but are not desirable, since they are neither autologous nor normal epithelial cells.

An intriguing variation of the organoid unit approach to small intestinal tissue engineering was recently reported by Lloyd and his group in Britain (Lloyd *et al.*, 2006). In this study, they implanted tubular poly-lactide-*co*-glycolide (PGLA) foam scaffolds subcutaneously into rats. The PLGA had a thickness of 2 mm and was supported by inner silicone tubing. After five weeks, the silicone tubing was removed, and the lumen was filled with intestinal organoid units suspended in a hydrogel containing HGF. All constructs that were examined four weeks later revealed spheroid cysts with evidence of intestinal mucosa within the lumen. A potential benefit of this approach is the apparent efficiency by which neomucosa can be generated from a given sample of donor tissue. Nevertheless, there are a number of confounding factors (e.g., hydrogel, growth factor administration) that may also explain the positive results in this study. It remains unclear whether any smooth muscle was present in these constructs at harvest. Furthermore, as we have already discussed, the overall spherical architecture of these engineered intestinal grafts remains less than ideal.

## Natural Scaffold Approaches in Intestinal Engineering

Some groups have explored different methods for creating tissue-engineeered intestine using acellular, natural scaffolds. Unfortunately, these results have generally been more mixed. Small intestinal submucosa (SIS), an extracellular matrix material derived from porcine small intestine, has been a biomaterial preferred by a number of investigators. For example, Chen and colleagues have repaired partial and full circumferential defects in the small intestine using either single-layer or multilayered SIS in a canine model (Chen and Badylak, 2001). None of these scaffolds was seeded with isolated cells prior to implantation. Evaluation of the partial circumferential repairs showed no luminal narrowing when followed for up to one year postoperatively. Histology at later time points demonstrated layers of remodeled wall containing a mucosal epithelium as well as varying amounts of smooth muscle and collagen that resembled native small intestine. Unfortunately, all dogs that received a tubular segment of SIS placed as an interposition graft had significant problems, including obstruction and anastomotic leak.

The use of SIS as an interposition graft in an ileal loop rodent model has been reported by Wang and colleagues (2003). This group used a four-ply layer of SIS without any cell seeding and experienced minimal complications with this material. However, up to 40% graft shrinkage was observed when rats were followed out to 24 weeks postoperatively. Histologically, mucosal ingrowth of intestinal epithelium was present, although the smooth muscle layer was less organized, and no neuronal cells have been demonstrated to date (Fig. 46.7) (Wang *et al.*, 2005). Recently, Dunn and his colleagues reported their experience with SIS after duodenotomy in a rodent model and observed complete epithelialization of a 6-mm-diameter defect by four weeks postoperatively (De Ugarte *et al.*, 2004). Unfortunately, the muscularis layer was notably absent in all examined specimens.

Collagen is another natural scaffold that has been used in small intestinal tissue engineering. In a canine model, Hori and colleagues resected a portion of jejunum and reconstructed the intestine using an acellular collagen sponge supported by a silicone stent (Hori *et al.*, 2001). Small intestinal epithelium was present overlying these grafts at harvest, but no muscle layer was present. In follow-up studies, cell seeding of the collagen sponge with gastric smooth muscle cells, but not bone marrow–derived mesenchymal stem cells, seemed to induce long-term formation of smooth muscle layer beneath the intestinal epithelia (Hori *et al.*, 2002; Nakase *et al.*, 2006).

## III. TISSUE-ENGINEERED ESOPHAGUS

The esophagus, which measures 25 cm in length in adult males, starts as a continuation of the pharynx and

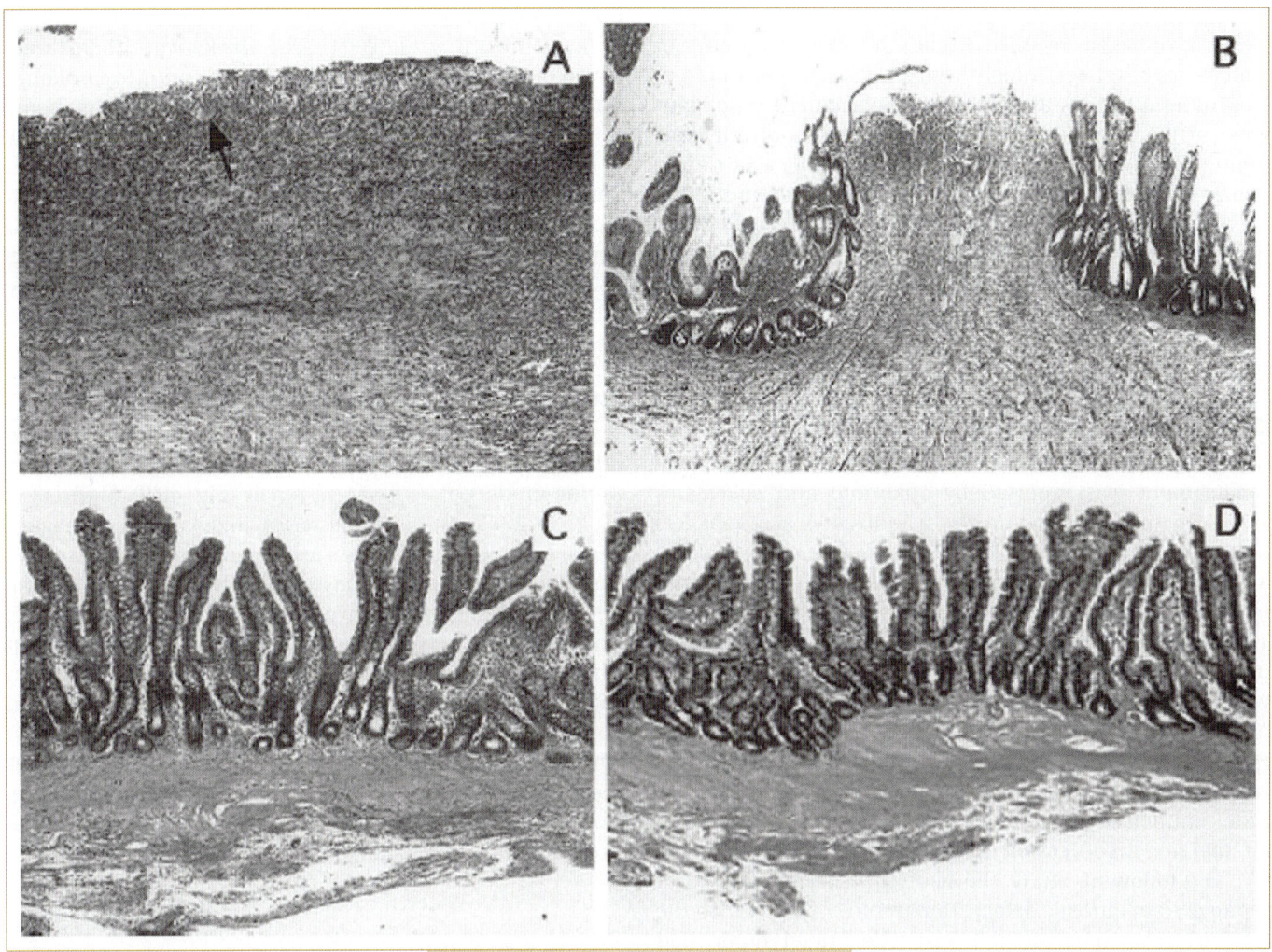

**FIG. 46.7.** Hematoxylin- and eosin-stained photomicrographs of rat small intestine generated on small intestinal submucosa (SIS). **(A)** No epithelialization at the luminal surface by two weeks. Arrows depict neovascularization. **(B)** Major portion of the epithelialized graft luminal surface by eight weeks. Inflammatory reaction was minimal. **(C)** A well-organized mucosal epithelial layer with smooth muscle cells regeneration has occurred at three months. **(D)** A well-developed mucosa, smooth muscle, and serosa of the regenerated bowel wall in another animal (original magnification, ×100). From Wang *et al.* (2003), p. 1598.

ends at the cardia of the stomach. Its blood supply is derived from other organs in the neck, chest, and abdomen. The esophagus resembles a muscular tube lined with nonkeratinizing squamous epithelium. Scattered submucosal mucus glands provide lubrication. The muscularis layer plays a vital role in the peristaltic propulsion of food, such that both solids and liquids can enter the stomach even when being turned upside-down. As with all portions of the alimentary tract, the nerve supply of the esophagus is both parasympathetic and sympathetic. A rich network of intrinsic ganglia within the submucosa and muscularis provides essential innervation. The upper one-third of the esophagus is composed predominantly of striated (voluntary) skeletal muscle. In the lower third, the esophagus contains only smooth (involuntary) muscle. At the junction of the esophagus and

stomach is a sphincter valve that provides a physiologic barrier between the two organs. As the food approaches the closed ring, the surrounding muscles relax and allow the passage of food.

There remains considerable interest in the creation of tissue-engineered esophagus for the treatment of a wide range of pathologies. Unlike the rest of the gastrointestinal tract, the esophagus has little redundancy, thereby leaving little autologous esophageal tissue available for reconstruction. In neonates born with long-gap esophageal atresia, the esophagus is commonly repaired by placing a segment of colon into the chest from the upper pouch to the lower pouch, based on its own vascular supply. Although this procedure has been successful, long-term complications, including anastomotic leak, stricture, dysmotility, and malignant transformation, are not uncommon (Ure *et al.*, 1995). Esophageal reconstruction after chemical injury or cancer has also been fraught with similar complications.

Initial attempts at creating a replacement esophagus used artificial materials, including rubber, polyethylene, Teflon (PTFE), and polypropylene, among others (Sato *et al.*, 1997). Most of these materials were eventually extruded and served as poor scaffolds to facilitate tissue ingrowth. Furthermore, major complications, such as anastomotic leak and stenosis, were a common occurrence. Autologous tissue and homografts, including skin, trachea, and fascia, have been used as esophageal substitutes, with minimal success (Sato *et al.*, 1997).

## Collagen-Based Scaffold Approaches

One of the early studies of bioprosthetic esophageal replacement was reported by Takimoto and colleagues (1993). In this study, they created a 5-cm cervical esophageal defect in a canine model. The defect was then repaired using a two-layered tube consisting of a collagen sponge matrix (composed of types I and III) and an inner silicone stent. The stent was subsequently removed endoscopically at two to four weeks postoperatively. After stent removal at four weeks, regenerated esophageal tissue was found at the site of the defect, and all dogs were able to tolerate oral feeding. Histologically, there was regeneration of host-derived tissue within the collagen sponge, as shown by stratified squamous epithelia and striated muscle, complete with an inner circular and an outer longitudinal layer.

In a follow-up study, the same prosthesis was used to repair a 5-cm thoracic defect (Yamamoto *et al.*, 1999). At the time of stent removal, the host-regenerated tissue had replaced the gap in all dogs, and the mucosa had fully regenerated at three months. Some mild stenosis and shrinkage did occur, but not after three months postoperatively. The muscularis mucosa was seen as islets of smooth muscle by 12 months. Unfortunately, many of these grafts were quite thin, and muscle did not extend into the middle of the regenerated segment. This observation may have been due to the relatively poor blood supply in the mediastinal region of the esophagus. However, increasing blood vessel ingrowth into these grafts using either an omental wrap or basic fibroblast growth factor has not led to substantially improved results (Yamamoto *et al.*, 2000; Hori *et al.*, 2003). More recently, this collagen sponge technique was used after circular myotomy of the esophagus and may promote healing by facilitating connective tissue ingrowth (Komuro *et al.*, 2002).

One cell-based strategy using collagen scaffolds has seeded these constructs with cultured keratinocytes (Sato *et al.*, 1994; Hayashi *et al.*, 2004). In one variation of this approach, Sato and colleagues isolated keratinocytes from rodent esophageal biopsies and expanded these cells *ex vivo* in growth medium. The cells were subsequently seeded onto collagen type I gels that were embedded with dermal fibroblasts and smooth muscle cells derived from the same donor (Hayashi *et al.*, 2004). The sheets were then transferred onto PGA mesh or porcine dermis prior to implantation into the latissimus dorsi muscle of athymic rats. Using this multilayered approach, the investigators have been able to demonstrate a fairly complex esophageal construct containing the major components of native esophagus, including smooth muscle (Fig. 46.8). The tubular architecture has been maintained *in vivo* using PGA. Unpublished data from Sato has shown that these grafts can be rolled over silicone tubes and anastomosed to rat jejunum in an end-to-end fashion.

## Submucosal Scaffold–Based Approaches

Acellular scaffolds such as porcine-derived SIS and urinary bladder submucosa have been used in several canine models of esophageal repair (Fig. 46.9) (Badylak *et al.*, 2000, 2005). In his initial series, Badylak and colleagues (2000) created full-thickness lesions in the cervical esophagus in dogs and repaired these defects with SIS. After several weeks, all of the scaffolds were replaced by skeletal muscle, which was oriented appropriately and was contiguous with adjacent normal esophageal skeletal muscle (Fig. 46.10). In addition, there was an intact squamous epithelium derived from host migration of epithelium. Unfortunately, a major limitation to the use of these acellular scaffolds in esophageal-tissue engineering has been the problem of stricture, which occurs at a much higher frequency after repair of full circumferential defects.

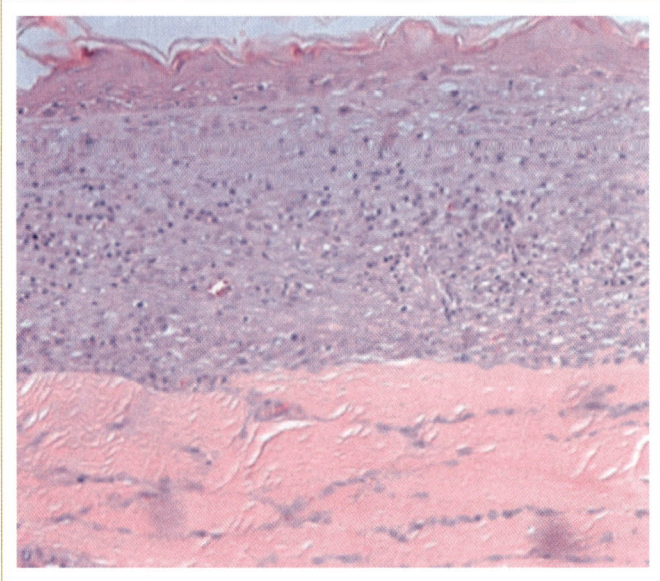

**FIG. 46.8.** Hematoxylin- and eosin-stained photomicrograph showing rat tissue–engineered esophagus using keratinocytes seeded onto a collagen gel containing fibroblasts (original magnification, ×100). From Fuchs *et al.* (2001), p. 586.

**FIG. 46.9.** Photograph of a single-layer sheet of decellularized porcine submucosa used for esophageal reconstruction in a canine model. From Badylak *et al.* (2005), p. 89.

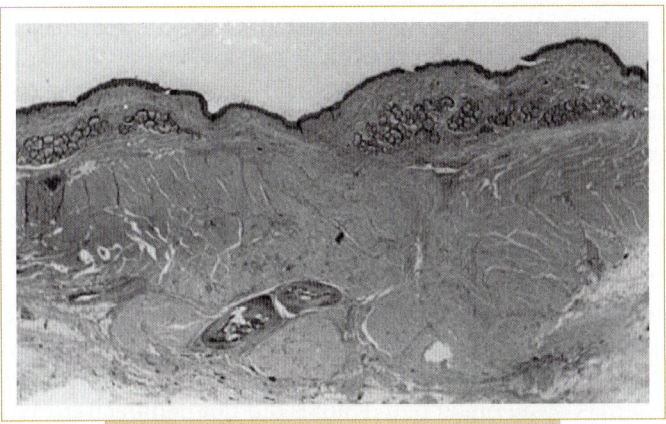

**FIG. 46.10.** Masson trichrome-stained photomicrograph demonstrating near-complete remodeling of the extracellular matrix (ECM) scaffold with neoesophageal tissue after five months in a canine model (original magnification, ×120). There is an intact squamous epithelium, a submucosal layer that is focally devoid of glandular elements, and bundles of partially organized skeletal muscle. From Badylak *et al.* (2000), p. 1100.

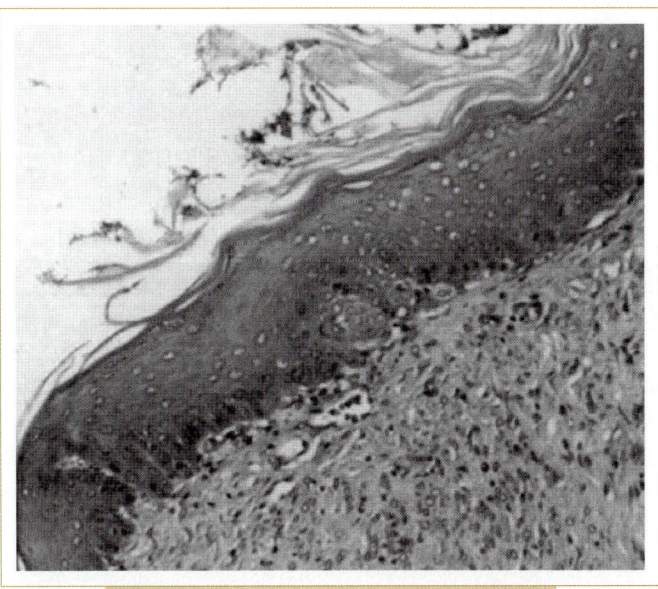

**FIG. 46.11.** Hematoxylin- and eosin-stained photomicrograph illustrating tissue-engineered esophagus in a rodent model based on esophageal organoid units (original magnification, ×100). From Grikscheit *et al.* (2003a), p. 539.

## Synthetic Scaffold–Based Approaches

Grikscheit, from our group, has demonstrated the feasibility of the organoid unit approach in esophageal-tissue engineering (Grikscheit *et al.*, 2003). Applying the same collagenase/dispase digestion technique used to produce small intestinal organoid units, esophageal organoid units were successfully isolated from neonatal esophagus and labeled overnight with green fluorescence protein (GFP) to follow their long-term fate *in vivo*. The organoid units were subsequently seeded onto PGA/PLLA tubes and implanted in the omentum of syngeneic rats. After four weeks *in vivo*, histological evaluation revealed the formation of a complete esophageal wall, including full-thickness stratified squamous mucosa, submucosa, and muscularis propria (Fig. 46.11). Neuronal innervation was not seen in any harvested specimens. The constructs were then used to replace the abdominal esophagus, either as an onlay patch or as an interposition graft. Postoperatively, all animals gained appropriate weight. Although gross inspection of these grafts at sacrifice seven weeks later revealed a massively dilated and cystic construct, there was no evidence of anastomotic leak or obstruction in any rat on fluoroscopy (Fig. 46.12). Long-term survival of the donor cells was demonstrated by maintenance of the GFP signal at harvest. Given the cystic characteristics of these esophageal constructs using PGA, ongoing studies are attempting to seed organoid units onto newer biomaterials with more favorable mechanical characteristics that would allow for maintenance of the tubelike architecture of the native esophagus.

## IV. TISSUE-ENGINEERED STOMACH

The stomach is a well-vascularized, short-term storage reservoir located in the left upper quadrant of the abdominal cavity. It receives ingested food from the esophagus and delivers it to the duodenum. Although the adult stomach has a volume of only 50 mL when empty, as the muscles of the upper portion of the stomach relax, approximately one liter of swallowed material can be accepted (Brunicardi *et al.*, 2004).

Vigorous contractions produced by the muscular wall of the stomach within an acidic (pH 1–3) environment enable the breakdown of food. The stomach is also an

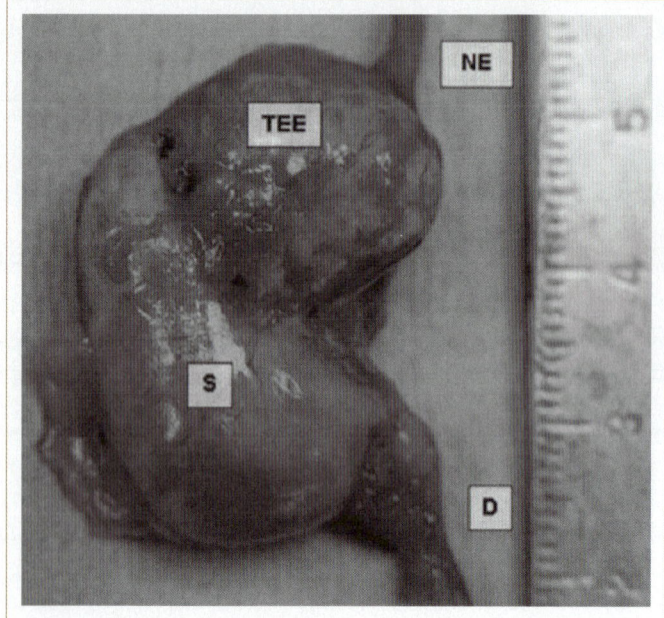

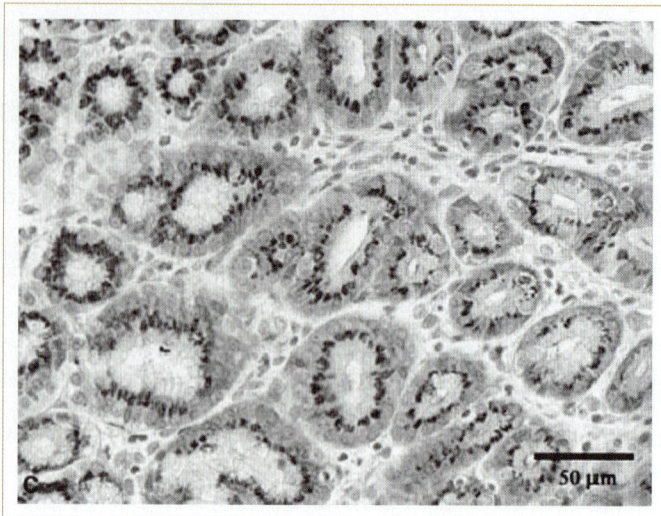

**FIG. 46.13.** Immunohistochemical staining for gastric mucin in rat tissue–engineered stomach (original magnification, ×200). From Maemura *et al.* (2003), p. 63.

**FIG. 46.12.** Gross appearance of rat tissue–engineered esophagus (TEE) at harvest following end-to-end anastomosis with the native alimentary tract. The native esophagus (NE) is shown above the TEE. The stomach (S) is located distally and joins the duodenum (D). From Grikscheit *et al.* (2003), p. 543.

important site for the digestion of proteins. Chief cells, which secrete pepsinogen to aid in protein digestion, as well as parietal cells, which secrete hydrochloric acid, help to form a densely packed mucosa composed of simple, columnar epithelium. Intrinsic factor production by parietal cells is important for the absorption of vitamin $B_{12}$ in the ileum. Food is delivered slowly into the small intestine upon relaxation of the pylorus, a ringlike valve composed of smooth muscle (Brunicardi *et al.*, 2004).

Gastric adenocarcinoma is one of the most common cancers worldwide. In Japan, gastric cancer afflicts approximately one in 1000 patients. For proximal gastric cancer, a total or near-total gastrectomy may be required for survival. Numerous gastric replacement techniques involving autologous jejunum have been used in an effort to improve the quality of life in these patients. Unfortunately, the morbidity associated with these procedures remains high, in part because intestinal substitutes cannot sufficiently recapitulate the specialized functions (i.e., acid secretion, enzymatic digestion, intrinsic factor secretion) of the native stomach. Significant morbidity is common and includes severe malnutrition, weight loss, reflux esophagitis, and anemia (Brunicardi *et al.*, 2004). A tissue-engineered stomach would represent a significant advance toward improving clinical outcomes in these patients.

## Collagen Scaffold–Based Approaches

The earliest efforts to create a tissue-engineered stomach attempted to reconstruct the stomach wall using a collagen sponge scaffold in a canine model. With this approach, Hori and his colleagues (2001a) resected a $4 \times 4$-cm portion of the anterior wall of the stomach and subsequently reconstructed the stomach using a collagen sponge scaffold reinforced with PGA felt. Omentum was sutured to the outer wall of the scaffold to facilitate vascular ingrowth, and a silicone sheet was left in place on the luminal side to protect the scaffold from degradation by digestive juices. After four weeks, the silicone sheet was removed. By 16 weeks, histological analyses revealed a highly organized layer of stomach mucosa overlying the collagen scaffold as well as evidence of a thin muscle layer. Unfortunately, a follow-up functional analysis using organ bath studies did not demonstrate any acetylcholine-induced contraction of the tissue-engineered stomach wall (Hori *et al.*, 2002b).

## Synthetic Scaffold–Based Approaches

Our group has employed the organoid unit approach to create tissue-engineered stomach in a rodent model (Maemura *et al.*, 2003, 2004; Grikscheit *et al.*, 2004). In these studies, gastric organoid units were isolated from both neonatal and adult donors and seeded onto PGA/PLLA tubes implanted into the omentum of syngeneic adult hosts. Three to four weeks later, the engineered constructs resembled cysts and showed well-developed gastric epithelium on histology. Well-formed basal gastric glands were also identified, and immunohistochemical staining demonstrated positive expression of gastrin and mucin within the mucosal lining (Fig. 46.13). A well-organized smooth muscle layer was also present. After anastomosis of tissue-engineered stomach to host native small intestine, the mucosal architecture within these constructs was maintained. In one of these studies, the native stomach was resected and the cephalad end of the tissue-engineered stomach was anasto-

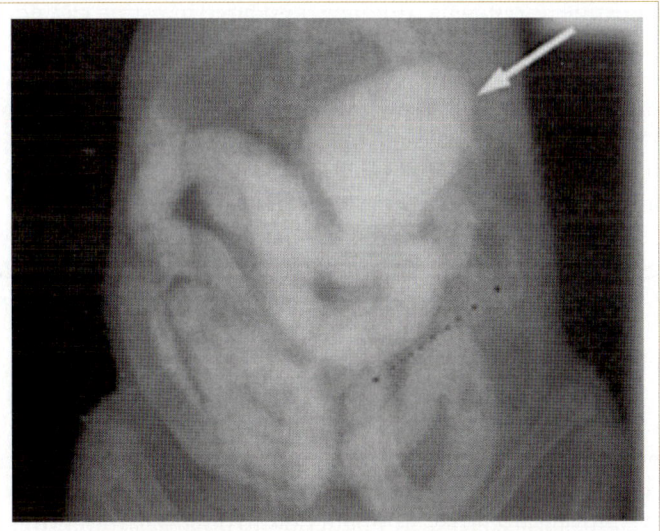

**FIG. 46.14.** Upper gastrointestinal contrast study 10 minutes after barium injection after replacement of the native stomach with tissue-engineered stomach in rats. The white arrow points to the location of the tissue-engineered stomach. From Maemura *et al.* (2003), p. 63.

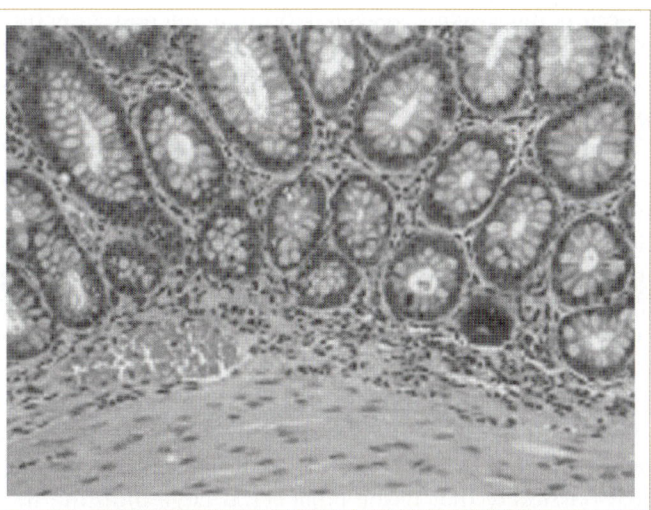

**FIG. 46.15.** Hematoxylin- and eosin-stained photomicrograph of rat tissue–engineered large intestine (original magnification, ×100). From Grikscheit *et al.* (2003b), p. 38.

mosed to the native esophagus (Maemura *et al.*, 2003). At follow-up, there was no evidence of obstruction or stenosis after an upper gastrointestinal study (Fig. 46.14). Unfortunately, preliminary evidence has not shown a significant short-term improvement in weight gain when compared to control rats that underwent total gastrectomy with a Roux-en-Y reconstruction. Thus, the functionality of the tissue-engineered stomach remains to be determined.

## V. TISSUE-ENGINEERED LARGE INTESTINE

The distal portion of the alimentary tract, known as the large intestine or colon, extends from the cecum to the rectum. The large intestine has a much larger caliber than the small intestine but measures only 1.5 meters in length. The primary function of the colon is to store waste and to reclaim any remaining water that is not absorbed by the small intestine. The ileum expels approximately 1500 mL of fluid per day, of which 1350 mL is absorbed by the colon. Digestion is an underrecognized activity of the large intestine. Primarily anaerobic bacteria within the colon have the ability to ferment proteins, dietary fiber, and carbohydrate. After the colon has processed all the material received by the ileum, its contents are propelled into the rectum as feces, where it remains until expelled during a bowel movement (Brunicardi *et al.*, 2004).

The large intestine is lined with simple columnar epithelium without villi or Paneth cells. Unlike the small intestine, the longitudinal muscle of the colon forms three dominant cables of muscle, called *taeniea coli*, that originate in the cecum and fuse together to form a circumferential coat at the junction of the sigmoid colon and

rectum. The major blood supply of the large intestine is derived from both the superior and inferior mesenteric arteries. As in other areas of the alimentary tract, the large intestine is innervated via the sympathetic and parasympathetic nervous systems.

There are some clinical conditions, such as ulcerative colitis, familial adenomatous polyposis, and long-segment Hirschsprung's disease, in which a total protocolectomy (i.e., removal of the entire colon and rectum) is required for survival. Although an ileal pouch can provide some restorative function in these patients, no reliable replacement therapy for the native large intestine currently exists. Postcolectomy morbidities, including high stool frequency, pouchitis, and pelvic sepsis as a result of pouch leak, can be significant and debilitating for patients (Meagher *et al.*, 1998). The cumulative frequency of pouchitis at major referral centers approaches 50% over 10 years. Therapy directed at relieving pouch morbidity includes revision or conversion to end-ileostomy, transrectal catheter drainage, antibiotics, and steroids (Fonkalsrud and Bustorff-Silva, 1999). Ileal mucosal adaptation in the pouch through colonic metaplasia and chronic inflammation has been postulated to increase long-term neoplasia.

To date, the organoid unit approach using PGA scaffolds has been the only reported strategy to create tissue-engineered large intestine. In our laboratory, Grikscheit used minced neonatal rat large intestine reliably to generate tissue-engineered large intestinal cysts (Grikscheit *et al.*, 2002). These grafts contained all of the histological elements of native large intestine, including long crypts of Lieberkuhn, goblet cells, and the presence of acetylcholinesterase within the lamina propria (Fig. 46.15). Ussing chamber data indicated *in vitro* function consistent with that of mature colonocytes, including a positive short-circuit current response to theophylline indicating intact ion transfer. Transmission electron microscopy revealed normal microarchitecture. After anastomosis to native small intestine, colon morphology was preserved, and gross fluid absorption was noted *in vivo*.

In a follow-up study, the physiologic function of tissue-engineered colon as a replacement therapy was evaluated (Grikscheit *et al.*, 2003b). Rats received either end-ileostomies or engineered colon constructs, which were anastomosed in a side-to-side fashion 1 cm from the stoma. Transit times, as measured by phenol red gavage, were more than doubled in rats with tissue-engineered intestine when compared to controls. In addition, animals with tissue-engineered colon had a lower degree of weight loss, lower stool moisture content, high stool short-chain fatty acids, and higher total serum bile acids. These results suggest the real possibility that tissue-engineered large intestine may someday be a useful treatment in patients requiring proctocolectomy.

## VI. CONCLUSIONS

Since its inception approximately two decades ago, alimentary-tract-tissue engineering has been a vibrant field of study within regenerative medicine. Through the combined efforts of physicians, engineers, cell biologists, and others, remarkable progress has been made with regard to the fabrication of these relatively complex tissues. Using the organoid unit approach, we have gained a better understanding of intestinal morphogenesis and of other basic biological phemonenon, including epithelial regeneration, epithelial–mesenchymal interactions, cell–matrix interactions, and the role of different growth factors in intestinal development. For the first time, tissue-engineered small intestine, esophagus, stomach, and large intestine can now be reliably created, with each of these engineered tissues containing all of the major histological components of full-thickness, native tissue. More importantly, preliminary studies of these tissue-engineered constructs are beginning to show evidence for physiological function in a number of animal models.

Nevertheless, we have also learned that re-creating the alimentary tract is not a simple task. Successful application of tissue-engineered gastrointestinal structures in humans requires overcoming many unique challenges. As a result, clinical application of engineered gastrointestinal tissues still remains a number of years away. Ongoing work must work to understand better how optimally to harvest the relevant cell populations and how to expand them either in monolayer culture or in three-dimensional culture systems. We must also learn more about how to build gastrointestinal structures composed of a dynamic and complex epithelial lining over a fully innervated muscular tube. It is likely that the resultant tissue also needs to be in the appropriate gross geometry in order to have adequate functional capacity for peristalsis. The use of more advanced biomaterials, as outlined in other chapters in this book, may be particularly helpful in this regard.

As with many scientific and medical endeavors, incremental and stepwise progress in alimentary-tract-tissue engineering is likely to continue, and we should remain excited and optimistic about its future. The increasing number of investigators worldwide that are currently engaging in this fascinating area of research is a testament to the promise of alimentary-tract-tissue engineering for the thousands of unfortunate patients in need.

## VII. REFERENCES

Abu-Elmagd, K. M. (2006). Intestinal transplantation for short-bowel syndrome and gastrointestinal failure: current consensus, rewarding outcomes, and practical guidelines. *Gastroenterology* **130**, S132–S137.

Avansino, J. R., Chen, D. C., Hoagland, V. D., Woolman, J. D., Haigh, W. G., and Stelzner, M. (2005). Treatment of bile acid malabsorption using ileal stem cell transplantation. *J. Am. Coll. Surg.* **201**, 710–720.

Badylak, S., Meurling, S., Chen, M., Spievack, A., and Simmons-Byrd, A. (2000). Resorbable bioscaffold for esophageal repair in a dog model. *J. Pediatr. Surg.* **35**, 1097–1103.

Badylak, S. F., Vorp, D. A., Spievack, A. R., Simmons-Byrd, A., Hanke, J., Freytes, D. O., Thapa, A., Gilbert, T. W., and Nieponice, A. (2005). Esophageal reconstruction with ECM and muscle tissue in a dog model. *J. Surg. Res.* **128**, 87–97.

Brunicardi, F. C., Andersen, D. K., Billiar, T. R., Dunn, D. L., Hunter, J. G., and Pollock, R. E. (2004). "Schwartz's Principles of Surgery," 8th ed. McGraw-Hill Professional, New York.

Chen, M. K., and Badylak, S. F. (2001). Small-bowel tissue engineering using small-intestinal submucosa as a scaffold. *J. Surg. Res.* **99**, 352–358.

Choi, R. S., and Vacanti, J. P. (1997). Preliminary studies of tissue-engineered intestine using isolated epithelial organoid units on tubular synthetic biodegradable scaffolds. *Transplant. Proc.* **29**, 848–851.

Choi, R. S., Riegler, M., Pothoulakis, C., Kim, B. S., Mooney, D., Vacanti, M., and Vacanti, J. P. (1998). Studies of brush border enzymes, basement membrane components, and electrophysiology of tissue-engineered neointestine. *J. Pediatr. Surg.* **33**, 991–996; discussion 996–997.

De Ugarte, D. A., Choi, E., Weitzbuch, H., Wulur, I., Caulkins, C., Wu, B., Fonkalsrud, E. W., Atkinson, J. B., and Dunn, J. C. (2004). Mucosal regeneration of a duodenal defect using small-intestine submucosa. *Am. Surg.* **70**, 49–51.

Duxbury, M. S., Grikscheit, T. C., Gardner-Thorpe, J., Rocha, F. G., Ito, H., Perez, A., Ashley, S. W., Vacanti, J. P., and Whang, E. E. (2004). Lymphangiogenesis in tissue-engineered small intestine. *Transplantation* **77**, 1162–1166.

Evans, G. S., Flint, N., Somers, A. S., Eyden, B., and Potten, C. S. (1992). The development of a method for the preparation of rat intestinal epithelial cell primary cultures. *J. Cell Sci.* **101** (Pt. 1), 219–231.

Fonkalsrud, E. W., and Bustorff-Silva, J. (1999). Reconstruction for chronic dysfunction of ileoanal pouches. *Ann. Surg.* **229**, 197–204.

Fuchs, J. R., Nasseri, B. A., and Vacanti, J. P. (2001). Tissue engineering: a 21st century solution to surgical reconstruction. *Ann. Thorac. Surg.* **72**, 577–591.

Gardner-Thorpe, J., Grikscheit, T. C., Ito, H., Perez, A., Ashley, S. W., Vacanti, J. P., and Whang, E. E. (2003). Angiogenesis in tissue-engineered small intestine. *Tissue Eng.* **9**, 1255–1261.

Grikscheit, T. C., Ogilvie, J. B., Ochoa, E. R., Alsberg, E., Mooney, D., and Vacanti, J. P. (2002). Tissue-engineered colon exhibits function *in vivo*. *Surgery* **132**, 200–204.

Grikscheit, T., Ochoa, E. R., Srinivasan, A., Gaissert, H., and Vacanti, J. P. (2003a). Tissue-engineered esophagus: experimental substitution by onlay patch or interposition. *J. Thorac. Cardiovasc. Surg.* **126**, 537–544.

Grikscheit, T. C., Ochoa, E. R., Ramsanahie, A., Alsberg, E., Mooney, D., Whang, E. E., and Vacanti, J. P. (2003b). Tissue-engineered large intestine resembles native colon with appropriate *in vitro* physiology and architecture. *Ann. Surg.* **238**, 35–41.

Grikscheit, T. C., Siddique, A., Ochoa, E. R., Srinivasan, A., Alsberg, E., Hodin, R. A., and Vacanti, J. P. (2004). Tissue-engineered small intestine improves recovery after massive small bowel resection. *Ann. Surg.* **240**, 748–754.

Gupte, G. L., Beath, S. V., Kelly, D. A., Millar, A. J., and Booth, I. W. (2006). Current issues in the management of intestinal failure. *Arch. Dis. Child* **91**, 259–264.

Hayashi, K., Ando, N., Ozawa, S., Kitagawa, Y., Miki, H., Sato, M., and Kitajima, M. (2004). A neo-esophagus reconstructed by cultured human esophageal epithelial cells, smooth muscle cells, fibroblasts, and collagen. *Asaio J.* **50**, 261–266.

Hori, Y., Nakamura, T., Matsumoto, K., Kurokawa, Y., Satomi, S., and Shimizu, Y. (2001a). Experimental study on *in situ* tissue engineering of the stomach by an acellular collagen sponge scaffold graft. *Asaio J.* **47**, 206–210.

Hori, Y., Nakamura, T., Matsumoto, K., Kurokawa, Y., Satomi, S., and Shimizu, Y. (2001b). Tissue engineering of the small intestine by acellular collagen sponge scaffold grafting. *Int. J. Artif. Organs* **24**, 50–54.

Hori, Y., Nakamura, T., Kimura, D., Kaino, K., Kurokawa, Y., Satomi, S., and Shimizu, Y. (2002a). Experimental study on tissue engineering of the small intestine by mesenchymal stem cell seeding. *J. Surg. Res.* **102**, 156–160.

Hori, Y., Nakamura, T., Kimura, D., Kaino, K., Kurokawa, Y., Satomi, S., and Shimizu, Y. (2002b). Functional analysis of the tissue-engineered stomach wall. *Artif. Organs* **26**, 868–872.

Hori, Y., Nakamura, T., Kimura, D., Kaino, K., Kurokawa, Y., Satomi, S., and Shimizu, Y. (2003). Effect of basic fibroblast growth factor on vascularization in esophagus tissue engineering. *Int. J. Artif. Organs* **26**, 241–244.

Kaihara, S., Kim, S., Benvenuto, M., Kim, B. S., Mooney, D. J., Tanaka, K., and Vacanti, J. P. (1999a). End-to-end anastomosis between tissue-engineered intestine and native small bowel. *Tissue Eng.* **5**, 339–346.

Kaihara, S., Kim, S. S., Benvenuto, M., Choi, R., Kim, B. S., Mooney, D., Tanaka, K., and Vacanti, J. P. (1999b). Successful anastomosis between tissue-engineered intestine and native small bowel. *Transplantation* **67**, 241–245.

Kaihara, S., Kim, S. S., Kim, B. S., Mooney, D., Tanaka, K., and Vacanti, J. P. (2000). Long-term follow-up of tissue-engineered intestine after anastomosis to native small bowel. *Transplantation* **69**, 1927–1932.

Kent Leach, J., Kaigler, D., Wang, Z., Krebsbach, P. H., and Mooney, D. J. (2006). Coating of VEGF-releasing scaffolds with bioactive glass for angiogenesis and bone regeneration. *Biomaterials* **27**, 3249–3255.

Kim, S. S., Kaihara, S., Benvenuto, M., Choi, R. S., Kim, B. S., Mooney, D. J., Taylor, G. A., and Vacanti, J. P. (1999a). Regenerative signals for tissue-engineered small intestine. *Transplant. Proc.* **31**, 657–660.

Kim, S. S., Kaihara, S., Benvenuto, M. S., Choi, R. S., Kim, B. S., Mooney, D. J., and Vacanti, J. P. (1999b). Effects of anastomosis of tissue-engineered neointestine to native small bowel. *J. Surg. Res.* **87**, 6–13.

Komuro, H., Nakamura, T., Kaneko, M., Nakanishi, Y., and Shimizu, Y. (2002). Application of collagen sponge scaffold to muscular defects of the esophagus: an experimental study in piglets. *J. Pediatr. Surg.* **37**, 1409–1413.

Lloyd, D. A., Ansari, T. I., Gundabolu, P., Shurey, S., Maquet, V., Sibbons, P. D., Boccaccini, A. R., and Gabe, S. M. (2006). A pilot study investigating a novel subcutaneously implanted precellularized scaffold for tissue engineering of intestinal mucosa. *Eur. Cell Mater.* **11**, 27–33; discussion 34.

Maemura, T., Shin, M., Sato, M., Mochizuki, H., and Vacanti, J. P. (2003). A tissue-engineered stomach as a replacement of the native stomach. *Transplantation* **76**, 61–65.

Maemura, T., Shin, M., Ishii, O., Mochizuki, H., and Vacanti, J. P. (2004). Initial assessment of a tissue-engineered stomach derived from syngeneic donors in a rat model. *Asaio J.* **50**, 468–472.

Meagher, A. P., Farouk, R., Dozois, R. R., Kelly, K. A., and Pemberton, J. H. (1998). J ileal pouch–anal anastomosis for chronic ulcerative colitis: complications and long-term outcome in 1310 patients. *Br. J. Surg.* **85**, 800–803.

Nakase, Y., Hagiwara, A., Nakamura, T., Kin, S., Nakashima, S., Yoshikawa, T., Fukuda, K., Kuriu, Y., Miyagawa, K., Sakakura, C., *et al.* (2006). Tissue engineering of small-intestinal tissue using collagen sponge scaffolds seeded with smooth muscle cells. *Tissue Eng.* **12**, 403–412.

Organ, G. M., Mooney, D. J., Hansen, L. K., Schloo, B., and Vacanti, J. P. (1992). Transplantation of enterocytes utilizing polymer-cell constructs to produce a neointestine. *Transplant. Proc.* **24**, 3009–3011.

Patel, H. R., Tait, I. S., Evans, G. S., and Campbell, F. C. (1996). Influence of cell interactions in a novel model of postnatal mucosal regeneration. *Gut* **38**, 679–686.

Perez, A., Grikscheit, T. C., Blumberg, R. S., Ashley, S. W., Vacanti, J. P., and Whang, E. E. (2002). Tissue-engineered small intestine: ontogeny of the immune system. *Transplantation* **74**, 619–623.

Ramsanahie, A., Duxbury, M. S., Grikscheit, T. C., Perez, A., Rhoads, D. B., Gardner-Thorpe, J., Ogilvie, J., Ashley, S. W., Vacanti, J. P., and Whang, E. E. (2003). Effect of GLP-2 on mucosal morphology and SGLT1 expression in tissue-engineered neointestine. *Am. J. Physiol. Gastrointest. Liver Physiol.* **285**, G1345–G1352.

Sato, M., Ando, N., Ozawa, S., Miki, H., and Kitajima, M. (1994). An artificial esophagus consisting of cultured human esophageal epithelial cells, polyglycolic acid mesh, and collagen. *Asaio J.* **40**, M389–M392.

Sato, M., Ando, N., Ozawa, S., Miki, H., Hayashi, K., and Kitajima, M. (1997). Artificial esophagus. *Mater. Sci. Forum* **250**, 105–114.

Tait, I. S., Flint, N., Campbell, F. C., and Evans, G. S. (1994a). Generation of neomucosa *in vivo* by transplantation of dissociated rat postnatal small-intestinal epithelium. *Differentiation* **56**, 91–100.

Tait, I. S., Evans, G. S., Flint, N., and Campbell, F. C. (1994b). Colonic mucosal replacement by syngeneic small intestinal stem cell transplantation. *Am. J. Surg.* **167**, 67–72.

Tait, I. S., Penny, J. I., and Campbell, F. C. (1995). Does neomucosa induced by small-bowel stem cell transplantation have adequate function? *Am. J. Surg.* **169**, 120–125.

Takimoto, Y., Okumura, N., Nakamura, T., Natsume, T., and Shimizu, Y. (1993). Long-term follow-up of the experimental replacement of the esophagus with a collagen–silicone composite tube. *Asaio J.* **39**, M736–M739.

Tavakkolizadeh, A., Berger, U. V., Stephen, A. E., Kim, B. S., Mooney, D., Hediger, M. A., Ashley, S. W., Vacanti, J. P., and Whang, E. E. (2003). Tissue-engineered neomucosa: morphology, enterocyte dynamics, and SGLT1 expression topography. *Transplantation* **75**, 181–185.

Thompson, J. S. (1987). Growth of neomucosa after intestinal resection. *Arch. Surg.* **122**, 316–319.

Ure, B. M., Slany, E., Eypasch, E. P., Gharib, M., Holschneider, A. M., and Troidl, H. (1995). Long-term functional results and quality of life after colon interposition for long-gap oesophageal atresia. *Eur. J. Pediatr. Surg.* **5**, 206–210.

Vacanti, J. P. (1988). Beyond transplantation. Third annual Samuel Jason Mixter lecture. *Arch. Surg.* **123**, 545–549.

Vacanti, J. P. (2003). Tissue and organ engineering: can we build intestine and vital organs? *J. Gastrointest. Surg.* **7**, 831–835.

Vacanti, J. P., Morse, M. A., Saltzman, W. M., Domb, A. J., Perez-Atayde, A., and Langer, R. (1988). Selective cell transplantation using bioabsorbable artificial polymers as matrices. *J. Pediatr. Surg.* **23**, 3–9.

Wang, Z. Q., Watanabe, Y., and Toki, A. (2003). Experimental assessment of small-intestinal submucosa as a small-bowel graft in a rat model. *J. Pediatr. Surg.* **38**, 1596–1601.

Wang, Z. Q., Watanabe, Y., Noda, T., Yoshida, A., Oyama, T., and Toki, A. (2005). Morphologic evaluation of regenerated small bowel by small-intestinal submucosa. *J. Pediatr. Surg.* **40**, 1898–1902.

Yamamoto, Y., Nakamura, T., Shimizu, Y., Matsumoto, K., Takimoto, Y., Kiyotani, T., Sekine, T., Ueda, H., Liu, Y., and Tamura, N. (1999). Intrathoracic esophageal replacement in the dog with the use of an artificial esophagus composed of a collagen sponge with a double-layered silicone tube. *J. Thorac. Cardiovasc. Surg.* **118**, 276–286.

Yamamoto, Y., Nakamura, T., Shimizu, Y., Matsumoto, K., Takimoto, Y., Liu, Y., Ueda, H., Sekine, T., and Tamura, N. (2000). Intrathoracic esophageal replacement with a collagen sponge–silicone double-layer tube: evaluation of omental-pedicle wrapping and prolonged placement of an inner stent. *Asaio J.* **46**, 734–739.

# Chapter Forty-Seven

# Liver Stem Cells

*Eric Lagasse*

## I. INTRODUCTION

The identification of liver stem cells would have two possible applications. In cell-based therapy, transplantation of liver stem cells could be an alternative to whole-liver transplantation for patients suffering liver dysfunctions. For liver cancers, the identification of liver cancer stem cells would result in novel anticancer therapies directed against cancer stem cells. The field of cell therapy and that of anti-cancer therapy are inextricably linked; understanding the biology of liver stem cells will benefit both fields greatly. This chapter is limited to liver-derived stem cells.

The issue of the presence of liver stem cells in adult liver remains controversial. This is partially due to the fact that the proliferation of adult hepatocytes, a well-differentiated cell, accomplishes regeneration of the liver after injury, with the capacity of self-renewal. Another reason why the issue of a liver stem cell is still contentious may be the lack of clear evidence of the liver stem cell's presence in the adult liver. This topic of liver stem cells has received more attention recently with the prospective identification of stem/progenitor cells in fetal liver. Here we review some recent evidence of the lineage organization of the liver and the value of the prospective isolation of a liver stem cell for regenerative medicine.

## II. DEFINITION OF A TISSUE-DERIVED STEM CELL

The classic definition of a tissue-derived stem cell comes from the study of hematopoietic stem cells. These stem cells have been defined as clonogenic cells capable of self-renewal and differentiation into multiple committed progenitor cells, which then proliferate and differentiate into functionally specialized mature cells (Bryder *et al.*, 2006). In the search for hematopoietic stem cells, which was driven by the need for bone marrow transplantation in patients, a hirarchy of blood-forming cells from primitive stem cells, committed progenies, and mature cells was uncovered (Till and McCulloch, 1961; Siminovitch *et al.*, 1963; Schofield, 1970, 1978; Spangrude *et al.*, 1988; Baum *et al.*, 1992). With the prospective isolation of hematopoietic stem cells (Spangrude *et al.*, 1988; Baum *et al.*, 1992), it was then possible to compare stem cells to all the other blood cells in the regenerative process of bone marrow transplantation. To the surprise of many, only the hematopoietic stem cells, and not the other more committed progenitor cells of the bone marrow, were able to sustain a long-term engraftment in animal models. Thus, in the context of bone marrow transplantation, only stem cells seemed to be important for the sustained regeneration of the hematopoietic system. This is in part due to the foremost property a stem cell must have, the capacity for self-renewal. Cells without self-renewing potential, such as committed progenitor cells, will still be capable of generating enough progeny for a short-term engraftment but will not be capable of sustaining a long-term engraftment over many years. Therefore, self-renewal is a key function for stem cells.

Another property of stem cells is their multipotentiality. Stem cells are capable of producing progeny in more than one lineage. Again, the best-known example of a pluripotent tissue-derived stem cell is hematopoietic stem cells. Hematopoietic stem cells have the capacity to give rise to many lineages, from short-term stem cells to multipotent progenitor cells, to oligolineage progenitors to differentiated progenies (Bryder *et al.*, 2006). In this process, hematopoietic stem cells lose their self-renewal with functionally irreversible maturation steps. Finally, clonality is probably the ultimate proof of the presence of stem cells, because it allows one to demonstrate that both self-renewal and multipotent potential are present at the single-cell level. However, for stem cells to maintain their stemness, at least some of the cells must undergo division without differentiation while others differentiate.

Thus, any claim for the isolation and identification of a liver stem cell has one minimal requirement: the demonstration that a liver stem cell can be prospectively isolated and is a clonogenic cell, capable of self-renewal and differentiation to the expected lineages of the liver, namely the hepatocyte and bile duct lineages (see Fig. 47.1). In this chapter, we present a compilation of experiments published recently that support these assumptions.

## III. CELLULAR ORGANIZATION OF THE ADULT LIVER

The liver is one of the largest organs of the body. It can be considered an exocrine gland with glandular acinus, composed of hepatocytes, with draining ducts lined by bile duct cells. Both the bile duct cells (also called *cholangiocytes*) and the hepatocytes (also called *parenchyma cells*) belong to the same hepatic lineage with a common stem cell, the liver or hepatic stem cells (see Fig. 47.1). Within the acinus, also called the *hepatic lobule*, two distinct vascular beds are found: A set of six portal triads are present at the periphery of the acinus, composed of hepatic arteriole, portal venule, and a bile duct. The central vein is in the middle of the hepatic lobule. Blood enters the liver from the portal venule or hepatic arteriole and flows between liver sinusoids that have fenestrated endothelium toward the central vein. Thus, contact between blood and hepatocytes allows filtration of the blood. A bile canaliculus forms a channel adjacent to the hepatocytes and serves to drain secreted bile toward the bile duct of the portal triad. In addition to hepatocytes and bile duct cells, stellate cells (formerly called *Ito cells*), Kupffer cells (the macrophage of the liver), endothelial cells, fibroblasts, and leukocytes can present in significant numbers in the liver. Hepatocytes along the hepatic plate, from the portal triad to the central vein, present a display of heterogeneous biochemical properties as well as a pattern of gene expression.

Different zonal locations have been distinguished. The periportal zone is where most of the primitive cells would reside. The canals of Hering, which represent the connection between the bile canaliculi and the bile duct, are in the periportal zone and have been proposed to be the site of liver stem cells. Then comes the mid-acinar zone and lastly the pericentral zone, adjacent to the central vein and presenting the more mature hepatocytes. Liver-specific gene expression with distinct patterns has been associated with different zones of the liver acinus. However, findings of alteration in gene expression after rerouting of the blood flow through the liver support the interpretation of regulated gene expression by microenvironment (Jungermann and Kietzmann, 1996).

## IV. HEPATOCYTES: THE FUNCTIONAL UNIT OF THE LIVER WITH STEM CELL PROPERTIES

### The Liver Is an Amazing Organ Because It Can Regenerate

Centuries ago, the Greek myth of Prometheus told of the mortal who stole fire from Zeus and was punished by having his liver devoured by an eagle at dusk, only to have it regenerated by next morning. Even though the myth exaggerated the regenerative potential of liver, this phenomenon was recognized much later in medicine, with some the early scientific reports in the literature coming from Germany and England. The classic experiment by Higgins and Anderson (1931) demonstrated experimentally that surgical removal of two-thirds of a rat liver was possible without significant mortality and, more importantly, resulted in regeneration of the remaining lobes of the liver in five to seven days by compensatory hyperplasia.

Serial surgical resection of the liver (hepatectomy) in experimental animals is possible, resulting each time in the extraordinary compensatory growth of the remaining liver lobes, suggesting that this extraordinary regenerative potential of the liver may be due to liver stem cells. This model of hepatectomy has been intensively studied and to the surprise of many is not dependent on liver stem cells. In contrast to other regenerative tissues, such as bone marrow and gut, mature cells, mostly hepatocytes, in response to stimuli like growth factors, will enter into a cell cycle and drive liver regeneration. Hepatocytes were shown to be the main proliferating cell responsible for the regrowth of the liver, because they are the most abundant epithelial cells of the liver (~60%), representing most of the mass of the organ due

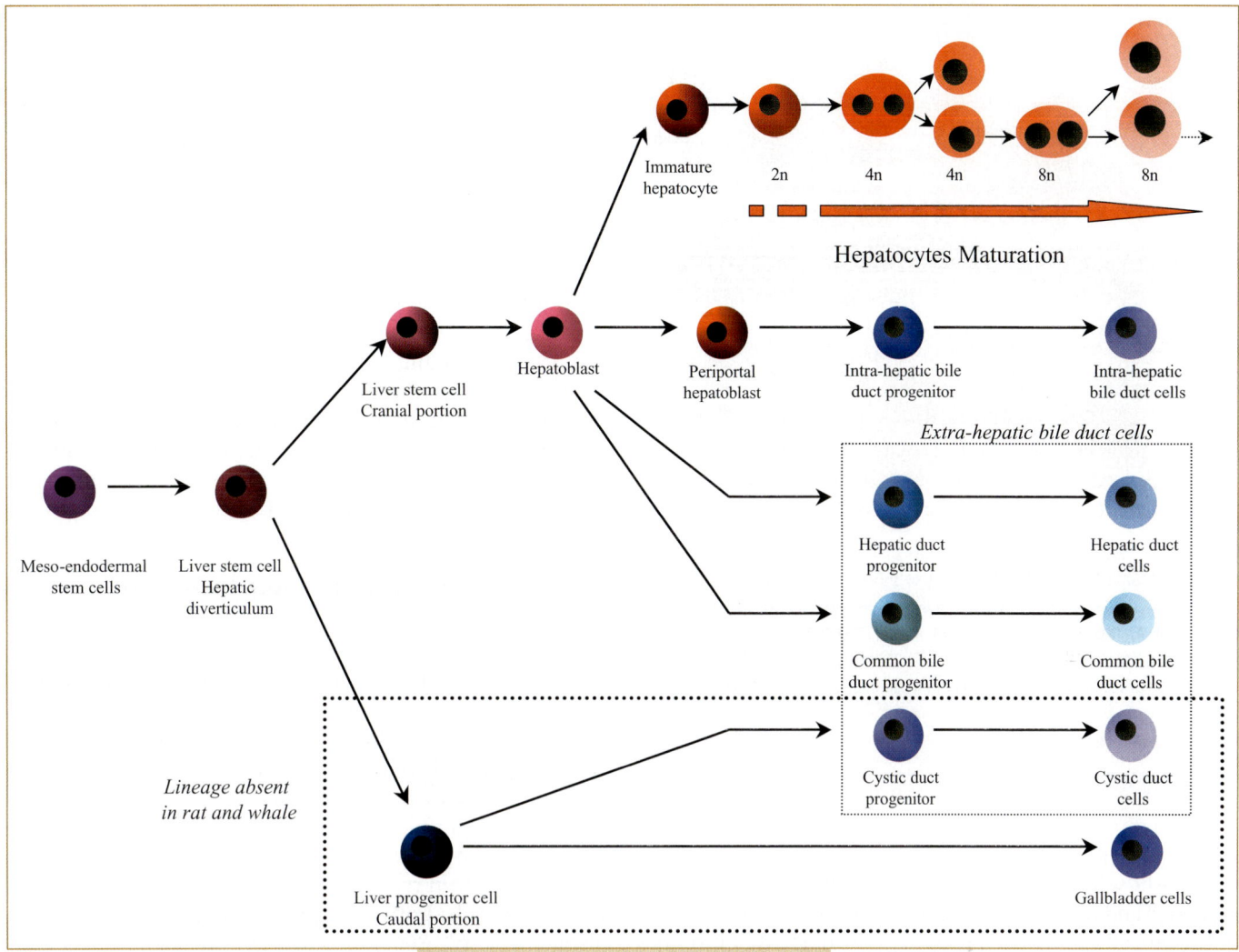

**FIG. 47.1.** Model of the hepatic lineage in the mammalian system. Based on the review by Shiojiri (1997). Gallbladder and cystic duct do not develop through hepatic development in the rat and whales.

to their large size. Hepatocyte proliferation will start in the periportal region and will spread to the centrilobular region. In general, the regeneration of the liver mass required each hepatocyte to undergo an average of 1.4 rounds of replication to restore the liver mass. However, other type of cells are also involved in the process of liver regeneration, including biliary epithelial cells, hepatic stellate cells (found in the space of Disse, juxtaposed between hepatocytes and liver sinusoids), endothelial cells lining sinusoids, fibroblasts, and Kupffer cells (see the review by Michalopoulos and DeFrances, 1997).

## Hepatocyte Transplantation: The "Proof-of-Concept" to Cell-Based Therapy for Liver Disease in Animal Models

Our understanding of the potential of cell-based therapy for liver disease has come from studies of transplanted liver cells in animal models. The first experiments described

transplantation and survival of hepatocytes in spleen (Mito *et al.*, 1979). This work was followed by the demonstration that these ectopic hepatocytes were functional and could proliferate (Darby *et al.*, 1986; Selden *et al.*, 1986; Gupta *et al.*, 1987). Hepatocyte transplantation has been used in experimental animal models with inherited metabolic liver diseases (Matas *et al.*, 1976; Groth *et al.*, 1977). These experiments have established that allotransplantation of normal hepatocytes and autotransplantation of genetically modified hepatocytes can correct metabolic defects.

More recently, the use of transgenic and knockout mice provided excellent models of liver regeneration from cell transplantation. In these models, inherited chronic hepatocellular toxicity induced massive necrosis of the host/endogenous hepatocytes and a quasi clonal proliferation of healthy donor hepatocytes transplanted.

The urokinase plasminogen activator (uPa) transgenic mouse, driven by an albumin promoter, was one of the first

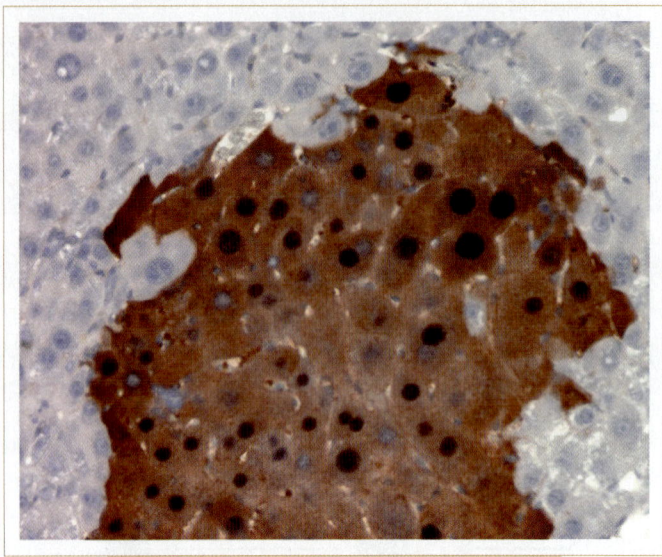

**FIG. 47.2.** Regenerative hepatic nodule in the FAH-mutant mouse. Liver cells suspension from a wild-type mouse (FAH$^{+/+}$) was transplanted into a tyrosinemic mouse (FAH$^{-/-}$-mutant mouse). A few weeks after cell transplantation, FAH$^{+/+}$ nodules were detected in the diseased liver (Peroxydase-positive hepatocytes stained with anti-FAH antibody). In general, two months after transplantation, most of the tyrosinemic liver is reconstituted by FAH$^{+/+}$ hepatocytes.

genetic models (Rhim *et al.*, 1994). In this model, massive engraftment of the donor liver cells was achieved because the transplanted cells had a proliferative growth advantage over the host liver cells and replicated when host cells disappeared. The Fah knockout mouse is another genetic model of hepatocellular toxicity, with the advantage that it reproduces a human lethal inherited liver disease, tyrosinemia type 1, by deleting the fumarylacetoacetate hydrolase (*Fah*) gene (Grompe *et al.*, 1995). The animals develop severe liver disease, which can be rescued by blocking the tyrosine pathway upstream using the drug 2-(2-nitro-4-trifluoro-methylbenzyol)-1,3 cyclohexanedione (NTBC). Thus, FAH mutants developed to adulthood can breed and remain healthy until NTBC is removed. After NTBC withdrawal, liver failure occurs, and the mice die within six to eight weeks. When FAH-mutant mice are transplanted intrasplenically with single-cell suspensions of syngeneic hepatocytes from wild-type mice, the FAH-positive cells migrate to the liver, invade the parenchyma, and divide until >95% of the liver cell mass is replaced by donor hepatocytes (see Fig. 47.2) (Overturf *et al.*, 1996). Repopulated animals are healthy, and as few as 100 donor hepatocytes are sufficient to rescue the animal from the lethal liver defect. Moreover, repopulating cells can be serially transplanted to measure their cell division capacity (Overturf *et al.*, 1997).

Because hepatocytes are terminally differentiated cells, which need only limited proliferation to restore liver mass, it was originally thought that hepatocytes have a limited

capacity of proliferation. However, both the urokinase plasminogen activator (uPa) transgenic mouse and the fulmarylacetoacetate hydrolase (FAH) knockout mouse demonstrated that hepatocytes have a high proliferative potential. With the uPa transgenic mouse, 12 or more cell divisions were necessary for liver regeneration; in the FAH mouse, Grompe and coworkers serially transplanted the mice with hepatocytes, resulting in an average of 69 divisions. This unique capacity of hepatocytes to regenerate the liver is based on two essential features: extensive and continuous liver injury, and a strong selective advantage for the transplanted cell to survive as compared to the host cells. The preconditioning for hepatocyte engraftment and liver regeneration is similar to what is needed for efficient bone marrow transplantation. Here again, through chemotherapy and/or irradiation, the recipient hematopoietic system is depleted and the donor cells have a strong selective advantage, which allowed engraftment and transplantation of the blood system.

## Transdifferentiation of Hepatocytes into Biliary Cells

Are hepatocytes liver stem cells? Hepatocytes do not fulfill one essential criterion of stem cells, multipotentiality. Hepatocytes may be considered more like what B- and T-cells would be for the immune system. Upon activation, T- or B-cells can self-renew and expand in large numbers to generate plasma cells or activated CD4+ or CD8+ T-cells. However, under normal circumstances, T- or B-cells do not generate any myeloid or erythroid cells. Interestingly, hepatocytes have been reported potentially to transdifferentiate into biliary cells, both *in vitro* (Michalopoulos *et al.*, 2002; Nishikawa *et al.*, 2005) and *in vivo* (Michalopoulos *et al.*, 2005). Although no clonal approaches or prospective isolation of hepatocytes had been proposed in these studies, these results, if confirmed, would suggest that hepatocytes can function as facultative stem cells, having both self-renewing capacity as well as possible lineage commitments. As seen in Figs. 47.3 and 47.4 and in the next paragraph, hepatocytes are not a homogeneous population of cells. Consequently, it could still be argued that only a distinct population of primitive hepatocytes may have this "stemness" capacity.

## Hepatocyte Polyploidy and Cell Aging

Hepatocytes are highly differentiated cells of the liver capable of multiple synthetic and metabolic functions. However, hepatocytes can be also viewed as a heterogeneous population of cells containing small and large cells, diploid to polyploidy, mononucleated to binucleated (see Figs. 47.3 and 47.4). The polyploidization of hepatocytes after birth is another interesting physiological process of this specialized cell. During growth and development, hepatocytes undergo dramatic changes, which are characterized by a gradual polyploidization, with the successive appear-

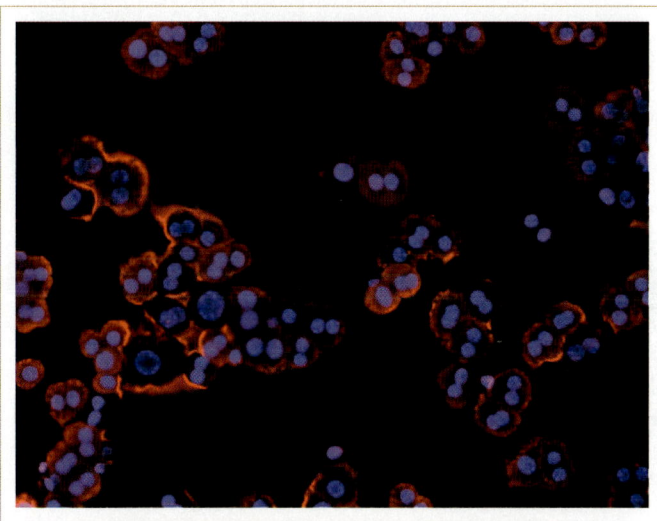

**FIG. 47.3.** Cytospin of mouse adult liver cells. Hepatocytes are a heterogeneous population of cells, small and large, mono- and binucleated. Cells were stained with propidium iodide and counterstained with Hoechst.

ance of tetraploid and octoploid cells with one or two nuclei. In this process, human or rat hepatocytes of a newborn are exclusively diploid ($2n$) but will subsequently generate $4n$, $8n$, and eventually $16n$ mono- and binucleated hepatocytes. The accumulation rate of binucleated and polyploidy cells is very slow in the human, with an intensification of the polyploidization after the age of 50 (Kudryavtsev *et al.*, 1993).

Polyploidy is a general physiological process found in many cellular systems (Anatskaya *et al.*, 1994). Normally, polyploidization is a strategy of cell growth that increases metabolic output and is viewed as an alternative to cell division that is indicative of terminal differentiation and senescence (Sigal *et al.*, 1995, 1999). The cellular mechanism that governs the passage from mononucleated $2n$ to binucleated $2 \times 2n$ or mononucleated $4n$ was recently unveiled and involved the abortion of cytokinesis, which induced the formation of binucleated hepatocytes (Guidotti *et al.*, 2003). In the proposed model, binucleated $2n$ hepatocytes could divide and generate binuclear tetraploid ($2 \times 2n$) daughter cells. This binucleated ($2 \times 2n$) tetraploid hepatocyte could itself generate two ($4n$) mononucleated cells (see Fig. 47.1). In rodents, studies of the transplanted hepatocytes isolated via flow cytometry using the DNA profile of diploid, tetraploid, and octoploid liver cells showed that all fractions could engraft, proliferate, and regenerate liver tissues (see Fig. 47.4) (Weglarz *et al.*, 2000). These data supported the conclusion that multiple hepatocyte ploidy classes can serve as progenitors for regenerating hepatocyte foci in damaged liver. However, isolation of hepatocytes by centrifugal elutriation into three hepatic fractions—small (16 microns), medium-sized (21 microns), and large (27 microns) hepatocytes—showed that the small fraction had a surprisingly

lower repopulation capacity during the first round of transplantation (Overturf *et al.*, 1999). Clearly, more studies need to be done to understand the self-renewal and commitment potential of these different classes of hepatocytes.

## V. LIVER STEM CELLS

### Liver Development

Mammalian organisms begin as a totipotent stem cell, the fertilized ovum. Then, through a process of specification and determination, cells develop specialized functions and grow into organs. The initiation of liver formation begins from the debut of ventral endoderm from the embryonic foregut. In the mouse, where most of the advances in liver development have been made, hepatic genes are first induced in a segment of definitive endoderm at about 8.5 days of gestation (7–8 somite pair stage, three weeks in the human) (Lemaigre and Zaret, 2004). This induction requires fibroblast growth factor (FGF) from adjacent cardiogenic medoderm as well as bone morphogenic protein (BMP) from nearby septum transversum mesenchyme. In addition, after hepatic endoderm epithelium becomes more proliferative, interactions with endothelial cells are crucial for this early phase of development. The hepatic diverticulum is divided into cranial and caudal portions. Hepatic cords from the cranial portion invade the septum transversum mesenchyme and develop into the liver parenchyma and intra- and extrahepatic ducts. Gallbladder and cystic duct derive from the caudal portion of the hepatic diverticulum, a region not formed through development in the rat embryo. This would explain why no gallbladder and cystic duct are found in the adult rat (Shiojiri, 1997). Once the liver bud emerges from the developing gut tube, hematopoietic cells migrate in by day 11 in the mouse, and the liver cells are referred to as *hepatoblasts*, which already express liver-specific genes such as albumin (ALB), alpha-fetoprotein (AFP), and cytokeratin 19 (ck19) (see Fig. 47.5) in the human. The hepatoblasts will give rise to definitive hepatocytes and bile duct cells (Fig. 47.1). Recent observations uncovered an unexpected relationship between extrahepatic bile duct morphogenesis and pancreas development, with common factors, like Pdx-1, influencing differentiation of both cell lineages.

### Prospective Identification of Liver Stem/ Progenitor Cells

Clonogenic rat cells were the first liver progenitor cells to be prospectively isolated (Kubota and Reid, 2000). Using an *in vitro* colony-forming assay and flow cytometry technology, a population of fetal liver cells, RT1A$^{1-}$/OX18$^{low}$/ICAM-1$^+$, was isolated at ED13. These cells were considered hepatoblasts and were shown to exhibit bipotential differentiation *in vitro* (hepatocyte and bile duct markers). This finding was followed by the identification of a similar population of hepatoblasts in mouse fetal liver (Tanimizu *et al.*,

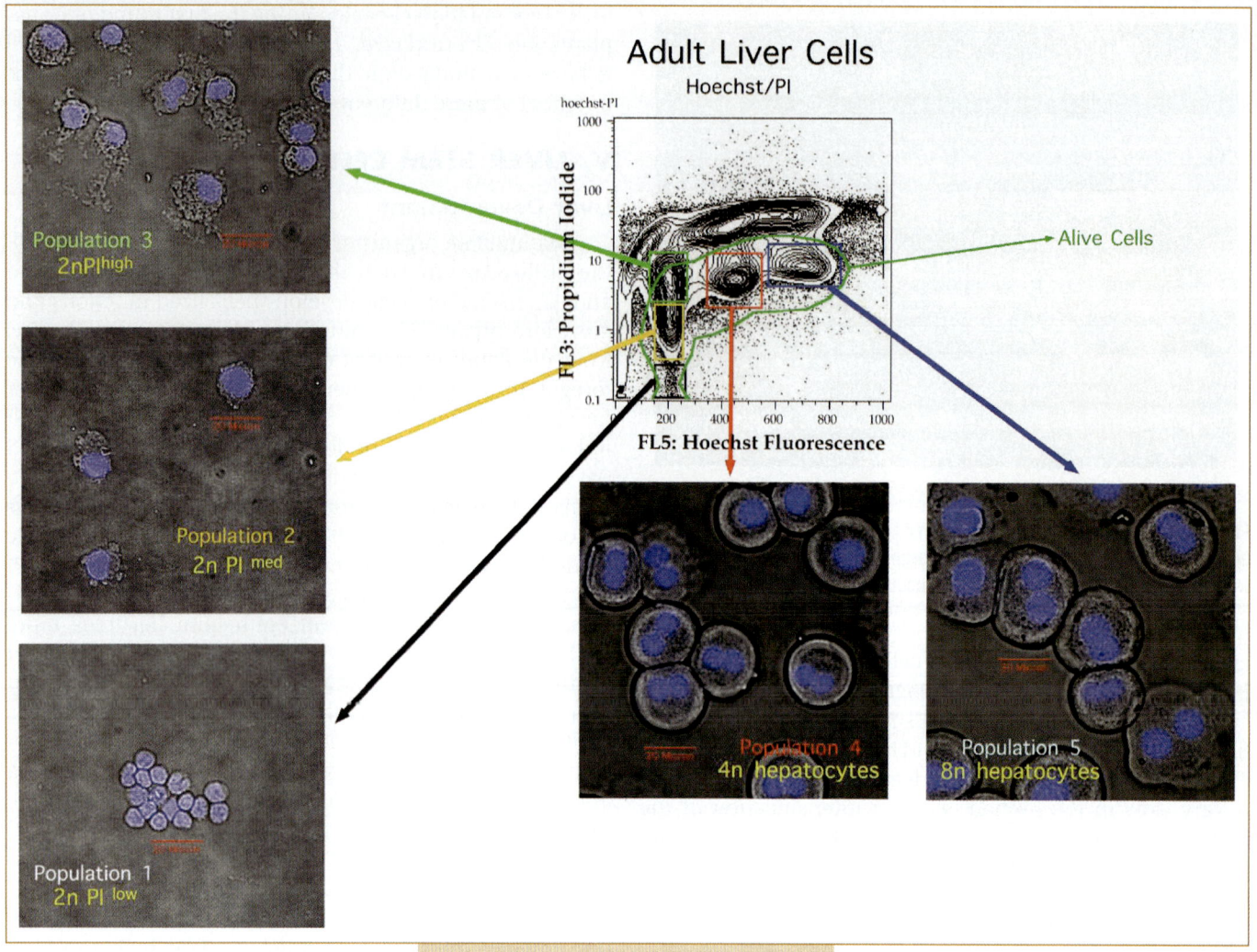

**FIG. 47.4.** Isolation of mouse adult liver cells using a fluorescent-activated cell sorter. Liver cells were purified based on their ploidy (DNA staining, Hoechst) and autofluorescence. Diploid cells have low levels of Hoechst and have a low-to-high level of autofluorescence (Populations 1–3). Tetraploid hepatocytes (Population 4) and octoploid hepatocytes (Population 5) have high levels of autofluorescence and are mono- and binucleated cells. Dead cells (propidium iodide–positive cells) were excluded.

2003; Nierhoff *et al.*, 2005). By combining a different culture system with flow cytometry, more primitive fetal liver cells capable of forming large colonies and providing their descendants for a relatively longer period were isolated from mouse fetal livers (Suzuki *et al.*, 2000, 2002). The Met$^+$/CD49f$^{+/low}$/c-kit$^-$/CD45$^-$/TER119$^-$ fetal liver cells were clonally propagated in culture and were shown continuously to produce hepatocytes and cholangiocytes while maintaining their primitiveness. When these cells clonally expanded *in vitro* were transplanted into mice, they morphologically and, to some extent, functionally differentiated into hepatocytes and cholangiocytes. Furthermore, these mouse fetal liver cells were capable of differentiating into pancreatic acinar cells and gastric as well as intestinal epithelial cells upon transplantation into these organs. Therefore, these

cells are regarded as endodermal stem cells (Suzuki *et al.*, 2002).

We know very little about the different progenitor populations in human fetal liver. Recently, several isolations and characterizations of human fetal liver stem and/or progenitor cells were described. Using EpCam and magnetic beads (magnetic activated cell sorting, MACS), Epcam$^+$ hepatoblasts and hepatic stem cells were isolated from fetal, neonatal, and adult liver and cultured in short- (10 days) and long-term (20 days) assays (Schmelzer *et al.*, 2006). EpCAM-negative liver cells proved to be more than 95% diploid and polyploid hepatocytes. Hepatic stem cells were Epcam$^+$, NCAM$^+$, CLDN-3$^+$, ck19$^+$, c-kit$^+$, aquaporin 4$^+$ Alb$^{low}$, and AFP$^-$, while hepatoblasts were EpCAM$^+$, NCAM$^-$, CLDN-3$^-$, ck19$^{low}$, and c-kit$^{low}$. No prospective isolation by flow cytom-

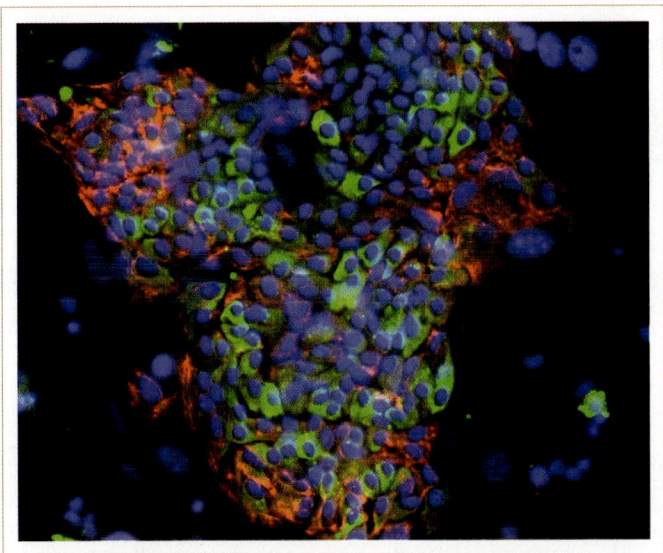

**FIG. 47.5.** *In vitro* culture of human fetal liver cells. Human fetal liver cells were cultured for several weeks on stroma. The colony was stained with AFP (alpha-fetoprotein, green), a marker for hepatoblasts, and ck19 (cytokeratin 19, red), a marker for bile duct cells. This colony contains a mix of cells positive for ck19 only (bile duct cells), AFP only (hepatoblats/small hepatocytes), and ck19/AFP bipotent progenitor cells.

etry, clonal approaches, or *in vivo* transplantation were presented in this work, but more extensive studies from this group are currently under review.

A multipotent progenitor population (hFLMPC) was reported very recently to be capable of differentiating into liver and mesenchymal lineages (Dan *et al.*, 2006). Human fetal liver cells were maintained in culture on feeder layers and the derived colony was clonally isolated by serial dilution. These progenitor cells have the immunophenotype CD34$^+$, CD90$^+$, c-kit$^+$, EPCAM$^+$, c-met$^+$, SSEA-4$^+$, CK18$^+$, CK19$^+$, Albumin$^-$, alpha-fetoprotein$^-$, CD44h$^+$, and vimentin$^+$. Placed in appropriate media, hFLMPC will differentiate into hepatocytes and bile ducts as well as into fat, bone, cartilage, and endothelial cells. Interestingly, hFLMPC survive and differentiate into hepatocytes *in vivo* when transplanted into animal models of liver disease. hFLMPC does not express hepatic markers such as albumin, alpha-fetoprotein, and primitive hepatic transcription factors (HNF1a, HNF3b, and HNF4a) in their undifferentiated state, and it is capable of mesenchymal properties. This result would suggest that hFLMPC is a mesenchymal–epithelial transitional cell, probably derived from a mesendodermal origin. Further analysis and characterization will be necessary to determine the true origin of this human liver progenitor cell.

## Liver Stem Cells and Adult Liver

Hepatocytes and cholangiocytes are the two mature epithelial cells in the adult liver. Both derive from the hepa-toblast population during embryogenesis of the liver. Unfortunately, very little is known about the identification of an adult liver stem/progenitor cell. There is, however, evidence of the presence of such stem/progenitor cells in adult liver. In rodent injury models, damage to the liver combined with an arrest in the proliferative capacity of hepatocytes leads to proliferation of *oval cells*. These cells are not small hepatocytes but a distinct population of liver cells, which differentiates into hepatocytes and restores the loss of hepatic parenchyma (Fausto and Campbell, 2003; Lowes *et al.*, 2003). A similar population of liver cells, called *ductular hepatocytes*, has been reported in patients suffering a variety of liver diseases (Vessey and de la Hall, 2001). These cells have a gene expression associated with both hepatocytes and cholangiocytes, and their presence has been reported in the histopathology of human diseases, from fulminant hepatitis to primary biliary cirrhosis.

In rodent models, experimental studies have shown that oval cells derive from cells of the biliary epithelium. The precursor of oval cells has been proposed to be the elusive adult liver stem cell and would reside in the canals of Hering. The canals of Hering were originally identified as luminal channels linking the hepatocyte canalicular system to the biliary tree. These channels are in direct physical contact with hepatocytes on the one side and bile duct cells on the other side. Recent studies have described the canals of Hering as the structures that penetrate deep inside the lobules (Saxena and Theise, 2004) and could be considered as a candidate *stem cell niche*. Studies in the rat treated with partial hepatectomy followed by the administration of *N*-2 acetyl-aminofluorene (AAF), a protocol used to generate large populations of oval cells, demonstrated that indeed these cells are located in the canals of Hering before differentiating into hepatocytes (Paku *et al.*, 2004).

## Liver Stem Cells and Cancer

Stem cell biology and cancer biology are inextricably linked. It is widely accepted that cancer from multistep events and stem cells may be the only cells that have a life span that is long enough to acquire the requisite number of genetic changes for neoplastic growth. There is overwhelming evidence that virtually all cancers are clonal; that is, they represent the offspring of a single cell, evolving from a series of sequential mutations due to genetic instability and/or environmental factors. Additionally, transplant experiments have demonstrated that tumor cells are functionally heterogeneous, in that only a limited number of tumor cells are able to initiate tumorigenicity. As the result, cancers can be viewed as newly formed abnormal tissues initiated by a few cancer-initiating cells or cancer stem cells that undergo an aberrant and poorly regulated process of organogenesis. This hierarchy theory for tumor cells in cancer predicts that the tumor would comprise rare cancer-initiating cells that are different from the vast majority of the cells that made

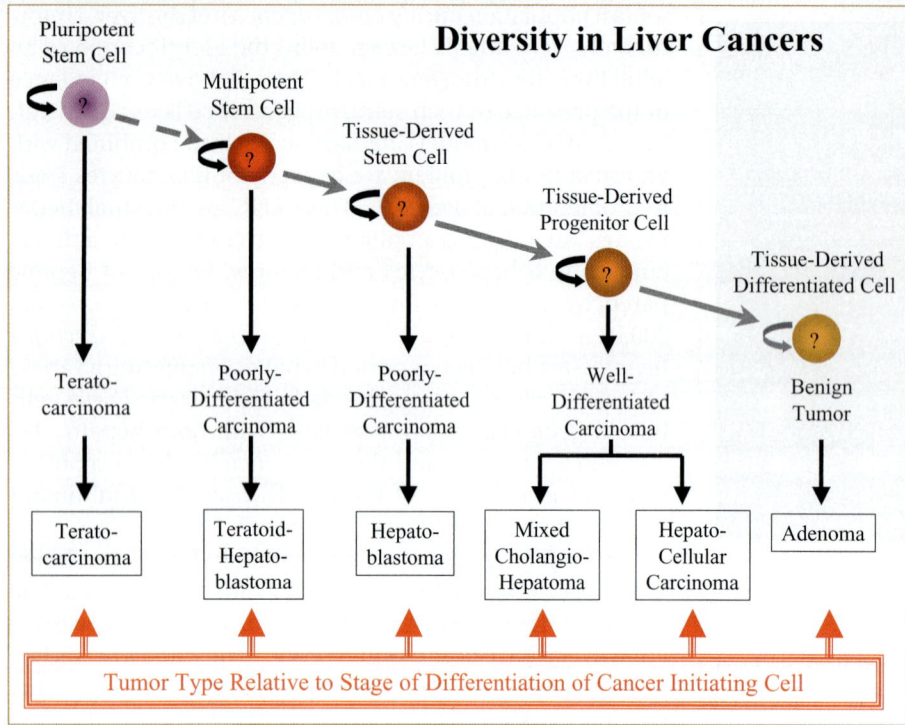

## Diversity in Liver Cancers

Pluripotent Stem Cell

Multipotent Stem Cell

Tissue-Derived Stem Cell

Tissue-Derived Progenitor Cell

Tissue-Derived Differentiated Cell

Terato-carcinoma

Poorly-Differentiated Carcinoma

Poorly-Differentiated Carcinoma

Well-Differentiated Carcinoma

Benign Tumor

| Terato-carcinoma | Teratoid-Hepato-blastoma | Hepato-blastoma | Mixed Cholangio-Hepatoma | Hepato-Cellular Carcinoma | Adenoma |

Tumor Type Relative to Stage of Differentiation of Cancer Initiating Cell

**FIG. 47.6.** Hepatic cell lineage and cancer type. Based on the model of maturation arrest of the liver cell lineage by Stuart Sell (Sell and Ilic, 1997). The diversity of liver cancers is based on cancer-initiating cells, which have acquired self-renewing capacity at a different stage of differentiation in the hepatic pathway. Each cancer-initiating cell can be regarded as a cancer stem cell, capable of both self-renewal and to some extent differentiation. A major challenge will be the isolation of the cancer-initiating cells in each tumor type.

the tumor and would have a function analogous to that of normal stem cells. A hallmark of all cancers is the capacity for unlimited self-renewal, which is also a defining characteristic of normal stem cells, as described earlier. Given these shared attributes between cancer-initiating cells and stem cells, it has been proposed that cancers may be initiated by transforming events that take place in normal stem cells, which would give rise to a cancer stem cell. Alternatively, cancer cells may acquire signaling pathways associated with stem cells functions (see, for review, Reya *et al.*, 2001).

An in-depth review of liver cancer etiology is beyond the scope of this section. Here we focus only on the possible role of stem cells and progenitor cells as the cellular origin for human liver cancers (see Fig. 47.6). During liver development, hepatoblastomas are the most common liver tumors in childhood. In these pediatric liver tumors, cells resembling progenitor cells have been observed. It has been postulated that hepatoblastomas have a stem cell origin, for they can be composed of both epithelial and mesenchymal tissue components, with possible intrahepatic bile duct–like formations or mixed type (teratoid hepatoblastoma). The presence of different cell populations bearing stem cell markers in human hepatoblastoma have been observed, with ductal cells coexpressing stem cell markers and hepatic lineage markers phenotypically resembling hepatic stem-like cells. These findings support the thesis that stem cells play a role in the histogenesis of hepatoblastoma (Fiegel *et al.*, 2004). An interesting parallel could be made between the recent isolation of a fetal mesenchymal–epithelial tran-

sitional cell, described earlier, and the tumor cells of origin in hepatoblastomas (Dan *et al.*, 2006). However, no conclusive information on the cellular origin of hepatoblastomas has been reported.

For adult liver, most of the chronic human liver diseases are characterized by progenitor cell activation, the ductular reaction, which comprises expansion of a transit-amplifying cell compartment of small "biliary" cells. These cells can differentiate into biliary and hepatic lineages. The ductular reaction is the human equivalent of rodent oval cell activation, a cellular response to carcinogenic agent inhibiting the proliferation of mature hepatocytes after a regenerative stimulus. Cirrhotic stages of a variety of chronic liver diseases are characterized by hepatocyte senescence, which should trigger similar progenitor activation. The degree of progenitor cell activation increases with the severity of the disease, making such cells very likely carcinogen targets.

However, in the adult liver the identity of the cancer-initiating cell is more problematic for the two major primary tumors, hepatocellular carcinoma (HCC), one of the most common cancers worldwide and the major consequence of chronic viral hepatitis, and cholangiocarcinoma (CC). Both cancers evolve from focal precursor lesions that reflect the accumulation of genetic and epigenetic alterations, contributing to multistep carcinogenesis. In HCC, small dysplastic foci reflecting the earliest premalignant lesions consist of progenitor cells and intermediate hepatocytes. HCCs express markers of progenitors and biliary cells such as ck19, ck7, and OV6, and ck19 expression has been associated with a worse prognosis after surgical treatment (Roskams, 2006). This observation is a very strong argument in favor of the progenitor cell origin of at least part of the HCCs. It also seems likely that mature polyploid hepatocytes are not the cells of origin of HCCs, a cancer of diploid cells in precursor lesions. Based on animal models of experimental hepatocarcinogenesis, four distinct cell lineages susceptible of neoplastic transformation have been anticipated (Sell, 2002). This hypothesis is based on the fact that there is a considerably different cellular response to injuries in the many different models of hepatocarcinogenesis. Out of the four cell lineages, hepatocytes would be implicated in the initiation of some models of HCC, while biliary epithelium cells would be the cell responsible in some other models of CC. Hepatic progenitor cell and oval cell activations would be responsible for initiating many liver cancers. Finally, periductular cells that respond to periportal injury would also be involved as the last candidate tumor cells (Sell, 2002).

# VI. LIVER STEM CELLS AND THERAPEUTIC APPROACHES

## Stem Cell Therapy for Liver Diseases

### An Urgent Need for Alternative Approaches to Orthotopic Liver Transplantation (OLT)

Liver diseases, such as infections, alcohol-related ailments, cancers, and autoimmune, metabolic, and drug-induced liver diseases represent a major health burden in modern times. Chronic hepatitis, currently the most common cause for orthopedic liver transplantation (OLT), afflicts an estimated 4 million people in the United States alone (350 million people worldwide), and acute liver failure is, in general, associated with unacceptable high mortality rates of up to 90%. Most of these conditions lead to acute or chronic liver failure if no definitive treatment can be offered. The only available routine treatment for hepatic failure is allogeneic, OLT, a procedure that cures some of these end-stage chronic liver diseases. But the procedure is costly and requires the use of lifelong immunosuppressive drugs. Plus, the availability of donor livers is limited to less than 5000 annually (United States alone). Patients on the waiting list for OLT, using fairly stringent criteria of acceptance, are severalfold greater, and the demand for liver transplantation is on the increase while the limited supply of organs for transplantation has not increased. With organ transplantation not available for a large fraction of patients with liver disease to liver failure, cell transplantation for the treatment of acute and chronic liver diseases is an increasingly attractive prospect. Lastly, treating conditions such as metabolic diseases, where overall liver functions generally are intact, with a surgical procedure like OLT is certainly quite extreme and should indeed be substituted by a less radical, less expensive, and potentially long-term curable procedure like cell therapy.

## Clinical State of the Art of Hepatocyte Transplantation in Patients with Inborn Errors of Metabolism

Due to the successes of hepatocyte transplantation in animals, several groups have been evaluating cell therapy in a variety of human diseases (see review by Gupta and Chowdhury, 2002; Selden and Hodgson, 2004). Several dozen of patients have undergone isolated hepatocyte transplantation for metabolic diseases. Cell-based transplantations have been applied in individual cases and in small, uncontrolled series of patients suffering from inborn errors of metabolism, mainly involving the liver (summarized by Burlina, 2004). Although the beneficial outcome in some of these liver cell transplantations has been very encouraging (Fox *et al.*, 1998; Strom *et al.*, 1999), the limited clinical experience in hepatocyte transplantation is a current problem. *Importantly, the single most outstanding issue preventing the use of hepatocyte transplantation is the limited availibility of hepatocytes.* The normal source of cells for hepatocyte transplantation is donor livers. Because hepatocyte transplantation is still experimental and, for ethical reasons, as many patients are dying and will still die waiting for a liver transplant, transplanted liver cells are obtained from organs unsuitable for transplantation. While a liver that has suffered extensive trauma and is unsuitable for transplantation can still be used for isolation of a few billion healthy hepatocytes, it nonetheless has a high risk of providing cells of poor quality and viability. The problem of cell availability would be solved by alternative sources of hepatocytes. The use of stem and progenitor cells, liver derived, to be applied in the treatment of inherited and acquired liver diseases could solve the problem of hepatocyte availability. In addition, the search for alternatives to liver cells (not liver derived) could also produce an adequate source of cells. The possibility of deriving mature liver cells from ES cells and other stem cells has elicited considerable interest. We need to extend our current knowledge and understanding of these various stem cell populations to make possible their future therapeutic uses in cell transplantation, liver regeneration, and bioartificial liver devices.

## Stem Cells and Anticancer Therapy

Because of the importance stem cells and progenitor cells may have in cancer, understanding what controls the maintenance of stem cells (self-renewal) and the differentiation signals (cell commitment) will give insights into the cellular mechanism involved in cancer (see review by Sell, 2006). One such signal that is important at the stem cell level for maintaining proliferation at an early stage of differentiation and that may have a similar role in cancer is the Wnt/β-catenin pathway. Wnt was discovered as a putative proto-oncogene before it was recognized as a homolog of the *Drosophila* windless (wnt-1) gene, a member of a family of genes highly conserved that secreted signaling molecules regulating cell-to-cell interaction during wing embryogenesis of the fly. The Wnt pathway has been also associated with colon cancer and the familial adenomatous polyposis, an identified genetic predisposition for colon cancer. Here the adenomatous polyposis coli (APC) gene codes for a tumor-suppressor protein, which binds to β-catenin and down-regulates its function. The APC gene is a major suppressor gene in colon cancer, and molecular lesions of the gene inhibit the phosphorylation and degradation of β-catenin. As a consequence, β-catenin concentration increases, resulting in colon cell proliferation and cancer. Mutations and overexpression of β-catenin is found in most human liver cancer, such as HCC and hepatoblastomas, and is postulated to be driving the proliferation of cancer cells. Treatment of liver cancers targeting β-catenin is one novel approach for therapeutic intervention. More cell signaling pathways, including Notch, BPM, and sonic hedgehog, may ultimately lead to new drugs for anticancer treatments (Sell, 2002).

## VII. CONCLUSION

Major progress has been made in the identification of liver stem and progenitor cells. However, more needs to be done. So far only a few laboratories have prospectively isolated fetal liver cells, but there is no clear consensus about the phenotype of fetal liver stem cells. For adult liver, the liver stem cells are still elusive.

A cell-based therapy with the transplantation of hepatocytes in small-animal models has demonstrated their efficacy, and clinical trials for human liver diseases may be achieved within the near future. But much remains to be learned. We still need to determine the best "engrafting cell" for cell transplantation: adult hepatocytes, fetal hepatocytes, or liver stem cells. Cell expansion will also be an issue for a viable therapeutic procedure. Finally, transplanted cells must have a selective advantage to engraft and repopulate the diseased liver. A starting point will be genetic metabolic disorders, which could provide the environment necessary for an efficient and therapeutic cell transplantation. But the prospects for the future of liver diseases, in general, will be a procedure similar to myeloablation, necessary to prepare the patient for bone marrow or hematopoietic stem cell engraftment.

For liver cancers, identification of cancer-initiating cells is in its infancy. As with normal stem and progenitor cells, cancer stem and progenitor cells will need to be prospectively identified. The study of the tumor cell hierarchy in liver cancer may help our understanding of the normal cell hierarchy in adult liver. Finally, the clinical application of the stem cell theory of cancer lies in the development of novel therapies to induce terminal differentiation of cancer cells. Of course, an understanding of the cellular mechanisms that control stem cell self-renewal and differentiation is key in the development of new approaches for both cancer- and cell-based therapies.

## VIII. REFERENCES

Anatskaya, O. V., Vinogradov, A. E., *et al.* (1994). Hepatocyte polyploidy and metabolism/life-history traits: hypotheses testing. *J. Theor. Biol.* **168**(2), 191–199.

Baum, C. M., Weissman, I. L., *et al.* (1992). Isolation of a candidate human hematopoietic stem-cell population. *Proc. Natl. Acad. Sci. U.S.A.* **89**(7), 2804–2808.

Bryder, D., Rossi, D. J., *et al.* (2006). Hematopoietic stem cells: the paradigmatic tissue-specific stem cell. *Am. J. Pathol.* **169**(2), 338–346.

Burlina, A. B. (2004). Hepatocyte transplantation for inborn errors of metabolism. *J. Inherit. Metab. Dis.* **27**(3), 373–383.

Dan, Y. Y., Riehle, K. J., *et al.* (2006). Isolation of multipotent progenitor cells from human fetal liver capable of differentiating into liver and mesenchymal lineages. *Proc. Natl. Acad. Sci. U.S.A.* **103**(26), 9912–9917. Epub 2006 Jun 16.

Darby, H., Gupta, S., *et al.* (1986). Observations on rat spleen reticulum during the development of syngeneic hepatocellular implants. *Br. J. Exp. Pathol.* **67**(3), 329–339.

Fausto, N., and Campbell, J. S. (2003). The role of hepatocytes and oval cells in liver regeneration and repopulation. *Mech. Dev.* **120**(1), 117–130.

Fiegel, H. C., Gluer, S., *et al.* (2004). Stem-like cells in human hepatoblastoma. *J. Histochem. Cytochem.* **52**(11), 1495–1501.

Fox, I. J., Chowdhury, J. R., *et al.* (1998). Treatment of the Crigler–Najjar syndrome type I with hepatocyte transplantation. *N. Engl. J. Med.* **338**(20), 1422–1426.

Grompe, M., Lindstedt, S., *et al.* (1995). Pharmacological correction of neonatal lethal hepatic dysfunction in a murine model of hereditary tyrosinaemia type I. *Nat. Genet.* **10**(4), 453–460.

Groth, C. G., Arborgh, B., *et al.* (1977). Correction of hyperbilirubinemia in the glucuronyltransferase-deficient rat by intraportal hepatocyte transplantation. *Transplant. Proc.* **9**(1), 313–316.

Guidotti, J. E., Bregerie, O., *et al.* (2003). Liver cell polyploidization: a pivotal role for binuclear hepatocytes. *J. Biol. Chem.* **278**(21), 19095–19101.

Gupta, S., and Chowdhury, J. R. (2002). Therapeutic potential of hepatocyte transplantation. *Semin. Cell Dev. Biol.* **13**(6), 439–446.

Higgins, G., and Anderson, R. (1931). Experimental pathology of the liver. I. Restoration of the liver of the white rat following partial surgical removal. *Arch. Pathol.* **12**, 187–202.

Jungermann, K., and Kietzmann, T. (1996). Zonation of parenchymal and nonparenchymal metabolism in liver. *Annu. Rev. Nutr.* **16**, 179–203.

Kubota, H., and Reid, L. M. (2000). Clonogenic hepatoblasts, common precursors for hepatocytic and biliary lineages, are lacking classical major histocompatibility complex class I antigen. *Proc. Natl. Acad. Sci. U.S.A.* **97**(22), 12132–12137.

Kudryavtsev, B. N., Kudryavtseva, M. V., *et al.* (1993). Human hepatocyte polyploidization kinetics in the course of life cycle. *Virchows Arch. B Cell. Pathol. Incl. Mol. Pathol.* **64**(6), 387–393.

Lemaigre, F., and Zaret, K. S. (2004). Liver development update: new embryo models, cell lineage control, and morphogenesis. *Curr. Opin. Genet. Dev.* **14**(5), 582–590.

Lowes, K. N., Croager, E. J., *et al.* (2003). Oval cell–mediated liver regeneration: role of cytokines and growth factors. *J. Gastroenterol. Hepatol.* **18**(1), 4–12.

Matas, A. J., Sutherland, D. E., *et al.* (1976). Hepatocellular transplantation for metabolic deficiencies: decrease of plasms bilirubin in Gunn rats. *Science* **192**(4242), 892–894.

Michalopoulos, G. K., and DeFrances, M. C. (1997). Liver regeneration. *Science* **276**(5309), 60–66.

Michalopoulos, G. K., Bowen, W. C., *et al.* (2002). Hepatocytes undergo phenotypic transformation to biliary epithelium in organoid cultures. *Hepatology* **36**(2), 278–283.

Michalopoulos, G. K., Barua, L., *et al.* (2005). Transdifferentiation of rat hepatocytes into biliary cells after bile duct ligation and toxic biliary injury. *Hepatology* **41**(3), 535–544.

Mito, M., Ebata, H., Kusano, M., Onishi, T., Hiratsuka, M., Saito, T. (1979). Studies on ectopic liver utilizing hepatocyte transplantation into the rat spleen. *Transplantation* **11**, 585–591.

Nierhoff, D., Ogawa, A., *et al.* (2005). Purification and characterization of mouse fetal liver epithelial cells with high *in vivo* repopulation capacity. *Hepatology* **42**(1), 130–139.

Nishikawa, Y., Doi, Y., *et al.* (2005). Transdifferentiation of mature rat hepatocytes into bile ductlike cells *in vitro*. *Am. J. Pathol.* **166**(4), 1077–1088.

Overturf, K., Al-Dhalimy, M., *et al.* (1996). Hepatocytes corrected by gene therapy are selected *in vivo* in a murine model of hereditary tyrosinaemia type I. *Nat. Genet.* **12**(3), 266–273.

Overturf, K., Al-Dhalimy, M., *et al.* (1997). Serial transplantation reveals the stem-cell-like regenerative potential of adult mouse hepatocytes. *Am. J. Pathol.* **151**(5), 1273–1280.

Overturf, K., Al-Dhalimy, M., *et al.* (1999). The repopulation potential of hepatocyte populations differing in size and prior mitotic expansion. *Am. J. Pathol.* **155**(6), 2135–2143.

Paku, S., Nagy, P., *et al.* (2004). 2-Acetylaminofluorene dose-dependent differentiation of rat oval cells into hepatocytes: confocal and electron microscopic studies. *Hepatology* **39**(5), 1353–1361.

Reya, T., Morrison, S. J., *et al.* (2001). Stem cells, cancer, and cancer stem cells. *Nature* **414**(6859), 105–111.

Rhim, J. A., Sandgren, E. P., *et al.* (1994). Replacement of diseased mouse liver by hepatic cell transplantation. *Science* **263**(5150), 1149–1152.

Roskams, T. (2006). Liver stem cells and their implication in hepatocellular and cholangiocarcinoma. *Oncogene* **25**(27), 3818–3822.

Saxena, R., and Theise, N. (2004). Canals of Hering: recent insights and current knowledge. *Semin. Liver Dis.* **24**(1), 43–48.

Schmelzer, E., Wauthier, E., *et al.* (2006). The phenotypes of pluripotent human hepatic progenitors. *Stem Cells* **24**(8), 1852–1858.

Selden, C., and Hodgson, H. (2004). Cellular therapies for liver replacement. *Transplant. Immunol.* **12**(3–4), 273–288.

Selden, C., Johnstone, R., *et al.* (1986). Human serum does contain a high-molecular-weight hepatocyte growth factor: studies pre- and post-hepatic resection. *Biochem. Biophys. Res. Commun.* **139**(1), 361–366.

Sell, S., and Ilic, S. (1997). "Liver Stem Cells." Chapman and Itall, New York.

Sell, S. (2002). Cellular origin of hepatocellular carcinomas. *Semin. Cell Dev. Biol.* **13**(6), 419–424.

Sell, S. (2006). Cancer stem cells and differentiation therapy. *Tumour Biol.* **27**(2), 59–70. Epub 2006 Mar. 24.

Shiojiri, N. (1997). Development and differentiation of bile ducts in the mammalian liver. *Microsc. Res. Tech.* **39**(4), 328–335.

Sigal, S. H., Gupta, S., *et al.* (1995). Evidence for a terminal differentiation process in the rat liver. *Differentiation* **59**(1), 35–42.

Sigal, S. H., Rajvanshi, P., *et al.* (1999). Partial hepatectomy–induced polyploidy attenuates hepatocyte replication and activates cell aging events. *Am. J. Physiol.* **276**(5 Pt. 1), G1260–G1272.

Spangrude, G. J., Heimfeld, S., *et al.* (1988). Purification and characterization of mouse hematopoietic stem cells. *Science* **241**(4861), 58–62.

Strom, S. C., Chowdhury, J. R., *et al.* (1999). Hepatocyte transplantation for the treatment of human disease. *Semin. Liver Dis.* **19**(1), 39–48.

Suzuki, A., Zheng, Y., *et al.* (2000). Flow-cytometric separation and enrichment of hepatic progenitor cells in the developing mouse liver. *Hepatology* **32**(6), 1230–1239.

Suzuki, A., Zheng, Y. W., *et al.* (2002). Clonal identification and characterization of self-renewing pluripotent stem cells in the developing liver. *J. Cell Biol.* **156**(1), 173–184.

Tanimizu, N., Nishikawa, M., *et al.* (2003). Isolation of hepatoblasts based on the expression of Dlk/Pref-1. *J. Cell Sci.* **116**(Pt. 9), 1775–1786.

Till, J. E., and McCulloch, E. A. (1961). A direct measurement of the radiation sensitivity of normal mouse bone marrow cells. *Radiat. Res.* **14**, 213–223.

Vessey, C. J., and de la Hall, P. M. (2001). Hepatic stem cells: a review. *Pathology* **33**(2), 130–141.

Weglarz, T. C., Degen, J. L., *et al.* (2000). Hepatocyte transplantation into diseased mouse liver. Kinetics of parenchymal repopulation and identification of the proliferative capacity of tetraploid and octaploid hepatocytes. *Am. J. Pathol.* **157**(6), 1963–1974.

# Chapter Forty-Eight

# Liver

*Gregory H. Underhill, Salman R. Khetani, Alice A. Chen, and Sangeeta N. Bhatia*

## I. INTRODUCTION

Cell-based therapies for liver failure offer the potential to augment or replace whole-organ transplantation. However, the development of such therapies poses unique challenges, stemming largely from the complexity of liver structure and function. The field of liver-tissue engineering encompasses several approaches collectively aimed at providing novel therapeutic options for liver disease patients as well as elucidating fundamental characteristics of liver biology. These approaches include the development of (1) *in vitro* model systems that recapitulate normal liver function, (2) extracorporeal bioartifical liver devices for the temporary support of liver failure patients, and (3) three-dimensional implantable therapeutic constructs. Advances in each of these aspects are reviewed in this chapter, within the context of current treatments for liver disease and additional clinical alternatives, such as surgical advancements for organ transplant and cell transplantation strategies. Critical issues relevant to all of these areas, including cell sourcing and animal models, are also discussed.

## II. LIVER FAILURE AND CURRENT TREATMENTS

Liver failure is a significant health problem, representing the cause of death of over 40,000 individuals in the United States every year, and it can generally be separated into two major categories: fulminant hepatic failure, also referred to as *acute liver failure*, and chronic hepatic failure resulting from chronic end-stage liver disorders. The term *fulminant hepatic failure* is utilized for cases in which hepatic encephalopathy develops within two weeks of the initial onset of jaundice, whereas *subfulminant hepatic failure* is applied to cases in which encephalopathy develops between two weeks and three months after the appearance of initial symptoms. Hepatic encephalopathy is a neuropsychiatric condition that can be divided into four stages, ranging from minor effects, such as mild confusion and sleep disorder, to deep coma. Although fulminant hepatic failure is relatively rare, with approximately 2000 cases in the United States per year, it exhibits a high mortality rate, approximately 28% (W. M. Lee, 2003). The major identified causes of fulminant hepatic failure include acetaminophen overdose, idiosyncratic drug reactions, and viral hepatitis A and B (W. M. Lee, 2003). Of note, a 2002 multicenter etiology study illustrated that 17% of fulminant hepatic failure cases remained of indeterminant origin (Ostapowicz *et al.*, 2002). In addition to hepatic encephalopathy, other clinical aspects of fulminant hepatic failure are infection, both bacterial and fungal, coagulopathy, as well as broad-spectrum metabolic, cardiorespiratory, and hemodynamic abnormalities. Though spontaneous recovery has been observed due to the regenerative capacity of the liver, this type of recovery is difficult to predict, and it rarely occurs in such etiologies as idiosyncratic drug toxicity and hepatitis B (W. M. Lee, 2003). Liver transplantation is the only therapy shown to alter mortality directly, and therefore it is the standard of care in most clinical settings. As a result, all fulminant hepatic failure patients who meet the criteria for orthotopic liver transplant are immediately listed as United Network for Organ Sharing Status 1 on presentation. Factors that preclude this designation are irreversible brain damage, unresponsive cerebral edema, extrahepatic uncontrollable sepsis and malignancy, and multisystem organ failure. Despite the effectiveness of liver transplant in improving short-term survival of fulminant hepatic failure patients, the utility of this approach remains limited, due to the scarcity of donor organs.

Liver failure due to chronic liver diseases, while exhibiting a longer time course of disease pathogenesis, is much more common than fulminant failure, with chronic liver disease and cirrhosis representing the 12th leading cause of death (1.1% of total deaths) in the United States in 2002 (Kochanek *et al.*, 2004). The most common causes of chronic liver disease are hepatitis C, alcohol-induced and nonalcoholic fatty liver disease (NAFLD), and hepatitis B (W. R. Kim *et al.*, 2002). Other etiologies include primary sclerosing cholangitis, primary biliary cirrhosis, $\alpha_1$-antitrypsin deficiency, autoimmune hepatitis, hereditary hemochromatosis, and liver cancer. The prevalence of hepatitis C infection in the United States population has been estimated at 1.8%, or nearly 4 million individuals (Alter *et al.*, 1999). Notably, cirrhosis initiated by hepatitis C infection is the most frequent cause for liver transplantation, accounting for 40–50% of individuals who have undergone transplant and those on the waiting list (R. S. Brown, 2005). In addition, the long-term inflammation precipitated by chronic hepatitis C infection can also promote the development of hepatocellular carcinoma. As a result, substantial efforts are focused on understanding hepatitis C viral pathogenesis and the development of approaches for the improved control of the virus before and after transplantation. Taken together, so-called fatty liver diseases also comprise a major proportion of chronic liver disease patients (W. R. Kim *et al.*, 2002). In particular, NAFLD is an increasingly prevalent condition in the United States, present in approximately 20% of adults,

of which a subset (2–3% of adults) exhibits nonalcoholic steatohepatitis (NASH), defined by the presence of characteristic injury and necroinflammatory changes in addition to excessive fat accumulation. NAFLD pathogenesis has been shown to be associated with risk factors and conditions such as obesity, type 2 diabetes mellitus, hyperlipidemia, hyperinsulinemia, and insulin resistance. Collectively, chronic liver disorders can progress toward the eventual development of dangerous conditions, including portal hypertension and hepatic encephalopathy. This progression, often referred to as *acute on chronic liver failure*, is analogous to fulminant hepatic failure cases; organ transplantation is the only effective therapy currently. As a means to determine organ allocation more accurately to patients on the liver transplant waiting list, the MELD (mathematical model for end-stage liver disease) system was implemented in 2002, which assigns a priority score based on three prognostic indicators: bilirubin level, creatine level, and INR, a measure of blood-clotting time (R. S. Brown, 2005). Although overall improvements in liver allocation have been achieved following the introduction of the MELD system, regional variations in MELD scores exist, and the ability of this model to accurately predict outcomes across the entire score distribution and for distinct pathologies is less clear.

In order to expand the supply of available livers, several surgical options have been pursued, including the use of non-heart-beating donors and split liver transplants from cadaveric or living donors (K. A. Brown, 2005). Partial liver transplants take advantage of the body's role in the regulation of liver mass and the significant capacity for regeneration exhibited by the mammalian liver. This regenerative process has been extensively examined utilizing experiments in rodent models, which demonstrate that partial hepatectomy or chemical injury induces the proliferation of the existing mature cell populations within the liver, including hepatocytes, bile duct epithelial cells, and others, resulting in the replacement of lost liver mass. However, liver regeneration is difficult to control clinically; and although partial liver transplants have demonstrated some effectiveness, biliary and vascular complications are major concerns in these procedures (K. A. Brown, 2005). Furthermore, in spite of these surgical advances and improvements in organ allocation, there is an increasing divergence between the number of patients awaiting transplantation and the number of available organs, suggesting that it is unlikely that liver transplantation procedures alone will meet the increasing demand.

Consequently, alternative approaches are needed and are being actively pursued, including several nonbiological extracorporeal support systems (discussed in more detail later in this chapter), such as plasma exchange, plasmapheresis, hemodialysis, and hemoperfusion over charcoal or various resins. These systems have shown limited success, likely due to the narrow range of functions inherent to these devices. The liver exhibits a complex array of over 500 functions, including detoxification, synthetic, and metabolic processes. Hence, recapitulation of a substantial number of liver functions will be required to offer sufficient liver support. As a means to provide the multitude of known as well as currently unidentified liver functions, cell-based therapies have been proposed as an alternative to both organ transplantation and the use of strictly nonbiological systems (Allen and Bhatia, 2002). These therapies encompass approaches aimed at providing temporary support, such as extracorporeal bioartificial liver (BAL) devices, as well as more permanent adjunct interventions, such as cell transplantation, transgenic xenografts, and implantable hepatocellular constructs (Fig. 48.1). Collectively, the development of these types of cell-based therapies for liver disease is a major aim of liver-tissue engineering, and the fundamental advances and current status of these approaches are reviewed in this chapter.

## III. CELL SOURCES FOR LIVER CELL–BASED THERAPIES

The choice of cell type is a critical parameter for any cell-based therapy. Table 48.1 highlights some of the key issues regarding the use of various cell sources. Immortalized hepatocyte cell lines, such as HepG2 (human hepatoblastoma) (Kelly and Darlington, 1989), the HepG2-derived line C3A (Ellis *et al.*, 1996), HepLiu (SV40 immortalized) (J. Liu *et al.*, 1999), and immortalized fetal human hepatocytes (Yoon *et al.*, 1999), have been utilized as readily available surrogates for hepatic tissue. However, it has been suggested that these cells lack the full functional capacity of primary adult hepatocytes, and for clinical applications there is a risk that oncogenic factors could be transmitted to the patient. Thus, the generation of conditionally immortalized lines and the incorporation of inducible suicide genes have been considered as potential precautionary measures. The use of primary hepatocyte–based systems could potentially eliminate these issues and provide the appropriate collection of liver functions. Primary porcine hepatocytes have been utilized in a range of BAL device configurations, with some encouraging results. However, the utility of xenogeneic

porcine cells for human liver therapies is restricted by immunogenicity and the potential for xenozoonotic transmission of infectious agents such as porcine endogenous retrovirus (PERV).

Primary human hepatocytes are ultimately the preferred cell type for cell-based therapies, and the development of primary hepatocyte–based approaches is the focus of substantial ongoing research. Yet progress has been hindered by the limited supply of primary human hepatocytes and certain aspects of hepatocyte physiology. Discussed in

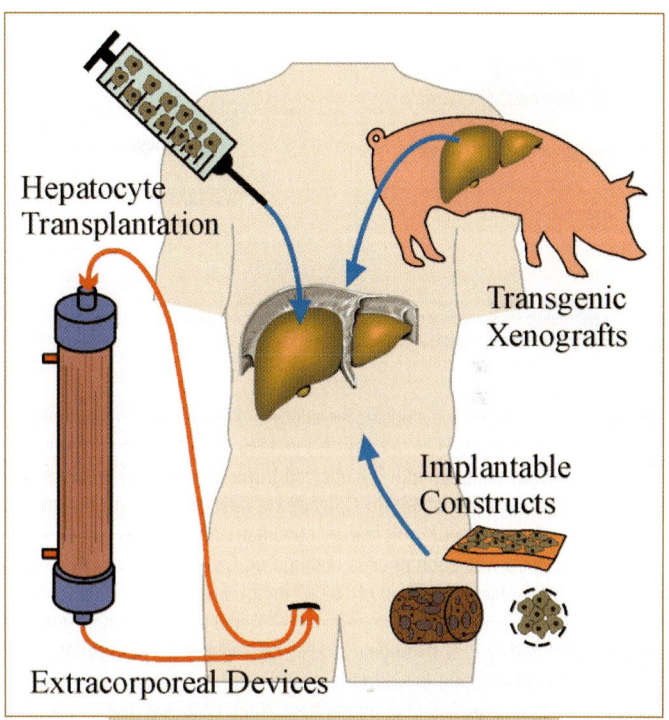

**FIG. 48.1.** Cell-based therapies for liver disease. Extracorporeal devices perfuse the patient's blood or plasma through bioreactors containing hepatocytes. Hepatocytes are transplanted directly or implanted on scaffolds. Transgenic animals are being raised in order to reduce complement-mediated damage of the endothelium. From Allen *et al.* (2001).

| Table 48.1. | Cell sources for liver therapies |
| --- | --- |
| *Cell source* | *Critical issues* |
| *Primary hepatocytes* | Sourcing, expansion, phenotypic instability, |
| Human, xenogeneic | immunogenicity, safety (xenozoonotic) |
| *Immortalized hepatocyte lines* | Range of functions, genomic instability, safety |
| Tumor-derived, SV40, telomerase, spontaneously immortalized | (tumorigenicity) |
| *Stem cells* | Sourcing, differentiation efficiency, phenotypic |
| Embryonic, liver progenitors (hepatoblasts, oval cells), other lineages (HSC, MAPC) | instability, immunogenicity, safety (tumorigenicity) |

*Abbreviations*: SV40, simian virus 40; HSC, hematopoietic stem cells; MAPC, multipotent adult progenitor cells.

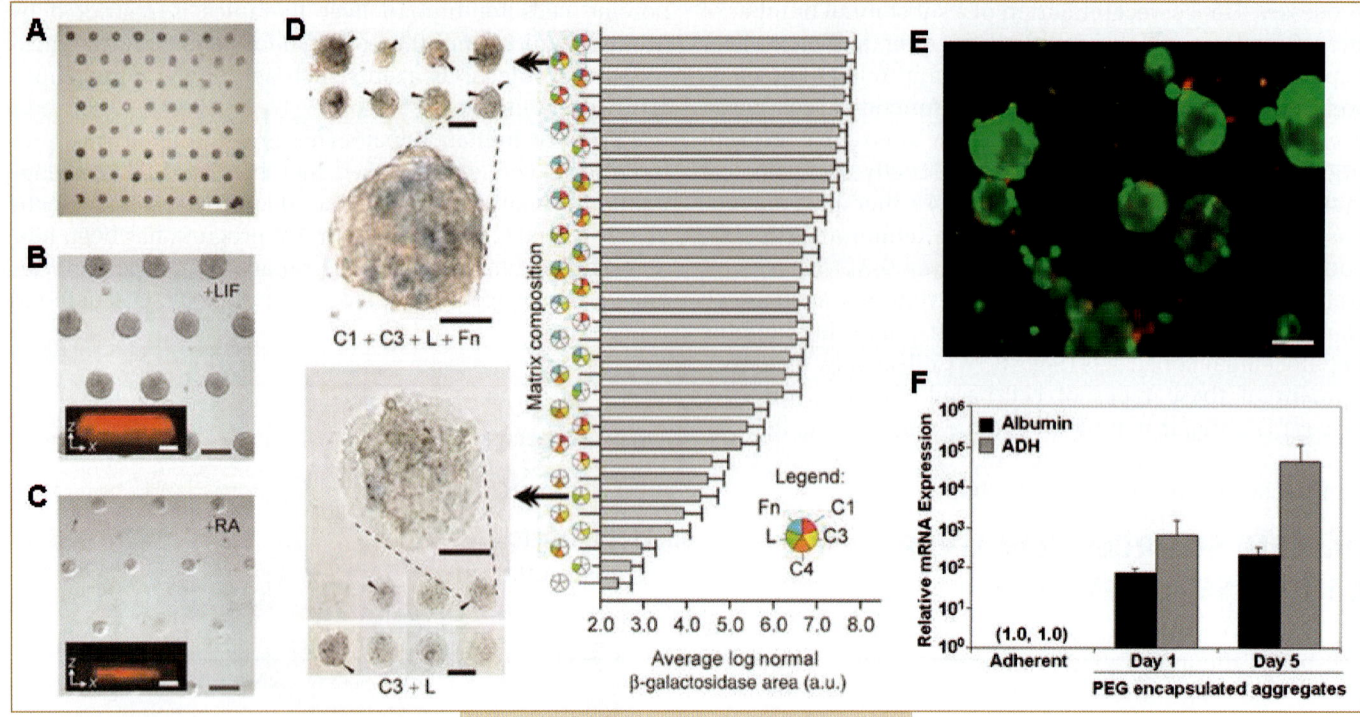

**FIG. 48.2.** Platforms for studying hepatocyte differentiation of embryonic stem cells and liver progenitors. **(A)** Bright field alkaline phosphatase staining of day 1 ES cultures on ECM microarrays in 15% serum medium (scale bar, 1 mm). **(B, C)** Phase-contrast images of day 3 arrays cultured with LIF (B) and with RA (C). Cells cultured with LIF showed three-dimensional features (in B, inset *x-z* confocal section, ~77-μm thickness). In contrast, RA-induced cells grew as a relatively thin sheet (in C, inset *x-z* section, ~25-μm thickness). Scale bars, 250 μm (inset scale bars, 50 μm). **(D)** Bright-field micrograph of selected X-gal-stained ECM microarray conditions after 3 d of culture in RA. C1 + C3 + L + Fn (top left images) induced higher *Ankrd17* reporter activity (arrowheads) than was seen in cells cultured on C3 + L (bottom left images). Scale bars, 250 μm. Magnified views of reporter activity: scale bars, 50 μm. *Bar graph:* hierarchical depiction of "blue" image area (pooled data from four microarrays) for each of the matrix mixtures. Error bars, s.e.m. (*n* = 32). The C1 + C3 + L + Fn culture condition induced ~27-fold more β-galactosidase image area than the C3 + L cultures (Flaim *et al.*, 2005). **(E)** BMEL cell aggregates encapsulated within PEG hydrogels exhibit high viability. Fluorescent labeling distinguished viable (green) from nonviable (red) cells. Scale bar, 100 μm. **(F)** Expression of albumin (black bars) and alcohol dehydrogenase (ADH, grey bars) mRNA was determined by real-time quantitative RT-PCR at day 1 and day 5 following encapsulation of aggregated BMEL cells. Expression for each gene is displayed relative to basal expression exhibited by adherent BMEL cells prior to aggregation. The housekeeping gene, HPRT, was utilized as a normalization control, and data are presented as the mean ± SD (*n* = 3, independent experiments).

more detail later, hepatocytes exhibit a loss of liver-specific functions under many conditions *in vitro*. In addition, particularly for human hepatocytes, despite the significant proliferative capacity during regenerative responses *in vivo*, mature hepatocyte proliferation in culture is limited (Mitaka, 1998). As a result, alternative cell sources for liver cell–based therapies are being investigated, such as diverse stem cell populations, which can retain significant proliferative ability *in vitro* and exhibit either pluripotency or multipotency, thereby constituting a possible source of hepatocytes as well as other liver cell types. Preliminary evidence suggests that pluripotent embryonic stem cells can be induced to differentiate toward the hepatocyte lineage in culture (Chinzei *et al.*, 2002; Hamazaki *et al.*, 2001; Ishizaka *et al.*, 2002; Yamada *et al.*, 2002), and methods to improve the range of acquired hepatocyte functions as well as differentiation efficiency continue to be explored (Heng *et al.*, 2005). Recently, an extracellular matrix microarray platform was utilized to

examine the influence of combinatorial mixtures of matrix molecules on embryonic stem cell differentiation toward an early hepatic fate (Fig. 48.2) (Flaim *et al.*, 2005). In this system, an approximately 140-fold difference in the induction of an early hepatic marker (Ankrd17) was observed between the least and the most efficient conditions (Fig. 48.2), underscoring the importance of high-throughput technologies in the elucidation of factors affecting stem cell differentiation.

In addition to embryonic stem cells, various fetal or adult stem/progenitor populations have similarly been investigated. For example, multipotent adult progenitor cells (MAPC) derived from bone marrow have been shown to exhibit hepatocyte differentiation potential (Schwartz *et al.*, 2002). Fetal hepatoblasts and oval cells are also intriguing possible cell types for liver cell–based therapies. Hepatic development proceeds through the differentiation of liver precursor cells, termed *hepatoblasts*, which exhibit bipotential differentiation capacity, defined by the ability to differentiate into both hepatocytes and bile duct epithelial cells (cholangiocytes) (Lemaigre and Zaret, 2004). Notably, in certain types of severe and chronic liver injury, an adult progenitor cell population, termed *oval cells*, which shares many phenotypic markers and functional properties with fetal hepatoblasts, mediates compensatory liver repair through a similar differentiation program (Sell, 2001). Exposure to toxins or carcinogens that block hepatocyte proliferation, and consequently normal liver regeneration, results in the proliferation of oval cells within the liver and subsequent liver repair due to the differentiation of these cells. Recent work by Weiss and colleagues demonstrates the development of bipotential mouse embryonic liver (BMEL) cell lines that are derived from mouse E14 embryos and exhibit characteristics reminiscent of fetal hepatoblasts and oval cells (Strick-Marchand and Weiss, 2002). In particular, these cell lines are nontransformed and proliferative, demonstrate up-regulation of hepatocyte or bile duct epithelial markers under distinct culture conditions *in vitro* (Strick-Marchand and Weiss, 2002), and exhibit the capacity to home to the liver and undergo bipotential differentiation *in vivo* within a regeneration context (Strick-Marchand *et al.*, 2004). This system highlights the potential of progenitor cell lines for tissue repair. As such, toward the eventual incorporation of BMEL cells into implantable tissue-engineered constructs, and discussed in more detail later in this chapter, BMEL cells have been successfully encapsulated within a biomaterial scaffold (PEG hydrogel), and their differentiation toward the hepatocyte lineage was demonstrated to proceed efficiently in this system (Fig. 48.2).

Although diverse stem and progenitor cell populations exhibit vast potential regarding integration into hepatic therapies, many challenges remain, including the ability to dictate and enhance differentiation, particularly within multicellular systems. Furthermore, regardless of the cell source, the stabilization of hepatocyte functions remains a primary issue. Microenvironmental signals, including soluble mediators, cell–extracellular matrix interactions, and cell–cell interactions, have been implicated in the regulation of hepatocyte function. Accordingly, the development of robust hepatocyte *in vitro* culture models that allow for the controlled reconstitution of these environmental factors is a fundamental prerequisite toward a thorough understanding of mechanisms regulating hepatocyte processes and the improved functionality of liver cell–based therapies.

## IV. *IN VITRO* HEPATIC CULTURE MODELS

An extensive range of liver model systems have been developed, some of which include: perfused whole-organs and wedge biopsies; precision-cut liver slices; isolated primary hepatocytes in suspension or cultured on extracellular matrix; immortalized liver cell lines; isolated organelles, membranes, or enzymes; and recombinant systems expressing specific drug metabolism enzymes (Guillouzo,

1998). While perfused whole organs, wedge biopsies, and liver slices maintain many aspects of the normal *in vivo* microenvironment and architecture, they typically suffer from short-term viability (<24 h) and limited nutrient/oxygen diffusion to inner cell layers. Purified liver fractions and single-enzyme systems are routinely used in high-throughput systems to identify enzymes involved in the metabolism of new pharmaceutical compounds, although they lack the complete spectrum of gene expression and cellular machinery required for liver-specific functions. In addition, cell lines derived from hepatoblastomas or from immortalization of primary hepatocytes are finding limited use as reproducible, inexpensive models of hepatic tissue. However, such lines are plagued by abnormal levels and a repertoire of hepatic functions (Sivaraman *et al.*, 2005), perhaps most notably the divergence of nuclear receptor–mediated regulation of cytochrome P450 enzymes (Kawamoto *et al.*, 1999). Though each of these models has found utility for focused questions in drug metabolism research, isolated primary hepatocytes are generally considered to be most suitable for constructing *in vitro* platforms for a multitude of applications. A major limitation in the use of primary hepatocytes is that they are notoriously difficult to maintain in culture, due to the precipitous decline in viability and liver-specific functions on isolation from the liver (Guillouzo, 1998). Accordingly, substantial research has been conducted since the mid-1980s toward elucidating the specific molecular stimuli that can maintain phenotypic functions in hepatocytes. In subsequent sections, we present examples of strategies that have been employed to improve the survival and liver-specific functions of primary hepatocytes in culture.

### *In Vivo* Microenvironment of the Liver

In order to engineer an optimal microenvironment for hepatocytes *in vitro*, one can utilize as a guide the precisely defined architecture of the liver, in which hepatocytes interact with diverse extracellular matrix molecules, nonparenchymal cells, and soluble factors (i.e., hormones, oxygen) (Fig. 48.3). Structurally, the two lobes of the liver contain repeating functional units called *lobules*, which are centered on a draining central vein. Portal triads at each corner of a lobule contain portal venules, arterioles, and bile ductules. The blood supply to the liver comes from two major blood vessels on its right lobe: the hepatic artery (one-third of the blood) and the portal vein (two-thirds). The intrahepatic circulation consists of sinusoids, which are small, tortuous vessels lined by a fenestrated basement membrane lacking endothelium that is separated from the hepatocyte compartment by a thin extracellular matrix region termed the *space of Disse*. The hepatocytes, constituting ~70% of the liver mass, are arranged in unicellular plates along the sinusoid, where they experience homotypic cell interactions. Several types of junctions (i.e., gap junctions, cadherins, and tight junctions) and bile canaliculi at the interface of

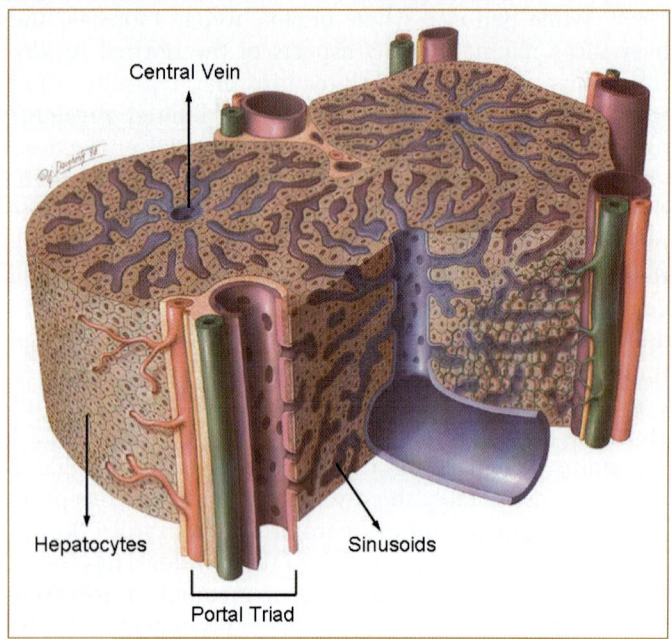

**FIG. 48.3.** The precisely defined architecture of the repeating unit of the liver, the lobule. Hepatocytes are arranged in cords along the length of the sinusoid, where they interact with extracellular matrix molecules, nonparenchymal cells, and gradients of soluble factors. Nutrient- and oxygen-rich blood from the intestine flows into the sinusoid via the portal vein. After being processed by the hepatocytes, the blood enters the central vein and into the systemic circulation. Reproduced with permission from J. Daugherty.

hepatocytes facilitate the coordinated excretion of bile to the bile duct and subsequently to the gall bladder. Nonparenchymal cells, including stellate cells, cholangiocytes (biliary ductal cells), sinusoidal endothelial cells, and Kupffer cells (macrophages), interact with hepatocytes to modulate their diverse functions. In the space of Disse, hepatocytes are sandwiched between layers of extracellular matrix (collagen types I–IV, laminin, fibronectin, heparin sulfate proteoglycans), the composition of which varies from the portal triad to the central vein (Reid *et al.*, 1992).

Within the liver lobule, hepatocytes are partitioned into three zones, based on morphological and functional variations along the length of the sinusoid (zonation). Zonal differences have been observed in virtually all hepatocyte functions. For instance, compartmentalization of gene expression is thought to underlie the capacity of the liver to operate as a "glucostat." Furthermore, zonal differences in expression of cytochrome-P450 enzymes have also been implicated in the zonal hepatotoxicity observed with some xenobiotics (Lindros, 1997). Possible modulators of zonation include blood-borne hormones, oxygen tension, pH levels, extracellular matrix composition, and innervation. Therefore, a precisely defined microarchitecture, coupled with specific cell–cell, cell-soluble factor, and cell–matrix interactions allows the liver to carry out its many diverse functions, which can be broadly categorized as protein

synthesis (i.e., albumin, clotting factors), cholesterol metabolism, bile production, glucose and fatty acid metabolism, and detoxification of endogenous (i.e., bilirubin, ammonia) and exogenous (drugs and environmental compounds) substances.

## Two-Dimensional Cultures

Numerous studies have illustrated improvements in the maintenance of primary hepatocyte morphology, survival, and liver-specific functions through modifications in culture conditions. For instance, supplementation of culture medium with low concentrations of hormones, corticosteroids, growth factors, vitamins, amino acids, or trace elements can stabilize the hepatic phenotype (Guillouzo, 1998). In addition, nonphysiologic factors, such as Phenobarbital (drug inducer) and dimethylsulfoxide (organic solvent), are also effective (Isom *et al.*, 1985; Miyazaki *et al.*, 1985). Some complex serum-free hormonally defined media preparations have also been proposed for culturing hepatocytes; however, these formulations often do not maintain liver-specific gene expression and differentiated functions beyond a few days (Guillouzo, 1998). A few mitogenic and comitogenic factors can induce proliferation in hepatocytes under standard culture conditions. For example, when plated at low densities and maintained in serum-free culture medium supplemented with nicotinamide and epidermal growth factor, about 15% of rat hepatocytes undergo three rounds of cell divisions (Mitaka *et al.*, 1991). Human hepatocytes have also displayed proliferation when stimulated with epidermal growth hormone, hepatocyte growth hormone, or human serum (Blanc *et al.*, 1992). However, such limited growth is unlikely to meet the demand for the several billion hepatocytes required for cell-based therapies.

The impact of extracellular matrix on the differentiated functions of hepatocytes has been widely studied. In general, the efficiency of hepatocyte attachment is enhanced by coating substrates with simple extracellular matrix proteins (typically collagen); however, in most cases a concomitant increase in hepatocyte spreading leads to a loss of liver-specific functions (LeCluyse *et al.*, 1996). Presentation of extracellular matrices of different compositions and topologies can stabilize hepatocyte morphology and a limited set of phenotypic functions. For instance, hepatocyte culture on biomatrix, a complex mixture of extracellular matrix components extracted from liver, has been shown to improve hepatocyte function as compared to monolayers on collagen (LeCluyse *et al.*, 1996; Lin *et al.*, 2004). In order to systematically probe the role of distinct extracellular matrix molecules, the matrix microarray platform for embryonic stem cells described earlier was also utilized for primary hepatocyte culture. These experiments similarly revealed a synergistic influence of extracellular matrix combinations on hepatocyte function (Flaim *et al.*, 2005). Furthermore, when sandwiched between two layers of gelled collagen, hepatocytes from a variety of species maintain a cuboidal shape, secrete albumin, and synthesize urea (marker of nitrogen metabolism) (Guillouzo, 1998). Rat hepatocytes, in particular, secrete albumin at a high rate for 40 days in sandwich cultures, exhibit improved cytochrome P450 induction, and form a contiguous, anastomosing network of bile canaliculi indicative of polarity (Dunn *et al.*, 1989; LeCluyse *et al.*, 1994). Even though two-dimensional sandwich modifications improve the longevity of hepatocyte cultures, an imbalance of phase I/II detoxification processes typically occurs over time (Richert *et al.*, 2002). Also, the presence of an overlaid layer of extracellular matrix may present diffusion barriers for molecular stimuli (i.e., drug candidates), and the fragility of the gelled matrix may hinder scale-up within bioartificial liver devices.

Heterotypic interactions between hepatocytes and their nonparenchymal neighbors are known to be important at multiple stages *in vivo*. Liver specification from the endodermal foregut and mesenchymal vasculature during development is believed to be mediated by heterotypic interactions (Houssaint, 1980; Matsumoto *et al.*, 2001). Similarly, nonparenchymal cells of several types modulate cell-fate processes of hepatocytes under both physiologic and pathophysiologic conditions within the adult liver (Michalopoulos and DeFrances, 1997; Olson *et al.*, 1990). *In vitro*, through extensive studies initially pioneered by Guguen-Guillouzo and colleagues (1983), the viability and liver-specific functions of hepatocytes from multiple species have been shown to be stabilized for several weeks on cocultivation with a wide range of nonparenchymal cell types from within and outside the liver (Bhatia *et al.*, 1999). Furthermore, induction has been reported by nonparenchyma (both primary and immortalized) across species barriers, suggesting possible conservation of underlying mechanisms. Hepatocytes in cocultures maintain for weeks the polygonal morphology, distinct nuclei and nucleoli, well-demarcated cell–cell borders, and a visible bile canaliculi network indicative of freshly isolated cells or cells *in vivo* (Fig. 48.4). Hepatic cocultures have been utilized to investigate various physiologic and pathophysiologic processes, including host response to sepsis, mutagenesis, xenobiotic toxicity, response to oxidative stress, lipid metabolism, and induction of the acute-phase response (Bhatia *et al.*, 1999).

### Three-Dimensional Cultures

Culture of hepatocytes on substrates that promote aggregation into three-dimensional spheroids has also been extensively explored. For example, when seeded on nonadhesive surfaces such as positively charged dishes and dishes coated with liver-derived proteoglycans, hepatocytes aggregate and within one to two days form spheroidal structures with a mean diameter of ~50 μm. Over the course of several weeks, small aggregates fuse to form larger spheroids that are 150–175 μm in diameter (LeCluyse *et al.*, 1996). Seeding of hepatocytes into spinner flasks or on polystyrene dishes incubated on a rotary shaker allows larger spheroids to be

formed within 24 hours. Under spheroidal culture conditions, hepatocyte survival and functions are improved over standard monolayers on collagen (Landry *et al.*, 1985). The underlying mechanisms for these improvements are potentially the retention of a three-dimensional cytoarchitecture, the presence of extracellular matrix surrounding the spheroids, and the establishment of homotypic cell–cell contacts. Accordingly, static cultures of spheroids on collagen-coated dishes have been shown to disassemble, spread, and subsequently lose differentiated functions in culture (Hsiao *et al.*, 1999; Powers *et al.*, 2002). Culture of hepatocytes on a substratum coated with Matrigel (laminin-rich basement membrane extract derived from a mouse sarcoma) also induces formation of adherent three-dimensional heterogeneous spheroids and leads to retention of hepatocyte functions including CYP450 activity (Bissell *et al.*, 1987; LeCluyse, 2001). However, these effects are difficult to interpret mechanistically, since contamination of Matrigel with proteins, hormones, and growth factors has been previously reported (Guillouzo, 1998; Vukicevic *et al.*, 1992). In addition, probably the most significant drawback of Matrigel-based platforms is an imbalance of phase I/II detoxification processes (i.e., decline in CYP450 activities) that occurs over the course of a few days in culture (Richert *et al.*, 2002).

Encapsulation of spheroids has been utilized as a means to control cell–cell interactions. For example, one strategy employs hepatocyte spheroids, first suspended in a methylated collagen solution and then extruded from a needle into a terpolymer solution to form microcapsules (Chia *et al.*, 2000). A variety of other synthetic and natural scaffolds have been explored for encapsulation of single hepatocytes or hepatocyte aggregates. The development and characterization of such scaffolds are described in a subsequent section of this chapter focused on implantable constructs. In particular, spheroid cultures have been shown to be useful as the basis for both small-scale and large-scale bioreactor systems. However, the utility of spheroidal cultures can be somewhat restricted by certain properties of this configuration, including the fusion of small heterogeneous spheroids into larger structures; cell death in the center of large aggregates due to insufficient transport of nutrients and oxygen or due to bile accumulation in size-controlled, compacted spheroids; and the limited *in situ* observation of cells in aggregates. Overall, the development of novel approaches and tools to identify the important three-dimensional signals underlying hepatocyte function in spheroid structures as well as robust platforms for spheroid maintenance are active areas of research.

### Bioreactor Cultures

The development of bioreactors for hepatocellular culture has been extensively studied in the last few decades. Several different bioreactor designs have been explored for use in extracorporeal liver devices, and these are discussed in a subsequent section. Here, our focus is primarily on

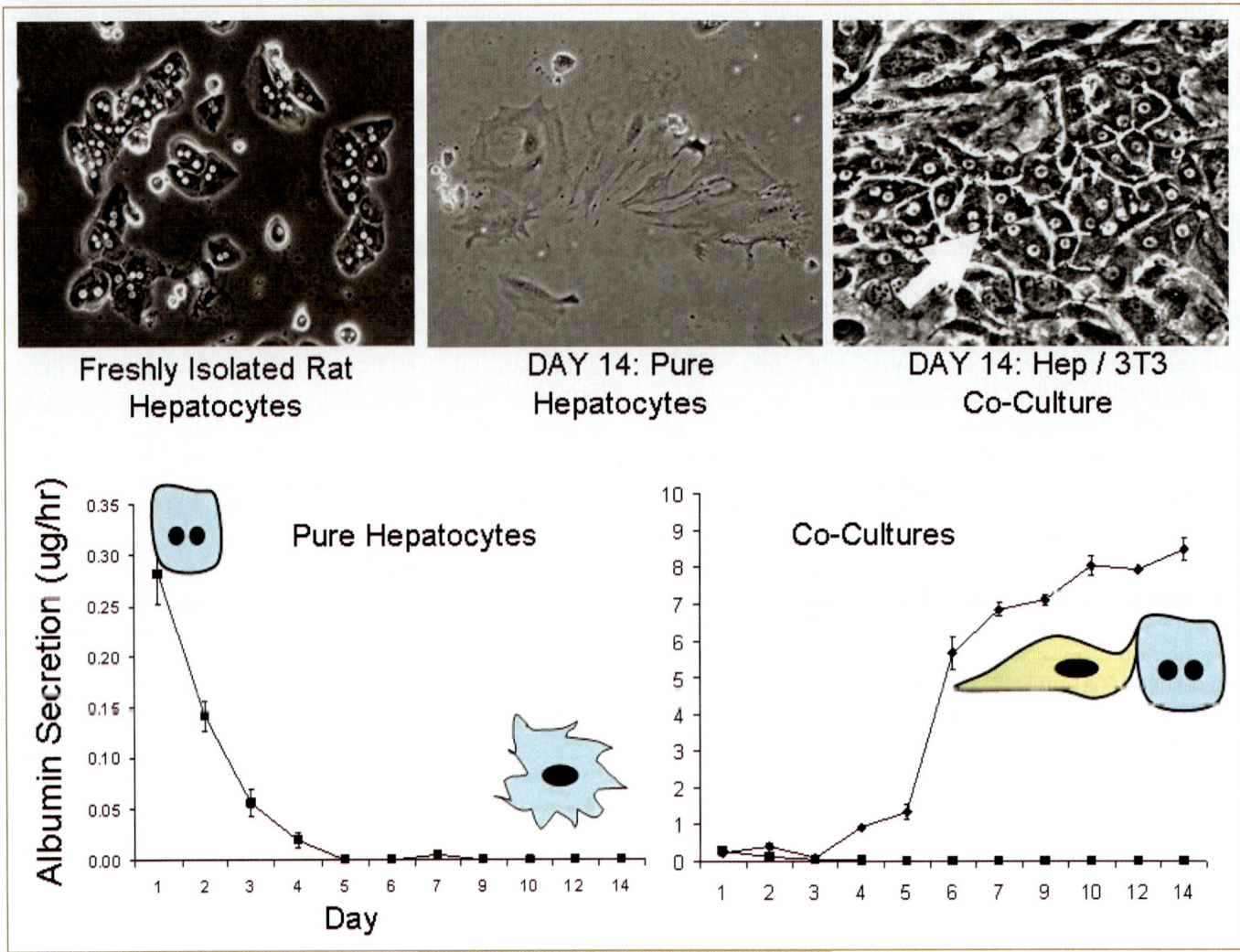

**FIG. 48.4.** The coculture effect. Isolated primary hepatocytes from a variety of species lose characteristic morphological features and liver-specific functions upon isolation from the liver. Cocultivation of hepatocytes with a wide range of nonparenchymal cells from both within and outside the liver can stabilize the hepatic phenotype. Shown here are primary rat hepatocytes cocultivated with 3T3 murine embryonic fibroblasts. Hepatocytes in cocultures maintain the polygonal morphology (see arrow in picture), visible bile canaliculi, and distinct nuclei and nucleoli as seen in freshly isolated cells and *in vivo*. On the other hand, hepatocytes in pure culture adopt a fibroblastic morphology. Furthermore, liver-specific functions (i.e., albumin secretion) decline rapidly in pure hepatocytes, whereas they are retained in cocultures for several weeks.

small-scale bioreactors that have been designed for fundamental investigations in hepatology and/or for use in drug development.

The distribution of hepatic functions along the length of the sinusoid is thought to be modulated by gradients of diverse factors such as oxygen, hormones, nutrients, and extracellular matrix molecules. Current *in vitro* models of the liver in static media conditions offer a relatively homogenous view of liver function. Recently, a small-scale, parallel-plate bioreactor was developed that allows exposure of hepatocyte-nonparenchymal cocultures to a continuous range of oxygen tensions in order to mimic *in vivo* conditions in the liver lobule (Fig. 48.5) (Allen *et al.*, 2005). Cocultures exposed to physiologic gradients of oxygen demonstrated regional heterogeneity of CYP2B and CYP3A expression as seen *in vivo*. Expression of these drug metabolism enzymes can be further modulated via chemical inducers and growth factors. Additionally, acetaminophen exposure induced maximal cell death in the downstream perivenous-like hepatocytes, which is consistent with the centrilobular patterns of cell death observed *in vivo*. Overall, the ability to decouple oxygen tension from gradients of other soluble stimuli and cell–cell interaction effects within this platform represents an important tool for the systematic investigation of the role of extracellular stimuli in zonation. Also, miniaturization of this bioreactor system utilizing microfluidics is currently being pursued for the high-throughput examination of drug metabolism and toxicity in

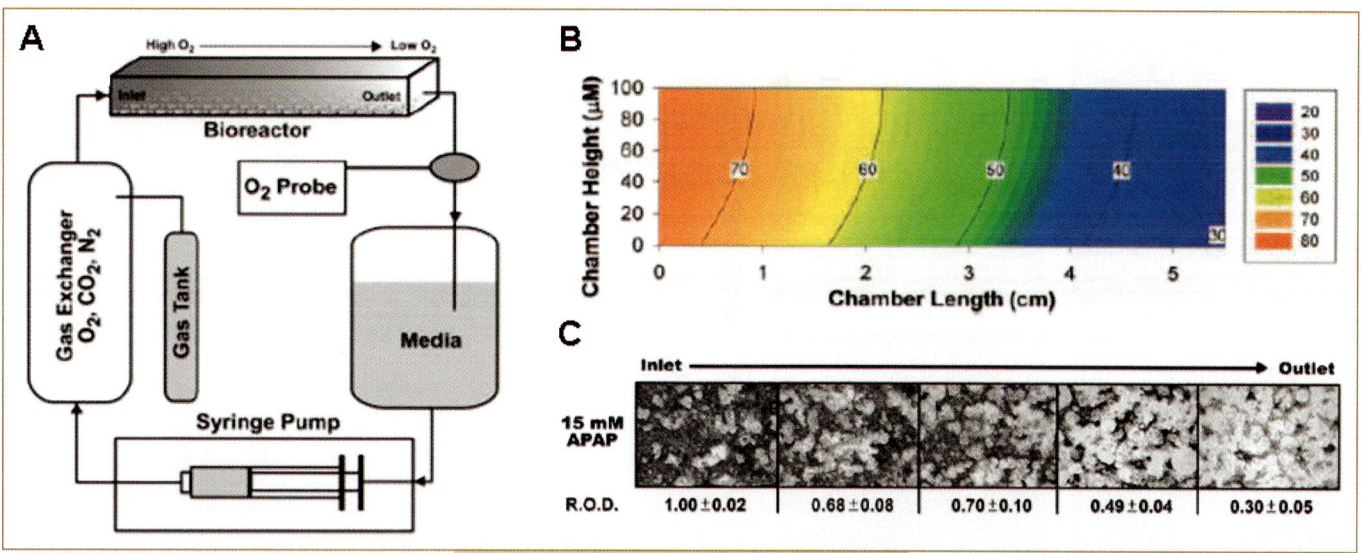

**FIG. 48.5.** Zonation and toxicity in a hepatocyte bioreactor. **(A)** Cocultures of hepatocytes and nonparenchymal cells are first created on collagen-coated glass slides and then placed in a bioreactor circuit, where the oxygen concentration at the inlet is held at a constant value. Depletion of oxygen by cells creates a gradient of oxygen tensions along the length of the chamber, similar to that observed *in vivo*. **(B)** Two-dimensional contour plot of modeled oxygen profile in the medial cross section of the reactor. Results depict the cell surface gradient formed with inlet $pO_2$ of 76 mmHg and flow rate of 0.3 mL/min. **(C)** Bright-field images of MTT-stained (measure of cell viability), perfused cultures were acquired from five regions along the length of the slide. Images from 15 mm acetaminophen (APAP) treatment are shown, with relative optical density (R.O.D) values representing mean and standard error of mean ($n = 3$). The zonal pattern of APAP toxicity seen here is consistent with that observed *in vivo*. From Allen *et al.* (2005).

addition to the further analysis of the combinatorial influence of microenvironmental factors on zonated hepatocyte functions.

Hepatic aggregate culture in microbioreactors has also been investigated. For instance, a 1-cm$^2$ planar polymer scaffold with 900 microcontainers has been shown to be suitable for culture of uniformly sized three-dimensional hepatic aggregates (Knedlitschek *et al.*, 1999). The bottom faces of the microcontainers are modified to contain laser-drilled pores for fluid perfusion through the cellular structures. Aggregates in microcontainers retain liver-specific morphology and functions for up to two weeks. In particular, the cells secrete albumin, display active drug metabolism enzymes, and are responsive to CYP450-inducing compounds. In another platform, hepatic spheroids are cultured inside an array of microchannels created using deep reaction ion etching of silicon wafers (Sivaraman *et al.*, 2005). The wafers are combined with a cell-retaining filter and support in a bioreactor system that delivers culture medium across the top of the array and through the spheroids in each channel. Characterization via gene expression profiling, protein expression, and activity of drug metabolism enzymes indicates that the spheroids in the microfabricated bioreactor retain liver-specific characteristics for two to three weeks in culture. Inducibility of specific CYP450 enzymes via prototypic chemical inducers has also been explored in this platform, and the simultaneous examination of drug responses in six parallel reactors has been proposed (Sivaraman *et al.*, 2005).

Several other coculture bioreactor systems have been developed. For instance, a flat-plate bioreactor was used specifically to evaluate the effects of oxygenation and shear stress on bulk markers of hepatocyte function toward the fabrication of a bioartificial liver support system (Tilles *et al.*, 2001). In another perfusion system, pure rat hepatocytes and cocultures seeded on collagen-coated slides display liver-specific functions for up to two weeks and are responsive to induction of CYP450 activity by pharmaceutical compounds (Gebhardt *et al.*, 2003). In order to improve oxygen delivery to cells, a 96-well perfused microbioreactor with a gas-permeable biocompatible membrane was developed to culture collagen sandwich cocultures of hepatocytes and liver-derived nonparenchymal cells (Gebhardt *et al.*, 2003). Preliminary studies indicate that hepatocytes in this system maintain numerous important functions, including albumin and urea secretion, constitutive expression of phase I and phase II detoxification enzymes, and the inducible expression of CYP1A1. Collectively, bioreactor platforms can provide substantial control of hepatocyte culture conditions, and thus hepatocyte functions, suggesting that these systems will be key components of drug development studies as well as fundamental investigations of hepatocyte biology.

## Microtechnology Tools to Optimize and Miniaturize Liver Cultures

The ability to control tissue architecture along with cell–cell and cell–matrix interactions on the order of single-

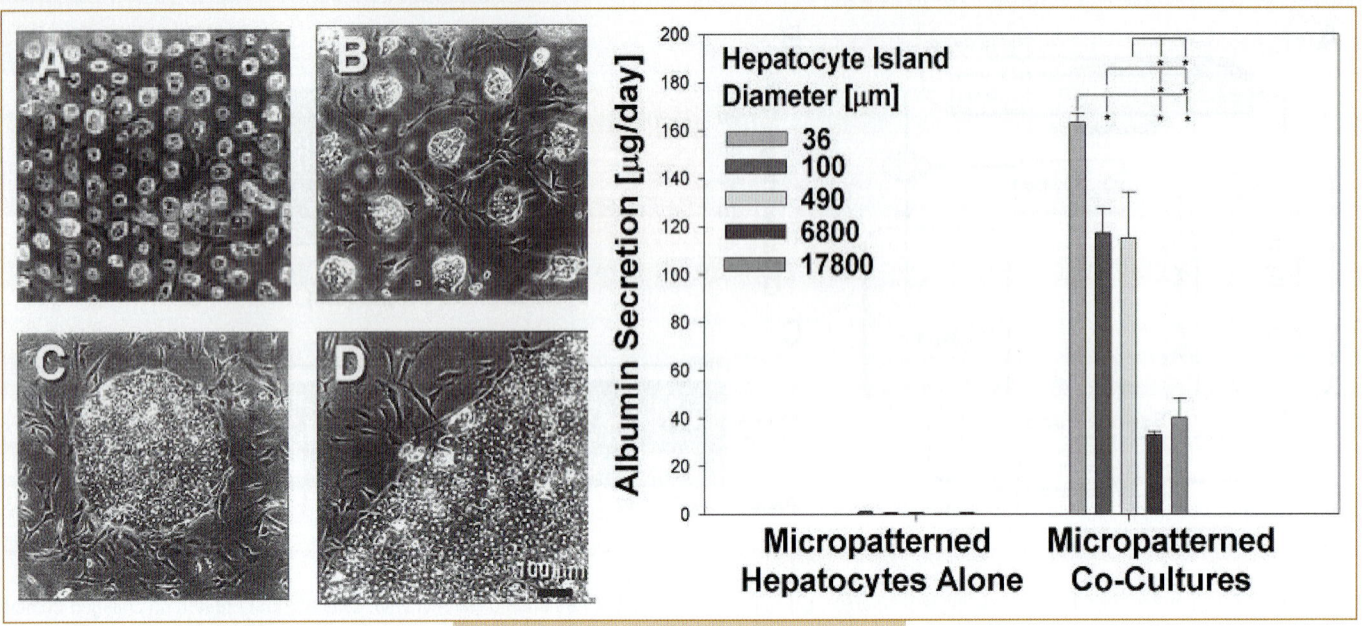

**FIG. 48.6.** Optimization of liver-specific functions in cocultures via photolithographic micropatterning. Shown here are phase-contrast micrographs of micropatterned primary rat hepatocytes surrounded by supportive 3T3-J2 fibroblasts. Ratio of cell populations and total cell numbers were kept constant across pattern geometries, which displayed a broad range of heterotypic interface. **(A)** 36-μm hepatocyte islands with 90 μm center to center spacing (36/90); **(B)** 100/250; **(C)** 490/1230; **(D)** 6800/16,900. Liver-specific functions (i.e., albumin secretion) were higher in cocultures than in pure hepatocyte cultures. Furthermore, the degree of up-regulation varied with micropatterned geometry (17,800 μm represents a single hepatocyte island surrounded by fibroblasts). From Bhatia *et al.* (1999).

cell dimensions represents another important tool in the investigation of mechanisms underlying tissue development and, ultimately, the realization of tissue-engineered systems. Semiconductor-driven microtechnology tools now offer an unprecedented micrometer-scale control over cell adhesion, shape, and multicellular interactions (Folch and Toner, 2000). Thus, since the mid-1990s, microtechnology tools have emerged both to probe biomedical phenomena at relevant length scales and to miniaturize and parallelize biomedical assays (e.g., DNA microarrays, microfluidics).

In the context of the liver, a photolithographic cell-patterning technique has enabled the optimization of liver-specific functions in cocultures via engineering of the balance between homotypic (hepatocyte–hepatocyte) and heterotypic (hepatocyte–nonparenchymal) cell–cell interactions (Bhatia *et al.*, 1999). Specifically, micropatterned cocultures were created in which hepatocyte islands of controlled diameters were surrounded by supporting nonparenchymal cells (Fig. 48.6). Results indicated that rat hepatocyte–fibroblast cocultures with a maximal initial heterotypic interface (i.e., single-hepatocyte islands surrounded by supporting cells) had the highest levels of liver-specific functions as compared to other micropatterned configurations with similar cell numbers and hepatocyte-to-nonparenchymal ratios (Fig. 48.6). More recent work with primary human hepatocytes indicates that these cells are more dependent on homotypic interactions than their rat

counterparts (manuscript in preparation). As such, human hepatocyte cocultures displayed optimal functions once the appropriate balance of homotypic and heterotypic interactions was achieved. Furthermore, miniaturization of micropatterned human cocultures into a multiwell format has been performed in order to create a microscale human liver tissue model for drug development. The utility of this platform for pharmaceutical applications has been demonstrated via characterization of global gene expression, phase I/II drug metabolism, secretion of liver-specific products, drug–drug interactions, and the susceptibility to a panel of putative hepatotoxins.

As described earlier, several culture techniques have been explored for the formation of hepatic spheroids. These methods exhibit certain limitations, including an inability to immobilize the spheroids at defined locations and the heterogeneity of the structures, which can result in cell necrosis at the core of large and coalesced spheroids due to depletion of oxygen and nutrients. Recently, microfabrication and microcontact printing techniques have been combined to develop a microarray that contains immobilized spheroids of a uniform size (Fukuda *et al.*, 2005). These spheroids retained a liver-specific phenotype as assessed by the expression of liver-enriched transcription factors, secretion of albumin, and the presence of urea cycle enzymes.

The field of hepatic-tissue engineering continues to evolve toward creating an optimal microenvironment for

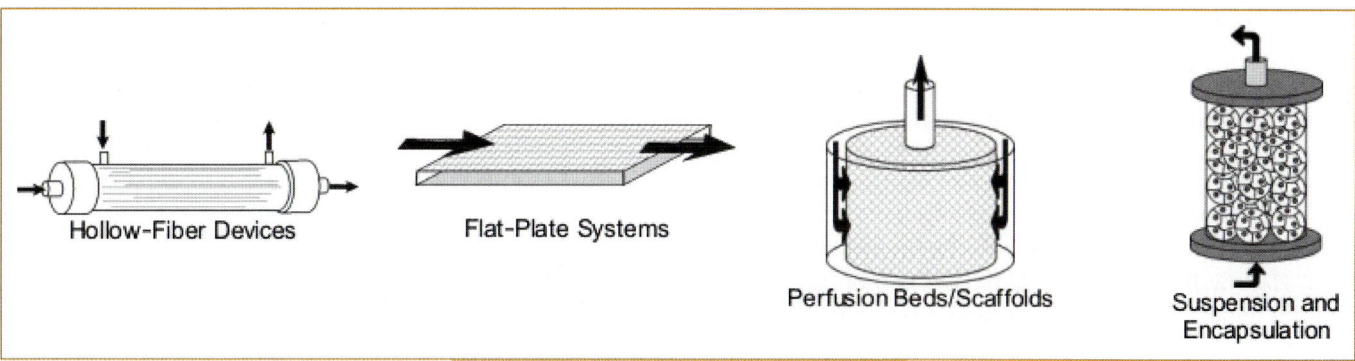

**FIG. 48.7.** Schematics of cell-based bioreactor designs. The majority of liver cell–based bioreactor designs fall into these four general categories, each with inherent advantages and disadvantages. From Allen and Bhatia (2002).

liver cells *in vitro*. Overall, it is evident that many different culture conditions can preserve at least some phenotypic features of fully functional hepatocytes. Since detailed differences can exist in these model systems, current strategies have focused on selecting a platform that is appropriate for a particular application. Although progress has been made, further work is required to obtain a more complete picture of the molecular signals that provide phenotypic stability of liver cultures. In particular, as a systems-level view of hepatocyte biology emerges, perhaps it will become evident how multiple distinct inputs influence hepatocyte function. The development of highly functional *in vitro* liver platforms will ultimately facilitate the clinical effectiveness of cell-based therapies and enable high-throughput screening of candidate drugs for liver-specific metabolism and toxicity earlier in drug discovery, which could potentially reduce development costs and help create safer drugs for patients.

# V. EXTRACORPOREAL BIOARTIFICIAL LIVER DEVICES

One of the most promising approaches for cell-based therapies for liver failure is the development of extracorporeal support devices, which, analogous to kidney dialysis systems, would process the blood or plasma of liver-failure patients. These devices are aimed principally at providing temporary support to liver-failure patients to enable sufficient regeneration of the host liver tissue or serve as a bridge to transplantation. Initial extracorporeal device designs utilized primarily nonbiological mechanisms, such as hemoperfusion, hemodialysis, plasmapheresis, and plasma exchange (Allen *et al.*, 2001). Hemoperfusion removes toxins, but also other useful metabolites, by passage of blood or plasma through a charcoal column. A modification of this approach, referred to as *hemodiadsorbtion* and utilized in the Liver Dialysis Unit developed by HemoCleanse, reduces direct contact with charcoal components through the

utilization of a flat-membrane dialyser containing charcoal and exchange resin particles (Ellis *et al.*, 1999). In general, charcoal perfusion systems have been the most extensively studied nonbiological configuration, including clinical evaluation in patients with acute liver failure, although no clear survival improvement has been observed (Yarmush *et al.*, 1992). An alternative strategy, the Molecular Adsorbent Recirculating System (MARS®), uses a method termed *albumin dialysis*, in which albumin-bound toxins are cleared by interaction with an albumin-impregnated dialysis membrane (Stange *et al.*, 1993). This device has been shown to be effective for the reduction of plasma bile acids, bilirubin, and other albumin-bound toxins; results demonstrate that its use can provide clinical benefit (Sen and Williams, 2003).

In order to provide the full array of synthetic, metabolic, and detoxification functions important for the effective treatment of liver failure, which is lacking in these non-biological schemes, biological approaches, including cross-hemodialysis, whole-liver perfusion, and liver-slice perfusion, have been explored (Yarmush *et al.*, 1992). While these methods have demonstrated some benefit, they are difficult to implement on a large scale clinically. Consequently, substantial efforts have been made toward the development of extracorporeal bioartificial liver (BAL) devices containing hepatic cells that would exhibit a myriad of critical liver functions and could be employed in a clinical setting. BAL devices that have been proposed and studied can be broadly categorized into four main types, summarized in Fig. 48.7: hollow-fiber devices, flat-plate and monolayer systems, perfusion bed or porous matrix devices, and suspension reactors, with each of these general designs exhibiting innate advantages and disadvantages. Overall, an effective BAL device would satisfy several important criteria, including maintenance of cell viability and hepatic functions, efficient bidirectional mass transfer, and scalability to therapeutic levels. We next discuss these issues and review recent advances in BAL device design.

## Cell Viability and Function

The components and characteristics of primary hepatocyte culture models have underscored the importance of microenvironmental signals and cellular configuration in the regulation of hepatocyte processes. Thus, the development of effective BAL systems is predicated on the appropriate incorporation of environmental and organizational cues that would enable stability of the hepatocellular component. Hollow-fiber devices, the most common BAL design, contain hepatic cells within cartridge units similar to those utilized in hemodialysis systems (J. K. Park and Lee, 2005). The hollow-fiber membranes serve as a scaffold for cell attachment and compartmentalization, although adequate nutrient transport and proper environmental stimuli are potentially limited in these configurations. Multiple modifications aimed at providing critical cellular cues have been explored. For instance, exposure of hepatocytes to plasma or blood of a sick patient may necessitate distinct alterations in hepatocyte culture conditions. Specifically, preconditioning with physiologic levels of insulin, lower than in normal culture medium, has been shown to prevent fat accumulation in hepatocytes on exposure to plasma (Chan et al., 2003). Additionally, the supplementation of plasma with amino acids has been demonstrated to increase albumin and urea synthesis (Washizu et al., 2000). An important role has also been suggested for homotypic cell–cell interactions. Single-cell suspensions of hepatocytes have been used in some devices because of their advantageous transport characteristics, although these systems quickly lose metabolic activity (Margulis et al., 1989). Due to the enhanced function of hepatocyte spheroids relative to dispersed cells, many device configurations contain either attached or encapsulated hepatocyte aggregates (J. K. Park and Lee, 2005). For example, the original HepatAssist system, previously developed by Circe Biomedical, was a hollow-fiber device containing microcarrier-attached porcine hepatocyte aggregates within the extracapillary space (Rozga et al., 1993). Collagen gel entrapment of hepatocyte aggregates has also been added to some hollow-fiber designs to improve function, and many perfusion-bed systems integrate hepatocyte aggregates within a polymeric network of pores or capillaries (van de Kerkhove et al., 2004). Furthermore, encapsulation of preformed hepatocyte spheroids within calcium-alginate beads has been explored as a means to promote hepatocyte stability while simultaneously providing an immunoisolation barrier (J. K. Park and Lee, 2005).

The inclusion of other stable in vitro model characteristics has similarly been investigated. For instance, a fundamental component of the modular extracorporeal liver support (MELS) system, developed by Gerlach and coworkers, is the spontaneous aggregation of hepatocytes in coculture with nonparenchymal cells, resulting in the formation of tissuelike organoid structures (Gerlach, 1996). Bile canaliculi and bile-duct structures as well as matrix deposition have been observed in this system, which utilized both porcine cells and primary human hepatocytes isolated from livers unsuitable for transplant (Sauer et al., 2002). Platforms based on collagen gel "sandwich culture," a stable hepatocyte-only culture configuration, have also been developed (De Bartolo et al., 2000).

## Mass Transfer

Bidirectional mass transfer is another primary consideration in the design of BAL systems and is required to provide vital nutrients to the incorporated cells and simultaneously allow export of therapeutic cellular products. Device configuration determines both the convective and the diffusive properties of the system, thereby dictating the exchange of soluble components. In particular, diffusion resistance is often a major constraint in BAL devices. Factors commonly limiting diffusive transport are membrane structures, collagen gels, and nonviable cells. For example, semipermeable membranes are often utilized in BAL devices in order to enable selectivity in the size of exchanged factors. Inherent to most hollow-fiber devices, but also utilized in some flat-plate and perfusion-bed systems, such membranes with a designated molecular weight cutoff act to prevent the transport of immunologic components and larger xenogeneic substances while maintaining transport of carrier proteins such as albumin. Although clearly not ideal, in order to maximize mass transfer, immunologic barriers have been eliminated from some device designs (van de Kerkhove et al., 2002), with the assumption that the short duration of contact with xenogeneic cells will result in minimal immunological complications. Perfusion-bed systems allow for enhanced mass transfer due to the direct contact with the perfusing media, although fluid flow distribution is highly contingent on the type of packing material. Encapsulation of dispersed or aggregated cells represents another strategy for immunoisolation but can increase diffusion resistance (Dixit and Gitnick, 1998; Yanagi et al., 1989). The development of novel methods for the microencapsulation of hepatocytes is an active research area and is discussed in more detail in the implantable applications section of this chapter.

Oxygen tension is an important mediator of hepatocyte function (Bhatia et al., 1996; Holzer and Maier, 1987; Nauck et al., 1981); thus, the improved regulation of oxygen delivery is a major goal of many recently developed BAL platforms. Strategies have included the incorporation of additional fiber components that carry oxygen directly into the device (Gerlach et al., 1994; Wolfe et al., 2002), and this approach has been employed in systems shown to be effective for the maintenance of certain hepatocyte functions (Gerlach et al., 1994). In contrast with other designs, flat-plate geometries can be perfused in a relatively uniform manner, although this configuration may result in the exposure of cells to shear stress, causing deleterious effects on cellular function (De Bartolo *et al.*, 2000; Kan *et al.*, 1998; Taguchi *et al.*, 1996; Tilles *et al.*, 2001). Several approaches for minimizing shear-stress exposure in the flat-plate geometry have been explored, including the fabrication of grooved substrates for shear protection as well as the integration of adjacent channels separated by a gas-permeable membrane as a means to decouple oxygen exchange and volumetric flow rate (J. Park *et al.*, 2005; Roy *et al.*, 2001; Tilles *et al.*, 2001). Notably, as discussed earlier, oxygen has been implicated in the heterogeneous distribution of hepatocyte functions along the liver sinusoid, and this distribution can be recapitulated *in vitro* within a bioreactor system (Allen and Bhatia, 2003). The eventual incorporation of oxygen gradients, as well as gradients of other, diverse stimuli, such as hormones and growth factors, into BAL designs could provide a means to simulate more closely the range of hepatocyte functions exhibited *in vivo* and to further enhance the effectiveness of BAL devices.

## Scale-Up

The successful clinical implementation of any BAL device is dependent on the ability to scale the device to a level that provides effective therapy. Hepatocyte transplantation experiments in humans as well as rat models have demonstrated some improvement in various blood parameters following the transplantation of cell numbers representing on the order of 1–10% of total liver mass (Bilir *et al.*, 2000; Eguchi *et al.*, 1997; Gupta and Chowdhury, 2002; Strom *et al.*, 1999). However, engraftment efficiency and potential indirect influences on the regeneration of the native liver are important variables in the transplantation setting and can make it difficult to determine the direct effect and functional capacity of transplanted cells. BAL devices that have been tested clinically have used approximately 0.5 $\times 10^9$ to $1 \times 10^{11}$ porcine hepatocytes or $4 \times 10^{10}$ C3A hepatoblastoma cells (van de Kerkhove *et al.*, 2004). The target capacity of most current BAL designs is approximately $1 \times 10^{10}$ hepatocytes, representing roughly 10% of total liver weight. Increasing cartridge size and the use of multiple cartridges have been utilized to scale-up hollow-fiber-based systems. For example, the recently developed HepatAssist-2 (http://www.arbios.com) system can contain $1.5 \times 10^{10}$ hepatocytes, three times the capacity of the original version. Perfusion systems or devices utilizing encapsulated cells can be scaled fairly easily to the desired size; however, such expanded configurations normally present a large priming or dead volume. Stacked-plate designs have been suggested as a means to scale-up flat-plate systems, although these modifications may introduce channeling effects and heterogeneous flow distribution (Allen *et al.*, 2001). Since hepatocytes exhibit substantial metabolic activity, oxygen delivery is a critical parameter in the scaling up of BAL devices. Accordingly, in addition to improvements in gas exchange, methods to prevent hepatocyte apoptosis resulting from hypoxic injury have also been investigated (Nyberg *et al.*, 2000).

Overall, the development of BAL devices exhibiting therapeutic levels of function is a major challenge, and developing modifications aimed at further improving the capacity and efficiency of these systems is a central goal in the field.

## Regulation and Safety

Similar to other tissue-engineering-based therapies, the regulation of BAL devices is complex, due to the hybrid nature of these systems. Currently, BAL devices are being regulated as drugs through the Center for Biologics and Evaluation Research of the Food and Drug Administration. The primary safety concerns for BAL systems are similar to those for other cell-based therapies, such as escape of tumorigenic cells, immune reactions to foreign antigens, and xenozoonosis. Other potential complications generally associated with extracorporeal blood treatment include hemodynamic, metabolic, and hypothermia-related abnormalities as well as problems linked with catheterization and anticoagulation (Sen and Williams, 2003). In order explicitly to prevent escape of tumorigenic cells, such as the C3A cell line used in the ELAD system, downstream filters have been added to BAL designs (Millis *et al.*, 2002), an approach that is generally acknowledged as an adequate precautionary measure.

With regard to the utilization of porcine hepatocytes, there is some evidence for the presence of antibodies directed against porcine antigens in the serum of patients treated with BAL devices (Baquerizo *et al.*, 1997). However, the clinical significance of these findings remains unclear, since high titers are not observed until one to three weeks, for IgM and IgG isotypes, respectively. These results suggest that immune rejection may not be a significant problem in the context of BAL therapy, except in cases of repetitive treatments. Still, appropriate modifications through cell sourcing or device design aimed at limiting immunologic complications would likely be important in the expansion of BAL treatment options to chronic-liver-disease patients and patients with repetitive episodes of acute decompensation. As mentioned earlier, exposure to porcine cells can also represent a risk of xenozoonotic transmission of PERV, ubiquitous in the genome of bred pigs. PERV has been shown to infect human cell lines *in vitro* (Patience *et al.*, 1997), although studies examining transmission to BAL-treated patients have not demonstrated any evidence of infection (Pitkin and Mullon, 1999). Although specific transmission of PERV may not occur during the course of BAL therapy, the use of xenogeneic cells in BAL devices will always incur a note of caution.

## Ongoing Clinical Trials

A number of BAL devices have been tested clinically. The characteristics of these systems are provided in Table 48.2. Important practical issues include the use of whole blood versus plasma, the type of anticoagulation regiment, and the utilization of fresh or cryopreserved cells. The use of whole blood exhibits the advantage of including oxygen-

**Table 48.2.  Characteristics of eight bioartificial liver systems**

| BAL system | Configuration | Cell source/amount | Trial phase | Comments |
|---|---|---|---|---|
| HepatAssist (Circe Biomedical, Lexington, MA) | Hollow fiber, polysulphone; microcarrier attached | Cryopreserved porcine $(5-7 \times 10^9)$ | II/III | Plasma, citrate anticoagulation, 0.15–0.20-μm pore size, 6 h/session for 1–5 days |
| ELAD (Vital Therapies, La Jolla, CA) | Hollow fiber, cellulose acetate; large aggregates | C3A human cell line (200–400 g) | II | Plasma, heparin anticoagulation, 120-kDa cutoff, continuous up to 107 h |
| MELS (Charité Virchow, Berlin, Germany) | Hollow fiber, interwoven, multicompartment; tissue organoids | Freshly isolated porcine or human (250–500 g) | I/II | Plasma, heparin anticoagulation, 400 kDa cutoff, continuous up to 3 d |
| BLSS (Excorp Medical, Oakdale, MN) | Hollow fiber, cellulose acetate; collagen gel entrapped | Freshly isolated porcine (70–120 g) | I/II | Whole blood, heparin anticoagulation, 100-kDa cutoff, 12 h/session for up to 2 sessions |
| RFB-BAL (Univ. of Ferrara, Italy) | Radial flow bioreactor; aggregates | Freshly isolated porcine (200–230 g) | I/II | Plasma, heparin/citrate anticoagulation, 1-μm polyester screen, 6–24-h treatments |
| AMC-BAL (Univ. of Amsterdam, the Netherlands) | Nonwoven polyester matrix, spirally wound; aggregates | Freshly isolated porcine $(10 \times 10^9)$ | I | Plasma, heparin anticoagulation, direct cell–plasma contact, up to 18 h/session for up to 2 sessions |
| LiverX-2000 (Algenix Inc., Minneapolis, MN) | Hollow fiber; collagen entrapped | Freshly isolated porcine (40–80 g) | I | Whole blood, heparin anticoagulation |
| HBAL (Nanjing Univ., Nanjing, China) | Hollow fiber; polysulfone; adsorption column | Freshly isolated porcine $(10 \times 10^9)$ | I | Plasma, 100-kDa cutoff, one to two 6-h treatments |

containing erythrocytes; however, undesirable leukocyte activation and cell damage may arise. In contrast, perfusion of plasma prevents hematopoietic cell injury, but the solubility of oxygen in plasma devoid of oxygen carriers is quite low. Furthermore, heparin coagulation is normally used in BAL systems, although deleterious effects of heparin exposure on hepatocyte morphology and function have been suggested (Stefanovich et al., 1996). Regarding cell source, freshly isolated hepatocytes represent the ideal condition, due to the documented decrease in hepatocyte function following cryopreservation (Chesne et al., 1993; De Loecker et al., 1990). Yet cryopreservation enables flexibility in the timing of therapies; thus, methods to enhance cryopreserved hepatocyte function are actively being explored (Sosef et al., 2005).

The design of clinical trials for BAL devices poses a significant challenge. In particular, liver failure progression is highly variable and etiology dependent. Additionally, one of the major manifestations of liver failure, hepatic encephalopathy, is difficult to quantify clinically. As a result, patients in clinical trials must be randomized while still controlling for the stage at which support was initiated as well as etiology. Similarly, the determination of the relevant control therapy can be difficult. Ideally, in order to minimize nonspecific effects of extracorporeal treatment, a nonbiological control such as veno-venous dialysis would be utilized. Another challenge is the choice of clinical endpoint. Most clinical trials utilize endpoints of 30-day survival and 30-day transplant-free survival; however, trials can often be confounded by the fact that acute-liver-failure patients receive transplants variably, depending on organ availability and the eligibility criteria in place at a given center. Furthermore, interpretation of the specific role of incorporated live, functional hepatocytes can be complicated by the presence of nonbiological adjuncts, such as charcoal perfusion, in some designs. A direct comparison of the effect of nonbiological systems alone, dead or nonhepatocyte cells, and live hepatocytes would provide substantial insight concerning the effectiveness of the cellular components, particularly given that dead hepatocytes and nonhepatocyte cells have been shown to offer some survival benefit in various animal models of acute liver failure (Makowka et al., 1980). Notably, the ability to assess more accurately the viability and function of cells during BAL treatment would be a major advance. Such information would be crucial in determining treatment time and the potential requirement for device replacement, both important considerations due to the instability of hepatocellular function in many contexts and the demonstrated detrimental effect of plasma from liver-failure patients on cultured hepatocytes (Sakai et al., 1996). Finally, even if clinical trials of current BAL devices do not prove efficacy, information obtained from these studies coupled with improvements in cell sourcing and functional stabilization will represent the foundation for the next generation of devices.

## VI. CELL TRANSPLANTATION

In addition to temporary extracorporeal support, the development of cell-based therapies for liver treatment aimed at the eventual replacement of damaged or diseased tissue is an active area of investigation. One potential cell-based approach is the transplantation of isolated mature hepatocytes. In experiments utilizing rodent models, transplanted hepatocytes were demonstrated to exhibit substantial proliferative capacity and the ability to replace diseased tissue under some limited conditions (Overturf et al., 1997; Rhim et al., 1994; Sokhi et al., 2000). The in vivo proliferation of transplanted hepatocytes is highly dependent on the presence of an adequate regenerative environment. In animal models, discussed in more detail later in this chapter, regenerative stimulation is provided by transgenic injury, partial hepatectomy, portocaval shunting, or the administration of hepatotoxic agents prior to cell transplantation. However, these approaches would be difficult to adapt to a clinical setting. For cell delivery, several distinct approaches have been explored, including intrasplenic or intraperitoneal injection and direct injection into peripheral veins, the portal vein, or the splenic artery. Major limitations of isolated cell transplantation include the inefficient engraftment and limited survival of transplanted hepatocytes, which has been collectively reported at only 20–30% of injected cells (Gupta et al., 1999). Several studies have demonstrated methods to improve engraftment and enhance the selective proliferation of transplanted hepatocytes (Guha et al., 1999; Laconi et al., 1998; Mignon et al., 1998), although, analogous to regeneration models, the clinical utility of these approaches remains to be determined. The time for engraftment and proliferation of transplanted hepatocytes also can create a substantial lag time [48 h in one study (Bilir et al., 2000)] before clinical benefit is observed, which could restrict the utility of cell transplantation for certain clinical conditions, such as fulminant hepatic failure. Consequently, at least in the immediate future, hepatocyte transplantation may provide the maximum clinical benefit for patients with hereditary metabolic disorders, which can be effectively treated with relatively little liver mass (2%–5%).

Similar to other cell-based approaches, the feasibility of hepatocyte transplantation is constrained by the availability of allogeneic human hepatocytes. Only a limited supply of human hepatocytes is currently available from collagenase perfusion of organs regarded as inappropriate for transplantation. Accordingly, many other cell source options are under investigation, as discussed earlier. For example, immortalization has been explored as a means to promote hepatocyte proliferation and survival, although tumor development is a major concern with this approach. Numerous stem cell or progenitor cell populations are also being examined as an alternative to mature hepatocytes in the cell transplantation context. Significant efforts have been focused on prospectively identifying intrahepatic stem cells that participate in liver regeneration, specifically, the possible role of bone marrow–derived cells. Although it has become clear that bone marrow–derived hepatocytes are rare under most circumstances and likely arise through cell fusion (Thorgeirsson and Grisham, 2006), if sufficiently optimized, the potential utility of bone marrow cells as a gene delivery vehicle for liver repair remains an intriguing option. Overall, elucidation of the mechanisms regulating the proliferation and hepatic differentiation of various stem and progenitor cell populations could lay the groundwork for the development of robust cell-based liver transplantation therapies.

# VII. THREE-DIMENSIONAL HEPATOCELLULAR SYSTEMS: DEVELOPMENT OF IMPLANTABLE THERAPEUTIC CONSTRUCTS

The development of implantable cellular platforms is another promising cell-based therapy for liver disease. Similar in principle to cell transplantation, in that hepatocytes are transplanted to perform liver functions, this approach is characterized by the immobilization or encapsulation of hepatocytes, using biomaterials in order to improve cellular survival and function. Notably, implantation of preformed hepatocellular constructs could potentially represent an approach to avoid many of the limitations inherent in cell transplantation strategies, namely inefficient engraftment, the required repopulation advantage of donor cells, and the resultant lag phase before clinical benefit. Implantable systems can generally be separated into two categories: (1) platforms based on acellular scaffolds, which serve as a substrate for cellular attachment, and (2) platforms in which hepatic cells are fully encapsulated within the biomaterial scaffold.

## Acellular Scaffolds for Cellular Attachment

Hepatocytes are known to be anchorage dependent, and initial experiments examining implantable approaches were formulated based on the hypothesis that providing a substrate for cellular attachment would enhance function. For example, hepatocytes were attached to collagen-coated dextran microcarriers and transplanted intraperitoneally into animals exhibiting two different model genetic disorders (Demetriou et al., 1986). Microcarriers were found to serve as an adequate platform for cell attachment and enhanced survival and function of the transplanted hepatocytes in both model systems. As a result of these studies, microcarriers were utilized to retain and promote the function of porcine hepatocytes in the HepatAssist BAL device described earlier. Various other microcarrier chemistries have been explored for hepatocyte culture, including cellulose (Kasai et al., 1992; Kino et al., 1998), gelatin (Tao et al., 2003), and gelatin/chitosan composite (Li et al., 2004). The potential effects of microcarrier architecture, such as a

**Table 48.3. Scaffolds utilized for hepatocellular constructs**

| Category | Compositions |
|---|---|
| Natural | Collagen, chitosan, collagen/chitosan composites, alginate, alginate composites, biomatrix, peptides, hyaluronic acid, fibrin, gelatin |
| Synthetic | PLLA, PLGA, PLLA/PLGA composite, PGA, PEG, PCL, PET, PVA |

*Abbreviations*: PLLA, poly(L-lactic acid); PLGA, poly(DL-lactide-*co*-glycolide); PGA, polyglycolic acid; PEG, poly(ethylene glycol); PCL, polycaprolactone; PET, poly(ethylene terephtalate); PVA, polyvinyl alcohol.

porous surface topography, on hepatocyte attachment and aggregation have also been investigated (Tao et al., 2003). Such systems share homology with acellular porous scaffolds, utilized in many fields of tissue engineering.

A central tenet of many implantable tissue-engineering approaches is the use of porous scaffolds that provide mechanical support, often in conjunction with cues for growth and morphogenesis. Collectively, the porosity and biodegradability of these constructs enables the integration of the implant with host tissue, particularly the host blood vessels. Proposed hepatocellular constructs have utilized scaffolds of diverse natural and synthetic compositions (Table 48.3). Collagen or various alginate and chitosan composites are the most commonly used natural/biologically derived scaffold materials for hepatocyte culture. Collagen sponge structures have been proposed as platforms for tissue-engineered constructs in a wide array of cellular systems as well as substrates for promoting wound repair. In a study examining hepatocyte interaction with collagen scaffolds exhibiting a range of pore sizes (10–82 µm), pore size was found to be an important factor regulating cell spreading and cell–cell interactions, both of which can influence hepatocyte functions (Ranucci et al., 2000). In addition, alginate scaffolds with pore sizes of approximately 100 µm have been shown to be well suited for encouraging spheroidal aggregation of hepatocytes, due to the weakly adhesive nature of the material (Glicklis et al., 2000). Spheroid formation in this platform acts to promote hepatocyte stabilization, an effect that was described in detail earlier. Adaptations on the pure alginate formulation, such as an alginate/galactosylated chitosan/heparin composite system (Seo et al., 2006), have been reported to enhance aggregation and cell viability further (Chung et al., 2002; Haque et al., 2005; Maguire et al., 2006; Seo et al., 2006). Although natural scaffold materials exhibit robust biocompatibility and some are capable of mediating substantial cell attachment (e.g., collagen), biodegradable synthetic polymers that offer flexibility in structure, mechanical properties, and degradation kinetics have also been investigated for use in hepatic-tissue-engineering applications.

Polyesters such as poly(L-lactic acid) (PLLA) and poly(DL-lactide-co-glycolide) (PLGA) are the most common synthetic polymers utilized in the generation of porous tissue-engineering constructs, due to their biocompatibility, biodegradability, and highly versatile properties. These systems have been shown to be supportive for hepatocyte culture and have been used as scaffolds for hepatocyte transplantation. For example, a composite PLLA/PLGA scaffold was demonstrated to support long-term engraftment of hepatocytes on mesenteric transplantation in a rodent injury model (Mooney et al., 1997). Of note, PLLA and PLGA exhibit distinct hydrolysis kinetics; thus, the adjustment of the relative contribution of these components has been illustrated as a means to regulate degradation time. Furthermore, modifications of PLGA scaffolds have been shown to improve functionality. Specifically, infiltration of hydrophilic polyvinyl alcohol (PVA) into PLGA scaffolds was noted to enhance hepatocyte seeding (Mooney et al., 1995), and alkali hydrolysis and extracellular matrix coating of PLGA constructs can improve hepatocyte attachment (Fiegel et al., 2004; Hasirci et al., 2001; Nam et al., 1999). Other modifications of polymer scaffold chemistries, such as the addition of biologically active factors, are discussed in more detail later. In order to obtain a homogeneous cellular distribution, porous, acellular scaffolds are normally seeded utilizing gravity, centrifugal forces, or convective flow or through the recruitment of cells with chemokines or growth factors. However, incorporation of hepatocytes into scaffolds is hindered by their relatively immotile nature and their limited proliferation ex vivo. Also, while porous scaffold systems continue to be explored for use in liver constructs, many of these scaffold architectures are essentially two-dimensional surfaces, from the hepatocyte perspective, perhaps limiting their utility toward the full recapitulation of three-dimensional cues, which will likely be important for certain hepatocyte functions, such as the coordination of bile drainage.

## Hepatocellular Encapsulation Platforms

The development of highly functional implantable liver systems is contingent on the generation of fully populated, thick constructs that concurrently promote three-dimensional conformations, hepatocyte stability, and the transport of nutrients and large macromolecules. Consequently, as alternatives to macroporous scaffold systems, approaches aimed at the efficient and homogeneous encapsulation of hepatocytes within a fully three-dimensional structure have been explored. In particular, hydrogels that exhibit high water content and thus mechanical properties similar to those of tissues are widely utilized for various tissue-engineering applications, including hepatocellular platforms (Drury and Mooney, 2003). Similar to preformed acellular scaffolds, encapsulation schemes have employed various natural and synthetic chemistries (Table 48.3). Natural platforms have been based primarily on biologically derived extracellular matrix, such as collagen gel (Dixit et al., 1990), peptides (Semino et al., 2003), and alginate and alginate-based composites (Haque et al., 2005; Hirai et al., 1993; Maguire et al., 2006; Miura et al., 1986). In general, microencapsulation of hepatocytes in these types of systems has been shown to facilitate hepatocyte aggregation and to promote function. For instance, alginate-based encapsulation platforms have been demonstrated to support hepatocyte spheroid culture while simultaneously providing an immunoisolation barrier (Haque et al., 2005; Hirai et al., 1993; Miura et al., 1986), and consequently they have been proposed for use in both implantable constructs and BAL devices.

Synthetic hydrogels, such as poly(ethylene glycol) (PEG)-based systems, have become increasingly utilized in tissue-engineering applications, due to their hydrophilicity and resistance to protein adsorption, their biocompatibility, and the ability to customize these gels through the modification of chain length and the addition of bioactive elements (Peppas et al., 2000). Also, such synthetic hydrogels can be polymerized in the presence of cells, thereby generating three-dimensional constructs with a uniform cellular distribution throughout the network. PEG-based hydrogels have been used for the encapsulation of a diverse array of cell types. Initial studies examining PEG hydrogel encapsulation of hepatic cells utilized immortalized hepatocytes or hepatoblastoma cell lines (Itle et al., 2005; V. A. Liu and Bhatia, 2002), demonstrating the compatibility of these cell types with this system. However, encapsulation of primary hepatocytes may necessitate distinct hydrogel conditions. Accordingly, recent experiments have begun to explore the potential utility of PEG hydrogels as a three-dimensional platform for primary hepatocytes, including the important parameters influencing encapsulated hepatocyte function. For example, following encapsulation in PEG hydrogels, primary hepatocytes exhibit a rapid and substantial decrease in viability (manuscript in preparation). However, analogous to strictly two-dimensional coculture systems, coculture with 3T3 fibroblasts greatly improved the maintenance of hepatocyte survival and function. In addition, PEG chain length determines the mesh size and diffusive characteristics of the hydrogel network, and culture within 10% (w/v) 20-kDa PEG hydrogels was shown to facilitate the release of secreted albumin, a critical parameter for an implantable construct as well as the accurate assessment of encapsulated hepatocyte function. Incorporation of liver progenitor cells into PEG hydrogel constructs has similarly been investigated. Specifically, homotypic cell–cell interactions were recently demonstrated to be important for the maintenance of BMEL cell viability in PEG hydrogels, and the efficient differentiation of these cells toward the hepatocyte lineage was compatible with this platform (Fig. 48.2). Overall, synthetic hydrogel systems represent promising platforms for implantable constructs as well as models for the in vitro investigation of hepatocellular responses within a three-dimensional context.

## Modifications in Scaffold Chemistry

Although synthetic polymer scaffolds offer many advantages, these systems are often limited in their capacity to mediate aspects of hepatocyte function, such as cell adhesion. Yet the inert nature of synthetic systems could also be viewed as an advantage, by facilitating the controlled incorporation of biologically active elements aimed at regulating distinct aspects of cellular function. Multiple diverse approaches have been explored for modulating hepatocyte interactions with synthetic platforms. These include addition of extracellular matrix molecule coatings (Fiegel *et al.*, 2004; Hasirci *et al.*, 2001) and the incorporation of various sugar residues, such as lactose and heparin (Gutsche *et al.*, 1996), and galactose (Chua *et al.*, 2005; T. G. Park, 2002), all of which have been implicated in improving hepatocyte adhesion and function within polymer scaffolds. Similarly, addition of poly(*N-p*-vinylbenzyl-4-*O*-β-D-galactopyranosyl-D-glucoamide) (PVLA) to PLLA scaffolds has been shown to improve hepatocyte adhesion due to its integral hydrophilic oligosaccharide residue (Karamuk *et al.*, 1999; J. S. Lee *et al.*, 2005; Mayer *et al.*, 2000). Furthermore, incorporation of epidermal growth factor (EGF) into PET fabric scaffolds was suggested to improve rates of hepatocyte aggregation (Mayer *et al.*, 2000). Taken together, these examples highlight modifications of the polymer scaffold backbone through either nonspecific adsorbtion or chemical conjugation of biological factors that subsequently act to influence hepatocyte processes. As an alternative to the integration of entire biomolecules, polymer scaffolds can also be conjugated with bioactive peptide sequences.

The addition of adhesive peptides that interact with integrin receptors has been extensively utilized to promote cell attachment in polymer systems. In particular, inclusion of RGD peptides within biomaterial scaffolds has been shown appreciably to influence the adhesion and function of a diverse array of cell lines and primary cell populations. Regarding hepatic cells, grafting of RGD peptides to PLLA scaffolds has been shown to enhance hepatocyte attachment (Carlisle *et al.*, 2000). The presence of RGD peptides has also been suggested to affect hepatocyte functions within hydrogel networks. For instance, RGD conjugation significantly improved the long-term stability of primary hepatocyte function within PEG hydrogels (manuscript in preparation). The additional incorporation of adhesive peptides that bind other integrins may serve as a way to modulate and enhance hepatocyte function further within synthetic polymer substrates. Moreover, although not yet applied to hepatocellular systems, the integration of matrix metalloproteinase-sensitive peptide sequences into hydrogel networks as degradable linkages has been shown to enable cell-mediated remodeling of the gel (Lutolf *et al.*, 2003; Mann *et al.*, 2001; Seliktar *et al.*, 2004). The capacity to

modify biomaterial scaffold chemistry through the introduction of biologically active factors will likely enable the finely tuned regulation of cell function and interactions with host tissues important for implantable systems.

## Controlling Three-Dimensional Architecture and Cellular Organization

In addition to alterations in scaffold chemistry, another approach for improving the functionality of tissue engineered constructs is based on the premise of more closely mimicking *in vivo* microarchitecture. The 3D architecture of native tissues influences not only cellular function, but also mechanical properties and the interfacing of tissue units with the microcirculation. The ability to fabricate cellular scaffolds with a highly defined structure would facilitate the recapitulation of the appropriate micro-scale environment for cell viability and function as well as the integration of macro-scale properties dictating mechanical characteristics, nutrient delivery, and the coordination of multicellular processes. Recent application of CAD-based manufacturing technologies to biomaterial scaffold fabrication has enabled multiscale regulation of scaffold features.

CAD-based rapid prototyping strategies that have been applied to polymer scaffold fabrication exploit multiple modes of assembly, including fabrication using heat, light, adhesives, or molding; these techniques have been extensively reviewed elsewhere (Tsang and Bhatia, 2004). For example, three-dimensional printing with adhesives combined with particulate leaching has been utilized to generate porous PLGA scaffolds for hepatocyte attachment (S. S. Kim *et al.*, 1998b), and microstructured ceramic (Petronis *et al.*, 2001) and silicon scaffolds (Kaihara *et al.*, 2000; Ogawa *et al.*, 2004) have been proposed as platforms for hepatocyte culture. Furthermore, molding and microsyringe deposition have been demonstrated to be robust methods for fabricating specified three-dimensional PLGA structures toward integration into implantable systems (Vozzi *et al.*, 2003).

Microfabrication techniques have similarly been employed for the generation of patterned cellular hydrogel constructs. For instance, microfluidic molding has been used to form biological gels containing cells into various patterns (Tan and Desai, 2003). In addition, syringe deposition in conjunction with micropositioning was recently illustrated as a means to generate patterned gelatin hydrogels containing hepatocytes (Wang *et al.*, 2006). Patterning of synthetic hydrogel systems has also recently been explored. Specifically, the photopolymerization property of PEG hydrogels enables the adaptation of photolithographic techniques to generate patterned hydrogel networks. In this process, patterned masks printed on transparencies act to localize the UV exposure of the prepolymer solution and, thus, to dictate the structure of the resultant hydrogel. The major advantages of photolithography-based techniques for patterning of hydrogel structures are its simplicity and flexibility. Photopatterning has been employed to surface pattern

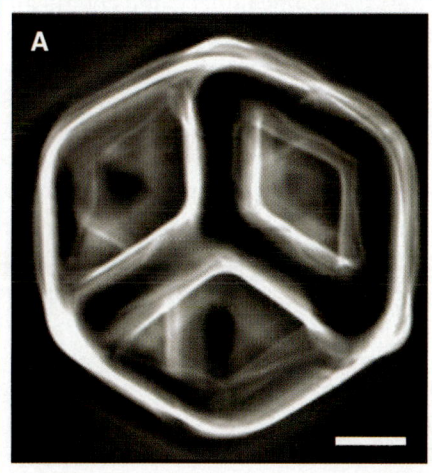
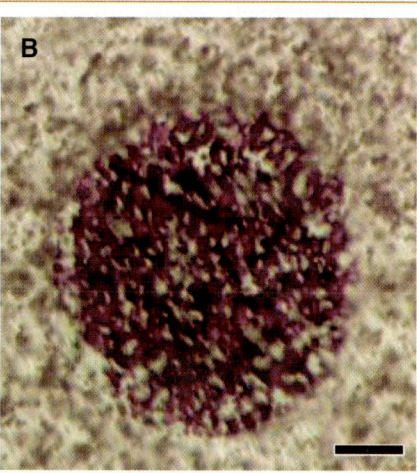
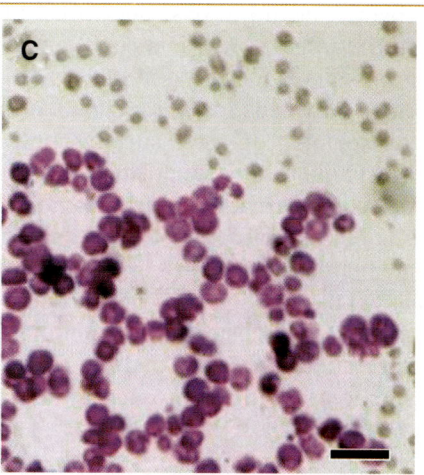

**FIG. 48.8.** Multiscale regulation of hydrogel structure and cellular organization. **(A)** Hydrogel photopatterning was utilized to generate a multilayer branched construct to facilitate adequate perfusion of encapsulated cells. Scale bar, 500 μm. **(B)** Encapsulated hepatocytes were initially patterned in approximately 1-mm-diameter islands, followed by the polymerization of a surrounding hydrogel network containing encapsulated fibroblasts. Hepatocyte-specific glycogen staining (purple-magenta) demonstrates the localization of these hepatocyte domains. Scale bar, 250 μm. **(C)** Hepatocytes organized via electropatterning were entrapped in islands by photopatterning and then surrounded by electropatterned encapsulated fibroblasts. Glycogen staining (purple-magenta) was performed on these dual patterned constructs. Scale bar, 50 μm.

biological factors (Hahn *et al.*, 2006), produce hydrogel structures with a range of sizes and shapes (Beebe *et al.*, 2000; Revzin *et al.*, 2001), as well as build multilayer cellular networks (V. A. Liu and Bhatia, 2002). Consequently, hydrogel photopatterning technology is ideally suited for the regulation of scaffold architecture at the multiple length scales required for implantable hepatocellular constructs. As a demonstration of these capabilities, photopatterning of PEG hydrogels was recently utilized to generate hepatocyte/fibroblast coculture hydrogels with a defined organization (Fig. 48.8) (manuscript in preparation). Also, the additional combination of photopatterning with dielectrophoresis-mediated cell patterning enabled the construction of hepatocellular hydrogel structures with an organization defined at the cellular scale (Fig. 48.8). Of note, perfusion of hepatocellular constructs has been implicated in the improvement of hepatocyte functions (S. S. Kim *et al.*, 1998a; T. G. Park, 2002). Photopatterning of PEG hydrogels containing hepatocyte/fibroblast cocultures into a three-dimensional branched network (Fig. 48.8) and culture under flow conditions enhanced encapsulated hepatocyte function (manuscript in preparation). Overall, the ability to dictate scaffold architecture coupled with other advances in scaffold material properties, chemistries, and the incorporation of bioactive elements will serve as the foundation for the future development of improved tissue-engineered liver constructs that can be customized spatially, physically, and chemically.

## Host Interactions

Further challenges in the design of therapeutic implantable liver devices are more specifically associated with the interactions with the host environment. These include issues related to vascularization, remodeling, the biliary system, and immunologic considerations. Notably, a significant challenge for the design of implantable liver constructs is the need to overcome transport limitations within the grafted construct due to the lack of functional vasculature. Within the normal liver environment, hepatocytes are supplied by an extensive sinusoidal vasculature with minimal extracellular matrix and a lining of fenestrated (sinusoidal) endothelial cells. Together these features allow for the efficient transport of nutrients to the highly metabolic hepatocytes. Strategies to incorporate vasculature into engineered constructs include the microfabrication of vascular units with accompanying surgical anastomosis during implantation (Griffith *et al.*, 1997; Kaihara *et al.*, 2000). For example, polymer molding using microetched silicon has been shown to generate extensive channel networks with capillary dimensions (Borenstein *et al.*, 2002). The incorporation of angiogenic factors within the implanted scaffolds has also been explored. Specifically, integration of cytokines important in angiogenesis, such as VEGF (Smith *et al.*, 2004). bFGF (H. Lee *et al.*, 2002), and VEGF in combination with PDGF (Richardson *et al.*, 2001), has been shown to promote the recruitment of host vasculature to implanted constructs. Furthermore, preimplantation of VEGF-releasing alginate scaffolds prior to hepatocyte seeding was demonstrated to enhance capillary density and to improve engraftment (Kedem *et al.*, 2005).

In addition to interactions with the vasculature, integration with other aspects of host tissue will constitute important future design parameters. For instance, incorporation of hydrolytic or protease-sensitive domains into hepatocellular hydrogel constructs could enable the degradation of these systems following implantation. Of note, liver regeneration proceeds in conjunction with a distinctive array of remodeling processes, such as protease expression and extracellular matrix deposition. Interfacing with these features could provide a mechanism for the efficient integration of implantable constructs. Similar to whole-liver or cell transplantation, the host immune response following transplant of tissue-engineered constructs is also a major consideration. Immunosuppressive treatments will likely play an important role in initial therapies, although stem cell–based approaches hold the promise of implantable systems with autologous cells. Furthermore, harnessing the liver's unique ability to induce antigen-specific tolerance (Racanelli and Rehermann, 2006) could potentially represent another means for improving the acceptance of engineered grafts. Finally, incorporation of excretory functions associated with the biliary system will ultimately be required in future designs. Toward this end, current studies are focused on the development of *in vitro* models that exhibit biliary morphogenesis and recapitulate appropriate polarization and bile canaliculi organization (Auth *et al.*, 2001; Ishida *et al.*, 2001; Sudo *et al.*, 2005) as well as platforms for the engineering of artificial bile duct structures (Miyazawa *et al.*, 2005).

## VIII. ANIMAL MODELS

An essential element in the development of clinically relevant cell-based liver therapies of all types is the use of animal models in order to examine therapeutic effects as well as safety considerations. A wide range of large-animal models have been used to examine the effectiveness of extracorporeal support systems for liver failure (van de Kerkhove *et al.*, 2004). These systems can generally be divided into four major categories: partial hepatectomy, total hepatectomy, toxic, and ischemia models. Several important criteria are utilized in the development of animal models of fulminant hepatic failure. These include reproducibility, reversibility, liver failure–induced cell death, and a sufficient time interval for diagnosis and therapeutic intervention (Rahman and Hodgson, 2000). A recently proposed rat model exhibiting these criteria utilized a combinatorial approach; specifically, endotoxin treatment was performed in addition to 70% liver resection in order to produce an inflammatory component (Skawran *et al.*, 2003).

Animal models have also been used extensively to investigate liver regeneration. These can broadly be separated into two categories: surgically induced and chemically induced (Palmes and Spiegel, 2004). Models in the chemical category include exposure to toxic doses of carbon tetrachloride or acetaminophen, both of which can induce localized centrilobular necrosis. Chemical-induced injury models are particularly useful for testing the efficacy of cell-based liver therapies, for these systems more closely mimic liver injuries commonly occurring in humans (i.e., drug toxicity). In the surgical category, the two-thirds partial hepatectomy model in the rat has been widely employed as a model in which the injury stimulus is well defined (Higgins and Anderson, 1931). Of note, parabiotic systems have demonstrated that after partial hepatectomy, factors regulating liver cell proliferation are present in the circulation (Fisher *et al.*, 1971; Moolten and Bucher, 1967). Consequently, although the hepatectomy model is less relevant clinically, it may serve as a well-controlled system to examine the importance of regenerative cues in the engraftment and proliferation of hepatic constructs implanted in extrahepatic sites. Extrahepatic implantation in subcutaneous or mesenteric sites is often utilized for tissue-engineered constructs due to the simplicity of access and improved imaging capabilities, although intrahepatic models that display relevant hepatotrophic stimuli continue to be explored. Overall, utilization of various animal models, both surgical and chemical, will likely be important in testing the effectiveness of cell-based liver therapies. Furthermore, knowledge of the mechanisms of liver injury and regeneration gained from these systems will provide an important foundation for the design of engineered liver tissue.

## IX. CONCLUSION

Although many challenges remain for the improvement of tissue-engineered liver therapies; substantial progress has been made toward a thorough understanding of the necessary components. The parallel development of highly functional *in vitro* systems as well as extracorporeal and implantable therapeutic devices is based on contributions from diverse disciplines, including regenerative medicine, developmental biology, transplant medicine, and bioengineering. In particular, novel technologies such as hydrogel chemistries, high-throughput platforms, and microfabrication techniques represent enabling tools for investigating the critical role of the microenvironment in liver function and, subsequently, the development of structurally complex and clinically effective engineered liver systems.

## X. REFERENCES

Allen, J. W., and Bhatia, S. N. (2002). Engineering liver therapies for the future. *Tissue Eng.* **8**, 725–737.

Allen, J. W., and Bhatia, S. N. (2003). Formation of steady-state oxygen gradients in vitro: application to liver zonation. *Biotechnol. Bioengi.* **82**, 253–262.

Allen, J. W., Hassanein, T., and Bhatia, S. N. (2001). Advances in bioartificial liver devices. *Hepatology* **34**, 447–455.

Allen, J. W., Khetani, S. R., and Bhatia, S. N. (2005). *In vitro* zonation and toxicity in a hepatocyte bioreactor. *Toxicol. Sci.* **84**, 110–119.

Alter, M. J., Kruszon-Moran, D., Nainan, O. V., McQuillan, G. M., Gao, F., Moyer, L. A., Kaslow, R. A., and Margolis, H. S. (1999). The prevalence of hepatitis C virus infection in the United States, 1988 through 1994. *N. Engl. J. Med.* **341**, 556–562.

Auth, M. K., Joplin, R. E., Okamoto, M., Ishida, Y., McMaster, P., Neuberger, J. M., Blaheta, R. A., Voit, T., and Strain, A. J. (2001). Morphogenesis of primary human biliary epithelial cells: induction in high-density culture or by coculture with autologous human hepatocytes. *Hepatology* **33**, 519–529.

Baquerizo, A., Mhoyan, A., Shirwan, H., Swensson, J., Busuttil, R. W., Demetriou, A. A., and Cramer, D. V. (1997). Xenoantibody response of patients with severe acute liver failure exposed to porcine antigens following treatment with a bioartificial liver. *Transplant. Proc.* **29**, 964–965.

Beebe, D. J., Moore, J. S., Bauer, J. M., Yu, Q., Liu, R. H., Devadoss, C., and Jo, B. H. (2000). Functional hydrogel structures for autonomous flow control inside microfluidic channels. *Nature* **404**, 588–590.

Bhatia, S., Toner, M., Foy, B., Rotem, A., O'Neil, K., Tompkins, R., and Yarmush, M. (1996). Zonal liver cell heterogeneity: effects of oxygen on metabolic functions of hepatocytes. *Cell. Eng.* **1**, 125–135.

Bhatia, S. N., Balis, U. J., Yarmush, M. L., and Toner, M. (1999). Effect of cell–cell interactions in preservation of cellular phenotype: cocultivation of hepatocytes and nonparenchymal cells. *FASEB J.* **13**, 1883–1900.

Bilir, B. M., Guinette, D., Karrer, F., Kumpe, D. A., Krysl, J., Stephens, J., McGavran, L., Ostrowska, A., and Durham, J. (2000). Hepatocyte transplantation in acute liver failure. *Liver Transpl.* **6**, 32–40.

Bissell, D. M., Arenson, D. M., Maher, J. J., and Roll, F. J. (1987). Support of cultured hepatocytes by a laminin-rich gel. Evidence for a functionally significant subendothelial matrix in normal rat liver. *J. Clin. Invest.* **79**, 801–812.

Blanc, P., Etienne, H., Daujat, M., Fabre, I., Zindy, F., Domergue, J., Astre, C., Saint Aubert, B., Michel, H., and Maurel, P. (1992). Mitotic responsiveness of cultured adult human hepatocytes to epidermal growth factor, transforming growth factor alpha, and human serum. *Gastroenterology* **102**, 1340–1350.

Borenstein, J. T., Terai, H., King, K. R., Weinberg, E. J., Kaazempur-Mofrad, M. R., and Vacanti, J. P. (2002). Microfabrication technology for vascularized tissue engineering. *Biomed. Microdevices* **4**, 167–175.

Brown, K. A. (2005). Liver transplantation. *Curr. Opin. Gastroenterol.* **21**, 331–336.

Brown, R. S. (2005). Hepatitis C and liver transplantation. *Nature* **436**, 973–978.

Carlisle, E. S., Mariappan, M. R., Nelson, K. D., Thomes, B. E., Timmons, R. B., Constantinescu, A., Eberhart, R. C., and Bankey, P. E. (2000). Enhancing hepatocyte adhesion by pulsed plasma deposition and polyethylene glycol coupling. *Tissue Eng.* **6**, 45–52.

Chan, C., Berthiaume, F., Lee, K., and Yarmush, M. L. (2003). Metabolic flux analysis of hepatocyte function in hormone- and amino acid–supplemented plasma. *Metab. Eng.* **5**, 1–15.

Chesne, C., Guyomard, C., Fautrel, A., Poullain, M. G., Fremond, B., De Jong, H., and Guillouzo, A. (1993). Viability and function in primary culture of adult hepatocytes from various animal species and human beings after cryopreservation. *Hepatology* **18**, 406–414.

Chia, S. M., Leong, K. W., Li, J., Xu, X., Zeng, K., Er, P. N., Gao, S., and Yu, H. (2000). Hepatocyte encapsulation for enhanced cellular functions. *Tissue Eng.* **6**, 481–495.

Chinzei, R., Tanaka, Y., Shimizu-Saito, K., Hara, Y., Kakinuma, S., Watanabe, M., Teramoto, K., Arii, S., Takase, K., Sato, C., Terada, N., and Teraoka, H. (2002). Embryoid-body cells derived from a mouse embryonic stem cell line show differentiation into functional hepatocytes. *Hepatology* **36**, 22–29.

Chua, K. N., Lim, W. S., Zhang, P., Lu, H., Wen, J., Ramakrishna, S., Leong, K. W., and Mao, H. Q. (2005). Stable immobilization of rat hepatocyte spheroids on galactosylated nanofiber scaffold. *Biomaterials* **26**, 2537–2547.

Chung, T. W., Yang, J., Akaike, T., Cho, K. Y., Nah, J. W., Kim, S. I., and Cho, C. S. (2002). Preparation of alginate/galactosylated chitosan scaffold for hepatocyte attachment. *Biomaterials* **23**, 2827–2834.

De Bartolo, L., Jarosch-Von Schweder, G., Haverich, A., and Bader, A. (2000). A novel full-scale flat-membrane bioreactor utilizing porcine hepatocytes: cell viability and tissue-specific functions. *Biotechnol. Prog.* **16**, 102–108.

De Loecker, R., Fuller, B. J., Gruwez, J., and De Loecker, W. (1990). The effects of cryopreservation on membrane integrity, membrane transport, and protein synthesis in rat hepatocytes. *Cryobiology* **27**, 143–152.

Demetriou, A. A., Whiting, J. F., Feldman, D., Levenson, S. M., Chowdhury, N. R., Moscioni, A. D., Kram, M., and Chowdhury, J. R. (1986). Replacement of liver function in rats by transplantation of microcarrier-attached hepatocytes. *Science* **233**, 1190–1192.

Dixit, V., and Gitnick, G. (1998). The bioartificial liver: state of the art. *Eur. J. Surg. Suppl.* **5**, 71–76.

Dixit, V., Darvasi, R., Arthur, M., Brezina, M., Lewin, K., and Gitnick, G. (1990). Restoration of liver function in Gunn rats without immunosuppression using transplanted microencapsulated hepatocytes. *Hepatology* **12**, 1342–1349.

Drury, J. L., and Mooney, D. J. (2003). Hydrogels for tissue engineering: scaffold design variables and applications. *Biomaterials* **24**, 4337–4351.

Dunn, J. C., Yarmush, M. L., Koebe, H. G., and Tompkins, R. G. (1989). Hepatocyte function and extracellular matrix geometry: long-term culture in a sandwich configuration [published erratum appears in *FASEB J.* 3(7), 1873]. *FASEB J.* **3**, 174–177.

Eguchi, S., Lilja, H., Hewitt, W. R., Middleton, Y., Demetriou, A. A., and Rozga, J. (1997). Loss and recovery of liver regeneration in rats with fulminant hepatic failure. *J. Surg. Res.* **72**, 112–122.

Ellis, A. J., Hughes, R. D., Wendon, J. A., Dunne, J., Langley, P. G., Kelly, J. H., Gislason, G. T., Sussman, N. L., and Williams, R. (1996). Pilot-controlled trial of the extracorporeal liver assist device in acute liver failure. *Hepatology* **24**, 1446–1451.

Ellis, A. J., Hughes, R. D., Nicholl, D., Langley, P. G., Wendon, J. A., O'Grady, J. G., and Williams, R. (1999). Temporary extracorporeal liver support for severe acute alcoholic hepatitis using the BioLogic-DT. *Int. J. Artif. Organs* **22**, 27–34.

Fiegel, H. C., Havers, J., Kneser, U., Smith, M. K., Moeller, T., Kluth, D., Mooney, D. J., Rogiers, X., and Kaufmann, P. M. (2004). Influence of flow conditions and matrix coatings on growth and differentiation of three-dimensionally cultured rat hepatocytes. *Tissue Eng.* **10**, 165–174.

Fisher, B., Szuch, P., Levine, M., and Fisher, E. R. (1971). A portal blood factor as the humoral agent in liver regeneration. *Science* **171**, 575–577.

Flaim, C. J., Chien, S., and Bhatia, S. N. (2005). An extracellular matrix microarray for probing cellular differentiation. *Nat. Methods* **2**, 119–125.

Folch, A., and Toner, M. (2000). Microengineering of cellular interactions. *Annu. Rev. Biomed. Eng.* **2**, 227–256.

Fukuda, J., Sakai, Y., and Nakazawa, K. (2006). Novel hepatocyte culture system developed using microfabrication and collagen/polyethylene glycol microcontact printing. *Biomaterials* **27**, 1061–1070.

Gebhardt, R., Hengstler, J. G., Muller, D., Glockner, R., Buenning, P., Laube, B., Schmelzer, E., Ullrich, M., Utesch, D., Hewitt, N., Ringel, M., Hilz, B. R., Bader, A., Langsch, A., Koose, T., Burger, H. J., Maas, J., and Oesch, F. (2003). New hepatocyte *in vitro* systems for drug metabolism: metabolic capacity and recommendations for application in basic research and drug development, standard operation procedures. *Drug. Metab. Rev.* **35**, 145–213.

Gerlach, J. C. (1996). Development of a hybrid liver support system: a review. *Int. J. Artif. Organs* **19**, 645–654.

Gerlach, J. C., Encke, J., Hole, O., Müller, C., Ryan, C. J., and Neuhaus, P. (1994). Bioreactor for a larger-scale hepatocyte *in vitro* perfusion. *Transplantation* **58**, 984–988.

Glicklis, R., Shapiro, L., Agbaria, R., Merchuk, J. C., and Cohen, S. (2000). Hepatocyte behavior within three-dimensional porous alginate scaffolds. *Biotechnol. Bioeng.* **67**, 344–353.

Griffith, L. G., Wu, B., Cima, M. J., Powers, M. J., Chaignaud, B., and Vacanti, J. P. (1997). *In vitro* organogenesis of liver tissue. *Ann. N.Y. Acad. Sci.* **831**, 382–397.

Guguen-Guillouzo, C., Clement, B., Baffet, G., Beaumont, C., Morel-Chany, E., Glaise, D., and Guillouzo, A. (1983). Maintenance and reversibility of active albumin secretion by adult rat hepatocytes co-cultured with another liver epithelial cell type. *Exp. Cell. Res.* **143**, 47–54.

Guha, C., Sharma, A., Gupta, S., Alfieri, A., Gorla, G. R., Gagandeep, S., Sokhi, R., Roy-Chowdhury, N., Tanaka, K. E., Vikram, B., and Roy-Chowdhury, J. (1999). Amelioration of radiation-induced liver damage in partially hepatectomized rats by hepatocyte transplantation. *Cancer Res.* **59**, 5871–5874.

Guillouzo, A. (1998). Liver cell models in *in vitro* toxicology. *Environ. Health Perspect.* **106**(Suppl. 2), 511–532.

Gupta, S., and Chowdhury, J. R. (2002). Therapeutic potential of hepatocyte transplantation. *Semin. Cell Dev. Biol.* **13**, 439–446.

Gupta, S., Gorla, G. R., and Irani, A. N. (1999). Hepatocyte transplantation: emerging insights into mechanisms of liver repopulation and their relevance to potential therapies. *J. Hepatol.* **30**, 162–170.

Gutsche, A. T., Lo, H., Zurlo, J., Yager, J., and Leong, K. W. (1996). Engineering of a sugar-derivatized porous network for hepatocyte culture. *Biomaterials* **17**, 387–393.

Hahn, M. S., Taite, L. J., Moon, J. J., Rowland, M. C., Ruffino, K. A., and West, J. L. (2006). Photolithographic patterning of polyethylene glycol hydrogels. *Biomaterials* **27**, 2519–2524.

Hamazaki, T., Iiboshi, Y., Oka, M., Papst, P. J., Meacham, A. M., Zon, L. I., and Terada, N. (2001). Hepatic maturation in differentiating embryonic stem cells *in vitro*. *FEBS Lett.* **497**, 15–19.

Haque, T., Chen, H., Ouyang, W., Martoni, C., Lawuyi, B., Urbanska, A. M., and Prakash, S. (2005). *In vitro* study of alginate-chitosan microcapsules: an alternative to liver cell transplants for the treatment of liver failure. *Biotechnol. Lett.* **27**, 317–322.

Hasirci, V., Berthiaume, F., Bondre, S. P., Gresser, J. D., Trantolo, D. J., Toner, M., and Wise, D. L. (2001). Expression of liver-specific functions by rat hepatocytes seeded in treated poly(lactic-*co*-glycolic) acid biodegradable foams. *Tissue Eng.* **7**, 385–394.

Heng, B. C., Yu, H., Yin, Y., Lim, S. G., and Cao, T. (2005). Factors influencing stem cell differentiation into the hepatic lineage *in vitro*. *J. Gastroenterol. Hepatol.* **20**, 975–987.

Higgins, G. M., and Anderson, R. M. (1931). Experimental pathology of the liver. I: restoration of the liver of the white rat following partial surgical removal. *Arch. Pathol.* **12**, 186–202.

Hirai, S., Kasai, S., and Mito, M. (1993). Encapsulated hepatocyte transplantation for the treatment of D-galactosamine-induced acute hepatic failure in rats. *Eur. Surg. Res.* **25**, 193–202.

Holzer, C., and Maier, P. (1987). Maintenance of periportal and pericentral oxygen tensions in primary rat hepatocyte cultures: influence on cellular DNA and protein content monitored by flow cytometry. *J. Cell. Physiol.* **133**, 297–304.

Houssaint, E. (1980). Differentiation of the mouse hepatic primordium. I. An analysis of tissue interactions in hepatocyte differentiation. *Cell Differ.* **9**, 269–279.

Hsiao, C. C., Wu, J. R., Wu, F. J., Ko, W. J., Remmel, R. P., and Hu, W. S. (1999). Receding cytochrome P450 activity in disassembling hepatocyte spheroids. *Tissue Eng.* **5**, 207–221.

Ishida, Y., Smith, S., Wallace, L., Sadamoto, T., Okamoto, M., Auth, M., Strazzabosco, M., Fabris, L., Medina, J., Prieto, J., Strain, A., Neuberger, J., and Joplin, R. (2001). Ductular morphogenesis and functional polarization of normal human biliary epithelial cells in three-dimensional culture. *J. Hepatol.* **35**, 2–9.

Ishizaka, S., Shiroi, A., Kanda, S., Yoshikawa, M., Tsujinoue, H., Kuriyama, S., Hasuma, T., Nakatani, K., and Takahashi, K. (2002). Development of hepatocytes from ES cells after transfection with the HNF-3beta gene. *FASEB J.* **16**, 1444–1446.

Isom, H. C., Secott, T., Georgoff, I., Woodworth, C., and Mummaw, J. (1985). Maintenance of differentiated rat hepatocytes in primary culture. *Proc. Nat. Acad. Sci. U.S.A.* **82**, 3252–3256.

Itle, L. J., Koh, W. G., and Pishko, M. V. (2005). Hepatocyte viability and protein expression within hydrogel microstructures. *Biotechnol. Prog.* **21**, 926–932.

Kaihara, S., Borenstein, J., Koka, R., Lalan, S., Ochoa, E. R., Ravens, M., Pien, H., Cunningham, B., and Vacanti, J. P. (2000). Silicon micromachining to tissue engineer branched vascular channels for liver fabrication. *Tissue Eng.* **6**, 105–117.

Kan, P., Miyoshi, H., Yanagi, K., and Ohshima, N. (1998). Effects of shear stress on metabolic function of the co-culture system of hepatocyte/nonparenchymal cells for a bioartificial liver. *Asaio J.* **44**, M441–M444.

Karamuk, E., Mayer, J., Wintermantel, E., and Akaike, T. (1999). Partially degradable film/fabric composites: textile scaffolds for liver cell culture. *Artif. Organs* **23**, 881–884.

Kasai, S., Sawa, M., Nishida, Y., Onodera, K., Hirai, S., Yamamoto, T., and Mito, M. (1992). Cellulose microcarrier for high-density culture of hepatocytes. *Transplant. Proc.* **24**, 2933–2934.

Kawamoto, T., Sueyoshi, T., Zelko, I., Moore, R., Washburn, K., and Negishi, M. (1999). Phenobarbital-responsive nuclear translocation of the receptor CAR in induction of the CYP2B gene. *Mol. Cell. Biol.* **19**, 6318–6322.

Kedem, A., Perets, A., Gamlieli-Bonshtein, I., Dvir-Ginzberg, M., Mizrahi, S., and Cohen, S. (2005). Vascular endothelial growth factor–releasing scaffolds enhance vascularization and engraftment of hepatocytes transplanted on liver lobes. *Tissue Eng.* **11**, 715–722.

Kelly, J. H., and Darlington, G. J. (1989). Modulation of the liver specific phenotype in the human hepatoblastoma line Hep G2. *In Vitro Cell. Dev. Biol.* **25**, 217–222.

Kim, S. S., Utsunomiya, H., Koski, J. A., Wu, B. M., Cima, M. J., Sohn, J., Mukai, K., Griffith, L. G., and Vacanti, J. P. (1998a). Survival and function of hepatocytes on a novel three-dimensional synthetic biodegradable polymer scaffold with an intrinsic network of channels. *Ann. Surg.* **228**, 8–13.

Kim, S. S., Utsunomiya, H., Koski, J. A., Wu, B. M., Cima, M. J., Sohn, J., Mukai, K., Griffith, L. G., and Vacanti, J. P. (1998b). Survival and function of hepatocytes on a novel three-dimensional synthetic biodegradable polymer scaffold with an intrinsic network of channels [see comments]. *Ann. Surg.* **228**, 8–13.

Kim, W. R., Brown, R. S., Jr., Terrault, N. A., and El-Serag, H. (2002). Burden of liver disease in the United States: summary of a workshop. *Hepatology* **36**, 227–242.

Kino, Y., Sawa, M., Kasai, S., and Mito, M. (1998). Multiporous cellulose microcarrier for the development of a hybrid artificial liver using isolated hepatocytes. *J. Surg. Res.* **79**, 71–76.

Knedlitschek, G., Schneider, F., Gottwald, E., Schaller, T., Eschbach, E., and Weibezahn, K. F. (1999). A tissue-like culture system using microstructures: influence of extracellular matrix material on cell adhesion and aggregation. *J. Biomech. Eng.* **121**, 35–39.

Kochanek, K. D., Murphy, S. L., Anderson, R. N., and Scott, C. (2004). Deaths: final data for 2002. *Natl. Vital Stat. Rep.* **53**, 1–115.

Laconi, E., Oren, R., Mukhopadhyay, D. K., Hurston, E., Laconi, S., Pani, P., Dabeva, M. D., and Shafritz, D. A. (1998). Long-term, near-total liver replacement by transplantation of isolated hepatocytes in rats treated with retrorsine. *Am. J. Pathol.* **153**, 319–329.

Landry, J., Bernier, D., Ouellet, C., Goyette, R., and Marceau, N. (1985). Spheroidal aggregate culture of rat liver cells: histotypic reorganization, biomatrix deposition, and maintenance of functional activities. *J. Cell Biol.* **101**, 914–923.

LeCluyse, E. L. (2001). Human hepatocyte culture systems for the *in vitro* evaluation of cytochrome P450 expression and regulation. *Eur. J. Pharm. Sci.* **13**, 343–368.

LeCluyse, E. L., Audus, K. L., and Hochman, J. H. (1994). Formation of extensive canalicular networks by rat hepatocytes cultured in collagen-sandwich configuration. *Am. J. Physiol.* **266**, C1764–C1774.

LeCluyse, E., Bullock, P. L., and Parkinson, A. (1996). Strategies for restoration and maintenance of normal hepatic structure and function in long-term cultures of rat hepatocytes. *Adv. Drug Delivery Rev.* **22**, 133–186.

Lee, H., Cusick, R. A., Browne, F., Ho Kim, T., Ma, P. X., Utsunomiya, H., Langer, R., and Vacanti, J. P. (2002). Local delivery of basic fibroblast growth factor increases both angiogenesis and engraftment of hepatocytes in tissue-engineered polymer devices. *Transplantation* **73**, 1589–1593.

Lee, J. S., Kim, S. H., Kim, Y. J., Akaike, T., and Kim, S. C. (2005). Hepatocyte adhesion on a poly[*N-p*-vinylbenzyl-4-*O*-beta-D-galacto-pyranosyl-D-glucoamide]-coated poly(L-lactic acid) surface. *Biomacromolecules* **6**, 1906–1911.

Lee, W. M. (2003). Acute liver failure in the United States. *Semin. Liver Dis.* **23**, 217–226.

Lemaigre, F., and Zaret, K. S. (2004). Liver development update: new embryo models, cell lineage control, and morphogenesis. *Curr. Opin. Genet. Dev.* **14**, 582–590.

Li, K., Wang, Y., Miao, Z., Xu, D., Tang, Y., and Feng, M. (2004). Chitosan/gelatin composite microcarrier for hepatocyte culture. *Biotechnol. Lett.* **26**, 879–883.

Lin, P., Chan, W. C., Badylak, S. F., and Bhatia, S. N. (2004). Assessing porcine liver–derived biomatrix for hepatic tissue engineering. *Tissue Eng.* **10**, 1046–1053.

Lindros, K. O. (1997). Zonation of cytochrome P450 expression, drug metabolism and toxicity in liver. *Gen. Pharmacol.* **28**, 191–196.

Liu, J., Pan, J., Naik, S., Santangini, H., Trenkler, D., Thompson, N., Rifai, A., Chowdhury, J. R., and Jauregui, H. O. (1999). Characterization and evaluation of detoxification functions of a nontumorigenic immortalized porcine hepatocyte cell line (HepLiu). *Cell Transpl.* **8**, 219–232.

Liu, V. A., and Bhatia, S. N. (2002). Three-dimensional photopatterning of hydrogels containing living cells. *Biomed. Microdevices* **4**, 257–266.

Lutolf, M. P., Lauer-Fields, J. L., Schmoekel, H. G., Metters, A. T., Weber, F. E., Fields, G. B., and Hubbell, J. A. (2003). Synthetic matrix metalloproteinase-sensitive hydrogels for the conduction of tissue regeneration: engineering cell-invasion characteristics. *Proc. Natl. Acad. Sci. U.S.A.* **100**, 5413–5418.

Maguire, T., Novik, E., Schloss, R., and Yarmush, M. (2006). Alginate-PLL microencapsulation: effect on the differentiation of embryonic stem cells into hepatocytes. *Biotechnol. Bioeng.* **93**, 581–591.

Makowka, L., Falk, R. E., Rotstein, L. E., Falk, J. A., Nossal, N., Langer, B., Blendis, L. M., and Phillips, M. J. (1980). Reversal of experimental acute hepatic failure in the rat. *J. Surg. Res.* **29**, 479–487.

Mann, B. K., Gobin, A. S., Tsai, A. T., Schmedlen, R. H., and West, J. L. (2001). Smooth muscle cell growth in photopolymerized hydrogels with cell adhesive and proteolytically degradable domains: synthetic ECM analogs for tissue engineering. *Biomaterials* **22**, 3045–3051.

Margulis, M. S., Erukhimov, E. A., Andreiman, L. A., and Viksna, L. M. (1989). Temporary organ substitution by hemoperfusion through suspension of active donor hepatocytes in a total complex of intensive therapy in patients with acute hepatic insufficiency. *Resuscitation* **18**, 85–94.

Matsumoto, K., Yoshitomi, H., Rossant, J., and Zaret, K. S. (2001). Liver organogenesis promoted by endothelial cells prior to vascular function. *Science* **294**, 559–563.

Mayer, J., Karamuk, E., Akaike, T., and Wintermantel, E. (2000). Matrices for tissue engineering — scaffold structure for a bioartificial liver support system. *J. Controlled Release* **64**, 81–90.

Michalopoulos, G. K., and DeFrances, M. C. (1997). Liver regeneration. *Science* **276**, 60–66.

Mignon, A., Guidotti, J. E., Mitchell, C., Fabre, M., Wernet, A., De La Coste, A., Soubrane, O., Gilgenkrantz, H., and Kahn, A. (1998). Selective repopulation of normal mouse liver by Fas/CD95-resistant hepatocytes. *Nat. Med.* **4**, 1185–1188.

Millis, J. M., Cronin, D. C., Johnson, R., Conjeevaram, H., Conlin, C., Trevino, S., and Maguire, P. (2002). Initial experience with the modified extracorporeal liver-assist device for patients with fulminant hepatic failure: system modifications and clinical impact. *Transplantation* **74**, 1735–1746.

Mitaka, T. (1998). The current status of primary hepatocyte culture. *Int. J. Exp. Pathol.* **79**, 393–409.

Mitaka, T., Sattler, C. A., Sattler, G. L., Sargent, L. M., and Pitot, H. C. (1991). Multiple cell cycles occur in rat hepatocytes cultured in the presence of nicotinamide and epidermal growth factor. *Hepatology* **13**, 21–30.

Miura, Y., Akimoto, T., Kanazawa, H., and Yagi, K. (1986). Synthesis and secretion of protein by hepatocytes entrapped within calcium alginate. *Artif. Organs* **10**, 460–465.

Miyazaki, M., Handa, Y., Oda, M., Yabe, T., Miyano, K., and Sato, J. (1985). Long-term survival of functional hepatocytes from adult rat in the presence of phenobarbital in primary culture. *Exp. Cell Res.* **159**, 176–190.

Miyazawa, M., Torii, T., Toshimitsu, Y., Okada, K., Koyama, I., and Ikada, Y. (2005). A tissue-engineered artificial bile duct grown to resemble the native bile duct. *Am. J. Transplant.* **5**, 1541–1547.

Moolten, F. L., and Bucher, N. L. (1967). Regeneration of rat liver: transfer of humoral agent by cross circulation. *Science* **158**, 272–274.

Mooney, D. J., Park, S., Kaufmann, P. M., Sano, K., McNamara, K., Vacanti, J. P., and Langer, R. (1995). Biodegradable sponges for hepatocyte transplantation. *J. Biomed. Mater. Res.* **29**, 959–965.

Mooney, D. J., Sano, K., Kaufmann, P. M., Majahod, K., Schloo, B., Vacanti, J. P., and Langer, R. (1997). Long-term engraftment of hepatocytes transplanted on biodegradable polymer sponges. *J. Biomed. Mater. Res.* **37**, 413–420.

Nam, Y. S., Yoon, J. J., Lee, J. G., and Park, T. G. (1999). Adhesion behaviors of hepatocytes cultured onto biodegradable polymer surface modified by alkali hydrolysis process. *J. Biomater. Sci. Polym. Ed.* **10**, 1145–1158.

Nauck, M., Wolfle, D., Katz, N., and Jungermann, K. (1981). Modulation of the glucagon-dependent induction of phosphoenolpyruvate carboxykinase and tyrosine aminotransferase by arterial and venous oxygen concentrations in hepatocyte cultures. *Eur. J. Biochem.* **119**, 657–661.

Nyberg, S. L., Hardin, J. A., Matos, L. E., Rivera, D. J., Misra, S. P., and Gores, G. J. (2000). Cytoprotective influence of ZVAD-fmk and glycine on gel-entrapped rat hepatocytes in a bioartificial liver. *Surgery* **127**, 447–455.

Ogawa, K., Ochoa, E. R., Borenstein, J., Tanaka, K., and Vacanti, J. P. (2004). The generation of functionally differentiated, three-dimensional hepatic tissue from two-dimensional sheets of progenitor small hepatocytes and nonparenchymal cells. *Transplantation* **77**, 1783–1789.

Olson, M., Mancini, M., Venkatachalam, M., and Roy, A. (1990). Hepatocyte cytodifferentiation and cell-to-cell communication. *In* "Cell Intercommunication" (W. De Mello, ed.), pp. 71–92. CRC Press, Boca Raton, FL.

Ostapowicz, G., Fontana, R. J., Schiodt, F. V., Larson, A., Davern, T. J., Han, S. H., McCashland, T. M., Shakil, A. O., Hay, J. E., Hynan, L., Crippin, J. S., Blei, A. T., Samuel, G., Reisch, J., and Lee, W. M. (2002). Results of a prospective study of acute liver failure at 17 tertiary care centers in the United States. *Ann. Intern. Med.* **137**, 947–954.

Overturf, K., al-Dhalimy, M., Ou, C. N., Finegold, M., and Grompe, M. (1997). Serial transplantation reveals the stem-cell-like regenerative potential of adult mouse hepatocytes. *Am. J. Pathol.* **151**, 1273–1280.

Palmes, D., and Spiegel, H. U. (2004). Animal models of liver regeneration. *Biomaterials* **25**, 1601–1611.

Park, J. K., and Lee, D. H. (2005). Bioartificial liver systems: current status and future perspective. *J. Biosci. Bioeng.* **99**, 311–319.

Park, J., Berthiaume, F., Toner, M., Yarmush, M. L., and Tilles, A. W. (2005). Microfabricated grooved substrates as platforms for bioartificial liver reactors. *Biotechnol. Bioeng.* **90**, 632–644.

Park, T. G. (2002). Perfusion culture of hepatocytes within galactose-derivatized biodegradable poly(lactide-*co*-glycolide) scaffolds prepared by gas foaming of effervescent salts. *J. Biomed. Mater. Res.* **59**, 127–135.

Patience, C., Takeuchi, Y., and Weiss, R. A. (1997). Infection of human cells by an endogenous retrovirus of pigs. *Nat. Med.* **3**, 282–286.

Peppas, N. A., Bures, P., Leobandung, W., and Ichikawa, H. (2000). Hydrogels in pharmaceutical formulations. *Eur. J. Pharm. Biopharm.* **50**, 27–46.

Petronis, S., Eckert, K. L., Gold, J., and Wintermantel, E. (2001). Microstructuring ceramic scaffolds for hepatocyte cell culture. *J. Mater. Sci. Mater. Med.* **12**, 523–528.

Pitkin, Z., and Mullon, C. (1999). Evidence of absence of porcine endogenous retrovirus (PERV) infection in patients treated with a bioartificial liver support system. *Artif. Organs* **23**, 829–833.

Powers, M. J., Domansky, K., Kaazempur-Mofrad, M. R., Kalezi, A., Capitano, A., Upadhyaya, A., Kurzawski, P., Wack, K. E., Stolz, D. B., Kamm, R., and Griffith, L. G. (2002). A microfabricated-array bioreactor for perfused 3D liver culture. *Biotechnol. Bioeng.* **78**, 257–269.

Racanelli, V., and Rehermann, B. (2006). The liver as an immunological organ. *Hepatology* **43**, S54–S62.

Rahman, T. M., and Hodgson, H. J. (2000). Animal models of acute hepatic failure. *Int. J. Exp. Pathol.* **81**, 145–157.

Ranucci, C. S., Kumar, A., Batra, S. P., and Moghe, P. V. (2000). Control of hepatocyte function on collagen foams: sizing matrix pores toward selective induction of 2D and 3D cellular morphogenesis. *Biomaterials* **21**, 783–793.

Reid, L. M., Fiorino, A. S., Sigal, S. H., Brill, S., and Holst, P. A. (1992). Extracellular matrix gradients in the space of Disse: relevance to liver biology [editorial]. *Hepatology* **15**, 1198–1203.

Revzin, A., Russell, R. J., Yadavalli, V. K., Koh, W. G., Deister, C., Hile, D. D., Mellott, M. B., and Pishko, M. V. (2001). Fabrication of poly(ethylene glycol) hydrogel microstructures using photolithography. *Langmuir* **17**, 5440–5447.

Rhim, J. A., Sandgren, E. P., Degen, J. L., Palmiter, R. D., and Brinster, R. L. (1994). Replacement of diseased mouse liver by hepatic cell transplantation. *Science* **263**, 1149–1152.

Richardson, T. P., Peters, M. C., Ennett, A. B., and Mooney, D. J. (2001). Polymeric system for dual growth factor delivery. *Nat. Biotechnol.* **19**, 1029–1034.

Richert, L., Binda, D., Hamilton, G., Viollon-Abadie, C., Alexandre, E., Bigot-Lasserre, D., Bars, R., Coassolo, P., and LeCluyse, E. (2002). Evaluation of the effect of culture configuration on morphology, survival time, antioxidant status and metabolic capacities of cultured rat hepatocytes. *Toxicol. In Vitro* **16**, 89–99.

Roy, P., Baskaran, H., Tilles, A. W., Yarmush, M. L., and Toner, M. (2001). Analysis of oxygen transport to hepatocytes in a flat-plate microchannel bioreactor. *Ann. Biomed. Eng.* **29**, 947–955.

Rozga, J., Holzman, M. D., Ro, M. S., Griffin, D. W., Neuzil, D. F., Giorgio, T., Moscioni, A. D., and Demetriou, A. A. (1993). Development of a hybrid bioartificial liver. *Ann. Surg.* **217**, 502–509; discussion 509–511.

Sakai, Y., Naruse, K., Nagashima, I., Muto, T., and Suzuki, M. (1996). *In vitro* function of porcine hepatocyte spheroids in 100% human plasma. *Cell Transplant.* **5**, S41–S43.

Sauer, I. M., Zeilinger, K., Obermayer, N., Pless, G., Grunwald, A., Pascher, A., Mieder, T., Roth, S., Goetz, M., Kardassis, D., Mas, A., Neuhaus, P., and Gerlach, J. C. (2002). Primary human liver cells as source for modular extracorporeal liver support — a preliminary report. *Int. J. Artif. Organs* **25**, 1001–1005.

Schwartz, R. E., Reyes, M., Koodie, L., Jiang, Y., Blackstad, M., Lund, T., Lenvik, T., Johnson, S., Hu, W. S., and Verfaillie, C. M. (2002). Multipotent adult progenitor cells from bone marrow differentiate into functional hepatocyte-like cells. *J. Clin. Invest.* **109**, 1291–1302.

Seliktar, D., Zisch, A. H., Lutolf, M. P., Wrana, J. L., and Hubbell, J. A. (2004). MMP-2 sensitive, VEGF-bearing bioactive hydrogels for promotion of vascular healing. *J. Biomed. Mater. Res.* **68A**, 704–716.

Sell, S. (2001). The role of progenitor cells in repair of liver injury and in liver transplantation. *Wound Repair Regen.* **9**, 467–482.

Semino, C. E., Merok, J. R., Crane, G. G., Panagiotakos, G., and Zhang, S. (2003). Functional differentiation of hepatocyte-like spheroid structures from putative liver progenitor cells in three-dimensional peptide scaffolds. *Differentiation* **71**, 262–270.

Sen, S., and Williams, R. (2003). New liver support devices in acute liver failure: a critical evaluation. *Semin. Liver Dis.* **23**, 283–294.

Seo, S. J., Choi, Y. J., Akaike, T., Higuchi, A., and Cho, C. S. (2006). Alginate/galactosylated chitosan/heparin scaffold as a new synthetic extracellular matrix for hepatocytes. *Tissue Eng.* **12**, 33–44.

Sivaraman, A., Leach, J. K., Townsend, S., Iida, T., Hogan, B. J., Stolz, D. B., Fry, R., Samson, L. D., Tannenbaum, S. R., and Griffith, L. G. (2005). A microscale *in vitro* physiological model of the liver: predictive screens for drug metabolism and enzyme induction. *Curr. Drug Metab.* **6**, 569–591.

Skawran, S., Palmes, D., Budny, T., Bahde, R., Stratmann, U., and Spiegel, H. U. (2003). Development and evaluation of an experimental model for investigating the pathogenesis and therapeutic strategies of acute liver failure. *Transplant. Proc.* **35**, 3142–3146.

Smith, M. K., Peters, M. C., Richardson, T. P., Garbern, J. C., and Mooney, D. J. (2004). Locally enhanced angiogenesis promotes transplanted cell survival. *Tissue Eng.* **10**, 63–71.

Sokhi, R. P., Rajvanshi, P., and Gupta, S. (2000). Transplanted reporter cells help in defining onset of hepatocyte proliferation during the life of F344 rats. *Am. J. Physiol. Gastrointest. Liver Physiol.* **279**, G631–G640.

Sosef, M. N., Baust, J. M., Sugimachi, K., Fowler, A., Tompkins, R. G., and Toner, M. (2005). Cryopreservation of isolated primary rat hepatocytes: enhanced survival and long-term hepatospecific function. *Ann. Surg.* **241**, 125–133.

Stange, J., Ramlow, W., Mitzner, S., Schmidt, R., and Klinkmann, H. (1993). Dialysis against a recycled albumin solution enables the removal of albumin-bound toxins. *Artif. Organs* **17**, 809–813.

Stefanovich, P., Matthew, H. W., Toner, M., Tompkins, R. G., and Yarmush, M. L. (1996). Extracorporeal plasma perfusion of cultured hepatocytes: effect of intermittent perfusion on hepatocyte function and morphology. *J. Surg. Res.* **66**, 57–63.

Strick-Marchand, H., and Weiss, M. C. (2002). Inducible differentiation and morphogenesis of bipotential liver cell lines from wild-type mouse embryos. *Hepatology* **36**, 794–804.

Strick-Marchand, H., Morosan, S., Charneau, P., Kremsdorf, D., and Weiss, M. C. (2004). Bipotential mouse embryonic liver stem cell lines contribute to liver regeneration and differentiate as bile ducts and hepatocytes. *Proc. Natl. Acad. Sci. U.S.A.* **101**, 8360–8365.

Strom, S. C., Chowdhury, J. R., and Fox, I. J. (1999). Hepatocyte transplantation for the treatment of human disease. *Semin. Liver Dis.* **19**, 39–48.

Sudo, R., Mitaka, T., Ikeda, M., and Tanishita, K. (2005). Reconstruction of 3D stacked-up structures by rat small hepatocytes on microporous membranes. *FASEB J.* **19**, 1695–1697.

Taguchi, K., Matsushita, M., Takahashi, M., and Uchino, J. (1996). Development of a bioartificial liver with sandwiched-cultured hepatocytes between two collagen gel layers. *Artif. Organs* **20**, 178–185.

Tan, W., and Desai, T. A. (2003). Microfluidic patterning of cells in extracellular matrix biopolymers: effects of channel size, cell type, and matrix composition on pattern integrity. *Tissue Eng.* **9**, 255–267.

Tao, X., Shaolin, L., and Yaoting, Y. (2003). Preparation and culture of hepatocyte on gelatin microcarriers. *J. Biomed. Mater. Res. A* **65**, 306–310.

Thorgeirsson, S. S., and Grisham, J. W. (2006). Hematopoietic cells as hepatocyte stem cells: a critical review of the evidence. *Hepatology* **43**, 2–8.

Tilles, A. W., Baskaran, H., Roy, P., Yarmush, M. L., and Toner, M. (2001). Effects of oxygenation and flow on the viability and function of rat hepatocytes cocultured in a microchannel flat-plate bioreactor. *Biotechnol. Bioeng.* **73**, 379–389.

Tsang, V. L., and Bhatia, S. N. (2004). Three-dimensional tissue fabrication. *Adv. Drug Deliv. Rev.* **56**, 1635–1647.

van de Kerkhove, M. P., Di Florio, E., Scuderi, V., Mancini, A., Belli, A., Bracco, A., Dauri, M., Tisone, G., Di Nicuolo, G., Amoroso, P., Spadari, A., Lombardi, G., Hoekstra, R., Calise, F., and Chamuleau, R. A. (2002). Phase I clinical trial with the AMC-bioartificial liver. *Int. J. Artif. Organs* **25**, 950–959.

van de Kerkhove, M. P., Hoekstra, R., Chamuleau, R. A., and van Gulik, T. M. (2004). Clinical application of bioartificial liver support systems. *Ann. Surg.* **240**, 216–230.

Vozzi, G., Flaim, C., Ahluwalia, A., and Bhatia, S. N. (2003). Fabrication of PLGA scaffolds using soft lithography and microsyringe deposition. *Biomaterials* **24**, 2533–2540.

Vukicevic, S., Kleinman, H. K., Luyten, F. P., Roberts, A. B., Roche, N. S., and Reddi, A. H. (1992). Identification of multiple active growth factors in basement membrane Matrigel suggests caution in interpretation of cellular activity related to extracellular matrix components. *Exp. Cell Res.* **202**, 1–8.

Wang, X., Yan, Y., Pan, Y., Xiong, Z., Liu, H., Cheng, J., Liu, F., Lin, F., Wu, R., Zhang, R., and Lu, Q. (2006). Generation of three-dimensional hepatocyte/gelatin structures with rapid prototyping system. *Tissue Eng.* **12**, 83–90.

Washizu, J., Berthiaume, F., Chan, C., Tompkins, R. G., Toner, M., and Yarmush, M. L. (2000). Optimization of rat hepatocyte culture in citrated human plasma. *J. Surg. Res.* **93**, 237–246.

Wolfe, S. P., Hsu, E., Reid, L. M., and Macdonald, J. M. (2002). A novel multi-coaxial hollow-fiber bioreactor for adherent cell types. Part 1: hydrodynamic studies. *Biotechnol. Bioeng.* **77**, 83–90.

Yamada, T., Yoshikawa, M., Kanda, S., Kato, Y., Nakajima, Y., Ishizaka, S., and Tsunoda, Y. (2002). *In vitro* differentiation of embryonic stem cells into hepatocyte-like cells identified by cellular uptake of indocyanine green. *Stem Cells* **20**, 146–154.

Yanagi, K., Ookawa, K., Mizuno, S., and Ohshima, N. (1989). Performance of a new hybrid artificial liver support system using hepatocytes entrapped within a hydrogel. *ASAIO Trans.* **35**, 570–572.

Yarmush, M. L., Dunn, J. C., and Tompkins, R. G. (1992). Assessment of artificial liver support technology. *Cell. Transplant.* **1**, 323–341.

Yoon, J. H., Lee, H. V., Lee, J. S., Park, J. B., and Kim, C. Y. (1999). Development of a nontransformed human liver cell line with differentiated-hepatocyte and urea-synthetic functions: applicable for bioartificial liver. *Int. J. Artif. Organs* **22**, 769–777.

# Part Thirteen

# Hematopoietic System

# Chapter Forty-Nine

# Hematopoietic Stem Cells

*Malcolm A. S. Moore*

## I. INTRODUCTION

Hematopoietic stem cells (HSC) are characterized by their extensive self-renewal capacity and pluripotency. The probabilities of asymmetric versus symmetric division of HSC can be stochastically determined or influenced by external signals. There are many *in vitro* systems involving either bone marrow stromal support or provision of a combination of recombinant hematopoietic growth factors or both that can maintain HSC proliferation and differentiation over many weeks. However, the goal of significant expansion of the HSC population *in vitro* has proved more elusive. There has been an explosive increase in knowledge of the cellular and molecular bases of HSC regulation with the identification of pathways implicating Notch, Wnt, and Hedgehog as well as the cytokine signaling through the c-Kit, Flt3, IL-6-R, mpl, and Tie-2 receptors and downstream pathways involving Jak/STAT and homeobox proteins. There is considerable redundancy in pathways regulating HSC, and both additive and synergistic interactions between different pathways determine the balance between self-renewal and differentiation. With the identification of specific niches within the bone marrow, including endosteal and endothelial, it is now recognized that intimate interactions between HSC and regulatory components of the marrow microenvironment (osteoblasts, osteoclasts, granulocytes, mesenchymeal cells, endothelium) determine HSC proliferative status, pool size, differentiation, and mobilization. The migration of HSC between different niches and the vascular compartment is regulated by the chemotactic action of stromal-derived chemokine SDF-1 acting through its receptor CXCR4, in combination with CD44 and hyaluronic acid. The release of various proteases within the marrow environment leads to cleavage of stromal and HSC-associated adhesion molecules, receptors, cytokines, and chemokines, providing a further level of regulation.

## II. HISTORICAL BACKGROUND

For more than a century, the nature and developmental potential of stem cells within the hematopoietic system has been debated. The monophyletic hypothesis proposed that there was a single type of stem cell generating all myeloid and lymphoid lineage cells, whereas the polyphyletic hypothesis proposed a number of lineage-restricted stem cells. The prevailing view is a monophyletic one, with a pluripotent hematopoietic stem cell compartment and a hierarchy of progressively more lineage-restricted progenitor cells. The concept of a self-renewing, pluripotential hematopoietic stem cell (HSC) achieved experimental validation from the pioneering work of Till and McCulloch in the early 1960s using the spleen colony forming assay (CFU-s) in irradiated mice (reviewed in McCulloch and Till, 2005). They undertook a secondary passage of individually excised spleen colonies and observed a high variability in the numbers of secondary colonies generated, which fitted a skewed (gamma) distribution. The probability that a single CFU-s on division would generate a new CFU-s (self-renewal) was calculated as 6, while production of a differentiated progenitor cell was 4. The conclusion was that this process was random, or stochastic. Other studies at this time indicated the existence of "hematopoietic inductive microenvironments" that influenced the differentiation of HSC. The relevance of stochastic processes versus external inductive signals in the regulation of HSC self-renewal and differentiation is still debated.

## III. PROPERTIES OF HEMATOPOIETIC STEM CELLS (HSC)

### Asymmetric Division

HSC can divide either symmetrically or asymmetrically, depending on environmental, developmental, and stochastic factors (Fig. 49.1). Asymmetric division can be governed by intrinsic partitioning of regulators of cell fate (e.g., inhibition of Notch signaling by Numb and inhibition of Numb by Musachi) or by asymmetric exposure to extrinsic cues (reviewed in Ho, 2005). Symmetric divisions occur during hematopoietic expansion during development and in hematopoietic regeneration following myelodepletion or HSC transplantation. In the latter situations, well-documented data show HSC expansion in murine models. For example, after successive serial transfers of HSC in irradiated mice, a cumulative 8400-fold increase in HSC was observed (reviewed in Sauvageau et al., 2004). Extensive expansion of human HSC was also observed following prolonged in vitro cultures of hematopoietic precursors from umbilical cord blood (Gammaitoni et al., 2004). There is less evidence of symmetrical HSC division either in vitro or in vivo in adult humans (reviewed in Moore, 2005).

### HSC Assays

In vitro or in vivo assays for HSC generally use a limiting-dilution strategy. Primitive, high-proliferative-potential progenitor (HPP-CFC) culture systems have been used as surrogate HSC assays (Muench et al., 1993). More recent assays involve the development of cobblestone areas of phase-dark cells that develop beneath marrow stroma following two to three weeks (murine) or more than five weeks (human) coculture with bone marrow or CD34+ cells (Jo et al., 2000) (Fig. 49.2). These cobblestone area–forming cells (CAFC) have self-renewal and pluripotent differentiation potential. A modification of this assay involves quantitation of secondary progenitor erythroid and myeloid colony formation (CFC), e.g., after five weeks of stromal or cytokine-dependent culture (long-term culture–initiating cell assay-LTC-IC). The in vivo assays for murine HSC involve limiting dilution, competitive (e.g., with the addition of a genetically distinct marrow population) repopulation assay in irradiated recipients, with evaluation after three to six months. The immunodeficient SCID or NOD/SCID mouse supports human hematopoiesis following intravenous injection of $2 \times 10^4$–$2 \times 10^5$ CD34+ cells and quantitation of engraftment by measure of human CD45+ cells in the murine femoral bone marrow at five weeks and beyond can be used under limiting dilution conditions to quantify human HSC (SCID repopulating cells — SRC) (reviewed in Dick and Lapidot, 2005). Human HSC with long-term engraftment of NOD/SCID mice are in the CD38−, CD34+ fraction (Ishikawa et al., 2003). Mazurier et al. (2003) identified a new class of human HSC by direct intrafemoral injection in NOD-SCID mice — these were identified as CD34+, CD38low, CD36− subpopulations termed rapid SCID repopulating cells (R-SRCs), which rapidly generate high levels of human myeloid and erythroid regeneration within the injected femur, migrate to the blood, and colonize individual bone within two weeks after transplantation. While the presence of CD34 and the absence of CD38 provide reliable markers for human HSC, murine HSC are routinely isolated as cKit+, Sca-1+, lineage negative (Lin−) populations, with additional features such as Hoechst dye exclusion. CD34+ is expressed on long-term repopulating HSC of the murine fetus and neonate but decreases with age so that HSC of 10-week-old mice are CD34− (Ogawa, 2002; Matsuoka et al., 2001). The

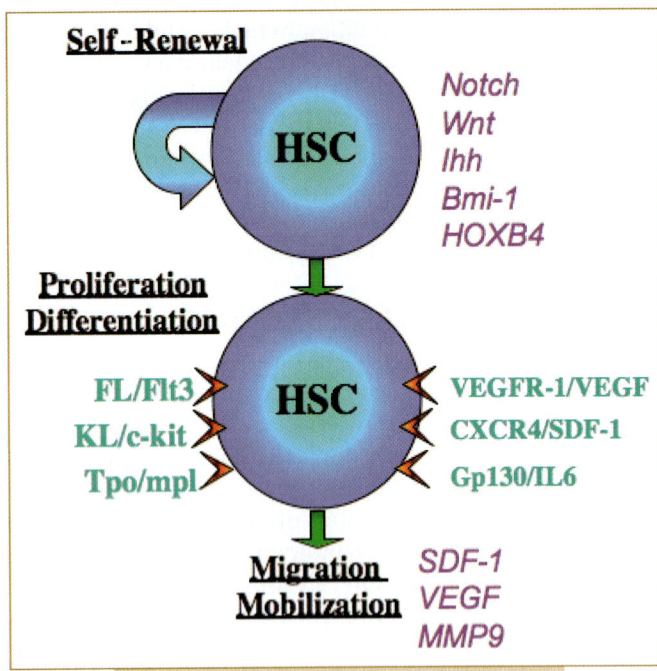

**FIG. 49.1.** Model of HSC self-renewal and differentiation illustrating the role of key cytokines — Flt3 ligand (FL), cKit ligand (KL, also termed *stem cell factor* or Steel factor), thrombopoietin (Tpo), and Interleukin-6 (IL-6). The chemokine receptor CXCR4 and its ligand SDF-1 and the VEGF Receptor-1 (VEGFR-1) and its ligands are implicated in HSC migration and mobilization. The Notch, Wnt, and Indian Hedgehog (Ihh) signaling pathways have also been implicated in HSC self-renewal, as have the Polycomb group transcriptional repressor Bmi-1 and the HOX homeodomain protein HOXB4.

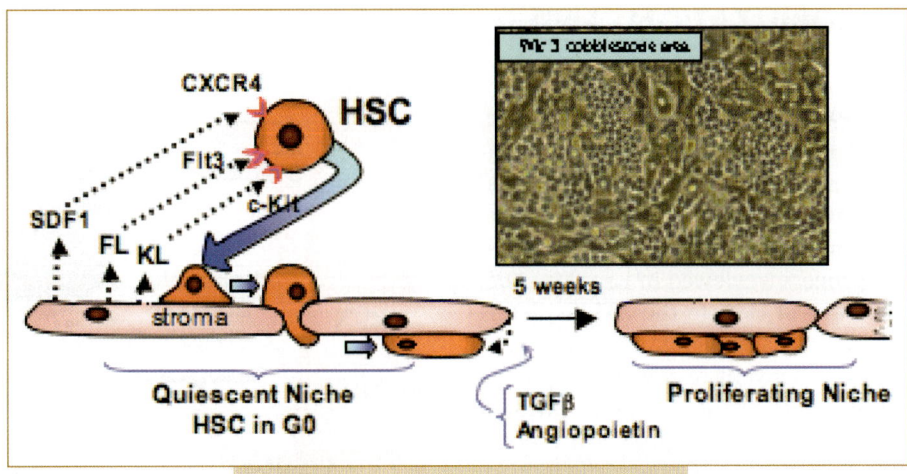

**FIG. 49.2.** A model illustrating the concept of the "late" cobblestone area–forming cell (CAFC) and its interaction with bone marrow stromal elements. SDF-1 produced by the marrow stroma establishes a gradient that attracts CXCR4+ HSC that adhere via VLA-4-VCAM-1–mediated adhesion and then migrate beneath the stroma. Stromal membrane–associated cKit ligand (KL) and Flt3 ligand (FL) provide survival and proliferation signals to the HSC, while negative regulatory influences from the stroma (TGFβ, angiopoietin) override proliferative signals and place the HSC in a G0 state that persists for some weeks. Escape from quiescence may be a stochastic process or driven by an alteration in the balance between proliferation stimulating and inhibiting factors. At this stage the HSC proliferates and differentiates, forming colonies of 5–100 phase-dark cells by week 5. At this stage each cobblestone area contains on average of 1–4 HSC and 5–20 progenitors (CFU-GM, BFU-E).

observations clearly demonstrate that CD34 expression reflects the activation/kinetic state of HSC and that expression is reversible. In the murine system, the majority of HSC are CD38+, and there is a reciprocal relationship of CD34 and CD38 expression (Ogawa, 2002). Cell surface receptors of the SLAM family are differentially expressed among functionally distinct HSC, and progenitor populations of the mouse and HSC are CD150+, CD244−, and CD48−, while multipotent progenitors are CD244+, CD150−, and CD48− (Kiel et al., 2005).

### HSC Numbers and Proliferative Status

It has been suggested that HSC numbers are conserved in mammals, with cats, mice, and rats having comparable numbers (~11,000–12,000) (Abkowitz et al., 2002; McCarthy, 2003). If extrapolated to humans, the frequency of HSC would be $0.7–1.5/10^8$ marrow cells, a frequency that is 20-fold less than estimated by the NOD/SCID repopulating assay. Measurements of granulocyte telomere length have been used to estimate human HSC kinetics, and results showed that replication is infrequent, on average once every 45 weeks, substantially slower than the average rates estimated for murine (once per 2.5 weeks) or cats (once per 8.3–10 weeks) (Shepherd et al., 2004).

## IV. ONTOGENY OF HSC

In the avian and mammalian systems, the ontogeny of hematopoiesis involves a sequence of vascular migration streams of primitive hematopoietic stem and progenitor cells (hematogenous metastases), initiating from the earliest sites of HSC development and sequentially colonizing developing myeloid (fetal liver, spleen, bone marrow) and primary lymphoid (thymus) organs (reviewed in Moore, 2006). In mammals, hematopoiesis is initiated in the blood islands of the yolk sac, where both primitive and definitive lineage hematopoietic precursors arise. Intraembryonic hematopoiesis is subsequently initiated in the aorta–gonad–mesonephros (AGM) region in or near the endothelium of the dorsal aorta, where HSC capable of repopulating adult animals arise (Moore, 2006). Bertrand et al. (2005) identified a population of CD45−ve, c-Kit+, CD31+, HSC arising in the AGM region in the splanchnic mesoderm underlying aortic endothelial cells, within GATA-3+, CD31+ cell clusters, and these cells did not colocalize with aortic endothelium as previously thought. The onset of definitive (adult engraftment potential) HSC activity in the AGM region was paralleled by that in the placenta starting in the mouse at E10.5–11.0 days of gestation (Gekas et al., 2005). There remains considerable debate as to the role of yolk sac hematopoiesis in lymphohematopoietic development. One view is that the yolk sac is a transient source of primitive hematopoietic progenitors, whereas definitive, pluripotent HSC arise exclusively within the AGM region. Alternatively, and more plausible, is the view that HSC arise independently in the AGM and yolk sac and migrate to the fetal liver and thymus — organs that become major developmental sites of hematopoiesis and lymphopoiesis (Moore, 2006). The subsequent migration of HSC from the fetal liver to bone marrow provides an opportunity for harvesting HSC from the circulation. The umbilical cord blood population is a rich source of HSC and has been used as an HSC transplant source in a wide range of hematological and oncologic disorders (reviewed in Broxmeyer, 2005).

# V. MIGRATION, MOBILIZATION, AND HOMING OF HSC

## Bone Marrow (BM) Homing of HSC

Large numbers of HSC move into and out of the circulation on a daily basis. Homing of HSC to the BM is rapid (hours), involving cells traversing across the BM sinusoidal endothelium, which presents adhesion molecules and chemokines facilitating the process (reviewed in Lapidot *et al.*, 2005). The seeding efficiency of HSC to the BM, as determined by secondary passaging, has generally been accepted as 10–20% by 24 h. However, Benveniste *et al.* (2003), using competitive repopulation and tracking of a single purified HSC in mice, concluded that HSC engraft with near-absolute efficiency. This view has recently been challenged, since individual HSC (SP cells) were only able to reconstitute up to 35% of mice (Camargo *et al.*, 2006). The transition from quiescence to active cell cycling of HSC was associated with significant loss of engraftment potential (Glimm *et al.*, 2000), which was not associated with loss of expression of the integrins VLA-4, VLA-5 or the chemokine receptor CXCR4.

## Role of the SDF-1/CXCR4 Chemokine Pathway in HSC Homing and Migration

The chemokine SDF-1 is expressed on bone marrow vascular endothelium, immature osteoblasts in the endosteal region, and marrow stromal cells, while its receptor, CXCR4, is expressed on HSC and progenitors (Figs. 49.2 and 49.3). The SDF-1/CXCR4 pathway plays a major role in regulating the mobilization, migration, and retention of HSC (reviewed in Lapidot *et al.*, 2005). This chemokine pathway is essential for HSC seeding from the fetal liver to the bone marrow during development; however, CXCR4-deficient HSC can, with reduced efficiency, engraft adult irradiated mice (reviewed in Moore, 2006). Elevation of plasma levels of SDF-1 occurs following intravenous injection of an adenovector expressing SDF-1, and the consequent reversal of the SDF-1 gradient from blood to marrow leads to mobilization of HSC and progenitors (Hattori *et al.*, 2001). Overexpression of CXCR4 on human CD34+ cells by gene transfer increased their proliferation, migration, and NOD/SCID engraftment potential (Lapidot *et al.*, 2005). CXCR4 neutralization abolished human intravenous or intrafemoral CD34+ engraftment in NOD/SCID mice, indicating the essential role of this receptor in BM seeding and colonization as well as in homing (Lapidot *et al.*, 2005).

The adhesion receptor CD44 and its major ligand, hyaluronic acid, are essential for HSC homing to marrow and spleen. HSC express CD44 and migrate on hyaluronic acid toward a gradient of SDF-1, acquiring a polarized morphology with CD44 concentrated at the leading edge of pseudopodia (Avigdor *et al.*, 2004). Since hyaluronic acid is expressed on both BM endothelium and endosteum, it is likely that

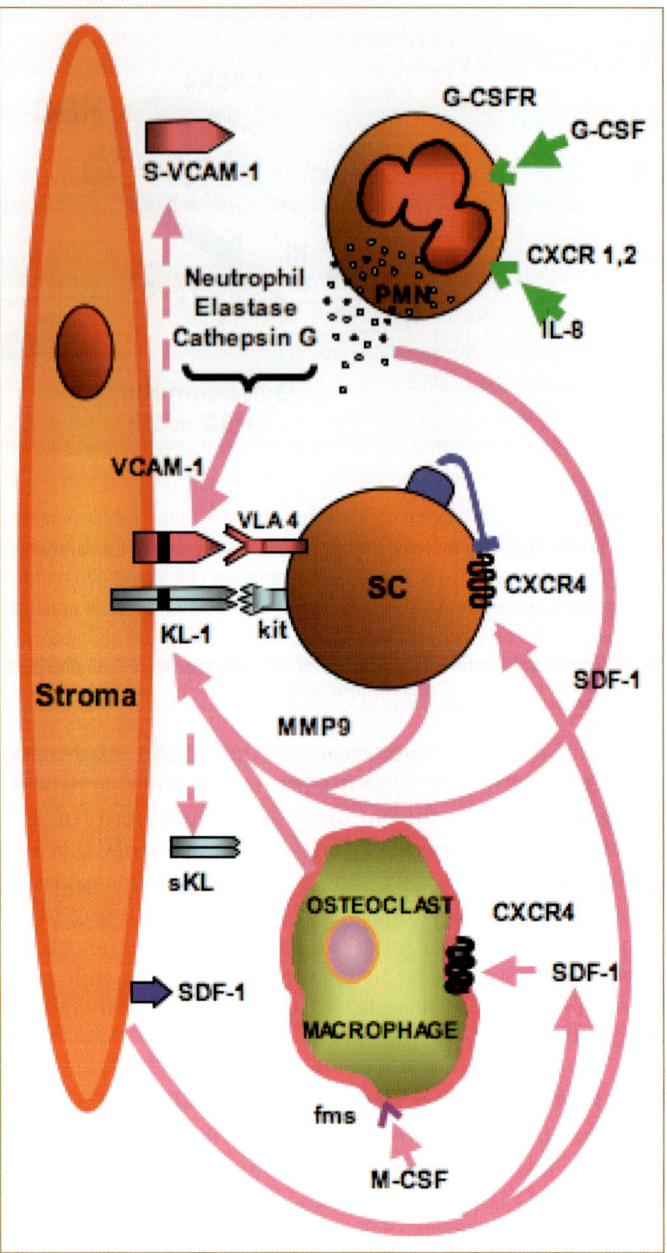

**FIG. 49.3.** Model illustrating HSC adherence to bone marrow stromal cells via VLA4 binding to VCAM-1 and cKit-binding transmembrane cKit ligand (KL-1). SDF-1 produced by the stroma induces matrix metalloproteinase-9 (MMP9) production by HSC and macrophages, and this in turn cleaves the transmembrane KL-1, releasing soluble active ligand (sKL). The HSC-mobilizing role of G-CSF is explained in part by increased numbers and activation of neutrophils within the bone marrow, with secretion of a number of proteases (elastase, cathepsin G) that cleave VCAM-1 and untether HSC. Protease activity is also implicated in the cleavage and inactivation of SDF-1. HSC express the membrane-bound ectopeptidase CD26, which removes dipeptides from the amino-terminus of proteins and acts as a negative regulator of the SDF-1-CXCR4 axis.

CD44 plays a key role in SDF-1, mediated transendothelial migration, and localization in specific BM niches.

## HSC Mobilization

Beginning in the mid-1980s, autologous peripheral blood was used for transplantation, particularly blood obtained after chemotherapy at the rebound in hematopoietic regeneration. The availability of two hematopoietic growth factors, G-CSF and GM-CSF, in recombinant form led to their use as HSC- and progenitor-mobilizing agents. G-CSF treatment for mobilization of HSC, peaking at five to six days, is currently the most efficient strategy for clinical HSC collection. A number of chemokines (IL-8, MCP-1, MIP-1α, Groβ, SDF-1) also induce rapid HSC mobilization in murine models, with peak HSC and progenitor mobilization in a few hours, in contrast to the days required with CSFs. Movement of HSC between endosteal, perivascular, and intravascular sites is influenced by G-CSF activation of marrow neutrophils, with resulting protease release and cleavage and inactivation of SDF-1 (Fig. 49.3). One mechanism proposed to explain the ability of G-CSF treatment to mobilize HSC involves a reported reduction in SDF-1⁻ possibly due to degradation by neutrophil protease activity and an increase in its receptor CXCR4 in the bone marrow, whereas their protein expression in blood was less affected (Petit *et al.*, 2002). Activated neutrophils release matrix metalloproteinase-9 (MMP-9) and lactoferrin from specific granules and neutrophil elastase, cathepsin G, and proteinase 3 from azurophilic granules (Fig. 49.3) (reviewed in Papayannopoulou, 2004). The release of multiple proteases within the BM can lead to cleavage or degradation of vascular cell adhesion molecule-1 (VCAM-1), membrane-bound cKit ligand (KL), SDF-1, and its receptor CXCR4 — all events that can untether HSC from their niches and facilitate their proliferation, differentiation, and mobilization (Fig. 49.3) (Heissig *et al.*, 2002). Studies in genetic models of protease deficiency, including models with multiple protease deficiencies, surprisingly did not show impaired G-CSF-induced HSC mobilization (Papayannopoulou, 2004). Thus, the contribution of a single protease to mobilization may not be critical, but multiple proteases acting in a cell context–specific manner appear important. Targeting and reduction of SDF-1 or its receptor may be common denominators of proteases implicated in HSC mobilization. HSC express the membrane-bound ectopeptidase CD26, which removes dipeptides from the amino-terminus of proteins, and CD26 null mice have an attenuated response to G-CSF. CD26 expression on HSC negatively regulated their homing and engraftment, and by inhibiting or deleting CD26 it was possible to increase greatly the efficiency of engraftment (Christopherson *et al.*, 2004). CD26 expression may regulate HSC engraftment potential by inactivating SDF-1 produced by marrow stroma. CXCR4 antagonists such as AMD3100 have also been shown to be effective in mobilized HSC with long-term repopulating capacity in nonhuman primates (Laro-

chelle *et al.*, 2006). Signals from the sympathetic nervous system also regulate HSC egress from bone marrow (Katayama *et al.*, 2006). Pharmacologic or genetic ablation of adrenergic neurotransmission resulted in the failure of HSC and progenitors to egress the marrow following G-CSF stimulation, while a β2 adrenergic agonist enhanced HSC mobilization in both control and norepinephrine-deficient mice. Kollet *et al.* (2006) have linked bone remodeling with regulation of stress-induced mobilization of HSC and progenitor cells. They showed that an increase in osteoclast numbers is associated with mobilization. RANK ligand (RANKL) is a member of the TNF family of cytokines, which is necessary for the production of osteoclasts, is a product of osteoblasts, and is responsible for the balance between bone production and destruction in skeletal remodeling. RANKL altered osteoclast expression of MMP-9 and cathepsin K, which cleave membrane-bound KL, resulting in decreased osteoblast KL and the production of osteopontin. Inhibition of osteoclasts with calcitonin reduced G-CSF stem/progenitor mobilization. Mice deficient in the protein tyrosine phosphatase-ε (PTP-ε) have dysfunctional osteoclasts, and HSC in these mice did not mobilize on treatment with either RANKL or G-CSF (Kollet *et al.*, 2006).

The Rho guanosine triphosphatases, Rac1 and Rac2, play distinct roles in actin organization, hematopoietic cell survival, and proliferation. Deletion of both Rac1 and Rac2 murine alleles led to a massive egress of HSC from the marrow to blood, and Rac1⁻/⁻ mice showed profound defects in engraftable HSC (Gu *et al.*, 2003). Rac2-deficient HSC showed defective adhesion, motility, and interaction with the hematopoietic microenvironment and had a significant competitive repopulation defect (Jansen *et al.*, 2005).

## VI. HSC PROLIFERATION AND EXPANSION *IN VITRO*

The *in vitro* maintenance and potential expansion of HSC has been the subject of extensive investigation since the mid-1970s, beginning with murine long-term bone marrow cocultures with murine bone marrow stroma (Dexter cultures) and subsequent adaptation of this system to humans, using both human and murine stroma. With the discovery of a number of hematopoietic growth factors (interleukins, colony stimulating factors) and their availability as recombinant proteins, *in vitro* culture systems were developed that supported extensive cell and progenitor expansion in the absence of stroma but in the presence of combinations of cytokines (Moore, 2005).

### Stromal-Based Cultures

Dexter and colleagues first introduced *in vitro* culture systems for long-term growth of murine BM on preestablished marrow stroma. Similar systems, but with the addition of corticosteroids, were developed for human (Moore *et al.*, 1980) and primate BM (Moore *et al.*, 1979). With adult human LTBMC, hematopoiesis was generally limited to five to eight weeks, and there was continuing decay in repopulating HSC and progenitors over time. A number of modifications to the long-term stromal coculture system have been developed to prevent the decrease in stem and progenitor output, for example, the use of different sources of stroma, including human vascular endothelium, human fetal spleen, and stroma from different species, including mouse, primate, and pig. Frequent readdition of fresh media and cytokines, ranging from one exchange per week up to total daily feeding, resulted in increased CFC and LTC-IC output and led to the development of clinical scale flat-plate perfusion bioreactors (Koller *et al.*, 1993). Unfortunately, the use of media exchange alone has not translated into expansion of repopulating HSC. Cord blood CD34$^+$ cells have been expanded in transwell culture systems (noncontact cultures) above AFT024 mouse fetal liver stromal cells with additional cytokines (Flt3 ligand [FL], KL, thrombopoietin [Tpo], IL7) (Lewis *et al.*, 2001). In stromal-supported cultures of cord blood, CD34$^+$ cells and cytokine- (FL, Tpo, KL $^{+/-}$IL6) enhanced production of CAFC (11–89-fold over six weeks) was noted, while NOD/SCID repopulating cells increased 10–14-fold by two weeks but were absent by week 5 (Kusadasi *et al.*, 2001). Expanded cells engrafted primary and secondary NOD/SCID mice, and day 7–expanded cells engrafted primary, secondary, and tertiary fetal sheep. The murine OP9 stromal cell line derived from osteopetrotic op/op mice has been used to support long-term murine hematopoiesis in the absence of added cytokines; however, it is not supportive of human CD34$^+$ cells. The addition of Tpo to the stromal coculture system or transduction of OP9 with an adenovector expressing Tpo permitted very efficient long-term (16+ weeks) human hematopoiesis initiated with cord blood CD34$^+$ cells, with sustained generation and extensive cumulative expansion of LTC-IC (Gammaitoni *et al.*, 2004). In a comparable system of coculture of OP9 transduced with adenovectors expressing Tpo, KL, and FL but using adult bone marrow or G-CSF-mobilized blood CD34$^+$ cells, hematopoiesis was sustained for 12 weeks, but a much lesser degree of LTC-IC expansion was noted (three- to four-fold) (Feugier *et al.*, 2005). In the latter system, addition of human (but not murine) M-CSF or CD14$^+$ monocyte/macrophages to the cultures substantially decreased progenitor and HSC expansion, indicating that accumulation of differentiated cells, particularly macrophages, exerted a negative influence on HSC proliferation.

## Cytokine-Supplemented Cultures

It was found that direct contact between stromal cells and hematopoietic cells was not necessary for long-term *in vitro* hematopoiesis (Lewis *et al.*, 2001), and hematopoiesis could be maintained in cultures fed with media conditioned by stromal feeders. This information suggested the feasibility of using stromal-free cytyokine-supplemented cultures for expanding HSC and progenitors *in vitro*. Over the years,

the use of phenotypic and functional assays has identified a number of cytokines that have distinct stimulatory effects on primitive hematopoietic cells, particularly when used in various combinations (Fig. 49.1). These include FL, KL, Tpo, IL-1, IL-3, IL-6, IL-11, IL-12, G-CSF, and GM-CSF (Haylock *et al.*, 1992; Muench *et al.*, 1993; reviewed in Moore, 2005). Using limiting-dilution assays for competitive repopulating cells, modest degrees of HSC expansion have been reported in murine *in vitro* expansion systems using combinations of KL, FL, and IL-11. Optimal growth factor combinations are required to achieve *in vitro* HSC expansion, and while KL and FL are sufficient to maintain survival and proliferation, retention of HSC function requires activation of additional pathways, for example, gp130, stimulated via IL-6 or IL-11 (reviewed in Sauvageau *et al.*, 2004). Haylock *et al.* (1992) were the first to report a CFC expansion of 66-fold in human CD34$^+$ cultures stimulated with IL-1, IL-3, IL-6, G-CSF, and GM-CSF. Subsequent studies showed the importance of specific cytokines for the expansion of defined progenitor cell types. In a human CD34$^+$ expansion system, IL-6 plus soluble IL-6 receptor (to maximize IL-6 signaling where IL-6 receptor density could be limiting), together with FL, KL, and Tpo, resulted in an approximate fourfold expansion of HSC, as determined by NOD/SCID limiting-dilution repopulation (Ueda *et al.*, 2000). Both FL and IL-6/sIL-6 are important for expanding cord blood CD34$^+$-derived LTC-IC, while the addition of KL to either of these factors enhances generation of CFC (Zandstra *et al.*, 1998). In contrast, FL, KL, and IL-3 most efficiently stimulated BM LTC-IC proliferation, whereas the addition of IL-6/sIL-6R or Tpo to this combination was required for expansion of CFC (Zandstra *et al.*, 1998). Regardless of the cytokine combination used, rigorous limiting-dilution engraftment studies in NOD/SCID mice have, in the majority of studies, shown that in cytokine-supplemented cultures there was generally a loss, maintenance, or modest (two- or fourfold) expansion of human SRC (Ueda *et al.*, 2000). In contrast, Piacibello *et al.* (1999) (Gammaitoni *et al.*, 2004; reviewed in Moore, 2005) reported a major expansion of LTC-IC and progenitors (3000-fold) and a 70-fold expansion of NOD/SCID engrafting HSC (SRC) in cultures of CB CD34$^+$ cells with FL, KL, Tpo, and IL-6. Under similar conditions but with CD34$^+$ cells from adult BM and mobilized peripheral blood (MPB), *ex vivo* expansion was shorter in duration and extent, with SRC expansion of only sixfold by three weeks (Gammaitoni *et al.*, 2004). In a modified stromal-free culture system with cord blood–derived CD34$^+$ cells and a cocktail of four cytokines (KL, FL, Tpo, and IL-6), with CD34$^+$ re-isolation at monthly intervals, continuous expansion of HSC was observed over 16–20 weeks by *in vitro* CAFC and LTC-IC assays and by NOD/SCID engraftment and secondary and tertiary passaging (Gammaitoni *et al.*, 2004) (Table 49.1). Despite extensive proliferation, the telomere length of cultured hematopoietic cells initially increased, and it was only at late stages of culture that telomere shortening was detected. Telomere

**Table 49.1.** Stem and progenitor cell production, telomere length, and telomerase activity of CB CD34$^+$ cells generated after repeated expansion-isolation procedures ("fractionated" LTC)

| | Input | 1st isolation, wk 4 | 2nd isolation, wk 8 | 3rd isolation, wk 12 | 4th isolation, wk 16 | 5th isolation, wk 20 |
|---|---|---|---|---|---|---|
| CD34$^+$ ($\times 10^3$) | 100 | 380 | 1820 | 6550 | 30,100 | 144,500 |
| CD34$^+$/38/Lin$^-$, $\times 10^3$ | 11 | 72 | 400 | 1560 | 7600 | 6700 |
| CFC, $\times 10^3$ | 7.9 | 15.7 | 20.7 | 222.7 | 1033.4 | 2408.3 |
| CFU-GEMM, $\times 10^3$ | 1.8 | 3.8 | 17 | 43.7 | 501.7 | 647.4 |
| LTC-IC, $\times 10^3$ | 14.3 | 40 | 73.5 | 791 | 1063.5 | 0.082 |
| Telomere length by flow-FISH, kb | 9.1 | 9.9 | 9.3 | 9.3 | 9.3 | 7.25 |
| Telomerase activity, TPG units | 41.95 | 27.06 | 16.56 | 11.1 | 10.1 | 4.8 |
| **Engraftment, no. injected mice/no. positive** | | | | | | |
| Primary, $3 \times 10^5$ CD34$^+$ infused | 6/6 | 11/11 | 5/5 | 7/7 | 7/7 | 5/5 |
| Secondary | 2/2 | 3/3 | 2/2 | 3/3 | 2/2 | 0/1 |
| Tertiary | 0/2 | 1/1 | 1/1 | 1/1 | 1/1 | ND |
| SRC/100,000 CD34$^+$ cells | 2.8 | 9.4 | 14.4 | 19 | 58.3 | 1.2 |

Modified from Gammaitoni *et al.* (2004).

length stabilization correlated with high telomerase levels (Table 49.1). In contrast, cytokine-stimulated adult CD34$^+$ cells from bone marrow or G-CSF–mobilized blood showed CD34$^+$ and NOD/SCID engraftment expansions of sixfold for only three to four weeks, with telomere shortening and low levels of telomerase activity. Human fetal liver–derived HSC with NOD/SCID repopulating capacity have been expanded 10- to 100-fold net expansion over 28 days of culture with FL, Tpo, KL, and IL-6 and 8% human AB plasma (Rollini *et al.*, 2004). The CD133$^+$ G0 cell subpopulation from cord blood is enriched for HSC and progenitors (CFC), with an LTC-IC incidence of 1 in 4.2 cells and a CFC incidence of 1 : 2.8 cells (Summers *et al.*, 2004). These cells could be expanded with cytokines in a serum-free, stromal-free culture system for up to 30 weeks, resulting in a 100-million-fold amplification of progenitors. Quantitation of HSC expansion *in vitro* is complicated by divergence in results when *in vitro* HSC assays (CAFC, LTC-IC) are used as an endpoint, versus *in vivo* (SCR). While this has led to considerable debate as to the validity of the assays, the issue is probably related to the efficiency of the respective systems. Clearly, the *in vivo* assay requires HSC survival and homing to the HSC niches of the mouse bone marrow, and there is evidence that *in vitro* culture or cycling HSC may be less efficient *in vivo* while retaining potential to self-renew and form CAFC/LTC-IC *in vitro*. In this context, Liu *et al.* (2003) cultured CB CD34$^+$ cells for up to five days with a cocktail of cytokines, labeled the cells with 111-indium, and deter-

mined recovery at 48 h following intravenous injection into NOD/SCID mice. Cultured, cycling HSC and progenitors showed a reduction of marrow homing (from 11.3% to 5.4%), with reduced homing to spleen, liver, and lung. Acute cytokine deprivation in the *in vivo* environment and Fas/CD95-mediated cell death may be responsible for the reduced engraftment efficiency of cultured HSC.

## Morphogens and HSC Regulation

In addition to the classic hematopoietic growth factors, signaling pathways implicated in general development have also been shown in various models to influence HSC proliferation and differentiation (Fig. 49.1).

### Notch

In differential gene display, Notch1 has consistently been shown to characterize the murine and human HSC. Notch1-null mouse embryos showed defective hematopoietic (and vascular) development and were devoid of HSC. Retrovirus-mediated expression of activated Notch1 enhanced HSC self-renewal, and a similar effect of differentiation inhibition and progenitor/HSC expansion was reported with activated Notch4 (Int3) (Ye *et al.*, 2003). The Bernstein group (Varnum-Finney *et al.*, 2003; Delaney *et al.*, 2005) showed that incubation of murine BM precursors with the Notch ligand Delta1 extracellular domain fused to the Fc portion of human IgG1, together with cytokines (KL, IL-6, IL-11, and Flt3L), inhibited myeloid differentiation and promoted several log increases in precursors capable of short-term lymphoid and myeloid repopulation. Addition of IL-7 promoted T-lymphocyte development, whereas GM-CSF induced myeloid differentiation. It has been shown that Notch signaling mediated by both Delta and Jagged ligands expands the HSC compartment while blocking or delaying terminal myeloid differentiation. The quantitative aspects of Notch signaling in determining hematopoietic precursor fate has been demonstrated by Delaney *et al.* (2005), who showed in CD34$^+$ cord blood cultures that low densities of the Notch ligand Delta1 enhanced *in vitro* generation of CD34$^+$ cells as well as CD14 and CD7 cells, consistent with myeloid and lymphoid differentiation, whereas higher concentrations induced apoptosis of CD34$^+$ cells but not CD7 T-cell precursors. A role for combinatorial effects of Jagged stimulation of Notch and cytokine-induced signaling pathways (Kit, Flt3, Mpl) was reported in murine HSC cultures, with a 10- to 20-fold expansion of HSC with long-term repopulating potential (Kertesz *et al.*, 2006). Inhibition of Notch signaling leads to accelerated differentiation of HSC *in vitro* and depletion of HSC *in vivo* (Duncan *et al.*, 2005). However, using a Cre-LoxP-mediated inactivation system, mice with simultaneous inactivation of Jagged and Notch1 survived normally, even following chemotherapy-induced myelosuppression (Mancini *et al.*, 2005). HSC regulation is complex, cell-context dependent, and plagued by potential compensation systems, and while the study excludes an essential role for Jagged1 and Notch1 during hematopoieis, there are four Notch receptors and five ligands, and the Notch pathway cross-talks with the Wnt pathway, with more than nine frizzled receptors and 12 ligands (Duncan *et al.*, 2005). The *Drosophila* hairy and enhancer of split (HES) 1 basic helix-loop-helix protein is a major downstream effector of the Notch pathway and is expressed at high levels in HSC-enriched CD34$^+$, CD38$^-$ subpopulations. Conditional expression of HES1 in murine and human HSC inhibited cell cycling *in vitro* and cell expansion *in vivo*, with preservation of long-term reconstituting function.

## Wnt

The Wnt pathway has been implicated in HSC proliferation (reviewed in Reya and Cleavers, 2005). Notch signaling was required for Wnt-mediated maintenance of undifferentiated HSC but not for their survival or entry into cell cycle (Duncan *et al.*, 2005). Wnt signaling represses glycogen synthase kinase-3β (GSK3β), leading to accumulation of β-integrin but also accumulation of intracellular fragments of Notch and activation of Notch target genes, such as Hes1. Hematopoietic repopulation by mouse or human stem cells has been augmented by administration of a GSK-3 inhibitor to recipient mice (Trowbridge *et al.*, 2006). The inhibitor improved neutrophil and megakaryocyte recovery and mouse survival and enhanced sustained long-term repopulation. The GSK-3 inhibitors modulated gene targets of Wnt, hedgehog, and Notch pathways in primitive hematopoietic

cells, without affecting mature cells. Wnt-5A injected *in vivo* into immunodeficient mice enhanced engraftment by human CD34$^+$ cells (Murdoch *et al.*, 2003). Reya *et al.* (2003) showed that overexpression of activated β-catenin expanded the pool of HSC in long-term culture. Activation of Wnt signaling in HSC increases expression of HOXB4 and Notch1, both implicated in HSC self-renewal. Ectopic expression of axin or the frizzled ligand-binding domain, both inhibitors of the Wnt signaling pathway, inhibited HSC growth *in vitro* and reduced *in vivo* engraftment. Wnt3a-palmitolylated protein has been purified and shown to induce self-renewal of HSC (Willert *et al.*, 2003). The essential role of β-catenin in hematopoiesis has been questioned in studies of mice with Cre-loxP-mediated inactivationed of β-catenin, since there was no impaired HSC ability to self-renew and differentiate (Cobas *et al.*, 2004).

## Hedgehog

Sonic hedgehog (SHH) treatment in CD34$^+$ cultures induced expansion of human HSC, and Noggin, a specific inhibitor of bone morphogenic protein-4 (BMP-4), inhibited SHH-induced proliferation, indicating that SHH regulates HSC via mechanisms that are dependent on downstream BMP signals (Bhardwaj *et al.*, 2001). Ihh-expressing human stromal cells supported CD34$^+$ cells, with markedly enhanced production of progenitor CFC above the stroma but no change in long-term cobblestone formation (Kobune *et al.*, 2004).

## Bone Morphogenic Protein

CD34$^+$, CD38$^-$ HSC express the bone morphogenic protein type 1 (BMP-1) receptors activin-like kinase (ALK-3 and ALK-6) and their downstream transducers SMAD-1, -4, and -5. Like TGFβ, high concentrations of BMP-2, -4, and -7 inhibited HSC proliferation but maintained their long-term survival and repopulating potential, whereas low concentrations of BMP-4 induced HSC proliferation and differentiation (Bhatia *et al.*, 1999).

## Molecular Pathways Implicated in HSC Self-Renewal

### Jak-STAT

Transduction of mouse HSC with constitutively activated STAT3 enhanced HSC self-renewal under stimulated but not homeostatic conditions, while a dominant negative form of STAT3 suppressed self-renewal (Chung *et al.*, 2006). The STAT5 pathway is activated strongly following ligand binding to the erythropoietin and IL-3 receptors and weakly following FL binding to Flt3. However, constitutively activating mutants of Flt3, found in ~25% of human acute myeloid leukemia, are associated with strong activation of STAT5 (reviewed in Moore, 2005). A constitutively activated double mutant of STAT5a [STAT5a(1*6)] transduced into CD34$^+$ cells promoted enhanced HSC self-renewal, as measured by

CAFC assay, and promoted enhanced erythroid differentiation relative to myeloid (Schuringa *et al.*, 2004). This was causally linked to down-regulation of C/EBPα, a transcription factor uniquely associated with granulocytic differentiation.

### Bmi-1

The Polycomb-group transcriptional repressor gene Bmi-1 has been implicated in HSC maintenance, and loss-of-function studies showed profound defects in HSC (Park *et al.*, 2003). Bmi-1 overexpression down-regulated expression of p16 and p19Arf, which are encoded by ink4a, and enhanced HSC symmetrical division, resulting in expansion of multipotent progenitors *in vitro* and enhanced HSC repopulation *in vivo* (Jacobs *et al.*, 1999).

### The HOX Family of Hematopoietic Regulators

The HOX homeodomain (HD) proteins are DNA-binding transcription factors that have long been recognized as key regulators of development and hematopoiesis. *HOX* genes are expressed at various stages during hematopoietic development. Mounting evidence suggests that *HOXA9* plays important roles in normal hematopoiesis. Targeted disruption of *HOXA9* in mice severely reduces the number of HSC and progenitor cells, while enforced expression of *HOXA9* promotes proliferative expansion of HSC and progenitor cells and subsequently inhibits their differentiation (reviewed in Sauvageau *et al.*, 2004; Moore, 2005). These data highlight the importance of precise control of HOXA9 protein levels during hematopoiesis. HOXB4 plays a critical role in promoting HSC self-renewal and engraftment (reviewed in Sauvageau *et al.*, 2004; Moore, 2005). Combinations of early-acting cytokines increased HOXB4 promoter activity in primitive hematopoietic cells, and Tpo acting via Mpl and p38MAPK increased HoxB4 expression two- to threefold in primitive hematopoietic cells (Kirito *et al.*, 2003). In Tpo$^{-/-}$ mice, hematopoietic HoxB4 expression was two- to fivefold lower than in wild-type animals (Kirito *et al.*, 2003). Wnt signaling in primitive hematopoietic cells also induces HOXB4 expression (Reya *et al.*, 2003). HOXB4-null mice had reduced HSC and progenitor cells, due to impaired proliferative capacity, but did not show perturbed lineage commitment. Retroviral-mediated ectopic expression of HOXB4 resulted in a rapid increase in proliferation of murine HSC both *in vivo* (1000-fold increase in transduced HSC in a murine transplant model) and *in vitro* (40-fold expansion of murine HSC), with retention of lymphomyeloid repopulating potential and enhanced regenerative capability in mice (reviewed in Sauvageau *et al.*, 2004). However, high levels of HOXB4 expression in human umbilical cord blood (CB) CD34$^+$ cells were recently reported either to increase proliferation of HSC and inhibit differentiation (Schiedlmeier *et al.*, 2003) or to direct the cells toward a myeloid differentiation program rather than increasing proliferation (Brun *et al.*, 2003). These studies suggest that in human hematopoietic progenitors, HOXB4 affects cell fate decisions (self-renewal, differentiation, or a differentiation block) in a concentration-dependent manner. Therefore, like HOXA9, the relative abundance of HOXB4 requires precise regulation. Molecular mechanisms and target genes responsible for HOXB4-induced HSC expansion remain to be elucidated. Overexpression of HOXB4 point mutations lacking the capacity to bind DNA (HOXB4(N$^{51}$→A)) failed to enhance proliferative activity of transduced BM populations, whereas mutants that blocked the capacity of HOX to cooperate with the transcription factor PBX in DNA-binding HOXB4 (W→G) conferred a pronounced proliferative advantage *in vitro* and *in vivo* to transduced BM populations (Beslu *et al.*, 2004). This mutant was comparable to wild-type HOXB4, in that its elevated level promoted a comparable degree of HSC expansion. This was distinct from results obtained by knocking down the expression of PBX in HOXB4-overexpressing BM cells using lentivector transduction with a PBX antisense construct. In this system the HSC were more than 20 times more competitive than HSC overexpressing HOXB4 with PBX levels intact, which were, in turn, 20–50 times more competitive that wild-type BM (Krosl *et al.*, 2003). The likely explanation for these observations is that HOXB4 and PBX genes act on distinct pathways in the HSC, the former promoting self-renewal and the latter inhibiting it. A developmentally important mechanism for regulating HOX gene expression was identified by Davidson *et al.* (2003). Cdx4 belongs to the caudal family of homeobox genes, which have been implicated in anteroposterior patterning of the axial skeleton and regulation of Hox gene expression. A Cxc4 mutation in zebra fish causes severe anemia, with complete absence of Runx1 in blood cells (Davidson *et al.*, 2003). Injection of mutants with Hoxb7a and Hox9a mRNA almost completely rescued Cdx4 mutants. Retroviral transduction of mouse ES–derived embryoid body cells with Cdx4 increased expression of HoxB4 (×30), HoxB3 (19×), HoxB8 (5×), and Hoxa9 (4×) — all implicated in HSC or progenitor expansion. And Cxc4 was more potent than HoxB4 in stimulating hematopoiesis and CFU-GEMM production (Davidson *et al.*, 2003).

The abundance of a given cellular protein is regulated by the interplay between its biosynthesis and degradation. During normal hematopoietic development, *HOXA9* is strongly expressed in the CD34$^+$ populations enriched in early myeloid progenitors and is turned off when cells exit the CD34$^+$ compartment and undergo terminal differentiation. In conjunction with decreased biosynthesis, rapid turnover of HOXA9 would ensure low steady-state levels, which are necessary for proper execution of differentiation into myeloid lineages. HOXB4 is also expressed at high levels in HSC/progenitor compartments and is down-regulated but maintains low-level expression during differentiation. The studies of the biochemical mechanisms controlling the activities of HOXA9 and HOXB4 thus far have been focused primarily on transcriptional regulation. Little is known about how their cellular abundance is controlled at the posttranslational level. Identification of proteins involved in the removal of HOXA9 and HOXB4 will be necessary for understanding the elaborate regulatory circuitry governing hematopoiesis. Posttranslational regulation of HOX protein levels has been linked to their ubiquitin-dependent proteolysis by the CUL-4A ubiquitin-ligase (Y. Zhang et al., 2003).

Some transcription factors travel between cells because they contain protein domains that allow them to do so. This is the case for the HIV transcription factor TAT and for several homeoproteins, such as Engrailed, HOXA5, HOXB4, HOXC8, EMX1, OTX2, and PAX6. Direct paracrine homeoprotein activity has not been considered, yet in theory it would represent a way for neighboring hematopoietic cells to exchange proliferative and differentiative signals. Bone marrow stromal cells (murine MS-5 stroma) have been lentivector transduced with a vector expressing HOXB4 linked to an immunoglobulin kappa chain leader sequence (signal peptide), which is cleaved during protein secretion. Coculture of human CB CD34$^+$ cells on this stroma resulted in a two- to threefold greater expansion of cells and progenitors (CFC), a four- to tenfold greater expansion of LTC-IC, and a 2.5-fold expansion of NOD/SCID repopulating cells relative to control over five weeks of culture (Amsellem et al., 2003). A biologically active TAT–HOXB4 fusion protein has been produced and expressed as a recombinant protein that on purification could be added to murine HSC cultures (Sauvageau et al., 2004). Since the half-life of intracellular HOXB4 was ~1 h, TAT–HOXB4 was added to cell culture every 3 h for four days together with cytokines (KL, IL-6, IL-3), resulting in a five- to sixfold expansion of HSC, as measured by competitive repopulation.

Ectopic overexpression of HOXB4 in murine bone marrow produced HSC that were ~40-fold more competitive than untransduced cells in mouse repopulation assay, and by three to five months the HSC pool size of reconstituted mice equaled, but never exceeded, that of untreated control mice (Sauvageau et al., 2004). High-level ectopic expression of HOXB4 in human cord blood CD34$^+$ cells had a selective growth advantage in NOD/SCID mice but with substantial impairment in myeloerythroid differentiation and B-cell development (Shojaei et al., 2005). In nonhuman primate competitive repopulating transplantation models, HOXB4-overexpressing CD34$^+$ cells had a 56-fold higher short-term, and fivefold higher long-term (6 mo) engraftment than control cells (X. Zhang et al., 2006).

## VII. NEGATIVE REGULATION OF HSC

TGFβ maintains HSC in a quiescent or slowly cycling state, and this cell-cycle arrest was linked to up-regulation of the cyclin-dependent kinase inhibitor p57KIP2 in primary human hematopoietic cells (Fig. 49.2) (Scandura et al., 2004). Fortunel et al. (2003) attributed the TGFβ affect in part to down-modulation of cell surface expression of tyrosine kinase receptors cKit, Flt3, and IL-6R and the Tpo receptor Mpl. This negative regulatory role of TGFβ has been challenged by Larsson et al. (2003), who showed that TGFβ-R1–null mice had normal in vivo hematopoiesis and a normal HSC cell-cycle distribution and did not differ in long-term HSC repopulating potential as compared to wild-type animals. The negative regulatory role of angiopoietin and osteopontin is discussed in the following section in the context of quiescence of HSC in osteoblasts niches.

Two cyclin-dependent kinase inhibitors (CKI), p21 (Cip1/Waf1) and p27kip1, have been shown to govern the pool size of HSC and progenitors, respectively (Cheng and Scadden, 2002). Enforced expression of the HOXB4 transcription factor and down-regulation of p21 (Cip1/Waf) can each independently increase murine HSC proliferation (Miyake et al., 2006). When p21 knockdown and HOXB4 overexpression were combined in HSC, long-term competitive repopulating cells expanded 100-fold in five days, compared to wild-type HSC, and threefold greater than HOXB4 alone. The transcription factor MEF (or EL4F) regulates quiescence of primitive hematopoietic cells (Lacorazza et al., 2006). MEF-null HSC display increased residence in G0, with reduced 5-bromodeoxyuridine incorporation in vivo and impaired cytokine proliferation in vitro. MEF-null mice are consequently relatively resistant to the myelosupressive effects of chemotherapy or radiation.

## VIII. HEMATOPOIETIC STEM CELL NICHES

Direct visualization of engrafted murine GFP$^+$ HSC in candidate niches of the mouse bone has been obtained (Yoshimoto et al., 2003). In myeloablative conditions, GFP$^+$ foci were first detected at the femoral epiphyses and some ribs and vertebra and thereafter spread to other bones. In nonmyeloablative transplants GFP$^+$ cells localized in femoral epiphyses and in some vertebra and ribs but remained quiescent for at least four months, presumably due to residence in quiescent niches (Fig. 49.2). Primitive hematopoietic cells, including HSC, are retained in the endosteal niche, whereas lineage-committed cells localize to the central marrow regions (Nilsson et al., 2003) (Fig. 49.4). These two populations display striking differences in the expression of cell surface adhesion molecules. Murine SLAM CD150-expressing HSC localize in various sites after in vivo injection, including endothelial niches provided by the sinusoids of spleen and marrow as well as endosteal niches (Fig. 49.4) (Kiel et al., 2005). The presence of glycosaminoglycan hyaluronic acid on HSC appears critical for the spatial distribution of transplanted stem cells in vivo. Furthermore, binding of hyaluronic acid by a surrogate ligand results in marked inhibition of HSC proliferation and granulocyte differentiation. A unique feature of the bone niche for HSC is the high concentration of calcium ions at the endosteal surface. HSC express a calcium-sensing receptor (CaR), and mice defi-

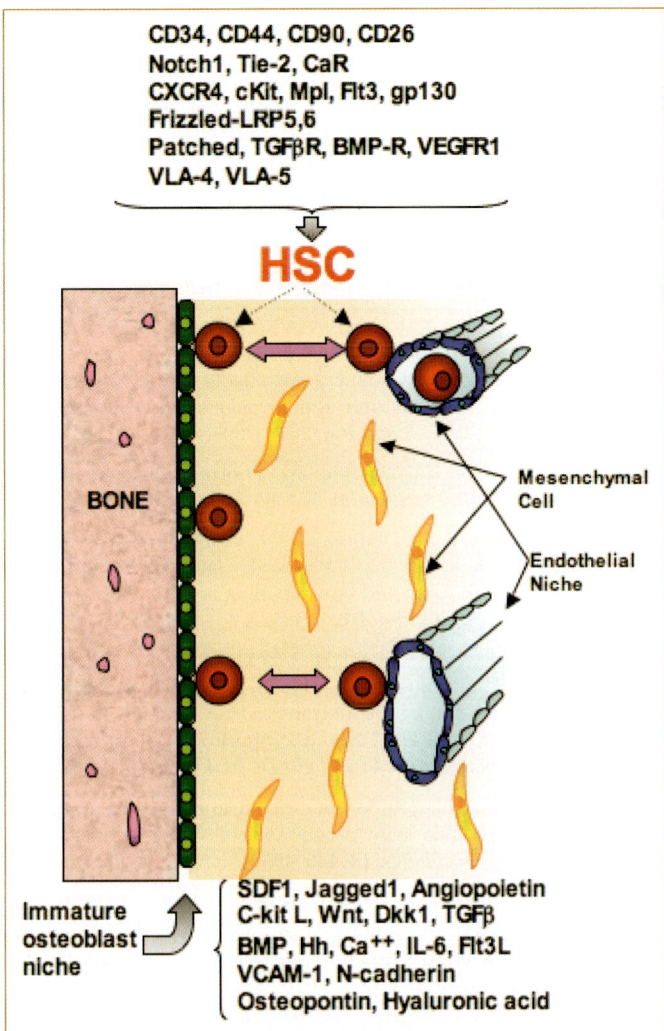

CD34, CD44, CD90, CD26
Notch1, Tie-2, CaR
CXCR4, cKit, Mpl, Flt3, gp130
Frizzled-LRP5,6
Patched, TGFβR, BMP-R, VEGFR1
VLA-4, VLA-5

**HSC**

BONE

Mesenchymal
Cell

Endothelial
Niche

Immature
osteoblast
niche

SDF1, Jagged1, Angiopoietin
C-kit L, Wnt, Dkk1, TGFβ
BMP, Hh, Ca++, IL-6, Flt3L
VCAM-1, N-cadherin
Osteopontin, Hyaluronic acid

**FIG. 49.4.** Human CD34+/CXCR4+ HSC from the circulation transit the BM sinusoids comprising the endothelial niches and home to the endosteal niche. Interactive molecules are shown that are expressed either on the HSC or on the immature osteoblasts.

cient in CaR had normal numbers of HSC in the fetal liver but were highly deficient in HSC localizing to the endosteal niche (Adams *et al.*, 2006).

Further evidence for the importance of the osteoblast niche has been provided in studies in mice with genetic modifications leading to increased numbers of osteoblasts. Mice with constitutively activated parathyroid receptors have enhanced numbers of osteoblasts associated with a doubling of BM Sca-1+c−Kit+Lin− cells and increased LTC-IC and long-term repopulating HSC (Calvi *et al.*, 2003). Osteoblasts with activated receptors produced high levels of the Notch ligand Jagged1, which may be responsible for HSC expansion, since HSC expansion was abrogated following treatment with a gamma secretase inhibitor that blocked Notch activation. Mutant mice with conditional inactivation of the BMP receptor 1A led to an increase in the number of

spindle-shaped N-cadherin+, CD45− osteoblasts (J. Zhang *et al.*, 2003). This was linked to a 2.4-fold increase in quiescent HSC and competitive repopulating cells in the marrow. Since β-catenin forms a complex with N-cadherin, this may play a role in retaining HSC in osteoblastic niches (Fig. 49.4).

A quiescent state is probably critical for maintenance of HSC. HSC expressing the Tie2 receptor are quiescent and antiapoptotic and comprise a side population that adheres to osteoblasts in the BM niche (Arai *et al.*, 2004). The Tie-2 ligand, angiopoietin-1, induced HSC quiescence, their adherence to the endosteal surface, cobblestone formation, and retention of long-term repopulating activity (Fig. 49.2). The matrix glycoprotein, osteopontin (OPN), acts as a negative regulatory restraining factor on HSC within the osteoblast niche (Stier *et al.*, 2005; Nilsson *et al.*, 2005). Ostoblasts at the endosteal bone surface produce varying amounts of osteopontin in response to stimulation and HSC specifically bind to OPN via β1 integrin. Exogenous OPN potently suppresses the proliferation of HSC *in vitro*, and OPN-deficient mice have increased numbers of HSC with markedly enhanced cycling.

c-Myc appears to control the balance between HSC self-renewal and differentiation, possibly by regulating the interaction between HSC and their niches (Wilson *et al.*, 2004). Conditional elimination of c-Myc activity in BM resulted in severe cytopenia and accumulation of self-renewing HSC *in situ*, with impaired differentiation. The c-Myc-deficient HSC appear trapped in stem cell niches, possibly due to up-regulation of N-cadherin and a number of adhesion molecules. Enforced c-Myc expression in HSC repressed N-cadherin and integrins, leading to loss of self-renewal at the expense of differentiation.

## IX. CONCLUSION

The current dogma recognizes that specific factors determine specific cell fates, ranging from stem cell (HSC) self-renewal through commitment to common myeloid (CMP) or common lymphoid progenitors (CLP) to lineage-restricted progenitor differentiation. A contrary hypothesis to current dogma recognizes that HSC fate may instead be governed by (small) quantitative shifts in the relative activation of known signaling pathways (Davey and Zandstra, 2004). Alterations in the relative levels of pathway activation may arise from dynamic shifts in the expression of signaling components (e.g., receptors, ligands). Because extracellular conditions vary over small ranges, cells are induced to overcome signaling threshold barriers and, as a consequence, adopt new, stable cell fates. *Ex vivo* HSC culture systems can be manipulated (1) quantititatively, e.g., by varying the magnitude of signaling pathway activation, influencing cell fate decisions; (2) temporally, e.g., by defining windows of opportunity for stimulation; and (3) spatially, e.g., by fixed location-dependent signaling (niches).

## X. REFERENCES

Abkowitz, J. L., Catlin, S. N., McCallie, M. T., and Guttorp, P. (2002). Evidence that the number of hematopoietic stem cells per animal is conserved in mammals. *Blood* **100**, 2665–2667.

Adams, G. B., Chabner, K. T., Alley, I. R., Olson, D. P., Szczepiorkowski, Z. M., Poznansky, M. C., Kos, C. H., Pollak, M. R., Brown, E. M., and Scadden, D. T. (2006). Stem cell engraftment at the endosteal niche is specified by the calcium-sensing receptor. *Nature* **439**, 599–603.

Amsellem, S., Pflumio, F., Bardinet, D., Izac, B., Charneau, P., Romeo, P. H., Dubart-Kupperschmitt, A., and Fichelson, S. (2003). *Ex vivo* expansion of human hematopoietic stem cells by direct delivery of the HOXB4 homeoprotein. *Nature Med.* **9**, 1423–1427.

Arai, F., Hirao, A., Ohmura, M., Sato, H., Matsuoka, S., Takubo, K., Ito, K., Koh, G. Y., and Suda, T. (2004). Tie2/Angiopoietin-1 signaling regulates hematopoietic stem cell quiescence in the bone marrow niche. *Cell* **118**, 149–161.

Avigdor, A., Goichberg, P., Shivtiel, S., Dar, A., Peled, A., Samira, S., Kollet, O., Hershkovitz, R., Alon, R., Hardan, I., Ben-Hur, H., Naor, D., Nagler, A., and Lapidot, T. (2004). CD44 and hyaluronoic acid cooperate with SDF-1 in the trafficking of human CD34$^+$ stem/progenitor cells to bone marrow. *Blood* **103**, 2981–2989.

Benveniste, P., Cantin, C., Hyam, D., and Iscove, N. N. (2003). Hematopoietic stem cells engraft in mice with absolute efficiency. *Nat. Immunol.* **4**, 708–713.

Bertrand, J. Y., Giroux, S., Golub, R., Klaine, M., Jalil, A., Boucontet, L., Godin, I., and Cumano, A. (2005). Characterization of purified intra-embryonic hematopoietic stem cells as a tool to define their site of origin. *Proc. Natl. Acad. Sci. U.S.A.* **102**, 134–139.

Beslu, N., Krosl, J., Laurin, M., Mayotte, N., Humphries, R. K., and Sauvageau, G. (2004). Molecular interactions involved in HOXB4-induced activation of HSC self-renewal. *Blood* **104**, 2307–2314.

Bhardwaj, G., Murdoch, B., Wu, D., Baker, D. P., Williams, K. P., Chadwick, K., Ling, L. E., Karanu, F. N., and Bhatia, M. (2001). Sonic hedgehog induces the proliferation of primitive human hematopoietic cells via BMP regulation. *Nat. Immunol.* **2**, 172–180.

Bhatia, M., Bonnet, D., Wu, D., Murdoch, B., Wrana, J., Gallacher, L., and Dick, J. E. (1999). Bone morphogenetic proteins regulate the developmental program of human hematopoietic stem cells. *J. Exp. Med.* **189**, 1139–1148.

Broxmeyer, H. (2005). Biology of cord blood cells and future prospects for enhanced clinical benefit. *Cytotherapy* **7**, 209–218.

Calvi, L. M., Adams, G. B., Weibrecht, K. W., Weber, J. M., Olson, D. P., Knight, M. C., Martin, R. P., Schipani, E., Divieti, P., Bringhurst, F. R., Milner, L. A., Kronenberg, H. M., and Scadden, D. T. (2003). Osteoblastic cells regulate the haematopoietic stem cell niche. *Nature* **425**, 841–846.

Camargo, F. D., Chambers, S. M., Drew, E., McNagny, K. M., and Goodell, M. A. (2006). Hematopoietic stem cells do not engraft with absolute efficiencies. *Blood* **107**, 501–507.

Cheng, T., and Scadden, D. T. (2002). Cell cycle entry of hematopoietic stem cells and progenitor cells controlled by distinct cyclin-dependent kinase inhibitors. *Int. J. Hematol.* **75**, 460–465.

Christopherson, K. W., 2nd, Hangoc, G., Mantel, C. R., and Broxmeyer, H. E. (2004). Modulation of hematopoietic stem cell homing and engraftment by CD26. *Science* **305**, 1000–1003.

Chung, Y. J., Park, B. B., Kang, Y. J., Kim, T. M., Eaves, C. J., and Oh, I. H. (2006). Unique effects of STAT3 on the early phase of hematopoietic stem cell regeneration. *Blood* (Apr. 13) [Epub ahead of print].

Cobas, M., Wilson, A., Ernst, B., Mancini, S. J., MacDonald, H. R., Kemler, R., and Radtke, F. (2004). Beta-catenin is dispensable for hematopoiesis and lymphopoiesis. *J. Exp. Med.* **199**, 221–229.

Davey, R. E., and Zandstra, P. W. (2004). Signal processing underlying extrinsic control of stem cell fate. *Curr. Opin. Hematol.* **11**, 95–101.

Davidson, A. J., Ernst, P., Wang, Y., Dekens, M. P., Kingsley, P. D., Palis, J., Korsmeyer, S. J., Daley, G. Q., and Zon, L. I. (2003) *Cdx4* mutants fail to specify blood progenitors and can be rescued by multiple *hox* genes. *Nature* **425**, 300–306.

Delaney, C., Varnum-Finney, B., Aoyama, K., Brashem-Stein, C., and Bernstein, I. D. (2005). Dose-dependent effects of the Notch ligand Delta1 on *ex vivo* differentiation and *in vivo* marrow repopulating ability of cord blood cells. *Blood* **106**, 2693–2699.

Dick, J. E., and Lapidot, T. (2005). Biology of normal and acute myeloid leukemia stem cells. *Int. J. Hematol.* **82**, 389–396.

Duncan, A. W., Rattis, F. M., DiMascio, L. N., Congdon, K. L., Pazianos, G., Zhao, C., Yoon, K., Cook, J. M., Willert, K., Gaiano, N., and Reya, T. (2005). Integration of Notch and Wnt signaling in hematopoietic stem cell maintenance. *Nat. Immunol.* **6**, 314–322.

Feugier, P., Li, N., Jo, D-Y., Shieh, J. H., MacKenzie, K. L., Lesesve, J. F., Latger-Cannard, V., Bensoussan, D., Crystal, R. G., Rafii, S., Stoltz, J. F., and Moore, M. A. S. (2005). Osteopetrotic mouse stroma with thrombopoietin, c-kit ligand, and Flk-2 ligand supports long-term mobilized CD34$^+$ hematopoiesis *in vitro*. *Stem Cells Devel.* **14**, 505–516.

Fortunel, N. O., Hatzfeld, J. A., Monier, M. N., and Hatzfeld, A. (2003). Control of hematopoietic stem/progenitor cell fate by transforming growth factor-beta. *Oncol. Res.* **13**, 445–453.

Gammaitoni, L., Weisel, K. C., Gunetti, M., Wu, K. D., Bruno, S., Pinelli, S., Bonati, A., Aglietta, M., Moore, M. A., and Piacibello, W. (2004). Elevated telomerase activity and minimal telomere loss in cord blood long-term cultures with extensive stem cell replication. *Blood* **103**, 4440–4448.

Gekas, C., Dieterlen-Lievre, F., Orkin, S. H., and Mikkola, H. K. (2005). The placenta is a niche for hematopoietic stem cells. *Dev. Cell* **8**, 365–375.

Glimm, H., Oh, I. H., and Eaves, C. J. (2000). Human hematopoietic stem cells stimulated to proliferate *in vitro* lose engraftment potential during their S/G(2)/M transit and do not reenter G(0). *Blood* **96**, 4185–4193.

Gu, Y., Filippi, M. D., Cancelas, J. A., Siefring, J. E., Williams, E. P., Jasti, A. C., Harris, C. E., Lee, A. W., Prabhakar, R., Atkinson, S. J., Kwiatkowski, D. J., and Williams, D. A. (2003). Hematopoietic cell regulation by Rac1 and Rac2 guanosine triphosphatases. *Science* **302**, 445–449.

Hattori, K., Heissig, B., Tashiro, K., Honjo, T., Tateno, M., Shieh, J-H., Hackett, N., Quitoriano, M. S., Crystal, R. G., Rafii, S., and Moore, M. A. S. (2001). Plasma elevation of stromal derived factor-1 induces mobilization of mature and immature hematopoietic progenitor and stem cells. *Blood* **97**, 3354–3360.

Haylock, D. N., To, L. B., Dowse, T. L., Juttner, C. A., and Simmons, P. J. (1992). *Ex vivo* expansion and maturation of peripheral blood CD34$^+$ cells into the myeloid lineage. *Blood* **80**, 1405–1412.

Heissig, B., Hattori, K., Dias, S., Friedrich, M., Ferris, B., Hackett, N., Crystal, R. G., Besmer, P., Lyden, D., Moore, M. A. S., Werb, Z., and Rafii, S. (2002). Recruitment of stem cells from the bone marrow niche requires MMP-9 mediated release of kit-ligand. *Cell* **109**, 625–637.

Ho, A. D. (2005). Kinetics and symmetry of division of hematopoietic stem cells. *Exp. Hematol.* **33**, 1–8.

Ishikawa, F., Livingston, A. G., Minamiguchi, H., Wingard, J. R., and Ogawa, M. (2003). Human cord blood long term engrafting cells are CD34⁺ CD38⁻. *Leukemia* **17**, 960–964.

Jacobs, J. J., Kieboom, K., Marino, S., DePinho, R. A., and van Lohuizen, M. (1999). The oncogene and Polycomb-group gene bmi-1 regulates cell proliferation and senescence through the ink4a locus. *Nature* **397**, 164–168.

Jansen, M., Yang, F. C., Cancelas, J. A., Bailey, J. R., and Williams, D. A. (2005). Rac2-deficient hematopoietic stem cells show defective interaction with the hematopoietic microenvironment and long-term engraftment failure. *Stem Cells* **23**, 335–346.

Jo, D-Y., Rafii, S., Hamada, T., and Moore, M. A. S. (2000) Chemotaxis of primitive hematopoietic cells in response to stromal cell-derived factor-1. *J. Clin Invest.* **105**, 101–111.

Katayama, Y., Battista, M., Kao, W. M., Hidalgo, A., Peired, A. J., Thomas, S. A., and Frenette, P. S. (2006). Signals from the sympathetic nervous system regulate hematopoietic stem cell egress from bone marrow. *Cell* **124**, 407–421.

Kertesz, Z., Vas, V., Kiss, J, Urban, V. S., Pozsonyi, E., Kozma, A., Paloczi, K., and Uher, F. (2006). *In vitro* expansion of long-term repopulating hematopoietic stem cells in the presence of immobilized Jagged-1 and early acting cytokines. *Cell Biol. Int.* (Apr. 18) [Epub ahead of print].

Kiel, M. J., Yilmaz, O. H., Iwashita, T., Yilmaz, O. H., Terhorst, C., and Morrison, S. J. (2005). SLAM family receptors distinguish hematopoietic stem and progenitor cells and reveal endothelial niches for stem cells. *Cell* **121**, 1109–1121.

Kirito, K., Fox, N., and Kaushansky, K. (2003). Thrombopoietin stimulates HoxB4 expression: an explanation for the favorable effects of Tpo on hematopoietic stem cells. *Blood* **102**, 3172–3180.

Kobune, M., Ito, Y., Kawano, Y., Sasaki, K., Uchida, H., Nakamura, K., Dehari, H., Chiba, H., Takimoto, R., Matsunaga, T., Terui, T., Kato, J., Niitsu, Y., and Hamada, H. (2004). Indian hedgehog gene transfer augments hematopoietic support of human stromal cells including NOD/SCID-β2m⁻/⁻ repopulating cells. *Blood* **104**, 1002–1009.

Koller, M. R., Emerson, S. G., and Palsson, B. O. (1993). Large-scale expansion of human stem and progenitor cells from bone marrow mononuclear cells in continuous perfusion cultures. *Blood* **82**, 378–384.

Kollet, O., Dar, A., Shivtiel, S., Kalinkovich, A., Lapid, K., Sztainberg, Y., Tesio, M., Samstein, R. M., Goichberg, P., Spiegel, A., Elson, A., and Lapidot, T. (2006). Osteoclasts degrade endosteal components and promote mobilization of hematopoietic progenitor cells. *Nat. Med.* **12**, 657–664.

Krosl, J., Beslu, N., Mayotte, N., Humphries, R. K., and Sauvageau, G. (2003). The competitive nature of HOXB4-transduced HSC is limited by PBX1: the generation of ultra-competitive stem cells retaining full differentiation potential. (2003). *Immunity* **18**, 561–571.

Kusadasi, N., Koevoet, J. L., van Soest, P. L., and Ploemacher, R. E. (2001). Stromal support augments extended long-term *ex vivo* expansion of hemopoietic progenitor cells. *Leukemia* **15**, 1347–1358.

Lacorazza, H. D., Yamada, T., Liu, Y., Miyata, Y., Sivina, M., Nunes, J., and Nimer, S. D. (2006). The transcription factor MEF/ELF4 regulates the quiescence of primitive hematopoietic cells. *Cancer Cell* **9**, 175–187.

Lapidot, T., Dar, A., and Kollet, O. (2005). How do stem cells find their way home? *Blood* **106**, 1901–1910. Review.

Larochelle, A., Krouse, A., Metzger, M., Orlic, D., Donahue, R. E., Fricker, S., Bridgere, G., Dunbar, C. E., and Hematti, P. (2006). AMD3100 mobilizes hematopoietic stem cells with long-term repopulating capacity in nonhuman primates. *Blood* **107**, 3772–3778.

Larsson, J., Blank, U., Helgadottir, H., Bjornsson, J. M., Ehinger, M., Goumans, M. J., Fan, X., Leveen, P., and Karlsson, S. (2003). TGF-beta signaling-deficient hematopoietic stem cells have normal self-renewal and regenerative ability *in vivo* despite increased proliferative capacity *in vitro*. *Blood* **102**, 3129–3135.

Lewis, I. D., Almeida-Porada, G., Du, J., Lemischka, I. R., Moore, K. A., Zanjani, E. D., and Verfaillie, C. M. (2001). Umbilical cord blood cells capable of engrafting in primary, secondary, and tertiary xenogeneic hosts are preserved after *ex vivo* culture in a noncontact system. *Blood* **97**, 3441–3449.

Liu, B., Buckley, S. M., Lewis, I. D., Goldman, A. I., Wagner, J. E., and van der Loo, J. C. (2003). Homing defect of cultured human hematopoietic cells in the NOD/SCID mouse is mediated by Fas/CD95. *Exp. Hematol.* **31**, 824–832.

Mancini, S. J., Mantei, N., Dumortier, A., Suter, U., Macdonald, H. R., and Radtke, F. (2005). Jagged1-dependent Notch signaling is dispensable for hematopoietic stem cell self-renewal and differentiation. *Blood* **105**, 2340–2342.

Matsuoka, S., Ebihara, Y., Xu, M., Ishii, T., Sugiyama, D., Yoshino, H., Ueda, T., Manabe, A., Tanaka, R., Ikeda, Y., Nakahata, T., and Tsuji, K. (2001). CD34 expression on long-term repopulating hematopoietic stem cells changes during developmental stages. *Blood* **97**, 419–425.

Mazurier, F., Doedens, M., Gan, O. I., and Dick, J. E. (2003). Rapid myelo-erythroid repopulation after intrafemoral transplantation of NOD-SCID mice reveals a new class of human stem cells. *Nat. Med.* **9**, 959–963.

McCarthy, K. F. (2003). Marrow frequency of rat long-term repopulating cells: evidence that marrow hematopoietic stem cell concentration may be inversely proportional to species body weight. *Blood* **101**, 3431–3435.

McCulloch, E. A., and Till, J. E. (2005). Perspectives on the properties of stem cells. *Nat. Med.* **11**, 1026–1028.

Miyake, N., Brun, A. C., Magnusson, M., Miyake, K., Scadden, D. T., and Karlsson, S. (2006). HOXB4-induced self-renewal of hematopoietic stem cells is significantly enhanced by p21 deficiency. *Stem Cells* **24**, 653–661.

Moore, M. A. S. (2005). Converging pathways in leukemogenesis and stem cell self-renewal. *Exp. Hematol.* **33**, 719–737.

Moore, M. A. S. (2006). The ontogeny of the hematopoietic system. *In* "Handbook of Stem Cells, Vol. 2 — Adult and Fetal" (R. P. Lanza, H. M. Blau, D. A. Melton, M. A. S. Moore, E. D. Thomas, C. M. Verfaillie, I. L. Weissman, and M. D. West), Ch. 15, pp. 159–174.

Moore, M. A. S., Sheridan, A. P. C., Allen, T. D., and Dexter, T. M. (1979). Prolonged hematopoiesis in a primate bone marrow culture system: characteristics of stem cell production and the hematopoietic microenvironment. *Blood* **54**, 775–793.

Moore, M. A. S., Broxmeyer, H. E., Sheridan, A. P. C., Meyers, P. A., Jacobsen. N., and Winchester, R. J. (1980). Continous human bone marrow culture: Ia antigen characterization of probable human pluripotential stem cells. *Blood* **55**, 682–690.

Muench, M. O., Firpo, M. T., and Moore. M. A. S. (1993). Bone marrow transplantation with Interleukin-1 plus kit-ligand *ex vivo* expanded bone marrow accelerates hematopoietic reconstitution in mice without the loss of stem cell lineage and proliferative potential. *Blood* **81**, 3463–3473.

Murdoch, B., Chadwick, K., Martin, M., Shojaei, F., Shah, K. V., Gallacher, L., Moon, R. T., and Bhatia, M. (2003). Wnt-5A augments repopulating capacity and primitive hematopoietic development of human blood stem cells *in vivo*. *Proc. Natl. Acad. Sci. U.S.A.* **100**, 3422–3427.

Nilsson, S. K., Haylock, D. N., Johnston, H. M., Occhiodoro, T., Brown, T. J., and Simmons, P. J. (2003). Hyaluronan is synthesized by primitive hemopoietic cells, participates in their lodgment at the endosteum following transplantation, and is involved in the regulation of their proliferation and differentiation *in vitro*. *Blood* **101**, 856–862.

Nilsson, S. K., Johnston, H. M., Whitty, G. A., Williams, B., Webb, R. J., Denhardt, D. T., Bertoncello, I., Bendall, L. J., Simmons, P. J., and Haylock, D. N. (2005). Osteopontin, a key component of the hematopoietic stem cell niche and regulator of primitive hematopoietic progenitor cells. *Blood* **106**, 1232–1239.

Ogawa, M. (2002). Changing phenotypes of hematopoietic stem cells. *Exp. Hematol.* **30**, 3–6.

Papayannopoulou, T. (2004). Current mechanistic scenarios in hematopoietic stem/progenitor cell mobilization. *Blood* **103**, 1580–1585.

Park, I-K., Qian, D., Kiel, M., Becker, M. W., Pihalja, M., Weissman, I. L., Morrison, S. J., and Clarke, M. F. (2003). Bmi-1 is required for maintenance of adult self-renewing haematopoietic stem cells. *Nature* **423**, 302–305.

Petit, I., Szyper-Kravitz, M., Nagler, A., Lahav, M., Peled, A., Habler, L., Ponomaryov, T., Taichman, R. S., Arenzana-Seisdedos, F., Fujii, N., Sandbank, J., Zipori, D., and Lapidot, T. (2002). G-CSF induces stem cell mobilization by decreasing bone marrow SDF-1 and up-regulating CXCR4. *Nat. Immunol.* **3**, 687–694.

Piacibello, W., Sanavio, F., Severino, A., *et al.* (1999). Engraftment in nonobese diabetic severe immunodeficient mice of human CD34⁺ cord blood cells after *ex vivo* expansion: evidence for the amplification and self-renewal of repopulating stem cells. *Blood* **93**, 3736–3749.

Reya, T., and Cleavers, H. (2005). Wnt signalling in stem cells and cancer. *Nature* **434**, 843–850. Review.

Reya, T., Duncan, A. W., Ailles, L., Domen, J., Scherer, D. C., Willert, K., Hintz, L., Nusse, R., and Weissman, I. L. (2003). A role for Wnt signalling in self-renewal of haematopoietic stem cells. *Nature* **423**, 409–414.

Rollini, P., Kaiser, S., Faes-van't, Hull, E., Kapp, U., and Leyvraz, S. (2004). Long-term expansion of transplantable human fetal liver hematopoietic stem cells. *Blood* **103**, 1166–1170.

Sauvageau, G., Iscove, N. N., and Humphries, R. (2004). *In vitro* and *in vivo* expansion of hematopoietic stem cells. *Oncogene* **23**, 7223–7232.

Scandura, J. M., Boccuni, P., Massague, J., and Nimer, S. D. (2004). Transforming growth factor beta-induced cell cycle arrest of human hematopoietic cells requires p57KIP2 up-regulation. *Proc. Natl. Acad. Sci. U.S.A.* **101**, 15231–15236.

Schiedlmeier, B., Klump, H., Will, E., Arman-Kalcek, G., Li, Z., Wang, Z., Rimek, A., Friel, J., Baum, C., and Ostertag, W. (2003). High-level ectopic HOXB4 expression confers a profound *in vivo* competitive growth advantage on human cord blood CD34⁺ cells, but impairs lymphomyeloid differentiation. *Blood* **101**, 1759–1768

Schuringa, J. J., Chung, K. Y., Morrone, G., and Moore, M. A. S. (2004). Constitutive activation of STAT5 promotes human hematopoietic stem cell self-renewal and erythroid differentiation. *J. Exp. Med.* **200**, 623–635.

Shepherd, B. E., Guttorp, P., Lansdorp, P. M., and Abkowitz, J. L. (2004). Estimating human hematopoietic stem cell kinetics using granulocyte telomere lengths. *Exp. Hematol.* **32**, 1040–1050.

Shojaei, F., Trowbridge, J., Gallacher, L., Yuefei, L., Goodale, D., Karanu, F., Levac, K., and Bhatia, M. (2005). Hierarchical and ontogenic positions serve to define the molecular basis of human hematopoietic stem cell behavior. *Dev. Cell* **8**, 651–663.

Stier, S., Ko, Y., Forkert, R., Lutz, C., Neuhaus, T., Grunewald, E., Cheng, T., Dombkowski, D., Calvi, L. M., Rittling, S. R., and Scadden, D. T. (2005). Osteopontin is a hematopoietic stem cell niche component that negatively regulates stem cell pool size. *J. Exp. Med.* **201**, 1781–1791.

Summers, Y. J, Heyworth, C. M., de Wynter, E. A., Hart, C. A., Chang, J., and Testa, N. G. (2004). AC133⁺ G0 cells from cord blood show a high incidence of long-term culture-initiating cells and a capacity for more than 100 million-fold amplification of colony-forming cells *in vitro*. *Stem Cells* **22**, 704–715.

Trowbridge, J. J., Xenocostas, A., Moon, R. T., and Bhatia, M. (2006). Glycogen synthase kinase-3 is an *in vivo* regulator of hematopoietic stem cell repopulation. *Nat. Med.* **12**, 89–98.

Ueda, T., Tsuji, K., Yoshino, H., Ebihara, Y., Yagasaki, H., Hisakawa, H., Mitsui, T., Manabe, A., Tanaka, R., Kobayashi, K., Ito, M., Yasukawa, K., and Nakahata, T. (2000). Expansion of human NOD/SCID-repopulating cells by stem cell factor, Flk2/Flt3 ligand, thrombopoietin, IL-6 and soluble IL-6 receptor. *J. Clin. Invest.* **105**, 1013–1021.

Varnum-Finney, B., Brashem-Stein, C., and Bernstein, I. D. (2003). Combined effects of Notch signaling and cytokines induce a multiple log increase in precursors with lymphoid and myeloid reconstituting ability. *Blood* **101**, 1784–1789.

Willert, K., Brown, J. D., Danenberg, E., Duncan, A. W, Weissman, I. L., Reya ,T., Yates, J. R., 3rd, and Nusse, R. (2003). Wnt proteins are lipid-modified and can act as stem cell growth factors. *Nature* **423**, 448–452.

Wilson. A., Murphy, M. J., Oskarsson, T., Kaloulis, K., Bettess, M. D., Oser, G. M., Pasche, A. C., Knabenhans, C., Macdonald, H. R., and Trumpp, A. (2004). c-Myc controls the balance between hematopoietic stem cell self-renewal and differentiation. *Genes Dev.* **18**, 2747–2763.

Yoshimoto, M., Shinohara, T., Heike, T., Shiota, M., Kanatsu-Shinohara, M., and Nakahata, T. (2003). Direct visualization of transplanted hematopoietic cell reconstitution in intact mouse organs indicates the presence of a niche. *Exp. Hematol.* **31**, 733–740.

Zandstra, P. W., Conneally, E., Piret, J. M., and Eaves, C. J. (1998). Ontogeny-associated changes in the cytokine responses of primitive human haematopoietic cells. *Br. J. Haematol.* **101**, 770–778.

Zhang, C. C., Kaba, M., Ge, G., Xie, K., Tong, W., Hug, C., and Lodish, H. F. (2006). Angiopoietin-like proteins stimulate *ex vivo* expansion of hematopoietic stem cells. *Nat. Med.* **12**, 240–245.

Zhang, J., Niu, C., Huang, H., He, X., Tong, W. G., Ross, J., Haug, J., Johnson, T., Feng, J. Q., Harris, S., Wiedemann, L. M., Mishina, Y., and Li, L. (2003). Identification of the haematopoietic stem cell niche and control of the niche size. *Nature* **425**, 836–841.

Zhang, X. B., Beard, B. C., Beebe, K., Storer, B., Humphries, R. K., and Kiem, H. P. (2006). Differential effects of HOXB4 on nonhuman primate short- and long-term repopulating cells. *Plos. Med.* **3**, e173.

Zhang, Y., Morrone, G., Zhang, J., Chen, X., Lu, X., Ma, L., Moore, M. A. S., and Zhou, P. (2003). CUL-4A stimulates ubiquitylation and degradation of the HOXA9 homeodomain protein. *Embo. J.* **22**, 6057–6067.

# Chapter Fifty

# Red Blood Cell Substitutes

*Thomas Ming Swi Chang*

## I. INTRODUCTION

Hemoglobin molecules extracted from red blood cells are modified by microencapsulation or cross-linking. This stabilizes the hemoglobin molecules and also allows the sterilization of the product to remove HIV and other microorganisms. A number of modified hemoglobin molecules have been developed, and extensive clinical trials have been carried out. These clinical studies have led to the recent finding that it is important to have nanodimension-modified hemoglobin in order to avoid problems related to vasoconstriction caused by the removal of nitric oxide. Although development toward routine clinical uses of blood substitutes is complex and slow, one type of nano-dimension polyhemoglobin has been approved for routine clinical uses in South Africa and in Phase III clinical trials in North America. New generations of modified hemoglobin are also being actively developed, including nano-dimension polyhemoglobin-catalase-superoxide dismutase, nano-artificial red blood cells of lipid or biodegradable polymeric membrane, and others.

In 1957, the author initiated the first research on artificial cells including modified hemoglobin (Chang, 1957). Artificial cells (Fig. 50.1) retain biologically active materials such as hemoglobin, enzymes, cells, and adsorbent and other materials (Chang, 1964, 1972, 2005, 2007). Their dimensions can vary in the macro, micro, nano, and molecular ranges (Fig. 50.2). The membrane of artificial cells allows permeable molecules such as oxygen and substrates to enter and products, peptides, and other materials to leave. In this way the enclosed materials are protected from immunological rejection and other materials from the external environment. Until 1989, most research was on artificial cells containing enzymes, cells, microorganisms, adsorbents, peptides, and others.

Concentrated research and development on modified hemoglobin started only after 1987 because of public concerns of HIV in donor blood. After extraction from red blood cells and before modification, hemoglobin can be sterilized by pasteurization, ultrafiltration, and chemical means. These procedures can remove microorganisms, including those responsible for AIDS, hepatitis, and other diseases. There are many situations where modified hemoglobin can potentially be used to substitute for red blood cells (Chang, 1997, 2003, 2004, 2005; Klein, 2000; Rudolph *et al.*, 1997; Tuschida, 1998; Kobayashi *et al.*, 2005; Winslow, 2003, 2006). These include trauma surgery, cardiopulmonary bypass surgery, cancer surgery, other elective surgery, and cardioplegia. Another area is in severe traumatic injuries, such as traffic accidents and other accidents that result in severe bleeding and hemorrhagic shock. The number of traumatic injuries requiring blood substitutes in civilian use is small when compared to the requirements in major disasters or wars. Modified hemoglobin is especially useful in emergency situations such as these. Since modified

*Principles of Tissue Engineering, 3rd Edition*
*ed. by Lanza, Langer, and Vacanti*

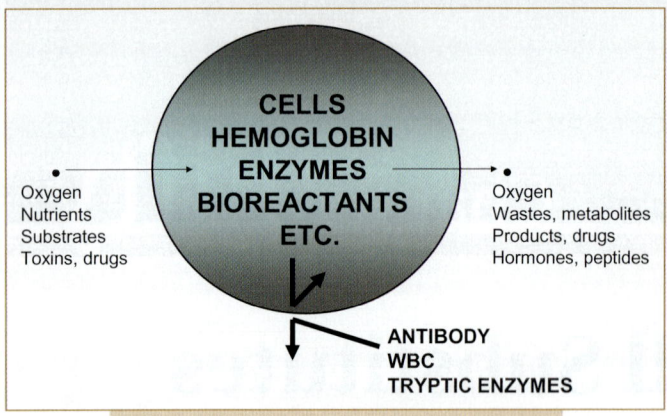

**FIG. 50.1.** Basic principle of artificial cells. From Chang (2004), with copyright permission from Marcel Dekker, Publisher.

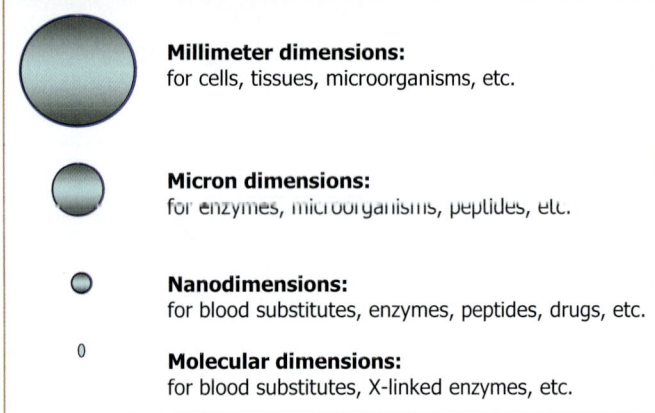

**FIG. 50.2.** Artificial cells in the macro, micro, nano, and molecular dimensions. From Chang (2004), with copyright permission from Marcel Dekker, Publisher.

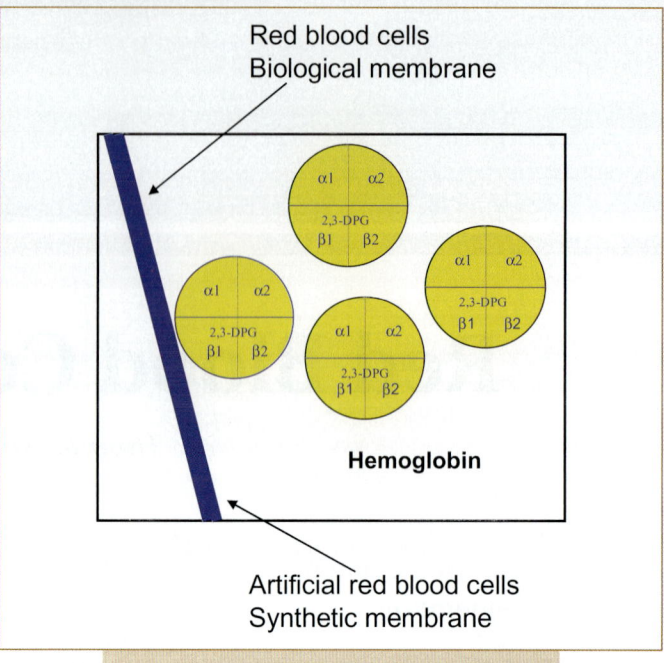

**FIG. 50.3.** Inside the red blood cell, each hemoglobin molecule is a tetramer of four subunits: two alpha subunits and two beta subunits. Cofactors 2,3-DPG, retained in the red blood cell, help the hemoglobin to release oxygen as required by the tissues. An artificial red blood cell has the same content as red blood cells, but the membrane is made of synthetic material. Modified from Chang (1992), with permision of Marcel Dekker, Publisher.

hemoglobin does not contain red blood cell membrane and therefore no blood group antigens, it can be used without the need for cross-matching or typing. This could save much time and facilities and would permit on-the-spot transfusion as required, similar to giving intravenous solution. This is further facilitated by the fact that modified hemoglobin can be stored as a solution at room temperature for more than two years. It can also be lyophilized and stored as a stable dried powder that can be reconstituted with the appropriate salt solution just before use. Another requirement is where the patients' religious belief does not allow them to use donor blood for transfusion.

## II. MODIFIED HEMOGLOBIN

### Why Do We Have to Modify Hemoglobin?

Hemoglobin in the red blood cell is responsible for carrying oxygen from the lung and delivering oxygen to tissues. It is a tetramer of four subunits: two alpha subunits and two beta subunits (Perutz, 1989) (Fig. 50.3). Hemoglobin is in the "oxy," relaxed, or "R" state when it is carrying oxygen. To release oxygen, the hemoglobin molecule undergoes conformational change with a 15° rotation. The molecule is then in the "deoxy," tensed, or "T" state. Red blood cells contain a cofactor, 2,3-DPG, which facilitates this conformational change. Thus, hemoglobin inside red blood cells has a high $P_{50}$, which allows it readily to release oxygen to the tissue at physiological oxygen tensions.

Hemoglobin can be extracted from red blood cells by removing the cell membranes to form stroma-free hemoglobin. Attempts in clinical trials in patients to use stroma-free hemoglobin as a blood substitute were not successful because of renal toxicity and other adverse effects (Savitsky *et al.*, 1978). When infused into the circulation, each hemoglobin molecule of four subunits (tetramer) is rapidly broken down into two subunits (dimers). These toxic dimers are the major causes of the adverse effects.

### Modified Hemoglobin

Biotechnological approaches can be used either to cross-link or to encapsulate the hemoglobin molecules, to prevent hemoglobin from dissociating into dimers and also to allow hemoglobin to have an acceptable $P_{50}$ (Chang, 1957, 1964, 1971, 1972). Many groups have since contributed to major progress, especially since around 1990. These include Abuchowski, Agishi, Alayash, Bakker, Benesch, Biro,

Bonhard, Bucci, Bund and Jandl, Chang, DeAngelo, D'Agnillo, DeVenuto, Estep, Faivre, Farmer, Feola, Fratantoni, Fronteicelli, Gould, Greenburg, Hedlund, Hess, Hori, Hsia, Iwashita, Jesch, Kobayashi, Keipert, Klugger, Liu, Manning, Messmer, Moss, Nose, Olson, Privalle, Pristoupil, Rudolph, Segal, Sekiguchi, Sideman, Shorr, Valeri, Winslow, Wong, and Yang, among many others. Microencapsulation of hemoglobin to prepare artificial red blood cells (Chang, 1957, 1964) is a rather ambitious approach (Fig. 50.3). Although this attempt to mimic red blood cells has resulted in a complete red blood cell substitute, it is rather complicated. Further research is needed. This approach, now considered a third-generation modified hemoglobin, is discussed later. Simpler cross-linked modified hemoglobin has been developed as a first-generation modified hemoglobin for more immediate clinical applications, as discussed next.

## III. FIRST-GENERATION MODIFIED HEMOGLOBIN

### Polyhemoglobin Based on Nanobiotechnology

Hemoglobin contains numerous amino groups, many of which are on the surface of the hemoglobin molecule. This author first reported the use of nanobiotechnology with a bifunctional agent (diacid) to cross-link and assemble hemoglobin molecules into polyhemoglobin (Chang, 1964, 1972) (Fig. 50.4). This was used first to form ultrathin polyhemoglobin membranes for artificial red blood cells. It was also found that by decreasing the size of artificial cells,

all the hemoglobin molecules become cross-linked into polyhemoglobin. Cross-linking prevents the breakdown of hemoglobin tetramers into dimers. The reaction is as follows (Chang, 1964, 1972):

$$Cl–CO–(CH_2)_8–CO–Cl + HB–NH_2$$
$$\text{Diacid} \qquad \text{Hemoglobin}$$
$$= HB–NH–CO–(CH_2)_8–CO–NH–HB$$
$$\text{Polyhemoglobin}$$

Glutaraldehyde is another bifunctional agent first used by this author in a similar nanobiotechnological approach to cross-link and assemble hemoglobin molecules into soluble nanodimension polyhemoglobin (Chang, 1971):

$$H–CO–(CH_2)_3–CO–H + HB–NH_2$$
$$\text{glutaraldehyde} \qquad \text{hemoglobin}$$
$$= HB–NH–CO–(CH_2)_3–CO–NH–HB$$
$$\text{Polyhemoglobin}$$

Since then, a number of other groups have independently carried out extensive studies using this principle of glutaraldehyde-cross-linked polyhemoglobin. The two most successful are glutaraldehyde-cross-linked human polyhemoglobin (Gould *et al.*, 1998, 2002) and glutaraldehyde-cross-linked bovine polyhemoglobin (Pearce and Gawryl, 1998; Sprung *et al.*, 2002). Another glutaraldehyde polyhemoglobin uses hemoglobin from red blood cells of placentas discarded after birth (T. Li *et al.*, 2006). Other cross-linkers have also been developed.

### Intramolecularly Cross-Linked Hemoglobin

Studies have been carried out specifically to cross-link hemoglobin molecules intramolecularly to form cross-linked tetrameric hemoglobin (Fig. 50.4). For example, a bifunctional agent, 2-Nor-2-formylpyridoxal 5-phosphate, which is also a 2,3-DPG analog, can intramolecularly cross-link the two beta subunits of the hemoglobin molecules (Bunn and Jandl, 1968). Another 2,3-DPG pocket modifier, bis(3,5-dibromosalicyl) fumarate (DBBF), intramolecularly cross-links the two alpha subunits of the hemoglobin molecule (Walder *et al.*, 1979; Nelson, 1998).

### Conjugated Hemoglobin

Conjugated hemoglobin involves the cross-linking of hemoglobin to polymers (Chang, 1964, 1972). The use of soluble polymers resulted in soluble conjugated hemoglobin with good circulation time (Iwashita, 1992; Wong, 1988). Two new nanodimension conjugated hemoglobin molecules are being tested in ongoing clinical trials (X. Z. Li *et al.*, 2005; Winslow, 2003, 2006).

### Sources of Hemoglobin

Where do we obtain all the hemoglobin needed for preparing modified hemoglobin? Red blood cells from outdated donor blood is one major source of human hemoglobin. Another source is human hemoglobin from red blood cells

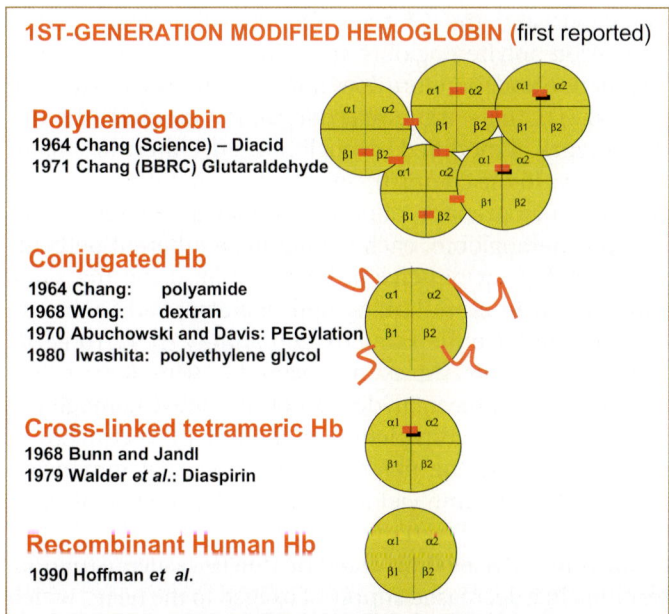

**1ST-GENERATION MODIFIED HEMOGLOBIN** (first reported)

**Polyhemoglobin**
1964 Chang (Science) – Diacid
1971 Chang (BBRC) Glutaraldehyde

**Conjugated Hb**
1964 Chang: polyamide
1968 Wong: dextran
1970 Abuchowski and Davis: PEGylation
1980 Iwashita: polyethylene glycol

**Cross-linked tetrameric Hb**
1968 Bunn and Jandl
1979 Walder *et al.*: Diaspirin

**Recombinant Human Hb**
1990 Hoffman *et al.*

**FIG. 50.4.** Molecular-dimension red blood cell substitutes in the form of polyhemoglobin, conjugated hemoglobin, intramolecularly cross-linked hemoglobin, and recombinant hemoglobin. Modified from Chang (2004), with copyright permission from Marcel Dekker, Publisher.

of placenta discarded after the delivery of babies (T. Li *et al.*, 2006). In addition to hemoglobin from human sources, another source is bovine hemoglobin (Pearce and Gawryl, 1998). Another promising source is human hemoglobin produced by recombinant technology in microorganisms (Frontecelli *et al.*, 2004; Freytag and Templeton, 1997; Hoffman *et al.*, 1990). One of these has been tested in clinical trials (Frytag and Templeton, 1997), and the result of clinical trials has led to the development of a second-generation recombinant human hemoglobin (Doherty *et al.*, 1998). Another potential source that is being developed is synthetic heme, which is linked to recombinant human albumin (Tsuchida, 1998; Tsuchida *et al.*, 2002).

## Efficacy and Safety of Modified Hemoglobin Blood Substitutes

Modified hemoglobin no longer has problems related to the breakdown of hemoglobin into toxic dimers. However, how safe is modified hemoglobin when injected into animals and humans? Animal studies using properly prepared modified hemoglobin have shown no adverse effects on coagulation factors, leucocytes, platelets, or complement activation (Chang, 1997, 2003, 2004, 2005; Klein, 2000; Rudolph *et al.*, 1997; Tsuchida, 1998; Kobayashi *et al.*, 2005; Winslow, 2003, 2006). Very severe and sensitive immunological studies have been carried out for polyhemoglobin, conjugated hemoglobin, and microencapsulated hemoglobin (Chang *et al.*, 1992; Chang, 1997). Results show that modified homologous polyhemoglobins (e.g., rat polyhemoglobin injected into rats) are not immunogenic, even with repeated injection. Heterologous polyhemoglobin (e.g., nonrat hemoglobin injected into rats) is not immunogenic initially but is immunogenic after repeated injections with Freund's adjuvant. Conjugation and microencapsulation markedly decreased the antigenicity of heterologous polyhemoglobin. Other studies show that without the use of Freund's adjuvant, repeated subcutaneous and intravenous injections of heterologous cross-linked hemoglobin are much less immunogenic (Estep *et al.*, 1992).

Another important area of safety study is the distribution of modified hemoglobin after infusion and also its effect on the reticuloendothelial system. Another factor is that the safety response in animals is not necessaily the same as in humans, especially in the case of immunological and hypersensitivity types of reactions. We have worked out a simple *in vitro* screening test (Chang, 1997), which consists of adding 0.1 mL of modified hemoglobin to a test tube containing 0.4 mL of human plasma or blood and then analyzing for complement activation. This has been used to detect trace contamination of blood group antigens, antibody–antigen complexes, endotoxin, trace fragments of microorganisms, impurities in polymers, some emulsifiers, and others. Detailed animal studies show that modified hemoglobins are effective in short-term applications such as hemorrhagic shock, exchange transfusion, hemodilution

in surgery, and some other conditions. However, despite all the detailed *in vitro* and *in vivo* safety and efficacy studies and screening tests, as just described, extensive clinical trials since the early 1990s show that some of the blood substitutes have problems related to increased vasoconstriction, as discussed next.

## Vasoactivity and Nitric Oxide

Not all modified hemoglobins have been successful in clinical trials, since a vasopressor effect has been observed in some of the first-generation modified hemoglobin blood substitutes. For example, intramolecularly cross-linked hemoglobin and first-generation recombinant human hemoglobin blood substitutes contain 100% molecular-dimension modified hemoglobin. Infusion causes vasopressor effects and also increased smooth muscle contractions. With another type of polyhemoglobin, which contains more than 30% molecular-dimension modified hemoglobin, significant vasoactivity and increased smooth muscle contractions can also be observed when using larger volumes. On the other hand, the use of nano-dimension polyhemoglobin with <1% molecular-dimension modified hemoglobin did not show vasopressor effects, even when large volumes of 10 liters were infused. This has led to the proposal that the intercellular junctions of the endothelial lining of vascular wall allow molecular-dimension hemoglobin to enter the interstitial space (Fig. 50.5). There, hemoglobin acts as a sink in binding and removing the nitric oxide needed for maintaining the normal tone of smooth muscles. This results in the constriction of blood vessels and other smooth muscles, especially those of the esophagus and the GI tract. On the other hand, nano-dimension polyhemoglobin would not cross the intercellular junction and therefore does not result in vasoconstriction (Fig. 50.5). Others argue that one cannot compare the different types of modified hemoglobin, since there are major differences in the chemistry involved and in oxygen affinity. We therefore prepare laboratory versions of nanodimension polyhemoglobin, each containing a different percentage of molecular-dimension hemoglobin, using the same glutaraldehyde cross-linking and characterized to ensure that they all have the same oxygen affinity (B. L. Yu *et al.*, 2006). The result shows that samples with a very low percentage of molecular-dimension modified hemoglobin do not cause vasoconstriction or changes in electrocardiogram. Increasing the percentage of molecular-dimension modified hemoglobin yielded an increasing degree of vasoconstriction and elevation of the ST segment of the electrocardiogram. ST elevation could be due to vasoconstriction, resulting in a decreased supply of oxygen to the heart, which may explain the observation of small myocardial lesions in some primates and swine after infusion with one type of molecular-dimension modified hemoglobin (Burhop and Estep, 2001). This theory is further supported by the lack of vasoconstriction using nonextravasating hemoglobin

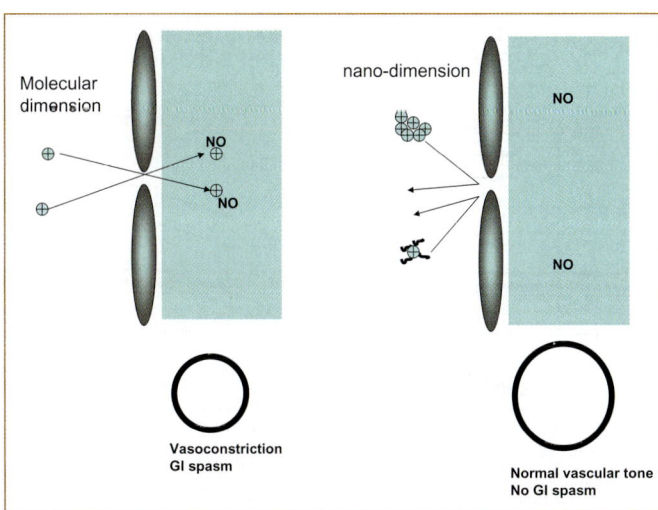

| | DONOR RBC | BLOOD SUBSTITUTES (Polyhemoglobin) |
|---|---|---|
| INFECTION | Possible | Sterilized |
| SOURCE | Limited | "Unlimited"? |
| BLOOD GROUPS | Yes | None |
| USAGE | Xmatch, typing | Immediately |
| STORAGE | 42 days | >1 year |
| FUNCTIONS | RBC | Oxygen carrier |

**FIG. 50.5.** Molecular-dimension modified hemoglobin can transverse the intercellular junction of the endothelial cells of the vascular wall. It will then bind with nitric oxide, thus removing it and resulting in increased vasoactivity, since removal of nitric oxide results in contraction of smooth muscles of the vascular wall and other smooth muscles. Nanodimension modified hemoglobin molecules do not contain any significant free molecular-dimension modified hemoglobin. They cannot transverse the intercellular junction and therefore do not cause vasoconstriction. From Chang (1997), with permission of the copyright holder.

**FIG. 50.6.** Comparison of polyhemoglobin with donor red blood cells. Polyhemoglobin has many advantages over red blood cells and is useful during surgery. However, it cannot be used in a number of other clinical conditions, because, unlike red blood cells (rbc), polyhemoglobin is only an oxygen carrier. It has no rbc enzymes needed for many functions, including the removal of oxygen radicals. Furthermore, its circulation time is much shorter than that of rbc. Modified from Chang (2004), with copyright permission from Marcel Dekker, Publisher.

polymer (Matheson *et al.*, 2002). Furthermore, a second-generation molecular-dimension recombinant human hemoglobin that did not bind nitric oxide did not cause vasoconstriction (Doherty *et al.*, 1998). Since recombinant hemoglobin crosses the intercellular junction and is removed quickly, the circulation time can be increased by cross-linking to form nanodimension polyhemoglobin (Fronticelli *et al.*, 2004). Another approach in an ongoing clinical trial is to prepare PEG-conjugated hemoglobin that, with its water of hydration, would result in a nanodimension modified hemoglobin, and this does not result in vasopressor effects (Winslow, 2006).

### Present Clinical Status of First-Generation Modified Hemoglobin

Two of the most successful ones are based on glutaraldehyde-cross-linked polyhemoglobin. These were developed independently by two groups based on Chang's basic principle of glutaraldehyde-cross-linked polyhemoglobin (Chang, 1971). One is pyridoxalated glutaraldehyde human polyhemoglobin (Gould *et al.*, 1998, 2002). The researchers show in Phase III clinical trials that this can successfully replace extensive blood loss in trauma surgery by maintaining the hemoglobin level, with no reported side effects (Gould *et al.*, 2002). They have infused up to 10 liters into individual trauma surgery patients. Another one is glutaraldehyde-cross-linked bovine polyhemoglobin, which has been extensively tested in clinical trials (Pearce and

Gawryl, 1998; Sprung *et al.*, 2002). This bovine polyhemoglobin has been approved for routine patient uses in South Africa. These two polyhemoglobins have been approved for compassionate use in humans, and they are awaiting regulatory approval for routine clinical use in humans in North America. Conjugated hemoglobin of sufficient nano-dimension is another promising approach that is in phase II clinical trials (Winslow, 2006).

Nanodimension polyhemoglobin and conjugated hemoglobin have a number of advantages when compared to donor red blood cells (Fig. 50.6). Unlike red blood cells, there is no blood group, and thus it can be given immediately on the spot without time-consuming typing and cross-matching. It can be sterilized and is free of infectious agents. Donor blood has to be stored at 4°C and is good for only 42 days. Polyhemoglobin can be stored at room temperature for more than a year. Thus, it has an important role in a number of clinical conditions. In some other conditions, related to severe and sustained ischemia, we may have to add antioxidant enzymes to oxygen carriers in order to avoid the potential for ischemia reperfusion injury (Fig. 50.7).

## IV. NEW GENERATIONS OF MODIFIED HEMOGLOBIN

### Polyhemoglobin Cross-Linked with RBC Antioxidant Enzymes

In sustained severe hemorrhagic shock or other ischemic conditions, oxygen carrier alone may result in ischemia reperfusion injuries, due to the production of oxygen

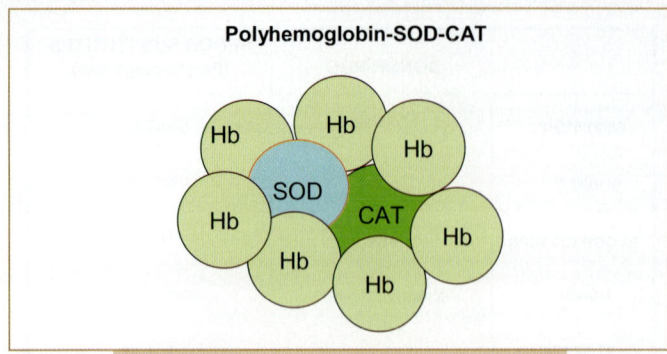

**FIG. 50.7.** Cross-linking of hemoglobin with two rbc enzymes to form polyhemoglobin-catalase-superoxide dismutase (PolyHb-CAT-SOD). Unlike polyhemoglobin, this has rbc enzymes that can remove oxygen radicals. Modified from Chang (2004), with copyright permission from Marcel Dekker, Publisher.

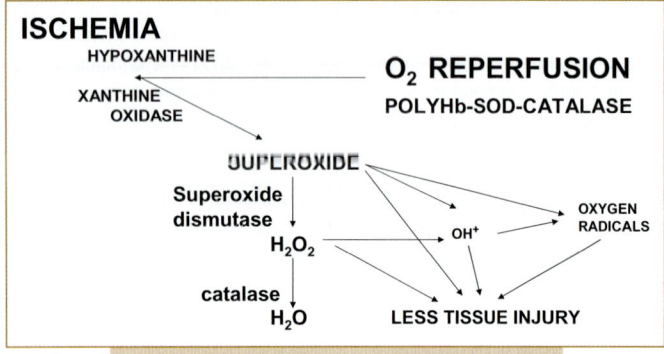

**FIG. 50.8.** In conditions such as severe sustained hemorrhagic shock, stroke, myocardial infarction, and organ transplantation, reperfusion with polyhemoglobin can sometimes result in oxygen radicals that cause tissue injury. PolyHb-CAT-SOD can supply oxygen and at the same time significantly lower any oxygen radicals formed. From Chang (1997), with copyright permission from the copyright holder.

radicals (Fig. 50.8) (Alayash, 2004; Chang, 2003). We are studying the use of a cross-linked polyhemoglobin-superoxide dismutase-catalase (PolyHb-SOD-CAT) (Fig. 50.7) (D'Agnillo and Chang, 1998). The intestine is one of the organs most likely to be injured in this type of ischemia reperfusion injury. We found that, unlike PolyHb, PolyHb-SOD-CAT did not cause a significant increase in oxygen radicals when used to reperfuse ischemic rat intestine (Chang, 1997). Polyhemoglobin is in solution and is therefore more likely to be able to perfuse partially obstructed vessels in stroke, myocardial infarction, and other conditions as compared to red blood cells. In a rat stroke model (Fig. 50.9), we found that after 60 minutes of ischemia, reperfusion with polyHb resulted in a significant increase in the breakdown of blood–brain barrier and an increase in brain edema (Powanda and Chang, 2002). On the other hand, polyHb-SOD-CAT did not result in these adverse changes (Powanda and Chang, 2002).

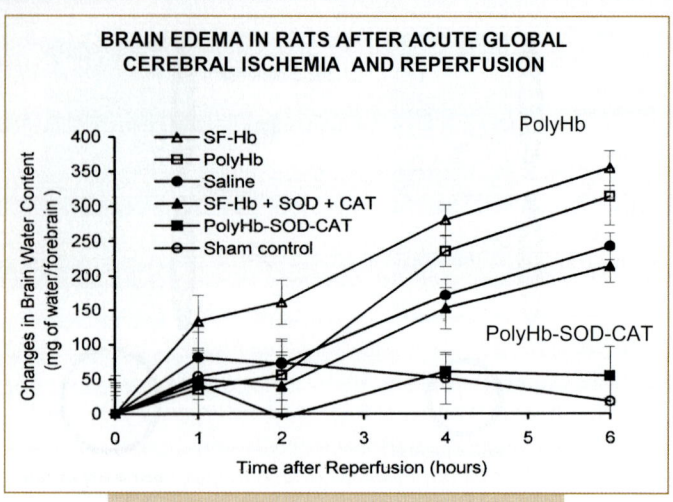

**FIG. 50.9.** This is a rat model of acute global cerebral ischemia followed by reperfusion with different oxygen-carrying solutions. Unlike polyhemoglobin, polyHb-CAT-SOD does not cause brain edema when used in this situation. From Powanda and Chang (2002), with copyright permission from Marcel Dekker, Publisher.

## Polyhemoglobin Cross-Linked with Tyrosinase

Tumors are not well perfused with blood, but radiation therapy and chemotherapy work better with better tissue oxygenation. Since polyhemoglobin is in solution, it can perfuse tumor more effectively than red blood cells and thus has been studied for potential use in radiation therapy and chemotherapy of tumors (Shorr *et al.*, 1996; Pearce and Gawryl, 1998). We have recently cross-linked tyrosinase with hemoglobin to form a soluble polyhemoglobin–tyrosinase complex (B. L. Yu and Chang, 2004). This has the potential dual function of supplying the needed oxygen for radiation therapy and in addition lowers the systemic tyrosine needed for the growth of melanoma. This preparation inhibits the growth of murine B16F10 melanoma culture. In mice, it significantly delays the growth of B16F10 melanoma as compared to the control group, without causing adverse effects or changes in the growth of the treated animals (B. L. Yu and Chang, 2004).

## Artificial Red Blood Cells

The first artificial red blood cells (Figs. 50.1 and 50.3) have an oxygen dissociation curve similar to that for red blood cells, since 2–3-DPG is retained inside (Chang, 1957). Hemoglobin also stays inside as tetramers, and red blood cell enzymes such as carbonic anhydrase and catalase retain their activities (Chang, 1964, 1972). These artificial red blood cells do not have blood group antigens on the membrane and therefore do not aggregate in the presence of blood group antibodies (Chang, 1972). However, the single major problem is the rapid removal of these artificial cells from the circulation. Most of the studies since that time aimed to improve survival in the circulation by decreasing uptake by the reticuloendothelial system. Since removal of sialic acid from biological red blood cells resulted in their rapid removal from the circulation, we started to modify the surface properties of artificial red blood cells (Chang, 1972). This included synthetic polymers, negatively charged polymers, cross-linked protein, lipid–protein, lipid–polymer, and others. Artificial red blood cells have since been extensively explored by many researchers around the world. These include Beissinger, Chang, Farmer, Hunt, Kobayashi, Lee, Mobed, Nishiyia, Rabinovic, Rudolph, Schmidt, Sinohara, Szebeni, Takeoka, Tsuchida, Takahashi, and Usuba, among many others.

One major area of progress is the preparation of submicron lipid membrane artificial red blood cells, resulting in significant improvements in circulation time (Djordjevich and Miller, 1980). By the addition of polyethylene glycol to the lipid membrane, the circulation time has been increased to a half-time of about 36 hours in rats (Philips *et al.*, 1999). These advances now make it possible to scale-up for detailed preclinical studies toward clinical trials (Tsuchida, 1998; Kobayashi et al., 2005). The uptake is mainly by the reticuloendothelial system. It is possible to replace 90% of the red blood cells in rats with these artificial red blood cells. The animals with this percentage of exchange transfusion remain viable. Studies also reported effectiveness in hemorrhagic shock. There are no changes in the histology of brain, heart, kidneys, and lungs of rats.

Using a modification of this author's earlier method of micron-dimension biodegradable polymeric membrane artificial cells (Chang, 1976), we recently prepared nanodimension artificial red blood cells (Chang, 1997, 2003; W. P. Yu and Chang, 1996). These nano-artificial rbc of 80–150 nanometers contain all the red blood cell enzymes and can convert methemoglobin to hemoglobin (Chang *et al.*, 2003) (Fig. 50.10). Our recent studies show that, using a polyethylene-glycol-polylactide copolymer membrane, we are able to increase the circulation time of these nano-artificial red blood cells to double that of polyhemoglobin (Fig. 50.11) (Chang *et al.*, 2003). Further studies show that one infusion with a volume of one-third the total blood volume resulted in no adverse effects.

## V. A CHEMICAL APPROACH BASED ON PERFLUOROCHEMICALS

There is another type of blood substitute, based on a chemical approach first shown by Clark and then Geyer using perfluorochemicals (Riess, 2001). This approach was developed by a number of investigators, including Chen, Clark, Geyer, Faithful, Keipert, Lowe, Mitsuno, Naito, Nose, Ohyanagi, Reiss, Sloviter, and Yokoyama, among many others. Perfluorochemicals (PFCs) are synthetic fluids in which oxygen can dissolve and that can be made into fine emulsions for use as oxygen carriers. Their greatest advantage is that they can be produced in large amounts. Further-

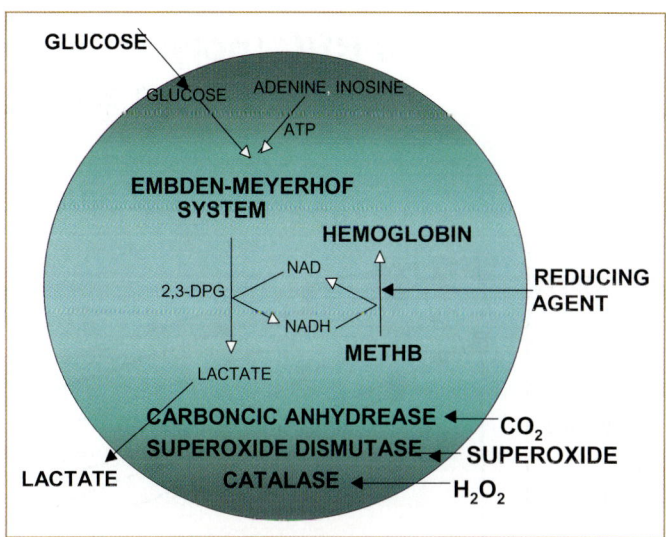

**FIG. 50.10.** Nanodimension artificial red blood cells with polyethylene-glyco-polylactide membrane. In addition to hemoglobin, this contains the same enzymes that are normally present in red blood cells. Thus, it has the complete function of red blood cells. From Chang *et al.* (2003), with copyright permission from Marcel Dekker, Publisher.

more, their purity can be more easily controlled. However, PFCs have a lower capacity for carrying oxygen than Hb. Improved perfluorochemicals have allowed a higher concentration of PFC to be used and a blood substitute based on perfluoro-octyl bromide ($C_8F_{17}Br$) with egg-yolk lecithin as the surfactant has been tested in clinical trials. The results are being carefully analyzed.

## VI. CONCLUSIONS

It was mistakenly thought that blood substitute would be a simple approach that could easily be developed for clinical applications when needed. Thus, the crisis of HIV-contaminated donor blood was followed by efforts to be able to quickly manufacture blood substitutes for clinical use. However, many years have passed with numerous failures, and no blood substitute is yet in routine clinical use in North America, and only one type of first-generation material is in the final phases of Phase III clinical trials. We now know that we have to approach blood substitutes as a highly interdisciplinary type of basic and applied research and development involving polymer chemistry, cell physiology, molecular biology, biotechnology, and other areas. The belated but ongoing emphasis on this interdisciplinary approach will accelerate the clinical realization of blood substitutes. This brief overview cannot include the numerous ongoing studies and research in this area. More details are available in the many detailed publications, books, and symposium volumes in this area (Chang, 1997, 2003, 2005, 2007; Klein, 2000; Rudolph *et al.*, 1997; Tuschida, 1998; Kobayashi *et al.*, 2005; Winslow, 2003, 2006).

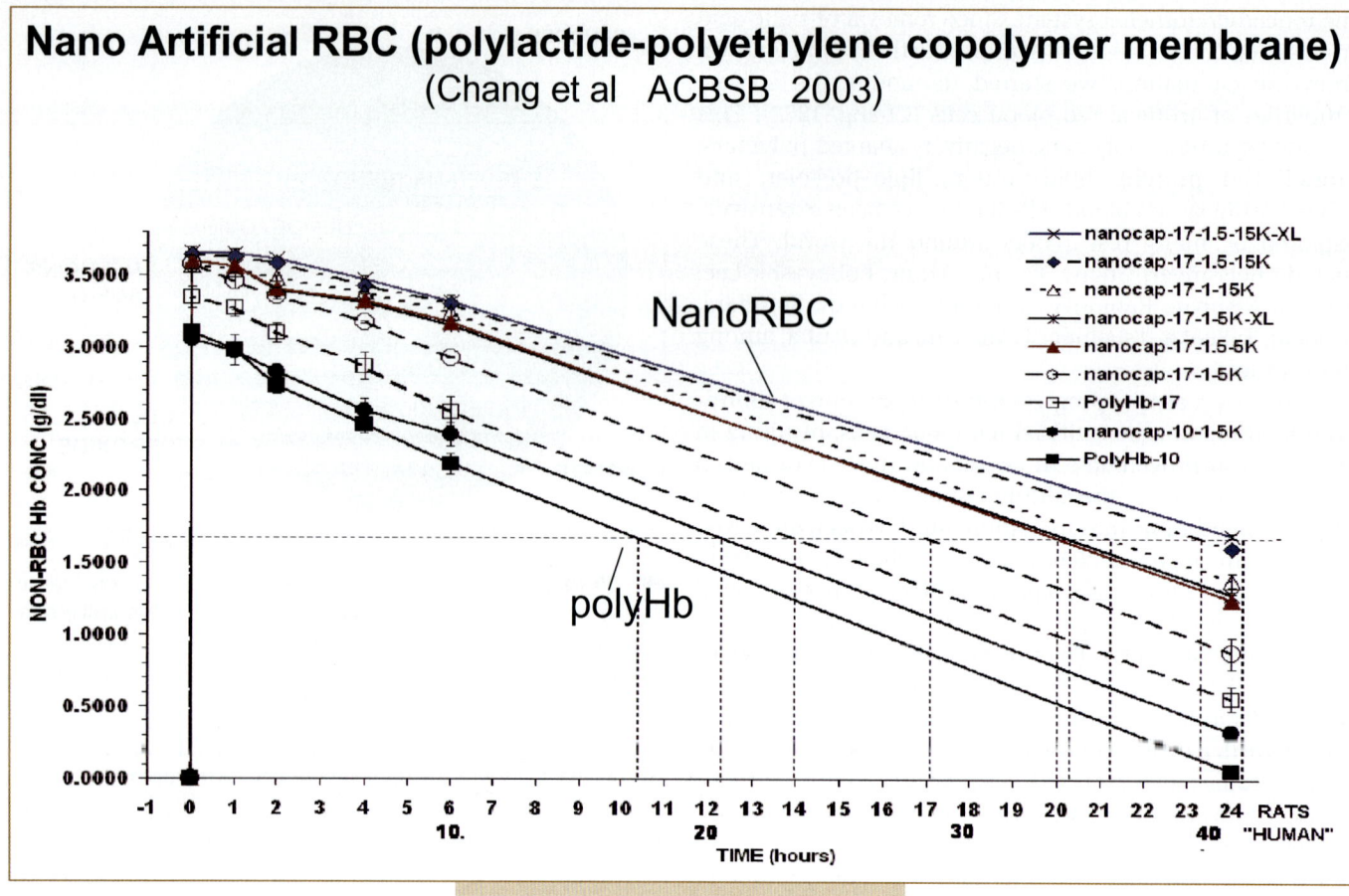

**FIG. 50.11.** Nanodimension artificial red blood cells with polyethylene-glyco-polylactide membrane. The circulation time is double that of polyhemoglobin. The circulation half-time of polyhemoglobin in human is about 24 hours. This means that the nanodimension artificial rbc may have a circulation time of about 48 hours in human From Chang *et al.* (2003), with copyright permission from the publisher, Marcel Dekker.

## VII. LINK TO WEBSITES

www.artcell.mcgill.ca or www.artificialcell.info is the website of the International Network for Artificial Cells, Blood Substitutes, and Biotechnology. It contains review articles and books (including Chang's 1972 and 1997 monographs) that can be accessed without cost. This website also links to other websites in the area of blood substitutes and artificial cells.

## VIII. ACKNOWLEDGMENTS

This author acknowledges the support of the Canadian Institutes of Health Research, the "Virage" Centre of Excellence in Biotechnology from the Quebec Ministry, the MSSS-FRSQ Research Group (d'equipe) award on "Blood Substitutes in Transfusion Medicine" from the Quebec Ministry of Health's "Haemovigillance and Transfusion Medicine Program," and an operating grant from the Research Fund of the Bayer/Canadian Blood Agency/Haema Quebec/Canadian Institutes of Health Research.

## IX. REFERENCES

Alayash, A. I. (2004). Oxygen therapeutics: can we tame Hb? *Nature Rev. Drug Discov.* **3,** 152–159.

Bunn, H. F., and Jandl, J. H. (1968). The renal handling of hemoglobin. *Trans. Assoc. Am. Physicians* **81,** 147.

Burhop, K. E., and Estep T. E. (2001). Hb-induced myocardial lesions. *Artif. Cells Blood Substit. Immob. Biotech.* **29,** 101–106.

Chang, T. M. S. (1957). Hemoglobin corpuscles. Report of a research project of B.Sc. Honours Physiology, McGill University, 1–25, Medical Library, McGill University (reprinted in *J. Biom. Artif. Cells Artif. Organs* **16,** 1–9, 1988).

Chang, T. M. S. (1964). Semipermeable microcapsules. *Science* 146(3643), 524.

Chang, T. M. S. (1971). Stabilization of enzyme by microencapsulation with a concentrated protein solution or by cross-linking with glutaraldehyde. *Biochem. Biophys. Res. Commun.* **44**, 1531–1533.

Chang, T. M. S. (1972). Artificial cells. Monograph. Charles C Thomas, Springfield, IL (full text available at www.artcell.mcgill.ca).

Chang, T. M. S. (1976). Biodegradable semipermeable microcapsules containing enzymes, hormones, vaccines, and other biologicals. *J. Bioeng.* **1**, 25–32.

Chang, T. M. S. (1997). "Blood Substitutes: Principles, Methods, Products and Clinical Trials," Vol. 1. Karger Basel, Switzerland (full text at www.artcell.mcgill.ca).

Chang, T. M. S. (2003). New generations of red blood cell substitutes. *J. Int. Med.* **253**, 527–535.

Chang, T. M. S. (2005). Therapeutic applications of polymeric artificial cells. *Nat. Revi. Drug Discov.* **4**, 221–235.

Chang, T. M. S. (2007). Artificial cells: biotechnology, nanomedicine, regenerative medicine, blood substitutes, bioencapsulation and cell/stem cell therapy. World Scientific Publisher, Singapore.

Chang, T. M. S., Lister, C., Nishiya, T., and Varma, R.(1992). Effects of different methods of administration and effects of modifications by microencapsulation, cross-linkage or PEG conjugation on the immunological effects of homologous and heterologous hemoglobin. *J. Biomater. Artif. Cells Immob. Biotechnol.* **20**, 611–618.

Chang, T.M.S., Powanda, D., and Yu, W. P. (2003). Ultrathin polyethylene-glycol-polylactide copolymer membrane nanocapsules containing polymerized Hb and enzymes as nano-dimension red blood cell substitutes. *Artif. Cells Blood Substit. Biotechnol.* **31**(3), 231–248.

D'Agnillo, F., and Chang, T. M. S. (1998). PolyHb-superoxide dismutase catalase as a blood substitute with antioxidant properties. *Nat. Biotechnol.* **16**(7), 667–671.

Djordjevich, L., and Miller, I. F. (1980). Synthetic erythrocytes from lipid encapsulated Hb. *Exp. Hematol.* **8**, 584.

Doherty, D. H., Doyle, M. P., Curry, S. R., Vali, R. J., Fattor, T. J., Olson, J. S., and Lemon, D. D. (1998). Rate of reaction with nitric oxide determines the hypertensive effect of cell-free hemoglobin. *Nat. Biotechnol.* **16**, 672–676.

Estep, T. N., Gonder, J., Bornstein, I., Young, S., and Johnson, R. C. (1992). Immunogenicity of diaspirin crosslinked hemoglobin solutions. *J. Biomater. Artif. Cells Immob. Biotechnol.* **20**, 603–610.

Freytag, J. W., and Templeton, D. (1997). Optro™ (recombinant human hemoglobin): a therapeutic for the delivery of oxygen and the restoration of blood volume in the treatment of acute blood loss in trauma and surgery. *In* "Red Cell Substitutes: Basic Principles and Clinical Application" (A. S. Rudolph, R. Rabinovici, and G. Z. Feuerstein, eds.), pp. 325–334. Marcel Dekker, New York.

Fronticelli, C., Bellelli, A., and Briniger, W. S. (2004). Approach to the engineering of hemoglobin-based oxygen carrier. *Transfusion Mediciner* **5**, 516–520.

Gould, S. A., Sehgal, L. R., Sehgal, H. L., DeWoskin, R., and Moss, G. S. (1998). The clinical development of human polymerized hemoglobin. *In* "Blood Substitutes: Principles, Methods, Products and Clinical Trials," Vol. 2 (T. M. S, Chang, ed.), pp. 12–28. Karger, Basel, Switzerland.

Gould, S. A., Moore, E. E., Hoyt, D. B., Ness, P. M., Norris, E. J., Carson, J. L., Hides, G. A., Freeman, I. H. G., DeWoskin R., and Moss, G. S. (2002). The life-sustaining capacity of human polymerized hemoglobin when red cells might be unavailable. *J. Am. Coll. Surg.* **195**, 445–452.

Hoffman, S. J., Looker, D. L., Roehrich, J. M., *et al.* (1990). Expression of fully functional tetrameric human hemoglobin in *escherichia coli. Proc. Natl. Acad. Sci. U.S.A.* **87**, 8521–8525.

Iwashita, Y. (1992). Relationship between chemical properties and biological properties of pyridoxalated hemoglobin-polyoxyethylene. *J. Biomater. Artif. Cells Immob. Biotechnol.* **20**, 299–308.

Kobayashi, K. Tsuchida, E., and Horinouchi, H. (eds.). (2005). "Artificial Oxygen Carriers." Spring-Verlag New York.

Klein, H. G. (2000). The prospects for red-cell substitutes. *N. Engl. J. Med.* **342**(22), 1666–1668.

Li, T., Yu, R., Zhang, H. H. Liang, W. G., Yang, X. M., and Yang, C. M. (2006). A method for purification and viral inactivation of human placenta hemoglobin. *Artfif. Cells Blood Substit. Biotechnol.* (in press).

Li, X. Z., Zhang, X. W., and Liu, Q. (2005). Determination of the molecular weight distribution of PEGylated bovine hemoglobin (PEG-bHb). *Artif. Cells Blood Substit. Biotechnol.* **33**, 13–28.

Matheson, B., Kwansa, H. E., Rebel, A., and Bucci, E. (2002). Vascular responses to infusions of a nonextravasating Hb polymer. *J. Appl. Physiol.* **93**, 1479–1486.

Nelson, D. J. (1998). Blood and HemAssist™ (DCLHb): potentially a complementary therapeutic team. *In* "Blood Substitutes: Principles, Methods, Products and Clinical Trials," Vol. 2 (T. M. S. Chang, ed.), pp. 39–57. Karger, Basel, Switzerland.

Pearce, L. B., and Gawryl, M. S. (1998). Overview of preclinical and clinical efficacy of Biopure's Hemopure. *In* "Blood Substitutes: Principles, Methods, Products and Clinical Trials," Vol. 2 (T. M. S. Chang, ed.), pp. 82–98. Karger, Basel, Switzerland.

Perutz, M. F. (1989). Myoglobin and hemoglobin: role of distal residues in reactions with haem ligands. *Trends Biochem. Sci.* **14**, 42–44.

Philips, W. T., Klpper, R. W., Awasthi, V. D., Rudolph, A. S., Cliff, R., Kwasiborski, V. V., and Goins, B. A. (1999). Polyethylene glycol-modified liposome-encapsulated Hb: a long circulating red cell substitute. *J. Pharm. Exp. Therapeutics* **288**, 665–670.

Powanda, D., and Chang, T. M. S. (2002). Cross-linked polyHb-superoxide dismutase-catalase supplies oxygen without causing blood–brain barrier disruption or brain edema in a rat model of transient global brain ischemia-reperfusion. *Artif. Cells Blood Substit. Immobil. Biotechnol.* **30**, 25–42.

Riess, J. G. (2001). Oxygen carriers. Chem. Rev. **101**, 2797–2919.

Rudolph, A. S., Rabinovici, R., and Feuerstein, G. Z. (eds.). (1997). "Red Blood Cell Substitutes: Basic Principles and Clinical Applications." Marcel Dekker, New York.

Savitsky, J. P., Doozi, J., Black, J., Arnold, J. D. (1978). A clinical safety trial of stroma-free hemoglobin. *Clin. Pharm. Ther.* **23**, 73.

Shorr, R. G., Viau, A. T., and Abuchowski, A. (1996). Phase 1B safety evaluation of PEG Hb as an adjuvant to radiation therapy in human cancer patients. *Artif. Cells Blood Substit. Immobil. Biotechnol.* **24**, 407.

Sprung J., Kindscher, J. D., Wahr, J. A., Levy, J. H., Monk T. G., Moritz, M. W., and O'Hara, P. J. (2002). The use of bovine hemoglobin glutamer-250 (Hemopure) in surgical patients: results of a multicenter, randomized, single-blinded trial. *Anesth. Analg.* **94**(4), 799–808.

Tsuchida, E (ed.). (1998). "Blood Substitutes: Present and Future Perspectives." Elsevier, Amsterdam.

Tsuchida, E., Komatsu, T., Yanagimoto, T., and Sakai H. (2002). Preservation stability and *in vivo* administration of albumin-heme hydrid solution as an entirely synthetic oxygen carrier. *Polymer Adv. Technol.* **13**, 845–850.

Walder, J. A., Zaugg, R. H., Walder, R. Y., Steele, J. M., and Klotz, I. M. (1979). Diaspirins that cross-link alpha chains of Hb: bis(3,5-dibromo-salicyl) succinate and bis(3,5-dibormosalicyl)fumarate. *Biochemistry* **18**, 4265–4270.

Winslow, R. (2003). Current status of blood substitute research: towards a new paradigm. *J. Int. Med.* **253**, 508–517.

Winslow, R. (ed.). (2006). "Blood Substitutes." Academic Press, San Diego, CA.

Wong, J. T. (1988). Rightshifted dextran-hemoglobin as blood substitute. *Biomater Artif. Cells Artif. Organs* **16**, 237–245.

Yu, B. L., and Chang, T. M. S. (2004). *In vitro* and *in vivo* effects of polyHb–tyrosinase on murine B16F10 melanoma. *Melanoma Res.* **14**, 197–202.

Yu, B. L., Liu, Z. C., and Chang, T. M. S. (1996). PolyHb with different percentage of tetrameric Hb and effects on vasoactivity and electrocardiogram. *Artif. Cells Blood Substit. Biotechnol.* **2** (in press).

Yu, W. P., and Chang, T. M. S. (1996). Submicron polymer membrane Hb nanocapsules as potential blood substitutes: preparation and characterization. *Artif. Cells Blood Substit. Immobil. Biotechnol.* **24**, 169–184.

# Lymphoid Cells

*Una Chen*

## I. INTRODUCTION

For researchers interested in cell-based therapy, it is a great challenge to learn how to expand normal lymphoid cells and their precursor cells *ex vivo* under defined conditions. Lymphocytes are defined by cell surface receptor-immunoglobulin (Ig) for B-cells and the T-cell receptor (TCR) for T-cells. Precursor B-cells bear a pre-B receptor, which contains lambda 5 in mice ("physi" in humans), and precursor T-cells bear a pre-TCR-alpha. No biochemically and genetically defined surface receptors for the progenitors of lymphoid precursors have been reported. It is not entirely clear how committed stem cells differentiate into lymphoid cells, which then mature into effector cells. Many theories have been postulated. Differentiation and maturation involve both antigen-independent and antigen-dependent processes. For the first process, I will explore the sequential commitment model proposed by Brown *et al.* (1985, 1988). For my treatment of the second process, I am indebted to many colleagues who helped me to summarize current views.

Ligand–receptor interactions must play an important role in activating lymphoid cells to proliferate and to differ-

entiate. The term *ligand* means all external signals, including those provided by stroma cells and cytokines during the antigen-independent stage and those provided by antigens (Ag) and antigen-presenting cells (APCs), or accessory cells during the antigen-dependent stage. The process of turning on and turning off of transcription factors at each stage of differentiation as a consequence of ligand–receptor interaction is the key element controlling both the phenotype and the function of cells. I will discuss transcription factors known to be important at various stages of lymphopoiesis due to the notion that the growth of lymphoid (precursor) cells can be manipulated at will by using genetic tools. There are a few examples of expansion of normal lymphoid (precursor) cells in culture. All are based on the support of stroma cell lines and cytokines. On removal of these elements, cells either differentiate toward terminal effector cells or die.

I have arbitrarily defined seven stages of lymphopoiesis to discuss in detail. For each one, I discuss the available cellular and molecular markers, the current understanding of cellular phenotype, the potential of these cells to expand *ex vivo*, and, if possible, what could be done in the future.

*Principles of Tissue Engineering, 3rd Edition*
ed. by Lanza, Langer, and Vacanti

Both somatic and genetic manipulations are possible in animal models. By combining modern cellular technology and available mutant mice, one should be able to grow normal lymphoid (precursor) cells at every stage of lymphopoiesis.

Despite the efforts of many, human stroma cell lines that can reproducibly support the growth of different stages of lymphoid (precursor) cells are still under development. To apply inducible regulation of stage-specific and lymphoid-specific transcription factors in controlling the growth of normal lymphoid cells seems to be the key element.

The aim of this chapter is to understand lymphocytes and their precursors. The purpose is to learn how these cells might be manipulated to make them useful in cell-based, somatic gene therapy. I attempt to address the possibility of growing normal lymphoid cells and their stem cells of mouse and human origin in a controllable manner. That is, cells should proliferate *ex vivo* without becoming malignantly transformed and with little or no differentiation.

Three main types of stem cells can be distinguished: totipotential, pluripotent, and multipotential. Totipotential and pluripotential stem cells usually divide symmetrically to give rise to two daughter stem cells with the same properties that are identical in phenotype to their parent. Under appropriate conditions, totipotent and pluripotent stem cells can differentiate into any other form of stem cell. The only known stem cell lines that seem to be close to totipotential are mouse embryonic stem cells (ESCs) and, more recently, human ESCs. Multipotential stem cells, which exist for the lifetime of an organism, undergo primarily asymmetric divisions. One daughter cell is another stem cell like its parent, whereas the other daughter cell is a more highly differentiated cell that performs a tissue-specific function. Due to this unique property, multipotential stem cells are ideal vehicles for cell-based, somatic gene therapy. They will carry the transgene for the lifetime of the organism, and they will maintain expression of the transgene in the differentiated cells that they spawn.

Both lymphoid and lymphoid precursor cells are the progenies of hematopoietic stem cells (HSCs). One of the main tasks here is to review whether lymphoid cells and their progenitors possess stem cell–like properties. Are they self-renewing? And with available culture conditions and technology, can they expand *ex vivo* under controllable growth conditions? Unlimited growth of lymphoid cells and their progenitors is well documented. Lymphoid leukemia cells develop either spontaneously or after infection with viruses. These cell lines are useful for studying lymphoid cells but not for cell-based therapy, because they develop tumors when reintroduced into the organism. Thus, this chapter is devoted to exploring the possibility of growing normal lymphoid cells that can be engineered in culture and reimplanted into syngeneic or autologous hosts for therapeutic purposes. Other normal cell types, expandable at will using genetic approaches, are available: tet-on (tetracycline-inducible) glial stem cells (Muth *et al.*, 2001; Elmshaeuser *et al.*, 2002) and tet-off (tetracycline-regulated) keratinocytic and lung stem cells (Chen, 2001, 2006a, 2006b).

## II. PROPERTIES OF LYMPHOCYTES

Lymphoid precursor cells are bipotential progenitors of pre-T- and pre-B-cells. The development of lymphoid precursor cells is independent of the presence of Ag. It is controversial whether lymphoid precursor cells divide asymmetrically with self-renewal. Lymphocyte types are defined by cell surface receptors-Ig for B-cells and the TCR for T-cells. Precursor B-cells bear a pre-B receptor, which contains lambda 5 in mice, and "physi" in humans, and precursor T-cells bear a pre-TCR-$\alpha$. So far, no biochemically and genetically defined surface receptors for the progenitors of lymphoid precursors have been reported. Lymphocytes and their precursors are similar to cells of other lineages, in that they proliferate, differentiate, communicate with other cells, age, and die. However, lymphocytes also possess unique properties that distinguish them from other cells: (1) formation of receptor genes by VDJ recombination, (2) requirement for a thymuslike environment to generate mature $CD4^+/8^+$ T-cells, (3) requirement for somatic education and selection by Ag, (4) somatic hypermutation to generate more diversity in B-cells, (5) immunologic memory, (6) Ig heavy-chain class switch recombination in B-cells, and (7) kappa light-chain editing somatically to generate new specificities of B-cell receptor (Chen, 2000). Development of techniques to expand *ex vivo*, to manipulate genetically, and to reimplant lymphocytes and their precursors into host animals requires lymphocyte engineers.

## III. LYMPHOCYTE ENGINEERING: REALITY AND POTENTIAL

Two models describe the sequence of cell commitment during lymphopoiesis: stochastic and inductive. In this section, I use an inductive model in an attempt to explain lymphocyte commitment during development of the organism and in the adult.

## IV. INDUCTIVE MODEL OF SEQUENTIAL CELL COMMITMENT OF HEMATOPOIESIS

This model is divided into two parts. The first part deals with the Ag-independent stage, starting with the differentiation from the null cells (fertilized egg to embryonic stem cells) to mesoderm and then to lymphoid precursor cells. A significant portion of the first part of this model was expanded from Bailey's hypothesis of developmental progress (1986) and modified according to Brown *et al.* (1985, 1988).

## Antigen-Independent Stage of Lymphopoiesis — Part 1

Because multipotential stem cells are committed to differentiate, it is hypothesized that the differentiation sequence is genetically determined. Cells within the sequence are precommitted in how they're determined. Cells within the sequence are precommitted in their ability to respond to various inducers of differentiation; and on encountering appropriate factors or a suitable microenvironment, they proliferate and mature along a particular pathway. It is proposed that multipotential stem cells that do not receive a signal for differentiation toward mature end cells progress to the next stage in the sequence of development. Alternatively, cells that do not receive a signal die. Only cells that receive proper signals will differentiate further toward the mature end cells. Because commitment is gradual, there is a continuous spectrum of multipotential stem cells. Thus, some of them may be able to respond to inducers of one sort or another. In terms of self-renewal, multipotential stem cell populations continually occupy each potential for proliferation at any given time and therefore respond to the requirement for each cell type. Some daughter cells will differentiate into the next stage. The continual development of stem cells may not be diverted entirely toward one cell type, even in the presence of differentiation factors for that type. Because stem cells divide as they respond, some cells are still able to progress to the next stage of commitment.

## Antigen-Dependent Stage of Lymphopoiesis — Part 2

The second part of this model deals with the Ag-dependent stage, after receptors are expressed on the surface of lymphocytes. There are two main lineages of lymphocytes: T-cells and B-cells. T-cells can mature into CD8$^+$ cytotoxic T-cells and CD4$^+$ helper T-cells. Based on their function and the spectrum of cytokines produced, helper T-cells can be divided into Th$_1$ and Th$_2$ subpopulations. The main functions of CD4$^+$ T-cells are to recognize cell-bound antigen and to communicate with and help cytotoxic T-cells and B-cells to perform their duties as effector cells. Both lineages generate memory cells. How and when lymphoid cells are committed to differentiate into effector cells or into memory cells remain a mystery.

## V. DIAGRAMS TO EXPLAIN THIS MODEL

Figure 51.1 is a simple diagram to explain this model. HSCs are multipotential stem cells derived from the mesoderm. Originally they are descended from common stem cells (null, or 0, cells), such as fertilized eggs, and from ESCs. The cells diagonally to the right (Fig. 51.1) represent cells committed to hematopoietic lineages, which gradually lose their multipotentiality during embryogenesis. This concept is expressed with numbers (5, 4, 3, and 2) inside the progeni-

tor cells on the diagonal. For example, HSCs (noncycling) enter the proliferating stage to become the hematopoietic progenitor cells (HPCs, cycling). The number 5 cell, for instance, means that the HPCs at this stage have the potential to differentiate into five different types of precursor cells.

The committed, monopotential precursor cells then differentiate toward mature end cells, such as megakaryocytes (Me), erythrocytes (E), neutrophils/granulocytes (G), monocytes (M), and lymphocytes (L). Lymphoid precursor cells then differentiate into monopotential precursors for B- and T-cells. After expression of Ag receptors on the cell surface and on encountering Ag, B-cells differentiate into plaque-forming cells (PFCs, i.e., plasma cells) or memory B-cells. T-cells become CD8$^+$ or CD4$^+$ after thymic education, maturation, and negative and positive selection. Memory T-cells are generated later during differentiation. (Fig. 51.4 shows the markers and events that occur during the development of pre-T-cells — the process of maturation, education, and apoptosis of mature CD4$^+$ or CD8$^+$ T-cells from CD4$^+$8$^+$ T-cells, also shown in figures derived from Kisielow and von Boehmer, 1995; Rodewald and Fehling, 1998, cited in Chen, 2000.) Genes encoding transcription factors and other important genes known to be expressed specifically in lymphopoiesis and that affect function, i.e., loss of function (knockout, down) or gain of function (ectrophic expression by transducing with retroviral vectors containing trasngene of interests), include *Brachyury; scl; Cdx4; HoxB4; c-myb; GATA1, -2, -3; rbtn2; pu.1/spi-l; Ikaros; TIS21; D3 cycline; Notch-Delta; Tcf1; E2A, pax5; BAFF; Oct-2, Blimp1; OBF1; G9a; Mad1, Bcl6; Bach2; E47; Id2, XBP1; IRF4; pTalpha; bcl-x;Dnmt1; foxp3; TLR; p53*, and more (Matthias and Rolink, 2005).

## VI. SOME COMMENTS ON THIS MODEL

Very little is known about how totipotent and pluripotent stem cells differentiate into ectoderm, mesoderm, and endoderm or how mesoderm differentiates into hematopoietic lineages. These are the central issues of embryogenesis, and this part of the model is intentionally very sketchy. Why and when multipotential, lineage-specific stem cells divide asymmetrically or progress to the next stage of commitment are unknown. The underlying rules may be stochastic or inductive. But the progression of the stage of commitment is more likely due to the turning on and off of genes encoding transcription factors.

Brown and coworkers originally formulated their model based on the data from the differentiation pattern of several myeloid cell lines. The scheme of the development of lymphoid lineages was purely hypothetical in the original model, and the development of mature lymphoid cells (in Part 2) was not included. It was proposed that precursor B-cells become committed before precursor T-cells. I have modified this step to the precommitment to lymphoid precursor cells. Also, I have changed the commitment of

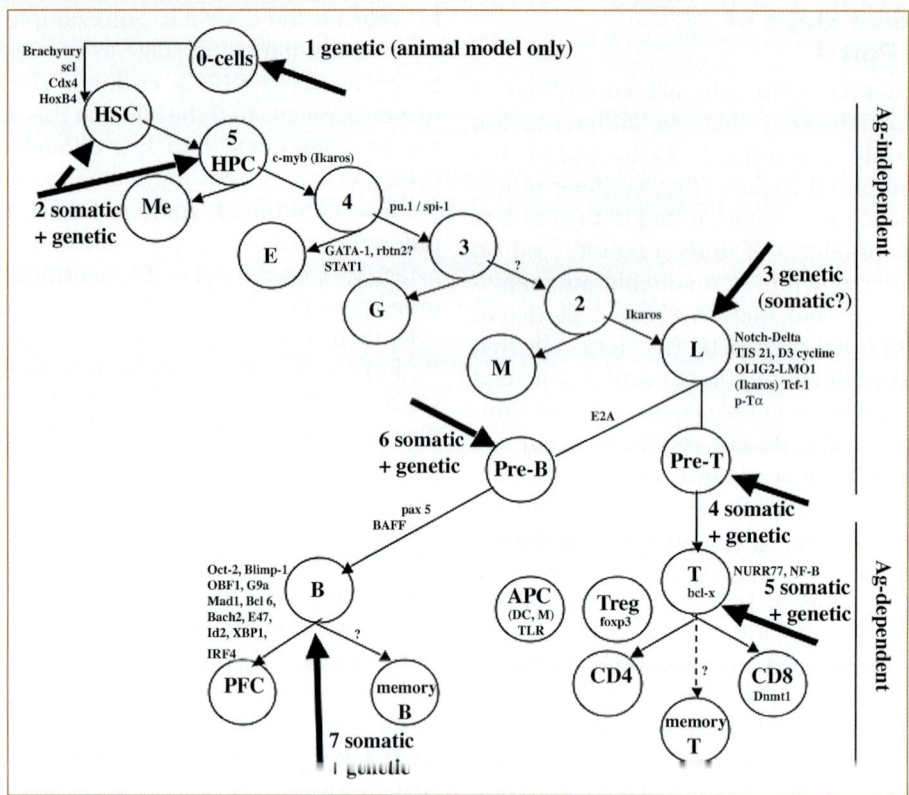

**FIG. 51.1.** Proposed sequence of cell commitment during lymphopoiesis and stages of hematopoiesis and lymphopoiesis at which somatic and/or genetic engineering of lymphocytes and their precursor cells might be feasible. Solid block arrows point to the cell stages that might be targeted. The first part of the theory is the sequential commitment model of hematopoiesis of Brown *et al.* (1985, 1988), modified to include the pattern of expression of transcription factors. The cells depicted diagonally to the right represent hematopoietic stem cells (HSCs) committed to multipotential precursor cells or hematopoietic progenitor cells (HPCs), which gradually lose pluripotency, as expressed by the numbers 5, 4, 3, and 2 inside the stem cells. Number 5, for example, means that the HPCs at this stage have the potential to differentiate into five different types of precursor cells. The committed monopotential precursor cells can then differentiate toward mature end cells, such as megakaryocytes (Me), erythrocytes (E), neutrophils/granulocytes (G), monocytes (M), and finally lympho-cytes (L). HSCs are derived from very primitive stem cells (null cells, or 0 cells), such as fertilized eggs and/or embryonic stem cells (ESCs). In the second part of the theory, lymphoid precursor cells differentiate into monopotential precursor B-cells (B) and monopotential precursor T-cells (T). After the expression of cell surface receptors and on encountering antigen (Ag), B-cells can differentiate to plaque-forming cells (PFCs) or memory B-cells. T-cells can become CD8+ or CD4+ cells after education and maturation in the thymus. Memory T-cells can be generated at a certain stage of differentiation. Regulatory T-cells are rediscovered and APCs (antigen processing cells), including dendritic cells and macrophage, seem to play a role in directing peptide-antigen recognition and activation of Th1 vs Th2. Lipid antigen is directed to TCR via the CD1 molecule pathway. The transcription factors are expressed and are known to affect differentiation. The data come mostly from phenotypic expression in loss/gain of function approaches: Brachyury, scl, c-myb, GATA-1, rbtn2, pu.1/spi-1, Ikaros, Tcf-1, E2A, pax 5, Oct-2, Blimp-1, pTalpha, bcl-x, and p53 (refs. cited in Chen, 2000). Recent additions of the orchester are Cdx4, HoxB4, STAT1 (Halupa *et al.*, 2005), Notch-Delta, TIS21, D3 cycline, foxp3, BAFF, OBF1, G9a, Mad1, Bcl6, Bach2, E47, Id2, XBP1, IRF4, and TLR (refs. cited in the text).

lymphoid precursor cells to occurring after the myeloid precursor cells. However, the committed lymphoid precursor cells should still be viewed as hypothetical, because there are many examples supporting the common lineage development of B-cells and myeloid cells and not of precursor T-cells. Committed lymphoid precursor cells are not clearly identified and isolated (details in Stage 3, later).

There is ambiguity in the original model. For example, the issue of the timing of each point of decision was not addressed. The original hypothesis implied that stem cells that do not receive the appropriate signals would proceed "spontaneously" to the next stage in the differentiation

sequence. Based on published data and on observations on the abortive development of lymphoid cells from blood islands derived from embryoid bodies (EBs) (Chen *et al.*, 1997; Chen, 2000) and on the essential requirement of cytokines to maintain and to propogate stem cells derived from transgenic fetal liver, I tend not to be in favor of spontaneous progression; appropriate inductive signals for the transcriptional activation of lineage-committed genes are stringently required. Moreover, because pro-grammed cell death plays an important role in the fate of cells during embryogenesis, I have included apoptosis as another important factor. This model has gained support from the phenotypes of mice in which genes encoding transcription factors have been deleted by knockout, knockdown, the so-called loss-of-function approach, and from the phenotype of cells in which transgenes are transduced using retroviral vectors, the so-called gain of function approaches. Transcription factors controlling the sequential commitment of HSCs have been summarized (Chen, 2000).

# VII. CRITERIA FOR ENGINEERING DEVELOPMENTAL STAGES OF LYMPHOPOIESIS

I attempt to explore the feasibility of engineering lymphoid cells and their stem cells at several defined stages of differentiation. The criteria for deciding on feasibility are (1) intrinsic self-renewal of HSCs, HPCs, and more differentiated cells; (2) the availability of cell surface markers for identification and purification; (3) the supply of recombinant growth factors and appropriate stroma cells; (4) the existence of favorable cell culture conditions that promote growth instead of differentiation; (5) the possibility of educating lymphocytes to become antigen-specific memory cells; and (6) the possibility of introducing, either genetically or somatically, genes of interest along with inducible promoter and regulatory sequences. Technical advances make the engineering of HSCs, HPCs, and lymphoid cells attractive for further manipulation, such as introducing new genes or reprogramming cells.

# VIII. STAGES OF LYMPHOPOIESIS FOR ENGINEERING

Figure 51.1 shows a schematic diagram of the stages at which lymphoid cells and lymphoid precursor cells might be engineered. The stages are next discussed individually.

## Stage 1: 0 Cells

Null (0) cells include the fertilized eggs, which are totipotent, and the blastocyst-derived embryonic stem (ES) cell lines, which are pluripotent. Mouse ES cell lines of different strain origin have been available for a long time, and they possess the property of differentiation into different somatic lineages *in vitro* and *in vivo*, homologous recombination, and germ-line transmitted through the crossing and breeding of progeny. ES cells from many animal species are also available. Establishment of human ES-like cell lines has been reported since 1998, and now many IFV units can derive such cell lines. Several lineage-committed cells, such as from bone and cartilage, have been shown to be obtainable through differentiation *in vitro*. However, strictly speaking, the only way to prove that an hES cell line is truly pluripotent is to demonstrate germ-line transmission in human beings — a task that should not even be considered. Similarly, genetic and cellular engineering at this level should be limited to study *in vitro*. ES cells are stem cell lines that divide symmetrically. Unlike lineage-committed stem

cells, which are rare, ES cells are almost an exception, being readily available in abundance. mES cells can proliferate without differentiation if feeder cells and/or cytokines such as LIF (leukemia inhibitory factor) are provided, while hES cell lines established so far are feeder cell dependent, and LIF has no supportive effect. They can be induced to differentiate by removing the feeder cells (and cytokines). A few surface molecules are available for characterizing ES cells. Antibodies against various components that have been used to characterize such cells, such as antibody against stage-specific embryonic antigen-l (SSEA-l), enable one to distinguish undifferentiated from differentiated mES cells. mES cells are SSEA1$^+$, SSEA3$^-$, SSEA4$^-$, TRA-1-60$^-$, TRA-1-81$^-$, Oct4$^-$, Oct6$^+$, AP (alkaline phosphatase)$^+$, telomerase$^+$; hES cells are SSEA1$^-$, SSEA3$^+$, SSEA4$^+$, TRA1-60$^+$, TRA1-81$^+$, Oct 4$^+$, AP$^+$, telomerase$^+$.

In recent years, stem cell genomics, or the systems biology approach to stem cells, has become popular. It considers the complex nature of cells and aims at a comprehensive understanding of stem cell mechanisms. Gene expression profiles of different stem cell types are being determined using DNA microarrays. These molecular signatures are stored and available in the stem cell database and allow a comparison of different stem cell types. On the basis of the obtained molecular signature, candidate genes, such as transcription factors, are subjected to the loss-of-function approach by knocked-out, knocked-in, or knocked-down by RNAi, and the effects on the expression of downstream genes are determined with DNA microarrays. In this way, gene regulatory networks could be identified rapidly.

For example, this approach is used to identify gene regulatory networks implicated in the self-renewal and differentiation of mES cells. Based on a microarray time course study of retinoic acid–induced ES cell differentiation, 169 genes that are down-regulated in the course of differentiation were found, and 40 of them were further investigated. ES cell cultures infected with lentiviral vectors containing the EGFP gene in addition to one of these 40 genes (RNAi knockdown or overexpression) were investigated in a competition assay, where respective infected cells were mixed with noninfected cells. In this way, higher or lower proliferation rates of the infected cells could be determined based on measured EGFP fluorescence. Changes in ES cell number could be due to a function of the respective gene in cell-cycle progression, cell survival, or prevention of lineage-specific differentiation. Downstream effects of changes of gene expression were determined with DNA microarrays. Such an approach also applies to the study of other cells, such as hES cells and hEC cells (Sperger *et al.*, 2003), HSCs (Ivanova *et al.*, 2002), pre-B-cells (Hoffmann *et al.*, 2003), and T-reg-cells (Bruder *et al.*, 2004). In addition to microarray studies for stem cells, the extent of posttranscriptional modifications within the stem cell transcription is under way. Thousands of genes expressed in HSCs and ES cells undergo alternative splicing. Using combined computational and experimental analyses, the frequency of alternative splicing has been found to be especially high in tissue-specific genes, as compared to ubiquitous genes. Negative regulation of constitutively active splicing sites can be a prevalent mode for generation of splicing variants, and alternative splicing is generally not conserved between orthologous genes in human and mouse (Pritsker *et al.*, 2005; Lemischka and Pritsker, 2006).

Many genes encoding lymphoid-specific functions have been knocked out, and more studies are under way to develop specific deletions or replacements of genes by various techniques (Fig. 51.1). Vectors with inducible promoters are being used to introduce genes in a controllable, tissue-specific manner.

## Growth Requirement for Differentiation of ES Cells to Lymphoid Cells

Under appropriate conditions, ES cells can differentiate *in vitro*. Many different culture systems are available for studying the development of lineage commitment. Because ES cells are pluripotential, optimal culture conditions would enable the formation of mature cells and tissues and even of organs in well-organized three-dimensional structures *in vitro*. That is, one should be able to generate an artificial fetus in culture without any external cells or other factors. In fact, of course, organogenesis from ES cell culture is but a dream. Nevertheless, with the culture conditions and system that we have developed, mature and embryonic cells in some form of tissuelike organization, though never organs, of certain orderly structures (gutlike, epidermallike, and thymuslike structures), can be observed.

Many genes have been identified that play important roles in the differentiation of ES cell to mesoderm and then to derivatives such as HSCs (see figure legends). Recently, more genes have been identified; for example, mMix (a homeobox gene Mix-like) is reported to play a critical role in gastrulation and may function early in the recruitment and/or expansion of mesodermal progenitors to the hemangioblastic and hematopoietic lineages (Willey *et al.*, 2006). Hematopoietic defects have been identified in differentiating mMix$^{-/-}$ ES cells. Conditional induction of mMix in ES-derived EBs results in the early activation of mesodermal markers prior to expression of Brachury/T and acceleration of the mesodermal developmental program. Increased numbers of mesodermal cells, hemangioblastic cells, and HSCs form in response to premature activation of mMix. Differentiation to embryonic and mature blood cells proceeds normally and without an apparent bias in the representation of different hematopoietic cell fates.

Many researchers have attempted but failed to study the differentiation of lymphopoietic precursor cells obtained from mES cells. Many observe the development of yolk sac blood islands in culture; these presumably contain HSCs. But only myeloid cells could be produced in the culture

systems reported. Only a few systems allow the generation of lymphoid precursor cells, mostly B-cells but also T- and B-cells, in a quantity sufficient for further manipulation (Chen *et al.*, 1997; Chen, 2000).

The reasons for these variations are unclear. It is partially due to the origin and pluripotentiality, hence the fragility and variability, of ES cells but partially due to experimental manipulations, different culture conditions, (Chen *et al.*, 1997, also recently reviewed by Dang *et al.*, 2002), and the transient expression of genes relevant to the developmental stage of hemato-lymphopoiesis from mesoderm equivalent at the yolk sac stage. HoxB4 (see later) is an example of such genes of recent discovery. Even in the hands of a single researcher, the potential of lymphopoiesis from the same ES cell lines varies greatly among different experiments, and why this is so remains a mystery. The culture conditions vary, depending on which lineages are chosen for study. For example, methylcellulose culture is preferred for myeloid lineages but is inferior for lymphoid lineages. Several culture conditions that favor the differentiation of ES cells to lymphoid lineages are available.

Through the work of many groups, it has become clear that lymphoid precursor cells can be obtained from ESCs in culture. However, the development of mature B plasma cells and mature T-cells requires a two-step procedure — a two-step culture procedure for plasma cells, and culture and animal implantation for mature B- and T-cells. When chimeric bone marrow cells from embryoid body–implanted nude mice, which contain cells of ESC origin, were injected into the host mice, the bone marrow cells were shown to repopulate the primary and secondary lymphoid organs, i.e., bone marrow, spleen, and lymph nodes. The data indicate that HSCs and HPCs of ESC origin can be transferred. However, when injecting into the host animal, such as lethally irradiated adult mice instead of unirradiated nude mice, dissociated cells from blood island of the EB, instead of the whole EB, killed the mice when intravenial injection was performed, and failed to repopulate the host immune system or to achieve a stable engraftment when intraperitoneal injection was performed (reviewed in Chen *et al.*, 1997).

The recent work on the expression of (transcriptional) molecules such as HoxB4, Cdx4 (of Cdx-Hox pathway), and especially the transduction of transgenic expression under the inducible tet-promoter using retroviral vector, has been shown to understand the mechanism and pathway. HoxB4 is a homeodomain-containing transcription factor with diverse roles in embryonic development and the regulation of adult stem cells. This gene has a double life and can act in opposite ways when expressed by different cells, promoting the proliferation of stem cells while activating the apoptotic pathway in some embryonic structures. It is implicated in self-renewal of definitive HSCs, transiently expressed in blood developmental stages somewhere between the yolk sac stage and the fetal body stage and also expressed in human cervical carcinoma (Lopez *et al.*, 2006). Persistent overexpression of HoxB4, however, inhibits the development of blood lineage.

It was examined to show its effect on hemopoiesis using a variety of tools, including retroviral transgenic expression (Sauvageau *et al.*, 1995; Helgason *et al.*, 1996; Antonchuk *et al.*, 2002; Kyba *et al.*, 2002). Expression of HoxB4 in primitive progenitors, and when combined with culture on stromal cells, induces a switch to the definitive HSC phenotype. These progenitors engraft lethally irradiated adult mice and contribute to hematopoiesis in primary and secondary hosts. Thus, HoxB4 expression promotes the transition from primitive cells to become HSCs. Overexpression, a high level of expression, of HoxB4, however, is shown to perturb the differentiation and to predispose the manipulated cells to leukemogenesis. HoxB4 may affect cell growth in a dose-dependent manner and may disturb the differentiation into lymphopoiesis (Will *et al.*, 2006; Pilat *et al.*, 2005; Abramovich *et al.*, 2005).

This approach involving the manipulation of essential genes has overcome, partially, the problems of variability of the development of hematopoietic cells from mesoderm-derived cells, such as yolk sac cells and EBs, and increased engraftment into irradiated adult mice (Wang *et al.*, 2005). Combining HoxB4 gain of function and Cdx4 gain of function has been shown to increase the multilineage haematopoietic, including lymphoid, engraftment of lethally irradiated adult mice.

Others that have been tried are: mMix (see earlier, Willey *et al.*, 2006); Stat 5, a signal transducer under similar inducible conditions to express in mES-derived HSC and to grow similarly on OP9 stroma cells, shown to increase the induction of preferable myeloid, rather than lymphoid, differentiation (Kyba *et al.*, 2003); BCR/ABL, a chronic myeloid leukemia–associated oncoprotein, is shown to transform a subset of HSC between erythryo- and lymphoid-myeloid lineages (Daley, 2003).

The combined *in vitro* and *in vivo* culture system described earlier using nude mice as the hosts can also be employed to study thymic stroma as well as well-organized, highly differentiated cells of many other lineages, including gut epithelia, skin with hair follicles, bone, muscle, neurons, and glia. One essential requirement for applying this combined system with success is that the cells differentiate first in culture, until the EB developed before the subcutaneous implantation. Otherwise, tumors may grow, due to the expression of the acylated dimeric iLRP (immature 32–44 kDa laminin receptor protein) and other, still-to-be-identified oncofetal antigen (OFA) positive in early to mid-gestation (Rohrer *et al.*, 2001). The development of endogenic thymic epithelial cells as well as stroma cells inside the ES-embryonic bodies also explain partially the findings of an efficient development of T cells in the absence of exogenous stroma cells.

Stroma cell line RP.0.10 is able to support both T- and B-lymphoid precursor cells derived from ESCs in some labs,

but this has not been reproducible elsewhere. Stroma cell lines OP9 and S17 were originally reported to support B-precursor cells and myeloid but not T-precursor cells derived from ESCs. Many other stroma cell lines do not support the differentiation of ESCs to lymphoid precursor cells. The S17 stroma cell line was compared to many stroma cell lines for the production of cytokines and stroma function, i.e., the stem cell support function (ref. cited in Chen, 2000) (see also later, Stage 2); there were no correlations in the assays. Thus a yet-to-be-identified extracellular matrix molecule of S17 cell lineage may play a role in supporting the function of HPCs.

Recently, with the insertion of genes involved in Notch-Delta signaling pathways, it was shown that the OP9-derived cell line, OP9-DL1 (Delta-like 1), increases efficient T-cells, which appeared from sources such as fetal liver (Taghon *et al.*, 2005), human cord blood (La Motte-Mohs *et al.*, 2005), and ES cells (de Pooter *et al.*, 2005). The study using switch cultures shows that while necessary to induce and sustain T-cell development, Notch/Delta signaling is not sufficient for T-lineage specification and commitment but, instead, can be permissive for the maintenance and proliferation of uncommitted progenitors that are omitted in binary-choice models. Moreover, it was demonstrated that a three-dimensional structure to allow the interaction of precursor-supporting cells plays an essential role for it to function well (see also later, Stage 4).

Other candidate molecules and cells that may support the lymphopoiesis include the following: dlk, an epidermal growth factor–like molecule of 66 kDa in stroma cell lines was reported to influence the requirement for IL (interleukin)-7 in supporting mouse pre-B lymphopoiesis, and dlk is suggested to play an important role in the bone marrow HPC microenvironment. Another high-molecular-weight CD166 (HCA/ALCAM) glycoprotein was shown to express in human HPCs as well as stroma cells. This molecule is involved in adhesive interaction between HPCs and stroma cells in the most primary blood-forming organs.

Rethinking the seemingly conflicting results and the inconsistent report of differentiation of ES cells to lymphocytes, I wonder if culturing ES-derived lymphoid progenitor cells on stroma cell lines derived from the most primitive stroma environment from the fetus (such as paraaortic cells) may be a better choice to obtain consistent results, rather than using the existing stroma cell lines. Indeed, stroma cell lines from such a source have been derived.

The field of hESCs (human embryonic stem cells), only in its infancy, represents a theoretically inexhaustible source of precursor cells that could be differentiated into any cell type to treat degenerative, malignant, or genetic diseases or injury due to inflammation, infection, and trauma. This pluripotent, endlessly dividing cell has been hailed as a possible means of treating various diseases. hESCs are also an invaluable research tool to study development, both normal and abnormal, and can serve as a platform to develop and test new therapies. They are also a potential source of HSCs for therapeutic transplantation and can provide a model for human hematopoiesis. The first stable hES cell line was reported in 1998 (Thomson *et al.*). Eight years later, many centers could establish hES cell lines. The current goal is to establish clinical grade hES cell lines for potential therapeutic purposes. At least two IND (introducing new drug) have been submitted for clinical trials, such as treating spinal cord injury repair. None of these studies is ready for treating blood disorders yet. The attempt to differentiate hESCs *in vitro* to HSCs has been reported (Kaufman *et al.*, 2001; Vodyanki *et al.*, 2005; Wang *et al.*, 2005; Zambidis *et al.*, 2005; Narayan *et al.*, 2006; Bowles *et al.*, 2006).

From such studies, hES cells are shown to differentiate to HSCs, using various protocols and different supportive stroma cell lines, with or without cytokines. It seems that the differentiation into mesoderm-derived lineages and then to mesodermal-hematoendothelial (MHE) lineage can occur, resembling human yolk sac development. The cellular and molecular kinetics of the stepwise differentiation of hESCs to primitive and definitive erythromyelopoiesis from hEBs in serum-free clonogenic assays have been demonstrated (Zambidis *et al.*, 2005). Hematopoiesis initiates from CD45+ hEB cells, with the emergence of MHE colonies. They can generate endothelium and form organized, yolk sac–like structures that secondarily generate multipotent primitive HSPCs, erythroblasts, and CD13+CD45+ macrophages. A first wave of hematopoiesis follows MHE colony emergence and is predominated by primitive erythropoiesis. A second wave of a definitive type of colonies of erythroid, GM, and multilineage CFCs follows. These stages of hematopoiesis proceed spontaneously from hEB-derived cells without requirement for supplemental growth factors during hEB differentiation. Gene expression analysis revealed that initiation of hematopoiesis correlated with increased levels of SCL/TAL1, GATA1, GATA2, CD34, CD31, and Cdx4. These data indicate that the earliest events of embryonic and definitive hematopoiesis can be differentiated from hESCs *in vitro*.

However, differentiating hEBs further into HSCs, and lymphohematopoietic lineages becomes variable and unpredictable, according to several reports. In one report, when cocultured with S17, a mouse stroma cell line, or C166, a yolk sac endothelial cell line, and in the presence of FCS, without cytokines, hES cell-derived HSCs are shown to be CD34+TAL1+LMO2+GATA2+. When cultured on semisolid media with growth factors, these HSCs form characteristic erythroid, myeloid, and megakaryocyte colonies. More terminally differentiated hematopoietic cells derived from hES under these conditions also express normal surface antigens of the corresponding lineages: glycophorin A+, CD15+, and CD41+. No lymphoid lineage was generated using such stroma cells and culture conditions. When cocultured with OP9, a mouse stroma cell, it is possible to obtain up to 20% of CD34+ cells and isolate up to 10(7) CD34+ cells with more than 95% purity from a similar number of initially plated hES cells after eight to nine days of culture. The hES-derived CD34+ cells were highly enriched with CFC (colony-forming cells), expressing GATA1+ GATA2+ SCL/TAL1+ Flk1+, and retained clonogenic potential after *in vitro* expansion. Such HSCs are shown to be CD34+CD90+CD117+CD164+ CD38− ADH (aldehyde dehydrogenase)+ rhodamine 123$^{flux+}$. When cocultured with hES-derived CD34+ cells on MS-5 stroma cells in the presence of SCF, Flt3-L, IL-7, and IL-3, they differentiated into B lymphoid cells, natural killer cells, and myeloid (macrophages and granulocytes) lineages. No T lymphoid lineage could be demonstrated to generate under such conditions. The OP9 DL1 cell line (see earlier) was not applied in this study. The derived cells can differentiate into other cells when implanted into fetal sheep intraperitoneally and differentiated in primary and secondary hosts (Narayan *et al.*, 2006).

The transit expression of transcription factor HoxB4 (see earlier) appears to play a role in determining cells to become lymphoid cells in the mouse system and would explain the occasional determination of precursor cells to commit to lymphoid lineages. It implies right away that it can apply to study the lymphohemopoiesis in the human system. Indeed, a recent report (Bowles *et al.*, 2006) has provided data on such an attempt. Culture of hES cells on mouse feeder cells or in cell-free culture conditions results in low levels of differentiated HPCs. Transgenic stable HoxB4-hESC clones were generated by the lipofectamin transfection method. *In vitro* differentiation of hESCs, as EBs, in serum containing medium without cytokines led to the sequential expansion of erythroid, myeloid, and monocytic progenitors from day 10 of culture. These cells retained the capacity to develop into formed blood elements during culture. Overexpression of HoxB4 considerably augments the development of three myeloid lineages of hESCs; however, no lymphoid progeny were reported. Since this approach is very different from the transduction using mouse retroviral vector and involves the human cell system, it is too early to draw any conclusions at this stage regarding the role of HoxB4 in human lymphohemopoiesis.

Theoretically, lentiviral vector introducing HoxB4 under a tet-on system should be a comparable approach for further study. However, another report does not support this notion or show a functional role of HoxB4 in HSCs in their combined *in vitro* and *in vivo* system, using SCID mice (Wang *et al.*, 2005). It is interesting that hES-derived HSCs failed to reconstitute i.v. transplanted host because of cellular aggregation causing fatal emboli formation, which is similar to our attempt using mEBs-derived cells. Direct femoral injection allowed host survival and resulted in multilineage hematopoietic repopulation. However, hES-derived HSCs had limited proliferative and migratory capacity as compared with somatic HSCs, which correlated with a distinct gene expression pattern of hESC-derived HSCs that included HoxA and HoxB gene clusters. Transduction of HoxB4 had no effect on the repopulating capacity of hES-derived cells.

A yet-to-be-identified molecular program is postulated to contribute to the atypical behavior of hES-derived cells *in vivo*.

Thus, conflicting reports continue to surface, giving the impression that the differentiation into lymphohemopoiesis seems to be, if not more rare, then as rare and as variable as the *in vitro* differentiation system using mES cells. The questions of the variability among different hES cell lines and clones, similar to the observation using mES cells, are also reported. The derived somatic tissue–committed stem precursor cells remain rare. How to study individual gene expression using hES cells and its relationship with the group of genes and the final outcome of the cell fate is a challenge and will require more time in the future. Combining multiple genes, such as mMix, Cdx4, and HoxB4, with a gain-of-function approach might be the next possibility to attempting to increase the lymphohematopoiesis derived from hES cells (also, with Bowles, personal communication).

## Stage 2: HSCs and HPCs

The terms *hematopoietic* stem cells, *progenitor* cells, and *precursor* cells are used loosely in the literature. Hematopoietic stem cells and hematopoietic progenitor cells are defined as two populations of stem cells sharing similar surface markers. HSCs are rather quiescent, noncycling stem cells, whereas HPCs are cycling cells. In mice, $c\text{-kit}^+$, $thy\text{-}1^{low}$, $lin^-$, and $scal^+$ bone marrow cells are defined as HSCs or HPCs. HSCs and HPCs in bone marrow are heterogeneous in size and in self-renewal capacity. Moreover, they have finite life spans. In human bone marrow, $CD34^+HLA^-lin^-thy\text{-}1^{low}rhodamine123^{low}$ stem cells are generally defined as HSCs and HPCs. However, CD34 is also expressed in other lineages, such as vascular endothelial cells and muscle precursor cells. It is well established that mouse HSCs and HPCs can be isolated, cultured, and transduced by retroviruses and used to repopulate animal hosts. In addition to bone marrow, they can be isolated from fetal liver, cord blood, fetal blood, yolk sac, and paraaortic spanchnopleura.

### Do HSCs and HPCs Self-Renew?

The self-renewal of HSCs and HPCs was carefully examined (ref. cited in Chen, 2000). By transferring bone marrow stem cells from one host to the other, it was found that the pool of donor bone marrow stem cells was smaller on transfer, and it was concluded that self-renewal is limited for somatic bone marrow stem cells. However, the results could be due to a dilution effect of the donor stem cells, if a certain absolute number of stem cells was required for the successful implantation of donor stem cells in the bone marrow. Using fluorescence *in situ* hybridization (FISH) and telomerase assays, it was shown that telomere DNA length predicts the age and replication capacity of human fibroblasts.

The telomere DNA length has been correlated with the replication potential of HSCs and HPCs. In human $CD34^+$, $CD71^{low}$, and $CD45RA^{low}$ bone marrow stem cells, the length of telomeric DNA in HSCs and HPCs is correlated with the age of the donor. The stem cells in adults have shorter telomere than found in fetal cells, but both have the same telomerase activity. When the HPCs were expanded with cytokines without the support of stroma cells, there was a loss of telomeric DNA in culture. The loss of telomeric DNA in HSCs and HPCs from older people and cultured stem cells was interpreted to mean that their replication potential is finite. An alternative interpretation would be that the culture conditions that allow stem cells to grow in the presence of cytokines may not favor the maintenance of telomerase activity and hence not the self-renewal of HSCs and HPCs.

Many culture systems for studying the biology of mouse HSCs and HPCs are available. Many mouse stroma cell lines support the growth of HPCs. Studies on human stem cells use mouse stroma cell lines, such as S17 and AFTO24 (Nolta *et al.*, 2002), among many others, or mixed primary stroma populations, recently called MSCs (mesenchymal stem cells), isolated from human bone marrow or other sources, to support purified HSCs and HPCs or unseparated bone marrow cells. The establishment of heterogeneous human stroma cell lines with the help of a plasmid-containing SV40 T antigen under the control of an inducible metallothionein promoter has been reported, but their ability to support lymphopoiesis could not be demonstrated. Mixed human stroma cells, or MSCs, and cell lines that support human long-term culture-initiating cells also exist. However, these human stroma cell lines could not be shown to support the development of human lymphocytes from bone marrow stem cells.

Recently, MSCs have become a promising field, and many clinical trials are under way. The expansion of human MSCs in culture presents a challenge. They proliferate poorly in medium supplemented with human serum, and they require the presence of selective lots of FCS. A recent report has indicated that prolonging the culture of MSCs runs the risk of becoming tumor cells (Rubio *et al.*, 2005).

Major problems in this field are the variability of culture conditions, the efficiency of differentiation of the cultured cells to lymphoid lineages, and the poor reproducibility in the hands of different investigators. Also contributing to problems is the failure to report specific reagents and ingredients used as well as undescribed procedures, stroma cell conditions, batches of serum, and growth factors required in each system. Of course, the intrinsic multipotentiality of HSCs and HPCs also contributes significantly to variability. Many have claimed that a cocktail of cytokines alone could promote the differentiation of myeloid lineages from HSCs and HPCs in mouse and human (ref. cited in Chen, 2000) systems. A few have claimed the development of B-cells at various stages of maturation, but no mature T-cells could be found in such culture conditions. The maintenance of long-term cultures of self-renewing stem cells with potency for lymphopoiesis, especially T lymphopoiesis, requires stroma cells, additional cytokines, a three-dimensional structure to support the precursor cell–stroma cell interaction (also see Stage 3), and other culture conditions yet to be defined. Several existing culture systems seem to be promising for expansion *ex vivo* and potential differentiation of human and mouse HSCs. Human-based supporting cell lines, such as an endothelial-like cell line, ECV304/T24, were used to generate split-function amphotropic packaging cell lines. This manipulated cell line, APEX, was used for transduction and for support of the growth of HSPs (Dando *et al.*, 2001). The development of supporting human cell lines for lymphopoiesis is essential for the production of large quantities of cultured cells for manipulation and reimplantation.

## Selected Examples of Clinical Applications of Human HSCs and HPCs

Using human HSCs and HPCs as the cell base for gene therapy is a complicated issue. In most clinical protocols, HPCs are transduced *ex vivo* with retroviral vectors and reimplanted into patients. One example undergoing clinical trial for a decade is the treatment of adenosine deaminase (ADA) deficiency. ADA deficiency exists in all cells examined. However, T-cells of these patients are selectively missing, causing a server combined immunodeficiency (SCID) symptom. Despite decades of study, the mechanism by which the ADA defect causes specific deficiency of T-cells remains unknown. Possible mechanisms have been postulated, such as the apoptotic pathway of CD95 (Fas/apo-1)-induced cell death.

Patients suffering from ADA deficiency have been treated via two protocols: One trial uses autologous cord blood CD34$^+$ cells in ADA-deficienct babies, and the combined cell and enzyme therapies seem to generate better immune function. Another trial uses ADA gene transfer therapy. Stem cells were expanded *in vitro* and transduced with retroviral vectors containing ADA genes. Several ADA-deficient children were subsequently followed up using the PCR (polymerase chain reaction) assay to assess the expression of circulating T-cells bearing the transduced ADA gene. Peripheral blood carrying the transduced ADA gene was shown to persist for seven months before disappearing. Bone marrow cells appeared later and became the major source of T-cells expressing the transduced ADA gene in the peripheral blood. Although initial reports of these studies claimed to be successful, careful evaluation of ADA gene therapy protocols shows that there is a major problem with the original protocols. During *ex vivo* expansion and manipulation for gene transfer, HPCs lose their capacity to proliferate in favor of differentiation and maturation. There was a need for repeated treatments of the same patient with retrovirus transduction, and continuous infusion of PEG (polyethylene glycol)-conjugated ADA enzyme. The differ-

ent results are due to the fact that very few mature, functional T-cells make enough enzyme to detoxify the serum in ADA patients. If stem cells are rare, small differences in technique and uncontrolled variables could make the difference between success and failure. Another, not mutually exclusive, interpretation of the lack of long-term reconstituting stem cells is that the cycling HPCs, transduced, are not the true stem cells. Human immunodeficiency virus (HIV)-based lentiviral vector has been shown to efficiently transfer foreign genes into postmitotic and noncycling cells, including HSCs. This vector and several package cell lines are still under development. Improved culture conditions for the growth and maintenance of human stem cells in large quantities are required. The establishment and use of reliable human stroma cell lines that could support the growth of lymphopoietic HPCs *in vitro* are urgently needed.

The current status of a phase 1/2 clinical trial of ADA-deficient SCID using gene-corrected HPCs has been reported. It has moved from the early trials of safety and feasibility to recent studies demonstrating efficacy and clinical benefit, an improved protocol running with four patients who have been enrolled for one to three years, without the infusion of PEG-ADA. It is a combination of low-dose chemotherapy with busulfan and a single transplantation with ADA-retroviral-transduced HPC. No patient has received a second dose of transduced HPC yet; no side effect has been reported to date, and they can return home and live normally. A protocol with IL-3 or IL-7 was included in another gene therapy clinical trial for ADA-SCID, resulting in long-term engraftment of stem/progenitor cells. The future direction of this clinical trial is to extend this improved protocol to treat more ADA patients (Aiuti *et al.*, 2002, 2003; Ficara *et al.*, 2004).

There is another recent study of the SCID-XI gene therapy clinical trials, in which HPCs are the target of murine leukemia virus (MLV)-derived retroviral vectors containing IL (interleukin) receptor gamma common chains. Immune function has been restored in 9 of 10 treated children with the transduction of gamma c gene transfer in CD34$^+$ cells. The distribution of both T-cell receptor (TCR) V beta family usage and TCR V beta CDR3 (complementarity-determining region 3) length revealed a broadly diversified T-cell repertoire. Further retroviral integration site analysis showed that insertion sites were shared by progenic T- and B-cells, granulocytes, monocytes, and the transduced CD34$^+$. This finding demonstrates the initial transduction of very primitive multipotent progenitor cells with self-renewal capacity. These results provide evidence in the setting of a clinical trial that CD34$^+$ cells maintain both lymphomyeloid potential as well as self-renewal capacity after *ex vivo* manipulation (Schmidt *et al.*, 2005). However, due to the developed leukemia of two treated children (Hacein-Bey-Abina *et al.*, 2003; Wood *et al.*, 2006), the trial has been on hold for a period of time, and the research of this group has turned to focus on the basic biology of the oncogenicity of the vector-related insertion site LMO2 of the affected T-cells and of using the Lmo2-TLL transgenic mice model. The integrations are found to reside within FRA11E, a common fragile site known to correlate with chromosomal break-points in tumors. The fragile sites attract a nonrandom number of MLV integrations. This explains the mechanism of two out of three leukemia cases. The other leukemia is currently attributed to the transplanted IL2RG gene (Wood et al., 2006). This clinical trial is running continuously in the UK (Thrasher et al., 2005), and clinical trials subsequently resumed in France and the United States based on case-by-case study (Christoph von Kalle, Fischer, 2005).

## Stage 3: Lymphoid Precursor Cells

The characterization of lymphoid lineage–committed precursor cells defined in terms of cell surface markers and functional assays of precursors and progenies is a challenge. One cell type that is described closest to the definition of lymphoid precursor cells is the CD4$^{low}$ precursor population isolated from adult thymus (ref. cited in Chen, 2000). Because this population of cells is restricted in its potential to differentiate into myeloid cells, it is preferentially committed to differentiate into T- and B-cells. These cells can also differentiate into thymic dendritic cells. The molecular markers of this population have not been well characterized. It would be interesting to see whether this population of cells can differentiate into myogenic cells, because Morgan and Partridge obtained differentiated myogenic cells from adult thymus. Thymus contains lymphoid and presumably mesenchymal precursor cells, recently called MSC (mesenchymal stem cells), which can develop into other lineages, such as bone, cartilage, and lung.

Little is known about the culture conditions for growing this population of cells. It is possible that stroma cell lines such as S17 could be used to expand the CD4$^{low}$ population of cells in culture, because it has been possible to use S17 and cytokines to maintain paraaortic splanchnopleural-derived precursor cells possessing the T-cell lineage potency in culture for some time (in Chen, 2000).

A lymphoid lineage–preferable transcriptional factor, lkaros, was thought to be an important molecular marker of lymphoid precursor cells at this specific stage of development. lkaros was subsequently shown to have multiple effects on HSCs, lymphoid cells, T precursor cells, and others.

Using the transgenic approach, it is possible to trace lymphopoiesis between common lymphoid precursor and alpha-beta T-cells. A transgenic mouse system is applied using pre-T-cell receptor alpha (pTalpha) promoter to drive the human CD25 (hCD25) surface marker as a reporter. It marked intra- and extrathymic lymphoid precursors but not myeloid cells. The extrathymic precursors were characterized as a common lymphoid precursor population (no. 1) expressing CD19$^-$B220$^+$Thy1$^+$CD4$^+$ cells using clonogenic assays. The earliest intrathymic precursors were CD4$^{low}$CD8$^-$CD25$^-$CD44$^+$c-Kit$^+$ cells and Notch-1 mRNA$^+$ (Gounari et al., 2002).

By using the regulatory sequences from the gene encoding pre T cell receptor alpha to drive hCD25 reporter and to produce transgenic mice, the same group (Martin et al., 2003) identified another common lymphoid precursor population (no. 2), which was B220$^+$c-Kit$^-$. In short-term culture, population no. 2 cells could be derived from the no. 1 subset and contained cells that in clonogenic assays were characterized as bipotent T- and B-precursors. Mature alpha-beta T-cells were produced when transgenic bone marrow cells were injected i.v.; thymocytes were cultured using a thymic organ culture system. The no. 2 subset may represent the most differentiated population with T-cell potential before commitment to the B-cell lineage.

The human counterpart of the CD4$^{low}$ population is unknown. Certain batches of FCS (fetal calf serum) might be supportive, because such batches are shown to support the growth and expansion of human MSC derived from cord blood. The expansion of CD34$^+$, CD31$^+$ human bone marrow B-progenitor cells and partial differentiation to B-precursors in serum-free culture medium in the presence of mixed primary human stroma cells and IL-7 for a limited duration has been shown to be possible (ref. cited in Chen, 2000). A few such CD34$^+$ precursor cells could differentiate into natural killer (NK) cells and T-precursors when subjected to a secondary culture condition, with the thymic environment provided. It is not clear whether T-precursor cells are derived from HPCs or are the precommited precursor cells migrating to bone marrow. The three-dimensional structure and the expression of Notch/Delta signaling of the thymic environment have been shown to play critical roles for supporting the differentiation of cells to the T-cell lineages (Mohtashami and Zuniga-Pflucker, 2006).

## Stage 4: Precursor T-Cells

Precursor T-cells are identified to be thy-l$^+$, CD117$^{low}$, DC3$^-$, pTα$^+$ (ref. cited in Chen, 2000). They can be derived from fetal and adult blood. Using markers such as the CD25 and CD44, they can be classified into subpopulations, as shown in the two figures by Rodewald and Fehling (1998, cited in Chen, 2000), also in Fig. 4 of Hoffmann et al. (2003) and more recent reports. TIS21(TPA-inducible sequence 21) is identified as an antiproliferative T-cell molecule of the BTG/TOB (B-cell translocation gene/transducer of ErbB-2) family member, also expressed in the quiescent T-precursors and in peripheral anergic and unstimulated T-cells. With the expression of TIS21, it is possible to divide such precursor T cells further into four populations. They are the quiescent CD44$^+$ CD25$^-$ TIS21$^+$ early cells and CD44$^-$ CD25$^+$ TIS21$^+$ cells prior to TCR beta selection. After the selection, the cells are the proliferating CD25$^+$ CD44$^+$ TIS-21$^{low}$precursor T-cells and CD25$^{low}$ CD44$^-$ TIS21$^{low}$cells. By transgenic overexpression of TIS21 in precursor T-cells and HPC, it can inhibit the expansion of thymocytes (Konrad and Zuniga-Pflucker, 2005). Thus, somatic and genetic manipulations would become easier with cloning and identifying such "quiencenting genes" and their ability to express at stage-specific matter.

Other genes of interests are the pre-T-cell receptor gene (preTCR) and the antiapoptotic A1 gene (BCL2A1) (Mandal *et al.*, 2005). PreTCR$^+$ cells are selected to survive and differentiate further, whereas preTCR$^-$ cells are selected to die. The mechanisms of preTCR-mediated survival are under investigation. The induction of the A1 gene will induce pre-T-cell survival by inhibiting activation of caspase-3. *In vitro* knockdown of A1 expression can compromise survival, even in the presence of a functional preTCR. However, the overexpression of preTCR-induced A1 can contribute to T-cell leukemia in both mice and humans. Whether it can contribute to the expansion of this population in culture is not known.

Another gene could be of interest: the cell-cycle control genes, such as the D3 cycline of D type (D1, D2, D3) (Sicinska *et al.*, 2003). It could play a role in immature T-cell expansion. The D3 gene has been demonstrated to overexpress in several human lymphoid malignancies. The D3$^{-/-}$ mice failed to undergo normal expansion of immature T-cells and showed greatly reduced susceptibility to T-cell malignancies triggered by specific oncogenic pathways. Knockdown of cyclin D3 inhibited proliferation of human acute lymphoblastic leukemias that were of immature T-cell origin.

Both OLIG2 and LMO1 were overexpressed with large thymic tumor masses (Y. W. Lin *et al.*, 2005). Gene expression profiling of thymic tumors that developed in OLIG2-LMO1 mice revealed up-regulation of Notch1. Two genes considered to be downstream of Notch1 — Deltex1 (Dtx1) and preTCR-alpha — are also up-regulated. In three of six primary thymic tumors, mutations in the Notch1 heterodimerization, or proline-, glutamic acid–, serine-, and threonine-rich domains, are found. The established OLIG2-LMO1 leukemic cell line was suppressed by a gamma-secretase inhibitor, suggesting that Notch1 up-regulation is important for the proliferation of OLIG2-LMO1 leukemic cells.

T-lymphocytes are generated in the thymus, where developing thymocytes must accept one of two fates: They either differentiate or they die. These fates are determined by signals that originate from the TCR. CD4$^+$CD8$^+$ thymocytes undergo one of three fates in the thymus: positive selection, negative selection (TCR-mediated apoptosis), or death by neglect. Only 5% of developing thymocytes are exported as mature T-cells. Negative selection of thymocytes that express TCRs with high affinity for self-peptide-MHC deletes potentially self-reactive thymocytes, generating a largely self-tolerant peripheral T-cell repertoire. Most negative selection is thought to occur in the thymic medulla, for this contains two types of specialized antigen-presenting cell–dendritic cells (DC) and thymic epithelial cells (TECs). Medullary TECs transcribe genes that are normally expressed in peripheral tissues. Negative selection can occur before or after, thus independent of, positive selection and

in thymocytes at all stages of development. Negative selection in response to high-affinity ligands might be due to increased TCR occupancy or a slower off rate (kinetic proofreading). Although discrepancies exist between blocking experiments and genetically deficient mice, TCR signal and a second costimulatory signal might be required for negative selection. The kinetics of mitogen-activated protein kinase (MAPK) signaling might determine positive- vs. negative-selection signals. Extracellular signal-regulated kinase (ERK) is induced more rapidly during negative selection, which might determine the triggering of transcriptional factors, NUR77 and NF-B (Werlen *et al.*, 2003; Palmer, 2003).

Precursor T-cells are not self-renewing. Apoptosis occurs easily, and only some cells of this compartment differentiate into double-positive mature T-cells. A scale-up expansion of this population presents a challenge, with the foregoing available regulatory genes newly identified and modern manipulation tools in hand. Fetal thymus organ culture systems and suspension cultures with dissociated thymic epithelial cells and, more recently, a cell line transgenic with various notch-ligand, for studying the microenvironment, the mechanism, and the differentiation and education of T-cells, are available (ref. cited in Chen, 2000). However, how to expand precursor T-cells *ex vivo* without differentiation and apoptosis remains a challenge. It is possible to expand transiently paraaortic splanchnopleural–derived precursor T-cells in culture and perish quickly. The physical environment of culturing, such as varying the O$_2$ level of the incubator where the fetal organ culture was set up, was shown to help the survival of (precursor) T-cells. Free radicals promote molecules that lead to the apoptotic pathway. Thus, reducing the concentration of local O$_2$ with a high level of N$_2$ in the incubator or supplying the cultured cells with a high density of stroma cells favors the survival of (precursor) T-cells.

## Stage 5: Mature T-Cells and Memory T-Cells

Using cultures of mature T-cells in the presence of APCs (antigen-presenting cells) and recombinant cytokines, one can study the growth and differentiation of T-cells from a variety of sources. Long-term mouse and human T-cell clones are also available. For unknown reasons, T-cells can be maintained as clones in culture far longer than B-cells. However, long-term human T-cell clones, even from healthy donors, have not been reimplanted into patients. Human T-cells (mainly natural killer cells) grown in the presence of cytokines in a short-term culture have been reimplanted into autologous cancer patients. No study was reported to evaluate their long-term repopulating potentiality in the human severe combined immunodeficient (SCID) model or the self-renewal property of long-term mouse T-cell clones *in vivo*. Most CD8$^+$ and CD4$^+$ T-cells have a short life span. Memory T-cells have been well studied, but in reality they do not exist in abundance and can be demonstrated *in vivo* only by repeated priming with antigen. They are very difficult to define at the cellular and molecular levels. There is no isotype 3′-end downstream of the TCR constant region for class switching to occur. Somatic hypermutation of the TCR-beta gene has never been claimed, and it is controversial whether the TCR-alpha gene is hypermutable. Memory T-cells are thus defined using criteria such as accelerated cellular responses, distinct pathways of lymphocyte recirculation *in vivo*, distinct DNA motifs of TCR genes, cytokine-producing pattern and diminishing expression of surface markers (ref. cited in Chen, 2000), and functional and antigen requirements. In mice, memory CD4$^+$ T-cells are CD45RO, L-selectin (MEL-14)$^{low}$. This is the equivalent of human CD45RO, keeping in mind that the CD45R family (A, B, C, and O) may not be the best marker to define naive versus memory T-cells. Because foreign antigens do not always quickly elicit unprimed T-cells, memory T-cells must exist, but the commitment, the mechanism of development, and the maintenance of these cells are unknown. Memory T-cells are thought to be generated either when T-cells acquire specificity to kill or to help during thymic education, or they are generated during the mature stage.

## Memory CD8$^+$ T-Cells

Because cell-bound antigen on APCs (see later, for details) and more than one signal are required for educating T-cells to perform effector functions instead of becoming tolerized, work on the generation of CD8$^+$ memory T-cells has been mainly *in vivo*, using viruses. Is it possible to prevent cytotoxic T-cells from performing their function by releasing granzymes. If so, what happens to committed CD8$^+$ T-cells? Do they die, or become anergic, or become memory cells? A study addressing the avoidance of granzyme B–induced apoptosis in target cells is interesting in this regard (ref. cited in Chen, 2000). Dephosphorylation of cdc2 was shown to be a critical step in granzyme B–induced apoptosis in the targets of cytotoxic T-cells. A nuclear kinase encoded by the *wee* 1 gene was transiently expressed and shown to induce phosphorylation of the tyrosine residues of cdc2 kinase, and that in turn provoked mitosis and the rescue of target cells. Because cytotoxic T-cells are subject to being killed by their colleagues, the apoptosis pathway in these cytotoxic T-cells could be similar to that of the target cells. The priming, clonal expansion, and differentiation into memory T-cells can be achieved. The expansion of memory T-cells can be demonstrated in culture. Besides viral peptide antigen, recent examples are the use of myco-bacterial glycolipids as the antigen, the antigen processed by CD1b, CD1e (Gilleron *et al.*, 2004; Sallet *et al.*, 2005). Primed T-cells are harvested from a human adult and stimulated *in vitro* with antigen with hIL-2. The T-cell clone can demontrate the killing of cells infected with *Mycobacterium tuberculosis* (discussed further later).

It is interesting to learn that DNA methylation may contribute to regulation of mouse T effector cell function (Makar and Wilson, 2004a). In Dnmt1$^{-/-}$ (the maintenance DNA methyltransferase), silencing of IL-4, IL-5, IL-13, and IL-10 in CD8 T-cells was abolished, and expression of these T$_h$2 cytokines increased drastically as compared with that of control CD8 T-cells. T$_h$2 cytokine expression also increased in Dnmt1$^{-/-}$ CD4 T-cells, but the increase was less than for CD8 T-cells. As a result, both Dnmt1$^{-/-}$ CD4 and CD8 T-cells expressed high and comparable amounts of T$_h$2 cytokines. Loss of Dnmt1 had more subtle effects on IL-2 and IFN expression and did not affect the normal bias for greater IL-2 expression by CD4 T-cells and greater IFN expression by CD8 T-cells or the exclusive expression of perforin and granzyme B by the CD8 T-cells. Dnmt1 and DNA methylation seem to be necessary to prevent cell-autonomous T$_h$2 cytokine expression in CD8 T-cells, but they are not essential for maintaining proper T-cell subset-specific expression of T$_h$1 or CTL effectors. Thus, transcription factors and DNA methylation are complementary and nonredundant mechanisms by which the T$_h$2 effector program is regulated.

## Memory CD4$^+$ T-Cells

Like memory CD8$^+$ T-cells, the generation of memory CD4$^+$ T-cells is studied *in vivo*. This population of cells has been well characterized in mice. Effector memory CD4$^+$ cells in mouse spleen against the soluble protein antigen keyhole limpet hemocyanin (KLH) and other protein antigen were found to be CD45RO, L-selectin$^-$, CD44$^+$ and to produce elevated levels of IL-4 (ref. cited in Chen, 2000). Whether this population of cells can be expanded *ex vivo* remains unknown. Because T-cells of IL-2 knockout mice survive much longer than those of conventional mice, they may be useful for studying the development of memory T-cells. Moreover, molecules such as Fas and Fas ligands of the apoptotic pathways play critical roles in determining the fate of cells after activation. The Bcl-2/bcl-x family seems to function by helping the survival of cells via action on the intermediate steps of apoptosis. In the future, genes controlling the cell death pathway may play critical roles in the development of, and the subsequent genetic manipulation of, memory T-cells. It is important to improve the conditions for growing CD8$^+$ T-cells *in vitro*, because for tumor therapy, the generation of tumor-specific CD8$^+$ T-cells is a critical step.

Although surface markers and the life span of a population of T-cells equivalent to those of the mouse system have been documented, ethical and safety considerations have prevented a systematic study of human memory T-cells *in vivo*. The possibility of producing human CD4$^+$ T-cells on a large scale has been reported. By stimulating peripheral, resting T-cells with cytokines such as IL-2, anti-CD28, and solid-phase anti-CD3, they reported that survival and proliferation of CD4$^+$, but not CD8$^+$, T-cells could be greatly promoted. What remains to be shown are the specificity and function of these cells and how they are related to the regulation of the Bcl-2/bcl-x family and CD95 (Fas/apo-1) ligand and whether they are candidate memory CD4$^+$ T-cells or abortive T-cells.

## Role of APC, Toll-Like Receptor, in the Activation of CD4⁺, T$_h$1 vs. T$_h$2 Pathways

The recognition and stimulation of antigen epitopes by T-cells requires that antigen molecules first be processed to become fragmented epitope and presented properly by APCs (antigen-presenting cells). The APCs identified so far are macropahges, dendritic cells, and B-cells when in the process of T–B-cell cooperation. T-cells will only recognize the antigenic epitope when it is embedded in MHC (major histocompatibility complexes, H2 in mouse and HLA in humans), which enables the immune system to distinguish its own cells from foreign cells.

Foreign antigens are great engineers. The nature of the antigen to be processed by the APCs decides which processed antigenic epitopes are to be presented by which kind of T-cells and how to influence the reaction of the immune system. For example, when cytosol enters the host, viral proteins will be digested by proteasomes to become fragmented epitopes and be presented on the surface of APCs in complex with class I MHC. Their interaction with the CD8⁺ T-cells will enable them to become cytotoxic T-cells to perform cell-mediated immunity.

The soluble bacterial (with exception, see next section) toxins, phagocytized into acidic vesicles in APCs, will be processed by the vesicular proteases, and the fragmented epitopes will be presented on the surface of APCs in complex with class II MHCs. In macrophages, foreign proteins within the acidic phagocytic vesicles will be digested and presented on their surface in complex with class II MHC, similar to APCs. These epitopes are recognized by the CD4⁺ helper T-cells and will cause immune responses such as the activation of phagocytes and antibody production by activated B-cells.

In recent years, two major subsets, T$_h$1–T$_h$2, have been recognized in the population of CD4⁺ T-cells. T$_h$1 is shown to secrete gamma interferon and to help cell-mediated responses. T$_h$1 stimulation results in local inflammatory responses, including the activation of macrophages, and the production of complement-fixing and -opsonizing antibodies. T$_h$2 is shown to secrete IL-4 and IL-3, IL-5, and IL-13 and to help antibody-mediated responses. T$_h$2 activation leads to the production of IgG and IgE and to the activation of basophil and eosinophils to fight against allergens and parasites.

It has been a puzzle as to how T$_h$1 vs. T$_h$2 cells recognize and distinguish the epitopes of foreign antigen on APCs, since both require the presence of class II MHCs. The discovery of the mammalian equivalent of TLRs (Toll-like receptors) derived from *Drosophila* "Toll" and the subsequent study of the signaling via such receptors are among the exciting reports to address such a question. TLRs are found on the surface of APCs (dendritic cells, macrophages); so far, 11 different receptors have been shown to bind to various cellular components of microorganisms.

As shown in Fig. 51.2 (O'Neill, 2004) and in Burdin *et al.* (2004), the binding of most of these components to the preferable TLRs produces a differential signaling cascade that stimulates immune response in favor of T$_h$1 over T$_h$2. Stimulation of TLR7 (and TLR8 in humans) (Heil *et al.*, 2004; Diebold *et al.*, 2004) by synthetic imidazoquinoline compounds leads to activation of the T$_h$2 pathway; while stimulation of TLR4 by MPL (monophosphoryl Lipid A) and LPS and stimulation of TLR9 by unmethylated CpG of bacterial DNA lead to activation of the T$_h$1 pathway. The activation, via TLR1, 2, 6, TLR3, TLR5, and the new addition of TLR11 (Zhang *et al.*, 2004) leading to which specific pathway remains open. Future understanding of the mechanism and the accompanying design of the tools will lead to the engineering of T$_h$1 versus T$_h$2 pathways (O'Neill, 2004; O'Neill *et al.*, 2003; Takeda *et al.*, 2003; Glazer and Nikaido, 2007).

## Role of CD1 in the T-Cell Recognition of Lipid Antigen

Besides peptide-protein antigen, which stimulates the Th1, Th2 pathway via TLR on APCs, lipid antigen activates T-cells by CD1 moleules, which is independent of TLR. Recent studies have shown that CD1 molecules are lipid antigen-presenting molecules that offer the lipid antigens to the TCR of T-cells, resembling MHC presentation of peptides, a MHC class III molecule. They have no structural homology with TLR, though they are also present on APCs. It may cause confusion, derived from the fact that some lipids are TLR ligands and therefore trigger the classical TLR-dependent responses on specific interaction. Some lipids are recognized by the immune system as classical antigens, resembling peptides associated with MHCs, whereas other lipids trigger the TLR innate receptors. Little is known about the delivery of lipid antigens from either extracellular compartments or CD1⁻ cells to CD1⁺ APCs. Apolipoprotein E is indicated to be involved in binding lipid antigens and delivering them to APC (Schumann and De Libero, 2006).

A panel of T-cell clones with different lipid specificities isolated from *Mycobacterium tuberculosis* was established. A novel lipid antigen belonging to the group of diacylated sulfoglycolipids, Ac2SGL (2-palmitoyl or 2-stearoyl-3-hydroxyphthioceranoyl-2′-sulfate-alpha-alpha′-D-trehalose), was identified. Ac2SGL is mainly presented by CD1b after internalization in a cellular compartment with low pH. Ac2SGL-specific T-cells release interferon-gamma, efficiently recognize *M. tuberculosis*–infected cells, and kill intracellular bacteria. The presence of Ac2SGL-responsive T-cells *in vivo* is strictly dependent on previous contact with *M. tuberculosis* but independent of the development of clinically overt disease. These properties identify Ac2SGL as a promising candidate to be tested in novel vaccines against tuberculosis (Gilleron *et al.*, 2004). Another isolated molecule, PIM6 (hexamannosylated phosphatidyl-myo-inositols), stimulates CD1b-restricted T-cells after partial digestion of the oligomannose moiety by lysosomal alpha-mannosidase, and soluble CD1e, one of the CD1 family

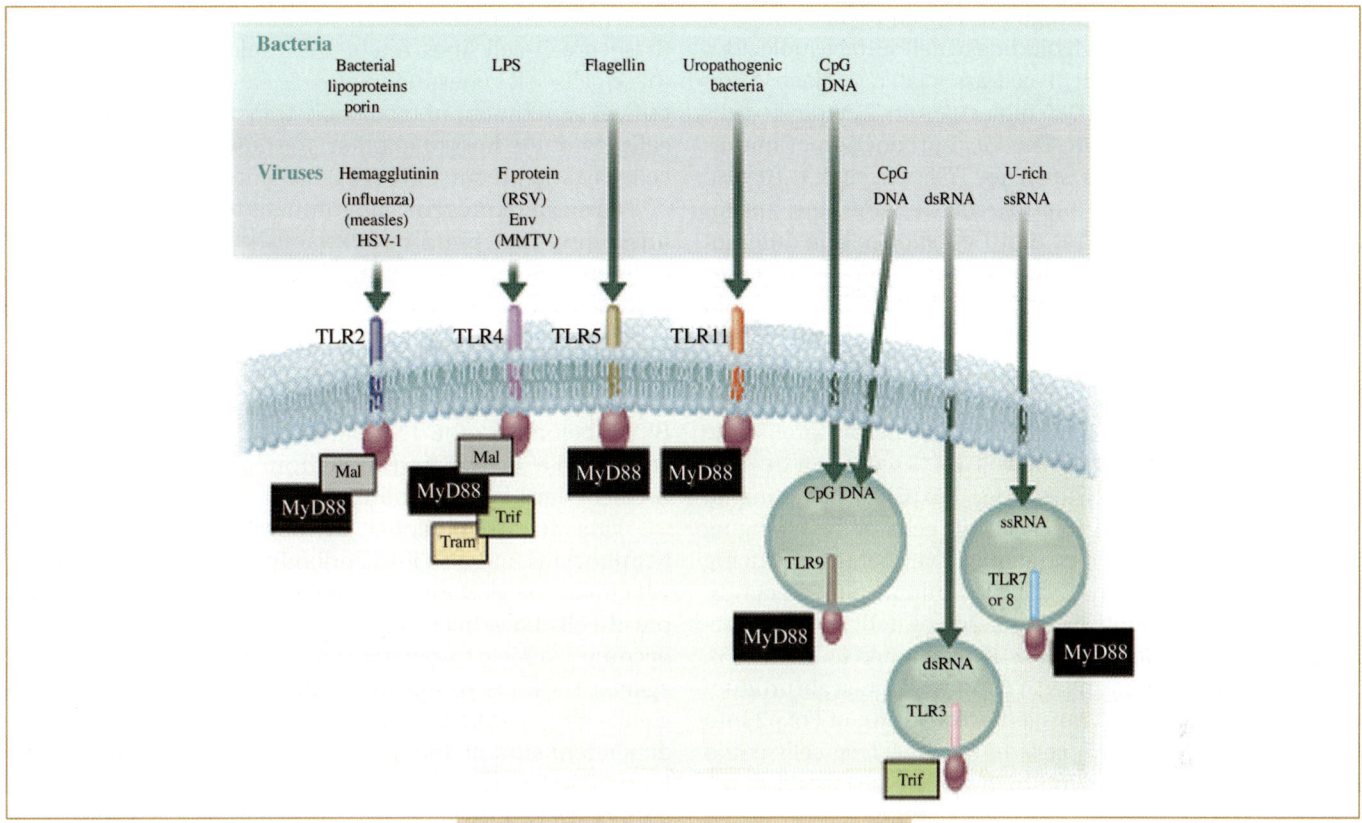

**FIG. 51.2.** Going down the Toll mine. Toll-like receptors (TLRs) provide a repertoire for sensing pathogen-derived molecules during the innate immune response. TLRs in endosomal membranes detect bacterial and viral nucleic acids. The relative contribution of each TLR to the innate immune response is not yet known because pathogens contain multiple ligands specific for several different TLRs. The signaling pathways associated with each TLR are different, although they share common components. Specific signals may emanate from the adapter proteins recruited by each TLR (MyD88, Mal, Trif, and Tram). One important question concerns how the immune response is tailored to each pathogen according to the activation of specific signaling pathways triggered by different pathogen products (O'Neill *et al.*, 2003). HSV-1, herpes simplex virus 1; LPS, lipopolysaccharide; RSV, respiratory syncytial virus (corrected following the erratum, *Science*, postdate 4 June 2004); MMTV, mouse mammary tumor virus; Porin (bacterium, not influenza). From O'Neill (2004) with permission. The figure has been modified following the erratum in *Science*, postdate 4 June 2004), i.e., Porin is a bacterial component and not viral and is shifted from the grey region upward to the green region.

members, is required for the processing. Recombinant CD1e was able to bind glycolipids and assist in the digestion of PIM6. It is proposed that CD1e helps expand the repertoire of glycolipidic T-cell antigens to optimize antimicrobial immune responses through glycolipid editing (de la Salle *et al.*, 2005). How CD1 present lipid antigens to T-cells has been reveiwed and graphically explained (De Libero and Mori, 2005).

### Innate Immunity and T Regulatory Cells

Recently, another heterogeneous population of T-cells has been re-identified, the so-called Treg (T regulatory) or suppressor cells. They can be innate or induced. The current literature is rather confusing. Here only a simplified overview is given.

The subset of Treg cells can be identified using different surface markers and the secreted cytokines. They include (1) naturally occurring T-cells possessing CD4$^+$ CD25$^+$ Nrp1$^+$ Foxp3$^+$ TGF beta$^+$ and (2) cells induced in the periphery following antigen exposure, such as Treg1 cells possessing CD4$^+$ CD25$^{-/+}$ Foxp3$^{-/+}$ IL10$^+$ IFN gamma$^+$ TGF beta$^+$ IL5$^+$, Th3 cells possessing CD4$^+$CD25$^+$Foxp3$^?$ TGF beta$^+$ IL-4$^{+/-}$ IL-10$^{+/-}$, and Treg cells possessing CD8$^+$ CD25$^{low}$ foxp3$^-$ IL-10$^+$ TGF beta$^{+/-}$ (Bruder *et al.*, 2004). The cells express Foxp3 (forkhead/winged helix transcription factor), which appears to play a key role in supporting their action (Fontenot *et al.*, 2003).

APCs are essential in the activation of Treg cells; the immature APCs can support the differentiation of Treg cells. Targeting of antigens to immature dendritic cells has been shown to result in antigen-specific T-cell tolerance *in vivo*, which implies that some autoimmune diseases, transplant rejection, and allergy can be treated. The mechanism of suppression could be multiple and is not entirely known; the tools for such study are under development. One possibility is to tolerate the APCs, and the other is via the activation of

Treg. Treg cells can down-regulate Th1 and Th2 responses. Direct cell contact through binding of cell surface molecules such as CTLA4 (cytotoxic T-cell-associated Antigen 4) on Treg cells to CD80 and CD86 molecules on T cytotoxic cells results in the suppression. The local production of immunosuppressive cytokines such as TGFbeta and IL10 also results in suppression. Suppression requires the appropriate colocalization of Treg and T effector cells in different tissue and may involve interference with the T-cell receptor signaling that triggers transcription factors important in regulating effector cell function, such as IL2 gene transcription (von Boehmer, 2005).

Treg are usually anergic (do not respond or proliferate to antigen stimulation) and IL2 dependent and proliferate poorly in culture. The truly naive CD4$^+$ T-cells can be converted into Treg cells expressing Foxp3 by targeting of peptide-agonist ligands to dendritic cells or by changing culture conditions, such as adding TGF-beta or reducing IL2. Treg cell populations induced in minute antigen conditions could subsequently be expanded by delivery of higher- or immunogenic-dose antigen (Kretschmer *et al.*, 2005). Since this field is young, how to conduct large-scale production is still developing. Transgenic expression of Foxp3 into antigenic primed CD4$^+$ T-cells to become Treg cells could be further developed by improving vectors (Jaeckel *et al.*, 2005).

## Stage 6: Precursor B-Cells

Mouse and human precursor B-cells have been characterized extensively. Surface markers and molecular events of precursor lymphoid to precursor B-cells at intermediate stages of development have been defined (Table 51.1, Fig. 51.3, Hoffmann *et al.*, 2003). The markers for peripheral B-cells at different stages are summarized in Table 51.2 (Matthias and Rolink, 2005).

### Mouse Pre-B-Cell Lines

In the mouse system, with the help of several stroma cell lines and recombinant growth factors such as IL7 (Chen, 2000), it becomes feasible to expand mouse pre-B-cells without gross differentiation. On release from the stroma cells and IL7 and in the presence of the bacterial mitogen LPS, some pre-B-cell lines differentiate into plasma cells. Pre-B-cell clones have been established from various lymphoid organs of wild-type mice, transgenic mice, and knock-out mice. Thus, both somatic and genetic manipulation of these cell lines becomes possible. Some pre-B-cell lines have been used to repopulate SCID mice and RAG-2 knock-out mice. Injected cells migrate to bone marrow, lymph nodes, peritoneal cavity, and spleen. Plasma cells, mature B-cells, and pre-B-cells were detected in the host. The percentage of cells that mature into various B compartments seems to vary from experiment to experiment.

However, several questions remain to be answered. Do these pre-B-cell lines retain the capacity to expand? Are these cells self-renewing *in vivo*, as stem cells must be? The critical experiment of repeatedly transferring donor pre-B-cells from one host to another, to show that the implanted cells are still precursor B-cells, has not yet been done.

Pax5 signatures the commitment to B-cell lineage. It is interesting that pre-B-cells established from pax5$^{-/-}$ mice differentiated into T-cells, myeloid cells, dendritic cells, and osteoclasts but not mature B-cells in SCID mice. The data indicate that the microenvironment, which plays a role in keeping pre-B-cells committed to the B-cell lineage, may be missing in the pax5$^{-/-}$ mice. BAFF (B-cell activation factor, BLYS) belongs to the TNF family, and its receptors have a critical role in the transition from immature to mature B-cells (Matthias and Rolink, 2005).

Due to its potential to develop into mature B-lymphocytes and also into antibody-secreting plasma cells, which can be performed either *in vitro* or *in vivo*, mouse pre-B-cells have been subjected to genetic modification to become a vehicle to generate and secrete human antibodies against a variety of infectious antigens in SCID mice. The genetically modified pre-B-cells can be subjected to T-cell-dependent stimuli, inducing somatic hypermutation in the human antibody constructs. In this way, affinity-matured antibodies can be generated, allowing the development of therapeutic antibodies with the pharmacological efficacies.

### Curiosities of Growing Human Precursor B-Cells

The question of whether human precursor B-cells can expand *ex vivo* is open. Several reports claim it is possible by using either primary mixed human or mouse stroma cells (ref. cited in Chen, 2000). When subjected to mixed stroma cells and cytokines such as IL-7, most human bone marrow cells expand for a limited period and then either perish or differentiate. The culture conditions established seem to be only for short-term expansion of pre-B-cells. However, no normal human pre-B-cell lines have been established. This distinguishes mouse and human precursor B-cells. The study of human pre-B-cells cannot be separated completely from the study of human HSCs and HPCs, given the complexity of the cell types involved and the difficulty in establishing human stroma cell lines. The establishment of human stroma cell lines supporting the growth of human pre-B-cells is critical. The failure lies partially in the inability of adherent cells derived from human bone marrow (stroma cells) to proliferate well under normal culture conditions with conventional sera.

Methods for immortalizing human cells include transfecting plasmids or retroviruses containing oncogenes such as SV40 T antigen. One group (J. Dando, personal communication) has done that with human fetal bone marrow, and many stroma cell lines have been established; a few cell lines are partially characterized for supporting the growth and expansion of cord blood stem cells. Cell proliferation is

**Table 51.1.** Expression of cellular and molecular markers during early stages of B-cell development

| | Progenitor B stage | | Precursor B stage | |
|---|---|---|---|---|
| *Human B-cell* | *I* | *II* | *III* | *Pre-B* |
| CD34 | + | + | + | − |
| CD38 | nd | + | + | + |
| 'l}rL | − | + | + | + |
| CD10 | nd | −+ | + | + |
| CD19 | − | − | + | + |
| lambda-like/Vpre-B | + | + | + | + |
| Rag 1 | + | + | + | + |
| TdT | + | + | + | − |
| VH/Cmu. | − | + | + | + |
| V-kappa/C-kappa | − | + | + | + |
| mbl/B29 | +/− | +/− | + | + |
| cyto-mu | − | − | +/− | + |

| | | | | | |
|---|---|---|---|---|---|
| *Mouse B-cell* | *A* | *B* | *C* | *C'* | *D* |
| B220 | + | + | + | + | + |
| CD43 | + | + | + | + | − |
| HSA | − | + | + | + | + |
| BP1 | − | − | + | + | + |
| lambda-5/Vpre-B | +/− | + | + | + | + |
| Rag 1–2 | +/− | + | + | + | + |
| TdT | +/− | + | + | +/− | − |
| mbl | +/− | + | + | + | + |
| D-JH | | + | + | | |
| VH-D-JH | | | +(out) | | +(on) |

| | *Pro-B* | *Pre-B I* | *Pre-B II* | |
|---|---|---|---|---|
| | | | *Large* | *Small* |
| CD43 | + | + | +/− | − |
| c-Kit | (+)/− | (+) | − | − |
| CD25 | − | − | + | + |
| IL-7Ralpha | + | + | + | (+) |
| CD19 | − | + | + | + |
| 'l}rL | + | + | + | + |
| lambda5/Vpre-B | + | + | + | + |
| Rag 1–2 | + | + | − | + |
| TdT | + | + | − | − |
| cyto-mu. | (+) | − | + | + |
| $D_H$–$J_H$ | − | + | + | + |
| $V_H$–$D$-$J_H$ | − | − | + | + |
| $V_L$–$J_L$ | − | − | − | + |

This table is modified from Chen (2000), Table 43.1, and data are partially derived from Martensson *et al.* (2002) and Matthias and Rolink (2005).

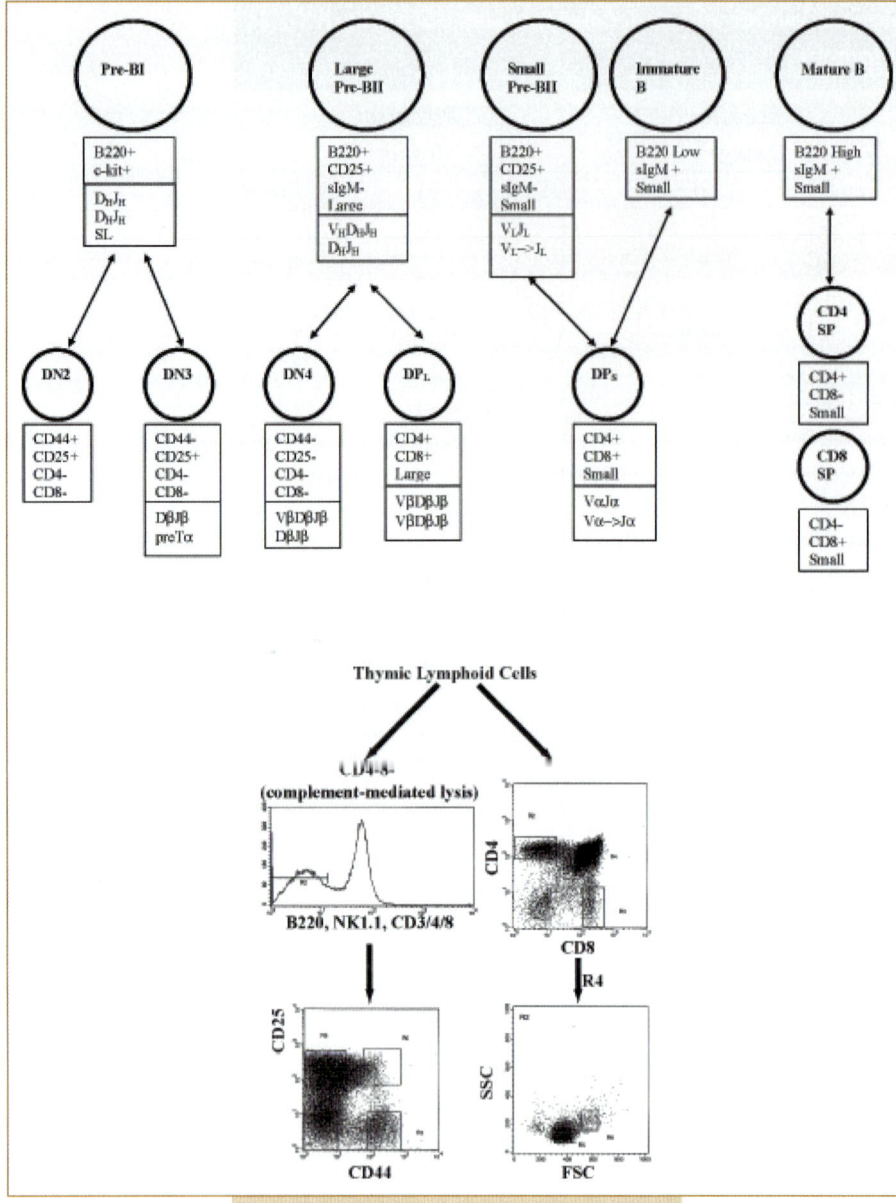

**FIG. 51.3.** Surface markers and sorting strategy for isolation of thymic subsets. **(A)** Synopsis of cellular populations and surface markers used for isolation of lymphocyte developmental stages. *Top:* B-cell development; *bottom:* T-cell development. Cellular stages are depicted as circles. Markers used for separation of the stages appear in the upper boxes; Ig and TCR gene loci rearrangement status appear in the lower boxes. Arrows connect corresponding stages between B- and T-cell development. SL, surrogate L chain. **(B)** Cell-sorting strategy for separation of T-cell precursors. To obtain DN thymocytes, CD4+ and CD8+ cells were removed by complement-mediated lysis from thymus single-cell suspension. Remaining lymphocytes were stained with a panel of lineage markers (B220, NK.1.1, CD3, CD4, CD8), and negative cells (R2 gate in top left panel) were gated and analyzed for CD25 and CD44 surface expression (bottom left). For the isolation of DP and SP subsets, single-cell suspension was stained with CD4 and CD8 mAbs (top right); SP cells were sorted according to gates R2 for CD4+ and R3 for CD8+. DP thymocytes (gate R4 in top right panel) were further resolved into large (R5) and small (R6) subsets (bottom right). From Hoffmann *et al.* (2003) with permission.

**Table 51.2.** Expression of cellular markers during periphery stages of B-cell development

| B-cell marker | Stage periphery | | | | |
| --- | --- | --- | --- | --- | --- |
| | Immature | | | Mature | |
| | t1B cell | t2B cell | t3B cell | MZB cell | Follicular B-cell |
| CD21 | – | + | + | + | + |
| CD23 | – | + | + | (+) | + |
| CD93 | + | + | + | – | – |
| IgM | + | + | + | + | + |
| IgD | (+) | + | + | (+) | + |

t: transitional; peripheral immature B-cells can be further classified into three transitional stages. MZ: marginal zone. Modified from Matthias and Rolink (2005).

a prerequisite for the stable integration of transgenes into chromosomes and for immortalization. A breakthrough for human HSCs and pre-B-cells would be to establish stroma cell lines for B lymphopoiesis and to optimize the conditions for the growth of stroma cells, as has been done for human MSCs (ref. cited in Chen, 2000). In recent years, MSCs have been shown to expand to large quantity when selected lots of FCS are used. Prolonged culturing of MSCs under such conditions generates abnormalities of chromosome and tumors when transplanted in immunoincompetent mice (Rubio *et al.*, 2005).

## Stage 7: Memory B-Cells

B-cells with surface Ig isotypes such as IgG, IgA, and IgE, which possess higher affinities for antigen, are generally defined as memory B-cells. In human beings, $CD38^+$, $CD20^+$ germinal center B-cells can be distinguished from memory $CD38^-$, $CD20^+$ B-cells and $CD38^+$, $CD20^-$ plasma cells. In mice, until recently there were no reliable surface markers to distinguish memory B-cells from plasma cells and mature primary B-cells.

The mechanisms of memory B-cell generation are unknown and have been the subject of debate for decades. Memory B-cells are mature B-cells that have encountered antigen that have been activated but not tolerized and have switched to higher-affinity isotypes. IgM-secreting plasma cells, Syndecan-1/$CD138^+$, are terminal cells destined to die. It is unknown how memory B-cells develop *in vivo*. *In vitro*, at least two systems may generate memory B-cells: the germinal center–like culture system and the suspension culture system, wherein resting B-cells are activated by LPS plus anti-mu. These B-cells are activated, alive, proliferating, and nonabortive, but they are not plasma cells.

### A System Stimulating Germinal Centers
### *In Vitro* to Culture B-Cells

In a culture environment that mimics the germinal center in lymphoid organs, B-cells survive better and live longer. A germinal center–like environment is a culture system that provides cytokines and supporting cells from either purified follicular dendritic cells (FDCs) from mouse spleen (ref. cited in Chen, 2000) or stroma cells (L-cells transfected with CD40 ligand called CD154, or fibroblasts). Both human and mouse B-cells survive for two weeks instead of three to four days, and absolute cell numbers increase two- to threefold.

The *in vitro* germinal center culture system developed was designed to study the differentiation of mouse B-cells into plasma cells rather than to maintain long-term B-cells *in vitro*. Mouse FDCs are nonproliferating, terminally differentiated cells of unknown origin. However, other studies reported that the primary mouse FDCs could be partially replaced by fibroblast cell lines expressing CD154. The cytokines, required to maintain mature B-cells in growth phase and differentiation, are controversial. For mouse B-cells,

combined IL-2, IL-4, and IL-5 induce differentiation into mature cells; for further differentiation into plasma cells, IL-6 seems to be essential.

For human B-cells, IL-2 plus IL-I0, combined IL-3, IL-6, IL-7, and combined IL-2, IL-6, IL-10 have been reported to play roles in plasma cell differentiation. Cellular interactions, including those mediated by CD40 and CD154, are critical in the generation of both memory B-cells and plasma cells. If CD154 is removed in the secondary culture, human B-cells differentiate into plasma cells. These memorylike cells are neither cell lines nor cell clones; rather, they are a mixed B-cell type with a limited life span (up to a few weeks), and they preferentially switch to certain Ig isotypes, such as IgG and IgA. The data suggest that down-regulation of the J chain may not be essential during the development of memory B-cells.

IL-21, a newly identified cytokine affecting T-cells, NK-cells, and B-cells, signaling through IL-21 receptor and the common cytokine receptor gamma-chain ($gamma_c$), has recently been shown to play a role in stimulating the differentiation of mouse (Ozaki *et al.*, 2004) and human B-cells (Ettinger *et al.*, 2005). In mouse, IL-21 is shown to promote differentiation of B-cells into postswitch and plasma cells, using IL-21-transgenic mice and hydrodynamics-based gene delivery of IL-21 plasmid DNA *in vivo* and *in vitro*. IL-21 induces expression of Blimp-1 and Bcl-6, which play a role in the development of autoimmune disease. When human B-cells were stimulated through the BCR (B-cell receptor, anti-IgM), IL-21 induced minimal proliferation, IgD down-modulation, and small numbers of plasma cells. In contrast, after anti-CD40 activation, IL-21 induced extensive proliferation, class switch recombination, and plasma cell differentiation. On cross-linking both BCR and CD40, IL-21 induced the largest numbers of plasma cells. IL-21 drove both $CD27^+$ memory cells and naive cord blood B-cells to differentiate into plasma cells. In the latter, the effect of IL-21 was more potent than the combination of IL-2 and IL-10. IL-21 costimulation induced the expression of Blimp-1 and AICD (activation-induced cytidine deaminase), required for CSR (class switch recombination), secreted IgG from B-cells, but did not induce somatic hypermutation. IL-2 enhanced the effects of IL-21, whereas IL-4 inhibited IL-21-induced plasma cell differentiation.

### A Suspension Culture System for Stimulating
### B-Cells with LPS Plus Anti-mu

Systems for the short-term culture of primary splenic lymphocytes have long existed (Chen, 2000). These systems are invaluable for studying the proliferation and differentiation of B-cells, T- and B-cell interaction, the priming of B-cells by antigen, and the mechanism of memory B-cell generation. If B cells could be kept alive and untolerized but could be prevented from becoming IgM secretory plasma cells, they might become memory B-cells. One example is the finding that when stimulated with a bacterial mitogen,

LPS, some B-cells die and some proliferate and become plasma cells (Chen, 2000). When stimulated with anti-mu, most B-cells die right away, some proliferate and exhibit growth arrest at $G_1$ phase and then die two days later, and none become plasma cells. When stimulated with LPS plus anti-mu, most B-cells proliferate and none become plasma cells. This non-plasma-cell phenomenon has been known as an antidifferentiation effect, and it was postulated to be a way to generate memory B-cells (Chen, 2000). Through the efforts of many, the molecular mechanism of this antidifferentiation phenomenon became clear. In the presence of the two stimuli, B-cells proliferate maximally, over 90% being in the cell cycle (Chen, 2000), but IgM secretion is turned off. The block has been shown to be primarily at the level of nuclear RNA processing of the $mu_m$-to-$mu_s$ switch. Inducible nuclear factors binding to the pre-mRNA secretory polyA site have been reported, though the nature of these factors remains unclear.

### Transcription Factors Active in the Development of Plasma Cells

On activation of B-cells, many transcription factors (Oct-2, OBF-1, Blimp-1, or PRDI-BF1 in human, Xbp1, IRF4, Bcl6, etc.) become engaged in the production and secretion of Ig genes. B-cell-preferential transcription factors include Oct-2 (Chen, 2000) and Blimp-l (ref. cited in Chen, 2000, and more recent publication). The transcription factors play a role in the decision to switch from $mu_m$ to $mu_s$. Blimp-l is described as a cofactor of transcription factor pu.l and was shown to bind to multiple Ig-enhancer motifs and the J-chain regulatory element. It is crucial for the transcription of mu, kappa, and J chains. Blimp-l is active in many cell lines and drives the maturation of B-cells into Ig-secreting cells. Blimp-l has been shown to down-regulate *c-myc* in B-cell lines.

Oct2 is POU-domain-containing transcription factor, binds to the mu intron enhancer octamer motif, on which they can form a ternary complex with the coactivator OBF1, and is essential in transcriptional activation of the mu chain. Oct2 and OBF-1 are mainly B-cell-specific protein. Mice that lack Oct2 are not viable, and mature B-cells are affected (Corcoran *et al.*, 1993). It was shown that Oct2 and the J chain are highly expressed in LPS-stimulated B-cells and are diminished in LPS + anti-mu-stimulated B cells. It has been shown that Blimp-l is highly expressed in LPS-stimulated B-cells and is diminished in LPS + anti-mu-stimulated B-cells. On the other hand, sterile gamma chain is highly expressed in the latter system. Transfection of *Blimp-l* into LPS + anti-mu-activated B cells provoked them to become IgM-secreting plasma cells. The data indicated that transcription factors such as Oct2 and Blimp-l are tightly regulated in plasma cell development. If one postulates that LPS + anti-mu-stimulated B-cells represent some stage in memory B-cell development, then the down-regulation of Oct2 and Blimp-l reflects the specific transcriptional regulation when B-cells make the commitment to the memory cell pathway instead of the plasma cell pathway. The identification of such candidate transcription factors that control memory B-cell commitment provided a powerful genetic tool to manipulate the turning on and off of these lineages at will.

To determine the role of Blimp-1 in the functioning of plasma cells, mice in which the gene encoding Blimp-1 could be deleted in an inducible manner have been produced (Shapiro-Shelef *et al.*, 2005). Deletion of Blimp-1 either *in vitro* or *in vivo* leads to loss of previously formed $B220^{low}$ Syndecan-1/$CD138^{high}$ plasma cells. Assays have been performed and show that Blimp-1 is required for the formation and maintenance of long-lived plasma cells.

Interestingly, Blimp-1 is also active in some differentiated cells, including a human osteosarcoma cell line (Gyory *et al.*, 2004), terminal differentiated myeloid cell lines, and freshly isolated bone marrow mononuclear cells. Its action is shown to be antiproliferative by inhibiting c-myc gene expression pathway (Tamura *et al.*, 2003). In B-cells, Blimp-1 is required for terminal differentiation of B-cells to plasma cells. The optimal DNA recognition sequence for Blimp-1 has been identified. The consensus is very similar to a subset of sites recognized by IRFs (IFN regulatory factors) that contain the sequence GAAAG. Competition experiments showed that Blimp-1, IRF-1, and IRF-2 have similar binding affinities for functionally im-portant regulatory sites containing this sequence. Blimp-1 does not bind to all IRF sites and does not recognize IRF-4/PU.1 or IRF-8 sites lacking the GAAAG sequence. Chromatin immunoprecipitation studies showed that Blimp-1, IRF-1, and IRF-2 all bind the IFN-beta promoter *in vivo*, as predicted by the *in vitro* binding parameters, and in cotransfections Blimp-1 inhibits IRF-1-dependent activation of the IFN-beta promoter. Thus, Blimp-1 competes *in vivo* with a subset of IRF proteins and helps predict the sites and IRF family members that may be affected (Kuo and Calame, 2004).

The human Blimp-1, PRDI-BF1, is a DNA-binding protein involved in postinduction repression of INF-beta gene transcription in response to viral infection. In terminal differentiation of B-cells, it has an essential function in driving differentiation and therein silences multiple genes. In the osteosarcoma cell line U2OS, PRDI-BF1 is shown to assemble silent chromatin over the INF-beta promoter through recruitment of the histone H3 lysine methyltransferase G9a. G9a is recruited only when in a complex with PRDI-BF1. G9a catalytic activity is required for the accu-mulation of methylated histone H3 and transcriptional silencing mediated by PRDI-BF1 *in vivo*. It establishes a mechanism for the recruitment of G9a, the main mam-malian euchromatic methyltransferase, and defines non-embryonic targets of G9a (Gyory *et al.*, 2004). G9a can target the specific silencing of IFN-beta gene expression and blocking of V(D)J recombination (Osipovich *et al.*, 2004). A drawing of the model of action is shown in Makar and Wilson (2004b).

Microarray experiments of the expression of Blimp-1 indicate that it regulates a large set of genes of plasma cell expression signature (Sciammas and Davis, 2005). Blimp-1 affects numerous aspects of plasma cell maturation, ranging from migration, adhesion, and homeostasis to antibody secretion. It regulates immunoglobulin secretion by affecting the nuclear processing of the mRNA transcript and by affecting protein trafficking by regulating genes that impact the activity of the endoplasmic reticulum. The differentiation events that Blimp-1 regulates appear to be modulated, depending on the activation state of the B-cell, and hint at the complexity of Blimp-1 and the genetic program that it initiates to produce a pool of plasma cells.

There is a concerted activation-suppression of transcriptional factors during B-cell differentiation; besides the foregoing examples, see also the review by Matthias and Rolink (2005). In (human) germinal center B-cells, a set of genes has been shown to be involved in the differentiation: down-regulation of Bcl-6, which is implied to transform B-cells, activation of Blimp-1/PRDI-BF1, modulation of Myc, and the up-regulation of the Mad1 and Mad4 transcription factors. Recent data have suggested that Mad1 acts as a transcriptional repressor of Bcl-6 (Lee *et al.*, 2006). In addition, IRF4/Mum1 is expressed in germinal center cells and CD138[+] plasma cells (Buettner *et al.*, 2005). Transcription factor E47 is required for CSR (class switch recombination), at least in part, via expression of AICD (activation-induced cytidine deaminase). Id2 has been identified as a negative regulator of E47 (Frasca *et al.*, 2004). Bach2 is a B-cell-specific transcription repressor interacting with the small Maf proteins, whose expression is high only before the plasma cell stage. It is critical for CSR and somatic hypermutation (SHM) of Ig genes. In Bach2[-/-] mice, B-cells produced IgM, as did wild-type cells, and abundantly expressed Blimp-1 and XBP-1, critical regulators of the plasmacytic differentiation, indicating that Bach2 was not required for the plasmacytic differentiation itself. However, they failed to undergo efficient CSR. These findings define Bach2 as a key regulator of antibody response and provide insight into the orchestration of CSR and SHM during plasma cell differentiation (Muto *et al.*, 2004). Mitf (microphthalmia-associated transcription factor) is highly expressed in naive B-cells, where it antagonizes the process of terminal differentiation through the repression of IRF-4. Defective Mitf activity results in spontaneous B-cell activation, antibody secretion, and autoantibody production. Conversely, ectopic Mitf expression suppresses the expression of IRF-4, CD138[+] plasma cells, and antibody secretion. Mitf regulates B-cell homeostasis by suppressing the antibody-secreting fate (L. Lin *et al.*, 2004).

The role of Xbp1 (X-box binding protein 1), another transcriptional factor (besides Blimp-1) essential for the differentiation of B-cells into plasma cells, has been studied (Shaffer *et al.*, 2004). By using microarray analyses, a set of genes have been defined whose induction during mouse plasmacytic differentiation is dependent on Blimp-1 and/or Xbp1. Blimp-1[-/-] B-cells failed to up-regulate most plasma cell–specific genes, including Xbp1. Differentiating Xbp1[-/-] B-cells induced Blimp-1 normally but failed to up-regulate genes encoding many secretory pathway components. Conversely, ectopic expression of Xbp1 induced a wide spectrum of secretory pathway genes and physically expanded the endoplasmic reticulum. Xbp1 increased cell size, lysosome content, mitochondrial mass and function, ribosome numbers, and total protein synthesis. Xbp1 is essential to coordinate diverse changes in cellular structure and function resulting in the characteristic phenotype of professional secretory cells.

## IX. CONCLUDING REMARKS AND PROSPECTS FOR LYMPHOCYTE ENGINEERING

I have discussed and summarized the current understanding of the expansion of lymphoid cells and their precursors *ex vivo* at certain stages of lymphopoiesis. I have tried to address the feasibility of expanding lymphoid cells under controlled growth conditions; that is, we need to expand untransformed, nonmalignant cells. In general, in order to maintain the status of cell survival and growth without apoptosis and differentiation, cytokines and cell contact with feeder cells are required. Fundamental questions regarding the engineering of lymphoid cells and their precursors for therapeutic purposes remain and can be traced to our current understanding of the immune system. Do we ask too much for the survival in culture of cells that are programmed to die? From extensive studies in gene-manipulated mice, it is possible to generate antigen-specific memory T-cells; it is still puzzling that there is no good systematic study of human memory cells in culture, although the surface markers are defined.

If there is massive programmed cell death during the development from HSCs to lymphoid precursors and from pre-T-cell to T-cell maturation, I wonder whether it is realistic to try to produce enough HSCs, HPCs, and precursor lymphoid cells for therapeutic purposes. In several clinical protocols, expansion of cells *ex vivo* for the purpose of reinfusion into patients is limited to as few passages as possible in order to avoid mutation and contamination *in vitro*. Bioreactors for large-scale production of cells in liquid suspension using cytokines are available. However, they are not designed for coculturing of stem cells with stroma cells. With advances in culture technology and bioreactors and with increased supply of recombinant cytokines, it becomes possible to obtain a quantity sufficient for reimplantation from 10 ml of bone marrow cells. However, under these conditions, very few cells engage in lymphopoiesis. Thus, to grow the HSCs and HPCs consistently and to favor lymphopoiesis, there is a great need for a better way to grow human stem cells using human stroma cell lines. The current study using transcripton factors such as HoxB4, Cdx4, and mMix to manipulate the behavior of hESCs *in vitro* may be a promising approach.

In view of the massive apoptosis at several stages of lymphopoiesis, it is amazing that mouse precursor B-cells can grow normally, become lines and clones, and retain the potential to differentiate *in vitro* and *in vivo*. To grow cells from other stages of lymphopoiesis, it might be advantageous to use cells from the many available mutated mice.

The future of cell-based immune therapy lies in the *ex vivo* expansion of cells. Because the techniques to establish ligand-regulatable vectors are available, the derived cell lines will become available and will become valuable resources for many purposes. Other areas remain to be improved, including the search for novel markers of true HSCs and precursor lymphocytes; better sources of HSCs and HPCs, such as cord blood and fetal tissues and hES cells; improved retroviral vectors; and large-scale culture systems for expansion of HSCs and HPCs and precursor lymphoid cells.

With advances in genetic tools and mutated-mice technology, it is possible to turn on or off the transcription factors that control the differentiation of cells. With the recent discovery and understanding of Toll-like molecules on APCs to stimulate $T_h1$ vs. $T_h2$, and the rediscovery of Treg cells, the approaches to fight against autoimmune diseases, allergy, and transplantation rejection will become accessible. Thus, immunology will continue to be a very exciting field.

## X. ACKNOWLEDGMENTS

The author wishes to thank Lucien Caro, Don Gerson, Anja Holtz, Raju Pareek, and all our other colleagues for fruitful discussions and critical reading and editing of this chapter. The author also wishes to thank Hiroshi Nikaido for sharing his work before publication and to Harald von Boehmer, Gerhard Hoffmann, Fritz Melchers, Christoph Wilson, Nicolas Burdin, Alexander Aiuti, Gennardo De Libero, Lucia Mori, Patrick Matthias, Ton Rolink, Kris Bowles, and Roger Pedersen for giving permission to use figures and tables and for conducting kind discussions and for providing inside information on their work. This work is supported by funds from third-party agencies: German DFG, the European Community, W-P, CMU.

## XI. REFERENCES

Abramovich, C., Pineault, N., Ohta, H., and Humphries, R. K. (2005). Hox genes: from leukemia to hematopoietic stem cell expansion. *Ann. N.Y. Acad. Sci.* **1044**, 109–116.

Aiuti, A., Slavin, S., Aker, M., Ficara, F., Deola, S., Mortellaro, A., Morecki, S., Andolfi, G., Tabucchi, A., Carlucci, F., Marinello, E., Cattan eo, F., Vai, S., Servida, P., Miniero, R., Roncarolo, M. G., and Bordignon, C. (2002). Correction of ADA-SCID by stem cell gene therapy combined with non-myeloablative conditioning. *Science* **296**, 2410–2413.

Aiuti, A., Ficara, F., Cattaneo, F., Bordignon, C., and Roncarolo, M. G. (2003). Gene therapy for adenosine deaminase deficiency. *Curr. Opin. Allergy Clin. Immunol.* **3**, 461–466.

Antonchuk, J., Sauvageau, G., and Humphries, R. K. (2002). HOXB4-induced expansion of adult hematopoietic stem cells *ex vivo. Cell* **109**, 39–45.

Bester, A. C., Schwartz, M., Schmidt, M., Garrigue, A., Hacein-Bey-Abina, S., Cavazzana-Calvo, M., Ben-Porat, N., Von Kalle, C., Fischer, A., and Kerem, B. (2006). Fragile sites are preferential targets for integrations of MLV vectors in gene therapy. *Gene Ther.* (Mar. 2, online publication).

Bowles, K. M., Vallier, L., Smith, J. R., Alexander, M. R., and Pedersen, R. A. (2006). HOXB4 overexpression promotes hematopoietic development by human embryonic stem cells. *Stem Cells* (Jan. 12, online publication).

Brown, G., Bunce, D. M., and Guy, G. R. (1985). Sequential derrermination of lineage potentials during haemopoiesis. *Br. J. Cancer* **52**, 681–686.

Brown, G., Bunce, D. M., Lord, J., M., and McConnell, E. M. (1988). The development of cell lineages: a sequential model. *Differentiation (Berlin)* **39**, 83–89.

Bruder, D., Probst-Kepper, M., Westendorf, A. M., Geffers, R., Beissert, S., Loser, K., von Boehmer, H., Buer, J., and Hansen, W. (2004). Neuropilin-1: a surface marker of regulatory T-cells. *Eur. J. Immunol.* **34**, 623–630.

Buettner, M., Greiner, A., Avramidou, A., Jack, H. M., and Niedobitek, G. (2005). Evidence of abortive plasma cell differentiation in Hodgkin and Reed–Sternberg cells of classical Hodgkin lymphoma. *Hematol. Oncol.* **23**, 127–132.

Burdin, N., Guy, B., and Moingeon, P. (2004). Immunological foundations to the quest for new vaccine adjuvants. *Biodrugs* **18**, 79–93.

Chen, U. (2000). Lymphoid cells. *In* "Principles of Tissue Engineering," (R. P. Lanza, R. Langer, and J. Vacanti, eds.) 2nd ed., pp. 611–629. Academic Press, New York.

Chen, U. (2001). Methods for growing stem cells. Patent Application no. WO01/14530.

Chen, U. (2006a). Methods of growing stem cells. Australian patent application no. 67031/00.

Chen, U. (2006b). Methods for growing stem cells. U.S. patent application no. 09/957,458.

Chen, U., Esser, R., Kotlenga, K., *et al.* (1997). Potential application of quasi-totipotent ES cells, a ten-year study of soft-tissue engineering with ES cells. *J. Tissue Eng.* **3**, 321–328.

Corcoran, I. M., Karvelas, M., Nossal, G. V., *et al.* (1993). Oct-2, although not required for early B-cell development, is critical for later B cell maturation and for postnatal survival. *Genes Dev.* **7**, 570–582.

Daley, G. Q. (2003). From embryos to embryoid bodies: generating blood from embryonic stem cells. *Ann. N.Y. Acad. Sci.* **996**, 122–131.

Dando, J. S., Roncarolo, M. G., Bordignon, C., and Aiuti, A. (2001). A novel human packaging cell line with hematopoietic supportive capacity increases gene transfer into early hematopoietic progenitors. *Hum. Gene Ther.* **12**, 1979–1988.

Dang, S. M., Kyba, M., Perlingeiro, R., Daley, G. Q., and Zandstra, P. W. (2002). Efficiency of embryoid body formation and hematopoietic development from embryonic stem cells in different culture systems. *Biotechnol. Bioeng.* **78**, 442–453.

de la Salle, H., Mariotti, S., Angenieux, C., Gilleron, M., Garcia-Alles, L. F., Malm, D., Berg, T., Paoletti, S., Maitre, B., Mourey, L., Salamero, J., Cazenave, J. P., Hanau, D., Mori, L., Puzo, G., and De Libero, G. (2005). Assistance of microbial glycolipid antigen processing by CD1e. *Science* **310**, 1321–1324.

De Libero, G., and Mori, L. (2005). Recognition of lipid antigens by T-cells. *Nat. Rev. Immunol.* **5**, 485–496.

de Pooter, R. F., Cho, S. K., and Zuniga-Pflucker, J. C. (2005). *In vitro* generation of lymphocytes from embryonic stem cells. *Methods Mol. Biol.* **290**, 135–147.

Diebold, S. S., Kaisho, T., Hemmi, H., Akira, S., and Reis e Sousa, C. (2004). Innate antiviral responses by means of TLR7-mediated recognition of single-stranded RNA. *Science* **303**, 1529–1531.

Elmshaeuser, C., Bechtel, J., Motta, I., Schipke, C., Kettenmann, H., Schmalbruch, H., Kann, M., Beck, E., and Chen, U. (2002). Characterization of a mouse tet-on glia precursor cell line *in vitro* and *in vivo* using the electrophysiological measurement. *J. Physiology Paris* **96**, 329–338.

Ettinger, R., Sims, G. P., Fairhurst, A. M., Robbins, R., da Silva, Y. S., Spolski, R., Leonard, W. J., and Lipsky, P. E. (2005). IL-21 induces differentiation of human naive and memory B-cells into antibody-secreting plasma cells. *J. Immunol.* **175**, 7867–7879.

Ficara, F., Superchi, D. B., Hernandez, R. J., Mocchetti, C., Carballido-Perrig, N., Andolfi, G., Deola, S., Colombo, A., Bordignon, C., Carballido, J. M., Roncarolo, M. G., and Aiuti, A. (2004). IL3 or IL7 increases *ex vivo* gene transfer efficiency in ADA-SCID BM CD34+ cells while maintaining *in vivo* lymphoid potential. *Mol. Ther.* **10**, 1096–1108.

Fontenot, J. D., Favin, M. A., and Rudensky, A. Y. (2003). Foxp3 programs the development and function of CD4+CD25+ regulatory T-cells. *Nat. Immunol.* **4**, 330–336.

Frasca, D., Riley, R. L., and Blomberg, B. B. (2004). Effect of age on the immunoglobulin class switch. *Crit. Rev. Immunol.* **24**, 297–320.

Gilleron, M., Stenger, S., Mazorra, Z., Wittke, F., Mariotti, S., Bohmer, G., Prandi, J., Mori, L., Puzo, G., and De Libero, G. (2004). Diacylated sulfoglycolipids are novel mycobacterial antigens stimulating CD1-restricted T cells during infection with Mycobacterium tuberculosis. *J. Exp. Med.* **199**, 649–659.

Glazer, A. N., and Nikaido, H. (2007). Recombinant and synthetic vaccines. *In* "Microbial Biotechnology: Fundamentals of Applied Microbiology," (A. Glazer, and H. Nikaido, eds.), 2nd ed., ch. 5. Cambridge University Press, New York.

Gounari, F., Aifantis, I., Martin, C., Fehling, H. J., Hoeflinger, S., Leder, P., von Boehmer, H., and Reizis, B. (2002). Tracing lymphopoiesis with the aid of a pTalpha-controlled reporter gene. *Nat. Immunol.* **3**, 489–496.

Gyory, I., Wu, J., Fejer, G., Seto, E., and Wright, K. L. (2004). PRDI-BF1 recruits the histone H3 methyltransferase G9a in transcriptional silencing. *Nat. Immunol.* **5**, 299–308.

Hacein-Bey-Abina, S., Von Kalle, C., Schmidt, M., McCormack, M. P., Wulffraat, N., Leboulch, P., Lim, A., Osborne, C. S., Pawliuk, R., Morillon,

E., Sorensen, R., Forster, A., Fraser, P., Cohen, J. I., de Saint Basile, G., Alexander, I., Wintergerst, U., Frebourg, T., Aurias, A., Stoppa-Lyonnet, D., Romana, S., Radford-Weiss, I., Gross, F., Valensi, F., Delabesse, E., Macintyre, E., Sigaux, F., Soulier, J., Leiva, L. E., Wissler, M., Prinz, C., Rabbitts, T. H., Le Deist, F., Fischer, A., and Cavazzana-Calvo, M. (2003). LMO2-associated clonal T-cell proliferation in two patients after gene therapy for SCID-X1. *Science* **302**, 400–401.

Halupa, A., Bailey, M. L., Huang, K., Iscove, N. N., Levy, D. E., and Barber, D. L. (2005). A novel role for STAT1 in regulating murine erythropoiesis: deletion of STAT1 results in overall reduction of erythroid progenitors and alters their distribution. *Blood* **105**, 552–561.

Heil, F., Hemmi, H., Hochrein, H., Ampenberger, F., Kirschning, C., Akira, S., Lipford, G., Wagner, H., and Bauer, S. (2004). Species-specific recognition of single-stranded RNA via toll-like receptor 7 and 8. *Science* **303**, 1526–1529.

Helgason, C. D., Sauvageau, G., Lawrence, H. J., Largman, C., and Humphries, R. K. (1996). Overexpression of HOXB4 enhances the hematopoietic potential of embryonic stem cells differentiated *in vitro*. *Blood* **87**, 2740–2749.

Ivanova, N. B., Dimos, J. T., Schaniel, C., Hackney, J. A., Moore, K. A., and Lemischka, I. R. (2002). A stem cell molecular signature. *Science* **298**, 601–604.

Jaeckel, E., von Boehmer, H., and Manns, M. P. (2005). Antigen-specific FoxP3-transduced T-cells can control established type 1 diabetes. *Diabetes* **54**, 306–310.

Kaufman, D. S., Hanson, E. T., Lewis, R. L., Auerbach, R., and Thomson, J. A. (2001). Hematopoietic colony-forming cells derived from human embryonic stem cells. *Proc. Natl. Acad. Sci. U.S.A.* **98**, 10716–10721.

Kisielow, P., and von Boehmer, H. (1995). Development and selection of T-cells: facts and puzzles. *Adv. Immunol.* **58**, 87–209.

Konrad, M. A., and Zuniga-Pflucker, J. C. (2005). The BTG/TOB family protein TIS21 regulates stage-specific proliferation of developing thymocytes. *Eur. J. Immunol.* **35**, 3030–3042.

Kretschmer, K., Apostolou, I., Hawiger, D., Khazaie, K., Nussenzweig, M. C., and von Boehmer, H. (2005). Inducing and expanding regulatory T-cell populations by foreign antigen. *Nat. Immunol.* **6**, 1219–1227.

Kuo, T. C., and Calame, K. L. (2004). B lymphocyte–induced maturation protein (Blimp)-1, IFN regulatory factor (IRF)-1, and IRF-2 can bind to the same regulatory sites. *J. Immunol.* **173**, 5556–5563.

Kyba, M., Perlingeiro, R. C., and Daley, G. Q. (2002). HoxB4 confers definitive lymphoid-myeloid engraftment potential on embryonic stem cell and yolk sac hematopoietic progenitors. *Cell* **109**, 29–37.

Kyba, M., Perlingeiro, R. C., Hoover, R. R., Lu, C. W., Pierce, J., and Daley, G. Q. (2003). Enhanced hematopoietic differentiation of embryonic stem cells conditionally expressing Stat5. *Proc. Natl. Acad. Sci. U.S.A.* **100**, 11904–11910.

La Motte-Mohs, R. N., Herer, E., and Zuniga-Pflucker, J. C. (2005). Induction of T-cell development from human cord blood hematopoietic stem cells by Delta-like 1 *in vitro*. *Blood* **105**, 1431–1439.

Lee, S. C., Bottaro, A., Chen, L., and Insel, R. A. (2006). Mad1 is a transcriptional repressor of Bcl-6. *Mol. Immunol.* (Jan. 16, online publication).

Lemischka, I. R., and Pritsker, M. (2006). Alternative splicing increases complexity of stem cell transcriptome. *Cell Cycle* **5**, 347–351.

Lin, L., Gerth, A. J., and Peng, S. L. (2004). Active inhibition of plasma cell development in resting B-cells by microphthalmia-associated transcription factor. *J. Exp. Med.* **200**, 115–122.

Lin, Y. W., Deveney, R., Barbara, M., Iscove, N. N., Nimer, S. D., and Slape, C., Aplan, P. D. (2005). OLIG2 (BHLHB1), a bHLH transcription factor, contributes to leukemogenesis in concert with LMO1. *Cancer Res.* **65**, 7151–7158.

Lopez, R., Garrido, E., Pina, P., Hidalgo, A., Lazos, M., Ochoa, R., and Salcedo, M. (2006). HOXB homeobox gene expression in cervical carcinoma. *Int. J. Gynecol. Cancer* **16**, 329–335.

Makar, K. W., and Wilson, C. B. (2004a). DNA methylation is a non-redundant repressor of the Th2 effector program. *J. Immunol.* **173**, 4402–4406.

Makar, K. W., and Wilson, C. B. (2004b). Sounds of a silent Blimp-1. *Nat. Immunol.* **5**, 241–242.

Mandal, M., Borowski, C., Palomero, T., Ferrando, A. A., Oberdoerffer, P., Meng, F., Ruiz-Vela, A., Ciofani, M., Zuniga-Pflucker, J. C., Screpanti, I., Look, A. T., Korsmeyer, S. J., Rajewsky, K., von Boehmer, H., and Aifantis, I. (2005). The BCL2A1 gene as a pre-T-cell receptor-induced regulator of thymocyte survival. *J. Exp. Med.* **201**, 603–614.

Martensson, I. L., Rolink, A., Melchers, F., Mundt, C., Licence, S., and Shimizu, T. (2002). The pre-B-cell receptor and its role in proliferation and Ig heavy-chain allelic exclusion. *Semin. Immunol.* **14**, 335–342.

Martin, C. H., Aifantis, I., Scimone, M. L., von Andrian, U. H., Reizis, B., von Boehmer, H., and Gounari, F. (2003). Efficient thymic immigration of B220+ lymphoid-restricted bone marrow cells with T precursor potential. *Nat. Immunol.* **4**, 866–873.

Matthias P., and Rolink, A. (2005). Transcriptional networks in developing and mature B-cells. *Nat. Immunol.* **5**, 497–508.

Mohtashami, M., and Zuniga-Pflucker, J. C. (2006). Cutting edge: three-dimensional architecture of the thymus is required to maintain delta-like expression necessary for inducing T-cell development. *J. Immunol.* **15**, 730–734.

Moingeon, P., Batard, T., Fadel, R., Frati, F., Sieber, J., and Overtvelt, L. V. (2006). Immune mechanisms of allergen-specific sublingual immunotherapy. *Allergy* **61**, 151–165.

Muth, H., Elmshauser, C., Broad, S., Schipke, C., Kettenmann, H., Beck, E., Kann, M., Motta, I., and Chen, U. (2001). Cell-based delivery of cytokines allows for the differentiation of a doxycycline inducible oligodendrocyte precursor cell line *in vitro*. *J. Gene Medicine* **3**, 585–598.

Muto, A., Tashiro, S., Nakajima, O., Hoshino, H., Takahashi, S., Sakoda, E., Ikebe, D., Yamamoto, M., and Igarashi, K. (2004). The transcriptional programme of antibody class switching involves the repressor Bach2. *Nature* **429**, 566–571.

Narayan, A. D., Chase, J. L., Lewis, R. L., Tian, X., Kaufman, D. S., Thomson, J. A., and Zanjani, E. D. (2006). Human embryonic stem cell–derived hematopoietic cells are capable of engrafting primary as well as secondary fetal sheep recipients. *Blood* **107**, 2180–2183.

Nolta, J. A., Thiemann, F. T., Arakawa-Hoyt, J., Dao, M. A., Barsky, L. W., Moore, K. A., Lemischka, I. R., and Crooks, G. M. (2002). The AFT024 stromal cell line supports long-term *ex vivo* maintenance of engrafting multipotent human hematopoietic progenitors. *Leukemia* **16**, 352–361.

O'Neill, L. A. (2004). After the Toll rush. *Science* **303**, 1481–1483.

O'Neill, L. A., Fritzferald, K., and Bowie, A. G. (2003). The Toll-IL-1 receptor adaptor family grows to five members. *Trends Immunol.* **24**, 286–290.

Osipovich, O., Milley, R., Meade, A., Tachibana, M., Shinkai, Y., Krangel, M. S., and Oltz, E. M. (2004). Targeted inhibition of V(D)J recombination by a histone methyltransferase. *Nat. Immunol.* **5**, 309–316.

Ozaki, K., Spolski, R., Ettinger, R., Kim, H. P., Wang, G., Qi, C. F., Hwu, P., Shaffer, D. J., Akilesh, S., Roopenian, D. C., Morse, H. C. 3rd, Lipsky, P. E., and Leonard, W. J. (2004). Regulation of B-cell differentiation and plasma cell generation by IL-21, a novel inducer of Blimp-1 and Bcl-6. *J. Immunol.* **173**, 5361–5371.

Palmer, E. (2003). Negative selection — clearing out the bad apples from the T-cell repertoire. *Nat. Rev. Immunol.* **3**, 383–391.

Pilat, S., Carotta, S., Schiedlmeier, B., Kamino, K., Mairhofer, A., Will, E., Modlich, U., Steinlein, P., Ostertag, W., Baum, C., Beug, H., and Klump, H. (2005). HOXB4 enforces equivalent fates of ES-cell-derived and adult hematopoietic cells. *Proc. Natl. Acad. Sci. U.S.A.* **102**, 12101–12106.

Pritsker, M., Doniger, T. T., Kramer, L. C., Westcot, S. E., and Lemischka, I. R. (2005). Diversification of stem cell molecular repertoire by alternative splicing. *Proc. Natl. Acad. Sci. U.S.A.* **102**, 14290–14295.

Rodewald, H.-R, and Fehling, H. J. (1998). Molecular and cellular events in early thymocyte development. *Adv. Immunol.* **69**, 1–112.

Rohrer, J. W., Barsoum, A. L., and Coggin, J. H. (2001). The development of a new universal tumor-rejection antigen expressed on human and rodent cancers for vaccination, prevention of cancer, and antitumor therapy. *Mod. Asp. Immunobiol.* **1**, 191–195.

Rubio, D., Garcia-Castro, J., Martin, M. C., de la Fuente, R., Cigudosa, J. C., Lloyd, A. C., and Bernad, A. (2005). Spontaneous human adult stem cell transformation. *Cancer Res.* **65**, 3035–3039.

Sauvageau, G., Thorsteinsdottir, U., Eaves, C. J., Lawrence, H. J., Largman, C., Lansdorp, P. M., and Humphries, R. K. (1995). Overexpression of HOXB4 in hematopoietic cells causes the selective expansion of more primitive populations *in vitro* and *in vivo*. *Genes Dev.* **9**, 1753–1765.

Schmidt, M., Hacein-Bey-Abina, S., Wissler, M., Carlier, F., Lim, A., Prinz, C., Glimm, H., Andre-Schmutz, I., Hue, C., Garrigue, A., Le Deist, F., Lagresle, C., Fischer, A., Cavazzana-Calvo, M., and von Kalle, C. (2005). Clonal evidence for the transduction of CD34+ cells with lymphomyeloid differentiation potential and self-renewal capacity in the SCID-X1 gene therapy trial. *Blood* **105**, 2699–2706.

Schumann, J., and De Libero, G. (2006). Serum lipoproteins: Trojan horses of the immune response? *Trends Immunol.* **27**, 57–59.

Sciammas, R., and Davis, M. M. (2005). Blimp-1: immunoglobulin secretion and the switch to plasma cells. *Curr. Top. Microbiol. Immunol.* **290**, 201–224.

Shaffer, A. L., Shapiro-Shelef, M., Iwakoshi, N. N., Lee, A. H., Qian, S. B., Zhao, H., Yu, X., Yang, L., Tan, B. K., Rosenwald, A., Hurt, E. M., Petroulakis, E., Sonenberg, N., Yewdell, J. W., Calame, K., Glimcher, L. H., and Staudt, L. M. (2004). XBP1, downstream of Blimp-1, expands the secretory apparatus and other organelles and increases protein synthesis in plasma cell differentiation. *Immunity* **21**, 81–93.

Shapiro-Shelef, M., Lin, K. I., Savitsky, D., Liao, J., and Calame, K. (2005). Blimp-1 is required for maintenance of long-lived plasma cells in the bone marrow. *J. Exp. Med.* **202**, 1471–1476.

Sicinska, E., Aifantis, I., Le Cam, L., Swat, W., Borowski, C., Yu, Q., Ferrando, A. A., Levin, S. D., Geng, Y., von Boehmer, H., and Sicinski, P. (2003). Requirement for cyclin D3 in lymphocyte development and T-cell leukemias. *Cancer Cell* **4**, 417–418.

Sperger, J. M., Chen, X., Draper, J. S., Antosiewicz, J. E., Chon, C. H., Jones, S. B., Brooks, J. D., Andrews, P. W., Brown, P. O., and Thomson, J. A. (2003). Gene expression patterns in human embryonic stem cells and human pluripotent germ cell tumors. *Proc. Natl. Acad. Sci. U.S.A.* **100**, 13350–13355.

Taghon, T. N., David, E. S., Zuniga-Pflucker, J. C., and Rothenberg, E. V. (2005). Delayed, asynchronous, and reversible T-lineage specification induced by Notch/Delta signaling. *Genes Dev.* **19**, 965–978.

Takeda, K., Kaisho, T., and Akira, S. (2003). Toll-like receptors. *Ann. Rev. Immunol.* **21**, 335–376.

Tamura, T., Kong, H. J., Tunyaplin, C., Tsujimura, H., Calame, K., and Ozato, K. (2003). ICSBP/IRF-8 inhibits mitogenic activity of p210 Bcr/Abl in differentiating myelod progenitor cells. *Blood* **102**, 4547–4554.

Thomson, J. A., Itskovitz-Eldor, J., Shapiro, S. S., *et al.* (1998). Embryonic stem cell lines derived from human blastocysts. *Science* **282**, 1145–1147.

Thrasher, A. J., Hacein-Bey-Abina, S., Gaspar, H. B., Blanche, S., Davies, E. G., Parsley, K., Gilmour, K., King, D., Howe, S., Sinclair, J., Hue, C., Carlier, F., von Kalle, C., de Saint Basile, G., le Deist, F., Fischer, A., and Cavazzana-Calvo, M. (2005). Failure of SCID-X1 gene therapy in older patients. *Blood* **105**, 4255–4257.

Vodyanik, M. A., Bork, J. A., Thomson, J. A., and Slukvin, I. I. (2005). Human embryonic stem cell-derived CD34$^+$ cells: efficient production in the coculture with OP9 stromal cells and analysis of lymphohematopoietic potential. *Blood* **105**, 617–626.

von Boehmer, H. (2005). Mechanisms of suppression by suppressor T-cells. *Nat. Immunol.* **6**, 338–344.

Wang, L., Menendez, P., Shojaei, F., Li, L., Mazurier, F., Dick, J. E., Cerdan, C., Levac, K., and Bhatia, M. (2005). Generation of hematopoietic repopulating cells from human embryonic stem cells independent of ectopic HOXB4 expression. *J. Exp. Med.* **201**, 1603–1614.

Wang, Y., Yates, F., Naveiras, O., Ernst, P., and Daley, G. Q. (2005). Embryonic stem cell–derived hematopoietic stem cells. *Proc. Natl. Acad. Sci. U.S.A.* **102**, 19081–19086.

Werlen, G., Hausmann, B., Naeher, D., and Palmer, E. (2003). Signaling life and death in the thymus: timing is everything. *Science* **299**, 1859–1863.

Will, E., Speidel, D., Wang, Z., Ghiaur, G., Rimek, A., Schiedlmeier, B., Williams, D. A., Baum, C., Ostertag, W., and Klump, H. (2006). HOXB4 inhibits cell growth in a dose-dependent manner and sensitizes cells towards extrinsic cues. *Cell Cycle* **5**, 14–22.

Willey, S., Ayuso-Sacido, A., Zhang, H., Fraser, S. T., Sahr, K. E., Adlam, M. J., Kyba, M., Daley, G. Q., Keller, G., and Baron, M. H. (2006). Acceleration of mesoderm development and expansion of hematopoietic progenitors in differentiating ES cells by the mouse mix-like homeodomain transcription factor. *Blood* (Jan. 10 online).

Zambidis, E. T., Peault, B., Park, T. S., Bunz, F., and Civin, C. I. (2005). Hematopoietic differentiation of human embryonic stem cells progresses through sequential hematoendothelial, primitive, and definitive stages resembling human yolk sac development. *Blood* **106**, 860–870.

Zhang, D., Zhang, G., Hayden, M. S., Greenblatt, M. B., Bussey, C., Flavell, R. A., and Ghosh, S. (2004). A Toll-like receptor that prevents infection by uropathogenic bacteria. *Science* **303**, 1522–1526.

# Kidney and Genitourinary System

Chapter Fifty-Two

# Stem Cells in Kidney Development and Regeneration

*Gregory R. Dressler*

## I. INTRODUCTION

Chronic and acute renal failure are significant clinical problems whose frequencies are expected to increase with the prevalence of diabetes, hypertension, and obesity. In the United States, more than 320,000 patients were undergoing dialysis treatment for end-stage renal disease in 2003, at a cost of more than $14 billion. This cohort is expected to increase to 650,000 by 2010 (Nolan, 2005). Yet the mortality associated with long-term dialysis remains unacceptably high. In the case of acute renal failure due to ischemia or nephrotoxicity, the frequency of mortality in the clinical setting has not dipped significantly below 50% (Liu, 2003), despite many advances in understanding the cell physiology of renal injury. Clearly, new approaches for the treatment of chronic and acute renal disease must be explored. Tissue engineering, cell replacement therapies, and novel growth and differentiation factors are being developed to address the limitations of current therapies. All of these approaches rely on the knowledge base obtained from basic developmental studies in the kidney, regarding the potency of renal stem cells and the genetic basis of epithelial cell differentiation and proliferation. This chapter outlines the basic elements of renal development and renal stem cell biology and discusses current and future applications of emerging technologies for renal disease.

## II. KIDNEY DEVELOPMENT

A detailed description of early renal development and nephron formation is covered in several recent reviews (Dressler, 2002; Kuure *et al.*, 2000). In brief, the mammalian kidney develops from a region of mesoderm, called the *intermediate mesoderm*, that lies between the axial and lateral plate mesoderm along the mediolateral axis of the embryo. The earliest morphological indication of unique derivatives arising from the intermediate mesoderm is the formation of the pronephric duct, or primary nephric duct. This single-cell-thick epithelial tube runs bilaterally, beginning from the mid-thoracic region, to a posterior cavity called the *cloaca*. As it grows, it induces a linear array of epithelial tubules, which extend medioventrally and are thought to derive from mesenchyme surrounding this duct (Fig. 52.1). The tubules are referred to as *pronephric* or *mesonephric*, depending on their position and degree of development, and represent an evolutionarily more primitive excretory system that forms transiently in mammals until it is replaced by the adult, or metanephric, kidney. The adult kidney, or metanephros, is formed at the caudal end of the nephric duct when an outgrowth, called the *ureteric bud* or *metanephric diverticulum*, extends into the surrounding metanephric mesenchyme. Outgrowth, or budding, of the epithelia requires signals emanating from the mesenchyme.

*Principles of Tissue Engineering, 3rd Edition*
*ed. by Lanza, Langer, and Vacanti*

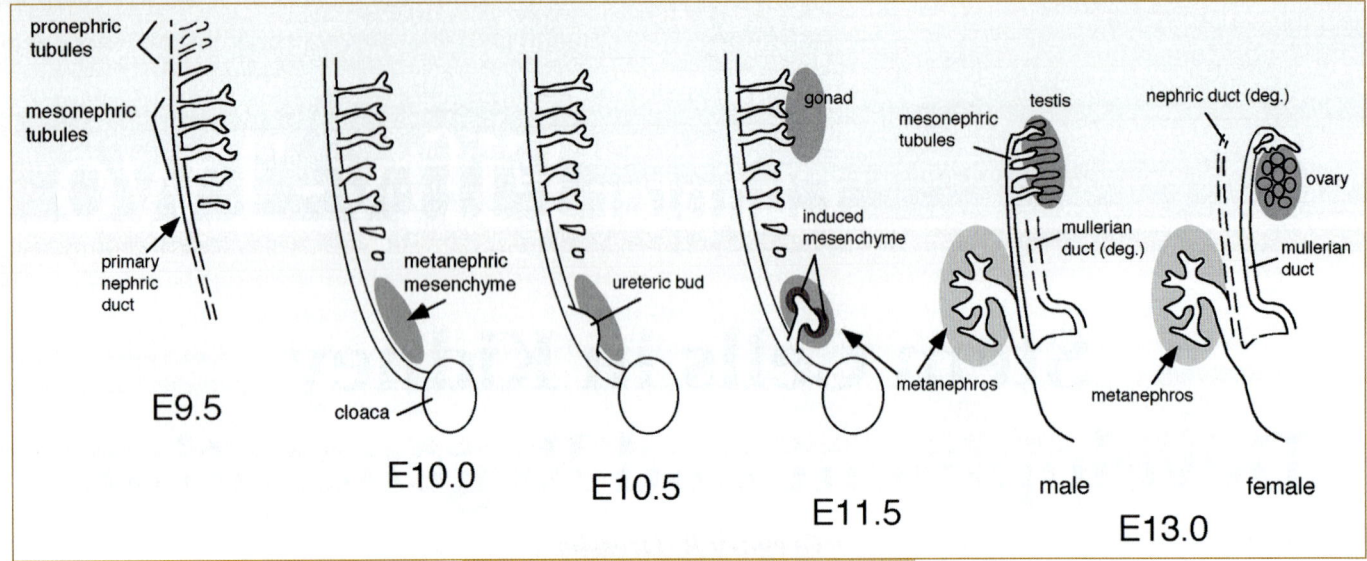

**FIG. 52.1.** Urogenital derivatives of the intermediate mesoderm. A schematic of the mammalian urogenital tract during development is shown. The embryonic stages (E9.5) are shown as days postfertilization in the mouse. The primary nephric duct extends anterior to posterior, inducing mesonephric tubules along the ventral aspect. At the posterior end, the metanephric mesenchyme is induced by the ureteric bud, an outgrowth of the nephric duct. Ureteric bud–derived signals induce condensation of the mesenchyme, whereas mesenchymal signals induce branching morphogenesis in the bud epithelia. By E13, male- and female-specific urogenital derivatives become apparent.

Genetic and biochemical studies indicate that outgrowth of the ureteric bud is mediated by the transmembrane tyrosine kinase RET, which is expressed in the nephric duct, and the secreted neurotrophin GDNF, which is expressed in the metanephric mesenchyme.

Once the ureteric bud has invaded the metanephric mesenchyme, inductive signals emanating from the bud initiate the conversion of the metanephric mesenchyme to epithelium (Fig. 52.2). The induced, condensing mesenchymal cells aggregate around the tips of the bud and will form a primitive polarized epithelium, the renal vesicle. Through a series of cleft formations, the renal vesicle forms first a comma and then an S-shaped body, whose most distal end remains in contact with the ureteric bud epithelium and fuses to form a continuous epithelial tubule. This S-shaped tubule begins to express genes specific for glomerular podocyte cells at its most proximal end, markers for more distal tubules near the fusion with the ureteric bud epithelia, and proximal tubule markers in between. Endothelial cells begin to infiltrate the most proximal cleft of the S-shaped body as the vasculature of the glomerular tuft takes shape. At this stage, the glomerular epithelium consists of a visceral and a parietal component, with the visceral cells becoming podocytes and the parietal cells the epithelia surrounding the urinary space. The capillary tuft consists of capillary endothelial cells and a specialized type of smooth muscle cell, termed the *mesangial cell*, whose origin remains unclear.

While these renal vesicles are generating much of the epithelia of the nephron, the ureteric bud epithelia continue to undergo branching morphogenesis in response to signals derived from the mesenchyme. Branching follows a stereotypical pattern and results in new mesenchymal aggregates induced at the tips of the branches, as new nephrons are sequentially induced. This repeated branching and induction results in the formation of nephrons along the radial axis of the kidney, with the oldest nephrons being more medullary and the younger nephrons located toward the periphery. However, not all cells of the mesenchyme become induced and convert to epithelia; some cells remain mesenchymal and migrate to the interstitium. These interstitial mesenchymal cells, or *stromal* cells, are essential for providing signals that maintain branching morphogenesis of the ureteric bud and survival of the mesenchyme.

Defining the population of cells that generate the kidney depends in part on which stage is considered. At the time of metanephric mesenchyme induction, there are at least two primary cell types, the mesenchyme and the ureteric bud epithelia. Though these cells are phenotypically distinguishable, they do express some common markers and share a common region of origin. As development progresses, it was thought that most of the epithelium of the nephron was derived from the metanephric mesenchyme, whereas the branching ureteric bud epithelium generates the collecting ducts and the most distal tubules. This view has been challenged by cell-lineage tracing methods *in vitro*, which indicate some plasticity at the tips of the ureteric bud epithelium such that the two populations may intermingle (Qiao *et al.*, 1995). Thus, at the time of induction, epithelial cells can convert to mesenchyme, just as the mesenchymal

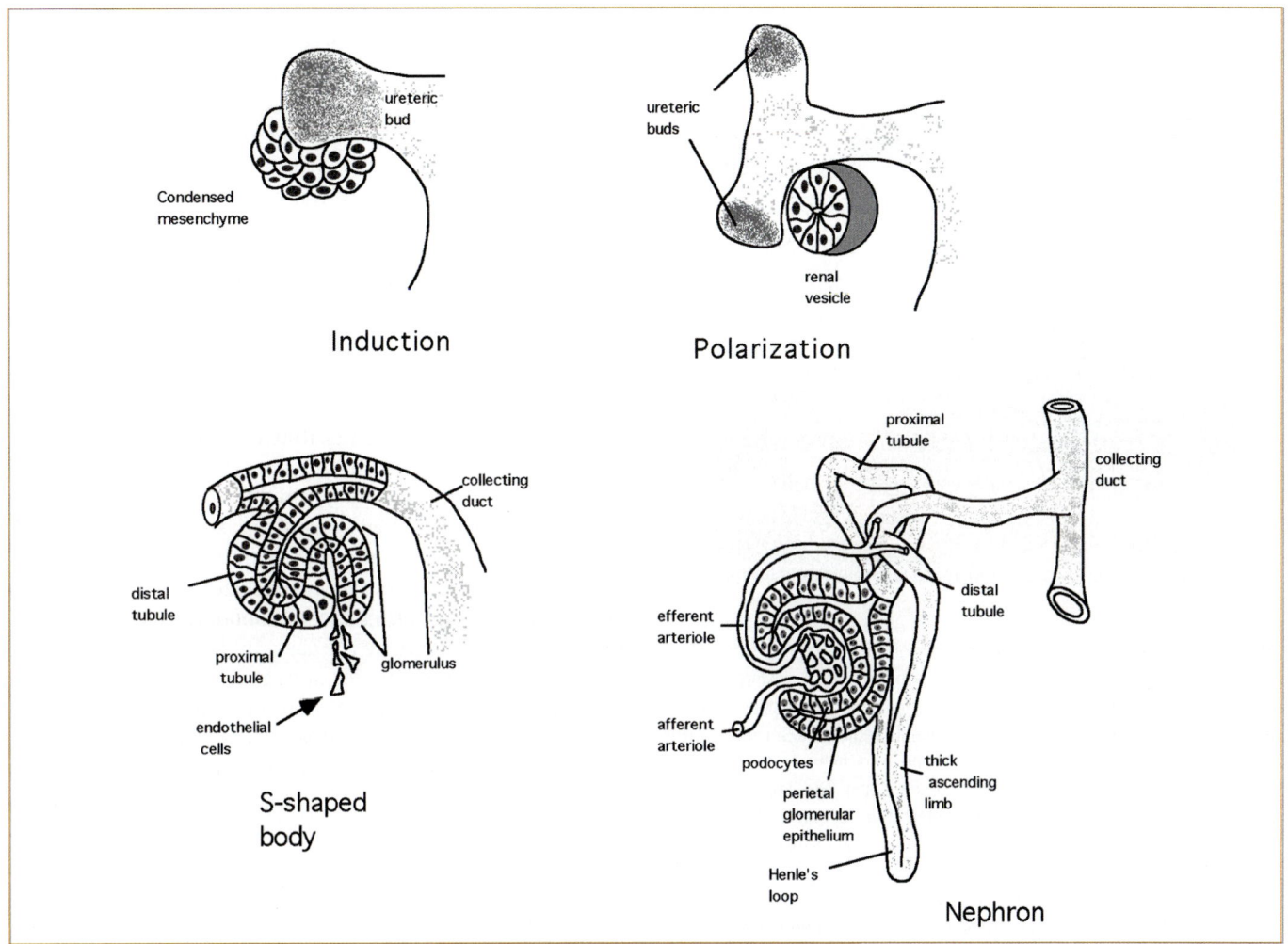

**FIG. 52.2.** Sequential conversion of the metanephric mesenchyme to renal epithalia. The metanephric mesenchymal cells are induced to condense around the tips of the invading ureteric bud. These condensations undergo conversion to a primitive epithelial vesicle, the renal vesicle. The renal vesicle is in close proximity to the branching ureteric bud epithelia and fuses to form a continuous epithelial tubule. At the S-shaped body stage, the most proximal cleft becomes vascularized by infiltrating endothelial precursors. The nephron takes shape as the podocyte precursors and endothelial cells intermingle to make the glomerular tuft and the tubular components proliferate and elongate.

aggregates can convert to epithelia. Regardless of how the mesenchyme is induced, the cells are predetermined to make renal epithelia. Thus, their potential as renal stem cells has begun to be explored. To understand the origin of the metanephric mesenchyme, we begin with the patterning of the intermediate mesoderm.

## III. GENES THAT SPECIFY EARLY KIDNEY CELL LINEAGES

The early events controlling the specification of the renal cell lineages may be common among the pro- and mesonephric regions. Indeed, many of the same genes expressed in the pronephric and mesonephric tubules are instrumental in early metanephric development. Some of these events that underlie regional specification have been studied in more amenable organisms, including fish and

amphibians, in which pronephric development is less transient and of functional significance.

### Regionalization of the Mesoderm

While formation of the nephric duct is the earliest morphological evidence of renal development, the expression of intermediate mesodermal-specific markers precedes nephric duct formation temporally and marks the intermediate mesoderm along much of the A–P body axis. The earliest markers specific for the intermediate mesoderm are two transcription factors of the Pax family (Fig. 52.3), Pax2 and Pax8, which appear to function redundantly in nephric duct formation and extension (Bouchard *et al.*, 2002). The homeobox gene lim1 is expressed in the intermediate mesoderm, but it is initially expressed in the lateral plate mesoderm before becoming more restricted (Tsang *et al.*, 2000).

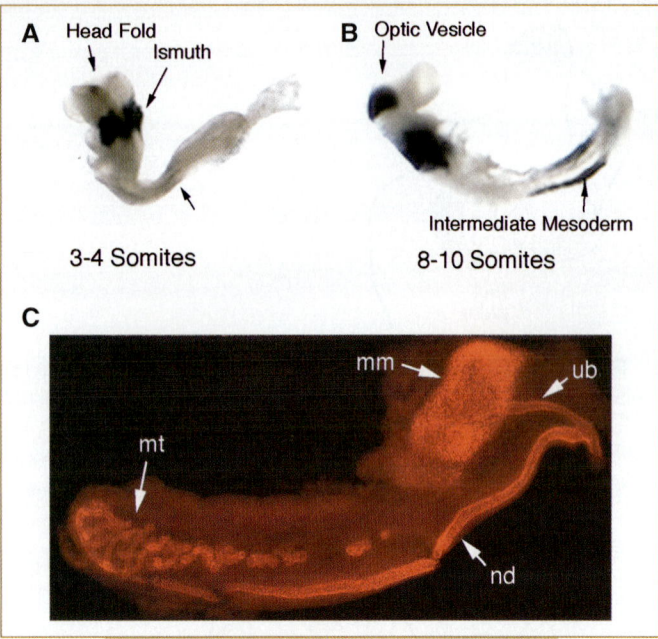

**FIG. 52.3.** The activation of Pax2 expression in the intermediate mesoderm. **(A & B)** The embryos shown carry a Pax2 promoter driving the LacZ gene, and expression is visualized by staining for beta-galactosidase activity. One of the earliest markers for the nephrogenic region, Pax2 expression is activated around the 4–5-somite stage in the cells between the axial and lateral plate mesoderm (A). By the 8-somite stage (B), Pax2 marks the growing intermediate mesoderm, even before the nephric duct is formed. Pax2 is also expressed in parts of the nervous system, especially the midbrain/hindbrain junction, known as the *rhombocephalic ismuth*, and in the optic placode and cup. **(C)** By E11, staining for Pax2 protein reveals expression in all epithelial derivatives of the intermediate mesoderm, including the nephric duct (nd), the mesonephric tubules (mt), the ureteric bud (ub) branching from the posterior duct, and the metanephric mesenchyme (mm).

Similarly, Odd-1 (Odd skipped related 1) is also expressed in a broad swath of lateral and intermediate mesoderm (James and Schultheiss, 2005). Genetics implicates all four genes in some aspects of regionalization along the intermediate mesoderm. In the mouse, Pax2 mutants begin nephric duct formation and extension but lack mesonephric tubules and the metanephros (Torres *et al.*, 1995). Pax2/Pax8 double mutants have no evidence of nephric duct formation and do not express the lim1 gene (Bouchard *et al.*, 2002). Null mutants in lim1 also lack the nephric duct and show reduced ability to differentiate into intermediate mesoderm-specific derivatives (Tsang *et al.*, 2000). Odd-1 mutants are able to generate a nephric duct that expresses Pax2 but do not have mesonephric tubules or evidence of a metanephros (Wang *et al.*, 2005). Since lim1 expression precedes Pax2 and Pax8 and is spread over a wider area in the pre- and postgastrulation embryo and is Pax independent, it seems likely that maintenance and restriction of lim1 expression within the intermediate mesoderm requires activation of the Pax2/8 genes at the 5–8-somite stage.

Are Pax2/8 sufficient, then, to specify the renal progenitors? This question was addressed in the chick embryo by Bouchard *et al.* (2002) using replication-competent retroviruses expressing a Pax2b cDNA. Ectopic nephric ducts were generated within the general area of the intermediate mesoderm on retrovirally driven Pax2b expression. These ectopic nephric ducts were not obtained with either lim1 or Pax8 alone. Strikingly, the ectopic nephric ducts paralleled the endogenous ducts and were not found in more paraxial or lateral plate mesoderm. This would suggest that Pax2's ability to induce duct formation does require some regional competence, perhaps only in the lim1-expressing domain.

If Pax2/8 and lim1 restriction in the intermediate mesoderm are the earliest events that distinguish the nephrogenic zone from surrounding paraxial and lateral plate mesoderm, the question then remains as to how these genes are activated. In the axial mesoderm, signals derived from the ventral notochord pattern the somites along the dorsal–ventral axis. Similar notochord-derived signals could also pattern mesoderm along the mediolateral axis. However, this does not appear to be the case. In the chick embryo, the notochord is dispensable for activation of the Pax2 gene in the intermediate mesoderm. Rather, signals derived from the somites, or paraxial mesoderm, are required for activation of Pax2 (Mauch *et al.*, 2000). More recently, James and Schultheiss (2005) used bone morphogenetic protein 2 (Bmp-2) to induce expression of intermediate mesodermal markers in a concentration-specific manner. *In vitro*, high levels of Bmp-2 induce a lateral plate mesodermal fate, whereas low concentrations generate more intermediate mesodermal tissue. Bmp expression is high in the dorsolateral overlying ectoderm. In an independent experiment, nephric duct formation can be inhibited if the overlying ectoderm is removed, yet duct formation is restored with exogenous Bmps (Obara-Ishihara *et al.*, 1999). The model proposed by James and Schultheiss suggests that factors within the axial mesoderm suppress the expression of intermediate mesodermal markers and that this suppression is lost at low concentrations of Bmps. In the *Xenopus* embryo, retinoic acid and activins are able to expand the pronephric region in animal cap assays, suggesting a positive role for these potential morphogens (Osafune *et al.*, 2002). Similarly, mouse embryonic stem cells respond to activin-A and retinoic acid by expressing many markers of the intermediate mesoderm and its derivatives (Kim and Dressler, 2005). Thus, the mediolateral patterning events appear to be driven by opposing signals to specify the intermediate mesoderm at the interface between lateral plate and axial mesoderm.

If Pax2/8 mark the entire nephric region (Fig. 52.3), there must be additional factors that specify the position of elements along the anterior–posterior (A–P) axis in the intermediate mesoderm. Such patterning genes could determine whether a mesonephric or metanephric kidney is formed within the Pax2 positive domain. Among the known A–P patterning genes are members of the HOX gene family. Indeed, mice that have deleted all genes of the Hox11 paralogous group have no metanephric kidneys (Wellik et al., 2002), though it is not clear whether this is truly a shift in A–P patterning or a lack of induction. A–P patterning of the intermediate mesoderm may also depend on the FoxC family of transcription factors. FoxC1 and FoxC2 have similar expression domains in the presomitic and intermediate mesoderm, as early as E8.5 (Kume et al., 2000). As nephric duct extension progresses, FoxC1 is expressed in a dorso-ventral gradient with the highest levels near the neural tube and lower levels in the BMP4 positive ventrolateral regions. In FoxC1 homozygous null mutants, the anterior boundary of the metanephric mesenchyme, as marked by GDNF expression, extends rostrally (Kume et al., 2000). This results in formation of a broader ureteric bud along the A–P axis and eventual duplication of ureters. Similar defects are observed in compound heterozygotes of FoxC1 and FoxC2, indicating some redundancy and gene dosage effects. Thus, FoxC1 and FoxC2 may set the anterior boundary of the metanephric mesenchyme, at the time of ureteric bud outgrowth, by suppressing genes at the transcriptional level.

## Genes That Function at the Time of Metanephric Induction

Induction and conversion of the metanephric mesenchyme to renal epithelia requires the concerted action of many genes. In Pax2 mutants, there is no evidence of ureteric bud outgrowth, despite the presence of a nephric duct. Ureteric bud outgrowth is controlled primarily by the receptor type tyrosine kinase RET, which is expressed on the nephric duct epithelia, the secreted signaling protein GDNF, which is expressed in the metanephric mesenchyme, and the GPi-linked protein GFRα1, which is expressed in both tissues. Pax2 mutants have no ureteric buds because they do not express GDNF in the mesenchyme and fail to maintain high levels of RET expression in the nephric duct (Brophy et al., 2001). Despite the lack of bud, the metanephric mesenchyme is morphologically distinguishable in Pax2 mutants. While lacking GDNF, it does express other markers of the mesenchyme, such as Six2 (Torres et al., 1995). In vitro recombination experiments using Pax2 mutant mesenchyme, surgically isolated from E11 mouse embryos, and heterologous inducing tissues indicate that Pax2 mutants are unable to respond to inductive signals (Brophy et al., 2001). Thus, Pax2 is necessary for specifying the region of intermediate mesoderm destined to undergo mesenchyme-to-epithelium conversion. In humans, the necessity of Pax2 function is further underscored because the loss of a single Pax2 allele is associated with renal-coloboma syndrome, which is characterized by hypoplastic kidneys with vesicoureteral reflux (Sanyanusin et al., 1995).

A second essential gene for conversion of the metanephric mesenchyme to epithelia is Eya1, a vertebrate homolog of the Drososphila eyes-absent gene. In mice homozygous for an Eya1 mutation, kidney development is arrested at E11 because ureteric bud growth is inhibited and the mesenchyme remains uninduced, though Pax2 and WT1 expression appears normal (Xu et al., 1999). However, two other markers of the metanephric mesenchyme, Six2 and GDNF expression, are lost in the Eya1 mutants. The loss of GDNF expression most probably underlies the failure of ureteric bud growth. However, it is not clear if the mesenchyme is competent to respond to inductive signals if a wild-type inducer were to be used in vitro. The eyes-absent gene family is part of a conserved network that underlies cell specification in several other developing tissues. Eya proteins share a conserved domain but lack DNA binding activity. The Eya proteins interact directly with the Six family of DNA-binding proteins. Mammalian Six genes are homologs of the Drosophila sina oculis homeobox gene. This cooperative interaction between Six and Eya proteins is necessary for nuclear translocation and transcriptional activation of Six target genes (Ohto et al., 1999). In humans, mutations in either Six1 (Ruf et al., 2004) or Eya1 (Abdelhak et al., 1997) are associated with branchio-oto-renal syndrome, further underscoring these genetic and biochemical interactions.

The Wilms' tumor suppressor gene, WT1, is another early marker of the metanephric mesenchyme and is essential for its survival. Wilms' tumor is an embryonic kidney neoplasia that consists of undifferentiated mesenchymal cells, poorly organized epithelium, and surrounding stromal cells. Expression of WT1 is regulated spatially and temporally in a variety of tissues and is further complicated by the presence of at least four isoforms, generated by alternative splicing. In the developing kidney, WT1 can be found in the uninduced metanephric mesenchyme and in differentiating epithelium after induction (Armstrong et al., 1992; Pritchard-Jones et al., 1990). Early expression of WT1 may be mediated by Pax2 (Dehbi et al., 1996). Initial expression levels are low in the metanephric mesenchyme, but they become up-regulated at the S-shaped body stage in the precursor cells of the glomerular epithelium, the podocytes. High WT1 levels persist in the adult podocytes. In the mouse, WT1-null mutants have complete renal agenesis (Kreidberg et al., 1993), because the metanephric mesenchyme undergoes apoptosis and the ureteric bud fails to grow out of the nephric duct. The arrest of ureteric bud growth is most probably due to lack of signaling by the WT1 mutant mesenchyme.

The transcription factor Gata-3 is also required for proper expression of RET in the nephric duct (Grote et al., 2006). Gata-3 mutations exhibit increased duct cell proliferation and misdirected ureteric buds (Grote et al., 2006; Lim et al., 2000). Early expression of Gata 3 in the intermediate mesoderm requires Pax2/8, though direct regulation has not been demonstrated. In humans, Gata-3 is associated with hypothyroidism, deafness, renal anomalies, and the so-called HDR syndrome, which is a result of haplo-insufficiency (Van Esch *et al.*, 2000).

## IV. ESTABLISHMENT OF ADDITIONAL CELL LINEAGES

At the time of ureteric bud invasion, at least two cell lineages appear to be established, the metanephric mesenchyme and the ureteric bud epithelia (Fig. 52.4). As branching morphogenesis and induction of the mesenchyme progress, additional cell lineages are evident. The early E11.5 mouse metanephros contains precursors for most all cell types, including endothelial, stromal, epithelial, and mesangial cells. Yet it is far from clear if these cell types share a common precursor or if the metanephric mesenchyme is a mixed population of precursors. The latter point may well be true for the endothelial lineage. While transplantation studies with lineage markers indicate that the vasculature can be derived from E11.5 metanephric kidneys, the Flk1 positive endothelial precursors have been observed closely associated with the ureteric bud epithelium, shortly after invasion, and are probably not derived from the metanephric mesenchyme. The lineages most likely to share a common origin within the metanephric mesenchyme are the stromal and epithelial. The maintenance of these two lineages is essential for renal development, for the ratio of stroma to epithelia is a critical factor for the renewal of mesenchyme and the continued induction of new nephrons.

### Epithelia versus Stroma

What are the early events in the induced mesenchyme that separate the stromal lineage from the epithelial lineage? The inducing signal is most likely to be Wnt9b, because it is expressed in the ureteric bud and Wnt9b mouse mutants fail to induce the mesenchyme while still undergoing early branching morphogenesis (Carroll *et al.*, 2005). Wnt genes encode a family of secreted peptides that are known to function in the development of many tissues. In response to these Wnt-inductive signals, Pax2-positive cells aggregate at the tips of the ureteric bud. Subsequently, activation of Wnt4 in these early aggregates appears critical to promote polarization. Mice homozygous for a wnt-4 mutation exhibit renal agenesis, due to growth arrest shortly after branching of the ureteric bud (Stark *et al.*, 1994). Although some mesenchymal aggregation has occurred, there is no evidence of cell differentiation into a polarized epithelial vesicle. Expression of Pax2 is maintained but reduced. Thus, wnt-4 may be a secondary inductive signal in the mesenchyme that propagates or that maintains the primary induction response in the epithelial lineage.

The transcription factor BF-2 is expressed in uninduced mesenchyme and becomes restricted to those cells not undergoing epithelial conversion after induction (Hatini *et al.*, 1996). BF-2 expression is found along the periphery of the kidney and in the interstitial mesenchyme, or stroma. After induction, there is little overlap between BF-2 and the Pax2 expression domain, prominent in the condensing pretubular aggregates. Clear lineage analysis is still lacking, though the expression patterns are consistent with the interpretation that mesenchyme cells may already be partitioned into a BF-2-positive stromal precursor and a Pax2-positive epithelial precursor prior to or shortly after induction. Mouse mutants in BF-2 exhibit severe developmental defects in the kidney that point to an essential role for BF-2 in maintaining growth and structure (Hatini *et al.*, 1996). Early ureteric bud growth and branching is unaffected, as is the formation of the first mesenchymal aggregates. However, at later stages (E13–14) these mesenchymal aggregates fail to differentiate into comma- and S-shaped bodies at a rate similar to wild type. Branching of the ureteric bud is greatly reduced at this stage, resulting in formation of fewer new mesenchymal aggregates. The fate of the initial aggregates is not fixed, for some are able to form epithelium and most all express the appropriate early markers, such as Pax2, wnt-4, and WT1. Nevertheless, it appears that

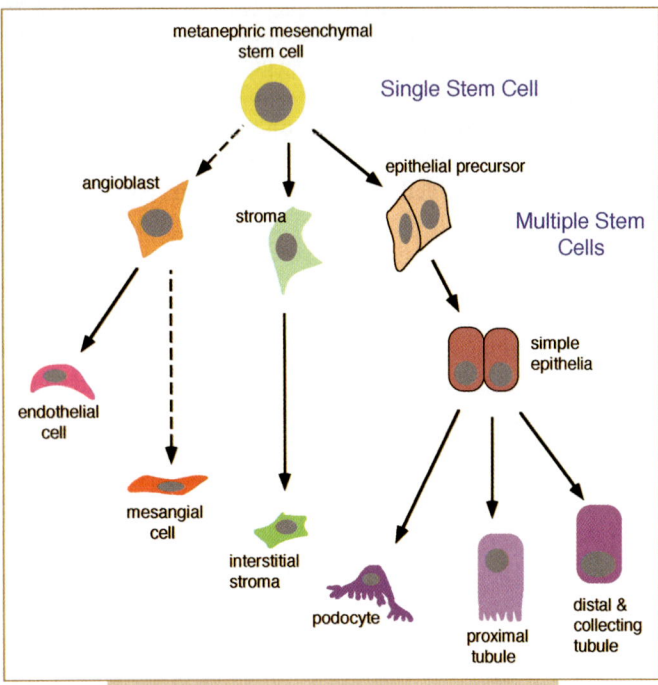

**FIG. 52.4.** Major cell lineages of the kidney. Whether the kidney arises from a single renal stem cell or from multiple independent lineages remains to be determined. However, the basic differentiation scheme is becoming clearer. The cell-lineage relationships are outlined schematically, with dotted lines reflecting ambiguity in terms of direct lineages. The metanephric mesenchyme contains angioblasts and stromal and epithelial precursors. Whether angioblasts arise from mesenchymal cells or are a separate lineage that surround the mesenchyme is not entirely clear. Similarly, the origin of the mesangial cell is not well defined. Stromal and epithelial cells may share a common precursor, the metanephric mesenchyme, but segregate at the time of induction. The epithelial cell precursors, found in the aggregates at the ureteric bud tips, generate most all of the epithelial cell types in the nephron.

the BF-2-expressing stromal lineage is necessary to maintain growth of both ureteric bud epithelium and mesenchymal aggregates. Perhaps factors secreted from the stroma provide survival or proliferation cues for the epithelial precursors, in the absence of which the non-self-renewing population of mesenchyme is exhausted.

Some survival factors that act on the mesenchyme have already been identified. The secreted TGF-β family member BMP7 and the fibroblast growth factor FGF2 in combination dramatically promote survival of uninduced metanephric mesenchyme *in vitro* (Dudley *et al.*, 1999). FGF2 is necessary to maintain the ability of the mesenchyme to respond to inductive signals *in vitro*. BMP7 alone inhibits apoptosis but is not sufficient to enable mesenchyme to undergo tubulogenesis at some later time. After induction, exogenously added FGF2 and BMP7 reduce the proportion of mesenchyme that undergoes tubulogenesis while increasing the population of BF-2 positive stromal cells (Dudley *et al.*, 1999). At least after induction occurs, there is a delicate balance between a self-renewing population of stromal and epithelial progenitor cells, the proportion of which must be well regulated by both autocrine and paracrine factors. Whether this lineage decision has already been made in the uninduced mesenchyme remains to be determined.

The role of stroma in regulating renal development is further underscored by studies with retinoic acid receptors. It is well documented that vitamin A deficiency results in severe renal defects (Lelievre-Pegorier *et al.*, 1998). In organ culture, retinoic acid stimulates expression of RET to increase dramatically the number of ureteric bud branch points, increasing the number of nephrons (Vilar *et al.*, 1996). Yet it is the stromal cell population that expresses the retinoic acid receptors, specifically *RARα* and *RARβ2*. Genetic studies with *RARα* and *RARβ2* homozygous mutant mice indicate no significant renal defects when either gene is deleted. However, double homozygotes mutant for both *RARα* and *RARβ2* exhibit severe growth retardation in the kidney (Mendelsohn *et al.*, 1999). These defects are due primarily to decreased expression of the RET protein in the ureteric bud epithelia and limited branching morphogenesis. Surprisingly, overexpression of RET with a HoxB7/RET transgene can completely rescue the double RAR mutants (Batourina *et al.*, 2001). These studies suggest that stromal cells provide paracrine signals for maintaining RET expression in the ureteric bud epithelia and that retinoids are required for stromal proliferation. Reduced expression of stromal cell marker BF-2, particularly in the interstitium of RAR double mutants, supports this hypothesis.

## Cells of the Glomerular Tuft

The unique structure of the glomerulus is intricately linked to its ability to retain large macromolecules within the circulating bloodstream while allowing for rapid diffusion of ions and small molecules into the urinary space. The glomerulus consists of four major cell types: the endothelial cells of the microvasculature, the mesangial cells, the podocyte cells of the visceral epithelium, and the parietal epithelium. The development of the glomerular architecture and the origin of the individual cell types are just beginning to be understood.

The podocyte is a highly specialized epithelial cell whose function is integral to maintaining the filtration barrier in the glomerulus. The glomerular basement membrane separates the endothelial cells of the capillary tufts from the urinary space. The outside of the glomerular basement membrane, which faces the urinary space, is covered with podocyte cells and their interdigitated foot processes. At the basement membrane, these interdigitations meet to form a highly specialized cell–cell junction, called the *slit diaphragm*. The slit diaphragm has a specific pore size to enable small molecules to cross the filtration barrier into the urinary space while retaining larger proteins in the bloodstream. The podocytes are derived from condensing metanephric mesenchyme and can be visualized with specific markers at the S-shaped body stage. While there are a number of genes expressed in the podocytes, there are only a few factors known to regulate podocyte differentiation. These include the WT1 gene, which is required early for metanephric mesenchyme survival but whose levels increase in podocyte precursors at the S-shaped body stage. In the mouse, complete WT1 null animals lack kidneys, but reduced gene dosage and expression of WT1 results in specific podocyte defects (Guo *et al.*, 2002). Thus, the high levels of WT1 expression in podocytes appear to be required and make these precursor cells more sensitive to gene dosage. The basic helix-loop-helix protein Pod1 is expressed in epithelial precursor cells and in more mature interstitial mesenchyme. At later developmental stages, Pod1 is restricted to the podocytes. In mice homozygous for a *Pod1*-null allele, podocyte development appears arrested (Quaggin *et al.*, 1999). Normal podocytes flatten and wrap their foot processes around the glomerular basement membrane. *Pod1*-mutant podocytes remain more columnar and fail to develop foot processes fully. Since Pod1 is expressed in epithelial precursors and in the interstitium, it is unclear whether these podocyte effects are due to a general developmental arrest because of the stromal environment or a cell autonomous defect within the *Pod1*-mutant podocyte precursor cells.

Within the glomerular tuft, the origin of the endothelium and the mesangium is less clear. At the S-shaped body stage, the glomerular cleft forms at the most proximal part of the S-shaped body, furthest from the ureteric bud epithelium. Vascularization of the developing kidney is first evident within this developing tuft. The origin of these invading endothelial cells has been studied in some detail. Under normal growth conditions, kidneys excised at the time of induction and cultured *in vitro* do not exhibit signs of vascularization, leading to the presumption that endothelial cells migrate to the kidney some time after induction.

However, hypoxygenation or treatment with vascular endothelial growth factor (VEGF) promotes survival or differentiation of endothelial precursors in these same cultures, suggesting that endothelial precursors are already present and require growth differentiation stimuli (Tufro *et al.*, 1999; Tufro-McReddie *et al.*, 1997). It is likely that VEGF secretion by podocyte precursor cells helps attract neighboring endothelial cells into the proximal cleft (Eremina *et al.*, 2003). *In vivo* transplantation experiments using lacZ-expressing donors or hosts also demonstrate that the E11.5 kidney rudiment has the potential to generate endothelial cells, although recruitment of endothelium is also observed from exogenous tissue, depending on the environment (Hyink *et al.*, 1996; Robert *et al.*, 1996, 1998).

The data are consistent with the idea that cells within the E11.5 kidney have the ability to differentiate along the endothelial lineage. Are these endothelial cells generated from the metanephric mesenchyme? Using a lacZ knockin allele for the endothelial specific receptor Flk-1, Robert *et al.* (1998) showed that presumptive angioblasts are dispersed along the periphery of the E12 kidney mesenchyme, with some positive cells invading the mesenchyme along the aspect of the growing ureteric bud. At later stages, Flk-1-positive angioblasts were localized to the nephrogenic zone, the developing glomerular cleft of the S-shaped bodies, and the more mature capillary loops, whereas VEGF localizes to the parietal and visceral glomerular epithelium. Injection of neutralizing VEGF antibodies into newborn mice, at a time when nephrogenesis is still ongoing, disturbs vascular growth and glomerular architecture (Kitamoto *et al.*, 1997). The data suggest that endothelial cells originate independent of the metanephric mesenchyme and invade the growing kidney from the periphery and along the ureteric bud.

The mesangial cells are located between the capillary loops of the glomerular tuft and have been referred to as specialized *pericytes*. The pericytes are found within the capillary basement membranes and have contractile abilities, much like a smooth muscle cell. Whether the mesangial cell is derived from the endothelial or epithelial lineage remains unclear. However, genetic and chimeric analyses in the mouse have revealed a clear role for the platelet-derived growth factor receptor (PDGFr) and its ligand (PDGF). In mice deficient for either PDGF or PDGFr (Leveen *et al.*, 1994; Soriano, 1994), a complete absence of mesangial cells results in glomerular defects, including the lack of microvasculature in the tuft. PDGF is expressed in the developing endothelial cells of the glomerular tuft, whereas the receptor is found in the presumptive mesangial cell precursors. Using ES cell chimeras of *Pdgfr*$^{-/-}$ and *Pdgfr*$^{+/+}$ genotypes, only the wild-type cells could contribute to the mesangial lineage (Lindahl *et al.*, 1998). This cell-autonomous effect indicates that signaling from the developing vasculature promotes proliferation and/or migration of the mesangial precursor cells. Also, expression of PDGFr and smooth muscle actin

supports a model where mesangial cells are derived from smooth muscle of the afferent and efferent arterioles during glomerular maturation (Lindahl *et al.*, 1998).

## V. REGENERATION AND RENAL STEM CELLS

The ability of renal epithelial cells to regenerate after injury has prompted much speculation regarding the source of adult renal stem cells and the potential for utilizing such stem cells to enhance recovery further from both acute and chronic renal failure. Acute tubular necrosis (ATN) is the most common cause of acute renal failure in a clinical setting. Often the result of hypotension, surgical cross-clamping of the aorta or renal arteries, or exposure to toxicants, ATN has been studied in humans and in a variety of well-characterized animal models. Despite many advances in understanding the cell physiology of ischemic and toxic renal injury, the high frequency of mortality in the clinical setting has not dipped significantly below 50% (Liu, 2003). In animal models, ATN is the result of sublethal damage that is characterized by ATP depletion, mitochondrial dysfunction, loss of epithelial cell polarity, and loss of adhesion to extracellular matrix (Nony and Schnellmann, 2003). Such changes can trigger the intrinsic apoptotic response, most likely through cytochrome C release from mitochondria. Cells within the S3 segment of the proximal tubules and the thick ascending limb seem to be particularly sensitive to damaging agents (Shanley *et al.*, 1986; Venkatachalam *et al.*, 1978; Witzgall *et al.*, 1994). Despite significant cell death after ischemia/reperfusion, folic acid, or mercury chloride treatment, renal function in animal models of ATN can recover and proximal tubule morphology is restored if the extent of damage is sublethal.

The process of proximal tubule regeneration in rodent models of ATN has been characterized in much detail. After clearing of cellular debris, cells begin to repopulate the damaged area within a few days. The proliferation of cells within the S2 and S3 segments is accompanied by the expression of many immediate-early genes, such as c-myc and AP1 (Safirstein, 1994), and the reexpression of developmental regulatory genes such as Pax2 (Imgrund *et al.*, 1999). This reentry into the cell cycle and recapitulation of a developmental program has lent credence to the idea that surviving cells may dedifferentiate, migrate toward, and repopulate the denuded basement membrane of the S3 segment (Safirstein, 1999). While it is clear that sublethal injury promotes cell proliferation and the expression of dedifferentiation markers along the damaged proximal tubules, the origin of these newly differentiating cells has been the subject of some debate. The conventional models presumed that a subpopulation of viable, remaining cells could repopulate the damaged tubules (Nony and Schnellmann, 2003; Wallin *et al.*, 1992). Alternatively, the existence of a small population of renal stem cells within the environment of the adult kidney was postulated as a potential source of renewable epithelia after injury (Al-Awqati and Oliver, 2002). More recently, a population of slowly dividing cells from the adult renal papilla was identified that could differentiate into multiple epithelial and parenchymal cell types in response to hypoxia (Oliver *et al.*, 2004). These findings support the idea that renal stem cells are intrinsic to the adult kidney.

In contrast, other investigators have pursued the potential of bone marrow–derived cells as a source of renal stem cells. Two early reports demonstrated bone marrow–derived proximal tubule cells in mice subject to ischemia and reperfusion (Kale *et al.*, 2003; Lin *et al.*, 2003). In these experiments, the majority of regenerated proximal tubule cells appeared to be derived from donor bone marrow stem cells either transplanted prior to or injected intravenously directly after injury. However, these findings lead to a great deal of skepticism and have since been disputed. The potential of whole bone marrow to contribute to proximal tubule regeneration many months after transplantation was examined in detail by Szczypka *et al.* (2005), who found little evidence for bone marrow–derived renal epithelial cells after induction of ATN by folic acid, despite numerous bone marrow–derived leukocytes within the renal interstitium. Similarly, Duffield *et al.* (2005) found no evidence for bone marrow–derived renal epithelial cells after injury, after careful examination of histological sections by deconvolusion microscopy. In a follow-up article, Lin *et al.* (2005) were not able to demonstrate a significant fraction of bone marrow–derived renal epithelial cells after injury, despite their promising initial findings (Lin *et al.*, 2003).

In addition to whole–bone marrow transplants, purified human mesenchymal stem cells have been used for testing renal replacement in animal models. Such cells have been recombined with developing rodent kidneys *in vitro* and *in vivo* and show the potential to differentiate into tubular epithelium (Yokoo *et al.*, 2005, 2006). Injection of mesenchymal stem cells in rodent models of ischemia reduces the severity of renal failure, although this does not appear to be due to transdifferentiation into epithelia (Duffield *et al.*, 2005; Lange *et al.*, 2005). Finally, bone marrow–derived cells were used to rescue a genetic defect in a collagen IV chain (Col4A3), a mouse model of Alport's syndrome, such that the glomerular basement membrane was partially restored (Sugimoto *et al.*, 2006). Again, this correction does not necessarily mean that bone marrow–derived cells contributed to epithelial structures; rather, replacement of capillary endothelial cells by genetically corrected bone marrow cells could help remodel the aberrant glomerular basement membrane. Regardless of the cellular mechanisms involved, the data do point to significant therapeutic potential for mesenchymal stem cells and other bone marrow derivatives in alleviating certain renal insults.

An alternative source of renal stem cells may be the true embryonic stem (ES) cells isolated from the inner cell mass of the blastocyst. On injection into developing kidney rudiments, ES cells can contribute to tubular epithelium and interstitium, suggesting that the local environment is conducive to differentiation of ES cells into renal derivatives (Steenhard *et al.*, 2005) (Fig. 52.5). Furthermore, the contribution of ES cells to renal epithelia can be enhanced if the cells are first differentiated *in vitro* by mesodermal factors known to be important for early renal development (Kim and Dressler, 2005). A combination of retinoic acid, activin, and BMP7 was shown to stimulate expression of early renal markers in embryoid bodies cultured *in vitro* (Fig. 52.5). Once these renal progenitor cells were injected into a developing kidney they were able to contribute exclusively to the renal proximal and distal tubules. Strikingly, ES cell derivatives were never found in the glomerulus, suggesting that an essential factor for podocyte differentiation was missing.

The above outlined experiments with ES cells highlight some of the difficulties regarding development and cell-lineage specification in the kidney. A prospective renal stem cell should be self-renewing and able to generate all of the cell types in the kidney, including epithelial, glomerular, mesangial, and stromal. Whether this can be achieved by a single cell or requires multiple progenitor cell types is still not clear. In simplest terms, all of the cells in the kidney could be generated by a single stem cell population, as outlined in Fig. 52.4. Yet even at the earliest stage of kidney development there are already two identifiable cell types, ureteric bud epithelia and metanephric mesenchyme. The presence of early endothelial precursors at the time of induction would make a third distinct cell type. If the three earliest cell types can actually by derived from the metanephric mesenchyme, a single stem cell may indeed exist and continue to proliferate as development progresses. At present, the data suggest that stroma and epithelia may share a common origin, whereas endothelial cells and their potential smooth muscle derivatives constitute a second lineage. However, even if there are three separate lineages already demarcated within the metanephric mesenchyme, the most relevant with respect to the repair of renal tissue is the epithelial lineage. Thus, if we consider the possibility of an epithelial stem cell, the following points would be among the criteria for selection: (1) The cells would most likely be a derivative of the intermediate mesoderm. (2) The cells would express a combination of markers specific for the metanephric mesenchyme. (3) The cells should be able to contribute to all epithelial components of the nephron, *in vitro* and *in vivo*.

The metanephric mesenchyme contains the renal progenitor cells but is essentially quiescent and does not proliferate in the absence of induction. However, growth conditions that mimic induction might be able to allow for the mesenchyme cells to proliferate while suppressing their differentiation into epithelium. If metanephric mesenchymal cells were able to proliferate *in vitro*, the markers they might express should include Pax2, lim1, WT1, GDNF, Six2,

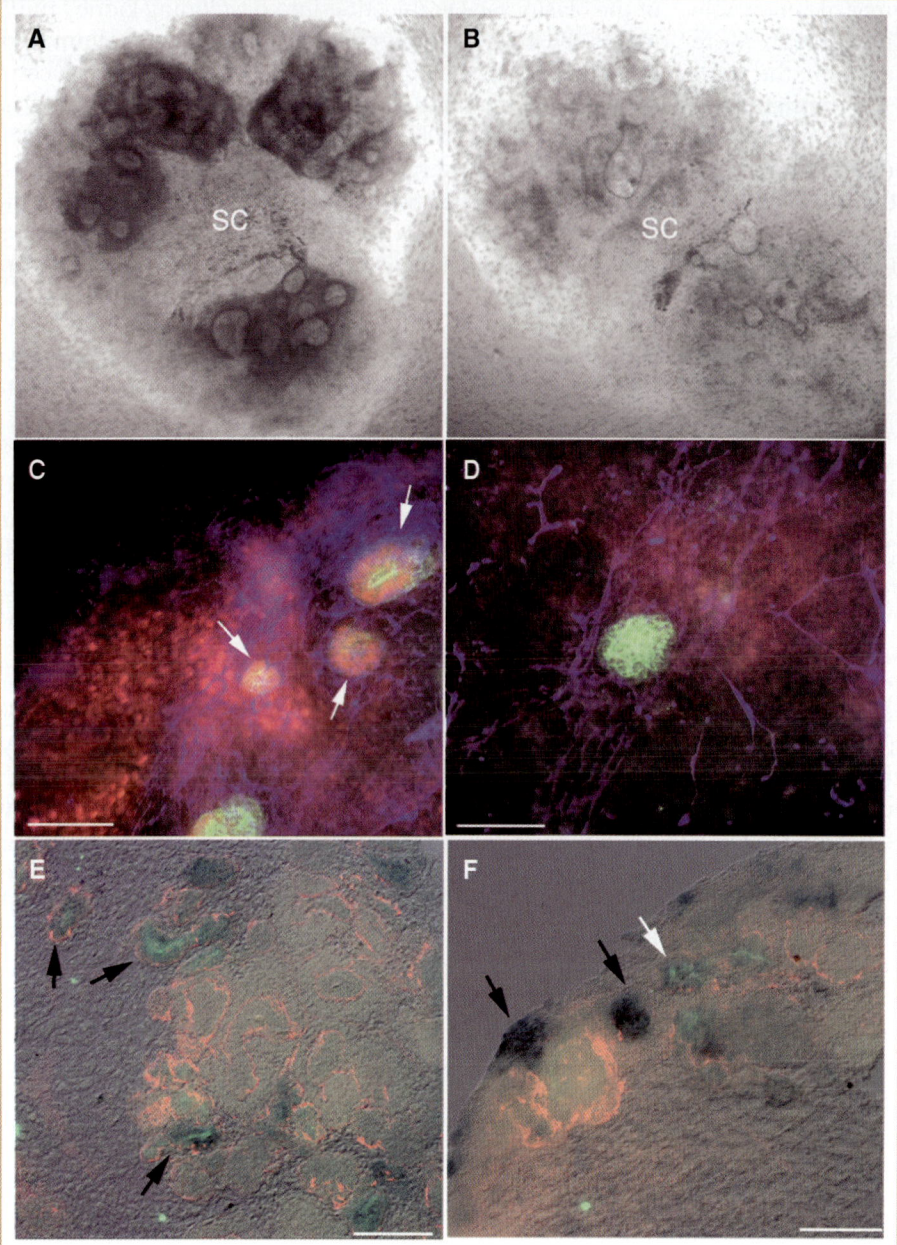

**FIG. 52.5.** Embryonic stem cells can differentiate into renal epithelia. ES cells were allowed to form embryoid bodies and cultured with retinoic acid, activin, and BMP7 (RA7, panels A, C, E) as described by Kim and Dressler (2005). Control ES cells were allowed to form embryoid bodies but were cultured in media without additional differentiation promoting factors (B, D, F). **(A & B)** Embryoid bodies cocultured with inducing tissue, i.e., dorsal spinal cord, show large three-dimensional epithelial structures when treated with RA7 but not with media alone. **(C & D)** Epithelial structures generated from RA7-treated ES cells *in vitro* stain positive for Pax2 (red), E-cadherin (green), and laminin (blue), whereas controls show few Pax2-positive epithelia. **(E & F)** RA7-treated lacZ-positive cells (blue) injected into a developing kidney rudiment integrate into tubules (arrows), as indicated by staining for laminin (orange) and LTA lectin (proximal tubules, green). Many control ES cells remain as aggregates of mesenchyme (black arrows), although occasional ES cells are seen within tubules (white arrow).

and BF-2. Expression of these markers would indicate a mesenchymal cell that had not decided between the stromal and epithelial lineages. If cells express Pax2, WT1, and Wnt-4 but not BF-2, they could be an epithelial stem cell. Such epithelial stem cells, when injected into or recombined with an *in vitro* cultured metanephric kidney, should be able to make all of the epithelial cells along the proximal–distal axis of the nephron. Such epithelial stem cells could prove significant in regenerating damaged tubules in acute and chronic renal injury.

At present, the complexity of the kidney still impedes progress in the area of tissue- and cell-based therapies. Not only must the right cells be made; they must be able to organize into a specialized three-dimensional tubular structure capable of fulfilling all of the physiological demands put on the nephrons. Developmental biology can provide a framework for understanding how these cells arise and what factors promote their differentiation and growth. While we may not be able to make a kidney from scratch, it seems within the realm of possibility to provide the injured adult kidney with cells or factors to facilitate its own regeneration. Given the high incidence and severity of acute and chronic renal insufficiency, such therapies would be most welcome indeed.

# VI. ACKNOWLEDGMENTS

I thank the members of my lab for valuable discussion regarding this topic, particularly Pat Brophy, Yi Cai, and Sanj Patel. G.R.D. is supported by NIH grants DK054740, DK062914, and DK069689.

# VII. REFERENCES

Abdelhak, S., Kalatzis, V., Heilig, R., Compain, S., Samson, D., Vincent, C., Weil, D., Cruaud, C., Sahly, I., Leibovici, M., *et al.* (1997). A human homologue of the *Drosophila* eyes-absent gene underlies branchio-oto-renal (BOR) syndrome and identifies a novel gene family. *Nat. Genet.* **15**, 157–164.

Al-Awqati, Q., and Oliver, J. A. (2002). Stem cells in the kidney. *Kidney Int.* **61**, 387–395.

Armstrong, J. F., Pritchard-Jones, K., Bickmore, W. A., Hastie, N. D., and Bard, J. B. L. (1992). The expression of the Wilms' tumor gene, *WT1*, in the developing mammalian embryo. *Mech. Dev.* **40**, 85–97.

Batourina, E., Gim, S., Bello, N., Shy, M., Clagett-Dame, M., Srinivas, S., Costantini, F., and Mendelsohn, C. (2001). Vitamin A controls epithelial/mesenchymal interactions through Ret expression. *Nat. Genet.* **27**, 74–78.

Bouchard, M., Souabni, A., Mandler, M., Neubuser, A., and Busslinger, M. (2002). Nephric lineage specification by Pax2 and Pax8. *Genes Dev.* **16**, 2958–2970.

Brophy, P. D., Ostrom, L., Lang, K. M., and Dressler, G. R. (2001). Regulation of ureteric bud outgrowth by Pax2-dependent activation of the glial derived neurotrophic factor gene. *Development* **128**, 4747–4756.

Carroll, T. J., Park, J. S., Hayashi, S., Majumdar, A., and McMahon, A. P. (2005). Wnt9b plays a central role in the regulation of mesenchymal to epithelial transitions underlying organogenesis of the mammalian urogenital system. *Dev. Cell* **9**, 283–292.

Dehbi, M., Ghahremani, M., Lechner, M., Dressler, G., and Pelletier, J. (1996). The paired-box transcription factor, PAX2, positively modulates expression of the Wilms' tumor suppressor gene (WT1). *Oncogene* **13**, 447–453.

Dressler, G. R. (2002). Development of the excretory system. *In* "Mouse Development: Patterning, Morphogenesis, and Organogenesis" (J. Rossant and P. T. Tam, eds.), pp. 395–420. Academic Press, San Diego, CA.

Dudley, A. T., Godin, R. E., and Robertson, E. J. (1999). Interaction between FGF and BMP signaling pathways regulates development of metanephric mesenchyme. *Genes Dev.* **13**, 1601–1613.

Duffield, J. S., Park, K. M., Hsiao, L. L., Kelley, V. R., Scadden, D. T., Ichimura, T., and Bonventre, J. V. (2005). Restoration of tubular epithelial cells during repair of the postischemic kidney occurs independently of bone marrow–derived stem cells. *J. Clin. Invest.* **115**, 1743–1755.

Eremina, V., Sood, M., Haigh, J., Nagy, A., Lajoie, G., Ferrara, N., Gerber, H. P., Kikkawa, Y., Miner, J. H., and Quaggin, S. E. (2003). Glomerular-specific alterations of VEGF-A expression lead to distinct congenital and acquired renal diseases. *J. Clin. Invest.* **111**, 707–716.

Grote, D., Souabni, A., Busslinger, M., and Bouchard, M. (2006). Pax 2/8-regulated Gata 3 expression is necessary for morphogenesis and guidance of the nephric duct in the developing kidney. *Development* **133**, 53–61.

Guo, J. K., Menke, A. L., Gubler, M. C., Clarke, A. R., Harrison, D., Hammes, A., Hastie, N. D., and Schedl, A. (2002). WT1 is a key regulator of podocyte function: reduced expression levels cause crescentic glomerulonephritis and mesangial sclerosis. *Hum. Mol. Genet.* **11**, 651–659.

Hatini, V., Huh, S. O., Herzlinger, D., Soares, V. C., and Lai, E. (1996). Essential role of stromal mesenchyme in kidney morphogenesis revealed by targeted disruption of Winged Helix transcription factor BF-2. *Genes Dev.* **10**, 1467–1478.

Hyink, D. P., Tucker, D. C., St John, P. L., Leardkamolkarn, V., Accavitti, M. A., Abrass, C. K., and Abrahamson, D. R. (1996). Endogenous origin of glomerular endothelial and mesangial cells in grafts of embryonic kidneys. *Am. J. Physiol.* **270**, F886–F899.

Imgrund, M., Grone, E., Grone, H. J., Kretzler, M., Holzman, L., Schlondorff, D., and Rothenpieler, U. W. (1999). Re-expression of the developmental gene Pax-2 during experimental acute tubular necrosis in mice 1. *Kidney Int.* **56**, 1423–1431.

James, R. G., and Schultheiss, T. M. (2005). Bmp signaling promotes intermediate mesoderm gene expression in a dose-dependent, cell-autonomous and translation-dependent manner. *Dev. Biol.* **288**, 113–125.

Kale, S., Karihaloo, A., Clark, P. R., Kashgarian, M., Krause, D. S., and Cantley, L. G. (2003). Bone marrow stem cells contribute to repair of the ischemically injured renal tubule. *J. Clin. Invest.* **112**, 42–49.

Kim, D., and Dressler, G. R. (2005). Nephrogenic factors promote differentiation of mouse embryonic stem cells into renal epithelia. *J. Am. Soc. Nephrol.* **16**, 3527–3534.

Kitamoto, Y., Tokunaga, H., and Tomita, K. (1997). Vascular endothelial growth factor is an essential molecule for mouse kidney development: glomerulogenesis and nephrogenesis. *J. Clin. Invest.* **99**, 2351–2357.

Kreidberg, J. A., Sariola, H., Loring, J. M., Maeda, M., Pelletier, J., Housman, D., and Jaenisch, R. (1993). *WT1* is required for early kidney development. *Cell* **74**, 679–691.

Kume, T., Deng, K., and Hogan, B. L. (2000). Murine forkhead/winged helix genes Foxc1 (Mf1) and Foxc2 (Mfh1) are required for the early organogenesis of the kidney and urinary tract. *Development* **127**, 1387–1395.

Kuure, S., Vuolteenaho, R., and Vainio, S. (2000). Kidney morphogenesis: cellular and molecular regulation. *Mech. Dev.* **92**, 31–45.

Lange, C., Togel, F., Ittrich, H., Clayton, F., Nolte-Ernsting, C., Zander, A. R., and Westenfelder, C. (2005). Administered mesenchymal stem cells enhance recovery from ischemia/reperfusion-induced acute renal failure in rats. *Kidney Int.* **68**, 1613–1617.

Lelievre-Pegorier, M., Vilar, J., Ferrier, M. L., Moreau, E., Freund, N., Gilbert, T., and Merlet-Benichou, C. (1998). Mild vitamin A deficiency leads to inborn nephron deficit in the rat. *Kidney Int.* **54**, 1455–1462.

Leveen, P., Pekny, M., Gebre-Medhin, S., Swolin, B., Larsson, E., and Betsholtz, C. (1994). Mice deficient for PDGF B show renal, cardiovascular, and hematological abnormalities. *Genes Dev.* **8**, 1875–1887.

Lim, K. C., Lakshmanan, G., Crawford, S. E., Gu, Y., Grosveld, F., and Engel, J. D. (2000). Gata3 loss leads to embryonic lethality due to noradrenaline deficiency of the sympathetic nervous system. *Nat. Genet.* **25**, 209–212.

Lin, F., Cordes, K., Li, L., Hood, L., Couser, W. G., Shankland, S. J., and Igarashi, P. (2003). Hematopoietic stem cells contribute to the regeneration of renal tubules after renal ischemia-reperfusion injury in mice. *J. Am. Soc. Nephrol.* **14**, 1188–1199.

Lin, F., Moran, A., and Igarashi, P. (2005). Intrarenal cells, not bone marrow–derived cells, are the major source for regeneration in post-ischemic kidney. *J. Clin. Invest.* **115**, 1756–1764.

Lindahl, P., Hellstrom, M., Kalen, M., Karlsson, L., Pekny, M., Pekna, M., Soriano, P., and Betsholtz, C. (1998). Paracrine PDGF-B/PDGF-Rbeta signaling controls mesangial cell development in kidney glomeruli. *Development* **125**, 3313–3322.

Liu, K. D. (2003). Molecular mechanisms of recovery from acute renal failure. *Crit. Care. Med.* **31**, S572–S581.

Mauch, T. J., Yang, G., Wright, M., Smith, D., and Schoenwolf, G. C. (2000). Signals from trunk paraxial mesoderm induce pronephros formation in chick intermediate mesoderm. *Dev. Biol.* **220**, 62–75.

Mendelsohn, C., Batourina, E., Fung, S., Gilbert, T., and Dodd, J. (1999). Stromal cells mediate retinoid-dependent functions essential for renal development. *Development* **126**, 1139–1148.

Nolan, C. R. (2005). Strategies for improving long-term survival in patients with ESRD. *J. Am. Soc. Nephrol.* **16** (Suppl. 2), S120–S127.

Nony, P. A., and Schnellmann, R. G. (2003). Mechanisms of renal cell repair and regeneration after acute renal failure. *J. Pharmacol. Exp. Ther.* **304**, 905–912.

Obara-Ishihara, T., Kuhlman, J., Niswander, L., and Herzlinger, D. (1999). The surface ectoderm is essential for nephric duct formation in intermediate mesoderm. *Development* **126**, 1103–1108.

Ohto, H., Kamada, S., Tago, K., Tominaga, S. I., Ozaki, H., Sato, S., and Kawakami, K. (1999). Cooperation of six and eya in activation of their target genes through nuclear translocation of Eya. *Mol. Cell. Biol.* **19**, 6815–6824.

Oliver, J. A., Maarouf, O., Cheema, F. H., Martens, T. P., and Al-Awqati, Q. (2004). The renal papilla is a niche for adult kidney stem cells. *J. Clin. Invest.* **114**, 795–804.

Osafune, K., Nishinakamura, R., Komazaki, S., and Asashima, M. (2002). *In vitro* induction of the pronephric duct in *Xenopus* explants. *Dev. Growth Differ.* **44**, 161–167.

Pritchard-Jones, K., Fleming, S., Davidson, D., Bickmore, W., Porteous, D., Gosden, C., Bard, J., Buckler, A., Pelletier, J., Housman, D., *et al.* (1990). The candidate Wilms' tumor gene is involved in genitourinary development. *Nature* **346**, 194–197.

Qiao, J., Cohen, D., and Herzlinger, D. (1995). The metanephric blastema differentiates into collecting system and nephron epithelia *in vitro*. *Development* **121**, 3207–3214.

Quaggin, S. E., Schwartz, L., Cui, S., Igarashi, P., Deimling, J., Post, M., and Rossant, J. (1999). The basic helix-loop-helix protein pod1 is critically important for kidney and lung organogenesis. *Development* **126**, 5771–5783.

Robert, B., St. John, P. L., Hyink, D. P., and Abrahamson, D. R. (1996). Evidence that embryonic kidney cells expressing flk-1 are intrinsic, vasculogenic angioblasts. *Am. J. Physiol.* **271**, F744–F753.

Robert, B., St. John, P. L., and Abrahamson, D. R. (1998). Direct visualization of renal vascular morphogenesis in Flk1 heterozygous mutant mice. *Am. J. Physiol.* **275**, F164–F172.

Ruf, R. G., Xu, P. X., Silvius, D., Otto, E. A., Beekmann, F., Muerb, U. T., Kumar, S., Neuhaus, T. J., Kemper, M. J., Raymond, R. M., Jr., *et al.* (2004). SIX1 mutations cause branchio-oto-renal syndrome by disruption of EYA1-SIX1-DNA complexes. *Proc. Natl. Acad. Sci. U.S.A.* **101**, 8090–8095.

Safirstein, R. (1994). Gene expression in nephrotoxic and ischemic acute renal failure. *J. Am. Soc. Nephrol.* **4**, 1387–1395.

Safirstein, R. (1999). Renal regeneration: reiterating a developmental paradigm. *Kidney Int.* **56**, 1599–1600.

Sanyanusin, P., Schimmenti, L. A., McNoe, L. A., Ward, T. A., Pierpont, M. E. M., Sullivan, M. J., Dobyns, W. B., and Eccles, M. R. (1995). Mutation of the *Pax2* gene in a family with optic nerve colobomas, renal anomalies and vesicoureteral reflux. *Nat. Genet.* **9**, 358–364.

Shanley, P. F., Rosen, M. D., Brezis, M., Silva, P., Epstein, F. H., and Rosen, S. (1986). Topography of focal proximal tubular necrosis after ischemia with reflow in the rat kidney. *Am. J. Pathol.* **122**, 462–468.

Soriano, P. (1994). Abnormal kidney development and hematological disorders in PDGF beta-receptor mutant mice. *Genes Dev.* **8**, 1888–1896.

Stark, K., Vainio, S., Vassileva, G., and McMahon, A. P. (1994). Epithelial transformation of metanephric mesenchyme in the developing kidney regulated by Wnt-4. *Nature* **372**, 679–683.

Steenhard, B. M., Isom, K. S., Cazcarro, P., Dunmore, J. H., Godwin, A. R., St John, P. L., and Abrahamson, D. R. (2005). Integration of embryonic stem cells in metanephric kidney organ culture. *J. Am. Soc. Nephrol.* **16**, 1623–1631.

Sugimoto, H., Mundel, T. M., Sund, M., Xie, L., Cosgrove, D., and Kalluri, R. (2006). Bone-marrow-derived stem cells repair basement membrane collagen defects and reverse genetic kidney disease. *Proc. Natl. Acad. Sci. U.S.A.* **103**, 7321–7326.

Szczypka, M. S., Westover, A. J., Clouthier, S. G., Ferrara, J. L., and Humes, H. D. (2005). Rare incorporation of bone marrow–derived cells into kidney after folic acid–induced injury. *Stem Cells* **23**, 44–54.

Torres, M., Gomez-Pardo, E., Dressler, G. R., and Gruss, P. (1995). Pax-2 controls multiple steps of urogenital development. *Development* **121**, 4057–4065.

Tsang, T. E., Shawlot, W., Kinder, S. J., Kobayashi, A., Kwan, K. M., Schughart, K., Kania, A., Jessell, T. M., Behringer, R. R., and Tam, P. P. (2000). Lim1 activity is required for intermediate mesoderm differentiation in the mouse embryo. *Dev. Biol.* **223**, 77–90.

Tufro, A., Norwood, V. F., Carey, R. M., and Gomez, R. A. (1999). Vascular endothelial growth factor induces nephrogenesis and vasculogenesis. *J. Am. Soc. Nephrol.* **10**, 2125–2134.

Tufro-McReddie, A., Norwood, V. F., Aylor, K. W., Botkin, S. J., Carey, R. M., and Gomez, R. A. (1997). Oxygen regulates vascular endothelial growth factor-mediated vasculogenesis and tubulogenesis. *Dev. Biol.* **183**, 139–149.

Van Esch, H., Groenen, P., Nesbit, M. A., Schuffenhauer, S., Lichtner, P., Vanderlinden, G., Harding, B., Beetz, R., Bilous, R. W., Holdaway, I., *et al.* (2000). GATA3 haplo-insufficiency causes human HDR syndrome. *Nature* **406**, 419–422.

Venkatachalam, M. A., Bernard, D. B., Donohoe, J. F., and Levinsky, N. G. (1978). Ischemic damage and repair in the rat proximal tubule: differences among the S1, S2, and S3 segments. *Kidney Int.* **14**, 31–49.

Vilar, J., Gilbert, T., Moreau, E., and Merlet-Benichou, C. (1996). Metanephros organogenesis is highly stimulated by vitamin A derivatives in organ culture. *Kidney Int.* **49**, 1478–1487.

Wallin, A., Zhang, G., Jones, T. W., Jaken, S., and Stevens, J. L. (1992). Mechanism of the nephrogenic repair response. Studies on proliferation and vimentin expression after 35S-1,2-dichlorovinyl-L-cysteine nephrotoxicity *in vivo* and in cultured proximal tubule epithelial cells. *Lab. Invest.* **66**, 474–484.

Wang, Q., Lan, Y., Cho, E. S., Maltby, K. M., and Jiang, R. (2005). Odd-skipped related 1 (Odd 1) is an essential regulator of heart and urogenital development. *Dev. Biol.* **288**, 582–594.

Wellik, D. M., Hawkes, P. J., and Capecchi, M. R. (2002). Hox11 paralogous genes are essential for metanephric kidney induction. *Genes Dev.* **16**, 1423–1432.

Witzgall, R., Brown, D., Schwarz, C., and Bonventre, J. V. (1994). Localization of proliferating cell nuclear antigen, vimentin, c-Fos, and clusterin in the postischemic kidney. Evidence for a heterogenous genetic response among nephron segments, and a large pool of mitotically active and dedifferentiated cells. *J. Clin. Invest.* **93**, 2175–2188.

Xu, P. X., Adams, J., Peters, H., Brown, M. C., Heaney, S., and Maas, R. (1999). Eya1-deficient mice lack ears and kidneys and show abnormal apoptosis of organ primordia. *Nat. Genet.* **23**, 113–117.

Yokoo, T., Ohashi, T., Shen, J. S., Sakurai, K., Miyazaki, Y., Utsunomiya, Y., Takahashi, M., Terada, Y., Eto, Y., Kawamura, T., *et al.* (2005). Human mesenchymal stem cells in rodent whole-embryo culture are reprogrammed to contribute to kidney tissues. *Proc. Natl. Acad. Sci. U.S.A.* **102**, 3296–3300.

Yokoo, T., Fukui, A., Ohashi, T., Miyazaki, Y., Utsunomiya, Y., Kawamura, T., Hosoya, T., Okabe, M., and Kobayashi, E. (2006). Xenobiotic kidney organogenesis from human mesenchymal stem cells using a growing rodent embryo. *J. Am. Soc. Nephrol.* **17**, 1026–1034.

## Chapter Fifty-Three

# Renal Replacement Devices

*H. David Humes*

I. Introduction
II. Basics of Kidney Function

III. Tissue-Engineering Approach to Renal Function Replacement
IV. References

## I. INTRODUCTION

The rapid understanding of the cellular and molecular bases of organ function and disease processes will be translated in the next decade into new therapeutic approaches to a wide range of clinical disorders, including acute and chronic renal failure. Central to these new therapies are the developing technologies of cell therapy and tissue engineering, which are based on the ability to expand stem or progenitor cells in tissue culture to perform differentiated tasks and to introduce these cells into the patient either via extracorporeal circuits or as implantable constructs. Cell therapy devices are currently being developed to replace the filtrative, metabolic, and endocrinologic functions of the kidney lost in both acute and chronic renal failure. This chapter summarizes the current state of development of a renal tubule–assist device and a bioartificial hemofilter. These devices have the promise to be combined to produce a wearable or implantable bioartificial kidney for full renal replacement therapy that may significantly diminish morbidity and mortality in patients with acute or chronic renal disease.

The kidney was the first solid organ whose function was approximated by a machine and a synthetic device. In fact, renal substitution therapy with hemodialysis or chronic ambulatory peritoneal dialysis (CAPD) has been the only successful long-term ex vivo organ substitution therapy to date (Iglehart, 1993). The kidney was also the first organ to be successfully transplanted from a donor individual to an autologous recipient patient. However, the lack of widespread availability of suitable transplantable organs has kept kidney transplantation from becoming a practical solution in most cases of chronic renal failure.

Although long-term chronic renal replacement therapy with either hemodialysis or CAPD has dramatically changed the prognosis of renal failure, it is not complete replacement therapy, because it provides only filtration function (usually on an intermittent basis) and does not replace the homeostatic, regulatory, metabolic, and endocrine functions of the kidney. Because of the nonphysiologic manner in which dialysis performs or does not perform the most critical renal functions, patients with end-stage renal disease (ESRD) on dialysis continue to have major medical, social, and economic problems (U.S. Renal Data System, 2004). Accordingly, dialysis should be considered as renal substitution rather than renal replacement therapy.

Tissue engineering of an implantable artificial kidney composed of both biologic and synthetic components could result in substantial benefits for patients by increasing life expectancy, mobility, and quality of life, with less risk of infection and with reduced costs. This approach could also be considered a cure rather than a treatment for patients. A successful tissue-engineering approach to the kidney depends on a thorough knowledge of the physiologic basis of kidney function.

## II. BASICS OF KIDNEY FUNCTION

The excretory function of the kidney is initiated by filtration of blood at the glomerulus, which is an enlargement of the proximal end of the tubule, incorporating a vascular

*Principles of Tissue Engineering, 3rd Edition*
ed. by Lanza, Langer, and Vacanti

tuft. The structure of the glomerulus is designed to provide efficient ultrafiltration of blood, to remove toxic wastes from the circulation yet retain important circulating components, such as albumin. The regulatory function of the kidney, especially with regard to fluid and electrolyte homeostasis, is provided by the tubular segments attached to the glomerulus. The functional unit of the kidney is therefore composed of the filtering unit (the glomerulus) and the regulatory unit (the tubule). Together they form the basic component of the kidney, the nephron. In addition to these excretory and regulatory functions, the kidney is an important metabolic and endocrine organ. Erythropoietin, active forms of vitamin D, renin, angiotensin, prostaglandins, leukotrienes, kallikrein-kinins, various cytokines, and complement components are some of the endocrinologic compounds produced by the kidney. Because of the efficiency inherent in the kidney as an excretory organ, life can be sustained with only 5–10% of normal renal excretory function. Accordingly, the approach to a tissue-engineered construct becomes easier to entertain, especially because only a fraction of normal renal excretory function is required to maintain life (Humes, 2000).

The process of urine formation begins within the capillary bed of the glomerulus (Brenner and Humes, 1977). The glomerular capillary wall has evolved into a structure with the property to separate as much as one-third of the plasma entering the glomerulus into a solution of a nearly ideal ultrafiltrate. This high rate of ultrafiltration across the glomerular capillary is a result of hydraulic pressure generated by the heart and vascular tone of the preglomerular and postglomerular vessels as well as by the high hydraulic permeability of the glomerular capillary wall. The hydraulic pressure and permeability of the glomerular capillary bed are at least two times and two orders of magnitude higher, respectively, than those of most other capillary networks within the body (Landis and Pappenheimer, 1965). Despite this high rate of water and solute flux across the glomerular capillary wall, this same structure retards the filtration of important circulating macromolecules, especially albumin, so all but the lower-molecular-weight plasma proteins are restricted in the passage across this filtration barrier (Anderson and Quinn, 1974; Chang et al., 1975; Brenner et al., 1978).

This ultrafiltration process of glomeruli in normal human kidneys forms approximately 100 mL of filtrate every minute. Because daily urinary volume is roughly two liters, more than 98% of the glomerular ultrafiltrate must be absorbed by the renal tubule. The bulk of reabsorption, 50–65%, occurs along the proximal tubule. Similar to glomerular filtration, fluid movement across the renal proximal tubule cell is governed by physical forces. Unlike the fluid transfer across the glomerular capillary wall, however, tubular fluid flux is driven principally by osmotic and oncotic pressures rather than hydraulic pressure. Renal proximal tubule fluid absorption is based on active Na$^+$ transport,

requiring the energy-dependent Na$^+$,K$^+$-ATPase located along the basolateral membrane of the renal tubule cell to promote a small degree of luminal hypotonicity (Andreoli and Schafer, 1978). This small degree of osmotic difference (2–3 mOsm/kg H$_2$O) across the renal tubule is sufficient to drive isotonic fluid reabsorption, due to the very high diffusive water permeability of the renal tubule cell membrane. Once across the renal proximal tubule cell, the transported fluid is taken up by the peritubular capillary bed, due to the favorable oncotic pressure gradient. This high oncotic pressure within the peritubular capillary is the result of the high rate of protein-free filtrate formed in the proximate glomerular capillary bed (Knox et al., 1983). As can be appreciated, an elegant system has evolved in the nephron to filter and reabsorb large amounts of fluid in bulk to attain high rates of metabolic product excretion while maintaining regulatory salt and water balance.

## III. TISSUE-ENGINEERING APPROACH TO RENAL FUNCTION REPLACEMENT

In designing an implantable bioartificial kidney for renal function replacement, essential features of kidney tissue must be utilized to direct the design of the tissue-engineering project. The critical elements of renal function must be replaced, including the excretory, regulatory transport, and endocrinologic functions. The functioning excretory unit of the kidney, as detailed previously, is composed of the filtering unit, the glomerulus, and the regulatory or transport unit, the tubule. Therefore, a bioartificial kidney requires two main units, the glomerulus and the tubule, to replace excretory function.

### Bioartificial Hemofilter

The potential for a bioartificial glomerulus has been achieved with the use of polysulfone fibers *ex vivo* with maintenance of ultrafiltration in humans for a few days with a single device (Golper, 1986; Kramer et al., 1977). The availability of hollow fibers with high hydraulic permeability has been an important advancement in biomaterials for replacement of glomerular ultrafiltration. Conventional hemodialysis for ESRD has used membranes in which solute removal is driven by a concentration gradient of the solute across the membranes and is, therefore, predominantly a diffusive process. Another type of solute transfer also occurs across the dialysis membrane via a process of ultrafiltration of water and solutes across the membrane. This convective transport is independent of the concentration gradient and depends predominantly on the hydraulic pressure gradient across the membrane. Both diffusive and convective processes occur during traditional hemodialysis, but diffusion is the main route of solute movement.

The development of synthetic membranes with high hydraulic permeability and solute-retention properties in convenient hollow-fiber form has promoted ESRD therapy

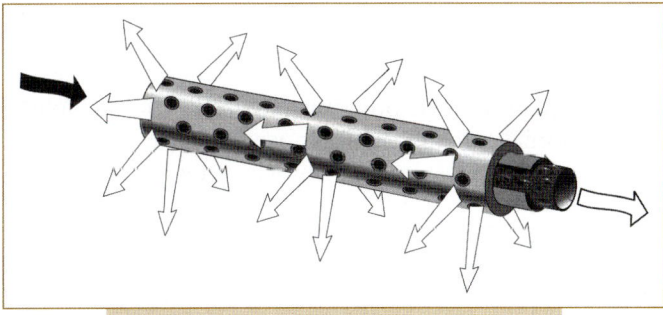

**FIG. 53.1.** Schematic of a tissue-engineered hemofilter composed of a microporous synthetic biocompatible hollow fiber, a preadhered extracellular matrix, and a confluent monolayer of autologous endothelial cells lining the luminal surface of the fiber. Arrows refer to vectorial ultrafiltrate formation.

based on convective hemofiltration rather than diffusive hemodialysis (Colton *et al.*, 1975; Henderson *et al.*, 1975). Removal of uremic toxins, predominantly by the convective process, has several distinct advantages, because it imitates the glomerular process of toxin removal with increased clearance of higher-molecular-weight solutes and removal of all solutes (up to a molecular-weight cutoff) at the same rate. Distinct differences are apparent between diffusive and convective transport across a semipermeable membrane. The clearance of a molecule by diffusion is negatively correlated with the size of the molecule. In contrast, clearance of a substance by convection is dependent on size, up to a certain molecular weight. The bulk movement of water carries passable solutes along with it in approximately the same concentration as in the fluid.

Development of an implantable device that mimics glomerular filtration will thus depend on convective transport (Fig. 53.1). This physiologic function has been achieved clinically with the use of polymeric hollow fibers *ex vivo*. Major limitations to the currently available technology for long-term replacement of filtration function include bleeding associated with required anticoagulation, diminution of filtration rate due either to protein deposition in the membrane over time or to thrombotic occlusion, and large amounts of fluid replacement required to replace the ultrafiltrate formed from the filtering unit. The use of autologous endothelial cell–seeded conduits along filtration surfaces may provide improved long-term hemocompatibility and hemofiltration *in vivo* (Kadletz *et al.*, 1992; Schnider *et al.*, 1988; Shepard *et al.*, 1986). A report detailing the capability of isolating angioblasts from peripheral blood makes this approach more readily achievable (Asahara *et al.*, 1997). Microfabricated capillary networks with microelectromechanical techniques and endothelial attachment and growth have made substantive advances recently (Shin *et al.*, 2004).

For differentiated endothelial cell morphology and function, an important role for various components of the extracellular matrix (ECM) has been demonstrated (Carey,

1991; Carley *et al.*, 1988). The ECM has clearly been shown to dictate phenotype and gene expression of endothelial cells, thereby modulating morphogenesis and growth. Various components of the ECM, including collagen type I, collagen type IV, laminin, and fibronectin, have been shown to affect endothelial cell adherence, growth, and differentiation. Of importance, ECM produced by Madin-Darby canine kidney (MDCK) cells, a permanent renal epithelial cell line, promotes endothelial cells to develop fenestrations. Fenestrations are large openings that act as channels or pores for convective transport through the endothelial monolayer and are important in the high hydraulic permeability and sieving characteristics of glomerular capillaries (Carley *et al.*, 1988; Milici *et al.*, 1985). Thus, the ECM component on which the endothelial cells attach and grow may be critical in the functional characteristics of the lining monolayer.

A second approach for a tissue-engineered device that may prove to have clinical utility is an implantable hemofilter with a capacity to produce 2–4 mL/min of ultrafiltrate. Although this rate of filtration is considerably less than the 10–15 mL/min rate required to correct the uremic state in ESRD patients, it is sufficient enough to improve solute clearance to lessen dialysis time and optimize clearance rates of urea and other uremic toxins. In addition, this rate of ultrafiltration from a stand-alone implanted hemofilter is small enough so that, with a urinary bladder capacity of 500–750 mL, a patient can tolerate the frequency of urination required from the filtrate's draining directly into the urinary tract system. Both of these intermediate devices will provide substantial experience to improve the durability, efficacy, and efficiency of an implantable hemofilter and a tubule-assist device to achieve the required hemofiltration rate of 15–20 mL/min and selective reabsorption for adequate clearance while avoiding volume depletion and intolerable urinary frequency.

An approach toward developing a durable and implantable bioartificial filtration device is to promote site-directed neovascularization *in vivo* (Thompson *et al.*, 1988). In this formulation, angiogenic factors (Folkman and Shing, 1992) can be delivered via endogenous or exogenous routes to induce targeted angiogenesis and neovascular beds among implanted biocompatible hollow fibers. These hollow follows are envisioned to act as collecting conduits for ultrafiltrate produced by the newly formed capillary network induced by the angiogenic factors. This formulation is based on the intrinsic property inherent in all capillary beds to produce ultrafiltrate (Landis and Pappenheimer, 1965). This filtrate, or *transudate*, will collect in the HF network rather than the usual physiologic sites consisting of the interstitial space and lymphatics. The vectorial filtrate flow will be from capillary through interstitium into hollow fiber and, ultimately, into an external drainage and collection system.

Our laboratory has just completed a series of studies in a rat model to promote via surgical techniques a tissue-engineered neovascular bed within an implant to permit capillary ultrafiltration with and without angiogenic factors as a critical step to produce a bioartificial hemofilter. The results demonstrated a proof of concept of producing a bioartificial hemofiltration device with a durability of at least three to four months, the longest time period tested (Tiranathanagul *et al.*, 2005). Scale-up in large-animal models is being planned as a next step in this tissue-engineering program.

### Bioartificial Tubule

As detailed earlier, the efficiency of reabsorption, even though dependent on natural physical forces governing fluid movement across biologic as well as synthetic membranes, requires specialized epithelial cells to perform vectorial solute transport. Fortunately, a population of cells residing in the adult mammalian kidney has retained the capacity to proliferate and differentiate morphogenically into tubule structures *in vitro* (Humes and Cieslinski, 1992; Humes *et al.*, 1996) and can be used as the key cellular element of a tissue-engineered renal tubule device.

The bioartificial renal tubule can readily be conceived as a combination of living cells supported by polymeric substrata, using epithelial progenitor cells cultured on water- and solute-permeable membranes seeded with various biometric materials so that expression of differentiated vectorial transport and metabolic and endocrine functions is attained (Fig. 53.2). With appropriate membranes and biomatrices, immunoprotection of cultured progenitor cells has been achieved concurrent with long-term functional performance as long as conditions support tubule cell viability. This bioartificial tubule has been shown to transport salt and water effectively along osmotic and oncotic gradients (MacKay *et al.*, 1998).

A bioartificial proximal tubule satisfies a major requirement of reabsorbing a large volume of filtrate to maintain salt and water balance within the body. The need for additional tubule segments to replace other nephronal functions, such as the loop of Henle, to perform more refined homeostatic elements of the kidney, including urine concentration or dilution, may not be necessary. Patients with moderate renal insufficiency lose the ability to regulate salt and water homeostasis finely, because they are unable to concentrate or dilute, yet are able to maintain reasonable fluid and electrolyte homeostasis due to redundant physiologic compensation via other mechanisms. Thus, a bioartificial proximal tubule, which reabsorbs isoosmotically the majority of the filtrate, may be sufficient to replace required tubular function to sustain fluid electrolyte balance in a patient with ESRD.

### Implantable Tissue-Engineered Kidney

The development of a bioartificial filtration device and a bioartificial tubule processing unit would lead to the possibility of an implantable bioartificial kidney consisting of a filtration device followed in series by the tubule unit, into which filtrate flows directly. This tubule unit should maintain viability, because metabolic substrates and low-molecular-weight growth factors are delivered to the tubule cells from the ultrafiltration unit (Nikolovski *et al.*, 1999). Furthermore, immunoprotection of the cells grown within the hollow fiber is achievable due to the impenetrability of immunologically competent cells through the hollow fiber (O'Neil *et al.*, 1997). Rejection of transplanted cells will, therefore, not occur. This arrangement thereby allows the filtrate to enter the hollow-fiber network internal compartments, which are lined with confluent monolayers of renal tubule cells for regulated transport function.

This device could be either used extracorporeally or implanted within a patient. The specific implant site for a bioartificial kidney will depend on the final configuration of both bioartificial filtration and tubule devices. As presently conceived, the endothelial-lined bioartificial filtration hollow fibers can be placed into an arteriovenous circuit using the common iliac artery and vein, similar to the surgical connection for a renal transplant. The filtrate is connected in series to a bioartificial proximal tubule, with the reabsorbate transported back into the systematic circulation. The processed filtrate exiting the tubule unit is then connected via tubing to the proximate ureter for drainage and urine excretion via the recipient's own urinary collecting system.

Although the ultimate goal of this approach is to construct a fully implantable bioartificial kidney with a hemofilter and tubular system, the pathway to achieve this goal will most likely occur via a staged developmental strategy, with each intermediate device providing both clinical therapeutic benefit and substantial experience in the use of more elementary tissue-engineered devices (Humes *et al.*, 1997). In this regard, the initial component of the bioartificial kidney to be developed for clinical evaluation will be an extracorporeal renal tubule–assist device (RAD) to optimize current hemofiltration approaches to treat the clinical disorder of acute renal failure (ARF).

### Development of a Renal Tubule–Assist Device

Replacement of the multivariate tubular functions of the kidney cannot be achieved with inanimate membrane

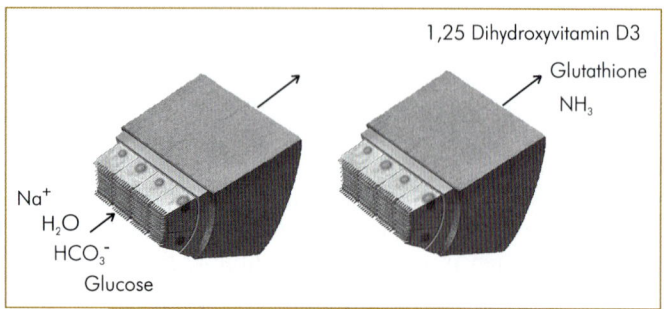

**FIG. 53.2.** Schematic of a tissue-engineered renal tubule. Renal epithelial cells form a confluent monolayer along the inner surface of a polysulfone hollow fiber with preadhered matrix molecules.

devices, as has been accomplished with the renal ultrafiltration process, but requires the use of the naturally evolved biologic membranes of the renal tubular epithelium. In this regard, the tissue engineering of a bioartificial renal tubule as a cell therapy device to replace these missing functions can be conceived as a combination of living cells supported on synthetic scaffolds (MacKay *et al.*, 1998; Nikolovski *et al.*, 1999). A bioartificial tubule can be constructed utilizing renal tubule progenitor cells (Humes and Cieslinski, 1992; Humes *et al.*, 1996), cultured on semipermeable hollow-fiber membranes on which extracellular matrix has been layered to enhance the attachment and growth of the epithelial cells (Timpl *et al.*, 1979). These hollow-fiber synthetic membranes not only provide the architectural scaffold for these cells but also provide immunoprotection, as has been observed in the long-term implantation of the bioartificial pancreas in a xenogeneic host (O'Neil *et al.*, 1997). The successful tissue engineering of a bioartificial tubule as a confluent monolayer has been achieved in a single hollow-fiber bioreactor system (Fig. 53.3) (MacKay *et al.*, 1998). The scale-up from a single-fiber system to a multifiber bioreactor and porcine renal proximal tubule cells has also been successfully demonstrated (Humes *et al.*, 1999b). Porcine cells were used initially because the pig is currently considered the best source of organs for both human xenotransplantation and immunoisolation cell therapy due to its anatomic and physiologic similarities to human tissue and the relative ease of breeding pigs in large numbers (Cozzi and White, 1995). The risk of viral transmission across species, however, requires human cells for clinical use.

The scale-up from a single hollow-fiber device to a multifiber bioartificial RAD device has proceeded with porcine renal proximal tubule cells grown as confluent monolayers along the inner surface of polysulfone immunoisolating hollow fibers (Humes *et al.*, 1999b). These hollow-fiber cartridges are packaged in bioreactor cartridges with membrane surface areas as large as 0.7 m², resulting in a device containing up to $2.5 \times 10^9$ cells. *In vitro* studies of these RADs have demonstrated their retention of differentiated active vectorial transport of sodium, bicarbonate, glucose, and organic anions (Humes *et al.*, 1999b). These transport properties were suppressed with specific transport inhibitors. Ammoniagenesis and glutathione metabolism, which are important differentiated metabolic processes of the kidney, were also demonstrated in these devices. Synthesis of 1,25-dihydroxyvitamin $D_3$ as a key endocrinologic metabolite was also documented. These metabolic processes were also shown to be regulated by important physiologic parameters and achieved rates comparable to those of a whole kidney.

The critical next step for clinical application of a renal tubule cell therapy device was to ascertain whether the RAD maintained differentiated renal functional performance, similar to that observed *in vitro*, and viability in an extracorporeal hemoperfusion circuit in an acutely uremic dog. The RAD was accordingly placed in series to a standard hollow-

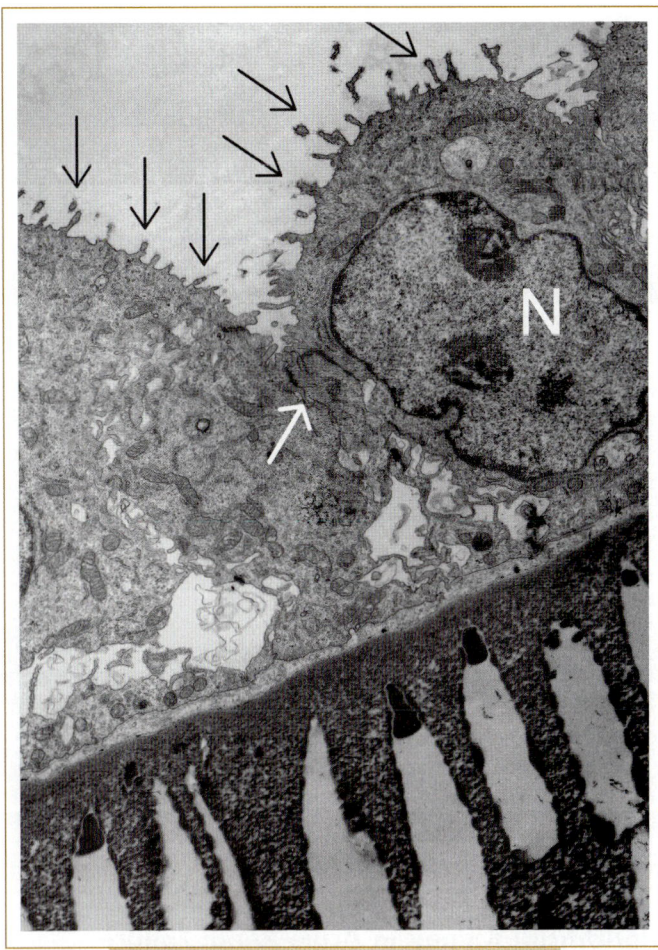

**FIG. 53.3.** Electron micrograph of a tissue-engineered bioartificial renal tubule. The nucleus (N) is indicated; black arrows delineate apical microvilli (a differentiated morphologic characteristic of proximal tubule cells), and the white arrow identifies the tight junctional complex of a transporting epithelium.

fiber hemofiltration cartridge; the two cartridges had ultrafiltrate and postfiltered blood connections to duplicate the structural anatomy of the nephron and to mimic the functional relationship between the renal glomerulus and tubule (see Fig. 53.4). The successful completion of this step has been reported (Humes *et al.*, 1999a). Data show that fluid and electrolyte balances in the animals, as reflected by plasma parameters, were adequately controlled with the bioartificial kidney. In fact, plasma potassium and blood urea nitrogen levels were more easily controlled during RAD treatment as compared with sham control conditions. The step-up of oncotic pressure in the postfiltrate blood, which was delivered to the antiluminal space of the bioartificial proximal tubules, allowed the fractional reabsorption of sodium and water to achieve 40–50% of the ultrafiltrate volume, an amount similar to that seen in the nephron *in vivo*. Active transport of potassium, bicarbonate, and glucose by the RAD was demonstrated. Successful metabolic activity was also observed.

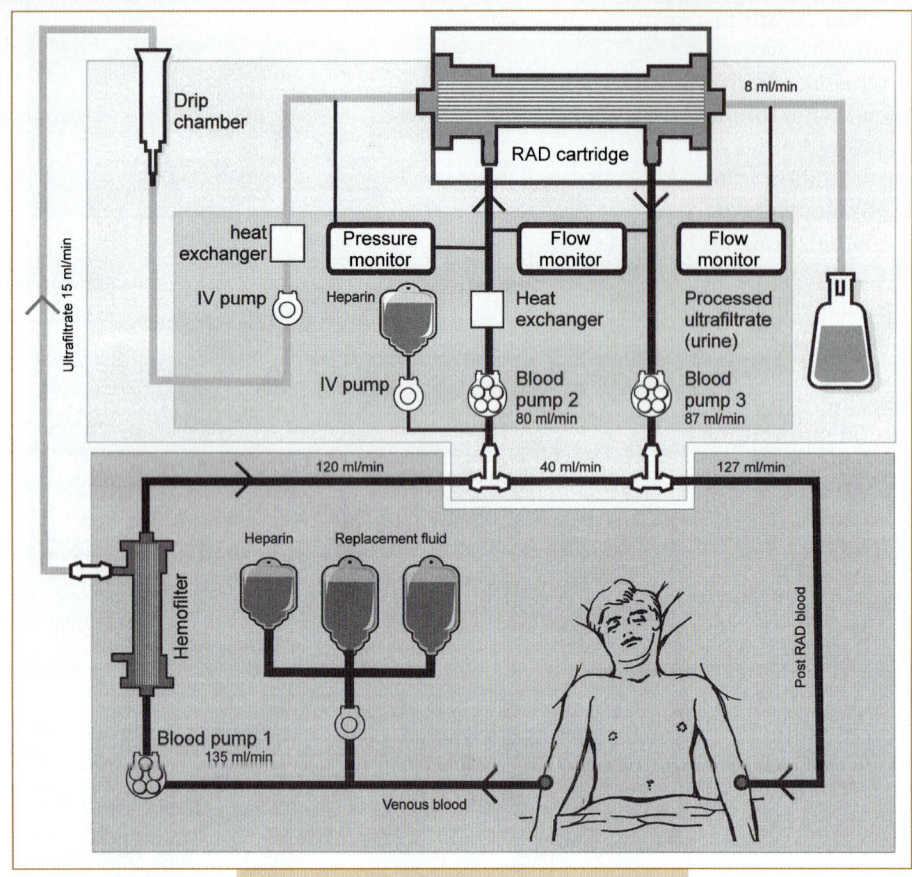

**FIG. 53.4.** The extracorporeal hemoperfusion circuit of a tissue-engineered kidney for the treatment of a patient with acute renal failure. The synthetic hemofilter is connected in series to the bioartificial renal tubule–assist device (RAD). Heat exchangers, flow and pressure monitors, and multiple pumps are required for optimal functioning of this extracorporeal device.

## Bioartificial Kidney in Acute Renal Failure

Current therapy for ischemic or toxic ARF or acute tubular necrosis (ATN) is predominantly supportive in nature. Uremia is treated with either intermittent hemodialysis or continuous hemofiltration, treatments that have had substantial impact on this disease process over the last 25 years. Patients with ATN, however, still have an exceedingly high mortality rate of greater than 50%, especially in sepsis (Schrier and Wang, 2004), despite maintenance of normal electrolyte balance and improvement in the uremic state. The high mortality is caused by the propensity of these patients to develop systemic inflammatory response syndrome followed by multiorgan dysfunction syndrome (Breen and Bihari, 1998). The sequential failure of organ systems apparently unrelated to the site of the initial insult has been correlated with altered plasma cytokine levels observed in sepsis (Donnelly and Robertson, 1994; Pinsky *et al.*, 1993; Marty *et al.*, 1994).

Because ARF secondary to ischemic or nephrotoxic insults arises from ATN, predominantly to renal proximal tubule cells, replacement of the functions of these cells during the episode of ATN and in conjunction with hemofiltration provides almost full renal replacement therapy. The addition of metabolic activity, such as ammoniagenesis and glutathione reclamation, endocrine activity, such as activation of vitamin $D_3$ (low levels of which seem to

correlate with high mortality rates in hospitalized patients; Thomas *et al.*, 1998), immunoregulatory support, and cytokine homeostasis may provide additional physiologic replacement activities to change the current natural history of this disease process (Humes, 2000).

Recent experiments investigated whether treatment with the bioartificial kidney would alter the course of ARF and ATN with sepsis in animal models. Mongrel dogs that underwent surgical nephrectomy–induced ARF were treated with continuous venovenous hemofiltration (CVVH) and either a RAD containing cells or an identically prepared sham noncell cartridge. After four hours of therapy, 2 mg/kg intravenous endotoxin was infused over 1 h to simulate gram-negative septic shock. Mean peak levels of an antiinflammatory cytokine, IL-10, and mean arterial pressures were found to be significantly higher in cell-treated animals (Fissell *et al.*, 2002).

To further assess the effect of the bioartificial kidney in ARF with bacterial sepsis, dogs were nephrectomized and 48 hours later administered intraperitoneally with $3 \times 10^{11}$ *Escherichia coli* cells per kilogram body weight (Fissell *et al.*, 2003). Immediately after bacteria administration, animals were placed in a CVVH circuit with either a RAD with cells or a sham RAD without cells. Cell RAD treatment maintained better cardiovascular performance, as determined by mean arterial blood pressure and cardiac output, for longer periods than sham RAD therapy. All sham RAD animals expired within 2–10 hours after bacteria administration, whereas all cell RAD-treated animals survived greater than 10 hours. Plasma cytokine levels in the bacteremic animals were assessed. Levels of IL-10, an antiinflammatory cytokine, were significantly elevated in the cell RAD group as compared with the control group. A significant correlation between the rise in plasma IL-10 levels and the decline in mean arterial pressure was observed. The cell RAD maintained renal metabolic activity throughout the septic period (Fissell *et al.*, 2003).

In another study, pigs with normal kidney function were administered intraperitoneally with $3 \times 10^{11}$ *E. coli* cells per kilogram body weight. One hour later, animals were placed in a CVVH circuit containing either a RAD with cells or a sham RAD without cells. All animals developed ARF with anuria within 2–4 h after bacteria administration. Cell RAD treatment maintained better cardiovascular performance, as determined by cardiac output and renal blood flow, for longer periods than sham RAD therapy. Consistently, the cell RAD group survived longer than the controls ($10 \pm 2$ hours versus $5 \pm 1$ hour, respectively). RAD treatment was associated with significantly lower plasma circulating levels of IL-6, a proinflammatory cytokine, and interferon-γ. These data demonstrate that septic shock results in early ARF and that RAD treatment in a bioartificial kidney circuit improves cardiovascular performance associated with changes in cytokine profiles and confers a significant survival advantage (Humes *et al.*, 2003).

## Phase I and II Clinical Experience with a Human Renal Tubule–Assist Device

With these encouraging preclinical data, the Food and Drug Administration approved an Investigational New Drug application to study the RAD containing human cells in patients with ATN receiving CVVH. Human kidney cells were isolated from kidneys donated for cadaveric transplantation but found unsuitable for such purpose because of anatomic or fibrotic defects. The initial results in the first 10 treated patients in this Phase I/II trial demonstrated that this experimental treatment can be delivered safely under study protocol guidelines in this critically ill patient population for up to 24 hours when used in conjunction with CVVH (Humes *et al.*, 2004). These data also indicate that the RAD maintains and exhibits viability, durability, and functionality in this *ex vivo* clinical setting. Cardiovascular stability of the patients was maintained, and increased native renal function, as determined by elevated urine outputs, temporally correlated with RAD treatment, a finding that requires additional study. The isolated and expanded human cells also demonstrated differentiated metabolic and endocrinologic activity in this *ex vivo* treatment. Glutathione degradation and endocrinologic conversion of 25-OH-$D_3$ to 1,25-(OH)$_2$-$D_3$ by the RAD tubule cells were demonstrated.

All 10 patients were critically ill with ARF and multiorgan dysfunction syndrome, with predicted hospital mortality rates between 80% and 95%. One patient expired within 12 hours after RAD treatment because of his family's request to withdraw ventilatory life support. Another patient expired after a surgical catastrophe (toxic megacolon) required discontinuation of RAD treatment after only 12 hours. Of the remaining eight patients, six survived past 28 days with renal function recovery. The other two patients died from nonrecoverable complications unrelated to RAD therapy and ARF, including fungal pericarditis and vancomycin-resistant enterococcus septicemia in one patient and ischemic colitis with bowel perforations in the other patient. Plasma cytokine levels suggest that RAD therapy produces dynamic and individualized responses in patients, depending on their unique pathophysiologic conditions. For the subset of patients who had excessive proinflammatory levels, RAD treatment resulted in significant declines in granulocyte-colony stimulating factor (G-CSF), IL-6, IL-10, and especially IL-6/IL-10 ratios, suggesting a greater decline in IL-6 relative to IL-10 levels and a less proinflammatory state. These results were encouraging and led to an FDA-approved, randomized, controlled, open-label Phase II investigation, which was recently completed at 10 clinical sites to assess the safety and early efficacy of this cell therapy approach. The results have been equally as compelling as the Phase I/II trial results. Renal cell therapy improved the 28-day mortality rate from 61% in the conventional CVVH-treated control group to 33% in the RAD-treated group (Tumlin *et al.*, 2005).

## Bioartificial Kidney in End-Stage Renal Disease

A bioartificial kidney for long-term use in ESRD, similar to short-term use in ARF, would integrate tubular cell therapy and the filtration function of a hemofilter. Although all ESRD patients could conceivably benefit from a bioartificial kidney, patients in the inflammatory stage of this disease would likely benefit most and will be the target population for clinical study in the near future. For this patient population, however, there are obvious limitations in using an extracorporeal RAD connected to a hemofiltration circuit. Whereas cell therapy, as in ARF, currently can be administered only intermittently, a bioartificial kidney suitable for long-term use in ESRD patients ideally would be capable of performing continuously, like the native kidney, to reduce risks from fluctuations in volume status, electrolytes, and solute concentrations and to maintain acid–base and uremic toxin regulation. These additional functions require the eventual design and manufacture of a compact implantable bioartificial kidney incorporating a RAD capable of self-monitoring and self-repair. The advent of silicon bulk and surface micromachining offers hope for the development of such a device. These technologies allow the fabrication of pores, beams, gears, and pressure sensors and the patterned deposition of cells in engineered microenvironments and nanoenvironments. Surface micromachining has been used to provide capillary-like conduits for blood flow in hepatocyte tissue cultures, and micropatterning has been used to control cell growth and differentiation (Kaihara *et al.*, 2000; Chen *et al.*, 1998). Clinicians look forward to the possible use of these emerging technologies in the manufacture of a bioartificial kidney for universal, optimal renal replacement therapy for ESRD patients.

## Cell Sourcing

A key critical hurdle for any bioartificial organ program, including kidney replacement, is identifying a robust cell source for the replacement device. In this regard, a specific cell-based therapy requires several important methodologic choices and the solution of a number of technological problems. For cell-based therapies, cells need to be expanded in large quantities while maintaining uniformity in activity and being pathogen free. Current approaches to ensure robust cell expansion and uniformity requirements are dependent on either stem/progenitor cells or transformed cells. The use of human embryonic stem (ES) cells versus adult stem cells is under rigorous societal debate, with the current political environment strongly favoring adult stem cell processes (Brower, 1999; Wertz, 2002). The plasticity of adult stem cells to transdifferentiate from one lineage pathway to another is also under careful scientific scrutiny. The early support for stem cell plasticity appears to be questioned by recent reports demonstrating stem cell fusion with tissue-specific differentiated cells, resulting in polyploidy rather than true stem cell transdifferentiation with normal diploid chromosomal numbers (Terada *et al.*, 2002; Wang *et al.*, 2003; Vassilopoulos *et al.*, 2003). The ability of bone marrow stem cells to differentiate into a variety of cell types within the kidney, including glomerular vascular and tubular elements, has been demonstrated (Brodie and Humes, 2005). These reports, however, demonstrate highly variable engraftment rates and inconsistent phenotypic differentiation. The issue of cell fusion in these experiments has not been addressed. Current cell-based approaches, therefore, are directed toward utilizing adult tissue-specific stem cell expansion, but the potential use of ES cells is also being aggressively pursued.

The utilization of transformed cells, including applications to deliver a gene product with gene therapy, has recently come under intense scrutiny due to safety concerns. The autologous transplantation of genetically modified hematopoietic stem cells in children with adenosine deaminase deficiency, which leads to severe immunodeficiency, resulted in the development of acute leukemia in some of the patients due to genetic integration of the vector in the hematopoietic stem cells (Hacein-Bey-Abina *et al.*, 2003). The ability to retrieve or deactivate these transformed cells following cell implantation is required to mitigate this high risk. Even the use of nontransformed cells may have safety concerns. Implantation of nerve cells in patients with Parkinson's disease leads to a high rate of severe and uncontrollable dyskinetic activity (Dunnett *et al.*, 2001); implantation of myoblasts into the heart has resulted in high rates of cardiac arrhythmias (Menasché *et al.*, 2003).

A choice between autologous and nonautologous human cells is also critical in the formulation of a cell-based application. Nonautologous cells must overcome natural host immunologic rejection processes. Since most indications preclude the use of immunosuppressant drugs to accommodate the discordant cell implant, immunoprotection of nonautologous cells has been approached with microencapsulation techniques using ultrathin synthetic membranes to prevent entry of antibodies and immunocompetent cells of the host. Implantation of cellular microcapsules has had limited success because of poor long-term functional performance secondary to progressive loss of cell viability (Orive *et al.*, 2003). Success with short-term cell therapy utilizing hollow-fiber bioreactors in an extracorporeal blood perfusion circuit for organ replacement therapy in acute disorders, including ATN, has been more forthcoming (Humes *et al.*, 1999a; Fissell *et al.*, 2003). The use of autologous cells, although overcoming the immunologic barrier, has its own set of problems. Autologous approaches require obtaining the patient's own cells, expanding them *in vitro* in large quantities over several weeks, and then reintroducing the cells in a site-specific manner. Thus, each treatment is an individualized and nonscalable process with substantial logistical and regulatory hurdles, including maintenance of the uniform quality of cells, avoidance of introduced pathogens during cell processing, and potential retrievability after implantation.

A final technologic hurdle for cell-based therapies is the maintenance of cell viability during long-term implantation. Maintenance of cell function depends on adequate nutrient and oxygen delivery to the cellular implant (Avgoustiniatos and Colton, 1997). Creative approaches to induce and maintain formation of a neovascular capillary bed in and around the cell implant with various drug-delivery and cell scaffold formulations have been demonstrated experimentally but have not yet been successfully translated to the clinic (Richardson *et al.*, 2001). An intravascular cell encapsulation implant approach looks promising but requires further experimentation to determine its successful implementation (Humes, 1998).

Although cell-based therapy has substantial technological, regulatory, and ethical barriers, the potential to develop innovative treatments for a large number of clinical disorders, including acute and chronic renal diseases, is expanding rapidly. Progress in this field is highly dependent on an interdisciplinary approach at the interface of a number of scientific disciplines. This approach is at times empiric rather than reductive in nature, without full understanding of the manner in which cells may alter the complex pathophysiology of a systemic disorder. This limitation should not preclude continued efforts in this field. In fact, this empiric approach may result in unanticipated insights into basic biology, similar to those obtained in immunology as the field of human solid organ transplantation evolved. Cell therapy has the potential creatively to leverage nature's ability to provide new and much needed treatments to patients with acute and chronic diseases.

# IV. REFERENCES

Anderson, J. L., and Quinn, J. A. (1974). Restricted transport in small pores. A model for steric exclusion and hindered particle motion. *Biophys. J.* **14**, 130.

Andreoli, T. E., and Schafer, J. A. (1978). Volume absorption in the pars recta: III. Luminal hypotonicity as a driving force for isotonic volume absorption. *Am J. Physiol.* **234**(4), F349.

Asahara, T., Murohara, T., Sullivan, A., Silver, M., van der Zee, R., Li, T., Witzenbichler, B., Schatteman, G., and Isner, J. M. (1997). Isolation of putative progenitor endothelial cells for angiogenesis. *Science* **275**, 964–967.

Avgoustiniatos, E. S., and Colton, C. K. (1997). Effect of external oxygen mass transfer resistances on viability of immunoisolated tissue. *Ann. N.Y. Acad. Sci.* **831**, 145–167.

Breen, D., and Bihari, D. (1998). Acute renal failure as a part of multiple organ failure: the slippery slope of critical illness. *Kidney Int.* **66**(Suppl.), S25–S33.

Brenner, B. M., and Humes, H. D. (1977). Mechanisms of glomerular ultra-filtration. *N. Engl. J. Med.* **297**, 148.

Brenner, B. M., Hostetter, T. H., and Humes, H. D. (1978). Molecular basis of proteinuria of glomerular origin. *N. Engl. J. Med.* **298**, 826.

Brodie, J. C., and Humes, H. D. (2005). Stem cell approaches for the treatment of renal failure. *Pharmacol. Rev.* **57**, 299–313.

Brower, V. (1999). Human ES cells: can you build a business around them? *Nat. Biotechnol.* **17**, 139–142.

Carey, D. J. (1991). Control of growth and differentiation of vascular cells by extracellular matrix proteins. *Annu. Rev. Physiol.* **53**, 161.

Carley, W. W., Milica, A. J., and Madri, J. A. (1988). Extracellular matrix specificity for the differentiation of capillary endothelial cells. *Exp. Cell Res.* **178**, 426.

Chang, R. L. S., Robertson, C. R., Deen, W. M., *et al.* (1975). Permselectivity of the glomerular capillary wall to macromolecules: I. Theoretical considerations. *Biophys. J.* **15**, 861.

Chen, C. S., Mrksich, M., Huang, S., *et al.* (1998). Micropatterned surfaces for control of cell shape, position, and function. *Biotechnol. Prog.* **14**, 356–363.

Colton, C. K., Henderson, L. W., Ford, C. A., *et al.* (1975). Kinetics of hemodiafiltration. *In vitro* transport characteristics of a hollow-fiber blood ultrafilter. *J. Lab. Clin. Med.* **85**, 855.

Cozzi, E., and White, D. (1995). The generation of transgenic pigs as potential organ donors for humans. *Nat. Med.* **1**, 965–966.

Donnelly, S. C., and Robertson, C. (1994). Mediators, mechanisms and mortality in major trauma. *Resuscitation* **28**, 87–92.

Dunnett, S. B., Bjorklund, A., and Lindvall, O. (2001). Cell therapy in Parkinson's disease — Stop or go? *Nat. Rev. Neurosci.* **2**, 365–369.

Fissell, W. H., Dyke, D. B., Weitzel, W. F., *et al.* (2002). Bioartificial kidney alters cytokine response and hemodynamics in endotoxin-challenged uremic animals. *Blood Purif.* **20**, 55–60.

Fissell, W. H., Lou, L., Abrishami, S., Buffington, D. A., and Humes, H. D. (2003). Bioartificial kidney ameliorates gram-negative bacteria-induced septic shock in uremic animals. *J. Am. Soc. Nephrol.* **14**, 454–461.

Folkman, J., and Shing, Y. (1992). Angiogenesis. *J. Biol. Chem.* **267**(16), 10931–10934.

Gage, F. H. (1998). Cell therapy. *Nature (London)* **392**, 518–524.

Golper, T. A. (1986). Continuous arteriovenous hemofiltration in acute renal failure. *Am J. Kidney Dis.* **6**, 373.

Hacein-Bey-Abina, S., von Kalle, C., Schmidt, M., Le Deist, F., Wulffraat, N., McIntyre, E., Radford, I., Villeval, J. L., Fraser, C. C., Cavazzana-Calvo, M., and Fischer, A. (2003). A serious adverse event after successful gene therapy for X-linked severe combined immunodeficiency. *N. Engl. J. Med.* **348**, 255–256.

Henderson, L. W., Colton, C. K., and Ford, C. A. (1975). Kinetics of hemodiafiltration: II. Clinical characterization of a new blood cleansing modality. *J. Lab. Clin. Med.* **85**, 372.

Humes, H. D. (1998). Implantable device and use thereof. U.S. Patent 5,704,910, January 6, 1998.

Humes, H. D. (2000). Bioartificial kidney for full renal replacement therapy. *Semin. Nephol.* **20**, 71–82.

Humes, H. D., and Cieslinski, D. A. (1992). Interaction between growth factors and retinoic acid in the induction of kidney tubulogenesis in tissue culture. *Exp. Cell Res.* **201**, 8–15.

Humes, H. D., Krauss, J. C., Cieslinski, D. A., and Funke, A. J. (1996). Tubulogenesis from isolated single cells of adult mammalian kidney: clonal analysis with a recombinant retrovirus. *Am. J. Physiol.* **271**(40), F42–F49.

Humes, H. D., MacKay, S. M., Funke, A. J., and Buffington, D. A. (1997). The bioartificial renal tubule assist device to enhance CRRT in acute renal failure. *Am. J. Kidney Dis.* **30**, S28–S31.

Humes, H. D., Buffington, D. A., MacKay, S. M., Funke, A. J., and Weitzel, W. F. (1999a). Replacement of renal function in uremic animals with a tissue-engineered kidney. *Nat. Biotechnol.* **17**, 451–455.

Humes, H. D., MacKay, S. M., Funke, A. J., and Buffington, D. A. (1999b). Tissue engineering of a bioartificial renal tubule assist device: *in vitro* transport and metabolic characteristics. *Kidney Int.* **55**, 2502–2514.

Humes, H. D., Buffington, D. A., Lou, L., *et al.* (2003). Cell therapy with a tissue-engineered kidney reduces the multiple-organ consequences of septic shock. *Crit. Care Med.* **31**, 2421–2428.

Humes, H. D., Weitzel, W. F., Bartlett, R. H., *et al.* (2004). Initial clinical results of the bioartificial kidney containing human cells in ICU patients with acute renal failure. *Kidney Int.* **66**, 1578–1588.

Iglehart, J. K. (1993). The American health care system: the End-Stage Renal Disease Program. *N. Engl. J. Med.* **328**, 366.

Kadletz, M., Magometshnigg, H., Minar, E., *et al.* (1992). Implantation of *in vitro* endothelialized polytetrafluoroethylene grafts in human beings. *J. Thorac. Cardiovasc. Surg.* **104**, 736.

Kaihara, S., Borenstein, J., Koka, R., *et al.* (2000). Silicon micromachining to tissue engineer branched vascular channels for liver fabrication. *Tissue Eng.* **6**, 105–117.

Knox, F. G., Mertz, J. I., Burnett, J. C., *et al.* (1983). Role of hydrostatic and oncotic pressures in renal sodium reabsorption. *Circ. Res.* **52**, 491.

Kramer, P., Wigger, W., Rieger, J., *et al.* (1977). Arteriovenous hemofiltration: a new and simple method for treatment of overhydrated patients resistant to diuretics. *Klin. Wochenschr.* **55**, 1121.

Landis, E. M., and Pappenheimer, J. R. (1963). Exchange of substances through the capillary walls. *In* "Handbook of Physiology" (W. F. Hamilton and P. Dow, eds.) Circulation, Sec. 2, Vol. 2, p. 962. American Physiological Society, Washington, DC.

Landis, E. M., and Pappenheimer, J. R. (1965). Exchange of substances through the capillary walls. *In* "Handbook of Physiology" (W. F. Hamilton and P. Dow, eds.), Sec. 2, Vol. 2, p. 961. American Physiological Society, Washington, DC.

MacKay, S. M., Funke, A. J., Buffington, D. A., and Humes, H. D. (1998). Tissue engineering of a bioartificial renal tubule. *ASAIO J.* **44**, 179–183.

Marty, D., Misset, B., Tamion, F., *et al.* (1994). Circulating interleukin-8 concentrations in patients with multiple organ failure of septic and nonseptic origin. *Crit. Care Med.* **22**, 673–679.

Menasché, P., Hagège, A. A., Vilquin, J. T., Desnos, M., Abergel, E., Pouzet, B., Bel, A., Sarateanu, S., Scorsin, M., Schwartz, K., *et al.* (2003). Autologous skeletal myoblast transplantation for severe postinfarction left ventricular dysfunction. *J. Am. Coll. Cardiol.* **41**, 1078–1083.

Milici, A. J., Furie, M. B., and Carley, W. W. (1985). The formation of fenestrations and channels by capillary endothelium *in vitro*. *Proc. Natl. Acad. Sci. U.S.A.* **82**, 6181.

Nikolovski, J., Gulari, E., and Humes, H. D. (1999). Design engineering of a bioartificial renal tubule cell therapy device. *Cell Transplant.* **8**(4), 351–364.

O'Neil, J. J., Stegemann, J. P., Nicholson, D. T., Mullon C. J.-P., Maki, T., Monaco, A. P., and Solomon, B. A. (1997). Immunoprotection provided by the bioartificial pancreas in a xenogeneic host. *Transplant. Proc.* **29**, 2116–2117.

Orive, G., Hernández, R. M., Gascon, A. R., Calafiore, R., Chang, T. M., De Vos, P., Hortelano, G., Hunkeler, D., Lacík, I., Shapiro, A. M., *et al.* (2003). Cell encapsulation: promise and progress. *Nat. Med.* **9**, 104–107.

Pinsky, M. R., Vincent, J. L., Deviere, J., *et al.* (1993). Serum cytokine levels in human septic shock: relation to multiple-system organ failure and mortality. *Chest* **103**, 565–575.

Richardson, T. P., Peters, M. C., Ennett, A. B., and Mooney, D. J. (2001). Polymeric system for dual growth factor delivery. *Nat. Biotechnol.* **19**, 1029–1034.

Schnider, P. A., Hanson, S. R., Price, T. M., *et al.* (1988). Durability of confluent endothelial cell monolayers of small-caliber vascular prostheses *in vitro*. *Surgery* **103**, 456.

Schrier, R. W., and Wang, W. (2004). Acute renal failure and sepsis. *N. Engl. J. Med.* **351**, 159–169.

Shepard, A. D., Eldrup-Jorgensen, J., Keough, E. M., *et al.* (1986). Endothelial cell seeding of small-caliber synthetic grafts in the baboon. *Surgery* **99**, 318.

Shin, M., Matsuda, K., Ishii, O., *et al.* (2004). Endothelialized networks with a vascular geometry in microfabricated poly(dimethyl siloxane). *Biomed. Microdev.* **6**, 269–278.

Terada, N., Hamazaki, T., Oka, M., Hoki, M., Mastalerz, D. M., Nakano, Y., Meyer, E. M., Morel, L., Petersen, B. E., and Scott, E. W. (2002). Bone marrow cells adopt the phenotype of other cells by spontaneous cell fusion. *Nature* **416**, 542–545.

Thomas, M. K., Lloyd-Jones, D. M., Thadhani, R. I., *et al.* (1998). Hypovitaminosis D in medical inpatients. *N. Engl. J. Med.* **338**, 777–783.

Thompson, J. A., Anderson, K. D., Dipietro, J. M., *et al.* (1988). Site-directed neovessel formation *in vivo*. *Science* **241**, 1349–1352.

Timpl, R., Rhode, H., Robey, P. G., Rennard, S. I., Foidart, J. M., and Martin, G. M. (1979). Laminin-A glycoprotein from basement membranes. *J. Biol. Chem.* **254**, 9933–9937.

Tiranathanagul, K., Dhawan, V., Lou, L., Borschel, G., Williams, J., Tziampazis, E., Buffington, D., Brown, D., and Humes, H. D. (2005). Tissue engineering of an implantable bioartificial hemofilter. *J. Am. Soc. Nephrol.* **16**(Abstracts), 38A.

Tumlin, J., Wali, R., Brennan, K., and Humes, H. D. (2005). Effect of the renal assist device (RAD) on mortality of dialysis-dependent acute renal failure: a randomized, open-labeled, mulitcenter, phase II trial. *J. Am. Soc. Nephrol.* **16**(Abstracts), 46A.

U.S. Renal Data System. (2004). USRDS 2004 Annual Data Report: Atlas of End-Stage Renal Disease in the United States. National Institutes of Health, National Institute of Diabetes and Digestive and Kidney Diseases, Bethesda, MD.

Vassilopoulos, G., Wang, P. R., and Russell, D. W. (2003). Transplanted bone marrow regenerates liver by cell fusion. *Nature* **422**, 901–904.

# Chapter Fifty-Four

# Genitourinary System

*Anthony Atala*

## I. INTRODUCTION

There is a severe shortage of donor tissues and organs that is worsening yearly, given the aging population. Currently, patients suffering from diseased and injured organs are treated with transplanted organs or cells. Tissue-engineering techniques are being investigated for the replacement of absent or deficient genitourinary structures, including urethra, bladder, male and female genital tissues, ureter, and renal structures. The goal of tissue engineering is to develop biologic substitutes that can restore and maintain normal function.

This chapter reviews recent advances in regenerative medicine and describes applications of new technologies to treat the genitourinary system, including reconstruction of the patient's organs with autologous cells. This approach allows lost tissue function to be restored or replaced with limited complications due to rejection or the inherently different functional parameters of replacement tissues currently used. Most of the effort expended to engineer genitourinary tissues has occurred within the last decade. Before these engineering techniques can be applied to humans, further studies need to be performed in many of the tissues described in this chapter.

The genitourinary system may be exposed to a variety of possible injuries from the time the fetus develops. Individuals may suffer from congenital disorders, cancer, trauma, infection, inflammation, iatrogenic injuries, or other conditions that may lead to genitourinary organ damage or loss and necessitate eventual reconstruction. Whenever there is a lack of native urologic tissue, reconstruction may be performed with native nonurologic tissues (skin, gastrointestinal segments, or mucosa from multiple body sites), homologous tissues (cadaver fascia, cadaver or donor kidney), heterologous tissues (bovine collagen), or artificial materials (silicone, polyurethane, Teflon). The tissues used for reconstruction may lead to complications because of their inherently different functional parameters. In most cases, the replacement of lost or deficient tissues with functionally equivalent tissues would improve the outcome for these patients. This goal may be attainable with the use of tissue-engineering techniques.

## II. RECONSTITUTION STRATEGIES

The goal of tissue engineering is to develop biologic substitutes that can restore and maintain normal function. Tissue engineering may involve matrices alone, wherein the body's natural ability to regenerate is used to orient or direct new tissue growth, or it may use matrices with cells. When cells are used for tissue engineering, donor tissue (heterologous, allogeneic, or autologous) is dissociated into individual cells, which are implanted directly into the host or expanded in culture, attached to a support matrix, and reimplanted after expansion. Ideally, this approach allows lost tissue function to be restored or replaced in toto and

*Principles of Tissue Engineering, 3rd Edition*
ed. by Lanza, Langer, and Vacanti

with limited complications (Atala *et al.*, 1993a, 1993b, 1994; Atala, 1997, 1999; Cilento *et al.*, 1994; Yoo and Atala, 1997; Fauza *et al.*, 1998; Machluf and Atala, 1998; Yoo *et al.*, 1998; Amiel and Atala, 1999; Kershen and Atala, 1999; Oberpenning *et al.*, 1999; Park *et al.*, 1999).

## III. ROLE OF BIOMATERIALS

Biomaterials in genitourinary–tissue engineering function as an artificial extracellular matrix (ECM) and elicit biologic and mechanical functions of native ECM found in body tissues. Biomaterials facilitate the localization and delivery of cells and/or bioactive factors (such as cell adhesion peptides and growth factors) to desired sites in the body; define a three-dimensional space for the formation of new tissues with appropriate structure; and guide the development of new tissues with appropriate function (Kim and Mooney, 1998). While direct injection of cell suspensions without biomaterial matrices has been used (Brittberg *et al.*, 1994), it is difficult to control the localization of transplanted cells.

The ideal biomaterial should be biocompatible, promote cellular interaction and tissue development, and possess proper mechanical and physical properties. Generally, three classes of biomaterials have been used for engineering of genitourinary tissues: naturally derived materials, such as collagen and alginate; acellular tissue matrices, such as bladder submucosa and small intestinal submucosa; and synthetic polymers, such as polyglycolic acid (PGA), polylactic acid (PLA), and poly(lactic-*co*-glycolic acid) (PLGA). While naturally derived materials and acellular tissue matrices have the potential advantage of biologic recognition, synthetic polymers can be produced reproducibly on a large scale with controlled properties of strength, degradation rate, and microstructure.

## IV. VASCULARIZATION

A restriction of tissue engineering is that cells cannot be implanted in volumes exceeding 3 mm$^3$ because of the limitations of nutrition and gas exchange (Folkman and Hochberg, 1973). To achieve the goals of engineering large, complex tissues and possibly internal organs, vascularization of the regenerating cells is essential. Three approaches have been used for vascularization of bioengineered tissue: (1) incorporation of angiogenic factors in the bioengineered tissue; (2) seeding EC with other cell types in the bioengineered tissue; and (3) prevascularization of the matrix prior to cell seeding. Angiogenic growth factors may be incorporated into the bioengineered tissue prior to implantation, in order to attract host capillaries and to enhance neovascularization of the implanted tissue. Many obstacles must be overcome before large entire tissue-engineered solid organs are produced. Recent developments in angiogenesis research may provide important knowledge and essential materials to accomplish this goal.

## V. PROGRESS IN TISSUE ENGINEERING OF UROLOGIC STRUCTURES

Tissue-engineering techniques are currently being investigated for the replacement of lost or deficient genitourinary structures, including urethra, bladder, male and female genital tissues, ureter, and renal structures.

### Urethra

Various strategies have been proposed to regenerate urethral tissue. Woven meshes of PGA, without cells, have been used to reconstruct urethras in dogs (Olsen *et al.*, 1992). PGA has been used as a cell transplantation vehicle to engineer tubular urothelium *in vivo*. Small intestinal submucosa (SIS) without cells was used as an onlay patch graft for urethroplasty in rabbits (Kropp *et al.*, 1998), and a homologous free graft of acellular urethral matrix was also used in a rabbit model (Sievert *et al.*, 2000).

Bladder-derived acellular collagen matrix has proven to be a suitable graft for repairing urethral defects in rabbits. The created neourethras demonstrated a normal urothelial luminal lining and organized muscle bundles (F. Chen *et al.*, 1999). Results were confirmed clinically in a series of patients with a history of failed hypospadias reconstruction whose urethral defects were repaired with human bladder acellular collagen matrices (Fig. 54.1) (Atala *et al.*, 1999). An advantage of this material over nongenital tissue grafts for urethroplasty is that it is "off the shelf," eliminating the need for additional surgical procedures for graft harvesting and decreasing operative time and potential morbidity from the harvest procedure.

The foregoing techniques, using nonseeded acellular matrices, were successfully applied experimentally and clinically for onlay urethral repairs. However, when tubularized repairs were attempted experimentally, adequate urethral tissue regeneration was not achieved, and complications, such as graft contracture and stricture formation, ensued (le Roux, 2005). Seeded tubularized collagen matrices have performed better than their nonseeded counterparts in animal studies. In a rabbit model, entire urethral segments were resected, and urethroplasties were performed with tubularized collagen matrices, either seeded or nonseeded. The tubularized collagen matrices seeded with autologous cells formed new tissue, which was histologically similar to native urethra. Those without cells led to poor tissue development, fibrosis, and stricture formation.

### Bladder

Gastrointestinal segments are commonly used as tissues for bladder replacement or repair. However, these tissues are designed to absorb specific solutes, and when they come in contact with the urinary tract, multiple complications may ensue, including infection, metabolic disturbances, urolithiasis, perforation, increased mucus production, and malignancy (McDougal, 1992). Because of these problems,

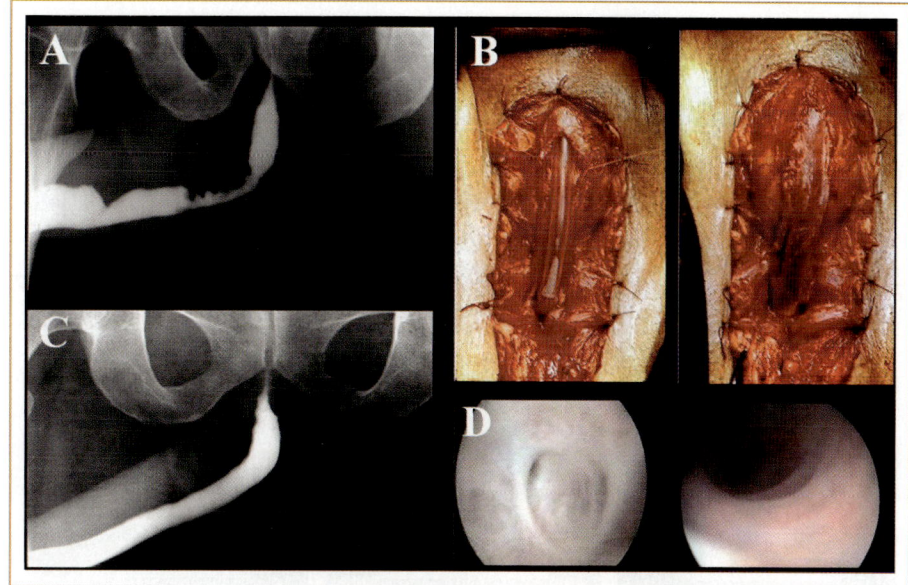

**FIG. 54.1.** Tissue-engineered urethra using a collagen matrix. **(A)** Representative case of a patient with a bulbar stricture. **(B)** Urethral repair. Strictured tissue is excised, preserving the urethral plate on the left side, and matrix is anastomosed to the urethral plate in an onlay fashion on the right. **(C)** Urethrogram six months after repair. **(D)** Cystoscopic view of urethra before surgery on the left side and four months after repair on the right side.

investigators have attempted alternative reconstructive procedures for bladder replacement or repair, such as the use of tissue expansion, seromuscular grafts, matrices for tissue regeneration, and tissue engineering with cell transplantation.

### Tissue Expansion

A system of progressive dilation for ureters and bladders has been proposed as a method of bladder augmentation but has not yet been attempted clinically (Lailas *et al.*, 1996; Satar *et al.*, 1999). Augmentation cystoplasty performed with the dilated ureteral segment in animals has resulted in increased bladder capacity ranging from 190% to 380% (Lailas *et al.*, 1996). A system progressively to expand native bladder tissue has also been used for augmenting bladder volumes in animals. Within 30 days after progressive dilation, neoreservoir volume was expanded at least 10-fold. Urodynamic studies showed normal compliance in all animals, and microscopic examination of the expanded neoreservoir tissue confirmed a normal histology. A series of immunocytochemical studies demonstrated that the dilated bladder tissue maintained normal phenotypic characteristics (Satar *et al.*, 1999).

### Seromuscular Grafts

Seromuscular grafts and deepithelialized bowel segments, either alone or over a native urothelium, have also been attempted (Blandy, 1961; Harada *et al.*, 1965). Keeping the urothelium intact avoids complications associated with the use of bowel in continuity with the urinary tract (Blandy, 1961; Harada *et al.*, 1965). An example of this strategy is to combine the techniques of autoaugmentation and enterocystoplasty. An autoaugmentation is performed, and the diverticulum is covered with a demucosalized gastric or intestinal segment.

### Matrices

Nonseeded allogeneic acellular bladder matrices have served as scaffolds for the ingrowth of host bladder wall components. The matrices are prepared by mechanically and chemically removing all cellular components from bladder tissue (Yoo *et al.*, 1998; Probst *et al.*, 1997). The matrices serve as vehicles for partial bladder regeneration, and relevant antigenicity is not evident. For example, SIS (a biodegradable, acellular, xenogeneic collagen-based tissue-matrix graft) was first used in the early 1980s as an acellular matrix for tissue replacement in the vascular field. It has been shown to promote regeneration of a variety of host tissues, including blood vessels and ligaments (Badylak *et al.*, 1989). Animal studies have shown that the nonseeded SIS matrix used for bladder augmentation can regenerate *in vivo* (Kropp *et al.*, 2004).

In multiple studies using various materials as nonseeded grafts for cystoplasty, the urothelial layer regenerated normally, but the muscle layer, although present, was not fully developed (Yoo *et al.*, 1998; Probst *et al.*, 1997). Often the grafts contracted to 60–70% of their original sizes (Portis *et al.*, 2000), with little increase in bladder capacity or compliance (Landman *et al.*, 2004).

Studies involving acellular matrices that may provide the necessary environment to promote cell migration, growth, and differentiation are being conducted. With continued research, these matrices may have a clinical role in bladder replacement in the future. Recently, bladder regeneration has been shown to be more reliable using SIS derived from the distal ileum (Kropp *et al.*, 2004).

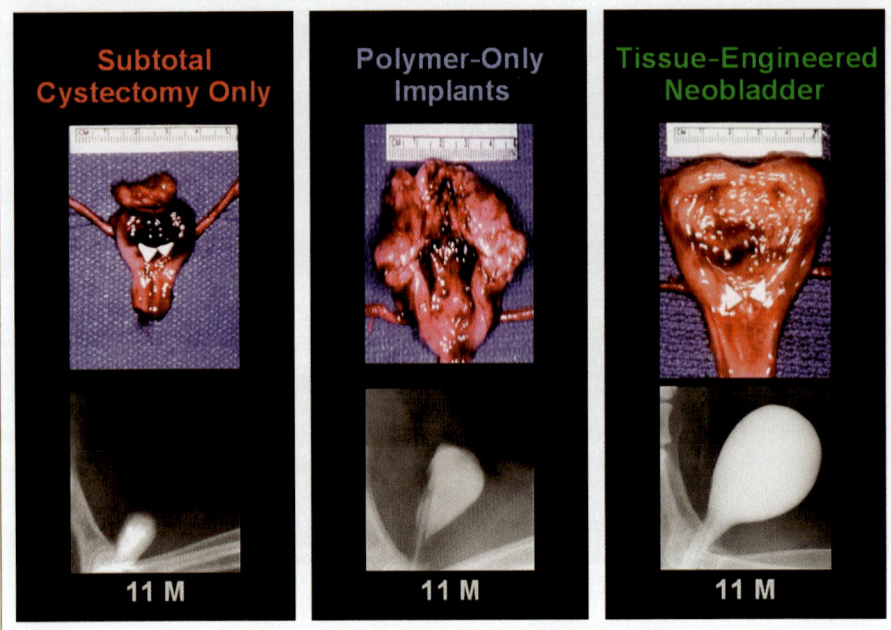

| Subtotal Cystectomy Only | Polymer-Only Implants | Tissue-Engineered Neobladder |
|---|---|---|
| 11 M | 11 M | 11 M |

**FIG. 54.2.** Comparison of tissue-engineered neobladders. Gross specimens and cystograms at 11 months of the cystectomy-only, nonseeded controls, and cell-seeded tissue-engineered bladder replacements. The cystectomy-only bladder had a capacity of 22% of the preoperative value and a decrease in bladder compliance to 10% of the preoperative value. The nonseeded controls showed significant scarring, with a capacity of 46% of the preoperative value and a decrease in bladder compliance to 42% of the preoperative value. An average bladder capacity of 95% of the original precystectomy volume was achieved in the cell-seeded tissue-engineered bladder replacements, and the compliance showed almost no difference from preoperative values, which were measured when the native bladder was present (106%).

## Cell Transplantation

Cell-seeded allogeneic acellular bladder matrices have been used for bladder augmentation in dogs. Trigone-sparing cystectomy was performed in dogs randomly assigned to one of three groups. One group underwent closure of the trigone without a reconstructive procedure; another underwent reconstruction with a nonseeded bladder-shaped biodegradable scaffold; and the last underwent reconstruction using a bladder-shaped biodegradable scaffold that delivered seeded autologous urothelial cells and smooth muscle cells (Oberpenning et al., 1999).

The cystectomy-only and nonseeded controls maintained average bladder capacities of 22% and 46% of preoperative values, respectively, compared with 95% in the cell-seeded tissue-engineered bladder replacements (Fig. 54.2). The subtotal cystectomy reservoirs that were not reconstructed and the polymer-only reconstructed bladders showed a marked decrease in bladder compliance (10% and 42% total compliance). The compliance of the cell-seeded tissue-engineered bladders showed almost no difference from preoperative values, which were measured when the native bladder was present (106%). Histologically, the nonseeded scaffold bladders presented a pattern of normal urothelial cells with a thickened fibrotic submucosa and a thin layer of muscle fibers (Fig. 54.2, middle parcel). The retrieved tissue-engineered bladders showed a normal cellular organization, consisting of a trilayer of urothelium, submucosa, and muscle (Fig. 54.2, right parcel) (Oberpenning et al., 1999). Preliminary clinical trials for the application of this technology have been performed and are under evaluation.

## Genital Tissues

Tissue-engineering techniques have been used to reconstruct male and female genital tissues.

### Corporal Smooth Muscle

Because one of the major components of the phallus is corporal smooth muscle, the creation of autologous functional and structural corporal tissue *de novo* would be beneficial. To examine functional parameters of engineered corpora, acellular corporal collagen matrices were obtained from donor rabbit penis, and autologous corpus cavernosal smooth muscle and endothelial cells were harvested, expanded, and seeded on the matrices. The entire rabbit corpora was removed and replaced with engineered scaffolds. The experimental corporal bodies demonstrated intact structural integrity by cavernosography and showed similar pressure by cavernosometry when compared with normal controls. The control rabbits (without cells) failed to show normal erectile function throughout the study. Mating activity in the animals with the engineered corpora appeared normal by one month after implantation. The presence of sperm was confirmed during mating and was present in all the rabbits with the engineered corpora. The female rabbits mated with the animals implanted with engineered corpora and also conceived and delivered healthy pups. Animals implanted with the matrix alone were unable to demonstrate normal mating activity and failed to ejaculate into the vagina (K. L. Chen et al., 2005).

### Engineered Penile Prostheses

Although silicone is an accepted biomaterial for penile prostheses, biocompatibility remains a concern (Thomalla *et al.*, 1987). The use of a natural prosthesis composed of autologous cells may be advantageous.

A recent study using an autologous system investigated the feasibility of applying the engineered cartilage rods *in situ* (Yoo *et al.*, 1999). Autologous chondrocytes harvested from rabbit ear were grown and expanded in culture. The cells were seeded onto biodegradable poly-L-lactic acid–coated PGA polymer rods and implanted into the corporal spaces of rabbits. Examination at retrieval one month later showed the presence of well-formed, milky-white cartilage structures within the corpora. All polymers were fully degraded by two months. There was no evidence of erosion or infection in any of the implantation sites.

Subsequent studies assessed the long-term functionality of the cartilage penile rods *in vivo* (Yoo *et al.*, 1999). To date, the animals have done well and can copulate and impregnate their female partners without problems.

## Female Genital Tissues

Congenital malformations of the uterus may have profound implications clinically. Patients with cloacal exstrophy or intersex disorders may not have sufficient uterine tissue for future reproduction.

We investigated the possibility of engineering functional uterine tissue using autologous cells (Wang *et al.*, 2003). Autologous rabbit uterine smooth muscle and epithelial cells were harvested and then grown and expanded in culture. These cells were seeded onto preconfigured uterine-shaped biodegradable polymer scaffolds, which were then used for subtotal uterine tissue replacement in the corresponding autologous animals. Upon retrieval six months after implantation, histological, immunocytochemical, and Western blot analyses confirmed the presence of normal uterine tissue components. Biomechanical analyses and organ bath studies showed that the functional characteristics of these tissues were similar to those of normal uterine tissue. Breeding studies using these engineered uteri are currently being performed.

Several pathologic conditions, including congenital malformations and malignancy, can adversely affect normal vaginal development or anatomy. Vaginal reconstruction has traditionally been challenging due to the paucity of available native tissue.

Vaginal epithelial and smooth muscle cells of female rabbits were harvested, grown, and expanded in culture. The cells were seeded onto biodegradable polymer scaffolds, which were then implanted into nude mice for up to six weeks. Immunocytochemical, histological, and Western blot analyses confirmed the presence of vaginal tissue phenotypes. Electrical field stimulation studies in the tissue-engineered constructs showed similar functional properties to those of normal vaginal tissue. When these constructs were used for autologous total vaginal replacement, patent vaginal structures were noted in the tissue-engineered specimens, while the nonseeded structures were noted to be stenotic.

## Ureter

Ureteral nonseeded matrices have been used as a scaffold for the ingrowth of ureteral tissue in rats. On implantation, the acellular matrices promoted the regeneration of the ureteral wall components (Dahms *et al.*, 1997). In a more recent study, nonseeded ureteral collagen acellular matrices were tabularized, but attempts to use them to replace 3-cm segments of canine ureters were unsuccessful (Osman *et al.*, 2004).

Cell-seeded biodegradable polymer scaffolds have been used with more success to reconstruct ureteral tissues. In one study, urothelial and smooth muscle cells isolated from bladders and expanded *in vitro* were seeded onto PGA scaffolds with tubular configurations and implanted subcutaneously into athymic mice. After implantation, the urothelial cells proliferated to form a multilayered luminal lining of tubular structures, while the smooth muscle cells organized into multilayered structures surrounding the urothelial cells. Abundant angiogenesis was evident. Polymer scaffold degradation resulted in the eventual formation of natural urothelial tissues. This approach has also been used to replace ureters in dogs (Yoo *et al.*, 1995).

## Renal Structures

Due to its complex structure and function, the kidney is possibly the most challenging organ in the genitourinary system to reconstruct using tissue-engineering techniques. However, concepts for a bioartificial kidney are emerging. Some investigators are pursuing the replacement of isolated kidney function parameters with the use of extracorporeal units, while others are working toward the replacement of total renal function by tissue-engineered bioartificial structures.

### Ex Vivo Renal Units

Although dialysis is currently the most prevalent form of renal replacement therapy, the relatively high rates of morbidity and mortality have spurred investigators to seek alternative solutions involving *ex vivo* systems.

To assess the viability and physiologic functionality of a cell-seeded device to replace the filtration, transport, metabolic, and endocrinologic functions of the kidney in acutely uremic dogs, researchers introduced a synthetic hemofiltration device combined with a renal tubular cell therapy device (containing porcine renal tubules in an extracorporeal perfusion circuit). Levels of potassium and blood urea nitrogen (BUN) were controlled during treatment with the device. The fractional reabsorption of sodium and water was possible, and active transport of potassium, bicarbonate, and glucose and a gradual ability to excrete ammonia were observed. These results demonstrated the technologic feasibility of an extracorporeal assist device that is

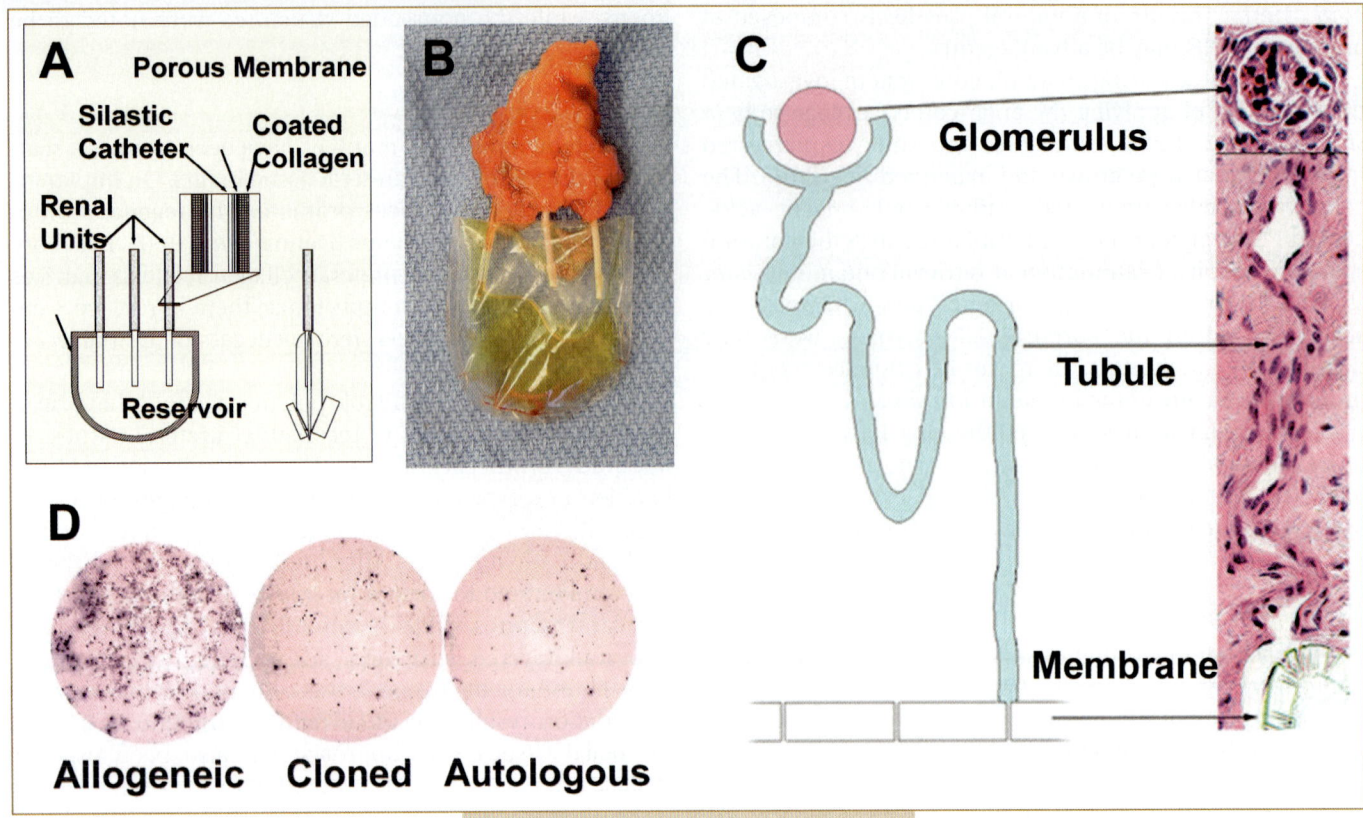

**FIG. 54.3.** Creation of kidney tissue from therapeutic cloning and tissue-engineering strategies. **(A)** Illustration of the tissue-engineered renal unit. **(B)** Renal unit seeded with cloned cells, three months after implantation, showing the accumulation of urinelike fluid. **(C)** There was a clear unidirectional continuity between the mature glomeruli, their tubules, and the polycarbonate membrane. **(D)** Elispot analyses of the frequencies of T-cells that secrete IFN-gamma after primary and secondary stimulation with allogeneic renal cells, cloned renal cells, or nuclear donor fibroblasts.

reinforced by the use of proximal tubular cells (Humes *et al.*, 1999).

Using similar techniques, a tissue-engineered bioartificial kidney — consisting of a conventional hemofiltration cartridge in series with a renal tubule–assist device containing human renal proximal tubule cells — was used in patients with acute renal failure in the intensive care unit. Initial clinical experience with the bioartificial kidney and the renal tubule–assist device suggests that such therapy may provide a dynamic and individualized treatment program as assessed by acute physiologic and biochemical indices (Humes *et al.*, 2003).

### In Vivo Renal Structures

Another method of improving renal function involves augmenting renal tissue with kidney cell expansion *in vitro* and subsequent autologous transplantation. The feasibility of achieving renal cell growth, expansion, and *in vivo* reconstitution with tissue-engineering techniques has been explored.

Most recently, an attempt was made to harness the reconstitution of renal epithelial cells to generate functional nephron units. Renal cells harvested and expanded in culture were seeded onto a tubular device constructed from a polycarbonate membrane connected at one end to a Silastic catheter terminating in a reservoir. The device was implanted into athymic mice. Histological examination of the implanted devices over time revealed extensive vascularization, with formation of glomeruli and highly organized tubulelike structures. Immunocytochemical staining confirmed the renal phenotype. Yellow fluid consistent with the makeup of dilute urine in its creatinine and uric acid concentrations was retrieved from inside the implant (Yoo *et al.*, 1996). Further studies using nuclear transfer techniques have been performed showing the formation of renal structures in cows (Fig. 54.3) (Lanza *et al.*, 2002). Challenges facing this technology include the expansion to larger, three-dimensional structures.

## VI. ADDITIONAL APPLICATIONS

Tissue engineering and cell therapy hold promise for a number of additional genitourinary applications.

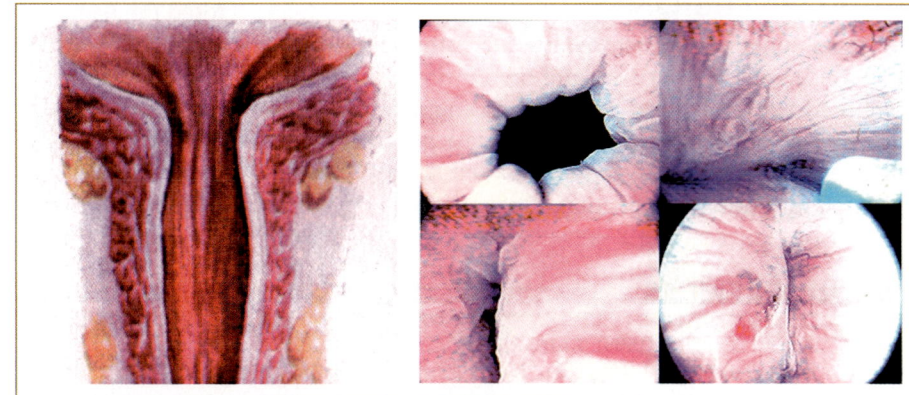

**FIG. 54.4.** Tissue-engineered bulking agent. Chondrocytes are harvested and combined with alginate *in vitro*, and the suspension is injected cystoscopically as a bulking agent to treat urinary incontinence.

## Fetal-Tissue Engineering

Improved prenatal diagnostic techniques have led to the use of intervention before birth to reverse potentially life-threatening processes. Several strategies may be pursued to facilitate the future prenatal management of urologic disease. Having a ready supply of urologic-associated tissue for surgical reconstruction at birth may be advantageous.

Theoretically, once the diagnosis of the pathologic condition is confirmed prenatally, a small tissue biopsy could be obtained under US guidance. These biopsy materials could then be processed and the various cell types expanded *in vitro*. Using tissue-engineering techniques, reconstituted structures *in vitro* could then be readily available at the time of birth for reconstruction (Fauza *et al.*, 1998).

## Injectable Therapies

Both urinary incontinence and vesicoureteral reflux (VUR) are common conditions affecting the genitourinary system that can be treated with injectable bulking agents. The ideal substance for endoscopic treatment of VUR and incontinence should be injectable, nonantigenic, nonmigratory, volume stable, and safe for human use.

Animal studies have shown that chondrocytes can easily be harvested and combined with alginate *in vitro*; the suspension can easily be injected cystoscopically; and the elastic cartilage tissue formed is able to correct VUR without any evidence of obstruction (Atala, 1994). The first human application of cell-based tissue-engineering technology for urologic applications occurred with the injection of chondrocytes for the correction of VUR in children and for urinary incontinence in adults (Fig. 54.4) (Bent *et al.*, 2001; Diamond and Caldamone, 1999).

With cell therapy techniques, the use of autologous smooth muscle cells was explored for both urinary incontinence and VUR applications (Cilento and Atala, 1995).

The potential use of injectable, cultured myoblasts for the treatment of stress urinary incontinence (SUI) has also been investigated. Use of injectable muscle precursor cells has also been studied for treatment of urinary incontinence due to irreversible urethral sphincter injury or maldevelopment (Yiou *et al.*, 2003). A clinical trial involving the use of muscle-derived stem cells (MDSC) to treat SUI has also been performed with good results. Biopsies of skeletal muscle were obtained, and autologous myoblasts and fibroblasts were cultured. Under US guidance, myoblasts were injected into the rhabdosphincter, and fibroblasts mixed with collagen were injected into the submucosa. One year following injection, the thickness and function of the rhabdosphincter had significantly increased, and all patients were continent (Strasser *et al.*, 2004). These are the first demonstrations of the replacement of both sphincter muscle tissue and its innervation by the injection of muscle precursor cells.

In addition, injectable muscle-based gene therapy and tissue engineering were combined to improve detrusor function in a bladder injury model and may have potential as a novel treatment option for urinary incontinence (Huard *et al.*, 2002).

## Testicular Hormone Replacement

Patients with testicular dysfunction require androgen replacement for somatic development. Conventional treatment consists of periodic intramuscular injections of chemically modified testosterone or, more recently, skin patch applications. However, long-term nonpulsatile testosterone therapy is not optimal and can cause multiple problems, including erythropoiesis and bone density changes.

A system was designed wherein Leydig cells were microencapsulated for controlled testosterone replacement. Purified Leydig cells were isolated and encapsulated in an alginate–poly-L-lysine solution. The encapsulated Leydig cells were injected into castrated animals, and serum testosterone was measured serially; the animals were able to maintain testosterone levels in the long term (Machluf *et al.*, 1998). These studies suggest that microencapsulated Leydig cells may be able to replace or supplement testosterone in situations where anorchia or testicular failure is present.

## VII. CONCLUSION

Tissue-engineering efforts are currently being undertaken for every type of tissue and organ within the urinary system. Most of the effort expended to engineer genitourinary tissues has occurred within the last decade. Tissue-engineering techniques require a cell culture facility designed for human application. Personnel who have mastered the techniques of cell harvest, culture, and expansion as well as polymer design are essential for the successful application of this technology. Before these engineering techniques can be applied to humans, further studies need to be performed in many of the tissues described. Recent progress suggests that engineered urologic tissues and cell therapy may have clinical applicability.

## VIII. REFERENCES

Amiel, G. E., and Atala, A. (1999). Current and future modalities for functional renal replacement. *Urol. Clin. North. Amer.* **26**, 235–246.

Atala, A. (1994). Use of nonautologous substances in VUR and incontinence treatment. *Dial. Pediatr. Urol.* **17**, 11–12.

Atala, A. (1997). Tissue engineering in the genitourinary system. *In* "Tissue Engineering" (A. Atala and D. Mooney, eds.), p 149. Birkhauser Press, Boston.

Atala, A. (1999). Future perspectives in reconstructive surgery using tissue engineering. *Urol. Clin. North. Amer.* **26**, 157–165.

Atala, A., Cima, L. G., Kim, W., Paige, K. T., Vacanti, J. P., Retik, A. B., and Vacanti, C. A. (1993a). Injectable alginate seeded with chondrocytes as a potential treatment for vesicoureteral reflux. *J. Urol.* **150**, 745–747.

Atala, A., Freeman, M. R., Vacanti, J. P., Shepard, J., and Retik, A. B. (1993b) Implantation *in vivo* and retrieval of artificial structures consisting of rabbit and human urothelium and human bladder muscle. *J. Urol.* **150**, 608–612.

Atala, A., Kim, W., Paige, K. T., Vacanti, C. A., and Retik, A. B. (1994). Endoscopic treatment of vesicoureteral reflux with chondrocyte-alginate suspension. *J. Urol.* **152**, 641–643.

Atala, A., Guzman, L., and Retik, A. (1999). A novel inert collagen matrix for hypospadias repair. *J. Urol.* **162**, 1148–1151.

Badylak, S. F., Lantz, G. C., Coffey, A., and Geddes, L. A. (1989). Small intestinal submucosa as a large-diameter vascular graft in the dog. *J. Surg. Res.* **47**, 74–80.

Bent, A., Tutrone, R., McLennan, M., Lloyd, L. K., Kennelly, M. J., and Badlani, G. (2001). Treatment of intrinsic sphincter deficiency using autologous ear chondrocytes as a bulking agent. *Neurourol. Urodynam.* **20**, 157–165.

Blandy, J. P. (1961). Neal pouch with transitional epithelium and anal sphincter as a continent urinary reservoir. *J. Urol.* **86**, 749–767.

Brittberg, M., Lindahl, A., Nilsson, A., Ohlsson, C., Isaksson, O., and Peterson, L. (1994). Treatment of deep cartilage defects in the knee with autologous chondrocyte transplantation. *N. Engl. J. Med.* **331**, 889–895.

Chen, F., Yoo, J. J., and Atala, A. (1999). Acellular collagen matrix as a possible "off-the-shelf" biomaterial for urethral repair. *Urology* **54**, 407–410.

Chen, K. L., Yoo, J. J., and Atala, A. (2005). Total penile corpora cavernosa replacement using tissue-engineering techniques. *Regenerate*, abstract.

Cilento, B. G., and Atala, A. (1995). Treatment of reflux and incontinence with autologous chondrocytes and bladder muscle cells. *Dial. Pediatr. Urol.* **18**, 11–15.

Cilento, B. G., Freeman, M. R., Schneck, F. X., Retik, A. B., and Atala, A. (1994). Phenotypic and cytogenetic characterization of human bladder urothelia expanded *in vitro*. *J. Urol.* **152**, 655–670.

Dahms, S. E., Piechota, H. J., Nunes, L., Dahiya, R., Lue, T. F., and Tanagho, E. A. (1997). Free ureteral replacement in rats: regeneration of ureteral wall components in the acellular matrix graft. *Urology* **50**, 818–825.

Diamond, D. A., and Caldamone, A. A. (1999). Endoscopic correction of vesicoureteral reflux in children using autologous chondrocytes: preliminary results. *J. Urol.* **162**, 1185–1188.

Fauza, D. O., Fishman, S., Mehegan, K., and Atala, A. (1998). Videofetoscopically assisted fetal tissue engineering: bladder augmentation. *J. Pediatr. Surg.* **33**, 7–12.

Folkman, J., and Hochberg, M. (1973). Self-regulation of growth in three dimensions. *J. Exp. Med.* **138**, 745–753.

Harada, N., Yano, H., Ohkawa, T., Misse, T., Kurita, T., and Nagahara, A. (1965). New surgical treatment of bladder tumors: mucosal denudation of the bladder. *Br. J. Urol.* **37**, 545–547.

Huard, J., Yokoyama, T., Pruchnic, R., Ou, A., Li, Y., Lee, J. Y., Somogyi, G. T., de Groat, W. C., and Chancellor, M. B. (2002). Muscle-derived cell-mediated *ex vivo* gene therapy for urological dysfunction. *Gene Ther.* **9**, 1617–1626.

Humes, H. D., Buffington, D. A., MacKay, S. M., Funke, A. J., and Weitzel, W. F. (1999). Replacement of renal function in uremic animals with a tissue engineered kidney. *Nat. Biotechnol.* **17**, 451–455.

Humes, H. D., Weitzel, W. F., Bartlett, R. H., Swaniker, F. C., and Paganini, E. P. (2003). Renal cell therapy is associated with dynamic and individualized responses in patients with acute renal failure. *Blood Purif.* **21**, 64–71.

Kershen, R. T., and Atala, A. (1999). Advances in injectable therapies for the treatment of incontinence and vesicoureteral reflux. *Urol. Clin. North. Amer.* **26**, 81–94.

Kim, B. S., and Mooney, D. J. (1998). Development of biocompatible synthetic extracellular matrices for tissue engineering. *Trends Biotechnol.* **16**, 224–230.

Kropp, B. P., Ludlow, J. K., Spicer, D., Rippy, M. K., Badylak, S. F., Adams, M. C., Keating, M. A., Rink, R. C., Birhle, R., and Thor, K. B. (1998). Rabbit urethral regeneration using small intestinal submucosa onlay grafts. *Urology* **52**, 138–142.

Kropp, B. P., Cheng, E. Y., Lin, H. K., and Zhang, Y. (2004). Reliable and reproducible bladder regeneration using unseeded distal small intestinal submucosa. *J. Urol.* **172**, 1710–1713.

Lailas, N. G., Cilento, B., and Atala, A. (1996). Progressive ureteral dilation for subsequent ureterocystoplasty. *J. Urol.* **156**, 1151–1153.

Landman, J., Olweny, E., Sundaram, C. P., Andreoni, C., Collyer, W. C., Rehman, J., Jerde, T. J., Lin, H. K., Lee, D. I., Nunlist, E. H., *et al.* (2004). Laparoscopic mid-sagittal hemicystectomy and bladder reconstruction with small intestinal submucosa and reimplantation of ureter into small intestinal submucosa: 1-year follow-up. *J. Urol.* **171**, 2450–2455.

Lanza, R. P., Chung, H. Y., Yoo, J. J., Wettstein, P. J., Blackwell, C., Borson, N., Hofmeister, E., Schuch, G., Soker, S., Moraes, C. T., *et al.* (2002). Generation of histocompatible tissues using nuclear transplantation. *Nat. Biotechnol.* **20**, 689–696.

le Roux, P. J. (2005). Endoscopic urethroplasty with unseeded small intestinal submucosa collagen matrix grafts: a pilot study. *J. Urol.* **173**, 140–143.

Machluf, M., and Atala, A. (1998). Emerging concepts for tissue and organ transplantation. *Graft* **1**, 31–37.

Machluf, M., Boorjian, S., Caffaratti, J., Kershen, R., Orsola, A., and Atala, A. (1998). Microencapsulation of Leydig cells: a new system for the therapeutic delivery of testosterone. *Pediatrics.* **102**(Suppl.), 32.

McDougal, W. S. (1992). Metabolic complications of urinary intestinal diversion. *J. Urol.* **147**, 1199–1208.

Oberpenning, F. O., Meng, J., Yoo, J., and Atala, A. (1999). *De novo* reconstitution of a functional urinary bladder by tissue engineering. *Nat. Biotechnol.* **17**, 149–155.

Olsen, L., Bowald, S., Busch, C., Carlsten, J., and Eriksson, I. (1992). Urethral reconstruction with a new synthetic absorbable device. *Scan. J. Urol. Nephrol.* **26**, 323–326.

Osman, Y., Shokeir, A., Gabr, M., El-Tabey, N., Mohsen, T., and El-Baz, M. (2004). Canine ureteral replacement with long acellular matrix tube: is it clinically applicable? *J. Urol.* **172**, 1151–1154.

Park, H. J., Kershen, R., Yoo, J., and Atala, A. (1999). Reconstitution of human corporal smooth muscle and endothelial cells *in vivo. J. Urol.* **162**, 1106–1109.

Portis, A. J., Elbahnasy, A. M., Shalhav, A. L., Brewer, A., Humphrey, P., McDougall, E. M., and Clayman, R. V. (2000). Laparoscopic augmentation cystoplasty with different biodegradable grafts in an animal model. *J. Urol.* **164**, 1405–1411.

Probst, M., Dahiya, R., Carrier, S., and Tanagho, E. A. (1997). Reproduction of functional smooth muscle tissue and partial bladder replacement. *Br. J. Urol.* **79**, 505–515.

Satar, N., Yoo, J., and Atala, A. (1999). Progressive bladder dilation for subsequent augmentation cystoplasty. *J. Urol.* **162**, 829–831.

Sievert, K. D., Bakircioglu, M. E., Nunes, L., Tu, R., Dahiya, R., and Tanagho, E. A. (2000). Homologous acellular matrix graft for urethral reconstruction in the rabbit: histological and functional evaluation. *J. Urol.* **163**, 1958–1965.

Strasser, H., Berjukow, S., Markstiner, R., Margreiter, E., Hering, S., Bartsch, G., and Hering, S. (2004). Stem cell therapy for urinary stress incontinence. *Exp. Gerontol.* **39**, 1259–1265.

Thomalla, J. V., Thompson, S. T., Rowland, R. G., and Mulcahy, J. J. (1987). Infectious complications of penile prosthetic implants. *J. Urol.* **138**, 65–67.

Wang, T., Koh, C. J., and Yoo, J. J. (2003). Creation of an engineered uterus for surgical reconstruction. Presented at the Proceedings of the American Academy of Pediatrics Section on Urology, New Orleans, LA.

Yiou, R., Yoo, J. Y., and Atala, A. (2003). Restoration of functional motor units in a rat model of sphincter injury by muscle precursor cell autografts. *Transplantation* **76**, 1053–1060.

Yoo, J. J., and Atala, A. (1997). A novel gene delivery system using urothelial tissue engineered neo-organs. *J. Urol.* **158**, 1066–1070.

Yoo, J. J., Satar, N., Retik, A. B., and Atala, A. (1995). Ureteral replacement using biodegradable polymer scaffolds seeded with urothelial and smooth muscle cells. *J. Urol.* **153**(Suppl.), 375A.

Yoo, J. J., Ashkar, S., and Atala, A. (1996). Creation of functional kidney structures with excretion of urine-like fluid *in vivo. Pediatrics.* **98**(Suppl.), 605.

Yoo, J. J., Meng, J., Oberpenning, F., and Atala, A. (1998). Bladder augmentation using allogenic bladder submucosa seeded with cells. *Urology* **51**, 221–225.

Yoo, J., Park, H. J., Lee, I., and Atala, A. (1999). Autologous engineered cartilage rods for penile reconstruction. *J. Urol.* **162**, 1119–1121.

Part Fifteen

# Musculoskeletal System

# Chapter Fifty-Five

# Mesenchymal Stem Cells

*Faye H. Chen, Lin Song, Robert L. Mauck, Wan-Ju Li, and Rocky S. Tuan*

## I. INTRODUCTION

Mesenchymal stem cells (MSCs) are nonhematopoietic progenitor cells found in adult tissues, and they are characterized by their extensive proliferative ability in an uncommitted state and the potential to differentiate along various lineages of mesenchymal origin in response to appropriate stimuli. Because of these properties and the ease of their isolation and *in vitro* expansion, MSCs are naturally considered an attractive candidate progenitor cell type for tissue-engineering applications. Since their discovery in the late 1960s, the burgeoning field of MSC investigation has yielded increasing knowledge about their biology, which has in turn provided tools for controlling their activity and cell fate, aspects critical for their use in tissue engineering. MSCs can easily be isolated from an increasing number of adult tissues, and they have been shown to differentiate into cells indicative of various mesenchymal lineages, including cellular phenotypes representative of the musculoskeletal tissues, such as cartilage, bone, muscle, ligament, and tendon. A plethora of factors control the extensive proliferative and multipotent differentiation capacity of MSCs, including growth factors, cytokines, hormones, various signaling molecules, and transcription factors. For tissue engineering of musculoskeletal tissues, their primary functions demand specific tissue form and mechanical strength, and the success of the outcome is dependent on the critical factors of scaffolds, cells, and environmental cues. Optimizing these three factors in musculoskeletal-tissue engineer-

ing, by exploiting our current knowledge on controlling MSC differentiation, has produced promising results in generating engineered tissue replacements for cartilage, bone, osteochondral, and other musculoskeletal tissues using MSCs, although there is clearly room for significant knowledge-based improvements. The current understanding, recent advances, and remaining challenges for the deployment of MSCs for musculoskeletal-tissue-engineering applications constitute the subject of this chapter.

## II. MSC BIOLOGY RELEVANT TO MUSCULOSKELETAL-TISSUE ENGINEERING

An ideal cell source for tissue engineering should exhibit the following characteristics: availability and easy access of the source cells, capacity for extensive self-renewal or expansion to generate sufficient quantity, capacity to differentiate readily into cell lineages of interest on instructive differentiation cues, and lack of or minimal immunogenic or tumorigenic ability. We discuss the intrinsic properties of MSCs that determine their behavior as well as external factors that affect the growth, differentiation, and the developmental outcomes of MSCs.

### MSC Identification

MSCs were first described by Friedenstein and colleagues in 1966 as a population of adherent, colony-forming,

*Principles of Tissue Engineering, 3rd Edition*
*ed. by Lanza, Langer, and Vacanti*

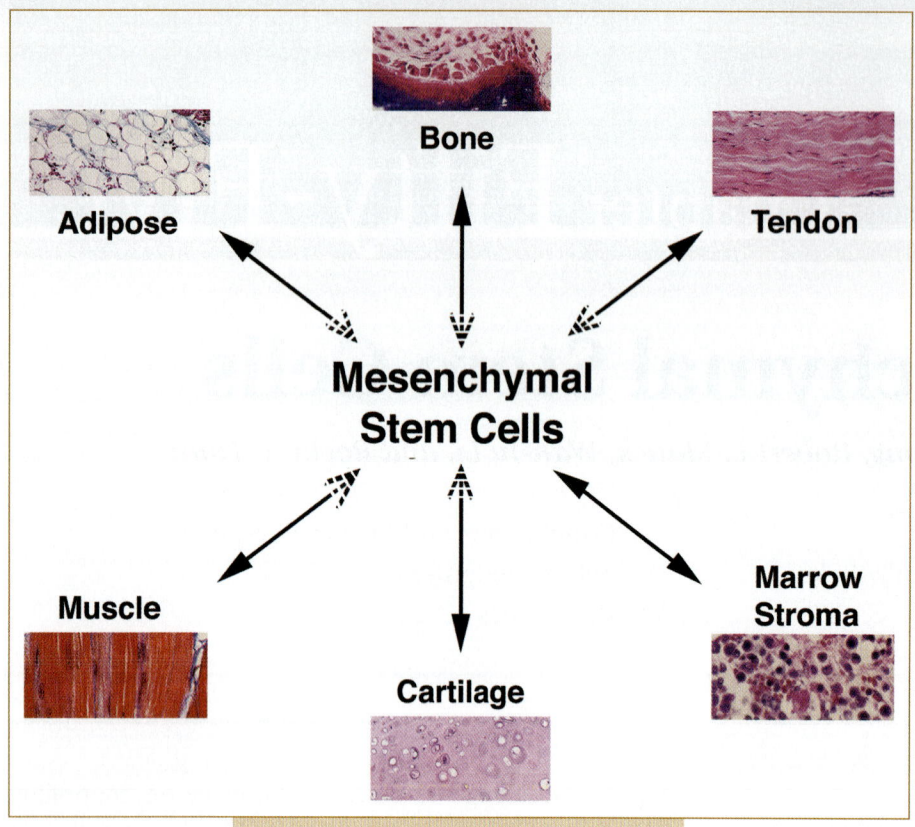

**FIG. 55.1.** Multilineage differentiation potential of mesenchymal stem cells. MSCs are able to differentiate into cell types of multiple lineages under appropriate conditions, including bone, cartilage, adipose, muscle, tendon, and stroma. The arrows are presented as bidirectional, suggesting that differentiated mesenchymal stem cells are capable of dedifferentiation and transdifferentiation (Song and Tuan, 2004). Adapted from Tuan *et al.* (2003).

fibroblast-like cells able to undergo osteogenic differentiation (Friedenstein *et al.*, 1966). MSCs have since been shown to possess the capacity to differentiate into cells characteristic of several mesenchymally derived tissues, including cartilage, bone, fat, muscle, tendon, and hematopoietic-supporting marrow stroma (Fig. 55.1). In addition to these mesenchymal lineages, MSCs can differentiate into other tissue types, including hepatocytes and neural tissues (Caplan, 1991; Jiang *et al.*, 2002; Pittenger *et al.*, 1999; Prockop, 1997; Suon *et al.*, 2004).

Experimentally, MSCs are identified by the expression of a number of surface markers, including STRO-1, SB-10, SH3, and SH4 antigens as well as Thy-1 (CD90), TGF-β receptor type III endoglin (CD105), hyaluronic acid receptor CD44, integrin α1 subunit CD29, CD133, p75LNGFR, and activated leukocyte-cell adhesion molecules (ALCAM, CD166). MSCs are negative for the hematopoietic markers, CD34, CD45, and CD14. SH3, SH4, and STRO-1 antibodies recognize antigens that are present on mesenchymal stem cells but not hematopoietic cells (for review, see Baksh *et al.*, 2004; Gregory *et al.*, 2005; Tuan *et al.*, 2003). However, these are not expressed exclusively by MSCs and are found on other cell types as well. To date, there is no one single marker or a combination of markers that has been shown to be specific and exclusive to MSCs. Therefore it remains a challenge to isolate MSCs specifically from a mixed cell population. Frequently, a combination of antibodies is used to characterize the MSC phenotype retrospectively. Inconsistency in the literature on the growth characteristics and differentiation potential of MSCs underscores the need for a functional definition of MSCs. At present, there is lack of a unifying definition as well as information on specific markers that define the cell types characterized as MSCs, with the sole definition being their ability (1) to proliferate to self-renew and remain in an undifferentiated state until provided the signal to do so, and (2) to differentiate along specific mesenchymal lineages when induced to do so (Fig. 55.1).

## Tissue Sources of MSCs

Bone marrow remains the best-studied tissue source for MSCs. In addition, MSC-like cells have been identified in a number of other tissues, including adipose, periosteum, synovial membrane, muscle, dermis, deciduous teeth, pericytes, trabecular bone, articular cartilage, immobilized peripheral blood, cord blood, and placenta (Baksh *et al.*, 2004; Gregory *et al.*, 2005; Tuan *et al.*, 2003) (Table 55.1). The endogenous functions of these stem cells in these various tissues remain unclear, although they are generally thought to participate in local tissue repair and regeneration. In addition, bone marrow MSCs can give rise to stromal cells that support hematopoiesis *in vitro* and *in vivo*, possibly through providing extracellular components as well as various growth factors and cytokines (Majumdar *et al.*, 2000).

| Table 55.1. | Tissue sources of mesenchymal stem cells |
|---|---|
| *Tissue source* | *Representative references* |
| Bone marrow | Friedenstein *et al.* (1966) |
| | Prockop (1997) |
| | Pittenger *et al.* (1999) |
| Trabecular bone | Tuli *et al.* (2003) |
| | Osyczka *et al.* (2002) |
| Muscle | Young *et al.* (1995) |
| | Bosch *et al.* (2000) |
| | Williams *et al.* (1999) |
| Adipose | Zuk *et al.* (2002) |
| | Erickson *et al.* (2002) |
| Periosteum | De Bari *et al.* (2001) |
| | Nakahara *et al.* (1991) |
| Synovial membrane | De Bari *et al.* (2001) |
| | Nishimura *et al.* (1999) |
| Articular cartilage | Alsalameh *et al.* (2004) |
| | Dowthwaite *et al.* (2004) |
| | Tallheden *et al.* (2003) |
| Skin | Young *et al.* (1995) |
| | Toma *et al.* (2001) |
| Pericyte | Brighton *et al.* (1992) |
| Peripheral blood | Kuznetsov *et al.* (2001) |
| | Zvaifler *et al.* (2000) |
| Deciduous teeth | Miura *et al.* (2003) |
| Periodontal ligament | Lekic and McCulloch (1996) |
| | Seo *et al.* (2005) |

Although bone marrow is relatively easily accessible, the potential donor site morbidity, pain, and low cell incidence have prompted researchers to find alternative sources for MSC isolation to meet the demand of cell therapy. In addition to easy accessibility, MSC yield, proliferation rate, multipotency, and differentiation potential determine the usefulness of MSCs from different tissue origins. Although MSCs from different tissues show similar phenotypic characteristics, it is not clear if these are the same MSCs, and they clearly show different propensities in proliferation and differentiation. For example, a recent study comparing human MSCs derived from bone marrow, umbilical cord blood, and adipose tissue showed that adipose tissue contained the highest number of hMSCs while umbilical cord blood contained the lowest. However, umbilical cord blood–derived MSCs could be cultured longest and showed the highest proliferation capacity, whereas bone marrow–derived MSCs showed the lowest proliferation capacity (Kern *et al.*, 2006). Sakaguchi *et al.* (2005) compared the properties of MSCs isolated from bone marrow, synovium, periosteum, skeletal muscle, and adipose tissue and found significant differences in their differentiation potentials.

MSCs from bone marrow, synovium, and periosteum showed greater chondrogenic activities, with synovium-derived MSCs exhibiting the highest capacity. With regard to osteogenesis, the rate of matrix mineralization was highest in bone marrow–, synovium-, and periosteum-derived MSCs. Therefore, the optimal source of MSCs for therapeutic use for the musculoskeletal system remains to be determined.

In addition to tissue source, donor age and disease stage may directly affect MSC yield, rate of proliferation, and multipotency. There seems to be decreasing MSC number and proliferation rate as well as differentiation potentials with increasing age (Quarto *et al.*, 1995). MSCs isolated from osteoporotic women showed significantly reduced proliferative response and osteogenic differentiation (Rodriguez *et al.*, 1999). With bone marrow MSCs, advanced osteoarthritis (OA) condition, regardless of age, has a significantly deleterious impact on proliferative capacity and chondrogenic and adipogenic activities when compared to healthy donors (Murphy *et al.*, 2002). The potential decrease of MSC quantity and quality with age and disease may limit the use of autologous MSCs in clinical settings, which warrants further study. This indicates the need for using allogeneic MSCs for tissue repair, which might carry the risk of immune rejection and allogeneic reaction by the host. This topic is discussed in detail later.

## MSC Isolation and *In Vitro* Culture

MSCs are rare in bone marrow, occurring with a frequency of 0.001–0.01% of the total nucleated cells (Pittenger *et al.*, 1999). Various protocols of MSC isolation are available, and some are species specific. Human bone marrow MSCs can be isolated by simple direct plating. Frequently, this is performed with the layer of mononuclear cells obtained after Percoll or Ficoll density gradient centrifugation. With time and medium changes, the nonadherent hematopoietic cells are washed away, and MSCs appear as small, adherent, spindle-shaped fibroblast-like cells. Further purification can be performed by either surface marker–based positive selection or negative selection using either magnetic activated cell sorting (MACS) or fluorescence activated cell sorting (FACS). Positive selections have been based on the selection of markers that are expressed by the MSCs (although not exclusively). These include STRO-1, CD105, CD133, and p75LNGFR. Negative selection is based on the exclusion of markers that are expressed by hematopoietic cells, using antibodies against CD34, CD45, and CD11b.

MSCs are normally cultured in Dulbecco's modified eagles medium (DMEM) or α-minimum essential medium (α-MEM) supplemented with 10% preselected batches of fetal bovine serum at 37°C with 5% $CO_2$ in a humidified chamber. After initial expansion, cultures are maintained at densities of 50–10,000 cells/$cm^2$ (Gronthos *et al.*, 1994; Jiang *et al.*, 2002; Sekiya *et al.*, 2002a). The extensive expansion capacity of MSCs depends on the harvesting techniques, the culture conditions, and the condition and age of the donor. The growth of MSCs exhibits a lag phase of growth, followed by an exponential growth phase and a plateau state when confluent. On subculturing, the same cycle resumes. Microscopically, MSCs exhibit a fibroblast-like morphology under light microscope (Fig. 55.2). There is a wide range in the number of population doublings the MSCs can undergo, with the average maximal value of 38 ±

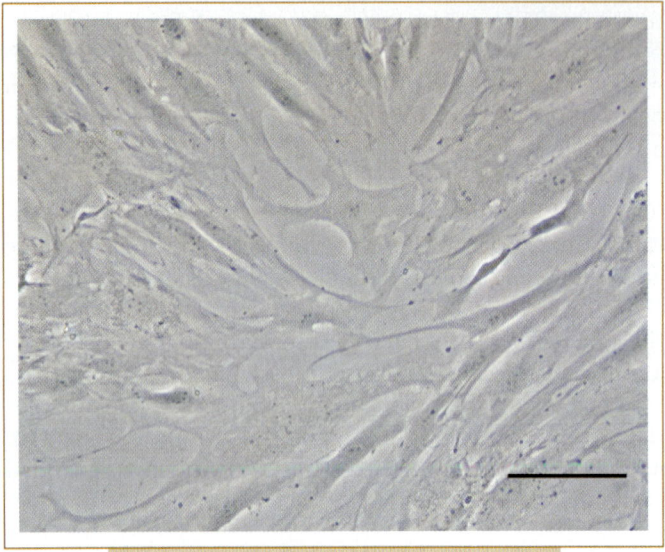

**FIG. 55.2.** Morphology of human bone marrow–derived mesenchymal stem cells. MSCs exhibit fibroblastic phenotype when maintained in monolayer culture, as observed by phase contrast microscopy. Bar, 100 μm.

4, at which point the cells become large and flatten and become senescent (Table 55.2). The osteogenic potential of MSCs appears to be conserved throughout extensive passage (Bruder *et al.*, 1997). Culture conditions, such as initial seeding density, also play a role in the expansion capacity of MSCs. For example, higher proliferation profiles of MSCs was seen when plated at low initial plating density (1.5–3 cells/cm²) but not at high density (12 cells/cm²), resulting in a dramatic increase in the fold expansion of total cells (2000-fold vs. 60-fold increase, respectively) (Colter *et al.*, 2000).

## MSC Self-Renewal and Proliferation Capacity

*In vivo*, MSCs remain in a mitotically quiescent state (G0 stage of the cell cycle), as demonstrated by analysis of fresh bone marrow cell harvests continuously exposed to tritiated thymidine labeling (Friedenstein *et al.*, 1974). BrdU labeling of cells in a human trabecular bone explant culture also reveals that cell proliferation, including that of Stro-1-positive mesenchymal stem/progenitor cells, was inhibited completely within the explant tissue milieu (Song *et al.*, 2005). Since most of the tissues/organs from which MSCs are derived exhibit a relatively slow turnover rate in the adult organism, it is not unreasonable to propose that self-renewal or proliferation of MSCs is normally suppressed *in vivo* in the course of tissue homeostasis, possibly regulated by intrinsic factors present in the tissue microenvironment and/or through direct or indirect interaction with neighboring cells.

*In vitro*, primary MSC cultures exhibit a lag phase of growth, ranging from 24 hours to five days, followed by an exponential phase of four to six days before reaching the

| Table 55.2. | *In vitro* population doubling potential of human mesenchymal stem cells | | |
|---|---|---|---|
| *Cell source* | *Culture condition* | *Population doublings* | *References* |
| Bone marrow | αMEM, 17% FBS, 2 mM L-glutamine | 10 | Pochampally *et al.* (2004) |
| Bone marrow | Coon's modified Ham's F12 medium, 10% FCS | 22–23 | Muraglia *et al.* (2000) |
| Bone marrow | αMEM, 10% FBS | 15 | Lee *et al.* (2004) |
| Bone marrow | αMEM, 10% FCS | 24 ± 11 to 41 ± 10 | Stenderup *et al.* (2003) |
| Bone marrow | αMEM, 20% FBS, 100 uM L-ascorbate-2-phosphate | 20 | Gronthos *et al.* (2003) |
| Bone marrow | Coon's modified Ham's F-12, 10% FCS, 1 ng/mL FGF-2 | 22–23 | Banfi *et al.* (2000) |
| Bone marrow | DMEM, 10% FBS | 38 ± 4 | Bruder *et al.* (1997) |
| Bone marrow | αMEM, 20% FBS | 30 | Colter *et al.* (2000) |
| Adipose tissue | αMEM, 10% FBS | 20 | Lee *et al.* (2004) |
| Pancreas | DMEM, 2% FBS, MCDB-201, 10 ng/ml EGF, 1× insulin-transferrin selenium, 1× linoleic acid BSA, 10 nM Dex, 10 mM ascorbic acid phosphate | 30 | Hu *et al.* (2003) |

plateau phase, after which cell division slows dramatically. Cell-cycle studies revealed that the majority of the MSCs (between 78.7% and 96.45%) are in the G0/G1 phase of the cell cycle, while a very small fraction of MSCs are engaged in active proliferation. Depending on the tissue source, donor age, and culture conditions, population doubling times of hMSCs vary to a large extent. During the log phase of the cell growth, human MSCs isolated from bone marrow and fetal pancreas exhibited a similar population doubling time, ranging from 10 to 30 hours. On the other hand, postnatal human MAPCs, a subset of MSCs, divide much slower, with a population doubling time of 48–72 hours.

Unlike embryonic stem cells, MSCs do not expand indefinitely *in vitro* when cultured in the presence of serum, which could be due to their intrinsic property as adult stem cells or due to suboptimal culture environment. After extensive propagation, MSCs change their phenotype from a fibroblastic shape to a more flattened morphology with extensive podia and actin stress fiber formation, a phenomenon usually referred to as *replicative senescence*. Consistent with their limited expansion capacity, hMSCs do not express telomerase activity, and their telomere length is reduced as the cells grow older. Furthermore, murine MSCs derived from telomerase knockout mice failed to differentiate into adipocytes and chondrocytes and lost telomere at late passage (Liu *et al.*, 2004). Therefore, telomerase activity and telomere maintenance appear to play a crucial role in maintaining the self-renewal and multidifferentiation potential of MSCs. Indeed, hMSCs stably expressing telomerase reverse transcriptase can undergo 80 (Okamoto *et al.*, 2002; Shi *et al.*, 2002) or 260 (Simonsen *et al.*, 2002) population doublings without growth arrest. In addition, growth factors play a role in the expansion capacity of MSCs. Basic fibroblast growth factor (bFGF) can prolong the replicative capacity of hMSCs and increase total cell numbers by several-fold when included in the basal culture medium.

## Multilineage Differentiation of MSCs

MSC differentiation is tightly controlled by several factors, such as growth factors, hormones, extracellular matrix molecules, and signaling pathways. Relevant to musculoskeletal tissue engineering, *in vitro* chondrogenesis of MSCs, or cartilage formation, is accomplished by culturing the cells in a three-dimensional culture condition, either as pellet culture, micromass cultures, or three-dimensional scaffold, in the presence of a member of the transforming growth factors (TGF-β1, TGF-β2, or TGF-β3 or bone morphogenetic proteins, BMPs). Supplementation with dexamethasone also promotes the expression of chondrocyte phenotype. Chondrogenesis is demonstrated histologically by the expression of genes encoding cartilage matrix components, including aggrecan, collagen types II and IX, and COMP (cartilage oligomeric matrix protein). Control of MSC chondrogenesis is discussed in more detail in the cartilage-tissue-engineering section of this chapter.

Osteogenesis, i.e., bone formation, can be induced *in vitro* by treating MSCs with the synthetic glucocorticoid dexamethasone, β-glycerophosphate, ascorbic acid, and 1,25-dihydroxyvitamin $D_3$. Alkaline phosphatase (ALP) activity and calcification of the extracellular matrix are typical markers used for detecting preosteoblasts and mature osteoblasts, respectively. Dexamethasone stimulates MSC proliferation and increases ALP activity of osteogenic hMSCs. However, matrix mineralization occurs only when dexamethasone functions together with β-glycerophosphate and ascorbic acid, and it is further enhanced by 1,25-dihydroxyvitamin $D_3$.

Growth factors also play certain osteoinductive roles. For instance, treatment of human MSCs with insulin like growth factor 1 (IGF-1), epidermal growth factors (EGF), and vascular endothelial growth factor (VEGF) increases the activity of ALP and mineralization *in vitro*. On the other hand, platelet-derived growth factor (PDGF) has little inductive effect on hMSC osteogenesis.

Members of the Wnt family of signaling proteins have recently been shown to impact human MSC osteogenesis (Boland *et al.*, 2004; Cho *et al.*, 2006; De Boer *et al.*, 2004). Wnts are a family of secreted cysteine-rich glycoproteins that have been implicated in the regulation of stem cell maintenance, proliferation, and differentiation during embryonic development. In humans, mutations in the Wnt coreceptor, LRP5, lead to defective bone formation. LRP5 gain of function mutation results in high bone mass, where loss of function causes an overall loss of bone mass and strength, indicating that Wnt signaling is positively involved in embryonic osteogenesis. *In vitro*, exposure of human MSCs to Wnt3a-conditioned medium or overexpression of ectopic Wnt3a inhibits osteogenesis. The expression of several osteoblast specific genes, e.g., ALP, bone sialoprotein, and osteocalcin, was dramatically reduced, while the expression of Cbfa1/Runx2, an early osteoinductive transcription factor, was not altered, implying that Wnt3a-mediated canonical signaling pathway is necessary but not sufficient to completely block hMSC osteogenesis. On the other hand, Wnt5a, a typical noncanonical Wnt member, has been shown to promote osteogenesis (Boland *et al.*, 2004). Since Wnt3a promotes MSC proliferation during early osteogenesis, it is very likely that canonical Wnt signaling functions in the initiation of early osteogenic commitment by increasing the osteoprogenitor reservoir, while noncanonical Wnt drives the progression of osteoprogenitor to mature functional osteoblasts. The identity and actions of intracellular mediators of Wnt signaling remain to be elucidated.

## Plasticity of MSCs

Plasticity/transdifferentiation is a process whereby a cell type committed to and progressing along a specific developmental lineage switches into another cell type of a different lineage through genetic reprogramming. As demonstrated by several studies, terminally differentiated cells can switch their phenotype under appropriate stimulations. For example, chondrocytes can become osteocytes (Gentili *et al.*, 1993; Kahn and Simmons, 1977), and adipocytes can convert to osteoblasts (Bennett *et al.*, 1991; Beresford *et al.*, 1992; Bianco *et al.*, 1988). Since all of these cells are the mature progenies of MSCs, the conversion from one to another might reflect the plastic property of MSCs. By using an *in vitro* differentiation strategy, we have demonstrated that human MSCs precommitted to a given mesenchyme cell lineage can transdifferentiate into other cell types in response to inductive extracellular cues (Song and Tuan, 2004). Understanding the molecular mechanisms that control the transdifferentiation potential of MSCs will facilitate the identification of regulatory factors, thus providing tools to manipulate adult stem cells for cell-based tissue-engineering applications.

## MSC Heterogeneity

Individual colonies derived from single MSC precursors exhibit a heterogeneous nature in terms of cell proliferation and multilineage differentiation potential. For instance, only a minor proportion of colonies (17%) derived from adult human bone marrow continued to grow beyond 20 population doublings, while the majority of the colonies exhibited early senescence (Gronthos *et al.*, 2003). There is also a marked difference in the differentiation capacity of MSC colonies. For example, Pittenger *et al.* (1999) reported that only one-third of the initial adherent bone marrow–derived MSC clones are multipotent to differentiate along the osteo-, chondro-, and adipogenic pathways (osteo/chondro/adipo). Furthermore, nonimmortalized cell clones examined by Muraglia *et al.* (2000) demonstrated that 30% of the *in vitro* derived MSC clones exhibited a trilineage (osteo/chondro/adipo) differentiation potential, while the remainder displayed a bilineage (osteo/chondro) or unilineage potential (osteo). These observations are consistent with other *in vitro* studies using conditionally immortalized clones. Even derived from a single colony, cells seem to be heterogeneous. Kuznetsov *et al.* (1997) demonstrated that only 58.8% of the single colony-derived clones had the ability to form bone within hydroxyapatite-tricalcium phosphate ceramic scaffolds after implantation in immunodeficient mice. Similar results were reported by using purer populations of MSCs maintained *in vitro* (Gronthos *et al.*, 2003). Taken together, these results suggest that MSCs are heterogeneous with respect to their developmental potential, placing a significant challenge on selecting the most potent cells for clinical application in tissue regeneration.

There exist a number of inconsistencies and discrepancies among currently published results on MSCs, perhaps reflecting the heterogeneity of the population, different isolation and culture methods, as well as the different stimuli used for the differentiation procedures. There has not been a single MSC specific marker unequivocally to identify MSCs, and there is heterogeneity in MSC populations. Most likely, distinct and unrelated cell types are being studied, and only a subpopulation of the MSCs are true stem cells.

## MSC Effect on Host Immunobiology

Ideally, autologous MSCs would have advantages over allogeneic MSCs for regenerative medicine because autologous MSCs pose few immunological complications. However, due to the rarity of MSCs and also to the possible decrease in quantity as well as quality of the MSCs with age and disease, it is more feasible to consider using allogeneic MSCs for replacing or repairing damaged tissues. Before this is considered acceptable, it is important to understand MSC-elicited host immunological reactions. Allogeneic cells are normally detected and deleted by the host immune systems. However, MSCs have been surprisingly different in this aspect and offer several advantages, including possible suppressing and modulatory effects on host immune response.

The first immunological advantage of MSCs is that they are hypoimmunogenic and can evade host immune system, as shown by several *in vitro* experiments. This makes MSCs attractive for allogeneic transplantation, whose major limitation is host immune rejection. MSCs express low (fetal) to intermediate (adult) major histocompatibility complex (MHC) class I molecules, and do not express MHC class II molecules on their cell surface (Gotherstrom *et al.*, 2004; Le Blanc *et al.*, 2003). The expression of MHC class I molecules helps to protect MSCs from deletion by natural killer cells. The lack of surface MHC class II expression gives the MSCs the potential to escape recognition by alloreactive CD4$^+$ T-cells. MSCs do contain an intracellular pool of MHC class II molecules that can be mobilized onto the cell surface by treatment with interferon-$\gamma$ (IFN-$\gamma$). However, induced surface expression of MHC class II still does not render the MSCs immunogenic (Le Blanc *et al.*, 2003). After differentiation, MSCs continue to express MHC class I but not class II molecules on their cell surface, and they continue to be nonimmunogenic (Gotherstrom *et al.*, 2004; Le Blanc *et al.*, 2003).

Experimental evidence suggests that MSCs can interact directly with immune cells and modulate and suppress alloreativity. MSCs inhibit T-cell proliferation *in vitro* (Bartholomew *et al.*, 2002; Di Nicola *et al.*, 2002; Tse *et al.*, 2003). MSCs do not seem to express costimulatory molecules, CD40, CD40 ligand, B7-1, and B7-2, and probably do not activate alloreative T-cells (Tse *et al.*, 2003). In fact, MSCs suppress T-cell activation and proliferation. For example, MSCs have been shown to suppress CD4$^+$ and CD8$^+$ T-cells in mixed lymphocyte cultures (Di Nicola *et al.*, 2002; Tse *et al.*, 2003). Even though T-cell proliferation can be induced by exogenous costimulation, when they are cocultured with MSCs in the presence of a stimulant, T-cell proliferation is not observed (Tse *et al.*, 2003). MSCs can also induce apoptosis of activated T-cells but not resting T-cells (Plumas *et al.*, 2005). Supporting these observations, Bartholomew (Bartholomew *et al.*, 2002) showed that allogeneic baboon MSCs inhibited the proliferation of lymphocytes *in vitro* and prolonged skin graft survival *in vivo*. In addition to the effect of MSCs on T-cells, MSCs can affect dendritic cell differentiation and maturation and interfere with their function (Beyth *et al.*, 2005; Zhang *et al.*, 2004). They also alter the phenotype of natural killer cells and can suppress the proliferation, cytokine secretion, and cytotoxicity of these cells against MHC class I targets (Sotiropoulou *et al.*, 2006).

In addition to cell–cell interaction mediated inhibition, MSCs are capable of secreting soluble factors to create a local immunosuppressive environment. These factors have been shown to include hepatocyte growth factor (HGF), TGF-β1, IL-10, and prostaglandin E2 (Aggarwal and Pittenger, 2005; Di Nicola *et al.*, 2002). For example, MSCs are shown constitutively to express HGF and TGF-β1. When anti-HGF and anti-TGF-β1 antibodies are included, MSC inhibition on T-cell proliferation is lifted (Di Nicola *et al.*, 2002). IL-10 is a cytokine for regulatory T-cells and can suppress inflammatory immune response. Similarly, IL-10 is shown to be produced either constitutively or increased in coculture, and MSCs can mediate the suppressive activities partially through IL-10 (Aggarwal and Pittenger, 2005; Beyth *et al.*, 2005; Rasmusson *et al.*, 2005). Collectively, these *in vitro* studies have suggested that MSCs can interact with the various subsets of cells of the immune system, alter the response of the immune cells, and shift the response from a pro-inflammatory response to an antiinflammatory response, possibly through inhibition of the pro-inflammatory cytokines, such as IFN-γ and tumor necrosis factor-α (TNF-α), and stimulation of the immunosuppressive cytokines, including IL-10 and prostaglandin E2 (Aggarwal and Pittenger, 2005). The role of secreted factors in the immunoregulatory action of MSCs remains an actively investigated area. Contradictory results exist as to which factors are important for this function. For example, although a role for IL-10, TGF-β, and prostaglandin E2 has been suggested, in other studies none of these factors was found to be responsible for the immunosuppressive action of MSCs (Tse *et al.*, 2003). Despite the discrepancies on the mechanism of action, the foregoing studies suggest that MSCs can be transplanted between MHC-incompatible individuals.

The *in vitro* studies showing that MSCs possess immunomodulatory and immunosuppressive activities suggest that MSCs can be potentially used *in vivo* for enhancing the engraftment of other tissues (e.g., hematopoietic stem cells) or for the prophylactic prevention and even possibly as a treatment of graft-versus-host disease, to prevent rejection and to promote transplant and patient survival. However, before MSC treatment can be used as standard therapy on humans, more *in vivo* animal studies need to be performed, and the biology and mechanism of MSC immunomodulation effect need to be better elucidated.

### Safety of Using MSCs for Transplantation

The ability of MSCs to undergo extensive self-renewal via proliferation raises some concern as to whether MSCs, after prolonged *in vitro* culture, can become tumorigenic. Although most MSC transplantations did not show obvious malignant transformation, there has been evidence suggesting otherwise. Rubio *et al.* (2005) showed that although standard short-term *in vitro* culture (six to eight weeks) seems safe, MSCs can undergo spontaneous transformation after long-term culture of four to five months.

The immunosuppressive properties, especially the potential systemic immunosuppressive ability of MSCs, bring caution to the use of MSCs under certain clinical conditions, such as cancer. Using a murine melanoma tumor model, it has been shown that cotransplantation of an MSC cell line (C3H10T1/2) favors tumor growth of subcutaneously injected B16 melanoma cells (Djouad *et al.*, 2003). However, this tumor-promoting effect was not observed in another study of coculture of a rat MSC line, MPC1cE, with rat colon carcinoma cells in a gelatin matrix. In this case, an inhibitory effect of MSCs on the outgrowth of the tumor cells was observed (Ohlsson *et al.*, 2003). The effect of MSCs on tumor growth requires further investigation to rule out the potential side effect of the therapeutic use of MSCs.

## III. MSCs IN MUSCULOSKELETAL TISSUE ENGINEERING

The musculoskeletal system of the human body is designed to sustain and maintain its form and function in the face of enormous load-bearing demands throughout a normally active life. Form and mechanical strength are vital for the function of this system. The mechanical and biochemical properties of the musculoskeletal system that define its structure and function also define the functional requirements of a tissue-engineered substitute. Successful tissue engineering based replacement of native tissues will likely require constructs that possess functional properties similar to those of the native tissues to minimize premature failure. There are two basic approaches in tissue engineering. The first one is *ex vivo* tissue engineering, in which the construct is cultivated *in vitro* to achieve appropriate functionality before implantation. The second approach is *in vivo* tissue engineering, in which the construct is allowed to mature *in vivo* for tissue repair and regeneration. For both approaches, three components govern the eventual outcome of tissue-engineered constructs: appropriate scaffold, instructive environment, and responsive cells.

### Cartilage Tissue Engineering

The need for engineered cartilage arises from the fact that while the tissue often functions well through a lifetime of use, 9% of the U.S. population age 30 and older have osteoarthritis (OA) of the hip or knee. The total cost of OA is estimated at $28.6 billion per year in the United States alone

(Felson and Zhang, 1998), with over 200,000 knee replacements performed each year. The intrinsic healing capacity of the native tissue is limited, and, given the increasing incidence of OA and the increasing life expectancy of the population, there is a growing demand for novel repair strategies. Effective treatment of cartilage injuries may eliminate or forestall the need for joint replacement, thus enhancing the quality of life.

## General Properties of Articular Cartilage

Articular cartilage is the dense white tissue that lines the surfaces of joints and functions to transmit the high stresses associated with joint motion. The tissue consists of both a solid extracellular matrix (ECM) component as well as a fluid phase (Mow et al., 2005). The solid ECM is composed of a dense network of specialized molecules that engender the unique mechanical properties of the tissue. Collagen content (predominantly collagen type II) of the tissue ranges from 5% to 30% by wet weight, while proteoglycan (PG) content ranges from 2% to 10%. In addition to these major elements, the ECM includes numerous minor collagens (e.g., types VI, IX, and XI) and linking molecules (COMP, hyaluronan, link protein, fibronectin). Together, these elements make up the fluid-filled fibrous network of the cartilage ECM, with larger collagen fibers interwoven throughout an array of large PG aggregates (aggrecan core protein with its covalently linked keratin and chondroitin sulfate moieties) attached to long hyaluronan chains. These core constituents change with age (resulting in the emergence of adult functional properties) and deteriorate with disease processes (such as OA).

The cellular component of cartilage, the chondrocytes, comprises less than 10% of the tissue volume and maintains the cartilage by synthesizing ECM constituents to balance their continual degradation and remodeling. Chondrocytes display a rich transcriptional profile in situ reflective of their specialized cartilaginous phenotype, which is lost when these cells dedifferentiate with expansion in monolayer culture. In situ, chondrocytes are tuned to the mechanical loading environment, altering rates and location of matrix ECM production with changes in physiologic loading (Guilak and Hung, 2005).

The mechanical properties of articular cartilage underlie its ability to act as a low friction bearing surface over a lifetime of use. The dense, negatively charged PG-rich ECM results in an equilibrium compression modulus of 0.2–1.4 MPa (Mow et al., 2005). The collagen content engenders a tensile modulus that is higher, ranging from 1 to 30 MPa (Woo et al., 1976). One unique feature of cartilage is that under transient or cyclical loading at physiologic frequencies (0.1–2 Hz), interstitial fluid pressure is maintained at elevated values. This fluid pressurization arises with contact and supports >90% of applied stress, shielding the matrix from excess deformation (Soltz and Ateshian, 1998). Fluid pressurization also results in a higher dynamic modulus than equilibrium modulus, effectively stiffening the tissue with higher frequency loading. Cartilage use arises from movement, with the resulting forces that are multiples of body weight acting on the cartilage, resulting in stresses at the surface of $\geq 2$ MPa (Brown and Shaw, 1983).

## Cartilage Tissue Engineering with Chondrocytes

When tissue engineering was originally proposed as a strategy for repairing diseased or damaged tissues, it was believed that articular cartilage would be one of the first successes in this new field, owing to its relatively simple composition (possessing a single cell type) and lack of neural and vascular supply. The first approaches focused on the chondrocyte itself, postulating that because these cells make and maintain matrix in vivo, exogenous cells should be able to reconstitute the tissue when implanted. Early in vivo successes demonstrated enhanced repair when high-density chondrocyte solutions were transplanted to focal defects beneath a periosteal flap (Brittberg et al., 1994), a procedure now commonly referred to as autologous chondrocyte transplantation/implantation (ACT/ACI) and commercialized as the Carticel® method by Genzyme Biosurgery. While questions remain regarding the efficacy and cost effectiveness of this approach, it nevertheless stands as the first clinically available tissue-engineering strategy to enhance cartilage repair.

In addition to using chondrocytes directly as therapeutics, there has been a growing interest in the combination of chondrocytes with a carrier scaffold for in vivo implantation or in vitro maturation in culture conditions permissive of neotissue growth. Experimental challenges include finding a source of chondrocytic cells, phenotype maintenance, and the provision of an appropriate three-dimensional matrix. Regardless of the chondrocyte transplantation carrier materials used, it should be noted that in nearly all cases, repair is significantly better with cell-seeded scaffolds than with acellular scaffolds alone. Scaffolding materials composed of woven and nonwoven macroporous meshes of biodegradable polymers such as poly-glycolic acid (PGA), poly-lactic acid (PLA), and their copolymers have shown great potential in cartilage-tissue engineering (Vunjak-Novakovic et al., 1999). Recent fabrication technologies have further improved the deployment of such polymers, either by changing the length scales of fiber components to produce nanofibrous scaffolds (Li et al., 2003) or by modifying the deposition methodology to define depth-dependent structural properties (Sherwood et al., 2002).

In addition to fibrous scaffolds, various biocompatible hydrogels have been applied for the in vitro and in vivo growth of cartilage constructs. These hydrogels include alginates (Paige et al., 1995; Rowley et al., 1999), fibrin gels (Fortier et al., 1997), agarose (Buschmann et al., 1992), collagen type I (Wakitani et al., 1998), collagen type II (Nehrer et al., 1997), photopolymerizing poly(ethylene glycol) (PEG) gels (Elisseeff et al., 2002), and, more recently, self-assembled peptide hydrogels (Kisiday *et al.*, 2002). Continued modifications of these gels, as with the recent modifications of PEG-based hydrogels to introduce a matrix metalloproteinase (MMP)-cleavable linker, may further enhance tissue formation by promoting remodeling of the carrier material as new tissue is deposited (Park *et al.*, 2004). Combinations of scaffold materials (e.g., fibrous and gel scaffolds) may also be valuable in this pursuit.

## Factors Influencing Outcomes of Tissue-Engineered Cartilage

The evolution of a tissue-engineered construct is a time-dependent process, controlled by numerous (and sometimes unanticipated) factors. This creates the difficult engineering problem of construct optimization, and numerous studies have addressed how the variation in such parameters influences the development of biochemical and mechanical properties of the tissue. For example, increasing the seeding density of cells within a scaffold typically results in a more rapid development of construct properties (S. C. Chang *et al.*, 2001). Variations in the amount and formulation of culture media can also have profound effects on chondrocyte biosynthetic activities. For example, increasing the amount of fetal bovine serum (FBS) in growth medium has been shown to enhance the mechanical properties of constructs (Mauck *et al.*, 2003). Specific supplementation with biologically active growth factors found in the synovial fluid can also influence chondrocyte biosynthetic activities and construct maturation. In particular, anabolic factors (those leading to increased production and retention of ECM), including IGF-1, TGF-β1 and -β3, BMP-2, 12, 13, and IL-4, have all been shown to positively modulate the growth of engineered cartilage constructs (Blunk *et al.*, 2002; Gooch *et al.*, 2001). More recently, novel reduced-serum or chemically defined serum-free formulations have been introduced that support the growth of cartilaginous constructs (Kisiday *et al.*, 2002), reducing variability among constructs and enhancing their potential for clinical application by defining media constituents and limiting the inclusion of animal products in the culture process.

In addition to changing cell densities and media formulation, recent cartilage-engineering efforts have focused on the native loading environment as a potential modulator of engineered construct growth. In this so called "functional tissue engineering" approach, bioreactors are designed to recapitulate some portion of the *in vivo* environment in the *in vitro* culture system (Butler *et al.*, 2000). Cartilage may be particularly amenable to this approach, for it is a tissue whose function and form are defined by its physical environment. One of the first bioreactors applied for cartilage-tissue engineering was the rotating-wall vessel (Freed *et al.*, 1998). In this system, constructs tumble in a fluid-filled annulus, enhancing nutrient transport while maintaining a low-shear environment, with long-term culture resulting in cartilage-like constructs with near-native material proper-

ties (Vunjak-Novakovic *et al.*, 1999). In addition to enhancing growth via improved nutrient transport, beneficial effects of direct application of physical forces on engineered articular cartilage constructs have been widely reported. For example, when a combination of hydrostatic pressure and perfusion was applied to chondrocyte-seeded polymer meshes, improved matrix deposition was observed (Carver and Heath, 1999). In other studies, dynamic deformational loading of constructs improved the biochemical and mechanical properties of constructs (Mauck *et al.*, 2000). Cyclical shear deformation of high-density scaffold-free chondrocyte cultures on a calcium phosphate scaffold improves properties as well (Waldman *et al.*, 2003). This applied mechanical environment may also interact synergistically with anabolic growth factors (Gooch *et al.*, 2001). As these systems become more advanced, the rate and quality of cartilaginous-tissue growth may be expedited and a functional tissue-engineered cartilage construct realized.

## MSCs and Cartilage Tissue Engineering

Although significant progress has been made in the production of engineered cartilage constructs with chondrocytes, several significant impediments exist that limit their clinical application. First, chondrocytes are present in limited supply and are often of an aged and/or diseased state in patients presenting with OA. While most tissue-engineering studies in the literature report on newborn- or juvenile-derived chondrocytes, a recent study has shown that aged chondrocytes are less able to form functional constructs in agarose culture (Tran-Khanh *et al.*, 2005). Finally, chondrocytes must be isolated from the joint tissue itself, a process that may further complicate an already damaged joint. One opportunity for overcoming such concerns is the use of adult MSCs. As described earlier, these cells are readily obtained from the adult bone marrow and other tissues and retain a multilineage potential. These cells are expandable in culture (Bruder *et al.*, 1997) and may be grown in sufficient numbers to populate engineered scaffolds.

## MSC Chondrogenesis Potential and Control

Control of chondrogenesis in MSCs is a complex and developing research area, with much of our understanding of the relevant molecules and processes stemming from a continuing elucidation of the events that control healthy cartilage homeostasis as well as cartilage formation in the developing limb. Elements including soluble factors such as growth factors, cytokines, hormones, various intracellular signaling pathways and transcription factors, environmental factors such as mechanical loading and oxygen levels, and seeding density all affect chondrogenic differentiation of MSCs.

The standard system of MSC chondrogenesis involves a three-dimensional culture of MSCs under the stimulation of a suitable chondrogenic stimulus. High-density cultures of pellet or micromass are frequently employed, reminiscent of early chondrogenesis in development, where condensation of the early progenitor cells initiates the cascade of events leading to cartilage formation (Tuan *et al.*, 2004). In addition, MSCs are also seeded on various three-dimensional scaffolds to induce chondrogenesis. In this system, cell-seeding density, similar to what has been observed with chondrocytes, also affects the extent of chondrogenesis. For example, MSCs seeded at a higher density in agarose gels exhibited more cartilage-specific gene expression (Huang *et al.*, 2004b). MSCs that have undergone chondrogenic differentiation assume a chondrocyte-like phenotype characterized by increases in PG deposition and expression of aggrecan, COMP, and collagen type II (Barry *et al.*, 2001). Microarray analysis has shown that numerous other cartilage ECM elements increase in their expression as well (Sekiya *et al.*, 2002b).

Growth factors that have regulatory effects on MSCs include members of the TGF-β superfamily, the IGFs, the FGFs, and the PDGFs. Among these growth factors, TGF-βs, including TGF-β1, TGF-β2, and TGF-β3, as well as BMPs are the most potent inducers to promote chondrogenesis of MSCs (Fig. 55.3). For human MSCs, TGF-β2 and TGF-β3 were shown to be more active than TGF-β1 in promoting chondrogenesis (Barry *et al.*, 2001). BMPs, known for their involvement in cartilage formation, act alone or in concert with other growth factors to induce or enhance MSC chondrogenic differentiation. For example, BMP2, BMP4, or BMP6, combined with TGF-β3, induced the chondrogenic phenotype in cultured bone marrow derived hMSC pellets,

with BMP2 seemingly the most effective (Sekiya *et al.*, 2005). Other growth factors, such as IGF, FGF, and PDGF, are important signaling molecules that mediate chondrocyte physiology, rather than promoting chondrogenesis of MSCs, and therefore commonly work with TGF-βs to promote chondrogenesis and enhance chondrocytic activities of differentiated MSCs. The promitotic activity of the FGFs has also been exploited for cell expansion purposes (Mastrogiacomo *et al.*, 2001). Interestingly, FGF2-supplemented hMSCs proliferated more rapidly and exhibited greater chondrogenic potential than untreated controls (Solchaga *et al.*, 2005). Wnt proteins have recently been implicated in the progression of rheumatoid arthritis and OA (Loughlin *et al.*, 2004). Canonical Wnt signaling has been shown to enhance MSC differentiation (Yano *et al.*, 2005), and Wnt signaling in chondrogenesis has been shown to cross-talk with TGF-β signaling (L. Fischer *et al.*, 2002; Tuli *et al.*, 2003; Zhou *et al.*, 2004).

Growth factors act on cells and induce various intracellular signaling pathways to coordinate transcription factors and change cellular phenotype. One of the most important molecules intrinsic to the assumption of the cartilaginous phenotype is the transcription factor sox9. The role of sox9 in cartilage formation was first observed in condensations in the developing mouse limb, and its presence is thought to be required for cartilage formation (Bi *et al.*, 1999). In human bone marrow derived MSCs, exogenous expression of sox9 led to increased PG deposition (Tsuchiya *et al.*, 2003).

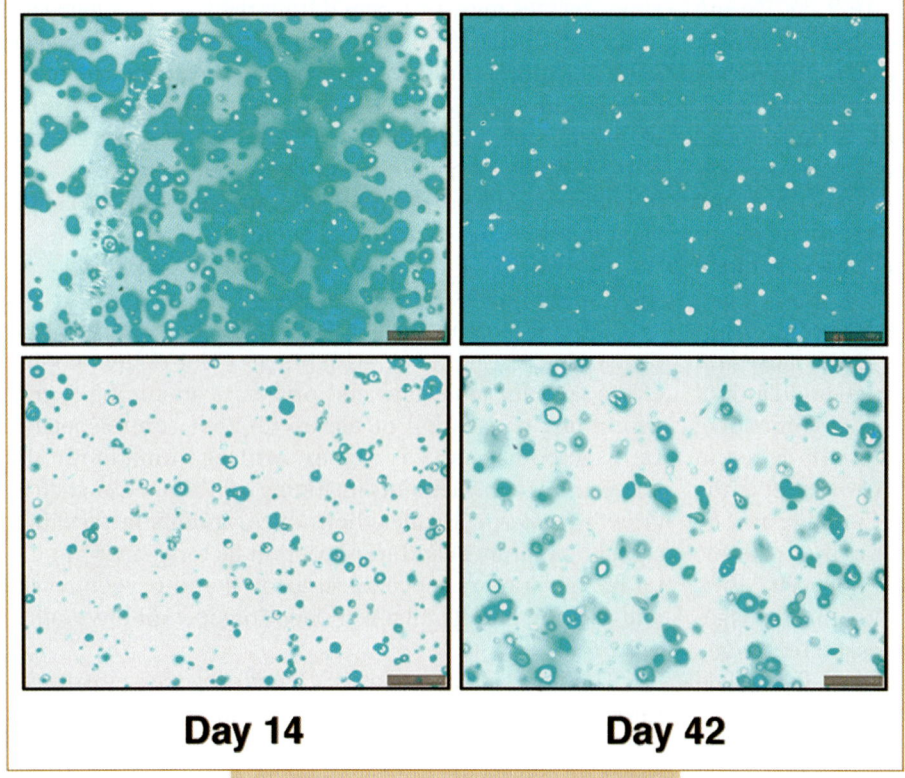

**Day 14**　　**Day 42**

**FIG. 55.3.** Chondrogenic differentiation of mesenchymal stem cells: effect of TGF-β3. Time-dependent increase in Alcian blue staining of matrix sulfated proteoglycan deposition by bovine MSCs maintained in three-dimensional agarose culture ($20 \times 10^6$ cells/mL) in a chemically defined chondrogenic medium with (top) and without (bottom) supplementation of 10 ng/mL TGF-β3. Bar, 100 μm.

In addition to sox9, other *cis* and *trans* acting factors and intracellular signaling pathways have been shown to modulate differentiation toward a chondrogenic phenotype. Factors such as AP2-α have been shown to negatively regulate chondrogenesis in ATDC5 cells, a prechondrogenic cell line, that have been induced with either insulin or TGF-β (Huang *et al.*, 2004d). Intracellular signaling cascades, particularly those involving the MAP kinases, p38 and ERK-1, are activated by TGF-β1 during cartilage-specific gene expression in MSCs, and specific inhibitors of these pathways block or limit chondrogenesis (Tuli *et al.*, 2003). In MSCs, it was also demonstrated that TGF-β1 and p38 MAP kinase signaling pathways converge on AP2 DNA binding during chondrogenesis, suggesting that one mechanism by which these signaling cascades may induce the cartilage phenotype is through interruption of the transcriptional repression exerted by AP2 (Tuli *et al.*, 2002). Taken together, these findings in MSCs suggest a complex interplay of extracellular growth factor molecules, signal transduction to the nucleus, and coordinated control of transcription factors for the induction of the chondrogenic phenotype. The current challenge in finding the most efficient growth factor(s) for MSC chondrogenesis is that the regulatory effects of signal molecules are dependent on property, dose, and timing of the molecules administered to the cells. That may explain some of the contradictory results regarding the effects of specific growth factors on chondrogenesis.

## Cartilage Tissue Engineering with MSCs

The use of MSCs for cartilage regeneration and tissue engineering has shown enormous potential and initial success. Of note, one current surgical strategy for enhancing the repair of cartilage defects is via microfracturing of the subchondral bone. This technique provides entrance for marrow elements, including MSCs, to the wound site, and has been shown to generate an enhanced, albeit fibrous, repair response. MSCs have been used as cellular therapeutics directly, via direct injection into the joint space (with and without a carrier matrix) in an undifferentiated state (Murphy *et al.*, 2003) and after differentiation down a cartilage lineage *in vitro* (Yoo *et al.*, 2000). These cells may be further transduced to express morphogenetic proteins that drive their own chondrogenesis after implantation (Palmer *et al.*, 2005).

In the context of tissue engineering, MSCs have been shown to undergo chondrogenesis in a variety of three-dimensional scaffolds. In one of the first studies of its kind, avian bone marrow stromal cells were shown to undergo chondrogenesis and osteogenesis in nonwoven PGA scaffolds with culture in inductive media (Martin *et al.*, 1998). Since then, positive chondrogenesis has been demonstrated in such scaffolds as agarose (Awad *et al.*, 2004; Huang *et al.*, 2004b), alginate (Awad *et al.*, 2004), fibrin (Worster *et al.*, 2001), gelatin (Awad *et al.*, 2004), collagen types I and II hydrogels (Bosnakovski *et al.*, 2006), polyethylene oxide

(Williams *et al.*, 2003), silk (Meinel *et al.*, 2004), and PLA (Noth *et al.*, 2002) as well as on biodegradable nanofibers (Li *et al.*, 2005b) (Table 55.3). Polymeric nanofibrous scaffolds composed of ultrafine fibers at the nanometer-level scale, by virtue of structural similarity to natural ECM, represent promising structures for cartilage tissue engineering. Recent findings indicate that the biological activities of chondrocytes (Li *et al.*, 2003) and mesenchymal stem cells (Li *et al.*, 2005b) are crucially dependent on the dimensionality of the extracellular scaffolds and that nanofibrous scaffold may be a biologically preferred scaffold/substrate for proliferation and phenotype maintenance of chondrocytes (Fig. 55.4) and chondrogenic differentiation of mesenchymal stem cells. In addition, for cartilage-tissue engineering, both *in vitro* and *in vivo* results have shown that nanofibrous scaffolds support chondrogenic differentiation of MSCs.

One remaining caveat in the MSC-based engineered cartilage tissue is the functional mechanical properties of the cartilaginous matrix formed. Recent studies have shown that MSCs in agarose and alginate hydrogels do undergo chondrogenesis and do deposit ECM with increasing mechanical properties with time in culture (Mauck *et al.*, 2006). However, the mechanical properties of the forming matrix were inferior to those of donor-matched chondrocytes similarly maintained. These findings suggest that, as was done with primary chondrocytes, a complete optimization regimen may be required to generate MSC-laden constructs that possess functional properties similar to those of the native tissue.

One such approach is to consider the role of mechanical signals in this process. Recent studies have shown that even fully differentiated tissues alter their phenotype with changes in the mechanical environment. Functional tissue engineering is likely to be dependent on similar physical forces that regulate development and maintenance of cellular phenotype. Indeed, some of the very signals that cause differentiation may be used to optimize construct growth. To date, few studies have directly examined mechanically induced differentiation in a tissue-engineering environment. For example, tensile stretching increased ligament differentiation of MSCs in collagen (Altman *et al.*, 2002), while cyclic substrate strain enhanced osteogenesis (Simmons *et al.*, 2003). Both static and dynamic loading have been shown to increase the chondrogenesis of limb bud cells in collagen and agarose (Elder *et al.*, 2001; Takahashi *et al.*, 1998), depending on the frequency and duration. Fluid pressure has recently been shown to enhance chondrogenesis of pellet cultures (Angele *et al.*, 2003), and studies in agarose hydrogels (Huang *et al.*, 2004a) have demonstrated enhanced MSC chondrogenesis with dynamic deformational loading. These findings motivate a continued exploration in the role of the mechanical loading environment in both inducing differentiation as well as enhancing subsequent construct maturation.

**Table 55.3.** Cartilage-tissue engineering with mesenchymal stem cells

| Multipotential cell source | Scaffold composition | Chondrogenic outcome parameters | References |
|---|---|---|---|
| Chicken bone marrow | PGA nonwoven mesh | HC, IHC, BC | Martin *et al.* (1998) |
| Bovine and rabbit bone marrow, human adipose tissue, chick limb bud | Agarose hydrogel | HC, IHC, GE, MT, RLI | Awad *et al.* (2004) Huang *et al.* (2004b) Li *et al.* (2005) Mauck *et al.* (2006) Elder *et al.* (2000) Elder *et al.* (2001) |
| Human bone marrow, human adipose tissue | Alginate hydrogel | IHC, GE, RLI, MT | Awad *et al.* (2004) Majumdar *et al.* (2001) Erickson *et al.* (2002) |
| Equine bone marrow | Fibrin | HC, ISH, IHC, RLI, BC | Worster *et al.* (2001) |
| Human adipose tissue | Gelatin | IHC, GE, RLI, MT | Awad *et al.* (2004) |
| Human adipose tissue | Elastin-like peptide hydrogel | HC, IHC, BC, GE | Betre *et al.* (2006) |
| Bovine bone marrow, mouse limb bud | Collagen I and II hydrogel | HC, IHC, GE, ISH | Bosnakovski *et al.* (2006) Takahashi *et al.* (1998) |
| Goat bone marrow | Poly(ethylene glycol)-diacrylate photo-cross-linkable hydrogel | HC, IHC, GE, BC | Williams *et al.* (2003) |
| Human bone marrow | Porous silk foams (with and without RGD-modification) | HC, IHC, SEM, GE, BC | Meinel *et al.* (2004) Wang *et al.* (2005) |
| Human bone marrow | Nanofibrous poly-ε-caprolactone mesh | SEM, GE, RLI, BC, HC, IHC | Li *et al.* (2005b) |

*Abbreviations*: GE, gene expression; BC, biochemical composition; ISH, *in situ* hybridization: HC, histochemical staining; IHC, immunohistochemistry; RLI, radiolabel incorporation; SEM, scanning electron microscopy; MT; mechanical testing.

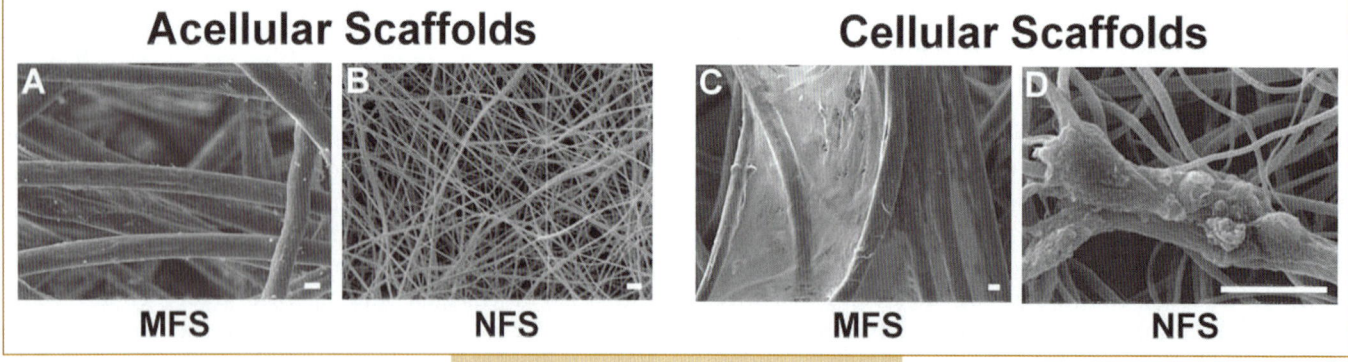

**FIG. 55.4.** Use of nanofibrous biomaterial scaffold for cartilage tissue engineering with mesenchymal stem cells: comparison with microfibrous scaffold. Ultrastructural morphology of acellular and cellular poly(L-lactic acid) (PLLA) microfibrous scaffolds and nanofibrous scaffolds were examined by scanning electron microscopy. **(A)** A PLLA microfibrous scaffold showed random orientation of microfibers, with diameters ranging from 15 to 20 μm. **(B)** A PLLA nanofibrous scaffold produced by the electrospinning process showed random orientation of ultrafine fibers, with diameters ranging from 500 to 900 nm, defining a matrix with interconnecting pores. **(C)** Spread cellular sheets composed of fibroblast-like cells spanned between microfibers in the microfibrous scaffold culture after 28 days. **(D)** Cellular aggregates composed of globular, chondrocyte-like cells growing on nanofibers in the nanofibrous scaffold culture after 28 days. Bar, 10 μm.

## Bone Tissue Engineering

Bone has a vigorous potential to regenerate itself after damage; however, the efficacious repair for large bone defects resulting from resection or trauma or nonunion fractures still requires the implantation of bone grafts. Natural bone grafts possessing excellent osteoconduction and mechanical stability have been used extensively in clinical settings; depending on the relationship between the donor and the recipient, bone grafts are categorized into autografts, allografts, and xenografts. Autografts, considered the gold standard for bone implantation, have the advantage of immunecompatibility over allografts and xenografts. However, problems such as donor site morbidity, risk of infection, and the availability of bone tissue of the correct size and shape limit the use of autografts in orthopedic applications.

Two hundred thousand surgical cases of bone-grafting procedures are performed annually, and thus the demand for bone grafts continues to rise and is expected to be even greater over the next decade as the population ages. One possible remedy for the shortage of bone grafts is a functional tissue-engineered bone graft. Both *in vivo* and *in vitro* tissue engineering approaches are employed for bone-tissue engineering using MSCs. The primary difference between these two approaches is the additional step of allowing the *ex vivo* formation of mature bone grafts, which provides a better control over the growth environment of bone tissue and could also ensure the production of mature, functional bone grafts before surgical implantation. However, this process requires longer times and skillful techniques to produce tissue-engineered bone grafts. In contrast, because the *in vivo* approach is more direct, development of this strategy is ahead of that of the other, and it has already been used in clinical applications.

MSCs are an attractive cell type for bone tissue engineering due to their potentials of extensive expansion and default osteogenic differentiation. Successful and complete osteogenic differentiation of MSCs is imperative for bone formation and still remains a challenge in bone-tissue engineering. Although it has been well known that the signals driving MSCs into osteogenic differentiation include biochemical cues from growth factors/cytokines and physical contacts from cell–cell or cell–matrix interactions, the molecular mechanisms regulating osteogenic differentiation of MSCs have yet to be fully described.

Normal or genetically modified MSCs as therapeutic agents injected into recipients have successfully induced ectopic bone formation. However, the injected cells require three-dimensional biomaterial scaffolds to secure them at the implantation site, provide physical protection, and maintain and direct tissue shape. In MSC based bone tissue engineering, various biomaterial scaffolds have been evaluated for their potential as cell carriers. These scaffolds may be made from natural (Chen *et al.*, 2002) or synthetic (E. M. Fischer *et al.*, 2003; Partridge *et al.*, 2002) materials that have been fashioned into structures with different shapes and sizes. In general, natural polymers, such as collagens, contain bioactive domains favorable for biological activities involved in tissue regeneration, whereas synthetic polymers, such as poly(α-hydroxyesters), feature controllable material properties that can approximate the physical properties of native tissue. Among the materials that have been used in bone tissue engineering, hydroxyapatite (HA) and their derivatives, such as β-tricalcium phosphate (β-TCP), are the most common scaffold materials for osteogenic induction of MSCs. Bioabsorbable TCP-based scaffolds, compared to HA-based ones, showed comparable results on ectopic bone formation (Boo *et al.*, 2002). Moreover, both synthetic ceramics support osteogenic differentiation more efficaciously than demineralized bone matrix (Kasten *et al.*, 2005). To improve the affinity of osteoconductive ceramics for cells, HA has been coated with bioactive peptides (Sawyer *et al.*, 2005) or proteins (Yang *et al.*, 2005), to enhance MSC attachment and osteogenic differentiation. Furthermore, it is well known that the chemistry and architecture of scaffolds are important factors in promoting MSC-based bone formation. Recently, a novel electrospun nanofibrous scaffold, morphologically similar to native ECM, has been applied to bone-tissue engineering. The ultrafine ECM-biomimetic nanofibers favorably supported ostegenic differentiation of MSCs *in vitro* (Li *et al.*, 2005a) and *in vivo* (Shin *et al.*, 2004).

A recent preferred trend is to combine multiple materials to produce a composite with the properties from each of the composed materials. Composite scaffolds are expected to be physically and biologically superior to single-material-based scaffolds, for the properties of a composite may be programmatically varied by mixing different materials in various ratios. Both the composition and the relative ratio of the constituent materials can affect bone formation. HA has been used as a primary material combined with another material, such as TCP (Shi *et al.*, 2005), PLGA (Turhani *et al.*, 2005), and chitin (Huang *et al.*, 2004c), to produce various composite scaffolds. It was reported that scaffolds with different ratios of HA/TCP loaded with MSCs showed different extents of bone formation *in vivo*. Composites in which the HA/TCP ratio was designed to coordinate scaffold degradation with tissue deposition seemed optimal in promoting the greatest ectopic bone formation (Arinzeh *et al.*, 2005).

Growth factors, cytokines, and other nonproteinaceous chemical factors are critical for osteogenic differentiation of MSCs, as discussed earlier. To augment bone formation successfully, it is necessary to continuously introduce osteo-inductive molecules, most of which have a short half-life, into the cell culture or the defect site. One strategy for enhancing bone formation is to use biomaterial scaffolds both as a cell carrier as well as a reservoir for the release of growth factors in a controllable manner. BMP2 is the most efficacious growth factor among the BMP family members and has been incorporated in various forms of biomaterial scaffolds to induce osteogenesis in *ex vivo* cultures (Huang *et al.*, 2004c) or to stimulate bone formation in *in vivo* models (Saito *et al.*, 2001).

Bioreactor systems, such as spinner flasks, rotating-wall vessels, and perfusion chambers, have also been demonstrated to be useful tools for enhancing bone formation of MSC-laden scaffolds *in vitro*. Such systems more closely resemble the *in vivo* environment of cells by providing a dynamic culture system that improves mass transport. Some bioreactor designs also apply mechanical loading to the cultured bone constructs, actively inducing osteoblast maturation and stimulating bone matrix production. These MSC-laden constructs with mechanical loading in culture became biologically more mature and mechanically stronger, compared to those without mechanical loading (Mauney *et al.*, 2004).

## Osteochondral Tissue Engineering

Severe joint defects often extend to damage or destruction of subchondral bone, leading to associated pain and mechanical instability of the joint. It is well known that the body attempts a spontaneous repair of osteochondral lesions that penetrate the subchondral bone by activating the healing mechanism mediated by MSCs in the exposed bone marrow. Clinically accepted treatments, such as microfracture, apply the same concept to encourage healing and regeneration of the lost cartilage. However, the density of MSCs released by the physical destruction of subchondral bone is low and often not enough for efficient repair. Instead of mechanically strong hyaline cartilage, the lesion is usually repaired by fibrocartilage that erodes over a short time period. In addition, due to the removal of subchondral bone, increased mechanical instability of the lesion site makes the repair even more difficult. Clinical results show that, even for a partial-thickness cartilage defect, there are beneficial effects in exposing subchondral bone by drilling as well as the incorporation of a bone layer as an anchor to integrate with host tissue securely. Therefore osteochondral grafts or plugs are used clinically in the treatment of both chondral and osteochondral defects.

A tissue-engineered osteochondral graft, composed of both cartilage and bone components, is a promising alternative to the use of autologous osteochondral grafts. The engineered construct is more controllable in size and shape, and donor site morbidity experienced when grafts are harvested for mosaicplasty procedures is eliminated. For osteochondral tissue engineering, a number of three-dimensional biomaterial scaffolds have been developed, with the aim of optimizing MSC attachment, proliferation, and osteogenic/chondrogenic differentiation. Such scaffolds, including hyaluronan (Radice *et al.*, 2000) and hyaluronan-gelatin (Angele *et al.*, 1999), can be used as delivery vehicles *in vivo* for a large number of *ex vivo*–expanded MSCs to the lesion site while providing physical protection to the resident cells. In this approach, although MSCs are retained locally at the osteochondral lesion for the repair, the actual regeneration of tissue is unpredictable and depends primarily on the host tissue. A more sophisticated approach for osteochondral tissue engineering is to take advantage of the multidifferentiation potential of MSCs to fabricate an osteochondral composite graft *in vitro* or *ex vivo* before implantation. In contrast to using merely three-dimensional biomaterial scaffolds as delivery vehicles without *in vitro* tissue preformation, this approach can produce a functional osteochondral graft that is biologically similar to the native host tissue. Angele *et al.* (1999) have shown that the repair of osteochondral lesion, using the precultured osteochondral graft, is indeed more effective.

Osteochondral grafts are single-unit biphasic composite materials consisting of two tissues with different properties, with cartilage on the top of bone, and as such have been technically challenging to engineer. Chondrogenesis and osteogenesis of MSCs require different physical and biochemical cues from matrices and soluble growth factors/cytokines, respectively. A commonly used approach is to fabricate cartilage and bone independently before integrating them together using sutures or glues (Gao *et al.*, 2001; Schaefer *et al.*, 2000). In one approach, MSCs loaded into a hyaluronan or a TCP ceramic scaffold were separately induced to undergo chondrogenic or osteogenic differentiation. Then the two components were integrated together with fibrin sealant, becoming a single unit of an osteochondral construct (Gao *et al.*, 2001; Schaefer *et al.*, 2000). One drawback with this method is the less-than-optimal integration between the chondral and osteo constructs; poor integration leads to discontinuous cell distribution at the interface and/or the possibility of eventual separation of the two components. Several groups have reported different alternatives to solve the problem of discontinuous interface. In one approach, MSCs after *in vitro* chondrogenic or osteogenic differentiation were loaded into two separate polyethylene glycol (PEG) hydrogel layers and the cell-loaded PEG gels were then photopolymerized together. Since the two components were combined before gel solidification, the osteochondral construct exhibited a less defined gap line (Alhadlaq and Mao, 2005). Another approach was to apply a press-coating process (Noth *et al.*, 2002) to fabricate osteochondral constructs. A PLA scaffold was press-coated with a high density pellet of MSCs previously induced in chondrogenic culture and subsequently seeded with preosteogenically differentiated MSCs at the other end of the scaffold. Macroscopically, the osteochondral composite consisted of a cartilage-like layer adherent to and overlying a dense bonelike component (Tuli *et al.*, 2004). Since both cartilage and bone were produced in a single unit, there existed no gap between the two tissues; instead, an interface resembling the native osteochondral junction was observed.

The ideal scenario for the fabrication of biphasic osteochondral constructs would be to differentiate MSCs cultured in a single-unit scaffold into chondrocytes on the top and osteoblasts on the bottom. To this end, the biomaterial scaffold should chemically and structurally support both chondrogenesis and osteogenesis. Novel nanofibrous scaffolds morphologically resembling natural ECMs have been shown successfully to support chondro- and osteogenesis of MSCs *in vitro* (Li *et al.*, 2005a). Another need for achieving this goal is a culture system in which an MSC-laden construct can differentiate into cartilage and bone simultaneously. A double-chamber bioreactor with a unique two-compartment design allowing the storage of different media was used to culture osteochondral constructs (C. H. Chang *et al.*, 2004). With this double-chamber bioreactor, it would be feasible to engineer an autologous osteochondral graft using MSCs by means of a nanofibrous scaffold.

## Engineering Other Skeletal Tissues with MSCs

In addition to their clear roles in engineering bone and cartilaginous tissues, MSCs may be readily used for the repair of numerous other musculoskeletal tissues. This is partially predicated on the belief that endogenous tissue repair of these tissues may be driven by progenitor cells located within or those that migrate into injury sites, proliferate, and deposit new matrix. Indeed, an increasing number of musculoskeletal tissues have been found to harbor cells that retain some multipotential characteristics, including articular cartilage, meniscus, trabecular bone, synovium, heart and skeletal muscle, and periodontal ligaments (Table 55.1). These findings suggest a wide distribution of these cells and consequently the potential for effecting repair of specific tissues via their exogenous application. If successful, such cell-based tissue engineering therapies would rectify what are otherwise untreatable and progressively debilitating disorders.

Tendon and ligament tissue engineering was one of the first areas in which MSCs have found application. In early studies, it was observed that rabbit MSCs contract collagen type I carrier gels and that the delivery of these constructs to patellar and Achilles tendon defects improved the biomechanical properties of the repair tissue as compared to acellular controls (Young *et al.*, 1998). Newly emerging silk scaffolds that have been modified with arginine-glycine-aspartic acid moieties and/or formed into nanofibrous scaffolds may also prove useful for tendon/ligament-tissue engineering with MSCs (Jin *et al.*, 2004). Recent studies, applying tensional and torsional mechanical stimulations to MSC-seeded collagen gels *in vitro*, have shown enhanced fibrous differentiation (Altman *et al.*, 2002). These studies suggest a role for MSCs in accelerating the repair of tendon and ligament defects that may be modulated by the mechanical loading environment.

Another tissue in which exogenous MSC application has been used to effect repair is the fibrocartilaginous knee meniscus. It has been reported that intraarticular injection of MSCs (absent a tissue-engineering delivery vehicle) can have an ameliorating effect on articular cartilage and meniscus degeneration. Injection of MSCs in hyaluronan in an anterior cruciate ligament resection/medial meniscectomy model led to some regeneration of the meniscus and protection of the underlying cartilage (Murphy *et al.*, 2003), with GFP-labeled MSCs colonizing the healing meniscus. Alternatively, direct delivery of MSCs to meniscal defects within a fibrin clot led to the maintenance of cells in the defect for up to eight weeks after implantation (Izuta *et al.*, 2005). Clearly, MSCs can play a role in the reparative process of the meniscus, although the mode of their administration and the nature of the carrier materials require further optimization.

In addition to these purely fibrous structures, MSC therapy is being widely investigated for treatment of other musculoskeletal tissues, including skeletal and cardiac muscle and intervertebral disc. Taken together, these findings suggest that MSCs will find wide application in the treatment of a number of different musculoskeletal disorders, greatly expanding the realm of regenerative medicine and ultimately enhancing the quality of human life.

## Gene Therapy in Musculoskeletal-Tissue Engineering

Gene therapy is a promising technique for disease treatment, in which the genetically modified cells are transferred into or generated within individuals for therapeutic purposes. The cells with modified genes can be programmed to differentiate into desired cells or produce proteins needed for tissue repair. Although some safety issues still remain, a likely future scenario is the merging of gene therapy, cell therapy, and tissue engineering for the treatment of musculoskeletal diseases.

Two strategies are currently used in gene therapy, viral transduction and nonviral transfection, and both strategies can be conducted *in vivo* and *ex vivo*. *In vivo* gene therapy is simple but often lacks tracking of the gene-modified cells and target site specificity. An attractive alternative is to combine gene therapy with a tissue-engineering approach to transduce or transfect cells of interest *ex vivo* and then to use a carrier to deliver these genetically modified cells to the target site. This approach offers the advantages of flexibility of target cell type and retaining of gene-modified cells at the site of interest. Gene transduction using viral vectors, such as retrovirus, adenovirus, and lentivirus, effectively modifies the host chromosome but raises concerns on mutagenesis and the immune response. In contrast, the nonviral approach is safer, but the efficiency of transfection is lower. More research efforts are needed to overcome the limitations associated with each approach. The ultimate goal is to develop a simple, safe, and effective approach to transfer genes into the cell of interest.

Currently, the most frequently applied example of gene therapy in bone tissue engineering is the transduction/ transfection of interested cells with BMP genes. BMPs play important roles in the regulation of osteogenic differentiation of MSCs and the production of bone matrix during bone formation. To overcome the short half-life limit of growth factors, MSCs are transduced or transfected with BMP genes for the continuous expression of the required BMPs *in vitro* and *in vivo*. The BMP2-transduced MSCs were induced to differentiate into osteoblasts producing bone matrix and to synthesize BMP2, attracting the host cells to migrate and differentiate. Compared to control MSCs, the BMP2-producing MSCs effectively enhance bone formation, even at large defect sites like segmental femoral defects (Lieberman *et al.*, 1999). Similar results have also been reported when BMP2- or BMP4-transduced MSCs were delivered to bone defects using different biomaterial scaffolds, such as demineralized bone matrix (Lieberman *et al.*, 1999) and gelatin (Gysin *et al.*, 2002). Runx2 is an early transcription factor regulating osteoblast differentiation. One recent study has shown that Runx2-modified MSCs cultured in three-dimensional polymeric scaffolds successfully created a bone construct with significantly up-regulated osteoblast differentiation and mineralization (Byers *et al.*, 2004). Other osteogenic transcription factors, such as Osterix, and Foxc1 have also been shown to promote MSC osteogenic differentiation.

The potential benefits of cartilage repair using gene therapy are also gaining recognition. For example, BMP7-transduced MSCs delivered by PGA scaffolds successfully regenerated cartilage, whereas the control nontransduced group did poorly (Mason *et al.*, 2000), and BMP4 transduction induced chondrogenesis and enhanced cartilage repair (Kuroda *et al.*, 2006). The level of chondrogenesis in MSCs transduced with various growth factors was dependent on the level of expression of the exogenous genes (Palmer *et al.*, 2005).

## IV. CONCLUSIONS AND FUTURE PERSPECTIVES

MSCs, isolated from various tissues, can be maintained in culture and self renew to yield a large number of cells in an undifferentiated stage and then, under appropriate conditions, to differentiate into cells of various mesenchymal lineages, including cell types of the musculoskeletal tissues. Due to their ease of isolation, their capacity for undergoing *in vitro* proliferation to achieve a cell number large enough for cell therapy, their ability to undergo lineage-specific differentiation into musculoskeletal cells, and their potential immunomodulatory advantages, MSCs present significant potential in musculoskeletal tissue engineering, which promises to bring hope to patients and surgeons alike for the generation of functional tissue substitutes. As discussed in this chapter, MSCs have been used in tissue engineering of a number of musculoskeletal tissues, including cartilage, bone, osteochondral constructs, ligament, and tendon, with varying degrees of success. Currently, tissue-engineered constructs have not been readily accepted for clinical use to treat skeletal tissue defects. There clearly is a need for further research that combines the concerted efforts of biologists as well as engineers and clinicians. Critical to the success of these approaches is a better understanding of MSC biology, which still poses great challenges. So far, there has not been a marker that can be used unequivocally to identify and prospectively select MSCs from various tissues. The factors that regulate phenotype transition, i.e., uncommitted vs. differentiated phenotype, remain to be identified and studied. Additionally, the various growth factors, signaling pathways, and transcription factors that can influence MSCs to differentiate completely and stably into a desirable lineage require further elucidation. Further complicating the matter is the heterogeneity and differences among various tissue sources of MSCs. Despite our limited understanding of MSC biology, the use of MSCs in musculoskeletal tissue engineering has shown great promise. Future efforts in this area should include the fabrication of optimized scaffolds, systematic studies of the interplay of MSC, scaffold and environmental factors, and the development of quantitative outcome measurements for tissue-engineered constructs. *In vivo* testing of the engineered construct serves as the gold standard for the long-term survival of the constructs and warrants much greater attention. However, the immunosuppressive and anti-inflammatory effects of MSCs offer encouragement in their use in allogeneic transplantation in the inflammatory environment of an injury site. With further research and development, the use of MSCs in tissue engineering is expected to bring to fruition a tissue substitute that is functionally suitable for implantation to improve the quality of life of patients with debilitating musculoskeletal injuries.

## V. ACKNOWLEDGMENT

This work is supported by the NIAMS Intramural Research Program (NIH Z01 41113).

## VI. REFERENCES

Aggarwal, S., and Pittenger, M. F. (2005). Human mesenchymal stem cells modulate allogeneic immune cell responses. *Blood* **105**, 1815–1822.

Alhadlaq, A., and Mao, J. J. (2005). Tissue-engineered osteochondral constructs in the shape of an articular condyle. *J. Bone Joint Surg. Amer.* **87**, 936–944.

Altman, G. H., Horan, R. L., Martin, I., Farhadi, J., Stark, P. R., Volloch, V., Richmond, J. C., Vunjak-Novakovic, G., and Kaplan, D. L. (2002). Cell differentiation by mechanical stress. *FASEB J.* **16**, 270–272.

Angele, P., Kujat, R., Nerlich, M., Yoo, J., Goldberg, V., and Johnstone, B. (1999). Engineering of osteochondral tissue with bone marrow mesenchymal progenitor cells in a derivatized hyaluronan-gelatin composite sponge. *Tissue Eng.* **5**, 545–554.

Angele, P., Yoo, J. U., Smith, C., Mansour, J., Jepsen, K. J., Nerlich, M., and Johnstone, B. (2003). Cyclic hydrostatic pressure enhances the chondrogenic phenotype of human mesenchymal progenitor cells differentiated *in vitro*. *J. Orthop. Res.* **21**, 451–457.

Arinzeh, T. L., Tran, T., McAlary, J., and Daculsi, G. (2005). A comparative study of biphasic calcium phosphate ceramics for human-mesenchymal-stem-cell-induced bone formation. *Biomaterials* **26**, 3631–3638.

Awad, H. A., Wickham, M. Q., Leddy, H. A., Gimble, J. M., and Guilak, F. (2004). Chondrogenic differentiation of adipose-derived adult stem cells in agarose, alginate, and gelatin scaffolds. *Biomaterials* **25**, 3211–3222.

Baksh, D., Song, L., and Tuan, R. S. (2004). Adult mesenchymal stem cells: characterization, differentiation, and application in cell and gene therapy. *J. Cell Mol. Med.* **8**, 301–316.

Barry, F., Boynton, R. E., Liu, B., and Murphy, J. M. (2001). Chondrogenic differentiation of mesenchymal stem cells from bone marrow: differentiation-dependent gene expression of matrix components. *Exp. Cell Res.* **268**, 189–200.

Bartholomew, A., Sturgeon, C., Siatskas, M., Ferrer, K., McIntosh, K., Patil, S., Hardy, W., Devine, S., Ucker, D., Deans, R., Moseley, A., and Hoffman, R. (2002). Mesenchymal stem cells suppress lymphocyte proliferation *in vitro* and prolong skin graft survival *in vivo*. *Exp. Hematol.* **30**, 42–48.

Bennett, J. H., Joyner, C. J., Triffitt, J. T., and Owen, M. E. (1991). Adipocytic cells cultured from marrow have osteogenic potential. *J. Cell Sci.* **99**(Pt. 1), 131–139.

Beresford, J. N., Bennett, J. H., Devlin, C., Leboy, P. S., and Owen, M. E. (1992). Evidence for an inverse relationship between the differentiation of adipocytic and osteogenic cells in rat marrow stromal cell cultures. *J. Cell Sci.* **102**(Pt. 2), 341–351.

Beyth, S., Borovsky, Z., Mevorach, D., Liebergall, M., Gazit, Z., Aslan, H., Galun, E., and Rachmilewitz, J. (2005). Human mesenchymal stem cells alter antigen-presenting cell maturation and induce T-cell unresponsiveness. *Blood* **105**, 2214–2219.

Bi, W., Deng, J. M., Zhang, Z., Behringer, R. R., and de Crombrugghe, B. (1999). Sox9 is required for cartilage formation. *Nat. Genet.* **22**, 85–89.

Bianco, P., Costantini, M., Dearden, L. C., and Bonucci, E. (1988). Alkaline phosphatase positive precursors of adipocytes in the human bone marrow. *Br. J. Hematol.* **68**, 401–403.

Blunk, T., Sieminski, A. L., Gooch, K. J., Courter, D. L., Hollander, A. P., Nahir, A. M., Langer, R., Vunjak-Novakovic, G., and Freed, L. E. (2002). Differential effects of growth factors on tissue-engineered cartilage. *Tissue Eng.* **8**, 73–84.

Boland, G. M., Perkins, G., Hall, D. J., and Tuan, R. S. (2004). Wnt 3a promotes proliferation and suppresses osteogenic differentiation of adult human mesenchymal stem cells. *J. Cell Biochem.* **93**, 1210–30.

Boo, J. S., Yamada, Y., Okazaki, Y., Hibino, Y., Okada, K., Hata, K., Yoshikawa, T., Sugiura, Y., and Ueda, M. (2002). Tissue-engineered bone using mesenchymal stem cells and a biodegradable scaffold. *J. Craniofac. Surg.* **13**, 231–239.

Bosnakovski, D., Mizuno, M., Kim, G., Takagi, S., Okumura, M., and Fujinaga, T. (2006). Chondrogenic differentiation of bovine bone marrow mesenchymal stem cells (MSCs) in different hydrogels: influence of collagen type II extracellular matrix on MSC chondrogenesis. *Biotechnol. Bioeng.* **93**, 1152–1163.

Brittberg, M., Lindahl, A., Nilsson, A., Ohlsson, C., Isaksson, O., and Peterson, L. (1994). Treatment of deep cartilage defects in the knee with autologous chondrocyte transplantation [see comments]. *N. Engl. J. Med.* **331**, 889–895.

Brown, T. D., and Shaw, D. T. (1983). *In vitro* contact stress distributions in the natural human hip. *J. Biomech.* **16**, 373–384.

Bruder, S. P., Jaiswal, N., and Haynesworth, S. E. (1997). Growth kinetics, self-renewal, and the osteogenic potential of purified human mesenchymal stem cells during extensive subcultivation and following cryopreservation. *J. Cell Biochem.* **64**, 278–294.

Buschmann, M. D., Gluzband, Y. A., Grodzinsky, A. J., Kimura, J. H., and Hunziker, E. B. (1992). Chondrocytes in agarose culture synthesize a mechanically functional extracellular matrix. *J. Orthop. Res.* **10**, 745–758.

Butler, D. L., Goldstein, S. A., and Guilak, F. (2000). Functional tissue engineering: the role of biomechanics. *J. Biomech. Eng.* **122**, 570–575.

Byers, B. A., Guldberg, R. E., and Garcia, A. J. (2004). Synergy between genetic and tissue engineering: Runx2 overexpression and *in vitro* construct development enhance *in vivo* mineralization. *Tissue Eng.* **10**, 1757–1766.

Caplan, A. I. (1991). Mesenchymal stem cells. *J. Orthop. Res.* **9**, 641–650.

Carver, S. E., and Heath, C. A. (1999). Semicontinuous perfusion system for delivering intermittent physiological pressure to regenerating cartilage. *Tissue Eng.* **5**, 1–11.

Chang, C. H., Lin, F. H., Lin, C. C., Chou, C. H., and Liu, H. C. (2004). Cartilage tissue engineering on the surface of a novel gelatin-calcium-phosphate biphasic scaffold in a double-chamber bioreactor. *J. Biomed. Mater. Res. B Appl. Biomater.* **71**, 313–321.

Chang, S. C., Rowley, J. A., Tobias, G., Genes, N. G., Roy, A. K., Mooney, D. J., Vacanti, C. A., and Bonassar, L. J. (2001). Injection molding of chondrocyte/alginate constructs in the shape of facial implants. *J. Biomed. Mater. Res.* **55**, 503–511.

Chen, F., Mao, T., Tao, K., Chen, S., Ding, G., and Gu, X. (2002). Bone graft in the shape of human mandibular condyle reconstruction via seeding marrow-derived osteoblasts into porous coral in a nude mice model. *J. Oral Maxillofac. Surg.* **60**, 1155–1159.

Cho, H. H., Kim, Y. J., Kim, S. J., Kim, J. H., Bae, Y. C., Ba, B., and Jung, J. S. (2006). Endogenous Wnt signaling promotes proliferation and suppresses osteogenic differentiation in human adipose derived stromal cells. *Tissue Eng.* **12**, 111–121.

Colter, D. C., Class, R., DiGirolamo, C. M., and Prockop, D. J. (2000). Rapid expansion of recycling stem cells in cultures of plastic-adherent cells from human bone marrow. *Proc. Natl. Acad. Sci. U.S.A.* **97**, 3213–3218.

De Boer, J., Wang, H. J., and Van Blitterswijk, C. (2004). Effects of Wnt signaling on proliferation and differentiation of human mesenchymal stem cells. *Tissue Eng.* **10**, 393–401.

Di Nicola, M., Carlo-Stella, C., Magni, M., Milanesi, M., Longoni, P. D., Matteucci, P., Grisanti, S., and Gianni, A. M. (2002). Human bone marrow stromal cells suppress T-lymphocyte proliferation induced by cellular or nonspecific mitogenic stimuli. *Blood* **99**, 3838–3843.

Djouad, F., Plence, P., Bony, C., Tropel, P., Apparailly, F., Sany, J., Noel, D., and Jorgensen, C. (2003). Immunosuppressive effect of mesenchymal stem cells favors tumor growth in allogeneic animals. *Blood* **102**, 3837–3844.

Elder, S. H., Goldstein, S. A., Kimura, J. H., Soslowsky, L. J., and Spengler, D. M. (2001). Chondrocyte differentiation is modulated by frequency and duration of cyclic compressive loading. *Ann. Biomed. Eng.* **29**, 476–482.

Elisseeff, J. H., Lee, A., Kleinman, H. K., and Yamada, Y. (2002). Biological response of chondrocytes to hydrogels. *Ann. N.Y. Acad. Sci.* **961**, 118–122.

Felson, D. T., and Zhang, Y. (1998). An update on the epidemiology of knee and hip osteoarthritis with a view to prevention. *Arthritis Rheum.* **41**, 1343–1355.

Fischer, E. M., Layrolle, P., Van Blitterswijk, C. A., and De Bruijn, J. D. (2003). Bone formation by mesenchymal progenitor cells cultured on dense and microporous hydroxyapatite particles. *Tissue Eng.* **9**, 1179–1188.

Fischer, L., Boland, G., and Tuan, R. S. (2002). Wnt-3A enhances bone morphogenetic protein-2-mediated chondrogenesis of murine C3H10T1/2 mesenchymal cells. *J. Biol. Chem.* **277**, 30870–30878.

Fortier, L. A., Nixon, A. J., Mohammed, H. O., and Lust, G. (1997). Altered biological activity of equine chondrocytes cultured in a three-dimensional fibrin matrix and supplemented with transforming growth factor beta-1. *Am. J. Vet. Res.* **58**, 66–70.

Freed, L. E., Hollander, A. P., Martin, I., Barry, J. R., Langer, R., and Vunjak-Novakovic, G. (1998). Chondrogenesis in a cell-polymer-bioreactor system. *Exp. Cell Res.* **240**, 58–65.

Friedenstein, A. J., Piatetzky-Shapiro, I., and Petrakova, K. V. (1966). Osteogenesis in transplants of bone marrow cells. *J. Embryol. Exp. Morphol.* **16**, 381–390.

Friedenstein, A. J., Chailakhyan, R. K., Latsinik, N. V., Panasyuk, A. F., and Keiliss-Borok, I. V. (1974). Stromal cells responsible for transferring the microenvironment of the hemopoietic tissues. Cloning *in vitro* and retransplantation *in vivo*. *Transplantation* **17**, 331–340.

Gao, J., Dennis, J. E., Solchaga, L. A., Awadallah, A. S., Goldberg, V. M., and Caplan, A. I. (2001). Tissue-engineered fabrication of an osteochondral composite graft using rat bone marrow–derived mesenchymal stem cells. *Tissue Eng.* **7**, 363–371.

Gentili, C., Bianco, P., Neri, M., Malpeli, M., Campanile, G., Castagnola, P., Cancedda, R., and Cancedda, F. D. (1993). Cell proliferation, extracellular matrix mineralization, and ovotransferrin transient expression during *in vitro* differentiation of chick hypertrophic chondrocytes into osteoblast-like cells. *J. Cell Biol.* **122**, 703–712.

Gooch, K. J., Blunk, T., Courter, D. L., Sieminski, A. L., Bursac, P. M., Vunjak-Novakovic, G., and Freed, L. E. (2001). IGF-I and mechanical environment interact to modulate engineered cartilage development. *Biochem. Biophys. Res. Commun.* **286**, 909–915.

Gotherstrom, C., Ringden, O., Tammik, C., Zetterberg, E., Westgren, M., and Le Blanc, K. (2004). Immunologic properties of human fetal mesenchymal stem cells. *Am. J. Obstet. Gynecol.* **190**, 239–245.

Gregory, C. A., Prockop, D. J., and Spees, J. L. (2005). Non-hematopoietic bone marrow stem cells: molecular control of expansion and differentiation. *Exp. Cell Res.* **306**, 330–335.

Gronthos, S., Graves, S. E., Ohta, S., and Simmons, P. J. (1994). The STRO-1+ fraction of adult human bone marrow contains the osteogenic precursors. *Blood* **84**, 4164–4173.

Gronthos, S., Zannettino, A. C., Hay, S. J., Shi, S., Graves, S. E., Kortesidis, A., and Simmons, P. J. (2003). Molecular and cellular characterization of highly purified stromal stem cells derived from human bone marrow. *J. Cell Sci.* **116**, 1827–1835.

Guilak, F., and Hung, C. T. (2005). Physical regulation of cartilage metabolism. *In* "Basic Orthopedic Biomechanics and Mechano-Biology" (V. C. Mow and R. Huiskes, eds.), pp. 259–300. Lippincott Williams & Wilkins, Philadelphia.

Gysin, R., Wergedal, J. E., Sheng, M. H., Kasukawa, Y., Miyakoshi, N., Chen, S. T., Peng, H., Lau, K. H., Mohan, S., and Baylink, D. J. (2002). *Ex vivo* gene therapy with stromal cells transduced with a retroviral vector containing the BMP4 gene completely heals critical size calvarial defect in rats. *Gene Ther.* **9**, 991–999.

Huang, C. Y., Hagar, K. L., Frost, L. E., Sun, Y., and Cheung, H. S. (2004a). Effects of cyclic compressive loading on chondrogenesis of rabbit-bone-marrow-derived mesenchymal stem cells. *Stem Cells* **22**, 313–323.

Huang, C. Y., Reuben, P. M., D'Ippolito, G., Schiller, P. C., and Cheung, H. S. (2004b). Chondrogenesis of human bone marrow-derived mesenchymal stem cells in agarose culture. *Anat. Rec. A Discov. Mol. Cell Evol. Biol.* **278**, 428–436.

Huang, W., Carlsen, B., Wulur, I., Rudkin, G., Ishida, K., Wu, B., Yamaguchi, D. T., and Miller, T. A. (2004c). BMP-2 exerts differential effects on differentiation of rabbit bone marrow stromal cells grown in two-dimensional and three-dimensional systems and is required for *in vitro* bone formation in a PLGA scaffold. *Exp. Cell Res.* **299**, 325–334.

Huang, Z., Xu, H., and Sandell, L. (2004d). Negative regulation of chondrocyte differentiation by transcription factor AP-2alpha. *J. Bone Miner. Res.* **19**, 245–255.

Izuta, Y., Ochi, M., Adachi, N., Deie, M., Yamasaki, T., and Shinomiya, R. (2005). Meniscal repair using bone marrow–derived mesenchymal stem cells: experimental study using green fluorescent protein transgenic rats. *Knee* **12**, 217–223.

Jiang, Y., Jahagirdar, B. N., Reinhardt, R. L., Schwartz, R. E., Keene, C. D., Ortiz-Gonzalez, X. R., Reyes, M., Lenvik, T., Lund, T., Blackstad, M., Du, J., Aldrich, S., Lisberg, A., Low, W. C., Largaespada, D. A., and Verfaillie, C. M. (2002). Pluripotency of mesenchymal stem cells derived from adult marrow. *Nature* **418**, 41–49.

Jin, H. J., Chen, J., Karageorgiou, V., Altman, G. H., and Kaplan, D. L. (2004). Human bone marrow stromal cell responses on electrospun silk fibroin mats. *Biomaterials* **25**, 1039–1047.

Kahn, A. J., and Simmons, D. J. (1977). Chondrocyte-to-osteocyte transformation in grafts of perichondrium-free epiphyseal cartilage. *Clin. Orthop. Relat. Res.* **129**, 299–304.

Kasten, P., Vogel, J., Luginbuhl, R., Niemeyer, P., Tonak, M., Lorenz, H., Helbig, L., Weiss, S., Fellenberg, J., Leo, A., Simank, H. G., and Richter, W. (2005). Ectopic bone formation associated with mesenchymal stem cells in a resorbable calcium deficient hydroxyapatite carrier. *Biomaterials* **26**, 5879–5889.

Kern, S., Eichler, H., Stoeve, J., Kluter, H., and Bieback, K. (2006). Comparative analysis of mesenchymal stem cells from bone marrow, umbilical cord blood or adipose tissue. *Stem Cells* **24**, 1294–301.

Kisiday, J., Jin, M., Kurz, B., Hung, H., Semino, C., Zhang, S., and Grodzinsky, A. J. (2002). Self-assembling peptide hydrogel fosters chondrocyte extracellular matrix production and cell division: implications for cartilage tissue repair. *Proc. Natl. Acad. Sci. U.S.A.* **99**, 9996–10001.

Kuroda, R., Usas, A., Kubo, S., Corsi, K., Peng, H., Rose, T., Cummins, J., Fu, F. H., and Huard, J. (2006). Cartilage repair using bone morphogenetic protein 4 and muscle-derived stem cells. *Arthritis Rheum.* **54**, 433–442.

Kuznetsov, S. A., Krebsbach, P. H., Satomura, K., Kerr, J., Riminucci, M., Benayahu, D., and Robey, P. G. (1997). Single-colony-derived strains of human marrow stromal fibroblasts form bone after transplantation *in vivo. J. Bone Miner. Res.* **12**, 1335–1347.

Le Blanc, K., Tammik, C., Rosendahl, K., Zetterberg, E., and Ringden, O. (2003). HLA expression and immunologic properties of differentiated and undifferentiated mesenchymal stem cells. *Exp. Hematol.* **31**, 890–896.

Li, W. J., Danielson, K. G., Alexander, P. G., and Tuan, R. S. (2003). Biological response of chondrocytes cultured in three-dimensional nanofibrous poly(ε-caprolactone) scaffolds. *J. Biomed Mater. Res.* **67A**, 1105–1114.

Li, W. J., Tuli, R., Huang, X., Laquerriere, P., and Tuan, R. S. (2005a). Multilineage differentiation of human mesenchymal stem cells in a three-dimensional nanofibrous scaffold. *Biomaterials* **26**, 5158–5766.

Li, W. J., Tuli, R., Okafor, C., Derfoul, A., Danielson, K. G., Hall, D. J., and Tuan, R. S. (2005b). A three-dimensional nanofibrous scaffold for cartilage tissue engineering using human mesenchymal stem cells. *Biomaterials* **26**, 599–609.

Lieberman, J. R., Daluiski, A., Stevenson, S., Wu, L., McAllister, P., Lee, Y. P., Kabo, J. M., Finerman, G. A., Berk, A. J., and Witte, O. N. (1999). The effect of regional gene therapy with bone morphogenetic protein-2-producing bone marrow cells on the repair of segmental femoral defects in rats. *J. Bone Joint Surg. Am.* **81**, 905–917.

Liu, L., DiGirolamo, C. M., Navarro, P. A., Blasco, M. A., and Keefe, D. L. (2004). Telomerase deficiency impairs differentiation of mesenchymal stem cells. *Exp. Cell Res.* **294**, 1–8.

Loughlin, J., Dowling, B., Chapman, K., Marcelline, L., Mustafa, Z., Southam, L., Ferreira, A., Ciesielski, C., Carson, D. A., and Corr, M. (2004). Functional variants within the secreted frizzled-related protein 3 gene are associated with hip osteoarthritis in females. *Proc. Natl. Acad. Sci. U.S.A.* **101**, 9757–9762.

Majumdar, M. K., Thiede, M. A., Haynesworth, S. E., Bruder, S. P., and Gerson, S. L. (2000). Human marrow-derived mesenchymal stem cells (MSCs) express hematopoietic cytokines and support long-term hematopoiesis when differentiated toward stromal and osteogenic lineages. *J. Hematother. Stem Cell Res.* **9**, 841–848.

Martin, I., Padera, R. F., Vunjak-Novakovic, G., and Freed, L. E. (1998). *In vitro* differentiation of chick embryo bone marrow stromal cells into cartilaginous and bonelike tissues. *J. Orthop. Res.* **16**, 181–189.

Mason, J. M., Breitbart, A. S., Barcia, M., Porti, D., Pergolizzi, R. G., and Grande, D. A. (2000). Cartilage and bone regeneration using gene-enhanced tissue engineering. *Clin. Orthop. Relat. Res.* **379**, S171–S178.

Mastrogiacomo, M., Cancedda, R., and Quarto, R. (2001). Effect of different growth factors on the chondrogenic potential of human bone marrow stromal cells. *Osteoarthritis Cartilage* **9** (Suppl. A), S36–S40.

Mauck, R. L., Soltz, M. A., Wang, C. C., Wong, D. D., Chao, P. H., Valhmu, W. B., Hung, C. T., and Ateshian, G. A. (2000). Functional tissue engineering of articular cartilage through dynamic loading of chondrocyte-seeded agarose gels. *J. Biomech. Eng.* **122**, 252–260.

Mauck, R. L., Wang, C. C., Oswald, E. S., Ateshian, G. A., and Hung, C. T. (2003). The role of cell seeding density and nutrient supply for articular cartilage tissue engineering with deformational loading. *Osteoarthritis Cartilage* **11**, 879–890.

Mauck, R. L., Yuan, X., and Tuan, R. S. (2006). Chondrogenic differentiation and functional maturation of bovine mesenchymal stem cells in long-term agarose culture. *Osteoarthritis Cartilage* **14**, 179–189.

Mauney, J. R., Sjostorm, S., Blumberg, J., Horan, R., O'Leary, J. P., Vunjak-Novakovic, G., Volloch, V., and Kaplan, D. L. (2004). Mechanical stimulation promotes osteogenic differentiation of human bone marrow stromal cells on 3D partially demineralized bone scaffolds *in vitro. Calcif. Tissue Int.* **74**, 458–468.

Meinel, L., Hofmann, S., Karageorgiou, V., Zichner, L., Langer, R., Kaplan, D., and Vunjak-Novakovic, G. (2004). Engineering cartilage-like tissue using human mesenchymal stem cells and silk protein scaffolds. *Biotechnol. Bioeng.* **88**, 379–391.

Mow, V. C., Gu, W. Y., and Chen, F. H. (2005). Structure and function of articular cartilage and meniscus. *In* "Basic Orthopedic Biomechanics and Mechano-Biology" (V. C. Mow and R. Huiskes, eds.), pp 181–258. Lippincott-Raven, Philadelphia.

Muraglia, A., Cancedda, R., and Quarto, R. (2000). Clonal mesenchymal progenitors from human bone marrow differentiate *in vitro* according to a hierarchical model. *J. Cell Sci.* **113**(Pt. 7), 1161–1146.

Murphy, J. M., Dixon, K., Beck, S., Fabian, D., Feldman, A., and Barry, F. (2002). Reduced chondrogenic and adipogenic activity of mesenchymal stem cells from patients with advanced osteoarthritis. *Arthritis Rheum.* **46**, 704–713.

Murphy, J. M., Fink, D. J., Hunziker, E. B., and Barry, F. P. (2003). Stem cell therapy in a caprine model of osteoarthritis. *Arthritis Rheum.* **48**, 3464–3474.

Nehrer, S., Breinan, H. A., Ramappa, A., Shortkroff, S., Young, G., Minas, T., Sledge, C. B., Yannas, I. V., and Spector, M. (1997). Canine chondrocytes seeded in type I and type II collagen implants investigated *in vitro. J. Biomed. Mater. Res.* **38**, 95–104.

Noth, U., Tuli, R., Osyczka, A. M., Danielson, K. G., and Tuan, R. S. (2002). *In vitro* engineered cartilage constructs produced by press-coating biodegradable polymer with human mesenchymal stem cells. *Tissue Eng.* **8**, 131–144.

Ohlsson, L. B., Varas, L., Kjellman, C., Edvardsen, K., and Lindvall, M. (2003). Mesenchymal progenitor cell–mediated inhibition of tumor growth *in vivo* and *in vitro* in gelatin matrix. *Exp. Mol. Pathol.* **75**, 248–255.

Okamoto, T., Aoyama, T., Nakayama, T., Nakamata, T., Hosaka, T., Nishijo, K., Nakamura, T., Kiyono, T., and Toguchida, J. (2002). Clonal heterogeneity in differentiation potential of immortalized human mesenchymal stem cells. *Biochem. Biophys. Res. Commun.* **295**, 354–361.

Paige, K. T., Cima, L. G., Yaremchuk, M. J., Vacanti, J. P., and Vacanti, C. A. (1995). Injectable cartilage. *Plast. Reconstr. Surg.* **96**, 1390–1398.

Palmer, G. D., Steinert, A., Pascher, A., Gouze, E., Gouze, J. N., Betz, O., Johnstone, B., Evans, C. H., and Ghivizzani, S. C. (2005). Gene-induced chondrogenesis of primary mesenchymal stem cells *in vitro. Mol. Ther.* **12**, 219–228.

Park, Y., Lutolf, M. P., Hubbell, J. A., Hunziker, E. B., and Wong, M. (2004). Bovine primary chondrocyte culture in synthetic matrix metalloproteinase-sensitive poly(ethylene glycol)-based hydrogels as a scaffold for cartilage repair. *Tissue Eng.* **10**, 515–522.

Partridge, K., Yang, X., Clarke, N. M., Okubo, Y., Bessho, K., Sebald, W., Howdle, S. M., Shakesheff, K. M., and Oreffo, R. O. (2002). Adenoviral BMP-2 gene transfer in mesenchymal stem cells: *in vitro* and *in vivo* bone formation on biodegradable polymer scaffolds. *Biochem. Biophys. Res. Commun.* **292**, 144–152.

Pittenger, M. F., Mackay, A. M., Beck, S. C., Jaiswal, R. K., Douglas, R., Mosca, J. D., Moorman, M. A., Simonetti, D. W., Craig, S., and Marshak, D. R. (1999). Multilineage potential of adult human mesenchymal stem cells. *Science* **284**, 143–147.

Plumas, J., Chaperot, L., Richard, M. J., Molens, J. P., Bensa, J. C., and Favrot, M. C. (2005). Mesenchymal stem cells induce apoptosis of activated T cells. *Leukemia* **19**, 1597–1604.

Prockop, D. J. (1997). Marrow stromal cells as stem cells for nonhematopoietic tissues. *Science* **276**, 71–74.

Quarto, R., Thomas, D., and Liang, C. T. (1995). Bone progenitor cell deficits and the age-associated decline in bone repair capacity. *Calcif. Tissue Int.* **56**, 123–129.

Radice, M., Brun, P., Cortivo, R., Scapinelli, R., Battaliard, C., and Abatangelo, G. (2000). Hyaluronan-based biopolymers as delivery vehicles for bone-marrow-derived mesenchymal progenitors. *J. Biomed. Mater. Res.* **50**, 101–109.

Rasmusson, I., Ringden, O., Sundberg, B., and Le Blanc, K. (2005). Mesenchymal stem cells inhibit lymphocyte proliferation by mitogens and alloantigens by different mechanisms. *Exp. Cell Res.* **305**, 33–41.

Rodriguez, J. P., Garat, S., Gajardo, H., Pino, A. M., and Seitz, G. (1999). Abnormal osteogenesis in osteoporotic patients is reflected by altered mesenchymal stem cells dynamics. *J. Cell Biochem.* **75**, 414–423.

Rowley, J. A., Madlambayan, G., and Mooney, D. J. (1999). Alginate hydrogels as synthetic extracellular matrix materials. *Biomaterials* **20**, 45–53.

Rubio, D., Garcia-Castro, J., Martin, M. C., de la Fuente, R., Cigudosa, J. C., Lloyd, A. C., and Bernad, A. (2005). Spontaneous human adult stem cell transformation. *Cancer Res.* **65**, 3035–3039.

Saito, N., Okada, T., Horiuchi, H., Murakami, N., Takahashi, J., Nawata, M., Ota, H., Nozaki, K., and Takaoka, K. (2001). A biodegradable polymer as a cytokine delivery system for inducing bone formation. *Nat. Biotechnol.* **19**, 332–335.

Sakaguchi, Y., Sekiya, I., Yagishita, K., and Muneta, T. (2005). Comparison of human stem cells derived from various mesenchymal tissues: superiority of synovium as a cell source. *Arthritis Rheum.* **52**, 2521–2529.

Sawyer, A. A., Hennessy, K. M., and Bellis, S. L. (2005). Regulation of mesenchymal stem cell attachment and spreading on hydroxyapatite by RGD peptides and adsorbed serum proteins. *Biomaterials* **26**, 1467–1475.

Schaefer, D., Martin, I., Shastri, P., Padera, R. F., Langer, R., Freed, L. E., and Vunjak-Novakovic, G. (2000). *In vitro* generation of osteochondral composites. *Biomaterials* **21**, 2599–2606.

Sekiya, I., Larson, B. L., Smith, J. R., Pochampally, R., Cui, J. G., and Prockop, D. J. (2002a). Expansion of human adult stem cells from bone marrow stroma: conditions that maximize the yields of early progenitors and evaluate their quality. *Stem Cells* **20**, 530–541.

Sekiya, I., Vuoristo, J. T., Larson, B. L., and Prockop, D. J. (2002b). *In vitro* cartilage formation by human adult stem cells from bone marrow stroma defines the sequence of cellular and molecular events during chondrogenesis. *Proc. Natl. Acad. Sci. U.S.A.* **99**, 4397–4402.

Sekiya, I., Larson, B. L., Vuoristo, J. T., Reger, R. L., and Prockop, D. J. (2005). Comparison of effect of BMP-2, -4, and -6 on *in vitro* cartilage formation of human adult stem cells from bone marrow stroma. *Cell Tissue Res.* **320**, 269–276.

Sherwood, J. K., Riley, S. L., Palazzolo, R., Brown, S. C., Monkhouse, D. C., Coates, M., Griffith, L. G., Landeen, L. K., and Ratcliffe, A. (2002b). A three-dimensional osteochondral composite scaffold for articular cartilage repair. *Biomaterials* **23**, 4739–4751.

Shi, S., Gronthos, S., Chen, S., Reddi, A., Counter, C. M., Robey, P. G., and Wang, C. Y. (2002). Bone formation by human postnatal bone marrow stromal stem cells is enhanced by telomerase expression. *Nat. Biotechnol.* **20**, 587–591.

Shi, S., Bartold, P. M., Miura, M., Seo, B. M., Robey, P. G., and Gronthos, S. (2005). The efficacy of mesenchymal stem cells to regenerate and repair dental structures. *Orthod. Craniofac. Res.* **8**, 191–199.

Shin, M., Yoshimoto, H., and Vacanti, J. P. (2004). *In vivo* bone tissue engineering using mesenchymal stem cells on a novel electrospun nanofibrous scaffold. *Tissue Eng.* **10**, 33–41.

Simmons, C. A., Matlis, S., Thornton, D. J., Chen, S., Wang, C. Y., and Mooney, D. J. (2003). Cyclic strain enhances matrix mineralization by adult human mesenchymal stem cells via the extracellular signal-regulated kinase (ERK1/2) signaling pathway. *J. Biomech.* **36**, 1087–1096.

Simonsen, J. L., Rosada, C., Serakinci, N., Justesen, J., Stenderup, K., Rattan, S. I., Jensen, T. G., and Kassem, M. (2002). Telomerase expression extends the proliferative life span and maintains the osteogenic potential of human bone marrow stromal cells. *Nat. Biotechnol.* **20**, 592–596.

Solchaga, L. A., Penick, K., Porter, J. D., Goldberg, V. M., Caplan, A. I., and Welter, J. F. (2005). FGF-2 enhances the mitotic and chondrogenic potentials of human adult bone marrow–derived mesenchymal stem cells. *J. Cell Physiol.* **203**, 398–409.

Soltz, M. A., and Ateshian, G. A. (1998). Experimental verification and theoretical prediction of cartilage interstitial fluid pressurization at an impermeable contact interface in confined compression. *J. Biomech.* **31**, 927–934.

Song, L., and Tuan, R. S. (2004). Transdifferentiation potential of human mesenchymal stem cells derived from bone marrow. *FASEB J.* **18**, 980–982.

Song, L., Young, N. J., Webb, N. E., and Tuan, R. S. (2005). Origin and characterization of multipotential mesenchymal stem cells derived from adult human trabecular bone. *Stem Cells Dev.* **14**, 712–721.

Sotiropoulou, P. A., Perez, S. A., Gritzapis, A. D., Baxevanis, C. N., and Papamichail, M. (2006). Interactions between human mesenchymal stem cells and natural killer cells. *Stem Cells* **24**, 74–85.

Suon, S., Jin, H., Donaldson, A. E., Caterson, E. J., Tuan, R. S., Deschennes, G., Marshall, C., and Iacovitti, L. (2004). Transient differentiation of adult human bone marrow cells into neuron-like cells in culture: development of morphological and biochemical traits is mediated by different molecular mechanisms. *Stem Cells Dev.* **13**, 625–635.

Takahashi, I., Nuckolls, G. H., Takahashi, K., Tanaka, O., Semba, I., Dashner, R., Shum, L., and Slavkin, H. C. (1998). Compressive force promotes sox9, type II collagen and aggrecan and inhibits IL-1beta expression resulting in chondrocgeneis in mouse embryonic limb bud mesenchymal cells. *J. Cell Sci.* **111**, 2067–2076.

Tran-Khanh, N., Hoemann, C. D., McKee, M. D., Henderson, J. E., and Buschmann, M. D. (2005). Aged bovine chondrocytes display a diminished capacity to produce a collagen-rich, mechanically functional cartilage extracellular matrix. *J. Orthop. Res.* **23**, 1354–1362.

Tse, W. T., Pendleton, J. D., Beyer, W. M., Egalka, M. C., and Guinan, E. C. (2003). Suppression of allogeneic T-cell proliferation by human marrow stromal cells: implications in transplantation. *Transplantation* **75**, 389–397.

Tsuchiya, H., Kitoh, H., Sugiura, F., and Ishiguro, N. (2003). Chondrogenesis enhanced by overexpression of sox9 gene in mouse bone marrow–derived mesenchymal stem cells. *Biochem. Biophys. Res. Commun.* **301**, 338–343.

Tuan, R. S., Boland, G., and Tuli, R. (2003). Adult mesenchymal stem cells and cell-based tissue engineering. *Arthritis Res. Ther.* **5**, 32–45.

Tuan, R. S. (2004). Biology of developmental and regenerative skeletogenesis. *Clin. Orthop.* **427**, S105–S117.

Tuli, R., Seghatoleslami, M. R., Tuli, S., Howard, M. S., Danielson, K. G., and Tuan, R. S. (2002). p38 MAP kinase regulation of AP-2 binding in TGF-beta1-stimulated chondrogenesis of human trabecular bone–derived cells. *Ann. N.Y. Acad. Sci.* **961**, 172–177.

Tuli, R., Tuli, S., Nandi, S., Huang, X., Manner, P. A., Hozack, W. J., Danielson, K. G., Hall, D. J., and Tuan, R. S. (2003). Transforming growth factor-β–mediated chondrogenesis of human mesenchymal progenitor cells involves N-cadherin and mitogen-activated protein kinase and Wnt signaling cross-talk. *J. Biol. Chem.* **278**, 41227–41236.

Tuli, R., Nandi, S., Li, W. J., Tuli, S., Huang, X., Manner, P. A., Laquerriere, P., Noth, U., Hall, D. J., and Tuan, R. S. (2004). Human mesenchymal progenitor cell–based tissue engineering of a single-unit osteochondral construct. *Tissue Eng.* **10**, 1169–1179.

Turhani, D., Watzinger, E., Weissenbock, M., Yerit, K., Cvikl, B., Thurnher, D., and Ewers, R. (2005). Three-dimensional composites manufactured with human mesenchymal cambial layer precursor cells as an alternative for sinus floor augmentation: an *in vitro* study. *Clin. Oral. Implants Res.* **16**, 417–424.

Vunjak-Novakovic, G., Martin, I., Obradovic, B., Treppo, S., Grodzinsky, A. J., Langer, R., and Freed, L. E. (1999). Bioreactor cultivation conditions modulate the composition and mechanical properties of tissue-engineered cartilage. *J. Orthop. Res.* **17**, 130–138.

Wakitani, S., Goto, T., Young, R. G., Mansour, J. M., Goldberg, V. M., and Caplan, A. I. (1998). Repair of large full-thickness articular cartilage defects with allograft articular chondrocytes embedded in a collagen gel. *Tissue Eng.* **4**, 429–444.

Waldman, S. D., Spiteri, C. G., Grynpas, M. D., Pilliar, R. M., and Kandel, R. A. (2003). Long-term intermittent shear deformation improves the quality of cartilaginous tissue formed *in vitro*. *J. Orthop. Res.* **21**, 590–596.

Williams, C. G., Kim, T. K., Taboas, A., Malik, A., Manson, P., and Elisseeff, J. (2003). *In vitro* chondrogenesis of bone marrow–derived mesenchymal stem cells in a photopolymerizing hydrogel. *Tissue Eng.* **9**, 679–688.

Woo, S. L., Akeson, W. H., and Jemmott, G. F. (1976). Measurements of nonhomogeneous, directional mechanical properties of articular cartilage in tension. *J. Biomech.* **9**, 785–791.

Worster, A. A., Brower-Toland, B. D., Fortier, L. A., Bent, S. J., Williams, J., and Nixon, A. J. (2001). Chondrocytic differentiation of mesenchymal stem cells sequentially exposed to transforming growth factor-beta1 in monolayer and insulin-like growth factor-I in a three-dimensional matrix. *J. Orthop. Res.* **19**, 738–749.

Yang, S. H., Hsu, C. K., Wang, K. C., Hou, S. M., and Lin, F. H. (2005). Tricalcium phosphate and glutaraldehyde crosslinked gelatin incorporating bone morphogenetic protein — a viable scaffold for bone-tissue engineering. *J. Biomed. Mater. Res. B Appl. Biomater.* **74**, 468–475.

Yano, F., Kugimiya, F., Ohba, S., Ikeda, T., Chikuda, H., Ogasawara, T., Ogata, N., Takato, T., Nakamura, K., Kawaguchi, H., and Chung, U. I. (2005). The canonical Wnt signaling pathway promotes chondrocyte differentiation in a Sox9-dependent manner. *Biochem. Biophys. Res. Commun.* **333**, 1300–1308.

Yoo, J. U., Mandell, I., Angele, P., and Johnstone, B. (2000). Chondrogenitor cells and gene therapy. *Clin. Orthop. Relat. Res*, **379**, S164–S170.

Young, R. G., Butler, D. L., Weber, W., Caplan, A. I., Gordon, S. L., and Fink, D. J. (1998). Use of mesenchymal stem cells in a collagen matrix for Achilles tendon repair. *J. Orthop. Res.* **16**, 406–413.

Zhang, W., Ge, W., Li, C., You, S., Liao, L., Han, Q., Deng, W., and Zhao, R. C. (2004). Effects of mesenchymal stem cells on differentiation, maturation, and function of human monocyte-derived dendritic cells. *Stem Cells Dev.* **13**, 263–271.

Zhou, S., Eid, K., and Glowacki, J. (2004). Cooperation between TFG-beta and Wnt pathways during chondrocyte and adipocyte differentiation of human marrow stromal cells. *J. Bone Miner. Res.* **19**, 463–470.

# Chapter Fifty-Six

# Bone Regeneration

*Chantal E. Holy, F. Jerry Volenec, Jeffrey Geesin, and Scott P. Bruder*

## I. INTRODUCTION

Research in bone regeneration truly took off in the 1960s when Burwell *et al.* confirmed bone marrow's osteogenicity and Urist discovered bone morphogens. Scientific advances in bone-tissue engineering was thus achieved through both in-depth understanding of bone cell biology as well as progress in the isolation, purification, and commercial-scale generation of bone-specific growth factors. Two schools of thought related to the engineering of bone tissue evolved from these findings: Bone repair was accomplished, on one hand, through the use of cell-based therapies; while, on the other hand, bone repair was addressed using an increasingly diverse palette of growth factors. In this chapter, currently available approaches as well as ongoing research paths to engineer bone tissue *in vivo* are described.

### Bone Engineering — Paradigm Shift from *In Vitro* to *In Vivo* Tissue Growth

The concept of regenerating bone tissue for orthopedic and spinal applications has spurred long-term research efforts and, despite significant progress, is still facing a number of clinical challenges. Autologous iliac crest bone graft (ICBG) is considered the gold standard to repair bone, despite its limited supply and the morbidities associated with harvesting this tissue, such as infections, seromas, hematomas, herniation, vascular and neurologic injuries, and iliac wing fractures (Arrington *et al.*, 1996). Although severe complications from harvesting autograft are rare, donor site pain is reported in the literature in ranges of 25–49%, with 19–27% of patients experiencing chronic site pain two years postoperatively (Fernyhough *et al.*, 1992; Heary *et al.*, 2002). Unfortunately, these morbidities are also found to be unrelated to surgical approach or patient demographics, such as age and gender, thus making it impossible for surgeons to predict or prevent pain within their patient population (A. R. Gupta *et al.*, 2001). Significant research therefore focuses on developing new bone regeneration options that would alleviate the need for autograft harvesting.

In the early 1990s, advances in bone cell biology, combined with expertise in material sciences, fueled the notion of engineering bone chips in petri dishes. The initial concept consisted of taking cells from a patient, seeding them *in vitro* to create bone chips that would look and feel very much like autologous bone chips, and then reimplanting them in the patient as needed. This approach was based on the idea that cells harvested from a patient's bone marrow would generate unlimited volumes of tissue, which would also be devoid of risks of rejection or disease transmission. These concepts were in part confirmed by advances in stem cell biology that demonstrated the potential of bone marrow aspirates to grow in culture and to generate significant numbers of stem cells (Bruder *et al.*, 1997).

While this design seemed to address some of the key limitations related to the use of autograft, it brought along its own challenges. From a purely logistical perspective, the *in vitro* phase of the tissue-engineering design was riddled with complications, for it would require a long and potentially complex regulatory path to market, carried risks of culture contamination, and was not easily optimized for large-scale applications. The economics of this approach could also prove particularly unfavorable since the majority of cases seen by orthopedic surgeries are known to respond with excellent clinical outcomes while employing fairly basic technology; the market for a true "tissue-engineered" approach would therefore be limited to a relatively small number of high-risk patients.

In addition to these logistical and financial considerations, parallel progress in medical devices, pharmacology, and cell biology further modified the bone regeneration landscape with the development of new point-of-care cell concentration devices, recombinant growth factor technologies, or allogeneic stem cell banks. The concept of an *in vitro* engineered bone chip was thus made obsolete. Instead, current approaches focus on creating the right biological environment *in vivo* to promote bone healing.

### Two Parallel Approaches to Bone Regeneration

A decade of bone-tissue-engineering endeavors and more than 50 years of intense bone biology research highlighted the three key elements required to create a biological environment suitable for bone formation: (1) presence of matrix material for cell attachment and proliferation (osteoconduction); (2) critical mass of osteogenic cells (stem cells or progenitors that are capable of forming new bone tissue); and (3) biological factors to promote chemotaxis, cell proliferation, and differentiation. Currently, two parallel paths are being explored to create this optimized bone bed. The first approach focuses primarily on the delivery of a critical mass of osteoprogenitor cells to the repair site — and relies on autologous growth factors present in the body in physiological concentrations — to form bone. The second approach is based on the delivery of high doses of biological factors — and relies on the availability and responsiveness of autologous stem cells in the body — to achieve bone formation.

### Cell-Based Therapies

The clinical use of cell-based therapies to aid in bone regeneration dates back to the 1940s, when the first reports were published on the use of bone marrow injections as a source of stem cells into nonunions (Phemister, 1947). In the 1960s, Burwell (1964) further defined the bone regeneration potential of the iliac crest autograft as a function of the crest's marrow content. Since that time, while plain bone marrow injections are still in some cases the therapeutic option du jour, scientific developments provided health care professionals with (1) optimized matrices to attach and maintain bone marrow cells; (2) point-of-care bone marrow cell concentration methods; and (3) allogeneic, culture-expanded stem cell concentrates. Review of these major approaches and published results are described further in this chapter.

### Growth Factor–Based Therapies

Concurrent with Burwell's bone marrow research was Urist's first description of bone morphogenetic proteins (BMPs), which were later characterized as a family of growth factors present in the bone matrix and capable of inducing local progenitor cells into osteoblastic development (Urist, 1965; Lindholm *et al.*, 1982). These findings were followed by hundreds of publications describing the role of BMPs, and more specifically BMP-2, in bone formation. A recombinant human form of BMP-2 (rhBMP-2) was thus bioengineered to be mass-produced and was the first bone morphogen to enter the clinical market, in 2003. A review of rhBMP-2 and other growth factors on the market or in development for bone repair is provided in this chapter.

## II. CELL-BASED APPROACH TO BONE-TISSUE ENGINEERING

As just defined, cell-based therapies are based on the delivery of cells and rely on the presence of autologous growth factors for biological stimulation. More specifically, cells required for cell-based bone regeneration therapies are so-called osteogenic cells. Osteogenic cells include all cells that are or will differentiate into osteoblasts capable of forming bone. Currently, only autogenic and allogeneic osteogenic cells are used clinically or evaluated for use. These human cells can be harvested from two key sources: (1) differentiated bone tissue, and (2) nonbony tissues containing mesenchymal stem cells (MSCs); such tissues include bone marrow as well as other connective tissues (Bruder and Caplan, 2000).

Differentiated bone tissues have recently been described as potential sources of osteogenic cells. These tissues consisted of mature bone fragments, obtained as debris from reaming procedures (Wenisch *et al.*, 2005). Bone fragments did not contain any bone marrow, and thus the osteogenicity of the bone fragments was described as specifically due to bone-lining cells. This source of osteogenic cells has not been widely investigated yet and thus will not be described further in this chapter.

The key sources for osteogenic cells therefore remains nonbony connective tissues and bone marrow. Various differentiated connective tissues have been shown to contain osteogenic cells that could form bone in specific culture conditions (Zuk *et al.*, 2002) and are currently being investigated for potential use in bone regeneration. Bone marrow, on the other hand, has a long track record of preclinical and clinical efficacy for bone repair and remains the most easily accessible and richest source of MSCs.

### Current Therapies

#### Use of Bone Marrow as a Source of Stem Cells

Of all human tissues, bone marrow was found to be particularly advantageous from a cell-sourcing perspective, for it can be obtained from the iliac crest using a

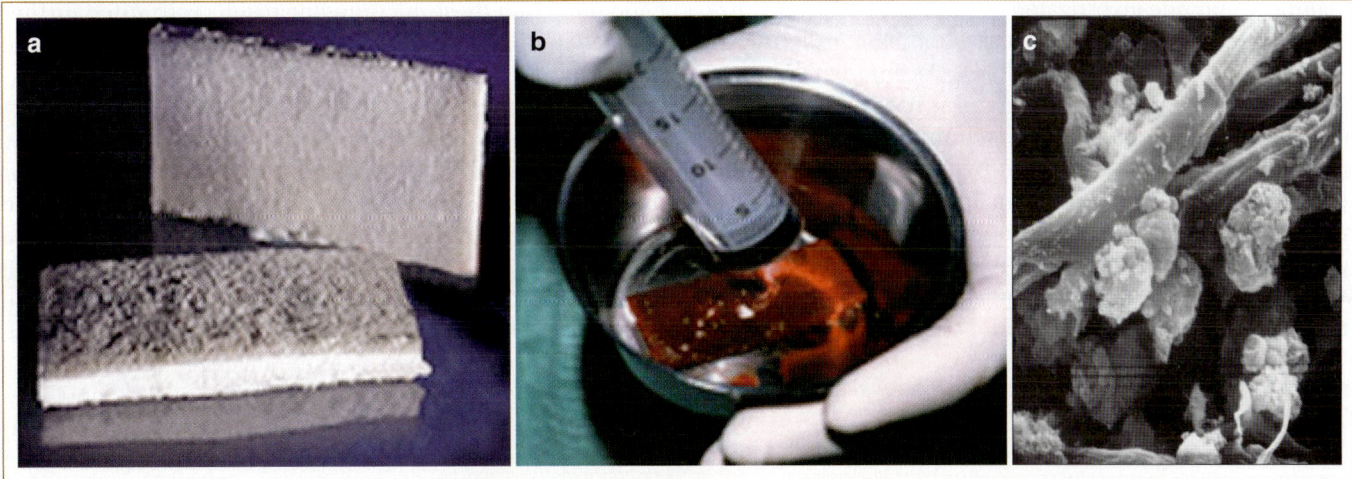

**FIG. 56.1.** Photographs of a hydroxyapatite-coated collagen type I sponge (HEALOS®) prior to use **(a)** and soaked with autologous bone marrow **(b)**. Retention of bone marrow–derived cells on the HEALOS strips was observed by scanning electron microscopy **(c)**.

noninvasive procedure with a simple needle aspiration. More importantly, it is self-renewing and was found to regenerate its own nucleated cellular content within 90 days (Stroncek *et al.*, 1991). Bone marrow has therefore been extensively evaluated, from a cell biology perspective as well as from an overall efficacy point of view, in culture dishes as well as in animal and clinical cases.

### Cellular Content of Bone Marrow

Bone marrow was shown in the mid-1960s to contain two different stem cell systems: hematopoietic and mesenchymal (Friedenstein *et al.*, 1968). The hematopoietic system was shown to ensure long-term formation and renewal of all hematopoietic tissues and contains cell types at all stages of hematopoietic cell maturation. Hematopoietic stem cells (HSCs) were also isolated from the marrow and shown to survive throughout life (Quesenberry *et al.*, 2005). The mesenchymal system was found to be responsible for the long-term formation and renewal of the "fibroblastic" stromal cells, which were involved in "mesenchymal" tissue formation, including bone, cartilage, muscle, ligament, and adipose tissue, as well as marrow stroma. The mesenchymal and hematopoietic systems were also found to be interdependent, because cytokines from mesenchymal stem cells could affect hematopoietic cell differentiation, and vice versa (Kuznetsov *et al.*, 1997).

Recently, a third type of precursor cell was identified within bone marrow: the "side population" (SP); these cells are defined by their ability to regenerate the hematopoietic compartment as well as to differentiate into osteoblasts through a mesenchymal intermediate (Olmsted-Davis *et al.*, 2002).

While bone marrow's cellular content was fairly extensively analyzed, little research was done on the plasma content of bone marrow. Thus, the noncellular phase of bone marrow is still mostly undefined. However, bone marrow, with its diverse cellularity and its mostly unknown plasma content, has been tested extensively *in vivo*, in many animal models as well as in clinical cases.

### In Vivo *Evidence of Bone Marrow Osteogenicity: Preclinical Data*

Bone marrow research *in vivo* has been strongly correlated with biomaterial research, for it was demonstrated early on that osteogenic cells required adequate surfaces for growth and proliferation. Ohgushi *et al.* (1989) utilized a porous three-dimensional hydroxyapatite carrier to demonstrate that bone marrow cells could heal critical-sized defects in rats. In addition, Tiedeman *et al.* (1991) used bone marrow aspirate on demineralized bone matrices to heal critical-sized tibial defects in dogs. Novel composite carriers that combine collagen and hydroxyapatite (HEALOS®) were also shown effectively to regenerate bone when combined with autologous bone marrow (Fig. 56.1). In this study, Tay *et al.* (1998) used a rabbit posterolateral fusion model to evaluate the effectiveness of bone marrow with or without heparin. Bone marrow on the carrier was found comparable to autograft.

In all published research, bone marrow osteogenicity was found to be critical to ensure efficacy. Poorly cellular bone marrow typically did not result in bone formation, while highly cellular bone marrow was found to be efficacious. An example of this concept was brought forward by two studies published in 2005 (Kraiwattanapong *et al.*, 2005; Minamide *et al.*, 2005). Both Kraiwattanapong and Minamide replicated the evaluation performed by Tay *et al.* in 1998, using bone marrow sources with lower cellularity. Tay *et al.* used bone marrow at 238 million cells/mL and observed healing in all cases. Kraiwattanapong utilized bone marrow at 30 million cells/mL and did not observe any healing. Minamide investigated two different cell concentrations, 1 million cells/mL as well as 100 million cells/mL. No healing was observed at 1 million cells/mL, while healing equivalent to autograft was observed at 100 million cells/mL, indicating a minimum cell mass requirement for efficacy.

This finding was supported by Andersen *et al.* (2003), who evaluated the relationship between the number of osteoblasts in an iliac crest bone graft and the resulting fusion mass in a posterolateral mini-pig-fusion model. The study found a statistically significant correlation between the number of cells implanted and successful fusion.

### In Vivo *Evidence of Bone Marrow Osteogenicity: Clinical Data*

While no long-term, multicenter, prospective, randomized, blinded clinical studies have been completed to evaluate the efficacy of bone marrow vs. autograft, multiple clinical reports have described bone marrow for use in bone healing. Salama and Weissman (1978) published a preliminary report on 28 patients undergoing long-bone repair under conditions covering a wide range of indications. Autologous bone marrow aspirates were used in all cases and provided very satisfactory results. Connolly *et al.* (1991) described a technique to inject bone marrow directly into bone grafting sites, thereby alleviating the need for open surgery. Comparing healing patterns in 100 patients, the study reported an 80% healing rate following marrow grafting. In a clinical study for a collagen–calcium phosphate graft material (Collagraft®), Chapman *et al.* (1997) described the efficacy of bone marrow with the carrier in the treatment of long-bone fractures. No significant difference between the autograft and the bone marrow carrier groups was observed. In a more challenging clinical application, Garg *et al.* (1993) reported the use of percutaneous autogenous bone marrow grafting in 20 cases of nonunited fracture. After five months, 17 cases progressed to healing. Delayed union and nonunion cases were also treated with bone marrow by Sim *et al.* (1993), who described 11 cases that healed within a median time of 10 weeks following injection of bone marrow.

Bone marrow with allograft was used in a variety of pediatric cases, including cysts, fibromas, long-bone nonunions, and tibia-lengthening procedures (Wientroub *et al.*, 1989). In this study, all cases showed good new bone formation, with no adverse reaction. In a subsequent pediatric tibial nonossifying fibromas case, Tiedeman *et al.* (1991) also reported successful healing after injection of demineralized bone powder with autologous bone marrow.

Grafting bone marrow was also proven effective in medically compromised patients, for example, cancer patients, with delayed union or nonunions. Healey *et al.* (1991) reported bone marrow injections in eight patients with primary sarcomas, and bone formation was observed in seven patients after marrow injection, while complete healing was observed in five patients. Healey *et al.*

concluded that these encouraging results warranted further clinical studies and that their findings suggested a useful technique for the treatment of delayed unions and nonunions in difficult clinical circumstances.

In clinical spinal fusion applications, bone marrow combined with either demineralized bone matrix (DBM) or osteoconductive substrates also resulted in improved bone fusion rates. Results similar to autograft were observed in a retrospective posterior spinal fusion study examining 88 consecutive patients and comparing (1) autologous iliac crest bone graft, (2) freeze-dried corticocancellous bone without marrow, and (3) demineralized bone matrix plus autologous bone marrow. Success rates of 88% and 89% were observed in the autologous iliac crest bone graft and the demineralized bone matrix plus autologous bone marrow groups, respectively. The highest failure rate (28%) was obtained in the freeze-dried corticocancellous bone without marrow group. The authors concluded that augmentation of demineralized bone with bone marrow resulted in fusion rates similar to those of iliac crest bone graft (Price *et al.*, 2003). Similar findings have been reported with the use of synthetic grafting materials. Bone marrow aspirate combined with a mineralized collagen matrix (HEALOS®) produced similar fusion rates to those observed with autograft in a posterior spine fusion study (Kitchel, 2006).

While bone marrow was consistently reported clinically and preclinically to be as effective as autograft for bone repair, indications in animal models, as described earlier, suggested a relationship between the efficacy of marrow and its inherent cellularity (Kraiwattanapong *et al.*, 2005; Minamide *et al.*, 2005; Tay *et al.*, 1998). Clinically, bone marrow cellularity, as a function of age and gender, was evaluated by multiple groups; consolidated findings from these studies seemed to indicate that cellularity decreased rapidly in early years and after puberty. A slower but continued decrease in cellularity was observed subsequently during adult life; this effect was also found to be greater in women than in men (Muschler *et al.*, 2001). Other factors, such as air pollution particulates and nicotine, were found to affect bone marrow cellularity and bone marrow cell responsiveness (Liu *et al.*, 2001). Thus, in order to maintain the efficacy of bone marrow to regenerate bone despite decreasing cell numbers and reduced cell sensitivity, significant effort was geared toward creating methods to concentrate cells at a given repair site.

### Use of Concentrated-Stem-Cell Therapies

#### Point-of-Care Concentration Techniques — Use of Autogenic Cells

Centrifugation was the first method utilized to concentrate osteogenic cells, initially based on protocols described by Friedenstein *et al.* in 1968. In 1989, Connolly *et al.* established a protocol to recover close to 100% of all nucleated cells within a bone marrow aspirate, centrifuging marrow samples at 400 times gravity for 10 minutes. Connolly was among the first to suggest that concentrating bone marrow–derived nucleated cells on bone grafts would increase the bone-forming efficacy of the grafts. These investigations demonstrated that using a four fold cell concentrate of rabbit bone marrow significantly improved the bone-forming rate in a rabbit intraperitoneal chamber model.

The concept of concentrating cells became particularly attractive in challenging clinical cases, as, for example, in nonunions. Connolly utilized his centrifugation methods in the clinic and was able to achieve 100% healing in 18 cases of nonunion (Connolly, 1998). More recently, Hernigou *et al.* (2005) used the same centrifugation methods to inject concentrated bone marrow cells in tibial nonunions of 60 patients. In this study, a 4.2-fold cell concentration ratio was typically achieved (from $612 \pm 134$ progenitors/cm$^3$ before concentration to an average of $2579 \pm 1121$ progenitors/cm$^3$ after concentration). Union was obtained in 88% of cases (53 patients), and the bone marrow that had been injected into the nonunions of those patients contained an average of $54{,}962 \pm 17{,}431$ progenitors. In contrast, the total number of osteoprogenitors injected in the seven patients who did not heal was significantly lower, at $19{,}324 \pm 6843$ progenitors ($p < 0.01$), than that of patients who healed. As seen in rabbits, the success of a bone graft in healing challenging defects was correlated directly with the number of cells implanted in that site. This study represented the first clinical attempt to quantify the required cell numbers for successful union.

To facilitate the cell concentration procedure in the operating room, Muschler *et al.* (2003) developed a point-of-care, disposable, cell-enrichment device. This method relied on the physical principles of an affinity column to populate a custom-designed graft matrix with a high proportion of osteoprogenitor cells found in bone marrow. The custom-design graft matrix for the cell-enrichment technology was engineered to be highly porous and to possess a high surface area for cell attachment. When marrow passed through this matrix at an optimized, controlled flow, nucleated cells attached while hematopoietic cells passed through. The result was a graft material that contained an increased concentration of nucleated and, more specifically, osteoprogenitor cells. Muschler *et al.* (2003) first tested the selective cell retention technology in a posterolateral fusion model in dogs. An established posterior spinal fusion model was used to compare (1) cell-enriched cancellous bone alone, (2) cancellous bone plus a bone marrow clot, and (3) cell-enriched cancellous bone plus a bone marrow clot. The hypothesis in this study was twofold: On one hand, cell enrichment would increase graft osteogenicity, and, in addition, adding a bone marrow clot to the graft would even further enhance its efficacy. Results from union score, quantitative computed tomography, and mechanical testing all demonstrated that the bone matrix plus enriched bone marrow clot was superior to all other groups. These data

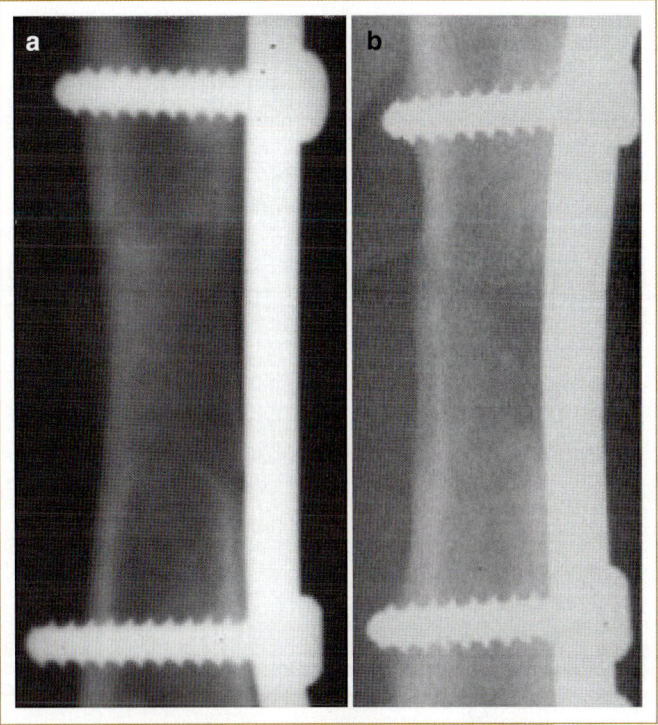

**FIG. 56.2.** Representative x-ray radiographs 16 weeks postoperatively of canine critical-sized femoral defects implanted with **(a)** autograft and **(b)** cell-enriched demineralized bone matrix with cancellous chips clotted with platelet gels.

also confirm that cell enrichment significantly improved graft performance.

In a subsequent study, the actual cell concentrate was compared directly to whole bone marrow. Groups included (1) matrix alone (demineralized cortical bone powder), (2) matrix plus marrow, and (3) matrix with enriched marrow cells. Enriched matrix grafts delivered a mean of 2.3 times more cells and approximately 5.6 times more progenitors than matrix mixed with bone marrow. Again, union scores and fusion volumes both confirmed that selective cell retention improved healing outcomes (Muschler *et al.*, 2005).

Using the same selective cell retention methodology, Brodke *et al.* (2006) used a canine critical-sized segmental defect to evaluate the healing efficacy of cell-enriched grafts versus autograft. Canine demineralized bone matrix (cDBM) and cancellous chips were enriched in osteoprogenitors and placed in 21-mm-long osteoperiosteal femoral defects; animals were sacrificed 16 weeks postoperatively, and the femurs were removed and analyzed. The results showed equivalency between both the cell-enriched grafts and autograft. Both resulted in 100% bridging bone across the length of the defects (Figs. 56.2 and 56.3 illustrate fusion via radiography and CT). Histology sections further demonstrated bone formation across the defects in all autograft and cell-enriched cases (Fig. 56.4).

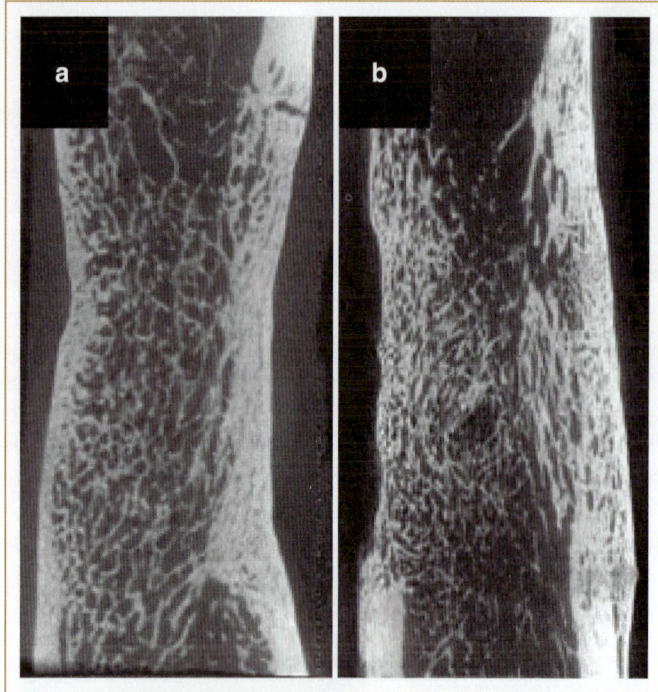

**FIG. 56.3.** Representative μCT scan of canine critical-sized femoral defects 16 weeks postoperatively implanted with **(a)** autograft and **(b)** cell-enriched demineralized bone matrix with cancellous chips.

Finally, a sheep model was also used to evaluate the healing potential of cell-enriched grafts using autologous bone marrow. In this model, the bone-grafting efficacy of tricalcium phosphate (CaP) grafts in three different configurations was evaluated: (1) CaP alone, (2) CaP saturated with whole bone marrow, and (3) CaP enriched with osteoprogenitor cells. (Graft placement in the gutter is shown in Fig. 56.5.) In this model, CaP enriched with osteoprogenitor cells reached 33% fusion, while autograft fused at only 25%. CaP alone did not result in any fusions, and CaP saturated with whole bone marrow reached only 8% fusion (three-dimensional CT reconstructions of fusions shown in Fig. 56.6) (M. C. Gupta *et al.*, in press). These results indicate once more that osteoprogenitor enrichment increased bone graft osteogenicity.

Human clinical evaluations of the cell-enriched grafts followed preclinical studies. A pilot clinical trial conducted at the Cleveland Clinic reviewed spinal fusion outcomes of 21 patients who received demineralized bone matrix enriched using Muschler's technique. Twenty out of 21 patients showed radiographic evidence of fusion at 12 months (Lieberman, 2004). This study was followed by a prospective, multicenter study. Fifty-one patients across five centers were included in the study. All underwent one- or two-level posterolateral fusions. Grafts were prepared using iliac crest bone marrow aspirate. Selective cell retention of the mar-row was prepared on demineralized bone matrix. After 12 months, VAS scores were decreased favorably by an

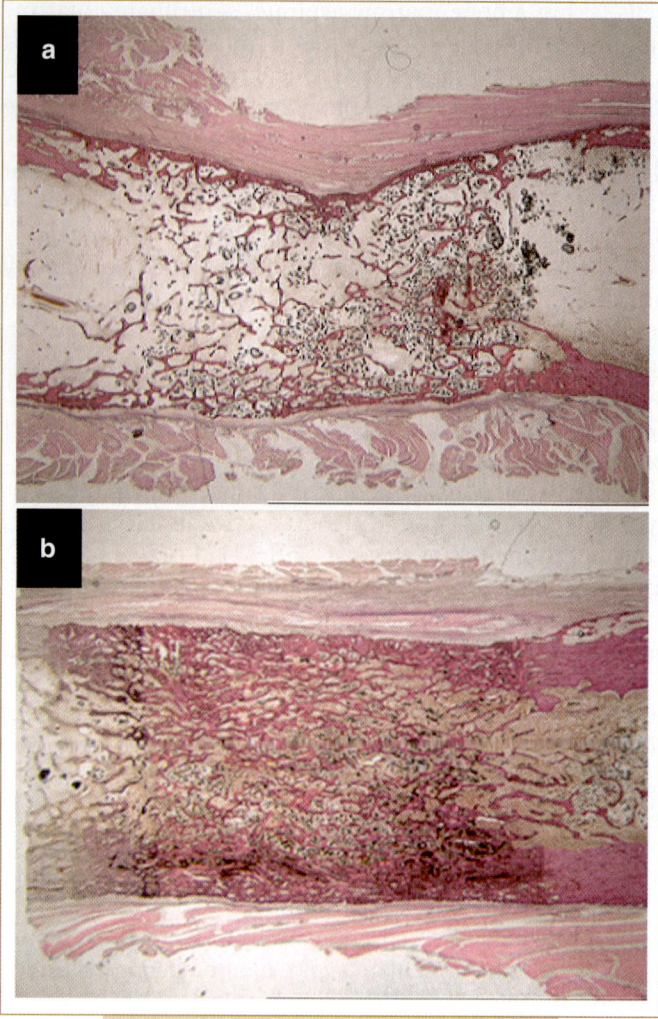

**FIG. 56.4.** Micrograph of undecalcified, hematoxylin- and eosin-stained histological section of canine critical-sized femoral defects 16 weeks postoperatively filled with **(a)** autograft and **(b)** cell-enriched demineralized bone matrix with cancellous chips, clotted with platelet gels.

average of 55% for back pain, 58.5% for right leg pain and 65.7% for left leg pain. Fusion rates were 84.2% (Wang *et al.*, 2005).

While fusion results using cell concentration techniques matched those obtained using autograft, a small percentage of patients, as seen with autograft, still did not show radiographic evidence of solid bone formation. It was hypothesized that cell numbers in these few patients was so drastically compromised that neither autograft — the gold standard — nor physiological cell-enrichment techniques could regenerate their bone. At the same time, recent evidence indicates that adult stem cells may not express immunologically relevant cell surface markers and, thus, could be harvested from an individual, grown in culture, and potentially implanted in another individual without being then rejected. A new approach was therefore envisioned for these high-risk patients: the use of allogeneic, culture-expanded

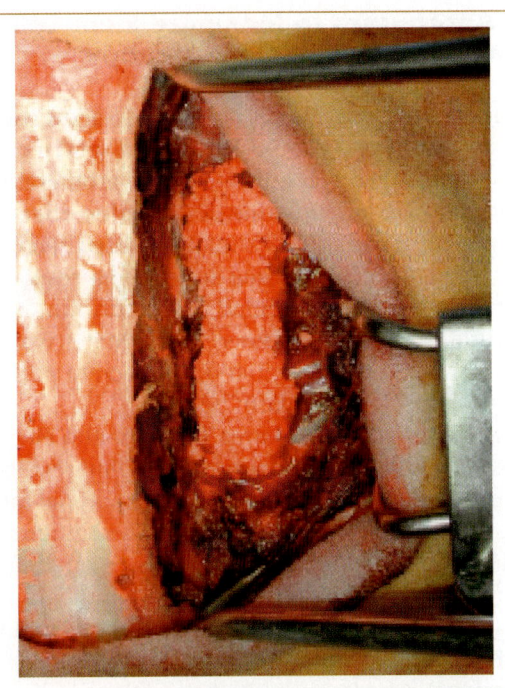

**FIG. 56.5.** Photograph of bone marrow–derived cell-enriched calcium phosphate grafts placed in the posterolateral gutters in sheep. The grafts, approximately 5 cc on each side, were placed in between decorticated transverse processes between L4 and L5.

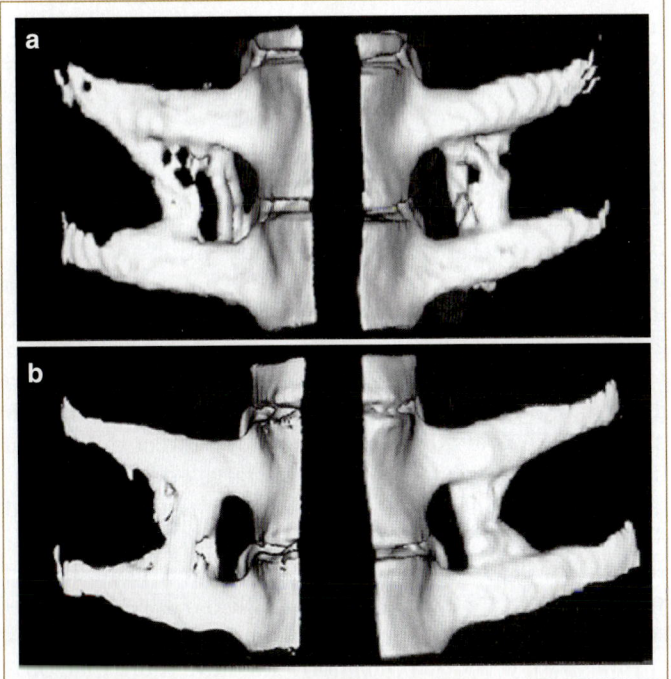

**FIG. 56.6.** Three-dimensional μCT reconstructions of fusion sites in sheep six months postoperatively, following implantation of **(a)** autograft and **(b)** cell-enriched calcium phosphate grafts clotted with platelet gels.

cells, stored in cell banks and ready for use at any time, such that concentration levels would now reach several hundred times native levels.

### Culture-Expanded Approaches — Use of Allogeneic Cells

Key culture-expansion methodologies for bone marrow–derived mesenchymal stem cells were published starting in the late 1980s. Maniatopoulos *et al.* (1988) first described a process to isolate osteoprogenitor cells from rat bone marrow. This technique involved explanting the entire femur of the rats and flushing the bone marrow in an osteogenic cell media. After 7–14 days, alkaline-phosphatase-positive colonies would become visible, indicating potential differentiation of osteogenic cells.

Cell-isolation methodologies were then developed to culture human bone marrow–derived cells, and Jaiswal *et al.* (1997) established a reproducible and standardized system for the *in vitro* osteogenic differentiation of human marrow–derived MSCs. Muschler *et al.* (1997) utilized similar cell-isolation techniques to quantify osteogenic precursors in bone marrow aspirates of patients.

As scientists deepened their understanding of osteoblastic cell growth and proliferation conditions, it was hypothesized that allogeneic MSCs might be applicable to bone repair and regeneration if one could successfully mute immunoreactive groups. Surprisingly, this turned out to be unnecessary, for *in vivo* evaluations of allogeneic MSC implantation consistently demonstrated a lack of immune response and rejection. In one instance, analysis of circulating antibody levels against MSCs nine weeks postimplantation in a canine cranial site supported the hypothesis that neither autologous nor allogeneic MSCs induced a systemic response by the host (Arinzeh *et al.*, 2003). The authors concluded that autologous and allogeneic MSCs had the capacity to regenerate bone within craniofacial defects (De Kok *et al.*, 2003). More recently, undifferentiated human MSCs were shown not to express immunologically relevant cell surface markers. They also seemed to inhibit the proliferation of allogeneic T-cells *in vitro*. Thus, MSCs could be described as immunoprivileged or immunomodulating cells (Niemeyer *et al.*, 2004). These findings confirmed that allogeneic stem cells could indeed become a possible therapeutic tool. As such, they are currently being developed for bone and soft tissue repair.

While no additional preclinical or clinical studies were conducted on the use of allogeneic cells, these cells became available in 2006 through a special regulation that allows human tissue to be commercialized in the United States with very limited regulatory scrutiny. Commercially-available cells are, however, not culture expanded but used at concentrations similar to those found in human bone.

While allogeneic culture-expanded stem cells could potentially be of value for high-risk patients, currently available preclinical evidence failed to demonstrate superiority of this technique over point-of-care cell-concentration methods. To address this shortfall, it was hypothesized that the undefined marrow stromal environment, which is maintained when using point-of-care concentration techniques but is washed away when using culture-expanded cells, may affect bone formation in ways that are not yet clearly understood. This fact was based on two publications (Brodke *et al.*, 2006; Muschler *et al.*, 2003) that both highlighted the importance of bone marrow stroma or bone marrow clots for bone regeneration. Clearly, more research is needed to address these issues and other unknown parameters in bone regeneration.

## Ongoing Research

While most research from the 1960s to the late 1990s focused on understanding osteoprogenitor cells from the bone marrow, research groups also investigated other tissues that may provide clues or cellular responses related to bone formation. Was there another repository of stem cells that may be easily accessible and may provide cells that, for unknown reasons, would be more responsive than those found in the marrow? One key focus thus turned to another tissue, widely available among Americans and easily accessible: fat. Fat is the most convenient tissue other than bone marrow for harvesting MSCs, for it is easily biopsied, cultured, expanded, and transduced.

### New Autologous Sources for Stem Cells

Not surprisingly, cells derived from connective tissues such as muscle and fat were shown to "behave" similarly to bone marrow–derived MSCs: These cells, when cultured in the right conditions, would differentiate into bone, as seen with bone marrow–derived cells (Betz *et al.*, 2005). Interestingly, but again not surprisingly, these cells were shown to differentiate in many different connective tissues, i.e., bone, fat, cartilage, and muscle tissues, with growth factors specific to each culture condition (Zuk *et al.*, 2002). Surprisingly, however, cells derived from adipose tissue were shown to contain proliferative properties that did not decline with age. In addition, this tissue was described as a richer and more effective source of osteoprogenitor cells than bone marrow, one that could easily be modified to express BMP-2 (Dragoo, 2003).

While all these *in vitro* considerations were very intriguing, only a bone regeneration model *in vivo* would address whether or not these cells were suitable for bone regeneration. Cowan *et al.* (2004) used these cells to investigate the *in vivo* osteogenic capability of adipose-derived stromal (ADAS) cells to heal critical-sized mouse calvarial defects. These ADAS cells were harvested from the subcutaneous anterior abdominal wall, yielding ~800 mg of fat tissue, and were shown to be multipotent and available in large numbers. *In vitro*, they were observed to attach and proliferate rapidly. The authors also reported a yield of cells that, by itself, was much higher than that of typical bone marrow.

## Novel Approaches to Maximize Stem Cell Survival and Growth: Gene Therapy

Gene therapy was initially designed to allow transfer of genetic material into cells, which in turn secreted specific proteins that could affect cell migration, proliferation, and differentiation via an autocrine or paracrine effect. In this model, growth factors, unlike manufactured recombinant proteins, would be synthesized *in situ* as a result of gene transfer and would be presented to the surrounding tissue in a natural, cell-based manner. Local MSCs would then undergo osteogenic differentiation or secrete appropriate tissue and thus effectively form bone.

Gene therapy involves three fundamental elements: a sequence of DNA encoding a protein of interest, a vector that would facilitate the entry of genetic material into cells, and target cells for that gene segment. Two different types of therapeutic conditions could be envisioned for gene therapy: (1) conditions that require continuous, sustained delivery of specific proteins; and (2) conditions that require a one-time bone regeneration agent (e.g., trauma cases and spinal fusions). A commonly cited example of a clinical case requiring continuous, sustained delivery of proteins is osteogenesis imperfecta. In this case, a patient would be implanted with MSCs genetically modified with a retrovirus containing the gene for normal type I collagen. These cells would reestablish themselves in the bone marrow and thus provide mesenchymal progenitors with the appropriate collagen type I building capabilities.

For one-time bone repair applications (e.g., bone fractures, spinal fusions), gene therapy would generate short-term gene expression. Thus, nonviral vectors, which are typically easier to generate, more stable than viruses, and less immunogenic, were considered.

In addition to short- and long-term protein expression, two types of gene therapy approaches could be envisioned: a so-called *in vivo* approach as well as an *ex vivo* approach. *In vivo* gene therapy involved the direct transfer of genes into patients. In preclinical studies, this method was shown to be easy to perform but had relatively poor transduction efficiencies and could cause inflammatory responses. An *ex vivo* gene therapy involved transducing cells *in vitro* and then implanting those cells at a specific site. This approach presented several advantages, including direct delivery of the osteogenic gene to a desired site, targeted cell transduction for gene delivery, a supply of cells that participated directly in the osteoinductive process, and the potential for controlling the rate or extent of gene expression by using known vectors (Nussenbaum *et al.*, 2004). However, as discussed earlier, would the market support such a complex approach for clinical applications that have more than 90% success rate with available technologies?

Both *in vivo* and *ex vivo* gene therapies were tested in animal studies. In a preliminary study, Viggeswarapu *et al.* (2001) investigated *in vivo* gene therapy involving the LMP-1 gene in an adenovirus construct, which was used to transduce bone marrow–derived buffy-coat cells. These cells were implanted in a posterolateral fusion model in rabbits. All animals showed solid fusion at four weeks postsurgery. In a subsequent study, H. S. Kim *et al.* (2003) used a similar gene for *ex vivo* gene therapy in rabbits. In this study, the authors researched conditions to minimize immune responses and maximize fusion and found that minor protocol modifications, including doubling the viral dose or number of cells infected and increasing the infection time, could overcome immune complications. Bone formation and regeneration was also recently achieved in several different animal models, including rat femurs, mouse skulls, and porcine anterior spinal fusion.

While successful in these preclinical trials for one-time bone repair applications, it is worth mentioning that simple bone marrow aspirates also fused these challenging defects (Tay *et al.*, 1998). Thus, as of today, these techniques do not show superiority to the most accessible of all approaches presented in this chapter. Current thinking seems to indicate that the use of gene therapy for these applications may be unnecessary, since simpler and more economical approaches can effectively treat bone conditions such as fractures and fusions. Gene therapy, however, may be better suited for conditions that, like osteogenesis imperfecta, require continuous protein expression, since there is no currently available satisfactory treatment option to heal these pathologies.

As cell-based research progressed over the last decades, a parallel research arm focused on understanding growth factors to manipulate cells responsible for bone formation.

## III. GROWTH FACTOR–BASED THERAPIES

While bone growth factor research seemed to belong primarily to the pharmaceutical world, it has really been championed by the "bone device" industry. This was due primarily to the fact that these stimulants rely heavily on matrices that would contain them and allow them to be delivered in a specific fusion site. From a purely practical perspective, it also seemed reasonable that those companies involved in distributing orthopedic devices would also distribute the biological components required to heal bone.

### Current Therapies

As with cellular approaches, the search for growth factors first focused on autologous factors that could easily be concentrated in the operating room. Growth factors were found to be readily stored in platelets, so platelet concentration devices to be used at point-of-care sites became available on the market. A review of the scientific findings around platelet concentrates for bone repair is included below.

In addition to autologous factors, biotechnology provided the bone device industry with recombinant technologies that allowed mass production of naturally occurring bone morphogens. In 2003, the first such morphogen, recombinant human bone morphogenetic protein-2 (rhBMP2) entered the market for specific indications in spine fusion.

### Autologous Sources: Platelet Concentrates

Marx *et al.* (1998) first evaluated the role of platelets in normal bone healing. His research led to the development of a whole new industry to develop and sell platelet concentrators for orthopedic and spinal applications.

Marx hypothesized that platelets, which were known to contribute to normal bone healing through the release of growth factors, could be concentrated and thus accelerate the healing cascade in a bone defect site. These platelet-derived growth factors had been shown to exhibit chemotactic activity as well as a mitogenic effect on osteoprogenitor cells. An important consideration in the use of concentrations of platelets [also called platelet gels or platelet-rich plasma (PRP)] was the fact that, by concentrating platelets instead of concentrating individual growth factors, an entire "cocktail" of factors would be concentrated and thus act in concert to increase bone fusion. Using clinical biopsy samples, Marx successfully demonstrated that these platelet gels could indeed accelerate the bone-healing process (Marx *et al.*, 1998; Marx, 2001).

As a result of Marx's observations, multiple clinical studies were initiated to evaluate PRP in craniofacial (Aghaloo *et al.*, 2004), long-bone (Barrow and Pomeroy, 2005), and spine applications (Weiner and Walker, 2003). While most publications confirmed Marx's observations, some manuscripts were controversial. Weiner *et al.*, in their posterolateral fusion study, even suggested an inhibitory effect due to PRP. This study prompted a series of investigations to understand exactly how the PRP used in these particular cases were prepared and whether anything in the preparation method could explain that unexpected outcome. Some critical clues were obtained from these observations: Premature platelet activation as well as plasma protein denaturation could result from inappropriate plasma preparation methods and could impact the efficacy of PRP (Kevy and Jacobson, 2004). Thus, critical care needed to be used in preparing these concentrates.

More recently, research focused on the use of platelet gels for diabetic patients. The fact that diabetics had lower base levels of growth factors — as low as one-third those of nondiabetics — had already been demonstrated in the early 2000s (Lu *et al.*, 2003). Interestingly, Gandhi *et al.* (2006) demonstrated that using PRP in a bony defect site of diabetic rats could play a significant role in promoting fracture healing.

While PRP was shown overall to promote bone healing and held promise for diabetic patients, it still maintained the stigma of a technology that, while helpful, did not *consistently* demonstrate efficacy. Thus, the idea of a purified recombinant human factor provided to surgeons at a fixed concentration and thus, consistent efficacy, became increasingly attractive.

### Recombinant Human Proteins

A by-product of research performed on bone mineralization mechanisms, the discovery of bone morphogens within human bone matrix certainly accounted for one of the major milestones in bone research (Urist, 1965). Urist not only discovered BMPs, but also performed the first implantations of BMPs in humans in the late 1980s (Johnson *et al.*, 1988). It would take more than 20 years to get the first BMPs on the market: by 2006, rhBMP-2 had become available for interbody fusion and tibial repair, while rhBMP-7 was granted exemptions for humanitarian use in the treatment of recalcitrant tibial nonunions.

### rhBMP-2

Countless preclinical and clinical studies were performed with rhBMP-2, of which only a few are selected for review in this chapter. Using a rabbit posterolateral fusion model, Schimandle *et al.* (1995) reported 100% fusion in rabbits implanted with rhBMP-2 on a type 1 fibrillar collagen sponge. In this model, only 42% of sites implanted with autogenous bone graft from the iliac crest achieved fusion. Similarly, Sandhu *et al.* (1996) reported 100% posterolateral fusion in beagle dogs within 12 weeks after the implantation of rhBMP-2 on a collagen sponge. In this study, no fusions were observed with autogenous bone graft from the iliac crest.

As investigations evolved from smaller to larger animal species, doses required to achieve healing with rhBMP-2 needed to be increased. It was hypothesized that more evolved species contained fewer responsive cells and thus required higher concentrations of growth factor. Martin *et al.* (1999) showed that rhBMP-2 concentrations effective in lower animals (0.43 mg/mL) were not effective in posterolateral fusions in primates, and thus effective doses were increased to reach 1.5 mg/mL doses for human use.

On July 2, 2002, the U.S. Food and Drug Administration (FDA) cleared rhBMP-2 for distribution on a type I collagen sponge for use in combination with an LT cage for treatment of degenerative lumbar disc disease. This clearance followed a 279-patient clinical trial comparing rhBMP-2 with autograft. Two years following surgery, 94.5% of rhBMP-2-treated patients and 88.7% of autograft-treated patients achieved radiographic fusion without revision surgery (McKay and Sandhur, 2002).

rhBMP-2 on type I collagen sponges was also investigated in a posterolateral fusion model (Alexander and Branch, 2002). In this study, rhBMP-2 was found to be at least as effective as autologous iliac bone for posterolateral reconstruction, in terms of promoting radiographic fusion. However, 58% of the rhBMP-2 patients demonstrated "greater than expected bone formation" dorsal to the cage, a finding seen in none of the controls. In 30% of the rhBMP-2 cases, this bone compromised the central canal, the neural foramen, or both. The study was halted and rhBMP-2 was deemed unsuitable for posterolateral construct. A following study was published in 2005, evaluating the use of rhBMP-2 in posterolateral applications with a compression-resistant matrix. In this study, 89% fusion was observed in patients implanted with a total dose of 40 mg rhBMP-2, a dose four times higher than that suggested for intervertebral fusion (Glassman *et al.*, 2005). This result was very similar to that observed in a comparable surgical setting using concentrated bone marrow aspirates (see earlier, Wang *et al.*, 2005), indicating that, whether using supraphysiological concentrations of growth factors or bone marrow cell concentrate, the percent healing rates for posterolateral fusions reached a plateau that these current technologies, when used separately, could still not overcome. rhBMP-7, also called rhOP-1, showed very comparable trends, as described later.

rhBMP-2 was also evaluated for efficacy in a long-bone-defect model. In a prospective, randomized, single-blind study, 450 patients with an open tibial fracture were randomized to receive either the standard of care (intramedullary nail fixation and routine soft-tissue management), the standard of care and an implant containing 6 mg of rhBMP-2, or the standard of care and an implant containing 12 mg of rhBMP-2. The primary outcome measure was the proportion of patients requiring secondary intervention because of delayed union or nonunion within 12 months postoperatively. The results indicated highest efficacy in the 12-mg group, and, in that group, superiority to the standard of care was achieved in reducing the frequency of secondary interventions. These results prompted the FDA to grant rhBMP-2 market approval for tibial fractures.

### rhOP-1

rhOP-1, following closely behind rhBMP-2 in terms of development timeline, demonstrated similar osteogenicity in preclinical and clinical models. Cook *et al.* (1994) demonstrated lumbar facet fusions in a dog model using rhOP-1 with a collagen-based carrier. All animals implanted with the rhOP-1 achieved stable facet fusion by 12 weeks, while animals implanted with autogenous bone graft achieved fusion only 26 weeks postoperative. Cunningham *et al.* (2001) studied posterolateral fusion in the dog using rhOP-1 and found that after 12 weeks, 27% of the posterolateral sites implanted with autograft had fused, 83% of the posterolateral sites implanted with autograft and rhBMP-7 had fused, and 72% of the sites implanted with rhBMP-7 alone had fused.

Interestingly, Den Boer *et al.* (2003) investigated the healing potential of ceramic grafts containing either bone marrow or rhOP-1 in a 3-cm segmental bone defect in sheep tibia. Five treatment groups were included: no implant, autograft, hydroxyapatite alone, hydroxyapatite loaded with rhOP-1, and hydroxyapatite loaded with autologous bone marrow. At 12 weeks, torsional strength and stiffness of the healing tibiae were about two to three times higher for autograft and hydroxyapatite plus rhOP-1 or bone marrow, compared to hydroxyapatite alone and empty defects. The mean values of both combination groups were comparable to those of autograft. Healing of bone defects, treated with porous hydroxyapatite, was enhanced by the addition of rhOP-1 or autologous bone marrow. The results of these composite biosynthetic grafts were equivalent to those of autograft.

So, while preclinical studies demonstrated the efficacy of rhOP-1 for bone repair, no evidence indicated an increased efficacy of the recombinant protein vs. native levels of bone marrow in animal models.

rhOP-1 was further investigated *in vivo*. Patel *et al.* (2001) presented the results of a safety and efficacy study using rhOP-1 in combination with autograft for posterolateral fusion. In this study, 16 patients with degenerative lumbar spondylolisthesis and stenosis were randomized to instrumented posterior spinal fusion and decompression with either autograft bone alone or a combination of autograft bone and rhBMP-7 with the collagen and CMC delivery vehicle. No instrumentation was used in this study. Six months following surgery, 9 of 12 (75%) patients in the autograft/rhBMP-7 cohort were graded fused, whereas 2 of 4 (50%) patients in the autograft alone group achieved fusion. While these results seem far less impressive than those obtained with rhBMP-2, it is important to remember that the indication that was used is significantly more challenging and, thus, may require a significantly stronger biological stimulant. Thus, a follow-up trial is currently ongoing for the evaluation and market approval of rhOP-1 in spinal applications.

For long-bone repair, rhOP-1 was compared to autogenous bone graft in a tibial nonunion study. This study successfully demonstrated that rhOP-1 was as efficacious as autograft, but it did not show any improved healing with rhOP-1 as compared to autograft (Friedlaender *et al.*, 2001). Accordingly, the FDA did not grant postmarketing approval (PMA) but instead approved a more limited humanitarian device exemption (HDE) for this application.

## Ongoing Research: New Growth Factors

Both rhBMP-2 and rhOP-1, as presented earlier, were shown to be equivalent to autograft in repairing bone. Therefore, the next generation of growth factors needed to present specific market edges or scientific improvements that would justify their use.

Key issues surrounding both rhBMP-2 and rhOP-1 were related to their long-term safety (Alexander and Branch, 2002;

Dawson, 2003). These long-term safety considerations really came down to these large quantities of growth factors needed in each case; for rhBMP-2, it was estimated that one dose for spinal surgery would equal about 100 times the native levels of BMP-2 in an entire skeleton. Such high doses could generate immune and/or tumorigenic responses. Therefore, the need for growth factors with similar osteoinductivity to both rhBMP-2 and rhOP-1 but with reduced potential for long-term risk emerged as a development opportunity.

In addition to safety concerns, both rhBMP-2 and rhOP-1 presented challenges in their delivery to the surgical site: An optimal carrier for growth factor still seemed to be missing in both cases. Thus, another development opportunity included improving the delivery of these factors through an optimized matrix material.

### Chrysalin

Chrysalin, a new growth factor currently in development by OrthoLogic Inc., has not been widely described in the literature. It is thus unclear whether Chrysalin would provide a clear scientific improvement over rhBMP-2 and rhOP-1. This growth factor may therefore only be developed to provide choices to surgeons.

Chrysalin was characterized as the 23-amino-acid fragment representing the natural amino acid sequence of the receptor-binding domain of human thrombin (prothrombin amino acids 508–530, TP508) (Glenn *et al.*, 1988). Used as a single injection, Chrysalin was shown to accelerate fracture repair in the mid-diaphysis of the rat tibias by about 30–40%. Histological examination of the healing fracture indicated that this single injection of Chrysalin stimulated endochondral ossification. Callus formation was increased, resulting in increased resistance to torsional stress. Recent studies also showed that Chrysalin promoted cartilage formation in rabbit osteochondral defects (Grande *et al.*, 2002). Chrysalin's mechanism of action on endochondral ossification was studied by Schwartz *et al.* (2005) using an *in vitro* model of growth-plate cartilage regulation. Results indicated that Chrysalin promoted the cartilage extracellular matrix synthesis and, to some extent, chondroblastic proliferation, but did not significantly affect endochondral differentiation.

Because preclinical results seemed promising, Chrysalin entered a Phase III clinical trial to assess its effectiveness to accelerate healing of unstable, displaced distal radius fractures. Unfortunately, in early 2006, OrthoLogic Inc. reported that treatment with 10 μg of Chrysalin did not demonstrate a statistically significant benefit as compared to placebo in the primary efficacy endpoint of "time to removal of immobilization."

With competitors such as rhBMP-2, OP-1, and recombinant human growth and differentiation factor 5 (rhGDF-5), it thus becomes unclear how Chrysalin would compete, yet alone gain marketing approval. A strategic repositioning

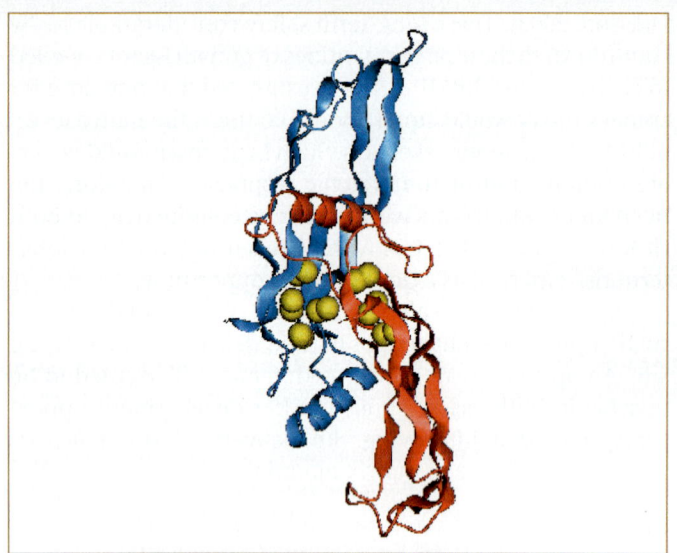

**FIG. 56.7.** Three-dimensional rendering of the recombinant human growth and differentiation factor-5 (rhGDF-5) structure. Circular dichroism of rhGDF-5 further confirmed presence of 2.4% α-turns, 41% β-sheets, 8% β-turns, and 48% random structures.

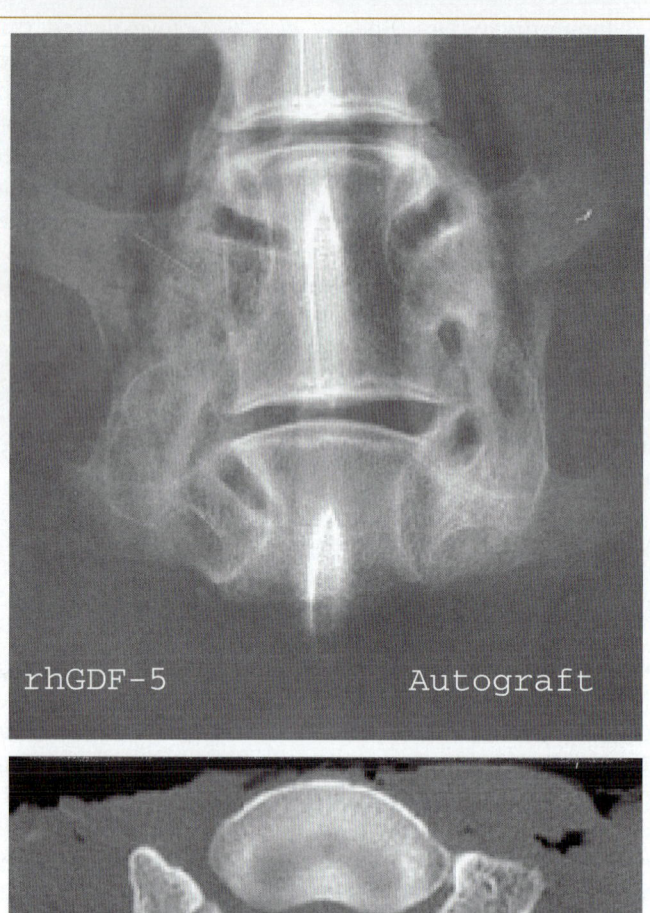

**FIG. 56.8.** Representative radiograph of a sheep L4–L5 spine segment fused with HEALOS/rhGDF-5 six months postoperatively. HEALOS/rhGDF-5 was implanted on one side, while autograft was used on the other side. Surgeries were performed endoscopically.

of Chrysalin's value proposition in terms of health/cost benefits seemed necessary to bring this molecule to market. While the clinical trial addressed only the effect of Chrysalin on bone regeneration, it is possible that this molecule may be found effective with other clinical endpoints or for the repair of other connective tissues.

### Recombinant Human Growth and Differentiation Factor-5

Recombinant human growth and differentiation factor-5 (rhGDF-5) was described as a subgroup-defining BMP family member with structural similarity to BMP-2 and BMP-7 (Fig. 56.7). Following its initial discovery in 1992, GDF-5 was further rediscovered in various tissues and by several unrelated research groups, all of which named the molecule differently. As a result, GDF-5 was called MP52, BMP-14, LAP-4, CDMP-1, and, more recently, radotermin (Triantafilou *et al.*, 2001; Hotten *et al.*, 1994). rhGDF-5's market "edges" over rhBMP-2 and rhOP-1 were twofold: (1) An optimized matrix (HEALOS®), on which rhGDF-5 is lyophilized, would improve the delivery of the growth factor in the surgical site, and (2) efficacy at lower doses would alleviate some of the long-term safety concerns.

HEALOS/rhGDF-5 was tested in a series of animal models, ranging from rats, rabbits, and sheep to nonhuman primates, and in long bones, flat bones, and spinal applications. Three key preclinical studies were most recently published in rabbits (Magit *et al.*, 2005) and sheep (D. H. Kim *et al.*, 2004; Jahng *et al.*, 2004).

Magit *et al.*, using the rabbit posterolateral fusion model, performed a randomized study using three different doses of HEALOS/rhGDF-5 bone graft (0.5 mg/cc, 1.0 mg/cc, and

1.5 mg/cc), as well as autograft and carrier-only controls. A total of 13 animals were included per arm (for a total of 65 rabbits) to ensure statistical relevance. The animals were maintained for eight weeks and then analyzed by manual palpation and histology. All animals implanted with HEALOS/rhGDF-5 fused at the eight-week endpoint. Less than 40% fusion was observed in the autograft arm, and no fusion was seen in the carrier-only controls. Histologically, the 0.5-mg/cc and 1.0-mg/cc doses resulted in the greatest bone volume formation.

HEALOS/rhGDF-5 was then evaluated in a variety of ovine spine fusion models. The first posterolateral fusion study in sheep was performed by D. H. Kim *et al.* (2004). This study compared HEALOS/rhGDF-5 to autograft in an instrumented lumbar fusion model. Eleven skeletally mature, female sheep underwent intertransverse process spine fusion at two noncontiguous fusion segments, L2–L3 and L5–L6. Each animal was treated with autograft at one level and 0.5 mg/cc HEALOS/rhGDF-5 bone graft at the other level. At 6 and 12 months, fusion was assessed by manual palpation, radiography, and histology. Manual palpation indicated fusion at all treated levels, at both time points, a result that was further confirmed by radiographic analyses and histology. Histomorphometry was also performed on all samples and showed similar *de novo* bone volumes among all study groups. This study demonstrated that HEALOS/rhGDF-5 induced bone formation similar to autograft in a posterolateral, large-animal fusion model (Fig. 56.8).

In a second posterolateral fusion study in sheep, Jahng *et al.* (2004) investigated the use of HEALOS/rhGDF-5 for endoscopic surgery. The efficacy of 0.5 mg/cc HEALOS/rhGDF-5 was thus compared to autograft in an instrumented L4–L5 fusion model performed using an endoscopic approach. Twelve mature sheep were implanted with autogenous iliac crest bone on one side of the vertebral bodies and 0.5 mg/cc HEALOS/rhGDF-5 on the other side. Groups of four sheep each were euthanized at two, four, and six months postoperatively for manual palpation, radiography, and histological evaluation. No complications or ectopic bone formation were observed, confirming previous observations that rhGDF-5 does not induce bone in heterotopic sites. At two months, radiographic evaluation of the fusion sites showed the greatest bone growth in sites implanted with HEALOS/rhGDF-5. At four and six months, solid posterolateral fusion was observed in both the autograft and the HEALOS/rhGDF-5 sites. These results reinforced previously recorded fusion rates and also highlighted the safety of HEALOS/rhGDF-5, since no ectopic bone formation was observed, a result that was particularly important in view of rhBMP-2's ectopic bone formation history in its aborted clinical trial. rhGDF-5 is currently undergoing clinical trials in the United States for use in the spine.

## IV. CONCLUSION AND FUTURE TRENDS

Significant progress has been achieved over the last several decades in bone-tissue regeneration. Specifically, in-depth understanding of bone marrow cells was critical to optimize the use of bone marrow in bone defects and develop methods of concentrating these osteogenic components to further enhance the healing potential of novel devices. In parallel, growth factors also significantly changed the therapeutic landscape and may continue to do so going forward.

As our understanding of the complex phenomenon of bone formation increases, there will be an increasing number of potent grafts and cell therapies available to help in bone-related surgical procedures. Short- and medium-term research may include new osteopromotive and osteo-inductive growth factors for bone repair and bone fusion, as well as improved delivery systems for existing growth factors. In the long term, focus may shift to injectable formulations that could form bone *in situ* and would altogether alleviate the need for invasive surgeries.

## V. REFERENCES

Aghaloo, T. L., Moy, P. K., and Freymiller, E. G. (2004). Evaluation of platelet-rich plasma in combination with anorganic bovine bone in the rabbit cranium: a pilot study. *Int. J. Oral Maxillofac. Impl.* **19**(1), 59–65.

Alexander, J., and Branch, C. (2002). Recombinant human bone morphogenetic protein-2 in a posterior lumbar interbody fusion construct: 2-year clinical and radiologic outcomes. *Spine J.* **2**, 48.

Andersen, T., Christensen, F. B., Laursen, M., Lund-Olesen, L., Gelineck, J., and Bunger, C. (2003). *In vitro* osteoblast proliferation as a predictor for spinal fusion mass. *Spine J.* **3**, 285–288.

Arinzeh, T. L., Peter, S. J., Archambault, M. P., Van Den Bos, C., Gordon, S., Kraus, K., Smith, A., and Kadiyala, S. (2003). Allogeneic mesenchymal stem cells regenerate bone in a critical-sized canine segmental defect. *JBJS* **85A**(10), 1927–1935.

Arrington, E. D., Smith, W. J., Chambers, H. G., Bucknell, A. L., and Davino, N. A. (1996). Complications of iliac crest bone graft harvesting. *Clin. Orthop. Rel. Res.* **329**, 300–309.

Barrow, C. R., and Pomeroy, G. C. (2005). Enhancement of syndesmotic fusion rates in total ankle arthroplasty with the use of autologous platelet concentrate. *Foot Ankle Int.* **26**(6), 458–461.

Betz, O., Vrahas, M., Baltzer, A., Lieberman, J. R., Robbins, P. D., and Evans, C. H. (2005). Gene transfer approaches to enhancing bone healing. *In* "Bone Regeneration and Repair" (J. R. Lieberman and G. E. Friedlaender, eds.), pp. 158–162. Humana Press, Totowa, NJ.

Brodke, D. S., Pedrozo, H. A., Kapur, T. A., Holy, C. E., Attawia, M., Kraus, K. H., Kadiyala, S., and Bruder, S. P. (2006). Bone grafts prepared with selective cell retention technology heal canine segmental defects as effectively as autograft. *J. Orthop. Res.* (in press).

Bruder, S. P., and Caplan, A. I. (2000). Bone regeneration through cellular engineering. *In* "Principles of Tissue Engineering" (R. P. Lanza, R. Langer, and J. Vacanti, eds.), pp. 683–693. Academic Press, San Diego, CA.

Bruder, S. P., Jaiswal, N., and Haynesworth, S. E. (1997). Growth kinetics, self-renewal, and the osteogenic potential of purified human mesenchymal stem cells during extensive subcultivation and following cryopreservation. *J. Cell Biochem.* **64**(2), 278–294.

Burwell, R. G. (1964). Studies in the transplantation of bone VII. The fresh composite homograft-autograft of cancellous bone: an analysis of factors leading to osteogenesis in marrow transplants and in marrow containing bone grafts. *J. Bone Joint Surg. Br.* **46**(1), 110–140.

Chapman, M. W., Bucholz, R., and Cornell, C. (1997). Treatment of acute fractures with a collagen–calcium phosphate graft material. *JBJS* **79A**(4), 495–502.

Connolly, J. F. (1998). Clinical use of marrow osteoprogenitor cells to stimulate osteogenesis. *Clini. Orthop. Rel. Res.* **355**(Suppl.), S257–S266.

Connolly, J. F., Guse, R., Lippiello, L., and Dehne, R. (1989). Development of an osteogenic bone-marrow preparation. *J. Bone Joint Surg. Am.* **71**, 684–691.

Connolly, J. F., Guse, R., Lippiello, L., and Dehne, R. (1991). Autologous marrow injection as a substitute for operative grafting of tibial nonunions. *Clin. Orthop. Rel. Res.* **266**, 259–270.

Cook, S. D., Dalton, J. E., Tan, E. H., Whitecloud, T. S., III, and Rueger, D. C. (1994). *In vivo* evaluation of recombinant human osteogenic protein (rhOP-1) implants as a bone graft substitute for spinal fusions. *Spine* **19**, 1655–1663.

Cowan, C. M., Shi, Y. Y., Aalami, O. O., Chou, Y. F., Carina, M., Thomas, R., Quarto, N., Contag, C. H., Wu, B., and Longaker, M. T. (2004). Adipose-derived adult stromal cells heal critical-size mouse calvarial defects. *Nat. Biotechnol.* **22**(5), 560–567.

Cunningham, B. W., Shimamoto, N., Sefter, J. C., et al. (2002). Osseointegration of autograft versus osteogenic protein-1 in posterolateral spinal arthrodesis:emphasis on the comparative mechanisms of bone induction. *Spine J.* **2**, 11–24.

Dawson, E. G. (2003). Letters to the editor. *Spine J.* **3**, 87–89.

De Kok, I. J., Peter, S. J., Archambault, M., Van den Bos, C., Kadiyala, S., Aukhil, I., and Cooper, L. F. (2003). Investigation of allogeneic mesenchymal stem cell–based alveolar bone formation: preliminary findings. *Clin. Oral Implants Res.* **14**(4), 481–489.

Den Boer, F. C., Wippermann, B. W., Blokhuis, T. J., Patka, P., Bakker, F. C., and Haarman, H. J. (2003). Healing of segmental bone defects with granular porous hydroxyapatite augmented with recombinant human osteogenic protein-1 or autologous bone marrow. *J. Orthop. Res.* **21**, 521–552.

Dragoo, J. L. (2003). Bone induction by BMP-2 transduced stem cells derived from human fat. *J. Orthop. Res.* **21**(4), 622–629.

Fernyhough, J. C., Schimandle, J. J., Weigel, M. C., Edwards, C. C., and Levine, A. M. (1992). Chronic donor site pain complicating bone graft harvesting from the posterior iliac crest for spinal fusion. *Spine* **17**(12), 1474–1480.

Friedenstein, A. J., Petrakova, K. V., Kurolesova, A. I., and Frolova, G. P. (1968). Heterotopic transplants of bone marrow. Analysis of precursor cells for osteogenic and hematopoietic tissues. *Transplantation* **6**(2), 230–247.

Friedlaender, G. E., Perry, C. R., Cole, J. D., Cook, S. D., Cierny, G., and Muschler, G. F., *et al.* (2001). Osteogenic protein-1 (bone morphogenetic protein-7) in the treatment of tibial nonunions. *J. Bone Joint Surg. Am.* **83A**(Suppl. 1), S151–S158.

Fritzell, P., Hagg, O., Wessberg, P., and Nordwall, A. (2001). Lumbar fusion versus nonsurgical treatment for chronic low back pain: a multicenter randomized controlled trial from the Swedish Lumbar Spine Study Group. *Spine* **26**(23), 2521–2532.

Gandhi, A., Dumas C., O'Connor, J. P., Parsons, R., and Lin, S. S. (2006). The effects of local platelet rich plasma delivery on diabetic fracture healing. *Bone* **38**, 540–546.

Garg, N. K., Gaur, S., and Sharma, S. (1993). Percutaneous autologenous bone marrow grafting in 20 cases of ununited fracture. *Acta Orthop. Scand.* **64**(6), 671–672.

Glassman, S. D., Dimar, J. R., Carreon, L. Y., Campbell, M. J., Puno, R. M., and Johnson, J. R. (2005). Initial fusion rates with rhBMP-2/ compression resistant matrix and a hydroxyapatite and tricalcium phosphate/collagen carrier in posterolateral fusion. *Spine* 1694–1698.

Glenn, K. C., Frost, G. H., Bergmann, J. S., and Carney, D. H. (1988). Synthetic peptides bind to high-affinity thrombin receptors and modulate thrombin mitogenesis. *Pept. Res.* **1**, 65–73.

Govender, S., Csimma, C., Genant, H. K., Valentin-Opran, A., Amit, Y., Arbel, R., Aro, H., Atar, D., Bishay, M., and Borner, M. G., et al. (2002). BMP-2 Evaluation in Surgery for Tibial Trauma (BESTT) Study Group. Recombinant human bone morphogenetic protein-2 for treatment of open tibial fractures: a prospective, controlled, randomized study of 450 patients. *J. Bone Joint Surg. Am.* **84A**(12), 2123–2134.

Grande, D., Karnaugh, R., Ryaby, J., Dines, D., Razzano, P., Crowther, R., Carney, D., and Wu, D. (2002). *In vivo* evaluation of the synthetic thrombin peptide, TP508, in articular cartilage repair. *Orthop. Res. Soc. Meeting*, 447.

Gupta, A. R., Shah, N. R., Patel, T. C., and Grauer, J. N. (2001). Perioperative and long-term complications of iliac crest bone graft harvesting for spinal surgery: a quantitative review of the literature. *Int. Med. J.* **8**(3), 163–166.

Gupta, M. C., Theerajunyaporn, T., Schmid, M. B., Holy, C. E., Kadiyala, S., and Bruder, S. P. (in press). Use of mesenchymal stem cells enriched grafts in an ovine posterolateral lumbar spine model. *Spine*.

Healey, J. H., Zimmerman, P. A., Jessop, A. B., McDonnel, M., and Lane, J. M. (1991). Percutaneous bone marrow grafting of delayed union and nonunion in cancer patients. *Clin. Orthop. Rel. Res.* **256**, 280–285.

Heary, R. F., Schlenk, R. P., Sacchieri, T. A., Barone, D., and Brotea, C. (2002). Persistant iliac crest donor site pain: independent outcome assessment. *Neurosurgery* **50**(3), 510–516.

Hernigou, Ph., Poignard, A., Beaujean, F., Rouard, H. (2005). Percutaneous autologous bone-marrow grafting for nonunions: influence of the number and concentration of progenitor cells. *J. Bone Joint Surg (Am.)* **87**, 1430–1437.

Hotten, G., Neidhardt, H., and Jacobowsky, B., Pohl, J. (1994). Cloning and expression of recombinant human growth/differentiation factor 5. *Biochem. Biophys. Res. Com.* **204**(2), 646–652.

Jahng, T. A., Fu, T. S., and Cunningham, B. W., Dmitriev, A. E., and Kim, D. H. (2004). Endoscopic instrumented posterolateral lumbar fusion with HEALOS and recombinant human growth/differentiation factor-5. *Neurosurgery* **54**(1), 171–181.

Jaiswal, N., Haynesworth, S. E., Caplan, A. I., and Bruder, S. P. (1997). Osteogenic differentiation of purified, culture-expanded human mesenchymal stem cells *in vitro*. *J. Cell. Biochem.* **64**(2), 295–312.

Johnson, E. E., Urist, M. R., and Finerman, G. A. (1988). Bone morphogenetic protein augmentation grafting of resistant femoral nonunions. A preliminary report. *Clin. Orthop.* **230**, 257–265.

Kevy, S. V., and Jacobson, M. S. (2004). Comparison of methods for point of care preparation of autologous platelet gel. *JECT* **36**, 28–35.

Kim, D. H., Jahng, T. A., and Fu, T. S., Zhang, H. Y. and Novak, S. A. (2004). Evaluation of HEALOS/MP52 osteoinductive bone graft for instrumented lumbar intertransverse process fusion in sheep. *Spine* **29**(24), 2800–2808.

Kim, H. S., Viggeswarapu, M., Boden, S. D., Liu, Y., Hair, G. A., Louis-Ugbo, J., Murakami, H., Minamide, A., Suh, D. Y., and Titus, L. (2003). Overcoming the immune response to permit *ex vivo* gene therapy for spine fusion with human type 5 adenoviral delivery of the LIM mineralization protein-1 cDNA. *Spine* **28**(3), 219–226.

Kitchel, S. H. (2006). A preliminary comparative study of radiographic results using mineralized collagen and bone marrow aspirate vs. autologous bone in the same patients undergoing posterior lumbar interbody fusion with instrumented posterolateral lumbar fusion. *The Spine Journal* **6**, 405–412.

Kraiwattanapong, C., Boden, S. D., Louis-Ugbo, J., Attallah, E., Barnes, B., and Hutton, W. C. (2005). Comparison of Healos/bone marrow to INFUSE(rhBMP-2/ACS) with a collagen-ceramic sponge bulking agent as graft substitutes for lumbar spine fusion. *Spine* **30**(9), 1001–1007.

Kuznetsov, S. A., Krebsbach, P. H., Satomura, K., Kerr, J., Riminucci, M., Benayahu, D., and Robey, P. G. (1997). Single-colony-derived strains of human marrow stromal fibroblasts form bone after transplantation *in vivo*. *J. Bone Min. Res.* **12**(9), 1335–1347.

Lieberman, I. (2004). Local cell delivery strategies. Bone Summit, Cleveland, OH.

Lindholm, T. S., Nilsson, O. S., and Lindholm, T. C. (1982). Extraskeletal and intraskeletal new bone formation induced by demineralized bone matrix combined with bone marrow cells. *Clin. Orthop. Rel. Res.* **171**, 251–255.

Liu, X. D., Zhu, Y. K., Umino, T., Spurzem, J. R., Romberger, D. J., Wang, H., Reed, E., and Rennard, S. I. (2001). Cigarette smoke inhibits osteogenic differentiation and proliferation of human osteoprogenitor cells in monolayer and three dimensional collagen gel culture. *J. Lab. Clin. Med.* **137**(3), 208–219.

Lu, H., Kraut, D., Gerstenfeld, L. C., and Graves, D. T. (2003). Diabetes interferes with the bone formation by affecting the expression of transcription factors that regulate osteoblast differentiation. *Endocrinology* **144**, 346–352.

Magit, D. P., Maak, T., Troiano, N., et al. (2006). Healos/recombinant human growth and differentiation Factor-5 induces posterolateral lumbar fusion in a New Zealand white rabbit model. *Spine* **31**, 2180–2188.

Maniatopoulos, C., Sodek, J., and Melcher, A. H. (1988). Bone formation *in vitro* by stromal cells obtained from bone marrow of young adult rats. *Cell Tissue Res.* **254**, 317–330.

Marx, R. E. (2001). Platelet-rich plasma (PRP): what is PRP and what is not PRP? *Implant Dentistry* **10**(4), 225–228.

Marx, R. E., Carlson, E. R., Eichstaedt, R. M., Schimmele, S. R., Strauss, J. E., and Georgeff, K. R. (1998). Platelet-rich plasma: growth factor enhancement for bone grafts. *Oral Surg. Oral Med. Oral Pathol. Oral Radiol. Endod.* **85**(6), 638–646.

McKay, B., and Sandhu, H. S. (2002). Use of recombinant human bone morphogenetic protein-2 in spinal fusion applications. *Spine* **27**(16 Suppl. 1), S66–S85.

Minamide, A., Yoshida, M., Kawakami, M., Yamasaki, S., Kojima, H., Hashizume, H., and Boden, S. D. (2005). The use of cultured bone marrow cells in type I collagen gel and porous hydroxyapatite for posterolateral lumbar spine fusion. *Spine* **30**(10), 1134–1138.

Muschler, G. F., Boehm, C., and Easley, K. (1997). Aspiration to obtain osteoblast progenitor cells from human bone marrow: the influence of aspiration volume. *J. Bone Joint Surg. Am.* **79**(11), 1699–1709.

Muschler, G. F., Nitto, H., Boehm, C. A., and Easley, K. A. (2001). Age- and gender-related changes in the cellularity of human bone marrow and the prevalence of osteoblastic progenitors. *J. Orthop. Res.* **19**(1), 117–125.

Muschler, G. F., Nitto, H., and Matsukura, Y., *et al.* (2003). Spine fusion using cell matrix composites enriched in bone marrow–derived cells. *Clin. Orthop. Rel. Res.* **407**, 102–118.

Muschler, G. F., Nitto, H., Matsukura, Y., Boehm, C., Valdevit, A., Kambic, H., Davros, W. Powell, K., and Easley, K. (2005). Selective retention of bone marrow–derived cells to enhance spinal fusion. *Clin. Orthop. Rel. Res.* **432**, 242–251.

Niemeyer, P., Seckinger, A., Simank, H. G., Kasten, P., Sudkamp, N., and Krause, U. (2004). Allogenic transplantation of human mesenchymal stem cells for tissue engineering purposes: an *in vitro* study. *Orthopade* **33**(12), 1346–1353.

Ohgushi, H., Goldberg, A. I., and Caplan, A. I. (1989). Repair of bone defects with marrow cells and porous ceramic. Experiments in rats. *Acta Orthop. Scand.* **60**, 334–339.

Olmsted-Davis, E. A., Gugala, Z., Gannon, F. H., Yotnda, P., McAlhany, R. E., Lindsey, R. W., and Davis A. R. (2002). Use of a chimeric adenovirus vector enhances BMP2 production and bone formation. *Hum. Gene Ther.* **13**(11), 1337–1347.

Phemister, D. (1947). Treatment of ununited fractures by onlay bone grafts without screw or tie fixation and without breaking of the fibrous union. *J. Bone Joint Surg.* **29**, 946–960.

Price, C. T., Connolly, J. F., Carantzas, A. C., and Ilyas, I. (2003). Comparison of bone grafts for posterior spinal fusion in adolescent idiopathic scoliosis. *Spine* **28**(8), 793–798.

Prockop, D. J. (1997). Marrow stromal cells as stem cells for nonhematopoietic tissues. *Science* **276**(5309), 71–74.

Quesenberry, P. J., Colvin, G. A., Abedi, M., Dooner, G., Dooner, M., Aliotta J., Keaney, P., Luo, L., Demers, D., Peterson, A., Foster, B., and Greer, D. (2005). The stem cell continuum. *Ann. N.Y. Acad. Sci.* **1044**, 228–235.

Salama, R., and Weissman, S. L. (1978). The clinical use of combined xenografts of bone and autologous red marrow. *JBJS* **60B**(1), 111–115.

Sandhu, H. S., Kanim, L. E., Kabo, J. M., Liu, D., Delamarter, R. B., and Dawson, E. G. (1996). Effective doses of recombinant human bone morphogenetic protein-2 in experimental spinal fusion. *Spine* **21**, 2115–2122.

Sandhu, H. S., Kanim, L. E., Toth, J. M., Kabo, J. M., Liu, D., Delamarter, R. B., and Dawson, E. G. (1997). Experimental spinal fusion with recombinant bone morphogenetic protein-2 without decortication of osseous elements. *Spine* **22**, 1171–1180.

Schimandle, J. H., Boden, S. D., and Huttonm, W. C. (1995). Experimental spinal fusion with recombinant human bone morphogenetic protein-2. *Spine* **20**, 1326–1337.

Schwartz, Z., Carney, D. H., Crowther, R. S., Ryaby, J. T., and Oyan, B. D. (2005). Thrombin peptide (tp508) treatment of rat growth plate cartilage cells promotes proliferation and retention of the chondrocytic phenotype while blocking terminal endochondral differentiation. *J. Cell Physiol.* **202**, 336–343.

Sim, R., Liang, T. S., and Tay, B. K. (1993). Autologous marrow injection in the treatment of delayed and nonunion in long bones. *Singapore Med. J.* **34**, 412–417.

Stroncek, D. F., McGlave, P., Ramsay, N., and McCullough, J. (1991). Effects on donors of second bone marrow collections. *Transfusion* **31**(9), 819–822.

Tay, B. K., Le, A. X., Heilman, M., Lotz, J., and Bradford, D. S. (1998). Use of a collagen-hydroxyapatite matrix in spinal fusion. *Spine* **23**(21), 2276–2281.

Tiedeman, J. J., Connolly, J. F., Strates, B. S., and Lippiello, L. (1991). Treatment of nonunion by percutaneous injection of bone marrow and demineralized bone matrix. An experimental study in dogs. *Clin. Orthop.* **268**, 294–302.

Triantafilou, K., Triantafilou, M., and Dedrick, R. L. (2001). A CD14-independent LPS receptor cluster. *Nat. Immunol.* **2**(4), 338–345.

Urist, M. R. (1965). Bone. Formation by autoinduction. *Science* **150**, 893–899.

Vaccaro, A. R., Patel, T., Fischgrund, J., et al. (2003). A pilot safety and efficacy study of OP-1 putty (rhBMP-7) as an adjunct to iliac crest autograft in posterolateral lumbar fusions. *Eur. Spine J.* **12**, 495–500.

Viggeswarapu, M., Boden, S. D., Liu, Y., Hair, G. A., Louis-Ugbo, J., Murakami, H., Kim, H. S., Mayr, M. T., Hutton, W. C., and Titus, L. (2001). Adenoviral delivery of LIM mineralization protein-1 induces new-bone formation *in vitro* and *in vivo*. *J. Bone Joint Surg. (Am.)* **83A**(3), 364–376.

Wang, J. C., Youssef, J. A., Lieberman, I. H., Brodke, D. S., Lauryssen, C., Haynesworth, S. E., and Muschler, G. F. (2005). A prospective, multicenter study of selective osteoprogenitor cell retention for enhancement of lumbar spinal fusion. *IMAST*, Banff, Canada.

Weiner, B. K., and Walker, M. (2003). Efficacy of autologous growth factors in lumbar intertransverse fusions. *Spine* **28**(17), 1968–1971.

Wenisch, S., Trinkaus, K., Hild, A., Hose, D., Herde K., Heiss, C., Kilian, O., Alt, V., and Schettler, R. (2005). Human reaming debris: a source of multipotent stem cells. *Bone* **36**(1), 74–83.

Wientroub, S., Goodwin, D., Khermosh, O., and Salama, R. (1989). The clinical use of autologous marrow to improve osteogenic potential of bone grafts in pediatric orthopedics. *J. Ped. Orthop.* **9**(2), 186–190.

Zuk, P. A., Zhu, M., Ashjian, P., De Ugarte, D. A., Huang, J. I., Mizuno, H., Alfonso, Z. C., Fraser, J. K., Benhaim, P., and Hedrick M. H. (2002). Human adipose tissue is a source of multipotent stem cells. *Mol. Biol. Cell* **13**(12), 4279–4295.

# Chapter Fifty-Seven

# Bone and Cartilage Reconstruction

*Wei Liu, Wenjie Zhang, and Yilin Cao*

## I. INTRODUCTION

Bone and cartilage reconstruction is an important area of tissue-engineering research. Tissue engineering offers a better approach than conventional surgery for bone and cartilage repair by reducing donor site morbidity. To achieve this goal, animal study of repairing clinically relevant tissue defects, especially in large animals, has become a crucial step.

Bone and cartilage tissue repair is an important task for plastic and orthopedic surgeons. One of the disadvantages of the traditional surgical repair procedure is the need to harvest autologous tissues from a donor site of the human body for the repair of a defect on another site, thus leaving a secondary tissue defect. Tissue engineering offers an approach to tissue repair and regeneration without the necessity of donor site morbidity. For practical applications, a series of experiments should be performed to prove the concept and the practical feasibility. In the initial phase, immunodeficient-animal models are usually used to screen all potential components of the proposed construct, such as matrix, cells, and signaling substances. In the second phase, an appropriate tissue-defect model in small animals is used to prove the principle of engineered tissue repair. In the preclinical phase, clinically relevant models are established in large animals to study further the tissue formation process and the outcomes of engineered tissue repair in order to safely translate animal studies into human clinical trials. Regarding bone and cartilage engineering, a large number of studies have been performed in the initial and the second phases, according to the published literature. Based on these and our own results of early-phase studies, we have focused on tissue engineering and repair in large-animal models in the past few years and have further conducted clinical trials of bone repair in patients on a small scale. This chapter introduces our experience in bone and cartilage reconstruction in large-animal models and our experience in clinical trials of engineered bone repair of patients' craniofacial defects.

## II. BONE RECONSTRUCTION

As early as 1993, Vacanti *et al.* applied tissue-engineering techniques to the construction of bone grafts in the subcutaneous tissue of nude mice using periosteum-derived osteoblasts and degradable polymer scaffold, which clearly proved the concept that the tissue-engineering approach can regenerate bone tissue *in vivo* (Vacanti *et al.*, 1993; Vacanti and Upton, 1994). Later, more papers reported success in repairing clinically relevant bone defects using the tissue-engineering approach.

Calvarial bone defect is a commonly used model to test the feasibility of bone engineering and repair in different animal studies and to define the role that seed cells or scaffold material plays in the bone repair process. Although relatively small, the mouse model is still being used for bone-engineering studies. Kaplan's group recently used silk as a scaffold material for bone engineering (Meinel *et al.*, 2005). In their study, human mesenchymal stem cells (hMSC) were first isolated and seeded on a porous silk fibroin scaffold, *in vitro* engineered for five weeks in a bioreactor, and then transplanted to repair the calvarial bone defects created in nude mice with a diameter of 4 mm. Their results showed that hMSC could undergo osteogenic differentiation on this novel scaffold and that *in vitro* engineered bone could achieve better reparative results than the cell–scaffold complex without *in vitro* engineering. In another study, Cowan *et al.* (2004) used adult male FBV mice to study engineered bone repair of a critical-sized calvarial defect using osteogenically induced adipose-derived stromal cells and PLGA scaffold. They created a defect in parietal bone with a diameter as large as 5 mm and then repaired the defect with PLGA seeded with adipose-derived stromal cells or osteoblasts. The results showed that apatite-coated PLGA seeded with either type of cells could completely heal the bone defects and thus demonstrated that adipose-derived stromal cells could serve as a cell source for bone regeneration as well as bone marrow stromal cells.

Compared to the mouse model, rabbit calvarial repair seems to be used more often. As early as 1999, Breitbart *et al.* used this model to determine the effect of BMP-7 gene on engineered bone formation. They transduced rabbit periosteal cells with retrovirus containing a human BMP-7 gene and then seeded them on polyglycolic acid scaffold for bone repair. After 12 weeks, better bone formation was observed both histologically and radiographically in the transduced cell group than in the nontransduced group. In a similar study, Chang *et al.* (2004) also demonstrated that BMP-2-gene-transfected bone marrow stem cells could repair the calvarial bone defect better than nontransfected stem cells when applied with biomaterial to the bone defect site. In several other studies, the rabbit calvarial model was used to test scaffold materials for bone formation. With this model, bioglass was proved not to be a proper scaffold material for repairing large-sized bone defects (Moreira-Gonzalez *et al.*, 2005). Dutta Roy *et al.*'s (2003) study showed that porous HA scaffolds with engineered macroscopic channels had a significantly higher percentage of new bone formation when compared with the scaffolds without channels in a rabbit calvarial defect model, which indicates that scaffold geometry can enhance the ability of a ceramic material to accelerate the healing of bone defects.

The bone-engineering studies in mouse and rabbit calvarial models provide convincing data that critical-sized defects of flat bone in small animals can be completely healed with a tissue-engineering approach. However, these results may not necessarily represent the bone formation and repairing process in human beings, because low-level mammals usually have a stronger ability than human beings for tissue regeneration. In contrast, large mammals, such as the sheep, dog, and pig, may reflect better than small animals the actual bone formation process in human beings, which will provide a guide to clinical application. Therefore, in the past few years our center has focused on engineered bone formation studies in large mammals. For example, in a study of sheep cranial bone defect repair (Shang *et al.*, 2001), bone marrow stromal cells (BMSCs) were isolated, expanded, and osteogenically induced with dexamethasone, β-glycerophosphate, and vitamin $D_3$. The induced cells were then mixed with calcium alginate to form a cell–scaffold construct for *in vivo* implantation. In created bilateral cranial defects (20 mm in diameter), the experimental side was repaired with a bone graft comprising *in vitro*-induced autologous BMSCs and calcium alginate. Histology demonstrated new bone formation at six weeks postrepair, which became more mature by 18 weeks (Fig. 57.1). Three-dimensional computed tomography (3D CT) scanning confirmed that the bone defects were almost completely repaired by engineered bone at 18 weeks. In contrast, the control defect, which was transplanted with calcium alginate alone, remained unrepaired (Fig. 57.2). Furthermore, chemical analysis showed that the engineered bone tissues contained a high level of calcium (71.6% of normal bone tissue), suggesting that engineered bone can achieve good mineralization.

As a gel form scaffold, calcium alginate might be more suitable for engineering injectable bone than flat bone. In addition, tight contact between host bone and implanted bone construct might be important for better regeneration of engineered bone *in vivo*, meaning a relatively solid scaffold material should be preferred. Furthermore, a scaffold with natural tissue structure and components should be more suitable for clinical trials. Considering these, we performed a similar study in a canine model by using demineralized bone matrix (DBM) and BMSCs (manuscript in preparation). First, bilateral bone defects with a size of 2 cm × 2 cm were created. Then a BMSC-DMB construct that had been preincubated in osteogenic medium for seven days was implanted on the experimental side. Similarly treated scaffold construct without cell loading was implanted in a defect on the other side as a control. As shown in Fig. 57.3, bone formation was observed on the experimental side and remained stable when examined by x-ray and 3D CT scanning at 12 months postimplantation, and the defect was well repaired by the engineered bone. Histology also verified the bone formation on the experimental side but not on the control side. Of note, the engineered bone formed cancellous bone structure instead of lamellar structure of natural calvarial bone (Fig. 57.4). As expected, the control

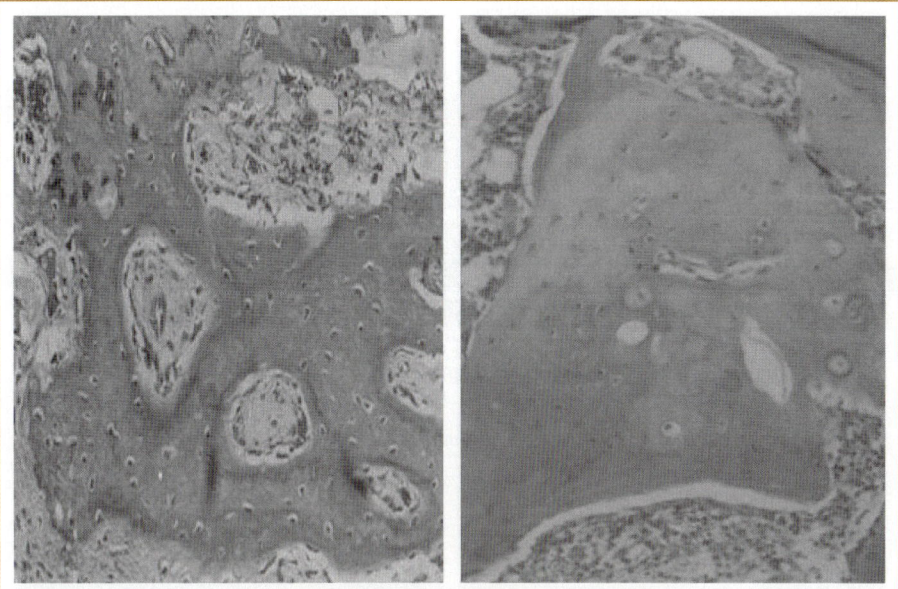

**FIG. 57.1.** Histology of tissue-engineered cranial bone at six weeks (*left*) and 18 weeks (*right*). Reprinted by permission of the *Journal of Craniofacial Surgery* from Shang *et al.* (2001).

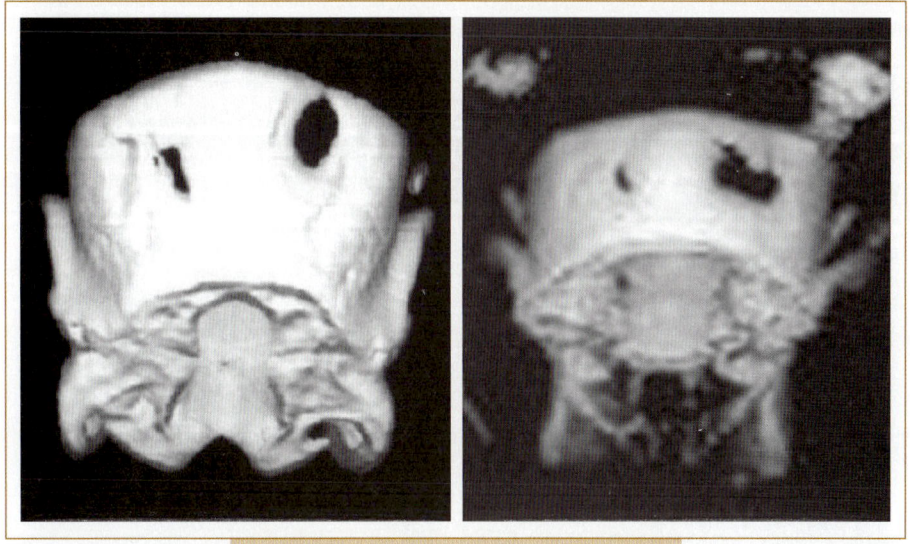

**FIG. 57.2.** Three-dimensional CT evaluation of tissue-engineered bone repair of cranial defect at 12 weeks (*left panel*) and 18 weeks (*right panel*). *Left side:* Experimental group. *Right side:* Control group. Reprinted by permission of the *Journal of Craniofacial Surgery* from Shang *et al.* (2001).

side remained unhealed when evaluated by X-ray and 3D CT scanning examination at 12 months. Only fibrotic tissue was observed in its histology (Figs. 57.3 and 57.4). This work demonstrated that DBM could serve as a suitable scaffold material for bone engineering in large animals, and it might be an appropriate material for clinical trials. Recently, Mauney *et al.* (2005) also demonstrated *de novo* bone formation after *in vivo* implantation of DBM seeded with human BMSCs. In addition, they found that the level of mineralization of DBM also has an influence on the osteogenic differentiation of seeded BMSCs.

Of interest is that engineered flat bone did not form a typical lamellar structure, as natural flat bone has. Instead, a cancerous bone structure was observed in engineered cranial bone. This phenomenon is probably due to the lack of proper mechanical loading during engineered bone formation *in vivo*. Unlike the bone-engineering process, during development and growth a natural calvarial bone may sustain a continuous and long-term pressure from growing brain tissue and eventually forms a lamellar structure.

Therefore, an *in vivo* model that allows for bone remodeling by mechanical loading is particularly important for generating a bone that has the structure and mechanical strength similar to those of natural bone counterpart. Different from flat bone, weight-bearing long bones usually sustain mechanical loading in their physiological activities during their lifetime. These kinds of bones, including femur, tibia, and metatarsus, might be fit for such a model for engineering weight-bearing bone. In 2000, Petite *et al.* succeeded

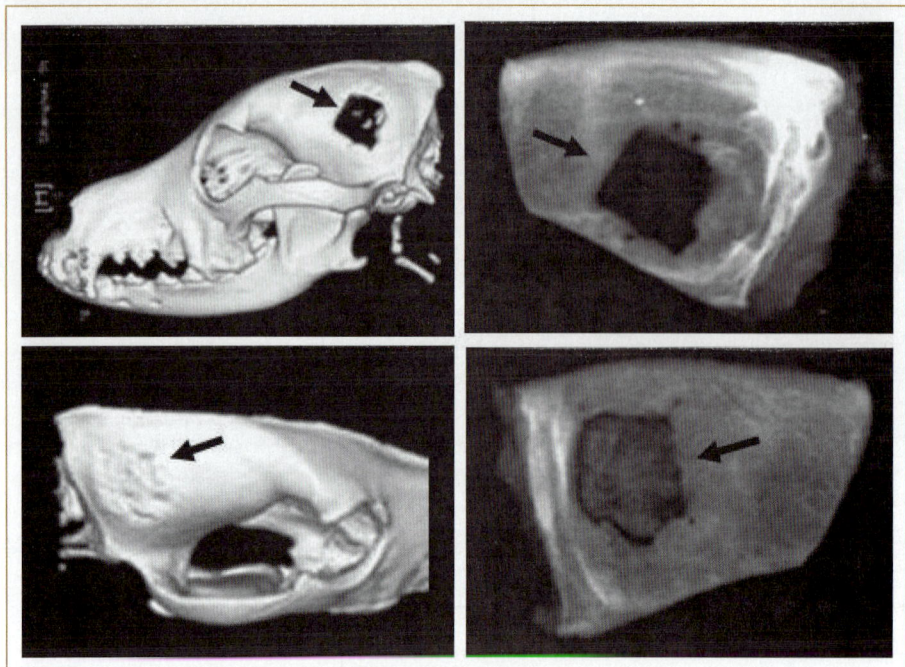

**FIG. 57.3.** Three-dimensional CT and x-ray evaluation of cranial bone engineered by DBM and BMSCs at 12 months. *Left panels:* CT scanning. *Right panels:* X-ray. *Top panels:* Control group. *Bottom panels:* Experimental group. Arrow indicates the location of created defect.

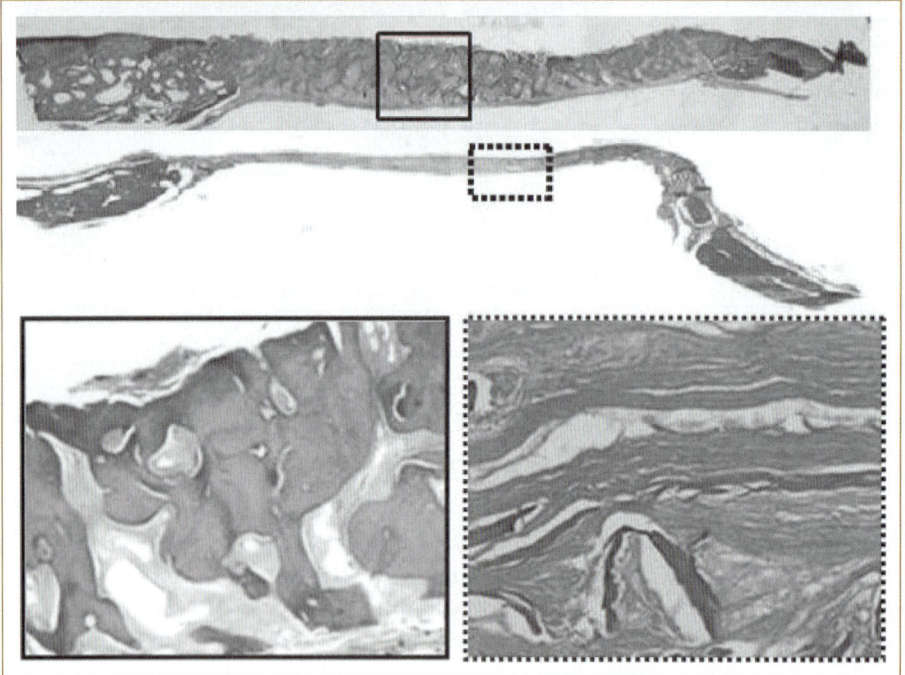

**FIG. 57.4.** Histological evaluation of cranial bone engineered by DBM and BMSCs at 12 months. *Top panel:* Histology of experimental repaired structure, including interface and central areas (merged picture). *Middle panel:* Histology of control repaired structure, including interface and central areas (merged picture). *Bottom panels:* Cancellous bone structure of experimental group at central area (solid border picture, *left*) and fibrotic tissue structure of control group at central area (dotted border picture, *right*).

in engineering a metatarsus in a sheep model using coral and bone marrow stromal cells and showed the feasibility of generating a weight-bearing bone in large animals.

Femur is the largest weight-bearing bone in animals and human beings. In a recently published study (Zhu *et al.*, 2006), we tested the possibility of engineering femoral bone in a goat model by using BMSCs and coral scaffold. As other studies reported, BMSCs were expanded *in vitro* and osteogenically induced and then seeded onto a natural coral scaffold and cultured *in vitro* for one week before implantation *in vivo*. To provide a mechanical loading for the remodeling of the engineered bone, a special internal fixation device was developed that contains an internal fixation rod, with a diameter a littler smaller than that of marrow cavity, and two sets of interlocking nails for fixing the rod at each end. After engineered bone formation, the interlocking nails can be removed, leaving the rod in the marrow cavity and thus allowing for mechanical loading by body weight.

In the surgical procedure, 22 adult goats, 12–14 months of age and weighing from 16.5 to 23 kg (an average of 19.6 kg), were included in this study and randomly divided into experimental group (implanted with cell–scaffold construct, $n = 10$), scaffold control group (implanted with scaffold alone, $n = 10$), and blank control group (without implantation, $n = 2$). After anesthesia, the aspect of the right femoral shaft was exposed and a 25-mm-long (20% of the femur's total length) osteoperiosteal segmental defect was created with proper fixation by using the device. The defects then were treated with either cell-loaded scaffold, or scaffold alone or were left untreated.

Animals were sacrificed at either four months or eight months, respectively, for gross observation and histological and biomechanical analyses. In order to enhance tissue remolding, interlocking nails were removed at six months in animals that would be sacrificed at eight months, to give the newly formed bone a mechanical loading.

As shown in Fig. 57.5, radiography examination demonstrated new bone formation one month after implantation at the site where BMSCs were implanted with coral. At six months, bone union was achieved at the defect site by engineered bone tissue, and so the interlocking nails were

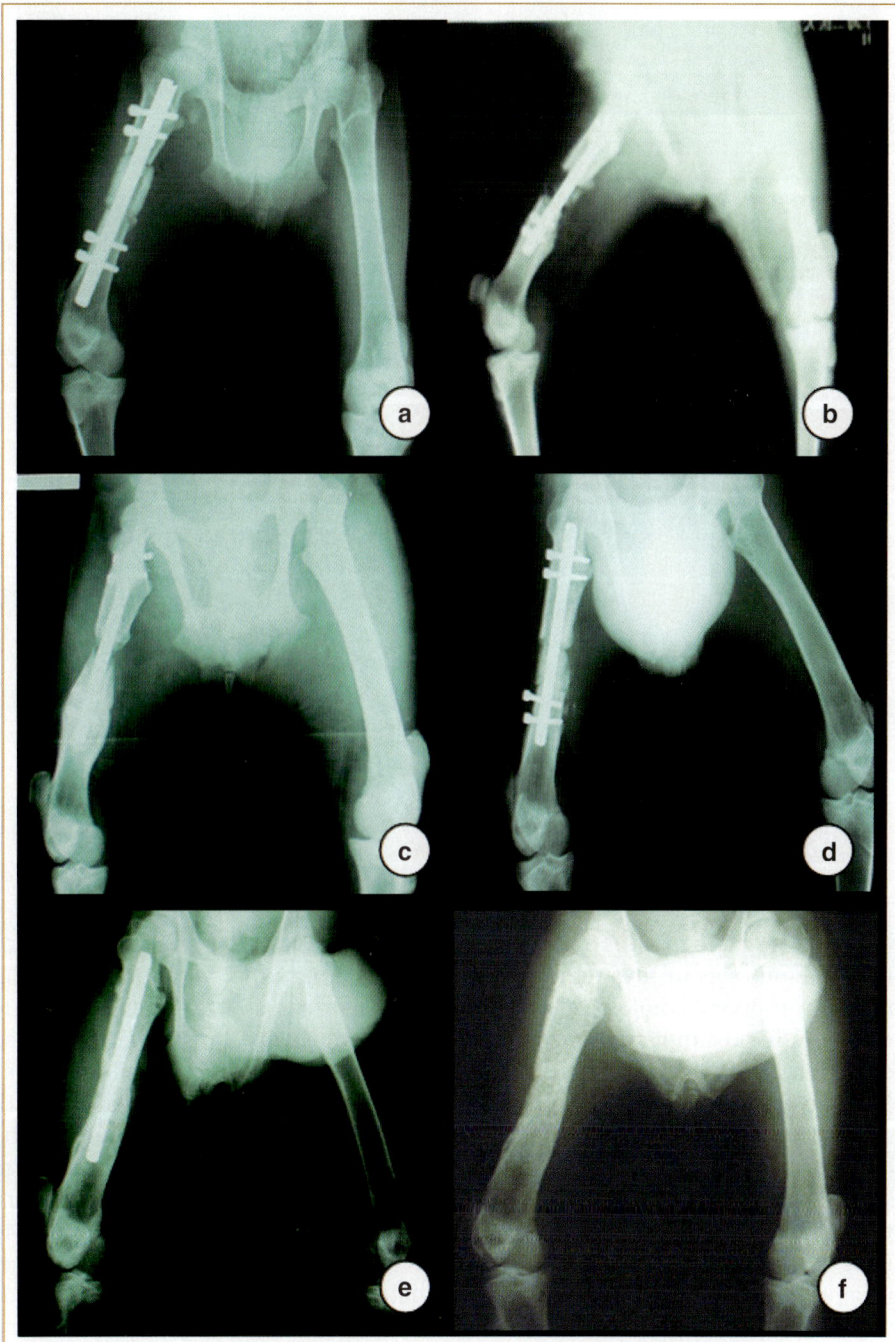

**FIG. 57.5.** Radiographs taken of treated goat's femur at different time frames. **(a–c)** Femur defects treated with coral alone. **(d–f)** Femur defects treated with coral loaded with culture-expanded and osteo-inducted BMSCs. **(a)** Femur defect treated with coral alone at day 1 after implantation. The cylinder with a density less than that of cortex can be found. **(b)** Coral density mostly disappeared at one month after implantation. **(c)** At eight months after implantation, minimal bone formation at the bone cut ends without bridging defect. **(d)** Femur defect treated with coral loaded with culture-expanded and osteo-inducted BMSCs at one month after implantation. The shape of the implant was maintained, with callus formation at the interface to the bone cut ends. **(e)** At six months after implantation, newly formed bone can afford weight bearing after removing the interlocking nails. **(f)** At eight months after implantation, the density of newly formed bone is equal to that of cortex. Reprinted by permission of *Tissue Engineering* from Zhu et al. (2006).

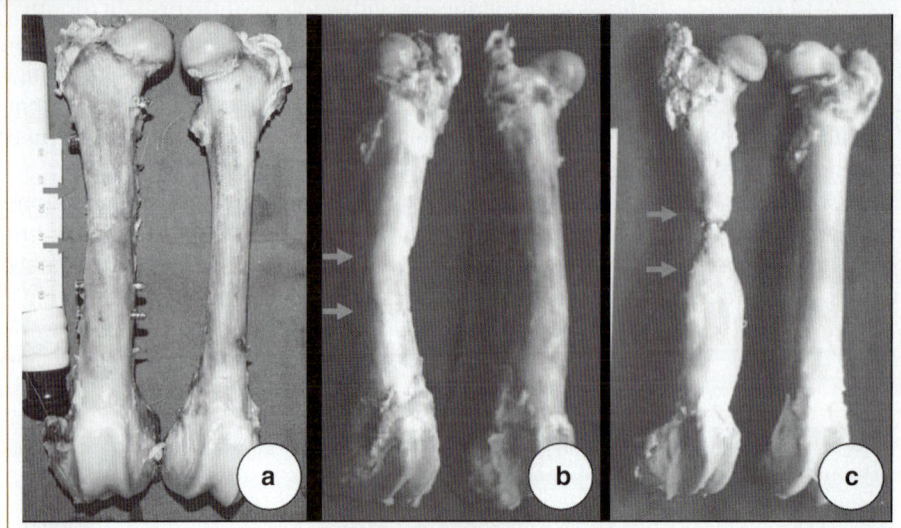

**FIG. 57.6.** Gross view of repaired defects in experimental and control groups. **(a)** Femur defect implanted with coral loaded with culture-expanded and osteo-inducted BMSCs at four months. Newly cancellous bone formed, with red color. **(b)** Femur defect implanted with coral loaded with culture-expanded and osteo-inducted BMSCs at eight months. White-colored, remodeled cortex formed. **(c)** Femur defect implanted with coral alone at eight months. Nonunion formed. **(a–c)** *Left,* experimental bone defects; *right,* normal bone serving as a positive control. Reprinted by permission of *Tissue Engineering* from Zhu *et al.* (2006).

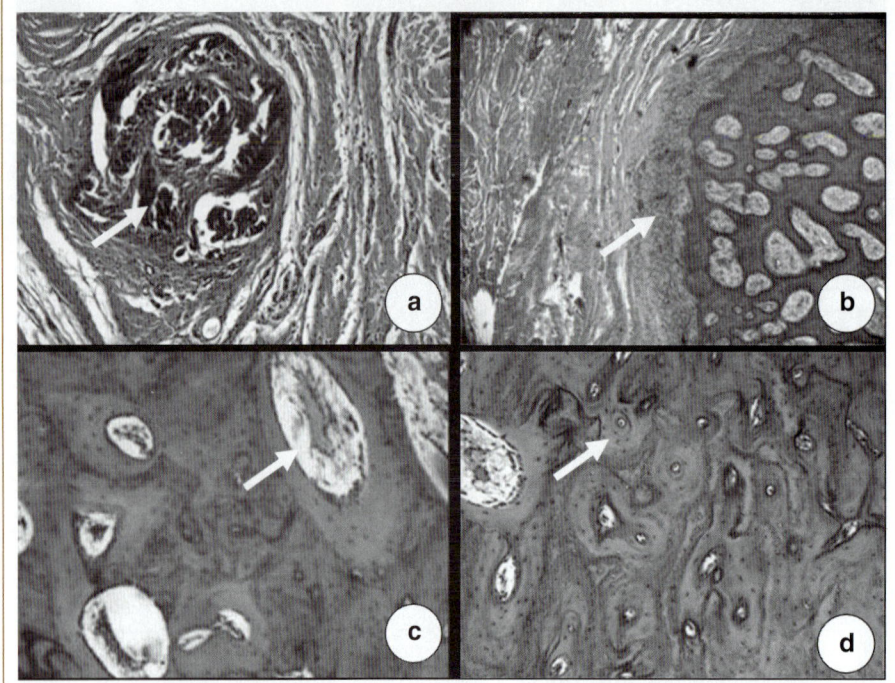

**FIG. 57.7.** H&E staining of repaired bone tissues. **(a)** Femur defect treated with coral alone at four months after implantation. Residual coral particle remained, surrounded by fibrous tissue (arrow, ×100). **(b)** Femur defect treated with coral alone at eight months after implantation. Coral particle disappeared completely, and a certain level of bone formed at bone cut end (arrow, ×100). **(c)** Femur defect implanted with coral loaded with culture-expanded and osteo-inducted BMSCs at four months. Coral completely disappeared and woven bone was formed. White regions indicated by arrows are newly formed medullary canal (×100). **(d)** Femur defect implanted with coral loaded with culture expanded and osteo-inducted BMSCs at eight months. Mature bone structure was formed and Haversian system became observable (arrow, ×100). Reprinted by permission of *Tissue Engineering* from Zhu *et al.* (2006).

removed. After two months of mechanical loading, the engineered femur was further remodeled to become a mature bone. As observed radiologically, cortex bone was formed at eight months postimplantation. Interestingly, marrow cavity formation inside the engineered bone was also observed at this time point. Grossly, the engineered bone also exhibited an appearance similar to that of natural femoral bone (Fig. 57.6). Histology demonstrated the formation of trabecular and woven bone at four months. At eight months, the engineered bone became obviously more mature. Although still being an irregular pattern, the formed Haversian system became observable, indicating that mechanical loading plays an important role in the remodeling of engineered

bone (Fig. 57.7). In contrast, implantation of scaffold alone without cell seeding could not generate a femoral bone tissue sufficiently to achieve a bone union both grossly and histologically (Figs. 57.5, 57.6, and 57.7).

In addition to structural restoration by engineered bone, functional recovery is an important requirement for engineered bone repair. In this study, a three-point bending test demonstrated that engineered femoral bones at eight months were much stronger than those of four months. In addition, no significant difference was found with respect to bending load strength and bend rigidity between the eight-month engineered bones and natural counterpart bones. However, the engineered bones were still inferior to natural

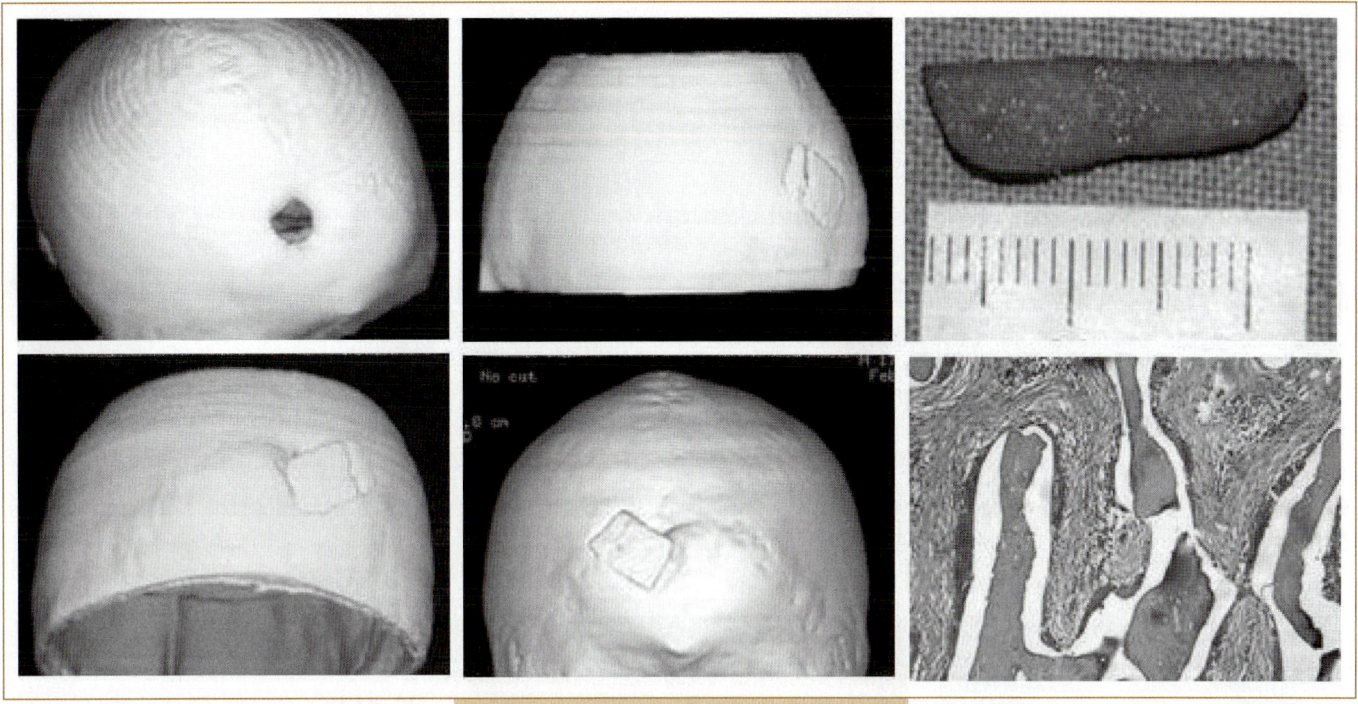

**FIG. 57.8.** Three-dimensional CT and histological evaluation of the clinical repair of a cranial defect with tissue-engineered bone. *Top left:* Before operation. *Top middle:* Three months postrepair. *Bottom left:* Twelve months postrepair. *Bottom middle:* Five years postrepair. *Top right:* Bone biopsy; *bottom right:* biopsy shows cancellous bone structure of engineered bone at 18 months postrepair (HE).

bones with regard to bend intention and bend displacement (Zhu *et al.*, 2006).

## III. CLINICAL APPLICATION OF ENGINEERED BONE

Vacanti *et al.* (2001) performed the first clinical trial of engineered bone in thumb reconstruction, using autologous osteogenic precursor cells isolated from periosteum and initially appeared to have succeeded. However, long-term follow-up was not reported. Recently, Warnke *et al.* (2004) reported successful clinical treatment of mandibular bone defect in a male patient with tissue-engineered bone. A titanium mesh cage with the mandibular shape of the patient was first filled with bone mineral blocks infiltrated with recombinant human BMP-7 and bone marrow mixture and then implanted into the patient's latissimus dorsi muscle for seven weeks to form a vascularized bone graft. The combined engineered bone graft was then transplanted functionally to repair the mandibular defect, and the bone remodeling and mineralization was also observed via scintigraphy examination.

Our center initiated a clinical trial of engineered bone around 2001, focused mainly on the repair of human cranio-maxillofacial bone defects using autologous BMSCs and demineralized bone matrix (Chai *et al.*, 2003). As a typical case, shown in Fig. 57.8, we started the clinical trial on a small-size cranial defect. Bone marrow was harvested from

the patient to isolate BMSCs. After *in vitro* expansion and osteogenic induction, cells were seeded on DBM scaffold and *in vitro* cultured for seven days and then implanted to repair the patient's defect. As illustrated by 3D CT scanning, bone formation was observed at three months postimplantation and maintained stable when followed up at 12 months and five years postoperation. During a secondary revision operation, a bone biopsy was taken at 18 months postimplantation, and the histology showed a typical structure of cancellous bone. Therefore, we proved that engineered bone could form and remain stable long term in the human being. Based on the success in small-size bone repair, we further challenged ourselves to repair a craniofacial bone defect of larger size and more complicated 3D structure. Figure 57.9 demonstrates another typical case. The patient had a craniofacial bone defect involving both frontal and orbital roof bones. To proceed with the repair, the patient's 3D CT scanning data were first obtained and then used to 3D-print a mold that was fit exactly into the defect site. According to the 3D shape of the mold, DBM materials were trimmed and assembled to form a 3D-shape scaffold. After cell seeding and *in vitro* culture for seven days, the cell-scaffold construct was implanted to repair the defect. The patient has been followed up for up to three years with 3D CT scanning, which demonstrated that engineered bone not only satisfactorily repaired the defect with complicated 3D structure, but also remained stable without absorption even after three years. Our clinical trial results indicate that tissue

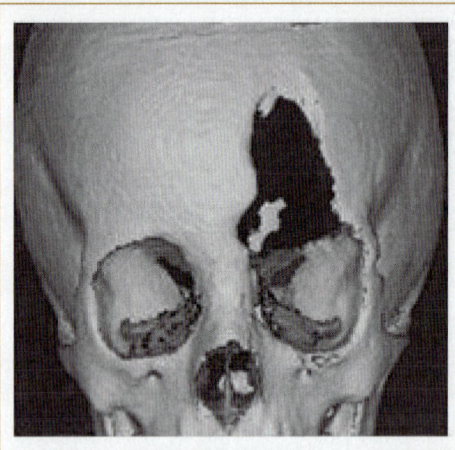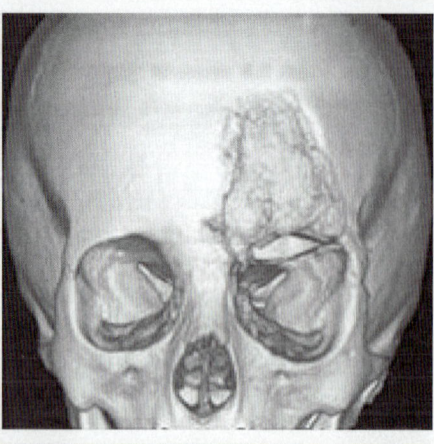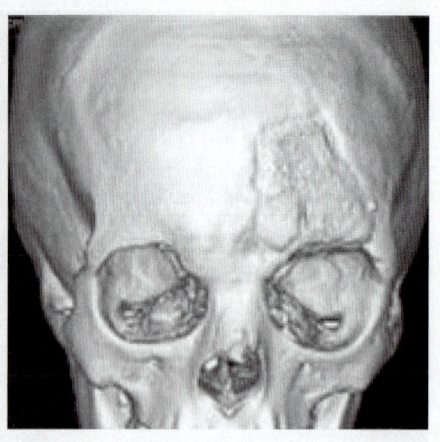

**FIG. 57.9.** Three-dimensional CT evaluation of clinical repair of a complicated craniofacial bone defect with tissue-engineered bone. *Left:* Before operation. *Middle:* One year postrepair. *Right:* Three years postrepair.

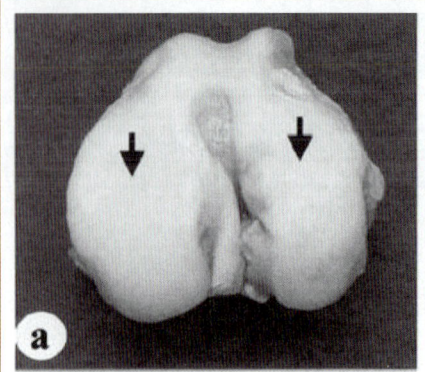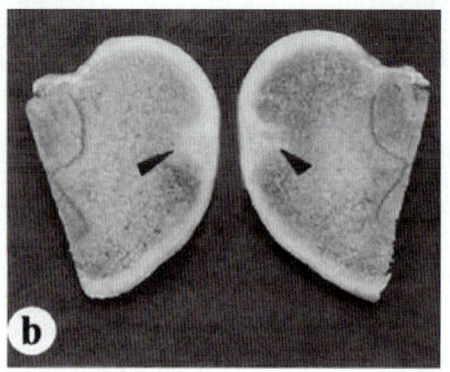

**FIG. 57.10.** Macrograph of a right knee joint harvested from the experimental group 24 weeks postrepair. **(a)** Gross examination shows that the defects are completely repaired by the engineered cartilage (arrows), leading to a smooth joint surface indistinguishable from the surrounding normal cartilage. **(b)** A cross section demonstrates ideal interface healing with the adjacent normal cartilage (arrowheads). Reprinted by permission of *Tissue Engineering* from Liu *et al.* (2002).

engineering will become a novel approach for bone regeneration and repair.

## IV. CARTILAGE RECONSTRUCTION

As with bone tissue, the proof of concept of cartilage engineering was performed in the subcutaneous tissue of the nude mouse (Vacanti *et al.*, 1993). In 1997, Cao *et al.* reported that cartilage with a complicated 3D structure, such as the human ear shape, could be engineered using nude mouse as an animal model, which reveals the great potential for the clinical application of engineered cartilage (Cao *et al.*, 1997). With regard to clinically relevant models for cartilage engineering and repair, articular cartilage repair is the most commonly involved subject; others include tracheal cartilage and meniscus.

Repairing an articular-cartilage defect using tissue engineering was reported many years ago (Wakitani *et al.*, 1989; Chu *et al.*, 1995; Grande *et al.*, 1997; Richardson *et al.*, 1999). While some studies focused on repairing a cartilage defect in small-animal models using allogenic chondrocyte–engineered hyaline cartilage (Wakitani *et al.*, 1989), our center has tried to repair the osteochondral defect in large

mammals using autologous chondrocyte–engineered cartilage (Liu *et al.*, 2002). In a porcine model, autologous articular cartilage at a non-weight-bearing area was harvested from the knee joint on one side. On the other side, an 8-mm full-thickness articular-cartilage defect deep to the underlying cancellous bone was created at both weight-bearing areas of the medial and lateral femoral condyles. The cell–scaffold construct containing polyglycolic acid (PGA), polyethylene–polypropylene hydrogel, and chondrocytes was then transplanted to repair the defects in the experimental group. In the control group, the defects were either transplanted with scaffold material alone or left unrepaired.

Grossly, cartilage tissues were formed in the defects of the experimental group as early as four weeks after transplantation. Histological examination demonstrated the presence of hyaline cartilage tissue. At 24 weeks postrepair, gross examination revealed a complete repair of the defects by engineered cartilage, shown by a smooth articular surface indistinguishable from nearby normal cartilage. A cross section revealed ideal interface healing between the engineered cartilage and the adjacent normal cartilage (Fig. 57.10). Histological analysis of the tissue harvested from repaired defects further revealed a typical structure of cartilage lacuna and ideal interface healing to adjacent normal cartilage as well as to underlying cancellous bone (Fig. 57.11). Moreover, the engineered cartilage exhibited an enhanced extracellular matrix production and improved biomechanical properties, indicating that engineered cartilage resembles the native articular cartilage not only in morphology and histology, but also in biochemical components and biomechanical properties as well. Unfortunately, the subchondral bone defect was also repaired by the engineered cartilage rather than by engineered bone (Fig. 57.10), suggesting that chondrocytes may not be an optimal cell source to achieve a physiological repair of both cartilage and subchondral bone defects for an articular osteochondral defect. Other cell sources for optimal repair need to be investigated. In recent years, a few clinical trials of matrix-induced autologous chondrocyte implantation have been conducted in the treatment of articular cartilage lesions (Bartlett *et al.*, 2005; Marcacci *et al.*, 2005). Positive outcomes have been achieved in short-term and long-term follow-up (up to five years). The encouraging clinical results obtained indicate that autologous chondrocyte implantation is a safe and effective therapeutic option for the treatment of articular-cartilage lesions.

Although chondrocytes are a useful seed cell source for cartilage engineering and are relatively safe for clinical treatment, immune rejection will still be the major obstacle to the use of allogenic chondrocytes (Malejczyk *et al.*, 1991; Moskalewski *et al.*, 2002), and the use of autologous chondrocytes will face the challenge of donor site morbidity and limited cell source due to their limited expansion capacity, cell aging, and dedifferentiation during *in vitro* expansion (Parsch *et al.*, 2002, 2004). In addition, our experience indicates that chondrocytes are not appropriate to bony repair of an osteochondral defect (Liu *et al.*, 2002). Therefore, chondrocyte-based therapy may not be optimal for clinical applications. In contrast, mesenchymal stem cell–based therapy may potentially avoid these disadvantages. Recent studies showed that BMSCs have become a reliable replacement for chondrocytes due to their capability for chondrogenic differentiation and rapid proliferation (Pittenger *et al.*, 1999; Gao and Caplan, 2003). There have been many reports in the past decade of *in vitro* chondrogenic induction of BMSCs, and the induced cells could express chondrocyte-specific molecules, such as type II collagen, aggrecan, and Sox 9 (Mastrogiacomo *et al.*, 2001; Winter *et al.*, 2003; Fukumoto *et al.*, 2003; Williams *et al.*, 2003; Evans *et al.*, 2004). Moreover, several reports have demonstrated that induced BMSCs actually formed cartilage tissue *in vivo* (Wakitani *et al.*, 1994; Im *et al.*, 2001; Fuchs *et al.*, 2003; Gao *et al.*, 2004; Wayne *et al.*, 2005).

The repair of articular cartilage with BMSCs has been reported for many years. However, most of these studies focused on small-animal models, such as rats (Oshima *et al.*,

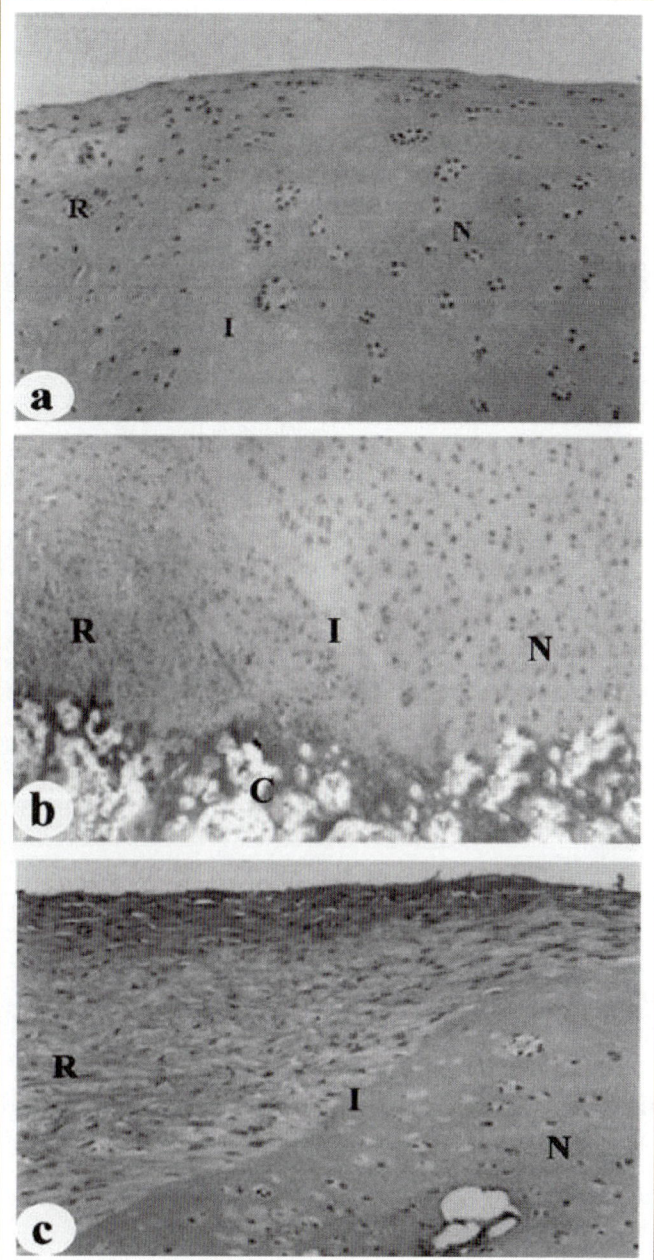

**FIG. 57.11.** Histology of tissues harvested at cartilage defects 24 weeks postrepair. In the experimental group, histological examination demonstrates a typical structure of hyaline cartilage lacunae **(a, b)**. In addition, engineered cartilage heals ideally at the interface with the surrounding normal cartilage **(a)** and the underlying cancellous bone **(b)**. In contrast, the defect in the control group is repaired by fibrous tissue and fibrocartilage. A poor interface healing to the adjacent cartilage tissue is also observed **(c)**. *Abbreviations:* N, normal cartilage; I, interface healing; R, repaired tissue; C, cancellous bone. (H&E staining; original magnification, ×100.) Reprinted by permission of *Tissue Engineering* from Y. Liu *et al.* (2002).

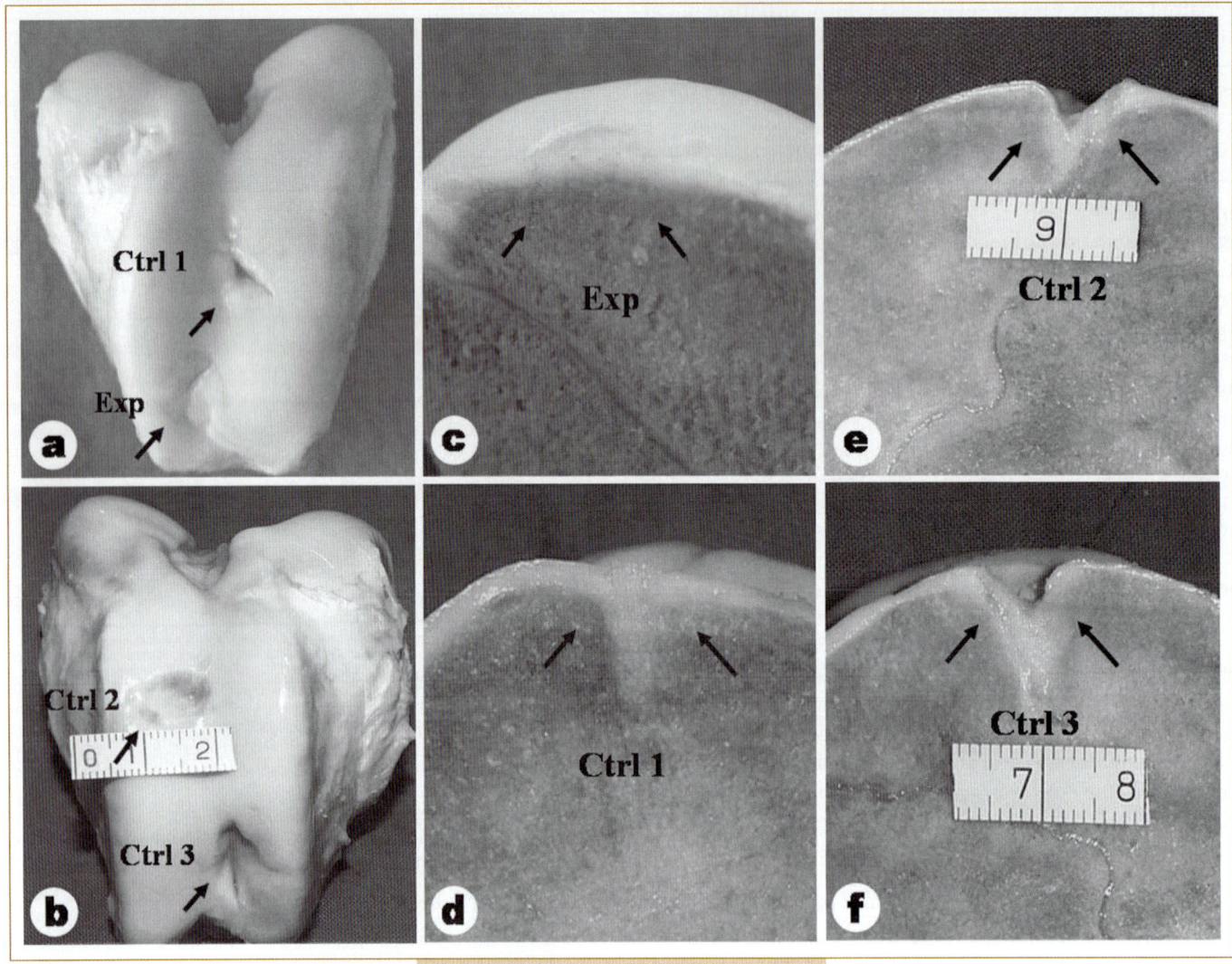

**FIG. 57.12.** Gross and cross-sectional view of repaired defects at six months postrepair. Arrows indicate the repaired regions. The experimental defect exhibits a relatively regular surface (Exp, **a**) and the osteochondral defect is completely repaired with both engineered cartilage and bone when observed at the cross section **(c)**. The repaired surface of control group 1 remains irregular (Ctrl 1, **a**), but the osteochondral defect is mostly repaired at the cross section **(d)**. The defects in control group 2 (Ctrl 2, **b, e**) and control group 3 (Ctrl 3, **b, f**) remain largely unrepaired at both cartilage and bony layers. Reprinted by permission of *Tissue Engineering* from Zhou *et al.* (2007).

2004, 2005) and rabbits (Im *et al.*, 2001; Gao *et al.*, 2004). Due to the differences in species and defect size, the data obtained from small animals might be inappropriate to elucidate the repair mechanism in human beings. In addition, there is still controversy regarding the true cell source (endogenous vs. exogenous) for the engineered repair of osteochondral defects, although some related studies have been performed in small animals using labeled seed cells (Oshima *et al.*, 2004, 2005). This is particularly true in large-animal model studies, in which no direct evidence has yet been provided. Therefore, two issues remain worthy to explore in this area: (1) discovering whether the approach is feasible or not in large animals for engineered repair of an articular osteochondral defect using autologous BMSCs,

especially for a long-term study; and (2) providing more convincing evidence to demonstrate the true cell source of repaired cartilage and its subchondral bone. To address these concerns, we recently performed a study of repairing articular osteochondral defect with autologous BMSCs in a total of 18 pigs using the previously established porcine model (Zhou *et al.*, 2006). BMSCs were isolated from hybrid pigs' marrow and treated with dexamethasone (40 ng/mL) alone or chondrogenically induced with dexamethasone and transforming growth factor-β1 (TGF-β1, 10 ng/mL). The cells were then seeded, respectively, onto polylactic acid (PLA)-coated PGA scaffolds to generate cell–scaffold constructs. Four osteochondral defects in each animal were created at non-weight-bearing areas of knee joints (two

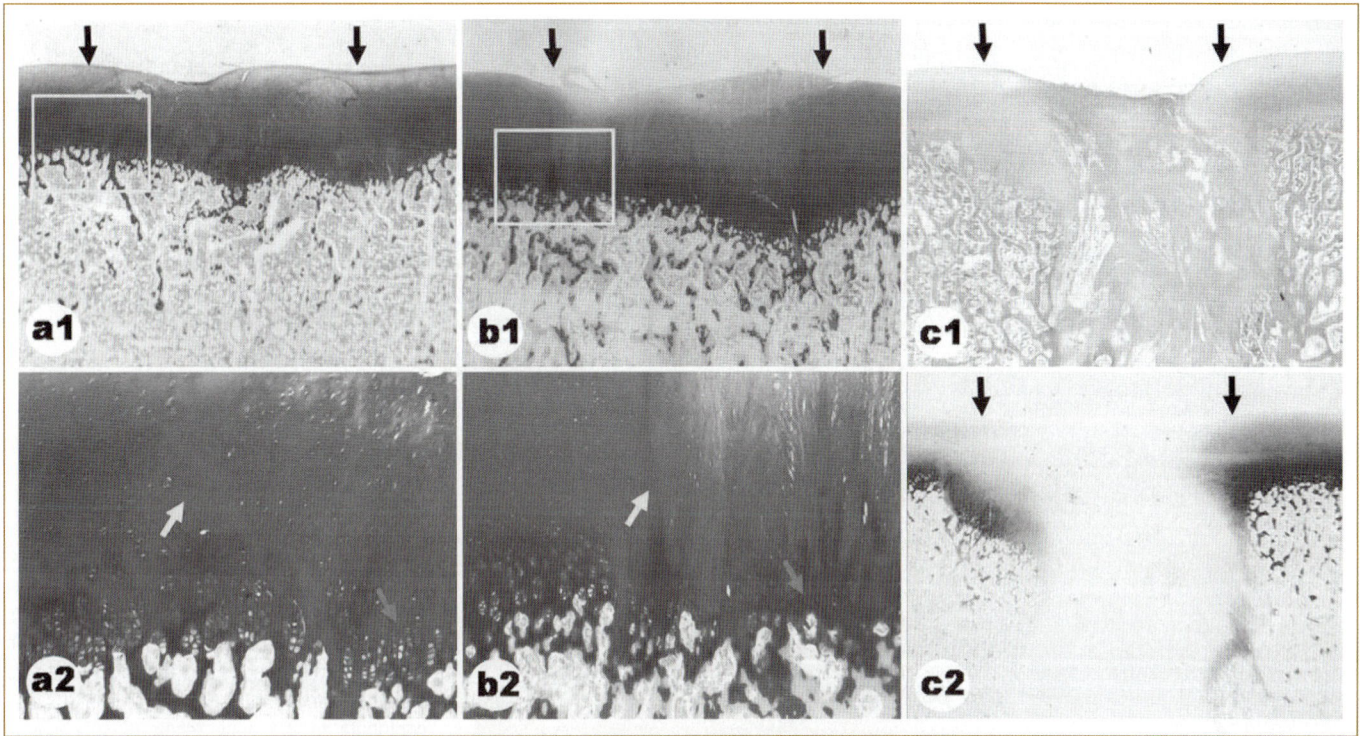

**FIG. 57.13.** Histology of repaired tissue at six months postrepair. The repaired regions are indicated between black arrows. **(a)** Histologic view of repaired tissue in the experimental group by Toluidine blue staining, a representative of *complete repair*. The rectangular region indicates the interface area (a1, ×10), which lies between the normal and engineered tissues. In the amplified view (a2, ×40), a white arrow indicates the interface line between native (left) and engineered (right) tissues. Interestingly, a nearly normal tidemark structure is observed in the engineered tissue, as indicated by the gray arrow. **(b)** Histologic view of repaired tissue in control group 1 by Toluidine blue staining, a representative of *incomplete repair*, which shows a relatively poor tissue structure in cartilage repair area (b1, ×10). In an amplified view (b2, ×40), normal articular tissue structure (hyaline cartilage and cancellous bone) is observed at a normal region (left side of the interface, as indicated by the white arrow) with an obvious tidemark structure. At the repaired area (right side of the interface), however, the repaired tissue is composed mainly of fibrocartilage and cancellous bone. Additionally, no mature tidemark structure is observed at the repaired osteochondral junction area (gray arrow). **(c)** Histologic view of repaired tissue in control group 2 by HE staining and Toluidine blue staining, a representative of *no repair*. The defect area is filled by fibrotic tissue (c1, HE, ×10) and shows no metachromatic matrix staining (c2, Toluidine blue, ×10). Reprinted by permission of *Tissue Engineering* from Zhou *et al.* (2007).

defects/side). Chondrogenically induced BMSC-PGA/PLA constructs were implanted in the defects of the experimental group (Exp). In control groups, the defects were repaired with dexamethasone-treated BMSC-PGA/PLA (Ctrl 1) constructs or PGA/PLA construct alone (Ctrl 2) or left unrepaired (Ctrl 3). To trace the implanted BMSCs, cells were retrovirally labeled with green fluorescent protein (GFP) before implantation.

The results showed that cartilage tissue was formed in the experimental group defects as early as three months after transplantation, and an improved result was achieved after six months. Additionally, the experimental group achieved the best reparative results among all tested groups. Grossly, most of the defects in this group were repaired by newly formed cartilage-like tissue with relatively regular surfaces (Fig. 57.12a). Cross sections showed fine interface healing between repaired tissue and normal osteochondral tissues. The thickness of repaired cartilage approached a

level similar to that of adjacent normal cartilage. Interestingly, both cartilage and bony defects were satisfactorily repaired by engineered cartilage and bone, respectively (Fig. 57.12c). However, in control group 1, only some defects achieved satisfactory repair with relatively smooth articular surfaces and relatively normal osteochondral structures, but most defects showed irregular surfaces in the central area of the defects and poorly repaired subchondral bone (Fig. 57.12a, d). As expected, defects of control group 2 and control group 3 healed poorly, showing unrepaired defect and partial formation of osteochondral tissue only at the interface between the defect areas and the native tissue areas (Fig. 57.12b, e, f).

Histology also showed that the experimental group achieved the best results. Most defects exhibited the structures of mature hyaline cartilage and cancellous bone, with cellular density and cartilage thickness similar to those of normal tissues (Fig. 57.13a1). Importantly, a nearly normal

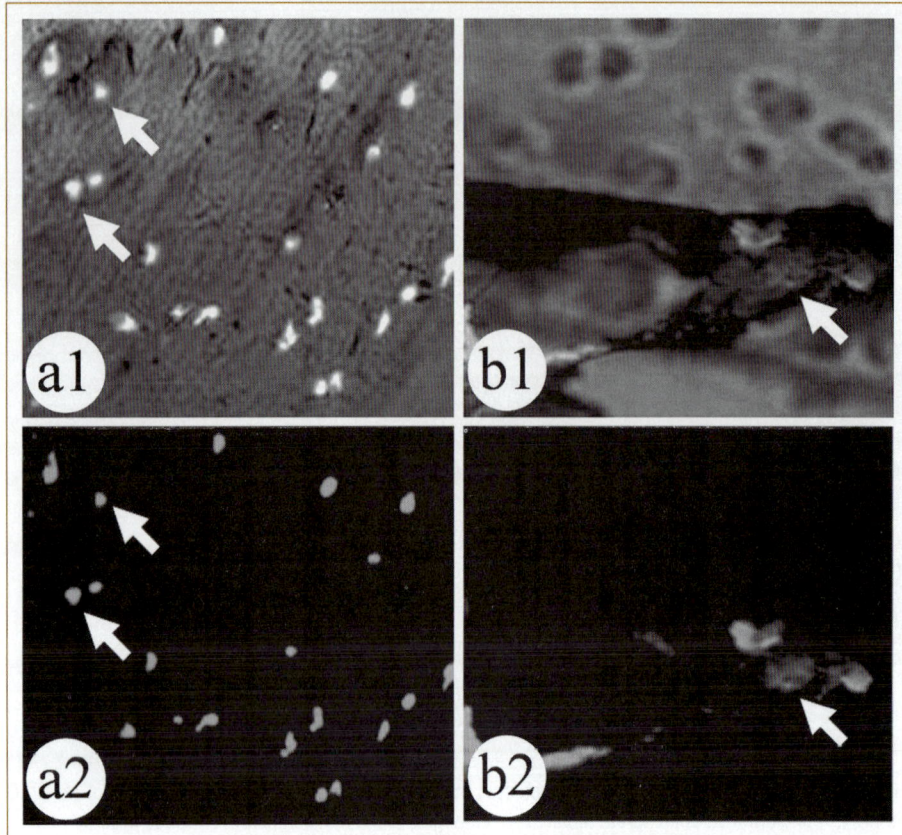

**FIG. 57.14.** GFP expression in the repaired cartilage and cancellous bone at seven months postrepair by laser confocal microscope (×200). **(a)** GFP-labeled BMSCs are evenly distributed in repaired cartilage area and form typical lacuna structures (arrows). **(b)** GFP-labeled BMSCs also contribute to the bony repair of subchondral bone. *Top panels:* Images under regular light (a1, b1); *bottom panels:* images under fluorescence light (a2, b2). Reprinted by permission of *Tissue Engineering* from Zhou *et al.* (2007).

tidemark could be observed between repaired cartilage and cancellous bone. Strong metachromatic matrix production and satisfactory interface healing were also observed, despite the fact that a tiny difference between repaired tissue and native articular tissue remains observable (Fig. 57.13a2). The results were less satisfactory in control group 1 because only some of the defects were repaired by the mature hyaline cartilage and cancellous bone. Yet the tissue quality remained inferior to that of the experimental group. In addition, a larger part of the defects in control group 1 showed fibrocartilage repair and the lack of an obvious tidemark at the osteochondral junction (Fig. 57.13b1, b2). In contrast to the experimental group and control group 1, the defects of control groups 2 and 3 showed mainly fibrotic repair, with poor interface healing and poor metachromatic matrix staining at the area with the most defects (Fig. 57.13c1, c2). All these results indicate that implanted cells play an important role in tissue regeneration and repair and that *in vitro* chondrogenic induction is beneficial for that *in vivo* cartilage regeneration.

To further determine the fate of implanted BMSCs and the cell source for osteochondral tissue formation, confocal microscopy was used to examine the tissue sample in which GFP-labeled BMSCs were implanted. As shown in Fig. 57.14,

GFP-labeled cells remained observable at both cartilage and subchondral bone layers, even at seven months postoperation. Importantly, these GFP-labeled BMSCs actually formed typical lacuna structures (Fig. 57.14a1), suggesting that the implanted BMSCs have differentiated into both chondrocytes and osteoblasts, respectively, in their respective *in vivo* environments and that implanted BMSCs are the vital cell source for the osteochondral tissue repair.

Biomechanical property analysis of the repaired cartilage showed that the compressive moduli increased with postrepair time. At three months, the moduli reached 43.82% and 30.37% of normal cartilage level, respectively, in the experimental group and control group 1, and they further increased to 80.27% and 62.69% at six months postrepair. In addition, the glycosaminoglycan content of the experimental group reached 93.25% of normal content, with no significant difference between the native cartilage group and the experimental groups ($p > 0.05$). Thus, BMSC-mediated repair not only restores the osteochondral tissue structure, but also generates the tissues with mechanical property and biochemical components similar to those of natural osteochondral tissues.

The foregoing study demonstrated that BMSCs are a better cell source than chondrocytes to achieve tissue-engineered repair of an articular osteochondral defect. In addition, it also proves that implanted BMSCs can differentiate into both chondrocytes and osteoblasts in their related *in vivo* niches. Encouragingly, there are already a few reports of clinical trials of repairing articular-cartilage defect using BMSC-engineered cartilage, although the therapeutic effect needs to be improved further (Wakitani *et al.*, 2002).

Besides articular-cartilage repair, we succeeded in the repair of meniscal defects in a porcine model (Liu *et al.*, 2003). The meniscus was first harvested from one side to isolate meniscal fibrochondrocytes. A full-thickness meniscal defect (1 cm long) was then created on the other side, and the two ends of the defect were bridged by an acellular intestinal submucosa membrane that wrapped around the ends. Inside the membrane tube, PGA fibers mixed with polyethylene–polypropylene hydrogel and fibrochondrocytes were transplanted. In the control group, biomaterial alone was used. Grossly, the engineered tissue resembled native meniscus in morphology, color, and texture at 25 weeks posttransplantation of the cell–scaffold construct. Histologically, the engineered meniscus displayed a structure typical of fibrocartilage tissue, similar to that of native meniscus (Fig. 57.15). As expected, no meniscus tissue was formed in the control group.

*In vitro* cartilage engineering is an important area to pursue. We have succeeded as well in engineering cartilage tissue in the culture dish by using chondrocytes and polyglycolic acid (PGA) unwoven fibers, such as tracheal tubular cartilage (Fig. 57.16, *left*). However, BMSCs might be a better cell source than chondrocytes for cartilage engineering, as already described. As shown in Fig. 57.16 (*right*), cartilage tissue could be generated using PGA unwoven fibers and chondrogenically induced BMSCs. A future direction will be to engineer cartilage tissue with a 3D structure, like the human ear shape, by using BMSCs so that engineered cartilage can be used in clinical application.

## V. CONCLUSION

Many studies, including ours, have demonstrated the feasibility of bone and cartilage reconstruction in small and large animals using tissue-engineering techniques. BMSCs have been shown to be a preferred cell source for tissue reconstruction. In addition, clinically relevant tissue defects could be repaired by engineered bone and cartilage in large mammals. Moreover, the success in clinical trials of repairing patients' craniofacial defects with engineered bone reveals the promising future of tissue engineering as a favored approach for bone and cartilage reconstruction.

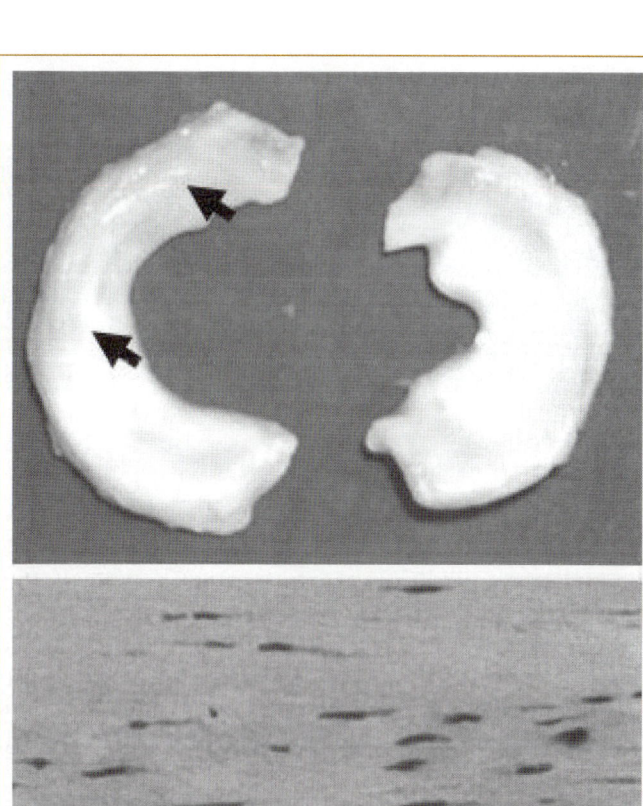
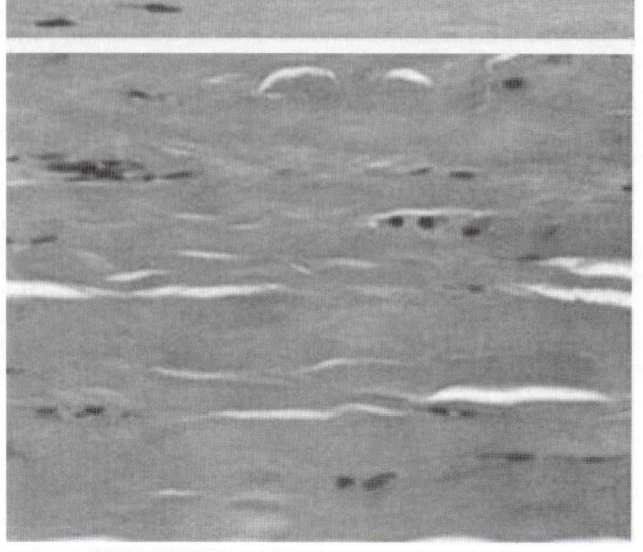

**FIG. 57.15.** Gross view (*top*) and histology of engineered meniscus at 25 weeks postrepair (*middle*) and natural meniscus (*bottom*) (arrows indicate the engineered meniscus). Reprinted by permission of *Tissue Engineering* from W. Liu *et al.* (2003).

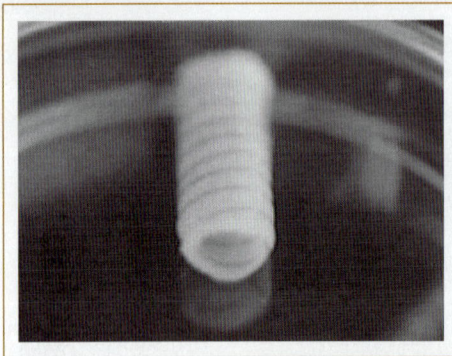

**FIG. 57.16.** *In vitro* cartilage engineering. *Left:* Tracheal tubular cartilage engineered by chondrocytes and PGA unwoven fibers. *Right:* Cartilage tissue engineered by chondrogenically induced BMSCs and PGA unwoven fibers.

## VI. ACKNOWLEDGEMENTS

The authors would like to thank Drs. Lei Cui, Yulai Weng, Ming Wang, Zhuo Wang, and Lian Zhu for their contributions in bone engineering and Dr. Gang Chai for clinical trials of engineered bone. We also appreciate the contribution to cartilage engineering from Drs. Guangdong Zhou, Yanchung Liu, Fuguo Chen, and Tianyi Liu. The research projects are supported by the National "973" (2005CB522700) and "863" Project Foundation, the Shanghai Science and Technology Development Foundation, and the Key Laboratory Foundation of the Shanghai Education Committee.

## VII. REFERENCES

Bartlett, W., Gooding, C. R., Carrington, R. W., Skinner, J. A., Briggs, T. W., and Bentley, G. (2005). Autologous chondrocyte implantation at the knee using a bilayer collagen membrane with bone graft. A preliminary report. *J. Bone Joint Surg. Br.* **87**, 330–332.

Breitbart, A. S., Grande, D. A., Mason, J. M., Barcia, M., James, T., and Grant, R. T. (1999). Gene-enhanced tissue engineering: applications for bone healing using cultured periosteal cells transduced retrovirally with the BMP-7 gene. *Ann. Plast. Surg.* **42**, 488–495.

Cao, Y., Vacanti, J. P., Paige, K. T., Upton, J., and Vacanti, C. A. (1997). Transplantation of chondrocytes utilizing a polymer cell construct to produce tissue engineered cartilage in the shape of a human ear. *Plast. Reconstr. Surg.* **100**, 297–304.

Chai, G., Zhang, Y., Liu, W., Cui, L., and Cao, Y. L. (2003). Clinical application of tissue-engineered bone repair of human craniomaxillofacial bone defects. *Zhonghua Yi Xue Za Zhi* **83**, 1676–1681.

Chang, S. C., Chuang, H., Chen, Y. R., Yang, L. C., Chen, J. K., Mardini, S., Chung, H. Y., Lu, Y. L., Ma, W. C., and Lou, J. (2004). Cranial repair using BMP-2 gene engineered bone marrow stromal cells. *J. Surg. Res.* **119**, 85–91.

Chu, C. R., Coutts, R. D., Yoshioka, M., Harwood, F. L., Monosov, A. Z., and Amiel, D. (1995). Articular cartilage repair using allogeneic perichondrocyte-seeded biodegradable porous polylactic acid (PLA): a tissue-engineering study. *J. Biomed. Mater. Res.* **29**, 1147–1154.

Cowan, C. M., Shi, Y. Y., Aalami, O. O., Chou, Y. F., Mari, C., Thomas, R., Quarto, N., Contag, C. H., Wu, B., and Longaker, M. T. (2004). Adipose-derived adult stromal cells heal critical-size mouse calvarial defects. *Nat. Biotechnol.* **22**, 560–567.

Dutta Roy, T., Simon, J. L., Ricci, J. L., Rekow, E. D., Thompson, V. P., and Parsons, J. R. (2003). Performance of hydroxyapatite bone repair scaffolds created via three-dimensional fabrication techniques. *J. Biomed. Mater. Res. A* **67**, 1228–1237.

Evans, J. F., Niu, Q. T., Canas, J. A., Shen, C. L., Aloia, J. F., and Yeh, J. K. (2004). ACTH enhances chondrogenesis in multipotential progenitor cells and matrix production in chondrocytes. *Bone* **35**, 96–107.

Fuchs, J. R., Hannouche, D., Terada, S., Vacanti, J. P., and Fauza, D. O. (2003). Fetal tracheal augmentation with cartilage engineered from bone marrow–derived mesenchymal progenitor cells. *J. Pediatr. Surg.* **38**, 984–987.

Fukumoto, T., Sperling, J. W., Sanyal, A., Fitzsimmons, J. S., Reinholz, G. G., Conover, C. A., and O'Driscoll, S. W. (2003). Combined effects of insulin-like growth factor-1 and transforming growth factor-beta1 on periosteal mesenchymal cells during chondrogenesis *in vitro*. *Osteoarthritis Cartilage* **11**, 55–64.

Gao, J., and Caplan, A. I. (2003). Mesenchymal stem cells and tissue engineering for orthopedic surgery. *Chir. Organi. Mov.* **88**, 305–316.

Gao, J., Knaack, D., Goldberg, V. M., and Caplan, A. I. (2004). Osteochondral defect repair by demineralized cortical bone matrix. *Clin. Orthop. Relat. Res.* **427**, S62–S66.

Grande, D. A., Halberstadt, C., Naughton, G., Schwartz, R., and Manji, R. (1997). Evaluation of matrix scaffolds for tissue engineering of articular cartilage grafts. *J. Biomed. Mater Res.* **34**, 211–220.

Im, G. I., Kim, D. Y., Shin, J. H., Hyun, C. W., and Cho, W. H. (2001). Repair of cartilage defect in the rabbit with cultured mesenchymal stem cells from bone marrow. *J. Bone Joint Surg. Br.* **83**, 289–294.

Liu, W., Cui, L., and Cao, Y. (2003). A closer view of tissue engineering in China: the experience of tissue construction in immunocompetent animals. *Tissue Eng.* **9**(Suppl. 1), S17–S30.

Liu, Y., Chen, F., Liu, W., Cui, L., Shang, Q., Xia, W., Wang, J., Cui, Y., Yang, G., Liu, D., Wu, J., Xu, R., Buonocore, S. D., and Cao, Y. (2002). Repairing large porcine full-thickness defects of articular cartilage using autologous chondrocyte-engineered cartilage. *Tissue Eng.* **8**, 709–721.

Malejczyk, J., Osiecka, A., Hyc, A., and Moskalewski, S. (1991). Effect of immunosuppression on rejection of cartilage formed by transplanted allogeneic rib chondrocytes in mice. *Clin. Orthop. Relat. Res.* **269**, 266–273.

Marcacci, M., Berruto, M., Brocchetta, D., Delcogliano, A., Ghinelli, D., Gobbi, A., Kon, E., Pederzini, L., Rosa, D., Sacchetti, G. L., Stefani, G., and Zanasi, S. (2005). Articular cartilage engineering with Hyalograft C: 3-year clinical results. *Clin. Orthop. Relat. Res.* **435**, 96–105.

Mastrogiacomo, M., Cancedda, R., and Quarto, R. (2001). Effect of different growth factors on the chondrogenic potential of human bone marrow stromal cells. *Osteoarthr. Cartilage* **9**, S36–S40.

Mauney, J. R., Jaquiery, C., Volloch, V., Heberer, M., Martin, I., and Kaplan, D. L. (2005). *In vitro* and *in vivo* evaluation of differentially demineralized cancellous bone scaffolds combined with human bone marrow stromal cells for tissue engineering. *Biomaterials* **26**, 3173–3185.

Meinel, L., Fajardo, R., Hofmann, S., Langer, R., Chen, J., Snyder, B., Vunjak-Novakovic, G., and Kaplan, D. (2005). Silk implants for the healing of critical size bone defects. *Bone* **37**, 688–698.

Moreira-Gonzalez, A., Lobocki, C., Barakat, K., Andrus, L., Bradford, M., Gilsdorf, M., and Jackson, I. T. (2005). Evaluation of 45S5 bioactive glass combined as a bone substitute in the reconstruction of critical size calvarial defects in rabbits. *J. Craniofac. Surg.* **16**, 63–70.

Moskalewski, S., Hyc, A., and Osiecka-Iwan, A. (2002). Immune response by host after allogeneic chondrocyte transplant to the cartilage. *Microsc. Res. Tech.* **58**, 3–13.

Oshima, Y., Watanabe, N., Matsuda, K., Takai, S., Kawata, M., and Kubo, T. (2004). Fate of transplanted bone-marrow-derived mesenchymal cells during osteochondral repair using transgenic rats to simulate autologous transplantation. *Osteoarthritis Cartilage* **12**, 811–817.

Oshima, Y., Watanabe, N., Matsuda, K., Takai, S., Kawata, M., and Kubo, T. (2005). Behavior of transplanted bone marrow–derived GFP mesenchymal cells in osteochondral defect as a simulation of autologous transplantation. *J. Histochem. Cytochem.* **53**, 207–216.

Parsch, D., Brummendorf, T. H., Richter, W., and Fellenberg, J. (2002). Replicative aging of human articular chondrocytes during ex vivo expansion. *Arthritis Rheum.* **46**, 2911–2916.

Parsch, D., Fellenberg, J., Brummendorf, T. H., Eschlbeck, A. M., and Richter, W. (2004). Telomere length and telomerase activity during expansion and differentiation of human mesenchymal stem cells and chondrocytes. *J. Mol. Med.* **82**, 49–55.

Petite, H., Viateau, V., Bensaid, W., Meunier, A., de Pollak, C., Bourgnignon, M., Oudina, K., Sedel, L., and Guillemin, G. (2002). Tissue-engineered bone regeneration. *Nat. Biotechnol.* **18**, 959–963.

Pittenger, M. F., Mackay, A. M., Beck, S. C., Jaiswal, R. K., Douglas, R., Mosca, J. D., Moorman, M. A., Simonetti, D. W., Craig, S., and Marshak, D. R. (1999). Multilineage potential of adult human mesenchymal stem cells. *Science* **284**, 143–147.

Richardson, J. B., Caterson, B., Evans, E. H., Ashton, B. A., and Roberts, S. (1999). Repair of human articular cartilage after implantation of autologous chondrocytes. *J. Bone Joint Surg. Br.* **81**, 1064–1068.

Shang, Q., Wang, Z., Liu, W., Shi, Y., Cui, L., and Cao, Y. (2001). Tissue-engineered bone repair of sheep cranial defects with autologous bone marrow stromal cells. *J. Craniofac. Surg.* **12**, 586–93.

Vacanti, C. A., and Upton, J. (1994). Tissue-engineered morphogenesis of cartilage and bone by means of cell transplantation using synthetic biodegradable polymer matrices. *Clin. Plast. Surg.* **21**, 445–462.

Vacanti, C. A., Kim, W., Upton, J., Vacanti, M. P., Mooney, D., Schloo, B., and Vacanti, J. P. (1993). Tissue-engineered growth of bone and cartilage. *Transplant. Proc.* **25**, 1019–1021.

Vacanti, C. A., Bonassar, L. J., Vacanti, M. P., and Shufflebarger, J. (2001). Replacement of an avulsed phalanx with tissue-engineered bone. *N. Engl. J. Med.* **344**, 1511–1514.

Wakitani, S., Kimura, T., Hirooka, A., Ochi, T., Yoneda, M., Yasui, N., Owaki, H., and Ono, K. (1989). Repair of rabbit articular surfaces with allograft chondrocytes embedded in collagen gel. *J. Bone Joint Surg. Br.* **71**, 74–80.

Wakitani, S., Goto, T., Pineda, S. J., Young, R. G., Mansour, J. M., Caplan, A. I., and Goldberg, V. M. (1994). Mesenchymal cell–based repair of large, full-thickness defects of articular cartilage. *J. Bone Joint Surg. Am.* **76**, 579–592.

Wakitani, S., Imoto, K., Yamamoto, T., Saito, M., Murata, N., and Yoneda, M. (2002). Human autologous culture expanded bone marrow mesenchymal cell transplantation for repair of cartilage defects in osteoarthritic knees. *Osteoarthritis Cartilage* **10**, 199–206.

Warnke, P. H., Springer, I. N., Wiltfang, J., Acil, Y., Eufinger, H., Wehmoller, M., Russo, P. A., Bolte, H., Sherry, E., Behrens, E., and Terheyden, H. (2004). Growth and transplantation of a custom vascularized bone graft in a man. *Lancet* **364**, 766–770.

Wayne, J. S., McDowell, C. L., Shields, K. J., and Tuan, R. S. (2005). *In vivo* response of polylactic acid–alginate scaffolds and bone marrow–derived cells for cartilage tissue engineering. *Tissue Eng.* **11**, 953–963.

Williams, C. G., Kim, T. K., Taboas, A., Malik, A., Manson, P., and Elisseeff, J. (2003). *In vitro* chondrogenesis of bone marrow–derived mesenchymal stem cells in a photopolymerizing hydrogel. *Tissue Eng.* **9**, 679–688.

Winter, A., Breit, S., Parsch, D., Benz, K., Steck, E., Hauner, H., Weber, R. M., Ewerbeck, V., and Richter, W. (2003). Cartilage-like gene expression in differentiated human stem cell spheroids: a comparison of bone marrow–derived and adipose tissue–derived stromal cells. *Arthritis Rheum.* **48**, 418–429.

Zhou, G. D., Liu, W., Cui, L., Wang, X., Liu, T., and Cao, Y. L. (2006). Repair of porcine articular osteochondral defects in non-weight-bearing areas with autologous bone marrow stromal cells. *Tissue Eng.* **12**, 3209–3221.

Zhu, L., Liu, W., Cui, L., and Cao, Y. (2006). Tissue-engineered bone repair of goat-femur defects with osteogenically induced bone marrow stromal cells. *Tissue Eng.* **12**, 423–433.

# Chapter Fifty-Eight

# Regeneration and Replacement of the Intervertebral Disc

*Lori A. Setton, Lawrence J. Bonassar, and Koichi Masuda*

## I. INTRODUCTION

The intervertebral disc (IVD) is the fibrocartilaginous part of a "three-joint complex" that governs motion, flexibility, and weight bearing in the spine (Fig. 58.1). As part of this complex, the disc undergoes a lifetime of "wear and tear" that contributes to multiple IVD disorders of enormous consequence for human disability and suffering. These IVD disorders are poorly understood musculoskeletal pathologies characterized by multiple anatomic features, including internal disc disruption, IVD tears, and herniated disc fragments (Andersson, 1996; Bodguk, 1988; Boos *et al.*, 2002). These anatomic features are believed to associate with nerve root compression or irritation, spinal canal narrowing (stenosis or spondylolisthesis), or loss of disc height, which contribute to symptoms of low-back pain, neurological deficits, and disability that affect between 4% and 33% of the U.S. population annually (Hurri and Karppinen, 2004; Praemer *et al.*, 1999; Woolf and Pfleger, 2003). Like most cartilaginous tissues, the IVD is an avascular and alymphatic structure that exhibits little to no capacity for repair following injury and experiences aging-related cell density losses that may further limit biologically mediated repair (Urban and Roberts, 2003). The extreme mechanical demands on the IVD may also contribute to tissue failure and degeneration, due to the high magnitudes of pressure, compressive, tensile, and shear stresses, and strains that result from joint loading, muscle activation, and spinal flexibility. As a result, strategies to intervene in the progression of IVD disorders are met with significant biological and mechanical challenges that frustrate success.

Numerous surgical procedures have been developed to treat IVD disorders, focused largely on bony fusion across the disc space to restore stability and eliminate symptomatic motions and weight bearing. The majority of these procedures have relied on fixation of devices to inhibit motion during the bony fusion process. In cases where the pathology permits, removal of extruded IVD fragments may be performed in a procedure termed *discectomy*. Together these procedures comprise more than 300,000 inpatient hospitalizations annually in the United States alone (DeFrances and Podgornik, 2006). More recently, intervertebral disc replacements have been approved as an alternative therapy to bony fusion for disc-related pathologies (McAfee, 2004). The concept for these "motion-preservation devices"

*Principles of Tissue Engineering, 3rd Edition*
*ed. by Lanza, Langer, and Vacanti*

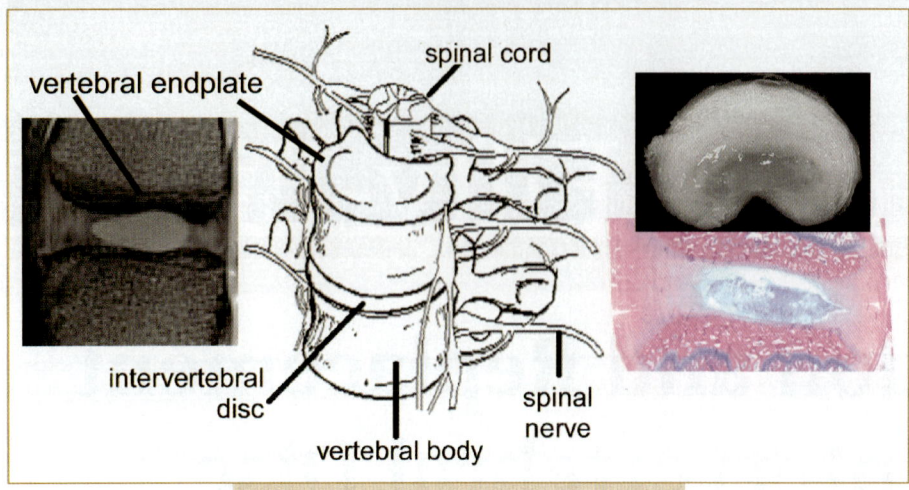

**FIG. 58.1.** Schema of spinal motion segment illustrating location of intervertebral disc between superior and inferior vertebral bodies. (*Left*) MRI appearance of immature lumbar disc with characteristic intense nucleus pulposus region. (*Right, top*) Gross appearance of nondegenerate lumbar disc and (*right, bottom*) histological appearance of immature disc in a stained section. Modified schema reprinted with permission from Columbia-Presbyterian Neurosurgery at www.cumc.columbia.edu/dept/nsg.

is that maintenance of load sharing across the IVD and a small range of motion are important to provide full range of spinal motion, to maintain IVD health, and to minimize IVD height loss, facet joint degeneration, stenosis, and related symptoms that may occur in progressive IVD pathology. These devices present all the risks associated with conventional joint replacements, such as subsidence, wear, and failure, and are indicated for only a small portion of symptomatic pathologies. There exists a very compelling need to develop alternative strategies not only to treat the consequences of IVD disorders, but also to detect and limit the progression of symptomatic IVD pathology.

Success with cellular therapies for articular-cartilage regeneration, gene therapy, and *in vitro* regeneration of cartilaginous tissue has raised hope for tissue-engineered treatments for IVD disorders. Tissue-engineered approaches to IVD regeneration have been focused on implantation of cell-supplemented or acellular biomaterials that may partially replace the IVD structure as well as on delivery of cells or bioactive factors designed to promote the natural repair process. In this chapter, a review of these tissue-engineering strategies is provided, along with evaluations of their adaptation and implementation for treatment of IVD disorders.

## II. IVD STRUCTURE AND FUNCTION

In all structures of the IVD, the extracellular matrix provides physical and biochemical cues that regulate cell-mediated repair or breakdown in mature or aging tissues (Oegema, 1990, 2002). The native matrix organization and interaction with the local IVD cell population will be important considerations in the design of any tissue-engineered regeneration strategy. The IVD is composed of a centrally situated and gelatinous tissue, the nucleus pulposus, which differs substantially from the more fibrocartilaginous anulus fibrosus, on the radial periphery (Fig. 58.2). On both superior and inferior faces is a cartilaginous endplate that provides an intimate mechanical and biophysical connection between the vascularized vertebral bone and the avascular IVD. Both

the anulus fibrosus, with a vascularized periphery, and the cartilaginous endplates are believed to be important routes of nutrient transport to all cells of the IVD (Nachemson *et al.*, 1970; Urban *et al.*, 1977). Given the very low cell density of the IVD, maintenance of both cellularity and a generous nutrient supply are often held to be critical to a successful biologically based regenerative strategy.

The immature nucleus pulposus is highly hydrated (>80% water), with extracellular matrix components that include randomly organized type II collagen fibers and multiple forms of negatively charged proteoglycans (Table 58.1, Roughley, 2004). A population of large and highly vacuolated cells is present in the nucleus pulposus during development and growth, with a shift toward a more chondrocyte-like cell population by age 7 (Meachim and Cornah, 1970; Taylor and Twomey, 1988; Trout *et al.*, 1982). Like all IVD regions, the nucleus pulposus contains multiple collagenous and noncollagenous proteins, including types III, V, VI, and IX collagens, elastin, fibronectin, and laminin (Hayes *et al.*, 2001; Oegema, 1993; Roberts *et al.*, 1991; J. Yu *et al.*, 2002). The nucleus pulposus is largely loaded in compression (Fig. 58.3) and experiences high interstitial swelling and fluid pressures, which arise from joint loading and a high density of osmotically active, proteoglycan-associated negative charges (Urban and McMullin, 1985, 1988). Nachemson and coworkers showed, as early as the 1960s, that this interstitial fluid pressure is greater than 0.5 MPa (or approximately five times atmospheric pressure) in the nucleus pulposus region (Andersson *et al.*, 1982; Nachemson, 1960, 1992). An early loss of hydration or tearing in the nucleus pulposus (Boos *et al.*, 2002), often detected as a loss of MR signal (Yu *et al.*, 1989), is believed to contribute to a loss of fluid pressurization in the IVD, which may lead to herniation or stenosis with aging (Buckwalter, 1995; McNally and Adams, 1992; Schultz *et al.*, 1982). With loss of fluid pressurization, the load distribution to the anulus fibrosus will shift from a characteristic outward "bulging" of the anulus to one of inward displacements (Adams *et al.*, 1996; Nachemson,

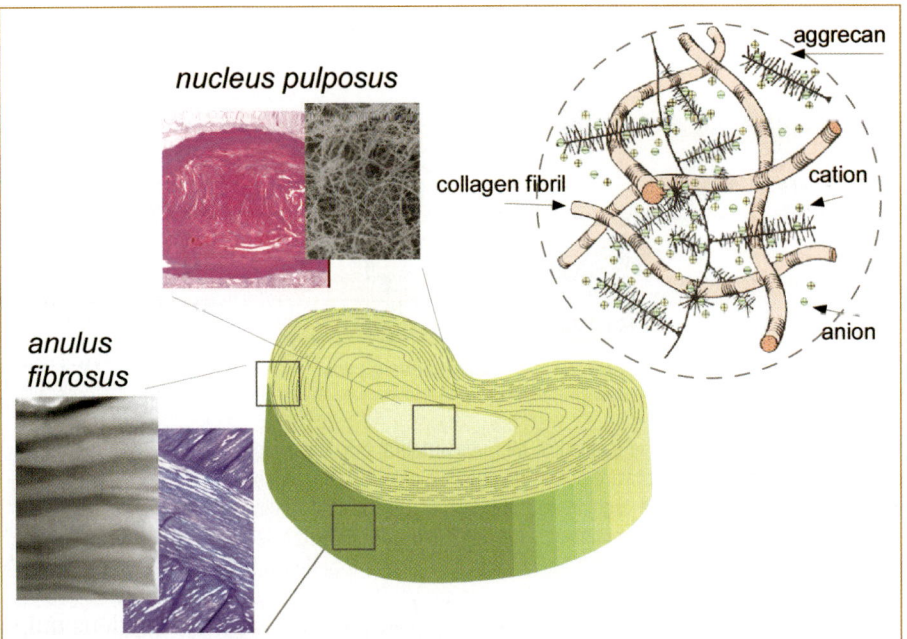

**FIG. 58.2.** Schema of different zones and microstructures comprising the intervertebral disc. Anulus fibrosus insets of macroscopic appearance and stained section illustrate the lamellar structure of the tissue. The lamellae comprise aligned collagen fiber bundles that are oriented with alternating angles of ±60°. Nucleus pulposus insets of a stained section and scanning electron micrograph illustrate the randomly organized network of fine collagen fibers and gelatinous nature of the tissue. Circular inset contains a schema of building blocks for these cartilaginous tissues that include banded type I and type II collagen fibrils, aggrecan and smaller proteoglycans, water, and multiple ionic species.

| Table 58.1. | Ranges reported for compositional features and mechanical properties for nucleus pulposus and anulus fibrosus tissue regions of the nondegenerate intervertebral disc[a] |
|---|---|

| | Water (% wt) | Collagen (% dry wt) | Proteoglycan (% dry wt) | Other proteins (% dry wt) | Compressive modulus (MPa) | Shear modulus (MPa) | Tensile modulus (MPa) | Interstitial pressure (MPa)[b] |
|---|---|---|---|---|---|---|---|---|
| **Nucleus pulposus** | 70–90 | 15–35 | 25–60 | 20–45 | 0.5–1.5 | 0.005–0.01 | NA | 0.5–3.0 |
| **Anulus fibrosus** | 65–80 | 10–65 | 10–35 | 15–40 | 0.5–1.5 | 0.08–0.40 | 20–50[c] ‖ circ | 0.1–1.0 |
| **Notes** | | Types I, II, VI, IX, XI | Aggrecan, decorin, biglycan, fibromodulin, versican, and more | | | | 0.5–5.0[c] ⊥ circ | |

[a]Both composition and mechanical properties of the disc vary substantially with region and with degeneration. Additional mechanical features important to tissue function, such as failure strength, are not shown here.
[b]Reported also as peak hydrostatic pressures, or swelling pressures.
[c]⊥, ‖ denote perpendicular and parallel to circumferential direction.

1992; Panjabi *et al.*, 1988; Shirazi-Adl, 1992). Partial or complete removal of the nucleus pulposus, occurring in some discectomy procedures, may lead to a loss of disc pressurization and disc height that will transfer loads to facet joints of the spine, increase segmental range of motion, and impact overall spinal stability. Restoration of this interstitial swelling pressure in the nucleus pulposus, or restoration of MR signal intensity, is an oft-cited criterion for restoration of a healthy functioning disc.

The anulus fibrosus is a lamellar, fibrocartilaginous structure that is highly organized into distinct lamellae (Coventry *et al.*, 1945a, 1945b) of highly oriented, and largely type I collagen–containing fiber bundles (Cassidy *et al.*, 1989; Hickey and Hukins, 1980). Type II collagen concentration increases toward the innermost region of the anulus fibrosus as the concentration of type I collagen is diminished. As with the nucleus pulposus, the anulus fibrosus contains proteoglycans within the collagenous extracellular matrix, although at lesser concentrations, which vary from outer to inner regions of the tissue. The collagen reinforcement within the anulus fibrosus resists the tensile loads, which arise during physiological joint motions, and the swelling effects, which give rise to significant anular bulging and deformation. Consequently, the anulus fibrosus has a

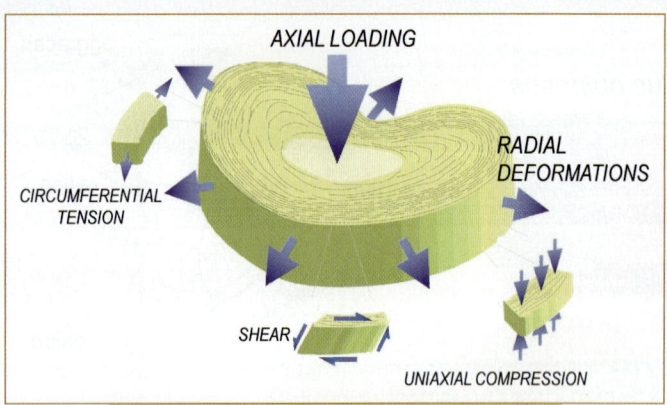

**FIG. 58.3.** Axial compressive loading of the intervertebral disc gives rise to a radial deformation, or "outward bulge," as the disc deforms in response to the compressive load. The high tensile stiffness of the healthy anulus fibrosus in the circumferential direction acts to restrict this outwardly directed deformation. Tissues of the disc will be variably loaded and experience a combination of compression, tension, and shear, as shown. Pressurization of the central and gelatinous nucleus pulposus is an important mechanism for load support and load transfer to the anulus fibrosus, which contributes to maintenance of disc height.

very high stiffness in tension, with moduli that vary with the angle of orientation along the principal collagen fiber direction (Table 58.1, Ebara *et al.*, 1996; Elliott and Setton, 2001; Fujita *et al.*, 1997; Galante, 1967; Holzapfel *et al.*, 2005; Skaggs *et al.*, 1994). Cells of the anulus fibrosus originate from the mesenchyme and exhibit many characteristics of fibroblasts and chondrocytes (Bayliss and Johnstone, 1992; Oegema, 2002; Postacchini *et al.*, 1984; Rufai *et al.*, 1995; Urban and Roberts, 1995). These cells are sparsely distributed in the mature IVD and exhibit very little intrinsic ability for self-repair. Disorders of the IVD that involve displacement or herniation of an IVD fragment are believed to arise from tears in the anulus fibrosus region, and discectomy procedures frequently involve removing a portion of this anulus tissue. Some tissue-engineering strategies are being developed around restoration of healthy anulus fibrosus function or composition, although the complexities of anulus structure and composition make this a very challenging goal.

The hyaline cartilage endplates of the IVD are important structures that transmit and distribute loads of the spinal column to the discs. Because of their direct contact with both the anulus fibrosus and the nucleus pulposus, the endplates are believed to be an important route of nutrient transport, particularly to cells of the nucleus pulposus (Antoniou *et al.*, 1996; Benneker *et al.*, 2005; Roberts *et al.*, 1996; Selard *et al.*, 2003; Urban *et al.*, 1977). With aging, the cartilage endplate will thin, as it undergoes mineralization and eventual replacement by bone. This mineralization of the endplate is thought to impede diffusion and nutrient flow to the disc, principally the nucleus pulposus, which is lacking in an alternate short-diffusion pathway. Endplate

changes, such as sclerosis, fracture, and modified vascularity, may be detected by MRI changes (Modic *et al.*, 1984) and are believed to contribute to symptomatic IVD degeneration (Bodguk, 1988; Kokkonen *et al.*, 2002; Weishaupt *et al.*, 2001). Thus, tissue-engineering strategies that preserve the health of the endplate without inducing additional damage are believed to be critical to restoring IVD function.

## III. BIOMATERIALS FOR NUCLEUS PULPOSUS REPLACEMENT

### *In Situ* Hydrating Polymers

The development of biomaterials and cellular therapies for tissue-engineered IVD replacement has a long history but has not often progressed past preclinical evaluations. The complexity of the IVD, with its three distinct substructures and multiple pathologies, together with very harsh loading conditions and mechanical requirements, has led to challenges for engineering tissue replacements. The concept that nucleus pulposus changes are an important contributor to IVD disorders has led to an initial focus on the use of acellular biomaterials for restoration of the nucleus pulposus tissue or function (Carl *et al.*, 2004; Di Martino *et al.*, 2005; Klara and Ray, 2002). In this section, attention is given to strategies developed around the concept of using *in situ* hydrating, synthetic polymers to restore nucleus pulposus hydration and, consequently, IVD disc pressure and disc height. The device with the longest clinical history is based on a copolymeric hydrogel encased in a polyethylene-fiber jacket (polyacrylonitrile and polyacrylamide, PDN™, Raymedica Inc., Fig. 58.4). When implanted in a dessicated state, the polymers absorb water while the polyethylene jacket restricts excessive swelling of the polymer. Similar concepts have been developed based on implantation of preformed devices constructed from semihydrated poly(vinyl) alcohol (Aquarelle™, Stryker Spine Inc.), a copolymer of poly(vinyl alcohol) (PVA) and poly(vinyl pyrrolidone) (PVP) (Thomas *et al.*, 2003), or modified poly(acrylonitrile) reinforced by a Dacron mesh (Bertagnoli *et al.*, 2005, NeuDisc™, Replication Medical). As shown for the PVA/PVP copolymeric implant (Fig. 58.5), the design goal is to exploit implant swelling pressure to restore the high compressive stiffness of the IVD, which is lost on dehydration or denucleation of the nucleus pulposus (Joshi *et al.*, 2006). The relevant stiffness is that measured after placement of the implant, with stiffness values reflecting both the material behaviors of the implant as well as the integration with the containing anulus fibrosus and endplates. An additional concept that has promoted development of these devices is an ability to maintain disc height.

As with many tissue replacements, there is a long list of requirements that must be satisfied for biomaterials to be used in this application, including the needs to achieve (1) favorable mechanical stiffness or mechanical properties matched to that of the native structure; in particular, the

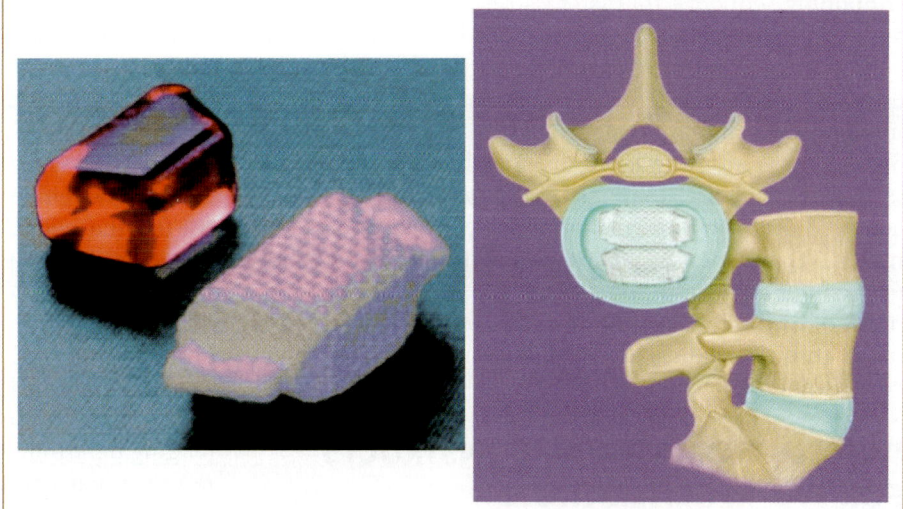

**FIG. 58.4.** (*left*) Prosthetic disc nucleus (PDN, Raymedica), shown on the left, is composed of an *in situ* swelling synthetic polymer. The pellet is encased in a polyethylene jacket that restricts water imbibition, shown adjacent to the pellet. (*right*) The PDN is designed to fit into either the anterior portion of the vacated disc nucleus or the posterior portion. Shown is a schematic of placement of both anterior and posterior components within the nucleus pulposus disc space. Modified from Klara and Ray (2002) with permission.

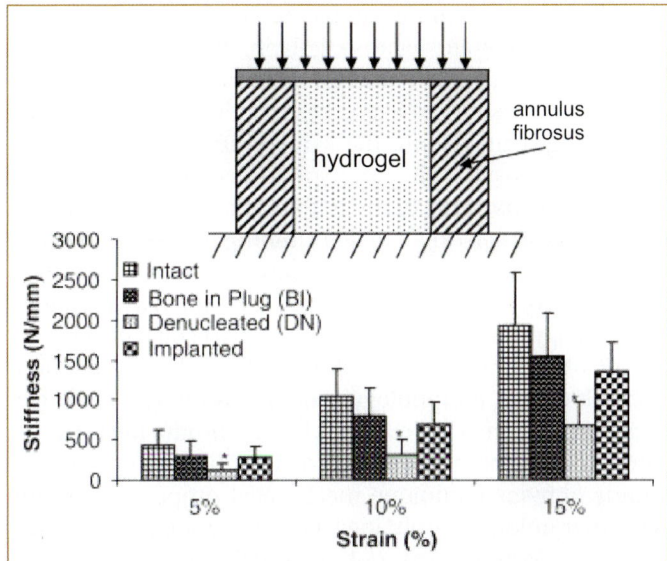

**FIG. 58.5.** (*top*) Schematic of polymeric hydrogel implants placed within the contained space of the intervertebral disc. The compressive stiffness of the disc implant can be measured as the relationship between axial deformation and applied compressive load, as shown. (*bottom*) Stiffness values were obtained for an intact human motion segment (devoid of posterior elements). When denucleated, the stiffness drops significantly. Implantation of a copolymeric hydrogel of PVA/PVP is capable of restoring stiffness for the motion segment to intact values. Modified from Joshi *et al.* (2006) with permission.

compressive stresses generated must not exceed the failure strength of the adjacent endplate, in order to avoid endplate fracture or subsidence of the device, IVD height loss, and associated problems; (2) integration with adjacent structures in order to promote load transfer, minimize device migration or extrusion, and restore stability for the motion segment; (3) durability, or an ability to maintain physical support over millions of cycles of loading; (4) minimal gen-

eration of wear debris, if appropriate; and (5) standards of biocompatibility without eliciting systemic, cellular, or immunotoxicity. Some of the polymeric devices for nucleus pulposus replacement have experienced device extrusion, endplate failure, and endplate sclerosis after implantation. These observations are thought to relate to a mismatch in mechanical stiffness that leads to excessive endplate loading and poor integration associated with device migration (Boyd *et al.*, 2004; Huang *et al.*, 2005). Controlling interstitial hydration of the polymer is a desirable feature for this class of polymers in general, because excessive swelling can cause implant stiffness and endplate overloading.

### *In Situ* Forming Polymers

Injectable polymer systems, such as polymers that will undergo a physical transition to a gel-like or solidlike form via cross-linking or thermal or pH-induced transitioning, have been evaluated for placement into a vacant nucleus pulposus space (Bao and Yuan, 2002; Boyd *et al.*, 2004; Temenoff and Mikos, 2000; Thomas *et al.*, 2003). Many of the requirements for success of this strategy are similar to those just described, with the additional requirements that the polymers must exhibit minimal leaching during the *in situ* forming procedure and must provide the benefit of minimally invasive insertion into the disc space. *In situ* curing polyurethane is one widely studied *in situ* curing polymer that has been delivered to the disc space through an inflatable polyurethane balloon in order to contain excessive swelling.

Two alternative injectable polymers evaluated for nucleus pulposus replacement have been developed from cross-linkable biopolymers. The BioDisc™ (Cryolife) is composed of bovine serum albumin that is cross-linked via glutaraldehyde at the time of injection to form a mechanically stiff implant. Similarly, Nucore™ (SpineWave Inc.) is a protein hydrogel developed from a silk and elastin peptide–

containing sequence that is cross-linked via di-isocyanate at the time of injection (Boyd *et al.*, 2004). The cross-linking confers an extra stiffness to protein polymers, which is necessary to achieve satisfactory stiffness values for a disc implant. Both implant systems have been able to maintain disc stiffness and to restore disc height when implanted, providing evidence that successful integration is being achieved on injection. Furthermore, systems composed of native IVD polymers, such as the elastin peptide sequences, may confer some additional benefit as recognized components of the body. These approaches are promising as they move through clinical trials of implant feasibility in the current period.

## IV. CELL-BIOMATERIAL CONSTRUCTS FOR IVD REGENERATION

A persistent limitation of materials-based replacements of the IVD is their biologically mediated or mechanically induced failure due to the harsh loading conditions and cellular responses within the disc space. These challenges are rooted in the fact that the materials used for such applications have no capacity for self-renewal or self-repair. This has led to increasing interest in tissue-engineering methods to regenerate new IVD *in situ* or to transplant IVD tissue that has been generated *ex vivo*. Such strategies have been employed to augment repair of other types of cartilage, most notably articular cartilage, and meniscus, which share some features of the harsh biologic and mechanical loading environment within the IVD.

### Scaffolds for Cell-Based Tissue Engineering in the IVD

As in other cartilage-tissue-engineering applications, a main strategy for IVD regeneration has been the inclusion of cells with biomaterials to enable production and long-term maintenance of newly generated tissue. Biomaterials that enable appropriate cellular phenotypes and matrix biosynthesis and that sometimes enable polymeric degradation or resorption have been proposed as alternative implantable biomaterials and have been studied largely *in vitro*. The goals for use of these scaffolds are similar to those for other biomaterial implants, with the added requirements that the biomaterial must generate no cytotoxic or immunogenic degradation or breakdown fragments and that new matrix formation is enabled. Studies of cell-biomaterial constructs cultured *in vitro* have demonstrated potential for many materials (Table 58.2), including thermosensitive gels such as chitosan, modified chitosans, and elastinlike polypeptides (Au *et al.*, 2003; Betre *et al.*, 2002; Mwale *et al.*, 2005; Roughley *et al.*, 2006); self-associating gels composed of agarose, collagen, and fibrin (Gruber *et al.*, 2004; Peretti *et al.*, 2006); or modified forms of these same materials, cross-linkable alginates, polyethylene glycol, poly(glycolic acids), and more (Baer *et al.*, 2001; Burkoth and Anseth, 2000;

Elisseeff, 2004; Masuda *et al.*, 2003; Mercier *et al.*, 2004; Sontjens *et al.*, 2006). *In vitro* studies with these materials are based on evaluating new matrix formation and sometimes degradation characteristics through culturing native disc or other cell types within these matrices. Hydrogels, such as alginate and gelatin, have been used most commonly for engineering nucleus pulposus tissue, likely due to the fact that such materials reasonably approximate the gel-like properties of the native tissue. Cells of different origin, including native IVD cells, stem cells, and chondrocytes, are capable of synthesizing and depositing collagen and glycosaminoglycans within these hydrogels, although there is little agreement on the targeted composition necessary to achieve a satisfactory tissue construct. This is a particularly challenging determination for the IVD because the matrix contains varying amounts of both types I and II collagen, so the exclusive presence of type II collagen does not serve as a phenotypic matrix marker, as is the case for articular cartilage.

Efforts to regenerate anulus fibrosus have also involved gels such as alginate, agarose, gelatin, and collagen as well as fibers or sponges made from materials such as poly(glycolic) acid, collagen, hyaluronic acid, and/or glycosaminoglycans. Often, the same scaffolds evaluated for nucleus pulposus cells are also studied with cells of the anulus fibrosus, with findings that generally illustrate the importance of cell origin in determining the resultant extracellular matrix synthesis. A common observation, however, is that cells of either origin that are maintained in a rounded morphology tend to generate more type II collagen, characteristic of hyaline cartilage, whereas those that are cultured in an elongated morphology generate more type I collagen (Fig. 58.6). A main challenge has been reproducing the intricate lamellar arrangement of collagen fibers that give the anulus fibrosus its unique mechanical properties and the cells their unique morphology. For this reason, the majority of anulus-tissue-engineered materials have been generated from scaffolds that lack any apparent lamellar microstructure. Some investigators have developed polymeric scaffolds with anisotropic features, such as an oriented honeycomb structure, demonstrating production of multiple collagen types as well as proteoglycan (Sato *et al.*, 2003). These results are indeed suggestive of the potential to engineer anisotropic collagenous tissues, although no results for generating new, functional lamellar anulus tissue are known to exist to date.

Assessment of the success of IVD-tissue-engineering efforts is critical to moving this technology toward clinical application. Most studies have focused on generating new IVD *in vitro*, with few documenting tissue formation and integration in preclinical evaluations *in vivo*. To date, the most common tool for assessment of newly generated IVD tissue has been histology as a method to evaluate cell and tissue morphology. Given the structural complexity of the tissue, analysis of gene expression and extracellular matrix

**Table 58.2.** Representative overview of studies involving cell scaffold–based tissue engineering of IVD using cells obtained from native tissues only

| Material | Cell type | Cell source | Cell density | In vitro/in vivo | Assessment |
|---|---|---|---|---|---|
| PCL[a] | AF, NP | Bovine | $5 \times 10^3/cm^2$ | In vitro | Histology, SEM, gene expression |
| Alginate[b] | AF, NP | Porcine | $4 \times 10^6/mL$ | In vitro | DNA, ECM analysis |
| Gelatin/C6S/HA[c] | NP | Human | $20 \times 10^6/mL$ | In vitro | Histology, DNA, ECM analysis, gene expression |
| CPP[d] | NP | Bovine | $16 \times 10^6/cm^2$ | In vitro | Histology, mechanical analysis |
| Agarose, collagen[e] | AF | Human | $0.2 \times 10^6/mL$ | In vitro | Histology, ECM analysis |
| Gelatin, PLA[f] | NP | Porcine | $5 \times 10^6/mL$ | In vitro | Histology, ECM analysis, gene expression |
| Collagen/GAG[g] | AF | Canine | $40 \times 10^6/mL$ | In vitro | Histology, ECM analysis |
| Collagen/HA[h] | AF, NP | Bovine | $13 \times 10^6/mL$ | In vitro | Histology DNA, ECM analysis, gene expression |
| Alginate[i] | AF, NP | Porcine | $1–10 \times 10^6/mL$ | In vitro | Histology, gene expression, mechanical analysis |
| Collagen[j] | AF | Lapine | $10 \times 10^6/mL$ | In vivo | Histology |
| PGA, alginate[k] | AF, NP | Ovine | $25–50 \times 10^6/mL$ | In vivo | Histology, DNA, ECM analysis, mechanical analysis |

*Abbreviations*: PCL (polycaprolactone); C6S (chondroitin-6-sulfate); HA (hyaluronan); CPP (calcium polyphosphate); PLA (polylactic acid); GAG (glycosaminoglycan); PGA (polyglycolic acid); AF (anulus fibrosus); NP (nucleus pulposus); SEM (scanning electron microscopy); DNA (deoxyribonucleic acid); ECM (extracellular matrix).
[a]Johnson *et al.*, *Eur Spine J*, 2006.
[b]Akeda *et al.*, *Spine*, 2006.
[c]Yang *et al.*, *Artific Organ*, 2005. Yang *et al.*, *J Biomed Mat Res B*, 2005.
[d]Hamilton *et al.*, *Biomaterials*, 2005; Seguin *et al.*, *Spine*, 2004.
[e]Gruber *et al.*, *Biomaterials*, 2006.
[f]Brown *et al.*, *J Biomed Mat Res A*, 2005.
[g]Saad and Spector, *J Biomed Mat Res A*, 2004; Rong *et al.*, *Tissue Eng*, 2002.
[h]Alini *et al.*, *Spine*, 2003.
[i]Baer *et al.*, *J Orthop Res*, 2001; Wang *et al.*, *Spine*, 2001.
[j]Sato *et al.*, *Spine*, 2003.
[k]Mizuno *et al.*, *Spine*, 2004. Mizuno *et al.*, *Biomaterials*, 2006.

composition has been commonly employed to confirm the appropriate phenotypic behavior in engineered IVD. Relatively few studies have documented mechanical analysis of engineered IVD tissue, but this will undoubtedly be critical as efforts to engineer functional tissue continue. Nevertheless, these *in vitro* studies begin to lay the foundation for necessary and/or sufficient characteristics of a successful scaffold for nucleus pulposus replacement. From these studies, for example, it is evident that a high starting cell density and a high degree of initial matrix stability are essential for promoting long-term construct stability and matrix accumulation, eventually to restore mechanical function and swelling pressure (Wilson *et al.*, 2002).

### Composite Cell-Biomaterial IVD Implants

Cell-based regeneration of the IVD *ex vivo* is complicated by the inherent multicomponent structure of the IVD, which includes two distinct regions, the anulus fibrosus and the nucleus pulposus. Given this added complication, it is not surprising that there are fewer examples of efforts to engineer integration of the multiple components of the IVD *ex vivo*. This can prove to be a critical limitation because integration is needed to ensure proper load transfer and to limit damaging motions during disc loading. In studies conducted by Bonassar and coworkers, IVD regeneration was attempted with a fully integrated scaffold combining poly(glycolic *co*-lactic acid) as a scaffold for anulus fibrosus and a cross-linked alginate hydrogel as a scaffold for nucleus pulposus tissue (Fig. 58.7) (Mizuno *et al.*, 2004, 2006). Primary cells for culture within each scaffold region were derived from the corresponding native IVD tissues, and the resultant cell-laden scaffolds were implanted subcutaneously into athymic mice for a period of 12 or 16 weeks. Results illustrate spatially directed matrix regeneration with extracellular matrix that exhibited distinct morphologies and contained both collagen and glycosaminoglycans (Fig.

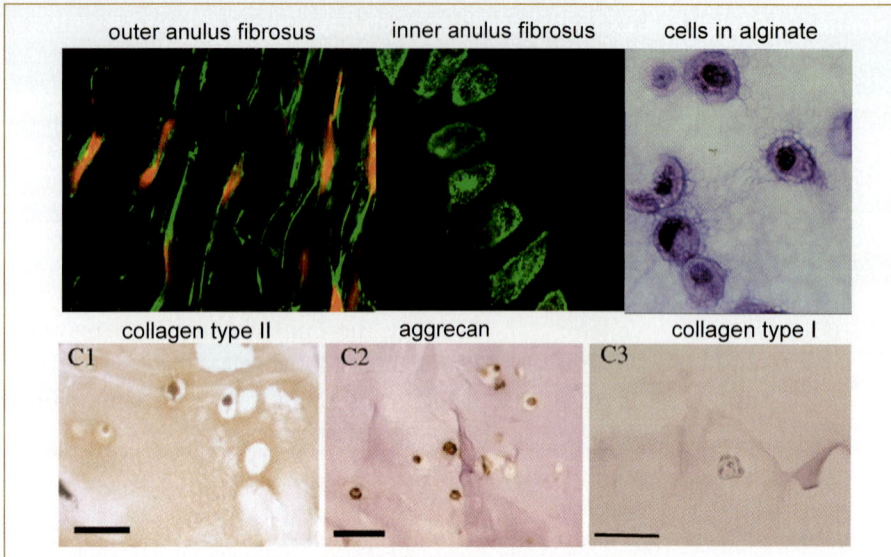

**FIG. 58.6.** **Top:** (*left*) Anulus fibrosus cells are highly elongated in the outermost regions of the tissue but adopt a more rounded morphology in the inner tissue regions, as shown (*middle*). (*right*) When anulus fibrosus cells are embedded in an alginate hydrogel, they assume a rounded morphology. **Bottom:** Cells in alginate will express matrix proteins associated with chondrocytes, such as type II collagen and aggrecan. Bottom figure modified from Le Maitre *et al.* (2005) with permission.

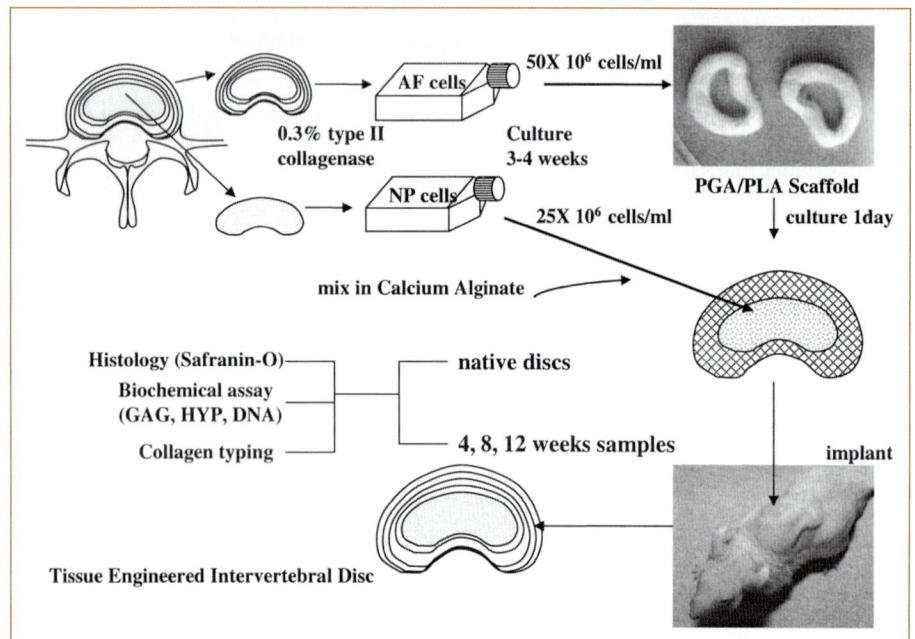

**FIG. 58.7.** Schema of tissue engineering for intervertebral discs. Anulus fibrosus and nucleus pulposus cells were isolated and cultured separately prior to seeding onto PGA/PLA scaffolds (anulus fibrosus) or suspension in a solution of 2% alginate (nucleus pulposus). The cell–alginate mixture was injected into the empty center of the PGA/PLA construct. The assembled tissue-engineered intervertebral disc constructs were implanted subcutaneously onto the dorsum of athymic mice and harvested at times up to 12 weeks. Reprinted with permission from Mizuno *et al.* (2004).

58.8). In biomechanical tests, the composite tissue-engineered disc was found to have a compressive modulus about one order of magnitude lower than that of the native tissue, with a permeability to fluid flow that fell between values for the nucleus pulposus and the anulus fibrosus. Thus, this approach illustrated an ability for cell-laden scaffolds to regenerate extracellular matrix with some of the functional and compositional features of the native tissue.

In another integrative tissue-engineering study of note, investigators Kandel and coworkers generated nucleus pulposus tissue *in vitro* by culturing primary bovine nucleus pulposus cells at high density on a calcium polyphosphate substrate, in order to mimic the natural integration of the nucleus pulposus against the vertebral endplate (Fig. 58.9, Seguin *et al.*, 2004). A similar strategy was employed to generate a calcified tissue–cartilaginous endplate–nucleus pulposus construct by first generating a hyaline cartilage tissue layer on the calcium polyphosphate substrate prior to seeding with nucleus pulposus cells (Fig. 58.9 Hamilton *et al.*, 2006). The nucleus pulposus cells formed tissue with a proteoglycan, but not collagen content matched to that of the native nucleus pulposus. Importantly, functional properties in some testing configurations approached that of the native tissue. Additional work will be required in adapting

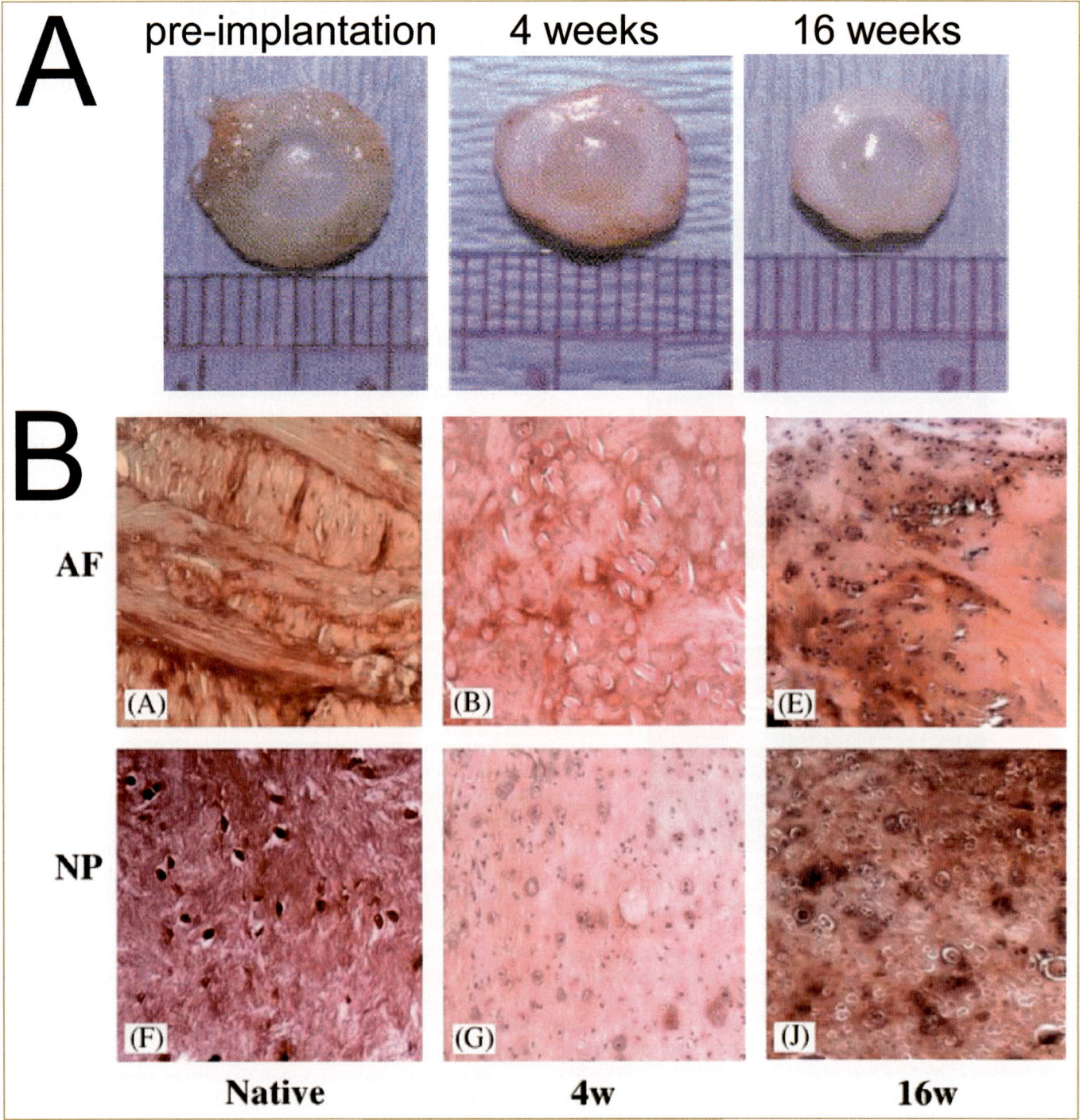

**FIG. 58.8.** **(A)** Gross morphology of tissue-engineered intervertebral disc constructs before implantation (*left*) and 16 weeks after implantation in the sub-cutaneous pouch of athymic mice. **(B)** Safranin O–stained sections of anulus fibrosus from (*left*) native ovine lumbar intervertebral disc and (*middle*) tissue-engineered intervertebral disc constructs at 4 weeks and (*right*) 16 weeks after implantation. **(C)** Safranin O–stained sections of nucleus pulposus from (*left*) native ovine lumbar intervertebral disc and (*middle*) tissue-engineered intervertebral disc constructs at four weeks and (*right*) 16 weeks after implantation. Modified from Mizuno *et al.* (2006) with permission.

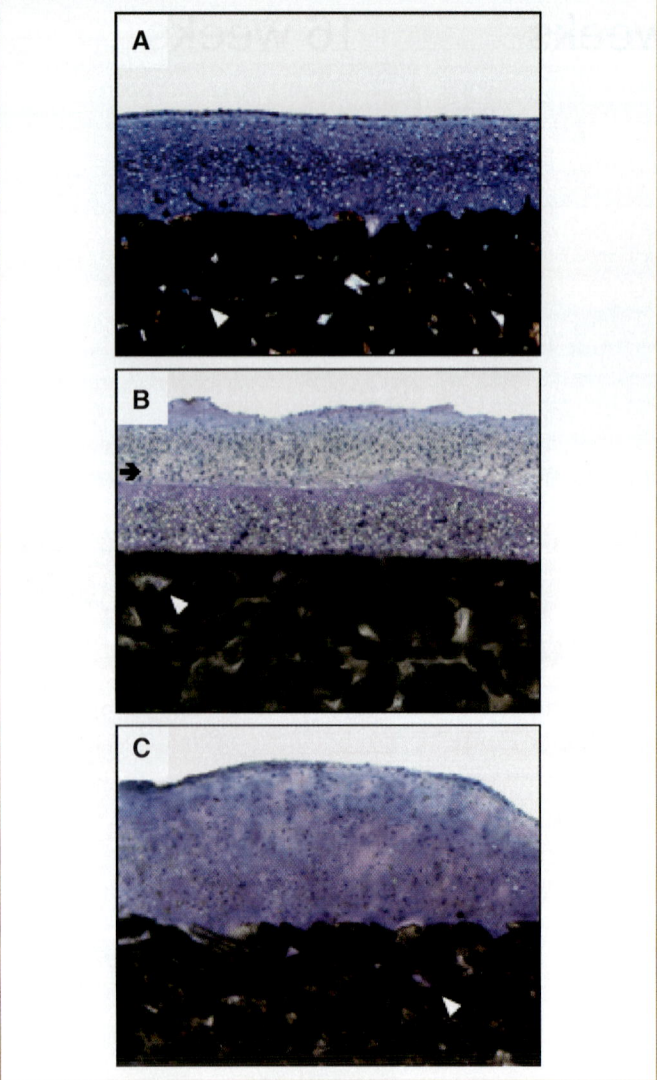

**FIG. 58.9.** Histological appearance of multicomponent tissue-engineered construct. Bovine articular chondrocytes were placed on the top surface of a porous calcium polyphosphate construct (CPP) and allowed to form cartilage *in vitro*. Nucleus pulposus cells were then placed onto the *in vitro*–formed hyaline cartilage and cultured for periods up to eight weeks. **(A)** *In vitro*–formed cartilage at two weeks (time at which the nucleus pulposus cells would be seeded); **(B)** *in vitro*–formed nucleus pulposus-cartilage–CPP composite (triphasic construct) at eight weeks following seeding of chondrocytes; and **(C)** *in vitro*–formed cartilage tissue alone (no nucleus pulposus cells) at eight weeks. Arrowheads indicate tissue growing within the pores of the CPP; arrow indicates interface between cartilage and nucleus pulposus tissue. (Toluidine blue stain; original magnification ×50). Reprinted from Hamilton and coworkers (2006) with permission.

these integrative tissue-engineering approaches to ensure that mechanical integration with adjacent tissues is adequate, but these studies focused on generating integrated nucleus–endplate or nucleus–anulus are an important step in illustrating feasibility for this approach.

# V. CELLULAR ENGINEERING FOR INTERVERTEBRAL DISC REGENERATION

Given the relatively small numbers of studies in the area of IVD-tissue engineering, there is a surprising amount of breadth not only to the biomaterials, but also to the cell sources utilized for regeneration. The question of cell source is of particular note for IVD-tissue engineering, given that the availability of autologous disc cells is extremely low in the adult and that the phenotype of cells varies so substantially with both spatial position and with age. In animals studied for IVD-tissue engineering *ex vivo*, the origin of cells in the nucleus pulposus may be partly notochordal or mesenchymal, depending on the age of the animal in question. As such, the choice of species used as a source of cells may be quite important. Due to the ease of availability, porcine and bovine cells are the most commonly used, with other efforts reporting the use of cells of canine, lapine, and ovine origin as well as human. However, cells derived from bovine tissues may be exclusively mesenchymal in origin, while those derived from porcine, lapine, or ovine sources may be largely notochordal. These phenotypic differences add an additional and unique complicating factor for investigators studying preclinical models for IVD-tissue regeneration.

Given the very limited availability of native IVD cells that can be effectively harvested for tissue engineering, there has been interest in using other cells as sources for these efforts. The primary target for other sources has been mesenchymal stem cells (MSCs) derived from sources such as bone marrow (Richardson *et al.*, 2006a) and adipose tissue (Li *et al.*, 2005). A major challenge in this approach has been the development of methods to guide the development of MSCs toward phenotypes found in the IVD (see next section). This has been attempted through manipulation of the culture medium and gas conditions (Risbud *et al.*, 2004) as well as through coculture with primary cells from the IVD (Richardson *et al.*, 2006b). In comparison to the use of adult primary disc cells derived from often-pathological or degenerated IVDs, the use of autologous or other MSCs or progenitor cells may be most promising to the future of *ex vivo* tissue-engineering strategies that rely on cell supplementation.

In addition to origin, cell density is known to have a profound effect on the efficacy of the tissue-engineering process. Here there has been a great deal of variability in protocols, with studies reporting densities of delivered cells ranging from $0.2 \times 10^6$ to $50 \times 10^6$ cells/mL. While the lower end of this scale is likely more reflective of the actual cell density in nucleus pulposus tissue, the densities at the higher end of the scale are more in line with those known to be effective in generating other types of cartilage (Puelacher *et al.*, 1994). A critical concern for disc-tissue regeneration, particularly in the case of strategies that employ high cell densities, is the issue of nutrient and gas supply necessary to maintain cell viability and health. The IVD is both avascular and alymphatic, meaning that the transport of nutrients and oxygen is driven largely by diffusion from the vascularized periphery and through the vertebral endplates (Holm *et al.*, 1981; Maroudas *et al.*, 1975; Nachemson *et al.*, 1970; Stairmand *et al.*, 1991; Urban *et al.*, 1977). It is noteworthy that calcification and endplate changes in the degenerating IVD can lead to impaired nutrient transport, which is presumably linked to decreased cell viability (Benneker *et al.*, 2005; Roberts *et al.*, 1996). Thus, supplementation of scaffolds with very high cell densities may not be optimized for long-term survival in the largely hypoxic and glucose-poor, lactate-rich environment within the IVD. This concern has been expressed for cell-laden IVD scaffolds, but it has not been directly addressed or investigated as an issue in IVD-tissue regeneration.

## Cellular Supplementation in the IVD

If the local environment within the IVD is conducive to the survival of cells, direct cell supplementation without biomaterial scaffolds may hold promise for IVD repair. This strategy has been pursued by several groups, using either IVD cells, chondrocyte-like cells, or progenitor cells. In the first reported work, Nishimura and Mochida (1998) inserted nucleus pulposus cells from the rabbit, following removal of nucleus tissue, and showed some beneficial effects in inhibiting the degenerative IVD changes of nucleotomy. Similar procedures have also shown the effectiveness of autologous disc cell implantation in both a sand rat model of spontaneous disc degeneration (Gruber *et al.*, 2002) as well as a canine model of disc degeneration (Ganey *et al.*, 2003). Furthermore, work by Nomura and coworkers (2001) has shown that supplementation with allogeneic nucleus pulposus cells did not induce any appreciable host-versus-graft rejection response and also retarded disc degeneration in a rabbit nucleotomy model. It is noteworthy that nucleus pulposus cell insertion resulted in a slightly poorer outcome than did insertion of the allograft nucleus pulposus tissue itself, indicating that inclusion of the extracellular matrix present in the allograft may be as important as or more important than the absolute number of cells inserted (Nomura *et al.*, 2001).

Limitations will always exist in obtaining sufficient numbers of autologous or allogeneic disc cells from a single site, as well as concerns about impaired cellular activity for the native cells. Some studies have thus focused on using a coculture of nucleus pulposus and anulus fibrosus cells to stimulate cell metabolism prior to reinsertion (Okuma *et al.*, 2000). These approaches were shown to be effective in delaying some degenerative features, such as the loss of disc architecture following reinsertion of the activated cells in a rabbit model. Still other studies have focused on delivery of cells through allograft tissues based on the concept that preservation of extracellular matrix is an equally important criteria for regeneration (Matsumoto *et al.*, 2001a; Sato *et al.*, 2003; Seguin *et al.*, 2004; Yung Lee *et al.*, 2001).

Since 2002, a prospective, controlled, multicenter study has been performed to compare autologous disc cell transplantation plus discectomy against discectomy alone (Meisel *et al.*, 2006). The interim analysis of the first 28 patients at two years showed a clinically significant reduction of low-back pain in the transplantation group as compared to the discectomy group, suggesting a potential benefit of the cell transplantation strategies described earlier. Little is understood about the mechanism by which the cell supplementation provides this benefit, although disc hydration but not disc height was found to be higher in the patients receiving the cell transplantation as compared to the discectomy group. This clinical study underscores the role of autologous cell–mediated biological factors in regulating symptoms with IVD pathology and illustrates a potentially important role for sustaining cell viability of the IVD in inhibiting this pathology.

Clinically, the autologous reinsertion of the nucleus pulposus cells into the degenerative disc remains challenging (Evans, 2006). As a source of cells for transplantation, MSCs that can be harvested from a patient's own bone marrow are a possible candidate (Leung *et al.*, 2006; Risbud *et al.*, 2004; Sakai *et al.*, 2005). *In vitro*, the differentiation of MSCs into nucleus pulposus–like or chondrocyte-like cells has been shown under hypoxic and high-osmotic conditions (Risbud *et al.*, 2004) and also with TGF-β stimulation (Steck *et al.*, 2005). Studies have followed injection of autologous MSCs embedded in atelocollagen gel as well as direct injection of MSCs into rabbit or rat models of IVD degeneration, and these have observed an ability for these cells to differentiate or regenerate matrix (Crevensten *et al.*, 2004; Sakai *et al.*, 2005, 2003). In one set of studies, transplantation of MSCs with an atelocollagen carrier into the rabbit discs effectively maintained disc height, MR signal intensity, and the histological appearance of the nucleus pulposus and anulus fibrosus regions at 24 weeks after transplantation (Fig. 58.10, Sakai *et al.*, 2003). Some of the positive outcomes observed for tissue regeneration in the animal models may arise from factors released from MSCs or direct contact with MSCs, which may enhance metabolism of native disc cells (Yamamoto *et al.*, 2004; Richardson *et al.*, 2006b).

Another approach to supplement cells without donor morbidity is the use of established cell lines, as shown recently following transplantation of a human nucleus pulposus cell line (Sakai *et al.*, 2005, 2004) into degenerated discs of the rabbit nucleotomy model (Iwashina *et al.*, 2006). Despite evidence that the cell supplementation was beneficial in retarding the progression of IVD degeneration, concerns about tumorigenesis and/or carcinogenesis associated with cell lines and the consequences of the use of a recombinant SV40 adenoviral vector need to be clarified before this strategy could be widely adopted. Given the increasing importance of MSCs and other progenitor cell therapies in articular-cartilage-, meniscus-, and other cartilage-regeneration strategies, expanded research on use of MSCs

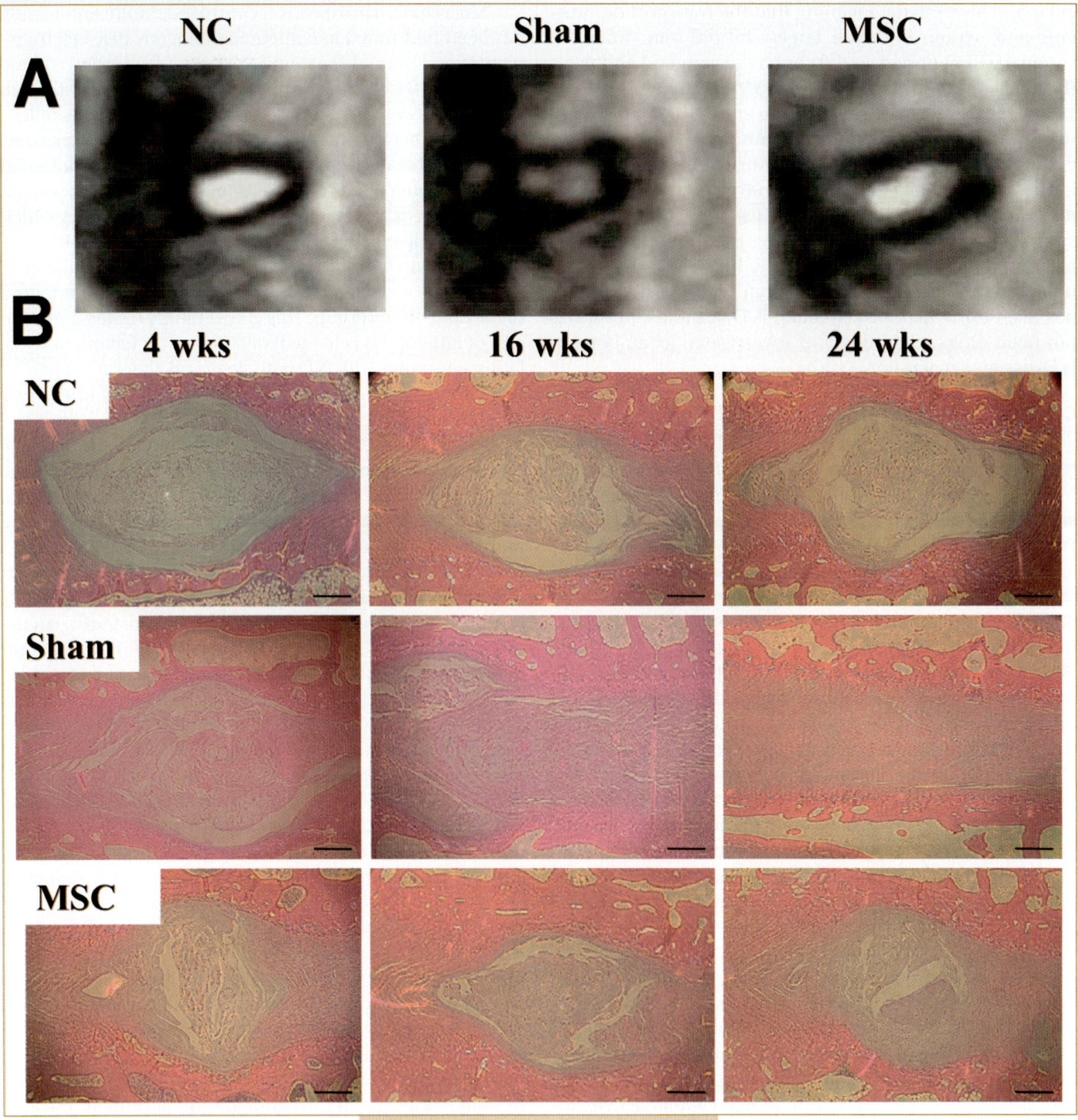

**FIG. 58.10.** Rabbit IVDs were subjected to annular puncture as a model to induce IVD degeneration in a study by Sakai and coworkers. At two weeks after puncture, subsets of IVDs were transplanted with MSCs or left as sham-operated controls. **(A)** MRI of normal control (NC), sham (disc-degeneration model), and MSC-transplanted rabbits taken 26 weeks postinduction of degeneration (24 weeks after MSC transplantation). Significant recovery of T2-weighted signal intensity is seen in discs of MSC-transplanted discs as compared to very low signal intensity in sham. **(B)** Histological changes seen after MSC transplantation in discs. Control group discs show oval-shaped nucleus, with no collapse of the inner annular structure. Sham operated discs show collapse of the inner anulus morphology from four weeks (six weeks after induction of degeneration) and fibrotic changes in the nucleus pulposus completed by 24 weeks. MSC group discs showed relatively preserved inner anulus structure, with minimal fibrosis in the nucleus region. Bar = 200 μm. Reprinted with permission from Sakai and coworkers (2006).

can be expected to comprise a significant effort in the future of IVD regeneration.

## VI. GROWTH FACTORS AND BIOLOGICS FOR INTERVERTEBRAL DISC REGENERATION

Disc cells modulate their activity by a variety of substances, including cytokines, growth factors, enzymes, and enzyme inhibitors, in a paracrine and/or autocrine fashion (Masuda and An, 2004). Tissue-engineering approaches to disc regeneration have been based on attempts to up-regulate important matrix proteins (e.g., aggrecan) or to down-regulate proinflammatory cytokines [e.g., interleukin-1 (IL-1), tumor necrosis factor-$\alpha$ (TNF-$\alpha$)] (Ahn et al., 2002; Burke et al., 2003; Igarashi et al., 2000; Kang et al., 1996; Le Maitre et al., 2005; Olmarker and Larsson, 1998; Seguin et al., 2005; Weiler et al., 2005), and matrix-degrading enzymes (e.g., metalloproteinases and aggrecanases) (Evans, 2006; Le Maitre et al., 2004; Roberts et al., 2000; Sztrolovics et al., 1997). Delivery of these modulating biologic agents, with and without scaffolds and/or through cell transplantation, has been the subject of many years of efforts in tissue engineering. In vitro studies have shown that the rate of matrix synthesis or gene expression for matrix proteins, principally proteoglycan or collagen, can be increased several-fold in IVD cells in the presence of supplemental transforming growth factor-$\beta$ (TGF-$\beta$), osteogenic protein-1 (OP-1) (Imai et al., 2002; Masuda et al., 2003), bone morphogenetic proteins (e.g., BMP-2) (Kim et al., 2003; Tim Yoon et al., 2003), growth and differentiation factor-5 (GDF-5) (Chujo et al., in print; Li et al., 2004), epidermal growth factor (EGF) (Gruber et al., 1997; Thompson et al., 1991), or insulinlike growth factor-1 (IGF-1) (Osada et al., 1996). Other studies have demonstrated the potential of these growth factors as well as platelet-derived growth factor (PDGF) to reduce cell apoptosis and to promote cell proliferation (Gruber et al., 1997, 2000). Autologous platelet-rich plasma, which contains a variable mixture of growth factors, has also been shown to be an effective stimulator of cell proliferation, proteoglycan and collagen synthesis, as well as proteoglycan accumulation, when added to IVD cell cultures in vitro (Akeda et al., 2006). In a different approach, supplementation of IVD cell cultures with a naturally occurring antiinflammatory molecule, interleukin-1 receptor antagonist (rhIL-1Ra), has been shown to inhibit the down-regulation of biosynthesis induced by the proinflammatory cytokine, IL-1 (Akeda et al., 2006, Pichika et al., 2006). This illustrates that stimulatory factors as well as antiinflammatory or anticatabolic factors may be considered for therapeutic purposes in IVD regeneration. Overall, these in vitro studies have illustrated the potential for biologics to assist in matrix regeneration through controlling both cell metabolism and cell number, and they have paved the way for more recent studies evaluating these biologics in vivo.

Protein injection into the disc space is relatively simple and practical and has been the most widely studied of all approaches for delivery of biologics for IVD regeneration. Walsh and coworkers (2001) reported the in vivo effect of a single injection of several growth factors, including basic fibroblast growth factor (bFGF), GDF-5, IGF-1, or TGF-$\beta$, in mouse caudal discs with degeneration induced by static compression. An increase in neomatrix was observed following the injection of GDF-5, while increases in the anulus fibrosus cell population were found under the influence of IGF-1 (Walsh et al., 2002). In separate studies by Masuda and coworkers, a single intradiscal administration of rhOP-1 in normal rabbit discs in vivo was shown to result in increased disc height and proteoglycan content in the nucleus pulposus regions, in comparison to a saline injection control group (An et al., 2005). In an animal model of disc degeneration caused by needle puncture of the anulus fibrosus, an injection of rhOP-1 (100 µg/disc) restored disc height, structural change, and mechanical properties (Fig. 58.11) (Masuda et al., 2006; Miyamoto et al., 2006). The effectiveness of direct protein delivery was also confirmed in experiments using rhGDF-5, where a single injection of rhGDF-5 resulted in a restoration of disc height and improvements in MRI and histological grading scores in this same animal model of disc degeneration (Chujo et al., 2006). These works are important for demonstrating that biologic manipulation of IVD cells in vitro may translate to an observed effect in vivo and that direct protein delivery may be useful for promoting new matrix formation in the absence of cell delivery. It is noteworthy that the latter studies using rhOP-1 and rhGDF-5 also provide documentation of preclinical measures, such as disc height, disc mechanics, and MRI appearance, that may be important for translating these technologies to noninvasive or minimally invasive clinical outcomes for patient treatment.

## VII. GENE THERAPY FOR INTERVERTEBRAL DISC REGENERATION

While the studies described earlier illustrate a range of proteins considered as possible therapies for IVD regeneration, it is important to consider the unavoidable limitations of protein delivery to the disc space. Issues such as protein half-life and solubility, the need for a proper carrier, the need to preserve mechanical environment or cell numbers, and/or the presence of inhibitors are all factors that can be expected to affect the therapeutic efficacy of protein delivery in vivo. A consideration for the use of recombinant protein therapies is also cost, because some disc pathologies and the need to inhibit disc degeneration may be chronic in nature or require multiple treatments. Gene therapy has been advocated as a therapeutic alternative for the delivery of biologics in disc regeneration (Nishida et al., 1999; Wehling et al., 1997; Yoon, 2005; Yoon et al., 2004). DNAs that encode specific proteins may be delivered into

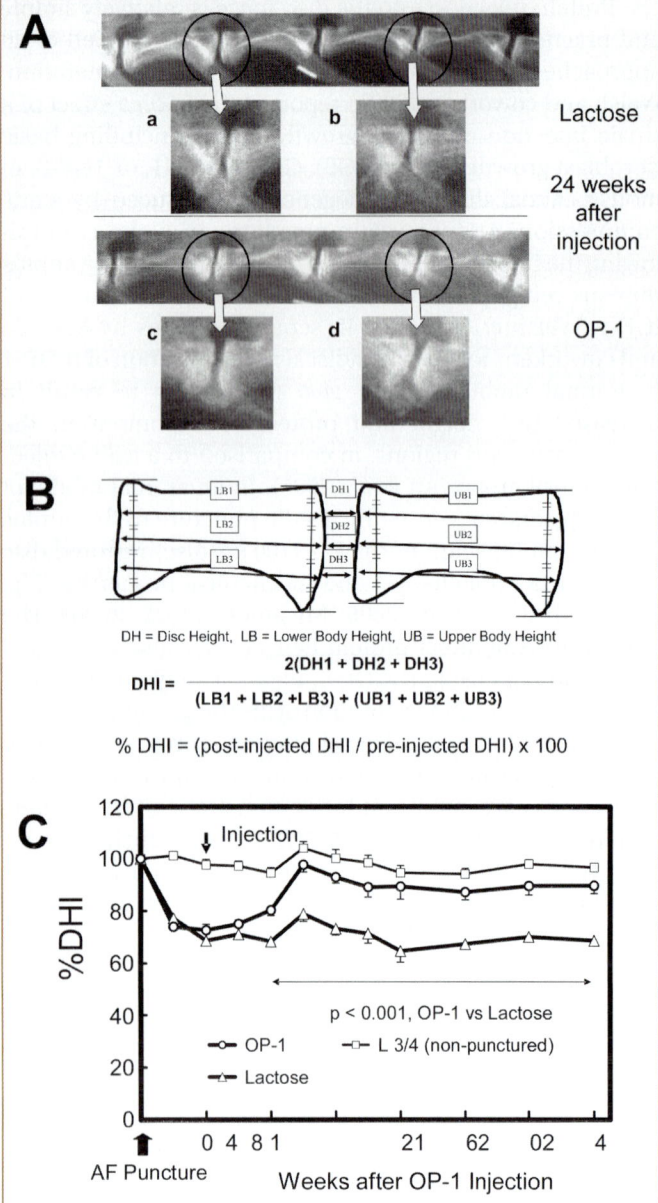

**FIG. 58.11.** IVDs of rabbit lumbar spines were subjected to annular puncture by an 18-gauge needle to induce disc degeneration. Four weeks after the initial puncture, the vehicle (5% lactose, 10 μL) or OP-1 in 5% lactose (100 μg/10 μL at each level) was injected into the center of the nucleus pulposus. **(A, top)** Radiographs at the 24-week time after injection with lactose. **(A, bottom)** Radiographs of rabbit IVDs following injection with OP-1 in the experimental group, shown at 24 weeks after injection. **(B)** Method for radiographic measurement of disc height index (DHI). The average IVD height (DHI) was calculated by measurements obtained from the anterior, middle, and posterior portions and divided by the average of adjacent vertebral body heights. LBH indicates lower body height; UBH, upper body height. **(C)** Changes in % DHI after the annular puncture and OP-1 injections. As shown, % DHI of injected discs in the OP-1 group was significantly higher than in the lactose-injected control group by four weeks (*p* < 0.001, repeated ANOVA). This difference in % DHI was maintained out to later periods.

the cells by viral or nonviral transfection, with the result that these cells produce proteins to, theoretically, prolong the duration of action. The first successful attempt for *in vitro* gene transfer was reported for chondrocyte-like cells from the endplate of the IVD using a retroviral-mediated technique (Wehling *et al.*, 1997). An *ex vivo* approach was used based on harvesting host cells, infecting these host cells *in vitro*, selecting and enriching infected cells, and finally returning these cells to the host. This approach avoids problems associated with low cell numbers and transfection efficiencies, but it is both challenging and costly to perform. Nevertheless, because a decrease of cells by apoptosis or necrosis is considered to be associated with advanced disc degeneration, cell supplementation with genetically manipulated cells will continue to hold promise for disc regeneration.

Adenoviral vectors often possess high titers and infectivity and are able to infect nondividing cells such as IVD cells. Adenoviral-mediated gene transfer to human IVD cells has been shown to be efficient and to produce transcripts across nondegenerative to degenerative cell types, using adenovirus carrying lacZ (Ad/CMV-lacZ) or luciferase marker genes (Ad/CMV-luciferase) as well as Sox9 (Ad/Sox9-GFP), GDF-5, and TGF-β1 adenoviral constructs (Moon *et al.*, 2000; Wang *et al.*, 2004; Paul *et al.*, 2003). Yoon and coworkers (2004) also used adenoviral vector to transfer LIM mineralization protein-1 (LMP-1) to rat IVD cells *in vitro* and observed an increase of BMP-2 and BMP-7 gene expression and protein production, and proteoglycan synthesis. The feasibility of using direct *in vivo* adenoviral-mediated gene transfer to disc cells has also been demonstrated using the lacZ, TGF-β, LMP-1, and Sox9 genes (Moon *et al.*, 2000; Nishida *et al.*, 1998; Yoon *et al.*, 2004), finding transgene expression to be present or to exert a biological effect on biosynthesis, often for several weeks.

In addition to up-regulation of anabolic factors, inhibition of catabolic processes has been studied using gene therapy for IVD regeneration. Wallach and coworkers (2003) reported that gene transfer of the tissue inhibitor of metalloproteinase-1 (TIMP-1), an inhibitor of catabolic enzymes, can increase proteoglycan accumulation within pellet cultures of human IVD cells. LeMaitre recently reported that human disc cells infected with Ad-IL-1 receptor antagonist (Ad/IL-1Ra) were resistant to IL-1 (Le Maitre *et al.*, 2006). When *in vitro* infected cells were injected into disc explants *in vitro*, IL-1 receptor antagonist protein expression was also increased and maintained for the two-week time period investigated.

There are significant concerns about adenoviral vector use clinically, however, which may include significant toxicity when used in spinal applications (Driesse *et al.*, 2000; Wallach *et al.*, 2006). These concerns have led investigators to begin consideration of adeno-associated viral vectors (Lattermann *et al.*, 2005) and baculoviral vectors, the latter of which may be nontoxic (Liu *et al.*, 2006). Both approaches may provide safe alternatives for future disc therapies, although much work remains. Also, to avoid safety concerns found with viral gene transfer, several nonviral methods for direct gene transfer to cells have been proposed. Preliminary reports using microbubble-enhanced ultrasound gene therapy (Nishida *et al.*, 2006) and a "gene-gun" method (Chang *et al.*, 2000; Matsumoto *et al.*, 2001b) have shown that introduction of a marker or growth factor gene could be accomplished and provide sustained gene expression without the need for viral vectors. Transfection efficacy with nonviral means is lower, however, than that in viral transfection, and further investigation will be needed to apply these in a clinical setting. Nevertheless, both viral and nonviral transfection methods have pros and cons. Safety and immunological reactions as well as the control of expression in viral-mediated gene therapy are potential problems. The comparatively low immunologic exposure of the healthy disc and its low cellularity seem to suggest that safety with gene delivery of therapeutic agents is a lesser concern in the IVD. Cells are needed to transduce the biological effect, however, so transplantation of *ex vivo* transfected cells may be an important part of the future potential for gene therapy in the IVD. During this time, work continues on identification of broader and more diverse molecular targets that can be useful for gene delivery in the treatment of IVD regeneration.

## VIII. CONCLUDING REMARKS

Efforts to regenerate and replace the tissues of the intervertebral disc have virtually exploded over the last two decades, although the field remains in its infancy. The complexity of the diverse degenerative and pathological processes that affect the IVD as well as the intrinsic complexity of the heterogeneous disc structures demand that multiple strategies be developed for treatment of the IVD. Partial IVD replacements using acellular, preformed, or *in situ*–formed biomaterials have the longest history of development and are important for defining procedural outcomes that will be relevant to long-term functional success. Development of strategies using cells, biologics, or gene therapy is often focused on the restoration of a single tissue source, such as nucleus pulposus or anulus fibrosus, and with or without biomaterial scaffolds. Only a few tissue-engineering solutions have been proposed to integrate two dissimilar tissues in the repair process, and additional work to promote integration among native, neogenerated, and implanted tissues will be critical to restoring IVD function. Many of the identified strategies derive largely from knowledge gained in cartilage-tissue engineering, although the differing cellular, functional, and structural requirements of the IVD suggest that custom approaches are needed. Advances in IVD cell biology are needed to enable the identification of novel therapeutic targets, to select for classes of biomaterials, and to suggest appropriate drug delivery strategies, for disc cell phenotype, cell-biomaterial interactions, and the biology of aging for these cells are still poorly understood. While a diverse array of molecules, cell sources, and materials is suggested as appropriate for IVD regeneration, additional work is needed to reveal some common and unique themes in human IVD cell responses that focus research on IVD-specific strategies. Currently underway clinical trials of autologous cell therapies or autologous protein products will pave the way for later generations of cellular and biologic-based therapies, because they are expected to illustrate the unique challenges of treating the pathologic and aged human IVD. The next decade promises great advances in the translation of basic and applied sciences to the clinical treatment of IVD regeneration and replacement.

## IX. ACKNOWLEDGMENTS

Supported in part by grants from the NIH (P01AR048152, R01AR47442, R01EB00263). Editorial contributions of Ms. C. M. Flahiff are greatly appreciated.

## X. REFERENCES

Adams, M. A., McNally, D. S., and Dolan, P. (1996). "Stress" distributions inside intervertebral discs. The effects of age and degeneration. *J. Bone Joint Surg. Br.* **78**(6), 965–972.

Ahn, S. H., Cho, Y. W., Ahn, M. W., Jang, S. H., Sohn, Y. K., and Kim, H. S. (2002). mRNA expression of cytokines and chemokines in herniated lumbar intervertebral discs. *Spine* **27**(9), 911–917.

Akeda, K., An, H. S., Pichika, R., Attawia, M., Thonar, E., Lenz, M. E., Uchida, A., and Masuda, K. (2006). Platelet-rich plasma (PRP) stimulates the extracellular matrix metabolism of porcine nucleus pulposus and anulus fibrosus cells cultured in alginate beads. *Spine* **31**(9), 959–966.

An, H. S., Takegami, K., Kamada, H., Nguyen, C. M., Thonar, E. J., Singh, K., Andersson, G. B., and Masuda, K. (2005). Intradiscal administration of osteogenic protein-1 increases intervertebral disc height and proteoglycan content in the nucleus pulposus in normal adolescent rabbits. *Spine* **30**(1), 25–31.

Andersson, G. B. J. (1996). Intervertebral disk herniation: epidemiology and natural history. *In* "Low Back Pain: A Scientific and Clinical Overview" (J. N. Weinstein and S. L. Gordon, eds.), pp. 7–21. American Academy of Orthopaedic Surgeons, Rosemont, IL.

Andersson, G. B., Ortengren, R., and Nachemeson, A. (1982). Disc pressure measurements when rising and sitting down on a chair. *Eng. Med.* **11**(4), 189–190.

Antoniou, J., Goudsouzian, N. M., Heathfield, T. F., Winterbottom, N., Steffen, T., Poole, A. R., Aebi, M., and Alini, M. (1996). The human lumbar endplate. Evidence of changes in biosynthesis and denaturation of the extracellular matrix with growth, maturation, aging, and degeneration. *Spine* **21**(10), 1153–1161.

Au, A., Ha, J., Polotsky, A., Krzyminski, K., Gutowska, A., Hungerford, D. S., and Frondoza, C. G. (2003). Thermally reversible polymer gel for chondrocyte culture. *J. Biomed. Mater. Res.* **67**(4), 1310–1319.

Baer, A. E., Wang, J. T., Kraus, V. B., and Setton, L. A. (2001). Collagen gene expression and mechanical properties of intervertebral disc cell-alginate cultures. *J. Orthop. Res.* **19**(1), 2–10.

Bao, Q. B., and Yuan, H. A. (2002). New technologies in spine: nucleus replacement. *Spine* **27**, 1245–1247.

Bayliss, M. T., and Johnstone, B. (1992). Biochemistry of the intervertebral disc. *In* "The Lumbar Spine and Back Pain" (M. I. V. Jayson, ed.), pp. 111–131. Churchill Livingstone, New York.

Benneker, L. M., Heini, P. F., Alini, M., Anderson, S. E., and Ito, K. (2005). 2004 Young investigator award winner: vertebral endplate marrow contact channel occlusions and intervertebral disc degeneration. *Spine* **30**(2), 167–173.

Bertagnoli, R., Sabatino, C. T., Edwards, J. T., Gontarz, G. A., Prewett, A., and Parsons, J. R. (2005). Mechanical testing of a novel hydrogel nucleus replacement implant. *Spine J.* **5**(6), 672–681.

Betre, H., Setton, L. A., Meyer, D., and Chilkoti, A. (2002). Characterization of a genetically engineered elastin-like polypeptide for cartilaginous tissue repair. *Biomacromolecules* **3**, 910–916.

Bodguk, N. (1988). "Clinical Anatomy of the Lumbar Spine and Sacrum." Churchill Livingstone, Edinburgh, UK.

Boos, N., Weissbach, S., Rohrbach, H., Weiler, C., Spratt, K. F., and Nerlich, A. G. (2002). Classification of age-related changes in lumbar intervertebral discs: 2002 Volvo Award in basic science. *Spine* **27**(23), 2631–2644.

Boyd, L. M., Chen, J., Kraus, V. B., and Setton, L. A. (2004). Conditioned medium differentially regulates matrix protein gene expression in cells of the intervertebral disc. *Spine* **29**(20), 2217–2222.

Buckwalter, J. A. (1995). Aging and degeneration of the human intervertebral disc. *Spine* **20**(11), 1307–1314.

Burke, J. G., RW, G. W., Conhyea, D., McCormack, D., Dowling, F. E., Walsh, M. G., and Fitzpatrick, J. M. (2003). Human nucleus pulposus can respond to a pro-inflammatory stimulus. *Spine* **28**(24), 2685–2693.

Burkoth, A. K., and Anseth, K. S. (2000). A review of photocrosslinked polyanhydrides: *in situ* forming degradable networks. *Biomaterials* **21**(23), 2395–2404.

Carl, A., Ledet, E., Yuan, H., and Sharan, A. (2004). New developments in nucleus pulposus replacement technology. *Spine J.* **4**(6 Suppl.), 325S–329S.

Cassidy, J. J., Hiltner, A., and Baer, E. (1989). Hierarchical structure of the intervertebral disc. *Connect. Tissue Res.* **23**, 75–88.

Chang, S., Masuda, K., Takegami, K., Sumner, D., Thonar, E. J.-M. A., Andersson, G., and An, H. (2000). Gene gun–mediated gene transfer to intervertebral disc cells. *Trans. Orthop. Res. Soc.* **25**, 231.

Chujo, T., An, H., Akeda, K., Miyamoto, K., Muehleman, C., Attawia, M., Andersson, G., and Masuda, K. (2006). Effects of growth differentiation factor-5 (GDF-5) on the intervertebral disc — *in vitro* bovine study and *in vivo* rabbit disc degeneration model study. *Spine* **31**(25), 2909–2917.

Coventry, N. B., Ghormley, R. K., and Kernohan, J. W. (1945a). The intervertebral disc — its microscopic anatomy and pathology. Part I:

anatomy, development and physiology. *J. Bone Joint Surg.* **27A**, 105–112.

Coventry, N. B., Ghormley, R. K., and Kernohan, J. W. (1945b). The intervertebral disc — its microscopic anatomy and pathology. Part II: changes in the intervertebral disc concomittant with age. *J. Bone Joint Surg.* **27A**, 233–247.

Crevensten, G., Walsh, A. J. L., Ananthakrishnan, D., Page, P., Wahba, G. M., Lotz, J. C., and Berven, S. (2004). Intervertebral disc cell therapy for regeneration: mesenchymal stem cell implantation in rat intervertebral discs. *Ann. Biomed. Eng.* **32**(3), 430–434.

DeFrances, C. J., and Podgornik, M. N. (2006). 2004 national hospital discharge survey. *In* "Advance Data from Vital and Health Statistics," May 4. National Center for Health Statistics, Centers for Disease Control.

Di Martino, A., Vaccaro, A. R., Lee, J. Y., Denaro, V., and Lim, M. R. (2005). Nucleus pulposus replacement: basic science and indications for clinical use. *Spine* **30**(16 Suppl.), S16–S22.

Driesse, M. J., Esandi, M. C., Kros, J. M., Avezaat, C. J., Vecht, C., Zurcher, C., van der Velde, I., Valerio, D., Bout, A., and Sillevis Smitt, P. A. (2000). Intra-CSF administered recombinant adenovirus causes an immune response–mediated toxicity. *Gene Ther.* **7**(16), 1401–1409.

Ebara, S., Iatridis, J. C., Setton, L. A., Mow, V. C., and Weidenbaum, M. (1996). The tensile properties of human lumbar anulus fibrosus and their variations with region in the intervertebral disc. *Spine* **21**, 452–461.

Elisseeff, J. (2004). Injectable cartilage tissue engineering. *Expert Opin. Biol. Ther.* **4**(12), 1849–1859.

Elliott, D. M., and Setton, L. A. (2001). Anisotropic and inhomogeneous tensile behavior of the human anulus fibrosus: experimental measurement and material model predictions. *J. Biomech. Eng.* **123**(3), 256–263.

Evans, C. (2006). Potential biologic therapies for the intervertebral disc. *J. Bone Joint Surg. Am.* **88**(Suppl. 2), 95–98.

Fujita, Y., Duncan, N. A., and Lotz, J. C. (1997). Radial tensile properties of the lumbar anulus fibrosus are site and degeneration dependent. *J. Orthop. Res.* **15**(6), 814–819.

Galante, J. O. (1967). Tensile properties of the human lumbar anulus fibrosus. *Acta Orthop. Scand.* **Suppl. 100**, 1–91.

Ganey, T., Libera, J., Moos, V., Alasevic, O., Fritsch, K. G., Meisel, H. J., and Hutton, W. C. (2003). Disc chondrocyte transplantation in a canine model: a treatment for degenerated or damaged intervertebral disc. *Spine* **28**(23), 2609–2620.

Gruber, H. E., Fisher, E. C., Jr., Desai, B., Stasky, A. A., Hoelscher, G., and Hanley, E. N., Jr. (1997). Human intervertebral disc cells from the anulus: three-dimensional culture in agarose or alginate and responsiveness to TGF-beta1. *Exp. Cell Res.* **235**(1), 13–21.

Gruber, H. E., Norton, H. J., and Hanley, E. N., Jr. (2000). Anti-apoptotic effects of IGF-1 and PDGF on human intervertebral disc cells *in vitro*. *Spine* **25**(17), 2153–2157.

Gruber, H. E., Johnson, T. L., Leslie, K., Ingram, J. A., Martin, D., Hoelscher, G., Banks, D., Phieffer, L., Coldham, G., and Hanley, E. N., Jr. (2002). Autologous intervertebral disc cell implantation: a model using *Psammomys obesus*, the sand rat. *Spine* **27**(15), 1626–1633.

Gruber, H. E., Leslie, K., Ingram, J., Norton, H. J., and Hanley, E. N. (2004). Cell-based tissue engineering for the intervertebral disc: *in vitro* studies of human disc cell gene expression and matrix production within selected cell carriers. *Spine J.* **4**(1), 44–55.

Hamilton, D. J., Seguin, C. A., Wang, J., Pilliar, R. M., and Kandel, R. A. (2006). Formation of a nucleus pulposus-cartilage endplate construct *in vitro. Biomaterials* **27**(3), 397–405.

Hayes, A. J., Benjamin, M., and Ralphs, J. R. (2001). Extracellular matrix in development of the intervertebral disc. *Matrix Biol.* **20**(2), 107–121.

Hickey, D. S., and Hukins, D. W. L. (1980). X-ray diffraction studies of the arrangement of collagenous fibers in human fetal intervertebral disc. *J. Anatomy* **131**, 81–90.

Holm, S., Maroudas, A., Urban, J. P., Selstam, G., and Nachemson, A. (1981). Nutrition of the intervertebral disc: solute transport and metabolism. *Connect. Tissue Res.* **8**, 101–119.

Holzapfel, G. A., Schulze-Bauer, C. A., Feigl, G., and Regitnig, P. (2005). Single lamellar mechanics of the human lumbar anulus fibrosus. *Biomech. Model Mechanobiol.* **3**(3), 125–140.

Huang, R. C., Wright, T. M., Panjabi, M. M., and Lipman, J. D. (2005). Biomechanics of nonfusion implants. *Orthop. Clin. North Am.* **36**(3), 271–280.

Hurri, H., and Karppinen, J. (2004). Discogenic pain. *Pain* **112**, 225–228.

Igarashi, T., Kikuchi, S., Shubayev, V., and Myers, R. R. (2000). 2000 Volvo Award winner in basic science studies: exogenous tumor necrosis factor-alpha mimics nucleus pulposus-induced neuropathology. Molecular, histologic, and behavioral comparisons in rats. *Spine* **25**(23), 2975–2980.

Imai, Y., An, H., Thonar, E., Andersson, G., and Masuda , K. (2002). Recombinant human osteogenic protein-1 (rhOP-1) up-regulates extracellular matrix metabolism by human anulus fibrosus and nucleus pulposus cells. *Spine J.* **2**, 101S.

Iwashina, T., Mochida, J., Sakai, D., Yamamoto, Y., Miyazaki, T., Ando, K., and Hotta, T. (2006). Feasibility of using a human nucleus pulposus cell line as a cell source in cell transplantation therapy for intervertebral disc degeneration. *Spine* **31**(11), 1177–1186.

Kang, J. D., Georgescu, H. I., McIntyre-Larkin, L., Stefanovic-Racic, M., Donaldson, W. F., 3rd, and Evans, C. H. (1996). Herniated lumbar intervertebral discs spontaneously produce matrix metalloproteinases, nitric oxide, interleukin-6, and prostaglandin E2. *Spine* **21**(3), 271–277.

Kim, D. J., Moon, S. H., Kim, H., Kwon, U. H., Park, M. S., Han, K. J., Hahn, S. B., and Lee, H. M. (2003). Bone morphogenetic protein-2 facilitates expression of chondrogenic, not osteogenic, phenotype of human intervertebral disc cells. *Spine* **28**(24), 2679–2684.

Klara, P. M., and Ray, C. D. (2002). Artificial nucleus replacement: clinical experience. *Spine* **27**(12), 1374–1377.

Kokkonen, S.-M., Kurunlahti, M., Tervonen, O., Iikko, E., and Vanharanta, H. (2002). Endplate degeneration observed on magnetic resonance imaging of the lumbar spine: correlation with pain provocation and disc changes observed on computed tomography diskography. *Spine* **27**(20), 2274–2278.

Lattermann, C., Oxner, W. M., Xiao, X., Li, J., Gilbertson, L. G., Robbins, P. D., and Kang, J. D. (2005). The adeno-associated viral vector as a strategy for intradiscal gene transfer in immune competent and pre-exposed rabbits. *Spine* **30**(5), 497–504.

Le Maitre, C. L., Freemont, A. J., and Hoyland, J. A. (2004). Localization of degradative enzymes and their inhibitors in the degenerate human intervertebral disc. *J. Pathol.* **204**(1), 47–54.

Le Maitre, C. L., Freemont, A. J., and Hoyland, J. A. (2005). The role of interleukin-1 in the pathogenesis of human intervertebral disc degeneration. *Arthritis Res. Ther.* **7**(4), R732–R745.

Le Maitre, C. L., Freemont, A. J., and Hoyland, J. A. (2006). A preliminary *in vitro* study into the use of IL-1Ra gene therapy for the inhibition of intervertebral disc degeneration. *Int. J. Exp. Pathol.* **87**(1), 17–28.

Leung, V. Y., Chan, D., and Cheung, K. M. (2006). Regeneration of intervertebral disc by mesenchymal stem cells: potentials, limitations, and future direction. *Eur. Spine J.* (in print).

Li, X., Leo, B. M., Beck, G., Balian, G., and Anderson, G. D. (2004). Collagen and proteoglycan abnormalities in the GDF-5-deficient mice and molecular changes when treating disk cells with recombinant growth factor. *Spine* **29**(20), 2229–2234.

Li, X. D., Lee, J. P., Balian, G., and Anderson, D. G. (2005). Modulation of chondrocytic properties of fat-derived mesenchymal cells in cocultures with nucleus pulposus. *Connect. Tissue Res.* **46**(2), 75–82.

Liu, X., Li, K., Song, J., Liang, C., Wang, X., and Chen, X. (2006). Efficient and stable gene expression in rabbit intervertebral disc cells transduced with a recombinant baculovirus vector. *Spine* **31**(7), 732–735.

Lotz, J. C., and Kim, A. J. (2005). Disc regeneration: why, when, and how. *Neurosurg. Clin. N. Am.* **16**(4), 657–663, vii.

Maroudas, A., Stockwell, R. A., Nachemson, A., and Urban, J. (1975). Factors involved in the nutrition of the human lumbar intervertebral disc: cellularity and diffusion of glucose *in vitro. J. Anat.* **120**, 113–130.

Masuda, K., and An, H. S. (2004). Growth factors and the intervertebral disc. *Spine J.* **4**(6 Suppl.), 330S–340S.

Masuda, K., Takegami, K., An, H., Kumano, F., Chiba, K., Andersson, G. B., Schmid, T., and Thonar, E. (2003). Recombinant osteogenic protein-1 upregulates extracellular matrix metabolism by rabbit anulus fibrosus and nucleus pulposus cells cultured in alginate beads. *J. Orthop. Res.* **21**(5), 922–930.

Masuda, K., Imai, Y., Okuma, M., Muehleman, C., Nakagawa, K., Akeda, K., Thonar, E., Andersson, G., and An, H. S. (2006). Osteogenic protein-1 injection into a degenerated disc induces the restoration of disc height and structural changes in the rabbit anular puncture model. *Spine* **31**(7), 742–754.

Matsumoto, T., Masuda, K., An, H., Chen, S., Williamson, A., Sah, R., Andersson, G. B. J., Rueger, D., and Thonar, E. J.-M. A. (2001a). Formation of transplantable disc-shaped tissues by nucleus pulposus and anulus fibrosus cells: biochemical and biomechanical properties. *Trans. Orthop. Res. Soc.* **26**, 897.

Matsumoto, T., Masuda, K., Chen, S., An, H., Andersson, G. B. J., Aota, Y., Horvath, E., and Thonar, E. J.-M. A. (2001b). Transfer of osteogenic protein-1 gene by gene gun system promotes matrix synthesis in bovine intervertebral disc and articular cartilage cells. *Trans. Orthop. Res. Soc.* **26**, 30.

McAfee, P. C. (2004). The indications for lumbar and cervical disc replacement. *Spine J.* **4**, 177S–181S.

McNally, D. S., and Adams, M. A. (1992). Internal intervertebral disc mechanics as revealed by stress profilometry. *Spine* **17**, 66–73.

Meachim, G., and Cornah, M. S. (1970). Fine structure of juvenile human nucleus pulposus. *J. Anat.* **107**(2), 337–350.

Meisel, H. J., Ganey, T., Hutton, W. C., Libera, J., Minkus, Y., and Alasevic, O. (2006). Clinical experience in cell-based therapeutics: intervention and outcome. *Eur. Spine J.* (in print).

Mercier, N. R., Costantino, H. R., Tracy, M. A., and Bonassar, L. J. (2004). A novel injectable approach for cartilage formation *in vivo* using PLG microspheres. *Ann. Biomed. Eng.* **32**(3), 418–429.

Miyamoto, K., Masuda, K., Kim, J., Inoue, N., Akeda, K., Andersson, G., and An, H. (2006). Intradiscal injections of osteogenic protein-1 restore the viscoelastic properties of degenerated intervertebral discs. *Spine J.* **6**(6), 692–703.

Mizuno, H., Roy, A. K., Vacanti, C. A., Kojima, K., Ueda, M., and Bonassar, L. J. (2004). Tissue-engineered composites of anulus fibrosus and nucleus pulposus for intervertebral disc replacement. *Spine* **29**(12), 1290–1297.

Mizuno, H., Roy, A. K., Zaporojan, V., Vacanti, C. A., Ueda, M., and Bonassar, L. J. (2006). Biomechanical and biochemical characterization of composite tissue-engineered intervertebral discs. *Biomaterials* **27**(3), 362–370.

Modic, M. T., Weinstein, M. A., Pavlicek, W., Dengel, F., Paushter, D., Boumphrey, F., Duchesneau, P. M., Ngo, F. Q. H., and Meaney, T. F. (1984). Magnetic resonance of intervertebral-disk disease. *Magn. Reson. Med.* **1**(2), 207–208.

Moon, S. H., Gilbertson, L. G., Nishida, K., Knaub, M., Muzzonigro, T., Robbins, P. D., Evans, C. H., and Kang, J. D. (2000). Human intervertebral disc cells are genetically modifiable by adenovirus-mediated gene transfer: implications for the clinical management of intervertebral disc disorders. *Spine* **25**(20), 2573–2579.

Mwale, F., Iordanova, M., Demers, C. N., Steffen, T., Roughley, P., and Antoniou, J. (2005). Biological evaluation of chitosan salts cross-linked to genipin as a cell scaffold for disk tissue engineering. *Tissue Eng.* **11**(1–2), 130–140.

Nachemson, A. (1960). Lumbar intradiscal pressure. Experimental studies on postmortem material. *Acta Orthop. Scand. Suppl.* **43**, 1–103.

Nachemson, A. (1992). Lumbar mechanics as revealed by lumbar intradiscal pressure measurements. *In* "The Lumbar Spine and Back Pain" (M. I. V. Jayson, ed.), 4th ed., pp. 157–171. Churchill Livingstone, New York.

Nachemson, A., Lewin, T., Maroudas, A., and Freeman, M. A. (1970). *In vitro* diffusion of dye through the endplates and the anulus fibrosus of human lumbar intervertebral discs. *Acta Orthop. Scand.* **41**(6), 589–607.

Nishida, K., Kang, J. D., Suh, J. K., Robbins, P. D., Evans, C. H., and Gilbertson, L. G. (1998). Adenovirus-mediated gene transfer to nucleus pulposus cells. Implications for the treatment of intervertebral disc degeneration. *Spine* **23**(22), 2437–2442.

Nishida, K., Kang, J. D., Gilbertson, L. G., Moon, S. H., Suh, J. K., Vogt, M. T., Robbins, P. D., and Evans, C. H. (1999). Modulation of the biologic activity of the rabbit intervertebral disc by gene therapy: an *in vivo* study of adenovirus-mediated transfer of the human transforming growth factor-beta 1 encoding gene. *Spine* **24**(23), 2419–2425.

Nishida, K., Doita, M., Takada, T., Kakutani, K., Miyamoto, H., Shimomura, T., Maeno, K., and Kurosaka, M. (2006). Sustained transgene expression in intervertebral disc cells *in vivo* mediated by microbubble-enhanced ultrasound gene therapy. *Spine* **31**(13), 1415–1419.

Nishimura, K., and Mochida, J. (1998). Percutaneous reinsertion of the nucleus pulposus. An experimental study. *Spine* **23**, 1531–1538.

Nomura, T., Mochida, J., Okuma, M., Nishimura, K., and Sakabe, K. (2001). Nucleus pulposus allograft retards intervertebral disc degeneration. *Clin. Orthop.* **389**, 94–101.

Oegema, T. R. (1990). Recent advances in understanding the biochemistry of the IVD. *In* "Disorders of the Lumbar Spine (Y. Floman, ed.), pp. 53–71. Aspen, Rockville, MD.

Oegema, T. R. J. (1993). Biochemistry of the intervertebral disc. *Clin. Sports Med.* **12**(3), 419–439.

Oegema, T. R. J. (2002). The role of disc cell heterogeneity in determining disc biochemistry: a speculation. *Biochem. Soc. Trans.* **30**(6), 839–844.

Okuma, M., Mochida, J., Nishimura, K., Sakabe, K., and Seiki, K. (2000). Reinsertion of stimulated nucleus pulposus cells retards intervertebral disc degeneration: an *in vitro* and *in vivo* experimental study. *J. Orthop. Res.* **18**(6), 988–997.

Olmarker, K., and Larsson, K. (1998). Tumor necrosis factor alpha and nucleus-pulposus-induced nerve root injury. *Spine* **23**(23), 2538–2544.

Osada, R., Ohshima, H., Ishihara, H., Yudoh, K., Sakai, K., Matsui, H., and Tsuji, H. (1996). Autocrine/paracrine mechanism of insulin-like growth factor-1 secretion, and the effect of insulin-like growth factor-1 on proteoglycan synthesis in bovine intervertebral discs. *J. Orthop. Res.* **14**(5), 690–699.

Panjabi, M., Brown, M., Lindahl, S., Irstam, L., and Hermens, M. (1988). Intrinsic disc pressure as a measure of integrity of the lumbar spine. *Spine* **13**(8), 913–917.

Paul, R., Haydon, R. C., Cheng, H., Ishikawa, A., Nenadovich, N., Jiang, W., Zhou, L., Breyer, B., Feng, T., Gupta, P., *et al.* (2003). Potential use of sox9 gene therapy for intervertebral degenerative disc disease. *Spine* **28**, 755–763.

Peretti, G. M., Xu, J. W., Bonassar, L. J., Kirchhoff, C. H., Yaremchuk, M. J., and Randolph, M. A. (2006). Review of injectable cartilage engineering using fibrin gel in mice and swine models. *Tissue Eng.* **12**(5), 1151–1168.

Pichika, R., Betre, H., Setton, L.A., An, H. S., and Masuda, K. (2006). Bioactivity of IL–1 receptor antagonist (IL1Ra) and a related fusion protein in inhibiting IL-1–mediated effects in human intervertebral disc cells. *Trans. Orthop. Res. Soc.* **52**,

Postacchini, F., Bellocci, M., and Massobrio, M. (1984). Morphologic changes in anulus fibrosus during aging: an ultrastructural study in rats. *Spine* **9**(6), 596–603.

Praemer, A., Furner, S., and Rice, D. P. (1999). Back and other joint pain. *In* "Musculoskeletal Conditions in the United States," pp. 27–33. American Academy of Orthopaedic Surgeons, Rosemont, IL.

Puelacher, W. C., Kim, S. W., Vacanti, J. P., Schloo, B., Mooney, D., and Vacanti, C. A. (1994). Tissue-engineered growth of cartilage — the effect of varying the concentration of chondrocytes seeded onto synthetic-polymer matrices. *Int. J. Oral Maxillofac. Surg.* **23**(1), 49–53.

Richardson, S. M., Curran, J. M., Chen, R., Vaughan-Thomas, A., Hunt, J. A., Freemont, A. J., and Hoyland, J. A. (2006a). The differentiation of bone marrow mesenchymal stem cells into chondrocyte-like cells on poly-L-lactic acid (PLLA) scaffolds. *Biomaterials* **27**(22), 4069–4078.

Richardson, S. M., Walker, R. V., Parker, S., Rhodes, N. P., Hunt, J. A., Freemont, A. J., and Hoyland, J. A. (2006b). Intervertebral disc cell-mediated mesenchymal stem cell differentiation. *Stem Cells* **24**(3), 707–716.

Risbud, M. V., Albert, T. J., Guttapalli, A., Vresilovic, E. J., Hillibrand, A. S., Vaccaro, A. R., and Shapiro, I. M. (2004). Differentiation of mesenchymal stem cells towards a nucleus pulposus-like phenotype *in vitro*: implications for cell-based transplantation therapy. *Spine* **29**(23), 2627–2632.

Roberts, S., Menage, J., Duance, V., Wotton, S., and Ayad, S. (1991). 1991 Volvo Award in basic sciences. Collagen types around the cells of the intervertebral disc and cartilage end plate: an immunolocalization study. *Spine* **16**(9), 1030–1038.

Roberts, S., Urban, J. P., Evans, H., and Eisenstein, S. M. (1996). Transport properties of the human cartilage endplate in relation to its composition and calcification. *Spine* **21**(4), 415–420.

Roberts, S., Caterson, B., Menage, J., Evans, E. H., Jaffray, D. C., and Eisenstein, S. M. (2000). Matrix metalloproteinases and aggrecanase: their role in disorders of the human intervertebral disc. *Spine* 25(23), 3005–3013.

Roughley, P. J. (2004). Biology of intervertebral disc aging and degeneration: involvement of the extracellular matrix. *Spine* 29(23), 2691–2699.

Roughley, P., Hoemann, C., DesRosiers, E., Mwale, F., Antoniou, J., and Alini, M. (2006). The potential of chitosan-based gels containing intervertebral disc cells for nucleus pulposus supplementation. *Biomaterials* 27(3), 388–396.

Rufai, A., Benjamin, M., and Ralphs, J. R. (1995). The development of fibrocartilage in the rat intervertebral disc. *Anat. Embryol.* **192**, 53–62.

Sakai, D., Mochida, J., Yamamoto, Y., Nomura, T., Okuma, M., Nishimura, K., Nakai, T., Ando, K., and Hotta, T. (2003). Transplantation of mesenchymal stem cells embedded in Atelocollagen gel to the intervertebral disc: a potential therapeutic model for disc degeneration. *Biomaterials* 24(20), 3531–3541.

Sakai, D., Mochida, J., Yamamoto, Y., Toh, E., Iwashina, T., Miyazaki, T., Inokuchi, S., Ando, K., and Hotta, T. (2004). Immortalization of human nucleus pulposus cells by a recombinant SV40 adenovirus vector: establishment of a novel cell line for the study of human nucleus pulposus cells. *Spine* 29(14), 1515–1523.

Sakai, D., Mochida, J., Iwashina, T., Watanabe, T., Nakai, T., Ando, K., and Hotta, T. (2005). Differentiation of mesenchymal stem cells transplanted to a rabbit degenerative disc model: potential and limitations for stem cell therapy in disc regeneration. *Spine* 30(21), 2379–2387.

Sato, M., Asazuma, T., Ishihara, M., Ishihara, M., Kikuchi, T., Kikuchi, M., and Fujikawa, K. (2003). An experimental study of the regeneration of the intervertebral disc with an allograft of cultured anulus fibrosus cells using a tissue-engineering method. *Spine* 28(6), 548–553.

Schultz, A., Andersson, G., Ortengren, R., Haderspeck, K., and Nachemson, A. (1982). Loads on the lumbar spine. Validation of a biomechanical analysis by measurements of intradiscal pressures and myoelectric signals. *J. Bone Joint Surg. Am.* **64**(5), 713–720.

Seguin, C. A., Grynpas, M. D., Pilliar, R. M., Waldman, S. D., and Kandel, R. A. (2004). Tissue–engineered nucleus pulposus tissue formed on a porous calcium polyphosphate substrate. *Spine* 29(12), 1299–1306.

Seguin, C. A., Pilliar, R. M., Roughley, P. J., and Kandel, R. A. (2005). Tumor necrosis factor-alpha modulates matrix production and catabolism in nucleus pulposus tissue. *Spine* 30(17), 1940–1948.

Selard, E., Shirazi-Adl, A., and Urban, J. P. (2003). Finite-element study of nutrient diffusion in the human intervertebral disc. *Spine* **28**(17), 1945–1953.

Shirazi-Adl, A. (1992). Finite-element simulation of changes in the fluid content of human lumbar discs. Mechanical and clinical implications. *Spine* 17, 206–212.

Skaggs, D. L., Weidenbaum, M., Iatridis, J. C., Ratcliffe, A., and Mow, V. C. (1994). Regional variation in tensile properties and biochemical composition of the human lumbar anulus fibrosus. *Spine* **19**, 1310–1319.

Sontjens, S. H., Nettles, D. L., Carnahan, M. A., Setton, L. A., and Grinstaff, M. W. (2006). Biodendrimer-based hydrogel scaffolds for cartilage tissue repair. *Biomacromolecules* 7(1), 310–316.

Stairmand, J. W., Holm, S., and Urban, J. P. (1991). Factors influencing oxygen concentration gradients in the intervertebral disc. A theoretical analysis. *Spine* 16, 444–449.

Steck, E., Bertram, H., Abel, R., Chen, B., Winter, A., and Richter, W. (2005). Induction of intervertebral disc–like cells from adult mesenchymal stem cells. *Stem Cells* 23(3), 403–411.

Sztrolovics, R., Alini, M., Roughley, P. J., and Mort, J. S. (1997). Aggrecan degradation in human intervertebral disc and articular cartilage. *Biochem. J.* **326**, 235–241.

Taylor, J. R., and Twomey, L. T. (1988). The development of the human intervertebral disc. *In* "The Biology of the Intervertebral Disc" (P. Ghosh, ed.), pp. 39–82. CRC Press, Boca Raton, FL.

Temenoff, J. S., and Mikos, A. G. (2000). Injectable biodegradable materials in orthopaedic tissue engineering. *Biomaterials* 21, 2405–2412.

Thomas, J., Lowman, A., and Marcolongo, M. (2003). Novel associated hydrogels for nucleus pulposus replacement. *J. Biomed. Mater. Res.* 67(4), 1329–1337.

Thompson, J. P., Oegema, T. J., and Bradford, D. S. (1991). Stimulation of mature canine intervertebral disc by growth factors. *Spine* **16**(3), 253–260.

Tim Yoon, S., Su Kim, K., Li, J., Soo Park, J., Akamaru, T., Elmer, W. A., and Hutton, W. C. (2003). The effect of bone morphogenetic protein-2 on rat intervertebral disc cells *in vitro*. *Spine* **28**(16), 1773–1780.

Trout, J. J., Buckwalter, J. A., Moore, K. C., and Landas, S. K. (1982). Ultrastructure of the human intervertebral disc: I. Changes in notochordal cells with age. *Tissue Cell* **14**, 359–369.

Urban, J. P., and McMullin, J. F. (1985). Swelling pressure of the intervertebral disc: influence of proteoglycan and collagen contents. *Biorheology* **22**(2), 145–157.

Urban, J. P., and McMullin, J. F. (1988). Swelling pressure of the lumbar intervertebral discs: influence of age, spinal level, composition, and degeneration. *Spine* **13**(2), 179–187.

Urban, J. P., and Roberts, S. (1995). Development and degeneration of the intervertebral discs. *Mol. Med. Today* **1**(7), 329–335.

Urban, J. P., and Roberts, S. (2003). Degeneration of the intervertebral disc. *Arthritis Res. Ther.* **5**(3), 120–130.

Urban, J. P., Holm, S., Maroudas, A., and Nachemson, A. (1977). Nutrition of the intervertebral disk. An *in vivo* study of solute transport. *Clin. Orthop.* **129**, 101–114.

Wallach, C. J., Sobajima, S., Watanabe, Y., Kim, J. S., Georgescu, H. I., Robbins, P., Gilbertson, L. G., and Kang, J. D. (2003). Gene transfer of the catabolic inhibitor TIMP-1 increases measured proteoglycans in cells from degenerated human intervertebral discs. *Spine* **28**(20), 2331–2337.

Wallach, C. J., Kim, J. S., Sobajima, S., Lattermann, C., Oxner, W. M., McFadden, K., Robbins, P. D., Gilbertson, L. G., and Kang, J. D. (2006). Safety assessment of intradiscal gene transfer: a pilot study. *Spine J.* 6(2), 107–112.

Walsh, A. L., Kleinstueek, F., Lotz, J., and Bradford, D. (2001). Growth factor treatment in degenerated intervertebral discs. *Trans. Orthop. Res. Soc.* **26**, 0892.

Walsh, A. L., Lotz, J., and Bradford, D. (2002). Single and multiple injections of GDF-5, IGF-1, or TGF-beta into degenerated intervertebral discs. *Trans. Orthop. Res. Soc.* **27**, 0820.

Wang, H., Kroeber, M., Hanke, M., Ries, R., Schmid, C., Poller, W., and Richter, W. (2004). Release of active and depot GDF 5 after adenovirus-mediated overexpression stimulates rabbit and human intervertebral disc cells. *J. Mol. Med.* **82**(2), 126–134.

Wehling, P., Schulitz, K. P., Robbins, P. D., Evans, C. H., and Reinecke, J. A. (1997). Transfer of genes to chondrocytic cells of the lumbar spine. Proposal for a treatment strategy of spinal disorders by local gene therapy. *Spine* **22**(10), 1092–1097.

Weiler, C., Nerlich, A. G., Bachmeier, B. E., and Boos, N. (2005). Expression and distribution of tumor necrosis factor alpha in human lumbar intervertebral discs: a study in surgical specimen and autopsy controls. *Spine* **30**(1), 44–53.

Weishaupt, D., Zanetti, M., Hodler, J., Min, K., Fuchs, B., Pfirrmann, C. W. A., and Boos, N. (2001). Painful lumbar disc derangement: relevance of endplate abnormalities at MR imaging. *Radiology* **218**(2), 420–427.

Wilson, C. G., Bonassar, L. J., and Kohles, S. S. (2002). Modeling the dynamic composition of engineererd cartilage. *Arch. Biochem. Biophys.* **408**, 246–254.

Woolf, A. D., and Pfleger, B. (2003). Burden of major musculoskeletal conditions. *Bull. World Health Organ.* **81**, 646–656.

Yamamoto, Y., Mochida, J., Sakai, D., Nakai, T., Nishimura, K., Kawada, H., and Hotta, T. (2004). Up-regulation of the viability of nucleus pulposus cells by bone marrow–derived stromal cells: significance of direct cell-to-cell contact in coculture system. *Spine* **29**(14), 1508–1514.

Yoon, S. T. (2005). Molecular therapy of the intervertebral disc. *Spine J.* **5**(6 Suppl.), 280S–286S.

Yoon, S. T., Park, J. S., Kim, K. S., Li, J., Attallah-Wasif, E. S., Hutton, W. C., and Boden, S. D. (2004). ISSLS prize winner: LMP-1 upregulates intervertebral disc cell production of proteoglycans and BMPs *in vitro* and *in vivo*. *Spine* **29**(23), 2603–2611.

Yu, J., Winlove, P. C., Roberts, S., and Urban, J. P. (2002). Elastic fiber organization in the intervertebral discs of the bovine tail. *J. Anat.* **201**(6), 465–475.

Yu, S., Haughton, V. M., Sether, L. A., Ho, K. C., and Wagner, M. (1989). Criteria for classifying normal and degenerated lumbar intervertebral disks. *Radiology* **170**, 523–526.

Yung Lee, J., Hall, R., Pelinkovic, D., Cassinelli, E., Usas, A., Gilbertson, L., Huard, J., and Kang, J. (2001). New use of a three-dimensional pellet culture system for human intervertebral disc cells: initial characterization and potential use for tissue engineering. *Spine* **26**(21), 2316–2322.

# Chapter Fifty-Nine

# Articular Cartilage Injury

*J. A. Buckwalter, J. L. Marsh, T. Brown, A. Amendola, and J. A. Martin*

## I. INTRODUCTION

In an effort to decrease the risk of posttraumatic osteoarthritis, surgeons and scientists have been seeking ways to prevent progressive joint degeneration following injury. Surgeons have used operative treatments, including penetrating subchondral bone, soft-tissue grafts, and cell transplants, to stimulate restoration of damaged articular surfaces, with variable results. This chapter covers recent work suggesting that several biologic agents, including capsase inhibitors and antioxidants, may minimize mechanical damage to chondrocytes and experimental studies indicating that use of artificial matrices, growth factors, and immature chondrocytes or stem cells may promote formation of a new joint surface.

Mechanical loading of articular surfaces that exceeds the tolerance of those surfaces damages chondrocytes and their matrix, damage that can cause joint degeneration leading to the clinical syndrome of posttraumatic osteoarthritis. The risk of posttraumatic osteoarthritis depends on the type and severity of the injury and also on the repair and remodeling of the damaged articular surfaces. Three classes of joint injuries can be identified, based on the type of articular surface damage: (1) chondral damage and in some cases subchondral bone damage that does not cause visible disruption of the articular cartilage, (2) mechanical disruption of the joint surface limited to articular cartilage (chondral ruptures or tears), and (3) mechanical disruption of articular cartilage and subchondral bone (articular surface fractures). In most instances, chondrocytes can repair damage that does not disrupt the articular surface if they are protected from further injury. Mechanical disruption of articular cartilage stimulates chondrocyte synthetic activity, but this response, with few if any exceptions, fails to repair the tissue damage. Disruption of articular cartilage and subchondral bone stimulates chondral and bony repair. The osteochondral repair response usually heals the bony injury, but the chondral repair tissue does not duplicate the properties of normal articular cartilage.

Normal pain-free movement depends on the unique properties of the articular cartilage that forms the bearing surfaces of synovial joints (Buckwalter and Mankin, 1997). Degeneration of this remarkable tissue causes osteoarthritis: joint pain and dysfunction that limits mobility (Buckwalter *et al.*, 2000, 2005). The mechanisms, frequency, and natural history of articular-surface injuries are poorly understood, but it is clear that these injuries can lead to posttraumatic osteoarthritis (Buckwalter, 1992, 1995, 2002; Buckwalter and Mankin, 1997). Limited awareness of chondral and osteochondral injuries and difficulty in diagnosing these injuries makes it impossible to determine accurately their incidence or their relationship to the development of joint degeneration (Buckwalter and Lane, 1996, 1997; Buckwalter and Mankin, 1997). However, arthroscopic examinations of injured knee joints suggest

*Principles of Tissue Engineering, 3rd Edition*
*ed. by Lanza, Langer, and Vacanti*

that closed articular-surface injuries occur frequently (Noyes *et al.*, 1980; Noyes and Stabler, 1989). One group of surgeons examined arthroscopically 85 knees with traumatic hemarthrosis but absent or negligible ligamentous instability (Noyes *et al.*, 1980). Twenty percent of these knees had articular-surface defects. In many patients cartilage injuries occur in association with injuries to other joint tissues, including menisci, ligaments, joint capsule, and synovium. In these people the cartilage injury may be overlooked; even when it is identified it is difficult to distinguish the effects of the cartilage injury from the effects of the injuries to the other tissues. Damage to articular surfaces that does not result in visible disruption of articular cartilage or subchondral bone is not easily detected, although it probably occurs far more frequently than chondral and osteochondral fractures (Buckwalter, 1992, 2002).

Recent advances in methods of diagnosing articular-surface injuries, including arthroscopy and magnetic resonance imaging (Spindler *et al.*, 1993; D. L. Johnson *et al.*, 1998b; Rubin *et al.*, 2000), combined with reports of new methods of stimulating cartilage repair or regeneration and osteochondral transplantation have increased interest in these injuries. Clinical evaluation of patients with articular-surface damage and determining the appropriate role of these new treatments or the need for any treatment requires an understanding of the mechanisms of these injuries and their natural history. This chapter discusses the relationship between articular-surface injury and joint degeneration, mechanisms of articular-surface injuries, the responses of articular surfaces to injury, and approaches to preventing joint degeneration following joint injury.

## II. ARTICULAR-CARTILAGE INJURY AND JOINT DEGENERATION

The end stage of posttraumatic osteoarthritis, the osteoarthritis that follows joint injury, is identical to that of primary osteoarthritis, but patients with posttraumatic osteoarthritis are often young or middle-aged adults and have a well-defined precipitating insult (Gelber *et al.*, 2000; Buckwalter and Brown, 2004). Clinical experience and epidemiologic studies show that meniscal, ligament, and joint-capsule tears, joint dislocations, and intraarticular fractures increase the risk of the progressive joint degeneration that causes posttraumatic osteoarthritis (Gelber *et al.*, 2000; Marsh *et al.*, 2002; Nelson *et al.*, 2006). Participation in sports that expose joints to high levels of impact or torsional loading also increases the risk of joint degeneration (Buckwalter and Lane, 1997; Buckwalter and Brown, 2004).

The risk of osteoarthritis following joint injury varies with the type of injury: Meniscal and ligamentous injuries have a lower risk than intraarticular fractures. Gelber *et al.* (2000) found that 13.9% of those who had a knee injury (including meniscal, ligamentous, or bone injuries) during adolescence or young adulthood developed knee osteoarthritis, as compared with 6.0% of those who did not have a

knee injury. A study of patients who suffered ligamentous and meniscal injuries of the knee reported that they had a tenfold increased risk of ostoearthritis as compared with patients who do not have joint injury (Gillquist and Messner, 1999). Intraarticular fractures have the greatest risk of osteoarthritis. Depending on the severity of the injury and on the joint, the risk ranges from about 25% to more than 50% of patients (Marsh *et al.*, 2002; Buckwalter and Brown, 2004).

The time interval between joint injury and the development of osteoarthritis varies from less than a year in patients with severe intraarticular fractures to a decade or more in some patients with ligamentous or meniscal injuries (Marsh *et al.*, 2002; Buckwalter and Brown, 2004). Because many joint injuries occur in young adults, the population of patients with posttraumatic osteoarthritis includes many individuals under 50 years of age. But older individuals may have an increased risk of osteoarthritis after joint injuries. Studies of patients with intraarticular fractures of the knee show that patients older than 50 years of age have a twofold-to-fourfold greater risk of developing osteoarthritis than younger patients; patients over 40 years who have acetabular fractures and patients over 50 years who have displaced ankle fractures may also have a greater risk of OA than younger patients who have similar injuries; and age increases the risk of knee-joint degeneration after anterior cruciate ligament injury (Buckwalter and Brown, 2004; Buckwalter *et al.*, 2004).

## III. MECHANISMS OF ARTICULAR-CARTILAGE INJURIES

Understanding the mechanisms of articular surface injuries requires appreciation of how loads and rate of loading affect articular cartilage (Buckwalter, 2002; Martin *et al.*, 2004). Slowly applied loads and suddenly applied loads differ considerably in their effects. The articular-cartilage extracellular matrix consists of water and a macromolcular framework formed primarily by collagens and large aggregating proteoglycans (Buckwalter and Mankin, 1997). The collagens give the tissue its form and tensile strength, and the interaction of aggregating proteoglycans with water gives the tissue its stiffness to compression, resilience, and probably its durability. Loading of articular surfaces causes movement of fluid within the articular-cartilage matrix that dampens and distributes loads within the cartilage and to the subchondral bone (Mow and Rosenwasser, 1988). When this occurs slowly, the fluid movement allows the cartilage to deform and decreases the force applied to the matrix macromolecular framework. When it occurs too rapidly for fluid movement through the matrix and deformation of the tissue, as with sudden impact or torsional joint loading of the joint surface, the matrix macromolecular framework sustains a greater share of the force. If this force is great enough, it ruptures the matrix macromolecular framework, damages cells, and exceeds the ability of articular cartilage to prevent subchondral-bone damage by dampening and distributing loads.

*In vivo*, expected and unexpected and slow and sudden movements or impacts may differ in the amount of force transmitted to joint surfaces. Muscle contractions absorb much of the energy and stabilize joints during slow, expected movements or impacts. Sudden or unexpected movements or impacts may occur too rapidly for muscle contractions to stabilize joints and decrease the forces on the articular surfaces. For this reason, sudden and unexpected movements or impacts can transmit greater forces to joint surfaces.

Acute or repetitive blunt joint trauma can damage articular cartilage and the calcified cartilage zone–subchondral-bone region while leaving the articular surface intact (Donohue *et al.*, 1983; Thompson *et al.*, 1991; Buckwalter, 1992; Zang *et al.*, 1999; Loening *et al.*, 2000). The intensity and type of joint loading that can cause chondral and subchondral damage without visible tissue disruption has not been well defined. Physiologic levels of joint loading do not appear to cause joint injury, but impact loading above that associated with normal activities but less than that necessary to produce cartilage disruption can cause alterations of the cartilage matrix and damage chondrocytes (Donohue *et al.*, 1983; Thompson *et al.*, 1991; Buckwalter, 1992; Jeffrey *et al.*, 1995, 1997; Loening *et al.*, 2000). Experimental evidence shows that loss of proteoglycans or alteration of their organization (in particular, a decrease in proteoglycan aggregation) occurs before other signs of cartilage injury following impact loading. The loss of proteoglycans may be due either to increased degradation of the molecules or to decreased synthesis. Significant loss of matrix proteoglycans decreases cartilage stiffness and increases its permeability. These alterations may cause greater loading of the remaining macromolecular framework, including the collagen fibrils, increasing the vulnerability of the tissue to further damage from loading. These injuries may cause other matrix abnormalities besides loss of proteoglycans, such as distortions of the collagen fibril meshwork and disruptions of the collagen fibril proteoglycan relationships and swelling of the matrix (Donohue *et al.*, 1983), and they may injure chondrocytes (Buckwalter, 1992; Loening *et al.*, 2000).

Currently there is no clinically applicable method of detecting alterations in cartilage matrix composition such as decreased proteoglycan concentration and increased water concentration; however, new imaging techniques may provide methods of assessing articular-cartilage composition. When probing the articular surface, surgeons sometimes find regions of apparent softening that may result from alterations in the matrix, and devices are being developed that will allow *in vivo* measurement of articular-surface stiffness. Combined with information about cartilage composition, these measurements may make it possible to better define injuries to the articular surface that do not result in visible tissue disruption.

Disrupting a normal articular surface with a single impact requires substantial force, presumably because of the ability of articular cartilage and subchondral bone to dampen and distribute loads. A transarticular load of 2170 newtons applied to canine patellofemoral joints caused fractures in the zone of calcified cartilage visible by light microscopy and articular cartilage fissures that extended from the articular surface to the transitional or superficial radial zone of the articular cartilage (Thompson *et al.*, 1991). A study of the response of human articular cartilage to blunt trauma showed that articular cartilage could withstand impact loads of up to 25 newtons per square millimeter (25 MPa) without apparent damage. Impact loads exceeding this level caused chondrocyte death and cartilage fissures (Repo and Finlay, 1977). The authors suggested that reaching a stress level that could cause cartilage damage required a force greater than that necessary to fracture the femur. Another study (Haut, 1989) measured the pressure on human patellofemoral articular cartilage during impact loading and found that impact loads less than the level necessary to fracture bone caused stresses greater than 25 MPa in some regions of the articular surface. With the knee flexed 90°, 50% of the load necessary to cause a bone fracture produced joint pressures greater than 25 MPa for nearly 20% of the patellofemoral joint. At 70% of the bone fracture load, nearly 35% of the contact area of the patellofemoral joint pressures exceeded 25 MPa, and at 100% of the bone fracture load 60% of the patellofemoral joint pressures exceeded 25 MPa. These latter results show that impact loads can disrupt cartilage without fracturing bone.

Other experimental investigations show that repetitive impact loads split articular-cartilage matrix and initiate progressive cartilage degeneration (Weightman *et al.*, 1973; Weightman, 1976; Dekel and Weissman, 1978). Cyclic loading of human cartilage samples *in vitro* caused surface fibrillation (Weightman, 1976), and periodic impact loading of bovine metacarpal phalangeal joints *in vitro* combined with joint motion caused degeneration of articular cartilage (Radin and Paul, 1971). Repeated overuse of rabbit joints *in vivo* combined with peak overloading caused articular cartilage damage, including formation of chondrocyte clusters, fibrillation of the matrix, thickening of subchondral bone, and penetration of subchondral capillaries into the calcified zone of articular cartilage (Dekel and Weissman, 1978). The extent of cartilage damage appeared to increase with longer periods of repetitive overloading, and deterioration of the cartilage continued following cessation of excessive loading. This latter finding suggests that some cartilage damage is not immediately visible.

An investigation of cartilage plugs also showed that repetitive loading disrupted the tissue and that the severity of the damage increased with increasing load and increasing number of loading cycles (Zimmerman *et al.*, 1988). Two hundred and fifty cycles of a 1000-pound-per-square-inch compression load caused surface abrasions. Five hundred cycles produced primary fissures penetrating to calcified cartilage, and 1000 cycles produced secondary fissures extending from the primary fissures. After 8000 cycles, the fissures coalesced and undermined cartilage fragments. Higher loads caused similar changes with fewer cycles. The experiments suggested that repetitive loading can propogate vertical cartilage fissures from the joint surface to calcified cartilage and cause extension of oblique fissures into areas of intact cartilage.

Clinical studies have identified articular-cartilage fissures, flaps, and free fragments and changes in subchondral bone similar to those produced experimentally by single and repetitive impact loads (Buckwalter, 1992; Levy et al., 1996). In at least some patients, acute impact loading of the articular surface or twisting movements of the joint apparently caused these injuries. In other patients, the cartilage damage may have resulted from repetitive loading. Magnetic resonance imaging of joints soon after an acute impact or torsional load occasionally shows changes in subchondral bone consistent with damage to the zone of calcified cartilage and subchondral bone, even when the articular surface is intact (Vellet et al., 1991; Spindler et al., 1993; D. L. Johnson et al., 1998; Rubin et al., 2000).

Clinical experience suggests that chondral ruptures or tears and osteochondral fractures result from similar impact and twisting-joint injuries, but they tend to occur in different age groups, and some individuals may have a greater risk of chondral tears. Chondral tears generally occur in skeletally mature people, while osteochondral fractures typically occur in skeletally immature people or young adults. This difference may result from age-related changes in the mechanical properties of the articular surface, including the uncalcified cartilage, the calcified cartilage zone, and the subchondral bone. That is, age-related alterations in the articular-cartilage matrix decrease the tensile stiffness and strength of the superficial zone; and the calcified cartilage zone subchondral bone region mineralizes fully following completion of skeletal growth, presumably creating a marked difference in mechanical properties between the uncalcified cartilage and the calcified cartilage subchondral bone region. Taken together these changes probably increase the risk of ruptures of the superficial-cartilage matrix and of these ruptures extending to the calcified cartilage subchondral bone region. Genetically determined abnormalities of the articular cartilage may also increase the risk of chondral ruptures from a given impact or torsional load, but the relationships between known genetic abnormalities of articular cartilage and cartilage properties have not been well defined.

# IV. RESPONSE OF ARTICULAR CARTILAGE TO INJURY

Articular-surface injuries can be classified based on the type of tissue damage and the repair response: (1) cartilage matrix and cell injuries, that is, damage to the joint surface that does not cause visible mechanical disruption of the articular surface; (2) chondral fissures; flap tears, or chondral defects, that is, visible mechanical disruption of articular cartilage limited to articular cartilage; and (3) osteochondral injuries, that is, visible mechanical disruption of articular cartilage and bone (Buckwalter et al., 1988b, 1990, 2003; Buckwalter, 1992; Buckwalter and Mankin, 1997) (Table 59.1).

## Matrix and Cell Injuries

Acute or repetitive blunt trauma, including excessive impact loading, can cause alterations in the articular-cartilage matrix, including a decrease in proteoglycan concentration and possibly disruptions of the collagen fibril framework. Injuries that do not cause an apparent articular-cartilage injury, including joint dislocations or ligament and joint capsule tears, may have associated damage to the articular-cartilage cells and matrix (Tiderius et al., 2005). The ability of chondrocytes to sense changes in matrix composition and to synthesize new molecules makes it possible for them to repair damage to the macromolecular framework (Martin and Buckwalter, 2000). It is not clear at what point this type of injury becomes irreversible and leads to progressive loss of articular cartilage. Presumably, the chondrocytes can restore the matrix as long as the loss of matrix proteoglycan does not exceed what the cells can rapidly produce, if the fibrillar collagen meshwork remains intact, and if enough chondrocytes remain viable. When these conditions are not met, the cells cannot restore the matrix, the chondrocytes will be exposed to excessive loads, and the tissue will degenerate.

## Chondral Injuries

Acute or repetitive trauma can cause focal mechanical disruption of articular cartilage, including fissures, chondral flaps or tears, and loss of a segment of articular cartilage (Buckwalter, 1992). The lack of blood vessels and lack of cells that can repair significant tissue defects limit the response of cartilage to injury (Buckwalter et al., 1990; Buckwalter and Mow, 1992). Chondrocytes respond to tissue injury by proliferating and increasing the synthesis of matrix macromolecules near the injury. But the newly synthesized matrix and proliferating cells do not fill the tissue defect, and soon after injury the increased proliferative and synthetic activity ceases.

## Osteochondral Injuries

Unlike injuries limited to cartilage, injuries that extend into subchondral bone cause hemorrhage and fibrin clot formation and activate the inflammatory response (Buckwalter et al., 1990, 2003; Buckwalter and Mow, 1992). Soon after injury, blood escaping from the damaged-bone blood vessels forms a hematoma that temporarily fills the injury site. Fibrin forms within the hematoma and platelets bind to fibrillar collagen. A continuous fibrin clot fills the bone defect and extends for a variable distance into the

| Table 59.1. | Chondral and osteochondral injuries | | |
|---|---|---|---|
| *Injury* | *Clinical presentation* | *Repair response* | *Potential for healing* |
| Damage to chondral matrix and/or cells and/or subchondral bone without visible disruption of the articular surface | No known symptoms, although subchondral-bone injury may cause pain<br><br>Inspection of the articular surface and current clinical imaging methods for articular cartilage cannot detect this type of injury<br><br>Imaging of subchondral bone may show abnormalities | Synthesis of new matrix macromolecules<br><br>Cell proliferation? | If the basic matrix structure remains intact and enough viable cells remain, the cells can restore the normal tissue composition<br><br>If the matrix and/or cell population sustains significant damage or if the tissue sustains further damage, the lesion may progress to cartilage degeneration |
| Cartilage disruption (chondral fractures or ruptures) | May cause mechanical symptoms, synovitis, pain, and joint effusions | No fibrin clot formation or inflammation<br><br>Synthesis of new matrix macromolecules and cell proliferation, but new tissue does not fill the cartilage defect | Depending on the location and size of the lesion and the structural integrity, stability, and alignment of the joint, the lesion may or may not progress to cartilage degeneration |
| Cartilage and bone disruption (osteochondral fractures) | May cause mechanical symptoms, synovitis, pain, and joint effusions | Formation of a fibrin clot, inflammation, invasion of new cells, and production of new chondral and osseous tissue | Depending on the location and size of the lesion and the structural integrity, stability, and alignment of the joint, the lesion may or may not progress |

cartilage defect. Platelets within the clot release vasoactive mediators and growth factors or cytokines (small proteins that influence multiple cell functions, including migration, proliferation, differentiation, and matrix synthesis), including transforming growth factor beta and platelet-derived growth factor. Bone matrix also contains growth factors, including transforming growth factor beta, bone morphogenic protein, platelet-derived growth factor, insulinlike growth factor I, insulinlike growth factor II, and possibly others. Release of these growth factors may have an important role in the repair of osteochondral defects. In particular, they probably stimulate vascular invasion and migration of undifferentiated cells into the clot and influence the proliferative and synthetic activities of the cells. Shortly after entering the tissue defect, the undifferentiated mesenchymal cells proliferate and synthesize a new matrix. Within two weeks of injury, some mesenchymal cells assume the rounded form of chondrocytes and begin to synthesize a matrix that contains type II collagen and a relatively high concentration of proteoglycans. These cells produce regions of hyalinelike cartilage in the chondral and bone portions of the defect. Six to eight weeks following injury, the repair tissue within the chondral region of osteochondral defects contains many chondrocytelike cells in a matrix consisting of type II collagen, proteoglycans, some type I collagen, and noncollagenous proteins. Unlike the cells in the chondral portion of the defect, the cells in the bony portion of the defect produce immature bone, fibrous tissue, and hyalinelike cartilage.

The chondral repair tissue typically has a composition and structure intermediate between that of hyaline cartilage and fibrocartilage; and it rarely, if ever, replicates the elaborate structure of normal articular cartilage (Buckwalter *et al.*, 1988a, 1988b; Buckwalter and Mankin, 1997a, 1997b). Occasionally, the cartilage repair tissue persists unchanged or progressively remodels to form a functional joint surface. But in most large osteochondral injuries, the chondral repair tissue begins to show evidence of depletion of matrix proteoglycans, fragmentation and fibrillation, increasing collagen content, and loss of cells with the appearance of chondrocytes within a year or less. The remaining cells often assume the appearance of fibroblasts as the surrounding matrix comes to consist primarily of densely packed collagen fibrils. This fibrous tissue usually fragments and often disintegrates, leaving areas of exposed bone. The inferior mechanical properties of chondral repair tissue may be responsible for its frequent deterioration (Buckwalter and Mow, 1992; Buckwalter and Mankin, 1997). Even repair tissue that successfully fills osteochondral defects is less stiff and more permeable than normal articular cartilage, and the orientation and organization of the collagen fibrils in even the most hyalinelike cartilage repair tissue does not follow the pattern seen in normal articular cartilage. In addition, the repair tissue cells may fail to establish the normal relationships between matrix macromolecules, in particular, the relationship between cartilage proteoglycans and the collagen fibril network. The decreased stiffness and increased permeability of repair cartilage matrix may increase loading of the macromolecular framework during joint use, resulting in progressive structural damage to the matrix collagen and proteoglycans, thereby exposing the repair chondrocytes to excessive loads and further compromising their ability to restore the matrix.

Clinical experience and experimental studies suggest that the success of chondral repair in osteochondral injuries may depend to some extent on the severity of the injury, as measured by the volume of tissue or surface area of cartilage injured and the age of the individual (Buckwalter et al., 1996). Smaller osteochondral defects that do not alter joint function heal more predictably than larger defects that may change the loading of the articular surface. Potential age-related differences in healing of chondral and osteochondral injuries have not been thoroughly investigated, but bone heals more rapidly in children than in adults, and the articular cartilage chondrocytes in skeletally immature animals show a better proliferative response to injury and synthesize larger proteoglycan molecules than those from mature animals (Buckwalter et al., 1993, 1994b; Martin and Buckwalter, 1996a, 1996b, 2000; Martin et al., 1997). Furthermore, a growing synovial joint has the potential to remodel the articular surface to decrease the mechanical abnormalities created by a chondral or osteochondral defect.

## V. PREVENTING JOINT DEGENERATION FOLLOWING INJURY

Orthopedic surgeons routinely perform extensive surgical procedures, some having substantial complication rates, in an effort to restore the alignment and congruity of articular surfaces following intraarticular fractures (Marsh et al., 2002). The purpose of these procedures is to decrease residual joint incongruity and thereby to decrease focal elevations of contact stress, presumed to be responsible for posttraumatic osteoarthritis. These widely accepted prac-

tices are based largely on the assumption that joints are less likely to develop osteoarthritis if the peak stresses on focal areas of the articular surface are reduced. However, there is little evidence to guide surgeons in determining how much stress the articular surface can tolerate, in the form of either acute impact or chronically increased stress, and the potential for human joints to repair and remodel the articular surface after injury is poorly understood.

With better understanding of the role of mechanical forces in the pathogenesis of osteoarthritis, clinical and basic research could be focused on developing new, minimally invasive, and nonsurgical treatments of joint injuries that have a high risk of posttraumatic osteoarthritis and on treatments that prevent or delay the development of joint degeneration in patients at high risk of posttraumatic osteoarthritis. There is increasing evidence that biologic interventions can decrease mechanical stress–induced chondrocyte damage. For example, the work of D'Lima and coworkers (2001a, 2001b, 2001c) shows that capsase inhibition can decrease mechanically induced chondrocyte and apoptosis, and Haut and colleagues have reported that P188 surfactant can limit chondrocyte necrosis following impact loading (Phillips and Haut, 2004; Rundell et al., 2005). Other investigations show that antioxidants can prevent mechanically induced chondrocyte damage (Kurz et al., 2004; Beecher et al., 2006).

## VI. PROMOTING ARTICULAR SURFACE REPAIR

Better understanding of articular-cartilage injuries and recognition of the limitations of the natural repair responses have contributed to the recent interest in cartilage repair and regeneration (Buckwalter et al., 1990; Buckwalter, 1992; Buckwalter and Mow, 1992; Buckwalter and Lohmander, 1994; Buckwalter and Martin, 1995; Buckwalter, 2004). In the last three decades clinical and basic scientific investigations have shown that implantation of artificial matrices, growth factors, perichondrium, periosteum, and transplanted chondrocytes and mesenchymal stem cells can stimulate formation of cartilaginous tissue in synovial joint osteochondral and chondral defects (Buckwalter and Lohmander, 1994; Buckwalter et al., 1994a; Buckwalter, 1996a, 1996b, 2004; Messner and Gillquist, 1996; Buckwalter and Mankin, 1997).

### Penetration of Subchondral Bone

Experimental and clinical investigations show that penetration of subchondral bone leads to formation of fibrocartilagenous repair tissue on the articular surfaces of synovial joints (Buckwalter and Lohmander, 1994; Buckwalter et al., 1996; L. L. Johnson, 1996; Mithoefer et al., 2005; Gudas et al., 2006). In regions with full-thickness loss or advanced degeneration of articular cartilage, penetration of the exposed subchondral bone disrupts subchondral blood vessels, leading to formation of a fibrin clot that fills the bone defect and usually covers the exposed bone surface (Buckwalter and Lohmander, 1994; Buckwalter *et al.*, 1996). If the surface is protected from excessive loading, undifferentiated mesenchymal cells migrate into the clot, proliferate, and differentiate into cells with the morphologic features of chondrocytes (Shapiro *et al.*, 1993). In most instances, over a period of six to eight weeks they form bone in the osseous portion of the defect and fibrocartilagenous tissue in the chondral portion (L. L. Johnson, 1986, 1990; Buckwalter *et al.*, 1990). Initially, the chondral repair tissue can closely resemble articular cartilage in gross and light-microscopic appearance, but it fails to duplicate fully the composition (especially the types and concentrations of collagens and proteoglycans), structure, and mechanical properties of normal articular cartilage, and in many instances it deteriorates with time (Buckwalter *et al.*, 1988, 1990; Buckwalter and Mow, 1992; Buckwalter and Mankin, 1997).

Surgeons currently use a variety of methods of penetrating subchondral bone to stimulate formation of a new cartilaginous surface, including arthroscopic drilling and abrasion of the articular surface and making multiple small-diameter defects or fractures with an awl or similar instrument, a method referred to as the *microfracture technique* (L. L. Johnson, 1986, 1990, 1996; Buckwalter and Lohmander, 1994; Levy *et al.*, 1996; Steadman *et al.*, 1997; Mithoefer *et al.*, 2005a). Multiple authors report that these procedures can decrease the symptoms due to isolated articular-cartilage defects of the knee in a majority of patients (Sprague, 1981; Friedman *et al.*, 1984; L. L. Johnson, 1986, 1990; Ewing, 1990; Levy *et al.*, 1996; Steadman *et al.*, 1997; Mithoefer *et al.*, 2005a).

### Periosteal and Perichondrial Grafts

The potential benefits of periosteal and perichondrial grafts include introduction of a new cell population along with an organic matrix and some protection of the graft or host cells from excessive loading. Animal experiments and clinical experience show that perichondrial and periosteal grafts placed in articular-cartilage defects can produce new cartilage (Salter, 1993; Buckwalter and Lohmander, 1994). O'Driscoll has described the use of periosteal grafts for the treatment of isolated chondral and osteochondral defects and demonstrated that these grafts can produce a new articular surface (O'Driscoll and Salter, 1986; O'Driscoll *et al.*, 1988; Buckwalter, 1996). Other investigators have reported encouraging results with perichondrial grafts (Engkvist and Johansson, 1980; Homminga *et al.*, 1990). However, one study suggests that increasing patient age adversely affects the results of soft-tissue grafts. Seradge *et al.* (1984) studied the results of rib perichondrial arthroplasties in 16 metacarpophalangeal joints and 20 proximal interphalangeal joints at a minimum of three years following surgery. Patient age was directly related to the results. One hundred percent of the patients in their twenties and 75% of the patients in their thirties had good results following metacarpophalangeal-joint arthroplasties. Seventy-five percent of the patients in their teens and 66% of the patients in their twenties had good results following proximal interphalangeal-joint arthroplasties. None of the patients older than 40 years had a good result with either type of arthroplasty. The clinical observation that perichondrial grafts produced the best results in younger patients (Seradge *et al.*, 1984) agrees with the concept that age may adversely affect the ability of undifferentiated cells or chondrocytes to form an articular surface or that with age the population of cells that can form an articular surface declines (Buckwalter *et al.*, 1993; Martin *et al.*, 1997; Martin and Buckwalter, 2000).

### Cell Transplantation

Transplantation of cells grown in culture provides another method of introducing a new cell population into chondral and osteochondral defects. Experimental work has shown that both chondrocytes and undifferentiated mesenchymal cells placed in articular-cartilage defects survive and produce a new cartilage matrix (Wakitani *et al.*, 1988, 1989, 1994a, 1994b, 1994c; Buckwalter and Mankin, 1997). In addition to these animal experiments, orthopedic surgeons have used autologous chondrocyte transplants for treatment of localized cartilage defects (Brittberg *et al.*, 1994, 1996; Peterson, 1996; Minas and Peterson, 1997; Fu *et al.*, 2005; Mithoefer *et al.*, 2005b). Proponents of this procedure report that it produces satisfactory results, including the ability to return to demanding physical activities, in as many as 90% of patients (Minas and Peterson, 1997; Mithoefer *et al.*, 2005b).

### Artificial Matrices

Treatment of chondral defects with growth factors or cell transplants requires a method of delivering and in most instances at least temporarily stabilizing the growth factors or cells in the defect. For these reasons, the success of these approaches often depends on an artificial matrix. In addition, artificial matrices may allow, and in some instances stimulate, ingrowth of host cells, matrix formation, and binding of new cells and matrix to host tissue (Paletta *et al.*, 1992; Hoemann *et al.*, 2005). Investigators have found that implants formed from a variety of biologic and nonbiologic materials, including treated cartilage and bone matrices, collagens, hyaluronan, chitosan, fibrin, carbon fiber, hydroxlyapatite, porous polylactic acid, polytetrafluoroethylene, polyester, and other synthetic polymers, facilitate restoration of an articular surface (Buckwalter and Lohmander, 1994; Buckwalter and Mankin, 1997; Dorotka *et al.*, 2005; Hoemann *et al.*, 2005; Nehrer *et al.*, 2006). Lack of studies that directly compare different types of artificial matrices makes it difficult to evaluate their relative merits, but the available reports show that this approach may contribute to restoration of an articular surface.

## Growth Factors

Growth factors influence a variety of cell activities, including proliferation, migration, matrix synthesis, and differentiation. Many of these factors, including the fibroblast growth factors, insulinlike growth factors, and transforming growth factor betas, have been shown to affect chondrocyte metabolism and chondrogenesis (Buckwalter and Lohmander, 1994; Buckwalter et al., 1996; Henson et al., 2005; Schmidt et al., 2006). Bone matrix contains a variety of these molecules, including transforming growth factor betas, insulinlike growth factors, bone morphogenic proteins, platelet-derived growth factors, and others (Buckwalter et al., 1995, 1996). In addition, mesenchymal cells, endothelial cells, and platelets produce many of these factors. Thus, osteochondral injuries and exposure of bone due to loss of articular cartilage may release these agents, which affect the formation of cartilage repair tissue, and they probably have an important role in the formation of new articular surfaces after currently used surgical procedures, including penetration of subchondral bone. Local treatment of chondral or osteochondral defects with growth factors has the potential to stimulate restoration of an articular surface (Hunziker and Rosenberg, 1994, 1996; Hunziker, 2001; Schmidt et al., 2006). Despite the promise of this approach, the wide variety of growth factors, their multiple effects, the interactions among them, and the possibility that the responsiveness of cells to growth factors may decline with age (Buckwalter et al., 1993; Pfeilschifter et al., 1993; Martin and Buckwalter, 1996; Martin et al., 1997) have made it difficult to develop a simple strategy for using these agents to treat articular-surface injuries. However, development of growth factor–based treatments for isolated chondral and osteochondral injuries in combination with other approaches, including use of artificial matrices and cell transplants, appears promising (Buckwalter, 2004; Fan et al., 2006).

## VII. CONCLUSION

Articular-surface injuries are a significant unsolved problem. They are common, the value of current treatments is uncertain, and many of these injuries initiate progressive joint degeneration, a condition recognized as posttraumatic osteoarthritis. Surgeons attempt to decrease the risk of posttraumatic osteoarthritis by restoring joint stability, congruity, and alignment following injury. In addition, to promote repair and remodeling of damaged articular surfaces, they penetrate subchondral bone and insert periosteal and perichondral grafts and autologous chondrocytes. The results of these procedures vary considerably among patients, and there is limited information concerning the long-term outcomes. However, the available studies of people who have suffered joint injuries indicate that even with optimal current treatment the risk of posttraumatic osteoarthritis is high following chondral tears and articular-surface fractures. For this reason there is a clear need to improve the treatment of joint injuries. Clinical and experimental studies show that chondrocyte and mesenchymal stem cell transplantation, synthetic matrices, growth factors, and combinations of these treatments have the potential to restore articular surfaces. Other investigations suggest that biologic interventions may minimize chondrocyte damage due to mechanical forces. The most dramatic improvements in the treatment of articular-surface injuries are likely to come from approaches that help maintain chondrocyte viability and function, restore joint stability and congruity, and promote articular-surface repair and remodeling following joint injury.

## VIII. REFERENCES

Beecher, B. R., Martin, J. A., et al. (2006). Vitamin E blocks shear stress-induced chondrocyte death in articular cartilage. Trans. 52nd Annual Meeting Orthopaedic Research Society: abstract 1517.

Brittberg, M., Lindahl, A., et al. (1994). Treatment of deep cartilage defects in the knee with autologous chondrocyte transplantation. New Eng. J. Med. **331**, 889–895.

Brittberg, M., Nilsson, A., et al. (1996). Rabbit articular cartilage defects treated with autologous cultured chondrocytes. Clin. Orthop. Rel. Res. **326**, 270–283.

Buckwalter, J. A. (1992). Mechanical injuries of articular cartilage. In "Biology and Biomechanics of the Traumatized Synovial Joint" (G. Finerman, ed.), pp. 83–96. American Academy of Orthopedic Surgeons, Park Ridge, IL.

Buckwalter, J. A. (1995). Osteoarthritis and articular cartilage use, disuse and abuse: experimental studies. J. Rheumatol. **22**(suppl. 43), 13–15.

Buckwalter, J. A. (1996a). Cartilage researchers tell progress: technologies hold promise, but caution urged. Am. Acad. Orthop. Surg. Bull. **44**(2), 24–26.

Buckwalter, J. A. (1996b). Regenerating articular cartilage: why the sudden interest? Orthopaedics Today **16**, 4–5.

Buckwalter, J. A. (2002). Articular cartilage injuries. Clin. Orthop. Rel. Res. **402**, 21–37.

Buckwalter, J. A. (2004). Can tissue engineering help orthopaedic patients? Clinical needs and criteria for success. In "Tissue Engineering in Musculoskeletal Clinical Practice" (L. J. Sandell and A. J. Grodizinsky, eds.), pp. 3–16. American Academy of Orthopedic Surgeons, Rosemont, IL.

Buckwalter, J. A., and Brown, T. D. (2004). Joint injury, repair, and remodeling: roles in posttraumatic osteoarthritis. Clin. Orthop. Rel. Res. **423**, 7–16.

Buckwalter, J. A., and Lane, N. E. (1996). Aging, sports, and osteoarthritis. Sports Med. Arth. Rev. **4**, 276–287.

Buckwalter, J. A., and Lane, N. E. (1997). Athletics and osteoarthritis. Am. J. Sports Med. **25**, 873–881.

Buckwalter, J. A., and Lohmander, S. (1994). Operative treatment of osteoarthrosis: current practice and future development. *J. Bone Joint Surg.* **76A**, 1405–1418.

Buckwalter, J. A., and Mankin, H. J. (1997a). Articular cartilage I. Tissue design and chondrocyte–matrix interactions. *J. Bone Joint Surg.* **79A**(4), 600–611.

Buckwalter, J. A., and Mankin, H. J. (1997b). Articular cartilage II. Degeneration and osteoarthrosis, repair, regeneration and transplantation. *J. Bone Joint Surg.* **79A**(4), 612–632.

Buckwalter, J. A., and Mow, V. C. (1992). Cartilage repair in osteoarthritis. *In* "Osteoarthritis: Diagnosis and Management" (R. W. Moskowitz, D. S. Howell, V. M. Goldberg, and H. J. Mankin, eds.), 2nd ed., pp. 71–107. Saunders, Philadelphia.

Buckwalter, J. A., and Martin, J. A. (1995). Degenerative joint disease. *Clinical Symposia.* Summit, NJ, Ciba Geigy. **47**, 2–32.

Buckwalter, J. A., Hunziker, E. B., *et al.* (1988a). Articular cartilage: composition and structure. *In* "Injury and Repair of the Musculoskeletal Soft Tissues" (S. L. Woo and J. A. Buckwalter, eds.), pp. 405–425. American Academy of Orthopedic Surgeons. Park Ridge, IL.

Buckwalter, J. A., Rosenberg, L. C., *et al.* (1988b). Articular cartilage: injury and repair. *In* "Injury and Repair of the Musculoskeletal Soft Tissues" (S. L. Woo and J. A. Buckwalter, eds.), pp. 465–482. American Academy of Orthopedic SurgeonsPark Ridge, IL.

Buckwalter, J. A., Rosenberg, L. A., *et al.* (1990). Articular cartilage: composition, structure, response to injury, and methods of facilitation repair. *In* "Articular Cartilage and Knee Joint Function: Basic Science and Arthroscopy" (J. W. Ewing, ed.), pp. 19–56. Raven Press, New York.

Buckwalter, J. A., Woo, S. L.-Y., *et al.* (1993). Soft tissue aging and musculoskeletal function. *J. Bone Joint Surg.* **75A**, 1533–1548.

Buckwalter, J. A., Mow, V. C., *et al.* (1994a). Restoration of injured or degenerated articular surfaces. *J. Am. Acad. Ortho. Surg.* **2**, 192–201.

Buckwalter, J. A., Roughley, P. J., *et al.* (1994b). Age-related changes in cartilage proteoglycans: quantitative electron microscopic studies. *Micros. Res. Tech.* **28**, 398–408.

Buckwalter, J. A., Glimcher, M. M., *et al.* (1995). Bone biology II. Formation, form, modeling and remodeling. *J. Bone Joint Surg.* **77A**, 1276–1289.

Buckwalter, J. A., Einhorn, T. A., *et al.* (1996). Healing of musculoskeletal tissues. *In* "Fractures" (C. A. Rockwood and D. Green, eds.), pp. 261–304. Lippincott, Philadelphia.

Buckwalter, J. A., Martin, J. A., *et al.* (2000). Synovial joint degeneration and the syndrome of osteoarthritis. *Instr. Course Lect.* **49**, 481–489.

Buckwalter, J. A., Martin, J. A., *et al.* (2003). Osteochondral repair of primate knee femoral and patellar articular surfaces: implications for preventing posttraumatic osteoarthritis. *Iowa Orthop. J.* **23**, 66–74.

Buckwalter, J. A., Saltzman, C., *et al.* (2004). The impact of osteoarthritis: implications for research. *Clin. Orthop. Rel. Res.* **427**(Suppl.), S6–S15.

Buckwalter, J. A., Mankin, H. J., *et al.* (2005). Articular cartilage and osteoarthritis. *Instr. Course Lect.* **54**, 465–480.

D'Lima, D. D., Hashimoto, S., *et al.* (2001a). Human chondrocyte apoptosis in response to mechanical injury. *Osteoarthritis Cartilage* **9**(8), 712–719.

D'Lima, D. D., Hashimoto, S., *et al.* (2001b). Impact of mechanical trauma on matrix and cells. *Clin. Orthop. Rel. Res.* **391**(Suppl.), S90–S99.

D'Lima, D. D., Hashimoto, S., *et al.* (2001c). Prevention of chondrocyte apoptosis. *J. Bone Joint Surg. Am.* **83A**(Suppl. 2), 25–26.

Dekel, S., and Weissman, S. L. (1978). Joint changes after overuse and peak overloading of rabbit knees *in vivo. Acta Orthop. Scand.* **49**, 519–528.

Donohue, J. M., Buss, D., *et al.* (1983). The effects of indirect blunt trauma on adult canine articular cartilage. *J. Bone Joint Surg.* **65A**(7), 948–956.

Dorotka, R., Bindreiter, U., *et al.* (2005). Marrow stimulation and chondrocyte transplantation using a collagen matrix for cartilage repair. *Osteoarthritis Cartilage* **13**(8), 655–664.

Engkvist, O., and Johansson, S. H. (1980). Perichondral arthroplasty: a clinical study in 26 patients. *Scand. J. Plast. Reconstr. Surg.* **14**, 71–87.

Ewing, J. W. (1990). Arthroscopic treatment of degenerative meniscal lesions and early degenerative arthritis of the knee. *In* "Articular Cartilage and Knee Joint Function. Basic Science and Arthroscopy" (J. W. Ewing, ed.), pp. 137–145. Raven Press, New York.

Fan, H., Hu, Y., *et al.* (2006). Porous gelatin-chondroitin-hyaluronate tri-copolymer scaffold containing microspheres loaded with TGF-beta1 induces differentiation of mesenchymal stem cells in vivo for enhancing cartilage repair. *J. Biomed. Mater. Res. A* **77**(4), 785–794.

Friedman, M. J., Berasi, D. O., *et al.* (1984). Preliminary results with abrasion arthroplasty in the osteoarthritic knee. *Clin. Orthop.* **182**, 200–205.

Fu, F. H., Zurakowski, D., *et al.* (2005). Autologous chondrocyte implantation versus debridement for treatment of full-thickness chondral defects of the knee: an observational cohort study with 3-year follow-up. *Am. J. Sports Med.* **33**(11), 1658–1666.

Gelber, A. C., Hochberg, M. C., *et al.* (2000). Joint injury in young adults and risk for subsequent knee and hip osteoarthritis. *Ann. Intern. Med.* **133**(5), 321–328.

Gillquist, J., and Messner, K. (1999). Anterior cruciate ligament reconstruction and the long-term incidence of gonarthrosis. *Sports Med.* **27**(3), 143–156.

Gudas, R., Stankevicius, E., *et al.* (2006). Osteochondral autologous transplantation versus microfracture for the treatment of articular cartilage defects in the knee joint in athletes. *Knee Surg. Sports Traumatol. Arthrosc.* **14**(9), 834–842.

Haut, R. C. (1989). Contact pressures in the patellofemoral joint during impact loading on the human flexed knee. *J. Ortho. Res.* **7**, 272–280.

Henson, F. M., Bowe, E. A., *et al.* (2005). Promotion of the intrinsic damage-repair response in articular cartilage by fibroblastic growth factor-2. *Osteoarthritis Cartilage* **13**(6), 537–544.

Hoemann, C. D., Hurtig, M., *et al.* (2005). Chitosan-glycerol phosphate/blood implants improve hyaline cartilage repair in ovine microfracture defects. *J. Bone Joint Surg. Am.* **87**(12), 2671–2686.

Homminga, G. N., Bulstra, S. K., *et al.* (1990). Perichondral grafting for cartilage lesions of the knee. *J. Bone Joint Surg.* **72B**, 1003–1007.

Hunziker, E. B. (2001). Growth-factor-induced healing of partial-thickness defects in adult articular cartilage. *Osteoarthritis Cartilage* **9**, 22–32.

Hunziker, E. B., and Rosenberg, L. C. (1996). Repair of partial-thickness defects in articular cartilage: cell recruitment from the synovial membrane. *J. Bone Joint Surg.* **78A**, 721–733.

Hunziker, E. B., and Rosenberg, R. (1994). Induction of repair partial thickness articular cartilage lesions by timed release of TGF-Beta. *Trans. Ortho. Res. Soc.* **19**, 236.

Jeffrey, J. E., Gregory, D. W., *et al.* (1995). Matrix damage and chondrocyte viability following a single impact load on articular cartilage. *Arch. Biochem. Biophys.* **10**, 87–96.

Jeffrey, J. E., Thomson, L. A., *et al.* (1997). Matrix loss and synthesis following a single impact load on articular cartilage *in vitro. Biochem. Biophys. Acta* **15**, 223–232.

Johnson, D. L., Urban, W. P., *et al.* (1998). Articular cartilage changes seen with magnetic resonance imaging–detected bone bruises associated with anterior cruciate ligament rupture. *Am. J. Sports Med.* **26**, 409–414.

Johnson, L. L. (1986). Arthroscopic abrasion arthroplasty. Historical and pathologic perspective: present status. *Arthroscopy* **2**, 54–59.

Johnson, L. L. (1990). The sclerotic lesion: pathology and the clinical response to arthroscopic abrasion arthroplasty. *In* "Articular Cartilage and Knee Joint Function. Basic Science and Arthroscopy" (J. W. Ewing, ed.), pp. 319–333. Raven Press, New York.

Johnson, L. L. (1996). Arthroscopic arbrasion arthroplasty. *In* "Operative Arthroscopy" (J. B. McGinty, ed.), pp. 427–446. Lippincott-Raven, Philadelphia.

Kurz, B., Lemke, A., *et al.* (2004). Influence of tissue maturation and antioxidants on the apoptotic response of articular cartilage after injurious compression. *Arthritis Rheum.* **50**(1), 123–130.

Levy, A. S., Lohnes, J., *et al.* (1996). Chondral delamination of the knee in soccer players. *Am. J. Sports Med.* **24**, 634–639.

Loening, A. M., James, I. E., *et al.* (2000). Injurious mechanical compression of bovine articular cartilage induces chondrocyte apoptosis. *Arch. Biochem. Biophys.* **381**, 205–212.

Martin, J. A., and Buckwalter, J. A. (1996a). Articular cartilage aging and degeneration. *Sports Med. Arth. Rev.* **4**, 263–275.

Martin, J. A., and Buckwalter, J. A. (1996b). Fibronectin and cell shape affect age-related decline in chondrocyte synthetic response to IGF-I. *Trans. Ortho. Res. Soc.* **21**, 306.

Martin, J. A., Ellerbroek, S. M., *et al.* (1997). The age-related decline in chondrocyte response to insulin-like growth factor-I: the role of growth factor binding proteins. *J. Ortho. Res.* **15**, 491–498.

Martin, J. A., and Buckwalter, J. A. (2000). The role of chondrocyte–matrix interactions in maintaining and repairing articular cartilage. *Biorheology* **37**, 129–140.

Marsh, J. L., Buckwalter, J., *et al.* (2002). Articular fractures: does an anatomic reduction really change the result? *J. Bone Joint Surg. Am.* **84A**(7), 1259–1271.

Martin, J. A., Brown, T., *et al.* (2004). Posttraumatic osteoarthritis: the role of accelerated chondrocyte senescence. *Biorheology* **41**(3–4), 479–491.

Messner, K., and Gillquist, J. (1996). Cartilage repair: a critical review. *Acta Orthop. Scand.* **67**(5), 523–529.

Minas, T., and Peterson, L. (1997). Chondrocyte transplantation. *Oper. Tech. Orthop.* **7**(4), 323–333.

Mithoefer, K., Williams 3rd, R. J., *et al.* (2005a). The microfracture technique for the treatment of articular cartilage lesions in the knee. A prospective cohort study. *J. Bone Joint Surg. Am.* **87**(9), 1911–1920.

Mithoefer, K., Peterson, L., *et al.* (2005b). Articular cartilage repair in soccer players with autologous chondrocyte transplantation: functional outcome and return to competition. *Am. J. Sports Med.* **33**(11), 1639–1646.

Mow, V. C., and Rosenwasser, M. P. (1988). Articular cartilage: biomechanics. *In* "Injury and Repair of the Musculoskeletal Soft Tissues" (S. L. Woo and J. A. Buckwalter, eds.), pp. 427–463. American Academy of Orthopaedic Surgeons, Park Ridge, IL.

Nehrer, S., Domayer, S., *et al.* (2006). Three-year clinical outcome after chondrocyte transplantation using a hyaluronan matrix for cartilage repair. *Eur. J. Radiol.* **57**(1), 3–8.

Nelson, F., Billinghurst, R. C., *et al.* (2006). Early posttraumatic osteoarthritis-like changes in human articular cartilage following rupture of the anterior cruciate ligament. *Osteoarthritis Cartilage* **14**(2), 114–119.

Noyes, F. R., and Stabler, C. L. (1989). A system for grading articular cartilage lesions at arthroscopy. *Am. J. Sports Med.* **17**(4), 505–513.

Noyes, F. R., Bassett, R. W., *et al.* (1980). Arthroscopy in acute traumatic hemarthrosis of the knee. *J. Bone Joint Surg.* **62A**(5), 687–695.

O'Driscoll, S. W., and Salter, R. B. (1986). The repair of major osteochondral defects in joint surfaces by neochondrogenesis with autogenous osteoperiosteal grafts stimulated by continuous passive motion: an experimental investigation in the rabbit. *Clin. Orthop.* **208**, 131.

O'Driscoll, S. W., Keeley, F. W., *et al.* (1988). Durability of regenerated articular cartilage produced by free autogenous periosteal grafts in major full-thickness defects in joint surfaces under the influence of continuous passive motion. *J. Bone Joint Surg.* **70A**, 595–606.

Paletta, G. A., Arnoczky, S. P., *et al.* (1992). The repair of osteochondral defects using an exogenous fibrin clot. An experimental study in dogs. *Am. J. Sports Med.* **20**, 725–731.

Peterson, L. (1996). Articular cartilage injuries treated with autologous chondrocyte transplantation in the human knee. *Acta Orthop. Belgia* **62**(Suppl. 1), 196–200.

Pfeilschifter, J., Diel, I., *et al.* (1993). Mitogenic responsiveness of human bone cells *in vitro* to hormones and growth factors decreases with age. *J. Bone Miner. Res.* **8**, 707–717.

Phillips, D. M., and Haut, R. C. (2004). The use of a nonionic surfactant (P188) to save chondrocytes from necrosis following impact loading of chondral explants. *J. Orthop. Res.* **22**(5), 1135–1142.

Radin, E. L., and Paul, I. L. (1971). Response of joints to impact loading. *In vitro* wear. *Arth. Rheum.* **14**(3), 356–362.

Repo, R. U., and Finlay, J. B. (1977). Survival of articular cartilage after controlled impact. *J. Bone Joint Surg.* **59A**(8), 1068–1075.

Rubin, D. A., Harner, C. D., *et al.* (2000). Treatable chondral injuries of the knee: frequency of associated focal subchondral edema. *Am. J. Roentgenol.* **174**, 1099–1106.

Rundell, S. A., Baars, D. C., *et al.* (2005). The limitation of acute necrosis in retro-patellar cartilage after severe blunt impact to the *in vivo* rabbit patello-femoral joint. *J. Ortho. Res.* **23** (on line).

Salter, R. B. (1993). "Continuous Passive Motion CPM: A Biological Concept for the Healing and Regeneration of Articular Cartilage, Ligaments and Tendons, from Original Research to Clinical Applications". Williams and Wilkins, Baltimore.

Schmidt, M. B., Chen, E. H., *et al.* (2006). A review of the effects of insulin-like growth factor and platelet derived growth factor on *in vivo* cartilage healing and repair. *Osteoarthritis Cartilage* **14**, 403–412.

Seradge, H., Kutz, J. A., *et al.* (1984). Perichondrial resurfacing arthroplasty in the hand. *J. Hand Surg.* **9A**, 880–886.

Shapiro, F., Koide, S., *et al.* (1993). Cell origin and differentiation in the repair of full-thickness defects of articular cartilage. *J. Bone Joint Surg.* **75A**, 532–553.

Spindler, K. P., Schils, J. P., *et al.* (1993). Prospective study of osseous, articular and meniscal lesions in recent anterior cruciate ligament tears by magnetic resonance imaging and arthroscopy. *Am. J. Sports Med.* **21**, 551–557.

Sprague, N. F. (1981). Arthroscopic debridement for degenerative knee joint disease. *Clin. Orthop.* **160**, 118–123.

Steadman, J. R., Rodkey, W. G., *et al.* (1997). Microfracture technique for full-thickness chondral defects: technique and clinical results. *Oper. Tech. Orthop.* **7**(300–304), 294–299.

Thompson, R. C., Oegema, T. R., *et al.* (1991). Osteoarthritic changes after acute transarticular load: an animal model. *J. Bone Joint Surg.* **73A**, 990–1001.

Tiderius, C. J., Olsson, L. E., *et al.* (2005). Cartilage glycosaminoglycan loss in the acute phase after an anterior cruciate ligament injury: delayed gadolinium-enhanced magnetic resonance imaging of cartilage and synovial fluid analysis. *Arthritis Rheum.* **52**(1), 120–127.

Vellet, A. D., Marks, P. H., *et al.* (1991). Occult posttraumatic osteochondral lesions of the knee: prevalence, classification, and short-term sequelae evaluated with MR imaging. *Radiology* **178**, 271–276.

Wakitani, S., Kimura, T., *et al.* (1988). Repair of rabbits' articular surfaces by allograft of chondrocytes embedded in collagen gels. *Trans. Orthop. Res. Soc.* **13**, 440.

Wakitani, S., Kimura, T., *et al.* (1989). Repair of rabbit articular surfaces with allograft chondrocytes embedded in collagen gel. *J. Bone Joint Surg.* **71B**, 74–80.

Wakitani, S., Goto, T., *et al.* (1994a). Mesenchymal stem cell–based repair of a large articular cartilage and bone defect. *Trans. Orthop. Res. Soc.* **19**, 481.

Wakitani, S., Goto, T., *et al.* (1994b). Mesenchymal cell–based repair of large, full-thickness defects of articular cartilage. *J. Bone Joint Surg.* **76A**, 579–592.

Wakitani, S., Ono, K., *et al.* (1994c). Repair of large cartilage defects in weight-bearing and partial weight-bearing articular surfaces with allograft articular chondrocytes embedded in collagen gels. *Trans. Orthop. Res. Soc.* **19**, 238.

Weightman, B. (1976). Tensile fatigue of human articular cartilage. *J. Biomech.* **9**, 193–200.

Weightman, B. O., Freeman, M. A. R., *et al.* (1973). Fatigue of articular cartilage. *Nature* **244**, 303–304.

Zang, H., Vrahas, M. S., *et al.* (1999). Damage to rabbit femoral articular cartilage following direct impacts of uniform stresses: an *in vitro* study. *Clin. Biomechanics* **14**, 543–548.

Zimmerman, N. B., Smith, D. G., *et al.* (1988). Mechanical disruption of human patellar cartilage by repetitive loading *in vitro. Clin. Orthop. Rel. Res.* **229**(April), 302–307.

# Chapter Sixty

# Tendons and Ligaments

*Francine Goulet, Lucie Germain, A. Robin Poole, and François A. Auger*

## I. INTRODUCTION

Ligament bioengineering involves some particular technological challenges. Synthetic, semisynthetic, or natural scaffolds are under development to establish optimal conditions for permanent replacement of torn ligaments. The anterior cruciate ligament (ACL) of the knee joint, one of the strongest ligaments of the body, is often the target of traumatic injuries. Unfortunately, its healing potential is limited, and the surgical options for its replacement are frequently associated with knee instability, pain, or muscular weaknesses. Several parameters must be taken into account to produce a graftable bioengineered ACL (bACL) for the permanent replacement of a torn ACL. First, biocompatibility of the bACL will determine its level of integration in the bone knee joint. Second, the ultrastucture of its scaffold is critical for cell colonization, before and/or after implantation. Third, the biomechanical features of the bACL must be adequate to sustain the physiological stress that will be applied on its structure postgrafting. Finally, the regeneration and remodeling of the bACL is expected to occur *in situ* postimplantation, leading to a functional ACL that plays its role as a knee joint stabilizer. Our research group has developed a bACL that seems to possess these various advantages and may eventually be used for torn-ACL replacement in humans. Our tissue-enginnering approach and concepts are presented in this chapter.

The need for human tissues and organs has increased over the last few years, mostly because of population growth.

The risks for transmission of contagious diseases have also imposed serious restrictions on the availability of human transplants. Recent advances in tissue and organ engineering have generated a lot of interest among clinicians and patients in various fields of medicine. Tissue engineering offers the opportunity to target and control various biophysiological parameters of tissue development to make it competent for implantation; for example, collagen fibers and cells can be aligned in culture under defined conditions (Black *et al.*, 1999; Huang *et al.*, 1993; Kanda and Matsuda 1994; Kim *et al.*, 1999; Goulet *et al.*, 2000; Paquette *et al.*, 2003). Perhaps most importantly, bioengineered tissue substitutes are expected to respond and adapt to the mechanical stresses that occur following implantation *in vivo* (Goulet *et al.*, 2004). This chapter is dedicated to ligament and tendon bioengineering. The technological approach developed to produce a tissue-engineered human anterior cruciate ligament substitute is described. A bioengineered ACL (bACL) was developed by seeding human ligament fibroblasts in a hydrated collagen matrix (Goulet *et al.*, 2000, 2004). This bACL is anchored with two bones, since bone-to-bone insertion is reported as the most secure method for ligament fixation (Amiel *et al.*, 1986; Arnoczky *et al.*, 1982). Such an ACL substitute is a good tool to study connective-tissue repair and the environmental and cellular factors that may affect collagen alignment, cross-linking, and remodeling *in vitro*. It may also become a therapeutic alternative for torn-ACL replacement, considering the promising outcomes

*Principles of Tissue Engineering, 3rd Edition*
*ed. by Lanza, Langer, and Vacanti*

observed *in vivo* postimplantation in the caprine model (Goulet *et al.*, 2004).

## II. NEED FOR BIOENGINEERED TENDON AND LIGAMENT SUBSTITUTES

In the United States alone, at least 120,000 patients per year undergo tendon or ligament repairs (Langer and Vacanti, 1993), including approximately 50% surgical ACL reconstructions requested, using various surgical approaches (Frank and Jackson, 1997). The rupture of an ACL of the knee joint frequently occurs during sports activities such as skiing, snowboarding, soccer, and hockey. Several mechanisms can lead to ACL rupture. An ACL injury may result from a violent twisting of the knee, which can happen when an athlete plants his or her foot and suddenly changes direction. The ACL can also tear if the knee is hyperextended. Almost any sport that involves jumping, cutting, or twisting has an inherent risk of an ACL rupture. The completely torn or disrupted ACL does not appear to heal very well, at least to the extent that healing leads to a return to function (Hart *et al.*, 2005). The therapeutic options for ACL reconstruction include three categories of ACL substitutes: (1) synthetic prostheses, (2) allogenic natural substitutes (ACL excised from cadavers), and (3) natural autografts (central portion of the patellar tendon or hamstring tendon of the patient). Unfortunately, none of these surgical alternatives provides a long-term optimal solution. Synthetic material implantation was a very popular surgical technique in the 1980s, but it frequently led to implant degeneration and failure (Olson *et al.*, 1988; Woods *et al.*, 1991). Joint laxity occurs with synthetic prostheses about 12 months postsurgery. The result of ACL reconstruction with nondegradable prostheses has shown many problems, including long-term failure due to synovitis, arthritis, and mechanical deterioration. Other issues include lack of supportive-tissue ingrowth or disorganized and weak tissue formation. Stress shielding is also a major concern because the host tissue will not remodel or mature if the mechanical load is maintained by the prosthesis. For these reasons, synthetic prostheses are not the first option chosen (Olson *et al.*, 1988; Woods *et al.*, 1991; Cloutier *et al.*, 1993; Frank and Jackson, 1997). However, some new biosynthetic materials are under investigation (Laurencin and Freeman, 2005).

The use of allograft substitutes (ACL taken from cadavers) overcomes the need for autologous tissues, avoiding donor-site morbidity. However, it has limitations, considering the factors of failure, including risks of disease transmission, graft rejection, and inflammation (Noyes *et al.*, 1994; Sterling *et al.*, 1995). Allogeneic grafting of ligaments is still at an experimental stage and may lead to immunologic reactions that prevent good wound healing (Jackson and Kurzweil, 1991; Sabiston *et al.*, 1990). In fact, the main problem associated with this option for ACL replacement is the decellularization of the allografts prior to implantation.

The dead cells present in these tissues must be eliminated to reduce considerably the risks of immune rejection postgrafting. Detergents are used to reach this goal, but they also alter the structure of the scaffold and the quality of the collagen fibers that support the tissues. Moreover, residual traces of detergents such as SDS cannot be totally eliminated from the allografts, hampering repopulation of the implants by host cells *in situ* postgrafting. Finally, cell debris can remain attached to the decellularized scaffold of the allografts and potentially induce inflammatory reactions postgrafting.

Assuming proper processing techniques, autografts using the patellar tendon (Fulkerson and Langeland, 1995; Stengel *et al.*, 2005), quadriceps tendon–patellar bone (Chen *et al.*, 1999), or the hamstring tendons (Puddu and Ippolito, 1983) avoid immune rejection problems. Permanent attachment of ACL prosthesis to bone is of critical importance. Since the mid-1990s, the best solution has been the bone–patellar tendon–bone grafts (Kragh and Branstetter, 2005). Unfortunately, ligament translation or laxity is observed postsurgery (Stengel *et al.*, 2005), adding to the morbidity associated with the partial loss of a healthy tendon, chronic knee pain, loss of motion, knee instability, quadriceps weakness, and patella rupture (Vergis and Gillquist, 1995). Little is known about the physiological and pathological events that regulate the histological organization of ACL *in vivo* (Frank, 2004).

This chapter summarizes the recent advances in the domain of tendon and ligament bioengineering. Tissue engineering has engendered considerable interest, notably for its potential applications in orthopedic surgery, since ligaments and tendons are frequently targets of sports and aging trauma.

## III. HISTOLOGICAL DESCRIPTION OF TENDONS AND LIGAMENTS

Tendons connect bones and muscles, while ligaments join one bone to another. The literature used to describe the histological structures of fascia, tendons, and ligaments as similar. They are dense connective tissues consisting of fibroblasts surrounded by type I collagen bundles (Amiel *et al.*, 1984). This may explain why patellar tendon and tensor fascia have frequently been used to replace the torn anterior cruciate ligament of the knee joint (Arnoczky *et al.*, 1982). The central quadriceps tendon autograft was reported as a promising alternative for ACL reconstruction (Fulkerson and Langeland, 1995). However, some pain, loss of motion, knee instability, and other problems can be associated with this approach (Tanzer and Lenczner, 1990; Tria *et al.*, 1994; Vergis and Gillquist, 1995), according to the type of trauma and depending on variables unique to each individual (Harner *et al.*, 1994; Johnson *et al.*, 1992; Parker *et al.*, 1994). The success of such grafts depends on the revascularization of the transplanted tissues progressively surrounded by a synovial membrane, rich in vessels (Arnoczky *et al.*, 1982). The tendon would finally gain some ligament properties, and the word *ligamentization* was applied by Amiel *et al.* (1986), to describe this physiological phenomenon studied *in vivo* postgrafting.

Both ligaments and tendons consist mainly of closely packed and thick collagen fibers (predominantly type I, with a small proportion of type III collagen) but also include small quantities of glycosaminoglycans (dermatan sulfate and hyaluronic acid) (Danylchuk *et al.*, 1978; Watanebe *et al.*, 1994). The composition of these biochemical constituents is modulated during tissue growth and development, also changing in response to functional requirements (Watanebe *et al.*, 1994).

Tendon collagen fibrillogenesis is initiated early in development by fibroblasts (Trelstad and Hayashi, 1979). Fibrils are embedded in an organized, hydrated proteoglycan matrix and cross-linked through aldol or Schiff base adducts between aldehydes on one or more of the α-chains of the collagen molecules and aldehydes or amino groups on adjacent chains or molecules (Davison, 1989). Such cross-links contribute to the tensile strength of the fibrils and, thus, to the tensile properties of the whole tendon (Davison, 1989). The aldehyde-derived cross-links are found in two forms: Some are unstable in dilute acids and others are stable (Davison, 1989). The ratio of one form to the other varies in different tissues, generally increasing with aging.

Data reported by Amiel *et al.* (1984) suggested that ligaments are more metabolically active than tendons, having plumper cellular nuclei, higher DNA content, larger amounts of reducible cross-links, and more type III collagen, as compared with tendons. They also contain slightly less total collagen than tendons and more glycosaminoglycans, particularly close to the joint (Amiel *et al.*, 1984).

Extracellular matrix fibers in ligaments are also arranged less regularly than in tendons (Puddu and Ippolito, 1983), although the fibers are oriented parallel to the longitudinal axis of both tissues (Amiel *et al.*, 1984). For example, the ACL is composed of multiple fascicles, the basic unit of which is type I collagen (Bessette and Hunter, 1990). The collagen fibrils are nonparallel but are themselves arranged into fibers oriented roughly along the long axis of the ligament in a wavy, undulating pattern ("crimp"), which slowly straightens out as small loads are applied to the ligament (Bessette and Hunter, 1990). To give an example of the forces applied on such tissues *in vivo*, the ACL usually supports loads of about 169 newtons and ruptures around 1730 newtons (Bessette and Hunter, 1990). An elongation of only 6% (about 2 mm) of human ACL (about 32 mm long) is reported to be the limit beyond which damage must be expected (Amis and Dawkins, 1991). Thus, the main challenge associated with the production of bioengineered tendons and ligaments is to obtain strong and functional tissues.

## IV. TISSUE-ENGINEERED ACL SUBSTITUTES

### Basic Features of Bioengineered ACLs

Several approaches are proposed to produce bioengineered tissues *in vitro*. However, the development of such a technology always involves the use of biocompatible and, preferably, biodegradable materials that (1) can provide mechanical resistance and (2) can be colonized and reorganized by living cells *in situ* postgrafting. Indeed, some tendons and ligaments are at the top of the list in terms of injury frequency and problems to repair, replace, or heal. As often reported, in contrast with the medial collateral ligament, ACL regeneration is hampered *in vivo* (Lyon *et al.*, 1991; Ross *et al.*, 1990). Moreover, certain anatomic factors may predispose people to ACL injury (Harner *et al.*, 1994). In the young and active population, reconstruction is often the best therapeutic option. However, to overcome the drawbacks associated with ACL repair and healing, several efforts have been made to produce a bioengineered ACL model over the last decade (Petrigliano *et al.*, 2006; Spindler *et al.*, 2006). Selecting the right scaffold for the starting point to build a bACL is going to impact the outcome (Petrigliano *et al.*, 2006).

In a composite material, the orientation and density of fibers in the matrix significantly affect its mechanical properties. From a macroscopic point of view, the ACL can also be considered a composite material, where the fibroblasts secrete collagen that undergoes fibrillogenesis. Hence, it is not surprising to see that human ligaments have a very low stiffness in traction for small strains, because of the cross-linked organization of the fibers (Strocchi *et al.*, 1992).

Collagen remains the basic protein of interest in the field of connective-tissue bioenginering. Dunn *et al.* (1992, 1993, 1994) reported successful replacement of ACL by cross-linked collagen prostheses implanted in rabbits. A similar approach was also reported for the regeneration of the Achilles tendon (Kato *et al.*, 1991). The use of native collagen for the production of tissue-engineered ligaments leads to a better understanding of the mechanisms responsible for its secretion by the cells, its assembly, and its remodelling. Thus, collagen alone or with cells isolated from the tissue of origin was used to produce bioengineered functional tissues *in vitro*. The main idea behind such attempts was first to produce *in vitro* models of ligaments or tendons by tissue engineering in order to implant them permanently *in vivo*, where their histological and functional properties should be improved by local and systemic stimuli.

### A New Living bACL

A new concept for the production of human bACL *in vitro* was developed in our laboratory. This approach combines the skills and knowledge associated with tissue culture and biomechanical principles. Such expertise has been successfully developed in our laboratory for other bioengineered tissues, such as human skin (Auger *et al.*, 1995; Black *et al.*, 1999), human bronchi (Goulet *et al.*, 1996; Paquette *et al.*, 2003, 2004), and human blood vessels (L'heureux *et al.*, 2001). Our research programs aim at producing bioengineered tissues that could serve both experimental and clinical interests.

The main differences between our bACL and other ligament models are the following advantages: (1) It contains living ACL fibroblasts that contract, synthesize, and initiate remodeling and organization of the extracellular matrix in which they are initially seeded in culture; (2) it could be a very useful tool to study connective tissue repair *in vitro*, since the cells can maintain their activities for at least two months in culture; (3) it is produced without chemical cross-linking agents and synthetic material; (4) it avoids healthy ligament or tendon donor-site morbidity; and (5) bones are used to anchor the bACL, being included in the bACL structure right from the start of its production, to facilitate its eventual transplantation *in vivo*. Such an ACL substitute would greatly reduce the risks of immune reactions and infections and promote the permanent graft integration posttransplantation. This approach also eliminates the risks of chemical cytotoxicity and the production of foreign molecules due to mechanical friction, which could occur *in vivo*, in the case of its eventual use as an ACL substitute (Olson *et al.*, 1988).

## Autologous Cell Seeding of bACLs

We preferentially have used autologous ACL fibroblasts to cellularize the bACLs grafted in goat knee joints. However, other sources of autologous fibroblasts, and possibly mesenchymal stem cells, could be seeded in bACLs *in vitro*, to assess their capacity to contract, synthesize, and remodel the collagen fiber network that composes the tissue scaffold. Animal experimentation could also be initiated to evaluate the histological and biomechanical properties of bACLs seeded with ACL fibroblasts as compared to grafts populated with other sources of cells. The phenotypic stability of stem cell–derived fibroblasts would have to be monitored during several passages in culture before considering cellularization of bACLs with these cells. Stem cell–derived fibroblasts, expected to show faster doubling times and a longer life span than ACL cells *in vitro*, may differentiate into various cellular phenotypes in response to the physiological environment and the biomechanical stimuli that characterize functional knee joints. Therefore, until animal experimentation is performed, it would be premature to use stem cells to populate bACLs.

## Matrix Remodeling in Response to Mechanical Stimuli

Various research groups have shown that fibroblasts seeded in collagen gels can degrade and reorganize the sur-

**FIG. 60.1.** bACL native collagen scaffold. Photomicrograph taken during scanning electron microscope analyses of an acellular bACL bovine type I collagen scaffold that shows the alignment of the fibers in response to static tension applied for 24 h on the tissue in culture (×6000).

rounding extracellular matrix and adopt a specific orientation in a contracted collagen lattice as a function of culture conditions and time (Auger *et al.*, 1995; López Valle *et al.*, 1992b; Bell *et al.*, 1981; Delvoye *et al.*, 1983; Grinnell and Lamke, 1984; Guidry and Grinnell, 1985, 1986; Kasugai *et al.*, 1990). Several studies have also shown that mechanical stimuli to biological tissues can produce ultrastructural and histological content variations. For instance, it is well known that ligament remodeling depends on the mechanical stress to which it is subjected *in vivo* (Amiel *et al.*, 1984). In addition, bACL so far produced have collagen and cells aligned in a parallel direction to the stress exerted (Goulet *et al.*, 2000, 2004). Finally, Huang *et al.* (1993) previously reported that the mechanical properties along with the ultrastructure of ligament equivalents change in response to mechanical strengthening *in vitro*. Similar observations were made following *in vitro* elongation studies on rabbit ACL (Matyas *et al.*, 1994). Thus, it is possible and structurally relevant to induce the parallel alignment of the matrix fibers that compose the scaffold of a tissue-engineered ACL substitute in culture (Fig. 60.1). This alignment can be achieved by subjecting the bACL to static or dynamic longitudinal stretching *in vitro* (Goulet *et al.*, 2000) (Fig. 60.2). We previously reported that fibroblasts isolated from human ACL biopsies and seeded in a collagen matrix can secrete and reorganize matricial fibers in response to static tension (Goulet *et al.*, 1997a, 1997b).

## Strength of the bACL at the Time of Implantation

*In vitro*, the strength of the bACL does not compare with the strength of native ACL. Therefore, when a bACL is

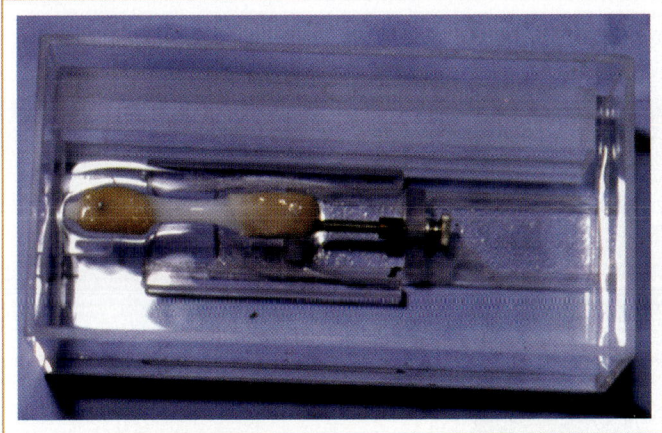

**FIG. 60.2.** A bACL maintained under tension *in vitro*. Macroscopic view of an acellular bACL subjected to minimal static tension for 24 h in culture.

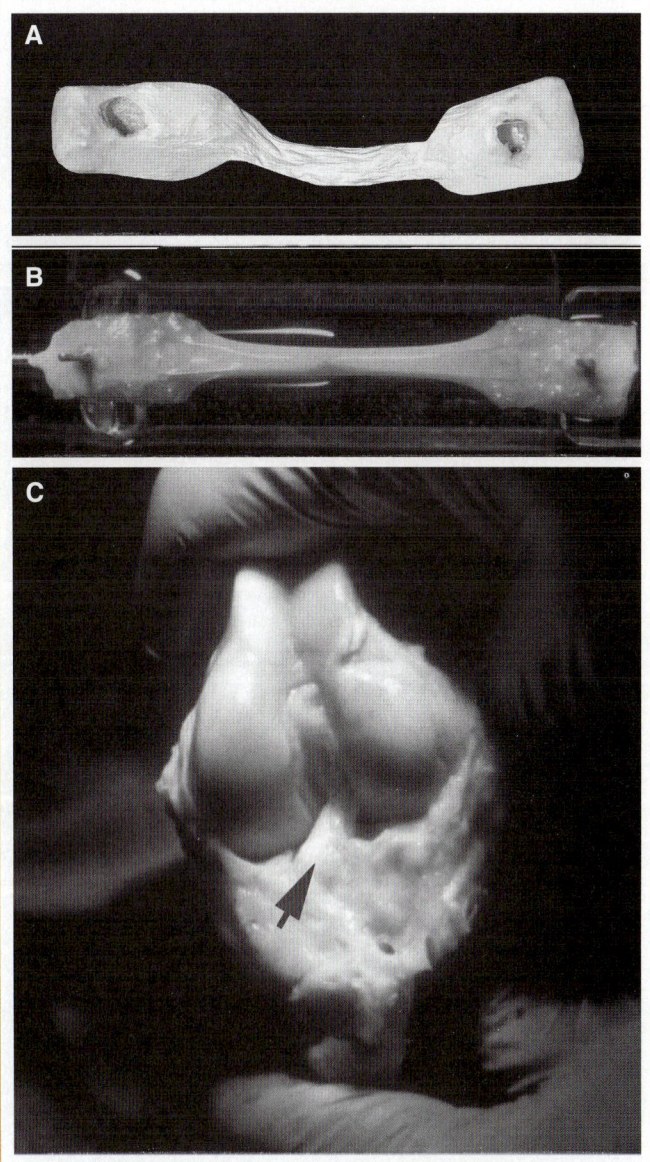

**FIG. 60.3.** Construction of bACL and postgrafting view *in situ*. **(A)** Macroscopic appearance of the initial collagen layer contracted onto two bone plugs and lyophilized. **(B)** A bACL after the second collagen coating and ready for grafting. **(C)** Macroscopic view of a bACL *in situ* (arrow) grafted for one year into a goat knee.

produced for knee joint implantation, a polyglyconate MAXON surgical thread (size 3–0), resorbable within six weeks postgrafting, can be used initially to reinforce tissue structure. The thread is passed through holes made in the two bone plugs anchoring the bACL and is then tied (Goulet *et al.*, 2004). The bone plugs and thread are counterrotated to provide a stronger link between the osseous anchors. For casting the bACLs, DMEM containing fetal calf serum (FCS) and 1.0 mg/mL of solubilized bovine type I collagen is quickly mixed with a suspension of autologous ACL fibroblasts ($2.5 \times 10^5$ cells/mL). The mixture is poured into tubes containing the bone plugs linked by the surgical thread. These bACLs are cultured for 24 h in DMEM supplemented with 10% FCS, 50 µg/mL ascorbic acid, and antibiotics, during which time the collagen is contracted by the cells onto and around the twisted surgical thread. The bACLs are frozen overnight at −70°C and subsequently lyophilized (Fig. 60.3A). In culture, in the absence of a lyophilized core and of the resorbable thread, all bACL matrix would break at the bone–collagen interface at forces varying from 0.2 to 0.5 N. When a lyophilized core is added, the bACL can support up to 2 N before rupture. This step aims at gathering the collagen fibers that are aligned in the tissue, to reinforce the resulting scaffold. In addition, the biodegradable thread, integrated between the bone plugs that anchor the lyophilized core, allows the bACL to sustain up to 25 N before rupture. Following rehydration of the lyophilized central core in fresh DMEM, a second coating of collagen and cells is applied as described earlier (Fig. 60.3B). A bilayered bACL is obtained with a rehydrated central core and a living, cell-populated outer layer. The bACLs are viable prior to implantation, since the fibroblasts progressively contract the outer collagen layer *in vitro*. The resultant bACLs are kept in culture for about a week until grafted into their respective hosts. After a year *in situ* postgrafting in a goat knee joint, the bACL shows macroscopic features comparable to the contralateral native ACL (Fig. 60.3C).

## Integration of the bACL in the Goat Knee Joint

Data reported from animal experimentation showed that vascularization, cell migration, tissue remodeling, and extracellular matrix deposition occured *in vivo* in acellular bACLs 12 weeks postgrafting in the rabbit knee joint (Dunn *et al.*, 1992, 1993, 1994) and in cellularized bACLs a year posttransplantation in the goat knee joint (Goulet *et al.*, 2004). Several research groups are currently involved in the

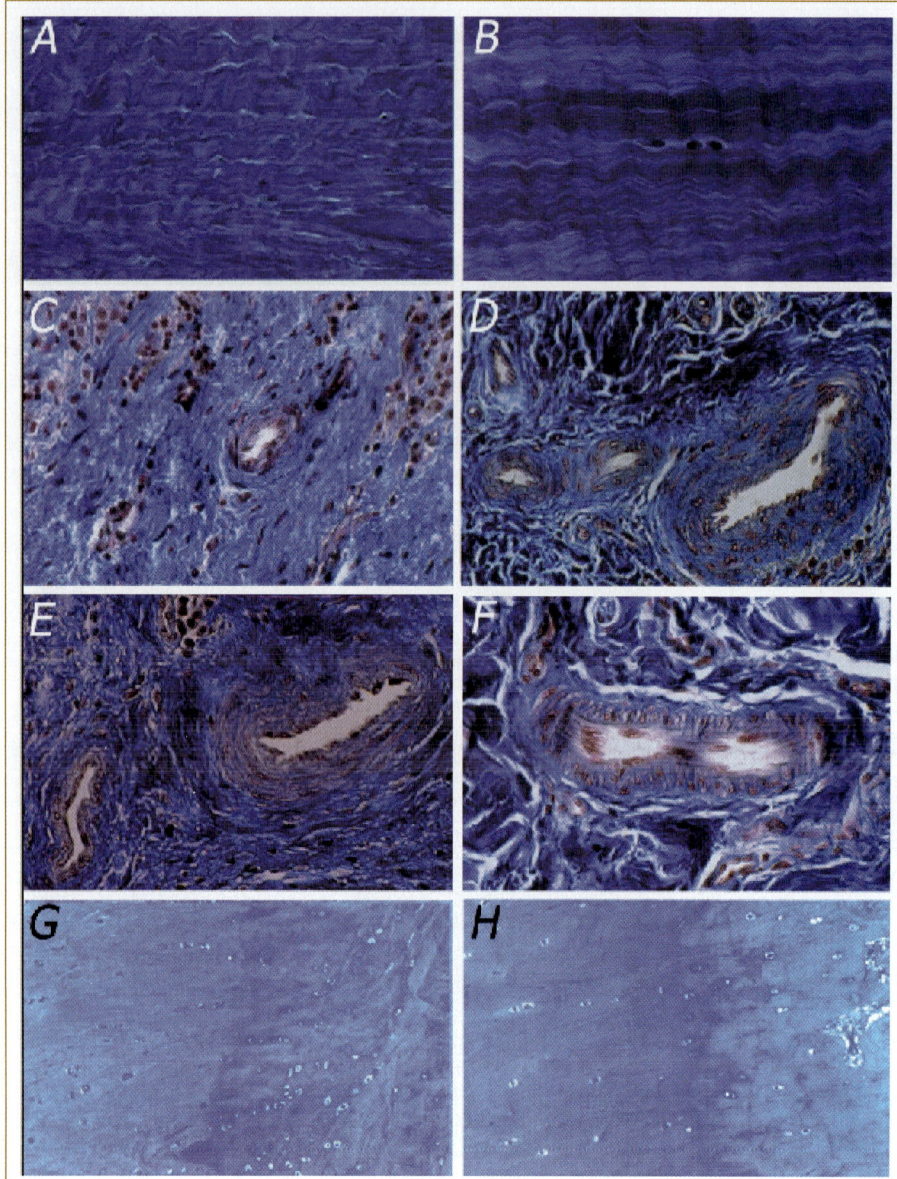

**FIG. 60.4.** Histological analyses *ex vivo* of a bACL, compared to a natural ACL. Light-microscope histology of paraffin sections of bACL grafted in goat knee and stained with Masson's trichrome method showed a dense network of collagen fibers organized in a typical crimp pattern in bACL grafted for 13 months **(A)** and native ACL **(B)** (×120). Vascularization observed in synovial membranes attached to bACLs grafted for one month **(C)**, six months **(D)**, and 11 months **(E)**, compared with blood vessel in contralateral native ACL **(F)**. Light-microscope histology of tissue sections embedded in LR white and stained with toluidine blue showed fibrocartilage and chondrocytes at the junction of the ligament proper and the bone in both the bACL grafted for six months **(G)** and native ACL **(H)** (×150).

development of human bACLs that would include living cells, to obtain tissues sharing several histological and functional properties with natural ACL (Laurencin and Freeman, 2005; Petrigliano *et al.*, 2006). When a bACL is implanted in a goat knee joint, it is slowly populated with host cells. These cells are distributed in the matrix fiber network of the graft and regenerate its collagen matrix (Fig. 60.4A). After a year, the implanted bACL shares several histological properties with a natural ACL (Fig. 60.4B). The presence of structure blood vessels is shown in the bACLs grafted for a month in the caprine model (Fig. 60.4C–E). Similarly, fibrocartilage containing chondrocytes is formed at the bone–ligament interface, contributing to the strengthening of the graft at its insertion sites (Fig. 60.4F). At the insertion sites of both bACL and natural ACL, fibrocartilaginous tissue is apparent,

as evidenced by the cartilage markers aggrecan and link protein, six months after grafting the bACLs (Goulet *et al.*, 2004). Since the bACL does not initially contain any chondrocyte, these data strongly suggest that the ultrastructure of the bovine type I collagen scaffold of the graft allows its colonization and regeneration by host cells *in situ*.

The organization and number of blood vessels in the graft are highly comparable to their distribution in natural ACLs (Fig. 60.4G). The vascularization of the bACL *in situ* postimplantation is certainly the first critical step for its permanent integration in a knee joint. The scaffold of the bACL allows endothelial cells to migrate and form such blood vessels. This phenomenon is dependent on the level of integration of the bone plugs of the bACL into the bones of the host knee joint. The vascularized graft being exposed to

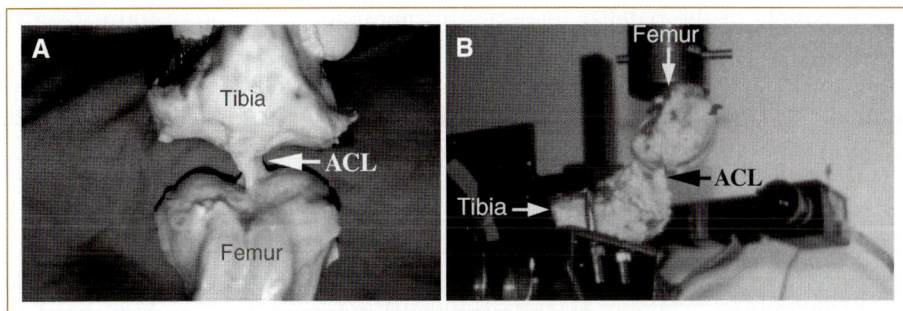

**FIG. 60.5.** Rupture assay of a bACL *ex vivo*. Macroscopic view of an 11-month-old graft *ex vivo* as prepared for biomechanical testing before **(A)** and during **(B)** application of tension prior to rupture.

nutriments, its matrix remodeling and strengthening can be initiated. Thus, the bACL seems to offer a promising alternative for torn-ACL replacement in a knee joint.

## Matrix Remodeling and Biomechanical Reinforcement of the bACL *In Vivo*

During the animal implantation phase, the reinforcement of the grafted bACL progresses slowly in response to the mechanical stimuli that occur in the host knee joint (Goulet *et al.*, 2004). The ultimate strength of each graft *ex vivo* is evaluated using an Instron traction machine (Fig. 60.5). The average bACL strength from seven grafted knees was 23% after 11 months and 36% after 13 months, compared with the average ultimate strengths of natural goat ACL corresponding to 100% (Goulet *et al.*, 2004). This strengthening can be compared to the evolution of the bone–patellar ligament–bone autograft currently used for ACL reconstruction in human knees. This type of autograft weakens *in situ* at about six weeks posttransplantation. Despite this decrease in functionality, when such a graft is successful, it is slowly revascularized after about six weeks. It gains most ACL histologic and functional properties after 30 weeks in humans. Such a physiological process was first called *ligamentization* by Amiel *et al.* (1986). Observations made in the goat knee joint suggest that a similar tissue remodeling occurs *in vivo* in tissue-engineered ACLs (Goulet *et al.*, 2004). Such remodeling is expected progressively to bring the bACL mechanical properties to an optimal level in the knee joint.

## V. CONCLUSION

Recent advances in bioengineering and cellular biology have led to revolutionary new therapeutical applications in many medical fields. The bioengineered ligament model described in this chapter presents features that can be modulated throughout culture conditions and time *in vitro* (our bACL) and very likely *in vivo* postgrafting in animals.

One of the major problems in the implantation of bACL is the fixation of the graft to the bone. Presently, one of the most popular options for reconstruction is using the patellar tendon (Arnoczky *et al.*, 1982) with a piece of spongious bone at each end. In the bACL we have developed *in vitro*, the spongious bone structure allows the penetration of the

collagen matrix among its trabiculae prior to its polymerization. This approach for bioengineered ligament fixation seems promising, since it facilitates surgical manipulations and promotes graft integration in the knee joint of the host. Synthetic bone plugs will be interesting alternatives to assess prior to clinical applications of the bACL as a torn-ACL substitute.

Nevertheless, the main advantages of our bACL over other ligament models stem from its histological organization, which is progressively modulated by the living ACL fibroblasts included in its structure *in vitro*. Because fibroblasts seeded in the bACL secrete and reorganize extracellular matrix constituents in culture, it makes it a promising tool to study the role of various cellular elements involved in the maintenance or induction of ligament repair *in vitro*.

This chapter was meant to give the reader an overview of the recent achievements and approaches that combine cell biology and tissue-engineering concepts to produce tendons and ligaments in culture. These bioengineered tissues will indeed have to be optimized prior to human transplantation. However, investigators involved in this domain have not yet started to exploit all the possibilities offered by such a new biotechnological concept. For instance, gene therapy might find some interesting applications by introducing living transfected cells within various types of bioengineered tissues.

Despite the fact that challenges lie ahead for those aiming at reproducing tissues and organs *in vitro*, we can all be hopeful, looking at the ongoing advances reached in various medical fields through tissue bioengineering (Langer and Vacanti, 1993). It is quite probable that in the near future, people will benefit from bioengineered tendon and ligament substitutes. Their production would start in a laboratory and reach its final completion in a human body. It is well recognized that a delay of six weeks after injury is required before ACL surgery is performed. In fact, it is better to perform surgical intervention of the ACL when the nearly normal range of motion of the knee has returned and inflammation has been mostly eliminated (Johnson *et al.*, 1992). This delay could be long enough to allow the production of a bACL *in vitro*. Engineering of autologous ACL would avoid the need for detrimental harvesting of the patient's tissues, like a sample of the patellar tendon. This approach may limit the needs for knee surgery, since it may be possible to graft a bACL anchored with bones simply by arthoscopy. In addition, bioengineered tissues cannot be rejected post-grafting when produced with autologous cells. That is the joint ultimate goal shared by researchers and clinicians in the various research teams around the world. Such a realization would certainly revolutionize the present therapeutic skills in tendon and ligament implantation.

## VI. ACKNOWLEDGMENTS

Dr. Albert Normand died early after the initiation of our work on human-ligament bioengineering. Our research team wants to express our sincere gratitude to this devoted orthopedic surgeon, who was the first to propose the concept of bACL development *in vitro*. The authors are also grateful to Drs. Jean Lamontagne and Réjean Cloutier for their contribution as orthopedic surgeons on the implantation of bACLs in the caprine model, to Richard Janvier for sample preparation for electron microscopy analyses, and to Julie Tremblay and Lina Robayo for technical assistance. This work was supported by the Medical Research Council of Canada, the Canadian Arthritis Society, the Canadian Orthopedic Association, the Foundation of the Enfant-Jesus and Saint-Sacrement Hospitals, and The Club Richelieu de Limoilou, Quebec.

## VII. REFERENCES

Amiel, D., Frank, C., Harwood, F., Fronek, J., and Akeson, W. (1984). Tendons and ligaments: a morphological and biochemical comparison. *J. Orthop. Res.* **1**, 257–265.

Amiel, D., Kleiner, J. B., Roux, R. D., Harwood, F. L., and Akeson, W. (1986). The phenomenon of "ligamentization": anterior cruciate ligament reconstruction with autogenous patellar tendon. *J. Orthop. Res.* **4**, 162–172.

Amis, A. A., and Dawkins, G. P. C. (1991). Functional anatomy of the anterior cruciate ligament: fiber bundle actions related to ligament replacements and injuries. *J. Bone Joint Surg.* **73B**, 260–267.

Andrews, M., Noyes, F. R., and Barber-Westin, S. D. (1994). Anterior cruciate ligament allograft reconstruction in the skeletally immature athlete. *Am. J. Sports Med.* **22**, 48–54.

Arnoczky, S. P., Tarvin, G. B., and Marshall, J. L. (1982). Anterior cruciate ligament replacement using patellar tendon. *J. Bone Joint Surg.* **64A**, 217–224.

Auger, F. A., López Valle, C. A., Guignard, R., Tremblay, N., Noël, B., Goulet, F., and Germain, L. (1995). Skin equivalents produced using human collagens. *In Vitro Cell. Dev. Biol.* **31**, 432–439.

Bell, E., Ehrlich, H. P., Buttle, D. J., and Nakatsuji, T. (1981). Living tissue formed *in vitro* and accepted as skin-equivalent tissue of full thickness. *Science* **211**, 1052–1054.

Bellows, C. G., Melcher, A. H., and Aubin, J. E. (1982). Association between tension and orientation of periodontal ligament fibroblasts and exogenous collagen fibers in collagen gels *in vitro*. *J. Cell Sci.* **58**, 125–138.

Bessette, G. C., and Hunter, R. E. (1990). The anterior cruciate ligament. *Orthopedics* **13**, 551–562.

Black, A., Berthod, F., L'Heureux, H., *et al.* (1999). *In vitro* reconstruction of a human capillary-like network in a tissue-engineered skin equivalent. *FASEB J.* **12**, 1331–1340.

Chen, C.-H., Chen, W.-J., and Shih, C.-H. (1999). Arthroscopic anterior cruciate ligament reconstruction with quadriceps tendon-patellar bone autograft. *J. Trauma* **46**, 678–682.

Cloutier, R., Lacasse, D., and Normand, A. (1993). ACL reconstruction with LAD: a five-year follow-up. *J. Bone Joint Surg. [Br.]* **74**(Suppl. III), 273–274.

Danylchuk, K. D., Finlay, J. B., and Krcek, J. P. (1978). Microstructural organization of human and bovine cruciate ligaments. *Clin. Orthop. Rel. Res.* **131**, 294–298.

Davison, P. F. (1989). The contribution of labile cross-links to the tensile behavior of tendons. *Connect. Tissue Res.* **18**, 293–305.

Delvoye, P., Nusgens, B., and Lapière, C. M. (1983). The capacity of retracting a collagen matrix is lost by dermatosparactic skin fibroblasts. *J. Invest. Dermatol.* **81**, 267–270.

Dunn, M. G., Tria, A. J., Kato, Y. P., Bechler, J. R., Ochner, R. S., Zawadsky, J. P., and Silver, F. (1992). Anterior cruciate ligament reconstruction using a composite collagenous prosthesis: a biomechanical and histologic study in rabbits. *Am. J. Sports Med.* **20**, 507–515.

Dunn, M. G., Avasarala, P. N., and Zawadsky, J. P. (1993). Optimization of extruded collagen fibers for ACL reconstruction. *J. Biomed. Mater. Res.* **27**, 1545–1552.

Dunn, M. G., Maxian, S. H., and Zawadsky, J. P. (1994). Intraosseous incorporation of composite collagen prostheses designed for ligament reconstruction. *J. Orthop. Res.* **12**, 128–137.

Fernyhough, W., Aukhil, I., and Link, T. (1987). Orientation of gingival fibroblasts in stimulated periodontal spaces *in vitro* containing collagen gel. *J. Periodontol.* **58**, 762–769.

Frank, C. B. (2004). Ligament structure, physiology and function. *J. Musculoskelet. Neuronal Interact.* **4**, 199–201.

Frank, C. B., and Jackson, D. W. (1997). The science of reconstruction of the anterior cruciate ligament. *J. Bone Joint Surg. Am.* **79**, 1556–1576.

Fulkerson, J. P., and Langeland, R. (1995). An alternative cruciate reconstruction graft: the central quadriceps tendon. *Arthroscopy* **11**, 252–254.

Goulet, F., Boulet, L.-P., Chakir, J., Tremblay, N., Dubé, J., Laviolette, M., and Auger, F. A. (1996). Morphological and functional properties of bronchial cells isolated from normal and asthmatic subjects. *Am. J. Resp. Cell Mol. Biol.* **15**, 312–318.

Goulet, F., Germain, L., Rancourt, D., Caron, C., Normand, A., and Auger, F. A. (1997a). Tendons and ligaments. *In* "Principles of Tissue Engineering" (R. Lanza, R. Langer, and W. L. Chick, eds.), pp. 633–644. Academic Press, San Diego, CA.

Goulet, F., Germain, L., Caron, C., Rancourt, D., Normand, A., and Auger, F. A. (1997b). Tissue-engineered ligament. *In* "Ligaments and Ligamentoplasties" (L. H. Yahia, ed.), pp. 367–377. Springer-Verlag, Berlin.

Goulet, F., Rancourt, D., Cloutier, R., Germain, L., Poole, A. R., and Auger, F. A. (2000). Tendons and ligaments. *In* "Principles of Tissue Engineering" (R. Lanza, R. Langer, and J. Vacanti, eds.), 2nd ed., pp. 711–722. Academic Press, San Diego, CA.

Goulet, F., Rancourt, D., Cloutier, R., Tremblay, P., Belzil, A. M., Lamontagne, J., Bouchard, M., Tremblay, J., Stevens, L.-M., Labrosse, J., Langelier, E., and McKee, M. D. (2004). Torn ACL: a new bioengineered substitute brought from the lab to the knee joint. *Appl. Bionics Biomech.* **1**, 115–121.

Green, H., Kehinde, O., and Thomas, J. (1979). Growth of cultured human epidermal cells into multiple epithelia suitable for grafting. *Proc. Natl. Acad. Sci. U.S.A.* **76**, 5665–5668.

Grinnell, F., and Lamke, C. R. (1984). Reorganization of hydrated collagen lattices by human skin fibroblasts. *J. Cell Sci.* **66**, 51–63.

Guidry, C., and Grinnell, F. (1985). Studies on the mechanisms of hydrated collagen gel reorganization by human skin fibroblasts. *J. Cell Sci.* **79**, 67–81.

Guidry, C., and Grinnell, F. (1986). Contraction of hydrated collagen gels by fibroblasts: evidence for two mechanisms by which collagen fibrils are stabilized. *Collagen Rel. Res.* **6**, 515–529.

Hansbrough, J. F., Boyce, S. T., Cooper, M. L., and Foreman, T. J. (1989). Burn wound closure with cultured autologous keratinocytes and fibroblasts attached to a collagen–glycosaminoglycan substrate. *JAMA* **262**, 2125–2130.

Harner, C. D., Paulos, L. E., Greenwald, A. E., Rosenberg, T. D., and Cooley, V. C. (1994). Detailed analysis of patients with bilateral anterior cruciate ligament injuries. *Am. J. Sports Med.* **22**, 37–43.

Hart, D. A., Shrive, N. G., and Goulet, F. (2005). Tissue engineering of ACL replacements. *Sports Med. Arthroscopy Rev.* **13**, 170–176.

Huang, D., Chang, T. R., Aggarwal, A., Lee, R. C., and Ehrlich, H. P. (1993). Mechanisms and dynamics of mechanical strengthening in ligament-equivalent fibroblast-populated collagen matrices. *Ann. Biomed. Eng.* **21**, 289–305.

Jackson, D. W., and Kurzweil, P. R. (1991). Allograft in knee ligament surgery. *In* "Ligament and Extensor Mechanism of the Knee: Diagnosis and Treatment" (W. N. Scott, ed.), pp. 349–360. Mosby Year Book, St. Louis, MO.

Johnson, R. J., Beynnon, B. D., Nichols, C. E., and Renstrom, A. F. H. (1992). Current concepts review: the treatment of injuries of the anterior cruciate ligament. *J. Bone Joint Surg.* **74A**, 140–151.

Kanda, K., and Matsuda, T. (1994). Mechanical stress–induced orientation and ultrastructural change of smooth muscle cells cultured in three-dimensional collagen lattices. *Cell. Transplant.* **3**, 481–492.

Kasugai, S., Susuki, S., Shibata, S., Yasui, S., Amano, H., and Ogura, H. (1990). Measurements of the isometric contractile forces generated by dog periodontal ligament fibroblasts *in vitro. Arch. Oral Biol.* **35**, 597–601.

Kato, Y. P., Dunn, M. G., Zawadsky, J. P., Tria, A. J., and Silver, F. H. (1991). Regeneration of achille tendon with a collagen tendon prosthesis. *J. Bone Joint Surg.* **73A**, 561–574.

Kim, B. S., Nikolovski, J., Bonadio, J., and Mooney, D. J. (1999). Cyclic mechanical strain regulates the development of engineered smooth muscle tissue. *Nat. Biotechnol.* **17**, 979–983.

Kragh, J. F., Jr., and Branstetter, J. G. (2005). Suture holes in anterior cruciate ligament bone–patellar tendon–bone grafts. *Arthroscopy.* **21**, 1011.e1–1011.e3.

Kurosaka, M., Yoshiya, S., and Andrish, J. T. (1987). A biomechanical comparison of different surgical techniques of graft fixation in anterior cruciate ligament reconstruction. *Am. J. Sports Med.* **15**, 225–229.

Langer, R., and Vacanti, J. P. (1993). Tissue engineering. *Science* **260**, 920–926.

Laurencin, C. T., and Freeman, J. W. (2005). Ligament tissue engineering: an evolutionary materials science approach. *Biomaterials* **26**, 7530–7536.

L'Heureux, N., Stoclet, J. C., Auger, F. A., Lagaud, G. J., Germain, L., and Andriantsitohaina, R. A. (2001). Human tissue-engineered vascular media: a new model for pharmacological studies of contractile responses. *FASEB J.* **15**, 515–524.

López Valle, C. A., Auger, F. A., Rompré, P., Bouvard, V., and Germain, L. (1992). Peripheral anchorage of dermal equivalents. *Br. J. Dermatol.* **127**, 365–371.

Lyon, R. M., Akeson, W. H., Amiel, D., Kitabayashi, L. R., and Woo, S. L. (1991). Ultrastructural differences between the cells of the medial collateral and the anterior cruciate ligaments. *Clin. Othop. Rel. Res.* **272**, 279–286.

Matyas, J., Edwards, P., Miniaci, A., Shrive, N., Wilson, J., Bray, R., and Frank, C. (1994). Ligament tension affects nuclear shape *in situ*: an *in vitro* study. *Connect. Tissue Res.* **31**, 45–53.

Noyes, F. R., Barber-Westin, S. D., and Roberts, C. S. (1994). Use of allografts after failed treatment of rupture of the anterior cruciate ligament. *J. Bone Joint Surg.* **76A**, 1019–1031.

Olson, E. J., Kang, J. D., Fu, F. H., Georgescu, H. I., Mason, G. C., and Evans, C. H. (1988). The biomechanical and histological effects of artificial ligament wear particles: *in vitro* and *in vivo* studies. *Am. J. Sports Med.* **16**, 558–570.

Paquette, J. S., Tremblay, P., Bernier, V., Auger, F. A., Laviolette, M., Germain, L., Boutet, M., Boulet, L. P., and Goulet, F. (2003). Production of tissue-engineered three-dimensional human bronchial models. *In Vitro Cell. Dev. Biol. Animal.* **39**, 213–220.

Paquette, J. S., Tremblay, P., Bernier, V., Tremblay, N., Laviolette, M., Boutet, M., Boulet, L.-P., and Goulet, F. (2004). A new cultured human bronchial equivalent that reproduce *in vitro* several cellular events associated with asthma. *Eur. Cells Mat.* **7**, 1–11.

Parker, A. W., Drez, D., and Cooper, J. L. (1994). Anterior cruciate ligament injuries in patients with open physes. *Am. J. Sports Med.* **22**, 44–47.

Petrigliano, F. A., McAllister, D. R., and Wu, B. M. (2006). Tissue engineering for anterior cruciate ligament reconstruction: a review of current strategies. *Arthroscopy* **22**, 441–451.

Puddu, G., and Ippolito, E. (1983). Reconstruction of the anterior cruciate ligament using the semitendinosus tendon. Histologic study of a case. *Am. J. Sports Med.* **11**, 14–16.

Ross, S. M., Joshi, R., and Frank, C. B. (1990). Establishment and comparison of fibroblast cell lines from the medial collateral and anterior cruciate ligaments of the rabbit. *In Vitro Cell Dev. Biol.* **26**, 579–584.

Sabiston, P., Frank, C. Y., Lam, T., and Shrive, N. (1990). Allograft ligament transplantation. A morphological and biochemical evaluation of a medial collateral ligament complex in a rabbit model. *Am. J. Sports Med.* **18**, 160–168.

Spindler, K. P., Murray, M. M., Devin, C., Nanney, L. B., and Davidson, J. M. (2006). The central ACL defect as a model for failure of intra-articular healing. *J. Orthop. Res.* **24**, 401–406.

Stengel, D., Matthes, G., Seifert, J., Tober, V., Mutze, S., Rademacher, G., Ekkernkamp, A., Bauwens, K., Wich, M., and Casper, D. (2005). Resorbable screws versus pins for optimal transplant fixation (SPOT) in anterior cruciate ligament replacement with autologous hamstring grafts: rationale and design of a randomized, controlled, patient and investigator blinded trial. *BMC Surg.* **5**, 1–9.

Sterling, J. C., Meyers, M. C., and Calvo, R. D. (1995). Allograft failure in cruciate ligament reconstruction: follow-up evaluation of eighteen patients. *Am. J. Sports Med.* **23**, 173–178.

Strocchi, R., De Pasquale, V., Gubellini, P., Facchini, A., Marcacci, M., Buda, R., Zaffaghini, S., and Ruggeri A. (1992). The human anterior cruciate ligament: histological and ultrastructural observations. *Anatomy* **180**, 515–51.

Tanzer, M., and Lenczner, E. (1990). The relationship of intercondylar notch size and content to notchplasty requirement in anterior cruciate ligament surgery. *Arthroscopy* **6**, 89–93.

Trelstad, R. L., and Hayashi, K. (1979). Tendon collagen fibrillogenesis: intracellular subassemblies and cell surface changes associated with fibril growth. *Develop. Biol.* **71**, 228–242.

Tria, A. J., Alicea, J. A., and Cody, R. P. (1994). Patella Baja in anterior cruciate ligament reconstruction of the knee. *Clin. Orthop. Rel. Res.* **299**, 229–234.

Vergis, A., and Gillquist, J. (1995). Graft failure in intra-articular anterior cruciate ligament reconstructions: a review of the literature. *Arthroscopy* **11**, 312–321.

Woods, G. A., Indelicato, P. A., and Prevot, T. J. (1991). The Gore-tex anterior cruciate ligament prosthesis. Two- versus three-year results. *Am. J. Sports Med.* **19**, 48–55.

# Chapter Sixty-One

# Mechanosensory Mechanisms in Bone

*Upma Sharma, Antonios G. Mikos, and Stephen C. Cowin*

## I. INTRODUCTION

The mechanosensory mechanisms in bone include (1) the cell system that is stimulated by external mechanical loading applied to the bone; (2) the system that transduces that mechanical loading to a communicable signal; and (3) the systems that transmit that signal to the effector cells for the maintenance of bone homeostasis and for strain adaptation of the bone structure. The effector cells are the osteoblasts and the osteoclasts. These systems and the mechanisms they employ have not yet been unambiguously identified. We review here candidate systems. In particular we summarize the current theoretical and experimental evidence suggesting that osteocytes are the principal mechanosensory cells of bone, that they are activated by strain amplificiation from fluid flowing through the osteocyte canaliculi, and that the electrically coupled three-dimensional network of osteocytes and lining cells is a communications system for the control of bone homeostasis and structural strain adaptation. Finally, the effect of mechanical forces on the differentiation of mesenchymal stem cells into bonelike cells is considered.

It has long been known that living adult mammalian bone tissue adapts its material properties and whole bones adapt their shape in response to altered mechanical loading (Frost, 1964; Wolff, 1870, 1892, 1986). Progress is being made in understanding the cellular mechanisms that accomplish the absorption and deposition of bone tissue. The physiological mechanism by which the mechanical loading applied to bone is sensed by the tissue and the mechanism by which the sensed signal is transmitted to the cells that accomplish the surface deposition, removal, and maintenance have not been identified. The purpose of this chapter is to review some background research on these mechanosensory mechanisms. See Burger *et al.* (1998) for an earlier review of similar literature.

All vital cells are "irritable," i.e., capable of perturbation by and response to alterations in their external environment. The mechanosensing process(es) of a cell enable it to sense the presence of and respond to extrinsic physical loadings. This property is widespread in uni- and multicellular animals (French, 1992), plants (Goldsmith, 1994; Wildon *et al.*, 1992), and bacteria (Olsson and Hanson, 1995). Tissue sensibility is a property of a connected set of cells, and it is accomplished by the intracellular processes of mechanoreception and mechanotransduction. *Mechanoreception* is the term used to describe the process that transmits the informational content of an extracellular mechanical stimulus to a receptor cell. *Mechanotransduction* is the term used to describe the process that transforms the mechanical stimulus' content into an *intra*cellular signal. We employ the term *mechanosensory* to mean both mechanoreception and mechanotransduction.

Additional processes of *inter*cellular transmission of transduced signals are required at tissue, organ, and organismal structural levels.

## II. THE CONNECTED CELLULAR NETWORK (CCN)

The bone cells that lie on all bony surfaces are osteoblasts, either active or inactive. Inactive osteoblasts are called *bone-lining cells*; they have the potential of becoming active osteoblasts. The bone cells that are buried in the extracellular bone matrix are the osteocytes. Each osteocyte, enclosed within its mineralized lacuna, has many (perhaps as many as 80) cytoplasmic processes. These processes are approximately 15 μm long and are arrayed three-dimensionally in a manner that permits them to interconnect with similar processes of up to as many as 12 neighboring cells. These processes lie within mineralized bone matrix channels called *canaliculi*. The small space between the cell process plasma membrane and the canalicular wall is filled with bone fluid and macromolecular complexes of unknown composition. All bone cells except osteoclasts are extensively interconnected by the cell process of the osteocytes, forming a *connected cellular network* (CCN) (Cowin *et al.*, 1991; Moss, 1991a, 1991b). The interconnectivity of the CCN is graphically illustrated in Fig. 61.1, a scanning electron micrograph showing the replicas of lacunae and canaliculi *in situ* in mandibular bone from a young subject aged 22 years. The inset shows enlarged lacunae, identified by a rectangle.

The touching cell processes of two neighboring bone cells contain gap junctions (Bennett and Goodenough, 1978; Civitelli, 1995; Doty, 1989; D. B. Jones and Bingmann, 1991; S. J. Jones *et al.*, 1993; Schirrmacher *et al.*, 1992). A *gap junction* is a channel connecting two cells. The walls of the channel consist of matching rings of proteins piercing the membrane of each cell; when the rings associated with two cells connect with each other, the cell-to-cell junction is formed. This junction allows ions and compounds of low molecular weight (less than 1 kDa) to pass between the two cells without passing into the extracellular space. The proteins making up a gap junction are called *connexins*; in bone, the protein is either connexin 43 or connexin 45, with 43 predominating (the number refers to the size of the proteins, calculated in kilodaltons) (Cheng *et al.*, 2001; Lecanda *et al.*, 1998; Minkoff *et al.*, 1999). A ring of connexins in one cell membrane is generally called a *connexon* or *hemichannel*. Connexin 43-null mice display inhibited ossification and exhibit increased craniofacial abnormalities, while osteoblasts deficient in this protein show a significant decrease in markers of the osteoblastic phenotype, suggesting that connexin 43 plays a pivotal role in normal osteogenesis (Lecanda *et al.*, 2000). In osteocytes, fluid shear stress causes a redistribution of connexin 43 and an opening of the gap junctions, suggesting that these channels may be used to transmit signals in response to mechanical loading (Cheng *et al.*, 2001).

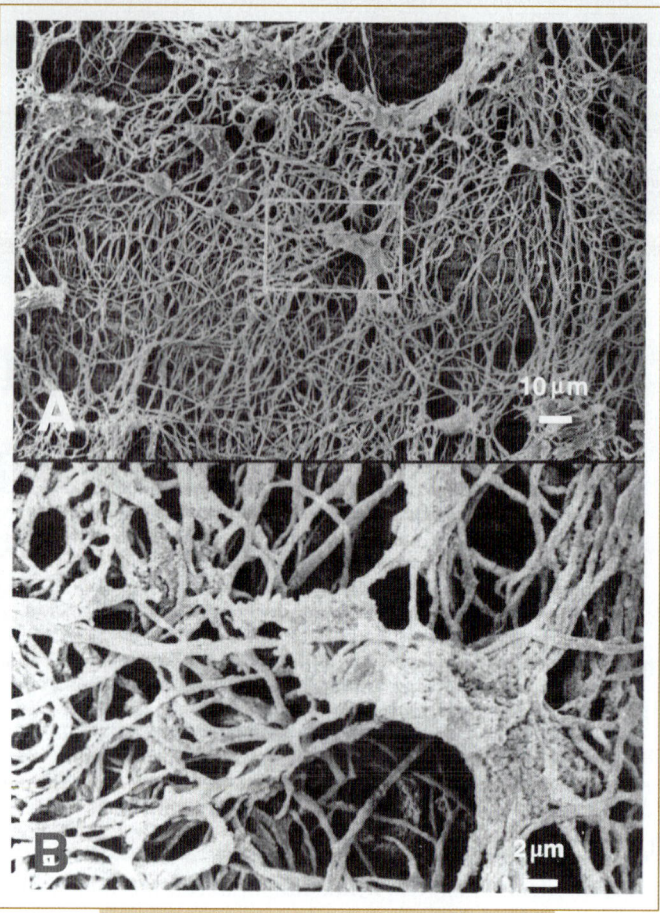

**FIG. 61.1.** Scanning electron micrograph showing the replicas of lacunae and canaliculi *in situ* in mandibular bone from a young subject age 22 years. The inset shows enlarged lacunae, identified by a rectangle. This micrograph illustrates the interconnectivity of the CCN. From Atkinson and Hallsworth (1983) with permission.

Furthermore, in bone, gap junctions connect superficial osteocytes to periosteal and endosteal osteoblasts. All osteoblasts are similarly interconnected laterally on a bony surface; perpendicular to the bony surface, gap junctions connect periosteal osteoblasts with preosteoblastic cells, and these, in turn, are similarly interconnected. Effectively, each CCN is a true syncytium (Duncan and Misler, 1989; S. J. Jones *et al.*, 1993; Schirrmacher *et al.*, 1992). Gap junctions are found where the plasma membranes of a pair of markedly lapping canalicular processes meet (Rodan, 1992). In compact bone, canaliculi cross the cement lines that form the outer boundary of osteons. Thus extensive communication exists between osteons and interstitial regions (Curtis *et al.*, 1985).

Bone cells are electrically active (Bingmann *et al.*, 1988, 1998; Chesnoy-Marchais and Fritsch, 1988; Massass *et al.*, 1990; Rubinacci *et al.*, 1998). In addition to permitting the intercellular transmission of ions and small molecules, gap junctions allow both electrical and fluorescent dye transmission (Jeansonne *et al.*, 1979; Lecanda *et al.*, 2000; Moreno *et al.*, 1994; Schirrmacher *et al.*, 1993; Spray, 1998). Gap junctions are electrical synapses, in contradistinction to interneuronal, chemical synapses; and, significantly, they permit bidirectional signal traffic (e.g., biochemical, ionic, electrical). In a physical sense, the CCN represents the hard wiring (Cowin *et al.*, 1991; Moss, 1991a, 1991b; Nowak, 1992) of bone tissue.

## III. MECHANOSENSATION ON THE CCN

### Stimuli

The *stimulus* for bone remodeling is defined as that particular aspect of the bone's stress or strain history that is employed by the bone to sense its mechanical load environment and to signal for the deposition, maintenance, or resorption of bone tissue. The bone tissue domain or region over which the stimulus is felt is called the *sensor* domain. When an appropriate stimulus parameter exceeds threshold values, loaded tissues respond by the triad of bone adaptation processes: deposition, resorption, and maintenance. The CCN is the site of intracellular stimulus reception, signal transduction, and intercellular signal transmission. It is thought that stimulus reception occurs in the osteocyte (Cowin *et al.*, 1991) and that the CCN transduces and transmits the signal to the surface lining or osteoblast. The osteoblasts alone regulate bone deposition and maintenance directly and regulate osteoclastic resorption indirectly (T. J. Martin and Ng, 1994). The possible role of the osteoblast as a stimulus receptor has not yet been thoroughly investigated (Owan *et al.*, 1997). Although it is reasonably presumed that initial mechanosensory events occur at the plasma membrane of the osteocytic soma and/or canalicular processes, the initial receptive and subsequent transductive processes are not well understood.

It follows that the true biological stimulus, though much discussed, is not precisely known. A variety of mechanical loading stimuli associated with ambulation (at a frequency of 1–2 Hz) have been considered for bone remodeling. The majority have followed Wolff (1892, 1986) in suggesting that some aspect of the mechanical loading of bone is the stimulus. The mechanical stimuli suggested include strain (Cowin and Hegedus, 1976), stress (Wolff, 1892, 1986), strain energy (Fyhrie and Carter, 1986; Huiskes *et al.*, 1987), strain rate (Fritton *et al.*, 2000; Goldstein *et al.*, 1991; Hert *et al.*, 1971, 1969, 1972; Lanyon, 1984; O'Connor *et al.*, 1982), and fatigue microdamage (Carter and Hayes, 1976; R. B. Martin and Burr, 1982). In some cases the time-averaged values of these quantities are suggested as the mechanical stimulus, and in others the amplitudes of the oscillatory components and/or peak values of these quantities are the candidates for the mechanical stimulus. Two dozen possible stimuli were compared in a combined experimental and analytical approach (Brown *et al.*, 1990). The data supported strain energy density, longitudinal shear stress, and tensile principal

stress, or strain as stimuli; no stimulus that could be described as rate dependent was among the two dozen possible stimuli considered in the study. For a consideration of the stimulus in microgravity, see Cowin (1998).

The case for strain rate as a remodeling stimulus has been building since the early 1970s. The animal studies of Hert and his coworkers (1971, 1969, 1972) suggested the importance of strain rate. Experiments have quantified the importance of strain rate over strain as a remodeling stimulus (Fritton *et al.*, 2000; Goldstein *et al.*, 1991; Lanyon, 1984). The studies directed at the understanding of the cellular mechanism for bone remodeling have suggested that the prime mover is the bone strain rate–driven motion of the bone fluid whose signal is transduced by osteocytes (Cowin *et al.*, 1995; Weinbaum *et al.*, 1991, 1994). In the model of Weinbaum *et al.* (1991, 1994) the shear stress from the bone fluid flow over the osteocytic processes in the canaliculi is a cellular mechanism–based model, suggesting strain rate as a stimulus. The study of Gross *et al.* (1997), which showed bone deposition to be related to strain gradients by the model of Weinbaum *et al.* (1991, 1994) also demonstrated a dependence on strain rate.

During *in vitro* culture, it has been shown that bone cells exposed to fluid flow respond by displaying increased prostaglandin, nitric oxide (NO), and intracellular $Ca^{2+}$ production (Chen *et al.*, 2000; Klein-Nulend *et al.*, 1995a, 1995b), decreased apoptosis (Bacabac *et al.*, 2004; Pavalko *et al.*, 2003a), and rearrangement of junctional proteins (Thi *et al.*, 2003). More specifically, it has been shown that osteocytes, but not periosteal fibroblasts, are extremely sensitive to fluid flow, resulting in increased prostaglandin as well as nitric oxide (NO) production (Klein-Nulend *et al.*, 1995a, 1995b). Three different cell populations — osteocytes, osteoblasts, and periosteal fibroblasts — were subjected to two stress regimes, pulsatile fluid flow (PFF) and intermittent hydrostatic compression (IHC) (Klein-Nulend *et al.*, 1995b). IHC was applied at 0.3 Hz with 13-kPa peak pressure. PFF was a fluid flow with a mean shear stress of 0.5 Pa with cyclic variations of 0.02 Pa at 5 Hz. The maximal hydrostatic pressure rate was 130 kPa/s, and the maximal fluid shear stress rate was 12 Pa/s. Under both stress regimes, osteocytes appeared more sensitive than osteoblasts, and osteoblasts more sensitive than periosteal fibroblasts. However, despite the large difference in peak stress and peak stress rate, PFF was more effective than IHC. Osteocytes, but not the other cell types, responded to 1-hour PFF treatment, with a sustained prostaglandin $E_2$ (PGE$_2$) upregulation lasting at least one hour after PFF was terminated. By comparison, IHC needed six hours of treatment before a response was found. These results suggested that osteocytes are more sensitive to mechanical stress than are osteoblasts, which are again more sensitive than periosteal fibroblasts. Furthermore, osteocytes appeared particularly sensitive to fluid shear stress, more so than to hydrostatic stress. These conclusions are in agreement with the theory that osteocytes are the main mechanosensory cells of bone and that they detect mechanical loading events by the canalicular flow of interstitial fluid that results from that loading event. Weinbaum et al. (1994) used Biot's porous-media theory to relate loads applied to a whole bone to the flow of canalicular interstitial fluid past the osteocytic processes. Their calculations predict fluid-induced shear stresses of 0.8–3 Pa, as a result of peak physiological loading regimes. The findings that bone cells in vitro actually respond to fluid shear stress of 0.2–6 Pa (Hung et al., 1995; Klein-Nulend et al., 1995a, 1995b; Reich et al., 1990; Williams et al., 1994) lend experimental support to the theory.

Prostaglandin is necessary for the new bone formation in response to mechanical loading (Forwood, 1996). It is known that fluid flow induces $PGE_2$ release, cyclooxygenase 2 (COX-2) expression, gap junction function, and connexin 43 expression in osteocyte-like cells (Cherian et al., 2003). When these cells are exposed to media containing $PGE_2$, an increase — similar to that observed in the presence of fluid flow — is observed in the gap junction function and connexin 43 expression. When $PGE_2$ is depleted from the culture media, this effect is significantly reduced. Therefore, the stimulatory effect of fluid flow on gap junctions is regulated, at least in part, by the release of $PGE_2$ (Cheng et al., 2001).

Osteocytes also rapidly release NO in response to stress (Klein-Nulend et al., 1995a, 1995b; Pitsillides et al., 1995), and this NO response seems to be required for the stress-related prostaglandin release (Klein-Nulend et al., 1995a). Therefore, the behavior of osteocytes compares to that of endothelial cells that regulate the flow of blood through the vascular system, and also respond to fluid flow of 0.5 Pa with increased prostaglandin and NO production (Hecker et al., 1993). The response of endothelial cells to shear stress is likely related to their role in mediating an adaptive remodeling of the vasculature so as to maintain constant endothelial fluid shear stress throughout the arterial site of the circulation (Kamiya et al., 1984).

Bakker et al. (2001) considered the effects of various stimuli (i.e., wall shear stress, streaming potentials, and chemotransport) on the production of both NO and $PGE_2$ by primary mouse bone cells. By adding dextran to the media, the viscosity, and thereby the shear stress, could be altered without affecting the streaming potentials or chemotransport. This study indicated that shear stress, not streaming potentials or chemotransport, was the stimulus yielding increased NO and $PGE_2$ production in mouse bone cells (Bakker et al., 2001).

Like osteocytes, osteoblasts release NO and $PGE_2$ in response to stress (Bacabac et al., 2004; McAllister and Frangos, 1999; Vance et al., 2005). When osteoblasts are subjected to 15 min of fluid shear stress, the resulting NO production is linearly dependent on the fluid shear stress rate, with peak production immediately after the onset of flow (Bacabac et al., 2004). The release of NO in osteoblasts is less sensitive to sustained flow and more sensitive to transients

(McAllister and Frangos, 1999). A rapid increase in intracellular $Ca^{2+}$ is also observed in response to shear stress; this intracellular $Ca^{2+}$ is thought to be connected to NO release in osteoblasts (Chen et al., 2000; McAllister and Frangos, 1999). Osteoblasts exposed to oscillatory fluid flow (1 Hz) at shear stresses of 1 and 2 Pa resulted in an increase in intracellular $Ca^{2+}$ and an increase in the overall percentage of cells responding when a 10- or 15-s rest was added after every 10 loading cycles (Batra et al., 2005).

Fluid shear stress also decreases the apoptosis of osteoblasts (Bakker et al., 2004; Pavalko et al., 2003a). Since these cells are responsible for bone formation, signals that influence osteoblastic apoptosis influence new bone deposition directly. Apotosis induced in osteoblasts using tumor necrosis factor-$\alpha$ (TNF-$\alpha$), a molecule known to stimulate bone resorption after menopause, is inhibited by fluid shear stress. TNF-$\alpha$-induced apoptosis of osteoblasts occurs through a mechanism involving the activation of PI3-kinase signaling and inhibition of caspace-3 (Pavalko et al., 2003a). Another observed effect of fluid shear stress in osteoblast- and osteocyte-like cells is a rearrangement of junctional proteins (e.g., connexin 43 and 45) causing a disruption in junctional communication (Thi et al., 2003).

Skeletal muscle contraction is a typical bone-loading event and has been suggested (Moss, 1968, 1978) as a stimulus. Frequency is one of the critical parameters of the muscle stimulus, and it serves to differentiate this stimulus from the direct mechanical loads of ambulation, which occur at a frequency of 1–2 Hz. The frequency of contracting muscle in tetanus is from 15 Hz to a maximum of 50–60 Hz in mammalian muscle (McMahon, 1984). It has been observed (McLeod and Rubin, 1992; Rodriquez et al., 1993) that these higher-order frequencies, significantly related to bone adaptational responses, are (Rubin et al., 1993) "present within the [muscle contraction] strain energy spectra regardless of animal or activity and implicate the dynamics of muscle contraction as the source of this energy band." The close similarity of muscle stimulus frequencies to bone tissue response frequencies is noted later.

## Reception and Transduction

The osteocyte has been suggested as the stimulus sensor, the receptor of the stimulus signal (Cowin et al., 1991); histologic and physiologic data are consistent with this suggestion (Aarden et al., 1994). The placement and distribution of osteocytes in the CCN three-dimensional array is architecturally well suited to sense deformation of the mineralized tissue encasing them (Lanyon, 1993). Since only a population of cells, and not an individual receptor (Edin and Trulsson, 1992), can code unambiguously, the osteocytes in the CCN are potential mechanoreceptors by virtue of their network organization.

Osteocytic mechanotransduction may involve a number of different processes or cellular systems. These processes include stretch- and voltage-activated ion channels, cytomatrix sensation-transduction processes, mechanosomes, cytosensation by fluid shear stresses, cytosensation by streaming potentials, and exogenous electric field strength. Each of these processes or cellular systems is discussed next.

### Stretch- and Voltage-Activated Ion Channels

The osteocytic plasma membrane contains stretch-activated ion channels (Duncan and Misler, 1989; Guggino *et al.*, 1989; Hamill and McBride, 1996; Keynes, 1994), which are also found in many other cell types (French, 1992; Ghazi *et al.*, 1998; Morris, 1990). When activated in strained osteocytes, they permit passage of certain ions (Sachs, 1986, 1988; Sackin, 1995), including $K^+$, $Ca^{2+}$, $Na^+$, and $Cs^+$. Such ionic flow may, in turn, initiate cellular electrical events; e.g., bone cell stretch-activated channels may modulate membrane potential as well as $Ca^{2+}$ ion flux (French, 1992; Harter *et al.*, 1995). Rough estimates of osteocytic mechanoreceptor strain sensitivity have been made (Cowin *et al.*, 1991), and the calculated values cover the morphogenetically significant strain range of 0.1–0.3% in the literature (Lanyon, 1984; Rubin and Lanyon, 1984, 1987). This appears to be too low a strain to open a stretch-activated ion channel. A model involving bone fluid flow has been suggested for the amplification of the 0.1–0.3% strain applied to a whole bone to a value an order of magnitude larger at the osteocytic cytoplasmic process membrane (Cowin and Weinbaum, 1998). Stretch-activated ion channels also occur in osteoblastic cells (Davidson *et al.*, 1990).

As in most cells, the osteocytic plasma membrane contains voltage-activated ion channels, and transmembrane ion flow may be a significant osseous mechanotransductive process (Chesnoy-Marchais and Fritsch, 1988; Ferrier *et al.*, 1991; Jan and Jan, 1992; Ravesloot *et al.*, 1991). It is also possible that such ionic flow generates osteocytic action potentials, capable of transmission through gap junctions (Schirrmacher *et al.*, 1993).

### Cytomatrix Sensation-Transduction Processes

The mineralized matrix of bone tissue is strained when loaded. Macromolecular mechanical connections between the extracellular matrix and the osteocytic cell membrane exist, and these connections may be capable of transmitting information from the strained extracellular matrix to the bone cell nuclear membrane. The basis of this mechanism is the physical continuity of the transmembrane integrin molecule, which is connected extracellularly with the macromolecular collagen of the organic matrix and intracellularly with the cytoskeletal actin. The latter, in turn, is connected to the nuclear membrane (Carvalho *et al.*, 1998; Dolce *et al.*, 1996; Hughes *et al.*, 1993; Ingber, 1998; Janmey, 1998; Meazzini *et al.*, 1998; Richardson and Parsons, 1995; Salter *et al.*, 1997; Shapiro *et al.*, 1995; Shyy and Chien, 1997; Watanabe *et al.*, 1993). It is suggested that such a cytoskeletal lever chain, connecting to the nuclear membrane, can

provide a physical stimulus able to activate the osteocytic genome (D. B. Jones and Bingmann, 1991), possibly by first stimulating the activity of such components as the *c-fos* genes (Dayhoff *et al.*, 1992; Haskin *et al.*, 1993; D. B. Jones and Bingmann, 1991; Machwate *et al.*, 1995; Pavalko *et al.*, 1998a; Petrov and Usherwood, 1994; Sadoshima *et al.*, 1992; Salter *et al.*, 1997; Uitto and Larjava, 1991; N. Wang *et al.*, 1993; Watanabe *et al.*, 1993; Yanagishita, 1993).

A number of studies have investigated the role of the cytoskeleton in fluid shear–induced signaling in bone cells (McGarry *et al.*, 2005; Norvell *et al.*, 2004; Pavalko *et al.*, 1998b; Ponik and Pavalko, 2004). It has been suggested that fluid shear stress leads to increased expression of COX-2 and c-*fos* (Pavalko *et al.*, 1998a) or intracellular $Ca^{2+}$ (Chen *et al.*, 2000) through mechanisms involving the reorganization of the cytoskeleton. Norvell *et al.* (2004) examined the effect of disruption of the major types of cytoskeletal networks, actin microfilaments, microtubules, and intermediate filaments on fluid shear–induced expression of COX-2 or release of $PGE_2$. Using a 90-min steady fluid shear stress, they found that none of the major networks were required for the fluid shear–induced release of $PGE_2$. Additionally, intact actin microfilaments and microtubules were not necessary for the fluid shear–induced increase in COX-2 expression (Norvell *et al.*, 2004).

However, the studies of McGarry *et al.* (2005), in contrast with those of Norvell *et al.* (2004), indicated that $PGE_2$ release in osteocytes exposed to pulsatile fluid flow was significantly decreased when actin microfilaments or microtubules were disrupted. Furthermore, the shear-induced release of $PGE_2$ from osteoblasts was significantly enhanced when actin microfilaments or microtubules were disrupted. The differences in the results between the studies of Norvell *et al.* (2004) and McGarry *et al.* (2005) were attributed to the differences in the flow profiles: a steady flow profile versus a pulsatile one. Also, the opposite effects of fluid shear on the release of $PGE_2$ in osteoblasts and osteocytes was hypothesized to be due to differences in the composition of the cytoskeleton of these cell types (osteocyte structure is largely dependent on actin) (McGarry *et al.*, 2005).

In another study, the idea that focal adhesions influence mechanotransduction in osteoblasts was considered (Ponik and Pavalko, 2004). To do so, methods were used to limit formation of focal adhesions: When integrin–fibronectin interactions were blocked, both fluid shear–induced COX-2 expression and $PGE_2$ release were reduced. These results indicated that focal adhesions were involved in the mechanotransduction of shear-induced signaling, resulting in the release of $PGE_2$ and the induction of COX-2 (Ponik and Pavalko, 2004).

### Mechanosomes

Pavalko *et al.* (2003b) proposed a model for skeletal mechanotransduction involving *mechanosomes*. This model hypothesizes that mechanotransduction is a combination of stretch-activated ion channels and cytomatrix sensation-transduction processes. In this model, deformation induced by mechanical loading causes flow through canalicular space, which applies strain to the tissue matrix. Additionally, this flow yields deformation of the membranes of bone cells, causing stretch-activated cation channels to open and resulting in an increase in $Ca^{2+}$ in the cytosol. $Ca^{2+}$-dependent kinase cascades are thereby initiated. Simultaneously, the deformation of the membranes causes a conformational change in the membrane proteins. These proteins are linked to a signaling scaffold that releases mechanosomes, protein complexes able to transport mechanical information, into the nucleus, where gene activity is altered (Pavalko et al., 2003b).

## Solitary Cilium as Mechanosensors

It was suggested in 2003 that the solitary cilium on osteoblasts and osteocytes might turn out to be a major mechanosensor in bone (Whitfield, 2003). Recent work has supported that suggestion (Andrews, 2006). Xiao et al. (2006) have demonstrated the presence of cilia on osteoblasts and osteocytes and has linked polycystin-1, a protein that associates with cilia and allows them to sense fluid flow in the kidney, with skeletal development in mutant mice. Malone et al. (2006) provided preliminary data that loss of cilia resulted in decreased sensitivity to flow. It has not been conclusively demonstrated that these cilia are mechanosensors for osteocytes. It will be important to determine how a single cilium on an osteocyte cell body can mediate the mechanosensory functions ascribed to the osteocyte.

## Cytosensation by Fluid Shear Stresses

A hypothesis concerning the mechanism by which the osteocytes housed in the lacunae of mechanically loaded bone sense the load applied to the bone by the detection of dynamic strains was suggested by Weinbaum et al. (1991, 1994). It was proposed that the osteocytes are stimulated by relatively small fluid shear stresses acting on the membranes of their osteocytic processes. A hierarchical model of bone tissue structure that related the cyclic mechanical loading applied to the whole bone to the fluid shear stress at the surface of the osteocytic cell process was presented (Weinbaum et al., 1994). However, an inconsistency in this model was discovered. Specifically, the strains in whole bone are usually less than 0.1% (Fritton et al., 2000; Rubin and Lanyon, 1984); however, mechanical deformation of bone cells in vitro requires strains one to two orders of magnitude higher than those measured in whole bone (You et al., 2000). You et al. (2001) proposed a model to account for this contradiction. Central to this model is the pericellular space, located between the osteocytic cell process membrane and the canalicular wall, through which bone fluid can flow (You et al., 2001, and references therein). This space is filled by an attached pericellular organic

matrix, supported by transverse fibrils. When whole bone is deformed, the induced pressure gradient will result in bone fluid's entering into this space, inducing a drag force on the pericellular matrix fibrils (You et al., 2001). This drag force is coupled to an intracellular actin cytoskeleton, resulting in strain amplification. For physiological loading ranges, this strain amplification can produce cellular strains that are two orders of magnitude greater than tissue-level strains (less than 0.1%), consistent with levels required for in vitro deformation of bone cells. These results suggest that the mechanism for cell stimulation is not direct fluid shear stress, as previously suggested, but, rather, tension from the fibers that traverse the space between the osteocytic cell process and the canalicular wall. The tension in the fibers is a reaction to the drag of the fluid flowing over and through the fibers (Han et al., 2004; You et al., 2001).

Several investigators (Johnson et al., 1982; Kufahl and Saha, 1990; Piekarski and Munro, 1977) have examined other aspects of the lacunar-canalicular porosity using simple circular-pore models and have attempted to analyze its possible physiological importance. These studies have emphasized primarily the importance of the convective flow in the canaliculi between the lacunae as a way of enhancing the supply of nutrients between neighboring osteocytes. Previous studies on the relaxation of the excess pore pressure have been closely tied to the strain-generated potentials (SGPs) associated with bone fluid motion. The SGP studies are briefly reviewed next.

## Cytosensation by Streaming Potentials

The fact that the extracellular bone matrix is negatively charged due to its proteins means that a fluid electrolyte bounded by the extracellular matrix will have a diffuse double layer of positive charges. When the fluid moves, the excess positive charge is convected, thereby developing streaming currents and streaming potentials. The cause of the fluid motion is the deformation of the extracellular matrix due to whole-bone mechanical loading. Pollack and coworkers (Pollack et al., 1984; Salzstein and Pollack, 1987; Salzstein et al., 1987) have laid an important foundation for explaining the origin of strain-generated potentials (SGPs). However the anatomical site in bone tissue that is the source of the experimentally observed SGPs is not agreed on. Salzstein and Pollack (1987) concluded that this site was the collagen-hydroxyapatite porosity of the bone mineral, because small pores of approximately 16-nm radius were consistent with their experimental data if a poroelastic-electrokinetic model with unobstructed and connected circular pores was assumed (Salzstein et al., 1987). Cowin et al. (1995), using the model of Weinbaum et al. (1994), have shown that the data of Salzstein and Pollack (1987), Scott and Korostoff (1990), and Otter et al. (1992) are also consistent with the larger pore space (100 nm) of the lacunar-canalicular porosity as the anatomical source site of the SGPs if the hydraulic drag and electrokinetic contribution associated with the passage of bone fluid through the surface matrix (glycocalyx) of the osteocytic process are accounted for (Otter *et al.*, 1992; Scott and Korostoff, 1990). The mathematical models of Salzstein *et al.* (1987) and Weinbaum *et al.* (1994) are similar, in that they combine poroelastic and electrokinetic theory to describe the phase and magnitude of the SGP. The two theories differ in the description of the interstitial fluid flow and streaming currents at the microstructural level and in the anatomical structures that determine the flow. In Weinbaum *et al.* (1994) this resistance resides in the fluid annulus that surrounds the osteocytic processes, i.e., the cell membrane of the osteocytic process, the walls of the canaliculi, and the glycocalyx (also called the *surface matrix* or *capsule*) that exists in this annular region. In Cowin *et al.* (1995) the presence of the glycocalyx increases the SGPs and the hydraulic resistance to the strain-driven flow. The increased SGP matches the phase and amplitude of the measured SGPs. In the Salzstein *et al.* (1987) model this fluid resistance and SGP are achieved by assuming that an open, continuous small pore structure ($\approx$16-nm radius) exists in the mineralized matrix. The poroelastic model of Weinbaum *et al.* (1994) for bone fluid flow has been developed further (Zhang *et al.*, 1997, 1998a, 1998b), and a review of the related poroelastic literature has been published (Cowin, 1999).

Experimental evidence indicating that the collagen-hydroxyapatite porosity of the bone mineral is unlikely to serve as the primary source of the SGP is obtained from several sources, including the estimates of the pore size in the collagen-hydroxyapatite porosity and permeability studies with different-size-labeled tracers in both mineralized and unmineralized bone. Such permeability studies clearly show time-dependent changes in the interstitial pathways as bone matures. At the earliest times, the unmineralized collagen–proteoglycan bone matrix is porous to large solutes. Studies with ferritin (10 nm in diameter) in two-day-old chick embryo show a continuous halo around primary osteons five minutes after the injection of this tracer (Dillaman, 1984). The halo passes right through the lacunar-canalicular system, suggesting that, before mineralization, pores of the size predicted in Salzstein and Pollack (1987) (radii $\approx$ 16 nm) can exist throughout the bone matrix. In contrast, Montgomery *et al.* (1988), using this same tracer in adult dogs, also found a fluorescent halo surrounding the Haversian canals (Montgomery *et al.*, 1988); however, this halo was not continuous, but formed by discrete lines, suggesting that the pathways were limited to discrete pores whose spacing was similar to that observed for canaliculi. This conclusion is supported by the studies of Tanaka and Sakano (1985) in the alveolar bone of five-day-old rats using the much smaller tracer microperoxidase (MP) (2 nm) (Tanaka and Sakano, 1985). These studies clearly showed that the MP penetrated only the unmineralized matrix surrounding the lacunae and the borders of the canaliculi (see Fig. 13 of this study) and was absent from the mineralized matrix. Using more mature rats, the failure of MP (2 nm) and reactive red (1 nm) tracers to penetrate the mineralized matrix tissue from the bone fluid compartments was confirmed (L. Y. Wang *et al.*, 2004).

### Exogenous Electric Field Strength

Bone responds to exogenous electrical fields (Otter *et al.*, 1998). While the extrinsic electrical parameters are unclear, field strength may play an important role (Brighton *et al.*, 1992; Otter *et al.*, 1998). A significant parallelism exists between the parameters of exogenous electrical fields and the endogenous fields produced by muscle activity. Bone responds to exogenous fields in an effective range of 1–10 µV/cm, strengths that are on the order of those endogenously produced in bone tissue during normal (muscle) activity (McLeod *et al.*, 1993). These exogenous fields are thought to affect the activity of hormones, growth factors, cytokines, and mechanical forces, which, in turn, alter the response of bone cells (Spadaro, 1997).

### Uniqueness of Osseous Mechanosensation

The difference between mechanosensation in bone tissue and mechanosensation in nonosseous processes is revealing and thought to be significant. First, most mechanosensory cells are cytologically specialized (e.g., rods and cones in the retina); bone cells are not. Second, one bone-loading stimulus can evoke three adaptational responses (deposition, maintenance, and resorption), while nonosseous process stimuli evoke just one (e.g., vision). Third, osseous signal transmission is aneural, while all other organismal mechanosensational signals (Moss-Salentijn, 1992) utilize afferent neural pathways (e.g., the visual pathway; Hackney and Furness, 1995). Fourth, the evoked bone adaptational responses are confined within each bone independently (e.g., within a femur); there is no organismal involvement (e.g., touch is a generalized somatic sensation). However, all afferent transductive processes, osseous and nonosseous, share many common mechanisms and processes (Gilbertson, 1998; Wilson and Sullivan, 1998).

### Signal Transmission

From a communications viewpoint, the CCN is multiply noded (each osteocyte is a node) and multiply connected. Each osteocytic process is a connection between two osteocytes, and each osteocyte is multiply connected to a number of osteocytes that are near neighbors (see Fig. 61.1). Cell-to-cell communication is considered first, and then we discuss the opening of hemichannels as a means of transmitting signaling factors. Finally, we speculate on the ability of the CCN to compute as well as signal. It is useful to note the possibility that bone cells, like neurons, may communicate intercellular information by volume transmission, a process that does not require direct cytological contact, but, rather, utilizes charges in the environment (Fuxe and Agnati, 1991; Marotti, 1996; Schaul, 1998).

## Cell-to-Cell Communication

In order to transmit a signal over the CCN, one osteocyte must be able to signal a neighboring osteocyte, which will then pass the signal on until it reaches an osteoblast on the bone surface. There are a variety of chemical and electrical cell-to-cell communication methods (De Mello, 1987). The passage of chemical signals, such as $Ca^{2+}$, from cell to cell appears to occur at a rate that would be too slow to respond to the approximately 30-Hz signal associated with muscle firing. We focus here on electrical cell-to-cell communication. Zhang et al. (1997, 1998a) have formulated a cable model for cell-to-cell communication in an osteon. The spatial distribution of intracellular electric potential and current from the cement line to the lumen of an osteon was estimated, because the frequency of the loading and conductance of the gap junction were altered. In this model the intracellular potential and current are driven by the mechanically induced strain-generated streaming potentials (SGPs) produced by the cyclic mechanical loading of bone. The model differs from earlier studies (Harrigan and Hamilton, 1993), in that it pursues a more physiological approach, in which the microanatomical dimensions of the connexon pores, the osteocytic processes, and the distribution of cellular membrane area and capacitance are used to estimate quantitatively the leakage of current through the osteoblast membrane, the time delay in signal transmission along the cable, and the relative resistance of the osteocytic processes and the connexons in their open and closed states. The model predicts that the cable demonstrates a strong resonant response when the cable coupling length approaches the osteonal radius. The theory also predicts that the pore pressure relaxation time for the draining of the bone fluid into the osteonal canal is of the same order as the characteristic diffusion time for the spread of current along the membrane of the osteocytic processes. This coincidence of characteristic times produced a spectral resonance in the cable at 30 Hz. These two resonances lead to a large amplification of the intracellular potential and current in the surface osteoblasts, which could serve as the initiating signal for osteoblasts to conduct remodeling.

## Opening of Hemichannels

As described earlier, it is known that the stimulatory effect of fluid flow on gap junctions and connexin-43 expression is at least partly regulated by the release of $PGE_2$ (Cheng et al., 2001). This observation raised the question of how the $PGE_2$ molecules (negatively charged at physiological pH) are transported to the cell's exterior. Recently, it has been shown that connexin-43 hemichannels, found in osteocytes and osteoblasts, can be opened by mechanical strain. The opening of these hemichannels was demonstrated to act as a direct route for exit of intracellular $PGE_2$ (Cherian et al., 2005, and references therein). This novel mechanism could have important implications for the rapid exit, and thereby the transmission, of other signaling molecules involved in bone adaptive processes.

## Signal Processing and Integration

When a physical representation of a CCN, such as Figure 61.1, is viewed by someone familiar with communications, there is often an intuitive response that the CCN may function as a neural network for processing the mechanical loading stimulus signals being felt over the network. That idea is explored here with no justification other than shared intuition. A CCN is operationally analogous to an artificial neural network in which massively parallel, or parallel distributed, signal processing occurs (Edin and Trulsson, 1992; Martino et al., 1994). Fortunately, the bases of connectionist theory are sufficiently secure to permit a biologically realistic CCN model (Dayhoff et al., 1992; Hinton and Anderson, 1989; McClelland and Rumelhart, 1987; Zorntzer et al., 1990).

A CCN consists of a number of relatively simple, densely interconnected processing elements (bone cells), with many more interconnections than cells. Operationally these cells are organized into layers: an initial input, a final output, and one or more intermediate or hidden layers. However, such networks need not be numerically complex to be operationally complex (Kupfermann, 1992).

The operational processes are identical, in principle, for all bone cells in all layers. All cells in any layer may simultaneously receive several weighted (i.e., some quantitative measure) inputs. In the initial layer these are the loading stimuli (mechanoreception). Within each cell independently, "all the weighted inputs are then summed" (Wasserman, 1989). This net sum is then compared, within the cell, against some threshold value. If this liminal value is exceeded, a signal is generated (mechanotransduction in input layer cells) that is then transmitted identically to all the hidden layer cells (adjacent osteocytes) to which each initial layer cell is connected. Similar processes of weighted signal summation, comparison, and transmission occur in these layers until final layer cells (osteoblasts) are reached. The outputs of these surface-situated cells regulate the specific adaptation process of each group of osteoblasts (Parfitt, 1994). All neighboring osteoblasts that carry out an identical bone adaptational process form a communication compartment, a cohort of structurally and operationally similar cells, since all these cells are interconnected by open, functional gap junctions. At the boundary between such compartments that are carrying out different adaptational processes, the intervening gap junctions are closed and are incapable of transmitting information. These boundaries are probably changing continuously because some of the cells have some downtime (Jeansonne et al., 1979; Spray, 1998).

Information is not stored discretely in a CCN, as is the case in conventional computers. Rather, it is distributed across all or part of the network, and several types of information may be stored simultaneously. The instantaneous state of a CCN is a property of all of its cells and of their connections. Accordingly its informational representation is redundant, ensuring that the network is fault, or error, tolerant; i.e., one or several inoperative cells causes little or no noticeable loss in network operations (Wasserman, 1989).

CCNs exhibit oscillation; i.e., iterative reciprocal (feedback) signaling between layers enables them to self-organize adjustively. This is related to the fact that CCNs are not preprogrammed; rather, they learn by unsupervised training (Fritzke, 1994), a process involving the adaptation of the CCNs to the responses of the cytoskeleton to physical activity (Dayhoff et al., 1992). In this way the CCN adjusts to the customary mechanical loading of the whole bone (Lanyon, 1984). In a CCN, structurally more complex attributes and behavior gradually self-organize and emerge during operation. These are not reducible; they are neither apparent nor predictable from a prior knowledge of the behavior of individual cells.

As noted earlier, gap junctions, as electrical synapses, permit bidirectional flow of information. This is the cytological basis for the oscillatory behavior of a CCN. The presence of sharp discontinuities between groups of phenotypically different osteoblasts is also related to an associated property of gap junctions, i.e., their ability to close and thus prevent the flow of information (Kam and Hodgins, 1992; Kodama and Eguchi, 1994). Significantly, informational networks can also transmit inhibitory signals, a matter beyond our present scope (Marotti et al., 1992).

It is suggested that a CCN displays the following attributes: Developmentally it is self-organized and self-adapting, and, in the sense that it is epigenetically regulated, it is untrained. Operationally it is a stable, dynamic system, whose oscillatory behavior permits feedback; in this regard, it is noted that a CCN operates in a noisy, nonstationary environment and that it also employs useful and necessary inhibitory inputs.

The CCN permits a triad of histological responses to a (seemingly) unitary loading event. Although in this chapter, as in almost all the related literature, the organization of bone cells is treated as if it existed in only two dimensions and as if bone tissue loadings occurred only at certain discrete loci, and that without consideration of loading gradients, the biological situation is otherwise. Given such a loading event, a three-dimensional bone volume, gradients of deformation must exist, and each osteocyte may sense correspondingly different strain properties. Moreover, it is probable that each osteocyte potentially is able to transmit three different signals in three different directions, some stimulatory and some inhibitory; such states have not yet been modeled (Moss, 1997a, 1997b).

### A Tentative Mechanotransduction Synthesis

The molecular lever mechanisms that permit muscle function to regulate directly the genomic activity of strained bone cells, including their phenotypic expression, when combined together with electric field effects and contraction frequency energetics, provides a biophysical basis for an earlier hypothesis of epigenetic regulation of skeletal tissue adaptation (Moss, 1962, 1968; Moss and Salentijn, 1969; Moss and Young, 1960).

It is probable (Moss, 1997c, 1997d) that electrical and mechanical transductive processes are neither exhaustive nor mutually exclusive. While utilizing differing intermediate membrane mechanisms and/or processes, they share a common final pathway (Bhalla and Iyengar, 1999); i.e., both mechanical and electrical transductions result in trans plasma membrane ionic flow(s), creating a signal(s) capable of intercellular transmission to neighboring bone cells via gap junctions (Cowin et al., 1995; Weinbaum et al., 1994; Zeng et al., 1994; Zhang et al., 1997, 1998a). These signals are inputs to a CCN, whose outputs regulate the bone adaptational processes.

The primacy of electrical signals is suggested here, since while bone cell transduction may also produce small biochemical molecules that can pass through gap junctions, the time course of mechanosensory processes is believed to be too rapid for the involvement of secondary messengers (Carvalho et al., 1994; French, 1992; Wildon et al., 1992). As we noted earlier, the passage of chemical signals, such as $Ca^{2+}$, from cell to cell appears to occur at a rate that would be too slow to respond to the approximately 30-Hz signal associated with muscle firing. The opening of hemichannels in response to mechanical strain could potentially provide a route for rapid release of intracellular signaling molecules.

## IV. MESENCHYMAL STEM CELLS

Osteoblasts and osteocytes are derived from the mesenchymal stem cells (MSCs), found mainly in bone marrow. While much research has focused on the effect of mechanical loading on osteocytes and osteoblasts, the response of these progenitor cells to mechanical stimulation has only recently been explored (Bancroft et al., 2002; Holtorf et al., 2005; Sikavitsas et al., 2003). A perfusion bioreactor was developed to examine the effect of mechanical forces on the differentiation of MSCs in 3D scaffolds for tissue-engineering applications (Bancroft et al., 2002; Sikavitsas et al., 2003). MSCs were cultured for periods of up to 16 days, and markers of osteoblastic differentiation were monitored (alkaline phosphatase activity, osteocalcin and osteopontin secretion, and calcium content); cells cultured in the presence of fluid flow exhibited enhanced osteoblastic differentiation as compared to static controls for all flow conditions tested (perfusion rates varied from 0.3 to 3 mL/min). The effect of the fluid flow on the deposited mineralized matrix was found to be dose dependent and due to fluid shear stress rather than chemotransport (Bancroft et al., 2002; Sikavitsas et al., 2003). Using a cylindrical-pore model, the shear stress experienced by MSCs was estimated to be less than 1 dyne/cm$^2$; when the pore size of the scaffold was decreased, thereby reducing the exerted shear stress, greater mineral deposition was observed after 16 days of culture

(Holtorf *et al.*, 2005). These results indicate that fluid shear stress influences the osteoblastic differentiation of MSCs during *in vitro* culture on 3D scaffolds. Additionally, it was demonstrated that MSCs cultured in the perfusion bioreactor deposited an extracellular matrix (ECM) rich in osteoblastic differentiation factors; fresh MSCs seeded on this ECM showed enhanced osteoblastic differentiation (Datta *et al.*, 2006). Further study is required to investigate the role of fluid flow on MSC differentiation *in vivo* and the signaling pathways affected by exposure to fluid flow.

## V. ACKNOWLEDGMENTS

In the earlier editions of this volume, this chapter was co-authored with Melvin L. Moss (January 3, 1923–June 25, 2006), a passionate scientist and teacher.

## VI. REFERENCES

Aarden, E. M., Burger, E. H., and Nijweide, P. J. (1994). Function of osteocytes in bone. *J. Cell Biochem.* **55**(3), 287–299.

Andrews, N. A. (2006). New results sharpen the focus on mechanosensory mechanisms. *BoneKEy-Osteovision.* **3**(12), 7–10.

Atkinson, P. J., and Hallsworth, A. S. (1983). The changing structure of aging human mandibular bone. *Gerodontology* **2**, 57–66.

Bacabac, R. G., Smit, T. H., Mullender, M. G., Dijcks, S. J., Van Loon, J. J., and Klein-Nulend, J. (2004). Nitric oxide production by bone cells is fluid shear stress rate dependent. *Biochem. Biophys. Res. Commun.* **315**(4), 823–829.

Bakker, A. D., Soejima, K., Klein-Nulend, J., and Burger, E. H. (2001). The production of nitric oxide and prostaglandin e(2) by primary bone cells is shear stress dependent. *J. Biomech.* **34**(5), 671–677.

Bakker, A., Klein-Nulend, J., and Burger, E. (2004). Shear stress inhibits while disuse promotes osteocyte apoptosis. *Biochem. Biophys. Res. Commun.* **320**(4), 1163–1168.

Bancroft, G. N., Sikavitsast, V. I., van den Dolder, J., Sheffield, T. L., Ambrose, C. G., Jansen, J. A., and Mikos, A. G. (2002). Fluid flow increases mineralized matrix deposition in 3d perfusion culture of marrow stromal osteloblasts in a dose-dependent manner. *Proc. Natl. Acad. Sci. U.S.A.* **99**(20), 12600–12605.

Batra, N. N., Li, Y. J., Yellowley, C. E., You, L., Malone, A. M., Kim, C. H., and Jacobs, C. R. (2005). Effects of short-term recovery periods on fluid-induced signaling in osteoblastic cells. *J. Biomech.* **38**(9), 1909–1917.

Bennett, M. V., and Goodenough, D. A. (1978). Gap junctions, electrotonic coupling, and intercellular communication. *Neurosci. Res. Program Bull.* **16**(3), 1–486.

Bhalla, U. S., and Iyengar, R. (1999). Emergent properties of networks of biological signaling pathways. *Science* **283**(5400), 381–387.

Bingmann, D., Tetsch, D., and Fritsch, J. (1988). Membrane properties of bone cells derived from calvaria of newborn rats (tissue culture). *Pfluger's Arch.* **S412**, R14–R15.

Bingmann, D., Tetsch, D., and Fritsch, J. (1998). Membraneigenschaften von zellen aus knochenexplantaten. *Z. Zahnartzl. Implantol.* **IV**, 277–281.

Brighton, C. T., Okereke, E., Pollack, S. R., and Clark, C. C. (1992). *In vitro* bone-cell response to a capacitively coupled electrical field. The role of field strength, pulse pattern, and duty cycle. *Clin. Orthop. Rel. Res.* **285**, 255–262.

Brown, T. D., Pedersen, D. R., Gray, M. L., Brand, R. A., and Rubin, C. T. (1990). Toward an identification of mechanical parameters initiating periosteal remodeling: a combined experimental and analytic approach. *J. Biomech.* **23**(9), 893–905.

Burger, E. L., Klein Nulend, J., and Cowin, S. C. (1998). Mechanotransduction in bone. In "Advances in Organ Biology" (M. Zaidi, ed.), pp. 107–118. JAI Press, London.

Carter, D. R., and Hayes, W. C. (1976). Bone compressive strength: the influence of density and strain rate. *Science* **194**(4270), 1174–1176.

Carvalho, R. S., Scott, J. E., Suga, D. M., and Yen, E. H. (1994). Stimulation of signal transduction pathways in osteoblasts by mechanical strain potentiated by parathyroid hormone. *J. Bone Miner. Res.* **9**(7), 999–1011.

Carvalho, R. S., Schaffer, J. L., and Gerstenfeld, L. C. (1998). Osteoblasts induce osteopontin expression in response to attachment on fibronectin: demonstration of a common role for integrin receptors in the signal transduction processes of cell attachment and mechanical stimulation. *J. Cell. Biochem.* **70**(3), 376–390.

Chen, N. X., Ryder, K. D., Pavalko, F. M., Turner, C. H., Burr, D. B., Qiu, J., and Duncan, R. L. (2000). $Ca^{2+}$ regulates fluid shear–induced cytoskeletal reorganization and gene expression in osteoblasts. *Am. J. Physiol. Cell Physiol.* **278**(5), C989–C997.

Cheng, B., Kato, Y., Zhao, S., Luo, J., Sprague, E., Bonewald, L. F., and Jiang, J. X. (2001). Pge(2) is essential for gap junction–mediated intercellular communication between osteocyte-like mlo-y4 cells in response to mechanical strain. *Endocrinology* **142**(8), 3464–3473.

Cherian, P. P., Cheng, B., Gu, S., Sprague, E., Bonewald, L. F., and Jiang, J. X. (2003). Effects of mechanical strain on the function of gap junctions in osteocytes are mediated through the prostaglandin ep2 receptor. *J. Biol. Chem.* **278**(44), 43146–43156.

Cherian, P. P., Siller-Jackson, A. J., Gu, S., Wang, X., Bonewald, L. F., Sprague, E., and Jiang, J. X. (2005). Mechanical strain opens connexin 43 hemichannels in osteocytes: a novel mechanism for the release of prostaglandin. *Mol. Biol. Cell* **16**(7), 3100–3106.

Chesnoy-Marchais, D., and Fritsch, J. (1988). Voltage-gated sodium and calcium currents in rat osteoblasts. *J. Physiol.* **398**, 291–311.

Civitelli, R. (1995). Cell–cell communication in bone. *Calcified Tissue Int.* **56**, S29–S31.

Cowin, S. C. (1998). On mechanosensation in bone under microgravity. *Bone* **22**(5 Suppl.), 119S–125S.

Cowin, S. C. (1999). Bone poroelasticity. *J. Biomech.* **32**(3), 217–238.

Cowin, S. C., and Hegedus, D. H. (1976). Bone remodeling. 1. Theory of adaptive elasticity. *J. Elasticity* **6**(3), 313–326.

Cowin, S. C., and Weinbaum, S. (1998). Strain amplification in the bone mechanosensory system. *Am. J. Med. Sci.* **316**(3), 184–188.

Cowin, S. C., Moss-Salentijn, L., and Moss, M. L. (1991). Candidates for the mechanosensory system in bone. *J. Biomech. Eng.* **113**(2), 191–197.

Cowin, S. C., Weinbaum, S., and Zeng, Y. (1995). A case for bone canaliculi as the anatomical site of strain generated potentials. *J. Biomech.* **28**(11), 1281–1297.

Curtis, T. A., Ashrafi, S. H., and Weber, D. F. (1985). Canalicular communication in the cortices of human long bones. *Anat. Rec.* **212**(4), 336–344.

Datta, N., Pham, Q. P., Sharma, U., Sikavitsas, V. I., Jansen, J. A., and Mikos, A. G. (2006). *In vitro* generated extracellular matrix and fluid shear stress synergistically enhance 3d osteoblastic differentiation. *Proc. Natl. Acad. Sci. U.S.A.* **103**(8), 2488–2493.

Davidson, R. M., Tatakis, D. W., and Auerbach, A. L. (1990). Multiple forms of mechanosensitive ion channels in osteoblast-like cells. *Pflugers Arch.* **416**(6), 646–651.

Dayhoff, J. E., Hameroff, S. R., and Lahoz-Beltra, R. (1992). Intracellular mechanisms in neuronal learning: adaptive models. *Int. Joint Conf. Neural Networks* **I73–I78**.

De Mello, W. C. (1987). The ways cells communicate. *In* "Cell-to-Cell Communication" (W. C. DeMello, ed.), pp. 1–20. Plenum Press, New York.

Dillaman, R. M. (1984). Movement of ferritin in the 2-day-old chick femur. *Anat. Rec.* **209**(4), 445–453.

Dolce, C., Kinniburgh, A. J., and Dziak, R. (1996). Immediate early-gene induction in rat osteoblastic cells after mechanical deformation. *Arch. Oral Biol.* **41**(12), 1101–1108.

Doty, S. B. (1989). Cell-to-cell communication in bone tissue. *In* "The Biological Mechanism of Tooth Eruption and Root Resorption" (Z. Davidovitch, ed.), pp. 61–69. EBSCO Media, Birmingham, AL.

Duncan, R., and Misler, S. (1989). Voltage-activated and stretch-activated Ba$^{2+}$-conducting channels in an osteoblast-like cell line (umr 106). *FEBS Lett.* **251**(1–2), 17–21.

Edin, B. B., and Trulsson, M. (1992). Neural network analysis of the information content in population responses from human periodontal receptors. *Sci. Neural Networks* **1710**, 257–266.

Ferrier, J., Grygorczyk, C., Grygorczyk, R., Kesthely, A., Lagan, E., and Xia, S. L. (1991). Ba$^{(2+)}$-induced action potentials in osteoblastic cells. *J. Membr. Biol.* **123**(3), 255–259.

Forwood, M. R. (1996). Inducible cyclo-oxygenase (cox-2) mediates the induction of bone formation by mechanical loading *in vivo*. *J. Bone Miner. Res.* **11**(11), 1688–1693.

French, A. S. (1992). Mechanotransduction. *Annu. Rev. Physiol.* **54**, 135–152.

Fritton, S. P., McLeod, K. J., and Rubin, C. T. (2000). Quantifying the strain history of bone: spatial uniformity and self-similarity of low-magnitude strains. *J. Biomech.* **33**(3), 317–325.

Fritzke, B. (1994). Growing cell structures — a self-organizing network for unsupervised and supervised learning. *Neural Networks* **7**(9), 1441–1460.

Frost, H. M. (1964). "The Laws of Bone Structure." Charles C. Thomas, Springfield, IL.

Fuxe, J., and Agnati, L. F., eds. (1991). "Volume Transmission in the Brain." Raven Press, New York.

Fyhrie, D. P., and Carter, D. R. (1986). A unifying principle relating stress to trabecular bone morphology. *J. Orthop. Res.* **4**(3), 304–317.

Ghazi, A., Berrier, C., Ajouz, B., and Besnard, M. (1998). Mechanosensitive ion channels and their mode of activation. *Biochimie* **80**(5–6), 357–362.

Gilbertson, T. A. (1998). Peripheral mechanisms of taste. *In* "The Scientific Basis of Eating" (R. W. A. Linden, ed.), pp. 1–28. Karger, Basel.

Goldsmith, P. (1994). Plant stems: a possible model system for the transduction of mechanical information in bone modeling. *Bone* **15**(3), 249–250.

Goldstein, S. A., Matthews, L. S., Kuhn, J. L., and Hollister, S. J. (1991). Trabecular bone remodeling: an experimental model. *J. Biomech.* **24**(Suppl. 1), 135–150.

Gross, T. S., Edwards, J. L., McLeod, K. J., and Rubin, C. T. (1997). Strain gradients correlate with sites of periosteal bone formation. *J. Bone Miner. Res.* **12**(6), 982–988.

Guggino, S. E., Lajeunesse, D., Wagner, J. A., and Snyder, S. H. (1989). Bone remodeling signaled by a dihydropyridine- and phenylalkyl-amine-sensitive calcium channel. *Proc. Natl. Acad. Sci. U.S.A.* **86**(8), 2957–2960.

Hackney, C. M., and Furness, D. N. (1995). Mechanotransduction in vertebrate hair cells: structure and function of the stereociliary bundle. *Am. J. Physiol.* **268**(1 Pt. 1), C1–C13.

Hamill, O. P., and McBride, D. W., Jr. (1996). The pharmacology of mechanogated membrane ion channels. *Pharmacol. Rev.* **48**(2), 231–252.

Han, Y., Cowin, S. C., Schaffler, M. B., and Weinbaum, S. (2004). Mechanotransduction and strain amplification in osteocyte cell processes. *Proc. Natl. Acad. Sci. U.S.A.* **101**(47), 16689–16694.

Harrigan, T. P., and Hamilton, J. J. (1993). Bone strain sensation via transmembrane potential changes in surface osteoblasts: loading rate and microstructural implications. *J. Biomech.* **26**(2), 183–200.

Harter, L. V., Hruska, K. A., and Duncan, R. L. (1995). Human osteoblast-like cells respond to mechanical strain with increased bone matrix protein production independent of hormonal regulation. *Endocrinology* **136**(2), 528–535.

Haskin, C., Cameron, I., and Athanasiou, K. (1993). Physiological levels of hydrostatic pressure alter morphology and organization of cytoskeletal and adhesion proteins in mg-63 osteosarcoma cells. *Biochem. Cell. Biol.* **71**(1–2), 27–35.

Hecker, M., Mulsch, A., Bassenge, E., and Busse, R. (1993). Vasoconstriction and increased flow: two principal mechanisms of shear stress–dependent endothelial autacoid release. *Am. J. Physiol.* **265**(3 Pt. 2), H828–H833.

Hert, J., Liskova, M., and Landrgot, B. (1969). Influence of the long-term, continuous bending on the bone. An experimental study on the tibia of the rabbit. *Folia Morphol. (Praha)* **17**(4), 389–399.

Hert, J., Liskova, M., and Landa, J. (1971). Reaction of bone to mechanical stimuli. 1. Continuous and intermittent loading of tibia in rabbit. *Folia Morphol. (Praha)* **19**(3), 290–300.

Hert, J., Pribylova, E., and Liskova, M. (1972). Reaction of bone to mechanical stimuli. 3. Microstructure of compact bone of rabbit tibia after intermittent loading. *Acta Anat. (Basel)* **82**(2), 218–230.

Hinton, G. E., and Anderson, J. A. (1989). "Parallel Models of Associative Memory." Lawrence Erlbaum, Hillsdale, NJ.

Holtorf, H. L., Datta, N., Jansen, J. A., and Mikos, A. G. (2005). Scaffold mesh size affects the osteoblastic differentiation of seeded marrow stromal cells cultured in a flow perfusion bioreactor. *J. Biomed. Mater. Res. A* **74**(2), 171–180.

Hughes, D. E., Salter, D. M., Dedhar, S., and Simpson, R. (1993). Integrin expression in human bone. *J. Bone Miner. Res.* **8**(5), 527–533.

Huiskes, R., Weinans, H., Grootenboer, H. J., Dalstra, M., Fudala, B., and Slooff, T. J. (1987). Adaptive bone-remodeling theory applied to prosthetic-design analysis. *J. Biomech.* **20**(11–12), 1135–1150.

Hung, C. T., Pollack, S. R., Reilly, T. M., and Brighton, C. T. (1995). Real-time calcium response of cultured bone cells to fluid flow. *Clin. Orthop. Rel. Res.* **313**, 256–269.

Ingber, D. E. (1998). Cellular basis of mechanotransduction. *Biol. Bull.* **194**(3), 323–325; discussion 325–327.

Jan, L. Y., and Jan, Y. N. (1992). Tracing the roots of ion channels. *Cell* **69**(5), 715–718.

Janmey, P. A. (1998). The cytoskeleton and cell signaling: component localization and mechanical coupling. *Physiol. Rev.* **78**(3), 763–781.

Jeansonne, B. G., Feagin, F. F., McMinn, R. W., Shoemaker, R. L., and Rehm, W. S. (1979). Cell-to-cell communication of osteoblasts. *J. Dent. Res.* **58**(4), 1415–1423.

Johnson, M. W., Chakkalakal, D. A., Harper, R. A., Katz, J. L., and Rouhana, S. W. (1982). Fluid flow in bone *in vitro*. *J. Biomech.* **15**(11), 881–885.

Jones, D. B., and Bingmann, D. (1991). How do osteoblasts respond to mechanical stimulation? *Cells Mater.* **1**, 329–340.

Jones, S. J., Gray, C., Sakamaki, H., Arora, M., Boyde, A., Gourdie, R., and Green, C. (1993). The incidence and size of gap junctions between the bone cells in rat calvaria. *Anat. Embryol. (Berl.)* **187**(4), 343–352.

Kam, E., and Hodgins, M. B. (1992). Communication compartments in hair follicles and their implication in differentiative control. *Development* **114**(2), 389–393.

Kamiya, A., Bukhari, R., and Togawa, T. (1984). Adaptive regulation of wall shear stress optimizing vascular tree function. *Bull. Math. Biol.* **46**(1), 127–137.

Keynes, R. D. (1994). The kinetics of voltage-gated ion channels. *Q. Rev. Biophys.* **27**(4), 339–434.

Klein-Nulend, J., Semeins, C. M., Ajubi, N. E., Nijweide, P. J., and Burger, E. H. (1995a). Pulsating fluid flow increases nitric oxide (no) synthesis by osteocytes but not periosteal fibroblasts — correlation with prostaglandin up-regulation. *Biochem. Biophys. Res. Comm.* **217**(2), 640–648.

Klein-Nulend, J., van der Plas, A., Semeins, C. M., Ajubi, N. E., Frangos, J. A., Nijweide, P. J., and Burger, E. H. (1995b). Sensitivity of osteocytes to biomechanical stress *in vitro*. *FASEB J.* **9**(5), 441–445.

Kodama, R., and Eguchi, G. (1994). The loss of gap junctional cell-to-cell communication is coupled with dedifferentiation of retinal pigmented epithelial cells in the course of transdifferentiation into the lens. *Int. J. Dev. Biol.* **38**(2), 357–364.

Kufahl, R. H., and Saha, S. (1990). A theoretical model for stress-generated fluid flow in the canaliculi-lacunae network in bone tissue. *J. Biomech.* **23**(2), 171–180.

Kupfermann, I. (1992). Neural networks — they do not have to be complex to be complex. *Behav. Brain Sci.* **15**(4), 767–768.

Lanyon, L. E. (1984). Functional strain as a determinant for bone remodeling. *Calcif. Tissue Int.* **36**(Suppl. 1), S56–S61.

Lanyon, L. E. (1993). Osteocytes, strain detection, bone modeling and remodeling. *Calcif. Tissue Int.* **53**(Suppl. 1), S102–S106; discussion S106–S107.

Lecanda, F., Towler, D. A., Ziambaras, K., Cheng, S. L., Koval, M., Steinberg, T. H., and Civitelli, R. (1998). Gap junctional communication modulates gene expression in osteoblastic cells. *Mol. Biol. Cell* **9**(8), 2249–2258.

Lecanda, F., Warlow, P. M., Sheikh, S., Furlan, F., Steinberg, T. H., and Civitelli, R. (2000). Connexin43 deficiency causes delayed ossification, craniofacial abnormalities, and osteoblast dysfunction. *J. Cell. Biol.* **151**(4), 931–944.

Machwate, M., Jullienne, A., Moukhtar, M., and Marie, P. J. (1995). Temporal variation of c-fos proto-oncogene expression during osteoblast differentiation and osteogenesis in developing rat bone. *J. Cell. Biochem.* **57**(1), 62–70.

Malone, A. M. D., Anderson, C. T., Temiyasathit, S., Tang, J., Tummala, P., Sterns, T., and Jacobs, C. R. (2006). Primary cilia: mechanosensory organelles in bone cells. *J. Bone Min. Res.* **21**(Suppl 1), S39.

Marotti, G. (1996). The structure of bone tissues and the cellular control of their deposition. *Ital. J. Anat. Embryol.* **101**(4), 25–79.

Marotti, G., Ferretti, M., Muglia, M. A., Palumbo, C., and Palazzini, S. (1992). A quantitative evaluation of osteoblast–osteocyte relationships on growing endosteal surface of rabbit tibiae. *Bone* **13**(5), 363–368.

Martin, R. B., and Burr, D. B. (1982). A hypothetical mechanism for the stimulation of osteonal remodelling by fatigue damage. *J. Biomech.* **15**(3), 137–139.

Martin, T. J., and Ng, K. W. (1994). Mechanisms by which cells of the osteoblast lineage control osteoclast formation and activity. *J. Cell Biochem.* **56**(3), 357–366.

Martino, R. L., Johnson, C. A., Suh, E. B., Trus, B. L., and Yap, T. K. (1994). Parallel computing in biomedical research. *Science* **265**(5174), 902–908.

Massass, R., Bingmann, D., Korenstein, R., and Tetsch, P. (1990). Membrane potential of rat calvaria bone cells: dependence on temperature. *J. Cell Physiol.* **144**(1), 1–11.

McAllister, T. N., and Frangos, J. A. (1999). Steady and transient fluid shear stress stimulate no release in osteoblasts through distinct biochemical pathways. *J. Bone Miner. Res.* **14**(6), 930–936.

McClelland, J. L., and Rumelhart, D. E. (1987). "Parallel Distributed Processing. Psychological and Biological Models." M.I.T. Press, Cambridge, MA.

McGarry, J. G., Klein-Nulend, J., and Prendergast, P. J. (2005). The effect of cytoskeletal disruption on pulsatile fluid flow–induced nitric oxide and prostaglandin e-2 release in osteocytes and osteoblasts. *Biochem. Biophys. Res. Comm.* **330**(1), 341–348.

McLeod, K. J., and Rubin, C. T. (1992). The effect of low-frequency electrical fields on osteogenesis. *J. Bone Joint Surg. Am.* **74**(6), 920–929.

McLeod, K. J., Donahue, H. J., Levin, P. E., Fontaine, M. A., and Rubin, C. T. (1993). Electric fields modulate bone cell function in a density-dependent manner. *J. Bone Miner. Res.* **8**(8), 977–984.

McMahon, T. A. (1984). "Muscles, Reflexes, and Locomotion." Princeton University Press, Princeton, NJ.

Meazzini, M. C., Toma, C. D., Schaffer, J. L., Gray, M. L., and Gerstenfeld, L. C. (1998). Osteoblast cytoskeletal modulation in response to mechanical strain *in vitro*. *J. Orthop. Res.* **16**(2), 170–180.

Minkoff, R., Bales, E. S., Kerr, C. A., and Struss, W. E. (1999). Antisense oligonucleotide blockade of connexin expression during embryonic bone formation: evidence of functional compensation within a multigene family. *Dev. Genet.* **24**(1–2), 43–56.

Montgomery, R. J., Sutker, B. D., Bronk, J. T., Smith, S. R., and Kelly, P. J. (1988). Interstitial fluid flow in cortical bone. *Microvasc. Res.* **35**(3), 295–307.

Moreno, A. P., Rook, M. B., Fishman, G. I., and Spray, D. C. (1994). Gap junction channels: distinct voltage-sensitive and -insensitive conductance states. *Biophys. J.* **67**(1), 113–119.

Morris, C. E. (1990). Mechanosensitive ion channels. *J. Membr. Biol.* **113**(2), 93–107.

Moss, M. L. (1962). The functional matrix. *In* "Vistas in Orthodontics" (B. Kraus and R. Reidel, eds.). Lea and Febiger, Philadelphia.

Moss, M. L. (1968). A theoretical analysis of the functional matrix. *Acta Biotheor.* **18**(1), 195–202.

Moss, M. L. (1978). "The Muscle–Bone Interface: An Analysis of a Morphological Boundary." Center for Human Growth and Development, Ann Arbor, MI.

Moss, M. L. (1991a). Alternate mechanisms of bone remodeling: their representation in a connected cellular network model. *Ann. Biomed. Eng.* **19**, 636.

Moss, M. L. (1991b). Bone as a connected cellular network: modeling and testing. *In* "Topics in Biomedical Engineering" (G. Ross, ed.), pp. 117–119. Pergamon Press, New York.

Moss, M. L. (1997a). The functional matrix hypothesis revisited. 1. The role of mechanotransduction. *Am. J. Orthod. Dentofacial. Orthop.* **112**(1), 8–11.

Moss, M. L. (1997b). The functional matrix hypothesis revisited. 2. The role of an osseous connected cellular network. *Am. J. Orthod. Dentofacial. Orthop.* **112**(2), 221–226.

Moss, M. L. (1997c). The functional matrix hypothesis revisited. 3. The genomic thesis. *Am. J. Orthod. Dentofacial. Orthop.* **112**(3), 338–342.

Moss, M. L. (1997d). The functional matrix hypothesis revisited. 4. The epigenetic antithesis and the resolving synthesis. *Am. J. Orthod. Dentofacial. Orthop.* **112**(4), 410–417.

Moss, M. L., and Salentijn, L. (1969). The capsular matrix. *Am. J. Orthod.* **56**(5), 474–490.

Moss, M. L., and Young, R. W. (1960). A functional approach to craniology. *Am. J. Phys. Anthropol.* **18**, 281–292.

Moss-Salentijn, L. (1992). The human tactile system. *In* "Advanced Tactile Sensing for Robotics" (H. R. Nicholls, ed.). World Scientific, Singapore.

Norvell, S. M., Ponik, S. M., Bowen, D. K., Gerard, R., and Pavalko, F. M. (2004). Fluid shear stress induction of cox-2 protein and prostaglandin release in cultured mc3t3-e1 osteoblasts does not require intact microfilaments or microtubules. *J. Appl. Physiol.* **96**(3), 957–966.

Nowak, R. (1992). Cells that fire together, wire together. *J. NIH Res.* **4**, 60–64.

O'Connor, J. A., Lanyon, L. E., and MacFie, H. (1982). The influence of strain rate on adaptive bone remodelling. *J. Biomech.* **15**(10), 767–781.

Olsson, S., and Hanson, B. S. (1995). Action potential-like activity found in fungal mycelia is sensitive to stimulation. *Naturwissch* **82**, 30–31.

Otter, M. W., Palmieri, V. R., Wu, D. D., Seiz, K. G., MacGinitie, L. A., and Cochran, G. V. (1992). A comparative analysis of streaming potentials *in vivo* and *in vitro*. *J. Orthop. Res.* **10**(5), 710–719.

Otter, M. W., McLeod, K. J., and Rubin, C. T. (1998). Effects of electromagnetic fields in experimental fracture repair. *Clin. Orthop. Rel. Res.* (355 Suppl.), S90–S104.

Owan, I., Burr, D. B., Turner, C. H., Qiu, J., Tu, Y., Onyia, J. E., and Duncan, R. L. (1997). Mechanotransduction in bone: osteoblasts are more responsive to fluid forces than mechanical strain. *Am. J. Physiol.* **273**(3 Pt. 1), C810–C815.

Parfitt, A. M. (1994). Osteonal and hemi-osteonal remodeling: the spatial and temporal framework for signal traffic in adult human bone. *J. Cell. Biochem.* **55**(3), 273–286.

Pavalko, F. M., Chen, N. X., Turner, C. H., Burr, D. B., Atkinson, S., Hsieh, Y. F., Qiu, J., and Duncan, R. L. (1998a). Cytoskeletal-integrin interactions are required for fluid shear–induced mechanical signaling in mc-3t3-e1 osteoblasts. *Mol. Biol. Cell* **9**, 163A–163A.

Pavalko, F. M., Chen, N. X., Turner, C. H., Burr, D. B., Atkinson, S., Hsieh, Y. F., Qiu, J., and Duncan, R. L. (1998b). Fluid shear–induced mechanical signaling in mc3t3-e1 osteoblasts requires cytoskeleton–integrin interactions. *Am. J. Physiol.* **275**(6 Pt. 1), C1591–C1601.

Pavalko, F. M., Gerard, R. L., Ponik, S. M., Gallagher, P. J., Jin, Y., and Norvell, S. M. (2003a). Fluid shear stress inhibits tnf-alpha-induced apoptosis in osteoblasts: a role for fluid shear stress–induced activation of pi3-kinase and inhibition of caspase-3. *J. Cell Physiol.* **194**(2), 194–205.

Pavalko, F. M., Norvell, S. M., Burr, D. B., Turner, C. H., Duncan, R. L., and Bidwell, J. P. (2003b). A model for mechanotransduction in bone cells: the load-bearing mechanosomes. *J. Cell. Biochem.* **88**(1), 104–112.

Petrov, A. G., and Usherwood, P. N. (1994). Mechanosensitivity of cell membranes. Ion channels, lipid matrix and cytoskeleton. *Eur. Biophys. J.* **23**(1), 1–19.

Piekarski, K., and Munro, M. (1977). Transport mechanism operating between blood supply and osteocytes in long bones. *Nature* **269**(5623), 80–82.

Pitsillides, A. A., Rawlinson, S. C., Suswillo, R. F., Bourrin, S., Zaman, G., and Lanyon, L. E. (1995). Mechanical strain-induced NO production by bone cells: a possible role in adaptive bone (re)modeling? *FASEB J.* **9**(15), 1614–1622.

Pollack, S. R., Petrov, N., Salzstein, R., Brankov, G., and Blagoeva, R. (1984). An anatomical model for streaming potentials in osteons. *J. Biomech.* **17**(8), 627–636.

Ponik, S. M., and Pavalko, F. M. (2004). Formation of focal adhesions on fibronectin promotes fluid shear stress induction of cox-2 and pge2 release in mc3t3-e1 osteoblasts. *J. Appl. Physiol.* **97**(1), 135–142.

Ravesloot, J. H., Van Houten, R. J., Ypey, D. L., and Nijweide, P. J. (1991). High-conductance anion channels in embryonic chick osteogenic cells. *J. Bone Miner. Res.* **6**(4), 355–363.

Reich, K. M., Gay, C. V., and Frangos, J. A. (1990). Fluid shear stress as a mediator of osteoblast cyclic adenosine monophosphate production. *J. Cell Physiol.* **143**(1), 100–104.

Richardson, A., and Parsons, J. T. (1995). Signal transduction through integrins: a central role for focal adhesion kinase? *Bioessays* **17**(3), 229–236.

Rodan, G. A. (1992). Introduction to bone biology. *Bone* **13**(Suppl. 1), S3–S6.

Rodriguez, A. A., Agre, J. C., Knudtson, E. R., Franke, T. M., and Ng, A. V. (1993). Acoustic myography compared to electromyography during isometric fatigue and recovery. *Muscle Nerve* **16**(2), 188–192.

Rubin, C. T., and Lanyon, L. E. (1984). Regulation of bone formation by applied dynamic loads. *J. Bone Joint Surg. Am.* **66**(3), 397–402.

Rubin, C. T., and Lanyon, L. E. (1987). Osteoregulatory nature of mechanical stimuli: function as a determinant for adaptive remodeling in bone. *J. Orthop. Res.* **5**(2), 300–310.

Rubin, C. T., Donahue, H. J., Rubin, J. E., and McLeod, K. J. (1993). Optimization of electric field parameters for the control of bone remodeling: exploitation of an indigenous mechanism for the prevention of osteopenia. *J. Bone Miner. Res.* **8**(Suppl. 2), S573–S581.

Rubinacci, A., Villa, I., Dondi Benelli, F., Borgo, E., Ferretti, M., Palumbo, C., and Marotti, G. (1998). Osteocyte-bone lining cell system at the origin of steady ionic current in damaged amphibian bone. *Calcif. Tissue Int.* **63**(4), 331–339.

Sachs, F. (1986). Biophysics of mechanoreception. *Membr. Biochem.* **6**(2), 173–195.

Sachs, F. (1988). Mechanical transduction in biological systems. *Crit. Rev. Biomed. Eng.* **16**(2), 141–169.

Sackin, H. (1995). Mechanosensitive channels. *Annu. Rev. Physiol.* **57**, 333–353.

Sadoshima, J., Takahashi, T., Jahn, L., and Izumo, S. (1992). Roles of mechano-sensitive ion channels, cytoskeleton, and contractile activity in stretch-induced immediate-early gene expression and hypertrophy of cardiac myocytes. *Proc. Natl. Acad. Sci. U.S.A.* **89**(20), 9905–9909.

Salter, D. M., Robb, J. E., and Wright, M. O. (1997). Electrophysiological responses of human bone cells to mechanical stimulation: evidence for specific integrin function in mechanotransduction. *J. Bone Miner. Res.* **12**(7), 1133–1141.

Salzstein, R. A., and Pollack, S. R. (1987). Electromechanical potentials in cortical bone — ii. Experimental analysis. *J. Biomech.* **20**(3), 271–280.

Salzstein, R. A., Pollack, S. R., Mak, A. F., and Petrov, N. (1987). Electromechanical potentials in cortical bone — i. A continuum approach. *J. Biomech.* **20**(3), 261–270.

Schaul, N. (1998). The fundamental neural mechanisms of electroencephalography. *Electroencephalogr. Clin. Neurophysiol.* **106**(2), 101–107.

Schirrmacher, K., Schmitz, I., Winterhager, E., Traub, O., Brummer, F., Jones, D., and Bingmann, D. (1992). Characterization of gap junctions between osteoblast-like cells in culture. *Calcif. Tissue Int.* **51**(4), 285–290.

Schirrmacher, K., Brummer, F., Dusing, R., and Bingmann, D. (1993). Dye and electric coupling between osteoblast-like cells in culture. *Calcif. Tissue Int.* **53**(1), 53–60.

Scott, G. C., and Korostoff, E. (1990). Oscillatory and step response electromechanical phenomena in human and bovine bone. *J. Biomech.* **23**(2), 127–143.

Shapiro, F., Cahill, C., Malatantis, G., and Nayak, R. C. (1995). Transmission electron microscopic demonstration of vimentin in rat osteoblast and osteocyte cell bodies and processes using the immunogold technique. *Anat. Rec.* **241**(1), 39–48.

Shyy, J. Y., and Chien, S. (1997). Role of integrins in cellular responses to mechanical stress and adhesion. *Curr. Opin. Cell. Biol.* **9**(5), 707–713.

Sikavitsas, V. I., Bancroft, G. N., Holtorf, H. L., Jansen, J. A., and Mikos, A. G. (2003). Mineralized matrix deposition by marrow stromal osteoblasts in 3d perfusion culture increases with increasing fluid shear forces. *Proc. Natl. Acad. Sci. U.S.A.* **100**(25), 14683–14688.

Spadaro, J. A. (1997). Mechanical and electrical interactions in bone remodeling. *Bioelectromagnetics* **18**(3), 193–202.

Spray, D. C. (1998). Gap junction proteins: where they live and how they die. *Circ. Res.* **83**(6), 679–681.

Tanaka, T., and Sakano, A. (1985). Differences in permeability of microperoxidase and horseradish peroxidase into the alveolar bone of developing rats. *J. Dent. Res.* **64**(6), 870–876.

Thi, M. M., Kojima, T., Cowin, S. C., Weinbaum, S., and Spray, D. C. (2003). Fluid shear stress remodels expression and function of junctional proteins in cultured bone cells. *Am. J. Physiol. Cell Physiol.* **284**(2), C389–C403.

Uitto, V. J., and Larjava, H. (1991). Extracellular matrix molecules and their receptors: an overview with special emphasis on periodontal tissues. *Crit. Rev. Oral Biol. Med.* **2**(3), 323–354.

Vance, J., Galley, S., Liu, D. F., and Donahue, S. W. (2005). Mechanical stimulation of mc3t3 osteoblastic cells in a bone tissue-engineering bioreactor enhances prostaglandin e2 release. *Tissue Eng.* **11**(11–12), 1832–1839.

Wang, L. Y., Ciani, C., Doty, S. B., and Fritton, S. P. (2004). Delineating bone's interstitial fluid pathway *in vivo*. *Bone* **34**(3), 499–509.

Wang, N., Butler, J. P., and Ingber, D. E. (1993). Mechanotransduction across the cell surface and through the cytoskeleton. *Science* **260**(5111), 1124–1127.

Wasserman, P. D. (1989). "Neural Computation. Theory and Practice." van Nostrand Reinhold, New York.

Watanabe, H., Miake, K., and Sasaki, J. (1993). Immunohistochemical study of the cytoskeleton of osteoblasts in the rat calvaria. Intermediate filaments and microfilaments as demonstrated by detergent perfusion. *Acta Anat. (Basel)* **147**(1), 14–23.

Weinbaum, S., Cowin, S. C., and Zeng, Y. (1991). A model for the fluid shear stress excitation of membrane ion channels in osteocytic processes due to bone strain. *In* "Advances in Bioengineering" (J. R. Vanderby, ed.). American Society of Mechanical Engineers, New York.

Weinbaum, S., Cowin, S. C., and Zeng, Y. (1994). A model for the excitation of osteocytes by mechanical loading-induced bone fluid shear stresses. *J. Biomech.* **27**(3), 339–360.

Whitfield, J. F. (2003). Primary cilium: Is it an osteocyte's strain-sensing flowmeter? *J. of Cell. Biochem.* **89**, 233–237.

Wildon, D. C., Thain, J. F., Minchin, P. E. H., Gubb, I. R., Reilly, A. J., Skipper, Y. D., Doherty, H. M., Odonnell, P. J., and Bowles, D. J. (1992). Electrical signaling and systemic proteinase-inhibitor induction in the wounded plant. *Nature* **360**(6399), 62–65.

Williams, J. L., Iannotti, J. P., Ham, A., Bleuit, J. and Chen, J. H. (1994). Effects of fluid shear stress on bone cells. *Biorheology* **31**(2), 163–170.

Wilson, D. A., and Sullivan, R. M. (1998). Peripheral mechanisms of smell. *In* "The Scientific Basis of Eating" (R. W. A. Linden, ed.). Karger, Basel, Switzerland.

Wolff, J. (1870). Uber der innere architektur der knochen und ihre bedeutung fur die frage vom knochenwachstum. *Arch. path. anat. physio. med.* **50**, 389–453.

Wolff, J. (1892). "Das Gesetz der Transformation der Knochen." Hirschwald, Berlin.

Wolff, J. (1986). "The Law of Bone Remodeling." Springer, Berlin.

Xiao, Z., Zhang, S., Mahlios, J., Zhou, G., Magenheimer, B. S., Guo, D., Dallas, S. L., Maser, R., Calvet, J. P., Bonewald, L., and Quarles, L. D. (2006). Cilia-like structures and polycystin-1 in osteoblasts/osteocytes and associated abnormalities in skeletogenesis and Runx2 expression. *J. Biol. Chem.* **281**(41), 30,884–30,895.

Yanagishita, M. (1993). Function of proteoglycans in the extracellular matrix. *Acta Pathol. Jpn.* **43**(6), 283–293.

You, J., Yellowley, C. E., Donahue, H. J., Zhang, Y., Chen, Q., and Jacobs, C. R. (2000). Substrate deformation levels associated with routine physical activity are less stimulatory to bone cells relative to loading-induced oscillatory fluid flow. *J. Biomech. Eng.* **122**(4), 387–393.

You, L., Cowin, S. C., Schaffler, M. B., and Weinbaum, S. (2001). A model for strain amplification in the actin cytoskeleton of osteocytes due to fluid drag on pericellular matrix. *J. Biomech.* **34**(11), 1375–1386.

Zeng, Y., Cowin, S. C., and Weinbaum, S. (1994). A fiber matrix model for fluid flow and streaming potentials in the canaliculi of an osteon. *Ann. Biomed. Eng.* **22**(3), 280–292.

Zhang, D., Cowin, S. C., and Weinbaum, S. (1997). Electrical signal transmission and gap junction regulation in a bone cell network: a cable model for an osteon. *Ann. Biomed. Eng.* **25**(2), 357–374.

Zhang, D., Weinbaum, S., and Cowin, S. C. (1998a). Electrical signal transmission in a bone cell network: the influence of a discrete gap junction. *Ann. Biomed. Eng.* **26**(4), 644–659.

Zhang, D., Weinbaum, S., and Cowin, S. C. (1998b). Estimates of the peak pressures in bone pore water. *J. Biomech. Eng.* **120**(6), 697–703.

Zorntzer, S. F., Davis, J., and Lau, C. (1990). "An Introduction to Neural and Electronic Networks." Academic Press, San Diego, CA.

# Chapter Sixty-Two

# Skeletal-Tissue Engineering

*Matthew D. Kwan, Derrick C. Wan, and Michael T. Longaker*

## I. INTRODUCTION

As the U.S. population ages, skeletal-tissue injury and loss from trauma or disease exact a burgeoning socioeconomic burden on the health care system. Recent data place U.S. health care expenditures at over $1.8 trillion (HCUP, 2006). Skeletal-related injuries and diseases represent a significant portion of this cost. According to 2003 data from the U.S. Health Cost and Utilization Project, there were approximately 1 million hospital admissions for appendicular skeletal fractures in the United States, with an aggregate cost of over $26 billion (HCUP, 2006). Costs for skull and facial fractures contributed an additional $1.3 billion. Of all fractures, approximately 10% are complicated with impaired healing (Logeart-Avramoglou *et al.*, 2005). Thus, regenerative medicine offers an exciting option for addressing the reconstruction of osseous defects. With the culmination of developments in materials science, bioengineering, stem cell biology, and molecular biology, the potential of tissue engineering to accelerate bone regeneration becomes a distinct possibility.

The clinician is confronted with an array of large bone defects resulting from trauma, resection of malignancies, nonunion of fractures, and congenital malformations. Autografts remain the best option for reconstructing bony defects because they provide the osteogenic cells, osteoinductive factors, and lattice needed for bone regeneration. They can be obtained from the patient's calvarium, ribs, iliac crest, distal femur, greater trochanter, or proximal tibia (Perry, 1999). Autografts, however, subject the patient to the inherent risks of surgery during harvest (Laurencin *et al.*, 2006). In a retrospective study of autografts harvested from the iliac crest, a complication rate nearing 10% was found, including infection, fracture, large hematomas, pain, paresthesia, and nerve injury (Younger and Chapman, 1989). Furthermore, autografts are a relatively scarce resource and are limited by the amount of bone that can be removed without causing undue morbidity to the donor site. In situations where autografts are not sufficient or possible, allogeneic bone, from donors or cadavers, may be used (Laurencin *et al.*, 2006). However, use of allografts is accompanied by the risk for disease transmission, immunologic rejection, and graft-versus-host disease.

Ceramics and metals are the other major categories of bone substitutes available to repair skeletal defects, each with its associated problems. Metals provide immediate mechanical support but fail to integrate with surrounding tissue (Salgado *et al.*, 2004). Metals are also at higher risk for infection and fatigue loading. Ceramics, mimicking the inorganic composition of bone, tend to integrate well but possess low tensile strength and are very brittle (Temenoff and Mikos, 2000). Numerous other synthetic materials have been developed to aid in bone reconstruction, including the use of plaster of Paris, polymer-based substitutes, and bioactive glass.

The wide range of options available for osseous reconstruction is indicative of the many advances made in bio-

*Principles of Tissue Engineering, 3rd Edition*
*ed. by Lanza, Langer, and Vacanti*

materials, but it also reflects the inadequacies of any single method. As such, tissue engineering portends great promise for being able to address more adequately the issue of repairing bone defects. A foretaste of tissue engineering is already in clinical application, in the form of delivering osteogenic cytokines, specifically bone morphogenetic proteins (BMPs), to induce bone formation. The translational goal is ultimately to deliver both osteogenic cells and cytokines via a biologically active scaffold, in order to heal a bony defect in an accelerated fashion.

This chapter begins with an overview of distraction osteogenesis, a modality that translates directed mechanical force into endogenous bone formation, to emphasize the important role of the mechanical environment on bone formation and to highlight the growth factors that play a role in inducing bone formation. This chapter then discusses the necessary components of bone-tissue engineering, cellular sources, osteogenic cytokines, and scaffolds, with an emphasis on recent developments.

## II. DISTRACTION OSTEOGENESIS

Distraction osteogenesis is relevant to tissue engineering because it provides a model for the study of endogenous bone formation in large skeletal defects. First described by Alexander Codivilla in 1905 for limb lengthening and later codified by Gavril Ilizarov in the 1950s, distraction osteogenesis represents an endogenous form of bone-tissue engineering (Codivilla, 1905; Ilizarov, 1989a; Ilizarov, 1989b). New bone is formed during the process of separating osteogenic fronts with gradual but constant mechanical force. As a first step, an osteotomy is usually performed on the bone of interest, followed by application of rigid fixation. This is followed by a latency period of several days, during which a soft callus forms between the two osteogenic fronts. Gradual distraction is then applied, followed by stable fixation, until a stable osseous regenerate is formed. It is thought that the mechanical forces applied by distraction can contribute to bone formation by induction of cytokines that guide mesenchymal cells in the bony gap to differentiate along an osteogenic lineage. Distraction osteogenesis not only triggers bone formation, but also stimulates local angiogenesis as part of the process (Aronson, 1994; Fang et al., 2005).

Much research has been devoted toward characterizing the association between stress and strain patterns with bone formation. Correlating tensile force measurements with histology, Loboa et al. (2004) demonstrated that the greatest amount of bone formation occurs during active distraction, the period of greatest strain. Loboa went on further to characterize the forces of distraction, using finite-element analysis models created from three-dimensional computed tomography image data of rat mandibles at different phases of distraction osteogenesis (Loboa et al., 2005). The models described patterns of moderate hydrostatic stress within the gap, predictive of intramembranous ossification, and patterns of mild compressive stress in the periphery, consistent

with endochondral ossification. These data derived from finite-element analysis were consistent with previous histological findings.

Great interest surrounds research characterizing how mechanical forces may be translated into molecular signals that promote bone regeneration. Tong et al. (2003) described the role of focal adhesion kinase (FAK), a regulator of the integrin-mediated signal transduction cascade, in distraction osteogenesis. In a rat model of mandible distraction osteogenesis, Tong and colleagues demonstrated immunolocalization of FAK in regions of new bone formation secondary to distraction, which was absent in the control groups, where new bone formation occurred without distraction. Similarly, recent work has also colocalized c-SRC, a kinase involved with activation of the mechanical transduction complex (p130), in regions of bone regeneration secondary to distraction osteogenesis (Rhee and Buchman, 2005). While signaling molecules involved with transduction of mechanical forces are being identified, further work elucidating the mechanisms of these messenger molecules is required to clarify the influence of the mechanical environment on skeletal-tissue engineering.

Distraction osteogenesis has also emphasized the multitude of growth factors that participate in regenerating bone in this environment. Elucidation of their roles and mechanisms in bone regeneration will be a key element in successful tissue engineering. Transforming growth factor β1 (TGF-β1), a potent growth factor known to stimulate osteoblast proliferation, has been demonstrated, via Northern blot analysis of distracted rat mandibles, to have increased expression after osteotomy, continuing through the distraction period (Mehrara et al., 1999). Immunohistochemistry revealed that TGF-β1 during the distraction phase is localized to osteoblasts, primitive mesenchymal cells, and the extracellular matrix. Data regarding the effects of exogenous TGF-β1, in general, suggest that it enhances osseous healing (Bouletreau et al., 2002).

Insulinlike growth factor-1 (IGF-1) has also been implicated to play a role in the early stage of distraction osteogenesis. IGF-1 has been shown in vitro and in vivo to stimulate osteoprogenitor cell mitosis and differentiation (Spencer et al., 1991). In a canine tibia-lengthening model, serum levels of IGF-1 increased initially during the early distraction period, followed by elevations in local IGF-1 levels at the region of distraction (Lammens et al., 1998). Schumacher et al. (1996) also presented evidence supporting the role of IGF-1 in early distraction, where periosteal IGF-1 levels in the rat tibia were increased only during active lengthening. Exogenous addition of IGF-1 in a rabbit model of mandible distraction osteogenesis resulted in increased bone formation as compared to control groups (Stewart et al., 1999).

Like TFG-β1 and IGF-1, studies have shown fibroblast growth factors (FGFs) to mediate bone formation in distraction osteogenesis. Immunohistochemical staining in a sheep mandible distraction model revealed FGF-2 staining in the region of distraction, with greater staining in animals with high-rate distraction (Farhadieh *et al.*, 1999). Exogenous administration of recombinant human FGF-2 to a rabbit model of long-bone lengthening, during the end of the distraction period, increased the mineral content of the callus (Okazaki *et al.*, 1999).

Lastly, bone morphogenetic proteins (BMPs), known mediators in bone formation, have been localized to regions of successful distraction osteogenesis. In examining femoral lengthening in rats, gene expression levels of BMP-2 and BMP-4 were found to be elevated during the period of distraction (Sato *et al.*, 1999). Likewise, in distraction osteogenesis of rabbit tibia, immunohistochemical staining revealed intense BMP-2, -4, and -7 staining in fibroblast-like cells and chondrocytes during the distraction phase (Rauch *et al.*, 2000). This was confirmed in membranous bone, where analysis of the bone generated from mandibular lengthening in rabbits revealed BMP-2 and BMP-4 to be highly expressed in osteoblasts during distraction and in chondrocytes during consolidation (Campisi *et al.*, 2003).

Distraction osteogenesis is thus a useful modality for translating discrete applications of mechanical force into molecular signals that induce skeletal regeneration. However, this process is not without its own morbidities, such as soft-tissue infection, osteomyelitis, pintract infection or loosening, and patient discomfort (Mofid *et al.*, 2002). Distraction osteogenesis is also limited by the large hardware necessary to accomplish distraction and by the length of time required for large defects. Given the foregoing, cell-based tissue engineering of bone represents an alternative for addressing large osseous defects.

## III. CRITICAL-SIZED DEFECTS

In the search for optimal elements of successful skeletal-tissue engineering, the critical-sized defect has proved to be an indispensable tool for evaluating the performance of various constructs in an *in vivo* setting. Critical-sized defects are bone defects that fail to heal without intervention. They allow the performance, dosages, and combinations of various cellular sources, osteogenic cytokines, and scaffolds to be evaluated.

Common types of defects include calvarial and long-bone or mandible segmental defects (Salgado *et al.*, 2004). The calvarial model is particularly useful in evaluating constructs for craniofacial defects, given the relative easy handling of the calvarial plate and the support provided by the surrounding intact bone, absolving the need for fixation. It is also a relatively unloaded model mechanically, minimizing the influence of exogenous forces in the investigation of specific cellular elements. Long-bone segmental defects can also be created in the radius, where the ulna provides endogenous fixation of that limb. In situations where load bearing of the construct is desired, femur defects can be used.

These models all allow serial radiographic examination through the use of modalities such as microcomputed tomography. Ultimately, however, histology remains the gold-standard tool in evaluating the efficacy of constructs. Histomorphometric analysis is used to evaluate the amount of new bone formation. Histology also allows for the evaluation of bone resorption, changes in endogenous structure, and quality of the bone regenerate (Salgado *et al.*, 2004). Serving as functional *in vivo* bone regeneration assays, critical-sized defects allow researchers to assess the rate and degree of bone healing for individual components of a bone construct as well as their effect in concert (Cowan *et al.*, 2005).

## IV. CELLULAR THERAPY

Arguably, cells are the most crucial component to skeletal-tissue engineering. The two broad categories of populations available as sources of osteoprogenitor cells include osteoblasts and multipotent cells. At first glance, it would seem that osteoblasts could make a reasonable candidate for bone-tissue engineering. These are cells already committed along a bone lineage and readily form mineralized matrix. However, given the limited number of cells available, their low expansion rate, and donor-site morbidity, the osteoblast population is not an ideal cellular source for skeletal-tissue engineering (Heath, 2000).

As such, stem cells appear to be the most promising option for regenerative medicine. Stem cells are defined as a population of cells with the ability for prolonged division while undifferentiated, as well as the ability to differentiate along multiple lineages with the proper biological cues (Thomson *et al.*, 1998). Stem cells of embryonic origin have the most plasticity along the totipotent hierarchy. By definition, they are able to differentiate along all three embryonic germ layers: the endoderm, the ectoderm, and the mesoderm (Thomson *et al.*, 1998). The ability of embryonic stem cells to differentiate into osteoblasts and bone was demonstrated *in vitro* with murine embryonic stem cells, in the presence of dexamethasone (Buttery *et al.*, 2001). However, because of the ethical and political controversy surrounding the use of embryonic stem cells, researchers have focused their attention on postnatal sources of multipotent cells (Bahadur, 2003; Chin, 2003; Weissman, 2002).

Postnatal stem cells have been identified in bone marrow, periosteum, muscle, fat, brain, skin, and umbilical cord blood (Salgado *et al.*, 2004). Although discussion continues as to the degree of their differentiation potential, research in this area has been spurred by the capacity of these postnatal progenitor cells to differentiate along multiple lineages. The most commonly studied source of postnatal stem cells is the bone marrow. The major components of bone marrow are hematopoietic and mesenchymal stromal cells. The hematopoietic component provides the progenitors that will differentiate into all the mature blood cell phenotypes. Bone marrow mesenchymal stem cells

(BMSCs) are easily isolated from the hematopoetic components because of their unique quality to adhere to polystyrene culture dishes. Substantial work has demonstrated the ability of BMSCs to be guided along multiple mesenchymal lineages, including bone, cartilage, muscle, ligament, tendon, and stroma (Kadiyala *et al.*, 1997; Pittenger *et al.*, 1999, 2000). BMSCs have been shown, *in vitro*, to differentiate along an osteogenic pathway when cultured in the presence of dexamethasone, ascorbic acid, and β-glycerophosphate (Kadiyala *et al.*, 1997; Pittenger *et al.*, 1999).

The use of BMSCs in skeletal-tissue engineering has been demonstrated in multiple animal models. Oghushi and colleagues (1989) implanted bone marrow cells loaded on calcium phosphate ceramics into critical-sized defects of rat femurs. They demonstrated a higher rate of complete bone union and accelerated healing in the bone marrow cell group as compared to controls. Similar results were shown by Bruder *et al.* (1998), where human BMSCs, seeded on a ceramic carrier, were implanted into femur defects of athymic rats. Defects with BMSC implants healed and were significantly stronger under mechanical loading conditions than defects implanted with cell-free ceramics. In a rabbit calvarial model, BMSCs seeded on a fibrin scaffold demonstrated osseous regeneration similar to control implants seeded with osteoblasts (Schantz *et al.*, 2003).

Although the concept of BMSCs is attractive for tissue engineering, it is limited by the low frequency of mesenchymal stem cells within bone marrow stroma. Analysis of human bone marrow aspirates have found that the incidence of BMSCs can be as low as 1 in 27,000 cells (Banfi *et al.*, 2000). Furthermore, the procedure of acquiring bone marrow is a painful one, requiring general or local anesthesia. Other limitations on the use of BMSCs include the need for sera and growth factor supplements for *in vitro* expansion prior to use (Baksh *et al.*, 2003). Although subject to debate, BMSCs from older animals have been observed *in vitro* to form less bone than that from young animals, although this may be a reflection of their differences in proliferative ability (Bergman *et al.*, 1996).

Recently, a mesenchymal cell population has been isolated from the stromal component of adipose tissue (AMC) and has been demonstrated to be multipotent (Lin *et al.*, 2006; Zuk *et al.*, 2001). Zuk *et al.* described isolating an adherent cell population from human adipose tissue and inducing these cells to differentiate along adipogenic, chondrogenic, osteogenic, and myogenic lineages. Comparison of CD marker antigens demonstrated a similar expression profile between AMCs and BMSCs (Zuk *et al.*, 2002). Osteogenic differentiation conditions were found to be identical to that for BMSCs, supplementing media with dexamethasone, β-glycerophosphate, and ascorbate-2-phosphate. Confirmation of osteogenic differentiation was obtained with alkaline phosphatase and von Kossa staining as well as expression of bone-related genes, including Cbfa-1, alkaline phosphatase, osteopontin, and osteocalcin (Zuk *et al.*, 2002).

This finding is significant because AMCs are an abundant multipotent cell population that can be procured from individuals with minimal morbidity, expanded fairly rapidly, and used in an autologous fashion for tissue regeneration or reconstruction (Aust *et al.*, 2004; De Ugarte *et al.*, 2003; Zuk *et al.*, 2002, 2001).

Lee *et al.* (2003) demonstrated the ability of AMCs to differentiate into bone, *in vivo*, by seeding polylactic glycolic acid (PLGA) scaffolds with predifferentiated AMCs from Lewis rats and implanting them subcutaneously. Bone formation was demonstrated eight weeks later by tissue morphology on hematoxylin-eosin staining and immunohistochemical staining for osteocalcin.

The efficacy of AMCs in healing osseous defects was demonstrated in a model of critical-sized calvarial defects in mice (Cowan *et al.*, 2004). Implanting AMCs seeded on hydroxyapatite-coated PLGA scaffolds into calvarial defects, significant bone formation was demonstrated by x-ray, histology, and molecular imaging. At two weeks, significant bone formation was seen in the defect, with complete bony bridging by 12 weeks. Comparison of defects implanted with AMC-seeded scaffolds versus BMSC-seeded scaffolds demonstrated similar rates of bone formation. This experiment established an important proof of principle that AMCs contributed to bone formation in a critical-sized, skeletal defect in mice.

Interestingly, although AMCs from young mice demonstrated a greater ability to proliferate, AMCs from young and old mice both exhibited robust osteogenic differentiation (Shi *et al.*, 2005). This observation may be attributed to the presence of BMP-2 and retinoic acid in the osteogenic differentiation media for mouse AMCs, accounting for the continued ability of AMCs from adult animals to display robust osteogenic differentiation. This finding is encouraging because it suggests that, with the proper cues, cells from aging animals are still able to maximize their potential for bone formation.

Thus, the mesencyhymal stem cell (MSC) population seems to be a very promising cellular candidate for promoting skeletal regeneration. MSCs derived from fat are especially attractive because of their ease of procurement, the abundance of MSCs derived from a harvest, and their rapid expansion. Whether MSCs are derived from bone marrow or adipose tissue, they still remain a heterogeneous population that remains to be fully characterized. Future direction in developing a cellular source for skeletal-tissue engineering includes the need to identify markers of a mesenchymal progenitor cell subpopulation that possesses the strongest proclivity for osteogenic differentiation.

## V. CYTOKINES

The study of bone formation in developmental biology and distraction osteogenesis has revealed a host of growth factors involved with this process. Research on these cytokines has provided much information on the effect of these signaling molecules on cell proliferation, differentiation, adhesion, migration, and ultimately bone formation. An understanding of these interactions will undoubtedly allow for their use in bone-tissue engineering. It is foreseeable that successful skeletal-tissue engineering will involve the incorporation of the appropriate cytokines with an osteoprogenitor population.

Already in use clinically, BMPs seem to be the most promising candidate cytokine in skeletal-tissue engineering. BMPs, members of the TGF-β superfamily, were originally isolated by Urist from bovine bone extracts and found to induce ectopic bone formation subcutaneously in rats (Urist, 1965). This large group of proteins, comprising nearly one-third of the TGF-β superfamily, has also been found to be involved in mesoderm induction, skeletal patterning, and limb development (Duprez et al., 1996). BMPs are known to control both intramembranous as well as endochondral ossification through chemotaxis and mitosis of mesenchymal cells, induction of mesenchymal commitment to osteogenic or chondrogenic differentiation, and programmed cell death (Reddi, 2001). BMPs transmit their signals via ligand binding to the heteromeric complex of types I and II serine/threonine kinase receptors on the cell surface (Shi and Massague, 2003). The ligand signal is then transduced intracellularly via activation of SMAD (signaling, mothers against decapentaplegic) proteins, which subsequently migrate to the nucleus to effect gene expression (Noth et al., 2003). BMP signaling has also been demonstrated to be transmitted via the MAP-kinase (mitogen-activated protein) pathway (Gallea et al., 2001).

BMP-2, -4, -6, -7, and -9 are known to be the most osteoinductive (Jadlowiec et al., 2003). It is thought that BMPs regulate osteoblast differentiation via increased transcription of core-binding factor-1/Runt–related family 2 (Cbfa1/Runx2), a molecule known to be necessary for commitment along an osteoblastic lineage (Ducy et al., 1997; Gu et al., 2004; Harada et al., 1999; Otto et al., 1997; Tou et al., 2003). These osteoinductive BMPs have all been demonstrated to stimulate osteogenic differentiation in multiple cell lines, including fibroblasts, chondrocytes, BMSCs, and AMCs. The effect of BMPs has also been noted to be concentration dependent. At low concentrations, they foster chemotaxis and cellular proliferation (Zimmerman et al., 1996). At high concentrations, BMPs induce bone formation (Urist, 1997). BMPs are more potent at inducing bone formation as heterodimers than as homodimers. In culture, BMP-2/6, BMP-2/7, and BMP-4/7 heterodimers have been shown to promote higher alkaline phosphatase levels than homodimer combinations (Aono et al., 1995). These data have also been corroborated in vivo (Israel et al., 1996).

FGFs are a highly conserved family of 24 peptides, which transmit their signal via a family of four transmembrane tyrosine kinases. FGF-2, the most abundant ligand of the family, is known to increase osteoblast proliferation and to enhance bone formation in vitro and in vivo (Kimoto et al., 1998; Nakamura et al., 1997). In FGF-2-happloinsufficient and -null mice, decreased bone mineral density was observed grossly and correlated with decreased expression of FGF receptor-2 and Runx2 at a molecular level (Naganawa et al., 2006). Exogenous FGF-2 was able to rescue decreased bone nodule formation in osteoblast cultures from these transgenic mice.

IFG-1 and IGF-2 are 7.6-kDa polypeptides that have been demonstrated to stimulate bone collagen synthesis as well as osteogenesis and chondrogenesis (Linkhart et al., 1996). In a transgenic mouse, where IGF expression was specifically up-regulated in osteoblasts, there was increased bone formation of the distal femur as compared to control mice (Zhao et al., 2000). Of interest, however, is that histology did not reveal an increase in the number of osteoblasts, suggesting that IGF-1 up-regulated the activity of existing bone-forming cells. Conversely, in an IGF-1-null, transgenic mouse model, the size and bone-formation rates of the knockout mice were significantly reduced as compared to their wild-type littermates (Bikle et al., 2002).

Platelet-derived growth factor (PDGF), a 30-kDa polypeptide, has also been demonstrated to be a potent stimulus for osteoblast proliferation, chemotaxis, as well as collagen activity. The utility of PDGF in vivo has also been shown, with enhanced bone formation after local application of PDGF to tibial osteotomies in rabbits and canine mandibles after periodontal surgery. PDGF, commonly used in chronic wounds, is now being studied clinically in periodontics. Nevins et al. (2005), in a multicenter, randomized, controlled trial, demonstrated that the application of recombinant human PDGF in a tricalcium phosphate matrix significantly increased periodontal bone formation.

While the bulk of tissue-engineering applications involve the utilization of BMPs, multiple other cytokines exist for potential use in skeletal-tissue regeneration. The intuitive approach for delivery of cytokines involves incorporation of these molecules into scaffolds. Recent work, however, has explored the possibility of utilizing microspheres and liposomes as delivery mechanisms. This system would allow the growth factor to be retained at the site of interest for an extended period while maintaining its biological activity (DiTizio et al., 2000). The process of creating microspheres involves combining the cytokine, which is dissolved in aqueous phase, with a polymer, which is dissolved in an appropriate solvent (Borden et al., 2002). The combination of these two solutions is then dispersed into another aqueous phase, which serves to create the microspheres. This emulsion is heated to evaporate the solvent. The microspheres are then isolated by centrifugation and sintered for scaffold formation.

The delivery of cytokines in this encapsulated form has been examined in several animal models. Woo et al. (2001) described encapsulating recombinant human BMP-2 within PLGA microspheres and implanting them in rat calvarial defects. Compared to a control group with an immediate-release formulation, the microsphere group demonstrated sustained release of BMP-2 and greater amounts of bone matrix formation. Similar results have been demonstrated in a long-bone model with the delivery of recombinant human BMP-7 in PLGA microspheres to a critical-sized, ulna defect in rabbits (Borden *et al.*, 2004).

The challenges of incorporating cytokines in skeletal-tissue engineering involve identifying molecules whose actions are specific to bone formation. Other issues requiring further work include the timing of delivery, effective dosages, and compatible carrier mechanisms. Study of these growth factors *in vitro* and in critical-sized defect models will help provide answers to these questions.

## VI. SCAFFOLDS

Paralleling research on an ideal osteoprogenitor cell population and the identification of potential cytokines for bone-tissue engineering, advances in materials science have provided an abundance of innovations on developing an appropriate carrier for these cells and molecules. In the selection of a biomimetic scaffold for engineering bone, the material should be osteoconductive, osteoinductive, biocompatible, and biodegradeable. *Osteoconductivity* refers to the ability of the graft to support the attachment of cells and allow new cell migration and vessel formation. The *osteoinductive* quality of scaffolds describes their ability to induce nondifferentiated stem cells or progenitor cells along an osteogenic lineage. The scaffold should not elicit an immune response. In examining the structure of scaffolds, maximizing the porosity of the scaffold, to promote cellular and neovascular ingrowth, must be balanced by respect for the structural integrity of the lattice (Salgado *et al.*, 2004). Given the foregoing, an overview of the scaffolds available currently ensues in relation to these criteria. These matrices can be generalized into three categories: natural, mineral based, and synthetic polymers.

Natural scaffolds include the use of collagen, hyaluronic acids, calcium alginate, and chitosan (Bumgardner *et al.*, 2003; Cho *et al.*, 2005; Saadeh *et al.*, 2001; Seol *et al.*, 2004; Solchaga *et al.*, 2002). Natural scaffolds are typically biodegradable. In many instances, they exhibit osteoinductive properties, exclusive of cells or cytokines. Implantation of collagen type I alone into critical-sized defects of rat mandibles resulted in healing (Saadeh *et al.*, 2001). The disadvantage of natural scaffolds is their lack of mechanical stability and, hence, ability to be used in load-bearing regions of the skeleton. Use of natural scaffolds in humans is also limited by the biochemical changes often induced by sterilization procedures (Zahraoui and Sharrock, 1999).

Mineral-based scaffolds include calcium phosphate ceramics and bioactive glass. Calcium phosphate ceramics are available as hydroxyapatite or β-tricalcium phosphate, with hydroxyapatite most closely mimicking the structural and chemical characteristics of the mineral component of bone (Nacamuli and Longaker, 2005). Tricalcium

phosphate is marked by a high dissolution rate that accelerates material resorption and elicits an immune response. In contrast, hydroxyapatite has high chemical stability. In general, mineral-based scaffolds provide an osteoinductive signal to encourage differentiation of progenitor cells along an osteogenic lineage. There is, however, great variability in the quality of calcium phosphate ceramics to support osteogenesis. This is due to the difficulty in reproducibly creating these scaffolds (El-Ghannam and Ning, 2006).

Polymer scaffolds, including polylactic acid, polyglycolic acid, polydoxanonone, and polycaprolactone, have been engineered to provide a greater ability to withstand mechanical forces. These lattices are marked by their incredible strength (Behravesh *et al.*, 1999; Ishaug *et al.*, 1997). They are also designed to be biodegradable via hydrolysis as a result of their local mileau. Polymer scaffolds allow ingrowth of bony tissue, but they lack osteoinductive properties. Thus, current work is focused on combining polymer formulations with hydroxyapatite or growth factors, which would encourage cellular osteogenic differentiation.

Hydrogels represent another class of polymer scaffolds. They are formed by polymerization and cross-linking of molecules such as acrylic acid, and *N*-isopropylacrylamide (Stile *et al.*, 1999). Hydrogels are an attractive option because of their temperature-dependent physical properties. They can be designed to be gelatinous at room temperature, but they take on more rigid qualities at body temperatures (Stile and Healy, 2002). This property could allow for the administration of tissue-engineering constructs via injection. Hydrogels also allow for relatively easy chemical manipulation of individual peptides. Incorporation of arginine-glycine-aparganine (RGD) peptide motifs on these polymers has been demonstrated to enhance osteoblast adhesion and proliferation (Rezania and Healy, 1999).

Recently, electrospinning has emerged as a novel method to construct scaffolds that more closely mimic natural extracellular matrix. Electrospinning involves creation of nanowoven fibers by shooting a jet of polymer solution through a high electric field (Li *et al.*, 2006). Adjusting the polymer solution and the electric field allows the fiber diameter and porosity to be controlled (Zong *et al.*, 2003). The advantage of this technique is not only that it mimics ECM, but that various pro-osteogenic materials or cytokines can be incorporated more conveniently into this scaffold. Li *et al.* (2006) incorporated BMP-2 and nanoparticles of hydroxyapatite into silk fibroin fiber scaffolds using this technique. They demonstrated, *in vitro*, higher mineralization of human MSCs cultured on these scaffolds than in controls.

## VII. ANGIOGENESIS

Another frontier of work that has emerged is providing the bone construct with a vascular supply. In hostile cellular microenvironments, such as large defects, irradiated regions, and contaminated wounds, blood flow to the implanted bone graft could increase its performance and survival. Fukui *et al.* (2005) provided bone-forming constructs, consisting of human MSCs and osteogenic cytokines in a gelatin carrier, with blood flow from an epigastric flap. Implants provided with a vascular source formed higher bone mineral density than those that were not.

Without a vascular supply, osseous constructs depend on diffusion to supply the necessary nutrients and to remove metabolic waste. Especially in large defects, the diffusion process is limited by an effective distance of 150–200 μm from blood vessels (Colton, 1995). Furthermore, hypoxic conditions are known to contribute to the significant attrition rates of MSCs *in vitro* (Logeart-Avramoglou *et al.*, 2005).

One avenue of research that may provide a solution to this problem is the delivery of angiogenic molecules that could accelerate vascular invasion of the implanted bone construct. Vascular endothelial growth factor (VEGF), a 34- to 46-kDa homodimeric glycoprotein known to be an essential regulator of angiogenesis, is a potential candidate (Ferrara, 2000; Ferrara *et al.*, 2003; Fong *et al.*, 1996). The VEGF family consists of placental growth factor (PLGF), VEGF-B, VEGF-C, and VEGF-D. They bind to three homologous high-affinity transmembrane receptors: VEGF-R1 (Flt-1), VEGF-R2 (Flk-1), and VEGF-R3 (Flt-4). The role of VEGF in endochondral bone formation was demonstrated in neonatal mice by the systemic application of mouse Flt(1-3)-IgG, a soluble, VEGF-receptor, chimeric protein that would competitively bind to VEGF (Gerber *et al.*, 1999). This experimental approach resulted in suppressed capillary invasion, impaired cartilage remodeling, and defective bone formation. VEGF expression is found in bone fracture–healing sites and growth plates and is thought to regulate the migration of endothelial cells to sites of osseous regeneration (Jadlowiec *et al.*, 2003).

Current experiments involving the delivery of VEGF to regions of osseous regeneration have met with mixed results. In a model of distraction osteogenesis of rabbit tibias, animals treated with VEGF at the bony defect demonstrated increased blood flow but did not demonstrate increased bone formation at the site (Eckardt *et al.*, 2003). In a recent experiment by Huang and colleagues (2005), bone formation of various PLGA constructs was evaluated following a period of subcutaneous implantation in SCID mice. PLGA implants contained combinations of human BMSCs and plasmid DNA–encoding BMP-4 or VEGF. Scaffolds containing VEGF-encoding plasmids demonstrated a significant increase in blood vessel formation. The combination of all three components in one scaffold exhibited the greatest amount of osseous regeneration. While this experiment contained many confounding variables, it suggests that increased blood vessel formation with other appropriate components may aid in more robust bone formation. Given the critical role of VEGF in endochondral ossification, further work could more appropriately examine the effect of adding VEGF to cell-seeded scaffolds, with respect to healing of skeletal defects of varying sizes.

## VIII. TISSUE ENGINEERING IN PRACTICE

While the goal of bone-tissue engineering is to incorporate an osteoprogenitor cell population with the necessary cytokines on a biomimetic matrix, current clinical applications of tissue engineering involve only the delivery of cytokines, namely recombinant human BMP-2 and BMP-7. A significant criticism of these applications is the high doses necessary to obtain clinical relevance and the resultant costs associated with manufacturing recombinant human BMPs. Concern also exists about the rare but reported incidences of high blood pressure and even myocardial infarctions in animal models, which could be explained by an immunologic response to the cytokines (Cowan *et al.*, 2005).

The Infuse® Bone Graft/LT Cage® fusion device is currently in use for skeletally mature patients with degenerative disc disease at one level from L4-S1 (Medtronic Sofamor Danek, Minneapolis, MN). This product involves an absorbable, bovine collagen implant soaked with recombinant human BMP-2 (rhBMP-2). Humans require supraphysiologic doses ranging from 0.4 to 1.5 mg/mL of rhBMP-2 to form new bone. Because of the high doses, there have been concerns of uncontrolled bone growth around the implantation site or cancer, given links of osteosarcoma with BMP activity (Raval *et al.*, 1996; Yoshikawa *et al.*, 1994). A recent clinical study of Infuse® in 74 consecutive patients revealed radiographic fusion was 100% at 12 and 24 months. No bone overgrowth or sarcomas were observed. The study was significantly weakened, however, by the lack of a control group (Villavicencio *et al.*, 2005).

rhBMP-2 has also been used in patients requiring maxillary sinus floor augmentation, in preparation for enodosseous dental implants. Recombinant human BMP-2, on an absorbable collage sponge, at 0.75 and 1.5 mg/mL, was applied to the maxillary sinus floor. Radiographic evaluation for increases in bone mass in the maxillary sinus floor revealed that rhBMP-2 was able to induce bone formation, but it was not as effective as autogenous bone grafting in fostering osseous formation. Notably, though, rates of successful placement of dental implants were similar between the bone graft treatment arm and those that received rhBMP-2 (Boyne, 2001).

The other product approved by the FDA for clinical use is recombinant human BMP-7 (rhBMP-7) (Stryker Orthopedics, Mahwah, NJ). It is also delivered via a bovine collagen matrix and has been approved for the treatment of nonunion in the tibia of at least nine months, secondary to trauma, in skeletally mature patients. Recombinant human BMP-7 has been approved only for use in cases where previous treatment with autograft failed. In a recent small-series report, the use of rhBMP-7 in five patients who had failed allografts did not result in any significant healing (Delloye *et al.*, 2004). Recombinant human BMP-7 has also been in clinical use to aid intertransverse process fusion in the lumbar spine (Vaccaro *et al.*, 2005). In a putty formulation consisting of rhBMP-7, bony fusion rates were comparable to groups treated with autograft. They did not report any adverse events, ectopic bone formation, or systemic toxicities associated with use of this product over a two-year period.

## IX. CONCLUSION

The intersection of advances in stem cell biology, molecular biology, biochemistry, bioengineering, and materials sciences has brought to the forefront the nearing reality of regenerative medicine in addressing problems of skeletal defects. Efforts are encouraged by the further understanding of the functions and interactions of osteogenic cytokines as well as by the identification of postnatal osteoprogenitor cells in bone marrow and adipose tissue. The influence of nanotechnology on scaffold design and the possibility of sustained-release formulations of growth factors via microspheres are promising developments.

Work remains to be performed on further elucidation of the mechanisms of the multitude of osteogenic cytokines as well as on developing effective delivery systems for these molecules to avoid the need for supraphysiologic doses. Efforts also need to be directed toward developing methods to provide vascular perfusion of large osseous constructs, potentially through the use of pro-angiogenic agents. In the realm of stem cell biology, the need for identifying characteristics of an enriched osteoprogenitor population also represents an area requiring further investigation. Given the immense biomedical burden of skeletal defects and the significant developments in skeletal-tissue engineering, osseous regeneration provides a promising and attainable goal that can be achieved only through interdisciplinary collaboration.

## X. REFERENCES

Aono, A., Hazama, M., Notoya, K., Taketomi, S., Yamasaki, H., Tsukuda, R., Sasaki, S., and Fujisawa, Y. (1995). Potent ectopic bone-inducing activity of bone morphogenetic protein-4/7 heterodimer. *Biochem. Biophys. Res. Commun.* **210**, 670–677.

Banfi, A., Muraglia, A., Dozin, B., Mastrogiacomo, M., Cancedda, R., and Quarto, R. (2000). Proliferation kinetics and differentiation potential of *ex vivo* expanded human bone marrow stromal cells: implications for their use in cell therapy. *Exp. Hematol.* **28**, 707–715.

Bikle, D. D., Sakata, T., Leary, C., Elalieh, H., Ginzinger, D., Rosen, C. J., Beamer, W., Majumdar, S., and Halloran, B. P. (2002). Insulin-like growth factor I is required for the anabolic actions of parathyroid hormone on mouse bone. *J. Bone Miner. Res.* **17**, 1570–1578.

Borden, M., Attawia, M., Khan, Y., El-Amin, S. F., and Laurencin, C. T. (2004). Tissue-engineered bone formation *in vivo* using a novel sintered polymeric microsphere matrix. *J. Bone Joint Surg. Br.* **86**, 1200–1208.

Boyne, P. J. (2001). Application of bone morphogenetic proteins in the treatment of clinical oral and maxillofacial osseous defects. *J. Bone Joint Surg. Am.* **83A(Suppl. 1)**, S146–S150.

Bruder, S. P., Kurth, A. A., Shea, M., Hayes, W. C., Jaiswal, N., and Kadiyala, S. (1998). Bone regeneration by implantation of purified, culture-expanded human mesenchymal stem cells. *J. Orthop. Res.* **16**, 155–162.

Buttery, L. D., Bourne, S., Xynos, J. D., Wood, H., Hughes, F. J., Hughes, S. P., Episkopou, V., and Polak, J. M. (2001). Differentiation of osteoblasts and *in vitro* bone formation from murine embryonic stem cells. *Tissue Eng.* **7**, 89–99.

Campisi, P., Hamdy, R. C., Lauzier, D., Amako, M., Rauch, F., and Lessard, M. L. (2003). Expression of bone morphogenetic proteins during mandibular distraction osteogenesis. *Plast. Reconstr. Surg.* **111**, 201–208; discussion 209–210.

Codivilla, A. (1905). On the means of lengthening in the lower limbs, the muscles and tissues which are shortened through deformity. *Am. J. Orthop. Surg.* **2**, 353–369.

Cowan, C. M., Shi, Y. Y., Aalami, O. O., Chou, Y. F., Mari, C., Thomas, R., Quarto, N., Contag, C. H., Wu, B., and Longaker, M. T. (2004). Adipose-derived adult stromal cells heal critical-size mouse calvarial defects. *Nat. Biotechnol.* **22**, 560–567.

Cowan, C. M., Soo, C., Ting, K., and Wu, B. (2005). Evolving concepts in bone tissue engineering. *Curr. Top. Dev. Biol.* **66**, 239–285.

Delloye, C., Suratwala, S. J., Cornu, O., and Lee, F. Y. (2004). Treatment of allograft nonunions with recombinant human bone morphogenetic proteins (rhBMP). *Acta Orthop. Belg.* **70**, 591–597.

Ducy, P., Zhang, R., Geoffroy, V., Ridall, A. L., and Karsenty, G. (1997). Osf2/Cbfa1: a transcriptional activator of osteoblast differentiation. *Cell* **89**, 747–754.

Eckardt, H., Bundgaard, K. G., Christensen, K. S., Lind, M., Hansen, E. S., and Hvid, I. (2003). Effects of locally applied vascular endothelial growth factor (VEGF) and VEGF-inhibitor to the rabbit tibia during distraction osteogenesis. *J. Orthop. Res.* **21**, 335–340.

El-Ghannam, A., and Ning, C. Q. (2006). Effect of bioactive ceramic dissolution on the mechanism of bone mineralization and guided tissue growth *in vitro*. *J. Biomed. Mater. Res. A* **76**, 386–397.

Fang, T. D., Salim, A., Xia, W., Nacamuli, R. P., Guccione, S., Song, H. M., Carano, R. A., Filvaroff, E. H., Bednarski, M. D., Giaccia, A. J., and Longaker, M. T. (2005). Angiogenesis is required for successful bone induction during distraction osteogenesis. *J. Bone Miner. Res.* **20**, 1114–1124.

Farhadieh, R. D., Dickinson, R., Yu, Y., Gianoutsos, M. P., and Walsh, W. R. (1999). The role of transforming growth factor-beta, insulin-like growth factor I, and basic fibroblast growth factor in distraction osteogenesis of the mandible. *J. Craniofac. Surg.* **10**, 80–86.

Ferrara, N., Gerber, H. P., and LeCouter, J. (2003). The biology of VEGF and its receptors. *Nat. Med.* **9**, 669–676.

Fukui, M., Akita, S., and Akino, K. (2005). Ectopic bone formation facilitated by human mesenchymal stem cells and osteogenic cytokines via nutrient vessel injection in a nude rat model. *Wound Repair Regen.* **13**, 332–340.

Gerber, H. P., Vu, T. H., Ryan, A. M., Kowalski, J., Werb, Z., and Ferrara, N. (1999). VEGF couples hypertrophic cartilage remodeling, ossification and angiogenesis during endochondral bone formation. *Nat. Med.* **5**, 623–628.

HCUP. (2006). Healthcare Cost and Utilization Project, Agency for Healthcare Research and Quality.

Huang, Y. C., Kaigler, D., Rice, K. G., Krebsbach, P. H., and Mooney, D. J. (2005). Combined angiogenic and osteogenic factor delivery enhances bone marrow stromal cell–driven bone regeneration. *J. Bone Miner. Res.* **20**, 848–857.

Ilizarov, G. A. (1989). The tension-stress effect on the genesis and growth of tissues: part II. The influence of the rate and frequency of distraction. *Clin. Orthop. Relat. Res.* **239**, 263–285.

Israel, D. I., Nove, J., Kerns, K. M., Kaufman, R. J., Rosen, V., Cox, K. A., and Wozney, J. M. (1996). Heterodimeric bone morphogenetic proteins show enhanced activity *in vitro* and *in vivo*. *Growth Factors* **13**, 291–300.

Jadlowiec, J. A., Celil, A. B., and Hollinger, J. O. (2003). Bone-tissue engineering: recent advances and promising therapeutic agents. *Expert Opin. Biol. Ther.* **3**, 409–423.

Lammens, J., Liu, Z., Aerssens, J., Dequeker, J., and Fabry, G. (1998). Distraction bone healing versus osteotomy healing: a comparative biochemical analysis. *J. Bone Miner. Res.* **13**, 279–286.

Lee, J. A., Parrett, B. M., Conejero, J. A., Laser, J., Chen, J., Kogon, A. J., Nanda, D., Grant, R. T., and Breitbart, A. S. (2003). Biological alchemy: engineering bone and fat from fat-derived stem cells. *Ann. Plast. Surg.* **50**, 610–617.

Li, C., Vepari, C., Jin, H. J., Kim, H. J., and Kaplan, D. L. (2006). Electrospun silk-BMP-2 scaffolds for bone tissue engineering. *Biomaterials* **27**, 3115–3124.

Loboa, E. G., Fang, T. D., Warren, S. M., Lindsey, D. P., Fong, K. D., Longaker, M. T., and Carter, D. R. (2004). Mechanobiology of mandibular distraction osteogenesis: experimental analyses with a rat model. *Bone* **34**, 336–343.

Loboa, E. G., Fang, T. D., Parker, D. W., Warren, S. M., Fong, K. D., Longaker, M. T., and Carter, D. R. (2005). Mechanobiology of mandibular distraction osteogenesis: finite element analyses with a rat model. *J. Orthop. Res.* **23**, 663–670.

Logeart-Avramoglou, D., Anagnostou, F., Bizios, R., and Petite, H. (2005). Engineering bone: challenges and obstacles. *J. Cell. Mol. Med.* **9**, 72–84.

Mehrara, B. J., Rowe, N. M., Steinbrech, D. S., Dudziak, M. E., Saadeh, P. B., McCarthy, J. G., Gittes, G. K., and Longaker, M. T. (1999). Rat mandibular distraction osteogenesis: II. Molecular analysis of transforming growth factor beta-1 and osteocalcin gene expression. *Plast. Reconstr. Surg.* **103**, 536–547.

Mofid, M. M., Inoue, N., Atabey, A., Marti, G., Chao, E. Y., Manson, P. N., and Vander Kolk, C. A. (2002). Callus stimulation in distraction osteogenesis. *Plast. Reconstr. Surg.* **109**, 1621–1629.

Naganawa, T., Xiao, L., Abogunde, E., Sobue, T., Kalajzic, I., Sabbieti, M., Agas, D., and Hurley, M. M. (2006). *In vivo* and *in vitro* comparison of the effects of FGF-2 null and haplo-insufficiency on bone formation in mice. *Biochem. Biophys. Res. Commun.* **339**, 490–498.

Nakamura, K., Kurokawa, T., Kawaguchi, H., Kato, T., Hanada, K., Hiyama, Y., Aoyama, I., Nakamura, T., Tamura, M., and Matsumoto, T. (1997). Stimulation of endosteal bone formation by local intraosseous application of basic fibroblast growth factor in rats. *Rev. Rhoum. Engl. Ed.* **64**, 101–105.

Nash, T. J., Howlett, C. R., Martin, C., Steele, J., Johnson, K. A., and Hicklin, D. J. (1994). Effect of platelet-derived growth factor on tibial osteotomies in rabbits. *Bone* **15**, 203–208.

Nevins, M., Giannobile, W. V., McGuire, M. K., Kao, R. T., Mellonig, J. T., Hinrichs, J. E., McAllister, B. S., Murphy, K. S., McClain, P. K., Nevins, M. L., Paquette, D. W., Han, T. J., Reddy, M. S., Lavin, P. T., Genco, R. J.,

and Lynch, S. E. (2005). Platelet-derived growth factor stimulates bone fill and rate of attachment level gain: results of a large multicenter randomized controlled trial. *J. Periodontol.* **76**, 2205–2215.

Ohgushi, H., Goldberg, V. M., and Caplan, A. I. (1989). Repair of bone defects with marrow cells and porous ceramic. Experiments in rats. *Acta Orthop. Scand.* **60**, 334–339.

Okazaki, H., Kurokawa, T., Nakamura, K., Matsushita, T., Mamada, K., and Kawaguchi, H. (1999). Stimulation of bone formation by recombinant fibroblast growth factor-2 in callotasis bone lengthening of rabbits. *Calcif. Tissue Int.* **64**, 542–546.

Pittenger, M. F., Mackay, A. M., Beck, S. C., Jaiswal, R. K., Douglas, R., Mosca, J. D., Moorman, M. A., Simonetti, D. W., Craig, S., and Marshak, D. R. (1999). Multilineage potential of adult human mesenchymal stem cells. *Science* **284**, 143–147.

Rauch, F., Lauzier, D., Croteau, S., Travers, R., Glorieux, F. H., and Hamdy, R. (2000). Temporal and spatial expression of bone morphogenetic protein-2, -4, and -7 during distraction osteogenesis in rabbits. *Bone* **26**, 611–617.

Reddi, A. H. (2001). Bone morphogenetic proteins: from basic science to clinical applications. *J. Bone Joint Surg. Am.* **83A(Suppl. 1)**, S1–S6.

Rezania, A., and Healy, K. E. (1999). Biomimetic peptide surfaces that regulate adhesion, spreading, cytoskeletal organization, and mineralization of the matrix deposited by osteoblast-like cells. *Biotechnol. Prog.* **15**, 19–32.

Rhee, S. T., and Buchman, S. R. (2005). Colocalization of c-Src (pp60src) and bone morphogenetic protein 2/4 expression during mandibular distraction osteogenesis: *in vivo* evidence of their role within an integrin-mediated mechanotransduction pathway. *Ann. Plast. Surg.* **55**, 207–215.

Saadeh, P. B., Khosla, R. K., Mehrara, B. J., Steinbrech, D. S., McCormick, S. A., DeVore, D. P., and Longaker, M. T. (2001). Repair of a critical size defect in the rat mandible using allogenic type I collagen. *J. Craniofac. Surg.* **12**, 573–579.

Salgado, A. J., Coutinho, O. P., and Reis, R. L. (2004). Bone-tissue engineering: state of the art and future trends. *Macromol. Biosci.* **4**, 743–765.

Sato, M., Ochi, T., Nakase, T., Hirota, S., Kitamura, Y., Nomura, S., and Yasui, N. (1999). Mechanical tension-stress induces expression of bone morphogenetic protein (BMP)-2 and BMP-4, but not BMP-6, BMP-7, and GDF-5 mRNA, during distraction osteogenesis. *J. Bone Miner. Res.* **14**, 1084–1095.

Schantz, J. T., Teoh, S. H., Lim, T. C., Endres, M., Lam, C. X., and Hutmacher, D. W. (2003). Repair of calvarial defects with customized tissue-engineered bone grafts I. Evaluation of osteogenesis in a three-dimensional culture system. *Tissue Eng.* **9(Suppl. 1)**, S113–S126.

Schumacher, B., Albrechtsen, J., Keller, J., Flyvbjerg, A., and Hvid, I. (1996). Periosteal insulin-like growth factor I and bone formation. Changes during tibial lengthening in rabbits. *Acta Orthop. Scand.* **67**, 237–241.

Shi, Y. Y., Nacamuli, R. P., Salim, A., and Longaker, M. T. (2005). The osteogenic potential of adipose-derived mesenchymal cells is maintained with aging. *Plast. Reconstr. Surg.* **116**, 1686–1696.

Stewart, K. J., Weyand, B., van't Hof, R. J., White, S. A., Lvoff, G. O., Maffulli, N., and Poole, M. D. (1999). A quantitative analysis of the effect of insulin-like growth factor-1 infusion during mandibular distraction osteogenesis in rabbits. *Br. J. Plast. Surg.* **52**, 343–350.

Stile, R. A., Burghardt, W. R., and Healy, K. E. (1999). Synthesis and characterization of injectable poly(*n*-isopropylacrylamide)-based hydrogels that support tissue formation *in vitro*. *Macromolecules* **32**, 7370–7379.

Thomson, J. A., Itskovitz-Eldor, J., Shapiro, S. S., Waknitz, M. A., Swiergiel, J. J., Marshall, V. S., and Jones, J. M. (1998). Embryonic stem cell lines derived from human blastocysts. *Science* **282**, 1145–1147.

Tong, L., Buchman, S. R., Ignelzi, M. A., Jr., Rhee, S., and Goldstein, S. A. (2003). Focal adhesion kinase expression during mandibular distraction osteogenesis: evidence for mechanotransduction. *Plast. Reconstr. Surg.* **111**, 211–222.

Urist, M. R. (1965). Bone: Formation by autoinduction. *Science* **150**, 893.

Urist, M. R. (1997). Bone morphogenetic protein: the molecularization of skeletal system development. *J. Bone Miner. Res.* **12**, 343–346.

Vaccaro, A. R., Anderson, D. G., Patel, T., Fischgrund, J., Truumees, E., Herkowitz, H. N., Phillips, F., Hilibrand, A., Albert, T. J., Wetzel, T., and McCulloch, J. A. (2005). Comparison of OP-1 Putty (rhBMP-7) to iliac crest autograft for posterolateral lumbar arthrodesis: a minimum 2-year follow-up pilot study. *Spine* **30**, 2709–2716.

Villavicencio, A. T., Burneikiene, S., Nelson, E. L., Bulsara, K. R., Favors, M., and Thramann, J. (2005). Safety of transforaminal lumbar interbody fusion and intervertebral recombinant human bone morphogenetic protein-2. *J. Neurosurg. Spine* **3**, 436–443.

Weissman, I. L. (2002). Stem cells — scientific, medical, and political issues. *N. Engl. J. Med.* **346**, 1576–1579.

Woo, B. H., Fink, B. F., Page, R., Schrier, J. A., Jo, Y. W., Jiang, G., DeLuca, M., Vasconez, H. C., and DeLuca, P. P. (2001). Enhancement of bone growth by sustained delivery of recombinant human bone morphogenetic protein-2 in a polymeric matrix. *Pharm. Res.* **18**, 1747–1753.

Yoshikawa, H., Rettig, W. J., Takaoka, K., Alderman, E., Rup, B., Rosen, V., Wozney, J. M., Lane, J. M., Huvos, A. G., and Garin-Chesa, P. (1994). Expression of bone morphogenetic proteins in human osteosarcoma. Immunohistochemical detection with monoclonal antibody. *Cancer* **73**, 85–91.

Younger, E. M., and Chapman, M. W. (1989). Morbidity at bone graft donor sites. *J. Orthop. Trauma* **3**, 192–195.

Zhao, G., Monier-Faugere, M. C., Langub, M. C., Geng, Z., Nakayama, T., Pike, J. W., Chernausek, S. D., Rosen, C. J., Donahue, L. R., Malluche, H. H., Fagin, J. A., and Clemens, T. L. (2000). Targeted overexpression of insulin-like growth factor I to osteoblasts of transgenic mice: increased trabecular bone volume without increased osteoblast proliferation. *Endocrinology* **141**, 2674–2682.

Zuk, P. A., Zhu, M., Ashjian, P., De Ugarte, D. A., Huang, J. I., Mizuno, H., Alfonso, Z. C., Fraser, J. K., Benhaim, P., and Hedrick, M. H. (2002). Human adipose tissue is a source of multipotent stem cells. *Mol. Biol. Cell* **13**, 4279–4295.

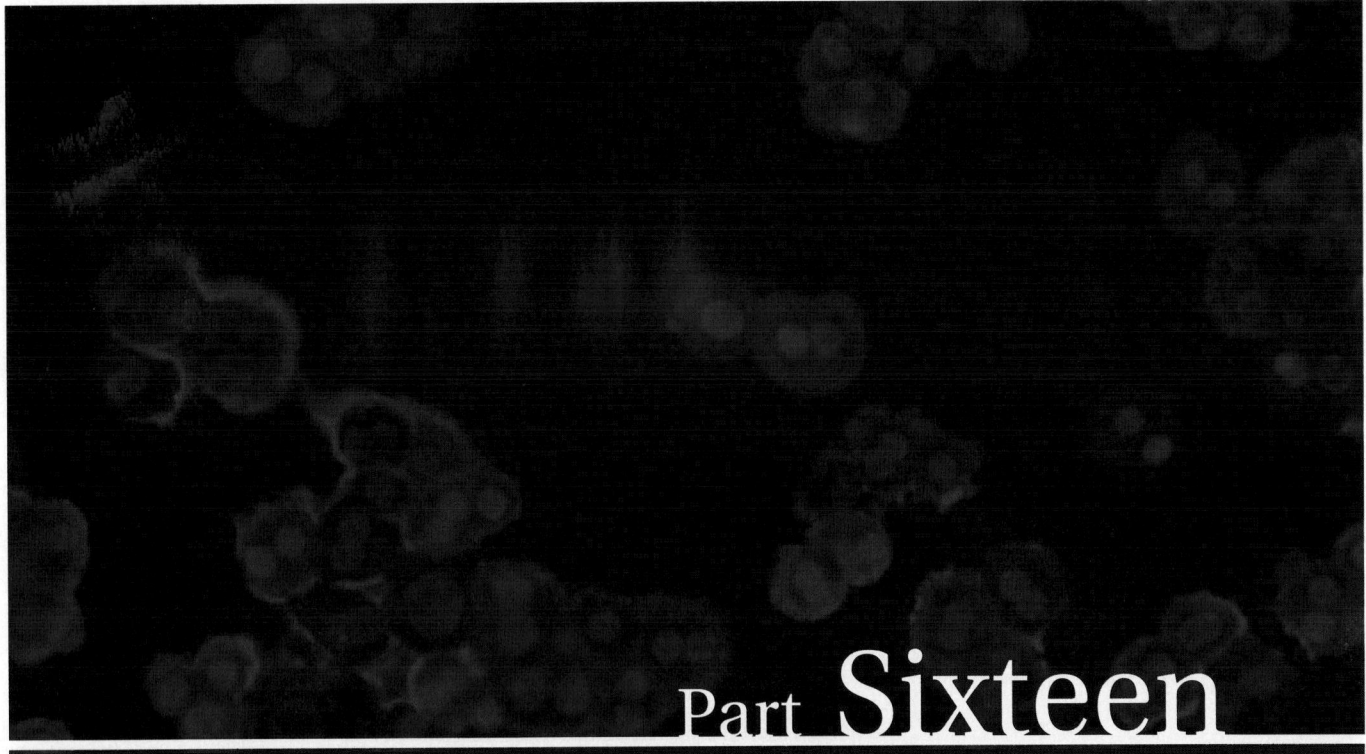

Part Sixteen

# Nervous System

# Chapter Sixty-Three

# Neural Stem Cells

*Lorenz Studer*

## I. INTRODUCTION

This chapter summarizes the current state of neural stem cell research, with a particular focus toward applications in regenerative medicine and tissue engineering. The text highlights the close relationship between developmental biology and neural stem cell biology and discusses techniques for isolation, characterization, and differentiation of neural stem cells. Protocols for the neural differentiation of mouse and human embryonic stem cells are also presented. Numerous examples are provided to direct ES cells toward specific neuronal or glial fates. Finally, current and future applications of neural stem cells are discussed, and the advantages of ES cell–derived primary and endogenous neural stem cells are compared. Neural stem cell biology has grown into a discipline with powerful novel tools and techniques. Neural stem cells will become an essential instrument to unraveling basic questions related to brain development, providing a unique cellular interface with tissue-engineering approaches and working toward the promise of medical breakthroughs.

For most of the last century the thought of applying tissue-regeneration and -engineering principles to the central nervous system was considered naive and unrealistic. The neuroscience community was under the strong influence of the famous neuroanatomist Ramon Y Cajal, who stated that (in the adult central nervous system) "everything may die, nothing may be regenerated." Interestingly, however, in the very same paragraph Ramon Y Cajal also wrote that "it is for the science of the future to change, if possible, this harsh decree." After nearly a century of failures, the time may have come to take on this challenge. Today a number of novel tools are available that have fundamentally changed our ability to approach CNS regeneration and disease, to apply principles of tissue engineering, and to make steps toward the ultimate dream of applying CNS repair to the diseased brain. Among the key developments that have made this possible is the discovery that the adult brain continues to harbor a reservoir of neural stem cells that may be harnessed for repair, the ability to culture primary neural stem and progenitor cells *in vitro*, and the ability systematically to direct cell fate from mouse and human embryonic stem cells.

We begin with a brief introduction to neural development and then cover neural stem cells, addressing the current advantages and limitations of these cells. Then we deal with the derivation of neural cells from embryonic stem cells, followed by a more general discussion of neural stem cell applications in tissue engineering, developmental biology, disease modeling, and regenerative medicine.

## II. NEURAL DEVELOPMENT

Neural induction occurs during gastrulation via "organizer" signals emanating from the node and notochord and acting on the overlying dorsal ectoderm. While the initial

*Principles of Tissue Engineering, 3rd Edition*
*ed. by Lanza, Langer, and Vacanti*

work by Speamann and Mangold (1924) suggested the presence of a direct neural inducer signal, experiments performed over half a century later led to the conclusion that neural induction occurs during early gastrulation by a default mechanism (Munoz-Sanjuan and Brivanlou, 2002). The default model suggests that key molecules capable of inducing neural fate act via inhibiting BMP signaling. These include Chordin (Sasai et al., 1995), Noggin (Lamb et al., 1993), Follistatin (Hemmati Brivanlou et al., 1994), and cerberus. Recent studies suggest that FGF signals emanating from precursors of the organizer prior to gastrulation contribute to a prepattern of neural induction via activation of Sox3 and ERNI (Streit et al., 2000). However, experiments in *Xenopus* suggest that FGF signaling may exert proneural activity, in part via activating MAPK and leading to inhibition of BMP signaling via SMAD1 phosphorylation (Kuroda et al., 2005; Pera et al., 2003). Other major pathways involved during neural induction are IGF (Pera et al., 2001, 2003) and Wnt signaling (Baker et al., 1999; Wilson et al., 2001).

Neural plate cells undergo a set of morphological and molecular changes leading to neural fold and neural tube formation and closure. These morphogenetic events are followed by orchestrated waves of neural proliferation and differentiation. Neural fate specification is highly dependent on regional signals specific to the anteroposterior (AP) and dorsoventral (DV) locations within the embryo, translating into distinct subdomains defined by the combinatorial expression of homeodomain and bHLH transcription factors. There is considerable evidence for the hypothesis that anterior neural fates are established by default while posterior fates require exposure to caudalizing signals such as FGF, Wnt, Nodal, and retinoids during neural induction (for review, see Lumsden and Krumlauf, 1996). The combinatorial action of these caudalizing signals may further subdivide the CNS into discrete AP domains (Nordstrom et al., 2006). DV patterning involves the dose-dependent antagonistic action of the secreted molecules sonic hedgehog (SHH) and BMP (for review, see Jessell, 2000). SHH is expressed ventrally in the notochord and floor plate, while BMPs are secreted dorsally from the roof plate. There is ample evidence from explant (Ericson et al., 1995; Roelink et al., 1995; Ye et al., 1998) and ES cell differentiation studies (Barberi et al., 2003; J. Y. Kim et al., 2002; S.-H. Lee et al., 2000; Mizuseki et al., 2003; Wichterle et al., 2002) supporting a concentration-dependent role for SHH for generating ventral CNS fates (Briscoe et al., 2000). However, genetic studies in SHH mutant embryos, including the SHH/Gli3 (Litingtung and Chiang, 2000; Persson et al., 2002; Rallu et al., 2002) or SHH/Rab23 mice (Eggenschwiler et al., 2001), suggest that DV patterning can occur in the absence of SHH.

The role for BMPs in driving dorsal CNS fates is supported by various *in vitro* studies in explant culture and by multiple ES cell differentiation protocols (Liem et al., 1995, 1997; Mizuseki et al., 2003). However, genetic studies again suggest a more complex picture. Loss-of-function studies

via the ablation of the roof plate (K. J. Lee et al., 2000), and gain-of-function studies via expression of the BMP receptor type 1a (BMPR1a/Alk3) under control of the nestin enhancer (Panchision et al., 2001) are compatible with a dose-dependent role for BMPs in dorsal patterning. However, loss of the BMP receptor (BMPR1a) results in the loss of choroid plexus fates without obviously affecting other dorsal cell fates (Hebert et al., 2002) and the loss of the dl1 and dl2 domains in the spinal cord (Wine-Lee et al., 2004). Similar to the ventral spinal cord, the dorsal spinal cord can be subdivided into distinct progenitor domains, defined by the combinatorial expression of specific bHLH factors, and homeodomain proteins, with dorsal precursor domains marked by Olig3 expression (Muller et al., 2005).

Region-specific neural fates are adopted in a sequential manner, leading to appropriate neuron subtypes and followed by astroglial and oligodendroglial differentiation. Neuronal differentiation is mediated by inhibition of the Notch pathway, repressing proneural bHLH genes (for review, see Gaiano and Fishell, 2002; Tanigaki et al., 2001). Subsequent astrocytic fates require the activation of the Jak/STAT pathway, exerting an instructive role on multipotent neural progenitors to drive astrocytic differentiation (Bonni et al., 1997; Rajan and McKay, 1998). However, recent insight into the neurogenic properties of radial glial (Noctor et al., 2002) cells and the identification of adult neural stem cells expressing astrocytic markers (Doetsch et al., 1999) suggest a more complex and dynamic interaction between neural stem cell and astrocytic fates. Similarly, our current understanding of the developmental origin of oligodendrocytes has undergone significant changes. Early studies suggested that oligodendrocytes are derived from bipotent glial precursors termed *O2A progenitors* (Ffrench-Constant and Raff, 1986) or from other glially committed precursors (Rao et al., 1998). However, more recent studies reported a direct lineage relationship between motoneuron and oligodendrocytes (He et al., 2001; Lu et al., 2002; Yung et al., 2002; Zhou and Anderson, 2002) based on Olig2 expression. Oligodendrocyte progenitor cells (OPCs) arise from the ventricular zone (VZ) of the embryonic spinal cord (for review, see Rowitch, 2004) and ganglionic eminences in the forebrain (Kessaris et al., 2006). While most Olig2$^+$ oligodendrocyte progenitors emanate from the ventral CNS, recent work has also demonstrated Olig2$^+$/PDGFRa$^+$ progenitors located with the dorsal spinal cord (Cai et al., 2005; Vallstedt et al., 2005).

## III. NEURAL STEM CELLS

### Isolation

The presence of neural stem cells in the CNS has been debated for decades. Persistent neural precursor cells in adult mammalian CNS were described by Altman and Das (1965) more than 40 years ago. However, the idea of a stem cell–like population in the mammalian CNS was largely ignored at that time. Subsequent work, in the early 1980s, provided convincing evidence of persistent neurogenic populations in the adult canary brain (Goldman and Nottebohm, 1983) and showed a clear link between dividing ventricular zone precursor cells and postmitotic neurons derived from these precursor populations using pulse chase $H^3$-thymidine labeling. Parallel studies were performed describing in detail all the dividing precursor populations in the developing rat CNS (Frederiksen and McKay, 1988). Technical limitations in manipulating stem cells *in vivo* required the development of *in vitro* assays. Early efforts demonstrated the isolation of precursor populations and early attempts aimed at manipulating cell fate (Cattaneo and McKay, 1990). A key step in the explosive growth of neural precursor cell studies was the development of the neurosphere culture system (Gritti *et al.*, 1995; Kilpatrick and Bartlett, 1993; Reynolds and Weiss, 1992). Under neurosphere conditions, neural precursors are proliferated free-floating in the presence of EGF or FGF2. Human neurosphere cultures are often supplemented with LIF in addition to EGF and FGF2 (Galli *et al.*, 2000). Neurospheres can be derived clonally, and the capacity for neurosphere formation has been used as an assay to determine neural stem cell identity. The prospective isolation of neural stem cells based on the expression of AC133 (Uchida *et al.*, 2000), Lex1 (Capela and Temple, 2002), or combinations of surface markers (Rietze *et al.*, 2001) was largely based on neurosphere formation capacity. However, such data need to be interpreted cautiously, for neurosphere formation is not a true test of stemness. Neurospheres are heterogeneous structures and in the adult SVZ are obtained at higher efficiencies from transient amplifying cells than from the presumptive SVZ stem cells (Doetsch *et al.*, 2002).

An alternative approach is the growth of neural stem–like cells as monolayer cultures in the presence of FGF2 on plates precoated with fibronectin or laminin (Davis and Temple, 1994; Johe *et al.*, 1996; Palmer *et al.*, 1995). These conditions are amenable to determining precise lineage relationships and manipulations at the single-cell level. Complete lineage trees for single cortical stem cells have been defined under such conditions (Qian *et al.*, 2000). More recent work suggests that the combination of FGF2 and EGF in monolayer cultures improves the ability of long-term expanded precursors to undergo neuronal fate specification. Such symmetrically dividing neural stem cell populations express markers of radial glia and are capable of differentiating into GABAergic neurons and glutamatergic neurons after more than 100 population doublings (Conti *et al.*, 2005).

## Characterization

### Clonal Assays, Self-Renewal, and Differentiation Potential

Over 10 years of intensive research has convincingly demonstrated self-renewal and multilineage differentiation of rodent neural stem cells. Clonal cultures have been a key assay for these studies. However, similar assays in human neural stem cells have been technically challenging, and

there are very few examples of clonal human neurosphere cultures from primary unmodified neural stem cells. Another major challenge in neural stem cell biology has been the inability to properly control neuronal subtype specification. While many tissue-specific stem cells, such as hematopoietic stem cells, are capable of differentiation into most if not all progeny within an organ system (Spangrude *et al.*, 1988), long-term expanded neural stem cells do not efficiently give rise to all neuronal subtypes but are largely limited to the production of GABA and glutamatergic neurons (Caldwell *et al.*, 2001; Jain *et al.*, 2003). The derivation of midbrain dopamine neurons has served as a model for these difficulties. Functional midbrain dopamine neurons can be derived from short-term expanded precursors isolated from the early rodent and human midbrain (Sanchez-Pernaute *et al.*, 2001; Studer *et al.*, 1998). However, long-term expansion causes a dramatic drop in the efficacy of midbrain dopamine neuron generation (J. Yan *et al.*, 2001). Several strategies based on intrinsic and extrinsic fate induction have been developed in an effort to overcome these problems (see later). However, none of these approaches has succeeded in deriving unlimited numbers of midbrain dopamine neurons that exhibit full functionality. The mechanisms responsible for the restricted neuronal subtype potential remain to be elucidated. The subtypes most difficult to generate are neurons born during limited temporal windows in early development. This suggests that competent precursors capable of dopaminergic differentiation may be lost during expansion or that required noncell autonomous cofactors are missing in expanded cultures. It is well known that mitogens such as FGF2 can deregulate neural patterning states in precursor cell cultures (Gabay *et al.*, 2003). Epigenetic factors such as methylation and histone modification also influence neural fate specification, such as neuronal and astrocytic fate choice (Hsieh *et al.*, 2004; Song and Ghosh, 2004). Many of the basic regulators of neural stem cell fate and differentiation, such as the NRSF complex and the polycomb gene BMI, involve epigenetic modifications (Ballas *et al.*, 2005; Valk-Lingbeek *et al.*, 2004). Future strategies to overcome neuronal subtype restrictions in neural stem cells may include isolating neural stem cells at an earlier developmental state under conditions that retain early competency, the induction of competence in later precursors using extrinsic cues, or direct manipulation of the epigenetic state. These challenges are major roadblocks to the use of neural stem cells in neurodegenerative disorders such as Parkinson's disease (PD) and amyotrophic lateral sclerosis (ALS; also called Lou Gehrig's disease).

## Differentiation and Plasticity

### Neurons

There is a clear stage-dependent bias in the ability of neural precursor cells to yield neuronal progeny. While precursors at early developmental stages produce neurons exclusively, later precursors show mixed neuronal/glial differentiation progressing toward a glial bias in the postnatal and adult brain. Various strategies have been developed to drive neuronal differentiation from multipotent precursors. Factors shown to enhance neuronal fate specification in embryonic rat neural precursors and in human neurosphere progeny are PDGF (Caldwell *et al.*, 2001; Johe *et al.*, 1996) and neurotrophic factors, including NT3 and NT-4/5 (Caldwell *et al.*, 2001; Vicario-Abejon *et al.*, 1995). Adult-derived hippocampal precursors show enhanced neuronal differentiation after exposure to a combination of retinoic acid and neurotrophins (Takahashi *et al.*, 1999), though overall neuronal yield in these cultures remains low. Interestingly, both PDGF and neurotrophic factors have also been shown to promote oligodendroglial fates under different circumstances (Calver *et al.*, 1998; Marmur *et al.*, 1998). A common requirement of neuronal differentiation is the activation of proneural bHLH factors. Indeed, the forced expression of bHLH factors in neural precursors drives neuronal specification and inhibits glial differentiation (Nieto *et al.*, 2001).

### Midbrain DA Neuron Differentiation

Dopamine neuron differentiation has been an obvious target for neural stem cell research, given the long history of experimental cell transplantation in Parkinson's disease (PD). One of the first examples of generating midbrain dopamine neurons *in vitro* came from studies based on the isolation of neural precursor cells from the developing midbrain. These cells can be propagated *in vitro* in serum-free medium in the presence of FGF2. Differentiation by growth factor withdrawal leads to the generation of large numbers of midbrain DA neurons capable of reversing behavioral deficits after transplantation into the striatum of PD rats (Studer *et al.*, 1998). While such studies were encouraging, they also showed clear limitations of the midbrain precursor system. While DA neuron differentiation was highly efficient in precursors proliferated for short *in vitro* culture periods, long-term expanded precursors rapidly lost the ability to generate midbrain DA neuron progeny, despite intact neuronal differentiation potential (J. Yan *et al.*, 2001). Many groups subsequently tried to overcome these limitations with modified culture protocols, such as the use of ascorbic acid during differentiation, culture of midbrain precursors under reduced oxygen levels (Studer *et al.*, 2000), cocktail of growth factors (Carvey *et al.*, 2001; Storch *et al.*, 2001) thought to enhance DA neuron differentiation potential, or, most recently, the use of various Wnt signaling molecules (Castelo-Branco *et al.*, 2003) or TGFβ-related molecules (Farkas *et al.*, 2003). Genetic strategies to drive DA neuron fates in neural precursors of nonmidbrain origin include expression of Nurr1 alone (J. Y. Kim *et al.*, 2003; Wagner *et al.*, 1999) or in combination with bHLH factors or SHH and BclXL (Park *et al.*, 2006), the latter strategy resulting in DA neurons capable of survival and function *in vivo*. While some of these strategies provide clear improvements in our ability to control DA neuron differentiation in neural stem cells, the levels of improvement remain modest, and cell identity may not always correspond to bona fide midbrain DA neuron.

### Motoneurons

Motoneurons are born during a brief developmental window early in development. There is limited evidence that motoneurons can be generated from neural stem cells. While there are data on the derivation of cholinergic neurons from human neurospheres via SHH priming (Wu *et al.*, 2002), these findings have not been confirmed in independent studies. Short-term expansion of embryonic spinal cord precursors followed by motoneuronal fate specification has been demonstrated (Mayer-Proschel *et al.*, 1997), but, similar to the midbrain DA neuron problem just discussed, it is limited to restricted precursor cells after short expansion rates only.

### Other Neuron Subtypes

Recent work comparing the properties of FGF2-expanded neural precursor cells isolated from the adult hippocampus and the adult hypothalamus suggested that both populations can yield comparable neuronal subtypes after differentiation (Markakis *et al.*, 2004). This study concluded that neuronal fates are at least in part plastic and controlled *in vivo* by environmental factors that restrict neuronal fate to region-appropriate subtypes, while such restricted fates are accessible *in vitro*. Similar plasticity has been suggested after expansion and *in vitro* proliferation of forebrain precursor cells when implanted into ectopic environments (Hitoshi *et al.*, 2002).

## Glia

### Astrocytes

Astrocytic differentiation of multipotent neural stem cells has served as one of the key paradigms of fate specification via an extrinsic signaling factor. It has been shown that similar to the O2A progenitors cells, multipotent neural stem cells undergo rapid and highly efficient astrocytic differentiation in response to CNTF (Johe *et al.*, 1996). It has been further shown that CNTF acts via activation of STAT3 and that the response can be blocked using a dominant negative inhibitor of STAT signaling (Bonni *et al.*, 1997; Rajan and McKay, 1998). However, CNTF astrocytic differentiation in response to CNTF is dependent on the appropriate developmental stage, and early neural precursors such as E12 rat forebrain cells do activate STATs in response to CNTF but do not yield astrocytic progeny (Molne *et al.*, 2000). The responsiveness to LIF seems to correlate with the expression of the EGFR (Viti *et al.*, 2003). Another signaling pathway that is capable of promoting astrocytic fates under appropriate conditions is the BMP pathway (Gross *et al.*, 1996).

## Oligodendrocytes

Thyroid hormone (T3) has been one of the key factors to regulate oligodendroglial fates from neural stem cells (Johe *et al.*, 1996). Other factors that promote oligodendroglial differentiation or proliferation include SHH (Lu *et al.*, 2000; Pringle *et al.*, 1996), PDGF (Barres *et al.*, 1993), and neurotrophic factors, particularly NT3 (Barres *et al.*, 1994).

### Non-CNS Fates

One example of neural stem cell plasticity is the ability to yield neural crest–like progeny after exposure to FGF2 and BMPs (Sailer *et al.*, 2005). The derivation of neural crest–like populations has also been reported from the adult SVZ. These cells exhibit a migratory behavior *in vivo* similar to that of neural crest precursors (Busch *et al.*, 2006). Examples of neural stem cell differentiation into nonneural progeny include the *in vitro* differentiation of skeletal muscle–like cells (Valtz *et al.*, 1991). Other reports on the differentiation of neural stem cells into nonectodermal derivatives such as blood (Bjornson *et al.*, 1999) and the pluripotent behavior of neural stem cells after transplantation into the early blastocyst (Clarke *et al.*, 2000) remain controversial and have not been independently confirmed.

## IV. EMBRYONIC STEM CELL–DERIVED NEURAL STEM CELLS AND NEURAL PROGENY

ES cells are capable of virtually unlimited *in vitro* proliferation and offer many important advantages for both basic and applied research. These include ease of genetic manipulation, access to the earliest stages of neural development, and the comprehensive differentiation potential. Neural differentiation has become one of the most utilized *in vitro* differentiation assays in ES cell research, due to the high efficiency of obtaining neural progeny and the large number of important basic and preclinical applications. The use of ES cells in regenerative medicine became a distinct possibility with the isolation of human ES and embryonic germ (EG) cells (Reubinoff *et al.*, 2000; Shamblott *et al.*, 1998; Thomson *et al.*, 1998).

### Neural Induction

The first step in generating neural subtypes from ES cells is neural induction. There are at least three basic strategies to achieve neural induction: embryoid-body–(EB) based protocols, stromal feeder–mediated neural induction, and protocols based on default differentiation into neural fates.

### Embryoid Body–Based Protocols

EBs are formed through aggregation of ES cells in suspension cultures. Cell interactions within the EB mimic the signaling events in normal development, particularly gastrulation. Therefore EB cultures typically result in derivatives of all three germ layers (Doetschman *et al.*, 1985; for review, see Weiss and Orkin, 1996), and EB formation is often used as evidence for pluripotency. While the derivation of neural progeny is inefficient under basic EB conditions, a number of protocols have been developed to enhance neural induction and to select and expand EB-derived neural precursors. The first EB-mediated neural differentiation protocol was based on exposure to retinoic acid (RA) for four days after four days of RA-free culture (4–/4+ protocol) (Bain *et al.*, 1995). Many variations on the basic protocol have been developed (Fraichard *et al.*, 1995; Renoncourt *et al.*, 1998; Strübing *et al.*, 1995). While no clear mechanistic understanding of the action of RA in EBs has emerged, there seems to be a requirement for Wnt inhibition during the process (Aubert *et al.*, 2002; Verani *et al.*, 2006). One major concern in RA-based neural induction protocols is the concomitant effect of RA on AP patterning mediated through activating the Hox gene cascade (Krumlauf, 1994). An alternative EB-based strategy is the exposure to conditioned medium derived from the hepatocarcinoma cell line (HepG2), which appears to induce neuroectodermal fate directly (Rathjen *et al.*, 2002). Accordingly, HepG2-treated aggregates do not express endodermal or mesodermal markers but give rise directly to neural progeny expressing progenitor markers such as Sox1, 2, and nestin. The active component within HepG2-conditioned medium remains to be isolated. Unlike the RA protocols, HepG2 conditions are not thought to bias neural subtype composition toward specific AP or DV fates. A third strategy for achieving neural induction is an RA-free EB-based protocol that subjects EB progeny to neural-selective growth conditions (Brustle *et al.*, 1999; S. H. Lee *et al.*, 2000; Okabe *et al.*, 1996). Neural selection from EB progeny is achieved under minimal-growth conditions in a serum-free medium supplemented with insulin, transferrin, and selenite (ITS medium). These conditions select for immature neural cells expressing the intermediate filament nestin. These nestin+ precursors can be replated and directed toward various neuronal and glial fates using a combination of patterning, survival, and lineage-promoting factors. Commercial kits have become available that provide all the necessary reagents for this approach. However, ES cell line–specific variability in neural induction efficiency has been observed, particularly at stage III of differentiation (selection of nestin+ precursors from EBs in ITS medium) (Barberi *et al.*, 2003; Wakayama *et al.*, 2001). Another strategy for neural induction is based on the serum-free EB protocols (SFEBs). This strategy is based on aggregating feeder-free cultures of ES cells under serum-free conditions. The technique is particularly useful for the derivation of forebrain fates (Watanabe *et al.*, 2005). Both classic EB and SFEB protocols have been adapted for use in primate and human ES cells (Schulz *et al.*, 2004; Zhang *et al.*, 2001).

### Stromal Feeder–Mediated Neural Induction

Bone marrow–derived stromal cell lines have been used for many years to support the growth of undifferentiated hematopoietic stem cells (Collins and Dorshkind, 1987; Croisille et al., 1994; Nakano et al., 1994; Sutherland et al., 1991), acting in part via expression of the membrane protein mKirre, a mammalian homolog of the gene kirre of Drosophila melanogaster (Ueno et al., 2003). More recently, it has been reported that many of the stromal cell lines that support hematopoietic stem cell growth exhibit neural-inducing properties in coculture with mouse ES cells (Barberi et al., 2003; Kawasaki et al., 2000, 2002). Stromal cell lines with the highest efficiencies of neural induction are typically at the preadipocytic stage of differentiation. While most of these cell lines were isolated from the bone marrow (e.g., PA-6, MS5, S17), stromal cells derived from the aorta–gonad–mesonephros (AGM) region were found to induce neural differentiation equally well in mouse ES and nuclear transfer ES cells (Barberi et al., 2003). While the molecular nature of this stromal-derived inducing activity remains largely unknown, the efficiency and robustness of neural induction via this approach is extremely high (Barberi et al., 2003). There is some controversy as to the extent that feeder-mediated induction biases regional specification or neuronal-versus-glial fate choice (Barberi et al., 2003; Mizuseki et al., 2003; Watanabe et al., 2005). However, the system can be readily adapted to the differentiation of human embryonic stem cells (Perrier et al., 2004) and modified to yield a number of specific neuronal or glial cell types (Perrier et al., 2004; Tabar et al., 2005).

### Neural Differentiation by Default

Coculture-free direct neural differentiation protocols are based on the default hypothesis proposing that in the absence of cell–cell signaling, particularly in the absence of BMPs, ectodermal cells will adopt a neural fate (Munoz-Sanjuan and Brivanlou, 2002). A number of ES cell studies demonstrated that under minimal conditions in the absence of BMP, neural induction does occur in nonadherent (Tropepe et al., 2001) or adherent (Ying et al., 2003) monocultures. Loss-of-function studies suggest an early role for FGFs during neural induction of ES cells, as suggested by studies in vivo (Streit et al., 2000). However, the efficacy of default monoculture approaches in comparison to EB and stromal feeder protocols has not been determined. A variation of this protocol was recently suggested to generate fairly homogeneous populations of radial glial-like cells capable of undergoing differentiation to presumptive cortical pyramidal-like cells (Bibel et al., 2004). Subsequent work demonstrated that ES cell–derived radial glial-like cells can be maintained long term in the presence of FGF2/EGF and that these symmetrically diving precursors retain neurogenic potential over extended in vitro cultures (Conti et al., 2005). Most of these cells appear to undergo GABAergic and glutamatergic differentiation compatible with forebrain characteristics. The ability of these cells to yield neuronal

cell types specific to other regions, such as midbrain DA neurons or spinal cord motoneurons, has not been demonstrated.

## Neural Stem Cells

An important issue relates to the question of whether ES cell–derived neural precursor cells exhibit properties of bona fide neural stem cells. Evidence on the multipotent nature of mouse ES cell–derived neural precursors was obtained in clonal studies following single-cell progeny after neural induction proliferation and differentiation of ES progeny. These studies demonstrated clonal derivation of neuronal, astrocytic, and oligodendrocytic progeny comparable to primary neural stem cells (Barberi et al., 2003). Neural stem cell behavior was also observed in long-term expanded ES–derived neural progeny maintained in the presence of FGF2/EGF (Conti et al., 2005). In human ES cells, neural stem cell–like behavior was illustrated in vivo after transplantation of human ES–derived neural progeny into the rat striatum or SVZ (Tabar et al., 2005). These cells were capable of integrating in the adult rodent SVZ, expressing markers of SVZ stem cells and transit-amplifying cells, and migrating along the rostral migratory stream to the olfactory bulb. Human-derived olfactory bulb neurons were detected expressing markers comparable to those derived from endogenous rat SVZ neural stem cells.

## Neuronal and Glial Differentiation

### ES-Derived Neurons

Neuronal differentiation occurs rapidly on neural induction of mouse ES cells. Neuronal subtype specification can be influenced by the neural induction strategy, such as the use of RA, stromal feeders, or SFEB protocol. Strategies to direct neuronal subtype differentiation are based on the application of signals that mimic early patterning events in the embryo to define AP and DV domains within the body axis of the developing embryo.

### Midbrain Dopaminergic Neurons

Protocols for the dopaminergic differentiation of mouse ES cells closely follow findings obtained in explant studies that identified FGF8 and SHH as critical factors in midbrain dopamine neuron specification (Hynes et al., 1995a, 1995b; Ye et al., 1998, 2001). The effect of SHH/FGF8 on ES-derived neural precursors was first described using an EB-based five-step differentiation protocol (S. H. Lee et al., 2000). Under these conditions up to 34% of all neurons expressed tyrosine hydroxylase (TH), the rate-limiting enzyme in the synthesis of dopamine. A further increase in dopamine neuron yield was obtained in ES cells overexpressing Nurr1 (J. H. Kim et al., 2002). Nearly 80% of all neurons express TH under these conditions, and expression of midbrain dopaminergic markers remains stable even after transgenic expression of Nurr1 has been silenced. Midbrain dopaminergic differentiation was also obtained using coculture of

ES cells on the stromal feeder cell line (PA6) (Kawasaki *et al.*, 2000). PA6 cell–mediated neural induction yielded dopaminergic differentiation in up to 16% of all neurons, without requiring any exposure to exogenous SHH and FGF8 but in the presence of AA. When combining stromal feeder–mediated neural induction with exposure to SHH/FGF8, about 50% of all neurons express TH (Barberi *et al.*, 2003), without requiring any transgene such as Nurr1 to further push dopamine neuron phenotype. Very similar results can be obtained with primate and human ES cells (Cibelli *et al.*, 2002; Kawasaki *et al.*, 2002; Perrier *et al.*, 2004; Y. Yan *et al.*, 2005).

The numbers of TH neurons need to be interpreted carefully in all *in vitro* differentiation studies because TH is an unreliable marker for identifying dopamine neurons. TH is also expressed in other catecholaminergic neurons, including noradrenergic and adrenergic cells, and it is induced under various extrinsic conditions, such as cell stress, hypoxia, and growth factors. It is therefore essential that studies reporting on the derivation of midbrain dopamine neurons provide additional markers and evidence of dopamine neuron function *in vitro* and *in vivo* (for review, see Perrier and Studer, 2003). The derivation of TH neurons has also been achieved using a monolayer default neural induction protocol (Ying *et al.*, 2003). However, efficiency of dopamine neuron induction, characterization of midbrain phenotype, and *in vitro* and *in vivo* functionality using the default neural induction protocol were not reported. Most recent studies suggest that transgenic expression of the transcription factor Lmx1A can increase DA neuron yield in such monolayer cultures. Lmx1A expression under control of the nestin enhancer elements in combination with SHH/FGF8 treatments leads to midbrain DA neuron conversion of most ES cell progeny (Andersson *et al.*, 2006). Future studies will be required to compare directly the properties of DA neurons induced genetically versus those induced by purely epigenetic means using extrinsic cues only.

### Serotonergic Neurons

The developmental origin of serotonergic neurons is closely related to that of midbrain dopamine neurons. Both neuronal subtypes are dependent on signals emanating from the isthmic organizer (Goridis and Rohrer, 2002). Accordingly, serotonergic neurons are a major "contaminating" neuronal subtype in protocols aimed at the derivation of midbrain dopaminergic cells. Protocols designed specifically to increase serotonergic versus dopaminergic differentiation have been based on exposure to FGF4. Application of exogenous FGF4 preceding FGF8 and SHH application ectopically induces serotonergic neurons in explant culture (Ye *et al.*, 1998). While application of FGF4 to neural precursors in the presence of FGF2 at stage IV (neural precursor cell proliferation) of the multistep EB differentiation protocol does not yield a significant increase in serotonergic differentiation (S. H. Lee *et al.*, 2000), FGF4 added in the absence of FGF2 does cause a dramatic shift from dopaminergic to

serotonergic differentiation (J. H. Kim *et al.*, 2002). Efficient derivation of hindbrain serotonergic neurons using stromal feeder–mediated neural induction also involves early FGF4 exposure in the absence of FGF2, followed by FGF2, FGF8, and SHH application (Barberi *et al.*, 2003). Novel strategies to refine serotonergic differentiation protocols might make use of novel basic developmental insights that demonstrated the importance of the transcription factor Lmx1b in serotonergic differentiation (Ding *et al.*, 2003).

### Motor Neurons

The development of spinal motor neurons has been studied in great detail (for review, see Jessell, 2000; S. K. Lee and Pfaff, 2001). Early studies in mouse ES cells demonstrated that cells expressing markers of motoneurons can be generated using an EB induction protocol in combination with RA exposure (2–/7+) (Renoncourt *et al.*, 1998). More systematic approaches using RA exposure in combination with exogenous SHH to promote ventral fates have yielded ES-derived motoneurons at high efficiency and provided a demonstration on how developmental pathways can be harnessed to direct ES cell fate *in vitro* (Wichterle *et al.*, 2002). By creating an ES cell line that expresses GFP under the control of the motor neuron–specific gene HB9, these ES cell–derived motor neurons could also be readily identified and purified. The *in vivo* properties of ES-derived motoneurons were demonstrated on transplantation into the spinal cord of early chick embryos. ES-derived motoneurons were detected in the ventral spinal cord, extended axons, and innervated nearby muscle targets (Wichterle *et al.*, 2002). Efficient derivation of motoneurons can also be achieved using stromal feeders (Barberi *et al.*, 2003; Mizuseki *et al.*, 2003) in combination with SHH and RA treatment. The next challenges for *in vitro* motoneuron differentiation protocols will be the selective generation of motoneurons of distinct AP and columnar identity. Insights from developmental biology studies will provide a good starting point toward that end (Dasen *et al.*, 2003, 2005; Sockanathan *et al.*, 2003). Recent studies in mouse ES cells suggested that the RA/SHH protocols yield primarily cervical and brachial phenotypes of medial motor column identity (Soundararajan *et al.*, 2006). Human ES cell derivation of motoneuron progeny has been demonstrated using an EB-based differentiation strategy followed by exposure to SHH/RA (X. J. Li *et al.*, 2005), and isolation of motoneuron progeny was achieved using transient transfection with an HB9 enhancer–driven GFP reporter construct (Singh *et al.*, 2005).

### GABA Neurons

GABA cells are the main inhibitor neuron type within the brain and are found at high densities in basal forebrain structures, particularly in the striatum. The presence of GABAergic neurons during ES cell differentiation *in vitro* has been reported under various conditions, including the classic 4–/4+ EB-based differentiation protocols (Bain *et al.*,

1995), which yield approximately 25% GABA neurons (Gottlieb, 2002). Interestingly, protocols with shorter periods of RA-free EB formation followed by extended RA treatment (2–/7+) select for motoneurons rather than GABA neurons (Renoncourt *et al.*, 1998). This suggests that the timing of RA application might be crucial for neural fate specification. The presence of GABA neurons has also been reported under default neural induction conditions (Ying *et al.*, 2003). Differentiation directed to GABA neurons has been achieved using a stromal feeder–based approach (Barberi *et al.*, 2003). Neural induction on MS5 is followed by neural precursor proliferation in FGF2 and subsequent exposure to SHH and FGF8. The delayed application of FGF8 and SHH promotes ventral forebrain identities, as determined by the expression of the forebrain-specific marker FOXG1B (BF-1) (Tao and Lai, 1992) and the increase in GABAergic differentiation (Barberi *et al.*, 2003). In addition to forebrain striatal and cortical GABA neurons there are many other types of GABA neurons in various brain regions, including the thalamus (Houser *et al.*, 1980), midbrain (Ribak *et al.*, 1976), and cerebellum (Hatten *et al.*, 1984). A recent example of the generation of a region-specific GABA neuron type is the derivation of forebrain GABA neurons (Watanabe *et al.*, 2005). GABAergic neuron subtypes of particular clinical relevance may be cortical interneurons for the treatment of intractable epilepsy and medium spiny GABA neurons in Huntington's disease.

### Glutamate Neurons

Glutamate neurons can be readily obtained from mouse ES cells. For example, in the Bain protocol (Bain *et al.*, 1995) approximately 70% of all neurons are glutamatergic, and ES-derived neurons with NMDA and non-NMDA receptor subtypes have been described (Finley *et al.*, 1996). Very similar neuron subtype compositions have been obtained with various related protocols (Strübing *et al.*, 1995; Wernig *et al.*, 2002). More detailed physiological data on glutamatergic neurons have been reported after coculture of ES-derived neurons on hippocampal brain slices (Benninger *et al.*, 2003). Interestingly, this study suggested a possible bias toward establishing AMPA- versus NMDA-type synaptic contacts (Benninger *et al.*, 2003). There is currently no detailed data on the derivation of glutamatergic neurons using stromal feeder protocols. However, default neural differentiation protocols appear efficient at yielding cortical-type glutamatergic neurons derived from radial glial-like precursors (Bibel *et al.*, 2004). Cerebellar granule cells have been induced in an SFEB culture–based system on exposure to BMP4 and Wnt3A (Su *et al.*, 2006).

### Other Neuronal and Neural Subtypes

The presence of about 5% glycinergic neurons has been reported using the classic 4–/4+ EB protocol (Bain *et al.*, 1995; Gottlieb, 2002). However, directed differentiation into this main inhibitory neuronal subtype in the spinal cord has not been demonstrated. Other interesting neural types generated from ES cells are precursors of the otic anlage (H. Li *et al.*, 2003). These precursors were obtained by culturing EBs for 10 days in EGF and IGF, followed by FGF2 expansion. After transplantation of these precursors, *in vivo* differentiation was observed into cells expressing markers of mature hair cells (H. Li *et al.*, 2003). Derivation of radial glial cells from mouse ES cells (Bibel *et al.*, 2004; Conti *et al.*, 2005; Liour and Yu, 2003) provides another interesting assay system to probe neuronal and glial lineage relationships in early neural development. Cell fates related to eye development, including lenslike precursors (Ooto *et al.*, 2003) and retinal ganglion cells (Ikeda *et al.*, 2005a), have been reported. The efficiency of generating eye-related fates is high in primate and human ES cells. For example, a recent study reported highly efficient differentiation toward retinal fates in human ES cells following simple exposure to factors that promote anterior CNS fates, including IGF1, Noggin, and Dkk (Lamba *et al.*, 2006).

### ES-Derived Glia

Neural precursors derived from mouse ES cells can be readily differentiated into astrocytic and oligodendroglial progeny under conditions similar to those described for primary neural precursors. The first reports on the glial differentiation of mouse ES cells were based on the 4–/4+ EB protocols (Bain *et al.*, 1995; Fraichard *et al.*, 1995; Strübing *et al.*, 1995) or multistep EB differentiation protocols (Okabe *et al.*, 1996). Most of the glial progeny under these conditions are astrocytes, with only a few immature oligodendrocytes present. However, subsequent studies have defined conditions for the selective generation of both astrocytes and oligodendrocytes.

### Oligodendrocytes

Highly efficient differentiation into oligodendrocytes was reported first using a modified multistep EB–based protocol (Brustle *et al.*, 1999). ES-derived neural precursors were expanded with FGF2, followed by FGF2 + EGF and FGF2 + PDGF. These conditions yielded a population of A2B5$^+$ glial precursors that are capable of differentiation into both astrocytic (~36% GFAP$^+$) and oligodendrocytic (~38% O4$^+$) progeny on mitogen withdrawal. The 4–/4+ EB RA induction protocol has recently been optimized for the production of oligodendrocytic progeny. This study (Billon *et al.*, 2002) demonstrated efficient selection of neural progeny by both positive (Sox1–eGFP) and negative (Oct4–HSV–thymidine kinase) selection, oligodendrocytic differentiation in RA-induced EBs after expansion in FGF2, followed by dissociation and replating in serum-free medium containing FGF2 and SHH. The final step involved SHH and FGF2 withdrawal and the addition of PDGF and thyroid hormone (T3). Under these conditions, ~50% of all cells express oligodendroglial markers. Optimized conditions for oligodendrocyte differentiation using HepG2 or default neural differentiation protocols have not yet been reported. However, stromal feeder–mediated induction, initially thought to bias toward neuronal progeny (Kawasaki *et al.*, 2000), can readily be adapted to derive oligodendrocytes at very high efficiencies and without requiring any genetic selection (Barberi *et al.*, 2003). In human ES cells, oligodendrocytic fates have been reported *in vitro* and *in vivo* (Keirstead *et al.*, 2005; Nistor *et al.*, 2005; Tabar *et al.*, 2005). The use of oligodendrocytic progeny in spinal cord injury has been proposed as one of the first clinical applications for human ES cells.

### Astrocytes

Highly efficient differentiation of ES cells into astrocytes has been reported recently using stromal feeder–mediated neural induction, followed by sequential exposure to FGF2, FGF2/EGF, EGF/CNTF, and CNTF. Over 90% of all cells expressed the astrocytic marker GFAP under these conditions (Barberi *et al.*, 2003). Significant numbers of GFAP cells were also obtained using HepG2-mediated neural differentiation (Rathjen *et al.*, 2002) or multistep EB protocols (Brustle *et al.*, 1999). Glial progenitors obtained with a multistep EB protocol were recently "transplanted" *in vitro* into hippocampal slices and revealed that full physiological maturation of ES-derived astrocytes can be achieved on interaction with an appropriate host environment (Scheffler *et al.*, 2003). Under these conditions, ES-derived astrocytes integrated seamlessly into host astrocytic networks tightly coupled via gap junctions. Human ES cells also readily give rise to astrocytic progeny (Tabar *et al.*, 2005; Zhang *et al.*, 2001).

### ES-Derived Neural Crest Progeny

ES cells provide a powerful assay to study neural crest development *in vitro*. The main strategy for deriving neural crest–like structures from ES cells is based on the exposure to BMPs following neural induction. The feasibility of this approach was demonstrated for mouse and nonhuman primate ES cells using the PA6 stromal feeder cell system (Mizuseki *et al.*, 2003). This study showed the development of both sensory as well as sympathetic neurons in a BMP dose-dependent manner. The derivation of smooth muscle cells required growth in chicken extract in combination with BMP withdrawal. No melanocytes or Schwann cells could be obtained under these conditions (Mizuseki *et al.*, 2003). Another recent study suggested efficient neural crest induction, including Schwann cell differentiation, using an EB-based multistep differentiation protocol in combination with BMP2 treatment (Gossrau *et al.*, 2003). Neural crest formation has also been reported in HepG2-mediated neural differentiation protocols on exposure to staurosporine (Rathjen *et al.*, 2002), previously reported to induce avian neural crest development (Newgreen and Minichiello, 1996). However, characterization of neural crest progeny was limited to morphological observations and expression of Sox10, a marker expressed during neural crest development but also during glial development in the CNS. Recent studies demonstrated neural crest progeny from human ES cells (Pomp *et al.*, 2005) and the ability to yield melanocyte-like cells (Motohashi *et al.*, 2006). Future studies will be required to systematically define conditions for the derivation of all neural crest lineages *in vitro* and to validate stability and function of neural crest phenotypes *in vivo*. One particularly interesting strategy for the future might be the isolation of neural crest stem cell populations from ES cells. Such ES cell–derived neural crest stem cells could be derived at various developmental stages and of various regional identities to re-create more fully the neural crest diversity *in vitro*. The derivation of unlimited numbers of neural crest stem cells derived from mouse or human embryonic stem cells will provide an important alternative to studying neural crest biology, given the limited availability and the lack of techniques to reliably expand primary neural crest stem cells *in vitro*.

## V. APPLICATIONS IN BIOLOGY AND DISEASE

### Tissue Engineering

Stem cell biology and tissue engineering are highly complementary approaches to achieve ultimate control over cell fate, cell behavior, and *in vivo* performance. Engineering approaches have been used to influence neuronal fate specification in neural progenitor cultures using high-epitope-density nanofibers. Nearly exclusive neuronal differentiation was achieved using such nanofibers presenting the laminin epitope IKVAV at the expense of astrocytic differentiation (Silva *et al.*, 2004). Matrix conditions that influence neuronal subtype specification have also been reported during neural differentiation of mouse ES cells. In this study, laminin-based attachment matrices promoted GABAergic fates in a multistep EB-based protocol (Goetz *et al.*, 2006). Three-dimensional fibrin matrices have been used to optimize neural differentiation in the classic 4−/4+ protocols (Willerth *et al.*, 2006). The combination of 3D matrices such as poly(lactic-*co*-glycolic acid)/poly(L-lactic acid) polymer scaffolds with defined-growth factors promoted the growth of various tissue types from human embryonic stem cells, including neural fates (Levenberg *et al.*, 2003).

Another important application for tissue engineering in neural stem cell biology is the ability to guide axonal growth using tools such as photofabricated gelatin-based nerve conduits (Gamez *et al.*, 2004) or alginate-based capillary hydrogels (Prang *et al.*, 2006). The combination of peripheral nerve conduits, growth factors, and fibrin glue has been used successfully in animal models of spinal cord injury (Cheng *et al.*, 1996). Other engineering approaches to spinal cord injury include the use of polymer scaffold in combination with neural stem cells that appear to promote endogenous axon regeneration (Teng *et al.*, 2002). Neural stem cells embedded in scaffolds have also been used as "biobridges" to fill cavities in the brain caused by hypoxic ischemic injury (K. I. Park *et al.*, 2002). Such implants were capable of rerouting axonal projections within the host brain, though it remains to be determined whether the induced plasticity can be sufficiently controlled to predict functional outcome. The progress in neural fate specification and ES cell biology will provide unprecedented opportunities for tissue engineering. These will be not limited to the use of engineering to enhance stem cell technologies but include the use of stem cell–derived products for engineering applications such as the development of neural prostheses, neural networks, and sophisticated neural computing.

## Disease Modeling

Embryonic stem cells offer a host of attractive features for use in developmental biology, disease modeling, and drug development. Some early examples of how ES cells could be harnessed for the study of neural disease are in the area of Parkinson's disease, Alzheimer's disease, and ALS. For each of these diseases, cell lines carrying specific genetic alterations associated with the disease have been generated. DJ1-mutant ES cells mimic a rare recessive form of Parkinson's disease and faithfully reproduce failures in oxidative stress response (Martinat *et al.*, 2004). SOD1G93A-mutant ES cells can be used to study cell intrinsic versus cell nonautonomous features in ALS (Bruijn *et al.*, 2004). There are also recent reports on the neural differentiation of ES cells that exhibit the disease-causing mutant of APP knocked into the endogenous APP locus (Abe *et al.*, 2003) and mutant APP expressing neural progenitors (Millet *et al.*, 2005). Other potentially interesting applications include the use of neural stem cells to model for prion infectivity and disease progression, allowing efficient matching of disease agent and genetic background (Giri *et al.*, 2006; Milhavet *et al.*, 2006). Such pilot studies suggest that ES cells can be used to model important aspects of neurodegenerative disease and may present a uniquely suitable system to develop for pharmacological screens and high-throughput drug discovery.

## Cell Repair

One of the driving forces behind deciphering the developmental program that controls cell fate specification is the hope that such insights could be harnessed for generating specialized cells for therapy. Despite the excitement about the potential of ES cells in neural repair, there are only a few examples where such approaches have been tested in animal models of disease.

### Cell Transplantation

#### Parkinson's Disease (PD)

One of the most widely discussed applications is the derivation of unlimited numbers of dopamine neurons from human ES cells for the treatment of Parkinson's disease. PD is particularly attractive for cell transplantation due to the relatively defined pathology that mainly affects midbrain dopamine neurons and the mostly unknown etiology that currently precludes causative treatment. At the onset of clinical symptoms, the majority of midbrain dopamine neurons have already died, providing further rationale for a cell-replacement approach. The first ES cell–based study that showed functional improvement in 6OHDA-lesioned rats, an animal model of Parkinson's disease, was based on the transplantation of low numbers of largely undifferentiated mouse ES cells isolated after short-term differentiation in EB cultures (Bjorklund *et al.*, 2002). Spontaneous differentiation into large numbers of neurons with midbrain dopamine characteristics was observed. However, the clinical relevance of this approach is rather limited due to the high rate of tumor formation (>50% of the animals with surviving grafts developed teratomas). At the same time, remarkable functional improvement was obtained after transplantation of dopamine neurons derived from mouse ES cells overexpressing Nurr1 (J. H. Kim *et al.*, 2002). This study was based on a multistep EB differentiation protocol. In addition to behavioral restoration in 6OHDA-lesioned rats, this study demonstrated *in vivo* functionality via electrophysiological recordings from grafted dopamine neurons in acute brain slices obtained from the grafted animals. However, the need for genetic modification raises safety concerns that may preclude clinical translation. Functional recovery with dopamine neurons derived from naive mouse ES cells was recently reported (Barberi *et al.*, 2003). This study demonstrated functional improvement *in vivo* after grafting dopamine neurons derived from ES and nuclear transfer ES (ntES) using stromal feeder–mediated differentiation. Successful grafting of ntES-derived dopamine neurons provided a first example of therapeutic cloning in neural disease (Barberi *et al.*, 2003). Differentiation of human ES cells into midbrain dopamine neurons has been reported following a stromal feeder–based strategy of neural induction, ventral midbrain specification by combined exposure to SHH/FGF8, and terminal differentiation in the presence of various differentiation and survival factors, including ascorbic acid, BDNF, GDNF, dbcAMP, and TGFb3 (Perrier *et al.*, 2004). A number of variations have been developed over the last few years, and alternative EB-based strategies have also emerged (Schulz *et al.*, 2004; Y. Yan *et al.*, 2005). However, unlike mouse ES–derived DA neurons, the survival of human and nonhuman primate ES–derived DA neurons has remained rather poor, with no or modest behavioral effects and with significant contamination of ES-derived forebrain precursors (Ferrari *et al.*, 2006; Martinat *et al.*, 2006; C. H. Park *et al.*, 2005; Sanchez-Pernaute *et al.*, 2005; Sonntag *et al.*, 2006; Takagi *et al.*, 2005). A recent study suggested that coculture with primary immortalized human midbrain astrocytes may significantly increase the survival of human ES cell–derived DA neurons *in vitro* and *in vivo*. However, the study found tumors in all the grafted animals, making interpretation of behavioral data problematic and raising considerable safety concerns for this approach. However, the large numbers of surviving TH$^+$ neurons reported in this study are a first step in the development of preclinical strategies for the treatment of PD.

### Huntington's Disease (HD)

The pathology in HD preferentially affects GABAergic medium spiny neurons, the main neuron subtype in the striatum. Similar to PD, fetal tissue transplantation trials in HD (Bachoud-Levi *et al.*, 2000; Freeman *et al.*, 2000; Gaura *et al.*, 2003) provide guidance on how to design stem cell–based approaches for HD. However, in the case of HD, the grafted cells need to project long distances from the striatum to the globus pallidus and the substantia nigra pars reticulata. ES-derived GABAergic neurons have not yet been tested in any animal model of HD, but the *in vitro* derivation of GABA neurons with forebrain characteristics has been achieved at high efficiencies (Barberi *et al.*, 2003; Watanabe *et al.*, 2005), though the proportion of medium spiny neurons among these GABA cells has not yet been determined. Another interesting avenue for stem cell research in HD is the recent derivation of human ES cells with HD mutations from embryos discarded after preimplantation diagnostics (Mateizel *et al.*, 2006). Such HD ES cell lines may provide invaluable insights into the selective vulnerability of the striatal GABAergic cell population.

### Spinal Cord Injury and Motoneuron Disorders

Traumatic or degenerative injuries to the spinal cord are often devastating and irreversible. Cell replacement using stem cells has been touted as a potential application for stem cell research. However, the complexity of cell therapy in spinal cord injury is enormous and far from resolved. Motor neurons are one of the main cell types affected by spinal cord injuries and by various degenerative diseases, such as ALS. The efficient derivation of motoneurons from mouse ES cells has been demonstrated with both EB-based (Wichterle *et al.*, 2002) and stromal feeder protocols (Barberi *et al.*, 2003; Mizuseki *et al.*, 2003). The functionality of ES-derived neurons *in vivo* has been addressed via xenografts into the developing chick spinal cord (Miles *et al.*, 2004; Soundararajan *et al.*, 2006; Wichterle *et al.*, 2002). However, a more recent study showed that mouse ES–derived motoneurons can survive in the adult spinal cord (Harper *et al.*, 2004) and that the application of a cocktail of growth factors can partly overcome white matter inhibition and improve target innervation (Deshpande *et al.*, 2006). Despite such progress, issues related to motoneuron subtype specification, survival in the diseased spinal cord, and efficiency of axonal projections remain formidable challenges for future translation. Functional improvement in a rodent spinal cord injury model have been reported after grafting mouse ES cells differentiated using a modified 4−/4+ protocol (McDonald *et al.*, 1999). The grafted cells differentiated *in vivo* into oligodendrocytes, neurons, and astrocytes and induced significant functional improvement in BBB scores as compared to sham injected animals. The mechanism by which functional improvement was obtained remains controversial (Privat *et al.*, 2000). Based on the ability for efficient differentiation into oligodendrocytes *in vivo*, it was suggested that remyelination of denuded axons might be a key factor (McDonald *et al.*, 2000). Functional improvement was also reported after grafting EB-derived cells obtained from human EG cells (Kerr *et al.*, 2003). EB-derived cell populations derived from human EG cells were implanted into the CSF of rats after virus-induced neuronopathy and motor neuron degeneration, a model of ALS. While a very small number of transplanted cells started to express markers compatible with motor neuron fate, most cells differentiated into neural progenitor or glial cells. It was concluded that the functional improvement was due to enhancing host neuron survival and function rather than to reestablishing functional connections from graft-derived motoneurons. The same hypothesis of protecting host neurons and providing myelin to surviving but denuded host axons led to the development of human ES cell–based grafting strategies using oligodendroglial precursors (Faulkner and Keirstead, 2005; Keirstead *et al.*, 2005; Nistor *et al.*, 2005).

### Stroke

Limited information is available on the performance of ES-derived progeny in animal models of stroke. A recent study showed that grafted ES cells can survive in a rat stroke model induced by transient ischemia via occlusion of the middle cerebral artery. The goal of this study was noninvasive imaging of the grafted cells using high-resolution MRI after transfection with ultrasmall superparamagnetic iron oxide particles (Hoehn *et al.*, 2002). Transplantation of neurally differentiated progeny was reported for mouse ES cells expressing Bcl-2 (Wei *et al.*, 2005) and for primate ES–derived neural progenitors (Hayashi *et al.*, 2006; Ikeda *et al.*, 2005b), with evidence of graft-derived projections into the host brain. Cell transplantation efforts for the treatment of stroke are complicated by the multiple cell types that are affected and the variability of the affected cell populations, depending on stroke location.

### Demyelination

There is significant evidence for the ability of primary and immortalized neural progenitor cells to achieve remyelination in a number of relevant animal models (Groves *et al.*, 1993; Windrem *et al.*, 2004; Yandava *et al.*, 1999). The capacity of mouse ES–derived progeny to remyelinate *in vivo* has been demonstrated after transplantation of highly purified ES-derived glial progenitors into the spinal cord of *md* rats (Brustle *et al.*, 1999) that lack the X-linked gene encoding myelin proteolipid protein (PLP), an animal model of Pelizaeus–Merzbacher syndrome. This study showed impressive *in vivo* differentiation results and yielded large grafts composed of myelinating oligodendrocytes. However, the grafted cells were not able to extend the short life span of these animals, precluding detailed functional analyses. A second study demonstrated remyelination after grafting purified oligosphere cultures derived from 4–/4+ EBs into the spinal cord of *shiverer* mice or into the chemically demyelinated spinal cord (Liu *et al.*, 2000).

The recent success in generating oligodendrocytes from human ES cells with evidence of *in vivo* remyelination capacity (Nistor *et al.*, 2005) offers a new dimension toward preclinical and clinical applications.

### Endogenous Recruitment

The persistence of proliferating neural precursors in the adult SVZ and hippocampus suggests recruitment of endogenous cells as a potential alternative to cell transplantation. Endogenous recruitment offers considerable advantages, given the noninvasive nature and lack of adverse immunological, inflammatory, or vascular response. However, it appears likely that these approaches will be limited to selective brain regions. Examples include the recruitment of SVZ-type neural precursors to the striatum (Arvidsson *et al.*, 2002) and differentiation of proliferating hippocampal precursors into CA1 pyramidal neurons (Nakatomi *et al.*, 2002). Increased efficiency of striatal neuron differentiation was achieved by ectopic expression of BDNF via adenoviral vectors (Chmielnicki *et al.*, 2004) or by the stimulation of the SVZ precursor population using FGF2 or Notch activation (Androutsellis-Theotokis *et al.*, 2006). Some of the preclinical models that may benefit most from these approaches for striatal repair are stroke and Huntington's disease.

One of the most intriguing reports regarding endogenous precursor cell potential was the replacement of cortical projection neurons through presumptive recruitment of SVZ precursors (Magavi *et al.*, 2000). The results of this study suggested that the minimally invasive nature of cell ablation via retrograde uptake of a phototoxic dye may have uncovered a broader differentiation potential inherent to SVZ precursors but suppressed in conventional injury models. However, the extremely low efficiency of the effect and the limited follow-up make it difficult to judge the full impact of this finding.

## VI. CONCLUSION

Embryonic stem cell–derived primary and endogenous neural stem cells provide unique tools for basic research and the development of new therapies. Embryonic stem cells offer unlimited cell numbers, full differentiation potential, and ease of introducing stable genetic modifications. Primary neural stem cells have a more limited proliferation and differentiation potential. However, for applications suitable to primary neural stem cells, they may offer a greater level of predictability and safety. Endogenous neural precursors can be harnessed without the invasive extraction of transplantation procedures, but our current understanding allows only limited control over cell fates, migration behavior, and functional integration. Beyond the role in regenerative medicine, neural stem cells, particularly ES cell–derived neural progeny, will become an essential tool for gene discovery and the study of basic development and disease. The generation reporter ES cells libraries and the establishment of cell lines with loss-of-function and gain-of-function mutations for most genes should become a reality in the near future. *In vitro* ES cell differentiation protocols will provide unlimited sources of defined neuronal and glial subtypes for pharmacological and genetic screens and for use in toxicology. However, one of the most important contributions of neural stem cell biology may be its role, as in basic biology, in unraveling the complex signals governing development from pluripotency all the way through neural specification and the formation of complex neuronal networks. Aspects of neural stem cell biology and tissue engineering will likely merge into a new, coherent discipline at the cutting edge of biology, tissue engineering, and modern medicine.

## VII. REFERENCES

Abe, Y., Kouyama, K., Tomita, T., Tomita, Y., Ban, N., Nawa, M., Matsuoka, M., Niikura, T., Aiso, S., Kita, Y., Iwatsubo, T., and Nishimoto, I. (2003). Analysis of neurons created from wild-type and Alzheimer's mutation knock-in embryonic stem cells by a highly efficient differentiation protocol. *J. Neurosci.* **23**, 8513–8525.

Altman, J., and Das, G. D. (1965). Autoradiographic and histological evidence of postnatal hippocampal neurogenesis in rats. *J. Comp. Neurol.* **124**, 319–335.

Andersson, E., Tryggvason, U., Deng, Q., Friling, S., Alekseenko, Z., Robert, B., Perlmann, T., and Ericson, J. (2006). Identification of intrinsic determinants of midbrain dopamine neurons. *Cell* **124**, 393–405.

Androutsellis-Theotokis, A., Leker, R. R., Soldner, F., Hoeppner, D. J., Ravin, R., Poser, S. W., Rueger, M. A., Bae, S. K., Kittappa, R., and McKay, R. D. (2006). Notch signaling regulates stem cell numbers *in vitro* and *in vivo*. *Nature* **442**, 823–826.

Arvidsson, A., Collin, T., Kirik, D., Kokaia, Z., and Lindvall, O. (2002). Neuronal replacement from endogenous precursors in the adult brain after stroke. *Nat. Med.* **8**, 963–970.

Aubert, J., Dunstan, H., Chambers, I., and Smith, A. (2002). Functional gene screening in embryonic stem cells implicates Wnt antagonism in neural differentiation. *Nat. Biotechnol.* **20**, 1240–1245.

Bachoud-Levi, A. C., Bourdet, C., Brugieres, P., Nguyen, J. P., Grandmougin, T., Haddad, B., Jeny, R., Bartolomeo, P., Boisse, M. F., Dalla Barba, G., Degos, J. D., Ergis, A. M., Lefaucheur, J. P., Lisovoski, F., Pailhous, E., Remy, P., Palfi, S., Defer, G. L., Cesaro, P., Hantraye, P., and Peschanski, M. (2000). Safety and tolerability assessment of intrastriatal neural allografts in five patients with Huntington's disease. *Exp. Neurol.* **161**, 194–202.

Bain, G., Kitchens, D., Yao, M., Huettner, J. E., and Gottlieb, D. I. (1995). Embryonic stem cells express neuronal properties *in vitro*. *Dev. Biol.* **168**, 342–357.

Baker, J. C., Beddington, R. S., and Harland, R. M. (1999). Wnt signaling in *Xenopus* embryos inhibits bmp4 expression and activates neural development. *Genes Dev.* **13**, 3149–3159.

Ballas, N., Grunseich, C., Lu, D. D., Speh, J. C., and Mandel, G. (2005). REST and its corepressors mediate plasticity of neuronal gene chromatin throughout neurogenesis. *Cell* **121**, 645–657.

Barberi, T., Klivenyi, P., Calingasan, N. Y., Lee, H., Kawamata, H., Loonam, K., Perrier, A. L., Bruses, J., Rubio, M. E., Topf, N., Tabar, V., Harrison, N. L., Beal, M. F., Moore, M. A., and Studer, L. (2003). Neural subtype specification of fertilization and nuclear transfer embryonic stem cells and application in parkinsonian mice. *Nat. Biotechnol.* **21**, 1200–1207.

Barres, B. A., Schmid, R., Sendtner, M., and Raff, M. C. (1993). Multiple extracellular signals are required for long-term oligodendrocyte survival. *Development* **118**, 283–295.

Barres, B. A., Raff, M. C., Gaese, F., Bartke, I., Dechant, G., and Barde, Y. A. (1994). A crucial role for neurotrophin-3 in oligodendrocyte development. *Nature* **367**, 371–375.

Benninger, F., Beck, H., Wernig, M., Tucker, K. L., Brustle, O., and Scheffler, B. (2003). Functional integration of embryonic stem cell–derived neurons in hippocampal slice cultures. *J. Neurosci.* **23**, 7075–7083.

Bibel, M., Richter, J., Schrenk, K., Tucker, K. L., Staiger, V., Korte, M., Goetz, M., and Barde, Y. A. (2004). Differentiation of mouse embryonic stem cells into a defined neuronal lineage. *Nat. Neurosci.* **7**, 1003–1009.

Billon, N., Jolicoeur, C., Ying, Q. L., Smith, A., and Raff, M. (2002). Normal timing of oligodendrocyte development from genetically engineered, lineage-selectable mouse ES cells. *J. Cell Sci.* **115**, 3657–3665.

Bjorklund, L. M., Sanchez-Pernaute, R., Chung, S., Andersson, T., Chen, I. Y., McNaught, K. S., Brownell, A. L., Jenkins, B. G., Wahlestedt, C., Kim, K. S., and Isacson, O. (2002). Embryonic stem cells develop into functional dopaminergic neurons after transplantation in a Parkinson rat model. *Proc. Natl. Acad. Sci. U.S.A.* **99**, 2344–2349.

Bjornson, C. R., Rietze, R. L., Reynolds, B. A., Magli, M. C., and Vescovi, A. L. (1999). Turning brain into blood: a hematopoietic fate adopted by adult neural stem cells *in vivo*. *Science* **283**, 534–537.

Bonni, A., Sun, Y., NadalVicens, M., Bhatt, A., Frank, D. A., Rozovsky, I., Stahl, N., Yancopoulos, G. D., and Greenberg, M. E. (1997). Regulation of gliogenesis in the central nervous system by the JAK-STAT signaling pathway. *Science* **278**, 477–483.

Briscoe, J., Pierani, A., Jessell, T. M., and Ericson, J. (2000). A homeodomain protein code specifies progenitor cell identity and neuronal fate in the ventral neural tube. *Cell* **101**, 435–445.

Bruijn, L. I., Miller, T. M., and Cleveland, D. W. (2004). Unraveling the mechanisms involved in motor neuron degeneration in ALS. *Annu. Rev. Neurosci.* **27**, 723–749.

Brustle, O., Jones, K. N., Learish, R. D., Karram, K., Choudhary, K., Wiestler, O. D., Duncan, I. D., and McKay, R. G. (1999). Embryonic stem cell–derived glial precursors: a source of myelinating transplants. *Science* **285**, 754–756.

Busch, C., Oppitz, M., Sailer, M. H., Just, L., Metzger, M., and Drews, U. (2006). BMP-2-dependent integration of adult mouse subventricular stem cells into the neural crest of chick and quail embryos. *J. Cell Sci.* **119**, 4467–4474.

Cai, J., QI, Y., Hu, X., Tan, M., Liu, Z., Zhang, J., Li, Q., Sander, M., and Qiu, M. (2005). Generation of oligodendrocyte precursor cells from

mouse dorsal spinal cord independent of Nkx6 regulation and Shh signaling. *Neuron* **45**, 41–53.

Caldwell, M. A., He, X. L., Wilkie, N., Pollack, S., Marshall, G., Wafford, K. A., and Svendsen, C. N. (2001). Growth factors regulate the survival and fate of cells derived from human neurospheres. *Nat. Biotechnol.* **19**, 475–479.

Calver, A. R., Hall, A. C., Yu, W. P., Walsh, F. S., Heath, J. K., Betsholtz, C., and Richardson, W. D. (1998). Oligodendrocyte population dynamics and the role of PDGF *in vivo*. *Neuron* **20**, 869–882.

Capela, A., and Temple, S. (2002). LeX/ssea-1 is expressed by adult mouse CNS stem cells, identifying them as nonependymal. *Neuron* **35**, 865–875.

Carvey, P. M., Ling, Z. D., Sortwell, C. E., Pitzer, M. R., McGuire, S. O., Storch, A., and Collier, T. J. (2001). A clonal line of mesencephalic progenitor cells converted to dopamine neurons by hematopoietic cytokines: a source of cells for transplantation in Parkinson's disease. *Exp. Neurol.* **171**, 98–108.

Castelo-Branco, G., Wagner, J., Rodriguez, F. J., Kele, J., Sousa, K., Rawal, N., Pasolli, H. A., Fuchs, E., Kitajewski, J., and Arenas, E. (2003). Differential regulation of midbrain dopaminergic neuron development by Wnt-1, Wnt-3a, and Wnt-5a. *Proc. Natl. Acad. Sci. U.S.A.* **100**, 12747–12752.

Cattaneo, E., and McKay, R. (1990). Proliferation and differentiation of neuronal stem cells regulated by nerve growth factor. *Nature* **347**(6295), 762–765.

Cheng, H., Cao, Y., and Olson, L. (1996). Spinal cord repair in adult paraplegic rats: partial restoration of hind limb function. *Science* **273**, 510–513.

Chmielnicki, E., Benraiss, A., Economides, A. N., and Goldman, S. A. (2004). Adenovirally expressed noggin and brain-derived neurotrophic factor cooperate to induce new medium spiny neurons from resident progenitor cells in the adult striatal ventricular zone. *J. Neurosci.* **24**, 2133–2142.

Cibelli, J. B., Grant, K. A., Chapman, K. B., Cunniff, K., Worst, T., Green, H. L., Walker, S. J., Gutin, P. H., Vilner, L., Tabar, V., Dominko, T., Kane, J., Wettstein, P. J., Lanza, R. P., Studer, L., Vrana, K. E., and West, M. D. (2002). Parthenogenetic stem cells in nonhuman primates. *Science* **295**, 819.

Clarke, D. L., Johansson, C. B., Wilbertz, J., Veress, B., Nilsson, E., Karlstrom, H., Lendahl, U., and Frisen, J. (2000). Generalized potential of adult neural stem cells [see comments]. *Science* **288**, 1660–1663.

Collins, L. S., and Dorshkind, K. (1987). A stromal cell line from myeloid long-term bone-marrow cultures can support myelopoiesis and B-lymphopoiesis. *J. Immunol.* **138**, 1082–1087.

Conti, L., Pollard, S. M., Gorba, T., Reitano, E., Toselli, M., Biella, G., Sun, Y. R., Sanzone, S., Ying, Q. L., Cattaneo, E., and Smith, A. (2005). Niche-independent symmetrical self-renewal of a mammalian tissue stem cell. *Plos. Biol.* **3**, 1594–1606.

Croisille, L., Auffray, I., Katz, A., Izac, B., Vainchenker, W., and Coulombel, L. (1994). Hydrocortisone differentially affects the ability of murine stromal cells and human marrow-derived adherent cells to promote the differentiation of CD34++/. *Blood* **84**, 4116–4124.

Dasen, J. S., Liu, J. P., and Jessell, T. M. (2003). Motor neuron columnar fate imposed by sequential phases of Hox-c activity. *Nature* **425**, 926–933.

Dasen, J. S., Tice, B. C., Brenner-Morton, S., and Jessell, T. M. (2005). A Hox regulatory network establishes motor neuron pool identity and target-muscle connectivity. *Cell* **123**, 477–491.

Davis, A. A., and Temple, S. (1994). A self-renewing multipotential stem cell in embryonic rat cerebral cortex. *Nature* **372**, 263–266.

Deshpande, D. M., Kim, Y. S., Martinez, T., Carmen, J., Dike, S., Shats, I., Rubin, L. L., Drummond, J., Krishnan, C., Hoke, A., Maragakis, N., Shefner, J., Rothstein, J. D., and Kerr, D. A. (2006). Recovery from paralysis in adult rats using embryonic stem cells. *Ann. Neurol.* **60**, 32–44.

Ding, Y. Q., Marklund, U., Yuan, W., Yin, J., Wegman, L., Ericson, J., Deneris, E., Johnson, R. L., and Chen, Z. F. (2003). Lmx1b is essential for the development of serotonergic neurons. *Nat. Neurosci.* **6**, 933–938.

Doetsch, F., Caille, I., Lim, D. A., Garcia-Verdugo, J. M., and Alvarez-Buylla, A. (1999). Subventricular zone astrocytes are neural stem cells in the adult mammalian brain. *Cell* **97**, 703–716.

Doetsch, F., Petreanu, L., Caille, I., Garcia-Verdugo, J. M., and Alvarez-Buylla, A. (2002). EGF converts transit-amplifying neurogenic precursors in the adult brain into multipotent stem cells. *Neuron* **36**, 1021–1034.

Eggenschwiler, J. T., Espinoza, E., and Anderson, K. V. (2001). Rab23 is an essential negative regulator of the mouse Sonic hedgehog signaling pathway. *Nature* **412**, 194–198.

Ericson, J., Muhr, J., Placzek, M., Lints, T., Jessell, T. M., and Edlund, T. (1995). Sonic hedgehog induces the differentiation of ventral forebrain neurons: a common signal for ventral patterning within the neural tube. *Cell* **81**, 747–756.

Farkas, L. M., Dunker, N., Roussa, E., Unsicker, K., and Krieglstein, K. (2003). Transforming growth factor-beta(s) are essential for the development of midbrain dopaminergic neurons *in vitro* and *in vivo*. *J. Neurosci.* **23**, 5178–5186.

Faulkner, J., and Keirstead, H. S. (2005). Human embryonic stem cell–derived oligodendrocyte progenitors for the treatment of spinal cord injury. *Transpl. Immunol.* **15**, 131–142.

Ferrari, D., Sanchez-Pernaute, R., Lee, H., Studer, L., and Isacson, O. (2006). Transplanted dopamine neurons derived from primate ES cells preferentially innervate DARPP-32 striatal progenitors within the graft. *Eur. J. Neurosci.* **24**, 1885–1896.

Ffrench-Constant, C., and Raff, M. C. (1986). Proliferating bipotential glial progenitor cells in adult rat optic nerve. *Nature* **319**, 499–502.

Finley, M. F., Kulkarni, N., and Huettner, J. E. (1996). Synapse formation and establishment of neuronal polarity by P19 embryonic carcinoma cells and embryonic stem cells. *J. Neurosci.* **16**, 1056–1065.

Fraichard, A., Chassande, O., Bilbaut, G., Dehay, C., Savatier, P., and Samarut, J. (1995). *In vitro* differentiation of embryonic stem cells into glial cells and functional neurons. *J. Cell Sci.* **108**, 3181–3188.

Frederiksen, K., and McKay, R. D. (1988). Proliferation and differentiation of rat neuroepithelial precursor cells *in vivo*. *J. Neurosci.* **8**, 1144–1151.

Freeman, T. B., Cicchetti, F., Hauser, R. A., Deacon, T. W., Li, X. J., Hersch, S. M., Nauert, G. M., Sanberg, P. R., Kordower, J. H., Saporta, S., and Isacson, O. (2000). Transplanted fetal striatum in Huntington's disease: phenotypic development and lack of pathology. *Proc. Natl. Acad. Sci. U.S.A.* **97**, 13877–13882.

Gabay, L., Lowell, S., Rubin, L., and Anderson, D. J. (2003). Deregulation of dorsoventral patterning by FGF confers trilineage differentiation capacity on CNS stem cells *in vitro*. *Neuron* **40**, 485–499.

Gaiano, N., and Fishell, G. (2002). The role of notch in promoting glial and neural stem cell fates. *Annu. Rev. Neurosci.* **25**, 471–490.

Galli, R., Pagano, S. F., Gritti, A., and Vescovi, A. L. (2000). Regulation of neuronal differentiation in human CNS stem cell progeny by leukemia inhibitory factor. *Dev. Neurosci.* **22**, 86–95.

Gamez, E., Goto, Y., Nagata, K., Iwaki, T., Sasaki, T., and Matsuda, T. (2004). Photofabricated gelatin-based nerve conduits: nerve tissue regeneration potentials. *Cell Transplant.* **13**, 549–564.

Gaura, V., Bachoud-Levi, A. C., Ribeiro, M. J., Nguyen, J. P., Frouin, V., Baudic, S., Brugieres, P., Mangin, J. F., Boisse, M. F., Palfi, S., Cesaro, P., Samson, Y., Hantraye, P., Peschanski, M., and Remy, P. (2003). Striatal neural grafting improves cortical metabolism in Huntington's disease patients. *Brain* DOI 10.1093/brain/awh003.

Giri, R. K., Young, R., Pitstick, R., DeArmond, S. J., Prusiner, S. B., and Carlson, G. A. (2006). Prion infection of mouse neurospheres. *Proc. Natl. Acad. Sci. U.S.A.* **103**, 3875–3880.

Goetz, A. K., Scheffler, B., Chen, H. X., Wang, S., Suslov, O., Xiang, H., Brustle, O., Roper, S. N., and Steindler, D. A. (2006). Temporally restricted substrate interactions direct fate and specification of neural precursors derived from embryonic stem cells. *Proc. Natl. Acad. Sci. U.S.A.* **103**, 11063–11068.

Goldman, S. A., and Nottebohm, F. (1983). Neuronal production, migration, and differentiation in a vocal control nucleus of the adult female canary brain. *Proc. Natl. Acad. Sci. U.S.A.* **80**, 2390–2394.

Goridis, C., and Rohrer, H. (2002). Specification of catecholaminergic and serotonergic neurons. *Nat. Rev. Neurosci.* **3**, 531–541.

Gossrau, G., Wernig, M., and Brustle, O. (2003). Neural crest fates in differentiating embryonic stem cell cultures. *Soc. Neurosci. Abstr.* **347**(8).

Gottlieb, D. I. (2002). Large-scale sources of neural stem cells. *Annu. Rev. Neurosci.* **25**, 381–407.

Gritti, A., Cova, L., Parati, E. A., Galli, R., and Vescovi, A. L. (1995). Basic fibroblast growth factor supports the proliferation of epidermal growth factor–generated neuronal precursor cells of the adult mouse CNS. *Neurosci. Lett.* **185**, 151–154.

Gross, R. E., Mehler, M. F., Mabie, P. C., Zang, Z., Santschi, L., and Kessler, J. A. (1996). Bone morphogenetic proteins promote astroglial lineage commitment by mammalian subventricular zone progenitor cells. *Neuron* **17**, 595–606.

Groves, A. K., Barnett, S. C., Franklin, R. J., Crang, A. J., Mayer, M., Blakemore, W. F., and Noble, M. (1993). Repair of demyelinated lesions by transplantation of purified o-2a progenitor cells (see comments). *Nature* **362**, 453–455.

Harper, J. M., Krishnan, C., Darman, J. S., Deshpande, D. M., Peck, S., Shats, I., Backovic, S., Rothstein, J. D., and Kerr, D. A. (2004). Axonal growth of embryonic stem cell–derived motoneurons *in vitro* and in motoneuron-injured adult rats. *Proc. Natl. Acad. Sci. U.S.A.* **101**, 7123–7128.

Hatten, M. E., Francois, A. M., Napolitano, E., and Roffler-Tarlov, S. (1984). Embryonic cerebellar neurons accumulate [3H]-gamma-aminobutyric acid: visualization of developing gamma-aminobutyric acid-utilizing neurons *in vitro* and *in vivo*. *J. Neurosci.* **4**, 1343–1353.

Hayashi, J., Takagi, Y., Fukuda, H., Imazato, T., Nishimura, M., Fujimoto, M., Takahashi, J., Hashimoto, N., and Nozaki, K. (2006). Primate embryonic stem cell–derived neuronal progenitors transplanted into ischemic brain. *J. Cereb. Blood Flow Metab.* **26**, 906–914.

He, W., Ingraham, C., Rising, L., Goderie, S., and Temple, S. (2001). Multipotent stem cells from the mouse basal forebrain contribute GABAergic neurons and oligodendrocytes to the cerebral cortex during embryogenesis. *J. Neurosci.* **21**, 8854–8862.

Hebert, J. M., Mishina, Y., and McConnell, S. K. (2002). BMP signaling is required locally to pattern the dorsal telencephalic midline. *Neuron* **35**, 1029–1041.

Hemmati Brivanlou, A., Kelly, O. G., and Melton, D. A. (1994). Follistatin, an antagonist of activin, is expressed in the Spemann organizer and displays direct neuralizing activity. *Cell* **77**, 283–295.

Hitoshi, S., Tropepe, V., Ekker, M., and van der Kooy, D. (2002). Neural stem cell lineages are regionally specified, but not committed, within distinct compartments of the developing brain. *Development* **129**, 233–244.

Hoehn, M., Kustermann, E., Blunk, J., Wiedermann, D., Trapp, T., Wecker, S., Focking, M., Arnold, H., Hescheler, J., Fleischmann, B. K., Schwindt, W., and Buhrle, C. (2002). Monitoring of implanted stem cell migration *in vivo*: a highly resolved *in vivo* magnetic resonance imaging investigation of experimental stroke in rat. *Proc. Natl. Acad. Sci. U.S.A.* **99**, 16267–16272.

Houser, C. R., Vaughn, J. E., Barber, R. P., and Roberts, E. (1980). GABA neurons are the major cell type of the nucleus reticularis thalami. *Brain Res.* **200**, 341–354.

Hsieh, J., Nakashima, K., Kuwabara, T., Mejia, E., and Gage, F. H. (2004). Histone deacetylase inhibition–mediated neuronal differentiation of multipotent adult neural progenitor cells. *Proc. Natl. Acad. Sci. U.S.A.* **101**, 16659–16664.

Hynes, M., Porter, J. A., Chiang, C., Chang, D., Tessier-Lavigne, M., Beachy, P. A., and Rosenthal, A. (1995a). Induction of midbrain dopaminergic neurons by sonic hedgehog. *Neuron* **15**, 35–44.

Hynes, M., Poulsen, K., Tessier-Lavigne, M., and Rosenthal, A. (1995b). Control of neuronal diversity by the floor plate: contact-mediated induction of midbrain dopaminergic neurons. *Cell* **80**, 95–101.

Ikeda, H., Osakada, F., Watanabe, K., Mizuseki, K., Haraguchi, T., Miyoshi, H., Kamiya, D., Honda, Y., Sasai, N., Yoshimura, N., Takahashi, M., and Sasai, Y. (2005a). Generation of Rx+/Pax6+ neural retinal precursors from embryonic stem cells. *Proc. Natl. Acad. Sci. U.S.A.* **102**, 11331–11336.

Ikeda, R., Kurokawa, M. S., Chiba, S., Yoshikawa, H., Ide, M., Tadokoro, M., Nito, S., Nakatsuji, N., Kondoh, Y., Nagata, K., Hashimoto, T., and Suzuki, N. (2005b). Transplantation of neural cells derived from retinoic acid-treated cynomolgus monkey embryonic stem cells successfully improved motor function of hemiplegic mice with experimental brain injury. *Neurobiol. Dis.* **20**, 38–48.

Jain, M., Armstrong, R. J., Tyers, P., Barker, R. A., and Rosser, A. E. (2003). GABAergic immunoreactivity is predominant in neurons derived from expanded human neural precursor cells *in vitro*. *Exp. Neurol.* **182**, 113–123.

Jessell, T. M. (2000). Neuronal specification in the spinal cord: inductive signals and transcriptional codes. *Nat. Rev. Genet.* **1**, 20–29.

Johe, K. K., Hazel, T. G., Müller, T., Dugich-Djordjevic, M. M., and McKay, R. D. G. (1996). Single factors direct the differentiation of stem cells from the fetal and adult central nervous system. *Genes Devel.* **10**, 3129–3140.

Kawasaki, H., Mizuseki, Nishikawa, S., Kaneko, S., Kuwana, Y., Nakanishi, S., Nishikawa, S., and Sasai, Y. (2000). Induction of midbrain dopaminergic neurons from ES cells by stromal cell–derived inducing activity. *Neuron* **28**, 31–40.

Kawasaki, H., Suemori, H., Mizuseki, K., Watanabe, K., Urano, F., Ichinose, H., Haruta, M., Takahashi, M., Yoshikawa, K., Nishikawa, S. I., Nakatsuji, N., and Sasai, K. (2002). Generation of dopaminergic neurons-sand pigmented epithelia from primate ES cells by stromal cell-derived inducing activity. *Proc. Natl. Acad. Sci. U.S.A.* **99**, 1580–1585.

Keirstead, H. S., Nistor, G., Bernal, G., Totoiu, M., Cloutier, F., Sharp, K., and Steward, O. (2005). Human embryonic stem cell–derived oligodendrocyte progenitor cell transplants remyelinate and restore locomotion after spinal cord injury. *J. Neurosci.* **25**, 4694–4705.

Kerr, D. A., Llado, J., Shamblott, M. J., Maragakis, N. J., Irani, D. N., Crawford, T. O., Krishnan, C., Dike, S., Gearhart, J. D., and Rothstein, J. D. (2003). Human embryonic germ cell derivatives facilitate motor recovery of rats with diffuse motor neuron injury. *J. Neurosci.* **23**, 5131–5140.

Kessaris, N., Fogarty, M., Iannarelli, P., Grist, M., Wegner, M., and Richardson, W. D. (2006). Competing waves of oligodendrocytes in the forebrain and postnatal elimination of an embryonic lineage. *Nat. Neurosci.* **9**, 173–179.

Kilpatrick, T. J., and Bartlett, P. F. (1993). Cloning and growth of multipotential neural precursors: requirements for proliferation and differentiation. *Neuron* **10**, 255–265.

Kim, J. H., Auerbach, J. M., Rodriguez-Gomez, J. A., Velasco, I., Gavin, D., Lumelsky, N., Lee, S. H., Nguyen, J., Sanchez-Pernaute, R., Bankiewicz, K., and McKay, R. (2002). Dopamine neurons derived from embryonic stem cells function in an animal model of Parkinson's disease. *Nature* **418**, 50–56.

Kim, J. Y., Koh, H. C., Lee, J. Y., Chang, M. Y., Kim, Y. C., Chung, H. Y., Son, H., Lee, Y. S., Studer, L., McKay, R., and Lee, S. H. (2003). Dopaminergic neuronal differentiation from rat embryonic neural precursors by Nurr1 overexpression. *J. Neurochem.* **85**, 1443–1454.

Krumlauf, R. (1994). Hox genes in vertebrate development. *Cell* **78**, 191–201.

Kuroda, H., Fuentealba, L., Ikeda, A., Reversade, B., and De Robertis, E. M. (2005). Default neural induction: neuralization of dissociated *Xenopus* cells is mediated by Ras/MAPK activation. *Genes Dev.* **19**, 1022–1027.

Lamb, T. M., Knecht, A. K., Smith, W. C., Stachel, S. E., Economides, A. N., Stahl, N., Yancopoulos, G. D., and Harland, R. M. (1993). Neural induction by the secreted polypeptide noggin. *Science* **262**, 713–718.

Lamba, D. A., Karl, M. O., Ware, C. B., and Reh, T. A. (2006). Efficient generation of retinal progenitor cells from human embryonic stem cells. *Proc. Natl. Acad. Sci. U.S.A.* **103**, 12769–12774.

Lee, K. J., Dietrich, P., and Jessell, T. M. (2000). Genetic ablation reveals that the roof plate is essential for dorsal interneuron specification. *Nature* **403**, 734–740.

Lee, S. K., and Pfaff, S. L. (2001). Transcriptional networks regulating neuronal identity in the developing spinal cord. *Nat. Neurosci.* **4**(Suppl.), 1183–1191.

Lee, S. -H., Lumelsky, N., Studer, L., Auerbach, J. M., and McKay, R. D. G. (2000). Efficient generation of midbrain and hindbrain neurons from mouse embryonic stem cells. *Nat. Biotechnol.* **18**, 675–679.

Levenberg, S., Huang, N. F., Lavik, E., Rogers, A. B., Itskovitz-Eldor, J., and Langer, R. (2003). Differentiation of human embryonic stem cells on three-dimensional polymer scaffolds. *Proc. Natl. Acad. Sci. U.S.A.* **100**, 12741–12746.

Li, H., Roblin, G., Liu, H., and Heller, S. (2003). Generation of hair cells by stepwise differentiation of embryonic stem cells. *Proc. Natl. Acad. Sci. U.S.A.* **100**, 13495–13500.

Li, X. J., Du, Z. W., Zarnowska, E. D., Pankratz, M., Hansen, L. O., Pearce, R. A., and Zhang, S. C. (2005). Specification of motoneurons from human embryonic stem cells. *Nat. Biotechnol.* **23**, 215–221.

Liem, K. F., Tremml, G., Roelink, H., and Jessell, T. M. (1995). Dorsal differentiation of neural plate cells induced by BMP-mediated signals from epidermal ectoderm. *Cell* **82**, 969–979.

Liem, K. F. J., Tremml, G., and Jessell, T. M. (1997). A role for the roof plate and its resident TGFbeta-related proteins in neuronal patterning in the dorsal spinal cord. *Cell* **91**, 127–138.

Liour, S. S., and Yu, R. K. (2003). Differentiation of radial glia-like cells from embryonic stem cells. *Glia* **42**, 109–117.

Litingtung, Y., and Chiang, C. (2000). Specification of ventral neuron types is mediated by an antagonistic interaction between shh and gli3. *Nat. Neurosci.* **3**, 979–985.

Liu, S., Qu, Y., Stewart, T. J., Howard, M. J., Chakrabortty, S., Holekamp, T. F., and McDonald, J. W. (2000). Embryonic stem cells differentiate into oligodendrocytes and myelinate in culture and after spinal cord transplantation. *Proc. Natl. Acad. Sci. U.S.A.* **97**, 6126–6131.

Lu, Q. R., Yuk, D., Alberta, J. A., Zhu, Z., Pawlitzky, I., Chan, J., McMahon, A. P., Stiles, C. D., and Rowitch, D. H. (2000). Sonic hedgehog–regulated oligodendrocyte lineage genes encoding bHLH proteins in the mammalian central nervous system. *Neuron* **25**, 317–329.

Lu, Q. R., Sun, T., Zhu, Z. M., Ma, N., Garcia, M., Stiles, C. D., and Rowitch, D. H. (2002). Common developmental requirement for Olig function indicates a motor neuron/oligodendrocyte connection. *Cell* **109**, 75–86.

Lumsden, A., and Krumlauf, R. (1996). Patterning the vertebrate neuraxis. *Science* **274**, 1109–1115.

Magavi, S. S., Leavitt, B. R., and Macklis, J. D. (2000). Induction of neurogenesis in the neocortex of adult mice. *Nature* **405**, 951–955.

Markakis, E. A., Palmer, T. D., Randolph-Moore, L., Rakic, P., and Gage, F. H. (2004). Novel neuronal phenotypes from neural progenitor cells. *J. Neurosci.* **24**, 2886–2897.

Marmur, R., Kessler, J. A., Zhu, G. F., Gokhan, S., and Mehler, M. F. (1998). Differentiation of oligodendroglial progenitors derived from cortical multipotent cells requires extrinsic signals including activation of gp130/LIF beta receptors. *J. Neurosci.* **18**, 9800–9811.

Martinat, C., Shendelman, S., Jonason, A., Leete, T., Beal, M. F., Yang, L., Floss, T., and Abeliovich, A. (2004). Sensitivity to oxidative stress in DJ-1-deficient dopamine neurons: an ES-derived cell model of primary parkinsonism. *PLoS Biol.* **2**, e327.

Martinat, C., Bacci, J. J., Leete, T., Kim, J., Vanti, W. B., Newman, A. H., Cha, J. H., Gether, U., Wang, H., and Abeliovich, A. (2006). Cooperative transcription activation by Nurr1 and Pitx3 induces embryonic stem cell maturation to the midbrain dopamine neuron phenotype. *Proc. Natl. Acad. Sci. U.S.A.* **103**, 2874–2879.

Mateizel, I., De, T. N., Ullmann, U., Cauffman, G., Sermon, K., Van, D., V, De, R. M., Degreef, E., Devroey, P., Liebaers, I., and Van, S. A. (2006). Derivation of human embryonic stem cell lines from embryos obtained after IVF and after PGD for monogenic disorders. *Hum. Reprod.* **21**, 503–511.

Mayer-Proschel, M., Kalyani, A. J., Mujtaba, T., and Rao, M. S. (1997). Isolation of lineage-restricted neuronal precursors from multipotent neuroepithelial stem cells. *Neuron* **19**, 773–785.

McDonald, J. W., Liu, X. Z., Qu, Y., Liu, S., Mickey, S. K., Turetsky, D., Gottlieb, D. I., and Choi, D. W. (1999). Transplanted embryonic stem cells survive, differentiate and promote recovery in injured rat spinal cord. *Nature Med.* **5**, 1410–1412.

McDonald, J. W., Gottlieb, D. I., and Choi, D. W. (2000). Reply to "What is a functional recovery after spinal cord injury?" *Nature Med.* **6**, 358.

Miles, G. B., Yohn, D. C., Wichterle, H., Jessell, T. M., Rafuse, V. F., and Brownstone, R. M. (2004). Functional properties of motoneurons derived from mouse embryonic stem cells. *J. Neurosci.* **24**, 7848–7858.

Milhavet, O., Casanova, D., Chevallier, N., McKay, R. D., and Lehmann, S. (2006). Neural stem cell model for prion propagation. *Stem Cells* **24**, 2284–2291.

Millet, P., Lages, C. S., Haik, S., Nowak, E., Allemand, I., Granotier, C., and Boussin, F. D. (2005). Amyloid-beta peptide triggers Fas-independent apoptosis and differentiation of neural progenitor cells. *Neurobiol. Dis.* **19**, 57–65.

Mizuseki, K., Sakamoto, T., Watanabe, K., Muguruma, K., Ikeya, M., Nishiyama, A., Arakawa, A., Suemori, H., Nakatsuji, N., Kawasaki, H., Murakami, F., and Sasai, Y. (2003). Generation of neural crest-derived peripheral neurons and floor plate cells from mouse and primate embryonic stem cells. *Proc. Natl. Acad. Sci. U.S.A.* **100**, 5828–5833.

Molne, M., Studer, L., Tabar, V., Ting, Y. T., Eiden, M. V., and McKay, R. D. (2000). Early cortical precursors do not undergo LIF-mediated astrocytic differentiation. *J. Neurosci. Res.* **59**, 301–311.

Motohashi, T., Aoki, H., Yoshimura, N., and Kunisada, T. (2006). Induction of melanocytes from embryonic stem cells and their therapeutic potential. *Pigment Cell Res.* **19**, 284–289.

Muller, T., Anlag, K., Wildner, H., Britsch, S., Treier, M., and Birchmeier, C. (2005). The bHLH factor Olig3 coordinates the specification of dorsal neurons in the spinal cord. *Genes Dev.* **19**, 733–743.

Munoz-Sanjuan, I., and Brivanlou, A. H. (2002). Neural induction, the default model and embryonic stem cells. *Nat. Rev. Neurosci.* **3**, 271–280.

Nakano, T., Kodama, H., and Honjo, T. (1994). Generation of lympho-hematopoietic cells from embryonic stem cells in culture. *Science* **265**, 1098–1101.

Nakatomi, H., Kuriu, T., Okabe, S., Yamamoto, S., Hatano, O., Kawahara, N., Tamura, A., Kirino, T., and Nakafuku, M. (2002). Regeneration of hippocampal pyramidal neurons after ischemic brain injury by recruitment of endogenous neural progenitors. *Cell* **110**, 429–441.

Newgreen, D. F., and Minichiello, J. (1996). Control of epitheliomesenchymal transformation. II. Cross-modulation of cell adhesion and cytoskeletal systems in embryonic neural cells. *Dev. Biol.* **176**, 300–312.

Nieto, M., Schuurmans, C., Britz, O., and Guillemot, F. (2001). Neural bHLH genes control the neuronal versus glial fate decision in cortical progenitors. *Neuron* **29**, 401–413.

Nistor, G. I., Totoiu, M. O., Haque, N., Carpenter, M. K., and Keirstead, H. S. (2005). Human embryonic stem cells differentiate into oligodendrocytes in high purity and myelinate after spinal cord transplantation. *Glia* **49**, 385–396.

Noctor, S. C., Flint, A. C., Weissman, T. A., Wong, W. S., Clinton, B. K., and Kriegstein, A. R. (2002). Dividing precursor cells of the embryonic cortical ventricular zone have morphological and molecular characteristics of radial glia. *J. Neurosci.* **22**, 3161–3173.

Nordstrom, U., Maier, E., Jessell, T. M., and Edlund, T. (2006). An early role for Wnt signaling in specifying neural patterns of Cdx and Hox gene expression and motor neuron subtype identity. *PLoS. Biol.* **4**, e252.

Okabe, S., Forsberg-Nilsson, K., Spiro, A. C., Segal, M., and McKay, R. D. G. (1996). Development of neuronal precursor cells and functional postmitotic neurons from embryonic stem cells *in vitro*. *Mech. Dev.* **59**, 89–102.

Ooto, S., Haruta, M., Honda, Y., Kawasaki, H., Sasai, Y., and Takahashi, M. (2003). Induction of the differentiation of lentoids from primate embryonic stem cells. *Invest. Ophthalmol. Vis. Sci.* **44**, 2689–2693.

Palmer, T. D., Ray, J., and Gage, F. H. (1995). FGF-2-responsive neuronal progenitors reside in proliferative and quiescent regions of the adult rodent brain. *Mol. Cell. Neurosci.* **6**, 474–486.

Panchision, D. M., Pickel, J. M., Studer, L., Lee, S. -H., Turner, P., Hazel, T. G., and McKay, R. D. (2001). Sequential actions of BMP receptors control neural precursor cell production and fate. *Genes Dev.* **15**, 2094–2110.

Park, C. H., Minn, Y. K., Lee, J. Y., Choi, D. H., Chang, M. Y., Shim, J. W., Ko, J. Y., Koh, H. C., Kang, M. J., Kang, J. S., Rhie, D. J., Lee, Y. S., Son, H., Moon, S. Y., Kim, K. S., and Lee, S. H. (2005). *In vitro* and *in vivo* analyses of human embryonic stem cell–derived dopamine neurons. *J. Neurochem.* **92**, 1265–1276.

Park, C. H., Kang, J. S., Shin, Y. H., Chang, M. Y., Chung, S., Koh, H. C., Zhu, M. H., Oh, S. B., Lee, Y. S., Panagiotakos, G., Tabar, V., Studer, L., and Lee, S. H. (2006). Acquisition of *in vitro* and *in vivo* functionality of Nurr1-induced dopamine neurons. *FASEB J.* **20**, 2553–2555.

Park, K. I., Teng, Y. D., and Snyder, E. Y. (2002). The injured brain interacts reciprocally with neural stem cells supported by scaffolds to reconstitute lost tissue. *Nat. Biotechnol.* **20**, 1111–1117.

Pera, E. M., Wessely, O., Li, S. Y., and De Robertis, E. M. (2001). Neural and head induction by insulin-like growth factor signals. *Dev. Cell* **1**, 655–665.

Pera, E. M., Ikeda, A., Eivers, E., and De Robertis, E. M. (2003). Integration of IGF, FGF, and anti-BMP signals via Smad1 phosphorylation in neural induction. *Genes Dev.* **17**, 3023–3028.

Perrier, A. L., and Studer, L. (2003). Making and repairing the mammalian brain — *in vitro* production of dopaminergic neurons. *Sem. Cell Develop. Biol.* **14**, 181–189.

Perrier, A. L., Tabar, V., Barberi, T., Rubio, M. E., Bruses, J., Topf, N., Harrison, N. L., and Studer, L. (2004). Derivation of midbrain dopamine neurons from human embryonic stem cells. *Proc. Natl. Acad. Sci. U.S.A.* **101**, 12543–12548.

Persson, M., Stamataki, D., te, W. P., Andersson, E., Bose, J., Ruther, U., Ericson, J., and Briscoe, J. (2002). Dorsal — ventral patterning of the spinal cord requires Gli3 transcriptional repressor activity. *Genes Dev.* **16**, 2865–2878.

Pomp, O., Brokhman, I., Ben-Dor, I., Reubinoff, B., and Goldstein, R. S. (2005). Generation of peripheral sensory and sympathetic neurons and neural crest cells from human embryonic stem cells. *Stem Cells* **23**, 923–930.

Prang, P., Muller, R., Eljaouhari, A., Heckmann, K., Kunz, W., Weber, T., Faber, C., Vroemen, M., Bogdahn, U., and Weidner, N. (2006). The promotion of oriented axonal regrowth in the injured spinal cord by alginate-based anisotropic capillary hydrogels. *Biomaterials* **27**, 3560–3569.

Pringle, N. P., Yu, W. P., Guthrie, S., Roelink, H., Lumsden, A., Peterson, A. C., and Richardson, W. D. (1996). Determination of neuroepithelial cell fate: induction of the oligodendrocyte lineage by ventral midline cells and sonic hedgehog. *Dev. Biol.* **177**, 30–42.

Privat, A., Ribotta, M. G., and Orsal, D. (2000). What is a functional recovery after spinal cord injury? *Nat. Med.* **6**, 358.

Qian, X. M., Shen, Q., Goderie, S. K., He, W. L., Capela, A., Davis, A. A., and Temple, S. (2000). Timing of CNS cell generation: a programmed sequence of neuron and glial cell production from isolated murine cortical stem cells. *Neuron* **28**, 69–80.

Rajan, P., and McKay, R. D. G. (1998). Multiple routes to astrocytic differentiation in the CNS. *J. Neurosci.* **18**, 3620–3629.

Rallu, M., Machold, R., Gaiano, N., Corbin, J. G., McMahon, A. P., and Fishell, G. (2002). Dorsoventral patterning is established in the telen-cephalon of mutants lacking both Gli3 and hedgehog signaling. *Development* **129**, 4963–4974.

Rao, M. S., Noble, M., and Mayer-Proschel, M. (1998). A tripotential glial precursor cell is present in the developing spinal cord. *Proc. Natl. Acad. Sci. U.S.A.* **95**, 3996–4001.

Rathjen, J., Haines, B. P., Hudson, K. M., Nesci, A., Dunn, S., and Rathjen, P. D. (2002). Directed differentiation of pluripotent cells to neural lineages: homogeneous formation and differentiation of a neurectoderm population. *Development* **129**, 2649–2661.

Renoncourt, Y., Carroll, P., Filippi, P., Arce, V., and Alonso, S. (1998). Neurons derived *in vitro* from ES cells express homeoproteins characteristic of motoneurons and interneurons. *Mech. Dev.* **79**, 185–197.

Reubinoff, B. E., Pera, M. F., Fong, C. Y., Trounson, A., and Bongso, A. (2000). Embryonic stem cell lines from human blastocysts: somatic differentiation *in vitro*. *Nat. Biotechnol.* **18**, 399–404.

Reynolds, B. A., and Weiss, S. (1992). Generation of neurons and astrocytes from isolated cells of the adult mammalian central nervous system (see comments). *Science* **255**, 1707–1710.

Ribak, C. E., Vaughn, J. E., Saito, K., Barber, R., and Roberts, E. (1976). Immunocytochemical localization of glutamate decarboxylase in rat substantia nigra. *Brain Res.* **116**, 287–298.

Rietze, R. L., Valcanis, H., Brooker, G. F., Thomas, T., Voss, A. K., and Bartlett, P. F. (2001). Purification of a pluripotent neural stem cell from the adult mouse brain. *Nature* **412**, 736–739.

Roelink, H., Porter, J. A., Chiang, C., Tanabe, Y., Chang, D. T., Beachy, P. A., and Jessell, T. M. (1995). Floor plate and motor neuron induction by different concentrations of the amino-terminal cleavage product of sonic hedgehog autoproteolysis. *Cell* **81**, 445–455.

Rowitch, D. H. (2004). Glial specification in the vertebrate neural tube. *Nat. Rev. Neurosci.* **5**, 409–419.

Sailer, M. H., Hazel, T. G., Panchision, D. M., Hoeppner, D. J., Schwab, M. E., and McKay, R. D. (2005). BMP2 and FGF2 cooperate to induce neural-crest-like fates from fetal and adult CNS stem cells. *J. Cell Sci.* **118**, 5849–5860.

Sanchez-Pernaute, R., Studer, L., Bankiewicz, K. S., Major, E. O., and McKay, R. D. (2001). *In vitro* generation and transplantation of precursor-derived human dopamine neurons. *J. Neurosci. Res.* **65**, 284–288.

Sanchez-Pernaute, R., Studer, L., Ferrari, D., Perrier, A. L., Lee, H., Vinuela, A., and Isacson, O. (2005). Long-term survival of dopamine neurons derived from parthenogenetic primate embryonic stem cells (Cyno1) in rat and primate striatum. *Stem Cells* **23**, 914–922.

Sasai, Y., Lu, B., Steinbeisser, H., and De Robertis, E. M. (1995). Regulation of neural induction by the Chd and Bmp-4 antagonistic patterning signals in *Xenopus*. *Nature* **377**, 757.

Scheffler, B., Schmandt, T., Schroder, W., Steinfarz, B., Husseini, L., Wellmer, J., Seifert, G., Karram, K., Beck, H., Blumcke, I., Wiestler, O. D., Steinhauser, C., and Brustle, O. (2003). Functional network integration of embryonic stem cell-derived astrocytes in hippocampal slice cultures. *Development* **130**, 5533–5541.

Schulz, T. C., Noggle, S. A., Palmarini, G. M., Weiler, D. A., Lyons, I. G., Pensa, K. A., Meedeniya, A. C., Davidson, B. P., Lambert, N. A., and Condie, B. G. (2004). Differentiation of human embryonic stem cells to dopaminergic neurons in serum-free suspension culture. *Stem Cells* **22**, 1218–1238.

Shamblott, M. J., Axelman, J., Wang, S., Bugg, E. M., Littlefield, J. W., Donovan, P. J., Blumenthal, P. D., Huggins, G. R., and Gearhart, J. D.

(1998). Derivation of pluripotent stem cells from cultured human primordial germ cells. *Proc. Natl. Acad. Sci. U.S.A.* **95**, 13726–13731.

Silva, G. A., Czeisler, C., Niece, K. L., Beniash, E., Harrington, D. A., Kessler, J. A., and Stupp, S. I. (2004). Selective differentiation of neural progenitor cells by high-epitope-density nanofibers. *Science* **303**, 1352–1355.

Singh, R. N., Nakano, T., Xuing, L., Kang, J., Nedergaard, M., and Goldman, S. A. (2005). Enhancer-specified GFP-based FACS purification of human spinal motor neurons from embryonic stem cells. *Exp. Neurol.* **196**, 224–234.

Sockanathan, S., Perlmann, T., and Jessell, T. M. (2003). Retinoid receptor signaling in postmitotic motor neurons regulates rostrocaudal positional identity and axonal projection pattern. *Neuron* **40**, 97–111.

Song, M. R., and Ghosh, A. (2004). FGF2-induced chromatin remodeling regulates CNTF-mediated gene expression and astrocyte differentiation. *Nat. Neurosci.* **7**, 229–235.

Sonntag, K. C., Pruszak, J., Yoshizaki, T., van, A. J., Sanchez-Pernaute, R., and Isacson, O. (2007). Enhanced yield of neuroepithelial precursors and midbrain-like dopaminergic neurons from human embryonic stem cells using the BMP antagonist noggin. *Stem Cells* **25**, 411–418.

Soundararajan, P., Miles, G. B., Rubin, L. L., Brownstone, R. M., and Rafuse, V. F. (2006). Motoneurons derived from embryonic stem cells express transcription factors and develop phenotypes characteristic of medial motor column neurons. *J. Neurosci.* **26**, 3256–3268.

Spangrude, G. J., Heimfeld, S., and Weissman, I. L. (1988). Purification and characterization of mouse hematopoietic stem cells. *Science* **241**, 58–62.

Speamann, H., and Mangold, H. (1924). Induktion von embryonanlagen durch implantation artfremder organisatoren. *Wilhelm Roux Arch. Entw. Mech. Organ.* **100**, 599–638.

Storch, A., Paul, G., Csete, M., Boehm, B. O., Carvey, P. M., Kupsch, A., and Schwarz, J. (2001). Long-term proliferation and dopaminergic differentiation of human mesencephalic neural precursor cells. *Exp. Neurol.* **170**, 317–325.

Streit, A., Berliner, A. J., Papanayotou, C., Sirulnik, A., and Stern, C. D. (2000). Initiation of neural induction by FGF signaling before gastrulation. *Nature* **406**, 74–78.

Strübing, C., Ahnert-Hilger, G., Shan, J., Wiedenmann, B., Hescheler, J., and Wobus, A. M. (1995). Differentiation of pluripotent embryonic stem cells into the neuronal lineage *in vitro* gives rise to mature inhibitory and excitatory neurons. *Mech. Dev.* **53**, 275–287.

Studer, L., Tabar, V., and McKay, R. D. (1998). Transplantation of expanded mesencephalic precursors leads to recovery in parkinsonian rats. *Nature Neurosci.* **1**, 290–295.

Studer, L., Csete, M., Lee, S. -H., Kabbani, N., McKay, R. D., Wold, B., and McKay, R. D. (2000). Enhanced proliferation, survival and dopaminergic differentiation of CNS precursors in lowered oxygen. *J. Neurosci.* **20**, 7377–7383.

Su, H. L., Muguruma, K., Matsuo-Takasaki, M., Kengaku, M., Watanabe, K., and Sasai, Y. (2006). Generation of cerebellar neuron precursors from embryonic stem cells. *Dev. Biol.* **290**, 287–296.

Sutherland, H. J., Eaves, C. J., Lansdorp, P. M., Thacker, J. D., and Hogge, D. E. (1991). Differential regulation of primitive human hematopoietic cells in long-term cultures maintained on genetically engineered murine stromal cells. *Blood* **78**, 666–672.

Tabar, V., Panagiotakos, G., Greenberg, E. D., Chan, B. K., Sadelain, M., Gutin, P. H., and Studer, L. (2005). Migration and differentiation of neural precursors derived from human embryonic stem cells in the rat brain. *Nat. Biotechnol.* **23**, 601–606.

Takagi, Y., Takahashi, J., Saiki, H., Morizane, A., Hayashi, T., Kishi, Y., Fukuda, H., Okamoto, Y., Koyanagi, M., Ideguchi, M., Hayashi, H., Imazato, T., Kawasaki, H., Suemori, H., Omachi, S., Iida, H., Itoh, N., Nakatsuji, N., Sasai, Y., and Hashimoto, N. (2005). Dopaminergic neurons generated from monkey embryonic stem cells function in a Parkinson primate model. *J. Clin. Invest.* **115**, 102–109.

Takahashi, J., Palmer, T. D., and Gage, F. H. (1999). Retinoic acid and neurotrophins collaborate to regulate neurogenesis in adult-derived neural stem cell cultures. *J. Neurobiol.* **38**, 65–81.

Tanigaki, K., Nogaki, F., Takahashi, J., Tashiro, K., Kurooka, H., and Honjo, T. (2001). Notch1 and Notch3 instructively restrict bFGF-responsive multipotent neural progenitor cells to an astroglial fate. *Neuron* **29**, 45–55.

Tao, W., and Lai, E. (1992). Telencephalon-restricted expression of BF-1, a new member of the HNF-3/fork head gene family, in the developing rat brain. *Neuron* **8**, 957–966.

Teng, Y. D., Lavik, E. B., Qu, X. L., Park, K. I., Ourednik, J., Zurakowski, D., Langer, R., and Snyder, E. Y. (2002). Functional recovery following traumatic spinal cord injury mediated by a unique polymer scaffold seeded with neural stem cells. *Proc. Natl. Acad. Sci. U.S.A.* **99**, 3024–3029.

Thomson, J. A., Itskovitz-Eldor, J., Shapiro, S. S., Waknitz, M. A., Swiergiel, J. J., Marshall, V. S., and Jones, J. M. (1998). Embryonic stem cell lines derived from human blastocysts. *Science* **282**, 1145–1147.

Tropepe, V., Hitoshi, S., Sirard, C., Mak, T. W., Rossant, J., and van der Kooy, D. (2001). Direct neural fate specification from embryonic stem cells: a primitive mammalian neural stem cell stage acquired through a default mechanism. *Neuron* **30**, 65–78.

Uchida, N., Buck, D. W., He, D. P., Reitsma, M. J., Masek, M., Phan, T. V., Tsukamoto, A. S., Gage, F. H., and Weissman, I. L. (2000). Direct isolation of human central nervous system stem cells. *Proc. Natl. Acad. Sci. U.S.A.* **97**, 14720–14725.

Ueno, H., Sakita-Ishikawa, M., Morikawa, Y., Nakano, T., Kitamura, T., and Saito, M. (2003). A stromal cell–derived membrane protein that supports hematopoietic stem cells. *Nat. Immunol.* **4**, 457–463.

Valk-Lingbeek, M. E., Bruggeman, S. W., and van, L. M. (2004). Stem cells and cancer: the polycomb connection. *Cell* **118**, 409–418.

Vallstedt, A., Klos, J. M., and Ericson, J. (2005). Multiple dorsoventral origins of oligodendrocyte generation in the spinal cord and hindbrain. *Neuron* **45**, 55–67.

Valtz, N. L., Hayes, T. E., Norregaard, T., Liu, S. M., and McKay, R. D. (1991). An embryonic origin for medulloblastoma. *New Biol.* **3**, 364–371.

Verani, R., Cappuccio, I., Spinsanti, P., Gradini, R., Caruso, A., Magnotti, M. C., Motolese, M., Nicoletti, F., and Melchiorri, D. (2007). Expression of the Wnt inhibitor Dickkopf-1 is required for the induction of neural markers in mouse embryonic stem cells differentiating in response to retinoic acid. *J. Neurochem.* **100**, 242–250.

Vicario-Abejon, C., Johe, K. K., Hazel, T. G., Collazo, D., and McKay, R. D. (1995). Functions of basic fibroblast growth factor and neurotrophins in the differentiation of hippocampal neurons [see comments]. *Neuron* **15**, 105–114.

Viti, J., Feathers, A., Phillips, J., and Lillien, L. (2003). Epidermal growth factor receptors control competence to interpret leukemia inhibitory factor as an astrocyte inducer in developing cortex. *J. Neurosci.* **23**, 3385–3393.

Wagner, J., Akerud, P., Castro, D. S., Holm, P. C., Canals, J. M., Snyder, E. Y., Perlmann, T., and Arenas, E. (1999). Induction of a midbrain dopaminergic phenotype in Nurr1-overexpressing neural stem cells by type 1 astrocytes. *Nat. Biotechnol.* **17**, 653–659.

Wakayama, T., Tabar, V., Rodriguez, I., Perry, A. C., Studer, L., and Mombaerts, P. (2001). Differentiation of embryonic stem cell lines generated from adult somatic cells by nuclear transfer. *Science* **292**, 740–743.

Watanabe, K., Kamiya, D., Nishiyama, A., Katayama, T., Nozaki, S., Kawasaki, H., Watanabe, Y., Mizuseki, K., and Sasai, Y. (2005). Directed differentiation of telencephalic precursors from embryonic stem cells. *Nat. Neurosci.* **8**, 288–296.

Wei, L., Cui, L., Snider, B. J., Rivkin, M., Yu, S. S., Lee, C. S., Adams, L. D., Gottlieb, D. I., Johnson, E. M., Jr., Yu, S. P., and Choi, D. W. (2005). Transplantation of embryonic stem cells overexpressing Bcl-2 promotes functional recovery after transient cerebral ischemia. *Neurobiol. Dis.* **19**, 183–193.

Wernig, M., Tucker, K. L., Gornik, V., Schneiders, A., Buschwald, R., Wiestler, O. D., Barde, Y. A., and Brustle, O. (2002). Tau EGFP embryonic stem cells: an efficient tool for neuronal lineage selection and transplantation. *J. Neurosci. Res.* **69**, 918–924.

Wichterle, H., Lieberam, I., Porter, J. A., and Jessell, T. M. (2002). Directed differentiation of embryonic stem cells into motor neurons. *Cell* **110**, 385–397.

Willerth, S. M., Arendas, K. J., Gottlieb, D. I., and Sakiyama-Elbert, S. E. (2006). Optimization of fibrin scaffolds for differentiation of murine embryonic stem cells into neural lineage cells. *Biomaterials* **27**, 5990–6003.

Wilson, S. I., Rydstrom, A., Trimborn, T., Willert, K., Nusse, R., Jessell, T. M., and Edlund, T. (2001). The status of Wnt signaling regulates neural and epidermal fates in the chick embryo. *Nature* **411**, 325–330.

Windrem, M. S., Nunes, M. C., Rashbaum, W. K., Schwartz, T. H., Goodman, R. A., McKhann, G., Roy, N. S., and Goldman, S. A. (2004). Fetal and adult human oligodendrocyte progenitor cell isolates myelinate the congenitally dysmyelinated brain. *Nature Med.* **10**, 93–97.

Wine-Lee, L., Ahn, K. J., Richardson, R. D., Mishina, Y., Lyons, K. M., and Crenshaw, E. B., III (2004). Signaling through BMP type 1 receptors is required for development of interneuron cell types in the dorsal spinal cord. *Development* **131**, 5393–5403.

Wu, P., Tarasenko, Y. I., Gu, Y., Huang, L. Y., Coggeshall, R. E., and Yu, Y. (2002). Region-specific generation of cholinergic neurons from fetal human neural stem cells grafted in adult rat. *Nat. Neurosci.* **5**, 1271–1278.

Yan, J., Studer, L., and McKay, R. D. G. (2001). Ascorbic acid increases the yield of dopaminergic neurons derived from basic fibroblast growth factor–expanded mesencephalic precursors. *J. Neurochem.* **76**, 307–311.

Yan, Y., Yang, D., Zarnowska, E. D., Du, Z., Werbel, B., Valliere, C., Pearce, R. A., Thomson, J. A., and Zhang, S. C. (2005). Directed differentiation of dopaminergic neuronal subtypes from human embryonic stem cells. *Stem Cells* **23**, 781–790.

Yandava, B. D., Billinghurst, L. L., and Snyder, E. Y. (1999). "Global" cell replacement is feasible via neural stem cell transplantation: evidence from the dysmyelinated shiverer mouse brain. *Proc. Natl. Acad. Sci. U.S.A.* **96**, 7029–7034.

Ye, W. L., Shimamura, K., Rubenstein, J. R., Hynes, M. A., and Rosenthal, A. (1998). FGF and Shh signals control dopaminergic and serotonergic cell fate in the anterior neural plate. *Cell* **93**, 755–766.

Ye, W. L., Bouchard, M., Stone, D., Liu, X. D., Vella, F., Lee, J., Nakamura, H., Ang, S. L., Busslinger, M., and Rosenthal, A. (2001). Distinct regulators control the expression of the mid-hindbrain organizer signal FGF8. *Nat. Neurosci.* **4**, 1175–1181.

Ying, Q. L., Stavridis, M., Griffiths, D., Li, M., and Smith, A. (2003). Conversion of embryonic stem cells into neuroectodermal precursors in adherent monoculture. *Nat. Biotechnol.* **21**, 183–186.

Yung, S. Y., Gokhan, S., Jurcsak, J., Molero, A. E., Abrajano, J. J., and Mehler, M. F. (2002). Differential modulation of BMP signaling promotes the elaboration of cerebral cortical GABAergic neurons or oligodendrocytes from a common sonic hedgehog-responsive ventral forebrain progenitor species. *Proc. Natl. Acad. Sci. U.S.A.* **99**, 16273–16278.

Zhang, S. C., Wernig, M., Duncan, I. D., Brustle, O., and Thomson, J. A. (2001). *In vitro* differentiation of transplantable neural precursors from human embryonic stem cells. *Nat. Biotechnol.* **19**, 1129–1133.

Zhou, Q., and Anderson, D. J. (2002). The bHLH transcription factors OLIG2 and OLIG1 couple neuronal and glial subtype specification. *Cell* **109**, 61–73.

# Brain Implants

*Lars U. Wahlberg*

## I. INTRODUCTION

Tissue-engineered brain implants have great potential to enable future treatments of many neurological disorders for which no disease-modifying treatments are available. Proof-of-concept clinical studies of cell replacement and restorative treatments have already been made in the clinic, with mixed but encouraging results. In particular, the human fetal tissue transplantation in Parkinson's disease (PD) has paved the way for more advanced tissue-engineering concepts.

A major obstacle to developing cellular implants for a large number of patients has been the inability to grow cells with the appropriate therapeutic potential in large numbers. The discovery of expandable stem cells from both embryonic and adult sources has therefore created exciting sources of cell lines potentially applicable to a large number of patients. In combination with an increasing knowledge of developmental biology, these expandable stem cell preparations can be increasingly pushed into relevant therapeutic cells for use in tissue-engineered brain implants.

In parallel with the cell replacement approaches, restorative and neuroprotective therapies have been developed that utilize the therapeutic effects of endogenous or genetically engineered protein factors. Encapsulated cell biodelivery devices have been made that combine advanced tissue engineering with genetic engineering to deliver regenerative growth factors to the brains of patients with neurological disorders such as PD. In combination with growth factors, artificial scaffoldings, and axonal outgrowth modulators, more sophisticated implants are being generated. In the future, implants like these are likely to restore more complex neural circuits and help a large number of patients in great medical need.

Experimental data strongly support the use of brain implants and tissue-engineering concepts in the treatment of brain disorders and promise new disease-modifying medical products for patients. Various cellular implants have already been applied in the clinic in proof-of-concept studies in Parkinson's disease (PD) (Lindvall and Björklund, 2004), Huntington's disease (HD) (Dunnett and Rosser, 2004), Alzheimer's disease (AD) (Blesch and Tuszynski, 2004), epilepsy (Björklund and Lindvall, 2000), and stroke (Savitz *et al.*, 2004). However, a cellular implant has yet to reach the market as a therapeutic product. PD has been a major target for brain implants and in this chapter is used as a disease example to illustrate the tissue-engineering concepts applied so far.

Despite the success of many drugs for PD, such as L-3,4-dihydroxyphenylalanine (L-dopa) therapy, introduced in the late 1960s, the current treatments of PD do not stop the progressive dopamine neuron dysfunction and cell death. Therefore, with time, patients on chronic L-dopa therapy develop both progressive symptoms and drug-induced side effects and require additional treatment options. In the early 1990s, surgical treatment strategies initially developed during the 1950s, such as the ventrolateral pallidotomy for Parkinsonian rigidity, were rejuvenated because new treatment strategies were desperately needed (Speelman and Bosch, 1998). With improvements in imaging and surgical techniques, ablative procedures have yielded excellent results in selected groups of PD patients. Implantable neural stimulators that inhibit neuronal transmission in local areas by high-frequency stimulation have more recently replaced much of the ablative therapies and can yield good therapeutic results without inducing permanent lesions in the brain (Deuschl *et al.*, 2006). Despite these successful applications of neurosurgical procedures to PD, these procedures are based on the inhibition or destruction of normal neurons in attempts to compensate for the disease damage. They do not address the biology of the underlying disease itself, and, albeit successfully applied in many patients, the destruction or inhibition of normal tissue is not an optimal treatment for neurological disorders. There is therefore a need for new treatment strategies that can address the pathology more directly and offer disease-modifying effects. Fortunately, the accumulated knowledge of the pathological processes, molecular and cell biology, biomaterials, imaging, and surgical procedures now make it possible to implement disease-modifying tissue-engineering concepts in the treatment of PD and other neurological diseases.

In many untreatable neurological disorders, the progressive loss of neurons and their associated function is the primary underlying cause for the symptoms. Therefore, various cell implant strategies have been designed either to *replace* the neurons or their function or to *protect or regenerate* the function and health of the diseased neurons or a combination of both (Fig. 64.1). Clinical applications to replace the dopaminergic function in patients with PD have so far utilized primary tissues or cells. Dopaminergic neurons derived from stem cells and more sophisticated tissue-engineered implants have not yet been applied in the clinic, but rapid progress in preclinical efforts is being made in this area. In neuroprotective or restorative strategies, cellular implants secreting either endogenous or engineered growth factors or cytokines have successfully been applied in animal models and even in the clinic for amyotrophic lateral sclerosis (ALS) (Aebischer *et al.*, 1996), HD (Bloch *et al.*, 2004), and AD (Tuszynski *et al.*, 2005).

This chapter reviews some of the cell replacement and regenerative brain implants applied in the clinic and what may be developed in the future.

## II. CELL REPLACEMENT IMPLANTS

### Primary Tissue Implants

As already mentioned, oral L-dopa therapy is the main treatment for PD. L-dopa is a precursor to dopamine that passes the blood–brain barrier and is taken up mainly by the dopaminergic neurons that in turn convert L-dopa to dopamine and increase their dopamine production and

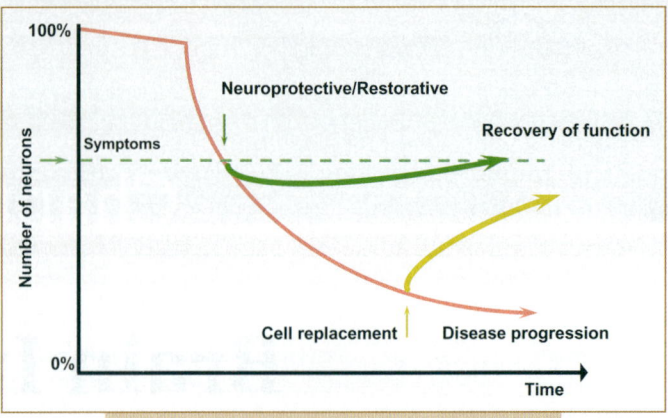

**FIG. 64.1.** Cell replacement and protective/regenerative brain implants in PD. The degeneration in PD is schematically depicted, showing an accelerated dopaminergic cell loss as seen in patients. At the time of symptoms, about 60% of the dopaminergic neurons remain and would be amenable to neuroprotective or restorative treatment with the application of protective and regenerative factors. In late-stage PD, the need for the replacement of lost dopaminergic neurons may be necessary. In future tissue-engineering strategies, a combination of cells, growth factors, and polymeric scaffoldings may be able to regenerate the dopaminergic pathways in a physiological manner and restore normal functioning. The relatively local degeneration in PD makes this disease a good initial target for brain implants, in comparison to stroke or other diseases with more unpredictable and diffuse injuries.

storage. However, with the progressive loss of dopaminergic neurons, the L-dopa therapy eventually becomes ineffective, and severe fluctuations in the ability to initiate movements occur. Because of the finding that L-dopa can increase the production of dopamine and alleviate the symptoms of PD, a reasonable therapeutic approach may be to implant dopamine- or L-dopa-secreting cells in the relevant areas of the brain (striatum).

Considering this idea, the first clinical transplantation for PD using a cellular brain implant was done at the Karolinska Hospital in Stockholm, Sweden, in the early 1980s (Backlund *et al.*, 1985). In this experimental procedure, autologous dopamine-secreting adrenal chromaffin cells were harvested from one of the adrenal glands from the patient and successfully transplanted to the striatum. The procedure was adopted very quickly by the medical community, and initial reports indicated good clinical results. However, with time, other studies showed poor survival of the cells and minimal positive clinical effects, resulting in cessation of the treatment. However, the concept of cellular brain implants had made its definite entry into the clinic, and many lessons were learned along the way.

At about the same time as the first chromaffin cell transplants were made in Stockholm, a promising cell transplantation strategy for PD was developed by Anders Björklund and coworkers at Lund University in southern Sweden (Björklund and Lindvall, 2000). They collected discarded aborted fetal tissue and dissected out the ventral mesencephalon to create cell suspensions containing developing dopaminergic neurons for transplantation experiments. After several years of extensive validation of the concept in animal models, cells were transplanted to the striatum of two patients (Lindvall *et al.*, 1988). The first results were relatively unimpressive but prompted modifications to various parts of the experimental procedure, and a second pair of patients transplanted about one year later with the modified techniques fared much better (Lindvall *et al.*, 1994). These patients showed positive clinical recovery starting about four months after the procedure. Positron emission tomography (PET) data indicated that the grafts were able to survive and take up and secrete dopamine. More than 10 years out from the procedure, one of the patients showed persistent graft viability on PET scanning, and the patient was maintained on minimal L-dopa therapy (Piccini *et al.*, 1999). To date, more than 300 patients have been transplanted with fetal ventral mesencephalic tissue at different centers around the world. However, the lack of suitable donor material, the heterogeneity of the tissues and preparations, and the inability to industrialize the process have all made it difficult to establish standardized medical trials and therapy geared to a large number of patients. In fact, two controlled trials with fetal transplantation have shown only minimal efficacy, and some patients have developed movement side effects (dyskinesias) that appear related to the grafting procedure (Freed *et al.*, 2001; Olanow *et al.*, 2003). Therefore, fetal transplantation as a therapy for PD has come to a halt, and cell replacement strategies for PD are awaiting alternative cells and implants that lend themselves to a reproducible industrial process applied in well-controlled trials.

It also should be noted that clinical trials with porcine-derived ventral mesencephalic cells also failed in clinical trials (A. Pollack, 2001). With concerns regarding the transmission of animal diseases to humans (zoonosis), few if any additional clinical applications of animal-derived cellular brain implants are currently ongoing. In fact, the use of animal-derived products in the manufacture of human brain implants should be avoided to the largest degree possible, because safety and regulatory concerns should be minimized.

Despite disappointing results in the clinical application of both chromaffin and fetal-derived primary tissues from both human and pig sources, the experimental work surrounding primary tissue transplantation in PD has demonstrated several important points that will facilitate the development of tissue-engineered implants in the treatment of PD and other neurological disorders.

1. Allogeneic cells can survive over many years in the brain after a relatively brief initial immunosuppression therapy (18 months) but do not survive well without it.
2. Grafted neurons can integrate, function, and interact with the host brain in a physiological and reciprocal manner.
3. Trial designs and outcome measures have been developed that facilitate safety and efficacy measures in the clinic.
4. Surgical techniques have been developed that allow for the injection and implantation of cells and tissue-engineered products in the brain.
5. Imaging techniques have been developed to evaluate the implantation and function of the brain implants.
6. Animal models have been developed to translate and scale-up experimental brain implants for the clinic.

## Cell Line Implants

As mentioned earlier, important drawbacks using primary tissues are its limited supply, its heterogeneity, and the difficulty and prohibitive costs to implement the necessary good manufacturing principles (GMP) for harvesting, manipulation, storage, and later use. For example, fetal transplantation experiments in PD required fresh tissue from four to eight donors, resulting in procedural difficulties and poor quality control, and may help explain the poor efficacy outcomes in the aforementioned controlled trials.

Therefore, the ability to expand and store cells in cell banks is paramount to create allogeneic cell alternatives to primary tissue sources. The expansion of normal and genetically unmodified cells derived from tissue donations can give rise to expandable primary cell lines potentially useful for brain implantation with or without further manipulation. The primary cell lines normally retain a limited number of cell cycles but allow for the proliferation of enough cells to transplant hundreds to thousands of patients from a single donor. Cultured primary cells can be expanded while retaining normal geno- and phenotypes, with contact inhibition and differentiation behaviors intact. These cells are therefore relatively safe to use, and the formation of tumors or other abnormal behaviors are relatively unlikely.

Retinal pigmented epithelial (RPE) cells derived from the retinas of organ donors can make primary cell lines with limited expansion capacity and are currently in clinical trials for PD. The RPE cells secrete L-dopa and are thought to function by increasing the intrastriatal L-dopa concentration, with subsequent conversion to dopamine by residual dopaminergic nerve endings and glia. These RPE cells are grown and transplanted on gelatin microcarriers to improve survival and prevent immune rejection. A report on a clinical pilot trial was recently published and showed that these implants were well tolerated and safe (Stover *et al.*, 2005). A Phase II trial is currently ongoing and is estimated to be finished in 2008.

The RPE cells do not work in a physiological manner and do not form neurons and associated neurites, and, even though they may help patients, it is unlikely that these implants will restore function to baseline. A more versatile source of expandable primary cell lines capable of making various replacement neurons or glia are mitogen-responsive human neural progenitor/stem cell cultures that can be isolated from various regions of aborted and adult human central nervous system (CNS) tissues and expanded for more than one year *in vitro*. These cells are often grown as *neurospheres* and can be expanded in defined and animal-free media containing growth factors (Carpenter *et al.*, 1999). These cells can form the three major phenotypes of the nervous system (neurons, astrocytes, and oligodendrocytes) *in vitro* and *in vivo* and show excellent survival, without the formation of tumors. These human neural stem cell–containing cultures have been transplanted to various regions in animals and have been shown to survive, integrate, migrate, differentiate, extend neurites, and arborize (Fricker *et al.*, 1999). Even though the cells tend to retain the markers consistent with the anatomical region from which they were isolated (Piao *et al.*, 2006), the cells can be manipulated with epigenetic and genetic factors to make specific cellular subtypes potentially useful for cell replacement implants. For a PD application, relevant nigral dopaminergic neurons need to be generated from these cells. However, so far only dopaminergic neurons with limited differentiation have been generated (Christophersen *et al.*, 2006). Reportedly, clinical applications in Batten's disease with unmanipulated human neurosphere cell lines are currently being explored (Stem-Cells, Inc., press release, 2006).

In addition to neurospheres, adherent primary cell lines of neurogenic glia have been made from both mouse and human developing neural tissues that display a glial phenotype during expansion but are capable of generating neurons during differentiating conditions (Skogh *et al.*, 2001). Similarly, so-called NS cells, expressing glia and stem cell markers, have been derived from ES cells and the developing CNS and appear to differentiate into neural phenotypes similar to those of the adherent glia and neurosphere cultures (Pollard *et al.*, 2006). The adherent cultures may have an advantage over neurospheres, from an industrial point of view, as adherent cells can be readily cloned and expanded and have been shown to retain their phenotype during expansion without differentiating into a heterogeneous mixture of progenitors. It is currently unclear if neurospheres or adherent NS cells can be made into dopaminergic neurons suitable for replacement therapy in PD.

In general, all primary cell lines derived from tissue stem cells have a large but limited expansion potential and show senescence (Ostenfeld *et al.*, 2000; Villa *et al.*, 2004). This may be due to the successive loss of immortal stem cells through the asymmetric division into progenitors (as seen in neurospheres), or, alternatively, the stem cells themselves have a limit to their proliferation. Both mouse and human ES cells defy the normal senescence of primary cells and can be expanded from a single clone indefinitely without loosing pluripotentiality (Miura *et al.*, 2004). From an industrial and tissue-engineering perspective, this feature is extremely attractive because a single donation could give rise to a cell line source with the capacity to make all organs of the body in unlimited numbers. One major drawback of ES cell–derived products, however, is that the ES cell itself cannot be implanted but needs to undergo the relevant development *in vitro* to make suitable cells for transplantation, e.g., dopaminergic neurons for PD or islet cells for diabetes mellitus. Because it is difficult to make pure cultures without retaining one or more undifferentiated ES cells, both the risk of heterogeneous cell preparations and the risk of tumorigenesis need to be overcome to develop ES cell–derived brain implants (Roy *et al.*, 2006). For neural applications, the recently described ES cell–derived NS cell may become a progenitor of choice for brain implant products because it is restricted to the neural lineage and is most likely nontumorigenic and can be expanded as a clonal cell line. It is currently unclear if the undifferentiated NS cell is immortal, like its parental ES cell, or if it displays senescence, like its neurosphere counterpart.

Cell replacement strategies using cell lines need to employ differentiation or selection methods in order to make the relevant replacement cell, e.g., dopaminergic neurons for a PD application. The generation of functioning human dopaminergic neurons *in vivo* akin to those derived from primary VM tissues has been difficult to achieve. Although neurons with the dopaminergic machinery can be made from both growth factor–expanded human neurospheres and genetically immortalized committed dopaminergic neuroblasts, it has been difficult to achieve survival and function *in vivo* (Christophersen *et al.*, 2006). Interestingly, even though ES cells would hypothetically need more steps to be differentiated into dopaminergic neurons for a PD application, relatively short-step protocols that use developmental signals involved in the rostrocaudal and ventrodorsal specification of the midbrain can push mouse ES cells into functional dopaminergic neurons in a rat model of PD (Kim *et al.*, 2002). More recent studies have also identified important transcription signals involving the Lmx1a and Msx1 homeobox genes that, when overexpressed in ES cells, can yield dopaminergic neurons with markers consistent with substantia nigra neurons (Andersson *et al.*, 2006). Transplantation of these dopaminergic neurons in a rat model of PD yields excellent survival, neurite extension, and function, consistent with results from primary VM tissues. These findings may pave the way forward to make relevant "nigral" dopaminergic neurons from human ES cells and neural stem cell cultures in the not-too-distant future. However, similar to other experiments with ES cells, these cells also form tumors *in vivo* (Björklund, personal communication), and this issue needs to be worked out before clinical applications.

## III. CELL PROTECTION AND REGENERATION IMPLANTS

The use of primary tissues or cell lines in cell replacement approaches is aimed at making transplantable mimics of the cells lost in the disease process. In PD, the use of chromaffin cells, dissected developing ventral mesencephalon, RPE cells, or stem cells has been aimed at replacing or augmenting the dopaminergic function. However, primary tissues, cell lines, and genetically modified cells also produce secreted factors that can influence the nearby host or transplanted cells in potentially beneficial ways. Several growth factors are endogenously made by cells, including fibroblast growth factors, transforming growth factors, and interleukins, that can have neuroprotective and regenerative effects on nearby nerve cells. Custom therapeutic cell lines can also be made by genetic engineering to secrete specific growth factors such as nerve growth factor (NGF), glial cell line–derived neurotrophic factor (GDNF), and ciliary neurotrophic factor (CNTF) that when implanted in relevant anatomical areas can affect specific neuronal populations in neuroprotective and regenerative ways (*ex vivo* gene therapy).

## Cell Implants Secreting Endogenous Factors

A carcinoma cell line derived from a human testicular teratocarcinoma isolated from a metastasis in a patient has been used in clinical applications to treat stroke (Nelson *et al.*, 2002). This immortal cell line is pluripotent and can be induced to stop dividing and to differentiate into a neuronal phenotype using retinoic acid. A preparation of this cell line was investigated in the clinic for the treatment of ischemic stroke based on animal data suggesting that the postinjury transplantation of this cell line into an infarcted area can improve recovery. The mechanisms surrounding this effect are unclear but are likely related to beneficial factors released from the cells. In a study in patients with lacunar stroke in randomized controlled Phase II trials at the University of Pittsburgh, the therapy with this cell line failed to meet the efficacy endpoints (Kondziolka *et al.*, 2005). The transplantation of a cell line derived from a human cancer has obvious risks associated with it. Importantly, the approval of this trial demonstrates that cell transplantation for severe neurological disorders is seen as a reasonable strategy by the regulatory agency, as long as safety and some efficacy can be demonstrated in animal models.

Other groups are investigating the transplantation of immortal cell lines but are using genetic engineering to immortalize cells. Advances in genetic engineering have made it possible to extend the number of doublings a primary cell line can go through by inserting various oncogenes and cell-cycle regulators. This allows for the selection, clonal expansion, and banking of a large number of cells. Besides the genetic modification, these cells retain otherwise-normal genotypic characteristics. ReNeuron, a British biotechnology company, has made immortal human neural stem cells that have shown regenerative effects in stroke models (K. Pollock *et al.*, 2006). Preparations for clinical trials are under way, and it is expected that clinical trials in stroke patients will be initiated in the near future.

Similarly, StemCells, Inc., is using growth factor–expanded human neurosphere cell lines in a strategy to treat Batten's disease, a rare neurometabolic disorder, with the

idea that the endogenous enzymes and factors made by the stem cells will have a therapeutic effect (StemCells, Inc., press release, 2006).

Last, implanted autologous mesenchymal stem cells are being studied in clinical trials for stroke (Bang *et al.*, 2005) and potentially in neurodegenerative disorders (Blondheim *et al.*, 2006).

The use of nonspecific neuroprotective and regenerative strategies based on the implantation of cells with unclear mechanisms may pose regulatory problems because the risk–benefit analyses become difficult to make. For example, even though positive results were inferred from the published trial with the human teratocarcinoma–derived cell line in stroke, the clinical data were not convincing enough to continue clinical development. In this trial, no significant adverse events or tumors were reported. But if they had occurred, a major setback for tissue-engineered brain implants could have been the result. The risk–benefit analysis is often difficult, and the predictive value and scale-up issues from the use of animal models are not straightforward. The regulatory agencies have therefore a real dilemma, and, similar to the setbacks experienced in the field of gene therapy, a push to do clinical trials with poorly characterized cell preparations and mechanisms in patients desperate for a treatment may cause significant adverse events that can create setbacks for the whole field of tissue engineering. On the other hand, a too-restrictive regulatory body may make the hurdle of bringing potentially beneficial but complex tissue-engineered products into the clinic too costly and difficult.

These regulatory issues are hard to resolve. But as experience with cell-containing implants build, it is likely that the decision making and risk–benefit analyses will improve. Some of the clinical trials using primary autologous cells, such as hematopoietic or mesenchymal stem cells derived from the bone marrow, can also bypass regulatory scrutiny and only need approval by a local ethics committee. Unfortunately, this has led to the initiation of clinical trials based on very little evidence of preclinical beneficial effects, causing potentially false hopes, personal expenses, and potentially harmful side effects to patients desperate for therapy.

## Cell Implants Secreting Engineered Factors (*Ex Vivo* Gene Therapy)

Because PD involves a slow and progressive degeneration of dopaminergic neurons, a protective and/or regenerative strategy could be applied. Many protein factors have been shown to protect fetal dopaminergic neurons both *in vitro* and *in vivo*, and one of the most powerful factors is glial cell line–derived neurotrophic factor (GDNF) (Lin *et al.*, 1993). This factor promotes the survival (neuroprotective effect) and neurite extension (regenerative effect) of dopaminergic neurons both *in vitro* and *in vivo*. Based on positive animal data, GDNF has been tried in humans in two separate randomized clinical trials (Lang *et al.*, 2006; Nutt

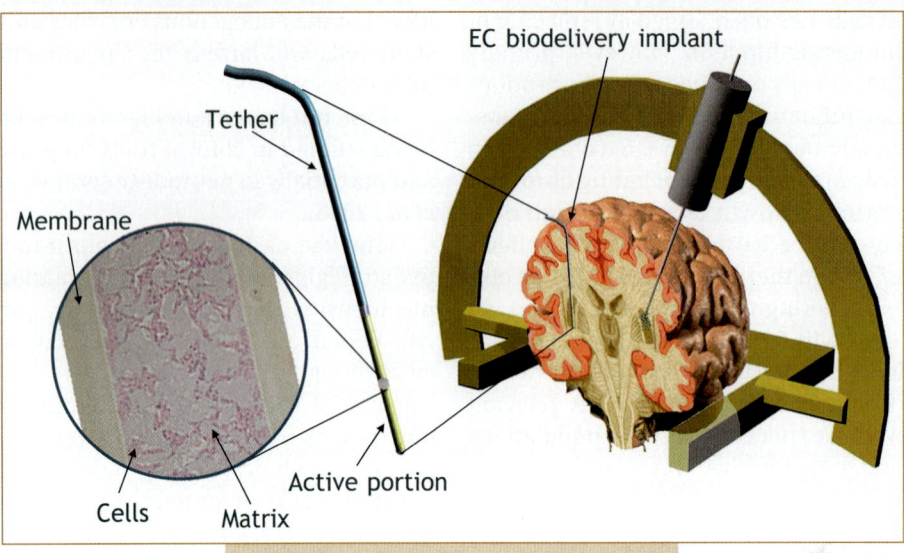

**FIG. 64.2.** Stereotactic surgery for cell replacement and EC biodelivery implants. On the right, an oblique frontal view of a cut section of the human brain shows a stereotactic injection of cells into the putamen, which is the standard method used for cell transplantation in PD. On the opposite side, the placement of a regenerative and/or protective EC biodelivery device is shown. On the left, an EC biodelivery device is shown illustrating the genetically modified human cell line attached to the polymeric matrix inside of a semipermeable hollow fiber. The cell-containing active tip is held in place by an inert tether, which is secured at the level of the skull under the skin for easy access and retrieval. The genetically modified cells housed in the tip receive nutrients from the surrounding brain and remain viable for more than a year *in vivo*. The membrane is also immunoprotective and allows for the therapeutic transgene product to be secreted into the brain interstitium.

*et al.*, 2003). The first trial used monthly intracerebroventricular bolus injections of GDNF and failed to meet both safety and efficacy endpoints. The second trial employed intraputaminal infusion of GDNF and, though safe, did not meet the efficacy endpoints. This is an unfortunate outcome for patients suffering from PD, and critics of the trial have indicated that inconsistencies in catheter design and other parameters may explain the poor results (Salvatore *et al.*, 2006).

From both the clinical and animal data, it appears necessary to deliver the GDNF in low but chronic doses to the dopaminergic nerve endings within the striatum in order to have a regenerative effect. It is currently the view of many that implantation of genetically modified cells or encapsulated cells or the use of viral vectors may be better at delivering GDNF than utilizing available pumps and catheters. Safety concerns with unregulated and unstoppable gene therapeutic approaches may favor the use of a tissue-engineered product based on an implantable and retrievable encapsulated cell biodelivery of GDNF.

It is beyond the scope of this chapter to expand on the topic of *ex vivo* and *in vivo* gene therapy, but it is important to mention that the delivery of growth factors to the brain via implants is an important cornerstone of tissue-engineering strategies for brain repair.

## Encapsulated Cell Biodelivery Implants

The implantation of naked cells has the advantage of allowing for migration, integration, and the formation of neurites and synapses in replacement strategies. The migration and honing mechanisms that neural stem cells display in models of stroke and glioma may also be utilized to deliver regenerative or tumoricidal agents in genetically modified cells. However, naked cells cannot readily be removed; and if a potent protein factor is being delivered to the brain, the inability to stop the treatment may pose a problem if untoward effects are noted or if the regenerative treatment is needed for only a limited amount of time. A device containing encapsulated cells that secrete the factor combines the advantages of cell and gene therapy with that of the safety of a retrievable device. One of the types of encapsulated cell (EC) biodelivery devices is depicted in Fig. 64.2 and consists of a hollow-fiber membrane surrounding a core of genetically modified cells seeded on a polymer matrix that in turn is attached to a tether. These encapsulated cell implants are true tissue-engineered devices that combine genetically modified cell lines with artificial scaffolding enclosed behind an immunoprotective membrane. The polymeric membrane excludes larger molecules and cells but allows for the bidirectional passage of nutrients and transgene products. The encapsulated cells can thus be protected from immune rejection, making allogeneic or even xenogeneic transplantation possible and immunosuppressive therapy unnecessary. The host is also protected from the implanted, genetically modified cells, and the risks of gene transfer or tumor formation are greatly diminished. The tether allows for handling, implantation, and removal. These devices can be implanted intraparenchymally, intracerebroventricularly, or intrathecally, depending on the application. Cellular survival and continuous production of factors have been demonstrated for at least 12 months in the brain, allowing for long-term delivery of therapeutic factors (Wahlberg, unpublished data). Encapsulated devices secreting GDNF have been studied in rodent models for PD and have shown both neuroprotective and neuroregenerative effects on dopaminergic cells (Sajadi *et al.*, 2006). Plans for clinical development with these devices in patients with PD have been announced (Michael J. Fox Foundation press release, 2004).

Recently, other investigators have published data on using encapsulated choroid plexus cells in the treatment of HD, an inherited neurodegenerative disorder (Emerich *et al.*, 2006). These cells are reportedly therapeutic because they secrete various endogenous factors that have neuroprotective and regenerative effects. The application of the choroid plexus cells in this study was made with an injectable microencapsulated-cell configuration. In this setting, the encapsulation provided immunoprotection for the porcine-derived primary choroid plexus cells. Unlike the macroencapsulation described above, these devices would not be retrievable and may be less suited for applications in which the treatment may need to be stopped.

## Controlled-Release Implants

Acellular synthetic polymeric brain implants that are able to deliver protein factors or other drugs to the CNS have also been developed (Fournier *et al.*, 2003). These systems normally release drugs by degradation- or diffusion-based mechanisms over an extended time (weeks) but cannot sustain release over a long time (months), which is possible with cellular-based systems.

Appropriately designed, polymeric controlled-release devices have several possible applications and could, for example, support the survival and integration of transplanted cells. Furthermore, a polymeric system can support the sequential release of growth factors that may be necessary to fully support the stepwise differentiation of immature cells. This concept could become applicable to neural stem cells, which may lack important embryonic developmental signals in the adult brain (Wahlberg, 1997).

## IV. COMBINED REPLACEMENT AND REGENERATION IMPLANTS

From a tissue-engineering point of view, the future goal is to make replacement organs for the body that can take over the function of a failed organ or structure in an anatomically and physiologically correct manner. And even though it would be difficult if not impossible to make entirely new brains, it should become possible not only to replace cells but also to make new axonal pathways and restore the correct connections.

The transplantation of fetal dopaminergic cells to the striatum is called *heterotopic transplantation*. This means that the dopaminergic cells are transplanted into an anatomical region different from their normal location, which is the substantia nigra (SN). The heterotopic implantation of dopaminergic neurons may result in the loss of important normal innervation and feedback loops. Many transplants for PD may thus work only as simple cellular pumps that increase the striatal dopamine levels. Although simplicity is desired, an ultimate strategy to treat PD could be to transplant the dopaminergic neurons to their anatomically correct position (homotopic), regenerate the nigrostriatal axonal pathway, and induce terminal sprouting and innervation of the striatal target neurons. This would regenerate the appropriate connections and represent a more physiologic strategy.

An initial approach may be to provide survival factors to the implanted cells. Even in the most optimal VM grafts applied heterotopically in PD, the total fraction of surviving dopaminergic neurons was only about 10–20% (Brundin and Björklund, 1987). This required a large number of donors (four to eight) to ensure enough surviving cells for a clinical effect. The combination of VM grafts with the neuroprotective and regenerative effect of GDNF is therefore a logical idea. Experimental data indeed show that the application of GDNF delivered by encapsulated cells in combination with either rat or human VM grafts increase the survival, neurite extension, and innervation of the striatum in a rat model of PD (Ahn *et al.*, 2005; Sautter *et al.*, 1998). Similarly, it would be expected that GDNF would have survival and regenerative effects on dopaminergic cells derived from ES or other stem cells when placed *in vivo*. A combined approach, with dopaminergic grafts and encapsulated cell implants secreting GDNF, may therefore be contemplated in future transplantation studies in PD.

A large challenge for tissue-engineering approaches in the treatment of neurological disorders is the regeneration of axonal pathways. Axons between the cell bodies and their targets often extend for several centimeters in the brain and close to one meter between the brain and the lumbar spinal cord neurons in an adult. Compared to the relatively short distances that the axons had to grow during development, the regeneration in the adult may pose a particular challenge. Fortunately, science has made progress in this area, and what was thought impossible only a few years ago now seems more feasible. Several molecules are now known to promote and to guide axonal outgrowth. As mentioned, GDNF is a strong promoter of axonal outgrowth of dopaminergic neurons. In addition, certain extracellular matrix (ECM) proteins, such as laminin, can guide axonal outgrowth (Zhou and Azmitia, 1988), and extensive nigrostriatal reconstruction has been accomplished using bridges of striatal tissues in combination with fetal mesencephalic grafts placed in the substantia nigra (Dunnett *et al.*, 1989). The finding that the central myelin and glial scars are inhibitory to axonal outgrowth has led to the identification of various inhibitory molecules that can be manipulated in various ways with inhibitors and enzymes (Liu *et al.*, 2006). From a tissue-engineering point of view, the combination of replacement cells with regeneration channels or scaffoldings capable of releasing survival- and neurite-promoting factors and coated with molecules that facilitate axonal outgrowth may thus become a future reality. Using nanotechnology, synthetic bridges can be made that promote extensive fiber regeneration and functional restoration (Ellis-Behnke *et al.*, 2006).

Even though mounting data show that axonal bridges can improve axonal growth in animal models, these methods have not been applied to humans. As the cellular building blocks become better refined, it it likely that more "true" tissue-engineered brain implants will enter the clinic. These types of implants could have great potential use for regeneration in many areas of the CNS, particularly the spinal cord.

## V. DISEASE TARGETS FOR BRAIN IMPLANTS

As mentioned, PD has been a major target for cellular brain implants. However, many other neurological disorders should become amenable to tissue-engineered implants.

In Huntington's disease (HD), several neuronal populations slowly degenerate and cause the clinical signs of choreiform movements and progressive dementia. HD is inherited as an autosomal dominant disease, and the responsible mutation has been located on chromosome 4. Carriers of the disease can therefore be screened for and identified before the onset of symptoms. This makes a neuroprotective strategy for HD an attractive possibility, where the delivery of neurotrophic factors could prolong the symptom-free interval. One such factor is ciliary neurotrophic factor (CNTF), which protects striatal neurons in both rodent and nonhuman primate models of HD (Mittoux *et al.*, 2000). Considering these results, a small clinical proof-of-concept trial using an intracerebroventricularly placed encapsulated-cell device secreting CNTF was completed (Bloch *et al.*, 2004). The study indicated that the placement was safe and would warrant additional trials. However, data also suggested that improved devices and an intrastriatal implantation approach may improve the efficacy. This approach may be taken in future trials.

Cell replacement strategies have also been tried in HD (Dunnett and Rosser, 2004). Primary fetal striatal tissue transplantation for HD has been performed at a handful of centers in the world. Long-term follow-up has described mild improvements in some of the implanted patients (Bachoud-Levi *et al.*, 2006). Because the number of implanted patients is few, it is currently difficult to draw any major conclusions regarding the clinical efficacy of transplantation for HD. One possible advantage over PD is that the transplantation for HD involves homotopic implantation, which, at least theoretically, should allow for the differentiation of the fetal tissue using normal environmental cues. However, in HD, multiple sets of neuronal populations degenerate, including both cortical and striatal neurons. The homotopic transplantation for HD may thus require more extensive regeneration of axonal pathways than in PD.

Other diseases that could be amenable to the implantation of cells within the brain are the myelin disorders. Animal experiments have shown the ability of various neural tissues, precursors, and stem cells to remyelinate areas of demyelination (Rice *et al.*, 2004).

One substitute for oligodendrocytes (the myelinating cells of the CNS) may be to transplant growth factor–expanded Schwann cells [the myelinating cell of the peripheral nervous system (PNS)]. This is based on the fact that patients with CNS myelin disorders do not display demyelination of the PNS. In fact, remyelination of central axons may occur spontaneously by ingrowth of Schwann cells from the periphery. An intriguing therapeutic possibility therefore is to grow Schwann cells from a nerve biopsy and to expand these cells in culture. In turn, these cells could be transplanted into demyelinated areas in the same patient. Physiologically, this may not be the best strategy, for Schwann cells myelinate only single axons, whereas an oligodendrocyte myelinates multiple axons. However, animal data indicate that, for a limited volume, Schwann cells can remyelinate and restore function to central demyelinated areas. Based on animal data, a small clinical study was performed. Reportedly, the Schwann cell transplants did not survive, but the procedure appears to be safe (Stangel and Hartung, 2002). It is currently unclear if new trials with Schwann cells are to be expected.

One of the most common neurological disorders is epilepsy, which affects about 1–2% of the population. Epilepsy is characterized by recurrent abnormal electrical discharges in the brain, affecting subparts of the brain or generalizing to deeper parts in the brain and resulting in unconsciousness. A subgroup of these patients has temporal lobe epilepsy, which is generated by a loss of neurons and an imbalance of inhibitory and excitatory neurotransmitters in the hippocampal formation. In medically intractable cases, this disease can sometimes be treated surgically with the removal of the medial hippocampus and the abnormal area. This procedure eliminates or reduces the frequency of seizures in selected patients but involves a major surgical procedure and the ablation of normal tissue. A less invasive procedure may be to implant inhibitory cells in the seizure focus, which would raise the seizure threshold (Björklund and Lindvall, 2000). This idea is supported by animal experimentation data that indicate that locus coeruleus grafts and the local delivery of inhibitory substances such as GABA and Galanin can increase the seizure threshold and treat established seizures.

Other disease indications that may benefit from brain implant strategies include stroke, brain injury from trauma, AD, and rare disorders such as cerebellar degeneration and inherited metabolic disorders. Besides the brain, the spinal cord and retina are potential targets for similar approaches.

## VI. SURGICAL CONSIDERATIONS

The surgical implantation of most brain implants involves the use of stereotactic techniques. The stereotactic method (stereotaxis) in brain surgery was established in the beginning of the 21st century and is now well established in neurosurgical practice (Speelman and Bosch, 1998). It involves attaching a rigid frame (stereotactic frame) to the skull, followed by imaging, such as MRI. Attached markers (fiducials) create a three-dimensional coordinate system in which any point in the brain can be defined and related to the frame with high precision. In the operating room, the markers used during imaging are replaced with holders that guide the instruments. It is a relatively simple and low-risk neurosurgical procedure often done under local anesthesia and mild sedation. The procedure is therefore safe and relatively painless. The patients are usually discharged from the hospital after an overnight observation.

## VII. CONCLUSIONS

In this chapter, various brain implants have been described that may have the potential to treat PD and other neurological disorders using tissue-engineering strategies. Most of the clinical applications to date involve the transplantation of primary cell suspensions, such as those derived from fetal tissue. The use of manipulated adult and embryonic stem cells have shown great promise in animal models, but embryonic stem cell–derived replacement cells carry risks of tumor formation. Adult neural stem cells do not tend to form tumors *in vivo*. However, these cell lines have been lacking the necessary phenotypes for clear applications in indications such as PD. Applications using growth factor support, genetic engineering, scaffolds, extracellular matrices, and encapsulation have all been able to improve survival and function of brain implants in experimental settings. Rapid development toward clinical applications is ongoing, and it is expected that cell replacement and restorative treatments using tissue-engineered brain implants will have a great impact on neurological disorders in the relatively near future. The ultimate implants are yet to be developed and may combine stem cells, genetically modified cells, controlled-delivery devices, axonal bridges, scaffolds, and encapsulated cells.

## VIII. REFERENCES

Aebischer, P., *et al.* (1996). Intrathecal delivery of CNTF using encapsulated genetically modified xenogeneic cells in amyotrophic lateral sclerosis patients. *Nat. Med.* **2**, 696–699.

Ahn, Y. H., *et al.* (2005). Increased fiber outgrowth from xeno-transplanted human embryonic dopaminergic neurons with co-implants of polymer-encapsulated genetically modified cells releasing glial cell line–derived neurotrophic factor. *Brain Res. Bull.* **66**, 135–142.

Andersson, E., *et al.* (2006). Identification of intrinsic determinants of midbrain dopamine neurons. *Cell* **124**, 393–405.

Bachoud-Levi, A. C., *et al.* (2006). Effect of fetal neural transplants in patients with Huntington's disease six years after surgery: a long-term follow-up study. *Lancet Neurol.* **5**, 303–309.

Backlund, E. O., *et al.* (1985). Transplantation of adrenal medullary tissue to striatum in parkinsonism. First clinical trials. *J. Neurosurg.* **62**, 169–173.

Bang, O. Y., *et al.* (2005). Autologous mesenchymal stem cell transplantation in stroke patients. *Ann. Neurol.* **57**, 874–882.

Björklund, A., and Lindvall, O. (2000). Cell replacement therapies for central nervous system disorders. *Nat. Neurosci.* **3**, 537–544.

Blesch, A., and Tuszynski, M. H. (2004). Gene therapy and cell transplantation for Alzheimer's disease and spinal cord injury. *Yonsei Med. J.* **45** (Suppl.), 28–31.

Bloch, J., *et al.* (2004). Neuroprotective gene therapy for Huntington's disease, using polymer-encapsulated cells engineered to secrete human ciliary neurotrophic factor: results of a phase I study. *Hum. Gene Ther.* **15**, 968–975.

Blondheim, N. R., *et al.* (2006). Human mesenchymal stem cells express neural genes, suggesting a neural predisposition. *Stem Cells Dev.* **15**, 141–164.

Brundin, P., and Björklund, A. (1987). Survival, growth and function of dopaminergic neurons grafted to the brain. *Prog. Brain Res.* **71**, 293–308.

Carpenter, M. K., *et al.* (1999). *In vitro* expansion of a multipotent population of human neural progenitor cells. *Exp. Neurol.* **158**, 265–278.

Christophersen, N. S., *et al.* (2006). Induction of dopaminergic neurons from growth factor–expanded neural stem/progenitor cell cultures derived from human first trimester forebrain. *Brain Res. Bull.* **70**, 457–466.

Deuschl, G., *et al.* (2006). A randomized trial of deep-brain stimulation for Parkinson's disease. *N. Engl. J. Med.* **355**, 896–908.

Dunnett, S. B., and Rosser, A. E. (2004). Cell therapy in Huntington's disease. *NeuroRx* **1**, 394–405.

Dunnett, S. B., *et al.* (1989). Nigrostriatal reconstruction after 6-OHDA lesions in rats: combination of dopamine-rich nigral grafts and nigrostriatal "bridge" grafts. *Exp. Brain Res.* **75**, 523–535.

Ellis-Behnke, R. G., *et al.* (2006). Nano neuro knitting: peptide nanofiber scaffold for brain repair and axon regeneration with functional return of vision. *Proc. Natl. Acad. Sci. U.S.A.* **103**, 5054–5059.

Emerich, D. F., *et al.* (2006). Extensive neuroprotection by choroid plexus transplants in excitotoxin lesioned monkeys. *Neurobiol. Dis.* **23**, 471–480.

Fournier, E., *et al.* (2003). Biocompatibility of implantable synthetic polymeric drug carriers: focus on brain biocompatibility. *Biomaterials* **24**, 3311–3331.

Freed, C. R., *et al.* (2001). Transplantation of embryonic dopamine neurons for severe Parkinson's disease. *N. Engl. J. Med.* **344**, 710–719.

Fricker, R. A., *et al.* (1999). Site-specific migration and neuronal differentiation of human neural progenitor cells after transplantation in the adult rat brain. *J. Neurosci.* **19**, 5990–6005.

Kim, J. H., *et al.* (2002). Dopamine neurons derived from embryonic stem cells function in an animal model of Parkinson's disease. *Nature* **418**, 50–56.

Kondziolka, D., *et al.* (2005). Neurotransplantation for patients with subcortical motor stroke: a phase 2 randomized trial. *J. Neurosurg.* **103**, 38–45.

Lang, A. E., *et al.* (2006). Randomized controlled trial of intraputamenal glial cell line–derived neurotrophic factor infusion in Parkinson disease. *Ann. Neurol.* **59**, 459–466.

Lin, L. F., *et al.* (1993). GDNF: a glial cell line–derived neurotrophic factor for midbrain dopaminergic neurons. *Science* **260**, 1130–1132.

Lindvall, O., and Bjorklund, A. (2004). Cell therapy in Parkinson's disease. *NeuroRx* **1**, 382–393.

Lindvall, O., *et al.* (1988). Fetal dopamine-rich mesencephalic grafts in Parkinson's disease. *Lancet* **2**, 1483–1484.

Lindvall, O., *et al.* (1994). Evidence for long-term survival and function of dopaminergic grafts in progressive Parkinson's disease. *Ann. Neurol.* **35**, 172–180.

Liu, B. P., *et al.* (2006). Extracellular regulators of axonal growth in the adult central nervous system. *Philos. Trans. R. Soc. Lond. B. Biol. Sci.* **361**, 1593–1610.

Mittoux, V., *et al.* (2000). Restoration of cognitive and motor functions by ciliary neurotrophic factor in a primate model of Huntington's disease. *Hum. Gene Ther.* **11**, 1177–1187.

Miura, T., *et al.* (2004). Cellular life span and senescence signaling in embryonic stem cells. *Aging Cell* **3**, 333–343.

Nelson, P. T., *et al.* (2002). Clonal human (hNT) neuron grafts for stroke therapy: neuropathology in a patient 27 months after implantation. *Am. J. Pathol.* **160**, 1201–1206.

Nutt, J. G., *et al.* (2003). Randomized, double-blind trial of glial cell line–derived neurotrophic factor (GDNF) in PD. *Neurology* **60**, 69–73.

Olanow, C. W., *et al.* (2003). A double-blind controlled trial of bilateral fetal nigral transplantation in Parkinson's disease. *Ann. Neurol.* **54**, 403–414.

Ostenfeld, T., *et al.* (2000). Human neural precursor cells express low levels of telomerase *in vitro* and show diminishing cell proliferation with extensive axonal outgrowth following transplantation. *Exp. Neurol.* **164**, 215–226.

Piao, J. H., *et al.* (2006). Cellular composition of long-term human spinal cord- and forebrain-derived neurosphere cultures. *J. Neurosci Res.* **84**, 471–482.

Piccini, P., *et al.* (1999). Dopamine release from nigral transplants visualized *in vivo* in a Parkinson's patient. *Nat. Neurosci.* **2**, 1137–1140.

Pollack, A. (2001). Companies announce setback in treatment for Parkinson's. *N.Y. Times.*

Pollard, S. M., *et al.* (2006). Adherent neural stem (NS) cells from fetal and adult forebrain. *Cereb. Cortex* **16** (Suppl. 1), i112–i120.

Pollock, K., *et al.* (2006). A conditionally immortal clonal stem cell line from human cortical neuroepithelium for the treatment of ischemic stroke. *Exp. Neurol.* **199**, 143–155.

Rice, C., *et al.* (2004). Cell therapy in demyelinating diseases. *NeuroRx* **1**, 415–423.

Roy, N. S., *et al.* (2006). Functional engraftment of human ES cell–derived dopaminergic neurons enriched by coculture with telomerase-immortalized midbrain astrocytes. *Nat. Med.* **12**, 1259–1268.

Sajadi, A., *et al.* (2006). Transient striatal delivery of GDNF via encapsulated cells leads to sustained behavioral improvement in a bilateral model of Parkinson disease. *Neurobiol. Dis.* **22**, 119–129.

Salvatore, M. F., *et al.* (2006). Point source concentration of GDNF may explain failure of phase II clinical trial. *Exp. Neurol.* **202**, 497–505.

Sautter, J., *et al.* (1998). Implants of polymer-encapsulated genetically modified cells releasing glial cell line–derived neurotrophic factor improve survival, growth, and function of fetal dopaminergic grafts. *Exp. Neurol.* **149**, 230–236.

Savitz, S. I., *et al.* (2004). Cell therapy for stroke. *NeuroRx* **1**, 406–414.

Skogh, C., *et al.* (2001). Generation of regionally specified neurons in expanded glial cultures derived from the mouse and human lateral ganglionic eminence. *Mol. Cell. Neurosci.* **17**, 811–820.

Speelman, J. D., and Bosch, D. A. (1998). Resurgence of functional neurosurgery for Parkinson's disease: a historical perspective. *Mov. Disord.* **13**, 582–588.

Stangel, M., and Hartung, H. P. (2002). Remyelinating strategies for the treatment of multiple sclerosis. *Prog. Neurobiol.* **68**, 361–376.

Stover, N. P., *et al.* (2005). Intrastriatal implantation of human retinal pigment epithelial cells attached to microcarriers in advanced Parkinson disease. *Arch. Neurol.* **62**, 1833–1837.

Tuszynski, M. H., *et al.* (2005). A phase 1 clinical trial of nerve growth factor gene therapy for Alzheimer disease. *Nat. Med.* **11**, 551–555.

Villa, A., *et al.* (2004). Long-term molecular and cellular stability of human neural stem cell lines. *Exp. Cell. Res.* **294**, 559–570.

Wahlberg, L. U. (1997). Implantable bioartificial hybrids for targeted therapy in the central nervous system. Department of Clinical Neurosciences, Ph.D. Thesis, Karolinska Institute, Stockholm.

Zhou, F. C., and Azmitia, E. C. (1988). Laminin facilitates and guides fiber growth of transplanted neurons in adult brain. *J. Chem. Neuroanat.* **1**, 133–146.

# Chapter Sixty-Five

# Spinal Cord

## John W. McDonald and Daniel Becker

## I. INTRODUCTION

Spinal cord injury is a major medical problem because there currently is no way to repair the central nervous system and restore function. In this chapter, we focus on embryonic stem cells as an important research tool and potential therapy. We quickly review the epidemiology, functional anatomy, and pathophysiology of spinal cord injury and then discuss spontaneous regeneration and current limitations on repair. We also summarize features of spinal cord development that might guide restoration strategies. We then review studies that have utilized embryonic stem cells in spinal cord repair. We conclude that progress has been good, that knowledge is still too limited, and that harnessing the potential of embryonic stem cells will be important for solving the problem of spinal cord injury.

## II. THE PROBLEM

Nearly 12,000 people in the United States suffer a traumatic spinal cord injury (SCI) every year, and about a quarter of a million Americans are living with this devastating condition (National Spinal Cord Injury Database, 2001). There are also four to five times as many spinal cord injuries caused by medical conditions, such as multiple sclerosis, a common disorder that destroys the myelin insulation on nerves in the cervical spinal cord. Other medical causes include amyotrophic lateral sclerosis (ALS; Lou Gehrig's disease), polio and postpolio syndrome, human T-lymphotropic virus type 1 (HTLV-1), human immunodeficiency virus type 1 (HIV-1), metabolic deficits such as of vitamin B-12, as well as causes of myelopathy, such as stenosis and disc herniation (McDonald and Sadowsky, 2002). The worldwide incidence of traumatic and nontraumatic SCI is greater per capita than in the United States.

The consequences of SCI depend on the level at which the cord is damaged. Generally, injuries in the neck produce tetraplegia, with loss of bowel and bladder function, whereas lesions in the thoracic or lumbar area may cause paraplegia, also with bowel and bladder dysfunction. In its severest form, SCI causes complete paralysis and loss of sensation throughout the body, inability to control bowel and bladder function, trouble controlling autonomic functions such as blood pressure, and inability to breathe or cough. The long-term disability from SCI results not only from the initial loss of function but also from the complications that accumulate. Up to 30% of individuals with SCI are hospitalized every year for complications (McKinley et al., 1999), such as severe spasticity, infections (lung, skin, bowel, bone, urinary tract), osteoporosis and pathologic bone fractures, autonomic dysreflexia, and heterotropic ossification (McDonald and Sadowsky, 2002; Sjolund, 2002). Another major long-term

*Principles of Tissue Engineering, 3rd Edition*
*ed. by Lanza, Langer, and Vacanti*

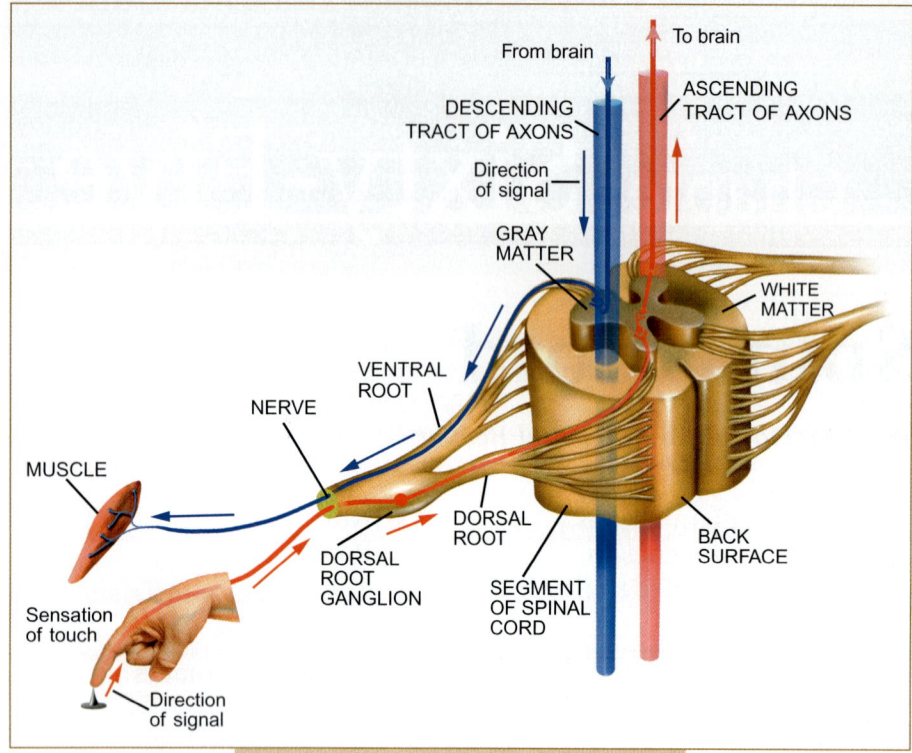

From brain

To brain

DESCENDING
TRACT OF AXONS

ASCENDING
TRACT OF AXONS

Direction
of signal

GRAY
MATTER

WHITE
MATTER

VENTRAL
ROOT

NERVE

MUSCLE

DORSAL
ROOT

DORSAL
ROOT
GANGLION

BACK
SURFACE

SEGMENT
OF SPINAL
CORD

Sensation
of touch

Direction
of signal

**FIG. 65.1.** Organization of the spinal cord. A segment of cord reveals the butterfly-shaped gray matter at the core and an outer ring of white matter. The main components of the gray matter are neuronal cell bodies and glial cells and blood vessels. The white matter also contains astrocytes and blood vessels, but it consists mostly of axons and oligodendrocytes (glial cells that wrap axons in white, insulated myelin). Axonal tracts that ascend in the cord, such as the one shown in red, convey sensory messages received from the body; the descending tracts (blue) carry motor commands to muscles. Reproduced with permission of Alexander and Turner Studio, FL. © 2002 Edmond Alexander.

complication is muscle wasting. This results from disuse and the absence of nerve impulses, which are critical for maintaining junctions between nerves and muscles (Auld and Robitaille, 2003; Rossi *et al.*, 2003). Therefore, patients who maintain their body in the best condition for nervous system repair will benefit most from future therapeutic strategies.

## III. SPINAL CORD ORGANIZATION

Unlike the brain, the spinal cord has its white matter (nerve axons) on the outside and gray matter (cell bodies) on the inside (Fig. 65.1). The gray matter houses neurons that project to their level of the periphery to control movement and receive sensory signals. The white matter carries axonal connections to and from the brain, and half of all those axons are myelinated. In general, the spinal cord has much less potential for regeneration than the peripheral nervous system. Moreover, most traumatic SCIs also injure the incoming and outgoing (*afferent* and *efferent*) peripheral nerves at the injury level. In most cases, however, the caudal cord remains intact beginning several segments below the injury level. Circuitry in those segments can produce reflexes by activating a ventral motor neuron to produce muscle contraction when it receives a sensory stimulus from the periphery of the body (Fig. 65.1).

The distal spinal cord also contains groups of nerve cells that generate the patterns of activity needed for walking and running. Finally, there are central pattern generators, which

govern particular motor functions in the periphery that are quite complex in nature and well controlled. Interested readers are referred to recent monographs on this subject (Burke, 2001; Dietz, 1995; Barbeau *et al.*, 1999). Because the upper spinal cord plays only a limited role in controlling these pattern generators, people with injury to the cervical spine can walk or ride a stationary bicycle if their muscles are stimulated appropriately. The existence of additional pattern generators has also been predicted. For example, the phrenic nucleus, housed at C3 through C5 in the upper neck, appears to generate the pattern of movements needed for respiration. Unfortunately, this part of the spine is a common site of traumatic injury.

## IV. INJURY

Traumatic injury occurs when broken fragments of bone and ligament impinge on the soft spinal cord, which is no wider than the thumb. The cord responds by swelling until it encounters resistance from the bony canal of the spinal column. The swelling that compounds the initial injury reduces venous blood flow, causing a secondary venous infarct in the central part of the cord. This initiates a cascade of events that injures neighboring tissue. During this secondary phase, which occurs in the first 24 hours after the primary phase, cells die by excitotoxic necrosis (caused by release of harmful concentrations of chemicals from other cells) as well as by apoptosis (cell suicide), causing the initial injury site to enlarge rapidly into a hole in the middle

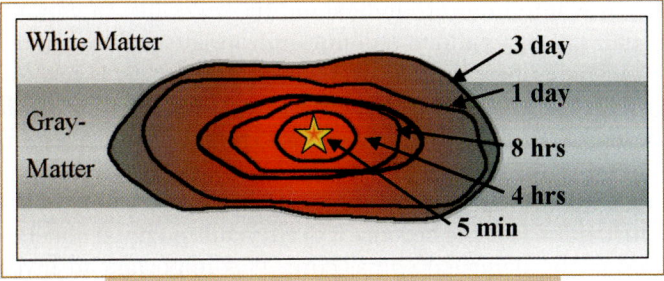

**FIG. 65.2.** Evolution of primary to secondary spinal cord injury. Following the initial trauma, the injury site enlarges rapidly as a consequence of secondary injury, particularly over the first 24 hours.

of the spinal cord (Fig. 65.2). Because a donutlike rim of viable tissue usually remains at the level of injury, SCI preferentially affects gray matter.

A second wave of very delayed cell death occurs during the weeks after an SCI (Crowe *et al.*, 1997). It removes mostly oligodendrocytes (the cells that wrap nerve axons in a fatty layer of myelin) from adjacent white matter tracts. Since each oligodendrocyte myelinates 10–40 different axons, loss of one oligodendrocyte leads to exponential loss of myelin. As a result, nerve axons cease to function because they cannot conduct electrical impulses when this insulating sheath is lost. This progressive phase of secondary injury is a good target for potential therapies, such as neuroprotective stem cell transplantation.

The problem does not stop with the secondary wave of cell death, however. An injured, underactive nervous system may be unable to adequately replace cells, particularly housekeeping cells called *glial cells*, which normally turn over. Therefore, individuals with SCI may experience a slow, progressive loss of neurological function over very long periods in addition to complications from their initial injury. It is important to consider this potential for further loss of function when designing therapeutic regimens.

## V. SPONTANEOUS REGENERATION

Not long ago, the adult nervous system was believed to have no capacity for repair or regeneration. Now a growing body of evidence indicates that the capacity for spontaneous regeneration may be much greater than previously perceived. Data obtained since the 1960s have demonstrated that new cells are added to the nervous system continually; these cells include neurons in at least two brain regions, the hippocampus and the olfactory bulb (Eriksson *et al.*, 1998; Winner *et al.*, 2002). Moreover, glial cells are frequently born and are capable of regeneration (Herndon *et al.*, 1977; Ludwin, 1984). Recent work has shown that severed or intact axons can sprout even long after injury (Weidner *et al.*, 2001).

Nonetheless, our ability to maximize spontaneous regeneration is currently limited. The recent detection of endogenous stem cells within the spinal cord raises hope

that such cells can be harnessed to repair the damaged spinal system. Although the birth of new neurons from these progenitors has never been demonstrated within the spinal cord (Horner *et al.*, 2000), new glial cells, including oligodendrocytes and astrocytes, are continually being added. In fact, injury stimulates cell birth. Our laboratory recently showed that nearly 2 million cells are born in the spinal cord each day, though most eventually die (Becker *et al.*, 2003). Collectively, these data suggest that cell turnover does occur in the nervous system, albeit much more slowly than in other organs of the body. Moreover, we recently showed that patterned neural activity, generated by repetitive tasks such as cycling, can stimulate cell birth, suggesting that behavioral modification could be important for maximizing cord regeneration and functional recovery (Becker *et al.*, 2003).

## VI. CURRENT LIMITATIONS AND APPROACHES TO REPAIR AND REDEFINING GOALS

Although some limited spontaneous regeneration is known to occur (Tuszynski *et al.*, 2002; Prewitt *et al.*, 1997), dramatic self-repair of the nervous system does not take place. A growing body of evidence suggests that factors in the nervous system actively inhibit regeneration. Such factors include inhibitory proteins in the cord, which guide regrowing connections, and scar tissue, which contains chondroitin sulfate and other proteoglycans (Morgenstern *et al.*, 2002). Reduced production of growth factors that stimulate regrowth also limits regeneration.

Figure 65.3 and 65.15 outline the general strategies for repairing the damaged spinal cord. Note that a complete cure of nervous system injury is not practical or required. Partial repair translates into proportionally greater recovery of function. For example, only about 10% of the functional connections are required to support locomotion in cats (Blight, 1983), and humans missing more than half the gray matter in the cervical spinal cord can still walk and run fairly normally.

One repair strategy is to transplant stem cells or other biomaterials to fill the cyst that forms at the eye of the injury. This cyst acts as a physical barrier to the growth of anatomically intact axons in the surrounding donut of white matter. The lost cells include segmental motor neurons, sensory neurons, oligodendrocytes, and astrocytes. Although it is possible for endogenous cells to give birth to new oligodendrocytes and astrocytes, this response is limited; moreover, it has not so far been possible to obtain new neurons in the spinal cord from endogenous cells (Horner *et al.*, 2000).

As well as filling in the cyst, it may be necessary to repair axons that no longer function properly because they have lost their myelin or have been inappropriately myelinated. Several approaches have been used to overcome this problem, which can manifest itself as a segmental

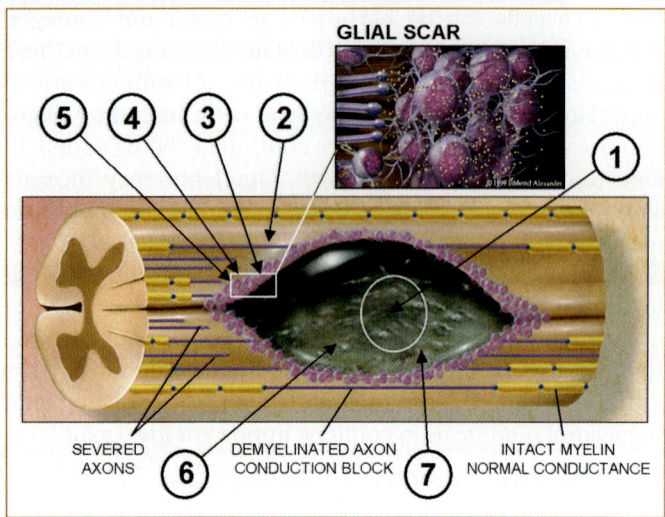

**GLIAL SCAR**

(5) (4) (3) (2)

(1)

SEVERED AXONS

(6) DEMYELINATED AXON CONDUCTION BLOCK

(7) INTACT MYELIN NORMAL CONDUCTANCE

**FIG. 65.3.** Common strategies toward regeneration of the damaged spinal cord. **(1)** *Prevention of progression of secondary injury:* Necrotic and apoptotic cell death would be prevented by antiexcitotoxic drugs and antiapoptotic treatments. **(2)** *Compensation for demyelination:* Chemicals that prevent conduction block in demyelinated areas and agents that encourage surviving oligodendrocytes to remyelinate axons would be provided. Lost oligodendrocytes would be replenished. **(3)** *Removal of inhibition:* Agents that block the actions of natural inhibitors of regeneration or drugs that down-regulate expression of inhibitory proteins would be provided. **(4)** *Promotion of axonal regeneration:* Growth factors that promote regeneration (sprouting) of new axons would be provided. **(5)** *Direction of axons to proper targets:* Guidance molecules would be provided or their expression would be increased in host cells. **(6)** *Creation of bridges:* Bridges would be implanted into the cyst to provide directional scaffolding that encouraged axonal growth. **(7)** *Replacement of lost cells:* Cells capable of generating all cell types (progenitor cells or embryonic stem cells) would be implanted. Substances that induce undifferentiated cells to replace dead cells would be provided. Also, cells that had been genetically engineered to deliver regenerative molecules would be transplanted. Reproduced with permission of Alexander and Turner Studio, FL. © 2002 Edmond Alexander.

conduction block. For example, potassium channels on dysfunctional axons can be blocked pharmacologically, and preexisting or transplanted oligodendrocytes and their progenitors can be encouraged to make new myelin (Nashmi and Fehlings, 2001; Novikova *et al.*, 2002).

Another strategy is to remove or mask the effects of proteins in the glial scar around the cyst that actively inhibit the regrowth of new connections (Tatagiba *et al.*, 1997). Antibodies that block the inhibitory effects of these proteins promote sprouting (Tatagiba *et al.*, 1997). Alternatively, cells might be delivered to the cyst to digest inhibitory proteins in the scar; or the initial expression of such proteins might be inhibited. Moreover, certain growth factors promote the self-repair of physically broken connections. These factors can be provided exogenously or by transplanting genetically altered cells that release them (Bregman *et al.*, 2002; Murray *et al.*, 2002; Tuszynski *et al.*, 2002).

It is important to understand the feasibility of various repair strategies and to redefine appropriate goals of repair. If one ranks the foregoing strategies in terms of likely success (McDonald and Sadowsky, 2002), it becomes clear that it will be very difficult to persuade axons to regrow across a lesion, extend all the way down the spinal cord, and make connections with the appropriate target cells. Remyelination seems much more feasible because it occurs continually at a steady rate in the damaged spinal cord. It is important to investigate all possible strategies, however, because some will materialize sooner and some later. Most importantly, we must understand that multiple strategies delivered over time will be more important than the elusive magic bullet.

When defining appropriate goals for therapy, improving patients' quality of life must surely rank first. To this end, strategies that limit complications are important. Moreover, individuals with SCI often value small gains in function more highly than larger gains, such as walking. Thus, the top goal for most patients is recovery of bowel, bladder, or sexual function. Distant seconds include recovery of respiratory function for those whose respiration is impaired and use of a single hand for those without hand function (McDonald and Sadowsky, 2002). Paradoxically, most animal models focus on recovering the ability to walk. This is both the least appropriate goal and the most difficult.

## VII. SPINAL CORD DEVELOPMENT

To understand spinal cord regeneration, it is necessary to understand spinal cord development. In humans, oligodendrocytes are found in cultures of fetal spinal cord at 7 and 12 weeks of gestation (Dickson *et al.*, 1985; Satoh and Kim, 1994). Myelination begins at 10–11 weeks of gestation (Gamble, 1969; Weidenheim *et al.*, 1992) and continues throughout the second year of life (Benes *et al.*, 1994; Puduslo and Jang, 1984). Signals derived from axons regulate the growth of progenitor cells and the survival of oligodendrocytes (Barres and Raff, 1996). For example, sonic hedgehog (SHH), a protein synthesized by notocord and floor plate cells, induces the differentiation of motor neurons in the ventral cord (Roelink *et al.*, 1995), and SHH signaling induces oligodendrocyte precursors (OP) to emerge in the embryonic spinal cord (Hajihosseini *et al.*, 1996). Platelet-derived growth factor (PDGF) is a potent regulator of OP migration and proliferation, while insulinlike growth factor (IGF-1) acts on both neurons and oligodendrocytes. Other locally synthesized growth factors appear to control the balance between OP proliferation and differentiation (Dubois-Dalcq and Murray, 2000).

Many different neuronal phenotypes arise from various progenitor pools during CNS development, but most of the pathways are poorly understood. One of the best described is the generation of spinal motor neurons (Jessell, 2000), which involves several steps. Through signaling pathways that enable cells to respond to external cues (pathways involving bone morphogenetic protein [BMP], fibroblast growth factor [FGF], and Wnt), ectodermal cells obtain a rostral neural character (Munoz-Sanjuan and Brivanlou, 2002). In response to caudalizing signals, such as retinoic acid (RA), these progenitors acquire a spinal positional identity (Durston *et al.*, 1998). Through the ventralizing action of SHH, spinal progenitor cells gain their motor neuron phenotype (Briscoe and Ericson, 2001).

These findings raise the question of whether embryonic stem (ES) cells can be directed along specific pathways to produce specific neural cell populations for CNS repair. In fact, very encouraging steps have already been taken. Early reports revealed the possibility of generating cells with motor neuron characteristics from ES cells (Renoncourt *et al.*, 1998). Later, Wichterle and colleagues (2002) demonstrated that signaling factors operating along the rostrocaudal and dorsoventral axes of the neural tube to specify motor neuron fate *in vivo* could be harnessed *in vitro* to direct the differentiation of mouse ES cells into functional motor neurons.

## VIII. EMBRYONIC STEM CELLS

ES cells have unique features that are important for spinal cord repair. They represent every cell type in the body and are the earliest stem cells capable of replicating indefinitely without aging, and their DNA can be modified easily, even in single cells. They also fulfill two criteria that are essential for nervous system repair through transplantation: (1) They normally belong in the spinal cord; and (2) they contain the recipient's own DNA, obviating the need for immunosuppression, which cannot be used in spinal cord patients because of the increased incidence of infections.

Several methods are available for obtaining ES cells. The most common is *in vitro* fertilization, which requires a fertilized egg and therefore produces cells that differ genetically from the host. A second strategy involves somatic nuclear transfer, which takes a nucleus from a somatic cell, such as a skin cell, and transfers it to a fertilized egg whose nucleus has been removed (Fig. 65.4A) (Sotomaru *et al.*, 1999). The subsequent ES cells therefore contain only the genetic material of the recipient. Another possibility for women is parthenogenesis (Kaufman *et al.*, 1983), which tricks an egg into thinking it is fertilized, allowing it to begin duplicating its DNA (Fig. 65.4B). Although this process is unable to create a viable embryo (because factors derived from sperm are necessary for preimplantation), it can produce normal cells.

Many studies of spinal cord repair have involved differentiated cells. The earliest studies used peripheral nerve grafts to demonstrate that the nervous system has the capacity to regenerate but that the environment in the CNS is not permissive (Aguayo *et al.*, 1981). Substantial progress has been made with transplantation of peripheral nervous system cells and non–nervous system cells, including genetically engineered fibroblasts, bone marrow cells, glial cells,

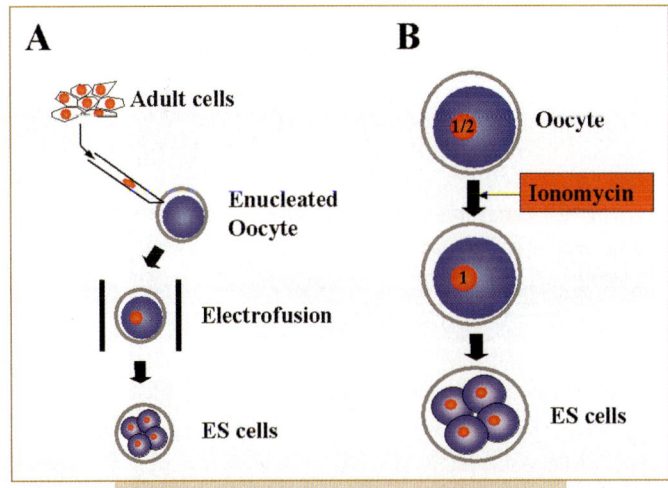

**FIG. 65.4.** ES cells can be made by *in vitro* fertilization and two additional methods: **(A)** *Somatic nuclear transfer:* Donor cells are placed under the zona pelucida into the perivitelline space of enucleated oocytes. The cell nucleus is introduced into the cytoplast by electrofusion, which also activates the oocyte. It can then be grown to the blastocyst stage to harvest the ES cells. **(B)** *Parthenogenic activation:* For activation, oocytes are briefly exposed to a calcium ionophore such as ionomycin. The resulting cell can mature for ES cell harvest.

neurons, and mixtures of glia and neurons (Bunge, 2000; Murray *et al.*, 2002; Hains *et al.*, 2003; Jiang *et al.*, 2003; Akiyama *et al.*, 2002). In general, fetal sources of cells have been the most successful because derivatives of postnatal and adult cells are less able to withstand neural transplantation. One recent exception has been neural stem cells derived from the adult CNS (Cao *et al.*, 2002a).

Another very active area of investigation is transdifferentiation (Kennea and Mehmet, 2002): isolating cells from organs other than the nervous system and transforming them into neural progenitor cells. Although initial progress suggested that this strategy might work with many types of tissue, the problem of cell-to-cell fusion may have dampened enthusiasm for this approach (Liu and Rao, 2003). Suffice it to say that transdifferentiation has not yet produced cells suitable for transplantation, although clearly the potential exists.

### Embryonic Stem Cells and the Neural Lineage

Several protocols are available for converting ES cells into neural lineage cells. However, protocols for mouse and human cells differ because the constraints of the human ES cell system have not yet been clearly defined (Fig. 65.5). Differentiating murine ES cells traditionally begins with floating spheres called *embryoid bodies*. These bodies are akin to the neural spheres of stem cells obtained from the adult CNS. Retinoic acid is a key induction agent for producing neural progenitors from mouse ES cells (Gottlieb and Huettner, 1999). Neural cells resembling anatomically

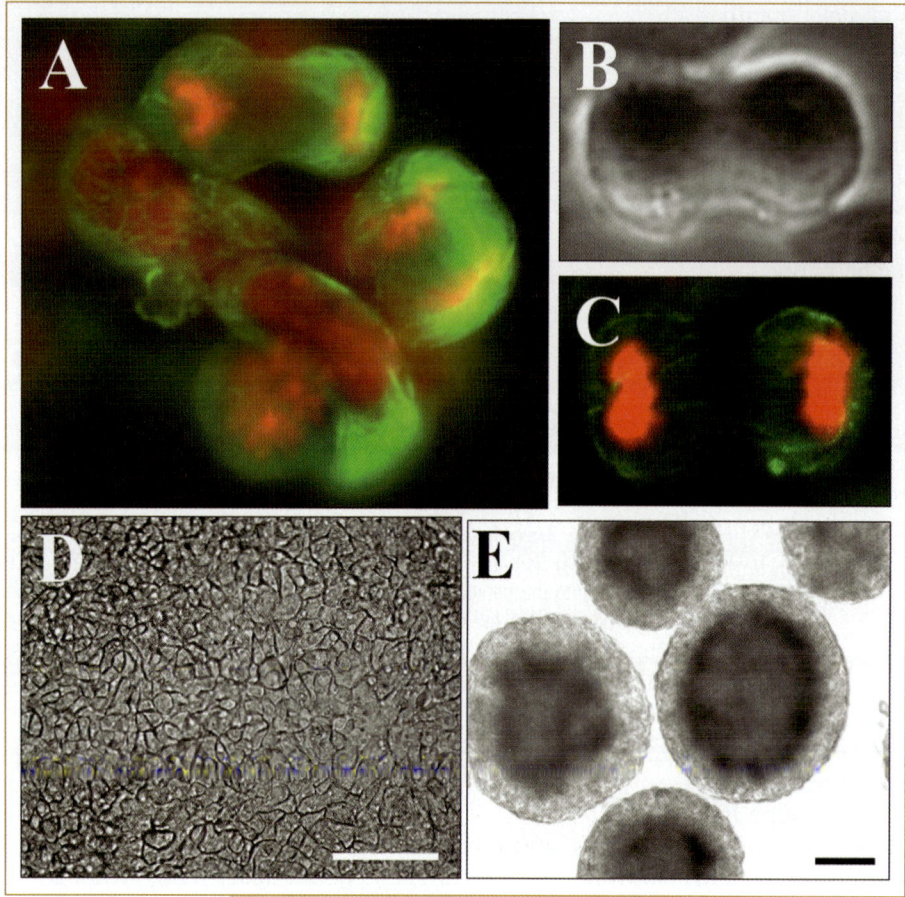

**FIG. 65.5.** **(A–C)** Undifferentiated mouse ES cells dividing in a culture dish. **(A, C)** Immunofluorescence images demonstrate anti-B-tubulin (green) and anti-DNA (Hoechst; red). The phase image (B) of the identical field (C) is shown in Panel B. **(D)** Undifferentiated human ES cells. **(E)** Embryoid bodies derived from human ES cells. Immunofluorescence (green), Hoechst 33342 (blue). Scale bar D–E = 100 μm.

normal neurons, astrocytes, and oligodendrocytes from the CNS can easily be derived from mouse ES cells using these protocols (Fig. 65.6).

## Embryonic Stem Cell Transplantation

Use of ES cells for neural transplantation is in its infancy, and only a very limited amount of work has been completed with the spinal cord. These early studies have relied on transplantation during the embryonic (Wichterle *et al.*, 2002), postnatal (Brustle *et al.*, 1999; McKay, 1997), and adult (McDonald *et al.*, 1999; Liu *et al.*, 2000; Brustle *et al.*, 1999) periods in the normal or injured spinal cord. Overall, they demonstrate that ES cells have a remarkable ability to integrate into the injured region of the cord and differentiate appropriately.

A recent study by Thomas Jessell and colleagues provides exceptional evidence that ES cells can participate in the normal development of spinal cord cells, including motor neurons (Wichterle *et al.*, 2002). It demonstrated that developmentally relevant signaling factors can induce mouse ES cells to differentiate into spinal progenitor cells and then motor neurons through the normal developmental pathway (Fig. 65.7). Thus, the signals that promote the

differentiation of neural stem cells *in vivo* are also effective when applied to ES cells. This group further demonstrated that motor neurons derived from ES cells can populate the embryonic spinal cord, extend axons, and form synapses with target muscles (Fig. 65.8). Therefore, they not only participate in normal development but also grow appropriately when transplanted into the embryonic spinal cord, targeting muscle. Thus, inductive signals in normal pathways of neurogenesis can direct ES cells to form specific classes of CNS neurons.

Oliver Brustle, Ian Duncan, and Ron McKay were the first to transplant progenitors derived from ES cells into the adult and embryonic spinal cord (Brustle *et al.*, 1999). They demonstrated that ES cells that are transplanted into brain and spinal cord of normal adult animals can differentiate into oligodendrocytes that can myelinate axons. They generated oligodendrocyte precursors efficiently by first supplementing cultures of ES cells with fibroblast growth factor (BFGF) and epidermal growth factor and later including platelet-derived growth factor (PDGF). About 38% of the cells in the resulting cultures were oligodendrocytes. To investigate whether these oligodendrocytes could myelinate *in vivo*, cells grown in the presence of BFGF and PDGF

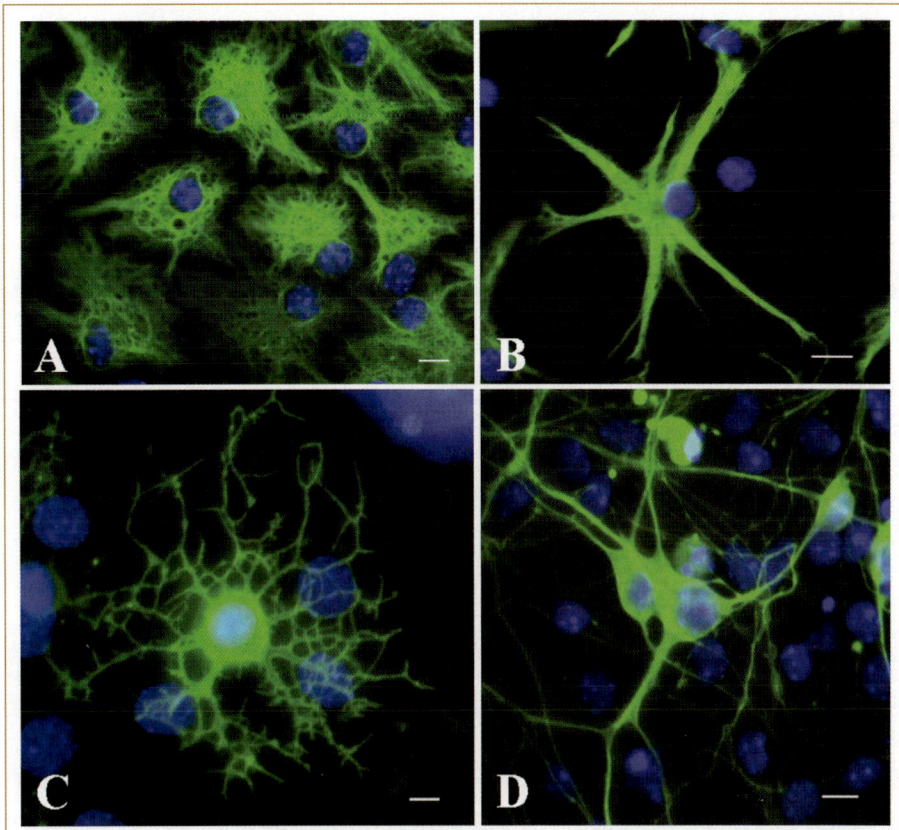

**FIG. 65.6.** ES cells differentiate into the principal types of neural cells. **(A)** Type I and **(B)** type II astrocytes (anti-GFAP). **(C)** Oligodendrocytes (anti-O1). **(D)** Neurons (anti–β tubulin). Scale bar = 10 μm. From Becker *et al.* (2003). Reproduced with permission.

were injected into the spinal cord of one-week-old myelin-deficient rats, a model for a human myelin disorder. Two weeks after transplantation, numerous myelin sheaths were detected in six of the nine affected rats. The original 100,000 cells had migrated widely, and they made myelin with appropriate ultrastructure. Therefore, glial precursors derived from ES cells and transplanted into the neonatal rat spinal cord migrated over several millimeters and differentiated into myelinating oligodendrocytes and astrocytes.

This group also transplanted precursors derived from ES cells into the cerebral ventricles of developing rodents (embryonic day 17) (Brustle *et al.*, 1999). Three weeks later, proteolipid protein (PLP)-positive myelin sheaths were evident in a variety of brain regions. The cells' exogenous origin was confirmed by *in situ* hybridization with a probe to mouse satellite DNA. Importantly, there was no observable evidence of abnormal cellular differentiation or tumor formation.

In the same year, John McDonald, David Gottlieb, and Dennis Choi demonstrated for the first time that ES cells that had been induced to become neural cell precursors could be successfully transplanted into the injured spinal cord (McDonald *et al.*, 1999). Examination of the spinal cord nine days after 1 million precursor cells were transplanted into a cyst caused by contusion injury showed that the cells had survived, grafted, migrated long distances, and differentiated into the three principle neural cell types: neurons, astrocytes, and oligodendrocytes (Figs. 65.9 and 65.10). This group also demonstrated that transplantation was associated with a significant and sizable improvement in function. Their study was the first demonstration that transplanted embryonic precursors can successfully repair the damaged adult nervous system, an important finding, given that conditions in the damaged adult nervous system are much less favorable for regeneration than those in the neonatal spinal cord. Subsequently, Liu and colleagues (2000) demonstrated that precursors derived from ES cells and transplanted into the injured adult nervous system can remyelinate neurons in an anatomically appropriate manner (Fig. 65.11). Furthermore, they showed that oligodendrocytes derived from ES cells functioned normally, myelinating multiple axons in culture, just as they do in the normal nervous system (Fig. 65.12). Using patch clamp analysis, Jim Huettner and John McDonald later demonstrated that the physiologic characteristics of such oligodendrocytes are similar to those of oligodendrocytes taken from the adult spinal cord (Fig. 65.12C and Fig. 65.13). Oligodendrocytes derived from ES cells represent the entire oligodendrocyte lineage, from

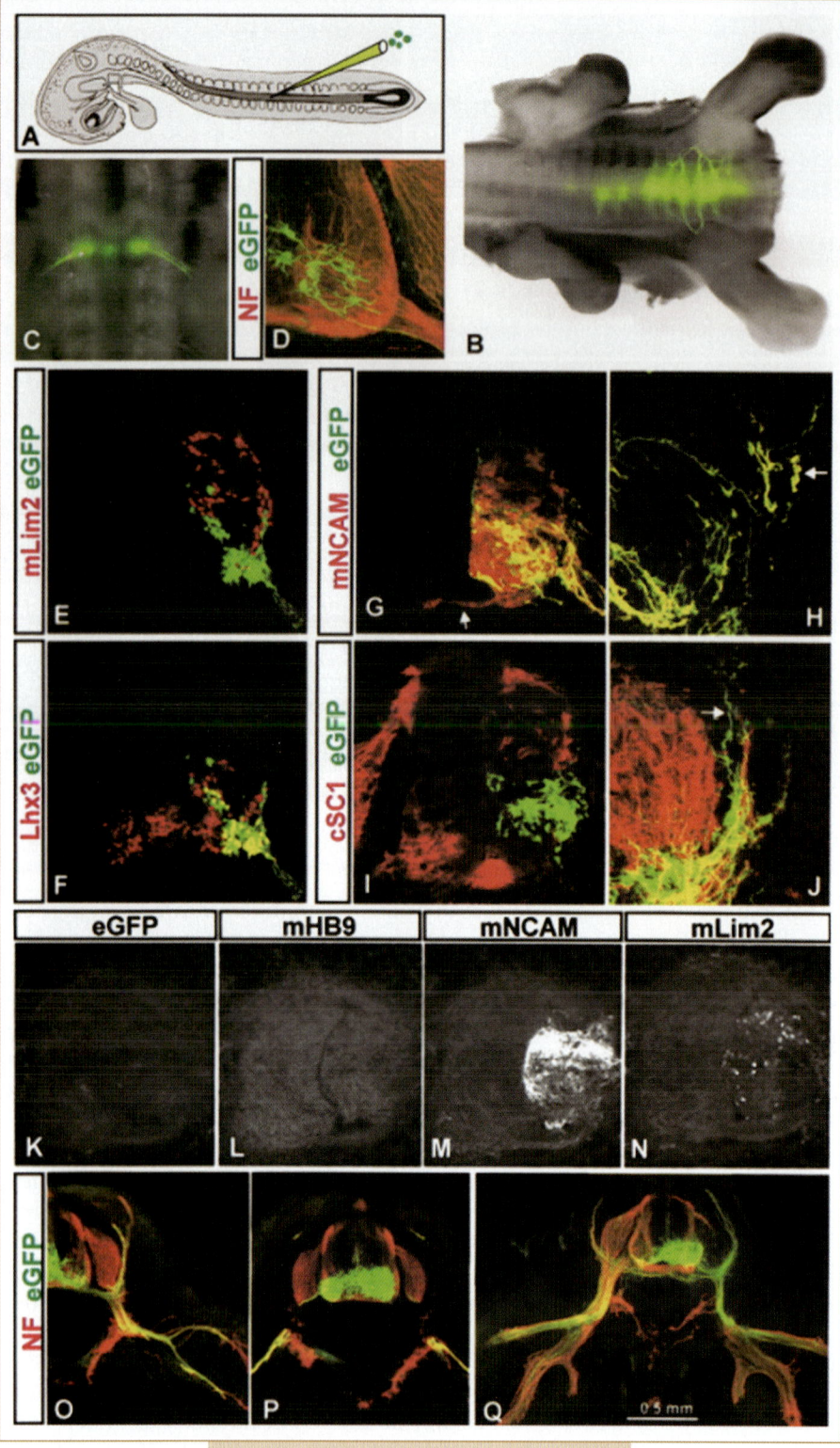

**FIG. 65.7.** Embryonic transplantation of motor neurons derived from mouse ES cells. Integration of transplanted ES cell–derived motor neurons into the spinal cord *in vivo*. **(A)** Implantation of HBG3 ES cell–derived motor neuron–enriched embryoid bodies into stage 15–17 chick spinal cord. **(B)** Bright-field/fluorescence image showing eGFP⁺ motor neurons in thoracic and lumbar spinal cord, assayed at stage 27 (ventral view). **(C, D)** Location of FACS-sorted ES cell–derived eGFP⁺ motor neurons in thoracic spinal cord, assayed at stage 27. eGFP⁺ motor neurons are clustered in the ventral spinal cord (D). **(E–J)** Transverse sections through stage 27 chick spinal cord at rostral cervical levels after transplantation of motor neuron-enriched embryoid bodies. Motor neurons are concentrated in the ventral spinal cord and are segregated from transplanted interneurons, labeled by a mouse-specific Lim2 antibody (E). Many ES cell–derived motor neurons coexpress eGFP and Lhx3 (F). ES cell–derived motor neurons (G) and axons (*arrow*, H) are labeled by rodent-specific anti-NCAM antibody but do not express the chick motor neuron marker protein SC1 (I, J). eGFP⁻, NCAM⁺ axons cross the floor plate but do not project out of the spinal cord (*arrows*, G, H). **(K–N)** Transverse sections of thoracic spinal cord at stage 27, after grafting of embryoid bodies grown with RA (2 µM) and anti-Hh antibody (5E1, 30 µg/mL). No mouse-derived motor neurons were detected by either eGFP (K) or a mouse-specific anti-HB9 antibody (L). In contrast, many mouse-derived NCAM⁺ (M) and Lim2⁺ (N) interneurons are present. **(O–Q)** Transverse sections through stage 27 spinal cord at thoracic (O, P) and lumbar (Q) levels after grafting embryoid bodies enriched with motor neurons. eGFP⁺ motor neurons are concentrated in the ventral spinal cord. Ectopic eGFP⁺ motor neurons are located within the lumen of the spinal cord. eGFP⁺ axons exit the spinal cord, primarily via the ventral root, and project along nerve branches that supply axial (O–Q), body wall (O, P), and dorsal and ventral limb (Q) muscles. The pathway of axons is detected by neurofilament (NF) expression. eGFP⁺ axons are not detected in motor nerves that project to sympathetic neuronal targets. From Wichterle *et al.* (2002). Reproduced with permission.

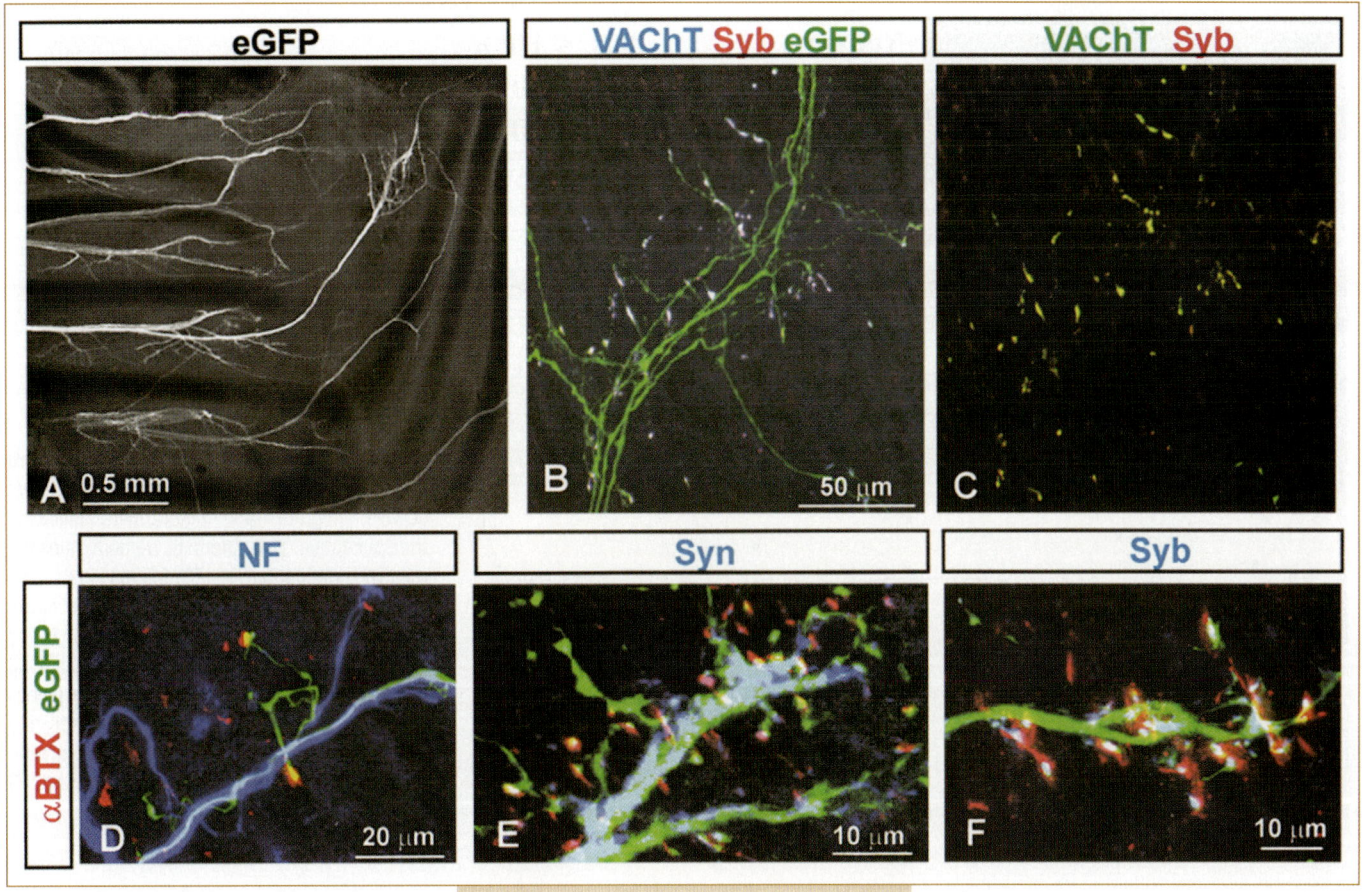

**FIG. 65.8.** Anatomic integration of motor neurons derived from transplanted mouse ES cells. Synaptic differentiation of ES cell–derived motor neurons *in vivo*. **(A)** Whole-mount preparation of stage-35 chick embryonic rib cage. ES cell–derived eGFP⁺ motor axons contact intercostal muscles. **(B, C)** Coexpression of synaptobrevin (Syb) and vesicular acetylcholine transporter (VAChT) in the terminals of eGFP⁺ motor axons at sites of nerve contact with muscle. The anti-Syb and -VAChT antibodies recognize mouse but not chick proteins. **(D)** Neurofilament (NF) and eGFP expression in motor axons that supply intercostal muscles. eGFP⁺ axons lack NF expression. The terminals of eGFP⁺ axons coincide with clusters of acetylcholine receptors, defined by α-bungarotoxin (αBTX) labeling. **(E)** Coincidence of synaptotagmin (Syn) expression in eGFP⁺ motor axon terminals and αBTX labeling. **(F)** Coincidence of synaptobrevin (Syb) expression in eGFP⁺ motor axon terminals and αBTX labeling. From Wichterle *et al.* (2002). Reproduced with permission.

early oligodendrocyte precursors to mature, myelinating oligodendrocytes.

In culture, neurons derived from ES cells differentiate rapidly and spontaneously create neural circuits with anatomical (Fig. 65.14) and physiological evidence of excitatory and inhibitory synapses (Strubing *et al.*, 1995; Finley *et al.*, 1996; Okabe *et al.*, 1996). Substantial neural differentiation is also evident *in vivo* in the model of contusion injury (McDonald *et al.*, 1999), where neurons show extensive axonal outgrowth with presumptive morphological evidence of synapse formation. Moreover, immunological evidence of cholinergic, serontonergic, GABAergic, glycinergic, and glutamatergic neurons has been obtained *in vitro* and *in vivo* (data not shown).

Implantation, survival, and migration of transplanted ES cell–derived precursors in the damaged spinal cord have been verified with real-time polymerase chain reaction (PCR)

and also by using magnetic resonance imaging (MRI) to track oligodendrocytes prelabeled with paramagnetic agents. In both cases, migration up to 1 cm from the transplantation site was evident (Bulte *et al.*, 2001; Lu *et al.*, 2003).

After other groups demonstrated that multiple rounds of transplantation can deliver different types of stem cells to the CNS, our group systematically evaluated intravenous, intraventricular, intrathecal, and interparanchemal transplantation of embryo bodies derived from ES cells using our 4–/4+ protocol (four days in culture without retinoic acid followed by four days with retinoic acid). Although some of the cells that were transplanted intravenously entered the CNS, they remained in blood vessels and showed only limited neural differentiation. Intracerebral ventricular administration enabled some transplanted ES cells to be incorporated into the CNS, but this approach had to be discontinued because the cells obstructed the third and

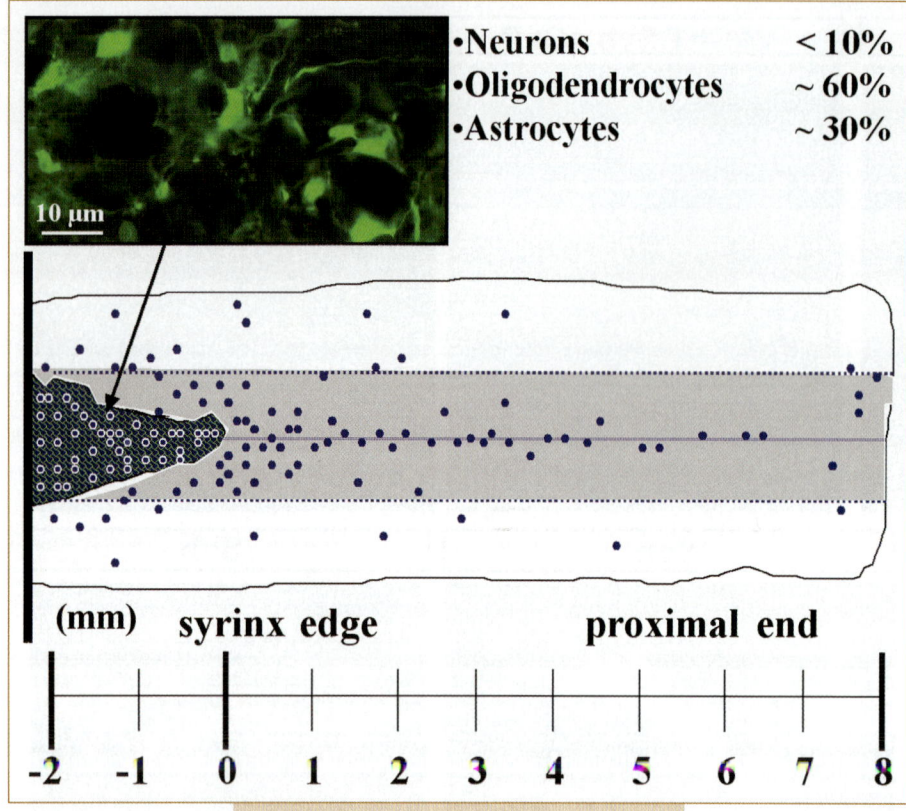

- Neurons          < 10%
- Oligodendrocytes ~ 60%
- Astrocytes       ~ 30%

10 μm

(mm)   syrinx edge          proximal end

-2  -1  0  1  2  3  4  5  6  7  8

**FIG. 65.9.** In the contusion-injured spinal cord of the rat, ES cell–derived neural precursors survive, migrate, and differentiate following transplantation. The schematic demonstrates the relative distribution of ES-derived cells two weeks after transplantation into the central cavity nine days after spinal contusion injury. The cavity is partially filled, and cells have migrated long distances. Most of the distant cells are identified as oligodendrocytes, while astrocytes and neurons remain restricted to the transplantation site. Left top inset shows gfp-expressing ES cell–derived neural cells.

fourth ventricles, causing hydrocephalus. However, we made a remarkable observation when we transplanted a large number of ES cell–derived neural precursors intrathecally into the lumbar spine (Ao and McDonald, 2002). Three to four months later, ES cells were observed in well-formed tissues surrounding the lumbar sacral nerve roots and extending down the sacral canal. Further evaluation clearly showed that the cells had differentiated into tissue that bore a remarkable resemblance to some components of the normal spinal cord. For example, extensive myelination and cells resembling cholinergic motor neurons were observed. Zones of the peripheral nervous system were also clearly demarcated from the CNS. These results were quite surprising because intraparenchymal transplantation into the spinal cord characteristically produces abnormal macro-organization of cells. Thus, this was the first demonstration that transplanted ES cells can organize themselves into CNS-like tissues. It will now be important to understand the constraints required to produce this degree of self-organization.

## IX. NOVEL APPROACHES TO CNS REPAIR (Fig. 65.15)

Most neural transplantation studies, including those using ES cell-derived neural progenitors, have used cells to replace cells lost after injury. Given that we do not know which types of differentiated cells are required for repair, neural precursors may be ideal for this purpose because environmental clues can decide their developmental fate (Liu *et al.*, 2000; Cao *et al.*, 2002; Murray and Fisher, 2001). Moreover, such precursors can serve as bridges to support axonal regrowth (Bunge, 2002).

To try to overcome the physical and chemical constraints imposed by the injured nervous system, ES cells are often genetically altered so they will deliver growth molecules, such as neurotrophin-3, after transplantation (Novikova *et al.*, 2002; Murray and Fisher, 2001; Tuszynski *et al.*, 2002), and they are particularly well suited to this task. McDonald and Silver recently adopted a novel approach to overcoming restraints by showing that early progenitors of ES cells can phagocytize key inhibitory molecules in glial scar tissue (Vadivelu *et al.*, 2003). These cells therefore created an inhibitor-free bridge over which axons rapidly sprouted from the transplant into normal cord. By nine days after transplantation, axons from the graft had grown up to 1 cm — a rate of more than 1 mm per day! This rate is similar to that seen in the normal embryo (Mouveroux *et al.*, 2000).

Another novel approach is to use stem cell transplantation to limit the secondary injury that occurs after nervous system injury (Teng *et al.*, 2002). For example, we recently demonstrated that transplanting ES cells can limit the delayed death of neurons and oligodendrocytes. Since most

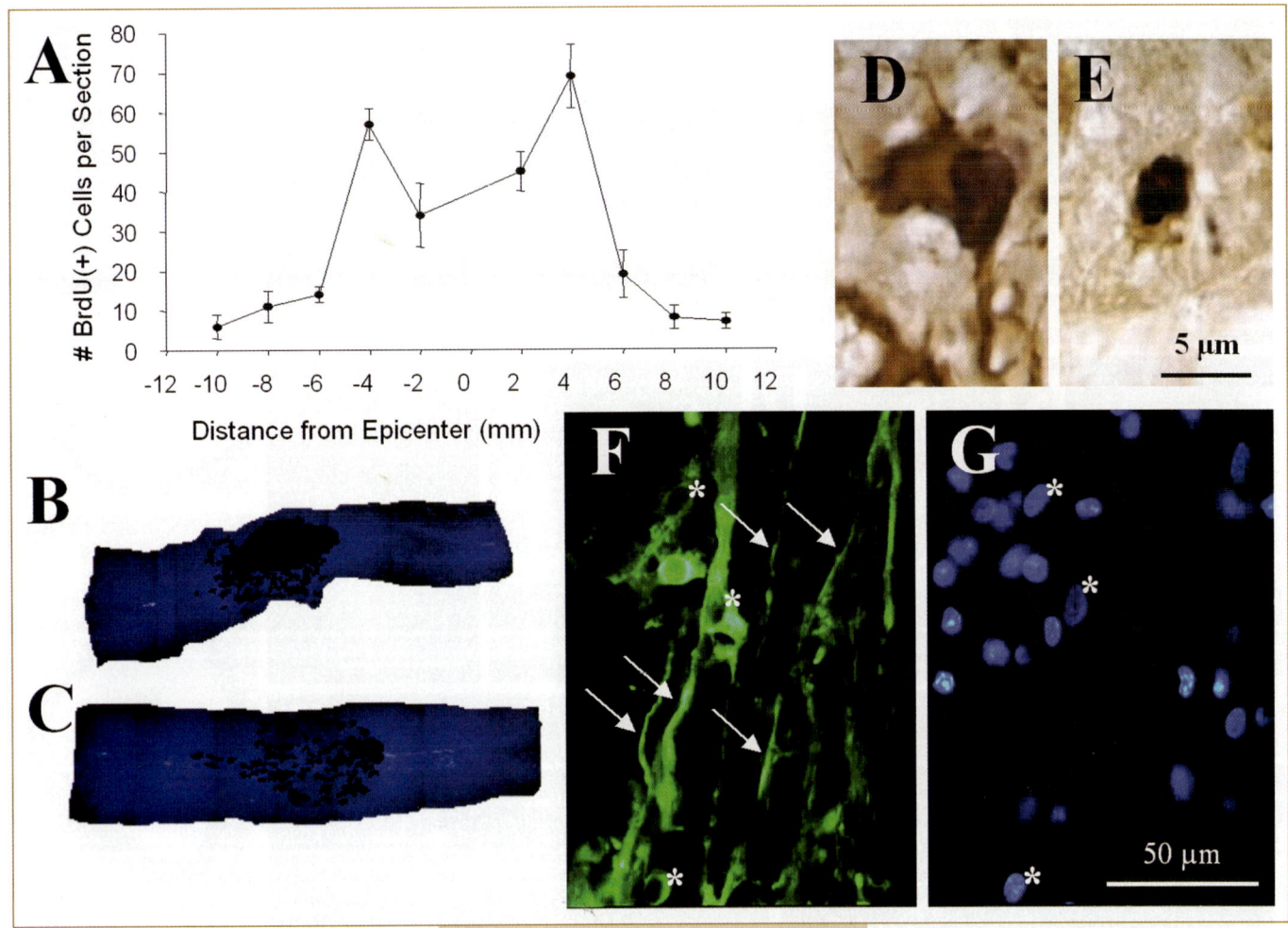

**FIG. 65.10.** BrdU-labeled ES cell–derived cells two weeks after transplantation. Mean ± SEM BrdU-labeled nuclei per 1-mm segment in longitudinal sections. **(A)**. Hoechst 33342-labeled sections 42 days after injury, transplanted with vehicle **(B)** or ES cells **(C)** nine days after injury. BrdU-positive cell colabeled with GFAP **(D)**. BrdU-labeled cell colabeled with APC CC-1 **(E)**. The mouse-specific marker EMA indicates processes (arrows) emanating from ES cells **(F)**. Corresponding nuclei are marked by asterisks **(G)**. Modified from McDonald *et al.* (1999) with permission.

transplantations are performed at the time of injury, it is possible that many of their results may be attributable to this neuroprotective role. It seems clear that even genetically unmodified ES cells release large quantities of growth factors.

Although replacing lost neurons is difficult, it might be possible to use neuronal replacements to create bridging circuits across the injury site. Descending axons could synapse onto ES cell–derived neurons that subsequently synapsed onto key pattern generators in the lower spinal cord. More global delivery of neurotransmitters via synaptic or nonsynaptic mechanisms might enhance functions such as locomotion. Previous work by others has demonstrated that release of noradrenergic and serotonergic neurotransmitters can stimulate and enhance the central pattern generator for locomotion (Rossignol *et al.*, 1998). More recent work indicates that release of brain-derived neurotrophic

factor and neurotrophin-3 can perform this function (Ankeny *et al.*, 2001).

Because ES cells are embryonic in nature, neural progenitors derived from them may be able to reprogram the adult CNS so that damage can be repaired (Tada *et al.*, 2001). ES cells are also unique in that they have the potential to replace cells derived from multiple embryonic germ layers. In the CNS, it will be important to replace lost nonneural cells as well as neural cells to regain normal CNS function, such as neovascularization. Growing recognition of the link between neovascularization and neurogenesis is strengthening this approach. We recently found that ES cells in transplanted 4–/4+ embryoid bodies differentiate into both neural cells and vascular endothelial cells (unpublished observations).

In the embryo, stem cells differentiate into motor neurons and grow toward muscle, where they form neuro-

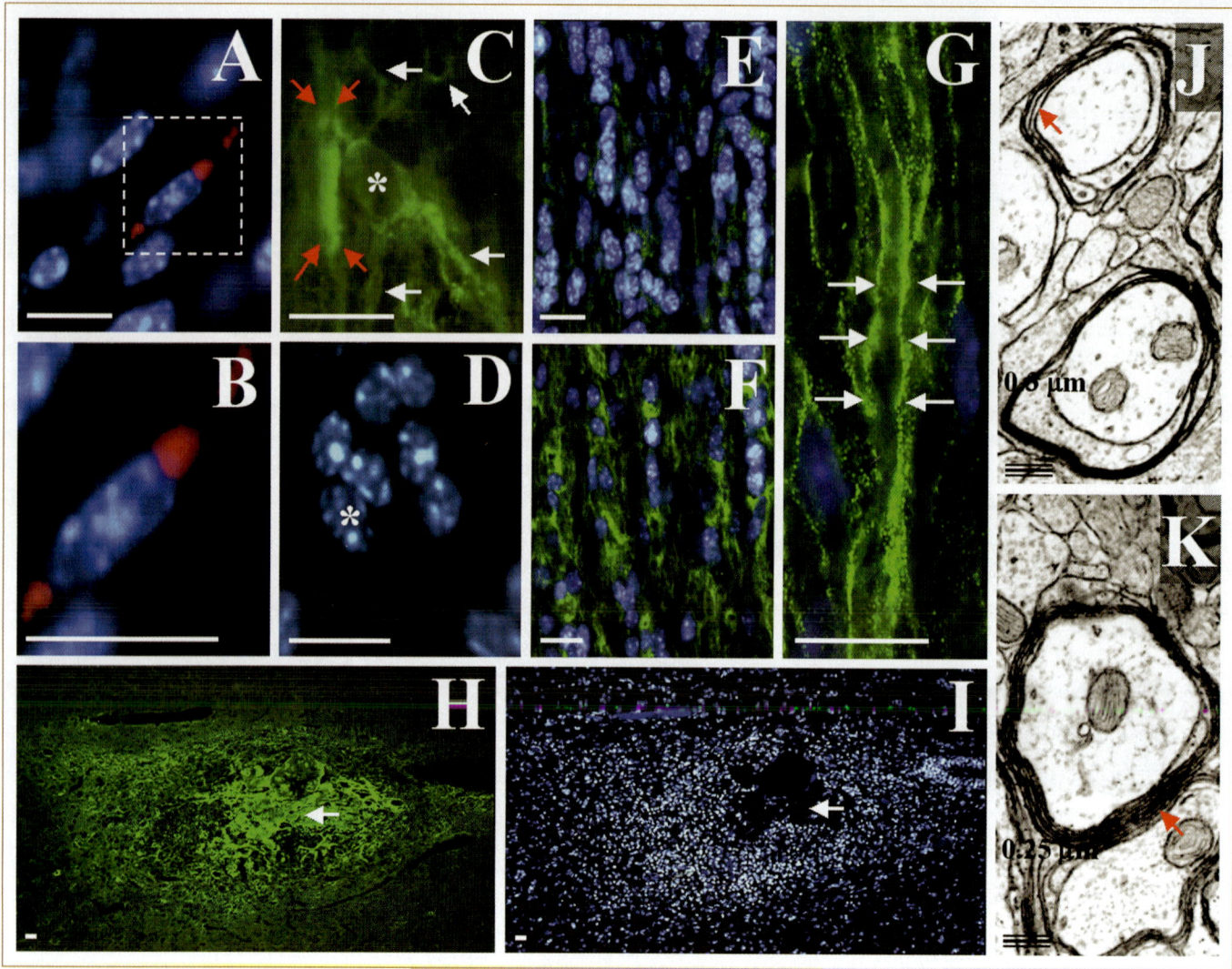

**FIG. 65.11.** Cells derived from ES oligospheres can migrate and myelinate axons when transplanted into dysmyelinated spinal cords of adult *shiverer* mice, which lack the gene for myelin basic protein (MBP). Transplanted cells were identified by cell tracker orange (CTO) epifluorescence (red) or immuno-reactivity for MBP (green). Hoechst 33342 (blue). CTO-labeled cells aligned with native intrafascicular oligodendrocytes in white matter **(A, B)**. An ES cell–derived (MBP⁺) oligodendrocyte (asterisk) with longitudinally oriented processes (white arrows) is shown in panels **C** and **D**. Red arrows mark probable myelination around an adjacent axon **(C)**. Little MBP immunoreactivity is seen in white matter in a longitudinal section of spinal cord from a mouse that received a sham transplantation **(E)**. A gradient of MBP immunoreactivity centers on the site of ES cell transplantation **(F)**. Panel **G** (high magnification) shows intrafascicular oligodendrocyte nuclei (blue) and MBP immunoreactivity (green), two indications of axonal myelination (white arrows), in white matter from a mouse transplanted with ES cells. The spatial distribution of MBP immunoreactivity one month after ES cell transplantation is shown at low magnification **(H)**, with corresponding Hoechst 33342 counterstaining **(I)**. White arrows indicate the center of the transplanted site. Transmission EM shows four loose wraps of myelin, the maximal number of layers typically seen around axons in control animals (red arrow, **J**), and nine or more compact wraps around axons from the transplanted area (red arrow, **K**). Shiverer-mutant mice lack a functional MBP gene that is required to form mature compact myelin; therefore, the presence of mature compact myelin is a standard for transplant oligodendrocyte–associated myelin. Scale bars: A–I = 10 μm; J, K = 0.3 μm. Reproduced with permission from Liu *et al.* (2000).

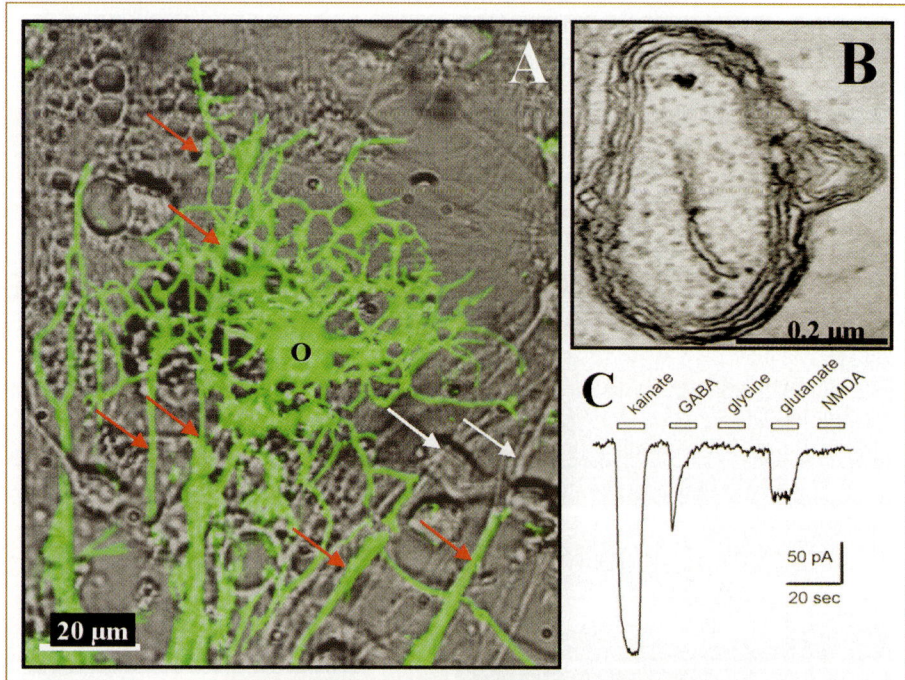

**FIG. 65.12.** ES cells produced mature oligo-dendrocytes with normal anatomical features of myelination and physiological response to neu-rotransmitters. Reproduced in adapted form with permission from Liu *et al.* (2000).

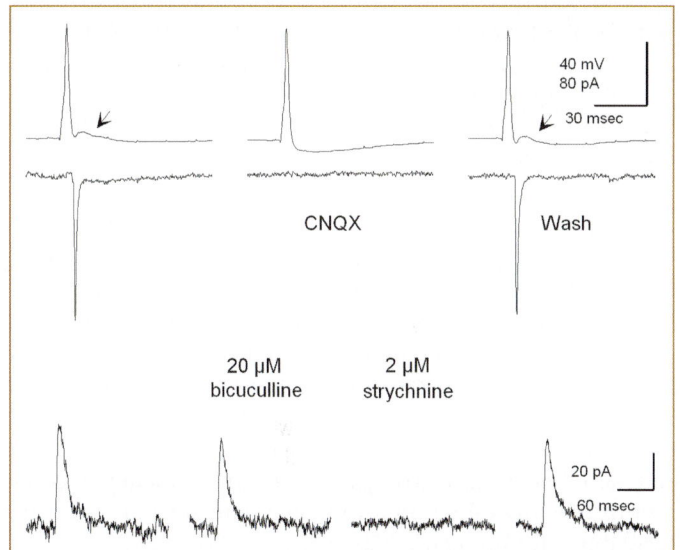

**FIG. 65.13.** Excitatory (*top*) and inhibitory (*bottom*) synaptic transmission between neurons derived *in vitro* from mouse ES cells. Action potentials were evoked in the presynaptic neuron by current injection. Excitatory postsynaptic currents, blocked by superfusion with the selective glutamate receptor antagonist 6-cyano-7-nitroquinoxaline-2,3-dione (CNQX; 10 micromolar), were recorded under voltage clamp. Arrows point to an autaptic excitatory synaptic potential that is also blocked by CNQX. Inhibitory synaptic currents were evoked in a different ES cell pair and tested sequentially with the antagonists bicuculline and strychnine, which are selective for gamma-aminobutyric acid (GABA) and glycine receptors, respectively. For this cell pair, only the post-synaptic currents are shown. Transmission was blocked by strychnine, indi-cating that glycine was the transmitter. Other presynaptic cells evoked bicuculline-sensitive synaptic currents, indicating that GABA mediated trans-mission (not shown).

muscular junctions. Using ES-derived motor neurons to replace lost motor neurons at injured segments will not enhance function because chronically denervated muscle is unable to reassemble functional neuromuscular junctions. However, transplantation might avoid the long-term muscle wasting that results from denervation. To maintain dener-vated muscle, for example, it might be possible to transplant ES cells into the distal nerve stump or directly into the muscle. Once in place, the transplanted cells could differen-tiate into motor neurons. If they were genetically altered, the motor neurons could be selectively removed once the repair was properly achieved.

More recent studies are beginning to combine stem cells, including ES cells, with biomaterials. The advent of nanotechnology should further enhance this approach (Lockman *et al.*, 2002).

Finally, remyelination is one of the most pragmatic approaches to restoring function to the damaged spinal cord. Because many potentially functional connections remain in the outer donut of surviving tissue, appropriate remyelination could substantially improve function. One must consider, however, that dysmyelination rather than demyelination is the biggest problem in the damaged nervous system. Often, inappropriate myelination is more harmful than no myelination at all.

## X. TOWARD HUMAN TRIALS

Some of the animal studies described previously have prompted early human safety trials of ES cell transplanta-tion. Most have focused on Parkinson's disease, where unforeseen exaggeration of dystonia has limited advances

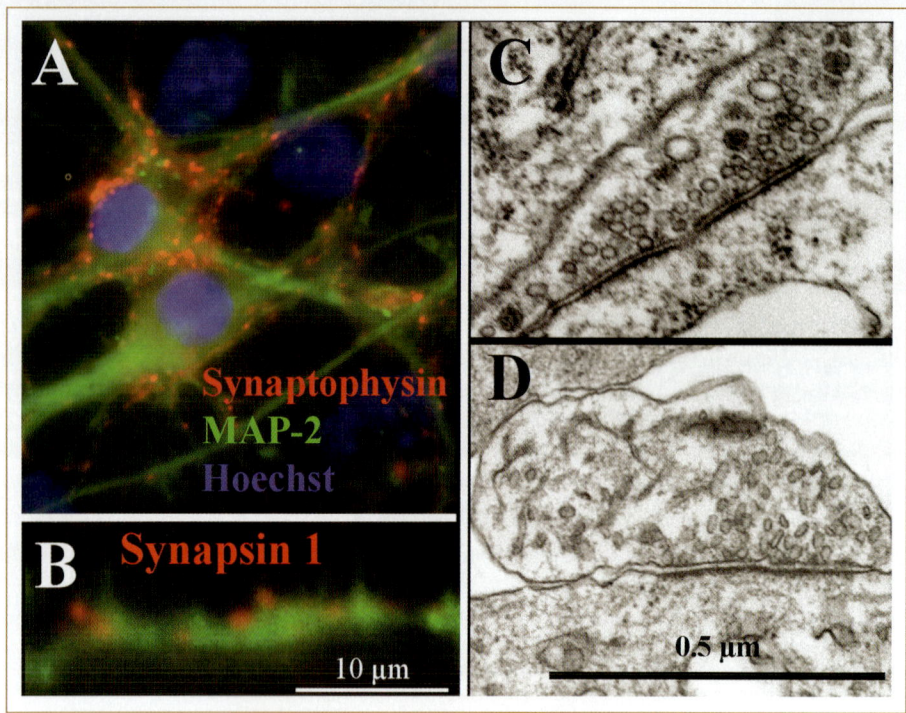

**FIG. 65.14.** ES cells differentiate into neurons that spontaneously create neural circuits. Presumptive presynaptic sites (red) oppose dendrites (green) in **(A)** and **(B)**. **(C)** and **(D)** demonstrate ultrastructural characteristics of synaptic profiles *in vitro*.

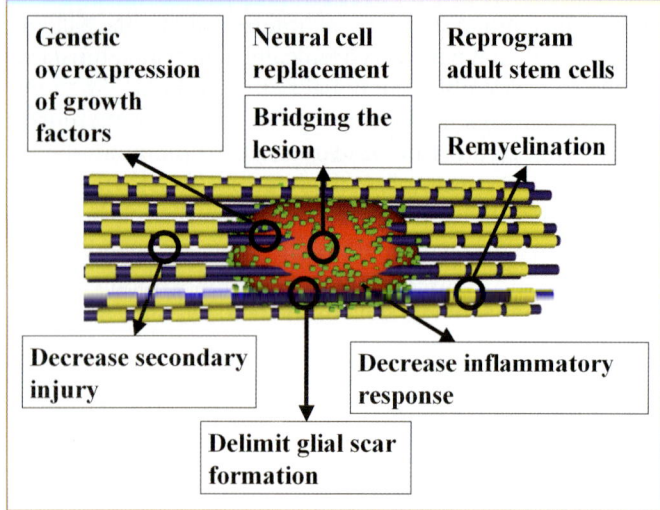

**FIG. 65.15.** Novel approaches to spinal cord repair using ES cells. Remyelination is one of the most pragmatic approaches to restoring function to the damaged spinal cord. ES cell–derived precursors can serve as bridges to support axonal regrowth. They can phagocytize key inhibitory molecules in glial scar tissue. These cells therefore created an inhibitor-free bridge over which axons sprouted rapidly from the transplant into normal cord. Newly generated neurons can be used to create bridging circuits across the injury site. ES cell transplantation can limit the secondary injury. To try to overcome the constraints of the injured nervous system, ES cells are often genetically altered so that they will deliver growth molecules, such as NT3, after transplantation. Because ES cells are embryonic in nature, their progenitors may also be able to reprogram the adult CNS to optimize spontaneous host regeneration. Replacement of nonneural cells as well as neural cells will be important for regaining normal CNS function, such as neovascularization. ES cells can have similar applications in the peripheral nervous system.

in transplantation therapy (Ma *et al.*, 2002). Phase I human trials for repairing the spinal cord are also under way. They include transplantation of porcine-derived stem cells for the purpose of remyelination (Diacrin; http://www.diacrin.com) as well as allogeneic transplantation of olfactory ensheathing glia (Carlos Lima; Lisbon, Portugal). Although safety data are not yet available, it is important to note that such trials are paving the way for further uses of ES cells. With the recent advent of human ES cells, human transplantation appears more feasible, though somatic transplantation will be required to ensure a genetic match. Such manipulation of cells raises many regulatory concerns that must be addressed before clinical trials can move forward. The recent demonstration of *in vitro* oocyte generation from mouse ES cells raises the possibility that ethically acceptable procedures for human nuclear transfer may soon be available (Hubner *et al.*, 2003).

## XI. CONCLUSIONS

Studies of neurotransplantation for repairing the damaged nervous system are making good progress. Early animal studies showed that mouse ES cells can replace neurons, astrocytes, and oligodendrocytes, instigate appropriate remyelination, and even improve locomotion. As is the case for all transplantation studies, however, the mechanisms underlying functional improvement remain unclear. Nevertheless, ES cells offer a novel approach to deciphering these mechanisms. For the first time, ES cell genetics is allowing investigators to track transplanted cells that have been engineered to express green fluorescent protein. This and other types of genetic modification permit one to track the integration and differentiation of ES cells, assess behavioral recovery, selectively remove the cells to see whether recovery wanes, and insert more cells to see if functional recovery occurs anew. This proof of principle will be required to identify mechanisms in neural repair.

Although it is impossible to predict the future of the field, it is clear that murine and human ES cells are tools that will revolutionize neurobiology and neural transplantation by providing the unprecedented ability to selectively deliver key growth and regulatory factors. Also, ES cell transplantation promises to be one of the greatest potential therapies for chronic nervous system disorders. Only the future will reveal its full potential for treating human disease and disability, but it is possible to say with confidence that ES cells will have a major impact on repairing the human CNS within our lifetimes.

## XII. ACKNOWLEDGMENTS

We would like to thank the people in our laboratory who have contributed to this work over the years and also our collaborators: Jeff Bulte at NIH; Dennis W. Choi, David I. Gottlieb, David Gutmann, Chung Hsu, Mark F. Jacquin, Gene Johnson, and Carl Lauryssen at Washington University in St. Louis. This team obtained a program project grant from the National Institutes of Health (NIH; NS39577) to evaluate the potential of embryonic stem cells to repair spinal cord injury. We would also like to thank our animal technicians, who express rodent bladders three times daily, seven days a week, including holidays (Joan Bonnot, Joseph Galvez, Brandy Jones, Amy Hansen) and our TC staff, who passage ES cells every other day, endlessly (Ashley Johnson, Laura Luecking, Becky Purcell). This work was supported by NINDS and NIDCR grants NS39577, NS40520, NS45023, and DE07734.

## XIII. REFERENCES

Aguayo, A. J., David, S., and Bray, G. M. (1981). Influences of the glial environment on the elongation of axons after injury: transplantation studies in adult rodents. *J. Exp. Biol.* **95**, 231–240.

Akiyama, Y., Radtke, C., Honmou, O., and Kocsis, J. D. (2002). Remyelination of the spinal cord following intravenous delivery of bone marrow cells. *Glia* **39**, 229–236.

Ankeny, D. P., McTigue, D. M., Guan, Z., Yan, Q., Kinstler, O., Stokes, B. T., and Jakeman, L. B. (2001). Pegylated brain-derived neurotrophic factor shows improved distribution into the spinal cord and stimulates locomotor activity and morphological changes after injury. *Exp. Neurol.* **170**, 85–100.

Ao, H., and McDonald, J. W. (2002). Self-assembly of ES cells into neural tissues with features resembling normal spinal cord. *Abstr. Soc. Neurosci. Program* No. 34.17.

Auld, D. S., and Robitaille, R. (2003). Perisynaptic Schwann cells at the neuromuscular junction: nerve- and activity-dependent contributions to synaptic efficacy, plasticity, and reinnervation. *Neuroscientist* **9**, 144–157.

Barbeau, H., McCrea, D. A., O'Donovan, M. J., Rossignol, S., Grill, W. M., and Lemay, M. A. (1999). Tapping into spinal circuits to restore motor function. *Brain Res. Brain Res. Rev.* **30**, 27–51.

Barres, B. A., and Raff, M. C. (1996). Axonal control of oligodendrocyte development. *In* "Glia Cell Development: Basic Principles and Clinical Relevance" (K. R. Jessen and W. D. Richardson, eds.), pp. 71–83. Bios Scientific, Oxford, UK.

Becker, D., Grill, W. M., and McDonald, J. W. (2003). Functional electrical stimulation helps replenish neural cells in the adult CNS after spinal cord injury. Paper presented at the 55th Annual Meeting of the American Academy of Neurology.

Benes, F. M., Turtle, M., Khan, Y., and Farol, P. (1994). Myelination of a key relay zone in the hippocampal formation occurs in the human brain during childhood, adolescence, and adulthood. *Arch. Gen. Psychiatry* **51**, 477–484.

Blight, A. R. (1983). Cellular morphology of chronic spinal cord injury in the cat: analysis of myelinated axons by line-sampling. *Neuroscience* **10**, 521–543.

Bregman, B. S., Coumans, J. V., Dai, H. N., Kuhn, P. L., Lynskey, J., McAtee, M., and Sandhu, F. (2002). Transplants and neurotrophic factors increase regeneration and recovery of function after spinal cord injury. *Prog. Brain Res.* **137**, 257–273.

Briscoe, J., and Ericson, J. (2001). Specification of neuronal fates in the ventral neural tube. *Curr. Opin. Neurobiol.* **11**, 43–49.

Brustle, O., Jones, K. N., Learish, R. D., Karram, K., Choudhary, K., Wiestler, O. D., Duncan, I. D., and McKay, R. D. (1999). Embryonic stem cell–derived glial precursors: a source of myelinating transplants. *Science* **285AB**, 754–756.

Bulte, J. W., Douglas, T., Witwer, B., Zhang, S. C., Strable, E., Lewis, B. K., Zywicke, H., Miller, B., van Gelderen, P., Moskowitz, B. M., Duncan, I. D., and Frank, J. A. (2001). Magnetodendrimers allow endosomal magnetic labeling and *in vivo* tracking of stem cells. *Nat. Biotechnol.* **19**, 1141–1147.

Bunge, M. B. (2002). Bridging the transected or contused adult rat spinal cord with Schwann cell and olfactory ensheathing glia transplants. *Prog. Brain Res.* **137**, 275–282.

Burke, R. E. (2001). The central pattern generator for locomotion in mammals. *Adv. Neurol.* **87**, 11–24.

Cao, Q., Benton, R. L., and Whittemore, S. R. (2002a). Stem cell repair of central nervous system injury. *J. Neurosci. Res.* **68**, 501–510.

Cao, Q. L., Howard, R. M., Dennison, J. B., and Whittemore, S. R. (2002b). Differentiation of engrafted neuronal-restricted precursor cells is inhibited in the traumatically injured spinal cord. *Exp. Neurol.* **177**, 349–359.

Crowe, M. J., Bresnahan, J. C., Shuman, S. L., Masters, J. N., and Beattie, M. S. (1997). Apoptosis and delayed degeneration after spinal cord injury in rats and monkeys. *Nat. Med.* **3**, 73–76.

Dickson, J. G., Kesselring, J., Walsh, F. S., and Davison, A. N. (1985). Cellular distribution of 04 antigen and galactocerebroside in primary cultures of human foetal spinal cord. *Acta Neuropathol. (Berl.)* **68**, 340–344.

Dietz, V. (1995). Central pattern generator. *Paraplegia* **33**, 739.

Dubois-Dalcq, M., and Murray, K. (2000). Why are growth factors important in oligodendrocyte physiology? *Pathol. Biol. (Paris)* **48**, 80–86.

Durston, A. J., van der, W. J., Pijnappel, W. W., and Godsave, S. F. (1998). Retinoids and related signals in early development of the vertebrate central nervous system. *Curr. Top. Dev. Biol.* **40**, 111–175.

Eriksson, P. S., Perfilieva, E., Bjork-Eriksson, T., Alborn, A. M., Nordborg, C., Peterson, D. A., and Gage, F. H. (1998). Neurogenesis in the adult human hippocampus. *Nat. Med.* **4**, 1313–1317.

Finley, M. F., Kulkarni, N., and Huettner, J. E. (1996). Synapse formation and establishment of neuronal polarity by P19 embryonic carcinoma cells and embryonic stem cells. *J. Neurosci.* **16**, 1056–1065.

Gamble, H. J. (1969). Electron microscope observations on the human foetal and embryonic spinal cord. *J. Anat.* **104**, 435–453.

Gottlieb, D. I., and Huettner, J. E. (1999). An *in vitro* pathway from embryonic stem cells to neurons and glia. *Cells Tissues Organs* **165**, 165–172.

Hains, B. C., Johnson, K. M., Eaton, M. J., Willis, W. D., and Hulsebosch, C. E. (2003). Serotonergic neural precursor cell grafts attenuate bilateral hyperexcitability of dorsal horn neurons after spinal hemisection in rat. *Neuroscience* **116**, 1097–1110.

Hajihosseini, M., Tham, T. N., and Dubois-Dalcq, M. (1996). Origin of oligodendrocytes within the human spinal cord. *J. Neurosci.* **16**, 7981–7994.

Herndon, R. M., Price, D. L., and Weiner, L. P. (1977). Regeneration of oligodendroglia during recovery from demyelinating disease. *Science* **195**, 693–694.

Hubner, K., Fuhrmann, G., Christenson, L. K., Kehler, J., Reinbold, R., De La, F. R., Wood, J., Strauss III, J. F., Boiani, M., and Scholer, H. R. (2003). Derivation of oocytes from mouse embryonic stem cells. *Science*

Horner, P. J., Power, A. E., Kempermann, G., Kuhn, H. G., Palmer, T. D., Winkler, J., Thal, L. J., and Gage, F. H. (2000). Proliferation and differentiation of progenitor cells throughout the intact adult rat spinal cord. *J. Neurosci.* **20**, 2218–2228.

Jessell, T. M. (2000). Neuronal specification in the spinal cord: inductive signals and transcriptional codes. *Nat. Rev. Genet.* **1**, 20–29.

Jiang, S., Wang, J., Khan, M. I., Middlemiss, P. J., Salgado-Ceballos, H., Werstiuk, E. S., Wickson, R., and Rathbone, M. P. (2003). Enteric glia promote regeneration of transected dorsal root axons into spinal cord of adult rats. *Exp. Neurol.* **181**, 79–83.

Kaufman, M. H., Robertson, E. J., Handyside, A. H., and Evans, M. J. (1983). Establishment of pluripotent cell lines from haploid mouse embryos. *J. Embryol. Exp. Morphol.* **73**, 249–261.

Kennea, N. L., and Mehmet, H. (2002). Neural stem cells. *J. Pathol.* **197**, 536–550.

Liu, Y., and Rao, M. S. (2003). Transdifferentiation — fact or artifact. *J. Cell Biochem.* **88**, 29–40.

Liu, S., Qu, Y., Stewart, T. J., Howard, M. J., Chakrabortty, S., Holekamp, T. F., and McDonald, J. W. (2000). Embryonic stem cells differentiate into oligodendrocytes and myelinate in culture and after spinal cord transplantation. *Proc. Natl. Acad. Sci. U.S.A.* **97**, 6126–6131.

Lockman, P. R., Mumper, R. J., Khan, M. A., and Allen, D. D. (2002). Nanoparticle technology for drug delivery across the blood–brain barrier. *Drug Dev. Ind. Pharm.* **28**, 1–13.

Lu, J., Bulte, J. W., Liu, S., Schottler, F., Howard, M. J., Zywicke, H., van Gelderen, P., Douglas, T., Frank, J. A., and McDonald, J. W. (2003). MR imaging of transplanted labeled embryonic stem (ES) cells in the contusion-injured spinal cord. Paper presented at the 31st Annual Meeting of the Society for Neuroscience.

Ludwin, S. K. (1984). Proliferation of mature oligodendrocytes after trauma to the central nervous system. *Nature* **308**, 274–275.

Ma, Y., Feigin, A., Dhawan, V., Fukuda, M., Shi, Q., Greene, P., Breeze, R., Fahn, S., Freed, C., and Eidelberg, D. (2002). Dyskinesia after fetal cell transplantation for parkinsonism: a PET study. *Ann. Neurol.* 628–634.

McDonald, J. W., and Sadowsky, C. (2002). Spinal-cord injury. *Lancet* **359**, 417–425.

McDonald, J. W., Liu, X. Z., Qu, Y., Liu, S., Mickey, S. K., Turetsky, D., Gottlieb, D. I., and Choi, D. W. (1999). Transplanted embryonic stem cells survive, differentiate and promote recovery in injured rat spinal cord. *Nat. Med.* **5**, 1410–1412.

McKay, R. (1997). Stem cells in the central nervous system. *Science* **276**, 66–71.

McKinley, W. O., Jackson, A. B., Cardenas, D. D., and DeVivo, M. J. (1999). Long-term medical complications after traumatic spinal cord injury: a regional model systems analysis. *Arch. Phys. Med. Rehabil.* **80**, 1402–1410.

Morgenstern, D. A., Asher, R. A., and Fawcett, J. W. (2002). Chondroitin sulphate proteoglycans in the CNS injury response. *Prog. Brain Res.* **137**, 313–332.

Mouveroux, J. M., Verkijk, M., Lakke, E. A., and Marani, E. (2000). Intrinsic properties of the developing motor cortex in the rat: *in vitro* axons from the medial somatomotor cortex grow faster than axons from the lateral somatomotor cortex. *Brain Res. Dev. Brain Res.* **122**, 59–66.

Munoz-Sanjuan, I., and Brivanlou, A. H. (2002). Neural induction, the default model and embryonic stem cells. *Nat. Rev. Neurosci.* **3**, 271–280.

Murray, M., and Fischer, I. (2001). Transplantation and gene therapy: combined approaches for repair of spinal cord injury. *Neuroscientist* **7**, 28–41.

Murray, M., Kim, D., Liu, Y., Tobias, C., Tessler, A., and Fischer, I. (2002). Transplantation of genetically modified cells contributes to repair and recovery from spinal injury. *Brain Res. Brain Res. Rev.* **40**, 292–300.

Nashmi, R., and Fehlings, M. G. (2001). Mechanisms of axonal dysfunction after spinal cord injury: with an emphasis on the role of voltage-gated potassium channels. *Brain Res. Brain Res. Rev.* **38**, 165–191.

National Spinal Cord Injury Database. (2001). Spinal Cord Injury, Facts and Figures.

Novikova, L. N., Novikov, L. N., and Kellerth, J. O. (2002). Differential effects of neurotrophins on neuronal survival and axonal regeneration after spinal cord injury in adult rats. *J. Comp. Neurol.* **452**, 255–263.

Okabe, S., Forsberg-Nilsson, K., Spiro, A. C., Segal, M., and McKay, R. D. (1996). Development of neuronal precursor cells and functional postmitotic neurons from embryonic stem cells *in vitro*. *Mech. Dev.* **59**, 89–102.

Poduslo, S. E., and Jang, Y. (1984). Myelin development in infant brain. *Neurochem. Res.* **9**, 1615–1626.

Prewitt, C. M., Niesman, I. R., Kane, C. J., and Houle, J. D. (1997). Activated macrophage/microglial cells can promote the regeneration of sensory axons into the injured spinal cord. *Exp. Neurol.* **148**, 433–443.

Renoncourt, Y., Carroll, P., Filippi, P., Arce, V., and Alonso, S. (1998). Neurons derived in vitro from ES cells express homeoproteins characteristic of motoneurons and interneurons. *Mech. Dev.* **79**, 185–197.

Roelink, H., Porter, J. A., Chiang, C., Tanabe, Y., Chang, D. T., Beachy, P. A., and Jessell, T. M. (1995). Floor plate and motor neuron induction by different concentrations of the amino-terminal cleavage product of sonic hedgehog autoproteolysis. *Cell* **81**, 445–455.

Rossi, S. G., Dickerson, I. M., and Rotundo, R. L. (2003). Localization of the CGRP receptor complex at the vertebrate neuromuscular junction and its role in regulating acetylcholinesterase expression. *J. Biol. Chem.*

Rossignol, S., Chau, C., Brustein, E., Giroux, N., Bouyer, L., Barbeau, H., and Reader, T. A. (1998). Pharmacological activation and modulation of the central pattern generator for locomotion in the cat. *Ann. NY. Acad. Sci.* **860**, 346–359.

Satoh, J., and Kim, S. U. (1994). Proliferation and differentiation of fetal human oligodendrocytes in culture. *J. Neurosci. Res.* **39**, 260–272.

Sjolund, B. H. (2002). Pain and rehabilitation after spinal cord injury: the case of sensory spasticity? *Brain Res. Brain Res. Rev.* **40**, 250–256.

Sotomaru, Y., Kato, Y., and Tsunoda, Y. (1999). Induction of pluripotency by injection of mouse trophectoderm cell nuclei into blastocysts following transplantation into enucleated oocytes. *Theriogenology* **52**, 213–220.

Strubing, C., Ahnert-Hilger, G., Shan, J., Wiedenmann, B., Hescheler, J., and Wobus, A. M. (1995). Differentiation of pluripotent embryonic stem cells into the neuronal lineage in vitro gives rise to mature inhibitory and excitatory neurons. *Mech. Dev.* **53**, 275–287.

Tada, M., Takahama, Y., Abe, K., Nakatsuji, N., and Tada, T. (2001). Nuclear reprogramming of somatic cells by in vitro hybridization with ES cells. *Curr. Biol.* **11**, 1553–1558.

Tatagiba, M., Brosamle, C., and Schwab, M. E. (1997). Regeneration of injured axons in the adult mammalian central nervous system. *Neurosurgery* **40**, 541–546.

Teng, Y. D., Lavik, E. B., Qu, X., Park, K. I., Ourednik, J., Zurakowski, D., Langer, R., and Snyder, E. Y. (2002). Functional recovery following traumatic spinal cord injury mediated by a unique polymer scaffold seeded with neural stem cells. *Proc. Natl. Acad. Sci. U.S.A.* **99**, 3024–3029.

Tuszynski, M. H., Grill, R., Jones, L. L., McKay, H. M., and Blesch, A. (2002). Spontaneous and augmented growth of axons in the primate spinal cord: effects of local injury and nerve growth factor–secreting cell grafts. *J. Comp. Neurol.* **449**, 88–101.

Vadivelu, S., Stewart, T. J., Miller, J. H., Tom, V., Liu, S., Li, Q., Howard, M. J., Silver, J., and McDonald, J. W. (2003). Embryonic stem cell transplantation: penetration of the glial scar and robust axonal regeneration into white matter in the injured adult rat spinal cord. Paper presented at the 33rd Annual Meeting of the Society for Neuroscience.

Weidenheim, K. M., Kress, Y., Epshteyn, I., Rashbaum, W. K., and Lyman, W. D. (1992). Early myelination in the human fetal lumbosacral spinal cord: characterization by light and electron microscopy. *J. Neuropathol. Exp. Neurol.* **51**, 142–149.

Weidner, N., Ner, A., Salimi, N., and Tuszynski, M. H. (2001). Spontaneous corticospinal axonal plasticity and functional recovery after adult central nervous system injury. *Proc. Natl. Acad. Sci. U.S.A.* **98**, 3513–3518.

Wichterle, H., Lieberam, I., Porter, J. A., and Jessell, T. M. (2002). Directed differentiation of embryonic stem cells into motor neurons. *Cell* **110**, 385–397.

Winner, B., Cooper-Kuhn, C. M., Aigner, R., Winkler, J., and Kuhn, H. G. (2002). Long-term survival and cell death of newly generated neurons in the adult rat olfactory bulb. *Eur. J. Neurosci.* **16**, 1681–1689.

# Chapter Sixty-Six

# Protection and Repair of Audition

*Richard A. Altschuler, Yehoash Raphael, David C. Martin, Jochen Schacht, David J. Anderson, and Josef M. Miller*

## I. INTRODUCTION

Deafness is this country's number-one disability, with hearing loss affecting more than 35 million individuals. There is an increasing ability to use tissue engineering to prevent acquired deafness and to restore hearing following deafness. Hearing loss and cochlear damage from noise, ototoxic drugs, trauma, disease, and perhaps even aging can be reduced by relieving oxidative stress to the cochlea and enhancing endogenous protective systems with agents such as neurotrophic factors, heat shock proteins, and antioxidants. Genetic modification to induce regeneration of hair cells following their loss has now been shown in animal models and may soon become clinically feasible. Stem cells are also being developed as a mechanism to replace lost hair cells or auditory nerve. Survival factors can be applied to the cochlea to improve auditory nerve survival following deafness and to induce regeneration of its peripheral processes. Major advances have also been made in region-specific delivery of therapeutics to the cochlea to allow these interventions, including gene transfer and microcannulation. Auditory prostheses also provide for return of hearing following deafness, with significant improvements occurring in cochlear prostheses as well as in development of central auditory prostheses. Prostheses can also provide a mechanism for application of therapeutics through microchannels. Biopolymers on and associated with prostheses not only can improve their function and histocompatability but also can provide for other mechanisms of therapeutics.

Hearing loss affects individuals in all age groups. It increases with age, affecting one in three over the age of 65 and 50% of individuals over the age of 75. With the increased longevity of our population, this disability is also increasing. Approximately one-half of all hearing losses are thought to be of genetic etiology. Acquired deafness affects one in four to five individuals and can occur from a variety of causes, including noise, drugs, trauma, disease, and aging. It is obviously important to develop interventions to help to prevent hearing loss and treatment to restore hearing after deafness. In the latter case, major progress has been made in inducing regeneration of the sensory hair cells, whose loss results in profound deafness. In addition, recent advances in our understanding of molecular mechanisms as well as new developments in cochlear prostheses also provide promise for new and effective treatments for protection from acquired and genetic deafness and better restoration of hearing. This chapter discusses the basis and methods for achieving such interventions.

*Principles of Tissue Engineering, 3rd Edition*
*ed. by Lanza, Langer, and Vacanti*

## II. INTERVENTIONS TO PREVENT HEARING LOSS/COCHLEAR DAMAGE

### Oxidative Stress

Oxidative stress, the overproduction of reactive oxygen species (ROS; free radicals) in the cell, now appears to be recognized as a common mechanism by which many traumatic insults, including noise, ototoxic drugs, and possibly aging, cause damage to the inner ear. The formation of ROS, such as superoxide, hydroxyl radical, and nitric oxide, may occur by different mechanisms for each stress, but in each it appears to occur early and throughout the traumatizing events. Normally contained within a physiological range by endogenous antioxidant systems, ROS in excess can cause cellular damage, and the resulting redox imbalance will trigger cell death pathways of apoptosis and necrosis.

### Prevention of Aminoglycoside-Induced Hearing Loss

Aminoglycoside antibiotics, a class of drugs that includes neomycin, kanamycin, tobramycin, amikacin, and gentamicin, can exert a plethora of effects on afflicted tissues, such as the block of calcium channels and interference with enzymatic mechanisms (Forge and Schacht, 2000). ROS formation has been causally linked to the adverse effects on the inner ear, both through studies of the mechanism of these drugs *in vitro* and *in vivo* and perhaps most compellingly through studies of protection from these adverse effects by means of antioxidants.

ROS can be formed by the ability of aminoglycosides to chelate iron and abstract electrons from donors such as polyunsaturated fatty acids (Priuska and Schacht, 1995; Lesniak *et al.*, 2005). The propensity to accelerate ROS formation is shared by all aminoglycoside antibiotics (Sha and Schacht, 1999a) and is also evident in organ culture of inner ear tissues and the cochlea *in vivo* (Clerici *et al.*, 1996; Hirose *et al.*, 1997; Jiang *et al.*, 2006a). In addition, aminoglycosides can activate redox-dependent molecular signaling pathways linked to Rho-GTPases, which can, in turn, activate NADPH oxidase, an enzyme responsible for the formation of superoxide radicals (Jiang *et al.*, 2006b).

On oxidative insult, complex signaling pathways of cell death and survival are initiated, a network of interfacing signaling systems involving a variety of transcription factors and proteases and intracellular organelles such as mitochondria and lysosomes. Indicative of this complexity, cell death by aminoglycosides includes both apoptosis and necrosis of hair cells in cochlear and vestibular organs (Nakagawa *et al.*, 1998; Ylikoski *et al.*, 2002; see also Forge and Schacht, 2000), and both caspase-dependent and caspase-independent pathways appear to contribute to hair cell pathology. Studies implicating caspase-3 in hair cell death have been performed, primarily *in vitro* with explant cultures and isolated cells or in the vestibular system (e.g.,

Matsui *et al.*, 2004). The JNK apoptotic pathway is also often implicated, since pharmacological inhibitors (e.g., CEP-1347) can offer some protection *in vitro* from aminoglycosides (Pirvola *et al.*, 2000; Bodmer *et al.*, 2002). Like the caspase pathway, involvement of the JNK pathway is reported mostly in cultured explants but may contribute to the overall pattern of cell death *in vivo*. A dominance of caspase-independent cell death emerges in a chronic drug treatment model, where the onset of cochlear deficit is delayed and continues to develop after the cessation of treatment, akin to the clinical situation. In this model, activated calpain and cathepsins are the major mediators of cell death (Jiang *et al.*, 2006a).

The establishment of oxidant stress as the mechanism underlying both apoptotic and necrotic cell death in ototoxicity provides a basis for a rational approach to prevention. For example, small synthetic molecules designed to inhibit one of the many steps in the apoptotic cascade are a potential approach to stave off cell death. For a clinical application, however, a systemic application of such powerful signaling molecules may have far-ranging physiological consequences, particularly when applied to drug treatment, which may last for weeks. Local gene therapy, the process of virally introducing a gene into a tissue, may be more suitable in such a situation (Kawamoto *et al.*, 2004). However, given the complexity of cell death mechanisms, targeting a single pathway may not be sufficient.

A potentially more feasible method of prevention would be antioxidant therapy, which has become a successful clinical approach to many pathologies that involve free radicals. This type of intervention would also act directly on the ROS upstream of the ensuing cell death pathways and therefore suppress toxic mechanisms at the very onset. A wide variety of antioxidant molecules can indeed attenuate hearing loss and cell death *in vivo* while suppressing free radical formation (Song and Schacht, 1996; Song *et al.*, 1997, 1998; Conlon and Smith, 1998; Conlon *et al.*, 1999; Sha and Schacht, 2000). The attenuation achieved in animal models was quite dramatic. For example, a gentamicin-induced hearing loss of 60–80 dB could be reduced to a negligible loss of 10 dB or less (Song *et al.*, 1997). Since neither the serum levels of the drugs nor their antibacterial efficacy were compromised, antioxidant therapy appeared promising for a clinical application.

A clinically feasible prophylactic therapy requires drugs that by themselves are nontoxic and easily administered to the patients. Among the drugs surveyed in animal experimentation, compounds such as lipoic acid or D-methionine appear useful because they are available for clinical application in other contexts (for a summary of therapeutic opportunities, see Rybak and Whitworth, 2005). However, a particularly promising compound was salicylate (Sha and Schacht, 1999b). Aspirin (acetyl salicylate) is hydrolyzed, within minutes after ingestion, to salicylate and might therefore serve as a protective medication.

The efficacy of aspirin was recently tested in a randomized double-blind and placebo-controlled study in patients receiving gentamicin for acute infections (Sha *et al.*, 2006). Fourteen of 106 patients (13%) met the criteria of hearing loss in the placebo group, while only 3 out of 89 (3%) sustained a hearing loss in the aspirin group. Aspirin did not influence gentamicin serum levels or the course of therapy. These results indicate that therapeutic protection from aminoglycoside ototoxicity may be extrapolated from animal models to the clinic. Furthermore, simple antioxidants and even medications as common as aspirin may hold significant promise to attenuate the risk of aminoglycoside-induced hearing loss.

## Prevention of Noise-Induced Hearing Loss

The mechanisms of noise-induced pathology are potentially more complex than those for drug-induced ototoxicity, including not only the formation of reactive oxygen species but also vasoconstriction and direct mechanical damage. In addition, the sources of noise trauma can be highly variable, ranging from a single high-energy impact, such as an explosion, to impulse noise, to chronic noise exposure at various levels of intensity. Each of these conditions may induce variations in the mechanism of ROS formation or variations on cell death and survival pathways. Furthermore, while a protective therapy can be timed with drug administration, patients may report noise-induced hearing loss days, months, or years into the trauma. To what extent a delayed intervention may be successful is yet another question that remains to be resolved.

As with aminoglycosides, there is good evidence that ROS are formed in the inner ear following intense sounds (Ohlemiller *et al.*, 1999), specifically in the stria vascularis (Yamane *et al.*, 1995) and the organ of Corti (Ohinata *et al.*, 2000a). Indicators of ROS formation can be detected at the onset of the noise exposure (Ohinata *et al.*, 2000a) and may persist for hours (Yamane *et al.*, 1995) or even days (Yamashita *et al.*, 2004). Attesting to the importance of endogenous antioxidants, decreased glutathione in the inner ear enhances noise trauma, while dietary supplementation with glutathione attenuates these effects (Yamasoba *et al.*, 1998; Ohinata *et al.*, 2000b; Henderson *et al.*, 2006).

While the precise origin of ROS in the cochlea following noise exposure remains somewhat speculative, there are several sources most likely to contribute. A surge of superoxide radicals followed by chain reaction formation of other free radicals (via Fenton-type reactions) may be due to increased mitochondrial activity, a prolonged tissue hypoxia due to vasoconstriction (Brown *et al.*, 2003), and a rebound hyperfusion (Thorne *et al.*, 1987). Reactive nitrogen species may arise from the activation or induction of the enzyme nitric oxide synthase. The different isoforms of this enzyme serve a variety of physiological processes in normal tissue physiology, and the product of the enzymatic reaction, nitric oxide, is an important second messenger molecule. When produced in excess, not only is nitric oxide potentially damaging as a free radical, but it can also combine with superoxide to produce the highly reactive and destructive peroxynitridc. Nitric oxidc lcvcls can rise after noise exposure (Shi *et al.*, 2002; Yamashita *et al.*, 2004a), perhaps as a consequence of an enhanced release of excitatory amino acids (Puel *et al.*, 1994, 1996), and may contribute to the overall pathophysiology of noise-induced hair cell loss.

Multiple cell death pathways — partly independent and partly overlapping — may be triggered by the spectrum of radicals generated in response to noise. The fact that both necrosis and apoptosis are observed in noise trauma attests to the existence of multiple cell death pathways (Henderson *et al.*, 2006). This notion is also supported by the emergence of caspase-dependent and caspase-independent cell death pathways and the involvement of transcription factors that code for different signaling cascades, such as AP-1, endonuclease G, and the calcineurin-related activation of BCL-2 family proteins (Hu *et al.*, 2002; Yamashita *et al.*, 2004b; Matsunobu *et al.*, 2004; Vicente-Torres and Schacht, 2006).

Restoring redox balance is therefore also one of the potential means of intervention in noise-induced hearing loss, the other being a manipulation and interference with cell death pathways. The latter has been done successfully in animal experimentations *in vivo* (Pirvola *et al.*, 2000; J. Wang *et al.*, 2003), but similar concerns may arise here as in the prevention of drug-induced hearing loss, in that such systemic intervention in important physiological processes may have unwanted consequences. Other approaches successful in animals may also be impractical for the clinic. For example, the up-regulation of antioxidant defenses by systemically administered R-phenyl isopropyl adenosine has adverse physiological effects (Hu *et al.*, 1997). A local application may bypass such a problem but is more invasive and affords less control over drug concentrations.

Nevertheless, antioxidant therapy does hold promise as an interventive strategy, with an appropriate choice of drugs (see Lynch and Kil, 2005). Noise-induced hearing loss can be attenuated effectively by scavengers such as superoxide dismutase and allopurinol (Seidman *et al.*, 1993), antioxidants, and iron chelators (Yamasoba *et al.*, 1999) and by antioxidant drugs that may have clinical appeal, such as ebselen (Pourbakht and Yamasoba, 2003) or *N*-acetylcysteine (Kopke *et al.*, 2000). Multidrug combinations are promising, for example, acetyl-ʟ carnitine with ᴅ-methionine or *N*-acetylcysteine (Kopke *et al.*, 2002, 2005). Alternative strategies aimed at vasodilation and providing adequate blood flow to the cochlea during noise exposure may also indirectly relate to ROS formation. Drugs such as pentoxifylline and sarthran improved blood flow to the cochlea and also reduced the temporary effects of noise exposure (Latoni *et al.*, 1996; Goldwin *et al.*, 1998). Magnesium supplementation likewise maintains cochlear blood flow under noise exposure and reduces the resulting hearing loss (Haupt and Scheibe, 2002).

The translation of these findings from animal models to practical situations in humans has been partially successful in tests on populations exposed to high levels of noise during military exercises and leisure activities. Daily supplementation with magnesium reduced the incidence of permanent threshold shifts in military recruits from 21.5% to 11.2% (Attias *et al.*, 1994) and, likewise, reduced temporary threshold shifts in a group of young volunteers (Attias *et al.*, 2004). In contrast, a group of nightclub visitors taking *N*-acetyl cysteine sustained the same temporary threshold shift as the control group (Kramer *et al.*, 2005, cited in Lynch and Kil, 2005). Ongoing trials on U.S. Army and Navy recruits currently explore the efficacy of *N*-acetyl cysteine and of ebselen (Lynch and Kil, 2005).

While most strategies for interventions rely on applying the protective treatment prior to or concomitant with noise exposure, recent evidence indicates that the formation of free radicals continues for days following noise exposure (Yamashita *et al.*, 2004a) and that, consequently, an antioxidant treatment may still be effective for a short period postexposure (Yamashita *et al.*, 2005).

### Heat Shock Proteins

Heat shock proteins (HSPs) provide another natural protective mechanism. HSPs can achieve their protective role by influencing the stress-related denaturation of proteins (either reducing the denaturation or enhancing renaturation and regaining the correct tertiary structure), as chaperones or through an influence on cell death cascades. Several families of HSPs (commonly grouped by their molecular weights) include HSP 25/27, HSP 32, and HSP 70/72. Expression of the inducible form of HSP 70 in the cochlea has been shown with heat (Dechesne *et al.*, 1992; Yoshida *et al.*, 1999), transient ischemia (Myers *et al.*, 1992), noise (Lim *et al.*, 1993), and ototoxic drugs (Oh *et al.*, 2000). This expression is generally found in the sensory hair cells and is transient, peaking three to seven hours following the stress. HSP 27 is found constitutively (without the need for induction by stress) and has a more widespread distribution, not only in sensory hair cells, but in supporting cells and the lateral wall tissues (Leonova *et al.*, 2002). Levels increase following stress. HSP 32 has a small constitutive expression and is up-regulated following stress in hair cells and stria vascularis (Fairfield *et al.*, 2004).

A protective role of HSPs in the inner ear is suggested by two types of studies. In one type of study, HSP levels are up-regulated by an initial stress, either low-level noise (Altschuler *et al.*, 1999a) or heat (Yoshida *et al.*, 1999), followed by a noise exposure that would normally cause a significant hearing loss. Either preexposure resulted in a significant (30-dB) reduction in noise-induced hearing loss and hair cell loss. Mikuriya *et al.* (2005) showed reduced hearing loss and less hair cell loss from noise in guinea pigs in which geranylgernylacetone was used to increase induction of heat shock proteins. A second type of study examines

heat shock factor 1 (HSF1), a major transcription factor for the HSPs, which has been shown to be present in hair cells and stria vascularis in the rodent cochlea (Fairfield *et al.*, 2002). In mice in which HSF1 has been knocked out, there is decreased protection from noise, with increased hearing loss and increased loss of hair cells (Fairfield *et al.*, 2005; Sugahara *et al.*, 2003).

### Neurotrophic Factors

Neurotrophic factors play an important role in the development of the cochlea, in regulating differentiation of sensory and neuronal elements, as well as in the formation of afferent and efferent connections (for reviews, see Fritzsch *et al.*, 2003, 2006). Many neurotrophic factors and/or their receptors remain in the mature cochlea, including GDNF, NT3, and BDNF (Ylikoski *et al.*, 1993, 1998; Fritzsch *et al.*, 2006; Stöver *et al.*, 2000, 2001), where they play a role in maintenance and protection. Neurotrophic factors can have a protective action through an influence on ROM or intervention in the cascade of events induced by ROM formation to block cell death. Removal of neurotrophic factors leads to an increase in reactive oxygen species, which can induce cell death through the apoptotic pathways discussed previously. While removal of neurotrophic factors is therefore pro-apoptotic, neurotrophic factors themselves can also induce pro-survival pathways and provide protection. Neurotrophic factors may reduce oxidative stress–driven increases in intracellular $Ca^{2+}$ (Hegarty *et al.*, 1997; Mattson *et al.*, 1995; Mattson and Furakawa, 1996) through induced expression of calcium-binding proteins (Collazo *et al.*, 1992; Cheng and Mattson, 1994) or antioxidant enzymes (Mattson *et al.*, 1995; Mattson and Furakawa, 1996).

Infusion of neurotrophins into the inner ear fluids prior to stress provides protection from hearing loss (for reviews, see Altschuler *et al.*, 1999a; Kopke *et al.*, 1999; LePrell *et al.*, 2003; Miller *et al.*, 1998; Miller *et al.*, 1999). When levels of glial line–derived neurotrophic factor (GDNF) or neurotrophin 3 (NT3) are increased in the inner ear prior to noise overstimulation or the administration of ototoxic drugs there is significant protection, with decreased hearing loss and hair cell loss (Altschuler *et al.*, 1999b, Keithley *et al.*, 1998; A. M. Miller *et al.*, 1999; Park *et al.*, 1998; Shoji *et al.*, 2000, 2002; Yagi *et al.*, 1999). This protection is less effective if the neurotrophic factors are provided after the stress rather than before, suggesting that they play a greater role in protection than in repair. BDNF and FGF were less effective in preventing hair cell loss (Shoji *et al.*, 2002, Yamasoba *et al.*, 2001).

## III. HAIR CELL REGENERATION

Once hair cells of the mammalian cochlea are lost, they will not regenerate spontaneously. The only possible way to restore hearing via biological means is by replacing the missing hair cells. Such therapy is being developed via several different approaches, including the use of stem cells, enhancement of cell proliferation, and reactivation of developmental signals that regulate hair cell development. Knowledge of the developmental genes that regulate the proliferation and differentiation of hair cells during development is essential for designing hair cell regeneration strategies. Technological means for manipulating gene expression in the mature ear are also necessary. Progress in these two fields has allowed for important recent breakthroughs in attempts to induce cochlear hair cell regeneration.

One important goal for treating the deaf cochlea is to increase the number of cells and to direct the differentiation of some of the new cells toward hair cell phenotype. The p27$^{Kip1}$ gene is expressed in nonsensory cells and keeps them quiescent (Lowenheim et al., 1999; P. Chen and Segil, 1999). Mice with deficient expression of the p27$^{Kip1}$ gene exhibit organ of Corti cells with an excessive number of hair cells. It appears that in the absence of cell-cycle arrest, cells continue to proliferate after they normally would have stopped had p27$^{Kip1}$ been active. Interestingly, mice with this induced mutation have a severe hearing loss, despite the large number of hair cells (P. Chen and Segil, 1999; Kanzaki et al., 2006). Nevertheless, the ability to use genetic manipulation for inducing growth of new hair cells is a considerable breakthrough that can lead to feasible clinical applications with further refining and advancing of the technology. Inhibition of the cell cycle can also be removed by blocking expression of other genes, including Ink4d and Rb1 (P. Chen et al., 2003; Sage et al., 2005), leading to extranumerary hair cells in the tissue. The genetic manipulations that are common in the mouse model cannot be done in humans, for both technical and ethical reasons. However, further development of the technology for gene transfer and manipulation in the somatic cells of a target tissue may help advance this approach for clinically relevant hair cell regeneration therapy.

Hair cells and supporting cells develop from common progenitor cells (Fekete and Wu, 2002). Their phenotype is determined by a set of genes expressed in a sequence and influencing neighboring cells in a cross-interactive way. Some of the genes that signal for the differentiation of hair cells have been identified (Fekete and Wu, 2002; Barald and Kelley, 2004; Woods et al., 2004; Zine, 2003; Bermingham et al., 1999; Bryant et al., 2002). Knowledge of these genes helped in the design of a strategy for inducing hair cell regeneration based on forced expression of the hair cell developmental genes in nonsensory cells that remain in the mature deaf cochlea. Thus, the goal of this genetic manipulation is to induce a phenotypic conversion in supporting cells and to turn them into hair cells. Birds and other non-mammalian vertebrates can regenerate hair cells spontaneously (Ryals and Rubel, 1988; Corwin and Cotanche, 1988) using a mechanism of transdifferentiation, with or without a proliferative state (Raphael, 1992; Stone and Cotanche, 1994; Adler and Raphael, 1996; Roberson et al., 1996).

To induce expression of developmental genes in the supporting cells of the deaf organ of Corti, genes were delivered via adenovirus vectors that were inoculated into the scala media (Ishimoto et al., 2002). Atoh1 is a gene encoding a transcription factor that is necessary and sufficient for hair cell differentiation in the developing epithelial ridge (Woods et al., 2004; Bermingham et al., 1999; Zine et al., 2001; Doetzlhofer et al., 2004). Overexpression of Atoh1 (or its homolog) resulted in the appearance of cells that had the surface morphology of hair cells in the auditory epithelium (Kawamoto et al., 2003; Izumikawa et al., 2005). These cells were presumed to be new hair cells and were found in the organ of Corti proper as well as in an ectopic location adjacent to the organ of Corti. Similar results were obtained in culture work on explants of developing (Zheng and Gao, 2000) and mature organ of Corti (Shou et al., 2003). Together, these studies have shown that the nonsensory cells in the deaf ear retain the competence to respond to developmental genes and to convert to new hair cells.

Electrophysiological thresholds of the deaf Atoh1-treated animals improved, but the morphology of the new hair cells often appeared abnormal, especially in the outer hair cell region (Izumikawa et al., 2005). It is possible that combining the transdifferentiation step with forced cell proliferation in the organ of Corti or its vicinity will yield better morphological and physiological restoration of the tissue. To advance such therapies toward clinical applications, it will be necessary to test and optimize efficiency and safety and to determine the ability of this approach to accomplish regeneration in ears that have been deafened for a long time.

## IV. AUDITORY NERVE SURVIVAL FOLLOWING DEAFNESS

Profound deafness is usually associated with a loss of inner hair cells, the sensory cells responsible for transduction and activation of the auditory nerve. If there is loss of inner hair cells, a series of pathophysiological changes follow, including scar formation, a loss of the peripheral processes of the auditory nerve, and, over time, a substantial loss of the auditory nerve itself. This loss of auditory nerve is related to the loss of survival/maintenance factors, including the deafferentation-associated loss of neural activity and the loss of neurotrophic factors that had been released by hair cells and other cochlear elements lost during the scar process. Loss of these survival/maintenance factors causes these auditory neurons to enter into the cell death cycle.

Cochlear prostheses depend on direct stimulation of the auditory nerve, so auditory nerve survival is of major importance. The auditory nerve loss can be blocked or reduced either by blocking the cell death pathway or by replacing the survival/maintenance factors. Thus, activity can be replaced by direct cochlear electrical stimulation with a cochlear prosthesis, and auditory nerve survival is enhanced (Lousteau, 1987; Hartshorn *et al.*, 1991; J. M. Miller *et al.*, 1995; Mitchell *et al.*, 1997; Leake *et al.*, 1991, 1992, 1995). Stimulation may serve as a survival factor through activation of voltage-gated ion channels and/or through up-regulation of autocrine factors, including neurotrophic factors. Neurotrophic factors can be infused into the cochlear fluids, to replace those that are lost, which is even more effective than electrical stimulation in enhancing auditory nerve survival following deafness (Ernfors *et al.*, 1996; J. M. Miller *et al.*, 1997; A. M. Miller *et al.*, 1999; Staecker *et al.*, 1996; Ylikoski *et al.*, 1998). The combination of electrical stimulation and application of neurotrophic factors is more effective than when either is applied singly (Kanzaki *et al.*, 2002; Shepherd *et al.*, 2005).

There is also interest in inducing regrowth of peripheral processes of the auditory nerve. Cochlear prostheses might have reduced thresholds if their target is closer, making them more energy efficient and reducing battery requirements. If regrowth can be directed toward specific sites on the cochlear prostheses, this might also allow a greater number of channels to used, resulting in more and better frequency separation. Studies have now shown that infusion of neurotrophic factors BDNF, NT-3, and FGF can induce regrowth of the peripheral process of the auditory nerve that regress following inner hair cell loss (Wise *et al.*, 2005; Altschuler *et al.*, 1999). Thus, neurotrophic factors can serve not only to enhance survival of the auditory nerve but also to restore its peripheral processes, although different factors may be necessary for most effective treatment(s). Directed regrowth remains a challenge.

## V. GENETIC DEAFNESS

One in 1000 newborns suffers from hearing impairment. In developed countries, more than half of the afflicted babies are deaf for genetic reasons. Genetic inner ear impairments can be nonsyndromic, affecting only hearing, or syndromic, with multiple defects occurring in several body systems. Many of the genes for deafness, as well as their products, have now been identified. Some of the affected proteins of the inner ear include cytoskeletal proteins such as myosin and a tectorial membrane protein (reviewed by Probst and Camper, 1999). Identification of the genes that are mutated in families with genetic inner ear impairments is important not only for diagnostic purposes, but also for the potential for prevention and cure. For example, Probst *et al.* (1999) identified a mutation in *Myo-15* as the cause for deafness and balance impairments in the shaker 2 deafness mouse. This mouse is a model for several families in which mutations in the very same gene cause genetic inner ear impairments (A. Wang *et al.*, 1998). Probst *et al.* (1998) demonstrated that in transgenic mice destined (genetically) to become deaf, insertion of the wild-type (correct) gene sequence into the fertilized egg corrected the genetic deficit, leading to normal inner ear structure and function in the adult mouse. This demonstrated that addition of the correct copy of the gene can rescue the inner ear from genetic deficits. To make such interventions viable for clinical applications, it is necessary to develop the means for delivering genes into the inner ear cells that are affected directly by the mutations.

## VI. METHODS OF THERAPEUTIC INTERVENTION

With effective treatments for protection from acquired deafness and treatment following deafness identified, the next challenge is developing the methods for providing these interventions.

### Cochlear Prostheses

Auditory prostheses can provide direct stimulation of the auditory pathways for return of hearing following deafness. In severe and profound sensorineural hearing loss, the sensory cells of the cochlea are destroyed. Cochlear prostheses bypass the damaged receptor epithelium and electrically excite the auditory nerve directly. The development of the cochlear prosthesis has been the success story of the field of neuroprostheses. It has provided a therapy for the profoundly deaf, where none previously existed. The first FDA-approved inner ear implants, approximately 25 years ago, were single-channel devices that provided crude input and were an aid to lip reading. Now multichannel devices, combined with many advances in signal-processing strategies, routinely provide intelligible speech perception (e.g., Rauschecker and Shannon, 2002; Skinner *et al.*, 2002) with open set speech discrimination and even use of the telephone in the majority of implant recipients. With the success of cochlear implant performance and the low risk associated with its placement, the age at which implantation is considered has decreased to where very young children receive implants and show benefit from early placements, taking advantage of critical periods for auditory system development. There is also increasing bilateral placement to allow binaural hearing, important for spatial localization (Laszig *et al.*, 2004, Van Hoesel *et al.*, 2002; Verschuur *et al.*, 2005), with the benefit also greatest for early implantation during or preceding critical periods (e.g., Litovsky *et al.*, 2006). Even with later placements, there is still some induced maturation of pathways and improvement in performance (e.g., Gordon *et al.*, 2003).

### Central Auditory Prostheses

Central auditory prostheses can be advantageous when the cochlea is not suitable for implantation, when there is insufficient auditory nerve survival, or when auditory nerve must be removed in patients with VIII nerve tumors (acoustic neuroma). While a peripheral cochlear implant allows for more normal auditory pathway processing, the central implants can take advantage of the tonotopic organization of auditory nuclei for frequency separation. Central auditory prostheses can also have reduced thresholds because of the proximity to neurons as well and increased dynamic range. Cochlear nucleus implants have been successfully applied in many subjects, with excellent performance recently reported (Lenarz *et al.*, 2002, 2004; Otto *et al.*, 2002; Colletti and Shannon, 2005) in nontumor patients; when there is a tumor (the most frequent need for central auditory prostheses), the complications of the tumor and its removal can compromise the long-term function of the cochlear nucleus implant (e.g., Otto *et al.*, 2002; Schwartz *et al.*, 2003). Auditory prostheses are therefore being developed for the inferior colliculus (Lenarz *et al.*, 2005) and in animal models show improved thresholds, increased dynamic range, and better frequency discrimination than cochlear implants (Lenarz *et al.*, 2006; Lim and Anderson, 2003, 2006, 2007).

## Gene Therapy

Gene therapy is used to manipulate levels of specific proteins in cells and tissues. It can be used to inactivate specific proteins or to overexpress them. In most cases, this is accomplished by introducing a foreign gene into these cells. Viral vectors are the most efficient vehicles for gene transfer. Adenovirus, herpes simplex virus, adeno-associated virus, and lentivirus can mediate gene transfer into cells of the inner ear (Patel *et al.*, 2004; Lalwani and Mhatre, 2003; Crumling and Raphael, 2006). When mature mammalian inner ears are inoculated with adenoviral vectors, fibroblasts and other cells of connective tissue origin are transduced with the highest efficiency (Raphael *et al.*, 1996; Yagi *et al.*, 1999). At present this limits the applicability of adenovirus-based gene therapy in the inner ear to genes encoding secreted and diffusible proteins, such as the neurotrophins. Overexpression of the human GDNF transgene with an adenovirus vector has been shown to protect hair cells from noise and ototoxic drug insults in guinea pigs (Yagi *et al.*, 1999). Similar treatment was also efficient in protecting spiral ganglion neurons from degeneration following hair cell loss (Yagi *et al.*, 2000). Several viral vectors have been shown to bestow protection on hair cells and spiral ganglion neurons by inserting transgenes encoding for neurotrophic factors and antioxidant genes (Cooper *et al.*, 2006; Kawamoto *et al.*, 2004; Nakaizumi *et al.*, 2004; Bowers *et al.*, 2002; X. Chen *et al.*, 2001; Staecker *et al.*, 1998).

## Biopolymers and Scaffolding

Biopolymers can be synthesized that contain both the protein sequence for silk, to provide strength and flexibility, along with sequences of functional domains that naturally occur in the extracellular matrix protein to confer specific biological properties (Cappello, 1990; Cappello *et al.*, 1990). These can be spun in microstructured thin films on the surface of prosthetic devices. Biopolymers can be produced that contain the sequence of fibronectin or laminin to provide a biological glue (for neurons and epidermal cells, respectively), which can then be coated on implants. This can, in turn, be used to influence and stabilize the neuro-probe–tissue interface or to attach transformed cells for *ex vivo* gene transfer (see later). The biopolymer can also be used for timed release of specific factors. Thus a prosthesis could be providing electrical stimulation while releasing one or more chemical factors through microchannels (see later) and other factors through diffusion out of the biopolymer. Specific placement of biopolymer and microchannels can also provide different factors to different regions.

## Microcannulation — Miniosmotic Pumps

The fluid spaces of the cochlea provide for a closed environment well suited to receive local delivery of chemicals. Local delivery provides improved access and avoids the side effects that systemic delivery could entail. The technique of miniosmotic pumps with cannula into the inner ear fluid of scala tympani or into the middle ear ear with access through the round window has been developed in animals (e.g., Brown *et al.*, 1993; Prieskorn and Miller, 2000; Park *et al.*, 1998) and effectively utilized for local application of chemicals that provide protection from acquired deafness or enhanced survival of auditory nerve (see earlier sections). While devices for fluid delivery to the inner ear can be developed for human application (Schwab *et al.*, 2004; Hoffer *et al.*, 2001; DeCicco *et al.*, 1998), most recent efforts have focused on combining fluid delivery with cochlear prostheses (see next section).

## Combinations: Prostheses with Microchannels

Studies now show that combining electrical stimulation and chemical delivery is more effective in enhancing auditory nerve survival following deafness than either applied by itself (Kanzaki *et al.*, 2002; Shepherd *et al.*, 2005). Moreover, more patients with some residual hearing have been shown to benefit from cochlear implants and more patients with surviving hair cells are now considered candidates for cochlear implants. These subjects would benefit from protection of these remaining hair cells from the trauma of cochlear prosthesis insertion. There is also the potential of inducing regrowth of peripheral processes toward the stimulation sites, which could lower thresholds and enhance selectivity and separation. Therefore recent efforts have been made by cochlear implant manufacturers and research groups to develop cochlear prostheses capable of both electrical stimulation of the auditory nerve and delivery of pharmaceuticals into the cochlear fluids (Altschuler *et al.*, 2005). These have been successfully applied in animal studies (Shepherd *et al.*, 2005), and clinical application is beginning.

## Polymer Coatings for Cochlear Electrodes

There is considerable interest in the development of polymer coatings that could improve the long-term performance of cochlear implants. By lowering the impedance of the electrode, this would reduce energy requirements and increase battery life. Also, the ability to attract the neurons closer to the electrode surface, or to bring the electrodes closer to the spiral ganglion neurons by growing conducting polymer filaments out from the electrode itself, should improve the ability to target individual neuronal populations. Recent work shows the potential for improving the sensitivity, selectivity, and lifetime of cochlear implants.

Conducting polymers such as polypyrrole (PPy) have been used to make electrically active coatings on biomedical devices (Schmidt, Shastri, Vacanti, and Langer, 1997). It has been found that electrochemically deposited coatings of polypyrrole can significantly reduce the impedance of electrodes, particularly at the 1 kHz frequency of interest for neural recording (Cui, Hetke, Wiler, Anderson, and Martin, 2001) (George *et al.*, 2005). One issue with PPy is that it is relatively unstable in a biological environment, due presumably to coupling defects in the chemistry along the backbone. The diethoxy-functionalized thiophene poly (3,4-ethylene dioxythiophene) (PEDOT) has shown to be much more stable, and is also able to readily deposited on conducting electrodes (Cui and Martin, 2003). PEDOT films can be tailored into a wide variety of geometries including porous templates with variable pore size (Yang and Martin, 2004b; Yang and Martin, 2004a) nanotubes (Abidian and Martin, 2006), and nanofibrils (Yang, Lipkin, and Martin, 2007).

PPy and PEDOT can also be used to incorporate and delivery pharmacological agents. It has been shown that bioactive molecules can be realized from the conducting polymers using an external bias (Pernault and Reynolds, 2000) (George *et al.*, 2006). Neurotrophins such as NGF (Kim, Richardson-Burns, Hendricks, Sequera, and Martin, 2006) (Gomez and Schmidt, 2007) and NT-3 (Richardson *et al.*, 2007) have been incorporated into PPy and PEDOT, and have shown to enhance neurite regrowth. Nanotubes of conducting polymers can be formed around filaments of degradable polymers such as PLGA containing anti-inflammatory agents or neurotrophins (Abidian and Martin, 2006).

Cochlear implants can also be coated with hydrogels, in order to provide attachments sites for cells, to provide a means for drug delivery, and for the delivery of genetically-transfected cells. These gels can also be used as matrices for polymerization of conducting polymer (Kim, Abidian, and Martin, 2004), and they provide a means for drug delivery from degradable nanoparticles (Kim and Martin, 2006). One useful strategy is likely to be the use of hydrogels loaded with cells transfected to deliver BDNF (Nakaizumi, Kawamoto, Minoda, and Raphael, 2004).

Recently it has been demonstrated that PEDOT can be directly grown in living tissue (Richardson-Burns, Hendricks, and Martin, 2007). Efforts are now actively underway to polymerize conducting polymers directly in the cochlea itself, with the intention of creating a more intimate, stable contact between the electrode and the spiral ganglion cells.

This could be part of a hybrid coating that also delivers neurotrophins, anti-inflammatory agents, or transfected cells.

## Stem Cells

The use of stem cells, exogenous or endogenous, provides the potential to treat deafness by replacing lost hair cells or lost auditory nerve. Embryonic stem cells or stem cells derived from bone marrow can be induced to reach a hair cell–like phenotype (Li *et al.*, 2003; Parker and Cotanche, 2004) or an auditory nerve–like phenotype (Kondo *et al.*, 2005) *in vitro*. *In vivo* studies have shown that mouse embryonic stem cells can be placed into the cochlear fluids of scala tympani or directly into the auditory nerve, survive, and differentiate into a neuronal phenotype (Okano *et al.*, 2005; Sakamoto *et al.*, 2004; Hu *et al.*, 2005; Hernandez *et al.*, 2006; Hildebrand *et al.*, 2005). Stem cells that reach a neuronal phenotype will send processes that reinnervate hair cells (Martinez-Monedero *et al.*, 2006; Matsumoto *et al.*, 2005) or integrate with auditory nerve processes. We found that differentiation into an auditory nerve–like phenotype typical of SGN was improved by applying factors that guide auditory nerve differentiation during normal development. The cochlear fluid compartments provide a means for applying factors *in vivo*. We used mouse embryonic stem cells with Ngn1, a proneural gene associated with auditory nerve fate determination, under control of a doxyxline-inducible promoter. We then induced transient *in vivo* expression of Neurogenin 1 and followed this with intrascalar infusion of growth factors GDNF and BDNF. We found increased numbers of stem cells reaching a glutamatergic phenotype and that these differentiated stem cells gave off processes that integrated with the remaining auditory nerve (Hernandez *et al.*, 2006).

Another exciting development is the finding that there are endogenous pluripotential cells that can be gathered from human (Rask-Andersen *et al.*, 2005) or animal (Rask-Andersen *et al.*, 2005; Zhai *et al.*, 2005) cochlea or vestibule (Li *et al.*, 2003). This provides the potential that these cells can be cultured and placed into the cochlea to replace hair cells or auditory nerve, or perhaps they can be induced to differentiate *in vivo* to provide replacement.

## VII. CONCLUSIONS

Major advances in our understanding of the molecular mechanisms underlying deafness and of the factors that influence and modulate its expression and progression have occurred since the mid-1990s. We can expect, as progress in these areas continues, that this knowledge will form the basis for *molecular otology*. Novel tissue engineering–based therapeutic interventions may become a major part of the practice of otolaryngology in the 21st century.

# VIII. REFERENCES

Abidian, M., and Martin, D. C. (2006). Conducting Polymer Nanotubes for Controlled Drug Release. *Advanced Materials* **18**, 405–409.

Adler, H. J., and Raphael, Y. (1996). New hair cells arise from supporting cell conversion in the acoustically damaged chick inner ear [published erratum appears in *Neurosci. Lett.* 1996 May 24; **210**(1), 73]. *Neurosci. Lett.* **205**, 17–20.

Altschuler, R. A., Cho, Y., Ylikoski, J., Pirvola, U., Magal, E., and Miller, J. M. (1999a). Rescue and regrowth of the auditory nerve by neurotrophic factors following deafferentation. *Ann. N. Y. Acad. Sci.* **884**, 305–311.

Altschuler, R. A., Miller, J. M., Raphael, Y., and Schacht, J. (1999b). Strategies for protection of the inner ear from noise induced hearing loss. *In* "Cochlear Pharmacology and Noise Trauma" (D. K. Prasher and B. Canlon, eds.), pp. 98–112. Noise Research Network Publications, London.

Altschuler, R. A., Hochmair, I., and Miller, J. M. (2005). Cochlear implantation: a path to inner ear pharmaceutics. *In* "Meniere's Disease and Inner Ear Homeostasis Disorders" (D. J. Lim, ed.), pp. 14–17. House Ear Institute.

Attias, J., Weisz, G., Almog, S., Shahar, A., Wiener, M., Joachims, Z., Netzer, A., Ising, H., Rebentisch, E., and Guenther, T. (1994). Oral magnesium intake reduces permanent hearing loss induced by noise exposure. *Am. J. Otolaryngol.* **15**, 26–32.

Attias, J., Sapir, S., Bresloff, I., Reshef-Haran, I., and Ising, H. (2004). Reduction in noise-induced temporary threshold shift in humans following oral magnesium intake. *Clin. Otolaryngol. Allied Sci.* **29**, 635–641.

Barald, K. F., and Kelley, M. W. (2004). From placode to polarization: new tunes in inner ear development. *Development* **131**, 4119–4130.

Bermingham, N. A., Hassan, B. A., Price, S. D., Vollrath, M. A., Ben-Arie, N., Eatock, R. A., *et al.* (1999). Math1: an essential gene for the generation of inner ear hair cells. *Science* **284**, 1837–1841.

Bodmer, D., Brors, D., Bodmer, M., and Ryan, A. F. (2002). Rescue of auditory hair cells from ototoxicity by CEP-11 004, an inhibitor of the JNK signaling pathway. *Laryngorhinootologie* **81**, 853–856.

Bowers, W. J., Chen, X., Guo, H., Frisina, D. R., Federoff, H. J., and Frisina, R. D. (2002). Neurotrophin-3 transduction attenuates Cisplatin spiral ganglion neuron ototoxicity in the cochlea. *Mol. Ther.* **6**, 12–18.

Brown, J. N., Miller, J. M., Altschuler, R. A., and Nuttall, A. L. (1993). Osmotic pump implants for chronic infusion of drugs into the inner ear. *Hearing Res.* **70**, 167–172.

Brown, J. N., Miller, J. M., and Schacht, J. (2003). 8-Iso-prostaglandin F$_{2\alpha}$, a product of noise exposure, reduces inner ear blood flow. *Audiol. Neuro-Otol.* **8**, 207–221.

Bryant, J., Goodyear, R. J., and Richardson, G. P. (2002). Sensory organ development in the inner ear: molecular and cellular mechanisms. *Br. Med. Bull.* **63**, 39–57.

Cappello, J. (1990). The biological production of protein polymers and their use. *Trends Biotechnol.* **8**, 309–311.

Cappello, J., Crissman, J., Dorman, M., Mikolajczak, M., Textor, G., Marquet, M., and Ferrari, F. (1990). Genetic engineering of structural protein polymers. *Biotechnol. Progr.* **6**, 198–202.

Chen, P., and Segil, N. (1999). p27(Kip1) links cell proliferation to morphogenesis in the developing organ of Corti. *Development* **126**, 1581–1590.

Chen, P., Zindy, F., Abdala, C., Liu, F., Li, X., Roussel, M. F., *et al.* (2003). Progressive hearing loss in mice lacking the cyclin-dependent kinase inhibitor Ink4d. *Nat. Cell. Biol.* **5**, 422–426.

Chen, X., Frisina, R. D., Bowers, W. J., Frisina, D. R., and Federoff, H. J. (2001). HSV amplicon-mediated neurotrophin-3 expression protects murine spiral ganglion neurons from cisplatin-induced damage. *Mol. Ther.* **3**, 958–963.

Cheng, B., and Mattson, M. P. (1992). NT-3 and BDNF protect CNS neurons against metabolic/excitotoxic insults. *Brain Res.* **640**, 56–67.

Clerici, W. J., Hensley, K., DiMartino, D. L., and Butterfield, D. A. (1996). Direct detection of ototoxicant-induced reactive oxygen species generation in cochlear explants. *Hearing Res.* **98**, 116–124.

Collazo, D., Takahashi, H., and McKay, R. D. (1992). Cellular targets and trophic functions of neurotrophin-3 in the developing rat hippocampus. *Neuron* **9**, 643–656.

Colletti, V., and Shannon, R. V. (2005). Open set speech perception with auditory brainstem implant? *Laryngoscope* **115**(11), 1974–1978.

Conlon, B. J., and Smith, D. W. (1998). Attenuation of neomycin ototoxicity by iron chelation. *Laryngoscope* **108**, 284–287.

Conlon, B. J., Aran, J.-M., Erre, J.-P., and Smith, D. W. (1999). Attenuation of aminoglycoside-induced cochlear damage with the metabolic antioxidant a-lipoic acid. *Hearing Res.* **128**, 40–44.

Cooper, L. B., Chan, D. K., Roediger, F. C., Shaffer, B. R., Fraser, J. F., Musatov, S., *et al.* (2006). AAV-mediated delivery of the caspase inhibitor XIAP protects against cisplatin ototoxicity. *Otol. Neurotol.* **27**, 484–490.

Corwin, J. T., and Cotanche, D. A. (1988). Regeneration of sensory hair cells after acoustic trauma. *Science* **240**, 1772–1774.

Crumling, M. A., and Raphael, Y. (2006). Manipulating gene expression in the mature inner ear. *Brain Res.* **1091**, 265–269.

Cui, X., Hetke, J. F., Wiler, J. A., Anderson, D. J., and Martin, D. C. (2001). Electrochemical deposition and characterization of conducting polymer polypyrrole/PSS on multichannel neural probes. *Sensors and Actuators A: Physical* **93**(1), 8–18.

Cui, X., and Martin, D. C. (2003). Electrochemical Deposition and Characterization of Poly(3,4-ethylenedioxythiophene) on Neural Microelectrode Arrays. *Sensors and Actuators B: Chemical* **89**, 92–102.

Dechesne, C. J., Kim, H. N., Nowak, T. S., and Wenthold, R. J. (1992). Expression of heat shock protein, HSP 72, in guinea pig and rat cochlea after hyperthermia: immunocytochemical and *in situ* hybridization analysis. *Hearing Res.* **59**, 195–204.

DeCicco, M. J., Hoffer, M. E., Kopke, R. D., Wester, D., Allen, K. A., Gottshall, K., and O'Leary, M. J. (1998). Round-window microcatheter-administered microdose gentamicin: results from treatment of tinnitus associated with Meniere's disease. *Int. Tinnitus J.* **4**(2), 141–143.

Doetzlhofer, A., White, P. M., Johnson, J. E., Segil, N., and Groves, A. K. (2004). *In vitro* growth and differentiation of mammalian sensory hair cell progenitors: a requirement for EGF and periotic mesenchyme. *Dev. Biol.* **272**, 432–447.

Ernfors, P., Duan, M. L., El Shamy, W. M., and Canlon, B. (1996). Protection of auditory neurons from aminoglycoside toxicity by NT-3. *Nat. Med.* **2**, 463–467.

Fairfield, D. A., Kanicki, A. C., Lomax, M. I., and Altschuler, R. A. (2002). Expression and localization of heat shock factor (HSF) 1 in the rodent cochlea. *Hearing Res.* **173**, 109–118.

Fairfield, D. A., Kanicki, A. C., Lomax, M. I., and Altschuler, R. A. (2004). Induction of heat shock protein 32 (Hsp32) in the rat cochlea following hyperthermia. *Hearing Res.* **188**, 1–11.

Fairfield, D. A., Margaret, I., Lomax, M. I., Dootz, G. A., Chen, S., Galecki, T. A., Benjamin, I. J., David, F., Dolan, D. F., and Altschuler, R. A. (2005). Heat shock factor 1 (Hsf1) deficient mice exhibit decreased recovery of hearing following noise overstimulation. *J. Neurosci Res.* **81**, 589–596.

Fekete, D. M., and Wu, D. K. (2002). Revisiting cell fate specification in the inner ear. *Curr. Opin. Neurobiol.* **12**, 35–42.

Forge, A., and Schacht, J. (2000). Aminoglycoside, antibiotics. *Audiol. Neuro-Otol.* **5**, 3–22.

Fritzsch, B. (2003). Development of inner ear afferent connections, forming primary neurons and connecting them to the developing sensory epithelium. *Brain Res. Bull.* **60**, 423–434.

Fritzsch, B., Pauley, S., and Beisel, K. W. (2006). Cells, molecules and morphogenesis: the making of the vertebrate ear. *Brain Res.* **1091**, 151–171.

George, P. M., LaVan, D. A., Burdick, J. A., Chen, C.-Y., Liang, E., and Langer, R. (2006). Electrically Controlled Drug Delivery from Biotin-Doped Conductive Polypyrrole. *Advanced Materials* **18**, 577–581.

George, P. M., Lyckman, A. W., LaVan, D. A., Hegde, D. A., Leung, Y., Avasare, R., *et al.* (2005). Fabrication and biocompatibility of polypyrrole implants suitable for neural prosthetics. *Biomaterials* **26**(17), 3511–3519.

Goldwin, B., Khan, M. J., Shivapuja, B., Seidman, M, D., and Quirk, W. S. (1998). Sarthran preserves cochlear microcirculation and reduces temporary threshold shifts after noise exposure. *Otolaryngol. Head Neck Surg.* **118**, 576–583.

Gomez, N., and Schmidt, C. E. (2007). Nerve growth factor-immobilized polypyrrole: Bioactive electrically conducting polymer for enhanced neurite extension. *Journal of Biomedical Materials Research*, in press.

Gordon, K. A., Papsin, B. C., and Harrison, R. V. (2003). Activity-dependent developmental plasticity of the auditory brain stem in children who use cochlear implants. *Ear Hearing* **24**(6): 485–500.

Hartshorn, D. O., Miller, J. M., and Altschuler, R. A. (1991). Protective effect of electrical stimulation on the deafened guinea pig cochlea. *Otolaryngol. Head Neck Surg.* **104**, 311–319.

Haupt, H., and Scheibe, F. (2002). Preventive magnesium supplement protects the inner ear against noise-induced impairment of blood flow and oxygenation in the guinea pig. *Magnes. Res.* **15**, 17–25.

Hegarty, J. L., Kay, A. R., and Green, S. H. (1997). Trophic support of cultured spiral ganglion neurons by depolarization exceeds and is additive to that by neurotrophins or cAMP and requires elevation of Ca++ with a set range. *J. Neurosci.* **17**, 1959–1970.

Henderson, D., Bielefeld, E. C., Harris, K. C., and Hu, V. H. (2006). The role of oxidative stress in noise-induced hearing loss. *Ear Hearing* **27**, 1–19.

Hernandez, J., Wys, N., Prieskorn, D., Velkey, M., O'Shea, K. S., Miller, J. M., and Altschuler, R. A. (2006). Stem cells for auditory nerve replacement [Abstract]. *Assoc. Res. Otolaryngol.* **29**, 1113.

Hildebrand, M. S., Dahl, H. H., Hardman, J., Coleman, B., Shepherd, R. K., and de Silva, M. G. (2005). Survival of partially differentiated mouse embryonic stem cells in the scala media of the guinea pig cochlea. *J. Assoc. Res. Otolaryngol.* **6**(4), 341–354.

Hirose, K., Hockenberry, D. N., and Rubel, E. W. (1997). Reactive oxygen species in chick hair cells after gentamicin exposure *in vitro. Hearing Res.* **104**, 1–14.

Hoffer, M. E., Kopke, R. D., Weisskopf, P., Gottshall, K., Allen, K., Wester, D., and Balaban, C. (2001). Use of the round-window microcatheter in the treatment of Meniere's disease. *Laryngoscope* **111**(11 Pt. 1), 2046–2049.

Hu, B. H., Zheng, X. Y., McFadden, S. L., Kopke, R. D., and Henderson, D. (1997). R-phenylisopropyladenosine attenuates noise-induced hearing loss in the chinchilla. *Hearing Res.* **113**, 198–206.

Hu, B. H., Henderson, D., and Nicotera, T. M. (2002). Involvement of apoptosis in progression of cochlear lesion following exposure to intense noise. *Hearing Res.* **166**, 62–71.

Hu, Z., Wei, D., Johansson, C. B., Holmstrom, N., Duan, M., Frisen, J., and Ulfendahl, M. (2005). Survival and neural differentiation of adult neural stem cells transplanted into the mature inner ear. *Exp. Cell. Res.* **302**(1), 40.

Izumikawa, M., Minoda, R., Kawamoto, K., Abrashkin, K. A., Swiderski, D. L., Dolan, D. F., *et al.* (2005). Auditory hair cell replacement and hearing improvement by Atoh1 gene therapy in deaf mammals. *Nat. Med.* **11**, 271–276.

Jiang, H., Sha, S.-H., Forge, A., and Schacht, J. (2006a). Caspase-independent pathways of hair cell death induced by kanamycin *in vivo. Cell Death Diff.* **13**, 20–30.

Jiang, H., Sha, S.-H., and Schacht, J. (2006b). Rac/Rho pathway regulates actin depolymerization induced by aminoglycoside antibiotics. *J. Neurosci. Res.* **83**, 1544–1551.

Jiang, H., Talaska, A. E., Schacht, J., and Sha, S.-H. (2006c). Oxidative imbalance in the aging inner ear. *Neurobiol. Aging.* Epub.

Kanzaki, S., Stover, T., Kawamoto, K., Prieskorn, D. M., Altschuler, R. A., Miller, J. M., and Raphael, Y. (2002). Glial cell line–derived neurotrophic factor and chronic electrical stimulation prevent VIII cranial nerve degeneration following denervation. *J. Comp. Neurol.* **454**(3), 350–360.

Kanzaki, S., Beyer, L. A., Swiderski, D. L., Izumikawa, M., Stover, T., Kawamoto, K., *et al.* (2006). p27(Kip1) deficiency causes organ of Corti pathology and hearing loss. *Hearing Res.* **214**, 28–36.

Kawamoto, K., Kanzaki, S., Yagi, M., Stover, T., Prieskorn, D. M., Dolan, D. F., Miller, J. M., and Raphael, Y. (2001). Gene-based therapy for inner ear disease. *Noise Health* **3**(11), 37–47.

Kawamoto, K., Ishimoto, S., Minoda, R., Brough, D. E., and Raphael, Y. (2003). Math1 gene transfer generates new cochlear hair cells in mature guinea pigs *in vivo. J. Neurosci,* **23**, 4395–4400.

Kawamoto, K., Sha, S. H., Minoda, R., Izumikawa, M., Kuriyama, H., Schacht, J., *et al.* (2004). Antioxidant gene therapy can protect hearing and hair cells from ototoxicity. *Mol. Ther.* **9**, 173–181.

Keithley, E. M., Ma, C. L., Ryan, A. F., Louis, J. C., and Magal, E. (1998). GDNF protects the cochlea against noise damage. *Neuroreport* **9**, 2183–2187.

Kim, D.-H., and Martin, D. C. (2006). Sustained release of dexamethasone from hydrophilic matrices using PLGA nanoparticles for neural drug delivery. *Biomaterials* **27**(15), 3031–3037.

Kim, D.-H., Richardson-Burns, S. M., Hendricks, J., Sequera, C., and Martin, D. C. (2006). Effect of Immobilized Nerve Growth Factor (NGF) on Conductive Polymers: Electrical Properties and Cellular Response. *Advanced Functional Materials* **17**, 79–86.

Kim, D., Abidian, M., and Martin, D. C. (2004). Conducting Polymers Grown in Hydrogel Scaffolds Coated on Neural Prosthetic Devices. *Journal of Biomedical Materials Research* **71A**(4), 577–585.

Kondo, T., Johnson, S. A., Yoder, M. C., Romand, R., and Hashino, E. (2005). Sonic hedgehog and retinoic acid synergistically promote sensory fate specification from bone-marrow-derived pluripotent stem cell. Proc. *Natl. Acad. Sci. U.S.A.* **102**, 4789–4794.

Kopke, R., Allen, K. A., Henderson, D., Hoffer, M., Frenz, D., and Van de Water, T. (1999). A radical demise. Toxins and trauma share common pathways in hair cell death. *Ann. N.Y. Acad. Sci.* **884**, 171–191.

Kopke, R. D., Weisskopf, P. A., Boone, J. L., Jackson, R. L., Wester, D. C., Hoffer, M. E., *et al.* (2000). Reduction of noise-induced hearing loss using L-NAC and salicylate in the chinchilla. *Hearing Res.* **149**: 138–146.

Kopke, R. D., Coleman, J. K., Liu, J., Campbell, K. C., and Riffenberg, R. H. (2002). Enhancing intrinsic cochlea stress differences to reduce noise-induced hearing loss. *Laryngoscope* **112**, 1515–1532.

Kopke, R., Bielefeld, E., Liu, J., Zheng, J., Jackson, R., Henderson, D., and Coleman, J. K. (2005). Prevention of impulse noise–induced hearing loss with antioxidants. *Acta Oto-Laryngol.* **125**, 235–243.

Lalwani, A. K., and Mhatre, A. N. (2003). Cochlear gene therapy. *Ear Hearing* **24**, 342–348.

Laszig, R., Aschendorff, A., Stecker, M., Muller-Deile, J., Maune, S., Dillier, N., Weber, B., Hey, M., Begall, K., Lenarz, T., Battmer, R. D., Bohm, M., Steffens, T., Strutz, J., Linder, T., Probst, R., Allum, J., Westhofen, M., and Doering, W. (2004). Benefits of bilateral electrical stimulation with the nucleus cochlear implant in adults: six-month postoperative results. *Otol. Neurotol.* **25**(6), 958–968.

Latoni, J., Shivapuja, B., Seidman, M., D, and Quirk, W. S. (1996). Pentoxifylline maintains cochlear microcirculation and attenuates temporary threshold shifts following acoustic overstimulation. *Acta Oto-laryngologica*, **116**, 388–394.

Leake, P. A., Hradek, G. T., Rebscher, S. J., and Snyder, R. L. (1991). Chronic intracochlear electrical stimulation induces selective survival of spiral ganglion neurons in neonatally deafened cats. *Hearing Res.* **54**, 251–271.

Leake, P. A., Snyder, R. L., Hradek, G. T., and Rebscher, S. J. (1992). Chronic intracochlear stimulation in neonatally deafened cats: effects of intensity and stimulating electrode position. *Hearing Res.* **64**, 99–117.

Leake, P. A., Snyder, R. L., Hradek, G. T., and Rebscher, S. J. (1995). Consequences of chronic electrical stimulation in neonatally deafened cats. *Hearing Res.* **82**, 65–80.

Lenarz, M., Matthies, C., Lesinski-Schiedat, A., Frohne, C., Rost, U., Illg, A., Battmer, R. D., Samii, M., and Lenarz, T. (2002). Auditory brainstem implant part II: subjective assessment of functional outcome. *Otol. Neurotol.* **23**(5), 694–697.

Lenarz, M., Lim, H. H., Anderson, D. J., Patrick, J., and Lenarz, T. (2005). Electrophysiological assessment and validation of the auditory midbrain implant (AMI) [Abstracts]. *Assoc. Res. Otolaryngol.* **28**, 1299.

Lenarz, M., Lenarz, T., Reuter, G., and Patrick, J. (2004). Auditory Midbrain Implant AMI: eine Alternative zum Hirnstammimplantat, Laryngo-Rhino-Otol Vol. 83.

Lenarz, M., Lim, H. H., Patrick, J. F., Anderson, D. J., and Lenarz, T. (2006). Electrophysiological validation of a human prototype auditory midbrain implant in a guinea pig model. *J. Assoc. Res. Otolaryngol.* **7**, 383–398.

LePrell, C., Dolan, D., Schacht, J., Miller, J. M., Lomax, M., and Altschuler, R. A. (2003). Pathways for protection from noise-induced hearing loss. *Noise Health* **5**, 1–17.

Leonova, E. V., Fairfield, D. A., Lomax, M. I., and Altschuler, R. A. (2002). Constitutive expression of HSP 27 in the rat cochlea. *Hearing Res.* **163**, 61–70.

Lesniak, W., Pecoraro, V.-L., and Schacht, J. (2005). Ternary complexes of gentamicin with iron and lipid catalyze formation of reactive oxygen species. *J. Chem. Res. Toxicol.* **18**, 357–364.

Li, H., Roblin, G., Liu, H., and Heller, S. (2003a). Generation of hair cells by stepwise differentiation of embryonic stem cells. *Proc. Natl. Acad. Sci. U.S.A.* **100**, 13495–13500.

Li, H., Liu, H., and Heller, S. (2003b). Pluripotent cells from the adult mouse inner ear. *Nat. Med.* **10**, 1293–1299.

Lim, H. H. (2005). Effects of electrical stimulation of the inferior colliculus on auditory cortical activity: Implications for an auditory midbrain implant (AMI). Ph. D., University of Michigan, Ann Arbor, MI.

Lim, H. H., and Anderson, D. J. (2003). Feasibility experiments for the development of a midbrain auditory prosthesis, Neural Engineering, 2003. Conference Proceedings. IEEE, Capri, IT. pp. 193–196.

Lim, H. H., and Anderson, D. J. (2006). Auditory cortical responses to electrical stimulation of the inferior colliculus: implications for an auditory midbrain implant. *J. Neurophysiol.* **96**, 975–988.

Lim, H. H., and Anderson, D. J. (2007). Antidromic activation reveals tonotopically organized projections from primary auditory cortex to the central nucleus of the inferior colliculus in guinea pig. *J. Neurophysiol.* **97**, 1413–1427.

Lim, H. H., Jenkins, O. H., Myers, M. W., Miller, J. M., and Altschuler, R. A. (1993). Detection of HSP72 synthesis after acoustic overstimulation in rat cochlea. *Hearing Res.* **69**, 146–150.

Litovsky, R. Y., Johnstone, P. M., Godar, S., Agrawal, S., Parkinson, A., Peters, R., and Lake, J. (2006). Bilateral cochlear implants in children: localization acuity measured with minimum audible angle. *Ear Hearing* **27**(1), 43–59.

Lousteau, R. J. (1987). Increased spiral ganglion cell survival in electrically stimulated, deafened guinea pig cochleae. *Laryngoscope* **97**, 837–842.

Lowenheim, H., Furness, D. N., Kil, J., Zinn, C., Gultig, K., Fero, M. L., *et al.* (1999). Gene disruption of p27(Kip1) allows cell proliferation in the postnatal and adult organ of corti. *Proc. Natl. Acad. Sci. U.S.A.* **96**, 4084–4088.

Lynch, E. D., and Kil, J. (2005). Compounds for the prevention and treatment of noise-induced hearing loss. *Drug Disc. Today* **10**, 1291–1298.

Martinez-Monedero, R., Corrales, C. E., Cuajungco, M. P., Heller, S., and Edge, A. S. (2006). Reinnervation of hair cells by auditory neurons after selective removal of spiral ganglion neurons. *J. Neurobiol.* **66**(4), 319–313.

Matsui, J. I., Gale, J. E., and Warhol, M. E. (2004). Critical signaling events during the aminoglycoside-induced death of sensory hair cells *in vitro*. *J. Neurobiol,* **61**, 250–266.

Matsumoto, M., Nakagawa, T., Higashi, T., Kim, T. S., Kojima, K., Kita, T., Sakamoto, T., and Ito, J. (2005). Innervation of stem cell–derived neurons into auditory epithelia of mice. *Neuroreport* **16**(8), 787–789.

Matsunobu, T., Ogita, K., and Schacht, J. (2004). Modulation of AP-1/DNA binding activity by acoustic overstimulation in the guinea pig cochlea. *Neuroscience* **123**, 1037–1043.

Mattson, M. P., and Furakawa, K. (1996). Programmed cell life: antiapoptotic singaling and therapeutic strategies for neurodegenerative disorders. *Restor. Neurol. Neurosci.* **9**, 191–205.

Mattson, M. P., Lovell, M. A., Furukawa, K., and Markesebery, W. R. (1995). Neurotrophic factors attenuate glutamate-induced accumulation of peroxides, elevation of intracellular $Ca^{2+}$ concentration, and neurotoxicity and increase antioxidant enzyme activities in hippocampal neurons. *J. Neurochem.* **65**, 1740–1751.

Mikuriya, T., Sugahara, K., Takemoto, T., Tanaka, K., Takeno, K., Shimogori, H., Nakai, A., and Yamashita, H. (2005). Geranylgeranylacetone, a heat shock protein inducer, prevents acoustic injury in the guinea pig. *Brain Res.* **1065**(1–2), 107–114.

Miller, A. M., Yamasoba, T., and Altschuler, R. A. (1999). Hair cell and spiral ganglion neuron preservation and regeneration — influence of growth factors. *Curr. Opin. Otolaryngol. Head Neck Surg.* **6**, 301–307.

Miller, J. M., and Altschuler, R. A. (1995). Effectiveness of different electrical stimulation conditions in preservation of spiral ganglion cells following deafness. *Ann. Otol. Rhinol. Laryngol.* **166**, 57–60.

Miller, J. M., Chi, D. H., Kruszka, P., Raphael, Y., and Altschuler, R. A. (1997). Neurotrophins can enhance spiral ganglion cell survival after inner hair cell loss. *Int. J. Develop. Neurosci.* **15**, 631–643.

Miller, J. M., Yamasoba, T., Shoji, F., and Altschuler, R. A. (1998). Neurotrophin and antioxidant protection of the inner ear from noise induced damage. *In* "Recent Advances in Inner Ear Research" (N. Suga and A. Furukawa, eds.), pp. 113–128. Kanehara Co., Ltd., Tokyo.

Mitchell, A., Miller, J. M., Finger, P., Heller, J., and Altschuler, R. A. (1997). Effects of chronic high-rate electrical stimulation on the cochlea and eighth nerve in the deafened guinea pig. *Hearing Res.* **105**, 30–43.

Myers, M. W., Quirk, W. S., Rizk, S. S., Miller, J. M., and Altschuler, R. A. (1992). Expression of the major mammalian stress protein in the cochlea following transient ischemia. *Laryngoscope* **102**, 981–987.

Nakagawa, T., Yamane, H., Takayama, M., Sunami, K., and Nakai, Y. (1998). Apoptosis of guinea pig cochlear hair cells following aminoglycoside treatment. *Eur. Arch. Otorhinolaryngol.* **255**, 127–131.

Nakaizumi, T., Kawamoto, K., Minoda, R., and Raphael, Y. (2004). Adenovirus-mediated expression of brain-derived neurotrophic factor protects spiral ganglion neurons from ototoxic damage. *Audiol. Neurootol.* **9**, 135–143.

Oh, S. H., Yu, W. S., Song, B. H., Lim, D., Koo, J. W., Chang, S. O., and Kim, C. S. (2000). Expression of heat shock protein 72 in rat cochlea with cisplatin-induced acute ototoxicity. *Acta. Otolaryngol.* **120**(2), 146–150.

Ohinata, Y., Miller, J. M., Altschuler, R. A., and Schacht, J. (2000a). Intense noise induces formation of vasoactive lipid peroxidation products in the cochlea. *Brain Res.* **878**: 163–173.

Ohinata, Y., Yamasoba, T., Schacht, J., and Miller, J. M. (2000b). Glutathione limits noise-induced hearing loss. *Hearing Res.* **146**: 28–34.

Ohlemiller, K. K., Wright, J. S., and Dugan, L. L. (1999). Early elevation of cochlear reactive oxygen species following noise exposure. *Audiol. Neurootol.* **4**(5), 229–236.

Okano, T., Nakagawa, T., Endo, T., Kim, T. S., Kita, T., Tamura, T., Matsumoto, M., Ohno, T., Sakamoto, T., Iguchi, F., and Ito, J. (2005). Engraftment of embryonic stem cell–derived neurons into the cochlear modiolus. *Neuroreport* **16**(17), 191.

Otto, S. R., Brackmann, D. E., Hitselberger, W. E., Shannon, R. V., and Kuchta, J. (2002). Multichannel auditory brainstem implant: update on performance in 61 patients. *J. Neurosurg.* **96**(6), 1063–1071.

Parker, M. A., and Cotanche, D. A. (2004). The potential use of stem cells for cochlear repair. *Audiol. Neurootol.* **9**(2), 72–80.

Patel, N. P., Mhatre, A. N., and Lalwani, A. K. (2004). Biological therapy for the inner ear. *Expert Opin. Biol. Ther.* **4**, 1811–1819.

Pernault, J.-M., and Reynolds, J. R. (2000). Use of Conducting Electroactive Polymers for Drug Delivery and Sensing of Bioactive Molecules. A Redox Chemistry Approach. *Journal of Physical Chemistry* **B104**, 4080–4090.

Pirvola, U., Xing-Qun, L., Virkkala, J., Saarma, M., Murakata, C., and Camoratto, A. M. (2000). Rescue of hearing, auditory hair cells, and neurons by CEP-1347/KT7515, an inhibitor of c-Jun N-terminal kinase activation. *J. Neurosci.* **20**: 43–50.

Pourbakht, A., and Yamasoba, T. (2003). Ebselen attenuates cochlear damage caused by acoustic trauma. *Hearing Res.* **181**, 100–108.

Prieskorn, D. M., and Miller, J. M. (2000). Technical report: chronic and acute intracochlear infusion in rodents. *Hearing Res.* **140**(1–2), 212–215.

Priuska, E., and Schacht, J. (1995). Formation of free radicals by gentamicin and iron and evidence for an iron/gentamicin complex. *Biochem. Pharmacol.* **50**, 1749–1752.

Probst, F. J., and Camper, S. A. (1999). The role of mouse mutants in the identification of human hereditary hearing loss genes. *Hearing Res.* **130**, 1–6.

Probst, F. J., Fridell, R. A., Raphael, Y., Saunders, T. L., Wang, A., Liang, Y., Morell, R, J., Touchman, J. W., Lyons, R. H., Noben-Trauth, K., Friedman, T. B., and Camper, S. A. (1998). Correction of deafness in shaker-2 mice by an unconventional myosin in a BAC transgene. *Science* **280**, 1444–1447.

Probst, F. J., Chen, K. S., Zhao, Q., Wang, A., Friedman, T. B., Lupski, J. R., and Camper, S. A. (1999). A physical map of the mouse shaker-2 region contains many of the genes commonly deleted in Smith–Magenis syndrome (del17p11.2p11.2). *Genomics* **55**, 348–352.

Puel, J. L., Pujol, R., Tribillac, F., Ladrech, S., and Eybalin, M. (1994). Excitatory amino acid antagonists protect cochlear auditory neurons from excitotoxicity. *J. Comp. Neurol.* **341**, 241–256.

Puel, J. L., d'Aldin, G., Saffiedine, S., Eybalin, M., and Pujol, R. (1996). Excitotoxicity and plasticity of IHC-auditory nerve contribrutes to both temporary and permanent threshold shift. *In* "Scientific Basis of Noise-Induced Hearing Loss" (A. Axelsson, H. Borchgrevink, R. P. Hamernik, P. A. Hellstrom, D. Henderson, and R. J. Salvi, eds.), pp. 36–42. Thieme Medical Publishers, New York.

Raphael, Y. (1992). Evidence for supporting cell mitosis in response to acoustic trauma in the avian inner ear. *J. Neurocytol.* **21**, 663–671.

Raphael, Y., and Yagi, M. (1998). Gene transfer and the inner ear. *Curr. Opin. Otolaryngol. Head Neck Surg.* **6**, 311–315.

Raphael, Y., Frisancho, J. C., and Roessier, B. J. (1996). Adenoviral-mediated gene transfer into guinea pig cochlear cells *in vivo. Neurosci. Lett.* **207**, 137–141.

Rask-Andersen, H., Bostrom, M., Gerdin, B., Kinnefors, A., Nyberg, G., Engstrand, T., Miller, J. M., and Lindholm, D. (2005). Regeneration of human auditory nerve, *in vitro/in vivo* demonstration of neural progenitor cells in adult human and guinea pig spiral ganglion. *Hearing Res.* **203**, 180–191.

Rauscheker, J. P., and Shannon, R. V. (2002). Sending sound to the brain. *Science* **295**, 1025–1029.

Richardson-Burns, S. M., Hendricks, J. L., and Martin, D. C. (2007). Electrochemical polymerization of conducting polymers in living neural tissue. *Journal of Neural Engineering* **4**, L6–L13.

Richardson, R. T., Thompson, B., Moulton, S., Newbold, C., Lum, M. G., Cameron, A., *et al.* (2007). The effect of polypyrrole with incorporated neurotrophin-3 on the promotion of neurite outgrowth from auditory neurons. *Biomaterials* **28**, 513–523.

Roberson, D. W., Kreig, C. S., and Rubel, E. W. (1996). Light-microscopic evidence that direct transdifferentiation gives rise to new hair cells in regenerating avian auditory epithelium. *Aud. Neurosci.* **2**, 195–205.

Ryals, B. M., and Rubel, E. W. (1988). Hair cell regeneration after acoustic trauma in adult *Coturnix* quail. *Science* **240**, 1774–1776,

Rybak, L. P., and Whitworth, T. A. (2005). Ototoxicity: therapeutic opportunities. *Drug Discovery Today* **10**(19), 1313–1321.

Sage, C., Huang, M., Karimi, K., Guticrrcz, G., Vollrath, M. A., Zhang, D. S., *et al.* (2005). Proliferation of functional hair cells *in vivo* in the absence of the retinoblastoma protein. *Science* **307**, 1114–1118.

Sakamoto, T., Nakagawa, T., Endo, T., Kim, T. S., Iguchi, F., Naito, Y., Sasai, Y., and Ito, J. (2004). Fates of mouse embryonic stem cells transplanted into the inner ears of adult mice and embryonic chickens. *Acta Otolaryngol, Suppl.* **551**, 48–52.

Schmidt, C. E., Shastri, V. R., Vacanti, J. P., and Langer, R. (1997). Stimulation of Neurite Outgrowth Using an Electrically Conducting Polymer. *Proc. Natl. Acad. Sci. USA*, **94**, 8948–8953.

Schwab, B., Lenarz, T., and Heermann, R. (2004). Use of the round-window micro cath for inner ear therapy — results of a placebo-controlled, prospective study on chronic tinnitus. *Laryngorhinootoly* **83**(3), 164–172.

Schwartz, M. S., Otto, S. R., Brackmann, D. E., Hitselberger, W. E., and Shannon, R. V. (2003). Use of a multichannel auditory brainstem implant for neurofibromatosis type 2. *Stereotact. Funct. Neurosurg.* **81**(1–4), 110–114.

Seidman, M. D., Shivapuja, B. G., and Quirk, W. S. (1993). The protective effects of allopurinol and superoxide dismutase on noise-induced cochlear damage. *Otolaryngol. Head Neck Surg.* **109**, 1052–1056.

Sha, S.-H., and Schacht, J. (1999a). Formation of free radicals by aminoglycoside antibiotics. *Hearing Res.* **128**, 112–118.

Sha, S.-H., and Schacht, J. (1999b). Salicylate attenuates gentamicin-induced ototoxicity. *Lab. Invest.* **79**, 807–813.

Sha, S.-H., and Schacht, J. (1999c). Formation of reactive oxygen species following bioactivation of gentamicin. *Free Radical Biol. Med.* **26**, 341–347.

Sha, S. H., and Schacht, J. (2000). Antioxidants attenuate gentamicin-induced free-radical formation *in vitro* and ototoxicity *in vivo*: D-methionine is a potential protectant. *Hearing Res.* **142**, 34–40.

Sha, S.-H., Qiu, J.-H., and Schacht, J. (2006). Aspirin to prevent gentamicin-induced hearing loss. *N. Engl. J. Med.* **354**, 1856–1857.

Shannon, R. V., Fayad, J., Moore, J., Lo, W. W., Otto, S., Nelson, R. A., and O'Leary, M. (1993). Auditory brainstem implant: II. Postsurgical issues and performance. *Otolaryngol. Head Neck Surg.* **108**, 634–642.

Shepherd, R. K., Coco, A., Epp, S. B., and Crook, J. M. (2005). Chronic depolarization enhances the trophic effects of brain-derived neurotrophic factor in rescuing auditory neurons following a sensorineural hearing loss. *J. Comp. Neurol.* **486**(2), 145–158.

Shi, X., Ren, T., and Nuttall, A. L. (2002). The electrochemical and fluorescence detection of nitric oxide in the cochlea and its increase following loud sound. *Hearing Res.* **164**, 49–58.

Shoji, F., Yamasoba, T., Magal, E., Dolan, D. F., Altschuler, R. A., and Miller, J. M. (2000). Glial cell line–derived neurotrophic factor has a dose-dependent influence on noise-induced hearing loss in the guinea pig cochlea. *Hearing Res.* **142**(1–2), 41–55.

Shoji, F., Miller, A. L., Mitchell, A., Yamasoba, T., Altschuler, R. A., and Miller, J. M. (2002). Differential protective effects of neurotrophins in

the attenuation of noise-induced hair cell loss. *Hearing Res.* 2000 Aug; **146**(1–2), 134–134.

Shou, J., Zheng, J. L., and Gao, W. Q. (2003). Robust generation of new hair cells in the mature mammalian inner ear by adenoviral expression of Hath1. *Mol. Cell. Neurosci.* **23**, 169–179.

Skinner, M. W., Arndt, P. L., and Stallker, S. J. (2002). Nucleu 24 advanced encoder conversion study: performance versus preference. *Ear Hearing* **23**, 2S–17S.

Song, B.-B., and Schacht, J. (1996). Variable efficacy of radical scavengers and iron chelators to attenuate gentamicin ototoxicity in guinea pig *in vivo*. *Hearing Res.* **94**, 87–93.

Song, B.-B., Anderson, D. J., and Schacht, J. (1997). Protection from gentamicin ototoxicity by iron chelators in guinea pig *in vivo*. *J. Pharmacol. Exp. Ther.* **282**, 369–377.

Song, B.-B., Sha, S.-H., and Schacht, J. (1998). Iron chelators protect from aminoglycoside-induced cochleo- and vestibulotoxicity in guinea pig. *Free Radical Biol. Med.* **25**, 189–195.

Staecker, H., Kopke, R., Malgrange, B., Lefebvre, P., and Van De Water, T. R. (1996). NT-3 and BDNF therapy prevents loss of auditory neurons following loss of hair cells. *NeuroReport* **7**, 889–894.

Staecker, H., Gabaizadeh, R., Federoff, H., and Van De Water, T. R. (1998). Brain-derived neurotrophic factor gene therapy prevents spiral ganglion degeneration after hair cell loss. *Otolaryngol. Head Neck Surg.* **119**, 7–13.

Stone, J. S., and Cotanche, D. A. (1994). Identification of the timing of S phase and the patterns of cell proliferation during hair cell regeneration in the chick cochlea. *J. Comp. Neurol.* **341**, 50–67.

Stover, T., Gong, T. L., Cho, Y., Altschuler, R. A., and Lomax, M. I. (2000). Expression of the GDNF family members and their receptors in the mature rat cochlea. *Brain Res. Mol. Brain Res.* **76**(1), 25–35.

Stover, T., Nam, Y., Gong, T. L., Lomax, M. I., and Altschuler, R. A. (2001). Glial cell line-derived neurotrophic factor (GDNF) and its receptor complex are expressed in the auditory nerve of the mature rat cochlea. *Hearing Res.* **155**(1–2), 143–151.

Sugahara, K., Inouye, S., Izu, H., Katoh, Y., Takemoto, T., Shimogori, H., Yamashita, H., and Nakai, A. (2003). Heat shock transcription factor HSF1 is required for survival of sensory hair cells against acoustic over-exposure. *Hearing Res.* **182**(1–2), 88–96.

Thorne, P. R., Nuttall, A. L., Scheibe, F., and Miller, J. M. (1987). Sound-induced artifact in cochlear blood flow measurements using the laser Doppler flowmeter. *Hearing Res.* **31**, 229–234.

Van Hoesel, R., Ramsden, R., and Odriscoll, M. (2002). Sound-direction identification, interaural time delay discrimination, and speech intelligibility advantages in noise for a bilateral cochlear implant user. *Ear Hearing* **23**(2), 137–149.

Verschuur, C. A., Lutman, M. E., Ramsden, R., Greenham, P., and O'Driscoll, M. (2005). Auditory localization abilities in bilateral cochlear implant recipients. *Otol. Neurotol.* **26**(5), 965–971.

Vicente-Torres, M. A., and Schacht, J. (2006). A BAD link to mitochondrial cell death in noise-induced hearing loss. *J. Neurosci. Res.* **83**, 1564–1572.

Wang, A., Liang, Y., Fridell, R. A., Probst, F. J., Wilcox, E. R., Touchman, J. W., Morton, C. C., Morell, R. J., Noben-Trauth, K., Camper, S. A., and Friedman, T. B. (1998). Association of unconventional myosin MYO15 mutations with human nonsyndromic deafness DFNB3. *Science* **280**, 1447–1451.

Wang, J., Van de Water, T. R., Bonny, C., de Ribaupierre, F., Puel, J. L., and Zine, A. (2003). A peptide inhibitor of c-Jun N-terminal kinase protects against both aminoglycoside and acoustic trauma–induced auditory hair cell death and hearing loss. *J. Neurosci.* **23**, 8596–8607.

Wise, A. K., Richardson, R., Hardman, J., Clark, G., and O'leary, S. (2005). Resprouting and survival of guinea pig cochlear neurons in response to the administration of the neurotrophins brain-derived neurotrophic factor and neurotrophin-3. *J. Comp. Neurol.* **487**(2), 147–165.

Woods, C., Montcouquiol, M., and Kelley, M. W. (2004). Math1 regulates development of the sensory epithelium in the mammalian cochlea. *Nat. Neurosci.* **7**, 1310–1318.

Yagi, M., Magal, E., Sheng, Z., Ang, K. A., and Raphael, Y. (1999). Hair cells are protected from aminoglycoside ototoxicity by adenoviral-mediated overexpression of GDNF. *Hum. Gene Ther.* **10**, 813–823.

Yagi, M., Kanzaki, S., Kawamoto, K., Shin, B., Shah, P. P., Magal, E., *et al.* (2000). Spiral ganglion neurons are protected from degeneration by GDNF gene therapy. *J. Assoc. Res. Otolaryngol.* **1**, 315–325.

Yamane, H., Nakai, Y., Takayama, M., Konishi, K., Iguchi, H., Nakagawa, T., Shibata, S., Kato, A., Sunami, K., and Kawakatsu, C. (1995). The emergence of free radicals after acoustic trauma and strial blood flow. *Acta Otolaryngol.* (Suppl.) **519**, 87–92.

Yamashita, D., Jiang, H.-Y., Schacht, J., and Miller, J. M. (2004a). Delayed production of free radicals following noise exposure. *Brain Res.* **1019**, 201–209.

Yamashita, D., Miller, J. M., Jiang, H., Minami, S. B., and Schacht, J. (2001b). AIF and EndoG in noise-induced hearing loss. *Neuroreport* **15**, 2719–2722.

Yamashita, D., Jiang, H. -Y., Schacht, J., and Miller, J. M. (2005). Postexposure treatment attenuates noise-induced hearing loss. *Neuroscience* **134**, 633–642.

Yamasoba, T., Nuttall, A. L., Harris, C., Raphael, Y., and Miller, J. M. (1998). Role of glutathione in protection against noise-induced hearing loss. *Brain Res.* **784**, 82–90.

Yamasoba, T., Schacht, J., Shoji, F., and Miller, J. M. (1999). Attenuation of cochlear damage from noise trauma by an iron chelator, a free radical scavenger and glial cell line–derived neurotrophic factor *in vivo*. *Brain Res.* **815**, 317–325.

Yamasoba, T., Altschuler, R. A., Raphael, Y., Miller, A. M., Shoji, F., and Miller, J. M. (2001). Absence of hair cell protection by exogenous FGF-1 and FGF-2 delivered to the guinea pig cochlea *in vivo*. *Noise Health* **3**, 65–78.

Yang, J., Lipkin, K., and Martin, D. C. (2007). Electrochemical fabrication of conducting polymer poly(3,4-ethylenedioxythiophene) (PEDOT) nanofibrils on microfabricated neural prosthetic devices. *Journal of Biomaterials Science — Polymer Edition*, in press.

Yang, J., and Martin, D. C. (2004a). Microporous Conducting Polymers on Neural Prosthetic Devices. I. Electrochemical Deposition. *Sensors and Actuators B: Chemical* **101**(1–2), 133–142.

Yang, J., and Martin, D. C. (2004b). Microporous Conducting Polymers on Neural Prosthetic Devices. II. Physical Characterization. *Sensors and Actuators A: Physical* **113**(2), 204–211.

Ylikoski, J., Pirvola, U., Moshnyakov, M., Palgi, J., Arumae, U., and Saarma, M. (1993). Expression patterns of neurotrophins and their receptor mRNAs in the rat inner ear. *Hearing Res.* **65**, 69–78.

Ylikoski, J., Pirvola, U., Suvanto, P., Liang, X.-Q., Virkkala, J., Magal, E., Altschuler, R. A., Miller, J. M., and Saarma, M. (1998). Guinea pig auditory neurons are protected by GDNF from degeneration after noise trauma. *Hearing Res.* **124**, 17–26.

Ylikoski, J., Xing-Qun, L., Virkkala, J., and Pirvola, U. (2002). Blockade of c-Jun N-terminal kinase pathway attenuates gentamicin-induced cochlear and vestibular hair cell death. *Hearing Res.* **163**, 71–61.

Yoshida, N., Kristiansen, A., and Liberman, M. C. (1999). Heat stress and protection from permanent acoustic injury in mice. *J. Neurosci.* **19**(22), 10116–10124.

Zhai, S., Shi, L., Wang, B. E., Song, W., Hu, Y., and Gao, W. G. (2005). Isolation and culture of human hair cell progenitors from postnatal rat cochleae. *J. Neurobiol.* **3**, 282–293.

Zheng, J. L., and Gao, W. Q. (2000). Overexpression of Math1 induces robust production of extra hair cells in postnatal rat inner ears. *Nat. Neurosci.* **3**, 580–586.

Zine, A. (2003). Molecular mechanisms that regulate auditory hair-cell differentiation in the mammalian cochlea. *Mol. Neurobiol.* **27**, 223–238.

Zine, A., Aubert, A., Qiu, J., Therianos, S., Guillemot, F., Kageyama, R., *et al.* (2001). Hes1 and Hes5 activities are required for the normal development of the hair cells in the mammalian inner ear. *J. Neurosci.* **21**, 4712–4720.

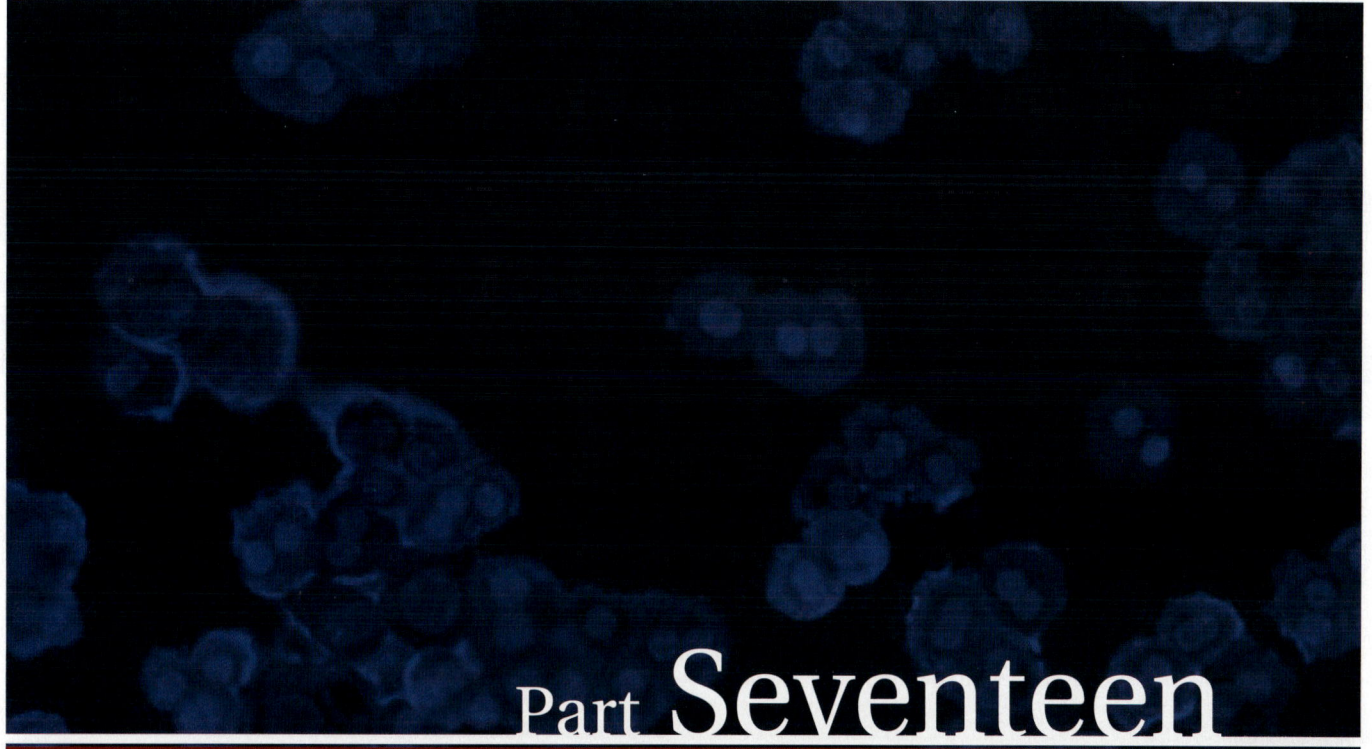

# Part Seventeen

# Ophthalmic Applications

# Chapter Sixty-Seven

# Stem Cells in the Eye

*Michael E. Boulton, Julie Albon, and Maria B. Grant*

## I. INTRODUCTION

The eye is a complex organ consisting of epithelial, mesenchymal, connective, and neural tissue. Vision is dependent on carefully regulated structural and functional integration of these tissues. Since the early 1980s it has become increasingly apparent that there are a number of stem cell niches in ocular tissues that are important in maintenance and repair. This chapter considers our current knowledge of ocular stem cells and reflects on their therapeutic potential in repairing damage to ocular tissue.

## II. CORNEAL EPITHELIAL STEM CELLS

Limbal stem cells (LSCs) are crucial to corneal epithelial tissue repair and regeneration throughout the life of the adult cornea (see Dua and Azuaro-Blanco, 2000a; Kinoshita *et al.*, 2001; Boulton and Albon, 2004). The ability of LSCs to self-renew and generate daughter cells that undergo progressive differentiation until they are shed from the surface as terminally differentiated cells supports the concept of "stemness" in the cornea.

The corneal epithelium is the outermost layer of the cornea, whose functions include transparency, refraction, and protection. It consists of a stratified, multilayered (five to seven layers) epithelium that is continually renewed from a population of relatively undifferentiated cells that reside in the basal cell layer of the corneal limbus. The *limbus* is defined as a narrow ring of tissue situated between the cornea and the conjunctiva, terminating anteriorly at Bowman's membrane. The limbus, as is appropriate for all stem cell niches, offers both physical protection and nourishment to the stem cells that reside there. The deep undulations of the Palisades Vogt of ensure that the limbal stem cells are hidden and protected from any hostile external environment, while the nearby limbal blood vessels provide nourishment. In addition, melanin in LSCs is thought to protect against ultraviolet light and reactive oxygen species. LSCs are slow-cycling cells that self-renew and produce transient amplifying cells (TACs). The TACs differentiate into postmitotic cells as they move centripetally toward the central cornea and upward from the basal corneal epithelial cell layer, before being shed from the epithelial surface as terminally differentiated, flattened superficial cells (Fig. 67.1) (Beebe and Masters, 1996; Thoft, 1984).

Interestingly, a recent study by Dua *et al.* (2005) suggests that the limbal stem cell niche may extend beyond that originally thought. They termed this novel anatomical structure, which formed cords of ABCG2-positive epithelial cell extending from the limbal palisades, the *limbal epithelial crypt.* Further studies are required to determine the "stemness" of cells derived from this structure.

### The Limbal Stem Cell Niche

Evidence in support of the limbus as a niche for stem cells that renew the corneal epithelium has been described

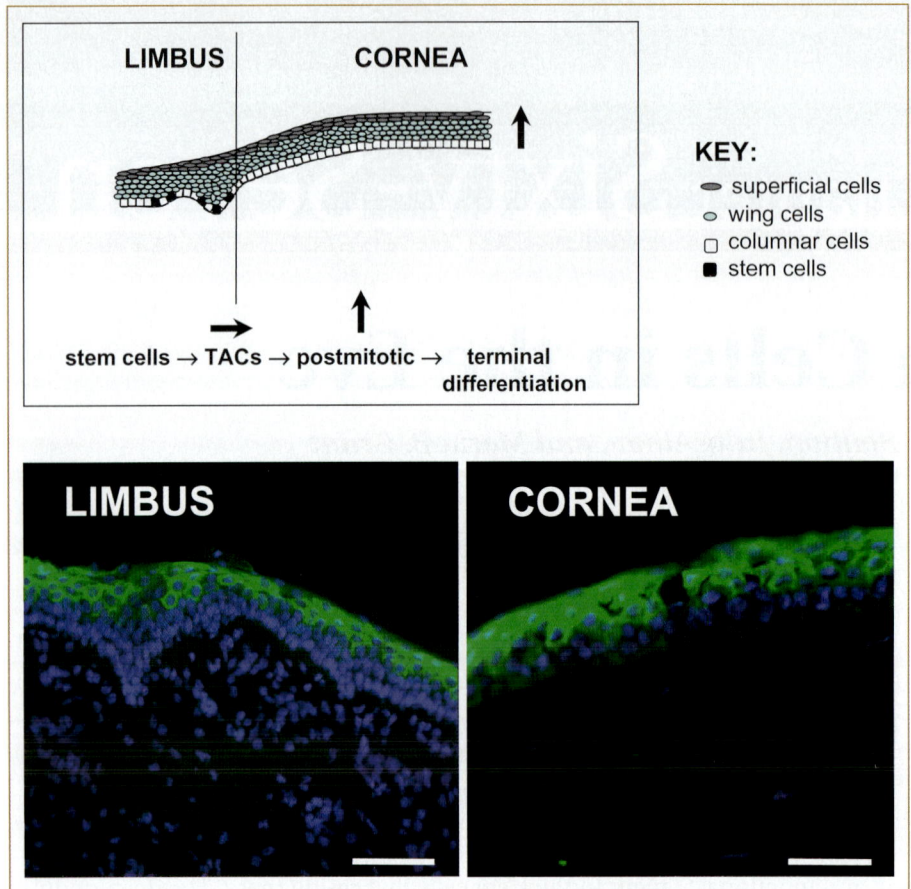

**FIG. 67.1.** Diagram of the architecture of the corneal epithelium and limbus (upper panel). Cytokeratin localizes to all cell layers of the human corneal epithelium, but only in the suprabasal layers in the limbal epithelium (lower panels). Bar = 50 μM.

in several studies. The migration of limbal pigmented cells toward the central cornea in a wound-healing response provided evidence for the existence of LSCs in the Palisades of Vogt (Davenger and Evenson, 1971). The fact that cells in the limbal niche are slow-cycling (Cotsarelis *et al.*, 1989), relatively undifferentiated cells (Schermer *et al.*, 1986; Chaloin-Dufau *et al.*, 1990) and, when cultured *in vitro*, have a higher proliferative potential than those of cultured central or peripheral corneal epithelial cells (Ebato *et al.*, 1998; Lindberg *et al.*, 1993; Pellegrini *et al.*, 1999) provided further evidence. The distribution of LSCs does not appear to be uniform; the population of LSCs differs according to region, with the greatest being in the superior and inferior cornea (Lauweryns *et al.*, 1993; Wiley, 1991). A single, unipotent LSC self-renews and produces a transient amplifying cell (TAC). The TACs proliferate to produce more TACs, dependent on requirement (e.g., proliferation will be increased during corneal regeneration following wounding) (Fig. 67.1). Each stem cell is capable of producing TACs throughout one's lifetime, but it is likely that a proportion of LSCs remain in the resting state of $G_0$ for at least part of their lifetime (Becker *et al.*, 1965; Lajtha, 1979; Cotsarelis *et al.*, 1989). Following wounding, LSC proliferation rate can be up-regulated as much as eight-to ninefold (Cotsarelis *et al.*, 1989; Lehrer *et al.*, 1998; Chung *et al.*, 1992) within 12 hours, compared to

a twofold increase in TACs following central epithelial injury (Cotsarelis *et al.*, 1989). Epithelial cell proliferation rate is highest in the peripheral cornea, in comparison to both central cornea and limbus (Ebato *et al.*, 1998; Ladage *et al.*, 2001). The change in the frequency and number of TAC division represents a shortening of cell-cycle time triggered by cell injury and loss on wounding (Cotsarelis *et al.*, 1989; Lehrer *et al.*, 1998).

As for other stem cell niches, the adjacent stromal environment plays a critical role in defining stem cell destiny. The precise molecular mechanism by which the stromal niche regulates limbal stem cell fate is unclear, although it is likely due to intricate interactions between the stem cell and its microenvironment (Morrison *et al.*, 1997; Watt and Hogan, 2000) as well as short- and long-range signals.

## Differential Regulation of Limbal Stem Cells and TACs

The regulation of stem cell homeostasis typically involves various short- and long-range internal and external factors. The environments that encompass limbal stem cells and TACs differ in various ways and are important in determining cell fate in each region (Suda *et al.*, 2006; Wolosin, 2000). Important definers include the differences that exist between the basal substrate that underlies these cells in the limbus and peripheral corneal peripheral cornea. Differences include distinct laminin isoforms, collagen IV α-chains (Ljubimov *et al.*, 1995; Tuori *et al.*, 1996), and the basement membrane–related AE27 BM antigen (Kolega *et al.*, 1989).

Cytokines are well known to modulate both corneal epithelial and stromal wound-healing responses (Wilson *et al.*, 2001) through a number of mesenchymal–epithelial interactions (Li and Tseng, 1995; Carrington and Boulton, 2005; Carrington *et al.*, 2006). It is almost certain that they play similar roles in the regulation of LSCs and TACs. Depending on the mesenchymal–epithelial interaction in which they are involved, cytokines can be assigned to groups. Epithelial cells secrete type I cytokines (TGFβ, IL1β, PDGF-BB) to modulate fibroblasts; mediators of both epithelial and fibroblasts are type II cytokines (IGF1, TGFβ1, TGFβ2, bFGF); and corneal fibroblasts produce type III cytokines (KGF, HGF). Interplay between epithelial cells and stromal fibroblasts in the cornea and limbus is likely to influence cell behavior and phenotype (Li and Tseng, 1995; Li *et al.*, 1999; Wilson *et al.*, 1994; Carrington and Boulton, 2005; Carrington *et al.*, 2006). A good example is the presence of TGFβ 1, 2, and 3 and receptors in the limbal epithelium. TGFβ inhibits LSC proliferation (Li *et al.*, 1999) and may therefore serve to influence stem cell maintenance in the limbus. Stem cell regulation by many cytokines/growth factors has been reviewed and reported by various investigators (Kruse and Volcker, 1997; Dua and Azuaro-Blanco, 2000a, 2000b; Daniels *et al.*, 2001; Kruse and Tseng, 1993; J. Imanishi *et al.*, 2000, Kruse, 1994). In addition we have recently shown that Notch/Jagged/Delta interactions play a key role in TAC differentiation (A. Ma *et al.*, 2007).

## Evidence of Corneal Epithelial Cell Plasticity

The plasticity of adult corneal epithelial cells is implied by their ability to alter phenotype in response to signals from the embryonic dermis (Ferraris *et al.*, 2000). These findings suggest that corneal epithelial cells can be reprogrammed. The screw model of stem cell behavior suggests that adult TACs, once differentiated and under certain signals, can be induced to revert to their original stemness (Loeffler and Potten, 1997; Potten and Loeffler, 1990). In the foregoing study, adult corneal epithelial cells appear to have matured (after dedifferentiation) into cells that express epidermal specific keratin 10. The corneal epithelial cells responded to signals and directions of their new environmental niche, such that their fate was redirected to allow participation in hair-follicle morphogenesis (Ferraris *et al.*, 2000; Pearton *et al.*, 2004). The mechanisms that control these processes are still to be determined.

The potential for epithelial cell differentiation into a neural cell phenotype has also been suggested. However, since the embryonic corneal epithelium is derived from the neural ectoderm, it is not surprising that a subset of epithelial cells expresses the neural marker nestin (Seigel *et al.*, 2003) or that Notch 1 expression can be induced in cultured limbal epithelial cells (Zhao *et al.*, 2002).

## Pursuit of a Corneal Stem Cell Marker

General stem cell characteristics that are exhibited by corneal LSCs define their stem cell status. The slow-cycling nature of LSCs was first recognized by the detection of label-retaining cells in the basal cell layer of the limbus (Cotsarelis *et al.*, 1989). The lack of differentiation of a subset of cells within the basal cell layer of the corneal limbus adds to their potential as being stem cells. The relative undifferentiated status of LSCs has been demonstrated by the lack of the epithelial differentiation markers, 64-kDa cytokeratin 3 (CK3), (Schermer *et al.*, 1986), and the cornea-specific cytokeratin 12 (Chaloin-Dufau *et al.*, 1990, Kurpakus *et al.*, 1990; R. Wu *et al.*, 1994) in the limbal basal epithelium, compared to localization to all layers of the corneal and suprabasal cells of limbal epithelium (Fig. 67.1). In addition, limbal stem cells are small and round and have features suggestive of a more primitive phenotype (reviewed by Schlotzer-Schrehardt and Kruse, 2005).

The TACs are rapid-cycling cells, with limited proliferative activity, which, like stem cells, can be up-regulated in corneal wound healing. At the corneal margin, basal cells coexpress vimentin and cytokeratin 19 (Kasper, 1992; Lauweryns *et al.*, 1993), and as the TAC migrates across the limbocorneal margin, the cells express differentiation features: cytokeratin 3 and 12 (Kurpakus *et al.*, 1990; Schermer *et al.*, 1986, Wolosin *et al.*, 2000). Lavker *et al.* (1991) identified peak proliferative activity in this zone.

As with many tissues derived from stem cells, to date a definitive marker that characterizes the limbal stem cell has yet to be identified. However, a number of negative differentiation markers and putative stem cell–specific markers have been investigated and are comprehensively reviewed by Wolosin *et al.* (2004) and Schlotzer-Schrehardt and Kruse (2005). Table 67.1 lists some of the markers described that may be used to discriminate basal limbal epithelial cells from basal corneal epithelial cells. Used in combination, these markers have potential in limbal stem cell isolation and in the characterization of the differentiation state.

**Table 67.1.** Positive and negative markers of limbal basal epithelial cells

| Negative markers of LSCs | Limbal basal cell markers |
| --- | --- |
| K12/K3 | ABCG2 |
| Connexin 43 | K19 |
| P cadherin | Vimentin |
| Involucrin | Integrin α9 |
| Integrins α2, α6, β4 | KGF-R |
| Nestin | Metallothionein |

A number of putative stem cell markers have been suggested, but most are not solely expressed in a subpopulation of cells in the basal cell layer of the limbus or are highly expressed in the limbal basal epithelium (i.e., positive for corneal basal cells also). Alpha-enolase is present in basal cells of the embryonic cornea (Chung *et al.*, 1992) and was observed in basal limbal epithelial cells (Zieske, 1994; Zieske *et al.*, 1992). However, its potential as a stem cell marker was unlikely, for alpha-enolase is expressed in basal cells (nonstem cells) of several stratified epithelia and was also detected in basal cells of the peripheral cornea. Pellegrini *et al.* (2001) indicated the transcription factor p63 as a potential keratinocyte stem cell marker. However, p63 did not appear to be exclusive to corneal stem cells, for low levels were occasionally observed in peripheral corneal basal cells. Also, p63 was identified at low levels in meroclones (young TACs) as compared to holoclones (stem cells) (Pellegrini *et al.*, 2001).

Epithelial tissues are suggested to consist of around 0.4% adult stem cells. In the limbus, estimations for the stem cell population that maintains the corneal epithelium indicate less than 10% of the basal limbal epithelial cell population (Lavker *et al.*, 1991), or 100 cells, in the rodent cornea (Collinson *et al.*, 2002). Like other stem cells, a subpopulation (known as side-population cells) of limbal epithelial cells, approximately 0.4%, are able to efflux Hoescht 33342, a property attributable to the ATP-binding cassette transporter G2 (Watanabe *et al.*, 2004; Wolosin *et al.*, 2004; de Paiva *et al.*, 2005; Umemoto *et al.*, 2006). Interestingly, de Paiva *et al.* (2005) and Umemoto *et al.* (2006) demonstrated conflicting results in the potential of side-population cells to proliferate and form colonies *in vitro*. Umemoto *et al.* (2006) reported that the discrepancy in findings was due to the non-colony-forming side-population cells being arrested in G0/G1, having no telomerase activity and thus representing a quiescent stem cell population in the basal cell layer of the limbus. A study of rabbit limbal epithelial side-population (SP) cells concurred: the SP cells represented 0.73% of the population, were small, undifferentiated, and noncycling, and could be induced to enter the cell cycle on wounding (Park *et al.*, 2006).

## Potential for Tissue Engineering of Limbal Stem Cells in Ocular Surface Disease

In corneal pathologies such as traumatic injuries (chemical and thermal burns), contact lens-induced keratopathy, Stevens Johnson syndrome, and ocular pemphigoid (Chiou *et al.*, 1998; Wagoner, 1997), the absence or depletion of LSCs results in failure of corneal epithelial formation. Instead, persistent epithelial defects, neovascularisation, scarring, ulceration, and in some instances corneal perforation result. In such cases, corneal transplantation that involves full-thickness transplant of a central corneal button (without inclusion of the limbal region) will not restore vision.

Several investigators have harnessed the proliferative potential of corneal and limbal epithelial cells for therapeutic purposes (Pellegrini *et al.*, 1997, 1999; Kruse and Tseng, 1993; Koizumi *et al.*, 2001a, 2001b). Limbal epithelial cells have a higher proliferative potential than cells in the peripheral and central cornea, and clonogenicity studies have confirmed that holoclones (derived from stem cells) have the greater capacity for clonal expansion than meroclones or paraclones (Pellegrini *et al.*, 1999). Clinically, cells derived from this region have led to successful limbal transplantation, with the formation of a stratified corneal epithelium (Dua and Azuaro-Blanco, 2000b; Koizumi *et al.*, 2001a, 2001b). In the absence of the limbal epithelium, conjunctivalization of the cornea results in an abnormal corneal surface (Huang and Tseng, 1991; Kruse *et al.*, 1990).

Initially, autologous conjunctival transplantation was performed to treat limbal deficiency–related ocular surface disease (Thoft, 1977; Herman *et al.*, 1983; Clinch *et al.*, 1992). However, since the discovery that "conjunctival transdifferentiation" does not occur (Dua, 1998; Kruse *et al.*, 1990; W. Chen *et al.*, 2003), other methods were investigated. Keratoepithelioplasty describes the transplant of lenticules of peripheral cornea (Thoft, 1984), and in 1989 the first human conjunctival limbal autograft was performed (Kenyon and Tseng, 1989). Kinoshita *et al.* (1991) then introduced the concept of transplantation of limbal stem cells in rabbits. Human limbal stem cell autografts utilize the limbal cells of the healthy contralateral eye, whereas bilateral disorders necessitate stem cell allografts isolated from living tissue-matched eyes or nonmatched cadaver eyes.

Next, the field progressed to *ex vivo* expansion of limbal stem cells. This approach minimized the risk of limbal stem cell deficiency in the donor eye (J. J. Chen and Tseng, 1990), since only a small biopsy was required and allograft rejection was reduced due to the elimination of Langerhan cells during culture (Lavker *et al.*, 2004). Pellegrini *et al.* (1997) reported the first successful autologous graft for unilateral severe ocular surface disease, using cultivated corneal epithelial stem cells. A number of substrates for the culture of epithelial stem cells have been introduced, including fibrin, amniotic membrane, and thermoresponsive plastic. The latter allows cultivated epithelial cells to be released on change of temperature so that only the epithelial sheets are transplanted (Nishida *et al.*, 2004a). The fibrin gel, used by Rama *et al.* (2001) as a carrier, is degraded following transplant, whereas the amniotic membrane appears to persist following transplantation (Connon *et al.*, 2006).

Preserved amniotic membrane has been used for several years in combination with limbal transplantation to promote ocular surface reconstruction in patients suffering from severe ocular-surface disease. The benefits of transplanting amniotic membrane in transplantation include provision of growth factors important in reepithelialization, antiinflammatory effects, inhibition of conjunctival fibrosis (Kinoshita *et al.*, 2004), and antibacterial properties. It is also postulated that the amniotic membrane (AM) provides the limbal stem cells with a new niche (Gruetrich *et al.*, 2003). A number of studies have evaluated the use of intact AM versus denuded AM (epithelial cells removed) as a substrate for limbal stem cell culture (Tsai *et al.*, 2000; Schwab *et al.*, 2000; Koizumi *et al.*, 2002). Limbal epithelial cells have been shown to demonstrate stem-like properties when cultured on intact AM, including slow cycling and lack of expression of the differentiation markers CK3, CK12, and connexin 43 (Gruetrich *et al.*, 2003). However, cultivation on a denuded AM showed increased limbal epithelial cell migration and produced a better-stratified epithelium (Koizumi *et al.*, 2002). If denuded AM is used to cultivate epithelial cells, growth-arrested 3T3 feeder layers are required to synthesize the growth factors that were previously produced by AM epithelial cells.

Bioengineered epithelial tissue equivalents have been derived from both limbal explants and cell suspensions isolated from limbal epithelium (Koiziumi *et al.*, 2002). Both techniques generate epithelial cell sheets that express CK3 and CK12, although more evident in superficial cells of the cell suspension–derived sheets. The latter also resulted in the formation of a greater number of desmosomes, smaller basal intercellular spaces, and secure attachment via hemidesmosomes to the underlying basement membrane. The use of the air-interface method of culture, whereby the level of media is reduced to below the epithelial cell surface, ensures that a well-differentiated, stratified epithelial tissue is available for transplantation.

*Ex vivo* expanded cells have been used, in both allografts and autografts, to promote repopulation of the recipient cornea by donor-derived cells to facilitate restoration of epithelial integrity and vision (Tsai *et al.*, 2000; Tseng *et al.*, 1998; Tsubota *et al.*, 1999; Schwab *et al.*, 2000; Koizumi *et al.*, 2001a, 2001b). In the case of allografts, immunosuppression is required to prevent epithelial rejection.

Donor epithelial cells have been identified in the recipient bed for up to 30 months after limbal allograft transplantation (Shimazaki *et al.*, 2001), but the long-term duration of donor-derived epithelial stem cell viability remains uncertain (Shimazaki *et al.*, 2001; Williams *et al.*, 1995; Swift *et al.*, 1996). Although successful long-term clinical outcome has been described in limbal allografts, donor epithelial-cell survival could not be detected (Henderson *et al.*, 2001). Whether donor limbal stem cell survival is sustained or activation of resident recipient stem cells occurs on reestablishment of niche signals has yet to be confirmed.

## Tissue-Engineered Stem Cells from Other Tissues as an Alternative to Limbal Epithelial Cells

### Oral Mucosa

Nakamura *et al.* (2003) pioneered the idea of using *ex vivo* expanded oral mucosal cells in autologous grafts to reconstruct the corneal epithelium of a rabbit. As with

limbal epithelial cells cultured *ex vivo*, oral epithelial cells were enzymatically isolated from oral mucosal biopsies and cultured on denuded AM using the air-lift culture system. The epithelial tissue equivalent, which was generated after two to three weeks culture, showed cornea-like properties with expression of CK3 (but not CK12) and the formation of a five- to six-layered stratified epithelium with desmosomes, hemidesmosomes, and tight junctions. Subsequently, the feasibility of using oral mucosa to treat corneal epithelial pathology in humans has been confirmed (Nishida *et al.*, 2004b; Nakamura *et al.*, 2004; Inatomi *et al.*, 2006).

### Bone Marrow Stem Cells (BMSCs)

Bone marrow–derived CD34$^+$ progenitor cells have been identified in the cornea, particularly in the stroma (Nakamura *et al.*, 2005). This suggests that BMSCs may play a role in stromal maintenance and repair. Transplantation of bone marrow–derived human mesenchymal stem cells (cultivated on human amniotic membrane) onto the ocular surface of a chemically damaged rat cornea demonstrated a novel idea for therapeutic reconstruction of the ocular surface. However, it appeared that the mesenchymal stem cells did not differentiate into corneal epithelial cells, but served to suppress inflammation and angiogenesis (Y. Ma *et al.*, 2006).

## The Bioengineered Cornea

The idea of a bioengineered cornea has arisen, with corneal equivalents being reconstructed from corneal cell lines. Griffith *et al.* (1999) reconstructed a human cornea from immortalized cells. The resultant corneal equivalent behaved similarly to a normal cornea with respect to morphology, transparency, ion and fluid transport, and gene expression. However, bioengineering for the production of a replacement implant for wounded or diseased human corneas is still at the investigatory stage and requires further development in order to produce a bioimplant with the same tensile strength, curvature and durability as a normal cornea. Perhaps using LSCs, BMSCs or combinations thereof in the future will overcome some of these problems and obviate the need for immortalized cell lines. The ideal artificial implant is likely to be a composite of biomaterials and cells, to provide a transparent, flexible, but strong biocompatible implant that can withstand surgical procedures as well as normal day-to-day mechanical stresses. To this purpose the chemically cross-linked collagen–glycosaminoglycan biomatrix (Pieper *et al.*, 2002) and naturally occurring biomaterials (e.g., coral or sponges) have been suggested.

## Stem Cells in the Corneal Endothelium and Conjunctival Epithelium

While the first part of this chapter concentrated on LSCs, there is evidence for other stem cell populations in the anterior segment of the eye. The conjunctival epithelium,

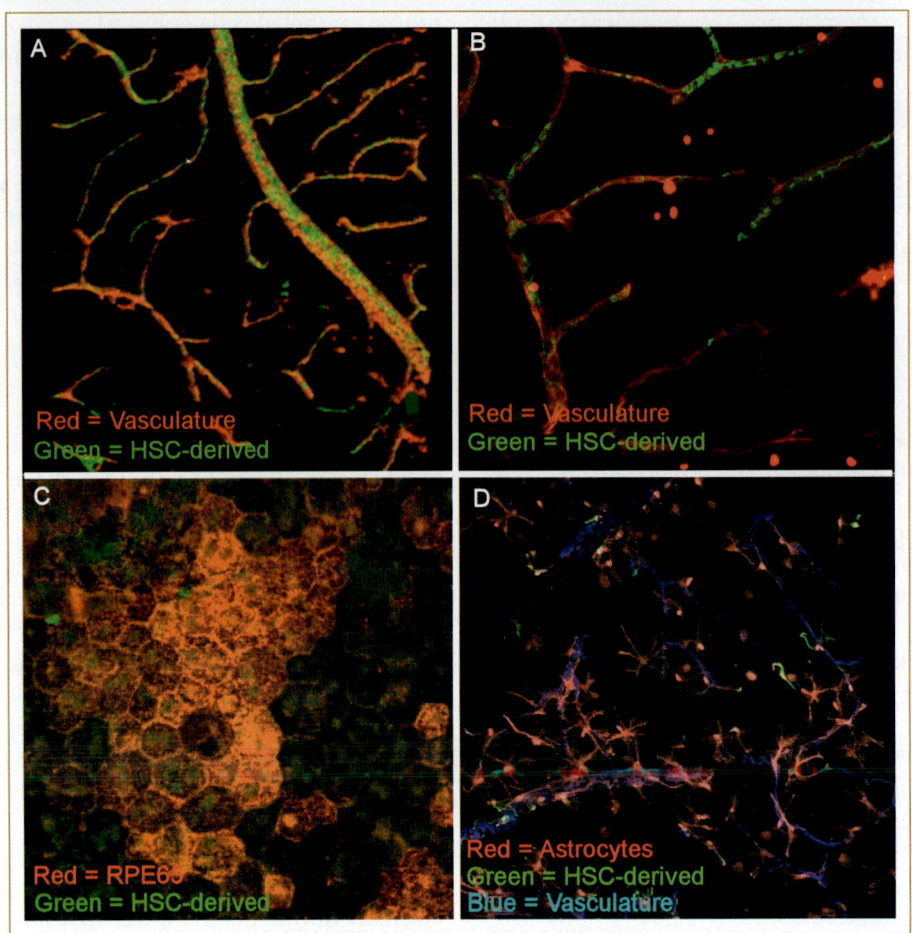

**FIG. 67.2.** Immunofluorescent localization of HSC-derived gfp$^+$ cells. All images were taken from chimeric animals whose bone marrow had been ablated by irradiation and then reconstituted with HSC from gfp homozygous donor mice. In panels **A** (z-projection of a confocal stack) and **B** (epifluorescence fluoromicrograph), the eyes of chimeric animals were first injected with a recombinant viral vector expressing VEGF, after which branch retinal vessel occlusion was performed by laser photocoagulation. The resulting combination of growth factor overexpression and ischemic injury leads to the recruitment and incorporation of bone marrow–derived endothelial progenitor cells to areas of vascular injury in the neural retina. The red fluorescence results from binding of rhodamine-conjugated *R. communis* agglutinin binding specifically to mature endothelial cells. Note the gfp$^+$ cells in these panels colocalizing with the vascular walls. The chimeric mouse in panel **C** had its RPE damaged by intraperitoneal injection of NaIO$_3$. Red fluorescence in this panel results from an antibody to RPE65, found only on the surface of mature RPE cells. Green fluorescence again results from gfp expression. Note that the colocalization of these markers (orange-yellow), along with the hexagonal morphology, indicates that damaged RPE can be repaired by recruitment and differentiation of bone marrow–derived cells. Finally, the chimeric animal in panel **D** was subjected to laser rupture of Bruch's membrane to induce choroidal neovascularization. Within the choroidal vascular lesion, astrocytes are identified by red fluorescence via antibody to S100β, vascular cells by blue fluorescence through *G. simplicifolia* isolectin B4, and bone marrow–derived cells by gfp. Note the gfp$^+$ cellular processes emanating from astrocytes, especially center right. Panels A, B, and D are 40× original magnification. Panel C originally 60× magnification. Images in panels A and B courtesy of Nilanjana Sengupta and Sergio Caballero, Pharmacology and Therapeutics, University of Florida, Gainesville, FL; images in panels C and D courtesy of Tailoi Can-Ling, Dept. of Anatomy, University of Sydney, Sydney, NSW, Australia.

like the corneal epithelium, undergoes constant renewal. The source of the population of stem cells is controversial, with different reports localizing stem cells to the fornix, palpebral, and mucocutaneous zones. Evidence demonstrating slow-cycling cells with higher proliferative capacity in these regions (Wei *et al.*, 1995; Wirstschafter *et al.*, 1999; W. Chen *et al.*, 2003), together with clinical observations, supports this concept. Unlike the stem cells of the corneal epithelium, conjunctival stem cells are not unipo-

tent; they differentiate into either a mucin-secreting goblet cell or an epithelial cell. Wolosin *et al.* (2004) suggests that the different epithelial phenotypes of the conjunctiva and cornea are under the control of genes present in only one of the two epithelia. In addition, Whikehart *et al.* (2005) proposed the existence of a stem cell niche in the posterior limbus, between the endothelium and trabecular meshwork that supplies both trabecular meshwork and corneal endothelium. Furthermore, Kawasaki *et al.* (2006) have described clusters of corneal epithelial cells residing ectopically in human conjunctival epithelium.

## III. RETINAL PROGENITOR CELLS

It has been extensively reported that stem cells exist in the retina of fish and amphibians and that these cells add to the retina throughout their lifetime (Perron and Harris, 2000). Furthermore, these cells, which are located to the ciliary-retinal interface, are also able to regenerate a complete retina, including retinal pigment epithelium (RPE), under appropriate experimental conditions (Reh and Levine, 1998). By contrast, the neural retina and RPE in mammals are largely developed by the early postnatal period and show no evidence of adult regeneration as observed in fish and amphibians (Perron and Harris, 2000; Fisher and Reh, 2003).

Surprisingly, retinal progenitor cells were not identified in mammals such as mouse, rat, cow, and human until 2000 (Tropepe *et al.*, 2000; Ahmad *et al.*, 2000). These retinal progenitor cells, which are located at the ciliary marginal zone, represent only about 0.2% of pigmented cells in the ciliary margin. These progenitors display many of the properties associated with stem cells: (1) They are multipotential, (2) they can self-renew, and (3) they are proliferative and express the neuroectodermal marker nestin. Other potential markers can include CD133, CD15, Notch, Numb, and FGFR4 (see Young, 2005). These retinal progenitor cells can clonally proliferate *in vitro* to form sphere colonies of cells that exhibit differentiation markers for a variety of cell types, including photoreceptors, intermediate neurons, and Muller glia (Tropepe *et al.*, 2000; Layer *et al.*, 2001). Thus it would appear that these progenitors have the potential, given the right environment, to be engineered into the morphological and functional layers associated with the retina.

The inability of these ciliary-margin progenitor cells to renew retinal cells in the postnatal period in mammals indicates that the progenitor cells find themselves in an inhibitory environment. Little is known about what regulates retinal progenitor cells, but transcription factors such as Pax6, Six3, Rx1, Chx10, and Hes1 (Tropepe *et al.*, 2000; Marquardt and Gruss, 2002) are strongly implicated, as is the cyclin-dependent kinase inhibitor protein, p27$^{Kip1}$ (Levine *et al.*, 2000). Growth factors such as FGF2, EGF, and IGF-I have all been reported to regulate proliferation, but their precise role remains equivocal (Tropepe *et al.*, 2000; Ahmad *et al.*, 2000; Fisher and Reh, 2003; Levine *et al.*, 2000). Furthermore, it appears from studies in the chick that Muller glia have the potential to become neurogenic retinal progenitor cells. Furthermore, Muller glia cells appear to play an important role in the generation of multipotent precursor cells from embryonic retinal cells (Layer *et al.*, 2001).

The identification of adult retinal progenitor cells opens the possibility that these cells can, given the appropriate cues, be engineered for transplantation in retinal degenerations such as retinitis pigmentosa and age-related macular degeneration (Ali and Sowden, 2003). Due to the limited number of retinal progenitor cells and the difficulty of isolating them, investigators have concentrated on investigating the incorporation and differentiation in the retina of transplanted brain-derived neural progenitor cells (Akita *et al.*, 2002; Sakaguchi *et al.*, 2003; Guo *et al.*, 2003), embryonic retinal progenitor cells (Yang *et al.*, 2002), bone marrow–derived stem cells (Tomita *et al.*, 2000), and embryonic stem cells from the inner mass of the mouse blastocyst (Schraermeyer *et al.*, 2001). While these studies, largely carried out in rodents, provide proof of principle for stem cell transplantation, differentiation, and integration in the host retina, they do emphasize that the correct inductive cues are essential for a successful outcome. Studies in the pig have shown that retinal progenitor cell xenografts to the pig retina can result in integration and differentiation (Warfvinge *et al.*, 2005). Furthermore, recent research has shown that retinal progenitor cells can be isolated and greatly expanded *ex vivo*. When grafted into the degenerating retina of dystrophic *mice*, a subset of retinal stem cells developed into mature neurons, including cells expressing photoreceptor markers (Klassen *et al.*, 2004; Young, 2005). Not only was rescue of photoreceptors observed, but there was also integration of donor cells, and animals showed improved light responsiveness. Despite these successes, cell delivery, survival of transplanted cells, and regulated differentiation of grafted cells need to be optimized if we are to regenerate a fully functional adult retina.

## IV. BONE MARROW STEM CELLS

Although not resident in the eye, it appears that bone marrow stem cells (BMSCs) play an important role in retinal homeostasis and repair. Adult BMSCs are mostly quiescent cells that comprise ~0.05–0.1% of the bone marrow (T. Imanishi *et al.*, 2004; Morrison *et al.*, 1995; Spangrude *et al.*, 1988). Adult hematopoietic stem cells (HSCs) expand and differentiate exclusively in the bone marrow, from which they can be mobilized into the bloodstream. At the same time, BMSCs may divide to yield undifferentiated progeny with identical BMSC characteristics, fulfilling the two criteria for a stem cell: self-renewal and the ability to give rise to differentiated progeny. Cell–cell and cell–matrix interactions are crucial to the proliferation and differentiation of HSCs in the bone marrow niche. Depending on the circumstances, HSCs can traffic from the marrow into the circulation in large numbers, a process termed *mobilization* (Cottler-Fox *et al.*, 2003). However, Kucia and Ratajczak (2005) propose that in addition to BMSCs, bone marrow harbors versatile subpopulations of tissue-committed stem cells (TCSC) and perhaps even more primitive pluripotent stem cells. Exogenous administration of granulocyte colony-stimulating factor (G-CSF) is the standard method of inducing stem and progenitor cell release into the circulation in a clinical setting (Cottler-Fox *et al.*, 2003). The chemokines interleukin 8 (IL-8) (Rothe *et al.*, 1998) and stromal-derived factor 1 (SDF-1)

(Hattori *et al.*, 2001) both participate in G-CSF-mediated stem cell mobilization. SDF-1 as well as VEGF and PlGF (vascular endothelial growth factor and placental growth factor, respectively) induce osteoclasts to secrete the metalloprotease MMP-9, whose action results in the shedding of the membrane-bound cytokine stem cell factor (SCF) from the bone marrow, releasing it into the circulation.

On activation, BMSCs proliferate and differentiate into progenitor cells that can be found in the peripheral blood (Ikuta *et al.*, 1992). These progenitor cells may then differentiate further as required. BMSCs have been reported to have the potential to differentiate into a variety of other tissues, including liver (Lagasse *et al.*, 2000), muscle (Ferrari *et al.*, 1998), and neuronal cells (Brazelton *et al.*, 2000; Eglitis and Mezey, 1997; Mezey *et al.*, 2000), thus demonstrating a role for stem cell plasticity in tissue maintenance and repair. Recent studies support the possibility that BMSCs may also have broad potential to differentiate into various ocular cell types (Chan-Ling *et al.*, 2006; Harris *et al.*, 2006). Lethally irradiated mice were reconstituted with HSCs from mice homozygous for green fluorescent protein (GFP+) and subjected to laser-induced Bruchs membrane rupture. GFP+ retinal pigment epithelial cells, astrocytes, macrophages/microglia, pericytes, and vascular endothelial cells were observed in or adjacent to the wound site, confirming that HSCs have the capacity to differentiate into cells expressing different ocular phenotypes.

Postnatal neovascularization has previously been considered synonymous with proliferation and migration of preexisting endothelial cells resident within parent vessels, i.e., angiogenesis (Folkman, 1971). However, Asahara *et al.* (1997) identified cells derived from bone marrow capable of differentiating into neovasculature as CD34+-expressing cells. Thus, these cells presumably are key mediators of endothelial repair. We and others have demonstrated that HSCs provide functional hemangioblast activity during retinal and choroidal neovascularization (Grant, 2002, 2003; Espinosa-Heidmann *et al.*, 2005; Sengupta *et al.*, 2005; Takahashi *et al.*, 2004) and that HSCs give rise to a variety of lineages, including the CD34+ EPCs. This mechanism of endothelial repair represents a significant source of cells for neoangiogenic vessels. Defects in EPC function appear to be responsible for the development and persistence of acellular capillaries in diabetic retinopathy, since lack of repair of acellular capillaries contributes to the development of retinal ischemia and represents an irreversible

step in the progression of this disease (Bresnick *et al.*, 1976; Fong *et al.*, 2004). EPCs isolated from patients with type 1 diabetes have a decreased rate of migration, and incubation with a nitric oxide (NO) donor alters the EPC cytoskeleton, normalizing their rate of migration (Segal *et al.*, 2006). EPC migration can be stimulated by activation of growth factor receptors (Herbrig *et al.*, 2004; T. Imanishi *et al.*, 2004; Kim *et al.*, 2005; Tepper *et al.*, 2005; Valgimigli *et al.*, 2004; Wu *et al.*, 2004), such as vascular endothelial growth factor 1 and 2 (VEGFR1 and VEGFR2), and cytokine receptors, such as CXCR4 (receptor for SDF-1) (De Falco *et al.*, 2004; Hiasa *et al.*, 2004; Valgimigli *et al.*, 2004; Yamaguchi *et al.*, 2003). Inhibiting SDF-1 has been shown to reduce the degree of stem cell involvement in induced neovascularization (Sengupta *et al.*, 2003). It also appears that autologous bone marrow–derived lineage negative HSCs containing endothelial progenitors stabilize and rescue retinal blood vessels in two mouse models of retinal degeneration (Otani *et al.*, 2004). Interestingly, this regeneration of retinal vessels was accompanied by neurotrophic rescue of the retina and preservation of cone vision.

# V. POTENTIAL FOR STEM CELLS IN OCULAR REPAIR AND TISSUE ENGINEERING

The realization of the existence of undifferentiated cells, which self-renew and show plasticity, in adult tissues offers great potential in the treatment of both ocular and nonocular diseases. It is now becoming evident that both intraocular and extraocular stem cells have the ability to play a critical role in the maintenance of ocular tissues. Furthermore, the plasticity of these cells means that they can be engineered to repair injured or diseased ocular tissues. Corneal limbal stem cells are now routinely used in the restoration of the injured ocular surface, and bone marrow stem cells hold great promise in the repair of retinal damage.

Stem cells are an ideal starting point for tissue engineering of ocular tissues and offer a means of therapeutic intervention in a variety of pathologies involving cell/tissue loss (e.g., age-related macular degeneration, retinitis pigmentosa, glaucoma, and diabetic retinopathy). The future will tell if these stem cells can be used to deliver normal genes to correct genetic abnormalities in specific ocular tissues.

# VI. REFERENCES

Ahmad, I., Tang, L., and Pham, H. (2000). Identification of neural progenitors in the adult mammalian retina. *Biochem. Biophys. Res. Comm.* **270**, 517–521.

Akita, J., Takahashi, M., Hojo, M., Nishida, A., Haruta, M., and Honda, Y. (2002). Neuronal differentiation of adult rat hippocampus–derived neural stem cells transplanted into embryonic rat explanted retinas with retinoic acid pretreatment. *Brain Res.* **954**, 286–293.

Ali, R., and Sowden, J. (2003). Therapy may yet stem from cells in the retina. *Br. J. Ophthalmol.* **87**, 1058–1059.

Asahara, T., Murohara, T., Sullivan, A., Silver, M., van der Zee, R., Li, T., Witzenbichler, B., Schatteman, G., and Isner, J. M. (1997). Isolation of putative progenitor endothelial cells for angiogenesis. *Science* **275**, 964–967.

Becker, A. J., McCulloch, E. A., Siminovitch, L., and Till, J. E. (1965). The effect of differing demands for blood cell production on DNA synthesis by haemopoitic colony-forming cells of mice. *Blood* **26**, 296–307.

Beebe, D. C., and Masters, B. R. (1996). Cell lineage and the differentiation of corneal epithelial cells. *Invest. Ophthalmol. Vis. Sci.* **37**, 1815–1825.

Boulton M., and Albon J. (2004). Stem cells in the eye. *Int. J. Biochem. Cell. Biol.* **36**, 643–657.

Brazelton, T. R., Rossi, F., Keshet, G. I., and Blau, H. M. (2000). From marrow to brain: expression of neuronal phenotypes in adult mice. *Science* **290**, 1775–1779.

Bresnick, G. H., Engerman, R., Davis, M. D., de Venecia, G., and Myers, F. L. (1976). Patterns of ischemia in diabetic retinopathy. *Trans. Sect. Ophthalmol. Am. Acad. Ophthalmol. Otolaryngol.* **81**, OP694–OP709.

Carrington, L. M., and Boulton, M. (2005). Hepatocyte growth factor and keratinocyte growth factor regulation of epithelial and stromal corneal wound healing. *J. Cataract Refract. Surg.* **31**, 412–423.

Carrington, L. M., Albon, J., Anderson, I., Kamma, C., and Boulton, M. (2006). Differential regulation of key stages in early corneal wound healing by TGF-beta isoforms and their inhibitors. *Invest. Ophthalmol. Vis. Sci.* **47**, 1886–1894.

Chaloin-Dufau, C., Sun, T. T., and Dhouailly, D. (1990). Appearance of the keratin pair K3/K12 during embryonic and adult corneal epithelial differentiation in the chick and rabbit. *Cell. Differ. Dev.* **32**, 97–108.

Chan-Ling, T., Baxter, L., Afzal, A., Sengupta, N., Caballero, S., Rosinova, E., and Grant, M. B. (2006). Hematopoietic stem cells provide repair functions after laser-induced Bruch's membrane rupture model of choroidal neovascularization. *Am. J. Pathol.* **168**, 1031–1044.

Chen, J. J., and Tseng, S. C. (1990). Corneal epithelial wound healing in partial limbal deficiency, *Invest. Ophthalmol. Vis. Sci.* **31**, 1301–1314.

Chen, W., Ishikawa, M., Yamaki, K., and Sakuragi, S. (2003). Wistar rat palpebral conjunctiva contains more slow-cycling stem cells that have larger proliferative capacity: implication for conjunctival epithelial cell homeostasis. *Jpn. J. Ophthalmol.* **47**, 119–128.

Chiou, A.G-Y., Florakis, G. J., and Kazim, M. (1998). Management of conjunctival cicatrizing diseases and severe ocular dysfunction. *Surv. Ophthalmol.* **43**, 19–46.

Chung, E. H., Bukusolglu, G., and Zieske, J. D. (1992). Localization of corneal epithelial stem cells in the developing rat. *Invest. Ophthalmol. Vis. Sci.* **33**, 121–128.

Clinch, T. E., Goins, K. M., and Cobo, L. M. (1992). Treatment of contact-lens-related ocular surface disorders with autologous conjunctival transplantation. *Ophthalmology* **99**, 634–638.

Collinson, J. M., Morris, L., Reid,. A. I., Ramaesh, T., Keighren, M. A., Flockhart, J. H., Hill, R. E., Tan, S. S., Ramaesh, K., Dhillon, B., *et al.* (2002). Clonal analysis patterns of growth, stemcell, activity, and cell movement during development and maintenance of the murine corneal epithelium. *Dev. Dyn.* **254**, 432–440.

Connon, C. J., Nakamura, T., Quantock, A. J., and Kinoshita, S. (2006). The persistence of transplanted amniotic membrane in corneal stroma. *Am. J. Ophthalmol.* **141**, 190–192.

Cotsarelis, G., Cheng, S., Dong, G., Sun, T. T., and Lavker, R. (1989). Existence of slow-cycling limbal epithelial basal cell that can be preferentially stimulated to proliferate: implications on epithelial stem cells. *Cell* **57**, 201–209.

Cottler-Fox, M. H., Lapidot, T., Petit, I., Kollet, O., DiPersio, J. F., Link, D., and Devine, S. (2003). Stem cell mobilization. *Hematology Am. Soc. Hematol. Educ. Program.*, 419–437.

Daniels, J. T., Dart, J. K. G., Tuft, S. J., and Khaw, P. T. (2001). Corneal stem cells in review. *Wound Rep. Reg.* **9**, 483–494.

Davenger, M., and Evenson, A. (1971). Role of the pericorneal structure in renewal of corneal epithelium. *Nature* **229**, 560–561.

De Falco, E., Porcelli, D., Torella, A. R., Straino, S., Iachininoto, M. G., Orlandi, A., Truffa, S., Biglioli, P., Napolitano, M., Capogrossi, M. C., and Pesce, M. (2004). SDF-1 involvement in endothelial phenotype and ischemia-induced recruitment of bone marrow progenitor cells. *Blood* **104**, 3472–3482.

de Paiva, C. S., Chen, Z., Corrales, R. M., Plugfelder, S. C., and Li, D. Q. (2005). ABCG2 transporter identifies a population of clonogenic human limbal epithelial cells. *Stem Cells* **23**, 63–73.

Dua, H. S. (1998). The conjunctiva in corneal epithelial wound healing. *Br. J. Ophthalmol.* 82, 1407–1411.

Dua, H. S., and Azuaro-Blanco A. (2000a). Limbal stem cells of the corneal epithelium. *Surv. Ophthalmol.* **44**, 415–425.

Dua, H. S., and Azuaro-Blanco A. (2000b). Autologous limbal transplantation patients with unilateral corneal stem cell deficiency. *Br. J. Ophthalmol.* **84**, 273–278.

Dua, H. S., Shanmuganathan, V. A., Powell-Richards, A. O., Tighe, P. J., and Joseph, A. (2005). Limbal epithelial crypts: a novel anatomical structure and a putative limbal stem cell niche. *Br. J. Ophthalmol.* **89**, 529–532.

Ebato, B., Friend, J., and Thoft, R. A. (1988). Comparison of limbal and peripheral human corneal epithelium in tissue culture. *Invest. Ophthalmol. Vis. Sci.* **29**, 1533–1537.

Eglitis, M. A., and Mezey, E. (1997). Hematopoietic cells differentiate into both microglia and macroglia in the brains of adult mice. *PNAS* **94**, 4080–4085.

Espinosa-Heidmann, D. G., Reinoso, M. A., Pina, Y., Csaky, K. G., Caicedo, A., and Cousins, S. W. (2005). Quantitative enumeration of vascular smooth muscle cells and endothelial cells derived from bone marrow precursors in experimental choroidal neovascularization. *Exp. Eye Res.* **80**, 369–378.

Ferrari, G., Cusella-De Angelis, G., Coletta, M., Paolucci, E., Stornaiuolo, A., Cossu, G., and Mavilio, F. (1998). Muscle regeneration by bone marrow–derived myogenic progenitors. *Science* **279**, 1528–1530.

Ferraris, C., Chevalier, G., Favier, B., Jahoda, C. A. B., and Dhouailly, D. (2000). Adult corneal epithelium basal cells possess the capacity to activate epidermal, pilosebaceous and sweat genetic programmes in response to embryonic dermal stimuli. *Development* **127**, 5487–5495.

Fisher, A., and Reh, T. (2003). Potential of Muller glia to become retinal progenitor cells. *Glia* **43**, 70–76.

Folkman, J. (1971). Tumor angiogenesis: therapeutic implications. *N. Engl. J. Med.* **285**, 1182–1186.

Fong, D. S., Aiello, L., Gardner, T. W., King, G. L., Blankenship, G., Cavallerano, J. D., Ferris, F. L., 3rd, and Klein, R. (2004). Retinopathy in diabetes. *Diabetes Care* **27**(Suppl. 1), S84–S87.

Grant, M. B., May, W. S., Caballero, S., Brown, G. A., Guthrie, S. M., Mames, R. N., Byrne, B. J., Vaught, T., Spoerri, P. E., Peck, A. B., and Scott, E. W. (2002). Adult hematopoietic stem cells provide functional hemangioblast activity during retinal neovascularization. *Nat. Med.* **8**, 607–612.

Grant, M. B., Caballero, S., Brown, G. A., Guthrie, S. M., Mames, R. N., Vaught, T., and Scott, E. W. (2003). The contribution of adult hematopoietic stem cells to retinal neovascularization. *Adv. Exp. Med. Biol.* **522**, 37–45.

Griffith, M., Osborne, R., Munger, R., and Xiong, X. (1999). Functional human corneal equivalents constructed from cell lines. *Science* **286**, 2169–2172.

Gruetrich, M., Espana, E. M., and Tseng, S. C. G. (2003). *Ex vivo* expansion of limbal stem cells: amniotic membrane serving as a stem cell niche. *Surv. Ophthalmol.* **48**, 631–646.

Guo, Y., Saloupis, P., Shaw, S., and Rickman, D. (2003). Engraftment of adult neural progenitor cells transplanted to rat retina injured in transient ischaemia. *Invest. Ophthalmol. Vis. Sci.* **44**, 3194–3201.

Harris, J. R., Brown, G. A., Jorgensen, M., Kaushal, S., Ellis, E. A., Grant, M. B., and Scott, E. W. (2006). Bone marrow–derived cells home to and regenerate retinal pigment epithelium after injury. *Invest. Ophthalmol. Vis. Sci.* **47**, 2108–2113.

Hattori, K., Heissig, B., Tashiro, K., Honjo, T., Tateno, M., Shieh, J. H., Hackett, N. R., Quitoriano, M. S., Crystal, R. G., Rafii, S., and Moore, M. A. (2001). Plasma elevation of stromal cell–derived factor-1 induces mobilization of mature and immature hematopoietic progenitor and stem cells. Blood **97**, 3354–3360.

Henderson, T. R., Coster, D. J., and Williams, K. A. (2001). The long-term outcome of limbal allografts: the search for surviving cells. *Br. J. Ophthalmol.* **85**, 604–609.

Herbrig, K., Pistrosch, F., Oelschlaegel, U., Wichmann, G., Wagner, A., Foerster, S., Richter, S., Gross, P., and Passauer, J. (2004). Increased total number but impaired migratory activity and adhesion of endothelial progenitor cells in patients on long-term hemodialysis. *Am. J. Kidney Dis.* **44**, 840–849.

Herman, W. K., Doughman, D. J., and Lindstrom, R. L. (1983). Conjunctival autograft transplantation for unilateral ocular surface diseases. *Ophthalmology* **90**, 1121–1123.

Hiasa, K., Ishibashi, M., Ohtani, K., Inoue, S., Zhao, Q., Kitamoto, S., Sata, M., Ichiki, T., Takeshita, A., and Egashira, K. (2004). Gene transfer of stromal cell–derived factor-1alpha enhances ischemic vasculogenesis and angiogenesis via vascular endothelial growth factor/endothelial nitric oxide synthase–related pathway: next-generation chemokine therapy for therapeutic neovascularization. *Circulation* **109**, 2454–2461.

Holland, E. J. (1996). Epithelial transplantation for severe ocular surface disease. *Trans. Am. Ophthalmol. Soc.* **95**, 677–6743.

Huang, A. J., and Tseng, S. C. (1991). Corneal epithelial wound healing in the absence of limbal epithelium. *Invest. Ophthalmol. Vis. Sci.* **32**, 96–105.

Ikuta, K., Uchida, N., Friedman, J., and Weissman, I. L. (1992). Lymphocyte development from stem cells. *Annu. Rev. Immunol.* **10**, 759–783.

Ikutam, K., and Weissmanm, I. L. (1992). Evidence that hematopoietic stem cells express mouse c-kit but do not depend on steel factor for their generation. *Proc. Natl. Acad. Sci. U.S.A* **89**, 1502–1506.

Imanishi, J., Kamiyama, K., Iguchi, I., Kita, M., Sotozono, C., and Kinoshita, S. (2000). Growth factors: importance in wound healing and maintenance of transparency of the cornea. *Prog. Ret. Eye Res.* **19**, 114–129.

Imanishi, T., Hano, T., and Nishio, I. (2004). Angiotensin II potentiates vascular endothelial growth factor–induced proliferation and network formation of endothelial progenitor cells. *Hypertens. Res.* **27**, 101–108.

Inatomi, T., Nakamura, T., Koizumi, N., Sotozono, C., Yokoi, N., and Kinoshita, S. (2006). Midterm results on ocular surface reconstruction using cultivated autologous oral mucosal epithelial transplantation. *Am. J. Ophthalmol.* **141**, 267–275.

Kasper, M. (1992). Patterns of cytokeratin and vimentin in guinea pig and mouse eye tissue: evidence or regional variations in intermediate filament expression in limbal epithelium. *Acta Histochem.* **93**, 319–332.

Kawasaki, S., Tanioka, H., Yamasaki, K., Yokoi, N., Komuro, A., and Kinoshita, S. (2006). Clusters of corneal epithelial cells reside ectopically in human conjunctival epithelium. *Invest. Ophthalmol. Vis. Sci.* **47**, 1359–1367.

Kenyon, K. R., and Tseng, S. C. (1989). Limbal autograft transplantation for ocular surface disorders. *Ophthalmology* **96**, 709–722.

Kim, S. Y., Park, S. Y., Kim, J. M., Kim, J. W., Kim, M. Y., Yang, J. H., Kim, J. O., Choi, K. H., Kim, S. B., and Ryu, H. M. (2005). Differentiation of endothelial cells from human umbilical cord blood AC133⁻CD14⁺ cells. *Ann. Hematol.* **84**, 417–422.

Kinoshita, S., Ohashi, Y., Ohji, M., and Manabe, R. (1991). Long-term results of keratoepithelioplasty in Mooren's ulcer. *Ophthalmology* **98**, 438–442.

Kinoshita, S., Adachi, W., Sotozono, C., Nishida, K., Yokoi, N., Quantock, A. J., and Okubo, K. (2001). Characteristics of the human ocular surface epithelium. *Prog. Ret. Eye Res.* **20**, 639–673.

Kinoshita, S., Koizumi, N., and Nakamura, T. (2004). Tranplantable cultivated mucosal epithelial sheet for ocular surface reconstruction. *Exp. Eye Res.* **78**, 483–491.

Klassen, H. J., Ng, T. F., Kurimoto, Y., Kirov, I., Shatos, M., Coffey, P., and Young, M. J. (2004). Multipotent retinal progenitors express developmental markers, differentiate into retinal neurons, and preserve light-mediated behavior. *Invest. Ophthalmol. Vis. Sci.* **45**, 4167–4173.

Koizumi, N., Inatomi, T., Suzuki, T., Sotozono, C., and Kinoshita, S. (2001a). Cultivated corneal epithelial transplantation for ocular surface reconstruction in acute phase of Stevens–Johnson syndrome. *Arch. Ophthalmol.* **119**, 298–300.

Koizumi, N., Inatomi, T., Suzuki, T., Sotozono, C., and Kinoshita, S. (2001b). Cultivated corneal epithelial stem cell transplantation in ocular surface disorders. *Ophthalmology* **108**, 1569–1574.

Koizumi, N., Fullwood, N. J., Bairaktaris, G., and Inatomi, T. (2002). Cultivation of corneal epithelial cells on intact and denuded human amniotic membrane. *Invest. Ophthalmol. Vis. Sci.* **41**, 2506–2513.

Kolega, J., Manabe, M., and Sun, T. T. (1989). Basement membrane heterogeneity and variation in corneal epithelial differentiation. *Differentiation* **42**, 54–63.

Kruse, F. E. (1994). Stem cells and corneal epithelial regeneration. *Eye* **8**, 170–183.

Kruse, F. E., and Tseng, S. C. G. (1993). Growth factors modulate clonal growth and differentiation of cultured rabbit limbal and corneal epithelium. *Invest. Ophthalmol. Vis. Sci.* **34**, 1963–1976.

Kruse, F. E., and Volcker, H. E. (1997). Stem cells, wound healing, growth factors and angiogenesis in the cornea. *Curr. Opin. Ophthalmol.* **8**, 46–54.

Kruse, F. E., Chen, J. J., Tsai, R. J., and Tseng, S. C. (1990). Conjunctival transdifferentiation is due to the incomplete removal of limbal basal epithelium. *Invest. Ophthalmol. Vis. Sci.* **31**, 1909–1913.

Kucia, M., Reca, R., Jala, V. R., Dawn, B., Ratajczak, J., and Ratajczak, M. Z. (2005). Bone marrow as a home of heterogenous populations of nonhematopoietic stem cells. *Leukemia* **19**, 1118–1127.

Kurpakus, M. A., Stock, E. L., and Jones, J. C. (1990). Expression of the 55-kDa/64-kDa corneal keratins in ocular surface epithelium. *Invest. Ophthalmol. Vis. Sci.* **31**, 448–456.

Ladage, P. M., Yanamoto, K., Ren, D. H., Li, L., Jester, J. V., Petroll, W. M., Bergmanson, J. P. G., and Cavanagh, H. D. (2001). Proliferation rate of rabbit corneal epithelium during overnight rigid contact lens wear. *Invest. Ophthalmol. Vis. Sci.* **42**, 2802–2812.

Lagasse, E., Connors, H., Al-Dhalimy, M., Reitsma, M., Dohse, M., Osborne, L., Wang, X., Finegold, M., Weissman, I. L., and Grompe, M. (2000). Purified hematopoietic stem cells can differentiate into hepatocytes *in vivo. Nat. Med.* **6**, 1229–1234.

Lajtha, L. G. (1979). Stem cell concepts. *Differentiation* **14**, 23.

Lauweryns, B., van der Oord, J. J., De Vos, R., and Missotten, L. (1993). A new epithelial cell type in the human cornea. *Invest. Ophthalmol. Vis. Sci.* **34**, 1983–1990.

Lavker, R. M., Dong, G., Cheng, S. Z., Kudoh, K., Cotsarelis, G., and Sun, T. T., (1991). Relative proliferative rates of limbal and corneal epithelia. *Invest. Ophthalmol. Vis. Sci.* **32**, 1864–1875.

Lavker, R. M., Tseng, S. C. G., and Sun, T. T. (2004). Corneal epithelial stem cells at the limbus: looking at some old problems from a new angle. *Exp. Eye Res.* **78**, 433–446.

Layer, P., Rothermel, A., and Willbold, E. (2001). From stem cells towards neural layers: a lesson from reaggregated embryonic retinal cells. *NeuroReport* **12**, 39–46.

Lehrer, M. S., Sun, T. T., and Lavker R. M. (1998). Strategies of epithelial repair: modulation of stem cell and transit amplifying cell proliferation. *J. Cell Sci.* **111**, 2867–2875.

Levine, E., Close, J., Fero, M., Ostrovsky, A., and Reh, T. (2000). p27$^{Kip1}$ regulates cell-cycle withdrawal of late multipotent progenitor cells in the mammalian retina. *Dev Biol.* **219**, 299–314.

Li, D. Q., and Tseng, S. C. (1995). Three patterns of cytokine expression potentially involved in epithelial–fibroblast interactions of human ocular surface. *J. Cell. Physiol.* **163**, 61–79.

Li, D. Q., Lee, S. B., and Tseng, S. C. (1999). Differential expression and regulation of TGF-B1, TGF-B2, TGF-B3, TGF-BRI, TGF-BRII, TGF-BRIII in cultured human corneal, limbal and conjunctival fibroblasts. *Curr. Eye Res.* **19**, 154–161.

Lindberg, K., Brown, M. E., Chaves, H. V., Kenyon, H. R., and Rheinwald, J. G. (1993). *In vitro* propagation of human ocular surface epithelial cells for transplantation. *Invest. Ophthalmol. Vis. Sci.* **34**, 2672–2679.

Ljubimov, A. V., Burgeson, R. E., Butowski, R. J., Michael, A. F., Sun, T. T., and Kenney, M. C. (1995). Human corneal basement heterogeneity: topographical differences in the expression of type IV collagen and laminin isoforms. *Lab. Invest.* **72**, 461–471.

Loeffler, M., and Potten, C. S. (1997). Stem cells and cellular pedigrees — a conceptual introduction. *In* "Stem Cells" (C. Potten, ed.), pp. 1–27, Academic Press, London.

Ma, A., Boulton, M., Zhao, B., Connon, C., Cai, J., and Albon, J. (2007). A role for notch signaling in human corneal epithelial cell differentiation and proliferation. *Invest. Ophthalmol. Vis. Sci.* (In press).

Ma, Y., Xu, Y., Xiao, Z., Yang, W., Zhang, C., Song, E., Du, Y., and Li, L. (2006). Reconstruction of chemically burned rat corneal surface by bone marrow–derived mesenchymal stem cells. *Stem Cells.* **24**, 315–321.

Marquardt, T., and Gruss, P. (2002). Generating neuronal diversity in the retina: one for nearly all. *Trends Neurosci.* **25**, 32–38.

Mezey, E., Chandross, K. J., Harta, G., Maki, R. A., and McKercher, S. R. (2000). Turning blood into brain: cells bearing neuronal antigens generated *in vivo* from bone marrow. *Science* **290**, 1779–1782.

Morrison, S. J., Uchida, N., and Weissman, I. L. (1995). The biology of hematopoietic stem cells. *Annu. Rev. Cell Dev. Biol.* **11**, 35–71.

Morrison, S. J., Shah, N. M., and Anderson, D. J. (1997). Regulatory mechanisms in stem cell biology. *Cell* **88**, 287–298.

Nakamura, T., Endo, K.-I., Cooper, L. J., Fullwood, N., Tanifuji, N., Tsuzuki, M., Koizumi, N., Inatomi, T., Sano, Y., and Kinoshita, S. (2003). The successful culture and autologous transplantation of rabbit oral mucosal epithelial cells on amniotic membrane. *Invest. Ophthalmol. Vis. Sci.* **44**, 106–116.

Nakamura, T., Inatomi, T., Sotozono, C., Amemiya, T., Kanamura, N., and Kinoshita, S. (2004). Transplantation of cultivated autologous oral mucosal epithelial cells in patients with severe ocular surface disorders. *Br. J. Ophthalmol.* **88**, 1280–1284.

Nakamura, T., Ishikawa, F., Sonoda, K. H., Hisatomi, T., Qiao, H., Yamada, J., Fukata, M., Ishibashi, T., Harada, M., and Kinoshita, S. (2005). Characterization and distribution of bone marrow–derived cells in mouse cornea. *Invest. Ophthalmol. Vis. Sci.* **46**, 497–503.

Nishida, K., Yamato, M., Hayashida, Y., *et al.* (2004a). Functional bio-engineered corneal epithelial sheet grafts from corneal stem cells expanded *ex vivo* on a temperature-responsive cell culture surface. *Transplantation* **77**, 379–385.

Nishida, K., Yamato, M., Hayashida, Y., *et al.* (2004b). Corneal reconstruction with tissue-engineered cell sheets composed of autologous oral mucosal epithelium. *N. Engl. J. Med.* **351**, 1187–1196.

Otani, A., Dorrell, M. I., Kinder, K., Moreno, S. K., Nusinowitz, S., Banin, E., Heckenlively, J., and Friedlander, M. (2004). Rescue of retinal degeneration by intravitreally injected adult bone marrow–derived lineage-negative hematopoietic stem cells. *J. Clin. Invest.* **114**, 765–774.

Park, K, Lim, C. H., Min B.-M., Lee, J. L., Chung, H. Y., Joo, C. K., Park, C. W., and Son, Y. (2006). The side population cells in the rabbit limbus sensitively increased in response to the central cornea wounding. *Invest. Ophthalmol. Vis. Sci.* **47**, 892–900.

Pearton, D. J., Ferraris, C., and Dhouailly, D. (2004). Transdifferentiation of corneal epithelium: evidence for a linkage between the segregation of epidermal stem cells and the induction of hair follicles during embryogenesis. *Int. J. Dev. Biol.* **48**, 197–210.

Pellegrini, G., Traverso, C. E., Franz, A. T., Zingirian, M., Cancedda, R., and De Luca, M. (1997). Long-term restoration of damaged corneal surfaces with autologous cultivated epithelium. *Lancet* **349**, 990–993.

Pellegrini, G., Golisano, O., Paterna, P., Lambiase, A., Bonini, S., De Rama, P., and De Luca, M. (1999). Location and clonal analysis of stem cells and their differentiated progeny in the human ocular surface. *J. Cell. Biol.* **145**, 769–782.

Pellegrini, G., Dellambra, E., Golisano, O., Martenelli, E., Fantozzi, I., Bondanza, S., Ponzin, D., McKeon, F., and De Luca, M. (2001). p63 identifies keratinocyte stem cells. *Proc. Natl. Acad. Sci.* **98**, 3156–3161.

Perron, M., and Harris, W. (2000). Retinal stem cells in vertebrates. *Bioessays* **22**, 685–688.

Pieper, J. S., Hafmans, T., Veerkamp, T. H., and van Kuppervelt, T. H. (2002). Development of tailor-made collagen–glycosaminoglycan matrices: EDC/NHS cross-linking, and ultrastructural aspects. *Biomaterials* **21**, 581–593.

Potten, C. S., and Loeffler, M. (1990). Stem cells: attributes, cycles, spirals, pitfalls and uncertainties. *Development* **110**, 1001–1020.

Rama, P., Bonini, S., Lambiase, A., Golosano, O., Paterna, P., De Luca, M., and Pellegrini, G. (2001). Autologous fibrin-cultured limbal stem cells permanently restore the corneal surface of patients with total limbal stem cell deficiency. *Transplantation* **72**, 1478–1485.

Reh, T., and Levine, E. (1998). Multipotential stem cells and progenitors in the vertebrate retina. *Int. J. Neurobiol.* **36**, 206–220.

Rothe, L., Collin-Osdoby, P., Chen, Y., Sunyer, T., Chaudhary, L., Tsay, A., Goldring, S., Avioli, L., and Osdoby, P. (1998). Human osteoclasts and osteoclast-like cells synthesize and release high basal and inflammatory stimulated levels of the potent chemokine interleukin-8. *Endocrinology* **139**, 4353–4363.

Sakaguchi, S., van Hoffelen, S., and Young, M. (2003). Differentiation and morphological integration of neural progenitor cells transplanted into the developing mammalian eye. *Ann. N.Y. Acad. Sci.* **995**, 127–139.

Schermer, A., Galvin, S., and Sun, T. T. (1986). Differentiation related expression of major 64K keratin *in vivo* and in culture suggests limbal location of corneal epithelial stem cells. *J. Cell Biol.* **103**, 49–62.

Schlotzer-Schrehardt, U., and Kruse, F. E. (2005). Identification and characterisation of limbal stem cells. *Exp. Eye Res.* **81**, 247–264.

Schraermeyer, U., Thumann, G., Luther, T., Kociok, N., Armhold, S., Krut, K., Andressen, C., Addicks, K., and Bartz-Schmidt, K. (2001). Subretinally transplanted embryonic stem cells rescue photoreceptor cells from degeneration in the RCS rat. *Cell Transplant.* **10**, 673–680.

Schwab, I. R., Reyes, M., and Isseroff, R. R. (2000). Successful transplantation of bioengineered tissue replacements in patients with ocular surface disease. *Cornea* **19**, 421–426.

Segal, M. S., Shah, R., Afzal, A., Perrault, C. M., Chang, K., Schuler, A., Beem, E., Shaw, L. C., LiCalzi, S., Harrison, J. K., and Tran-Son-Tay, R. (2006). Nitric oxide cytoskeletal-induced alterations reverse the endothelial progenitor cell migratory defect associated with diabetes, *Diabetes.* **55**, 102–109.

Seigel, G. M., Sun, W., Salvi, R., Campbell, L. M., Sullivan, S., and Reddy, J. J. (2003). Human corneal stem cells display functional neuronal properties. *Mol. Vis.* **9**, 159–163.

Sengupta, N., Caballero, S., Mames, R. N., Butler, J. M., Scott, E. W., and Grant, M. B. (2003). The role of adult bone marrow–derived stem cells in choroidal neovascularization. *Invest. Ophthalmol. Vis. Sci.* **44**, 4908–4913.

Sengupta, N., Caballero, S., Mames, R. N., Timmers, A. M., Saban, D., and Grant, M. B. (2005). Preventing stem cell incorporation into choroidal neovascularization by targeting homing and attachment factors. *Invest. Ophthalmol. Vis. Sci.* **46**, 343–348.

Shimazaki, J., Kinado, M., Shinozaki, N., Shimmura, S., *et al.* (2001). Evidence of long-term survival of donor-derived cells after limbal allograft transplantation. *Invest. Ophthalmol. Vis. Sci.* **40**, 1664–1668.

Spangrude, G. J., Aihara, Y., Weissman, I. L., and Klein, J. (1988). The stem cell antigens Sca-1 and Sca-2 subdivide thymic and peripheral T-lymphocytes into unique subsets. *J. Immunol.* **141**, 3697–3707.

Suda, T., Fumio, A., and Shimmura, S. (2006). Regulation of stem cells in the niche. *Cornea* **24**, S12–S17.

Swift, G., Aggarwal, R., Davis, G., Coster, D., and Williams, K. (1996). Survival of rabbit limbal stem cell allografts. *Transplantation* **62**, 568–574.

Takahashi, H., Yanagi, Y., Tamaki, Y., Muranaka, K., Usui, T., and Sata, M. (2004). Contribution of bone-marrow-derived cells to choroidal neovascularization. *Biochem. Biophys. Res. Commun.* **320**, 372–375.

Tepper, O. M., Capla, J. M., Galiano, R. D., Ceradini, D. J., Callaghan, M. J., Kleinman, M. E., and Gurtner, G. C. (2005). Adult vasculogenesis occurs through *in situ* recruitment, proliferation, and tubulization of circulating bone marrow–derived cells. *Blood* **105**, 1068–1077.

Thoft, R. A. (1977). Conjunctival transplantation. *Arch. Ophthalmol.* **95**, 1425–1427.

Thoft, R. A. (1984). Keratoepithelioplasty. *Am. J. Ophthalmol.* **97**, 1–6.

Thoft, R. A., and Friend, J. (1983). The XYZ hypothesis of corneal epithelial maintenance. *Invest. Ophthamol. Vis. Sci.* **24**, 1442.

Tomita, M., Adachi, Y., Yamada, H., Takahashi, K., Kiuchi, K., Oyaizu, H., Ikebukuro, K., Kaneda, H., Matsumura, M., and Ikehara, S. (2000). Bone marrow–derived stem cells can differentiate into retinal cells in injured rat retina. *Stem Cells* **20**, 279–283.

Tropepe, V., Coles, B., Chiasson, B., Horsford, D., Elia, A., McInnes, R., and van der Kooy, D. (2000). Retinal stem cells in the adult mammalian eye. *Science* **287**, 2032–2036.

Tsai, R. J. F., and Tseng, S. C. G. (1994). Human allograft limbal transplantation for corneal surface reconstruction. *Cornea* **13**, 389–400.

Tsai, R. J. F., Li, L. M., and Chen, J. K. (2000). Reconstruction of damaged corneas by transplantation of autologous limbal epithelial cells. *N. Engl. J. Med.* **343**, 86–93.

Tseng, S. C., Prabhasawat, P., Barton, K., Gray, T., and Meller, D. (1998). Amniotic membrane transplantation with or without limbal allografts for corneal surface reconstruction in patients with limbal stem cell deficiency. *Arch. Ophthalmol.* **116**, 431–441.

Tsubota, K., Satake, Y., and Kaido, M. (1999). Treatment of severe ocular surface disorders with corneal epithelial stem–cell transplantation. *N. Engl. J. Med.* **340**, 1697–1703.

Tuori, A., Uusitalo, H., Burgeson, R. E., Tertunen, J., and Virtanen, I. (1996). The immunohistochemical composition of the human corneal basement membrane. *Cornea* **15**, 286–294.

Umemoto, T., Yamato, M., Nishida, K., Yang, J., Tano, Y., and Okano, T. (2006). Limbal epithelial side population cells have stem cell–like properties, including quiescent state. *Stem Cells* **24**, 86–94.

Valgimigli, M., Rigolin, G. M., Fucili, A., Porta, M. D., Soukhomovskaia, O., Malagutti, P., Bugli, A. M., Bragotti, L. Z., Francolini, G., Mauro, E., Castoldi, G., and Ferrari, R. (2004). CD34⁺ and endothelial progenitor cells in patients with various degrees of congestive heart failure. *Circulation* **110**, 1209–1212.

Wagoner, M. D. (1997). Chemical injuries of the eye: current concepts in pathophysiology and therapy. *Surv. Ophthalmol.* **41**, 275–313.

Warfvinge, K., Kiilgaard, J. F., Lavik, E. B., Scherfig, E., Langer, R., Klassen, H. J., and Young, M. J. (2005). Retinal progenitor cell xenografts to the pig retina: morphologic integration and cytochemical differentiation. *Arch. Ophthalmol.* **123**, 1385–1393.

Watanabe, K., Nishida, K., Yamato, M., *et al.* (2004). Human limbal epithelium contains side population cells expressing the ATP-binding cassette transporter ABCG2. *FEBS Lett.* **565**, 6–10.

Watt, F., and Hogan, A. (2000). Out of Eden: stem cells and their niches. *Science* **287**, 1427–1431.

Wei, Z. G., Cotsarelis, G., Sun, T. T., and Lavker, R. M. (1995). Label-retaining cells are preferentially located in fornical epithelium: implications on conjunctival epithelial homeostasis. *Invest. Ophthalmol. Vis. Sci.* **36**, 236–246.

Whikehart, D. R., Parikh, C. H., Vaughn, A. V., Mishler, K., and Edelhauser, H. F. (2005). Evidence suggesting the existence of stem cells for the human corneal endothelium. *Mol. Vis.* **11**, 816–824.

Williams, K. A., Brereton, H. M., Aggarwal, R., Sykes, P. J., Turner, D. R., Russ, G. R., and Coster, D. J. (1995). Use of DNA polymorphisms and the polymerase chain reaction to examine the survival of a human limbal stem cell allograft. *Am. J. Ophthalmol.* **120**, 342–350.

Wilson, S. E., He, Y. G., Weng, J, Zieske, J. D., Jester, J. V., and Schultz, G. S. (1994). Effect of epidermal growth factor, hepatocyte growth factor, keratinocyte growth factor on the motility, differentiation of human corneal epithelial cells. *Exp. Eye. Res.* **59**, 665–678.

Wilson, S. E., Mohan, R. R., Ambrosio, R., Hong, J. W., and Lee, J. S. (2001). The corneal wound healing response: cytokine-mediated interaction of the epithelium, stroma, and inflammatory cells. *Prog. Ret. Eye Res.* **20**, 625–637.

Wirstschafter, J. D., Ketcham, J. M., Weinstock, R. J., Tabesh, T., McCloon, M. K. (1999). Mucocutaneous junction as the major source of replacement palpebral conjunctival epithelial cells. *Invest. Ophthalmol. Vis. Sci.* **40**, 3138–3146.

Wolosin, J. M., Xiong, X., Schutte, M., Stegman, Z., and Tieng, A. (2000). Stem cells and differentiation stages in the limbocorneal epithelium. *Prog. Ret. Eye Res.* **19**, 223–255.

Wolosin, J. M., Budak, M. T., and Akinci, M. A. M. (2004). Ocular surface and stem cell development. *Int. J. Dev. Biol.* **48**, 981–991.

Wu, R., Zhu, G., Galvin, S., Xu, C., Haseba, T., Chaloin-Dufou, C., Dhouailly, D., Wei, Z. G., Lavker, R. M., Kao, W. Y., and Sun, T. T. (1994). Lineage-specific and differentiation-dependent expression of K12 keratin in rabbit corneal/limbal epithelial cells: cDNA cloning and Northern blot analysis. *Differentiation* 55, 137–144.

Wu, X., Rabkin-Aikawa, E., Guleserian, K. J., Perry, T. E., Masuda, Y., Sutherland, F. W., Schoen, F. J., Mayer, J. E., Jr., and Bischoff, J. (2004). Tissue-engineered microvessels on three-dimensional biodegradable scaffolds using human endothelial progenitor cells. *Am. J. Physiol. Heart Circ. Physiol.* **287**, H480–H487.

Yamaguchi, J., Kusano, K. F., Masuo, O., Kawamoto, A., Silver, M., Murasawa, S., Bosch-Marce, M., Masuda, H., Losordo, D. W., Isner, J. M., and Asahara, T. (2003). Stromal cell–derived factor-1 effects on *ex vivo* expanded endothelial progenitor cell recruitment for ischemic neovascularization. *Circulation* **107**, 1322–1328

Yang, P., Seiler, M., Aramant, R., and Whittemore, S. (2002). Differential lineage restriction of rat retinal progenitor cells *in vitro* and *in vivo*. *J. Neurosci. Res.* **69**, 466–476.

Young, M. J. (2005). Stem cells in the mammalian eye: a tool for retinal repair. *APMIS* **113**, 845–857.

Zhao, X., Das, A. V., Thoreson, W. B., James, J., Wattman, T. E., Rodriguez-Sierra, J., and Ahmed, I. (2002). Adult corneal limbal epithelium: a model for studying neural potential of non-neural stem cells/progenitors. *Dev. Biol.* **250**, 317–331.

Zieske, J. D. (1994). Perpetuation of stem cells in the eye. *Eye* **8**, 163–169.

Zieske, J. D., Bukusolglu, G., and Yankauckas, M. A. (1992). Characterisation of a potential marker of corneal stem cells. *Invest. Ophthalmol. Vis. Sci.* **33**, 143–152.

# Chapter Sixty-Eight

# Corneal-Tissue Replacement

*Jeffrey W. Ruberti, James D. Zieske, and Vickery Trinkaus-Randall*

## I. INTRODUCTION

Synthetic corneal replacements have been under speculative design for over 200 years. Recently developed advanced cell culture methods and the success of other engineered connective tissues (e.g., skin) have spurred numerous efforts since the early 1990s to produce engineered corneal tissue. Though there has been significant progress toward the development of synthetic corneal replacements and in the culturing of human corneal cells onto and within supporting natural substrates, in general the net results of these efforts has been somewhat disappointing. Currently two clinically viable synthetic cornea replacements and no clinically viable tissue-engineered corneas are approved for use in end-stage corneal disease. Given the relative success of tissue engineers in their approach to creating skin substitutes, the cornea would seem an attractive alternative next target. The lack of success is not surprising, for the cornea, unlike skin, comprises a highly organized stroma populated by three distinct cell types. Two particularly difficult problems must be resolved before suitable tissue-engineered corneal analogs can be generated: (1) replacement or replication of the ultrastructural arrangement of the stromal extracellular matrix and (2) generation of a reliable source of untransformed, proliferative corneal endothelial cells. In this chapter we review efforts to develop synthetic corneal analogs, examine the short history of corneal-tissue engineering, and present the latest methods that have shown promising results toward the realization of functional tissue-engineered corneal replacements.

Corneal disease and injury are the second leading causes of vision loss, affecting over 10 million people worldwide (Whitcher *et al.*, 2001). The leading cause of corneal blindness in the Third World is trachoma (4.5 million people), followed by corneal injury (1.5 million people). The vast majority of these cases would benefit from a suitable corneal replacement. In developed countries such as the United States, conditions that indicate the use of donor corneal grafts include Fuch's dystrophy, keratoconus, pseudo and aphakic bullous keratopathy, corneal stromal dystrophies, corneal scaring, and herpes simplex virus. As of 2000, there were approximately 33,000 corneal transplants performed each year in the United States (Aiken-O'Neill and Mannis, 2002). Though the donor cornea *potential* far outstrips the need, there are real cultural, logistical, and technical difficulties associated with procurement of donor graft material. In addition, donor grafts are typically variable in quality and can fail due to immunological rejection or endothelial decompensation, resulting in a failure rate for initial grafts of approximately 18% (Thompson *et al.*, 2003). Finally, there is a rising threat to the number of viable donor corneas in the form of LASIK procedures, which automatically disqualify donor tissue for transplantation. It is thus of significant clinical interest to develop a suitable corneal replacement for donor graft material. There is also a need for corneal replacements that are insensitive to the loss of limbal stem

*Principles of Tissue Engineering, 3rd Edition*
*ed. by Lanza, Langer, and Vacanti*

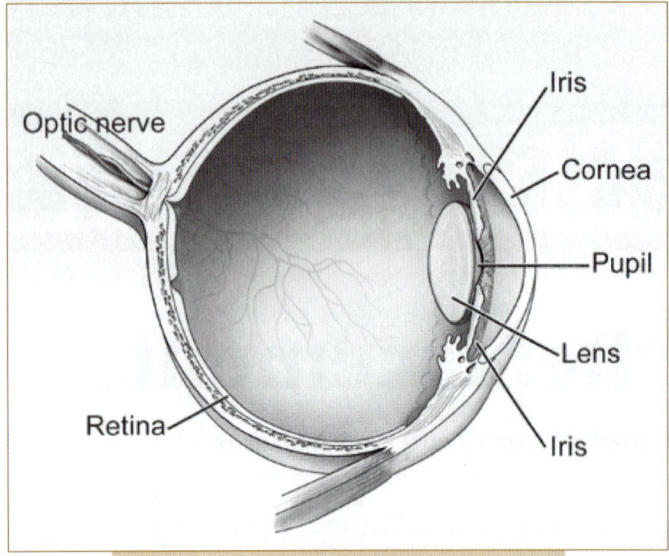

**FIG. 68.1.** Anatomy of the eye. The cornea is the transparent portion of the ocular tunic, which must protect the fragile intraocular contents, allow incident light to pass through to the retina, and provide a smooth optical interface for refraction. Figure from the National Eye Institute, National Institutes of Health.

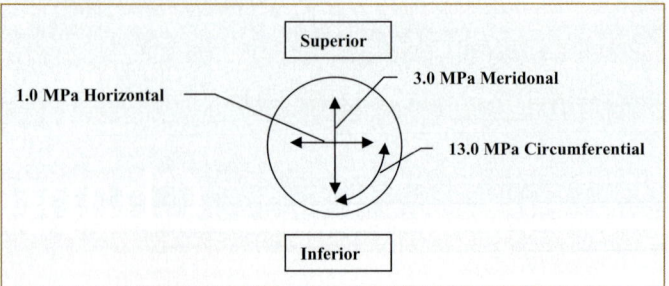

**FIG. 68.2.** Variation in corneal tensile strength as a function of direction. Though the cornea is typically considered a simple nematic stack of lamellae comprising aligned collagen that alternate in orientation by 90°, recent investigations demonstrate an array of fibrils that run circumferentially around the periphery. Preferred fibril orientation and aligned fibril concentration are reflected in the tensile strength, or tensile modulus. The figure shows the tensile modulus found in test strips excised and loaded in the direction of the arrows.

cells. A constellation of diseases and injuries can destroy the stem niches or disrupt the delicate systems that maintain the ocular surface (severe chemical burns, Stevens-Johnson, Sjogren's, and ocular pemphigoid, etc). For these conditions, donor grafts are contraindicated and an alternative approach such as autologous stem cell transplantation (Schwab *et al.*, 2000) or an all synthetic optic (Yaghouti and Dohlman, 1999) is required. Nonetheless, design of any tissue replacement requires a detailed understanding of the function that was performed by the native tissue. It is instructive, then, to review the functional role that the cornea plays in the visual system. There are three principal design requirements for the natural cornea: (1) protection of the fragile intraocular contents, (2) transparency to visible light, and (3) formation of a nearly perfect optical interface to refract light onto the retina. Nature has evolved an extremely elegant structure capable of meeting these critical design criteria simultaneously.

## Gross Corneal Structure and Function

The cornea is the anteriormost, clear portion of the tough ocular tunic (see Fig. 68.1). Corneal tissue comprises one-sixth of the total ocular globe and is a highly structured, relatively acellular, transparent collagenous tissue that joins the more disorganized and opaque sclera at the limbus. Though not perfectly circular, the diameter of the human cornea is about 12 mm. The average radius of curvature of the central anterior surface is 7.8 mm, and it is roughly 520 microns thick at the center and 650 microns thick in the periphery.

## Protection

The tensile mechanical strength of the cornea is a critical design criterion, in that the ocular tunic must survive significant traumatic impacts without rupture. The overall biomechanical properties of the cornea are complex because the tissue is highly anisotropic (different properties in different directions), heterogeneous, and viscoelastic (reviewed in Ethier *et al.*, 2004). The preferred collagen fibril orientation is reflected in the measured tensile strength of test specimens. Figure 68.2 shows the tensile modulus of the cornea as a function of location and orientation of the excised test strip. The tensile moduli shown in the figure were obtained from test specimens that were aligned with the principal collagen fibril directions, as determined by x-ray analysis (Aghamohammadzadeh *et al.*, 2004) and SEM (Radner *et al.*, 1998; Radner and Mallinger, 2002).

The compressive properties of the cornea depend on its hydration and are primarily the result of the fixed negative charges bound to the stromal proteoglycans (Hedbys, 1961). At normal hydration there are approximately 40 mEq (Kostyuk *et al.*, 2002) of fixed negative charge, which imbues the cornea with a swelling pressure of approximately 60 mmHg (Hedbys *et al.*, 1963).

## Transmission of Light

At normal hydration, the corneal can transmit up 97% of the incident visible light (Cox *et al.*, 1970). This value depends on the relative water content and distribution in the stroma (Goldman *et al.*, 1968). It is important to note that the water content (and transparency) of the cornea is maintained by a complex transport system that requires a functioning endothelium (pump) and epithelium (barrier) to operate (reviewed in Ruberti and Klyce, 2002).

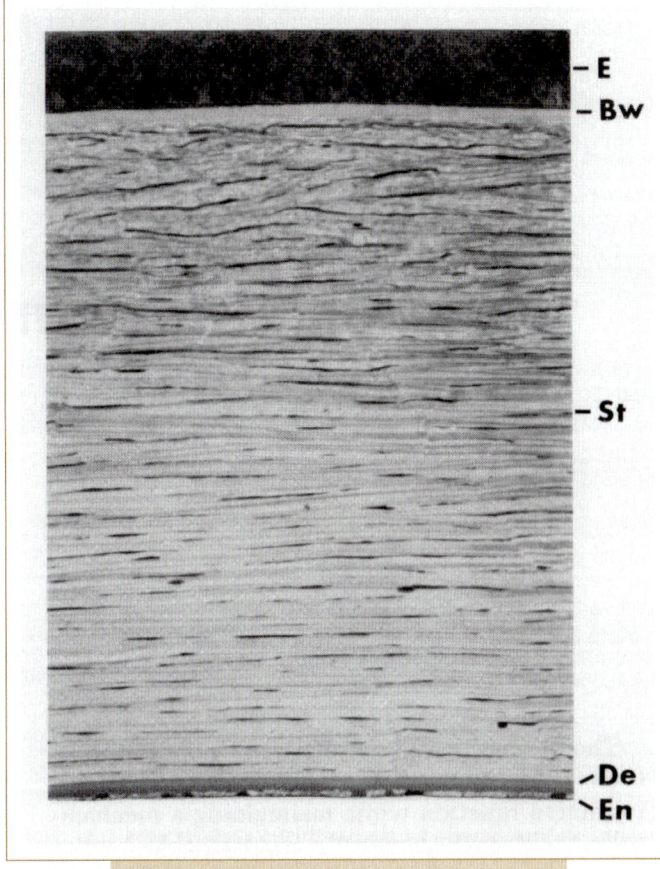

**FIG. 68.3.** The cornea (approximately 500 microns thick in humans) is a layered structure comprising the epithelium (E), Bowman's layer (Bw), stroma (St), Descemet's membrane (De), and the endothelium (En). Reproduced, with permission, from Klyce and Beuerman (1997).

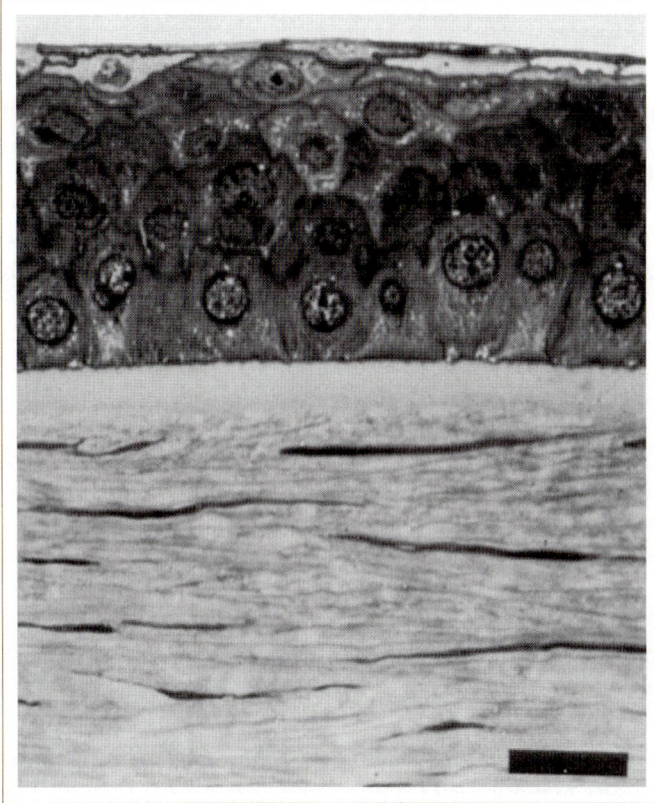

**FIG. 68.4.** Light micrograph of corneal epithelium and anterior stroma. Note stratified cell layer with basal cells (B), wing cells (W), and desquamating surface cells (D). Thin basal lamina (BL) resides at stromal epithelial interface. Bar is 20 microns. Reproduced, with permission, from Klyce and Beuerman (1997).

## Refraction of Light

The combination of the nearly perfectly spherical corneal anterior surface and the index of refraction change at the air/tear film interface generates approximately 80% (42 diopters) of the total refractive power of the human visual system.

## Corneal Microscopic Structure and Function

Three distinct morphological layers are important to the function of the cornea: the epithelium, the stroma, and the endothelium. Figure 68.3 shows the orderly structure of the tissue and the relative dimensions of the three major layers.

## Epithelium

The corneal epithelium is a 50-μm-thick multicellular "tight" stratified squamous epithelium comprising three distinct functional cellular strata. The deepest cellular layer, the stratum germinatum, is the only layer capable of undergoing mitosis (Hanna and O'Brien, 1960). The middle layer comprises the daughter cells (wing cells) of the basal layer, which are pushed anteriorly. The surface layer comprises the squamous cells that form the complete tight junctions that generate the primary *protective* barrier in the cornea. Like most squamous epithelia, the corneal epithelium is continually turned over (once every five to seven days; Hanna and O'Brien, 1960). This process depends on a stable supply of stem cells, which reside in niches at the junction between the cornea and the sclera (limbus) (Schermer *et al.*, 1986). A healthy epithelium has five to seven layers of cells and a basement membrane comprising laminin, type IV collagen, identifiable hemidesmosomes, and anchoring fibrils (Fig. 68.4).

Pathologies that chronically disrupt the ocular surface mucosa (ocular cicatricial pemphigoid, Stevens Johnson syndrome, etc.) or disrupt tear production (Sjogren's) or injuries that destroy the limbal stem cell niches (severe chemical or alkali burns) are contraindications for donor grafts (and any future engineered cornea) because the ocular surface will not epithelialize or wet properly. Epithelial cells are derived from surface ectoderm during embryogenesis.

## Stroma

The stroma is putatively the secretory product of invading mesenchymal cells during development that populate and then expand the extracellular matrix in the potential corneal space between the lens and the nascent epithelium (Cintron *et al.*, 1983). In adults, the stroma is a relatively acellular (3–10% quiescent corneal keratocytes by volume) 500-μm thick, extracellular matrix comprising hydrated type I/V heterotypic collagen fibrils (15% wet weight) of uniform diameter (~32 nm) in humans (Meek and Leonard, 1993); glycosaminoglycans (GAGs) keratan sulfate and dermatan sulfate (1% wet weight; Anseth, 1961); various proteoglycan (PG) core proteins (Axelsson and Heinegard, 1975); and other protein constituents, including fibronectin, laminin, and type VI collagen. The collagen fibrils are packed in 300–500 parallel arrays (lamellae) tangent to the corneal surface (Hamada *et al.*, 1972) and are principally responsible for the observed tensile mechanical properties of the cornea. The PGs and their associated GAGs contribute to the cornea's compressive and swelling material properties (Hedbys, 1961) and to the uniform spacing of the collagen fibrils.

## Endothelium

During development of the corneal endothelium, mesenchymal cells of neural crest origin migrate between the developing lens and the corneal epithelium. Endothelial cells are initially morphologically indistinct from stromal cells but begin to distinguish themselves by the formation of sinuous lateral borders, which interdigitate with adjacent cells to ultimately form the endothelial monolayer (Cintron *et al.*, 1988). In adults, the corneal endothelium is a transporting monolayer of approximately 400,000 hexagonal cells 20 microns in diameter and 4–6 microns in height (Fig. 68.5). This membrane is of critical importance to the maintenance of corneal transparency (reviewed in Ruberti and Klyce, 2002). It has long been known that the cornea is normally maintained in a state of relative deturgescence and that its tendency to imbibe fluid can cause swelling, opacity, and blindness. However, it was not until 1972 that Maurice discovered that the active mechanism that keeps the cornea dehydrated is a fluid pump located in the endothelium. This dehydrating mechanism is part of a sophisticated corneal transport system, which is described in detail in Ruberti and Klyce (2002). Nonetheless, the untransformed endothelial cells, which are critical to normal corneal function and which will be an important part of any tissue-engineered cornea, have, until recently (Engelmann *et al.*, 1999, 2004; Joyce and Zhu, 2004; McAlister *et al.*, 2005), been refractory to attempts at multiple expansions in culture.

## Stromal Ultrastructure

The corneal stroma comprises 90% of the total corneal thickness. It is thus not surprising that it plays a major role in providing the principal functions of the cornea (protec-

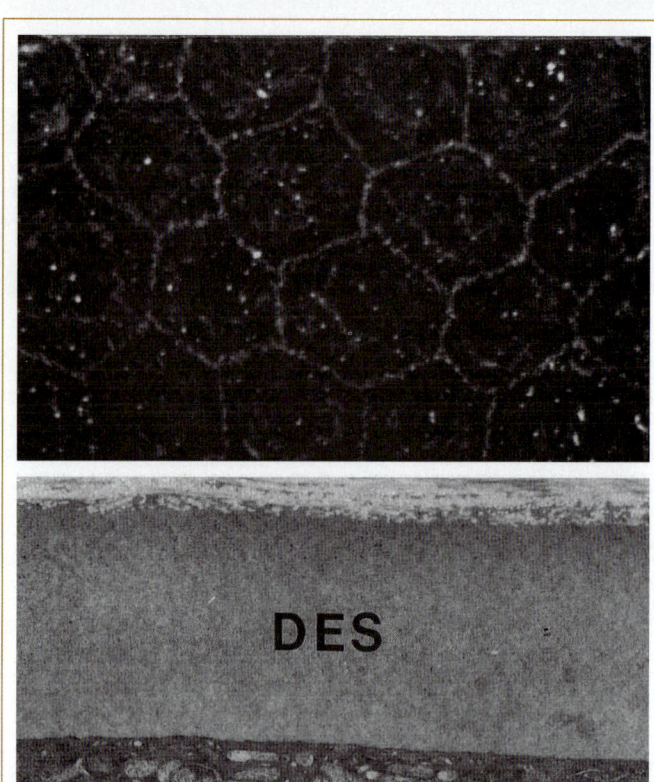

**FIG. 68.5.** Transverse (*top:* ×3150) and en face (*bottom:* ×900) electron micrographic views of the corneal endothelium. The en face view illustrates the hexagonal packing of the endothelial cells. Reproduced, with permission, from Beuerman and Pedroza (1996).

tion, transmission, and refraction). On close inspection, it is the exquisite *ultrastructural* organization of the stromal ECM that allows the cornea to meet these difficult design requirements. Figure 68.6 shows the organization of the corneal stromal collagen fibrils, which have a virtually monodisperse diameter distribution, uniform local interfibrillar spacing, and no interfibrillar covalent cross-links and are arranged in parallel arrays tangential to the corneal surface. This remarkable nanoscale arrangement of fibrils, which persists throughout the cornea, is *directly* responsible for corneal strength (Ethier *et al.*, 2004), clarity (Goldman *et al.*, 1968) and to some extent for its ability to refract light. The exquisite stromal organization is also the primary reason that attempts to produce a viable cornea via tissue engineering have, thus far, been unsuccessful. Any corneal analog comprising a collagenous extracellular matrix (ECM) must reproduce this arrangement or risk being too weak,

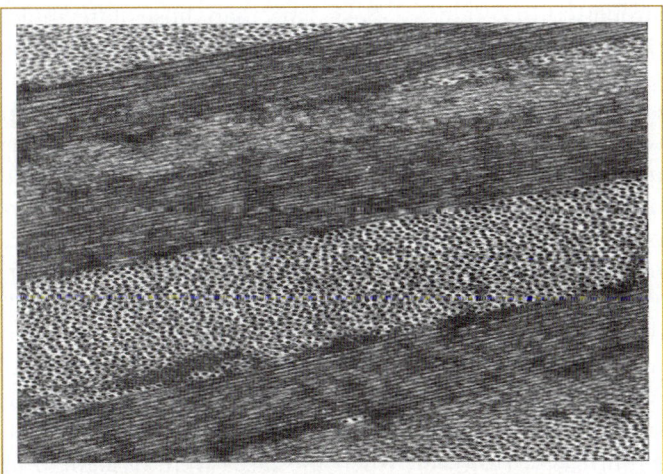

**FIG. 68.6.** Transmission electron micrograph of collagen fibril arrangement in cornea. Corneal collagen fibrils have a highly monodisperse diameter distribution and are arranged in oriented arrays within lamellae approximately 1–2 microns thick. (Section is normal to tangent plane.)

too opaque, or too irregular to form an appropriate refractive surface.

## II. SYNTHETIC CORNEAL REPLACEMENTS

Nondegradable synthetic corneal replacements, if ultimately successful, could obviate the need for donor corneas or tissue-engineered corneal constructs. As early as the 1700s, synthetic corneas were designed and implanted to give persons who had bilateral loss of eyesight finger-counting vision. Currently, synthetic corneas are indicated for persons affected by specific conditions, such as severe chemical burns, ocular pemphigoid, Stevens-Johnson syndrome, and recurrent graft rejections. In the late 1700s, one of the first documented efforts to develop a synthetic cornea was performed by Pellier de Quengsy (1789). In these trials he inserted a glass plate into an opaque eye. The plate was not stable and was rapidly extruded. Subsequently, efforts were made to stabilize the central optic with celluloid plates and egg membranes (Dimmer, 1891), plexiglass, and acrylate (Binder and Binder, 1956; Gyorffy, 1956) or supporting plates of solid or fenestrated poly(methylmethacrylate) (PMMA) (Castroviejo et al., 1969). No effort was made either to achieve the correct refractive index or to optimize the corneal curvature. The constructs were not biocompatible, and as a consequence cellular ingrowth did not occur and the device was not anchored. These initial prostheses led to the development by Cardona (1964) of "through and through devices."

Cardona was a pioneer in the development of keratoprostheses and demonstrated that highly purified and polymerized materials could be maintained in the cornea. The materials first used by him were rigid, impermeable, and secured with sutures, and neither keratocyte ingrowth nor deposition of matrix proteins was reported. A number of the devices protruded through the eyelid and used the surrounding tissue for stabilization. Devices of this style were used where lubrication of the cornea was minimal and often gave the person finger-counting vision. The more current Cardona devices include a PMMA optic surrounded by a perforated skirt of poly(tetrafluoroethylene) (PTFE) reinforced with a mesh composed of poly(ethylene) terephthalate (Dacron) and autologous tissue. The complexity of the device was matched only by the surgery, for the device is inserted through the eyelid, often requiring the removal of the lens, iris, and ocular muscles. Because there was no UV-absorbing tissue remaining, the optic was tinted to absorb UV irradiation. The final device often gave no more than 30° of visual field. These devices continued to be extruded because the material chosen for the periphery did not permit cellular ingrowth and the synthesis of matrix proteins. In contrast, the free edge permitted necrosis and allowed for epithelial downgrowth. Attempts to achieve interpenetration or anchoring of the material led to the development of the osteo-odontoprosthesis (OOKP) (Strampelli and Marchi, 1970). The premise of the Italian group was that a biological material would be more easily tolerated than a synthetic material. To manufacture their device, an acrylic optical core was embedded within the core of a tooth. The insertion of the device was complicated and was often accompanied by infiltration with vascularization, abscess formation, and extrusion. Blencke et al. (1978) modified the osteo-odonto prosthesis by replacing the tooth portion with a glass ceramic, which was again modified by Caiazza et al. (1990), who implanted a modified device using bone. These devices have been maintained for more than 12 years and continue to be used when surgery for penetrating keratoplasty is not suitable (Hille et al., 2005).

By the 1980s, investigators proposed that biological acceptance and retention of through-and-through devices depended on the ability of stromal fibroblasts to migrate into the device and deposit a matrix. To allow tissue integration, White and Gona (1988) devised a device with a central optical cylinder of poly(methyl methacrylate) (PMMA) and a periphery of Proplast with pore sizes ranging from 80 to 400 microns. The device had a high extrusion rate and induced signficant inflammatory responses. The next generation of porous skirt materials included the expanded poly(tetrafluoroethylene) (Impra) (White and Gona, 1988), carbon fibers that projected from the optic haptic as rays (Kain, 1990), and a melt-blown polybutylene:polypropylene web (Trinkaus-Randall et al., 1991).

The development of the olyofinic webs (3M, St. Paul, MN) (Trinkaus-Randall et al., 1990) was based on surgical mesh fabrics used in skin, muscle, and arterial wall. The optimal pore size determined in those tissues (Bobyn et al., 1982; Cook et al., 1989; Pourdeyhimi, 1989) was used in the development of the pore size for the cornea. A fiber diameter that ranged from 2 to 12 microns with a void volume of 88% was found to be optimal in *in vitro* and *in vivo* studies

(Trinkaus-Randall *et al.*, 1990, 1991). Legeais *et al.* (1994, 1995) used a similar model and fabricated a device from skirt porous PTFE (IMPRA) using a pore size of 18–22 microns, which also permitted fibroplasia.

The biological response (deposition of matrix components) of the cornea to devices based on synthetic biomaterials is of paramount importance if they are to remain stable *in vivo*. Trinkaus-Randall *et al.* (1991) found that the deposition of types I and III collagen in synthetic discs did not increase significantly until day 21. More recently, Drubaix's observations (Drubaix *et al.*, 1996) supported the earlier data that there are changes in the synthesis of collagen and that they resemble that of an injured cornea (deposition of types I and III collagen). These early studies demonstrated that the tensile strength of the implant interface was not sufficient unless cellular ingrowth occurs (Chirila *et al.*, 1993). When the deposition and synthesis of glycosaminoglycans were evaluated, the migration of cells into the discs mimicked the scenario of wound repair (Funderburgh and Chandler, 1989; Cintron *et al.*, 1990; Brown *et al.*, 1995). Heparan sulfate, absent in control corneas, was detected, and Trinkaus-Randall and Nugent (1998) demonstrated that FGF-2 was present transiently at the margin between the synthetic discs and the stroma. Trinkaus-Randall *et al.* (1994) found that pretreatment (by plasma cleaning or by coating with biologically active molecules) enhanced collagen deposition fivefold, which was associated with increased fibroplasia and decreased edema.

The development of implantable synthetic optic materials has had a long and laborious history and seeks to produce a material that permits light transmission in the visible range and protects the retina from ultraviolet damage. The latter requires a modification to absorb light at wavelengths below 300 nm and is critical in devices where the lens has been removed. It should have tensile properties similar to that of a native cornea and allow for continuous reshaping of both anterior and posterior curvatures. Epithelialization is a point of controversy, and most investigators now support Dohlman's premise that it only makes care more difficult and can cause more difficulties, such as epithelial downgrowth (Dohlman *et al.*, 1974).

The most commonly used material, PMMA, has been made in ultraviolet-absorbing forms for intraocular lenses, can be shaped and retains clarity over time, exhibits low toxicity, and is well tolerated. However, it is both rigid and hydrophobic and has poor diffusivity.

Hydrogels were evaluated because they are elastic, can be manufactured to have high tensile strength and maintain transparency, and can be modified to support epithelial cell growth (Laroche *et al.*, 1989; Trinkaus-Randall *et al.*, 1994; Wu *et al.*, 1994; Latkany *et al.*, 1997). A poly(vinyl alcohol) hydrogel (refractive index of 1.42) was developed for use as an optic by Trinkaus-Randall *et al.* (1988). The hydrogel was prepared from a poly(vinyl trifluoroacetate) precursor and possessed a higher elasticity (436% vs. 15%) and tensile strength (110 kg/cm$^2$ vs. 35 kg/cm$^2$) than those of the intact cornea. It possessed a water content around 70%, which is similar to that of the cornea (79%). When ultraviolet absorbers (phenyl-benzotraizole and a benzophenome derivative) were copolymerized, transmission of ultraviolet light was not detected below 340 nm (Tsuk *et al.*, 1997). The surface of the poly(vinyl alcohol) hydrogel was modified to support epithelial cell growth and maintenance (Latkany *et al.*, 1997) by coating with ECM components or by plasma treatment. The epithelial cells became confluent and multilayered, displayed typical cytoarchitecture, showed a basal–apical polarity, and synthesized a typical repertoire of adhesion proteins *in vitro*. This was also detected *in vitro* when the surfaces were modified with the tripeptide RGD (Laroche *et al.*, 1989). However, when the discs were implanted *in vivo*, the apical surface of the cultured cells was altered, with the shearing force associated with blinking and the cells sloughed off the surface of the optic. Epithelialization has continued to be one of the most challenging tasks in the development of keratoprostheses.

Two synthetic corneas have been tested in humans. The first "biocolonizable" was implanted in 24 human patients, and five failures were reported in the first six months (Legeais *et al.*, 1995). In a second model, the optic was silicon coated with PVP and the surface remained hydrophilic. These devices were implanted in eyes with dystrophy's, pseudophakic edema, pemphigus, and burns. Unfortunately, signs of decolonization after 9–12 months were noted; however, they were used successfully as temporary devices, which could help patients with retinal and corneal pathologies (Dong *et al.*, 2006) awaiting a transplant.

Dohlman, who was instrumental in the development of keratoprostheses (Dohlman *et al.*, 1974), has reevaluated conditions having the best prognosis and what factors can be used as predictors of outcome (Dohlman, 1997). He has continued to use a multipiece plastic device that is not colonized; however, he has demonstrated that with meticulous follow-up the patients can have an excellent outcome. Eyes with graft failure had the best outcome, while those with chemical burn and Stevens-Johnson syndrome had the worst outcomes.

Chang (Chang *et al.*, 1995) used a silicone membrane altered by plasma-induced graft copolymerization–treated HEMA. It was sutured as a penetrating keratoplasty; however, the graft did not remain in the plane of the host cornea and extruded by three months. It is likely that stromal downgrowth occurred where the fibroblasts grew under and essentially formed a second cornea, gradually extruding the cornea.

A number of colonizable devices ranging in size (4.5–9 mm) and design were evaluated in rabbits using both penetrating keratoplasty and lamellar keratoplasty models (Trinkaus-Randall *et al.*, 1997). The time course of fibroplasia depended on the model and on the size of the porous skirt. The devices that were the most successful were those

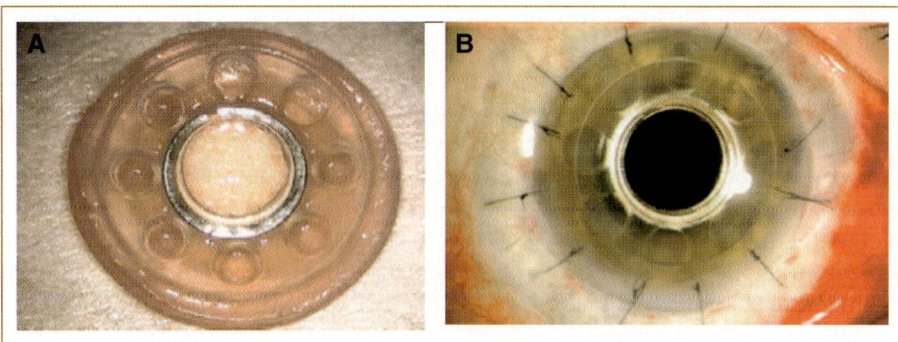

**FIG. 68.7.** Dohlman–Doane keratoprosthesis. **(A)** Prosthesis shown with large-diameter donor tissue carrier and central optical cylinder comprising PMMA. **(B)** Prosthesis one day after implantation. Reproduced, with permission, from Aquavella *et al.* (2005).

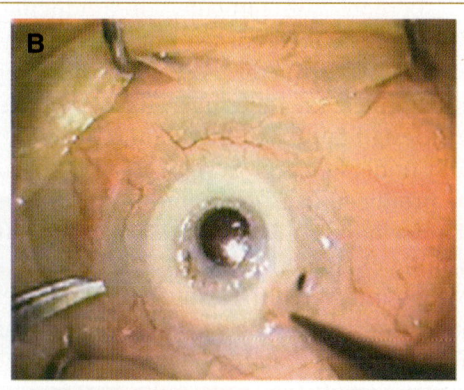

**FIG. 68.8.** Alphacor™ keratoprosthesis. **(A)** PHEMA keratoprosthesis with central optical zone porous skirt. **(B)** Alphacor at 12 weeks immediately following second stage of procedure, during which the tissue anterior to the optical zone was removed. Reproduced, with permission, from Hicks *et al.* (2003).

where Descemet's membrane was left intact. They were placed as posterior as possible in the stromal lamellar bed, and the flanges were inserted into the stromal lamellae. Fibroplasia was detected between two and three weeks. Devices with larger skirts were not supported by the cornea in the long term, presumably because there was less diffusivity of nutrients. Smaller models (7 mm in diameter) were retained in rabbit corneas for a minimum of one year, the duration of the trial. Collagen types III and VI were detected after six weeks. Heparan sulfate, a GAG, absent in the unwounded stroma but present in injured stromas, was detected (Trinkaus-Randall and Nugent, 1998). In all of these models the anterior chamber remained intact and no sign of leakage was detected. Parel's group also used a rabbit lamellar keratoplasty model to assess their nonperforating device after eyes were injured and scarring and vascularization induced (Stoiber *et al.*, 2005). The most optimal model using synthetic materials is probably that of Chirila, where both the peripheral rim and the central optic are cross-linked polymers of HEMA (Chirila *et al.*, 1993). In this model the periphery was made from a hydrogel sponge. The sponge has a high water content (80% w/w), facilitating diffusivity through the material. Fibroplasia was detected when the devices were pretreated with type I collagen (Chirila *et al.*, 1993). The central optic was an HEMA-2 ethoxyethyl methacrylate acrylamide. The entire device, given the name AlphaCor™, has been implanted in human patients (Fig. 68.8). The change in visual acuity pre- and postop is remarkable, with preoperative acuity being finger counting and postoperative (20/200) (Hicks *et al.*, 2003). The studies have now focused on the role of the environment (i.e., medications, cigarette smoke) and on the long-term stability and integrity of the device. Calcium was detected as a white opacity, indicating that topical medications can modify the haptics (Morrison *et al.*, 2006). In addition, brown deposits were found to be a result of smoking (Hicks *et al.*, 2004). Both of these indicate that to maintain the stability of the devices, patient management is paramount.

## III. CORNEAL-TISSUE ENGINEERING

The cornea is a fairly simple, sparsely populated, avascular, multilaminar structure that, at first glance, would seem to be an attractive target for tissue engineers. The corneal cell population comprises three different cell types (epithelium, fibroblasts/keratocytes, and endothelium) that require successful culturing techniques if a cornea is to be constructed by tissue-engineering methods. The history of corneal tissue engineering begins with culturing epithelial cells onto collagen mats (Friend *et al.*, 1982; Geggel *et al.*, 1985) and has continued to the production of full-thickness, multilaminar, populated constructs grown on collagen/polymer scaffolds, complete with innervation (Li *et al.*, 2003). In spite of the success in populating collagen-based gels with epithelium, fibroblasts/keratocytes, and endothelium, no corneal construct has been produced that is suitable for clinical use. There are two principal barriers to the production of a clinically viable cornea: (1) reproduction of a suitable corneal stromal analog and (2) the ability to populate the construct with viable human corneal endothelial cells. There has been significant progress on the latter problem (Joyce and Zhu, 2004), while the former has met with little documented success. Thus, at the moment it appears that the most critical barrier to successful tissue engineering of a cornea lies in the construction of a suitably strong, clear stromal matrix.

## A Note on Engineering and Tissue Engineering

Perhaps the difficulty of producing an appropriate tissue-engineered cornea might be discerned if the difference between the terms *tissue engineering* and *engineering* is examined. Engineering by its very nature employs a rational design approach that requires three things: (1) that we fully understand the design requirements of the structure to be produced (e.g., loads on the stanchions of a bridge); (2) that we have available at the time of construction a toolbox of fully characterized materials, devices, and systems that will be integrated to produce our engineered product (i.e., structural steel, concrete, etc.); and (3) that we have available methods that allow us to synthesize the components into the whole (i.e., nuts, bolts, hammers, etc.). In contrast, classical tissue-engineering methods are not rational, in that we often do not fully know the function of the tissue that is being engineered, we do not understand the properties of the materials (ECM components) and devices (cells) from which the final structure will be derived, and we have little idea about how the whole is assembled from the component parts. Why proceed without rationality, then? The answer, of course, is that we expect that the cells, which are typically seeded into/onto a degradable scaffold or "synthetic scar," will appropriately remodel the scaffold such that intended tissue function is reproduced. This approach has been reasonably effective for tissues that do not depend *critically* on highly organized and often highly anisotropic tissue ultrastructure, such as skin. However, for the engineering of tissues whose principal function depends critically on the precise arrangement of extracellular matrix components (i.e., cornea, ligament, tendon etc.), the results have been less than satisfactory. We must surmise, then, that there is a limit to the ability of seeded fibroblasts to remodel scaffolds, which is similar to the remodeling of a scar and can be extremely slow (Cintron *et al.*, 1982). For the corneal stroma, whose function depends on the exquisite organization of the ECM, perhaps it is time to pursue a more circumspect course that seeks to take a rational design approach rather than relying solely on the cells "to put things right." This can be accomplished either by attempting to construct the tissue ultrastructure *de novo* before cell seeding or by giving the seeded cells as many cues (mechanical, chemical, etc.) and as much support as possible to allow them to construct an appropriate ECM. Nonetheless, it is clear that more information about fibroblastic cell-behavior modulators (i.e., strain, growth factors, etc.), ECM component properties, and ECM component assembly kinetics is required.

## Engineering Corneal Epithelium

The epithelium is responsible for the protection of the stroma and the maintenance of the tear film. Though it is not necessary to produce corneal constructs with adherent epithelia to replace failed corneas (donor corneas are debrided of their epithelium prior to implantation), it is important that they support the attachment, spreading, and, ultimately, growth to confluence of epithelial cells to form a robust anterior corneal barrier. In addition, there have been numerous recent attempts to reengineer the surface of stem cell–deficient corneas via autologous stem cell transplantation on carriers such as amniotic membrane (Gomes *et al.*, 2003). Finally, organotypic corneal culture systems for *in vitro* research do require an intact functional corneal epithelium. Stratified squamous epithelial cells possess minimal intrinsic differentiation differences (Sun and Green, 1977) and may alter their behavior/differentiation as a function of the culture system environment. It has also been demonstrated that epithelial cells, which are moderately difficult to culture, may be supported by soluble factors produced by cocultured fibroblastic cells (Green *et al.*, 1977) or by treatment with exogenous factors, including EGF (Savage and Cohen, 1973). These observations suggest that environmental stimuli and cell-derived soluble factors are responsible for the final differentiation of the cells, which implies that culturing systems may be "tuned" to elicit appropriate cell behavior. An ideal corneal epithelium would form a confluent multilayer (five to seven cell layers) stratified structure with complete tight junctions at the surface, desmosomes between the cells, and hemidesmosomes with anchoring fibrils projecting into an adherent BM.

### Corneal Epithelial Cell Culture on Substrates

Corneal epithelial cells were first isolated from rabbit corneas and reseeded directly onto tissue culture dishes to form a confluent monolayer in long-term investigations by Sundar-Raj *et al.* (1980). However, to engineer a suitable cornea for replacement purposes, it is important to grow the cells on an implantable substrate. Numerous studies have investigated the behavior of corneal epithelial cells that have been isolated and then reseeded onto various natural substrates (Savage and Cohen, 1973; Friend *et al.*, 1982; Geggel *et al.*, 1985; Trinkaus-Randall *et al.*, 1988, 1988). One of the first investigations of corneal epithelial cell behavior following isolation and seeding onto substrates (denuded rabbit corneal stroma with or without basement membrane (BM)) was conducted by Friend *et al.* (1982). The results demonstrated that isolated corneal epithelial cells will adhere, grow to confluence, and stratify (to three layers *maximum*) on stromal carriers (with or without BM). In spite of the low number of cellular strata, the cells displayed microplicae, intracellular desmosomes, and normal-looking organelles. The major difference between cells seeded onto denuded stroma and those seeded onto residual basement membrane was that the cells seeded onto BM formed hemidesmosomes within one week. Cells seeded onto denuded stroma produced a BM within four days but did not generate hemidesmosomes for the period of the investigation. The next advance shifted the culturing of epithelium to reconstituted collagen instead of native corneal stromal tissue. This is a critical step in the development of an engineered cornea. In the study by Geggel *et al.* (1985), epithelial cells were lifted as an intact sheet from rabbit corneas and draped onto collagen gels, with the specific intent to *engineer* the ocular surface. The cells, which were grown on vitrogen gels in culture using SHEM (Jumblatt and Neufeld, 1983), maintained their confluence and apposition but did not stratify fully (only three layers) and lacked basement membrane components. Notably, the vitrogen disc was well tolerated for six weeks *in vivo*, with little induced inflammation or degradation of the onlay.

In 1988, Trinkaus-Randall *et al.* (1988) demonstrated that corneal basal epithelial cells (Trinkaus-Randall and Gipson, 1985) respond differently depending on the nature of the substrate on which they were plated: fibronectin, laminin type IV, or type I collagen. Their results suggest that laminin and type I collagen significantly enhance cell proliferation and support the growth and attachment (via sites of adhesion) of a normal-*looking* epithelium *in vitro*.

*Airlifting* is one of the more significant cell culture developments and was adapted from methods used to culture skin keratinocytes (Prunieras *et al.*, 1983) by Minami *et al.* (1993). Minami cultured bovine epithelial cells to subconfluence on a reconstituted type I collagen gel and then aspirated the medium until the cells on the gel surface were just exposed to (fully humidified) air. By 21 days, the epithelium formed a confluent multilaminar (six layers of cells) that stratified into three morphologically distinct cellular layers (basal, wing, and superficial). Distinct desmosomes were observed between cells, as was a fragmentary basal lamina on the collagen gel side of the epithelium. It is important to note that Minami performed this investigation as part of a full corneal reconstruction, with both keratocytes and endothelial cells present in the constructs. The paper noted that the *presence of the other cell lines enhanced epithelial proliferation and differentiation* as compared to controls on bare collagen gels. In a more comprehensive investigation, Zieske *et al.* (1994) cultured untransformed rabbit epithelium on untransformed rabbit keratocyte–populated collagen gels at a wet, dry, or moist interface, with or without mouse corneal endothelium (immortalized cell line). Zieske noted a profound influence of the endothelial cells (which were plated directly onto the reconstituted collagenous stroma) on the assembly of epithelial basement

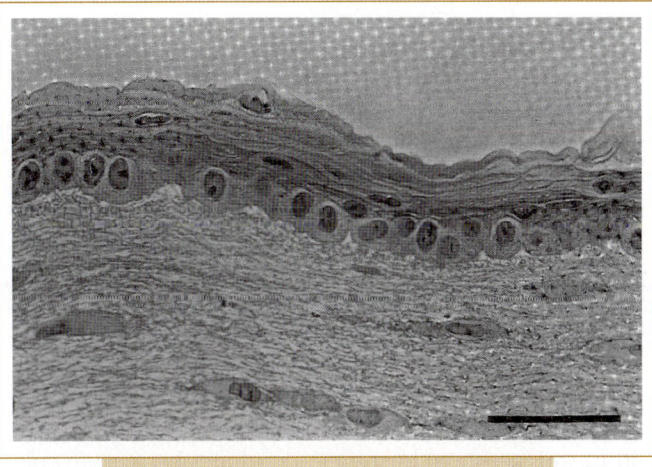

**FIG. 68.9.** Light micrograph of stratified corneal epithelium grown using airlift technique and in organotypic, coculture with endothelium and keratocytes. Reproduced, with permission, from Zieske *et al.* (1994). Bar = 50 microns.

membrane and epithelial cell differentiation. In the presence of the endothelium, (1) the epithelium forms an identifiable basement membrane, (2) the epithelial basal cells become more columnar, (3) α-enolase is restricted to the cells in the basal layer, (4) collagen IV and laminin localize to a continuous layer at the stromal–epithelium interface, and (5) α6 integrin becomes localized to the stromal side of the basal cells. Figure 68.9 shows the morphology of the epithelium grown at the moist air interface and in the presence of the endothelial cells.

### Engineering Human Corneal Epithelium

Culture of human corneal epithelial cells onto collagen-based substrates for the purposes of corneal tissue engineering has been performed with reasonable success (Ohji *et al.*, 1994; Germain *et al.*, 1999; Griffith *et al.*, 1999; Li *et al.*, 2003). Ohji *et al.* (1994) demonstrated that primary cultures of human epithelial cells produce basement membrane components (laminin, collagen type IV, collagen type VII, and perlecan) when cultured on stromal blocks or on collagen gels (with or without fibroblasts present). This study demonstrated the feasibility of culturing human epithelial cells on collagen gels. (See Protocol 68.1.)

In her seminal paper in which the three types of human corneal cells were used to generate an artificial cornea, May Griffith (Griffith *et al.*, 1999) obtained the "best results" through the use of immortalized epithelial cell lines that had been screened for biochemical, electrophysiological, and morphogical similarities to low-passage human primary cells. Unfortunately, the parameters that critically define the patency of healthy corneal epithelium (basement membrane components, desmosomes, hemidesmosomes, tight junctions, etc.) were not described in the paper. It should be noted that Griffith's group has focused on the task of

---

**Protocol 68.1**   Cell culture protocol of Ohji *et al.* (1994).

**Cell isolation:** *Epithelial cells.* Human corneal epithelial cell sheets were derived from $2 \times 2$-mm limbal explants of donor corneas. Explants were placed epithelial side up in 35-mm culture dishes using modified SHEM and were incubated at 37°C and 5% $CO_2$ 95% air for 12–15 days to gain confluency of epithelial cells (explants were removed at 10 days). Sheets of cells were lifted using dispase II. *Keratocytes.* Human keratocytes were derived from the central cornea of donor eyes. Explants were cultured in the modified SHEM for eight days and then moved, prior to keratocyte outgrowth, to 35-mm dishes and bathed in modified Eagle's medium containing 15% FBS and 40-µg/mL gentamicin sulfate. Confluent keratocyte cultures were lifted using trypsin (0.05%) and EDTA (0.02%). Media was changed once per week, and passages four to seven were used.

**Medium:** The medium used to support the cells in culture was a modified version of the SHEM (supplemented hormonal epithelial medium) of Jumblatt and Neufeld (1983), which contained DMEM-HEPES, Hams F-12, 15% FCS, 0.5% DMSO, 20 µg/mL gentamicin, 10 µg/mL EGF, 5 µg/mL insulin, 5 mM L-glutamine, and 0.1 mg/mL cholera toxin.

**Substrate preparation:** Debrided stromal collagen blocks were used for one study, and collagen gels derived from acid-soluble type I collagen were used for another. Vitrogen-100® was gelled (by neutralization and warming) in the presence of FBS with or without isolated human keratocytes. The total gel volume was 1.5 mL, and the keratocyte seeding count was $5 \times 10^5$ cells.

**Seeding/culturing protocol:** Epithelial cell sheets were placed onto stromal blocks or onto the collagen gels and covered with 1 mL of medium. One day 22 mL of medium was added to the dish, and the medium was changed every two to three days. The cell/gel constructs were harvested at one, two, three, and four weeks.

---

**Protocol 68.2**   Cell culture protocol of Germain *et al.* (1999).

**Cell isolation:** *Epithelial cells.* Human corneal epithelial sheets were teased from excised limbal annuli following exposure to dispase II neutral in 10 mM HEPES calcium chloride (1 mM) for 18 h at 4°C. The sheets were cut into small pieces and centrifuged for 10 min at $200 \times g$ at 20°C. The cells were seeded onto 25-cm² culture flasks with irradiated 3T3 fibroblasts in DME-HAM. *Keratocytes.* Human keratocytes were obtained from a mixed culture of fibroblasts and epithelial cells after culturing in fibroblast medium. Passages two to nine were used.

**Medium:** Epithelial medium (DME-HAM): Combination of Dulbecco–Vogt modified Eagle's medium (DME) with Ham's F12 in 3 : 1 ratio, supplemented with 24.3 µg/mL adenine, 5 µg/mL insulin, $2 \times 10^{-9}$ M 3,3′5′-triiodo-L-thyronine, 5 µg/mL human transferrin, 0.4 µg/mL hydrocortisone, $10^{-10}$ M cholera toxin, 10% fetal clone II newborn calf serum, 10 ng/mL human EGF, 100 IU/mL penicillin G, and 25 µg/mL gentamycin sulfate. Keratocyte medium: DME with 10% FCS antibiotics.

**Substrate preparation:** 9.5 mL of 2.8 mg/mL acetic acid extracted bovine dermal collagen (neutralized) and 1.0 mL DME containing $2.5–3.5 \times 10^5$ human corneal fibroblasts were mixed rapidly at 4°C with 5.6 mL DME and 3.7 mL FCS. 2.0 mL of the solution was poured into a 35-mm petri dish containing an anchorage (filter paper). The collagen was gelled at 37°C and covered with the keratocyte medium.

**Seeding/culturing protocol:** $3.8 \times 10^6$ epithelial cells in 0.4 mL of DME-HAM were added to the keratocyte/collagen sustrate (after four days of culture). The cells were allowed to attach for 2 h before adding 2 mL of additional DME-HAM. The medium was changed daily, and biopsies were taken after three days.

---

constructing stromal analogs that promote neurite ingrowth (Li *et al.*, 2003). Their motivation is derived from the well-known fact that loss of sensory innervation affects the corneal epithelium and, in particular, slows epithelial healing rates (Araki *et al.*, 1994). It is possible that improving neurite population rates may improve outcomes; however, current donor corneal graft neurons are necessarily severed from their roots, and epithelial failures are not a prominent cause of graft loss.

Though immortalized cells are useful tools for studying behavior *in vitro* (culture methods for immortalized cells are not reviewed), untransformed cells hold the greatest promise for production of an engineered corneal replacement. The location of epithelial harvesting of cells is important because there are spatial differences in the proliferative capacity of human corneal epithelial cells (Lindberg *et al.*, 1993). Limbally derived explants show the greatest ability to undergo multiple passages in culture. Two investigations have used primary human corneal epithelial cells for the purpose of corneal-tissue engineering (Germain *et al.*, 1999; Orwin and Hubel, 2000). In the first one (see Protocol 68.2), Germain *et al.* (1999) cultured primary human limbal epithelial cells onto collagen gels populated with primary human keratocytes. The cocultured epithelium stratified within three days with cuboidal basal cells and elongated suprabasal cells. Epithelial basal cells expressed β1, α3, α5, and α6 integrin subunits within 10 days (three submerged and seven at air–liquid interface). Laminin and type VII collagen was localized to the basal cell cytoplasm and to the junction of the epithelium and stroma within 10 days.

Orwin and Hubel cultured untransformed human epithelial cells, keratocytes, and endothelial cells alone or in combination onto specially constructed collagen sponges

**Protocol 68.3** Cell culture protocol of Orwin and Hubel (2000).

**Cell isolation:** *Epithelial cells.* Human epithelial cells were isolated from excised human corneas by gentle scraping following incubation overnight at 4°C in dispase (1.5 mg/mL) in Hanks balanced salt solution. The cells were seeded into a tissue culture flask at $2 \times 10^4$ cells/cm$^2$ in the epithelial medium. *Keratocytes.* Human keratocytes were obtained following collagenase digestion (37°C overnight) of debrided corneas. The keratocytes were washed and resuspended in the keratocyte medium. *Endothelial cells.* Human endothelial cells were trypsinized from the cornea, swabbed off, centrifuged, and seeded into tissue culture flasks at 8500 cells/cm$^2$ in endothelial medium.

**Media:** *Epithelial medium*: Keratinocyte growth medium (modified MCDB 153, Clonetics, Walkersville, MD) supplemented with 5 µg/mL insulin, 0.18 µg/mL hydrocortisone, 0.4% bovine pituitary extract, and 1 ng/mL EGF. *Keratocyte medium:* DMEM-F12/HAM with 10% FCS. *Endothelium medium:* MEM supplemented with 10% FBS, 0.292 mg/mL L-glutamine, 0.014 µl/mL 2-mercaptoethanol, 13.13 mg/mL chondroitin sulfate, 1 mM sodium pyruvate, and 0.1 mM nonessential amino acids.

**Substrate preparation:** Collagen sponges were prepared with acid-treated bovine hyde. Washed bovine corium was frozen, lyophilized, and ground to produce particles 83–125 microns in size. Collagen dispersions 0.5% wt/vol were blended with HCL to pH 3.0 at 4°C for 1 min. The dispersion was the lyophilized at –30°C and dehydrothermally cross-linked in a vacuum at 110°C/2.5

torr for five days. The sponges were gamma irradiated (17,500 rads). Prior to seeding, the sponges were cut into 14-mm-dia $\times$ 1–2-mm-thick disks and hydrated with 150 µL of media.

**Seeding/culturing protocol:** *Single cells:* Epithelial, keratocyte, or endothelial cells were seeded onto 14-mm-diameter $\times$ 1–2-mm thick sponges at 118,000 cells/cm$^2$ and allowed to sit at 37°C for one to two hours. Following attachment, 1 mL of media was applied to the culture dish. Every other day, 4 mL of fresh media were added to the dish for a duration of 21 days, *Cocultures:* (1) Epithelium/endothelium on same sponge. Endothelium was seeded on the center of the smooth side of the sponge and allowed to attach at 37°C in 5% CO$_2$ for 1–2 h. The dish was then supplemented with media and incubated overnight. The sponges were inverted and epithelium was seeded on the porous side in 50 µL of media. Following 1–2 h for attachment, the dish was supplemented every other day with equal parts of endothelial and epithelial media for 14 days. (2) Epithelium on sponge, endothelium cultured on other side of transwell membrane. The same protocol as before, but the epithelial cells were cultured on the bottom of a flask and the epithelial cells were cultured on the sponge, on top of a transwell insert. (3) Keratocytes in sponge, epithelium on sponge. Keratocytes were seeded onto the porous side of the sponge and cultured for three days. On day 4, epithelial cells were seeded onto the sponge surface and allowed to attach. Sponges were supplemented every other day with equal parts epithelial and keratocyte media for 15 days.

---

(Orwin and Hubel, 2000) (see Protocol 68.3). They found that coculture of epithelial cells and endothelial cells (cultured on the sponge or separated by a transwell insert) produced the highest degree of epithelial cell layering (two to four layers of cells). However, airlift techniques were not used in this study, and the specific markers of basement membrane formation, tight junction formation, or desmosome formation were not investigated.

## Summary of Epithelium in Culture

In summary, there is a long history of culturing primary corneal epithelial cells onto substrates with the intent of producing a morphologically appropriate epithelial membrane. The net results indicate that a reasonably mimetic (albeit thin) corneal epithelium can be produced *in vitro* if the substrate has appropriate components (type I collagen or laminin), if there are other cells (preferably endothelium) cocultured along with epithelial cells, if the medium is suitable, and if the culture is airlifted. Under these conditions, the epithelial cells will stratify to produce a multilaminar

architecture and will generate basement membrane components spontaneously. However, the history of culturing *human* epithelial cells is shorter, and there have been fewer assessments on the nature of the epithelium that has been produced. Specifically, there have been no studies that have shown formation of a reasonably continuous basement membrane with anchoring fibrils and associated hemidesmosomes. There have also been no long-term studies of human epithelial cultures. Nonetheless, progress to date has been good, and it is likely that further attempts to epithelialize tissue constructs with human cells await the production of a more appropriate stromal analog.

## Engineering the Corneal Stroma

The corneal stroma is an exquisitely organized collagen proteoglycan composite matrix populated sparsely by quiescent keratocytes. The stromal architecture provides high mechanical strength (Ethier *et al.*, 2004) and optimal curvature for refraction and allows light to pass virtually unimpeded (Goldman *et al.*, 1968). The keratocyte population also plays an important role in maintaining corneal clarity by index matching through expression of water-soluble corneal crystallin proteins (Jester *et al.*, 1999). Any tissue-engineered corneal construct must reproduce the function of the corneal stroma, or, as has been the case thus far, it will fail to achieve utility in the clinic. As with epithelial cell culturing, there has been some success in populating natural substrates with corneal fibroblasts/keratocytes. On the other hand, very little progress has been made toward producing a suitably strong, clear corneal stromal matrix into which corneal fibroblasts may be seeded.

## Keratocyte Cultures

In the human cornea *in vivo*, the stroma contains a sparse, quiescent (for an alternative view, see Muller *et al.*, 1995) network of index-matched, dendritic cells termed *keratocytes*. Human corneal keratocytes (HCKs) are distributed throughout the tissue and are thought to maintain the stroma through the continual production of stromal proteoglycans (Hassell *et al.*, 1992) and possibly proteins (Muller *et al.*, 1995). Upon corneal wounding, keratocytes can dedifferentiate into or are replaced by fibroblasts and myofibroblasts (Garana *et al.*, 1992), which participate in the normal wound-healing process. Corneal myofibroblasts are thought to play an important role in wound contraction (Jester *et al.*, 1987; Garana *et al.*, 1992), while corneal fibroblasts are synthetically active and can be induced to produce matrix components *in vitro* (Newsome *et al.*, 1974; Stoesser *et al.*, 1978). It was noted in the early 1970s that human corneal keratocytes are more proliferative under certain culture conditions (see Protocol 68.4) than either corneal epithelium or endothelium (Newsome *et al.*, 1974). Newsome *et al.* found that keratocytes (they actually had a fibroblastic morphology) migrated from explants after four days and then overran the epithelial cell layers that had emerged from the explant earlier. This was one of the first indications that HCKs could be grown successfully in large numbers in culture. However, it is important to understand the mutability of the keratocyte phenotype if stromal tissue is to be engineered. Normal stromas contain quiescent keratocytes, while wounded stromas contain fibroblastic and myofibroblastic phenotypes as well. In culture, the conversion of keratocytes from their quiescent phenotype to a fibroblastic phenotype is thought to be induced by the presence of serum in the medium (Beales *et al.*, 1999), while the differentiation to a myofibroblastic phenotype can be induced by adding TGF-β (Jester *et al.*, 1996). There is also evidence of the partial reversibility of fibroblasts to keratocyte morphology following the removal of the serum in the medium (Berryhill *et al.*, 2002). Reversion of fibroblasts to keratocyte phenotype, which possess corneal crystallins (aldehyde dehydrogenase) in large quantities, will ultimately be important for restoration of stromal transparency (Jester *et al.*, 1999). Most stromal analogs in tissue-engineered corneas include serum

---

> **Protocol 68.4** *Cell culture protocol of Newsome et al. (1974).*
>
> **Cell isolation:** *Epithelial cells, endothelial cells, and keratocytes.* Human corneal cells were allowed to migrate from the explants of cadavers between the ages of 18 and 68. Eyes were enucleated within eight hours of death and stored at 4°C in 0.85% NaCl for no more than 32 hours. Explants of $1 \times 2 \times 5$-mm of central corneal epithelium, Bowman's, and superficial stroma were used for the epithelial and keratocyte culture, while strips of Descemet's were used for the endothelial culture. The explants were rinsed twice in saline G (Puck's N-15 minus the amino acids) and placed into culture dishes with a sterile coverslip on top of them. Primary outgrowths of the cells between 10 and 20 mm in diameter were propagated by washing in $Ca^{++}$- and $Mg^{++}$-free PBS followed by incubation for 20 min at 37.5°C in 1.0 mL of Coon's enzyme (six units collagenase per mL, 0.1% trypsin, and 2.0% chick serum) with 4 mM EDTA for epithelium and endothelium and without EDTA for keratocytes. The suspended cells were pipetted onto new dishes containing growth medium.
>
> **Media:** *Growth medium:* Modified Ham's F12 with 5% FCS and 3 μg/mL penicillin.
>
> **Seeding/culturing protocol:** Cells were pipetted onto 60-mm plastic culture dishes or into 30-mL plastic tissue-culture flask. Media was changed every three days.

---

in the keratocyte medium (Germain *et al.*, 1999; Griffith *et al.*, 1999; Orwin and Hubel, 2000), which should induce the transformation to a fibroblastic phenotype and can induce contraction of collagen gel–based stromal substrates (Borderie *et al.*, 1999) as well.

### Keratocytes in Resorbable Scaffolds

To date, with the exception of some data reported in conference abstracts (Guo *et al.*, 2002; Germain *et al.*, 2004), stromal analogs for tissue-engineered corneas are created by seeding keratocytes/fibroblasts into collagen-based resorbable scaffoldings (Borderie *et al.*, 1999; Germain *et al.*, 1999; Griffith *et al.*, 1999; Orwin and Hubel, 2000). Unfortunately, there are very few data on the phenotype, synthetic, and catabolic behavior of corneal fibroblasts in such systems. However, it is known that corneal fibroblasts in three-dimensional collagen gels will organize into orthogonal arrays (Doane and Birk, 1991) and secrete type I, type V, and type VI collagen (Borderie *et al.*, 1999). Without long-term studies of the remodeling of such stromal analogs, it is not possible to comment on the potential of the classical resorbable scaffolding approach to corneal-tissue engineering. Nonetheless, the process of keratocyte-driven remodeling of a degradable scaffolding is likely to resemble stromal scar resolution, which, even under ideal conditions *in vivo*, can take years (Cintron *et al.*, 1982). If this fact is coupled with the difficult requirement that an implanted resorbing scaffold retain its clarity and mechanical strength during the remodeling process, then the prospect of developing a suitable cell-seeded resorbable system appears to be remote.

## Keratocytes in Scaffold-Free Systems

Development of the mammalian corneal stroma takes place under specialized conditions in a very thin layer between the epithelium and the nascent lens (Cintron *et al.*, 1983). It is possible that the rapid development of highly organized matrix can take place only under conditions that include a high density of synthetic cells (Cintron *et al.*, 1983) in a confined space (Martin *et al.*, 2000). Under appropriate conditions, human corneal fibroblasts cultured on plates at high density form multiple layers of cells that alternate in the orientation of their long axes by 90° in adjacent layers (Newsome *et al.*, 1974). Newsome also found that high-density cultures of human corneal fibroblasts deposit large quantities of extracellular fibrils (10–15 nm in diameter). Stimulation of collagen production by the addition of ascorbic acid to corneal fibroblast culture media has been demonstrated by Stoesser *et al.* (1978) and in general fibroblast cultures by Grinnell *et al.* (1989). Two laboratories have attempted to produce corneal stromal analogs in scaffold-free systems by stimulation of matrix production using ascorbic acid derivatives (Guo *et al.*, 2002; Germain *et al.*, 2004). In the investigation by Guo *et al.*, untransformed human corneal fibroblasts produced stroma-like arrays of alternating collagen fibrils embedded in a stratified cell–matrix construct (see Fig. 68.10). The most striking aspect of the cell-assembled constructs is their resemblance to the matrix produced during development *in vivo*. In particular, the alternating arrays of collagen, the stratification of the active keratocytes, and the high cellular density are all found in developing stroma as well (Cintron *et al.*, 1983). The ability of untransformed human corneal fibroblasts to produce organized arrays of collagen when stimulated is a very tantalizing development in the pursuit of an engineered cornea. Further details on this approach, detailed methods, and results await publication in the open literature.

## Generation of Stromal Analogs

Corneal stromal–tissue engineering is derived primarily from cell culture techniques that began with the seeding of corneal epithelial cells on denuded stromal carriers in the early 1980s (Friend *et al.*, 1982). This was quickly followed by the seeding of cells onto reconstituted type I collagen gels, which formed an ersatz stroma, with the specific intent of reengineering the corneal surface (Geggel *et al.*, 1985). Simple, reconstituted type I collagen derived from bovine skin or from rat tail tendon was used by many investigators interested in constructing stromal analogs (Geggel *et al.*,

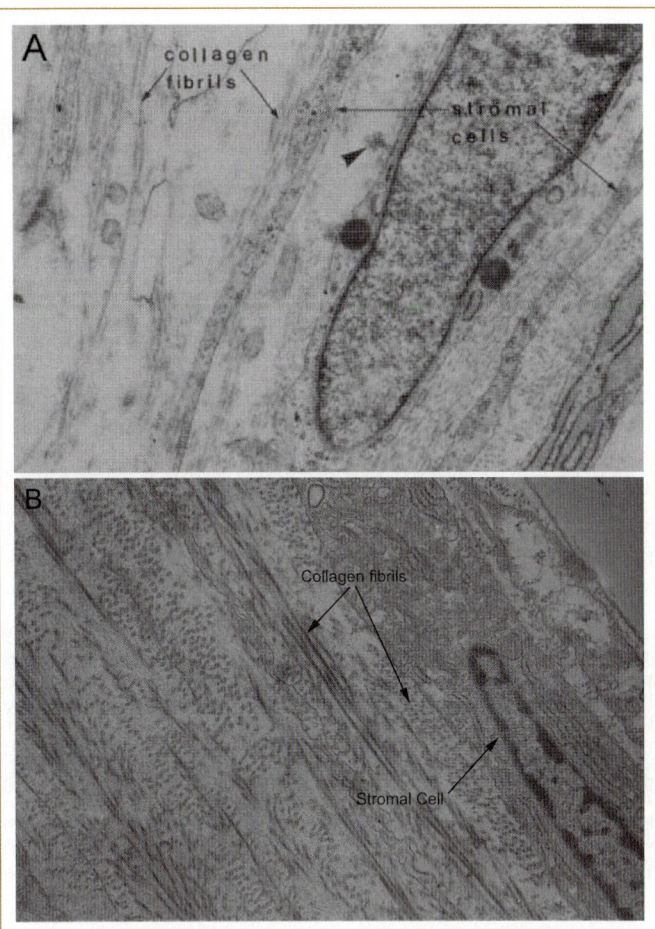

**FIG. 68.10.** Transmission electron micrographs of developing cell-assembled stroma. **(A)** Developing rabbit stroma in 21-day-old fetus. Image from Cintron *et al.* (1983). **(B)** Five-week scaffold-free stromal construct produced by untransformed human corneal fibroblasts in culture. Note synthetically active stromal cells and orthogonal arrangement of fibrils in lamellae.

1985; Minami *et al.*, 1993; Ohji *et al.*, 1994; Zieske *et al.*, 1994; Germain *et al.*, 1999). Acid-soluble collagen monomers such as Vitrogen-100® (which claims 80% type I collagen in monomeric form) can be purchased in quantity and form gels reliably by neutralization with NaOH followed by warming to 37°C. The resulting gels are weak, possess low density, are disorganized, and comprise long, native-looking type I collagen fibrils. Reconstituted type I collagen gels are highly biocompatible and can be gently gelled around a population of corneal fibroblasts (Germain *et al.*, 1999) but are not known to be particularly stable *in vivo* or *in vitro* without some form of cross-linking (Charulatha and Rajaram, 1997). To improve the stability of their reconstituted type I collagen stromal analog, Griffith *et al.* (1999) exposed the gel to 0.03–0.04% glutaraldehyde to cross-link it and then treated the gel with glycine to scavenge the unbound glutaraldehyde. Because of the known importance of glycosaminoglycans as

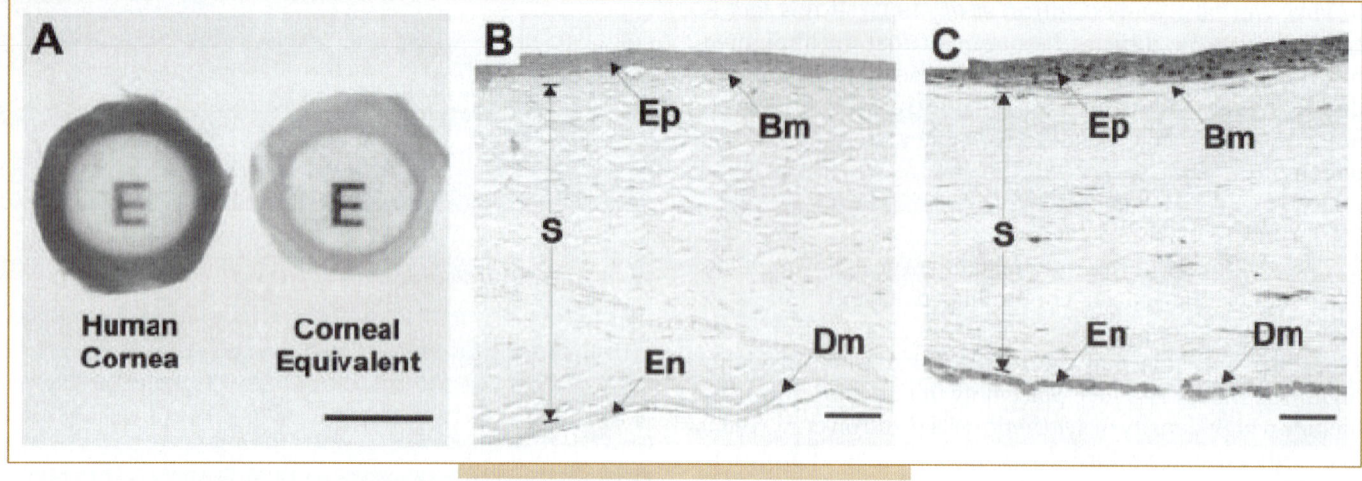

**FIG. 68.11.** Corneal construct of Griffith *et al.* **(A)** Demonstration of construct clarity compared to postmortem human cornea. Scale bar 10 mm. **(B)** Cultured eye bank cornea. **(C)** Cultured corneal equivalent. Scale bar 100 microns. Epithelium – EP; Bowman's membrane – Bm; stroma – S; Descemet's – Dm; endothelium – En. Reproduced, with permission, from Griffith *et al.* (1999).

structural molecules (Hedbys, 1961) in the cornea, chondroitin sulfate was also added to the collagen-based stromal analogs of Griffith (Griffith *et al.*, 1999). The net result of the cross-linking and addition of GAGs to the simple type I collagen gel, followed by human fibroblast seeding, produced a clear, populated corneal stroma (Fig. 68.11). Unfortunately, the details of the resulting stromal architecture were not thoroughly investigated, nor were the mechanical properties, so our ability to evaluate the success of this formulation is limited.

Reconstituted type I collagen gels typically comprise very low-density, randomly arranged arrays of collagen fibrils and possess difficult-to-control porosity and poor mechanical properties. In an effort to manipulate the mechanical properties and pore size of collagen constructs, Orwin and Hubel (2000) constructed stromal analogs that began with collagen fibrils (bovine corium) formed into sponges by grinding, lyophilization, and dehydrothermal (DHT) cross-linking. The sponges supported colonization by human primary epithelium, endothelium, and keratocytes alone or in combination (epithelium/endothelium or epithelium/keratocytes) *in vitro*. The details of the stromal matrix/keratocyte interaction was investigated to a limited extent and demonstrated that the keratocytes colonized and proliferated on the sponges (Fig. 68.12). The keratocytes also appeared to align with fibrils and to synthesize matrix components (collagen and proteoglycans).

In a separate study, Orwin *et al.* (2003) added exogenous chondroitin sulfate and/or hyaluronic acid to their collagen sponges and assessed the resulting mechanical and optical properties. Ultimately, the collagen sponges were better mechanically and optically than collagen gels but were substantially inferior to native corneas (equilibrium compressive modulus was three orders of magnitude smaller than native cornea; transparency was, at best, half of native

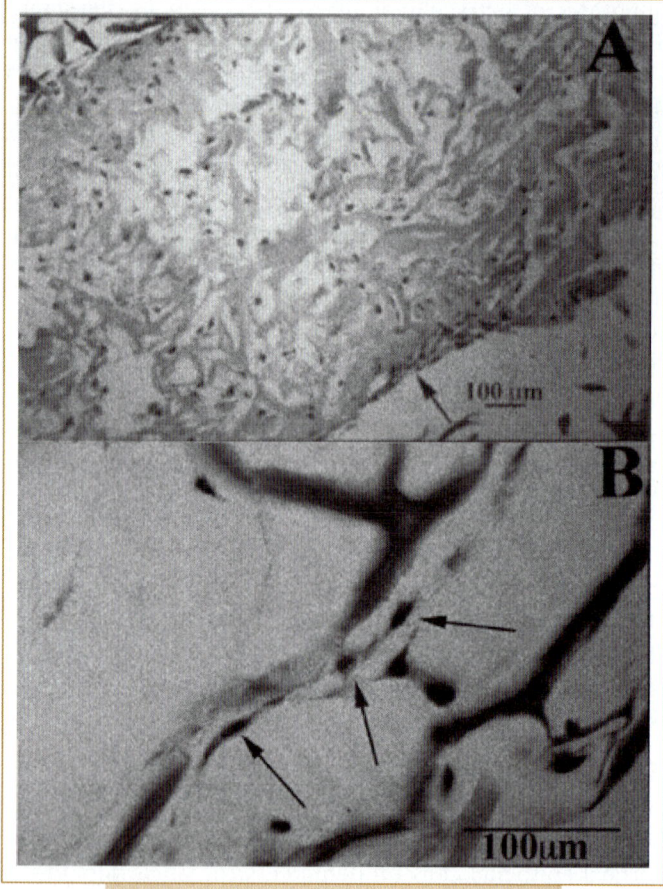

**FIG. 68.12.** H&E-stained stromal analog of Orwin and Hubel. **(A)** By day 10 of culture, the cells have migrated through the stromal sponge. **(B)** By day 22 of culture, cells have elongated along the collagen fibers. Reproduced, with permission, from Orwin and Hubel (2000).

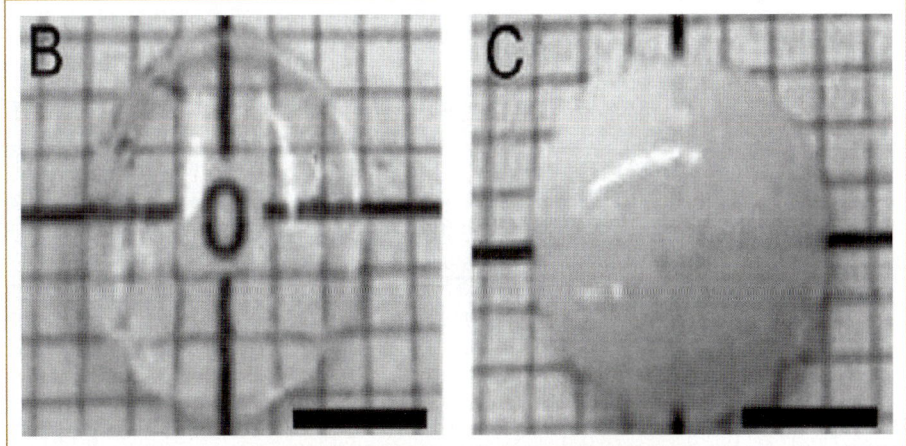

**FIG. 68.13.** Photographic comparison of the TERP5–collagen composite of Li *et al.* (2003) **(B)** and a reconstituted type I collagen gel **(C)**. Reproduced, with permission, from Li *et al.* (2003).

corneal transparency). Clearly, the difficulty of constructing a suitably strong and clear corneal stromal analog has hampered attempts to produce a full-thickness engineered cornea. In response to this difficulty, Griffith's research group abandoned natural scaffolds and investigated hybrid collagen–synthetic copolymers as potential stromal analogs (Li *et al.*, 2003). The copolymer [poly(*N*-isopropylacrylamide-coacrylic acid-coacryloxysuccinimide), PNiPAAm-coAAc-coASI, or TERP] was used because it spontaneously cross-links proteins and anchors peptides through primary amines. The TERP and a second formulation, TERP5, in which some of the acrylosuccinimide groups were reacted with YIGSR (to facilitate neurite extension and epithelial cell growth), were mixed with bovine collagen and molded to produce a controlled-thickness cornea. The resulting stromal analogs were clear (Fig. 68.13) and epithelialized well *in vitro*, with significantly more layers of cells (five) on the collagen–TERP5 constructs. The TERP5 and TERP matrices both promoted neurite ingrowth and were strong enough to be sutured (unlike the collagen-only gels). Implantation of the collagen–TERP5 stromal construct into micropigs (by LKP, leaving the posterior cornea and endothelium intact) demonstrated quiet integration of the copolymer matrix, rapid epithelialization (including the deposition of type VII collagen at the epithelial–matrix interface) by the host, and an enhanced rate of reinnervation over allografts. However, though the stromal constructs were suturable, care had to be taken to prevent cracking and damage during implantation. The authors state outright that their stromal construct is weak (but did not quantify the strength directly) as compared to native corneal stroma. This is probably due to the random arrangement and low density (3.4% wt/vol) of the collagen–polymer matrix. Unfortunately, the ability of neurites to populate the collagen–TERP5 constructs is quite possibly dependent on the density of the matrix, and any improvement in strength (by increasing the density) could be offset by a reduction in the rate of neurite migration.

## Summary of Stromal-Tissue Engineering

As has been demonstrated, the ability to culture human primary stromal keratocytes has advanced rapidly. The control of their differentiation state, contractile behavior, and ability to promote the secretion of matrix components by stimulation with adjutants like TGF-β and derivatives of ascorbic acid are all promising. However, the true difficulty associated with producing a corneal stromal analog is the exerting of control over the assembly and/or remodeling of the ECM. Understandably, the focus of most investigators has been primarily the behavior of the cellular constituents. The extracellular matrix, in spite of its dominant role in corneal function, has been, until recently, largely disregarded. Reconstituted type I collagen gels or fibrils have performed a useful but temporary role to aid in the study of primary human corneal cell behavior in culture. Their use is derived from their known biocompatibility and success in the tissue engineering of skin (Auger *et al.*, 1998). However, ultrastructurally, a cell-populated, reconstituted type I collagen gel (with or without cross-links or glycosaminoglycans) bears little resemblance to a real corneal stroma (except for the fact that both contain type I collagen and keratocytes). Since the principal functions of the cornea (mechanical integrity and clarity) are *directly* tied to stromal nanoscale structure, it is not surprising that the tissue-engineered corneas generated thus far are typically either not acceptably clear, not acceptably strong, or neither clear nor strong enough for use in the clinic. The problem with using self-assembled random arrays of collagen as a starting point for corneal-tissue engineering can be appreciated in Fig. 68.14, which shows SEMs of a self-assembled collagen gel and the cell-synthesized collagen in human cornea.

The ability of fibroblasts to remodel a randomly arranged collagen gel (as in Fig. 68.14B), either *in vitro* or *in vivo*, appears to be limited. Support for this point of view was provided by Cintron in his seminal work on the remodeling of rabbit trephination wounds (Cintron *et al.*, 1982). Cintron

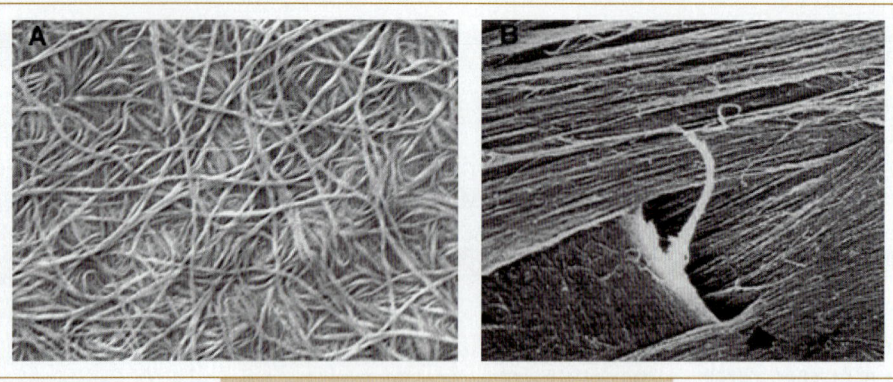

**FIG. 68.14.** Scanning electron micrographs of a reconstituted type I collagen gel **(A)** and the cell-assembled collagen in a native human cornea **(B)**. In the native cornea, arrays of highly organized fibrils split and pass above and below lamellae with different orientations. Current tissue-engineering methods expect seeded cells to resorb and replace randomly arranged collagen with highly organized collagenous matrix. The ability to exert control over the entropic randomization of self-assembled collagen in order to produce a more organized stromal analog may reduce the burden on seeded cells. The image in (B) is reproduced, with permission, from Radner *et al.* (1998).

found that fibrin-filled, full-thickness wounds in young rabbits were not successfully remodeled after two years! The process of primary fibroblast–driven remodeling of a scaffolding or synthetic scar (regardless of its composition) is likely to occur on a similar time scale. For tissues that are highly organized and anisotropic, such as the corneal stroma or tendons, it is thus unlikely that remodeling (either *in vitro* or *in vivo*) can reproduce the native structure in a reasonable period of time. The exquisite organization of highly oriented connective tissue such as cornea was produced in a completely different environment during morphogenesis (Cintron *et al.*, 1983). If a natural corneal stromal analog is to be engineered, we must either provide a similar environment to actively synthetic fibroblasts or focus on ways to control the organization of the collagen *de novo* on the benchtop. Indeed, recent work has demonstrated that untransformed human corneal fibroblasts can be induced to produce a stroma-like matrix (Guo *et al.*, 2002). As discussed, the ability to generate suitable natural stromal scaffold *de novo* (on the benchtop) is in its infancy, but methods such as spin-coating (Ruberti *et al.*, 2003) of collagen already offer some control over collagen assembly (see Fig. 68.15).

Tight control over collagen fibrillogenesis could make the production of suitably strong, clear, natural stromal possible in the near future. In addition, aligned collagen films or simply substrates with nano-scale aligned features may be used to guide the migration of fibroblasts to control the deposition of collagen arrays in scaffold-free systems similar to that of Guo *et al.* (2002).

### Engineering the Corneal Endothelium

Because of the large swelling pressure in the normal corneal stroma (Hedbys *et al.*, 1963) and the lack of interfibrillar cross-linking, corneal tissue has a tendency to imbibe fluid. In 1972, David Maurice determined that the mechanism responsible for countering the swelling pressure and

**FIG. 68.15.** Optical differential interference contrast image of collagen aligned on glass substrate. Collagen fibrils were self-assembled in a flowing confined film to control their orientation. Bar is 10 microns.

keeping the cornea in a state of relative deturgescence (and thus transparent) is an active transporter located in the corneal endothelium. A functioning confluent endothelial layer is not only a critical component of the corneal transport system; it has also been shown to influence positively the structure and behavior of corneal epithelium when cocultured (Zieske *et al.*, 1994; Orwin and Hubel, 2000). Thus, even if the stromal analog comprises a material that does not require "pumping down" to become transparent, the unknown signaling molecules derived from a functioning endothelium may still be required to generate a functional tissue-engineered cornea. However, the difficult task of supplying enough vital, differentiated, and functional untransformed human corneal endothelial cells (HCECs) to populate the construct requires solution. Fortunately, good progress has been made recently in this area.

## Endothelial Proliferative Capacity

Unlike other mammals, such as rabbit and pig, human corneal endothelial cells do not replicate *in vivo*; instead, they change their shape and spread to cover exposed Descemet's membrane following the loss of neighboring cells (Yee *et al.*, 1985). As a matter of course, corneal endothelial cell density declines in normal eyes from a postnatal value of 3000 cells/mm$^2$ at a rate of 0.5% per year (Bourne and Kaufman, 1976), yet adequate pump function is maintained throughout life. However, acceleration of the rate of normal endothelial cell loss through dystrophy, disease, or trauma (cataract surgery) can lead to declining visual acuity if the cell count drops below 500 cells/mm$^2$. Because the endothelial decompensation secondary to cell loss is irreversible *in vivo* (which is a leading indicator for corneal transplantation), there has been significant effort expended to understand why the cells do not replicate and to develop methods to induce their proliferation (reviewed in Engelmann *et al.*, 2004; Joyce and Zhu, 2004). Naturally, these studies have led to attempts to culture primary human corneal endothelial cells *in vitro* and have provided an excellent set of tools that may be extended to induce the colonization of engineered stromal constructs (Joyce and Zhu, 2004). Investigators have demonstrated that, *in vivo*, human corneal endothelial cells are actively maintained in a nonreplicative state but have found that they *do* retain their proliferative capacity and can be successfully cultured onto human stromas (Bohnke *et al.*, 1999; Engelmann *et al.*, 1999; Chen *et al.*, 2001). Transfection with SV40 large-T antigen has been used to generate higher numbers of HCECs (Feldman *et al.*, 1993), which are functional when cultured on denuded human stromas (Aboalchamat *et al.*, 1999). However, the use of immortalized cell lines to produce tissue-engineered corneas for the clinic is controversial and will not be discussed further. Instead, methods used to enhance the proliferative capacity of untransformed HCECs show the most promise for corneal regeneration (Joyce and Zhu, 2004; Engelmann *et al.*, 2004) and are discussed next.

## Expansion and Culture of Corneal Endothelial Cells

Though cell culture techniques have been developed for animal endothelium (Perlman and Baum, 1974; Perlman *et al.*, 1974; Gospodarowicz *et al.*, 1977; Giguere *et al.*, 1982; Savion *et al.*, 1982), they are largely ineffective when applied to adult human corneal endothelium (Newsome *et al.*, 1974). HCECs that are contact inhibited and arrested in the G1 phase of the cell cycle (Joyce *et al.*, 1996a, 1996b) are perniciously nonproliferative, and unique culture methods have evolved specifically for them (Engelmann *et al.*, 1999, Chen *et al.*, 2001). Adult HCECs have been shown to proliferate to some extent *in vitro* when stimulated by appropriate mitogens (such as FBS, FGF, EGF) (Nayak and Binder, 1984)

and/or by growth on appropriate matrix components (such as fibronectin, laminin, bovine endothelial elaborated ECM) (Samples *et al.*, 1991; Blake *et al.*, 1997). Additionally, disruption of the cell–cell junctions with EDTA can elicit a proliferative response (Senoo *et al.*, 2000), as can mechanical wounding (Senoo and Joyce, 2000). Laboratories working on the expansion of HCECs in culture have reported that cells from younger eyes (<30) can be cultured more readily than cells from older eyes (Baum *et al.*, 1979; Nayak and Binder, 1984; Samples *et al.*, 1991). Attempts to induce the proliferation of HCECs in culture have produced several multiparametric studies that have tested the effectiveness of various media components (Nayak and Binder, 1984; Engelmann *et al.*, 1988; Engelmann and Friedl, 1989, 1995; Yue *et al.*, 1989). However, the results of these investigations have been somewhat equivocal and have not demonstrated the ability to culture adult HCECs for multiple passages or at high densities (reviewed in Engelmann *et al.*, 2004; Joyce and Zhu, 2004). Only recently has the culture, expansion, and transplantation of HCECs from both young and old donor eyes been reasonably successful (Engelmann *et al.*, 1999; Chen *et al.*, 2001). Engelmann *et al.* (1999) attempted to seed adult donor HCECs onto the denuded stromas from different donors in culture. Their corneal endothelial isolation procedure is complex (see Protocol 68.5) and employs the use of selection medium which substitutes L-valine for D-valine to inhibit the growth of contaminating fibroblastic cells, which can overrun the HCEC culture. The potential for fibroblast contamination arises from the use of concentrated collagenase (type IV) to release the endothelial cells, because it may also release fibroblasts from the underlying stroma. Engelmann found optimal seeding densities of 2–5 $\times 10^5$ cells per 0.3–0.5 mL of fluid and that gentle centrifugation immediately following seeding aided in the attachment of the seeded cells to the recipient stroma. In spite of the careful protocol, a certain percentage of the cultured cells (not given) exhibited fibroblastic morphology and did not transfer well to the recipient stromas. Of the cultures that did exhibit polygonal morphology, the highest cell density was achieved at 11–12 days and was 1850 cells/mm$^2$ (mean 865, range 200–1850 cells/mm$^2$). Final density counts were independent of donor age, length of time of recipient cornea culturing (before seeding), the number of passages of seeded cells, or the presence of FGF (though FGF did improve the cell morphology). Recipient corneas with seeded endothelium exhibiting a polygonal monolayer deswelled at rates comparable to controls, indicating that pump function was established (however, the sample number was small). In spite of the ability to culture endothelial cells for multiple passages and for long periods, the mean endothelial density on the recipient stromas was still low as compared to normals.

Recently the laboratory of Dr. Joyce has been successful in inducing the proliferation of HCECs in culture (Chen *et al.*, 2001; Joyce and Zhu, 2004; Zhu and Joyce, 2004; Konomi

---

**Protocol 68.5** Endothelial cell culture protocol of Engelmann *et al.* (1999).

**Cell isolation:** Corneas were removed from donor eyes (age 30–70) and stored in MEM medium supplemented with 10% FCS and 1% chondroitin sulfate at 37°C for up to three months. To extract the HCECs, Engelmann first disrupted the cell–cell connections with trypsin EDTA and then disrupted the ECM under the cells with collagenase (type IV). Specifically, the corneas were placed on plastic dishes, endothelium side up, and a few drops of enzyme solution (EDTA/Trypsin, 0.02%/0.8% in HEPES buffered saline) was added to the endothelium and left for 90 minutes at 37°C. Collagenase (type IV) in high concentration (0.5%) in a 1 : 1 mixture of Ham's F12 and Iscove's DME (IF) was then applied to the endothelium for 90 minutes at 37°C to undercut the cells. This was sometimes followed by collagenase treatment at low concentrations (0.04% or 0.025%) for 17 hours. The HCECs were dislodged with 20 mL of IF injected through a small needle. Cell suspensions were concentrated at $900 \times g$ for 10 min and resuspended in 0.5 mL of the selection medium. They were then seeded in one well of a 24-well plat and maintained at 37°C and in 5% $CO_2$ to generate confluent primary cultures.

**Media:** *Selection medium:* The L-valine-free "selection medium" comprised IF plus 7.5% predialyzed, CM-preadsorbed, NCS and FCS (se paper for details of dialysis method for NCS and FCS), 0.08% chondroitin sulfate, 0.4 µg/mL FGF, and 50 mg/mL gentamycin. *F99:* 1 : 1 mixture of M199 and Ham's F12), supplemented with 10% FCS.

**Preparation of recipient cornea:** Human donor corneas were cultured in MEM supplemented with 2% FCS, 6% dextran, and penicillin [100 E/mL/streptomycin (100 µg/mL)] for 24 hours prior to denuding. Endothelial cells were denuded mechanically with a cotton swab, chemically by incubation with 10 mL of 0.04 M/L ammonium hydroxide for 5–10 min or physically by freezing to −80°C in 5 mL of MEM supplemented with 20% FCS. Typan blue was used to confirm full removal of the recipient HCECs. Freeze–thaw devitalization worked best and was the method of choice for the investigation.

**Seeding protocol:** Cultured monolayers of cells were harvested by exposure to 0.05% trypsin/0.02% EDTA. Cell counts from $5 \times 10^4$ to $5 \times 10^5$ were suspended in varying volumes of F99 (0.2–0.5 mL). Cell suspensions were "dropped" via pipette onto the denuded Descemet's membrane of a supported recipient cornea. The seeded supported corneas were placed in 50-mL Falcon tubes and centrifuged at $33 \times g$ for 5 min. The corneal discs were then transferred to 2-cm² dishes for organ culture.

**Corneal organ culture protocol:** Recipient corneas were cultured in F99 for 7–20 days with decreasing concentrations of FCS (days 1–2, 10%; days 3–5, 5%; days 5–20, 2%).

---

*et al.*, 2005). In 2001, her laboratory successfully cultured HCECs from donors 50–80 years old onto denuded Descemet's membrane of recipient corneas (see Protocol 68.6). Transplanted endothelial cell densities averaged 1900 cells/mm² at 14 days, which approaches cell densities achieved using transformed cells (Aboalchamat *et al.*, 1999) and is double the value obtained by Engelmann *et al.* (1999). Earlier passages (primary through P3) averaged greater than 1500 cells/mm², but later passages decreased the yield significantly. None of the cultures exhibited fibroblast morphology, but some cells did appear enlarged and exhibited chrraracteristics of senescent cells. The key differences in the culturing technique over previous methods were associated with the isolation technique for the HCECs, the lower concentration of FGF, and the inclusion of EGF, NGF, and bovine pituitary extract. The cultures exhibited enhanced longevity if the EGF, NGF, and pituitary extract were removed from the medium and the FBS concentration was reduced to 4%. Comparison of the effect of growth-promoting agents on young and old donors suggests that older cells in any medium have slower doubling times and generally produce a lower cell density. To obtain successful HCEC cultures, the Joyce laboratory provides some selection criteria and methodological tips: (1) Donors with traumatic injury or acute illness produced more successful cultures than donors will chronic illnesses. (2) Donors should have high endothelial cell density (>1500 cells/mm²). (3) Corneas should be preserved within 12 hours of death in optisol-GF and immediately cultured on arrival in the laboratory from the supplier. (4) Maximal numbers of healthy cells for culture are more readily obtained following overnight incubation of Descemet's membrane strips in FBS (8%). (5) Attachment of cells overnight to FNC-coated culture dishes indicates healthy cells. (6) Cells that are moderately elongated when attached represent healthy cells.

## Summary for Engineering the Corneal Endothelium

Recently, extremely encouraging results have been realized for the culturing of untransformed human corneal endothelial cells, which have generally resisted expansion in culture (Engelmann *et al.*, 2004; Joyce and Zhu, 2004). For the first time, the possibility of transplanting untransformed HCECs into patients or seeding HCECs in high density on artificial corneal constructs appears plausible. For tissue-engineered corneas, the ability to culture endothelial cells onto the stromal construct will likely enhance the chemical milieu and promote better epithelial morphological development (Zieske *et al.*, 1994). If the stromal construct exhibits

---

**Protocol 68.6** Endothelial cell culture protocol of Chen *et al.* (2000).

**Cell isolation:** Donor corneas (age 50–80 yr) were preserved in Optisol-GS and shipped to the laboratory on wet ice. Corneas were washed with M199 with gentamicin (50 μg/mL). Descemet's membrane with intact endothelium was excised in small strips and incubated in OptiMEM-I with 8% FBS overnight to stabilize the cells. The strips were centrifuged, washed in Hank's balanced salt solution, and incubated in 0.02% EDTA at 37°C for 1 h to loosen cell–cell junctions. Mechanical isolation of the cells was accomplished by forcing tissue and medium multiple times though the narrow opening in a flame-polished pipette. The cells and Descemet's strips were pelleted and resuspended in culture medium.

**Media:** *Culture medium:* Optimen-I, 8% FBS, FGF 40 ng/mL, EGF 5 ng/mL, NGF 20 ng/mL, human lipids 0.005%, RPMI-1640 multiple vitamin solution, ascorbic acid 20 μg/mL, calcium chloride 200 mg/L, 0.08% chondroition sulfate, gentamicin 50 μg/mL, and antibiotic/antimycotic solution diluted 1 : 100. Pituitary extract was added and human lipids and RPMI-1640 multiple vitamin solution were removed in later studies (Joyce and Zhu, 2004).

**Preparation of recipient cornea:** Descemet's membrane of the recipient cornea was denuded by treatment with 0.02 N of ammonium hydroxide and 0.5% sodium deoxycholate, followed by washing five times with 5 mmol/L EDTA. The cornea was washed one time with culture medium and placed, Descemet's up, in a six-well tissue culture plate using a sterile concave plastic support.

**Seeding protocol:** Cultured monolayers of cells in primary or passage 1 were harvested by trypsinization and centrifuged and resuspended to a concentration of $25 \times 10^5$ to $5.0 \times 10^5$. Cell suspensions were gently applied in small aliquots onto the denuded Descemet's membrane of a supported recipient cornea. Each aliquot was incubated at 37°C for 1 h, the medium was aspirated, and the seeding step was repeated until all of the cells were applied. Corneas were incubated overnight at 37°C.

**Corneal organ culture protocol:** Recipient corneas were cultured in culture medium for one week and then in culture medium without growth factors (but with serum for an additional week).

---

a tendency to swell and becomes opaque (similar to a native corneal stroma), then a functional endothelium will be necessary to reduce edema and promote transparency. The ability to culture untransformed HCECs is a major advance toward the goal of producing a tissue-engineered cornea.

## IV. CURRENT STATE OF CORNEAL-TISSUE ENGINEERING AND FUTURE DIRECTIONS

As a multilaminar structure, the cornea affords opportunities for reconstruction of each of the layers separately or in combination. Thus, diseases of each layer may be addressed without disruption of the entire structure. There have already been attempts to resurface the cornea using autologous limbal stem cell transplants alone or on amniotic membrane (Tsai *et al.*, 2000; Gomes *et al.*, 2003), and methods can now be envisioned that may allow the replacement of the corneal endothelium (Chen *et al.*, 2001; Zhu and

Joyce, 2004). Stromal replacement via synthetic optics is currently possible as well, but these implants remove the need for the limiting cell layers as a matter of course (Yaghouti and Dohlman, 1999; Chirila, 2001). A truly tissue-engineered corneal construct, which includes all three human cell types (untransformed) incorporated into and onto a natural stroma, has been an elusive target. Only Orwin and Hubel (2000) have cultured all three untransformed human cell types onto a collagen matrix ("sponge"). The result was too weak and too opaque for clinical consideration, and the colonization of the constructs did not produce convincing cellular strata. In general, the ability to culture the three corneal cell types *ex vivo* has improved to the point where the focus must now shift to the generation of a suitable stromal scaffold that will be functional at the time of implantation. Because of the long time scales associated with scar resolution *in vivo*, it is probably unreasonable to expect rapid resorption and remodeling of a degradable scaffolding that really is analogous to an artificial scar.

## V. REFERENCES

Aboalchamat, B., Engelmann, K., Bohnke, M., Eggli, P., and Bednarz, J. (1999). Morphological and functional analysis of immortalized human corneal endothelial cells after transplantation. *Exp. Eye Res.* **60**(5), 547–553.

Aghamohammadzadeh, H., Newton, R. H., and Meek, K. M. (2004). X-ray scattering used to map the preferred collagen orientation in the human cornea and limbus. *Structure* **12**(2), 249–256.

Aiken-O'Neill, P., and Mannis, M. J. (2002). Summary of corneal transplant activity Eye Bank Association of America. *Cornea* **21**(1), 1–3.

Anseth, A. (1961). Studies on corneal polysaccharides. III. Topographic and comparative biochemistry. *Exp. Eye Res.* **1**, 100–115.

Aquavella, J. V., Qian, Y., McCormick, G. J., and Palakuru, J. R. (2005). Keratoprosthesis: the Dohlman–Doane device. *Am. J. Ophthalmol.* **140**(6), 1032–1038.

Araki, K., Ohashi, Y., Kinoshita, S., Hayashi, K., Kuwayama, Y., and Tano, Y. (1994). Epithelial wound healing in the denervated cornea. *Curr. Eye Res.* **13**(3), 203–211.

Auger, F. A., Rouabhia, M., Goulet, F., Berthod, F., Moulin, V., and Germain, L. (1998). Tissue-engineered human skin substitutes developed from collagen-populated hydrated gels: clinical and fundamental applications. *Med. Biol. Eng. Comput.* **36**(6), 801–812.

Axelsson, I., and Heinegard, D. (1975). Fractionation of proteoglycans from bovine corneal stroma. *Biochem. J.* **145**(3), 491–500.

Baum, J. L., Niedra, R., Davis, C., and Yue, B. Y. (1979). Mass culture of human corneal endothelial cells. *Arch. Ophthalmol.* **97**(6), 1136–1140.

Beales, M. P., Funderburgh, J. L., Jester, J. V., and Hassell, J. R. (1999). Proteoglycan synthesis by bovine keratocytes and corneal fibroblasts: maintenance of the keratocyte phenotype in culture. *Invest. Ophthalmol. Vis. Sci.* **40**(8), 1658–1663.

Berryhill, B. L., Kader, R., Kane, B., Birk, D. E., Feng, J., and Hassell, J. R. (2002). Partial restoration of the keratocyte phenotype to bovine keratocytes made fibroblastic by serum. *Invest. Ophthalmol. Vis. Sci.* **43**(11), 3416–3421.

Beuerman, R. W., and Pedroza, L. (1996). Ultrastructure of the human cornea. *Microsc. Res. Tech.* **33**(4), 320–335.

Binder, G. B., and Binder, R. F. (1956). Experiments on plexiglas corneal implants. *Am. J. Ophthalmol.* **41**, 793–797.

Blake, D. A., Yu, H., Young, D. L., and Caldwell, D. R. (1997). Matrix stimulates the proliferation of human corneal endothelial cells in culture. *Invest. Ophthalmol. Vis. Sci.* **38**(6), 1119–1129.

Blencke, B. A., Hagan, P., Bromer, H., and Deurscher, K. (1978). Study on the use of glass ceramics in osteo-odonto-keratoplasty. *Ophthalmologica* **176**, 105–112.

Bobyn, J. D., Wilson, G. J., MacGregor, D. C., Pilliar, R. M., and Weatherly, G. C. (1982). Effect of pore size on the peel strength of attachment of fibrous tissue to porous-surfaced implants. *J. Biomed. Mater. Res.* **16**(5), 571–584.

Bohnke, M., Eggli, P., and Engelmann, K. (1999). Transplantation of cultured adult human or porcine corneal endothelial cells onto human recipients *in vitro*. Part II: evaluation in the scanning electron microscope. *Cornea* **18**(2), 207–213.

Borderie, V. M., Mourra, N., and Laroche, L. (1999). Influence of fetal calf serum, fibroblast growth factors, and hepatocyte growth factor on three-dimensional cultures of human keratocytes in collagen gel matrix. *Graefes. Arch. Clin. Exp. Ophthalmol.* **237**(10), 861–869.

Bourne, W. M., and Kaufman, H. E. (1976). Specular microscopy of human corneal endothelium *in vivo*. *Am. J. Ophthalmol.* **81**(3), 319–323.

Brown, C. T., Applebaum, E., Banwatt, R., and Trinkaus-Randall, V. (1995). Synthesis of stromal glycosaminoglycans in response to injury. *J. Cell Biochem.* **59**(1), 57–68.

Caiazza, S., Falcinelli, G., and Pintucci, S. (1990). Exceptional case of bone resorption in an osteo-odonto-keratoprosthesis. A scanning electron microscopy and x-ray microanalysis study. *Cornea* **9**(1), 23–27.

Cardona, H. (1964). Plastic keratoprostheses: a description of the plastic material and comparative histologic study of recipient corneas. *Am. J. Ophthalmol.* **58**, 247–252.

Castroviejo, R., Cardona, H., and DeVoe, A. G. (1969). The present status of prosthokeratoplasty. *Trans. Am. Ophthalmol. Soc.* **67**, 207–234.

Chang, P. C. T., Lee, S. D., and Huang, J. D. (1995). Biocompatibility of an artificial corneal membrane; *in vivo* animal study. *IOVS* (suppl.) **36**(4), 1467.

Charulatha, V., and Rajaram, A. (1997). Cross-linking density and resorption of dimethyl suberimidate–treated collagen. *J. Biomed. Mater. Res.* **36**(4), 478–486.

Chen, K. H., Azar, D., and Joyce, N. C. (2001). Transplantation of adult human corneal endothelium *ex vivo*: a morphologic study. *Cornea* **20**(7), 731–737.

Chirila, T. V. (2001). An overview of the development of artificial corneas with porous skirts and the use of PHEMA for such an application. *Biomaterials* **22**(24), 3311–3317.

Chirila, T. V., Constable, I. J., Crawford, G. J., Vijayasekaran, S., Thompson, D. E., Chen, Y. C., Fletcher, W. A., and Griffin, B. J. (1993). Poly(2-hydroxyethyl methacrylate) sponges as implant materials: *in vivo* and *in vitro* evaluation of cellular invasion. *Biomaterials* **14**(1), 26–38.

Cintron, C., Szamier, R. B., Hassinger, L. C., and Kublin, C. L. (1982). Scanning electron microscopy of rabbit corneal scars. *Invest. Ophthalmol. Vis. Sci.* **23**(1), 50–63.

Cintron, C., Covington, H., and Kublin, C. L. (1983). Morphogenesis of rabbit corneal stroma. *Invest. Ophthalmol. Vis. Sci.* **24**(5), 543–556.

Cintron, C., Covington, H. I., and Kublin, C. L. (1988). Morphogenesis of rabbit corneal endothelium. *Curr. Eye Res.* **7**(9), 913–929.

Cintron, C., Gregory, J. D., Damle, S. P., and Kublin, C. L. (1990). Biochemical analyses of proteoglycans in rabbit corneal scars. *Invest. Ophthalmol. Vis. Sci.* **31**(10), 1975–1981.

Cook, S. D., Barrack, R. L., Thomas, K. A., and Haddad, R. J., Jr. (1989). Quantitative histologic analysis of tissue growth into porous total knee components. *J. Arthroplasty* **4**(Suppl.), S33–S43.

Cox, J. L., Farrell, R. A., Hart, R. W., and Langham, M. E. (1970). The transparency of the mammalian cornea. *J. Physiol.* **210**(3), 601–616.

Dimmer, F. (1891). Zwie falle von celluloidplatten der homher. *Klin. Monarsbl. Augenheilkd.* **29**, 104–105.

Doane, K. J., and Birk, D. E. (1991). Fibroblasts retain their tissue phenotype when grown in three-dimensional collagen gels. *Exp. Cell Res.* **195**(2), 432–442.

Dohlman, C. H. (1997). Keratoprostheses. *In* "Cornea" (J. H. Krachmer, M. J. Mannis, and E. J. Holland, eds.), pp. 1855–1863. Mosby Year Book, Vol. III, St. Louis, MO.

Dohlman, C. H., Schneider, H. A., and Doane, M. G. (1974). Prosthokeratoplasty. *Am. J. Ophthalmol.* **77**(5), 694–670.

Dong, X., Wang, W., Xie, L., and Chiu, A. M. (2006). Long-term outcome of combined penetrating keratoplasty and vitreoretinal surgery using temporary keratoprosthesis. *Eye* **20**(1), 59–63.

Drubaix, I., Legeais, J. M., Malek-Chehire, N., Savoldelli, M., Menasche, M., Robert, L., Renard, G., and Pouliquen. Y. (1996). Collagen synthesized in fluorocarbon polymer implant in the rabbit cornea. *Exp. Eye Res.* **62**(4), 367–376.

Engelmann, K., and Friedl, P. (1989). Optimization of culture conditions for human corneal endothelial cells. *In Vitro Cell Dev. Biol.* **25**(11), 1065–1072.

Engelmann, K., and Friedl, P. (1995). Growth of human corneal endothelial cells in a serum-reduced medium. *Cornea* **14**(1), 62–70.

Engelmann, K., Bohnke, M., and Friedl, P. (1988). Isolation and long-term cultivation of human corneal endothelial cells. *Invest. Ophthalmol. Vis. Sci.* **29**(11), 1656–1662.

Engelmann, K., Drexler, D., and Bohnke, M. (1999). Transplantation of adult human or porcine corneal endothelial cells onto human recipients *in vitro*. Part I: cell culturing and transplantation procedure. *Cornea* **18**(2), 199–206.

Engelmann, K., Bednarz, J., and Valtink, M. (2004). Prospects for endothelial transplantation. *Exp. Eye Res.* **78**(3), 573–578.

Ethier, C. R., Johnson, M., and Ruberti, J. (2004). Ocular biomechanics and biotransport. *Annu. Rev. Biomed. Eng.* **6**, 249–273.

Feldman, S. T., Gjerset, R., Gately, D., Chien, K. R., and Feramisco, J. R. (1993). Expression of SV40 virus large T antigen by recombinant adenoviruses activates proliferation of corneal endothelium *in vitro*. *J. Clin. Invest.* **91**(4), 1713–1720.

Friend, J., Kinoshita, S., Thoft, R. A., and Eliason, J. A. (1982). Corneal epithelial cell cultures on stromal carriers. *Invest. Ophthalmol. Vis. Sci.* **23**(1), 41–49.

Funderburgh, J. L., and Chandler, J. W. (1989). Proteoglycans of rabbit corneas with nonperforating wounds. *Invest. Ophthalmol. Vis. Sci.* **30**(3), 435–442.

Garana, R. M., Petroll, W. M., Chen, W. T., Herman, I. M., Barry, P., Andrews, P., Cavanagh, H. D., and Jester, J. V. (1992). Radial keratotomy. II. Role of the myofibroblast in corneal wound contraction. *Invest. Ophthalmol. Vis. Sci.* **33**(12), 3271–3282.

Geggel, H. S., Friend, J., and Thoft, R. A. (1985). Collagen gel for ocular surface. *Invest. Ophthalmol. Vis. Sci.* **26**(6), 901–905.

Germain, L., Auger, F. A., Grandbois, E., Guignard, R., Giasson, M., Boisjoly, H., and Guerin, S. L. (1999). Reconstructed human cornea produced *in vitro* by tissue engineering. *Pathobiology* **67**(3), 140–147.

Germain, L., Carrier, P., Deschambeault, A., Talbot, M., Guerin, S. L., and Auger, F. A. (2004). "Fibroblasts Modulate Differentiation and Stratification of Epithelial Cells on Reconstructed Human Cornea Through the Production of Soluble Factors." ARVO, Fort Lauderdale, FL.

Giguere, L., Cheng, J., and Gospodarowicz, D. (1982). Factors involved in the control of proliferation of bovine corneal endothelial cells maintained in serum-free medium. *J. Cell Physiol.* **110**(1), 72–80.

Goldman, J. N., Benedek, G. B., Dohlman, C. H., and Kravitt, B. (1968). Structural alterations affecting transparency in swollen human corneas. *Invest. Ophthalmol.* **7**(5), 501–519.

Gomes, J. A., dos Santos, M. S., Cunha, M. C., Mascaro, V. L., Barros Jde N., and de Sousa, L. B. (2003). Amniotic membrane transplantation for partial and total limbal stem cell deficiency secondary to chemical burn. *Ophthalmology* **110**(3), 466–473.

Gospodarowicz, D., Mescher, A. L., and Birdwell, C. R. (1977). Stimulation of corneal endothelial cell proliferations *in vitro* by fibroblast and epidermal growth factors. *Exp. Eye Res.* **25**(1), 75–89.

Green, H., Rheinwald, J. G., and Sun, T. T. (1977). Properties of an epithelial cell type in culture: the epidermal keratinocyte and its dependence on products of the fibroblast. *Prog. Clin. Biol. Res.* **17**, 493–500.

Griffith, M., Osborne, R., Munger, R., Xiong, X., Doillon, C. J., Laycock, N. L., Hakim, M., Song, Y., and Watsky, M. A. (1999). Functional human corneal equivalents constructed from cell lines. *Science* **286**(5447), 2169–2172.

Grinnell, F., Fukamizu, H., Pawelek, P., and Nakagawa, S. (1989). Collagen processing, crosslinking, and fibril bundle assembly in matrix produced by fibroblasts in long-term cultures supplemented with ascorbic acid. *Exp. Cell Res.* **181**(2), 483–491.

Guo, X. Q., Hutcheon, A. E. K., Ruberti, J. W., and Zieske, J. D. (2002). Human Corneal Organotypic Cultures Using Untransformed Cells ARVO. *Invest. Opthalmol. Vis. Sci.* **43**.

Gyorffy, L. (1956). Acrylic corneal implants in keratoplasty. *Am. J. Ophthalmol.* **34**, 757–760.

Hamada, R., Giraud, J. P., Graf, B., and Pouliquen, Y. (1972). Etude analytique et statistique des lamelles, des keratocytes, des fibrillesdel collagene de la region centrale de la cornee humaine normale. *Arch. Ophtalmol.* **32**, 563–570.

Hanna, C., and O'Brien, J. E. (1960). Cell production and migration in the epithelial layer of the cornea. *Arch. Ophthalmol.* **64**, 536–539.

Hassell, J. R., Schrecengost, P. K., Rada, J. A., SundarRaj, N., Sossi, G., and Thoft, R. A. (1992). Biosynthesis of stromal matrix proteoglycans and basement membrane components by human corneal fibroblasts. *Invest. Ophthalmol. Vis. Sci.* **33**(3), 547–557.

Hedbys, B. O. (1961). The role of polysaccharides in corneal swelling. *Exp. Eye Res.* **1**, 81.

Hedbys, B. O., Mishima, S., and Maurice, D. M. (1963). The imbibition pressure of the corneal stroma. *Exp. Eye Res.* **2**, 99–111.

Hicks, C. R., Crawford, G. J., Lou, X., Tan, D. T., Snibson, G. R., Sutton, G., Downie, N., Werner, L., Chirila, T. V., and Constable, I. J. (2003). Corneal replacement using a synthetic hydrogel cornea, AlphaCor: device, preliminary outcomes and complications. *Eye* **17**(3), 385–392.

Hicks, C. R., Chirila, T. V., Werner, L., Crawford, G. J., Apple, D. J., and Constable, I. J. (2004). Deposits in artificial corneas: risk factors and prevention. *Clin. Exp. Ophthalmol.* **32**(2), 185–191.

Hille, K., Grabner, G., Liu, C., Colliardo, P., Falcinelli, G., Taloni, M., and Falcinelli, G. (2005). Standards for modified osteoodontokeratoprosthesis (OOKP) surgery according to Strampelli and Falcinelli: the Rome–Vienna Protocol. *Cornea* **24**(8), 895–908.

Jester, J. V., Rodrigues, M. M., and Herman, I. M. (1987). Characterization of avascular corneal wound-healing fibroblasts. New insights into the myofibroblast. *Am. J. Pathol.* **127**(1), 140–148.

Jester, J. V., Barry-Lane, P. A., Cavanagh, H. D., and Petroll, W. M. (1996). Induction of alpha-smooth muscle actin expression and myofibroblast transformation in cultured corneal keratocytes. *Cornea* **15**(5), 505–516.

Jester, J. V., Moller-Pedersen, T., Huang, J., Sax, C. M., Kays, W. T., Cavangh, H. D., Petroll, W. M., and Piatigorsky, J. (1999). The cellular basis of corneal transparency: evidence for "corneal crystallins." *J. Cell Sci.* **112**(Pt. 5), 613–622.

Joyce, N. C., and Zhu, C. C. (2004). Human corneal endothelial cell proliferation: potential for use in regenerative medicine. *Cornea* **23**(8 Suppl.), S8–S19.

Joyce, N. C., Meklir, B., Joyce, S. J., and Zieske, J. D. (1996a). Cell-cycle protein expression and proliferative status in human corneal cells. *Invest. Ophthalmol. Vis. Sci.* **37**(4), 645–655.

Joyce, N. C., Navon, S. E., Roy, S., and Zieske, J. D. (1996b). Expression of cell cycle–associated proteins in human and rabbit corneal endothelium *in situ*. *Invest. Ophthalmol. Vis. Sci.* **37**(8), 1566–1575.

Jumblatt, M. M., and Neufeld, A. H. (1983). Beta adrenergic and sero tonergic responsiveness of rabbit corneal epithelial cells in culture. *Invest. Ophthalmol. Vis. Sci.* **24**(8), 1139–1143.

Kain, H. L. (1990). [A new concept for keratoprosthesis]. *Klin. Monatsbl. Augenheilkd.* **197**(5), 386–392.

Klyce, S. D., and Beuerman, R. W. (1997). Structure and function of the cornea. *In* "The Cornea" (H. E. Kaufman, M. B. McDonald, and B. A. Barron, eds.), pp. 3–50. Butterworth-Heineman Medical, London.

Konomi, K., Zhu, C., Harris, D., and Joyce, N. C. (2005). Comparison of the proliferative capacity of human corneal endothelial cells from the central and peripheral areas. *Invest. Ophthalmol. Vis. Sci.* **46**(11), 4086–4091.

Kostyuk, O., Nalovina, O., Mubard, T. M., Regini, J. W., Meek, K. M., Quantock, A. J., Elliott, G. F., and Hodson, S. A. (2002). Transparency of the bovine corneal stroma at physiological hydration and its dependence on concentration of the ambient anion. *J. Physiol.* **543**(Pt. 2), 633–642.

Laroche, L., Honiger, J., Thenot, J. C., Scarano, M., and Assouline, M. (1989). Keratophakie Alloplastique. *Ophthalmologica* **3**, 227–228.

Latkany, R., Tsuk, A., Sheu, M. S., Loh, I. H., and Trinkaus-Randall, V. (1997). Plasma surface modification of artificial corneas for optimal epithelialization. *J. Biomed. Mater. Res.* **36**(1), 29–37.

Legeais, J. M., Renard, G., Parel, J. M., Serdarevic, O., Mei-Mui, M., and Pouliquen, Y. (1994). Expanded fluorocarbon for keratoprosthesis cellular ingrowth and transparency. *Exp. Eye Res.* **58**(1), 41–51.

Legeais, J. M., Renard, G., Parel, J. M., Savoldelli, M., and Pouliquen, Y. (1995). Keratoprosthesis with biocolonizable microporous fluorocarbon haptic. Preliminary results in a 24-patient study. *Arch. Ophthalmol.* **113**(6), 757–763.

Li, F., Carlsson, D., Lohmann, C., Suuronen, E., Vascotto, S., Kobuch, K., Sheardown, H., Munger, R., Nakamura, M., and Griffith, M. (2003). Cellular and nerve regeneration within a biosynthetic extracellular matrix for corneal transplantation. *Proc. Natl. Acad. Sci. U.S.A.* **100**(26), 15346–15351.

Lindberg, K., Brown, M. E., Chaves, H. V., Kenyon, K. R., and Rheinwald, J. G. (1993). *In vitro* propagation of human ocular surface epithelial cells for transplantation. *Invest. Ophthalmol. Vis. Sci.* **34**(9), 2672–2679.

Martin, R., Farjanel, J., Eichenberger, D., Colige, A., Kessler, E., Hulmes, D. J., and Giraud-Guille, M. M. (2000). Liquid crystalline ordering of procollagen as a determinant of three-dimensional extracellular matrix architecture. *J. Mol. Biol.* **301**(1), 11–17.

Maurice, D. M. (1972). The location of the fluid pump in the cornea. *J. Physiol.* **221**, 43–54.

McAlister, J. C., Joyce, N. C., Harris, D. L., Ali, R. R., and Larkin, D. F. (2005). Induction of replication in human corneal endothelial cells by E2F2 transcription factor cDNA transfer. *Invest. Ophthalmol. Vis. Sci.* **46**(10), 3597–3603.

Meek, K. M., and Leonard, D. W. (1993). Ultrastructure of the corneal stroma: a comparative study. *Biophys. J.* **64**(1), 273–280.

Minami, Y., Sugihara, H., and Oono, S. (1993). Reconstruction of cornea in three-dimensional collagen gel matrix culture. *Invest. Ophthalmol. Vis. Sci.* **34**(7), 2316–2324.

Morrison, D. A., Gridneva, Z., Chirila, T. V., and Hicks, C. R. (2006). Screening for drug-induced spoliation of the hydrogel optic of the AlphaCor artificial cornea. *Cont. Lens Anterior Eye* **29**(2), 93–100.

Muller, L. J., Pels, L., and Vrensen, G. F. (1995). Novel aspects of the ultrastructural organization of human corneal keratocytes. *Invest. Ophthalmol. Vis. Sci.* **36**(13), 2557–2567.

Nayak, S. K., and Binder, P. S. (1984). The growth of endothelium from human corneal rims in tissue culture. *Invest. Ophthalmol. Vis. Sci.* **25**(10), 1213–1216.

Newsome, D. A., Takasugi, M., Kenyon, K. R., Stark, W. F., and Opelz, G. (1974). Human corneal cells *in vitro*: morphology and histocompatibility (HL-A) antigens of pure cell populations. *Invest. Ophthalmol.* **13**(1), 23–32.

Ohji, M., SundarRaj, N., Hassell, J. R., and Thoft, R. A. (1994). Basement membrane synthesis by human corneal epithelial cells *in vitro*. *Invest. Ophthalmol. Vis. Sci.* **35**(2), 479–485.

Orwin, E. J., and Hubel, A. (2000). *In vitro* culture characteristics of corneal epithelial, endothelial, and keratocyte cells in a native collagen matrix. *Tissue Eng.* **6**(4), 307–319.

Orwin, E. J., Borene, M. L., and Hubel, A. (2003). Biomechanical and optical characteristics of a corneal stromal equivalent. *J. Biomech. Eng.* **125**(4), 439–444.

Pellier de Quengsy, G. (1789). "Precis ou cours d'opérations sur la chirurgie des yeux," Chez Didot and Mequignon, Paris.

Perlman, M., and Baum, J. L. (1974). The mass culture of rabbit corneal endothelial cells. *Arch. Ophthalmol.* **92**(3), 235–237.

Perlman, M., Baum, J. L., and Kaye, G. I. (1974). Fine structure and collagen synthesis activity of monolayer cultures of rabbit corneal endothelium. *J. Cell Biol.* **63**(1), 306–311.

Pourdeyhimi, B. (1989). Porosity of surgical mesh fabrics: new technology. *J. Biomed. Mater. Res.* **23**(A1 Suppl.), 145–152.

Prunieras, M., Regnier, M., and Woodley, D. (1983). Methods for cultivation of keratinocytes with an air-liquid interface. *J. Invest. Dermatol.* **81**(1 Suppl), 28s–33s.

Radner, W., and Mallinger, R. (2002). Interlacing of collagen lamellae in the midstroma of the human cornea. *Cornea* **21**(6), 598–601.

Radner, W., Zehetmayer, M., Aufreiter, R., and Mallinger, R. (1998). Interlacing and cross-angle distribution of collagen lamellae in the human cornea. *Cornea* **17**(5), 537–543.

Ruberti, J. W., and Klyce, S. D. (2002). Physiological system models of the cornea. *In* "Model's of the Visual System" (G. K. Hung and K. J. Ciuffreda, eds), pp. 3–55. Kluwer Academic/Plenum Publishers, New York.

Ruberti, J. W., Melotti, S. A., and Braithwaite, G. J. C. (2003). "Nanoscale Engineering of Type I Collagen to Mimic the Multiple Layers of Aligned Lamellae in Cornea." ARVO, Ft. Lauderdale, FL.

Samples, J. R., Binder, P. S., and Nayak, S. K. (1991). Propagation of human corneal endothelium *in vitro* effect of growth factors. *Exp. Eye Res.* **52**(2), 121–128.

Savage, C. R., Jr., and Cohen, S. (1973). Proliferation of corneal epithelium induced by epidermal growth factor. *Exp. Eye Res.* **15**(3), 361–366.

Savion, N., Isaacs, J. D., Shuman, M. A., and Gospodarowicz, D. (1982). Proliferation and differentiation of bovine corneal endothelial cells in culture. *Metab. Pediatr. Syst. Ophthalmol.* **6**(3–4), 305–320.

Schermer, A., Galvin, S., and Sun, T. T. (1986). Differentiation-related expression of a major 64K corneal keratin *in vivo* and in culture suggests limbal location of corneal epithelial stem cells. *J. Cell Biol.* **103**(1), 49–62.

Schwab, I. R., Reyes, M., and Isseroff, R. R. (2000). Successful transplantation of bioengineered tissue replacements in patients with ocular surface disease. *Cornea* **19**(4), 421–426.

Senoo, T., and Joyce, N. C. (2000). Cell-cycle kinetics in corneal endothelium from old and young donors. *Invest. Ophthalmol. Vis. Sci.* **41**(3), 660–667.

Senoo, T., Obara, Y., and Joyce, N. C. (2000). EDTA: a promoter of proliferation in human corneal endothelium. *Invest. Ophthalmol. Vis. Sci.* **41**(10), 2930–2935.

Stoesser, T. R., Church, R. L., and Brown, S. I. (1978). Partial characterization of human collagen and procollagen secreted by human corneal stromal fibroblasts in cell culture. *Invest. Ophthalmol. Vis. Sci.* **17**(3), 264–271.

Stoiber, J., Fernandez, V., Lamar, P. D., Kaminski, S., Acosta, A. C., Dubovy, S., Alfonso, E., and Parel, J. M. (2005). Biocompatibility of a nonpenetrating synthetic cornea in vascularized rabbit corneas. *Cornea* **24**(4), 467–473.

Strampelli, B., and Marchi, V. (1970). [Osteo-odonto-keratoprosthesis]. *Ann. Ottalmol. Clin. Ocul.* **96**(1), 1–57.

Sun, T. T., and Green, H. (1977). Cultured epithelial cells of cornea, conjunctiva and skin: absence of marked intrinsic divergence of their differentiated states. *Nature* **269**(5628), 489–493.

Sundar-Raj, C. V., Freeman, I. L., and Brown, S. I. (1980). Selective growth of rabbit corneal epithelial cells in culture and basement membrane collagen synthesis. *Invest. Ophthalmol. Vis. Sci.* **19**(10), 1222–1230.

Thompson, R. W., Jr., Price, M. O., Bowers, P. J., and Price, F. W., Jr. (2003). Long-term graft survival after penetrating keratoplasty. *Ophthalmology* **110**(7), 1396–1402.

Trinkaus-Randall, V., and Gipson, I. K. (1985). A technique for obtaining basal corneal epithelial cells. *Invest. Ophthalmol. Vis. Sci.* **26**(2), 233–237.

Trinkaus-Randall, V., and Nugent, M. A. (1998). Biological response to a synthetic cornea. *J. Controlled Release* **53**(1–3), 205–214.

Trinkaus-Randall, V., Capecchi, J., Newton, A., Vadasz, A., Leibowitz, H., and Franzblau, C. (1988a). Development of a biopolymeric keratoprosthetic material. Evaluation *in vitro* and *in vivo. Invest. Ophthalmol. Vis. Sci.* **29**(3), 393–400.

Trinkaus-Randall, V., Newton, A., and Franzblau, C. (1988b). Influence of substratum on corneal epithelial cell growth and protein synthesis. *Invest. Ophthalmol. Vis. Sci.* **29**(12), 1800–1809.

Trinkaus-Randall, V., Newton, A. W., Gipson, I. K., and Franzblau, C. (1988c). Carbohydrate moieties of the basal lamina: their role in attachment and spreading of basal corneal epithelial cells. *Cell Tissue Res.* **251**(2), 315–323.

Trinkaus-Randall, V., Capecchi, J., Sammon, L., Gibbons, D., Leibowitz, H. M., and Franzblau, C. (1990). *In vitro* evaluation of fibroplasia in a porous polymer. *Invest. Ophthalmol. Vis. Sci.* **31**(7), 1321–1326.

Trinkaus-Randall, V., Banwatt, R., Capecchi, J., Leibowitz, H. M., and Franzblau, C. (1991). *In vivo* fibroplasia of a porous polymer in the cornea. *Invest. Ophthalmol. Vis. Sci.* **32**(13), 3245–3251.

Trinkaus-Randall, V., Banwatt, R., Wu, X. Y., Leibowitz, H. M., and Franzblau, C. (1994). Effect of pretreating porous webs on stromal fibroplasia *in vivo. J. Biomed. Mater. Res.* **28**(2), 195–202.

Trinkaus-Randall, V., Wu, X. Y., Tablante, R., and Tsuk, A. (1997). Implantation of a synthetic cornea: design, development and biological response. *Artif. Organs* **21**(11), 1185–1191.

Tsai, R. J., Li, L. M., and Chen, J. K. (2000). Reconstruction of damaged corneas by transplantation of autologous limbal epithelial cells. *N. Engl. J. Med.* **343**(2), 86–93.

Tsuk, A. G., Trinkaus-Randall, V., and Leibowitz, H. M. (1997). Advances in polyvinyl alcohol hydrogel keratoprostheses: protection against ultraviolet light and fabrication by a molding process. *J. Biomed. Mater. Res.* **34**(3), 299–304.

Whitcher, J. P., Srinivasan, M., and Upadhyay, M. P. (2001). Corneal blindness: a global perspective. *Bull. World Health Organ.* **79**(3), 214–221.

White, J. H., and Gona, O. (1988). "Proplast" for keratoprosthesis. *Ophthalmic. Surg.* **19**(5), 331–333.

Wu, X. Y., Cornell-Bell, A., Davies, T. A., Simons, E. R., and Trinkaus-Randall, V. (1994). Expression of integrin and organization of F-actin in epithelial cells depends on the underlying surface. *Invest. Ophthalmol. Vis. Sci.* **35**(3), 878–890.

Yaghouti, F., and Dohlman, C. H. (1999). Innovations in keratoprosthesis: proved and unproved. *Int. Ophthalmol. Clin.* **39**(1), 27–36.

Yee, R. W., Matsuda, M., Schultz, R. O., and Edelhauser, H. F. (1985). Changes in the normal corneal endothelial cellular pattern as a function of age. *Curr. Eye Res.* **4**(6), 671–678.

Yue, B. Y., Sugar, J., Gilboy, J. E., and Elvart, J. L. (1989). Growth of human corneal endothelial cells in culture. *Invest. Ophthalmol. Vis. Sci.* **30**(2), 248–253.

Zhu, C., and Joyce, N. C. (2004). Proliferative response of corneal endothelial cells from young and older donors. *Invest. Ophthalmol. Vis. Sci.* **45**(6), 1743–1751.

Zieske, J. D., Mason, V. S., Wasson, M. E., Meunier, S. F., Nolte, C. J., Fukai, N., Olsen, B. R., and Parenteau, N. L. (1994). Basement membrane assembly and differentiation of cultured corneal cells: importance of culture environment and endothelial cell interaction. *Exp. Cell. Res.* **214**(2), 621–633.

# Chapter Sixty-Nine

# Vision Enhancement Systems

*Gislin Dagnelie*

## I. INTRODUCTION

This chapter provides an overview of visual system properties in health and disease, with an emphasis on the stages in visual-signal processing that are most commonly affected by disease and dysfunction. It then provides an overview of current approaches to vision restoration and of approaches that promise to migrate from the lab to the clinic in the coming years. In a third section the potential roles for engineered cells and tissues in this restoration process are discussed. The chapter closes with an estimate of the developments that can be expected to occur in the next 5–10 years.

Twenty years ago, the best hope vision researchers had for the enhancement of impaired vision was to build a better magnifier by integrating optics and electronic image processing. Since then, restoration of vision to functionally blind individuals has become not just an engineering target, but a realistic goal for which engineering milestones are being set and reached. The role of tissue engineering in this field may as yet be modest, but it has great promise, as will become clear in the following pages.

In this chapter's overview of approaches to the restoration and enhancement of impaired vision in human patients, inspired by tissue-engineering principles and related technologies, we argue that the nature of eye disease and the fragility and complexity of ocular tissues such as the neural retina do not (yet) lend themselves to application of the techniques emerging in the repair of other tissues. Thus, while some of the approaches presented here can function independent of tissue engineering as presented elsewhere in this volume, they at least complement those engineering approaches and lend themselves to future integration. This is particularly true in the areas of cell transplantation and neural prosthesis development, presented in the later part of this chapter, but to a lesser extent also for the optical and optoelectronic aids presented.

## II. VISUAL SYSTEM, ARCHITECTURE, AND (DYS)FUNCTION

Human vision is mediated by one of the most highly developed sensory systems found anywhere in nature. Its capacity to combine high spatial resolution near the center of fixation with a wide peripheral field of view, accurate depth perception, color discrimination, and light–dark adaptation over 12 orders of magnitude is unparalleled. Every stage in the system is organized to accomplish this: The photoreceptor layer in the retina provides the high signal amplification of the rods required for night vision and the dense packing of three cone types required for detailed

*Principles of Tissue Engineering, 3rd Edition*
*ed. by Lanza, Langer, and Vacanti*

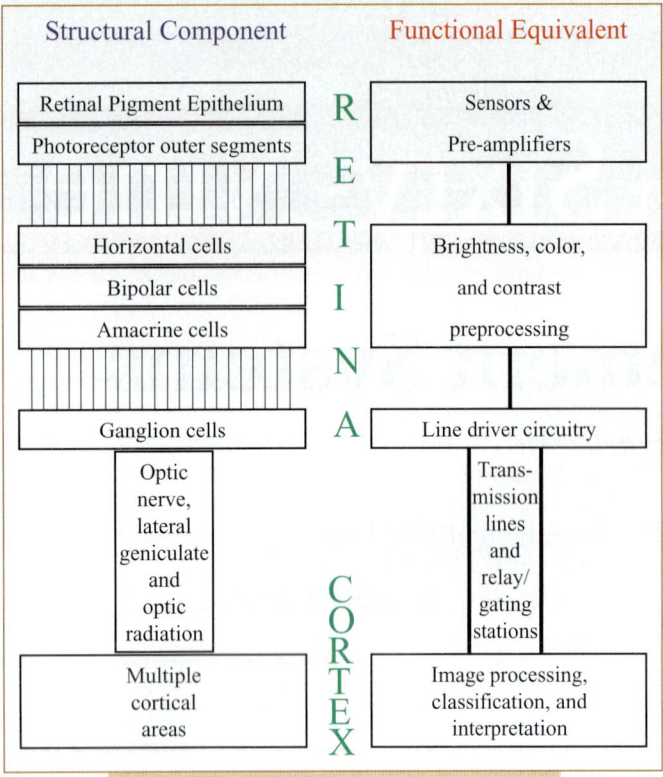

**FIG. 69.1.** Schematic representation of the visual system. The outer retina (RPE and photoreceptor layers) forms the sensor array, followed by several inner retinal preprocessing stages, the ganglion cell transmission stages, and further central processing stages in subcortical and cortical brain centers.

central and color vision. The intricate local preprocessing performed by subsequent retinal-cell layers augments these functions by performing brightness and color comparisons and helps to reduce the information stream acquired by over 100 million photoreceptors, to allow transport across a mere 1 million fibers in the optic nerve to the visual centers in the brain, where further parsing and interpretation of the image takes place.

From a functional point of view, the visual system can be understood as depicted in Fig. 69.1: with sensors, pre-amplifiers, preprocessors, transmission lines, and several central processor stages. The two most crucial stages in this process — the conversion from light into chemical and electrical signals, and the signal transmission from the eyes toward the brain — are also the most vulnerable ones. Light conversion and signal amplification in the photoreceptors require a highly complex interplay between the molecules inside these cells and ion channels and other permeable structures in the cell membrane. These components participate in interlocking cycles of conversion and regeneration and are assisted by surrounding cells — in particular the retinal pigmented epithelium (RPE) cells, which provide nutrients to, and digest cell membrane discarded from, the

photoreceptors. Any step in this intricate process can easily be disrupted by nutritional deficits, overexposure to short-wavelength light (presumably causing oxidative changes), attacks by pathogens, and especially genetic miscoding of one or more participating molecules. Since the mid-1980s, the list of mapped (>160) and identified (>120) genes has grown at an ever-increasing rate. For each identified gene there are multiple known mutations, often leading to distinct disease phenotypes; for an up-to-date list of the known mutations leading to loss of outer retinal function, see the Retnet website (Daiger *et al.*, 1996–2006). Not only do such mutations directly cause impaired signal transduction, but the additional energy demand, presence of abnormal molecules, and excess shedded cell membrane may exceed the RPE cells' support capacity, which inexorably leads to degeneration of both photoreceptors and RPE cells. The most common group of disorders caused by genetic miscoding of molecules involved in the phototransduction cycle is jointly known as *retinitis pigmentosa* (RP), while another group is caused by a breakdown of RPE function in the central retina due to either a genetic defect (Stargardt macular dystrophy) or a combination of genetic predisposition and environmental factors (age-related macular degeneration, AMD). Jointly, all these disorders are known as *retinal degenerations* (Fig. 69.2A and B).

Retinal neurons communicate in a variety of ways, with ion-gating channels fulfilling a role in reaching and maintaining an operating level (also known as the adaptation state) and neurotransmitters and gap junctions fulfilling the principal messenger roles in carrying information in chemical and electrical form, respectively. The loss of outer retinal function, and therefore of neurochemical and electrical signal transmission to the inner retina, will affect secondary retinal cells (horizontal, bipolar, amacrine, and ganglion cells) but not necessarily threaten their survival. Until a few years ago, morphometric studies of donor retinal tissue such as that shown in Fig. 69.2B suggested survival of bipolar and ganglion cells at rates of 80% and 30% in the macula (Santos *et al.*, 1997) and 40% and 20% at eccentricities up to 25° (Humayun *et al.*, 1999a), respectively, in retinas practically devoid of photoreceptors. In recent years, a series of highly detailed and elegant microanatomical studies using a technique called *computational molecular phenotyping* (CMP; Jones and Marc, 2005) has demonstrated that the connectivity patterns of these inner retinal-cell populations are fundamentally altered by the degeneration process (Fig. 69.3): Survival of these cells is predicated on their continued activity, and this they accomplish through the formation of new axonal and dendritic branches and through self-organization into clusters called *microneuromas*. These new and functionally random connections allow spontaneous oscillations to occur, in the absence of meaningful input signals from the erstwhile photoreceptors; RP patients may experience such spontaneous activities as photopsias, or "light shows" (Heckenlively, 1988).

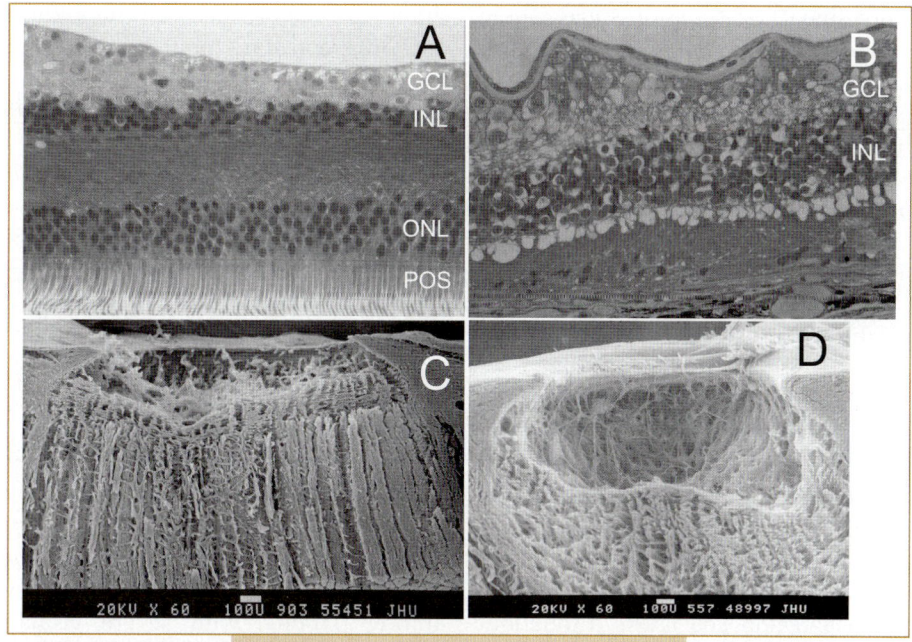

**FIG. 69.2.** Representative samples of ocular morphology in healthy and diseased conditions. **(A)** Cross section through the retina near its center (fovea), showing healthy photoreceptor outer segments (POS), multiple layers of photoreceptor cell nuclei in the outer nuclear layer (ONL; labeled black), bipolar cell nuclei in the inner nuclear layer (INL), and ganglion cell bodies in the ganglion cell layer (GCL). **(B)** Retina of a patient with a long history of retinal degeneration and bare light perception in the last years of life, showing lack of photoreceptor outer segments and cell bodies in comparison with micrograph A. **(C)** Scanning electron microscope cross section of the optic nerve head, showing healthy appearance of the support structure, the lamina cribrosa. **(D)** Optic nerve head from a patient with long history of glaucoma, showing compression of the lamina cribrosa and embedded nerve fibers (RGC axons). Micrographs C and D courtesy of Harry A. Quigley, M.D., the Johns Hopkins University, Baltimore, MD.

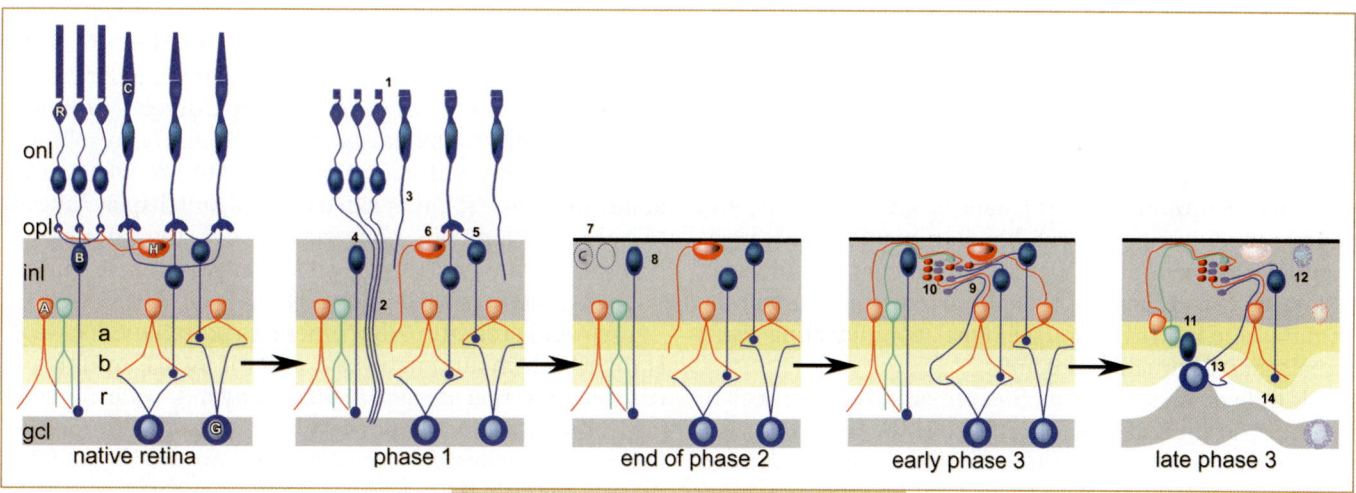

**FIG. 69.3.** Simplified schematic of possible alterations during retinal remodeling. The first panel shows a representative "native" mammalian retina with its basic complement of rods (R) and cones (C) driving bipolar (B) and horizontal (H) cells. In turn, bipolar cells drive amacrine (A) and ganglion cells (G) of the proximal retina to form the major cone-OFF (sublayer *a*), cone-ON (sublayer *b*), and rod-ON (sublayer *r*) zones of the inner plexiform layer (IPL). Remodeling occurs in phases, as indicated under each panel. In phase 1, rod and cone stress leads to truncation of outer segments (1) and, in some instances, extension of rod (2) and cone (3) axons deep into the inner nuclear (inl), inner plexiform, and even ganglion cell (gcl) layers. Both rod and cone bipolar cells truncate their dendrites (4, 5), and some rod bipolar cells may transiently switch to surviving cones (not shown). Horizontal cells also send axons into the inner plexiform layer (6). Phase 2 is a complex period of cell death, ablation of the outer nuclear layer (onl), and resolution of the distal margin of the neural retina into a largely confluent glial seal (7) formed by the distal processes of Müller cells. Surviving neurons may continue to alter their phenotypes by changing receptor expression patterns (8), and some cone cells may even escape cell death (C). Neural remodeling becomes even more extensive during phase 3, with formation of complex axon fascicles (9) and new synaptic complexes termed *microneuromas* (10). As remodeling continues throughout life, some neurons begin to migrate along glial columns (11), others die (12), and the inner plexiform layer becomes transformed through new synapse formation (13) and laminar deformation (14). © 2005 Robert E. Marc. Used by permission.

At the retinal output signal transmission stage, where retinal ganglion cells (RGCs) convert the neurotransmitter signals into electrical spike trains that travel along RGC axons toward the thalamic relay nuclei and other brain areas, the system is vulnerable to mechanical insults rather than to genetic dysfunction: injury to the optic nerve (trauma), increased pressure inside the eye [crushing the fragile axon fibers at the optic nerve head (glaucoma)], and inflammation (optic neuritis) or impaired blood supply (ischemic neuropathy) of the optic nerve itself. Each of these can impair or interrupt the signal-carrying capacity of optic nerve fibers (Fig. 69.2C/D).

In addition to damage occurring at these distinct stages, more generalized damage to the retina can occur. Common mechanisms for this are the leakage of capillaries in diabetic retinopathy and the interruption of the blood supply to the inner retina (retinal vascular occlusive disease), especially in individuals with a predisposition toward the development of blood clots. Each of these can lead to widespread cell death in the inner nuclear and ganglion cell layers throughout the affected area; vascular occlusive disease provokes angiogenesis; the new retinal vessels tend to leak and damage the retina's fragile structure. Diabetic retinopathy is a major cause of preventible blindness in the developed world and increasingly in developing countries as well.

## III. CURRENT AND NEAR-TERM APPROACHES TO VISION RESTORATION

As in all biomedical engineering, approaches to restore function can be based on reversal of the disease process and regeneration of autologous tissue (on tissue grafts) or on hybrid techniques involving tissue as well as synthetic materials and devices. To support the cells affected by disease or injury, neuroprotective substances, growth factors, and genetic modifications of cell function may be used, postponing or preventing further loss of function. In addition, as long as useful function remains, one can strengthen the stimulus and internal response signal, counteracting visual impairment as much as possible.

If little or no sensor function remains, one can seek to restore it in limited form through newly grown or transplanted photoreceptors and/or RPE cells, through a prosthetic device that will electrically stimulate the remaining secondary retinal cells, or through man-made tissues that mimic photoreceptor function. At the RGC level, substitution for lost signal transmission may be sought through protection of any remaining cells and administration of factors promoting axon regeneration. This may require the use of synthetic tissues and factors enabling reconnection of axons with central structures — or through *in situ* growth of new cells, promoted to differentiate into RGCs and to send new axons through the optic nerve.

In preliminary form, some of these approaches exist in the laboratory and even begin to reach the clinic, while others are merely ideas. In the following sections, existing and prospective methods are presented.

### Enhancing the Stimulus Through Optoelectronic and Optical Means

Most of these devices are based on a combination of optical and electronic image-enhancement techniques. Their common principle of operation is to improve visibility of the image to the diseased retina, through magnification, contrast and/or color enhancement, and filtering or feature extraction. Early incarnations, such as the low-vision enhancement system — LVES; no longer in production; see Fig. 69.4A and Massof (1999) — used optical magnification, zoom, automatic focus, and — for applications limited by available contrast, such as face recognition — analog image enhancement. These techniques have been adopted by more lightweight and cosmetically attractive video visors, such as the Jordy (Fig. 69.4B; Enhanced Vision Systems, Huntington Beach, CA) that enjoy considerable popularity among patients with severe visual impairment. More advanced image-processing techniques, such as digital feature enhancement and gaze-contingent remapping, will allow wearers to avoid losing crucial information in blind areas of the visual field. Such image processing techniques have been implemented in the laboratory but are not currently available in portable format due to the need for eye tracking and real-time image transformations. However, with the rapidly increasing power of chip-based image-processing technology, such wearable optoelectronic vision enhancers cannot be far off. Another limitation of current versions is the product of field of view and angular resolution, i.e., the number of pixels across the screen. Commercially available systems have standard video/VGA resolution, but this is likely to increase sharply with the wider availability of high-resolution video camera and display formats (e.g., HDTV). Similarly, compactness and color representation will be improved as smaller, lighter, and more luminous flat panel displays become available.

A useful property of the cameras in portable video visors is their built-in automatic gain control (AGC), resulting in constant internal image brightness over a wide range of environmental illumination levels. Among other properties, this makes the image-acquisition and -processing stages of these systems highly suitable as sensors and preprocessors, respectively, for prosthetic and tissue-based image enhancement systems.

Incidentally, many optical systems increase the visibility of the stimulus. One of the latest, aimed at patients with a large central scotoma (blind spot) in both eyes due to advanced AMD, uses an implantable miniature telescope (IMT™, Fig. 69.4C; VisionCare Ophthalmic Technologies, Saratoga, CA), placed behind the iris in the location of the crystalline lens. It provides one eye with a magnified view of the central visual field, while the other eye is not implanted, to preserve the wide peripheral view of the fellow eye.

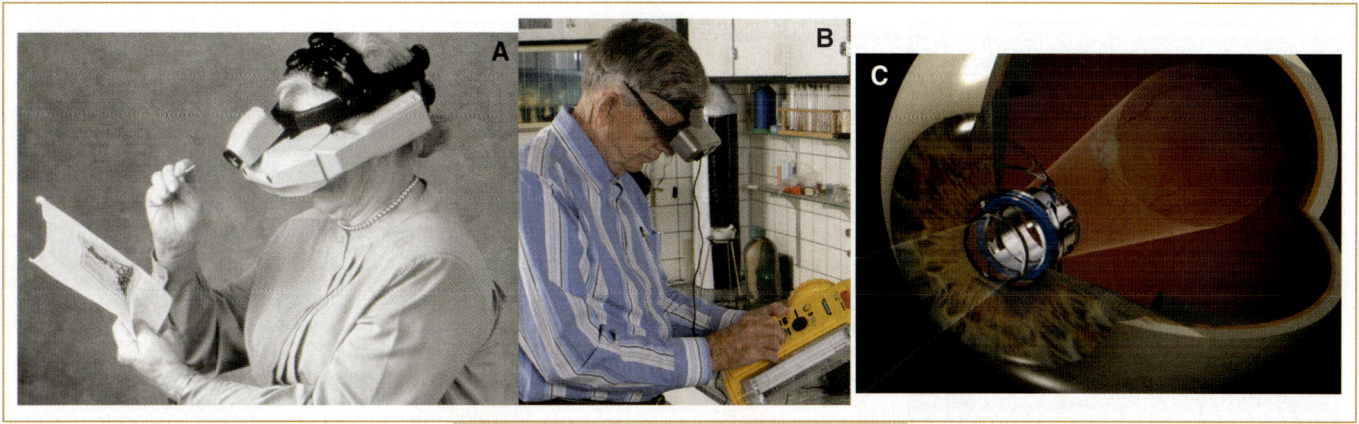

**FIG. 69.4.** **(A)** The low-vision enhancement system (LVES), developed at the Johns Hopkins University's Lions Vision Center, with support from NASA and the Department of Veterans Affairs, is a characteristic example of optoelectronic vision-enhancement systems. This system features binocular orientation cameras, a centrally placed 1.5–10× zoom camera with automatic focus, and a binocular projection path with 36° × 48° field of view and built-in refractive correction and alignment for the wearer. **(B)** The Jordy video visor has a narrower field of view but is much lighter, thanks to its LCD screens, and has a full color display. Photo courtesy of Enhanced Vision Systems. **(C)** The Implantable Miniature Telescope (IMT™) provides monocular 3× magnification, to benefit patients with large central scotomas. Diagram courtesy of VisionCare Ophthalmic Technologies.

Following an adaptation and rehabilitation training period of three to six months, most wearers have no difficulty switching between the magnified view in the implanted eye and the unenhanced view in the fellow eye. Note that this is a true prosthetic device, since it is permanent.

## Visual Prostheses Based on Electrical Tissue Stimulation

As was already mentioned, retinas with severe degeneration of the sensor layer retain high numbers of secondary cells. In analogy with the principle of operation of the cochlear prosthesis, i.e., restoring limited hearing through stimulation of spiral ganglion cells in the cochlea (Clark *et al.*, 1990), this provides the opportunity to convey pixelized images to the degenerated retina by a prosthetic device stimulating remaining secondary cells with a two-dimensional array of microelectrodes, not unlike the rastered images provided by a stadium scoreboard (Dagnelie and Massof, 1996). From an engineering point of view, one can envisage a range of possible approaches (stimulation electrodes under vs. over the retina; fully integrated photosensor, image-processor and -stimulator systems vs. external image capture and processing linked to an intraocular stimulating matrix, to name a few) but also a host of questions concerning biocompatibility, signal processing, and power management.

The most pressing question regarding the feasibility of conveying visual imagery to patients blind from retinal degeneration has been answered affirmatively. In a series of experiments started at Duke University in 1992, continued from 1993 until 2000 at Johns Hopkins, and since then confirmed through similar experiments at Harvard University and in several German university clinics, volunteers

with end-stage RP — whose remaining vision was limited to, at best, light perception — as well as several patients with advanced AMD have participated in tests where, during a surgical procedure under local anesthesia, the inner surface of the retina was electrically stimulated with small and brief biphasic current pulses applied through a single electrode or multiple electrodes (Humayun *et al.*, 1999b; Rizzo *et al.*, 2003; Hornig *et al.*, 2005). Among the most salient findings are the subjects' ability to see small punctate light flashes (phosphenes), whose perceived location corresponds exactly to that of the stimulation, and the ability to see simple patterns of multiple phosphenes when multiple electrodes are activated simultaneously. Stimulation at rates greater than 40–60 pulses per second is perceived as continuous stimulation, and perceived stimulus intensity increases with pulse duration and amplitude as well as repetition rate. Independent tests in blind volunteers and in amphibian retina (Greenberg, 1999) have demonstrated that stimulus pulses 1 ms or longer in duration preferentially stimulate the (deeper) bipolar cells rather than the (more superficial) RGCs.

Over the past several years, research in this field has moved from acute experiments to chronic implant models, and academic research has joined forces with start-up companies and in some cases government labs to develop dedicated technologies enabling chronic implants. A good example is the relocation of the implant group formerly at Johns Hopkins University to the Doheny Retina Institute at the University of Southern California, the founding of Second Sight Medical Products LLC (Sylmar, CA) as a corporate partner, and the collaboration of these two entities with laboratories at state universities (e.g., UC Santa Cruz and UC Berkeley) and in the U.S. Department of Energy

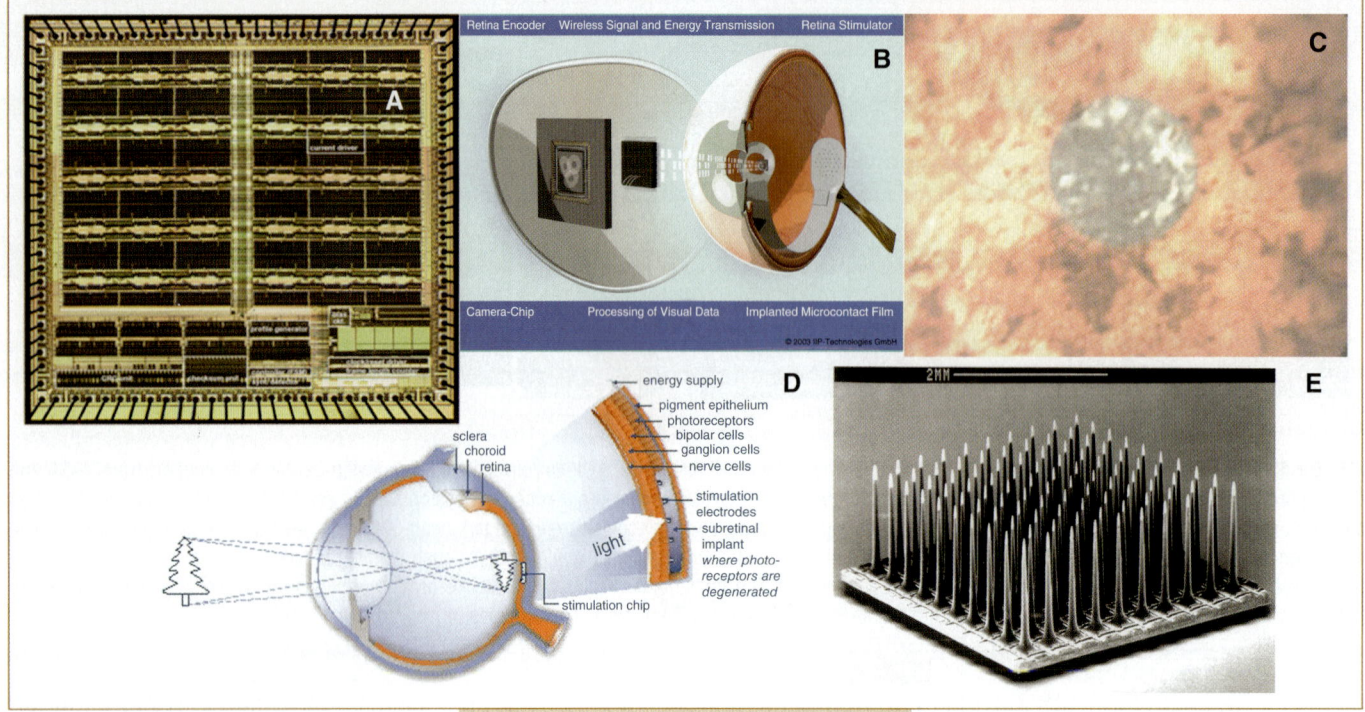

**FIG. 69.5.** **(A)** Photomicrograph of Die3 chip, developed at the University of California Santa Cruz for reception, decoding, and driving of 60 intraocular multielectrode signals. Each driver unit contains circuitry to drive five electrodes. For design considerations and other information, see Sivaprakasam *et al.* (2005). Photo courtesy of Wentai Liu, Ph.D., UCSC, Santa Cruz, CA. **(B)** Schematic epiretinal implant with transmission of electrode control signals (IR) and energy (RF) to the intraocular decoder chip. Diagram courtesy of IIP Technologies GmbH. **(C)** Optobionics Artificial Silicon Retina (ASR™) chip *in situ* under the retina of a late-stage RP patient. Photo courtesy of Optobionics Corporation. **(D)** Schematic subretinal implant with external energy supply and on-chip image capture and signal amplification and conditioning. Diagram courtesy of Retina Implant AG. **(E)** Silicon multielectrode designed by Normann *et al.* (1996) for penetrating stimulation of the visual cortex or retina. Interelectrode distances between 200 and 400 μm can be manufactured. The 1- to 2-mm-long electrode shafts are insulated, and the tips can be metallized with Pt black or other suitable metals. Image courtesy of Richard A. Normann, Ph.D., University of Utah.

(Oak Ridge, Sandia, Livermore). A sample product of this collaboration is the chipset (RF encoder/transmitter and decoder/demultiplexer/drivers) developed at UCSC to control an intraocular electrode array (Fig. 69.5A). Such devices are currently being tested in experimental animals with naturally occurring retinal degenerations.

At the present time, several other groups are pursuing intraocular prosthetic devices, in two variants: subretinal-stimulating photodiode arrays, and epiretinal-stimulating electrode arrays with external image capture and preprocessing. The latter, the more common prosthesis type, is being pursued by two groups in the United States — USC/Second Sight (Humayun *et al.*, 2003) and Harvard/MIT (Rizzo *et al.*, 2003) — and by two groups in Germany: IIP Technologies (2006) (Fig. 69.5B; IIP Technologies, 2006) and a consortium of German universities (Walter, 2006). The USC/Second Sight group has performed six 16-electrode implants based on cochlear implant technology, and all four groups are preparing prototypes with 50 or more electrodes and wireless transmission to the freely moving eye; the IIP group was the first to perform two such implants since

late 2005. A number of other groups in the United States, Germany, Switzerland, Japan, Korea, and Australia are working with similar implant designs. Some of these groups seek to stimulate the retina from outside the eyeball, to avoid the risk associated with retinal surgery, while accepting the reduced resolution inherent in the greater separation between electrodes and target cells. Others are experimenting with penetrating electrodes through the scleral wall of the eye, in order to bring the stimulus charge as close as possible to the target cells.

The second type of retinal prosthesis, a subretinal chip that uses photovoltaic elements for image capture in the location once occupied by the native photoreceptors, is conceptually simpler but in practice more challenging. In the oldest version (Fig. 69.5C; Chow *et al.*, 2004) of the subretinal array, a 2-mm-diameter silicon chip with approximately 5000 small photodiode units is placed under the retina, near the vascular arcades, typically in patients retaining a small central area of functional photoreceptors. The photodiodes convert incident light into small DC electrical currents and, while the precise mechanism of action of these currents remains to be elucidated, it appears that they lead to secondary release of neurotrophic factors in the surrounding retina, which in turn exert a beneficial effect on the remaining photoreceptors. This implant may therefore preserve some remaining vision, but it will not substitute an image by stimulating the secondary neurons.

An advanced version of the subretinal implant contains additional layers to provide signal amplification and pulse generation (Fig. 69.5D; Gekeler *et al.*, 2006). When placed at the level of the missing photoreceptor/RPE layer, it collects a sharp image of the outside world, mediated by the eye's optics, and generates localized electrical impulses that stimulate nearby (bipolar) cells in the overlying retina. Such an integrated prosthesis has great simplicity and elegance, but its feasibility hinges on three important premises: that nutrients and oxygen from the inner retinal vasculature will suffice to nourish the retina overlying the implant; that external energy can be provided to drive the amplifier and pulse generation stages; and that the heat generated by the implant's electronics can be safely dissipated into the underlying choroid without appreciably heating the overlying retina. Developers of such systems are concentrating on finding solutions for these problems as well as on the design of surgical methods to safely insert large arrays of such units under the retina (Volker *et al.*, 2004).

In addition to retinal stimulation, two other approaches to electrical stimulation of the visual system are being pursued. One of these, direct stimulation of optic nerve fibers through a set of electrodes mounted on the inside of a cuff placed around the optic nerve (Veraart *et al.*, 2003), is surgically less invasive than implantation of devices inside the eye, but it has the drawback that large numbers of fibers are stimulated simultaneously. Selective stimulation of individual optic nerve fibers will require microelectrode arrays much finer than those currently available and a sophisticated mapping system to establish the correspondence between visual field locations and individual electrodes.

The oldest attempts at vision restoration involved stimulation of the visual cortex through intracranial electrodes, controlled by external image-acquisition and signal-conditioning circuitry. Several decades ago, Brindley and Lewin (1968) took the first steps in this direction by implanting a set of electrodes over the visual cortex of a blind volunteer; the phosphenes described by this volunteer were similar to those elicited by intraocular stimulation. More recently, a team at the National Institute for Neurological Disorders and Stroke (NINDS) performed tests with over 30 electrodes penetrating the cortical surface (Schmidt *et al.*, 1996), and scientists at the University of Utah experimented with implantation of denser intracortical electrode arrays in cat and monkey visual cortex (Fig. 69.5E; Warren and Normann, 2005). The same group was also the first to perform simulation studies to establish minimum requirements for prosthetic vision (Cha *et al.*, 1992), an approach followed by several other groups (for a review, see Dagnelie,

2006). The cortical work that originated at NINDS is being continued through primate studies at the Illinois Institute of Technology (Bradley *et al.*, 2005), in preparation for future human implants.

The cortical prosthesis bypasses both retinal and optic nerve problems and might therefore be seen as a universal approach to vision restoration. Admittedly, it may be the only viable approach for patients whose inner retina and/or optic nerves have been destroyed by glaucoma or trauma. However, cortical stimulation requires complex surgery of an otherwise-healthy brain; moreover, the convoluted layout of the visual cortex complicates mapping of objects and locations in the outside world into a pixelized image that can be understood by the prosthesis wearer. Algorithms to establish such a map have been tested in primates (Bradley *et al.*, 2005) and through simulations in sighted volunteers (Dagnelie and Vogelstein, 1999) but thus far only for small sets of simulated phosphenes. The complexity of such algorithms for hundreds of phosphenes will be a significant burden on cortical implant research.

A final point of consideration for all visual prostheses for which the input image is acquired outside the eye is the altered role of eye movements. In natural vision, the intent of most eye movements is to bring an object of interest to the center of the retina, where it can be resolved with greater detail and receive directed attention. If a head-mounted or handheld camera is used for prosthetic image acquisition, movement of the camera rather than eye movements will recenter the image. To compensate for this difference, either a visual prosthesis can be equipped with eye movement monitoring and processing capability to shift the image accordingly, or the prosthesis wearer must learn to suppress eye movements as a means to acquire visual information, using camera (i.e., head or hand) movements instead.

A detailed review of developments in the areas of visual prosthesis development and simulation can be found in Dagnelie (2006).

### Retinal-Cell Transplantation

Cell transplantation in the neurosensory system can, in principle, restore function through two mechanisms: *rescue* of threatened cells (either by restoring a failing support system or through trophic factors secreted by the transplanted cells) and *replacement* of degenerating cells and *functional integration* of the transplanted cells into the host tissue. Both mechanisms may play a role in retinal cell transplantation, and the distinction is not always obvious, as will become clear.

There are two important distinctions in the choice of transplant modality. First, there is the distinction between *autologous* cells — harvested from the same individual who is to receive them, typically following amplification and/or differentiation in tissue culture — and *allografts* — harvested from a different donor, who is typically unaffected and immunologically matched. Second, the developmental stage of the transplanted cells may range from pluripotential stem cells to organ-specific undifferentiated cells to postmitotic (e.g., photoreceptor precursor) cells. Since the range of possible approaches far exceeds the scope of this chapter, only major developments will be indicated; a recent review of stem cell approaches to retinal degeneration can be found in Young (2005).

Some important categories of retinal disease are mediated by loss of RPE function, and both transplantation of autologous pigment epithelium and allografts have shown success in rescuing photoreceptor function. In the RCS rat, rescue of photoreceptor function can be accomplished by replacement of the degenerating RPE with a wide range of transplants, from RPE xenografts (harvested in a different species) (McGill et al., 2004) to autologous iris pigment epithelium (Semkova et al., 2002). In human retinal disease, macular translocation surgery in AMD has demonstrated the feasibility of rescuing threatened vision by relocating the neural retina over a relatively healthy area of RPE (de Juan and Fujii, 2001). Immunologically mismatched RPE allografts and xenografts may not provide viable treatments in human retinal disease: Contrary to the neural retina, where immune response appears to be muted or absent, a slow inflammatory response to mismatched transplanted RPE tissue is commonly observed (Zhang and Bok, 1998).

Retinal diseases that originate in a photoreceptor-cell defect do not lend themselves to treatment with autologous cell transplantation, since the same defect is likely to plague the transplanted cells. Since the mid-1990s, transplants with cell suspensions and organized sheets containing mature photoreceptors or photoreceptor precursors (with or without RPE) have been used therapeutically, with varying degrees of claimed efficacy. In a light-damaged rat model, morphological evidence of synapse formation (del Cerro et al., 1989) and behavioral evidence of regained function (DiLoreto et al., 1996) were among the earliest indicators that transplanted photoreceptors may be capable of assuming visual function. More recent evidence in the same direction from other laboratories suggests that functional synapse formation is possible in some animal models, while in others a gliotic seal under the retina effectively inhibits integration of the transplanted tissue with the host retina (Seiler and Aramant, 2005). The mechanisms responsible for such widely different results remain to be elucidated but are most likely associated with varying patterns of reorganization occurring in response to different retinal degeneration genotypes (Jones and Marc, 2005).

Fetal tissue has thus far proven more successful than fully developed retina or stem cells in forming synaptic connections and retinal morphology resembling that of intact retina, and it has become the tissue of choice in most transplantation attempts. This tissue also carries a lower risk of rejection by the host immune system, at least in photoreceptor grafts; indeed, rejection does not appear to occur in rat or mouse photoreceptor transplantation (del Cerro et al.,

1997). Admittedly, some caution in extrapolating this finding to other species is warranted, for immune reactions in rodents tend to be less severe than in humans. The use of fetal tissue for research and transplantation purposes is subject to legal and ethical concerns in a number of countries, and this is an important reason for continued efforts to develop transplantation techniques that use adult donor tissue.

Preliminary attempts at retinal-cell transplantation in blind volunteers, performed primarily to demonstrate safety, have yielded mixed, and at best modest, results. Allografts of cultured RPE appear to provide protection to functional AMD through trophic factors, but in exudative AMD the graft is quickly overwhelmed by an inflammatory reaction (Gouras and Algvere, 1996). Photoreceptor transplants appear to convey an improvement in vision in some cases (Humayun et al., 2000), but whether this is mediated by graft–host synapse formation or trophic support to the few remaining photoreceptors cannot be distinguished (del Cerro et al., 2000).

The ciliary body in the human eye contains pluripotential stem cells that will, under appropriate conditions, differentiate into retinal neuron precursors (Tropepe et al., 2000); hence, the potential for photoreceptor cell replacement from tissue culture does exist. At this time, however, any attempts to use bone marrow and retinal stem cells to replace degenerating photoreceptors in human patients through injection into the vitreous or subretinal space, as reported in the press and on the Internet, seem highly premature. It is unlikely that such attempts, without thorough research in animal models of retinal degeneration, will lead to effective and reliable vision restoration or even to long-term survival of the remaining photoreceptors.

## Optic Nerve Protection and Regeneration

Retinal ganglion cells (RGCs) behave as central nervous system neurons in their inability to recover from severe injury: RGCs whose axons are damaged by cutting or crushing the optic nerve undergo apoptosis within a few days following injury, although this particular mechanism of cell death may be restricted to certain classes of RGC (Kurimoto et al., 2003). And while the axons of peripheral (motor or sensory) neurons can regenerate on surgical reconnection of nerve fascia, CNS neurons appear to lose plasticity once their original outgrowth during the organism's development is completed, due to changes in both their internal makeup and their environment. Oligodendrocytes that form the protective myelin sheath around optic nerve axons and astrocytes appear to play a major role in preventing axon regeneration following injury (Dezawa and Adachi-Usami, 2000). It has been since the 1980s, however, that this environment can be effectively modified: Cut optic nerve axons in rat and cat will not form new neurites in their natural optic nerve environment yet can be made to regenerate into a peripheral nerve graft and reach the superior colliculus

(Aguayo *et al.*, 1987; Vidal-Sanz *et al.*, 2002; Watanabe and Fukuda, 2002). Two basic steps are thus necessary for therapeutic intervention in advanced optic nerve disease to become a reality: protection of RGCs from apoptosis following severe damage to the optic nerve, and axon regeneration, including formation of functional connections at the thalamus and other midbrain structures.

Studies in the last few years have provided encouraging indications in both areas. Intravitreal injections of anti-apoptotic drugs and neurotrophic factors can limit the loss of RGCs and prolong the window for application of other therapeutic modalities (Watanabe and Fukuda, 2002). Most of these therapies are aimed at preserving the RGC soma so that it can sustain a partially damaged axon, but there is still a need for targeted therapies that can either repair damaged axons or stimulate new axon growth (Levin, 2005). Some progress is being made in stimulating axonal growth: A recent report identified oncomodulin as a factor that is particularly effective in promoting growth following optic-nerve crush (Yin *et al.*, 2006).

## Drug Delivery

Since 1995, the understanding of photoreceptor and RGC death has changed dramatically. While it was previously thought that these cells die because a functional part — the photoreceptor outer segment and the axon, respectively — becomes dysfunctional, both events are now widely understood to trigger cell death through apoptosis and similar mechanisms. Better understanding of these mechanisms has led to the search for pharmacological interventions that may prevent cell death.

The process of RGC degeneration following transsection or crushing of the optic nerve can be halted, at least in animal models, with the use of neuroprotective agents such as neurotrophin-4/5 (NT-4/5; Sawai *et al.*, 1996). It turns out, however, that this protection has only a limited duration. Most RGCs die within two to four weeks, even with sustained administration of NT-4/5 (Clarke *et al.*, 1998). Various other neuroprotective strategies have been attempted in animal models, but their long-term efficacy as well as the feasibility of long-term delivery near the threatened site remain to be demonstrated.

In the case of photoreceptor degenerations, a variety of nerve growth factors and neuroprotective agents have been used, both *in vitro* and in animal studies, and several of these have shown promise (LaVail *et al.*, 1998). What remains to be worked out for most of these substances is the optimal delivery route. Systemic administration is not an option, since many of these agents do not easily pass the blood–retina barrier and/or have unwanted systemic side effects. Administration as an eye drop is not effective, because the active substance would need to diffuse through the cornea or sclera, and most of it would be washed away in the tear film. Repeated injection into the vitreous is unattractive to the patient, and it is not clear how long the active substance would persist in the eye and reach the outer retina.

Three relatively recent delivery techniques have all been successfully applied to ocular drugs in recent years: delivery by macromolecules, slow-release implants, and encapsulated-cell technology. The first of these approaches caused a revolution in the treatment of exudative AMD about 10 years ago, now widely known as photodynamic therapy (PDT): Liposomes with an embedded antiangiogenic substance (verteporfin) are injected into the bloodstream, and irradiation of the retina with a low-energy laser beam is used to release verteporfin in the choroidal space under the retina. The resulting oxygen free radicals attack vascular endothelial cells and stop angiogenesis (Asrani *et al.*, 1997). Unfortunately, this endothelial-cell cytotoxicity is not an effective long-term treatment, so, following the demonstrated safety of locally administered vascular endothelial growth factor (VEGF) inhibitors, PDT has over the last few years been replaced as the treatment of choice by intravitreal injections of such substances (Macugen, Avastin, and Lucentis; Rosenfeld *et al.*, 2006), even though injections have to be repeated once every six weeks. The longer-term outlook for these substances is very good. They can be embedded in poly(lactic-glycolic) acid (PLGA) microspheres to create slow-release implants that could be inserted into the eye through a small incision and attached to its inside wall, or they may even be effective when placed on the outside of the eyeball (Zhou and Wang, 2006). The third delivery method, encapsulated-cell technology (ECT), is elegantly exemplified by the NT-501-permeable-membrane delivery system developed by Neurotech SA (Lincoln, RI). When loaded with a transgenic cell line producing ciliary neurotrophic factor (CNTF) and implanted in the vitreous cavity, it will release a flow of CNTF for many months (Thanos *et al.*, 2004). Applications are sought in the area of photoreceptor protection in retinal degenerations; a phase I trial in retinitis pigmentosa has been completed successfully (Sieving *et al.*, 2006), and efficacy trials in AMD and RP were being planned for 2007.

In some diseases, even the simplest of drug delivery may be effective. In retinal ischemia, just as for other vascular blockages, clotbuster drugs, such as tissue plasminogen activator, can be intravenously injected. If this is done soon (hours or even days) after the ischemic event, much of the damage may be prevented; experimental therapy with these drugs has had promising results (Elman *et al.*, 2001).

## Genetic Interventions

Delivery issues also play a role in the introduction of new genes into the degenerating retina. Most retinal degenerations are genetic in nature, either inherited from one or both parents or caused by a new mutation. The premise of gene therapy is that intervention at an early stage may restore normal function to the cell and prevent the degeneration process that might otherwise ensue. Several strategies are needed to combat inherited retinal degenerations.

- In recessively inherited diseases, both copies of a gene are defective, and introduction of a third copy may be sufficient to achieve adequate production of a functional protein to replace the defective one encoded by the mutated gene. An inactivated (retro)virus is used to introduce a healthy copy of the gene. Following a number of demonstrations of successful gene transfection into RPE cells and photoreceptors *in vitro* and feasibility studies in rodents, a highly publicized study of vision rescue in a canine model of Leber's congenital amaurosis (Acland *et al.*, 2001) provided proof that gene therapy for some human retinal degenerations may soon be a reality

- In X-linked disease, males carry only one copy of the X-chromosome, including the defective gene; here, too, introduction of a healthy copy of the gene may be enough to achieve normal function.

- In dominant disease, a single bad copy of the gene suffices to poison the delicate balance of the cellular machinery or, at the very least, to prevent its proper function. To reduce or prevent this, a therapeutic intervention must block a step along the transcription pathway from the defective gene to its product protein but not inhibit expression of the good copy of the gene; a gene for such a blocking agent could, in principle, permanently neutralize the defect. This technique has been used in a transgenic rat model, using ribozymes to destroy mRNA produced by the P23H mutant gene defect responsible for one form of autosomal dominant RP (Drenser *et al.*, 1998). In a rat model with a naturally occurring degeneration, a gene for ribozyme production was successfully introduced into photoreceptors and proved effective even when introduced at a time when many photoreceptors had already succumbed to the effect of the defective rhodopsin gene (LaVail *et al.*, 2000), which bodes well for therapeutic interventions in dominant RP. These forms of RP tend to preserve substantial levels of vision into middle age, and this remaining vision might thus be rescued.

- An entirely separate class of retinal degenerations, typically asymptomatic until middle age and often associated with multisystem disease, is caused by mutations of the maternally inherited mitochondrial DNA, responsible primarily for cellular energy supply. Mitochondrial DNA has multiple copies, and it remains unclear whether single or multiple copies are responsible for these disorders. Moreover, the mitochondrial DNA does not lend itself to normal viral transfection techniques, which adds a level of complexity to potential gene therapeutic approaches. Recent reports suggest that proteins may be used to deliver expression-blocking factors or new functional genes into the mitochondrial milieu (Khan and Bennett, 2004).

Vision-related gene therapy research is concentrated around RP, Stargardt macular dystrophy, and related diseases with known inheritance patterns. Due to the multiple genes that can lead to these disorders and the enormous number of specific mutations that have already been identified, many gene therapeutic variants will be needed to address a substantial proportion of these disorders. This is an extremely active area of research; excellent reviews can be found in Delyfer *et al.* (2004) and Dinculescu *et al.* (2005),

Still, this means that a true cure may become available for these retinal degenerations. AMD and optic nerve diseases like glaucoma and ischemia, on the other hand, have no known genetic causes, though some genes associated with predisposition for such diseases have been identified. Genetic interventions may assist in preventing some forms of these diseases, but the prospects of other therapeutic approaches are more promising than those of gene therapy.

## IV. EMERGING APPLICATION AREAS FOR ENGINEERED CELLS AND TISSUES

Of the current approaches to vision enhancement considered earlier, strictly speaking only the use of stem and transgenic cells falls within the realm of tissue engineering. The complexity of ocular structures such as the retina and optic nerve poses daunting challenges to anyone seeking to re-create their function. While this may explain the lag in progress as compared to other organ systems, it should not keep researchers working to restore vision from drawing on the remarkable progress of tissue-engineering approaches. We next briefly consider four application areas, corresponding to the aforementioned three processing stages in the early visual system where severe vision loss may occur and the supporting retinal infrastructure.

### Photosensitive Structures

Current efforts in the areas of retinal-cell transplantation and intraocular prosthesis design, while promising, are in no way guaranteed to lead to reliable — i.e., long-term stable — restoration of useful (let alone high-quality) vision. As was explained earlier, each of the approaches currently being explored has inherent drawbacks: Epiretinal electrode arrays require external image acquisition and preprocessing, which may necessitate real-time eye movement tracking and compensation. Subretinal implants may cause atrophy of the overlying retina and are limited by the low yield of semiconductor photoelectric conversion, the need for signal amplification and thus external energy supply, and considerable heat production. Both implant types may be limited in resolution to ambulatory vision due to the 100- to 200-μm distance separating the electrodes from the target cells. As a rule of thumb, resolution in the underlying tissue will be no better than this distance. Retinal cell transplanta-

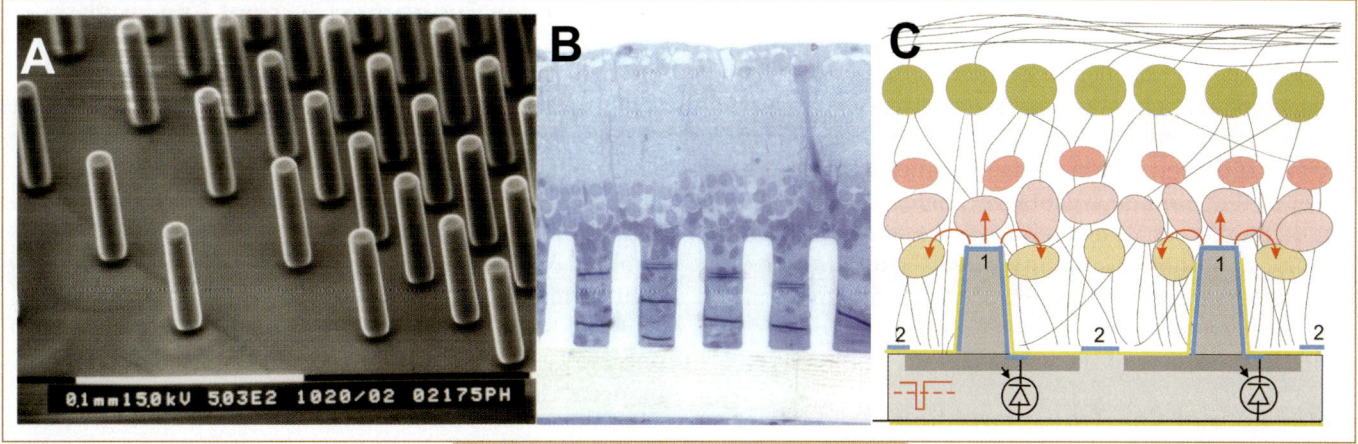

**FIG. 69.6.** Pillar electrodes penetrating into the middle of the inner nuclear layer of degenerated retina. **(A)** Electron micrograph of an array having pillars of 10 μm in diameter and 70 μm in height. **(B)** Histology of the RCS rat retina with a pillar array six weeks after implantation. **(C)** Schematic of a retinal prosthetic: retinal cells in the inner nuclear layer stimulated by pulsed electric fields emanating from the exposed tops of the pillars. Images provided by Daniel Palanker, Ph.D., Stanford University.

tion efforts are likely to be limited by the spotty record of transplanted cells in making functional contacts with the native inner retinal circuitry. And all three approaches are limited by microneuroma formation, destroying the functional diversity of the inner retina and limiting its resolution to 50–100 μm, i.e., at best 20/200 visual acuity — legal blindness.

For improved retinal prosthetics, it may be possible to develop electrodes that make more intimate contact with the target cells. If successful, this would improve resolution to the limit imposed by the repatterned inner retinal neuronal circuitry. One can envisage such a penetrating array as the Normann microelectrode in Fig. 69.5E, but the damage such an array might do to the delicate microphysiology of the retina is a distinct disadvantage. As an alternative, one might envisage inserting or growing, *in situ*, an array of parallel neurites. These might branch out of cells grown on the surface of implanted stimulating chips, penetrating the retina until they reach a specific target environment, e.g., the inner nuclear layer, where they would make synaptic contacts with the native cell population; these neurites would act as "tube electrodes," releasing either electrical charge or a neurotransmitter that would activate the target cells. An alternative approach would be an electrode structure that stimulates nearby retinal cells to form new neurites, spurred on by favorable growth conditions and a coating on the implant. Such coatings promoting cell growth, formed by microcontact printing, are under investigation (Mehenti *et al.*, 2006).

The distance between implant and target cells could also be reduced by enticing native retinal cells to migrate toward the electrode surface, adding a new twist to the cell migration seen in naturally occurring retinal repatterning (Jones and Marc, 2005). Such approaches are not as farfetched as they may sound. Several groups are experimenting with surface modifications on semiconductor chips that will allow cells to adhere to these surfaces and be activated by neurotransmitters (Atencia and Beebe, 2005), while researchers at Stanford have reported the tendency of retinal neurons to migrate through a perforated membrane placed under the photoreceptor layer (Fig. 69.6; Palanker *et al.*, 2004). Combinations of such technologies may lead to improved implant–tissue interfaces.

A fundamentally different approach to improving (sub)retinal prostheses may be the use of high-yield photoconversion systems. Such high-yield conversions are known to be performed by Photosystem I (PSI), a macromolecule present in the membrane of thylakoids, which can be found inside the chloroplasts that give the green color to plant leaves and algae. Recent experiments with thylakoids (J. W. Lee *et al.*, 1996) have shown that it is possible to anchor these structures onto a metal surface and use them as miniature photovoltaic elements. I. Lee *et al.* (1997) have also demonstrated that it is possible chemically to modify thylakoid surface membranes to create charge displacement in a specific direction. Kuritz *et al.* (2005) demonstrated the ability of PSI to impart photosensitive polarization and $Ca^{2+}$ ion movement to retinoblastoma cells in tissue culture, as a first example of cellular engineering that may eventually lead to artificial photoreceptors.

The engineering successes of thylakoids and PSI open the opportunity to create cells that assume a dipole charge distribution or, with the help of some form of intracellular electronics, produce a biphasic pulse between the poles; in a subsequent stage of development, automatic gain control could be incorporated as a limited form of light/dark adaptation. If these cells can be made to attach to the outer retinal layers or to migrate into the inner nuclear layer, one could achieve microscopic local current sources with sufficient conversion efficiency to ensure vision at a broad range of (day)light levels. To achieve true night vision, an external device, such as a night vision scope, could be used.

Note that in this idealized situation, the synthetic photoreceptors might be small and sensitive enough to provide good vision without external image preprocessing (e.g., magnification and contrast/edge enhancement). It is to be expected that preliminary forms of such light-sensing units will be neither small nor sensitive enough to provide the dynamic range and resolution required; at this intermediate solution level, external image processing with an advanced portable low-vision visor would be a necessary complement to this intraocular light-conversion array.

## Outer Retinal-Cell Transplantation

Despite almost two decades of careful investigation, reliable and widespread formation of synapses between transplanted photoreceptor precursors and native inner retinal cells — the gold standard for full integration of the transplanted cells into the host retina — remains elusive. Also, the effects of the host immune response on graft survival — especially of transplanted RPE — cannot be ignored, even if it appears to be mild, and this host reaction will have to be effectively controlled without long-term systemic immune suppression. Both areas can profit from tissue-engineering approaches. A complex set of conditions — cells in the proper stage of development, properly matched to the host retina, and spurred on by a proper combination of neuroprotective and neurotrophic modifications to the host environment — needs to be met for successful graft–host integration. Recent reports indicate that even widely held beliefs regarding the proper developmental stage of the transplanted cells may need to be revisited. In a mouse model, early postnatal photoreceptors showed widespread synapse formation with native retinal cells (Ali *et al.*, 2006), whereas most previous research has been directed at the use of postmitotic photoreceptor precursors in a much earlier stage of development.

Even if the cell population affected by the primary degeneration process is successfully replaced, secondary degeneration stages — of RPE cells in photoreceptor degenerative disease and of photoreceptors in RPE degeneration — may still prevent restoration of visual function. This problem can be addressed adequately only by performing a combined graft of RPE and photoreceptors, presumably prepared in a tissue culture environment and stimulated toward integration through a carefully tuned combination of neuroprotective and neurotrophic factors. Culturing stem cells or differentiated cell lines might allow the creation of heterogeneous structures such as RPE/photoreceptor double-sheets. Culture conditions and the growth stage of these sheets should be modulated to prepare the cells for integration with the host retina. In order to provide sufficient structural support to these fragile sheets during growth and transplantation, a resorbable polymer layer could be used as a substrate.

## Cell Matrices Supporting Axonal Regrowth

Cell culture conditions and retinal integration alone will not be sufficient in the case of ganglion cell transplantation with the objective of repopulating the optic nerve with axons carrying visual information to the brain. As discussed earlier, this will require the (re)creation of conditions favorable for axonal growth over distances of many centimeters and the ability of sprouting axons to contact target cells with the correct retinotopic mapping. This feat is accomplished effortlessly by RGC axons in the developing embryo, but to recreate the conditions for this to occur in an adult organism will be a considerable challenge.

As was noted earlier, an RGC axon damaged by glaucoma, optic nerve disease, or injury can only form new neurites under optimal environmental conditions, essentially mimicking those in a developing organism. At the same time, neuroprotective factors are required to sustain the RGC long enough for the axon to assume this support function. If the glaucoma diagnosis is made early and safe and effective neuroprotective agents become available, it may be possible to save most of the threatened axons and thus spare most RGCs and the patient's vision. In practice, however, many axons may have been lost by the time the diagnosis is made, and protection of the cell somata and stimulation of axon regrowth become the treatment objectives (Quigley and Iglesia, 2004). The experimental conditions employed thus far, using peripheral nerve sheaths to create a substrate for axonal regrowth, are less than ideal, because they do not provide an integrated environment in which regrowing axons combine with intact remaining axons and in which protection of the threatened RGCs is built into the environment. This is a very active field of research, offering great potential for tissue-engineering approaches, and one may expect such approaches to play a major role in achieving these objectives. Understanding the necessary conditions to create an integrated environment for axonal regrowth and creating novel synthetic materials to provide these conditions should provide wonderful challenges to tissue engineers.

One approach to such an integrated solution might be that engineered cells could be grown *in situ*, as a loose skeleton of supporting tissue, to follow the course of the optic nerve to the chiasm and optic tract; these cells would be programmed to exude the necessary factors promoting axon growth and RGC protection. Alternatively, it might be possible to modify the normal environment of the optic nerve (temporarily) to allow or even stimulate axonal growth. Assuming that it would be possible to guide outgrowing axons through the optic chiasm toward the appropriate structures and retinotopic projections in the midbrain, RGCs may restore their connections to the target cells and and resume functional visual processing.

## Repopulating Ischemic or Diabetic Retina

As mentioned earlier, new capillaries — formed under the influence of an angiogenic tissue response — tend to be poorly organized and fragile, causing leakage and thus a great deal of damage to the already stressed retinal tissue. Therefore, the prospects of restoring vision in such retinal areas are, at the present time, poor.

This may change, however, if cell populations can be grown *in vitro* and introduced into the retina under physiological conditions mimicking those during embryonal development. In that case the formation of new blood vessels could follow a more orderly pattern, and implanted cells would have a much better chance at forming functional connections. Whether and when it will be possible to re-create embryonal conditions and grow such integrated retinal tissue required to restore vision to an ischemic portion of the retina, from RPE to RGC axons, is difficult to predict. It is a challenge whose magnitude exceeds that of RPE/photoreceptor transplants, functional stimulation of inner retinal cells, and RGC protection/axonal regrowth combined.

## V. CONCLUSION: TOWARD 20/20 VISION

The potential applications of tissue engineering sketched in the previous sections pose enormous challenges, well exceeding the competency of any single group or institution. Concerted research efforts by multidisciplinary groups may allow the implementation of the complex systems required to restore and enhance vision. As researchers improve, on the one hand, their fundamental understanding of processes such as photoconversion, graft integration, immune regulation, and neuroprotection and, on the other, the engineering ability to control tissue properties and neurite growth, both *in vitro* and *in situ*, crude but functional vision restoration at the RPE/photoreceptor level and at the RGC level will advance to the level of experimental and even clinical therapy. Integration of all these areas to re-create the full range of retinal processing is a much more distant goal, for which the header of this section may well be too ambitious.

To accomplish any of these forms of vision restoration, however, funding mechanisms for multidisciplinary research and interest from the corporate sector will have to rise well beyond their current levels. While the number of severely visually impaired individuals and the economic impact of vision restoration alone may not justify that these approaches receive priority over treatment of life-threatening conditions, the investment required is relatively modest, and the improvement in quality of life for (nearly) blind patients can be very significant.

## VI. ACKNOWLEDGMENTS

Supported by the National Eye Institute (R01 EY12843).

## VII. REFERENCES

Acland, G. M., Aguirre, G. D., Ray, J., Zhang, Q., Aleman, T. S., Cideciyan, A. V., Pearce-Kelling, S. E., Anand, V., Zeng, Y., Maguire, A. M., Jacobson, S. G., Hauswirth, W. W., and Bennett, J. (2001). Gene therapy restores vision in a canine model of childhood blindness. *Nat. Genet.* **28**, 92–95.

Aguayo, A. J., Vidal-Sanz, M., Villegas-Perez, M. P., and Bray, G. M. (1987). Growth and connectivity of axotomized retinal neurons in adult rats with optic nerves substituted by PNS grafts linking the eye and the midbrain. *Ann. N.Y. Acad. Sci.* **495**, 1–9.

Ali, R. R., MacLaren, R. E., Pearson, R. A., MacNeil, A., and Sowden, J. C. (2006). Successful transplantation of photoreceptors into the retina of an adult mouse model of human retinitis pigmentosa. *Inv. Ophthalmol. Vis. Sci.* **47**, ARVO E-abstr. 5763.

Asrani, S., Zou, S., D'Anna, S., Lutty, G., Vinores, S. A., Goldberg, M. F., and Zeimer, R. (1997). Feasibility of laser-targeted photoocclusion of the choriocapillary layer in rats. *Inv. Ophthalmol. Vis. Sci.* **38**, 2702–2710.

Atencia, J., and Beebe, D. J. (2005). Controlled microfluidic interfaces. *Nature* **437**, 648–655.

Bradley, D. C., Troyk, P. R., Berg, J. A., Bak, M., Cogan, S., Erickson, R., Kufta, C., Mascaro, M., McCreery, D., Schmidt, E. M., Towle, V. L., and Xu, H. (2005). Visuotopic mapping through a multichannel stimulating implant in primate V1. *J. Neurophysiol.* **93**, 1659–1670.

Brindley, G. S., and Lewin, W. S. (1968). The sensations produced by electrical stimulation of the visual cortex. *J. Physiol.* **196**, 479–493.

Cha, K., Horch, K. W., and Normann, R. A. (1992). Reading speed with a pixelized vision system. *J. Opt. Soc. Amer.* **A9**, 673–677.

Chow, A. Y., Chow, V. Y., Packo, K. H., Pollack, J. S., Peyman, G. A., and Schuchard, R. (2004). The artificial silicon retina microchip for the treatment of vision loss from retinitis pigmentosa. *Arch. Ophthalmol.* **122**, 460–469.

Clark, G. M., Tong, Y. C., and Patrick, J. F. (1990). Introduction. *In* "Cochlear Prostheses" (G. M. Clark, Y. C. Tong, and J. F. Patrick, eds.), pp. 1–14. Churchill Livingstone, Melbourne, Australia.

Clarke, D. B., Bray, G. M., and Aguayo, A. J. (1998). Prolonged administration of NT-4/5 fails to rescue most axotomized retinal ganglion cells in adult rats. *Vision Res.* **38**, 1517–1524.

Dagnelie, G. (2006). Visual prosthetics 2006 — assessment and expectations. *Expert Rev. Med. Dev.* **3**, 315–325.

Dagnelie, G., and Massof, R. W. (1996). Towards an artificial eye. *IEEE Spectrum* **33**(5), 20–29.

Dagnelie, G., and Vogelstein, J. V. (1999). Phosphene mapping procedures for prosthetic vision. *In* "Vision Science and Its Applications," *OSA Technical Digest*, pp. 294–297. Optical Society of America, Washington, DC.

Daiger, S. P., Sullivan, L. S., and Bowne, S. J. (1996–2006). Retnet Retinal Information Network. http://www.retnet.org.

de Juan, E., Jr., and Fujii, G. Y. (2001). Limited macular translocation. *Eye* **15**, 413–423.

del Cerro, M., Notter, M. F., del Cerro, C., Wiegand, S. J., Grover, D. A., and Lazar, E. (1989). Intraretinal transplantation for rod-cell replacement in light-damaged retinas. *J. Neural Transplant.* **1**, 1–10.

del Cerro, M., Lazar, E. S., and DiLoreto, D., Jr. (1997). The first decade of continuous progress in retinal transplantation. *Microscopy Res. Techn.* **36**, 130–141.

del Cerro, M., Humayun, M. S., *et al.* (2000). Histologic correlation of human neural retinal transplantation. *Invest. Ophthalmol. Vis. Sci.* **41**, 3142–3148.

Delyfer, M. N., Leveillard, T., Mohand-Said, S., Hicks, D., Picaud, S., and Sahel, J. A. (2004). Inherited retinal degenerations: therapeutic prospects. *Biol. Cell* **96**, 261–269.

Dezawa, M., and Adachi-Usami, E. (2000). Role of Schwann cells in retinal ganglion cell axon regeneration. *Prog. Retin. Eye Res.* **19**, 171–204.

DiLoreto, Jr., D., del Cerro, M., Reddy, S. V., Janardhan, S., Cox, C., Wyatt, J., and Balkema, G. W. (1996). Water escape performance of adult RCS dystrophic and congenic rats: a functional and histomorphometric study. *Brain Res.* **717**, 165–172.

Dinculescu, A., Glushakova, L., Min, S. H., and Hauswirth, W. W. (2005). Adeno-associated virus-vectored gene therapy for retinal disease. *Hum. Gene Ther.* **16**, 649–663.

Drenser, K. A., Timmers, A. M., Hauswirth, W. W., and Lewin, A. S. (1998). Ribozyme-targeted destruction of RNA associated with autosomal-dominant retinitis pigmentosa. *Invest. Ophthalmol. Vis. Sci.* **39**, 681–689.

Elman, M. J., Raden, R. Z., and Carrigan, A. (2001). Intravitreal injection of tissue plasminogen activator for central retinal vein occlusion. *Trans. Am. Ophthalmol. Soc.* **99**, 219–221.

Gekeler, F., Szurman, P., Grisanti, S., Weiler, U., Claus, R., Greiner, T. O., Volker, M., Kohler, K., Zrenner, E., and Bartz-Schmidt, K. U. (2006). Compound subretinal prostheses with extraocular parts designed for human trials: successful long-term implantation in pigs. *Graefes. Arch. Clin. Exp. Ophthalmol.* In press.

Gouras, P., and Algvere, P (1996). Retinal cell transplantation in the macula: new techniques. *Vis. Res.* **36**, 4121–4126.

Greenberg, R. J., Velte, T. J., Humayun, M. S., Scarlatis, G. N., and de Juan, E., Jr. (1999). A computational model of electrical stimulation of the retinal ganglion cell. *IEEE Trans. Biomed. Eng.* **46**, 505–514.

Heckenlively, J. R., Yoser, S. L., Friedman, L. H., and Oversier, J. J. (1988). Clinical findings and common symptoms in retinitis pigmentosa. *Am. J. Ophthalmol.* **105**, 504–511.

Hornig, R., Laube, T., Walter, P., Velikay-Parel, M., Bornfeld, N., Feucht, M., Akguel, H., Rossler, G., Alteheld, N., Lutke Notarp, D., Wyatt, J., and Richard, G. (2005). A method and technical equipment for an acute human trial to evaluate retinal implant technology. *J. Neural. Eng.* **2**, S129–S134.

Humayun, M. S., Prince, M., de Juan, E., Jr., Barron, Y., Moskowitz, M. T., Klock, I. B., and Milam, A. H. (1999a). Morphometric analysis of the extramacular retina from postmortem eyes with retinitis pigmentosa. *Invest. Ophthalmol Vis. Sci.* **40**, 143–148.

Humayun, M. S., de Juan, Jr., E., Weiland, J. D., Dagnelie, G., Katona, S., Greenberg, R. J., and Suzuki, S. (1999b). Pattern electrical stimulation of the human retina. *Vision Res.* **39**, 2569–2576.

Humayun, M. S., de Juan, E., Jr., del Cerro, M., Dagnelie, G., Radner, W., and del Cerro, C. (2000). Human neural retinal transplantation. *Invest. Ophthalmol. Vis. Sci.*, **41**, 3100–3106.

Humayun, M. S., Weiland, J. D., Fujii, G. Y., Greenberg, R., Williamson, R., Little, J., Mech, B., Cimmarusti, V., Van Boemel, G., Dagnelie, G., and de Juan, E., Jr. (2003). Visual perception in a blind subject with a chronic microelectronic retinal prosthesis. *Vision Res.* **43**, 2573–2581.

IIP Technologies. (2006). IIP Tec home page: http://www.iip-tec.com/english/index.php4.

Jones, B. W., and Marc, R. E. (2005). Retinal remodeling during retinal degeneration. *Exp. Eye Res.* **81**, 123–137.

Khan, S. M., and Bennett, J. P., Jr. (2004). Development of mitochondrial gene replacement therapy. *J. Bioenerg. Biomembr.* **36**, 387–393.

Kurimoto, T., Miyoshi, T., Suzuki, A., Yakura, T., Watanabe, M., Mimura, O., and Fukuda, Y. (2003). Apoptotic death of beta cells after optic nerve transection in adult cats. *J. Neurosci.* **23**, 4023–4028.

Kuritz, T., Lee, I., Owens, E. T., Humayun, M., and Greenbaum, E. (2005). Molecular photovoltaics and the photoactivation of mammalian cells. *IEEE Trans. Nanobiosci.* **4**, 196–200.

LaVail, M. W., Yasumura, D., Matthes, M. T., Lau-Villacorta C., Unoki, K., Sung, C. H., and Steinberg, R. H. (1998). Protection of mouse photoreceptors by survival factors in retinal degenerations. *Invest. Ophthalmol. Vis. Sci.* **39**, 592–602.

LaVail, M. M., Yasumura, D., Matthes, M. T., Drenser, K. A., Flannery, J. G., Lewin, A. S., and Hauswirth, W. W. (2000). Ribozyme rescue of photoreceptor cells in P23H transgenic rats: long-term survival and late-stage therapy. *Proc. Natl. Acad. Sci. U.S.A.* **97**, 11488–11493.

Lee, J. W., Lee, I., and Greenbaum, E. (1996). Platinization: a novel technique to anchor photosystem I reaction centers onto a metal surface at biological temperature and pH. *Biosens. Bioel.* **11**, 375–387.

Lee, I., Lee, J. W., and Greenbaum, E. (1997). Biomolecular electronics: vectorial arrays of photosynthetic reaction centers. *Phys. Rev. Ltrs.* **79**, 3294–3297.

Levin, L. A. (2005). Neuroprotection and regeneration in glaucoma. *Ophthalmol. Clin. North Amer.* **18**, 585–596, vii.

Massof, R. W. (1999). Electro-optical head-mounted low-vision enhancement. *Pract. Optom.* **9**, 214–220.

McGill, T. J., Lund, R. D., Douglas, R. M., Wang, S., Lu, B., and Prusky, G. T. (2004). Preservation of vision following cell-based therapies in a model of retinal degenerative disease. *Vision Res.* **44**, 2559–2566.

Mehenti, N. Z., Tsien, G. S., Leng, T., Fishman, H. A., and Bent, S. F. (2006) A model retinal interface based on directed neuronal growth for single cell stimulation. *Biomed. Microdev.* **8**, 141–150.

Palanker, D., Huie, P., Vankov, A., Aramant, R., Seiler, M., Fishman, H., Marmor, M., and Blumenkranz, M. (2004). Migration of retinal cells through a perforated membrane: implications for a high-resolution prosthesis. *Invest. Ophthalmol. Vis. Sci.* **45**, 3266–3270.

Quigley, H. A., and Iglesia, D. S. (2004). Stem cells to replace the optic nerve. *Eye* **18**, 1085–1088.

Rizzo, J. F., 3rd, Wyatt, J., Loewenstein, J., Kelly, S., and Shire, D. (2003). Perceptual efficacy of electrical stimulation of human retina with a microelectrode array during short-term surgical trials. *Invest. Ophthalmol. Vis. Sci.* **44**, 5362–5369.

Rosenfeld, P. J., Heier, J. S., Hantsbarger, G., and Shams, N. (2006). Tolerability and efficacy of multiple escalating doses of ranibizumab (Lucentis) for neovascular age-related macular degeneration. *Ophthalmology* **113**, 632–641.

Santos, A., Humayun, M. S., de Juan, E., Jr., Greenberg, R. J., Marsh, M. J., Klock, I. B., and Milam, A. H. (1997). Preservation of the inner retina in retinitis pigmentosa. A morphometric analysis. *Arch. Ophthalmol.* **115**, 511–515.

Sawai, H., Clarke, D. B., Kittlerova, P., Bray, G. M., and Aguayo, A. J. (1996). Brain-derived neurotrophic factor and neurotrophin-4/5 stimulate growth of axonal branches from regenerating retinal ganglion cells. *J. Neurosci.* **16**, 3887–3894.

Schmidt, E. M., Bak, M. J., Hambrecht, F. T., Kufta, C. V., O'Rourke, D. K., and Vallabhnath, P. (1996). Feasibility of a visual prosthesis for the blind based on intracortical microstimulation of the visual cortex. *Brain.* **119**, 507–522.

Seiler, M. J., and Aramant, R. B. (2005). Transplantation of neuroblastic progenitor cells as a sheet preserves and restores retinal function. *Semin. Ophthalmol.* **20**, 31–42.

Semkova, I., Kreppel, F., Welsandt, G., Luther, T., Kozlowski, J., Janicki, H., Kochanek, S., and Schraermeyer, U. (2002). Autologous transplantation of genetically modified iris pigment epithelial cells: a promising concept for the treatment of age-related macular degeneration and other disorders of the eye. *Proc. Natl. Acad. Sci. U.S.A.* **99**, 13090–13095.

Sieving, P. A., Caruso, R. C., Tao, W., Coleman, H. R., Thompson, D. J., Fullmer, K. R., and Bush, R. A. (2006). Ciliary neurotrophic factor (CNTF) for human retinal degeneration: phase I trial of CNTF delivered by encapsulated cell intraocular implants. *Proc. Natl. Acad. Sci. U.S.A.* **103**, 3896–3901.

Sivaprakasam, M., Liu, W. T., Wang, G. X., Weiland, J. D., and Humayun, M. S. (2005). Architecture tradeoffs in high-density microstimulators for retinal prosthesis. IEEE Trans. Circuits Systems I — Regular Papers **52**, 2629–2641.

Thanos, C. G., Bell, W. J., O'Rourke, P., Kauper, K., Sherman, S., Stabila, P., and Tao, W. (2004). Sustained secretion of ciliary neurotrophic factor to the vitreous, using the encapsulated cell therapy-based NT-501 intraocular device. *Tissue Eng* **10**, 1617–1622.

Tropepe, V., Coles, B. L., Chiasson, B. J., Horsford, D. J., Elia, A. J., McInnes, R. R., and van der Kooy, D. (2000). Retinal stem cells in the adult mammalian eye. *Science* **287**, 2032–2036.

Veraart, C., Wanet-Defalque, M. C., Gerard, B., Vanlierde, A., and Delbeke, J. (2003). Pattern recognition with the optic nerve visual prosthesis. *Artif. Organs.* **27**, 996–1004.

Vidal-Sanz, M., Aviles-Trigueros, M., Whiteley, S. J., Sauve, Y., and Lund, R. D. (2002). Reinnervation of the pretectum in adult rats by regenerated retinal ganglion cell axons: anatomical and functional studies. *Prog. Brain Res.* **137**, 443–452.

Volker, M., Shinoda, K., Sachs, H., Gmeiner, H., Schwarz, T., Kohler, K., Inhoffen, W., Bartz-Schmidt, K. U., Zrenner, E., and Gekeler, F. (2004). *In vivo* assessment of subretinally implanted microphotodiode arrays in cats by optical coherence tomography and fluorescein angiography. *Graefes. Arch. Clin. Exp. Ophthalmol.* **242**, 792–799.

Walter, P. (2006). University Eye Clinic of the RWTH Aachen home page: http://www.eyenet-aachen.de/05-07-1-implants.html#epi_ret.

Watanabe, M., and Fukuda, Y. (2002). Survival and axonal regeneration of retinal ganglion cells in adult cats. *Prog. Retin. Eye Res.* **21**, 529–553.

Warren, D. J., and Normann, R. A. (2005). Functional reorganization of primary visual cortex induced by electrical stimulation in the cat. *Vision Res.* **45**, 551–565.

Yin, Y., Henzl, M. T., Lorber, B., Nazakawa, T., Thomas, T. T., Jiang, F., Langer, R., and Benowitz, L. I. (2006). Oncomodulin is a macrophage-derived signal for axon regeneration in retinal ganglion cells. *Nat. Neurosci.* **9**, 843–852.

Young, M. J. (2005). Stem cells in the mammalian eye: a tool for retinal repair. *Apmis* **113**, 845–857.

Zhang, X., and Bok, D. (1998) Transplantation of retinal pigment epithelial cells and immune response in the subretinal space. *Invest. Ophthalmol. Vis. Sci.* **39**, 1021–1027.

Zhou, B., and Wang, B. (2006). Pegaptanib for the treatment of age-related macular degeneration. *Exp. Eye. Res.* **83**, 615–619.

# Part Eighteen

# Oral/Dental Applications

# Chapter Seventy

# Biological Tooth Replacement and Repair

*Anthony J. (Tony) Smith and Paul T. Sharpe*

I. Introduction
II. Tooth Development
III. Whole-Tooth Tissue Engineering

IV. Dental-Tissue Regeneration
V. Conclusions
VI. References

## I. INTRODUCTION

Teeth might not at first seem an obvious organ to develop biological methods for replacement or repair. Teeth are nonessential organs, and modern dental treatments enable most dental problems to be treated using traditional nonbiological approaches. Teeth do, however, offer unique opportunities to develop tissue-engineering–based approaches on an organ that is not only easily accessible and non-life-threatening but one where there are very large patient numbers and a significant clinical problem. The challenges of developing methods needed to restore and repair complex organs are immense, but the biggest challenge will be the testing of any such organs on patients. Patients requiring repair or replacement of major internal organs (heart, liver, lungs, pancreas, etc.) are by definition likely to have a serious illness. Access to these organs requires major surgery; should the treatment fail, the consequences will be severe and life threatening. Teeth do not present any such problems and thus provide an opportunity for the proof of concept to be tested safely, with little chance of danger to the patients.

Although there are an increasing number of alternative treatments in dentistry, most are nonbiological and many are based on techniques that have been practiced for thousands of years. Dental implants, for example, involve the replacement of missing teeth with metal rods that are screwed into holes drilled into the jawbone. The practice of replacing missing teeth with metal implants can be traced back to the ancient Romans and Egyptians (Crubez *et al.*, 1998; Westbroek and Martin, 1998). The number of dental implants is increasing each year, and thus there is a need to develop new biologically based approaches. Similarly, many tooth fillings still use mercury-based amalgams, and the possibility that repair of dental hard tissues (dentine and enamel) might involve the use of cells to remineralize the damage to teeth naturally is an exciting prospect. This chapter explores the current status of research directed toward whole-tooth replacement and repair of dental disease.

## II. TOOTH DEVELOPMENT

Teeth are ectodermal appendages (hair, sweat glands, salivary glands, etc.) that develop from an increasingly well-understood series of reciprocal epithelial mesenchymal interactions (Figs. 70.1 and 70.2). In mammals, these interactions take place in the developing oral cavity between the oral ectoderm (epithelium) and cranial neural crest–derived mesenchyme (ectomesenchyme). The cellular origins of the signals that initiate tooth development have been the subject of much controversy over the years, but modern molecular and animal techniques have confirmed that the oral (pre-dental) epithelium is the source of the signals that initiate

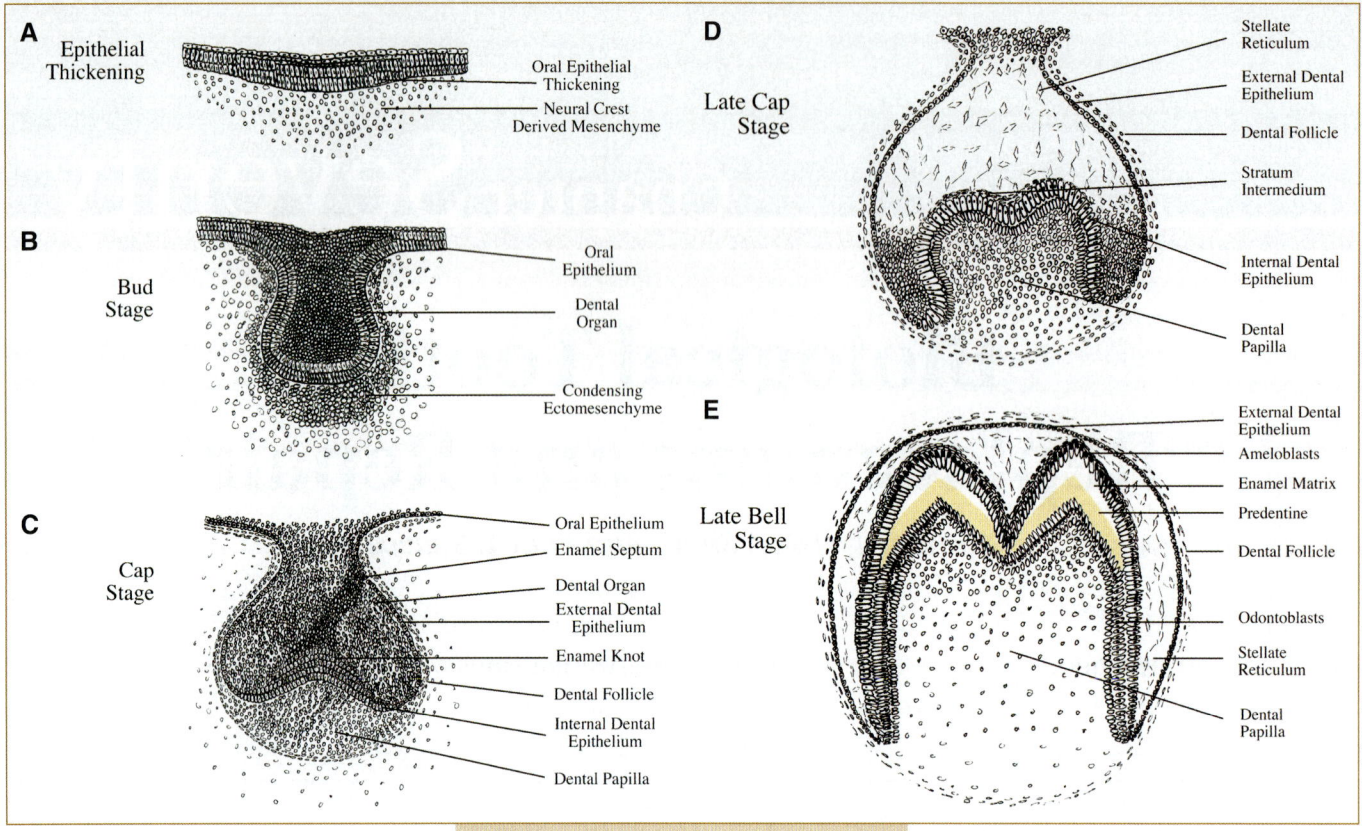

**FIG. 70.1.** Drawings of histological sections of mammalian first-molar-tooth development.

odontogenesis (reviewed in Thesleff and Sharpe 1997; Tucker and Sharpe, 2004). Recombination experiments between dental cells and nondental cells have identified temporal changes in the direction of inductive signals. Following the initial epithelial-to-mesenchymal inductive signals, a long series of temporally and spatially controlled exchange of signals governs each step in the increasingly complex development of a tooth. With the exception of the nerve supply, all the cells of a mature tooth originate from the oral epithelium and ectomesenchyme. The epithelial cells form only one functional cell type, the ameloblasts that are responsible for enamel formation. All of the other cell types, including the odontoblasts that are responsible for dentine formation, periodontal ligament cells, the pulp cells, etc., are derived from ectomesenchyme. The ability of ectomesenchyme to differentiate into these different cell types illustrates the stem cell–like properties of these cells. Classic recombination experiments in the 1980s established that non–neural crest cell populations of mesenchymal cells cannot respond to tooth-inductive signals, and, more recently, the requirement for dental mesenchymal cells to have stem cell properties has been established (see later). An understanding of the properties of predental epithelium, mesenchymal cells, and the interactions and molecules involved is the key to developing biological approaches for replacement and repair.

## III. WHOLE TOOTH-TISSUE ENGINEERING

### Stem Cell–Based Tissue Engineering of Teeth

The aim of this approach is to reproduce an embryonic tooth primordium from cultured epithelium and mesenchymal cells. Since it is established that embryonic tooth primordia are able to continue their embryonic development in ectopic adult sites such as the kidney, the expectation is that the artificial primordia will develop into teeth following implantation as tooth rudiments into the adult mouth. Thus, rather than a metal dental implant, a cell-based biological implant will be transplanted. Since the ectomesenchymal cells of the tooth primordium form the majority of tooth cells and regulate tooth shape, it is finding replacements for these cells that has been the main initial focus. The early inductive interactions that take place *in vivo* show that the mesenchymal cells must have stem cell–like properties. Cultured populations of stem cells have thus been used to replace these ectomesenchymal cells in recombinant tooth explants. Cultures of mouse ES and neural and adult bone marrow cells were aggregated to form a semi-solid mass, on which embryonic oral epithelium was placed (Ohazama *et al.*, 2004). When cultured for two to three days, the initiation of an odontogeneic response in the stem cell "mesenchyme" cells could be visualized with molecular markers. Transfer of the explants in the kidney capsules of

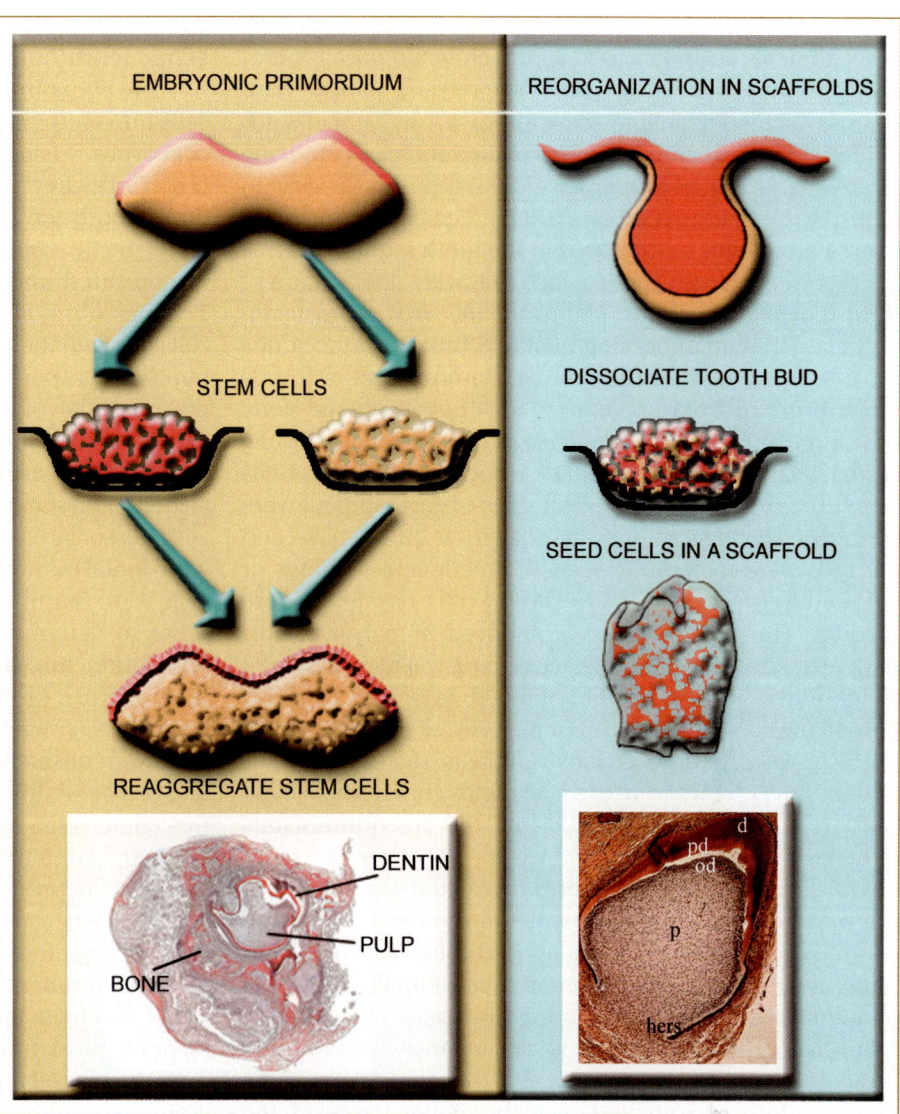

**FIG. 70.2.** Diagrammatic representation of two methods currently being explored for producing biological tooth replacement. Figure kindly drawn by Rachel Sartaj.

adult mice was then used to assay for tooth formation. "Bioteeth" explants made with mesenchyme derived from ES and neural stem cells proved difficult to transfer intact into kidneys. Explants made from adult bone marrow stromal cells, however, were substantially more robust and survived transfer. Following incubation of explants in kidneys for 10–14 days, clearly identifiable tooth crowns were formed, surrounded by bone. By using cells from genetically distinct transgenic mouse lines (GFP, LacZ), the stem cell origins of dental mesenchyme cells and bone cells could be confirmed. Cells derived from adult bone marrow stromal were thus capable of contributing to tooth formation is a way identical to ectomesencyme cells. Moreover, development of the tooth crowns appeared to follow the normal pathway of embryogenesis, with the formation of new bone completely surrounding the tooth. New bone formation is an essential (but often overlooked) component of biotooth formation. A biotooth has to be able to anchor itself to the jaw

bone with roots and a periodontal ligament. Concomitant bone formation must therefore occur during tooth development. Although the oral epithelium gives rise to only one major functional cell type in teeth (the ameloblasts) and these cells are lost from mature teeth, the oral epithelium is required for tooth primordia induction and further development. Sources of epithelium that can induce and participate in biotooth formation thus need to be identified. Currently, several different potential sources are under investigation (Odontis Ltd. unpublished information).

### Bioteeth from Cell-Seeded Scaffolds

In the early 1950s, Shirley Glasstone-Hughes demonstrated that when early-stage embryonic tooth primordia are physically divided into two halves, each half develops into a normal-size tooth (Glasstone-Hughes, 1952). This pioneering experiment demonstrated the developmental plasticity and regenerative capacity of embryonic tooth germs. This regenerative capacity has been exploited to devise a simple, biodegradable scaffold-based approach to biotooth generation. The early pioneering work of Vacanti and coworkers on the use of biodegradable scaffolds to act as supports for guided tissue regeneration has provided the basis for using the reorganizational properties of dental primordia cells to reform teeth *in vitro* (Cao *et al.*, 1997). The basic principle is to create scaffolds in the shape of the tooth required. These are then seeded with cells isolated following dissociation of third-molar-tooth germs. Third molars, or wisdom teeth, erupt late in human development and thus in young adults are present as dormant primordia. Both pig and rat third-molar primordia have been used with essentially the same procedure, utilizing biodegradable polymer scaffolds (Young *et al.*, 2002; Duailibi *et al.*, 2004; Honda *et al.*, 2005). Cells from third molars at the late bud stage of development were dissociated by incubation with collagenase and dispase, and the cells were then either seeded directly into scaffolds or cultured for up to six days in DMEM plus 10% FBS before seeding. The scaffolds were composed of polyglycolate/poly-L-lactate and poly-L-lactate-*co*-glycolide prepared using polyvinyl/sioxane and molded into the shape of human incisors and molars. For pig cells, scaffolds of $1 \times 0.5 \times 0.5$ cm were used; whereas for rat cells, rectangular scaffolds of $1 \times 5 \times 5$ mm were used. Scaffolds containing seeded cells were surgically implanted into the omentum of rats (athymic rats for pig cells) and left for 12–30 weeks. Histological sectioning of the explants revealed the formation of tiny tooth-like structures. The structures were between 1 and 2 mm in size and showed many of the features of molar-tooth crowns, including differentiation of ameloblasts and odontoblasts. Experiments using pig cells detected the tooth-like structures after 20 weeks, whereas with rat cells these formed after 12 weeks. In all cases, the shapes of the toothlets formed were independent of the shape of the scaffold, and, unlike natural tooth formation, no bone was formed in association with the teeth. Based on the previous results of Glasstone-Hughes, the most likely explanation for these results is one of reorganization rather than *de novo* formation. The small size of the teeth indicates that small numbers of dental epithelium and mesenchyme cells reassociated to form tooth germs. The fact that cells could be cultured for up to six days indicates that either the epithelial and/or mesenchymal cells are able to retain their odontogenic properties or the possibility exists that stem cells were present in the cell populations, but this remains to be demonstrated. A functional biotooth must be able to form roots; in order to do this, new bone must form at the same time as the tooth. In the scaffold approach, no new bone is formed. In order to address this, scaffold toothlike structures have been generated together with bone implants produced from osteoblasts induced from bone marrow progenitor cells seeded onto polyglycolide-*co*-lactide–fused wafer scaffolds (Young *et al.*, 2005). The codevelopment of bone and tooth-like structures permitted the early formation of roots

and thus demonstrated the possibility of utilizing a form of hybrid tooth/bone-tissue-engineering approach.

The phenomenon of reorganization of dental primordia cells into teeth has been investigated in detail by Lesot and colleagues. Using cap-stage-tooth germs (E14 — mouse) (Fig. 70.1), they showed that following complete dissociation of both the epithelium and mesenchymal cell derivatives, teeth could be produced when the cells were reaggregated and reassociated (Hu *et al.*, 2005a, 2005b). Thus, as observed with third-molar-tooth primordial, the cells retained their odontogenic capacity following dissociation. This property was utilized as a method of investigating the potential of non-dental cells to participate in tooth development. The experiments of Ohazama *et al.* (2004) showed that bone marrow stromal cells were able to replace all dental mesenchyme cells. Bone marrow cells were mixed with dissociated dental epithelial cells and reassociated with dental mesenchyme (Hu *et al.*, 2006). The bone marrow cells were found to contribute to ameloblast formation. In order to determine which cells within the crude marrow population might give rise to epithelial ameloblasts, cell sorting with c-kit was used to separate haematopoietic progenitor cells. When mixed with dental epithelial cells, these cells were observed to form both ameloblasts and odontoblasts. The ability of bone marrow mesenchymal progenitors (stem cells) to contribute to ameloblast formation was not assessed in these experiments. However, the principle, as originally proposed by Ohazama *et al.* (2004), that bone marrow cells can be a potential source of cells for tissue-engineering bioteeth is further supported by these reaggregation experiments. In all the experiments reported by Lesot and colleagues, the ability of nondental cells to contribute to tooth formation occurs only when the cells are mixed with dental cells from cap stage tooth germs. The experiments reported by Ohazama *et al.* (2004) did not require a mixed population. It remains to be seen therefore whether any commercial biotooth procedure could feature the use of dissociated embryonic third-molar dental cells. Since bone marrow can be used totally to replace dental mesenchyme in the absence of any dental mesenchyme cells, the prospect exists that if a source of epithelial cells can be identified that can replace dental epithelial cells in the absence of any epithelial cells, then bioteeth could be formed totally from adult cells. Dental epithelial cells produce the signals that initiate the whole process of tooth formation, and, thus, a nondental source of these cells must in part be capable of reproducing these signals.

One signal that has been identified is BMP4, which is specifically expressed in early predental epithelium before bud formation. When exogenous BMP4 is added to nondental embryonic explants, odontoblast and ameloblast differentiation can be detected (Ohazama *et al.*, 2005). Although these cells are organized together in cell layers as they appear *in vivo*, they are not present in a recognizable tooth structure. Nevertheless, the fact that one molecule applied to cells early can stimulate the differentiation of nondental cells into both epithelium and mesenchymal dental cells offers the possibility of engineering a dental epithelium from nondental epithelial cells.

## Root Formation

In order to be functional, bioteeth must develop roots. Root formation is a complex and little-understood process. However, although little is known of the molecules that stimulate and coordinate root formation, the absolute requirement for bone is established. If teeth form in the absence of bone, they cannot form roots, as observed in the seeded scaffold toothlike structures. When these scaffold teeth develop alongside new bone, root formation is stimulated (Young *et al.*, 2005). Tooth formation is most often studied experimentally by transplantation of tooth primordia to ectopic sites in adult rodents. The anterior chamber of the eye, the kidney capsule, the omentum, and the ear have all been used as sites that permit tooth formation. Root formation can occur at all of these sites if accompanied by bone development (Hu *et al.*, 2006). A prerequisite of any method of biotooth formation therefore is that tooth primordia be able to form teeth and roots integrated into jaw bone in the mouth. The mouth is, however, not routinely used as an experimental ectopic site for tooth development because of the difficulty of the surgery involved. In order to determine whether the adult mouth can support tooth formation, embryonic tooth primordia have been surgically transplanted into the mouths of adult mice. Not only do these develop into teeth in the mouth, but they also produce functional roots and erupt (Ohazama *et al.*, in press).

## IV. DENTAL-TISSUE REGENERATION

An important long-term goal for dental-tissue engineering is developing strategies for biotooth formation, thereby addressing many of the clinical problems arising from developmental anomalies and dental disease. However, in the shorter term there are many opportunities to be exploited for partial reconstruction of the tooth organ by tissue regeneration. Such approaches may be more readily achieved and still provide the potential for very significant impact on delivery of oral health care. Natural tissue regeneration, i.e., wound healing in the dental environment, is well recognized but represents a rather serendipitous event, and the approaches used to encourage such processes are somewhat empirical. A key factor in encouraging natural tissue regeneration is facilitating a conducive tissue environment in which the regenerative processes can take place, including moderation of chronic inflammatory events, control of bacterial infection, and minimizing tissue injury during any restorative surgical intervention. A good understanding of how to control these influences on tissue regeneration is fundamental to clinical success. There are also significant opportunities for the application of agents, whether directly through bioactive molecules or indirectly through agents that can release tissue-sequestered pools of these molecules locally, to promote regenerative processes and tip the balance between tissue degeneration and regeneration.

## Natural Tissue Regeneration

Natural tissue regeneration implies a cellular basis to regeneration, and the nonvital nature of mature dental enamel provides major hurdles to its regeneration by any but physicochemical remineralization processes. However, the vitality of the dentin–pulp complex provides significant opportunities for regeneration. Regeneration in the dentin–pulp complex will only take place, however, if there is a conducive tissue environment. Early studies highlighted the causal link between bacteria and inflammatory events in the dental pulp, and the presence of pulpal inflammation provides an effective barrier to initiation of regeneration therein (Rutherford and Gu, 2000). Thus, if natural tissue regeneration is to be facilitated, it is essential that bacterial infection and the consequent inflammation in the tooth are controlled. Traditionally, the former has often been achieved through extensive surgical removal of infected dental hard tissues, although the trend toward minimal intervention therapy during tooth restoration potentially places the tooth at risk through incomplete bacterial control. Sealing or "entombing" the bacteria within the restoration may help to compromise their viability, as may some of the chemical agents used in the placement of dental materials. Nevertheless, control of pulpal inflammation may be desirable in the context of tissue regeneration, and an improved understanding of the inflammatory mediators involved (McLachlan *et al.*, 2004, 2005) may allow specific targeting with novel antiinflammatory molecules.

The concept of tissue regeneration in the dentin–pulp complex has been recognized since the first report of tertiary dentinogenesis in response to injury from caries by Hunter in the 18th century, and dentistry has long been a pioneer in regenerative medicine through the use of calcium hydroxide to stimulate reparative dentinogenesis to bridge pulpal exposures in the dentin (Zander, 1939). Tertiary dentinogenesis (reactionary and reparative dentinogenesis are subvariants) represents the up-regulation of the dentin-secreting cells, the odontoblasts, in teeth after completion of tooth formation to initiate tissue regeneration in response to injury. With mild injury, the odontoblasts underlying the injury survive and respond by up-regulation of their secretory activity to form reactionary dentin, while with injury of greater intensity a number of these odontoblasts undergo necrosis and may be replaced by a new generation of odontoblast-like cells secreting a reparative dentin matrix (Smith *et al.*, 1995) (Fig. 70.3).

### Importance of the Injury–Regeneration Balance

Dental caries is one of the most widespread infectious diseases globally and continues to be a major health care

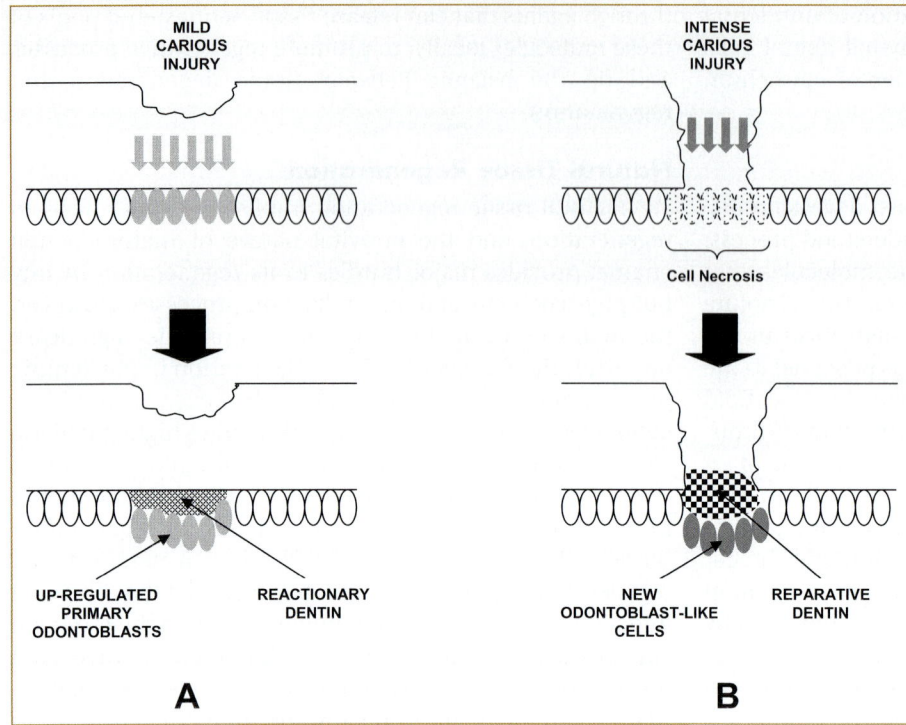

**FIG. 70.3.** Schematic diagrams of reactionary **(A)** and reparative **(B)** dentinogenesis. **(A)** Mild carious or other injury signals up-regulation of the primary odontoblasts underlying the injury site, leading to secretion of reactionary dentin. **(B)** More intense injury causes necrosis of the odontoblasts underlying the injury site, and progenitor cells are recruited from the pulp to differentiate into a new generation of odontoblast-like cells, which secrete reparative dentin. Dentin bridge formation at sites of pulpal exposure arises from reparative dentinogenesis.

problem. Initially, colonization of bacteria on the tooth surface within the dental plaque biofilm leads to demineralization and proteolytic degradation of the dental hard tissues following diffusion of bacterial acids and metabolites. As the disease progresses, both the hard and soft tissues of the tooth become infected, and a sustained bacterial challenge ensues, with consequent host inflammatory responses. The dynamics of the disease process determine the opportunities for tissue regeneration through tertiary dentinogenesis — with a rapidly progressing lesion, little tertiary dentinogenesis or regeneration is seen, but in a more slowly progressing lesion, induction of tertiary dentin secretion occurs immediately beneath the lesion (Bjørndahl, 2001). Classically, this has been assumed to reflect the intensity of microbial challenge, but more recent data indicate that the response is in part due to local release of dentin matrix components. During intense carious injury, appreciable levels of dentin matrix components are released, and these compromise survival of the odontoblasts (Smith *et al.*, 2005). It is possible that growth factors contained within the dentin matrix, particularly those from the TGF-β family, contribute to this loss of odontoblast survival, since levels of TGF-β similar to those released from the dentin matrix have a comparable effect on odontoblast survival *in vitro* (He *et al.*, 2005). At lower doses, however, these dentin matrix components and the cocktail of growth factors contained therein can stimulate regenerative events.

It has long been recognized that particles of dentin displaced into the pulp during surgery can act as a nidus for regeneration. Implantation of solubilized dentin matrix

components in the base of either unexposed (Smith *et al.*, 1995) or exposed cavities (Smith *et al.*, 1990) induces a regenerative response of tertiary dentinogenesis. Such responses mirror those seen beneath more slowly progressing carious lesions (Bjørndahl, 2001) and suggest that the natural regeneration seen during carious and other dental-tissue injury is induced by local release of bioactive tissue matrix components. If such processes are to be successfully mimicked for development of novel regenerative therapies, it is important that we understand the nature of the signaling molecules involved.

### Signaling Events in Dental Regeneration

It seems probable that the signaling events during tertiary dentinogenesis and regeneration in the dentin–pulp complex recapitulate many of those occurring during embryonic development. During tooth development, the cells of the inner enamel epithelium of the enamel organ induce the ectomesenchymal cells at the periphery of the dental papilla to differentiate into odontoblasts through the mediation of the dental basement membrane, which may function in the immobilization and presentation of the signaling molecules (Ruch *et al.*, 1995). Growth factors, especially those of the TGF-β family, may be key signaling molecules during induction of odontoblast differentiation (Begue-Kirn *et al.*, 1992; Thesleff and Vaahtokari, 1992). While an epithelial source of these molecules is not available in the mature tooth for signaling regeneration, sequestration of these molecules in the dentin matrix (Cassidy *et al.*, 1997; Roberts-Clark and Smith, 2000) following secretion by the odontoblasts may provide such a source. Their release during carious demineralization (Smith *et al.*, 2005) by lactic and other bacterial acid metabolites would allow their diffusion through the dentinal tubules to the odontoblasts and pulp cells, which express receptors for these growth factors. Application of recombinant members of the TGF-β family, either *in vitro* (Sloan *et al.*, 1999, 2000; Melin *et al.*, 2000) or *in vivo* (Rutherford *et al.*, 1994; Nakashima *et al.*, 1994; Tzaifas *et al.*, 1998), has been demonstrated to induce a regenerative response of tertiary dentinogenesis.

However, while this mirrors some of the signaling events of embryonic development, there is likely to be much reduced control over the signaling, since the rate of release of these molecular signals from the dentin matrix will be of variable intensity and the spectrum of cells with which they can interact will differ from embryogenesis.

## Control of Specificity of Dental-Tissue Regeneration

During wound healing in the dentin–pulp complex, a broad spectrum of tissue responses may be observed, ranging from regeneration of dentin tissue virtually indistinguishable from primary physiological dentin in terms of its tubularity and structure to secretion of atubular tissues with many of the structural features of bone. While terms such as *tertiary dentin*, *reparative dentin*, and *irritant dentin* have been used to encompass all of these responses, the dentinogenic specificity of some of the responses remains to be demonstrated. It is unclear, though, as to what the determinants are of the specificity of the response during natural tissue regeneration. Is it control of the molecular signaling processes, is it heterogeneity in the types of cells participating in the regeneration, or is it some other factor? Answering these questions will be fundamental to future strategies for exploiting tissue regeneration in the tooth in a controlled manner. The ability to determine whether tissue regeneration gives rise to tubular or atubular dentin matrix could be of great value in developing designer regenerative clinical approaches. For example, there may be significant benefit in directing secretion of a tubular dentin matrix in the tooth crown during regeneration to restore the normal physiological tissue architecture and function. However, it could be advantageous in possible exploitation of tissue regeneration for endodontic applications, such as root canal therapy, to generate an atubular dentin matrix, thereby providing an effective seal to the periapical region of the tooth.

## Dental Postnatal Stem Cells

The cells of the dental pulp have traditionally been considered to be neural crest–derived ectomesenchymal cells, although it is clear that during embryogenesis the migrating neural crest cells will intermingle with mesenchyme in the first branchial arch. Thus, cells of the pulp may not all share the same lineage, although those at the periphery, including the odontoblasts, appear to be of neural crest origin. In reparative situations, there are potentially a variety of possible progenitors in the pulp for differentiation of a new generation of odontoblast-like cells, including undifferentiated mesenchymal cells in the cell-rich layer of Höhl adjacent to the odontoblasts, perivascular cells, undifferentiated mesenchymal cells, and fibroblasts. This highlights the diversity in reparative response, which may occur after injury to the tooth. While the tissues that regenerate in such situations are all referred to as tertiary dentin (reactionary or reparative variants), in reality they represent a spectrum of tissue responses, dependent on the origin of the formative cells.

During normal tooth development, it has been suggested, cells destined to differentiate into odontoblasts have to achieve a level of competence before they can respond to an inductive signal for terminal differentiation (Ruch *et al.*, 1995). If this is the case, then it may be a very restricted population of cells able specifically to give rise to odontoblasts during regeneration in the mature tooth. Just prior to terminal differentiation during tooth development, the pre-odontoblasts align perpendicular to the dental basement membrane, and after the final cell division, one daughter receives the inductive signal to differentiate into an odontoblast, while the other daughter shares a similar developmental history, with the exception of this final inductive step. It has been presumed that this latter group of cells resides in the cell-rich layer of Höhl just beneath the odontoblast layer in the mature tooth and as such, these cells would be prime candidates as progenitors for odontoblast-like cells during regeneration. Certainly, there is a decline in the numbers of cells in this subodontoblastic site with age (Murray *et al.*, 2002), which mirrors anecdotal clinical reports of impaired tissue regeneration with age. Undifferentiated mesenchymal cells are also found within the central core of the pulp, which could contribute to regenerative processes. A specific population of postnatal dental pulp stem cells (DPSCs) has also been described (Gronthos *et al.*, 2002), which show many characteristics of postnatal stem cells, although it is possible that this population of cells may be heterogeneous in its phenotype. Transplantation of these cells subcutaneously into immunocompromised mice in association with hydroxyapatite/tricalcium phosphate powder generated ectopic deposits of reparative dentin with expression of dentin sialoprotein (Batouli *et al.*, 2003).

As interest in dental regeneration increases, it is important that detailed characterization be performed of the progenitor cells involved. Primary cultures of dental pulp cells give rise to many cells with myofibroblastic characteristics, and similar results have been observed during extended serial expansion of such pulp cultures (Smith *et al.*, 2005). Whether this represents asymmetric growth of these cells or the myofibroblast phenotype represents a default phenotype is unclear, but it is probable that such cells are not true postnatal stem cells, and any attempt to label them as such should be resisted in the absence of fuller characterization. Clearly, there is potential for a spectrum of derivations for the cells potentially involved in regeneration of the dentin–pulp complex, and the consequences of this may be a variety of phenotypic responses. It seems probable that only some of these responses may involve true postnatal stem cells. However, the diversity of cellular responses highlights the potential benefits of achieving control over these regenerative processes and the opportunities for developing strategies for directed tissue regeneration.

## Directed Tissue Regeneration

The concept of directing tissue regeneration will be successful only if it builds on the foundation of our knowledge of natural tissue regeneration, which in turn exploits our appreciation of the molecular cellular signaling events during normal tooth development. Focus on both cell- and signaling-based approaches has provided many interesting initial avenues for directed tissue regeneration in the tooth, but it is probable that the combination of these two approaches will prove the most effective in providing us with novel solutions for regenerating tooth structures.

### Signaling-Based Strategies

Signaling-based strategies have generally aimed to mimic those signaling events responsible for cell differentiation and secretion during embryonic tooth development and natural tissue regeneration. Thus, there has been a strong focus on application of growth factors, as well as various matrix-derived molecules, which also appear to stimulate regenerative events. Approaches have been used to up-regulate secretion by existing cells (reactionary dentinogenesis) as well as to induce the differentiation of new odontoblast-like cells (reparative dentinogenesis) for tissue regeneration, although the former has been constrained by our lack of understanding about the physiological control of odontoblast secretion during primary dentinogenesis.

Both *in vitro* (Sloan *et al.*, 1999, 2000; Melin *et al.*, 2000) and *in vivo* (Rutherford *et al.*, 1994; Nakashima *et al.*, 1994; Tzaifas *et al.*, 1998) application of growth factors, particularly of the TGF-β family, to the exposed pulp in pulp-capping situations induces regeneration, although the tissue formed has shown a range of appearances, from osteodentin to tubular dentin-like. There have also been several reports of the application of matrix molecules (Decup *et al.*, 2000; Six *et al.*, 2002; Liu *et al.*, 2004), promoting dentin regeneration, and a phase II clinical trial with a synthetic peptide derived from MEPE is presently in progress. While all of these approaches provide the foundation of a new era of biologically based regenerative therapies for the teeth, there are a number of limitations. The half-life of free growth factors is generally short; for instance, that of free TGF-β1 is estimated to be of the order of 2–3 min. In fact, when the growth factors are sequestered within the dentin matrix, they are remarkably well protected from degradation by association with extracellular matrix components. This protective mechanism can allow the growth factors to remain

"fossilized" for the life of the tooth, thereby providing an exquisite life-long potential for natural regeneration. Also, the consequences of short half-life for these molecules and the ubiquitous presence of proteolytic enzymes in the tissue milieu at sites of injury require high concentrations to be applied for regenerative signaling. As for all tissue-regenerative and -engineering strategies, the mode of delivery of signaling molecules is key to their effective action. To date, most signaling molecules have been simply applied as lyophilized powders or aqueous solutions, but considerable potential exists for development of novel delivery systems. An alginate hydrogel has shown encouraging results for delivery of TGF-βs for induction of *de novo* dentinogenesis on cut surfaces of pulp tissue (Dobie *et al.*, 2002), and innovative approaches are needed to deliver such molecules through the dentin matrix via the dentinal tubules. While passive diffusion through the tubules will have some effect, histomorphometry of restored human teeth suggests that natural regeneration occurs optimally when there is only a very limited residual dentin thickness, and a challenge will be to develop systems for delivery across greater distances in the dentin matrix. Gene therapy (see later) offers interesting opportunities for targeting signaling molecules to the regenerative site, although safe and suitable delivery systems are still required.

Calcium hydroxide has long been used to stimulate dentin bridge formation for dental regeneration, although its mechanism of action has remained largely elusive. Recent studies, however, have indicated that it probably acts by release of stores of endogenous bioactive molecules and growth factors sequestered within the dentin matrix, which are responsible for the cellular signaling (Graham *et al.*, 2006). The limited control on growth factor dissolution from the dentin matrix by calcium hydroxide may in part help to account for its rather variable activity during regeneration. However, its action does point to an interesting approach for regeneration where agents target release of endogenous signaling molecules from the tissue, thereby obviating some of the issues associated with the application of exogenous sources of these molecules.

### Cell- and Gene-Based Strategies

Cell- and gene-based strategies offer exciting opportunities for dentin–pulp regeneration, although many hurdles have to be overcome before these can become a clinical reality. Using electroporation, *Gdf11* gene transfer to an exposed pulp *in vivo* failed to provide effective regeneration (Nakashima *et al.*, 2002), although use of ultrasound for transfer of the *Gdf11* plasmid provided an osteodentinogenic regenerative response (Nakashima *et al.*, 2003). *In vivo* gene transfer of a recombinant adenovirus containing a full-length cDNA encoding mouse bone morphogenetic protein (BMP)-7 failed to induce reparative dentinogenesis in inflamed ferret dental pulps, while *ex vivo* transduced dermal fibroblasts induced regeneration in the same model, although the dentinogenic specificity of this response requires further characterization (Rutherford, 2001). These studies highlight the negative impact on regeneration of a nonconducive tissue environment, due to inflammation and, also, the limited numbers of stem/progenitor cells in the pulp for recruitment to participate in regeneration. The latter issue might be overcome by transplantation of suitable stem/progenitor cells for regeneration at sites of injury, especially if transduced with suitable signaling molecules. *Ex vivo* gene therapy by transplantation of *Gdf11* (Nakashima *et al.*, 2004) or *Bmp2* — electrotransfected pulp stem/progenitor cells (Iohara *et al.*, 2004) — shows promise for induction of regeneration, although the tissue formed was osteodentin-like in appearance. It is unclear whether the lack of dentinogenic specificity in these responses can be ascribed to the nature of the cells transplanted or to the inductive signaling molecules. However, the results highlight the importance of considering not only the cells and signaling molecules involved, but also the delivery mechanisms for effective therapies to be developed.

The anticipated low numbers of stem/progenitor cells for regeneration in the dental pulp indicate the need to optimize recruitment strategies for these cells, whether locally recruited from within the tissue or transplanted from without. The former approach avoids possible issues of immune reaction to nonautologous cellular material. However, local recruitment requires selection of those cells of appropriate lineage and phenotype. Cell markers for potential selection stem cells from embryonic first branchial arch tissue have been investigated (Deng *et al.*, 2004), and low-affinity NGF receptor is being targeted for selection of odontoblast-like cell progenitors from the mature dental pulp (Smith *et al.*, 2005). A combination of c-kit, CD34, and STR-1 has been reported for selection of stromal stem cells

from human deciduous dental pulps, which can be induced to differentiate into osteoblasts for bone regeneration (Laino *et al.*, 2006), while c-kit has been used for selection of odontoblast progenitor cells from bone marrow (Hu *et al.*, 2006). Use of such cell selection approaches may overcome problems associated with recruitment of stem/progenitor cells for odontoblast differentiation from a relatively small niche within the pulp during dental regeneration. Future strategies might include *ex vivo* selection of these cells for transplantation or immobilization of antibodies or other capture molecules for selection at regenerative sites for recruitment of the cells. A combination of these approaches with novel delivery of signaling molecules may prove effective in optimizing the regenerative response.

## V. CONCLUSIONS

Tissue-regenerative strategies offer exciting possibilities for development of novel clinical solutions for treatment of dental developmental anomalies and disease. The exquisite natural tissue-regenerative capacity of the dentin–pulp complex in the tooth provides an invaluable foundation on which to develop novel biologically based regenerative therapies. Particular attention will need to be focussed on both the control and the specificity of the regenerative processes, and a combination of natural and directed tissue regeneration offers many exciting opportunities for the future.

Progress in the identification, isolation, and understanding of the differentiation of embryonic and adult stem cells, together with a continuing understanding of the control of tooth development, will undoubtedly aid the production and refinement of approaches for biotooth formation. Although there remain many potential problems and pitfalls, biological tooth replacement is now a realistic possibility.

## VI. REFERENCES

Batouli, S., Miura, M., Brahim, J., Tsutsui, T. W., Fisher, L. W., Gronthos, S., Gehron Robey, P., and Shi, S. (2003). Comparison of stem-cell-mediated osteogenesis and dentinogenesis. *J. Dent. Res.* **82**, 976–981.

Begue-Kirn, C., Smith, A. J., Ruch, J. V., Wozney, J. M., Purchio, A., Hartmann, D., and Lesot, H. (1992). Effects of dentin proteins, transforming growth factor β1 (TGFβ 1) and bone morphogenetic protein 2 (BMP2) on the differentiation of odontoblasts *in vitro*. *Int. J. Dev. Biol.* **36**, 491–503.

Bjørndahl, L. (2001). Presence or absence of tertiary dentinogenesis in relation to caries progression. *Adv. Dent. Res.* **15**, 80–83.

Cao, Y., Vacanti, J. P., and Paige, K. T., *et al.* (1997). Transplantation of chondrocytes utilizing a polymer-cell construct to produce tissue-engineered cartilage in the shape of a human ear. *Plast. Recontr. Surg.* **100**(2), 297–302.

Cassidy, N., Fahey, M., Prime, S. S., and Smith, A. J. (1997). Comparative analysis of transforming growth factor-beta (TGF-β) isoforms 1–3 in human and rabbit dentine matrices. *Arch. Oral. Biol.* **42**, 219–223.

Crubez, E., Murail, P., Girard, L., *et al.* (1998). False teeth of the Roman world. *Nature* **391**, 29.

Decup, F., Six, N., Palmier, B., Buch, D., Lasfargues, J.-J., Salih, E., and Goldberg, M. (2000). Bone sialoprotein–induced reparative dentinogenesis in the pulp of rat's molar. *Clin. Oral Invest.* **4**, 110–119.

Deng, M. J., Jin, Y., Shi, J. N., Lu, H. B., Liu, Y., He, D. W., Nie, X., and Smith, A. J. (2004). Multilineage differentiation of ecto-mesenchymal cells isolated from the first branchial arch. *Tissue Eng.* **10**, 1597–1606.

Dobie, K., Smith, G., Sloan, A. J., and Smith, A. J. (2002). Effects of alginate hydrogels and TGF-β1 on human dental pulp repair *in vitro*. *Connect. Tissue Res.* **43**, 387–390.

Duailibi, M. T., Duailibi, S. E., and Young, C. S., *et al.* (2004). Bioengineered teeth from cultured rat tooth bud cells. *J. Dent. Res.* **83**(7), 532–528.

Glasstone-Hughes, S. (1952). The development of halved tooth germs: a study in experimental morphology. *J. Anat.* **86**, 12–25.

Graham, L., Cooper, P. R., Cassidy, N., Nor, J. E., Sloan, A. J., and Smith, A. J. (2006). The effect of calcium hydroxide on solubilisation of bioactive dentine matrix components. *Biomaterials* **27**, 2865–2873.

Gronthos, S., Brahim, J., Li, W., Fisher, L. W., Cherman, N., Boyde, A., DenBesten, P., Robey, P. G., and Shi, S. (2002). Stem cell properties of human dental pulp stem cells. *J. Dent. Res.* **81**, 531–535.

Honda, M. J., Sumita, Y., Kagami, H., and Ueda, M. (2005). Histological and immunohistochemical studies of tissue-engineered odontogenesis. *Arch. Histol. Cytol.* **68**(2), 89–101.

He, W.-X., Niu, Z.-Y., Zhao, S., and Smith, A. J. (2005). Smad protein-mediated transforming growth factor beta1 induction of apoptosis in the MDPC-23 odontoblast-like cell line. *Arch. Oral Biol.* **50**, 929–936.

Hu, B., Nadiri, A., Bopp-Kuchler, S., Perrin-Schmitt, F., Wang S., and Lesot, H. (2005a). Dental epithelial histo-morphogenesis in the mouse: positional information versus cell history. *Arch. Oral Biol.* **50**, 131–136.

Hu, B., Nadiri, A., Bopp-Kuchler, S., Perrin-Schmitt. F., and Lesot, H. (2005b). Dental epithelial histomorphogenesis *in vitro*. *J. Dent. Res.* **84**(6), 521–525.

Hu, B., Unda, F., Bopp-Kuchler, S., Jumenez, L., Wang, X. J., Haikel, Y., Wang, S. L., and Lesot, H. (2006). Bone marrow cells can give rise to ameloblast-like cells. *J. Dent. Res.* **85**(5), 416–421.

Iohara, K., Nakashima, M., Ito, M., Ishikawa, M., Nakashima, A., and Akamine, A. (2004). Dentin regeneration by dental pulp stem cell therapy with recombinant human bone morphogenetic protein 1. *J. Dent. Res.* **83**, 590–595.

Laino, G., Graziano, A., d'Aquino, R., Pirozzi, G., Lanza, V., Valiante, S., De Rosa, A., Naro, F., Vivarelli, E., and Papaccio, G. (2006). An approachable human adult stem cell source for hard-tissue engineering. *J. Cell. Physiol.* **206**, 693–701.

Liu, H., Li, W., Gao, C., Kumagai, Y., Blacher, R. W., and DenBesten, P. K. (2004). Dentonin, a fragment of MEPE, enhanced dental pulp stem cell proliferation. *J. Dent. Res.* **83**, 496–499.

McLachlan, J. L., Sloan, A. J., Smith, A. J., Landini, G., and Cooper, P. R. (2004). S100 and cytokine expression in caries. *Infect. Immunol.* **72**, 4102–4108.

McLachlan, J. L., Smith, A. J., Bujalska, I. J., and Cooper, P. R. (2005). Gene expression profiling of pulpal tissue reveals the molecular complexity of dental caries. *Biochim. Biophys. Acta.* **1741**, 271–281.

Melin, M., Joffre-Romeas, A., Farges, J.-C., Couble, M. L., Magloire, H., and Bleicher, D. (2000). Effects of TGFβ1 on dental pulp cells in cultured human tooth slices. *J. Dent. Res.* **79**, 1689–1696.

Murray, P. E., Stanley, H. R., Matthews, J. B., Sloan, A. J., and Smith, A. J. (2002). Aging human odontometric analysis. *Oral Surg. Oral Med. Oral Pathol. Oral Radiol.* **93**, 474–482.

Nakashima, M., Nagasawa, H., Yamada, Y., and Reddi, A. H. (1994). Regulatory role of transforming growth factor-beta, bone morphogenetic protein-2 and protein-4 on gene expression of extracellular matrix proteins and differentiation of pulp cells. *Dev. Biol.* **162**, 8–28.

Nakashima, M., Mizunuma, K., Murakami, T., and Akamine, A. (2002). Induction of dental pulp stem cell differentiation into odontoblasts by electroporation-mediated gene delivery of growth/differentiation factor 11 (Gdf11). *Gene Ther.* **9**, 814–818.

Nakashima, M., Tachibana, K., Iohara, K., Ito, M., Ishikawa, M., and Akamine, A. (2003). Induction of reparative dentin formation by ultra-sound-mediated gene delivery of growth/differentiation factor 11. *Hum. Gene Ther.* **14**, 591–597.

Nakashima, M., Iohara, K., Ishikawa, M., Ito, M., Tomokiyo, A., Tanaka, T., *et al.* (2004). Stimulation of reparative dentin formation by *ex vivo* gene therapy using dental pulp stem cells electrotransfected with growth/differentiation factor11(Gdf11). *Hum. Gene Ther.* **15**, 1045–1053.

Ohazama, A., Modino, S. A. C., Miletich, I., *et al.* (2004). Stem-cell-based tissue engineering of murine teeth. *J. Dent. Res.* **83**(7), 518–522.

Ohazama, A., Tucker, A. S., and Sharpe, P. T. (2005). Organized tooth-specific cellular differentiation stimulated by BMP4. *J. Dent. Res.* **84**, 603–606.

Roberts-Clark, D., and Smith, A. J. (2000). Angiogenic growth factors in human dentine matrix. *Arch. Oral Biol.* **45**, 1013–1016.

Ruch, J. V., and Lesot, H., and Begue-Kirn, C. (1995). Odontoblast differentiation. *Int. J. Dev. Biol.* **39**, 51–68.

Rutherford, R. B. (2001). BMP-7 gene transfer to inflamed ferret dental pulps. *Eur. J. Oral Sci.* **109**, 422–424.

Rutherford, R. B., and Gu, K. (2000). Treatment of inflamed ferret dental pulps with recombinant bone morphogenetic protein-7. *Eur. J. Oral Sci.* **108**, 202–206.

Rutherford, R. B., Spanberg, L., Tucker, M., Rueger, D., and Charette, M. (1994). The time course of the induction of reparative dentine formation in monkeys by recombinant human osteogenic protein-1. *Arch. Oral Biol.* **39**, 833–838.

Six, N., Tompkins, K., Lasfargues, J.-J., Veis, A., and Goldberg, M. (2002). Bioengineering of reparative dentin and pulp mineralization. *In* "Dentin/Pulp Complex. Proceedings of the International Conference on Dentin/Pulp Complex 2001" (T. Ishikawa, K. Takahashi, T. Maeda, H. Suda, M. Shimono, and T. Inoue, eds.), pp. 52–59. Quintessence, Tokyo.

Sloan, A. J., and Smith, A. J. (1996). Stimulation of the dentine-pulp complex of rat incisor teeth by transforming growth factor-β isoforms 1–3 *in vitro*. *Arch. Oral Biol.* **44**, 149–156.

Sloan, A. J., Rutherford, R. B., and Smith, A. J. (2000). Stimulation of the rat dentine-pulp complex by BMP7 *in vitro*. *Arch. Oral Biol.* **45**, 173–177.

Smith, A. J., Tobias, R. S., Plant, C. G., Browne, R. M., Lesot, H., and Ruch, J. V. (1990). *In vivo* morphogenetic activity of dentine matrix proteins. *J. Biol. Buccale* **18**, 123–129.

Smith, A. J., Cassidy, N., Perry, H., Begue-Kirn, C., Ruch, J. V., and Lesot, H. (1995). Reactionary dentinogenesis. *Int. J. Dev. Biol.* **39**, 273–280.

Smith, A. J., Patel, M., Graham, L., Sloan, A. J., and Cooper, P. R. (2005). Dentine regeneration: the role of stem cells and molecular signaling. *Oral Biosci. Med.* **2**, 127–132.

Thesleff, I., and Sharpe, P. (1997). Signaling networks regulating dental development. *Mech. Dev.* **67**, 111–123.

Thesleff, I., and Vaahtokari, A. (1992). The role of growth factors in determination and differentiation of the odontoblastic cell lineage. *Proc. Finn. Dent. Soc.* **88**, 357–368.

Tzaifas, D., Alvanou, A., Papadimitriou, S., Gasic, J., and Komnenou, A. (1998). Effects of recombinant fibroblast growth factor, insulin-like growth factor-II and transforming growth factor-β1 on dog dental pulp cells *in vivo*. *Arch. Oral Biol.* **43**, 431–444.

Tucker, A. S., and Sharpe, P. T. (2004). The cutting edge of mammalian development: how the embryo makes teeth. *Nat. Rev. Genet.* **5**, 499–508.

Tucker, A. S., and Matthews, K. L., Sharpe, P. T. (1998). Transformation of tooth type induced by inhibition of BMP signaling. *Science* **282**, 1136–1138.

Westbroek, P., and Martin, F. A. (1998). A marriage of bone and nacre. *Nature* **392**, 861–862.

Young, C. S., Terada, S., Vacanti, J. P., *et al.* (2002). Tissue engineering of complex tooth structures on biodegradable polymer scaffolds. *J. Dent. Res.* **10**, 695–700.

Young, C. S., Abukawa, H., Asrican, R., Ravens, M., Troulis, M. J., Kaban, L. B., Vacanti, J. P., and Yelick, P. (2005). Tissue-engineered hybrid tooth and bone. *Tissue Eng.* **11**(9/10), 1599–1610.

# Chapter Seventy-One

# Oral and Maxillofacial Surgery

*Simon Young, Kyriacos A. Athanasiou, Antonios G. Mikos, and Mark Eu-Kien Wong*

## I. INTRODUCTION

The oral and maxillofacial region is defined as the mouth and surrounding structures, bounded superiorly by the cranial base and inferiorly by the lower jaw. It is a highly complex area that includes morphologically intricate skeletal elements (Fig. 71.1), organs responsible for the special senses, lining and covering tissue (i.e., skin and subcutaneous fat), a rich neural and vascular network, and teeth. The prominent position adopted by this area makes it particularly vulnerable to injury, while the complicated interplay of events during embryogenesis of the craniofacial skeleton increases the possibility of developmental aberrations. Regular exposure to various disease-inducing agents, including complex carbohydrates (e.g., sugar), ultraviolet rays, thermal insults, and carcinogens (e.g., tobacco products), produces a variety of common pathological conditions, such as dental caries, burns, and malignant neoplasms. The loss of tissue integrity and continuity resulting from trauma, developmental deformities, and pathology imposes on physicians and dentists considerable reconstructive challenges. Meeting these requirements in the 21st century will include strategies based on tissue-engineering principles. While the discipline itself is several decades old, clinical applications for the reconstruction of the oral and maxillofacial region are currently absent. The purpose of this chapter, therefore, is to describe the special challenges posed by oral and maxillofacial reconstruction, outline existing reconstructive techniques, and review the available literature on tissue-engineering protocols that may be relevant. Providing tissue engineers with a description of current reconstructive modalities and, in particular, their shortcomings creates a reasonable starting point on which to base the development of revolutionary, new methods. We have specifically excluded descriptions of tissue engineering of teeth and special sensory organs, because these subjects are covered elsewhere in this book.

## II. SPECIAL CHALLENGES IN ORAL AND MAXILLOFACIAL RECONSTRUCTION

Virtually all tissue types of ectodermal, mesodermal, and endodermal origin are candidates for tissue-engineering strategies and are present in the oral and maxillofacial region. However, certain structures are more commonly impacted by disease, trauma, and developmental failures and constitute the focus of our discussion, though the reconstructive methods described can be applied to rarer conditions. Common *pathological* entities include benign and malignant cystic and neoplastic processes affecting the upper (maxilla) and lower (mandible) jaws as well as degenerative conditions involving the mandibular articulation (temporomandibular joints). These diseases, or the subsequent removal of pathologically involved tissue, can produce continuity defects of the jaws requiring the replacement of

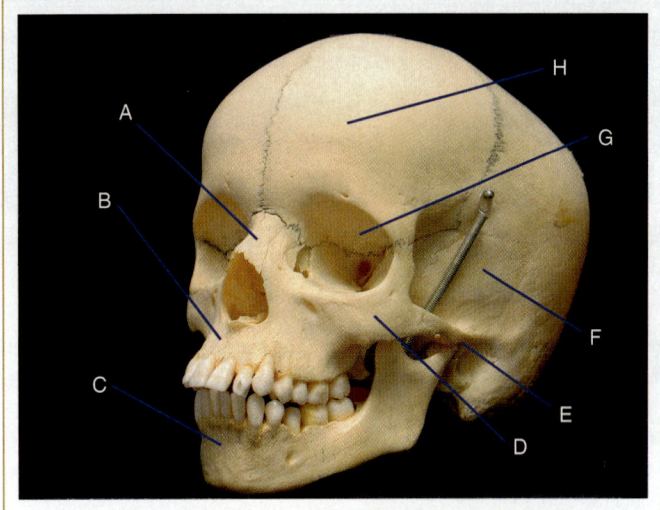

**FIG. 71.1.** The human skull is a complex region composed of many bones. Several key structures include the (A) nasal bone, (B) maxilla, (C) mandible, (D) zygomatic bone, (E) temporomandibular joint, (F) temporal bone, (G) orbital cavity, and (H) frontal bone.

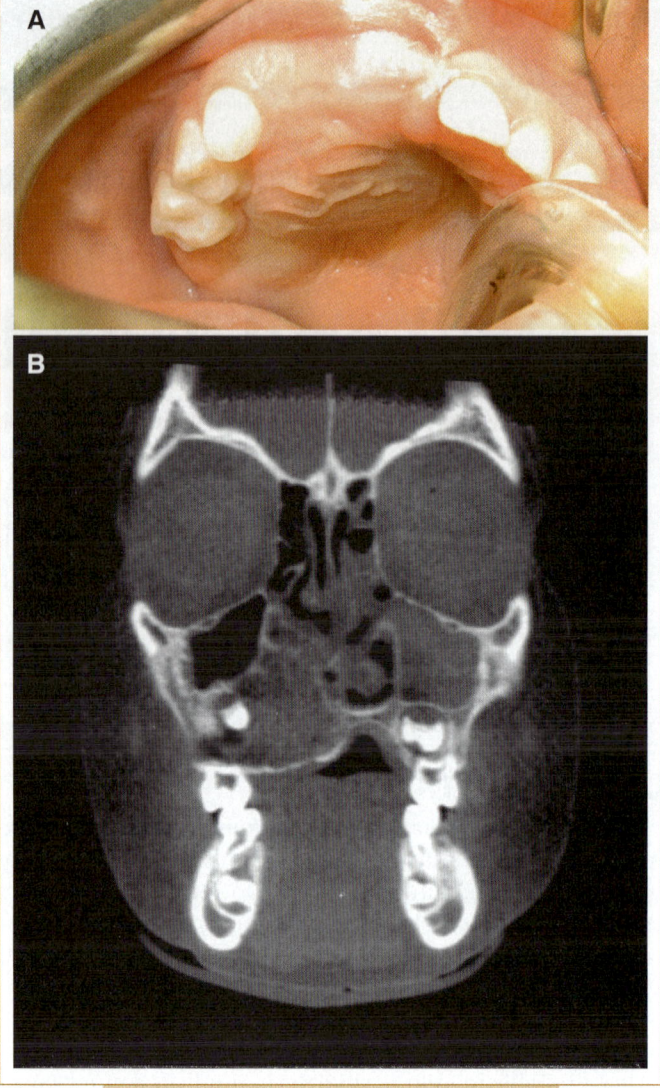

**FIG. 71.2.** **(A)** Intraoral photo of a patient with a giant cell tumor of the maxilla, which presents as a swelling of the palate and erosion of the supporting bone resulting in the loss of adjacent teeth. **(B)** Coronal CT scan of the same patient with a giant cell lesion of the right maxilla that has produced marked bone destruction and displacement of developing teeth. Sinusitis of the left maxillary sinus is an incidental finding.

bone, cartilage, and lining epithelium (Fig. 71.2). Since many of these conditions are frequently silent and the dimensions of the structures involved relatively small, their initial presentation is usually associated with significant tissue involvement. In addition to disease, nonphysiological loading of bone can produce loss of skeletal tissue, affecting the jaws and joints. *Edentulous bone loss* involves the resorption of the alveolar processes of the jaws (i.e., that portion of the jaw bone surrounding the tooth roots) following tooth removal (Fig. 71.3). This phenomenon is believed to be the result of direct loading of bone during mastication and the loss of physiological maintenance forces transmitted by the teeth. Over time, the loss of bone produces severely atrophied alveolar ridges, posing significant challenges to prosthetic reconstruction of the dentition and a predisposition to pathological fracture of the mandible (Seper *et al.*, 2004). In postmenopausal females, an increased incidence of alveolar resorption is observed following reductions in levels of bone-protective hormones (Bras *et al.*, 1983).

Degenerative diseases of the temporomandibular joints (TMJs) are commonly the result of nonphysiological mechanical forces produced by excessive ranges of motion of the joints or chronic microtrauma from various parafunctional habits (de Bont and Stegenga, 1993) (Fig. 71.4). The habits in question include jaw clenching and tooth grinding (nocturnal bruxism) and repetitive motion habits such as nail biting and gum chewing. Injury to the TMJs commonly affects the articulating surfaces of the condylar head and the glenoid fossa as well as the interpositional disc, producing a spectrum of disease from chondromalacia to severe osteoarthritis. As synovial joints, the TMJs can also fall victim to various immune-mediated disease processes, such as rheu-

matoid or psoriatic arthritis (Fig. 71.5). The inflammatory component is responsible for progressive structural tissue loss leading to changes in skeletal relationships and malocclusion (Helenius *et al.*, 2005). In advanced forms of disease, replacement of both cartilage and bone as a total joint reconstruction may be necessary to restore function or skeletal support to the mandible.

Maxillofacial trauma constitutes another group of conditions providing opportunities for tissue-engineering reconstruction. Whereas most forms of blunt trauma result in fractures where tissue loss is minimal, penetrating injuries produced by high-velocity missiles and projectiles often

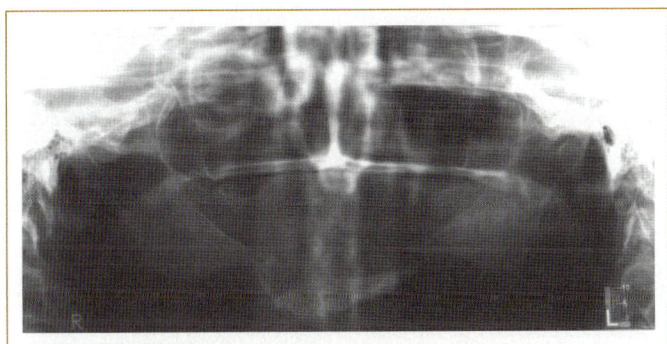

**FIG. 71.3.** Panoramic radiograph of a patient with a fractured atrophic mandible. Courtesy of Kamal Busaidy, D.D.S.

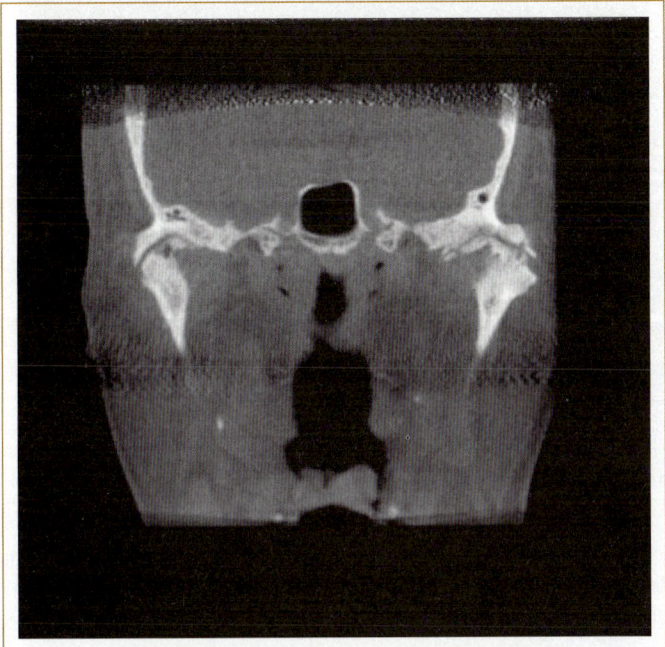

**FIG. 71.4.** Coronal CT scan demonstrating severe osteoarthritic degeneration of the temporomandibular joints.

create significant loss of bone and overlying soft tissue (Fig. 71.6). Finally, consideration should be given to the various forms of congenital facial clefts that commonly affect the oral and maxillofacial region. In a limited form, failure of the maxillary processes to fuse unilaterally or bilaterally produces alveolar clefts (Fig. 71.7). When the upper lip, maxilla, and palate are involved, a constellation of deformities associated with patients with unilateral or bilateral cleft lip and palate is present.

In the reconstruction of anatomical defects, the causative events must be taken into account to ensure long-term success. Defects produced by traumatic, developmental, and pathological conditions are associated with a defined endpoint. Assuming that pathology has been completely eradicated or that further traumatic insults do not occur,

defects produced by these mechanisms can be fully characterized with respect to size and missing tissue types. In contrast, tissue loss as a result of parafunctional habits, nonphysiological loading patterns, and immunologically mediated degeneration often continue following reconstruction. This set of circumstances will adversely affect any biological constructs produced by tissue-engineering techniques and impose an important limitation on the clinical application of their use. Before biological, rather than alloplastic, materials can be employed, correction of the underlying etiology is of paramount importance.

A special concern in oral and maxillofacial reconstruction is the potential exposure of grafted tissue to the external environment. Constructs used to restore defects involving the jaws, orbits, nose, and ears are potentially in direct contact with the mouth, sinuses (maxillary, ethmoidal, and frontal), nasal passages, and external environment. These areas are characterized by high moisture content, significant bacterial populations, and functional loads imposed by physiological activities such as chewing. If biological (i.e., tissue-engineered) constructs are to survive under these conditions, modifications to account for the dilutional effects of moisture, presence of infective organisms, and mechanical loads must be provided by the engineered tissue. For example, when *in vivo* polymerization of materials is intended, the presence of fluid must be considered. Alternatively, preformed constructs can be used. Colonization of constructs with a mixed population of aerobic and anaerobic bacteria is expected with reconstructions involving oral, nasal, and sinus-related structures. Porous constructs, capable of harboring potentially pathological organisms, might be modified to reduce either bacterial attachment or replication until lining tissue develops over the implant, forming a barrier to the external environment. In addition to contaminated wound sites, tissue constructs may be exposed to complicated mechanical loads before anisotropy is restored with the regeneration of biological tissue. Both the mandible and temporomandibular joints are subject to a combination of compressive, shear, and tensile loads, depending on the type and degree of function (Christensen and Mckay, 2000; van Eijden, 2000).

Another special feature of the maxillofacial region is the number of tissue types within a relatively small region. As a result of this proximity, traumatic, pathological, and developmental events often lead to the creation of composite defects requiring reconstruction of multiple tissue types. This results in a special challenge, not only to engineer composite tissues, but also to attach the various constructs to each other in their normal anatomical relationship.

Facial symmetry is an important consideration in oral and maxillofacial reconstruction. Since most structures are paired or contiguous (e.g., the orbits, zygomas, left and right maxillae, and mandible), accurate reproduction of the external form is an important aspect to preserve facial aesthetics. The paucity of overlying soft tissue as camou-

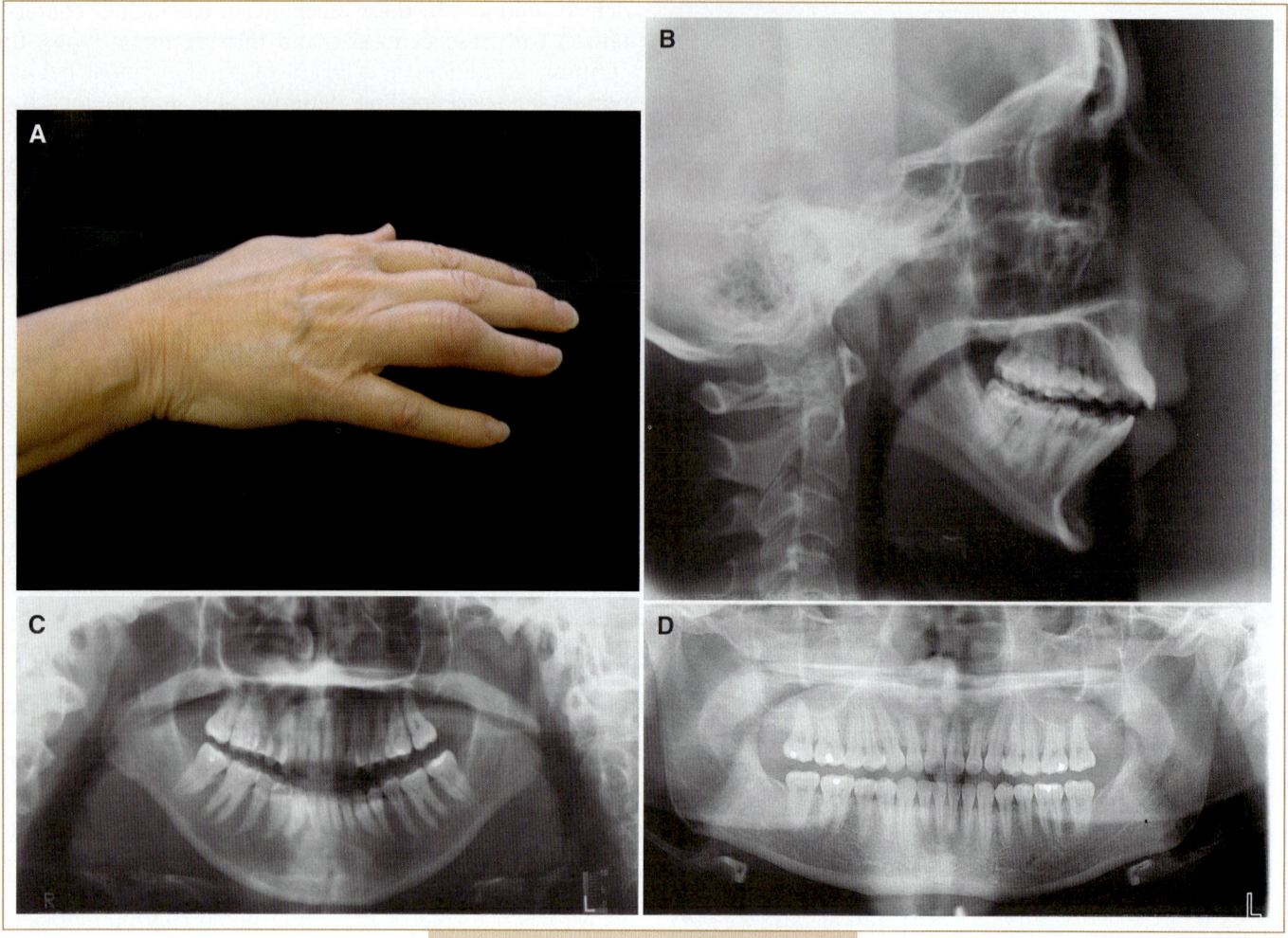

**FIG. 71.5.** **(A)** Rheumatoid deformation of the proximal interphalangeal joint of the fourth finger. **(B)** Lateral cephalometric radiograph of a patient with rheumatoid arthritis. Rheumatoid degeneration of the TMJs reduces posterior vertical support of the mandible, producing an anterior open bite. **(C)** Panoramic radiograph of the same patient with rheumatoid arthritis. Compare the appearance of the bilaterally resorbed condyles with normal condyles, seen in (D). **(D)** Panoramic radiograph of a healthy patient.

flage contributes to the exacting nature of oral and maxillofacial reconstruction, and these requirements impose on tissue-engineering methods the ability to compose and maintain accurate morphology. The advent of new, three-dimensional imaging techniques with the capacity to produce stereolithographic skeletal models that mirror both the normal anatomy and defect is a valuable adjunctive tool (Fig. 71.8). These models assist in the fabrication of scaffolds to support the reconstruction of missing tissue.

## III. CURRENT METHODS OF ORAL AND MAXILLOFACIAL RECONSTRUCTION

Several methods are used for oral and maxillofacial reconstruction, and the selection of a particular modality takes into account a number of important issues. Major factors that guide this process include the presence (or absence) of associated soft tissue, the vascularity and vas-

cular pattern present, a multidimensional characterization of the defect size, the types of missing tissue, availability of tissue for transfer, and both patient and surgeon preference. Most reconstructive techniques can be categorized into four categories:

1. Soft-tissue pedicled flaps
2. Nonvascularized soft- and hard-tissue grafts (the graft establishes, in a delayed fashion, a vascular network following implantation, relying on tissue diffusion to preserve the viability of the transplant)
3. Soft- and hard-tissue vascularized grafts (the graft is immediately perfused through an existing arterial–venous system)
4. Alloplastic reconstructions with prosthetic appliances

On occasion, composite techniques can be used, such as the staged reconstruction of a defect where soft tissue is first added to a defect site, followed by bone at a later time.

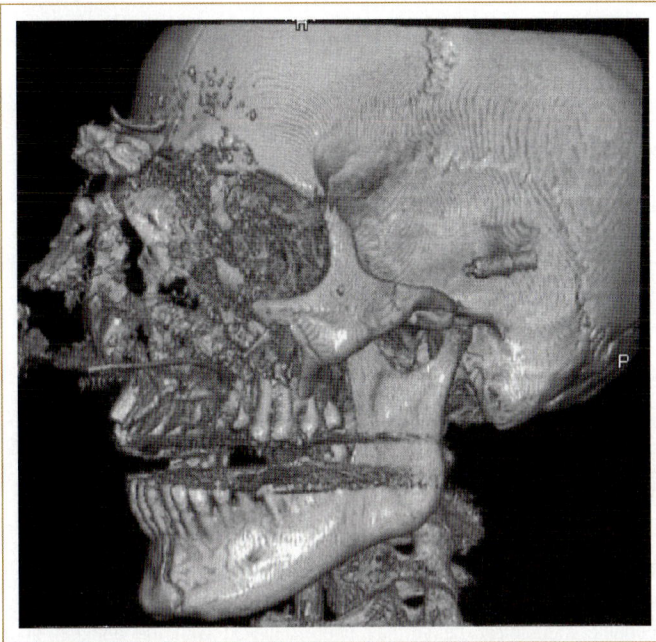

**FIG. 71.6.** Three-dimensional reconstructed radiograph of a patient with a self-inflicted gunshot wound, demonstrating the significant disruption and loss of maxillofacial skeletal structures.

## Mandibular Defects

Reconstruction of the lower jaw is indicated following removal of tissue during surgical excision of a pathological lesion or following loss of tissue from a traumatic injury. When malignant disease is present, not only is more radical removal of tissue required, but postoperative radiation therapy produces lasting compromise to both the cellularity and the vascularity of the remaining tissue. Blast effects from missile injuries can also produce significant composite injuries, with loss of bone soft tissue and diminished vascularity of the tissue bed. Two techniques are commonly employed for the reconstruction of mandibular defects. Vascularized grafts are indicated when the vascularity of the tissue bed is compromised by radiation or excessive scarring. They are also valuable when there is a requirement to replace both hard and soft tissue at the same time (Hidalgo, 1994). Hard- and soft-tissue defects can also be reconstructed using nonvascularized grafts, but their success relies on an adequately vascularized tissue bed to support the survival of transplanted cells before a new supply is established.

As composite structures, vascularized grafts either contain soft tissue alone (muscle, subcutaneous tissue with or without epithelium) or include hard- and soft-tissue components (bone and soft tissue). Since the vascular supply to bone is contained within a periosseous cuff of muscle and fibrous tissue, it is not possible to transplant only bone. The additional tissue transferred into the site of a bony defect often produces a bulky graft. While this can

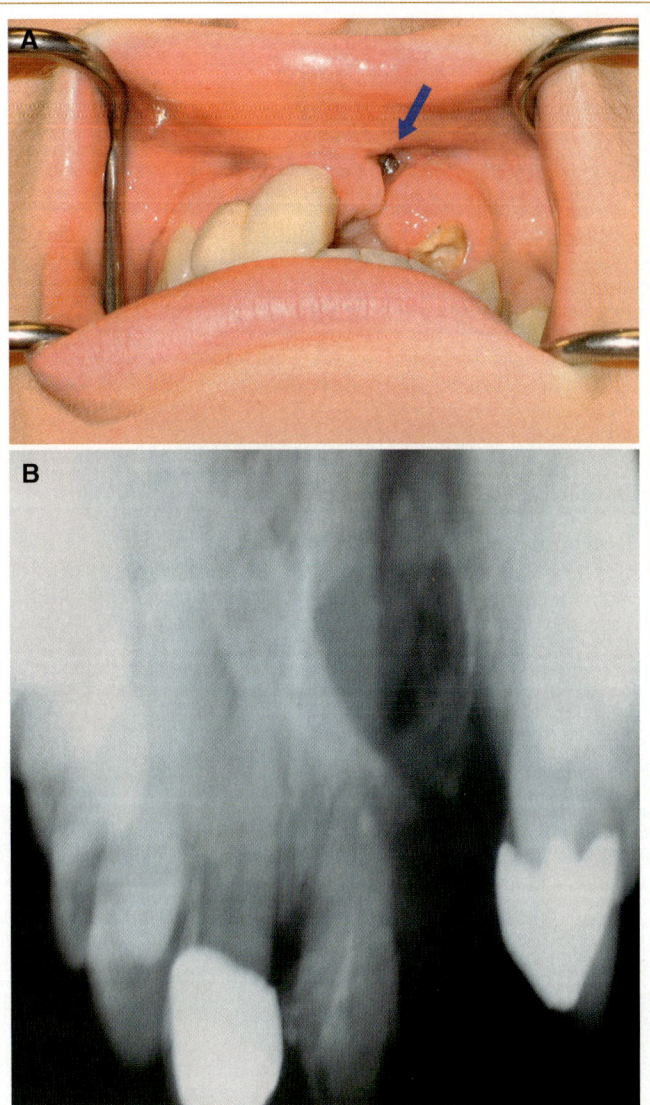

**FIG. 71.7.** **(A)** Intraoral view of a left maxillary alveolar cleft. An oronasal fistula is present at the superiormost aspect of the cleft (see arrow). **(B)** Radiograph of a cleft of the maxillary alveolar bone.

be easily excised once a new vascular network is established, a second procedure performed several months after the initial transplantation is required. Another potential limitation to the use of vascularized bone grafts for the reconstruction of mandibular defects is the amount of bone available, since the dimensions of the graft are determined by the morphology of the donor site and not the size of the defect (Fig. 71.9). Special techniques, such as osteotomizing the graft and folding it on itself, have been described, but this can compromise the blood supply to the graft. Vascularized grafts are harvested from a limited number of anatomical sites characterized by a dominant arterial supply venous drainage system. In addition, the en bloc harvesting of the graft must not compromise either the function of the donor

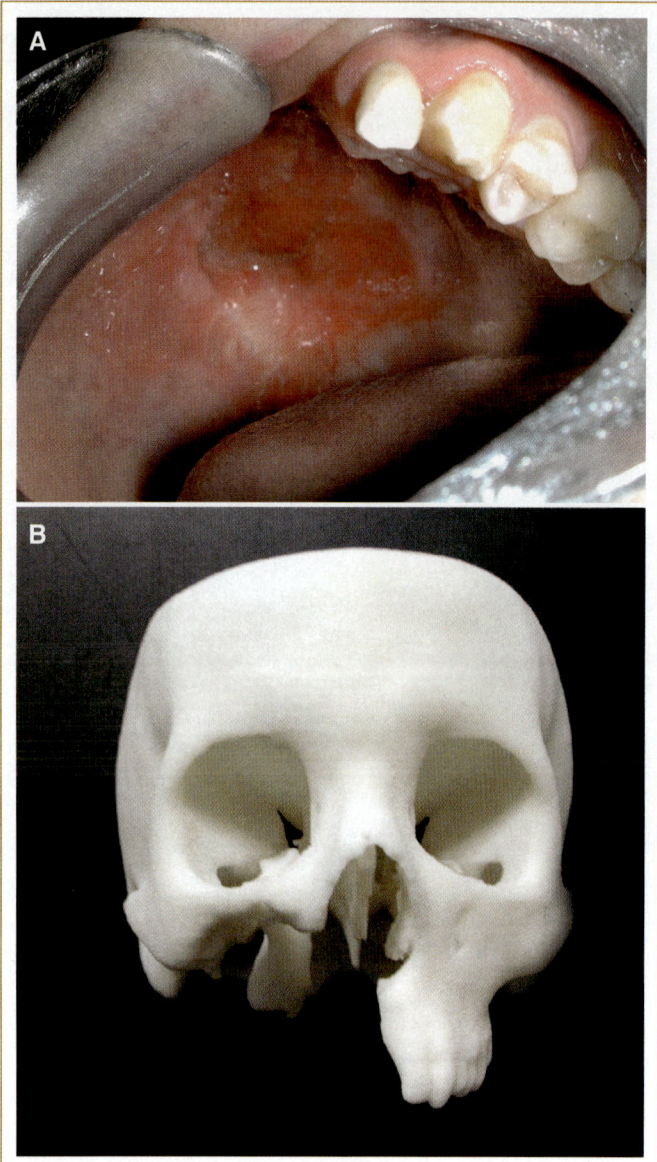

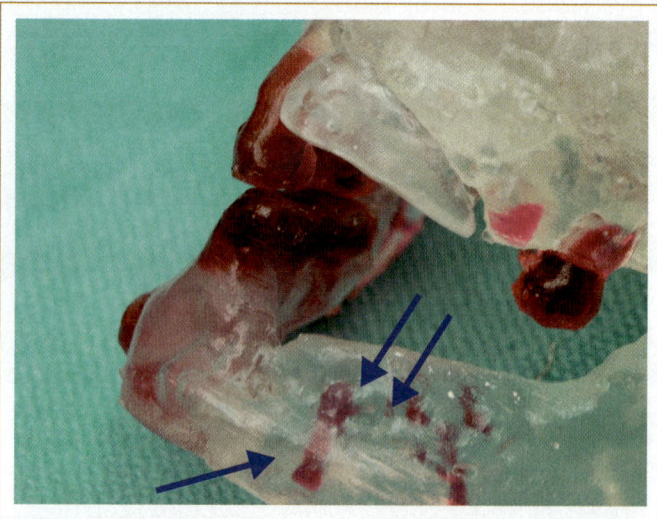

**FIG. 71.9.** Stereolithographic model illustrating a segmental defect reconstructed with a vascularized fibula graft (single arrow). Restoration of adequate mandibular height has been achieved with the addition of a nonvascularized ilium block graft (double arrows).

Mandibular defects can also be reconstructed using nonvascularized transplantations of autologous bone from various sites. Successful bone grafts rely on adequate cellularity and a sufficiently cellular and vascular recipient bed. When the soft-tissue bed is deficient or lacks a decent blood supply, addition of well-vascularized soft tissue is achieved by the rotation of a muscle flap (with or without skin) into the mandibular defect. The pectoralis major, latissimus dorsi, and deltopectoral flaps have all been described for this purpose. The bony reconstruction is delayed for a period of three to six months until the soft-tissue flap has healed. In patients whose soft tissue is adequate but avascular as a result of radiation therapy, hyperbaric oxygen therapy can improve the quality of the vascular supply in a course of treatments lasting between four and six weeks, where repeated exposures to pressurized room air promote tissue angiogenesis. This process adds both time and considerable expense to the reconstructive process, but it has been shown to be effective in improving the quality of the recipient bed. Once the soft tissue in a mandibular defect has been optimized with respect to quantity, cellularity, and vascularity, autologous bone is transferred from a donor site and molded to fit the dimensions of the defect. The bone graft can be retained with screws fixed to a rigid bone plate or held in position with the aid of cribs fashioned from either processed allogeneic bone or alloplastic materials. Depending on the size of the defect, bone can be harvested from the anterior ilium (suitable for defects up to 4 cm in length), the posterior ilium (defects up to 8 cm in length), the tibia or mandibular symphysis and rami (defects of less than 2 cm in length). Nonvascularized grafts, especially those combined with allogeneic bone, are susceptible to infection,

**FIG. 71.8.** **(A)** Patient following a right maxillectomy for removal of a benign odontogenic neoplasm. Defect has filled in with fibrous tissue stimulated by grafting the site with an allogeneic dermal matrix. **(B)** Stereolithographic model of the same patient, demonstrating the extent of the maxillary hard-tissue defect.

site or the vascular and neural supply of structures distal to the harvest. Commonly used donor sites that meet these requirements include the fibula, ilium, scapula, and distal radius. Vascularized grafts transplanted to mandibular defects are anastomosed to patent vessels adjacent to the mandible, such as the facial, lingual, or superior thyroid arteries and veins. This reconstructive approach is highly technique sensitive; and, while experienced microvascular surgeons achieve successful outcomes in over 90% of cases, less experienced surgeons or patients with underlying vascular disease (e.g., diabetes) enjoy less success.

particularly following exposure to the intraoral environment. When a nonvascularized graft is colonized by organisms, infection often ensues, and the graft fails to survive or integrate with the host bone. Aside from the potential for infection, nonvascularized grafts are less technique sensitive, allow complete reconstruction of a defect by customizing the volume of bone harvested, and are associated with less donor site morbidity (Marx, 1994).

## Maxillary Defects

Defects of the upper jaw pose difficult reconstructive challenges, from several perspectives, but must be undertaken to preserve speech, prevent the escape of food and fluids during eating, and maintain aesthetics. Unlike the mandible, which is related to the oral cavity alone, the maxilla is bounded both inferiorly by the mouth and superiorly by the nasal cavity and maxillary sinuses. Even when present, the thin lining epithelium does not provide a sufficiently cellular or vascular bed to support the transplantation of sufficient quantities of nonvascularized bone, and the potential for exposure of the graft to oral and nasal environments is high. Postoperative or posttraumatic scarring reduces the tissue envelope even more, further complicating reconstructive efforts. Staged reconstructions have been described involving the initial transfer of vascularized soft tissue with a pedicled flap, such as the temporalis muscle or temporoparietal flap, followed by the addition of bone several months later (Marx, 1994). As an alternative, vascularized flaps have been used because of their ability to transfer both hard and soft tissue at the same time (Bernhart *et al.*, 2003). However, the accompanying soft tissue and vascular pedicle may not be accommodated by the smaller dimensions of a maxillary defect, and this has limited their use to hemi- or total maxillary reconstructions. The simultaneous transfer of a large bulk of overlying soft tissue also results in a postoperative recovery period of several months, where the flap can prevent mouth closure and compromise eating.

As a result of these challenges, prosthetic appliances have become the most commonly used method to reconstruct maxillary defects. These devices incorporate teeth and a fitted base to separate the mouth from the superior defect (Sharma and Beumer, 2005). Excellent restoration of aesthetics as well as function can be achieved, but the fact that these devices are not permanently fixed in place and, in fact, require daily removal and cleaning reduces their acceptability by patients. In addition, adjustments are required periodically to account for remodeling of the underlying tissue bed.

The reconstruction of maxillary alveolar clefts is one exception to the use of prosthetic appliances as a primary reconstruction technique, even though they can be used very effectively to restore missing teeth in the cleft site or obturate an oral–nasal communication. When teeth are present in a cleft site, provision of bone is essential for erup-

tion and support (Eppley and Sadove, 2000). Alveolar grafting is therefore timed according to the presence and stage of development of adjacent teeth and is usually performed between the ages of 8 and 11 years. The procedure involves the development of soft-tissue flaps to isolate the mouth from the nasal cavity and placing autogenous bone between the cleft segments to restore maxillary continuity. Loss of the graft from infection, insufficient vascularity, or lack of functional stimulus is not an infrequent occurrence, and opportunities for tissue-engineering alternatives exist. This would be especially true if new interventions minimized the extent of surgery, since postsurgical scarring has been associated with restricted growth and development of the maxilla.

# IV. RELEVANT STRATEGIES IN ORAL- AND MAXILLOFACIAL-TISSUE ENGINEERING

Driven by the limited supply and inherent shortcomings of various autogenous, allogeneic, and prosthetic materials currently used for the reconstruction of oral and maxillofacial tissues, the potential for tissue-engineered biomaterials as alternatives is under serious investigation, with the hope that significantly improved therapies will result. While a diverse number of strategies are presently under development, the fundamental tenets of tissue engineering (TE) remain the same. These include consideration of the biological and mechanical properties of the scaffold material and its interactions with relevant bioactive molecules and cell populations.

Although multiple tissue types exist in the oral and maxillofacial region, TE research in this field has focused primarily on the regeneration of single tissues: the bony craniofacial skeleton, lining epithelium, the cartilages of the temporomandibular joint, auricle, and nose, and the teeth and surrounding periodontal tissue (please refer to Chapters 70 and 72 for teeth and periodontal regenerative initiatives, respectively).

## Bone Applications

An ideal biodegradable TE bone construct should combine the biocompatibility and osteoinductive potential of autologous bone with the availability and structural characteristics of allogeneic bone. Additional scaffold design considerations include porosity, pore interconnectivity, surface chemistry, and the ability to reproduce complex three-dimensional defects.

Scaffolds are responsible for a construct's initial mechanical integrity and provide surface area for cell attachment. Several biocompatible scaffold materials have been tested using *in vivo* models for craniofacial bone. Naturally derived materials (e.g., collagen, chitosan, alginate, gelatin, hyaluronic acid) (Arosarena and Collins, 2005; Hong *et al.*, 2000; Mukherjee *et al.*, 2003; Shang *et al.*, 2001; Zheng *et al.*,

2004), synthetic polymers [e.g., poly(lactic acid), poly(glycolic acid), poly(propylene fumarate), poly(ε-caprolactone)] (Chim and Schantz, 2006; Dean et al., 2005; Meyer et al., 2005; Schantz et al., 2003), metals (e.g., titanium mesh) (Blum et al., 2003; Sikavitsas et al., 2003), and calcium phosphate ceramics (Ripamonti et al., 2001) have all been tried. These materials are typically processed as porous structures or hydrogels that guide the morphology of regenerated tissue, allow for tissue ingrowth, and control the release of bioactive molecules, such as growth factors and nucleic acids. Bioactive molecules, incorporated into the scaffold from the wound itself, contribute significantly to the osteoinductive potential of the TE bone construct.

Other approaches to tissue engineering use novel biomaterials capable of implantation through minimally invasive surgery using transcutaneous photopolymerization to achieve the final scaffold form (Poshusta and Anseth, 2001). In this study, viability of osteoblasts in a photopolymerized poly(ethylene oxide) dimethacrylate scaffold was demonstrated. However, corresponding experiments employing TMJ disc cells encapsulated in alginate revealed a significant decrease in cell numbers, with no ECM production at any time point, suggesting that TMJ disc cells may not survive an encapsulated environment (Almarza and Athanasiou, 2004).

Growth factors act as mediators of cellular growth and differentiation during tissue regeneration and play an important role in extracellular matrix synthesis. Utilized as recombinant proteins in TE strategies, growth factors require a local population of target cells capable of effecting the desired response (Lieberman et al., 2002). This constituency of cells may be naturally present at the wound site or added to the scaffold at the time of fabrication (Seto et al., 2006) prior to implantation.

Factors that have been used for the regeneration of in vivo TE craniofacial bone include the bone morphogenetic proteins (BMPs) (Lutolf et al., 2003; Seto et al., 2006), transforming growth factor-beta (TGF-β) (Dean et al., 2005; Hong et al., 2000; Vehof et al., 2002), fibroblast growth factors (FGFs) (Eppley et al., 1991; Kawaguchi et al., 1994), insulin-like growth factors (IGFs) (Srouji et al., 2005), and platelet-derived growth factor (PDGF) (Vikjaer et al., 1997). The bulk of experience concerning the use of growth factors for bone repair has involved BMPs (Nussenbaum et al., 2004), and this popularity has been extended into clinical investigations using recombinant human BMP-2 (rhBMP-2) for alveolar ridge augmentation (Cochran et al., 2000; Fiorellini et al., 2005; Jung et al., 2003), maxillary sinus floor augmentation (Boyne et al., 2005), mandibular reconstruction following tumor resection (Moghadam et al., 2001), and distraction-assisted alveolar cleft repair (Carstens et al., 2005; Chin et al., 2005).

Jung et al. (2003) examined the effect of combining recombinant human BMP-2 with a xenogeneic bone substi-

tute in order to improve membrane-guided bone regeneration therapy of osseous defects in areas of dental implant placement. Although there was no statistically significant difference in percentage of newly formed bone at the rhBMP-2-treated site versus the control site at six months, a larger fraction of mature lamellar bone (76% vs. 56%) as well as increased graft to bone contact (57% vs. 29.5%) were present in the experimental sites.

In addition, Boyne et al. (2005) completed a phase II randomized controlled study investigating the safety and efficacy of rhBMP-2 combined with an absorbable collagen sponge (ACS) versus bone graft for staged maxillary sinus floor augmentation. It was concluded that rhBMP-2 had a similar safety profile to bone graft, with the added benefit of lacking donor site morbidity. In addition, the rhBMP-2/ACS treatment induced similar amounts of bone to the bone graft group, allowing for the placement and long-term functional loading of dental implants in approximately 75–80% of the patients treated.

The clinical use of rhBMP-2 to regenerate much larger bone defects has also been reported in the literature. Carstens et al. (2005) described the use of "distraction-assisted in situ osteogenesis" (DISO) to treat a severe facial cleft, in which rhBMP-2/ACS implantation was combined with distraction osteogenesis to create the patient's ramus and condyle as part of the surgical reconstruction.

A similarly spectacular application of rhBMP-7 has been described by Warnke et al. (2004) for the reconstruction of a 7-cm mandibular continuity defect in a patient who had received ablative tumor surgery and subsequent radiation treatment. A bone-muscle-flap prefabrication technique was utilized, in which computed tomography and computer-aided design techniques were used to fabricate a custom titanium mesh cage replicating the contours of the missing mandible. Within this cage, a combination of xenogeneic bone mineral blocks coated with rhBMP-7 and autologous bone marrow was placed, prior to implantation of the entire construct within the latissimus dorsi muscle of the patient. Following seven weeks of implantation within this "in vivo bioreactor," the viable mandibular replacement was harvested from the patient, along with part of the muscle containing a major artery and vein, which were subsequently anastomosed with vessels at the recipient site using microsurgical techniques. Four weeks after this transplantation surgery, the patient was able to undertake a small amount of mastication and enjoy more solid foods.

A significant drawback to growth factor strategies in tissue engineering is the shortage of naturally derived factors isolated from biological tissue. This deficiency has been addressed with the development of techniques to produce biologically active proteins using recombinant-engineering techniques. However, the use of recombinant proteins is not without concern (Nussenbaum et al., 2004). Compared to animal models, bone regeneration in humans does not appear to be as robust. In order to overcome this species recalcitrance, administration of factors in excess of naturally occurring concentrations appears to be necessary. The augmented administration of exogenous factors may potentially stimulate harmful biological effects, such as malignant transformation of cells, and also prove to be too expensive when compared to alternative techniques for tissue regeneration.

Attempts to address the shortcomings of recombinant protein-based strategies have spurred investigation into the use of gene delivery for tissue engineering. By delivering the gene for the expression of a protein with specific effects on a target cell population, successfully transfected cells will elaborate the protein constitutively. This results in higher and more constant levels of protein production (Hannallah et al., 2003). However, while both viral (Alden et al., 2000; Dunn et al., 2005; Lindsey, 2001) and nonviral (Huang et al., 2005) gene delivery vectors have been utilized for bone regeneration in cranial-defect animal models, compromises must be made with each. Adenoviral constructs have commonly been used as viral vectors to transfect craniofacial tissues and have the advantage of efficiently transfecting both replicating and quiescent cells (Warren et al., 2003). In addition, adenoviruses are easily manipulated and can be produced in high titers, and large amounts of genetic information can be inserted into them. However, concerns related to viral vectors include in vivo homologous recombination and the possibility of an immune response from the expression of viral antigens on the surfaces of transfected cells. These concerns have led to the development of nonviral vector agents (Hannallah et al., 2003).

While numerous nonviral gene delivery systems exist, a common problem is their low in vivo transfer efficiency (Warren et al., 2003). Nonetheless, such systems are able to deliver much larger genes with minimal immunogenicity. One promising modality of nonviral gene delivery for craniofacial applications is the use of cationic liposomes, which have been used to regenerate cranial-bone defects in rabbits by delivering BMP-2 plasmid cDNA (Ono et al., 2004). The low transfection efficiency of uncondensed, naked plasmid DNA has also been addressed by the use of the cationic macromer poly(ethylene imine), which has been used to condense BMP-4 plasmid DNA and deliver it in a sustained and localized manner from poly(lactic-co-glycolic acid) scaffolds within critical-size cranial defects (Huang et al., 2005).

Gene transfection can take place directly within the defect site by releasing the delivery vector in vivo from the TE scaffold (Alden et al., 2000; Dunn et al., 2005). Indirect delivery methods have also been described utilizing a target cell population harvested from the patient, performing in vitro transfection of the cells, and then reimplanting the transfected cells into the defect along with the TE scaffold material (Blum et al., 2003; Edwards et al., 2005; Nussenbaum et al., 2003). While the direct technique may

be simpler, it has a lower transfection efficiency and targets cells in a nonspecific manner (Nussenbaum et al., 2004). The indirect ex vivo approach, on the other hand, requires additional harvesting and culturing procedures but avoids the risks associated with placing viral vectors directly into the patient and disturbing the host genome. Ex vivo transfected cells are not immunologically privileged and may still express viral antigens on their surface, which can lead to a host response following implantation.

Alternative TE approaches to craniofacial reconstruction employ cell-seeded scaffolds as implants. These have potential benefits for regenerating tissues with compromised healing capacity, such as those affected by radiation therapy (Nussenbaum et al., 2003). Adult-derived mesenchymal stem cells (MSCs) retain the ability to form several of the tissues present in the craniofacial region, including bone, cartilage, muscle, tendon, and adipose tissue, making them ideal candidates for the seeding of bone TE constructs (Caplan and Bruder, 2001). However, the relative scarcity of MSCs derived from adult bone marrow (1 per 100,000 cells in adults, 1 per 2 million cells in the elderly) requires significant expansion of cell numbers in vitro before their use in cell-seeded TE scaffolds (Fong et al., 2003). Autologous, culture-expanded MSCs have been utilized in combination with alginate hydrogels for the treatment of large cranial-bone defects in sheep (Shang et al., 2001). Production of osteoblasts has been promoted by culturing MSCs in flow-perfusion bioreactors. This technique uses the inherent osteoinductive potential to elaborate bonelike extracellular matrix and fluid shear stresses synergistically to enhance the differentiation of MSCs toward an osteoblastic phenotype (Datta et al., 2006). These in vitro fabricated cellular/ECM constructs have tremendous potential in bone TE applications.

Aside from the biological components of tissue-engineering constructs, scaffold properties are also extremely important to the overall success of any particular strategy. A common misconception is that bone TE scaffolds for craniofacial applications do not require substantial strength, since the craniofacial skeleton is not subjected to heavy loading. However, in vivo studies demonstrate that many craniofacial bones undergo levels of strain similar to that experienced by the appendicular skeleton (Herring and Ochareon, 2005), substantiating the need for mechanical strength of potential bone TE scaffolds. Ideal scaffold design must therefore reconcile the need for high porosity and interconnectivity, which promotes tissue ingrowth and scaffold degradability, with a requirement for mechanical strength. While computational methods for designing and fabricating scaffold architectures to optimize both pore interconnectivity and load-bearing characteristics have been performed (Hollister et al., 2005), the final constructs must still be tested in animal models to determine their suitability in clinical applications.

The surface characteristics of bone-tissue-engineering scaffolds also determine their ability to regenerate tissue in the wound-healing environment. Surface chemistry has a significant effect on the interactions between the cell populations present in the defect and the biomaterial. Hydrophilic synthetic polymers such as oligo[poly(ethylene glycol) fumarate] (OPF) have been shown to impede bone healing in extraction sockets as compared to the hydrophobic polymer poly(propylene fumarate), based on the OPF macromer's prevention of protein adsorption and hence cell adhesion (Fisher et al., 2004). Interestingly, the resistance of OPF hydrogels to generalized cell adhesion has been used to advantage in the fabrication of biomimetic scaffolds, which are able to selectively encourage the migration of osteoblasts in vitro through the addition of specific binding peptides to their surfaces, such as osteopontin-derived peptide (Shin et al., 2004).

## Cartilage Applications

Compared to bone, far less work has been conducted into tissue engineering cartilage within the craniofacial region. This can be attributed partly to the disparate roles cartilage plays in different sites. For example, a tissue-engineered TMJ disc must be able to move in association with the joint while improving congruity between the articulative surfaces, while the cartilages of the auricle and nose provide skeletal support to the overlying soft tissue.

While in vitro expansion and seeding of harvested chondrocytes into degradable polymer scaffolds is capable of producing cartilage following implantation in immunocompetent-animal models (Chang et al., 2003; Kamil et al., 2004; Shieh et al., 2004), the clinical experience has been somewhat disappointing with resorption of tissue-engineered auricular cartilage after several months (Rotter et al., 2005). To date, engineered implants have yet to achieve the aesthetic appearance and stability of silicone-based maxillofacial prosthetic devices (Nussenbaum et al., 2004). In contrast to efforts with nasal and auricular reconstruction, tissue-engineered cartilage constructs have enjoyed greater success in orthopedic applications when used to regenerate osteochondral defects within the articulating surfaces. This difference is believed to be the result of the relatively immunoprivileged properties of the synovial joint environment.

Replacement of fibrocartilage to restore the articulative surfaces or interpositional disc of the TMJs constitutes a special challenge, including identifying a suitable source of healthy fibrochondrocytes. To date, the most commonly used cells have been derived either from the disc itself (Almarza and Athanasiou, 2004, 2005, 2006a, 2006b; Bean et al., 2005; Detamore and Athanasiou 2004, 2005b; Springer et al., 2001; Thomas et al., 1991) or from articular cartilage (Girdler, 1998; Puelacher et al., 1994; Springer et al., 2001). In order to create a sufficient number of cells to populate a construct, multiple passages are required. Unfortunately,

chondrocytes have been found to dedifferentiate into a more fibroblastic phenotype after only a couple of passages (Darling and Athanasiou, 2005), and cells from later passages also exhibit decreased extracellular matrix (ECM) protein expression (Allen et al., 2006a). Consequently, before TMJ disc engineering becomes a viable technique, a cell source that can yield a large population of TMJ disc cells, or a population of cells that rapidly fills a scaffold, must first be identified.

The earliest tissue-engineering study directed at reconstruction of the TMJ disc utilized a porous collagen scaffold seeded with articular cartilage cells. After two weeks, the construct appeared similar to a disc with regard to gross morphology and cell shape (Thomas et al., 1991). Later efforts tested fibers of poly(glycolic acid) (PGA) and poly(lactic acid) (PLA) and concluded that both materials were able to support cell attachment and matrix production and exhibited acceptable mechanical properties after 12 weeks (Puelacher et al., 1994). Another study, comparing PGA, polyamide filaments, expanded polytetrafluoroethylene (ePTFE) filaments, and bone blocks (Springer et al., 2001), demonstrated cell attachment and limited collagen production, but neochondrogenesis was not observed after four and eight weeks. While PGA is an acceptable scaffold substrate, the material degrades exceedingly rapidly, leaving constructs with limited mechanical integrity after only a few weeks. As an alternative, PLA nonwoven mesh has been tried, and initial results show promise, with retention of tensile and compressive integrity over a similar time scale (Allen and Athanasiou, 2006b; Almarza and Athanasiou, 2005).

In vivo investigations of implanted tissue-engineered TMJ discs are limited (Puelacher et al., 1994), although extensive mechanical, histological, and biochemical testing has been conducted as a preliminary step to characterize the behavior of the natural disc and constructs fabricated in vitro (Almarza and Athanasiou, 2004, 2005; Detamore and Athanasiou, 2003, 2005a, 2005b). For instance, the disc has been shown under conditions of unconfined compression to exhibit regional variations with regard to stiffness (Allen and Athanasiou, 2006c). The medial and anterior regions of the disc are more stiff than with other sites. Variations are also present between the superior and inferior surfaces of the disc, with the inferomedial and superoanterior regions demonstrating greater stiffness. Appreciating the regional characteristics of the disc becomes important as dynamic loading patterns of the joint during function are considered. Attempts at correlating this behavior to the distribution of glycosaminoglycans throughout the disc have not been uniformly successful, and other factors, such as collagen fiber size and orientation, may play a greater role in affecting the material properties.

Several studies have demonstrated the potential of growth factors for TMJ cartilage-tissue engineering. This potential was first observed in an experiment comparing the effects of transforming growth factor-$\beta_1$ (TGF-$\beta_1$) with prostaglandin $E_2$ ($PGE_2$) on bovine TMJ disc cells in mono-layer. TGF-$\beta_1$ increased cell proliferation 2.5-fold, while $PGE_2$ had no significant effect (Landesberg *et al.*, 1996). The effects of platelet-derived growth factor (PDGF), insulin-like growth factor (IGF), and basic fibroblast growth factor (bFGF) have also been assayed, using monolayer cultures of porcine TMJ disc cells. The results of these studies suggest that lower concentrations favor biosynthesis while higher concentrations favor proliferation (Detamore and Athanasiou, 2004). The most beneficial growth factors appear to be IGF-I and bFGF, both of which produce significant increases in collagen synthesis and cell proliferation. Since native tissue is exposed to a variety of growth factors, combination strategies are likely to prove more beneficial than single-factor therapy. To explore this hypothesis, IGF-I and bFGF in low concentration were combined with bFGF and TGF-$\beta_1$. This cocktail successfully demonstrated increased collagen production when applied to porcine TMJ disc cells seeded on PGA scaffolds (Detamore and Athanasiou, 2005a). However, while constructs exposed to growth factor combinations improved structural integrity and overall cellularity, a statistically significant improvement in biochemical or mechanical properties was not demonstrated (Almarza and Athanasiou, 2006b).

Although growth factors have received the most attention, positive biochemical stimulation is also likely to come from culture conditions and cellular interactions. An ascorbic acid concentration of 25 µg/mL has been shown to produce constructs with higher total collagen content and higher aggregate modulus relative to concentrations of 0 µg/mL or 50 µg/mL (Bean *et al.*, 2005). This was likely associated with improved seeding observed for the constructs cultured in 25 µg/mL of ascorbic acid. Initial cell seeding is another important consideration in any tissue-engineering construct, due to cell-to-cell interactions and signaling. It has been shown that PGA scaffolds seeded at saturation increased cellularity and ECM content relative to scaffolds seeded below saturation (Almarza and Athanasiou, 2005).

The native TMJ disc undergoes significant compression, tension, and shear (Tanaka *et al.*, 2003). While cells proliferate and produce ECM in static culture, mechanical stimuli may be required to produce an optimal tissue-engineered construct. A variety of mechanical stimuli may be beneficial, including compression, tension, hydrostatic pressure, and fluid shear stress. An extensive review of the mechanical bioreactors that have been used in engineering cartilaginous tissues has been published on the subject (Darling and Athanasiou, 2003).

Three recent studies have investigated the effects of mechanical stimulation on TMJ disc constructs. A low-shear fluid environment by means of a rotating wall bioreactor created constructs with dense matrix and cell composition (Detamore and Athanasiou, 2005b); however, when the biochemical content of these constructs was compared to that of those grown in static culture, no clear benefit of the bioreactor was observed. When disc cells were exposed to hydrostatic pressure in monolayer or PGA scaffolds, constant hydrostatic pressure at 10 MPa increased collagen production as compared to static culture (Almarza and Athanasiou, 2006a). In contrast, intermittent hydrostatic pressure from 0 to 10 MPa at 1-Hz frequency was detrimental to the constructs, producing less collagen and GAGs than unloaded controls. These results were consistent in both two- and three-dimensional cultures. In another recent study, dynamic tensile strain significantly reduced interleukin-1β-induced up-regulation of matrix metalloproteinase (Deschner *et al.*, 2006). This may have implications on future tissue-engineering studies, since MMPs play an important role in ECM degradation and remodeling.

## Composite-Tissue Applications

Success with the regeneration of single tissue types, such as bone and cartilage, has encouraged investigators to attempt the reconstruction of structures composed of multiple tissue types. Such anatomic structures may exist as composites of hard and soft tissues, which differ in their cellular composition and mechanical properties yet perform as a single functional unit (Rahaman and Mao, 2005).

The temporomandibular joint condyle serves as an example of a maxillofacial composite structure consisting of articular cartilage and subchondral bone and provides an excellent opportunity for composite osteochondral-tissue engineering. A study performed by Alhadlaq *et al.* (2004), used adult bone marrow MSCs, expanded in culture and induced to differentiate into separate osteogenic and chondrogenic lineages *in vitro*. The resultant cells were then encapsulated in poly(ethylene glycol)-based hydrogels and the cell–polymer solutions cross-linked in a mold that provided the correct stratified organization of the osteogenic and chondrogenic layers. Finally, the osteochondral constructs were implanted into the dorsum of immunodeficient mice for up to eight weeks. Histological and immunohistological analyses revealed both structural and immunohistochemical differences between the osteogenic and chondrogenic layers, which served as a primitive proof of concept of the potential for composite-tissue-engineered constructs in the craniofacial region.

## Animal Models

As tissue-engineering strategies become more sophisticated, the complex interactions between multiple cell populations, growth factors, and scaffold biomaterials require testing within clinically relevant *in vivo* healing environments. While computational models and *in vitro* testing provide proof-of-concept verification of a particular tissue-engineering strategy, more challenging test beds utilizing animal models are ultimately required, to allow discrimination of the healing potential of different biomaterial constructs (Guldberg *et al.*, 2004).

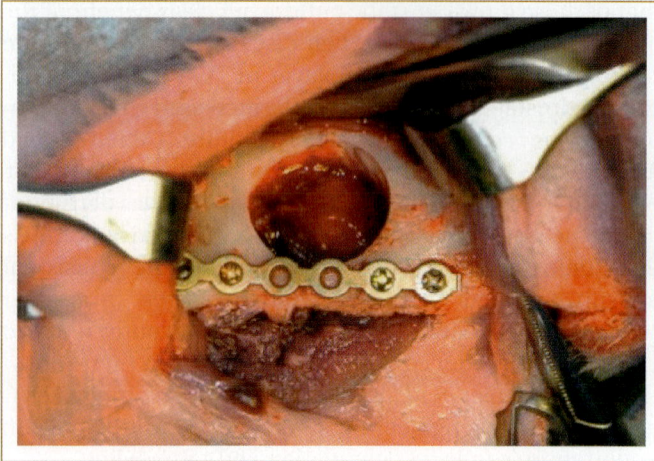

**FIG. 71.10.** Intraoperative photo of the rabbit mandibular critical-size bone defect. Reprinted from Young *et al.* (2007) with permission from Wiley & Sons, Inc. Copyright 2007.

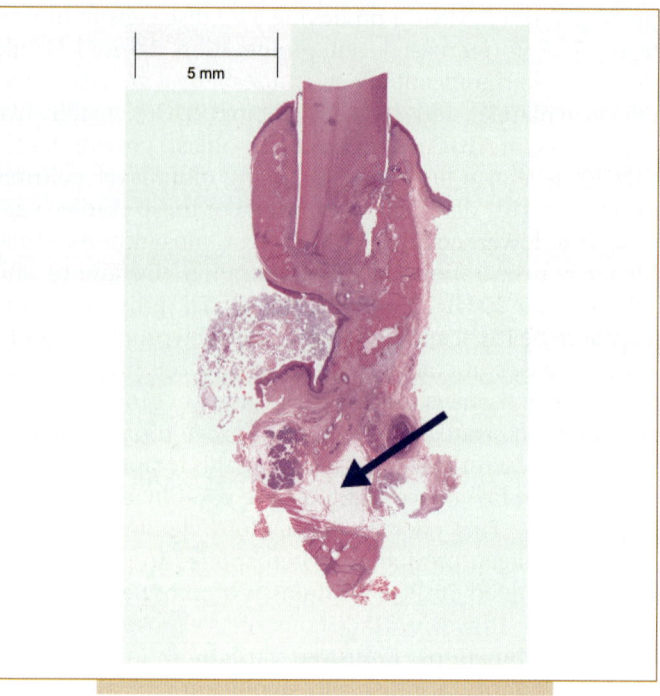

**FIG. 71.11.** Coronal section (stained with H&E) showing representative 16-week tissue repair within the rabbit mandibular critical-size defect. Scale bar represents 5 mm. As shown, the full-thickness defect at 16 weeks displays a lack of bony bridging across the center (arrow). Note that the image is oriented with the buccal surface on the left.

The ideal model for testing the validity of tissue-engineered bone constructs is the *critical size defect* (CSD): namely, an intraosseous defect that will not heal by bone formation during the lifetime of the animal (Hollinger and Kleinschmidt, 1990). Several CSD models have been described in the literature, including the cranial cavitational defect (Blum *et al.*, 2003; Dean *et al.*, 2005) and long-bone segmental defect (Hedberg *et al.*, 2005a, 2005b). These models do not accurately describe the oral and maxillofacial environment, because they do not simulate the unique masticatory stresses and cell populations seen in mandibular or maxillary wound healing. Mandibular defects have been described in rats, but they have the disadvantages of poor surgical access and a tendency for implanted materials to fall into the fascial spaces (Hollinger and Kleinschmidt, 1990). Conversely, larger animals, such as dogs (Huh *et al.*, 2005), minipigs (Henkel *et al.*, 2005), goats (Fennis *et al.*, 2005), and nonhuman primates (Fritz *et al.*, 2000), offer the advantages of easy surgical access and the ability to create large defects, but they are expensive to maintain.

To address the lack of an easily accessible, readily reproducible CSD model in an animal that was inexpensive and easy to maintain, a 10-mm, cylindrical, bicortical mandibular defect model in the New Zealand white rabbit was created (Young *et al.*, 2007) (Fig. 71.10). Both quantitative and qualitative outcome measures using histology (Fig. 71.11) and micro-computed tomography (Figs. 71.12 and 71.13) were used to evaluate the bone healing within this novel mandibular-defect model, which was shown to lack bone bridging at four months. It has also been shown that vascular ingrowth can be quantified in this model, making it an ideal testbed for angiogenic-tissue-engineering strategies as well.

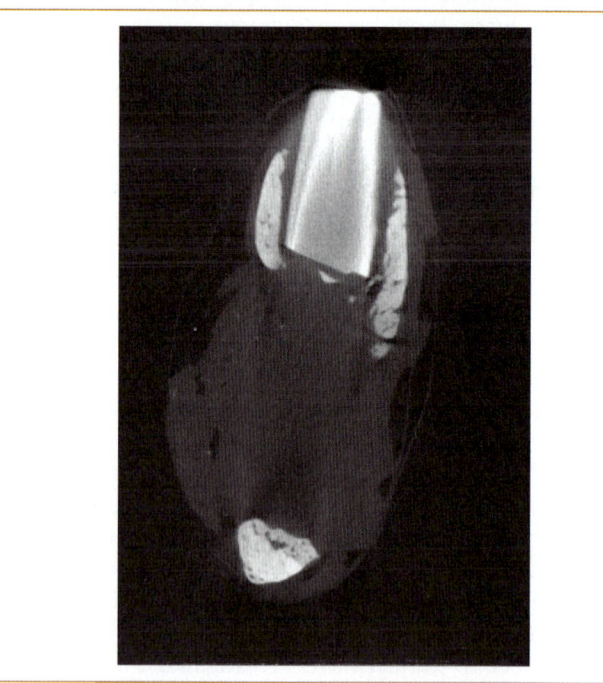

**FIG. 71.12.** Micro-CT-generated coronal tomogram showing representative 16-week defect healing of the critical-size defect. Qualitative micro-CT findings in terms of bone formation at 16 weeks were similar to the histological findings seen in Fig. 71.11.

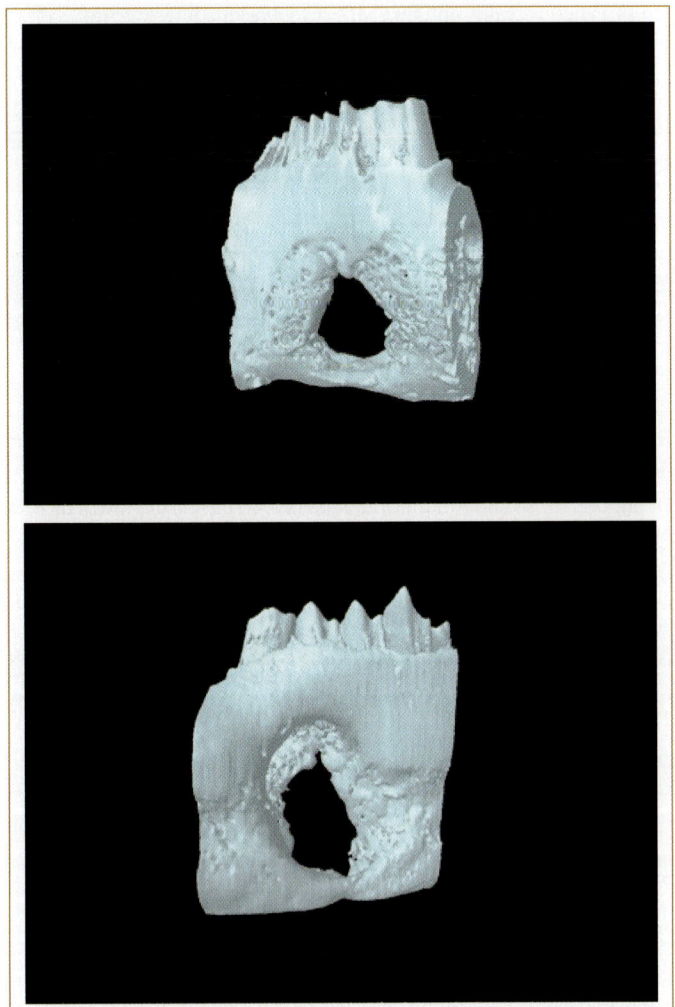

**FIG. 71.13.** Micro-CT-generated representative three-dimensional reconstructions. Note the buccal aspect of the specimen is shown on the top, with the lingual aspect on the bottom. The critical-size defect at 16 weeks shows a through-and-through defect and a lack of bony bridging across its center. Reprinted from Young *et al.* (2007) with permission from Wiley & Sons, Inc. Copyright 2007.

## V. FUTURE OF ORAL- AND MAXILLOFACIAL-TISSUE ENGINEERING

While significant progress has been made toward our ability to fabricate tissue in the laboratory for the reconstruction of defects in the oral and maxillofacial skeleton, considerable challenges remain before these techniques are embraced as a feasible clinical modality. Some of the issues pertain to tissue engineering itself, such as the ability to identify and harvest a suitable population of cells capable of fulfilling the functions of the desired tissue, supporting cellular differentiation and reproduction through physiological levels of growth and attachment factors, promotion of vasculogenesis, and the development of matrices that fulfill the physical requirements of a skeletal site. Since many of the structures in the oral and maxillofacial region are composed of multiple tissue types, the ability to engineer composite structures is also important, as is consideration of the unique environment to which tissue-engineered constructs are exposed. Even as these demands are met, other factors must be considered before TE tissue can be adopted as part of a reconstructive surgeon's armamentarium. Existing reconstructive techniques were originally based on macroscopic concerns for restoring the shape and size of a missing structure. Since TE is essentially based on cells, a microscopic appreciation of the function of a particular tissue type must first be derived. At this time, characterization of the function and pathological degradation of many of the structures in the oral and maxillofacial region is unfortunately deficient. In other words, before the replacement for a structure can be appropriately engineered, a better understanding of the function and local environment must first be achieved. This is particularly true of defects produced by ongoing pathological processes, where correction of the condition must precede replacement with yet another biological substrate. The future of oral and maxillofacial TE therefore lies in the hands of close collaborations between engineers and clinicians together.

## VI. REFERENCES

Alden, T. D., Beres, E. J., Laurent, J. S., Engh, J. A., Das, S., London, S. D., Jane, J. A., Jr., Hudson, S. B., and Helm, G. A. (2000). The use of bone morphogenetic protein gene therapy in craniofacial bone repair. *J. Craniofac. Surg.* **11**, 24–30.

Alhadlaq, A., Elisseeff, J. H., Hong, L., Williams, C. G., Caplan, A. I., Sharma, B., Kopher, R. A., Tomkoria, S., Lennon, D. P., Lopez, A., *et al.* (2004). Adult stem cell–driven genesis of human-shaped articular condyle. *Ann. Biomed. Eng.* **32**, 911–923.

Allen, K. D., and Athanasiou, K. A. (2006a). Gene expression changes in passaged cells from TMJ fibrocartilage. (Submitted).

Allen, K. D., and Athanasiou, K. A. (2006b). TMJ disc-tissue engineering on PLLA scaffolds: growth factor effects. *Osteoarthritis Cartilage.* (Submitted).

Allen, K. D., and Athanasiou, K. A. (2006c). Viscoelastic characterization of the porcine temporomandibular joint disc under unconfined compression. *J. Biomech.* **39**, 312–322.

Almarza, A. J., and Athanasiou, K. A. (2004). Seeding techniques and scaffolding choice for tissue engineering of the temporomandibular joint disk. *Tissue Eng.* **10**, 1787–1795.

Almarza, A. J., and Athanasiou, K. A. (2005). Effects of initial cell seeding density for the tissue engineering of the temporomandibular joint disc. *Ann. Biomed. Eng.* **33**, 943–950.

Almarza, A. J., and Athanasiou, K. A. (2006a). Effects of hydrostatic pressure on TMJ disc cells. *Tissue Eng.* **12**, 1285–1294.

Almarza, A. J., and Athanasiou, K. A. (2006b). Evaluation of three growth factors in combinations of two for temporomandibular joint disc-tissue engineering. *Arch. Oral Biol.* **51**, 215–221.

Arosarena, O. A., and Collins, W. L. (2005). Bone regeneration in the rat mandible with bone morphogenetic protein-2: a comparison of two carriers. *Otolaryngol. Head Neck Surg.* **132**, 592–597.

Bean, A. C., Almarza, A. J., and Athanasiou, K. A. (2005). Effects of ascorbic acid concentration for the tissue engineering of the temporomandibular joint disc. *J. Eng. Med.* (In press).

Bernhart, B. J., Huryn, J. M., Disa, J., Shah, J. P., and Zlotolow, I. M. (2003). Hard palate resection, microvascular reconstruction, and prosthetic restoration: a 14-year retrospective analysis. *Head Neck* **25**, 671–680.

Blum, J. S., Barry, M. A., Mikos, A. G., and Jansen, J. A. (2003). *In vivo* evaluation of gene therapy vectors in *ex vivo*-derived marrow stromal cells for bone regeneration in a rat critical-size calvarial defect model. *Hum. Gene Ther.* **14**, 1689–1701.

Boyne, P. J., Lilly, L. C., Marx, R. E., Moy, P. K., Nevins, M., Spagnoli, D. B., and Triplett, R. G. (2005). *De novo* bone induction by recombinant human bone morphogenetic protein-2 (rhBMP-2) in maxillary sinus floor augmentation. *J. Oral Maxillofac. Surg.* **63**, 1693–1707.

Bras, J., van Ooij, C. P., Duns, J. Y., Wansink, H. M., Driessen, R. M., and van den Akker, H. P. (1983). Mandibular atrophy and metabolic bone loss. A radiologic analysis of 126 edentulous patients. *Int. J. Oral Surg.* **12**, 309–313.

Caplan, A. I., and Bruder, S. P. (2001). Mesenchymal stem cells: building blocks for molecular medicine in the 21st century. *Trends Mol. Med.* **7**, 259–264.

Carstens, M. H., Chin, M., Ng, T., and Tom, W. K. (2005). Reconstruction of #7 facial cleft with distraction-assisted *in situ* osteogenesis (DISO): role of recombinant human bone morphogenetic protein-2 with Helistat-activated collagen implant. *J. Craniofac. Surg.* **16**, 1023–1032.

Chang, S. C., Tobias, G., Roy, A. K., Vacanti, C. A., and Bonassar, L. J. (2003). Tissue engineering of autologous cartilage for craniofacial reconstruction by injection molding. *Plast. Reconstr. Surg.* **112**, 793–799; discussion 800–801.

Chim, H., and Schantz, J. T. (2006). Human circulating peripheral blood mononuclear cells for calvarial bone-tissue engineering. *Plast. Reconstr. Surg.* **117**, 468–478.

Chin, M., Ng, T., Tom, W. K., and Carstens, M. (2005). Repair of alveolar clefts with recombinant human bone morphogenetic protein (rhBMP-2) in patients with clefts. *J. Craniofac. Surg.* **16**, 778–789.

Christensen, L. V., and McKay, D. C. (2000). Rotational and translational loading of the temporomandibular joint. *Cranio* **18**, 47–57.

Cochran, D. L., Jones, A. A., Lilly, L. C., Fiorellini, J. P., and Howell, H. (2000). Evaluation of recombinant human bone morphogenetic protein-2 in oral applications including the use of endosseous implants: 3-year results of a pilot study in humans. *J. Periodontol.* **71**, 1241–1257.

Darling, E. M., and Athanasiou, K. A. (2003). Biomechanical strategies for articular cartilage regeneration. *Ann. Biomed. Eng.* **31**, 1114–1124.

Darling, E. M., and Athanasiou, K. A. (2005). Rapid phenotypic changes in passaged articular chondrocyte subpopulations. *J. Orthop. Res.* **23**, 425–432.

Datta, N., Q, P. P., Sharma, U., Sikavitsas, V. I., Jansen, J. A., and Mikos, A. G. (2006). *In vitro*-generated extracellular matrix and fluid shear stress synergistically enhance 3D osteoblastic differentiation. *Proc. Natl. Acad. Sci. U.S.A.* **103**, 2488–2493.

Dean, D., Wolfe, M. S., Ahmad, Y., Totonchi, A., Chen, J. E., Fisher, J. P., Cooke, M. N., Rimnac, C. M., Lennon, D. P., Caplan, A. I., *et al.* (2005). Effect of transforming growth factor beta2 on marrow-infused foam poly(propylene fumarate) tissue-engineered constructs for the repair of critical-size cranial defects in rabbits. *Tissue Eng.* **11**, 923–939.

de Bont, L. G., and Stegenga, B. (1993). Pathology of temporomandibular joint internal derangement and osteoarthrosis. *Int. J. Oral Maxillofac. Surg.* **22**, 71–74.

Deschner, J., Rath-Deschner, B., and Agarwal, S. (2006). Regulation of matrix metalloproteinase expression by dynamic tensile strain in rat fibrochondrocytes. *Osteoarthritis Cartilage* **14**, 264–272.

Detamore, M. S., and Athanasiou, K. A. (2003). Motivation, characterization, and strategy for tissue engineering the temporomandibular joint disc. *Tissue Eng.* **9**, 1065–1087.

Detamore, M. S., and Athanasiou, K. A. (2004). Effects of growth factors on temporomandibular joint disc cells. *Arch. Oral Biol.* **49**, 577–583.

Detamore, M. S., and Athanasiou, K. A. (2005a). Evaluation of three growth factors for TMJ disc-tissue engineering. *Ann. Biomed. Eng.* **33**, 383–390.

Detamore, M. S., and Athanasiou, K. A. (2005b). Use of a rotating bioreactor toward tissue engineering the temporomandibular joint disc. *Tissue Eng.* **11**, 1188–1197.

Dunn, C. A., Jin, Q., Taba, M., Jr., Franceschi, R. T., Bruce Rutherford, R., and Giannobile, W. V. (2005). BMP gene delivery for alveolar bone engineering at dental implant defects. *Mol. Ther.* **11**, 294–299.

Edwards, P. C., Ruggiero, S., Fantasia, J., Burakoff, R., Moorji, S. M., Paric, E., Razzano, P., Grande, D. A., and Mason, J. M. (2005). Sonic hedgehog gene-enhanced tissue engineering for bone regeneration. *Gene Ther.* **12**, 75–86.

Eppley, B. L., and Sadove, A. M. (2000). Management of alveolar cleft bone grafting — state of the art. *Cleft Palate Craniofac. J.* **37**, 229–233.

Eppley, B. L., Connolly, D. T., Winkelmann, T., Sadove, A. M., Heuvelman, D., and Feder, J. (1991). Free bone graft reconstruction of irradiated facial tissue: experimental effects of basic fibroblast growth factor stimulation. *Plast. Reconstr. Surg.* **88**, 1–11.

Fennis, J. P., Stoelinga, P. J., and Jansen, J. A. (2005). Reconstruction of the mandible with an autogenous irradiated cortical scaffold, autogenous corticocancellous bone-graft and autogenous platelet-rich-plasma: an animal experiment. *Int. J. Oral Maxillofac. Surg.* **34**, 158–166.

Fiorellini, J. P., Howell, T. H., Cochran, D., Malmquist, J., Lilly, L. C., Spagnoli, D., Toljanic, J., Jones, A., and Nevins, M. (2005). Randomized study evaluating recombinant human bone morphogenetic protein-2 for extraction socket augmentation. *J. Periodontol.* **76**, 605–613.

Fisher, J. P., Lalani, Z., Bossano, C. M., Brey, E. M., Demian, N., Johnston, C. M., Dean, D., Jansen, J. A., Wong, M. E., and Mikos, A. G. (2004). Effect of biomaterial properties on bone healing in a rabbit tooth extraction socket model. *J. Biomed. Mater. Res. A* **68**, 428–438.

Fong, K. D., Nacamuli, R. P., Song, H. M., Warren, S. M., Lorenz, H. P., and Longaker, M. T. (2003). New strategies for craniofacial repair and replacement: a brief review. *J. Craniofac. Surg.* **14**, 333–339.

Fritz, M. E., Jeffcoat, M. K., Reddy, M., Koth, D., Braswell, L. D., Malmquist, J., and Lemons, J. (2000). Guided bone regeneration of large mandibular defects in a primate model. *J. Periodontol.* **71**, 1484–1491.

Girdler, N. M. (1998). *In vitro* synthesis and characterization of a cartilaginous meniscus grown from isolated temporomandibular chondroprogenitor cells. *Scand. J. Rheumatol.* **27**, 446–453.

Guldberg, R. E., Oest, M., Lin, A. S., Ito, H., Chao, X., Gromov, K., Goater, J. J., Koefoed, M., Schwarz, E. M., O'Keefe, R. J., *et al.* (2004). Functional integration of tissue-engineered bone constructs. *J. Musculoskelet. Neuronal Interact.* **4**, 399–400.

Hannallah, D., Peterson, B., Lieberman, J. R., Fu, F. H., and Huard, J. (2003). Gene therapy in orthopaedic surgery. *J. Bone Joint Surg. Am.* **84A**, 1046–1061.

Hedberg, E. L., Kroese-Deutman, H. C., Shih, C. K., Crowther, R. S., Carney, D. H., Mikos, A. G., and Jansen, J. A. (2005a). Effect of varied release kinetics of the osteogenic thrombin peptide TP508 from biodegradable, polymeric scaffolds on bone formation *in vivo. J. Biomed. Mater. Res. A* **72**, 343–353.

Hedberg, E. L., Kroese-Deutman, H. C., Shih, C. K., Crowther, R. S., Carney, D. H., Mikos, A. G., and Jansen, J. A. (2005b). *In vivo* degradation of porous poly(propylene fumarate)/poly(DL-lactic-*co*-glycolic acid)-composite scaffolds. *Biomaterials* **26**, 4616–4623.

Helenius, L. M., Hallikainen, D., Helenius, I., Meurman, J. H., Kononen, M., Leirisalo-Repo, M., and Lindqvist, C. (2005). Clinical and radiographic findings of the temporomandibular joint in patients with various rheumatic diseases. A case-control study. *Oral Surg. Oral Med. Oral Pathol. Oral Radiol. Endod.* **99**, 455–463.

Henkel, K. O., Gerber, T., Dorfling, P., Gundlach, K. K., and Bienengraber, V. (2005). Repair of bone defects by applying biomatrices with and without autologous osteoblasts. *J. Craniomaxillofac. Surg.* **33**, 45–49.

Herring, S. W., and Ochareon, P. (2005). Bone — special problems of the craniofacial region. *Orthod. Craniofac. Res.* **8**, 174–182.

Hidalgo, D. A. (1994). Fibula free flap mandibular reconstruction. *Clin. Plast. Surg.* **21**, 25–35.

Hollinger, J. O., and Kleinschmidt, J. C. (1990). The critical size defect as an experimental model to test bone repair materials. *J. Craniofac. Surg.* **1**, 60–68.

Hollister, S. J., Lin, C. Y., Saito, E., Lin, C. Y., Schek, R. D., Taboas, J. M., Williams, J. M., Partee, B., Flanagan, C. L., Diggs, A., *et al.* (2005). Engineering craniofacial scaffolds. *Orthod. Craniofac. Res.* **8**, 162–173.

Hong, L., Miyamoto, S., Hashimoto, N., and Tabata, Y. (2000). Synergistic effect of gelatin microspheres incorporating TGF-beta1 and a physical barrier for fibrous tissue infiltration on skull bone formation. *J. Biomater. Sci. Polym. Ed.* **11**, 1357–1369.

Huang, Y. C., Simmons, C., Kaigler, D., Rice, K. G., and Mooney, D. J. (2005). Bone regeneration in a rat cranial defect with delivery of PEI-condensed plasmid DNA encoding for bone morphogenetic protein-4 (BMP-4). *Gene Ther.* **12**, 418–426.

Huh, J. Y., Choi, B. H., Kim, B. Y., Lee, S. H., Zhu, S. J., and Jung, J. H. (2005). Critical size defect in the canine mandible. *Oral Surg. Oral Med. Oral Pathol. Oral Radiol. Endod.* **100**, 296–301.

Jung, R. E., Glauser, R., Scharer, P., Hammerle, C. H., Sailor, H. F., and Weber, F. E. (2003). Effect of rhBMP-2 on guided bone regeneration in humans. *Clin. Oral Implants Res.* **14**, 556–568.

Kamil, S. H., Vacanti, M. P., Aminuddin, B. S., Jackson, M. J., Vacanti, C. A., and Eavey, R. D. (2004). Tissue engineering of a human-sized and shaped auricle using a mold. *Laryngoscope* **114**, 867–870.

Kawaguchi, H., Kurokawa, T., Hanada, K., Hiyama, Y., Tamura, M., Ogata, E., and Matsumoto, T. (1994). Stimulation of fracture repair by recombinant human basic fibroblast growth factor in normal and streptozotocin-diabetic rats. *Endocrinology* **135**, 774–781.

Landesberg, R., Takeuchi, E., and Puzas, J. E. (1996). Cellular, biochemical and molecular characterization of the bovine temporomandibular joint disc. *Arch. Oral Biol.* **41**, 761–767.

Lieberman, J. R., Daluiski, A., and Einhorn, T. A. (2002). The role of growth factors in the repair of bone. Biology and clinical applications. *J. Bone Joint Surg. Am.* **84A**, 1032–1044.

Lindsey, W. H. (2001). Osseous-tissue engineering with gene therapy for facial bone reconstruction. *Laryngoscope* **111**, 1128–1136.

Lutolf, M. P., Weber, F. E., Schmoekel, H. G., Schense, J. C., Kohler, T., Muller, R., and Hubbell, J. A. (2003). Repair of bone defects using synthetic mimetics of collagenous extracellular matrices. *Nat. Biotechnol.* **21**, 513–518.

Marx, R. E. (1994). Clinical application of bone biology to mandibular and maxillary reconstruction. *Clin. Plast. Surg.* **21**, 377–392.

Meyer, U., Buchter, A., Hohoff, A., Stoffels, E., Szuwart, T., Runte, C., Dirksen, D., and Wiesmann, H. P. (2005). Image-based extracorporeal-tissue engineering of individualized bone constructs. *Int. J. Oral Maxillofac. Implants* **20**, 882–890.

Moghadam, H. G., Urist, M. R., Sandor, G. K., and Clokie, C. M. (2001). Successful mandibular reconstruction using a BMP bioimplant. *J. Craniofac. Surg.* **12**, 119–127; discussion 128.

Mukherjee, D. P., Tunkle, A. S., Roberts, R. A., Clavenna, A., Rogers, S., and Smith, D. (2003). An animal evaluation of a paste of chitosan glutamate and hydroxyapatite as a synthetic bone graft material. *J. Biomed. Mater. Res. B. Appl. Biomater.* **67**, 603–639.

Nussenbaum, B., Rutherford, R. B., Teknos, T. N., Dornfeld, K. J., and Krebsbach, P. H. (2003). *Ex vivo* gene therapy for skeletal regeneration in cranial defects compromised by postoperative radiotherapy. *Hum. Gene Ther.* **14**, 1107–1115.

Nussenbaum, B., Teknos, T. N., and Chepeha, D. B. (2004). Tissue engineering: the current status of this futuristic modality in head–neck reconstruction. *Curr. Opin. Otolaryngol. Head Neck Surg.* **12**, 311–315.

Ono, I., Yamashita, T., Jin, H. Y., Ito, Y., Hamada, H., Akasaka, Y., Nakasu, M., Ogawa, T., and Jimbow, K. (2004). Combination of porous hydroxyapatite and cationic liposomes as a vector for BMP-2 gene therapy. *Biomaterials* **25**, 4709–4718.

Poshusta, A. K., and Anseth, K. S. (2001). Photopolymerized biomaterials for application in the temporomandibular joint. *Cells Tissues Organs* **169**, 272–278.

Puelacher, W. C., Wisser, J., Vacanti, C. A., Ferraro, N. F., Jaramillo, D., and Vacanti, J. P. (1994). Temporomandibular joint disc replacement made by tissue-engineered growth of cartilage. *J. Oral Maxillofac. Surg.* **52**, 1172–1177; discussion 1177–1178.

Rahaman, M. N., and Mao, J. J. (2005). Stem cell–based composite tissue constructs for regenerative medicine. *Biotechnol. Bioeng.* **91**, 261–284.

Ripamonti, U., Crooks, J., and Rueger, D. C. (2001). Induction of bone formation by recombinant human osteogenic protein-1 and sintered porous hydroxyapatite in adult primates. *Plast. Reconstr. Surg.* **107**, 977–988.

Rotter, N., Haisch, A., and Bucheler, M. (2005). Cartilage- and bone-tissue engineering for reconstructive head and neck surgery. *Eur. Arch. Otorhinolaryngol.* **262**, 539–545.

Schantz, J. T., Hutmacher, D. W., Lam, C. X., Brinkmann, M., Wong, K. M., Lim, T. C., Chou, N., Guldberg, R. E., and Teoh, S. H. (2003). Repair of calvarial defects with customized tissue-engineered bone grafts II.

Evaluation of cellular efficiency and efficacy *in vivo*. *Tissue Eng.* **9**(Suppl. 1), S127–S139.

Seper, L., Piffko, J., Joos, U., and Meyer, U. (2004). Treatment of fractures of the atrophic mandible in the elderly. *J. Am. Geriatr. Soc.* **52**, 1583–1584.

Seto, I., Marukawa, E., and Asahina, I. (2006). Mandibular reconstruction using a combination graft of rhBMP-2 with bone marrow cells expanded *in vitro*. *Plast. Reconstr. Surg.* **117**, 902–908.

Shang, Q., Wang, Z., Liu, W., Shi, Y., Cui, L., and Cao, Y. (2001). Tissue-engineered bone repair of sheep cranial defects with autologous bone marrow stromal cells. *J. Craniofac. Surg.* **12**, 586–593; discussion 594–595.

Sharma, A. B., and Beumer, J., 3rd. (2005). Reconstruction of maxillary defects: the case for prosthetic rehabilitation. *J. Oral Maxillofac. Surg.* **63**, 1770–1773.

Shieh, S. J., Terada, S., and Vacanti, J. P. (2004). Tissue engineering auricular reconstruction: *in vitro* and *in vivo* studies. *Biomaterials* **25**, 1545–1557.

Shin, H., Zygourakis, K., Farach-Carson, M. C., Yaszemski, M. J., and Mikos, A. G. (2004). Attachment, proliferation, and migration of marrow stromal osteoblasts cultured on biomimetic hydrogels modified with an osteopontin-derived peptide. *Biomaterials* **25**, 895–906.

Sikavitsas, V. I., van den Dolder, J., Bancroft, G. N., Jansen, J. A., and Mikos, A. G. (2003). Influence of the *in vitro* culture period on the *in vivo* performance of cell/titanium bone tissue-engineered constructs using a rat cranial critical size defect model. *J. Biomed. Mater. Res. A* **67**, 944–951.

Springer, I. N., Fleiner, B., Jepsen, S., and Acil, Y. (2001). Culture of cells gained from temporomandibular joint cartilage on nonabsorbable scaffolds. *Biomaterials* **22**, 2569–2577.

Srouji, S., Rachmiel, A., Blumenfeld, I., and Livne, E. (2005). Mandibular defect repair by TGF-beta and IGF-1 released from a biodegradable osteoconductive hydrogel. *J. Craniomaxillofac. Surg.* **33**, 79–84.

Tanaka, E., Hanaoka, K., van Eijden, T., Tanaka, M., Watanabe, M., Nishi, M., Kawai, N., Murata, H., Hamada, T., and Tanne, K. (2003). Dynamic-shear properties of the temporomandibular joint disc. *J. Dent. Res.* **82**, 228–231.

Thomas, M., Grande, D., and Haug, R. H. (1991). Development of an *in vitro* temporomandibular joint cartilage analog. *J. Oral Maxillofac. Surg.* **49**, 854–856; discussion 857.

van Eijden, T. M. (2000). Biomechanics of the mandible. *Crit. Rev. Oral Biol. Med.* **11**, 123–136.

Vehof, J. W., Haus, M. T., de Ruijter, A. E., Spauwen, P. H., and Jansen, J. A. (2002). Bone formation in transforming growth factor beta-I-loaded titanium fiber mesh implants. *Clin. Oral Implants Res.* **13**, 94–102.

Vikjaer, D., Blom, S., Hjorting-Hansen, E., and Pinholt, E. M. (1997). Effect of platelet-derived growth factor-BB on bone formation in calvarial defects: an experimental study in rabbits. *Eur. J. Oral Sci.* **105**, 59–66.

Warnke, P. H., Springer, I. N., Wiltfang, J., Acil, Y., Eufinger, H., Wehmoller, M., Russo, P. A., Bolte, H., Sherry, E., Behrens, E., *et al.* (2004). Growth and transplantation of a custom vascularized bone graft in a man. *Lancet* **364**, 766–770.

Warren, S. M., Fong, K. D., Chen, C. M., Loboa, E. G., Cowan, C. M., Lorenz, H. P., and Longaker, M. T. (2003). Tools and techniques for craniofacial tissue engineering. *Tissue Eng.* **9**, 187–200.

Young, S., Bashoura, A. G., Borden, T., Baggett, L. S., Jansen, J. A., Wong, M. E., and Mikos, A. G. (2007). Development and characterization of a critical size alveolar bone defect model. *J. Biomed. Mater. Res. A.* (Submitted).

Zheng, H., Guo, Z., Ma, Q., Jia, H., and Dang, G. (2004). Cbfa1/osf2-transduced bone marrow stromal cells facilitate bone formation *in vitro* and *in vivo*. *Calcif. Tissue Int.* **74**, 194–203.

# Chapter Seventy-Two

# Periodontal-Tissue Engineering

*Hai Zhang, Hanson K. Fong, William V. Giannobile, and Martha J. Somerman*

I. Introduction

II. Factors for Periodontal Tissue Engineering and Regenerative Medicine

III. Current Approaches in Periodontal Tissue Engineering

IV. Future Directions

V. Acknowledgments

VI. References

## I. INTRODUCTION

Existing data from the National Institute of Dental and Craniofacial Research, National Institutes of Health, indicate that 86% of adults over the age of 70 have at least moderate periodontal disease, with approximately 25% having lost their teeth (Albandar *et al.*, 1999). Oral infections may have a marked negative impact on health and quality of life. Interestingly, but not surprisingly, millions of dental-oral-craniofacial procedures, ranging from tooth restoration to reconstruction of facial hard and soft tissues, are performed annually, compared with roughly 20,000 organ transplants and 500,000 joint replacements. The recognition that periodontal tissues, i.e., bone, cementum, and periodontal ligament, possess the capacity to repair/regenerate has led to substantial efforts focused on understanding the biologic basis for this activity and applying the knowledge gained to designing therapies that predictably promote periodontal tissue regeneration. This chapter covers some of the approaches that have been used.

In this first section a basic review of current therapies used to treat periodontal diseases is presented and a figure depicting current treatment modalities for damaged periodontal tissues and future approaches for regenerating lost periodontal tissues is provided (Fig. 72.1). The next section gives an overview of factors, known to and/or with strong evidence that they affect the behavior of cells associated with the regeneration of periodontal tissues. The following section discusses current approaches and future directions

being investigated for regenerating periodontal tissues, with a focus on cells considered to play a role in the formation of periodontal tissues, both during development and regeneration, and on delivery of factors and cells to local sites. The final section offers future directions/areas in need of research in order to design predictable regenerative treatment modalities for our patients.

The periodontium consists of the tissues supporting and investing the tooth and include the gingiva, periodontal ligament, cementum, and alveolar bone. Periodontal diseases are chronic inflammatory diseases caused by bacteria but also linked to trauma, environmental insults, genetic disorders, and systemic diseases. Periodontal diseases are manifested in the loss of both hard and soft tissues of the periodontium and eventual tooth loss if left untreated.

Researchers and clinicians in the field of dental-oral-craniofacial biology are among the leaders in applying bioengineering principles to regenerate tissues/organs. Dental materials have been and continue to be a major and successful area of focus, with advances in materials used to restore tooth structure (crown) in terms of adherence, aesthetics, and endurance; and in impression materials for assisting in the fabrication of crowns, dentures, partials, and other oral-craniofacial parts (Bayne, 2005). Knowledge gained from research targeted at improving the treatment modalities for replacing lost dental-oral-craniofacial tissues has resulted in enhanced bridge design for aesthetics and superior tissue integrity and tooth support, in more aes-

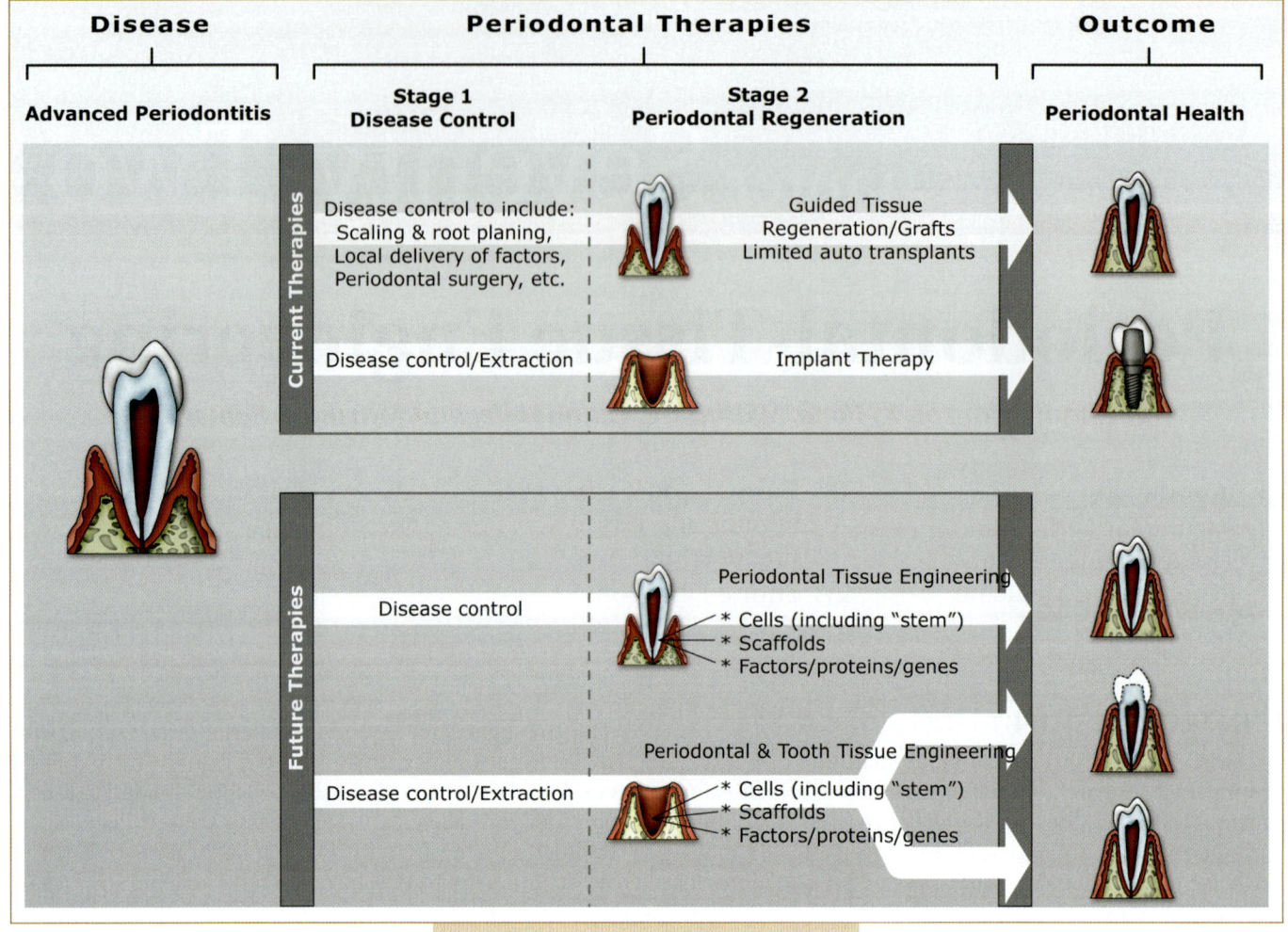

**FIG. 72.1.** Role of tissue engineering in periodontal therapies: Current status and future directions.

thetic replacement materials, and in an improved/increased variety of implants to select from for use as single and multiple tooth abutments.

Alternative approaches for the replacement of a single tooth include autotransplantation; but beyond the limitation that an individual must have a tooth/teeth for transplantation (e.g., third molars), reports of success have been limited (Gault and Warocquier-Clerout, 2002; Kallu *et al.*, 2005). In addition, several advances have been made in the application of graft materials and barrier membranes, often coupled with grafts, polymers, factors, and local delivery of factors. For example, antimicrobial agents are available for local delivery to periodontal pockets as a means of decreasing microbes locally without having to use a systemic approach for drug delivery. Further, oral-dose doxycycline (a tetracycline) is marketed as metalloproteinase inhibitor, but without antimicrobial activity. Low-dose doxycycline inhibits collagenase activity via decreasing availability of calcium, a requirement for activating collagenase, thereby decreasing collagen degradation normally associated with

the inflammatory response to bacteria. Unfortunately, while there have been some positive results with all of these approaches, the results are often not optimal and also not predictable (Wang *et al.*, 2005).

Fortunately, research approaches have provided insights into agents with potential for use in regenerating several tissues, including periodontal tissues. Areas of research focus that have had positive outcomes in the field of periodontal bioengineering include (a) defining the regulators controlling development of the periodontium; (b) understanding how healthy tissues are maintained with age; (c) recognizing the mechanisms that allow certain tissues, e.g., bone, to regenerate post trauma/disease; (d) unraveling the complexities of wound healing; (e) establishing stem cell properties within periodontal tissue biology; and (f) establishing the requirements for designing scaffolds/delivery systems for use in treating lost periodontal tissues. With the growing body of knowledge as to the cellular and molecular events controlling development and regeneration of tooth structures, the potential for designing predictable

| Table 72.1. | Effects of putative periodontal regenerative factors on periodontal tissues |
|---|---|
| *Growth factor* | *Effects: known and/or suggested* |
| Platelet-derived growth factor (PDGF) | Promotes migration, proliferation, and noncollagenous matrix synthesis of mesenchymal cells |
| Insulin-like growth factor-1 (IGF-1) | Promotes cell migration, proliferation, differentiation, and matrix synthesis of mesenchymal cells |
| Fibroblast growth factor-2 (FGF-2) | Promotes proliferation and attachment of endothelial cells and PDL cells and wound healing |
| Transforming growth factor-beta (TGF-β) | Promotes proliferation of cementoblasts and PDL fibroblasts |
| Selected bone morphogenetic proteins (BMPs) | Promotes proliferation, differentiation of mesenchymal cells, e.g., PDL fibroblasts, follicle cells, fibroblasts, toward a cell with capacity to form mineral matrix, e.g., osteoblasts, cementoblasts |
| Selected products of ameloblasts/epithelial cells, e.g., Emdogain (EMD); amelogenins (LRAP/TRAP); ameloblastin | Suggested roles in enhancing proliferation, protein synthesis, cell maturation, and mineral nodule formation in follicle cells, PDL cells, odontoblasts, osteoblasts, and cementoblasts and in wound healing |
| Selected anabolic factors, e.g., parathyroid hormone (PTH) | Intermittent PTH has been shown to promote bone formation *in vitro* and *in vivo* |
| Phosphates | Existing data suggest that cells within the periodontium are highly sensitive to changes in phosphate/pyrophosphate levels and further that phosphates, beyond being a physical-chemical regulator of crystal growth, may directly control expression of genes associated with maturation of cementoblasts and osteoblasts |

bioengineering approaches to replace/regenerate lost periodontal tissues is great. The next section is devoted to discussing some of the factors that hold promise for use in reconstructing periodontal tissues.

## II. FACTORS FOR PERIODONTAL TISSUE ENGINEERING AND REGENERATIVE MEDICINE

Following periodontal injury, a fibrin clot is formed at the wound site, releasing tissue growth factors locally, such as platelet-derived growth factor (PDGF) and transforming growth factor-beta (TGF-beta), from degranulating platelets (Okuda *et al.*, 2003; Tozum and Demiralp, 2003). These mitogenic polypeptides attract mesenchymal cells and fibroblasts to migrate into the periodontal wound and stimulate their repopulation in the defect (Marcopoulou *et al.*, 2003). The continuing process of periodontal tissue repair is followed by the formation of granulation tissue as a source for future periodontal connective tissue cells, such as osteoblasts, periodontal ligament fibroblasts, and cementoblasts (Bosshardt, 2005). One requirement for success in regenerating periodontal tissues is to have sufficient concentrations of ideal factors at the healing site. Some factors that have been considered as attractive candidates for delivery to these sites are discussed next and outlined in Table 72.1.

## Growth Factors

Growth factors have been used to target restoration of tooth-supporting bone, PDL, and cementum. A major focus of periodontal research has evaluated the impact of growth factor applications on periodontal tissue regeneration (reviewed in Giannobile, 1996; Cochran and Wozney, 1999; Anusaksathien and Giannobile, 2002; Nakashima and Reddi, 2003). These reviews describe various delivery systems and applications of growth factors, which are highlighted in Table 72.1. Advances in molecular cloning have made available unlimited quantities of recombinant growth factors for applications in tissue engineering. Recombinant growth factors known to promote skin and bone wound healing, such as PDGFs, insulin-like growth factors (IGFs), fibroblast growth factors (FGFs), and BMPs, have been used in preclinical and clinical trials for the treatment of periodontal osseous defects as well as alveolar bone-supporting endosseous dental implants.

### Platelet-Derived Growth Factor (PDGF)

PDGF is a member of a multifunctional polypeptide family that binds to two cell membrane tyrosine kinase receptors (PDGF Rα and PDGF Rβ) and subsequently exerts its biological effects on DNA replication, chemotaxis, extracellular matrix biosynthesis, and blocking growth arrest

(Kaplan *et al.*, 1979; Seppa *et al.*, 1982; Heldin *et al.*, 1989; Rosenkranz and Kazlaushas, 1999). PDGF-α and -β receptors are expressed in regenerating periodontal soft and hard tissues (Parkar *et al.*, 2001). In addition, PDGF initiates tooth-supporting periodontal ligament (PDL) cell chemotaxis (Nishimura and Terranova, 1996), mitogenesis (Oates *et al.*, 1993), matrix synthesis (Haase *et al.*, 1998), and attachment to tooth dentinal surfaces (Zaman *et al.*, 1999). Local topical application of PDGF alone or in combination with insulin-like growth factor-1 (IGF-1) results in repair of periodontal tissues, including bone and PDL (Lynch *et al.*, 1989, 1991; Rutherford *et al.*, 1992; Giannobile *et al.*, 1994, 1996).

## Bone Morphogenetic Proteins (BMPs)

The human genome encodes at least 20 BMPs (Reddi, 1998) that are multifunctional polypeptides belonging to the TGF-β superfamily of proteins (Wozney *et al.*, 1988). BMPs bind to type I and II receptors that function as serine-threonine kinases. The type I receptor protein kinase phosphorylates intracellular signaling substrates. The phosphorylated BMP-signaling Smads enter the nucleus and initiate the production of bone matrix proteins, leading to bone morphogenesis. The most remarkable feature of BMPs is the ability to induce *de novo* bone formation (Urist, 1965). BMPs are not only powerful regulators of cartilage and bone formation during embryonic development and regeneration in postnatal life, but also participate in the development and repair of other organs, such as the kidney and brain (Reddi, 2001). Studies have demonstrated the expression of BMPs during tooth development and periodontal repair, including in alveolar bone (Aberg *et al.*, 1997; Amar *et al.*, 1997). Investigations in animal models have shown the potential repair of alveolar bony defects around teeth and dental implants using BMP-2, -4, -7, and -12 (Lutolf *et al.*, 2003; Wikesjo *et al.*, 2003, 2004). In a recent clinical trial, BMP-2 delivered via a bioabsorbable collagen sponge promoted bone formation in buccal wall extraction socket defects when compared to collagen sponge alone (Fiorellini *et al.*, 2005). Furthermore, BMP7, also known as osteogenic protein-1, stimulates bone regeneration around teeth and endosseous dental implants and in maxillary sinus-floor-augmentation procedures (Rutherford *et al.*, 1992; Giannobile *et al.*, 1998; van den Bergh *et al.*, 2000).

## Factors Being Considered Based on Their Suggested Role During Development of Periodontal Tissues

### Epithelial-Like Products

It is well established that epithelial–mesenchymal (E–M) interactions are required for development of the tooth crown (enamel–ameloblasts; dentin–odontoblasts). The reader is referred to the website http://bite-it.helsinki.fi/ for a comprehensive list of signaling molecules and target genes expressed during crown development. Based on this knowledge, some success has been achieved toward regeneration of tooth structures in animal models, predominantly crowns, using a stem cell approach, mixing epithelial cells with mesenchymal cells at various stages of differentiation (Fong *et al.*, 2005; Zhang *et al.*, 2005). Accumulating data suggest that E–M interactions may also be involved in formation of periodontal tissues. In fact, a product currently marketed for use in regenerating periodontal tissues, Emdogain® (EMD, Straumann Biologics, Waltham, MA), is based on this concept. EMD is an extract of low-molecular-weight proteins derived from porcine tooth germs. The predominant protein in this extract is amelogenin, although it has been reported to contain several other factors, including proteolytically cleaved and alternatively spliced products of the amelogenin gene, i.e., leucine-rich amelogenin (LRAP) and tyrosine-rich amelogenin (TRAP), and ameloblastin and TGF-β (transforming growth factor β). *In vitro*, EMD, LRAP, TRAP, amelogenin, ameloblastin, and TGF-β have been shown to affect the behavior of mesenchymal cells, including reports of enhancing cell proliferation and cell adhesion (EMD) and altering genes associated with mineralization (EMD, LRAP, TRAP, amelogenin, ameloblastin, and TGF-β), while *in vivo* EMD has been shown to have positive clinical outcomes when used for regenerating lost periodontal tissues, although the results are often limited and not predictable (see reviews Esposito *et al.*, 2003; Giannobile and Somerman, 2003; Venezia *et al.*, 2004; Bartlett *et al.*, 2006; Heden and Wennstrom, 2006).

In spite of the fact that a product is used clinically based on the concept that E–M interactions are involved in development of periodontal tissues, it is still unclear as to whether or not E–M communications are required for development of periodontal tissues and, hence, their function in regeneration of periodontal tissues. Much of this doubt is based on the fact that the known roles for amelogenin and ameloblastin, secreted by ameloblasts, are as regulators of enamel formation via controlling crystal growth and structure. This has been highlighted by the murine phenotypes reported in overexpression and/or null for amelogenin or ameloblastin (Gibson *et al.*, 2001; Paine *et al.*, 2003; Paine and Snead, 2005). Mice null for amelogenin form enamel, but the crystal structure of the formed enamel lacks structure. An interesting finding was the noted increased root resorption in amelogenin-null mice, leading Kulkarni and his group to propose that amelogenin-like products, perhaps LRAP, have a protective effect on cementoblasts by decreasing the expression of RANKL, an activator of osteoclasts (Gibson *et al.*, 2001; Hatakeyama *et al.*, 2003, 2006).

One critical piece of information required is the identification of binding sites for amelogenin-like molecules and/or other epithelial derived factors on mesenchymal cells. Toward that end, Tompkins *et al.* (2006) reported the identification of a receptor for LRAP on the cell surface of myoblasts, LAMP-1, lysosomal-associated membrane protein–1, while Paine *et al.* reported that amelogenin bound to LAMP-3 in association with lysosomes (Wang *et al.*, 2005). The latter group proposed that LAMP-3 functions to degrade amelogenin. In contrast, Tompkins *et al.* propose that LRAP interacts with LAMP-1 on the cell surface and that this interaction results in downstream signaling events. Further studies, ongoing, will assist in clarifying the significance of amelogenin-like molecules and other epithelial-like molecules in regulating periodontal/root formation and, subsequently, periodontal regeneration.

### Phosphates

Several reports have demonstrated that alterations in phosphate/pyrophosphate regulation in both humans and animals have profound effects on bone homeostasis (Terkeltaub, 2001). Further, there is mounting evidence that inorganic phosphate (Pi), beyond its known role as a physiochemical regulator of mineralization, may control mineralization via signaling pathways. For example, mice null for tissue nonspecific alkaline phosphatase (TNAP), an enzyme that hydrolyzes pyrophosphate (PPi) to phosphate (Pi), are osteopenic and exhibit minimal or no cementum formation. The lack of cementum formation prevents the insertion of PDL fibers into cementum and surrounding bone, resulting in exfoliation of teeth. This tooth phenotype is noted in both animals and humans with mutations in TNAP. Importantly, there is a dental-specific form, odontohypophosphatasia, marked by abnormalities limited to teeth (Whyte, 2002). In contrast to TNAP, there are two genes and their products, PC-1 (nucleoside triphosphate pyrophosphohydrolase) and ANK, a protein of a mouse progressive ankylosis gene (ANKH — human homolog), that regulate PPi levels. ANK is a transmembrane protein regulating the transport of PPi from intracellular to extracellular locations. PC-1 regulates PPi levels in matrix vesicles and in extracellular compartments.

Mice with mutations in PC-1 or ANK have low levels of PPi, and this results in pathological ectopic calcification, including aortic valves and joints, mimicking an arthritic condition. In fact, mice with PC-1 mutation have been called tiptoe-walking (ttw) mice, since the severe ectopic calcification of joints results in an inability to walk. Mutated PC-1, with decreased extracellular pyrophosphate (ePPi) levels, has been linked to individuals affected with severe periarticular and vascular calcification (Rutsch *et al.*, 2001). Mutations in the human ANK gene have been identified in patients with a rare disorder, autosomal-dominant craniometaphyseal dysplasia. Affected individuals have marked osteosclerosis of craniofacial bones, often with neurological defects and flaring of metaphyses of long bones; but often, extracranial skeleton and joints are otherwise not affected (Reichenberger *et al.*, 2001). In terms of a tooth phenotype, Nociti *et al.* (2002) reported a marked increase in cementum formation in mice with either ANK/ANK or PC-1 mutations vs. wild-type control (greater that 10× increase); however, this was not associated with ectopic calcification or

decreased PDL space, and the surrounding alveolar bone appeared normal. To date, it is not established whether a tooth phenotype is noted in individuals with mutations in PC-1 or ANKH although a recent case report suggests a tooth phenotype (Zhang *et al.*, 2007).

The specific tooth phenotypes noted in murine models and human with mutations in TNAP have led us to hypothesize that periodontal tissues are phosphate-sensitive tissues, relative to several other tissues within the body. It is our belief that research targeted at understanding the genes/proteins and signaling pathways involved in phosphate-mediated cementum formation will result in the development of novel factors for use in delivery to sites for regeneration of lost periodontal support.

### Anabolic Factors

*Parathyroid hormone*, PTH, has been shown to have both catabolic and anabolic effects on osteoblastic-/cementoblastic-like cells (Fiaschi-Taesch and Stewart, 2003; Boabaid *et al.*, 2004). One of the mechanisms contributing to these activities include the ability of PTH to enhance the expression of RANKL in both osteoblasts and cementoblasts. RANKL is a key activator of osteoclasts via its binding to RANK on the cell surface osteoclasts. PTH given intermittently *in vitro* or *in vivo* (animal models and humans) results in enhanced bone formation. This knowledge has resulted in the use of PTHC (Forteo®) for treatment of severe osteoporosis. Forteo® [teriparatide (rhDNA origin) injection] contains human PTH (1-34) (Eli Lilly & Co.). It is tempting to speculate that controlled (intermittent) release of PTH or other anabolic factors at sites of periodontal destruction in humans would have very positive outcomes. Toward that end, Barros *et al.* (2003), using ligatures to induce periodontitis around first molars in rats, demonstrated that intermittent PTH, delivered subcutaneously over a 30-day period, protected the tooth site from periodontitis-induced bone resorption, noted in vehicle-treated rats. Currently, several groups are working on local delivery of PTH as well as other anabolic agents.

## III. CURRENT APPROACHES IN PERIODONTAL TISSUE ENGINEERING

### Role of Cells Within the Local Region

It has been know for decades that cells within the periodontium have the potential, when appropriately stimulated, to promote formation of new bone, cementum, and functional periodontium. This knowledge has led to considerable basic, translational, and clinical research focused on defining the signals and identifying the cells required to promote regeneration of periodontal tissues, with the ultimate goal of developing successful and predictable therapies.

Once the disease is under control, a viable approach is to promote the cells remaining at the site, in a manner to regenerate lost structures. Cells within the local region considered to play key roles in regenerating lost tissues include inflammatory cells, cells within the vasculature, PDL fibroblasts, osteoblasts, cementoblasts, and epithelial cells (see reviews: Gould *et al.*, 1980; McCulloch, 1985, 1995; Melcher, 1985; Bartold *et al.*, 2000; Nanci and Bosshardt, 2006). Factors released by inflammatory cells provide the motivation for fibroblasts and cells from the vasculature to begin the wound-healing process. Beyond the ability of a small population of cells within the PDL region to function as osteoblasts and/or cementoblasts (see later), these cells must control the balance between soft (PDL formation) and hard tissues (bone/cementum). Thus, a targeted area of research has emerged toward identifying both the promoters of mineralization as well as the negative regulators, allowing for formation of a functional PDL.

In addition, as also discussed later, understanding the factors and cells involved during development of periodontal tissues has provided clues to the cell types needed to promote regeneration of these tissues. For example, follicle cells, associated with the developing tooth and of mesenchymal origin, are considered to have the capacity to become cementoblasts, PDL fibroblasts, or osteoblasts, depending on the stimuli within the local region (Hakki *et al.*, 2001; Wise *et al.*, 2002; Zhao *et al.*, 2002; Zeichner-David *et al.*, 2003). However, when cementoblasts and follicle cells in PLGA polymer sponges were delivered to periodontal defects in rats, only cementoblasts provided an environment that promoted formation of new bone, cementum, and a PDL. In contrast, follicle cells under these same conditions inhibited mineral formation when compared with the vehicle control or cementoblasts. These results led M. Zhao *et al.* (2004) to hypothesize that follicle cells produce factors that control the extent of mineral formed and thereby allow for development of soft tissue–hard tissue interfaces during formation of the root/periodontium. Thus, defining the factors that trigger follicle cells to mature toward an osteoblast/cementoblast phenotype vs. PDL fibroblasts should provide information as to the regulators required for promoting PDL cells and other cells in the local environment to repair lost structures. As discussed earlier, some of the viable factors being considered include epithelial-like products produced by ameloblasts and/or Hertwig's epithelial root sheath (HERS), a doubled-layered sheath formed after the crown develops. The cells in this latter region proliferate apically and function to direct root morphogenesis. Products secreted by HERS, including BMPs, have been thought to stimulate follicle cells to develop into cementoblasts. An alternative suggestion is that HERS cells undergo transformation from an epithelial cell type to a mesenchymal cell type, i.e., cementoblast (for discussions on this topic, see Slavkin *et al.*, 1989; MacNeil and Somerman, 1999; Bosshardt and Nanci, 2004). Another aspect of the HERS that needs some attention is that as the root/periodontal region matures, remnant epithelial cells from HERS, called

*epithelial rest cells of Malassez* (ERCM), are noted within the PDL region. It has been suggested that these remnant cells have a role in wound healing, and thus some investigators have been working on ascertaining the factors secreted by ERCM (Hasegawa *et al.*, 2003).

## Stem Cells for Periodontal Tissue Engineering

The self-renewing stem cells (also called *mesenchymal stem cells*, MSCs) are capable of promoting tissue repair and regeneration. Recent research has shown that MSCs are present in mature dental tissues (Gronthos *et al.*, 2000; Miura *et al.*, 2003; Seo *et al.*, 2005). This stimulated some investigators to test the possibility of using stem cell–based therapies to regenerate the damaged periodontal tissues as a result of severe periodontitis (Miura *et al.*, 2004; Seo *et al.*, 2004).

MSCs have been identified in various human dental tissues, including adult third molars (dental pulp stem cells, DPSCs) (Gronthos *et al.*, 2000), exfoliated deciduous teeth (stem cells from human exfoliated deciduous teeth, SHED) (Miura *et al.*, 2003), and adult periodontal ligament (periodontal ligament stem cells, PDLSCs) (Seo *et al.*, 2004). Similar to the characterization of bone marrow stromal stem cells (BMSSC), these cells were identified by their capacity to form colony-forming units-fibroblast, (CFU-F) (Castro-Malaspina *et al.*, 1980). In addition, STRO-1, the putative stem cell marker, was also expressed in these dental-derived stem cells. Unlike DPSCs or SHED, which are derived from dental pulps and therefore could be characterized by the expression of odontoblast-specific protein, dentin sialophosphoprotein (DSPP) (MacDougall *et al.*, 1997; Feng *et al.*, 1998), there are no specific markers to characterize PDLSC. To date, scleraxis, a tendon-specific transcription factor, has been used to characterize PDLSC because of the similarity in collagen composition between PDL and tendon (Brent *et al.*, 2003).

Under proper inductive conditions, PDLSC has the capacity to differentiate toward different lineages, such as lipid-containing fat cells, and to form mineralized nodules as well as cementum-/PDL-like structures (Seo *et al.*, 2004). Sharpey's fibers were also observed to connect newly formed cementum to type I collagen-positive PDL-like tissue within the transplant.

Animal models and preliminary clinical trials have demonstrated great potential for the use of BM SSCs in treating various diseases (Azizi *et al.*, 1998; Hofstetter *et al.*, 2002; Toma *et al.*, 2002; Zhao *et al.*, 2002; Fuchs *et al.*, 2003; Gojo *et al.*, 2003), including cranial defects (Krebsbach *et al.*, 1998; Schantz *et al.*, 2003). It has been reported that porcine bone marrow progenitor cells seeded onto polyglycolide-*co*-lactide (PLGA) scaffolds can be used to repair critical defects in the pig mandible (Abukawa *et al.*, 2004). Toothlike structures were also generated in the omenta of rats using pig- and rat-tooth bud cells seeded onto polyglycolide (PGA) and PLGA scaffold materials (Young *et al.*,

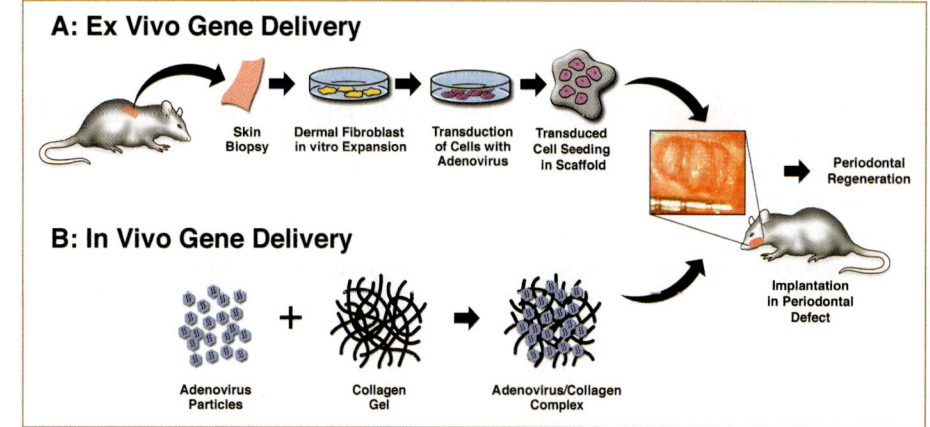

**FIG. 72.2.** Gene delivery approaches for periodontal-tissue engineering. **(A)** *Ex vivo* gene delivery involves the harvesting of tissue biopsies, expansion of cell populations, genetic manipulations of cells, and subsequent transplantation to periodontal osseous defects. **(B)** *In vivo* gene transfer involves the direct delivery of growth factor transgenes to the periodontal osseous defects. Reproduced from Ramseier *et al.* (2006) with the permission of Elsevier, Inc.

2002). By coordinating the timing of tooth and bone formation, a tooth–bone hybrid tissue joined by periodontal ligament was bioengineered recently (Young *et al.*, 2005). All of these have facilitated the idea of using MSCs to regenerate cementum, PDL, and bone lost as a result of advanced periodontitis. Kawaguchi *et al.* (2004) used *ex vivo* expanded BMSSC mixed with atelocollagen to regenerate experimental class III furcation defects in beagle dogs. One month after transplantation, the defects were almost completely covered with new cementum, regenerated periodontal ligament, and alveolar bone. In another study, cultured human PDLSCs were transplanted into surgical defects at the periodontal region of mandibular molars in immunocompromised rats (Seo *et al.*, 2004). PDL-like tissue was formed connecting alveolar bone and cementum surfaces. Thus it is promising that periodontal tissue could be regenerated by obtaining MSCs, which are then expanded *ex vivo* and transplanted to the areas where periodontal reconstruction is indicated.

## Limitations of Topical Growth Factor Administration

Topical delivery of growth factors to periodontal wounds has shown promising results, yet insufficient for the stimulation of consistent periodontal tissue engineering (Cochran and Wozney, 1999; Camelo *et al.*, 2003). Growth factor proteins, once delivered to the target site, tend to suffer from labiality, due primarily to proteolytic breakdown, receptor-mediated endocytosis, and solubility of the carrier (Anusaksathien and Giannobile, 2002). Because the half-lives of GFs are transient, the period of exposure may not be sufficient to act on osteoblasts, cementoblasts, or PDL cells. Therefore, different methods of growth factor delivery need to be considered (Anusaksathien *et al.*, 2005).

Investigations for periodontal bioengineering have examined a variety of methods combining delivery vehicles, such as scaffolds, with growth factors to target the defect site in order to optimize bioavailability (Lutolf and Hubbell, 2005). The scaffolds are designed to optimize the dosage of the growth factor and to control its release pattern, which may be pulsatile, constant, or time programmed (Babensee *et al.*, 2000). Additionally, the kinetics of the release and the duration of the exposure of the growth factor may be controlled (Hutmacher *et al.*, 2000).

To address transient bioactivity of GF peptides *in vivo*, gene therapy using a vector encoding the growth factor has been utilized to stimulate tissue regeneration. So far, two main strategies of gene vector delivery have been applied to periodontal tissue engineering. Gene vectors can be introduced directly to the target site (*in vivo* approach) (Jin *et al.*, 2004), or selected cells can be harvested, expanded, genetically transduced, and then reimplanted (*ex vivo* technique) (Jin *et al.*, 2003) (Fig. 72.2). *In vivo* gene transfer involves the insertion of the gene of interest directly into the defect, promoting genetic modification of the target cell. *Ex vivo* gene transfer includes the incorporation of genetic material into cells exposed from a tissue biopsy, with subsequent reimplantation into the recipient.

## Gene Delivery for Periodontal Tissue Engineering

Gene transfer strategies for tissue engineering have demonstrated success in healing soft-tissue wounds such as skin lesions (Crombleholme, 2000). Both PDGF plasmid (Hijjawi *et al.*, 2004) and Adenovirus (Ad)/PDGF gene delivery (Printz *et al.*, 2000) has been evaluated in preclinical and human trials. However, the latter has been able to exhibit more robust results favorable for clinical use (Gu *et al.*, 2004).

Early studies in dental applications using recombinant adenoviral vectors encoding PDGF showed that vector constructs can transduce cells derived from the periodontium (osteoblasts, cementoblasts, PDL cells, and gingival fibroblasts) (Giannobile *et al.*, 2001; Zhu *et al.*, 2001). Additionally, Chen and Giannobile (2002) were able to demonstrate the sustained effects of adenoviral delivery of PDGF for a better understanding of extended PDGF signaling. In an *ex vivo* investigation by Anusaksathien *et al.* (2003), it was shown that the expression of PDGF genes was prolonged for up to 10 days in gingival wounds. Adenovirus encoding PDGF-B (Ad/PDGF-B) transduced gingival fibroblasts and enhanced defect fill by induction of human gingival fibroblast migration and proliferation. However, in another study, continuous exposure of cementoblasts to PDGF-A had an inhibitory effect on cementum mineralization, possibly via the up-regulation of osteopontin (OPN) and subsequent enhancement of multinucleated giant cells (MNGCs) in cementum-engineered scaffolds. Furthermore, Ad/PDGF-1308 (a dominant-negative mutant of PDGF) inhibited mineralization of tissue-engineered cementum, possibly due to down-regulation of bone sialoprotein (BSP) and osteocalcin (OC), with a persistence of stimulation of MNGCs. These findings suggest that continuous exogenous delivery of PDGF-A may delay mineral formation induced by cementoblasts, while PDGF is clearly required for mineral neogenesis (Anusaksathien *et al.*, 2004).

Jin *et al.* (2004) demonstrated that direct *in vivo* gene transfer of PDGF-B stimulated tissue regeneration in large periodontal defects. Descriptive histology and histomorphometry revealed that human PDGF-B gene delivery promotes the regeneration of both cementum and alveolar bone, while PDGF-1308, has minimal effects on periodontal-tissue regeneration.

Jin *et al.* (2003) demonstrated that *ex vivo* delivery of BMP7-transduced fibroblasts in a gelatin matrix promoted periodontal tissue repair in large mandibular periodontal bone defects. BMP-7 gene transfer not only enhanced alveolar bone repair, it also stimulated cementogenesis and PDL fiber formation. Of interest, the alveolar bone formation was found to occur via a cartilage intermediate. However, when genes encoding the BMP antagonist noggin were delivered, inhibition of periodontal tissue formation resulted (Jin *et al.*, 2004). Dunn *et al.* (2005) recently demonstrated that direct *in vivo* gene delivery of BMP-7 in a collagen matrix promoted regeneration of extraction socket bone defects around dental implants. These experiments provide promising evidence showing the feasibility of both *in vivo* and *ex vivo* gene therapy for periodontal tissue regeneration and peri-implant osseointegration.

## Scaffold Materials for Tissue Engineering of Periodontal Tissues

Scaffolds are three-dimensional structures that provide an environment for healing and/or regeneration of damaged tissues and therefore have gained attention as attractive candidates for use in repair/regeneration of tissues (Nakashima and Akamine, 2005). Today, the choice of scaffold material has become wide-ranging in terms of chemical formulations and material properties (Garg, 1999; Reynolds *et al.*, 2003; Branfoot, 2005; Taba *et al.*, 2005). Regardless of the scaffold material selected, there are basic requirements these materials must satisfy in order to achieve successful tissue repair. They must be biocompatible, able to support

cell adhesion/differentiation/proliferation, and resorbable by the body. Those designed for delivery of biological mediators, cells/genes/proteins, to the wound site must be sustained at the site for a sufficient time to interact with cells in the local environment. Scaffold materials that have applicability to periodontal tissue repair include human bone derivatives, bioceramics, synthetic polymers, and composites thereof (Caffesse *et al.*, 2002; Reynolds *et al.*, 2003; Wang *et al.*, 2005). Please refer to Table 72.2 for a summary of current applications of scaffold materials in periodontal treatment therapies. For the most part, clinically accepted scaffold materials for treating periodontal diseases are those that perform well in promoting bone regeneration. However, the success of these materials in regenerating periodontal ligaments, bone, and cementum remains limited; furthermore, outcomes are not predictable (Caffesse *et al.*, 2002; Wang *et al.*, 2005). The next generation of scaffold materials for regenerating the entire periodontium may require that these materials have the capacity to deliver appropriate biological mediators in a specified temporal-spatial fashion.

### Human Bone Derivatives

Bone derivatives include autografts and allografts. Autografts, obtained from the same individual, are used to augment bone as needed for implants and also at sites of periodontal defects. While good results are achieved, especially for bone/ridge augmentation, the procedures can be invasive and require a healing phase and a second surgery (Garg, 1999; Wang *et al.*, 2005). Allografts are human bones that may come in powdered form or in fragments of cortical, cancellous, and cortical-cancellous bone of various sizes. Because allografts are osteoconductive and, properly sterilized, pose few adverse effects on immuno-response, they are widely used for regenerating lost bone tissues resulting from periodontal diseases. Clinical applications using commercial allografts include fenestrations, periodontal defects, and ridge augmentations.

Allografts can also be used in demineralized form, also known as demineralized bone matrix, where the mineral is dissolved away, leaving behind collagen, BMPs, and noncollagenous proteins. New bone is generated around demineralized allografts by osteoinduction, considered to be due to the presence of BMPs. While allografts have gained clinical acceptance, the results are neither consistent nor sufficient for the replacement of large defects.

### Bioceramics

Calcium phosphate, mainly as tricalcium phosphate (TCP) and hydroxyapatite (HA), have been used for bone regeneration applications because they are highly osteoconductive materials. Tricalcium phosphate is more readily resorbed than hydroxyapatite (Garg, 1999). Studies have established pure TCP or biphasic TCP/HA as viable graft

| Table 72.2. | Summary of scaffold materials being used/evaluated for treating periodontal defects | | |
|---|---|---|---|
| *Scaffold material* | *Chemical constituents* | *Characteristics favorable for regeneration therapy* | *Treatments* |
| Autografts | Calcium phosphate (HA), collagen, growth factors (e.g., BMP), etc. | Highly osteogenic | Osseous defects |
| Allografts (in combinations) | | | |
| Freeze-dried bone | May contain calcium phosphate (HA), collagen, growth factors (e.g., BMP), etc. | Osteoconductive, osteoinductive | Osseous defects |
| Demineralized freeze-dried bone | May contain collagen, growth factors (e.g., BMP), etc. | Osteoconductive | Osseous defects |
| Xenografts | Calcium phosphate (HA), some with added factors (e.g., synthetic cell-binding peptides) | Osteoconductive, high porosity, allowing tissue ingrowth | Osseous defects |
| Tricalcium phosphate (TCP) or biphasic TCP/HA (BCP) | Calcium phosphate: TCP (and HA) | Osteoconductive, resorbable | Osseous defects |
| Coral-derived IIA | Calcium phosphate (IIA) | Osteoconductive, high porosity, allowing tissue ingrowth | Osseous defects |
| Coral | Calcium carbonate (aragonite) | Osteoconductive, resorbable, high porosity, allowing tissue ingrowth | Osseous defects |
| Bioglass | Calcium, phosphorous, sodium, silicon, oxygen | Osteoconductive, resorbable | Osseous defects, ridge augmentation |
| Resorbable polymers | Polyglycolic acid, polylactic acid, copolymers thereof, some with added factors (e.g., growth factors, BMPs) | Resorbable, can tailor porosity and morphology such as nanofibers | Osseous defects and soft-tissue grafts |

materials for use in treating periodontal osseous defects (Garg, 1999). Bovine-derived hydroxyapatite can be fabricated into low-porosity blocks and granules by sintering, or into high-porosity structure by a low-heat alkaline extraction process, also known as xenografts (Caffesse *et al.*, 2002). Since the low-porosity bulk structure poses poor attachment and poor mechanical properties (brittleness), it is not widely applied for treating periodontal defects. Even though the issue of embrittlement is less severe in the granule analog and can be used for ridge augmentation, more porous structures are preferred for periodontal regenerative treatment, including ridge augmentation and osseous defects. The high porosity appears to be the key to an effective medium for cell/bone ingrowth and integration. In addition, porous hydroxyapatite has been studied as a scaffold material for use in guided tissue regeneration (to exclude epithelial downgrowth).

Using collagen and other barrier membranes in conjunction with porous xenografts and other graft materials

has yielded some success, but with variable results, depending on the type of defect (Caffesse *et al.*, 2002; Reynolds *et al.*, 2003; Eppley *et al.*, 2005; Wang *et al.*, 2005). Further, there are mixed reports as to whether or not results are improved when both a graft and a membrane are used vs. the use of either one alone (Reynolds *et al.*, 2003; Hartman *et al.*, 2004; Sculean *et al.*, 2005; Wang *et al.*, 2005).

Porous scaffolds have also been fabricated from corals. Corals skeletons, which are made of highly porous calcium carbonate, can be converted to hydroxyapatite via high-temperature treatment. The conversion process retains the porous structure, which enables bone ingrowth similar to that of porous xenograft. Alternatively, corals can be used in their native calcium carbonate form as a scaffold for use in bone regeneration. The calcium carbonate is highly resorbable and is readily osseointegrated. As a result, coral scaffolds have broad clinical applications for use in bone regeneration such as spinal disc fusions and regeneration of femur bone, as well as periodontal osseous defects (Eppley *et al.*, 2005). Results for periodontal defects are considered comparable to those for other allograft/alloplast materials (Garg, 1999). Further, applying coral scaffolds to guided tissue regeneration indicated enhancement of new bone formation, though it is unclear if this enhancement was due to the osteoconductivity of the coral scaffold (Reynolds *et al.*, 2003; Wikesjo *et al.*, 2003; Koo *et al.*, 2005).

Another class of bioceramics is bioactive glass, which is composed of calcium, phosphate, sodium, and silicon. Bioactive glass is not porous and therefore is not susceptible to any tissue ingrowth. However, its high-bioactivity surface is favorable for cell and tissue attachment. Commonly introduced to the wound site as particulates, the glass surface is readily colonized by osteogenic cells and then by a collagen network attaching the particulate to the surrounding tissue, enabling further bone growth and attachment on the particulates. Bioactive glass has been reported to be of value in treating intrabony defects (Garg, 1999; Park *et al.*, 2001).

### Synthetic Polymers

The most common polymer scaffold materials are polyglycolic acid (PGA), polylactic acid (PLA), and their copolymers (PLGA). These polymers are resorbable, with the resorption rate controllable by varying the molecular weight (Wu, 1995). Because of the ability to alter properties associated with biodegradability and porosity, polymers have a vast range of clinical applications, from use as barrier membranes in guided tissue regeneration to use as delivery systems for release of factors and/or cells to periodontal defects (Abukawa *et al.*, 2006). *In vitro* and *in vivo* studies using SCID mice have shown that PLGA is a viable scaffold material to deliver cementoblast cells as a means to regenerate periodontal tissues (Jin *et al.*, 2003; Zhao *et al.*, 2004). Similarly, stem cell studies such as those described earlier have used PLGA exclusively as the scaffold on which cells were seeded (Abukawa *et al.*, 2004; Young *et al.*, 2005). Furthermore, approaches to utilize PLGA as time-controlled release of growth factors are in active investigation. A study by Wei and coworkers (2006) demonstrated that the release of PDGF-BB could be controlled temporally by encapsulating PDGF-BB in PLGA microspheres, which in turn entrapped in a nanofibrous poly(L-lactic acid) (PLLA) scaffold. Sustained release of bioactive PDGF-BB was shown to be controllable from days to months by varying the molecular weight of the microspheres (Wei *et al.*, 2006).

In addition to PLA-PGA polymer systems, other formulations have shown potential as scaffold systems for use in periodontal tissue regeneration. A study by Nakahara *et al.* (2003) demonstrated successful fabrication of a collagen scaffold construct that sandwiched basic fibroblast growth factor (bFGF)-containing gelatin microspheres within the collagen scaffold and applied this to periodontal defects of beagle dogs. The results indicated formation of new bone, cementum, and PDL. In another approach, Hasegawa *et al.* (2005), using an athymic rat periodontal defect model, reported success in regenerating new bone, cementum, and a functional PDL by delivery of PDL cell sheets to the defective sites. The cell sheets were developed *in vitro* by applying a unique method to control cell adhesion by modifying culture temperature conditions and using a surface-grafted temperature-response polymer, poly(*N*-isopropylacrylamide) (PIPAAm).

The future of successful periodontal tissue engineering may very well lie in the ability to utilize effectively these biodegradable polymers as delivery vehicles for appropriate cells and/or factors. Improved delivery systems coupled with enhanced understanding of the mechanistic/biological regulators of tissue formation should result in better and more predictable treatment therapies within the next decade.

## IV. FUTURE DIRECTIONS

The emerging new technologies and major advances in our understanding of the cell/molecular factors controlling the function of cells within the periodontium, linked with improvements in scaffold design for local controlled delivery, and a new appreciation for the capacity of adult cells to function as stem cells have resulted in improvements in our approach to treating periodontal diseases. Nevertheless, predictable regenerative therapies have not kept up with the advances in implant therapy. As a consequence, it is often more attractive to replace severely damaged periodontal tissues with implants than to attempt to regenerate lost structures. Implants are not always the best treatment, for many reasons, among them location, inadequate bone, and lack of a PDL. Thus, ideally, one would like to be able to restore the "natural" tooth.

This chapter has focused on restoring periodontal tissues as one aspect of the long-term goal of restoring the whole tooth. The goal of predictably engineering periodontal tissues requires a strong collaboration amongst a multitude of disciplines, e.g., cell/molecular biologists, computational scientists, bioengineers, geneticists, chemists, and clinicians. The various pieces of information that are missing include (a) detailed information on the stem cell properties of periodontal tissues; (b) identifying the trigger factors, likely to be a well-orchestrated delivery of several factors over a period of time; (c) designing scaffolds that can deliver these factors over time; and (d) recognizing variability in host response to factors/cells. Advances in all of these areas are moving forward and have already resulted in improved approaches for restoring lost periodontal tissues.

# V. ACKNOWLEDGMENTS

The authors would like to thank Luke Wallace for assisting in preparation of the figures and critical review of the chapter by Brian Foster and Eugenie Fyfe. Reviews are cited instead of individual papers in some areas due to the limited space. The studies were supported by the following NIH/NIDCR grants: DE13047(MJS), DE15109(MJS), DE016619(WVG), DE09532(MJS and WVG), and DE13397 (MJS and WVG).

# VI. REFERENCES

Aberg, T., Wozney, J., and Thesleff, I. (1997). Expression patterns of bone morphogenetic proteins (Bmps) in the developing mouse tooth suggest roles in morphogenesis and cell differentiation. *Dev. Dyn.* **210**(4), 383–396.

Abukawa, H., Shin, M., Williams, W. B., Vacanti, J. P., Kaban L. B., and Troulis, M. J. (2004). Reconstruction of mandibular defects with autologous tissue-engineered bone. *J. Oral Maxillofac. Surg.* **62**(5), 601–606.

Abukawa, H., Papadaki, M., Abulikemu, M., Leaf, J., Vacanti, J. P., Kaban, L. B., and Troulis, M. J. (2006). The engineering of craniofacial tissues in the laboratory: a review of biomaterials for scaffolds and implant coatings. *Dent. Clin. North. Am.* **50**(2), 205–216.

Albandar, J. M., Brunelle, J. A., and Kingman, A. (1999). Destructive periodontal disease in adults 30 years of age and older in the United States, 1988–1994. *J. Periodontol.* **70**(1), 13–29.

Amar, S., Chung, K. M., Nam, S. H., Karatzas, S., Myokai, F., and Van Dyke, T. E. (1997). Markers of bone and cementum formation accumulate in tissues regenerated in periodontal defects treated with expanded polytetrafluoroethylene membranes. *J. Periodontal. Res.* **32**(1 Pt. 2), 148–158.

Anusaksathien, O., and Giannobile, W. V. (2002). Growth factor delivery to re-engineer periodontal tissues. *Curr. Pharm. Biotechnol.* **3**(2), 129–139.

Anusaksathien, O., Webb, S. A., Jin, Q. M., and Giannobile, W. V. (2003). Platelet-derived growth factor gene delivery stimulates *ex vivo* gingival repair. *Tissue Eng.* **9**(4), 745–756.

Anusaksathien, O., Jin, Q., Zhao, M., Somerman, M. J., and Giannobile, W. V. (2004). Effect of sustained gene delivery of platelet-derived growth factor or its antagonist (PDGF-1308) on tissue-engineered cementum. *J. Periodontol.* **75**(3), 429–440.

Anusaksathien, O., Jin, Q., and Ma, P. X. (2005). Scaffolding in periodontal engineering. *In* "Scaffolding in Tissue Engineering" (P. X. Ma and J. Eliseeff, eds.), p. 427. CRC Press, Boca Raton, FL.

Azizi, S. A., Stokes, D., Augelli, B. J., DiGirolamo, C., and Prockop, D. J. (1998). Engraftment and migration of human bone marrow stromal cells implanted in the brains of albino rats — similarities to astrocyte grafts. *Proc. Natl. Acad. Sci. U.S.A.* **95**(7), 3908–3913.

Babensee, J. E., McIntire, L. V., and Mikos, A. G. (2000). Growth factor delivery for tissue engineering. *Pharm. Res.* **17**(5), 497–504.

Barros, S. P., Silva, M. A., Somerman, M. J., and Nociti, Jr., F. H. (2003). Parathyroid hormone protects against periodontitis-associated bone loss. *J. Dent. Res.* **82**(10), 791–795.

Bartlett, J. D., Ganss, B., Goldberg, M., Moradian-Oldak, J., Paine, M. L., Snead, M. L., Wen, X., White, S. N., and Zhou, Y. L. (2006). Protein–protein interactions of the developing enamel matrix. *Curr. Top. Dev. Biol.* **74**, 57–115.

Bartold, P. M., McCulloch, C. A., Narayanan, A. S., and Pitaru, S. (2000). Tissue engineering: a new paradigm for periodontal regeneration based on molecular and cell biology. *Periodontol 2000* **24**, 253–269.

Bayne, S. C. (2005). Dental biomaterials: where are we and where are we going? *J. Dent. Educ.* **69**(5), 571–585.

Boabaid, F., Berry, J. E., Koh, A. J., Somerman, M. J., and McCauley, L. K. (2004). The role of parathyroid hormone–related protein in the regulation of osteoclastogenesis by cementoblasts. *J. Periodontol.* **75**(9), 1247–1254.

Bosshardt, D. D. (2005). Are cementoblasts a subpopulation of osteoblasts or a unique phenotype? *J. Dent. Res.* **84**(5), 390–406.

Bosshardt, D. D., and Nanci, A. (2004). Hertwig's epithelial root sheath, enamel matrix proteins, and initiation of cementogenesis in porcine teeth. *J. Clin. Periodontol.* **31**(3), 184–192.

Branfoot, T. (2005). Research directions for bone healing. *Injury Int. J. Care Injured.* **36S**, S51–S54.

Brent, A. E., Schweitzer, R., and Tabin, C. J. (2003). A somitic compartment of tendon progenitors. *Cell* **113**(2), 235–248.

Caffesse, R. G., de la Rosa, M., and Mota, L. F. (2002). Regeneration of soft- and hard-tissue periodontal defects. *Am. J. Dent.* **15**(5), 339–345.

Camelo, M., Nevins, M. L., Schenk, R. K., Lynch, S. E., and Nevins, M. (2003). Periodontal regeneration in human Class II furcations using purified recombinant human platelet-derived growth factor-BB (rhPDGF-BB) with bone allograft. *Int. J. Periodontics Restorative Dent.* **23**(3), 213–225.

Castro-Malaspina, H., Gay, R. E., Resnick, G., Kapoor, N., Meyers, P., Chiarieri, D., McKenzie, S., Broxmeyer, H. E., and Moore, M. A. (1980). Characterization of human bone marrow fibroblast colony-forming cells (CFU-F) and their progeny. *Blood* **56**(2), 289–301.

Chen, Q. P., and Giannobile, W. V. (2002). Adenoviral gene transfer of PDGF downregulates gas gene product PDGFalphaR and prolongs ERK and Akt/PKB activation. *Am. J. Physiol. Cell. Physiol.* **282**(3), C538–C544.

Cochran, D. L., and Wozney, J. M. (1999). Biological mediators for periodontal regeneration. *Periodontol 2000* **19**, 40–58.

Crombleholme, T. M. (2000). Adenoviral-mediated gene transfer in wound healing. *Wound Repair Regen* **8**(6), 460–472.

Dunn, C. A., Jin, Q., Taba, Jr., M., Franceschi, R. T., Bruce Rutherford, R., and Giannobile, W. V. (2005). BMP gene delivery for alveolar bone engineering at dental implant defects. *Mol. Ther.* **11**(2), 294–299.

Eppley, B. L., Pietrzak, W. S., and Blanton, M. W. (2005). Allograft and alloplastic bone substitutes: a review of science and technology for the craniomaxillofacial surgeon. *J. Craniofac. Surg.* **16**(6), 981–989.

Esposito, M., Coulthard, P., and Worthington, H. V. (2003). Enamel matrix derivative (Emdogain) for periodontal-tissue regeneration in intrabony defects. *Cochrane Database Syst. Rev.* (2), CD003875.

Feng, J. Q., Luan, X., Wallace, J., Jing, D., Ohshima, T., Kulkarni, A. B., D'Souza, R. N., Kozak, C. A., and MacDougall, M. (1998). Genomic organization, chromosomal mapping, and promoter analysis of the mouse dentin sialophosphoprotein (Dspp) gene, which codes for both dentin sialoprotein and dentin phosphoprotein. *J. Biol. Chem.* **273**(16), 9457–9464.

Fiaschi-Taesch, N. M., and Stewart, A. F. (2003). Minireview: parathyroid hormone–related protein as an intracrine factor — trafficking mechanisms and functional consequences. *Endocrinology* **144**(2), 407–411.

Fiorellini, J. P., Howell, T. H., Cochran, D., Malmquist, J., Lilly, L. C., Spagnoli, D., Toljanic, J., Jones, A., and Nevins, M. (2005). Randomized study evaluating recombinant human bone morphogenetic protein-2 for extraction socket augmentation. *J. Periodontol.* **76**(4), 605–613.

Fong, H. K., Foster, B. L., Popowics, T. E., and Somerman, M. J. (2005). The crowning achievement: getting to the root of the problem. *J. Dent. Educ.* **69**(5), 555–570.

Fuchs, J. R., Hannouche, D., Terada, S., Vacanti, J. P., and Fauza, D. O. (2003). Fetal tracheal augmentation with cartilage engineered from bone marrow–derived mesenchymal progenitor cells. *J. Pediatr. Surg.* **38**(6), 984–987.

Garg, A. K. (1999). Grafting materials in repair and restoration. *In* "Tissue Engineering: Applications in Maxillofacial Surgery and Periodontics" (S. E. Lynch, R. J. Genco, and R. E. Marx, eds.), pp. 83–101. Quintessence, Chicago.

Gault, P. C., and Warocquier-Clerout, R. (2002). Tooth autotransplantation with double periodontal ligament stimulation to replace periodontally compromised teeth. *J. Periodontol.* **73**(5), 575–583.

Giannobile, W. V. (1996). Periodontal tissue engineering by growth factors. *Bone* **19**(1 Suppl.), 23S–37S.

Giannobile, W. V., Finkelman, R. D., and Lynch, S. E. (1994). Comparison of canine and nonhuman primate animal models for periodontal regenerative therapy: results following a single administration of PDGF/IGF-I. *J. Periodontol.* **65**(12), 1158–1168.

Giannobile, W. V., Hernandez, R. A., Finkelman, R. D., Ryan, S., Kiritsy, C. P., D'Andrea, M., and Lynch, S. E. (1996). Comparative effects of platelet-derived growth factor-BB and insulin-like growth factor-I, individually and in combination, on periodontal regeneration in *Macaca fascicularis*. *J. Periodontal. Res.* **31**(5), 301–312.

Giannobile, W. V., Ryan, S., Shih, M. S., Su, D. L., Kaplan, P. L., and Chan, T. C. (1998). Recombinant human osteogenic protein-1 (OP-1) stimulates periodontal wound healing in class III furcation defects. *J. Periodontol.* **69**(2), 129–137.

Giannobile, W. V., Lee, C. S., Tomala, M. P., Tejeda, K. M., and Zhu, Z. (2001). Platelet-derived growth factor (PDGF) gene delivery for application in periodontal-tissue engineering. *J. Periodontol.* **72**(6), 815–823.

Giannobile, W. V., and Somerman, M. J. (2003). Growth and amelogenin-like factors in periodontal wound healing. A systematic review. *Ann. Periodontol.* **8**(1), 193–204.

Gibson, C. W., Yuan, Z. A., Hall, B., Longenecker, G., Chen, E., Thyagarajan, T., Sreenath, T., Wright, J. T., Decker, S., Piddington, R., *et al.* (2001). Amelogenin-deficient mice display an amelogenesis imperfecta phenotype. *J. Biol. Chem.* **276**(34), 31871–31875.

Gojo, S., Gojo, N., Takeda, Y., Mori, T., Abe, H., Kyo, S., Hata, J., and Umezawa, A. (2003). *In vivo* cardiovasculogenesis by direct injection of isolated adult mesenchymal stem cells. *Exp. Cell. Res.* **288**(1), 51–59.

Gould, T. R., Melcher, A. H., and Brunette, D. M. (1980). Migration and division of progenitor cell populations in periodontal ligament after wounding. *J. Periodontal Res.* **15**(1), 20–42.

Gronthos, S., Mankani, M., Brahim, J., Robey, P. G., and Shi, S. (2000). Postnatal human dental pulp stem cells (DPSCs) *in vitro* and *in vivo*. *Proc. Natl. Acad. Sci. U.S.A.* **97**(25), 13625–13630.

Gu, D. L., Nguyen, T., Gonzalez, A. M., Printz, M. A., Pierce, G. F., Sosnowski, B. A., Phillips, M. L., and Chandler, L. A. (2004). Adenovirus encoding human platelet-derived growth factor-B delivered in collagen exhibits safety, biodistribution, and immunogenicity profiles favorable for clinical use. *Mol. Ther.* **9**(5), 699–711.

Haase, H. R., Clarkson, R. W., Waters, M. J., and Bartold, P. M. (1998). Growth factor modulation of mitogenic responses and proteoglycan synthesis by human periodontal fibroblasts. *J. Cell Physiol.* **174**(3), 353–361.

Hakki, S. S., Berry, J. E., and Somerman, M. J. (2001). The effect of enamel matrix protein derivative on follicle cells *in vitro*. *J. Periodontol.* **72**(5), 679–687.

Hartman, G. A., Arnold, R. M., Mills, M. P., Cochran, D. L., and Mellonig, J. T. (2004). Clinical and histologic evaluation of an organic bovine bone collagen with or without a collagen barrier. *Int. J. Periodontics. Restorative Dent.* **24**(2), 127–135.

Hasegawa, N., Kawaguchi, H., Ogawa, T., Uchida, T., and Kurihara, H. (2003). Immunohistochemical characteristics of epithelial cell rests of Malassez during cementum repair. *J. Periodontal Res.* **38**(1), 51–56.

Hasegawa, M., Yamato, M., Kikuchi, A., Okano, T., and Ishikawa, I. (2005). Human periodontal ligament cell sheets can regenerate periodontal ligament tissue in an athymic rat model. *Tissue Eng.* **11**(3–4), 469–478.

Hatakeyama, J., Sreenath, T., Hatakeyama, Y., Thyagarajan, T., Shum, L., Gibson, C. W., Wright, J. T., and Kulkarni, A. B. (2003). The receptor activator of nuclear factor-kappa B ligand-mediated osteoclastogenic pathway is elevated in amelogenin-null mice. *J. Biol. Chem.* **278**(37), 35743–35748.

Hatakeyama, J., Philp, D., Hatakeyama, Y., Haruyama, N., Shum, L., Aragon, M. A., Yuan, Z., Gibson, C. W., Sreenath, T., Kleinman, H. K., *et al.* (2006). Amelogenin-mediated regulation of osteoclastogenesis, and periodontal cell proliferation and migration. *J. Dent. Res.* **85**(2), 144–149.

Heden, G., and Wennstrom, J. L. (2006). Five-year follow-up of regenerative periodontal therapy with enamel matrix derivative at sites with angular bone defects. *J. Periodontol.* **77**(2), 295–301.

Heldin, P., Laurent, T. C., and Heldin, C. H. (1989). Effect of growth factors on hyaluronan synthesis in cultured human fibroblasts. *Biochem. J.* **258**(3), 919–922.

Hijjawi, J., Mogford, J. E., Chandler, L. A., Cross, K. J., Said, H., Sosnowski, B. A., and Mustoe, T. A. (2004). Platelet-derived growth factor B, but not fibroblast growth factor 2, plasmid DNA improves survival of ischemic myocutaneous flaps. *Arch. Surg.* **139**(2), 142–147.

Hofstetter, C. P., Schwarz, E. J., Hess, D., Widenfalk, J., El Manira, A., Prockop, D. J., and Olson, L. (2002). Marrow stromal cells form guiding strands in the injured spinal cord and promote recovery. *Proc. Natl. Acad. Sci. U.S.A.* **99**(4), 2199–2204.

Hutmacher, D. W., Teoh, S. H., Zein, I., Ranawake, M., and Lau, S. (2000). Tissue-engineering research: the engineer's role. *Med. Device Technol.* **11**(1), 33–39.

Jin, Q. M., Anusaksathien, O., Webb, S. A., Rutherford, R. B., and Giannobile, W. V., (2003a). Gene therapy of bone morphogenetic protein for periodontal-tissue engineering. *J. Periodontol.* **74**(2), 202–213.

Jin, Q. M., Zhao, M., Webb, S. A., Berry, J. E., Somerman, M. J., and Giannobile, W. V., (2003b). Cementum engineering with three-dimensional polymer scaffolds. *J. Biomed. Mater. Res. A* **67**(1), 54–60.

Jin, Q. M., Anusaksathien, O., Webb, S. A., Printz, M. A., and Giannobile, W. V., (2004a). Engineering of tooth-supporting structures by delivery of PDGF gene therapy vectors. *Mol. Ther.* **9**(4), 519–526.

Jin, Q. M., Zhao, M., Economides, A. N., Somerman, M. J., and Giannobile, W. V., (2004b). Noggin gene delivery inhibits cementoblast-induced mineralization. *Connect. Tissue Res.* **45**(1), 50–59.

Kallu, R., Vinckier, F., Politis, C., Mwalili, S., and Willems, G. (2005). Tooth transplantations: a descriptive retrospective study. *Int. J. Oral Maxillofac. Surg.* **34**(7), 745–755.

Kaplan, D. R., Chao, F. C., Stiles, C. D., Antoniades, H. N., and Scher, C. D. (1979). Platelet alpha granules contain a growth factor for fibroblasts. *Blood* **53**(6), 1043–1052.

Kawaguchi, H., Hirachi, A., Hasegawa, N., Iwata, T., Hamaguchi, H., Shiba, H., Takata, T., Kato, Y., and Kurihara, H. (2004). Enhancement of periodontal-tissue regeneration by transplantation of bone marrow mesenchymal stem cells. *J. Periodontol.* **75**(9), 1281–1287.

Koo, K. T., Polimeni, G., Qahash, M., Kim, C. K., and Wikesjo, U. M. E. (2005). Periodontal repair in dogs: guided tissue regeneration enhances bone formation in sites implanted with a coral-derived calcium carbonate biomaterial. *J. Clin. Periodontol.* **32**(1), 104–110.

Krebsbach, P. H., Mankani, M. H., Satomura, K., Kuznetsov, S. A., and Robey, P. G. (1998). Repair of craniotomy defects using bone marrow stromal cells. *Transplantation* **66**(10), 1272–1278.

Lutolf, M. P., and Hubbell, J. A. (2005). Synthetic biomaterials as instructive extracellular microenvironments for morphogenesis in tissue engineering. *Nat. Biotechnol.* **23**(1), 47–55.

Lutolf, M. P., Weber, F. E., Schmoekel, H. G., Schense, J. C., Kohler, T., Muller, R., and Hubbell, J. A. (2003). Repair of bone defects using synthetic mimetics of collagenous extracellular matrices. *Nat. Biotechnol.* **21**(5), 513–518.

Lynch, S. E., Williams, R. C., Polson, A. M., Howell, T. H., Reddy, M. S., Zappa, U. E., and Antoniades, H. N. (1989). A combination of platelet-derived and insulin-like growth factors enhances periodontal regeneration. *J. Clin. Periodontol.* **16**(8), 545–548.

Lynch, S. E., de Castilla, G. R., Williams, R. C., Kiritsy, C. P., Howell, T. H., Reddy, M. S., and Antoniades, H. N. (1991). The effects of short-term application of a combination of platelet-derived and insulin-like growth factors on periodontal wound healing. *J. Periodontol.* **62**(7), 458–467.

MacDougall, M., Simmons, D., Luan, X., Nydegger, J., Feng, J., and Gu, T. T. (1997). Dentin phosphoprotein and dentin sialoprotein are cleavage products expressed from a single transcript coded by a gene on human chromosome 4. Dentin phosphoprotein DNA sequence determination. *J. Biol. Chem.* **272**(2), 835–842.

MacNeil, R. L., and Somerman, M. J. (1999). Development and regeneration of the periodontium: parallels and contrasts. *Periodontol. 2000* **19**, 8–20.

Marcopoulou, C. E., Vavouraki, H. N., Dereka, X. E., and Vrotsos, I. A. (2003). Proliferative effect of growth factors TGF-beta1, PDGF-BB and rhBMP-2 on human gingival fibroblasts and periodontal ligament cells. *J. Int. Acad. Periodontol.* **5**(3), 63–70.

McCulloch, C. A. (1985). Progenitor cell populations in the periodontal ligament of mice. *Anat. Rec.* **211**(3), 258–262.

McCulloch, C. A. (1995). Origins and functions of cells essential for periodontal repair: the role of fibroblasts in tissue homeostasis. *Oral Dis.* **1**(4), 271–278.

Melcher, A. H. (1985). Cells of periodontium: their role in the healing of wounds. *Ann. R. Coll. Surg. Engl.* **67**(2), 130–131.

Miura, M., Gronthos, S., Zhao, M., Lu, B. Fisher, L. W., Robey, P. G., and Shi, S. (2003). SHED: stem cells from human exfoliated deciduous teeth. *Proc. Natl. Acad. Sci. U.S.A.* **100**(10), 5807–5812.

Miura, M., Chen, X. D., Allen, M. R., Bi, Y., Gronthos, S., Seo, B. M., Lakhani, S., Flavell, R. A., Feng, X. H., Robey, P. G., *et al.* (2004). A crucial role of caspase-3 in osteogenic differentiation of bone marrow stromal stem cells. *J. Clin. Invest.* **114**(12), 1704–1713.

Nakahara, T., Nakamura, T., Kobayashi, E., Inoue, M., Shigeno, K., Tabata, Y., Eto, K., and Shimizu, Y. (2003). Novel approach to regeneration of periodontal tissues based on *in situ* tissue engineering: effects of controlled release of basic fibroblast growth factor from a sandwich membrane. *Tissue Eng.* **9**(1), 153–162.

Nakashima, M., and Akamine, A. (2005). The application of tissue engineering to regeneration of pulp and dentin in endodontics. *J. Endod.* **31**(10), 711–718.

Nakashima, M., and Reddi, A. H. (2003). The application of bone morphogenetic proteins to dental-tissue engineering. *Nat. Biotechnol.* **21**(9), 1025–1032.

Nanci, A., and Bosshardt, D. D. (2006). Structure of periodontal tissues in health and disease. *Periodontol. 2000* **40**, 11–28.

Nishimura, F., and Terranova, V. P. (1996). Comparative study of the chemotactic responses of periodontal ligament cells and gingival fibroblasts to polypeptide growth factors. *J. Dent. Res.* **75**(4), 986–992.

Nociti, F. H., Jr., Berry, J. E., Foster, B. L., Gurley, K. A., Kingsley, D. M., Takata, T., Miyauchi, M., and Somerman, M. J. (2002). Cementum: a phosphate-sensitive tissue. *J. Dent. Res.* **81**(12), 817–821.

Oates, T. W., Rouse, C. A., and Cochran, D. L. (1993). Mitogenic effects of growth factors on human periodontal ligament cells *in vitro*. *J. Periodontol.* **64**(2), 142–148.

Okuda, K., Kawase, T., Momose, M., Murata, M., Saito, Y., Suzuki, H. Wolff, L. F., and Yoshie, H. (2003). Platelet-rich plasma contains high levels of platelet-derived growth factor and transforming growth factor-beta and modulates the proliferation of periodontally related cells *in vitro*. *J. Periodontol.* **74**(6), 849–857.

Paine, M. L., and Snead, M. L. (2005). Tooth developmental biology: disruptions to enamel-matrix assembly and its impact on biomineralization. *Orthod. Craniofac. Res.* **8**(4), 239–251.

Paine, M. L., Wang, H. J., Luo, W., Krebsbach, P. H., and Snead, M. L. (2003). A transgenic animal model resembling amelogenesis imperfecta related to ameloblastin overexpression. *J. Biol. Chem.* **278**(21), 19447–19452.

Park, J. S., Suh, J. J., Choi, S. H., Moon, I. S., Cho, K. S., Kim, C. K., and Chai, J. K. (2001). Effects of pretreatment clinical parameters on bioactive glass implantation in intrabony periodontal defects. *J. Periodontol.* **72**(6), 730–740.

Parkar, M. H., Kuru, L., Giouzeli, M., and Olsen, I. (2001). Expression of growth-factor receptors in normal and regenerating human periodontal cells. *Arch. Oral Biol.* **46**(3), 275–284.

Printz, M. A., Gonzalez, A. M., Cunningham, M., Gu, D. L., Ong, M. Pierce, G. F., and Aukerman, S. L. (2000). Fibroblast growth factor 2–retargeted adenoviral vectors exhibit a modified biolocalization pattern and display reduced toxicity relative to native adenoviral vectors. *Hum. Gene Ther.* **11**(1), 191–204.

Ramseier, C. A., Abramson, Z. R., Jin, Q., and Giannobile, W. V. (2006). Gene therapeutics for periodontal regenerative medicine. *Dent. Clin. North Am.* **50**(2), 245–263.

Reddi, A. H. (1998). Role of morphogenetic proteins in skeletal-tissue engineering and regeneration. *Nat. Biotechnol.* **16**(3), 247–252.

Reddi, A. H. (2001). Bone morphogenetic proteins: from basic science to clinical applications. *J. Bone Joint Surg. Am.* **83A**(Suppl. 1, Pt. 1), S1–S6.

Reichenberger, E., Tiziani, V., Watanabe, S., Park, L., Ueki, Y., Santanna, C., Baur, S. T., Shiang, R., Grange, D. K., Beighton, P., et al. (2001). Autosomal dominant craniometaphyseal dysplasia is caused by mutations in the transmembrane protein ANK. *Am. J. Hum. Genet.* **68**(6), 1321–1326.

Reynolds, M. A., Aichelmann-Reidy, M. E., Branch-Mays, G. L., and Gunsolley, J. C. (2003). The efficacy of bone replacement grafts in the treatment of periodontal osseous defects. A systematic review. *Ann. Periodontol.* **8**(1), 227–265.

Rosenkranz, S., and Kazlauskas, A. (1999). Evidence for distinct signaling properties and biological responses induced by the PDGF receptor alpha and beta subtypes. *Growth Factors* **16**(3), 201–216.

Rutherford, R. B., Niekrash, C. E., Kennedy, J. E., and Charette, M. F. (1992a). Platelet-derived and insulin-like growth factors stimulate regeneration of periodontal attachment in monkeys. *J. Periodontal. Res.* **27**(4 Pt. 1), 285–290.

Rutherford, R. B., Sampath, T. K., Rueger, D. C., and Taylor, T. D. (1992b). Use of bovine osteogenic protein to promote rapid osseointegration of endosseous dental implants. *Int. J. Oral Maxillofac. Implants* **7**(3), 297–301.

Rutsch, F., Vaingankar, S., Johnson, K., Goldfine, I., Maddux, B., Schauerte, P., Kalhoff, H., Sano, K., Boisvert, W. A., Superti-Furga, A., et al. (2001). PC-1 nucleoside triphosphate pyrophosphohydrolase deficiency in idiopathic infantile arterial calcification. *Am. J. Pathol.* **158**(2), 543–554.

Schantz, J. T., Hutmacher, D. W., Lam, C. X., Brinkmann, M., Wong, K. M., Lim, T. C., Chou, N., Guldberg, R. E., and Teoh, S. H. (2003). Repair of calvarial defects with customized tissue-engineered bone grafts II. Evaluation of cellular efficiency and efficacy *in vivo*. *Tissue Eng.* **9**(Suppl. 1), S127–S139.

Sculean, A., Stavropoulos, A., Berakdar, M., Windisch, P., Karring, T., and Brecx, M. (2005). Formation of human cementum following different modalities of regenerative therapy. *Clin. Oral Invest.* **9**(1), 58–64.

Seo, B. M., Miura, M., Gronthos, S., Bartold, P. M., Batouli, S., Brahim, J., Young, M. Robey, P. G., Wang, C. Y., and Shi, S. (2004). Investigation of multipotent postnatal stem cells from human periodontal ligament. *Lancet* **364**(9429), 149–155.

Seo, B. M., Miura, M., Sonoyama, W., Coppe, C., Stanyon, R., and Shi, S. (2005). Recovery of stem cells from cryopreserved periodontal ligament. *J. Dent. Res.* **84**(10), 907–912.

Seppa, H., Grotendorst, G., Seppa, S., Schiffmann, E. and Martin, G. R. (1982). Platelet-derived growth factor in chemotactic for fibroblasts. *J. Cell Biol.* **92**(2), 584–588.

Slavkin, H. C., Bringas, Jr., P., Bessem, C., Santos, V., Nakamura, M., Hsu, M. Y., Snead, M. L., Zeichner-David, M., and Fincham, A. G. (1989). Hertwig's epithelial root sheath differentiation and initial cementum and bone formation during long-term organ culture of mouse mandibular first molars using serumless, chemically defined medium. *J. Periodontal Res.* **24**(1), 28–40.

Taba, M., Jr., Jin, Q. Sugai, J. V., and Giannobile, W. V. (2005). Current concepts in periodontal bioengineering. *Orthod. Craniofac. Res.* **8**(4), 292–302.

Terkeltaub, R. A. (2001). Inorganic pyrophosphate generation and disposition in pathophysiology. *Am. J. Physiol. Cell Physiol.* **281**(1), C1–C11.

Toma, C., Pittenger, M. F., Cahill, K. S., Byrne, B. J., and Kessler, P. D. (2002). Human mesenchymal stem cells differentiate to a cardiomyocyte phenotype in the adult murine heart. *Circulation* **105**(1), 93–98.

Tompkins, K., George, A., and Veis, A. (2006). Characterization of a mouse amelogenin [A-4]/M59 cell surface receptor. *Bone* **38**(2), 172–180.

Tozum, T. F., and Demiralp B. (2003). Platelet-rich plasma: a promising innovation in dentistry. *J. Can. Dent. Assoc.* **69**(10), 664.

Urist, M. R. (1965). Bone: formation by autoinduction. *Science* **150**(698), 893–899.

van den Bergh, J. P., ten Bruggenkate, C. M., Groeneveld, H. H., Burger, E. H., and Tuinzing, D. B. (2000). Recombinant human bone morphogenetic protein-7 in maxillary sinus floor elevation surgery in three patients compared to autogenous bone grafts. A clinical pilot study. *J. Clin. Periodontol.* **27**(9), 627–636.

Venezia, E., Goldstein, M., Boyan, B. D., and Schwartz, Z. (2004). The use of enamel matrix derivative in the treatment of periodontal defects: a literature review and meta-analysis. *Crit. Rev. Oral Biol. Med.* **15**(6), 382–402.

Wang, H. L., and Cooke, J. (2005). Periodontal regeneration techniques for treatment of periodontal diseases. *Dent. Clin. North Am.* **49**(3), 637–659.

Wang, H., Tannukit, S., Zhu, D., Snead, M. L., and Paine, M. L. (2005a). Enamel matrix protein interactions. *J. Bone Miner. Res.* **20**(6), 1032–1040.

Wang, H. L., Greenwell, H., Fiorellini, J., Giannobile, W., Offenbacher, S., Salkin, L., Townsend, C., Sheridan, P., and Genco, R. J. (2005a). Periodontal regeneration. *J. Periodontol.* **76**(9), 1601–1622.

Wei, G., Jin, Q., Giannobile, W. V., and Ma, P. X. (2006). Nanofibrous scaffold for controlled delivery of recombinant human PDGF-BB. *J. Controlled Release*.

Whyte, M. P. (2002). Hypophosphatasia: nature's window to alkaline phosphatase in man. *In* "Principles of Bone Biology" (2nd, ed.), pp. 1229–1248. Academic Press, San Diego.

Wikesjo, U. M., Lim, W. H., Razi, S. S., Sigurdsson, T. J., Lee, M. B., Tatakis, D. N., and Hardwick, W. R. (2003a). Periodontal repair in dogs: a bioabsorbable calcium carbonate coral implant enhances space provision for alveolar bone regeneration in conjunction with guided tissue regeneration. *J. Periodontol.* **74**(7), 957–964.

Wikesjo, U. M., Xiropaidis, A. V., Thomson, R. C., Cook, A. D., Selvig, K. A., and Hardwick, W. R. (2003b). Periodontal repair in dogs: rhBMP-2 significantly enhances bone formation under provisions for guided tissue regeneration. *J. Clin. Periodontol.* **30**(8), 705–714.

Wikesjo, U. M., Sorensen, R. G., Kinoshita, A., Jian Li, X., and Wozney, J. M. (2004). Periodontal repair in dogs: effect of recombinant human bone morphogenetic protein-12 (rhBMP-12) on regeneration of alveolar bone and periodontal attachment." *J. Clin. Periodontol.* **31**(8), 662–670.

Wise, G. E., Frazier-Bowers, S., and D'Souza, R. N. (2002). Cellular, molecular, and genetic determinants of tooth eruption. *Crit. Rev. Oral Biol. Med.* **13**(4), 323–334.

Wozney, J. M., Rosen, V., Celeste, A. J., Mitsock, L. M., Whitters, M. J., Kriz, R. W., Hewick, R. M., and Wang, E. A. (1988). Novel regulators of bone formation: molecular clones and activities. *Science* **242**(4885), 1528–1534.

Wu, X. S. (1995). Synthesis, properties of biodegradable lactic/glycolic acid polymers. *In* "Encyclopedic Handbook of Biomaterials and Bioengineering" (D. L. Wise, ed.), pp. 1015–1054. Marcel Dekker, New York.

Young, C. S., Terada, S., Vacanti, J. P., Honda, M., Bartlett, J. D., and Yelick, P. C. (2002). Tissue engineering of complex tooth structures on biodegradable polymer scaffolds. *J. Dent. Res.* **81**(10), 695–700.

Young, C. S., Abukawa, H., Asrican, R., Ravens, M., Troulis, M. J., Kaban, L. B., Vacanti, J. P., and Yelick, P. C. (2005). Tissue-engineered hybrid tooth and bone." *Tissue Eng.* **11**(9–10), 1599–1610.

Zaman, K. U., Sugaya, T., and Kato, H. (1999). Effect of recombinant human platelet–derived growth factor-BB and bone morphogenetic protein-2 application to demineralized dentin on early periodontal ligament cell response. *J. Periodontal. Res.* **34**(5), 244–250.

Zeichner-David, M., Oishi, K., Su, Z., Zakartchenko, V., Chen, L. S., Arzate, H., and Bringas, Jr., P. (2003). Role of Hertwig's epithelial root sheath cells in tooth root development. *Dev. Dyn.* **228**(4), 651–663.

Zhang, Y. D., Chen, Z., Song, Y. Q., Liu, C., and Chen, Y. P. (2005). Making a tooth: growth factors, transcription factors, and stem cells. *Cell Res.* **15**(5), 301–316.

Zhang, H., Somerman, M. J., Berg, J., Williams, B., and Cunningham, M. L. (2007). Dental anomalies in a child with craniometaphyseal dysplasia. *Pediatric Dentistry.* (In press).

Zhao, L. R., Duan, W. M., Reyes, M., Keene, C. D., Verfaillie, C. M., and Low, W. C. (2002). Human bone marrow stem cells exhibit neural phenotypes and ameliorate neurological deficits after grafting into the ischemic brain of rats. *Exp. Neurol.* **174**(1), 11 20.

Zhao, M., Xiao, G., Berry, J. E., Franceschi, R. T., Reddi, A., and Somerman, M. J. (2002). Bone morphogenetic protein 2 induces dental follicle cells to differentiate toward a cementoblast/osteoblast phenotype. *J. Bone Miner. Res.* **17**(8), 1441–1451.

Zhao, M., Jin, Q., Berry, J. E., Nociti, Jr., F. H., Giannobile, W. V., and Somerman, M. J. (2004). Cementoblast delivery for periodontal-tissue engineering. *J. Periodontol.* **75**(1), 154–161.

Zhu, Z., Lee, C. S., Tejeda, K. M., and Giannobile, W. V. (2001). Gene transfer and expression of platelet-derived growth factors modulate periodontal cellular activity. *J. Dent. Res.* **80**(3), 892–897.

Part Nineteen

# Respiratory System

# Chapter Seventy-Three

# Progenitor Cells in the Respiratory System

*Valérie Besnard and Jeffrey A. Whitsett*

## I. INTRODUCTION: LUNG BIOLOGY AND OPPORTUNITIES FOR REGENERATION OF THE LUNG

Acute and chronic pulmonary disorders are common causes of morbidity and mortality throughout life. Under normal circumstances, the human lung maintains respiration under a wide range of environmental and exertional conditions. The lung is capable of remarkable repair following injury caused by infection, inhaled toxicants, or resection. Since respiration is solely dependent on lung structure and function, failure to form adequate lung tissue during morphogenesis and loss of lung tissue related to acquired or inherited pulmonary disorders are common clinical occurrences that are often life threatening. Considerable interest has been generated in understanding the biological events underlying the repair or regeneration of lung tissue. Understanding the principles determining normal lung formation and repair will provide insights into the pathogenesis of pulmonary disease and open new avenues for therapy designed to enhance normal repair processes in the lung. Progress has been made in identifying critical

processes in stem cell and progenitor cell biology in general as well as those active during lung formation, differentiation, and repair. In this chapter, the identity and molecular processes regulating progenitor cells during lung morphogenesis and repair are discussed in relationship to research supporting opportunities for the application of endogenous and exogenous cells for repair and regeneration of the lung.

## II. COMPLEXITY OF LUNG STRUCTURE PRESENTS A CHALLENGE FOR TISSUE ENGINEERING

The formation of the lung occurred relatively late in evolution, being required for adaptation to air breathing needed to support the respiratory requirements of vertebrates of diverse sizes. The lung is not required for normal organogenesis of other organs. There is no evidence that it produces substances required for development or function of other organs. Its function is inherent in its structure. The lung is a remarkably complex organ, consisting of multiple cell types that are precisely arranged on complex scaffold support of conducting airways (trachea, bronchi, and bronchioles) that lead to alveoli, wherein an elastin-rich network supports the epithelial-lined saccules, perfused by an extensive vascular bed. Alveolar walls are lined by type I and type II epithelial cells that are uniquely suited for gas exchange and surfactant production, respectively. Since epithelial surfaces of the lung are exposed directly to environmental gases and particles, effective innate and adaptive immunity and mucociliary clearance systems have evolved to maintain patency and sterility of the lung. The adult human lung contains 300 million to 500 million alveoli, creating an extensive gas exchange region, the area of which changes dynamically during the respiratory cycle. The hydrated epithelial surfaces of the lung are in direct contact with environmental gases, creating surface tension at the air–liquid interface that is mitigated by the continuous presence of pulmonary surfactant. Surfactant, consisting primarily of phospholipid- and surfactant-associated proteins, is required for maintenance of lung volumes during the cyclical contraction and expansion of the lung during ventilation. Maintenance of lung structure, mucociliary clearance, host defense, and surfactant homeostasis as well as the pulmonary circulation are dependent on distinct cell types. The lung contains a number of endogenous pulmonary cell types and marrow-derived cells, including dendritic cells, macrophages, polymorphonuclear leukocytes, lymphocytes, eosinophils, and mast. The remarkable complexity of structure and the ongoing requirements for lung function greatly complicate the application of strategies for cell and organ replacement other than by the transplantation of normal lung tissue. While it is presently not possible to engineer tissue replacements capable of gas exchange, knowledge regarding the cells and molecular events controlling lung organogenesis and function will likely enhance strategies to control cellular processes that might be used for future cell- or tissue-based therapies for lung disorders.

## III. LUNG MORPHOGENESIS

### Commitment and Restriction of Pulmonary Cells in the Foregut Endoderm

Progenitor cells forming the respiratory epithelium are derived from foregut endoderm. Their precursors are first identified by the expression of thyroid transcription factor-1 (TTF-1) along the anterior–ventral region of the foregut tube, being recognized between E9 and E10 in the developing mouse (four to five weeks of gestation in the human) as the lung primordia form (Lazzaro et al., 1991). Progenitor cells lining the pharyngeal diverticulum form the tracheal-esophageal region. The epithelial cells form the trachea and bronchi and extend into the splanchnic mesenchyme. Regions of foregut endoderm are prespecified before the appearance of distinct organs along the foregut tube. The lung buds first appear between the thyroid and liver buds at approximately E9.5 in mouse development. These endoder-

mally derived progenitor cells proliferate. Lung tubules undergo stereotypic branching and budding to form the conducting airways of the fetal lung (Fig. 73.1). The tubules extend into the underlying splanchnic mesenchyme, driven by migrational and proliferative signals from mesenchymal cells in a process that is dependent on production of FGFs (fibroblast growth factors) that bind and activate receptors on the underlying endodermal cells (Cardoso et al., 1997; Warburton et al., 2000; Shannon and Hyatt, 2004). Lung formation has been divided into distinct structural epochs that are based on anatomic changes shared during vertebrate lung morphogenesis. The general processes involved are highly conserved, but they vary temporally and spatially among diverse species. These epochs are useful in understanding the genes and processes that occur sequentially during lung morphogenesis and are also useful in understanding the pathogenesis of congenital malformations that lead to lung hypoplasia or malformations commonly seen in newborn infants. Early defects in progenitor cell differentiation, survival, and proliferation can profoundly disrupt subsequent morphogenesis, leading to severe and life-threatening pulmonary malformations, including lung agenesis-hypoplasia, diaphragmatic hernia, tracheal–esophageal fistula, and cystadenomatoid malformations, among others. Figure 73.1 describes the epochs based on events occurring during mouse lung morphogenesis that are shared in mouse and human. These events serve to emphasize the complexity of the cellular processes involved in the formation of the lung, some of which are recapitulated during regeneration/engineering of new lung tissue.

TTF-1 is an early marker of the commitment of endodermal cells to pulmonary and thyroid cell lineages, appearing prior to formation of the definitive lung (Lazzaro et al., 1991). Outgrowth and branching of TTF-1 expressing endodermal cells that form the lung requires FGF signaling. Recent studies support the concept that FGF signaling centers from heart and/or from the splanchnic mesenchyme (duration and intensity) along the foregut are critical for lung formation (Serls et al., 2005). The dose and timing of exposure of endodermal precursors to FGF signaling from the cardiac mesenchyme influence the commitment of endoderm to pulmonary lineages (Serls et al., 2005). Cells receiving less FGF stimulation become committed to other organs, including the gastrointestinal tract, liver, and pancreas (Zaret, 2002). In contrast to the lung, thyroid morphogenesis is marked by coexpression of TTF-1 and PAX-8, whereas pulmonary precursors are marked by TTF-1 and FOXA2 (Bohinski et al., 1994; Di Palma et al., 2003). A lung-specific transcription factor has not been identified; thus, at present, the process of lung formation appears to be dependent on direct and indirect interactions of TTF-1 with other transcription factors. Understanding the events critical for restricting cells of the foregut endoderm to pulmonary lineage may be useful in the future in engineering potential progenitor cells into useful pulmonary cells.

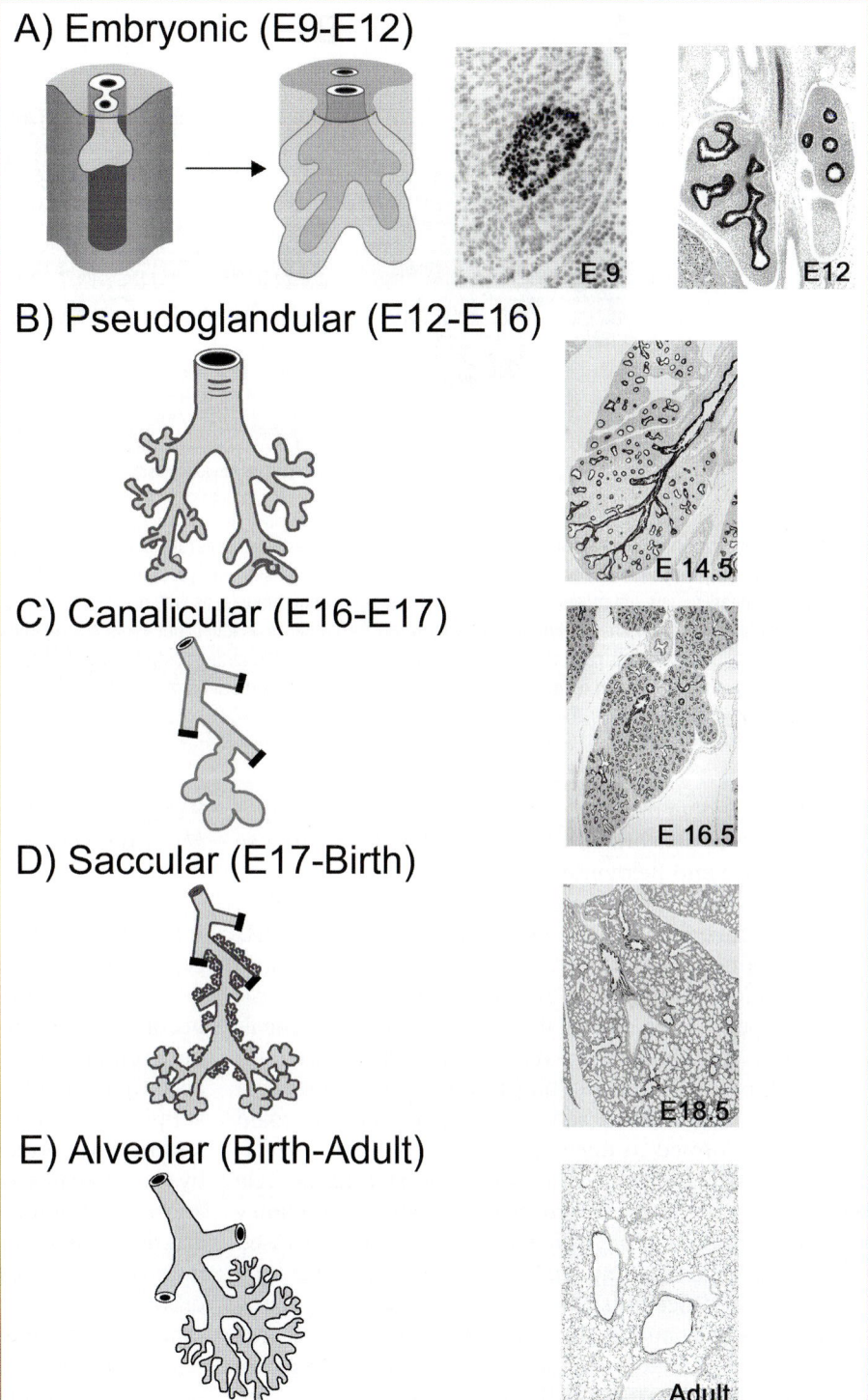

**FIG. 73.1.** Challenges in tissue engineering of the lung — the complexity of lung morphogenesis. The lung is a large organ, consisting of numerous cell types. Lung morphogenesis has been subdivided into epochs based on anatomic changes that are generally shared among various species during vertebrate morphogenesis. The lung buds emerge from the ventral-medial esophageal region of the foregut endoderm during the embryonic period of development **(A)**. The main airways and trachea extend into the splanchnic mesenchyme as branching morphogenesis proceeds during the pseudoglandular period of development **(B)**. During the canalicular period **(C)**, airway cell differentiation begins in multiple cell types, as evident in the respiratory epithelium, that differ along the cephalo-caudal axis. During the saccular-alveolar period **(D, E)**, the peripheral lung grows rapidly and peripheral saccules dilate and septate as they are subdivided into alveoli that comprise the gas exchange region of the lung required for survival after birth. Immunostaining (black cells) for Foxa1 marks the developing pulmonary epithelium or respiratory tubules. Illustrations shown are for mouse lung development.

## The Embryonic and Pseudoglandular Period: Branching Morphogenesis

Lung bud appears on the ventral–medial aspect of the mouse foregut at E9–9.5 and evaginates into the splanchnic mesenchyme as the primordial trachea and two main stem bronchi are formed. The tracheal and bronchial stalks extend, the latter forming the main bronchi, a process completed by approximately E10.5 in the mouse. Most vertebrate lungs display right–left asymmetry. For example, there are three main lobes on the right and two on the left in

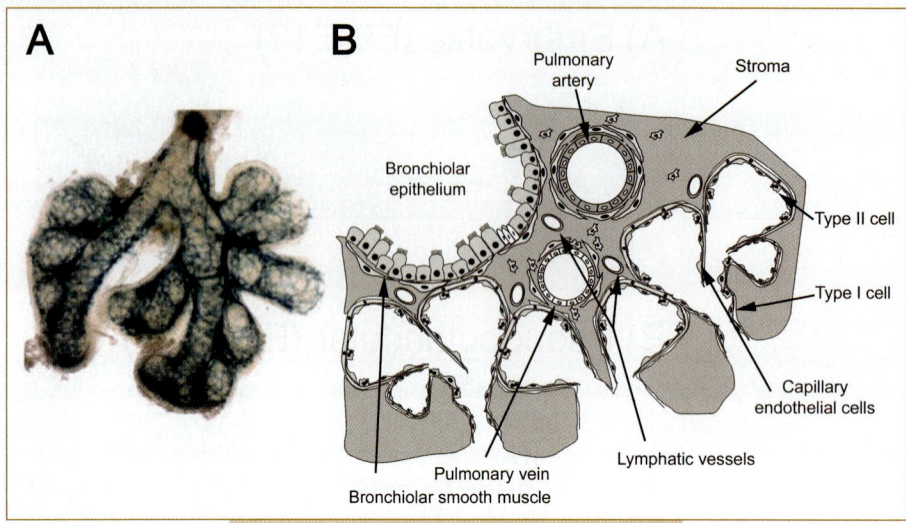

**FIG. 73.2.** Maturation of the pulmonary mesenchyme — vasculogenesis. Generally considered a relatively undifferentiated group of mesodermally derived cells, the pulmonary mesenchyme provides the progenitor cells for multiple cell types, including pulmonary and bronchial blood vessels (arteries, capillaries, veins, and lymphatics), smooth muscle cells surrounding the bronchial tubules, as well as stromal cells supporting peripheral lung saccules and distal regions of the conducting airways. Panel **A** demonstrates the early formation of vascular networks surrounding the lung tubules at E12.5 in the mouse, which are marked by the expression of β-galactosidase in pulmonary vascular cells under control of the flk-lacZ construct. Panel **B** demonstrates the organization of pulmonary vessels and stroma in relationship to bronchial and saccular structures in the late prenatal period of mouse lung development, demonstrating the complexity of organogenesis near the time of birth.

humans. A number of transcription factors have been implicated in the establishment of *situs inversus* in the mouse embryo (Raya and Belmonte, 2006).

FGF-10 from the mesenchyme (Deacon *et al.*, 1998), FGF-R2 in the endoderm (De Moerlooze *et al.*, 2000), SHH/ GLI 2,3 (Litingtung *et al.*, 1998; Pepicelli *et al.*, 1998), and retinoic acid receptors (RARs) (Mendelsohn *et al.*, 1994) play important roles during this early period of tracheal-pulmonary morphogenesis. While TTF-1 is required for formation of the peripheral lung and for complete separation of esophagus and trachea, the upper trachea, main bronchi are formed in mice lacking TTF-1 (Kimura *et al.*, 1996), FGF-10 (Deacon *et al.*, 1998), or FGF-R2IIIb (De Moerlooze *et al.*, 2000). There is considerable redundancy in RAR receptor function. Deletion of multiple RARs caused severe tracheal and lung malformations, including lung hypoplasia and agenesis (Mendelsohn *et al.*, 1994). Lung formation is also dependent on sonic hedgehog (SHH), produced by the endodermally derived cells, that is secreted, activating Ptch/Smo/Gli on the pulmonary mesenchyme (Pepicelli *et al.*, 1998). $Shh^{-/-}$ mice have tracheoesophageal fistula and simple cystlike lung sacs that fail to branch. As exemplified by the SHH and FGF pathways, multiple signaling centers are likely to control autocrine-paracrine signaling among and between endodermal and mesenchymally derived cells during formation of the lung that drive transcriptional processes influencing gene expression and cell behavior.

## Reciprocal Interactions Between Epithelial and Mesenchymal Cells

The development of the pulmonary mesenchyme, from which blood vessels and supportive tissue are formed, is carefully synchronized with that of the endodermally derived lung tubules. Stereotypic branching and budding occur as the bronchial and bronchiolar tubules form between approximately E11 and E16 in the mouse and are accompanied by the development of an extensive blood supply from both bronchial and pulmonary circulations. During this period, pulmonary blood vessels are produced by vasculogenesis and angiogenesis. Blood vessels coalesce and extend along the lung tubules to produce pulmonary arteries, veins, and lymphatics. Smooth muscle cells also differentiate, proliferate, and migrate along the bronchial tubules. Nerves are observed along vascular structures, innervating primarily the conducting regions of the lung. Cartilage appears along the ventral region of the trachea and bronchi in a precisely spaced manner. The lung tubules gradually taper along the cephalocaudal axis. Precartilaginous regions of the airway are well established by E11–E12 in the mouse, being marked by expression of Sox9 in precartilaginous tracheal–bronchial mesenchyme. Deletion or mutation in a number of transcription factors and signaling molecules, including SHH (Miller *et al.*, 2004), Sox9 (Mori-Akiyama *et al.*, 2003), RARs (Mendelsohn *et al.*, 1994), FGF signaling (Davis *et al.*, 1992), and FGF-18 (Whitsett *et al.*,

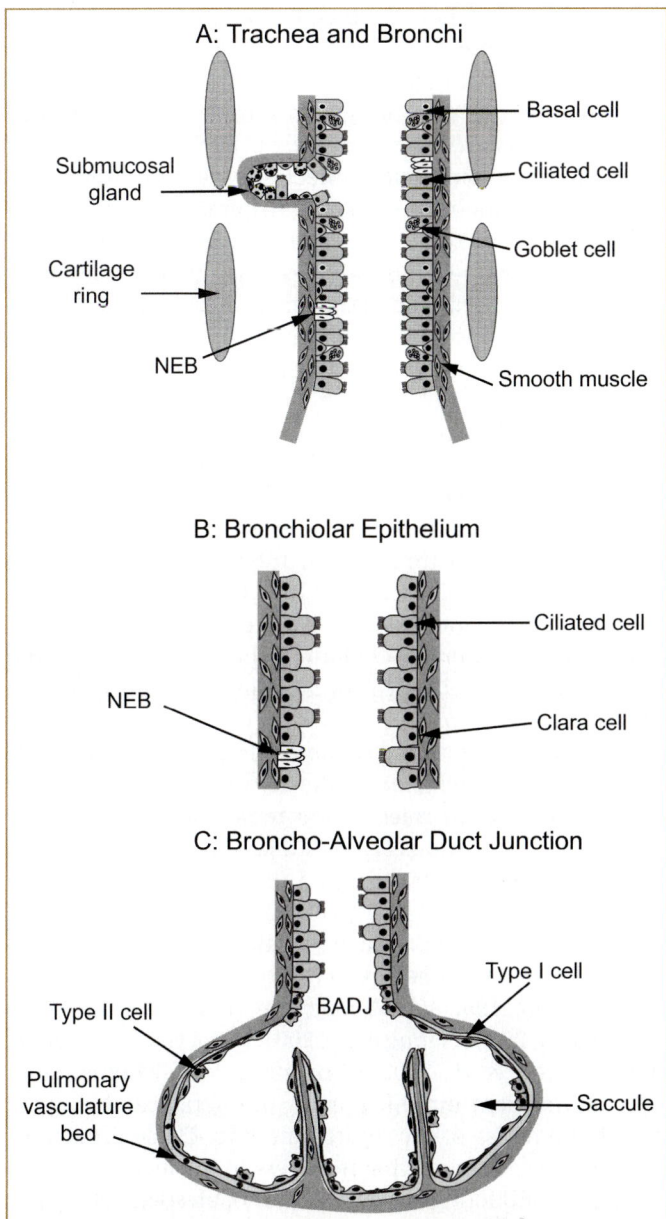

## A: Trachea and Bronchi

Submucosal gland

Cartilage ring

NEB

Basal cell

Ciliated cell

Goblet cell

Smooth muscle

## B: Bronchiolar Epithelium

NEB

Ciliated cell

Clara cell

## C: Broncho-Alveolar Duct Junction

Type II cell

Pulmonary vasculature bed

BADJ

Type I cell

Saccule

**FIG. 73.3.** Airway differentiation during the canalicular-saccular period. Structures and cell types differ along the cephalocaudal axis of the lung. A complex pseudostratified epithelium lines the larger cartilaginous, conducting airways. In many species, submucosal glands are located between the cartilaginous rings and in the submucosal layers. The epithelium consists primarily of ciliated, nonciliated, goblet, neuroendocrine, and basal cells **(A)**. The bronchi and bronchioles are lined primarily by a simple columnar epithelium consisting of ciliated and nonciliated epithelial cells as well as a rare subset of neuroendocrine cells that are found as isolated cells or in clusters termed neuroendocrine bodies (NEBs) **(B)**. The bronchoalveolar ducts are lined by cuboidal cells, leading to the saccules and alveoli that make up the peripheral gas exchange region of the lung **(C)**.

2002), disrupt tracheal cartilage formation in the developing mouse lung. Thus by E16, the general structure of the lung is well established along multiple axes, providing the scaffolding on which region-specific differentiation of various pulmonary epithelial cells begins.

Lung formation is highly dependent on reciprocal interactions between numerous cell types of both endodermal and mesenchymal origins. Removal of splanchnic mesenchyme that surrounds the developing lung buds arrests cell proliferation and branching morphogenesis (Wessells, 1970). The proliferation and differentiation of endodermal cells is strongly influenced by the underlying mesenchyme. For example, peripheral mesenchyme from the tips of lung buds to tracheal endoderm results in ectopic budding, branching, and peripheral lung specification of the upper airway (trachea), demonstrating remarkable plasticity of cell type during formation of the lung (Deterding and Shannon, 1995). It is likely that these interactions are controlled by cell–cell as well as paracrine and autocrine signals among various pulmonary cells. Ligands and receptors mediating these complex events are incompletely known. As seen in early embryonic patterning of the foregut, FGF signaling also plays an important role during branching morphogenesis. Evidence from gene targeting in the mouse suggests the important roles of FGF-10 (Deacon *et al.*, 1998), FGF-18 (Usui *et al.*, 2004), and FGF-9 (Colvin *et al.*, 2001), each playing an important and distinct role in lung formation. FGF-10 is sufficient to enhance cell migration, budding, and branching of the embryonic lung (Bellusci *et al.*, 1997). Conversely, the underlying endodermal cells synthesize and secrete a number of molecules critical for survival, proliferation, and differentiation of the pulmonary mesenchyme, including sonic hedgehog (SHH), vascular endothelial growth factors (VEGFs), and bone morphogenic proteins (BMPs).

## Epithelial Cell Differentiation During the Canalicular-Saccular Period

While respiratory epithelial cells lining the lung tubules consist of a relatively undifferentiated columnar-cuboidal cell layer in the embryonic period, a number of morphologically and biochemically distinct cell types become apparent during the canalicular-saccular period in the latter third of gestation. Epithelial surfaces of the larger cartilaginous airways are generally lined by a pseudostratified columnar epithelium consisting of basal, ciliated, goblet, Clara (nonciliated secretory cells), intermediate cells, and others. Many species, including the human, have large numbers of tracheal–bronchial submucosal glands, which also consist of numerous cell types. In most species, smaller airways, including bronchioles and respiratory acinar ducts, are lined by a single columnar or cuboidal epithelium, which consists primarily of ciliated and nonciliated secretory cells, as well as pulmonary neuroendocrine cells (PNECs), the latter a relatively rare cell type that expresses various neuropeptides. In the alveoli, peripheral lung epithelial cells become cuboidal (type II) or squamous (type I) as the peripheral saccules form. The numbers and characteristics of cells lining the respiratory tract vary along both dorsal–ventral and cephalocaudal axes and change dynamically during lung development. Numbers and differentiation of various epithelial cell types also vary during injury and repair in the postnatal period. The cellular and molecular mechanisms underlying the differentiation and maintenance of these diverse cell populations are being actively studied at the present, providing a conceptual framework useful in exploring lung regeneration and repair.

### Sacculation and Alveolarization

As if the processes of lung formation during branching morphogenesis were not complex enough, dramatic changes also occur in the saccular-alveolar period of lung development, as the structural transitions required for perinatal lung function occur. During this period, the distal fluid-filled lung saccules dilate. Epithelial cells proliferate and differentiate and the lung mesenchyme thins, as pulmonary blood capillaries come into close association with the walls of the lung saccules. At birth, the lungs expand, lung liquid is cleared, pulmonary blood flow increases, and surfactant is secreted into the alveoli to reduce surface tension as required for expansion. During the perinatal and postnatal period of lung growth, until maturity, the alveoli proliferate and septate, increasing the surface area available for gas exchange. The sacculation and alveolarization process is strongly influenced by stretch, requiring space for growth and expansion. Pressure from the amniotic fluid and production of lung liquid regulate growth of the peripheral lung. Various growth factors and hormones, including FGFs, EGF, PDGF, TGF-β, VEGFA, retinoids, and glucocorticoids, influence processes in the saccular-alveolar period of lung morphogenesis.

## IV. ENDOGENOUS PROGENITOR CELLS PLAY CRITICAL ROLES IN REPAIR OF THE RESPIRATORY EPITHELIUM AFTER BIRTH

After birth, the respiratory tract provides a remarkably large surface area that comes in direct contact with pathogens, particles, and toxicants present in inhaled air. In spite of the constant exposure to millions of particles each day, the respiratory tract is generally maintained in a sterile condition and is able to respond rapidly and robustly to various pulmonary injuries and infections. Repair of the respiratory epithelium depends primarily on endogenous cells lining the distinct regions of conducting and peripheral airways. Since respiration must be maintained at all times for survival, cellular responses mediating repair are initiated immediately following injury, resulting in rapid repair of the epithelium while maintaining alveolar-capillary permeabil-

ity and pulmonary homeostasis. Failure of repair, as seen in overwhelming infections, sepsis, or inhalation (e.g., following burns or smoke inhalation), causes loss of alveolar capillary integrity, often resulting in respiratory failure, termed *adult respiratory distress syndrome* (ARDS). Since the lung adapts to many pulmonary infections via both innate and acquired immune responses, the respiratory tract initiates remarkable protective and proliferative responses. Complex innate and acquired immune systems mediate the clearance of infected cells and pathogens. Since injury and infection often involve extensive regions of the lung, maintenance of epithelial barrier function must be maintained during the process of clearance of apoptotic or necrotic cells. It is therefore not surprising that the lung is capable of robust proliferative responses, which can be initiated throughout the tissue. In general, cell proliferation occurs at very low rates in the mature lung, in sharp contrast to findings in some organs, like the gastrointestinal tract, in which ongoing, rapid turnover of cells occurs. While the mature respiratory epithelium does not undergo rapid turnover (Breuer *et al.*, 1990), the respiratory epithelium responds rapidly to acute injury, and many cell types are capable of reentering the cell cycle during the repair process.

In conducting airways, there is strong evidence that basal cells and nonciliated epithelial cells proliferate rapidly and play important roles in the repair of the respiratory epithelium after various types of injury. *In vitro* and *in vivo* experiments have demonstrated the ability of basal cells, Clara cells, or other secretory cells, including epithelial cells in the bronchio-alveolar–duct junctions, to proliferate and differentiate into other respiratory epithelial cell types (Plopper *et al.*, 1992; Van Winkle *et al.*, 1995, 1999; Reynolds *et al.*, 2000a, 2000b; Hong *et al.*, 2001, 2004; Giangreco *et al.*, 2002; Lawson *et al.*, 2002). Progenitor cells, whether stem cells or transient amplifier cells, proliferate and migrate during the repair of the injured airways. These progenitor cells are present in varying numbers in distinct cell niches along the conducting airways. Repair of alveolar cells is initiated by the proliferation of type II epithelial cells and their subsequent differentiation into alveolar type I cells (Evans *et al.*, 1975, 1976).

### Models for Study of Lung Repair

Evidence from multiple experimental models used to study lung repair support the concept that endogenous cells provide the major, if not sole, cellular substrates for the repair of respiratory epithelium following acute injury (Plopper *et al.*, 1992; Bongso *et al.*, 1994; Reynolds *et al.*, 2000a, 2000b; Hong *et al.*, 2001, 2004; Giangreco *et al.*, 2002; Lawson *et al.*, 2002; Schoch *et al.*, 2004; Park *et al.*, 2006). A number of experimental models have been used to elucidate the cellular mechanisms involved in lung repair, including exposure to toxicants, infectious agents, and immunological challenges. Repair of the conducting airway epithelium has been studied using a variety of chemical and toxicant challenges selectively to injure specific cell types. Nonciliated cells (Clara cells) lining conducting airways of the mouse can be selectively killed with the toxicant naphthalene, which is metabolized to highly toxic metabolites, a process controlled by the expression of cytochrome P450, which is highly expressed in nonciliated respiratory epithelial cells (Plopper *et al.*, 1992; Van Winkle *et al.*, 1995, 1999; Reynolds *et al.*, 2000a, 2000b; Hong *et al.*, 2001, 2004; Giangreco *et al.*, 2002). Metabolites of naphthalene cause rapid and extensive injury of nonciliated cells in conducting airways. Basal cells play an important role during repair of the pseudostratified epithelium in the upper airways after naphthalene exposure (Bongso *et al.*, 1994; Borthwick *et al.*, 2001; Hong *et al.*, 2004). In the bronchioles, repair is initiated within hours after injury by the transdifferentiation and squamous metaplasia of ciliated or other toxicant-resistant cells in smaller conducting airways (Plopper *et al.*, 1992; Lawson *et al.*, 2002; Park *et al.*, 2006). These cells migrate and redifferentiate to contribute to the normal, heterogeneous cellularity of the airways within several days following injury. In the mouse, toxicant resistant or variant Clara cells and cells located near neuroendocrine bodies and in the bronchio-alveolar–duct region proliferate most actively following naphthalene exposure (Hong *et al.*, 2001; Kim *et al.*, 2005). Epithelial repair is accompanied by increased expression of a number of genes that are normally expressed during lung morphogenesis, including Foxa1, Foxa2, TTF-1, Sox2, Sox17, and Foxj1 (Park *et al.*, 2006). Thus, squamous metaplasia and transdifferentiation of existing airway epithelia play an important role in maintaining and repairing the respiratory epithelium after injury. The precise nature and state of differentiation of cells undergoing proliferation are not known with clarity, although a number of experiments support the concept that basal cells and nonciliated cells, as well as cells in distinct cell niches, are capable of proliferation, migration, and differentiation to restore the respiratory epithelium. Whether these cells are tissue stem cells or simply represent diverse populations of rapidly amplifying cells is unclear at present. Various toxicants have been used to study lung injury, including exposure to sulfur dioxide, ozone, detergents, and nitric oxide, which in general support the concept that both basal cells and nonciliated respiratory epithelial cells are the major source of progenitors that repair conducting airways after injury (Evans *et al.*, 1986).

Various *in vitro* model systems using isolated cells, xenografts, and tracheal–epithelial cells cultured at an air–liquid interface demonstrate the ability of various lung cells to proliferate and differentiate into multiple cell types. For example, purified basal cells and/or Clara cells can reform the complex epithelia consisting of numerous cell types (Bongso *et al.*, 1994; Schoch *et al.*, 2004). *In vitro* labeling of clones with retroviral vectors demonstrates the varying clonal potential for progenitor cells as the respiratory epithelium is regenerated during the reepithelialization of tra-

cheal grafts (Engelhardt *et al.*, 1991). Ciliated, nonciliated epithelial, and goblet cells are produced from these progenitors. In the peripheral lung, exposure to hyperoxia (80–95%) causes acute cell injury, inducing proliferation in subsets of epithelial cells. Alveolar type I cells are most susceptible to oxygen injury. Alveolar repair is initiated by proliferation of alveolar type II epithelial cells, which occurs two to three days following the injury. Subsets of these cells then differentiate into type I cells to complete the repair (Evans *et al.*, 1975).

## Evidence for Specialized Stem Cell Niches

The identity and properties of putative stem cells in the lung remain of considerable interest and are subject of active investigation. At present, specific markers and characteristics of the proposed stem cells have not been identified. Nevertheless, there is considerable evidence for pulmonary cells with stem cell characteristics. Tissue stem cells capable of continued self-renewal and proliferation have been proposed to play a role in the repair of multiple organs and tissues, including the lung. Stem cells are generally considered to represent a rare subset of nonproliferating cells that are normally maintained in an undifferentiated state. Such cells may be identified by prolonged retention of BrdU or $^3$H thymidine, which is incorporated into DNA during recurrent injury and proliferation of cells, termed *label-retaining cells* (LRCs). LRCs turn over slowly, are capable of self-renewal, and are able to provide progenitor cells that proliferate and differentiate. A number of experiments provide evidence for LRCs in specified niches within the respiratory tract (Reynolds *et al.*, 2000a, 2000b; Borthwick *et al.*, 2001; Engelhardt, 2001). LRCs were identified in the lung of mice exposed to naphthalene and after repeated exposure to detergent or $SO_2$, where they were identified in relatively restricted sites along conducting airways. LRCs were enriched in the necks of submucosal glands and in tracheal–bronchial folds in noncartilaginous regions of the conducting airways. After injury, these cells proliferated and migrated to repair the respiratory epithelium. The sites enriched in these LRCs are perhaps strategically located to minimize their direct injury from toxicants, or infection. Distinct surfaces of various cell types lining the respiratory epithelium may also determine their susceptibility to injury, toxicants, or infection. For example, the diverse cell surfaces present on various cells may influence attachment and/or uptake of pulmonary pathogens by the respiratory epithelium. Subsets of cells located near neuroepithelial bodies and those located near branch points of conducting airways were associated with repair following severe injury (Hong *et al.*, 2001). Cells located in bronchoalveolar ducts may represent another subset of cells with unique proliferative and differentiation capacities (Giangreco *et al.*, 2002; Kim *et al.*, 2005). Thus the lung is repaired primarily by the proliferation of endogenous progenitor cells, whose numbers and locations vary along the cephalocaudal axis of the airways. The cellular and molecular mechanisms involved in proliferation and redifferentiation is of considerable relevance in efforts to apply lung progenitor cells to tissue engineering.

## Lessons from Postpneumonectomy Regeneration

The remarkable capacity of the mammalian lung to regenerate following ipsilateral pneumonectomy provides evidence for the potential of tissue reengineering of the lung. This phenomenon contrasts sharply with the loss of lung function associated with many chronic pulmonary disorders, in which severe tissue loss and remodeling permanently impair lung function. In most mammals, ipsilateral pneumonectomy initiates rapid cell proliferation capable of substantially restoring lung functions and reserve capacity, a phenomenon that is well known clinically and experimentally. Many cellular components of the lung are involved in the regeneration process. Regrowth is strongly influenced by tissue caused by expansion of the remaining lung following removal of tissue and, to a lesser extent, by hypoxemia and endocrine factors. Regeneration is species and age dependent, the less mature lung being capable of more robust responses. In many animal models, growth of both airway and alveolar epithelium can occur in a balanced fashion, the growth of pulmonary vessels accompanying that of the respiratory tubules and maintaining ventilation/perfusion ratios. Under some experimental conditions, epithelial expansion occurs in a nonsynchronous fashion with that of the vascular system, resulting in limited improvement in gas exchange. As in other models of lung repair, proliferation and remodeling are mediated by the proliferation of endogenous pulmonary cells, and there is no evidence that exogenous or marrow-derived progenitor cells contribute substantially or biologically to the growth process. A number of peptides and hormones, including epidermal growth factor, platelet-derived growth factor, fibroblast growth factors, and retinoic acid, influence the repair process following pneumonectomy (Hsia, 2004). Although retinoids can enhance lung repair in experimental emphysema (Massaro and Massaro, 1997, 2000), the administration of retinoic acid alone does not enhance lung regeneration in dogs (Yan *et al.*, 2005). The ability of the mature lung to undergo remarkable compensatory growth, leading to enhanced pulmonary function, is a remarkable observation that provides a foundation for further investigation of tissue engineering of the lung.

## V. EVIDENCE FOR NONPULMONARY STEM CELLS IN THE LUNG

A number of controversial experiments provides support for the ability of marrow-derived or mesenchymal progenitor cells to contribute significantly to repair of the lung. Several studies support the concept that bone marrow stromal cells or mesenchymal-derived progenitor cells can migrate into various organs, where they express features of endogenous cells, including respiratory epithelial, myocardial, and gastrointestinal cells (Krause *et al.*, 2001; Kotton *et al.*, 2005). While engraftment and differentiation of cells in the lung rarely occur following bone marrow transplantation in normal mice, lung injury enhances the migration of bone marrow–derived cells into the lung. However, recent studies suggest that some of the engrafted cells may represent the fusion of bone marrow–derived cells with endogenous cells rather than the reprogramming and redifferentiation of the cells (Terada *et al.*, 2002; Chang *et al.*, 2005; Kotton *et al.*, 2005). Artifactual colocalization of cell-specific markers with other cell-labeling markers was demonstrated after bone marrow transplantation, further complicating interpretations of previous findings. The use of careful optical sectioning/imaging systems to discriminate colocalization from overlying or superimposing images was recommended (Kotton *et al.*, 2005). Thus, at the present time there is little evidence that bone marrow–derived cells play biologically relevant roles in lung repair. Nevertheless, the potential for the application of multipotent progenitor cells for use in cell replacement or tissue repair remains intriguing and merits further investigation.

The potential utility of embryonic stem cells for tissue regeneration is being actively explored for therapies of many diseases and multiple organs. Embryonic stem cells exhibit varied abilities to differentiate along distinct cell lineages, which can be further influenced by growth in selective media, by the addition of growth factors or extracellular matrix components, and/or by coculture with other cell types. Recent studies demonstrate that embryonic stem cells can differentiate into cells with characteristics of respiratory epithelial cells expressing lung-specific markers, including pro SP-C or Clara cell secretory protein (CCSP) (Coraux *et al.*, 2005; Van Vranken *et al.*, 2005). Coculture of embryonic lung mesenchyme with ES cells demonstrated the ability of the mesenchymal cells to enhance differentiation toward respiratory epithelial cell types (Van Vranken *et al.*, 2005). Nuclear extracts obtained from MLE-12 cells (an SV40 large T-antigen immortalized cell line) enhanced pulmonary selective differentiation of embryonic stem cells *in vitro* (Qin *et al.*, 2005). SV40 large T-antigen ES cells differentiate into multiple cell types, including those that express a variety of endodermal cell markers, when transplanted under renal or testicular capsules. Culture of ES cells at an air–liquid interface produced a complex, highly differentiated epithelium, with cell types characteristic of conducting airways, including ciliated, basal, and secretory cells (Coraux *et al.*, 2005). Taken together, ES cells may provide a source of cells that can be cultured under various conditions to enhance their differentiation into respiratory epithelial cell types. ES cells will be useful in identifying factors and processes regulating growth, differentiation, and specification of respiratory epithelial cells and will enable the study of their interactions with other pulmonary cell types. The utility of ES cells for regenerating pulmonary tissues that would be of therapeutic value remains to be explored.

## VI. MODELS FOR STUDY OF LUNG REGENERATION *IN VITRO*

While no single system has been developed for regeneration of functional lung tissue capable of engraftment into the lung, a number of model systems have been developed that allow the study of various aspects of the processes that will be required for lung regeneration. The remarkable complexity of the lung and its requirement for bronchial pulmonary arterial and lymphatic circulations as well as its connection to the conducting airways and the cardiovascular system present a considerable engineering hurdle. Various pulmonary cell types can be isolated and grown under conditions that retain their capacity to reorganize into lunglike structures *in vitro*. However, pulmonary cells cannot be readily infused or explanted into airways or into the pulmonary circulation to induce formation of functional tissue. Nevertheless, useful model systems for the study of lung regeneration have been developed for *in vitro* and *in vivo* studies. Various epithelial cell types, including basal, ciliated, and nonciliated airway cells, as well as alveolar type I and type II cells, can be substantially purified by differential protease treatments, cell sorting, differential attachment, and/or density gradient separation strategies. In general, respiratory epithelial cells rapidly lose their differentiated cell characteristics unless placed on specialized matrices or grown as organoids in which multiple cell types are cocultured. Various epithelial cell types can be immortalized from using transforming viruses or after expression of transforming genes in transgenic mice (Wikenheiser *et al.*, 1993). Such cells have been highly useful for the study of lung cell biology.

Respiratory epithelial cells differentiate in response to various culture conditions. Highly differentiated lung cells maintaining some of the features of lung parenchyma can be produced by coculture of dissociated fetal lung cells on collagen sponges (Gelfoam) (Douglas and Teel, 1976) or on other biomatrices, including Matrigel, a collagen-glycosaminoglycan rich matrix. Cell polarity, epithelial–mesenchymal cell interactions, and epithelial cell differentiation can be maintained under these conditions. Fetal lung has been explanted and maintained in culture for several days, during which epithelial cells undergo differentiation. Highly differentiated cell cultures of conducting airway epithelial cells have been produced by culture of mature respiratory epithelial cells on denuded trachea (Liu *et al.*, 1994). Tracheal grafts can be seeded with lung cells and then implanted and cultured under the skin of nude mice for prolonged periods of time. A highly differentiated pseudostratified respiratory epithelium consisting of multiple respiratory cell types forms under these conditions. Similarly, a complex epithelium can be generated by culturing isolated conducting airway cells on various biomatrices at an air–liquid interface (Coraux *et al.*, 2005; Vaughan *et al.*, 2006). However, in these models, growth and differentiation of lung epithelial cells occur without connection to blood vessels. The long-term viability of lung explanted tissue has been demonstrated in a xenograft model, in which explanted fetal lung tissue develops a vascular system that connects to the host circulation (Vu *et al.*, 2003). Both fetal lung explants, and embryonic stem cells transplanted under the testicular or renal capsules of SCID mice develop blood vessels derived by vasculogenesis from the explanted tissue, which connect with the host vascular system (Schwarz *et al.*, 2000; Vu *et al.*, 2003). Thus, a number of useful model systems, including those with isolated cells, ES cells, xenografts, and fetal lung explants, have been developed that are useful in exploring the complex biology underlying lung regeneration and repair. Fruitful lines of investigation support the concept that pulmonary cells can be isolated and cultured under conditions that was not in the growth of highly differentiated cells with many of the features of normal lung.

## VII. SUMMARY AND CONCLUSIONS

Disorders of the lung contribute substantially to morbidity and mortality throughout life. Lung malformations, genetic disorders affecting the lung, and diseases related to prematurity and other perinatal lung diseases represent the most common cause of illness in the perinatal period. Likewise, acute lung disease and chronic lung disease are encountered throughout postnatal life. Thus, the ability to provide lung tissue to enhance lung function would have diverse applications in medicine in the future. Knowledge regarding the fundamental principles underlying lung formation will provide the conceptual framework required to enhance normal lung repair and allow regeneration of lung tissue for treatment of inherited and acquired lung diseases. Because of its complexity, tissue engineering of the lung capable of enhancing respiration represents a high hurdle in the field of regenerative medicine. The study of lung formation, repair, and tissue engineering will likely provide knowledge regarding the cellular and molecular processes that allow for future cell- and tissue-based therapies for the treatment of lung disease.

## VIII. REFERENCES

Bellusci, S., Grindley, J., Emoto, H., Itoh, N., and Hogan, B. L. (1997). Fibroblast growth factor 10 (FGF10) and branching morphogenesis in the embryonic mouse lung. *Development* **124**, 4867–4878.

Bohinski, R. J., Di Lauro, R., and Whitsett, J. A. (1994). The lung-specific surfactant protein B gene promoter is a target for thyroid transcription factor 1 and hepatocyte nuclear factor 3, indicating common factors for organ-specific gene expression along the foregut axis. *Mol. Cell. Biol.* **14**, 5671–5681.

Bongso, A., Fong, C. Y., Ng, S. C., and Ratnam, S. (1994). Isolation and culture of inner cell mass cells from human blastocysts. *Hum. Reprod.* **9**, 2110–2117.

Borthwick, D. W., Shahbazian, M., Krantz, Q. T., Dorin, J. R., and Randell, S. H. (2001). Evidence for stem-cell niches in the tracheal epithelium. *Am. J. Respir. Cell. Mol. Biol.* **24**, 662–670.

Breuer, R., Zajicek, G., Christensen, T. G., Lucey, E. C., and Snider, G. L. (1990). Cell kinetics of normal adult hamster bronchial epithelium in the steady state. *Am. J. Respir. Cell. Mol. Biol.* **2**, 51–58.

Cardoso, W. V., Itoh, A., Nogawa, H., Mason, I., and Brody, J. S. (1997). FGF-1 and FGF-7 induce distinct patterns of growth and differentiation in embryonic lung epithelium. *Dev. Dyn.* **208**, 398–405.

Chang, J. C., Summer, R., Sun, X., Fitzsimmons, K., and Fine, A. (2005). Evidence that bone marrow cells do not contribute to the alveolar epithelium. *Am. J. Respir. Cell Mol. Biol.* **33**, 335–342.

Colvin, J. S., White, A. C., Pratt, S. J., and Ornitz, D. M. (2001). Lung hypoplasia and neonatal death in Fgf9-null mice identify this gene as an essential regulator of lung mesenchyme. *Development* **128**, 2095–2106.

Coraux, C., Nawrocki-Raby, B., Hinnrasky, J., Kileztky, C., Gaillard, D., Dani, C., and Puchelle, E. (2005). Embryonic stem cells generate airway epithelial tissue. *Am. J. Respir. Cell. Mol. Biol.* **32**, 87–92.

Davis, S., Bove, K. E., Wells, T. R., Hartsell, B., Weinberg, A., and Gilbert, E. (1992). Tracheal cartilaginous sleeve. *Pediatr. Pathol.* **12**, 349–364.

Deacon, T., Dinsmore, J., Costantini, L. C., Ratliff, J., and Isacson, O. (1998). Blastula-stage stem cells can differentiate into dopaminergic and serotonergic neurons after transplantation. *Exp. Neurol.* **149**, 28–41.

De Moerlooze, L., Spencer-Dene, B., Revest, J., Hajihosseini, M., Rosewell, I., and Dickson, C. (2000). An important role for the IIIb isoform of fibroblast growth factor receptor 2 (FGFR2) in mesenchymal-epithelial signalling during mouse organogenesis. *Development* **127**, 483–492.

Deterding, R. R., and Shannon, J. M. (1995). Proliferation and differentiation of fetal rat pulmonary epithelium in the absence of mesenchyme. *J. Clin. Invest.* **95**, 2963–2972.

Di Palma, T., Nitsch, R., Mascia, A., Nitsch, L., Di Lauro, R., and Zannini, M. (2003). The paired domain-containing factor Pax8 and the homeodomain-containing factor TTF-1 directly interact and synergistically activate transcription. *J. Biol. Chem.* **278**, 3395–3402.

Douglas, W. H., and Teel, R. W. (1976). An organotypic *in vitro* model system for studying pulmonary surfactant production by type II alveolar pneumocytes. *Am. Rev. Respir. Dis.* **113**, 17–23.

Engelhardt, J. F. (2001). Stem cell niches in the mouse airway. *Am. J. Respir. Cell. Mol. Biol.* **24**, 649–652.

Engelhardt, J. F., Allen, E. D., and Wilson, J. M. (1991). Reconstitution of tracheal grafts with a genetically modified epithelium. *Proc. Natl. Acad. Sci. U.S.A.* **88**, 11192–11196.

Evans, M. J., Cabral, L. J., Stephens, R. J., and Freeman, G. (1975). Transformation of alveolar type 2 cells to type 1 cells following exposure to NO2. *Exp. Mol. Pathol.* **22**, 142–150.

Evans, M. J., Johnson, L. V., Stephens, R. J., and Freeman, G. (1976). Cell renewal in the lungs of rats exposed to low levels of ozone. *Exp. Mol. Pathol.* **24**, 70–83.

Evans, M. J., Shami, S. G., Cabral-Anderson, L. J., and Dekker, N. P. (1986). Role of nonciliated cells in renewal of the bronchial epithelium of rats exposed to NO2. *Am. J. Pathol.* **123**, 126–133.

Giangreco, A., Reynolds, S. D., and Stripp, B. R. (2002). Terminal bronchioles harbor a unique airway stem cell population that localizes to the bronchoalveolar duct junction. *Am. J. Pathol.* **161**, 173–182.

Hong, K. U., Reynolds, S. D., Giangreco, A., Hurley, C. M., and Stripp, B. R. (2001). Clara cell secretory protein–expressing cells of the airway neuroepithelial body microenvironment include a label-retaining subset and are critical for epithelial renewal after progenitor cell depletion. *Am. J. Respir. Cell Mol. Biol.* **24**, 671–681.

Hong, K. U., Reynolds, S. D., Watkins, S., Fuchs, E., and Stripp, B. R. (2004). Basal cells are a multipotent progenitor capable of renewing the bronchial epithelium. *Am. J. Pathol.* **164**, 577–588.

Hsia, C. C. (2004). Signals and mechanisms of compensatory lung growth. *J. Appl. Physiol.* **97**, 1992–1998.

Kim, C. F., Jackson, E. L., Woolfenden, A. E., Lawrence, S., Babar, I., Vogel, S., Crowley, D., Bronson, R. T., and Jacks, T. (2005). Identification of bronchioalveolar stem cells in normal lung and lung cancer. *Cell* **121**, 823–835.

Kimura, S., Hara, Y., Pineau, T., Fernandez-Salguero, P., Fox, C. H., Ward, J. M., and Gonzalez, F. J. (1996). The T/ebp-null mouse: thyroid-specific enhancer-binding protein is essential for the organogenesis of the thyroid, lung, ventral forebrain, and pituitary. *Genes Dev.* **10**, 60–69.

Kotton, D. N., Fabian, A. J., and Mulligan, R. C. (2005). Failure of bone marrow to reconstitute lung epithelium. *Am. J. Respir. Cell. Mol. Biol.* **33**, 328–334.

Krause, D. S., Theise, N. D., Collector, M. I., Henegariu, O., Hwang, S., Gardner, R., Neutzel, S., and Sharkis, S. J. (2001). Multiorgan, multilineage engraftment by a single bone marrow–derived stem cell. *Cell* **105**, 369–377.

Lawson, G. W., Van Winkle, L. S., Toskala, E., Senior, R. M., Parks, W. C., and Plopper, C. G. (2002). Mouse strain modulates the role of the ciliated cell in acute tracheobronchial airway injury-distal airways. *Am. J. Pathol.* **160**, 315–327.

Lazzaro, D., Price, M., de Felice, M., and Di Lauro, R. (1991). The transcription factor TTF-1 is expressed at the onset of thyroid and lung morphogenesis and in restricted regions of the fetal brain. *Development* **113**, 1093–1104.

Litingtung, Y., Lei, L., Westphal, H., and Chiang, C. (1998). Sonic hedgehog is essential to foregut development. *Nat. Genet.* **20**, 58–61.

Liu, J. Y., Nettesheim, P., and Randell, S. H. (1994). Growth and differentiation of tracheal epithelial progenitor cells. *Am. J. Physiol.* **266**, L296–L307.

Massaro, G. D., and Massaro, D. (1997). Retinoic acid treatment abrogates elastase-induced pulmonary emphysema in rats. *Nat. Med.* **3**, 675–677.

Massaro, G. D., and Massaro, D. (2000). Retinoic acid treatment partially rescues failed septation in rats and in mice. *Am. J. Physiol.* **278**, L955–L960.

Mendelsohn, C., Lohnes, D., Decimo, D., Lufkin, T., LeMeur, M., Chambon, P., and Mark, M. (1994). Function of the retinoic acid receptors (RARs) during development (II). Multiple abnormalities at various stages of organogenesis in RAR double mutants. *Development* **120**, 2749–2771.

Miller, L. A., Wert, S. E., Clark, J. C., Xu, Y., Perl, A. K., and Whitsett, J. A. (2004). Role of Sonic hedgehog in patterning of tracheal-bronchial cartilage and the peripheral lung. *Dev. Dyn.* **231**, 57–71.

Mori-Akiyama, Y., Akiyama, H., Rowitch, D. H., and de Crombrugghe, B. (2003). Sox9 is required for determination of the chondrogenic cell lineage in the cranial neural crest. *Proc. Natl. Acad. Sci. U.S.A.* **100**, 9360–9365.

Park, K. S., Wells, J. M., Zorn, A. M., Wert, S. E., Laubach, V. E., Fernandez, L. G., and Whitsett, J. A. (2006). Transdifferentiation of ciliated cells during repair of the respiratory epithelium. *Am. J. Respir. Cell. Mol. Biol.* **34**, 151–157.

Pepicelli, C. V., Lewis, P. M., and McMahon, A. P. (1998). Sonic hedgehog regulates branching morphogenesis in the mammalian lung. *Curr. Biol.* **8**, 1083–1086.

Plopper, C. G., Suverkropp, C., Morin, D., Nishio, S., and Buckpitt, A. (1992). Relationship of cytochrome P-450 activity to Clara cell cytotoxicity. I. Histopathologic comparison of the respiratory tract of mice, rats and hamsters after parenteral administration of naphthalene. *J. Pharmacol. Exp. Ther.* **261**, 353–363.

Qin, M., Tai, G., Collas, P., Polak, J. M., and Bishop, A. E. (2005). Cell extract–derived differentiation of embryonic stem cells. *Stem Cells* **23**, 712–718.

Raya, A., and Belmonte, J. C. (2006). Left–right asymmetry in the vertebrate embryo: from early information to higher-level integration. *Nat. Rev. Genet.* **7**, 283–293.

Reynolds, S. D., Giangreco, A., Power, J. H., and Stripp, B. R. (2000a). Neuroepithelial bodies of pulmonary airways serve as a reservoir of progenitor cells capable of epithelial regeneration. *Am. J. Pathol.* **156**, 269–278.

Reynolds, S. D., Hong, K. U., Giangreco, A., Mango, G. W., Guron, C., Morimoto, Y., and Stripp, B. R. (2000b). Conditional clara cell ablation reveals a self-renewing progenitor function of pulmonary neuroendocrine cells. *Am. J. Physiol.* **278**, L1256–L1263.

Schoch, K. G., Lori, A., Burns, K. A., Eldred, T., Olsen, J. C., and Randell, S. H. (2004). A subset of mouse tracheal epithelial basal cells generates large colonies *in vitro. Am. J. Physiol.* **286**, L631–L642.

Schwarz, M. A., Zhang, F., Lane, J. E., Schachtner, S., Deutsch, G., Starnes, V., and Pitt, B. R. (2000). Angiogenesis and morphogenesis of murine fetal distal lung in an allograft model. *Am. J. Physiol.* **278**, L1000–L1007.

Serls, A. E., Doherty, S., Parvatiyar, P., Wells, J. M., and Deutsch, G. H. (2005). Different thresholds of fibroblast growth factors pattern the ventral foregut into liver and lung. *Development* **132**, 35–47.

Shannon, J. M., and Hyatt, B. A. (2004). Epithelial–mesenchymal interactions in the developing lung. *Annu. Rev. Physiol.* **66**, 625–645.

Terada, N., Hamazaki, T., Oka, M., Hoki, M., Mastalerz, D. M., Nakano, Y., Meyer, E. M., Morel, L., Petersen, B. E., and Scott, E. W. (2002). Bone marrow cells adopt the phenotype of other cells by spontaneous cell fusion. *Nature* **416**, 542–545.

Usui, H., Shibayama, M., Ohbayashi, N., Konishi, M., Takada, S., and Itoh, N. (2004). Fgf18 is required for embryonic lung alveolar development. *Biochem. Biophys. Res. Commun.* **322**, 887–892.

Van Vranken, B. E., Romanska, H. M., Polak, J. M., Rippon, H. J., Shannon, J. M., and Bishop, A. E. (2005). Coculture of embryonic stem cells with pulmonary mesenchyme: a microenvironment that promotes differentiation of pulmonary epithelium. *Tissue Eng.* **11**, 1177–1187.

Van Winkle, L. S., Buckpitt, A. R., Nishio, S. J., Isaac, J. M., and Plopper, C. G. (1995). Cellular response in naphthalene-induced Clara cell injury and bronchiolar epithelial repair in mice. *Am. J. Physiol.* **269**, L800–L818.

Van Winkle, L. S., Johnson, Z. A., Nishio, S. J., Brown, C. D., and Plopper, C. G. (1999). Early events in naphthalene-induced acute Clara cell toxicity: comparison of membrane permeability and ultrastructure. *Am. J. Respir. Cell. Mol. Biol.* **21**, 44–53.

Vaughan, M. B., Ramirez, R. D., Wright, W. E., Minna, J. D., and Shay, J. W. (2006). A three-dimensional model of differentiation of immortalized human bronchial epithelial cells. *Differentiation* **74**, 141–148.

Vu, T. H., Alemayehu, Y., and Werb, Z. (2003). New insights into saccular development and vascular formation in lung allografts under the renal capsule. *Mech. Dev.* **120**, 305–313.

Warburton, D., Schwarz, M., Tefft, D., Flores-Delgado, G., Anderson, K. D., and Cardoso, W. V. (2000). The molecular basis of lung morphogenesis. *Mech. Dev.* **92**, 55–81.

Wessells, N. K. (1970). Mammalian lung development: interactions in formation and morphogenesis of tracheal buds. *J. Exp. Zool.* **175**, 455–466.

Whitsett, J. A., Clark, J. C., Picard, L., Tichelaar, J. W., Wert, S. E., Itoh, N., Perl, A. K., and Stahlman, M. T. (2002). Fibroblast growth factor 18 influences proximal programming during lung morphogenesis. *J. Biol. Chem.* **277**, 22743–22749.

Wikenheiser, K. A., Vorbroker, D. K., Rice, W. R., Clark, J. C., Bachurski, C. J., Oie, H. K., and Whitsett, J. A. (1993). Production of immortalized distal respiratory epithelial cell lines from surfactant protein C/simian virus 40 large tumor antigen transgenic mice. *Proc. Natl. Acad. Sci. U. S.A.* **90**, 11029–11033.

Yan, X., Bellotto, D. J., Dane, D. M., Elmore, R. G., Johnson, R. L., Jr., Estrera, A. S., and Hsia, C. C. (2005). Lack of response to all-trans retinoic acid supplementation in adult dogs following left pneumonectomy. *J. Appl. Physiol.* **99**, 1681–1688.

Zaret, K. S. (2002). Regulatory phases of early liver development: paradigms of organogenesis. *Nat. Rev. Genet.* **3**, 499–512.

# Chapter Seventy-Four

# Lungs

*Anne E. Bishop and Julia M. Polak*

## I. INTRODUCTION

Repair or regeneration of defective lung cells or tissue would be of huge therapeutic potential. Cellular sources for the regeneration of lung tissue *in vivo* or lung-tissue engineering *in vitro* include endogenous pulmonary epithelial stem cells, extrapulmonary circulating stem cells, and embryonic stem cells. This chapter discusses the progress that has been made toward creating regenerative-medicine approaches to the treatment of lung injury and disease.

There is no doubt that a readily available means to repair, replace, or regenerate human lung tissue would have an enormous impact on health care. A huge variety of lung diseases exist, and most are debilitating and widespread and present a significant biomedical problem. Chronic obstructive pulmonary disease (COPD, or smoker's lung) alone affected more than 52 million individuals around the world in 1994, with approximately 16.2 million of these living in the United States, and is predicted to become the third-leading cause of death worldwide by 2020 (Pauwels *et al.*, 2001). In addition to the human suffering, the attendant financial burden is enormous: in 1993, $14.7 billion were spent on treatment of COPD, with the overall direct and indirect costs estimated at more than $30 billion (Sullivan *et al.*, 2000). Despite the significant burden lung diseases place on society, research funding has historically been disproportionately low, and the biological processes underlying many lung diseases remain poorly understood.

The ultimate aim of any regenerative-medicine strategy is to restore normal cellular architecture and function where tissues have been compromised by injury or disease. The lung is a particularly difficult target for any reparative or regenerative strategy for three main reasons: it has a highly complex structure, there is a wide range of cellular diversity and the pulmonary epithelium shows slow cell turnover rates. However, during recent years there has been a series of scientific advances leading toward targeted lung repair and/or regeneration. For example, our understanding of fetal lung morphogenesis has increased, and previously unknown regenerative pathways have been identified in the adult lung. Some of the most pivotal advances have been made in the area of stem cell biology, which has become big news in medical research since the mid-1990s. As the basis of natural pathways for tissue maintenance and regeneration, stem cells represent a key target for mediating repair *in vivo*. The targeted activation of endogenous pools of airway epithelial cells with the properties of stem cells could exploit existing repair mechanisms and augment the lung's innate regenerative capability. Implantation of pulmonary-epithelium or even pulmonary-tissue constructs produced *in vitro* has yet to be achieved but progress is being made in this direction too. In addition to the direct clinical benefits of such laboratory-created lung epithelium and/or tissue, they could be used as *in vitro* models of human lung development and disease for further investigation and

*Principles of Tissue Engineering, 3rd Edition*
ed. by Lanza, Langer, and Vacanti

manipulation and for drug discovery and toxicological testing.

The degree to which the lung can repair itself through the proliferation and differentiation of intrinsic stem cells is discussed in Chapter 73. In this chapter, we concentrate on the basic research being done with a view to repairing damaged lung and eventual engineering of the gas exchange component of the lung.

## II. LUNG STRUCTURE

In brief, the mammalian lung comprises branching airways that become narrower at each divergence until they form the small alveolar sacs at the distal ends that act as the gas exchange unit. There are about 300 million alveoli in the adult lung, providing the surface area in the order of 60–80 square meters that is needed for human respiration. Replicating such a structure in the laboratory is a truly daunting prospect for the tissue engineer. The pulmonary tree contains distinct anatomical regions, each lined by a confluent layer of different types of epithelial cells that forms the interface with the air space. Underlying the epithelium is the vasculature lined by endothelial cells, which, like the epithelium, form a continuum, in this case running from the main pulmonary artery, through capillaries, to the pulmonary vein and on to the left atrium. In between the epithelial and endothelial layers lies the pulmonary interstitium, which comprises a range of connective tissue cell types providing a scaffold for cell layers within the lung. The following summarizes the structure of these three main components of the lung.

### Epithelium

The proximal airways, the trachea and major bronchi, are lined by pseudostratified epithelium containing ciliated and mucous secretory (or goblet) cells on the luminal surface. Neuroendocrine cells first appear around eight weeks of gestation in human lung and are relatively frequent during development, where they play a major role in airway growth and development and form in the adult around 1% of the pulmonary epithelium (Lauwreyns and Cokelaere, 1973; Cutz and Orange, 1977). The lining of the smaller airways also includes ciliated epithelial cells, but, rather than goblet cells, the bronchioles possess the different, cuboidal, nonciliated Clara cells. The alveoli are lined by two epithelial phenotypes: the flattened squamous (type I) and the cuboidal (type II) pneumocytes. The type I pneumocyte is the cell type across which gas exchange occurs, and it is characteristically thin, with cytoplasmic extensions, a morphology that presents as little a barrier as possible to the diffusion of oxygen while maintaining the integrity of the alveolar wall. Type II pneumocytes are critical for maintaining alveolar homeostasis, clearing the alveolar air space of edema and secreting pulmonary surfactant that lowers surface tension and prevents airway collapse.

### Endothelium

The vasculature of the lung is different from that of the systemic circulation in a variety of ways: the vessels are of low resistance, blood in arteries is deoxygenated whereas that in veins is oxygenated, and specialized structural modifications of the alveolar capillaries promote gas diffusion. Throughout the pulmonary vasculature, the endothelial cells are continuous, bathed on one side by blood and on the other by interstitial fluid, and, unlike the epithelium, they share most characteristics, although there are some differences related to the function of each segment. For example, arterial endothelium is composed of relatively thick, elongated cells with numerous cytoplasmic organelles. Venous endothelial cells are thinner and polygonal and have fewer organelles, while those of the alveolar capillaries have an avesicular zone, the basal laminar of which fuses with that of type I pneumocytes to form the air–blood barrier.

### Interstitium

The interstitium, the space interposed between the airspace epithelium, the vascular endothelium and the pleural mesothelium, is the remains of the splanchnopleuric mesenchymal bed into which the airway tubes and blood vessels grew during lung morphogenesis. The cells of the interstitium are a spectrum, from smooth muscle to fibroblasts. Stops in between include myofibroblasts and pericytes, which are the particular myofibroblasts associated with the alveolar capillaries. Immune cells, in the form of macrophages and lymphocytes, mainly T-cells, are present in the interstitial fluid. A web of elastin and collagen fibers, interwoven with fibronectin fibrils and proteoglycan molecules, permeates the interstitial space and forms the matrix that supports the structural integrity of the lung while allowing its plastic deformation during respiration. The epithelial and endothelial basal laminae are considered the outer borders of the pulmonary interstitium.

## III. CELL SOURCES FOR LUNG REPAIR AND LUNG-TISSUE ENGINEERING

Primary cells, derived from body fluid, tissue outgrowth, or disaggregation, can be used for tissue repair and engineering but present problems of access, limited life span and failure to maintain phenotype *in vitro*. Primary cells have been used successfully, however, in the construction of *in vitro* models of lung. Stem cells, in contrast, can be isolated relatively easily from embryonic, fetal, or adult tissue, can self-renew (by definition) and can be encouraged to form a required cell phenotype, although all stem cell types are not equivalent because the range of cell types to which they can differentiate varies according to origin. For regenerative-medicine purposes, stem cells could provide a virtually inexhaustible cell source. Large-scale production of regenerative-medicine products, including cell therapies and engineered tissue constructs, could be envisaged if undifferentiated stem cells were expanded in culture and driven in batches to differentiate into different therapeutic cell types. Consequently, current research is focused on developing means to expand stem cells, to promote their efficient differentiation to required lineages, and to purify and manipulate the resulting cells into a form suitable for implantation (see Polak and Bishop, 2006, for review). For each potential target cell type or tissue, the initial step in developing a tissue-engineering strategy is the selection of the most appropriate stem cell on which to base it. As mentioned earlier, the structural complexity and cellular heterogeneity of the lung make the study of lung-specific stem cell biology difficult, so the isolation of sufficient numbers of pulmonary stem cells for potential therapeutic applications is simply unrealistic at this time. For this reason, attempts have been made to produce pulmonary epithelium from extrapulmonary stem cell types, which can be isolated and grown in greater numbers.

## Circulating Stem Cells

One possible source of stem cells for use in lung epithelial repair could be the circulation. Our understanding of the regeneration of many organs has been revised in recent years with the discovery that various tissues can be repaired by stem cells recruited from the circulation. Adult bone marrow has long been known to contain stem cells that can give rise to hematopoietic and mesenchymal cell lineages (Pittenger et al., 1999). However, these stem cells can differentiate not only toward a variety of mesenchymal cell types, such as adipocytes (Pittenger et al., 1999), osteocytes (Prockop 1997; Perreira et al., 1998), myocytes (Ferrari et al., 1998), and cardiomyocytes (Orlic et al., 2001), but also toward ectoderm, e.g., neurons (Mezey et al., 2000) and endoderm, e.g., hepatocytes (Petersen et al., 1999; Alison et al., 2000; Lagasse et al., 2000; Theise et al., 2000), and renal parenchymal cells (Poulsom et al., 2001). These observations have been taken to suggest that there is a subpopulation of stem cells in adult bone marrow that retains pluripotency throughout development and can act as a universal repair pathway. Consistent with this, multipotent adult progenitor cells (MAPCs) have been described in murine bone marrow that can differentiate in vitro at the single-cell level to all three germ layers (Jiang et al., 2002). Although the mechanism by which bone marrow–derived stem cells are recruited to repair tissue is not known, engraftment has been reported to be enhanced by tissue injury (Ferrari et al., 1998; Kotton et al., 2001; Okamoto et al., 2002; Theise et al., 2002; Ortiz et al., 2003; Abedi et al., 2004).

For the lung, there is some controversy as to whether stem cells are recruited from the circulation to repair the epithelium and, if so, whether or not this occurs by fusion with cells in situ. A variety of cell-tracing experiments has been carried out, using labeled or marker-mismatched cells derived from bone marrow fractions in murine models.

Marrow-derived cells were identified in most cases in the pulmonary epithelium of recipient mice, and the rate of engraftment was generally found to be improved in areas of overt lung damage (Kotton et al., 2001; Theise et al., 2002; Ortiz et al., 2003; Krause et al., 2001; Grove et al., 2002; Harris et al., 2004; Ishizawa et al., 2004; Mattsson et al., 2004; Yamada et al., 2004; Beckett et al., 2005; Loi et al., 2006). In one of these studies, bone marrow–derived pulmonary epithelium was seen 11 months after cell implantation, suggesting that this is a mechanism for effecting long-term repair (Grove et al., 2002). Apparent engraftment and transdifferentiation could be explained by fusion of bone marrow stem cells with endogenous somatic cells. This has been demonstrated in vitro (Terada et al., 2002; Ying et al., 2002; Alvarez-Delado et al., 2003) and also in vivo in cells known to form heterokaryons in certain pathologies, such as hepatocytes and cardiomyocytes (Alvarez-Delado et al., 2003; Vassilopoulous et al., 2003; Deb et al., 2003). Tests with pulmonary epithelium have given mixed results, with fusion with bone marrow stem cells being clearly demonstrated in vitro (Spees et al., 2003) but not seen in vivo (Grove et al., 2002; Alvarez-Delado et al., 2003; Harris et al., 2004). However, more recently, the results of all these studies have been questioned because of their reliance on colocalization of the bone marrow–specific label with pulmonary epithelial markers by microscopy. Two studies in animal models have been published arguing that these findings are actually a consequence of separate overlapping cells or autofluorescence, not true colocalization of markers, as originally suggested (Chang et al., 2005; Kotton et al., 2005).

There is another aspect of bone marrow stem cell engraftment in lung that concerns the discovery of lung-specific side population (SP) cells within pulmonary tissue. SP cells, identified by their unique flow cytometry profile when stained with the DNA dye Hoescht 33342, were originally shown in the bone marrow as being highly enriched for hematopoietic stem cell activity (Goodell et al., 1996). But recently, nonhematopoietic SP cells have been identified that seem to act as tissue-specific stem cells (Alvi et al., 2002; Wulf et al., 2003; Poliakova et al., 2004). SP cells, with the characteristics of airway epithelium and mesenchyme, have since been isolated from proximal and distal lung (Summer et al., 2003, 2004; Giangreco et al., 2004). Originally thought to represent another tissue-specific stem cell population, the most recent data show that lung-specific SP cells have a bone marrow origin, although they are phenotypically distinct from bone marrow SP cells (Summer et al., 2004). It is not known whether these cells colonize the lung during development only or are continually renewed throughout life by the bone marrow.

So far, investigation of human lung either has found no engraftment of bone marrow cells within the pulmonary epithelium or has shown that it occurs only infrequently (Kubit et al., 1994; Bittmann et al., 2001; Kleeberger et al., 2003; Surratt et al., 2003; Albera et al., 2005; Zander et al.,

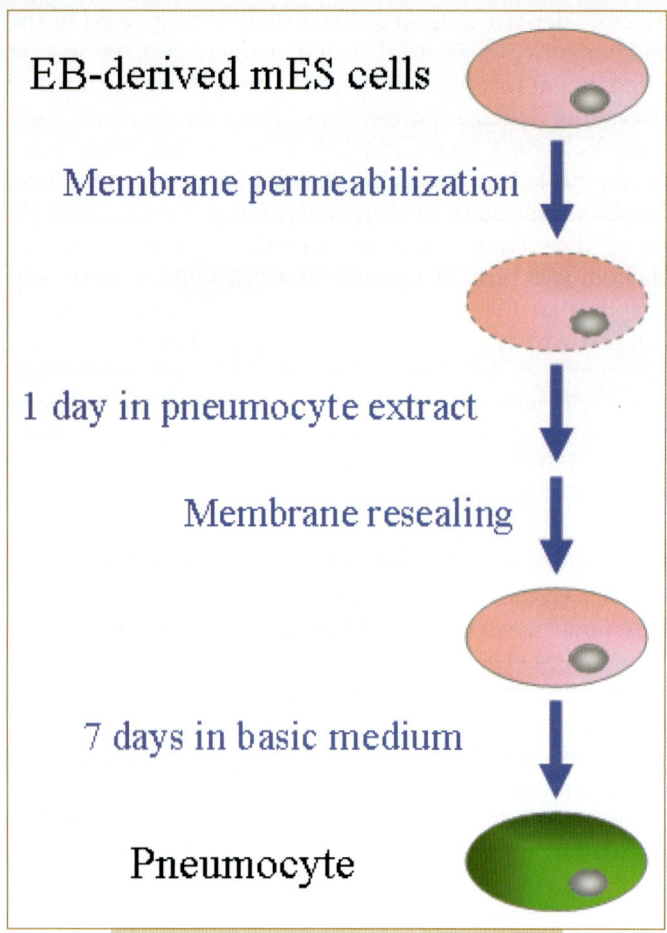

EB-derived mES cells

Membrane permeabilization

1 day in pneumocyte extract

Membrane resealing

7 days in basic medium

Pneumocyte

**FIG. 74.1.** Diagrammatic representation of cell extract–based reprogramming of embryonic stem cells to derive type II pneumocytes as described in Qin *et al.* (2005). EB = embryoid body; mES = murine embryonic stem.

regulate the natural repair process in a similar manner. If exogenous bone marrow or circulating stem cells could be delivered to the lung for repair, this could also provide a novel means for gene therapy. Isolated stem cells from the bone marrow or the circulation may be a source for the engineering of lung constructs, if they can be purified and expanded in sufficient numbers.

**Embryonic Stem Cells**

Embryonic stem cells (ESC) are particularly suited to tissue-engineering research, in view of their relative availability, known provenance, pluripotency and proliferative capacity (see Rippon and Bishop, 2004, for review). For derivation of lung cells, preliminary findings established the possibility of deriving alveolar airway epithelium, specifically type II pneumocytes, from murine ESC by supplementation of the culture medium with factors that maintain differentiated characteristics in primary lung epithelial cell cultures (Ali *et al.*, 2002; Rippon *et al.*, 2004). These ESC-derived cells express surfactant proteins, have an ultrastructure comparable to that of primary type II pneumocytes (i.e., they have microvili and lamellar bodies) and differentiate *in vitro* on plastic to type I pneumocytes. The original protocol has recently been refined and found to give rise to progenitor cells thought to be representative of those present in the early branching lung, around E10–E11 of murine development, which differentiate to form Clara cells and type I and type II pneumocytes (Rippon *et al.*, 2006).

It has proven difficult to produce yields of lung epithelial cells from ESC of more than a few percent by the manipulation of cell culture medium alone, an observation made for other endodermal cell lineages. Subsequently, other means have been tested to drive the formation of pulmonary epithelium from ESC. For example, it has been reported that combining medium supplementation with cell culture at an air–liquid interface can induce the formation of fully differentiated tracheobronchial airway epithelium from murine ESC (Coraux *et al.*, 2005). Alternatively, a method originally used to convert fibroblasts into T-cells using T-cell extracts (Håklien *et al.*, 2002) has been adapted and applied to murine ESC. This method involved permeabilizing stem cells and exposing them for one hour to crude lysates of a type II pneumocyte cell line. Following membrane resealing and culture in unsupplemented medium, the cells differentiated to type II pneumocytes with an efficiency of approximately 10%, well in excess of that observed with most other methods (Qin *et al.*, 2005) (Fig. 74.1).

In view of the critical role that underlying mesenchymal cells play in regulating lung epithelial differentiation *in vivo*, the inductive power of fetal lung mesenchyme on ESC has been investigated. Early differentiating embryoid bodies were wrapped in microdissected lung mesenchyme taken from E11.5 mouse embryos. After only five days, mesenchyme and ESC had coalesced and small channels had

2005). Unfortunately, these studies have been fraught with technical problems due to their reliance on the questionable colocalization of histological markers and the lack of appropriate human lung tissue samples of adequate quality. Nevertheless, if it is true that stem cells can be recruited by damaged pulmonary epithelium to effect repair, then there could be major clinical implications. As a precedent, the observation that bone marrow cells engraft in murine and human heart was rapidly translated to the clinic, where administration of autologous bone marrow, directly to the cardiac wall or via the vasculature, is being used to treat myocardial infarction and heart failure (see Wollert and Drexter, 2005, for review). It has also been reported recently that delivery of a specific fraction of stem cells in mobilized peripheral blood, which shows early markers for cells of all germ layers, can be used to reduce liver insufficiency in human patients (Gordon *et al.*, 2006). The complexity of the lung may make similar therapy more difficult to achieve for respiratory diseases, but it may at least be possible to up-

formed that were lined by cells expressing markers of alveolar epithelium (Van Vranken *et al.*, 2005). A similar approach has recently shown that the inductive properties of murine lung mesenchyme persist for only a short developmental window around embryonic day 11.5, the stage at which distal lung epithelial differentiation is initiated *in vivo* (Denham *et al.*, 2006). This approach may prove too limited to provide lung tissue for implantation, but it could provide a system for modeling lung development *in vitro.*

The basic research with murine cells obviously needs transfer to human cells before any firm steps can be made toward clinical applications. For ESC, this exercise is really just beginning, for human lines were isolated only in 1998 (Thomson *et al.*, 1998), and it is only within the last several years that they have become readily available to laboratories lacking the highly specialized expertise and facilities to isolate their own. Recently, type II pneumocytes were successfully differentiated from human ESC (Samadikuchaksaraei *et al.*, 2006) (Fig. 74.2), the first crucial step in the development of ESC-based treatments for lung disease. Whether ESC-derived pulmonary epithelial cells or engineered tissue will be used in the clinic remains to be seen because there are safety issues, principally regarding the possibility of accidental implantation of undifferentiated cells, that have yet to be addressed. In the meantime, there is no doubt that, at the very least, these cells have a range of indirect therapeutic applications, such as in disease modeling, drug discovery and toxicological screening.

## IV. LUNG-TISSUE CONSTRUCTS

Lung tissue has been constructed for *in vitro* use. For example, EpiAirway™ (www.Matteck.com) comprises normal, nonimmortalized human tracheal and bronchial epithelial cells cultured to form a pseudostratified structure that resembles pulmonary epithelial tissue. These constructs have been used in drug delivery research and toxicological testing. Similar models have been made of bronchus, specifically for the *in vitro* study of the mechanisms underlying asthma (Chakir *et al.*, 2001; Paquette *et al.*, 2003, 2004). For these, bronchial fibroblasts were seeded into a collagen gel and bronchial epithelial cells were grown on the surface. There have been attempts to engineer trachea for implantation (Kojima and Vacanti, 2004), and the first successful human case was recently reported (Omori *et al.*, 2005). In addition to direct implantation of an engineered construct, there has been early work on the engineering of trachea by aerosol delivery of cells (Roberts *et al.*, 2005). Although not yet tested in an animal model, an atomizer has been developed that successfully sprays mammalian chondrocytes and tracheal epithelial cells, with more than 70% viability and no effect on growth rate. An interesting feature of this system was the placing of the cells in a mixture of saline and the polymer Pluronic F-127, which is liquid at room temperature but gels in the body. This means that, on

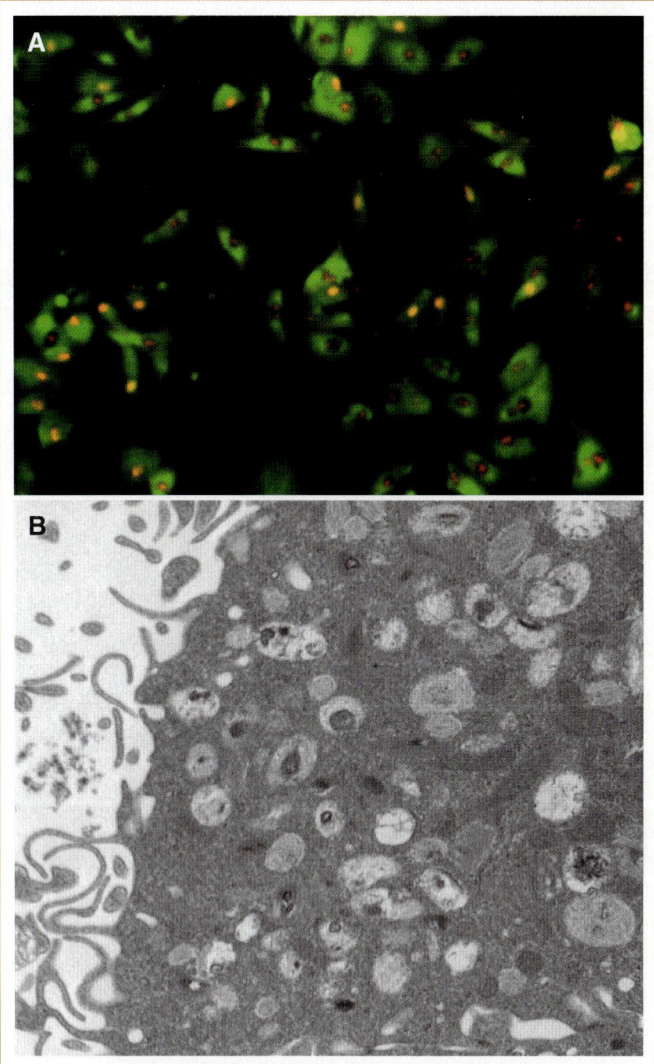

**FIG. 74.2.** **(A)** Human embryonic stem cell–derived type II pneumocytes identified by their immunoreactivity for the specific marker surfactant protein C (surfactant protein C immunoreactivity = green; propidium iodide nuclear counterstain = red). **(B)** Transmission electron micrograph of a human embryonic stem cell–derived type II pneumocyte showing the typical ultrastructure with lamellar bodies and microvilli.

instillation, the aerosol will gel in contact with the airway surface, thereby promoting adherence and engraftment of the cells.

The generation in the laboratory of the complex gas exchange component of the lung for human implantation, however, remains an elusive goal for tissue engineers. Pneumocytes attach fairly well to a range of polymer or bioactive scaffolds, particularly if an appropriate extracellar matrix protein is coated on the scaffold surface, e.g., laminin (Tan *et al.*, 2002) (Fig. 74.3). Attempts to create distal lung tissue can be traced back as far as 1976, when dispersed rat fetal pneumocytes grown on a collagen sponge reaggregated

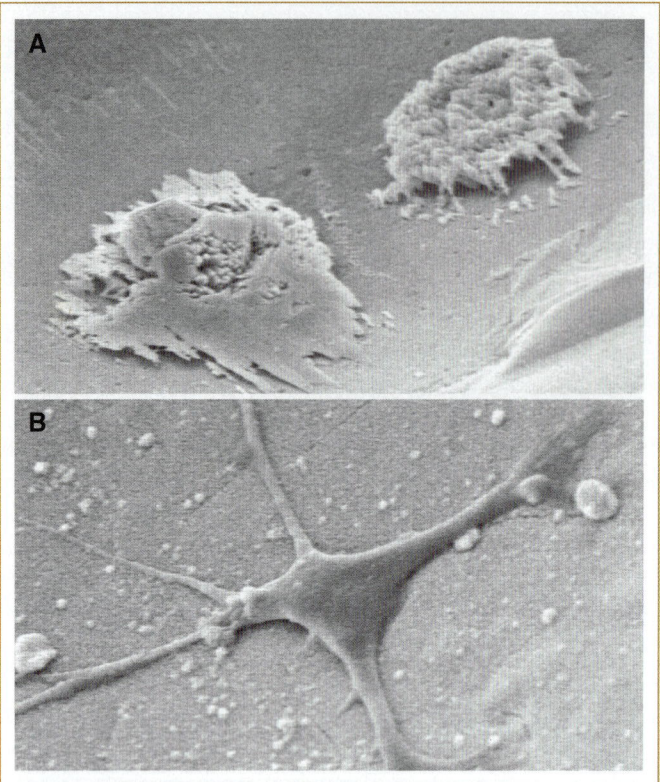

**FIG. 74.3.** **(A)** Scanning electron micrograph showing a single murine type II pneumocyte growing on the surface of Bioglass™. **(B)** A similar cell to that shown in **(A)** following coating of the surface of Bioglass™ with the extracellular matrix protein laminin. The cell shows increased spreading and contact with the scaffold surface.

around a lumen to form something that mimicked alveoli (Douglas and Teel, 1976; Douglas *et al.*, 1976a, 1976b). These alveolar-like structures, which formed within two days of culture and could be maintained for up to six weeks, were found to be composed mainly of type II pneumocytes, which were functional in that they secreted surfactant (Douglas *et al.*, 1983). Similar apparent reconstitution of alveolar structures *in vitro* has been reported for neonatal (Saito *et al.*, 1985) and adult rat lung cells grown on collagen gel (Sugihara *et al.*, 1993) or basement membrane (Shannon *et al.*, 1987; Kalina *et al.*, 1993) and for fetal rabbit lung cells grown on basement membrane (Blau *et al.*, 1988; Chinoy *et al.*, 1994). More recent developments in this area include the growth of dissociated rat lung cells on a collagen–glycosaminoglycan (GAG) tissue-engineering scaffold (Chen *et al.*, 2005). Not only did alveolar-like structures form, but the construct exhibited cell-mediated contraction, and smooth muscle actin was found in some of the cells in the scaffold, suggesting the formation of elastic units, as happens *in vivo*. The work also indicates the

potential usefulness in lung-tissue engineering of a scaffold, collagen-GAG, that is already well established in tissue engineering, e.g., for skin (Yannas *et al.*, 1989). Implantation of such a scaffold, seeded with cells possibly of a stem cell origin, as described earlier, might form the basis of a future engineering strategy. Another group reports the growth of mixed embryonic murine lung cells (epithelium, endothelium, and mesenchyme) in 3D on Matrigel and synthetic polymers (poly(lactic-*co*-glycotic) acid (PLGA), poly(L-lactic) acid (PLLA) formed into either foams or nanofibrous scaffolds (Mondrinos *et al.*, 2006). When grown on the hydrogel, it was possible to up-regulate the branching and sacculation of the resulting alveolar-like tissue by adding bFGF, FGF-7/10 to the culture medium. An exciting recent development has been the engineering of alveolar tissue by growing isolated lung progenitor cells on synthetic polymers (Cortiella *et al.*, 2006).

Artificial lungs, in the form of ECMOs (extracorporeal membrane oxygenators) or ECLS (life support systems), are being used in the clinic, but these are relatively inefficient and entirely mechanical at present. These systems are used mainly in cardiopulmonary bypass during heart surgery, but they are also used, to a lesser extent, to support patients in respiratory failure. Work is in progress to improve these machines, for example by increasing efficiency and/or making them intrathoracic, with a view to creating systems designed specifically for providing extended respiratory support (see Federspel and Henchir, 2004, for review). One means being investigated is to make the devices into biohybrids. A major step on the long, rocky road to making lung tissue for implantation could be the incorporation of epithelial and/or endothelial cells in extracorporeal gas exchange devices. For this latter possibility, the researchers at Imperial College London are collaborating with a company, Novalung™ (http://www.novalung.com), to try to increase the efficiency and life span of their interventional lung-assist device that is already in the clinic by incorporating stem cell–derived epithelium to make a biohybrid system.

## V. CONCLUSIONS

Tissue engineering promises to revolutionize our approach to the treatment of injury and disease. Lung disease is extremely prevalent worldwide, yet many lung diseases remain effectively incurable without a transplant, and the numbers of donor organs continues to dwindle. The latest research suggests that controlled and targeted lung repair could be effected by at least three methods: activating the body's existing repair pathways and stem cell pools, implanting healthy cells or tissue grown from stem cells in the laboratory or a combination of elements of both strategies. Exciting advances have been made in both areas in recent years, but the time frame for getting any of these approaches into the clinic remains unknown.

# VI. REFERENCES

Abedi, M., Greer, D. A., Colvon, G. A., Dembers, D. A., Dooner, M. S., Harpel, J. A., Pimentel, J., Menon, M. K., and Quesenberry, P. J. (2004). Tissue injury in marrow transdifferentiation. *Blood Cells Mol. Dis.* **32**, 42–46.

Albera, C., Polak, J. M., Janes, S., Albera C., Polak, J. M., Janes, S., Griffiths, M. J., Alison, M. R., Wright, N. A., Navaratnarasah, S., Poulsom, R., Jeffery, R., Fisher, C., *et al.* (2005). Repopulation of human pulmonary epithelium by bone marrow cells: a potential means to promote repair. *Tissue Eng.* **11**, 1115–1121.

Ali, N. N., Edgar, A. J., Samadikuchaksaraei, A., Timson, C. M., Romanska, H. M., Polak, J. M., and Bishop, A. E. (2002). Derivation of type II alveolar epithelial cells from murine embryonic stem cells. *Tissue Eng.* **8**, 541–550.

Alison, M. R., Poulsom, R., Jeffrey, R., *et al.* (2000). Hepatocytes from nonhepatic adult stem cells. *Nature* **406**, 257.

Alvarez-Dolado, M., Pardal, R., Garcia-Verdugo, J. M., Fike, J. R., Lee, H. O., Pfeffer, K., Lois, C., Morrison, S. J., and Alvarez-Buylla, A. (2003). Fusion of bone marrow–derived cells with Purkinje neurons, cardiomyocytes and hepatocytes. *Nature* **425**, 968–973.

Alvi, A., Clayton, H., Joshi, C., *et al.* (2002). Functional and molecular characterization of mammary side population cells. *Breast Cancer Res.* **5**, R1–R8.

Beckett, T., Loi, R., Prenovitz, R., Poynter, M., Goncz, K. K., Suratt, B. T., and Weiss, D. J. (2005). Acute lung injury with endotoxin or NO2 does not enhance development of airway epithelium from bone marrow. *Mol. Ther.* **12**, 680–686.

Bittmann, I., Dose, T., Baretton, G. B., *et al.* (2001). Cellular chimerism of the lung after transplantation. An interphase cytogenetic study. *Am. J. Clin. Pathol.* **115**, 525–533.

Blau, H., Guzowski, D. E., Siddiqi, Z. A., Scarpelli, E. M., and Bienkowski, R. S. (1988). Fetal type 2 pneumocytes form alveolar-like structures and maintain long-term differentiation on extracellular matrix. *J. Cell Physiol.* **136**, 203–214.

Chakir, J., Page, N., Hamid, Q., Laviolette, M., Boulet, L. P., and Rouabhia, M. (2001). Bronchial mucosa produced by tissue engineering: a new tool to study cellular interactions in asthma. *J. Allergy Clin. Immunol.* **107**, 36–40.

Chang, J. C., Summer, R., Sun, X., Fitzsimmons, K., and Fine, A. (2005). Evidence that bone marrow cells do not contribute to the alveolar epithelium. *Am. J. Respir. Cell Mol. Biol.* **33**, 335–342.

Chen, P., Marsilio, E., Goldstein, R. H., Yannas, I. V., and Spector, M. (2005). Formation of lung alveolar-like structures in collagen-glycosaminoglycan scaffolds *in vitro*. *Tissue Eng.* **11**, 1436–1448.

Chinoy, M. R., Antonio-Santiago, M. T., and Scarpelli, E. M. (2004). Maturation of undifferentiated lung epithelial cells into type II cells *in vitro*: a temporal process that parallels cell differentiation in vivo. *Anat. Rec.* **240**, 545–554.

Clara, M. (1937). Zur Histobiologie des Bronchialepithels. *Z. Mikrosk. Anat. Forsch.* **41**, 321.

Coraux, C., Nawrocki-Raby, B., Hinnrasky, J., *et al.* (2005). Embryonic stem cells generate airway epithelial tissue. *Am. J. Respir. Cell Mol. Biol.* **32**, 87–92.

Cortiella, J., Nichols, J. E., Kojima, K., Bonassar, L. J., Dargon, P., Roy, A. K., Vacant, M. P., Niles, J. A., and Vacanti, C. A. (2006). Tissue-engineered lung: an *in vivo* and *in vitro* comparison of polyglycolic acid and pluronic F-127 hydrogel/somatic lung progenitor cell constructs to support tissue growth. *Tissue Eng.* **12**, 1213–1225.

Cutz, E., and Orange, R. P. (1977). Mast cells and endocrine (APUD) cells of the lung. *In* "Asthma: Physiology, Immunopharmacology and Treatment" (L. M. Lichtenstein and K. F. Atisten, eds.), pp. 51–72. Academic Press, New York.

Deb, A., Wang, S., Skelding, K. A., Miller, D., Simper, D., and Caplice, N. M. (2003). Bone marrow–derived cardiomyocytes are present in adult human heart: a study of gender-mismatched bone marrow transplantation patients. *Circulation* **107**, 1247–1249.

Denham, M., Cole, T. J., and Mollard, R. (2006). Embryonic stem cells form glandular structures and express surfactant protein-C following culture with dissociated fetal respiratory tissue. *Am. J. Physiol. Lung Cell Mol. Physiol.* **290**, 1210–1215.

Douglas, W. H. J., and Teel, R. W. (1976). An organotypic *in vitro* model system for studying pulmonary surfactant production by type II alveolar pneumocytes. *Am. Rev. Resp. Dis.* **113**, 17.

Douglas, W. H. J., DelVecchio, P., Teel, R. W., Jones, R. M., and Farrell, P. M. (1976a). Culture of type II alveolar lung cells. *In* "Lung Cells in Disease" (A. Bouhuys, ed.), pp. 53–68. Elsevier, New York.

Douglas, W. H. J., Moorman, G. W., and Teel, R. W. (1976b). The formation of histotypic structures from monodisperse fetal rat lung cells cultured on a three-dimensional substrate. *In vitro* **12**, 373–381.

Douglas, W. H. J., Sommers-Smith, S. K., and Johnston, J. M. (1983). Phosphotidate phosphohydrolase activity as a marker for surfactant synthesis in organotypic cultures of type II alveolar pneumocytes. *J. Cell Sci.* **60**, 199–207.

Federspiel, W. J., and Henchir, K. A. (2004). Artificial lungs: basic principles and current applications. *In* "Encyclopedia of Biomaterials and Biomedical Engineering" (G. E. Wnek and G. L. Bowlin, eds.), pp. 910–921. Marcel Dekker, New York.

Ferrari, G., Cusella-De Angelis, G., Coletta, M., *et al.* (1998). Muscle regeneration by bone marrow–derived myogenic progenitors. *Science* **279**, 1528–1530.

Giangreco, A., Shen, H., Reynolds, S. D., and Stripp, B. R. (2004). Molecular phenotype of airway side population cells. *Am. J. Physiol. Lung Cell Mol. Physiol.* **286**, L642–L630.

Goodell, M. A., Brose, K., Paradis, G., Conner, A. S., and Mulligan, R. C. (1996). Isolation and functional properties of murine hematopoietic stem cells that are replicating *in vivo*. *J. Exp. Med.* **183**, 1797–1806.

Gordon, M. Y., Levicar, N., Pai, M., Bachellier, P., Dimarakis, I., Al-Allaf, F., M'hamdi, H., Thalji, T., Welsh, J. P., Marley, S. B., *et al.*, (2006). Characterization and clinical application of human CD34⁺ stem/progenitor cell populations mobilized into the blood by G-CSF. *Stem Cells* online. **24**, 1822–1830.

Grove, J. E., Lutzko, C., Priller, J. Henegariu, O., Theise, N. D., Kohn, D. B., and Krause, D. S. (2002). Marrow-derived cells as vehicles for delivery of gene therapy to pulmonary epithelium. *Am. J. Respir. Cell Mol. Biol.* **27**, 645–651.

Håkelien, A. M., Landsverk, H. B., Robl, J. M., Skalhegg, B. S., and Collas, P. (2002). Reprogramming fibroblasts to express T-cell functions using cell extracts. *Nat. Biotechnol.* **20**, 460–466.

Harris, R. G., Herzog, E. L., Bruscia, E. M., Grove, J. E., Van Arnam, J. S., and Krause, D. S. (2004). Lack of a fusion requirement of development of bone marrow–derived epithelia. *Science* **305**, 90–93.

Ishizawa, K., Kubo, H., Yamada, M., *et al.* (2004). Bone marrow–derived cells contribute to lung regeneration after elastase-induced pulmonary emphysema. *FEBS Lett.* **556**, 249–252.

Jiang, Y., Jahagirdar, B. N., Reinhardt, R. L. Blackstad, M., Reyes, M., and Verfaillie, C. M. (2002). Pluripotency of mesenchymal stem cells derived from adult marrow. *Nature* **418**, 41–49.

Kalina, M., Rikklis, S., and Blau, H. (1993). Pulmonary epithelial cell proliferation in primary culture of alveolar type II cells. *Exp. Lung Res.* **19**, 153–175.

Kleeberger, W., Versmold, A., Rothamel, T., *et al.* (2003). Increased chimerism of bronchial and alveolar epithelium in human lung allografts undergoing chronic injury. *Am. J. Pathol.* **162**, 1487–1494.

Kojima, J., and Vacanti, C. A. (2004). Generation of a tissue-engineered tracheal equivalent. *Biotechnol. Appl. Biochem.* **39**, 257–262.

Kotton, D. N., Ma, B. Y., Cardoso, W. V., Sanderson, E. A., Summer, R. S., Williams, M. C., and Fine, A. (2001). Bone marrow–derived cells as progenitors of lung alveolar epithelium. *Development* **128**, 5181–5188.

Kotton, D. N., Fabian, A. J., and Mulligan, R. C. (2005). Failure of bone marrow to reconstitute lung epithelium. *Am. J. Respir. Cell Mol. Biol.* **33**, 328–334.

Krause, D. S., Theise, N. D., Collector, M. I., Henegariu, O., Hwang, S., Gardner, R., Neutzel, S., and Sharkis, S. J. (2001). Multiorgan, multilineage engraftment by a single bone marrow–derived stem cell. *Cell* **105**, 369–377.

Kubit, V., Sonmez Alpan, E., Zeevi, A., Paradis, I., Dauber, J. H., Iacono, A., Keenan, R., Griffith, B. P., and Yousem, S. A. (1994). Mixed allogeneic chimerism in lung allograft recipients. *Hum. Pathol.* **25**, 408–412.

Lagasse, E., Connors, H., and Al-Dhalimy, M., (2000). Purified hematopoietic stem cells can differentiate into hepatocytes *in vivo*. *Nat. Med.* **6**, 1229–1234.

Lauweryns, J. M., and Cokelaere, M. (1973). Hypoxia-sensitive neuroepithelial bodies: intrapulmonary secretory neuroreceptors modulated by the CNS. *Z. Zellforsch. Milrosk. Anat.* **145**, 521–540.

Loi, R., Beckett, T., Goncz, K. K., Suratt, B. T., and Weiss, D. J. (2006). Limited restoration of cystic fibrosis lung epithelium *in vivo* with adult bone marrow–derived cells. *Am. J. Respir. Crit. Care Med.* **173**, 171–179.

Mattsson, J., Jansson, M., Wernerson, A., and Hassan, M. (2004). Lung epithelial cells and type II pneumocytes of donor origin after allogeneic hematopoietic stem cell transplantation. *Transplantation* **78**, 154–157.

Mezey, E., Chandross, K. J., Harta, G., Maki, R. A., and McKercher, S. R. (2000). Turning blood into brain: cells bearing neuronal antigens generated *in vivo* from bone marrow. *Science* **290**, 1779–1782.

Mondrinos, M. J., Koutzaki, S., Jiwanmall, E., Li, M., Dechadarevian, J. P., Lelkes, P. I., and Finck, C. M. (2006). Engineering three-dimensional pulmonary tissue constructs. *Tissue Eng.* **12**, 717–728.

Okamoto, R., Yajima, T., Yamazaki, M., Kanai, T., Mukai, M., Okamoto, S., Ikeda, Y., Hibi, T., Inazawa, J., and Watanabe, M. (2002). Damaged epithelia regenerated by bone marrow-derived cells in the human gastrointestinal tract. *Nat. Med.* **8**, 1011–1017.

Omori, K., Nakamura, T., Kanemaru, S., Asato, R., Yamashita, M., Tanaka, S., Magrufov, A., Ito, J., and Shimizu, Y. (2005). Regenerative medicine of the trachea: the first human case. *Ann. Otol. Rhinol. Laryngol.* **114**, 429–433.

Orlic, D., Kajstura, J., Chimenti, S., Orlic, D., Kajstura, J., Chimenti, S., Jakoniuk, I., Anderson, S. M., Li, B., Pickel, J., McKay, R., Nadal-Ginard, B., Bodine, D. M., *et al.* (2001). Bone marrow cells generate infarcted myocardium. *Nature* **410**, 701–705.

Ortiz, L. A., Gambelli, F., McBride, C., Gaupp, D., Baddoo, M., Kaminski, N., and Phinney, D. G. (2003). Mesenchymal stem cell engraftment in lung is enhanced in response to bleomycin exposure and ameliorates its fibrotic effects. *Proc. Natl. Acad. Sci. U.S.A.* **100**, 8407–8411.

Paquette, J. S., Tremblay, P., Bernier, V., Auger, F. A., Laviolette, M., Germain, L., Boutet, M., Boulet, L. P., and Goulet, F. (2003). Production of tissue-engineered three-dimensional human bronchial models. *In Vitro Cell Dev. Biol. Anim.* **39**, 213–220.

Paquette, J. S., Moulin, V., Tremblay, P., Bernier, V., Boutet, M., Laviolette, M., Auger, F. A., Boulet, L. P., and Goulet, F. (2004). Tissue-engineered human asthmatic bronchial equivalents. *Eur. Cell Mater.* **10**, 1–11.

Pauwels, R. A., Buist, A. S., Calverley, P. M., Jenkins, C. R., and Hurd, S. S. (2001). Global strategy for the diagnosis, management, and prevention of chronic obstructive pulmonary disease: NHLBI/WHO Global Initiative for Chronic Obstructive Lung Disease (GOLD) Workshop summary. *Am. J. Respir. Crit. Care Med.* **163**, 1256–1276.

Perreira, R. F., O'Hara, M. D., Laptev, A. V., Halford, K. W., Pollard, M. D., Class, R., Simon, D., Livezey, K., and Prockop, D. J. (1998). Marrow stromal cells as a source of progenitor cells for nonhemapoietic tissues in transgenic mice with a phenotype of osteogenesis imperfecta. *Proc. Natl. Acad. Sci. U.S.A.* **95**, 1142–1147.

Petersen, B. E., Bowen, W. C., Patrene, K. D., Mars, W. M., *et al.* (1999). Bone marrow as a potential source of hepatic oval cells. *Science* **284**, 1168–1170.

Pittenger, M. F., MacKay, A. M., Beck, S. C., Jaiswal, R. K., Douglas, R., Mosca, J. D., Moorman, M. A., Simonetti, D. W., Craig, S., and Marshak, D. R. (1999). Multilineage potential of adult human mesenchymal stem cells. *Science* **284**, 143–147.

Polak, J. M., and Bishop, A. E. (2006). Stem cells and tissue engineering: past, present and future. *Ann. New York Acad. Sci.* **1068**, 352–366.

Poliakova, L., Pirone, A., Farese, A., MacVittie, T., and Farney, A. (2004). Presence of nonhematopoietic side population cells in the adult human and nonhuman pancreas. *Transplant. Proc.* **36**, 1166–1168.

Poulsom, R., Forbes, S. J., Hodivala-Dilke, K., Ryan, E., Wyles, S., Navaratnarasah, S., Jeffery, R., Hunt, T., Alison, M., Cook, T., Pusey, C., and Wright, N. A. (2001). Bone marrow contributes to renal parenchymal turnover and regeneration. *J. Pathol.* **195**, 229–235.

Prockop, D. J. (1997). Marrow stromal cells for nonhemapoietic tissues. *Science* **276**, 71–74.

Qin, M. D., Tai, G. P., Collas, P., Polak, J. M., and Bishop, A. E. (2005). Cell extract–derived differentiation of embryonic stem cells. *Stem Cells* **23**, 712–718.

Rippon, H. J., and Bishop, A. E. (2004). Embryonic stem cells. *Cell Proliferation* **27**, 23–34.

Rippon, H. J., Ali, N. N., Polak, J. M., and Bishop, A. E. (2004). Initial observations on the effect of medium composition on the differentiation of murine embryonic stem cells to alveolar type II cells. *Cloning Stem Cells* **6**, 49–56.

Rippon, H. J., Polak, J. M., Qin, M., and Bishop, A. E. (2006). Derivation of distal lung epithelial progenitors from murine embryonic stem cells using a novel 3-step differentiation protocol. *Stem Cells.* **24**, 1389–1398.

Roberts, A., Wyslouzil, B. E., and Bonassar, L. (2005). Aerosol delivery of mammalian cells for tissue engineering. *Biotechnol. Bioeng.* **91**, 801–807.

Saito, K., Lwebuga-Mukasa, J., Barrett, C., Light, D., and Warshaw, J. B. (1985). Characteristics of primary isolates of alveolar type II cells from neonatal rats. *Exp. Lung Res.* **8**, 213–225.

Samadikuchaksaraei, A., Cohen, S., Isaac, K., Rippon, H. J., Polak, J. M., Bielby, R. C., and Bishop, A. E. (2006). Derivation of type II pneumocytes from human embryonic stem cells. *Tissue Eng.* **12**, 867–875.

Shannon, J. M., Mason, R. J., and Jennings, S. D. (1987). Functional differentiation of alveolar type II epithelial cells *in vitro*: effects of cell shape, cell–matrix interactions and cell–cell interactions. *Biochim. Biophys. Acta* **931**, 143–156.

Spees, J. L., Olson, S. D., Ylostalo, J., *et al.* (2003). Differentiation, cell fusion, and nuclear fusion during *ex vivo* repair of epithelium by human adult stem cells from bone marrow stroma. *Proc. Natl. Acad. Sci. U.S.A.* **100**, 2397–2402.

Sugihara, H., Toda, S., Miyabara, S., Fujiyama, C., and Yonemitsu, N. (1993). Reconstruction of alveolus-like structure from alveolar type II epithelial cells in three dimensional collagen gel matrix culture. *Am. J. Pathol.* **142**, 783–792.

Sullivan, S. D., Ramsey, S. D., and Lee, T. A. (2000). The economic burden of COPD. *Chest* **117**, 5S–9S.

Summer, R., Kotton, D. N., Sun, X., Ma, B., Fitzsimmons, K., and Fine, A. (2003). SP (side population) cells and Bcrp1 expression in lung. *Am. J. Physiol. Lung Cell Mol. Physiol.* **285**, L97–L104.

Summer, R., Kotton, D. N., Sin, X., Fitzsimmons, K., and Fine, A. (2004). Origin and phenotype of lung side population cells. *Am. J. Physiol.* **287**, 477–483.

Suratt, B. T., Cool, C. D., Serls, A. E., Chen, L., Varella-Garcia, M., Shpall, E. J., Brown, K. K., and Worthen, G. S. (2003). Human pulmonary chimerism after hematopoietic stem cell transplantation. *Am. J. Respir. Crit. Care Med.* **168**, 318–322.

Tan, A., Romanska, H. M., Lenza, R., Jones, J., Hench, L. L., Polak, J. M., and Bishop, A. E. (2002). The effect of 58S bioactive sol-gel-derived foams on the growth of murine lung epithelial cells. *Bioceramics* **15**, 719–724.

Terada, N., Hamazaki, T., Oka, M., Hoki, M., Mastalerz, D. M., Nakano, Y., Meyer, E. M., Morel, L., Petersen, B. E., and Scott, E. W. (2002). Bone marrow cells adopt the phenotype of other cells by spontaneous cell fusion. *Nature* **416**, 542–545.

Theise, N. D., Nimmakayalu, M., Gardner, R., Illei, P. B., Morgan, G., Teperman, L., Henegariu, O., and Krause, D. S. (2000). Liver from bone marrow in humans. *Hepatology* **32**, 11–16.

Theise, N. D., Henegariu, O., Grove, J., Jagirdar, J., Kao, P. N., Crawford, J. M., Badve, S., Saxena, R., and Krause, D. S. (2002). Radiation pneumonitis in mice: a severe-injury model for pneumocyte engraftment from bone marrow. *Exp. Hematol.* **30**, 1333–1338.

Thomson, J. A., Itskovitz-Eldor, J., Shapiro, S. S., Wakintz, M. A., Swiergiel, J. J., Maeshall, V. S., and Jones, J. M. (1998). Embryonic stem cells derived from human blastocysts. *Science* **282**, 1145.

Van Vranken, B., Romanska, H. M., Polak, J. M., Rippon, H. J., Shannon, J. M., and Bishop, A. E. (2005). Coculture of embryonic stem cells with pulmonary mesenchyme: a microenvironment that promotes differentiation of pulmonary epithelium. *Tissue Eng.* **11**, 1177–1187.

Vassilopoulos, G., Wang, P. R., and Russell, D. W. (2003). Transplanted bone marrow regenerates liver by cell fusion. *Nature* **422**, 901–904.

Wollert, K. C., and Drexler, H. (2005). Clinical applications of stem cells for the heart. *Circ. Res.* **96**, 151–163.

Wulf, G., Jackson, K., Brenner, M., and Goodell, M. (2003). Cells of the hepatic side population contribute to liver regeneration and can be replaced with BM stem cells. *Haematologica* **88**, 368–378.

Yamada, M., Kubo, H., Kobayashi, S., Ishizawa, K., Numasaki, M., Ueda, S., Suzuki, T., and Sasaki, H. (2004). Bone marrow–derived progenitor cells are important for lung repair after lipopolysaccharide-induced lung injury. *J. Immunol.* **172**, 1266–1272.

Yannas, I. V., Lee, E., Orgill, D. P., Skrabut, E. M., and Murphy, G. F. (1989). Synthesis and characterization of a model extracellular matrix that induces partial regeneration of adult mammalian skin. *Proc. Natl. Acad. Sci.* **86**, 933–937.

Ying, Q. L., Nichols, J., Evans, E. P., and Smith, A. G. (2002). Changing potency by spontaneous fusion. *Nature* **416**, 545–548.

Zander, D. S., Baz, M. A., Cogle, C. R., *et al.* (2005). Bone marrow–derived stem-cell repopulation contributes minimally to the Type II pneumocyte pool in transplanted human lungs. *Transplantation* **80**, 206–212.

# Part Twenty

## Skin

# Chapter Seventy-Five

# Cutaneous Stem Cells

*George Cotsarelis*

## I. INTRODUCTION

The field of epithelial stem cells is progressing rapidly, largely because of technical advances in molecular and cellular biology. Many astute observations over the last century led to predictions about epithelial stem cells that only recently have been definitively addressed through new techniques. Originally, putative epithelial stem cells were identified in the hair follicle bulge as quiescent "label-retaining cells." The study of these cells was hindered until the identification of bulge cell molecular markers, such as CD34 expression and K15 promoter activity. This allowed for the isolation and characterization of bulge cells from mouse follicles. Bulge cells possess stem cell characteristics, including multipotency, high proliferative potential, and their cardinal feature of quiescence. Lineage analysis demonstrated that all epithelial layers within the adult follicle and hair originated from bulge cells. Bulge cells only contribute to the interfollicular epidermis during wound healing. But after isolation, when combined with neonatal dermal cells, they regenerate new hair follicles, epidermis, and sebaceous glands. Bulge cells maintain their stem cell characteristics after propagation *in vitro*, so ultimately they may be useful for tissue-engineering applications. The isolation, characterization, and manipulation of stem cells in the skin holds promise for their future use in clinical applications for treatment of a variety of skin disorders, including hair loss (alopecia), wounds, pigmentary disorders, and skin aging. Here, we review the evolution of current thinking on cutaneous epithelial stem cells and delineate future challenges for applying our knowledge to reap clinical benefits.

The epidermis consists of multiple cell types and layers. The outermost layer, called the *stratum corneum*, is composed of dead corneocytes, which adhere tightly to each other to form a hydrophobic barrier that protects us from the environment and prevents water loss. The innermost layer at the base of the epidermis generates new cells that migrate to the surface while terminally differentiating and eventually forming the stratum corneum. In addition to keratinocytes, which produce intermediate filament proteins called *keratins* and constitute the majority of the cells, the epidermis houses melanocytes, which produce pigment, and Langerhans cells, which present antigen.

The epidermal surface is interrupted by orifices arising from adnexal structures, such as the hair follicle and sweat

*Principles of Tissue Engineering, 3rd Edition*
ed. by Lanza, Langer, and Vacanti

glands. The epidermis continues down into these structures, and cells from hair follicles and sweat ducts can move out and repopulate the epidermis after wounding. The adnexal structures possess a greater degree of tissue complexity than the epidermis. For example, in contrast to the stratified squamous epithelium of the epidermis, the hair follicle consists of at least eight different concentric layers of epithelia, which undergo degeneration and regeneration with each hair follicle cycle.

## II. WHAT ARE EPITHELIAL STEM CELLS?

Because self-renewing tissues, such as the epidermis and the hair follicle, continuously generate new cells to replenish the dead corneocytes and hairs, which are sloughed into the environment, their homeostasis is thought to be dependent on epithelial stem cells. Therefore, perhaps the simplest definition of an epithelial stem cell is based on lineage: *A stem cell is the cell of origin for terminally differentiated cells in adult tissues.* For example, tracing the lineage of a corneocyte or hair cell back to its ultimate source in adult skin leads to a stem cell. However, because the tools required to perform lineage analysis have not been available until recently, investigators have principally adopted definitions from the hematopoietic system. In particular, stem cells were felt to be self-renewing, multipotent, and clonogenic, similar to stem cells in the hematopoietic system, which can regenerate all of the blood lineages from one cell. In contrast to hematopoietic stem cells, cutaneous epithelial stem cell biologists also relied heavily on quiescence as a major stem cell characteristic. This can be attributed to the pioneering work of Bickenbach and Mackenzie (1984), who devised "label-retaining cell" methods for detecting quiescent cells in the epidermis, and of Morris and Potten, who showed that these cells retained carcinogens and possessed proliferative characteristics of stem cells (reviewed in Morris, 2000).

## III. LOCALIZATION OF EPITHELIAL STEM CELLS

With respect to the epidermis (which in this chapter refers to the interfollicular epidermis), we know that cells are generated through proliferation that occurs only in the basal layer; therefore, stem cells must be located there. Within the basal layer, however, keratinocytes display heterogeneous proliferative characteristics. In mouse skin, individual basal cells divide less frequently than the surrounding cells. The slower-cycling central cell and more rapidly proliferating surrounding cells constitute approximately 10 basal cells and are roughly organized into a hexagonal unit, which lies beneath a single keratinocyte (reviewed in Kaur, 2006). Based on these proliferative and morphological characteristics, the term *epidermal proliferative unit* (EPU) was coined to describe this architecture (Morris, 2000; Kaur, 2006). Without the benefit of direct

lineage analysis, it was assumed that the central cell within the EPU generates the rapidly proliferating cells, termed *transient* or *transit amplifying* (TA) cells, which move laterally and then differentiate and move upward. Thus, within the epidermis, the main source of cells, i.e., the stem cells, responsible for continual epidermal renewal appear to reside in the center of the EPU, and testing of this concept in unmanipulated epidermis on the back of the mouse supports this view (Ito *et al.*, 2005).

Similar to the epidermis, the hair follicle generates a terminally differentiated keratinized end product, the hair shaft, which is eventually shed. Tracing back a hair shaft cell to its origin in adult skin is not straightforward. In contrast to epidermis, the follicle undergoes cyclical regeneration and has a more complicated proliferative profile and architecture, with at least eight different epithelial lineages (Fig. 75.1). Hair is formed by rapidly proliferating matrix keratinocytes in the bulb located at the base of the growing (anagen) follicle. The duration of anagen varies drastically between hairs of differing lengths. For example, mouse hair and human eyebrow hair follicles stay in anagen for only two to four weeks, while scalp follicles can remain in anagen for many years. Nevertheless, matrix cells eventually stop proliferating, and hair growth ceases at catagen when the lower follicle regresses to reach a stage of rest (telogen). After telogen, the lower hair-producing portion of the follicle regenerates, marking the new anagen phase.

Because the lower portion of the follicle cyclically regenerates, hair follicle stem cells were thought to govern this growth. Historically, hair follicle stem cells were assumed to reside exclusively in the secondary germ (Fig. 75.2), which is located at the base of the telogen hair follicle. It was felt that the secondary germ moved downward to the hair bulb during anagen and provided new cells for production of the hair. At the end of anagen, the secondary germ was thought to move upward with the dermal papilla during catagen to come to rest at the base of the telogen follicle. This scenario of stem cell movement during follicle cycling was brought into question when a population of long-lived presumptive stem cells was identified using label-retaining cell methods in an area of the follicle surrounding the telogen club hair and not in the hair bulb (Cotsarelis *et al.*, 1990). That the presumptive stem cells localized to a previously defined area called the *bulge* was not appreciated until I read a description of the human embryonic follicle by Hermann Pinkus (1958):

> This *bulge*, often the most conspicuous detail of the young germ, is as large as the bulb. . . . The function of the bulge is obscure. While it serves as the point of insertion of the arrector muscle later in life, it develops much earlier than the muscle and the latter seems to originate quite independently in the skin near the sebaceous gland, and in many instances streaks by the bulge before approaching the lower follicle below it. Unna (1876) named the bulge area of the adult follicle the *hair bed* (Haarbett), believing that the club hair became implanted there and derived

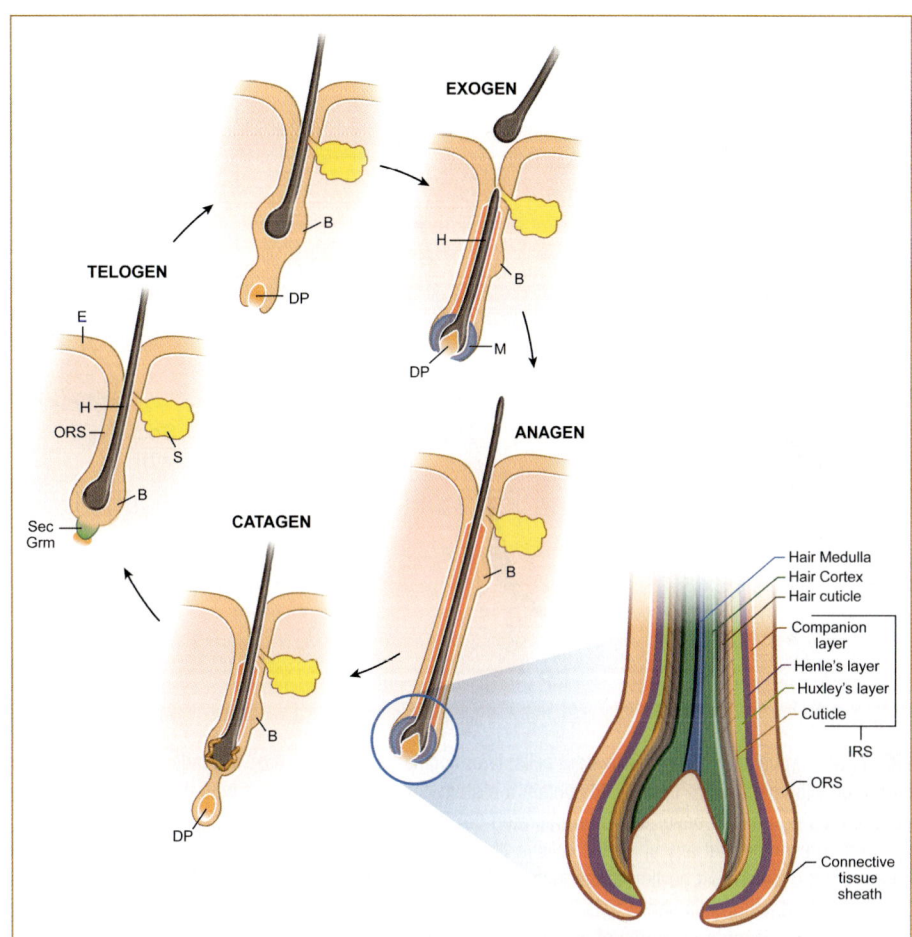

**FIG. 75.1.** Hair cycle and anatomy. The hair follicle cycle consists of stages of rest (telogen), hair growth (anagen), follicle regression (catagen), and hair shedding (exogen). The entire lower epithelial structure is formed during anagen and regresses during catagen. The transient portion of the follicle consists of matrix cells in the bulb that generate seven different cell lineages, three in the hair shaft and four in the inner root sheath (IRS).

additional growth from it. Stöhr gave it the neutral name "Wulst" (bulge or swelling). Some texts state that this is an area of marked proliferative activity, but no mitotic figures were observed in the bulge even if other parts of the follicle contained them. Whatever its function, the bulge marks the lower end of the "permanent follicle" later in life. Everything below it is expendable during the hair change [cycle].

This remarkable description, based purely on morphological observations, portended the characterization of the bulge cells in both human and mouse follicles as an area containing quiescent cells important for hair follicle cycling (Cotsarelis *et al.*, 1990; Lyle *et al.*, 1998; Morris *et al.*, 2004). In the mouse pelage follicle, the area analogous to the human bulge becomes morphologically apparent in the postnatal period during the first telogen stage at the site of arrector pili muscle attachment. The shape of the bulge in the mouse follicle results from displacement of the outer root sheath (ORS) by the club hair. In the human follicle, the bulge appears as a true thickening of the ORS, but it generally becomes much less apparent with age (Fig. 75.3).

The lack of markers for bulge cells hindered the study of this area. To date, probably the best (most specific) marker for mouse hair follicle bulge cells is CD34 expression, as first

defined by Trempus *et al.* (2003). CD34, which interestingly is also a hematopoietic stem cell marker in the human but not the mouse, highlights the bulge cells specifically within the cutaneous epithelium. Although it is also expressed by cells in the dermis, it is a cell surface protein, and antibodies recognizing CD34 can be used to collect viable bulge cells by fluorescent activated cell sorting. This allows for the isolation and molecular characterization of the bulge cells.

Keratin 15 expression in human bulge cells was first described by Lyle *et al.* (1998) (Fig. 75.3). K15 mRNA and protein are reliably expressed at high levels in the bulge, but lower levels of expression can be present in the basal layers of the lower follicle outer root sheath and the epidermis. Thus the use of K15 expression as the sole criterion for defining a bulge cell is not advisable. However, a K15 promoter used for generation of transgenic mice possesses a pattern of activity restricted to the bulge in the adult mouse (Liu *et al.*, 2003). This proved to be a powerful tool for studying bulge cells and is discussed later.

A salient feature of the bulge cells is their quiescence. In both adult mouse and human skin grafted to immunodeficient mice, the administration of nucleoside analogs, such as tritiated thymidine or bromodeoxyuridine, which are taken up by cells in S-phase, does not result in labeling of

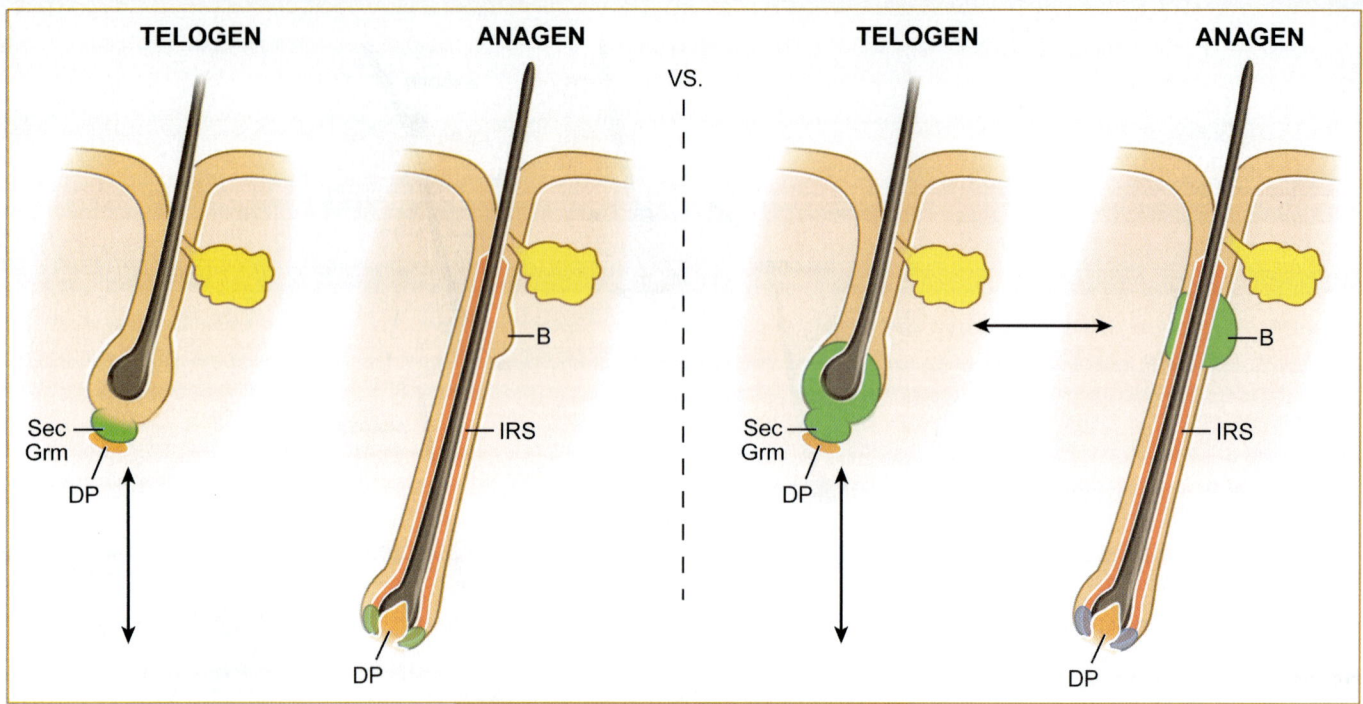

**FIG. 75.2.** Location of hair follicle stem cells: two models. In one view, the "secondary germ cells," named for their similarity to primary germ cells present during development, were thought to contain the stem cells for the follicle. It was thought that these cells migrated from the base of the telogen follicle to the bulb during anagen onset and then migrated back up during catagen. To date, no direct evidence (e.g., label-retaining cell studies or lineage analysis) has been presented that any epithelial cells in the bulb survive during catagen. The secondary germ cells found at the base of the telogen follicle appear to arise from the lowermost portion of the bulge at the end of catagen (Ito *et al.*, 2004). Based on morphology, the telogen secondary germ generates the new hair and IRS, although this needs to be addressed experimentally. The origin and fate of the matrix keratinocytes in the bulb, which possess an undifferentiated phenotype, is an area worthy of exploration. It is not known whether bulge cells migrate down the follicle during anagen to supply the matrix cells continuously with new cells or whether the matrix cells self-renew throughout anagen.

the bulge cells, except at anagen onset (Ito *et al.*, 2002; Ito *et al.*, 2004). Once labeled as either neonates or during anagen onset, when stem cells are proliferating, bulge cells can remain labeled for 14 months in the mouse and at least four months (the longest period examined) in the human (Lyle *et al.*, 1998). This prolonged quiescence is remarkable, given that the surrounding cells proliferate at a much higher rate, which suggests that bulge cells persist for the lifetime of the organism.

Once the bulge cells were identified as presumptive epithelial stem cells, based on their ability to retain label, it was necessary to evaluate whether these cells possessed other expected characteristics important for stem cells. In particular, did they exhibit a high proliferative potential? Were they multipotent, generating all epithelial cell types within the follicle as well as epidermis and sebaceous gland? Did they have a molecular signature of stemness shared with other stem cell populations? These questions were important not only from a biological perspective, but also from a therapeutic one. A better understanding of the role of bulge cells in renewal of the epithelium should provide insights into wound healing, gene therapy, aging, and carcinogenesis.

## IV. *IN VITRO* ASSESSMENT OF PROLIFERATIVE POTENTIAL

Proliferative capability can be assayed *in vitro* by examining the clonogenicity of individual keratinocytes through serial passage. Approximately 5% of adult epidermal basal cells possess a "holoclone" phenotype characterized by high reproductive capacity and low level of terminal differentiation. These cells are thought to represent stem cells (Barrandon and Green, 1987). Colony-forming efficiency (CFE; colonies per number of cells plated) serves as another indicator of proliferative capability and is thought to correlate with the number of stem cells in a tissue (Kobayashi *et al.*, 1993). These kinds of analyses revealed that the epidermis possesses three major types of cells, based on their proliferative capacity *in vitro*. Terminally differentiated cells have no proliferative capacity, TA cells have a limited capacity (forming abortive small colonies), and epidermal progenitor cells have clonogenic capacity (Jones *et al.*, 1995). For the epidermis, the number of colony-forming units is likely representative of the stem cell population, although direct proof, through lineage analysis, that all central cells within the EPU are holoclones is lacking.

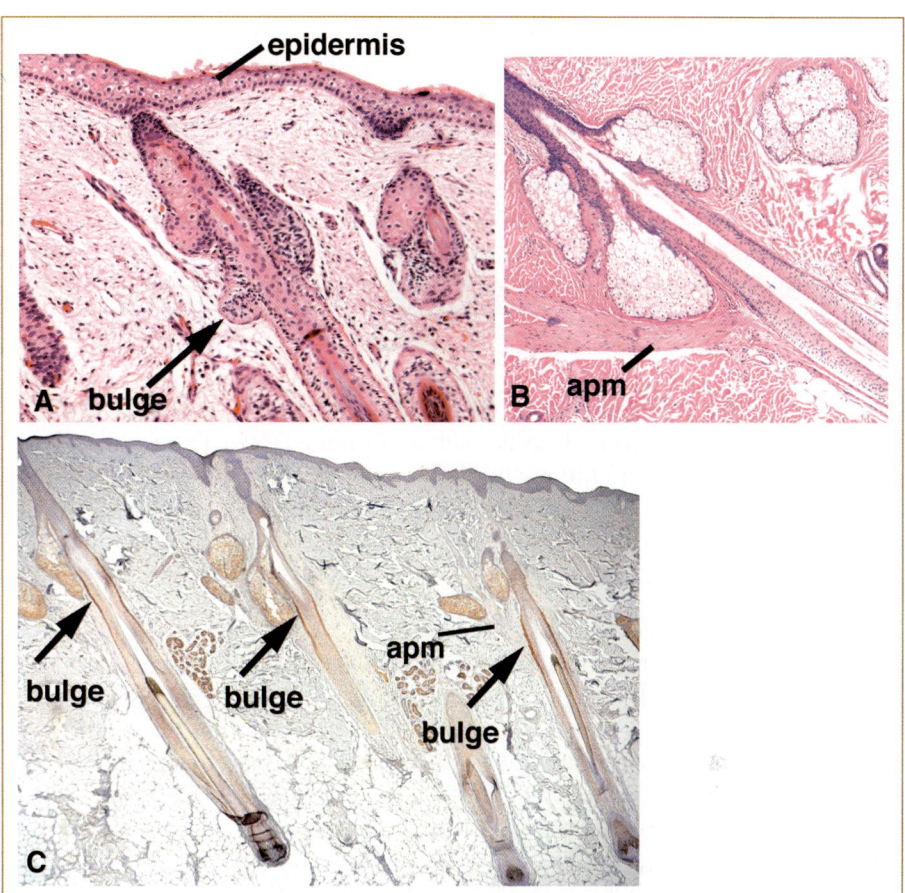

**FIG. 75.3.** The bulge is a prominent structure in fetal skin **(A)**, but generally is not morphologically distinct in the adult **(B)**. Immunostaining for keratin 15 in scalp preferentially detects bulge cells **(C)**. apm, arrector pili muscle.

Similar analyses have also been applied to the hair follicle epithelium, with mixed and sometimes contradictory results. In rat vibrissae (whisker follicles), CFE analysis localizes the large majority (>95%) of holoclones to the hair follicle bulge (Kobayashi *et al.*, 1993). However, in human hair follicles, CFE was reported greatest outside of the bulge region, in the lower outer root sheath (Rochat *et al.*, 1994). This was surprising, since this portion of the follicle undergoes degeneration during catagen, and these data are not in line with *in vivo*, label-retaining studies that localize LRC to the human hair follicle bulge (Lyle *et al.*, 1998). Possible explanations for the culture results are that current culture conditions actually support propagation of progenitor cells in addition to hair follicle stem cells or that culturing keratinocytes leads to "reprogramming" events that overestimate the number of true stem cells within a tissue.

Some of the discrepancies are explained by studies on vibrissa follicles, which show that the stage of the follicle at the time of isolation influences how well cells grow in culture (Oshima *et al.*, 2001). Since human hair follicles are not synchronized in their growth relative to one another, and since the anagen stage in human follicles lasts for many years, each isolated anagen follicle that was analyzed for CFE may have yielded very different results (Rochat *et al.*, 1994). For example, a follicle that has been in anagen for only one

month might produce a greater number of colonies from the bulge than one in anagen for six years, which may have more colonies from lower portions of the follicle.

Other types of *in vitro* studies based on epithelial outgrowths from explanted follicles have also produced conflicting results. Whole follicle explant cultures give rise to epithelial growths from the ORS (Yang *et al.*, 1993; Moll, 1996). In one study, the outgrowths were seen in the area believed to represent the bulge of human hair follicles (Yang *et al.*, 1993). In another, many outgrowths were seen in the upper central ORS, which was believed to be lower than the bulge but which likely includes the lower portion of what is the bulge (Moll, 1996). Again, these differences may be related to the heterogeneity of anagen in human follicles. More recently, however, Roh *et al.* (2005) decreased the influence of the hair follicle cycle by concentrating on telogen follicles, which possess a well-defined bulge area. They demonstrated major differences between bulge cells and matrix cells *in vitro*. The bulge cells possessed a higher proliferative potential than the matrix cells, while the matrix cells possessed a very mobile phenotype, demonstrated by time-lapse photography (matrix keratinocytes migrated 0.7 μm/min, compared to 0.1 μm/min for bulge cells). These data support the idea that bulb matrix keratinocytes are TA cells and possess a finite proliferative potential. This concept was originally proposed as an explanation for why anagen follicles eventually enter catagen (Cotsarelis *et al.*, 1990), although this needs further testing.

Recent data using CFE and holoclone analysis remain mixed, although most of the data support the idea that the bulge cells possess a high proliferative potential. Isolated bulge cells grown on feeder cells *in vitro* formed more numerous colonies than nonbulge keratinocytes (Trempus *et al.*, 2003). Morris showed that isolated bulge cells formed larger and more numerous colonies (Morris *et al.*, 2004). Blanpain *et al.* (2004) reported that the large majority of holoclones originate from the bulge in the adult mouse, while basal epidermis did not form any holoclones, although both of these populations had similar colony-forming efficiency. Given that similar culture conditions were used for all three studies, these findings may reflect the inherent variability and difficulty in interpreting *in vitro* proliferation studies (Joseph and Morrison, 2005).

## V. MULTIPOTENT BULGE CELLS?

If the stem cells of the hair follicle are located in the bulge, then these cells should give rise to all of the lower hair follicle epithelial cell layers. Early evidence supporting the concept that bulge cells generate the lower follicle includes proliferation studies showing that bulge cells proliferate preferentially at anagen onset (Wilson *et al.*, 1994; Ito *et al.*, 2004). More convincing evidence suggesting that the lower follicle originates from bulge cells came from *in vivo* labeling studies and transplantation studies. Taylor used a double-labeling technique to trace the progeny of bulge cells in intact pelage follicles (Taylor *et al.*, 2000). Faint labeling in a speckled pattern was observed in some cells of the lower follicle, suggesting that these cells had indeed originated in the bulge. Similarly, Tumbar *et al.* (2004) used persistence of GFP label as an indication that lower epithelial cells were progeny of the bulge cells. Neither study provided convincing evidence that all hair matrix keratinocytes in the bulb originated from bulge cells, and both suffered from an inability to mark bulge cells and their progeny permanently. Oshima *et al.* (2001) took a different approach and transplanted bulge regions from vibrissa follicles isolated by dissection from ROSA26 mice into non-ROSA follicles that were then grafted under the kidney capsule of an immunocompromised mouse. ROSA26 mice express *lacZ* under the control of the ubiquitous ROSA promoter, so the fate of the transplanted cells could be followed. After several weeks, they found labeled cells in the lower follicle, indicating that bulge cells or their progeny had migrated down the vibrissa follicle. At later time points, some follicles expressed *lacZ* in all epithelial cell layers of the lower follicle, suggesting that bulge cells do generate all of the cells of the lower follicle. These elegant studies were limited because of the unclear starting cell population, the manipulation performed during grafting, and the use of vibrissa follicles, which are markedly different than other mouse and human follicles.

More definitive evidence for bulge cell multipotency *in vivo* was reported using the K15 promoter to target these cells with an inducible Cre (CrePR1) construct (Morris *et al.*, 2004). CrePR1 is a fusion protein consisting of Cre-recombinase and a truncated progesterone receptor that binds the progesterone antagonist RU486 (Berton *et al.*, 2000). In *K15-CrePR1* transgenic mice, CrePR1 remains inactive in the cytoplasm of the K15-positive cells, except during RU486 treatment, which permits CrePR1 to enter the nucleus and catalyze recombination. We crossed the *K15-CrePR1* mice with *R26R* reporter mice (Berton *et al.*, 2000) that express *LacZ* under the control of a ubiquitous promoter after Cre-mediated removal of an inactivating sequence. Transient treatment of adult *K15-CrePR1;R26R* mice with RU486 results in permanent expression of *LacZ* in the bulge cells and in all progeny of the labeled bulge cells. From this approach, it is clear that cells originating in the bulge generated all epithelial cell types in the lower hair follicle.

Although these studies convincingly showed that the bulge cells as a whole gave rise to the lower follicle, none of these approaches addressed the question of how many stem cells participate in the formation of the new anagen follicle or at what point bulge cells or their progeny are committed to specific lineages (Kamimura *et al.*, 1997; Ghazizadeh and Taichman, 2001; Kopan *et al.*, 2002). Several studies have examined lineage within the follicle using reconstitution assays in which isolated keratinocytes are combined with neonatal dermal cells and grafted onto a nude mouse skin (Weinberg *et al.*, 1993). The mixture then forms hair follicles. By using a combination of keratinocytes labeled with a retrovirus encoding alkaline phosphatase plus nonlabeled keratinocytes for the epithelial component of the assay and then examining the proportion of labeled cells within the follicle, an indication of the number of clones within a follicle can be inferred (Kamimura *et al.*, 1997; Topley *et al.*, 1999). Dotto and Cotsarelis (2005) suggested that the majority of reconstituted follicles arose from three different clone types: one each for the ORS, the IRS, and the hair shaft. Other combinations were also present, including some follicles with partial labeling within one of the layers. Only rarely were entire follicles labeled, indicating that they may have arisen from single cells. The limitations of the reconstitution studies include the manipulation of the cells required for the reconstitution of the follicle, which likely mimics a wound-healing environment that is not normal; the resulting follicles lack normal orientation and often grow at right angles to each other — this limits the ability to accurately orient the follicles in tissue sections and makes analysis of clones difficult; the growth cycle of the follicles is not synchronized, making the evaluation of the hair cycle almost impossible. This especially limits evaluation of the bulge, which is most evident during telogen. Lastly, silencing of the reporter construct or its inadequate expression in follicle subpopulations may complicate matters as well.

Kopan, using chimeric mice, presented evidence for approximately four different clones within a follicle, but each with the ability to generate all epithelial cell types within the follicle (Kopan *et al.*, 2002). These chimeric studies were also limited by manipulation during embryogenesis. Related studies by the same group looked at X-inactivation in adult mice, and the findings support the concept that hair follicle stem cells are multipotent. But these studies were limited by the use of vibrissa follicles analyzed during the neonatal time period. Another recent study, examining Cre mice exhibiting recombination events of low frequency within their cells, also recognized several different types of clones responsible for different follicle layers (Legue and Nicolas, 2005). None of these three studies assessed clones within the bulge during telogen, so the question of how many and which bulge cells participate in hair follicle formation at anagen onset remains unanswered.

Ultimately, formal lineage analysis will be needed to elucidate the exact contribution of individual bulge cells to the different lineages within the follicle. Questions remaining include: At what point, if any, are stem cells set aside during development? When are hair follicle lineages established? That is, are individual bulge cells committed to a specific lineage, or is each bulge cell multipotent? How many bulge cells are required to generate a new lower anagen follicle, and is the number different based on the size and type of follicle? The answers may be important for designing cell-based treatments for alopecia (Stenn and Cotsarelis, 2005).

## VI. PLASTICITY OF BULGE CELLS?

Intriguing results suggest that cells isolated from the mouse hair follicle bulge area have the potential to differentiate into multiple different nonepithelial tissues, including Schwann cells and neurons (Amoh *et al.*, 2005). These cells were isolated from nestin-eGFP-transgenic mice (Li *et al.*, 2003). Nestin is a neural stem cell marker. The exact origin of the nestin positive cells and whether they represent a single or mixed population of cells need to be addressed. Because nestin was not differentially expressed in mouse or human bulge cells (Morris *et al.*. 1986; Tumbar *et al.*, 2004; Ohyama *et al.*, 2006), it is unlikely that these cells represent keratinocytes. However, regardless of their origin, nestin-positive cells from the skin possess remarkable regenerative capabilities when used in a nerve injury model (Amoh *et al.*, 2005). Similar *in vitro* plasticity findings using cells isolated from human follicles were also reported (Yu *et al.*, 2006); thus, the follicle could serve as a source of cells for a variety of tissue-engineering applications in the future.

## VII. BULGE CELLS: THE ULTIMATE CUTANEOUS EPITHELIAL STEM CELLS?

In addition to the role of the bulge cells in the formation of the hair follicle at anagen onset is a question of whether bulge cells are necessary for the homeostasis of the epidermis and sebaceous gland. One view is that bulge cells continuously provide progeny that repopulate these tissues (Lavker and Sun, 2000; Taylor *et al.*, 2000). If true, then ablation of the bulge should lead to failure of epidermal renewal. Recent experiments show that loss of bulge cells does not lead to loss of the epidermis, suggesting that the epidermis possesses stem cells capable of renewing themselves for long periods. Other studies have shown long-term persistence of clones within the epidermis. These studies utilized reconstitution assays (Kamimura *et al.*, 1997; Ghazizadeh and Taichman, 2001). Some evidence was provided that hair follicle–derived cells did move into the epidermis, especially adjacent to the hair follicle, but it was not clear that these cells were derived from the bulge. Morris *et al.* (2004) and Tumbar *et al.* (2004) did find that bulge cells can move into the epidermis, but this was not systematically studied using lineage analysis. Recent work in which isolated bulge keratinocytes were injected into neonatal mice and then followed showed that bulge cells contributed to the epidermis after engrafting, but eventually no cells arising from the injected cells were found in the epidermis, implying lack of movement of bulge cells to the epidermis (Claudinot *et al.*, 2005).

Studies addressing directly the questions of whether and when bulge cell progeny migrate to the epidermis have recently been accomplished. Using the K15CrePR;R26R-transgenic mouse, bulge cells were labeled in three-week-old mice (Morris *et al.*, 2004). Mice were followed for six months to determine the movement of bulge-derived cells into the epidermis. At the end of the experiment there was no evidence that bulge-derived cells had migrated to epidermis. Thus, under homeostatic conditions, the epidermis alone is responsible for its continual renewal. These findings are important for gene therapy since it will be important to target stem cells within the epidermis to correct genetic disorders such as epidermolysis bullosa (a congenital blistering disorder).

## VIII. ROLE OF BULGE CELLS IN WOUND HEALING

The contribution of the hair follicle to healing of the epidermis following wounding has been appreciated for decades by investigators working with mice and rabbits (Argyris, 1976). Clinicians are also well aware that keratinocytes emerge from the follicle to repopulate wounds. However, the role of the bulge cells in wound healing has only recently been characterized.

Bulge cell progeny migrate to the epidermis after different types of wounding. Using the K15CrePR;R26R-transgenic mouse, bulge cells were labeled in adult mice (Ito *et al.*, 2005). Excisional wounding with a 4-mm (punch) trephine resulted in the migration of bulge cell progeny into the healing epidermis. At least 25% of the newly formed epidermis originated from the bulge cells. Bulge cells were also stimulated to move into the epidermis following incisional wounds and after tape stripping, indicating that bulge cell activation plays a role in replenishing lost cells from the epidermis after wounding. Surprisingly, however, despite the presence of bulge-derived cells in the basal layer of the reepithelialized epidermis, the majority of the bulge-derived cells did not persist in the regenerated epidermis. This suggests that bulge cells and epidermal stem cells are intrinsically different, in that epidermal-derived cells seem better suited for establishing long-term epidermal proliferative units.

## IX. ROLE OF BULGE CELLS IN TUMORIGENESIS

Since epithelial stem cells are thought to have a life span at least as long as that of the organism, they are thought to be susceptible to multiple genetic "hits" that cumulatively may result in tumor formation. A great deal of evidence in the mouse system points to hair follicles and stem cells as the origin of many skin tumors (Morris, 2000). Keratinocyte stem cells appear to be the target of cutaneous carcinogens, and, since they are the slowest-cycling cells, hair follicle stem cells also retain carcinogens for extended periods and are thus more susceptible to tumor promotion (Morris, 2000).

Since human basal cell carcinomas (BCCs) are slowly growing tumors, composed of poorly differentiated cells with the ability to differentiate into various adnexal structures, it has been suggested that the cell of origin is a multipotent stem cell that is slowly cycling and has a high proliferative potential (Cotsarelis et al., 1990). There is considerable evidence that BCCs may arise from stem cells within the hair follicle (Hutchin et al., 2005). BCCs have been shown to express bcl-2 (Verhaegh et al., 1997), a marker of the permanent portion of the hair follicle, including the bulge (Stenn et al., 1994). Furthermore, the overexpression of sonic hedgehog (SHH) in mouse epidermis causes BCC-like tumors from invaginating hair follicles but has little effect on interfollicular epidermis (Oro et al., 1997), again suggesting a follicular origin for BCCs. The SHH signaling pathway, including ptc, is active in hair follicle morphogenesis during fetal development (Millar, 2002).

Because trichoepitheliomas (TEs), a benign type of hair follicle tumor, and BCCs have a similar clinical and histological appearance and may develop concurrently in some patients, it has been proposed that these tumors arise from a common precursor cell type within the hair follicle (Headington, 1976). Biochemically, both TEs and BCCs express cytokeratins 5, 6, 14, 17, and 19 (Schirren et al., 1997), a profile found in the infrainfundibular outer root sheath cells of the normal hair follicle, in fetal germ cells, and in the fetal hair follicle bulge. In addition, in one study,

13 of 13 TEs and about one-third of BCCs expressed K15, suggesting they originate from bulge cells (Jih et al., 1999). The findings that both sporadic TEs and BCCs contain mutations of the *patched* (*ptc*) gene (Johnson et al., 1996) provide compelling evidence that these tumors share a common tumorigenic mechanism and are linked to the hair follicle. The normal expression of *ptc* in the developing hair follicle and the presence of *ptc* mutations in sporadic TEs as well as both hereditary and sporadic BCCs suggest that these tumors arise from a common cell type within the hair follicle. The examination of TEs, BCCs, and other hair follicle tumors with additional markers for hair follicle stem cells should lead to insights into the role stem cells play in the formation of these tumors.

## X. STEM CELLS AND ALOPECIA

Alopecias can be classified into scarring and nonscarring types (Olsen et al., 2003). The localization of hair follicle stem cells to the bulge area may explain why some types of inflammatory alopecias cause permanent follicle loss (such as lichen planopilaris and discoid lupus erythematosis), while others (such as alopecia areata) are reversible (Paus and Cotsarelis, 1999). In cicatricial alopecias, inflammation involves the superficial portion of the follicle, including the bulge area, suggesting that the stem cells necessary for follicle regeneration are damaged. The inflammatory injury of alopecia areata, however, especially in early lesions, involves the bulbar region of the hair follicle, which is composed of bulge cell progeny. Because this area is immediately responsible for hair shaft production, its destruction leads to hair loss. However, the bulge area remains intact, and a new lower anagen follicle and subsequent hair shaft can be produced. Even patients with alopecia areata for many years can regrow their hair either spontaneously or in response to immunomodulation.

The bulge may actually be targeted by inflammation in androgenetic alopecia (common baldness) as well. Jaworsky et al. (1992) showed that in patients with early androgenetic alopecia, inflammatory cells localize to the bulge. Over time, this damage could contribute to the irreversible nature of androgenetic alopecia as well. The bulge area appears specifically attacked in early graft-versus-host disease (GVHD), which can cause alopecia (Murphy et al., 1991). These findings are also consistent with the idea that the bulge and lower hair follicle are "immune privileged," since they express low levels of MHCI.

## XI. BULGE CELLS FOR TISSUE ENGINEERING

An exciting approach for the use of hair follicle stem cells in the treatment of alopecia includes tissue engineering (Stenn and Cotsarelis, 2005). In one scenario, isolated hair follicle stem cells could be used for generating new follicles in bald scalp. For this to occur, isolated bulge cells must be capable of generating new hair follicles. At least two groups have shown that freshly isolated bulge cells from adult mice, when combined with neonatal dermal cells, formed hair follicles after injection into immunodeficient mice (Blanpain *et al.*, 2004; Morris *et al.*, 2004). These studies provided proof of concept that isolated stem cells could be a part of tissue-engineering approaches for treating alopecia.

A goal for treating alopecia with cell therapy approaches includes increasing the number of follicles, for example, by amplifying the number of stem cells *in vitro* prior to transplantation. Cultured keratinocytes from neonatal epidermis have been used for many years to generate hair follicles in reconstitution assays. Freshly isolated bulge cells from adult mice were shown to form hair follicles in skin-reconstitution assays (Morris *et al.*, 2004). Importantly, cultured, individually cloned bulge cells from adult mice also were shown to form hair follicles in skin-reconstitution assays (Blanpain *et al.*, 2004). However, the ratio of new follicles formed from the number of donor follicles and whether nonbulge kcratinocytes also possessed these properties were not analyzed.

The use of hair follicle stem cells for tissue-engineering approaches will depend on the isolation and characterization of human hair follicle stem cells. A major advance in this direction was reported by Ohyama *et al.* (2006). In this work, cell surface markers, including CD200 and FRIZZLED receptor, were identified on human bulge cells by using laser capture microdissection and microarray analysis for gene expression. A cocktail of antibodies against cell surface proteins was devised allowing for isolation of living hair follicle bulge stem cells; thus the stage is set for isolating human hair follicle stem cells to address biological questions relevant to diseases of the follicle and eventually for using these cells for therapeutic purposes.

## XII. MOLECULAR PROFILE — STEM CELL PHENOTYPE

Defining the stem cell phenotype at the molecular level is important for several reasons. Quiescence is a hallmark of these cells. Understanding the genes that distinguish bulge cells from proliferating TA cells as well as the genes that convert resting bulge cells to growing cells bodes well for gaining insights into the uncontrolled proliferation in cancer cells as well as the precisely orchestrated events of hair follicle formation at anagen onset. With the advent of microarrays, large-scale comparisons of gene expression in bulge cells versus nonbulge basal keratinocytes could be performed (Morris *et al.*, 2004; Tumbar *et al.*, 2004). These studies, in which bulge cells were isolated by two very different techniques, resulted in reassuringly similar results. At least 60% of the genes reported as differentially expressed by the two groups were the same. Both studies found that genes involved in activation of the WNT pathway were

generally decreased, while inhibitors of this pathway were decreased relative to nonbulge basal keratinocytes. These findings are in line with studies indicating the importance of WNT activation for anagen onset (Van Mater *et al.*, 2003). It is worth noting that large-scale gene-expression studies have been successful in transgenic mice; however, the relevance of the findings to human hair and stem cell biology has only recently been tested (Ohyama *et al.*, 2006). These investigators found many similarities to the mouse studies, thus validating the mouse as a useful model for studying human hair growth. However, very important differences were also described. In particular, CD34, which serves as a mouse bulge cell marker, is not expressed by human hair follicle bulge cells. CD200, a cell surface marker, may be useful for isolating bulge cells. These studies are a major step in our understanding of human hair follicle bulge cells.

Regarding genes that may maintain the stem cell phenotype, functional studies demonstrate that Rac-1 plays an important role in the self-renewal of the epidermis and the hair follicle (Benitah *et al.*, 2005; Dotto and Cotsarelis, 2005). Loss of Rac-1 causes a burst of proliferation in epidermal keratinocytes and then synchronized differentiation and loss of proliferative capability, resulting in thinning of the cutaneous epithelium. Thus, this gene suppresses proliferation and differentiation and seems important for the switch from stem cell to TA cell.

## XIII. THE BULGE AS STEM CELL NICHE

Keratinocytes within the bulge area may depend on their environment or niche for maintaining their stem cell characteristics. The bulge area also houses melanocyte stem cells, which are normally quiescent but proliferate at anagen onset to repopulate the new lower anagen hair follicle with melanocytes that generate melanin, leading to pigmentation of the hair (Nishimura *et al.*, 2002). Intriguingly, there is evidence that the follicle, if not the bulge, serves as a reservoir for immature Langerhans cells as well as other immunocytes (Kumamoto *et al.*, 2003). The role of the bulge environment in maintaining cells of different lineages in a relatively undifferentiated state or instructing (educating?) immune cells needs further study.

## XIV. CONCLUSION

The bulge begins as a prominent structure during human fetal development but then becomes less obvious, thus hindering its study. Markers for bulge cells, such as CD34 expression and K15 promoter activity, allowed for the isolation and characterization of these cells in mouse follicles. Lineage analysis demonstrated that all of the epithelial layers within the adult follicle and hair originate from bulge cells. Bulge cells possess stem cell characteristics, including multipotency, high proliferative potential, and their cardinal feature of quiescence. Although bulge cells only contribute to the epidermis during wound healing, after isolation, when combined with neonatal dermal cells, they regenerate new hair follicles, epidermis, and sebaceous glands. Propagation of bulge cells *in vitro* maintains their stem cell characteristics, although whether this is an exclusive property of bulge cells remains a question. Under normal homeostatic conditions, epidermal stem cells and hair follicle stem cells constitute two distinct populations. Understanding the differences between these two types of stem cells as well as for the signals important for directing differentiation of these cells into different lineages will be important for developing treatments based on stem cells as well as for clarifying their role in skin disease.

## XV. ACKNOWLEDGMENTS

The majority of the text and figures here are reproduced from Cotsarelis (2006).

## XVI. REFERENCES

Amoh, Y., Li, L., *et al.* (2005). Implanted hair follicle stem cells form Schwann cells that support repair of severed peripheral nerves. *Proc. Natl. Acad. Sci. U.S.A.* **102**(49), 17734–17738.

Argyris, T. (1976). Kinetics of epidermal production during epidermal regeneration following abrasion in mice. *Am. J. Pathol.* **83**(2), 329–340.

Barrandon, Y., and Green, H. (1987). Three clonal types of keratinocyte with different capacities for multiplication. *Proc. Natl. Acad. Sci. U.S.A.* **84**, 2302–2306.

Benitah, S. A., Frye, M., *et al.* (2005). Stem cell depletion through epidermal deletion of Rac1. *Science* **309**(5736), 933–935.

Berton, T. R., Wang, X. J., *et al.* (2000). Characterization of an inducible, epidermal-specific knockout system: differential expression of lacZ in different Cre reporter mouse strains. *Genesis* **26**(2), 160–161.

Bickenbach, J., and Mackenzie, I. (1984). Identification and localization of label-retaining cells in hamster epithelium. *J. Invest. Dermatol.* **82**, 618–622.

Blanpain, C., Lowry, W. E., *et al.* (2004). Self-renewal, multipotency, and the existence of two cell populations within an epithelial stem cell niche. *Cell* **118**(5), 635–648.

Claudinot, S., Nicolas, M., *et al.* (2005). Long-term renewal of hair follicles from clonogenic multipotent stem cells. *Proc. Natl. Acad. Sci. U.S.A.* **102**(41), 14677–14682.

Cotsarelis, G. (2006). Epithelial stem cells: a folliculocentric view. *J. Invest. Dermatol.* **126**, 1459–1468.

Cotsarelis, G., Sun, T. T., *et al.* (1990). Label-retaining cells reside in the bulge area of pilosebaceous unit: implications for follicular stem cells, hair cycle, and skin carcinogenesis. *Cell* **61**(7), 1329–1337.

Dotto, G. P., and Cotsarelis, G. (2005). Developmental biology. Rac1 up for epidermal stem cells. *Science* **309**(5736), 890–891.

Ghazizadeh, S., and Taichman, L. B. (2001). Multiple classes of stem cells in cutaneous epithelium: a lineage analysis of adult mouse skin. *Embo J.* **20**(6), 1215–1222.

Headington, J. T. (1976). Tumors of the hair follicle. A review. *Am. J. Pathol.* **85**, 479–514.

Hutchin, M. E., Kariapper, M. S., *et al.* (2005). Sustained hedgehog signaling is required for basal cell carcinoma proliferation and survival: conditional skin tumorigenesis recapitulates the hair growth cycle. *Genes Dev.* **19**(2), 214–223.

Ito, M., Kizawa, K., *et al.* (2002). Label-retaining cells in the bulge region are directed to cell death after plucking, followed by healing from the surviving hair germ. *J. Invest. Dermatol.* **119**(6), 1310–1316.

Ito, M., Kizawa, K., *et al.* (2004). Hair follicle stem cells in the lower bulge form the secondary germ, a biochemically distinct but functionally equivalent progenitor cell population, at the termination of catagen. *Differentiation* **72**(9–10), 548–557.

Ito, M., Liu, Y., *et al.* (2005). Stem cells in the hair follicle bulge contribute to wound healing but not to homeostasis of the epidermis. *Nat. Med.* (In press).

Jaworsky, C., Kligman, A. M., *et al.* (1992). Characterization of inflammatory infiltrates in male pattern alopecia: implications for pathogenesis. *Br. J. Dermatol.* **127**(3), 239–246.

Jih, D., Lyle, S., *et al.* (1999). Cytokeratin 15 expression in trichoepitheliomas and a subset of basal cell carcinomas suggests they originate from hair follicle stem cells. *J. Cutan. Pathol.* **26**, 113–118.

Johnson, R. L., Rothman, A. L., *et al.* (1996). Human homolog of patched, a candidate gene for the basal cell nevus syndrome. *Science* **272**(5268), 1668–1671.

Jones, P. H., Harper, S., *et al.* (1995). Stem cell patterning and fate in human epidermis. *Cell* **80**(1), 83–93.

Joseph, N. M., and Morrison, S. J. (2005). Toward an understanding of the physiological function of mammalian stem cells. *Dev. Cell* **9**, 173–183.

Kamimura, J., Lee, D., *et al.* (1997). Primary mouse keratinocyte cultures contain hair follicle progenitor cells with multiple differentiation potential. *J. Invest. Dermatol.* **109**, 534–540.

Kaur, P. (2006). Interfollicular epidermal stem cells: identification, challenges, potential. *J. Invest. Dermatol.* **126**(7), 1450–1458.

Kobayashi, K., Rochat, A., *et al.* (1993). Segregation of keratinocyte colony-forming cells in the bulge of the rat vibrissa. *Proc. Nat. Acad. Sci. U.S.A.* **90**(15), 7391–7395.

Kopan, R., Lee, J., *et al.* (2002). Genetic mosaic analysis indicates that the bulb region of coat hair follicles contains a resident population of several active multipotent epithelial lineage progenitors. *Dev. Biol.* **242**(1), 44–57.

Kumamoto, T., Shalhevet, D., *et al.* (2003). Hair follicles serve as local reservoirs of skin mast cell precursors. *Blood* **102**(5), 1654–1660.

Lavker, R. M. and Sun, T. T. (2000). Epidermal stem cells: properties, markers, and location. *Proc. Natl. Acad. Sci. U.S.A.* **97**(25), 13473–13475.

Legue, E. and Nicolas, J. F. (2005). Hair follicle renewal: organization of stem cells in the matrix and the role of stereotyped lineages and behaviors. *Development* **132**(18), 4143–4154.

Li, L., Mignone, J., *et al.* (2003). Nestin expression in hair follicle sheath progenitor cells. *Proc. Natl. Acad. Sci. U.S.A.* **100**(17), 9958–9961.

Liu, Y., Lyle, S., *et al.* (2003). Keratin 15 promoter targets putative epithelial stem cells in the hair follicle bulge. *J. Invest. Dermatol.* **121**(5), 963–968.

Lyle, S., Christofidou-Solomidou, M., *et al.* (1998). The C8/144B monoclonal antibody recognizes cytokeratin 15 and defines the location of human hair follicle stem cells. *J. Cell. Sci.* **111**(Pt. 21), 3179–3188.

Millar, S. E. (2002). Molecular mechanisms regulating hair follicle development. *J. Invest. Dermatol.* **118**(2), 216–225.

Moll, I. (1996). Differential epithelial outgrowth of plucked and micro-dissected human hair follicles in explant culture. *Arch. Dermatol. Res.* **288**, 604–610.

Morris, R. J. (2000). Keratinocyte stem cells: targets for cutaneous carcinogens. *J. Clin. Invest.* **106**(1), 3–8.

Morris, R., Fischer, S., *et al.* (1986). Evidence that a slowly cycling sub-population of adult murine epidermal cells retains carcinogen. *Cancer Res.* **46**, 3061–3066.

Morris, R. J., Liu, Y., *et al.* (2004). Capturing and profiling adult hair follicle stem cells. *Nat. Biotechnol.* **22**(4), 411–417.

Murphy, G. F., Lavker, R. M., *et al.* (1991). Cytotoxic folliculitis in GvHD. Evidence of follicular stem cell injury and recovery. *J. Cutan. Pathol.* **18**(5), 309–314.

Nishimura, E. K., Jordan, S. A., *et al.* (2002). Dominant role of the niche in melanocyte stem-cell fate determination. *Nature* **416**(6883), 854–860.

Ohyama, M., Terunuma, A., *et al.* (2006). Characterization and isolation of stem cell–enriched human hair follicle bulge cells. *J. Clin. Invest.* **116**(1), 249–260.

Olsen, E. A., Bergfeld, W. F., *et al.* (2003). Summary of North American Hair Research Society (NAHRS)–sponsored Workshop on Cicatricial Alopecia, Duke University Medical Center, February 10 and 11, 2001. *J. Am. Acad. Dermatol.* **48**(1), 103–110.

Oro, A. E., Higgins, K. M., *et al.* (1997). Basal cell carcinomas in mice overexpressing sonic hedgehog. *Science* **276**(5313), 817–821.

Oshima, H., Rochat, A., *et al.* (2001). Morphogenesis and renewal of hair follicles from adult multipotent stem cells. *Cell* **104**(2), 233–245.

Paus, R., and Cotsarelis G. (1999). The biology of hair follicles. *N. Engl. J. Med.* **341**(7), 491–497.

Pinkus, H. (1958). Embryology of hair. *In* "The Biology of Hair Growth" (W. Montagna and R. A. Ellis, eds.), pp. 1–32. Academic Press, New York.

Rochat, A., Kobayashi, K., *et al.* (1994). Location of stem cells of human hair follicles by clonal analysis. *Cell* **76**(6), 1063–1073.

Roh, C., Tao, Q., *et al.* (2005). *In vitro* differences between keratinocyte stem cells and transit-amplifying cells of the human hair follicle. *J. Invest. Dermatol.* **125**(6), 1099–1105.

Schirren, C. G., Burgdorf, W. H. C., *et al.* (1997). Fetal and adult hair follicle: an immunohistochemical study of anticytokeratin antibodies in formalin-fixed and paraffin-embedded tissue. *Am. J. Dermatopathol.* **19**(4), 334–340.

Stenn, K. S., and Cotsarelis, G. (2005). Bioengineering the hair follicle: fringe benefits of stem cell technology. *Curr. Opin. Biotechnol.* **16**(5), 493–497.

Stenn, K. S., Lawrence, L., *et al.* (1994). Expression of the bcl-2 proto-oncogene in the cycling adult mouse hair follicle. *J. Invest. Dermatol.* **103**(1), 107–111.

Taylor, G., Lehrer, M. S., *et al.* (2000). Involvement of follicular stem cells in forming not only the follicle but also the epidermis. *Cell* **102**(4), 451–461.

Topley, G. I., Okuyama, R., *et al.* (1999). p21(WAF1/Cip1) functions as a suppressor of malignant skin tumor formation and a determinant of keratinocyte stem-cell potential. *Proc. Natl. Acad. Sci. U.S.A.* **96**(16), 9089–9094.

Trempus, C., Morris, R., *et al.* (2003). Enrichment for living murine keratinocytes from the hair follicle bulge with the cell surface marker CD34. *J. Invest. Dermatol.* **120**, 501–511.

Tumbar, T., Guasch, G., *et al.* (2004). Defining the epithelial stem cell niche in skin. *Science* **303**(5656), 359–363.

Van Mater, D., Kolligs, F. T., *et al.* (2003). Transient activation of beta-catenin signaling in cutaneous keratinocytes is sufficient to trigger the active growth phase of the hair cycle in mice. *Genes Dev.* **17**(10), 1219–1224.

Verhaegh, M., Arends, J., *et al.* (1997). Transforming growth factor-beta and bcl-2 distribution patterns distinguish trichoepithelioma from basal cell carcinoma. *Dermatol. Surg.* **23**, 695.

Weinberg, W. C., Goodman, L. V., *et al.* (1993). Reconstitution of hair follicle development *in vivo*: determination of follicle formation, hair growth, and hair quality by dermal cells. *J. Invest. Dermatol.* **100**(3), 229–236.

Wilson, C., Cotsarelis, G., *et al.* (1994). Cells within the bulge region of mouse hair follicle transiently proliferate during early anagen: heterogeneity and functional differences of various hair cycles. *Differentiation* **55**(2), 127–136.

Yang, J. S., Lavker, R. M., *et al.* (1993). Upper human hair follicle contains a subpopulation of keratinocytes with superior in vitro proliferative potential. *J. Invest. Dermatol.* **101**(5), 652–659.

Yu, H., Fang, D., *et al.* (2006). Isolation of a novel population of multipotent adult stem cells from human hair follicles. *Am. J. Pathol.* **168**(6), 1879–1888.

# Chapter Seventy-Six

# Wound Repair

*Kaustabh Ghosh and Richard A. F. Clark*

## I. INTRODUCTION

The skin is the largest organ in the body, and its primary function is to serve as a protective barrier against the environment. Other important functions of the skin include fluid homeostasis, thermoregulation, immune surveillance, sensory detection, and self-healing. Loss of the integrity of large portions of the skin due to injury or illness may result in significant disability or even death. It is estimated that in 1992, there were 35.2 million cases of significant skin loss (Wound Care, 1993) that required major therapeutic intervention. Of these, approximately 7 million wounds become chronic.

The most common single cause of significant skin loss is thermal injury, which accounts for an estimated 1 million emergency room visits per year (American Burn Association, 2005). Other causes of skin loss include trauma (35 million wounds/yr in the United States require therapeutic intervention) and chronic ulcerations secondary to diabetes mellitus, pressure, and venous stasis. Every year in the United States there are approximately 2 million cases of chronic diabetic ulcers, many of which eventually necessitate amputation. Pressure ulcers and leg ulcers, including venous ulcers, affect another 2 million people in the United States, with treatment costs as high as $8 billion annually (Supp and Boyce, 2005). Of these approximately 600,000 patients suffer from venous ulcers, at an average cost of $9685 per patient (Olin *et al.*, 1999). In 2003, a survey estimated the U.S. market for advanced wound care products,

including biological and synthetic dressings, to be greater than $1.7 billion, which is expected to increase significantly as the population ages, becoming more susceptible to underlying causes of chronic wounds. The quality-of-life tolls of chronic wounds are extremely high.

Over the past two decades, extraordinary advances in cellular and molecular biology have greatly expanded our comprehension of the basic biological processes involved in acute-wound healing and the pathobiology of chronic wounds (Singer and Clark, 1999; Mustoe, 2004). These strides in basic knowledge have led to advancements in wound care resulting in more rapid closure of ulcers and normal wounds. Furthermore, since tumor microenvironments have many similarities to wound healing (Dvorak, 1986), increased knowledge in wound healing (Ashcroft *et al.*, 2003) has led to advances in tumor therapy (Basu *et al.*, 2006). For wound therapy, one recombinant growth factor, platelet-derived growth factor-BB (rPDGF-BB) (Regranex, Ortho-McNeil), and several skin substitutes (Dermagraft and TransCyte, Advance Tissue Sciences; Apligraf, Organogenesis; Integra Matrix Wound Dressings, Integra Life-Sciences Holding; OrCel, Ortec International; AlloDerm, LifeCell) have reached the marketplace for second-line therapy of recalcitrant ulcers (Singer and Clark, 1999; Balasubramani *et al.*, 2001). These therapeutic interventions can successfully improve the rate of wound closure. Unfortunately, rPDGF-BB must be applied daily, forcing patients to change their bandages daily, and two companies producing skin substitutes have undergone bankruptcy and reorganization. Regardless of the advanced wound care product, the ideal goal would be to regenerate tissues where both the structural and functional properties of the wounded tissue are restored to their levels prior to injury.

In contrast to adult wounds, embryonic wounds undergo complete regeneration, terminating in a scarless repair (Martin and Parkhurst, 2004; Redd *et al.*, 2004). Morphogenetic cues from this embryonic phenotype should be utilized to develop engineered constructs capable of tissue regeneration. Furthermore, the cellular response to biological stimuli depends on the mechanical strength of the extracellular matrix (ECM) (Grinnell, 2003; Vogel and Sheetz, 2006). Therefore, the therapeutic successes of tissue-engineered constructs will depend not only on their bioactivity but also on their optimal mechanical properties.

This chapter begins with an overview of the basic biology of wound repair, followed by a discussion of established practices and novel approaches to engineering tissue constructs for effective wound repair.

## II. BASIC BIOLOGY OF WOUND REPAIR

Wound repair is not a simple linear process in which growth factors released by phylogistic events activate parenchymal cell proliferation and migration; rather, it is an integration of dynamic interactive processes involving soluble mediators, formed blood elements, extracellular matrix, and parenchymal cells. Unencumbered, these wound repair processes follow a specific time sequence and can be temporally categorized into three major groups: inflammation, tissue formation, and tissue remodeling. The three phases of wound repair, however, are not mutually exclusive but, rather, overlap in time.

### Inflammation

Severe tissue injury causes blood vessel disruption, with concomitant extravasation of blood constituents. Blood coagulation and platelet aggregation generate a fibrin-rich clot that plugs severed vessels and fills any discontinuity in the wounded tissue. While the blood clot within vessel lumen reestablishes homeostasis, the clot within wound space acts as a growth factor reservoir and provides a provisional matrix for cell migration.

The primary cell types involved in the overall process of inflammation are platelets, neutrophils, and monocytes. Upon injury, successful reestablishment of homeostasis depends on platelet adhesion to interstitial connective tissue, which leads to their aggregation, coagulation, and activation. Activated platelets release several adhesive proteins to facilitate their aggregation, chemotactic factors for blood leukocytes, and multiple growth factors to promote new-tissue formation.

Of the two primary phagocytotic leukocytes (neutrophils and monocytes), neutrophils arrive first in large numbers due to their abundance in circulation. Infiltrating neutrophils cleanse the wounded area of foreign particles, including bacteria. If excessive microorganisms or indigestible particles have lodged in the wound site, neutrophils will probably cause further tissue damage as they attempt to clear these contaminants through the release of enzymes and toxic oxygen products. When particle clearance has been completed, generation of granulocyte chemoattractants ceases and the remaining neutrophils become effete.

### Transition Between Inflammation and Repair

Whether neutrophil infiltrates resolve or persist, monocyte accumulation continues, stimulated by selective monocyte chemoattractants (Clark, 1996). Besides promoting phagocytosis and debridement, adherence to extracellular matrix stimulates monocytes to undergo metamorphosis into inflammatory or reparative macrophages. Since cultured macrophages produce and secrete the peptide growth factors interleukin-1 (IL-1), platelet-derived growth factor-BB (PDGF-BB), transforming growth factor alpha (TGF-α), transforming growth factor beta (TGF-β), and fibroblast growth factor (FGF), presumably wound macrophages also synthesize these protein products (Clark, 1996). Although neutrophils and macrophages have a critical role in fighting infection and macrophages can contribute growth factors to the wound, it has become increasingly clear that too much of a good thing can be bad (Martin and Leibovich, 2005). In fact, from knockout and knockdown experiments, it is evident that wounds in some situations heal faster with fewer inflammatory cells, especially if microorganism invasion is avoided by some other means (Ashcroft *et al.*, 1999).

### Re-epithelialization

Re-epithelialization of a wound begins within hours after injury by the movement of epithelial cells from the surrounding epidermis over the denuded surface. Rapid reestablishment of the epidermal surface and its permeability barrier prevents excessive water loss and time of exposure to bacterial infections, which decreases the morbidity and mortality of patients who have lost a substantial amount of skin surface. If a wide expanse of the epidermis is lost, epidermal cells regenerate from stem cells in pilosebaceous follicles (Cotsarelis, 2006). Migrating epithelial cells markedly alter their phenotype by retracting their intracellular filaments, dissolving most of their desmosomes, and forming peripheral actin filaments (which facilitate cell movement) (Clark, 1996). These migrating cells also undergo dissolution of their hemidesmosomal links between the epidermis and the dermis. All these phenotypic alterations provide epithelial cells with the needed lateral mobility for migration over the wound site. Migrating epidermal cells possess a unique phenotype that is distinct from both the terminally differ-

| Table 76.1. | Integrin superfamily | | |
|---|---|---|---|
| *Integrins* | *Ligand* | *Integrins* | *Ligand* |
| **β1 Integrins** | | **αv Integrins** | |
| α1β1 | Fibrillar collagen, laminin-1 | αvβ1 | Fibronectin (RGD), vitronectin |
| α2β1 | Fibrillar collagen, laminin-1 | αvβ3 | Vitronectin (RGD), fibronectin, fibrinogen, von |
| α3β1 | Fibronectin (RGD), laminin-5, entactin, denatured collagens | | Willebrand factor, thrombospondin, denatured collagen |
| α4β1 | Fibronectin (LEDV), VCAM-1 | αvβ5 | Fibronectin (RGD), vitronectin |
| α5β1 | Fibronectin (RGD+PHSRN, the synergy site) | αvβ6 | Fibronectin, tenascin |
| α6β1 | Laminin | | |
| α7β1 | Laminin | **β2 Integrins** | |
| α8β1 | Fibronectin, vitronectin | αMβ2 | ICAM-1, iC3b, fibronogen, factor X |
| α9β1 | Tenasin | αLβ2 | ICAM-1, -2, and -3 |
| | | αXβ2 | IC3b, fibrinogen |
| **Other ECM integrins** | | | |
| αllbβ3 | Same as αvβ3 | | |
| α6β4 | Laminin | | |

entiated keratinocytes of normal (stratified) epidermis and the basal cells of stratified epidermis. It is now appreciated that the signals that control wound healing in the adult animal are similar to those that control epithelial fusion during embryogenesis (Jacinto *et al.*, 2001).

If the basement membrane is destroyed by injury, epidermal cells migrate over a provisional matrix of fibrin(ogen), fibronectin, tenascin, and vitronectin as well as stromal type I collagen (Clark, 1996). Wound keratinocytes express cell surface receptors for fibronectin, tenasin, and vitronectin, which belong to the integrin superfamily (Table 76.1) (Yamada *et al.*, 1996). In addition, α2β1 collagen receptors, which are normally disposed along the lateral sides of basal keratinocytes, redistribute to the basal membrane of wound keratinocytes as they come in contact with type 1 collagen fibers of the dermis. Whereas β1 integrins are clearly essential for normal re-epithelialization (Grose *et al.*, 2002), it is not clear which subtype is essential. It is most likely that there is a redundancy in the requirement for α5β1 and α2β1 in re-epithelialization.

The migrating wound epidermis does not simply transit over a wound coated with a fibrin/fibronectin provisional matrix but, rather, dissects through the wound, separating the clot and desiccated or otherwise nonviable tissue from underlying viable tissue (Kubo *et al.*, 2001). The path of dissection appears to be determined by the array of integrins that the migrating epidermal cells express on their cell membranes, as described previously. In addition, the αvβ3 receptor for fibrinogen/fibrin and denatured collagen are not expressed on keratinocytes *in vitro* or *in vivo*. Therefore, keratinocytes do not have the capacity to interact with these matrix proteins. Furthermore, in the presence of fibrinogen or fibrin, epidermal cells fail to bind fibronectin yet can bind

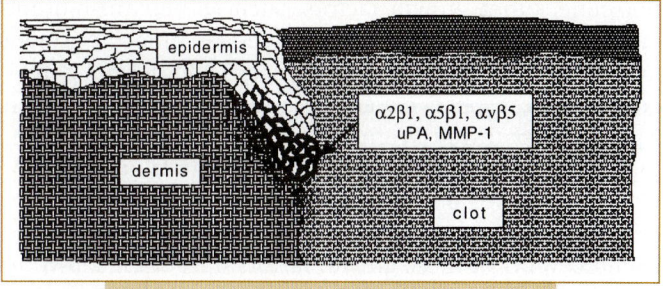

**FIG. 76.1.** Integrins in early (three-dimensional) reepithelialization. The migrating epidermis in contact with type I dermal collagen expresses a gene set including the collagen receptor α2β1, urokinase plasminogen activator (uPA), and interstitial collagenase (MMP-1). With this armamentarium, the epidermis can dissect a path between viable dermis and nonviable clot and denature stroma, ultimately leading to slough of the eschar.

type 1 collagen. Hence, migrating wound epidermis avoids the fibrin-rich, fibronectin-rich clot while migrating over a bed of fibronectin (via the α5β1 receptor) and type 1 collagen (via the α2β1 receptor) (Fig. 76.1).

Extracellular matrix degradation is clearly required for the dissection of migrating wound epidermis between the collagenous dermis and the fibrin eschar (Bugge *et al.*, 1996) and probably depends on epidermal cell production of collagenase, plasminogen activator, and stromelysin. Plasminogen activator activates collagenase (MMP-1) as well as plasminogen (Mignatti *et al.*, 1996) and therefore facilitates the degradation of interstitial collagen as well as provisional matrix proteins. Interestingly, keratinocytes in direct contact with collagen greatly increase the amount of MMP-1 they produce as compared to that produced when they reside on laminin-rich basement membrane or purified laminin

(Petersen *et al.*, 1990). The migrating epidermis of superficial skin ulcers and burn wounds, in fact, express high levels of MMP-1 mRNA in areas where it presumably comes in direct contact with dermal collagen (Saarialho-Kere *et al.*, 1992; Stricklin and Nanney, 1994).

One to two days after injury, epithelial cells at the wound margin begin to proliferate. Although the exact mechanism is still not clear, both proliferation and migration of epithelial cells may be triggered by the absence of neighboring cells at the wound margin (or the "free-edge effect"). The free edge effect in the wound epidermis may be secondary to modulation of cadherin junctions, as described for V-cadherins during angiogenesis (Dejana, 1996). In fact, recent studies indicate that epidermal desmosomes lose their hyperadhesiveness and cadherins switch from E-cadherins to P-cadherins at the wound edge (Garrod *et al.*, 2005; Koizumi *et al.*, 2005). Other possibilities, not exclusive of the former, are release of autocrine or paracrine growth factors that induce epidermal migration and proliferation and/or increased expression of growth factor receptors. Although some growth factors, such as insulin-like growth factor (IGF), may derive from the circulation and thereby act as a hormone, other growth factors, such as heparin-binding epidermal growth factor (HB EGF) and keratinocyte growth factor (KGF), derive from macrophages and dermal parenchymal cells, respectively, and act on epidermal cells through a paracrine pathway (Werner, 1998). In contrast, transforming growth factor (TGF)-α and -β originate from keratinocytes themselves and act directly on the producer cell or adjacent epidermal cells in an autocrine or juxtacrine fashion. Many of these growth factors have been shown to stimulate re-epithelialization in animal models (Brown *et al.*, 1989; Lynch *et al.*, 1989; Staiano-Coico *et al.*, 1993). Furthermore, lack of growth factors or their receptors in knockout mice support the hypothesis that growth factor activation of keratinocytes is required for optimal epidermal migration and/or proliferation during normal wound healing (Grose and Werner, 2004). Recently it has been demonstrated that JNK is a key signal transduction factor responsible for "resetting" the epidermal program from differentiation to proliferation and possibly migration (Gazel *et al.*, 2006).

As re-epithelialization progresses, basement membrane proteins reappear in a very ordered sequence from the margin of the wound inward, in a zipperlike fashion (Clark, 1996). Epidermal cells revert to their normal phenotype, once again firmly attaching to re-established basement membrane through hemidesmosomal proteins α6β4 integrin and 180-kDa bullous pemphigoid antigen (Litjens *et al.*, 2006), and to the underlying neodermis through type VII collagen fibrils (El Ghalbzouri *et al.*, 2004).

## Granulation Tissue

New stroma, often called *granulation tissue*, begins to form approximately four days after injury. The name derives from the granular appearance of newly forming tissue when it is incised and visually examined. Numerous new capillaries endow the neostroma with its granular appearance. Macrophages, fibroblasts, and blood vessels move into the wound space as a unit (Hunt, 1980), which correlates well with the proposed biologic interdependence of these cells during tissue repair. Macrophages and ingrowing parenchymal cells provide a continuing source of cytokines necessary to stimulate fibroplasia and angiogenesis, fibroblasts construct new extracellular matrix necessary to support cell ingrowth, and blood vessels carry oxygen and nutrients necessary to sustain cell metabolism. Recently the importance of oxygenation has been re-emphasized (Hopf *et al.*, 2005; Ueno *et al.*, 2006). The quantity and quality of granulation tissue depends on the biologic modifiers present, the activity level of target cells, and the extracellular matrix environment (Juliano and Haskill, 1993). As mentioned in the section on inflammation, the arrival of peripheral blood monocytes and their activation to macrophages establish conditions for continual synthesis and release of growth factors. In addition, and perhaps more importantly, injured and activated parenchymal cells can synthesize and secrete growth factors. For example, migrating wound epidermal cells produce vascular endothelial cell growth factor (VEGF) (Kishimoto *et al.*, 2000), TGF-β (Frank *et al.*, 1996), and PDGF BB (Ansel *et al.*, 1993), to which endothelial cells and fibroblasts respond, respectively. The provisional ECM also promotes granulation-tissue formation by positive feedback regulation of integrin ECM receptor expression (Xu and Clark, 1996). Once fibroblasts and endothelial cells express the appropriate integrin receptors, they invade the fibrin-rich/fibronectin-rich wound space (Fig. 76.2). Although it has been recognized for many years that ECM modulates cell differentiation via signal transduction from ligation of ECM receptors (Lin and Bissell, 1993; Nelson and Bissell, 2005), more recently it has become evident that the force and geometry of the ECM influence cell behavior and differentiation (Ingber, 2003; Discher *et al.*, 2005; Vogel and Sheetz, 2006).

### Fibroplasia

Components of granulation tissue derived from fibroblasts, including the cells themselves and the extracellular matrix, are collectively known as *fibroplasia*. Growth factors, especially PDGF and TGF-β (Heldin, 1996; Roberts and Sporn, 1996), in concert with the provisional matrix molecules (Xu and Clark, 1996), presumably stimulate fibroblasts of the periwound tissue to proliferate, express appropriate integrin receptors, and migrate into the wound space. Many of these growth factors are released from macrophages (Clark, 1996) or other tissue cells (Ansel *et al.*, 1993; Frank *et al.*, 1996); however, fibroblasts themselves can produce growth factors to which they respond in an autocrine fashion (Pardoux and Derynck, 2004). Multiple complex interactive biologic phenomena occur within fibroblasts as they respond to wound cytokines, including the induction of additional cytokines and modulation of cytokine receptor number or affinity. *In vivo* studies support the hypothesis

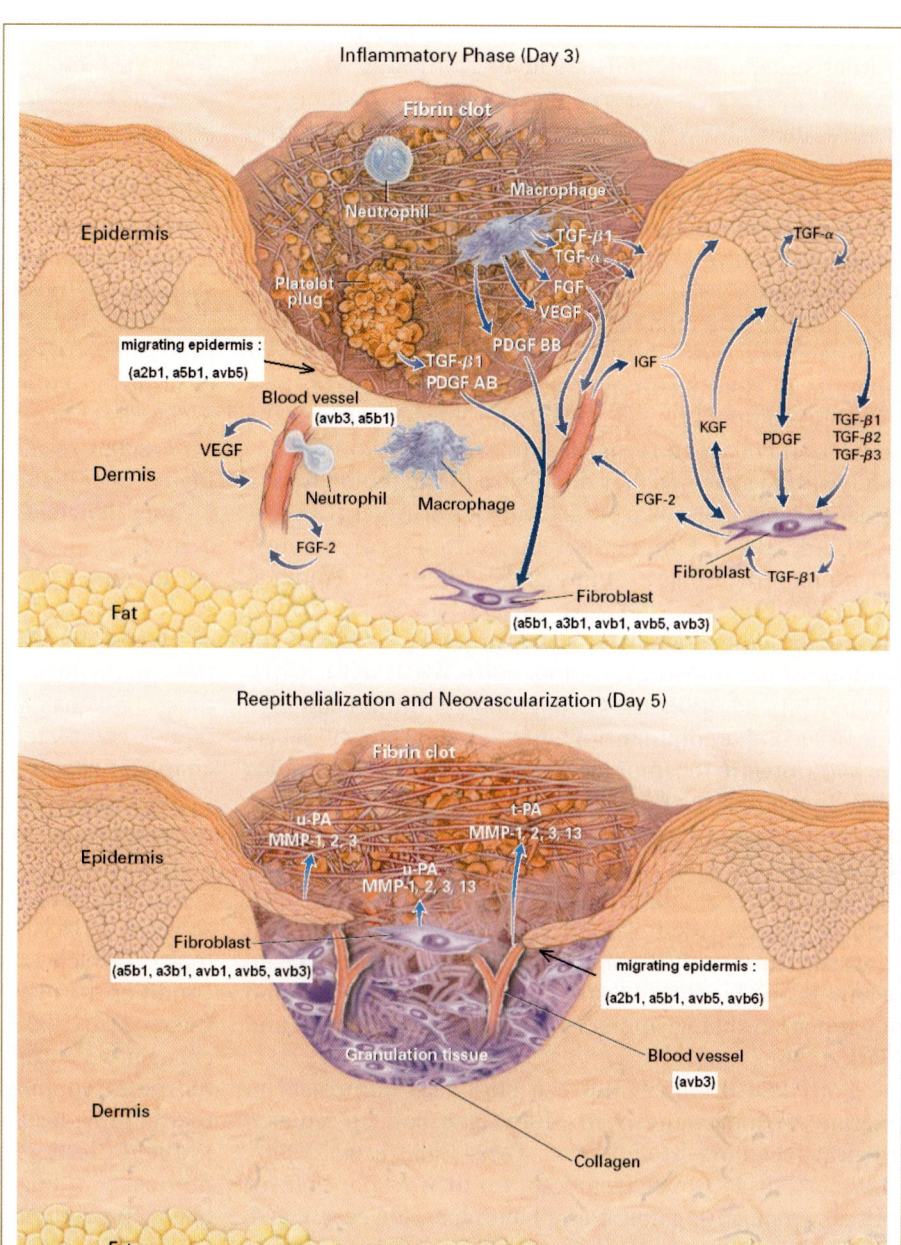

**FIG. 76.2.** After injury, platelets, inflammatory cells, and tissue cells (i.e., fibroblasts and endothelial and epidermal cells) secrete abundant quantities of multiple growth factors thought to be necessary for cell movement into the wound. However, it is the provisional matrix and integrin expression that acts as the rate-limiting step in granulation-tissue induction. Once the appropriate integrins are expressed on periwound endothelial cells and fibroblasts on day 3, the cells invade the wound space shortly thereafter (days 4 and 5). Fibroblasts and endothelial cells express the fibrinogen/fibrin receptor $\alpha V\beta 3$ and therefore are able to invade the fibrin clot; the epidermis, however, does not express $\alpha V\beta 3$ and therefore dissects under the clot. Proteinases play an important role during granulation-tissue formation by clearing the path for migrating tissue cells. Ultimately, the clot that has not been transformed into granulation tissue by invading fibroblasts and endothelial cells is dissected free of the wound by the migrating epidermis and sloughed as eschar. Modified from Singer, A. J., and Clark, R. A. F. (1999). Mechanisms of disease: cutaneous wound healing. *New Eng. J. Med.* **341**, 738–746, with permission.

that growth factors are active in wound repair fibroplasia. Several studies have demonstrated that PDGF, connective tissue factor (CTGF), TGF-$\alpha$, TGF-$\beta$, HB-EGF, and FGF family members are present at sites of tissue repair (Pierce *et al.*, 1992; Werner *et al.*, 1992; Ansel *et al.*, 1993; Levine *et al.*, 1993; Marikovsky *et al.*, 1993; Frank *et al.*, 1996; Grotendorst and Duncan, 2005). Furthermore, purified and recombinant-derived growth factors have been shown to stimulate wound granulation tissue in normal and compromised animals (Greenhalgh *et al.*, 1990; Mustoe *et al.*, 1991; Steed *et al.*, 1992; Shah *et al.*, 1995), and a single growth factor may work both directly and indirectly by inducing the production of other growth factors *in situ* (Mustoe *et al.*, 1991).

Structural molecules of the early ECM, coined *provisional matrix* (Clark *et al.*, 1982), contribute to tissue formation by providing a scaffold or conduit for cell migration (fibronectin) (Greiling and Clark, 1997), low impedance for cell mobility (hyaluronic acid) (Toole, 1991), a reservoir for cytokines (Wijelath *et al.*, 2002; Sahni *et al.*, 2006), and direct signals to the cells through integrin receptors (Juliano *et al.*, 2004; Ginsberg *et al.*, 2005; Walker and Assoian, 2005). Fibronectin appearance in the periwound environment as well as the expression of fibronectin receptors appear to be critical rate-limiting steps in granulation-tissue formation (Clark *et al.*, 1996; McClain *et al.*, 1996; Xu and Clark, 1996; Greiling and Clark, 1997). In addition, a dynamic reciprocity between fibroblasts and their surrounding ECM creates further complexity. That is, fibroblasts affect the ECM through new synthesis, deposition, and remodeling of the ECM, while the ECM affects fibroblasts by regulating their function, including their ability to synthesize, deposit, remodel, and generally interact with the ECM (Clark *et al.*, 1995; Xu and Clark, 1996; Grinnell, 2003; Even-Ram and Yamada, 2005). Thus, the reciprocal interactions between ECM and fibroblasts evolve dynamically during granulation-tissue development.

As fibroblasts migrate into the wound space, they initially penetrate the blood clot, composed of fibrin and lesser amounts of fibronectin and vitronectin. Fibroblasts presumably require fibronectin *in vivo* for movement from the periwound collagenous matrix into the fibrin/fibronectin-laden wound space, as they do *in vitro* for migration from a three-dimensional collagen gel into a fibrin gel (Greiling and Clark, 1997). Fibroblasts bind to fibronectin through receptors of the integrin superfamily (Table 76.1) (Yamada *et al.*, 1996). The Arg-Gly-Asp-Ser (RGDS) tetrapeptide within the cell-binding domain of these proteins is critical for binding to the integrin receptors $\alpha3\beta1$, $\alpha5\beta1$, $\alpha v\beta1$, $\alpha v\beta3$, and $\alpha v\beta5$. In addition, the CSIII domain of fibronectin provides a second binding site for human dermal fibroblasts via the $\alpha4\beta1$ integrin receptor (Gailit *et al.*, 1993). *In vivo* studies have shown that the RGD-dependent fibronectin receptors $\alpha5\beta1$ and $\alpha v\beta3$ are up-regulated on periwound fibroblasts the day prior to granulation tissue formation and on early granulation-tissue fibroblasts as they infiltrate the provisional matrix-laden wound (Xu and Clark, 1996). In contrast, the non-RGD-binding $\alpha1\beta1$ and $\alpha2\beta1$ collagen receptors on these fibroblasts either were suppressed or did not appear to change appreciably (Gailit *et al.*, 1996; Xu and Clark, 1996).

Both PDGF and TGF-$\beta$ can stimulate fibroblasts to migrate (Postlethwaite *et al.*, 1987) and can up-regulate integrin receptors (Heino *et al.*, 1989; Gailit *et al.*, 1996). Therefore, these growth factors appear to be responsible for inducing a migrating fibroblast phenotype. However, PDGF increases $\alpha5\beta1$ and $\alpha3\beta1$ while decreasing $\alpha1\beta1$ in cultured human dermal fibroblasts surrounded by fibronectin (Gailit *et al.*, 1996). Furthermore, fibronectin- or fibrin-rich environments promote the ability of PDGF to increase $\alpha5\beta1$ and $\alpha3\beta1$, but not $\alpha2\beta1$, mRNA steady-state levels by increasing the stability of these mRNA moieties (Xu and Clark, 1996). These data strongly suggest that the type of integrin increased by PDGF stimulation appears to depend on the ECM context and suggests a positive feedback between ECM and ECM receptors.

Movement into a cross-linked fibrin blood clot or any tightly woven extracellular matrix may also necessitate active proteolysis to cleave a path for migration. A variety of fibroblast-derived enzymes in conjunction with serum-derived plasmin are potential candidates for this task, including plasminogen activator, interstitial collagenase-1 and -3 (MMP-1 and MMP-13, respectively), the 72-kDa gela-

tinase A (MMP-2), and stromelysin (MMP-3) (Mignatti *et al.*, 1996; Vaalamo *et al.*, 1997). In fact, high levels of immuno-reactive MMP-1 have been localized to fibroblasts at the interface of granulation tissue with eschar in burn wounds (Stricklin and Nanney, 1994), and many stromal cells stain for MMP-1 and MMP-13 in chronic ulcers (Vaalamo *et al.*, 1997). While TGF-$\beta$ down-regulates proteinase activity (Overall *et al.*, 1989), PDGF stimulates the production and secretion of these proteinases (Circolo *et al.*, 1991).

From elegant knockout mice studies it is clear that the plasminogen activator–plasminogen system is critical for clearing the wound clot (Bugge *et al.*, 1996). Whereas MMP-9-deficient mice demonstrate altered fracture repair (Colnot *et al.*, 2003), results of MMP deficiency in genetically manipulated mice has not been so clear for soft-tissue healing. Neither deficiency of membrane-type-1 MMP, an activator of MMP-1, -3, -2, -9, and -13, nor deficiency of MMP-13 itself led to abnormal cutaneous wound healing (Mirastschijski *et al.*, 2004; Hartenstein *et al.*, 2006).

Once the fibroblasts have migrated into the wound, they gradually switch their major function to protein synthesis (Welch *et al.*, 1990). Ultimately the migratory phenotype is completely supplanted by a profibrotic phenotype characterized by decreased $\alpha3\beta1$ and $\alpha5\beta1$ provisional matrix receptor expression, increased $\alpha2\beta1$ collagen receptor expression, abundant rough endoplasmic recticulum, and Golgi apparatus filled with new collagen protein (Welch *et al.*, 1990; Xu and Clark, 1996). The fibronectin-rich provisional matrix is gradually supplanted with a collagenous matrix (Welch *et al.*, 1990; Clark *et al.*, 1995) Under these conditions, PDGF, which is still abundant in these wounds (Pierce *et al.*, 1995), stimulates extremely high levels of $\alpha2\beta1$ collagen receptor, but not $\alpha3\beta1$ or $\alpha5\beta1$ provisional matrix receptors, supporting the contention that the extracellular matrix provides a positive feedback for integrin expression (Xu and Clark, 1996). TGF-$\beta$, which is observed in wound fibroblasts at this time (Clark *et al.*, 1995), can induce fibroblasts to produce great quantities of collagen (Ignotz and Massague, 1986; Roberts *et al.*, 1986). IL-4 also can induce a modest increase in types I and III collagen production (Postlethwaite *et al.*, 1992). Since IL-4-producing mast cells are present in healing wounds as well as in fibrotic tissue, they may contribute to collagenous matrix accumulation in these sites.

Once an abundant collagen matrix is deposited in the wound, fibroblasts cease collagen production, despite the continuing presence of TGF-$\beta$ (Clark *et al.*, 1995). The stimuli responsible for fibroblast proliferation and matrix synthesis during wound repair were originally extrapolated from many *in vitro* investigations over the past quarter-century and then confirmed by *in vivo* manipulation of wounds within the last 15 years. Less attention had been directed toward elucidating the signals responsible for down-regulating fibroblast proliferation and matrix synthesis until more recently. Both *in vitro* and *in vivo* studies suggest that gamma-interferon may be one such factor (Duncan and Berman, 1985; Granstein *et al.*, 1987). In addition, collagen matrix can suppress both fibroblast proliferation and fibroblast collagen synthesis (Grinnell, 1994; Clark *et al.*, 1995). In contrast, a fibrin or fibronectin matrix has little or no suppressive effect on the mitogenic or synthetic potential of fibroblasts (Clark *et al.*, 1995; Tuan *et al.*, 1996).

Although the attenuated fibroblast activity in collagen gels is not associated with cell death, many fibroblasts in day-10 healing wounds develop pyknotic nuclei (Desmouliere *et al.*, 1995), a cytological marker for apoptosis, or programmed cell death, as well as other signs of apoptosis. These results in cutaneous wounds and other results in lungs and kidney suggest that apoptosis is the mechanism responsible for the transition from a fibroblast-rich granulation tissue to a relatively acellular scar (Desmouliere *et al.*, 1997). The signal(s) for wound fibroblast apoptosis have not been elucidated. Thus fibroplasia in wound repair is tightly regulated, whereas in fibrotic diseases such as keloid formation, morphea, and scleroderma these processes become dysregulated. Recent evidence suggests that the fibroblast apoptotic signals in keloids are disrupted either directly or indirectly (Moulin *et al.*, 2004; Linge *et al.*, 2005).

## Neovascularization

Fibroplasia would halt if neovascularization failed to accompany the newly forming complex of fibroblasts and extracellular matrix. The process of new blood vessel formation is called *angiogenesis*. The soluble factors that can stimulate angiogenesis in wound repair are gradually being elucidated. Angiogenic activity can be recovered from activated macrophages as well as various tissues, including the epidermis, soft-tissue wounds, and solid tumors. Molecules that appear to have angiogenic activity include vascular endothelial growth factor (VEGF), TGF-β, TGF-α, TNF-α, platelet factor-4 (PF-4), angiogenin, angiotropin, angiopoietin, interleukin-8 (IL-8), PDGF, thrombospondin, low-molecular-weight substances, including the peptide KGHK, low oxygen tension, biogenic amines, and lactic acid and nitric oxide (NO). Some of these factors, however, are intermediaries in a single angiogenesis pathway; for example, TNF-α induces PF-4, which stimulates angiogenesis through NO (Montrucchio *et al.*, 1997). To complicate matters further, not all growth factors within a family stimulate angiogenesis equally. For example, there are four isotypes of VEGF (VEGF-A, -B, -C, and -D) and three receptors (VEGFR1/Flt-1, VEGFR2/KDR/Flk-1, and VEGFR3) (Olsson *et al.*, 2006). VEGF-A does not interact with VEGFR1 and VEGFR2 equally, and the signal transduction stimulated is not the same. Furthermore, VEGF-C and -D stimulate lymphangiogenesis through VEGFR3 during development and solid tumor formation. Another complexity is that different growth factors affect blood vessel development at different stages. For example, VEGF-A

stimulates nascent sprout angiogenesis, while angiopoietin induces blood vessel maturation (Eklund and Olsen, 2006) and protects mature blood vessels (Brindle *et al.*, 2006).

Angiogenesis cannot be related directly to proliferation of cultured endothelial cells. In fact, Folkman (Folkman and Shing, 1992) has postulated that endothelial cell migration can induce proliferation. If this is true, endothelial cell chemotactic factors may be critical for angiogenesis. Some factors, of course, may have both mitogenic and chemotactic activities. For example, PDGF and EGF can be both chemotactic and mitogenic for dermal fibroblasts, while VEGF can be both motogenic and mitogenic for endothelial cells (Nissen *et al.*, 1998; Zachary, 2003; Gallicchio *et al.*, 2005).

Besides growth factors and chemotactic factors, an appropriate extracellular matrix is necessary for angiogenesis. Three-dimensional gels of extracellular matrix proteins have more pronounced effects on cultured endothelial cells than do monolayer protein coats (Tonnesen *et al.*, 2000), as has been observed with smooth muscle cells (Koyama *et al.*, 1996). Rat epididymal microvascular cells cultured in type I collagen gels with TGF-β produce capillary-like structures within one week (Madri *et al.*, 1996). Omission of TGF-β markedly reduces the effect. In contrast, laminin-containing gels in the absence of growth factors induce human umbilical vein and dermal microvascular cells to produce capillary-like structures within 24 hours of plating (Kubota *et al.*, 1988). Matrix-bound thrombospondin also promotes angiogenesis (Nicosia and Tuszynski, 1994), possibly through its ability to activate TGF-β. Recently it has been observed that type 1 collagen can protect newly formed blood vessels from apoptotic effects of angiostatic agents (Addison *et al.*, 2005). Together, these studies support the hypothesis that the extracellular matrix plays an important role in angiogenesis. Consonant with this hypothesis, angiogenesis in the chick chorioallantoic membrane is dependent on the expression of αvβ3, an integrin that recognizes fibrin and fibronectin as well as vitronectin (Brooks, Clark *et al.*, 1994). Furthermore, in porcine cutaneous wounds, αvβ3 is expressed only on capillary sprouts as they invade the fibrin clot (Clark *et al.*, 1996). *In vitro* studies in fact demonstrate that αvβ3 can promote endothelial cell migration on provisional matrix proteins (Leavesley *et al.*, 1993).

Given the information just outlined, a series of events leading to angiogenesis can be hypothesized. Substantial injury causes tissue-cell destruction and hypoxia. Potent angiogenesis factors such as FGF-1 and FGF-2 are released secondary to cell disruption, while VEGF is induced by hypoxia. Proteolytic enzymes released into the connective tissue degrade extracellular matrix proteins. Specific fragments from collagen, fibronectin, and elastin as well as many phlogistic agents recruit peripheral blood monocytes to the injured site, where these cells become activated macrophages that release more angiogenesis factors. Certain angiogenic factors, such as FGF-2, stimulate endothelial cells to release plasminogen activator and procollagenase. Plasminogen activator converts plasminogen to plasmin and procollagenase to active collagenase, and in concert these two proteases digest basement membrane constituents.

The fragmentation of the basement membrane allows endothelial cells to migrate into the injured site in response to FGF, fibronectin fragments, heparin released from disrupted mast cells, and other endothelial cell chemoattractants. To migrate into the fibrin/fibronectin-rich wound, endothelial cells express αvβ3 integrin. The newly forming blood vessels first deposit a provisional matrix containing fibronectin and proteoglycans but ultimately form basement membrane. TGF-β may induce endothelial cells to produce the fibronectin and proteoglycan provisional matrix as well as to assume the correct phenotype for capillary-tube formation. FGF and other mitogens, such as VEGF, stimulate endothelial cell proliferation, resulting in a continual supply of endothelial cells for capillary extension. Capillary sprouts eventually branch at their tips and join to form capillary loops through which blood flow begins. New sprouts then extend from these loops to form a capillary plexus.

Within a day or two after removal of angiogenic stimuli, capillaries undergo regression, as characterized by mitochondrial swelling in the endothelial cells at the distal tips of the capillaries, platelet adherence to degenerating endothelial cells, vascular stasis, endothelial cell necrosis, and ingestion of the effete capillaries by macrophages. Although αvβ3 has been shown to regulate apoptosis of endothelial cells in culture and in tumors (Brooks *et al.*, 1994), αvβ3 is not present on wound endothelial cells as they undergo programmed cell death, indicating another pathway of apoptosis in healing-wound blood vessels (X. Feng, R. A. F. Clark, and M. G. Tonnesen, unpublished observations). Thrombospondin appears to be a good candidate for this phenomenon (Koch *et al.*, 1992).

## Wound Contraction and Extracellular Matrix Organization

During the second and third weeks of healing, fibroblasts begin to assume a myofibroblast phenotype characterized by large bundles of actin-containing microfilaments disposed along the cytoplasmic face of the plasma membrane and the establishment of cell–cell and cell–matrix linkages (Welch *et al.*, 1990; Hinz *et al.*, 2004). In some (Desmouliere *et al.*, 1995) but not all (Welch *et al.*, 1990) wound situations, myofibroblasts express smooth muscle actin. Importantly, TGF-β can induce cultured human fibroblasts to express smooth muscle actin (Desmouliere *et al.*, 1993) and may also be responsible for its expression *in vivo* (Gabbiani, 2003).

The appearance of the myofibroblasts corresponds to the commencement of connective tissue compaction and the contraction of the wound. Fibroblasts link to the extracellular fibronectin matrix through α5β1 (Welch *et al.*, 1990); to collagen matrix through α1β1 and α2β1 collagen receptors (Ignatius *et al.*, 1990); and to each other through direct adherens junctions (Welch *et al.*, 1990). Fibroblast α2β1 receptors are markedly up-regulated in seven-day wounds (Xu and Clark, 1996), a time when new collagenous matrix is accumulating and fibroblasts are beginning to align with collagenous fibrils through cell–matrix connections (Welch *et al.*, 1990). New collagen bundles in turn have the capacity to join end-to-end with collagen bundles at the wound edge and ultimately to form covalent cross-links among themselves and with the collagen bundles of the adjacent dermis (Yamauchi *et al.*, 1987; Birk *et al.*, 1989). These cell–cell, cell–matrix, and matrix–matrix links provide a network across the wound whereby the traction of myofibroblasts on their pericellular matrix can be transmitted across the wound (Hinz, 2006).

Cultured fibroblasts dispersed within a hydrated collagen gel provide a functional *in vitro* model of tissue contraction (Bell *et al.*, 1979; Carlson and Longaker, 2004). When serum is added to the admixture, contraction of the collagen matrix occurs over the course of a few days. When observed with time-lapse microphotography, collagen condensation appears to result from a "collection of collagen bundles" executed by fibroblasts as they extend and retract pseudopodia attached to collagen fibers (Bell *et al.*, 1979). More recent elegant studies have further defined fibroblast–collagen interactions (Grinnell *et al.*, 2006). The transmission of these traction forces across the *in vitro* collagen matrix depends on two linkage events: fibroblast attachment to the collagen matrix through the α2β1 integrin receptors (Schiro *et al.*, 1991) and cross-links between the individual collagen bundles (Woodley *et al.*, 1991). This linkage system probably plays a significant role in the *in vivo* situation of wound contraction as well. In addition, cell–cell adhesions appear to provide an additional means by which the traction forces of the myofibroblast may be transmitted across the wound matrix (Hinz *et al.*, 2004).

F-actin bundle arrays, cell–cell and cell–matrix linkages, and collagen cross-links are all facets of the biomechanics of extracellular matrix contraction. The contraction process, however, needs a cytokine signal. For example, cultured fibroblasts mixed in a collagen gel contract the collagen matrix in the presence of serum (Bell *et al.*, 1979), PDGF (Clark *et al.*, 1989), or TGF-β (Werner *et al.*, 1994). Since TGF-β (Clark *et al.*, 1995), but not PDGF (Marikovsky *et al.*, 1993), persists in dermal wounds during the time of tissue contraction, it is the most likely candidate for the stimulus of contraction. Nevertheless, it is possible that both PDGF and TGF-β signal wound contraction, one more example of the many redundancies observed in the critical processes of wound healing. In summary, wound contraction represents a complex and masterfully orchestrated interaction of cells, extracellular matrix, and cytokines.

Collagen remodeling during the transition from granulation tissue to scar is dependent on continued collagen synthesis and collagen catabolism. The degradation of wound collagen is controlled by a variety of collagenase enzymes from macrophages, epidermal cells, and fibroblasts. These collagenases are specific for particular types of collagens, but most cells probably contain two or more different types of these enzymes (Mott and Werb, 2004). Three metalloproteinases (MMPs) have been described that have the ability to cleave native collagen: MMP-1, or classic interstitial collagenase, which cleaves types I, II, III, XIII, and X collagens; neutrophil collagenase (MMP-8); and a novel collagenase produced by breast carcinomas that is prominent in chronic wounds (MMP-13) (Vaalamo et al., 1997). Currently it is not known which *interstitial* collagenases are critical in the remodeling stage of human wound repair. For example, no wound-healing defect was observed in mice deficient for MMP-13 (Hartenstein et al., 2006). However, this fact may be attributable to the redundancy of nature.

Other MMPs potentially important in wound repair include two gelatinases, 72-kDa gelatinase A (MMP-2) and 92-kDa gelatinase B (MMP-9), with identical substrate specificity to degrade denatured collagens of all types as well as native types V and XI collagens; and stromelysin-1, -2, and -3, which degrade a wide variety of substrates, including types III, IV, V, VII, and IX collagens as well as proteoglycans and glycoproteins. Although MMP-2 and MMP-9 are expressed in early wounds (Larjava et al., 1993), they may play a lesser role in later tissue remodeling. MMP-2 is produced constitutively by many cells and only up-regulated by TGF-β (Overall et al., 1989), whereas MMP-9 is subject to regulation by a variety of physiologic signals, since its promoter contains two AP-1 regulatory elements (Huhtala et al., 1991). Two other metalloproteinases that do not belong to the aforementioned subgroups are macrophage metalloelastase and a novel transmembrane metalloproteinase (MT1-MMP), which can activate secreted progelatinase-A (proMMP-2) and other proenzymes of the MMP family. However, recent data suggest that this activator of MMPs may not be critical for normal cutaneous-wound healing (Mirastschijski et al., 2004). MMP enzymatic activities are controlled by various inhibitor counterparts called *tissue inhibitors of metalloproteinases* (TIMPs), which are finely regulated during wound repair (Frank et al., 1996).

Cytokines such as TGF-β, PDGF, and IL-1 and the extracellular matrix itself clearly play an important role in the modulation of collagenase and TIMP expression *in vivo*. Interestingly, type 1 collagen induces MMP-1 expression through the α2β1 collagen receptor while suppressing collagen synthesis through the α1β1 collagen receptor (Langholz et al., 1995). Type 1 collagen also induces expression of α2β1 receptors (Klein et al., 1991; Xu and Clark, 1996); thus collagen can induce the receptor that signals a collagen degradation–remodeling phenotype. Such dynamic, reciprocal cell–matrix interactions appear to occur generally during tissue formation and remodeling processes such as morphogenesis, tumor growth, and wound healing, to name a few (Nelson and Bissell, 2006).

Wounds gain only about 20% of their final strength by the third week, during which time fibrillar collagen has accumulated relatively rapidly and has been remodeled by myofibroblast contraction of the wound. Thereafter the rate at which wounds gain tensile strength is slow, reflecting a much slower rate of collagen accumulation. In fact, the gradual gain in tensile strength has less to do with new collagen deposition than further collagen remodeling, with formation of larger collagen bundles and an accumulation of intermolecular cross-links (Bailey et al., 1975). Nevertheless, wounds fail to attain the same breaking strength as uninjured skin. At maximum strength, a scar is only 70% as strong as intact skin (Levenson et al., 1965).

## III. CHRONIC WOUNDS

Acute wounds are those that heal through the routine processes of inflammation, tissue formation, and remodeling, which occur in a timely fashion. As discussed earlier, these processes may overlap temporally. However, the prolonged delay in any of these reparative processes may result in the formation of a chronic wound. Chronic wounds are often associated with underlying pathological conditions that contribute to an impaired healing. Venous leg ulcers and diabetic foot ulcers are common examples of chronic wounds caused or accentuated by an underlying disorder. While the former is induced by insufficient venous flow that results in increased blood pressure in the lower limb and, therefore, increased vascular permeability, the latter is caused by peripheral neuropathy that leads to abnormal load distribution on the foot surface and decreased sensation (Mustoe et al., 2006). Subsequently, these abnormalities cause a loss of tissue viability, suboptimal local tissue permeability, and an elevated and sustained inflammatory response.

## IV. TISSUE-ENGINEERED THERAPY: ESTABLISHED PRACTICE

Initial attempts to speed up wound repair and improve the quality of healing in chronic or burn wounds involved the use of synthetic, composite synthetic, or biological dressings (Bromberg et al., 1965; Robson et al., 1973; Purna and Babu, 2000). Although effective, these dressings did not offer any permanent treatment, since eventually an autograft had to be implanted to achieve complete healing, which is often undesirable due to donor site morbidity.

The advent of tissue-engineered constructs has, however, revolutionized the wound-healing practice (Rosso et al., 2005; Simpson, 2006). These constructs could be classified into two main categories: cellular and acellular. Regardless of whether they are cellular or acellular, the basic building block of these constructs is composed of a biomimetic and a scaffolding material. While the biomimetic functions to stimulate cells to perform their physiological functions, the scaffold typically provides a mechanical support for the cells to spread on and to proliferate within, to produce new tissue. However, scaffolds prepared from naturally occurring biopolymers may provide additional biological stimuli to support cell and tissue function (Lutolf and Hubbell, 2005). After implantation, these constructs promote faster healing, resulting in the development of a new tissue that bears a close structural and functional resemblance to the uninjured, host tissue.

These tissue-engineered constructs are typically targeted toward a specific application, i.e., for only epidermal repair, for only dermal repair, or both.

## Engineered Epidermal Constructs

Immediate wound coverage, whether permanent or temporary, is one of the cornerstones of wound management. Engineered epidermal constructs with qualities similar to those of autologous skin have been used to facilitate repair of split-thickness wounds.

Autologous cultured keratinocyte grafts were first used in humans by O'Conner *et al.* (1981). Subsequently there has been extensive experience with cultured epidermal grafts for the treatment of burns as well as other acute and chronic wounds (Butler and Orgill, 2005; Matouskova *et al.*, 2006). Epicel™ (Genzyme Tissue Repair) is an example of an engineered epidermal autograft. The major advantage of this technique is the ability to provide autologous grafts capable of covering large areas, with reasonable cosmetic results. Another significant advantage of autologous grafts is their ability to serve as permanent wound coverage, since the host does not reject them. Disadvantages include the two- to three-week time interval required before sufficient quantities of keratinocytes are available, the need for an invasive and painful procedure to obtain autologous donor cells, and the large costs, estimated at $13,000 per 1% total body surface area covered (Rue *et al.*, 1993). Furthermore, graft take is widely variable, based on wound status, patient age, general host status, and operator experience (Carsin *et al.*, 2000).

Cultured keratinocyte allografts were developed to help overcome the need for biopsy and separate cultivation for each patient to produce autologous grafts and the long lag period between epidermal harvest and graft product. Cultured epidermal cells from both cadavers and unrelated adult donors have been used for the treatment of burns (Yanaga *et al.*, 2001; Brychta *et al.*, 2002) and chronic leg ulcers (Navratilova *et al.*, 2004; Paquet *et al.*, 2005). Although a previous study showed that allografts made from neonatal foreskin keratinocytes were more responsive to mitogens than those from adult (cadaver) cells (Phillips and Gilchrest, 1991) and, therefore, were the preferred candidate for cultured allografts, a current investigation has revealed that such allografts are also more

immunogenic than the regular cultured skin substitutes (Erdag and Morgan, 2004).

To facilitate mass allograft production and wide availability, cryopreserved allografts have been developed (Navratilova *et al.*, 2004). These frozen constructs give fairly comparable results to fresh allografts (Navratilova *et al.*, 2004). In spite of these advances, the cultured epidermal grafts have failed to produce satisfactory responses. The primary reasons include the lack of mechanical strength and susceptibility to contractures. As an alternative, keratinocyte delivery systems were developed, where the cells were delivered to the injury site via a biodegradable scaffold. For example, laser skin, produced by FIDIA Advanced Biopolymer, Italy, is used to deliver keratinocytes via a chemically modified hyaluronan membrane, which is perforated with micron-sized holes that allow cells to grow to confluence.

## Engineered Dermal Constructs

While the use of cultured keratinocytes to enhance wound healing has met with some success, it lacks a dermal component that, if present, would help prevent wound contraction and provide greater mechanical stability. Allografts (containing dermis) from other sources have been used for many years, although they provide only temporary coverage due to their tendency to induce acute inflammation (Lamme *et al.*, 2002). However, this skin can be chemically treated to remove the antigenic epidermal cellular elements (Alloderm, Life Cell Corporation, Woodlans, TX) and has been used alone or in combination with cultured autologous keratinocytes for closure of various chronic wounds and burns (Gustafson and Kratz, 1999). In spite of these modifications, allogenic grafts, when compared with autologous grafts, have been shown to promote lower percent re-epithelialization and excessive wound contraction (Morimoto *et al.*, 2005). Furthermore, the overall therapeutic outcome of a skin graft depends also on the site from which the cells were isolated (during biopsy) (Wang *et al.*, 2004).

In 1981, Burke *et al.*, developed an acellular composite skin graft made of a collagen-based dermal lattice (containing bovine collagen and chondroitin-6-sulfate) with an outer silicone covering. After placement on the wound, the acellular dermal component recruits the host dermal fibroblasts while undergoing simultaneous degradation. About two to three weeks later, the silicone sheet is removed and covered with an autograft. This composite graft has been used successfully to treat burns and recently received FDA approval for this indication as well as for reconstructive surgery (Integra, Integra Life Sciences Corporation, Plainsboro, NJ) (Dantzer *et al.*, 2003). However, these constructs cannot be used in patients who are allergic to bovine products.

Another rendition of a dermal substitute is Transcyte (Dermagraft-TC) (Advanced Tissue Sciences Inc., La Jolla, CA). This product consists of an inner nylon mesh in which human fibroblasts are embedded, together with an outer silicone layer to limit evaporation. The fibroblasts are lysed in the final product by freeze-thawing. Prior to that time, the fibroblasts had manufactured collagen, matrix proteins, and cytokines, all of which promote wound healing by the host. Transcyte has been used successfully as a temporary wound coverage after excision of burn wounds (Purdue, 1997; Lukish *et al.*, 2001) and has recently been approved by the FDA for this indication. Dermagraft is a modification of this product in which a biodegradable polyglactin mesh is used instead of the nylon mesh, and the fibroblasts remain viable (Marston, 2004). Use of this dermal substitute has had limited success in the treatment of diabetic foot ulcers (Gentzkow *et al.*, 1996), owing largely to its inability to form stable adhesions with the final epidermal graft (Rennekampff *et al.*, 1996).

## Engineered Skin Substitutes

Full-thickness wounds involve the loss of both the epidermal and dermal layers of the skin. To treat such extensive wounds, Bell *et al.* (1981) first described a bilayered skin composite consisting of a collagen lattice with dermal fibroblasts that was covered with epidermal cells. Modification of this composite consisting of type I bovine collagen and live allogeneic human skin fibroblasts and keratinocytes has been developed (HSE, Apligraf, Organogenesis, Canton, MA). It has been used successfully in surgical wounds (Griffiths *et al.*, 2004) and venous ulcers (Curran and Plosker, 2002). In a large multicenter trial, this product resulted in accelerated healing of chronic nonhealing venous stasis ulcers when compared to standard compressive therapy (Falanga *et al.*, 1998). Both Apligraf (for venous stasis ulcers) and Dermagraft (for diabetic neurotrophic ulcerations) have been available in Canada since 1997 and more recently in the United States. Several other composite skin substitutes combining dermal and epidermal elements have been developed. Composite cultured skin (CCS, Ortec International Inc., New York, NY) is composed of an overlay of stratified neonatal keratinocytes on fibroblasts embedded in distinct layers of bovine type I collagen. This product is currently being evaluated in clinical trials for the treatment of burns and in patients with epidermolysis bullosa. More extensive comparative data of the various biologic dressings have been published elsewhere (Balasubramani *et al.*, 2001).

## V. TISSUE-ENGINEERED THERAPY: NEW APPROACHES

Although the increased healing rates observed with the use of these engineered constructs show promise for the treatment of burns and/or chronic wounds, they have several intrinsic disadvantages that limit their use: (1) The epidermal grafts are very fragile and therefore difficult to handle; (2) it is difficult to quality-control the large-scale production of any cell-populated matrix; and (3) while auto-grafts require skin biopsy, allografts may experience early rejection. Moreover, these constructs in general are only about 25% efficient, implying that they must be applied on at least four patients before their effect can be seen. These limitations suggest that further improvements be made so that tissue-engineered constructs are not only more effective but also less complex.

A general perception of novel engineering approaches emerges from the foregoing discussions, one that suggests the development of easy-to-handle, user-friendly, cost-effective, acellular, and potent constructs that are based on nonanimal products. Since tissue cells are the primary source of the various ECM molecules that facilitate and synchronize tissue repair, any acellular product must, therefore, be conductive to recruiting the host tissue cells rapidly and inductive to stimulating the invading cells to proliferate, synthesize new ECM, and, if required, differentiate. This design is nonetheless challenging. Cues learned from em-bryonic, morphogenetic, and wound repair events should be implemented to engineer such tissue constructs.

Over the years, various engineered skin constructs have used collagen as the preferred scaffolding material for cell seeding (Balasubramani *et al.*, 2001). The huge popularity of collagen can be attributed to its abundance in skin and, therefore, its recognition by cell surface receptors, its multiple biological roles during both cutaneous homeostasis and wound repair, and its ability to cross-link and thereby impart mechanical strength to the final construct (Ruszczak, 2003). However, during wound repair, collagen appears only during the later stages, after the invading and proliferating fibroblasts have filled the wound space (Clark, 1996). Therefore, collagen may not be optimal for initial cell migration. Instead, en masse fibroblast migration into the wound clot is accompanied by the synthesis and secretion of hyaluronan, which promotes migration by regulating cell adhesion to the ECM (Chen and Abatangelo, 1999). In fact, hyaluronan also appears coincident with tissue cell migration during embryogenesis and morphogenesis (Toole, 1991).

Hyaluronan is a nonsulfated glycosaminoglycan present in most human tissues. During wound repair, it serves multiple important functions, ranging from regulating inflammation to promoting fibroblast migration and proliferation (Chen and Abatangelo, 1999). Interestingly, hyaluronan has been implicated in the scarless or minimally scarred repair of fetal wounds, perhaps owing to its role in regulating the inflammatory response (Wisniewski and Vilcek, 1997) and collagen deposition (Longaker *et al.*, 1991). Furthermore, similar to synthetic polymers, hyaluronan can be chemically modified to obtain a variety of stable derivatives (Prestwich *et al.*, 1998). Therefore, by offering the advantages of both natural and synthetic materials, hyaluronan promises to be a more suitable scaffolding material for acellular matrices than is collagen. Indeed, chemically modified hyaluronan scaffolds have been successfully used for various tissue-engineering applications, including wound repair (Campoccia *et al.*, 1998; Kirker *et al.*, 2002).

Our laboratory is developing an acellular, hyaluronan-based matrix to promote dermal repair, where the intent is to provide a reparative dressing or packing material for acute and chronic wounds and, in addition, to facilitate granulation-tissue formation. Since fibroblast migration is the rate-limiting step in granulation-tissue formation (McClain *et al.*, 1996), the matrix must use a biomimetic that supports maximal fibroblast migration. Fibronectin turns out to be a favorable candidate since (1) it appears together with hyaluronan at times of cell migration during embryogenesis, morphogenesis, and wound repair (Toole, 1991; Clark, 1996); (2) fibroblast migration on hyaluronan/fibronectin gels is far greater (approximately fourfold) than on fibrin/fibronectin gels (D. Greiling and R. A. F. Clark, unpublished observation); (3) fibronectin has been shown to be required for fibroblast transmigration from a collagen gel to a fibrin gel (Greiling and Clark, 1997); and (4) fibronectin is absent in chronic wounds, where it is produced normally (Herrick *et al.*, 1996) but eliminated rapidly by the abundant proteases present in chronic wound fluid (Grinnell and Zhu, 1996). Although fibronectin appears to be an ideal biomimetic for use in hyaluronan scaffolds, its stability in the proteolytic environment of chronic wounds is a major concern.

Alternatively, the proteolytically stable arginine-glycine-asparatic acid (RGD) peptide sequence, the smallest cell recognition sequence in the 10th module of type III repeat of fibronectin, can be used to support key cell functions (Pierschbacher and Ruoslahti, 1984). RGD has been widely used to promote cell attachment and spreading in various tissue-engineering applications in general (Hersel *et al.*, 2003) and in wound-healing applications in particular (Pierschbacher *et al.*, 1994). This is perhaps because the RGD sequence is found in a variety of ECM molecules and, therefore, is recognized by the transmembrane integrin receptors of multiple cell types, including dermal and epidermal tissue cells. Our previous study has shown that hyaluronan hydrogels decorated with RGDS (arg-gly-asp-ser) support NIH 3T3 fibroblast functions *in vitro* (Shu *et al.*, 2004) and, when seeded with 3T3 fibroblasts and implanted in nude mice, produce granulation-like tissue in four weeks. Therefore, RGD-modified hyaluronan hydrogels possess great inductive properties. However, these hydrogels neither support optimal dermal fibroblast functions nor demonstrate conductive properties required of any acellular scaffold (Shu *et al.*, 2004).

To impart our acellular matrix with both inductive and conductive properties, we selected, as the biomimetics, three FN functional domains (FNIII$_{(8-11)}$, FNIII$_{(12-15)}$, and FNIII$_{(12-V-15)}$) that are necessary and sufficient for optimal dermal fibroblast migration *in vitro* (Clark *et al.*, 2003). Indeed, our hydrogels have been successful in supporting dermal fibroblast functions *in vitro* and in promoting wound repair *in vivo* (Ghosh *et al.*, 2006). Since these hydrogels can

be formulated at room temperature and physiological pH, they are compatible with both cells and the incorporated biological molecules. In addition, their rapid gelation (<10 min) advocates their possible injectable use. Furthermore, since these hydrogels are derived from components that can easily be produced by bacterial fermentation, they are very cost effective and, unlike animal-derived products, immunocompatible. Another important feature of this tissue-engineered construct is that it can be formulated into myriad physical forms, ranging from hydrogels to electrospun nanofibrous scaffolds (Ji *et al.*, 2006), the latter form mimicking the natural ECM architecture.

With a similar objective, several other groups have also developed "intelligent" scaffolds for tissue repair (Rosso *et al.*, 2005). These approaches commonly employ synthetic materials to build scaffolds, since they allow great flexibility during formulation. To impart bioactivity, these scaffolds contain potent biomimetics that can be recognized by tissue cells. However, as discussed earlier, cell invasion during granulation-tissue formation occurs concurrently with matrix degradation, which can typically be observed when using naturally derived materials. To elicit a similar response in synthetic materials, protease-sensitive sequences are incorporated within the scaffold that are cleaved on contact with the cell-secreted proteolytic enzymes (Lutolf *et al.*, 2003). Therefore, similar to our engineered matrix, these intelligent scaffolds combine the advantages of both natural and synthetic biomaterials to facilitate wound repair.

Traditionally, the structural component of the tissue-engineered constructs has been viewed as providing only a passive mechanical structure. The design proposed here reflects our understanding that cells respond primarily to biological signals. However, recent reports have shown that mechanical forces alone can govern cell and tissue phenotype in ways similar to biological stimuli (Ingber, 2003). Further studies have revealed that cells use an active tactile sensing mechanism to feel and respond to substrate mechanics (Discher *et al.*, 2005; Vogel and Sheetz, 2006). In particular, dermal fibroblasts have been shown to respond to substrate mechanics by regulating levels of gene transcription that eventually lead to differential ECM synthesis and their transformation into myofibroblasts (Chiquet *et al.*, 2003; Hinz, 2006). Since these processes are critical during wound repair, effective tissue-engineering approaches for wound repair would require optimization of both biological and mechanical effectors. In fact, our acellular matrix has been "tuned" to have optimal mechanical properties for skin wound healing.

The acellular tissue-engineered constructs discussed so far utilize scaffolding materials to provide mechanical support for tissue ingrowth and biomimetics to induce key cell functions. The primary goal of these novel approaches is to mimic the attributes of fibrin clot for parenchymal cell migration. However, a fibrin clot and the second provisional matrix composed of fibronectin and hyaluronan is composed not only of a fibrin/fibronectin scaffold and an array of clotting and fibrinolytic enzymes, but also a plethora of growth factors that had been released during platelet aggregation (Mosesson, 2005). Growth factors play a crucial role in the overall healing response, where they function to stimulate cell migration, proliferation, differentiation, and angiogenesis. Furthermore, growth factor deficiency often leads to impaired wound repair (Peters *et al.*, 2005). As a result, several groups have investigated the use of tissue-engineered constructs for local growth factor delivery, where the release of appropriate growth factors produced an increase in angiogenic activity (Richardson *et al.*, 2001; Cai *et al.*, 2005).

It is interesting to note that in spite of the release of growth factors immediately after wounding, there remains a three-day lag before granulation tissue is formed. This suggests that the growth factors may be retained and functional within the clot. This form of solid-state biochemistry may be unconventional, but it is backed by data from recent studies, which have shown that IGF and VEGF bound to specific molecular domains of FN retain or accentuate their bioactivity (Gui and Murphy, 2001; Wijelath *et al.*, 2002). In fact, studies from our own laboratory have shown that PDGF, when preloaded onto our engineered matrix containing specific domains of FN, retains its activity at the level typically observed in soluble state (K. Ghosh and R. A. F. Clark, unpublished observations). Therefore, by incorporating the appropriate growth factor–binding sequences/domains, tissue-engineered construct can be used as a growth factor repository, causing an increase in local concentration that may ultimately accentuate cell functions.

In conclusion, wound healing is a dynamic and fine-tuned cellular response aimed at reinstating tissue homeostasis after an insult. Vigorous cellular activities observed during wound repair are similar to those occurring during embryogenesis and morphogenesis, which indicates the enormous complexity of this physiological reparative process. That may also explain why, despite over two decades of intense research and development, we have still not identified an ideal therapy. However, novel tissue-engineering approaches are showing tremendous promise and are aiming to push the limits of human expectations of wound therapy.

## VI. REFERENCES

Markets for advanced wound care technologies. http://www.the-infoshop.com/study/bc25780_wound_care.html.

Addison, C. L., Nor, J. E., *et al.* (2005). The response of VEGF-stimulated endothelial cells to angiostatic molecules is substrate-dependent. *BMC Cell Biol.* **6**, 38.

American Burn Association. (2005). Burn incidence and treatment in the U.S.: 2000 Fact Sheet. http://www.ameriburn.org/pub/BurnIncidenceFactSheet.htm.

Ansel, J. C., Tiesman, J. P., *et al.* (1993). Human keratinocytes are a major source of cutaneous platelet-derived growth factor. *J. Clin. Invest.* **92**(2), 671–678.

Ashcroft, G. S., Yang, X., *et al.* (1999). Mice lacking Smad3 show accelerated wound healing and an impaired local inflammatory response [see comments]. *Nat. Cell Biol.* **1**(5), 260–266.

Ashcroft, G. S., Mills, S. J., *et al.* (2003). Estrogen modulates cutaneous wound healing by down-regulating macrophage migration inhibitory factor. *J. Clin. Invest.* **111**(9), 1309–1318.

Bailey, A. J., Bazin, S., *et al.* (1975). Characterization of the collagen of human hypertrophic and normal scars. *Biochim. Biophys. Acta* **405**(2), 412–421.

Balasubramani, M., Kumar, T. R., *et al.* (2001). Skin substitutes: a review. *Burns* **27**(5), 534–544.

Basu, A., Castle, V. P., *et al.* (2006). Cross-talk between extrinsic and intrinsic cell death pathways in pancreatic cancer: synergistic action of estrogen metabolite and ligands of death receptor family. *Cancer Res.* **66**(8), 4309–4318.

Bell, E., Ivarsson, B., *et al.* (1979). Production of a tissue-like structure by contraction of collagen lattices by human fibroblasts of different proliferative potential *in vitro*. *Proc. Natl. Acad. Sci. U.S.A.* **76**(3), 1274–1278.

Bell, E., Ehrlich, H. P., *et al.* (1981). Living tissue formed *in vitro* and accepted as skin-equivalent tissue of full thickness. *Science* **211**(4486), 1052–1054.

Birk, D. E., Zycband, E. I., *et al.* (1989). Collagen fibrillogenesis *in situ*: fibril segments are intermediates in matrix assembly. *Proc. Natl. Acad. Sci. U.S.A.* **86**(12), 4549–4553.

Brindle, N. P., Saharinen, P., *et al.* (2006). Signaling and functions of angiopoietin-1 in vascular protection. *Circ. Res.* **98**(8), 1014–1023.

Bromberg, B. E., Song, I. C., *et al.* (1965). The use of pig skin as a temporary biological dressing. *Plast. Reconstr. Surg.* **36**, 80–90.

Brooks, P. C., Clark, R. A., *et al.* (1994). Requirement of vascular integrin alpha v beta 3 for angiogenesis. *Science* **264**(5158), 569–571.

Brown, G. L., Nanney, L. B., *et al.* (1989). Enhancement of wound healing by topical treatment with epidermal growth factor. *N. Engl. J. Med.* **321**(2), 76–79.

Brychta, P., Adler, J., *et al.* (2002). Cultured epidermal allografts: quantitative evaluation of their healing effect in deep dermal burns. *Cell Tissue Bank* **3**(1), 15–23.

Bugge, T. H., Kombrinck, K. W., *et al.* (1996). Loss of fibrinogen rescues mice from the pleiotropic effects of plasminogen deficiency. *Cell* **87**(4), 709–719.

Burke, J. F., Yannas, I. V., *et al.* (1981). Successful use of a physiologically acceptable artificial skin in the treatment of extensive burn injury. *Ann. Surg.* **194**(4), 413–428.

Butler, C. E., and Orgill, D. P. (2005). Simultaneous *in vivo* regeneration of neodermis, epidermis, and basement membrane. *Adv. Biochem. Eng. Biotechnol.* **94**, 23–41.

Cai, S., Liu, Y., *et al.* (2005). Injectable glycosaminoglycan hydrogels for controlled release of human basic fibroblast growth factor. *Biomaterials* **26**(30), 6054–6067.

Campoccia, D., Doherty, P., *et al.* (1998). Semisynthetic resorbable materials from hyaluronan esterification. *Biomaterials* **19**(23), 2101–2127.

Carlson, M. A., and Longaker, M. T. (2004). The fibroblast-populated collagen matrix as a model of wound healing: a review of the evidence. *Wound Repair Regen.* **12**(2), 134–147.

Carsin, H., Ainaud, P., *et al.* (2000). Cultured epithelial autografts in extensive burn coverage of severely traumatized patients: a five year single-center experience with 30 patients. *Burns* **26**(4), 379–387.

Chen, W. Y., and Abatangelo, G. (1999). Functions of hyaluronan in wound repair. *Wound Repair Regen.* **7**(2), 79–89.

Chiquet, M., Renedo, A. S., *et al.* (2003). How do fibroblasts translate mechanical signals into changes in extracellular matrix production? *Matrix Biol.* **22**(1), 73–80.

Circolo, A., Welgus, H. G., *et al.* (1991). Differential regulation of the expression of proteinases/antiproteinases in fibroblasts. Effects of interleukin-1 and platelet-derived growth factor. *J. Biol. Chem.* **266**(19), 12283–12288.

Clark, R. A. F. (1996). Wound repair: overview and general considerations. *In* "The Molecular and Cellular Biology of Wound Repair" (R. A. F. Clark, ed.), pp. 3–50. Plenum Press, New York.

Clark, R. A., Lanigan, J. M., *et al.* (1982). Fibronectin and fibrin provide a provisional matrix for epidermal cell migration during wound reepithelialization. *J. Invest. Dermatol.* **79**(5), 264–269.

Clark, R. A., Folkvord, J. M., *et al.* (1989). Platelet isoforms of platelet-derived growth factor stimulate fibroblasts to contract collagen matrices. *J. Clin. Invest.* **84**(3), 1036–1040.

Clark, R. A., Nielsen, L. D., *et al.* (1995). Collagen matrices attenuate the collagen-synthetic response of cultured fibroblasts to TGF-beta. *J. Cell Sci.* **108**(Pt. 3), 1251–1261.

Clark, R. A., Tonnesen, M. G., *et al.* (1996). Transient functional expression of alphaVbeta 3 on vascular cells during wound repair. *Am. J. Pathol.* **148**(5), 1407–1421.

Clark, R. A., An, J. Q., *et al.* (2003). Fibroblast migration on fibronectin requires three distinct functional domains. *J. Invest. Dermatol.* **121**(4), 695–705.

Colnot, C., Thompson, Z., *et al.* (2003). Altered fracture repair in the absence of MMP9. *Development* **130**(17), 4123–4133.

Cotsarelis, G. (2006). Epithelial stem cells: a folliculocentric view. *J. Invest. Dermatol.* **126**(7), 1459–1468.

Curran, M. P., and Plosker, G. L. (2002). Bilayered bioengineered skin substitute (Apligraf): a review of its use in the treatment of venous leg ulcers and diabetic foot ulcers. *BioDrugs* **16**(6), 439–455.

Dantzer, E., Queruel, P., *et al.* (2003). Dermal regeneration template for deep hand burns: clinical utility for both early grafting and reconstructive surgery. *Br. J. Plast. Surg.* **56**(8), 764–774.

Dejana, E. (1996). Endothelial adherens junctions: implications in the control of vascular permeability and angiogenesis. *J. Clin. Invest.* **98**(9), 1949–1953.

Desmouliere, A., Geinoz, A., *et al.* (1993). Transforming growth factor-beta 1 induces alpha–smooth muscle actin expression in granulation tissue myofibroblasts and in quiescent and growing cultured fibroblasts. *J. Cell. Biol.* **122**(1), 103–111.

Desmouliere, A., Redard, M., *et al.* (1995). Apoptosis mediates the decrease in cellularity during the transition between granulation tissue and scar. *Am. J. Pathol.* **146**(1), 56–66.

Desmouliere, A., Badid, C., *et al.* (1997). Apoptosis during wound healing, fibrocontractive diseases and vascular wall injury. *Int. J. Biochem. Cell Biol.* **29**(1), 19–30.

Discher, D. E., Janmey, P., *et al.* (2005). Tissue cells feel and respond to the stiffness of their substrate. *Science* **310**(5751), 1139–1143.

Duncan, M. R., and Berman, B. (1985). Gamma interferon is the lymphokine and beta interferon the monokine responsible for inhibition of fibroblast collagen production and late but not early fibroblast proliferation. *J. Exp. Med.* **162**(2), 516–527.

Dvorak, H. F. (1986). Tumors: wounds that do not heal. Similarities between tumor stroma generation and wound healing. *N. Engl. J. Med.* **315**(26), 1650–1659.

Eklund, L., and Olsen, B. R. (2006). Tie receptors and their angiopoietin ligands are context-dependent regulators of vascular remodeling. *Exp. Cell Res.* **312**(5), 630–641.

El Ghalbzouri, A., Hensbergen, P., *et al.* (2004). Fibroblasts facilitate reepithelialization in wounded human skin equivalents. *Lab. Invest.* **84**(1), 102–112.

Erdag, G., and Morgan, J. R. (2004). Allogeneic versus xenogeneic immune reaction to bioengineered skin grafts. *Cell. Transplant.* **13**(6), 701–712.

Even-Ram, S., and Yamada, K. M. (2005). Cell migration in 3D matrix. *Curr. Opin. Cell Biol.* **17**(5), 524–532.

Falanga, V., Margolis, D., *et al.* (1998). Rapid healing of venous ulcers and lack of clinical rejection with an allogeneic cultured human skin equivalent. Human Skin Equivalent Investigators Group. *Arch. Dermatol.* **134**(3), 293–300.

Folkman, J., and Shing, Y. (1992). Angiogenesis. *J. Biol. Chem.* **267**(16), 10931–10934.

Frank, S., Madlener, M., *et al.* (1996). Transforming growth factors beta1, beta2, and beta3 and their receptors are differentially regulated during normal and impaired wound healing. *J. Biol. Chem.* **271**(17), 10188–10193.

Gabbiani, G. (2003). The myofibroblast in wound healing and fibrocontractive diseases. *J. Pathol.* **200**(4), 500–503.

Gailit, J., Pierschbacher, M., *et al.* (1993). Expression of functional alpha 4 beta 1 integrin by human dermal fibroblasts. *J. Invest. Dermatol.* **100**(3), 323–328.

Gailit, J., Xu, J., *et al.* (1996). Platelet-derived growth factor and inflammatory cytokines have differential effects on the expression of integrins alpha 1 beta 1 and alpha 5 beta 1 by human dermal fibroblasts *in vitro*. *J. Cell Physiol.* **169**(2), 281–289.

Gallicchio, M., Mitola, S., *et al.* (2005). Inhibition of vascular endothelial growth factor receptor 2-mediated endothelial cell activation by Axl tyrosine kinase receptor. *Blood* **105**(5), 1970–1976.

Garrod, D. R., Berika, M. Y., *et al.* (2005). Hyper-adhesion in desmosomes: its regulation in wound healing and possible relationship to cadherin crystal structure. *J. Cell Sci.* **118**(Pt. 24), 5743–5754.

Gazel, A., Banno, T., *et al.* (2006). Inhibition of JNK promotes differentiation of epidermal keratinocytes. *J. Biol. Chem.* **281**, 20530–20541.

Gentzkow, G. D., Iwasaki, S. D., *et al.* (1996). Use of dermagraft, a cultured human dermis, to treat diabetic foot ulcers. *Diabetes Care* **19**(4), 350–354.

Ghosh, K., Ren, X. D., *et al.* (2006). Fibronectin functional domains coupled to hyaluronan stimulate adult human dermal fibroblast responses critical for wound healing. *Tissue Eng.* **12**(3), 601–613.

Ginsberg, M. H., Partridge, A., *et al.* (2005). Integrin regulation. *Curr. Opin. Cell. Biol.* **17**(5), 509–516.

Granstein, R. D., Murphy, G. F., *et al.* (1987). Gamma-interferon inhibits collagen synthesis *in vivo* in the mouse. *J. Clin. Invest.* **79**(4), 1254–1258.

Greenhalgh, D. G., Sprugel, K. H., *et al.* (1990). PDGF and FGF stimulate wound healing in the genetically diabetic mouse. *Am. J. Pathol.* **136**(6), 1235–1246.

Greiling, D., and Clark, R. A. (1997). Fibronectin provides a conduit for fibroblast transmigration from collagenous stroma into fibrin clot provisional matrix. *J. Cell Sci.* **110**(Pt. 7), 861–870.

Griffiths, M., Ojeh, N., *et al.* (2004). Survival of Apligraf in acute human wounds. *Tissue Eng.* **10**(7–8), 1180–1195.

Grinnell, F. (1994). Fibroblasts, myofibroblasts, and wound contraction. *J. Cell Biol.* **124**(4), 401–404.

Grinnell, F. (2003). Fibroblast biology in three-dimensional collagen matrices. *Trends Cell Biol.* **13**(5), 264–269.

Grinnell, F., and Zhu, M. (1996). Fibronectin degradation in chronic wounds depends on the relative levels of elastase, alpha1-proteinase inhibitor, and alpha2-macroglobulin. *J. Invest. Dermatol.* **106**(2), 335–341.

Grinnell, F., Rocha, L. B., *et al.* (2006). Nested collagen matrices: a new model to study migration of human fibroblast populations in three dimensions. *Exp. Cell. Res.* **312**(1), 86–94.

Grose, R., and Werner, S. (2004). Wound-healing studies in transgenic and knockout mice. *Mol. Biotechnol.* **28**(2), 147–166.

Grose, R., Hutter, C., *et al.* (2002). A crucial role of beta 1 integrins for keratinocyte migration *in vitro* and during cutaneous wound repair. *Development* **129**(9), 2303–2315.

Grotendorst, G. R., and Duncan, M. R. (2005). Individual domains of connective tissue growth factor regulate fibroblast proliferation and myofibroblast differentiation. *FASEB J.* **19**(7), 729–738.

Gui, Y., and Murphy, L. J. (2001). Insulin-like growth factor (IGF)–binding protein-3 (IGFBP-3) binds to fibronectin (FN): demonstration of IGF-I/IGFBP-3/fn ternary complexes in human plasma. *J. Clin. Endocrinol. Metab.* **86**(5), 2104–2110.

Gustafson, C. J., and Kratz, G. (1999). Cultured autologous keratinocytes on a cell-free dermis in the treatment of full-thickness wounds. *Burns* **25**(4), 331–335.

Hartenstein, B., Dittrich, B. T., *et al.* (2006). Epidermal development and wound healing in matrix metalloproteinase 13–deficient mice. *J. Invest. Dermatol.* **126**(2), 486–496.

Heino, J., Ignotz, R. A., *et al.* (1989). Regulation of cell adhesion receptors by transforming growth factor-beta. Concomitant regulation of integrins that share a common beta 1 subunit. *J. Biol. Chem.* **264**(1), 380–388.

Heldin, C.-H. a. W., B. (1996). Role of platelet-derived growth factor *in vivo*. In "The Molecular and Cellular Biology of Wound Repair" (R. A. F. Clark ed.), pp. 249–274. Plenum Press, New York.

Herrick, S. E., Ireland, G. W., *et al.* (1996). Venous ulcer fibroblasts compared with normal fibroblasts show differences in collagen but not fibronectin production under both normal and hypoxic conditions. *J. Invest. Dermatol.* **106**(1), 187–193.

Hersel, U., Dahmen, C., *et al.* (2003). RGD-modified polymers: biomaterials for stimulated cell adhesion and beyond. *Biomaterials* **24**(24), 4385–4415.

Hinz, B. (2006). Masters and servants of the force: the role of matrix adhesions in myofibroblast force perception and transmission. *Eur J. Cell Biol.* **85**(3–4), 175–181.

Hinz, B., Pittet, P., *et al.* (2004). Myofibroblast development is characterized by specific cell–cell adherens junctions. *Mol. Biol. Cell* **15**(9), 4310–4320.

Hopf, H. W., Gibson, J. J., *et al.* (2005). Hyperoxia and angiogenesis. *Wound Repair Regen.* **13**(6), 558–564.

Huhtala, P., Tuuttila, A., *et al.* (1991). Complete structure of the human gene for 92-kDa type IV collagenase. Divergent regulation of expression for the 92- and 72-kilodalton enzyme genes in HT-1080 cells. *J. Biol. Chem.* **266**(25), 16485–16490.

Hunt, T. K. (1980). "Wound Healing and Wound Infection: Theory and Surgical Practice." Appleton-Century-Crofts, New York.

Ignatius, M. J., Large, T. H., *et al.* (1990). Molecular cloning of the rat integrin alpha 1-subunit: a receptor for laminin and collagen. *J. Cell Biol.* **111**(2), 709–720.

Ignotz, R. A., and Massague, J. (1986). Transforming growth factor-beta stimulates the expression of fibronectin and collagen and their incorporation into the extracellular matrix. *J. Biol. Chem.* **261**(9), 4337–4345.

Ingber, D. E. (2003). Tensegrity II. How structural networks influence cellular information–processing networks. *J. Cell Sci.* **116**(Pt. 8), 1397–1408.

Jacinto, A., Martinez-Arias, A., *et al.* (2001). Mechanisms of epithelial fusion and repair. *Nat. Cell Biol.* **3**(5), E117–E123.

Ji, Y., Ghosh, K., *et al.* (2006). Electrospun three-dimensional hyaluronic acid nanofibrous scaffolds. *Biomaterials* **27**(20), 3782–3792.

Juliano, R. L., and Haskill, S. (1993). Signal transduction from the extracellular matrix. *J. Cell Biol.* **120**(3), 577–585.

Juliano, R. L., Reddig, P., *et al.* (2004). Integrin regulation of cell signalling and motility. *Biochem. Soc. Trans.* **32**(Pt. 3), 443–446.

Kirker, K. R., Luo, Y., *et al.* (2002). Glycosaminoglycan hydrogel films as bio-interactive dressings for wound healing. *Biomaterials* **23**(17), 3661–3671.

Kishimoto, J., Ehama, R., *et al.* (2000). *In vivo* detection of human vascular endothelial growth factor promoter activity in transgenic mouse skin [see comments]. *Am. J. Pathol.* **157**(1), 103–110.

Klein, C. E., Dressel, D., *et al.* (1991). Integrin alpha 2 beta 1 is upregulated in fibroblasts and highly aggressive melanoma cells in three-dimensional collagen lattices and mediates the reorganization of collagen I fibrils. *J. Cell Biol.* **115**(5), 1427–1436.

Koch, A. E., Polverini, P. J., *et al.* (1992). Interleukin-8 as a macrophage-derived mediator of angiogenesis. *Science* **258**(5089), 1798–1801.

Koizumi, M., Matsuzaki, T., *et al.* (2005). Expression of P-cadherin distinct from that of E-cadherin in re-epithelialization in neonatal rat skin. *Dev. Growth Differ.* **47**(2), 75–85.

Koyama, H., Raines, E. W., *et al.* (1996). Fibrillar collagen inhibits arterial smooth muscle proliferation through regulation of Cdk2 inhibitors. *Cell* **87**(6), 1069–1078.

Kubo, M., Van de Water, L., *et al.* (2001). Fibrinogen and fibrin are anti-adhesive for keratinocytes: a mechanism for fibrin eschar slough during wound repair. *J. Invest. Dermatol.* **117**(6), 1369–1381.

Kubota, Y., Kleinman, H. K., *et al.* (1988). Role of laminin and basement membrane in the morphological differentiation of human endothelial cells into capillary-like structures. *J. Cell Biol.* **107**(4), 1589–1598.

Lamme, E. N., van Leeuwen, R. T., *et al.* (2002). Allogeneic fibroblasts in dermal substitutes induce inflammation and scar formation. *Wound Repair Regen.* **10**(3), 152–160.

Langholz, O., Rockel, D., *et al.* (1995). Collagen and collagenase gene expression in three-dimensional collagen lattices are differentially regulated by alpha 1 beta 1 and alpha 2 beta 1 integrins. *J. Cell Biol.* **131**(6 Pt. 2), 1903–1915.

Larjava, H., Salo, T., *et al.* (1993). Expression of integrins and basement membrane components by wound keratinocytes. *J. Clin. Invest.* **92**(3), 1425–1435.

Leavesley, D. I., Schwartz, M. A., *et al.* (1993). Integrin beta 1- and beta 3-mediated endothelial cell migration is triggered through distinct signaling mechanisms. *J. Cell Biol.* **121**(1), 163–170.

Levenson, S. M., Geever, E. F., *et al.* (1965). The healing of rat skin wounds. *Ann. Surg.* **161**, 293–308.

Levine, J. H., Moses, H. L., *et al.* (1993). Spatial and temporal patterns of immunoreactive transforming growth factor beta 1, beta 2, and beta 3 during excisional wound repair. *Am. J. Pathol.* **143**(2), 368–380.

Lin, C. Q., and Bissell, M. J. (1993). Multifaceted regulation of cell differentiation by extracellular matrix. *FASEB J.* **7**, 737–743.

Linge, C., Richardson, J., *et al.* (2005). Hypertrophic scar cells fail to undergo a form of apoptosis specific to contractile collagen — the role of tissue transglutaminase. *J. Invest. Dermatol.* **125**(1), 72–82.

Litjens, S. H., de Pereda, J. M., *et al.* (2006). Current insights into the formation and breakdown of hemidesmosomes. *Trends Cell. Biol.* **16**, 376–383.

Longaker, M. T., Chiu, E. S., *et al.* (1991). Studies in fetal wound healing. V. A prolonged presence of hyaluronic acid characterizes fetal wound fluid. *Ann. Surg.* **213**(4), 292–296.

Lukish, J. R., Eichelberger, M. R., *et al.* (2001). The use of a bioactive skin substitute decreases length of stay for pediatric burn patients. *J. Pediatr. Surg.* **36**(8), 1118–1121.

Lutolf, M. P., Lauer-Fields, J. L., *et al.* (2003). Synthetic matrix metalloproteinase-sensitive hydrogels for the conduction of tissue regeneration: engineering cell-invasion characteristics. *Proc. Natl. Acad. Sci. U.S.A.* **100**(9), 5413–5418.

Lutolf, M. P., and Hubbell, J. A. (2005). Synthetic biomaterials as instructive extracellular microenvironments for morphogenesis in tissue engineering. *Nat. Biotechnol.* **23**(1), 47–55.

Lynch, S. E., Colvin, R. B., *et al.* (1989). Growth factors in wound healing. Single and synergistic effects on partial-thickness porcine skin wounds. *J. Clin. Invest.* **84**(2), 640–646.

Madri, J. A., Sankar, S., and Romanic, A. M. (1996). Angiogenesis. *In* "The Molecular and Cellular Biology of Wound Repair" (R. A. F. Clark, ed.), pp. 355–372. Plenum Press, New York.

Marikovsky, M., Breuing, K., *et al.* (1993). Appearance of heparin-binding EGF-like growth factor in wound fluid as a response to injury. *Proc. Natl. Acad. Sci. U.S.A.* **90**(9), 3889–3893.

Marston, W. A. (2004). Dermagraft, a bioengineered human dermal equivalent for the treatment of chronic nonhealing diabetic foot ulcer. *Expert Rev. Med. Devices* **1**(1), 21–31.

Martin, P., and Parkhurst, S. M. (2004). Parallels between tissue repair and embryo morphogenesis. *Development* **131**(13), 3021–3034.

Martin, P., and Leibovich, S. J. (2005). Inflammatory cells during wound repair: the good, the bad and the ugly. *Trends Cell. Biol.* **15**(11), 599–607.

Matouskova, E., Broz, L., *et al.* (2006). Human allogeneic keratinocytes cultured on acellular xenodermis: the use in healing of burns and other skin defects. *Biomed. Mater. Eng.* **16**(Suppl. 4), S63–S71.

McClain, S. A., Simon, M., *et al.* (1996). Mesenchymal cell activation is the rate-limiting step of granulation-tissue induction. *Am. J. Pathol.* **149**(4), 1257–1270.

Mignatti, P., Rifkin, D. B., Welgus, H. G., and Parks, W. C. (1996). Proteinases and tissue remodeling. *In* "The Molecular and Cellular Biology of Wound Repair" (R. A. F. Clark, ed.), pp. 427–474. Plenum Press, New York.

Mirastschijski, U., Zhou, Z., *et al.* (2004). Wound healing in membrane-type-1 matrix metalloproteinase–deficient mice. *J. Invest. Dermatol.* **123**(3), 600–602.

Montrucchio, G., Lupia, E., *et al.* (1997). Nitric oxide mediates angiogenesis induced *in vivo* by platelet-activating factor and tumor necrosis factor-alpha. *Am. J. Pathol.* **151**(2), 557–563.

Morimoto, N., Saso, Y., *et al.* (2005). Viability and function of autologous and allogeneic fibroblasts seeded in dermal substitutes after implantation. *J. Surg. Res.* **125**(1), 56–67.

Mosesson, M. W. (2005). Fibrinogen and fibrin structure and functions. *J. Thromb. Haemost.* **3**(8), 1894–1904.

Mott, J. D., and Werb, Z. (2004). Regulation of matrix biology by matrix metalloproteinases. *Curr. Opin. Cell Biol.* **16**(5), 558–564.

Moulin, V., Larochelle, S., *et al.* (2004). Normal skin wound and hypertrophic scar myofibroblasts have differential responses to apoptotic inductors. *J. Cell. Physiol.* **198**(3), 350–358.

Mustoe, T. (2004). Understanding chronic wounds: a unifying hypothesis on their pathogenesis and implications for therapy. *Am. J. Surg.* **187**(5A), 65S–70S.

Mustoe, T. A., Pierce, G. F., *et al.* (1991). Growth factor–induced acceleration of tissue repair through direct and inductive activities in a rabbit dermal ulcer model. *J. Clin. Invest.* **87**(2), 694–703.

Mustoe, T. A., O'Shaughnessy, K., *et al.* (2006). Chronic wound pathogenesis and current treatment strategies: a unifying hypothesis. *Plast. Reconstr. Surg.* **117**(7 Suppl.), 35S–41S.

Navratilova, Z., Slonkova, V., *et al.* (2004). Cryopreserved and lyophilized cultured epidermal allografts in the treatment of leg ulcers: a pilot study. *J. Eur. Acad. Dermatol. Venereol.* **18**(2), 173–179.

Nelson, C. M., and Bissell, M. (2006). Of extracellular matrix, scaffolds, and signaling: tissue architecture regulates development, homeostasis, and cancer. *Annu. Rev. Cell. Dev. Biol.* **22**, 287–309.

Nicosia, R. F., and Tuszynski, G. P. (1994). Matrix-bound thrombospondin promotes angiogenesis *in vitro*. *J. Cell Biol.* **124**(1–2), 183–193.

Nissen, N. N., Polverini, P. J., *et al.* (1998). Vascular endothelial growth factor mediates angiogenic activity during the proliferative phase of wound healing. *Am. J. Pathol.* **152**(6), 1445–1452.

O'Conner, N. E., Mulliken, J. B., *et al.* (1981). Grafting of burns with cultured epithelium prepared from autologous epidermal cells. *Lancet* **1**(8211), 75–78.

Olin, J. W., Beusterien, K. M., *et al.* (1999). Medical costs of treating venous stasis ulcers: evidence from a retrospective cohort study. *Vasc. Med.* **4**(1), 1–7.

Olsson, A. K., Dimberg, A., *et al.* (2006). VEGF receptor signalling — in control of vascular function. *Nat. Rev. Mol. Cell. Biol.* **7**(5), 359–371.

Overall, C. M., Wrana, J. L., *et al.* (1989). Independent regulation of collagenase, 72-kDa progelatinase, and metalloendoproteinase inhibitor expression in human fibroblasts by transforming growth factor-beta. *J. Biol. Chem.* **264**(3), 1860–1869.

Paquet, P., Quatresooz, P., *et al.* (2005). Tapping into the influence of keratinocyte allografts and biocenosis on healing of chronic leg ulcers: split-ulcer controlled pilot study. *Dermatol. Surg.* **31**(4), 431–435.

Pardoux, C., and Derynck, R. (2004). JNK regulates expression and autocrine signaling of TGF-beta1. *Mol. Cell* **15**(2), 170–171.

Peters, T., Sindrilaru, A., *et al.* (2005). Wound-healing defect of CD18(–/–) mice due to a decrease in TGF-beta1 and myofibroblast differentiation. *EMBO J.* **24**(19), 3400–3410.

Petersen, M. J., Woodley, D. T., *et al.* (1990). Enhanced synthesis of collagenase by human keratinocytes cultured on type I or type IV collagen. *J. Invest. Dermatol.* **94**(3), 341–346.

Phillips, T. J., and Gilchrest, B. A. (1991). Cultured epidermal allografts as biological wound dressings. *Prog. Clin. Biol. Res.* **365**, 77–94.

Pierce, G. F., Tarpley, J. E., *et al.* (1992). Platelet-derived growth factor (BB homodimer), transforming growth factor-beta 1, and basic fibroblast growth factor in dermal wound healing. Neovessel and matrix formation and cessation of repair. *Am. J. Pathol.* **140**(6), 1375–1388.

Pierce, G. F., Tarpley, J. E., *et al.* (1995). Detection of platelet-derived growth factor (PDGF)-AA in actively healing human wounds treated with recombinant PDGF-BB and absence of PDGF in chronic nonhealing wounds. *J. Clin. Invest.* **96**(3), 1336–1350.

Pierschbacher, M. D., and Ruoslahti, E. (1984). Cell attachment activity of fibronectin can be duplicated by small synthetic fragments of the molecule. *Nature* **309**, 30–33.

Pierschbacher, M. D., Polarek, J. W., *et al.* (1994). Manipulation of cellular interactions with biomaterials toward a therapeutic outcome: a perspective. *J. Cell. Biochem.* **56**(2), 150–154.

Postlethwaite, A. E., Keski-Oja, J., *et al.* (1987). Stimulation of the chemotactic migration of human fibroblasts by transforming growth factor beta. *J. Exp. Med.* **165**(1), 251–256.

Postlethwaite, A. E., Holness, M. A., *et al.* (1992). Human fibroblasts synthesize elevated levels of extracellular matrix proteins in response to interleukin 4. *J. Clin. Invest.* **90**(4), 1479–1485.

Prestwich, G. D., Marecak, D. M., *et al.* (1998). Controlled chemical modification of hyaluronic acid: synthesis, applications, and biodegradation of hydrazide derivatives. *J. Controlled Release* **53**(1–3), 93–103.

Purdue, G. F. (1997). Dermagraft-TC pivotal efficacy and safety study. *J. Burn Care Rehabil.* **18**(1 Pt. 2), S13–S14.

Purna, S. K., and Babu, M. (2000). Collagen-based dressings — a review. *Burns* **26**(1), 54–62.

Redd, M. J., Cooper, L., *et al.* (2004). Wound healing and inflammation: embryos reveal the way to perfect repair. *Philos. Trans. R. Soc. Lond. B Biol. Sci.* **359**(1445), 777–784.

Rennekampff, H. O., Hansbrough, J. F., *et al.* (1996). Integrin and matrix molecule expression in cultured skin replacements. *J. Burn Care Rehabil.* **17**(3), 213–221.

Richardson, T. P., Peters, M. C., *et al.* (2001). Polymeric system for dual growth factor delivery. *Nat. Biotechnol.* **19**(11), 1029–1034.

Roberts, A. B., and Sporn, M. B. (1996). Transforming growth factor-b (TGF-b). *In* "The Molecular and Cellular Biology of Wound Repair" (R. A. F. Clark, ed.), pp. 275–310. Plenum Press, New York.

Roberts, A. B., Sporn, M. B., *et al.* (1986). Transforming growth factor type beta: rapid induction of fibrosis and angiogenesis *in vivo* and stimulation of collagen formation *in vitro*. *Proc. Natl. Acad. Sci. U.S.A.* **83**(12), 4167–4171.

Robson, M. C., Krizek, T. J., *et al.* (1973). Amniotic membranes as a temporary wound dressing. *Surg. Gynecol. Obstet.* **136**(6), 904–906.

Rosso, F., Marino, G., *et al.* (2005). Smart materials as scaffolds for tissue engineering. *J. Cell. Physiol.* **203**(3), 465–470.

Rue, L. W., 3rd, Cioffi, W. G., *et al.* (1993). Wound closure and outcome in extensively burned patients treated with cultured autologous keratinocytes. *J. Trauma* **34**(5), 662–667; discussion 667–668.

Ruszczak, Z. (2003). Effect of collagen matrices on dermal wound healing. *Adv. Drug Deliv. Rev.* **55**(12), 1595–1611.

Saarialho-Kere, U. K., Chang, E. S., *et al.* (1992). Distinct localization of collagenase and tissue inhibitor of metalloproteinases expression in wound healing associated with ulcerative pyogenic granuloma. *J. Clin. Invest.* **90**(5), 1952–1957.

Sahni, A., Khorana, A. A., *et al.* (2006). FGF-2 binding to fibrin(ogen) is required for augmented angiogenesis. *Blood* **107**(1), 126–131.

Schiro, J. A., Chan, B. M., *et al.* (1991). Integrin alpha 2 beta 1 (VLA-2) mediates reorganization and contraction of collagen matrices by human cells. *Cell* **67**(2), 403–410.

Shah, M., Foreman, D. M., *et al.* (1995). Neutralization of TGF-beta 1 and TGF-beta 2 or exogenous addition of TGF-beta 3 to cutaneous rat wounds reduces scarring. *J. Cell Sci.* **108**(Pt. 3), 985–1002.

Shu, X. Z., Ghosh, K., *et al.* (2004). Attachment and spreading of fibroblasts on an RGD peptide-modified injectable hyaluronan hydrogel. *J. Biomed. Mater. Res. A* **68**(2), 365–375.

Simpson, D. G. (2006). Dermal templates and the wound-healing paradigm: the promise of tissue regeneration. *Expert Rev. Med. Devices* **3**(4), 471–484.

Singer, A. J., and Clark, R. A. (1999). Cutaneous wound healing. *N. Engl. J. Med.* **341**(10), 738–746.

Staiano-Coico, L., Krueger, J. G., *et al.* (1993). Human keratinocyte growth factor effects in a porcine model of epidermal wound healing. *J. Exp. Med.* **178**(3), 865–878.

Steed, D. L., Goslen, J. B., *et al.* (1992). Randomized prospective double-blind trial in healing chronic diabetic foot ulcers. CT-102-activated platelet supernatant, topical versus placebo. *Diabetes Care* **15**(11), 1598–1604.

Stricklin, G. P., and Nanney, L. B. (1994). Immunolocalization of collagenase and TIMP in healing human burn wounds. *J. Invest. Dermatol.* **103**(4), 488–492.

Supp, D. M., and Boyce, S. T. (2005). Engineered skin substitutes: practices and potentials. *Clin. Dermatol.* **23**(4), 403–412.

Tonnesen, M. G., Feng, X., *et al.* (2000). Angiogenesis in wound healing. *J. Investig. Dermatol. Symp. Proc.* **5**(1), 40–46.

Toole, B. P. (1991). Proteoglycans and hyaluronan in morphogenesis and differentiation. *In* "Cell Biology of the Extracellular Matrix" (E. D. Hay, ed.). Plenum Press, New York.

Tuan, T. L., Song, A., *et al.* (1996). *In vitro* fibroplasia: matrix contraction, cell growth, and collagen production of fibroblasts cultured in fibrin gels. *Exp. Cell Res.* **223**(1), 127–134.

Ueno, C., Hunt, T. K., *et al.* (2006). Using physiology to improve surgical wound outcomes. *Plast. Reconstr. Surg.* **117**(7 Suppl.), 59S–71S.

Vaalamo, M., Mattila, L., *et al.* (1997). Distinct populations of stromal cells express collagenase 3 (MMP-13) and collagenase 1 (MMP-1) in chronic ulcers but not in normally healing wounds. *J. Invest. Dermatol.* **109**(1), 96–101.

Vogel, V., and Sheetz, M. (2006). Local force and geometry sensing regulate cell functions. *Nat. Rev. Mol. Cell. Biol.* **7**(4), 265–275.

Walker, J. L., and Assoian, R. K. (2005). Integrin-dependent signal transduction regulating cyclin D1 expression and G1 phase cell cycle progression. *Cancer Metastasis Rev.* **24**(3), 383–393.

Wang, H. J., Pieper, J., *et al.* (2004). Stimulation of skin repair is dependent on fibroblast source and presence of extracellular matrix. *Tissue Eng.* **10**(7–8), 1054–1064.

Welch, M. P., Odland, G. F., *et al.* (1990). Temporal relationships of F-actin bundle formation, collagen and fibronectin matrix assembly, and fibronectin receptor expression to wound contraction. *J. Cell Biol.* **110**(1), 133–145.

Werner, S. (1998). Keratinocyte growth factor: a unique player in epithelial repair processes. *Cytokine Growth Factor Rev.* **9**(2), 153–165.

Werner, S., Peters, K. G., *et al.* (1992). Large induction of keratinocyte growth factor expression in the dermis during wound healing. *Proc. Natl. Acad. Sci. U.S.A.* **89**(15), 6896–6900.

Werner, S., Breeden, M., *et al.* (1994). Induction of keratinocyte growth factor expression is reduced and delayed during wound healing in the genetically diabetic mouse. *J. Invest. Dermatol.* **103**(4), 469–473.

Wijelath, E. S., Murray, J., *et al.* (2002). Novel vascular endothelial growth factor binding domains of fibronectin enhance vascular endothelial growth factor biological activity. *Circ. Res.* **91**(1), 25–31.

Wisniewski, H. G., and Vilcek, J. (1997). TSG-6: an IL-1/TNF-inducible protein with antiinflammatory activity. *Cytokine Growth Factor Rev.* **8**(2), 143–156.

Woodley, D. T., Yamauchi, M., *et al.* (1991). Collagen telopeptides (cross-linking sites) play a role in collagen gel lattice contraction. *J. Invest. Dermatol.* **97**(3), 580–585.

Wound Care in the U.S.: Emerging Trends, Management and New Product Development. Medical Data International. (1993).

Xu, J., and Clark, R. A. (1996). Extracellular matrix alters PDGF regulation of fibroblast integrins. *J. Cell Biol.* **132**(1–2), 239–249.

Yamada, K. M., Gailit, J., and Clark, R. A. F (1996). Integrins in wound repair. *In* "The Molecular and Cellular Biology of Wound Repair" (R. A. F. Clark, ed.), pp. 311–338. Plenum Press, New York.

Yamauchi, M., London, R. E., *et al.* (1987). Structure and formation of a stable histidine-based trifunctional cross-link in skin collagen. *J. Biol. Chem.* **262**(24), 11428–11434.

Yanaga, H., Udoh, Y., *et al.* (2001). Cryopreserved cultured epidermal allografts achieved early closure of wounds and reduced scar formation in deep partial-thickness burn wounds (DDB) and split-thickness skin donor sites of pediatric patients. *Burns* **27**(7), 689–698.

Zachary, I. (2003). VEGF signalling: integration and multitasking in endothelial cell biology. *Biochem. Soc. Trans.* **31**(Pt. 6), 1171–1177.

# Chapter Seventy-Seven

# Bioengineered Skin Constructs

*Vincent Falanga and Katie Faria*

## I. INTRODUCTION

The engineering of skin tissue has been at the forefront of tissue engineering for many years and has now yielded some of the first medical products to come from tissue engineering. Today, over 200,000 patients have been treated with tissue-engineered skin products. This is the result of work over the last 30 years in the areas of skin cell biology, extracellular matrix biology, collagen scaffolds, polymer scaffolds, and tissue equivalents. The repair of skin using tissue engineering has taken on many forms, from simple to complex. The various strategies are discussed in this chapter, with an emphasis on the biological relevance of each to normal skin structure and function and the healing of wounds.

Skin wounds normally heal by formation of epithelialized scar tissue rather than regeneration of full-thickness skin. Consequently, strategies for the clinical management of wound healing have depended historically on providing a passive cover to the site of the wound while allowing the reparative mechanisms of wound healing: re-epithelialization, remodeling of granulation tissue, and formation of scar tissue. Therapy could do little more than facilitate these little-understood processes. Advances in our understanding of wound healing, wound assessment, the concerted action of several growth factors, the role of the extracellular matrix

in regulating the healing process, and the demonstrated ability of tissue-engineered constructs to promote wound healing highlight the potential for intervening therapeutically in tissue repair by providing lost epithelium, stimulating dermal regeneration, and reconstituting full-thickness skin.

Tissue-engineered skin substitutes are classified into two general categories; cellular-based products, which actively stimulate wound healing; and acellular products, which provide a substrate or covering to facilitate wound healing. Cellular substitutes include autologous epidermal cell sheets (Epicel®, Genzyme Tissue Repair, Cambridge, MA), allogeneic dermal substrates (Dermagraft™, Advanced BioHealing, La Jolla, CA), and human skin equivalent (HSE), composed of both living dermal and epidermal components (Apligraf®,[1] Organogenesis Inc., Canton, MA, Orcel®, Ortec International, New York, NY). Acellular products include Transcyte® (Advanced BioHealing), Integra® Artificial Skin (Integra Life Sciences, Plainsboro, NJ), and Alloderm® (Lifecell Corporation, Branchburg, NJ). These products represent the first of their kind and are the result of basic research in the biology of skin and wound healing and clinical experience with skin grafts, cultured keratinocyte grafts, acellular

---

[1] Apligraf is a registered trademark of Novartis.

*Principles of Tissue Engineering, 3rd Edition*
*ed. by Lanza, Langer, and Vacanti*

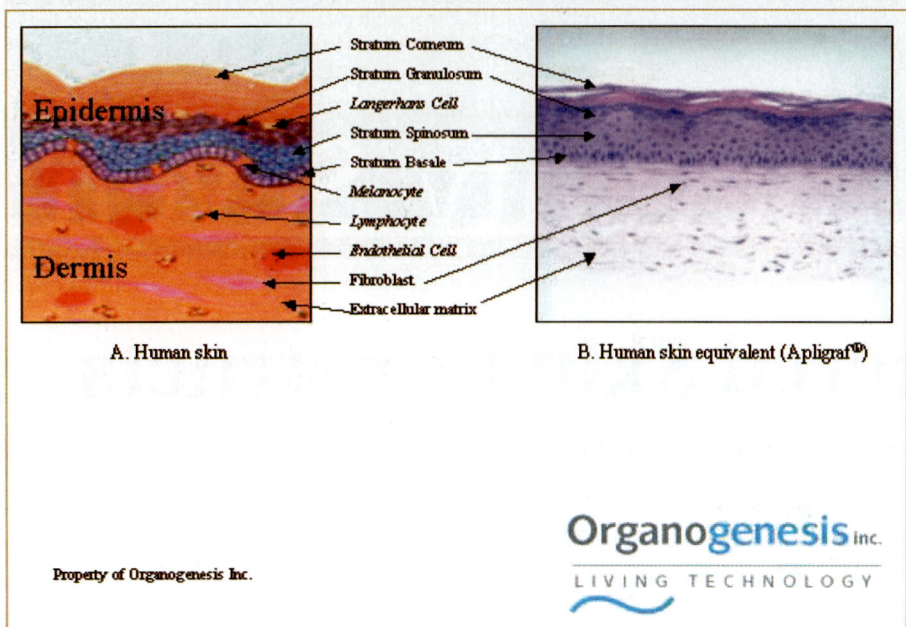

A. Human skin

Stratum Corneum
Stratum Granulosum
*Langerhans Cell*
Stratum Spinosum
Stratum Basale
*Melanocyte*
*Lymphocyte*
*Endothelial Cell*
*Fibroblast*
*Extracellular matrix*

B. Human skin equivalent (Apligraf®)

Property of Organogenesis Inc.

**Organogenesis** inc.
LIVING TECHNOLOGY

**FIG. 77.1.** The basic components of skin and engineered skin equivalent. **(A)** Diagram showing the major cell types of skin and their organization. Note that stratified keratinocytes make up the epidermis and display distinct morphological phenotypes. **(B)** A histological section of Apligraf HSE (hematoxylin and eosin, 142 X). Italics indicate cell types present in real skin but not in the engineered skin equivalent.

collagen matrices, cellular matrices, and cultured composite grafts.

## II. SKIN STRUCTURE AND FUNCTION

The passive and active functions of skin are carried out by specialized cells and structures located in the two main layers of skin, the epidermis and the dermis (Fig. 77.1A). Complex functional relationships between these two anatomic structures of skin maintain the normal properties of skin. Tissue-engineering applications in skin depend on an understanding of the structural components of skin, their spatial organization, and their functional relationships.

### The Epidermis

The skin is a physical barrier between the body and the external environment. The outermost layer of skin, the epidermis, must therefore be tough and impermeable to toxic substances and harmful organisms. It must also control the loss of water from the body to the relatively drier external environment.

The epidermis is composed primarily of keratinocytes, which form a stratified squamous epithelium (Fig. 77.1). Proliferating cells in the basal layer of the epidermis anchor the epidermis to the dermis and replenish the terminally differentiated epithelial cells lost through normal sloughing from the surface of the skin. These basal cells stop proliferating and terminally differentiate as they move from the basal layer through the suprabasal layers to the surface of the epidermis. Keratin filaments and desmosomes contribute physical strength in the living layers and maintain the integrity of the epidermis. The cornified envelopes serve as the bricks and the lipids as mortar.

The most superficial keratinocytes in the epidermis form the stratum corneum, the dead outermost structure that provides the physical barrier of the skin. In the last stages of differentiation, epithelial cells extrude lipids into the intercellular space to form the permeability barrier. The cells break down their nuclei and other organelles and form a highly cross-linked protein envelope immediately beneath their cell membranes. The physically and chemically resilient protein envelope connects to a dense network of intracellular keratin filaments to provide further physical strength to the epidermis.

Additional cells and structures in the epidermis perform specialized functions (Fig. 77.1A). Skin plays a major role in alerting the immune system to potential environmental dangers. The interacting cells in skin comprise a dynamic network capable of sensing a variety of perturbations (trauma, ultraviolet irradiation, toxic chemicals, and pathogenic organisms) in the cutaneous environment and rapidly sending appropriate signals that alert and recruit other branches of the immune system. To restore homeostasis in the skin immune system, the multiple proinflammatory signals generated by skin cells must eventually be counterbalanced by mechanisms capable of promoting resolution of a cutaneous inflammatory response. Dendritic cells of the immune system (Langerhans cells) reside in the epidermis and form a network of dendrites through which they interact with adjacent keratinocytes and nerves (Streilein and Bergstresser, 1984). Melanocytes distribute melanin to keratinocytes in the form of melanosomes. Melanin protects the epidermis and underlying dermis from ultraviolet radiation. Sweat glands help to regulate body temperature through evaporation of sweat secreted onto the skin surface. Sebaceous glands associated with hair follicles secrete sebum, an oily substance that lubricates and moisturizes hair and epidermis. Hair keeps the body warm in many mammals, although maintaining body temperature is not an important role for hair in humans. Hair follicles, however, are an important source of proliferating keratinocytes during re-epithelialization after severe wounds.

## The Dermis

The dermis underlies the epidermis. The dermis is divided into two regions: the papillary dermis, which lies immediately beneath the epidermis, and the deeper reticular dermis. The reticular dermis is more acellular and has a denser meshwork of thicker collagen and elastic fibers than the papillary dermis. The reticular dermis provides skin with most of its strength, flexibility, and elasticity. Loss of reticular dermis can often lead to excessive scarring and wound contraction.

The dermis provides physical strength and flexibility to skin as well as the connective tissue scaffolding that supports the extensive vasculature, lymphatic system, and nerve bundles. The dermis is relatively acellular, being composed predominantly of an extracellular matrix of interwoven collagen fibrils. Interspersed among the collagen fibrils are elastic fibers, proteoglycans, and glycoproteins.

Fibroblasts, the major cell type of the dermis, produce and maintain most of the extracellular matrix. Endothelial cells line the blood vessels and play a critical role in the skin immune system by controlling the extravasation of leukocytes. Cells of hematopoietic origin in the dermis (e.g., macrophages, lymphocytes) contribute to the surveillance function. A network of nerve fibers extends throughout the dermis, which serves the sensory role in the skin (and, to a more limited extent, a motor function). These nerve fibers also secrete neuropeptides that influence immune and inflammatory responses in skin through their effects on endothelial cells, leukocytes, and keratinocytes (Williams and Kupper, 1996).

## Immunology and the Skin

The first stage in the induction of a primary immune response in skin is the processing of antigen by dendritic cells (Langerhans cells with dermal dendritic cells), the antigen-presenting cells in skin. These cells process antigen and migrate out of the skin to the draining lymph node, where they can recruit and activate T-cells.

The interaction of keratinocytes and fibroblasts with T-cells has important implications for the use of allogeneic cells in tissue engineering. Under normal conditions, keratinocytes and fibroblasts do not express MHC–class II antigens; but they can be induced by interferon-γ to express MHC–class II molecules and thereby acquire the ability to present antigen to T-cells. Since the keratinocytes and fibroblasts are deficient in the necessary costimulatory molecules (Nickoloff and Turka, 1994; Phipps *et al.*, 1989), antigen presentation by keratinocytes and fibroblasts does not result in T-cell activation. Instead, this antigen presentation can result in T-cell nonresponsiveness (Gasparia and Katz, 1988; Bal *et al.*, 1990) or T-cell anergy (Gaspari and Katz, 1991). Therefore, the primary mode of skin rejection is likely mediated via an attack on the vasculature present in a normal skin graft (Pober *et al.*, 1986; Moulon *et al.*, 1999).

Autologous skin grafts avoid issues of immunogenicity, of course, but autologous grafts have significant limitations. Growing graft tissues from biopsy takes several weeks. The donor site creates another wound, and in some patients (e.g., severe-burn patients) there may be no appropriate donor site. Reproducibly making complex human skin equivalent (HSE) constructs to order from autologous cells would be technically difficult, time consuming, and very costly. Therefore, our ability to use allogeneic human cells effectively is a key element in the success of engineered-skin therapies.

## Cell Communication and Regulation in the Skin

Regulation of its own function is an essential role of skin. Epidermis, for example, produces parathyroid hormone–related protein (PTHrP) and plays a role in the regulation of keratinocyte growth and differentiation (Blomme *et al.*, 1998). Keratinocytes produce a large variety of polypeptide growth factors and cytokines, which act as signals between cells, regulate keratinocyte migration and proliferation, for example, and stimulate dermal cells in various ways (e.g., to promote matrix deposition and neovascularization) (McKay and Leigh, 1991; Parenteau *et al.*, 1997). Cytokines produced by keratinocytes are thought to regulate Langerhans cell migration and differentiation (Lappin *et al.*, 1996). Keratinocytes also translate a variety of stimuli into cytokine signals, which are transmitted to the other cells of the skin's immune system. Dermal cells produce and respond to cytokines and growth factors to regulate numerous processes critical to skin function (Williams and Kupper, 1996).

Approaches to dermal repair and regeneration center around control of fibroblast repopulation and collagen biosynthesis to limit scar-tissue formation. One of the keys to improving dermal repair is control, or redirection, of the wound-healing response so that scar tissue does not form. One promising observation is that fetal wounds heal without scarring. TGF-β1, which is not expressed in the fetus, is a potent stimulator of collagen biosynthesis by fibroblasts in the adult and is thought to be an important inducer of scar formation. The matrix composition of the fetal dermis is also significantly different than adult dermis or granulation tissue. Therefore, providing a better matrix environment for dermal cell repopulation may also be key in controlling scar formation.

## The Process of Wound Healing

The immediate tissue response to wounding is clot formation to stop bleeding. Simultaneously, there is a release of inflammatory cytokines that regulate blood flow to the area, recruit lymphocytes and macrophages to fight infection, and later stimulate angiogenesis and collagen deposition (Williams and Kupper, 1996). These latter processes result in the formation of granulation tissue, a highly vascularized and cellular wound connective tissue. Fibroblasts rich in actin, called *myofibroblasts* (Desmouliere and Gabbiani, 1996), are recruited through the action of factors such as platelet-derived growth factor (PDGF) and transforming growth factor-β (TGF-β). Granulation tissue forms in the wound bed, stimulated by factors such as PDGF. This tissue is gradually replaced by scar tissue through the action of the myofibroblasts and factors such as TGF-β. Keratinocytes are stimulated to proliferate and to migrate into the wound bed to restore epidermal coverage. From our preclinical observations using both full-thickness HSE and dermal matrices, coverage by the epidermis appears to play a key role in the regulation of the underlying inflammatory response. Providing a noninflammatory living connective tissue implant in the wound defect also appears to be beneficial in directing the granulation response.

Why should living tissue be different in its response? The epidermal and dermal response is regulated by inflammatory cytokines and by autocrine and paracrine factors produced by the dermal fibroblasts and epidermal keratinocytes (Ansel *et al.*, 1990; McKay and Leigh, 1991). These factors regulate growth and differentiation of keratinocytes, proinflammatory reactions, angiogenesis, and deposition of extracellular matrix. Living tissue created through tissue engineering can provide complex temporal control of factor delivery and effect and can be used to provide the needed combination of chemical, structural, and, last but not least, normal cellular elements (Sabolinski *et al.*, 1996).

As this brief description shows, wound healing involves the interaction of many tissue factors and elements. The poor healing response in chronic wounds has been attributed to an imbalance of factors rather than to an insufficiency of any particular factor (Parenteau *et al.*, 1997). However, most, if not all, factor-based approaches have had marginal success. Identification of putative wound-healing factors has led to several attempts to speed wound healing by local application of one or more factors that promote cell attachment and migration. Transforming growth factor-β (TGF-β), epidermal growth factor (EGF), vascular endothelial growth factor (VEGF), and platelet-derived growth factor (PDGF) have been candidates for this purpose (McKay and Leigh, 1991; Abraham and Klagsbrun, 1996). Of these, only PDGF has shown efficacy in clinical trials and is approved for clinical use (Regranex®, Ortho McNeil, Inc., Raritan, NJ 08869-0602). The arginine-glycine-aspartic acid (RGD) matrix peptide sequence has been found to promote the migration of connective tissue cells and, thus, to stimulate production of a dermal scaffold within the wound bed. This approach has been shown to accelerate healing of sickle-cell leg ulcer and diabetic ulcers (Steed *et al.*, 1995), as compared

with placebo but not when compared with standard care. In addition, complex cell extracts have been used in hopes of providing the appropriate mixture of elements. These include the use of platelet extract to provide primarily platelet-derived growth factor (PDGF) and the use of keratinocyte extracts to provide a complex mixture of elements of rapidly growing keratinocytes — again, with marginal effect, in part due to the complex nature of the wound healing response. In addition, the use of factors is not a sufficient approach, in and of itself, in situations where there is severe or massive loss of skin tissue.

## Impaired Healing and Its Mechanisms

Although there have been many recent advances in the scientific basis for tissue repair, the treatment of chronic wounds and intervening in situations associated with impaired healing have been very challenging problems. A number of reasons account for the impaired healing of chronic wounds, but in general the primary causes have to do with our inability to correct completely their fundamental pathophysiological abnormalities (Falanga, 2005). Preparatory steps before the use of advanced treatments seem to be required, and this preparatory phase may not have received proper attention in the past. For example, failure to prepare the wound properly may lie at the basis of the relative lack of success in the use of topically applied growth factors. This situation is slowly being corrected, but much more needs to be implemented in preparing the wound for the optimal success of advanced wound-healing products and devices. A few years ago, we proposed the notion of "wound bed preparation" as a series of steps to improve the wound before advanced products are used (Falanga, 2001). This concept has gained acceptance.

One of the basic differences between acute and chronic wounds is that in the former the sequence of steps and phases involved (coagulation, inflammation, migration and proliferation, and remodeling) occurs in a very orderly and linear fashion. Such is not the case in chronic wounds, where there is a fundamental asynchrony of the healing process. Within the chronic wound, the various phases of wound repair may be occurring at the same time or not in the appropriate sequence. Wound bed preparation is a way to get the wound to behave more like the acute wound. Often, surgical debridement is all that is required. At other times, treatment of bacterial infection, removal of edema, etc., are essential additional steps (Falanga, 2001).

Bacterial and, in some cases, fungal colonization or infection is a fundamental problem with nonhealing wounds. Some of the causes that foster this colonization and the development of occult infection have already been addressed. These factors include absent epithelium and its barrier properties, the constant wound exudate resulting from bacterial products and inflammation, and poor blood flow and hypoxia (Falanga, 2005). It is well established that wounds have a "bacterial burden" that interferes with healing. Thus, there is evidence that, regardless of the type of bacteria present, a level greater than or equal to $10^6$ organisms per gram of tissue is associated with serious healing impairment (Robson *et al.*, 1990). The configuration of the bacterial sheets or colonies is also important. Therefore, there is a great deal of interest at this time in the role played by biofilms, which represent bacterial colonies surrounded by a protective coat of polysaccharides. Biofilms develop mechanisms for antibiotic resistance (Siroky, 2002).

Bioengineered skin constructs may function, at least in part, through the timely and concerted release of stimulatory polypeptides. In this context, one must take into account that simple topical delivery of growth factors into the wound may not be adequate. The nonhealing wound microenvironment can best be described as hostile. Wound exudate in general; the vast abnormalities in the release, activation, and persistence of matrix metalloproteinases (MMPs); and lack of cell adhesion to substrates within the wound bed may all render the growth factors and cytokines unavailable to the healing process. A concept that takes into account these various components has been called the "trap hypothesis" (Falanga and Eaglstein, 1993). It has been hypothesized that nonhealing wounds, particularly in response to bacterial antigens, are characterized by chronic leakage of macromolecules into the wound. These macromolecules may impair healing by trapping cytokines and growth factors. The trap hypothesis suggests that, in spite of achieving adequate levels and even the orchestrated release of these growth factors, the polypeptides are quickly bound and unavailable to the healing process. Common macromolecules that might be involved in trapping include albumin, fibrinogen, and α-2-macroglobulin. The last is particularly important because it is an established scavenger for growth factors. Fibrinogen can bind to fibronectin, providing a mechanism for the trapping of transforming growth factor-β1 (TGF-β1) (Falanga, 1993).

Moist wound healing has been shown experimentally to accelerate re-epithelialization of acute wounds (Winter, 1962), and these observations have led to a number of moisture-retentive dressings (Ovington, 2001). For chronic wounds, moist wound healing has not been clearly shown to improve epidermal healing, but we do know that moist wound healing helps in the formation of granulation tissue and in relieving pain. Painless debridement, too, is another important property of moisture-retentive dressings (Ovington, 2001). The properties of moist wound healing are important in the field of bioengineered skin constructs, because such constructs lead to increased moisture in the wound bed. Proposed advantages of moist wound healing include retention of cytokines within the wound, facilitation of keratinocyte migration, prevention of bacterial entry, and even poorly understood but favorable electrical gradients. For example, acute wound fluid stimulates the *in vitro* proliferation of fibroblasts, keratinocytes, and endothelial cells. However, fluid and exudate from chronic wounds appear to

have a definite adverse effect on cellular proliferation and to contain excessive amounts of matrix metalloproteinases (MMPs), which can break down key matrix proteins critical to cell migration, such as fibronectin and vitronectin (Falanga, 2005). There is a great deal of information we still need about MMPs and their inappropriate activation in chronic wounds. Some of the information is often contradictory. For example, we do know that MMPs are critical to cell migration. Interstitial collagenase (MMP-1) is a case in point when it comes to keratinocyte migration. However, other enzymes (MMP-2, MMP-9) may prevent or interfere with healing (Falanga, 2005).

An ultimate goal for tissue-engineering constructs would be to offset very fundamental abnormalities that lead to impaired or slow healing. We have been discussing components of impaired healing that, in some way or another, can be approached or at least partially corrected by already available means. However, one important component of impaired healing is ischemia, due to poor arterial supply because of narrowing of blood vessels (i.e., atherosclerosis) or, indirectly, because of pressure on those blood vessels (i.e., pressure ulcers, diabetic neuropathic ulcers). The ischemia, of course, has important consequences for the other components of impaired healing we have discussed, such as bacterial colonization and infection. A challenge is how to use available tissue-engineering products or to modify and develop new ones to correct the problem of ischemia. There are some interesting possibilities that one can use as proof of principle. An example is the role of oxygen tension. Thus, there is no debate over the fact that long-term hypoxia is detrimental to the healing process. This has been readily shown with diabetic ulcers, where low levels of transcutaneous oxygen tension ($TcPO_2$) correlate with inability to heal (Fife *et al.*, 2002). However, and this makes sense even from a teleological point of view, short-term hypoxia actually stimulates healing. It has been shown that hypoxia can increase fibroblast proliferation, fibroblast clonal growth, and the synthesis of several growth factors, including PDGF, transforming growth factor-β, and vascular endothelial growth factor (VEGF), among others (Falanga, 2005). Therefore, modulation of the oxygen environment within the wound may offer the possibility of accelerating the healing process. One might even hypothesize that the application of living skin constructs may use up oxygen in the wound bed and that this temporary and relative hypoxia may provide a mechanism of action for these devices.

A very important mechanism for impaired healing is the phenotypic makeup of wound cells. This has critical implications for the use of bioengineered skin construct, in that these constructs may offset cellular phenotypic abnormalities. There is increasing evidence that the resident cells of a chronic wound have developed phenotypic changes that interfere with their response to growth factors and cytokines. Such abnormalities may affect cellular proliferation, locomotion, and the overall capacity to heal (Falanga, 2005).

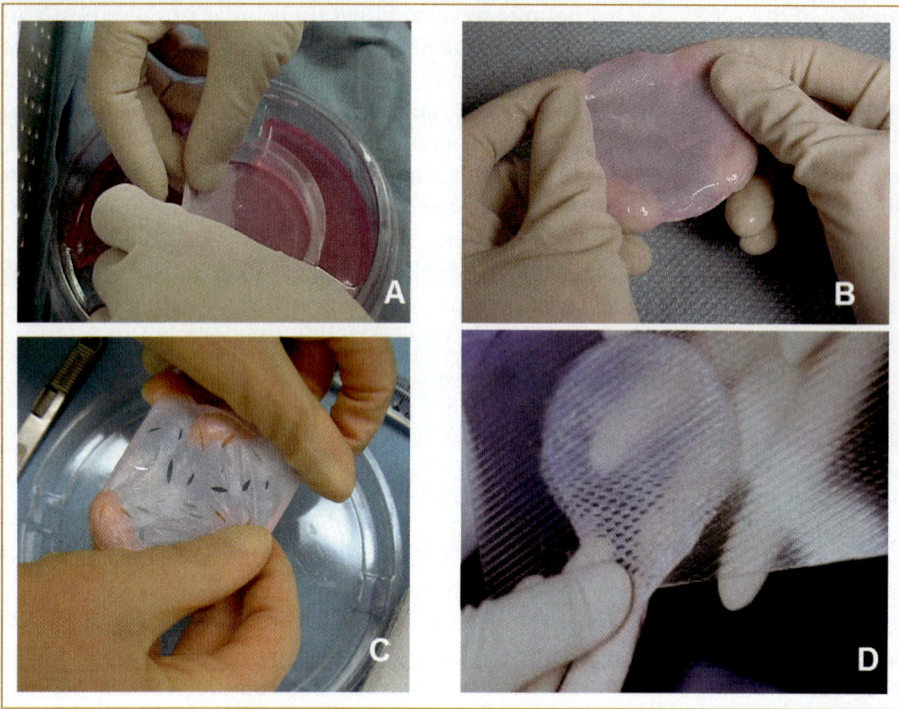

**FIG. 77.2.** Appearance of Apligraf. **(A)** Being removed from its pink nutrient agar. **(B)** Being held just prior to fenestration or meshing. **(C)** After fenestration with a scalpel. **(D)** After meshing at a 1.5 : 1 ratio. Copyright V. Falanga, 2006.

Unresponsiveness of chronic-wound fibroblasts to the action of transforming growth factor-β and platelet-derived growth factor (PDGF) have been found. Also, the signaling mechanism, which is so critical to the action of cytokines, may be impaired. For example, at least in venous ulcers, there is decreased phosphorylation of Smad2/3 and MAPK p44/42 (Kim *et al.*, 2003).

## III. ENGINEERING SKIN TISSUE

Although the epidermis has an enormous capacity to heal, there are situations in which it is necessary to replace large areas of epidermis or in which normal regeneration is deficient. The dermis has very little capacity to regenerate. The scar tissue that forms in the absence of dermis lacks the elasticity, flexibility, and strength of normal dermis. Consequently, scar tissue limits movement, causes pain, and is cosmetically undesirable. Engineered tissues that not only close wounds but also stimulate the regeneration of dermis would provide a significant benefit in human wound healing.

An engineered skin graft should incorporate as many of these factors as possible: (1) the extracellular matrix; (2) dermal fibroblasts; (3) the epidermis; and (4) a naturally occurring semipermeable membrane, the stratum corneum (Fig. 77.2). These components may act alone, but, more importantly, they act synergistically as part of a fully integrated tissue to protect the underlying tissues of a wound bed and to direct healing of the wound (Sabolinski *et al.*, 1996). Dermis containing fibroblasts may be necessary for the maintenance of the epidermal cell population. In turn, the epidermis is necessary for the formation of the so-called

neodermis, in the absence of a dermal layer and can dramatically influence underlying connective-tissue response. The formation of the epidermal barrier also likely influences these processes through control of water loss and its influence on epidermal physiology (Parenteau *et al.*, 1996).

### Design Considerations

Tissue engineering has not focused on regenerating certain skin structures, such as hair follicles and sebaceous glands, whose loss is clinically less significant than the loss of dermis and epidermis needed to cover and protect the underlying tissues. There is some preliminary evidence that sebaceous glands may be possible in the HSE (Wilkins *et al.*, 1994), although the development of functioning adnexal structures is likely to be years away. There has also been little need extraneously to stimulate regeneration of other dermal components (e.g., blood vessels and cells of the immune system) through tissue-engineering methods because these components have the ability to repopulate quickly and to normalize the area of a wound. Langerhans cells, for example, have been shown to migrate and repopulate effectively within months (Desmouliere and Gabbiani, 1996). Control of vascularization is dependent on the makeup of the extracellular matrix and the degree of inflammation present in the wound. Whether modification of vascularization through the use of exogenous factors will be of additional benefit for certain wounds remains to be determined. While pigmentation is not critical for wound healing, clinical studies using Apligraf HSE, which lacks melanocytes, has shown repigmentation of the grafted areas through repopulation of the area with host melanocytes, resulting in normal

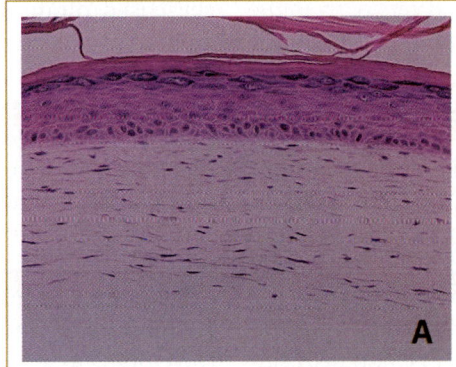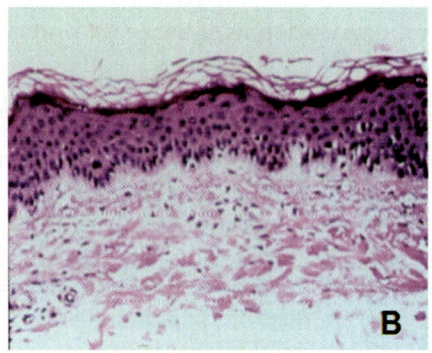

**FIG. 77.3.** Histological appearance of Apligraf **(A)** compared to normal human skin **(B)**. A well-defined epidermal and dermal layer is observed in both. Copyright V. Falanga, 2006.

skin color for each individual. The constructs should have sufficient mechanical strength to allow for clinical manipulations (Fig. 77.3). The approach tissue engineering has taken has been to focus primarily on providing or imitating structural and biological characteristics of dermis, epidermis, or both.

The key features to be replicated in an engineered skin construct are:

1.  A dermal or mesenchymal element capable of aiding appropriate dermal repair and epidermal support
2.  An epidermis capable of easily achieving biologic wound closure
3.  An epidermis capable of rapid re-establishment of barrier properties
4.  A permissive milieu for the components of the immune system, nervous system, and vasculature
5.  A tissue capable of achieving normalization of structure and additional function, such as reduction of long-term scarring and re-establishment of pigmentation
6.  Active cellular component(s) capable of responding to different wound types and conditions
7.  Sufficient mechanical strength to allow for clinical manipulation
8.  Persistence of cells in the wound for multiple weeks to stimulate the healing process through delivery of cytokines and matrix proteins

## Commercial Considerations

Tissue engineering is a unique and emerging industry. Engineered skin constructs, by virtue of being the first products commercialized, have been at the forefront, blazing trails in science, industry, and regulations. In recent years, more attention has been given to the subject of commercialization. In 2004, Lysaght and Hazlehurst published a review of tissue-engineering companies. The review highlighted that while the number of companies had increased from 2000 to 2002, annual spending and the number of full-time employees had declined. In addition, it showed a trend away from U.S.-based companies. Lysaght concluded that the industry is viable, characterized by small-scale companies heavily technology based, which is indicative of an industry

in transition from small-scale discovery-and-research companies to larger, more well-established profitable companies. The transition from start-up to well-established profitable companies is dependent on having a viable business model.

Apligraf, Dermagraft, and Orcel initially struggled with the task of commercializing their technologies. Significant restructuring of business models has occurred, with varying success. Dermagraft had been removed from the market by Smith & Nephew and reintroduced by Advanced BioHealing. In the case of Apligraf, Organogenesis was restructured. Much of the restructuring was centered on understanding the cost of manufacturing and working to reduce those costs while increasing efficiency and maintaining high quality. In addition to the work done to reduce costs and increase efficiencies, the company, Organogenesis, incorporated the functions formerly performed by their marketing partner: sales and marketing, customer service, reimbursement, and distribution. Having all of these functions in the same company is proving to be of benefit by allowing for a more integrated strategic company. As of 2006, Apligraf had been applied to over 130,000 patients.

There are many components to a business model. In general, the intent is to understand the relationship between the cost of getting the product to market and the revenues expected to be realized once the product is commercialized. If the cost of development, clinical testing, and manufacture of the product is much higher than what you can expect to receive in revenue after product launch, then the business model should be reconsidered. This may seem a simple principle, but in practice it can be difficult to capture the proper data for this analysis and properly anticipate changes in markets as the product evolves. It can take years to get to the point where you are ready to launch a product, making it necessary to continually revisit and adjust the business model accordingly. Alternate sources of revenue, different or additional clinical indications for the product, and modifications to the manufacturing process to reduce costs are just some of the components that could be reexamined in the business model. The costs associated with effective manufacturing have a major impact on the business model and are one element that can be considered very early in the product development life cycle. Understanding the unique skill sets required to market and sell these cell-based products is another area that requires significant consideration early in the process. However, we will not discuss this aspect of a business plan because it goes beyond the scope of this chapter.

Early in the product concept stage, consideration should be given to the design of the manufacturing process. Elements such as critical process parameters, quality control assays, production components and materials, process equipment, production facility requirements, and distribution methods need to be developed in some detail. From this design, the cost of manufacturing and strategies for scale-up and automation can be developed. The requirements for the manufacturing facility are directly related to the design of the manufacturing process. For example, a highly manual process requires a large, highly specialized staff with a relatively large manufacturing floor space, whereas a more automated process requires a less specialized staff and potentially less manufacturing floor space. An open process requires tighter environmental controls; developing a closed process reduces the need for tight environmental controls, thus reducing facility costs. All of these elements impact the cost of manufacturing and the potential for scale-up. These pioneering lessons learned in scale-up apply to the broader field of tissue engineering.

Through this process, strategies for scale-up and automation can be evaluated and planned for. More importantly, process components that will pose significant barriers to scale-up or automation can be identified and addressed while the product is still in research and development. This is critical when you consider that the further along in the development process, the more committed you become to the process design. The cost of making a significant process change increases the closer you get to manufacturing and becomes exponentially more expensive and time consuming post–product launch. Along these same lines, materials used in the process that are ill-defined, cell-based, single sources or of limited supply can contribute to a manufacturing process that is difficult to scale-up and control from a quality standpoint. These factors will influence the yield from the manufacturing process or, stated another way, the scrap rate for the product. High scrap rates increase manufacturing costs and create issues for inventory management.

Increasingly, agency's like the U.S. Medicaid and Medicare program as well as foreign government reimbursement agencies are requiring convincing data on the value and effectiveness of therapy as compared to standard of care or competing technologies (Archer and Williams 2005). The Center for Devices at the the FDA (21 CFR 820.30) requires products to be developed following design control procedures. Cell-based HSEs are classified and regulated as medical devices, though the trend at FDA is to regulate cell-based products as biologics. These design guidelines are in place to ensure that products that are developed will be safe and effective and have a sound business model. Establishing a design control system early will help refine the business model throughout the development life cycle. Not only is design control a regulatory requirement, but it makes good business sense.

There are critical issues and barriers to commercialization all along the continuum from product concept, through development, clinical evaluation, FDA approval, product launch, and reimbursement, to steady-state manufacturing. Developing a dynamic business model that takes all of these factors into account will increase the likelihood for a profit-generating product. Good science alone is not enough to ensure success. Understanding the challenges and working to incorporate process designs that are forward-looking, will be amenable to scale-up, address regulatory hurdles, and are supported by a viable business model are necessary for the industry to grow.

## IV. EPIDERMAL REGENERATION

Re-epithelialization of the wound is a paramount concern. Without epithelial coverage, no defense exists against contamination of the exposed underlying tissue or loss of fluid. The approaches to re-establishing epidermis are numerous, ranging from the use of cell suspensions to full-thickness skin equivalents possessing a differentiated epidermis. Silicone membranes have been used as temporary coverings in conjunction with dermal templates (Heimbach *et al.*, 1988). Regardless of approach, living epidermal keratinocytes are necessary to achieve permanent, biologic wound closure.

Green *et al.* (Phillips *et al.*, 1989; Green *et al.*, 1979) developed techniques for growing human epidermal keratinocytes from small patient-biopsy samples using coculture methods (Rheinwald and Green, 1975). The mouse 3T3 fibroblast feeder cell system allows substantial expansion of epidermal keratinocytes and can be used to generate enough thin, multilayered epidermal sheets to resurface the body of a severely burned patient (Gallico *et al.*, 1984). Once transplanted, the epidermal sheets quickly form epidermis and re-establish epidermal coverage. With time, the cultured epithelial autograft (CEA) stimulates formation of new connective tissue (*neodermis*) immediately beneath the epidermis, but scarring and wound contraction remain significant problems. Studies have shown that grafting of CEA onto pregrafted cadaver dermis greatly improves graft take (Odessey, 1992). Cultured epithelial autografts (Epicel) have been available since the late 1980s.

## V. DERMAL REPLACEMENT

Human cadaver allograft skin has been used when autologous skin grafts are not possible. Problems associated with human cadaver allografts include the possibility of an immune rejection reaction, potential for infection, and problems of supply and variability in the quality of the material. Decellularized dermal tissue has also been used in an attempt to recapitulate as much of the normal architecture as possible while providing a natural scaffold for re-epithelialization (Middelkoop *et al.*, 1995; Langdon *et al.*, 1988). Cadaver allograft dermis can be processed to make an immunologically inert, acellular dermal matrix with an intact basement membrane to aid the take and healing of ultrathin autografts (AlloDerm) (Lattari *et al.*, 1997). Currently, only the upper papillary layer of dermis is used clinically (AlloDerm). One limitation to this approach is that deep dermis and the more superficial papillary layer differ in architecture. The deep reticular dermis is needed to prevent wound contraction. Work is being done to develop a similar type of implant derived from deeper reticular dermis. Providing an appropriate scaffold for deep dermal repair remains a challenge for groups investigating native as well as synthetic matrices.

Tissue engineers have investigated the possibility of redirecting granulation-tissue formation through the use of scaffolds and livings cells. In one of the earliest tissue-engineering approaches to improving dermal healing, Yannas *et al.* (1982) designed a collagen–glycosaminoglycan sponge to serve as a scaffold or template for dermal extracellular matrix. The goal was to promote fibroblast repopulation in a controlled way that would decrease scarring and wound contraction. A commercial version of this material composed of bovine collagen and chondroitin sulfate with a silicone membrane covering (Integra™, Integra Life Sciences, Plainsboro, NJ) is currently approved for use in burns (Heimbach *et al.*, 1988). The dermal layer is slowly resorbed, and the silicone membrane is eventually removed, to be replaced by a thin autograft.

Several variations on the collagen sponge have been studied. Efforts have been made to improve fibroblast infiltration and collagen persistence by collagen cross-linking (Middelkoop *et al.*, 1995; Cooper *et al.*, 1996), by inclusion of other matrix proteins (Hansbrough *et al.*, 1989; Ansel *et al.*, 1990; deVries *et al.*, 1993) and hyaluronic acid (Cooper *et al.*, 1996), and by modifying the porosity of the scaffold (Hansbrough *et al.*, 1989; Yannas *et al.*, 1989).

Although matrix scaffolds have shown some improvement in scar morphology, no acellular matrix has yet been shown to lead to true dermal regeneration. This may be due in part to limits in cell repopulation, the type of fibroblast repopulating the graft, and control of the inflammatory and remodeling processes (i.e., the ability of the cells to degrade old matrix while synthesizing new matrix). The inflammatory response must be controlled in dermal repair in order to avoid the formation of scar tissue. Therefore, dermal scaffolds must not be inflammatory and must not stimulate a foreign-body reaction. This has been a problem in the past for some glutaraldehyde cross-linked collagen substrates (deVries *et al.*, 1993). The ability of the matrix to persist long enough to redirect tissue formation must be balanced with the effects of the matrix on inflammatory processes. One way to achieve this is to form a biological tissue that is recognized as living tissue, not a foreign substance.

There have been advances in the design of artificially grown dermal tissues using human neonatal fibroblasts grown on rectangular sheets of biodegradable mesh (Dermagraft). The fibroblasts propagate among the degrading fibers, producing extracellular matrix in the interstices of the mesh. A related product was a nonviable temporary covering for burns. In this case a nylon mesh coated with porcine collagen and layered with a nonpermeable silicone membrane (Biobrane®, Dow Hickam, Sugarland, TX) serves as a platform for deposition of human matrix proteins and associated factors by the human dermal fibroblasts (Transcyte, Dermagraft-TC). The material was then frozen to preserve the matrix and factors produced by the fibroblasts. The temporary covering would be removed prior to autografting.

## VI. COMPOSITE SKIN GRAFTS

Human skin autograft has been the gold standard for resurfacing the body and closing wounds that are difficult to heal. Cultured epidermal grafts are more likely to take when the dermal bed is relatively intact, probably because dermal factors influence epithelial migration, differentiation, attachment, and growth (Clark, 1993; Greiling and Clark, 1997). The epidermis and dermis act synergistically to maintain homeostasis (Parenteau *et al.*, 1997).

Boyce *et al.* have modified the approach first proposed by Yannas *et al.* to form a bilayered composite skin made using a modified collagen–glycosaminoglycan substrate seeded with fibroblasts and overlaid with epidermal keratinocytes (Boyce and Hansbrough, 1988). An autologous form of this composite skin construct has been used to treat severe burns, with some success (Hansbrough *et al.*, 1989). An allogeneic form of the construct showed improved healing in a pilot study in chronic wounds (Boyce *et al.*, 1995). A similar technology has been studied in the treatment of patients with genetic blistering diseases.

One of the first attempts to replicate a full-thickness skin graft was by Bell *et al.* (1981), who described a bilayered skin equivalent. The dermal component consisted of a lattice of type I collagen contracted by tractional forces of rat dermal fibroblasts trapped within the gelled collagen. This contracted lattice was then used alone or as a substrate for rat epidermal keratinocytes. They demonstrated the ability of these primitive skin equivalents to take as a skin graft in rats. This technology has now advanced to enable the production of large amounts of human Apligraf HSE from a single donor (Wilkins *et al.*, 1994). Using methods of organotypic culture that provides a three-dimensional culture environment permissive for proper tissue differentiation, the resulting HSE develops many of the structural, biochemical, and functional properties of human skin (Boyce and Hansbrough, 1988).

The process for formation of HSE has been covered in detail many times (Boyce and Hansbrough, 1988; Wilkins *et al.*, 1994) and will not be detailed here. However, there are points to be made about the approach to these procedures. The culture of HSE proceeds best with minimal intervention. Normal cell populations seem to have an intrinsic ability to re-express their differentiated program *in vitro*. A medium that supplies adequate amounts of nutrients, lipid precursors, vitamins, and minerals may be all that is required (Wilkins *et al.*, 1994). Another element is the environmental stimulus provided by culture at the air–liquid interface, which promotes differentiation and formation of the epidermal barrier (Boyce and Hansbrough, 1988).

The immunology of allogeneic tissue-engineered skin grafts is poorly understood. Inconsistencies in the literature are due, in part, to the complexity of biological and immunologic factors, which determine the ability of a graft to take and to persist over time. Among the properties that determine the immunogenicity of an engineered allogeneic graft are the purity of cell populations, the antigen-presenting capabilities of graft cells, and the vascularity of the graft.

The purity of cell populations is critical. Differences in cell purity between laboratories could contribute to the conflicting results found in murine and human studies (Hefton *et al.*, 1983; Thivolet *et al.*, 1986; Cairns *et al.*, 1993). Both culture condition and passage number affects the purity of cell populations. Very early passages of keratinocyte and fibroblast cultures might be expected to contain contaminating cell populations.

The antigen-presenting capabilities of keratinocytes and fibroblasts are also critical to determining the immunogenicity of the HSE grafts. Fibroblasts and keratinocytes are not professional antigen-presenting cells and fail to stimulate the proliferation of allogeneic T-cells (Nickoloff *et al.*, 1986; Niederwieser *et al.*, 1988; Gaspari and Katz, 1991). Both of these cell populations inherently do not express HLA class II molecules or costimulatory molecules such as B7-1 (Nickoloff *et al.*, 1993). The inability of keratinocytes and fibroblasts to induce proliferation of allogeneic T-cells is due primarily to the lack of expression of costimulatory molecules, even though aberrant antigen processing and invariant chain expression may also contribute (Nickoloff and Turka, 1994).

The ability to utilize allogeneic cells rather than autologous cells as in CEA therapy enables the reproducible manufacture of consistent Apligraf HSE (Wilkins *et al.*, 1994). The inability of epidermal keratinocytes and dermal fibroblasts to stimulate a T-cell response, discussed earlier, permits their use in allogeneic applications. Studies in athymic mice indicate that the use of a differentiated tissue such as the Apligraf HSE also beneficially affects its ability to engraft successfully (Nolte *et al.*, 1994; Parenteau *et al.*, 1996). Severe combined immunodeficient mice (SCID), who lack a functioning immune system, can be successfully transplanted with a functioning human immune system

without risk of rejection. SCID mice transplanted with human leukocytes were used as an *in vivo* model to assess the immunogenicity of allogeneic skin grafts (Moulon *et al.*, 1999). In these studies, human skin was rejected, while allogeneic tissue-engineered skin graft survival was 100%.

## VII. BIOENGINEERED SKIN: FDA-APPROVED INDICATIONS

The rather formidable obstacles to healing present in chronic wounds have made it difficult for the simple topical application of growth factors and cytokines to have a successful outcome. It can be argued that tissue engineering, particularly with living cells, has an advantage, in that cells may be capable of responding to the microenvironment and thus behave in a "smart" way, from an engineering point of view. Because of these considerations, cell therapy with bioengineered skin has been tested in difficult-to-heal wounds. Figures 77.2 and 77.3 show Apligraf before and after meshing and fenestration as well as its histological appearance. Venous and neuropathic diabetic ulcers have received the greatest attention. Two main types of living bioengineered skin have been proven effective in diabetic neuropathic foot ulcers and have received regulatory approval from the FDA. In a randomized 12-week trial of 208 patients with neuropathic ulcers, Apligraf led to complete wound closure in 56% of patients, compared to 38% in the control group (offloading alone) (Veves *et al.*, 2001). The Kaplan–Meier median time to complete wound closure was 65 days for Apligraf and 90 days for control. Figure 77.4 shows a diabetic neuropathic foot ulcer successfully treated with Apligraf. It is important to note, as detailed in the FDA-approved indication, that patients in the active arm of that study showed decreased incidence of osteomyelitis and amputation. These results on key complications of diabetic foot ulcers may have been due to faster healing. Another living cell product, Dermagraft, was shown in a 12-week randomized study to heal neuropathic foot ulcers, with an incidence of closure of 30% and 18% for the active and control arms, respectively (Marston *et al.*, 2003).

In addition to its indication for diabetic ulcers, Apligraf also remains the only approved tissue-engineered product for venous ulcers. In a pivotal multicenter randomized study of 293 patients with venous leg ulcers, Apligraf was more effective than leg compression alone (control) in the percentage of patients who healed by six months (63% vs. 49%) (Falanga *et al.*, 1998). That study also showed that Apligraf healed ulcers that were larger and deeper than those in the control group. No evidence of sensitization was observed with the bioengineered skin product. Interestingly, a subsequent reanalysis of the data from this pivotal trial, limited to 120 patients with venous ulcers longer than one year in duration, showed that Apligraf owed its overall effectiveness to these hard-to-heal ulcers. In this subgroup

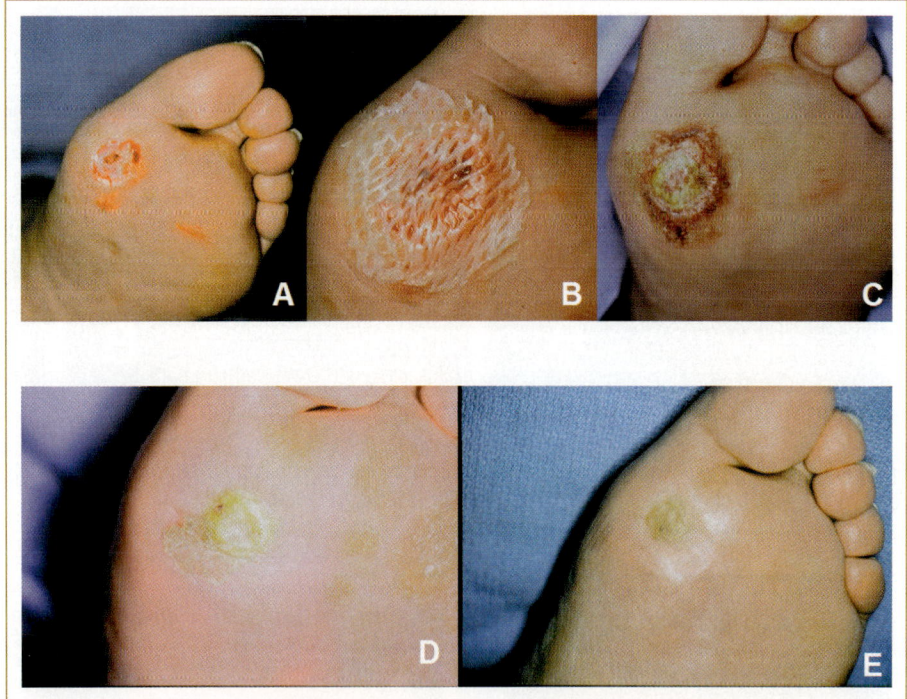

**FIG. 77.4.** Diabetic neuropathic foot ulcer successfully treated with Apligraf. **(A)** Just after surgical debridement to remove the necrotic wound bed and surrounding callus. **(B)** Wound covered with meshed Apligraf. **(C)** Appearance of construct a week later. **(D)** At three weeks. **(E)** Complete wound closure at five weeks. Copyright V. Falanga, 2006.

of the study population, Apligraf healed 47% of patients, compared to 19% in the control group (Falanga and Sabolinski, 1999). This observation, taken together with the mechanisms underlying impaired healing, suggests that cell therapy might have a greater effect when there are more substantial abnormalities leading to a prolonged failure to heal.

## Apligraf: Off-Label Uses

Off-label use is the practice of using products for purposes not approved by the FDA. Physicians can prescribe and use products as they see fit because they are not regulated and are recognized as experts in their fields. It is illegal for the parent companies of the products to promote off-label use. However, data generated by off-label use can be very valuable and lead to new insights into how the products function and, potentially, clinical trials for new indications.

From both a therapeutic and purely scientific standpoint, of critical importance in the context of the clinical trials just detailed is that, for the very first time, one has been able to show a beneficial effect of tissue-engineered products in situations of impaired healing. Understandably, clinicians have also used Apligraf off-label to accelerate healing in wounds not due to venous disease or diabetes (Shen and Falanga, 2000). Wounds that have been treated off-label with Apligraf include acute wounds after extensive skin surgery, donor sites after split-thickness skin harvesting, traumatic wounds, burns, inflammatory ulcers such as pyoderma gangrenosum and vasculitis, scleroderma digital ulcers, wounds after keloid removal, genetic conditions

such as epidermolysis bullosa, and a variety of situations that defy proper diagnosis because of their complexity. In these cases, bioengineered-skin treatment with Apligraf seems to offer a viable alternative to stimulate wound healing and relieve unacceptable pain and suffering. Some examples suffice. Pyoderma gangrenosum is an inflammatory ulcer often associated with rheumatoid arthritis, inflammatory bowel disease, and IgA gammopathy, among other predisposing factors. A peculiar feature of pyoderma gangrenosum is that it worsens with surgical manipulation and even develops at distant sites that are traumatized; thus, autologous grafting is a contraindication and bioengineered skin is an attractive therapeutic modality (DeImus et al., 2001). In a patient with multiple myeloma undergoing conditioning for bone marrow transplantation, extravasation of the chemotherapeutic drug adriamycin occurred into his chest and created a very large and deep painful wound. Apligraf proved to play a major role in healing the full-thickness wound (Fig. 77.5) and led to a dramatic relief in pain. Pressure ulcers, too, have been treated with Apligraf. This is of particular interest because, based on the engineering aspects of the construct, one would have expected that very deep wounds with loss of a great deal of tissue would not respond to Apligraf. Yet it appears that in such cases Apligraf is able to stimulate tissue regrowth in deep ulcers and not just re-epithelialization from the edges of the wound (Fig. 77.6). Scleroderma digital ulcers are ischemic in nature and have no known accepted treatment. Yet Apligraf appears promising in those wounds (Fig. 77.7). There is no evidence that Apligraf stimulates malignancy within the wound bed, possibly due to its stimulatory action. An interesting case in

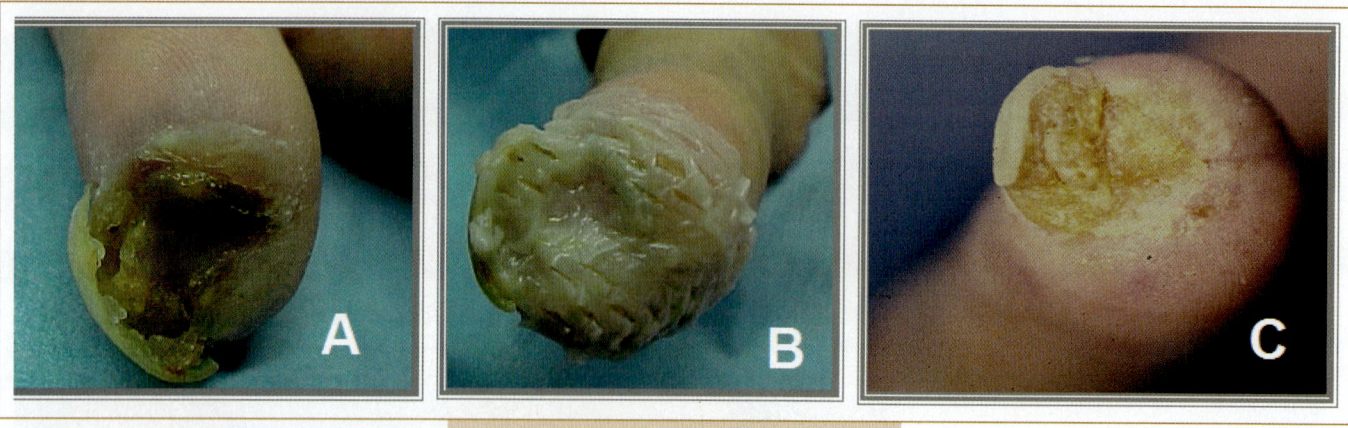

**FIG. 77.5.** Systemic sclerosis (scleroderma) digital ulcer treated with Apligraf. **(A)** Baseline ulcer with fibrinous wound bed. **(B)** Wound covered with meshed Apligraf. **(C)** Wound closure four weeks later. Copyright V. Falanga, 2006.

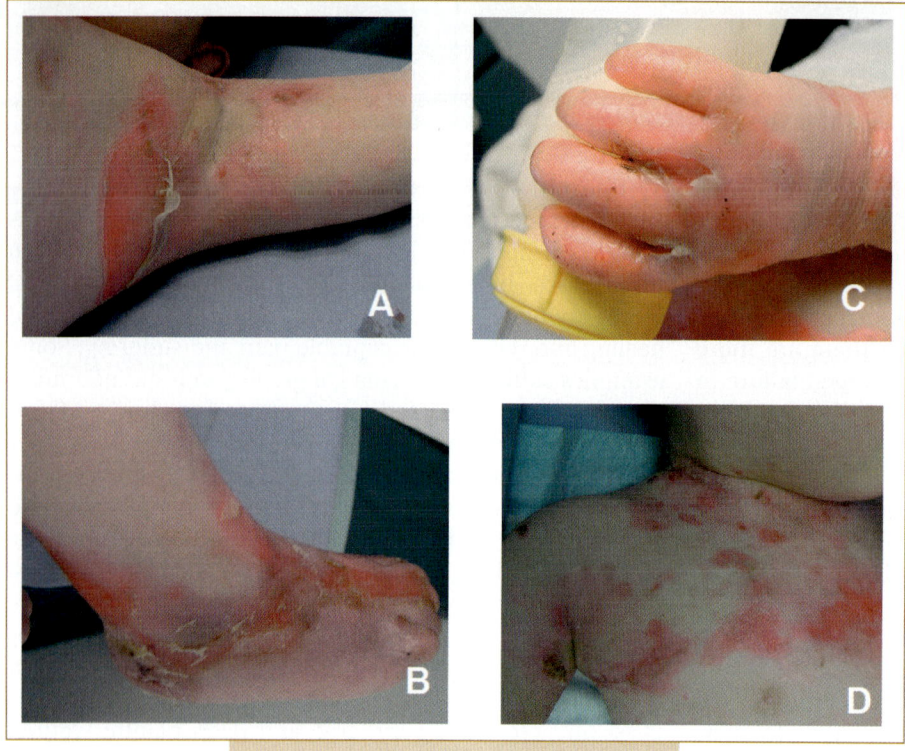

**FIG. 77.6.** Epidermolysis bullosa (dystrophic type). **(A)** Denuded area in axilla, with milia formation from repeated breakdown and healing. **(B)** Foot that has experienced many episodes of wounding, resulting in scarring and loss of toe webs. **(C)** Mitten deformity of the hand well under way due to constant injury and attempts at repair. **(D)** Several denuded areas on the chest, showing evidence of active epithelialization. The child had been treated with Apligraf. Copyright V. Falanga, 2006.

point is shown in Fig. 77.8, where multiple and ulcerated basal cell carcinomas in the foot of a patient with the hereditary form of basal cell nevus syndrome were treated with Apligraf.

The clinical results from the off-label use of Apligraf to treat epidermolysis bullosa, or EB, have prompted the initiation of clinical trials. Epidermolysis bullosa is a genetic skin disorder characterized by blistering of the skin and mucosae following mild mechanical trauma (Falabella *et al.*, 2000). Falabella *et al.* performed an open-label uncontrolled study of 15 patents with EB treated with Apligraf. The conclusion was that Apligraf induced rapid healing; the wounds remained healed for some weeks, and there was no acute rejection reaction or other adverse effects related to the treatment. Currently there is no specific treatment for EB, and the nature of the condition is that these wounds tend to heal slowly and in some instances fail to heal, becoming chronic wounds. Most wound treatment is supportive, such as sterile wound dressings and analgesics. As a result of the work done by this group as well as by others, Organogenesis has partnered with the epidermolysis bullosa patient advocate group (DebRA) to launch jointly a randomized, controlled, multicenter trial to offer evidence that Apligraf is efficacious in the treatment of EB.

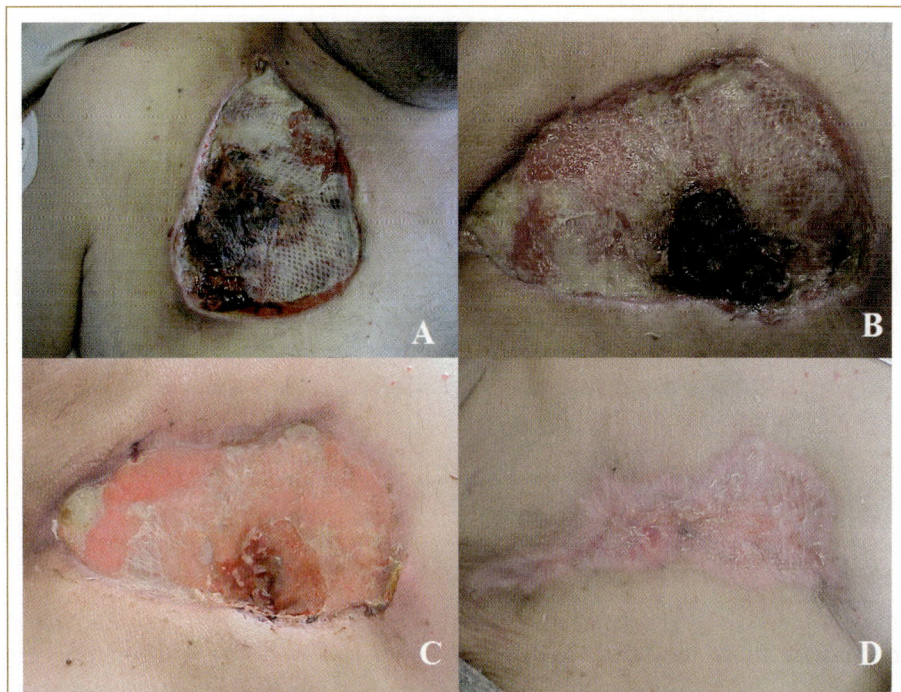

**FIG. 77.7.** Large and deep wound from extravasation of adriamycin into the chest wall during treatment for multiple myeloma and bone marrow conditioning. **(A)** Wound still partially covered with meshed Apligraf, two weeks after treatment with construct. **(B)** Wound at three weeks. **(C)** Wound at eight weeks. **(D)** Complete closure after ten weeks. Copyright V. Falanga, 2006.

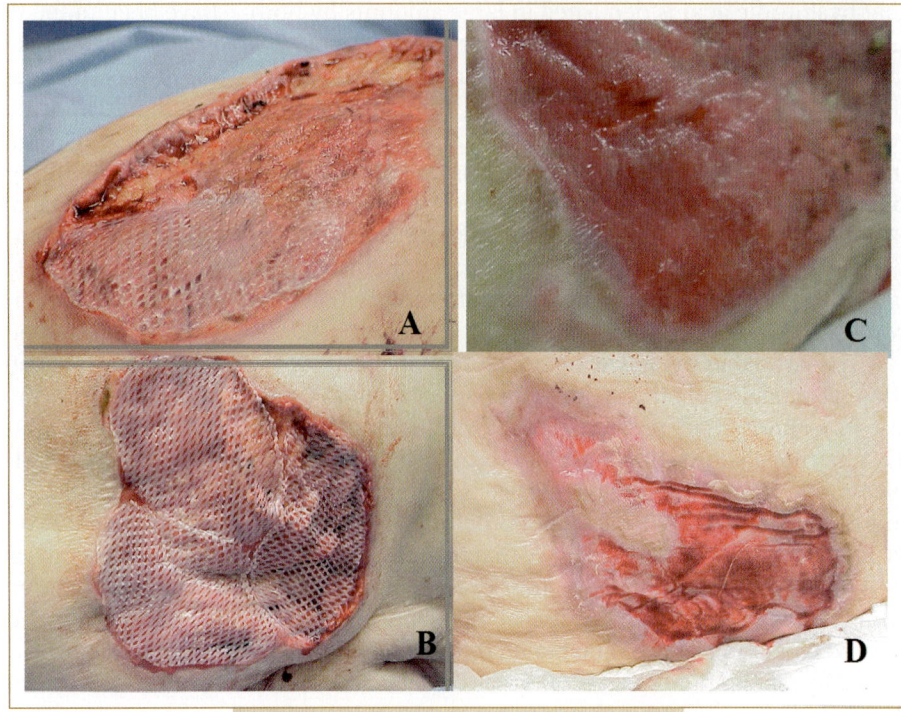

**FIG. 77.8.** Large sacral pressure ulcer. **(A)** Immediately after extensive surgical debridement and beginning of Apligraf application. **(B)** Wound fully covered with meshed Apligraf right after surgical debridement. **(C)** After three weeks the wound bed is now flush with the surrounding skin. **(D)** At week 5 the wound is largely epithelialized. The edges of the wound are rapidly advancing toward the center of the wound. Copyright V. Falanga, 2006.

# VIII. MECHANISMS OF ACTION OF BIOENGINEERED SKIN

The mechanisms of action by which bioengineered skin, in particular Apligraf, is effective in accelerating healing remain unknown. It has been stated that the delivery of living cells is associated with the release of growth factors and cytokines (Mansbridge *et al.*, 1998; Falanga *et al.*, 2003). Indeed, we have previously shown that, when wounded *in vitro*, Apligraf undergoes a staged expression of inflammatory cytokines and, later, growth factors, which is very much reminiscent of the normal process of wound healing. A host of cytokines are produced by Apligraf, and there is evidence

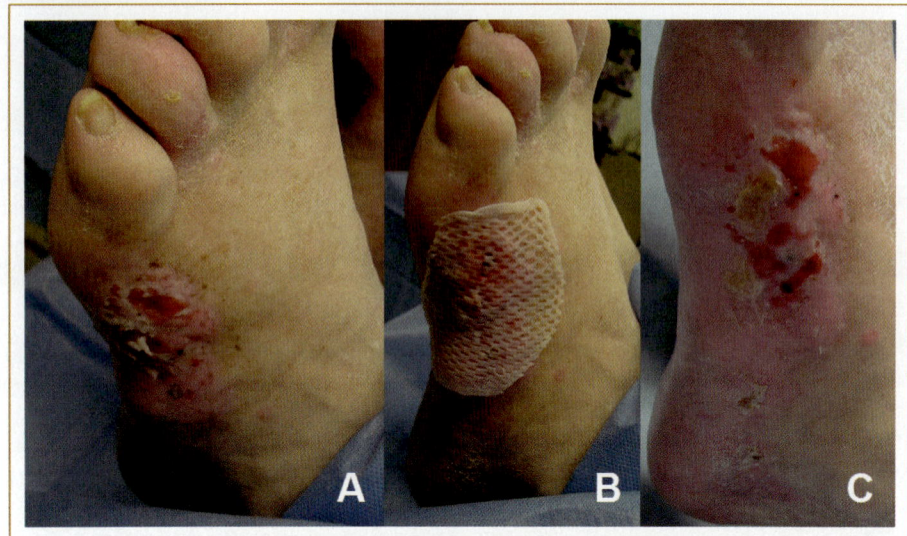

**FIG. 77.9.** Patient with a hereditary form of basal cell carcinomas (the basal cell nevus syndrome). His entire body was covered with these skin cancers. **(A)** Nonoperative ulcerated basal cell carcinomas (proven histologically) on the lateral side of the foot. **(B)** Ulcerated cancers treated with Apligraf. **(C)** Appearance of same site several months after Apligraf application and without obvious clinical evidence of extension of the skin cancers. Copyright V. Falanga, 2006.

that its epidermal and dermal components work in concert to produce these mediators that would not normally be detected with the epidermal and dermal component alone (Falanga *et al.*, 2003). In this regard, it is notable that there is a close interaction between Apligraf and the wound and that hyaluronic acid is deposited in large amounts by both the construct and the wound. Hyaluronic acid deposition is a hallmark of fetal wound healing, and it is possible that the Apligraf promotes a more fetal-like stimulation of the wound-healing process (Badiavas *et al.*, 2002). Another interesting property of Apligraf, which is due to its unique bilayered configuration, is that it undergoes epiboly *in vitro* (Nahm *et al.*, 2004). This ability to migrate over its own dermis suggests a very viable epidermal component, capable of allowing the construct to heal itself after injury (Falanga *et al.*, 2003).

The available evidence indicates that the cells from these allogeneic constructs, including Apligraf, do not persist in wounds (Phillips *et al.*, 2002; Griffiths *et al.*, 2004). This has been shown in acute wounds as well as in venous ulcers, where the donor allogeneic cells do not survive for more than four to six weeks, based on HLA typing using PCR. Thus, there is strong evidence that true engraftment or prolonged persistence of cells from these allogeneic constructs does not occur. The survival of cells from Apligraf may be longer in the neonatal period or in children, as suggested by reports that Apligraf cells may be detectable many more weeks or even months when the construct is applied in the wounds of young patients with epidermolysis bullosa, a condition characterized by a variety of genetic defects in molecules responsible for anchoring the epidermis to the dermis (i.e., laminin 5, type VII collagen) (Fivenson *et al.*, 2003). Still, in spite of a lack of longevity of the construct, the clinical trial results indicate that even a few days or weeks of exposure of the wound to Apligraf has beneficial effects. Possible mechanisms include a more orderly and orchestrated release of cytokines, deposition of extracellular matrix material important for early migration of mesenchymal cells and keratinocytes, and even the attraction of progenitor or stem cells from deep in the tissue or from the circulation (Falanga, 2005). For example, myoblast-like cells have been noted deep in the subcutaneous tissue.

Although it is tempting to think of Apligraf as a graft, available evidence indicates that this construct behaves like cell therapy. We have already discussed the issue of allogeneic cell longevity, which in itself speaks against the idea of Apligraf's being a graft. Moreover, there is no evidence that wound bed blood vessels grow into Apligraf, as is commonly observed with autologous grafts. Perhaps most telling is the fact that, uniformly, Apligraf seems to stimulate the edges of the wound to migrate toward the center. This "edge effect," which has also been reported with the application of living keratinocyte sheets, is the predominant observation with successful Apligraf treatment, although a variety of clinical presentations are seen that, at least at times, suggest that the construct is still present in the wound bed (Fig. 77.9). The edge effect strongly suggests that Apligraf does not act as a tissue replacement but, rather, as cell therapy. Indeed, the rate at which the edges of a wound migrate toward the center has been used to determine whether eventual wound closure will occur. The data appear to indicate that an advancement of the edge of more than 0.7 mm per week is about 80% sensitive and specific for ultimate wound closure. Apligraf seems to increase this healing rate, up to approximately 1.5 mm per week (Falanga and Sabolinski, 2000).

An interesting issue with regard to Apligraf, other than cytokine release and matrix deposition (but perhaps related to it), is its physical configuration and timing of contact with the wound bed. For example, we have shown that contact of Apligraf with the wound must ideally be for at least two weeks for a successful outcome to occur (Saap *et al.*, 2004).

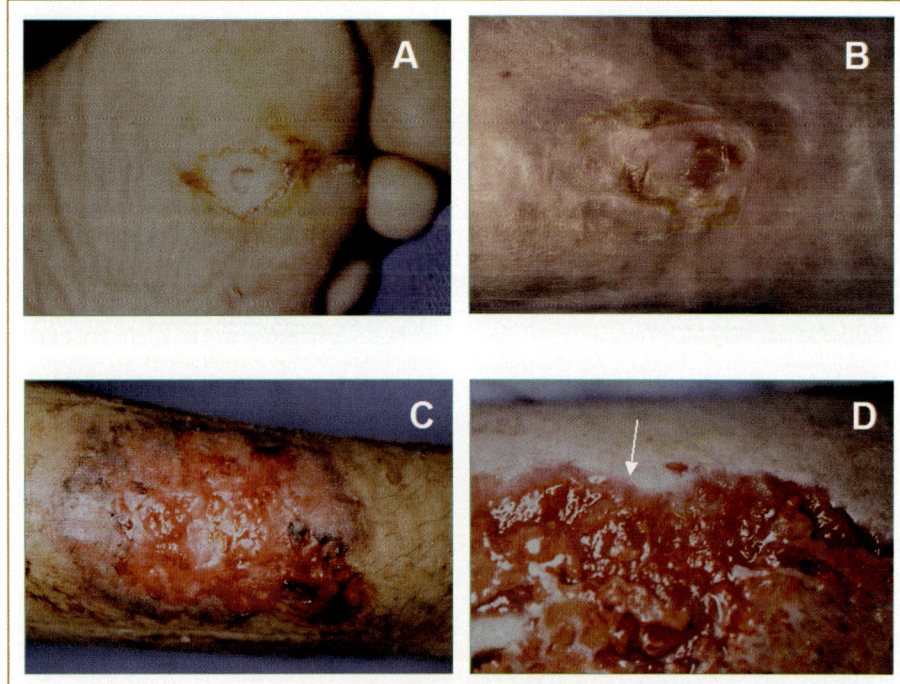

**FIG. 77.10.** Different clinical appearances of wounds after Apligraf application. **(A)** Diabetic foot ulcer, with white center representing Apligraf three weeks after application. **(B)** Venous ulcer still showing an Apligraf-like material six weeks later. **(C)** Pyoderma gangrenosum only two weeks after Apligraf application and with thin layer of epidermis completely covering the wound bed. **(D)** Classical "edge effect" (arrow) at the margins of a venous ulcer treated with Apligraf. The yellow material at the bottom of the pictures represents what is left of meshed Apligraf applied three weeks earlier. Copyright V. Falanga, 2006.

Another observation has been that the migrating tongue of the wound edge seems to "walk" over the Apligraf, even when the latter has the appearance of being broken down or certainly not intact (Fig. 77.10). Of most interest, defying the idea of architectural tissue replacement, is the observation that Apligraf is effective even when placed in the center of the wound, approximately 1 cm away from the wound edges, and that the beneficial clinical outcome occurs even when the construct is placed over the wound with its dermal side up (Fig. 77.11). Taken together, these observations and considerations lead to the notion that Apligraf acts as a tissue unit on its own, perhaps stimulated by the wound environment or by the meshing and fenestration of the construct that are usually done to allow for wound fluid to escape and to expand the coverage ability of the construct (Brem *et al.*, 2004; Cavorsi *et al.*, In press).

Cell therapy appears to be the primary mechanism, but the work done with epidermolysis bullosa patients suggests additional or alternative mechanisms. As discussed previously, Apligraf has been used off-label for the treatment of epidermolysis bullosa, or EB. Inherited EB is a heterogeneous group of rare genetic skin diseases. The diseases are characterized by blistering of the skin and mucosae following mild mechanical trauma; most forms result in a lifetime of blister and wound formation. The blistering is a result of the mechanical fragility of the basal cells of the epidermis, and this fragility has been linked to mutation in the keratin K5 and K14 genes (Falabella *et al.*, 1999). Although some forms of EB are associated with normal longevity, several severe forms have significant morbidity and increased mortality, as either a direct or an indirect result of the disease

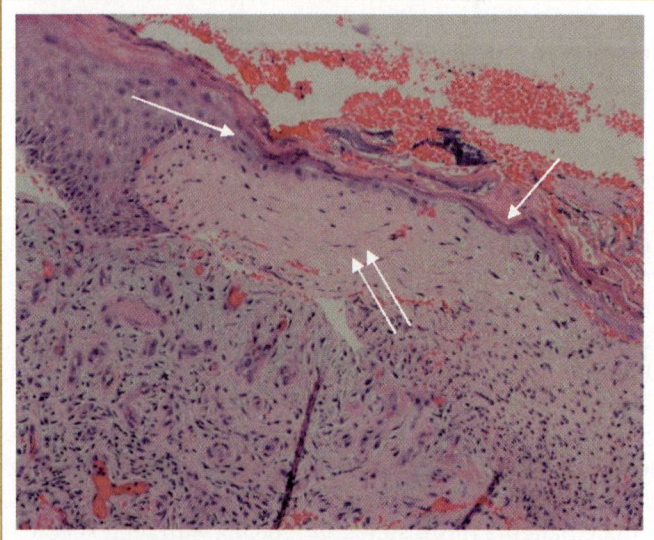

**FIG. 77.11.** Histological appearance and intimate interaction of Apligraf and an ulcer due to pyoderma gangrenosum several days after construct application. The edges of the wound are extending a thin layer of epidermis (single arrows) over the remnant of the construct (double arrows). It would appear that the native keratinocytes are using Apligraf as a substrate on which to migrate. Copyright V. Falanga, 2006.

(Falabella *et al.*, 1999, 2000). The work done with these patients has produced dramatic clinical results, where patients experience more rapid healing and a reduction in the frequency of blister formation in the treated areas as well as greater function and range of motion when used to

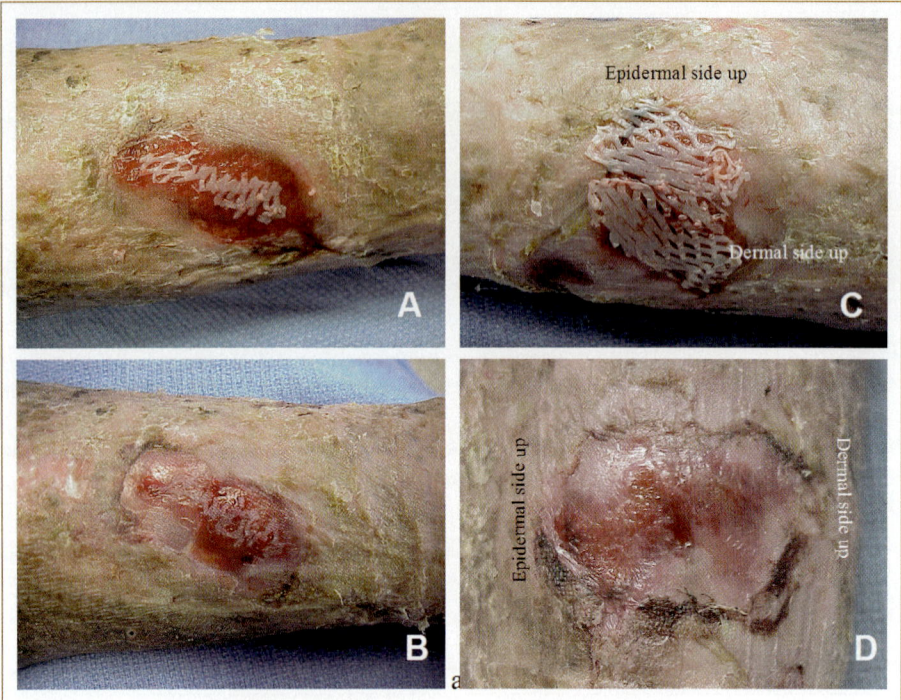

**FIG. 77.12.** Unusual applications of Apligraf in two patients with venous ulcers. **(A)** Apligraf applied in the center, about 1 cm away from the wound edge. **(B)** Almost-complete wound closure a week later. Apligraf is still visible in the center of the wound. **(C)** Apligraf applied in the normal way (epidermal side up) or with dermal side up. **(D)** Regardless of Apligraf orientation, there is definite evidence of reepithelialization from the wound edges. Copyright V. Falanga, 2006.

treat severe mitten deformity (Fivenson *et al.*, 2003a, 2003b; Falabella *et al.*, 2000).

With EB patients, the decrease in reblistering in treated sites as compared to untreated sites suggests that the patient's skin "learns" from the allogeneic tissue. In the EB patients, donor DNA has been found up to six months postoperatively (Fivenson *et al.*, 2003a, 2003b), in contrast to venous leg ulcers, where donor DNA was detected one month postoperatively but not detected at two months (Phillips *et al.*, 2002). The DNA analysis showed specific donor DNA sequences (HLADQb10201 and Short Tandem Repeat) at the treated sites six months after receiving Apligraft (Graftskin). The presence of this DNA is significant because it is unlikely that intact cells were present, due to the lack of Y-chromosome-specific sequences in any of the female patients (Fivenson *et al.*, 2003a). Although speculative, these findings would seem to support the idea that some degree of gene transfer has occurred, bearing some similarities to gene therapy, in that the diseased tissue is exposed to a source of normal genes and their proteins (Fivenson *et al.*, 2003a; Fivenson and Falabella, 2000).

## IX. CONCLUSION

Tissue engineering is an emerging field, with engineered skin constructs playing a lead role. Like the natural tissue these constructs are designed to mimic, the engineered tissues are proving to be dynamic in the way they respond to environmental stimuli. The understanding of how these products induce healing is evolving. Greater understanding into how engineered tissues promote wound healing is being achieved through the research being done with Apligraf in different wound types. Initially the action was believed to be similar to skin grafts, where the patient's native tissue was replaced with the graft tissue. It is now believed that the mode of action is more complex. Cell therapy appears to be the primary mode of action; however, there may be other secondary mechanisms at work as well. The work being done in this area is very exciting, and researchers have only begun to scratch the surface in terms of understanding the mechanisms and environmental stimuli involved.

## X. REFERENCES

Abraham, J. A., and Klagsbrun, M. (1996). Modulation of wound repair by members of the fibroblast growth factor family. *In* "The Molecular and Cellular Biology of Wound Repair" (R. A. F. Clark, Ed.), 2nd ed., pp. 195–248. Plenum Press, New York.

Ansel, J., Perry, P., and Brown, J. (1990). Cytokine modulation of keratinocyte cytokines. *J. Investi. Dermatol.* **94**(suppl), 101–107.

Archer, R., and Williams, D. J. (2005). Why tissue engineering needs process engineering. *Nat. Biotechnol.* **23**, 1353–1355.

Badiavas, E. V., Paquette, D., Carson, P., and Falanga, V. (2002). Human chronic wounds treated with bioengineered skin: histological evidence of host graft interactions. *J. Am. Acad. Dermatol.* **46**, 524–530.

Bal, V., McIndoe, A., Denton, G., Hudson, D., Lombardi, G., Lamb, J., and Lechler, R. (1990). Antigen presentation by keratinocytes induces tolerance in human T-cells. *Eur. J. Immunol.* **20**, 1893–1897.

Bell, E., Ehrlich, P., Buttle, D. J., and Nakatsuji, T. (1981). Living tissue formed *in vitro* and accepted as skin-equivalent of full thickness. *Science* **221**, 1052–2054.

Blomme, E. A., Werkmeister, J. R., Zhou, H., Kartsogiannis, V., Capen, C. C., and Rosol, T. J. (1998). Parathyroid hormone–related protein expression and secretion in a skin organotypic culture system. *Endocrine* **8**, 143–151.

Boyce, S. T., and Hansbrough, J. F. (1988). Biologic attachment, growth, and differentiation of cultured human keratinocytes on a graftable collagen and chondroitin-6-sulfate substrate. *Surgery* **103**, 421–431.

Boyce, S. T., Glatter, R., and Kitsmiller, J. (1995). Treatment of chronic wounds with cultured skin substitutes: a pilot study. *Wounds* **7**, 24–29.

Brem, H., Kirsner, R. S., and Falanga, V. (2004). Protocol for the successful treatment of venous ulcers. *Am. J. Surg.* **188**, 1881–1888.

Cairns, B. A., deSerres, S., Kilpatrick, K., Frelinger, J. A., and Meyer, A. A. (1993). Cultured keratinocyte allografts fail to induce sensitization in vivo. *Surgery* **114**(2), 416–422.

Cavorsi, J., Vicari, F., Wirthlin, D. J., Ennis, W., Kirsner, R. S., O'Connell, S. M., Steinberg, J., and Falanga, V. (In press). Best-practice algorithms for the use of a bilayered living cell therapy (Apligraf) in the treatment of lower extremity ulcers. *Wound Rep. Regen.*

Clark, R. A. F. (1993). Basics of cutaneous wound repair. *J. Dermatol. Surg. Oncol.* **19**, 639–706.

Cooper, M. L., Hansbrough, J. F., and Polareck, J. W. (1996). The effect of an arginine-glycine-aspartic acid peptide and hyaluronate synthetic matrix on epithelialization of meshed skin graft interstices. *J. Burn Care Rehabil.* **17**, 108–116.

DeImus, G., Tsoukas, M., Golomb, C., Wilkel, C., Nowak, M., and Falanga, V. (2001). Accelerated healing of pyoderma gangrenosum treated with bioengineered skin (Graftskin) and concomitant immunosuppression. *J. Am. Acad. Dermatol.* **44**, 61–66.

Desmouliere, A., and Gabbiani, G. (1996). The role of the myofibroblast in wound healing and fibrocontractive diseases. *In* "The Molecular and Cellular Biology of Wound Repair" (R. A. F. Clark, ed.), 2nd ed., pp. 321–323. Plenum Press, New York.

deVries, H. J. C., Mekkes, J. R., Middelkoop, E., Hinrichs, W. L. J., Wildevuur, C. H. R., and Westerhof, W. (1993). Dermal substitutes for full-thickness wounds in a one-stage grafting model. *Wound Rep. Regen.* **1**, 244–252.

Falabella, A. F., Schachner, L. A., Valencia, I. C., and Eaglstein, W. H. (1999). The use of tissue-engineered skin (Apligraf) to treat a newborn with epidermolysis bullosa. *Arch. Dermatol.* **135**, 1219–1222.

Falabella, A. F., Valencia, I. C., Eaglstein, W. H., and Schachner, L. A. (2000), Tissue-engineered sking (Apligraf) in the healing of patients wth epidermolysis bullosa wounds. *Arch. Dermatol.* **136**, 1225–1230.

Falanga, V. (1993). Chronic wounds: pathophysiological and experimental considerations. *J. Invest. Dermatol.* **100**, 751–755.

Falanga, V. (2001). Classifications for wound preparation and stimulation of chronic wounds. *Wound Rep. Regen.* **8**, 347–352.

Falanga, V. (2005). Wound healing and its impairment in the diabetic foot. *Lancet* 366, 1736–1743.

Falanga, V., and Eaglstein. W. H. (1993). The trap hypothesis of venous ulceration. *Lancet* **341**, 1006–1008.

Falanga, V., and Sabolinski, M. (1999). A bilayered living skin construct (Apligraf) accelerates complete closure of hard-to-heal venous ulcers. *Wound Rep. Reg.* **7**, 201–207.

Falanga, V., and Sabolinski, M. L. (2000). Prognostic factors for healing of venous and diabetic ulcers. *Wounds* **12**, 42A–46A.

Falanga, V., Margolis, D., Alvarez, O., Auletta, M., Maggiacomo, F., Altman, M., Jensen, J., Sabolinski, M., and Hardin-Young, J. (1998). Human Skin Equivalent Investigators Group. Rapid healing of venous ulcers and lack of clinical rejection with an allogeneic cultured human skin equivalent. *Arch. Dermatol.* **134**, 293–300.

Falanga, V., Isaacs, C., Paquette, D., Downing, G., Kouttab, N., Butmarc, J., Badiavas, E., and Hardin-Young, J. (2003). Wounding of bioengineered skin: cellular and molecular aspects after injury. *J. Invest. Dermatol.* **119**, 653–660.

Fife, C. E., Buyukcakir, C., Otto, G. H., *et al.* (2002). The predictive value of transcutaneous oxygen tension measurement in diabetic lower-extremity ulcers treated with hyperbaric oxygen therapy: a retrospective study of 1144 patients. *Wound Rep. Regen.* **10**, 198–207.

Fivenson, D. P., and Falabella, A. F. (2000). The use of Graftskin (Apligraft) for the treatment of epidermolysis bullosa. *Wounds* **12**, 64A–71A.

Fivenson, D. P., Schershun, L., Choucair, M. D., KuKuruga, D., Young, J., and Shwayder, T. (2003a). Graftskin therapy in epidermolysis bullosa. *J. Am. Acad. Dermatol.* **48**, 886–892.

Fivenson, D. P., Scherschun, L., and Cohen, L. V. (2003b). Apligraf in the treatment of severe mitten deformity associated with recessive dystophic epidermolysis bullosa. *Plast. Reconstruc. Surg.* **112**, 584–588.

Gallico, G. C., III, O'Connor, N. E., Compton, C. C., Kehinde, O., and Green, H. (1984). Permanent coverage of large burn wounds with autologous cultured human epithelium. *N. Eng. J. Med.* **311**, 448–451.

Gaspari, A. A., and Katz, S. I. (1988). Induction and functional characterization of class II MHC (Ia) antigens on murine keratinocytes. *J. Immunol.* **140**, 2956–2963.

Gaspari, A. A., and Katz, S. I. (1991). Induction of *in vivo* hyporesponsiveness to contact allergens by hapten-modified Ia+ keratinocytes. *J. Immunol.* **147**(12), 4155–4161.

Green, H., Kehinde, O., and Thomas, J. (1979). Growth of cultured human epidermal cells into multiple epithelia suitable for grafting. *Proc. Nat. Acad. Sci. U.S.A.* **76**, 5665–5668.

Greiling, D., and Clark, R. A. (1997). Fibronectin provides a conduit for fibroblast transmigration from collagenous stroma into fibrin clot provisional matrix. *J. Cell Sci.* **110**(Pt. 7), 861–870.

Griffiths, M., Ojeh, N., Livingstone, R., Price, R., and Navsaria, H. (2004). Survival of Apligraf in acute human wounds. *Tissue Eng.* **10**, 1180–1195.

Hansbrough, J. F., Boyce, S. T., Cooper, M. L., and Foreman, T. J. (1989). Burn wound closure with cultured autologous keratinocytes and fibroblasts attached to a collagen–glycosaminoglycan substrate. *J. Am. Med. Assoc.* **262**, 2125–2130.

Hefton, J. M., Madden, M. R., Finkelstein, J. L., and Shires, G. T. (1983). Grafting of burn patients with allografts of cultured epidermal cells. *Lancet* **2**(8347), 428–430.

Heimbach, D., Luterman, A., Burke, J. F., Cram, A., Herndon, D., Hunt, J., Jordan, M., McManus, W., Solem, L, and Warden, N. G. (1988). Artificial dermis for major burns: a multicenter randomized clinical trial. *Ann. Surg.* **208**, 313–320.

Kim, B-C., Kim, H. T., Park, S. H., Cha, J-S., Yufit, T., Kim, S-J., and Falanga, V. (2003). Fibroblasts from chronic wounds show altered TGF-β type II receptor expression. *J. Cell. Physiol.* **195**, 331–336.

Langdon, R. C., Cuono, C. B., Birchall, N., Madri, J. A., Kuklinska, E., McGuire, J., and Moellmann, G. E. (1988). Reconstitution of structure and cell function in human skin grafts derived from cryopreserved allogeneic dermis and autologous cultured keratinocytes. *J. Invest. Dermatol.* **91**, 478–485.

Lappin, M. B., Kimnber, I., and Norval, M. (1996). The role of dendritic cells in cutaneous immunity. *Arch. Dermatol. Res.* **288**, 109–121.

Lattari, V., Jones, L. M., Varcelotti, J. R., Latenser, B. A., Sherman, H. F., and Barrette, R. R. (1997). The use of a permanent dermal allograft in full-thickness burns of the hand and foot: a report of three cases. *J. Burn Care Rehabil.* **18**, 147–155.

Lysaght, M. J., and Hazlehurst, A. L. (2004). Tissue engineering: the end of the beginning. *Tissue Eng.* **10**, 309–320.

Mansbridge, J., Liu, K., Patch, R., Symons, K., and Pinney, E. (1998). Three-dimensional fibroblast culture implant for the treatment of diabetic foot ulcers: metabolic activity and therapeutic range. *Tissue Eng.* **4**, 403–414.

Marston, W. A., Hanft, J., Norwood, P., and Pollak, R. (2003). The efficacy and safety of Dermagraft in improving the healing of chronic diabetic foot ulcers: results of a prospective randomized trial. *Diabetes Care* **26**, 1701–1705.

McKay, I. A., and Leigh, I. M. (1991). Epidermal cytokines and their roles in cutaneous wound healing. *Br. J. Dermatol.* **124**, 513–518.

Middelkoop, E., deVries, H. J. C., Ruuls, L., Everts, V., Wildevuur, C. H. R., and Westerhof, W. (1995). Adherence, proliferation, and collagen turnover by human fibroblasts seeded into different types of collagen sponges. *Cell Tissue Res.* **280**, 447–453.

Moulon, K. S., Melder, R. J., Dharmidharka, V., Hardin-Young, J., Jain, R. K., and Briscoe, D. M. (1999). Angiogenesis in the huPBL-SCID model of human transplantation rejection. *Tranplantion.* **67**, 1626–1631.

Nahm, W. K., Philpot, B. D., Adams, M. M., Badiavas, E. V., Zhou, L. H., Butmarc, J., and Falanga, V. (2004). Significance of *N*-methyl-D-aspartate (NMDA) receptor–mediated signaling in human keratinocytes. *J. Cell. Physiol.* **200**, 309–317.

Nickoloff, B. J., and Turka, L. A. (1994). Immunological functions of nonprofessional antigen-presenting cells: new insights from studies of T-cell interactions with keratinocytes. *Immunol. Today* **15**(10), 464–469.

Nickoloff, B. J., *et al.* (1986). Human keratinocyte–lymphocyte reactions *in vitro. J. Invest. Dermatol.* **87**(1), 11–18.

Nickoloff, B. J., Mitra, R. S., Lee, K., Turka. L. A., Green, J., Thompson, C., and Shimizu, Y. (1993). Discordant expression of CD29 ligands, BB-1 and B7 on keratinocytes *in vitro* and psoriatic cells *in vivo. Am. J. Pathol.* **142**(4), 1029–1040.

Niederwieser, D., Aubock, J., Troppmair, J., Herold, M., Schuler, G., Boeck, G., Lotz, J., Fritsch, P., and Huber, C. (1988). IFN-mediated induction of MHC antigen expression on human keratinocytes and its influence on *in vitro* alloimmune responses. *J. Immunol.* **140**(8), 2556–2564.

Nolte, C. J., Oleson, M. A., Hansbrough, J. F., Morgan, J., Greenleaf, G., and Wilkins, L. (1994). Ultrastructural features of composite skin cultures grafted onto athymic mice. *J. Anat.* **185**, 325–333.

Odessey, R. (1992). Addendum: multicenter experience with cultured epidermal autograft for the treatment of burns. *J. Burn Care Rehab.* **13**, 174–180.

Ovington, L. G. (2001). Wound care products: how to choose. *Adv Wound Manage.* **40**, 259–264.

Parenteau, N. L., Sabolinski, M., Prosky, S., Nolte, C., Oleson, M., Kriwet, K., and Bilbo, P. (1996). Biological and physical factors influencing the successful engraftment of a cultured human skin substitute. *Biotechnol. Bioeng.* **52**, 3–14.

Parenteau, N. L., Sabolinski, M. L., Mulder, G., and Rovee, D. T. (1997). Wound research. *In* "Chronic Wound Care: A Clinical Source for Healthcare Professionals" (D. Krasner and D. Kane, eds.), 2nd ed., pp. 389–395. Health Management Publications, Wayne, PA.

Phillips, T. J., Kehinde, O., Green, H., and Gilchrest, B. A. (1989). Treatment of skin ulcers with cultured epidermal allografts. *J. Am. Acad. Dermatol.* **21**, 191–199.

Phillips, T. J., Manzoor, J., Rojas, A., Isaacs, C., Carson, P., Sabolinski, M., Young, J., and Falanga, V. (2002). The longevity of a bilayered skin substitute after application to venous ulcers. *Arch. Dermatol.* **138**, 1079–1081.

Phipps, R. P., Roper, R. L., and Stein, S. H. (1989). Alternative antigen presentation pathways: accessory cells which down-regulate immune responses. *Regional Immunol.* **2**, 326–339.

Pober, J. S., Collins, T., Bimmbrone, Jr., M. A., Libby, P., and Reiss, C. S. (1986). Inducible expression of class II major histocompatibility complex antigens and the immunogenicity of vascular endothelium. *Transplantation* **41**(2), 141–146.

Rheinwald, J. G., and Green, H. (1975). Serial cultivation of strains of human epidermal keratinocytes: the formation of keratinizing colonies from single cells. *Cell* **6**, 331–344.

Robson, M. C., Stenberg, B. D., and Heggers, J. P. (1990). Wound-healing alterations caused by infection. *Clin. Plast. Surg.* **17**, 485–492.

Saap, L., Donohue, K., and Falanga, V. (2004). Clinical classification of bioengineered skin use and its correlation with healing of diabetic and venous ulcers. *Dermatol. Surg.* **30**, 1095–1100.

Sabolinski, M. L., Alvarez, O., Auletta, M., Mulder, G., and Parenteau, N. L. (1996). Cultured skin as a "smart material" for healing wounds: experience in venous ulcers. *Biomaterials* **17**, 311–320.

Steed, D. L., Ricotta, J. J., Prendergast, J. J., Kaplan, R. J., Webster, M. W., McGill, J. B., and Schwartz, S. L. (1995). Promotion and acceleration of diabetic ulcer healing by arginine-glycine-aspartic acid (RGD) peptide matrix. RGD Study Group. *Diabetes Care,* **18**, 39–46.

Shen, J. T., and Falanga, V. (2000). Treatment of venous ulcers using a bilayered living skin construct. *Surg. Technol. Int.* **IX**, 77–80.

Siroky, M. B. (2002). Pathogenesis of bacteriuria and infection in the spinal cord–injured patient. *Am. J. Med.* **113**, 67S–79S.

Streilien, J. W., and Bergstresser, P. R. (1984). Langerhans cells: antigen-presenting cells of the epidermis. *Immunobiology* **168**, 285–300.

Thivolet, J., Faure, M., Demide, A., and Mauduit, G. (1986). Long-term survival and immunological tolerance of human epidermal allografts produced in culture. *Transplantation* **42**(3), 274–280.

Veves, A., Falanga, V., Armstrong, D. G., and Sabolinski, M. L. (2001). Graftskin, a human skin equivalent, is effective in the management of noninfected neuropathic diabetic foot ulcers: a prospective randomized trial. *Diabetes Care* **24**, 290–295.

Wilkins, L. M., Watson, S. R., Prosky, S. J., Meunier, S. F., and Parenteau, N. L. (1994). Development of a bilayered living skin construct for clinical applications. *Biotechnol. Bioeng.* **43**, 747–756.

Williams, I. R., and Kupper, T. S. (1996). Immunity at the surface: homeostatic mechanisms of the skin immune system. *Life Sci.* **58**(18), 1485–1507.

Winter, G. (1962). Formation of scab and the rate of epithelialization of superficial wounds in the skin of the young domestic pig. *Nature* **193**, 293–294.

Yannas, I. V., Burke, J. F., Orgill, D. P., and Skrabut, E. M. (1982). Wound tissue can utilize a polymeric template to synthesize a functional extension of skin. *Science* **215**, 174–176.

Yannas, I. V., Lee, E., Orgill, D. P., and Skrabut, E. M., and Murphy, G. F. (1989). Synthesis and characterization of a model extracellular matrix that induces partial regeneration of adult mammalian skin. *Proc. Nat. Acad. Sci. U.S.A.* **86**, 933–937.

# Clinical Experience

# Chapter Seventy-Eight

# Current State of Clinical Application

*Shaun M. Kunisaki and Dario O. Fauza*

I. Introduction
II. Current Challenges
III. Clinical Applications

IV. Conclusion
V. References

## I. INTRODUCTION

The overall impact of tissue engineering, as a multidisciplinary field, to clinical practice has been fairly diverse in nature. Directly or indirectly, it has enhanced our understanding of the structure–function relationships within normal and pathological tissues and of three-dimensional physiological processes. It has also broadened the possibilities for testing pharmacological therapies (Griffith and Naughton, 2002). Still, the benchmark for success remains, ultimately, its ability to generate new and more effective therapies for patients afflicted with severe tissue loss and/or organ failure (J. P. Vacanti, 1988). Only a handful of tissue-engineered products has reached the bedside, to date. Most of the clinically validated tissue-engineering therapies thus far have yielded modest benefits, often in small, well-defined patient populations. Besides, a long list of tissue-engineered products has either been abandoned during phase I/II trials or has failed in phase III testing (Lysaght and Hazlehurst, 2004). As with many other novel medical technologies, translation from the bench to the bedside has not been easy, and many hurdles remain to be surmounted. Perhaps this should not come as a surprise, in that the field is still maturing. The purpose of this chapter is to offer a general outlook on the present state of clinical application of tissue-engineering technology and its challenges for

further development. More detailed reviews of certain specific clinical applications are given in other chapters.

## II. CURRENT CHALLENGES

### Practical Considerations

Timing is an inherent limitation of many tissue-engineering concepts. Except for tissue-indicing substances, autologous cell–based approaches generally involve weeks, if not months, of processing, for example, for the culture of a sufficient number of cells for a given composite construct. Yet time is often an unaffordable luxury for many patients. Another practical barrier has been the cost of this technology. Elaborate and expensive infrastructures are necessary for the development and manufacture of engineered tissue. So-called good manufacturing practice (GMP) facilities are a prerequisite for FDA approval of cell-based therapies, which, in turn, cannot be pursued without a critical mass of highly trained personnel. Furthermore, certain tissues require preconditioning in complex bioreactors, which may not be readily amendable to scaled-up manufacturing and shipping (Griffith and Naughton, 2002). All of these issues translate into a chronic difficulty in establishing multicenter clinical trials, which are essential for the widespread application of this technology.

## Regulatory Matters

Regulatory constraints have significantly hampered the clinical translation of many tissue-engineering therapies (Ahsan and Nerem, 2005). In the United States in particular, the FDA has been notoriously slow in initiating and conducting product-approval processes. This problem has been attributed, at least in part, to the lack of clear regulatory frameworks and to uncertainties regarding whether tissue-engineered products should be classified as mechanical implants, biological materials, or both.

Regardless, for many companies devoted mostly to tissue-engineering technologies, these regulatory delays, combined with strict reimbursement policies and poor business models, have often been commercially unsustainable. Indeed, many firms with considerable interest in tissue engineering have exited the market since the turn of century. Other American biotechnology companies have begun collecting clinical data overseas at lower costs, given that some European Union and Asian countries have less stringent regulatory procedures with regard to the marketing and clinical application of novel medical devices.

## Engineering Limitations

Perhaps surprisingly for some, the ideal cell type for many clinical applications of tissue engineering has not necessarily been determined yet. In many cases, while differentiated autologous cells would be ideal, their use simply may not be a viable option, either because of current isolation and expansion limitations (e.g., hepatocytes, neurons, cardiomyocytes, pancreatic islets cells) or because they tend to dedifferentiate over time (e.g., chondrocytes). Furthermore, even autologous cells in culture may not be completely free of pathogens, since the growth medium often requires xenogeneic growth factors, such as fetal bovine serum, or the cells can only propagate on murine feeder layers, so infectious risks cannot be completely eliminated. Of course, adult and embryonic stem cells, along with therapeutic cloning, could provide solutions to all of these problems. This is discussed in depth in other chapters.

The ideal biomaterial for a given clinical application often has also yet to be determined. Many of the currently available synthetic scaffolds are still limited by foreign-body reaction when placed in an immunocompetent environment. These conditions can lead to a reduction in the diffusion of nutrients and waste products, fibrosis, and other complications. Additionally, the cytotoxic effects of macrophage-generated nitric oxide can reach and destroy the transplanted cells. Thus, it is not surprising that most of the scaffolds that have been implanted in humans to date are derived from natural sources (e.g., bone, dermis, intestinal submucosa). However, natural scaffolds have been associated with unfavorable mechanical properties (e.g., rapid or inconsistent degradation, low tensile strength). For these reasons, there remains a continued interest in the development of novel biocompatible synthetic biomaterials, such as elastomers. Scaffolds impregnated with growth factors or specific peptide sequences may also allow for better control of the surrounding microenvironment. Undoubtedly, these newer synthetic materials, further discussed in other chapters, will be instrumental in helping to broaden the type of tissues that can be engineered for human use.

### Blood Supply

Typically, any tissue greater than 1 cm in thickness cannot rely solely on vascular ingrowth from the host's vascular bed in order to remain viable *in vivo*. Thus, a major challenge for clinical application of large engineered tissues and engineered organs has been how to optimize the blood supply to the graft at the time of implantation. One interesting, now-already-preclinical approach to address this problem has been to create a preformed microcirculation within the engineered scaffold itself. Such a strategy has been employed by Vacanti and his colleagues using micro-electromechanical systems (MEMS) technology (Fidkowski et al., 2005). Preliminary work has enabled his group to develop a robust computational model of the vascular circulation, which includes the fractal nature of network topology, the rheology of blood flow through this computational system, and the mass transfer of oxygen and nutrients across the vascular bed. More recently, this method has allowed this group to etch vascular channels onto silicon wafers, which can then be transferred to biodegradable polymer systems. Multiple monolayers of this architecture can be stacked to form three-dimensional structures. Two other approaches to address the problem of vascularization — gene therapy and the scaffold-based encapsulation of angiogenic growth factors (e.g., vascular endothelial growth factor) — have also been pursued in preclinical models. Conceivably, these approaches might be applied in facilitating the formation of other complex accessory networks (e.g., neural, lymphatic, biliary) as well, within engineered tissue.

## III. CLINICAL APPLICATIONS

A proper overview of the current state of clinical application of tissue engineering demands a proper understanding of the extent of the field (Fauza, 2003). Although some still link the term *tissue engineering* solely to the paradigm of cell delivery within biocompatible scaffolds, the theme is certainly much broader (J. P. Vacanti and Langer, 1999). In essence, all tissue-engineering technologies fall into one of four main strategies: (1) delivery of isolated cells (cell transfers), (2) tissue-inducing substances, (3) extracorporeal and encapsulation techniques (closed systems), and (4) transplantation of cell within matrices (open systems). A summary list of selected tissue-engineering products representing all these strategies, which either have reached mainstream clinical practice or were undergoing clinical trials as of early 2006, is given in Table 78.1. Select general discussions follow, while more detailed information on certain specific applications can be found in other chapters.

## Cell Transfers

This approach involves simple injection or implantation of cells previously manipulated *ex vivo*, in the absence of matrices. The cells, whether autologous, allogeneic, or xenogeneic, may be altered by genetic, chemical, mechanical, or electrical stimuli prior to delivery. The aim of all cell transfers is to facilitate long-term cell engraftment, with the subsequent formation of new functional tissue, typically through the migration of host stroma and blood vessels. Perhaps the best and most successful example of this methodology is hematopoietic stem cell transplantation, which was first performed in humans in the 1950s, has long been well established clinically, and will not be further explored here. Other, more recent cell transfer–based therapies are briefly examined next.

## Knee Cartilage Repair

Full-thickness injuries to articular joint cartilage often do not heal, at least in part due to their avascular nature. Various orthopedic procedures, including removal of damaged meniscal tissue, microfracture, and articular resurfacing, are performed to treat these often-debilitating problems. While these techniques can be effective in terms of restoring a smooth articular surface, the functional results can be suboptimal because none of these methods induces regeneration of resilient and durable articular cartilage.

In 1997, Carticel®, an expanded chondrocyte preparation produced by Genzyme Biosurgery, became the first commercially available tissue-engineered product for the management of cartilage defects. To date, over 10,000

| Table 78.1. | Selected tissue-engineered products currently approved for human use or in clinical trials | |
|---|---|---|
| *Product/company* | *Website* | *Target disease* |
| Alloderm® | www.lifecell.com | Soft-tissue repair |
| Apligraf™ | www.apligraf.com | Diabetic/venous ulcers |
| Organogenesis | | |
| Carticel® | www.carticel.com | Knee cartilage defects |
| Genzyme Biosurgery | | |
| ELAD® | www.vitaltherapies.com | Acute hepatic failure |
| Vitagen | | |
| Encapsulated islets | www.amcyte.com | Diabetes (type I) |
| Amcyte | | |
| Epicel® | www.genzymebiosurgery.com | Burns |
| Genzyme Biosurgery | | |
| Grafton® DBM | www.osteotech.com | Bone nonunions |
| Osteotech | | |
| HepatAssist-2® | www.arbios.com | Acute hepatic failure |
| Abios Systems | | |
| Integra® | www.intregra-ls.com | Burns |
| Integra Life Sciences | | |
| Myocell® | www.bioheartinc.com | Ischemic heart disease |
| Bioheart, Inc. | | |
| Neural stem cells | www.neurogeneration.com | Parkinson's disease |
| NeuroGeneration | | |
| Provacel® | www.orisistx.com | Ischemic heart disease |
| Orisis Therapeutics | | |
| RAD | www.nephrostherapeutics.com | Acute renal failure |
| RenaMed Biologics | | |
| SIS® | www.cookbiotech.com | Soft-tissue repair |
| Cook Biotech | | |
| Tissue Repair Cell | www.aastrom.com | Bone nonunions |
| Aastrom Biosciences | | |

patients have been treated with it worldwide. From a small cartilage biopsy obtained from a less weight-bearing area, approximately 12 million chondrocytes are grown *in vitro* over a five-week time period. The cells are then injected into the area of articular cartilage damage beneath a periosteal flap at a density of 1.6 million cells/cm$^2$.

This autologous chondrocyte therapy is currently approved for use in selected patients with localized traumatic injury to the femoral condyle. The procedure can be particularly useful for the treatment of large single defects of up to 15 cm$^2$ or for the repair of multiple small defects. The original FDA approval of Carticel® was based on promising observational data in patients with full-thickness cartilage defects of the knee. In the seminal case series published by a Swedish group in 1994, 14 of the 16 patients with severe cartilage defects of the knee had good-to-excellent functional results (Brittberg *et al.*, 1994).

Nonetheless, in spite of the positive results using this product since 1997, the actual effectiveness and role of this therapy in the management of knee cartilage defects have remained somewhat unclear. In more recent, controlled studies, Carticel® has not been shown to be significantly better when compared with alternative approaches for joint-cartilage resurfacing (Bentley *et al.*, 2003; Horas *et al.*, 2003; Knutsen *et al.*, 2004). In one of these studies, autologous chondrocyte transplantation has demonstrated some good short-term results, as judged histologically and by the Short Form-36 score, but was shown not to be significantly different from microfracture after 24-month follow-up (Knutsen *et al.*, 2004). Clearly, long-term data will be required to determine the ultimate value of autologous chondrocyte transplantation among the various other treatments for traumatic knee injuries.

## Cellular Cardiomyoplasty

Every year, approximately 1.3 million individuals suffer a nonfatal myocardial infarction in the United States. Unfortunately, the myocardium cannot repair itself after these episodes, seemingly because of the absence of sufficient myocardial stem cells. The end result is therefore fibrotic scarring, resulting in a decrease in the ventricular ejection fraction and overall myocardial performance.

Recently, the local implantation of autologous skeletal muscle satellite cells or bone marrow–derived stem cells, a procedure now known as *cellular cardiomyoplasty*, has been evaluated as an experimental therapy in patients with ischemic myocardium. To date, over 200 patients have received this therapy worldwide. This approach aims to prevent the deleterious fibrotic remodeling of the injured myocardium by populating the damaged area with myogenic precursor cells. Support for this hypothesis has been based largely on animal data, which have shown that implanted cells can differentiate into multinucleated myocytelike cells, resulting in improved global ventricular function without requiring electromechanical coupling between the implanted cells and host cardiomyocytes (Murry *et al.*, 2005).

At this time, the clinical feasibility of the cell transfer approach in the management of ischemic heart disease has only been described in multiple descriptive pilot studies (Murry *et al.*, 2005). It has been shown that hundreds of millions of myoblasts can be grown from a small muscle biopsy under GMP conditions using either autologous or fetal bovine serum. Several uncontrolled studies have demonstrated modest improvements in ejection fraction and regional wall activity after skeletal myoblast transplantation (Herreros *et al.*, 2003; Menasche *et al.*, 2003; Siminiak *et al.*, 2004). At least two small randomized trials using autologous bone marrow mesenchymal cells have also demonstrated improvements in cardiac performance, including increased myocardial fluorodeoxyglucose uptake, enhanced wall motion, a reduction in ventricular end-systolic and end-diastolic volumes, and a net increase of 14% in ejection fraction when compared with a saline-infused control group (Chen *et al.*, 2004; Murry *et al.*, 2005).

The mechanisms for the observed benefits after clinical cellular cardiomyoplasty remain unknown. Because cell survival after the procedure appears to be quite low, proponents have speculated that the improved myocardial indices may be secondary to other factors, including increased angiogenesis, minimization of deleterious ventricular remodeling, and/or enhanced cytokine-mediated resident cell survival. Regardless of the mechanisms involved, additional studies are clearly needed in order to adequately assess the risks and benefits of cellular cardiomyoplasty. Whether or not direct myocardial injection of cells may also be associated with an increased risk for malignant arrhythmias warrants further scrutiny as well.

Several products based on the principle of cellular cardiomyoplasty are currently being tested in controlled trials, both in the United States and in Europe. One such therapy, marketed as MyoCell™ (Bioheart, Inc.), involves a collaboration of over 30 centers in Europe and has enrolled over 60 patients to date. In this approach, autologous skeletal myoblasts are expanded *ex vivo* and supplied as a cell suspension in a buffered salt solution. The product is then delivered directly into the epicardium during coronary artery bypass grafting surgery. In the United States, a donor-derived bone marrow preparation, marketed as Provacel™ (Osiris Therapeutics), is currently being evaluated, with preliminary unpublished data suggesting that the product is safe when injected intravenously within seven days of a myocardial infarction.

## Ocular Resurfacing

Damage to the corneal-limbal epithelium can occur after chemical/thermal burns, Stevens–Johnson syndrome, and ocular cicatricial pemphigoid. In severe cases, the underlying bulbar conjunctival cells cannot sufficiently regenerate a normal ocular surface, leading to fibrovascular scarring and subsequent corneal opacification and blindness. Corneal transplantation from an allogeneic donor has long been an accepted therapy in selected cases. However, all transplanted patients require chronic immunosuppressive therapy in order to prevent graft rejection.

Recently, an alternative approach has emerged using transplanted epithelial sheets derived from expanded autologous cells. This technique was first reported in 1997 in two patients with complete loss of the corneal-limbal epithelium (Pellegrini *et al.*, 1997). They had near-total blindness in one eye after an alkali burn injury and underwent ocular resurfacing using autologous sheet grafts (approximately 1 $\times$ $10^5$ cells each) cultured from a 1-mm$^2$ biopsy obtained from contralateral healthy limbus. A contact lens was placed over the graft to facilitate engraftment. After more than two years of follow-up, both patients reportedly maintained dramatic improvements in visual acuity. Clinical case reports by other groups have since confirmed limbal cell transplantation as a viable approach to ocular resurfacing (Tsai and Chen, 2000). A phase I clinical trial using cells cultured under FDA guidelines has recently been initiated.

Taking this concept one step further, ocular resurfacing using autologous epithelial cells has also been described in the setting of bilateral disease. Recently, several groups have reported their clinical experience in a small series of patients using oral mucosal epithelium as a cell source for reconstruction (Nishida *et al.*, 2004; Inatomi *et al.*, 2006). The cells were obtained from autologous buccal mucosa and were cultured on mouse fibroblast feeder layers. Various carriers, including amniotic membrane and temperature-responsive culture dishes made from poly(*N*-isopropylacrylamide), were used to facilitate expansion and transfer of these cells onto the diseased corneal bed (Fig. 78.1). Thus far, mid-term results at two years postoperatively have shown improved visual outcomes in many, but not all, patients (Inatomi *et al.*, 2006).

## Dopaminergic Cell Transplantation

Parkinson's disease, a disorder caused by a selective loss of mesencephalic dopaminergic neurons within the substantia nigra, is a common and debilitating neurodegenerative disorder. Affected patients have characteristic symptoms, including slowness in voluntary movements (bradykinesia), resting tremors, muscle rigidity, and gait disturbance. Current medical therapy, including the exogenous administration of dopamine (levodopa), can be effective in many patients but in some cases is associated with numerous secondary motor complications.

One type of cell transfer, the transplantation of human fetal dopaminergic cells, has been studied in recent years as a potential treatment for Parkinson's disease. Initial observational studies suggested prolonged survival of the

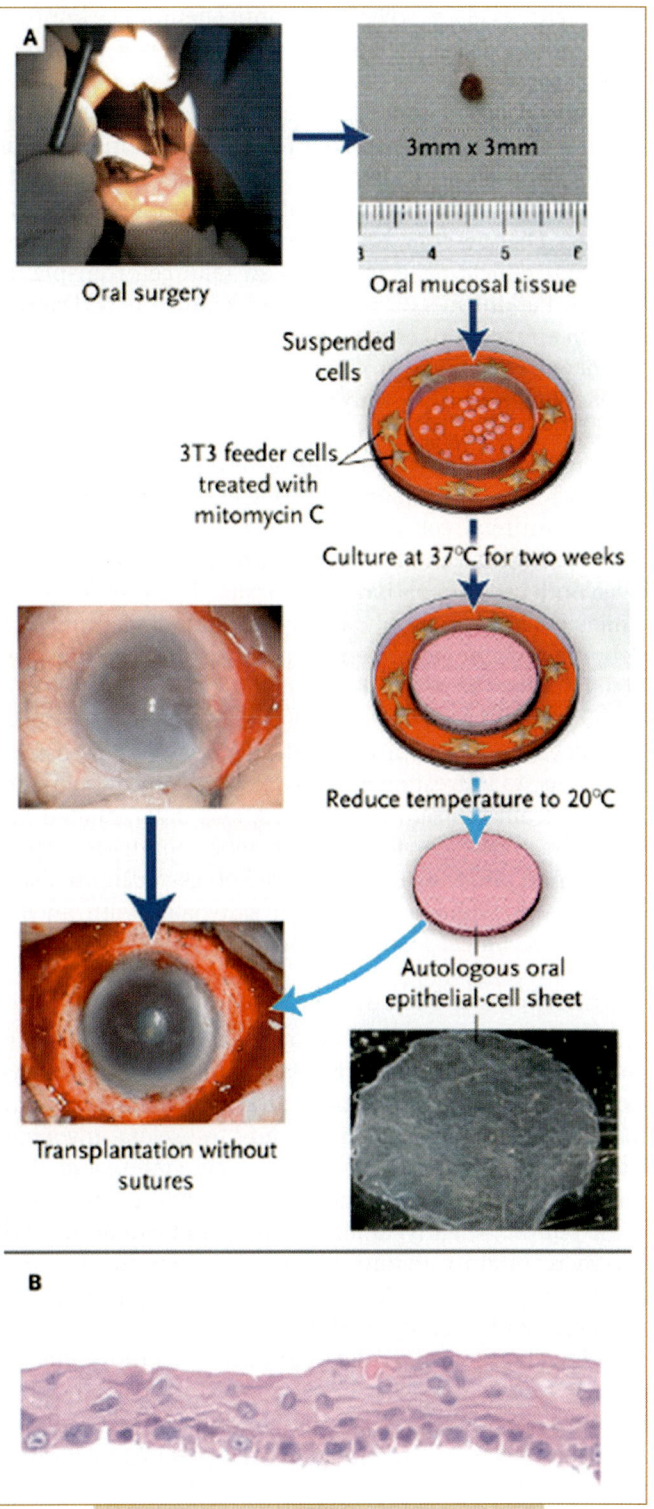

**FIG. 78.1.** Transplantation of autologous engineered cell sheets fabricated from oral mucosal epithelium. **(A)** Oral mucosa is removed, grown to form into cell sheets in culture, and transplanted directly onto the diseased eye. **(B)** Photomicrograph showing multilayered cell sheets that do not resemble the original oral mucosa. From Nishida *et al.* (2004).

transplanted cells as well as substantial neurologic improvements in a number of patients, in the absence of medication. Nevertheless, at least two published, randomized controlled trials comparing the effectiveness of human fetal cell transplantation with sham surgery demonstrated no significant differences between the two groups, except for some modest gains in a cohort of younger patients (Freed *et al.*, 2001; Olanow *et al.*, 2003). Furthermore, both the impractical and ethical aspects of human fetal cell transplantation for Parkinson's disease have precluded its widespread clinical application. Porcine fetal nigral cell transplantation has been reported to provide some benefit for Parkinson's disease in one phase I study (Fink *et al.*, 2000). However, this procedure ultimately proved no better than placebo in a subsequent, unpublished controlled trial.

Currently, a number of investigators have shifted their focus from fetal cell transplantation to other cell transfer approaches that employ either embryonic or neural stem cells. Both cell sources have a high capacity for self-renewal, which could conceivably supply an abundant number of neurons for the treatment of Parkinson's disease. One biotechnology firm, NeuroGeneration, has recently initiated some early-phase clinical trials using autologous neural stem cells. In their approach, a needle biopsy of the brain is performed to harvest neural cells. Neural stem cells are then isolated from the sample, cultured, and differentiated into dopaminergic neurons over a six- to nine-month period prior to injection into the putamen. Based on as-yet-unpublished studies, this technique has been associated with an 83% increase in clinical scores after one-year follow-up.

## Myoblast Transfer Therapy

Duchenne muscular dystrophy (DMD) is an X-linked genetic disorder caused by a defect in the encoding for dystrophin, a muscle-fiber-stabilizing protein. The disease results in chronic mechanical injury to skeletal myocytes, leading to a vicious cycle of myocyte degradation and fibrosis. By early adolescence, most affected children are wheelchair-bound; by early adulthood, most individuals develop severe respiratory failure and cardiomyopathy. Unfortunately, in spite of significant advances in our understanding of the pathophysiology of DMD, its management continues to be largely supportive, and there has yet to be any effective therapy for this disease.

One experimental approach to the treatment of DMD has been the transplantation of muscle precursor cells from normal donors into dystrophic host muscle, a procedure known as *myoblast transfer therapy* (MTT). The goal of MTT is to enable relocalization and expression of normal functional dystrophin transcripts within the host muscle. This approach was originally shown to have promising results in animal models using *mdx* mice. Clinically, there was also evidence of dystrophin transcript expression in DMD patients by reverse-transcriptase polymerase chain reaction (Law *et al.*, 1992).

Unfortunately, multiple clinical trials to date have not shown any objective benefit in DMD patients injected with donor myoblasts. In one of the initial human studies, dystrophin-positive fibers made up to 36% of the injected muscles after one month (Gussoni *et al.*, 1992). However, the levels of expression have generally been undetectable by six months postinjection. In a subsequent American study employing serial injections over a six-month time period, normal myoblasts harvested from the skeletal muscle of unaffected relatives also did not show any clinical benefit when transplanted into the skeletal muscles of affected boys (Mendell *et al.*, 1995).

The disappointing clinical results to date have likely to do with multiple factors, including poor cell survival, immune rejection, and limited cell distribution after injection. To our knowledge, no clinical trials of MTT are currently ongoing. Nonetheless, the past several years have witnessed a renaissance in preclinical efforts to enable MTT. These studies are attempting to address some of the shortcomings of the previous human studies, using better immunosuppressive regimens and alternative stem/progenitor cell sources (e.g., muscle-derived stem cells), among other approaches (Urish *et al.*, 2005).

## Other Cell-Transfer Applications

Other clinical applications of cell transfers have also been reported. Examples include either eminently anecdotal experiences, such as the intraportal injection of gene-transfected hepatocytes previously isolated from the recipient's own liver, as a means to reduce cholesterol levels in homozygous familial hypercholesterolemia, or better-established methods, employed in larger patient populations, such as pancreatic islet transplantation for type I diabetes. Recently, a Canadian group has reported an approximately 85% insulin-free rate at long-term follow-up after pancreatic islet cell transplantation (Shapiro *et al.*, 2000). Unfortunately, these clinical results have been difficult to replicate elsewhere, and more widespread application of this procedure continues to be hampered by difficulties in harvesting large numbers of islets and by the inability to expand these isolated cells *ex vivo* without senescence or dedifferentiation. Moreover, the procedure requires immunosuppressive therapy. In recent years, enthusiasm for a xenogeneic approach using porcine islet cells has dwindled somewhat because of increasing concerns regarding the introduction of porcine endogenous retroviruses (PERV) into humans. Further discussion of these other applications of cell transfer–based therapies are not within the scope of this chapter.

## **Tissue-Inducing Substances**

Tissue-inducing substances are injectable or implantable materials, devoid of any cells, that can stimulate, modulate, or guide the formation of new tissue *in situ*. This can be achieved via different mechanisms. These substances may provide the appropriate substrata for host cells to reorganize and develop into three-dimensional tissue. They may maximize gas exchange as well as nutrient and waste diffusion. Additionally, they may boast bioactive signals, such as cell-adhesion peptides and growth factors, which can be loaded or embedded within these materials to enhance tissue remodeling.

One of the major advantages of this approach has been its practicality as a biological substitute. Unlike cell-based therapies, they can be applied without extensive training. Moreover, they have enjoyed greater appeal among clinicians due to their true off-the-shelf availability. To date, a number of commercially available tissue-inducing substances have been employed across many surgical disciplines, most of them directed at skin or bone repair.

## Dermal Substitutes

Full-thickness burn injuries destroy both the epidermis and the dermis, leaving the skin unable to regenerate on its own. The use of meshed, split-thickness autologous skin grafts (autografting) has been the traditional method for covering full-thickness burn injuries. Unfortunately, this technique is also accompanied by high donor site morbidity and can produce undesirable cosmetic and functional results secondary to hypertrophic scarring. When full-thickness burn injuries are extensive (e.g., greater than 50% of total body surface area), achieving adequate burn wound coverage can be considerably more challenging because of the lack of autogolous donor sites. Although alternative materials for burn wound coverage, including allograft, xenograft, and synthetic dressings, can be used, these products generally lead to suboptimal outcomes.

The Integra® dermal regeneration template (Integra Life Sciences) is one of many off-the-shelf acellular products currently available for skin regeneration. It was originally described by Yannis and Burke over 20 years ago (Burke *et al.*, 1981) and remains the most widely accepted engineered product for patients with large burns. Integra has a bilaminar structure, consisting of cross-linked bovine collagen and chondroitin sulfate on the lower surface and an upper silicone membrane that serves an epidermal function. The pore size of the lower layer has been designed specifically to provide a template for the regeneration of dermis, as opposed to scar, by facilitating the controlled migration of host endothelium and fibroblasts. Once the neodermis is established (usually after three to six weeks), the silicone layer is removed, and an ultrathin (e.g., 0.15 mm) split-thickness skin graft can be applied.

Integra has been one of the best-studied tissue-engineered products to date. In a large multicenter, randomized trial, burn wounds treated with Integra healed faster and had less subjective hypertrophic scarring when compared to other materials (Heimbach *et al.*, 1988). Engraftment rates consistently range between 80% and 90%, and donor site morbidity is decreased, in part because of

faster healing at the donor sites. The long-term results also tend to be quite favorable (Heimbach *et al.*, 2003). Almost 10,000 patients have received this product for the management of burns worldwide, up to now.

## Demineralized Bone

The gold standard for reconstruction of major bone defects resulting from trauma, tumors, or infection is autologous cancellous bone. However, donor site complications, such as infection, hematoma, severe pain, and iliac wing fracture, are known to occur with some frequency. In addition, not infrequently there is a limited supply of autologous tissue, particularly in geriatric and pediatric patients. Cadaveric allografts can be used to circumvent some of the problems associated with autografts, but allograft supplies continue to be outstripped by the ever-growing demand for them. Moreover, concerns remain about possible disease transmission with allograft use. These limitations of conventional bone reconstruction techniques have spurred an ongoing interest in alternative tissue-inducing matrices, commonly referred to as *osteobiologics*. These materials are essentially graft substitutes with osteoconductive and osteoinductive properties that enable them to induce new bone formation.

Over the past decade, demineralized bone matrix (DBM) has been one of the most commonly employed osteobiologic materials. DBM is produced by a special demineralization processing of human cadaveric bone using hydrochloric acid. The resultant matrix is composed of 90% type I collagen and 10% noncollagenous proteins. DBM is also rich in bone morphogenic proteins (BMPs) and other osteogenic factors. Currently over half a dozen commercially available DBM products are on the market in the United States. A variety of formulations, including a powder and an injectable substrate, have been used clinically. DBM is currently used in combination with autologous bone or bone marrow, and it is approved for use in a number of settings in periodontal and orthopedic surgery. Although there appears to be some variability in the osteoinductive capacities of DBM, depending on its processing method and donor source, this material has been shown to facilitate bone healing when implanted into a fracture, arthrodesis, and other contained bone defects.

Still, in spite of its fairly widespread clinical acceptance, there has been a striking paucity of well-designed clinical studies documenting its effectiveness when compared to other reconstructive approaches. In one of the few prospective trials to date, a combination of DBM and an autograft led to similar healing rates after spinal fusions when compared to autografts alone (Cammisa *et al.*, 2004). Controlled data in other reconstructive settings are lacking. Further developments in osteoinduction driven by native insoluble bone morphogenetic protein and noncollagenous protein aggregates remain a continuing interest in the field.

## Closed Systems

In closed systems, construct cells are sequestered from the host, usually via a semipermeable membrane that allows for the bidirectional transport of nutrients and wastes. By manipulating the membrane pores, closed systems can be specially designed to prevent the diffusion of certain macromolecules, such as antibodies and complement, as well as lymphocytes, thereby preventing them from destroying the cells within the construct at the same time that beneficial/therapeutic donor cell products can be released into the host.

Closed systems can be subdivided into intracorporeal and extracorporeal. In intracorporeal systems, the transplanted cells are isolated from the host by either microencapsulation or macroencapsulation technologies. Microcapsules generally offer better flow of nutrients and wastes than macrocapsules and can be administered by injection. Macrocapsules, on the other hand, are physically more stable and easier to retrieve, should complications occur. In extracorporeal systems, the patient's blood or plasma is pumped *ex vivo* into a device housing living cells, which are separated from the blood/plasma by a semipermeable membrane prior to infusion back into the patient.

### Encapsulated Pancreatic Islet Cells

For decades there has been considerable interest in alternative cell-based therapies to pancreatic islet transplantation for type I diabetes. More recently, interest in islet cell encapsulation, a concept first introduced in the 1970s, has been renewed at multiple centers. Unfortunately, consistent clinical success with this approach has proven to be a formidable task, partly because of biocompatibility issues with the encapsulation membrane (Scharp *et al.*, 1994; Soon-Shiong *et al.*, 1994). Nevertheless, some progress has been made. For example, a recent uncontrolled trial using encapsulated porcine islets has suggested a decrease in insulin requirements in a subset of patients with diabetes (Valdes-Gonzalez *et al.*, 2005). In addition, at least one prospective phase I/II clinical trial is currently under way in Canada evaluating the effectiveness and safety of intraperitoneal injections of alginate-encapsulated human islet cells and low-dose, short-term immunosuppression in diabetic patients. Should these methods eventually be validated, they could have a major impact on the treatment for this disease.

### Bioartificial Liver

Acute liver failure can stem from a number of conditions and has been associated with mortality rates ranging from 60–90%. Its treatment remains largely supportive, with the only definitive therapy still being liver transplantation. Due to the chronic shortage of donor livers, a considerable number of patients die while waiting for an available organ.

A device that could reliably serve as either a bridge to liver transplantation, or as a temporary therapy until the native liver function is recovered, would represent an important breakthrough in this setting.

Recent clinical experience has been reported with a number of extracorporeal liver-support devices. For example, the ELAD® (Extracorporeal Liver Assist Device, Vitagen) uses a dual-pump dialysis system and hollow-fiber cartridges containing C3A cells derived from a human hepatoblastoma cell line. Plasma is ultrafiltrated through the fibers into the extracapillary space of the cartridge, where it comes in direct contact with the hepatocytes. Although the ELAD has been shown to be reasonably safe and to lead to improvements in clinical as well as biochemical parameters, no survival benefit was observed in an initial controlled study (Millis *et al.*, 2002). Subsequent refinements in the design of the apparatus have prompted its re-evaluation in currently ongoing clinical trials.

Another liver-support system, the HepatAssist® (Circe Biomedical), has also been tested in clinical trials. In contrast to the ELAD, the HepatAssist uses porcine hepatocytes attached to collagen-coated microcarriers (Fig. 78.2). Similar to the experience with the ELAD, however, unfortunately the original data from a large, randomized controlled trial

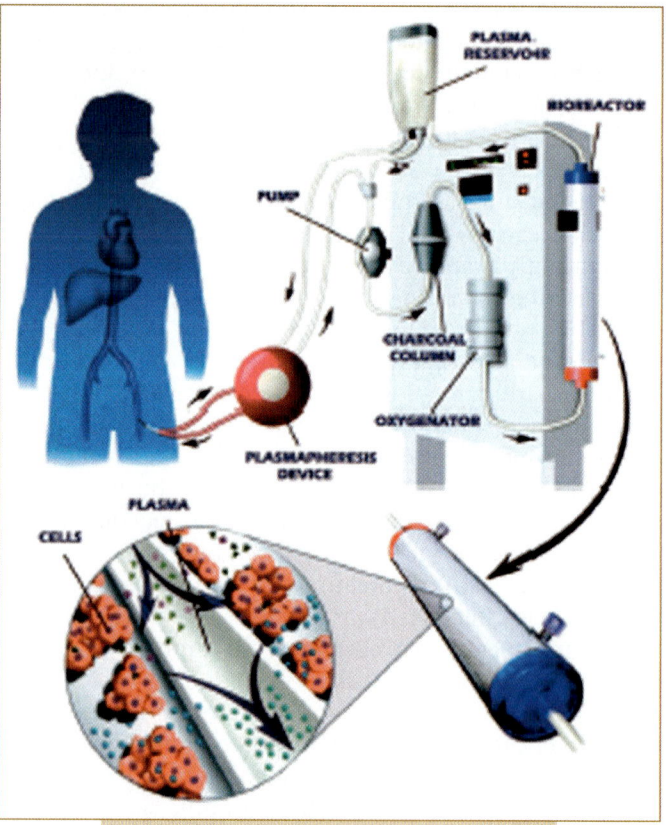

**FIG. 78.2.** Schematic outline of the HepatAssist® device loaded with microcarrier-attached porcine hepatocytes. From Demetriou *et al.* (2004).

did not show an increased survival in the majority of patients exposed to this device. Although a small survival benefit was subsequently reported in a subgroup of patients with fulminant/subfulminant hepatic failure (Demetriou et al., 2004), given the overall modest results, this device has yet to be approved for clinical use by the FDA. A newer-generation device, termed the HepatAssist-2® (Abios Systems), contains a number of design upgrades, including a threefold increase in hepatocyte mass. It is currently being tested in phase III clinical trials.

### Bioartificial Kidney

Renal failure is currently managed by hemodialysis, peritoneal dialysis, hemofiltration, or kidney transplantation. The roles of these therapies in providing adequate fluid clearance and in removing small and middle-sized solutes from the blood in patients with chronic renal failure are well established. Unfortunately, these treatment modalities have been shown to be less effective in critically ill patients with acute renal failure (ARF), with mortality rates in excess of 70%.

The bioengineered kidney has been proposed as a more effective therapy for the management of ARF. Since the death of most ARF patients is associated with a systemic inflammatory response syndrome, it has been postulated that a more complete device, offering filtration, metabolic, endocrine, and immunoregulatory functions similar to those performed by the native kidney, might lead to improved outcomes. To test this hypothesis, a bioengineered renal replacement system has been developed and tested in early phase trials by Humes and his colleagues at the University of Michigan. This renal tubule–assist device (RAD) consists of a conventional high-flux hemofilter followed in series by human renal tubular cells that have been grown to confluence along the inner surface of hollow fibers. In a small, uncontrolled clinical study in critically ill patients, these devices were used for up to 24 hours and demonstrated increased survival as well as differentiated metabolic and endocrine activity, as shown by glutathione degradation and the conversion of 25-OH-$D_3$ to 1,25-$(OH)_2$-$D_3$, respectively (Humes et al., 2004). Based on these data, a randomized, phase II trial, initiated in conjunction with RenaMed Biologics, is currently ongoing for patients with ARF secondary to acute tubular necrosis.

### Open Systems

In open systems, cells are seeded onto scaffolds that help guide their organization into the appropriate architecture (J. P. Vacanti and Langer, 1999). Also here, clinical experience has been largely anectodal. For instance, engineered bone based on a natural coral scaffold has been successfully implanted in a patient who sustained a traumatic left thumb avulsion (C. A. Vacanti et al., 2001). In another report, the native pulmonary valve was replaced with a decellularized pulmonary artery allograft that had been seeded with autol-

ogous vascular endothelial cells (Dohmen et al., 2002). Other investigators have recently described case reports ranging from the engineering of bone for craniofacial reconstruction to the creation of engineered respiratory airway tissue for a bronchopleural fistula (Macchiarini et al., 2004; Warnke et al., 2004). A case series from Italy has reported on the efficacy of a hyaluronan-based scaffold seeded with autologous chondrocytes for the repair of knee cartilage defects (Marcacci et al., 2005). Many isolated cases also remain unpublished (e.g., of chest wall repair using autologous chondrocytes). Albeit not yet disclosed, early reports on other applications are expected, such as engineered blood vessels for renal dialysis access, bone marrow–based constructs for bony nonunions, and different engineered constructs for the treatment of various urological diseases.

A few applications of open-system tissue engineering, however, have involved more sizable patient populations, even though the relevance of that experience has often been curbed by trial design limitations.

### Composite Skin Grafts

Foot ulcers are common complications of diabetes. Although most of these wounds eventually resolve under conventional wound management, many do not heal within three weeks. In some of these cases, serious infections can arise, and amputation may be the only option for definitive management. In an attempt to improve this scenario, a number of engineered skin constructs have been introduced clinically in recent years, as an adjunct to conventional wound management.

At this time, Apligraf™ (Organogenesis), which consists of human allogeneic neonatal keratinocytes and fibroblasts within a bovine dermal matrix, is one of the most widely used of these constructs. Its application in diabetic ulcer wounds has been validated in a randomized clinical trial (Veves et al., 2001). In that study, Apligraf™ plus optimal wound care demonstrated statistically faster rates of wound healing when compared to optimal wound care alone. Despite the allogeneic nature of the product, no signs of clinical rejection were observed. Apligraf™ is also approved for the management of long-standing venous stasis ulcers (Falanga et al., 1998). Experiments on the effectiveness of Apligraf™ in other settings, such as in the management of acute surgical wounds, burn wounds, and select severe congenital skin disorders, are currently under way.

At the same time, several important questions remain. For example, it is not yet known whether patients with a high risk for a poor outcome (e.g., due to the size and duration of the ulcer) might benefit from an earlier application of these products. In addition, the long-term benefit of a composite allogeneic skin construct has been questioned by some investigators, for at least one study has shown a lack of significant lasting engraftment of the Apligraf™ derived cells (Griffiths et al., 2004). These findings suggest that this product may essentially act more as a temporary (i.e.,

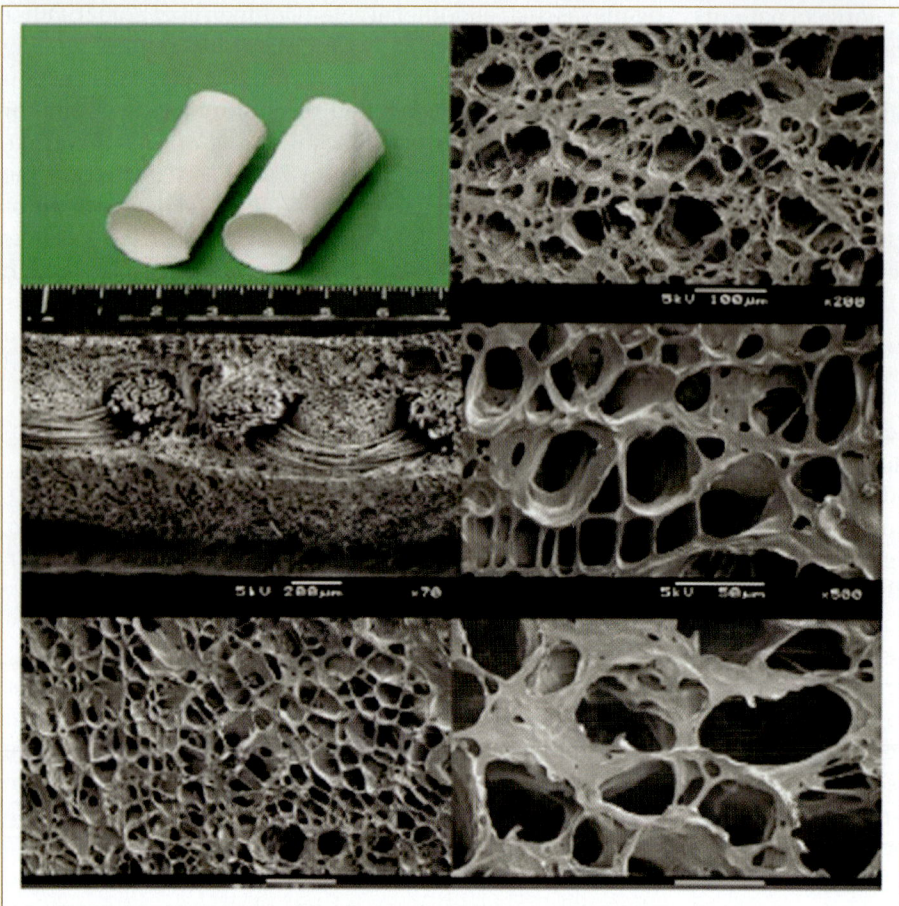

**FIG. 78.3.** Macroscopic and scanning electro-microscopic appearance of biodegradable scaffolds used for cardiovascular tissue–engineered conduits. The scaffolds are based on a copolymer of lactide acid and polycaprolactone. From Shin'oka *et al.* (2005).

four-weeks' duration) biological dressing, rather than as a bona fide permanent skin substitute.

## Pediatric Cardiovascular Conduits

An engineered vascular conduit would be highly desirable for infants and children born with different forms of severe cardiovascular anomalies. Currently available homografts and synthetic prostheses are hampered by numerous problems, including thrombosis, limited durability, susceptibility to infection, and calcification. In addition, because most of these conduits essentially cannot grow, the need for one or more reoperative procedures later in life is the rule.

In recent years, a Japanese group led by Shin'oka has amassed-substantial clinical experience with tissue-engineered conduits as vascular replacements in low-pressure systems. In 2001, they described their initial case report, a 4-year-old girl who developed total occlusion of the right intermediate pulmonary artery after a pulmonary artery angioplasty and a Fontan procedure and subsequently underwent reconstruction with an autologous vessel graft engineered from cells obtained from a peripheral vein (Shin'oka *et al.*, 2001). The construct remained patent for over seven months after implantation. That group has since reported their experience using a similar approach in over

40 children with varying forms of complex congenital anomalies (Fig. 78.3) (Matsumura *et al.*, 2003). Current methods are focusing on the use of uncultured, autologous bone marrow as the cell source for seeding scaffold tubes based on a copolymer of polylactic–polycaprolactone reinforced with woven polyglycolic acid (Hibino *et al.*, 2005; Shin'oka *et al.*, 2005).

Thus far, the midterm outcomes of engineered vascular grafts in children have shown no significant complications, as well as continued patency when followed up to a median time of 16.7 months postoperatively (Inatomi *et al.*, 2006). However, balloon angioplasty has been required in some cases because of tissue overgrowth at anastomotic sites. The long-term growth, remodeling, and durability patterns of these grafts remain unknown. Most importantly, there have been no well-controlled human trials comparing tissue-engineered conduits to those made only from synthetic materials. Without these data, the ultimate role of tissue engineering in this setting will justifiably continue to be questioned.

## IV. CONCLUSION

Tissue engineering may be at the inflection point of a steep upward slope on its developmental curve. As evidenced in many examples discussed in this chapter and others, concrete clinical benefits from this technology have accumulated in the past several years, albeit still at a much slower pace than what we can realistically expect for the future. Although the widespread availability of off-the-shelf tissues and organs has yet to become a reality, the field is both exceedingly young and vibrant. Much remains to be learned and developed, not only by scientists, clinicians, and entrepreneurs, but also by regulatory governmental agencies. Nevertheless, given its scientific premises, the potential magnitude of its impact to society, and what has been achieved thus far, many believe that it should only be a matter of time until tissue engineering reaches the mainstream of medical practice.

# V. REFERENCES

Ahsan, T., and Nerem, R. M. (2005). Bioengineered tissues: the science, the technology, and the industry. *Orthod. Craniofac. Res.* **8**, 134–140.

Bentley, G., Biant, L. C., Carrington, R. W., Akmal, M., Goldberg, A., Williams, A. M., Skinner, J. A., and Pringle, J. (2003). A prospective, randomized comparison of autologous chondrocyte implantation versus mosaicplasty for osteochondral defects in the knee. *J. Bone Joint Surg. Br.* **85**, 223–230.

Brittberg, M., Lindahl, A., Nilsson, A., Ohlsson, C., Isaksson, O., and Peterson, L. (1994). Treatment of deep cartilage defects in the knee with autologous chondrocyte transplantation. *N. Engl. J. Med.* **331**, 889–895.

Burke, J. F., Yannas, I. V., Quinby, W. C., Jr., Bondoc, C. C., and Jung, W. K. (1981). Successful use of a physiologically acceptable artificial skin in the treatment of extensive burn injury. *Ann. Surg.* **194**, 413–428.

Cammisa, F. P., Jr., Lowery, G., Garfin, S. R., Geisler, F. H., Klara, P. M., McGuire, R. A., Sassard, W. R., Stubbs, H., and Block, J. E. (2004). Two-year fusion rate equivalency between Grafton DBM gel and autograft in posterolateral spine fusion: a prospective controlled trial employing a side-by-side comparison in the same patient. *Spine* **29**, 660–666.

Chen, S. L., Fang, W. W., Ye, F., Liu, Y. H., Qian, J., Shan, S. J., Zhang, J. J., Chunhua, R. Z., Liao, L. M., Lin, S., *et al.* (2004). Effect on left ventricular function of intracoronary transplantation of autologous bone marrow mesenchymal stem cell in patients with acute myocardial infarction. *Am. J. Cardiol.* **94**, 92–95.

Demetriou, A. A., Brown, R. S., Jr., Busuttil, R. W., Fair, J., McGuire, B. M., Rosenthal, P., Am Esch, J. S., 2nd, Lerut, J., Nyberg, S. L., Salizzoni, M., *et al.* (2004). Prospective, randomized, multicenter, controlled trial of a bioartificial liver in treating acute liver failure. *Ann. Surg.* **239**, 660–667; discussion 667–670.

Dohmen, P. M., Lembcke, A., Hotz, H., Kivelitz, D., and Konertz, W. F. (2002). Ross operation with a tissue-engineered heart valve. *Ann. Thorac. Surg.* **74**, 1438–1442.

Falanga, V., Margolis, D., Alvarez, O., Auletta, M., Maggiacomo, F., Altman, M., Jensen, J., Sabolinski, M., and Hardin-Young, J. (1998). Rapid healing of venous ulcers and lack of clinical rejection with an allogeneic cultured human skin equivalent. Human Skin Equivalent Investigators Group. *Arch Dermatol.* **134**, 293–300.

Fauza, D. O. (2003). Tissue engineering: current state of clinical application. *Curr. Opin. Pediatr.* **15**, 267–271.

Fidkowski, C., Kaazempur-Mofrad, M. R., Borenstein, J., Vacanti, J. P., Langer, R., and Wang, Y. (2005). Endothelialized microvasculature based on a biodegradable elastomer. *Tissue Eng.* **11**, 302–309.

Fink, J. S., Schumacher, J. M., Ellias, S. L., Palmer, E. P., Saint-Hilaire, M., Shannon, K., Penn, R., Starr, P., VanHorne, C., Kott, H. S., *et al.* (2000). Porcine xenografts in Parkinson's disease and Huntington's disease patients: preliminary results. *Cell Transplant.* **9**, 273–278.

Freed, C. R., Greene, P. E., Breeze, R. E., Tsai, W. Y., DuMouchel, W., Kao, R., Dillon, S., Winfield, H., Culver, S., Trojanowski, J. Q., *et al.* (2001). Transplantation of embryonic dopamine neurons for severe Parkinson's disease. *N. Engl. J. Med.* **344**, 710–719.

Griffith, L. G., and Naughton, G. (2002). Tissue engineering — current challenges and expanding opportunities. *Science* **295**, 1009–1014.

Griffiths, M., Ojeh, N., Livingstone, R., Price, R., and Navsaria, H. (2004). Survival of Apligraf in acute human wounds. *Tissue Eng.* **10**, 1180–1195.

Gussoni, E., Pavlath, G. K., Lanctot, A. M., Sharma, K. R., Miller, R. G., Steinman, L., and Blau, H. M. (1992). Normal dystrophin transcripts detected in Duchenne muscular dystrophy patients after myoblast transplantation. *Nature* **356**, 435–438.

Heimbach, D., Luterman, A., Burke, J., Cram, A., Herndon, D., Hunt, J., Jordan, M., McManus, W., Solem, L., Warden, G., *et al.* (1988). Artificial dermis for major burns. A multicenter randomized clinical trial. *Ann Surg.* **208**, 313–320.

Heimbach, D. M., Warden, G. D., Luterman, A., Jordan, M. H., Ozobia, N., Ryan, C. M., Voigt, D. W., Hickerson, W. L., Saffle, J. R., DeClement, F. A., *et al.* (2003). Multicenter postapproval clinical trial of Integra dermal regeneration template for burn treatment. *J. Burn Care Rehabil.* **24**, 42–48.

Herreros, J., Prosper, F., Perez, A., Gavira, J. J., Garcia-Velloso, M. J., Barba, J., Sanchez, P. L., Canizo, C., Rabago, G., Marti-Climent, J. M., *et al.* (2003). Autologous intramyocardial injection of cultured skeletal muscle–derived stem cells in patients with nonacute myocardial infarction. *Eur. Heart J.* **24**, 2012–2020.

Hibino, N., Shin'oka, T., Matsumura, G., Ikada, Y., and Kurosawa, H. (2005). The tissue-engineered vascular graft using bone marrow without culture. *J. Thorac. Cardiovasc. Surg.* **129**, 1064–1070.

Horas, U., Pelinkovic, D., Herr, G., Aigner, T., and Schnettler, R. (2003). Autologous chondrocyte implantation and osteochondral cylinder transplantation in cartilage repair of the knee joint. A prospective, comparative trial. *J. Bone Joint Surg. Am.* **85A**, 185–192.

Humes, H. D., Weitzel, W. F., Bartlett, R. H., Swaniker, F. C., Paganini, E. P., Luderer, J. R., and Sobota, J. (2004). Initial clinical results of the bioartificial kidney containing human cells in ICU patients with acute renal failure. *Kidney Int.* **66**, 1578–1588.

Inatomi, T., Nakamura, T., Koizumi, N., Sotozono, C., Yokoi, N., and Kinoshita, S. (2006). Midterm results on ocular surface reconstruction using cultivated autologous oral mucosal epithelial transplantation. *Am J Ophthalmol* **141**, 267–275.

Knutsen, G., Engebretsen, L., Ludvigsen, T. C., Drogset, J. O., Grontvedt, T., Solheim, E., Strand, T., Roberts, S., Isaksen, V., and Johansen, O. (2004). Autologous chondrocyte implantation compared with microfracture in the knee. A randomized trial. *J. Bone Joint Surg. Am.* **86A**, 455–464.

Law, P. K., Goodwin, T. G., Fang, Q., Duggirala, V., Larkin, C., Florendo, J. A., Kirby, D. S., Deering, M. B., Li, H. J., Chen, M., *et al.* (1992). Feasibility, safety, and efficacy of myoblast transfer therapy on Duchenne muscular dystrophy boys. *Cell Transplant.* **1**, 235–244.

Lysaght, M. J., and Hazlehurst, A. L. (2004). Tissue engineering: the end of the beginning. *Tissue Eng.* **10**, 309–320.

Macchiarini, P., Walles, T., Biancosino, C., and Mertsching, H. (2004). First human transplantation of a bioengineered airway tissue. *J. Thorac. Cardiovasc. Surg.* **128**, 638–641.

Marcacci, M., Berruto, M., Brocchetta, D., Delcogliano, A., Ghinelli, D., Gobbi, A., Kon, E., Pederzini, L., Rosa, D., Sacchetti, G. L., *et al.* (2005). Articular cartilage engineering with Hyalograft C: 3-year clinical results. *Clin. Orthop. Relat. Res.* 96–105.

Matsumura, G., Hibino, N., Ikada, Y., Kurosawa, H., and Shin'oka, T. (2003). Successful application of tissue-engineered vascular autografts: clinical experience. *Biomaterials* **24**, 2303–2308.

Menasche, P., Hagege, A. A., Vilquin, J. T., Desnos, M., Abergel, E., Pouzet, B., Bel, A., Sarateanu, S., Scorsin, M., Schwartz, K., *et al.* (2003). Autologous skeletal myoblast transplantation for severe postinfarction left ventricular dysfunction. *J. Am. Coll. Cardiol.* **41**, 1078–1083.

Mendell, J. R., Kissel, J. T., Amato, A. A., King, W., Signore, L., Prior, T. W., Sahenk, Z., Benson, S., McAndrew, P. E., Rice, R., *et al.* (1995). Myoblast transfer in the treatment of Duchenne's muscular dystrophy. *N. Engl. J. Med.* **333**, 832–838.

Millis, J. M., Cronin, D. C., Johnson, R., Conjeevaram, H., Conlin, C., Trevino, S., and Maguire, P. (2002). Initial experience with the modified extracorporeal liver-assist device for patients with fulminant hepatic failure: system modifications and clinical impact. *Transplantation* **74**, 1735–1746.

Murry, C. E., Field, L. J., and Menasche, P. (2005). Cell-based cardiac repair: reflections at the 10-year point. *Circulation* **112**, 3174–3183.

Nishida, K., Yamato, M., Hayashida, Y., Watanabe, K., Yamamoto, K., Adachi, E., Nagai, S., Kikuchi, A., Maeda, N., Watanabe, H., *et al.* (2004). Corneal reconstruction with tissue-engineered cell sheets composed of autologous oral mucosal epithelium. *N. Engl. J. Med.* **351**, 1187–1196.

Olanow, C. W., Goetz, C. G., Kordower, J. H., Stoessl, A. J., Sossi, V., Brin, M. F., Shannon, K. M., Nauert, G. M., Perl, D. P., Godbold, J., *et al.* (2003). A double-blind controlled trial of bilateral fetal nigral transplantation in Parkinson's disease. *Ann. Neurol.* **54**, 403–414.

Pellegrini, G., Traverso, C. E., Franzi, A. T., Zingirian, M., Cancedda, R., and De Luca, M. (1997). Long-term restoration of damaged corneal surfaces with autologous cultivated corneal epithelium. *Lancet* **349**, 990–993.

Scharp, D. W., Swanson, C. J., Olack, B. J., Latta, P. P., Hegre, O. D., Doherty, E. J., Gentile, F. T., Flavin, K. S., Ansara, M. F., and Lacy, P. E. (1994). Protection of encapsulated human islets implanted without immunosuppression in patients with type I or type II diabetes and in nondiabetic control subjects. *Diabetes* **43**, 1167–1170.

Shapiro, A. M., Lakey, J. R., Ryan, E. A., Korbutt, G. S., Toth, E., Warnock, G. L., Kneteman, N. M., and Rajotte, R. V. (2000). Islet transplantation in seven patients with type 1 diabetes mellitus using a glucocorticoid-free immunosuppressive regimen. *N. Engl. J. Med.* **343**, 230–238.

Shin'oka, T., Imai, Y., and Ikada, Y. (2001). Transplantation of a tissue-engineered pulmonary artery. *N. Engl. J. Med.* **344**, 532–533.

Shin'oka, T., Matsumura, G., Hibino, N., Naito, Y., Watanabe, M., Konuma, T., Sakamoto, T., Nagatsu, M., and Kurosawa, H. (2005). Midterm clinical result of tissue-engineered vascular autografts seeded with autologous bone marrow cells. *J. Thorac. Cardiovasc. Surg.* **129**, 1330–1338.

Siminiak, T., Kalawski, R., Fiszer, D., Jerzykowska, O., Rzezniczak, J., Rozwadowska, N., and Kurpisz, M. (2004). Autologous skeletal myoblast transplantation for the treatment of postinfarction myocardial injury: phase I clinical study with 12 months of follow-up. *Am. Heart J.* **148**, 531–537.

Soon-Shiong, P., Heintz, R. E., Merideth, N., Yao, Q. X., Yao, Z., Zheng, T., Murphy, M., Moloney, M. K., Schmehl, M., Harris, M., *et al.* (1994). Insulin independence in a type 1 diabetic patient after encapsulated islet transplantation. *Lancet* **343**, 950–951.

Tsai, R. J., Li, L., and Chen, J. (2000). Reconstruction of damaged corneas by transplantation of autologous limbal epithelial cells(1). *Am. J. Ophthalmol.* **130**, 543.

Urish, K., Kanda, Y., and Huard, J. (2005). Initial failure in myoblast transplantation therapy has led the way toward the isolation of muscle stem cells: potential for tissue regeneration. *Curr. Top. Dev. Biol.* **68**, 263–280.

Vacanti, C. A., Bonassar, L. J., Vacanti, M. P., and Shufflebarger, J. (2001). Replacement of an avulsed phalanx with tissue-engineered bone. *N. Engl. J. Med.* **344**, 1511–1514.

Vacanti, J. P. (1988). Beyond transplantation. Third annual Samuel Jason Mixter lecture. *Arch. Surg.* **123**, 545–549.

Vacanti, J. P., and Langer, R. (1999). Tissue engineering: the design and fabrication of living replacement devices for surgical reconstruction and transplantation. *Lancet* **354**(Suppl. 1), SI32–SI34.

Valdes-Gonzalez, R. A., Dorantes, L. M., Garibay, G. N., Bracho-Blanchet, E., Mendez, A. J., Davila-Perez, R., Elliott, R. B., Teran, L., and White, D. J. (2005). Xenotransplantation of porcine neonatal islets of Langerhans and Sertoli cells: a 4-year study. *Eur. J. Endocrinol.* **153**, 419–427.

Veves, A., Falanga, V., Armstrong, D. G., and Sabolinski, M. L. (2001). Graftskin, a human skin equivalent, is effective in the management of noninfected neuropathic diabetic foot ulcers: a prospective randomized multicenter clinical trial. *Diabetes Care* **24**, 290–295.

Warnke, P. H., Springer, I. N., Wiltfang, J., Acil, Y., Eufinger, H., Wehmoller, M., Russo, P. A., Bolte, H., Sherry, E., Behrens, E., *et al.* (2004). Growth and transplantation of a custom vascularized bone graft in a man. *Lancet* **364**, 766–770.

Chapter Seventy-Nine

# Tissue-Engineered Skin Products

*Jonathan Mansbridge*

## I. INTRODUCTION

The culture of skin cells in organotypic systems has a comparatively long history, and such constructs have been seen as the simplest tissue-engineered products to fabricate. Accordingly, they are among the few tissue-engineered products that have reached the commercial market. This has brought the realization of many other factors, apart from materials and culture considerations, that are involved in the production of a successful commercial product. These include choice of scaffold cell sources, particularly the use of allogeneic cells, design of bioreactor, automation of the manufacturing process, cryopreservation, and distribution. Manufacturing takes place in a highly regulated environment, which imposes many constraints on design and strategic planning. These concepts are illustrated with the process involved in the manufacture of Dermagraft and TransCyte, taken as specific examples, including brief discussion of clinical trial experience and commercialization. Commercialization at this point involves, among other factors, satisfactory clinical trial results, demonstrating efficacy, an adequate rationale for the mechanism of action of the product, and an established reimbursement system from third-party payers. The principles covered in this discussion are summarized. To this point, therapeutic tissue-engineered products have involved fibroblasts alone or fibroblasts and keratinocytes. In the future we may expect to see products involving adnexal structures, such as hair follicles, which represent one of the few examples of organogenesis from adult cells, and the development of alternative methods of stasis preservation, such as desiccation.

Skin-tissue engineering was perceived as the simplest tissue-engineering application and was the earliest to be explored. The culture of fibroblasts, derived from skin biopsies, was established early, and the cells were comparatively easy to grow and were used to develop many of the early systems of tissue culture. The culture of keratinocytes was achieved by Rheinwald and Green in the late 1970s (Rheinwald and Green, 1975) and, using a different system, by Karasek (S. C. Liu *et al.*, 1978). Skin biology was second only to immunology in the application of the techniques of molecular and cell biology to medical problems, which has led to great advances in the understanding of the physiology of skin. Examples of methods that have been applied include the use of classical techniques of molecular and cell biology and transgenic animals.

The skin acts as a barrier between internal structures and the external environment. In the 1980s it was realized that the skin, in addition to acting as a barrier, has significant interaction with both the innate and the adaptive immune systems (Stingl *et al.*, 1978; Streilein, 1983; Stingl, 1989). It is, thus, much more than a passive barrier and actively recruits defensive mechanisms to protect against infection, colonization by microorganisms, and other external noxious entities.

With its long history of development, tissue-engineered skin substitutes have achieved market exposure and have some of the most extensive experience with the application of such products in a therapeutic setting. From its inception in 1979 (Bell *et al.*, 1979), engineered skin was seen as potentially a clinical product, and trials started in the 1980s. Preclinical testing was inaugurated and demonstrated that fibroblasts were apparently not rejected and persisted for a considerable period in experimental animals (Hull *et al.*, 1981, 1983; Sher *et al.*, 1983). This led to human clinical trials in the late 1980s and early 1990s. Tissue-engineered skin products first gained regulatory approval and appeared on the market in 1997. The first was TransCyte (Advanced Tissue Sciences), which was a nonviable burn product produced by tissue-engineering techniques. It was followed by Apligraf (Organogenesis) in 1998, the first tissue-engineered, viable, organotypic product, and Dermagraft in 2000. At the same time, many applications of such products for *in vitro* testing applications and models of skin physiology were developed.

Much of the discussion in this chapter is based on experience with Dermagraft and TransCyte. Both products have a similar basis in the growth of fibroblasts on three-dimensional scaffolds. The systems, however, differ substantially in detail. Dermagraft is grown on 2-inch × 3-inch sheets of a knitted polylactide/glycolide scaffold under static conditions in a bag bioreactor. Following seeding, the cultures are refed with medium every few days until harvest after about two weeks. Initially, the fibroblasts proliferate rapidly, much as in monolayer, but as they become confluent, at about eight to ten days, they lay down increasing amounts of extracellular matrix. At harvest, medium is replaced with cryoprotectant and the bioreactor is welded closed without exposure of the product to the environment, boxed, and frozen under controlled conditions. Dermagraft is stored below –65°C for up to six months. During the first month, sterility and analytical testing is performed and the paperwork completed to allow release of the product by the Quality Assurance Department. On thawing, which is performed by the end user, the cells of Dermagraft show 50–80% viability.

TransCyte is grown, using a similar process, on two 5-inch by 7-inch sheets of a knitted nylon (nondegradable) scaffold with a silastic backing in a continuous-flow, hard bioreactor. This system has advantages in the ease of taking medium samples and in slightly superior cell growth, but it adds considerably to the complexity of the system. At harvest, the cultures are rinsed and the bioreactors are sealed from the system by welding, packaged, and frozen without any care taken to maintain viability.

The three-dimensional structures formed by the cells under these conditions consist of cells embedded in extracellular matrix that they, themselves, have secreted. It comprises a complex series of molecules. The process of secretion may be thought of as formation of a foreign body capsule *in vitro*.

## II. TYPES OF THERAPEUTIC TISSUE-ENGINEERED SKIN PRODUCTS

The development of tissue-engineered skin has taken two fundamentally different approaches using alternative methods for the three-dimensional culture of fibroblasts and a third that has intermediate characteristics. One is the use of fibroblasts suspended in collagen gel, in which the gel acts as a substrate for the growth of keratinocytes and forms a structure that has some resemblance to human skin. The second involves culture on a three-dimensional nonbiologically derived, polymeric scaffold. Fibroblasts are responsive to signals from their environment and respond in quite different ways to the two types of culture. In a third system, fibroblasts are grown on a collagen sponge in serum containing medium. Since collagen adsorbs fibronectin and vitronectin, this system initially shows cell adhesion to collagen, fibronectin, and vitronectin, and the cultures show intermediate properties.

In collagen gel suspension, fibroblasts are surrounded by collagen, and respond through $\alpha_1\beta_1$ and $\alpha_2\beta_1$ integrins, which are collagen receptors, by becoming quiescent and nonproliferative (Langholz *et al.*, 1995). Characteristically, the collagen gel is contracted by the fibroblasts to less than 10% of its original volume in about 20 h, through the $\alpha_2\beta_1$ integrin. At the same time, many genes, notably those for collagen type I and proliferation, are repressed (Eckes *et al.*, 1993), and the cells show mechanosensitivity (Kessler *et al.*, 2001). The collagen repression is mediated through the $\alpha_1\beta_1$ integrin (Langholz *et al.*, 1995). If the gels are mechanically stressed, usually by preventing contraction with a ring, many genes, including collagen type I and proliferation-related genes, are somewhat induced relative to contracting gel fibroblasts, although not the level of monolayer fibroblasts (Kessler *et al.*, 2001). The cells also show a substantial up-regulation of genes associated with inflammation, such as the genes for IL-1, IL-6, and cyclooxygenase (Kessler-Becker *et al.*, 2004).

When grown on scaffolds, fibroblasts proliferate rapidly and then deposit large amounts of extracellular matrix. Initially, the conditions on the scaffold are not dissimilar from monolayer culture. The scaffold adsorbs proteins from the serum, which is a component of the medium still used for fibroblast culture. These include fibronectin and vitronectin, as in monolayer culture (Steele *et al.*, 1995). However, as the cells reach higher densities and the fibroblasts cease to proliferate, they proceed to lay down extracellular matrix at rates approaching their own weight per day. The matrix is loose connective tissue, much like the provisional matrix of granulation tissue, and may be thought of as foreign-body capsule formation occurring *in vitro*. The cells are clearly very active and induce several genes, among the most notable of which are neutrophil chemoattractant chemokines CXCL1, CXCL5, CXCL6, and CXCL8. They also induce the expression of IL-6. The conclusion from these observations is that the fibroblasts act to recruit neutrophil granulocyte components of the innate immune system, which is important in responding to the most likely type of environmental insult to which the dermis is liable, physical injury followed by bacterial colonization.

## III. COMPONENTS OF TISSUE-ENGINEERED SKIN GRAFTS AS RELATED TO FUNCTION

### Scaffold

Several types of scaffold have been employed for tissue-engineered skin implants. They include collagen gels, knitted polylactate/glycolate fabrics (PLGA), and collagen sponges. As discussed earlier, the scaffold has profound effects on gene expression and the function of the fibroblasts.

The Dermagraft and TransCyte processes used scaffolds obtained as finished commercial products from other companies, which already had regulatory approval. While this simplified the development of the products, it ultimately proved to be an expensive decision. A solution to this is to obtain a second source of raw material that is not a finished product and to establish its equivalence.

A further point here is that, while the use of already-approved products as scaffolds may be an easy initial path, the available materials may not be ideal for the particular application envisaged. There is a need to expand the range of scaffolds available for commercial tissue engineering. An example is enzymatically, rather than spontaneously, degrading scaffolds.

### Keratinocytes

Keratinocytes generate the impervious surface of the skin (the stratum corneum) and also have been thought of as having a major role in defense against microbial colonization as activators of immune responses. As a consequence, keratinocytes have been included, either alone or in combination, with other components in many skin implants (Ansel *et al.*, 1990). Their role in forming a physical barrier is important, but they also have the ability to produce antimicrobial peptides, such as β-defensin, psoriasin, and cathelicidin (Nizet *et al.*, 2001; Harder and Schroder, 2005) and a wide variety of cytokines capable of activating immune responses (Pivarcsi *et al.*, 2004; Albanesi *et al.*, 2005). The epidermis contains antigen-presenting cells that are capable of activating T-lymphocytes, under suitable conditions, to both cell-mediated and humoral adaptive responses. This is

largely a function of Langerhans cells, but it is possible that the keratinocytes also take part under special conditions. It has been argued that the application of a bag containing keratinocytes to a wound, which would permeable to proteins but not to cells, may be beneficial without incorporation of the cells into the patient (U.S. Patent 5972332).

It is notable that the keratinocytes show a higher expression of the fibroblast-stimulating growth factor PDGF-A chain gene than do fibroblasts. This may constitute part of the secretion of reciprocal paracrine growth factors by the epidermis and the dermis (Sato *et al.*, 1997). Such a pathway has been described by Fusenig, involving IL-1α secretion by keratinocytes, which stimulates secretion of FGF-7 (keratinocyte growth factor 1), and GMCSF, which promotes keratinocyte proliferation (Smola *et al.*, 1993). Such interactions may well be of importance during wound repair, when proliferation of both types of cells is important (Smola *et al.*, 1998).

In practice, allogeneic keratinocytes in skin implants appear to remain for some weeks and then disappear. This has been attributed to immunological rejection. But this is not clear because keratinocytes, cultured to the numbers required for implant, consist almost entirely of transiently amplifying cells that follow a differentiation pathway that leads to a modified form of apoptosis. They may, thus, be expected to be lost from the implant without the intervention of the immune system.

An approach that has been explored for increasing the coverage and survival of grafted epidermis is to mix allogeneic with autologous keratinocytes. This permits a comparatively small number of autologous keratinocytes, without extensive expansion, to cover a large area, by adding allogeneic keratinocytes (Rouabhia *et al.*, 1994). The allogeneic cells provide initial coverage, while the autologous population, which retains stemlike cells, provides the persistence. Most of the allogeneic cells are eventually lost, but the epidermis remains.

### Fibroblasts

Fibroblasts are the major producers of extracellular matrix (discussed next). They are not antigen-presenting cells and have been regarded as not having a major role in interactions with the immune system. However, their ability to secrete neutrophil attractant chemokines CXCL-1, CXCL-4, CXCL-5, and CXCL-8 and the cytokines IL-6 and IL-11 (discussed later) suggests that they may have a more important role in activating innate immune responses.

Fibroblasts do not appear to produce significant quantities of defensins, cathelicidin, or other antimicrobial peptides. Their major antimicrobial activity appears to be through recruiting neutrophils.

### Extracellular Matrix

Fibroblasts express a large array of extracellular matrix genes. Expressed genes represent collagens (types 1, 3, 5, 6), proteoglycans (decorin, lumican), thrombospondins, fibulins, fibronectin, tenascin C, metalloprotease inhibitors, lysyl oxidase, tissue factor inhibitor, and SPARC. It also involves bound growth factors, such as FGF-2 and TGF-$\beta$. The matrix involves multiple kinds of integrin ligand, including $\alpha_1\beta_1$, $\alpha_2\beta_1$, $\alpha_3\beta_1$, $\alpha_4\beta_1$, $\alpha_5\beta_1$, $\alpha_6\beta_1$, $\alpha_v\beta_1$, $\alpha_7\beta_1$, and $\alpha_6\beta_4$ and proteins known to modify adhesion. It provides complex stimulation that, since it is secreted by fibroblasts, is presumably appropriate for cell migration.

In systems where fibroblasts are cast in collagen gels, the initial signal is through the $\alpha_1\beta_1$ and $\alpha_2\beta_1$ integrin, and that remains the dominant stimulus until the cells elaborate other signaling molecules.

## Subcutaneous Fat

Subcutaneous fat has received little attention thus far in the engineering of skin. However, it has been shown to undergo major structural changes during the hair cycle in the mouse and is a highly vascular organ.

## Components of the Immune System

Inclusion of Langerhans cells has been accomplished in test systems but has not been applied to therapeutic products.

## Melanocytes

Other cells of the epidermis have been incorporated into test systems and have also been considered for transplant into epidermal tissue–engineered therapeutic products. An example is the inclusion of melanocytes as a treatment for vitiligo, which is a clinical problem in countries such as India. Systems including melanocytes have been used extensively in test systems for UV-protective preparations.

## Adnexal Structures

The self-assembly of hair follicles and their use as a therapeutic product is discussed later in this chapter. Inclusion of sweat glands has been explored, but they have not yet been included in tissue-engineered constructs. The early fear that patients treated with keratinocyte allografts might have problems with temperature control has not eventuated.

## IV. COMMERCIAL PRODUCTION OF TISSUE-ENGINEERED SKIN PRODUCTS

Commercial production takes place in a highly regulated system, which has many constraints if it is to be successful. These are discussed under a series of headings, dealing with physical constraints, such as bioreactor design, growth system design, and cell sources, and with broader considerations of the nature of preclinical and clinical assessment, regulation, reimbursement, and the marketplace. It is central to a commercial enterprise that it ultimately make a profit if it is to be successful. Since tissue engineering is inherently an expensive undertaking, minimizing costs at all stages is essential.

## Regulation

Regulation of tissue-engineered products by the FDA or other regulatory authorities is developing. Since commercial products have only been available since 1997, and since they are very different from other medical products, establishing a regulatory pathway has been complex. Tissue-engineered skin substitutes do not fit readily into any of the standard FDA categories: drugs (CDER), biologics (CBER), and devices (CDRH). Classification is important because requirements, criteria, and guidelines differ substantially between the centers. Skin implants were originally considered dermal or skin replacements and classified as devices. As it became evident that growth factor activity played a part in their mode of action, considerations appropriate to biologics played an increasing role. There is now within the FDA an Office of Combination Products responsible for assembling reviewers from different FDA centers to form a team to review such products. While tissue-engineered skin substitutes generally have more biochemical activity than would be expected of a pure device, they do not fit easily as a biologic. For instance, the concept of "dose" is not straightforward. If a component of the activity of these materials is through secretion of growth factors, growth factor production would seem to be a reasonable release criterion. However, unlike a purified growth factor preparation, these products are living, and the secretion of growth factors may vary depending on conditions. The cells adjust their cytokine output to the environment in the wound bed. It may be possible to devise conditions under which growth factor (for instance, VEGF) secretion is maximal (with optimal PDGF stimulation) and to determine output under those conditions. To avoid such problems, a "dose" of Dermagraft has been defined as a piece.

Also, at this stage the composition of tissue-engineered skin can only be determined to a limited extent. Usually, it is desirable that a biologic contain well-defined constituents, as in an active ingredient, at a known dose, and excipients. Thus far, studied components of skin implants have included those that are known and, to some degree understood, from other work. There is no comprehensive identification of constituents, and several that appear on expression arrays, such as lumican and collagen type 6, which appear prominently on arrays, have received inadequate attention.

## Product Development

Development of a commercial product should start with a *product concept*, from which a *design requirements* document is developed. As an example, Table 79.1 shows the product concepts for Dermagraft and TransCyte. The product is then developed to meet the design requirements

| Table 79.1. | Design concepts for Dermagraft and TransCyte |
|---|---|

**Dermagraft**
- Dermal replacement, later as a cytokine factory
- Allogeneic
- Viable product
- No sterile fill; i.e., bioreactor would be package
- Long shelf life (~6 months)
- Suitable size
- User-friendly presentation — modified bag

**TransCyte**
- Transitional covering for third-degree burns to replace cadaveric skin
- Allogeneic
- Nonviable
- No sterile fill; i.e., bioreactor would be package
- Long shelf life
- Suitable size (from Biobrane size)
- User-friendly presentation — clamshell bioreactor

by a series of hierarchical, more and more detailed design processes that ultimately comprise the *design master file*. The requirements can be divided into groups. In each case, the first statement is concerned with the therapeutic purpose of the product, the next three describe intrinsic characteristics of the product, and the last three are concerned with practical issues of importance to the final user. All these factors need to be considered from the earliest stages of development.

## Overall Concept

The initial application for Dermagraft was a dermal replacement for burn patients. Initial clinical application indicated that it provided little benefit and was slightly deleterious. As trials for chronic-wound applications progressed and it became evident that it was efficacious, the concept of Dermagraft as a cytokine source replaced the notion of a dermal replacement.

TransCyte, a replacement for cadaveric skin for covering third-degree burns after debridement, was seen as a means to overcome problems with rejection and potentially with disease transmission.

## Allogeneic Cell Source

The second item in both product concepts was that the products be allogeneic. The decision whether a product will be allogeneic or autologous is fundamental to the entire development process. While not impossible, the development of a commercially successful (profitable), autologous tissue-engineered product is very difficult. It requires separate, independent tissue culture suites (hood, incubator,

centrifuge, microscope, and air conditioning), which will be occupied by a product for a single individual for the entire period of culture. It is a service industry, with complex logistics and high cost. As an alternative, it would be possible, in principle, to construct an automatic machine that would accept a patient's tissue and produce the tissue-engineered construct in the hospital, with minimal intervention. An advantage to an autologous process is the comparatively low requirement for safety testing and the lack of immunological rejection by the patient.

Allogeneic tissue engineering has advantages in using only one tissue culture facility per product type, allowing economies of scale and straightforward logistics. The major disadvantages are the very large amount of safety testing required for the master cell banks and the potential for immunological rejection. The first of these disadvantages is ameliorated by the infrequency of making master cell banks. In the case of Dermagraft and TransCyte, one master cell was used for 15 years, so the investment could be amortized over a long period. Testing is determined by regulatory authorities from time to time, but it includes testing of the donor and, in the case of fibroblasts, the donor's mother for major human pathogens (HIV, hepatitis, CMV, etc.), xenogeneic pathogens, karyotype, identity (isoenzyme, VNTR, SNP, etc.), tumorigenicity, latent viruses, etc. This is tested both on the initial Master Cell Bank and again on cells reisolated from final product (the *end-of-production cell bank*).

## Viability of Product and No Sterile Fill

The viability of the product depends on its intended function and differs for Dermagraft and TransCyte. As discussed later, to package sterile material is a matter of some difficulty, and it was decided to avoid the issue. The bioreactor in which the tissue was grown was designed to form part of the final packaging.

## Shelf Life

Extended stasis preservation is important for commercial products, including tissue-engineered products. It allows time for quality control testing, for off-the-shelf availability, and for distribution. In addition, the Dermagraft and TransCyte processes take 12 weeks, from thawing a vial from the *manufacturer's working cell bank* (MWCB) to cell expansion, three-dimensional growth, and QC testing. The difficulty of predicting the market so far ahead so as to minimize waste is greatly aided by extended shelf life. This was achieved through cryopreservation at −70°C.

## Size, User Convenience

The remaining items in the product concept are related to convenience of use by the physician applying the product. This is an important item for a commercial product and needs much work in focus groups and thought. Thawing the product is already a major issue, requiring time and, in the case of TransCyte, careful estimation of the amount required, to avoid wasting expensive material or having to wait for a piece to be thawed. The Dermagraft bioreactor was designed so that it would open easily and the tissue would be positioned (through the use of a Z-weld that attached opposite sides of the scaffold to opposite walls of the bioreactor) so as to be easily rinsed prior to use. The bioreactor is translucent, to aid in marking the piece to be cut out for implantation.

## V. MANUFACTURE OF DERMAGRAFT AND TRANSCYTE

Overall schemata for the manufacture of Dermagraft and TransCyte are illustrated in Fig. 79.1. As can be seen, as much of the two processes as possible are identical. This economizes resources and reduces cost. In this case, the MCB, MWCB, the cell expansion process, and much of the release testing are all in common. In addition, the clean-room facility, all its ancillary equipment, and monitoring are the same.

The process overall takes about three months, from MWCB vial to releasable product. This makes the availability of shelf life through cryopreservation of particular importance to provide a buffer to minimize wastage or loss of supply.

### Cells

Several times per week, fibroblasts used in manufacture are recovered from liquid nitrogen storage (the MWCB) and expanded in roller bottles from the fifth to the eighth passage. This is a conventional procedure but may not be ideal. Fibroblasts for Apligraf are expanded on beads in a mixed bioreactor. It has become evident that the proliferation of fibroblasts is improved at low oxygen (2–6%) (Falanga and Kirsner, 1993), and, at this oxygen concentration, the development of a senescent phenotype may be reduced (Busuttil *et al.*, 2003; Parrinello *et al.*, 2003; Chen *et al.*, 2004). A tank bioreactor lends itself to active control of parameters

such as oxygen tension and pH and is worth exploring before a system gets fixed by regulation. The cells ideally should be grown just to the beginning of stationary phase. Further incubation seems to be slightly deleterious.

### Medium

The Dermagraft and TransCyte systems are grown in Dulbecco's-modified Eagle medium, supplemented with 10% calf serum, nonessential amino acids, and glutamine. All tissue culture beyond initial cell isolation from the foreskin is free of antimicrobial agents. Antimicrobial agents might hide otherwise significant contamination, which might only become evident in therapeutic application.

While serum-free medium is available for keratinocytes, it has been found difficult to achieve such systems for fibroblasts. With the increased concern about the transmission of prion diseases, there has been great interest in developing serum-free systems. Serum-free fibroblast media are commercially available, but while these have been successful in the research laboratory, transferring them to scaled-up systems and manufacturing has proved elusive. The major issue is that the cells do not grow to the extent required, apparently because, early on, they acquire a senescent phenotype. It was early established that serial subculture of fibroblasts leads to senescence after about 60–80 population doublings. While this can be attained in serum-containing medium, experience with serum-free media have indicated a life span of about nine doublings. Since Dermagraft is made at about the 28th generation, about 10 doublings from the MWCB, 19 from the MCB, and implanted fibroblasts should not be too close to senescence, serum-free media do not provide sufficient proliferation potential.

The alternative solution to serum-free medium is to take extreme caution in the source and treatment of serum. By regulation, all materials used in manufacture must be traceable to their sources. But, in addition, bovine serum is obtained only from countries that are free from bovine spongiform encephalopathy (BSE), which includes Australia and New Zealand, from closed herds with a known pedigree history that includes no animals from outside the country. Serum is also irradiated sufficiently to eliminate possible viral contamination.

### Bioreactor Design

Many of the considerations of bioreactor design involve questions of mass transport, conditions under which cells are able to grow, and so forth, much as in noncommercial tissue-engineering systems. A second series of considerations deals with the large-scale production process. These include ease of scale-up, ease of handling, minimal footprint, and maximal automation. In addition, the use of the product by a community not experienced with products of this type is also important. This leads to features to make the product easy for the end user to work with and apply.

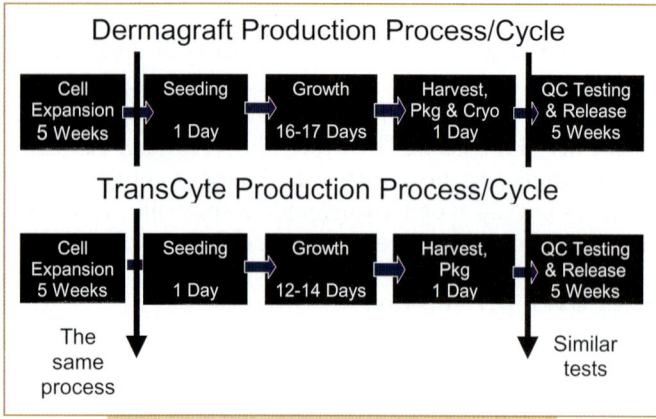

**FIG. 79.1.** Overall comparison of the Dermagraft and TransCyte processes.

The bioreactor used for the growth of Dermagraft consists of a bag with eight cavities welded into it, each of which contains a sheet of the scaffold. Such bags are attached in groups of 12 to manifolds, with a solid support, for cell, medium, or cryopreservative addition, to form a system. Several systems may be connected and all fed at the same time to form a lot, which is the unit of manufacture. The bag reactors, systems, and even lots may be assembled under clean but not necessarily sterile conditions and sterilized by 25–43 grays gamma-radiation prior to seeding. From this point, the process requires only nine sterile connections and is substantially automatic.

The TransCyte bioreactor systems are similarly assembled under clean conditions, complete with all associated tubing, and attached in groups of 12 to a manifold to form a system. This entire assembly is then sterilized by gamma radiation, as just described. Several systems can then be connected to form a lot. In this case, the inlets to each bioreactor are individually passed through a peristaltic pump to provide even flow. (Attempts to use constrictions to control flow was found to be inferior.) Again, as with the Dermagraft system, once set up and seeded, the TransCyte system is substantially automatic.

Automation in tissue-engineering systems is an important feature. Apart from reducing labor costs, automation greatly reduces the number of errors that have to be investigated, requiring time, effort, and cost. While operator errors are generally less serious than machine failures, they are much more numerous and still require individual investigation. Backup systems can be installed to minimize the impact of machine failure.

# VI. DERMAGRAFT AND TRANSCYTE PRODUCTION PROCESSES

The production of Dermagraft and TransCyte started with the cryopreserved master cell bank, which was stored at passage 3 in liquid nitrogen. About four times per year, cells were taken from this bank and expanded in roller bottles to fifth passage, when they were stored again as the MWCB. This was tested to ensure identity to the master cell bank. Several times per week, cells are taken from the MWCB and expanded to eighth passage in roller bottles, when they are seeded to the three-dimensional bioreactors. This process is conventional, large-scale tissue culture. It uses batch feeding and exposes the cells frequently to atmospheric oxygen. As discussed elsewhere in this chapter, fibroblasts grow better and senesce less rapidly in a lower-oxygen environment. Experiments have been conducted to expand the fibroblasts on beads in tank bioreactors, which allow both control of such variables as oxygen tension and pH as well as continuous feeding.

On seeding, the bioreactors are manipulated to ensure even distribution of the cells. In the case of Dermagraft, the bioreactors are rolled; in the case of TransCyte, each side is seeded successively under static conditions. The medium is replaced one day after seeding, and then the cells are fed every two to four days until harvest.

The time of harvest is determined from the glucose metabolism of the cultures. In the case of TransCyte, glucose consumption and lactate production are determined daily and interpreted as notional ATP turnover, assuming a P : O ratio for mitochondrial oxidative phosphorylation of 3. The resulting values are correlated with the performance of the system relative to release specifications and appropriate target values selected. In the case of Dermagraft, there was a concern that lactate release from the degradation of the scaffold might interfere with the ATP turnover estimation, so the time of harvest was determined on cumulative glucose utilization alone.

At harvest of Dermagraft, medium is replaced by cryopreservative solution, which consisted of 10% dimethyl-sulphoxide in phosphate-buffered saline, supplemented with calf serum, and the individual pieces of Dermagraft are sealed, removed from the surrounding bag, packaged, cooled slowly at a rate that fell within design parameters, and frozen. It is then stored in a freezer set to −75°C. The time between cryopreservative addition is at least 4 h and is validated to a 12-h delay. This time interval did not appear to affect the viability of the cells, and it was conjectured that it allowed induction and synthesis of stress proteins in response to the high osmotic pressure.

## Release Specifications

The release specifications include sterility for bacterial, fungi, mycoplasma, and endotoxin as well as for analytical criteria that ensure consistency of the product. Dermagraft uses four analytical release criteria: MTT reductase activity before freezing (in-process MTT), and after thawing, collagen content by Direct Red binding (Bedossa *et al.*, 1989) and DNA content. MTT reductase was used to determine the metabolic activity of the product, collagen as a surrogate for the amount of extracellular matrix, and DNA content was used to establish adequate cellular content. MTT reduction was an assay established early in the development of Dermagraft. Since then, several other substrates measuring essentially the same activity have been developed, several of which are less toxic to the cells than MTT and can be used repeatedly. It would be wise in the development of a new product to examine these early and select a method carefully. MTT reductase is frequently viewed as measuring mitochondrial activity. However, 70% of the activity is extra-mitochondrial (Musser and Oseroff, 1994), and it is better regarded as a general measure of cellular activity. The MTT reductase assay was originally developed to evaluate toxicity or cell proliferation, where values obtained after treatment were compared with control values. In determining the metabolic activity of tissue-engineered products, this is not possible, and a great deal of fruitless effort has been expended trying to obtain a standard for the MTT assay.

MTT determination before and after cryopreservation might be interpreted this way. However, changes occur in the viability and metabolic activity of cells for up to three days after cryopreservation, due to necrosis, apoptosis, and repair, so the MTT values are better regarded as independent absolute determinations than as relative values.

In the case of TransCyte, the microbiological criteria and the collagen and DNA determinations are similar to those for Dermagraft. However, because the cells are nonviable in the final TransCyte product, there is no MTT determination. Determination of MTT reductase prior to final freezing (in process MTT) was replaced by release specifications applied to ATP consumption, as described earlier.

Definition of specifications was initially within three standard deviations of the process performance. This was refined by clinical experience. In the case of MTT, it was evident during the clinical trial that the specifications needed to be refined, requiring a midtrial correction and ultimately an additional clinical trial.

### Distribution and Cryopreservation

The use of −70° cryostorage for Dermagraft was a compromise between an ideal solution and practicality and turned out to have several benefits. If the goal was 100% cell viability and indefinite shelf life, the ideal solution would probably have been storage at liquid nitrogen temperatures, probably using a vitrification procedure in which high concentrations of cryoprotectant and very rapid cooling are used to achieve the formation of a glass with minimal ice crystallization. While the technical difficulties of achieving sufficiently rapid cooling of a therapeutically useful structure and problems with the brittleness of plastics at low temperatures can probably be overcome, the method would still entail transport at liquid nitrogen temperatures and safety issues. Cryopreservation at dry ice temperatures provided adequate shelf life and allowed shipping in a container capable of maintaining −65°C for four days. This allowed distribution throughout the world and would generally permit two to three days of leeway for patient and physician scheduling at the site of use. If a center had a −70°C freezer, storage at this temperature would result in much longer shelf life.

In addition, the induction of stress proteins by the cryopreservation procedure (K. Liu *et al.*, 1998), combined with the suboptimal viability recovery, meant that necrotic cell debris, including heat shock proteins, were released from lysed cells. Such proteins activate macrophages through CD91 and possibly toll-like receptors (Curry *et al.*, 2003). While such necrotic products are likely not scarce in a chronic wound, this property may have contributed to the efficacy of Dermagraft.

While discussing cryopreservation, an important point that has received little attention is the thawing process. In the case of Apligraf, the thawing process involved solutions that had to be replaced, requiring a procedure of several hours, which was performed at the company. This provided shelf life but not off-the-shelf convenience and required additional resources from the company. Therefore, the procedure was not continued. In the case of Dermagraft, thawing is performed in the physician's office, out of the control of the company.

During the controlled freezing process, after the majority of the osmotic water removal has occurred, the cell contents reach a eutectic and freeze. While the majority of the material forms a glassy state, small ice nuclei may be formed. The slow growth of these nuclei during −70°C storage is probably what limits shelf life. During thawing, conditions arise that allow the growth of ice crystals, the nuclei enlarge and, if they become big enough, will kill the cells. It is important to take the cells through this stage as rapidly as possible to minimize growth. So rapid thawing, straight from the −70°C storage, is critical. The procedure that was adopted involved a thawing tub with a thermometer, which could be filled with warm water from the tap or a water bath at 37°C. A few seconds unprotected in ambient air was sufficient to reduce viability drastically.

### Problems with Commercial Culture for Tissue Engineering

In large-scale tissue culture, procedures are followed that differ from those found in a research laboratory. As a result, problems have arisen from time to time that have no counterpart in culture on a smaller scale. Frequently, it is difficult to establish the cause unequivocally because of the scale and cost of the experiments that would be required. When such a problem occurs, it is usual to assemble all those with potentially valuable ideas, analyze the incident with respect to changes in procedure (new batches of medium or other raw materials, newly implemented changes, etc.), and prioritize possible solutions. Usually the most likely and cheapest suggestions are explored. Such suggestions are frequently implemented more or less at the same time and immediately. When the problem ceases, as it usually does, it is very difficult to determine which suggestion solved it. At best, there is a reasonable rationale.

Examples of such occurrences include light damage to medium and deterioration of cell banks. It has been known for many years that medium is light sensitive and will degrade if left exposed to daylight or fluorescent light. However, such an event is never seen in the research laboratory, because medium is normally stored in a refrigerator in the dark. In commercial operation, it is sometimes necessary to make up medium a few weeks in advance and to store it in a cold room. If the lights in the cold room are fluorescent and are left on, deterioration of medium will be observed. Indeed, the Dermagraft plant was illuminated with yellow light to avoid such problems.

In a research laboratory, cells stocks frozen in liquid nitrogen are manipulated rarely, a few times per year. In commercial-scale tissue engineering, vials from the MWCB are removed several times per week. While one or two vials are identified and removed, the remainder is out of cryostorage and warming up. We noticed that, after some years of this, the fibroblasts could be shown to demonstrate a decreased life span. This was attributed to ice recrystallization during periodic warming to perhaps –120°C to –90°C during recovery of other vials. It is, thus, necessary to store the MCB independent of the MWCB and to devise a method of recovering selected vials with minimal disturbance to other vials.

# VII. CLINICAL TRIALS

The design and implementation of clinical trials is specialized and beyond the scope of this discussion. The process requires a great deal of time and expense. In the case of the CDRH division of the FDA, the initial trial is a pilot, or feasibility, trial, equivalent to a phase 1 trial in the CBE or CDER. This consists of a small number of patients (6–20) per treatment regime and is concerned primarily with safety. Any information obtained on efficacy in a trial of this magnitude is unlikely to be statistically significant but may provide useful information on the size and variability of the therapeutic effect that is valuable in design of the subsequent, pivotal trial.

The pivotal trials (equivalent to phase 2 trials in biologics and devices) involve much larger numbers of patients (50–500 in the case of devices), which are determined by a sample-size calculation, based on the results from the pilot trial, to give about 80% probability of a significant result ($p < 0.05$) if one exists.

Trials with chronic wounds take about two years or more to complete. The design phase — deciding on protocols, obtaining institutional approval for multiple centers, and preliminaries to the recruitment of patients — is likely to take at least six months. The actual trial involves screening many more patients than will actually take part in the trial, for many fail to conform to exclusion criteria, which reduce irrelevant and interfering factors, established in the design phase. With a trial involving a 12-week follow-up period, most should be completed within 18 months to two years. There is then a period of six months while the data are checked, discrepancies resolved, and the results of the trial evaluated. This period could be shortened substantially by logging the patient outcomes and checking them online as they are generated.

The design of the clinical trial should be conducted in close consultation with the relevant regulatory authorities. It is critically important that all aspects be thoroughly discussed and agreed on beforehand. In the case of Dermagraft, failure to follow advice from the FDA precisely led to considerable delay and additional expense.

Considerations in the selection of clinical trials include the size and value of the potential market, the expected efficacy of the product, and the difficulty of completing the trials. Five major areas of chronic wounds include venous stasis ulcers, diabetic foot ulcers, arterial ulcers, pressure (decubitus) ulcers, and all other chronic wounds. Of these, venous stasis ulcers represent the largest market. Diabetic ulcers are a valuable market, because, while smaller in number, the consequence of failure to heal such an ulcer is limb amputation, an expensive and seriously debilitating procedure. It is very difficult to perform a trial on decubitus ulcers, because the patients are frequently elderly and very infirm and may need approval from a guardian. Such trials are arduous, may be difficult to complete, and are best undertaken by a specialized group with well-developed access to such patients. In no cases does the treatment of chronic wounds with skin implants address the underlying cause of the ulcer, which lies in venous insufficiency, metabolic abnormalities, reduced arterial supply, or repeated ischemia. This is particularly evident in ulcers caused by arterial insufficiency. While the angiogenic activity implant may be capable of improving vascularity in the region of the wound, little may be gained if blood supply to the limb is inadequate.

The pivotal clinical trial for Dermagraft for diabetic ulcers was initiated in 1994. Dermagraft gave a statistically significant increase in healing of 19.1% that was similar to results obtained in similar trials with other products. These data were used in combination with mechanism of action and clinical instruction to support the clinical use of Dermagraft commercially.

The indications for which the product may be marketed (its labeling) are strictly limited by the clinical trials, and new trials are required to extend them. However, in certain categories, there may be an insufficient number of patients to support both a clinical trial and a commercial market. Examples are monogenetic congenital diseases, where, although the conditions may be very distressing, the entire population may number in the hundreds or a few thousand. An example in the chronic-wound field is epidermolysis bullosa (EB). For such diseases, the FDA has a category, Humanitarian Device Exemption (HDE), which allows much reduced clinical testing. It does, however, require Institutional Review Board (IRB) approval from each institution where it will be used. In the case of dermal implants, such a route has been used for Apligraf and Orcell for all genetic forms of EB and for Dermagraft for the dystrophic forms of EB. The category of "all other chronic wounds" includes many miscellaneous conditions, such as pyoderma gangrenosum and necrobiosis lipoidica diabeticorum, for which there is too small a patient population to perform a trial but that would be expected to benefit from treatment with Dermagraft or TransCyte. The HDE pathway may not be appropriate or available. However, since the FDA does not regulate the practice of medicine, some approach to those patients may be obtained through off-label use, although this cannot be formally promoted by the manufacturer.

## VIII. IMMUNOLOGICAL PROPERTIES OF TISSUE-ENGINEERED SKIN

Immunological rejection of tissue-engineered products is, at the time of writing, a controversial question. As was clearly established by Medawar in the 1940s, allogeneic transplants of whole live organs invariably cause immunological reactions that lead to rejection (Medawar, 1957). However, experience with some cultured cells and tissue-engineered constructs indicates, in some cases, no clinically significant rejection. At this point, some 90,000 pieces of Dermagraft have been implanted, in many cases several times to the same patient, without a single example of immunological rejection, and experience with Apligraf has been similar. The fibroblasts from Dermagraft have been found to persist in the wound site for six months, and survival for about a month has been obtained with Apligraf (Phillips *et al.*, 2002).

There are two major pathways of transplant rejection, direct and indirect. The direct pathway involves recognition of donor histocompatability antigens (HLA) by the host immune system, leading to acute rejection over about two weeks. This involves both HLA, of which Class II molecules, such as HLA-DR, are the most important, and costimulatory molecules on the transplant, the CD40 and the CD80 groups. These molecules are not normally expressed by fibroblasts. However, in monolayer tissue culture, the presence of γ-interferon, which may be present in chronic wounds, will cause induction of HLA-DR, CD-40, and genes involved with the physiological function of class II HLA (antigen presentation). In three-dimensional culture, many of the cells show a selective response to γ-interferon, which does not include induction of these molecules (Kern *et al.*, 2001). The indirect pathway of rejection involves display of transplant antigenic peptides by host antigen-presenting cells (macrophages, tissue dendritic cells, and endothelial cells). This gives rise to chronic rejection, which may cause destruction of the transplant over many months. While it cannot be excluded that this might occur, Dermagraft fibroblasts have been detected at six months from the time of implantation, and no clinical evidence for chronic rejection has been observed.

Acute rejection is primarily an attack on the endothelial cells of the vascular system, which are antigen-presenting cells and do express the component of the antigen presentation system (HLA Class II, CD40, CD80). It is possible that the lack of rejection of tissue-engineered skin products is related to the absence of such cells. Indeed, adding antigen-presenting cells (in this case, B-lymphocytes) to a tissue-engineered construct has restored susceptibility to immunological rejection (Rouabhia, 1996).

## IX. COMMERCIALIZATION

There are many contributors to the commercial success of a product. They include a satisfactory clinical trial, a good rationale for the performance of the product, convenience in use, a reasonable price, adequate reimbursement from third-party payers, sensitivity to complaints, and efficient distribution. Some of these require long-term effort and must be initiated early in the development process.

New products frequently find themselves selling into a conservative and skeptical market that arises from the hype involved in developing and funding the product. The first requirement in overcoming this view is a satisfactory clinical trial. This process is determined by regulatory bodies and is outside the scope of this chapter. However, the major criteria for the American authorities (FDA) are safety and efficacy. Unless the product is safe and efficacy is extraordinary, clinical results need to be backed up by a mechanism of action that, while it may not be fully established, must provide a logical reason for using the product. In the case of Dermagraft, this has led to a great deal of fundamental wound-healing research.

Equally important is the convenience of use, the cost, and the process for reimbursement of the physician applying it. Convenience of use and cost are functions of bioreactor and process design and company structure and are discussed elsewhere. Third-party reimbursement and health economics are major and specialized subjects that are outside the scope of this chapter. However, a major component of establishing satisfactory reimbursement in the United States involves the Center for Medicare and Medicaid Services (CMS). Government authorities with similar responsibilities are involved in other countries. Their criteria, in contrast to those of the FDA, are reasonableness and effectiveness/cost ratio. Both the FDA-type and CMS-type criteria have to be met for a product to be successful.

An activity that would prove very valuable in establishing the performance of a product is a system for tracking the clinical performance subsequent to regulatory approval to market the product. While complaints and adverse events are carefully tracked and reported, performance has not been. It may, however, become a regulatory requirement to confirm safety, efficacy, and cost effectiveness with a larger patient population than possible in a clinical trial. It is also valuable to the manufacturer because, while anecdotes of positive clinical experience may be heartening, they do not provide objective quantitative data. This was not performed in the cases of Dermagraft or TransCyte, making it difficult to assess the actual performance of the product in clinical use.

### Mechanism of Action

Understanding the mechanism of action of a product requires some understanding of the underlying pathology of the disease the product is intended to treat. Wound healing, and particularly the etiology and maintenance of chronic wounds, was poorly understood during the development of Dermagraft and TransCyte. Many phenomena have been observed, particularly in chronic wounds. The problem is sorting out how they relate to one another, i.e., which are etiological and which symptomatic, the difference between cause and response. The problem was less acute in the case of TransCyte because the product was designed as only a temporary covering for third-degree burns.

Changes from normal skin observed in chronic wounds include lack of keratinocyte migration, capillary cuffing, protease activity, abnormalities in macrophage activation, diminished vascular supply, and senescence in fibroblasts (Agren *et al.*, 2000). In developing a rationale for the action of Dermagraft, the first aspects considered were vascular supply and keratinocyte migration. Much study was undertaken on the angiogenic activities of Dermagraft, and it was shown that the tissue produced VEGF and hepatocyte growth factor. It was also shown to be angiogenic in the chick chorioallantoic membrane assay and to increase blood supply to diabetic ulcers in patients *in vivo* using laser-Doppler techniques. Thus, a case could be made that one mode of action was through its angiogenic effect.

The high protease activity of chronic wounds was combined with the lack of keratinocyte migration to hypothesize that the extracellular matrix in a chronic wound was degraded to the extent that the migration substrate was lacking. Indeed it was demonstrated that fibronectin was degraded by neutrophil elastase (Grinnell and Zhu, 1996). This led to work on the properties of Dermagraft and TransCyte as substrates for keratinocyte migration.

During comprehensive surveys of genes induced by three-dimensional culture, it was found that the most highly up-regulated gene was IL-8. Other ELR CXC chemokines were also highly induced. This led to studies of the secretion of these proteins and to changes in their production that might relate to the etiology of chronic wounds and the possible role of Dermagraft. It was found that Dermagraft secretes IL-8, sometimes in large quantities. The secretion is extremely variable. Monolayer fibroblasts also secrete CXCL-1, CXCL-5, and CXCL-6 (gro-$\alpha$, ENA-78, and GCP-2, respectively), which are all chemoattractive for neutrophils and activate them to a bactericidal phenotype. At the same time, it was observed that bacterial products (lipopolysaccharide) inhibit keratinocyte migration. This led to the hypothesis that a major function of fibroblasts in a wound context is the recruitment of neutrophil leukocytes to destroy colonizing bacteria and that failure of this system in chronic wounds results in the establishment of bacterial contamination that leads to failure of re-epithelialization. It is known that fibroblasts in chronic wounds display a senescencelike phenotype (Raffetto *et al.*, 1999) that grows slowly and is unresponsive to growth factors. The condition is probably related to stress-induced premature senescence caused by ischemia reperfusion injury, metabolic abnormality, extravasated red cells, and the inflammatory conditions of the chronic wound. It has been found that the production of CXCL-1, CXCL-5, CXCL-6, and CXCL-8 declines in senescent fibroblasts. Thus, in chronic wounds, a decline in the ability of the fibroblasts to recruit and activate neutrophils allows wound colonization by bacteria and failure of keratinocyte migration to close the wound. Chronic wounds seem to involve an arrest of normal healing at about the stage of neutrophil immigration. On this hypothesis, a major role of Dermagraft is to provide nonsenescent fibroblasts that are able to respond to the presence of bacteria with appropriate secretion of neutrophil chemoattractant chemokines.

In the case of TransCyte, the formation of a foreign-body capsulelike material *in vitro* leads to lack of ingrowth of the scaffold into the wound through a host foreign-body reaction and, thus, comparative ease of removal. The ability of TransCyte to reduce the pain of second-degree burns is common to many occlusive dressings.

A remarkable feature of dermal implants is that they do not appear to have led to an increase in infection. Indeed the general experience has been a slight reduction in infections. Preparations with keratinocytes included secrete antimicrobial peptides, but this is not known to be the case for fibroblasts. The most likely explanation is a combination of fibroblast secretion of neutrophil chemoattractant CXC chemokines and the provision of an extracellular matrix substrate for leukocyte migration.

## X. FUTURE DEVELOPMENTS

The major direction of development in skin products is toward simpler, possibly nonviable systems. Inclusion of live cells in a product entails many issues, such as the use of allogeneic cells, cryopreservation, distribution, inconvenience to the customer (for instance, the thawing procedure), and many manufacturing problems. Hence, studies were initiated to explore the possibility that a somewhat less effective but substantially less expensive product could be developed. This is a direction that has also been explored by other companies in the field.

It is a general principle that value is likely to be obtained by exploiting a technology base as far as possible. A major product of fibroblasts is collagen, in this case human. As a nonviable product, human collagen was extracted from the extracellular matrix laid down by fibroblasts and used, by injection, for the treatment of wrinkles.

An alternative to cryopreservation that has been explored is the possibility of storage by desiccation. Many natural systems, such as plant seeds, tardigrades, *Artemia*, and yeast, have developed the ability to survive desiccation and, in that state, may be able to survive extreme conditions of temperature, vacuum, radiation, and time (Crowe and Crowe, 2000). Many factors appear to be involved in this ability (Mattimore and Battista, 1996), but one that has attracted attention is trehalose, which is a disaccharide (1,1,$\alpha$,$\alpha$ diglucose) that is produced in large quantities by many desiccation-resistant organisms. It was found that TransCyte could be dried at room temperature in the presence of trehalose and irradiated to 25–40 grays to provide terminal sterilization, and it would recover its structure and wound adhesion on rehydration. Similar results could be obtained by careful lyophilization. While some success has been obtained in the retention of viability of human cells in the desiccated state, it is not yet possible to obtain the shelf life (at least one month and preferably at least six months) required for a tissue-engineering application. It is also questionable whether the radiation treatment possible with a nonviable product would be appropriate with living cells, and an aseptic drying-and-packaging system would have to be devised.

Thus far, the therapeutic applications of skin-tissue engineering have encompassed acute and chronic wounds and comprise dermal components alone or dermal components with epidermal structures. In the future, it is likely that adnexal structures will be added. Remarkable success in this direction has been achieved with hair follicles (Zheng *et al.*, 2005), where it has been found that hair follicle–inducing cells can be cultured so that they retain their properties and, mixed with hair follicle–derived keratinocytes, will, on injection, form a hair. The remarkable feature of this system is that the entire organ is formed: hair follicle, sebaceous gland, and erector pili muscle. This constitutes one of the few examples of true organogenesis observed using adult cells.

## XI. CONCLUSION

Experience with the commercial production of tissue-engineered skin has developed many principles that are important beyond the simple development of a tissue-engineered product. Most of these are based on well-known concepts used in other manufacturing processes but applied to tissue engineering. They include:

1. Optimized cryostorage of master cell banks with minimal access
2. Intermediate storage of expanded cell banks (MWCB)
3. Use of allogeneic cells
4. Bioreactors that permit scale-up and minimize footprint
5. Minimal aseptic connections
6. Maximal automation, to minimize errors
7. Avoidance of aseptic fill so that the bioreactor forms part of the final package
8. Bioreactor designed for end-user convenience
9. Cryostorage

While it may not be possible to incorporate all of these principles, as many as possible should be used. The major problem with tissue-engineered products in the marketplace is cost, and all these principles are directed toward reducing cost. Some, such as the use of allogeneic cells, may be difficult to implement. While experience with fibroblasts indicates that immunological rejection is not a major issue, this is not known for other cells. The possible use of allogeneic cells should, however, be checked, preferably in animals other than mice, because the advantages for a commercial product are great. Optimal cryostorage is a compromise between ideal tissue survival and other factors, such as ease of distribution.

## XII. REFERENCES

Agren, M. S., Eaglstein, W. H., Ferguson, M. W., Harding, K. G., Moore, K., Saarialho-Kere, U. K., and Schultz, G. S. (2000). Causes and effects of the chronic inflammation in venous leg ulcers. *Acta Derm. Venereol. Suppl. (Stockh.)* **210**, 3–17.

Albanesi, C., Scarponi, C., Giustizieri, M. L., and Girolomoni, G. (2005). Keratinocytes in inflammatory skin diseases. *Curr. Drug Targets Inflamm. Allergy* **4**, 329–334.

Ansel, J., Perry, P., Brown, J., Damm, D., Phan, T., Hart, C., Luger, T., and Hefeneider, S. (1990). Cytokine modulation of keratinocyte cytokines. *J. Invest. Dermatol.* **94**, 101S–107S.

Bedossa, P., Lemaigre, G., and Bacci, J. M. (1989). Quantitative estimation of the collagen content in normal and pathologic pancreas tissue. *Digestion* **44**, 7–13.

Bell, E., Ivarsson, G., and Merrill, C. (1979). Production of a tissue-like structure by contraction of collagen lattices by human fibroblasts of different proliferative potential in vitro. *Proc. Natl. Acad. Sci. U.S.A.* **76**, 1274–1278.

Busuttil, R. A., Rubio, M., Dolle, M. E., Campisi, J., and Vijg, J. (2003). Oxygen accelerates the accumulation of mutations during the senescence and immortalization of murine cells in culture. *Aging Cell* **2**, 287–294.

Chen, J. H., Stoeber, K., Kingsbury, S., Ozanne, S. E., Williams, G. H., and Hales, C. N. (2004). Loss of proliferative capacity and induction of senescence in oxidatively stressed human fibroblasts. *J. Biol. Chem.* **279**, 49439–49446. Epub 42004 Sep. 49416.

Crowe, J. H., and Crowe, L. M. (2000). Preservation of mammalian cells — learning nature's tricks. *Nat. Biotechnol.* **18**, 145–146.

Curry, J. L., Qin, J. Z., Bonish, B., Carrick, R., Bacon, P., Panella, J., Robinson, J., and Nickoloff, B. J. (2003). Innate immune-related receptors in normal and psoriatic skin. *Arch. Pathol. Lab. Med.* **127**, 178–186.

Eckes, B., Mauch, C., Huppe, G., and Krieg, T. (1993). Down-regulation of collagen synthesis in fibroblasts within three-dimensional collagen lattices involves transcriptional and posttranscriptional mechanisms. *FEBS Lett.* **318**, 129–133.

Falanga, V., and Kirsner, R. S. (1993). Low oxygen stimulates proliferation of fibroblasts seeded as single cells. *J. Cell. Physiol.* **154**, 506–510.

Grinnell, F., and Zhu, M. (1996). Fibronectin degradation in chronic wounds depends on the relative levels of elastase, alpha1-proteinase inhibitor, and alpha2-macroglobulin. *J. Invest. Dermatol.* **106**, 335–341.

Harder, J., and Schroder, J. M. (2005). Psoriatic scales: a promising source for the isolation of human skin–derived antimicrobial proteins. *J. Leukoc. Biol.* **77**, 476–486.

Hull, B., Sher, S., Friedman, L., Church, D., and Bell, E. (1981). Fibroblasts in a skin equivalent constructed *in vitro* persist after grafting. *J. Cell Biol.* **91**, 51a.

Hull, B. E., Sher, S. E., Rosen, S., Church, D., and Bell, E. (1983). Fibroblasts in isogeneic skin equivalents persist for long periods after grafting. *J. Invest. Dermatol.* **81**, 436–438.

Kern, A., Liu, K., and Mansbridge, J. N. (2001). Expression HLADR and CD40 in three-dimensional fibroblast culture and the persistence of allogeneic fibroblasts. *J. Invest. Dermatol.* **117**, 112–118.

Kessler, D., Dethlefsen, S., Haase, I., Plomann, M., Hirche, F., Krieg, T., and Eckes, B. (2001). Fibroblasts in mechanically stressed collagen lattices assume a "synthetic" phenotype. *J. Biol. Chem.* **276**, 36575–36585.

Kessler-Becker, D., Krieg, T., and Eckes, B. (2004). Expression of pro-inflammatory markers by human dermal fibroblasts in a three-dimensional culture model is mediated by an autocrine interleukin-1 loop. *Biochem. J.* **379**, 351–358.

Langholz, O., Rockel, D., Mauch, C., Kozlowska, E., Bank, I., Krieg, T., and Eckes, B. (1995). Collagen and collagenase gene expression in three-dimensional collagen lattices are differentially regulated by alpha 1 beta 1 and alpha 2 beta 1 integrins. *J. Cell Biol.* **131**, 1903–1915.

Liu, K., Yang, Y., Pinney, E., and Mansbridge, J. (1998). Cryopreservation of the three-dimensional fibroblast-derived tissue results in a stress response including induction of growth factors. *Tissue Eng.* **4**, 477.

Liu, S. C., and Karasek, M. (1978). Isolation and growth of adult human epidermal keratinocytes in cell culture. *J. Invest. Dermatol.* **71**, 157–162.

Mattimore, V., and Battista, J. R. (1996). Radioresistance of *Deinococcus radiodurans*: functions necessary to survive ionizing radiation are also necessary to survive prolonged desiccation. *J. Bacteriol.* **178**, 633–637.

Medawar, P. B. (1957). *In* "The Uniqueness of the Individual," Basic Books, New York.

Musser, D. A., and Oseroff, A. R. (1994). The use of tetrazolium salts to determine sites of damage to the mitochondrial electron transport chain in intact cells following *in vitro* photodynamic therapy with photofrin II. *Photochem. Photobiol.* **59**, 621–626.

Nizet, V., Ohtake, T., Lauth, X., Trowbridge, J., Rudisill, J., Dorschner, R. A., Pestonjamasp, V., Piraino, J., Huttner. K., and Gallo, R. L. (2001). Innate antimicrobial peptide protects the skin from invasive bacterial infection. *Nature* **414**, 454–457.

Parrinello, S., Samper, E., Krtolica, A., Goldstein, J., Melov, S., and Campisi, J. (2003). Oxygen sensitivity severely limits the replicative life span of murine fibroblasts. *Nat. Cell. Biol.* **5**, 741–747.

Phillips, T. J., Manzoor, J., Rojas, A., Isaacs, C., Carson, P., Sabolinski, M., Young, J., and Falanga, V. (2002). The longevity of a bilayered skin substitute after application to venous ulcers. *Arch. Dermatol.* **138**, 1079–1081.

Pivarcsi, A., Kemeny, L., and Dobozy, A. (2004). Innate immune functions of the keratinocytes. A review. *Acta Microbiol. Immunol. Hung.* **51**, 303–310.

Raffetto, J. D., Mendez, M. V., Phillips, T. J., Park, H. Y., and Menzoian, J. O. (1999). The effect of passage number on fibroblast cellular senescence in patients with chronic venous insufficiency with and without ulcer. *Am. J. Surg.* **178**, 107–112.

Rheinwald, J. G., and Green, H. (1975). Serial cultivation of strains of human epidermal keratinocytes: the formation of keratinizing colonies from single cells. *Cell* **6**, 331–343.

Rouabhia, M. (1996). *In vitro* production and transplantation of immunologically active skin equivalents. *Lab. Invest.* **75**, 503–517.

Rouabhia, M., Germain, L., Bergeron, J., and Auger, F. A. (1994). Successful transplantation of chimeric allogeneic-autologous cultured epithelium. *Transplant. Proc.* **26**, 3361–3362.

Sato, T., Kirimura, Y., and Mori, Y. (1997). The coculture of dermal fibroblasts with human epidermal keratinocytes induces increased prostaglandin E2 production and cyclooxygenase 2 activity in fibroblasts. *J. Invest. Dermatol.* **109**, 334–339.

Sher, A. E., Hull, B. E., Rosen, S., Church, D., Friedman, L., and Bell, E. (1983). Acceptance of allogeneic fibroblasts in skin-equivalent transplants. *Transplantation* **36**, 552–557.

Smola, H., Thiekotter, G., and Fusenig, N. E. (1993). Mutual induction of growth factor gene expression by epidermal–dermal cell interaction. *J. Cell Biol.* **122**, 417–429.

Smola, H., Stark, H. J., Thiekotter, G., Mirancea, N., Krieg, T., and Fusenig, N. E. (1998). Dynamics of basement membrane formation by keratinocyte–fibroblast interactions in organotypic skin culture. *Exp. Cell Res.* **239**, 399–410.

Steele, J. G., Dalton, B. A., Johnson, G., and Underwood, P. A. (1995). Adsorption of fibronectin and vitronectin onto *Primaria* and tissue culture polystyrene and relationship to the mechanism of initial attachment of human vein endothelial cells and BHK-21 fibroblasts. *Biomaterials* **16**, 1057–1067.

Stingl, G., Katz, S. I., Shevach, E. M., Rosenthal, A. S., and Green, I. (1978). Analogous functions of macrophages and Langerhans cells in the initiation in the immune response. *J. Invest. Dermatol.* **71**, 59–64.

Streilein, J. W. (1983). Skin-associated lymphoid tissues (SALT): origins and functions. *J. Invest. Dermatol.* **80**(Suppl), 12S–16S.

Zheng, Y., Du, X., Wang, W., Boucher, M., Parimoo, S., and Stenn, K. (2005). Organogenesis from dissociated cells: generation of mature cycling hair follicles from skin-derived cells. *J. Invest. Dermatol.* **124**, 867–876.

# Chapter Eighty

# Tissue-Engineered Cartilage Products

*David W. Levine*

## I. INTRODUCTION

Repair of damaged articular cartilage has been a long-standing clinical challenge. This is an area in which we have made significant clinical progress since the second edition of this textbook. This progress includes an improved understanding of the spectrum of cartilage injury and comorbid conditions, development of treatment algorithms, long-term experience with first-generation autologous cultured chondrocyte cell–based repair, early experience with second-generation autologous cultured chondrocyte cell–based repair, and improved methods for clinical evaluation both in practice and research settings.

The challenges of articular-cartilage repair relate to both the biologic characteristics of the tissue and the epidemiology of cartilage injury. Articular cartilage lacks intrinsic capacity for repair. The clinical strategy for autologous cultured chondrocyte cell therapy derived from the observations that articular cartilage is avascular and has a limited supply of cells adjacent to injuries to mediate a repair process (McPherson and Tubo, 2000). The articular chondrocytes that maintain normal cartilage homeostasis are embedded in an extracelluar matrix that limits a cell's ability to migrate to a zone of injury and that repair it with normally functioning tissue.

Articular-cartilage injuries are common, especially in the knee, and are often associated with damage to other joint structures. There is a spectrum of clinical presentations, ranging from some lesions that may be asymptomatic on initial diagnosis, often in the setting of an incidental finding during an arthroscopy initially indicated for another more evident injury, to lesions that may cause disabling symptoms. The public health impact of these latter types of lesions is magnified because they often occur in relatively young adults and may progress to end-stage arthritis. There clearly is a subset of patients with articular-cartilage injury that are very disabled. For example, preoperative baseline overall condition in a series of patients, mean age 37 years, treated with autologous chondrocyte implantation (ACI) was noted to be a mean of 3.2 on a 10-point scale, indicating poor function with "significant limitations that affect activities of daily living" (Browne *et al.*, 2005).

Although the incidence and prevalence of symptomatic full-thickness chondral injuries in the general population have not been established, it is known that cartilage lesions are a common finding at knee arthroscopy. Curl *et al.* (1997) reviewed records from 31,516 knee arthroscopies and reported Outerbridge Grade III cartilage lesions in 41% and Outerbridge Grade IV lesions in 19%. They reported an

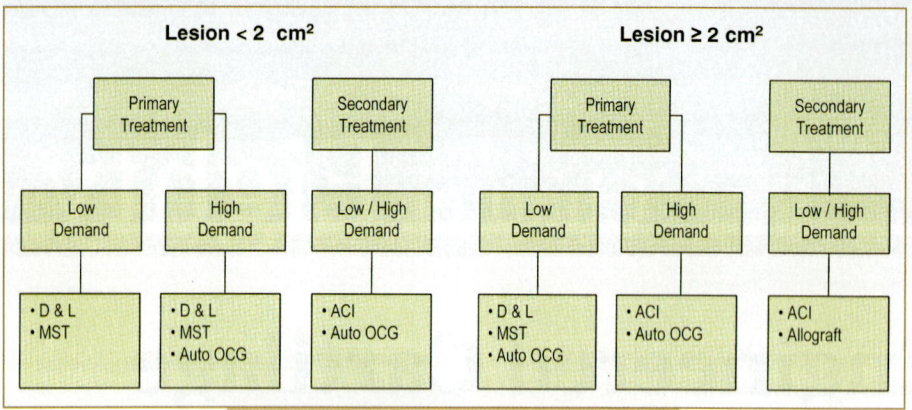

**FIG. 80.1.** Illustrative example of current treatment algorithms for repair of symptomatic cartilage lesions. The algorithm gives an overview of treatment options based on lesion size, patient functional demands, and prior surgical history. Given other factors and the complexity of decision making for individual patients, the authors of this algorithm caution against a mechanistic "menu-driven decision-making process." D&L = debridement and lavage; MST = marrow stimulation technique; Auto OCG = autologous osteochondral graft. Reproduced with permission from Cole and Farr (2001).

| Patient Factors | Co-Morbidity Factors | Lesion Factors | Procedure Factors |
|---|---|---|---|
| • Symptoms | • Ligaments | • Size | • Arthroscopy |
| • Age | • Meniscus | • Location | • Mini-arthrotomy |
| • Expectations | • Alignment | • Containment | • Arthrotomy |
| • Demand | • Bone | • Onset & duration | |
| • Prior treatments | | • Etiology | |
| • General health, systemic disorders | | | |

**FIG. 80.2.** Choosing the most appropriate and optimal cartilage repair treatment for an individual patient depends on multiple factors.

incidence of 5% Grade IV lesions when considering only lesions on the femoral condyle in patients less than 40 years old. Other authors have reported a similar incidence of full-thickness cartilage lesions in other arthroscopy series (Hjelle et al., 2002; Aroen et al., 2004).

Cartilage injury in the knee is often associated with other injuries, especially anterior cruciate ligament (ACL) rupture, meniscus tears, and patella dislocation (Maffulli et al., 2003; Tandogan et al., 2004; Nomura et al., 2003; Nomura and Inoue, 2004 ). For example, Maffulli et al. reported that 42% of patents in a series of 378 patients with ACL tears had associated articular cartilage injuries (Maffulli et al., 2003). In a multicenter registry of 1095 knee cartilage injuries, the etiology of the injury was associated with falls (25.9%), sports (27.4%), activities of daily living (21.2%), and motor vehicle accidents (5.5%), and the etiology could not be identified in 20% of cases (*Cartilage Repair Registry Volume 7*, 2001). In this same study, the onset of symptoms was noted to be acute in 64.6% of cases and gradual in 35.4% of cases. Another distinct etiology for articular cartilage lesions is osteochondritis desssicans (OCD) (Wall and Von Stein, 2003; Micheli et al., 2006).

Since the mid-1990s, treatment algorithms have evolved that reflect the range of cartilage lesion presentations, the different indications for different treatment modalities, and the outcome data (Minas, 1999; Cole and Farr, 2001; Scopp

and Mandelbaum, 2005). While these algorithms may differ in specific detail, they are directionally similar and base treatment guidelines on such factors as lesion size and location, prior surgical history, and patient expectations and functional demands. Figure 80.1 is provided as an illustrative example of current treatment algorithms (Cole and Farr, 2001). Clinical decision making for a given individual patient can be complex, and the multiple factors to evaluate and consider are outlined in Fig. 80.2. In decision making for treatment of articular-cartilage lesions, it is critical to consider the whole joint and the whole patient and not just the hole in the articular cartilage surface.

While it is beyond the scope of this chapter to review the outcome literature for all articular-cartilage treatment modalities, it is helpful to understand where traditional first-line arthroscopic treatments and the cell-based tissue-engineering experience fit in current treatment algorithms. In general, first-line arthroscopic treatments are used in less severe lesions, and autologous chondrocyte implantation is used in more severe lesions (Fig. 80.3). The impetus for developing ACI came from the recognition that while traditional treatments such as debridement and marrow stimulation techniques (Priddie drilling, abrasion arthroplasty, microfracture) may provide some benefit in smaller and/or less symptomatic lesions, they tend to be less effective and less durable in larger and/or more symptomatic lesions (McPherson and Tubo, 2000).

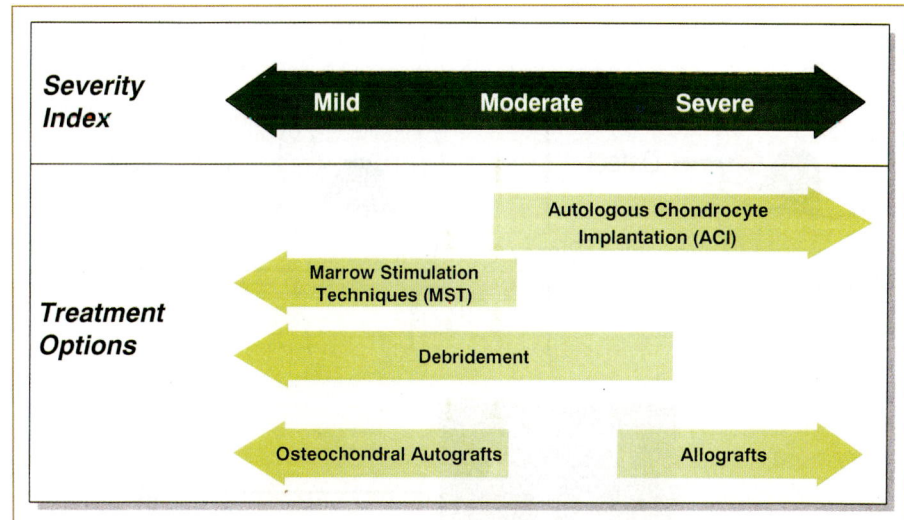

**FIG. 80.3.** There is a wide range of clinical presentations of articular-cartilage injury in the knee. Treatment options reflect severity, based on a combination of patient, lesion, and comorbity factors. The clinical experience with ACI has been at the more severe end of the spectrum of full-thickness articular-cartilage injury.

## II. CLINICAL EXPERIENCE WITH FIRST-GENERATION ACI

### Initial Experience

Lars Peterson and colleagues, in Gothenburg, Sweden, pioneered the clinical application of autologous cultured chondrocytes to repair symptomatic articular-cartilage lesions in the knee joint. The initial results in 23 patients with a mean follow-up of 39 months were published in a landmark paper in 1994 (Brittberg *et al.*, 1994). Results were reported for clinical outcome scores, macroscopic appearance during second look investigational arthroscopies, and histology evaluation of repair tissue biopsies (Fig. 80.4). Clinical results were good to excellent in 14 of 16 lesions (87.5%) on the femoral condyles. Eleven of 15 biopsies (73%) on the femoral condyles had the appearance of hyaline cartilage. Initial results in the seven patella lesions were less promising. Later experience with patella implants, in which patella alignment was addressed concurrently with cell implantation, yielded better results (Peterson *et al.*, 2000; Minas and Bryant, 2005).

### Cell Processing and Surgical Technique

The basic techniques and promising results reported in this initial study became the basis for a larger long-term experience at Peterson's center as well as the commercialization and multicenter experience with ACI. In 1995, Genzyme Corporation established (and subsequently obtained an FDA license for) a commercial GMP facility in the United States to perform the isolation and expansion of chondrocytes from cartilage biopsies. Also in 1995, a multicenter registry was established to track outcomes of ACI and other cartilage procedures. In 1997, the FDA approved Carticel® (autlogous cultured chondrocytes) as

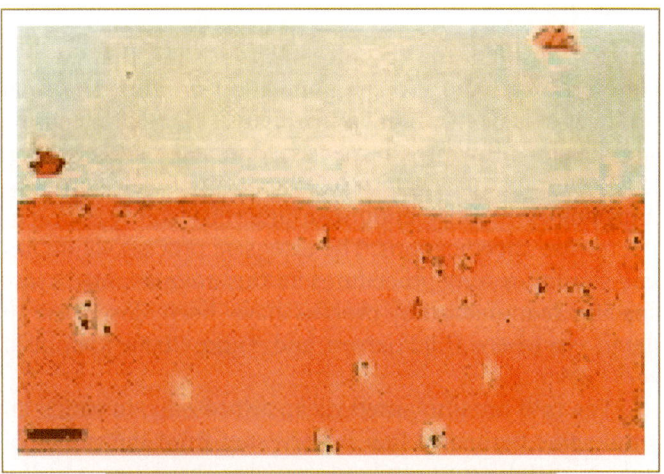

**FIG. 80.4.** Histologic section from a biopsy of cartilage repair tissue 36 months after treatment with autologous chondrocyte implantation of an articular-cartilage defect on the femoral condyle. From Brittberg *et al.* (1994). Copyright © 1994 Massachusetts Medical Society. All rights reserved.

the first cell therapy in orthopedics. Over 12,000 patients have now been treated with Carticel®. Other commercial cell-culturing facilities have been established outside the United States.

ACI is a two-stage procedure. In the first stage, a 200 to 300-mg specimen of cartilage is harvested arthroscopically from a lesser weight-bearing area of the trochlea ridge or the intercondylar notch. In the United States, the procedure is to send this sample to the Genzyme facility (Cambridge, MA), where the sample is enzymatically digested and the chondroycytes are isolated, expanded through primary and secondary culture, and then cryopreserved. Once the facility has been notified of an implantation date by the surgeon

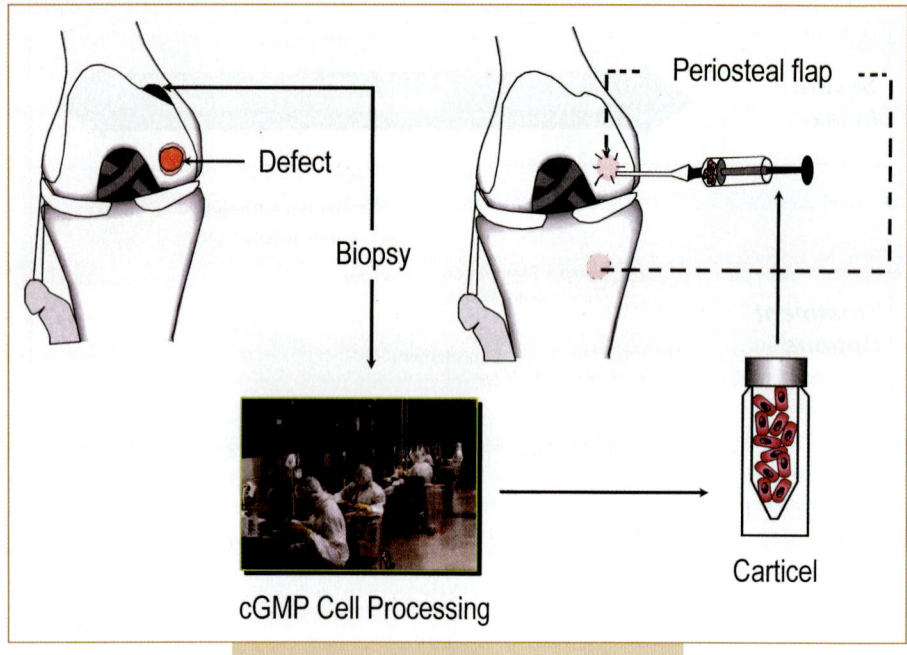

and the hospital, the cryopreserved cells are thawed and cultured to expand the cell population *in vitro*. The final product, a suspension of approximately 12 million cells, is shipped for implantation in a second-stage surgery (Fig. 80.5). All implants are tested for appropriate chondrocytic morphology, sterility, endotoxin, and cell viability as part of the release specifications. The extensive process validation and sterility-testing procedures for this unique, patient lot–specific manufacturing process have been described in detail elsewhere (McPherson and Tubo, 2000; Kielpinski *et al.*, 2005).

The cells are implanted via an arthrotomy, in which an autologous periosteal flap is harvested from the proximal tibia or distal femur, and sutured over the defect and the cells are injected into the defect under the patch after testing for a watertight seal. Fibrin glue can be used if needed to ensure a watertight seal (Fig. 80.6.)

## Long-Term Outcomes in the Swedish Series

Peterson *et al.* (2000) reported outcomes for 94 patients implanted between 1987 and 1995, with only one patient lost to follow-up. The patients were divided into five treatment groups based on defect classification and concurrent ACL procedures. Baseline characteristics and mean follow-up varied by defect category. In general these patients had large, chronic lesions and multiple prior surgeries. The mean defect size ranged from a 4.2 cm² in isolated femoral lesions to 4.7 cm² in OCD lesions. Mean follow-up ranged from 2.7 years in the multiple-lesion group to 4.2 years (range 2–9 years) in the isolated femoral condyle group. Outcome results based on patient assessment varied by defect category, with isolated femoral condyles 92% improved, femoral condyles with ACL reconstruction 75%

improved, OCD lesions 89% improved, multiple lesions 67% improved, and patella defects 68% improved.

Peterson *et al.* (2002) assessed durability of ACI in 61 patients with a mean follow-up of 7.4 years (range 5–11). Durability was assessed by comparing two-year follow-up results with longer-term follow-up. In addition to clinical outcome scores, 11 patients were assessed with second look arthroscopy, indentation probe, and biopsy. The key finding of this study is the durability of results from two years to longer follow-up, at a mean of 7.4 years. At two years, 50 of 61 patients (82%) had results graded as good or excellent; at 5–11 years' follow-up, 51 of 61 patients (84%) had results graded as good or excellent. Eight of 12 biopsies (67%) were graded as hyaline, and four of 12 (33%) were graded as fibrous. There were 10 treatment failures overall (16%), with the rate varying by defect treatment group. The lowest failure rate was in the isolated femoral condyle group (11%) and highest in the patella group (24%). Seven of the 10 treatment failures occurred in the first two years post-ACI, and the three other failures were rated fair to poor at two years. No patient rated as good or excellent at two years failed between two and 11 years.

## Multicenter Registry Studies

Micheli *et al.* (2001) reported results for the first 50 Cartilage Repair Registry patients with a minimum of three years of follow-up. The mean lesion size was 4.2 cm², and the median baseline modified Cincinnati score was 3. Thirty-nine patients (78%) had undergone one or more articular-cartilage procedures in the five years prior to ACI, and 13 patients (26%) had undergone three or more such procedures.

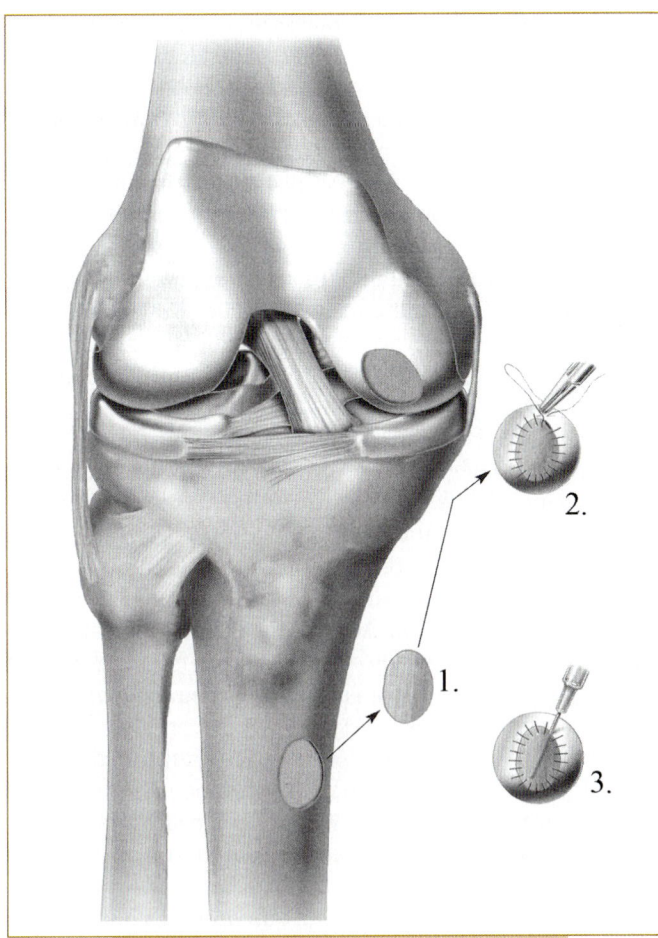

**FIG. 80.6.** Surgical implantation of first-generation ACI. The figure illustrates (1) harvest of periosteal patch, (2) suturing of periosteal patch over cartilage defect on femoral condyle, and (3) catheter in place for injection of autologous cultured chondroctyes under sutured patch. After chondrocyte injection, the catheter is removed and the catheter entry site is sealed with additional suture. Copyright © 2006 Genzyme Corporation. All rights reserved.

This study reported both patient and clinician evaluations. There was a close correlation between patient-reported scores and clinician evaluations. Improvement was seen in 85% of patients, 2% were unchanged, and 13% declined. Patients reported a median improvement of 5 points on the overall 10-point modified Cincinnati score from a median baseline of "3" ("poor" — "significant limitations affect activities of daily living") to a median score of "8" ("very good" — "only a few limitations with sports") at three-year follow-up. Eleven patients, including three treatment failures, underwent subsequent operations. The most common cause of symptoms requiring further intervention was hypertrophy of the periosteal patch, which occurred in six patients.

Browne *et al.* (2005) reported results for the first 100 Cartilage Repair Registry patients with femoral lesions to reach the five-year follow-up time frame. The majority of patients (70%) had at least one cartilage repair procedure prior to the index arthroscopy. Eighty-five percent had a single defect with a mean size of 4.2 cm², and 15% of patients had multiple defects with a mean lesion size of 9 cm².

The mean baseline score was 3.2. Eighty-seven patients completed full follow-up. Overall, mean Cincinnati scores improved by 2.6 points ($p < .0001$). Of the 87 patients who completed follow-up, 62 (71%) improved, 6 (7%) reported no change, and 19 (22%) worsened. Of the 62 who improved, the overall Cincinnati score improved 4.1 points from baseline ($p < 0.0001$). The authors concluded that "Patients in this study represent a clinical challenge with few treatment options. Despite the large lesions, prior surgical history, and poor function of patients at baseline, 71% of patients reported improvement in their overall condition score five years after implantation."

Fu *et al.* (2005) directly compared two Cartilage Registry Repair cohorts at three-year follow-up: a cohort of 58 debridement patients and a cohort of 58 ACI patients. The two cohorts had similar demographics and defect pathologies at baseline. However, more ACI patients failed a previous debridement or marrow stimulation procedure than the debridement patients. The mean total defect area for the debridment cohort was 4.5 cm², and for the ACI cohort it was 5 cm².

Fifty-six ACI patients and 42 debridement patients completed follow-up outcome assessments. Eighty-one percent of the ACI patients and 60% of the debridement patients reported median improvements of 5 and 2 points, respectively, in the overall condition score. ACI patients also reported greater improvements in the median pain and swelling scores than debridement patients. The treatment failure rate was the same for both ACI and debridement patients. Based on the *a priori* definition of treatment failure, there were three (5.2%) failures in the debridement group and four (6.9%) in the ACI group. The authors concluded that "Although patients treated with debridement for symptomatic, large, focal, chondral defects of the distal femur had some functional improvement at follow-up, ACI patients obtained higher levels of knee function and had greater relief from pain and swelling at three years."

Micheli *et al.* (2006) reported on ACI outcomes in children and adolescents. Thirty-seven registry patients with a mean lesion size of 5.2 cm² and a mean age of 16 at implant were in the cohort. Despite the young age of these patients, they were similar to adult ACI cohorts, in that the majority (70%) had undergone at least one prior knee procedure before the index arthroscopy for cartilage specimen harvest. Thirty-two patients completed follow-up assessments at a mean follow-up of 4.3 years. Twenty-eight patients (88%) reported a favorable outcome at a minimum of two-year follow-up.

## Other Studies of ACI

Knutsen *et al.* (2004) in Norway conducted a four-center, 80-patient randomized controlled trial comparing ACI to microfracture at a two-year follow-up. Endpoints were clinical outcome scores (Lysholm, VAS pain score, and SF-36) and a histology score. Both groups were improved at two years, with no statistically significant difference between the ACI and microfracture groups in Lysholm, VAS pain score, or SF-36 mental health subscale (results reported in aggregate, i.e., not stratified by lesion size). The authors do report a statistically significant difference in the SF-36 physical component score in favor of the microfracture arm; however, the authors also report different SF-36 physical component mean scores at baseline between the arms.

The authors report that microfracture outcomes varied by lesion size:

> In the microfracture group, patients with a lesion smaller than 4 cm$^2$ had significantly better clinical results (according to the Lysholm score, VAS pain score, and SF-36 physical component score) than those with a bigger defect ($p < 0.003$). We did not find this association between the size of the defect and the clinical outcome in the autologous chondrocyte implantation group ($p > 0.89$).

With regard to the histology comparison, the authors conclude, "There was a tendency in our study for the autologous chondrocyte implantation procedure to result in more hyaline repair cartilage than the microfracture procedure, but this was not a significant finding with the numbers available."

Study strengths include that it is a prospective, randomized trial. Limitations include that there were no trochlea lesions in the study, and the patient baseline condition appears to be less severe than the U.S. Carticel patient population. The authors concluded that "because microfracture is a relatively simple one-stage procedure, it may be more suitable for a primary first-line cartilage repair of a local contained defect. In patients in whom microfracture has failed and in those with bigger, noncontained defects, autologous chondrocyte implantation may be a better option."

Bentley *et al.* (2003) in the United Kingdom conducted a single-center, 100-patient randomized controlled trial comparing two ACI arms (ACI with periosteal flap coverage and ACI with type I/III collagen membrane coverage) to mosaicplasty, with a mean follow-up of 1.7 years. Outcome measures include clinical scores, second look arthroscopy with ICRS scoring, and histology.

The results overall favor ACI as compared to mosaicplasty in all measured parameters, including clinical outcome scores, histology results, and ICRS scoring on follow-up arthroscopy. Overall, 88% of the ACI patients vs. 69% of the mosaicplasty patients were graded as "excellent or good result."

Study strengths include that it is a prospective, randomized design with follow-up arthroscopy evaluation as well as clinical outcome scores. In addition, the patients in this study do appear to have some comparability to U.S. Carticel patients in terms of mean lesion size, baseline symptoms,

and prior surgical history. Mean lesion size is 4.66 cm$^2$. Mean duration of symptoms was 7.2 years, and all but six patients had undergone previous surgical interventions. Limitations of the study include that while the study showed better scores for ACI vs. mosaicplasty in all parameters, the differences in clinical scores were statiscally significant for only one anatomic location, the medial femoral condyle.

The authors conclude, "There is definite evidence, however, that ACI is valuable for selected patients in that it gives healing of hyaline cartilage in chondral and osteochondral defects. It can also dramatically reduce the symptoms of pain and disability. The continued use of mosaicplasty, however, appears to be dubious."

There are several other published studies on ACI from other centers. They are too numerous to review in detail here. In general, the results are similar to the studies just discussed. A few specific additional points are worth noting with regard to special populations or indications treated with ACI and health economics of this technology. Mithofer *et al.* (2005) reported on ACI to treat adolescent athletes. Overall, they report a 96% rate of return to high-impact sports and 60% achieving an athletic level equal to or higher than the preinjury level. Return rate to preinjury sports rate correlated with shorter preoperative symptom duration and a lower number of prior operations. This suggests that earlier intervention may optimize cell-therapy outcomes, an especially important consideration in younger patients.

Minas has reported successful outcomes in young adults with complex joint pathology for which ACI was performed as an alternative to an artificial joint arthroplasty (Minas, 2001; Minas and Bryant, 2005). Further, while the vast majority of ACI cases are in the knee, there are limited reports of successful use in other joints, especially the ankle. While the patient lot–specific manufacture of a live cellular product with stringent quality controls is inherently expensive, favorable health economics were demonstrated in three studies that assessed economic parameters (Minas, 1998; Lindahl *et al.*, 2001; Yates, 2003).

## Complications of ACI

Patients who undergo ACI may require subsequent surgical procedures on the treated knee. The majority of these procedures are arthroscopic interventions. Such procedures frequently have no findings attributable to ACI. The most common findings assessed related to ACI is graft overgrowth or tissue hypertrophy. This complication can present with symptoms of "catching" or "clicking" and generally responds well to arthroscopic debridement. The next-most-common findings at arthroscopy assessed as related to ACI treatment are joint adhesions or arthrofibrosis. These are treated with manipulation under anesthesia or arthroscopic lysis of adhesions. Graft complications, such as partial delamination of the graft and periosteal patch, have also been reported as findings at subsequent surgical procedures.

Multiple prior surgeries and a history of ligament reconstruction or meniscal procedures concomitant with ACI appear to be risk factors for a post-ACI subsequent procedure. The frequency of arthroscopic intervention after ACI is likely due to a combination of factors, including the associated arthrotomy necessary for implantation, complex knee pathology, including comorbid conditions, concurrent procedures and prior surgical history, and a hypertrophy of the graft or periosteal patch some patients.

## Summary of First-Generation ACI Clinical Experience

The data sources just summarized all have strengths and limitations, and no single source is definitive. Multiple factors make direct comparisons or data synthesis of these varied studies challenging. These factors include differences in study design, data collection instruments, length of follow-up, evaluation criteria, treatment failure definitions, and patient populations. However, bearing these factors in mind, these different studies are generally consistent and complementary in supporting the following overall conclusions regarding the efficacy of first-generation ACI:

- ACI has demonstrated successful clinical outcomes in challenging large, symptomatic cartilage lesions where other treatment modalities, such as debridement, microfracture, and osteochondral autografts, have been less successful.
- ACI has demonstrated successful clinical outcomes in patients who have failed prior cartilage repair procedures.
- ACI has demonstrated the ability to improve patients with very severe disability at baseline, such as pain and dysfunction with activities of daily living.

While these are important advances, first-generation ACI has limitations. These include the harvest and suturing of a periosteal flap to secure the implanted chondrocytes in the defect. Not only does this prolong the operative time and require a more extensive arthrotomy, but some authors have postulated that the periostal flap may contribute to the formation of hypertrophic tissue, the most common complication of first-generation ACI (Bentley *et al.*, 2003; Willers *et al.*, 2005). These limitations have led to efforts to develop improved technology for chondrocyte implantation. Goals include maintaining the benefits of achieving biological cartilage resurfacing and functional repair tissue while improving surgical delivery and reducing complications.

## III. CLINICAL EXPERIENCE WITH SECOND-GENERATION ACI

A significant degree of early clinical experience with second-generation ACI has been obtained in Europe and Australia. The first clinical step was to replace the autologous periosteal flap with a porcine type I/III collagen patch.

**FIG. 80.7.** Second-generation autologous cultured chondrocyte graft technology. In matrix-induced autologous chondrocyte implantation (MACI®), the cultured chondrocytes are seeded on 4 × 5-cm type I/III collagen membrane. This eliminates the need for the periosteal patch and extensive suturing used in first-generation ACI and facilitates less invasive surgical implantation. Copyright© 2005 Genzyme Corporation. All rights reserved.

This has been termed CACI, or collagen-covered ACI. Early results suggested clinical and histological results similar to periosteum-covered ACI, but with a lower incidence of post-implant hypertrophy (Briggs *et al.*, 2003; Bentley *et al.*, 2003).

In CACI, the chondrocytes are still delivered to the defect in a liquid suspension that requires suturing a collagen patch in place with a watertight seal. A further innovation was to combine the chondrocytes with an implantable membrane at the end stages of the culturing process. The chondrocytes adhere to the membrane and can be implanted as a construct that eliminates extensive suturing and facilitates less invasive surgical approaches (Fig. 80.7). Two different types of membrane are used in this manner and marketed outside the United States as second-generation ACI products. In matrix-induced autologous chondrocyte implantation (MACI®) (Genzyme Biosurgery, Cambridge, MA) the chondrocytes are seeded on a 4 × 5-cm porcine type I/III collagen membrane. In Hyalograft C® (FAB, Abano, Italy) chondrocytes are seeded on a 2 × 2-cm Hyaff 11 membrane, an esterified derivative of hyaluronan. Other biomaterial approaches to delivering chondrocytes for cartilage repair are being pursued by several entities, but they do not have extensive clinical experience to date.

Over 4000 patients have been implanted with second-generation ACI. The early clinical experience is promising, in that it meets the goals for a second-generation ACI technology (Bartlett *et al.*, 2005; Ronga *et al.*, 2005 Marlovits *et al.*, 2004, 2006; Marcacci *et al.*, 2005; D'Anchise *et al.*, 2005).

This approach eliminates the periostial patch and can be implanted less invasively with sutureless fixation. The next step for clinical development of second-generation ACI will be further controlled clinical trials that meet the requirements for product registration (BLA) in the United States and other jurisdictions that are developing similar regulations for cell-based therapies.

## IV. CONCLUSIONS

Further progress in treating articular-cartilage lesions is likely to come from several approaches, including the incremental improvement of autologous cultured chondrocyte implantation. Given the range of clinical presentations as outlined in the chapter introduction, it is unlikely that a single technology will be applicable to the entire spectrum of articular-cartilage pathology. Instead, one can envision an array of diagnostic and therapeutic tools to enable orthopedists to treat effectively each stage of cartilage injury and comorbid injury. These could include further development of cell-based therapy as well as other approaches, such as biomaterial implants alone or combined with repair-stimulating growth factors, which may be more applicable to different stage diseases.

There is a continuum from early chondral injury, identified when it may still be relatively asymptomatic, to symptomatic focal cartilage injury to end-stage arthritis. At the early end of this continuum is a need to better identify and mitigate the risk of progression of disease. Second-generation ACI focuses on improved delivery of autologous cultured chondrocytes for treatment of symptomatic cartilage lesions. This is well along in clinical development. Additional work is needed to assess the applicability of cell-therapy approaches to arthritis. This may include synthetic osteochondral constructs and is likely to focus on indications where traditional arthroplasty has the most limitations, e.g., younger, athletic patients. Synthetic osteochondral constructs may also be applicable for other etiologies, such as OCD lesions with extensive bone deficiency.

The clinical applications of cell-therapy tissue engineering to date have focused on autologous cultured chondrocytes. As discussed elsewhere in this book, there is extensive preclinical work on a wide range of other cell sources, such as adipose cells, mesenchymal and embryonic stem cells, and allogeneic chondrocytes, that have the potential to expand the clinical applications to other cell types. Cell modification strategies using gene therapy or growth factors to mediate cartilage repair with different cell types are also under preclinical investigation and have the potential to broaden the scope of cell-therapy clinical applications.

Developing and producing surgical implants with live cells poses scientific, logistical, commercial, and clinical challenges that are very different from the challenges with the development and production of more traditional orthopedic implants. To the extent that future cell-therapy applications in cartilage repair require *in vitro* cell processing on a manufacturing scale, the experience with Carticel® (autologous cultured chondrocytes) demonstrates the feasibility of manufacturing a high-quality cell product for surgical implantation. The clinical success with first-generation ACI in a patient population with very symptomatic and challenging cartilage lesions has laid the foundation for further progress in cell-based orthopedic-tissue engineering in general and cartilage repair in particular.

## V. REFERENCES

Anderson, A. F., Browne, J. E., Erggelet, C. E., Fu, F., Mandelbaum, B. R., Micheli L. J., and Moseley, J. B. (2001). *Cartilage Repair Registry Periodic Report 7*.

Aroen, A., Loken, S., Heir, S., Alvik, E., Ekeland, A., Granlund, O. G., and Engebretsen, L. (2004). Articular cartilage lesions in 993 consecutive knee arthroscopies. *Am. J. Sports Med.* **32**(1), 211–215.

Bartlett, W., Skinner, J. A., Gooding, C. R., Carrington, R. W. J., Flanagan, A. M., Briggs, T. W. R., and Bentley, G. (2005). Autologous chondrocyte implantation versus matrix-induced autologous chondrocyte implantation for osteochondral defects of the knee. *J. Bone Joint Surg. (Br.)* **87B**, 640–645.

Bentley, G., Biant, L. C., and Carrington, R. (2003). A prospective, randomised comparison of autologous chondrocyte implantation versus mosaicplasty for osteochondral defects in the knee. *J. Bone Joint Surg. (Br.)* **85**, 223–230.

Briggs, T. W. R., Mahroof, S., and David, L. A. (2003). Histological evaluation of chondral defects after autologous chondrocyte implantation of the knee. *J. Bone Joint Surg. (Br.)* **85**, 1077–1083.

Brittberg, M., Lindahl, A., Nilsson, A., Ohlsson, C., Isaksson, O., and Peterson, L. (1994). Treatment of deep cartilage defects in the knee with autologous chondrocyte transplantation. *N. Engl. J. Med.* **331**(14), 889–895.

Browne, J. E., Anderson, A. F., Arciero, R., Mandelbaum B., Moseley J. B., Micheli, L. J., Fu, F., and Erggelet, C. (2005). Clinical outcome of autologous chondrocyte implantation at 5 years in U.S. subjects. *Clin. Orthop. Rel. Res.* **436**, 237–245.

Cole, B. J., and Farr, J. (2001). Putting it all together. *Oper. Tech. Orthop.* **11**(2), 151–154.

Curl, W. W., Krome, J., Gordon, S. E., Rushing, J., Paterson Smith, B., and Poehling, G. G. (1997). Cartilage injuries: a review of 31,516 knee arthroscopies. *J. Arthrosc. Rel. Surg.* **13**(4), 456–460.

D'Anchise R., Manta N., Prospero E., Bevilacqua, C., and Gigante, A. (2005). Autologous implantation of chondrocytes on a solid collagen scaffold: clinical and histological outcomes after two years of follow-up. *J. Orthopaed. Traumatol.* **6**, 36–43.

Fu, F., Zurakowski, D., Browne, J. E., Mandelbaum B., Erggelet, C., Moseley, J. B., Anderson, A. F., and Micheli, L. J. (2005). Autologous chondrocyte implantation versus debridement for treatment of full-thickness chondral defects of the knee: an observational cohort study with 3-year follow-up. *Am. J. Sports Med.* **33**(11), 1658–1666.

Hjelle, K., Solheim, E., Strand, T., Muri, R., and Brittberg, M. (2002). Articular cartilage defects in 1000 knee arthroscopies. *J. Arthrosc. Rel. Surg.* **18**(7), 730–734.

Kielpinski, G., Prinzi, S., Duguid, J., and du Moulin, G. (2005). Roadmap to approval: use of an automated sterility test method as a lot release test for Carticel®, autologous cultured chondrocytes. *Cytotherapy* **7**(6), 531–541.

Knutsen, G., Engebretsen, L., and Ludvigsen, T. C. (2004). Autologous chondrocyte implantation versus microfracture: a prospective randomized Norwegian multicenter trial. *J. Bone Joint Surg. (Am.)* **86**, 455–464.

Lindahl, A., Brittberg, M., and Peterson, L. (2001). Health economics benefits following autologous chondrocyte transplantation for patients with focal chondral lesions of the knee. *Knee Surg. Sports Traumatol. Arthrosc.* **9**(6), 358–363.

Maffulli, N., Binfield, P. M., and King, J. B. (2003). Articular cartilage lesions in the symptomatic anterior cruciate ligament–deficient knee. *J. Arthrosc. Rel. Surg.* **19**(7), 685–690.

Marcacci, M., Massimo, B., Domenico, B., Delcogliano, A., Ghinelli, D., Gobbi, A., Kon, E., Pederzini, L., Rosa, D., Sacchetti, *et al.* (2005). Articular cartilage engineering with Hyalograft® C, three-year clinical results. *Clin. Orthop. Rel. Res.* **435**, 96–105.

Marlovits, S., Striessnig, G., Kutscha-Lissberg, F., Resinger, C., Aldrian, S. M., Vescei, V., and Trattnig, S. (2004). Early postoperative adherence of matrix-induced autologous chondrocyte implantation for the treatment of full-thickness cartilage defects of the femoral condyle. *Knee Surg. Sports Traumatol. Arthrosc.*, **13**, 451–457.

Marlovits, S., Zeller, P., Singer, P., Resinger, C., and Vécsei, V. (2006). Cartilage repair: generations of autologous chondrocyte transplantation. *Eur. J. Radiol.* **57**, 24–31.

McPherson, J. M., and Tubo, R. (2000). Articular cartilage injury. *In* "Principles of Tissue Engineering" (R. Lanza, R. Langer, and J. Vacanti, eds.), 2nd ed., pp. 697–609. Academic Press, San Diego, CA.

Micheli, L. J., Browne, J. E., Erggelet, C., Fu, F., Mandelbaum, B. R., Moseley, J. B., and Zurakowski, D. (2001). Autologous chondrocyte implantation of the knee: multicenter experience and minimum three-year follow-up. *Clin J Sports Med.* **11**, 223–228.

Micheli, L. J., Moseley, J. B., Anderson, A. F., Browne, J. E., Erggelet, C., Arciero, R., Fu, F., and Mandelbaum, B. R. (2006). Articular cartilage defects in children and adolescents: treatment with autologous chondrocyte implantation. *J. Pediatric. Orthop.* **26**(4), 455–460.

Minas, T. (1998). Chondrocyte implantation in the repair of chondral lesions of the knee: economics and quality of life. *Am. J. Orthop.* **27**(11), 739–744.

Minas, T. (1999). The role of cartilage repair techniques, including chondrocyte transplantation, in focal chondral knee damage. *AAOS Instr. Course Lect.* **48**, 629–643.

Minas, T. (2001). Autologous chondrocyte implantation for focal chondral defects of the knee. *Clin. Orthop. Relat. Res.* **391** (Suppl.), 349–361.

Minas, T., and Bryant, T. (2005). The role of autologous chondrocyte implantation in the patellofemoral joint. *Clin. Orthop. Rel. Res.* **436**, 30–39.

Mithofer, K., Minas, T., Peterson, L., Yeon, H., and Micheli, L. (2005). Functional outcome of knee articular cartilage repair in adolescent athletes. *Am. J. Sports Med.* **33**(8), 1147–1153.

Nomura, E., and Inoue, M. (2004). Cartilage lesions of the patella in recurrent patellar dislocation. *Am. J. Sports Med.* **32**(2), 498–502.

Nomura, E., Inoue, M., and Kurimura, M. (2003). Chondral and osteochondral injuries associated with acute patellar dislocation. *J. Arthrosc. Rel. Surg.* **19**(7), 717–721.

Peterson, L., Minas, T., and Brittberg, M. (2000). Two- to nine-year outcome after autologous chondrocyte transplantation of the knee. *Clin. Orthop. Rel. Res.* **374**, 212–234.

Peterson, L., Brittberg, M., and Kiviranta, I. (2002). Autologous chondrocyte transplantation. Biomechanics and long-term durability. *Am. J. Sports Med.* **30**(1), 2–12.

Ronga M., Grassi, F. A., Montoli, C., Bulgheroni, P., Genovese, E., and Cherubino, P. (2005). Treatment of deep cartilage defects of the ankle with matrix-induced autologous chondrocyte implantation. *Foot Ankle Surg.* **11**, 29–33.

Scopp J. M., and Mandelbaum, B. R. (2005). A treatment algorithm for the management of articular cartilage defects. *Orthop. Clin. N. Am.* **36**, 419–426.

Tandogan, R. N., Taser, O., Kayaalp, A., Taskiran, E., Pinar, H., Alparslan, B., and Alturfan, A. (2004). Analysis of meniscal and chondral lesions accompanying anterior cruciate ligament tears: relationship with age, time from injury, and level of sport. *Knee Surg. Sports Traumatol. Arthrosc.* **12**, 262–270.

Wall, E., and Von Stein, D. (2003). Juvenile osteochondritis dissecans. *Orthop. Clin. of North Am.* **34**, 341–353.

Willers, C., Chen, J., Wood, D., Xu, J., and Zheng, M. H. (2005). Autologous chondrocyte implantation with collagen bioscaffold for the treatment of osteochondral defects in rabbits. *Tissue Eng.* **11**(7–8), 1065–1076.

Yates, J. (2003). The effectiveness of ACI for treatment of full-thickness articular cartilage lesions in workers' compensation patients. *Orthopedics* **26**(3), 295–299.

# Chapter Eighty-One

# Tissue-Engineered Bone Products

*John F. Kay*

## I. INTRODUCTION

This chapter addresses the history and evolution of the three major constituents of bone that make up the necessary elements for current and future tissue-engineered bone replacements and adjuncts to bone repair. Today's clinical usage of various biomaterials has evolved from advances in the understanding and development of these three elements: those that are osteogenic, osteoconductive, and osteoinductive. One particular focus addresses the osteoinductive element, for it is within this class of materials that the increases in understanding, development, and manufacture have been the greatest. The reason the market for contemporary bone grafting has grown and why there has been measurable progress toward the goal of minimizing or eliminating autogenous iliac crest graft harvesting is due to the products' mimicking the osteoinductive element. The clinical use of composite grafting as a method to accomplish tissue engineering today is growing, and the developments of the last decade or so have set the stage for even more effective bone graft substances and the surgical techniques to use them effectively.

For centuries, one goal of medical specialists has been the creation of a viable substitute to repair bone. Through the ages, substances such as ivory, leather, noble metals, plaster of Paris, directly transplanted hard tissues from other species, and other hard substances have been used in an attempt to heal or repair bone tissues; all attempts were met with either short-term failures, long-term failures, or non-reproducible results. These earliest attempts were essentially empirical, for it was thought bone was a hard, lifeless structure that merely served to keep the body erect. They did, however, document the interest of individuals who dealt with addressing maladies of the body in repair and/or replacing damaged parts. Using the best knowledge and insight at hand at the time, the body tissues were attempted to be engineered. In the most recent decades, and armed with the perfect vision of hindsight, researchers and clinicians have made more noble attempts to develop substances to aid in the repair and regeneration of bone. The evolution of more discriminating analytical techniques and progress in materials science have allowed a more educated, targeted approach toward the quest, and advances in surgical technique and instrumentation have made attempts more successful. Just as with many other compounds and structures, engineering various constituents to perform together in a specific fashion now dominates product development of most biomaterials, including body tissues such as bone. Effectively no different than electrical or civil engineering, *tissue engineering* has become a common term in everyday biomaterials and hard-tissue clinical practice.

That bone was composed of several constituents capable of being separated and analyzed was realized in the early 1950s (Carlstrom and Engstrom, 1956; Eanes and Posner, 1970; Eastoe and Eastoe, 1954; Eastoe, 1956; Glimscher, 1959). Bone, now well studied, is known to be a complex, vital, and necessary composite tissue in the body, containing many individual constituents, and the analysis of bone has provided the basis for trying to mimic each piece and the whole (Recker, 2002). Over time, developmental efforts have attempted to subdivide the elements of bone and to mimic them on a sequentially smaller and smaller scale. With this methodology, the number of elements and the complexity of the duplication had increased (Davies, 2000). To engineer bone properly in today's world, one would have to split the material apart, create the best analogs to each constituent, and then recombine them in a manner that provides function in an orthopedic procedure. This will, no doubt, change, with the future offering replacement bones and bone segments "grown" *in situ*.

Finally, the advances in load-bearing medical device procedures, such as total hip replacement, total knee replacement, internal and external fracture fixation, and spinal instrumentation, have given rise to a need to develop and expand bone graft technology. For example, the use of structural allograft spacers and metallic spinal fusion cages has fostered the development of many bone graft substances to fill the spaces in their interiors, to aid bone adaptation at and in their surfaces, and to support the components designed to support patient loading.

## II. BONE HEALING

Without an understanding of the bone-healing process, development of bone repair materials and tissue augmentation would not be possible. The bone regeneration and repair process is not completely understood, but greater secrets have been unlocked on a seemingly monthly basis, and with this ever-increasing knowledge, more effective materials have been engineered. Traditionally, bone healing has been separated into four stages of healing: inflammation and clot formation, cellular infiltration and soft-callus formation, hard-callus formation, and remodeling. In terms of the actions that initiate the mechanism of repair leading to successful healing, the first stage is the most important, because both a favorable biological environment and mechanical stability need to be established to maximize the opportunity for a successful clinical outcome. In the following sections, various classes of materials are discussed. These advances in the elements of tissue engineering are designed to aid the normal bone-healing sequence, not make up for shortcomings in establishing the environment or stability for healing; in that way, they can increase the

chances for a successful outcome. No bone-grafting material is designed to overcome factors that may impede bone healing, such as smoking, alcohol, NSAIDS, an immuno-compromised host, soft-tissue disruption, and infection. While very complex from a biochemistry standpoint, the ability to regenerate bone and the most effective tissue-engineered bone grafts used today employ three basic elements of bone repair: osteogenic materials, osteoconductive materials, and osteoinductive materials (Helm, 2001; Kalfas, 2001; Pilitsas, 2002; Urist, 1965). The body employs all three to a greater or lesser degree, depending on the site of regeneration or repair and/or the severity of the repair problem. The three have also been summarized as cells, scaffolds, and signals. In contemporary bone-grafting language, the use of two or more such elements is called *composite bone grafting*.

## III. OSTEOGENIC GRAFTING MATERIALS

Osteogenic substances are ideally suited to perform as constituents of a bone graft because they have cellular constituents like those present at the host site. They are transplanted from one place in the body to another and may be transplanted directly, but they may also be treated in one way or another in between harvesting and implantation. The key feature of any osteogenic component is the presence of vital cells and a spectrum of naturally occurring proteins that can differentiate pluripotent stem cells down osteoblastic pathways (Cheng *et al.*, 2003).

Autograft bone, considered the gold standard, is the ideal bone graft if its vitality is maintained and it is securely affixed within the implant bed. The transplanted cells within an autograft are provided with a biologically familiar environment in which to perform as they would prior to transplantation, without any immunogenic response triggered by the host (Muschler *et al.*, 2003). The life span of these vital cells, however, is not known once removed from the body, nor is the optimal storage medium to preserve vitality for the time between harvest and use (Betz, 2002).

Bulk autogenous grafts can sustain functional loads if they are securely affixed and incorporated; morselized or minced-up autogenous bone has the advantage of containing vital cells and being formable and mixable with other graft materials to provide an osteogenic element to an osteoconductive and/or osteoinductive substance (Fleming, 2000). Division of the bone into small pieces has the advantage of providing a greater surface area and access to the vital cellular elements. For general bone-void filling of defects and posterolateral spinal fusions, minced, finely divided autogenous bone provides maximum release of osteogenic elements through high surface area; the trade-offs of fine mincing are most likely shorter vitality time, longer preparation time, and a greater need for containment. The pain, morbidity, and disadvantages of harvesting iliac crest autograft are well established, including long-term pain and the inability to serve as a future harvest site.

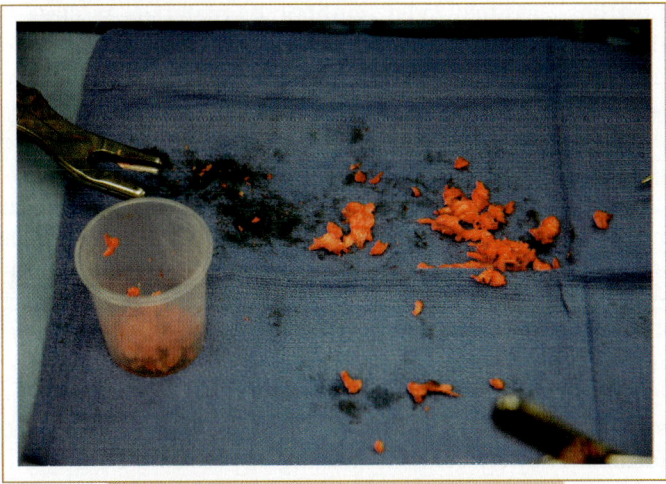

**FIG. 81.1.** Reducing autograft into smaller pieces is an effective way to provide greater access of vital cellular elements, because the surface area is increased. Depending on the application, mincing by rongeurs and creating even smaller pieces in a bone mill are effective ways to add osteogenic elements into a bone graft.

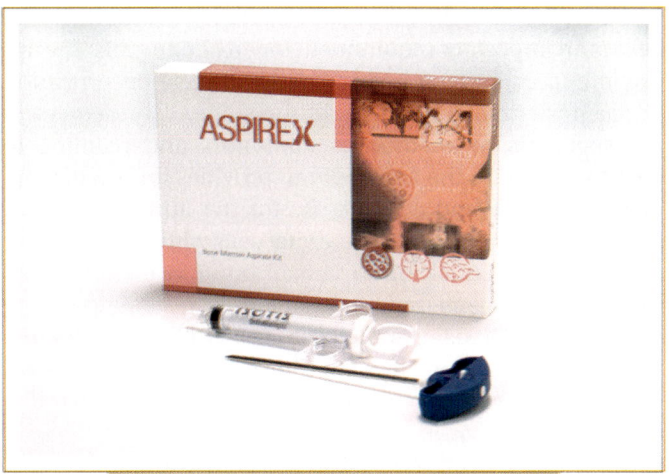

**FIG. 81.2.** A device for obtaining bone marrow aspirate employs a less invasive approach to marrow-rich sources such as the iliac crest, without the second-site surgery, pain, and/or longer-term morbidities and complications of larger autograft harvests.

And the advantages are documented as well: Notwithstanding its gold standard status, it is associated with an estimated 15–25% nonunion rate. One report estimates that approximately one-third of the harvest patients experience long-term (over two years) pain and complications (Burkus *et al.*, 2002b) (Fig. 81.1). Nonetheless, autograft is routinely used in orthopedic surgical procedures and it serves as the control for most preclinical and clinical evaluations, either required or strongly recommended by such regulatory agencies as the U.S. Food and Drug Administration (FDA) or for corporate marketing purposes; it has the most extensive historical database and is accepted as an appropriate positive control for performance comparisons.

A less invasive source of vital osteogenic cells is through the harvesting of bone marrow aspirate from the iliac crest or other marrow cavities. Such harvesting has become popular in recent years as the concept of composite grafting, the intentional combining of the essential elements previously described, grew in popularity and the desire to minimize or preclude iliac crest autograft harvests became a greater focus. Harvesting of bone-marrow aspirate is very technique sensitive, in that diligence must be exercised in order to yield a high volume of aspirate while minimizing excessive blood draw, and it is controversial in its reports of efficacy (Muschler, 2003; Hernigou, 2005; Fleming, 2000; Lieberman, 2006; Muschler, 2006). A relatively simple, inexpensive, disposable needle-type device (Fig. 81.2) can be used for bone-marrow aspirate harvesting. The osteogenic nature of the aspirate is acknowledged, but it may be that concentration is the only way to provide an efficacious, vital composite graft constituent (Hernigou, 2005). Bone marrow aspirated from the iliac crest, for example, is often substantially diluted by peripheral blood; studies indicate that

maintenance of a higher concentration of the osteogenic cells can be achieved by limiting the aspiration volume to 1–2 mL (Fleming, 2000). In concentrating the aspirate, one must be careful to protect the cells from mechanical damage, which would decrease the vitality of the aspirate. Nevertheless, bone-marrow aspirate can provide an osteogenic cell–containing addition to a composite graft that helps to create a moldable, packable graft or a surface coating on a porous implant such as a structural allograft.

The motivation to avoid harvesting bone from the iliac crest, along with the advent of composite grafting, has fostered the use of a series of blood-derived or "blood spin-down" products to enhance graft performance further. The concept is founded on the recognition that certain portions of the blood contain growth regulatory factors that may stimulate bone formation. Referred to as *platelet-rich plasma* (PRP) and/or autologous growth factors (AGF®), and on occasion improperly represented as similar to bone-marrow aspirate, these are a source of endogenous growth factors that are believed to act directly on reparative cells. Originally developed as an adjunct for wound healing, PRP contains a mixture of growth factors, primarily cytokines, that were later believed to enhance healing of osseous tissue, especially when combined with autograft. The reports in the literature are mixed, though mostly negative, regarding the efficacy of these materials when used as a stand-alone bone graft material (Anitua, 1999; Froum, 2002; Marx, 1998). However, these blood-derived products are effective in containing the multiple components used in composite grafts. Typically, a modestly priced, disposable blood-contact harvesting and container system is provided for single-patient use, with the centrifuge unit a fixed, nonfluid contact piece of operating room equipment.

The presence of osteogenic material containing vital cells is an important requirement for any bone graft. A properly prepared implant bed, such as complete decortication in the posteriolateral gutters of the spine, debridement of the segments of an osseous nonunion, and reaming to bleeding bone in the acetabulum, provides an osteogenic-element source and a viable site for the initiation of new bone formation; any additional vital cells within a bone graft add to the potential for successful bone formation. Used properly, osteogenic material can aid in hastening healing as part of a composite graft; its absence will not necessarily preclude healing but will reduce the chance for a successful outcome, especially in compromised sites and demanding applications.

## IV. OSTEOCONDUCTIVE BONE GRAFT MATERIALS

A critical part of natural bone during normal healing is its function as a scaffold for the newly apposing bone to grow on. Without that function as an osteoconductive substrate, differentiated cells would have no place to attach and build to effect the repair. Over 80 years ago, it was recognized that bone contained a mineral constituent composed primarily of a calcium phosphate (CaP) apatitic structure and that biological apatite exists as minute crystals on and within the bone collagen fibrils (deJong, 1926; Carlstrom and Engstrom, 1956; Eanes and Posner, 1970). In an effort to create a synthetic analog to bone mineral and in recognition that biocompatibility is a necessary element for any bone substrate, researchers created calcium phosphate–containing materials that were shown to exhibit the desired scaffolding characteristics. Hydroxylapatite (HA) and tricalcium phosphate (TCP) ceramics as well as calcium phosphate–based glasses were extensively characterized and found their way into clinical use in dental, oral and maxillofacial, and small-bone-void-filling applications (Laurencin, 2003). Their function as biocompatible substrates was further demonstrated by their extensive use as coatings for metallic dental and orthopedic prostheses, where they served as substrates for bone growth and adaptation (Geesink and Manley, 1993). Porous forms of ceramic HA and/or TCP have survived the rather empirical approach to substrate development and are the most utilized physical forms for bone-void filling; the porous structure provides a favorable microenvironment for cellular attachment. The CaPs have the additional advantages of relatively small production costs, no risk of disease transmission, and unlimited supply. Years of clinical use have demonstrated that they are purely conductive substrates, and care should be taken that they not be relied on as the sole graft material in larger defects. Another limitation of the CaPs stems from their brittle ceramic nature, so they are not greatly load bearing in block form and cannot be conveniently be cut and shaped by the clinician without risk of fracture. The FDA has classified most synthetic biomaterials as class II devices, requiring 510(k) submission and approval before commercialization; other regulatory agencies around the world, for the most part, classify the CaPs the same.

The most clinically utilized osteoconductive substrate is a natural one, allograft bone. Allograft bone, tissue transplanted from one human to another, has been processed in many ways over the years. Allograft can be fresh frozen or alternatively processed to minimize the potential of immunogenic response that may be elicited when used for filling a bone defect, acting as a void filler and scaffold for conducting new bone proliferation. Allograft can be cortical or cancellous bone. Structural allografts have enjoyed widespread use as a weight-bearing spacer for anterior spinal fusions. They are made by milling or precision-cutting the bone to create functional shapes in much the same function as metal or polymer materials are fabricated into space-preserving configurations. Cancellous bone, while not structural or capable of bearing significant loads, has become the most widely used form of bone-void filler and scaffold for new bone formation (Fig. 81.3). Cancellous bone chips are processed from cadaveric bone by tissue banks into various shapes and sizes. The greatest advantages of allograft are its availability, incorporation by bone, the ability to be fused with natural bone, and the modulus of elasticity. Processed allograft tissues with no additives are currently considered tissue for transplantation by the FDA and other regulatory bodies.

Over the last 30 years, interconnecting biocompatible porous metal coatings have been used to promote the fixation of hard-tissue implants through bone ingrowth (Bobyn et al., 1980; Cook et al., 1985). The success of biological ingrowth into a surface of a porous metal is based on the biocompatibility of the specific metal, titanium, cobalt-chrome alloy, and tantalum, for example. Since the mid-1990s, certain porous metals, most notably titanium and tantalum, have been shown to provide the function of an osteoconductive scaffold by mimicking in metal the interconnecting nature and mechanical properties of trabecular bone. They provide structural support, can be shaped and fixated with screws like bone, and thereby provide strength and ductility with a favorable environment for bone adaptation and ingrowth without the potential long-term liability of collapse or preferential resorption (Fig. 81.4) (Bobyn et al., 1999, 2004).

Over the last ten years or so, the use of allograft bone has grown as processing techniques have become more sophisticated, the safety of such grafts has become less an issue, and the limitations of metallic load-bearing structures (modulus of elasticity, nonresorbable nature, etc.) become more important. Organizations such as the American Association of Tissue Banks (AATB) have helped in establishing guidelines for processed tissue, with the FDA and other regulatory bodies using such guidelines as part of the basis for approval for marketing and widespread clinical use.

**FIG. 81.3.** **(A)** Cancellous chips are the most common form of allograft material used, due to their price, availability, biocompatibility, and track record as an osteoconductive bone graft extender. **(B)** Synthetic calcium phosphate ceramics based on tricalcium phosphate (TCP), hydroxylapatite (HA), or combinations (HA/TCP) are bone graft extenders that provide a biocompatible chemistry, favorable microenvironment, and porosity for bone adaptation. IsoTis OrthoBiologics, Inc. OsSatura TCP®.

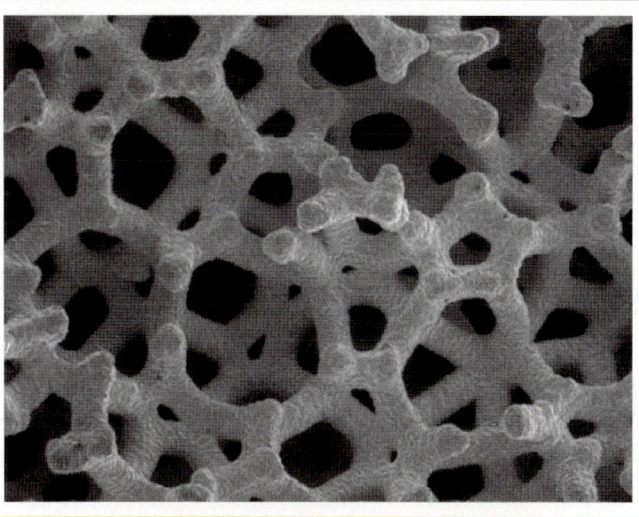

**FIG. 81.4.** Porous tantalum structure with high (75–80%) porosity and ~430-μm interconnected pore size used as a substitute for structural allograft and as an implant ingrowth surface. Zimmer, Inc. — Trabecular Metal®.

# V. OSTEOINDUCTIVE BONE GRAFT MATERIALS

Dr. Marshall Urist was the first to publish in a peer-reviewed journal a paper identifying the osteoinductive potential of demineralized bone matrix (DBM) (Urist, 1965). The first clinical use of a DBM-like substance goes back much further, when demineralized bovine xenograft was used successfully to repair long-bone and cranial defects in dogs and tibial defects in humans (Senn, 1889). Bone repairs itself by a complex series of steps and reactions that are mitigated by a series of naturally occurring proteins and growth factors that orchestrate the cellular recruitment, differentiation, and proliferation in the bone regeneration and repair process. The exact sequence and process the body uses are, at this writing, not entirely known. The DBM process is designed essentially to concentrate and make more readily available the elements responsible for participation in the natural sequence while removing those constituents that may get in the way of it.

Today, DBM is made from human bone by removing the natural mineral of ground cortical bone, leaving the bone collagen and native bone morphogenic proteins access to the bone repair environment. In making DBM, soft-tissue and cellular elements are removed through physical stripping, cutting, and a series of alcohol and washing treatments, and cancellous bone is removed. The diaphyseal region of the long bones provide the greatest yield and are the donor bones sought for DBM production; the cortical bone is ground to between 125 and 850 microns. Several demineralization processes can be used, but typically hydrochloric acid is used to leach out the mineral constituent. Other washing and cleansing steps render a material of granular, powdery consistency comprising primarily bone collagen and several noncollagenous proteins, notably the bone morphogenetic proteins. It took roughly 15 years for the concept of DBM to start to be used clinically, but containment of the small particles in a bone void was difficult, even with the use of blood or other material. In the early 1990s, the realization that DBM efficacy could be improved through the use of a containing carrier revolutionized the concept of osteoinductivity. The first carrier used to deliver DBM particles was glycerol, a water-soluble and biocompatible substance that, when combined with DBM, has a flowable consistency that allows delivery to the surgical site through an open-bore syringe. Subsequently, other inactive (noninductive) carriers were developed and used to deliver the active DBM particles, thus establishing a new and growing material concept in the clinical bone repair and augmentation industry (Fig. 81.5). The procurement of donor tissues and the processing of

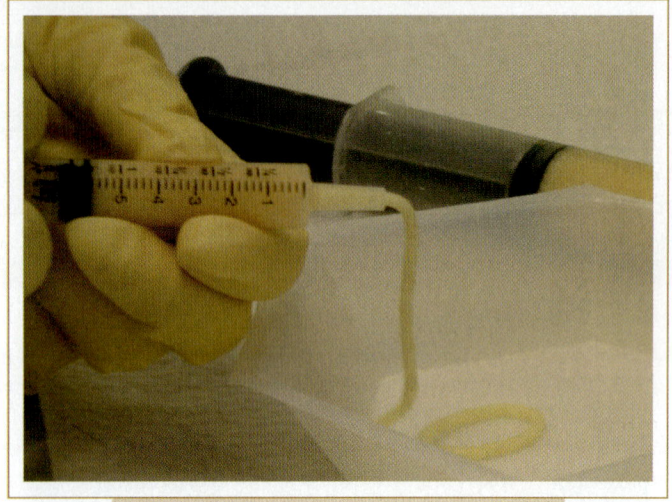

**FIG. 81.5.** A flowable putty form of inductive DBM particles contained in an inert (noninductive) carrier to facilitate convenient delivery and maintenance of the particles in the surgical site and to provide containment when the bone graft is extended with autograft, local bone, and/or osteoconductor. Isotis OrthoBiologics, Inc.

DBM is an established industry, with standards, controls, and demonstrations of efficacy in preclinical studies in support of regulatory applications. There are many suitable carriers used in commercially available DBM-based bone graft materials: sodium hyaluronate, glycerol, calcium sulfate and carboxymethylcellulose, lecithin, reverse-phase poloxamer, and porcine collagen. All are inert, in that they are not demonstrated to be capable of fostering cell differentiation down osteoblastic pathways as determined by validated *in vitro* or *in vivo* methods. The American Association of Tissue Banks (AATB, 2006) has established a series of guidelines for the preparation and quality maintenance of transplanted human tissues, and the FDA and other regulatory bodies around the world have, or are creating, governmental guidelines or requirements for the manufacture and use of these inductive substances (FDA, 2006). The advantages of such combinations are containment of the inductive particles in the surgical site and surgeon-friendly delivery. Such nonactive carriers must be removed to reveal the active BMPs and GFs to the bone repair environment, and they represent a limitation in use, since the inductivity of the combination is limited in part by the percentage of particles and the ability of the body to metabolize the carrier quickly. A balance must exist between carrier technology and biological function: If the carrier is metabolized too quickly, the active DBM particles may not be contained within the site long enough to aid in the differentiation and repair process, while if too slow or if locked in a mass of slowly resorbing material, the active DBM may not have access to the bone repair milieu in the optimal time to aid in the early stages of bone repair.

Until November 15, 2005, the DBM/carrier substances were commercialized with no prerequisite FDA approval to market; on that date, the combination of active DBM particles and any inactive carrier or ingredient were categorized as medical devices requiring application and FDA approval to market, before commercialization. The acceptance of the concepts of osteoinductivity by the clinical community fostered the development of other-than-putty and gel formulations to matrix, sponge, or fleece forms. The three-dimensional matrices are not generally bone derived, with dermal or tendon collagen used most frequently. The advantages of DBM-based materials are their long-term clinical use, their relatively low cost, and their graft-containment characteristics; they can form the basis of a composite graft of inductive, conductive, and osteogenic elements and have found increasing use in the bone repair arena. The disadvantages are the donor-based variability of DBM, the lower volume of inductive elements for some substances, the lack of any standardized measurement technique for DBM inductivity, and the inert carriers.

A recent innovation in the development and use of osteoinductive biomaterials came with the recognition of the limitations of an inert carrier for DBM particles, a further treatment of DBM particles to expose more effectively the naturally inductive elements found in bone. These natural human BMP-containing materials, nhBMP™, use the natural bone constituents themselves to form a flowable, inductive carrier for active DBM particles, thus providing a putty form of inductive graft material that does not suffer the liability of an inert carrier that must be removed before exposure of inductive particles becomes available to the bone repair site. The providing of greater inductivity within such materials is supported by the quantitation of various BMPs and GFs using contemporary analytical techniques (Kay *et al.*, 2005). The bone response is faster and more robust based on the carrier entering into the differentiation cascade and more effectively triggering the bone repair cascade. A spectrum of naturally occurring BMPs and growth factors is preserved in both the carrier and particles and is responsible for a more concentrated delivery of differentiating agents to the surgical site. Both flowable putty and lyophilized matrix forms are available for clinical use and have no inert ingredient additives (Fig. 81.6). The matrix form provides a spectrum of BMPs and GFs in a highly porous structure that allows for immediate mobilization and delivery of an inductive stimulus to the site, a distinct advantage over first-generation DBM-inert carrier combinations. The advantages of this class of substances are their moderate cost, high osteoinductivity, and lack of an inert carrier, while the disadvantage is the DBM-based nature, including the potential donor-based variability.

Genetic engineering has provided yet another source of osteoinductive bone graft materials. Since the mid-1980s, researchers have sought to identify the single BMP that can best trigger and accelerate bone healing. The individual BMPs from DBM were isolated and genetically engineered to increase yields that allowed implantation and assessment of individual capability to induce bone formation. Two recombinant BMPs, BMP-2 and BMP-7, are commercially available for several specific FDA-approved indications. These rhBMPs evolved from early studies on DBM and the specific BMPs contained therein. BMP-7 and BMP-2 were among those identified and were chosen by two commercial entities for development (Wozney, 2002; Rengachary, 2003; Sandberg, 1993). Preclinical studies showed acceleration of bone healing in a variety of defects, leading to the human clinical evaluations in application-specific studies (Boden *et al.*, 1996; Lovell, 1989; Schimandle, 1995). The recombinant BMPs provided for reduced blood loss, shorter operating times, and elimination of iliac crest harvesting but nonstatistically different fusion rates, assessed clinically and radiographically, over autograft. One recombinant form combined with autograft had no clinical advantage over autograft (Vaccaro, 2003).

Today, each commercially available recombinant material is provided in highly concentrated form, with a small quantity of an inert, osteoconductive carrier substance, dosed specifically for certain clinically approved indications, and each has been shown to be safe and effective. This class of inductive biomaterial appears to be most effective when on-label indications are followed, based on establishment of dosage, delivery, and efficacy (Burkus *et al.*, 2002a, 2002b). The clinical indications will increase based on additional ongoing clinical evaluations to determine dosage, delivery, and efficacy.

The advantages of the rhBMPs are their demonstrated ability to elicit a significant osseous response in the approved indications and the extensive clinical studies available to document the usage, while the disadvantages are the high cost, limited reimbursement, and potential side effects, associated primarily with off-label use, which can be serious in nature (Pradhan *et al.*, 2006; Smucker *et al.*, 2006; Treharne, 2004). Much more in-depth information on the recombinant materials can be obtained in other chapters of this book.

Perhaps more than the other two elements, advances in the development of osteoinductive agents and their demonstrated ability to influence the pathway of cellular differentiation have become the basis for contemporary concepts in tissue engineering. Advances in the understanding of bone induction and the introduction of substances designed specifically to address this mechanism are signs that treatment of bone healing is being pursued from a more advanced, mechanistic approach, representing truly relevant progress in our understanding. In the desire to use osteoinduction to aid in bone healing, the surgeon has had an expanded selection of materials to choose from since the turn of the century with the advent of higher-potency rhBMP and nhBMP™ technologies.

With the changes in the FDA posture to classify such osteoinductive materials as devices, the burden of proof has risen. There are no peer-reviewed publications that describe

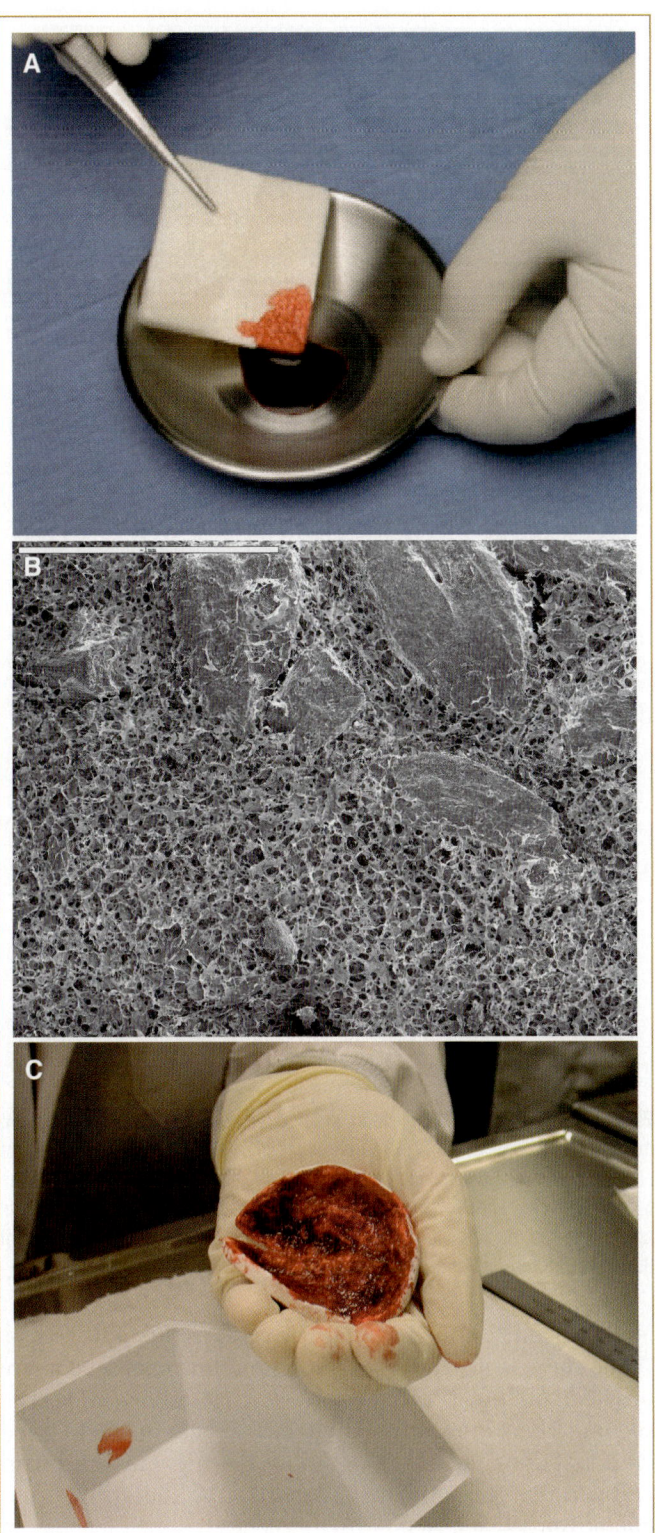

**FIG. 81.6.** **(A)** A matrix form of inductive bone graft material composed of inductive DBM particles contained in an inductive DBM-derived carrier. This material is made by lyophilizing a flowable DBM particle/DBM-based carrier substance to remove the moisture, resulting in the open, porous structure. **(B)** Scanning electron micrograph showing matrix and included DBM particles. **(C)** The use of an inductive matrix form to hold vital bone-marrow aspirate, as a composite graft. IsoTis OrthoBiologics, Inc. SEM magnification 50×, scale 1 mm.

well-conducted comparisons between any two osteoinductive materials. There are two primary reasons for this. First, the industry is relatively young and such studies are expensive, and to be published today, they would have to have been initiated some five to six years ago; this would have been too costly at that time. Second, because the substances were considered tissues for transplantation, there was no regulatory need. Today and going forward, specific orthopedic applications will necessarily be supported by clinical studies.

Osteoinductivity has been demonstrated in several *in vivo* and *in vitro* assays whose heritage goes back to the early work of Urist. He implanted test substances into the soft tissue of athymic rodents, and, if osteoinductive, they differentiated cells down osteoblastic pathways and resulted in bone-nodule formation; similar differentiation has been observed in a laboratory culture model and validated against the classic athymic rodent model (Urist, 1965; Urist and Strates, 1971; Katagiri *et al.*, 1994; Han *et al.*, 2003; Honsawek *et al.*, 2005).

The advantage of the osteoinductive materials as a class is the providing of nonautogenous triggering agents for cell differentiation down osteoblastic lines, thus aiding the initiation of the bone repair process. Dosage, concentration, delivery, and carrier substance appear to be the major determinants of clinical utility, while cost and regulatory status have effects on usage in certain clinical applications and institutions. Demineralized bone matrix particles contained in noninductive carriers, DBM particles in inductive DBM-based carriers, and recombinant biomaterials technologies are the currently available subdivisions within the class of osteoinductors, each delivering some level of inductive constituent to aid in the bone repair process.

## VI. COMPOSITE BONE GRAFTING

Composite bone grafting has become a preferred grafting philosophy and is the treatment modality of choice for many surgeons. The concept employs a selected combination of osteoconductive, osteoinductive, and osteogenic materials a clinician believes will achieve the best clinical result for that patient. As noted previously, the body employs all three mechanisms to some degree, depending on the site of regeneration or repair and/or the severity of the repair situation. Many options for cells, scaffolds, and signals are available to the clinician from each category, and the selection should be based on what clinical result is desired and the data to support each constituent's use, cost, and personal surgical techniques. Each constituent need not necessarily be harvested or obtained specifically for use as a graft material in every procedure; for example, a completely decorticated posteriorlateral spine or reamed acetabulum will often provide enough bleeding vital bone to preclude any additional autologous material, even bone-marrow aspirate (Fig. 81.7). The surgeon is faced with a continuum of different bone defects, each with a different intrinsic

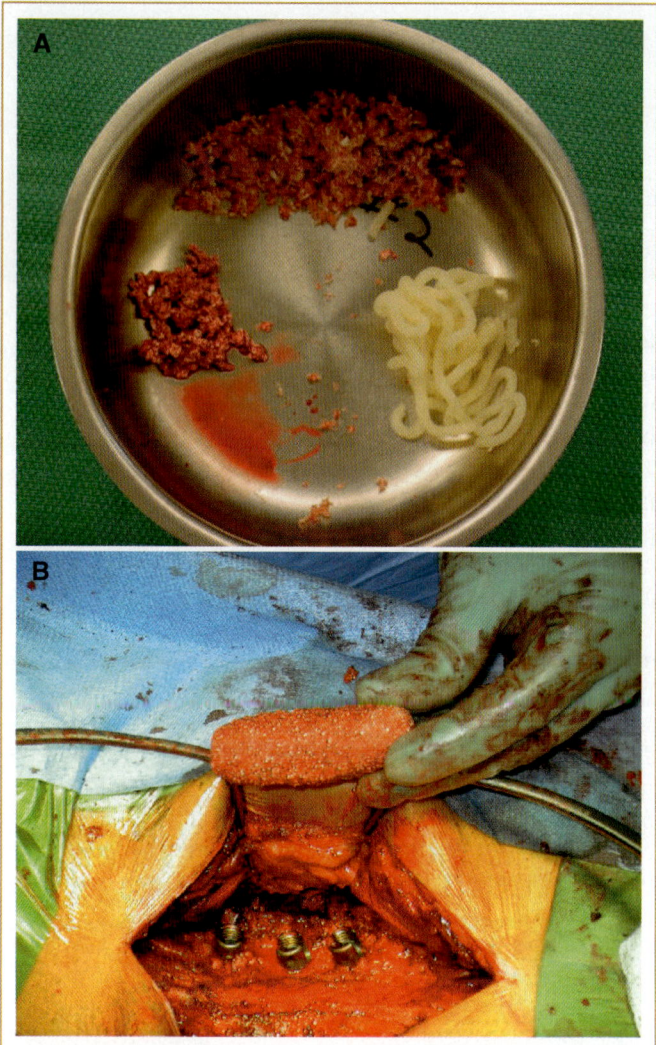

**FIG. 81.7.** A composite graft to preclude iliac crest harvest employing minced local bone to provide the osteogenic component, synthetic TCP to supply the conductive component, and a DBM-based putty form to provide osteoinduction, handling, and containment. Modest amounts of the synthetic are used, and the relative proportions are chosen to facilitate containment during and after placement.

ability to aid in bone healing. The options are there to use. While there is no universally accepted formula, several are provided in the clinical examples that follow in a later section. There are many sources of information about bone graft materials and their use; among them is one distributed and updated regularly by the American Academy of Orthopaedic Surgeons (AAOS) (AAOS, 2006).

## VII. REGULATORY ISSUES

The FDA regulates and provides the approval for use of medical devices and drugs after the demonstration of safety and efficacy. Many regulatory agencies of countries outside the United States utilize similar, if not identical, requirements and criteria for controlling and regulating bone graft materials. Currently, the materials described in this summary are all considered either medical devices or tissues for transplantation, requiring various degrees of approval, or none, at this writing. Cancellous bone chips and DBM-based materials with no inert ingredients added are considered tissues for transplantation and are presently governed by 21 CFR1270. Other substances are medical devices requiring either approval to market via the 510(k) process (21CFR-807, Subsection E), or the premarket approval (PMA) process pursuant to 21CFR-812 and 21CFR-814. The package insert for each material is important because it contains instructions for use, approved indications, contraindications, warnings, and other information important to understand before selecting the substances for clinical use in patients. The warnings and contraindications are particularly important because they are generally derived from specifically identified instances, complications, or test results. Specifically approved indications for the product are termed *on-label* uses, whereas all others are termed *off-label*. Additional information may be obtained from manufacturers' websites and product descriptions, in addition to promotional literature, published journal articles, clinical forums, and the like. In selecting constituents for composite bone grafting, the surgeon should consider the status of the biological and mechanical environments for bone healing and patient comorbidities and issues and then choose the most efficacious combination of the three essential elements to be used in consideration of the indications, contraindications, warnings, and prior clinical experience available.

## VIII. EXAMPLES OF TISSUE ENGINEERING IN CONTEMPORARY CLINICAL ORTHOPEDICS

Bone graft materials are used in a variety of clinical orthopedic procedures, including spinal fusion, total joint replacement, trauma, nonunions, oral and craniofacial, and other void- and defect-filling applications. The 2005 market for grafting and graft materials in the United States was estimated at approximately $1.8 billion and worldwide at almost $2.3 billion, with growth at an approximate 13.5% per year (Knowledge Enterprises, 2006). Worldwide, over 2.5 million procedures account for the substantial and growing market. The predominant use of graft materials is in the spine. This is a particularly demanding and often high-volume use of bone graft materials, where bone is used to fuse the unstable or damaged spinal vertebrae. The use of graft materials in the spine requires growth of bone to perform a function where it does not occur naturally. Almost all other grafting procedures involve the regeneration of bone where it would normally occur; such is the case in trauma, total joint arthroplasty, and other orthopedic applications of bone graft materials.

The growing market opportunity is the driving force for investment in tissue engineering and the development of better and more efficacious grafting materials and techniques. The ultimate reason for the increase in bone grafting is more consistent, predictable clinical results. This trend is occurring. Aside from the new materials, the ability to create composite grafts from several different constituents is an advance in clinical practice. The clinical community has become more sophisticated as well, with clinical professionals sometimes aiding in the discovery and development of new materials, concepts, and clinical applications.

The uses of bone graft materials have expanded greatly since the mid-1990s, and to provide individual, specific examples of that evolution is beyond the scope of this chapter. Figures 81.8 and 81.9 present two individual cases, however, where state-of-the art bone-grafting techniques involving contemporary, cost-effective, and accepted concepts of tissue engineering have been used to treat several common orthopedic conditions.

## IX. THE FUTURE

The future of bone-tissue engineering is bright. Engineers and scientists are constantly unlocking the secrets to the process of bone formation and healing. Biochemists and materials scientists are understanding and copying nature's codes, regulators, and biological substances more effectively, so the combinations of building blocks, triggers, and controls that will be tested and mass produced are greater and more effective. Advances in the development of each of the three elements of bone regeneration are taking place. Better osteoconductive scaffolds with greater affinity for new bone and more predictable bioresorption profiles are under development. The combinations of biopolyers, bioceramics, and naturally derived (collagen, for example) materials better mimics the existing natural scaffolds of bone in callus, vital autograft, and cancellous bone. More potent and/or optimally dosed natural, synthetic, or genetically engineered bone inductors are being evaluated for site-specific applicability and in combination with other carriers. Autogenous materials are being looked at for vitality preservation, *ex vivo* expansion, and optimization of potential with minimization of harvest insult. With the advances in osteoinductive biomaterials and the demonstration of the efficacy of composite grafting, it is doubtful that autograft use will increase. Future technologies, such as stem cell therapy and autogenous organ (bone) regeneration, are within the realm of possibility.

Several approaches in tissue engineering are being taken, so it is unclear whether in the future substances will be placed in a site, perhaps bear load, and become quickly incorporated and bioresorbed to leave a natural architecture and chemistry, or whether a fast-acting substance will be placed in a bone repair site to "supertrigger" the mobilization of natural body elements that heal the site using a body-derived series of materials in a natural sequence but at a highly rate-accelerated process. Both pathways can be envisioned, and both are being pursued within the

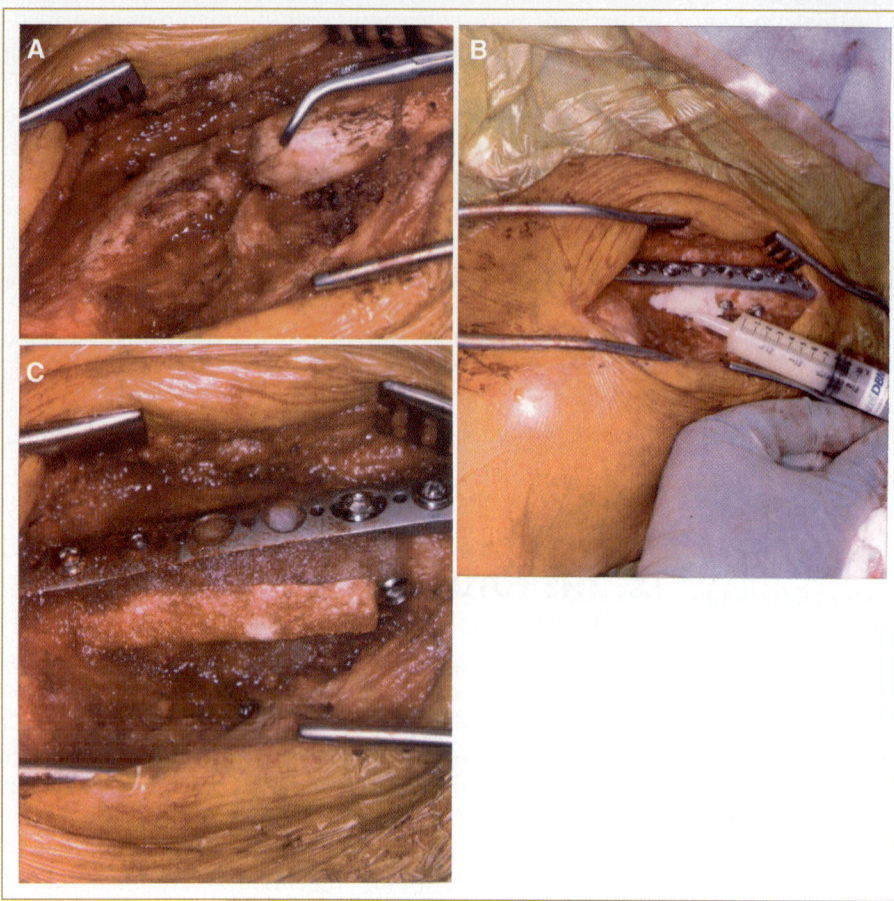

**FIG. 81.8.** Clavicle nonunion treated with a combination of a high-inductivity putty and inductive matrix form after thorough debridement and application of internal fixation. Gaps are filled with the putty, and the matrix is form cut and placed adjacent to the plate, with final contouring with the putty form. IsoTis OrthoBiologics, Inc.

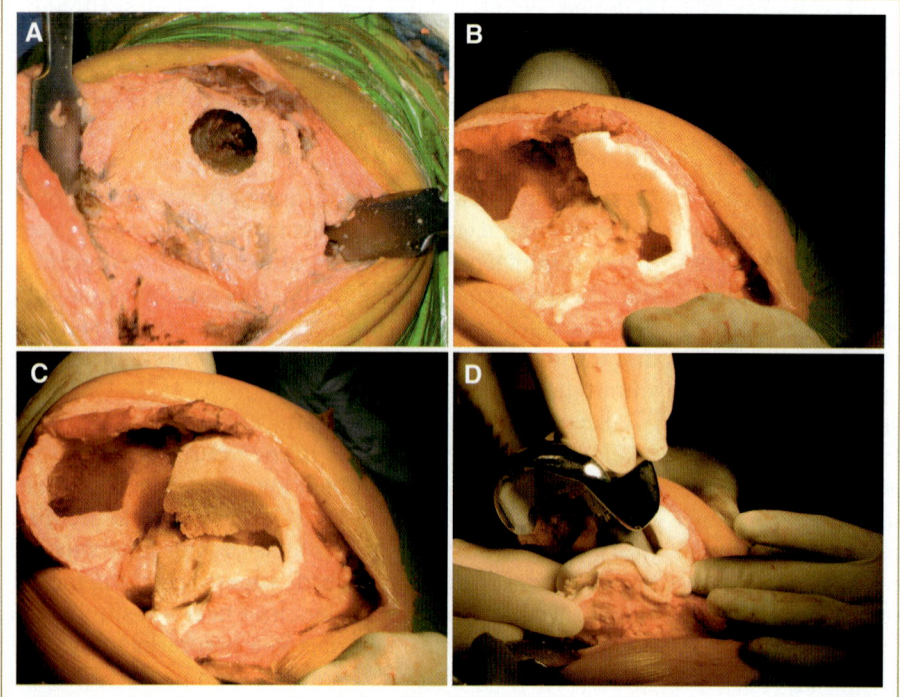

**FIG. 81.9.** Distal femoral allograft reconstruction. An inductive matrix form was used as an intermediary between the supporting, contoured structural allograft, eliminating gaps at the allograft–host bone interface. After impaction, a long-stemmed revision prosthesis is cemented into position. IsoTis OrthoBiologics, Inc.

academic and institutional research environments. Each pathway has its advantages and developmental challenges, but the work being done and the materials and processes being used and tested today increase the knowledge base and the art going forward. Clearly, orthopedic surgeons and their patients will benefit from the advances in biomaterials tissue engineering and the demonstration of their efficacy.

## X. CONCLUSIONS

Bone is a complex tissue. An engineered bone replacement has been a noble challenge, but advances are being made. The next edition of this volume will undoubtedly discuss further advances. This chapter has focused on the materials used in contemporary tissue-engineered bone grafts. These more potent and efficacious new materials, however, do not reduce the need for application of sound orthopedic principles of bone healing or understanding the effects of the comorbidities that can have negative effects on bone healing. The establishment of an ideal environment for hard-tissue healing and mechanical stability is essential. Three elements are required for bone repair: osteogenic cells, an osteoconductive growth substrate, and a substance that aids in cell differentiation. Many alternatives are available to the orthopedic surgeon today. It is the surgeon's educated judgment as to the most appropriate treatment for an individual patient, so a thorough understanding of the bone-healing process, the indications, contraindications, warnings, and handling characteristics of all materials at their disposal is most important in bone healing. This information is found on the device package insert and on the Internet, among other places. Finally, access to these new technologies will be limited or slowed by regulatory processes in place to ensure safe and effective products for clinical use.

## XI. ACKNOWLEDGMENTS

The author would like to acknowledge constructive discussions with and/or contributions by Robert Bess, M.D., Clemens von Blitterswick, Ph.D., Barbara Boyan, Ph.D., Jan P. Ertl, M.D., Susan Harding, M.D., IsoTis OrthoBiologics, Inc., Norman Johansen, M.D., Louis Keppler, M.D., Mark Munro, M.D., James Poser, Ph.D., Bruce van Dam, M.D., Jeffrey Wingate, M.D., and Robin Young, Zimmer, Inc.

## XII. REFERENCES

AAOS. (2006). American Academy of Orthopaedic Surgeons. http://www.aaos.org.

AATB. (2006). American Association of Tissue Banks. http://www.aatb.org.

Anitua, E. (1999). Plasma rich in growth factors: preliminary results of use in the preparation of future sites for implants. *Int. J. Oral Max. Impl.* **14**, 529–535.

Aro, H. T., Sandberg, M. M., and Vuorio, E. I. (1993). Gene expression during bone repair. *Clin. Orthop.* **289**, 292–312.

Bargar-Lux, J., and Recker, R. R. (2002). Embriology, anatomy, and microstructure of bone. *In* "Disorders of Bone and Mineral Metabolism" (F. L. Coe and M. J. Favus, eds.), 2nd ed., pp. 177–198. Lippencott, Williams, and Wilkins, Philadelphia.

Beujean, F., Hernigou, P., Poignard, A., and Rouard, H. (2006). Percutaneous autologous bone marrow grafting for nonunions. *J. Bone Joint Surg. Am.* **88**, 322–327.

Betz, R. R. (2002). Limitations of autograft and allografts: new synthetic solutions. *Orthopaedics* **25**, 561–570.

Bobyn, J. D., Cameron, H. U., Pilliar, R. M., and Weatherly, G. C. (1980). The optimum pore size for the fixation of porous-surfaced metal implants by the ingrowth of bone. *Clin. Orthop. Rel. Res.* **50**, 263–270.

Bobyn, J. D., Hacking, S. A., Krygier, J. J., Tanzer, M., and Stackpool, G. J. (1999). Characteristics of bone ingrowth and interface mechanics of a new porous tantalum biomaterial. *J. Bone Joint Surg. Br.* **81**, 907–914.

Bobyn, J. D., Hanssen, A. D., Poggie, R. A., *et al.* (2004). Clinical validation of a structural porous tantalum biomaterial for adult reconstruction. *J. Bone Joint Surg. Am.* **86A**, 123–129.

Boden, S. D., Hutton, W. C., and Schimandle, J. H. (1995). Experimental spinal fusion with recombinant human bone morphogenetic protein-2. *Spine* **20**, 1326–1337.

Boden, S. D., Morone, M. A., Moskovitz, P. A., *et al.* (1996). Video-assisted lateral intertransverse process arthrodesis. Validation of a new minimally invasive lumbar spinal fusion technique in the rabbit and non-human primate (*Rhesus*) models. *Spine* **21**, 2689–2697.

Burkus, K. J., Kitchel, S. H., Transfeldt, E. E., *et al.* (2002a). Clinical and radiographic outcomes of anterior lumbar interbody fusion using recombinant human bone morphogenetic protein-2. *Spine* **27**, 2396–2408.

Burkus, K. J., Dickman, C. A., Gornet, M. F., *et al.* (2002b). Anterior lumbar interbody fusion using rh-BMP-2 with tapered interbody cages. *J. Spinal Disord. Tech.* **15**, 337–349.

Cheng, H., Jiang, N., *et al.* (2003). Osteogenic activity of the 14 types of BMPs. *JBJS Am.* **55A**, 1544–1552.

Cook, S. D., Haddad, A. J., and Walsh, K. A. (1985). Interface mechanics and bone growth into porous Co-Cr-Mo alloy implants. *Clin. Orthop. Rel. Res.* **193**, 271–280.

Cornell, G., Fleming, J., and Muschler, G. (2000). Bone cells and matrices in orthopedic surgery. *Tissue Eng. Orthop. Surg.* **31**, 357–374.

Davies, J. E. (2000). "Bone Engineering." Em Squared, Toronto.

Dawson, E. G., Lovell, T. P., Nilsson, O. S., *et al.* (1989). Augmentation of spinal fusion with bone morphogenetic protein in dogs. *Clin. Orthop.* **243**, 266–274.

Dayoub, H., Holm, G. A., and Jane, J. A., Jr. (2001). Gene-based therapies for the induction of spinal fusion. *Neurosurg. Focus* **10**, 1–5.

DeJong, W. F. (1926). La substance minerale dans les os. *Rec. Trav. Chim.* **45**, 445–448.

Eanes, E. D., and Posner, A. S. (1970). Structure and chemistry of bone mineral. *In* "Biological Calcification" (H. Schraer, ed.), pp. 1–26. Appleton-Century-Crafts, New York.

Eastoe, J. E. (1956). The organic matrix of bone. *In* "The Biochemistry and Physiology of Bone" (G. H. Bourne, ed.), pp. 81–105. Academic Press, New York.

Eastoe, J. E., and Eastoe, B. (1954). The organic constituents of mammalian compact bone. *Biochem. J.* **57**, 453–459.

FDA. (2006). United States Food and Drug Administration. http://www.fda.gov.

Fischgrund, J., Patel, T., Vaccaro, A. R., *et al.* (2003). A pilot safety and efficacy study of OP-1 putty (rh-BMP-7) as an adjunct to iliac crest autograft in posterolateral lumbar fusions. *Eur. Spine J.* **12**, 495–500.

Froum, S. J., Tarnow, D. P., Wallace, S. S., *et al.* (2002). Effect of platelet-rich plasma on bone growth and osseointegration in human maxillary sinus grafts: three bilateral case reports. *Int. J. Perio. Rest. Dent.* **22**, 45–53.

Geesink, R. G. T., and Manley, M. T., eds. (1993). "Hydroxylapatite Coatings in Orthopaedic Surgery." Raven Press, New York.

Glimscher, M. J. (1959). Molecular biology of mineralized tissues with particular reference to bone. *Rev. Mod. Phys.* **31**, 359–393.

Han, B., Nimni, M., and Tang, B. (2003). Quantitative and sensitive *in vitro* assay for osteoinductive activity of demineralized bone matrix. *J. Orthop. Res.* **21**, 648–654.

Honsawek, S., *et al.* (2005). Extractable bone morphogenetic protein and correlation with induced new bone formation in an *in vivo* assay in the athymic mouse model. *Cell Tissue Banking* **6**, 13–23.

Kalfas, I. H. (2001). Principles of bone healing. *Neurosurg. Focus* **10**, 1–4.

Katagiri, T., Komacki, M., Yamaguchi, A., *et al.* (1994). Bone morphogenetic protein-2 converts the differentiation pathway of $C_2C_{12}$ myoblasts into the osteoblast lineage. *J. Cell Biol.* **127**, 1755–1766.

Kay, J. F., Khaliq, S. A., King, E., Murray, S. S., and Brochmann (2005). Quantification of BMP-2, BMP-4, and TGF-β1 in DBM and a DBM-based carrier. *Proc. Soc. Biomat. 30th Annu. Mtg.* **28**, 430.

Knowledge Enterprises. (2006). The worldwide orthopaedic market 2005–2006. Knowledge Enterprises, Chagrin Falls, OH, December 2006. http://www.theorthopeople.com.

Laurencin, C. T. (2003). Section III: polymers, ceramics, and other synthetic materials for bone graft substitutes. *In* "Bone Graft Substitutes" (C. T. Laurencin, ed.), pp. 227–308. ASTM, West Conshohocken, PA.

Lieberman, I. H. (2006). Osteoprogenitor cells from bone marrow to augment spinal fusions. http://www.spineuniverse.com/displayarticle.php/article2138.html.

Lucas, D. R., Pilitsas, J. G., and Rengachary, S. R. (2002). Bone healing and spinal fusion. *Neurosurg. Focus* **13**, 124–129.

Marx, R. E. (1998). Platelet-rich plasma: growth-factor enhancement for bone grafts. *Oral Surg. Med. Pathol.* **85**, 638–646.

Matsukara, Y., Muschler, G. F., Nitto, H., *et al.* (2003). Spine fusion using cell matrix composites enriched in bone marrow–derived cells. *Clin. Orthop. Rel. Res.* **407**, 102–118.

Muschler, G. F. (2006). Bone biology, skeletal reconstruction, aging, and osteoporosis. http://www.lerner.ccf.org/bme/muschler/lab/.

NASS. (2006). North American Spine Society. http://www.nass.org.

Pradhan, B., Bae, H. W., Dawson, E. G., Patal, V. V., and R. B. Delamarter (2006). Graft resorption with the use of bone morphogenetic protein: lessons from anterior lumbar interbody fusion using femoral ring allografts and recombinant human bone morphogenetic protein-2. *Spine* **31**, 2807–2814.

Rengachary, S. S. (2003). Bone morphogenetic proteins: basic concepts. *Neurosurg. Focus* **13**, 2992.

Senn, N. (1889). On the healing of aseptic bone cavities by implantation of antiseptic decalcified bone. *Am. J. Med. Sci.* **98**, 219–243.

Smucker, J. D., Rhee, J. M., Singh, K., Yoon, T., and Heller, J. G. (2006). Increased swelling complications associated with off-label usage of rh-BMP-2 in the anterior cervical spine. *Spine* **31**, 2813–2819.

Treharne, R. W. (2004). "Safety Alert: INFUSE Bone Graft." Medtronic Sofamor Danek, Memphis, TN.

Urist, M. (1965). Bone formation by autoinduction. *Science* **150**, 893–899.

Urist, M., and Strates, B. (1971). Bone morphogenetic protein. *J. Dent. Res.* **50**(Suppl. 6), 1392.

Wozney, J. M. (2002). Overview of bone morphogenetic proteins. *Spine* **27**(16S), S2–S8.

# Chapter Eighty-Two

# Tissue-Engineered Cardiovascular Products

*Thomas Eschenhagen, Herrmann Reichenspurner, and Wolfram-Hubertus Zimmermann*

## I. INTRODUCTION

Cardiovascular disease is by far the most common reason for morbidity and mortality in the Western world. It comprises malfunction of heart valves, blood vessels, and myocardium, all of which are subject to tissue-engineering programs. The development of tissue-engineered heart valves is fueled by the idea to create a growing valve for children with congenital cardiac malformations and a totally autologous biological valve for adults with acquired valve defects. Here the tissue-engineered product has to compare favorably to the existing spectrum of quite sophisticated and long-term approved biological and mechanical prostheses. Blood vessel engineering aims predominantly at generating small-diameter (<6 mm; e.g., coronary arteries) rather than large-diameter vessels (e.g., aorta, large mesenteric or femoral arteries). Dacron and other polymers are being used in the latter, with satisfying results. The driving forces in this field are the low long-term patency rates of classical saphenous vein grafts and the paucity of suitable arterial vessels (such as the internal mammaria artery) in coronary bypass surgery. Other potential applications include arteriovenous shunts for patients on chronic hemodialysis. Finally, myocardial-tissue engineering aims at creating force-developing cardiac tissue patches to correct cardiac congenital malformations in children ("the growing patch") and replace or support heart muscle function after myocardial infarction or in patients with cardiomyopathies. The total bioartificial heart, as publicly announced in the mid-1990s, would comprise all of the three tissue-engineering fields and would be the straight alternative to heart transplantation. But even with some very encouraging progress in many facets of the field, the total bioartificial heart remains a dream rather than a reasonable clinical perspective in the next 20 years.

If one searches Medline for "tissue engineering" and limits the search to clinical trials, only 26 papers pop up, none of them dealing with cardiovascular products. Instead, skin, cartilage, bone, and some urological applications have been tested in patients. This reflects, on the one hand, the differing advancements of various tissue-engineering fields, but it certainly also points at a problem: Testing skin is technically simple and bears only a limited risk to the patient, even in case of complete failure. Failure of a heart valve, a coronary artery conduit, or a load-carrying heart muscle patch, in contrast, would have disastrous consequences, as sadly proven in a study with decellularized porcine heart valves produced by CryoLife and implanted in children in Austria (Simon *et al.*, 2003). From the reported data, the valve had been implanted in children, and several died or had to be reoperated on. This disaster did not completely stop the development of decellularized xenogeneic valves, but it clearly raised the bar for further clinical trials in this field.

This chapter gives an overview on the current development of the three major tissue-engineered cardiovascular products (valves, vessels, and myocardium) but focuses mainly on myocardial-tissue patch engineering, being the field the authors have been actively involved in. For more thorough reviews on valves and vessels, the reader is referred to excellent recent papers (Isenberg *et al.*, 2006; Vesely, 2005).

## II. CLINICAL NEED FOR TISSUE-ENGINEERED CARDIOVASCULAR PRODUCTS?

Cardiovascular medicine has been among the most successful fields since the mid-1980s, and the progress made via improved drug therapy, invasive cardiology, cardiac surgery, and cardiac devices is impressive. Yet mortality and morbidity remain high, and it is unquestionable that there is room for improvement.

### Heart Valves

Malfunctioning heart valves are generally replaced surgically by either mechanical or biological prostheses, a procedure performed in more than 100,000 patients per year in the United States and more than 300,000 worldwide (Vesely, 2005). Whereas valve replacement is often life-saving and has a tremendous impact on quality of life, drawbacks do exist. Mechanical prostheses require life-long anticoagulation, with its potential hazards, and are still associated with a markedly increased rate of thromboembolism (4% per year) (Hammermeister *et al.*, 1993). Biological prostheses do not share these problems, but they have a limited half-life. None of the existing models grows, with the consequence that children with congenital valvular malformation have to be reoperated on with age. The ideal tissue-engineered living valves would overcome all of these problems. Yet the strength of these arguments has to be critically weighed. In particular, the best biological prostheses, such as the aortic homografts and the Edwards pericardial valve, have a durability close to 20 years (Biglioli *et al.*, 2004). In fact, this will often be lifelong, given that the majority of patients undergoing valve replacement are over 50 years old. Moreover, surgical techniques have improved and the high mortality associated with repeated valve replacements (formerly approximately 20%) has gone down considerably to approximately 4% (Vesely, 2005), allowing the option of repeated use of bioprosthetic valves rather than implanting a mechan-

ical valve. Thus, progress through tissue engineering appears most likely to be relevant in children with congenital malformations, where the availability of growing valves would constitute a major advancement.

### Blood Vessels

Considering that the market for heart valves is large, it becomes clearly evident, by looking at the patient numbers, that the need for blood vessels is even larger. Every year, 500,000 procedures are being performed that involve small-caliber blood vessels just in the United States (Isenberg *et al.*, 2006). For many years, poly(ethylene terephthalate) (PET; Dacron) or poly(tetrafluorethylene) (PTFE; Teflon) grafts have been in clinical use for the replacement or repair of large-diameter blood vessels, such as the aorta and the carotid arteries, with good results (Naylor *et al.*, 2004; Post *et al.*, 2001). In contrast, replacement of small-diameter vessels, such as coronary arteries, still requires venous or arterial autografts. The standard saphenous vein graft as used in coronary artery bypass grafting (CABG) has a high risk of acute or chronic vessel occlusion and requires a second surgical procedure. This does not apply to arterial autografts, such as the internal mammary artery, but this graft is often not available in sufficient number or quality. Thus, there is a clear need for tissue-engineered small arteries.

### Myocardial Patches

Compared to current treatment options (drugs, devices, transplantation, prostheses, cell therapy), cardiac repair with tissue-engineered myocardial patches appears complicated, expensive, and still very ambitious. Yet there are at least two areas where tissue engineering could prove to be valuable. The first, and maybe the most obvious, is reconstructing malformed hearts in children; the other, quantitatively much more important, is repairing infarcted or failing hearts in adults.

Heart malformations that might benefit from restoration using biological, autologous muscle patches include several forms of ventricular dysplasia, such as the hypoplastic left heart syndrome or tricuspid atresia with right ventricular dysplasia. Right now most of these infants undergo palliative surgery (a univentricular correction) ("Fontan-surgery"). Biologic active muscle patches might allow biventricular correction and true ventricular restoration. In addition, indications for a right ventricular outflow tract patch plasty are quite frequent in patients with pulmonary atresia or "Fallot"-type malformations, who might also benefit from a biological solution.

In contrast to cardiac malformations, myocardial infarction and its consequences belong to the most common diseases with huge socioeconomic impact. For the reason discussed earlier, namely the availability of conservative therapeutic options, only a minor fraction of patients surviving an acute myocardial infarction will be candidates for biological cardiac repair. Present cell-based clinical studies employ bone marrow–derived cells in patients with relatively well-preserved left ventricular (LV) function after acute myocardial infarction (Assmus *et al.*, 2002; Janssens *et al.*, 2006; Stamm *et al.*, 2003; Strauer *et al.*, 2002; Wollert *et al.*, 2004). These patients are very common and easy to study, but several arguments indicate that they are not the best-suited population for tissue-engineering strategies. First, if LV function is only mildly affected, the effect of an intervention aiming at providing the heart with new contractile myocardium is necessarily small. Second, current treatment after myocardial infarction with beta-blockers, statins, aspirin, and potentially ACE inhibitors is simple, safe, and quite effective, with a well-documented reduction of morbidity and mortality (Hippisley-Cox and Coupland, 2005). Third, spontaneous recovery, e.g., by improved perfusion and structural adaptation of the heart, namely by hypertrophy, occurs and obscures the potential effect of tissue patches. Finally, implanting tissue patches requires open heart surgery, which will not be justified in mild forms of cardiac disease. More likely, implantation of tissue patches could be considered in patients with chronic heart failure resulting from myocardial infarction. These patients might in fact benefit from a combination of tissue-engineered patch grafting, ventricular restoration surgery ("Dor-plasty"), surgical revascularization, and valve repair (Vitali *et al.*, 2003). A combination of the Dor-plasty and subsequent implantation of a contractile patch seems especially promising and would ideally enable complete functional regeneration of the ventricle (Eschenhagen and Zimmermann, 2005).

Whereas it is hardly disputable that numerous patients would profit from a direct support of contractile function, the question is whether tissue-engineered patches could play a significant role in this scenario when compared to current alternatives: (1) Mechanical left ventricular–assist devices (LVADs) acutely support the pumping activity of the left ventricle and thereby allow the heart to recover, a process leading to "reverse remodeling" (Wohlschlaeger *et al.*, 2005). Yet it seems unlikely that LVADs can completely and stably reverse a severe cardiac disease. Therefore, the place of LV-assist devices is bridging to transplantation rather than in long-term therapy. (2) Implantable defibrillators/cardioverters protect against lethal arrhythmias and thereby prolong survival (Bardy *et al.*, 2005) but in themselves have no effect on contractile function and heart failure symptoms. (3) Cardiac resynchronization therapy by means of biventricular pacing, in contrast, leads to improved heart function and has a documented effect on survival (Cleland *et al.*, 2005). Yet only a limited number of patients (those with conduction abnormalities) profit from this new therapy. (4) The ACORN device (Starling and Jessup, 2004) may be valuable in a limited subset of patients with large ventricles, but it is also hampered by severe complications, e.g., constrictive cardiomyopathy–like symptoms. Finally it seems clear that heart transplantation remains the only curative

therapy for terminal heart failure. However, the number of candidates largely exceeds the number of suitable donor organs, and this ratio has not improved over the years (Cai and Terasaki, 2004).

When considering a tissue-engineering strategy for patients with terminal heart failure, two different scenarios can be envisioned. The simple one is to add patches of tissue-engineered heart muscle on the diseased heart. A more elaborate one is to excise nonfunctioning myocardium and replace it with an engineered heart muscle patch. Obviously, the latter approach requires high mechanical stability of the patch to withstand systolic pressure, and this may be quite difficult to achieve. Thus, adding tissue-engineered patches to the heart ("biological ACORN") rather than replacing diseased parts is the more realistic perspective at this point.

## III. REQUIREMENTS FOR CLINICAL APPLICATION

The specific requirements for heart valves and blood vessels differ substantially from those for myocardial patches and are discussed only briefly here. The main part of this section focuses on the requirements for a clinical introduction of myocardial patches. The function of a heart valve depends mainly on its extracellular matrix component. In fact, the biological heart valves used today contain no living cell (e.g., the glutaraldehyd-fixed porcine xenograft or the bovine pericardial valve) or lose all their cellular components over time, as observed in case of the homografts (Schoen *et al.*, 1995). Thus, the currently used heart prostheses are essentially passivated structures in the bloodstream (Vesely, 2005). The hope of heart valve–tissue engineering is that, in the end, endothelialization of the prostheses reduces the risk of thromboembolism and immune responses and improves longevity, though this assumption remains unproven yet. In contrast, small-diameter blood vessels and heart muscle patches clearly depend on their cellular constituent to maintain function and longevity. Blood vessels need not only a high tensile strength as measured by the burst strength and an elastic component to provide recoil but also an intact endothelial cell layer to prevent thrombosis and to regulate vascular tone (Isenberg *et al.*, 2006). Myocardial patches need to generate systolic force, exhibit diastolic compliance, and, in case they were intended to withstand systolic blood pressure, high tensile strength. Given these features it becomes evident that the cellular component in myocardial patches plays a more prominent role than its matrix components. Finally, an important difference between the requirement for blood vessels and heart valves on the one hand and myocardial patches on the other is that the former can be engineered at a sufficient size without their own vascularization. In the following subsections, the conceptional requirements for a tissue-engineered myocardial patch are discussed in detail.

## A Tissue-Engineered Heart Muscle Patch Should Be Composed of Heart Cells

Engineered heart muscle should consist of bona fide myocardial tissue, including cardiac myocytes, blood vessels, and other cells that make up normal heart tissue. Whereas this criterion may appear trivial, it is not undisputed, and evidence exists that implantation of tissue-engineered constructs with nonmyocytes (Miyahara *et al.*, 2006) or without cells (Kofidis *et al.*, 2004) can have favorable effects on heart function, at least in animal models. Moreover, cardiac reconstructions with Dacron patches or other nonbiological materials is already a clinical reality. Thus, the necessity to create naturelike patches of heart muscles in the culture dish, with all the problems discussed herein, can well be challenged. Arguments for it are, at this point, partly derived experimentally, partly theoretical. Important experimental evidence comes from the failure of the "skeletal myoplasty approach" in heart failure. It was based on the observation that chronic electrical stimulation of skeletal muscle induces the expression of slow, cardiac-like myosin isoforms, which led to the hope that skeletal myocytes have the capacity to transdifferentiate into cardiac myocytes or at least into electrically coupling cardiac myocytelike cells. One early skeletal myoplasty procedure was the *latissimus dorsi plastic*, in which a vascularized piece of this muscle was surgically wrapped around the heart and electrically stimulated (Carpentier and Chachques, 1985). The other skeletal muscle–based procedure was the injection of autologous skeletal myoblasts (Menasche *et al.*, 2001). Neither the first nor the second approach turned out to be suitable to generate new myocardium in failing hearts. In fact, the first large randomized, placebo-controlled study with skeletal myoblasts was prematurely terminated in early 2006 due to a lack of efficacy and safety (personal communication). Though difficult to deduce from clinical studies, it is very likely that the absence of cardiac differentiation and the paucity of electrical coupling between the implanted skeletal myocytes and the host cardiac myocytes (Reinecke *et al.*, 2002) can induce life-threatening reentry tachyarrhythmias (Abraham *et al.*, 2005).

## A Tissue-Engineered Heart Muscle Patch Should Contain a Natural Extracellular Matrix

It is quite clear that the ideal engineered heart muscle should resemble natural heart tissue and therefore contain also natural extracellular matrix materials such as collagen type I/III and other basal membrane constituents. The question in the field is, rather, how to approach this aim. One strategy is to use the decellularized biomatrix of a heart and repopulate it with cardiac myocytes and the other cells that constitute heart muscle. Here the challenge is to employ chemical decellularization procedures that do not alter the matrix and do not interfere with the reseeding process. The second strategy is to let the cells produce their own matrix and just keep them in a three-dimensional space of the desired form for a given time until they have produced their own matrix. The third one is to design synthetic polymers that mimic the natural matrix as faithfully as possible and seed cells onto it. This ideal scaffold for cardiac-tissue engineering should promote cardiac-tissue formation, should be moldable in any desired three-dimensional form, and be biodegradable, nonimmunogenic, and nontoxic. They must, on the one hand, be highly compliant not to decrease diastolic properties and, on the other hand, exhibit mechanical strength to allow manual handling and systolic loading, especially if applied in the left ventricle.

## The Patch Should Have a Myocardiumlike Structure

Immature cells from embryonic chick and neonatal rat hearts exhibit a striking degree of remodeling capacity and, under suitable conditions, such as gyratory shaking, form beating, three-dimensional heartlike tissues, as already shown in the early 1950s (Moscona and Moscona, 1952). Similarly, embryonic stem cells easily form three-dimensional embryoid bodies that frequently contain beating cardiomyocytes (Martin and Evans, 1975). Histologically and functionally, these structures resemble immature cardiac tissues. Yet the usefulness of these structures for cardiac repair appears limited, for the following reasons: (1) They are relatively small (a few 10–100 µm); (2) they are spheric and do not exhibit the propensity to fuse spontaneously to larger, beating aggregates; (3) they contain randomly oriented rather than aligned cardiomyocytes. Thus, twitch force development of such aggregates is small and necessarily undirected, making simple cell aggregates unlikely to be suited for cardiac repair strategies. The three-dimensional structure and the muscle strand orientation of the ideal patch, in contrast, should be guided toward a physiological muscle architecture. In general, a longitudinal, parallel orientation of cardiac muscle strands will be desirable, because this is a major requisite for directed force development. Yet the normal heart has a complex macro-architecture with differently oriented muscle layers to support a wringing motion, to squeeze the blood forcefully into the circulation, rather than a straight contraction (Davis *et al.*, 2001). These biophysical aspects need to be taken into consideration when muscle patches are to be applied to replace complex myocardial structures *in vivo*.

## The Patch Should Develop Relevant Contractile Force

Maximal forces of intact cardiac preparations are in the range of 50 mN/mm² (Hasenfuss *et al.*, 1991). The force an engineered tissue patch develops can be taken as a good indicator of the cardiac tissue quality. It depends on several factors, including the fraction of the tissue occupied by cardiac myocytes (approximately 80% in normal cardiac tissue), the fraction of cardiac myocytes occupied by sarcomeres (normally 45%), the maturation of the sarcomeres (normally a semicrystalline, highly organized structure), and the isoforms of myofilament proteins (expression of adult rather than fetal isoforms; e.g., fast α-myosin heavy chain rather than the slow β-myosin heavy chain in rodents) as well as the orientation of muscle strands in parallel.

## The Patch Needs a Minimal Critical Size

The patch needs to be manufactured at a size large enough to affect realistically the function of an entire left ventricle or to be useful for cardiac reconstruction in case of malformations in childhood. It has been estimated that the human heart, which weighs around 300 g, contains 5 billion myocytes (Beltrami *et al.*, 1994). If an infarction was critical when >30% of the left ventricular mass was affected and one assumes the weight of the left ventricle to be half of that of a total heart including the atria then an ideal therapy would have to replace approximately 750 million cardiac myocytes. Such a patch would have a weight of 45 g or a size, for example, of $7 \times 7 \times 0.92$ cm.

## The Patch Should Be Vascularized

Present tissue-engineering experiences as well as studies of embryonic development, comparative anatomical studies in different species, and the biology of tumor formation show that a critical size of a tissue cannot be surpassed without perfusion of nutrients and oxygen, generally by vascularization. For example, in the absence of capillarization and perfusion, tumors implanted into nude mice did not reach a diameter of more than 2–3 mm (Folkman, 1971). These values, however, depend on both the metabolic demand and the cellular density of the respective tissue. Beating cardiac myocytes obviously have a very high metabolic activity. Accordingly, the density of capillaries in the adult heart is very high, amounting to approximately 2400–3300/mm² in rat and human at different postnatal stages (intercapillary distance ~19–20 μm) (Korecky *et al.*, 1982; Rakusan *et al.*, 1992). On the other hand, the early embryonic rat heart (until embryonic day 16) as well as the adult frog heart are avascular and nourished exclusively from the lumen by blood circulating within the trabecular system (Ratajska *et al.*, 2003; Sys *et al.*, 1997). Thus, large heart muscles develop either through vascularization or through intense trabecularization, with a single trabecel diameter being smaller than 50–75 μm. Given that an implanted patch has to survive in a nonperfused environment (the surface of the heart), sufficiently large patches for the human heart are likely to be possible only when the patch contains a functioning vasculature that can be surgically connected to the host circulation.

## The Patch Should Be Electrically Stable and Not Induce Arrhythmias

Sudden cardiac death, most commonly resulting from fatal arrhythmia, accounts for approximately 50% of cardiac death (one-third of total deaths). The two major arrhythmogenic mechanisms are reentry due to conduction abnormalities and abnormal automaticity, often as the consequence of cardiac repolarization defects ("heart failure is an acquired LQT syndrome"). Thus, the potential proarrhythmic risks of any tissue-engineering approach in heart failure has to be rigorously evaluated. The ideal patch should have no pacemaker potential that might act as a focus of automaticity. Even more importantly, the ideal patch should integrate robustly into the electrical syncytium of the host myocardium, on the one hand to be excited by the naturally propagated sinus rhythm and develop synchronized force and on the other hand not to cause arrhythmia. Suboptimal coupling could be even worse than no coupling because delayed conduction inevitably favors reentry circuits.

## The Patch Should Be Autologous

Organ transplantation represents major progress in modern medicine and justifies the need for life-long immunosuppression, with its unavoidable side effects. At first glance, it appears unrealistic to assume that patients and physicians would accept life-long immunosuppression to maintain a tissue-engineered patch. Therefore, the search for an autologous cell source and the design of totally autologous culture conditions is a major aim in the field. However, if cardiac-tissue engineering was to offer a curative treatment to patients who lack the alternative of conservative treatment or classical heart transplantation, then immunosuppression, with its side affects, appears acceptable.

## Tissue-Engineered Heart Muscle Must Be Generated Under GMP and GLP Guidelines

If a tissue-engineering strategy is to enter the clinical arena one day, it has to be developed under good manufacturing and good laboratory practice procedures (GMP and GLP). This means, for example, that xenogeneic compounds such as animal sera in the culture media have to be avoided, sterility must be ensured, and reproducibility is granted. This may actually make first-generation tissue-engineering products exceedingly expensive. Yet this is unlikely to be the real limiting factor for its further exploration because improvements in manufacturing processes, including large-scale production, are likely to reduce overall costs.

# IV. CURRENT CONCEPTS AND ACHIEVEMENTS IN ENGINEERING CARDIOVASCULAR PRODUCTS

The principal strategies in tissue engineering of valves, vessels, and myocardium are quite similar and have been developed in parallel, partly independent from each other. They can be broadly categorized into four approaches: (1) cell seeding onto decellularized native tissues, (2) cell seeding onto biodegradable polymer scaffolds, (3) cell entrapment in naturally occurring biogels such as collagen I and fibrin, and (4) the cell sheet technique. Consistent evidence exists that tissue formation can be significantly improved by chronic mechanical loading and culture under nonstatic conditions in a bioreactor.

## Decellularized Tissue

One of two strategies that actually made the way to clinical application so far is the unfixed, decellularized porcine valve, both with and without subsequent cell seeding. Decellularization of native tissues is the simplest technique and, in fact, does not require any specific tissue-engineering or cell biology procedure. Conceptionally, the natural tissue provides the ideal three-dimensional form, and decellularization is expected to reduce immune reactions inherent in the cellular component of the valve. However, the process of decellularization, usually performed with hyper- or hypotonic solutions, mild detergents, and proteolytic enzymes, exerts detrimental effects on the matrix that can give rise to thromboembolic complications and mechanical instability of the graft. Indeed, shortly after the first children received the acellular CryoLife valve in Vienna, Austria, serious valvular complications occurred and at least three children died (Simon *et al.*, 2003). The exact number of patients that received these grafts has never been reported. Despite these disastrous results, decellularized unseeded porcine valves, obtained via a different procedure, are still being implanted. A series of 50 patients in Berlin received such a graft during a Ross operation (replacement of the defective aortic valve by the endogenous pulmonary artery valve and implantation of the porcine valve in the pulmonary, low-pressure position), with apparently good results (Dohmen *et al.*, 2006; Konertz *et al.*, 2005). However, there is unpublished evidence that a number of patients has indeed died from valve-related complications (Vesely, 2005), casting serious doubts on this approach. The alternative approach that actually applies principles of tissue engineering is to repopulate the decellularized porcine valve with autologous, patient-derived cells *in vitro*. The expectation is that this procedure restores the biology of the graft, reduces thromboem-bolic complications, immune responses, and mechanical deterioration over time, and ultimately allows the valve to grow with the recipient. Indeed, positive results have been reported from experiments in sheep (Lichtenberg *et al.*, 2006) as well as from first implantations in children in the pulmonary position (A. Haverich, conference talk in Bad Reichenhall, Germany, February 2006). However, it appears too early to decide whether or not this approach can compare or even be superior to existing prostheses, such as the homograft of glutaraldehyde-fixed porcine or bovine pericardial valves.

Decellularized blood vessels and vessel-like structures made from small intestine submucosa (SIS) have also been tested, but overall the success appears limited (reviewed in Isenberg *et al.*, 2006). Similar to the valves, the process of decellularization apparently has detrimental effects on the tissue, leading to reduced tensile strength and compliance and shrinkage. Consequently, the decellularized blood vessels undergo aneurysm formation, infection, and thrombosis (Teebken and Haverich, 2002).

Similar to valves and vessels, there has been discussion about decellularizing whole hearts and repopulating the matrix with heart cells. Despite the apparent appeal of the idea, such a total bioartificial heart appears a rather unlikely option for therapy, given that one would have to reconstitute not only the heart muscle, but also the valves and the vessels, and this on the scale of a human heart. Thus, one could think of heart muscle patches derived from decellularized heart pieces, potentially with a perfusing vessel. Such experiments are ongoing in different laboratories but have not been published.

## Biodegradable Polymer Scaffolds

The second approach is to seed the respective cardiovascular cell type onto porous, biodegradable polymer scaffolds. Conceptionally the polymer serves as a three-dimensional template for the desired organ, and the cells are expected to degrade the polymer and produce their own extracellular matrix. This conventional tissue-engineering approach has the great advantage that the scaffold can theoretically be manufactured at any desired size, form, or microstructure and with various chemical and biological properties. In practice, however, the success in the cardiovascular field is limited. As Vesely puts it in his review with regard to heart valves: "The failure of this approach is much less dramatic [than the decellularized valve], as apparently none of these scaffolds have been tried in patients." This is actually not true for the blood vessel field, since a Japanese group reported a series of apparently successful implantations of tissue-engineered vascular grafts in the pulmonary artery position (Shin'oka *et al.*, 2003, 2005). They used a polyglycolic acid (PGA) sheet for mechanical reinforcement and a polycaprolactone-polylactic acid copolymer and seeded it with endothelial cells from a peripheral vein or, more recently, directly at the time of surgery with bone marrow cells. Since these experiments were done directly in patients, mechanical details and the long-term performance of these grafts remain unknown.

A principal problem of the polymer approach is that it is difficult to find the right balance between premature polymer degradation, with the risk of rupture (of a blood vessel or valve), and too little degradation, preventing tissue formation. For example, the early experiments with cardiac myocytes seeded on gelatine (Li *et al.*, 1999), alginate (Leor *et al.*, 2000), or PGA (Bursac *et al.*, 1999) showed survival of cells just on the surface of the matrix, suggesting that these scaffold materials prevented rather than promoted tissue formation. Much effort is directed toward newer matrix materials that may overcome the current limitations. Material sciences and particularly nanotechnology are progressing rapidly, and it may be that in a collaboration between engineers, physicists, biotechnologists, biologists, and morphologists the ideal scaffold material for creating truly artificial heart tissues will be designed (Buxton *et al.*, 2003; M. E. Davis *et al.*, 2005). Recent examples are tubular scaffolds for the engineering of blood vessels that were generated by electrospinning of collagen and elastin (Boland *et al.*, 2004) and collagen tubes for cardiac-tissue engineering (Yost *et al.*, 2004). Another innovative approach is to use MRI-derived three-dimensional images of an organ as a blueprint for the fully automated fabrication of a synthetic copy (Sodian *et al.*, 2005).

### Biopolymers

The third approach is based on the entrapment of the desired cardiovascular cell type in biopolymers such as collagen I and fibrin. These polymers form a gel in which the cells are trapped in a given three-dimensional form. The main advantage of this approach is that both collagen I and fibrin are approved materials in clinical use and apparently promote rather than prevent tissue formation. With this technique Weinberg and Bell (1969) generated blood vessels more than 20 years ago, and in the meantime many groups have adopted a similar approach using fibroblasts or smooth muscle cells to create blood vessels (L'Heureux *et al.*, 1993; Niklason *et al.*, 1999) or heart valves (Vesely, 2005) and cardiac myocytes for myocardial patches (Eschenhagen *et al.*, 1997; van Luyn *et al.*, 2002; Zimmermann *et al.*, 2000). A disadvantage of the biopolymer approach is that, until now, the tensile strength of the resulting tissues has been relatively low. None of it has been clinically applied yet.

### Cell Sheets

The fourth strategy differs from all others in that it goes without any scaffold material. In this approach, first described by Auger's group (L'Heureux *et al.*, 1998), vascular smooth muscle cells were grown on plastic culture dishes as normal monolayers until confluency and were then removed from the culture dish as a cellular sheet and wrapped around a porous mandrel to form the media, nourished from the inside through the mandrel. In the same manner, a second layer of fibroblast sheets was wrapped around the media to form the adventitia. Finally, the mandrel was removed and the tube populated with endothelial cells by installation and rotating overnight. These sheet-based blood vessels exhibit a very good burst pressure, exceeding that of human saphenous veins and vasoactive responses to several known receptor ligands (Laflamme *et al.*, 2005). Recently, sheet-based blood vessels made exclusively from human fibroblasts and human endothelial cells have been tested successfully in nonhuman primates (L'Heureux *et al.*, 2006). In a similar strategy, cardiac myocyte sheets can be stapled to form a contracting myocardial tissue (Shimizu *et al.*, 2002).

## V. STATE OF MYOCARDIAL-TISSUE ENGINEERING

### Heartlike Tissue Structure

Early experiments with biodegradable polymer scaffolds such as polyglycolic acid, alginate, and preformed collagen sponges showed that cardiac cells survive on these preformed three-dimensional structures but do not form synchronously beating, force-developing, heartlike structures (Bursac *et al.*, 1999; Leor *et al.*, 2000; Li *et al.*, 1999), likely because the scaffolds prevented the intense cell–cell contacts necessary for tissue formation. In contrast, tissue-engineering approaches based on biopolymers such as liquid collagen +/– Matrigel hydrogels or prefabricated collagen sponges + Matrigel as well as a cell sheet technique yielded spontaneously contracting myocardial tissues that exhibit a characteristic cardiac musclelike structure, both on the cellular and subcellular levels (Eschenhagen *et al.*, 1997; Radisic *et al.*, 2004; Shimizu *et al.*, 2002; van Luyn *et al.*, 2002) (Fig. 82.1). All of these approaches make use of the intrinsic capacity of immature cardiac cells to form heartlike muscle structures spontaneously (Akins *et al.*, 1999; Moscona and Moscona, 1952) that can be directed and enhanced by mechanical load (Eschenhagen *et al.*, 1997; Fink *et al.*, 2000), electrical stimulation (Radisic *et al.*, 2004), oxygenation and perfusion (Radisic *et al.*, 2005; Zimmermann *et al.*, 2006), growth factors (Zimmermann *et al.*, 2002b), and a native heart cell mix (Naito *et al.*, 2006). Thus, on a small scale, true heart muscle patches can be engineered from primary heart cells.

### Contractile Function

Such tissue-engineered heart muscle patches beat spontaneously and develop contractile force. Reported values range between 0.05 and 2 $mN/mm^2$ (Eschenhagen and Zimmermann, 2005), which is still considerably lower than the biological limit of 50 $mN/mm^2$. Under optimized culture conditions, including high oxygen, insulin, and auxotonic stretch plus electrical stimulation, EHTs develop up to 5 $mN/mm^2$ (own unpublished data). The remaining difference from native heart muscle is likely due to several factors: (1) The occupancy of the EHT tissue by muscle tissue is relatively low (<30% vs. almost 100%); (2) cardiac myocytes in EHTs are thinner when compared to the thick, bricklike adult cardiac myocytes of the ventricle; (3) orientation is not always perfectly parallel; and (4) the developmental state of the myocytes does not reach the level of terminal differention, as seen from the relatively high fraction of slow β-myosin heavy chain when compared to fast α-myosin heavy chain. Whereas such features clearly distinguish tissue-engineered patches from native myocardium, it is questionable whether, for the purpose of cardiac repair, a higher degree of maturation would be helpful. Immature cells have a higher remodeling capacity and resistance to

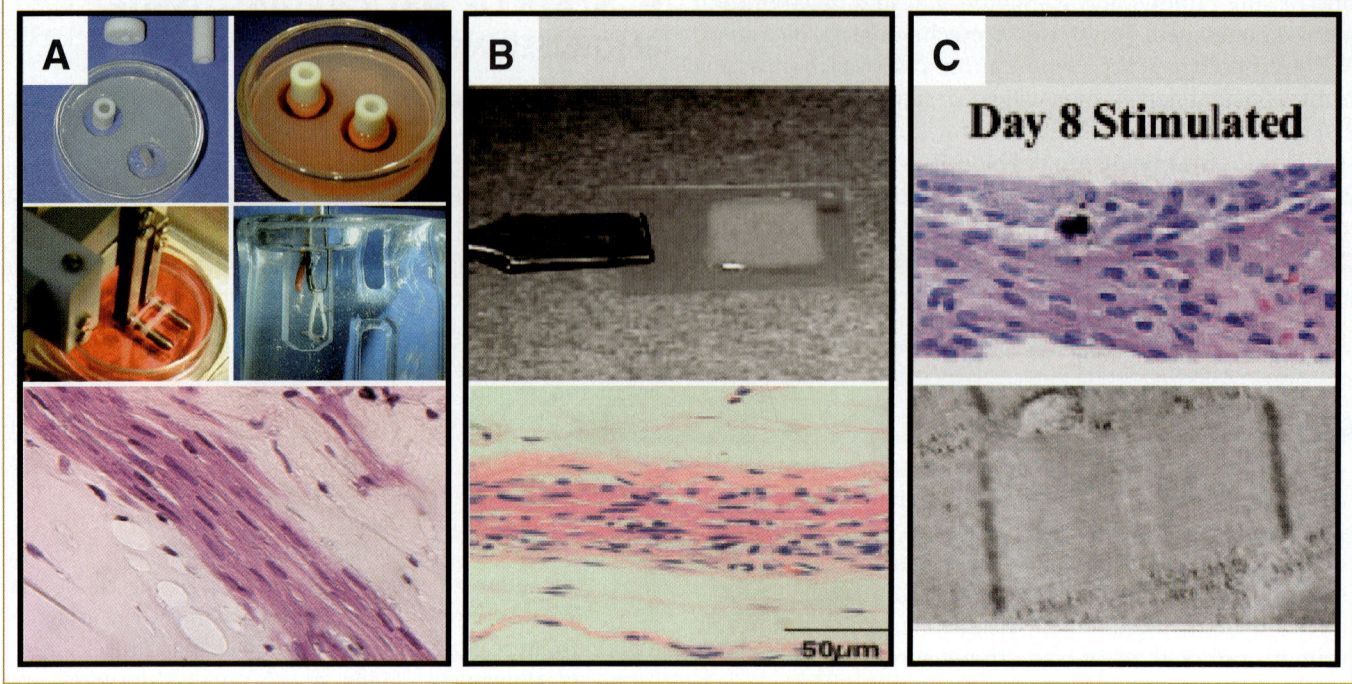

**FIG. 82.1.** Current approaches to myocardial-tissue engineering. **(A)** Macroscopic view of the setup to cast, culture, phasically stretch, and measure the contractile force of engineered heart tissue (EHT) based on collagen I/Matrigel hydrogels (Zimmermann *et al.*, 2002b). **(B)** Macroscopic (top) and microscopic view (H&E stain bottom) of stacked monolayers forming a beating three-dimensional tissue (Shimizu *et al.*, 2002). **(C)** Tissue formation (H&E stain, top) and sarcomere structure (bottom) in electrically stimulated collagen I + Matrigel-based constructs (Ultrafoam sponges) seeded with neonatal rat heart cells (Radisic *et al.*, 2004).

hypoxia [$Po_2$ in the fetus 20–25 mmHg; (Hollenberg *et al.*, 1976)], both could be critical parameters for integration into the host myocardium. Moreover, there is experimental evidence that myocytes undergo further maturation after implantation (Zimmermann *et al.*, 2002a). The heart of a neonatal rat weighs approximately 30 mg and that of an adult rat 1000 mg. This increase is due almost entirely to an increase in size and not number, meaning that the volume of myocytes increases 30-fold from neonatal to adult. If the same applies to cardiac myocytes in EHTs before and after implantation, the aforementioned ideal patch of 45 g could develop from an implanted patch of 1.5 g, a much more realistic number. Thus, it appears possible that the loose network of myocytes in EHTs at time of implantation form a much larger, compact muscle tissue with an adult structural and functional phenotype over time. Recent findings in rats four weeks after implantation support this assumption (Zimmermann *et al.*, 2006).

## Directed Process of Manufacturing

Currently, the most defined techniques to produce engineered myocardial patches for cardiac repair are (1) cell seeding on collagen/Matrigel sponges (Radisic *et al.*, 2004), (2) generation of collagen I/Matrigel-based EHTs (Zimmermann *et al.*, 2002b), and (3) the cell sheet technique (Shimizu *et al.*, 2002). All three techniques can be scaled up

in terms of width and length but, for the reasons discussed earlier, not yet in thickness. Scale-up and clinical application will also require some automation, but this appears possible with either technique and has actually already been established for the cell sheet technology (personal communication, T. Shimizu). Others generate cardiospheres (Akins *et al.*, 1999) or spheric microtissues (Kelm *et al.*, 2006) and implant them, but here cells are not under directed mechanical load and therefore are not well oriented and cannot be handled as a tissue patch. Thus, it is not clear whether these spheres have an advantage as compared to cells implanted in the heart as a suspension.

## Critical Size and Vascularization

The usual size of currently developed engineered myocardial tissue is in the range of centimeters in width/length and 0.1 to 1–4 millimeters in thickness (reviewed in Eschenhagen and Zimmermann, 2005). While the width and length can be expanded indefinitely, the thickness of compact myocardial tissue is limited in all models to 50–100 µm. This value may be enhanced by perfusion and high oxygen to a maximal theoretical thickness of 200 µm (Radisic *et al.*, 2005), but more seems unlikely to be achievable in the absence of vascularization and internal perfusion. A recent study demonstrated that multiloop EHTs of an initial size of approximately 15 × 15 × 4 mm formed a compact new layer of myocardium at a mean thickness of 0.5 mm when implanted onto infarcted rat hearts (Zimmermann *et al.*, 2006). Yet this is still far away from the thickness of the walls of the normal human left ventricle (mean 9 mm), and it is very likely that true vascularization is necessary to approach such values. One idea to vascularize tissue-engineered myocardium may be to support vascular ingrowth from adjacent or embedded vessels, either *in vitro* (Kofidis *et al.*, 2003) or *in vivo* (Shimizu *et al.*, 2006). Finally, some hope is related to the observation that primitive vascular structures develop spontaneously in EHTs (Zimmermann *et al.*, 2002b) (as in skeletal muscle constructs; Levenberg *et al.*, 2005). These vessels appear to find contact with the host vasculature after implantation (Zimmermann *et al.*, 2006). The angiogenic process could be further enhanced by other biological means, including angiogenic growth factors, but it is not clear at this point whether it helps to form thicker cardiac tissues *in vitro*.

## Electrical Integration and Improvement of Function After Implantation

Available data suggest that engineered sandwich constructs (Furuta *et al.*, 2006) and EHTs (Zimmermann *et al.*, 2006) find electrical contact with the host myocardium after implantation. This is in accordance with very careful studies of isolated cardiac myocytes injected directly into the myocardium (Rubart *et al.*, 2003). However, studying the question in engineered constructs implanted not in but onto the heart surface is not trivial and is difficult to substantiate (Eschenhagen *et al.*, 2006). Optical imaging with voltage-sensitive dyes and electrical mapping experiments suggest undelayed impulse propagation through the graft, but it is not entirely clear how the electrical contact is actually made between graft and host. Histological analyses did not unequivocally reveal cell–cell contacts in the respective zones (Furuta *et al.*, 2006; Zimmermann *et al.*, 2006), and high-resolution two-photon imaging (Rubart *et al.*, 2003) has not yet been performed in these models. Thus, at present it cannot be excluded that some of the apparently normal impulse propagation is due to electrotonic contact or even via fibroblasts. Present experiments with engineered heart tissues revealed no evidence for proarrhythmic effects (Zimmermann *et al.*, 2006). Yet the grafts can depolarize spontaneously, at least at the time of implantation, and this represents principally an arrhythmogenic risk. One reason this was not observed may be that, in the moment the engineered tissue couples to the host myocardium, it is overpaced by the higher frequency of the rat sinus rhythm. Finally, the rodent heart is not the most suited arrhythmia model, and more data are necessary to clarify this issue.

Several animal and human studies have suggested that cells of various origins when injected into the heart after myocardial infarction can have beneficial effects on contractile function (Dimmeler *et al.*, 2005; Laflamme and Murry, 2005; Murry *et al.*, 2002; Nadal-Ginard *et al.*, 2003).

The underlying mechanisms remain largely unknown but most likely do not include the formation of new contracting myocardium. Tissue-engineered myocardial patches, in contrast, are true heart muscles and should add new contracting units to the injured heart when implanted. Thus, engineered patches should have an effect on myocardial contractility that exceeds that of cell injections. Early studies actually reported some beneficial effects of implanting alginate (Leor *et al.*, 2000) or gelatine-based constructs (Li *et al.*, 1999) onto the heart, but these studies were not powered to quantify functional changes. We have recently performed a large-scale study in rats in which multiloop EHTs were implanted above the infarct scar 14 days after ligation of the left anterior descending (LAD) coronary artery (Fig. 82.2) (Zimmermann *et al.*, 2006). This study showed for the first time that tissue-engineered heart muscles can indeed restore contraction of an infarcted ventricular wall, improve ejection fraction in those animals with a poor function before implantation, and prevent further dilatation. Despite these encouraging results, it should be noted that even the implantation of a relatively large (surface approximately 1 cm$^2$) and thick (~1–4 mm) patch with nominally 12 million heart cells for a 1-g rat heart did not restore normal function. Importantly, however, control patches with noncontractile or fixed cells induced no major beneficial effects, supporting the notion that the improvement of systolic and diastolic function indeed derived from the cardiac myocytes in the EHT. Thus, the study may serve as a proof of principle for this approach.

## Autologous Patch

Tissue engineering at present utilizes almost exclusively allogeneic cardiac cells from neonatal rat or mouse or embryonic chick hearts. Accordingly, implantation studies used either immune-deficient models such as nude rats or mice (Furuta *et al.*, 2006; Shimizu *et al.*, 2002) or pharmacological immunosuppression (Zimmermann *et al.*, 2002a, 2006). The impact of these conditions on graft integration is unknown. A recent report describes a technique to generate EHTs from mouse embryonic stem cell–derived cardiomyocytes (Guo *et al.*, 2006), but this is also an allogeneic approach. The reason for the lack of autologous tissue-engineering approaches is simply that, until today, no one has identified a suitable adult stem cell source that allows allocation of sufficient numbers of true cardiac myocytes. Several reports have indicated that such stem cells exist in the heart (A. P. Beltrami *et al.*, 2003; Laugwitz *et al.*, 2005; Oh *et al.*, 2003), but these exciting new data have not yet been translated into tissue engineering.

## Scale-Up of GMP and GLP

Clinical application of tissue-engineered myocardial patches requires means to scale up the procedure under the strict regulations of good manufacturing (GMP) and laboratory practice (GLP). Although this question may appear far

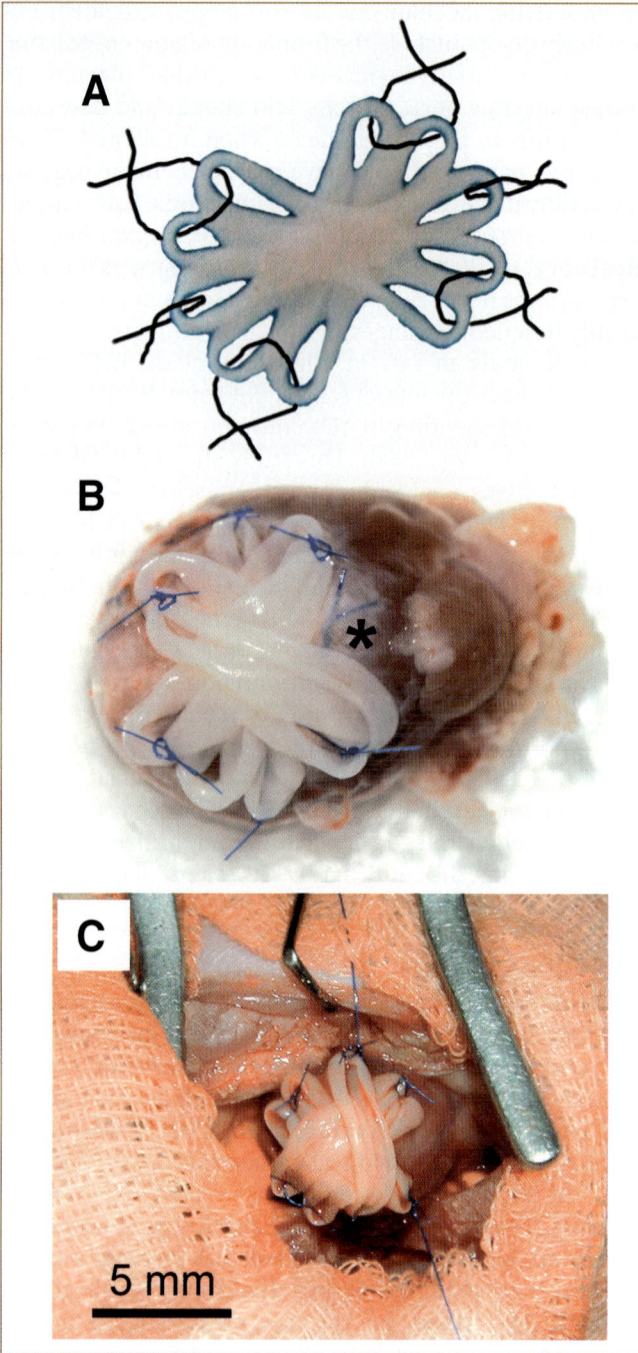

**FIG. 82.2.** EHT grafting onto infarcted rat heart. **(A)** Multiloop EHT with sutures before implantation. **(B)** Rat heart with multiloop EHTs sutured onto an infarcted rat heart *ex vivo*. **(C)** Rat heart situs showing a heart with a multiloop EHT fixed with six single-knot sutures placed in healthy myocardium just adjacent to the infarct scar through a left lateral thoracotomy.

away from the current state of the field, the technical effort and thus the costs will be important for the clinical success of a given technique. The cell sheet technique, seeding of collagen sponges, and the EHT technologies can be automated and performed in GMP/GLP facilities without

principal problems. However, the regulations of GMP/GLP do not permit the use of animal sera and growth supplements at any given point of the manufacturing process. Some progress has recently been made in defining serum-free and essentially Matrigel-free conditions for EHTs (Naito *et al.*, 2006), but more has to be done.

## VI. BOTTLENECKS

The critical bottlenecks in tissue engineering of heart valves, vessels, and myocardial patches differ. For a detailed review of the two former fields, the reader is referred to recent papers (Isenberg *et al.*, 2006; Vesely, 2005). It appears that the main hurdle for tissue-engineered heart valves is the relatively good performance of conventional prostheses and the disastrous consequences of any failure. Very encouraging progress has been made with tissue-engineered blood vessels, particularly with the sheet-based technique. In contrast to the myocardial patch, neither a suitable autologous cell source nor a sufficient size is a critical issue in this area. Recent sheet-based blood vessels exhibited excellent burst strength and compliance and, in rats and nonhuman primates, excellent long-term patency rates (L'Heureux *et al.*, 2006). Whether this also applies in humans is currently being tested. The long generation time (28 weeks) presently precludes urgent clinical use, but this is not a major drawback to these exciting new data.

Myocardial-tissue patches require less mechanical strength than vessels and valves, because they will be, with the exception of congenital malformations, likely applied as an add-on to an infarcted or failing heart. This also makes potential failure less dramatic. On the other hand, the lack of a suitable human cardiomyocyte source and the problem of critical size are the reasons why myocardial-tissue engineering is still far from any clinical application. Other problems, including scale-up, GMP/GLP- and serum-free culture conditions, and optimization of scaffolds can likely be solved by technical means. Electrical integration is subject to intense research, and increasing evidence suggests that myocardial grafts integrate well into the host myocardium.

### Cell Source

The most obvious problem is the identification of a suitable cell source. Since cardiac myocytes do not divide at a relevant rate, only two sources can be envisioned: (1) human adult stem cells (Beltrami *et al.*, 2003; Laugwitz *et al.*, 2005; Messina *et al.*, 2004; Oh *et al.*, 2003) and (2) human embryonic stem cells (Kehat *et al.*, 2001). The main advantage of the former is its autologous character, i.e., that the cells could ideally be taken from patient biopsies, amplified *ex vivo*, differentiated into cardiac myocytes, and used to engineer a myocardial-tissue patch that would be recognized by the patient's immune system as autologous. The crucial open questions are whether such cells exist in the body at a rate that realistically allows harvesting and amplification, whether they can generate a relevant number of true

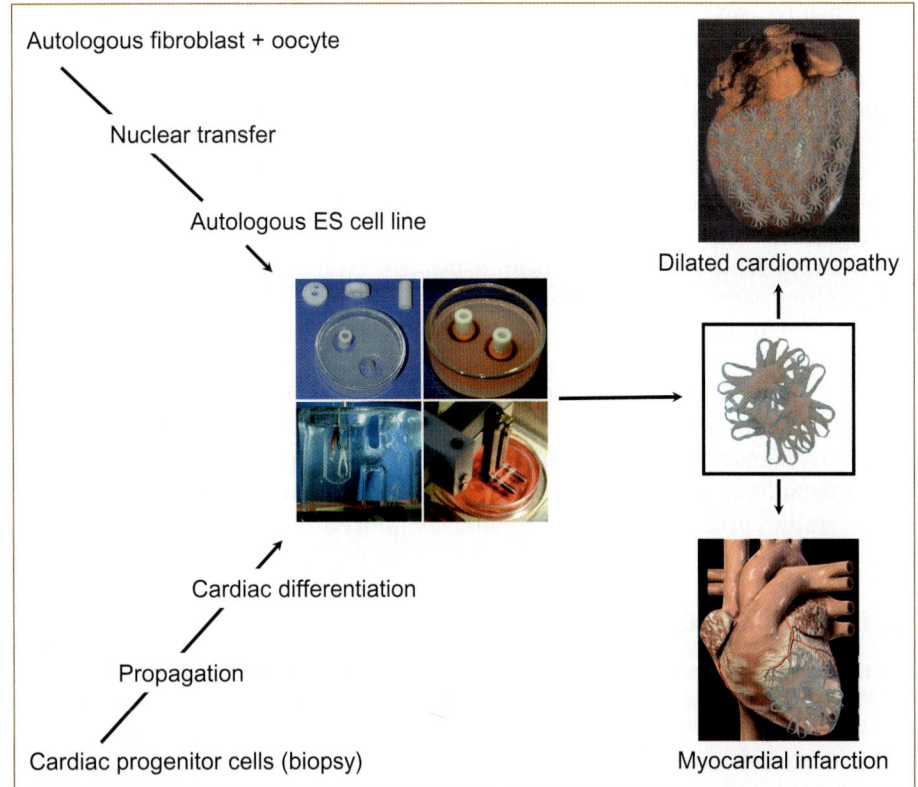

Autologous fibroblast + oocyte

Nuclear transfer

Autologous ES cell line

Dilated cardiomyopathy

Cardiac differentiation

Propagation

Cardiac progenitor cells (biopsy)

Myocardial infarction

**FIG. 82.3.** Perspectives of cardiac-tissue engineering exemplified with the EHT technology. Constructs will be made from ES cells that are ideally autologous, and derived from a nuclear transfer strategy or from adult cardiac progenitors that may be obtained from a heart muscle biopsy. For cardiac regeneration, several circular constructs can be woven together to form a multiloop construct. Several multiloop constructs together can form large "chain mail"-like networks for the treatment of dilated cardiomyopathy or a patch for the treatment of myocardial infarction.

cardiac myocytes, and whether one can produce engineered tissues from such cells. Extensive research is currently directed to solve these questions. The main advantage of human ES cells is that they unambiguously generate true cardiac myocytes (and many other cell types). Thus, improving the efficacy of cardiac myocyte differentiation will be essential to render ES cell–based tissue engineering feasible. The main disadvantage, however, is that ES cell–derived engineered tissues are allogeneic; i.e., their implantation would require life-long immunosuppression. Moreover, undifferentiated ES cells can form teratomas (Laflamme and Murry, 2005). Finally, legislative restrictions exist in many countries that, at present, represent hurdles in this part of research. The recent identification of multipotent adult germ-line stem cells from mouse testis with properties of ES cells (Guan *et al.*, 2006) opens an interesting perspective for an autologous cell source for myocardial-tissue engineering, at least for male patients. Finally, it is difficult to predict today which cell type will finally work, and therefore research to this end should be as broad and unprejudiced as possible.

### Size/Vascularization

Size truly matters, and it is possible that a sufficient size of an engineered tissue patch turns out to be the second most critical bottleneck. As outlined earlier, a realistic patch to improve heart function after myocardial infarction needs a surface of several square centimeters and a thickness of

millimeters and consists of several hundred million cardiac myocytes. This number of cells can only be provided by scale-up of stem cell cultures and directing stem cell differentiation into the cardiac lineage. But even with these cell numbers at hand, it will be a major task to generate myocardial tissues with a thickness of several millimeters. A combination of cardiac-muscle engineering principles and means to induce vascularization *in vitro* will be necessary to solve this problem. A possible scenario is to create a loose network of patches and to implant them on the heart as a biological ACORN device (Fig. 82.3).

## VII. SUMMARY

Tissue engineering of heart valves, small-diameter blood vessels, and myocardial patches is an exciting field with great clinical potential. The concerted research activity of several disciplines in the biomedical and engineering fields has yielded significant progress, but it still remains a major challenge to reproduce or at least approach the exquisite natural design of any of the cardiovascular structures and to translate such concepts to the clinic. With the exception of some rather premature implantations of unfixed, decellularized heart valves and cell-coated synthetic polymer conduits in children, tissue-engineered cardiovascular products have not entered the clinical arena yet. This seems wise, given the catastrophic consequences of failure in the case of a heart valve or a coronary artery.

But how can cardiovascular products make the step from the bench to the clinic? Rigorous *in vitro* testing and comprehensive small- and large-animal experiments are certainly necessary. Yet it is questionable whether this can prevent clinical failure. Once a tissue-engineering product has advanced to the large-animal testing stage, it becomes a matter of using the right model that ideally simulates human physiology and responsiveness to the respective intervention. This problem has been especially evident in large-animal valve trials. Sheep tend to develop massive fibrosis, which can be misleading in two respects — overestimation of the risk of valve degeneration, but also underestimation of mechanical instability because the fibrosis has in fact stabilizing effects on the valve (Schoen, 1998; Vesely, 2005). Pigs show a high level of calcification. Eventually, primates seem to be an ideal test bed. However, investigator-driven studies originating from academia are unlikely to be able to pay for these ultimate trials. On the other hand, industry will invest only if they sense a relevant profit based on successful preclinical data, leaving the field in a deadlocked situation. Thus, the only possible way to advance tissue engineering is to rigorously test available and new concepts in small-animal models and to advance the most promising into large-animal trials. Once a successful transition has been made into the clinic, it will be more likely that industry will finally be attracted by the field.

From all of the available data on cardiovascular-tissue engineering, it seems that tissue-engineered blood vessels have a good chance to be the first tissue-engineered products with a real practical value. The perspective of tissue-engineered valves will likely be restricted to children. Myocardial patches have a long way to go before they will be applied in humans, and their success will depend mainly on advances in stem cell technology.

# VIII. REFERENCES

Abraham, M. R., Henrikson, C. A., Tung, L., Chang, M. G., Aon, M., Xue, T., Li, R. A., B, O. R., and Marban, E. (2005). Antiarrhythmic engineering of skeletal myoblasts for cardiac transplantation. *Circ. Res.* **97**, 159–167.

Akins, R. E., Boyce, R. A., Madonna, M. L., Schroedl, N. A., Gonda, S. R., McLaughlin, T. A., and Hartzell, C. R. (1999). Cardiac organogenesis *in vitro*: reestablishment of three-dimensional tissue architecture by dissociated neonatal rat ventricular cells. *Tissue Eng.* **5**, 103–118.

Assmus, B., Schachinger, V., Teupe, C., Britten, M., Lehmann, R., Dobert, N., Grunwald, F., Aicher, A., Urbich, C., Martin, H., Hoelzer, D., Dimmeler, S., and Zeiher, A. M. (2002). Transplantation of progenitor cells and regeneration enhancement in acute myocardial infarction (TOPCARE-AMI). *Circulation* **106**, 3009–3017.

Bardy, G. H., Lee, K. L., Mark, D. B., Poole, J. E., Packer, D. L., Boineau, R., Domanski, M., Troutman, C., Anderson, J., Johnson, G., McNulty, S. E., Clapp-Channing, N., Davidson-Ray, L. D., Fraulo, E. S., Fishbein, D. P., Luceri, R. M., and Ip, J. H. (2005). Amiodarone or an implantable cardioverter-defibrillator for congestive heart failure. *N. Engl. J. Med.* **352**, 225–237.

Beltrami, A. P., Barlucchi, L., Torella, D., Baker, M., Limana, F., Chimenti, S., Kasahara, H., Rota, M., Musso, E., Urbanek, K., Leri, A., Kajstura, J., Nadal-Ginard, B., and Anversa, P. (2003). Adult cardiac stem cells are multipotent and support myocardial regeneration. *Cell* **114**, 763–776.

Beltrami, C. A., Finato, N., Rocco, M., Feruglio, G. A., Puricelli, C., Cigola, E., Quaini, F., Sonnenblick, E. H., Olivetti, G., and Anversa, P. (1994). Structural basis of end-stage failure in ischemic cardiomyopathy in humans. *Circulation* **89**, 151–163.

Biglioli, P., Spampinato, N., Cannata, A., Musumeci, A., Parolari, A., Gagliardi, C., and Alamanni, F. (2004). Long-term outcomes of the Carpentier–Edwards pericardial valve prosthesis in the aortic position: effect of patient age. *J. Heart Valve Dis.* **13(Suppl. 1)**, S49–S51.

Boland, E. D., Matthews, J. A., Pawlowski, K. J., Simpson, D. G., Wnek, G. E., and Bowlin, G. L. (2004). Electrospinning collagen and elastin: preliminary vascular-tissue engineering. *Front. Biosci.* **9**, 1422–1432.

Bursac, N., Papadaki, M., Cohen, R. J., Schoen, F. J., Eisenberg, S. R., Carrier, R., Vunjak-Novakovic, G., and Freed, L. E. (1999). Cardiac muscle–tissue engineering: toward an *in vitro* model for electrophysiological studies. *Am. J. Physiol.* **277**, H433–H444.

Buxton, D. B., Lee, S. C., Wickline, S. A., and Ferrari, M. (2003). Recommendations of the National Heart, Lung, and Blood Institute Nanotechnology Working Group. *Circulation* **108**, 2737–2742.

Cai, J., and Terasaki, P. I. (2004). Heart transplantation in the United States 2004. *Clin. Transpl.* 331–344.

Carpentier, A., and Chachques, J. C. (1985). Myocardial substitution with a stimulated skeletal muscle: first successful clinical case. *Lancet* **1**, 1267.

Cleland, J. G., Daubert, J. C., Erdmann, E., Freemantle, N., Gras, D., Kappenberger, L., and Tavazzi, L. (2005). The effect of cardiac resynchronization on morbidity and mortality in heart failure. *N. Engl. J. Med.* **352**, 1539–1549.

Davis, J. S., Hassanzadeh, S., Winitsky, S., Lin, H., Satorius, C., Vemuri, R., Aletras, A. H., Wen, H., and Epstein, N. D. (2001). The overall pattern of cardiac contraction depends on a spatial gradient of myosin regulatory light-chain phosphorylation. *Cell* **107**, 631–641.

Davis, M. E., Hsieh, P. C., Grodzinsky, A. J., and Lee, R. T. (2005). Custom design of the cardiac microenvironment with biomaterials. *Circ. Res.* **97**, 8–15.

Dimmeler, S., Zeiher, A. M., and Schneider, M. D. (2005). Unchain my heart: the scientific foundations of cardiac repair. *J. Clin. Invest.* **115**, 572–583.

Dohmen, P. M., da Costa, F., Holinski, S., Lopes, S. V., Yoshi, S., Reichert, L. H., Villani, R., Posner, S., and Konertz, W. (2006). Is there a possibility for a glutaraldehyde-free porcine heart valve to grow? *Eur. Surg. Res.* **38**, 54–61.

Eschenhagen, T., and Zimmermann, W. H. (2005). Engineering myocardial tissue. *Circ. Res.* **97**, 1220–1231.

Eschenhagen, T., Fink, C., Remmers, U., Scholz, H., Wattchow, J., Weil, J., Zimmermann, W., Dohmen, H. H., Schafer, H., Bishopric, N., Wakatsuki, T., and Elson, E. L. (1997). Three-dimensional reconstitution of embryonic cardiomyocytes in a collagen matrix: a new heart muscle model system. *FASEB J.* **11**, 683–694.

Eschenhagen, T., Zimmermann, W. H., and Kleber, A. G. (2006). Electrical coupling of cardiac myocyte cell sheets to the heart. *Circ. Res.* **98**, 573–575.

Fink, C., Ergun, S., Kralisch, D., Remmers, U., Weil, J., and Eschenhagen, T. (2000). Chronic stretch of engineered heart tissue induces hypertrophy and functional improvement. *FASEB J.* **14**, 669–679.

Folkman, J. (1971). Tumor angiogenesis: therapeutic implications. *N. Engl. J. Med.* **285**, 1182–1186.

Furuta, A., Miyoshi, S., Itabashi, Y., Shimizu, T., Kira, S., Hayakawa, K., Nishiyama, N., Tanimoto, K., Hagiwara, Y., Satoh, T., Fukuda, K., Okano, T., and Ogawa, S. (2006). Pulsatile cardiac tissue grafts using a novel three-dimensional cell sheet manipulation technique functionally integrates with the host heart, *in vivo. Circ. Res.* **98**, 705–712.

Guan, K., Nayernia, K., Maier, L. S., Wagner, S., Dressel, R., Lee, J. H., Nolte, J., Wolf, F., Li, M., Engel, W., and Hasenfuss, G. (2006). Pluripotency of spermatogonial stem cells from adult mouse testis. *Nature* **440**, 1199–1203.

Guo, X. M., Zhao, Y. S., Chang, H. X., Wang, C. Y., E, L. L., Zhang, X. A., Duan, C. M., Dong, L. Z., Jiang, H., Li, J., Song, Y., and Yang, X. J. (2006). Creation of engineered cardiac tissue *in vitro* from mouse embryonic stem cells. *Circulation* **113**, 2229–2237.

Hammermeister, K. E., Sethi, G. K., Henderson, W. G., Oprian, C., Kim, T., and Rahimtoola, S. (1993). A comparison of outcomes in men 11 years after heart-valve replacement with a mechanical valve or bioprosthesis. Veterans Affairs Cooperative Study on Valvular Heart Disease. *N. Engl. J. Med.* **328**, 1289–1296.

Hasenfuss, G., Mulieri, L. A., Blanchard, E. M., Holubarsch, C., Leavitt, B. J., Ittleman, F., and Alpert, N. R. (1991). Energetics of isometric force development in control and volume-overload of human myocardium. Comparison with animal species. *Circ. Res.* **68**, 836–846.

Hippisley-Cox, J., and Coupland, C. (2005). Effect of combinations of drugs on all cause mortality in patients with ischaemic heart disease: nested case-control analysis. *BMJ* **330**, 1059–1063.

Hollenberg, M., Honbo, N., and Samorodin, A. J. (1976). Effects of hypoxia on cardiac growth in neonatal rat. *Am. J. Physiol.* **231**, 1445–1450.

Isenberg, B. C., Williams, C., and Tranquillo, R. T. (2006). Small-diameter artificial arteries engineered *in vitro. Circ. Res.* **98**, 25–35.

Janssens, S., Theunissen, K., Boogaerts, M., and Van de Werf, F. (2006). Bone marrow cell transfer in acute myocardial infarction. *Nat. Clin. Pract. Cardiovasc. Med.* **3(Suppl. 1)**, S69–S72.

Kehat, I., Kenyagin-Karsenti, D., Snir, M., Segev, H., Amit, M., Gepstein, A., Livne, E., Binah, O., Itskovitz-Eldor, J., and Gepstein, L. (2001). Human embryonic stem cells can differentiate into myocytes with structural and functional properties of cardiomyocytes. *J. Clin. Invest.* **108**, 407–414.

Kelm, J. M., Djonov, V., Ittner, L. M., Fluri, D., Born, W., Hoerstrup, S. P., and Fussenegger, M. (2006). Design of custom-shaped vascularized tissues using microtissue spheroids as minimal building units. *Tissue Eng.* **12**, 2151–2160.

Klug, M. G., Soonpaa, M. H., Koh, G. Y., and Field, L. J. (1996). Genetically selected cardiomyocytes from differentiating embronic stem cells form stable intracardiac grafts. *J. Clin. Invest.* **98**, 216–224.

Kofidis, T., Lenz, A., Boublik, J., Akhyari, P., Wachsmann, B., Stahl, K. M., Haverich, A., and Leyh, R. G. (2003). Bioartificial grafts for transmural myocardial restoration: a new cardiovascular tissue culture concept. *Eur. J. Cardiothorac. Surg.* **24**, 906–911.

Kofidis, T., de Bruin, J. L., Hoyt, G., Lebl, D. R., Tanaka, M., Yamane, T., Chang, C. P., and Robbins, R. C. (2004). Injectable bioartificial myocardial tissue for large-scale intramural cell transfer and functional recovery of injured heart muscle. *J. Thorac. Cardiovasc. Surg.* **128**, 571–578.

Konertz, W., Dohmen, P. M., Liu, J., Beholz, S., Dushe, S., Posner, S., Lembcke, A., and Erdbrugger, W. (2005). Hemodynamic characteristics of the Matrix P decellularized xenograft for pulmonary valve replacement during the Ross operation. *J. Heart. Valve. Dis.* **14**, 78–81.

Korecky, B., Hai, C. M., and Rakusan, K. (1982). Functional capillary density in normal and transplanted rat hearts. *Can. J. Physiol. Pharmacol.* **60**, 23–32.

Laflamme, K., Roberge, C. J., Grenier, G., Remy-Zolghadri, M., Pouliot, S., Baker, K., Labbe, R., D'Orleans-Juste, P., Auger, F. A., and Germain, L. (2006). Adventitia contribution in vascular tone: insights from adventitia-derived cells in a tissue-engineered human blood vessel. *FASEB J.* **20**, 1245–1247.

Laflamme, M. A., and Murry, C. E. (2005). Regenerating the heart. *Nat. Biotechnol.* **23**, 845–856.

Laugwitz, K. L., Moretti, A., Lam, J., Gruber, P., Chen, Y., Woodard, S., Lin, L. Z., Cai, C. L., Lu, M. M., Reth, M., Platoshyn, O., Yuan, J. X., Evans, S., and Chien, K. R. (2005). Postnatal isl1[+] cardioblasts enter fully differentiated cardiomyocyte lineages. *Nature* **433**, 647–653.

Leor, J., Aboulafia-Etzion, S., Dar, A., Shapiro, L., Barbash, I. M., Battler, A., Granot, Y., and Cohen, S. (2000). Bioengineered cardiac grafts: a new approach to repair the infarcted myocardium? *Circulation* **102**, III56–III61.

Levenberg, S., Rouwkema, J., Macdonald, M., Garfein, E. S., Kohane, D. S., Darland, D. C., Marini, R., van Blitterswijk, C. A., Mulligan, R. C., D'Amore, P. A., and Langer, R. (2005). Engineering vascularized skeletal muscle tissue. *Nat. Biotechnol.* **23**, 879–884.

L'Heureux, N., Germain, L., Labbe, R., and Auger, F. A. (1993). *In vitro* construction of a human blood vessel from cultured vascular cells: a morphologic study. *J. Vasc. Surg.* **17**, 499–509.

L'Heureux, N., Paquet, S., Labbe, R., Germain, L., and Auger, F. A. (1998). A completely biological tissue-engineered human blood vessel. *FASEB J.* **12**, 47–56.

L'Heureux, N., Dusserre, N., Konig, G., Victor, B., Keire, P., Wight, T. N., Chronos, N. A., Kyles, A. E., Gregory, C. R., Hoyt, G., Robbins, R. C., and McAllister, T. N. (2006). Human tissue–engineered blood vessels for adult arterial revascularization. *Nat. Med.* **12**, 361–365.

Li, R. K., Jia, Z. Q., Weisel, R. D., Mickle, D. A., Choi, A., and Yau, T. M. (1999). Survival and function of bioengineered cardiac grafts. *Circulation* **100**, II63–II69.

Lichtenberg, A., Tudorache, I., Cebotari, S., Suprunov, M., Tudorache, G., Goerler, H., Park, J. K., Hilfiker-Kleiner, D., Ringes-Lichtenberg, S., Karck, M., Brandes, G., Hilfiker, A., and Haverich, A. (2006). Preclinical testing of tissue-engineered heart valves reendothelialized under simulated physiological conditions. *Circulation* **114**, I559–I565.

Martin, G. R., and Evans, M. J. (1975). Differentiation of clonal lines of teratocarcinoma cells: formation of embryoid bodies *in vitro. Proc. Natl. Acad. Sci. U.S.A.* **72**, 1441–1445.

Menasche, P., Hagege, A. A., Scorsin, M., Pouzet, B., Desnos, M., Duboc, D., Schwartz, K., Vilquin, J. T., and Marolleau, J. P. (2001). Myoblast transplantation for heart failure. *Lancet* **357**, 279–280.

Messina, E., De Angelis, L., Frati, G., Morrone, S., Chimenti, S., Fiordaliso, F., Salio, M., Battaglia, M., Latronico, M. V., Coletta, M., Vivarelli, E., Frati, L., Cossu, G., and Giacomello, A. (2004). Isolation and expansion of adult cardiac stem cells from human and murine heart. *Circ. Res.* **95**, 911–921.

Miyahara, Y., Nagaya, N., Kataoka, M., Yanagawa, B., Tanaka, K., Hao, H., Ishino, K., Ishida, H., Shimizu, T., Kangawa, K., Sano, S., Okano, T., Kitamura, S., and Mori, H. (2006). Monolayered mesenchymal stem cells repair scarred myocardium after myocardial infarction. *Nat. Med.* **12**, 459–465.

Moscona, A., and Moscona, H. (1952). The dissociation and aggregation of cells from organ rudiments of the early chick embryo. *J. Anat.* **86**, 287–301.

Murry, C. E., Whitney, M. L., Laflamme, M. A., Reinecke, H., and Field, L. J. (2002). Cellular therapies for myocardial infarct repair. *Cold Spring Harb. Symp. Quant. Biol.* **67**, 519–526.

Nadal-Ginard, B., Kajstura, J., Leri, A., and Anversa, P. (2003). Myocyte death, growth, and regeneration in cardiac hypertrophy and failure. *Circ. Res.* **92**, 139–150.

Naito, H., Melnychenko, I., Didie, M., Schneiderbanger, K., Schubert, P., Rosenkranz, S., Eschenhagen, T., and Zimmermann, W. H. (2006). Optimizing engineered heart tissue for therapeutic applications as surrogate heart muscle. *Circulation* **114**, I72–I78.

Naylor, R., Hayes, P. D., Payne, D. A., Allroggen, H., Steel, S., Thompson, M. M., London, N. J., and Bell, P. R. (2004). Randomized trial of vein versus dacron patching during carotid endarterectomy: long-term results. *J. Vasc. Surg.* **39**, 985–993; discussion 993.

Niklason, L. E., Gao, J., Abbott, W. M., Hirschi, K. K., Houser, S., Marini, R., and Langer, R. (1999). Functional arteries grown *in vitro. Science* **284**, 489–493.

Oh, H., Bradfute, S. B., Gallardo, T. D., Nakamura, T., Gaussin, V., Mishina, Y., Pocius, J., Michael, L. H., Behringer, R. R., Garry, D. J., Entman, M. L., and Schneider, M. D. (2003). Cardiac progenitor cells from adult myocardium: homing, differentiation, and fusion after infarction. *Proc. Natl. Acad. Sci. U.S.A.* **100**, 12313–12318.

Post, S., Kraus, T., Muller-Reinartz, U., Weiss, C., Kortmann, H., Quentmeier, A., Winkler, M., Husfeldt, K. J., and Allenberg, J. R. (2001). Dacron vs. polytetrafluoroethylene grafts for femoropopliteal bypass: a prospective randomized multicenter trial. *Eur. J. Vasc. Endovasc. Surg.* **22**, 226–231.

Radisic, M., Park, H., Shing, H., Consi, T., Schoen, F. J., Langer, R., Freed, L. E., and Vunjak-Novakovic, G. (2004). Functional assembly of engineered myocardium by electrical stimulation of cardiac myocytes cultured on scaffolds. *Proc. Natl. Acad. Sci. U.S.A.* **101**, 18129–18134.

Radisic, M., Deen, W., Langer, R., and Vunjak-Novakovic, G. (2005). Mathematical model of oxygen distribution in engineered cardiac tissue with parallel channel array perfused with culture medium containing oxygen carriers. *Am. J. Physiol. Heart Circ. Physiol.* **288**, H1278–H1289.

Rakusan, K., Flanagan, M. F., Geva, T., Southern, J., and Van Praagh, R. (1992). Morphometry of human coronary capillaries during normal growth and the effect of age in left ventricular pressure-overload hypertrophy. *Circulation* **86**, 38–46.

Ratajska, A., Ciszek, B., and Sowinska, A. (2003). Embryonic development of coronary vasculature in rats: corrosion casting studies. *Anat. Rec. A Discov. Mol. Cell. Evol. Biol.* **270**, 109–116.

Reinecke, H., Poppa, V., and Murry, C. E. (2002). Skeletal muscle stem cells do not transdifferentiate into cardiomyocytes after cardiac grafting. *J. Mol. Cell. Cardiol.* **34**, 241–249.

Rubart, M., Pasumarthi, K. B., Nakajima, H., Soonpaa, M. H., Nakajima, H. O., and Field, L. J. (2003). Physiological coupling of donor and host cardiomyocytes after cellular transplantation. *Circ. Res.* **92**, 1217–1224.

Schoen, F. J. (1998). Pathologic findings in explanted clinical bioprosthetic valves fabricated from photooxidized bovine pericardium. *J. Heart Valve Dis.* **7**, 174–179.

Schoen, F. J., Mitchell, R. N., and Jonas, R. A. (1995). Pathological considerations in cryopreserved allograft heart valves. *J. Heart Valve Dis.* **4**(Suppl. 1), S72–S75; discussion S75–S76.

Shimizu, T., Yamato, M., Isoi, Y., Akutsu, T., Setomaru, T., Abe, K., Kikuchi, A., Umezu, M., and Okano, T. (2002). Fabrication of pulsatile cardiac tissue grafts using a novel 3-dimensional cell sheet manipulation technique and temperature-responsive cell culture surfaces. *Circ. Res.* **90**, e40.

Shimizu, T., Sekine, H., Yang, J., Isoi, Y., Yamato, M., Kikuchi, A., Kobayashi, E., and Okano, T. (2006). Polysurgery of cell sheet grafts overcomes diffusion limits to produce thick, vascularized myocardial tissues. *FASEB J.* **20**, 708–710.

Shin'oka, T., Matsumura, K., Hibino, N., Naito, Y., Murata, A., Kosaka, Y., and Kurosawa, H. (2003). [Clinical practice of transplantation of regenerated blood vessels using bone marrow cells]. *Nippon Naika Gakkai Zasshi* **92**, 1776–1780.

Shin'oka, T., Matsumura, G., Hibino, N., Naito, Y., Watanabe, M., Konuma, T., Sakamoto, T., Nagatsu, M., and Kurosawa, H. (2005). Midterm clinical result of tissue-engineered vascular autografts seeded with autologous bone marrow cells. *J. Thorac. Cardiovasc. Surg.* **129**, 1330–1338.

Simon, P., Kasimir, M. T., Seebacher, G., Weigel, G., Ullrich, R., Salzer-Muhar, U., Rieder, E., and Wolner, E. (2003). Early failure of the tissue-engineered porcine heart valve SYNERGRAFT in pediatric patients. *Eur. J. Cardiothorac. Surg.* **23**, 1002–1006; discussion 1006.

Sodian, R., Fu, P., Lueders, C., Szymanski, D., Fritsche, C., Gutberlet, M., Hoerstrup, S. P., Hausmann, H., Lueth, T., and Hetzer, R. (2005). Tissue engineering of vascular conduits: fabrication of custom-made scaffolds using rapid prototyping techniques. *Thorac. Cardiovasc. Surg.* **53**, 144–149.

Stamm, C., Westphal, B., Kleine, H. D., Petzsch, M., Kittner, C., Klinge, H., Schumichen, C., Nienaber, C. A., Freund, M., and Steinhoff, G. (2003). Autologous bone-marrow stem-cell transplantation for myocardial regeneration. *Lancet* **361**, 45–46.

Starling, R. C., and Jessup, M. (2004). Worldwide clinical experience with the CorCap Cardiac Support Device. *J. Card. Fail.* **10**, S225–S233.

Strauer, B. E., Brehm, M., Zeus, T., Kostering, M., Hernandez, A., Sorg, R. V., Kogler, G., and Wernet, P. (2002). Repair of infarcted myocardium by autologous intracoronary mononuclear bone marrow cell transplantation in humans. *Circulation* **106**, 1913–1918.

Sys, S. U., Pellegrino, D., Mazza, R., Gattuso, A., Andries, L. J., and Tota, L. (1997). Endocardial endothelium in the avascular heart of the frog: morphology and role of nitric oxide. *J. Exp. Biol.* **200**, 3109–3118.

Teebken, O. E., and Haverich, A. (2002). Tissue engineering of small-diameter vascular grafts. *Eur. J. Vasc. Endovasc. Surg.* **23**, 475–485.

van Luyn, M. J., Tio, R. A., Gallego y van Seijen, X. J., Plantinga, J. A., de Leij, L. F., DeJongste, M. J., and van Wachem, P. B. (2002). Cardiac-tissue engineering: characteristics of in-unison-contracting two- and three-dimensional neonatal rat ventricle cell (co)-cultures. *Biomaterials* **23**, 4793–4801.

Vesely, I. (2005). Heart valve–tissue engineering. *Circ. Res.* **97**, 743–755.

Vitali, E., Colombo, T., Fratto, P., Russo, C., Bruschi, G., and Frigerio, M. (2003). Surgical therapy in advanced heart failure. *Am. J. Cardiol.* **91**, 88F–94F.

Weinberg, C. B., and Bell, E. (1986). A blood vessel model constructed from collagen and cultured vascular cells. *Science* **231**, 397–400.

Wohlschlaeger, J., Schmitz, K. J., Schmid, C., Schmid, K. W., Keul, P., Takeda, A., Weis, S., Levkau, B., and Baba, H. A. (2005). Reverse remodeling following insertion of left ventricular assist devices (LVAD): a review of the morphological and molecular changes. *Cardiovasc. Res.* **68**, 376–386.

Wollert, K. C., Meyer, G. P., Lotz, J., Ringes-Lichtenberg, S., Lippolt, P., Breidenbach, C., Fichtner, S., Korte, T., Hornig, B., Messinger, D., Arseniev, L., Hertenstein, B., Ganser, A., and Drexler, H. (2004). Intra-coronary autologous bone-marrow cell transfer after myocardial infarction: the BOOST randomized controlled clinical trial. *Lancet* **364**, 141–148.

Yost, M. J., Baicu, C. F., Stonerock, C. E., Goodwin, R. L., Price, R. L., Davis, J. M., Evans, H., Watson, P. D., Gore, C. M., Sweet, J., Creech, L., Zile, M. R., and Terracio, L. (2004). A novel tubular scaffold for cardiovascular-tissue engineering. *Tissue Eng.* **10**, 273–284.

Zimmermann, W. H., Didie, M., Wasmeier, G. H., Nixdorff, U., Hess, A., Melnychenko, I., Boy, O., Neuhuber, W. L., Weyand, M., and Eschenhagen, T. (2002a). Cardiac grafting of engineered heart tissue in syngenic rats. *Circulation* **106**, I151–I157.

Zimmermann, W. H., Schneiderbanger, K., Schubert, P., Didie, M., Munzel, F., Heubach, J. F., Kostin, S., Neuhuber, W. L., and Eschenhagen, T. (2002b). Tissue engineering of a differentiated cardiac muscle construct. *Circ. Res.* **90**, 223–230.

Zimmermann, W. H., Melnychenko, I., Wasmeier, G., Didie, M., Naito, H., Nixdorff, U., Hess, A., Budinsky, L., Brune, K., Michaelis, B., Dhein, S., Schwoerer, A., Ehmke, H., and Eschenhagen, T. (2006). Engineered heart tissue grafts improve systolic and diastolic function in infarcted rat hearts. *Nat. Med.* **12**, 452–458.

# Chapter Eighty-Three

# Tissue-Engineered Organs

*Steve J. Hodges and Anthony Atala*

## I. INTRODUCTION

Human organs are exposed to a variety of possible injuries from the time the fetus develops. Individuals may suffer from congenital disorders, cancer, trauma, infection, inflammation, iatrogenic injuries, or other conditions that may lead to organ damage or loss and necessitate eventual reconstruction or replacement. Whenever there is deficient organ function, replacement may be achieved with artificial organs or organ transplantation. These are flawed therapies, however, for artificial organs are not able to replace all physiologic functions and cannot fully integrate functionally into body systems, while organ transplantation is marred by donor shortages and immunologic rejection and requires chronic and morbid medical therapy. In most cases, the replacement of lost or deficient tissues with functionally equivalent autologous tissues would improve the outcome for patients immensely. Tissue-engineering therapies have been developed in order to achieve this goal.

The development of functional organ systems from the building blocks of cells, scaffolding, and growth factors is a Herculean task, however, as evidenced by the few successful tissue-engineering technologies currently in clinical use. An examination of the building blocks of tissue-engineering strategies as well as of the successful applications of these techniques in human clinical applications provides insight into the difficulties facing scientists as they seek to develop more complex bioengineered organs.

## II. TISSUE ENGINEERING: STRATEGIES FOR TISSUE RECONSTITUTION

The field of regenerative medicine encompasses various areas of technology, such as tissue engineering, stem cells, and cloning. One of the major components of regenerative medicine, tissue engineering, follows the principles of cell transplantation, materials science, and engineering toward the development of biological substitutes that can restore and maintain normal function. Tissue-engineering strategies generally fall into three categories: the use of acellular matrices (where matrices are used alone and which depend on the body's natural ability to regenerate for proper orientation and direction of new tissue growth), the use of simple cells, and the use of matrices combined with cells. The main obstacles that must be overcome for the clinical application of these strategies are localizing reliable, immunocompatible cell sources, developing biocompatible and functional scaffolding, and providing growth factors and vascularization for bioengineered organs.

Acellular tissue matrices are usually prepared by manufacturing artificial scaffolds or by removing cellular components from tissues via mechanical and chemical manipulation to produce collagen-rich matrices (Dahms *et al.*, 1998; Piechota *et al.*, 1998; Yoo *et al.*, 1998b; Chen *et al.*, 1999). These matrices tend to degrade slowly on implantation and are generally replaced by the extracellular matrix (ECM) proteins that are secreted by the ingrowing cells. Cells can be used in combination with scaffolds or injected as simple cell therapy, either with carriers (such as hydrogels) or alone (Atala, 2006).

## III. CELL SOURCES

When cells are used for tissue engineering, a small piece of donor tissue must be harvested and dissociated into individual cells. These cells are either implanted directly into the host or expanded in culture, attached to a support matrix, and then reimplanted into the host after expansion. The source of donor tissue can be heterologous (such as bovine), allogeneic (same species, different individual), or autologous. Ideally, both structural and functional tissue replacement will occur with minimal complications. The most preferred cells to use are autologous cells, where a biopsy of tissue is obtained from the host, the cells are dissociated and expanded in culture, and the expanded cells are implanted into the same host (Atala *et al.*, 1993, 1994; Cilento *et al.*, 1994; Yoo and Atala, 1997; Fauza *et al.*, 1998a, 1998b; Yoo *et al.*, 1998a, 1998b; Amiel and Atala, 1999; Kershen and Atala, 1999; Oberpenning *et al.*, 1999; Park *et al.*, 1999; Godbey and Atala, 2002). The use of autologous cells, although it may cause an inflammatory response, avoids rejection, and thus the deleterious side effects of immunosuppressive medications can be avoided.

One of the limitations of applying cell-based regenerative medicine techniques toward organ replacement has been the inherent difficulty of growing specific cell types in large quantities. Even when some organs, such as the liver, have a high regenerative capacity *in vivo*, cell growth and expansion *in vitro* may be difficult. By studying the privileged sites for committed precursor cells in specific organs as well as by exploring the conditions that promote differentiation, one may be able to overcome the obstacles that could lead to cell expansion *in vitro*. For example, urothelial cells could be grown in the laboratory setting in the past, but only with limited expansion. Several protocols were developed since the mid-1980s that identified the undifferentiated cells and kept them undifferentiated during their growth phase (Cilento *et al.*, 1994; Liebert *et al.*, 1997; Scriven *et al.*, 1997; Puthenveettil *et al.*, 1999). Using these methods of cell culture, it is now possible to expand a urothelial strain from a single specimen that initially covers a surface area of 1 cm$^2$ to one covering a surface area of 4202 m$^2$ (the equivalent area of one football field) within eight weeks (Cilento *et al.*, 1994). These studies indicated that it should be possible to collect autologous bladder cells from human patients, expand them in culture, and return them to the human donor in sufficient quantities for reconstructive purposes (Cilento *et al.*, 1994; Harriss, 1995; Freeman *et al.*, 1997; Liebert *et al.*, 1997; Lobban *et al.*, 1998; Nguyen *et al.*, 1999; Rackley *et al.*, 1999). Major advances have been achieved since the mid-1990s on the possible expansion of a variety of primary human cells, with specific techniques that make the use of autologous cells possible for clinical application.

An area of concern in the field of tissue engineering in the past was the source of cells for regeneration. The concept of creating engineered constructs involves initially obtaining cells for expansion from the diseased organ. A study recently showed that cultured neuropathic bladder smooth muscle cells possess and maintain different characteristics than normal smooth muscle cells *in vitro*, as demonstrated by growth assays, contractility, and adherence tests *in vitro* (Lin *et al.*, 2004). However, when neuropathic smooth muscle cells were cultured *in vitro* and seeded onto matrices and implanted *in vivo*, the tissue-engineered constructs showed the same properties as those tissues engineered with normal cells (Lai *et al.*, 2002). It is known that genetically normal progenitor cells, which are the reservoirs for new cell formation, are programmed to give rise to normal tissue, regardless of whether the niche resides in normal or diseased tissues. The stem cell niche and its role in normal tissue regeneration remain a fertile area of ongoing investigation.

## IV. ALTERNATE CELL SOURCES

### Stem Cells

Most current strategies for tissue engineering depend on a sample of autologous cells from the diseased organ of the host. However, for many patients with extensive end-stage organ failure, a tissue biopsy may not yield enough normal cells for expansion and transplantation. In other instances, primary autologous human cells cannot be expanded from a particular organ, such as the pancreas. In these situations, pluripotent human embryonic stem cells are envisioned as a viable source of cells, because they can serve as an alternative source of cells from which the desired tissue can be derived.

Human embryonic stem cells exhibit two remarkable properties: the ability to proliferate in an undifferentiated but pluripotent state (self-renew), and the ability to differentiate into many specialized cell types (Brivanlou *et al.*, 2003). They can be isolated by immunosurgery from the inner cell mass of the embryo during the blastocyst stage (five days postfertilization) and are usually grown on feeder layers consisting of mouse embryonic fibroblasts or human feeder cells (Richards *et al.*, 2002). More recent reports have shown that these cells can be grown without the use of a feeder layer (Amit *et al.*, 2003), thus avoiding the exposure of these human cells to mouse viruses and proteins. These cells have demonstrated longevity in culture by maintaining their undifferentiated state for at least 80 passages when grown using current published protocols (Thomson *et al.*, 1998; Reubinoff *et al.*, 2000).

Human embryonic stem cells have been shown to differentiate into cells from all three embryonic germ layers *in vitro*. Skin and neurons have been formed, indicating ectodermal differentiation (Schuldiner *et al.*, 2000; Reubinoff *et al.*, 2001; Schuldiner *et al.*, 2001). Blood, cardiac cells, cartilage, endothelial cells, and muscle have been formed, indicating mesodermal differentiation (Kaufman *et al.*, 2001; Kehat *et al.*, 2001; Levenberg *et al.*, 2002). Pancreatic cells have also been formed, indicating endodermal differentiation (Assady *et al.*, 2001). In addition, as further evidence of their pluripotency, embryonic stem cells can form embryoid bodies, which are cell aggregations that contain all three embryonic germ layers, while in culture and can form teratomas *in vivo* (Itskovitz-Eldor *et al.*, 2000).

Adult stem cells have the advantage of avoiding the ethical issues associated with embryonic cells, and they also do not transdifferentiate spontaneously to a malignant phenotype. The use of adult stem cells is limited for clinical therapy due to the difficulties encountered with cell expansion to large quantities.

Fetal stem cells derived from amniotic fluid and placentas have recently been described and represent a novel source of stem cells. The cells express markers consistent with human embryonic stem cells, such as OCT4 and SSEA-4, but do not form teratomas and do not spontaneously transdifferentiate to a malignant phenotype. The cells are multipotent and are able to differentiate into all three germ layers. In addition, the cells have a high replicative potential and could be stored for future self-use, without the risks of rejection and without ethical concerns (Lanza *et al.*, 1999).

## V. THERAPEUTIC CLONING

While there has been tremendous interest in the field of nuclear cloning since the birth of Dolly in 1997, the first successful nuclear transfer was reported over 50 years ago by Briggs and King (Briggs and King, 1952), cloned frogs which were the first vertebrates derived from nuclear transfer, were subsequently reported by Gurdon in 1962, but the nuclei were derived from nonadult sources (Gordon, 1962). In the past several years, tremendous advances in nuclear cloning technology have been reported, indicating the relative immaturity of the field. Dolly was not the first cloned mammal to be produced from adult cells; in fact, live lambs were produced in 1996 using nuclear transfer and differentiated epithelial cells derived from embryonic discs (Campbell *et al.*, 1996). The significance of Dolly was that she was the first mammal to be derived from an adult somatic cell using nuclear transfer (Wilmut *et al.*, 1997). Since then, animals from several species have been grown using nuclear transfer technology, including cattle (Cibelli *et al.*, 1998), goats (Baguisi *et al.*, 1999; Keefer *et al.*, 2002), mice (Wakayama *et al.*, 1998), and pigs (De Sousa *et al.*, 2002).

Two types of nuclear cloning, reproductive cloning and therapeutic cloning, have been described, and a better understanding of the differences between the two types may help to alleviate some of the controversy that surrounds these technologies. Banned in most countries for human applications, reproductive cloning is used to generate an embryo that has the identical genetic material as its cell source. This embryo can then be implanted into the uterus of a female to give rise to an infant that is a clone of the donor. On the other hand, therapeutic cloning is used to generate early-stage embryos that are explanted in culture to produce embryonic stem cell lines whose genetic material is identical to that of its source. These autologous stem cells have the potential to become almost any type of cell in the adult body and thus would be useful in tissue- and organ-replacement applications (Hochedlinger and Jaenisch, 2003). Therefore, therapeutic cloning, which has also been called *somatic cell nuclear transfer*, may provide an alternative source of transplantable cells. According to data from the Centers for Disease Control, it has been estimated that approximately 3000 Americans die every day of diseases that could have been treated with stem cell–derived tissues (Thomson *et al.*, 1998). With current allogeneic tissue transplantation protocols, rejection is a frequent complication because of immunologic incompatibility, and immunosuppressive drugs must be administered to treat and hopefully prevent host-versus-graft disease (Lanza *et al.*, 2001). The use of transplantable tissue and organs derived from therapeutic cloning may potentially lead to the avoidance of immune responses that typically are associated with transplantation of nonautologous tissues (Lanza *et al.*, 1999).

While promising, somatic cell nuclear transfer technology has certain limitations that require further improvements before therapeutic cloning can be applied widely in replacement therapy. Currently, the efficiency of the overall cloning process is low. The majority of embryos derived from animal cloning do not survive after implantation (Solter, 2000; Rideout *et al.*, 2001; Hochedlinger and Jaenisch, 2002). In practical terms, multiple nuclear transfers must be performed in order to produce one live offspring for animal-cloning applications. The potential for cloned embryos to grow into live offspring is between 0.5 and 18% for sheep, bovine, pigs, and mice (Tsunoda and Kato, 2002). However, greater success (80%) has been reported in cattle (Kato, *et al.*, 1998), which may be due in part to the availability of advanced bovine-supporting technologies, such as *in vitro* embryo production and embryo transfer, which have been developed for this species for agricultural purposes. To improve cloning efficiencies, further improvements are required in the multiple complex steps of nuclear transfer, such as enucleation and reconstruction, activation of oocytes, and cell-cycle synchronization between donor cells and recipient oocytes, that will more readily produce viable sources of cells.

Furthermore, common abnormalities have been found in newborn clones if they survive to birth, including enlarged size with an enlarged placenta (*large-offspring syndrome*) (Young *et al.*, 1998), respiratory distress and defects of the kidney, liver, heart, and brain (Cibelli *et al.*, 2002), obesity

(Tamashiro *et al.*, 2002), and premature death (Ogonuki *et al.*, 2002). These may be related to the epigenetics of the cloned cells, which involve the reversible modifications of the DNA or chromatin while the original DNA (genetic) sequences remain intact. Faulty epigenetic reprogramming in clones, where the DNA methylation patterns, histone modifications, and overall chromatin structure of the somatic nuclei are not being reprogrammed to an embryonic pattern of expression, may explain the aforementioned abnormalities (Hochedlinger and Jaenisch, 2003). Reactivation of key embryonic genes at the blastocyst stage is usually not present in embryos cloned from somatic cells, while embryos cloned from embryos consistently express early embryonic genes (Boiani *et al.*, 2002; Bortvin *et al.*, 2003). Proper epigenetic reprogramming to an embryonic state may help to improve the cloning efficiency and reduce the incidence of abnormal cloned cells.

## VI. BIOMATERIALS

For cell-based tissue engineering, once cells have been cultured and expanded, they are seeded onto a scaffold synthesized using the appropriate biomaterial. In tissue engineering, biomaterials replicate the biologic and mechanical functions of the native extracellular matrix (ECM) found in tissues in the body by serving as an artificial ECM. As a result, biomaterials provide a three-dimensional space for the cells to form into new tissues with appropriate structure and function and also can allow for the delivery of cells and appropriate bioactive factors (e.g., cell adhesion peptides, growth factors) to desired sites in the body (Kim and Mooney, 1998). Because the majority of mammalian cell types are anchorage dependent and will die if no cell adhesion substrate is available, biomaterials provide a cell adhesion substrate that can deliver cells to specific sites in the body with high loading efficiency. Biomaterials can also provide mechanical support against *in vivo* forces such that the predefined three-dimensional structure is maintained during tissue development. Furthermore, bioactive signals, such as cell adhesion peptides and growth factors, can be loaded along with cells to help regulate cellular function.

The ideal biomaterial should be biocompatible, in that it is biodegradable and bioresorbable to support the replacement of normal tissue without inflammation. Incompatible materials are destined for an inflammatory or foreign-body response that eventually leads to rejection and/or necrosis. In addition, the degradation products, if produced, should be removed from the body via metabolic pathways at an adequate rate that keeps the concentration of these degradation products in the tissues at a tolerable level (Bergsma *et al.*, 1995). Furthermore, the biomaterial should provide an environment in which appropriate regulation of cell behavior (e.g., adhesion, proliferation, migration, and differentiation) can occur such that functional tissue can form. Cell behavior in the newly formed tissue has been shown to

be regulated by multiple interactions of the cells with their microenvironment, including interactions with cell adhesion ligands (Hynes, 1992) and with soluble growth factors. In addition, biomaterials provide temporary mechanical support that allows the tissue to grow in three dimensions while the cells undergo spatial tissue reorganization. The properly chosen biomaterial should allow the engineered tissue to maintain sufficient mechanical integrity to support itself in early development, while in late development it should begin degradation such that it does not hinder further tissue growth (Kim and Mooney, 1998).

Generally, three classes of biomaterials have been utilized for engineering tissues: naturally derived materials (e.g., collagen and alginate), acellular tissue matrices (e.g., bladder submucosa and small intestinal submucosa), and synthetic polymers [e.g., polyglycolic acid (PGA), polylactic acid (PLA), and poly(lactic-*co*-glycolic acid) (PLGA)]. These classes of biomaterials have been tested with respect to their biocompatibility (Pariente *et al.*, 2001, 2002). Naturally derived materials and acellular tissue matrices have the potential advantage of biological recognition. However, synthetic polymers can be produced reproducibly on a large scale with controlled properties of their strength, degradation rate, and microstructure.

Collagen is the most abundant and ubiquitous structural protein in the body and can be readily purified from both animal and human tissues with an enzyme treatment and salt/acid extraction. Collagen implants degrade through a sequential attack by lysosomal enzymes. Controlling the density of the implant and the extent of intermolecular cross-linking can regulate the *in vivo* resorption rate. The lower the density, the greater the interstitial space and generally the larger the pores for cell infiltration, leading to a higher rate of implant degradation. Collagen contains cell-adhesion-domain sequences (e.g., RGD) that exhibit specific cellular interactions. This may assist in retaining the phenotype and activity of many types of cells, including fibroblasts (Silver and Pins, 1992) and chondrocytes (Sams and Nixon, 1995).

Alginate, a polysaccharide isolated from seaweed, has been used as an injectable cell-delivery vehicle (Smidsrod and Skjak-Braek, 1990) and a cell-immobilization matrix (Lim and Sun, 1980), owing to its gentle gelling properties in the presence of divalent ions such as calcium. Alginate is relatively biocompatible and approved by the FDA for human use as wound dressing material. Alginate is a family of copolymers of D-mannuronate and L-guluronate. The physical and mechanical properties of alginate gel are strongly correlated with the proportion and length of the polyguluronate block in the alginate chains (Smidsrod and Skjak-Braek, 1990).

Acellular tissue matrices are collagen-rich matrices prepared by removing cellular components from tissues (Dahms *et al.*, 1998; Piechota *et al.*, 1998; Yoo *et al.*, 1998b; Chen *et al.*, 1999). The matrices are often prepared by mechanical and chemical manipulation of a segment of tissue. The matrices degrade slowly on implantation and are replaced and remodeled by ECM proteins synthesized and secreted by transplanted or ingrown cells.

Polyesters of naturally occurring α-hydroxy acids, including PGA, PLA, and PLGA, are widely used in tissue engineering. These polymers have gained FDA approval for human use in a variety of applications, including sutures. The ester bonds in these polymers are hydrolytically labile, and these polymers degrade by nonenzymatic hydrolysis. The degradation products of PGA, PLA, and PLGA are nontoxic, natural metabolites and are eventually eliminated from the body in the form of carbon dioxide and water. The degradation rate of these polymers can be tailored from several weeks to several years by altering crystallinity, initial molecular weight, and the copolymer ratio of lactic to glycolic acid. Since these polymers are thermoplastics, they can easily be formed into a three-dimensional scaffold with a desired microstructure, gross shape, and dimension by various techniques, including molding, extrusion (Freed *et al.*, 1994), solvent casting, phase separation techniques, and gas-foaming techniques (Harris *et al.*, 1998). Many applications in tissue engineering often require a scaffold with high porosity and ratio of surface area to volume. Other biodegradable synthetic polymers, including poly(anhydrides) and poly(ortho-esters), can also be used to fabricate scaffolds for tissue engineering with controlled properties (Peppas and Langer, 1994).

## VII. GROWTH FACTORS

A key component to scaffold function is the presence of the appropriate growth factors for cell growth and maturation. Growth factors may be produced by cells themselves (autocrine) or by the surrounding environment (paracrine). Scientists have employed various strategies to enhance desirable growth factor production, in order to promote the cell growth and maturation needed for particular organ systems. Some investigators have attempted to influence cell growth factor production with gene therapy, by incorporating genes for growth factor transcription in the transplanted cells. Others have sought to simply inject the growth factors into the cells milieu, but this method often results in rapid dissipation of the agents, due to diffusion (Ikada, 2006).

A novel method of growth factor use in tissue engineering involves the incorporation of growth factors into carriers or cell scaffolds. The goals of these techniques are to provide localized and controlled release of growth factors necessary for cell development. Extensive research is under way to define the exact combination of growth factors needed for specific organ growth and maturation, with an emphasis on the use of these factors to induce more rapid cell development and differentiation, appropriate cell function and orientation, and extracellular matrix development and vascularization (Rosso *et al.*, 2005).

## VIII. VASCULARIZATION

One of the restrictions of the engineering of tissues is that cells cannot be implanted in volumes exceeding $3\,mm^3$ because of the limitations of nutrition and gas exchange (Folkman and Hochberg, 1973). To achieve the goals of engineering large, complex tissues and possibly internal organs, vascularization of the regenerating cells is essential.

Three approaches have been used for vascularization of bioengineered tissue: (1) incorporation of angiogenic factors in the bioengineered tissue, (2) seeding EC with other cell types in the bioengineered tissue, and (3) prevascularization of the matrix prior to cell seeding. Angiogenic growth factors may be incorporated into the bioengineered tissue prior to implantation, in order to attract host capillaries and to enhance neovascularization of the implanted tissue. There are many obstacles to overcome before large, entirely tissue-engineered solid organs are produced. Recent developments in angiogenesis research may provide important knowledge and essential materials to accomplish this goal.

## IX. CLINICAL APPLICATIONS

The assimilation of all these complex technologies into functional bioengineered organs is a complex task. By examining the early attempts at the application of tissue engineering for clinical use, the difficulties and challenges facing scientists in the future are more clearly delineated. The application of bioengineered organs for human use has been addressed in a logical fashion, beginning with the most simple organ system (e.g., skin) and extending to hollow organs (e.g., bladder) and finally to complex solid organs (e.g., kidneys).

### Skin

The concept of the clinical application of skin-tissue engineering had its origins in Massachusetts around 1980, when Yannas and Bell developed the technologies necessary for the growth of skinlike structures from single cells. Bell was able to create dermal skin tissue by culturing human neonatal foreskin fibroblasts in collagen gels (Bell *et al.*, 1979). Yannas and associates expanded this approach, and were able to engineer skin by applying the modern principles of biomaterial scaffolding and molecular cues to influence the growth of skin cells (Yannas and Burke, 1980; Yannas *et al.*, 1980). These efforts resulted in several commercially available bioengineered skin-replacement therapies. Yannas and coworkers developed Integra, a collagen template for dermal regeneration. This was followed shortly by Bell's development of a full-thickness skin substitute with keratinized epidermis and dermis, Apligraft. The most recent development in the skin substitute market is Dermagraft, a composite of dermal fibroblasts and collagen developed by culturing neonatal foreskin fibroblasts on PLGA scaffolds (Shastri, 2006).

Although all these products function well, clinical use has not been widespread. The failure of any of these excellent technologies to dominate the skin-replacement market emphasizes the somewhat-ignored factors of tissue engineering. Mainly, for a technology to be widely accepted as an organ-replacement technology, it must function better than all available options, be easy to use, and not be cost prohibitive. Indeed, a bioengineered organ that functions better than available therapies may never be accepted into widespread clinical use if it is difficult to use or far too expensive.

## Bladder

Currently, gastrointestinal segments are commonly used as tissues for bladder replacement or repair (McDougal, 1992; Kaefer *et al.*, 1997). Because of the problems encountered with the use of gastrointestinal segments, numerous investigators have attempted alternative reconstructive procedures for bladder replacement or repair, such as the use of tissue expansion, seromuscular grafts, matrices for tissue regeneration, and tissue engineering with cell transplantation (McDougal, 1992).

Cell-seeded allogeneic acellular bladder matrices are one of the few strategies of bioengineered bladder replacement with human clinical applications. The early work with this technology was performed in dogs, where a group of experimental canines underwent a trigone-sparing cystectomy with closure of the trigone without a reconstructive procedure or reconstruction with a nonseeded bladder-shaped biodegradable scaffold or reconstruction using a bladder-shaped biodegradable scaffold that delivered seeded autologous urothelial cells and smooth muscle cells (Oberpenning *et al.*, 1999).

The results were promising, with the cystectomy-only and nonseeded controls demonstrating decreased capacities and compliance as compared to preoperative values, while the cell-seeded tissue-engineered bladders showed increased capacity and compliance, with histology depicting normal cellular organization, consisting of a trilayer of urothelium, submucosa, and muscle (Oberpenning *et al.*, 1999).

Preliminary clinical trials for the application of this technology have been performed and have had promising results. Recently published data demonstrated that engineered autologous bladder tissue may be used safely to augment the bladder in children with end-stage bladder disease secondary to myelomeningocele (Atala *et al.*, 2006). In this study, seven patients aged 4–19 with high pressure or poorly compliant bladders were selected as candidates for autologous augmentation cystoplasty. Bladder biopsies were obtained from each patient, and smooth muscle and urothelial cells were cultured, expanded, and seeded on a composite scaffold. Approximately seven weeks after the biopsy, the autologously generated neobladders were used

for bladder reconstruction with or without omental wrapping. At a mean follow-up of 46 months, the increase in bladder volume and compliance was greatest in the engineered bladders coated in the omental wrap. Bowel function returned promptly after surgery, there were no metabolic complications, no stone formation or mucous production, and renal function was preserved. In addition, engineered-bladder biopsies demonstrated normal structural architecture and phenotype. Thus, this study provided good evidence in support of the use of engineered bladder tissue (created with autologous cells seeded on a composite scaffold and wrapped in omentum following implantation) for patients in need of cystoplasty.

## Kidney

Solid organs are the most challenging to reconstruct by tissue-engineering techniques because of their complex structure and function. Due to these complexities, some scientists approach the replacement of solid-organ function, such as liver or kidney, with bioartificial approaches, combining the function of the organ's predominant cells with an artificial structure and organization.

An example of this strategy is extracorporeal renal replacement. Although dialysis is currently the most prevalent form of renal-replacement therapy, the relatively high morbidity and mortality have spurred investigators to seek alternative solutions involving *ex vivo* systems. In an attempt to assess the viability and physiologic functionality of a cell-seeded device to replace the filtration, transport, metabolic, and endocrinologic functions of the kidney in acutely uremic dogs, a combination of a synthetic hemofiltration device and a renal tubular cell therapy device containing porcine renal tubules in an extracorporeal perfusion circuit was introduced. It was shown that levels of potassium and blood urea nitrogen (BUN) were controlled during treatment with the device. The fractional reabsorption of sodium and water was possible. Active transport of potassium, bicarbonate, and glucose and a gradual ability to excrete ammonia were observed. These results showed the technologic feasibility of an extracorporeal assist device that is reinforced by the use of proximal tubular cells (Humes *et al.*, 1999).

With similar techniques, a tissue-engineered bioartificial kidney consisting of a conventional hemofiltration cartridge in series with a renal tubule–assist device containing human renal proximal tubule cells was used in patients with acute renal failure (ARF) in the intensive care unit. The initial clinical experience with the bioartificial kidney and the renal tubule–assist device suggests that renal tubule cell therapy may provide a dynamic and individualized treatment program as assessed by acute physiologic and biochemical indices (Humes *et al.*, 2003).

The ultimate goal of tissue engineering, however, is the growth of fully functional and fully integrated replacement organs. Thus, another approach toward the achievement of improved renal function involves the augmentation of renal tissue with kidney cell expansion *in vitro* and subsequent autologous transplantation. The feasibility of achieving renal cell growth, expansion, and *in vivo* reconstitution via tissue-engineering techniques has been explored.

Most recently, an attempt was made to harness the reconstitution of renal epithelial cells for the generation of functional nephron units. Renal cells were harvested and expanded in culture. The cells were seeded onto a tubular device constructed from a polycarbonate membrane, connected at one end with a Silastic catheter that terminated in a reservoir. The device was implanted in athymic mice. Histologic examination of the implanted devices over time revealed extensive vascularization, with formation of glomeruli and highly organized tubulelike structures. Immunocytochemical staining confirmed the renal phenotype. Yellow fluid was collected from inside the implant, and the fluid retrieved was consistent with the makeup of dilute urine in its creatinine and uric acid concentrations (Atala, 2006). Further studies have been performed showing the formation of renal structures in cows via nuclear transfer techniques (Lanza *et al.*, 2002). Challenges await this technology, including the expansion of this system to larger, three-dimensional structures.

## X. CONCLUSION

Tissue-engineering efforts are currently being undertaken for almost every major type of tissue and organ within the body. Most of the effort expended to engineer these organs has occurred since the mid-1990s. Tissue-engineering techniques require a cell culture facility designed for human application. Personnel who have mastered the techniques of cell harvest, culture, and expansion as well as polymer design are essential for the successful application of this technology. Before these engineering techniques can be applied to humans, however, further studies need to be performed in many tissues.

In addition to the challenges already mentioned, numerous other hurdles face the tissue engineers of tomorrow. For example, physiologically accurate bioreactors must be developed and refined if complex organs are to be grown and functionalized prior to implantation in humans. Also, a refinement in the approaches to tissue engineering must be achieved, as scientists explore whether pre-engineered, functional bioengineered replacement organs must be produced to repair tissue function or whether the goal of tissue engineering should be to use simple biomaterials, cells, or growth factors to induce the repair or regrowth of these damaged organs.

## XI. REFERENCES

Amiel, G. E., and Atala, A. (1999). Current and future modalities for functional renal replacement. *Urol. Clin. North Am.* **26**(1), 235–246, xi.

Amit, M., Margulets, V., *et al.* (2003). Human feeder layers for human embryonic stem cells. *Biol. Reprod.* **68**(6), 2150–2156.

Assady, S., Maor, G., *et al.* (2001). Insulin production by human embryonic stem cells. *Diabetes* **50**(8), 1691–1697.

Atala, A. (2006). Recent applications of regenerative medicine to urologic structures and related tissues. *Curr. Opin. Urol.* **16**(4), 305–309.

Atala, A., Cima, L. G., *et al.* (1993). Injectable alginate seeded with chondrocytes as a potential treatment for vesicoureteral reflux. *J. Urol.* **150**(2 Pt. 2), 745–747.

Atala, A., Kim, W., *et al.* (1994). Endoscopic treatment of vesicoureteral reflux with a chondrocyte-alginate suspension. *J. Urol.* **152**(2 Pt. 2), 641–643; discussion 644.

Atala, A., Bauer, S. B., *et al.* (2006). Tissue-engineered autologous bladders for patients needing cystoplasty. *Lancet* **367**(9518), 1241–1246.

Baguisi, A., Behboodi, E., *et al.* (1999). Production of goats by somatic cell nuclear transfer. *Nat. Biotechnol.* **17**(5), 456–461.

Bell, E., Ivarsson, B., *et al.* (1979). Production of a tissuelike structure by contraction of collagen lattices by human fibroblasts of different proliferative potential *in vitro. Proc. Natl. Acad. Sci. U.S.A.* **76**(3), 1274–1278.

Bergsma, J. E., Rozema, F. R., *et al.* (1995). *In vivo* degradation and biocompatibility study of *in vitro* predegraded as polymerized polyactide particles. *Biomaterials* **16**(4), 267–274.

Boiani, M., Eckardt, S., *et al.* (2002). Oct4 distribution and level in mouse clones: consequences for pluripotency. *Genes Dev.* **16**(10), 1209–1219.

Bortvin, A., Eggan, K., *et al.* (2003). Incomplete reactivation of Oct4-related genes in mouse embryos cloned from somatic nuclei. *Development* **130**(8), 1673–1680.

Briggs, R., and King, T. J. (1952). Transplantation of living nuclei from blastula cells into enucleated frogs' eggs. *Proc. Natl. Acad. Sci. U.S.A.* **38**(5), 455–463.

Brivanlou, A. H., Gage, F. H., *et al.* (2003). Stem cells. setting standards for human embryonic stem cells. *Science* **300**(5621), 913–916.

Campbell, K. H., McWhir, J., *et al.* (1996). Sheep cloned by nuclear transfer from a cultured cell line. *Nature* **380**(6569), 64–66.

Chen, F., Yoo, J. J., *et al.* (1999). Acellular collagen matrix as a possible "off-the-shelf" biomaterial for urethral repair. *Urology* **54**(3), 407–410.

Cibelli, J. B., Stice, S. L., *et al.* (1998). Cloned transgenic calves produced from nonquiescent fetal fibroblasts. *Science* **280**(5367), 1256–1258.

Cibelli, J. B., Campbell, K. H., *et al.* (2002). The health profile of cloned animals. *Nat. Biotechnol.* **20**(1), 13–14.

Cilento, B. G., Freeman, M. R., *et al.* (1994). Phenotypic and cytogenetic characterization of human bladder urothelia expanded *in vitro. J. Urol.* **152**(2 Pt. 2), 665–670.

Dahms, S. E., Piechota, H. J., *et al.* (1998). Composition and biomechanical properties of the bladder acellular matrix graft: comparative analysis in rat, pig and human. *Br. J. Urol.* **82**(3), 411–419.

De Sousa, P. A., Dobrinsky, J. R., *et al.* (2002). Somatic cell nuclear transfer in the pig: control of pronuclear formation and integration with improved methods for activation and maintenance of pregnancy. *Biol. Reprod.* **66**(3), 642–650.

Fauza, D. O., Fishman, S. J., *et al.* (1998a). Videofetoscopically assisted fetal tissue engineering: bladder augmentation. *J. Pediatr. Surg.* **33**(1), 7–12.

Fauza, D. O., Fishman, S. J., *et al.* (1998a). Videofetoscopically assisted fetal tissue engineering: skin replacement. *J. Pediatr. Surg.* **33**(2), 357–361.

Folkman, J., and Hochberg, M. (1973). Self-regulation of growth in three dimensions. *J. Exp. Med.* **138**(4), 745–753.

Freed, L. E., Vunjak-Novakovic, G., *et al.* (1994). Biodegradable polymer scaffolds for tissue engineering. *Biotechnology (N.Y.)* **12**(7), 689–693.

Freeman, M. R., Yoo, J. J., *et al.* (1997). Heparin-binding EGF-like growth factor is an autocrine growth factor for human urothelial cells and is synthesized by epithelial and smooth muscle cells in the human bladder. *J. Clin. Invest.* **99**(5), 1028–1036.

Godbey, W. T., and Atala, A. (2002). *In vitro* systems for tissue engineering. *Ann. N.Y. Acad. Sci.* **961**, 10–26.

Gurdon, J. B. (1962). Adult frogs derived from the nuclei of single somatic cells. *Dev. Biol.* **4**, 256–273.

Harris, L. D., Kim, B. S., *et al.* (1998). Open-pore biodegradable matrices formed with gas foaming. *J. Biomed. Mater. Res.* **42**(3), 396–402.

Harriss, D. R. (1995). Smooth muscle cell culture: a new approach to the study of human detrusor physiology and pathophysiology. *Br. J. Urol.* **75**(Suppl. 1), 18–26.

Hochedlinger, K., and Jaenisch, R. (2002). Nuclear transplantation: lessons from frogs and mice. *Curr. Opin. Cell Biol.* **14**(6), 741–748.

Hochedlinger, K., and Jaenisch, R. (2003). Nuclear transplantation, embryonic stem cells, and the potential for cell therapy. *N. Engl. J. Med.* **349**(3), 275–286.

Humes, H. D., Buffington, D. A., *et al.* (1999). Replacement of renal function in uremic animals with a tissue-engineered kidney. *Nat. Biotechnol.* **17**(5), 451–455.

Humes, H. D., Weitzel, W. F., *et al.* (2003). Renal cell therapy is associated with dynamic and individualized responses in patients with acute renal failure. *Blood Purif.* **21**(1), 64–71.

Hynes, R. O. (1992). Integrins: versatility, modulation, and signaling in cell adhesion. *Cell* **69**(1), 11–25.

Ikada, Y. (2006). Challenges in tissue engineering. *J. R. Soc. Interface* **3**(10), 589–601.

Itskovitz-Eldor, J., Schuldiner, M., *et al.* (2000). Differentiation of human embryonic stem cells into embryoid bodies compromising the three embryonic germ layers. *Mol. Med.* **6**(2), 88–95.

Kaefer, M., Tobin, M. S., *et al.* (1997). Continent urinary diversion: the Children's Hospital experience. *J. Urol.* **157**(4), 1394–1399.

Kato, Y., Tani, T., *et al.* (1998). Eight calves cloned from somatic cells of a single adult. *Science* **282**(5396), 2095–2098.

Kaufman, D. S., Hanson, E. T., *et al.* (2001). Hematopoietic colony-forming cells derived from human embryonic stem cells. *Proc. Natl. Acad. Sci. U.S.A.* **98**(19), 10716–10721.

Keefer, C. L., Keyston, R., *et al.* (2002). Production of cloned goats after nuclear transfer using adult somatic cells. *Biol. Reprod.* **66**(1), 199–203.

Kehat, I., Kenyagin-Karsenti, D., *et al.* (2001). Human embryonic stem cells can differentiate into myocytes with structural and functional properties of cardiomyocytes. *J. Clin. Invest.* **108**(3), 407–414.

Kershen, R. T., and Atala, A. (1999). New advances in injectable therapies for the treatment of incontinence and vesicoureteral reflux. *Urol. Clin. North Am.* **26**(1), 81–94, viii.

Kim, B. S., and Mooney, D. J. (1998). Development of biocompatible synthetic extracellular matrices for tissue engineering. *Trends Biotechnol.* **16**(5), 224–230.

Lai, J. Y., Yoon, C. Y., *et al.* (2002). Phenotypic and functional characterization of *in vivo* tissue-engineered smooth muscle from normal and pathological bladders. *J. Urol.* **168**(4 Pt. 2), 1853–1857; discussion 1858.

Lanza, R. P., Cibelli, J. B., *et al.* (1999). Prospects for the use of nuclear transfer in human transplantation. *Nat. Biotechnol.* **17**(12), 1171–1174.

Lanza, R. P., Cibelli, J. B., *et al.* (2001). The ethical reasons for stem cell research. *Science* **292**(5520), 1299.

Lanza, R. P., Chung, H. Y., *et al.* (2002). Generation of histocompatible tissues using nuclear transplantation. *Nat. Biotechnol.* **20**(7), 689–696.

Levenberg, S., Golub, J. S., *et al.* (2002). Endothelial cells derived from human embryonic stem cells. *Proc. Natl. Acad. Sci. U.S.A.* **99**(7), 4391–4396.

Liebert, M., Hubbel, A., *et al.* (1997). Expression of mal is associated with urothelial differentiation *in vitro*: identification by differential display reverse-transcriptase polymerase chain reaction. *Differentiation* **61**(3), 177–185.

Lim, F., and Sun, A. M. (1980). Microencapsulated islets as bioartificial endocrine pancreas. *Science* **210**(4472), 908–910.

Lin, H. K., Cowan, R., *et al.* (2004). Characterization of neuropathic bladder smooth muscle cells in culture. *J. Urol.* **171**(3), 1348–1352.

Lobban, E. D., Smith, B. A., *et al.* (1998). Uroplakin gene expression by normal and neoplastic human urothelium. *Am. J. Pathol.* **153**(6), 1957–1967.

McDougal, W. S. (1992). Metabolic complications of urinary intestinal diversion. *J. Urol.* **147**(5), 1199–1208.

Nguyen, H. T., Park, J. M., *et al.* (1999). Cell-specific activation of the HB-EGF and ErbB1 genes by stretch in primary human bladder cells. *In Vitro Cell Dev. Biol. Anim.* **35**(7), 371–375.

Oberpenning, F., Meng, J., *et al.* (1999). *De novo* reconstitution of a functional mammalian urinary bladder by tissue engineering. *Nat. Biotechnol.* **17**(2), 149–155.

Ogonuki, N., Inoue, K., *et al.* (2002). Early death of mice cloned from somatic cells. *Nat. Genet.* **30**(3), 253–254.

Pariente, J. L., Kim, B. S., *et al.* (2001). *In vitro* biocompatibility assessment of naturally derived and synthetic biomaterials using normal human urothelial cells. *J. Biomed. Mater. Res.* **55**(1), 33–39.

Pariente, J. L., Kim, B. S., *et al.* (2002). *In vitro* biocompatibility evaluation of naturally derived and synthetic biomaterials using normal human bladder smooth muscle cells. *J. Urol.* **167**(4), 1867–1871.

Park, H. J., Yoo, J. J., *et al.* (1999). Reconstitution of human corporal smooth muscle and endothelial cells *in vivo*. *J. Urol.* **162**(3 Pt. 2), 1106–1109.

Peppas, N. A., and Langer, R. (1994). New challenges in biomaterials. *Science* **263**(5154), 1715–1720.

Piechota, H. J., Dahms, S. E., *et al.* (1998). *In vitro* functional properties of the rat bladder regenerated by the bladder acellular matrix graft. *J. Urol.* **159**(5), 1717–1724.

Puthenveettil, J. A., Burger, M. S., *et al.* (1999). Replicative senescence in human uroepithelial cells. *Adv. Exp. Med. Biol.* **462**, 83–91.

Rackley, R. R., Bandyopadhyay, S. K., *et al.* (1999). Immunoregulatory potential of urothelium: characterization of NF-kappaB signal transduction. *J. Urol.* **162**(5), 1812–1816.

Reubinoff, B. E., Pera, M. F., *et al.* (2000). Embryonic stem cell lines from human blastocysts: somatic differentiation *in vitro. Nat. Biotechnol.* **18**(4), 399–404.

Reubinoff, B. E., Itsykson, P., *et al.* (2001). Neural progenitors from human embryonic stem cells. *Nat. Biotechnol.* **19**(12), 1134–1140.

Richards, M., Fong, C. Y., *et al.* (2002). Human feeders support prolonged undifferentiated growth of human inner cell masses and embryonic stem cells. *Nat. Biotechnol.* **20**(9), 933–936.

Rideout, W. M., 3rd, Eggan, K., *et al.* (2001). Nuclear cloning and epigenetic reprogramming of the genome. *Science* **293**(5532), 1093–1098.

Rosso, F., Marino, G., *et al.* (2005). Smart materials as scaffolds for tissue engineering. *J. Cell. Physiol.* **203**(3), 465–470.

Sams, A. E., and Nixon, A. J. (1995). Chondrocyte-laden collagen scaffolds for resurfacing extensive articular cartilage defects. *Osteoarthritis Cartilage* **3**(1), 47–59.

Schuldiner, M., Yanuka, O., *et al.* (2000). Effects of eight growth factors on the differentiation of cells derived from human embryonic stem cells. *Proc. Natl. Acad. Sci. U.S.A.* **97**(21), 11307–11312.

Schuldiner, M., Eiges, R., *et al.* (2001). Induced neuronal differentiation of human embryonic stem cells. *Brain Res.* **913**(2), 201–205.

Scriven, S. D., Booth, C., *et al.* (1997). Reconstitution of human urothelium from monolayer cultures. *J. Urol.* **158**(3 Pt. 2), 1147–1152.

Shastri, V. P. (2006). Future of regenerative medicine: challenges and hurdles. *Artif. Organs* **30**(10), 828–834.

Silver, F. H., and Pins, G. (1992). Cell growth on collagen: a review of tissue engineering using scaffolds containing extracellular matrix. *J. Long Term Eff. Med. Implants* **2**(1), 67–80.

Smidsrod, O., and Skjak-Braek, G. (1990). Alginate as immobilization matrix for cells. *Trends Biotechnol.* **8**(3), 71–78.

Solter, D. (2000). Mammalian cloning: advances and limitations. *Nat. Rev. Genet.* **1**(3), 199–207.

Tamashiro, K. L., Wakayama, T., *et al.* (2002). Cloned mice have an obese phenotype not transmitted to their offspring. *Nat. Med.* **8**(3), 262–267.

Thomson, J. A., Itskovitz-Eldor, J., *et al.* (1998). Embryonic stem cell lines derived from human blastocysts. *Science* **282**(5391), 1145–1147.

Tsunoda, Y., and Kato, Y. (2002). Recent progress and problems in animal cloning. *Differentiation* **69**(4–5), 158–161.

Wakayama, T., Perry, A. C., *et al.* (1998). Full-term development of mice from enucleated oocytes injected with cumulus cell nuclei. *Nature* **394**(6691), 369–374.

Wilmut, I., Schnieke, A. E., *et al.* (1997). Viable offspring derived from fetal and adult mammalian cells. *Nature* **385**(6619), 810–813.

Yannas, I. V., and Burke, J. F. (1980). Design of an artificial skin. I. Basic design principles. *J. Biomed. Mater. Res.* **14**(1), 65–81.

Yannas, I. V., Burke, J. F., *et al.* (1980). Design of an artificial skin. II. Control of chemical composition. *J. Biomed. Mater. Res.* **14**(2), 107–132.

Yoo, J. J., and Atala, A. (1997). A novel gene delivery system using urothelial tissue-engineered neo-organs. *J. Urol.* **158**(3 Pt. 2): 1066–1070.

Yoo, J. J., Lee, I., *et al.* (1998a). Cartilage rods as a potential material for penile reconstruction. *J. Urol.* **160**(3 Pt. 2): 1164–1168; discussion 1178.

Yoo, J. J., Meng, J., *et al.* (1998b). Bladder augmentation using allogenic bladder submucosa seeded with cells. *Urology* **51**(2), 221–225.

Young, L. E., Sinclair, K. D., *et al.* (1998). Large-offspring syndrome in cattle and sheep. *Rev. Reprod.* **3**(3), 155–163.

# Regulation and Ethics

# Chapter Eighty-Four

# The Tissue-Engineering Industry

*Michael J. Lysaght, Elizabeth Deweerd, and Ana Jaklenec*

I. Introduction
II. The Age of Innocence
III. The Perfect Storm

IV. The Present Era
V. Concluding Perspectives
VI. References

## I. INTRODUCTION

The enterprise that has come to be known as tissue engineering had its genesis in the mid-1970s with the pioneering work of Bill Chick on a cell-based bioartificial pancreas (Chick *et al.*, 1977) and of Eugene Bell and Ioannis Yannas with the first fibroblast cell matrix and membrane living-skin surrogates (Bell *et al.*, 1981; Yannas *et al.*, 1982). Significantly, and even before the biotech boom, this work formed the basis for two subsequent early-stage commercial enterprises: Biohybrid and Organogenesis. In the 1980s other scientists extended the grasp of these approaches to the bioartificial liver (Wolf and Munkelt, 1975), kidney (Aebischer *et al.*, 1987), bone (Green, 1977), vascular grafts (Stanley *et al.*, 1982), and the like. Investigators worked in interdisciplinary teams and were funded by the agencies with jurisdiction over the pathologies being investigated. There were no meetings, or even sessions of society meetings, focused on shared themes and common technologies among the various efforts. The term tissue engineering did not exist (Viola *et al.*, 2004), and investigators affiliated with their core disciplines (e.g., endocrinology, engineering) rather than with each other. Commercial activity was low key and fragmented.

Beginning in the late 1980s and early 1990s, a very different landscape began to emerge and take form. Several start-up companies were formed with "deep pocket" venture backing, among them Advanced Tissue Sciences (1986), Organogenesis (1986), and CytoTherapeutics (1989). Baxter and WR Grace began well-funded internal programs related to the artificial pancreas. Others followed, and the early 1990s were marked by the proliferation of companies in a recognized biotech sector increasingly identified as *tissue engineering*, a term that first appeared in the literature in 1984 (Wolter and Meyer, 1985). The NSF had held a seminal workshop at Granlabbaken (Lake Tahoe) in 1987. Later, under the influence of Robert Nerem, Keystone Conferences on cell biology began to hold sessions on this new field and eventually commissioned fully dedicated conferences. A 1993 *Science* article by Langer and Vacanti (1993) became something of a technological manifesto for involved scientists. Commercial biotechnology conferences began to invite firms to come and tell their story, and firms, ever on the search for more development funds, were delighted to oblige. In 1995, the first issue of the journal *Tissue Engineering* was published. From that time forward, coherent records of the field were maintained that provided adequate documentation of three distinct stages of growth, which will be elaborated and analyzed in this chapter: (1) The age of innocence — the lustrum from 1995 to 2000 saw expansion and growth, unbridled optimism, abundant financing, exuberance, and perhaps a touch of hubris. (2) The perfect storm — between 2000 and 2002, when almost everything that could go wrong in fact did go wrong, leaving a ravaged industrial community and a poorer, sadder, and hopefully wiser cast of players. (3) The present era — from 2003 to the present time, when tissue engineering reinvented itself and has reemerged as a vibrant preclinical developmental industry with ascendant academic overtones. Much

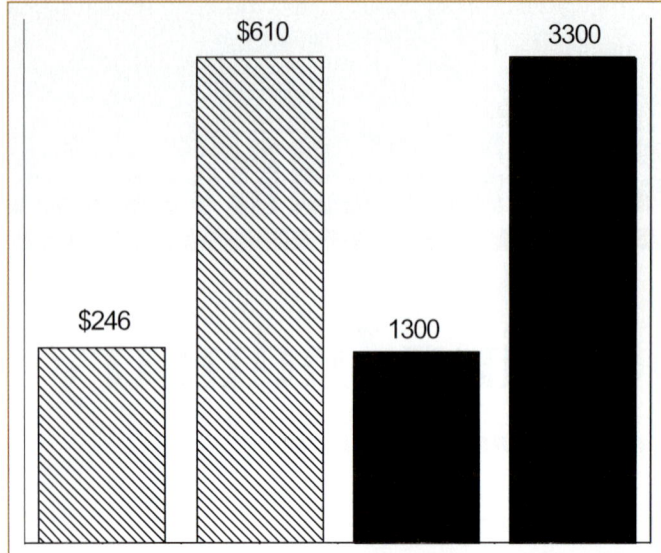

**FIG. 84.1.** The growth of tissue engineering: 1995–2000. During this six-year period, tissue engineering enjoyed a compound annual growth of greater than 15% per year. Cumulative investment exceeded $3 billion, and nearly all these funds came from the private sector. Left is total annual spending in the field; right is the number of employees.

of the data on which this chapter is based come from a series of our earlier reports in the field of tissue engineering (Lysaght, 1995; Lysaght *et al.*, 1998; Lysaght and Reyes, 2001), some as-yet-unpublished data for the next article in this series, and a very comprehensive NSF-commissioned report on the field (Viola *et al.*, 2004), as well as a comprehensive study describing the course of development in Europe, where things progressed along different paths than in the States (Bock *et al.*, 2003). Methods of data collection and analysis are fully described in these sources.

## II. THE AGE OF INNOCENCE

Figure 84.1 compares the tissue-engineering field in 1995 to that in 2000, using a consistent set of metrics. The number of companies increased from 73 to 89, the number of employees from ~1300 to ~3300, and the annual spending from $246 million to $610 million. Most companies were focused on either a specific product or a narrow range of products rather than on platform technologies. Broadly, companies were focused on structural applications (skin, bone, matrices) or on metabolic applications (bioartificial pancreas, bioartificial liver) or, in a few cases, on pure-play cellular applications (certain AIDS and cancer therapies). Virtually all the research and development was conducted in the context of start-ups, especially after Baxter and Grace spun their activities into Novocell and Circe, respectively. Companies could be very small, with fewer than 20 employees or, in some cases, as large as 200 full-time equivalents (FTEs). The smaller enterprises were essentially independent development groups structured as free-standing organizations; their activities were related mostly to early-stage

product development. The larger enterprises were usually structured as full-fledged organizations with all the trappings of an integrated company: separate departments for human resources, law, finance, investor relations, regulatory affairs, and the like. The bulk of activity in these larger organizations was centered on obtaining FDA approval for products believed ready for the market. The executive teams of the larger firms usually involved both scientists and general managers. Two products were approved by 2000: (1) Carticel® cartilage replacement, developed by Applied Biosciences and acquired by Genzyme; and (2) Apligraf®, a living-skin equivalent from Organogenesis. Both products were selling poorly, but this was generally regarded as the "going in" phase, with large-scale revenue soon to follow. Epicel, another skin equivalent, was available from Genzyme but only for compassionate use.

In the absence of robust product sales, from where was to come the revenue to pay for this bustling field? Financially, tissue engineering was modeled on the biotechnology industry, which had begun in the late 1970s and had grown of age in the 1990s, with companies like Amgen, Genentech, and Biogen providing enormous returns to early investors. Although every story was different, the most usual course was for an aspiring tissue-engineering firm to begin with seed money of ~$500,000 from venture capitalists, which would be followed by a larger investment, usually ~$5 million from a consortium of venture groups when the firm had recruited a credible scientific advisory board, a strong CEO, and an "investable" scientific staff. Subsequent rounds of financing from "mezzanine" investors would raise another $10 million to $20 million, usually on the basis of "anouncable" preclinical results or, better still, early "proof of principle" human trials. Two additional milestones were usually required in order for the firm to go public and raise a significant amount of capital by selling shares on the stock exchange (usually NASDAQ). First, companies needed a corporate partnership, in the form of a marketing and development contract with an established pharmaceutical or medical device company. Examples include ATS partnership with Smith & Nephew, CytoTherapeutics with ASTRA Pharmaceutical, and Organogenesis with Novartis. In these relationships, the corporate partner funded development of a product in return for defined marketing rights. Serious partnerships involved $20 million to $50 million, doled out over several years based on the achievement of defined development objectives. In addition to operating cash, the corporate partnership gave a firm considerable credibility in the investment community. The second major prerequisite prior to going public was having a product approved for FDA-level human trials. There were, and are, no uniform regulatory pathways for tissue-engineered products, and the difficulty of initiating FDA trials varied considerably from product to product. In any event, with a corporate partner and a "product in the clinic," companies were able to go public, raising another $20 million to $50 million. Once public, firms could return to the stock market, when favorable conditions prevailed, to raise more funds to finance further activities. This process was astonishingly successful: By 2000, a total of 16 firms were publicly traded, with a net capital value (number of shares multiplied by price per share) exceeding $2.5 billion (Lysaght and Reyes, 2001), all this for a field that had yet to produce a single profitable product.

At this juncture, the field of tissue engineering was a uniquely private sector and uniquely American. It is doubtful that government funding ever exceeded 10% of the total spending in any year, probably much less (Hellman, 2000). Academic laboratories operated more as partners with the private sector than as bastions of basic research and fundamental engineering. America had the field to itself — several factors explain the lack of a transatlantic or transpacific component to the field. Europe and the Far East did not have a start-up culture, and individual scientists were uncomfortable with the risk associated with the field. Minimal government support may have been even more of an impediment overseas, where the private sector is more enmeshed with the government than in the United States. And, at least until the close of the era, financing mechanisms for the field (venture capital, mezzanine investment, and public offerings for companies without products) were simply not part of the European or Asian financial rubric.

By the turn of the century, tissue engineering had enjoyed eight years of continuous growth. Sixteen companies were public, and investors with a good sense of timing enjoyed solid returns. Nearly 10 products were in some phase of regulatory-level clinical trials. In an exuberant report on TV's *Good Morning America*, science reporter Michael Gillian opined, "When historians look back, the greatest achievement of the 20th century will not be nuclear power, will not be space travel, will not be microelectronics, but, rather, will be in the fields of tissue engineering and genetic medicine" (Hellman, 2000). Further, *Time* magazine put tissue engineering at the top of its list of "Jobs for the New Millennium" (Atala *et al.*, 2006).

## III. THE PERFECT STORM

The period from 2000 to 2002 was not kind to the field of tissue engineering or to its investors and associated employees. First came the burst of the dot-com bubble and the decline of the NASDAQ. Investors at all stages lost their appetite for risk and sought safe harbors for their funds. Money was scarce; investors demanded higher-risk premiums, and performance was scrutinized closely. Unfortunately it was during this very time frame that the limited potential and flawed market performance of the first three tissue-engineering products were fully discerned. Dermagraft from ATS was introduced to the U.S. market in 2001 and thus joined Dermagraft (1997) and Carticel (1997). The peak sales of Carticel and Dermagraft were never more than $20 million per year; such a sales volume does not even cover the fixed cost of a GMP facility to produce a cell-based product, let alone provide a return on the several hundred million dollars in capital that had been required to bring them to market. Dermagraft sold only a few million dollars worth after its launch. In the face of high operating costs and diminutive sales, both Organogenesis and ATS filed for bankruptcy protection. Together, they had employed over 500 FTEs. Carticel, which had been a separate business unit (Genzyme Tissue Engineering) within Genzyme, was absorbed by Genzyme Surgical Products. The bankruptcy of two of the flagship firms in the field and the downgrading of the third further diminished investor appetite for tissue engineering.

But more bad news was to come. Nine tissue-engineering products had passed phase I clinical trials by 2000; three years later, all nine had either failed in later-phase trials or simply been abandoned (Lysaght and Hazlehurst, 2004). In phase III clinical trials, neither Circe's bioartificial liver nor CytoTherapeutic's cerecrib pain implant had reached statistically significant levels of efficacy, though this is as much a matter of the limited size of the trial population as one of technical device failure. Diacrin's implant for Parkinson's disease was put on clinical hold following incidences of bleeding. Three additional products (two related to bladder leakage and one for improved outcomes following heart bypass) had originally been developed by Reprogenesis and then acquired by Curos. All three products were to be regulated as drugs, and all were abandoned after phase I by Curos, who reckoned that product sales would not be high enough to allow recovery of the cost of clinical trials.

Even in the best of times, the bankruptcies of three flagship companies and nine (of nine) failed pipeline products would have severely challenged the industry. Given the already bad investment climate, the effect was devastating. Nineteen of 89 companies simply shut down, another 27 downsized drastically. About 1800 of 3000 workers in the field had to find new jobs, many in the neighboring field of stem cell research, which by then was beginning to acquire momentum.

Recently it has become fashionable to pronounce that the difficulties of early commercialization resulted from insufficient basic science, but this perspective simply does not stand up to rigorous examination. Rather, the firms failed for a combination of reasons: (1) excessively high manufacturing costs, (2) insufficient attention to reimbursement, (3) poorly structured agreements with marketing partners, (4) unrealistic assessments of the cost-to-benefit ratio of the early products, (5) unrealistic sales forecasts leading to excessive fixed costs of sales, and (6) financial exhaustion after protracted regulatory approval processes. More concisely, it could just be that these firms were good at developing technology but poor at other aspects of the medical products business.

Multiple factors also attend the failure and abandonment of products that had successfully passed the phase I stage. The small size of the study population made statistical evidence of efficacy a difficult achievement. Circe's bioartificial liver was evaluated on fewer than 100 patients, and CytoTherapeutic's neurocrib implant on only 60 patients. In comparison, 3900 patients received Pfizer's Viagra in its clinical trial and over 5000 patients received J&J's drug-eluting stent. Many argue that larger patient populations for early tissue-engineered products would have led to a statistical demonstration of efficacy, but the companies simply were unable to afford the larger trials. It costs about a billion dollars to bring a drug to market, whereas the development funds available for tissue-engineered products ran at most 15% of this ($150 million). A second issue is the regulatory pathway: The FDA is on a very steep learning curve with tissue engineering. Some products are treated as devices, including all that eventually reached the market. Some were required to follow the drug pathway, with its higher threshold for approval and a requirement for randomized, blinded trials; no tissue-engineered product assigned to this pathway has yet been approved. Still a third factor is availability of personnel skilled in designing and conducting clinical trials. This is a high-stakes enterprise for pharmaceutical companies, who treat their staff quite generously. It can be difficult for start-ups to acquire the levels of expertise and experience necessary to excel in this field.

In retrospect, one general concern is the appropriateness of the venture capital biotech model to the field of tissue engineering. VCs impose a very Darwinian structure on firms they control and aim for the occasional blockbuster rather than for a string of modest successes. For this reason, diversification of risk by individual start-ups is discouraged. Firms are motivated to grow very large very quickly because large teams have a better chance for the proverbial "home run." Quick growth can mean rushing milestones, such as beginning phase I trials at too early a stage. Finally, VCs are used to the biotech pharmaceutical industry and thus uncomfortable with a slow, cautious, incremental approach, which characterizes successful device development. It is somewhat churlish to be critical of a profession that raised billions of dollars for an unproven technology, but the VCs, like wealthy dysfunctional parents, may have spoiled their offspring.

## IV. THE PRESENT ERA

Tissue engineering has staged a remarkable comeback and transformation since the dark days of 2002 and 2003. The exact size of the current (mid-2006) industrial sector has not been reported, but our best estimate is that today's firms at least exceed the number from five years previously, with about the same total employment and slightly less overall spending.

Stated differently, there are more companies today than before, but they are smaller and have a lower burn rate per employee. Angels and SBA grants, more so than VCs, provide starting funds. Criteria for series-B or other mezzanine funds have become more rigorous. Although there are

exceptions, investors seem more willing to support early-stage research than to engage in the higher-risk clinical trial phase.

Smith & Nephew retained the marketing rights to ATS' Dermagraft, but sales were diminutive and the product was eventually discontinued. Genzyme's Carticel is placing around 2000–3000 units per year but claims to have reduced manufacturing costs to the point that the operation is profitable at this level. Impressively, Organogenesis, which had fallen to eight employees while under Chapter 11 protection, has rebounded and now claims to be profitable, with ~175 employees and with annual sales approaching 40,000 units. Cumulatively, over 120,000 patients have been treated. Organogenesis has thus become the first TE startup to reach profitability and also the first to reach sustained sales growth. Factors responsible for the turnaround include termination of the original marketing agreement with Norvartis, reduced manufacturing costs, enhanced emphasis on sales and marketing, improved packaging, focus on reimbursement, and a clearer appreciation of factors that will predispose dermatologists to adopt the new technology. The product itself has not changed.

Several firms have products in clinical trials. Renamed, which now has over 200 employees, has completed a phase II clinical trial of its bioartificial kidney for the treatment of acute renal failure and has announced plans for a larger, phase III trial. Considerable attention is being paid to the statistical power of the phase III trial, calculable from success rates in the phase I and II trials. Neurotech, a daughter company of CytoTherapeutics, has completed a successful phase II trial of an intraocular implant for the treatment of dry macular degeneration, with a relatively small staff of about 25 employees. Their strategy seems to be to let the staff size dictate the pace of clinical trials; their success certainly contravenes the common wisdom that very large staffs (hundreds or more) are a necessity for FDA trials. The firm has also benefited from clinical trial support from the NIH. Novato has developed vascular grafts from autologous fibroblasts, and these are being tested in a small FDA-approved trial in hemodialysis patients. Intercytex is developing a tissue-engineering lifestyle product to treat male pattern baldness and female alopecia by injecting an expanded population of hair-inductive dermal papilla cells. ReNeuron is developing a cell-carrier system for the treatment of Parkinson's disease. Finally, Aastrom Biosciences has developed a GMP process for collection, expansion, and characterization of bone marrow and is testing the process in a trial application of increasing vascularization in the diabetic foot. Each of these firms is taking a different approach to the clinic, but, importantly, they are refilling the pipeline that became so depleted earlier in the decade.

For completeness, we note the recent attention and publicity that attended the publication of Atala's paper reporting five-year survival of a bioartificial bladder in a series of patients whose bladders had been physically mal-

**Table 84.1.** Tissue-engineering/regenerative-medicine meetings in 2006

| Meeting title | Location | Date (2006) |
|---|---|---|
| Conference on Epithelial Technologies and Tissue Engineering | Washington, DC | April 2–4 |
| First Chinese European Symposium on Biomaterials in Regenerative Medicine | Suzhou, China | April 3–7 |
| Sixth REGENERATE conference on Tissue Engineering and Regenerative Medicine | Pittsburgh, PA | April 25–27 |
| Third Annual Tissue Engineering Symposium | Tampere, Finland | May 3–5 |
| New developments on polymers for tissue engineering | Maderia, Portugal | June 1–5 |
| Second International Conference "Strategies in Tissue Engineering" | Wurzburg, Germany | May 31–June 2 |
| Advances in Biomaterials for Drug Delivery and Regenerative Medicine | Naples, Italy | June 11–16 |
| Seventh Advanced Summer Course in Cell–Materials Interaction: Regenerative Medicine | Porto, Portugal | June 19–23 |
| Tissue and Cell Engineering Society 2006 — Annual Meeting | Sheffield, England | July 3–4 |
| Thirteenth CIRMIB Biomaterials School "Strategies and Technologies for Tissue Engineering and Reparative Medicine" | Ischia, Italy | July 10–14 |
| Fourteenth annual short course "Advances in Tissue Engineering" | Houston, TX | August 16–19 |
| Annual TERMIS–EU meeting | Rotterdam, Netherlands | October 8–11 |
| Second International Congress on Regenerative Biology | Stuttgart, Germany | October 9–11 |
| Fourth International Symposium "The Science and Technology of Skin Engineering" | Shanghai, China | October 15–17 |
| Sixth Annual Stem Cells and Regenerative Medicine Conference | Pittsburgh, PA | October 16–17 |
| Biomaterials in Regenerative Medicine | Vienna, Austria | October 22–25 |
| ESF-EMBO Symposium on Stem Cells in Tissue Engineering | Sant Feliu de Guixos, Spain | October 28–November 2 |

formed as a consequence of secondary diseases such as spina bifada (Atala *et al.*, 2006). This achievement was covered as front page news in the *New York Times*, *Wall Street Journal*, and *USA Today*. Major periodicals (*The Economist*, *Time*) also summarized the findings. The publicity was almost universally favorable and uncritical and certainly stands to remove some of the tarnish from the reputation of the field. Commercial rights to this technology are owned by Tengion, which currently has about 40 employees.

Several larger firms have begun to take tissue engineering seriously. Genzyme expanded beyond Carticel to cardiovascular applications (since abandoned) and to a major corporate partnership with Renamed. J&J has established a tissue-engineering division as a business unit to unify and drive all of its activities in the area. Medtronics is reportedly active in the field as well. These are major corporations with deep pockets, regulatory resources, manufacturing capability, and marketing skills. Unlike start-ups, they are not wed to a single product or application and can move from one technology to another as the situation dictates. They will likely have a very positive impact on the future course of commercial tissue engineering.

Globalization is another important trend. Up until the late 1990s, the United States could regard tissue engineering as "mare nostrum"; no longer. A report issued in 2003 and using a very broad definition of tissue engineering lists over 190 business units in the European Union alone (Bock *et al.*, 2003). Some of these are involved in autologous cell transplantation and are quasi-public; others are bona fide private-sector start-ups. England and Germany have established major tissue-engineering centers of excellence, and Holland has a national program in tissue engineering. In the Asian Pacific, Singapore and China are both active, and China, in fact, has two major tissue-engineering centers and Australia has one.

An important trend is the increased academization of the field. Whereas university laboratories used to represent a steadying tail on the dizzying kite of industrial tissue engineering, they now have emerged as its center of gravity. The NIH and NSF do not provide tallies of the cumulative extent of their funding in this area, but one need only look at the abstracts at any of the major tissue-engineering meetings to identify the size and scope of the academic contribution. Table 84.1 is an illustrative list of the number of tissue-engineering meetings held in 2006 (and this does not even count tissue-engineering sessions in more general meetings like ASAI2O and Biomaterials). Nothing could be more indicative of the increased role of academia than this proliferation of meetings, scheduled predominantly during the summer, and their symbiotic societies. Tissue engineering seems to have reversed its earlier practice and is now following the more normal paradigm of early-stage development in academia followed by commercialization by industry.

## V. CONCLUDING PERSPECTIVES

The difficulties and dislocations in tissue engineering early in the first decade of the 21st century were clearly unsettling and unnerving. But a rocky road to the clinic following a brilliant period of discovery and preclinical development is not atypical. Consider the early days of the pharmaceutical industry. Penicillin was first used clinically in the 1920s but did not become a successful commercial product until 25 years later, after World War II. Early difficulties with production of cortisone in the late 1940s nearly bankrupted Merck and forced its merger with Sharpe and Doehme. The polio vaccine survived its early trials and tribulations only because of the finiteness of the alternatives. And then came thalidomide, marking a far lower point than has ever been seen by tissue engineering. But by the mid-1960s, the industry was on its way, and it has become one of the great success stories of contemporary health care. Another example is dialysis for the treatment of end-stage renal disease: The extracorporeal technology was available from the mid-1940s, with workable chronic access available from 1962. But hemodialysis became a widely used therapy for the renal failure population only around 1975. Similar histories attend drug delivery, monoclonal antibody therapy, and the like. Things take time, development is messy, and setbacks and failures are a part of the cycle. But in the end, tissue engineering, with its strong ontologic rationale, still seems predestined to become an important, even dominant, segment of 21st-century medicine.

## VI. REFERENCES

Aebischer, P., Ip, T. K., Panol, G., and Galletti, P. M. (1987). The bioartificial kidney: progress towards an ultrafiltration device with renal epithelial cells processing. *Life Support Syst.* **5**, 159–168.

Atala, A., Bauer, S. B., Soker, S., Yoo, J. J., and Retik, A. B. (2006). Tissue-engineered autologousbladders for patients needing cytoplasty. *Lancet* **367**, 1241–1246.

Bell, E., Ehrlich, H. P., Buttle, D. J., and Nakatsuji, T. (1981). Living tissue formed *in vitro* and accepted as skin-equivalent tissue of full thickness. *Science* **211**, 1052–1054.

Bock, A. K., Ibarrate, D., and Rodriguez-Cerenzo, E. (2003). Human tissue-engineered products — today's markets and future prospects. "Report for European Science and Technology Observatory (ISTO)." Available online at http://esto.jrc.es/detailshort.cfm?ID_report=1127.

Chick, W. L., Perna, J. J., Lauris, V., Low, D., Galletti, P., Panol, G., Whittemore, A. D., Like, A., Colton, C. K., and Lysaght, M. J. (1977). Artificial pancreas using living beta cells: effect on glucose homeostasis in diabetic rats. *Science* **197**, 780–782.

Green, W. T. (1977). Articular cartilage repair: behavior of rabbit chondocytes during tissue culture and subsequent allografting. *Clin. Orthop. Rel. Res.* **124**, 237–250.

Hellman, K. (2000). "Presentation at the World Technology Evaluation Center (WTEC) Conference on Tissue Engineering, November 1–3, 2000." Previously available online at http://itri.loyola.edu/te/views/Hellman/sd001.html.

Langer, R., and Vacanti, J. P. (1993). Tissue engineering. *Science* **260**, 920–926.

Lysaght, M. J. (1995). Product development in tissue engineering. *Tissue Eng.* **1**, 221.

Lysaght, M. J., and Hazlehurst, A. L. (2004). Tissue engineering: the end of the beginning. *Tissue Eng.* **10**(1), 309–320.

Lysaght, M. J., and Reyes, J. (2001). The growth of tissue engineering. *Tissue Eng.* **7**, 485–493.

Lysaght, M. J., Nguy, N. A., and Sullivan, K. (1998). An economic survey of the emerging tissue-engineering industry. *Tissue Eng.* **4**, 231–238.

Stanley, J. C., Burkel, W. E., Ford, J. W., Vinter, D. W., Kahn, R. H., Whitehouse, W. M., and Graham, L. M. (1982). Enhanced patency of small-diameter, externally supported Dacron iliofemoral grafts seeded with endothelial cells. *Surgery* **92**, 994–1005.

Viola, J., Lal, B., and Grad, O. (2004). The emergence of tissue engineering as a research field. Chapter 3. Available online at http://www.nsf.gov/pubs/2004/nsf0450/start.htm.

Wolf, C. F., and Munkelt, B. E. (1975). Bilirubin conjugation by an artificial liver composed of cultured cells and synthetic capillaries. *Trans. Am. Soc. Artif. Intern. Organs* **21**, 16–27.

Wolter, J. R., and Meyer, R. F. (1985). Sessile macrophages forming clear endothelium-like membrane on the inside of successful keratoprosthesis. *Graefes Arch. Clin. Exp. Ophthalmol.* **222**, 109–117.

Yannas, I. V., Burke, J. F., Orgill, D. P., and Skrabut, E. M. (1982). Wound tissue can utilize a polymeric template to synthesize a functional extension of skin. *Science* **215**, 174–176.

# Chapter Eighty-Five

# The Regulatory Path From Concept to Market

*Kiki B. Hellman*

## I. INTRODUCTION

Therapeutic approaches for replacement, repair, restoration, or regeneration of diseased or damaged human organs or tissues have evolved over the last several years, from human donor organ and tissue transplants and implants of synthetic materials to *in vitro* engineered tissue constructs composed of autologous, allogeneic, or xenogeneic cells coupled with synthetic or natural matrix materials and/or pharmacological agents for either *in vivo* implantation or *ex vivo* use and cell therapies using either native, stem, or progenitor autologous or allogeneic cells for *in vivo* delivery. While organ/tissue transplantation and synthetic material implants continue as the standard of care, donor-organ shortages and indications where such approaches may not be feasible have led to a search for alternatives utilizing living tissue, which, in turn, has provided the impetus for engineered tissue solutions.

Engineered tissues can provide either a structural/mechanical or a metabolic function (Hellman *et al.*, 1998, 2000). Examples published in the scientific literature include, among others: artificial skin constructs; musculoskeletal applications, such as autologous cells for cartilage regenera-tion, engineered ligament and tendon, and bone graft substitutes; approaches for repair and regeneration of the cardiovascular system, including the myocardium, valves, and vessels; periodontal tissue repair; engineered cornea and lens; spinal cord repair and nerve regeneration; repair of the urogenital system; and approaches for functional restoration of vital metabolic organs, such as the pancreas, liver, and kidney, through either biohybrid organ implants or *ex vivo* support systems.

The promise of engineered-tissue therapies has been realized (Smith and Hellman, 2003; Hellman and Smith, 2006). Skin and musculoskeletal substitutes have been approved for use in the United States by the Food and Drug Administration (FDA). Other applications cited earlier are under either preclinical investigation or regulatory evaluation. Recent advances in stem cell and cytokine biology, materials science, engineering, and computer-assisted modeling and design, among others, are contributing to the development of second-generation engineered-tissue therapies.

In addition to therapeutic applications, *in vitro* engineered-tissue constructs are being applied as biosensors in

*Principles of Tissue Engineering, 3rd Edition*
ed. by Lanza, Langer, and Vacanti

diagnostic systems and as test models for toxicity assessment of pharmacological and other agents. Development of enabling technologies provides promising avenues for establishment of a service industry, e.g., cell banks/repositories, scaffold/matrix materials and reference material libraries, and customized tissue-specific bioreactors.

Integrity of the science, together with other critical determinants, is basic to the successful translation of tissue-engineering research into applications for the clinic and the marketplace (Hellman, 2004). Of these determinants, understanding the strategies developed by government entities for providing appropriate product regulatory oversight is key. Since a primary goal is the establishment of a global industry enabling companies to market products across national boundaries, a harmonized international regulatory approach, such as the International Conference on Harmonization for pharmaceutical products, would be ideal. While groups work toward that goal, and recognizing that the public's perception and subsequent market acceptance can be influenced by local social, political, legal, and ethical concerns, it is important to understand the approaches of regulatory entities where the research has moved successfully through product development to the marketplace (Hellman and Smith, 2006). Although the science is now worldwide and regulatory approaches are being developed in Europe and the Far East, among others, this discussion is limited to the regulatory strategies and evolving initiatives in the United States.

The FDA has recognized that an important segment of the products that it regulates results from application of novel technology, such as tissue engineering, and that product applications often pose new and complex issues. Thus, the agency has worked since the early 1990s on developing appropriate strategies for the regulatory oversight of human cells, tissues, and cellular- and tissue-based products. Most, if not all, engineered tissues fall into these categories.

## II. LEGISLATIVE AUTHORITY

### Laws

Since approaches for organ and tissue replacement, repair, restoration, and regeneration and their source materials span a broad spectrum of potential clinical applications, the responsibility for overseeing their development and commercialization within the U.S. federal government has been divided among different regulatory agencies, centers, and offices. The Health Resources Services Administration (HRSA) oversees the National Organ Transplant Program and the National Marrow Donor Program. The remaining products are regulated by the FDA.

The FDA is a science-based regulatory agency in the U.S. Public Health Service (PHS). The agency's legislative authority for product oversight, premarket approval, and postmarket surveillance and enforcement is derived principally from the Federal Food, Drug, and Cosmetic (FD&C) Act and the Public Health Service (PHS) Act. Under these authorities, the FDA evaluates and approves products for the marketplace, inspects manufacturing facilities, sometimes before and routinely during commercial distribution, and takes corrective action to remove products from commerce when they are unsafe, misbranded, or adulterated.

### FDA Mission and Organization

The FDA's mission is to promote and protect the public health through regulation of a broad range of products by ensuring the safety of foods, cosmetics, and radiation-emitting electronic products as well as ensuring the safety and effectiveness of human and veterinary pharmaceuticals, biologicals, and medical devices. The FDA's six centers are staffed with individuals expert in the science and regulations appropriate to a center's mission. The centers with regulatory oversight for human medical products are the Center for Drug Evaluation and Research (CDER), which regulates drugs; the Center for Biologics Evaluation and Research (CBER), which regulates biological products; and the Center for Devices and Radiological Health (CDRH), which regulates medical devices and radiation-emitting electronic products. However, each center can apply any of the statutory authorities to regulate its products. For example, many products reviewed by the CBER are regulated under the medical device authority. In addition to the centers, other offices, such as the Office of Regulatory Affairs (ORA) and the Office of Orphan Products (OOP), provide assistance to the centers on regulatory procedures and facility inspections, when necessary. The Office of Combination Products (OCP) is responsible for the regulatory oversight of combination products. Although the office does not perform product reviews for market approval or clearance, it assigns the combination product to the appropriate FDA center, ensures timely and effective premarket review and appropriate postmarket regulation, and serves as a resource to industry and the FDA centers' review staff (Smith and Hellman, 2003; Hellman and Smith, 2006). The OCP serves a very important function for the regulation of engineered tissue products, since many are combination products.

## III. PRODUCT REGULATORY PROCESS
### Classification of Products

Under federal law, a human medical product is classified as either a drug, a biological drug (biologic), a device, or a combination product, e.g., a combination of a drug, a biologic, and/or a device. The product's classification determines the premarket regulatory review and approval process for demonstration of safety and effectiveness utilized by the FDA and the FDA center with lead responsibility and jurisdiction for the product. For example, a *drug* is an article intended for use in the diagnosis, cure, mitigation, treatment, or prevention of disease in humans or other animals and an article (other than food) and other articles intended to affect the structure or any function of the body of humans or other animals [21USC321(g)]. A *biologic* is defined as a virus, therapeutic serum, toxin, antitoxin, vaccine, blood, blood component or derivative, allergenic product, or analogous product, . . . applicable to the prevention, treatment, or cure of diseases or injuries of man [42USC262(a)]. A device is an instrument, apparatus, . . . implant, *in vitro* reagent or other similar or related article which is intended for use in diagnosis of disease or conditions or in the cure, mitigation, treatment, or prevention of disease, in humans or other animals, or intended to affect the structure or any function of the body . . . and which does not achieve any of its principal intended purposes through chemical action within or on the body, . . . and which is not dependent on being metabolized for the achievement of any of its principal intended purposes [21USC201(h)].

## Combination Products

Advances in biomedical technology have generated products not readily classifiable as drugs, biologics, or devices as these terms are defined by federal law. To provide for the expanding varieties of products expressing features of more than one of these classifications, the FDA has been authorized to recognize combination products. These products constitute a growing category of innovative medical products. Examples include a drug with an implantable delivery device and autologous cells coupled with a scaffold for orthopedic use. While these products contribute to advancing medical care, they also pose a challenge for the FDA, since they straddle existing statutory classifications of regulated products, complicating the determination of the appropriate regulatory process (Hellman and Smith, 2006).

Congress recognized the existence of combination products when it enacted the Safe Medical Device Act of 1990 and established that the FDA shall classify a combination product according to its primary mode of action [Section 503(g) of the FD&C Act, 21USC353(g)]. From its determination of the product's primary mode of action, the agency could assign jurisdiction over the product to one of its established centers. For example, if the primary mode of action is that of a drug, the product is assigned to CDER, if that of a device to CDRH, and biologics to CBER. The FDA issued a final rule in 1991 establishing the process, i.e., Request for Designation (RFD), by which a product sponsor could petition the agency to make such an assignment [21CFR3.7].

The Medical Device User Fee and Modernization Act of 2002 (MDUFMA) modified Section 503(g) of the FD&C Act to require the FDA to establish an office with the primary responsibility for providing regulatory oversight of combination products. The Office of Combination Products in the Office of the Commissioner assigns the product to the appropriate FDA center, resolves any disputes over a prod-

uct's regulation, and is the focal point for both the FDA staff and industry regarding combination products.

There has been much progress since the office was established in making this somewhat-complex regulatory area more efficient, transparent, and better understood. Because of its role, the office has become the focus and, often, the primary point of entry for sponsors of combination products. The office encourages informal as well as formal interactions, i.e., through the RFD process, with sponsors regarding product jurisdictional questions.

While the FDA has traditionally required sponsors of an RFD to identify the product's primary mode of action and to recommend the lead center for product premarket review and regulation, there is no statutory definition of what constitutes a primary mode of action. To address concerns that, without a statutory codified definition, the assignment process has appeared arbitrary at times, the office published a proposed rule to amend the regulations and to define and codify both mode of action and primary mode of action, "Definition of Primary Mode of Action of a Combination Product: Proposed Rule" (69FR25527, May 7, 2004) (the PMOA Proposed Rule). Almost all comments received from the stakeholder community supported the Proposed Rule in whole or in part. The Final Rule was published on August 25, 2005 (70FR 49848-49862). *Mode of action* is defined as the means by which a product achieves its intended therapeutic effect, i.e., drug, biologic, or device mode of action. Since combination products have more than one identifiable mode of action, the *primary mode of action* is the single mode that provides the most important therapeutic action of the combination product. The *most therapeutic action* is the mode of action expected to make the greatest contribution to the overall intended therapeutic effects of the combination product. The Final Rule also describes an algorithm that the agency would use to assign a product to a center when it cannot determine with reasonable certainty which mode of action provides the most important therapeutic effect. The Final Rule would require a sponsor to base its recommendation of the center with primary jurisdiction for its product by using the definition and, if appropriate, the assignment algorithm. This framework is based on an assessment of the product as a whole, its intended use and effect, consistency with assignment of similarly situated products, and safety and effectiveness issues (Hellman and Smith, 2005c). Of note is that most, if not all, engineered tissue products are combination products.

# IV. PRODUCT PREMARKET SUBMISSIONS

## Investigational Studies

The FD&C Act requires demonstration of safety and effectiveness for new drugs and devices prior to introduction into interstate commerce. The PHS Act requires demonstration of safety, purity, and potency for biological products before introduction into interstate commerce. Consequently, premarket clinical studies must be performed under exemptions from these laws. For drugs and biologics, which are considered drugs under the FD&C Act, the application for the exemption is an Investigational New Drug (IND) application, (21CFR312). The application for exemption of a device is an Investigational Device Exemption (IDE) (21CFR812).

The contents of IND and IDE applications are similar (Hellman *et al.*, 2000). Applications will include a description of the product and manufacturing processes sufficient for an evaluation of product safety and preclinical studies that have been designed to assess risks and potential benefits of the product. The IND and IDE applications contain a proposal for a clinical protocol, which describes the indication being treated, proposed patient population, patient inclusion and exclusion criteria, treatment regimen, study endpoints, patient follow-up methods, and clinical trial-stopping rules. Both IND and IDE investigations require Institutional Review Board (IRB) approval before they may commence. Although IND and IDE requirements are somewhat different (e.g., in cost-recovery and device risk-assessment areas), the FDA applies comparable standards of safety and effectiveness for either type of application. When the FDA determines that there is sufficient information to allow a clinical investigation to proceed, the IND or IDE exemptions are approved.

The first clinical studies conducted under the IND or IDE applications are often clinical trials involving a small number of individuals (e.g., phase 1/feasibility studies) designed primarily to assess product safety. If these earlier studies indicate reasonable safety, phase 2 studies may be developed to investigate proper and safe dosing and potential efficacy. Phase3/pivotal studies utilize well-controlled clinical trial designs that support a determination of safety and effectiveness and lead to an application to the FDA for premarket approval of the product.

There may be situations in which the first study under an IND or IDE will not be a phase1/feasibility study (Hellman *et al.*, 2000). For example, this may occur when there is sufficient clinical experience to establish the safety of a product after use outside the United States or in a different patient population. The FDA may review data from clinical studies performed outside the United States in the IND/IDE process and/or in an application for marketing approval. The agency strongly recommends that the sponsor meet with FDA staff to discuss the clinical protocol, study results, statistical analyses, and applicability of the data to a U.S. population before submitting the premarket submission, i.e., Biologics License or Premarket Approval application (BLA/PMA).

## Premarket Submissions

According to the laws and regulations governing commercial distribution of human medical products, there are several different types of product premarket submissions, determined by the product's FDA classification. In general, the type of submission will depend on the type of product, i.e., drug, biologic, or device.

Engineered tissue products regulated as biologics will require review and approval of a BLA that demonstrates the safety and effectiveness of the product before it may be marketed commercially. If it is regulated as a device, a PMA demonstrating safety and effectiveness must be approved or a premarket notification [510(k)] must receive clearance. In order to obtain 510(k) premarket clearance, the sponsor must demonstrate substantial equivalence of the device to a legally marketed predicate device.

## Special Product Designations and Submissions

The FD&C Act recognized that there may be situations where the demand for new medical products may be such that the cost of obtaining marketing approval for a product may be prohibitive in view of the small size of the intended population (Hellman and Smith, 2006). To reduce the possibility that a cost–benefit analysis applied to product development for rare diseases will result in no available therapy, the FDA is authorized to grant special consideration and exceptions to reduce the economic burdens on product developers of products under such conditions. As a result, the FDA may be petitioned to grant a Humanitarian Device Exemption (HDE) for certain devices (FD&C Act, 520m) or to recognize certain drugs or biologics as orphan drugs (FD&C Act, 525, et seq.).

A Humanitarian Use Device (HUD) is a product that may be marketed under an exemption for treatment or diagnosis of a disease or condition that affects fewer than 4000 individuals per year in the United States. An HDE exempts an HUD from the effectiveness requirements for devices if certain criteria are met (FD&C Act, 529(m)(1), as amended February 1998). Several engineered skin constructs have been approved for market under the HUD designation.

Orphan drugs are those intended to treat a disease or condition affecting fewer than 200,000 individuals in the United States for which there is little likelihood that the cost of developing and distributing it in the United States will be recovered from sales of the drug in the United States. The orphan drug designation was established through an amendment of the FD&C Act by the 1982 Orphan Drug Act. An orphan drug is defined to include biologics licensed under Section 351 of the PHS Act. Under certain conditions, the FDA has authority to grant marketing exclusivity for an orphan drug in the United States for a period of seven years from the date the drug is approved for clinical use. Other benefits to sponsors include grant support for clinical trials, tax credits for clinical research expenses, and waiver of the prescription drug filing fee (Hellman and Smith, 2006). A sponsor must file a petition for orphan drug designation before any application for marketing approval.

### Postmarket Surveillance

Postmarket surveillance for therapeutic engineered tissue products is an important area of consideration. Manufacturers, user facilities, and health care professionals should report adverse events through the FDA MedWatch process. Postmarketing studies may be necessary when a sponsor seeks a change in product labeling, when studies are a condition of the FDA approval, or when such studies are necessary to protect the public health or to provide safety and effectiveness data (Hellman *et al.*, 2000). Additionally, postmarket surveillance of a device introduced into interstate commerce after January 1, 1991, may be required if it is intended for use in supporting or sustaining human life, presents a potential risk to human health, or is a permanent implant whose failure may cause serious, adverse health consequences or death (Section 522, FD&C Act).

# V. REVIEW OF PRODUCT PREMARKET SUBMISSIONS

Advances in tissue-engineering research have led to potential therapeutic products for many different medical conditions characterized by organ and/or tissue damage. As indicated previously, the products may provide either a structural/mechanical or metabolic function. To date, products have been developed either as *in vitro* engineered tissue constructs for implantation, cell therapies for *in vivo* delivery, or *ex vivo* systems. Representatives of these products are in different stages of development. First-generation products have been approved for use in the United States (Table 85.1), while many others are under either preclinical investigation or regulatory evaluation.

Since many of the products may consist of more than one component, i.e., biomolecule, cell/tissue, and/or biomaterial, they are considered combination products. A determination of the product's primary mode of action dictates the jurisdictional authority for the product and the primary reviewing center, i.e., CDER, CBER, or CDRH. However, regardless of the product's designation, review of any regulated product considers four basic elements [product manufacture, preclinical (laboratory and animal model) testing, clinical performance, and product labeling] in order to determine safety and effectiveness in support of the manufacturer's claim of intended use.

Regulatory evaluation is conducted on a case-by-case basis, and the manufacturer/sponsor is responsible for providing evidence of the product's safety and effectiveness. As indicated, product safety and effectiveness are evaluated with respect to the product's manufacture and clinical performance, as applicable, as well as the manufacturer's claim of intended use, i.e., the patient population to be treated and the product's role in the diagnosis, prevention, monitoring, treatment, or cure of a disease or condition. For engineered tissue products as well as other human medical products, issues of product manufacture include, among others, cell/tissue, biomaterial, and/or biomolecule sourcing, processing, and characterization; detection and avoidance of adventitious agents; product consistency and stability; and quality control/quality assurance procedures. Other important considerations include evaluation of the preclinical data, e.g., toxicity and immunogenicity testing for local/systemic and acute/chronic responses, as well as assessment of *in vivo* remodeling. Collecting data on product performance in humans requires insight into clinical trial design, e.g., patient entry criteria and study endpoints, study conduct, and subsequent data analyses.

At the request of the sponsor of a new drug, the FDA will facilitate the development and expedite the review of such a drug if it is intended for the treatment of a serious or life-threatening condition and it demonstrates the potential to address unmet medical needs for such a condition. The development program for such a drug or biologic may be designated a fast-track development program and may apply special procedures, such as accelerated approval based on surrogate endpoints, submission and review of portions of an application, and priority review to facilitate its development and expedite its review (Hellman *et al.*, 2000).

For devices, PMAs, PMA Supplements, and 510(k) applications may also undergo expedited review (Hellman *et al.*, 2000). In general, applications dealing with the treatment or diagnosis of life-threatening or irreversibly debilitating diseases or conditions may be candidates for expedited review if the device represents a clear, clinically meaningful advantage over existing technology, the device is a diagnostic or therapeutic modality for which no approved alternative exists, the device offers a significant advantage over existing approved alternatives, or availability of the device is in the best interests of patients. Granted expedited review status means that the marketing application will receive priority review before other applications. When multiple applications for the same type of device have also been granted expedited review, the applications will be reviewed with the priority according to their respective submission due dates.

# VI. HUMAN CELLS, TISSUES, AND CELLULAR- AND TISSUE-BASED PRODUCTS

With the recognition that an important segment of the products that it regulates often arises from applications of new technology, such as tissue engineering, and that the product applications may pose unique and complex questions, the FDA has devoted considerable resources since the early 1990s to the regulatory considerations of what have been termed human cellular- and tissue-based products (HCT/Ps). In February 1997, the FDA proposed a comprehensive tier-based approach for regulation of these prod-

**Table 85.1.** FDA-Approved Human Cellular- and Tissue-Based Products (HCT/Ps)

*Skin Applications*

| Product | Sponsor | Intended Use | Approval |
|---|---|---|---|
| Apligraf (Viable Allogeneic Fibroblasts/Keratinocytes On Type-1 Bovine Collagen) | Organogenesis Inc. | Standard therapeutic compression for treatment of non-infected partial and full-thickness skin ulcers | 1998-Device (PMA) |
| Dermagraft (Cryopreserved Dermal Substitute; Allogeneic Fibroblasts, Extracellular Matrix, Bioabsorbable Scaffold) | Advanced Tissue Sciences, Inc. | Treatment of full-thickness diabetic foot ulcers | 2001-Device (PMA) |
| Composite Cultured Skin (Viable Allogeneic Fibroblasts/ Keratinocytes On Collagen Matrix) | Ortec International, Inc. | Adjunct to standard autograft procedures for covering wounds and donor sites after surgical release of hand contractions in Recessive Dystrophic Epidermolysis Bullosa patients | 2001-Device (HDE) |
| Dermagraft (Cryopreserved Dermal Substitute; Allogeneic Fibroblasts, Extracellular Matrix, Bioabsorbable Scaffold) | Smith and Nephew Wound Management | Treatment of wounds related to Recessive Dystrophic Epidermolysis Bullosa | 2003-Device (HDE) |

*Musculoskeletal Applications*

| Product | Sponsor | Intended Use | Approval |
|---|---|---|---|
| Carticel (Autologous Cultured Chondrocytes) | Genzyme Corporation | Repair of femoral condyle caused by acute or repetitive fracture | 1997-Biologic (BLA) |
| OP-1 Implant (Recombinant Human Osteogenic Protein (rh OP-1), Type-1 Bovine Bone Collagen Matrix) | Stryker Biotech | Alternative to autograft in recalcitrant long bone non-unions | 2002-Device (PMA) |
| In FUSE Bone Graft/LT-Cage Lumbar Tapered Fusion Device (Recombinant Human Bone Morphogenetic Protein-2, Type-1 Bovine Bone Collagen, Titanium Alloy Cage) | Medtronic | Spinal fusion for degenerative disc disease | 2002-Device (PMA) |
| OP-1 Putty (Recombinant Human Osteogenic Protein (rh OP-1), Type-1 Bovine Bone Collagen Matrix, Putty Additive — Carboxymethyl Cellulose Sodium) | Stryker Biotech | Alternative to autograft in compromised patients requiring revision posterolateral lumbar spinal fusion for whom autologous bone and bone marrow-harvest are not feasible or expected to promote fusion | 2004-Device (HDE) |
| GEM 21S™ (Growth Factor Enhanced Matrix) (Recombinant Human Platelet Derived Growth Factor, Synthetic Beta Tricalcium Phosphate) | Biomimetic, Pharmaceuticals, Inc. | Treatment for periodontally-related defects; intrabony defects; furcation defects; gingival recession associated with periodontal defects. | 2005-Device (PMA) |

ucts, with the level of product review proportional to the degree of risk. On May 25, 2005, the final piece of this regulatory framework was put in place when the Current Good Tissue Practice for Human Cell, Tissues, and Cellular- and Tissue-Based Product Establishments: Final Rule (the CGTP Rule) became effective (Hellman and Smith, 2005a). Two earlier final rules, one providing for establishment registration and the other establishing processes for donor screening, had already set out significant portions of this framework. Publication of the CGTP Rule completes the set of regulations proposed in 1997 and issued in proposed or interim form since 2001 to implement the FDA's framework for regulation of HCT/Ps.

Defined as articles containing or consisting of human cells or tissues that are intended for implantation, transplantation, infusion, or transfer into a human recipient, HCT/Ps include skin, musculoskeletal tissue (bone and ligaments), ocular tissue (especially cornea), heart valve allografts, dura mater, hematopoietic stem and progenitor cells derived from peripheral and cord blood, reproductive tissue, cellular therapies, and combination products consisting of cells/tissue with a device and/or drug (such as cells on a natural or synthetic matrix).

The agency recognized the need for regulatory oversight of these products in the late 1980s and early 1990s because of a number of concerns. First, documented evidence of communicable disease transmission to recipients from infected donor tissue presented a primary public health concern. Second, the rapid growth of the industry with development of new applications and technologies for processing human cells and tissues, coupled with increased demand and international commerce, presented different issues. Finally, voluntary standards established by certain organizations had not been followed uniformly, because they are not legally enforceable. These factors, together with public demand for safe products, compelled the agency to effect appropriate solutions.

The tenets of the tiered risk–based approach initially outlined by the agency have been maintained in the CGTP Rule. Essentially, products meeting certain criteria, so-called *kick-down* factors, would be regulated solely under provisions of Section 361 of the U.S. PHS Act (361 Products) and would not be required to undergo premarket review. All others not meeting the kick-down factors would be regulated under existing drug, biologics, and device regulations, in addition to the new regulations addressing the incorporation of living biological materials into the finished product (Fig. 85.1).

The kick-down factors include minimal manipulation of the source tissue through the processing stage, homologous use, freedom from combination with another article, except a sterilizing, preserving, or storage agent, water, and crystalloids, and absence of intended systemic effect or dependence on the metabolic activity of living cells (except in cases of autologous use, use in first- or second-degree

**Tiered Approach**

- Regulated solely under Section 361 (PHS Act) if all "kick-down" criteria apply:
  - Minimally manipulated
  - Homologous use only
  - Not combined with another article (except sterilizing, preserving, or storage agent, water, or crystalloids)
  - Does not have systemic effect and is not dependent on metabolic activity of living cells (except autologous/reproductive use, use in first-/second-degree blood relatives)
  - Examples: "Banked Human Tissue" – cornea, skin, umbilical cord blood stem cells, cartilage, bone
  - Premarket application not required
- Regulated under Section 361 and biologic (IND/BLA) or device (IDE/PMA) regulations if HCT/P does not meet all "kick-down" criteria.

**FIG. 85.1.** FDA regulatory oversight for cells, tissues, and human cellular- and tissue-based products (HCT/Ps). From Current Good Tissue Practice Final Rule, Published 1/24/2004; Effective 5/25/2005; www.fda.gov/cber/rules/gtp.htm.

blood relatives, or reproductive use). Those HCT/Ps not meeting these criteria would be regulated under the FD&C Act as drugs, biologics, or devices. The risk-based approach is tiered, i.e., stratified, to provide the appropriate type and level of regulation based on a product's characteristics, with a platform of minimal requirements for all cells and tissues and additional requirements when necessary for product safety and effectiveness.

The CGTP requirements cover all aspects of production, including cell and tissue recovery, donor screening and testing, processing and process controls, supplies and reagents, equipment and facilities, environmental and labeling controls, storage conditions, product receipt, predistribution shipment and distribution, advertisement and deviation reporting, and tracking form donor to product consignee. Each establishment of the affected industry must develop and maintain a quality program covering all these requirements and take measures to report and track any product-related adverse event. The CGTP Rule also grants additional provisions to the FDA, including inspection authority, control of imports, and enforcement authority.

Thus, predictable regulatory requirements serve to support innovation in technology and the industry and to minimize elements of uncertainty in the product development process. Since many, if not all, engineered tissues are human cell- or tissue-based and/or combination products, the CGTP Rule and PMOA Final Rule serve to clarify the regulatory requirements for such products and to demonstrate the FDA's commitment to facilitating the development process for these products while, at the same time, maintaining the public confidence in safe, effective medical products for the marketplace.

## VII. SCIENCE AND PRODUCT DEVELOPMENT

Federal investment in basic biomedical research is expected to lead to an overall improvement in public health.

However, as observed and reported by the FDA in its March 2004 report "Challenges and Opportunity on the Critical Path to New Medical Products" (Critical Path Report), that expectation is not being fulfilled, and there is a discontinuity from basic research to application (Hellman and Smith, 2005b).

Data, based on 10-year trends, show that, while there has been an increase in research spending by federal government agencies, such as the National Institutes of Health, and industry, there has been a concomitant decrease in major drug and biological product submissions to the FDA. This is also true for devices, although not to as great an extent.

The FDA's analysis of this pipeline problem has led to the conclusion that the current medical product development path is becoming increasingly challenging, inefficient, and costly. To address these concerns, the FDA launched the Critical Path Initiative to identify the most pressing obstacles in the path and in technology translation. With publication of the Critical Path Report, the FDA framed the challenge as the shortage of modern tools to enable effective and efficient assessment of the safety and effectiveness of new medical products. Since then, the FDA has worked with FDA staff and external stakeholders to identify the most important challenges and to create the Critical Path Opportunities List as an outline of its strategy to overcome them.

While a number of issues and opportunities have been identified, certain common themes have emerged. The primary concerns are clinical trials and biomarker development. There is a need to improve clinical trials and outcome assessment generally. Accelerating the development and regulatory acceptance of biomarkers or other surrogate markers is perceived as an approach for their use in characterizing the product as well as in measuring outcome(s) for both preclinical and clinical studies. Other areas identified include bioinformatics, manufacturing and scale-up generally, i.e., moving from laboratory bench studies to a manufacturing process with appropriate system design controls to ensure a consistently reproducible, stable product, and progress in evaluating products developed through tissue-engineering approaches. In addition, development of therapies for specific at-risk populations, such as pediatrics, with better extrapolation methods and best practices in clinical trial design was felt to be especially important. It was noted that a key hurdle inhibiting innovation in tissue engineering is the difficulty in sufficiently characterizing the finished product to enable development of meaningful quality controls and product release specifications. For example, conventional techniques for evaluating cell characteristics cannot be applied to these products, since they may also include matrix materials and other components. Consensus on how to assess engineered tissue products and ensure manufacturing consistency would provide developers the predictability needed to fulfill the technology's full potential.

The initiative will continue as a formal process for continued input from all stakeholders and will be helpful to those engaged in engineered tissues research and product development. The FDA (2006) published the Critical Path Opportunities List and the full Critical Path Opportunities Report on May 16, 2006 (http://www.fda.gov/oc/initiative/criticalpath).

## VIII. CONCLUSION AND FUTURE PERSPECTIVES

Strategic investment in science, engineering, and allied disciplines is a critical determinant for advancing both basic and translational research in organ/tissue replacement, repair, restoration, and regeneration toward products for the clinic and the marketplace. However, to achieve successful product commercialization and market penetration, research strategies must be based on sound market analysis and demonstrated clinical need and with a product development plan in place to attract the needed funding support from the financial communities and approval from product regulatory and reimbursement authorities. Understanding the product regulatory process and specific points to consider for engineered human cellular- and tissue-based products will help companies in the development of their overall commercialization strategy. Moreover, since low reimbursement rates can often be the single greatest impediment to product acceptance by end users in the health care environment, attention to cost-recovery issues and their relationship to clinical and economic outcomes is equally important (Norton, 2004). All these determinants are interdependent and must be considered by companies when developing a sound business plan, since uncertainties in any one determinant can have a profound effect on the entire commercialization pathway.

The FDA's approach to regulation of human cellular- and tissue-based products and combination products as well as other evolving initiatives are indicative of the agency's commitment to providing the appropriate regulatory oversight for products generated from novel technology, such as tissue engineering. It is expected that the FDA will continue to build on these initiatives and on the cooperative approaches across the appropriate FDA centers and the Office of the Commissioner in its regulatory oversight for engineered tissues so that questions from manufacturers/sponsors are addressed early on in product development, product regulatory jurisdiction questions are addressed in a timely manner, and the product premarket review process becomes more transparent and simplified. This is especially important because the pursuit of new and different research directions focused on tissue and organ regeneration, such as the apparent shift toward the use of stem cell technology (Lysaght and Hazelhurst, 2004), will lead to the development of new and different products, posing unique product-specific issues (Fig. 85.2).

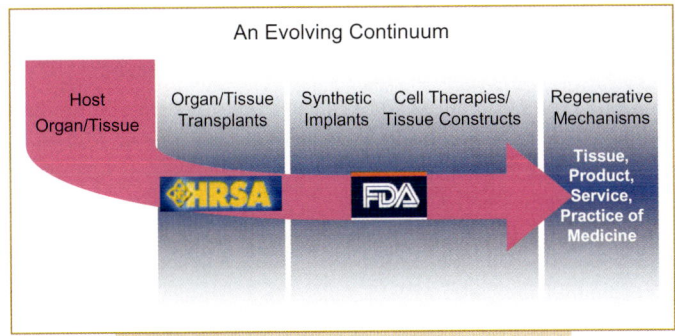

**FIG. 85.2.** Approaches for regulatory oversight of tissue repair, restoration, replacement, or regeneration: an evolving continuum.

The challenges for the tissue-engineering community are similarly multifold. For example, sponsors should consider the important determinants in the product regulatory path, such as the nature of the product, its manufacture and classification, i.e., tissue or product, its mode of action or primary mode of action if a combination product and overall therapeutic approach, and preclinical (animal models) and clinical strategies to assess safety and effectiveness, such as selection of appropriate outcome measures and assessment tools/methods. The sponsor's claim of intended use and whether the product will provide incremental or substantive therapeutic benefit compared to the standard of care will be important for end-user and market acceptance and, ultimately, cost reimbursement. The time to clinic and market will be dependent on the product's classification and subsequent submission and review of sponsor-generated data. For example, an orphan drug or humanitarian-use device will have a relatively shorter regulatory timeline than a product regulated under existing authorities as a drug, biologic, or device.

To advance the science and minimize the variables in engineered tissue systems, understanding the mechanisms and control processes in normal as well as diseased or damaged human organs and tissues will continue to be a necessary prerequisite for design of novel research strategies focused on applications for tissue repair, restoration, regeneration, and replacement. In this context and to advance the science, the following should continue to be examined: operative mechanisms in cell and developmental biology, interactions of engineered tissue constructs with the host and remodeling by the *in vivo* environment, and acute/chronic as well as local/systemic sequelae of either reparative or regenerative approaches through appropriate preclinical large-animal and clinical-monitoring studies. Progress in biomaterials science, such as the development of matrix materials customized for the cell(s) and application of interest, advances in manufacturing and scale-up techniques, such as development of tissue-customized bioreactors and process system design, and outcomes-assessment tools, such as noninvasive *in vivo* monitoring of implanted engineered tissues, will be important for translating research to applications (Hellman, 2006).

Ultimately, the challenge for the tissue-engineering community is to continue advances in the science while maintaining awareness of the product regulatory environment in the United States and abroad and to be an active voice for articulating the important issues in order to maintain a productive dialogue with the regulatory agencies and consumers so that engineered tissues find their proper place in the clinic and the marketplace.

## IX. ACKNOWLEDGMENTS

This material was originally contained in *Engineered tissues: the regulatory path from concept to market* from the journal "Advances in Experimental Medicine and Biology," Volume 585, pp. 363–376. It is used with the kind permission of Springer Science and Business Media.

## X. REFERENCES

FDA. (2006). Critical path opportunities list and the full critical path opportunities report. http://www.fda.gov/oc/initiative/criticalpath.

Hellman, K. (2004). Introduction, presentation at Inaugural Engineering Tissue Growth Executive Forum, Seattle, Washington.

Hellman, K. (2006). Engineered tissues: the regulatory path from concept to market. *Adv. Exp. Med. Biol.* **585**, 363–376.

Hellman, K., and Smith, D. (2005a). FDA final "Current Good Tissue Practice" rule. *Genet. Eng. News* **25**(6).

Hellman, K., and Smith, D. (2005b). Taking a look at the FDA's critical path initiative. *Genet. Eng. News* **25**(2).

Hellman, K., and Smith, D. (2005c). FDA regulation of combination products: evolving initiatives. *Genet. Eng. News* **25**(10).

Hellman, K., and Smith, D. (2006). The regulation of engineered tissues: emerging approaches, In "The Biomedical Engineering Handbook" (J. P. Fisher and A. G. Mikos, eds.), 4th ed., Springer, New York.

Hellman, K., Knight, D., and Durfor, C. N. (1998). Tissue engineering product applications and regulatory issues. In "Frontiers in Tissue Engineering" (C. W. Patrick, A. G. Mikos, and L. V. McIntire, eds.), pp. 341–366. Elsevier Science, Amsterdam.

Hellman, K., Solomon, R., Gaffey, C., Durfor, C., and Bishop, III, J. (2000). Regulatory considerations. *In* "Principles of Tissue Engineering" (R. Lanza, R. Langer, and J. P. Vacanti, eds.), 2nd ed., pp. 915–928. Academic Press, San Diego, CA.

Lysaght, M., and Hazelhurst, A. (2004). Tissue engineering: the end of the beginning. *Tissue Eng.* **10**, 309–321.

Norton, K. (2004). Regulation reimbursement — efficacy vs. effectiveness: what determines coverage? Presentation at the Inaugural Engineering Tissue Growth Executive Forum, Seattle, Washington.

Smith, D., and Hellman, K. (2003). Regulatory oversight and product development: charting the path, regulatory page. *Tissue Eng.* **9**(5), 1057–1058.

# Ethical Issues

## *Laurie Zoloth*

## I. INTRODUCTION

*Case Study:* The small black box holds a perfectly shaped ear. The scientist at the front of the room explains how it was made. A scaffold of nanoparticles supports fibroblast cells that grew over the form, and the ear now looks and feels actual: It can be transplanted to tissue. For burn patients, this represents an enormous chance and change. It is the prototype of a new genre of medicine, one that uses powerful technologies and methods of bioengineering and cellular biology to transform the matter of the world.

The ear in the box is not a freak example of a new technique. It is, in fact, one of a number of new devices that utilize the convergent technologies of several different fields of science and engineering to create tissue that can mimic the structure and function of the natural world. Other examples include the creation of skin grafts, corneas, bone, cartilage, and, in some pilot studies, bladders. Based on the new technologies of genomics, informatics, nanoscale engineering, molecular biology, and stem cell research, tissue engineering can be said to alter the concept of medicine itself. Instead of treating ailing tissues or organs with drugs intended to repair their structure or function, tissue engineering aims at replacing the diseased or injured or aging

tissues of the body with new ones entirely, made from component parts of the material of the world, both naturally occurring and synthetic.

Such an advance heralds a remarkable ability to heal, a long-awaited solution to several intractable problems, and a serious alternative to cadaveric or living-donor whole organ transplants, which have long been an ethically challenged sector of medicine (Caplan *et al.*, 1994; Fox and Swazey, 1992). Yet such a remarkable construction of the human body asks a great deal of any social world into which it is introduced, for it is the body that is the place of the self, the location of the acts of the sacred, and the sensory arbiter of the real. In fact, tissue engineering queries two of the very aspects of our humanity that we consider distinctive: our integral embodiment and our finitude. If we are indeed a collection of replaceable and adaptable parts, some people reason, what is it that separates us from any other engineered machine? If we can engineer, for example, a synthetic and improved lymphatic system, might we improve our chances to adapt to and overcome infectious disease? What other capacities for healing or alteration of our bodies might be prudent? How do we ensure that such changes are indeed ethical?

*Principles of Tissue Engineering, 3rd Edition*
ed. by Lanza, Langer, and Vacanti

It is this query that has greeted the new biotechnologies of the body, one based, this chapter argues, in social reactions largely shaped by culture both ancient and contemporary. We then ask: What are the ethical challenges to the field of tissue engineering? Does tissue engineering raise new ethical issues, or is it a description of one of the modalities enabled by the convergence of other technologies that have been understood to be individually ethically freighted?

In this chapter, I suggest using an established ethical framework that was suggested in 1999 by a committee of the American Association for the Advancement of Science on inheritable genetic germline modification and used far more widely by the field of bioethics to assess new technologies. We then review the responses given to new technologies in the past from a variety of sources in bioethics, philosophy, and theology. Finally, we reflect on how the legal and regulatory structure for tissue engineering has an impact on our reflections on ethical norms (Chapman *et al.*, 2003).

### Research Evaluation (Adapted from the AAAS Working Group on Human Inheritable Genetic Modifications 1998–2000 (Chapman *et al.*, 2003))

- Are there reasons, in principle, why performing the basic research should be impermissible?
- What contextual factors should be taken into account, and do any of these prevent the development and use of the technology?
- What purposes, techniques, or applications would be permissible and under what circumstances?
- On what procedures and structures, involving what policies, should decisions on appropriate techniques and uses be based?

## II. ARE THERE REASONS, IN PRINCIPLE, WHY PERFORMING THE BASIC RESEARCH SHOULD BE IMPERMISSIBLE?

Principled reasons for objections to basic research are extremely difficult to conceive in research that is, by its very nature, intended to be translational and clinical. Yet ethical objections to the manipulation, replacement, and engineering of human tissue can be seen as part of a long continuum of dissent about medical technology that began to assume full voice in the 1970s, when successful genetic manipulation of bacterial genomes became possible (Walters and Palmer, 1996).

All new technology raises new challenges — in particular, technology that refashions the embodied self, becomes a part of the "self" and the identity of the subject, and seems to raise the deepest anxieties. Even tissue engineering, an emerging field with clear targets, clinical successes, and patient needs, will raise familiar concerns.

First among these is the argument that human's possess an essential nature and live within an essential natural order that cannot be altered without harm. For C. S. Lewis (1998,

p. 274), this is expressed as a concern that the very acts of rational science — dissection, analysis, and quantification — are a violation of the sacred integrity that lies behind all of nature:

> Now I take it that when we understand a thing analytically, and then dominate and use it for our own convenience, we reduce it to the level of "nature," we suspend our judgments of value about it, ignore its final cause (if any), and treat it in terms of quantity. This repression of elements in what would otherwise be our total reaction to it is sometimes very noticeable and even painful: Something has to be overcome before we can cut up a dead man or a live animal in a dissecting room.

For Lewis, the understanding of the body as replaceable is disturbing: "The real objection," he says, "is that if man chooses to treat himself as raw material, raw material he will be, not raw material to be manipulated by himself as he fondly imagined, but by mere appetite" (p. 274). (Lewis imagines that new transformative technology will be manipulated by "controllers" who will eventually transform man into mere matter.)

Callahan (1973) echoes Lewis' concern, both in the sense that limits need to be placed on what is decent to do to nature and in the sense that such action is a part of a larger danger — that power in the hands of medicine to heal is really power in the hands of the elite, or the state, to manipulate and control. He argues:

> The word No perfectly sums up what I mean by a limit — a boundary point beyond which one should not go.... There are at least two reasons why a science of technological limits is needed. First, limits need to be set to the boundless hopes and expectations, constantly escalating, which technology has engendered. Advanced technology has promised transcendence of the human condition. That is a false promise, incapable of fulfillment.... Second, ... limits (are) necessary in order that the social pathologies resulting from technologies can be controlled.... [W]hile it can and does care, save, and free, it can also become the vehicle for the introduction of new repressions in society.

These objections, made over 30 years ago, are still made (despite, one may note with some irony, 30 years of medicine that have indeed seen rapid and successful advances, without their being used by the state for repression and without any fundamental change in the capacities for intellectual and spiritual self-possession). Nevertheless, the powerful arguments of opposition to the manipulation and replacement of tissues and organs continue, with some people worried that perfection itself is sought when healing is the goal. Such critics, many from the disability community, raise principled objections to the use of tissue engineering if the goal is to alter the disability. Activists in the deaf community, for example, defend their disability as a culture and a language exchange, not as a loss of function. Others are concerned that our society's focus on "fixing things" will allow a devaluation of the persons that currently bear the broken bodies and parts. Adrienne Asche suggests that there is a "troubling side to every cure, that those of us who are uncured are seen as less valuable, perhaps even expendible." For Gerald McKenney (1997), a community needs to embrace brokenness and to "deny that the worth of one's life is determined to how closely one conforms to societal standards of bodily perfection." McKinney is also concerned that if medicine is successful, it will create a social and economic system that "virtually demands that we be independent of the need to care for others."

## Duty and Healing: Natural Makers in a Broken World

While the opposition to medical technology has indeed been persistent, it has not been unchallenged. For many, the response lies in the nature of brokenness and the human duty to respond to the need of the suffering other (Freedman, 1998; Zoloth, 2003). The principle at stake in the assessment of tissue engineering as an ethical act is not how its use might potentially violate an abstract community in the future, but the actual problem of what one must do as a moral being when one's neighbor is in need. In this important sense, the duty to heal cannot be overridden by a "sense" of discomfort (as Lewis notes in the earlier quote).

It is the nature, goal, and meaning of science to address the human condition in all its yearning and capacity for defeat and failure. In this sense, there is no principle objection to the science of tissue engineering, and in fact there may well be a strong moral imperative to develop the technology.

Joseph Fletcher (1970) noted, in speaking of an earlier generation of medical technology and answering the critics of science, wrote: "The belief that God is at work directly or indirectly in all natural phenomena is a form of animism or simple pantheism. If we took it really seriously, all science, including medicine, would die away because we would be afraid to 'dissect God' or tamper with His activity". "Every widening and deepening of our knowledge of reality and of our control of its forces are the ingredients of both freedom and responsibility".

McKinney, Childress, and Lewis use an argument that is rooted in Christian moral theology: that since human persons are fallen creatures in a fallen world, we cannot really be counted on to know the right and the good. God, who is transcendent, from this fallenness, has set us in this place, not essentially to alter it toward our own transonic, but to find its meaning and purpose. Yet other faith traditions differ. For Jewish and Islamic theorists, the world is morally neutral. Human persons may — and will — fail in their aspirations but can be trusted to have the capacity for moral behavior and moral yearnings. Finding meaning in suffering is not the core task. The task is to alleviate suffering, which is understood as chaotic, meaningless, and agonistic. Hence, many of the core objects in principle are rooted in religious constructions and understandings.

## To Make Is to Know: Notes on an Old Problem About Knowledge

The classic debates of the 1970s are not the only set of problems engendered in the history of ethical responses to the technological gesture at the heart of tissue engineering. At stake as well is the special kind of knowledge that such making implies. For Aristotle and the Hellenists, useful knowledge, "practical wisdom," was phronesis. Phronesis implied actually doing an act, making, in order to know. The act of making, not the act of perception or contemplation alone, how wisdom and, indeed, rationality and power were achieved. Hence, making new tissue is a somewhat different moral gesture than curing the body by altering it with drugs that essentially allow the body to heal itself.

Second, the use of technology within the body of the patient is a different matter than the use of technology essentially to enhance the body of the practitioner. For all earlier technology, the thing that was changed or enhanced was the sense perceptions of the doctor. Stethoscopes and otoscopes allowed the sounds of the body to be more audible. X-rays, CT scans, and MRIs allow the inner vistas of the body to be revealed. EEGs and EKGs allow the electrical currents that animate the central and peripheral nervous systems to be charted in quantifiable units. Microscopes allow invasive bacteria to be seen at the microscopic and, increasingly, molecular levels. These earlier technologies extended the reach of what Bacon increasingly trusted and that the Greeks did not: the perception and observation of the phenomena of the world and the perception of the outcome of its deliberate perturbation.

> Bacon's method presupposes a double empirical and rational starting point. True knowledge is acquired if we proceed from lower certainty to higher liberty and from lower liberty to higher certainty. The rule of certainty and liberty in Bacon converges.... For Bacon, making is knowing and knowing is making (cf. Bacon IV [1901], 109–110). Following the maxim "command nature... by obeying her" (Sessions, 1999, p. 136; cf. Gaukroger, 2001, pp. 139 ff), the exclusion of superstition, imposture, error, and confusion are obligatory. Bacon introduces variations into "the maker's knowledge tradition" when the discovery of the forms of a given nature provide him with the task of developing his method for acquiring factual and proven knowledge (Stanford Encyclopedia of Philosophy website, 2003).

Thus, the world is known by understanding the parts of the world and, from that, theorizing (knowing), by induction, to principles or axioms or laws of nature, physics, and chemistry. In contemporary science, knowing is done largely by "unmaking," by deconstruction of the component parts in ways scientists of Bacon's era were unable to imagine. Many of these "unmaking" techniques, such as the splicing of alternative DNA and the manipulation of cellular structures, allow a sense of inherent interchangeability, as if the real and the person were merely a set of Lego parts awaiting clever recombination.

## What Is a Thing? The Perils of Deconstruction

Making actual tissue in mimesis of the real tissue of the actual body extends the Baconian act in radical ways. Here, the experimental perturbation is the unmaking of tissue and the remaking of tissue, only in more controllable form. This cannot help but excite concern about the nearly infinite possibilities for technological shaping of the self. Heidegger asks: "What is a thing?" and in so reflecting, understands a thing as an object separate from the self. But what of a made thing, an object that becomes the self?

The technology of the alteration of the patient is distinctive. Devices for altering the functioning of the body that become a part of the body and are actually a tissue of the body are a step beyond the idea of a device held within the body. This is important in any ethical assessment of the technology because the patient's consent and participation are needed for the final act of the technology to be completed. Such an event only happens in a specific context, for technologies, patients, and practioners operate in a social, religious, and economic context. Here we turn to the second ethical consideration.

## III. WHAT CONTEXTUAL FACTORS SHOULD BE TAKEN INTO ACCOUNT, AND DO ANY OF THESE PREVENT THE DEVELOPMENT AND USE OF THE TECHNOLOGY?

Tissue engineering is a complex procedure still in experimental stages. Yet to be an ethical technology, it must be directed toward accessibility, just distribution, and efficacy. Hence, the troubling context of widespread health care disparity is a problem not only for this advanced technology but for all newly emerging technologies. Emergence into an unjust world asks certain moral questions of new technological advances. First among these is the query about burdensomeness versus benefit in a context in which the vast majority of the world's people suffer from easily treatable infectious diseases such as tuberculosis, malaria, AIDS, and infant diarrhea. How can tissue engineering be justly promoted in the face of other, pressing needs?

This objection can typically be met by noting that it would be deeply inappropriate to withhold medical knowledge until the world is entirely perfected and that applications not only will be increasingly available to the poor (as in vaccines, once rare) but also that the very process of research has typically uncovered new and useful ways to understand disease.

The goal of tissue engineering is the widespread use of the technique. Unlike solid-organ transplants, which would always require significant resources far outside the capacities of developing-world clinics, tissue replacement, stem cell therapies, and other transportable therapies are designed for widespread use. The possibility to create a method for allografts that uses the patient's own cells and the possibility for allologous cells to provide an "off-the-shelf" source of tissue may provide the basis for access (but only if research priorities are discussed in advance of design), a process, as we describe later, that will need careful support and monitoring. The question of how to achieve this and how to enable a more just use of each technology has not yet been solved.

A second contextual factor for tissue engineering is that all human tissue is marked by its genomic identity. It is the very nature of cells, what allows them to copy and reproduce, to carry identifying markers linked to some person somewhere. In the past, such use has raised serious objections. Such tissue can be traced and known, which may have implications for the person who is the source of the tissue, raising significant new issues in genetic privacy for the donor.

Further, whose is the tissue that is derived from the cells of a particular body? Who should have the rights to, and a fair share of, the profits derived from its use? In the seminal case in the field, *Moore v. the Regents of the University of California (51 Cal 3d 120 271 Cal Reporter, 146, P2d 479, 1990)*, the issue of ownership was addressed. In this case, Mr. Moore had his T-cell lymphocytes taken from his spleen during the course of treatment for hairy-cell leukemia, cells that proved effective in deriving resistant cell cultures. Patented after manipulation to make a new "product," the cells were indeed profitable. Mr. Moore's complaint was that he was not informed of, much less a part of, the scientific enterprise and the lucrative payout for his cells. The case was decided in favor of the research labs. But in the ensuing decades, alert patients with unique cell types or unusual cancers sought for research have been selling their materials as personal possessions to the lab that wishes to procure them. Ownership is limited, however, by the common law of the United States and the European Union which constrains this ability to claim tissue as property. The goals of such restraints were put into place to prohibit the buying and selling of human tissue and organs, for fear, given the desperation of the poor, that selling the bodies of the poor would become permissible and lead to their exploitation. Thus, the entire process that allows for the derivation of tissue sources needs to be noted. The current context for tissue donation is a mixed system. Organs, tissues such as blood, corneas, and marrow are donated or exchanged without compensation. Gametes, however, are another matter entirely. Because the use of human sperm and eggs emerged in the context of fertility treatment and because this treatment was largely conducted in stand-alone, private clinics that functioned without public oversight or regulation, the marketplace standards prevailed. What originally began as a compassionate exchange between family members of gametes when an infertile couple could not conceive quickly changed into a robust marketplace in human gametes. As of this writing, international standards prohibit the use of marketplace incentives for gametes or embryos (International Society for Stem Cell Research, 2006).

A final context for the debate about the ethics of tissue engineering in general is the special case of human stem cells to make tissues. Because some applications of tissue engineering use stem cells as a part of the method of treatment (Egan, 2006), the debate about the ethics of the use of human stem cells is directly adjacent to this technology. For many, the origins of tissue matter a great deal. For some Christians, many Roman Catholics, and some Hindu sects, the destruction of the human embryo, even at the blastocyst stage, is tantamount to killing. For these faith traditions, the derivation of stem cells from embryos is always impermissible. For many other faith traditions, such as Judaism, Islam, Jainism, Buddhism, Confucian philosophy, and Daoism, the use of these cells is permissible within certain constraints, as we will see later. For all faith traditions, however, the manipulation of adult somatic cells in their precursor form is completely sanctioned. Precursor cells are not as flexible as pluripotent cells, and it is that very pluripotency and immortality that are important in tissue engineering. These factors raise concern. Yet the contextual factors alone do not prohibit entirely the use of this technology, for justice in distribution, the possibility of the loss of genetic privacy, and the controversy over stem cell research when pluripotent embryonic cells are used affect many aspects of the new techniques in medicine. Hence, we turn to the third major issue.

## IV. WHAT PURPOSES, TECHNIQUES, OR APPLICATIONS WOULD BE PERMISSIBLE AND UNDER WHAT CIRCUMSTANCES?

Many of the salient, justifying arguments for the use of tissue engineering hinge on the telos, or goal, of the treatment: If the goal is to cure or treat human disease, then the benefits will outweigh the burdens of the work — controversy, cost, and difficulty. Clearly, then, tissue engineering ought not to be used in a trivial or wasteful fashion. Human tissue is understood by many as deserving a special sort of "respect" (Geron Ethics Advisory Board, 1999; Nelson, 2002).

This proviso may not be so simple, for a core problem in genetic engineering has been the use of the technique for "enhancement" of human characteristics or traits. The initial ethical discussions about therapeutic uses of medicine versus cosmetic ones imagined ethical bright lines that would define the boundary between the use of such technology to restore "species normal functions" (Daniels, 1996) for each tissue and for the person as a whole. Yet medical practice has long gone beyond these lines, using surgery, for example, for cosmetic purposes. Will it be possible to restrict tissue replacement to burn victims, spinal cord injury, and diabetics? How can such a distinction be made?

Some tissue-replacement therapies, such as the use of skin grafts for full-face transplant, may also raise questions about the nature of identity. Indeed, the notion of a full-face transplant alerted us to the depth of resistance to identity-altering tissue replacements. (Could persons use any face? What if persons in need of facial transplants wished to change ethnicity? Should faces "match," and why?) Like many other aspects of this technology, this tension about identity was not new, only heightened. For example, the first years of organ transplant raised the same issues for recipients of hearts, a key aspect of identity in many cultures. If the face is our key determinant of the self in modernity and, even more so, if the brain is such, then how are we to understand the use of tissue engineering to transform identity?

Hence, linked inexorably to this technology are larger considerations of the use of tissue engineering for neuroscience, both for therapy and for enhancement. The applications of tissue transplant in Parkinson's disease are important. Yet will there be concern about this use of the neurons of a stranger in the brain of the self? Of all the possible uses of tissue engineering, the ones that may alter consciousness and memory are the most troubling. (What capacities or memories could neurons store?) Here, the need for restrictions on applications may be the clearest, yet it is not clear who ought to decide and who ought to ensure that the restrictions on unethical applications are maintained. By what criteria will such limits be set?

New research possibilities also offer applications to engineer gametes for use and storage. Engineered follicles may now be saved, frozen, matured, and used in animal models to create the possibility of human fertility after cancer chemotherapy or other environmental risk (Woodruff *et al.*, 2006).

With this, as with all such technology, there will have to be careful attention to how the market may drive technology toward specific research goals rather than others or to whether research goals will be framed only by the values of profit and efficacy and not by ones of more general interest: compassion, healing, and solidarity. The powerful applications and the potential for widespread use itself create the possibility for serious conflicts of interest, for serious market forces may be the core drivers of technology, especially in an aging population with increasing needs for all manner of new tissues and organs. This turns us to the consideration of our final set of issues.

## V. ON WHAT PROCEDURES AND STRUCTURES, INVOLVING WHAT POLICIES, SHOULD DECISIONS ON APPROPRIATE TECHNIQUES AND USES BE BASED?

Much of the first reviews of the ethical issues in tissue engineering have in fact focused on the issues of policy:

safety, patents, and gating. Products and drugs are typically controlled via four levels of restraints. The first is elaborate premarket gating, first involving animal models and then typically done for pharmaceuticals in a decade-long series of tests, phased to test the drug on an increasing but controllable number of human subjects. Such trials must be gender balanced, and subjects must give full, informed consent and be able to leave the trial at any time (which could be difficult in cases of implanted tissues).

The next gating is the system of intellectual property. Patents and licensing control the use of the products, even the replication of the experiments. The next gating is that of financial backing. To perform the enormous clinical trials, to do premarket investigaton, and, of course, actually to make and sell the product require a production apparatus, which must be assembled and supported. Finally, each drug or device must be approved for use by the insurers. As David Smith (2004) notes, tissue engineering faces a gauntlet of issues and a "new order of magnitude in interactions and science patents." Additionally, notes Smith, the "things" engineered are hybrids of two jurisdictions, that of drugs and that of devices.

Are genetically engineered insulin cells a drug, like insulin, a device, like a stent, or a biologic? Unlike stents, which are entirely synthetic, tissue engineering uses actual human cells, only manipulated in *de nova* ways.

Standards will need to be set for safety, efficacy, and fair use. Standards for clinical use, standards for clinical trial, and standards for tissue stability and purity will be needed for the research and application to be safe.

Getting informed consent in this case will present significant challenges. Patients in need of organs, for example, are particularly desperate, and their consent may be deeply affected by their utter lack of options. Eight percent of the medical system is already devoted to organ transplantation, and the lack of organs is an overwhelming problem for nearly half of the patients hoping for transplants (Lysaught and O'Leagh, 2001). Yet the first year of the use of engineered tissue will be experimental and will need to be conducted under the strongest possible set of NIH guidelines. How the first trials of engineered tissue are conducted will set the tone and the future for all subsequent use.

The question of policy and the regulation of policy are manifested in many of the first documents that evaluate the ethical and legal implications of tissue engineering. While, as Smith notes, the United States faces a complex regulatory system, the European Union has regulated such research products as medical products, and these will fall under the regulatory gaze of the European Medical Evaluation Agency (EMEA). In both the United States and the EU, the synthetic nature of tissue engineering, the very *de novo* quality of the work, and the uneasy greeting that met genetically modified food has created serious political opposition. Policies need to be crafted with transparency and full public participation, for such research needs not only public funding but public understanding of complex theory and practice of tissue engineering — what promises it can hold and what cautions need to be applied prior to use. Policymakers will need to attend to calls for justice in distribution, as was noted earlier, and will need to set in place structures for regulation.

How can new technologies best be regulated? I contend that a full array of regulatory structures can be employed. First among these are local committees, IRBS, and local review boards. The National Academies have played a large role in policy writing for both recombinant DNA and for stem cell research and, in both instances, called for special, national, ongoing oversight on such research. It would be prudent to reflect on the need for such a process for tissue engineering, for established structures largely address issues involving the use of donated tissue, not engineered tissue. Structures that protect human subjects also need strong enforcement, as noted earlier, both for donors and for recipients of tissues.

But regulation, government oversight, and market forces can only go so far in shaping just research goals and commitments. The goal of ethics is to develop moral agents who are aware of a constancy of duty toward subject and to humanity, who not only follow rules correctly but who, given the chance and grace to work at the frontiers of science, act with courage and decency in their research.

## VI. CONCLUSION

Tissue engineering suggests that an old dream — the replacement of human body parts — may be realized. While any sober and reflective scientist understands the long way to success of this idea, the science described in this volume clearly suggests that our society is on the road to the enactment of the possibility.

## VII. REFERENCES

Callahan, D. (1973). Science, limits and prohibitions. *Hastings Cent. Rep.* **3**, 5–7.

Caplan, A. L. (1994). "If I Were a Rich Man Could I Buy a Pancreas?: And Other Essays on the Ethics of Health Care," Indiana University Press, Bloomington, IN.

Chapman, A. R., and Frankel, M. S. (2003). "Designing Our Descendants: The Promises and Perils of Genetic Modification," The Johns Hopkins University Press, Baltimore, MD & London.

Daniels, N. (1996). "Justice and Justification: Reflective Equilibrium in Theory and Practice," Cambridge University Press, Cambridge, UK.

Eggan, K., Jurga, S., Gosden, R., Min, I. M., and Wagers, A. J. (2006). Ovulated oocytes in adult mice derive from non-circulating germ cell. *Nature* **441**, 1109–1114.

Fletcher, J. (1970) "Technological Devices in Medical Care." *In* "Who Shall Live" (K. Vaux, ed.), pp. 115–142. Fortress Press, Philadelphia, PA.

Fox, R., and Swazey, J.P (1992). Leaving the field. *Hastings Cent. Rep.* **22**, 9–15.

Freedman, B. (1999). "The Duty of Healing: Foundations of a Jewish Bioethic." Routledge, NY.

Gaukroger, S. (2001). "Francis Bacon and the Transformation of Early Modern Philosophy." Cambridge University Press, Cambridge, UK.

Geron Ethics Advisory Board (1999). Research with human embryonic stem cells: ethical considerations. *Hastings Cent. Rep.* **29**, 31–36.

Helman, K., and Smith, D. (2004). "Legal and regulatory issues in tissue engineering." Elsevier Academic Press, New York.

(2005) International Society for Stem Cell Research, Guidelines for Human Embryonic Stem Cell Research. http://www.nap.edu/books/0309096537/html.

Klein, J. (2003a). Bacon's quarrel with the Aristotelians. *Zeitsprünge* **7**, 19–31.

Klein, J. (2003b). "Francis Bacon." Stanford Encyclopedia of Philosophy website. http://plato.stanford.edu/entries/francis-bacon/

Lebacqz, K. (2001). On the elusive nature of respect. *In* "The Human Embryonic Stem Cell Debate: Science, Ethics, and Public Policy" (S. Holland, K. Lebacqz, and L. Zoloth, eds.), pp. 149–162. MIT Press, Cambridge, MA.

Lewis, C. S. (1998). The abolition of man. *In* "On Moral Medicine: Theological Perspectives in Medical Ethics" (S. E. Lammers and A. Verhey, eds.), pp. 247. Wm. B. Eerdmans Publishing Company, Grand Rapids, MI.

Lysaught, M. J., and O'Leagh, J. A. (2001). The growth of tissue engineering. *Tissue Eng.* **7**, 485–493.

McKenney, G. P. (1997). "To Relieve the Human Condition: Bioethics, Technology and the Body." State University of New York Press, Albany, NY.

*Moore v. The Regents of the University of California*, 51 Cal. 3d 120, 793 P.2d 479, 271 Cal. Rptr. 146 (1990).

Sessions, W. A. (2003). "Francis Bacon." Stanford Encyclopedia of Philosophy website. http://plato.stanford.edu/entries/francis-bacon/

Walters, L., and Palmer, J. G. (1996). "The Ethics of Human Gene Therapy." Oxford University Press, Oxford.

Xu, M., Kreeger, P. K., Shea, L. D., and Woodruff, T. K. (2006). Tissue-engineered follicles produce live, fertile offspring. *Tissue Eng.* June 1. Epub.

Zoloth, L. (2003). Freedoms, duties, and limits: the ethics of research in human stem cells. *In* "God and the Embryo: Religious Voices on Stem Cells and Cloning" (B. Waters and R. Cole-Turner, eds.), pp. 141–151. Georgetown University Press, Washington, D.C.

# Epilogue

This volume represents a current snapshot of tissue engineering, a single frame in a video sequence of ideas, science, and technological development spanning 30 years. The script evolves in real time and now includes substantive work in stem cell biology. Terms evolve as well. *Regenerative medicine* has been added to the nomenclature. Also included is a catalogue of clinical work, not only commercial prod- ucts but clinical trials. Business models have been tested and will continue to be adapted to maximize translation of innovation into patient care. New updates will continue as long as progress in the field continues to excite young and creative minds.

*Joseph Vacanti*

# INDEX